PETROLEUM PROCESSING HANDBOOK

OTHER McGRAW-HILL HANDBOOKS OF INTEREST

BAUMEISTER AND MARKS · Standard Handbook for Mechanical Engineers
BEEMAN · Industrial Power Systems Handbook
BELL · Petroleum Transportation Handbook
BLATZ · Radiation Hygiene Handbook
CARROLL · Industrial Instrument Servicing Handbook
COCKRELL · Industrial Electronics Handbook
CONSIDINE · Process Instruments and Controls Handbook
CONSIDINE AND ROSS · Handbook of Applied Instrumentation
CROCKER AND KING · Piping Handbook
ETHERINGTON · Nuclear Engineering Handbook
FACTORY MUTUAL ENGINEERING DIVISION · Handbook of Industrial Loss Prevention
FRICK · Petroleum Production Handbook
GRANT · Hackh's Chemical Dictionary
GUTHRIE · Petroleum Products Handbook
HEYEL · The Foreman's Handbook
HUSKEY AND KORN · Computer Handbook
JURAN · Quality Control Handbook
KAELBLE · Handbook of X-rays
KALLEN · Handbook of Instrumentation and Controls
KATZ · Handbook of Natural Gas Engineering
KING AND BRATER · Handbook of Hydraulics
KLERER AND KORN · Digital Computer User's Handbook
KNOWLTON · Standard Handbook for Electrical Engineers
KORN AND KORN · Mathematical Handbook for Scientists and Engineers
LANGE · Handbook of Chemistry
LASSER · Business Management Handbook
LEE AND NEVILLE · Handbook of Epoxy Resins
MAGILL, HOLDEN, AND ACKLEY · Air Pollution Handbook
MANAS · National Plumbing Code Handbook
MANTELL · Engineering Materials Handbook
MAYNARD · Industrial Engineering Handbook
MAYNARD · Top Management Handbook
MEITES · Handbook of Analytical Chemistry
MOODY · Petroleum Exploration Handbook
MORROW · Maintenance Engineering Handbook
PERRY · Chemical Engineers' Handbook
PERRY · Engineering Manual
RICHEY, HALL, AND JACOBSEN · Agricultural Engineers Handbook
SHAND · Glass Engineering Handbook
STANIAR · Plant Engineering Handbook
STREETER · Handbook of Fluid Dynamics
TOULOUKIAN · Retrieval Guide to Thermophysical Properties Research Literature
TRUXAL · Control Engineers' Handbook

PETROLEUM PROCESSING HANDBOOK

Co-Editors

WILLIAM F. BLAND

Editor and Publisher, PetroChemical News
Editorial Director, Petro/Chem Engineer

ROBERT L. DAVIDSON

Senior Editor, Chemical Engineering

McGRAW-HILL BOOK COMPANY

New York   San Francisco   Toronto   London   Sydney

CONTRIBUTING EDITORS

Harold S. Bell, * Consulting Engineer, Summit, N.J. (*Sec. 4, Equipment*)

Dr. A. S. Brunjes, Manager, Technical Information Group, The Lummus Co., Newark, N.J. (*Sec. 12, Physical Properties of Hydrocarbons*)

Vittorio de Nora, Oronzio de Nora Impianti Elettrochimico, Milano, Italy. (*Sec. 6, Chemicals and Catalysts*)

Matthew J. de Pasquale, Senior Instrument Engineer, The M. W. Kellogg Co., New York, N.Y. (*Sec. 10, Process Control and Instrumentation*)

Virgil B. Guthrie, Editor, *Petroleum Products Handbook*, Danbury, Conn. (*Sec. 11, Petroleum Products*)

James H. Herbert, Consultant, Accident and Fire Prevention, Christmas Cove, Maine. (*Sec. 9, Personnel and Plant Safety*)

Everette Kerns, Mechanical Engineer, Amanda, Ohio. (*Sec. 7, Maintenance and Construction*)

John A. King, Oronzio de Nora Impianti Elettrochimico, Milano, Italy. (*Sec. 6, Chemicals and Catalysts*)

Lawrence Lowy, Consulting Engineer, New York, N.Y. (*Sec. 4, Equipment*)

C. M. McKinney, Analytical Chemist, Bureau of Mines Petroleum Research Center, Bartlesville, Okla. (*Sec. 2, Evaluation of Crude Oils*)

Robert B. Norden, Senior Editor, *Chemical Engineering,* New York, N.Y. (*Sec. 5, Materials of Construction*)

Runne C. Ohrberg, Instrument Analytical Section Engineer, The M. W. Kellogg Co., New York, N.Y. (*Sec. 10, Process Control and Instrumentation*)

George C. Patterson, The M. W. Kellogg Co., New York, N.Y. (*Sec. 8, Offsite Facilities and Utilities*)

Morris D. Schoengold, Technical Information Division, Esso Research and Engineering Co., Linden, N.J. (*Sec. 13, Sources of Information*)

William C. Uhl, Editorial Director and Processing Editor, *World Petroleum*, New York, N.Y. (*Sec. 1, Introduction*)

George H. Unzelman, Chief Refinery Technologist, Western Region, Ethyl Corp., Los Angeles, Calif. (*Sec. 3, Processes*)

Charles J. Wolf, Chief Refinery Technologist, Eastern Region, Ethyl Corp., New York, N.Y. (*Sec. 3, Processes*)

* Deceased.

v

PREFACE

This *Petroleum Processing Handbook* is a compilation of practical information related to the refining (processing) of crude oil (petroleum) to convert it to useful petroleum products—primarily fuels and lubricants.

When the editors of this Handbook undertook the assignment of putting it together, they agreed that what was needed was a comprehensive collection of data that concentrated on the "what" and the "where" of petroleum processing: what processes are used in petroleum refining, what are they used for, where are they applied; what kinds of equipment and materials are employed in the construction and maintenance of petroleum processing plants; what problems are encountered in the safe operation of such plants; what products do they make, etc.

For that reason, this Handbook is truly a reference book. It is not a textbook, it does not include a lot of theory, it does not explain in any great detail the "how" of a process, a piece of equipment, or a practice.

It is intended for the engineers (experienced and otherwise) who want a ready reference where they can find basic data, for the technicians and plant operators who need to know more about the various elements that make up this complex industry, and for the laymen who of necessity deal with the industry and must understand its operations but don't want to get bogged down in the minute details of its technology.

We have tried to bring together a finely balanced collection of material, written to our specifications by experts in their individual specialized fields, that goes a long way toward satisfying the needs of all these groups. If some material is not included, it is only because the practical limitations of space precluded the use of everything available.

The heart of any such Handbook as this one, of course, is its index. The user of this Handbook will not "read" it in the sense that he reads a novel or a textbook. He will usually pick it up to find a particular piece of information, to answer a specific question. And he needs and expects to be able to find his subject with a minimum of effort. Considerable care has been taken, therefore, to prepare as comprehensive an index as is practical, with frequent use of cross references. The index includes nearly 3,000 individual entries.

This Handbook is obviously not the result of the work of only its two editors. Grateful acknowledgement is made to the many contributors who wrote specific parts of it, to the petroleum companies and the engineering firms which made it possible for men on their staffs to write individual sections and even provided technical data from their own files, to the various associations and societies (and particularly the American Petroleum Institute) who permitted us to use freely of their resources, and to the editorial and production staffs of the McGraw-Hill Book Company—who have exercised extreme patience and understanding in helping us bring to completion the long, drawnout process of producing this volume.

William F. Bland

Robert L. Davidson

CONTENTS

PETROLEUM PROCESSING HANDBOOK

Section 1

INTRODUCTION

By WILLIAM C. UHL

Editorial Director and Processing Editor
World Petroleum
New York, N.Y.

Petroleum, a mixture of organic compounds and primary hydrocarbons, comes from underground rock formations ranging in age from ten to several hundred million years. The process by which it formed and developed is not yet completely known.

Studies indicate that petroleum is formed mainly from microscopic-sized marine animals and plants. When these organisms died in water of low oxygen content, they did not decompose. Thus their remains sank to the bottom to be buried under accumulations of sediment. Their conversion to petroleum remains a subject of research today. The theory held generally is that bacteria converted the fats of the marine life into fatty acids. These in turn were changed, by mechanisms still unknown, to the asphaltic material called kerogen. Then, over millions of years, heat and pressure, plus probably catalytic agents in the rock, changed the kerogen to crude oil and gas.

Petroleum has been known and used by man for centuries. Explorers have found evidence of asphalt in buildings constructed more than 6,000 years ago in what today is Iraq. Historical records indicate use of mineral oil in various forms down through the ages. However, it is only a little over 100 years since the beginnings of the oil industry as it is known today, with the 1857 discovery of crude oil in western Ontario and the boom which followed the Drake discovery in Pennsylvania 2 years later.

Crude-oil deposits occur throughout the world, some of them in extremely remote and hard-to-reach areas. The major producing regions include Texas and California in the United States, Alberta in Canada, Venezuela, the Middle East, Indonesia, Russia, and North Africa. Areas of potential production exist on every continent (see Fig. 1-1). As a result, the petroleum industry has become truly international and requires huge capital investments and operating expenditures. It has grown to be the dominant source of the world's energy and organic chemicals.

In 1965, an estimated two-thirds of the Free World's energy came from oil. The Free World consumed petroleum at an average rate of 26.55 million bbl daily in 1965, or 1.9 times the consumption rate of 14.04 million B/D 10 years earlier. Total proved crude-oil reserves in the Free World at the start of 1966 were 324 billion bbl.

During the 1955–1964 decade, the over-all increases in consumption, crude reserves, tanker tonnage, and refinery throughput were not quite twofold, while the gross fixed assets expanded 2.2 times, from $63 billion to $136 billion.

Capital spending by the Free World's oil industry in 1964 alone was over $12 billion, of which 45 per cent went into production, 16 per cent into transportation, 18 per cent into refining and petrochemical plants, and 21 per cent into marketing and other facilities.

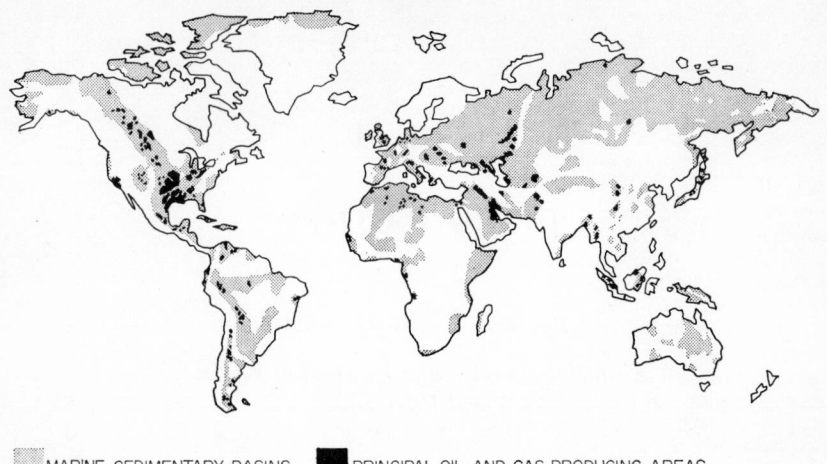

░ MARINE SEDIMENTARY BASINS   ██ PRINCIPAL OIL AND GAS PRODUCING AREAS

FIG. 1-1. Principal oil- and gas-producing areas of the world and the major sedimentary basins with producing potential. (*Source: Science in the Petroleum Industry, American Petroleum Institute.*)

CHEMISTRY OF PETROLEUM HYDROCARBONS

Crude petroleum is primarily a liquid of widely varying physical and chemical properties. Common colors are green, brown, and black and occasionally almost white or straw color. Specific gravity can range from 0.73 to 1.02; however, most crudes are between 0.80 and 0.95. Viscosity varies, too. Data for a large number of crudes indicate kinematic viscosities from 0.007 to 13 stokes at 100°F, though most of them range from 0.023 to 0.23 stoke.

Principal elements in crude petroleum are carbon and hydrogen, usually in a carbon-hydrogen ratio between 6 and 8. The hydrocarbons are mainly liquids and gases, with some solids in dispersion or solution. Among the many other materials usually present are small amounts of sulfur, nitrogen, and oxygen in the form of hydrocarbon derivatives; traces of such metals as nickel, vanadium, and iron; and water (emulsified in the oil and sometimes as high as 30 per cent). The water is generally in the form of saturated solutions of calcium and magnesium sulfates and sodium and magnesium chlorides.

The hydrocarbons are of three major types: paraffins, naphthenes, and aromatics. The very large range in their proportions and the ratios in which specific series of hydrocarbons appear determine the *physical* operations to be used in separating these components, such as by distillation. The presence and extent of some of the other materials—say sulfur and oxygen—determine the *chemical* operations to be used, i.e., catalytic conversions, hydrogen processing, and the like.

Natural gas, the gaseous component of crude petroleum, may also be found some distance away from an oil pool, in separate wells, having been separated from the liquid by natural processes under the ground. It is composed mainly of light paraffins—methane, ethane, propane, butane—plus some higher-boiling paraffins, nitrogen, carbon dioxide, and hydrogen sulfide.

A detailed study of the composition of petroleum, the work of American Petroleum Institute Research Project 6, was begun at the U.S. Bureau of Standards in 1928 and continued at Carnegie Institute of Technology since 1950. So far, 151 individual hydrocarbons have been isolated from one Ponca City, Okla., crude oil, covering a boiling range of -161 (methane) to 389°C (normal tetracosane). This is a small part of the total theoretically possible hydrocarbons. Apparently, the simpler molecular structures are more abundant than highly branched types in the lower-boiling ranges.

Therefore, the simpler paraffins, naphthenes, and aromatics predominate, and the isomers occur only in small percentages. At higher-boiling ranges, the content of polynuclear aromatics and polycycloparaffins increases while that of normal paraffins, branched-chain paraffins, and monocycloparaffins drops.

The refiner uses paraffins first for producing high-octane motor fuels and second for conversion to chemical intermediates. Very stable at atmospheric conditions, paraffins react mainly at elevated temperatures and pressures, for example, thermal cracking.

The main reactions are (1) rupture of the paraffin chain, generally toward the end, and (2) to a lesser extent, dehydrogenation to olefins, which may subsequently polymerize or form cyclic compounds. If the reaction takes place in the presence of a suitable catalyst, the chain tends to rupture closer to its mid-point and less gas is formed. Furthermore, the molecules which are adsorbed on the catalyst will isomerize, cyclize, and lose hydrogen. Catalytic cracking converts the paraffins mainly to isoparaffins, aromatics, and olefins.

Normal paraffins can be isomerized in the presence of aluminum chloride, boron trifluoride, and promoters such as hydrogen or alkyl halides or with platinum-containing catalysts. Higher-boiling isomers can be obtained by alkylation of lower-boiling isoparaffins with olefins, for example, the sulfuric acid alkylation of isobutane with butylene to produce isooctane.

The naphthenes dehydrogenate to aromatics in the presence of suitable catalysts, for example, catalytic reforming with a platinum catalyst.

The aromatics, besides providing a high-octane component for gasoline, can react to produce other valuable materials. One example is the alkylation of benzene with an olefin to make isopropyl benzene for high-octane aviation fuel.

REFINING

A petroleum refinery is an organized and coordinated arrangement of manufacturing processes designed to provide both physical and chemical change of crude petroleum into salable products with the qualities required and in the volumes demanded by the market.

A complete refining installation will include all necessary nonprocessing facilities: adequate tankage for storing crude oil, intermediate, and finished products, a dependable source of electric power; material-handling equipment; workshops and supplies for maintaining continuous 24-hr/day, 7-day/week operation; waste-disposal and water-treating equipment; and product-blending facilities.

Types of Refineries

Depending on the processes used, refineries can be classified as simple, complex, or fully integrated (see Fig. 1-2). A simple refinery will include crude-oil distillation, catalytic reforming, and treating. Its range of products is relatively limited: LP gases, motor fuels, kerosine, gas oil, diesel fuel, and fuel oil.

A more complex refinery will make a greater variety of products and require the following additional processes: vacuum distillation, catalytic cracking with gas recovery, polymerization, alkylation, asphalt oxidation. Cracked gases will feed polymerization and alkylation units to produce high-octane gasolines (motor and aviation) and may also be feedstock for petrochemical processing. Residue from the vacuum still will go to asphalt manufacture.

The fully integrated refinery makes a full range of products. A number of complex units will have been added to those already mentioned just to produce a line of finished lubricating oils, greases, and waxes. These will include high vacuum fractionation, solvent extraction, deasphalting, dewaxing, and treating.

Depending on the type of crude oil used, the refinery may have to include one or more hydrogen processes. If sour crude oil is used, and if sulfur can be recovered in profitable volumes, a unit for this purpose will be installed. If a coke market exists, some type of coking unit will be made a part of the refinery.

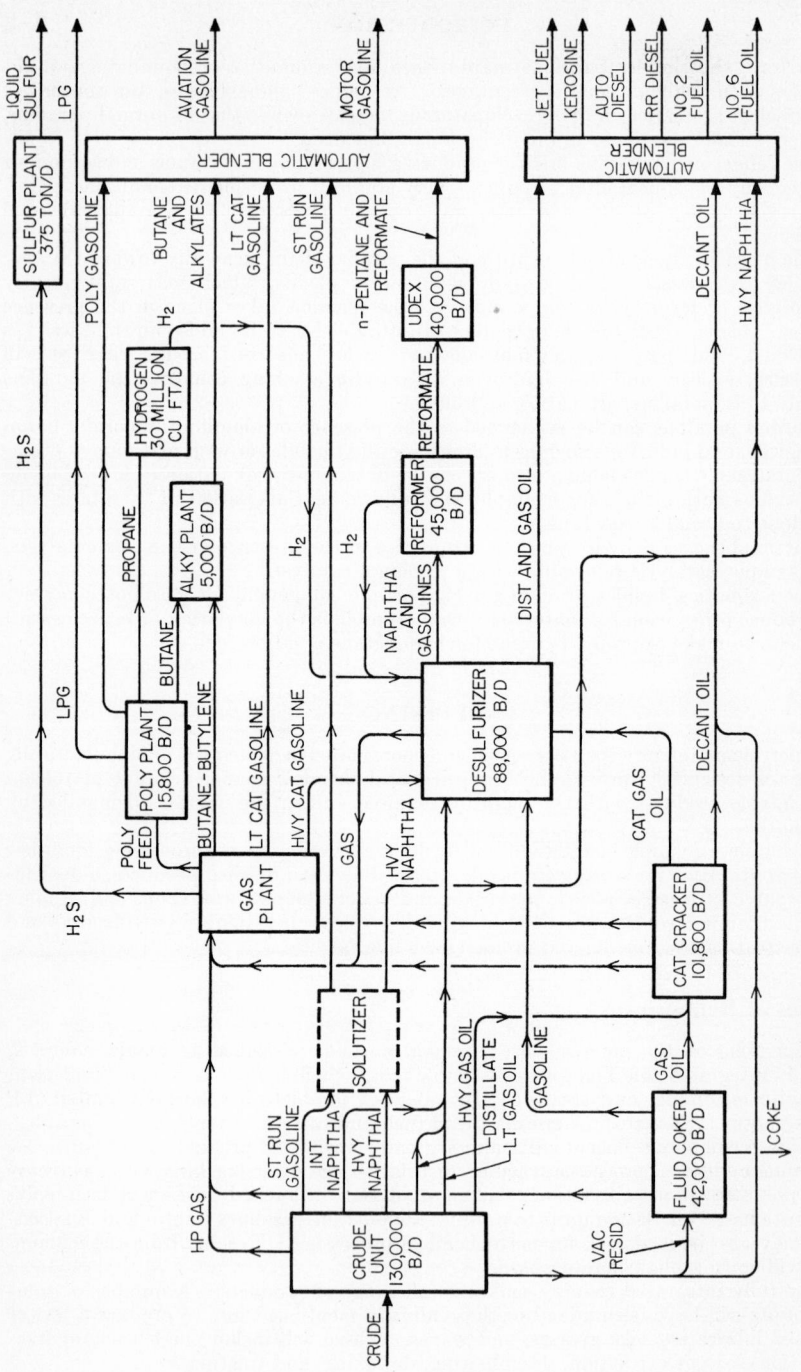

Fig. 1-2. Schematic flow chart of a modern refinery. (*Source: Petroleum Processing, July, 1957, pp. 102–103.*)

1–4

Process Selection

Processes are chosen, arranged, and interrelated according to the market the refinery serves and the products it manufactures.   A refinery must have flexibility; that is, its management must be able to alter operations as needed to maintain a balanced output.

For example, a refinery cannot produce a given volume of light oils (gasoline and the like) without simultaneously producing a certain quantity of higher-boiling materials, say heavy fuel oil.   If the gasoline moves to market faster than the fuel oil, the latter will accumulate in the storage tanks and eventually force the plant to shut down.   A flexible refinery will help solve this problem.   It would include a cracking unit to change the excess heavy fuel oil into more gasoline along with coke as the residual product, or it might include a vacuum still to separate the heavy fuel oil into lube-oil stocks and asphalt.

Types of Crude Oil

A significant controlling factor in the processes to be used is the type of crude oil to be run.   Crudes are commonly classified according to the residue from their distillation, this depending on their relative contents of three basic hydrocarbons: paraffins, naphthenes, and aromatics.

Table 1-1. General Properties of Crude Oils

Property	Paraffin base	Asphalt base
API gravity..........................	High	Low
Naphtha content.....................	High	Low
Naphtha octane number...............	Low	High
Naphtha odor........................	Sweet	Sour
Kerosine smoking tendency............	Low	High
Diesel-fuel knocking tendency..........	Low	High
Lube-oil pour point...................	High	Low
Lube-oil content.....................	High	Low
Lube-oil viscosity index...............	High	Low

Source: Purdy, G. A., *Petroleum, Prehistoric to Petrochemicals*, Copp Clark Publishing Co., Toronto, Canada, 1957, p. 72.

About 85 per cent of all crude oils fall into the following three classifications:

1. **Asphalt-base,** containing very little paraffin wax and a residue primarily asphaltic (predominantly condensed aromatics).   Sulfur, oxygen, and nitrogen contents are often relatively high.   Light and intermediate fractions have high percentages of naphthenes.   These crude oils are particularly suitable for making high-quality gasoline, machine lubricating oils, and asphalt.

2. **Paraffin-base,** containing little or no asphaltic materials, are good sources of paraffin wax, quality motor lube oils, and high-grade kerosine.   They usually have lower nonhydrocarbon content than do the asphalt-base crudes.

3. **Mixed-base,** containing considerable amounts of both wax and asphalt.   Virtually all products can be obtained, although at lower yields than from the other two classes.

Because of the variations in hydrocarbon fractions in different crude oils, there are several differences in general properties, as indicated in Table 1-1.

Crude-oil Fractions and Refinery Products

Refining starts with crude-oil distillation.   The usual fractions and their major uses are as follows:

1. Light gases, methane, ethane, and some propane (boiling range, -259 to $-44°F$), used as refinery fuel gas and/or petrochemical processing feedstock.

2. Propane (boiling point, $-44°F$), source of liquefied petroleum gas, feed for petrochemical processing.

3. Butanes (boiling range, 11 to 31°F), blended with motor gasoline to raise its volatility, feed for petrochemical processing.

4. Light naphtha (boiling range, 30 to 300°F), component of gasoline. When depentanized, it can be blended with heavy naphtha and serve as feed for catalytic reforming.

5. Heavy naphtha (boiling range, 300 to 400°F), feed for catalytic reforming, blended with light gas oil to make jet aircraft fuel.

6. Kerosine (boiling range, 400 to 500°F), jet fuel component, heating oil, solvent, illuminant.

7. Light gas oil (boiling range, 400 to 600°F), used in fuel oil and diesel fuels, may be blended with lighter oils to reduce pour point.

8. Heavy gas oil (boiling range, 600 to 800°F), combined with vacuum gas oil for catalytic cracking feed.

9. Vacuum gas oils (boiling range, 800 to 1100°F), feed for catalytic cracking, source of lube-oil distillates.

10. Residue (boiling range, 1100°F and up), source of heavy fuel oil when blended with gas oils, source of asphalts or waxes.

The naphthas are commonly referred to as the light oils; the kerosine and light gas oils are the middle distillates; the vacuum gas oils and residuals together represent reduced crude. Table 1-2 gives an indication of the effects on chemical composition of the major fractions by the type of crude oil.

Table 1-2. Chemical Composition of Petroleum Fractions

Fraction	Boiling range 50% ASTM distillation, °F	Paraffin-base crude, wt %			Asphaltic-base crude, wt %			
		Paraf.	Naph.	Arom.	Paraf.	Naph.	Arom.	Unsat.
Gasoline........	280	65	30	5	35	55	10	
Kerosine........	450	60	30	10	25	50	25	
Gas oil.........	600	35	55	15	...	65	33	2
Heavy distillate.	750	20	65	15	...	55	43	2

Source: Sachanen, A. N., *Conversion of Petroleum*, Reinhold Publishing Corporation, New York, 1948, 2d ed., p. 2.

Seldom are the fractions or even combinations of them suitable as finished products. The treating and conversion of them to salable materials are performed by the so-called downstream processes.

The product distribution pattern depends on the type of crude oil run and the process units used and is geared to market requirements. In the United States, the biggest demand is for motor fuels. Average product yield pattern in 1965 was gasolines, 45 per cent; kerosine, 6 per cent; gas oils and distillates, 23 per cent; residual fuel oil, 8 per cent; other products, 18 per cent. In Europe, on the other hand, greatest demand is for industrial fuels, and the 1965 pattern was gasoline, 21 per cent; kerosine, 3 per cent; distillates, 27 per cent; residuals, 35 per cent; others, 14 per cent. All yields are based on volumes.

The dominant use for petroleum products is as a source of energy. Over 85 per cent of oil serves as fuel. The balance serves a tremendous variety of applications: lubrication, paving, roof coatings, paint solvents, electrical insulation, wax coatings, to name just a few. Petroleum today is also the source of over one-third of all chemicals produced in the United States, a share amounting to more than 90 billion lb per year. Yet only about 4 per cent of the crude oil being processed is made into petrochemicals.

Refinery Operations

Operation of a refinery follows carefully planned programs of several types. One is the long-term program, usually covering a period of 5 years. This concerns itself primarily with the expected market and what products should be manufactured. It aids management in determining what processes will be needed and at what time. The long-term outlook is important because the design, engineering, and construction of new process units usually takes 2 years or more.

Another program is one covering about 1 year and is concerned mainly with (1) seasonal variations in product demand and inventory requirements and (2) planned shutdowns for inspection and maintenance.

A third program is the operating plan for periods of 1 to 3 months ahead. It is concerned with such variables as crude-oil supply, delivery schedules, and product flow requirements in considerable detail. It serves as the refinery manager's guide for day-to-day operations and helps prevent process bottlenecks or other upsets in the routine.

Because of the great importance of quantity and quality control, a modern refinery is one of the most highly instrumented operations of any kind of manufacturing today. All the more usual process variables—temperature, pressure, liquid level, and flow—are recorded and controlled, and much of this automatically. In addition, recent years have seen the development and addition of many instruments which record and control certain end-product qualities: viscosity, gravity, pH, color, content of key components, and the like. In many cases, such controls augment or even supplement the traditional laboratory tests and analyses: distillations, flash point, gum content, carbon residue, sulfur, vapor pressure, corrosivity, and the like.

Besides manufacturing, refinery management is concerned with and responsible for all the auxiliary operations. These include utilities (fuel, water, steam, electric power), waste disposal (both air- and water-pollution prevention), cooling water and cooling towers, compressed air, tank-farm operations, corrosion, evaporation losses, fire and accident prevention and protection, plant security (guards), maintenance, and technical and engineering services.

Historical

Today's refinery is a far different thing from the little so-called "teapots" of the early days of the industry. The average capacity of the 661 refineries in operation in the Free World by the end of 1965 was 46,500 B/D, and most of them, particularly the newest, contain many downstream processes.

The very term *refining* is by today's technological standards a complete misnomer. Words like *processing, manufacturing,* or *conversion* are much more apt. Once, a refinery was nothing more than a crude still, physically separating oil by heat. Today, almost 85 per cent of the finished products are the result of chemical change (see Table 1-3).

The early petroleum still of the 1860's consisted of a cast-iron kettle, perhaps 7 ft in diameter, with a tight cover, and set in firebrick over a wood-burning firebox. A gooseneck led vapors from its top through a coil of pipe immersed in running water in a wooden tank alongside. The first material to come over was naphtha, but this was usually combined with the remaining condensate, which was chiefly kerosine, since that was the only product wanted at that time. When a batch was finished, the still was cooled, cleaned, and refilled and the operation repeated. It usually took about 3 days to run a batch. If the kerosine was not pure enough, it had to be rerun in the same still, further naphtha and heavier ends being removed, both of which were discarded.

The first improvement was the development about 1870 of the shell still, a horizontal cylinder made of steel boiler plate instead of cast iron and holding 200 to 400 bbl of oil at a time. Shell stills were used on crude and certain crude fractions until well into the 1930's. They were arranged in batteries, however, and operated continuously.

The first significant improvement was the development of continuous distillation, the first unit being credited to Samuel Van Syckle in the 1870's in Pennsylvania.

A better method of separating fractions was attained in 1904 with the selective, or partial, condensation technique developed by Van Dyke and Irish. It was basically a steel tower added to the batch shell still, permitting withdrawal of additional fractions from the same batch. Its success prevented batch shell stills from being displaced by continuous shell stills for several years.

The problem of foaming in the shell still was overcome when M. J. Trumble of California introduced the first pipe still in 1911. In this, oil flowed continually through a long pipe which lined the inner walls of a firebox. It then entered a chamber

Table 1-3. Important Dates in the Development of Processes

1861—Batch still; Titusville, Penna., William Barnsdall and William H. Abbott.

1870's—Continuous still; Samuel Van Syckle, Pennsylvania.

1904—Selective condensation; J. W. Van Dyke and W. M. Irish, Atlantic Refining Co.

1911—Continuous pipe still; M. J. Trumble, California.

1913—Pressure cracking still; Dr. William M. Burton and associates, Standard Oil Co. (Indiana)

1914–1915—Continuous thermal cracking; C. P. Dubbs. Also, Dr. Walter M. Cross and Roy Cross. Also, the Holmes-Manley development of The Texas Co. (patents of Joseph H. Adams)

1930—Delayed coking; Standard Oil Co. (Indiana)

1934—Catalytic polymerization (phosphoric acid); Universal Oil Products Co.

1936—Catalytic cracking (continuous, fixed-bed); Eugene Houdry, backed by Socony-Vacuum Oil Co. and Sun Oil Co.

1939—Alkylation (sulfuric acid); Anglo-Iranian Oil Co., Humble Oil & Refining Co., Shell Development Co., Standard Oil Development Co., and The Texas Co.

1940—Hydrogen reforming; Standard Oil Development Co., Standard Oil Co. (Indiana), and M. W. Kellogg Co.

Early 1940's—Butane isomerization; Shell Development Co., Universal Oil Products Co., and Phillips Petroleum Co.

1941—Continuous catalytic cracking (moving-bed); Houdry Process Corp.

1941—Continuous catalytic cracking (fluid); Standard Oil Development Co.

1942—Alkylation (hydrofluoric acid); Universal Oil Products Co. and Phillips Petroleum Co.

1949—Catalytic reforming (platinum catalyst); Universal Oil Products Co.

1954—Fluid coking; Esso Research and Engineering Co.

where the vapors separated from the liquid, the vapors going on through a complex system of partial condensers, heat exchangers, and condensers for continuous fractionation.

Today's crude distillation process still uses the basic principles of the pipe still, but fractionation takes place in towers containing a series of horizontal trays where liquid condenses, collects, and is withdrawn.

Cracking by heat was discovered at almost the same time that distillation went into use, about 1861. Records have it that the discovery was quite accidental. A stillman reportedly left a still unattended with a strong fire under it while he visited a nearby saloon. He returned later than planned and found that the overhead product was a light distillate instead of the expected heavy oil. Investigation showed that the heavy oil had kept falling back into the pot because of the higher temperatures, and there it had partly decomposed to lower-boiling hydrocarbons. Called cracking distillation, the method was used for a long time to increase kerosine production. With a Pennsylvania crude, kerosine yield could be raised from about 50 per cent to 75 or 80 per cent, although at a lower quality.

During the early part of this century, the problem became one of increasing the yield of gasoline. William Burton and his associates at Standard Oil Co. in Whiting, Ind., developed pressure cracking, whereby gas oils were heated under pressure. A patent was obtained from the United States in 1912, and the first pressure stills—a batch operation—were installed at Whiting in 1913.

During and after World War I, several continuous gas-oil thermal cracking processes were developed and put into use. Some were called tube and tank processes. Hot gas oil, preheated by exchange with product streams, flowed through cracking coils (similar to the pipe-still furnace), then to soakers—huge drums where high pressures and temperatures were maintained—and finally to a tar separator and a fractionator.

Continued increases in automobile engine compression ratios and their rising octane requirements led to the development of cracking in the presence of a catalyst. The first commercially successful process, introduced by Eugene Houdry in 1936, was a fixed-bed, cyclic operation. Five years later, two continuous catalytic cracking processes came along: the moving-bed Thermofor process and the Fluid process of Standard Oil Development Co., a subsidiary of Standard Oil Co. (N.J.), and now Esso Research and Engineering Co.

Further needs for more high-octane materials led also to another process, alkylation, which came into commercial operation in 1939 in the sulfuric acid process. This represented the combined efforts of five companies: Anglo-Iranian Oil Co. Ltd., Humble Oil & Refining Co., Shell Development Co., Standard Oil Development Co., and Texaco. Each arrived independently at the same basic process; they then pooled their knowledge and made a joint announcement. Three years later, the hydrofluoric acid alkylation process was patented by Universal Oil Products Co., and Phillips Petroleum Co. developed, engineered, and built the first commercial plant.

When readily available supplies of isobutane in refineries were unable to keep up with alkylation requirements in the late 1930's and early 1940's, several variations of a butane isomerization process came along. Universal Oil Products, Shell Development, and Phillips Petroleum were instrumental in the development.

A major process development near the end of the 1940's was Universal Oil Products Co.'s Platforming, a catalytic reforming process using platinum-based catalyst, which did not require regenerating. The first commercial unit went into operation in October, 1949. The especial attraction of the technique was its value to smaller refiners. Adoption of Platforming and of some 12 more variations of platinum catalytic reforming processes made available for licensing during the ensuing few years was one of the more spectacular chapters in the history of petroleum processing.

The platinum technique, however, was preceded by two earlier ideas in reforming, both of which were probably major contributors to its development. The first was Hydroforming, a 1940 development by Standard Oil Development, Indiana Standard, and M. W. Kellogg from the basic hydrogen process rights Standard Oil Development had acquired from I. G. Farben in 1929. The second was Phillips Petroleum's Cycloversion process, first announced in 1943. These used nonplatinum catalysts which required regeneration.

Refiners were already familiar with the many uses to which hydrogen could be put in the plant, desulfurizing and removing various contaminants. It was the advent, however, of platinum reforming, producing excess hydrogen as a byproduct, that created the greatest impetus to the development and adoption of many new hydrogen treating processes in the 1950's. Today, hydrotreating is used to purify and to upgrade quality of many products—motor fuel, lube oils, and fuel oils. It is a must whenever sour crude is to be processed.

In the mid 1960's the use of hydrogen was extended to the hydrocracking process, which has given refiners still further flexibility in meeting a variety of product requirements from a wide range of raw materials.

PETROCHEMICALS AND THE REFINERY

"A petrochemical is a chemical compound or element recovered from petroleum or natural gas, or derived in whole or in part from petroleum or natural gas hydrocarbons, and intended for chemical markets" (*Petroleum Processing*, April, 1952, p. 490).

Examples range from ammonia made from natural gas to synthetic rubber, which is a mixture of hydrocarbon polymers. In addition to these two, the major petrochemicals include carbon black, sulfur, olefins, polyolefins, olefin derivatives, aromatics (benzene, toluene, xylene) and their derivatives, acetylene, phenol, alcohols, ketones, acrylonitrile, acetic anhydride, phthalic anhydride, maleic anhydride, and many others. Minor products number into the thousands.

Most petrochemicals are organic, but a few are inorganic. Most are manufactured from refinery gases or liquids; consequently many petrochemical plants are an integral part of a large refining complex or located very near by, perhaps drawing their raw materials through a short pipeline.

Process Development

Petrochemical processing was a natural development of the greater attention the industry started giving to more precise processing of conventional products. It was probably a direct result of the cracking process to produce gasolines. In early days the gases from cracking were simply burned, even though they contained a considerable amount of cracked gasoline. This wasteful practice led to the development of gas-recovery units accompanying cracking units and designed to separate propane and lighter gases from the heavier hydrocarbons. This equipment was the father of equipment used today to separate many petrochemical raw materials. Similar units were developed to obtain dry, salable gas from field natural gas by separating the condensates or natural-gas liquids.

Several forms of highly efficient fractionators evolved from the refiner's need to obtain more precise cuts. Three of these are useful to petrochemical processing because they can separate single compounds or small groups of compounds. These are superfractionation, azeotropic distillation, and extractive distillation.

The polymerization processes, developed in 1934, were aimed then at converting propylenes and butylenes in cracked gases to polymer gasolines to add to the total finished motor-fuel blend. The demand for still higher octanes led to separating the butylenes, polymerizing and then hydrogenating them into a material largely isooctane. Later, alkylation made possible a one-step combination of butylenes and isobutane to make isooctanes. Supplementing the supply of isobutane, isomerization was developed. These various fuel components are actually petrochemicals.

Still later, catalytic reforming provided another source of aromatics, primarily for use as high-octane gasoline components. Yet the aromatics, like the olefins derived from methane, propane, and butane, are also basic building blocks for petrochemicals. Hence, here too was the beginning of petrochemical processing.

Major Chemical Reactions

Among the many organic chemical reactions used in the manufacture of petrochemicals, some of which also play roles in conventional refining, the following are just a very few of the more important ones.

Oxidation. Completely oxidized hydrocarbons convert to carbon dioxide and water. **Partial oxidation** of some paraffins produces such petrochemicals as alcohols, organic acids, and ketones. Adipic acid, an intermediate in nylon manufacture, can be produced by oxidizing cyclohexane to cyclohexanol and cyclohexanol to adipic acid. Oxidation of benzene in the presence of vanadium pentoxide yields maleic anhydride; with naphthalene the product is phthalic anhydride.

Hydrogenation is adding one or more hydrogen atoms to an unsaturated or an olefinic compound. In nylon manufacture, the cyclohexanol comes from hydrogenation of benzene. Hydrogenation also occurs in the polymerization of isobutylene to di-isobutylene with dilute sulfuric acid, the hydrogenation taking place over a tungsten–nickel sulfide type of catalyst. The use of hydrogen in the catalytic desulfurization of fuels and lubes is not true hydrogenation (hydrogen combines with sulfur to form hydrogen sulfide). However, some saturation of olefinic bonds does occur.

Dehydrogenation is the removal of hydrogen atoms. Among examples that can be cited are the following: butane to butylene, butylene to butadiene, secondary butanol to methyl ethyl ketone. Paraffins are dehydrogenated to olefins in thermal cracking, and the reaction is reversible. In catalytic reforming, it goes the other direction.

Aromatization is conversion of paraffins or naphthenes to aromatics. In catalytic reforming, hexane is first cyclized to cyclohexane and then aromatized to benzene.

Chlorination can be *addition* of chlorine, for example, ethylene to its chloride. It can also be *substitution* of chlorine for hydrogen in a saturated compound, for example, methane to carbon tetrachloride and hydrogen chloride.   On the other hand, methane to methyl chloride, methylene chloride, and chloroform is an addition.

Alkylation is the addition of a straight- or branched-chain hydrocarbon (an alkyl) to any other compound.   Useful in manufacture of aviation gasoline, it also serves in petrochemicals, for example, tetrapropylene added to benzene to produce dodecyl benzene or methanol with benzene to produce toluene.

Hydration is the chemical combination of water with a compound, for example, ethylene and water to produce ethanol or ethylene oxide and water to make ethylene glycol.

Dehydration is removing chemically bound water from a compound, for example, water from ethylene cyanohydrin to produce acrylonitrile.

Hydrolysis is the use of water to split a compound, for example, isopropyl sulfate to isopropanol and sulfuric acid.

Polymerization is a very widely used process today, such as in the manufacture of polyethylene, polypropylene, or polybutadiene.   Another example is propylene polymerized to its tetramer for subsequent alkylation to benzene.

OTHER PHASES OF THE OIL INDUSTRY

The preceding provides background on the nature of petroleum, the refining of petroleum, and the manufacture of petrochemicals.   The following will give a broad description of the other major phases of the petroleum industry to indicate the interrelationship of exploration, production, and transportation with petroleum processing.

The less important phases of the industry are touched on only lightly or omitted altogether.   It is intended that the following will give an over-all orientation of the entire petroleum industry for those persons directly concerned with processing only.

EXPLORATION

Petroleum is found in pockets of porous rock, such as sandstone, surrounded by nonporous rock.   Gas, oil, and water have seeped into such pockets and been trapped. There are three types of traps: the anticline, the fault, and the stratigraphic trap (see Fig. 1-3).   The **anticline** is essentially an arch or bulge in the layers of rock. A **fault** is where earth movements have caused rock layers to crack and shift, so that a nonporous layer forms a part of the barrier trapping the oil.   A **stratigraphic trap** forms when two nonporous layers have joined forming a cul-de-sac.   There are many variations and combinations of the different types of traps.

Oil exploration requires (1) knowledge of underground structures and which types favor formation of oil pools and (2) methods of obtaining data about underground formations, primarily the field of geophysics.

Geophysical Exploration

Geophysical exploration obtains data on variations in the physical properties of subsurface rocks, including density, elasticity, magnetic and electrical properties.   It provides a means for determining the location of various types of rocks.   It does not, as yet, give direct evidence of the occurrence of oil; this still necessitates drilling a hole.

Up to about 30 years ago, when the science of geophysics was developed and its application began, oil exploration had to be carried out mainly on what could be observed *on the surface* of the earth.   Geophysics, which uses principles of physics in the study of geology, provided a method for seeing *below the surface*.   The principles used are basically magnetism, gravity, and sound waves.   The instruments are the magnetometer, gravimeter, and seismograph.

Magnetic Method

Oil tends to be found in layers of sedimentary rock, which lie on top of igneous, or basement, rock.   This is much denser than the sedimentary layers and contains a higher concentration of iron and other magnetic materials.   A **magnetometer,** fundamentally a specially designed magnetic compass, detects minute differences in the magnetic properties of various rock formations, thus helping to find structures which might contain oil.

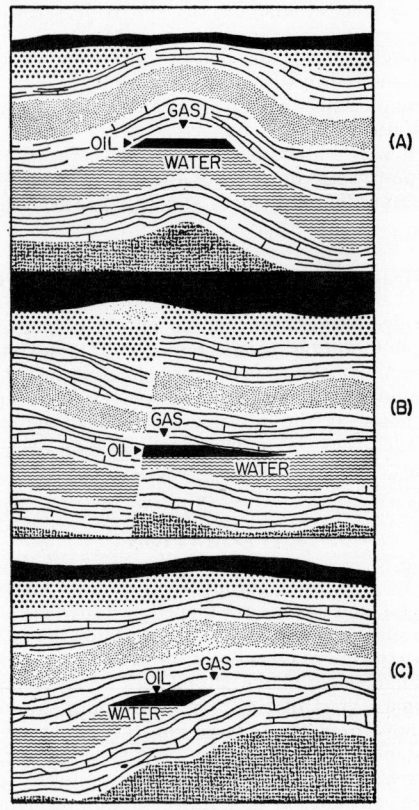

Fig. 1-3. Three types of geological formations in which oil tends to occur: (*A*) the anticline, (*B*) the fault, and (*C*) the stratigraphic trap.   (*Source: Science in the Petroleum Industry, American Petroleum Institute.*)

The magnetometer is carried over a plotted course in a selected area, recording on a strip chart the variations in magnetic intensity of the rock over which it passes.   The record indicates the depth of the basement rock, provided allowances are made for ground surface irregularities.

These data give a clue to places which might conceal anticlines or other oil-favorable structures.   Perhaps even more valuable is that the approximate total thickness of the sedimentary rock can be obtained.   This can save unwarranted expenditures later on more costly geophysics or even the drilling of a wildcat well where the sediment might not contain sufficient oil to pay for the investment.

Most magnetometer surveys used now are by air.   Not only does this method permit very large scale surveys to be made rapidly, but it also is highly accurate and permits surveys over regions where ground work would be impossible because of inaccessibility or, at least, very long and costly. In addition, such a survey is not affected by extraneous magnetic influences on the ground or just below the surface (see Fig. 1-4).

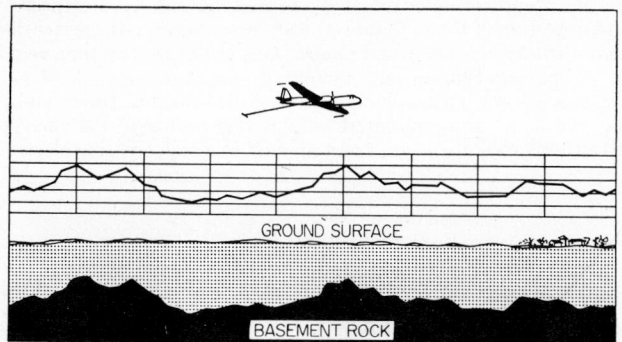

Fɪɢ. 1-4. In an aerial magnetic survey, the magnetometer is towed behind an airplane, its chart recording the depth of basement rock. (*Based on Science in the Petroleum Industry, American Petroleum Institute.*)

The instrument itself consists of a horizontally disposed magnet pivoted near, but not at, its center of gravity, so that the earth's magnetic field creates a torque opposed by that of its gravitation pull. The inclination of the system depends on the strength of the magnetic field, and the degree of change can be recorded by most conventional mechanisms.

Gravitational Method

The magnitude of gravitational pull of underground rock varies directly with its density, although the actual variations are extremely small. The **gravimeter,** a sensitive instrument which detects differences in the pull of gravity, therefore gives an indication of the location and density of underground rock formations.

Most gravimeters operate on about the same principle. A weight is suspended on a sensitive spring, and changes in gravity cause tiny changes in length or in twisting of the spring, which can be magnified optically or electrically and observed.

The instruments are used on land, over water-covered areas (by submerging them in special diving bells or occasionally in submarines), and they can be employed on aircraft.

The geophysicist must correlate gravimeter readings with known "normal" values for gravity at any given location on the earth's surface. Differences from the normal can be caused by geological and other influences, and it is such differences which provide an indication of subsurface structural formations.

Seismic Method

This technique is based fundamentally on the science of seismology which was developed for the study of earthquakes. The geophysicist creates his own "earthquake" by setting off a small and controlled charge of explosive in the bottom of a shallow hole in the ground. The **seismograph** measures the shock waves from the explosion (see Fig. 1-5).

The velocity of travel of a disturbance through the ground is in mathematical relationship to the elasticity and the density of the ground and its components. Furthermore, the sound or shock waves are reflected and refracted by materials of various densities in accordance with the same principles that apply in the reflection and refraction of light waves.

In practice, a series of detectors will be placed in a predetermined pattern surrounding the location for the explosion. The acoustic waves travel outwardly in all directions, like a rapidly expanding sphere. Some are reflected back to the ground surface by denser rock formations below. The wave velocity varies with the type of wave and the type of rock.

The formation depth is determined by measuring the time elapsed between the explosion and detection of the *reflected* wave at the surface. At greater distances from the explosion, *refracted* waves provide more data based on the time required to pass through the underground formation.

A shock wave detector, or geophone, is a modified form of the original earthquake seismograph. In it, a spring-mounted weight, responding to the effect of the shock wave, moves either a coil in a magnet or a magnet in a coil, creating electrical impulses.

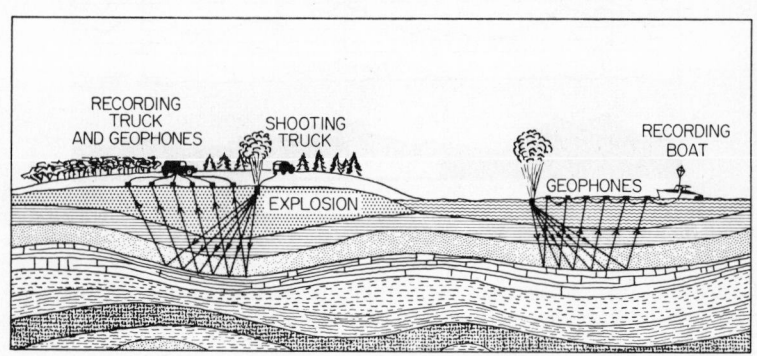

Fig. 1-5. Seismic exploration can be carried out both on land and on water. (*Source: Physics and Petroleum, American Petroleum Institute.*)

These are amplified and transmitted to a recorder, along with the signals from the other geophones.

Signals from each geophone control the movement of a tiny mirror in the recorder, which reflects a narrow light beam onto a moving strip of photosensitive film. When developed, the film shows the paths of the light beams that recorded vibrations picked up by the geophones. From the resulting trace, or graph, can be determined the depth and angle of repose of the rock layers from which the waves were reflected. Many trace records are required in surveying a potential oil area.

Seismic geophysical work is also carried out on the water, greatly aiding the search for oil on the continental shelves and other areas covered by water. A marine seismic project moves continually, with detectors being towed behind the boat at a constant speed and a fairly constant depth. Explosive charges are detonated at a position and time determined by the speed of the boat, so that a continuous survey of the reflecting horizons is obtained.

Several alternatives to the explosive have been developed for marine studies. One is a high-voltage electrical spark (10,000 volts) discharged under water, which transmits an acoustic pulse at short intervals. The device has permitted surveys at depths of 400 ft below the sea bed and in unusual cases as much as 1,200 ft.

Another device is the gas gun—an open-bottom combustion chamber towed behind the boat at about 5-ft depth. A mixture of propane and oxygen is fed into it and ignited by a small electric spark. Much deeper measurements can be attained, depending on the nature of the sea bed and the number of guns used. Average limit of penetration is 3,000 to 4,000 ft, although 8,000 ft has been achieved.

Exploratory Drilling and Well Logging

Another valuable exploration method is geophysical borehole logging. It involves drilling a well and then employing instruments to log or make measurements at various levels in the hole by such means as electrical resistivity, radioactivity, acoustics, or density. In addition, formation, or core, samples are taken for physical and chemical tests.

Electrical Logging

The use of electrical logging is based on the fact that the resistivity of a rock layer is a function of its fluid content. Oil-filled sand has a very high resistivity. The method consists of passing a current between an electrode at the surface and one which is lowered into the hole, the latter being uncased and filled with drilling mud. Any change in the resistivity conditions around the moving electrode will affect the flow of current and the voltage distribution around it. These voltage fluctuations can be measured by a pair of electrodes placed near the moving electrode and lowered with it. Several improvements on the original idea, developed in the 1920's by the Schlumberger brothers, have gone into use.

Radioactive Logging

The natural radioactive properties of many constituents of rock have made it possible to develop and use nuclear radiation detectors in the borehole or even in holes which have already been cased. Two commonly used methods are gamma-ray and neutron logging. In the first case, the natural radiations from the rock are used. With the second, a neutron source is employed to excite the release of radiation from the rock. The neutron source is usually a mixture of beryllium and radium but can be a miniature van de Graff particle accelerator. The neutron method is a means for determining the relative porosities of rock formations; the gamma-ray log helps define shale.

Acoustic Logging

This method is quite similar to surface seismic work. But instead of explosives (which cannot be used), an acoustic pulse generator operated electrically is used. In one instrument, the generator is separated from the receiver by a 5-ft-long acoustic insulator. The design permits automatic selection and recording of the travel times of the onsets of the pulses which travel through the rock wall of the hole as instrument moves down or up. Signals are recorded continuously at the surface, being transmitted through a special cable on which the instrument is suspended. The velocity log provided by the instrument helps to define beds and evaluate formation porosities.

Density Logging

Density can now be logged with a new development utilizing radioactivity. The instrument consists of a radioactive cobalt source of gamma rays and a Geiger counter as a detector which is shielded from the source. The rock formation is bombarded with the gamma rays, some of them being scattered back from the formation and entering the detector. The degree to which the original radiation is absorbed is a function of the density of the rock.

Core Sampling

Test well sampling is another important method used in the search for oil. Well data obtained from the examination of formation samples taken from various depths in the borehole are of considerable value in deciding further exploratory work. These samples can be (1) so-called "cores," which have been taken from the hole by a special coring device, or (2) drill cuttings screened from the circulating drilling mud.

The primary goal of sample examination is to identify the various strata in the borehole and compare their positions with the standard stratigraphic sequence of all the sedimentary rocks occurring in the specific basin in which the hole has been drilled. Examination is both physical and chemical.

A discussion of petroleum exploration cannot possibly cover all the phases of this complex activity within so few pages. Many scientific disciplines are used in the search for oil and gas, the most notable including general geology, stratigraphy, paleontology, palynology, petrology, marine geology and biology, geomorphology, photogeology, geophysics, and geochemistry.

DRILLING AND PRODUCTION

More then 80 per cent of the world's oil wells are drilled today by the rotary method, in which an abrasive bit is revolved at the end of a drill stem. The older cable-tool method, used almost exclusively until 1900, involves raising and dropping a heavy bit and drill stem attached by cable to a cantilever arm at the surface. It pulverizes the rock and earth, gradually forming the hole. The cable-tool system is generally preferred only for penetrating great thicknesses of hard rock at shallow depths and when the producing horizons are expected at the shallow depths.

In the rotary rig, the bit is at the bottom of a column of steel pipe through which is pumped a stream of drilling mud, the purpose of which is to bring cuttings back to the surface and help keep the bit cool. The mud flows out through tiny openings in the bit and returns to the surface in the annular space between the drill pipe and the borehole wall. The weight of the column is usually enough to attain penetration, but it can be augmented by a hydraulic pressure cylinder at the surface.

The Drilling Rig

There are many variations in design. However, all rigs have essentially the same components: a power plant, hoisting and rotating machinery, the drill column, a mud-circulation system, and auxiliaries (see Fig. 1-6).

Internal-combustion engines—gasoline, diesel, or gas—are the most widely used power plants, largely replacing steam, which was used more in the early days. However, where economics dictate, electricity is also used. Power requirements are related mainly to the depth to be drilled. A light rig designed for around 3,000 ft may use 250 to 300 brake horsepower; one designed for 12,000 ft or more may require as much as 2,000 bhp, and of this, 700 bhp might be needed to operate a heavy-duty mud pump.

The hoisting, or draw works, raises and lowers the drill pipe and casing. In a deep well this load can be as much as 200 tons. The height of the derrick depends on the number of joints of drill pipe to be withdrawn as a unit before being unscrewed. Common practice is to use three-joint units of 30 ft each, requiring a rig height of about 130 ft.

The drill-stem assembly is handled by block and tackle, usually five sheaves per block. A separate horizontal shaft on the winch has capstans, or "catheads," on its ends, used with heavy manila ropes for various pulling and moving jobs around the floor of the derrick.

The drill column is rotated by the rotary table, located in the middle of the rig floor above the borehole. The table also grips the drill stem when the hoist is disconnected and pipe sections are inserted or removed. It imparts rotary motion to the drill stem through the "kelly" attached to the upper end of the column, the kelly being a hollow bar of hexagonal or square cross section that fits into a similar-shaped hole in the center of the rotary table.

Couplings between pipe sections, called tool joints, are generally made to standardized specifications of the American Petroleum Institute. Drill pipe is usually $4\frac{1}{2}$ or $5\frac{1}{2}$ in. O.D. However, $2\frac{7}{8}$ and $3\frac{1}{2}$ in. are sometimes used, and for large holes (more rare), $6\frac{5}{8}$ and $8\frac{5}{8}$ in. may be used.

The drilling bit is connected to drill collars at the bottom of the stem. These are thick steel cylinders, 20 to 30 ft long; as many as 10 may be screwed together. They concentrate weight at the bottom of the column and exert tension on the more flexible pipe above, reducing the tendency of the hole to go off line and the drill pipe to fracture.

Bits have many designs, depending on the types of formation to be drilled. Variations include the number of blades, type of metal, and shape of cutting or abrading components. The hardest formations frequently require bits studded with diamonds.

A good mud-circulating system is probably one of the most important parts of a

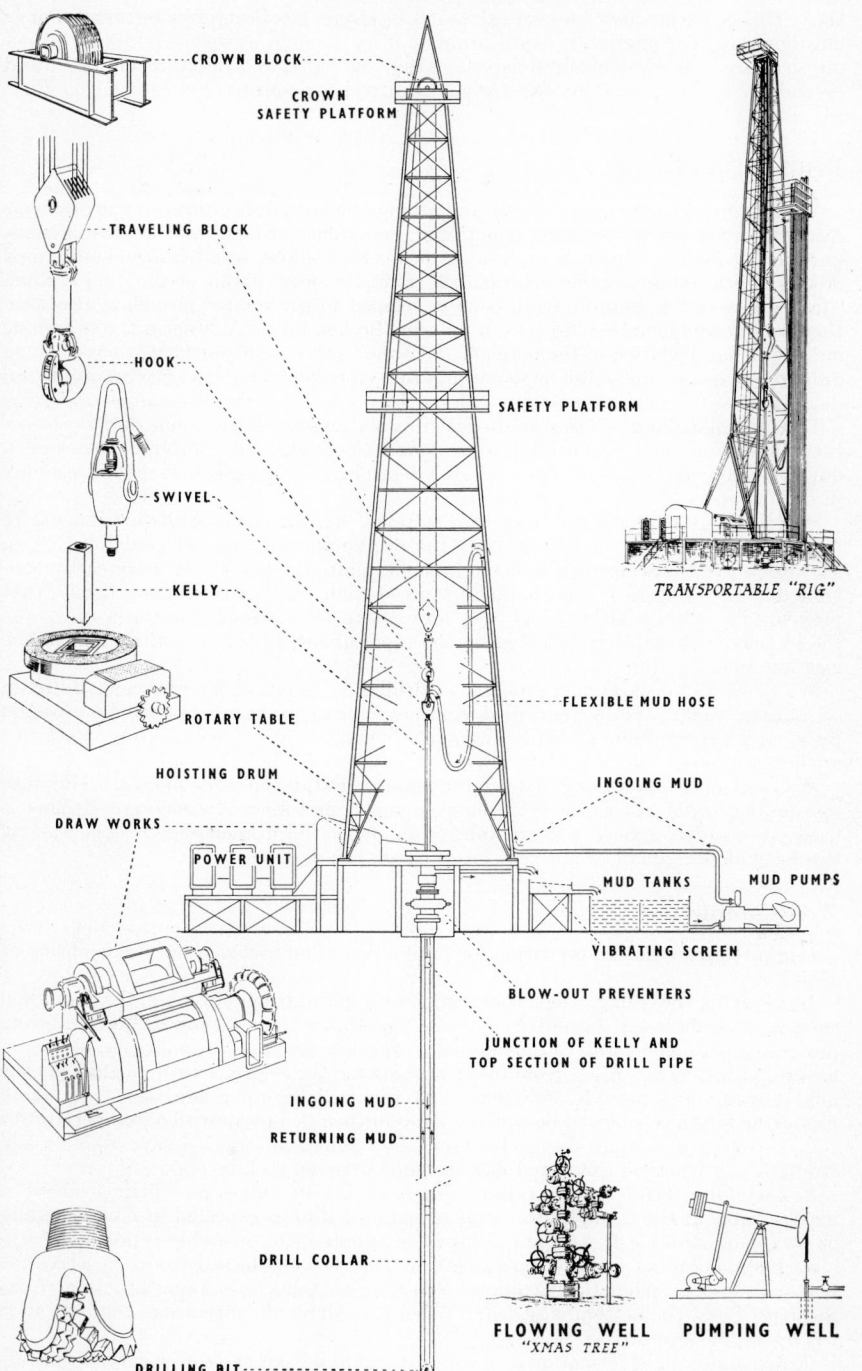

CROWN BLOCK

CROWN
SAFETY PLATFORM

TRAVELING BLOCK

HOOK

SWIVEL

KELLY

ROTARY TABLE

HOISTING DRUM

DRAW WORKS

POWER UNIT

TRANSPORTABLE "RIG"

SAFETY PLATFORM

FLEXIBLE MUD HOSE

INGOING MUD

MUD TANKS   MUD PUMPS

VIBRATING SCREEN

BLOW-OUT PREVENTERS

JUNCTION OF KELLY AND
TOP SECTION OF DRILL PIPE

INGOING MUD

RETURNING MUD

DRILL COLLAR

FLOWING WELL   PUMPING WELL
"XMAS TREE"

DRILLING BIT

FIG. 1-6. Simplified diagram of a rotary drilling rig (not to scale). (*Source: Modern Petroleum Technology, 3d ed., p. 141, The Institute of Petroleum, London, 1962.*)

rig. This is a system which keeps the mud in proper condition, free of rock cuttings or other abrasive materials it might bring up from the hole, as well as retaining proper physical and chemical characteristics. Mud, basically, is a mixture of water and special clays, but it may include various additives for control of viscosity, colloidity, pH, and the like.

Drilling Operations

Drilling an oil well is long, tedious, and often beset with difficulties. Some problems result from the formation being penetrated, occurrence of high-pressure gas, fissures, unexpected high pressures in permeable rock, etc. Others result from metallurgical or mechanical failures in the bit, the drill stem, the draw works, or the mud system. Many tools and techniques have been developed to solve most problems, probably the best known being the "fish" for recovering broken bits. Another is the technique of intentional deviation of the borehole, perhaps to avoid difficult formations or to go around an unrecoverable fish or sometimes to restore hole direction after an accidental deviation.

Directional drilling is also used to reach formations and targets not directly below the penetration point, i.e., drilling from shore to locations under water. A controlled deviation may also be used from a selected depth in an existing hole to attain economy in drilling costs.

Several kinds of tools are used in directional drilling along with instruments to help orient their position and measure the degree and direction of deviation. Two such tools are the whipstock and the knuckle joint, the first being more easily controlled. The whipstock is a gradually tapered wedge with a chisel-shaped base that prevents its rotation after it has been forced into the bottom of an open hole. As the bit moves down, it is deflected by the taper about 5° from the alignment of the existing hole.

Water offers no particular problem in drilling. Today many wells are drilled in offshore locations. Some are drilled from permanent platforms supported on caisson piles, and others from specially designed boats or from huge floating platforms anchored in place for the specific project.

Wells are drilled in water 500 ft deep and more with this type of equipment. Because marine platforms cost so much to build, operate, and service, directional drilling is used on a widely spaced pattern from a single platform anchorage. Also, the rig can be skidded around to different locations on the platform.

Well Completion

Completing a well and preparing for production of oil involve four major phases of activity:

1. Inserting the casing, which comprises one or more strings of tubing, the additional tubing being disposed concentrically. Part of this is done during drilling. Casing provides a permanent wall to the borehole; prevents cave-ins of formations; blocks off unwanted water, oil, or gas from other formations; provides a return passage for the mud stream; and provides for control of the well during production. Access to producing strata is achieved by making holes in the casing wall with a gun perforator.

2. Injecting a cement sheath between the casing and the borehole wall to add strength and block off unwanted flow of fluids between rock layers.

3. Installing the "Christmas tree," an assembly of valves and fittings above a master valve at the casinghead. This is installed if oil is expected to flow naturally or by gas or air lift. If mechanical lift is anticipated, the assembly is not used.

4. Installing a bottom-hole casing setting, the type of which depends on where the bottom is to be. When the bottom is below the producing formation, the gun perforator is used and the bottom is sealed. When it is above the formation, the bottom is left open, but screened.

Occasionally, tight formations are encountered and it becomes necessary to encourage flow. Several methods are used, one of which is to set off small explosions to fracture the rock. If the formation is mainly limestone, hydrochloric acid is sent down the hole to dissolve channels in the rock.   The acid is inhibited to protect the steel casing.   In sandstone, the preferred method is hydraulic fracturing.   A fluid with a viscosity high enough to hold coarse sand in suspension is pumped at very high pressure into the formation, fracturing the rock.   The grains of sand remain, helping to hold cracks open.

In the original version of the fracturing process, the fluid used was a gel.   Today, many fluids are used, including crude oil.   The technique can also be used in limestone, in which case the fluid is hydrochloric acid.

Production

Recovery of oil when a well is first brought in is usually by natural flow, forced by the pressure of gas that occurs with the oil.   A well is always carefully controlled in its "flush" stage of production in order to conserve oil.   This eliminates—except in case of an accident—the "gusher," which exemplifies the industry's early days when much oil was actually wasted.

Once natural pressure is insufficient, artificial lift must be used.   In some low-pressure reservoirs this may be required at the very start.   A mechanical pump is generally used, commonly a reciprocating plunger in a cylinder at the lower end of the tubing.   Power is transmitted to the plunger by a string of solid "sucker rods," operated by a motor-driven eccentric at the wellhead.

A second method of artificial lift is pressure maintenance.   Gas produced with the oil is processed in a field plant to remove condensates—natural gasoline, propane, butane—and the dry gas is returned to the formation through service, or input, wells, maintaining pressure in the reservoir (see Fig. 1-7).

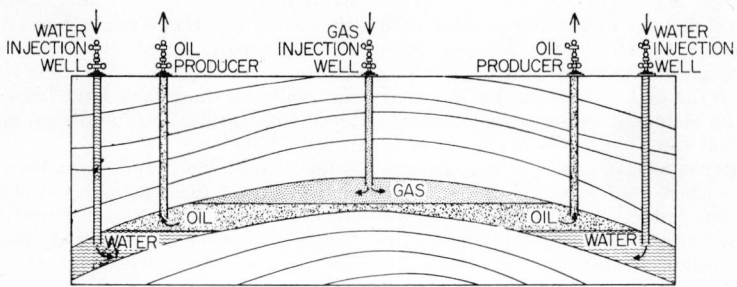

Fig. 1-7. How old oil fields are rejuvenated.   First gas, then later water, is injected to restore pressure and force oil to the surface.   (*Source: Science in the Petroleum Industry, American Petroleum Institute.*)

Secondary Recovery

Several methods have been developed to obtain oil when ordinary production systems are no longer economical.   A most frequently used secondary recovery technique is **water flooding** (Fig. 1-7).   Water is pumped under pressure into the reservoir by means of input wells.   Some passages in the formation are larger than others, however, and the water tends to flow freely through these, bypassing smaller passages where the oil remains.

A partial solution to this problem is possible with another secondary recovery technique, **miscible flooding.**   Liquid butane and propane are pumped into the ground under considerable pressure, dissolving the oil and carrying it out of the smaller passages.   Additional pressure is obtained by using natural gas.

A more recent development, *in situ* **combustion,** is regarded as having high potential, though still in an experimental stage.   It involves injecting air through a service well into the formation and igniting the oil at the sand face (see Fig. 1-8).   Expanding gases help push oil out of the producing well.   Further design studies and control methods are needed before *in situ* combustion can be used on a large scale.

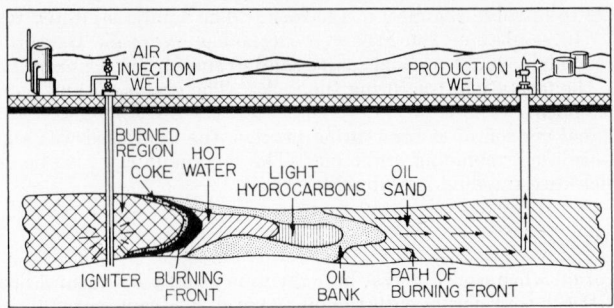

FIG. 1-8. *In situ* combustion involves burning a portion of the oil in place, supplying the energy needed to force the remaining oil to the surface. (*Source: Physics and Petroleum, American Petroleum Institute.*)

Oil-Water Separation

Fluids produced from a well are seldom pure crude oil; they are mixtures of oil, gas, and water which must be separated. The separation units at the surface are quite similar to those in a refinery. Usual practice is to have a single field processing plant serve a number of wells.

Gas is usually separated at as high a pressure as possible, reducing compression costs where the gas is to be used for gas lift or delivered to a pipeline. After gas removal lighter hydrocarbons and hydrogen sulfide are removed as necessary to obtain a crude oil of suitable vapor pressure for transport, yet retaining most of the natural gasoline constituents.

Once the materials which have no particular value to the refiner have been taken out, the volume of crude to be moved is reduced considerably. In addition, possible problems from hydrogen sulfide corrosion are localized.

Water removal must be as complete as possible. With light crudes, simple settling is sometimes sufficient. However, stable emulsions can form with any crude, and even free water will not settle out in a reasonable length of time from more viscous crudes. Dewatering and desalting processes include chemical treating, electrical precipitation, heating, and washing with fresh water or combinations of these.

OTHER SOURCES OF "PETROLEUM"

The term petroleum actually covers many materials other then ordinary crude oil. Some of these may eventually become large, supplementary supplies of petroleum hydrocarbons. Some occur in very large quantities, and economic methods for recovery have been the goal of extensive research for some years. Among the more significant are natural asphalt, oil shale, tar sands, and coal.

Natural asphalt formed by the evaporation of light constituents when crude petroleum seeped to the earth's surface. A number of large deposits have been found, including the famous asphalt lake in Trinidad. Sometimes the asphalt becomes almost as hard as coal and can be mistaken for it. Albertite and gilsonite are such asphalts. A large gilsonite deposit in Utah has been worked for some time.

Oil shale occurs in enormous deposits in Colorado, Utah, and Wyoming, on which much experimental work has been conducted by the United States government. However, no economical recovery process has yet been developed. Oil shales will yield a heavy oil in quantities up to 2 bbl per ton of shale. Sometimes called kerogen, shale oil is unlike either normal crude or asphalt, though it can be converted into a number of conventional refinery products.

The **tar sands** at Athabasca, in northern Alberta, Canada, are one of the major reserves of petroleum in the world, estimated at some 300 billion bbl. Yield is about 1 bbl of liquid from a ton of the sands, which lie very close to the surface. Fairly straightforward processing produces a material similar to normal crude oil. Close to $500 million is scheduled for investment by several groups of companies to develop production of around 135,000 B/D of "crude oil" from the tar sands by the 1970's.

Coal. Probably the best known source of petroleum hydrocarbons is the Fischer-Tropsch process, used by Germany in World War II to synthesize gasoline and other products from coal when the country's crude supply was cut off.

Also called coal hydrogenation, the process is as follows: Synthesis gas (a mixture of carbon monoxide and hydrogen), made by the reaction of steam and incandescent coal, is then used with more hydrogen as the feed gas to the Fischer-Tropsch unit. There, the mixture passes over a cobalt-thoria catalyst at atmospheric pressure and 450°F, producing a mixture of straight-chain olefins and paraffins with a boiling range similar to crude oil. Further processing produces gasolines, lube oils (polymerized olefins), and paraffin waxes. Synthesis gas can also be made from methane and serves as the starting point in petrochemicals manufacture.

High cost has prevented the Fischer-Tropsch process from displacing ordinary crude oil with coal as an economical source of petroleum hydrocarbons. A large plant has been operating for years in South Africa, the so-called SASOL plant; however it has never been expanded because of the economic obstacles.

TRANSPORTATION

Crude oil moves to refineries by pipeline, railroad tank cars, and tankers or barges. Products move from the refinery by all these and by tank trucks as well. Packaged, or less-than-bulk, movement of products is by whatever is most economical or fastest, depending on requirements.

The least expensive method of transporting crude oil is by water; transportation by pipeline costs about three times as much as by tanker, and rail is about three times pipeline. Choice of a method is usually based on both economics and the location of the source of crude with respect to the refinery. Conversely, in the selection of a site for a new refinery, the cost of crude transportation is frequently an important determining factor.

Pipelines

A large part of new pipeline construction in the United States in 1966 was for products, although crude-oil lines still dominate the total mileage of existing systems. In western Europe, however, the pattern is the reverse. Refineries are being built in major market areas, and crude lines will connect them to ocean ports, to which the oil will come by tanker from the Middle East and Africa.

A crude-oil pipeline system starts with the network of pipes in the field, the gathering system, which generally uses 2-in.-diameter pipe. Spur, or feeder, lines connect one or more field systems to the main, or trunk, line. Trunk lines range up to 30 in. (or more, occasionally) and transport oil directly to refineries or to marine terminals for transfer to tankers.

Line pipe is made of steel to standard specifications. Wall thickness varies between ¼ and ⅜ in., seldom exceeding 7⁄16 in. Pipe is seamless or welded. Sections are connected by welding throughout; fittings and valves are also welded.

Oil flows by pump or gravity or a combination of the two. Virtually all pumps are diesel-engine- or electric-motor-driven centrifugals. Early pipelines used steam-driven reciprocating pumps, and these are sometimes still used for very high viscosity oils. Pressures of about 1,200 psi are used for lines up to 12 in. For 24- to 30-in. pipe, pressures of 500 to 800 psi are more common. Among the largest pumps built for this service are those on the Trans-Arabian line, nominally 30 in., each driven by a 6,000-hp turbine and pumping up to 340,000 B/D at 800 psi. Long pipelines require frequent booster stations to restore pressure lost to line friction or gradient variations. A booster is usually installed wherever the line pressure drops to about 50 psi.

Economics determines choice of pipe diameter and number of pumping stations. A given throughput can be attained with a small-diameter line and high head loss or with a large-diameter line and low head loss. Normally, capital cost will be lower for the small diameter, but operating costs will be greater because more pumps and horsepower will be needed.

Pipelines are usually underground in most inhabited regions but are laid on the surface (on concrete blocks or metal supports) in remote areas, such as deserts. Underground pipe requires protection against external corrosion (mostly obtained by coatings and/or cathodic protection), whereas surface lines normally need only paint. On the other hand, underground pipe has no problem of expansion and contraction whereas surface lines do, and allowance for this movement must be made in designing and setting the supports.

Good practice is to design a line to allow for raising its capacity as growth in supply or demand requires. This can be accomplished by starting operations with reduced pump heads at all stations and adding series pumps later or by looping new sections of line into the system later.

Heavy, viscous oils are transported by reducing their viscosity. A variety of methods are used: (1) heating the oil at the input terminal and, as needed, at stations along the line; (2) thermally insulating the line; (3) emulsifying the oil with water and disposing of the water at the outlet terminal; (4) diluting the oil with a solvent, which is separated and returned to the input station through a small line.

Underwater lines are starting to find increased use with the growth in offshore production and the rising use of large tankers which cannot tie up at shore berths. Such lines are coated and wrapped as is buried pipe and then covered with a few inches of concrete, which not only protects the pipe but also provides weight needed to anchor it against tidal currents.

A pipeline can carry separate shipments of oil, one after another, with no barrier between. The batches are controlled by a dispatcher much like trains on a railroad. By knowing the volume of the batch, the line diameter, and the speed the oil is moving, the dispatcher can easily calculate when the head and tail of any given batch will reach any specific location along the line.

Samples taken from the line and tested for color and/or gravity enable operators to keep track of progress of batches. Another method is to insert a small quantity of a radioactive material in the line and detect its movement with a Geiger counter or similar instrument.

Refined products are handled in much the same way as crude. A small amount of mixing between batches does no harm to either as long as batches follow a proper sequence, normally the heavier down through the lighter products. If a product must be out of sequence, a buffer material with a composition permitting its being mixed with products on both sides can be used for the separating "barrier."

Pipeline throughput is usually measured by positive-displacement flowmeters. It is regulated to prevent low pressure on the suction side of the pump or high pressure on the station discharge. Flow is controlled several ways: (1) a control valve in the bypass of a positive displacement pump, (2) a control valve in the discharge of a constant-speed centrifugal pump, or (3) a speed controller on the prime mover of a variable-speed machine.

Instrumentation includes warning or shutoff devices to function when danger limits are reached by such variables as bearing temperatures, crude temperature, vibration, lube-oil temperature and pressure, cooling-water flow and temperature, and gland leakage.

Large pipeline systems are moving rapidly toward more use of automation and unattended station operation. Information is transmitted by telephone cable or radio waves to a central control point, and control signals transmitted back the same way to the individual pumping stations.

Tankers

Intercontinental flow of crude oil is by seagoing tanker. So is coastal trade; i.e., United States Gulf to East Coast. Ocean tankers also carry a considerable volume of refined products.

A tanker today is essentially a conventional ship with a conventional hull form. Inside, however, it is something else again. One authority compared it with a floating tank farm, the tanks crowded together, a powerhouse immediately next to it, and operating personnel living on top of it.

Tanker propulsion is diesel-engine, steam-turbine, or turboelectric drive. Horsepower ranges up to 20,000 have been attained in diesel engines in service now, and engines above 25,000 hp are being considered. A typical 40,000- to 50,000-dwt (deadweight tons) tanker, steam-turbine driven, has turbines of about 16,000 hp and uses steam at 600 psi and 900°F. Tanker speeds are 15 to 16½ knots, particularly ships in the 32,000- to 65,000-dwt range.

The amount of oil carried varies with the tanker tonnage. The famed T2 tankers of World War II, 16,500-dwt vessels, held almost 140,000 bbl. A 7,000-dwt ship carries some 56,000 bbl. A supertanker, in the 106,500-dwt class, will transport about 900,000 bbl. Several 200,000-dwt ships being built in Japan are designed for well over a million barrels, and tankers of up to 300,000-dwt, which would carry 2.2 billion barrels, were being considered in 1966.

The world tanker fleet, based on vessels 500 gross tons and over, totaled about 90 million dwt in 1965. Average tanker size has grown to 67,175 dwt for new ships on order. Larger vessels have lower operating costs per ton-mile, and their use will grow as more adequate port facilities become available.

Tankers are loaded through flexible hose connected to shore pipelines or by means of loading booms made of conventional pipe connected by swivel joints. Loading rates are limited by line size and by venting devices on the tanker. Some large ships can load at 30,000-bbl/hr rates or higher by pumps on the shore. Discharge is usually by ships' pumps, many larger ones having capacities as high as 15,000 bbl/hr. Some big tankers can be unloaded in less than 12 hr.

Tankers are generally divided into two groups: (1) the "dirty" or "black" oil ships, which transport crude oil, fuel oils, and diesel fuels, and (2) "clean" or "white" oil ships, which carry highly refined products. Wherever possible, a ship is kept in one service or the other. However, dirty ships must be cleaned occasionally, especially after many cargoes of certain crudes. Also, interior maintenance and repairs necessitate their being cleaned.

Mechanical washing machines are used; these are essentially rotating nozzles that spray water in virtually all directions at pressures up to 180 psi. After washing, tanks are ventilated and tested carefully to make sure they are free of combustible and/or hazardous gases.

A major problem in tanker operation is corrosion inside cargo tanks, caused by the repeated use of sea water for cleaning and as ballast on return voyages. A most effective technique to combat corrosion is cathodic protection, with permanent systems installed within the ship.

Some specialized tanker designs have been developed in recent years, the most significant of which is very probably the liquefied gas tanker. Equipped with insulated and refrigerated tanks, it will transport such materials as liquid natural gas, LP gases, or petrochemicals such as ammonia.

Another form of water transport is the barge, usually used on long, wide, and navigable rivers such as the Mississippi, the Thames, the Rhine, and the like. Barges usually carry products rather than crude.

Railroads and Highways

Petroleum products move to market in relatively small volume or in nonbulk form in tank cars or tank trucks or in drums or cans loaded into trucks. The transportation of products by pipeline and tanker has been discussed previously.

There are a number of differences between a tank car and a tank truck. Tank-car capacities are 6,000, 8,000, and 10,000 gal (U.S.), with special service cars ranging up to 12,000 gal. Tank trucks are 2,000 or 4,000 and occasionally 6,000 gal. Both may have compartments for smaller shipments of different products. Tank cars usually carry crude oil; tank trucks seldom do. Tank cars may be fitted with pipe coils for steam heating highly viscous materials. Some are also thermally insulated. Trucks seldom have coils but may be insulated. Tank cars are primarily for long-distance hauling to large-volume purchasers or distributing terminals for subsequent transfer of product to road vehicles.　Or they may serve industrial customers with rail sidings at their plant sites.

Trucks handle distribution in local marketing areas.　One type hauls gasoline to service stations; another carries fuel oil to industrial customers; another home-heating oil to domestic users.　Still another, of special design, has been developed for fueling planes at airports.

A recent development finding wider application is a road-rail combination, called "piggy-back."　It involves putting one or more tank-truck trailers on a railroad flat car, sending several such cars to various destinations by rail, and then hauling the individual trailers by a truck tractor to local customers from that point.

The distribution by vehicle of small packages—cartons, drums, cans, and the like—falls more into the field of materials handling.　It requires palletizing, the use of off-the-road vehicles such as fork-lift trucks, mechanized and automated filling machinery, roller and other types of conveyors, and a whole host of special equipment. Palletized containers, or large cartons of cans, are loaded into vans or railroad boxcars or similar vehicles for shipment to the customer.

Choice of transport is governed by several factors, only one of which is economics. Packaged products like greases, waxes, lube oils, and other specialties require truck or rail or both.　Larger volume products such as gasoline moving to service stations use tank trucks.　Domestic heating oil also moves from refinery or from terminal to the homeowner by tank truck.

REFERENCES

1. Ball, Douglas, and Turner, Dan S., *This Fascinating Oil Business*, The Bobbs-Merrill Company, Inc., New York, 1965.
2. Bell, H. S., *American Petroleum Refining*, D. Van Nostrand Company, Inc., Princeton, N.J., 1959, 4th ed.
3. Brantly, J. E., ed., *Rotary Drilling Handbook*, Palmer Publications, New York, 1961, 6th ed.
4. Dunstan, A. E., and Brooks, B. T., eds., *The Science of Petroleum*, Oxford University Press, London, 1950, vol. V, Part I (Crude Oils; Chemical and Physical Properties).
5. Institute of Petroleum, *Modern Petroleum Technology*, the Institute, London, 1962, 3d ed.
6. Institute of Petroleum, *The Oil Industry Tomorrow*, the Institute, London, 1962.
7. Kalichevsky, V. A., *The Amazing Petroleum Industry*, Reinhold Publishing Corporation, New York, 1943.
8. Purdy, G. A., *Petroleum, Prehistoric to Petrochemicals*, Copp Clark Publishing Co., Toronto, Canada, 1957.
9. Sachanen, A. N., *Conversion of Petroleum*, Reinhold Publishing Corporation, New York, 1948, 2d ed.

Section 2

EVALUATION OF CRUDE OILS

By C. M. McKINNEY

Analytical Chemist
Bureau of Mines Petroleum Research Center
Bartlesville, Okla.

INTRODUCTION

Crude oils are exceedingly complex mixtures, consisting predominantly of hydrocarbons and containing sulfur, nitrogen, oxygen, and metals as minor constituents. Although sulfur and at least one of the metals (mercury) have been reported as occurring in elemental form in some crude oils, most of the minor constituents occur in combination with carbon and hydrogen. The physical and chemical characteristics of crude oils and the yields and properties of products or fractions prepared from them vary considerably and are dependent upon the concentration of the various types of hydrocarbons and minor constituents present.

Knowledge of the characteristics of individual crude oils is essential to the achievement of maximum efficiency in refining. Application of such knowledge and the great flexibility of modern processing methods can prevent or minimize variations in the characteristics of finished products that can result from differences in crude-oil feedstocks.

Some types of crude oil have economic advantages as sources of fuels and lubricants with highly restrictive characteristics because they require less specialized processing than that needed for production of the same products from other types of crude oil. Also, some crude oils contain unusually low concentrations of components that are desirable fuel or lubricant constituents, and the production of these products from such crude oils may be economically infeasible.

Knowledge of the quantity and type of sulfur, nitrogen, and metals in crude oils is desirable. Raw products from high-sulfur crude oils often require extensive treatment for removal or alteration of sulfur compounds to improve corrosion characteristics and odor and, for gasolines, to improve combustion characteristics. Nitrogen compounds and metallo-organic compounds may cause deleterious effects to refinery processing catalysts if they are not removed prior to catalytic processing. Nitrogen compounds usually are present in lesser quantity than sulfur compounds; metallo-organic compounds are concentrated in the high-boiling or residual portion of the crude oil.

Proper interpretation of data resulting from crude-oil analyses requires knowledge of how these data are obtained and understanding of their significance. The remainder of this chapter includes brief discussions, with references that provide elaboration, on the data most commonly used in characterizing crude oils, how the data are obtained, certain calculated functions derived from the primary data, interpretation of crude-oil analysis data, and major uses of the data.

2–1

DISCUSSION

Crude-oil analytical methods usually may be classified as either preliminary or comprehensive. The most widely used of the preliminary type is the Bureau of Mines routine method.

Because of the extreme variation in characteristics of crude oils, the multitude of products manufactured from them, and variations in processing and conversion techniques employed by different refiners, development of a single, standardized, comprehensive analytical procedure has been considered infeasible. To satisfy specific needs with regard to type of products being produced and processing techniques being employed, most refiners have developed their own methods of crude-oil analysis and evaluation. Such methods are considered proprietary and are not widely applicable; consequently, detailed descriptions of the methods and correlations of the analytical

Table 2-1. Test Procedures Used in Bureau of Mines Routine Crude-oil Analysis

Material analyzed	Property determined	Test procedures and conditions*
Crude oil	Specific gravity	Bottle pycnometer,† at 60°/60°F
	Color	Opaque oils—comparison with color standards. Transparent or translucent oils—ASTM D 1500
	Sulfur content	ASTM D 129 (Bomb Combustion)
	Nitrogen content	Kjeldahl‡
	Viscosity	ASTM D 88 (Saybolt). ASTM D 445 (Cannon-Fenske viscometer—reverse-flow type for opaque liquids). If the Saybolt Universal viscosity at 100°F is less than 50 sec, no other viscosity is determined. If the viscosity at 100°F is 50 to 99 sec, viscosity is also determined at 77°F. If the viscosity at 100°F is 100 sec or more, the viscosity is also determined at 130°F
	Pour point	ASTM D 97
Fractions distilled at atmospheric pressure	Specific gravity	Westphal balance,† at 60°/60°F
	Refractive index	ASTM D 1218, n_D and n_g at 20°C
Fractions distilled at 40 mm Hg pressure	Specific gravity	Pipette pycnometer,† at 100°/60°F
	Refractive index	ASTM D 1218, n_D and n_g at 20°C of transparent and light-colored fractions with cloud points above 20°C
	Viscosity	ASTM D 445 (Cannon-Fenske viscometer for transparent liquids)
	Cloud point	ASTM D 97
Residuum	Specific gravity	Bottle pycnometer,† at 60°/60°F
	Carbon residue	ASTM D 524 (Ramsbottom)

* ASTM procedure described in reference 19.
† Described in reference 2.
‡ Described in reference 20.

data with yields and properties of products obtained during refinery processing are not generally available in the technical literature.

Both preliminary and comprehensive crude-oil analysis methods consist of (1) determination of the more important characteristics of the crude oil including API gravity, sulfur content, nitrogen content, viscosity, and pour point; (2) analytical distillation of the crude oil; and (3) determination of characteristics of the distillate fractions, blends of fractions, and the residuum. The major differences between the preliminary and comprehensive methods of analyses are in the quantity of crude oil distilled, separation efficiency of the distillation apparatus, minimum distillation pressure attained, maximum distillation temperature attained, number of fractions and blends prepared, and the amount and type of analytical data obtained for the distillate fractions, blends, and residuum.

BUREAU OF MINES CRUDE-OIL ANALYTICAL METHOD

The Bureau of Mines method of crude-oil analysis was developed in the period 1915 to 1920,[1] and since that time only minor changes have been made in the proce-

Table 2-2. Calculation Procedures Used in Bureau of Mines Routine Crude-oil Analysis

Determined property	Calculation
Specific gravity, 60°/60°F	Conversion to °API $$°API = \frac{141.5}{sp\ gr,\ 60°/60°F} - 131.5$$ Correlation index. Calculated as described in reference 5
Specific gravity, 100°/60°F	Conversion to specific gravity, 60°/60°F. Use appropriate factor from tables in ASTM-IP Petroleum Measurement Tables. ASTM D 1250
Viscosity, cs	Conversion from kinematic viscosity, centistokes to Saybolt Universal viscosity, sec. ASTM D 446
Refractive index, n_D and n_g, at 20°C	Specific dispersion $= \dfrac{\text{refractive index, } n_g - \text{refractive index, } n_D}{sp\ gr,\ 60°/60°F - 0.004} \times 10^4$ Note: Specific gravity, 60°/60°F − 0.004, is approximately equivalent to density at 20°/4°C for most hydrocarbon mixtures
Carbon residue, wt %	Conversion from wt % by Ramsbottom method (ASTM D 524) to wt % by Conradson method (ASTM D 189) ASTM D 524

Volume %	Volume % and certain properties of each fraction are used to calculate an approximate summary for each analysis as follows:
	Light gasoline: Vol % in fractions 1–3
	Gasoline and naphtha: Vol % in fractions 4–7
	Kerosine: Vol % in fractions 8–10 with specific gravities 0.825 or lower
	Gas oil: Vol % in fractions 8–10 with specific gravities 0.826 or higher plus distillate in fractions 11 and higher with viscosity below 50 sec at 100°F
	Lubricating distillates:
	  Nonviscous: Vol % with viscosity between 50 and 100 sec at 100°F
	  Medium: Vol % with viscosity between 100 and 200 sec at 100°F
	  Viscous: Vol % with viscosity above 200 sec at 100°F

Notes: All specific gravities and combination of fractions except for the lubricating distillates are weighted averages. Specific gravities for the lubricating distillates indicate the range and are obtained from graphical plots of specific gravity and viscosity vs. midpoint sum per cent of fractions 11 through 15.

dure.[2] During the intervening years some tests have been added, improvements have been made in analytical procedures, and the form of presentation has been changed slightly, but the early analyses can be compared directly with those made recently.

An outline of the procedure for obtaining a Bureau of Mines routine crude-oil analysis is shown in Fig. 2-1, test methods used during the procedure are shown in Table 2-1, and calculations used are summarized in Table 2-2.

Examination of Crude Oil

Gravity, color, sulfur content, nitrogen content, viscosity, pour point

Distillation at Atmospheric Pressure

10 distillate fractions cut at 25°C intervals from 50 to 275°C
Volume, gravity, refractive index—determined for each fraction

Distillation at 40 Mm Hg Pressure

5 distillate fractions obtained from atmospheric-distillation residuum, cut at 25°C intervals from 200 to 300°C
Volume, gravity, refractive index, viscosity, cloud point—determined for each fraction

Analysis of Residuum

Weight, gravity, carbon residue

Calculations

Volume and gravity of light gasoline, gasoline and naphtha, kerosine distillate, gas oil, lubricating distillates, residuum
Correlation index of fractions
Specific dispersion of fractions

Fig. 2-1. Outline of Bureau of Mines routine crude-oil analysis procedure.

Approximately 8,000 analyses of both United States and foreign crude oils have been made during the past 40 years by the Bureau of Mines, and a bibliography of published analyses has been prepared.[3] The latest comprehensive compilation includes 492 analyses of crude oils from 470 important oil fields in the United States.[4] Summaries of Bureau of Mines analyses of 401 crude oils from large United States oil fields and 96 foreign oil fields are included in Tables 2-3 and 2-4, respectively. Following Table 2-4, Tables 2-5 and 2-6 contain complete analyses of crude oils from 63 major United States and foreign oil fields.

It must be borne in mind that the routine analysis is not adapted for and was not specifically designed for the evaluation of a crude oil in terms of the products an individual refiner can manufacture. It is more a "snapshot" of the crude oil and indicates possibilities that may be explored by more complete analysis.

Comparison of routine analyses of samples probably has greatest utility, particularly when refining data are available for at least one of the samples. To aid in such comparisons the correlation index (CI),[5] based upon the specific gravity and boiling point of hydrocarbons in each of the distillate fractions, was developed. The CI of a fraction composed essentially of paraffins will be small, whereas fractions containing larger percentages of naphthenes or aromatics or both will have increasingly greater CI values. Graphical plots of CI vs. fraction number for two or more oils serve as a simple means of comparing the approximate composition of the distillates.[6,7]

Specific dispersion data are included in the analyses to provide information regarding the aromatic content of the distillates. Methods of calculating the aromatic content of petroleum-distillate fractions using specific dispersion data have been described.[8,9] These methods are based on the fact that the specific dispersions for naphthenes and paraffins are nearly equal and constant throughout the gasoline and kerosine boiling range whereas the specific dispersions of aromatics, although not constant, are considerably higher. A simplified application of the method for calculating the aromatic content of fractions 2 through 12 has been reported.[4]

By comparison of data in the approximate summaries of the analyses at the conclusion of this chapter, basic similarities or differences in yields and physical characteristics of comparable distillates may be observed. Similar comparisons may be made of the data in Tables 2-3 and 2-4.

Utilization of data to estimate the content of various products has been described in a series of journal articles.[6,10-13] Compilations of hydrocarbon-type data calculated from routine crude-oil analyses for fields in various geographical areas of the United States also have been published.[14] Published sulfur-content data[15-17] show how much of the crude-oil production in several geographical areas in the United States falls in each of several sulfur-content ranges. Analogous data regarding nitrogen content of United States crude oils have been published.[18]

REFERENCES

1. Dean, E. W., H. H. Hill, N. A. C. Smith, and W. A. Jacobs, "The Analytical Distillation of Petroleum and Its Products," *Bureau of Mines Bull.* 207, 82 pp., 1922.
2. Smith, N. A. C., H. M. Smith, O. C. Blade, and E. L. Garton, "The Bureau of Mines Routine Method for the Analysis of Crude Petroleum," *Bureau of Mines Bull.* 490, 82 pp., 1951.
3. Blade, O. C., "Bibliography of Reports Containing Analyses of Crude Oils by the Bureau of Mines Routine Method," *Bureau of Mines Inform. Circ.* 7921, 181 pp., 1959.
4. McKinney, C. M., and E. L. Garton, "Analyses of Crude Oils from 470 Important Oilfields in the United States," *Bureau of Mines Rept. Invest.* 5376, 276 pp., 1957.
5. Smith, H. M., "Correlation Index to Aid in Interpreting Crude-Oil Analyses," *Bureau of Mines Tech. Paper* 610, 34 pp., 1940.
6. Smith, N. A. C., H. M. Smith, and C. M. McKinney, "Refining Properties of New Crudes. Part 1—Significance and Interpretation of Bureau of Mines Routine Crudeoil Analysis," *Pet. Proc.*, vol. 5, pp. 609–614, 1950.
7. Smith, H. M., and John S. Ball, "Crude Oil Characterization by the Bureau of Mines Routine Method," *Pet. Engr.*, vol. 26:2, pp. C12–C14, 1954.
8. Grosse, A. V., and R. C. Wackker, "Quantitative Determinations of Aromatic Hydrocarbons by New Methods," *Ind. Engrg. Chem., Anal. Ed.*, vol. 11, pp. 614–624, 1939.
9. Gooding, R. M., N. G. Adams, and H. T. Rall, "Determination of Aromatics, Naphthenes and Paraffins by Refractometric Methods," *Ind. Engrg. Chem., Anal. Ed.*, vol. 18, pp. 2–13, 1946.
10. Smith, N. A. C., H. M. Smith, and C. M. McKinney, "Refining Properties of New Crudes. Part 2—Characteristics of Crude Oils from the Scurry County, Texas, Area," *Pet. Proc.*, vol. 5, pp. 730–734, 1950.
11. Ball, John S., W. J. Wenger, and M. L. Whisman, "Refining Properties of New Crudes. Part 3—Oils from 20 New Discoveries in the Rocky Mountain Region Analyzed," *Pet. Proc.*, vol. 5, pp. 842–846, 1950.
12. Smith, N. A. C., H. M. Smith, and C. M. McKinney, "Refining Properties of New Crudes. Part 4—Crude Oils from Recent Discoveries in Kansas," *Pet. Proc.*, vol. 5, pp. 960–964, 1950.
13. Smith, N. A. C., H. M. Smith, and C. M. McKinney, "Refining Properties of New Crudes. Part 5—Characteristics of Crude Oils from the 'Golden Trend' in Oklahoma," *Pet. Proc.*, vol. 6, pp. 980–990, 1951.
14. Smith, Harold M., "Composition of United States Crude Oils," *Ind. Engrg. Chem.*, vol. 44, pp. 2577–2585, 1952.
15. Smith, H. M., and O. C. Blade, "Trends in Supply of High-sulfur Crudes in the U.S.," *Oil & Gas J.*, vol. 48, pp. 73–78, 1947.
16. Blade, O. C., "High-sulfur Crude Oils—Trends in Supply. II. U.S. Data of 1945–47," *Pet. Ref.*, vol. 28, pp. 151–152, 1949.
17. Shelton, E. M., C. M. McKinney, and O. C. Blade, "Domestic Crudes Contain Less Sulfur," *Pet. Ref.*, vol. 36, pp. 257–260, 1957.
18. Ball, John S., and W. J. Wenger, "How Much Nitrogen in Crudes from Your Area?" *Pet. Ref.*, vol. 37, pp. 207–209, 1958.
19. American Society for Testing and Materials, *ASTM Standards on Petroleum Products and Lubricants (with Related Information)*, 1961, 38th ed., vol. 1.
20. Lake, G. R., Phillip McCutchan, Robin Van Meter, and J. C. Neel, "Effects of Digestion Temperature on Kjeldahl Analyses," *Anal. Chem.*, vol. 23, pp. 1634–1638, 1951.

Table 2-3. Properties of United States Crude Oils

Item No.	State / Field (Formation, age)[1]	Gravity, °API	Sulfur, wt. per cent	Viscosity, SUS at 100°F	Carbon residue of residuum, wt. per cent	Gasoline and naphtha		Kerosine distillate		Gas oil distillate		Lubricating distillate		Residuum	
						Per cent	Grav-ity, °API	Per cent	Grav-ity, °API	Per cent	Grav-ity, °API	Per cent	Gravity, °API	Per cent	Grav-ity, °API
	Alabama														
1	Citronelle (Rodessa, L. Cre.)	43.6	0.38	40	6.7	34.2	65.6	20.7	47.2	9.6	38.2	17.1	36.6-28.4	16.7	16.5
	Alaska														
2	Swanson River (Hemlock, Eoc.)	29.7	0.16	61	22.3	27.4	58.7	9.1	42.1	15.4	34.6	16.7	31.1-22.5	31.4	7.8
	Arkansas														
3	Magnolia (Reynolds-Smackover, Jur.)	38.4	0.90	38	6.7	32.2	59.2	10.5	43.8	22.2	35.2	11.4	31.5-25.9	20.5	17.3
4	Midway (Smackover, Jur.)	36.6	1.36	42	11.3	30.8	62.6	10.0	43.4	16.0	35.2	16.2	31.0-24.2	24.9	13.6
5	Schuler (Jones & Cotton Valley, Jur.)	32.8	1.55	52	12.0	26.4	60.2	9.5	43.2	15.5	35.8	16.3	31.3-24.5	31.7	13.2
6	Smackover (U. Cre.)	22.5	2.10	220	8.5	11.2	49.0	.....	.....	20.6	33.0	20.2	27.9-21.3	47.0	12.2
	California														
7	Belgian Anticline (Oceanic, Olig.)	35.0	0.59	40	8.3	35.5	56.4	5.5	41.1	21.2	35.4	15.5	31.3-22.5	22.2	11.9
8	Belridge, South (Tulare, Plio.-Pleist.)	15.0	0.23	2,440	11.3	2.1	44.3	.....	.....	17.2	30.4	29.6	24.3-12.0	49.4	7.1
9	Brea Olinda (Mio.)	24.0	0.75	135	14.2	19.4	51.3	.....	.....	20.7	33.8	20.4	28.9-16.8	37.7	7.8
10	Buena Vista (27-B Basal Etchegoin, Plio.)	30.6	0.59	46	12.1	33.9	54.2	.....	.....	21.2	32.5	15.0	27.1-18.4	28.0	11.0
11	Castaic Junction (Zone 10, Mohnian, Mio.)	19.0	3.40	1,230	9.4	17.1	57.4	.....	.....	16.0	33.2	13.1	27.9-19.2	53.3	5.6
12	Cat Canyon, West (Los Flores, Mio.)	17.5	5.07	3,000	13.8	13.3	58.9	2.6	42.1	14.0	33.2	13.5	27.5-20.0	55.7	4.8
13	Coalinga, East (Main Gatchell, Eoc.)	28.8	0.31	67	9.9	23.2	52.3	.....	.....	26.0	34.4	21.6	28.2-22.5	28.4	11.6
14	Coalinga Nose (Gatchell, Eoc.)	31.5	0.25	48	8.0	25.1	53.0	.....	.....	32.2	33.8	16.2	28.4-22.5	24.8	13.5
15	Coalinga, West (Temblor, Mio.)	20.2	0.55	195	10.9	6.9	45.2	.....	.....	26.5	29.7	28.9	24.5-14.4	36.5	10.3
16	Coles Levee, North (Mio.)	34.0	0.39	43	10.3	35.0	56.2	.....	.....	21.0	35.0	16.3	29.7-19.7	25.3	11.7
17	Coyote, West (Emery, Repetto, Plio.)	32.3	0.82	50	11.6	29.9	53.7	4.6	40.9	18.8	34.4	16.9	30.2-21.5	29.2	11.4
18	Cuyama, South (Dibblee, Mio.)	32.5	0.42	49	6.5	30.1	56.4	4.6	40.0	17.9	34.4	16.6	29.7-21.8	27.6	10.4
19	Cymric (McKittrick Group, Tulare, Plio.-Pleist.)	12.7	1.16	6,000	11.0	.....	.....	1.8	44.7	11.2	31.0	26.6	25.4-15.0	59.4	6.8
20	Dominguez (Plio.-Mio.)	29.9	0.40	60	12.2	26.4	52.7	5.5	40.7	19.5	34.6	17.9	29.9-21.5	29.9	11.0
21	Edison (Chanac, Jur.)	25.2	0.20	115	11.2	20.8	51.3	.....	.....	21.5	32.7	19.5	28.6-19.5	38.0	10.1

No.	Field														
22	Elk Hills (Shallow, U. Plio.)	22.8	0.68	135	4.6	11.1	49.9	...	...	28.7	31.7	22.0	25.7-17.8	37.4	11.1
23	Fruitvale (Chanac, Plio.-Mio.)	17.5	0.93	1,750	9.8	0.7	43.4	...	...	19.0	30.8	25.4	26.6-18.1	54.4	11.3
24	Gosford, East (Middle & Lower-Stevens, Mio.)	34.0	0.57	51	8.0	34.6	57.7	5.1	40.2	19.6	35.6	16.6	30.0-20.2	25.0	10.0
25	Greeley (Rio Bravo-Vedder, Mio.)	37.2	0.31	41	11.3	37.3	57.4	4.4	40.6	18.5	34.6	16.5	30.6-22.0	19.9	12.0
26	Guijarral Hills (Leda, Olig.)	36.8	0.63	40	10.9	37.7	58.4			20.3	35.0	16.1	30.8-22.6	20.7	12.0
27	Honor Rancho (Wayside, U. Mio.)	37.6	0.40	37	8.0	37.5	54.9			27.0	35.0	14.7	30.6-23.1	17.8	13.9
28	Huntington Beach (S. Main area, Mio.)	22.6	1.57	210	6.2	20.0	52.3			18.6	33.4	17.9	28.4-17.0	43.1	7.9
29	Inglewood (Vickers, L. Mio.)	18.1	2.50	680	12.3	11.7	48.5			19.3	31.9	17.2	26.4-17.9	51.6	7.6
30	Kern Front (Chanac, Plio.-Mio.)	14.8	0.85	5,100	10.0					13.7	31.0	29.1	25.4-15.0	55.8	8.9
31	Kern River (Kern River, Plio.-Pleist.)	12.6	1.19	6,000	10.0					9.6	29.9	23.3	24.5-15.1	65.7	8.2
32	Kettleman North Dome (Temblor, Mio.)	34.0	0.40	44	12.3	33.5	54.2	6.1	40.6	20.6	34.8	20.5	25.6-19.7	19.1	11.7
33	Long Beach (Alamitos, Repetto, Plio.)	22.6	1.29	208	6.3	13.7	51.3	4.6	40.0	17.9	34.2	24.1	29.1-17.8	38.7	8.7
34	Midway-Sunset (Plio.-Pleist.)	21.6	0.89	210	5.7	14.8	51.1			23.5	32.1	20.7	25.9-16.0	40.6	9.2
35	Montalvo, West (Colonia, Sespe, Olig.)	17.3	4.10	7,648	7.9	14.6	55.7	2.7	40.6	12.6	34.6	19.2	28.2-19.4	50.0	2.8
36	Mount Poso (Vedders, L. Mio.)	16.0	0.68	1,900	10.9					13.6	30.6	33.8	26.4-15.3	52.0	10.3
37	Newhall-Potrero (Modelo, Mio.)	32.7	0.56	46	11.4	33.9	57.2	4.5	40.4	16.8	34.4	14.9	30.4-22.1	28.4	10.1
38	Oxnard (McInnes, Sespe, Olig.)	25.7	1.72	95	7.6	20.3	56.2	3.8	41.1	17.3	35.0	19.6	29.3-19.0	36.0	7.1
39	Richfield (Kraemer, Mio.)	22.6	1.86	230	11.5	18.9	52.0			20.7	34.8	17.3	28.8-18.9	42.9	7.8
40	Rincon (Plio.)	28.2	1.40	80	11.0	26.3	56.7	3.8	41.1	16.4	34.8	18.4	30.6-20.3	34.1	8.5
41	Rio Bravo (Rio Bravo, Mio.)	38.6	0.35	38	7.1	40.3	56.9	4.6	40.2	19.2	35.2	14.9	31.3-24.2	18.6	15.0
42	Russell Ranch (Dibblee, Vaqueros, Mio.)	35.2	0.35	43	13.1	32.9	58.4			24.7	34.6	17.4	28.0-22.8	23.0	11.4
43	San Ardo (Lombardi, Mio.)	11.1	2.25	6,000	4.6					11.7	30.2	14.7	25.6-17.3	72.0	6.6
44	San Emidio Nose (Reef Ridge, Mio.)	29.7	0.83	59	6.8	27.3	56.2	4.4	41.3	18.2	35.0	19.1	29.9-20.5	30.3	10.6
45	Sansinena (Mio.)	28.6	0.87	63	9.8	28.1	52.5			22.5	34.0	18.3	29.5-20.0	30.9	10.0
46	Santa Fe Springs (Buckbee, Plio.)	32.8	0.33	47	10.1	28.5	52.5	6.3	40.6	22.7	34.4	17.5	30.4-23.1	23.2	12.0
47	Santa Maria Valley (Monterey, Mio.)	14.7	4.99	6,000	14.8	11.3	53.0			15.3	32.5	11.3	25.4-18.2	61.3	4.8
48	Seal Beach (McGrath, Mio.)	31.7	0.55	52	10.8	29.0	55.9	4.9	41.5	19.5	34.8	18.1	30.6-20.7	28.1	11.3
49	South Mountain (Sespe, Olig.)	23.3	2.79	220	13.5	20.4	58.2	3.4	41.1	15.5	33.6	14.1	28.9-20.2	46.0	7.5
50	Tejon, North (Zemorrian, 1-R, Mio.)	40.0	0.16	35	9.8	47.8	54.2	5.7	39.6	15.5	34.0	15.5	28.9-21.8	8.9	12.0
51	Torrance (Del Amo, Mio.)	23.8	1.84	160	13.2	17.9	52.5			19.4	34.0	18.4	28.4-20.0	41.9	9.4
52	Ventura (Pico-Repetto, Plio.)	31.3	0.94	56	13.5	30.0	57.4	4.1	40.4	21.0	34.6	16.3	31.1-20.7	31.5	10.9
53	Wheeler Ridge (Eoc.)	37.0	0.29	38	7.8	34.9	55.4			16.3	34.4	17.2	29.9-23.5	17.3	15.0
54	Wilmington (Harbor area, Terminal, Mio.)	22.3	1.33	210	6.3	16.7	52.5			28.1	33.0	20.7	27.1-17.1	42.1	8.7
	Colorado														
55	Adena (Dakota "J", Cre.)	44.7	<0.10	36	2.9	37.1	60.5	18.5	42.8	12.8	37.6	12.5	36.8-31.1	15.9	21.8
56	Rangely (Weber, Penn.)	34.8	0.56	48	7.6	26.1	59.5	10.3	41.5	15.3	34.6	20.3	32.1-24.5	26.5	15.6
57	Wilson Creek (Morrison, Jur.)	48.1	0.12	33	4.8	49.9	64.8	11.0	43.6	15.0	37.4	12.3	34.6-28.4	10.8	20.8

Table 2-3. Properties of United States Crude Oils (Continued)

Item No.	State / Field (Formation, age)[1]	Gravity, °API	Sulfur, wt. per cent	Viscosity, SUS at 100°F	Carbon residue of residuum, wt. per cent	Gasoline and naphtha Per cent	Gasoline and naphtha Gravity, °API	Kerosine distillate Per cent	Kerosine distillate Gravity, °API	Gas oil distillate Per cent	Gas oil distillate Gravity, °API	Lubricating distillate Per cent	Lubricating distillate Gravity, °API	Residuum Per cent	Residuum Gravity, °API
	Illinois														
58	Clay City (Miss.)..........	38.6	0.19	43	8.2	32.8	59.2	10.1	42.3	14.0	34.2	17.8	30.2-21.5	21.6	18.9
59	Dale (Aux Vases, Miss.)........	36.4	0.15	49	4.3	29.1	58.2	10.5	42.8	13.7	36.2	15.9	33.2-25.9	30.0	20.3
60	Lawrence (Bridgeport, Penn.)......	35.6	0.21	46	8.4	29.8	60.0	9.5	41.9	13.7	35.6	16.3	32.8-24.2	28.3	15.9
61	Loudon (Bethel, Miss.)............	36.2	0.22	45	7.6	30.2	59.5	10.2	42.6	14.2	36.4	15.7	33.6-25.9	29.0	16.4
62	New Harmony (McClosky, Miss.)	36.0	0.23	45	7.7	31.2	58.4	10.2	41.9	13.4	35.8	16.3	32.3-24.5	25.4	14.8
63	Robinson-Stoy (Robinson, Penn.)	37.4	0.20	44	9.0	32.6	60.5	10.3	41.9	13.3	35.8	16.3	32.8-25.6	25.4	16.8
64	Salem (Aux Vases, Miss.).........	37.2	0.17	43	11.4	32.5	59.7	9.7	42.1	14.5	35.4	15.9	33.2-24.9	25.3	16.0
	Indiana														
65	Griffin (Cypress, Miss.)...........	35.2	0.20	47	11.2	30.9	59.7	10.5	42.6	14.1	42.3	16.7	33.6-24.9	27.5	14.8
	Kansas														
66	Bemis-Shutts (U. Arbuckle, Ord.)......	34.6	0.57	52	11.9	28.3	60.5	9.6	42.3	14.2	36.6	16.5	32.8-25.6	29.5	14.1
67	Chase-Silica (Kansas City, Penn.)......	38.8	0.44	42	6.7	38.7	60.0	11.7	43.4	13.5	36.8	13.7	33.4-25.4	21.3	14.1
68	El Dorado (Admire, Perm.)........	36.8	0.18	43	8.3	32.1	59.5	13.3	43.4	16.6	36.0	17.8	32.5-24.9	19.5	12.9
69	Hall-Gurney (Kansas City, Penn.)......	39.4	0.34	43	8.6	36.2	60.0	11.0	43.2	15.3	37.0	14.9	33.6-26.3	21.1	15.7
70	Kraft-Prusa (Pre-Cambrian)........	43.0	0.27	38	12.7	40.9	64.5	11.8	44.1	14.9	37.4	13.7	33.8-26.1	17.6	15.9
71	Seely-Wick (Bartlesville, Penn.)......	41.1	0.23	38	9.6	35.3	61.8	10.5	42.8	14.3	36.4	16.3	33.4-26.8	20.8	17.9
72	Spivy-Grabs (Miss.)...............	23.5	0.93	84	14.3	22.8	56.4	9.1	41.7	11.8	34.6	18.5	31.3-23.1	36.7	1.6
73	Trapp (Arbuckle, Ord.)...........	39.2	0.41	41	11.2	34.3	62.9	9.9	42.8	15.8	37.2	15.2	33.6-26.3	23.4	16.5
	Louisiana														
74	Avery Island (U. Mio.)............	34.4	0.12	46	3.1	18.2	55.4	4.6	42.3	29.0	34.8	25.9	31.1-26.3	20.3	20.5
75	Bateman Lake (9900', U. Mio.)......	38.2	0.15	41	3.3	22.5	53.2	31.6	42.1	14.0	35.8	16.2	34.6-26.4	15.2	20.7
76	Bay de Chene (Mio.).............	33.6	0.27	52	5.0	19.3	55.7	11.5	41.3	20.9	35.2	22.1	31.5-24.5	25.2	18.1
77	Bay Marchand (3900' Mio.)........	20.2	0.46	270	7.5	2.5	45.8			22.2	30.8	38.8	25.2-16.2	34.8	11.7
78	Bay St. Elaine (U. Mio.)...........	33.6	0.39	49	4.3	21.2	52.0	6.7	41.3	25.4	35.6	21.5	31.7-25.6	24.5	19.2

No.	Name													
79	Bayou Sale (Morin, Mio.)	36.2	0.16	44	2.2	18.5	53.5	15.5	41.9	25.4	36.2	20.4	33.6-27.7	18.1	21.0
80	Black Bay, West (7300', Mio.)	30.0	0.27	57	6.3	15.2	54.2	5.5	42.1	24.8	35.0	21.7	30.6-23.7	29.8	15.6
81	Black Bay, West (8050', Mio.)	23.0	0.36	140	9.0	6.3	48.5			25.0	32.8	35.1	26.8-17.5	31.9	14.1
82	Black Bay, West (8300', Mio.)	30.6	0.26	58	7.1	16.2	54.2	4.9	43.0	23.9	35.8	21.7	30.6-22.6	30.7	15.7
83	Black Bay, West (8650', Mio.)	34.4	0.18	45	5.6	19.8	54.2	13.8	42.1	21.7	35.8	20.2	32.8-25.0	22.8	18.4
84	Black Bay, West (9100', Mio.)	35.2	0.17	43	5.1	21.7	55.2	13.0	42.1	20.3	36.2	20.5	33.2-25.4	22.4	18.6
85	Black Bay, West (9200', Mio.)	34.4	0.19	46	4.6	19.4	53.5	12.5	42.1	21.9	36.4	20.2	33.4-26.1	24.5	19.0
86	Caddo (Annona Chalk, U. Cre.)	36.8	0.37	44	6.0	27.0	54.7	12.4	41.7	20.4	36.8	18.5	34.6-28.9	21.5	20.7
87	Caillou Island (No. 70, Mio.)	35.4	0.23	45	6.3	28.1	56.7	12.1	41.9	16.5	35.4	19.4	32.7-24.0	23.1	17.1
88	Cameron, West (Block 45, U. Mio.)	39.2	<0.10	41	2.3	18.2	51.6	21.9	42.6	39.6	37.4	12.9	34.8-30.0	6.7	20.7
89	Cote Blanche Bay, West (U. Mio.)	33.6	0.16	49	2.9	16.6	51.1	5.5	41.5	30.8	38.4	23.7	31.9-26.1	21.9	20.2
90	Cotton Valley (Bodeaw, Jur.)	40.6	<0.10	46	0.6	19.1	60.0	15.8	44.7	15.7	35.6	22.3	37.2-34.4	25.9	31.9
91	Cox Bay (Mio.)	31.9	0.38	56	5.7	20.4	58.2	4.9	43.2	24.1	37.6	22.6	30.8-23.5	27.2	15.9
92	Delhi (Tuscaloosa, U. Cre.)	41.7	<0.10	39	5.1	33.1	61.0	10.4	44.3	18.3	35.2	19.0	35.0-27.9	16.8	19.0
93	Delta Farms (Mio.)	35.6	0.26	44	4.9	26.0	55.4	14.1	41.7	18.5	36.8	19.5	35.1-25.9	20.6	17.8
94	Duck Lake (U. Mio.)	36.4	0.14	44	2.0	11.7	52.0	13.1	41.5	39.4	33.6	21.3	35.0-27.1	14.3	22.1
95	Erath (Mio.)	31.0	0.20	54	2.5	15.0	54.4			36.6	36.6	27.7	29.5-24.0	20.5	19.0
96	Eugene Island (Block 32, 7500', Mio.)	39.2	<0.10	39	1.3	13.0	52.0	40.4	42.1	19.5	35.4	18.3	35.4-29.7	8.4	24.2
97	Eugene Island (Block 126, Mio.)	36.2	0.19	52	2.9	17.6	57.9	20.5	42.8	12.6	36.8	23.5	34.6-27.5	25.0	21.1
98	Eugene Island (Block 188, 9080', Mio.)	27.1	0.35	91	4.7	4.1	45.8			32.7	35.8	35.0	28.8-22.3	20.5	17.9
99	Garden Island Bay (Mio.)	34.8	0.22	49	7.4	23.6	55.4	5.3	42.3	24.0	36.2	23.8	32.7-24.3	20.9	16.2
100	Golden Meadow (Mio.)	37.6	0.18	44	1.9	19.1	54.4	23.0	42.6	17.7	36.8	22.0	36.0-28.6	17.4	22.3
101	Grand Bay (Mio.)	35.0	0.31	48	3.0	20.5	55.2	12.5	41.1	19.9	35.8	22.1	33.6-27.1	22.6	19.8
102	Grand Isle (Block 16, B-1, Seg. E, Plio.)	36.4	0.18	40	3.7	25.8	54.7	15.0	41.7	31.1	36.2	19.0	32.3-25.4	18.6	18.9
103	Grand Isle (Block 18, B-2, Plio.)	34.6	0.22	46	5.1	21.0	53.2	6.9	43.0	20.9	35.8	22.1	31.3-24.0	17.6	17.9
104	Grand Isle (Block 47, Plio.-Mio.)	33.6	0.23	56	5.7	18.0	56.2	12.7	41.9	27.5	32.1	22.7	32.1-25.0	24.6	18.2
105	Hackberry, West (U. Mio.)	31.3	0.29	45	5.2	22.0	49.5	7.0	40.0	10.6	36.4	23.0	28.4-22.8	20.2	18.2
106	Krotz Springs (Frio, Olig.)	54.9	<0.10	34	Nil	67.0	62.6	10.7	43.0	22.4	37.2	2.1	33.8-32.5	5.7	22.8
107	Lafitte (Mio.)	36.2	0.30	45	3.3	20.8	55.2	13.8	43.4	16.3	36.0	21.4	34.4-27.5	21.3	20.3
108	Lake Barre (R-1, Mio.)	40.4	0.14	39	2.2	36.2	57.2	14.4	42.8	33.1	36.2	18.3	33.4-25.9	14.0	20.3
109	Lake Pelto (U. Mio.)	34.6	0.21	45	3.9	17.9	57.4	7.5	42.1	25.3	36.8	21.1	32.7-25.6	19.3	18.9
110	Lake Salvador (Mio.)	35.4	0.14	47	5.6	14.6	52.5	13.8	42.8	19.3	36.8	30.7	34.2-25.7	14.3	18.4
111	Lake Washington (Mio.)	28.2	0.37	82	9.2	20.0	57.7	3.7	42.1	31.1	33.8	22.5	28.2-20.2	32.5	12.2
112	Leeville (U. Mio.)	35.4	0.20	45	4.7	26.8	52.4	15.7	42.3	16.8	36.2	18.8	32.3-24.0	21.7	18.4
113	Little Lake (Eggerella 2, Mio.)	32.1	0.27	58	5.3	18.3	55.9	12.2	41.9	18.8	34.8	23.8	31.9-24.7	26.1	16.7
114	Little Lake (Eggerella 4, Mio.)	32.5	0.28	61	7.1	18.2	56.2	13.1	42.1	17.7	35.2	22.6	31.9-24.3	26.7	14.8
115	Little Lake (Textularia Panamensis 1, Mio.)	31.7	0.27	59	5.3	17.8	54.0	13.2	41.5	18.9	34.6	22.1	30.4-24.5	26.7	16.4
116	Little Lake (Textularia Panamensis 2, Mio.)	36.2	0.15	46	3.9	17.3	55.4	13.7	43.6	26.6	37.4	24.6	34.8-27.7	17.6	20.2
117	Little Lake (Textularia Panamensis 6, Mio.)	46.3	<0.10	34	4.8	45.9	59.7	24.8	43.6	10.8	37.0	12.8	35.6-27.0	5.1	19.5

Table 2-3. Properties of United States Crude Oils (Continued)

Item No.	State / Field (Formation, age)¹	Gravity, °API	Sulfur, wt. per cent	Viscosity, SUS at 100°F	Carbon residue of residuum, wt. per cent	Gasoline and naphtha Per cent	Gasoline and naphtha Gravity, °API	Kerosene distillate Per cent	Kerosene distillate Gravity, °API	Gas oil distillate Per cent	Gas oil distillate Gravity, °API	Lubricating distillate Per cent	Lubricating distillate Gravity, °API	Residuum Per cent	Residuum Gravity, °API
118	Little Lake, South (Textularia Panamensis 1 "D", Mio.)	34.8	0.26	49	9.9	23.6	56.2	13.0	44.1	19.0	36.8	23.9	32.5-23.5	20.1	16.7
119	Main Pass (Block 69, Mio.)	30.6	0.25	61	6.2	16.0	53.2	4.6	41.1	25.4	35.2	23.6	31.7-24.7	29.6	17.0
120	Paradis (Paradis, Mio.)	36.0	0.23	41	4.4	29.4	53.0	7.2	43.6	26.1	36.0	18.2	31.7-24.5	18.9	19.0
121	Quarantine Bay (Mio.)	31.9	0.27	52	6.1	19.6	54.0	5.6	41.5	26.0	35.6	21.4	31.3-23.3	26.8	17.1
122	Romere Pass (Mio.)	37.4	0.30	44	7.1	26.4	60.0	11.0	43.8	19.3	37.2	19.4	34.0-27.5	22.9	17.9
123	Ship Shoal (Block 154, Mio.)	29.1	0.36	78	3.5	8.7	51.1	5.5	41.9	26.6	34.8	30.1	30.6-23.0	27.7	17.8
124	South Pass (Block 24, Mio.)	32.3	0.26	51	5.2	18.7	54.9	5.1	40.6	25.4	34.6	23.2	31.0-23.5	26.3	17.3
125	Timbalier Bay (Mio.)	34.4	0.33	43	7.7	31.6	57.4	5.5	42.6	19.9	34.2	19.6	30.2-21.1	21.8	14.5
126	Venice (Mio.)	37.6	0.24	41	6.0	30.4	55.9	13.3	42.6	20.2	36.2	16.7	33.0-26.6	19.2	18.1
127	Weeks Island (Mio.)	33.2	0.19	51	4.8	20.1	54.9	4.7	42.8	28.6	35.4	24.1	31.5-24.7	22.4	18.7
128	West Bay (Mio.)	32.1	0.27	54	6.8	17.3	53.7	5.3	42.6	27.5	36.4	21.9	32.7-24.7	27.3	17.9
129	West Delta (Block 30, Mio.)	27.0	0.33	92	5.7	9.5	50.9	4.7	40.0	24.4	33.8	28.5	29.1-22.1	32.7	16.2
130	West Delta (Block 53, KE, U. Mio.)	32.3	0.43	66	6.7	17.6	56.7	11.4	42.8	18.6	35.8	22.7	32.8-23.7	28.2	16.4
131	West Delta (Block 83, KE, U. Mio.)	35.0	0.37	48	6.7	22.1	58.7	12.6	43.4	18.0	36.6	18.9	32.7-26.3	26.8	17.3
	Michigan														
132	Albion (Trenton-Black River, Ord.)	41.9	0.10	44	3.5	28.9	63.1	17.3	45.8	9.4	36.6	13.3	35.0-28.9	25.7	20.8
	Mississippi														
133	Baxterville (L. Tuscaloosa, U. Cre.)	17.1	2.71	1,480	16.6	5.2	55.4	2.1	40.4	14.4	33.2	24.3	29.3-19.7	52.8	7.0
134	Brookhaven (L. Tuscaloosa, L. Cre.)	35.0	0.43	50	8.5	25.7	60.0	10.8	41.9	16.7	32.7	17.8	32.8-26.3	28.0	15.1
135	Bryan (Rodessa, L. Cre.)	37.2	1.47	47	6.3	32.3	67.0	15.3	47.2	10.0	36.4	13.7	34.0-27.1	27.5	10.7
136	Heidelberg (U. Tuscaloosa, U. Cre.)	23.3	3.75	370	10.0	19.2	64.5	6.3	43.5	11.1	35.0	16.1	30.6-20.2	45.4	5.0
137	Little Creek (L. Tuscaloosa, U. Cre.)	38.0	0.16	43	5.4	33.4	58.9	10.9	42.8	16.3	36.4	17.1	33.4-27.9	21.6	17.6
138	Raleigh (Hosston, L. Cre.)	45.8	0.43	58	5.7	40.9	64.2	18.1	45.2	10.6	37.0	13.9	34.4-28.9	14.7	19.8
139	Soso (11,701 Bailey, Rodessa, L. Cre.)	41.1	0.89	41	9.9	36.5	63.7	18.1	45.2	10.0	36.8	15.4	34.0-26.6	19.4	15.3
140	Tinsley (Selma, U. Cre.)	30.4	1.02	79	8.5	20.9	63.1	11.1	44.7	11.4	35.0	18.6	31.7-22.0	37.8	13.8

	C1	C2	C3	C4	C5	C6	C7	C8	C9	C10	C11	C12	C13	C14
Montana														
141 Cabin Creek (Mission Canyon, Miss.)	33.4	0.60	47	15.7	25.1	61.5	18.1	43.6	12.9	34.6	12.3	31.7-23.8	31.0	12.0
142 Cut Bank (Cut Bank, L. Cre.)	39.0	0.85	38	9.9	34.2	60.5	10.6	42.3	15.8	34.6	19.3	30.8-24.2	17.0	14.8
143 Pine (Dev.)	33.8	0.36	55	18.9	24.7	64.2	19.8	46.0	11.9	34.2	10.6	31.5-23.5	31.2	9.7
144 Poplar East (Madison, Miss.)	39.6	0.32	38	4.4	33.8	60.2	12.3	43.8	17.5	36.0	17.5	31.9-26.3	18.7	19.2
145 Sumatra (Amsden, Penn.)	29.5	0.65	72	11.2	16.2	53.5	10.9	41.9	16.8	35.2	22.5	32.5-24.7	32.5	14.4
New Mexico														
146 Bisti (Gallup, Cre.)	37.6	0.18	40	5.6	31.4	57.9	4.0	41.7	19.4	36.4	16.5	32.5-24.9	24.2	17.3
147 Caprock, East (Wolfcamp, Perm.)	43.2	0.17	35	3.3	46.0	56.7	12.2	42.3	17.3	35.8	12.4	32.7-27.3	11.7	21.3
148 Denton (Dev.)	46.0	0.17	35	4.0	44.3	61.0	11.3	42.8	16.3	36.0	13.3	33.0-28.2	11.5	21.6
149 Eunice-Monument (Grayburg, Perm.)	28.8	0.97	54	10.9	27.9	56.2	5.0	40.2	18.3	32.7	17.9	27.7-21.5	27.6	13.3
150 Gladiola (Wolfcamp, Perm.)	42.1	0.10	35	4.6	46.6	57.7	5.7	41.1	21.3	34.6	12.8	31.0-25.7	12.4	18.1
151 Hobbs (San Andres, Perm.)	37.4	1.41	41	9.6	35.5	60.8	4.5	41.3	17.0	34.0	19.6	28.6-20.7	17.9	11.1
152 Jalmat (Yates, Perm.)	36.2	1.22	47	6.0	32.8	58.2	10.5	42.8	13.4	36.0	19.0	31.5-23.8	11.6	12.6
153 Kemnitz (Wolfcamp, Perm.)	36.4	0.12	35	5.8	44.8	49.9	4.9	41.1	19.9	32.3	16.3	27.0-21.1	13.1	16.0
154 Langlie-Mattix (Queen, Perm.)	28.9	1.65	64	9.2	20.4	50.9	6.2	41.3	22.6	35.0	20.4	28.8-20.8	29.5	13.6
155 Lovington (Abo, Perm.)	39.4	0.36	36	4.6	41.0	55.7	5.0	41.5	19.0	34.0	13.6	29.3-24.0	18.0	17.9
156 Maljamar (Perm.)	38.6	0.70	37	8.8	37.5	57.7	4.7	42.8	19.7	35.6	17.3	30.8-24.2	15.7	15.7
157 Saunders (Wolfcamp, Perm.)	41.7	0.11	34	4.0	46.9	56.4	5.2	41.9	19.8	34.6	12.8	30.6-25.4	13.6	18.6
158 Vacuum (San Andres, Perm.)	35.0	0.95	42	9.2	33.5	54.0	4.7	42.6	21.3	35.0	17.1	30.0-22.5	22.2	14.4
159 Verde-Gallup (Mancos, Cre.)	39.6	0.12	39	3.4	31.6	59.2	10.7	42.6	16.0	37.4	19.1	34.2-25.9	20.7	20.3
North Dakota														
160 Antelope (Miss.)	42.8	<0.10	32	2.7	41.9	60.0	10.8	42.3	15.8	35.2	17.0	32.7-23.5	12.8	20.0
161 Beaver Lodge-Tioga (Mission-Canyon, Ord.)	46.0	0.23	34	2.7	46.6	60.8	9.8	42.1	13.1	35.0	13.9	31.0-26.3	12.2	20.3
162 Blue Butte (Madison, Miss.)	41.1	0.52	34	2.6	41.0	57.4	10.0	42.8	17.0	34.6	14.0	31.1-26.4	16.8	19.4
163 Tioga (Madison, Miss.)	41.3	0.31	35	2.8	40.7	59.5	11.3	43.0	15.3	35.0	15.9	31.1-25.9	15.2	20.0
Oklahoma														
164 Bradley (Springer, Penn. & Cunningham, Miss.)	35.0	0.22	56	6.7	24.3	57.4	15.6	43.0	9.6	36.6	19.5	34.8-27.1	30.4	18.6
165 Burbank (Layton, Penn.)	39.6	0.24	43	4.4	30.2	60.0	10.1	42.8	14.8	36.4	15.8	33.2-27.5	26.0	21.5
166 Cement (U. Melton, Penn.)	33.2	0.47	56	5.3	28.9	53.2	9.7	42.1	13.0	36.0	18.3	33.0-26.3	29.8	15.9
167 Cushing (Bartlesville, Penn.)	42.1	0.22	38	5.6	41.2	61.3	11.8	42.8	14.5	36.6	14.4	33.4-27.0	17.0	18.6
168 Eola-Robberson (Bromide, M. Ord.)	39.8	0.35	41	3.9	33.1	60.8	10.1	42.3	14.0	35.8	15.3	32.7-25.6	22.7	18.9
169 Eola-Robberson (Oil Creek, L. Ord.)	38.0	0.27	42	3.5	32.6	58.4	10.4	42.8	15.7	36.0	14.7	32.7-27.3	24.6	19.4
170 Glennpool (Glenn, Penn.)	37.4	0.31	42	6.5	32.3	56.9	12.0	42.3	16.4	36.0	15.7	32.5-26.1	23.0	18.1
Golden Trend														
171 Antioch, Southwest (Gibson, Miss.)	42.1	0.11	39	2.7	34.6	62.9	17.1	42.8	8.1	36.0	15.3	34.4-27.5	22.3	23.3

Table 2-3. Properties of United States Crude Oils (Continued)

Item No.	State / Field (Formation, age)[1]	Gravity, °API	Sulfur, wt. per cent	Viscosity, SUS at 100°F	Carbon residue of residuum, wt. per cent	Gasoline and naphtha Per cent	Gasoline and naphtha Gravity, °API	Kerosine distillate Per cent	Kerosine distillate Gravity, °API	Gas oil distillate Per cent	Gas oil distillate Gravity, °API	Lubricating distillate Per cent	Lubricating distillate Gravity, °API	Residuum Per cent	Residuum Gravity, °API
	State														
172	Elmore, Northeast (Gibson, Miss.)	42.1	0.14	38	3.0	37.0	62.1	17.4	43.0	8.8	36.4	15.9	34.4–28.4	20.2	22.6
173	New Hope, Southeast (Gibson, Miss.)	41.1	0.19	41	4.3	32.7	62.1	11.1	43.8	15.1	37.6	20.0	34.0–26.8	19.0	21.1
174	Panther Creek (Penn.)	46.5	0.14	37	3.8	51.5	61.5	11.6	43.8	12.1	37.2	11.3	34.0–27.7	13.3	20.8
175	Healdton (Healdton, Penn.)	28.9	0.92	110	10.7	17.2	54.2	3.9	42.8	20.6	37.0	21.3	32.5–25.2	36.7	15.0
176	Hewitt (Lone Grove, Pre-Cam.)	37.0	0.65	49	8.3	28.6	61.0	9.7	43.6	13.5	37.0	16.4	33.2–24.9	29.0	16.8
177	Joiner City (Bois D'Arc, Sil.-Dev.)	40.4	0.47	40	4.0	34.9	63.7	9.9	43.6	12.2	37.0	15.7	33.4–25.2	24.3	18.9
178	Knox (Dornick Hills, Penn.)	43.0	<0.10	37	3.6	37.3	60.0	11.9	43.4	14.0	37.6	18.8	35.4–28.4	16.3	22.8
179	Naval Reserve (Burbank, Penn.)	39.8	0.25	42	4.0	31.2	60.5	10.3	42.8	14.8	36.8	16.4	33.6–28.0	25.6	22.3
180	Oklahoma City (Wilcox, Ord.)	37.6	0.16	45	4.2	27.5	58.2	10.6	42.8	15.3	36.8	18.8	34.4–27.5	26.0	21.0
	Sho-Vel-Tum District														
181	Camp (Springer, Penn.)	28.0	1.41	115	11.4	22.2	58.2	4.1	42.3	16.0	36.2	15.8	34.2–26.8	40.9	12.6
182	Fox-Graham (Springer, Penn.)	36.0	0.57	49	6.6	27.0	57.4	11.5	43.6	15.5	37.2	15.8	34.0–26.3	29.0	17.9
183	Milroy (Deese, Penn.)	30.0	1.73	100	11.8	22.6	59.7	8.8	42.3	13.0	35.4	14.8	31.7–23.8	39.8	13.2
184	Sholem Alechem (Springer, Penn.)	26.8	1.44	150	7.9	21.5	58.7	3.3	42.6	14.8	35.4	15.5	30.4–21.8	43.2	10.6
185	Tatums (Deese, Penn.)	21.0	1.68	550	8.2	14.5	55.9	3.1	41.7	13.6	34.2	10.1	29.3–22.5	55.9	8.3
186	Velma (L. Dornick Hills, Springer, Penn.)	29.1	1.36	87	10.1	21.2	59.5	4.3	42.8	15.9	34.8	16.6	30.0–22.8	40.4	13.9
	Pennsylvania														
187	Bradford (U. Dev.)	41.1	0.11	44	1.6	30.7	56.2	18.1	44.1	8.7	38.4	15.9	36.8–31.1	25.0	25.7
	Texas														
188	Anahuac (Marg. No. 1, Olig.)	33.2	0.23	48	4.0	17.7	51.8	7.0	42.1	32.4	35.2	21.1	31.9–25.9	21.6	19.4
189	Andector (Ellen, Cam.-Ord.)	43.2	0.22	38	4.9	34.9	64.2	18.7	44.9	9.8	38.6	15.1	36.6–28.0	20.0	20.7
190	Andrews (Penn.)	39.0	0.11	41	5.6	33.8	59.2	10.1	42.3	14.1	36.2	16.4	34.0–26.8	24.2	19.8
191	Andrews (Wolfcamp, Perm.)	36.8	0.78	40	7.6	37.0	56.4	4.5	41.9	19.3	34.6	14.1	30.2–24.0	23.6	16.0
192	Andrews, North (Dev.)	44.3	0.30	37	3.7	39.3	62.9	10.7	44.3	14.1	37.8	15.1	34.4–27.7	17.6	22.1
193	Andrews, North (Ellen, Cam.-Ord.)	45.2	0.11	37	7.8	37.0	68.1	19.6	47.6	13.0	37.0	12.8	34.4–26.4	16.0	18.6
194	Andrews, South (Dev.)	44.7	<0.10	36	9.1	42.2	60.8	17.1	43.0	8.7	37.2	16.1	36.0–29.1	14.8	24.9

No.	Field (Formation, Age)														
195	Andrews, South (Wolfcamp, Perm.)	36.8	0.10	49	5.6	32.0	55.9	10.5	42.3	15.9	37.2	16.3	34.6–28.0	25.0	19.5
196	Bakke (Dev.)	44.7	0.16	37	1.1	42.4	61.3	10.7	43.8	15.0	38.0	15.0	35.4–29.9	15.0	24.3
197	Bakke (Ellen, Cam.-Ord.)	45.6	0.21	<32	6.3	36.7	65.7	19.5	45.8	13.1	38.0	11.1	35.0–27.7	18.2	25.7
198	Bakke (Penn.)	39.4	<0.10	40	2.2	33.9	58.9	10.7	43.4	14.2	36.0	17.7	35.2–28.6	22.7	21.8
199	Bakke (Wolfcamp, Perm.)	37.4	0.41	40	4.7	35.8	55.2	4.3	43.0	20.3	36.0	15.0	32.1–26.3	22.4	18.2
200	Bethany (4300' Glen Rose, L. Cre.)	41.5	0.23	41	6.8	30.9	61.5	21.6	47.4	10.0	37.8	16.5	36.2–29.1	19.4	19.0
201	Block 31 (Dev.)	44.5	0.18	35	3.9	43.3	61.8	6.0	43.0	13.1	37.2	15.5	34.4–27.1	15.2	21.6
202	Borregos (F-5, Frio, Olig.)	42.1	<0.10	35	3.3	42.3	55.9	6.6	42.6	32.6	35.6	12.3	31.7–22.0	4.3	12.5
203	Borregos (L-5, Frio, Olig.)	40.6	<0.10	36	3.9	38.2	53.5	17.3	42.3	21.4	34.2	13.0	31.5–19.0	5.7	11.3
204	Borregos (N-21, Frio, Olig.)	42.3	<0.10	35	4.0	31.6	54.0	25.6	42.1	32.2	36.0	7.5	33.2–23.0	2.0	13.2
205	Borregos (R-13, Vicksburg, Olig.)	38.2	<0.10	37	3.5	23.3	54.4	5.5	43.6	48.6	36.0	17.1	32.8–23.3	5.0	13.6
206	Borregos, South (R-5, Frio, Olig.)	39.0	<0.10	35	3.9	29.9	52.7	17.8	42.3	30.3	35.6	15.9	32.5–23.1	5.1	13.6
207	Cogdell area (Canyon Reef, Penn.)	41.7	0.38	37	7.6	38.2	62.6	9.9	42.1	14.4	35.8	14.4	32.7–26.8	19.9	17.5
208	Conroe (Cockfield, Eoc.)	37.0	<0.10	36	4.9	32.8	48.8			43.4	34.4	15.9	31.3–26.3	7.2	17.5
209	Cowden, North (Grayburg, Perm.)	30.4	1.89	51	10.9	27.7		<5.0	43.0	19.5	35.0	18.1	30.4–22.5	29.7	12.3
210	Cowden Deep (San Andres, Perm.)	36.6	0.96	42	6.7	35.0	58.7	10.4	41.7	15.0	34.8	15.0	31.7–24.7	24.6	17.1
211	Cowden, South (Grayburg, Perm.)	34.6	1.77	44	3.5	32.6	57.9	3.6	42.1	18.5	35.0	15.5	30.2–23.0	26.8	13.5
212	Darst Creek (Buda, L. Cre.)	36.6	0.76	49	7.8	23.5	58.9	22.5	44.1	12.2	36.4	18.8	34.6–28.6	22.5	17.5
213	Darst Creek (Edwards, L. Cre.)	36.8	0.78	46	6.8	25.7	57.9	19.4	43.4	13.1	36.4	17.9	34.6–28.6	23.2	17.8
214	Diamond "M" (Canyon Reef, Perm.)	45.4	0.20	35	4.9	43.0	61.5	5.0	42.3	17.5	36.6	13.5	32.8–27.7	16.1	20.2
215	Dollarhide (Clear Fork, Perm.)	37.4	0.42	43	5.8	33.4	60.5	11.1	42.6	13.8	35.8	17.5	32.7–24.9	24.0	17.9
216	Dollarhide (Dev.)	38.2	0.57	41	6.4	35.8	61.3	9.8	41.9	13.3	34.8	15.0	31.3–25.9	24.7	17.3
217	Dollarhide (Ellen., Cam.-Ord.)	41.5	0.23	40	6.0	31.6	65.0	20.9	46.7	14.5	37.8	11.5	34.8–27.7	21.4	19.7
218	Dollarhide (Sil.)	41.3	0.36	39	8.3	32.4	65.9	20.0	45.6	13.1	37.2	12.0	34.4–26.1	22.2	17.8
219	Dollarhide, East (Ellen., Cambro-Ord.)	42.3	0.10	40	4.3	35.7	62.9	17.2	44.5	9.8	36.4	13.6	30.4–24.3	20.9	21.0
220	Dune (Grayburg, Perm.)	29.7	3.11	57	7.4	24.4	55.9	4.8	42.1	20.7	34.3	14.6	34.2–25.9	35.0	13.2
221	East Texas (Woodbine, U. Cre.)	37.4	0.25	42	6.1	33.9	58.2	5.0	42.8	17.7	37.2	20.3	35.2–28.4	22.2	16.4
222	Emma (Dev.)	45.6	<0.10	36	2.9	39.9	62.9	11.9	44.5	14.8	38.2	13.5	36.4–27.5	16.7	23.3
223	Emma (Ellen., Cam.-Ord.)	49.2	<0.10	35	5.6	42.0	65.6	20.8	46.0	12.6	39.4	13.4	35.8–31.5	8.8	21.8
224	Emma (Grayburg-San Andres, Perm.)	49.0	<0.10	35	3.4	39.6	67.5	21.3	47.4	13.4	38.6	9.3		13.2	23.8
225	Emperor, Deep (Seven Rivers, Queen, Perm.)	35.6	1.11	45	7.1	34.0	57.7	10.8	42.3	14.0	34.6	16.1	30.0–23.5		15.7
226	Fairway (James, L. Cre.)	45.6	0.24	33	2.0	36.1	62.7	18.8	45.4	10.4	36.3	16.4	35.2–28.1	24.4	24.0
227	Fort Chadbourne (Odem, Penn.)	44.1	0.24	37	3.4	39.9	61.0	11.2	43.6	15.3	37.8	15.1	35.2–29.3	16.7	23.0
228	Foster (Grayburg, Perm.)	34.2	1.53	44	7.4	31.6	57.4	4.7	42.1	19.9	35.2	16.8	30.6–24.3	26.5	15.3
229	Fuhrman-Mascho (Grayburg, Perm.)	31.3	2.06	47	8.4	33.1	55.2	5.5	41.7	17.7	33.4	15.8	33.0–26.8	27.7	10.0
230	Fullerton (Clear Fork, Perm.)	39.6	0.47	40	5.8	35.1	58.7	10.9	43.2	14.2	36.0	16.3	33.0–26.4	21.0	19.4
231	Fullerton (Dev.)	41.5	0.32	38	4.4	37.4	61.3	11.4	43.6	15.6	37.4	15.8	34.6–28.8	19.6	22.0
232	Fullerton, South (Wolfcamp, Perm.)	43.4	0.17	36	3.0	39.5	61.3	10.6	43.8	13.7	37.2	14.8	33.6–27.9	18.3	22.1

Table 2-3. Properties of United States Crude Oils (Continued)

Item No.	State Field (Formation, age)[1]	Gravity, °API	Sulfur, wt. per cent	Viscosity, SUS at 100°F	Carbon residue of residuum, wt. per cent	Gasoline and naphtha Per cent	Gasoline and naphtha Gravity, °API	Kerosine distillate Per cent	Kerosine distillate Gravity, °API	Gas oil distillate Per cent	Gas oil distillate Grav., °API	Lubricating distillate Per cent	Lubricating distillate Gravity, °API	Residuum Per cent	Residuum Gravity, °API
233	Gillock (Hudgings, Frio, Olig.)	45.2	<0.10	34	3.5	38.2	60.2	16.5	43.2	21.6	37.0	14.2	34.6–26.1	6.6	19.8
234	Gillock, South (Frio, Olig.)	38.0	0.11	38	6.0	28.3	56.9	7.3	43.0	32.8	36.2	19.1	32.8–27.1	12.2	17.9
235	Goldsmith (5600', U. Clear Fork, Perm.)	38.0	0.52	46	5.7	30.4	58.7	10.7	43.6	15.0	36.6	16.7	33.4–23.5	24.1	18.2
236	Goldsmith (Clear Fork-Tubb, Perm.)	38.0	0.57	44	5.1	30.2	58.9	11.7	43.8	15.5	36.8	16.3	34.6–26.3	24.6	19.4
237	Goldsmith (Dev.)	40.9	0.16	40	6.7	35.4	61.8	10.5	43.4	13.2	36.4	16.0	33.6–27.3	21.9	19.7
238	Goldsmith (San Andres, Perm.)	38.4	1.16	40	6.7	35.4	58.7	4.5	42.8	18.1	35.8	15.6	31.0–26.8	21.7	15.3
239	Goldsmith, East (Holt, Perm.)	36.4	0.15	59	4.8	29.5	54.9	5.3	42.6	19.7	36.8	17.9	33.0–26.8	25.6	19.2
240	Goldsmith, North (Ellen, Cam.-Ord.)	37.0	0.58	44	5.6	29.6	58.9	10.4	43.8	15.6	36.2	17.8	33.0–25.0	24.7	17.5
241	Goldsmith, West (U. Clear Fork, Perm.)	37.4	0.53	44	5.2	31.7	60.2	10.2	43.4	14.6	36.4	17.6	32.8–25.9	26.4	18.2
242	Goldsmith, West (Ellen., Cam.-Ord.)	42.6	0.32	39	3.7	33.1	64.5	18.7	45.6	13.2	38.4	12.6	35.4–27.7	21.0	21.3
243	Goldsmith, West (Fusselman, Sil.)	37.4	0.96	43	6.1	31.9	60.0	9.8	42.6	15.8	35.0	15.3	31.3–25.2	25.4	16.8
244	Goldsmith, West (San Andres, Perm.)	34.4	1.38	43	9.1	33.2	56.7	11.0	41.7	16.2	31.3	14.1	30.0–25.0	24.8	13.6
245	Goose Creek (Frio, Olig.)	35.0	0.13	42	4.3	22.1	54.3	6.1	42.6	32.6	32.6	22.7	30.6–25.0	17.7	19.7
246	Hastings (Marginulina, Olig.)	31.5	0.15	48	4.7	20.1	52.0			36.9	33.4	23.1	29.5–21.6	20.3	18.2
247	Hastings, East (Frio, Olig.)	31.0	0.15	55	4.3	15.8	49.2			35.6	34.0	23.0	29.3–23.5	23.0	18.7
248	Hastings, West (Frio, Olig.)	30.2	0.17	58	5.8	16.4	50.6			35.6	33.6	23.0	28.6–22.2	22.4	17.0
249	Hawkins (Eagle Ford, U. Cre.)	26.8	2.19	135	6.0	20.7	63.1	7.3	44.5	12.7	35.4	15.4	30.6–22.0	43.1	9.3
250	Headlee (Dev.)	47.4	<0.10	37	0.4	45.7	62.1	17.6	43.8	9.8	37.6	13.2	37.0–31.0	12.2	25.4
251	Headlee (Ellen., Cam.-Ord.)	51.1	<0.10	35	1.5	43.0	67.0	22.3	48.5	14.7	40.2	8.8	38.0–32.8	9.8	26.3
252	High Island (Mio.)	27.3	0.26	79	6.2	10.1	47.2			38.6	32.1	30.1	28.0–20.5	21.0	15.7
253	Howard-Glasscock (Yates, Perm.)	30.6	1.18	61	7.6	21.8	53.2			24.5	35.0	19.5	31.5–25.4	33.3	17.3
254	Hull (Caprock, Mio.)	31.1	0.35	41	3.7	33.4	51.8			29.5	29.3	19.1	22.1–16.8	17.5	16.5
255	Jameson (Strawn, Penn.)	40.9	<0.10	34	1.6	36.3	56.9	13.5	43.0	15.0	38.6	17.1	34.4–29.9	17.1	23.7
256	Jameson (Strawn Reef, Penn.)	44.3	<0.10	36	1.3	41.7	60.0	18.0	42.6	8.6	36.8	14.3	35.4–30.0	15.6	25.0
257	Jo-Mill (Spraberry, Perm.)	37.4	0.11	43	3.7	32.6	57.7	4.5	41.7	17.6	35.8	15.2	31.9–27.5	27.8	19.8
258	Jordan (Ellen., Cam.-Ord.)	43.1	0.28	38	5.2	34.5	65.3	19.9	54.7	14.3	38.0	10.4	35.4–29.1	20.0	20.8
259	Jordan (San Andres, Perm.)	33.2	1.48	46	9.4	29.7	55.9	5.1	43.0	20.1	36.0	18.6	30.4–23.5	26.4	14.5
260	Kelly-Snyder (Canyon Reef, Penn.)	39.8	0.22	37	4.8	42.2	57.4	4.9	41.9	19.2	35.8	14.9	32.1–26.3	18.4	19.2

No.	Field (Formation)	1	2	3	4	5	6	7	8	9	10	11	12	13	14
261	Kelsey (Frio, Olig.)	40.6	0.13	35	8.9	38.5	57.4	6.3	43.6	33.4	35.4	14.1	32.1–17.6	6.8	9.9
262	Kelsey, South (18-A, Frio, Olig.)	43.4	<0.10	34	2.9	45.3	55.4	18.8	42.6	19.8	34.8	5.5	31.3–22.8	9.0	16.7
263	Kermit (Ellen., Cam.-Ord.)	41.9	0.19	39	2.3	30.0	61.8	19.2	44.9	13.1	37.2	16.8	35.2–29.9	20.3	24.3
264	Kermit (Yates and Seven Rivers, Perm.)	36.6	0.94	42	7.2	34.8	59.5	5.0	41.9	18.5	35.0	17.5	29.9–23.1	22.8	16.2
265	Kermit, South (Dev.)	32.3	0.79	81	4.8	17.0	58.7	8.6	43.8	13.1	36.8	22.6	32.7–26.8	37.5	19.7
266	Keystone-Colby (Queen, Perm.)	34.2	0.95	48	5.7	26.1	54.2	12.6	42.1	16.7	35.4	16.9	32.1–25.9	26.7	17.6
267	Keystone-Devonian (Dev.)	32.7	0.69	68	5.2	18.6	57.7	8.2	43.0	13.0	36.4	22.6	33.2–26.8	34.5	20.3
268	Keystone-Ellenburger (Ellen., Cam.-Ord.)	42.1	0.13	40	2.3	30.0	62.9	20.1	45.4	15.0	37.8	14.4	35.6–29.1	19.1	22.8
269	Keystone-Holt (San Angelo, Perm.)	37.8	0.63	43	5.1	31.7	57.7	11.8	43.6	17.1	36.6	15.4	33.2–26.6	23.3	18.4
270	Keystone-Silurian (Sil.)	35.4	0.49	51	5.2	21.9	57.9	19.5	43.4	12.8	36.2	17.3	34.0–27.3	28.5	19.5
271	KMA (Strawn, Penn.)	40.0	0.31	39	6.3	36.9	57.9	11.4	42.1	15.7	35.4	15.5	32.7–25.7	18.7	18.2
272	Lake Pasture (FT-560, Frio-Sinton, Olig.)	37.2	0.13	35	4.7	32.0	52.5	9.0	42.1	37.6	34.2	15.1	28.9–17.5	5.4	11.0
273	Lake Pasture (H-440, Greta, Olig.)	23.7	0.20	60	5.3	2.8	43.2			50.3	30.0	34.3	25.0–13.9	12.1	9.9
274	Levelland (San Andres, Perm.)	31.1	2.12	48	8.6	31.0	54.2	4.2	42.3	17.8	34.0	17.1	28.9–21.8	28.3	10.4
275	Liberty, South (EY, Olig.)	36.4	0.14	40	2.5	30.6	53.7	5.6	42.6	26.8	34.6	19.9	31.1–25.2	16.5	19.8
276	Luling-Branyon (Edwards, L. Cre.)	28.6	0.86	90	6.6	12.7	51.6	4.8	41.3	24.4	34.8	22.7	30.8–23.7	35.4	17.9
277	Magutex (Dev.)	40.2	0.30	38	4.3	38.5	58.4	11.0	42.3	15.6	36.6	15.8	34.0–28.0	18.9	20.8
278	Magutex (Ellen., Cam.-Ord.)	46.9	0.12	39	4.0	36.4	65.3	22.1	47.2	13.4	39.4	13.5	37.0–28.4	13.1	23.0
279	McElroy (Grayburg, Perm.)	31.5	2.37	53	10.5	24.7	57.4	4.4	41.9	20.2	35.0	19.5	29.7–22.8	29.6	13.3
280	Means (Grayburg, Perm.)	30.0	2.40	54	8.5	29.7	56.7	4.3	42.3	18.6	33.6	15.6	28.0–21.5	30.9	9.9
281	Means (Queen, Perm.)	35.6	1.11	46	11.8	31.3	59.7	10.2	43.2	12.6	35.6	15.1	31.9–25.4	28.5	13.9
282	Means, East (Strawn, Penn.)	43.0	0.10	35	3.0	45.2	58.4	5.3	43.2	17.5	36.4	14.6	32.3–26.8	15.4	21.5
283	Midland Farms (Ellen., Cam.-Ord.)	50.6	<0.10	34	2.2	42.8	66.1	22.5	48.3	14.0	39.6	9.6	37.0–32.7	9.4	27.1
284	Midland Farms (Grayburg, Perm.)	31.7	2.04	46	11.3	29.5	57.9	4.7	42.1	19.2	34.0	16.6	28.6–22.1	28.8	11.6
285	Midland Farms (Wolfcamp, Perm.)	39.6	0.13	40	3.1	36.6	56.9	10.3	41.3	13.5	35.6	15.5	33.2–27.3	22.1	21.3
286	Midland Farms, North (Grayburg, Perm.)	30.0	2.37	53	6.9	28.3	57.4	4.0	42.3	17.7	33.6	16.2	28.2–21.1	31.7	10.1
287	Midland Farms, Northeast (Ellen., Cam.-Ord.)	49.2	<0.10	38	3.0	41.3	64.8	23.5	47.8	14.4	39.8	11.6	37.8–30.2	8.7	23.5
288	Old Ocean (Armstrong, Frio, Olig.)	36.8	0.14	43	4.6	24.8	54.7	13.8	41.7	23.0	35.0	18.2	32.8–26.3	19.6	19.5
289	Old Ocean (Chenault, Frio, Olig.)	25.4	0.21	71	4.0	3.6	44.3	8.7	43.6	43.9	30.0	27.5	26.3–20.8	24.7	17.3
290	Panhandle (Moore County, Perm.)	40.4	0.55	47	5.6	31.1	61.5			13.6	38.4	16.1	36.0–31.1	27.2	20.0
291	Pegasus (Ellen., Cam.-Ord.)	53.0	<0.10	33	1.6	46.3	67.5	21.2	48.5	12.9	40.9	8.4	38.0–33.4	8.3	27.9
292	Pegasus (Penn.)	45.4	<0.10	36	1.3	42.9	61.0	9.4	43.2	12.7	36.4	16.3	33.3–24.7	17.1	23.5
293	Pegasus (Spraberry, Perm.)	35.6	0.17	48	8.5	31.9	59.2	9.4	41.9	12.4	36.4	13.8	32.3–24.7	28.4	16.5
294	Penwell (Ellen., Cam.-Ord.)	41.7	0.24	40	8.2	32.3	65.9	19.5	45.8	12.7	38.0	12.1	35.0–27.0	22.0	17.8
295	Penwell (San Andres, Perm.)	33.2	1.69	45	8.3	31.0	55.9	4.9	43.0	19.8	35.6	16.4	30.6–22.3	27.7	15.1
296	Pierce Junction (Frio, Olig.)	37.2	0.12	39	5.6	40.0	53.7	6.7	40.2	21.5	33.0	17.5	29.9–20.5	14.3	17.5
297	Plymouth (6100', Frio, Olig.)	42.3	0.12	34	6.1	44.2	57.4	8.0	44.3	30.8	34.6	10.9	28.2–15.1	5.0	9.2
298	Plymouth (Frio, Olig.)	40.6	0.13	37	5.4	47.0	56.9	7.6	42.1	26.3	32.8	12.1	27.0–14.8	6.0	8.9

Table 2-3. Properties of United States Crude Oils (Continued)

Item No.	State / Field (Formation, age)[1]	Gravity, °API	Sulfur, wt. per cent	Viscosity, SUS at 100°F	Carbon residue of residuum, wt. per cent	Gasoline and naphtha Per cent	Gasoline and naphtha Gravity, °API	Kerosine distillate Per cent	Kerosine distillate Gravity, °API	Gas oil distillate Per cent	Gas oil distillate Gravity, °API	Lubricating distillate Per cent	Lubricating distillate Gravity, °API	Residuum Per cent	Residuum Gravity, °API
299	Plymouth (Greta, Frio, Olig.)	23.5	0.19	55	4.7	6.1	54.7		....	47.4	28.9	31.0	23.8-12.0	15.1	9.3
300	Plymouth (Main Greta, Frio, Olig.)	28.8	0.15	44	5.8	19.8	51.1		....	42.3	30.6	26.4	24.7-11.7	9.8	9.2
301	Portilla (7100', Frio, Olig.)	40.4	<0.10	34	4.3	39.6	55.9	9.1	43.2	34.4	33.8	11.6	28.2-15.3	4.3	9.7
302	Portilla (7300', Frio, Olig.)	39.8	<0.10	36	4.3	38.6	55.2	7.4	41.7	32.8	34.8	12.4	29.3-19.0	6.6	10.7
303	Portilla (7400', Frio, Olig.)	39.6	0.12	35	4.6	37.8	54.9	7.1	42.3	35.9	34.6	13.2	30.0-19.5	5.3	10.7
304	Portilla (8100', Frio, Olig.)	39.0	0.12	35	4.8	30.5	53.7	7.9	42.6	42.8	34.6	14.6	31.5-20.3	4.0	10.9
305	Prentice (6700', Clear Fork, Perm.)	25.9	2.60	54	9.4	29.4	52.3	4.7	41.1	19.1	31.7	16.4	24.7-17.8	30.3	5.8
306	Prentice (Glorieta, Perm.)	28.6	2.68	47	5.8	31.9	54.2	4.5	41.1	19.4	31.0	15.8	24.9-18.1	27.2	5.9
307	Quitman (Eagle Ford, U. Cre.)	26.3	2.06	145	5.8	14.4	59.7	7.4	42.3	14.9	34.6	15.9	31.3-25.7	45.8	12.5
308	Quitman (Sub-Clarksville, U. Cre.)	16.2	3.64	3,700	12.0	8.6	58.9	2.2	42.8	13.7	34.4	19.6	27.5-17.0	55.0	4.2
309	Quitman (Trinity, L. Cre.)	43.8	0.92	39	8.5	43.8	66.4	16.9	44.1	8.6	35.4	12.0	32.8-26.6	18.4	15.4
310	Robertson (L. Clear Fork, Perm.)	34.0	1.31	44	9.2	32.4	56.7	4.7	42.1	19.2	35.4	17.0	30.6-24.0	25.5	13.5
311	Robertson (San Angelo-Clear Fork, Perm.)	29.9	1.95	49	6.1	29.1	53.7	4.0	41.7	18.6	33.8	16.3	28.0-22.1	31.4	12.9
312	Robertson, North (7100', Clear Fork, Perm.)	34.8	0.79	44	8.4	31.5	56.2	10.3	41.9	14.8	35.2	17.8	31.9-25.0	25.0	15.9
313	Russell (6100' Glorieta, Perm.)	32.7	1.20	40	11.7	35.7	53.7	5.1	41.9	20.2	32.3	15.0	26.6-21.5	22.8	10.7
314	Ryssell (7000' Clear Fork, Perm.)	34.6	1.23	39	10.6	38.2	55.4	4.6	42.1	19.5	32.3	14.3	26.6-21.6	21.7	11.4
315	Russell, North (Dev.)	40.2	0.31	37	7.8	35.9	61.3	10.3	42.1	14.9	35.0	13.8	31.7-25.2	22.1	16.4
316	Salt Creek (Canyon, Penn.)	36.8	0.63	41	10.3	33.4	59.5	10.5	41.7	15.0	35.2	15.4	32.5-26.3	24.3	15.6
317	Sand Hills (Ellen., Cam.-Ord.)	37.0	0.73	45	5.4	27.7	60.2	19.2	43.4	12.7	35.0	12.3	32.5-25.9	26.2	17.0
318	Sand Hills (McKnight, San Andres, Perm.)	31.7	3.33	45	9.9	24.3	57.2	4.4	42.1	18.4	34.2	17.5	28.2-22.5	23.9	4.7
319	Sand Hills (Tubb, Perm.)	36.8	0.92	42	7.6	33.6	59.5	10.6	43.4	14.9	36.4	15.5	32.7-26.1	25.0	16.8
320	Scarborough (Yates, Perm.)	34.0	1.00	47	8.2	30.2	57.2	6.0	42.6	20.2	35.4	19.5	30.2-22.8	23.7	15.1
321	Seeligson (Zone 14-B, Frio, Olig.)	41.5	<0.10	34	3.9	40.2	54.7	8.0	42.1	33.3	35.8	12.1	31.7-22.0	5.2	13.5
322	Seeligson (Zone 19-B, Frio, Olig.)	43.6	<0.10	35	6.9	39.0	54.4	18.8	41.5	22.7	35.8	12.7	31.9-20.0	5.4	11.4
323	Seeligson (Zone 19-C, Frio, Olig.)	41.9	<0.10	34	4.4	38.8	55.2	9.5	44.3	34.5	36.2	12.2	31.5-20.5	4.0	12.6
324	Seeligson (Zone 20, Frio, Olig.)	40.2	<0.10	34	3.4	40.2	53.2	8.9	43.2	31.7	34.8	13.2	30.6-19.4	5.8	12.3
325	Seeligson (Zone 21-D, Frio, Olig.)	41.5	0.12	36	4.4	28.8	55.2	21.0	43.2	28.1	36.2	14.9	34.0-21.5	5.2	13.6

326	Seminole (San Andres, Perm.)........	33.6	1.86	43	9.7	32.7	57.7	5.0	42.6	19.4	34.4	16.1	29.1-22.5	26.0	12.5
327	Seminole, West (San Andres, Perm.).....	31.7	2.27	45	12.2	32.9	55.9	5.0	42.8	19.0	34.4	15.1	28.9-22.0	27.8	10.6
328	Shafter Lake (Dev.).........	38.6	0.77	40	7.6	33.7	61.0	4.5	42.6	18.8	36.4	16.4	32.7-26.3	24.4	17.0
329	Shafter Lake (San Andres, Perm.).....	37.4	0.25	46	6.0	26.5	56.2	20.5	43.4	10.9	37.0	16.6	35.6-28.2	25.2	19.7
330	Sharon Ridge (1700' San Andres, Perm.)...	27.1	2.04	58	6.6	26.9	54.7	4.3	41.7	18.1	32.5	16.5	26.6-21.0	33.0	7.5
331	Sharon Ridge (2400' San Angelo, Perm.)...	28.2	1.71	49	13.4	23.3	51.1	4.2	41.9	17.8	33.0	16.5	27.3-21.0	32.1	9.2
332	Sharon Ridge (Clear Fork, Perm.).....	29.1	1.67	49	13.5	29.1	55.4	4.4	42.3	19.2	33.8	16.4	27.3-21.5	30.8	9.6
333	Shipley (Queen, Perm.)........	36.0	1.34	41	8.9	34.7	59.5	9.7	41.7	14.4	34.4	16.2	30.6-24.0	23.9	14.2
334	Slaughter (San Andres, Perm.)......	31.1	2.04	48	12.1	31.1	56.4	4.4	45.2	18.6	34.4	16.1	28.4-21.6	29.3	10.7
335	Spraberry Trend area (Spraberry, Perm.)...	35.0	0.18	48	7.1	31.3	55.9	4.5	41.3	17.5	36.2	15.0	32.3-23.7	28.4	17.0
336	Taft (Frio, Olig.)...........	21.6	0.21	85	3.9					43.9	28.4	36.0	24.2-14.5	20.0	11.1
337	Talco (Trinity, L. Cre.)......	20.5	3.00	520	17.6	10.7	58.9	7.0	42.3	13.7	34.2	17.9	33.8-24.7	50.7	9.7
338	Thompson (3600', Mio.).......	23.8	0.25	140	4.2	23.4	54.7	13.2	42.3	32.4	29.9	33.4	27.3-22.1	33.5	18.2
339	Thompson, North (Vicksburg, Olig.)...	36.4	0.11	46	3.3	7.2	45.2	17.5	43.0	20.9	35.6	19.6	33.2-25.2	22.1	21.0
340	Thompson, South (Mio.).......	25.7	0.20	64	4.4	37.7	54.9	9.7	42.1	42.4	28.4	27.7	24.7-20.5	22.6	18.1
341	Tijerina-Canales-Blucher (Frio, Olig.)...	40.5	<0.10	35	7.7	30.6	55.2	19.7	45.8	23.8	35.6	14.5	33.4-19.7	6.4	12.2
342	Tom O'Connor (Frio, Olig.).....	34.8	0.17	39	4.4	32.8	59.7	9.7	42.1	38.5	34.5	19.6	27.5-14.5	9.9	9.9
343	TXL (Dev.)................	38.6	0.50	41	6.0	33.7	64.5	9.8	42.3	13.7	36.0	17.7	32.8-25.9	23.5	18.2
344	TXL (Ellen, Cam.-Ord.).......	42.3	0.21	39	3.9	28.9	52.0	17.1	43.2	12.5	39.4	12.1	39.6-27.7	21.5	21.1
345	TXL (San Andres, Perm.).......	30.8	1.93	49	9.8	30.3	59.2	11.0	42.3	20.3	34.4	17.4	29.3-22.0	26.8	11.9
346	TXL (Tubb, Perm.)...........	36.4	0.54	47	5.2	39.1	62.3	5.2	42.1	14.1	36.0	15.8	32.8-26.1	28.5	17.9
347	University-Block 9 (Dev.)......	44.7	<0.10	36	2.7	29.5	56.7	4.7	41.9	9.1	35.6	14.0	34.0-29.3	17.4	22.8
348	University-Block 9 (Penn.).....	36.4	0.12	45	5.3	36.4	56.4	16.5	43.4	14.7	36.4	17.1	33.6-28.9	26.7	19.5
349	University-Block 9 (Wolfcamp, Perm.)...	37.0	0.57	39	6.2	26.5	64.8	16.5	44.1	19.5	35.0	15.9	30.2-25.7	21.0	16.0
350	Van (Woodbine-Dexter, U. Cre.)....	35.4	0.82	51	8.3	30.9	58.4	8.7	42.1	9.5	32.5	17.5	32.3-24.5	30.1	16.0
351	Waddell (Grayburg, Perm.)......	33.6	1.69	46	9.8	38.3	64.5	5.4	41.5	19.7	35.2	16.9	30.4-24.3	26.8	12.9
352	Walnut Bend (Hudspeth, Strawn, Penn.)...	46.0	0.23	38	3.3	37.5	64.2	5.3	42.6	7.3	37.0	14.0	35.6-23.1	20.8	26.6
353	Walnut Bend (U. Strawn, Penn.).....	44.1	0.17	38	3.5	24.5	59.2	4.5	42.6	9.3	37.0	14.5	35.8-29.1	20.6	23.3
354	Walnut Bend (Winger, L. Strawn, Penn.)...	31.0	0.86	77	8.3	31.6	59.7	4.6	42.3	12.6	35.2	18.2	31.1-22.0	35.7	14.2
355	Ward-Estes, North (Yates, Perm.)....	34.0	1.17	45	7.7	33.8	58.2	4.9	42.6	18.8	33.8	16.4	29.1-22.0	26.8	15.1
356	Ward, South, (Yates, Perm.)......	35.8	1.12	42	8.0	33.3	57.2	5.3	42.6	19.5	35.2	17.2	30.4-23.0	23.8	15.7
357	Wasson (San Andres, Perm.)......	32.8	1.76	43	5.2	33.9	54.9	4.5	42.3	19.6	34.2	14.3	29.3-22.8	27.8	11.9
358	Wasson 66 (Clear Fork, Perm.).....	31.9	1.40	44	13.6	33.3	54.4	4.6	42.6	18.8	33.6	16.1	26.4-20.2	24.1	9.3
359	Wasson 72 (Clear Fork, Perm.).....	33.2	1.01	42	11.9	33.9	54.9	4.9	41.7	19.6	33.4	16.6	27.3-21.0	21.7	11.0
360	Webster (Marginulina, Frio, Olig.)...	29.3	0.21	64	4.6	35.9	49.7		41.3	31.4	32.8	25.2	28.9-22.5	27.6	18.4
361	Welch (San Andres, Perm.).......	32.3	2.14	45	11.7	14.5	56.2	4.9	42.6	19.0	34.2	16.4	28.9-22.6	21.0	12.5
362	West Columbia ("Z," Frio, Olig.)...	28.0	0.23	65	4.2	31.7	51.3	5.3	41.7	22.7	33.8	29.4	29.1-21.3	28.6	16.4
363	West Columbia New (Frio, Olig.)....	28.6	0.19	63	6.2	13.7	50.1		41.3	27.7	32.7	25.3	28.4-21.5	26.9	16.2
364	West Ranch (41-A, Frio, Olig.)....	31.5	0.17	41	3.5	25.7	51.1	5.7		41.4	31.0	21.0	26.1-14.4	9.7	11.9

Table 2-3. Properties of United States Crude Oils (Continued)

Item No.	State / Field (Formation, age)[1]	Gravity, °API	Sulfur, wt. per cent	Viscosity, SUS at 100°F	Carbon residue of residuum, wt. per cent	Gasoline and naphtha Per cent	Gasoline and naphtha Gravity, °API	Kerosine distillate Per cent	Kerosine distillate Gravity, °API	Gas oil distillate Per cent	Gas oil distillate Gravity, °API	Lubricating distillate Per cent	Lubricating distillate Gravity, °API	Residuum Per cent	Residuum Gravity, °API
	State														
365	West Ranch (98-A, Frio, Olig.)	39.8	0.11	35	3.9	35.2	55.7	8.8	44.1	34.2	36.0	14.9	31.5-18.7	5.6	11.1
366	West Ranch (Glasscock, Frio, Olig.)	31.0	0.13	41	3.8	23.5	50.1	......	......	40.6	31.3	23.5	26.8-16.7	11.5	14.1
367	West Ranch (Greta, Olig.)	24.9	0.16	57	2.7	4.3	44.7	......	......	48.8	29.7	28.2	24.7-16.5	18.1	15.1
368	West Ranch (Ward, Frio, Olig.)	30.8	0.15	40	4.3	23.7	49.9	......	......	42.0	31.0	24.2	26.3-13.9	8.8	11.6
369	White Point, East (5900' Greta, Olig.)	27.3	0.13	44	4.3	15.4	47.2	......	......	45.7	30.6	26.0	23.1-14.7	12.2	12.3
370	White Point, East (5600' Brigham, Frio, Olig.)	38.4	0.13	35	3.5	40.5	54.7	7.0	42.0	30.8	32.3	14.3	26.6-16.5	6.6	12.9
371	Yates (San Andres, Perm.)	30.2	1.54	59	9.9	24.1	56.4	4.4	40.2	17.9	33.0	20.3	28.8-21.6	31.6	14.5
	Utah														
372	Aneth (Hermosa, Penn.)	40.4	0.20	38	3.5	35.0	59.5	10.2	42.1	16.9	36.8	15.7	34.2-31.0	21.2	23.1
373	McElmo Creek (Paradox, Penn.)	40.0	<0.10	37	3.0	34.0	57.9	11.5	42.6	9.5	37.8	19.9	35.8-28.9	24.0	22.8
374	Ratherford (Paradox, Penn.)	41.3	<0.10	37	2.9	34.5	59.7	10.0	42.8	16.8	36.8	15.1	34.4-28.2	21.3	22.3
375	White Mesa (Paradox, Penn.)	41.1	0.10	36	2.9	35.2	59.5	11.3	42.6	14.9	36.6	14.8	34.4-27.3	21.5	22.3
	Wyoming														
376	Beaver Creek (Steele, U. Cre.)	33.8	0.20	48	5.6	24.4	54.2	11.7	42.8	18.1	36.6	18.1	33.8-25.9	26.8	16.7
377	Big Muddy (Frontier, U. Cre.)	35.8	0.12	47	4.7	26.3	58.2	3.5	42.3	20.5	36.1	16.5	33.0-26.4	31.2	19.8
378	Big Sand Draw (Tensleep, Penn.)	34.2	1.35	43	12.4	26.6	60.0	12.8	43.6	17.8	34.8	19.7	30.4-21.5	21.8	11.6
379	Bonanza (Thermopolis, U. Cre.)	35.8	1.87	37	9.7	36.1	60.5	5.8	41.5	21.5	32.8	17.8	26.4-18.4	17.9	10.7
380	Byron (Tensleep, Penn.)	24.3	2.50	140	12.2	12.3	58.7	7.2	42.1	13.4	32.8	20.3	29.3-20.0	45.9	11.7
381	Cottonwood (Phosphoria, Perm.)	28.6	2.52	63	13.9	20.2	58.7	8.5	41.7	16.3	33.2	23.9	29.1-19.4	28.9	10.1
382	Coyote Creek (Minnelusa, Penn.)	40.9	<0.10	36	4.6	36.5	59.7	9.2	43.0	18.6	36.6	13.5	34.0-27.7	21.2	20.7
383	Donkey Creek (Dakota, L. Cre.)	39.4	0.12	37	4.0	34.4	59.5	10.8	41.5	14.9	36.0	13.8	33.8-27.7	24.3	19.5
384	Elk Basin (Frontier, U. Cre.)	43.2	<0.10	35	3.3	46.1	57.7	5.2	41.3	20.8	36.4	14.4	32.8-27.7	11.7	19.5
385	Four Bear (Madison, Miss.)	13.8	3.58	6,000	22.7	2.1	52.7	5.4	43.2	12.8	35.6	26.9	28.6-14.8	52.9	4.0
386	Frannie (Tensleep, Penn.)	27.5	2.43	66	11.6	19.0	58.9	4.3	43.4	18.6	34.4	21.7	27.5-19.4	35.2	10.9
387	Garland (Amsden, Penn.)	22.0	2.88	180	18.3	13.3	56.4	8.4	42.8	14.1	35.2	21.1	28.6-17.9	42.8	7.1

No.															
388	Glenrock (Dakota, L. Cre.)	34.4	0.16	55	6.5	24.9	55.7	6.7	43.2	17.0	37.0	15.1	34.6–28.0	34.7	18.7
389	Grass Creek (Frontier, U. Cre.)	44.5	<0.10	35	2.3	45.8	60.2	10.8	42.3	15.8	36.4	13.1	33.6–27.5	13.2	20.7
390	Grieve (L. Cre.)	38.2	<0.10	42	4.0	28.8	58.9	4.9	42.6	19.8	37.0	17.7	33.6–26.8	25.8	19.2
391	Hamilton Dome (Tensleep, Penn.)	22.6	2.98	230	15.2	11.2	58.9	8.7	44.1	12.2	34.2	22.4	29.5–19.7	45.1	9.0
392	Little Buffalo Basin (Phosphoria, Perm.)	20.7	3.31	340	15.3	12.1	60.2	6.3	41.7	12.0	32.7	19.4	27.5–18.1	49.3	8.2
393	Lost Soldier (Cam.)	35.2	1.23	41	11.4	29.1	61.3	10.7	42.8	15.5	34.2	19.8	30.2–22.1	23.5	13.2
394	Meadow Creek (Sussex, U. Cre.)	38.8	0.12	39	4.4	32.5	60.2	10.6	42.8	15.9	36.0	17.4	33.2–25.4	21.9	18.4
395	Murphy Dome (Tensleep, Penn.)	34.0	1.70	43	10.8	29.3	61.5	11.3	42.5	18.8	33.6	17.8	29.1–21.0	22.6	11.3
396	Oregon Basin (Embar-Tensleep-Madison, Perm.-Penn.-Miss.)	20.5	3.25	360	20.5	14.3	59.5	3.3	42.6	14.8	33.8	19.6	27.7–17.9	47.5	6.8
397	Salt Creek (Wall Creek, U. Cre.)	36.6	0.12	43	4.4	27.6	56.9	10.1	42.8	16.4	36.8	18.9	34.0–27.1	26.1	19.2
398	Steamboat Butte (Tensleep, Penn.)	28.2	2.18	66	16.1	19.3	61.0	9.1	44.1	17.0	33.4	20.6	29.5–20.8	33.1	9.4
399	Sussex (Lakota, L. Cre.)	39.0	0.37	38	6.4	31.7	60.0	10.4	42.8	13.6	36.8	19.6	34.2–26.6	21.9	18.2
400	Wertz (Tensleep, Penn.)	33.6	1.32	43	10.1	28.5	59.7	10.8	42.5	16.1	33.6	19.9	29.8–19.2	24.3	13.5
401	Winkleman Dome (Phosphoria, Perm.)	25.7	2.59	93	16.8	16.0	57.7	8.4	42.3	15.2	32.8	21.5	28.7–19.5	37.3	9.2

[1] Geologic age names are abbreviated as follows: Cambrian, Cam.; Cambro-Ordovician, Cam.-Ord.; Cretaceous, Cre.; Lower Cretaceous, L. Cre.; Upper Cretaceous, U. Cre.; Devonian, Dev.; Upper Devonian, U. Dev.; Eocene, Eoc.; Jurassic, Jur.; Miocene, Mio.; Lower Miocene, L. Mio.; Upper Miocene, U. Mio.; Mississippian, Miss.; Oligocene, Olig.; Ordovician, Ord.; Lower Ordovician, L. Ord.; Middle Ordovician, M. Ord.; Pennsylvanian, Penn.; Permian, Perm.; Pliocene, Plio.; Pliocene-Miocene, Plio.-Mio.; Pliocene-Pleistocene, Plio.-Pleist.; Upper Pliocene, U. Plio.; Silurian, Sil.; Pre-Cambrian, Pre-Cam.

Table 2-4. Properties of Foreign Crude Oils

Item No.	Continent / Country / Field (Formation, age)[1]	Gravity, °API	Sulfur, wt. per cent	Viscosity, SUS at 100°F	Carbon residue of residuum, wt. per cent	Gasoline and naphtha Per cent	Gasoline and naphtha Gravity, °API	Kerosine distillate Per cent	Kerosine distillate Gravity, °API	Gas oil distillate Per cent	Gas oil distillate Gravity, °API	Lubricating distillate Per cent	Lubricating distillate Gravity, °API	Residuum Per cent	Residuum Gravity, °API
	AFRICA														
	Egypt														
1	Asl (Eoc.)	22.1	2.05	410	18.4	8.2	55.7	13.3	44.7	10.7	35.8	18.7	32.7–23.8	48.7	8.6
2	Sudr (Eoc.)	22.6	2.06	300	21.6	13.8	58.7	7.7	44.1	14.1	35.6	18.5	31.3–22.6	45.2	7.2
	Libya														
3	Dahra (PL-5, Paleocene)	39.2	0.33	40	7.6	36.6	59.5	12.2	43.4	16.7	36.8	14.7	32.7–25.4	18.6	15.1
4	Dahra (PL-7, Paleocene)	33.0	0.61	54	6.5	20.4	56.7	5.2	41.7	27.3	37.4	20.1	32.3–25.7	25.2	14.8
5	Hofra (Paleocene)	38.0	0.33	41	7.7	33.8	56.2	13.8	42.6	19.0	38.4	14.1	32.8–26.6	18.6	16.0
6	Zelten (Paleocene)	39.2	0.23	38	5.8	31.0	60.0	10.9	43.6	6.3	39.8	26.0	36.2–25.6	22.3	17.6
	Morocco														
7	Bled Khatara	40.0	<0.10	43	8.5	27.1	60.2	22.1	49.9	12.8	37.6	19.5	36.6–29.0	17.9	17.9
	ASIA														
	Bahrein														
8	Bahrein	35.2	1.42	44	10.8	29.4	60.2	11.3	44.5	16.4	36.0	16.8	31.9–25.2	25.1	14.4
	Indonesia														
	Borneo														
9	Seria	36.8	<0.10	35	3.8	37.1	52.5	......	......	39.9	32.8	15.5	29.1–25.0	7.4	11.4
	Sumatra (Central)														
10	Minas	32.8	0.10	92	3.7	18.6	61.8	14.7	43.8	10.8	38.0	18.1	......	37.3	17.1
	Iran														
11	Agha Jari	34.6	1.43	46	9.1	28.8	60.8	10.2	43.2	14.6	34.6	15.2	30.6–23.0	28.9	13.0
12	Central area	38.0	1.20	40	11.2	32.2	61.0	10.8	43.8	14.8	35.8	16.8	31.7–23.7	21.5	12.9
13	Gach Saran (Asmari, Olig.-L. Mio.)	33.0	1.66	55	15.0	27.2	61.5	8.3	42.3	13.0	34.8	15.2	31.0–22.8	32.1	10.1

No.	Name															
	Iraq															
14	Kirkuk	10.4	23.3	32.5–23.5	16.3	36.6	14.9	44.5	9.8	63.7	35.5	14.6	42	1.93	36.6
	Kuwait															
15	Burgan	10.0	34.8	31.1–21.5	15.1	35.6	13.9	44.7	8.2	63.9	25.5	13.6	68	2.62	31.5
	Saudi Arabia															
16	Abqaiq	15.3	24.0	32.1–24.7	16.5	36.4	14.9	44.3	10.5	62.3	32.0	10.3	40	1.30	38.0
17	Abu Hadriya	13.0	28.6	31.0–24.7	18.3	35.8	17.9	44.7	11.3	57.4	23.3	12.8	50	1.69	32.3
18	Dammam	14.1	23.2	31.1–24.0	17.2	35.8	18.3	43.8	11.3	58.4	29.0	11.7	42	1.54	35.6
19	Ghawar-Ain Dar (Arab D, Clastic, Jur.)	13.3	29.8	31.3–24.2	15.8	35.8	16.6	44.7	9.9	62.3	27.8	11.3	49	1.66	33.6
20	Ghawar-Haradh (Arab D, Clastic, Jur.)	10.6	30.9	30.6–22.0	16.0	35.2	13.2	43.8	9.5	62.3	28.1	7.1	52	2.14	32.8
21	Ghawar-Huiya (Arab D, Clastic, Jur.)	10.7	31.4	30.6–21.5	15.9	35.4	13.7	43.6	9.3	61.5	28.4	10.0	55	2.14	32.3
22	Ghawar-Shedgum (Arab D, Clastic, Jur.)	12.0	29.0	31.1–22.8	16.5	34.4	10.1	43.2	15.4	63.1	28.0	7.7	51	1.85	34.0
23	Ghawar-Uthmaniyah (Arab D, Clastic, Jur.)	11.9	31.9	31.0–23.8	13.5	35.2	15.0	44.5	9.6	63.7	27.4	12.7	50	1.94	33.8
24	Khursaniyah (Arab A, Clastic, Jur.)	11.6	34.3	30.8–22.3	14.8	35.6	15.1	44.9	8.7	62.9	25.2	9.4	55	2.38	32.1
25	Khursaniyah (Arab B, Clastic, Jur.)	11.1	34.3	30.2–22.1	15.9	35.4	14.3	44.3	8.6	61.8	25.8	11.2	59	2.49	31.1
26	Khursaniyah (Arab C, Clastic, Jur.)	11.0	34.8	30.8–22.5	16.3	35.6	14.2	44.7	8.9	62.1	24.8	5.1	57	2.69	31.3
27	Khursaniyah (Arab D, Clastic, Jur.)	9.0	33.4	31.1–21.6	16.4	35.6	14.1	44.5	8.3	62.6	25.8	9.7	63	2.54	30.8
28	Qatif (Arab C, Clastic, Jur.)	10.0	36.1	30.2–21.5	17.5	35.2	17.2	44.7	9.7	58.4	19.3	16.6	72	2.57	27.7
29	Qatif (Arab D, Clastic, Jur.)	14.4	22.4	31.3–23.7	16.5	36.0	16.3	44.7	10.1	59.7	31.4	10.4	40	1.59	38.2
30	Safaniya-Bahrain (Cre.)	7.1	39.3	30.4–21.8	15.9	35.8	12.5	44.7	8.6	64.8	22.2	11.3	110	3.03	27.7
31	Safaniya-Zubair (Cre.)	9.3	33.6	31.1–22.5	14.8	36.0	14.1	45.4	9.5	65.0	26.6	11.0	56	2.63	31.9
	CENTRAL AMERICA															
	Cuba															
32	Bacuranao-Cruz Verde	11.4	31.8	27.0–21.1	18.5	32.5	21.9			51.8	27.7	6.9	61	0.87	28.2
33	Jatibonico (Jiquima, U. Cre.)	7.4	48.1	22.5–12.2	30.7	27.9	20.7					11.4	1,340	1.87	14.8
	EUROPE															
	France															
34	Lacq	3.8	47.1	30.8–20.3	15.7	35.8	12.8	45.2	7.1	68.6	16.0	15.7	430	4.34	21.8
	Germany															
35	Georgsdorf	16.2	62.7	29.3–24.2	10.6	35.0	10.9	44.1	2.8	60.5	11.8	11.5	2,340	0.94	24.3
36	Nienhagen	14.5	37.8	32.1–24.2	19.0	35.6	16.7	43.0	8.6	60.0	17.8	9.8	92	0.87	29.3
37	Suderbruch	14.5	34.4	32.8–24.7	19.6	36.6	17.1	43.6	8.1	62.6	19.9	9.9	71	0.71	31.5

Table 2-4. Properties of Foreign Crude Oils (Continued)

Item No.	Continent / Country / Field (Formation, age)[1]	Gravity, °API	Sulfur, wt. per cent	Viscosity, SUS at 100°F	Carbon residue of residuum, wt. per cent	Gasoline and naphtha		Kerosine distillate		Gas oil distillate		Lubricating distillate		Residuum	
						Per cent	Grav- ity, °API	Per cent	Grav- ity, °API	Per cent	Grav- ity, °API	Per cent	Gravity, °API	Per cent	Grav- ity, °API
	Netherlands														
38	Ijsselmonde (Barremian, L. Cre.)......	19.2	0.53	780	4.6	2.9	46.3	......	......	17.3	30.8	19.6	26.3–21.1	59.4	13.8
39	Schoonebeck (Valanginian, L. Cre.).....	24.0	0.96	1,200	4.4	8.0	53.7	6.4	41.9	9.8	36.0	9.5	32.3–27.0	65.9	17.8
	Yugoslavia														
40	Lendava..................	35.6	0.14	38	3.4	26.0	47.2	......		46.2	35.0	18.7	31.1–26.4	8.2	19.2
	NORTH AMERICA														
	Canada														
	Alberta														
41	Acheson (D-2, Dev.).............	39.8	0.36	36	6.9	36.0	59.5	9.6	42.1	16.0	36.0	15.6	32.7–27.0	20.8	17.9
42	Bonnie Glen (Leduc D-3, Dev.)......	42.1	0.25	35	6.5	38.8	58.9	13.1	41.5	13.9	35.6	17.5	32.5–25.2	13.3	17.8
43	Fenn-Big Valley (D-2, Dev.)........	33.4	1.09	52	11.2	26.3	59.7	4.0	41.5	15.9	35.8	18.2	31.7–22.6	30.5	12.9
44	Fenn-Big Valley (D-3, Dev.)........	34.2	0.71	49	8.7	27.1	58.4	3.9	41.5	17.9	36.4	18.7	33.0–25.9	29.0	14.8
45	Golden Spike (D-3, Dev.)..........	35.2	0.23	44	8.0	32.0	58.4	4.6	41.1	19.5	36.0	17.3	32.3–24.9	25.9	15.7
46	Harmattan, East (Rundle, Miss.).....	38.0	0.37	38	3.6	33.6	55.9	6.1	41.5	26.3	33.6	16.8	29.1–23.1	14.4	17.3
47	Innisfail (D-3, Dev.).............	43.8	0.58	35	2.8	40.8	58.9	12.7	42.1	16.1	35.8	16.5	33.0–28.0	11.6	21.5
48	Joarcam (Viking, L. Cre.).........	36.0	0.13	45	4.9	25.3	57.9	9.6	43.2	18.8	37.0	17.5	33.8–27.7	28.0	19.5
49	Joffre (D-2, Dev.)...............	42.1	0.17	36	6.6	35.9	59.5	9.9	41.9	16.0	35.8	16.4	32.8–27.5	16.6	19.0
50	Joffre (Viking, L. Cre.)...........	41.3	0.04	38	3.0	31.8	61.5	16.0	42.8	8.4	36.0	14.6	34.4–29.5	25.4	21.8
51	Kaybob (Beaverhill Lake, Dev.)......	43.2	0.04	36	3.0	39.0	57.9	11.4	42.1	16.5	36.8	19.0	34.2–26.6	12.2	21.0
52	Leduc (D-2, U. Dev.).............	39.8	0.30	38	7.6	36.7	59.2	10.2	41.5	14.0	35.4	16.2	32.5–25.6	19.9	17.3
53	Leduc (D-3, U. Dev.).............	40.9	0.30	37	7.1	36.0	59.5	9.6	41.7	16.0	35.6	15.7	32.8–25.9	18.7	17.9
54	Pembina Crown (Cardium, U. Cre.)...	36.8	0.24	46	6.3	28.0	60.0	8.6	41.3	12.7	37.0	17.8	34.0–28.2	30.1	17.6
55	Redwater (D-3, U. Dev.)...........	34.8	0.55	45	12.7	29.2	59.2	4.3	42.3	18.6	36.2	16.5	32.1–24.5	28.1	13.3
56	Stettler (Dev.).................	30.2	1.46	60	13.9	24.6	60.8	4.1	41.5	16.5	34.4	17.6	30.0–21.8	34.7	10.9
57	Sturgeon Lake, South (D-2, Nisku, Dev.)............	36.2	0.34	38	7.1	32.7	57.2	5.0	41.5	21.4	35.2	16.2	31.1–24.2	23.4	15.6

Table (rotated on page; field/reservoir data, rows 58–82). Column headers are not shown on this page.

No.	Field														
58	Sturgeon Lake, South (D-3, Dev.)	36.0	0.34	39	6.1	31.9	55.2	41.3	4.7	21.7	35.0	16.5	31.3–24.7	23.1	16.2
59	Sturgeon Lake, South (Leduc D-3, U. Dev.)	38.2	0.37	37	6.4	33.7	57.7	42.1	4.7	20.5	36.0	17.0	31.5–24.3	20.6	16.8
60	Swan Hills (Beaverhill Lake, Dev.)	40.8	0.80	35	5.0	36.9	58.4	42.3	9.7	16.8	36.6	16.2	32.8–26.6	17.9	16.8
61	Wizard Lake (D-3, Dev.)	37.0	0.24	41	7.6	33.1	58.7	41.7	4.8	18.3	35.8	17.4	32.1–24.7	23.9	16.2
	Manitoba														
62	North Virden (Miss.)	32.8	1.47	46	12.0	26.4	59.2	42.3	5.0	20.5	34.2	20.7	28.9–22.0	25.5	13.3
	Saskatchewan														
63	Coleville (L. Cre.)	14.1	3.38	3,450	19.4	6.3	48.8			15.8	29.9	21.4	23.5–16.0	56.1	5.6
64	Dollard (Upper Shaunavon, Jur.)	22.5	2.87	188	19.3	18.1	58.9	41.7	3.1	15.8	33.0	16.0	27.3–20.5	45.6	5.9
65	Fosterton (Roseray, Jur.)	22.5	2.76	247	16.1	17.2	58.7	42.8	3.5	16.5	34.0	18.1	28.2–20.3	44.1	7.0
66	Midale (Mission Canyon, Miss.)	31.0	1.89	49	13.4	24.1	57.2	43.8	9.5	15.7	33.8	16.3	29.7–23.6	32.2	10.6
67	Nottingham (Mission Canyon, Miss.)	39.4	0.82	36	7.6	34.2	58.9	42.3	10.1	15.9	34.6	14.9	30.8–25.6	21.0	17.0
68	Steelman (Charles, Miss.)	38.6	0.73	36	9.8	33.2	57.2	42.3	10.1	16.7	33.8	16.2	30.0–24.8	20.5	18.2
69	Steelman-Frobisher (Mission Canyon, Miss.)	40.6	0.56	35	5.2	37.1	58.4	41.7	10.8	16.0	34.6	15.0	30.8–25.7	18.3	18.1
70	Weyburn (Charles, Miss.)	31.7	2.12	46	17.0	28.4	57.9	41.9	8.9	14.3	33.2	14.3	28.6–21.5	29.4	6.7
71	Weyburn (Mission Canyon, Miss.)	35.4	1.66	39	15.4	34.4	59.5	42.1	4.4	18.4	35.2	14.9	29.8–22.3	23.7	9.2
	Mexico														
72	Cacalilao (Aqua Nueva)	12.3	5.23	6,000	14.1	7.3	50.4	40.0	3.2	14.4	31.9	10.1	24.9–18.4	67.9	
73	El Plan (Lignitic Cedral)	19.0	3.50	650	12.8	7.3	53.7	34.8	6.4	13.8	33.2	25.2	28.6–18.2	50.3	9.6
74	Ezequiel Ordonez	19.4	3.23	920	15.0	12.9	59.5	43.0	7.4	10.7	34.0	16.3	29.9–22.0	53.4	6.3
75	Naranjos (El Abra)	20.8	3.80	835	12.0	15.3	58.9	41.5	9.2	11.9	34.4	15.5	29.3–20.5	49.9	5.5
76	Poza Rica (L. Tamabra)	29.1	1.77	80	12.9	19.3	58.2	42.1	6.3	14.6	35.8	17.6	31.7–24.7	39.3	14.8
77	Santa Agueda (El Abra)	16.0	3.98	4,880	15.7	12.0	59.5			10.6	34.0	13.2	28.0–19.2	57.0	2.4
	SOUTH AMERICA														
	Chile														
78	Manantiales	40.9	<0.10	39	2.4	27.3	56.7	43.0	13.3	26.3	37.2	18.8	35.0–31.0	13.3	24.0
	Colombia														
79	Casabe ("A" Argillaceous, Colorado)	20.8	1.08	860	16.6	5.8	51.1			18.7	34.0	27.1	29.3–21.0	47.9	11.6
80	Casabe ("B2" Argillaceous, Mugrosa, L. Olig.)	20.5	1.06	660	12.6	6.1	48.5			17.9	33.0	30.1	28.0–19.7	45.1	12.5
81	Infantas	29.5	0.88	96	12.9	19.1	59.5	41.9	8.1	12.8	35.2	19.3	31.7–24.2	39.5	14.8
82	La Cira	24.3	0.96	220	13.9	12.1	54.9	40.2	2.9	16.2	34.0	24.4	29.5–22.0	43.0	13.5

Table 2-4. Properties of Foreign Crude Oils (Continued)

Item No.	Continent / Country / Field (Formation, age)[1]	Gravity, °API	Sulfur, wt. per cent	Viscosity, SUS at 100°F	Carbon residue of residuum, wt. per cent	Gasoline and naphtha Per cent	Gasoline and naphtha Gravity, °API	Kerosine distillate Per cent	Kerosine distillate Gravity, °API	Gas oil distillate Per cent	Gas oil distillate Gravity, °API	Lubricating distillate Per cent	Lubricating distillate Gravity, °API	Residuum Per cent	Residuum Gravity, °API
83	Tibu	33.0	0.99	64	11.3	22.8	61.3	8.9	43.4	13.5	36.8	17.7	33.6–26.4	35.7	15.3
84	Velasquez	28.2	0.85	110	12.3	16.6	59.5	8.1	42.6	14.0	35.2	24.6	31.5–22.5	35.4	13.2
	Peru														
85	La Brea-Parinas (High cold test)	37.6	<0.10	40	6.6	33.7	55.4	12.6	41.1	16.0	36.6	19.7	33.4–25.9	17.8	15.9
86	La Brea-Parinas (Low cold test)	34.8	0.12	41	7.4	38.0	55.4			20.1	34.0	20.5	29.7–20.7	20.7	14.1
	Venezuela														
87	Bachaquero	14.7	2.62	3,310	9.6	5.7	45.6			13.0	31.0	14.7	25.9–15.9	65.0	7.8
88	Boscan	11.3	5.53	6,000	7.2	2.2	52.5	2.0	40.9	11.5	32.7	12.6	24.3–18.4	68.7	4.0
89	Jusepin	32.5	0.89	50	7.8	24.9	57.7	5.0	41.7	18.9	34.6	18.8	30.8–24.2	29.4	15.4
90	Lagunillas	17.1	2.18	1,550	19.9	7.5	52.0			16.3	32.8	23.8	26.8–18.7	51.0	7.5
91	Mercedes (Heavy, C. Tert.)	25.0	1.76	120	17.2	13.7	48.8	4.9	42.1	22.4	35.2	20.6	30.8–24.5	37.6	9.9
92	Mercedes (Light, A, Tert.)	34.6	0.66	44	9.4	24.9	52.5	14.8	41.3	20.0	35.4	18.7	29.7–27.9	21.3	16.4
93	Oficina	38.0	0.59	38	12.7	35.6	58.2	5.2	42.3	22.2	35.0	16.4	30.8–24.5	18.5	13.3
94	Tia Juana	27.0	1.49	148	8.7	16.7	55.2	3.8	42.6	15.9	34.8	15.2	30.0–23.3	44.1	12.0
95	Tucupita	16.5	1.05	2,500	19.5	2.6	50.6	1.4	40.0	11.9	31.7	26.1	28.0–19.0	56.8	8.7
	WEST INDIES														
	Trinidad														
96	(Field name unknown)	33.0	0.40	46	9.7	36.2	56.7			21.6	32.5	16.8	26.6–20.5	23.2	11.1

[1] Geologic age names are abbreviated as follows: Cretaceous, Cre.; Lower Cretaceous, L. Cre.; Upper Cretaceous, U. Cre.; Devonian, Dev.; Upper Devonian, U. Dev.; Eocene, Eoc.; Jurassic, Jur.; Lower Miocene, L. Mio; Mississippian, Miss.; Oligocene, Olig.; Lower Oligocene, L. Olig.; Tertiary, Tert.

Table 2-5.1. U.S. Crude-petroleum Analysis

CITRONELLE FIELD (Mobile County, Alabama)
Rodessa, Lower Glenrose, Lower Cretaceous
10,708–11,200 ft

GENERAL CHARACTERISTICS

Gravity, specific, 0.808   Gravity, ° API, 43.6   Pour point, ° F., 30
Sulfur, percent, 0.38   Color, brownish black
Viscosity, Saybolt Universal at 100° F., 40 sec.   Nitrogen, percent, 0.02

DISTILLATION, BUREAU OF MINES ROUTINE METHOD

STAGE 1—Distillation at atmospheric pressure, 744 mm. Hg
First drop, 77 ° F.

Fraction No.	Cut temp. ° F.	Percent	Sum, percent	Sp. gr., 60/60° F.	° API, 60° F.	C. I.	Refractive index, n, at 20° C.	Specific dispersion	S. U. visc., 100° F.	Cloud test, ° F.
1	122	2.6	2.6	0.666	81.0					
2	167	3.5	6.1	.674	78.4	9.4	1.37257	127.7		
3	212	4.6	10.7	.689	73.9	6.6	1.38750	127.2		
4	257	5.7	16.4	.710	67.8	8.0	1.39815	127.0		
5	302	5.8	22.2	.728	62.9	8.5	1.40734	128.6		
6	347	5.8	28.0	.745	58.4	9.7	1.41645	127.8		
7	392	6.2	34.2	.762	54.2	12	1.42487	128.0		
8	437	6.0	40.2	.776	50.9	13	1.43241	130.3		
9	482	7.1	47.3	.793	46.9	15	1.43978	131.2		
10	527	7.6	54.9	.805	44.3	16	1.44764	137.3		

STAGE 2—Distillation continued at 40 mm. Hg

Fraction No.	Cut temp. ° F.	Percent	Sum, percent	Sp. gr., 60/60° F.	° API, 60° F.	C. I.	Refractive index, n, at 20° C.	Specific dispersion	S. U. visc., 100° F.	Cloud test, ° F.
11	392	3.8	58.7	0.829	39.2	24	1.45865	142.9	40	30
12	437	6.3	65.0	.837	37.6	24	1.46297	140.9	45	40
13	482	5.3	70.3	.849	35.2	26	1.46985	145.9	56	65
14	527	6.4	76.7	.866	31.9	31			80	80
15	572	4.9	81.6	.879	29.5	34			150	95
Residuum		16.7	98.3	.956	16.5					

Carbon residue, Conradson: Residuum, 6.7 percent; crude, 1.3 percent.

APPROXIMATE SUMMARY

	Percent	Sp. gr.	° API	Viscosity
Light gasoline	10.7	0.676	77.8	
Total gasoline and naphtha	34.2	0.718	65.6	
Kerosine distillate	20.7	.792	47.2	
Gas oil	9.6	.834	38.2	
Nonviscous lubricating distillate	10.7	.842-.870	36.6-31.1	50-100
Medium lubricating distillate	6.4	.870-.885	31.1-28.4	100-200
Viscous lubricating distillate	--	--	--	Above 200
Residuum	16.7	.956	16.5	
Distillation loss	1.7			

Table 2-5.2. U.S. Crude-petroleum Analysis

MAGNOLIA FIELD (Columbia County, Arkansas)
Reynolds-Smackover, Jurassic
7,392–7,570 ft

GENERAL CHARACTERISTICS

Gravity, specific, 0.833   Gravity, °API, 38.4   Pour point, °F, below 5
Sulfur, percent, 0.90   Color, brownish green
Viscosity, Saybolt Universal at 100° F., 38 sec.   Nitrogen, percent, 0.02

DISTILLATION, BUREAU OF MINES ROUTINE METHOD

STAGE 1—Distillation at atmospheric pressure, 749 mm. Hg
First drop, 82 °F.

Fraction No.	Cut temp. °F.	Percent	Sum. percent	Sp.gr. 60/60° F.	°API, 60° F.	C.L.	Refractive Index, n_D at 20° C.	Specific dispersion	S.U. vis. 100° F.	Cloud test, ° F.
1	122	1.3	1.3	0.661	82.6					
2	167	3.0	4.3	.694	81.6	4.6				
3	212	4.3	8.6	.699	70.9	11	1.37422	123.7		
4	257	5.7	14.3	.730	62.1	17	1.39287	126.9		
5	302	6.3	20.6	.757	55.4	22	1.41147	139.9		
6	347	6.6	27.2	.778	50.4	25	1.42443	144.5		
7	392	5.0	32.2	.791	47.4	25	1.43654	154.4		
8	437	5.3	37.5	.801	45.2	24	1.44388	154.2		
9	482	5.2	42.7	.813	42.6	25	1.44853	156.3		
10	527	7.6	50.3	.831	38.8	29	1.46475	162.5		

STAGE 2—Distillation continued at 40 mm. Hg

Fraction No.	Cut temp. °F.	Percent	Sum. percent	Sp.gr. 60/60° F.	°API, 60° F.	C.L.	Refractive Index, n_D at 20° C.	Specific dispersion	S.U. vis. 100° F.	Cloud test, ° F.
11	392	3.6	53.9	.851	34.8	35	1.47540	167.4	40	20
12	437	5.7	59.6	.861	32.8	35	1.48025	169.1	45	40
13	482	6.5	66.1	.881	29.1	41	1.49130	—	55	55
14	527	4.3	70.4	.895	26.7	45	1.49747	—	85	70
15	572	5.9	76.3	.901	25.6	45	—	—	150	85
Residuum		20.5	96.8	.951	17.3					

Carbon residue, Conradson: Residuum, 6.7 percent; crude, 1.6 percent.

APPROXIMATE SUMMARY

	Percent	Sp. gr.	°API	Viscosity
Light gasoline	8.6	0.681	76.3	
Total gasoline and naphtha	32.2	0.742	59.2	
Kerosene distillate	10.5	.807	43.8	
Gas oil	22.2	.849	35.2	
Nonviscous lubricating distillate	9.2	.868–.896	31.5–26.4	50-100
Medium lubricating distillate	2.2	.896–.899	26.4–25.9	100-200
Viscous lubricating distillate				Above 200
Residuum	20.5	.951	17.3	
Distillation loss	3.2			

Table 2-5.3. U.S. Crude-petroleum Analysis

SMACKOVER (ASPHALT) FIELD (Union County, Arkansas)

GENERAL CHARACTERISTICS

Gravity, specific, 0.919   Gravity, °API, 22.5   Pour point, °F, below 5
Sulfur, percent, 2.10   Color, brownish black
Viscosity, Saybolt Universal at 100° F., 220 sec.;   130° F., 125 sec.   Nitrogen, percent, 0.08

DISTILLATION, BUREAU OF MINES ROUTINE METHOD

STAGE 1—Distillation at atmospheric pressure, 742 mm. Hg
First drop, 160 °F.

Fraction No.	Cut temp. °F.	Percent	Sum. percent	Sp.gr. 60/60° F.	°API, 60° F.	C.L.	Refractive Index, n_D at 20° C.	Specific dispersion	S.U. vis. 100° F.	Cloud test, ° F.
1	122									
2	167									
3	212	1.7	1.7	0.741	59.5	26				
4	257	1.8	3.5	.748	57.7	32				
5	302	1.7	5.2	.777	50.6	35	1.42713	137.0		
6	347	2.9	8.1	.799	45.6	38	1.45638	153.8		
7	392	3.1	11.2	.818	41.5	42	1.46379	156.5		
8	437	3.3	14.5	.839	37.2	41	1.47093	155.6		
9	482	4.2	18.7	.847	35.6	41	1.47957	171.5		
10	527	5.4	24.1	.859	33.2	42				

STAGE 2—Distillation continued at 40 mm. Hg

Fraction No.	Cut temp. °F.	Percent	Sum. percent	Sp.gr. 60/60° F.	°API, 60° F.	C.L.	Refractive Index, n_D at 20° C.	Specific dispersion	S.U. vis. 100° F.	Cloud test, ° F.
11	392	3.7	27.8	0.873	30.6	45	1.48768	174.3	41	Below 5
12	437	5.2	33.0	.885	28.4	46	1.49127	—	47	do.
13	482	6.4	39.4	.899	25.9	50	1.49992	—	61	10
14	527	5.9	45.3	.914	23.3	54	1.50771	—	93	30
15	572	6.7	52.0	.922	22.0	55	1.51153	—	175	55
Residuum		47.0	99.0	.985	12.2					

Carbon residue, Conradson: Residuum, 8.5 percent; crude, 4.3 percent.

APPROXIMATE SUMMARY

	Percent	Sp. gr.	°API	Viscosity
Light gasoline	1.7	0.741	59.5	
Total gasoline and naphtha	11.2	0.784	49.0	
Kerosene distillate				
Gas oil	20.6	.860	33.0	
Nonviscous lubricating distillate	11.1	.888–.915	27.9–23.1	50-100
Medium lubricating distillate	7.8	.915–.925	23.1–21.5	100-200
Viscous lubricating distillate	2.5	.925–.926	21.5–21.3	Above 200
Residuum	47.0	.985	12.2	
Distillation loss	1.0			

Table 2-5.4. U.S. Crude-petroleum Analysis
COALINGA NOSE FIELD (Fresno County, California)
Gatchell Zone, Eocene
7,686–8,217 ft

GENERAL CHARACTERISTICS

Gravity, specific, 0.868   Gravity, ° API, 31.5   Pour point, °F., 25
Sulfur, percent, 0.25   Color, brownish green
Viscosity, Saybolt Universal at 77° F., 58 sec.;   Nitrogen, percent, 0.19

DISTILLATION, BUREAU OF MINES ROUTINE METHOD
Stage 1—Distillation at atmospheric pressure, 746 mm. Hg
First drop, 82 °F.

Fraction No.	Cut temp. °F.	Percent	Sum, percent	Sp.gr. 60/60° F.	° API 60° F.	C.I.	Refractive index, n_D at 20° C.	Specific dispersion	S.U. visc., 100° F.	Cloud test, °F.
1	122	1.3	1.3	0.659	83.2	20				
2	167	0.7	2.0	.696	71.8	27				
3	212	4.2	6.2	.731	62.1	27	1.39658	129.2		
4	257	5.7	11.9	.762	54.2	32	1.42148	141.7		
5	302	4.7	16.6	.780	49.9	33	1.43330	140.2		
6	347	4.2	20.8	.795	46.5	33	1.44190	145.5		
7	392	4.3	25.1	.810	43.2	34	1.44933	144.3		
8	437	4.6	29.7	.826	39.8	36	1.45832	146.1		
9	482	6.2	35.9	.844	36.2	40	1.46932	163.9		
10	527	6.8	42.7	.857	33.6	41	1.47813	170.0		

Stage 2—Distillation continued at 40 mm. Hg

Fraction No.	Cut temp. °F.	Percent	Sum, percent	Sp.gr. 60/60° F.	° API 60° F.	C.I.	Refractive index, n_D at 20° C.	Specific dispersion	S.U. visc., 100° F.	Cloud test, °F.
11	392	4.1	46.8	0.860	33.0	38	1.48172	173.5	.39	10
12	437	6.8	53.6	.872	30.8	40	1.48535	164.7	.44	40
13	482	7.3	60.9	.885	28.4	43	1.49259	174.2	.55	56
14	527	6.0	66.9	.897	26.3	46	1.49290	..	.80	70
15	572	6.6	73.5	.912	23.7	50			150	86
Residuum		24.8	98.3	.976	13.5					

Carbon residue, Conradson: Residuum, 8.0 percent; crude, 2.2 percent.

APPROXIMATE SUMMARY

	Percent	Sp. gr.	° API	Viscosity
Light gasoline	6.2	0.712	67.2	
Total gasoline and naphtha	25.1	0.767	53.0	
Kerosine distillate				
Gas oil	32.4	.856	33.8	
Nonviscous lubricating distillate	8.4	.885-.900	28.4-25.7	50-100
Medium lubricating distillate	7.8	.900-.919	25.7-22.5	100-200
Viscous lubricating distillate				
Residuum	24.8	.976	13.5	Above 200
Distillation loss	1.7			

Table 2-5.5. U.S. Crude-petroleum Analysis
CUYAMA, SOUTH FIELD (Santa Barbara County, California)
Dibblee, Miocene
4,195–4,540 ft

GENERAL CHARACTERISTICS

Gravity, specific, 0.863   Gravity, ° API, 32.5   Pour point, °F., 20
Sulfur, percent, 0.42   Color, brownish black
Viscosity, Saybolt Universal at 100° F., 49 sec.   Nitrogen, percent, 0.34

DISTILLATION, BUREAU OF MINES ROUTINE METHOD
Stage 1—Distillation at atmospheric pressure, 748 mm. Hg
First drop, 81 °F.

Fraction No.	Cut temp. °F.	Percent	Sum, percent	Sp.gr. 60/60° F.	° API 60° F.	C.I.	Refractive index, n_D at 20° C.	Specific dispersion	S.U. visc., 100° F.	Cloud test, °F.
1	122	1.9	1.9	0.642	88.9	11	1.37297	122.3		
2	167	2.2	4.1	.677	77.5		1.40222	124.9		
3	212	5.3	9.4	.726	63.4	24	1.41587	126.6		
4	257	6.2	15.6	.753	56.4	28	1.42765	134.5		
5	302	5.5	21.1	.773	51.6	30	1.43888	137.2		
6	347	4.9	26.0	.792	47.2	32	1.44900	138.7		
7	392	4.1	30.1	.810	43.2	34	1.45706	143.2		
8	437	4.6	34.7	.825	40.0	36	1.46568	151.0		
9	482	5.0	39.7	.838	37.4	38	1.47397	161.2		
10	527	5.7	45.4	.851	34.8	38				

Stage 2—Distillation continued at 40 mm. Hg

Fraction No.	Cut temp. °F.	Percent	Sum, percent	Sp.gr. 60/60° F.	° API 60° F.	C.I.	Refractive index, n_D at 20° C.	Specific dispersion	S.U. visc., 100° F.	Cloud test, °F.
11	392	3.3	48.7	0.860	33.0	38	1.48078	167.2	.40	14
12	437	5.7	54.4	.876	30.9	42	1.48373	161.8	.47	34
13	482	4.6	59.0	.888	27.9	45	1.49372	..	.63	54
14	527	4.2	63.2	.904	25.0	49	1.49224	..	.98	70
15	572	6.0	69.2	.916	23.0	52			220	80
Residuum		27.6	96.8	.897	10.4					

Carbon residue, Conradson: Residuum, 6.5 percent; crude, 1.9 percent.

APPROXIMATE SUMMARY

	Percent	Sp. gr.	° API	Viscosity
Light gasoline	9.4	0.698	71.2	
Total gasoline and naphtha	30.1	0.753	56.4	
Kerosine distillate	4.6	.825	40.0	
Gas oil	17.9	.853	34.4	
Nonviscous lubricating distillate	8.6	.878-.904	29.7-25.0	50-100
Medium lubricating distillate	4.2	.904-.914	25.0-23.3	100-200
Viscous lubricating distillate	3.8	.914-.823	23.3-21.8	Above 200
Residuum	27.6	.997	10.4	
Distillation loss	3.2			

Table 2-5.6. U.S. Crude-petroleum Analysis
ELK HILLS FIELD (Kern County, California)
Main Area, Shallow Zone, Pliocene
3,000–3,200 ft

GENERAL CHARACTERISTICS

Gravity, specific, 0.917   Gravity, °API, 22.8   Pour point, °F., below 5
Sulfur, percent, 0.68   Color, brownish black
Viscosity, Saybolt Universal at 100° F., 135 sec.;   Nitrogen, percent, 0.47

DISTILLATION, BUREAU OF MINES ROUTINE METHOD

STAGE 1—Distillation at atmospheric pressure, 737 mm. Hg
First drop, 171 °F.

Fraction No.	Cut temp. °F.	Percent	Sum. percent	Sp. gr. 60/60° F.	°API 60° F.	C.I.	Refractive index, n at 20° C.	Specific dispersion	S.U. visc. 100° F.	Cloud test, °F.
1	122									
2	167									
3	212	1.2	1.2	0.724	63.9		1.40106	124.4		
4	257	2.6	3.8	.759	54.9	31	1.41654	125.8		
5	302	3.6	7.4	.781	49.7	34	1.43116	130.2		
6	347	3.7	11.1	.813	42.6	42	1.44560	138.4		
7	392	4.2	15.3	.836	37.8	47	1.45870	143.4		
8	437	5.6	20.9	.852	34.6	49	1.46832	148.0		
9	482	6.9	27.8	.865	32.1	50	1.47806	163.5		
10	527	7.6	35.4	.881	29.1	52	1.48833	172.4		

STAGE 2—Distillation continued at 40 mm. Hg

Fraction No.	Cut temp. °F.	Percent	Sum. percent	Sp. gr. 60/60° F.	°API 60° F.	C.I.	Refractive index, n at 20° C.	Specific dispersion	S.U. visc. 100° F.	Cloud test, °F.
11	392	5.0	40.4	0.897	26.3	56	1.49786	165.8	46	Below 5
12	437	5.2	45.6	.905	24.6	56	1.50326	-	58	do.
13	482	5.2	57.8	.924	21.6	62	1.51153	-	97	do.
14	527	5.0	57.8	.938	19.4	65	1.51876	-	230	do.
15	572	4.0	61.8	.945	18.2	65			Over 400	do.
Residuum		37.4	99.2	.992	11.1					

Carbon residue, Conradson: Residuum, 4.6 percent; crude, 2.1 percent.

APPROXIMATE SUMMARY

	Percent	°API	Viscosity
Light gasoline	1.2	63.9	
Total gasoline and naphtha	11.1	49.9	
Kerosine distillate	28.7	31.7	
Gas oil	9.8	25.7–21.5	50-100
Nonviscous lubricating distillate	4.4	21.5–19.8	100-200
Medium lubricating distillate	7.8	19.8–17.8	Above 200
Viscous lubricating distillate			
Residuum	37.4	11.1	
Distillation loss	.8		

Table 2-5.7. U.S. Crude-petroleum Analysis
HUNTINGTON BEACH FIELD (Orange County, California)
South Main Area, Miocene
6,616 ft

GENERAL CHARACTERISTICS

Gravity, specific, 0.918   Gravity, °API, 22.6   Pour point, °F., below 5
Sulfur, percent, 1.57   Color, brownish black
Viscosity, Saybolt Universal at 100° F., 210 sec.;   Nitrogen, percent, 0.65
130° F., 140 sec.

DISTILLATION, BUREAU OF MINES ROUTINE METHOD

STAGE 1—Distillation at atmospheric pressure, 754 mm. Hg
First drop, 102 °F.

Fraction No.	Cut temp. °F.	Percent	Sum. percent	Sp. gr. 60/60° F.	°API 60° F.	C.I.	Refractive index, n at 20° C.	Specific dispersion	S.U. visc. 100° F.	Cloud test, °F.
1	122	1.8	1.8	0.680	76.6	-				
2	167	5.7	7.5	.747	57.9	34				
3	212	1.7	9.2	.768	52.7	35	1.40535	124.2		
4	257	3.9	13.1	.778	50.4	32	1.42567	126.5		
5	302	3.9	17.0	.800	45.4	36	1.43913	131.3		
6	347	3.0	20.0	.821	40.9	40	1.45053	142.1		
7	392	3.8	23.8	.837	37.6	42	1.46041	145.5		
8	437	3.7	27.5	.849	35.2	42	1.46780	148.9		
9	482	5.5	33.0	.860	33.0	42	1.47564	149.3		

STAGE 2—Distillation continued at 40 mm. Hg

Fraction No.	Cut temp. °F.	Percent	Sum. percent	Sp. gr. 60/60° F.	°API 60° F.	C.I.	Refractive index, n at 20° C.	Specific dispersion	S.U. visc. 100° F.	Cloud test, °F.
11	392	2.6	35.6	0.873	30.6	45	1.48527	159.8	43	Below 5
12	437	5.3	40.9	.884	28.6	46	1.48886	161.5	49	20
13	482	4.5	45.4	.898	26.1	49	1.49701	164.3	66	34
14	527	4.5	49.9	.917	22.8	55	1.50749	-	120	54
15	572	6.6	56.5	.936	19.7	-	1.51572	-	310	70
Residuum		43.1	99.6	1.015	7.9					

Carbon residue, Conradson: Residuum, 6.2 percent; crude, 3.0 percent.

APPROXIMATE SUMMARY

	Percent	Sp. gr.	°API	Viscosity
Light gasoline	7.5	0.731	62.1	
Total gasoline and naphtha	20.0	0.770	57.3	
Kerosine distillate	18.6			
Gas oil	7.4	.858	33.4	
Nonviscous lubricating distillate	3.6	.885–.910	28.4–24.0	50-100
Medium lubricating distillate	6.9	.910–.925	24.0–21.5	100-200
Viscous lubricating distillate		.925–.943	21.5–17.0	Above 200
Residuum	43.1	1.015	7.9	
Distillation loss	.4			

1/ Distillation discontinued at 568° F.

u. s. government printing office 16-57848-3

Table 2-5.8. U.S. Crude-petroleum Analysis

MIDWAY-SUNSET FIELD (Kern County, California)
Pleistocene-Pliocene
3,718 ft

GENERAL CHARACTERISTICS

Gravity, specific, 0.924 Gravity, °API, 21.6 Pour point, °F, below 5
Sulfur, percent, 0.89 Color, brownish green
Viscosity, Saybolt Universal at 100° F.: 210 sec.; Nitrogen, percent,
130° F.: 110 sec.

DISTILLATION, BUREAU OF MINES ROUTINE METHOD

STAGE 1—Distillation at atmospheric pressure, 751 mm. Hg
First drop, 147 °F.

Fraction No.	Cut temp. °F.	Percent	Sum. percent	Sp. gr. 60/60° F.	°API 60° F.	C.I.	Refractive index, n_D at 20° C.	Specific dispersion	S.U. visc. 100° F.	Cloud test, °F.
1	122	0.2	0.2	0.724	63.9	–				
2	167	3.0	3.2	0.740	59.7	31	1.41759	120.0		
3	212	3.5	6.7	.763	54.0	33	1.42839	124.0		
4	257	3.7	10.4	.783	49.2	36	1.43440	128.4		
5	302	4.4	14.8	.804	44.5	38	1.44031	135.8		
6	347	3.6	18.4	.831	38.8	44	1.45440	141.4		
7	392	3.9	22.3	.847	35.6	46	1.46502	152.5		
8	437	5.8	28.1	.863	32.5	49	1.47545	152.5		
9	482	6.4	34.5	.880	29.3	52	1.48647			
10	527									

STAGE 2—Distillation continued at 40 mm. Hg

Fraction No.	Cut temp. °F.	Percent	Sum. percent	Sp. gr. 60/60° F.	°API 60° F.	C.I.	Refractive index, n_D at 20° C.	Specific dispersion	S.U. visc. 100° F.	Cloud test, °F.
11	392	2.4	36.9	0.892	27.1	54	1.49443	168.8	44	Below 5
12	437	4.5	41.4	.901	25.1	54	1.49931	171.1	52	do.
13	482	5.8	47.2	.918	22.6	59	1.50820	168.6	78	do.
14	527	4.8	52.0	.934	20.0	63	1.51647	–	170	do.
15	572	7.0	59.0	.950	17.5	–	1.52344	–	Over 400	do.
Residuum		40.6	99.6	1.006	9.2					

Carbon residue, Conradson: Residuum, 5.7 percent; crude, 2.5 percent.

APPROXIMATE SUMMARY

	Percent	Sp. gr.	°API	Viscosity
Light gasoline	3.2	0.737	60.0	
Total gasoline and naphtha	14.8	0.775	51.1	
Kerosene distillate				
Gas oil	23.5	.865	32.1	
Nonviscous lubricating distillate	4.3	.899–.921	25.9–22.1	50-100
Medium lubricating distillate	4.8	.921–.935	22.1–19.8	100-200
Viscous lubricating distillate	8.8	.935–.946	19.8–16.0	Above 200
Residuum	40.6	1.006	9.2	
Distillation loss	.4			

U.S. GOVERNMENT PRINTING OFFICE 16-57688-3

Table 2-5.9. U.S. Crude-petroleum Analysis

SAN ARDO FIELD (Monterey County, California)
Lombardi, Miocene
2,100–2,250 ft

GENERAL CHARACTERISTICS

Gravity, specific, 0.992 Gravity, °API, 11.1 Pour point, °F, 80
Sulfur, percent, 2.25 Color, brownish black
Viscosity, Saybolt Universal at 100° F.: over 6,000 sec.; Nitrogen, percent, 0.91
130° F.: over 6,000 sec.

DISTILLATION, BUREAU OF MINES ROUTINE METHOD

STAGE 1—Distillation at atmospheric pressure, 749 mm. Hg
First drop, 352 °F.

Fraction No.	Cut temp. °F.	Percent	Sum. percent	Sp. gr. 60/60° F.	°API 60° F.	C.I.	Refractive index, n_D at 20° C.	Specific dispersion	S.U. visc. 100° F.	Cloud test, °F.
1	122									
2	167									
3	212									
4	257									
5	302									
6	347									
7	392	1.0	1.0	0.832	38.6	–				
8	437	1.4	2.4	.861	32.8	53				
1/ 9	482	4.1	6.5	.867	31.7	–	1.45473	125.6		
10	527									

STAGE 2—Distillation continued at 40 mm. Hg

Fraction No.	Cut temp. °F.	Percent	Sum. percent	Sp. gr. 60/60° F.	°API 60° F.	C.I.	Refractive index, n_D at 20° C.	Specific dispersion	S.U. visc. 100° F.	Cloud test, °F.
11	392	4.9	11.4	0.892	27.1	–	1.48846	141.1	43	Below 5
12	437	5.5	16.9	.901	24.2	58	1.49896	155.4	57	do.
13	482	6.3	23.2	.930	20.7	64	1.51060	161.6	99	do.
2/ 14	527	3.2	26.4	.946	18.1	–	1.51865	–	200	do.
15	572									
Residuum		72.0	98.4	1.025	6.6					

Carbon residue, Conradson: Residuum, 4.6 percent; crude, 3.4 percent.

APPROXIMATE SUMMARY

	Percent	Sp. gr.	°API	Viscosity
Light gasoline				
Total gasoline and naphtha				
Kerosene distillate				
Gas oil	11.7	0.875	30.2	
Nonviscous lubricating distillate	8.5	.901–.931	25.6–20.5	50-100
Medium lubricating distillate	4.6	.931–.946	20.5–18.1	100-200
Viscous lubricating distillate	1.6	.946–.951	18.1–17.3	Above 200
Residuum	72.0	1.025	6.6	
Distillation loss	1.6			

1/ Distillation discontinued at 482° F.
2/ Distillation discontinued at 509° F.

U.S. GOVERNMENT PRINTING OFFICE 16-57688-3

Table 2-5.10. U.S. Crude-petroleum Analysis

VENTURA FIELD (Ventura County, California)
Pico-Pliocene and Repetto-Pliocene
2,500–12,000 ft

GENERAL CHARACTERISTICS

Gravity, specific, 0.869 Gravity, °API, 31.3 Pour point, °F, 15
Sulfur, percent, 0.94 Color, brownish black
Viscosity, Saybolt Universal at 77° F., 69 sec.; Nitrogen, percent, 0.41
100° F., 56 sec.;

DISTILLATION, BUREAU OF MINES ROUTINE METHOD

STAGE 1—Distillation at atmospheric pressure, 753 mm. Hg
First drop, 82 °F.

Fraction No.	Cut temp. °F.	Percent	Sum. percent	Sp. gr. 60/60° F.	°API 60° F.	C.I.	Refractive index n_D at 20° C.	Specific dispersion	S.U. visc. 100° F.	Cloud test, °F.
1	122	1.8	1.8	0.651	85.9	11	1.37737	124.1		
2	167	2.5	4.3	.677	77.5	23	1.40031	126.1		
3	212	5.6	9.9	.723	64.2	23	1.41353	126.5		
4	257	6.6	16.5	.767	57.2	27	1.42455	126.0		
5	302	5.3	21.8	.788	53.0	27	1.43517	136.6		
6	347	4.3	26.1	.806	48.1	32	1.44552	140.6		
7	392	3.9	30.0	.823	44.1	32	1.45481	142.2		
8	437	4.1	34.1	.835	40.4	35	1.45481	142.2		
9	482	4.4	38.5	.835	38.0	35	1.46287	142.5		
10	527	5.4	43.9	.850	35.0	38	1.47177	153.4		

STAGE 2—Distillation continued at 40 mm. Hg

Fraction No.	Cut temp. °F.	Percent	Sum. percent	Sp. gr. 60/60° F.	°API 60° F.	C.I.	Refractive index n_D at 20° C.	Specific dispersion	S.U. visc. 100° F.	Cloud test, °F.
11	392	3.6	47.5	0.861	32.8	39	1.47952	165.2	42	14
12	437	3.9	51.4	.867	31.7	38	1.48282	165.0	47	34
13	482	4.9	56.3	.884	28.6	43	1.49085	157.3	62	50
14	527	5.1	61.4	.905	24.9	49			100	70
1/15	572	5.3	66.7	.920	22.3	-			230	86
Residuum		31.5	98.2	.994	10.9					

Carbon residue, Conradson: Residuum, 13.5 percent; crude, 4.9 percent.

APPROXIMATE SUMMARY

	Percent	Sp. gr.	°API	Viscosity
Light gasoline	9.9	0.698	71.2	
Total gasoline and naphtha	30.0	0.749	57.4	
Kerosine distillate	4.1	.823	40.4	
Gas oil	16.3	.852	34.6	
Nonviscous lubricating distillate	8.5	.870-.905	31.1-24.9	50-100
Medium lubricating distillate	4.5	.905-.919	24.9-22.5	100-200
Viscous lubricating distillate	3.7	.919-.930	22.5-20.7	Above 200
Residuum	31.5	.994	10.9	
Distillation loss	1.8			

1/ Distillation discontinued at 545° F.

Table 2-5.11. U.S. Crude-petroleum Analysis

WILMINGTON FIELD (Los Angeles County, California)
Terminal, Miocene
3,143–3,590 ft

GENERAL CHARACTERISTICS

Gravity, specific, 0.920 Gravity, °API, 22.3 Pour point, °F, below 5
Sulfur, percent, 1.33 Color, brownish black
Viscosity, Saybolt Universal at 100° F., 210 sec.; Nitrogen, percent,
130° F., 125 sec.;

DISTILLATION, BUREAU OF MINES ROUTINE METHOD

STAGE 1—Distillation at atmospheric pressure, 751 mm. Hg
First drop, 97 °F.

Fraction No.	Cut temp. °F.	Percent	Sum. percent	Sp. gr. 60/60° F.	°API 60° F.	C.I.	Refractive index n_D at 20° C.	Specific dispersion	S.U. visc. 100° F.	Cloud test, °F.
1	122	0.1	0.1			-				
2	167	1.0	1.0	0.672	79.1					
3	212	3.5	3.5	.723	64.2	23	1.39495	121.4		
4	257	3.5	7.0	.752	56.7	27	1.41476	127.7		
5	302	3.1	10.1	.774	51.3	30	1.42659	126.9		
6	347	3.2	13.3	.798	45.8	35	1.43718	145.7		
7	392	3.4	16.7	.817	41.7	38	1.45003	136.9		
8	437	3.4	20.1	.835	38.0	41	1.45978	140.2		
9	482	4.7	24.8	.850	35.0	42	1.46856	145.7		
10	527	5.9	30.7	.861	32.8	43	1.47890	154.7		

STAGE 2—Distillation continued at 40 mm. Hg

Fraction No.	Cut temp. °F.	Percent	Sum. percent	Sp. gr. 60/60° F.	°API 60° F.	C.I.	Refractive index n_D at 20° C.	Specific dispersion	S.U. visc. 100° F.	Cloud test, °F.
11	392	2.4	33.1	0.877	29.9	46	1.48726	155.2	42	Below 5
12	437	5.2	38.3	.891	27.3	49	1.49303	162.0	49	10
13	482	5.6	43.9	.908	24.3	54	1.50269	157.0	70	30
14	527	5.3	49.2	.924	21.6	58			135	50
15	572	7.6	56.8	.942	18.7				Over 400	68
Residuum		42.1	98.4	1.009	8.7					

Carbon residue, Conradson: Residuum, 6.3 percent; crude, 2.9 percent.

APPROXIMATE SUMMARY

	Percent	Sp. gr.	°API	Viscosity
Light gasoline	3.5	0.707	68.6	
Total gasoline and naphtha	16.7	0.769	52.5	
Kerosine distillate				
Gas oil	19.4	.860	33.0	50-100
Nonviscous lubricating distillate	4.5	.892-.915	27.1-23.1	100-200
Medium lubricating distillate		.915-.929	23.1-20.1	
Viscous lubricating distillate	8.6	.929-.952	20.8-17.1	Above 200
Residuum	42.1	1.009	8.7	
Distillation loss	1.1			

Table 2-5.12. U.S. Crude-petroleum Analysis

ADENA FIELD (Morgan County, Colorado)
Dakota "J," Cretaceous
5,592–5,612 ft

GENERAL CHARACTERISTICS

Gravity, specific, 0.803   Gravity, °API, 44.7   Pour point, °F, 10
Sulfur, percent, less than 0.10   Color, green
Viscosity, Saybolt Universal at 100° F., 36 sec.   Nitrogen, percent, 0.03

DISTILLATION, BUREAU OF MINES ROUTINE METHOD

STAGE 1—Distillation at atmospheric pressure, 746 mm. Hg
First drop, 80 °F.

Fraction No.	Cut temp. °F.	Percent	Sum. percent	Sp. gr. 60/60°F.	°API 60°F.	C.I.	Refractive index, n_D at 20° C.	Specific dispersion	S.U. visc. 100°F.	Cloud test, °F.
1	122	4.0	4.0	0.640	89.6					
2	167	3.2	7.2	.671	79.4	8.0	1.37718	124.5		
3	212	5.5	12.7	.716	66.1	19	1.39921	125.4		
4	257	7.1	19.8	.742	59.2	22	1.41351	132.2		
5	302	6.5	26.3	.763	54.0	25	1.42496	138.1		
6	347	5.5	31.8	.778	50.4	25	1.43385	141.1		
7	392	5.3	37.1	.789	47.8	24	1.43973	140.9		
8	437	5.3	42.4	.800	45.4	24	1.44573	140.2		
9	482	6.0	48.4	.811	43.0	24	1.45229	145.6		
10	527	7.2	55.6	.822	40.6	24	1.45855	147.6		

STAGE 2—Distillation continued at 40 mm. Hg

11	392	3.8	59.4	0.835	38.0	27			39	25
12	437	6.2	65.6	.840	37.0	25			43	40
13	482	5.5	71.1	.847	35.6	25			50	55
14	527	4.7	75.8	.855	34.0	26			64	70
15	572	5.1	80.9	.865	32.1	31			93	85
Residuum		15.9	96.8	.923	21.8					

Carbon residue, Conradson: Residuum, 2.9 percent; crude, 0.5 percent.

APPROXIMATE SUMMARY

	Percent	Sp. gr.	°API	Viscosity
Light gasoline	12.7	0.681	76.3	
Total gasoline and naphtha	37.7	0.731	60.5	
Kerosine distillate	18.5	.812	42.8	
Gas oil	12.8	.837	37.6	
Nonviscous lubricating distillate	11.2	.841–.867	36.8–31.7	50-100
Medium lubricating distillate	1.3	.867–.870	31.7–31.1	100-200
Viscous lubricating distillate				Above 200
Residuum	15.9	.923	21.8	
Distillation loss	3.2			

16-57858-8

Table 2-5.13. U.S. Crude-petroleum Analysis

RANGELY FIELD (Rio Blanco County, Colorado)
Weber, Pennsylvanian
5,960–6,459 ft

GENERAL CHARACTERISTICS

Gravity, specific, 0.851   Gravity, °API, 34.8   Pour point, °F, 10
Sulfur, percent, 0.56   Color, greenish black
Viscosity, Saybolt Universal at 100° F., 48 sec.   Nitrogen, percent, 0.07

DISTILLATION, BUREAU OF MINES ROUTINE METHOD

STAGE 1—Distillation at atmospheric pressure, 754 mm. Hg
First drop, 88 °F.

Fraction No.	Cut temp. °F.	Percent	Sum. percent	Sp. gr. 60/60°F.	°API 60°F.	C.I.	Refractive index, n_D at 20° C.	Specific dispersion	S.U. visc. 100°F.	Cloud test, °F.
1	122	1.4	1.4	0.647	87.2					
2	167	2.6	4.0	.670	79.7	7.5				
3	212	2.9	6.9	.709	68.1	.16	1.37406	125.6		
4	257	5.2	12.1	.731	62.1	18	1.39574	128.0		
5	302	3.9	16.0	.752	56.7	20	1.40817	128.2		
6	347	5.3	21.3	.772	51.8	23	1.41885	131.4		
7	392	4.8	26.1	.792	47.2	26	1.42916	135.2		
8	437	4.7	30.8	.810	43.2	29	1.43936	137.1		
9	482	5.6	36.4	.824	40.2	30	1.45630	140.5		
10	527	5.6	43.0	.843	36.4	34	1.45581	147.3		

STAGE 2—Distillation continued at 40 mm. Hg

11	392	2.9	45.9	0.854	34.2	36	1.47399	150.5	40	10
12	437	7.6	53.5	.861	32.8	35	1.47765	152.0	46	24
13	482	6.9	60.4	.880	29.3	41	1.48676	156.8	61	50
14	527	5.8	66.2	.891	27.3	43	1.49293	-	86	60
15	572	5.8	72.0	.902	25.4	45			160	74
Residuum		26.5	98.5	.962	15.6					

Carbon residue, Conradson: Residuum, 7.6 percent; crude, 2.3 percent.

APPROXIMATE SUMMARY

	Percent	Sp. gr.	°API	Viscosity
Light gasoline	6.9	0.682	76.0	
Total gasoline and naphtha	26.1	0.741	59.5	
Kerosine distillate	10.3	.818	41.5	
Gas oil	12.7	.852	34.6	
Nonviscous lubricating distillate	12.7	.865–.893	32.1–27.0	50-100
Medium lubricating distillate	7.6	.893–.907	27.0–24.5	100-200
Viscous lubricating distillate				Above 200
Residuum	26.5	.962	15.6	
Distillation loss	1.5			

16-57858-9

Table 2-5.14. U.S. Crude-petroleum Analysis
CLAY CITY, CONSOLIDATED, FIELD (Clay, Richland, Jasper, and Wayne Counties, Illinois)

Mississippian
2,174–3,534 ft

GENERAL CHARACTERISTICS

Gravity, specific, 0.832	Gravity, °API, 38.6
Sulfur, percent, 0.19	Color, brownish green
Viscosity, Saybolt Universal at 100° F., 43 sec.	Pour point, °F., 10
	Nitrogen, percent, 0.08

DISTILLATION, BUREAU OF MINES ROUTINE METHOD

Stage 1—Distillation at atmospheric pressure, 750 mm. Hg
First drop, 84 °F.

Fraction No.	Cut temp. °F.	Percent	Sum, percent	Sp. gr. 60/60° F.	°API, 60° F.	C.I.	Refractive index, n_D at 20° C.	Specific dispersion	S.U. visc., 100° F.	Cloud test, °F.
1	122	2.0	2.0	0.649	86.5	8.4				
2	167	2.8	4.8	.672	79.1	19	1.37821	122.0		
3	212	5.5	10.3	.715	66.4	21	1.39818	125.5		
4	257	6.3	16.6	.739	60.0	24	1.41088	129.9		
5	302	5.9	22.5	.760	54.7	25	1.42175	132.7		
6	347	5.5	28.0	.777	50.6	26	1.43201	131.6		
7	392	4.8	32.8	.793	46.9	27	1.44031	139.3		
8	437	4.9	37.7	.807	43.8	28	1.44798	139.7		
9	482	5.2	42.9	.820	41.1	28	1.45600	144.5		
10	527	5.6	48.5	.835	38.0	31	1.46401	146.5		

Stage 2—Distillation continued at 40 mm. Hg

Fraction No.	Cut temp. °F.	Percent	Sum, percent	Sp. gr. 60/60° F.	°API, 60° F.	C.I.	Refractive index, n_D at 20° C.	Specific dispersion	S.U. visc., 100° F.	Cloud test, °F.
11	392	4.3	52.8	0.861	32.8	39	1.47194	149.5	41	14
12	437	4.7	57.5	.871	31.0	40	1.47561	144.5	46	30
13	482	5.0	62.5	.884	28.6	43	1.48195	151.0	59	44
14	527	4.5	67.0	.903	25.2	49	1.48738	151.9	86	64
15	572	7.7	74.7	.916	23.0				190	80
Residuum		21.6	96.3	.941	18.9					

Carbon residue, Conradson: Residuum, 8.2 percent; crude, 2.0 percent.

APPROXIMATE SUMMARY

	Percent	Sp. gr.	°API	Viscosity
Light gasoline	10.3	0.690	73.6	
Total gasoline and naphtha	32.8	.742	59.2	
Kerosine distillate	10.1	.814	42.3	
Gas oil	14.0	.854	34.2	
Nonviscous lubricating distillate	8.8	.875–.905	30.2–24.9	50–100
Medium lubricating distillate	5.9	.905–.917	24.9–22.8	100–200
Viscous lubricating distillate	3.1	.917–.925	22.8–21.5	Above 200
Residuum	21.6	.941	18.9	
Distillation loss	3.7			

1/ Distillation discontinued at 554° F.

U. S. GOVERNMENT PRINTING OFFICE 16—57213-3

Table 2-5.15. U.S. Crude-petroleum Analysis
LOUDON FIELD (Fayette County, Illinois)

Bethel, Mississippian
1,550–1,566 ft

GENERAL CHARACTERISTICS

Gravity, specific, 0.844	Gravity, °API, 36.2
Sulfur, percent, 0.22	Color, brownish black
Viscosity, Saybolt Universal at 100° F., 45 sec.	Pour point, °F., 15
	Nitrogen, percent,

DISTILLATION, BUREAU OF MINES ROUTINE METHOD

Stage 1—Distillation at atmospheric pressure, 751 mm. Hg
First drop, 89 °F.

Fraction No.	Cut temp. °F.	Percent	Sum, percent	Sp. gr. 60/60° F.	°API, 60° F.	C.I.	Refractive index, n_D at 20° C.	Specific dispersion	S.U. visc., 100° F.	Cloud test, °F.
1	122	1.8	1.8	0.628	93.8	4.6				
2	167	2.0	3.8	.664	81.6	17	1.39701	123.8		
3	212	4.5	8.3	.711	67.5	21	1.40896	121.9		
4	257	6.2	14.5	.738	60.2	22	1.41980	126.7		
5	302	5.2	19.7	.756	55.7	23	1.43002	131.2		
6	347	5.5	25.2	.774	51.3	25	1.43967	133.9		
7	392	5.0	30.2	.791	47.4	25	1.44795	140.3		
8	437	5.0	35.2	.806	44.1	27	1.45522	142.3		
9	482	5.2	40.4	.819	40.4	28	1.46632	146.1		
10	527	6.0	46.4	.832	38.6	29				

Stage 2—Distillation continued at 40 mm. Hg

Fraction No.	Cut temp. °F.	Percent	Sum, percent	Sp. gr. 60/60° F.	°API, 60° F.	C.I.	Refractive index, n_D at 20° C.	Specific dispersion	S.U. visc., 100° F.	Cloud test, °F.
11	392	3.2	49.6	0.847	35.6	32	1.47145	149.2	40	10
12	437	5.2	54.8	.884	34.2	32	1.47493	150.6	45	30
13	482	4.8	59.6	.864	32.3	33	1.48090	153.5	56	50
14	527	5.4	65.0	.879	29.5	37			80	60
15	572	5.3	70.3	.893	27.0	41			145	75
Residuum		29.0	99.3	.957	16.4					

Carbon residue, Conradson: Residuum, 7.6 percent; crude, 2.5 percent.

APPROXIMATE SUMMARY

	Percent	Sp. gr.	°API	Viscosity
Light gasoline	8.3	0.682	76.0	
Total gasoline and naphtha	30.2	0.741	59.5	
Kerosine distillate	10.2	.813	42.6	
Gas oil	14.2	.843	36.4	
Nonviscous lubricating distillate	9.4	.857–.883	33.6–28.8	50–100
Medium lubricating distillate	6.3	.883–.899	28.8–25.9	100–200
Viscous lubricating distillate				Above 200
Residuum	29.0	.957	16.4	
Distillation loss	.7			

U. S. GOVERNMENT PRINTING OFFICE 16—57213-3

Table 2-5.16. U.S. Crude-petroleum Analysis

BEMIS-SHUTTS FIELD (Ellis County, Kansas)
Upper Arbuckle, Ordovician
3,592–3,598 ft

GENERAL CHARACTERISTICS

Gravity, specific, 0.852   Gravity, °API, 34.6   Pour point, °F, below 5
Sulfur, percent, 0.57   Color, brownish black
Viscosity, Saybolt Universal at 77° F., 68 sec.   Nitrogen, percent, 0.16
    100° F., 52 sec.

DISTILLATION, BUREAU OF MINES ROUTINE METHOD

STAGE 1—Distillation at atmospheric pressure, 743 mm. Hg
First drop, 84 °F.

Fraction No.	Cut temp. °F.	Percent	Sum, percent	Sp. gr. 60/60° F.	°API, 60° F.	C.I.	Refractive index, n_D at 20° C.	Specific dispersion	S.U. visc. 100° F.	Cloud test, °F.
1	122	1.6	1.6	0.650	86.2	7.0	1.37650	123.8		
2	167	2.7	4.3	.669	80.0		1.39426	123.1		
3	212	4.0	8.3	.706	68.9	15	1.40674	123.7		
4	257	6.2	14.5	.734	61.3	19	1.40674	121.7		
5	302	4.8	19.3	.753	56.4	20	1.42230	125.4		
6	347	5.0	24.3	.770	52.3	22	1.42971	126.3		
7	392	4.0	28.3	.790	47.6	25	1.43705	129.9		
8	437	4.5	32.8	.809	43.4	28	1.44580	131.4		
9	482	5.1	37.9	.818	41.5	27	1.45345	133.9		
10	527	6.0	43.9	.832	38.6	29	1.46158	139.1		

STAGE 2—Distillation continued at 40 mm. Hg

Fraction No.	Cut temp. °F.	Percent	Sum, percent	Sp. gr. 60/60° F.	°API, 60° F.	C.I.	Refractive index, n_D at 20° C.	Specific dispersion	S.U. visc. 100° F.	Cloud test, °F.
11	392	3.4	47.3	0.842	36.6	30	1.47022	148.2	41	Below 5
12	437	5.8	53.1	.856	33.8	33	1.47460	146.6	46	20
13	482	5.3	58.4	.870	31.1	36	1.48136	146.5	58	34
14	527	4.8	63.2	.882	28.9	39	1.48818	153.2	84	56
15	572	5.4	68.6	.894	26.8	41			145	80
Residuum		29.5	98.1	.972	14.1					

Carbon residue, Conradson; Residuum, 11.9 percent; crude, 4.0 percent.

APPROXIMATE SUMMARY

	Percent	Sp. gr.	°API	Viscosity
Light gasoline	8.3	0.683	75.7	
Total gasoline and naphtha	28.3	0.737	60.5	
Kerosine distillate	9.6	.814	42.3	
Gas oil	14.2	.842	36.6	
Nonviscous lubricating distillate	10.1	.861–.885	32.8–28.4	50–100
Medium lubricating distillate	6.4	.885–.901	28.4–25.6	100–200
Viscous lubricating distillate				Above 200
Residuum	29.5	.972	14.1	
Distillation loss	1.9			

Table 2-5.17. U.S. Crude-petroleum Analysis

EL DORADO FIELD (Butler County, Kansas)
El Dorado Shallow (Admire), Permian
650–668 ft

GENERAL CHARACTERISTICS

Gravity, specific, 0.841   Gravity, °API, 36.8   Pour point, °F, below 5
Sulfur, percent, 0.18   Color, brownish black
Viscosity, Saybolt Universal at 100° F., 43 sec.   Nitrogen, percent, 0.09

DISTILLATION, BUREAU OF MINES ROUTINE METHOD

STAGE 1—Distillation at atmospheric pressure, 748 mm. Hg
First drop, 88 °F.

Fraction No.	Cut temp. °F.	Percent	Sum, percent	Sp. gr. 60/60° F.	°API, 60° F.	C.I.	Refractive index, n_D at 20° C.	Specific dispersion	S.U. visc. 100° F.	Cloud test, °F.
1	122	1.1	1.1	0.638	90.3		1.37272	120.7		
2	167	2.0	3.1	.666	81.0	5.6	1.39360	122.7		
3	212	3.1	6.2	.704	69.5	14	1.40458	121.2		
4	257	6.4	12.6	.726	63.4	15	1.40458	121.5		
5	302	6.5	19.1	.746	58.2	17	1.41499	122.5		
6	347	7.2	26.3	.766	53.2	20	1.42492	124.3		
7	392	5.8	32.1	.785	48.8	23	1.43528	127.9		
8	437	6.7	38.8	.802	44.9	25	1.44474	124.6		
9	482	6.6	45.4	.817	41.7	27	1.45332	134.3		
10	527	7.7	53.1	.833	38.4	30	1.46230	142.2		

STAGE 2—Distillation continued at 40 mm. Hg

Fraction No.	Cut temp. °F.	Percent	Sum, percent	Sp. gr. 60/60° F.	°API, 60° F.	C.I.	Refractive index, n_D at 20° C.	Specific dispersion	S.U. visc. 100° F.	Cloud test, °F.
11	392	3.6	56.7	0.851	34.8	34	1.47221	145.8	41	20
12	437	6.4	63.1	.859	33.2	34	1.47612	148.0	46	35
13	482	5.1	73.6	.871	31.0	34	1.48312	—	58	30
14	527	6.2	79.8	.886	28.2	40	1.49305	—	85	65
15	572	6.2	79.8	.899	25.9	44			160	85
Residuum		19.5	99.3	.980	12.9					

Carbon residue, Conradson; Residuum, 8.3 percent; crude, 1.9 percent.

APPROXIMATE SUMMARY

	Percent	Sp. gr.	°API	Viscosity
Light gasoline	6.2	0.680	76.6	
Total gasoline and naphtha	32.1	0.741	59.5	
Kerosine distillate	13.3	.809	43.4	
Gas oil	16.6	.845	36.0	
Nonviscous lubricating distillate	10.2	.865–.889	32.5–27.7	50–100
Medium lubricating distillate	7.6	.889–.905	27.7–24.9	100–200
Viscous lubricating distillate				Above 200
Residuum	19.5	.980	12.9	
Distillation loss	.7			

Table 2-5.18. U.S. Crude-petroleum Analysis

BAY MARCHAND FIELD (Lafourche Parish, Louisiana)
3,900-ft Sand, Miocene
3,894–3,907 ft

GENERAL CHARACTERISTICS

Gravity, specific, 0.933   Gravity, °API, 20.2   Pour point, °F., below 5
Sulfur, percent, 0.46   Color, brownish green
Viscosity, Saybolt Universal at 100° F., 270 sec.; Nitrogen, percent,
    130° F., 125 sec.

DISTILLATION, BUREAU OF MINES ROUTINE METHOD

STAGE 1—Distillation at atmospheric pressure, 746 mm. Hg
First drop, 122 °F.

Fraction No.	Cut temp. °F.	Percent	Sum. percent	Sp. gr. 60/60° F.	°API 60° F.	C.I.	Refractive index n_D at 20° C.	Specific dispersion	S.U. visc. 100° F.	Cloud test, °F.
1	122									
2	167	1.2	1.2							
3	212	-	1.2							
4	257	-	1.2							
5	302	1.3	2.5							
6	347	3.7	6.2	0.798	45.8					
7	392			.846	35.8	46	1.43591	148.2		
8	437	6.2	12.4	.862	32.7	48	1.45969	128.9		
9	482			.880	29.3	52	1.46877	131.5		
10	527	9.4	21.8				1.47942	139.7		

STAGE 2—Distillation continued at 40 mm. Hg

Fraction No.	Cut temp. °F.	Percent	Sum. percent	Sp. gr. 60/60° F.	°API 60° F.	C.I.	Refractive index	Specific dispersion	S.U. visc. 100° F.	Cloud test, °F.
11	392	3.3	25.1	0.901	25.6	58			48	Below 5
12	437	10.2	35.3	.915	23.1	61			60	do.
13	482	9.6	44.9	.930	20.7	64			92	do.
14	527	8.5	53.4	.944	18.4	68			200	do.
15	572	10.1	63.5	.953	17.0	69			Over 400	do.
Residuum		34.8	98.3	.988	11.7					

Carbon residue, Conradson: Residuum, 7.5 percent; crude, 2.8 percent.

APPROXIMATE SUMMARY

	Percent	Sp. gr.	°API	Viscosity
Light gasoline	2.5	0.798	45.8	
Total gasoline and naphtha	22.2	.872	30.8	
Kerosine distillate	16.1	.903–.931	25.2–20.5	
Gas oil	8.4	.931–.944	20.5–18.4	
Nonviscous lubricating distillate	14.3	.944–.958	18.4–16.2	50–100
Medium lubricating distillate				100–200
Viscous lubricating distillate	34.8	.988	11.7	Above 200
Residuum	1.7			
Distillation loss				

Table 2-5.19. U.S. Crude-petroleum Analysis

CAILLOU ISLAND FIELD (Terrebonne Parish, Louisiana)
No. 70 Sand Series, Miocene
12,640–12,880 ft

GENERAL CHARACTERISTICS

Gravity, specific, 0.848   Gravity, °API, 35.4   Pour point, °F., below 5
Sulfur, percent, 0.23   Color, greenish black
Viscosity, Saybolt Universal at 100° F., 45 sec. Nitrogen, percent, 0.04

DISTILLATION, BUREAU OF MINES ROUTINE METHOD

STAGE 1—Distillation at atmospheric pressure, 748 mm. Hg
First drop, 9? °F.

Fraction No.	Cut temp. °F.	Percent	Sum. percent	Sp. gr. 60/60° F.	°API 60° F.	C.I.	Refractive index n_D at 20° C.	Specific dispersion	S.U. visc. 100° F.	Cloud test, °F.
1	122	0.4	0.4	0.668	80.3					
2	167	1.7	2.1	.670	79.7	7.5				
3	212	4.3	6.4	.712	67.2	18	1.39645	128.0		
4	257	4.9	11.3	.739	60.1	21	1.41197	133.9		
5	302	5.9	17.2	.761	54.4	24	1.42279	135.0		
6	347	5.6	22.8	.778	50.4	25	1.43199	136.0		
7	392	5.3	28.1	.791	47.4	25	1.43950	135.5		
8	437	5.5	33.6	.807	43.8	27	1.44745	137.6		
9	482	6.6	40.2	.824	40.2	30	1.45641	140.7		
10	527	7.0	47.2	.840	37.0	33	1.46481	144.1		

STAGE 2—Distillation continued at 40 mm. Hg

Fraction No.	Cut temp. °F.	Percent	Sum. percent	Sp. gr. 60/60° F.	°API 60° F.	C.I.	Refractive index	Specific dispersion	S.U. visc. 100° F.	Cloud test, °F.
11	392	5.8	53.0	0.851	34.8	34	1.47257	148.9	42	Below 5
12	437	6.5	59.5	.861	32.8	35	1.47719	148.9	49	30
13	482	5.4	64.9	.875	30.2	38	1.48426	-	64	50
14	527	5.7	70.4	.890	27.5	42	1.49006	-	100	64
15	572	5.7	76.1	.903	25.2	46	1.49701	-	210	74
Residuum		23.1	99.2	.952	17.1					

Carbon residue, Conradson: Residuum, 6.3 percent; crude, 1.6 percent.

APPROXIMATE SUMMARY

	Percent	Sp. gr.	°API	Viscosity
Light gasoline	6.4	0.698	71.2	
Total gasoline and naphtha	28.1	0.752	56.7	
Kerosine distillate	12.1	.816	41.9	
Gas oil	16.5	.848	35.4	
Nonviscous lubricating distillate	11.0	.862–.890	32.7–27.5	50–100
Medium lubricating distillate	5.0	.890–.901	27.5–25.6	100–200
Viscous lubricating distillate	3.4	.901–.910	25.6–24.0	Above 200
Residuum	23.1	.952	17.1	
Distillation loss	.8			

Table 2-5.20. U. S. Crude-petroleum Analysis

LAKE WASHINGTON FIELD (Plaquemines Parish, Louisiana)
Miocene
8,000–13,500 ft

GENERAL CHARACTERISTICS

Gravity, specific, 0.886   Gravity, °API, 28.2   Pour point, °F., below 5
Sulfur, percent, 0.37   Color, brownish black
Viscosity, Saybolt Universal at 100° F., 82 sec.; Nitrogen, percent, 0.15
130° F., 56 sec.

DISTILLATION, BUREAU OF MINES ROUTINE METHOD

STAGE 1—Distillation at atmospheric pressure, 737 mm. Hg
First drop, 84 °F.

Fraction No.	Cut temp. °F.	Percent	Sum, percent	Sp. gr. 60/60°F.	°API 60°F.	C.I.	Refractive index n_D at 20° C.	Specific dispersion	S.U. visc. 100°F.	Cloud test, °F.
1	122	1.2		0.664	81.6		1.37188	127.0		
2	167	1.3	2.5	.668	80.3	6.5	1.39451	130.0		
3	212	2.8	5.3	.708	68.4	16	1.41241	134.6		
4	257	3.7	9.0	.742	59.2	23	1.42314	134.7		
5	302	3.4	12.4	.761	54.4	24	1.43304	137.4		
6	347	3.7	16.1	.776	50.9	24	1.44146	136.3		
7	392	3.9	20.0	.797	46.0	28	1.45025	137.2		
8	437	3.7	23.7	.815	42.1	31	1.45399	137.2		
9	482	4.7	28.4	.830	39.0	33	1.45908	144.6		
10	527	6.6	35.0	.851	34.8	38	1.47024	148.9		

STAGE 2—Distillation continued at 40 mm. Hg

Fraction No.	Cut temp. °F.	Percent	Sum, percent	Sp. gr. 60/60°F.	°API 60°F.	C.I.	Refractive index n_D at 20° C.	Specific dispersion	S.U. visc. 100°F.	Cloud test, °F.
11	392	5.1	40.1	.872	30.8	44	1.48264	157.1	42	10
12	437	5.7	45.8	.886	28.2	47	1.48876	158.7	50	20
13	482	5.7	51.5	.900	25.7	50	1.49617	169.1	68	34
14	527	7.1	58.6	.917	22.8	55	1.50397		120	50
1/15	572	6.9	65.5	.928	21.0	57			300	60
Residuum		32.5	98.0	.985	12.2					

Carbon residue, Conradson: Residuum, 9.2 percent; crude, 3.3 percent.

APPROXIMATE SUMMARY

	Percent	Sp. gr.	°API	Viscosity
Light gasoline	5.3	0.688	74.2	
Total gasoline and naphtha	20.0	0.748	57.7	
Kerosine distillate	3.7	.815	42.1	
Gas oil	19.3	.856	33.8	
Nonviscous lubricating distillate	9.7	.886–.911	28.2–23.8	50–100
Medium lubricating distillate	5.5	.911–.925	23.8–21.5	100–200
Viscous lubricating distillate	7.3	.925–.933	21.5–20.2	Above 200
Residuum	32.5	.985	12.2	
Distillation loss	2.0			

1/ Distillation discontinued at 565° F.

Table 2-5.21. U. S. Crude-petroleum Analysis

SOUTH PASS (BLOCK 24) FIELD (Plaquemines Parish, Louisiana)
Various Members of Miocene
6,000–9,000 ft

GENERAL CHARACTERISTICS

Gravity, specific, 0.864   Gravity, °API, 32.3   Pour point, °F., below 5
Sulfur, percent, 0.26   Color, brownish green
Viscosity, Saybolt Universal at 77° F., 64 sec.; Nitrogen, percent, 0.07
100° F., 51 sec.

DISTILLATION, BUREAU OF MINES ROUTINE METHOD

STAGE 1—Distillation at atmospheric pressure, 754 mm. Hg
First drop, 91 °F.

Fraction No.	Cut temp. °F.	Percent	Sum, percent	Sp. gr. 60/60°F.	°API 60°F.	C.I.	Refractive index n_D at 20° C.	Specific dispersion	S.U. visc. 100°F.	Cloud test, °F.
1	122	1.0		0.662	82.2	7.0				
2	167	1.2	2.2	.669	80.0		1.38683	126.5		
3	212	2.5	4.7	.715	66.4	19	1.41490	131.7		
4	257	3.3	8.0	.748	57.7	26	1.42831	138.1		
5	302	3.3	11.3	.771	52.0	29	1.43709	144.2		
6	347	3.7	15.0	.791	47.4	32	1.44733	139.6		
7	392	3.7	18.7	.809	43.4	34	1.45399	144.0		
8	437	5.1	23.8	.822	40.6	34	1.46201	135.4		
9	482	6.2	30.0	.837	37.6	36	1.46819	159.5		
10	527	8.2	38.2	.849	35.2	37				

STAGE 2—Distillation continued at 40 mm. Hg

Fraction No.	Cut temp. °F.	Percent	Sum, percent	Sp. gr. 60/60°F.	°API 60°F.	C.I.	Refractive index n_D at 20° C.	Specific dispersion	S.U. visc. 100°F.	Cloud test, °F.
11	392	4.6	42.8	.858	33.4	37	1.47731		41	Below 5
12	437	9.2	52.0	.870	31.1	39	1.48103		47	14
13	482	7.1	59.1	.885	28.4	43	1.48786		61	40
14	527	6.3	65.4	.895	26.6	45	1.49409		90	66
15	572	7.0	72.4	.907	24.5	47	1.49936		175	74
Residuum		26.3	98.7	.951	17.3					

Carbon residue, Conradson: Residuum, 5.2 percent; crude, 1.5 percent.

APPROXIMATE SUMMARY

	Percent	Sp. gr.	°API	Viscosity
Light gasoline	4.7	0.692	73.0	
Total gasoline and naphtha	18.7	0.759	54.9	
Kerosine distillate	5.1	.822	40.6	
Gas oil	25.4	.852	34.6	
Nonviscous lubricating distillate	14.9	.871–.898	31.0–26.1	50–100
Medium lubricating distillate	6.8	.898–.910	26.1–24.0	100–200
Viscous lubricating distillate	1.5	.910–.913	24.0–23.5	Above 200
Residuum	26.3	.951	17.3	
Distillation loss	1.3			

Table 2-5.22. U.S. Crude-petroleum Analysis

BAXTERVILLE FIELD (Lamar and Marion Counties, Mississippi)
Lower Tuscaloosa, Upper Cretaceous
8,734–8,744 ft

GENERAL CHARACTERISTICS

Gravity, specific, 0.952 Gravity, °API, 17.1 Pour point, °F, 20
Sulfur, percent, 2.71 Color, brownish black
Viscosity, Saybolt Universal at 100° F., 1480 sec.; 130° F., 590 sec. Nitrogen, percent, 0.11

DISTILLATION, BUREAU OF MINES ROUTINE METHOD

STAGE 1—Distillation at atmospheric pressure, 747 mm. Hg
First drop, 171 °F.

Fraction No.	Cut temp. °F	Percent	Sum. percent	Sp. gr. 60/60° F.	°API 60° F.	C.I.	Refractive index n_D at 20° C.	Specific dispersion	S.U. visc. 100° F.	Cloud test, °F.
1	122	0.8	0.8	0.711	67.5					
2	167	1.2	2.0	0.735	61.0					
3	212	.7	2.7	.736	60.8	19				
4	257	1.1	3.8	.779	50.1	12				
5	302	1.4	5.2	.796	46.3	26				
6	347	2.1	7.3	.823	40.4	35				
7	392	3.1	10.4	.841	36.8	38				
8	437					41	1.42962	132.9		
9	482						1.46249	137.3		
10	527	5.5	15.9	.857	33.6		1.47127	145.3		

STAGE 2—Distillation continued at 40 mm. Hg

Fraction No.	Cut temp. °F	Percent	Sum. percent	Sp. gr. 60/60° F.	°API 60° F.	C.I.	Refractive index n_D at 20° C.	Specific dispersion	S.U. visc. 100° F.	Cloud test, °F.
11	392	2.0	17.9	0.866	31.9	41	1.47993	157.1	42	Below 5
12	437	6.5	24.4	.879	29.5	44	1.48509	156.7	49	20
13	482	6.9	31.3	.897	26.3	49	1.49241	163.6	65	34
14	527	6.9	38.2	.912	23.7	53	1.50358	173.8	100	50
16 [1]	572	7.8	46.0	.928	21.0		1.51058	175.6	210	66
Residuum		52.8	98.8	1.022	7.0					

Carbon residue, Conradson: Residuum, 16.6 percent; crude, 9.4 percent.

APPROXIMATE SUMMARY

	Percent	Sp. gr.	°API	Viscosity
Light gasoline	0.8	0.711	67.5	
Total gasoline and naphtha	5.2	0.757	55.4	
Kerosine distillate	2.1	.823	40.4	
Gas oil	14.4	.859	33.2–29.7	
Nonviscous lubricating distillate	8.1	.880–.902	29.7–23.7	50–100
Medium lubricating distillate	6.9	.910–.927	23.7–21.1	100–200
Viscous lubricating distillate	4.3	.922–.936	21.1–19.7	Above 200
Residuum	52.8	1.022	7.0	
Distillation loss	1.2			

[1] Distillation discontinued at 568° F.

U. S. GOVERNMENT PRINTING OFFICE 16—57838-3

Table 2-5.23. U.S. Crude-petroleum Analysis

CABIN CREEK FIELD (Fallon County, Montana)
Mission Canyon Dolomite—Mississippian
7,332–7,356 ft

GENERAL CHARACTERISTICS

Gravity, specific, 0.858 Gravity, °API, 33.4 Pour point, °F, below 5
Sulfur, percent, 0.60 Color, brownish black
Viscosity, Saybolt Universal at 100° F., 47 sec. Nitrogen, percent, 0.11

DISTILLATION, BUREAU OF MINES ROUTINE METHOD

STAGE 1—Distillation at atmospheric pressure, 760 mm. Hg
First drop, 82 °F.

Fraction No.	Cut temp. °F	Percent	Sum. percent	Sp. gr. 60/60° F.	°API 60° F.	C.I.	Refractive index n_D at 20° C.	Specific dispersion	S.U. visc. 100° F.	Cloud test, °F.
1	122	1.6	1.6	0.640	89.6		1.37159	129.0		
2	167	3.0	4.6	.666	81.0	5.6	1.39648	131.1		
3	212	3.4	8.0	.710	67.8	17	1.40812	133.7		
4	257	3.6	11.6	.734	61.3	19	1.41780	137.9		
5	302	4.1	15.7	.750	57.2	19	1.42619	139.3		
6	347	4.4	20.1	.765	53.5	19	1.43308	140.4		
7	392	5.0	25.1	.777	50.6	19	1.44152	141.7		
8	437	6.6	31.7	.793	46.9	21	1.45043	150.2		
9	482	4.7	36.4	.809	43.4	23	1.45819	151.0		
10	527	6.8	43.2	.822	40.6	24				

STAGE 2—Distillation continued at 40 mm. Hg

Fraction No.	Cut temp. °F	Percent	Sum. percent	Sp. gr. 60/60° F.	°API 60° F.	C.I.	Refractive index n_D at 20° C.	Specific dispersion	S.U. visc. 100° F.	Cloud test, °F.
11	392	4.1	47.3	0.841	36.8	29	1.46773	155.2	37	10
12	437	6.1	53.4	.850	35.0	30	1.47286	159.1	41	20
13	482	5.5	58.9	.867	31.7	35	1.48258	172.8	50	30
14	527	3.9	62.8	.890	27.5	42			64	55
15	572	5.6	68.4	.903	25.2	46			110	75
Residuum		31.0	99.4	.986	12.0					

Carbon residue, Conradson: Residuum, 15.7 percent; crude, 5.6 percent.

APPROXIMATE SUMMARY

	Percent	Sp. gr.	°API	Viscosity
Light gasoline	8.0	0.679	76.9	
Total gasoline and naphtha	25.1	0.733	61.5	
Kerosine distillate	18.1	.808	43.6	
Gas oil	12.9	.850	34.6	
Nonviscous lubricating distillate	3.9	.867–.900	31.4–25.7	50–100
Medium lubricating distillate		.900–.911	25.7–23.8	100–200
Viscous lubricating distillate				Above 200
Residuum	31.0	.986	12.0	
Distillation loss	.6			

U. S. GOVERNMENT PRINTING OFFICE 16—57838-3

Table 2-5.24. U.S. Crude-petroleum Analysis

PINE FIELD (Wilbaux County, Montana)
Devonian
8,910–8,970 ft

GENERAL CHARACTERISTICS

Gravity, specific, 0.856 Gravity, °API, 33.8 Pour point, °F., below 5
Sulfur, percent, 0.36 Color, black
Viscosity, Saybolt Universal at 77 °F., 67 sec.; 100 °F., 55 sec. Nitrogen, percent, _____

DISTILLATION, BUREAU OF MINES ROUTINE METHOD
STAGE 1—Distillation at atmospheric pressure, 760 mm. Hg
First drop, 82 °F.

Fraction No.	Cut temp. °F.	Percent	Sum, percent	Sp. gr. 60/60°F.	°API 60°F.	C.I.	Refractive index n_D at 20°C.	Specific dispersion	S.U. visc. 100°F.	Cloud test, °F.
1	122	1.9	1.9	0.648	86.9					
2	167	1.9	3.8	.660	82.9	2.7	1.37089	127.9		
3	212	2.9	6.7	.696	71.8	10	1.38984	125.6		
4	257	3.8	10.5	.713	67.0	8.9	1.39918	128.9		
5	302	4.0	14.5	.731	62.1	9.9	1.40865	130.1		
6	347	4.5	19.0	.748	57.7	11	1.41728	133.3		
7	392	5.7	24.7	.763	54.0	12	1.42538	131.8		
8	437	5.7	30.4	.780	49.9	15	1.43461	138.0		
9	482	6.6	37.0	.797	46.0	17	1.44475	142.4		
10	527	7.5	44.5	.810	43.2	19	1.45168	146.8		
STAGE 2—Distillation continued at 40 mm. Hg									
11	392	2.9	47.4	.830	39.0	24			36	20
12	437	5.9	53.3	.839	37.2	25			42	25
13	482	4.9	58.2	.864	32.3	33			48	35
14	527	3.9	62.1	.886	28.2	40			64	50
15	572	4.9	67.0	.903	25.2	46			110	70
Residuum		31.2	98.2	1.002	9.7					

Carbon residue, Conradson: Residuum, 18.9 percent; crude, 6.9 percent.

APPROXIMATE SUMMARY

	Percent	Sp. gr.	°API	Viscosity
Light gasoline	6.7	0.672	79.1	
Total gasoline and naphtha	24.7	0.723	64.2	
Kerosene distillate	19.8	.797	46.0	
Gas oil	11.9	.844	36.2	
Nonviscous lubricating distillate	7.1	.868–.899	31.5–25.9	50–100
Medium lubricating distillate	3.5	.899–.913	25.9–23.5	100–200
Viscous lubricating distillate				Above 200
Residuum	31.2	1.002	9.7	
Distillation loss	1.8			

Table 2-5.25. U.S. Crude-petroleum Analysis

EUNICE-MONUMENT FIELD (Lea County, New Mexico)
Grayburg, Permian
3,800–3,900 ft

GENERAL CHARACTERISTICS

Gravity, specific, 0.883 Gravity, °API, 28.8 Pour point, °F., below 5
Sulfur, percent, 0.97 Color, brownish green
Viscosity, Saybolt Universal at 100 °F., 54 sec.; 130 °F., 47 sec. Nitrogen, percent, 0.07

DISTILLATION, BUREAU OF MINES ROUTINE METHOD
STAGE 1—Distillation at atmospheric pressure, 746 mm. Hg
First drop, 126 °F.

Fraction No.	Cut temp. °F.	Percent	Sum, percent	Sp. gr. 60/60°F.	°API 60°F.	C.I.	Refractive index n_D at 20°C.	Specific dispersion	S.U. visc. 100°F.	Cloud test, °F.
1	122	4.0	4.0	0.673	78.8	-	1.37796	121.0		
2	167	4.1	8.1	.726	63.4	24	1.40240	122.9		
3	212	4.3	12.4	.744	58.7	24	1.41112	122.2		
4	257	5.5	17.9	.764	53.7	26	1.42247	127.1		
5	302	5.3	23.2	.786	48.5	29	1.43457	130.6		
6	347	4.7	27.9	.806	44.1	32	1.44543	139.7		
7	392	5.0	32.9	.824	40.2	35	1.45519	144.6		
8	437	5.1	38.0	.841	36.8	38	1.46478	149.9		
10	527	6.4	44.4	.860	33.0	42	1.47628	161.9		
STAGE 2—Distillation continued at 40 mm. Hg									
11	392	2.8	47.2	0.874	30.4	45	1.48771	–	42	Below 5
12	437	5.7	52.9	.886	28.2	47	1.49117	–	47	do.
13	482	4.4	57.3	.898	26.1	49	1.49898	–	61	do.
14	527	4.7	62.0	.912	23.7	53	1.50615		98	10
15	572	7.1	69.1	.920	22.3	54	1.51039		210	14
Residuum		27.6	96.7	.977	13.3					

Carbon residue, Conradson: Residuum, 10.9 percent; crude, 3.3 percent.

APPROXIMATE SUMMARY

	Percent	Sp. gr.	°API	Viscosity
Light gasoline	8.1	0.700	70.6	
Total gasoline and naphtha	27.9	0.754	56.2	
Kerosene distillate	5.0	.824	40.2	
Gas oil	18.3	.862	32.7	
Nonviscous lubricating distillate	8.6	.889–.912	27.7–23.7	50–100
Medium lubricating distillate	5.3	.912–.919	23.7–22.5	100–200
Viscous lubricating distillate	4.0	.919–.925	22.5–21.5	Above 200
Residuum	27.6	.977	13.3	
Distillation loss	3.3			

Table 2-5.26. U.S. Crude-petroleum Analysis

ALLEGANY FIELD (Allegany County, New York)
Caneadea—Richburg, Devonian
1,288–1,334 ft

GENERAL CHARACTERISTICS

Gravity, specific, 0.816   Gravity, °API, 41.9   Pour point, °F, below 5
Sulfur, percent, 0.12   Color, green
Viscosity, Saybolt Universal at 100° F., 42 sec.   Nitrogen, percent, 0.03

DISTILLATION, BUREAU OF MINES ROUTINE METHOD

STAGE 1—Distillation at atmospheric pressure, 749 mm. Hg
First drop, 84 °F.

Fraction No.	Cut temp., °F.	Percent	Sum, percent	Sp. gr., 60/60° F.	°API, 60° F.	C.I.	Refractive index, n_D at 20° C.	Specific dispersion	S.U. visc., 100° F.	Cloud test, °F.
1	122	2.4	2.4	0.637	90.6	5.6	1.37554	120.0		
2	167	2.7	5.1	.666	81.0	17	1.39713	126.8		
3	212	5.0	10.1	.710	67.8	21	1.41079	133.4		
4	257	6.1	16.2	.738	60.2	21	1.42256	126.9		
5	302	5.4	21.6	.758	55.2	23	1.43076	127.3		
6	347	4.7	26.3	.773	51.6	23	1.43643	136.5		
7	392	4.7	31.0	.783	49.2	22	1.44203	133.0		
8	437	4.9	35.9	.795	46.5	22	1.44889	142.2		
9	482	5.2	41.1	.808	43.6	22	1.45571	142.8		
10	527	6.1	47.2	.821	40.9	24				

STAGE 2—Distillation continued at 40 mm. Hg

Fraction No.	Cut temp., °F.	Percent	Sum, percent	Sp. gr., 60/60° F.	°API, 60° F.	C.I.	Refractive index, n_D at 20° C.	Specific dispersion	S.U. visc., 100° F.	Cloud test, °F.
11	392	3.3	50.5	0.831	38.8	25	1.46377	143.3	40	14
12	437	4.8	55.3	.837	37.6	24	1.46616	144.2	44	36
13	482	5.3	60.6	.848	35.4	26	1.47040	145.3	53	50
14	527	4.5	65.1	.859	33.2	28			70	74
15	572	5.6	70.7	.868	31.5	29			110	90
Residuum		27.0	97.7	.904	25.0					

Carbon residue, Conradson: Residuum, 2.2 percent; crude, 0.7 percent.

APPROXIMATE SUMMARY

	Percent	Sp. gr.	°API	Viscosity
Light gasoline	10.1	0.681	76.3	
Total gasoline and naphtha	31.0	0.735	61.0	
Kerosene distillate	16.2	.809	43.4	
Gas oil	9.2	.836	37.8	
Nonviscous lubricating distillate	10.3	.844–.865	36.2–32.1	50-100
Medium lubricating distillate	4.0	.865–.874	32.1–30.4	100-200
Viscous lubricating distillate				Above 200
Residuum	27.0	.904	25.0	
Distillation loss	2.3			

U. S. GOVERNMENT PRINTING OFFICE 16—57620-3

Table 2-5.27. U.S. Crude-petroleum Analysis

BEAVER LODGE-TIOGA FIELD (Williams County, North Dakota)
Mission-Canyon, Ordovician
10,495–14,066 ft

GENERAL CHARACTERISTICS

Gravity, specific, 0.797   Gravity, °API, 46.0   Pour point, °F, below 5
Sulfur, percent, 0.23   Color, green
Viscosity, Saybolt Universal at 100° F., 34 sec.   Nitrogen, percent, 0.02

DISTILLATION, BUREAU OF MINES ROUTINE METHOD

STAGE 1—Distillation at atmospheric pressure, 752 mm. Hg
First drop, 81 °F.

Fraction No.	Cut temp., °F.	Percent	Sum, percent	Sp. gr., 60/60° F.	°API, 60° F.	C.I.	Refractive index, n_D at 20° C.	Specific dispersion	S.U. visc., 100° F.	Cloud test, °F.
1	122	6.3	6.3	0.641	89.3	12	1.38156	125.0		
2	167	4.5	10.8	.680	76.6	12	1.40258	131.2		
3	212	8.6	19.4	.722	64.5	22	1.41803	135.5		
4	257	8.9	28.3	.751	56.9	27	1.42902	147.6		
5	302	6.7	35.0	.769	52.5	28	1.43694	148.2		
6	347	6.3	41.3	.783	49.2	28	1.44257	142.4		
7	392	5.3	46.6	.795	46.5	27	1.44864	143.2		
8	437	4.7	51.3	.808	43.6	28	1.45711	148.2		
9	482	5.1	56.4	.822	40.6	29	1.46701	154.6		
10	527	5.6	62.0	.839	37.2	32				

STAGE 2—Distillation continued at 40 mm. Hg

Fraction No.	Cut temp., °F.	Percent	Sum, percent	Sp. gr., 60/60° F.	°API, 60° F.	C.I.	Refractive index, n_D at 20° C.	Specific dispersion	S.U. visc., 100° F.	Cloud test, °F.
11	392	3.7	65.7	0.852	39.6	35	1.47597	157.4	40	10
12	437	4.8	70.5	.869	31.3	39	1.47988	159.0	46	34
13	482	4.0	74.5	.875	30.2	38	1.48547		58	54
14	527	4.4	78.9	.884	28.6	40	1.48968		81	70
15	572	4.5	83.4	.893	27.0	41			145	84
Residuum		12.2	95.6	.932	20.3					

Carbon residue, Conradson: Residuum, 2.7 percent; crude, 0.4 percent.

APPROXIMATE SUMMARY

	Percent	Sp. gr.	°API	Viscosity
Light gasoline	19.4	0.686	74.8	
Total gasoline and naphtha	46.6	0.736	60.8	
Kerosene distillate	9.8	.815	42.1	
Gas oil	13.1	.850	35.0	
Nonviscous lubricating distillate	8.6	.871–.886	31.0–28.2	50-100
Medium lubricating distillate	5.3	.886–.897	28.2–26.3	100-200
Viscous lubricating distillate				Above 200
Residuum	12.2	.932	20.3	
Distillation loss	4.4			

U. S. GOVERNMENT PRINTING OFFICE 16—57620-2

Table 2-5.28. U.S. Crude-petroleum Analysis

BURBANK FIELD (Osage County, Oklahoma)
Layton, Burbank, Pennsylvanian
2,700–2,760 ft

GENERAL CHARACTERISTICS

Gravity, specific, 0.827   Gravity, °API, 39.6   Pour point, °F., below 5
Sulfur, percent, 0.24   Color, brownish green
Viscosity, Saybolt Universal at 100° F., 43 sec.   Nitrogen, percent, 0.05

DISTILLATION, BUREAU OF MINES ROUTINE METHOD

STAGE 1—Distillation at atmospheric pressure, 753 mm. Hg
First drop, 82 °F.

Fraction No.	Cut temp. °F.	Percent	Sum. percent	Sp. gr. 60/60° F.	°API 60° F.	C.I.	Refractive index n_D at 20° C.	Specific dispersion	S.U. visc. 100° F.	Cloud test, °F.
1	122	2.3	2.3	0.639	89.9					
2	167	2.6	4.9	.667	80.6	6.1	1.37554	127.1		
3	212	4.8	9.7	.713	67.0	18	1.36992	125.5		
4	257	5.4	15.1	.739	60.0	21	1.41002	128.2		
5	302	4.5	19.6	.756	55.7	22	1.41934	131.3		
6	347	5.7	25.3	.775	51.1	24	1.42925	132.7		
7	392	4.9	30.2	.790	47.6	25	1.43735	145.4		
8	437	5.1	35.3	.806	44.1	27	1.44580	135.4		
9	482	5.0	40.3	.819	41.3	28	1.45318	134.6		
10	527	6.5	46.8	.834	38.2	30	1.46088	140.0		

STAGE 2—Distillation continued at 40 mm. Hg

Fraction No.	Cut temp. °F.	Percent	Sum. percent	Sp. gr. 60/60° F.	°API 60° F.	C.I.	Refractive index n_D at 20° C.	Specific dispersion	S.U. visc. 100° F.	Cloud test, °F.
11	392	2.4	49.2	0.844	36.2	31	1.46886	149.6	40	14
12	437	5.1	54.3	.854	34.2	32	1.47160	147.8	44	24
13	482	5.7	60.0	.863	32.5	33	1.47731	153.3	54	54
14	527	6.5	66.5	.880	29.3	38	1.48403	154.1	76	70
15	572	4.4	70.9	.887	28.0	38			130	84
Residuum		26.0	96.9	.925	21.5					

Carbon residue, Conradson: Residuum, 4.4 percent; crude, 1.3 percent.

APPROXIMATE SUMMARY

	Percent	Sp. gr.	°API	Viscosity
Light gasoline	9.7	0.683	75.7	
Total gasoline and naphtha	30.2	0.739	60.0	
Kerosene distillate	10.1	.812	42.8	
Gas oil	14.8	.843	36.4	
Nonviscous lubricating distillate	10.6	.859–.883	33.0–28.8	50–100
Medium lubricating distillate	5.2	.883–.890	28.8–27.5	100–200
Viscous lubricating distillate				Above 200
Residuum	26.0	.925	21.5	
Distillation loss	3.1			

Table 2-5.29. U.S. Crude-petroleum Analysis

CEMENT FIELD (Caddo County, Oklahoma)
Upper Melton, Pennsylvanian
5,790–5,824 ft

GENERAL CHARACTERISTICS

Gravity, specific, 0.859   Gravity, °API, 33.2   Pour point, °F., 25
Sulfur, percent, 0.47   Color, greenish black
Viscosity, Saybolt Universal at 100° F., 56 sec.; 130° F., 48 sec.   Nitrogen, percent, 0.15

DISTILLATION, BUREAU OF MINES ROUTINE METHOD

STAGE 1—Distillation at atmospheric pressure, 751 mm. Hg
First drop, 115 °F.

Fraction No.	Cut temp. °F.	Percent	Sum. percent	Sp. gr. 60/60° F.	°API 60° F.	C.I.	Refractive index n_D at 20° C.	Specific dispersion	S.U. visc. 100° F.	Cloud test, °F.
1	122	1.5	1.5	0.709	68.1	(-)	1.40961			
2	167	4.3	5.8	.744	58.7	24	1.42198	126.3		
3	212	7.7	13.5	.760	54.7	24	1.43059	125.3		
4	257	9.2	22.7	.775	51.1	24	1.43792	133.7		
5	302	6.2	28.9	.790	47.6	25	1.44624	137.4		
6	347	5.0	33.9	.809	43.4	28	1.45495	138.9		
7	392	4.7	38.6	.822	40.6	29	1.46281	135.7		
8	437	5.2	43.8	.836	37.8	31		143.8		

STAGE 2—Distillation continued at 40 mm. Hg

Fraction No.	Cut temp. °F.	Percent	Sum. percent	Sp. gr. 60/60° F.	°API 60° F.	C.I.	Refractive index n_D at 20° C.	Specific dispersion	S.U. visc. 100° F.	Cloud test, °F.
11	392	1.9	45.7	0.845	36.0	31	1.46896	158.7	39	16
12	437	6.2	51.9	.857	33.6	33	1.47194	147.4	45	26
13	482	6.5	58.4	.864	32.3	33	1.47729	-	57	44
14	527	5.2	63.6	.878	29.7	37	1.48334	-	79	66
15	572	6.3	69.9	.890	27.5	39			140	80
Residuum		29.8	99.7	.960	15.9					

Carbon residue, Conradson: Residuum, 5.3 percent; crude, 1.8 percent.

APPROXIMATE SUMMARY

	Percent	Sp. gr.	°API	Viscosity
Light gasoline	1.5	0.709	68.1	
Total gasoline and naphtha	28.9	0.766	53.2	
Kerosene distillate	9.7	.815	42.1	
Gas oil	13.0	.845	36.0	
Nonviscous lubricating distillate	11.4	.860–.881	33.0–29.1	50–100
Medium lubricating distillate	6.9	.881–.897	29.1–26.3	100–200
Viscous lubricating distillate				Above 200
Residuum	29.8	.960	15.9	
Distillation loss	.3			

Table 2-5.30. U.S. Crude-petroleum Analysis

SHO-VEL-TUM FIELD (Stephens County, Oklahoma)
Sholem Alechem Area
Hox-Deese, Springer, Pennsylvanian

GENERAL CHARACTERISTICS

Gravity, specific, 0.894   Gravity, °API, 26.8   Pour point, °F., below 5
Sulfur, percent, 1.44   Color, brownish black
Viscosity, Saybolt Universal at 100° F., 150 sec.; 130°F., 105 sec.   Nitrogen, percent, 0.37

DISTILLATION, BUREAU OF MINES ROUTINE METHOD

STAGE 1—Distillation at atmospheric pressure, 736 mm. Hg
First drop, 91 °F.

Fraction No.	Cut temp. °F.	Percent	Sum, percent	Sp. gr. 60/60°F.	°API 60°F.	C.I.	Refractive index, n_D at 20° C.	Specific dispersion	S.U. visc. 100°F.	Cloud test, °F.
1	122	1.6	1.6	0.646	87.5	8.9	1.37753	123.7		
2	167	1.4	3.0	.713	78.8	18	1.39622	123.4		
3	212	3.3	6.3	.713	67.0	18	1.40955	124.8		
4	257	4.0	10.3	.739	60.0	21	1.42111	130.3		
5	302	4.0	14.3	.759	54.9	23	1.43178	132.6		
6	347	3.4	17.7	.780	49.9	26	1.44187	136.4		
7	392	3.8	21.5	.797	46.0	28	1.45127	139.6		
8	437	3.3	24.8	.813	42.6	30	1.45994	145.8		
9	482	4.9	27.7	.830	39.0	33	1.46880	153.8		
10	527	4.6	34.3	.846	35.8	36				

STAGE 2—Distillation continued at 40 mm. Hg

11	392	2.7	37.0	0.862	32.7	39	1.47829	154.9	41	10
12	437	3.9	40.9	.872	30.8	40	1.48314	156.8	48	20
13	482	4.1	45.0	.884	28.6	43	1.49018		61	40
14	527	4.4	49.4	.900	25.7	47	1.49757	168.3	93	64
1/15	572	5.7	55.1	.915	23.1		1.50436		190	84
Residuum		43.2	98.3	.996	10.6					

Carbon residue, Conradson: Residuum, 7.9 percent; crude, 3.8 percent.

APPROXIMATE SUMMARY

	Percent	Sp. gr.	°API	Viscosity
Light gasoline	6.3	0.687	74.5	
Total gasoline and naphtha	21.5	0.744	58.7	
Kerosine distillate	3.3	.813	42.6	
Gas oil	14.8	.848	35.4	
Nonviscous lubricating distillate	8.0	.874–.901	30.4–25.6	50-100
Medium lubricating distillate	5.2	.901–.916	25.6–23.0	100-200
Viscous lubricating distillate	2.3	.916–.923	23.0–21.8	Above 200
Residuum	43.2	.996	10.6	
Distillation loss	1.7			

1/ Distillation discontinued at 559° F.

Table 2-5.31. U.S. Crude-petroleum Analysis

BRADFORD FIELD (McKean County, Pennsylvania)
Upper Devonian

GENERAL CHARACTERISTICS

Gravity, specific, 0.820   Gravity, °API, 41.1   Pour point, °F., below 5
Sulfur, percent, 0.11   Color, green
Viscosity, Saybolt Universal at 100° F., 44 sec.;   Nitrogen, percent, 0.01

DISTILLATION, BUREAU OF MINES ROUTINE METHOD

STAGE 1—Distillation at atmospheric pressure, 753 mm. Hg
First drop, 90 °F.

Fraction No.	Cut temp. °F.	Percent	Sum, percent	Sp. gr. 60/60°F.	°API 60°F.	C.I.	Refractive index, n_D at 20° C.	Specific dispersion	S.U. visc. 100°F.	Cloud test, °F.
1	122	2.0	2.0	0.671	79.4	(-)	1.39133	127.6		
2	167	4.5	6.5	.713	67.0	18	1.41120	126.6		
3	212	6.9	13.4	.738	60.2	21	1.42140	136.0		
4	257	5.8	19.2	.756	55.7	22	1.42893	136.9		
5	302	5.9	25.1	.770	52.3	22	1.43483	134.7		
6	347	5.6	30.7	.781	49.7	21	1.44056	134.6		
7	392	5.7	36.4	.793	46.9	21	1.44635	134.0		
8	437	5.7	42.1	.805	44.3	21	1.45097			
9	482	6.7	48.8	.817	41.7	22				
10	527									

STAGE 2—Distillation continued at 40 mm. Hg

11	392	1.9	50.7	0.827	39.6	23	1.46003	-	39	24
12	437	5.1	55.8	.835	38.0	23	1.46241	-	43	40
13	482	5.7	61.5	.843	36.4	23	1.46648	-	52	60
14	527	5.3	66.8	.854	34.2	25	1.47197	-	66	70
15	572	6.6	73.4	.864	32.3	27			100	80
Residuum		25.0	98.4	.900	25.7					

Carbon residue, Conradson: Residuum, 1.6 percent; crude, 0.4 percent.

APPROXIMATE SUMMARY

	Percent	Sp. gr.	°API	Viscosity
Light gasoline	6.5	0.700	70.6	
Total gasoline and naphtha	30.7	0.754	56.2	
Kerosine distillate	18.1	.806	44.1	
Gas oil	8.7	.833	38.4	
Nonviscous lubricating distillate	12.6	.841–.864	36.8–32.3	50-100
Medium lubricating distillate	3.3	.864–.870	32.3–31.1	100-200
Viscous lubricating distillate				
Residuum	25.0	.900	25.7	Above 200
Distillation loss	1.6			

U. S. GOVERNMENT PRINTING OFFICE 16-27333-3

Table 2-5.32. U.S. Crude-petroleum Analysis

CONROE FIELD (Montgomery County, Texas)
Cockfield, Eocene
5,000–5,200 ft

GENERAL CHARACTERISTICS

Gravity, specific, 0.840   Gravity, ° API, 37.0   Pour point, ° F., below 5
Sulfur, percent, less than 0.10   Color, green
Viscosity, Saybolt Universal at 100° F., 36 sec.   Nitrogen, percent, ...

DISTILLATION, BUREAU OF MINES ROUTINE METHOD

STAGE 1—Distillation at atmospheric pressure, 759 mm. Hg
First drop, 140 ° F.

Fraction No.	Cut temp. °F.	Percent	Sum. percent	Sp. gr. 60/60° F.	° API 60° F.	Refractive index, n_D at 20° C.	C. I.	Specific dispersion	S. U. vis. 100° F.	Cloud test, ° F.
1	122	1.1	1.1	0.680	76.6	—	—			
2	167	4.4	5.5	.745	58.4					
3	212	8.0	13.5	.775	51.1	1.40902	33	137.8		
4	257	7.3	20.8	.794	46.7	1.43218	38	155.6		
5	302	6.2	27.0	.814	44.5	1.44526	40	162.8		
6	347	5.8	32.8	.827	42.3	1.45072	38	163.4	40	25
7	392	7.4	40.2	.844	39.6	1.45479	36	158.1	45	40
8	437	10.2	50.4	.861	36.2	1.46281	37	161.1	57	60
9	482	11.4	61.8		32.8	1.47425	40	172.9	77	80
10	527					1.48231	43	176.8	130	95

STAGE 2—Distillation continued at 40 mm. Hg

Fraction No.	Cut temp. °F.	Percent	Sum. percent	Sp. gr. 60/60° F.	° API 60° F.	Refractive index, n_D at 20° C.	C. I.	Specific dispersion	S. U. vis. 100° F.	Cloud test, ° F.
11	392	7.5	69.3	0.865	32.1	1.41481	41	172.4		
12	437	8.0	77.3	.866	31.9	1.48251	37	—		
13	482	5.7	83.0	.873	30.6	1.48501	37	—		
14	527	4.5	87.5	.880	29.3		38			
15	572	4.6	92.1	.891	27.3		40			
Residuum		7.2	99.3	.950	17.5					

Carbon residue, Conradson: Residuum, 4.9 percent; crude, 0.4 percent.

APPROXIMATE SUMMARY

	Percent	Sp. gr.	° API	Viscosity
Light gasoline	5.5	0.732	61.8	
Total gasoline and naphtha	32.8	0.785	48.8	
Kerosine distillate				
Gas oil	43.4	.853	34.4	
Nonviscous lubricating distillate	11.1	.869–.884	31.3–28.6	50-100
Medium lubricating distillate	4.8	.884–.897	28.6–26.3	100-200
Viscous lubricating distillate				Above 200
Residuum	7.2	.950	17.5	
Distillation loss	.7			

U. S. GOVERNMENT PRINTING OFFICE   16—57835-2

Table 2-5.33. U.S. Crude-petroleum Analysis

EAST TEXAS FIELD (Rusk County, Texas)
Woodbine, Upper Cretaceous
3,650 ft

GENERAL CHARACTERISTICS

Gravity, specific, 0.838   Gravity, ° API, 37.4   Pour point, ° F., 20
Sulfur, percent, 0.25   Color, brownish black
Viscosity, Saybolt Universal at 100° F., 42 sec.   Nitrogen, percent, ...

DISTILLATION, BUREAU OF MINES ROUTINE METHOD

STAGE 1—Distillation at atmospheric pressure, 759 mm. Hg
First drop, 84 ° F.

Fraction No.	Cut temp. °F.	Percent	Sum. percent	Sp. gr. 60/60° F.	° API 60° F.	Refractive index, n_D at 20° C.	C. I.	Specific dispersion	S. U. vis. 100° F.	Cloud test, ° F.
1	122	1.3	1.3	0.634	91.7					
2	167	3.0	4.3	.674	78.4	1.37281	9.4	120.0		
3	212	6.2	10.5	.716	66.1	1.39884	19	122.1		
4	257	7.3	17.8	.743	58.9	1.41265	23	124.4		
5	302	5.8	23.6	.764	53.7	1.42487	26	131.1		
6	347	5.7	29.3	.783	49.2	1.43528	28	137.1	40	20
7	392	4.6	33.9	.798	45.8	1.44327	29	137.9	50	35
8	437	5.0	38.9	.812	42.8	1.45149	30	140.5	56	55
9	482	5.6	44.5	.827	39.6	1.45981	31	146.9	80	70
10	527	6.4	50.9	.839	37.2	1.46724	32	156.4	145	90

STAGE 2—Distillation continued at 40 mm. Hg

Fraction No.	Cut temp. °F.	Percent	Sum. percent	Sp. gr. 60/60° F.	° API 60° F.	Refractive index, n_D at 20° C.	C. I.	Specific dispersion	S. U. vis. 100° F.	Cloud test, ° F.
11	392	2.9	53.8	0.848	35.4	1.47371	33	159.6		
12	437	5.5	59.3	.854	34.2	1.47564	32	156.0		
13	482	6.2	65.5	.868	31.5	1.48207	35	152.8		
14	527	4.8	70.3	.883	28.8		39			
15	572	6.6	76.9	.893	27.0		41			
Residuum		22.2	99.1	.957	16.4					

Carbon residue, Conradson: Residuum, 6.1 percent; crude, 1.5 percent.

APPROXIMATE SUMMARY

	Percent	Sp. gr.	° API	Viscosity
Light gasoline	10.5	0.694	72.4	
Total gasoline and naphtha	33.9	0.746	58.2	
Kerosine distillate	5.0	.812	42.8	
Gas oil	17.7	.839	37.2	
Nonviscous lubricating distillate	13.1	.854–.886	34.2–28.2	50-100
Medium lubricating distillate	7.2	.886–.897	28.2–25.9	100-200
Viscous lubricating distillate				Above 200
Residuum	22.2	.957	16.4	
Distillation loss	.9			

U. S. GOVERNMENT PRINTING OFFICE   16—57835-2

Table 2-5.34. U.S. Crude-petroleum Analysis

GOLDSMITH FIELD (Ector County, Texas)
5,600 ft, Upper Clear Fork, Permian
5,622–5,722 ft

GENERAL CHARACTERISTICS

Gravity, specific, 0.835   Gravity, ° API, 38.0   Pour point, ° F., below 5
Sulfur, percent, 0.52   100° F., 46 sec.   Color, greenish black
Viscosity, Saybolt Universal at   Nitrogen, percent,

DISTILLATION, BUREAU OF MINES ROUTINE METHOD

STAGE 1—Distillation at atmospheric pressure, 748 mm. Hg   First drop, 84 ° F.

Fraction No.	Cut temp. °F.	Percent	Sum, percent	Sp. gr. 60/60° F.	° API, 60° F.	C.I.	Refractive index, n_D at 20° C.	Specific dispersion	S.U. visc, 100° F.	Cloud test, °F.
1	122	0.8	0.8	0.675	78.1	–				
2	167	2.2	3.0	.711	67.5	17	1.38060	126.3		
3	212	5.0	8.0	.737	60.5	20	1.39774	126.4		
4	257	6.7	14.7	.756	55.7	22	1.41009	129.6		
5	302	6.2	20.9	.773	51.6	23	1.42088	135.7		
6	347	4.9	25.8	.787	48.3	23	1.43024	140.3		
7	392	4.6	30.4	.801	45.2	24	1.43794	139.7		
8	437	5.4	35.8	.815	42.1	26	1.44564	139.8		
9	482	5.5	41.1	.829	39.2	28	1.45365	141.4		
10	527	5.2	47.3				1.46179	149.2		

STAGE 2—Distillation continued at 40 mm. Hg

Fraction No.	Cut temp. °F.	Percent	Sum, percent	Sp. gr. 60/60° F.	° API, 60° F.	C.I.	Refractive index, n_D at 20° C.	Specific dispersion	S.U. visc, 100° F.	Cloud test, °F.
11	392	3.0	50.3	0.848	35.4	33			39	10
12	437	6.2	56.8	.853	34.4	31			45	30
13	482	4.9	61.4	.864	32.3	33			56	50
14	527	5.3	66.7	.878	30.4	37			80	65
15	572	6.1	72.8	.903	25.2	46			150	80
Residuum		24.1	96.9	.944	18.2					

Carbon residue, Conradson: Residuum, 5.7 percent; crude, 1.6 percent.

APPROXIMATE SUMMARY

	Percent	Sp. gr.	° API	Viscosity
Light gasoline	8.0	0.697	71.5	
Total gasoline and naphtha	30.4	0.744	58.7	
Kerosene distillate	10.7	.808	43.6	
Gas oil	15.0	.842	36.6	
Nonviscous lubricating distillate	10.1	.858–.884	33.4–28.6	50–100
Medium lubricating distillate	6.6	.884–.913	28.1–23.5	100–200
Viscous lubricating distillate				Above 200
Residuum	24.1	.944	18.2	
Distillation loss	3.1			

Table 2-5.35. U.S. Crude-petroleum Analysis

GOLDSMITH FIELD (Ector County, Texas)
San Andres, Permian
4,234–4,290 ft

GENERAL CHARACTERISTICS

Gravity, specific, 0.833   Gravity, ° API, 38.4   Pour point, ° F., below 5
Sulfur, percent, 1.16   100° F., 40 sec.   Color, greenish black
Viscosity, Saybolt Universal at   Nitrogen, percent, 0.08

DISTILLATION, BUREAU OF MINES ROUTINE METHOD

STAGE 1—Distillation at atmospheric pressure, 746 mm. Hg   First drop, 84 ° F.

Fraction No.	Cut temp. °F.	Percent	Sum, percent	Sp. gr. 60/60° F.	° API, 60° F.	C.I.	Refractive index, n_D at 20° C.	Specific dispersion	S.U. visc, 100° F.	Cloud test, °F.
1	122	3.7	3.7	0.642	88.9	11	1.38190	126.5		
2	167	3.1	6.8	.677	77.5	24	1.40452	134.5		
3	212	6.8	12.9	.726	63.4	28	1.41885	139.9		
4	257	5.7	19.7	.753	56.4	29	1.42931	144.8		
5	302	5.1	25.4	.772	51.8	29	1.43686	142.6		
6	347	5.1	30.5	.785	48.8	28	1.44355	140.7		
7	392	4.5	35.4	.797	46.0	30	1.45088	142.1		
8	437	4.9	39.9	.812	42.8	30	1.45088	142.1		
9	482	4.8	44.7	.827	39.6	30	1.45933	148.7		
10	527	6.0	50.7	.844	36.2	35	1.46877	154.5		

STAGE 2—Distillation continued at 40 mm. Hg

Fraction No.	Cut temp. °F.	Percent	Sum, percent	Sp. gr. 60/60° F.	° API, 60° F.	C.I.	Refractive index, n_D at 20° C.	Specific dispersion	S.U. visc, 100° F.	Cloud test, °F.
11	392	3.2	53.9	0.855	34.0	36	1.47834	160.3	41	20
12	437	4.5	58.4	.867	31.7	38	1.48165	164.4	46	34
13	482	5.1	63.5	.879	29.5	40	1.48883	172.6	58	58
14	527	4.0	67.5	.892	27.1	43	1.49546	171.8	84	70
15	572	6.1	73.6	.915	23.1	51			160	80
Residuum		21.7	95.3	.964	15.3					

Carbon residue, Conradson: Residuum, 6.7 percent; crude, 1.7 percent.

APPROXIMATE SUMMARY

	Percent	Sp. gr.	° API	Viscosity
Light gasoline	12.9	0.690	73.6	
Total gasoline and naphtha	35.4	0.744	58.7	
Kerosene distillate	18.6	.812	42.8	
Gas oil	8.6	.848	35.8	
Nonviscous lubricating distillate	6.7	.871–.896	31.0–26.4	50–100
Medium lubricating distillate	6.7	.896–.927	26.4–21.1	100–200
Viscous lubricating distillate	2.1	.927–.929	21.1–20.8	Above 200
Residuum	21.7	.964	15.3	
Distillation loss	4.7			

Table 2-5.36. U.S. Crude-petroleum Analysis

HASTINGS FIELD (Brazoria County, Texas)
Marginulina, Oligocene
5,500–6,200 ft

GENERAL CHARACTERISTICS

Gravity, specific, 0.868   Gravity, °API, 31.5   Pour point, °F, below 5
Sulfur, percent, 0.15   Color, brownish green   Nitrogen, percent,
Viscosity, Saybolt Universal at 100° F., 48 sec.

DISTILLATION, BUREAU OF MINES ROUTINE METHOD

STAGE 1—Distillation at atmospheric pressure, 751 mm. Hg
First drop, 84 °F.

Fraction No.	Cut temp. °F.	Percent	Sum. percent	Sp. gr. 60/60° F.	°API, 60° F.	C.I.	Refractive index, n_D at 20° C.	Specific dispersion	S.U. visc., 100° F.	Cloud test, °F.
1	122	0.8	0.8	0.673	78.8	15				
2	167	1.0	1.8	.685	75.1					
3	212	3.0	4.8	.725	63.7	24	1.39374	127.7		
4	257	3.4	8.2	.755	55.9	29	1.41756	128.6		
5	302	3.1	11.3	.777	50.6	32	1.42985	135.4		
6	347	3.9	15.2	.798	45.8	35	1.44192	137.8		
7	392	4.9	20.1	.817	41.7	38	1.45217	139.9		
8	437	6.8	26.9	.833	38.4	40	1.46057	140.3		
9	482	8.0	34.9	.848	35.4	41	1.46875	148.0		
10	527	10.9	45.8	.864	32.3	44	1.47679	149.8		

STAGE 2—Distillation continued at 40 mm. Hg

Fraction No.	Cut temp. °F.	Percent	Sum. percent	Sp. gr. 60/60° F.	°API, 60° F.	C.I.	Refractive index, n_D at 20° C.	Specific dispersion	S.U. visc., 100° F.	Cloud test, °F.
11	392	7.3	53.1	0.873	30.6	45	1.48274	155.2	42	Below 5
12	437	7.8	60.9	.879	29.5	44	1.48474	156.2	50	do.
13	482	6.2	67.1	.889	27.7	45	1.49058	152.7	71	do.
14	527	5.7	72.8	.901	25.6	48			125	10
15	572	6.9	79.7	.916	28.0	52			280	20
Residuum		20.3	100.0	.945	18.2					

Carbon residue, Conradson: Residuum, 4.7 percent; crude, 1.0 percent.

APPROXIMATE SUMMARY

	Percent	Sp. gr.	°API	Viscosity
Light gasoline	4.8	0.708	68.4	
Total gasoline and naphtha	20.1	0.771	52.0	
Kerosine distillate	-	-	-	
Gas oil	36.9	.858	33.4	
Nonviscous lubricating distillate	10.2	.879–.895	29.5–26.6	50-100
Medium lubricating distillate	5.8	.895–.908	26.6–24.3	100-200
Viscous lubricating distillate	6.7	.908–.924	24.3–21.5	Above 200
Residuum	20.3	.945	18.2	
Distillation loss	.0			

Table 2-5.37. U.S. Crude-petroleum Analysis

HAWKINS FIELD (Wood County, Texas)
Sub-Clarksville (Eagle Ford), Upper Cretaceous
4,100–4,180 ft

GENERAL CHARACTERISTICS

Gravity, specific, 0.894   Gravity, °API, 26.8   Pour point, °F, below 5
Sulfur, percent, 2.19   Color, brownish black   Nitrogen, percent, 0.08
Viscosity, Saybolt Universal at 100° F., 135 sed.; 130° F., 84 sec.

DISTILLATION, BUREAU OF MINES ROUTINE METHOD

STAGE 1—Distillation at atmospheric pressure, 744 mm. Hg
First drop, 86 °F.

Fraction No.	Cut temp. °F.	Percent	Sum. percent	Sp. gr. 60/60° F.	°API, 60° F.	C.I.	Refractive index, n_D at 20° C.	Specific dispersion	S.U. visc., 100° F.	Cloud test, °F.
1	122	1.3	1.3	0.634	91.7	4.2				
2	167	2.1	3.4	.663	81.9		1.37978	126.8		
3	212	3.1	6.5	.698	71.2	11	1.40384	128.0		
4	257	3.3	9.8	.721	64.8	13	1.41473	130.2		
5	302	3.7	13.5	.742	59.2	15	1.42504	129.9		
6	347	3.7	17.2	.760	54.7	17	1.43383	131.4		
7	392	3.5	20.7	.778	50.4	19	1.44218	136.2		
8	437	3.3	24.0	.795	46.3	22	1.45141	137.1		
9	482	4.0	28.0	.811	43.0	24	1.46255	144.0		
10	527	5.4	33.4	.832	38.6	29				

STAGE 2—Distillation continued at 40 mm. Hg

Fraction No.	Cut temp. °F.	Percent	Sum. percent	Sp. gr. 60/60° F.	°API, 60° F.	C.I.	Refractive index, n_D at 20° C.	Specific dispersion	S.U. visc., 100° F.	Cloud test, °F.
11	392	2.8	36.2	0.851	34.8	34	1.47420		40	10
12	437	5.0	41.2	.866	31.9	37	1.48070		45	24
13	482	4.3	45.5	.883	28.8	42	1.49013		57	44
14	527	4.6	50.1	.902	25.4	48	1.50043		85	56
15	572	6.0	56.1	.913	23.5	50	1.50885		155	70
Residuum		43.1	99.2	1.005	9.3					

Carbon residue, Conradson: Residuum, 6.0 percent; crude, 2.9 percent.

APPROXIMATE SUMMARY

	Percent	Sp. gr.	°API	Viscosity
Light gasoline	6.5	0.674	78.4	
Total gasoline and naphtha	20.7	0.727	63.1	
Kerosine distillate	7.3	.804	44.5	
Gas oil	12.7	.848	35.4	
Nonviscous lubricating distillate	8.3	.873–.905	30.6–24.9	50-100
Medium lubricating distillate	7.1	.905–.922	24.9–22.0	100-200
Viscous lubricating distillate				Above 200
Residuum	43.1	1.005	9.3	
Distillation loss	.8			

Table 2-5.38. U.S. Crude-petroleum Analysis

KELLY-SNYDER FIELD (Scurry County, Texas)
Canyon Reef, Pennsylvanian
6,823 ft

GENERAL CHARACTERISTICS

Gravity, specific, 0.826 Gravity, °API, 39.8 Pour point, °F., below 5
Sulfur, percent, 0.22 Color, brownish green
Viscosity, Saybolt Universal at 100° F., 37 sec. Nitrogen, percent,

DISTILLATION, BUREAU OF MINES ROUTINE METHOD

STAGE 1—Distillation at atmospheric pressure, 759 mm. Hg
First drop, 82 °F.

Fraction No.	Cut temp. °F.	Percent	Sum, percent	Sp. gr. 60/60° F.	°API, 60° F.	C.I.	Refractive index, n_D at 20° C.	Specific dispersion	S.U. visc., 100° F.	Cloud test, °F.
1	122	2.1	2.1	0.647	86.9	16	1.38444	122.3		
2	167	4.8	6.9	.687	74.5	26	1.40428	125.2		
3	212	9.4	16.3	.729	62.6	27	1.41686	125.5		
4	257	8.2	24.5	.752	56.7	29	1.42759	125.5		
5	302	6.2	30.7	.772	51.8	30	1.43708	134.2		
6	347	5.6	36.3	.788	48.1	31	1.44580	134.2		
7	392	5.9	42.2	.803	44.7	31	1.45335	139.4		
8	437	4.9	47.1	.816	41.9	32	1.46107	138.2		
9	482	5.8	52.9	.832	38.6	34	1.46701	145.8		
10	527	5.6	58.5	.842	36.6	34		152.0		
STAGE 2—Distillation continued at 40 mm. Hg										
11	392	2.5	61.0	.854	34.2	36	1.47514	153.1	40	10
12	437	5.6	66.6	.859	33.2	34	1.48306	157.5	45	30
13	482	5.2	71.8	.871	31.0	37	1.48408	143.7	56	45
14	527	4.3	76.1	.885	28.4	40			79	65
15	572	5.1	81.2	.893	27.0	41			135	80
Residuum		18.4	99.6	.937	19.2					

Carbon residue, Conradson: Residuum, 4.8 percent; crude, 1.0 percent.

APPROXIMATE SUMMARY

	Percent	Sp. gr.	°API	Viscosity
Light gasoline	16.3	0.706	68.9	
Total gasoline and naphtha	42.2	0.749	57.4	
Kerosine distillate	4.9	.816	41.9	
Gas oil	19.2	.846	35.8	
Nonviscous lubricating distillate	9.4	.865-.888	32.1-27.9	50-100
Medium lubricating distillate	5.5	.888-.897	27.9-26.3	100-200
Viscous lubricating distillate				Above 200
Residuum	18.4	.939	19.2	
Distillation loss	.4			

U. S. GOVERNMENT PRINTING OFFICE 16—57530-3

Table 2-5.39. U.S. Crude-petroleum Analysis

McELROY FIELD (Crane and Upton Counties, Texas)
Grayburg, Permian
2,750-3,050 ft

GENERAL CHARACTERISTICS

Gravity, specific, 0.868 Gravity, °API, 31.5 Pour point, °F., 20
Sulfur, percent, 2.37 Color, brownish green
Viscosity, Saybolt Universal at 77° F., 62 sec.; Nitrogen, percent, 0.08
100° F., 53 sec.

DISTILLATION, BUREAU OF MINES ROUTINE METHOD

STAGE 1—Distillation at atmospheric pressure, 746 mm. Hg
First drop, 86 °F.

Fraction No.	Cut temp. °F.	Percent	Sum, percent	Sp. gr. 60/60° F.	°API, 60° F.	C.I.	Refractive index, n_D at 20° C.	Specific dispersion	S.U. visc., 100° F.	Cloud test, °F.
1	122	2.0	2.0	0.645	87.9	6.1		127.4		
2	167	1.8	3.8	.667	80.6	19	1.38530	129.7		
3	212	3.3	7.1	.715	66.4	24	1.41297	129.7		
4	257	4.6	11.7	.745	58.4	28	1.42707	137.0		
5	302	5.0	16.7	.769	52.5	30	1.43646	139.0		
6	347	4.0	20.7	.788	48.1	30	1.44485	139.9		
7	392	4.0	24.7	.804	44.5	32	1.45247	140.8		
8	437	4.4	29.1	.816	41.9	32	1.45984	147.5		
9	482	5.1	34.2	.829	39.2	32	1.46943	157.1		
10	527	7.1	41.3	.846	35.8	36				
STAGE 2—Distillation continued at 40 mm. Hg										
11	392	3.4	44.7	.860	33.0	38	1.47986	163.3	41	20
12	437	6.4	51.1	.876	30.0	42	1.48395	166.9	47	40
13	482	6.6	57.7	.886	28.2	44	1.49246	174.9	62	54
14	527	4.9	62.6	.899	25.9	47			95	70
15	572	6.2	68.8	.911	23.8	49			175	90
Residuum		29.6	98.4	.977	13.3					

Carbon residue, Conradson: Residuum, 10.5 percent; crude, 3.5 percent.

APPROXIMATE SUMMARY

	Percent	Sp. gr.	°API	Viscosity
Light gasoline	7.1	0.683	75.7	
Total gasoline and naphtha	24.7	0.749	57.4	
Kerosine distillate	4.4	.816	41.9	
Gas oil	20.2	.850	35.0	
Nonviscous lubricating distillate	11.3	.878-.890	29.7-27.5	50-100
Medium lubricating distillate	1.1	.890-.915	27.5-23.1	100-200
Viscous lubricating distillate	1.1	.915-.917	23.1-22.8	Above 200
Residuum	29.6	.977	13.3	
Distillation loss	1.6			

U. S. GOVERNMENT PRINTING OFFICE 16—57530-3

Table 2-5.40. U.S. Crude-petroleum Analysis

PANHANDLE FIELD (Hutchinson County, Texas)
Moore County, Permian
2,991–3,060 ft

GENERAL CHARACTERISTICS

Gravity, specific, 0.823 Gravity, °API, 40.4 Pour point, °F, 50
Sulfur, percent, 0.55 Color, dark green
Viscosity, Saybolt Universal at 100° F., 47 sec. Nitrogen, percent, 0.07

DISTILLATION, BUREAU OF MINES ROUTINE METHOD

STAGE 1—Distillation at atmospheric pressure, 746 mm. Hg
First drop, 86 °F.

Fraction No.	Cut temp. °F.	Percent	Sum, percent	Sp. gr. 60/60° F.	°API 60° F.	C.I.	Refractive index, n_D at 20° C.	Specific dispersion	S.U. visc. 100° F.	Cloud test, °F.
1	122	2.2	2.2	0.646	87.5					
2	167	3.1	5.3	.666	81.0	5.6	1.37493	114.1		
3	212	5.8	11.1	.717	65.5	17	1.39523	120.8		
4	257	6.8	17.9	.737	60.5	21	1.40837	125.2		
5	302	4.9	22.8	.755	55.9	21	1.41946	134.5		
6	347	5.6	28.4	.796	50.9	24	1.43104	138.7		
7	392	2.6	31.0	.797	47.6	25	1.43953	137.8		
8	437	4.3	35.4	.802	44.9	25	1.44532	138.6		
9	482	4.4	39.8	.814	42.3	25	1.45277	142.5		
10	527	5.5	45.3	.826	39.8	26	1.45943	142.3		

STAGE 2—Distillation continued at 40 mm. Hg

Fraction No.	Cut temp. °F.	Percent	Sum, percent	Sp. gr. 60/60° F.	°API 60° F.	C.I.	Refractive index, n_D at 20° C.	Specific dispersion	S.U. visc. 100° F.	Cloud test, °F.
11	392	3.6	48.9	0.834	38.2	26	1.46643	145.5	40	30
12	437	4.6	53.5	.840	37.0	25	1.46806	151.6	46	44
13	482	5.6	59.1	.851	34.8	27	1.47252	144.4	56	64
14	527	5.5	64.0	.864	32.3	30			78	80
15	572	5.5	69.5	.868	31.5	29			120	94
Residuum		27.2	96.7	.934	20.0					

Carbon residue, Conradson: Residuum, 5.6 percent; crude, 1.7 percent.

APPROXIMATE SUMMARY

	Percent	Sp. gr.	°API	Viscosity
Light gasoline	11.1	0.686	74.8	
Total gasoline and naphtha	31.1	0.733	61.5	
Kerosine distillate	8.7	.808	43.6	
Gas oil	13.6	.833	38.1	
Nonviscous lubricating distillate	10.9	.845–.866	36.0–31.9	50–100
Medium lubricating distillate	5.2	.866–.870	31.9–31.1	100–200
Viscous lubricating distillate				Above 200
Residuum	27.2	.934	20.0	
Distillation loss	3.3			

U. S. GOVERNMENT PRINTING OFFICE 16-57833-3

Table 2-5.41. U.S. Crude-petroleum Analysis

SLAUGHTER FIELD (Hockley County, Texas)
San Andres, Permian
5,000 ft

GENERAL CHARACTERISTICS

Gravity, specific, 0.870 Gravity, °API, 31.1 Pour point, °F, 25
Sulfur, percent, 2.04 Color, brownish black
Viscosity, Saybolt Universal at 100° F., 48 sec. Nitrogen, percent,

DISTILLATION, BUREAU OF MINES ROUTINE METHOD

STAGE 1—Distillation at atmospheric pressure, 751 mm. Hg
First drop, 93 °F.

Fraction No.	Cut temp. °F.	Percent	Sum, percent	Sp. gr. 60/60° F.	°API 60° F.	C.I.	Refractive index, n_D at 20° C.	Specific dispersion	S.U. visc. 100° F.	Cloud test, °F.
1	122	1.5	1.5	0.657	83.9					
2	167	2.8	4.3	.686	74.8	15	1.38356	133.8		
3	212	5.8	10.1	.728	62.9	25	1.40893	144.8		
4	257	6.4	16.5	.757	55.4	30	1.42388	147.4		
5	302	4.7	21.2	.772	51.8	29	1.43298	149.5		
6	347	5.7	26.9	.785	48.8	29	1.43944	144.8		
7	392	4.2	31.1	.797	46.0	28	1.44549	143.5		
8	437	4.4	35.5	.811	43.0	29	1.45222	144.7		
9	482	4.9	40.4	.828	39.4	32	1.46155	146.5		
10	527	6.2	46.6	.846	35.8	36	1.47218	159.1		

STAGE 2—Distillation continued at 40 mm. Hg

Fraction No.	Cut temp. °F.	Percent	Sum, percent	Sp. gr. 60/60° F.	°API 60° F.	C.I.	Refractive index, n_D at 20° C.	Specific dispersion	S.U. visc. 100° F.	Cloud test, °F.
11	392	3.1	49.7	0.868	31.5	42	1.48416	168.3	42	20
12	437	7.1	56.8	.883	28.8	46	1.48951	169.5	48	35
13	482	3.9	60.7	.894	26.8	47	1.49727	181.8	64	50
14	527	3.6	64.3	.907	24.5	50	1.50283	175.2	89	70
15	572	5.9	70.2	.919	22.5	53			160	80
Residuum		29.3	99.5	.995	10.7					

Carbon residue, Conradson: Residuum, 12.1 percent; crude, 4.1 percent.

APPROXIMATE SUMMARY

	Percent	Sp. gr.	°API	Viscosity
Light gasoline	10.1	0.706	68.9	
Total gasoline and naphtha	31.1	0.753	56.4	
Kerosine distillate	4.4	.813	45.2	
Gas oil	18.6	.853	34.4	
Nonviscous lubricating distillate	9.2	.885–.909	28.4–24.2	50–100
Medium lubricating distillate	6.7	.909–.923	24.2–21.8	100–200
Viscous lubricating distillate		.923–.924	21.8–21.6	Above 200
Residuum	29.3	.995	10.7	
Distillation loss	.5			

U. S. GOVERNMENT PRINTING OFFICE 16-57833-3

Table 2-5.42. U.S. Crude-petroleum Analysis

SPRABERRY TREND (Reagan County, Texas)
Spraberry, Permian
7,128–7,228 ft

GENERAL CHARACTERISTICS

Gravity, specific, 0.850   Gravity, °API, 35.0   Pour point, °F., 10
Sulfur, percent, 0.18   Color, brownish green
Viscosity, Saybolt Universal at 100° F., 48 sec.   Nitrogen, percent, 0.18

DISTILLATION, BUREAU OF MINES ROUTINE METHOD

STAGE 1—Distillation at atmospheric pressure, 750 mm. Hg
First drop, 82 °F.

Fraction No.	Cut temp. °F.	Percent	Sum. percent	Sp. gr. 60/60° F.	°API, 60° F.	C.I.	Refractive index, n_D at 20° C.	Specific dispersion	S.U. visc. 100° F.	Cloud test, °F.
1	122	1.4	1.4	0.658	83.6	20				
2	167	1.1	2.5	.697	71.5	23				
3	212	6.3	8.8	.724	63.9	23	1.39423	120.8		
4	257	7.9	16.7	.750	57.2	27	1.41347	120.5		
5	302	5.7	22.4	.769	52.5	28	1.42411	127.6		
6	347	4.8	27.2	.787	48.3	30	1.43462	127.6		
7	392	4.1	31.3	.805	44.3	32	1.44415	129.3		
8	437	3.5	35.8	.819	41.3	33	1.45258	138.2		
9	482	4.5	40.3	.831	38.8	33	1.45994	140.3		
10	527	5.4	45.7	.841	36.8	33	1.46621	148.3		

STAGE 2—Distillation continued at 40 mm. Hg

Fraction No.	Cut temp. °F.	Percent	Sum. percent	Sp. gr. 60/60° F.	°API, 60° F.	C.I.	Refractive index, n_D at 20° C.	Specific dispersion	S.U. visc. 100° F.	Cloud test, °F.
11	392	3.7	49.4	0.849	35.2	33	1.47319	–	41	10
12	437	5.0	54.4	.861	32.8	35	1.47610	–	46	30
13	482	4.3	58.7	.871	31.0	37	1.48236	–	60	56
14	527	4.8	63.5	.885	28.4	40	1.48986		90	70
15	572	4.8	68.3	.903	25.2	46			185	80
Residuum		28.4	96.7	.953	17.0					

Carbon residue, Conradson: Residuum, 7.1 percent; crude, 2.3 percent.

APPROXIMATE SUMMARY

	Percent	Sp. gr.	°API	Viscosity
Light gasoline	8.8	0.710	67.8	
Total gasoline and naphtha	31.3	0.755	55.9	
Kerosine distillate	4.5	.819	41.3	
Gas oil	17.5	.844	36.2	
Nonviscous lubricating distillate	8.3	.864–.876	32.3–30.0	50–100
Medium lubricating distillate	5.1	.876–.906	30.0–24.7	100–200
Viscous lubricating distillate	1.6	.906–.912	24.7–23.7	Above 200
Residuum	28.4	.953	17.0	
Distillation loss	3.3			

U. S. GOVERNMENT PRINTING OFFICE 16—37835–2

Table 2-5.43. U.S. Crude-petroleum Analysis

WARD ESTES, NORTH FIELD (Ward County, Texas)
Yates, Permian
2,696–2,756 ft

GENERAL CHARACTERISTICS

Gravity, specific, 0.855   Gravity, °API, 34.0   Pour point, °F., below 5
Sulfur, percent, 1.17   Color, greenish black
Viscosity, Saybolt Universal at 100° F., 45 sec.   Nitrogen, percent, 0.11

DISTILLATION, BUREAU OF MINES ROUTINE METHOD

STAGE 1—Distillation at atmospheric pressure, 737 mm. Hg
First drop, 82 °F.

Fraction No.	Cut temp. °F.	Percent	Sum. percent	Sp. gr. 60/60° F.	°API, 60° F.	C.I.	Refractive index, n_D at 20° C.	Specific dispersion	S.U. visc. 100° F.	Cloud test, °F.
1	122	2.8	2.8	0.649	86.5	6.1	1.37526	123.6		
2	167	2.7	5.5	.667	80.6		1.39550	121.1		
3	212	4.9	10.4	.710	67.8	17	1.40988	124.8		
4	257	5.9	16.3	.739	60.0	21	1.42163	131.3		
5	302	5.3	21.6	.761	54.4	24	1.43391	135.1		
6	347	5.5	27.1	.781	49.7	27	1.44410	140.7		
7	392	4.5	31.6	.800	45.4	30	1.45250	141.8		
8	437	5.4	37.0	.818	41.5	33	1.46214	146.3		
9	482	6.0	43.0	.836	37.8	36	1.47342	154.8		
10	527	5.4	48.4	.853	34.4	39				

STAGE 2—Distillation continued at 40 mm. Hg

Fraction No.	Cut temp. °F.	Percent	Sum. percent	Sp. gr. 60/60° F.	°API, 60° F.	C.I.	Refractive index, n_D at 20° C.	Specific dispersion	S.U. visc. 100° F.	Cloud test, °F.
11	392	3.2	51.6	0.872	30.8	44	1.48454	165.6	41	10
12	437	5.2	56.8	.879	29.5	44	1.48923	171.8	48	20
13	482	4.7	61.5	.896	26.4	48	1.49708	176.8	65	30
14	527	4.9	66.4	.912	23.7	53	1.50368	181.5	105	40
15	572	5.8	72.2	.918	22.6	53	1.50829	–	200	54
Residuum		26.8	99.0	.965	15.1					

Carbon residue, Conradson: Residuum, 7.7 percent; crude, 2.3 percent.

APPROXIMATE SUMMARY

	Percent	Sp. gr.	°API	Viscosity
Light gasoline	10.4	0.682	76.0	
Total gasoline and naphtha	31.6	0.740	59.7	
Kerosine distillate	5.4	.818	41.5	
Gas oil	18.8	.856	33.8	
Nonviscous lubricating distillate	7.6	.881–.876	29.1–24.0	50–100
Medium lubricating distillate	5.9	.910–.918	24.0–22.6	100–200
Viscous lubricating distillate	2.9	.918–.922	22.6–22.0	Above 200
Residuum	26.8	.965	15.1	
Distillation loss	1.0			

U. S. GOVERNMENT PRINTING OFFICE 16—37835–2

Table 2-5.44. U.S. Crude-petroleum Analysis
WASSON FIELD (Gaines County, Texas)
San Andres, Permian
4,900–5,150 ft

GENERAL CHARACTERISTICS

Gravity, specific, 0.861   Gravity, °API, 32.8   Pour point, °F, below 5
Sulfur, percent, 1.76   Color, brownish black
Viscosity, Saybolt Universal at 100° F., 43 sec.   Nitrogen, percent, 0.08

DISTILLATION, BUREAU OF MINES ROUTINE METHOD

STAGE 1—Distillation at atmospheric pressure, 750 mm. Hg
First drop, 82 °F.

Fraction No.	Cut temp. °F.	Percent	Sum, percent	Sp. gr. 60/60° F.	°API, 60° F.	C.I.	Refractive index, n_D at 20° C.	Specific dispersion	S.U. visc. 100° F.	Cloud test, °F.
1	122	2.4	2.4	0.645	87.9	12	1.38199	128.5		
2	167	2.6	5.0	.679	76.9		1.40351	137.4		
3	212	5.3	10.3	.723	64.2	23	1.42009	144.8		
4	257	7.0	17.3	.754	56.2	28	1.43256	146.2		
5	302	6.4	23.7	.775	51.1	31	1.43924	147.0		
6	347	5.1	28.8	.787	48.3	30	1.44477	143.3		
7	392	4.5	33.3	.798	45.8	29	1.44477	138.9		
8	437	4.5	37.8	.813	42.6	33	1.46015	139.0		
9	482	5.0	42.8	.830	39.0	39	1.46015	146.8		
10	527	5.6	48.3	.848	35.4	37	1.47087	161.3		

STAGE 2—Distillation continued at 40 mm. Hg

11	392	2.6	50.9	0.860	33.0	38	1.47973	-	39	10
12	437	6.4	57.3	.872	30.8	40	1.48630	-	44	30
13	482	4.3	61.6	.886	28.2	44	1.49372	-	54	54
14	527	5.3	66.9	.904	25.0	49	1.50091		77	70
15	572	4.8	71.7	.914	23.3	51			140	90
Residuum		27.8	99.5	.987	11.9					

Carbon residue, Conradson: Residuum, 5.2 percent; crude, 1.7 percent.

APPROXIMATE SUMMARY

	Percent	Sp. gr.	°API	Viscosity
Light gasoline	10.3	0.694	72.4	
Total gasoline and naphtha	33.3	0.750	57.2	
Kerosene distillate	14.5	.813	42.6	
Gas oil	19.6	.854	34.0	
Nonviscous lubricating distillate	8.7	.880–.907	29.3–24.5	50-100
Medium lubricating distillate	5.6	.907–.917	24.5–22.8	100-200
Viscous lubricating distillate				Above 200
Residuum	27.5	.987	11.9	
Distillation loss	.8			

16–37835–3

Table 2-5.45. U.S. Crude-petroleum Analysis
WEBSTER FIELD (Harris County, Texas)
Marginulina, Frio, Oligocene
5,090–5,110 ft

GENERAL CHARACTERISTICS

Gravity, specific, 0.880   Gravity, °API, 29.3   Pour point, °F, below 5
Sulfur, percent, 0.21   Color, greenish black
Viscosity, Saybolt Universal at 100° F., 64 sec.;   Nitrogen, percent, 0.05

DISTILLATION, BUREAU OF MINES ROUTINE METHOD

STAGE 1—Distillation at atmospheric pressure, 746 mm. Hg
First drop, 127 °F.

Fraction No.	Cut temp. °F.	Percent	Sum, percent	Sp. gr. 60/60° F.	°API, 60° F.	C.I.	Refractive index, n_D at 20° C.	Specific dispersion	S.U. visc. 100° F.	Cloud test, °F.
1	122	0.7	0.7	0.705	69.2	(-)	1.40393	125.6		
2	167	2.2	2.9	.723	60.2	30	1.41829	124.7		
3	212	2.9	5.8	.761	54.4	32	1.42868	129.0		
4	257	2.0	7.8	.780	49.9	33	1.44020	135.3		
5	302	3.0	10.8	.801	45.2	36	1.45103	149.5		
6	347	3.7	14.5	.821	40.9	40	1.45997	138.5		
7	392	5.5	20.0	.837	37.6	42	1.46819	146.8		
8	437	6.8	26.8	.854	34.2	44	1.47643	151.2		
9	482	10.1	36.9	.867	31.7	46				

STAGE 2—Distillation continued at 40 mm. Hg

11	392	6.4	43.3	0.872	30.8	44	1.48213	-	43	below 5
12	437	7.3	50.6	.884	28.6	46	1.48456	-	51	do.
13	482	6.6	57.2	.890	27.5	46	1.48953	-	73	do.
14	527	6.3	63.5	.907	24.5	50	1.49551		125	do.
15	572	7.6	71.1	.915	23.1	51	1.50021	-	290	do.
Residuum		27.6	98.7	.944	18.4					

Carbon residue, Conradson: Residuum, 4.6 percent; crude, 1.4 percent.

APPROXIMATE SUMMARY

	Percent	Sp. gr.	°API	Viscosity
Light gasoline	2.9	0.730	62.3	
Total gasoline and naphtha	14.5	0.781	49.7	
Kerosene distillate				
Gas oil	31.4	.861	32.8	
Nonviscous lubricating distillate	11.4	.882–.899	28.9–25.9	50-100
Medium lubricating distillate	5.9	.899–.911	25.9–23.8	100-200
Viscous lubricating distillate	7.9	.911–.919	23.8–22.5	Above 200
Residuum	27.6	.944	18.4	
Distillation loss	1.3			

U. S. GOVERNMENT PRINTING OFFICE   16–37835–3

Table 2-5.46. U.S. Crude-petroleum Analysis

YATES FIELD (Pecos County, Texas)
San Andres, Permian
973–1,038 ft

GENERAL CHARACTERISTICS

Gravity, specific, 0.875   Gravity, °API, 30.2   Pour point, °F., below 5
Sulfur, percent, 1.54   Color, brownish black
Viscosity, Saybolt Universal at 77° F., 66 sec.; 100° F., 59 sec.   Nitrogen, percent, 0.15

DISTILLATION, BUREAU OF MINES ROUTINE METHOD

Stage 1—Distillation at atmospheric pressure, 747 mm. Hg
First drop, 90 °F.

Fraction No.	Cut temp., °F.	Percent	Sum, percent	Sp. gr. 60/60° F.	°API 60° F.	C. I.	Refractive index, n_D at 20° C.	Specific dispersion	S. U. visc. 100° F.	Cloud test, °F.
1	122	1.3	1.3	0.642	88.9	7.0	1.37474	123.0		
2	167	1.4	2.7	.669	80.0	14	1.39764	126.6		
3	212	2.6	5.3	.704	69.5	20	1.40961	124.5		
4	257	4.3	9.6	.736	60.8	24	1.42114	126.4		
5	302	4.4	14.0	.761	54.4	24	1.43324	129.9		
6	347	4.9	18.9	.783	49.2	28	1.44521	134.5		
7	392	5.2	24.1	.805	44.3	32	1.45484	139.4		
8	437	4.4	28.5	.824	40.2	35	1.46428	144.8		
9	482	5.5	34.0	.841	36.8	38	1.47493	154.6		
10	527	7.0	41.0	.859	33.2	42				

Stage 2—Distillation continued at 40 mm. Hg

Fraction No.	Cut temp., °F.	Percent	Sum, percent	Sp. gr. 60/60° F.	°API 60° F.	C. I.	Refractive index, n_D at 20° C.	Specific dispersion	S. U. visc. 100° F.	Cloud test, °F.
11	392	2.5	43.5	0.876	30.0	46	1.48054	162.4	43	Below 5
12	437	5.4	48.9	.882	28.9	45	1.49011	169.4	49	do.
13	482	5.7	54.6	.900	25.7	50	1.49854	176.3	65	do.
14	527	5.2	59.8	.918	22.6	56	1.50512	181.6	110	do.
15	572	6.9	66.7	.922	22.0	51	1.51039	-	230	do.
Residuum		31.6	98.3	.969	14.5					

Carbon residue, Conradson: Residuum, 9.9 percent; crude, 3.5 percent.

APPROXIMATE SUMMARY

	Percent	Sp. gr.	°API	Viscosity
Light gasoline	5.3	0.680	76.6	
Total gasoline and naphtha	24.1	.753	56.4	
Kerosine distillate	4.4	.824	40.2	
Gas oil	17.9	.860	33.0	
Nonviscous lubricating distillate	9.2	.883-.914	36.2-23.3	50-100
Medium lubricating distillate	4.4	.914-.921	23.3-22.1	100-200
Viscous lubricating distillate	4.9	.921-.924	23.3-21.6	Above 200
Residuum	31.6	.969	14.5	
Distillation loss	1.7			

Table 2-5.47. U.S. Crude-petroleum Analysis

ANETH FIELD (San Juan County, Utah)
Upper Paradox, Hermosa, Pennsylvanian
5,700–5,840 ft

GENERAL CHARACTERISTICS

Gravity, specific, 0.823   Gravity, °API, 40.4   Pour point, °F., 25
Sulfur, percent, 0.20   Color, dark green
Viscosity, Saybolt Universal at 100° F., 38 sec.   Nitrogen, percent, 0.06

DISTILLATION, BUREAU OF MINES ROUTINE METHOD

Stage 1—Distillation at atmospheric pressure, 741 mm. Hg
First drop, 88 °F.

Fraction No.	Cut temp., °F.	Percent	Sum, percent	Sp. gr. 60/60° F.	°API 60° F.	C. I.	Refractive index, n_D at 20° C.	Specific dispersion	S. U. visc. 100° F.	Cloud test, °F.
1	122	3.1	3.1	0.648	86.9	10	1.38064	126.5		
2	167	3.7	6.8	.676	77.8	22	1.40165	128.6		
3	212	5.1	11.9	.721	64.8	24	1.41505	135.9		
4	257	6.9	18.8	.745	58.4	24	1.42544	138.0		
5	302	5.0	23.8	.761	54.4	26	1.43494	135.7		
6	347	5.8	29.6	.780	49.9	27	1.44247	140.9		
7	392	5.4	35.0	.794	46.7	28	1.44941	142.7		
8	437	4.9	39.9	.808	43.6	29	1.45657	147.5		
9	482	5.3	45.2	.821	40.9	29	1.46374	153.3		
10	527	6.9	52.1	.833	38.4	30				

Stage 2—Distillation continued at 40 mm. Hg

Fraction No.	Cut temp., °F.	Percent	Sum, percent	Sp. gr. 60/60° F.	°API 60° F.	C. I.	Refractive index, n_D at 20° C.	Specific dispersion	S. U. visc. 100° F.	Cloud test, °F.
11	392	3.0	55.1	0.840	37.0	29	1.46993	154.9	40	20
12	437	6.2	61.3	.849	35.2	29	1.47179	154.3	44	34
13	482	6.4	67.7	.858	33.4	30	1.47752	159.1	54	50
14	527	5.0	72.7	.868	31.5	32	1.48274	-	74	70
15	572	5.1	77.8	.870	31.1	30			115	80
Residuum		21.2	99.0	.915	23.1					

Carbon residue, Conradson: Residuum, 3.5 percent; crude, 0.8 percent.

APPROXIMATE SUMMARY

	Percent	Sp. gr.	°API	Viscosity
Light gasoline	11.9	0.688	74.2	
Total gasoline and naphtha	35.0	.741	59.5	
Kerosine distillate	10.2	.815	42.1	
Gas oil	16.9	.841	36.8	
Nonviscous lubricating distillate	11.3	.854-.869	34.2-31.3	50-100
Medium lubricating distillate	4.4	.869-.871	31.1-31.0	100-200
Viscous lubricating distillate				Above 200
Residuum	21.2	.915	23.1	
Distillation loss	1.0			

Table 2-5.48. U.S. Crude-petroleum Analysis

ELK BASIN FIELD (Park and Carbon Counties, Wyoming and Montana)
Frontier, Upper Cretaceous
1,500 ft

GENERAL CHARACTERISTICS

Gravity, specific, 0.810   Gravity, ° API, 43.2   Pour point, ° F., below 5
Sulfur, percent, less than 0.10   Color, brownish green
Viscosity, Saybolt Universal at 100° F., 35 sec.   Nitrogen, percent, 0.08

DISTILLATION, BUREAU OF MINES ROUTINE METHOD

STAGE 1—Distillation at atmospheric pressure, 745 mm. Hg
First drop, 86 ° F.

Fraction No.	Cut temp. °F.	Percent	Sum, percent	Sp. gr. 60/60° F.	°API 60° F.	Refractive index, n_D at 20° C.	C.I.	Specific dispersion	S.U. visc. 100° F.	Cloud test, °F.
1	122	2.0	2.7	0.638	90.3	1.38698	9.8	124.2		
2	167	3.6	6.3	.675	78.1	1.40231	23	123.8		
3	212	9.0	15.3	.723	64.2	1.41616	27	131.9		
4	257	11.0	26.3	.751	56.9	1.42834	29	139.9		
5	302	7.3	33.6	.771	52.0	1.43739	30	135.1		
6	347	6.6	40.2	.787	48.3	1.44488	30	142.5		
7	392	5.9	46.1	.801	45.2	1.45310	30	152.0		
8	437	5.2	51.3	.817	41.7	1.46238	34	148.9		
9	482	6.1	57.4	.832	38.6	1.46927	33	150.4		
10	527	6.0	63.4	.841	36.8					

STAGE 2—Distillation continued at 40 mm. Hg

Fraction No.	Cut temp. °F.	Percent	Sum, percent	Sp. gr. 60/60° F.	°API 60° F.	Refractive index, n_D at 20° C.	C.I.	Specific dispersion	S.U. visc. 100° F.	Cloud test, °F.
11	392	5.0	68.4	0.850	35.0	1.47493	34	155.1	41	20
12	437	4.5	72.9	.859	33.2	1.47679	34	160.7	47	40
13	482	5.1	78.0	.868	31.5	1.48192	35	160.0	58	60
14	527	3.7	81.7	.877	29.9		36		77	80
15	572	4.8	86.5	.885	28.4		37		125	90
Residuum		11.7	98.2	.937	19.5					

Carbon residue, Conradson: Residuum, 3.3 percent; crude, 4.5 percent.

APPROXIMATE SUMMARY

	Percent	Sp. gr.	° API	Viscosity
Light gasoline	15.3	0.697	71.5	
Total gasoline and naphtha	46.1	0.748	57.7	
Kerosine distillate	5.2	.817	41.3	
Gas oil	20.8	.843	36.4	
Nonviscous lubricating distillate	9.9	.861-.880	32.8-29.3	50-100
Medium lubricating distillate	4.5	.880-.887	29.3-27.7	100-200
Viscous lubricating distillate				Above 200
Residuum	11.7	.937	19.5	
Distillation loss	1.8			

U. S. GOVERNMENT PRINTING OFFICE   16-57630-3

Table 2-5.49. U.S. Crude-petroleum Analysis

HAMILTON DOME FIELD (Hot Springs, Wyoming)
Tensleep, Pennsylvanian
2,602-2,773 ft

GENERAL CHARACTERISTICS

Gravity, specific, 0.918   Gravity, ° API, 22.6   Pour point, ° F., below 5
Sulfur, percent, 2.98   Color, brownish black
Viscosity, Saybolt Universal at 100° F., 230 sec.;   Nitrogen, percent,
130° F., 135 sec.

DISTILLATION, BUREAU OF MINES ROUTINE METHOD

STAGE 1—Distillation at atmospheric pressure, 760 mm. Hg
First drop, 153 ° F.

Fraction No.	Cut temp. °F.	Percent	Sum, percent	Sp. gr. 60/60° F.	°API 60° F.	Refractive index, n_D at 20° C.	C.I.	Specific dispersion	S.U. visc. 100° F.	Cloud test, °F.
1	122	0.3	0.3							
2	167	0.7	1.0	0.699	70.9	1.39154	-	121.9		
3	212	1.6	2.6	.705	69.2	1.39341	5.2	122.3		
4	257	2.5	5.1	.731	62.1	1.40759	9.9	126.7		
5	302	3.0	8.1	.754	56.2	1.41896	14	129.6		
6	347	3.1	11.2	.775	51.1	1.42979	18	130.2		
7	392	3.9	15.1	.795	46.5	1.43990	22	132.4		
8	437	4.8	19.9	.815	42.1	1.45081	26	135.9		
9	482	6.6	26.5	.838	37.4	1.46365	32	146.3		

STAGE 2—Distillation continued at 40 mm. Hg

Fraction No.	Cut temp. °F.	Percent	Sum, percent	Sp. gr. 60/60° F.	°API 60° F.	Refractive index, n_D at 20° C.	C.I.	Specific dispersion	S.U. visc. 100° F.	Cloud test, °F.
11	392	6.8	33.3	0.874	30.4		-			
12	437	6.8	40.1	.889	27.7		45		46	20
13	482	6.4	46.5	.913	23.5		53		59	40
14	527	8.0	54.5	.928	21.0		57		90	60
15	572								185	75
Residuum		45.1	99.6	1.007	9.0					

Carbon residue, Conradson: Residuum, 15.2 percent; crude, 7.5 percent.

APPROXIMATE SUMMARY

	Percent	Sp. gr.	° API	Viscosity
Light gasoline	1.0	0.699	70.9	
Total gasoline and naphtha	11.2	0.743	58.9	
Kerosine distillate	8.7	.806	44.1	
Gas oil	12.2	.854	34.2	
Nonviscous lubricating distillate	12.0	.879-.915	29.5-23.1	50-100
Medium lubricating distillate	7.6	.915-.930	23.1-20.7	100-200
Viscous lubricating distillate	2.8	.930-.936	20.7-19.7	Above 200
Residuum	45.1	1.007	9.0	
Distillation loss	.4			

U. S. GOVERNMENT PRINTING OFFICE   16-57630-3

Table 2-5.50. U.S. Crude-petroleum Analysis

SALT CREEK FIELD (Natrona County, Wyoming)
Third Wall Creek, Upper Cretaceous
1,683–1,693 ft

GENERAL CHARACTERISTICS

Gravity, specific, 0.842   Gravity, °API, 36.6   Pour point, °F, 55
Sulfur, percent, 0.12   Color, brownish green
Viscosity, Saybolt Universal at 100° F., 43 sec.   Nitrogen, percent,

DISTILLATION, BUREAU OF MINES ROUTINE METHOD

STAGE 1—Distillation at atmospheric pressure, 77 mm. Hg
First drop, °F.

Fraction No.	Cut temp. °F.	Percent	Sum, percent	Sp. gr. 60/60° F.	°API 60° F.	C.I.	Refractive index n_D at 20° C.	Specific dispersion	S. U. visc. 100° F.	Cloud test, °F.
1	122	2.0	2.0	0.660	82.9	-	1.37089	127.2		
2	167	4.2	6.2	.714	66.7	18	1.39648	125.6		
3	212	6.3	12.5	.742	59.2	23	1.41089	131.3		
4	257	5.4	17.9	.762	54.2	25	1.42233	135.9		
5	302	5.5	23.4	.779	50.1	26	1.43218	137.4		
6	347	4.2	27.6	.793	46.9	26	1.43996	140.0		
7	392	5.5	33.1	.807	43.8	27	1.44757	140.6		
8	437	4.6	37.7	.819	41.3	28	1.45501	146.5		
9	482	7.1	44.8	.830	39.0	28	1.46173	149.8		
10	527									
					STAGE 2—Distillation continued at 40 mm. Hg					
11	392	3.7	48.5	0.846	35.8	32			40	15
12	437	6.6	55.1	.850	35.0	30			46	30
13	482	6.1	61.2	.863	32.5	33			57	60
14	527	5.3	66.5	.875	30.2	35			81	80
15	572	6.5	73.0	.886	28.2	37			140	95
Residuum		26.1	99.1	.939	19.2					

Carbon residue, Conradson: Residuum, 4.4 percent; crude, 1.3 percent.

APPROXIMATE SUMMARY

	Percent	Sp. gr.	°API	Viscosity
Light gasoline	6.2	0.697	71.5	
Total gasoline and naphtha	27.6	0.751	56.9	
Kerosine distillate	10.1	.812	42.8	
Gas oil	16.4	.841	36.8	
Nonviscous lubricating distillate	11.6	.855–.878	34.0–29.7	50–100
Medium lubricating distillate	7.3	.878–.892	29.7–27.1	100–200
Viscous lubricating distillate				Above 200
Residuum	26.1	.939	19.2	
Distillation loss	.9			

U. S. GOVERNMENT PRINTING OFFICE 16-37823-3

Table 2-6.1. Foreign Crude-petroleum Analysis

BAHREIN FIELD (Bahrein, Asia)

GENERAL CHARACTERISTICS

Gravity, specific, 0.849   Gravity, °API, 35.2   Pour point, °F,
Sulfur, percent, 1.42   Color, greenish black
Viscosity, Saybolt Universal at 100° F., 44 sec.   Nitrogen, percent,

DISTILLATION, BUREAU OF MINES ROUTINE METHOD

STAGE 1—Distillation at atmospheric pressure, 743 mm. Hg
First drop, 99 °F.

Fraction No.	Cut temp. °F.	Percent	Sum, percent	Sp. gr. 60/60° F.	°API 60° F.	C.I.	Refractive index n_D at 20° C.	Specific dispersion	S. U. visc. 100° F.	Cloud test, °F.
1	122	5.8	5.8	0.667	80.6	16	1.37647	127.0		
2	167	2.6	8.4	.709	68.1	17	1.39720	135.5		
3	212	4.1	12.5	.730	62.3	19	1.40900	135.1		
4	257	5.7	18.2	.751	56.9	19	1.42036	141.1		
5	302	6.1	24.3	.771	52.0	22	1.43201	144.5		
6	347	5.1	29.4	.785	48.8	22	1.43943	144.4		
7	392	6.2	35.6	.799	46.0	23	1.44529	143.4		
8	437	5.1	40.7	.812	42.8	23	1.45283	146.5		
9	482	6.8	47.5	.829	39.2	28	1.46680	155.3		
10	527									
					STAGE 2—Distillation continued at 40 mm. Hg					
11	392	3.0	50.5	0.848	35.4	33			41	20
12	437	7.1	57.6	.857	33.6	33			45	35
13	482	6.4	64.0	.877	29.9	39			57	55
14	527	4.6	68.6	.895	26.6	45			87	75
15	572	5.3	73.9	.900	25.7	44			135	85
Residuum		25.1	99.0	.970	14.4					

Carbon residue, Conradson: Residuum, 8 percent; crude, 3.1 percent.

APPROXIMATE SUMMARY

	Percent	Sp. gr.	°API	Viscosity
Light gasoline	8.4	0.680	76.6	
Total gasoline and naphtha	29.4	0.738	60.2	
Kerosine distillate	11.3	.804	44.5	
Gas oil	16.4	.845	36.0	
Nonviscous lubricating distillate	10.6	.866–.896	31.9–26.4	50–100
Medium lubricating distillate	6.2	.896–.903	26.4–25.2	100–200
Viscous lubricating distillate				Above 200
Residuum	25.1	.970	14.4	
Distillation loss	1.0			

U. S. GOVERNMENT PRINTING OFFICE 16-37823-3

Table 2-6.2. Foreign Crude-petroleum Analysis
AGHA JARI FIELD (Iran)

GENERAL CHARACTERISTICS

Gravity, specific, 0.852 — Gravity, °API, 34.6 — Pour point, °F., 20
Sulfur, percent, 1.43 — Color, brownish black
Viscosity, Saybolt Universal at 77° F., 53 sec.; 100° F., 46 sec. — Nitrogen, percent, 0.15

DISTILLATION, BUREAU OF MINES ROUTINE METHOD

STAGE 1—Distillation at atmospheric pressure, 743 mm. Hg
First drop, 82 °F.

Fraction No.	Cut temp. °F.	Percent	Sum, percent	Sp. gr. 60/60° F.	°API 60° F.	Refractive index, n_D at 20° C.	C.I.	Specific dispersion	S.U. visc. 100° F.	Cloud test, °F.
1	122	5.0	5.0	0.648	86.9		-	128.6		
2	167	3.7	8.7	.705	69.2	1.37312	14	130.7		
3	212	4.8	13.5	.735	61.0	1.39701	19	138.4		
4	257	4.9	18.4	.755	55.7	1.41094	22	140.2		
5	302	5.5	23.9	.774	51.3	1.42285	23	144.4		
6	347	4.9	28.8	.788	48.1	1.43227	24	144.4		
7	392	4.5	33.3	.801	45.2	1.43969	24	144.0		
8	437	5.7	39.0	.817	41.8	1.44637	24	141.8		
9	482					1.45490	27	146.0		
10	527	6.0	45.0	.836	37.8	1.46484	31	154.3		

STAGE 2—Distillation continued at 40 mm. Hg

Fraction No.	Cut temp. °F.	Percent	Sum, percent	Sp. gr. 60/60° F.	°API 60° F.	Refractive index, n_D at 20° C.	C.I.	Specific dispersion	S.U. visc. 100° F.	Cloud test, °F.
11	392	2.1	47.1	0.855	34.0	1.47460	36	155.9	40	10
12	437	5.8	52.9	.864	32.3	1.47832	37	154.5	44	30
13	482	5.6	58.5	.878	29.7	1.48620	40	154.5	54	45
14	527	4.8	63.3	.892	27.1		43		75	60
15	572	5.5	68.8	.908	24.3		48		130	75
Residuum		28.9	97.3	.979	13.0					

Carbon residue, Conradson: Residuum, 9.1 percent; crude, 3.0 percent.

APPROXIMATE SUMMARY

	Percent	Sp. gr.	°API	Viscosity
Light gasoline	8.7	0.672	79.1	
Total gasoline and naphtha	28.8	0.736	60.8	
Kerosine distillate	10.2	.810	43.2	
Gas oil	14.6	.852	34.6	
Nonviscous lubricating distillate	9.6	.873-.899	30.6-25.9	50-100
Medium lubricating distillate	5.6	.899-.916	25.9-23.0	100-200
Viscous lubricating distillate				Above 200
Residuum	28.9	.979	13.0	
Distillation loss	2.3			

Table 2-6.3. Foreign Crude-petroleum Analysis
CENTRAL AREA (Khuzistan, Iran)

GENERAL CHARACTERISTICS

Gravity, specific, 0.835 — Gravity, °API, 38.0 — Pour point, °F.,
Sulfur, percent, 1.20 — Color, greenish black
Viscosity, Saybolt Universal at 100° F., 40 sec. — Nitrogen, percent,

DISTILLATION, BUREAU OF MINES ROUTINE METHOD

STAGE 1—Distillation at atmospheric pressure, 744 mm. Hg
First drop, 84 °F.

Fraction No.	Cut temp. °F.	Percent	Sum, percent	Sp. gr. 60/60° F.	°API 60° F.	Refractive index, n_D at 20° C.	C.I.	Specific dispersion	S.U. visc. 100° F.	Cloud test, °F.
1	122	2.7	2.7	0.637	90.6	1.37309	3.2	127.6		
2	167	3.0	5.7	.661	82.6	1.39552	15	131.6		
3	212	4.9	10.6	.707	68.6	1.41121	20	139.3		
4	257	5.1	15.7	.737	60.5	1.42331	23	145.4		
5	302	6.1	21.8	.758	55.2	1.43277	24	145.4		
6	347	5.3	27.1	.776	50.9	1.43895	24	144.1		
7	392	5.1	32.2	.788	48.1	1.44498	24	144.1		
8	437	5.5	37.5	.800	45.4	1.45286	25	142.1		
9	482	5.5	43.0	.814	42.3	1.45286	25	145.6		
10	527	6.3	49.3	.830	39.0	1.46213	28	153.3		

STAGE 2—Distillation continued at 40 mm. Hg

Fraction No.	Cut temp. °F.	Percent	Sum, percent	Sp. gr. 60/60° F.	°API 60° F.	C.I.	S.U. visc. 100° F.	Viscosity
11	392	2.8	52.1	0.850	35.0	34	40	15
12	437	6.3	58.4	.859	33.3	34	45	30
13	482	5.5	63.9	.877	29.9	39	57	50
14	527	4.9	68.8	.893	27.0	44	82	70
15	572	5.8	74.6	.905	24.9	46	170	90
Residuum		21.5	96.1	.980	12.9			

Carbon residue, Conradson: Residuum, 11.2 percent; crude, 2.8 percent.

APPROXIMATE SUMMARY

	Percent	Sp. gr.	°API	Viscosity
Light gasoline	10.6	0.676	77.8	
Total gasoline and naphtha	32.2	0.735	61.0	
Kerosine distillate	10.8	.807	43.8	
Gas oil	14.8	.846	35.8	
Nonviscous lubricating distillate	9.7	.867-.895	31.7-26.6	50-100
Medium lubricating distillate	6.0	.895-.909	26.6-24.2	100-200
Viscous lubricating distillate	11.1	.909-.912	24.2-23.7	Above 200
Residuum	21.5	.980	12.9	
Distillation loss	3.9			

1/ Fields
Masjid-i-Sulaiman - 22.2%   White Oil Springs (Naft Safid) - 4.9%
Lali - 4.9%   Haft Kel - 68.0%

Table 2-6.4. Foreign Crude-petroleum Analysis
KIRKUK FIELD (Iraq)

GENERAL CHARACTERISTICS

Gravity, specific, 0.842	Gravity, °API, 36.6	Pour point, °F,	
Sulfur, percent, 1.93		Color, brownish black	
Viscosity, Saybolt Universal at 100° F., 42 sec.		Nitrogen, percent,	

DISTILLATION, BUREAU OF MINES ROUTINE METHOD

STAGE 1—Distillation at atmospheric pressure, 743 mm. Hg
First drop, 88 °F.

Fraction No.	Cut temp., °F.	Percent	Sum, percent	Sp. gr., 60/60° F.	°API, 60° F.	C.I.	Refractive index, n_D at 20° C.	Specific dispersion	S.U. visc., 100° F.	Cloud test, °F.
1	122	3.2	3.2	0.628	93.8					
2	167	4.6	7.8	.659	83.2	2.3	1.37157	119.9		
3	212	4.5	12.3	.696	71.8	10	1.38911	121.1		
4	257	5.9	18.2	.724	63.9	14	1.40426	128.9		
5	302	6.3	24.5	.751	56.9	19	1.41902	136.4		
6	347	5.5	30.0	.773	51.6	23	1.43121	140.6		
7	392	5.5	35.5	.786	48.5	23	1.43839	143.2		
8	437	4.7	40.2	.797	46.0	23	1.44370	140.7		
9	482	5.1	45.3	.810	43.2	23	1.45070	142.9		
10	527	6.0	51.3	.827	39.6	27	1.45991	148.5		

STAGE 2—Distillation continued at 40 mm. Hg

Fraction No.	Cut temp., °F.	Percent	Sum, percent	Sp. gr., 60/60° F.	°API, 60° F.	C.I.	Refractive index	Specific dispersion	S.U. visc., 100° F.	Cloud test, °F.
11	392	4.3	55.6	0.845	36.0	31			40	20
12	437	5.2	60.8	.855	34.0	32			45	35
13	482	5.4	66.2	.877	29.9	39			55	55
14	527	4.7	70.9	.894	26.8	44			85	70
15	572	5.6	76.5	.907	24.5	47			180	95
Residuum		23.3	99.8	.997	10.4					

Carbon residue, Conradson: Residuum, 14.6 percent; crude, 4.0 percent.

APPROXIMATE SUMMARY

	Percent	Sp. gr.	°API	Viscosity
Light gasoline	12.3	0.664	81.6	
Total gasoline and naphtha	35.5	0.725	63.7	
Kerosine distillate	9.8	.804	44.5	
Gas oil	14.9	.842	36.6	
Nonviscous lubricating distillate	11.2	.863-.901	32.5-25.6	50-100
Medium lubricating distillate	5.6	.901-.909	25.6-24.2	100-200
Viscous lubricating distillate	1.7	.909-.913	24.2-23.5	Above 200
Residuum	23.3	.997	10.4	
Distillation loss	.2			

Table 2-6.5. Foreign Crude-petroleum Analysis
ABQAIQ FIELD (Nejd Sultanate, Province of Al Hasa, Saudi Arabia)

GENERAL CHARACTERISTICS

Gravity, specific, 0.835	Gravity, °API, 38.0	Pour point, °F,	
Sulfur, percent, 1.30		Color, greenish black	
Viscosity, Saybolt Universal at 100° F., 40 sec.		Nitrogen, percent,	

DISTILLATION, BUREAU OF MINES ROUTINE METHOD

STAGE 1—Distillation at atmospheric pressure, 743 mm. Hg
First drop, 81 °F.

Fraction No.	Cut temp., °F.	Percent	Sum, percent	Sp. gr., 60/60° F.	°API, 60° F.	C.I.	Refractive index, n_D at 20° C.	Specific dispersion	S.U. visc., 100° F.	Cloud test, °F.
1	122	3.1	3.1	0.633	92.0					
2	167	3.1	6.2	.663	81.9		1.37387	126.3		
3	212	4.1	10.3	.698	71.2	11	1.39214	129.4		
4	257	5.0	15.3	.727	63.1	16	1.40723	134.9		
5	302	5.5	20.8	.751	56.9	19	1.42117	139.7		
6	347	5.9	26.7	.772	51.8	23	1.43251	145.7		
7	392	5.3	32.0	.787	48.3	23	1.44032	144.3		
8	437	5.6	37.6	.798	45.8	23	1.44570	143.7		
9	482	4.9	42.5	.812	42.8	24	1.45324	147.2		
10	527	6.3	48.8	.830	39.0	28	1.46328	156.1		

STAGE 2—Distillation continued at 40 mm. Hg

Fraction No.	Cut temp., °F.	Percent	Sum, percent	Sp. gr., 60/60° F.	°API, 60° F.	C.I.	Refractive index	Specific dispersion	S.U. visc., 100° F.	Cloud test, °F.
11	392	2.9	51.7	0.845	36.0	31			40	15
12	437	6.1	57.8	.856	33.8	33			44	30
13	482	4.6		.876	30.0	39			58	55
14	527		68.5	.892	27.1	43			76	70
15	572		73.9	.901	25.6	45			135	85
Residuum		24.0	97.9	.964	15.3					

Carbon residue, Conradson: Residuum, 10.3 percent; crude, 2.9 percent.

APPROXIMATE SUMMARY

	Percent	Sp. gr.	°API	Viscosity
Light gasoline	10.3	0.668	80.3	
Total gasoline and naphtha	32.0	0.730	62.3	
Kerosine distillate	10.5	.805	44.3	
Gas oil	14.9	.843	36.4	
Nonviscous lubricating distillate	10.9	.865-.896	32.1-26.4	50-100
Medium lubricating distillate	5.6	.896-.906	26.4-24.7	100-200
Viscous lubricating distillate				Above 200
Residuum	24.0	.964	15.3	
Distillation loss	2.1			

Table 2-6.6. Foreign Crude-petroleum Analysis

DAMMAM FIELD (Nejd Sultanate, Province of Al Hasa, Saudi Arabia)

GENERAL CHARACTERISTICS

Gravity, specific, 0.847   Gravity, °API, 35.6   Pour point, °F,
Sulfur, percent, 1.54   Color, greenish black
Viscosity, Saybolt Universal at 100° F., 42 sec.   Nitrogen, percent,

DISTILLATION, BUREAU OF MINES ROUTINE METHOD

STAGE 1—Distillation at atmospheric pressure, 741 mm. Hg
First drop, 84 °F.

Fraction No.	Cut temp. °F.	Percent	Sum, percent	Sp. gr. 60/60°F.	°API 60°F.	C.I.	Refractive index, n_D at 20° C.	Specific dispersion	S.U. visc. 100°F.	Cloud test, °F.
1	122	1.6	1.6							
2	167	2.0	3.6	0.653	85.2	-	1.37114	119.8		
3	212	3.3	6.9	.704	69.5	14	1.39522	132.5		
4	257	5.0	11.9	.735	61.0	15	1.41148	142.1		
5	302	5.7	17.6	.759	54.9	23	1.42530	148.2		
6	347	5.8	23.4	.777	50.6	25	1.43534	149.2		
7	392	5.6	29.0	.789	47.8	24	1.44158	148.3		
8	437	5.7	34.7	.801	45.2	24	1.44680	145.7		
9	482	5.6	40.3	.813	42.6	25	1.45359	149.7		
10	527	7.7	48.0	.830	39.0	28	1.46365	156.3		

STAGE 2—Distillation continued at 40 mm. Hg

Fraction No.	Cut temp. °F.	Percent	Sum, percent	Sp. gr. 60/60°F.	°API 60°F.	C.I.	Refractive index, n_D at 20° C.	Specific dispersion	S.U. visc. 100°F.	Cloud test, °F.
11	392	3.0	51.0	0.848	35.4	33			40	20
12	437	7.0	58.0	.859	33.2	34			44	35
13	482	6.3	64.3	.877	29.9	39			54	55
14	527	6.3	70.6	.893	27.0	44			79	75
15	572	5.2	75.8	.905	24.9	46			140	90
Residuum		23.2	99.0	.972	14.1					

Carbon residue, Conradson: Residuum, 11.7 percent; crude, 3.1 percent.

APPROXIMATE SUMMARY

	Percent	Sp. gr.	°API	Viscosity
Light gasoline	6.9	0.677	77.5	
Total gasoline and naphtha	29.0	.745	58.4	
Kerosene distillate	11.3	.807	43.8	
Gas oil	18.3	.846	35.8	
Nonviscous lubricating distillate	10.9	.870-.897	31.1-26.3	50-100
Medium lubricating distillate	6.3	.897-.910	26.3-24.0	100-200
Viscous lubricating distillate				Above 200
Residuum	23.2	.972	14.1	
Distillation loss	1.0			

Table 2-6.7. Foreign Crude-petroleum Analysis

LEDUC FIELD (Alberta Province, Canada)
D-2 (Leduc) Zone, Upper Devonian
5,050 ft

GENERAL CHARACTERISTICS

Gravity, specific, 0.826   Gravity, °API, 39.8   Pour point, °F, below 5
Sulfur, percent, 0.30   Color, brownish green
Viscosity, Saybolt Universal at 100° F., 38 sec.   Nitrogen, percent,

DISTILLATION, BUREAU OF MINES ROUTINE METHOD

STAGE 1—Distillation at atmospheric pressure, 746 mm. Hg
First drop, 86 °F.

Fraction No.	Cut temp. °F.	Percent	Sum, percent	Sp. gr. 60/60°F.	°API 60°F.	C.I.	Refractive index, n_D at 20° C.	Specific dispersion	S.U. visc. 100°F.	Cloud test, °F.
1	122	3.5	3.5	0.642	88.9	6.1	1.37486	125.8		
2	167	3.0	6.5	.667	80.6	20	1.39939	129.4		
3	212	5.9	12.4	.718	65.6	25	1.41366	133.6		
4	257	7.0	19.4	.746	58.2	27	1.42596	139.2		
5	302	6.6	26.0	.767	53.0	29	1.43611	142.6		
6	347	5.7	31.7	.785	48.8	30	1.44412	142.6		
7	392	5.0	36.7	.800	45.0	30	1.45084	142.7		
8	437	5.1	41.8	.812	42.8	30	1.45792	142.9		
9	482	5.1	46.9	.824	40.2	30	1.46614	147.2		
10	527	6.2	53.1	.840	37.0	33		153.7		

STAGE 2—Distillation continued at 40 mm. Hg

Fraction No.	Cut temp. °F.	Percent	Sum, percent	Sp. gr. 60/60°F.	°API 60°F.	C.I.	Refractive index, n_D at 20° C.	Specific dispersion	S.U. visc. 100°F.	Cloud test, °F.
11	392	2.9	56.0	0.850	35.0	34			41	20
12	437	5.8	61.8	.859	33.2	34			46	35
13	482	5.3	67.1	.872	30.8	37			58	55
14	527	4.6	71.7	.883	28.8	39			82	75
15	572	5.4	77.1	.895	26.6	42			155	90
Residuum		19.9	97.0	.951	17.3					

Carbon residue, Conradson: Residuum, 7.6 percent; crude, 1.7 percent.

APPROXIMATE SUMMARY

	Percent	Sp. gr.	°API	Viscosity
Light gasoline	12.4	0.684	75.4	
Total gasoline and naphtha	36.7	.742	59.2	
Kerosene distillate	10.2	.818	41.5	
Gas oil	14.0	.848	35.4	
Nonviscous lubricating distillate	9.8	.863-.886	32.5-28.2	50-100
Medium lubricating distillate	6.4	.886-.901	28.2-25.6	100-200
Viscous lubricating distillate				Above 200
Residuum	19.9	.951	17.3	
Distillation loss	3.0			

Table 2-6.8. Foreign Crude-petroleum Analysis

LEDUC FIELD (Alberta Province, Canada)
D-3 Zone, Upper Devonian
5,300 ft

GENERAL CHARACTERISTICS

Gravity, specific, 0.821   Gravity, °API, 40.9   Pour point, °F, below 5
Sulfur, percent, 0.30   Color, brownish green
Viscosity, Saybolt Universal at 100° F., 37 sec.   Nitrogen, percent,

DISTILLATION, BUREAU OF MINES ROUTINE METHOD

Stage 1—Distillation at atmospheric pressure, 746 mm. Hg
First drop, 84 °F.

Fraction No.	Cut temp. °F.	Percent	Sum. percent	Sp. gr. 60/60° F.	°API 60° F.	C.I.	Refractive index, n_d at 20° C.	Specific dispersion	S.U. visc. 100° F.	Cloud test, °F.
1	122	3.1	3.1	0.642	88.9					
2	167	3.0	6.1	.662	82.2					
3	212	5.3	11.4	.712	67.2	3.7	1.37387	125.0		
4	257	7.6	19.0	.744	58.7	18	1.39711	127.0		
5	302	5.7	24.7	.764	53.7	24	1.41272	133.4		
6	347	6.0	30.7	.783	49.2	26	1.42473	138.5		
7	392	5.3	36.0	.799	45.6	28	1.43526	144.1		
8	437	4.3	40.3	.810	43.2	29	1.44390	144.1		
9	482	5.3	45.6	.822	40.6	29	1.44952	149.5		
10	527	6.1	51.7	.836	37.8	31	1.45646	146.4		

Stage 2—Distillation continued at 40 mm. Hg

Fraction No.	Cut temp. °F.	Percent	Sum. percent	Sp. gr. 60/60° F.	°API 60° F.	C.I.	Refractive index, n_d at 20° C.	Specific dispersion	S.U. visc. 100° F.	Cloud test, °F.
11	392	4.5	56.2	0.848	35.4	33	1.47200	151.6	39	20
12	437	5.6	61.8	.857	33.6	33	1.47680		45	35
13	482	5.0	66.8	.870	31.1	36	1.48522		56	55
14	527	5.6	72.4	.886	28.2	40			82	75
15	572	4.9	77.3	.895	26.6	42			155	90
Residuum		18.7	96.0	.947	17.9					

Carbon residue, Conradson: Residuum, 7.1 percent; crude, 1.5 percent.

APPROXIMATE SUMMARY

	Percent	Sp. gr.	°API	Viscosity
Light gasoline	11.4	0.680	76.6	
Total gasoline and naphtha	36.0	0.741	59.5	
Kerosene distillate	9.6	.817	41.7	
Gas oil	16.0	.847	35.6	
Nonviscous lubricating distillate	9.3	.861–.888	32.8–27.0	**50-100**
Medium lubricating distillate	6.4	.888–.899	27.9–25.9	**100-200**
Viscous lubricating distillate				**Above 200**
Residuum	18.7	.947	17.9	
Distillation loss	4.0			

Table 2-6.9. Foreign Crude-petroleum Analysis

POZA RICA FIELD (Veracruz, Mexico)
Lower Tamabra
7,218

GENERAL CHARACTERISTICS

Gravity, specific, 0.881   Gravity, °API, 29.1   Pour point, °F, 20
Sulfur, percent, 1.77   Color, brownish black
Viscosity, Saybolt Universal at 77° F., 165 sec.;   Nitrogen, percent, 0.17

DISTILLATION, BUREAU OF MINES ROUTINE METHOD

Stage 1—Distillation at atmospheric pressure, 750 mm. Hg
First drop, 154 °F.

Fraction No.	Cut temp. °F.	Percent	Sum. percent	Sp. gr. 60/60° F.	°API 60° F.	C.I.	Refractive index, n_d at 20° C.	Specific dispersion	S.U. visc. 100° F.	Cloud test, °F.
1	122									
2	167	3.8	3.8	0.689	73.9	-	1.39789	132.0		
3	212	3.3	7.1	.726	63.4	15	1.40613	132.4		
4	257	4.0	11.1	.750	57.2	19	1.41882	133.4		
5	302	4.3	15.4	.771	52.0	22	1.42978	138.5		
6	347	3.9	19.3	.786	48.5	23	1.43822	140.5		
7	392	3.9	23.5	.800	45.4	24	1.44537	143.5		
8	437	4.2	28.5	.815	42.1	26	1.45332	145.1		
9	482	5.0	34.8	.831	38.8	29	1.46344	155.3		
10	527	6.3								

Stage 2—Distillation continued at 40 mm. Hg

Fraction No.	Cut temp. °F.	Percent	Sum. percent	Sp. gr. 60/60° F.	°API 60° F.	C.I.	Refractive index, n_d at 20° C.	Specific dispersion	S.U. visc. 100° F.	Cloud test, °F.
11	392	1.9	36.7	0.850	35.0	34	1.47200	154.8	40	15
12	437	6.1	42.7	.860	33.0	35	1.47680	157.9	44	30
13	482	6.0	48.8	.873	30.6	37	1.48522	159.6	55	50
14	527	5.4	54.2	.888	27.9	41			76	65
15	572	6.5	60.7	.900	25.7	44			130	80
Residuum		39.3	100.0	.967	14.8					

Carbon residue, Conradson: Residuum, 12.9 percent; crude, 5.6 percent.

APPROXIMATE SUMMARY

	Percent	Sp. gr.	°API	Viscosity
Light gasoline	3.8	0.689	47.8	
Total gasoline and naphtha	9.3	0.745	58.2	
Kerosene distillate	9.2	.808	41.5	
Gas oil	14.5	.846	35.8	
Nonviscous lubricating distillate	6.5	.867–.893	31.7–27.0	**50-100**
Medium lubricating distillate		.893–.906	27.0–24.7	**100-200**
Viscous lubricating distillate				**Above 200**
Residuum	39.3	.967	14.8	
Distillation loss	0.0			

Table 2-6.10. Foreign Crude-petroleum Analysis

LA CIRA FIELD (Santander Department, Colombia)
C Zone, La Paz
3,038–3,372 ft

GENERAL CHARACTERISTICS

Gravity, specific, 0.908   Gravity, °API, 24.3   Pour point, °F, _____
Sulfur, percent, 0.96   Color, brownish black
Viscosity, Saybolt Universal at 100° F., 220 sec.; 130° F., 125 sec   Nitrogen, percent, _____

DISTILLATION, BUREAU OF MINES ROUTINE METHOD

STAGE 1—Distillation at atmospheric pressure, 761 mm. Hg
First drop, 95 °F.

Fraction No.	Cut temp. °F.	Percent	Sum, percent	Sp. gr. 60/60° F.	°API 60° F.	C. I.	Refractive index, n_D at 20° C.	Specific dispersion	S. U. visc., 100° F.	Cloud test, °F.
1	122	0.1	0.1							
2	167	1.9	2.0	0.693	72.7	-	1.38911	121.7		
3	212	0.9	2.9	.745	58.4	24	1.41331	131.6		
4	257	2.5	5.4	.770	52.3	28	1.42670	137.1		
5	302	2.3	7.7	.790	47.6	31	1.43746	139.9		
6	347	1.8	9.5	.809	43.4	34	1.44724	140.1		
7	392	2.6	12.1	.824	40.2	35	1.45437	140.9		
8	437	2.9	15.0	.838	37.4	37	1.46205			
9	482	4.5	19.5	.854	34.2	40	1.47182	148.8		
10	527	6.9	26.4							

STAGE 2—Distillation continued at 40 mm. Hg

Fraction No.	Cut temp. °F.	Percent	Sum, percent	Sp. gr. 60/60° F.	°API 60° F.	C. I.	Refractive index, n_D at 20° C.	Specific dispersion	S. U. visc., 100° F.	Cloud test, °F.
11	392	3.2	29.6	0.871	31.0	44			45	Below 5
12	437	5.2	34.8	.882	28.9	45			52	do.
13	482	5.9	40.7	.897	26.3	49			73	do.
14	527	6.1	46.8	.911	23.8	52			135	do.
15	572	8.8	55.6	.918	22.6	53			290	do.
Residuum		43.0	98.6	.976	13.5					

Carbon residue, Conradson: Residuum, 13.9 percent; crude, 6.4 percent.

APPROXIMATE SUMMARY

	Percent	Sp. gr.	°API	Viscosity
Light gasoline	2.9	0.693	72.7	
Total gasoline and naphtha	12.1	0.759	54.9	
Kerosine distillate	6.9	.824	40.2	
Gas oil	16.2	.855	34.0	
Nonviscous lubricating distillate	9.2	.879-.903	29.5-25.2	50-100
Medium lubricating distillate	6.5	.903-.914	25.2-23.3	100-200
Viscous lubricating distillate	8.7	.914-.922	23.3-22.0	Above 200
Residuum	43.0	.976	13.5	
Distillation loss	1.4			

Table 2-6.11. Foreign Crude-petroleum Analysis

LA BREA—PARINAS FIELD (Piura Department, Peru)
(High Cold Test)
Salina

GENERAL CHARACTERISTICS

Gravity, specific, 0.837   Gravity, °API, 37.6   Pour point, °F, _____
Sulfur, percent, less than 0.10   Color, brownish green
Viscosity, Saybolt Universal at 100° F., 40 sec.   Nitrogen, percent, _____

DISTILLATION, BUREAU OF MINES ROUTINE METHOD

STAGE 1—Distillation at atmospheric pressure, 744 mm. Hg
First drop, 81 °F.

Fraction No.	Cut temp. °F.	Percent	Sum, percent	Sp. gr. 60/60° F.	°API 60° F.	C. I.	Refractive index, n_D at 20° C.	Specific dispersion	S. U. visc., 100° F.	Cloud test, °F.
1	122	1.0	1.0	0.653	85.2	-				
2	167	1.6	2.6	.719	65.3	21	1.37058	-		15
3	212	4.2	6.8	.748	57.7	26	1.39816	125.9		35
4	257	8.1	14.9	.767	53.0	27	1.41278	126.6		55
5	302	6.9	21.8	.785	48.8	29	1.42320	129.4		70
6	347	6.1	27.9	.801	45.2	30	1.43308	132.7		90
7	392	5.8	33.7	.815	42.1	31	1.44136	133.0		
8	437	6.2	39.9	.825	40.0	31	1.44832	132.1		
9	482	6.4	46.3	.835	38.0	31	1.45406	134.0		
10	527	6.7	53.0			31	1.46015	135.3		

STAGE 2—Distillation continued at 40 mm. Hg

Fraction No.	Cut temp. °F.	Percent	Sum, percent	Sp. gr. 60/60° F.	°API 60° F.	C. I.	Refractive index, n_D at 20° C.	Specific dispersion	S. U. visc., 100° F.	Cloud test, °F.
11	392	4.8	57.8	0.847	35.6	32			43	
12	437	7.1	64.9	.856	33.8	33			48	
13	482	5.4	70.3	.872	30.8	37			62	
14	527	5.8	76.1	.881	29.1	38			97	
15	572	5.9	82.0	.893	27.0	41			185	
Residuum		17.8	99.8	.960	15.9					

Carbon residue, Conradson: Residuum, 6.6 percent; crude, 1.3 percent.

APPROXIMATE SUMMARY

	Percent	Sp. gr.	°API	Viscosity
Light gasoline	6.8	0.694	72.4	
Total gasoline and naphtha	33.7	0.757	55.4	
Kerosine distillate	12.6	.820	41.1	
Gas oil	16.0	.842	36.6	
Nonviscous lubricating distillate	11.1	.858-.881	33.4-29.1	50-100
Medium lubricating distillate	6.2	.881-.895	29.4-26.6	100-200
Viscous lubricating distillate	1.9	.895-.899	26.5-25.9	Above 200
Residuum	17.8	.960	15.9	
Distillation loss	.2			

Table 2-6.12. Foreign Crude-petroleum Analysis
LA BREA—PARINAS FIELD (Piura Department, Peru)
(Low Cold Test)

GENERAL CHARACTERISTICS

Gravity, specific, 0.851 Gravity, °API, 34.8
Sulfur, percent, 0.12 Pour point, °F,
Viscosity, Saybolt Universal at 100° F., 41 sec. Color, brownish green
Nitrogen, percent,

DISTILLATION, BUREAU OF MINES ROUTINE METHOD

Stage 1—Distillation at atmospheric pressure, 745 mm. Hg
First drop, 79 °F.

Fraction No.	Cut temp. °F.	Percent	Sum, percent	Sp. gr. 60/60°F.	°API 60°F.	C.I.	Refractive index n_D at 20° C.	Specific dispersion	S.U. visc. 100°F.	Cloud test, °F.
1	122	1.6	1.6	0.632	92.4	6.1	1.37452	123.3		
2	167	2.2	3.8	.667	80.6	22	1.39858	123.2		
3	212	5.7	9.5	.721	64.8	27	1.41386	125.8		
4	257	9.9	19.4	.752	56.7	30	1.42376	131.0		
5	302	7.8	27.2	.774	51.3	34	1.43797	131.3		
6	347	6.3	33.5	.797	46.0	38	1.44834	133.3		
7	392	4.5	38.0	.818	41.5	38	1.45260	135.4		
8	437	5.5	43.5	.835	38.0	41	1.45760	135.1		
9	482	5.2	48.7	.852	34.6	43	1.46569	139.0		
10	527	6.2	54.9	.867	31.7	46	1.47398	143.4		

Stage 2—Distillation continued at 40 mm. Hg

Fraction No.	Cut temp. °F.	Percent	Sum, percent	Sp. gr. 60/60°F.	°API 60°F.	C.I.	Refractive index n_D at 20° C.	Specific dispersion	S.U. visc. 100°F.	Cloud test, °F.
11	392	2.6	57.5	0.871	31.0	44			46	Below 5
12	437	5.8	63.3	.888	27.9	48			55	do.
13	482	5.1	68.4	.901	25.6	51			84	do.
14	527	3.9	72.3	.911	23.8	52			170	do.
15	572	6.3	78.6	.923	21.8	55			380	do.
Residuum		20.7	99.3	.972	14.1					

Carbon residue, Conradson: Residuum, 7.4 percent; crude, 1.7 percent.

APPROXIMATE SUMMARY

	Percent	Sp. gr.	°API	Viscosity
Light gasoline	9.5	0.694	72.4	
Total gasoline and naphtha	38.0	0.757	55.4	
Kerosine distillate				
Gas oil	20.1	.855	34.0	
Nonviscous lubricating distillate	8.6	.878-.903	29.7-25.2	50-100
Medium lubricating distillate	4.9	.903-.914	25.2-23.3	100-200
Viscous lubricating distillate	7.0	.914-.930	23.3-20.7	Above 200
Residuum	20.7	.972	14.1	
Distillation loss	.7			

U. S. GOVERNMENT PRINTING OFFICE 16—57885-3

Table 2-6.13. Foreign Crude-petroleum Analysis
LAGUNILLAS FIELD (Zulia Department, Venezuela)

GENERAL CHARACTERISTICS

Gravity, specific, 0.952 Gravity, °API, 17.1
Sulfur, percent, 2.18 Pour point, °F,
Viscosity, Saybolt Universal at 100° F., 1550 sec. Color, brownish black
Nitrogen, percent,

DISTILLATION, BUREAU OF MINES ROUTINE METHOD

Stage 1—Distillation at atmospheric pressure, 752 mm. Hg
First drop, 172 °F.

Fraction No.	Cut temp. °F.	Percent	Sum, percent	Sp. gr. 60/60°F.	°API 60°F.	C.I.	Refractive index n_D at 20° C.	Specific dispersion	S.U. visc. 100°F.	Cloud test, °F.
1	122	1.9	1.9							
2	167	1.4	3.3							
3	212	.9	4.2	0.719	65.3	—	1.40050	123.9		
4	257	1.4	5.6							
5	302	1.9	7.5	.775	51.1	—	1.42773	133.1		
6	347	2.6	10.1	.815	42.1	37	1.44960	141.8		
7	392	1.9								
8	437	4.6	14.7	.835	38.0	41	1.46018	144.9		
9	482			.853	34.4	44	1.47056	152.5		
10	527	5.4	20.1	.863	32.5	44	1.47818	162.7		

Stage 2—Distillation continued at 40 mm. Hg

Fraction No.	Cut temp. °F.	Percent	Sum, percent	Sp. gr. 60/60°F.	°API 60°F.	C.I.	Refractive index n_D at 20° C.	Specific dispersion	S.U. visc. 100°F.	Cloud test, °F.
11	392	1.7	21.8	0.881	29.1	48			42	Below 5
12	437	5.3	27.1	.897	26.3	52			52	do.
13	482	6.0	33.4	.910	24.0	55			72	do.
14	527	9.2	42.6	.924	21.6	58			130	do.
1/ 15	572		47.6	.935	19.8				280	do.
Residuum		51.0	98.6	1.018	7.5					

Carbon residue, Conradson: Residuum, 19.9 percent; crude, 11.2 percent.

APPROXIMATE SUMMARY

	Percent	Sp. gr.	°API	Viscosity
Light gasoline	1.9	0.719	65.3	
Total gasoline and naphtha	7.5	0.771	52.0	
Kerosine distillate				
Gas oil	16.3	.861	32.8	
Nonviscous lubricating distillate	8.8	.894-.917	26.8-22.8	50-100
Medium lubricating distillate	6.4	.917-.929	22.8-20.8	100-200
Viscous lubricating distillate	8.6	.929-.942	20.8-18.7	Above 200
Residuum	51.0	1.018	7.5	
Distillation loss	1.4			

1/ Distillation discontinued at 565° F.

U. S. GOVERNMENT PRINTING OFFICE 16—57885-3

Section 3

PROCESSES

By GEORGE H. UNZELMAN

Chief Refinery Technologist, Western Region
Ethyl Corporation
Los Angeles, Calif.

and

CHARLES J. WOLF

Chief Refinery Technologist, Eastern Region
Ethyl Corporation
New York, N.Y.

PART 1. CATALYTIC PROCESSES

CATALYTIC CRACKING

Catalytic cracking is a refinery process to convert high-boiling nongasoline hydrocarbons into lower-boiling gasoline components. The catalyst may be in a fixed bed, moving bed, or fluid bed. Activated natural or synthetic catalysts—primarily silica alumina or silica magnesia—are employed in bead, pellet, or microspherical form. For similar catalyst composition and cracking conditions, natural catalyst results in a higher gasoline yield with lower antiknock quality than synthetic catalyst.

Feedstocks may range from naphtha cuts (included in normal heavier feed for upgrading) to reduced crudes. Usually feed preparation—to remove salts and heavy asphalts—is carried out through any one of the following ways: coking, propane deasphalting, furfural extraction, vacuum distillation, viscosity breaking, thermal cracking, and hydrodesulfurization.

The major process variables are temperature, pressure, catalyst-oil ratio (ratio of the weight of catalyst entering the reactor per hour to the weight of oil charged per hour), and space velocity (weight or volume of the oil charged per hour per weight or volume of catalyst in the reaction zone). Wide flexibility in product distribution and quality is possible through control of these variables along with the extent of internal recycle. Increased conversion can be obtained by (1) a higher temperature, (2) a higher pressure, (3) a lower space velocity, and (4) a higher catalyst-oil ratio.

Catalytic cracking as a commercial process came into use in the United States in 1936. In principle—conversion of fuel oil to gasoline—it is in direct competition with thermal cracking, and cracking construction today is essentially for catalytic usage.

The first catalytic cracker was of the fixed-bed variety.   However, today most cracking processing is carried out in moving or fluid beds.   On a total feed basis, about 99% of reported catalytic cracking capacity in 1962 was either the moving-bed (20%) or fluid-bed (79%) type.   Operating costs for these two process types are pretty much the same.   Direct operating costs for a 10,000-B/D unit run at 65% conversion range from 20 to 32 cents per bbl.   Total operating costs (including gas recovery) are 35 to 45 cents per bbl.

Though the phenomenal growth period for catalytic cracking installation has passed, it is still one of the most important processes in U.S. refineries.   In 1965, total catalytic cracking feed capacity was about 5,560,800 B/D (70% fresh feed, 30% recycle), representing over 50% of total U.S. crude running capacity.   This ratio of catalytic cracking to crude running capacity is not expected to increase markedly in the near future.   However, recent developments in catalyst manufacture and catalyst treating have given the industry catalysts which have longer "lives" and are more selective and, hence, give more economical product yields.

General catalytic cracking references include:

Bozeman, H. C., "Cat Cracking Begins Second 20 Years," *Oil & Gas J.*, pp. 67–78, July 23' 1962.

Berg, C., "Refining Capacity Up," *Petro/Chem Engr.*, pp. 45–48, January, 1962.

Carney, W., and A. W. Hoge, "Lion Oil Co. Catalytically Processes Sour Gas Oil," reviewed in *Pet. Proc.*, pp. 1207–1209, December, 1948.

Conn, A. L., and C. W. Brackin, "Cracking of High Sulfur Stocks, Use of Steam with Natural Catalyst," *Ind. Engrg. Chem.*, p. 1717, August, 1949.

Crocoll, J. F., and R. D. Jaquay, "What You Should Know about Catalytic Cracking," *Petro/Chem Engr.*, pp. C24–51, November, 1960.

DeBaun, R. M., "New Cracking Catalysts for More Dollars per Barrel," WPRA Meeting, San Antonio, Tex., March, 1960.

Flanders, R. L., et al., "How Feed Boiling Range Affects Cat Cracker Yields and Qualities," *Oil & Gas J.*, pp. 98–102, Mar. 7, 1960.

Goldtrap, W. A., and B. Skinner, "Refinery Conversion for Premium Motor Fuels," *Pet. Engr.*, p. 174, April, 1954.

Lawson, J. E., et al., "Get More Cracker Feed from Residua," *Pet. Ref.*, pp. 177–180, March, 1959.

Nelson, W. L., "Heat of Decomposition in Catalytic Cracking," *Oil & Gas J.*, p. 376, Nov. 10, 1949.

Nelson, W. L., "Catalytic Cracking Operating Costs," *Oil & Gas J.*, p. 129, Feb. 1, 1960.

Nelson, W. L., "Cat Cracking at 100% Conversion," *Oil & Gas J.*, p. 92, Jan. 15, 1962.

Nelson, W. L., "Paraffinic Crudes are Superior for Cat Cracking," *Oil & Gas J.*, p. 161, June 11, 1962.

Nelson, W. L., "Cost of Cat Cracking Plants Has Increased Little," *Oil & Gas J.*, pp. 114–116, July 2, 1962.

Noll, H. D., et al., "Commercial TCC Operations on Partially Vaporized Charge Stocks," NPA Meeting, Atlantic City, N.J., September, 1946, reviewed in *Pet. Proc.*, pp. 211–218, November, 1946.

Oden, E. C., and J. J. Perry, "17 Feeds for Cat Crackers," *Pet. Ref.*, pp. 191–193, March, 1954.

Pet. Proc., "8000 B/D Refinery Converts Thermal Unit to Catalytic Cracking for $200–250/ Bbl," pp. 411–416, June, 1947.

Pet. Ref., "More Catalytic Cracking Feed," p. 160, November, 1956.

Porter, R. W., "Unusual Techniques Feature the Production of Synthetic Bead Catalyst," *Chem. & Met. Engrg.*, p. 94, April, 1946.

Sittig, M., "Catalytic Cracking," *Pet. Proc.*, pp. 274–280, March, 1949.

Sittig, M., "Catalytic Cracking," *Pet. Ref.*, pp. 91–95, June, 1950.

Sittig, M., "Catalytic Cracking Techniques in Review," *Pet. Ref.*, pp. 263–316, September, 1952.

Sittig, M., "Catalytic Process Mechanics," *Pet. Proc.*, pp. 1048–1055, July, 1954.

Fluid-bed Catalytic Cracking

Fluidized-bed catalytic cracking processes use microspherical or powdered catalysts of either natural or synthetic origin.   The customary wide range of conventional catalytic charge stocks can be employed to yield gasoline, middle distillates, light olefins and isoparaffins, coke and fuel oil, and gas.   Product qualities and distributions are functions of the charge composition and boiling range, as well as of the operating conditions of the unit—amount of recycle, catalyst type, catalyst activity, conversion level, etc.

Normal fluid catalytic cracking operating conditions in the reactor are in the ranges of 880 to 980°F, 10 to 16 psig, 1.0 to 3.0/1.0 (w/hr/w) space velocity, and 8.0 to 12.0/1.0 catalyst-oil ratio. Regeneration temperatures and pressures are 1050 to 1100°F and 8 to 10 psig.

Cracking a gas-oil feed at about 60% conversion (based on 100% total possible disappearance of gas oil with zero recycle) can result in 35 to 45% debutanized gasoline, 40% cycle oil, 12% butane-butylene (50% olefin), and 8% C_3 hydrocarbons (60% olefin). Gasoline quality is 94 to 98 Research octane at 3.0 ml of TEL per gal. Feed pretreatment of vacuum distillation, decarbonizing or other solvent treating, visbreaking, coking, or hydrotreating might be used.

The several fluid catalytic cracking processes licensed differ primarily in mechanical design. Side-by-side, reactor-regenerator construction along with unitary vessel construction (the reactor either above or below the regenerator) are the two main mechanical variations. From a flow standpoint, all fluid catalytic cracking processes contact hydrocarbon feed and recycle with the finely divided catalyst in the reactor.

Catalyst circulates continuously between the reaction and regeneration zones. Heat of reaction is obtained from the regenerator through the catalyst flow. The catalyst rate is regulated to maintain the required temperature.

Hydrocarbon vapors are separated from the catalyst in the reactor and are sent to conventional fractionation facilities for separation.

Fluid catalytic cracking was first employed commercially in 1942. Since that time continual improvements in design have made the process the most widely used conversion process in the petroleum industry.

References include:

Andrews, J. M., "Cracking Characteristics of Cat Cracking Units," *Ind. Engrg. Chem.*, pp. 507–509, April, 1959.

Bergman, D. J., "How to Instrument a Fluid Catalytic Cracking Unit," *Pet. Ref.*, pp. 185–188, April, 1953.

Blazek, J. J., et al., "Catalytic Cracking with SM-30," API Refining Meeting, San Francisco, Calif., May 17, 1962.

Crocoll, J. F., and R. D. Jaquay, "Catalytic Cracking—Fluidization," *Petro/Chem Engr.*, pp. C26–29, December, 1960.

Ivey, F. E., "Flexible Refinery Yields through a New Fluid Cracking Catalyst," WPRA Meeting, San Antonio, Tex., June, 1960.

Krebs, R. W., "Applications of the Fluid Catalyst Technique to Catalytic Cracking and Hydrocarbon Synthesis," WPRA Meeting, Galveston, Tex., April, 1948.

Loper, B. H., "FCC Catalyst Containing 25% Alumina," WPRA Meeting, San Antonio, Tex., May, 1955.

MacDonald, M., "Automatic Control of a Fluid Catalytic Cracking Unit," *Pet. Ref.*, p. 87, October, 1946.

Olsen, C. R., and M. L. Sterba, "Effect of Reactor Temperature on Product Distribution and Product Quality in Fluid Catalytic Cracking," *Chem. Engrg. Prog.*, pp. 692–700, November, 1949.

Richardson, R. W., F. B. Johnson, and L. V. Robbins, Jr., "Fluid Catalyst Cracking with Silica-Magnesia," *Ind. Engrg. Chem.*, p. 1729, August, 1949.

Roquemore, R. W., and C. D. Strickland, "Cracking with High Alumina Catalyst," *Pet. Ref.*, pp. 231–233, May, 1957.

Thomas, E. J., "Fluid Catalytic Cracking of High Sulfur Stocks with Natural Catalysts," *Oil & Gas J.*, pp. 221–228, Mar. 23, 1950.

Thornton, Jr., D. P., "Fluid 'Cat Cracker' Designed to Use New MS Catalyst," *Pet. Proc.*, pp. 173–176, March, 1947.

Ullrich W., "How to Start up a Fluid Cat Cracker," *Pet. Ref.*, pp. 214–216, September, 1955.

VanDornick, E., "Modern Fluid Catalytic Cracking," *Pet. Engr.*, p. 149, April. 1947.

Wunderlich, O. A., and F. E. Ivey, "Regenerator Heat Balance Calculation," *Oil & Gas J.*, pp. 121–128, Jan. 20, 1958.

Model IV Fluid Catalytic Cracking

Licensed by the Esso Research and Engineering Company, the Model IV design (Fig. 3-1) incorporates the most recent engineering modifications.

Catalyst is transferred between the reactor and regenerator by means of U bends. Catalyst flow rate is varied by changing the amount of air injected into the spent-catalyst U bend. This feature eliminates the need for slide-valve control, a common source of erosion problems which limit run length. Regeneration air other than that used to control circulation enters the regenerator through a grid. Since no catalyst flows through the grid, it is possible to use a carbon-steel grid without concern for erosion or warpage.

Reactor and regenerator are mounted side by side, thus eliminating expansion and rotation joints and resulting in minimum overall height. Consequently, structural steel and foundation costs are lower and mechanical construction is greatly simplified.

The Model IV low-elevation design was preceded by the Model III (1947) balanced-pressure design, the Model II (1944) downflow design, and the original Model I (1941)

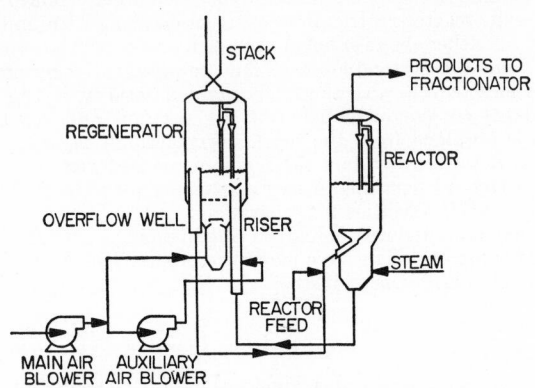

FIG. 3-1. Model IV fluid catalytic cracking.

upflow design. The first commercial Model IV installation in the United States was placed on-stream in 1952 at the Pan-Am Southern Corporation's Destrehan, La., refinery.

References include:

Hyd. Proc., "Fluid Catalytic Cracking—Model IV," p. 151, September, 1964.
McWhirter, Jr., W. E., et al., "Destrehan Model IV Fluid Cat Cracker," Pet. Ref., pp. 201–205, April, 1956.
Murphree, E. V., et al., "Fluid Catalyst Cracking for Premium Fuels," API Meeting, Chicago, Ill., November, 1943.
Murphree, E. V., "Progress in Petroleum Technology," ACS Meeting, New York, September, 1951.
Oil & Gas J., "Model IV Catalytic Cracking," p. 147, Apr. 5, 1965.
Pet. Proc., "First Cat Cracker in New England," pp. 1689–1691, November, 1953.
Pet. Ref., "Around the World with Model IV," p. 190, March, 1954.
Pet. Ref., "Fluid Catalytic Cracking," pp. 230–231, September, 1956.
Resen, F. L., "Model IV Cracker Proves Itself," Oil & Gas J., p. 64, Mar. 8, 1954.
Resen, F. L., "Model IV Cat Cracker Brings Lower Investment, Plus More Efficient Operation to Pan Am Southern at Destrehan," Oil & Gas J., pp. 214–222, Mar. 22, 1954.

Orthoflow Fluid Catalytic Cracking

Licensed by M. W. Kellogg Co., this process uses the unitary vessel design, which provides straight-line flow of catalyst and thereby minimizes erosion encountered in pipe bends.

Commercial Orthoflow designs are of three types: Models A and C, with the regenerator beneath the reactor, and Model B, with the regenerator above the reactor. In all cases, the catalyst stripping section is located between the reactor and the regenerator. Figure 3-2 shows a typical Model B design, and Fig. 3-3, a Model C design.

All these designs employ the heat-balanced principle incorporating fresh feed-recycle feed cracking. Local cost factors usually dictate the choice of model.

Orthoflow operating conditions generally fall within the following ranges: reactor, 885 to 950°F, 8 to 20 psig; regenerator, 1050 to 1200°F, 15 to 30 psig; catalyst-to-oil ratio, 6 to 20/1; space velocity, 1.0 to 16.0/1.0.

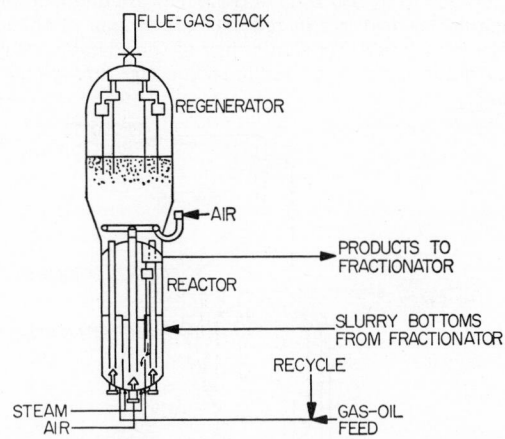

FIG. 3-2. Orthoflow fluid catalytic cracking—Model B.

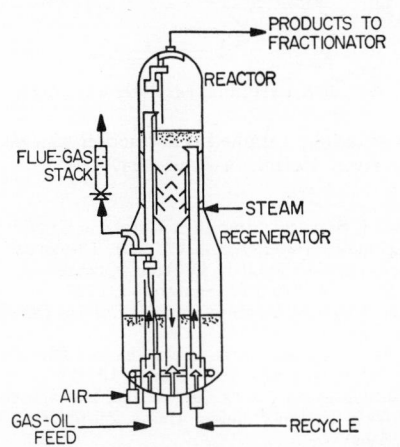

FIG. 3-3. Orthoflow fluid catalytic cracking—Model C.

References include:

Atteridg, P. T., and C. E. Slyngstad, "Orthoflow C," *Oil & Gas J.*, pp. 146–151, May 15, 1961.

Hyd. Proc., "Orthoflow Fluid Catalytic Cracking," p. 152, September, 1964.

Knaus, J. A., et al., "Orthoflow Catalytic Cracking of Reduced Crudes," AIChE Meeting, New York, December, 1961.

Oil & Gas J., "Orthoflow," p. 148, Apr. 5, 1965.

Pet. Ref., "Fluid Catalytic Cracking," pp. 232–233, September, 1956.

Reidel, J. C., "Cities Service Puts First Orthoflow Cat Cracker on Stream in U.S.," *Oil & Gas J.*, pp. 200–203, Mar. 24, 1952.

Weber, G., "Pure Adds Octanes with New Model (B) Orthoflow," *Oil & Gas J.*, pp. 99–100, Oct. 18, 1954.

UOP Fluid Catalytic Cracking

Licensed by Universal Oil Products Co., this is a fluid process of a unitary reactor-over-regenerator design, shown in Fig. 3-4.

This process is adaptable to the needs of both large and small refineries. Units having capacities of 1,200 to 53,500 B/D of fresh feed are in operation.

Important distinguishing features include (1) elimination of the air riser with its attendant large expansion joints, (2) elimination of considerable structural-steel supports, and (3) reduction in regenerator and in air-line size through use of 15 to 18 psig pressure operation.

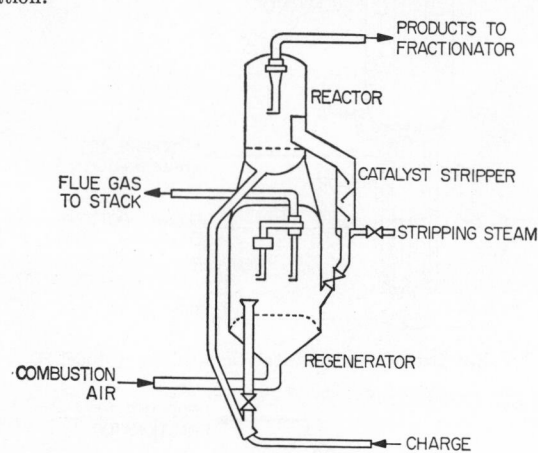

FIG. 3-4. UOP fluid catalytic cracking.

The first such unit employing a unified construction was placed on-stream by the Aurora Gasoline Co., Detroit, Mich., in June, 1947.

References include:

Anderson, N. K., and M. J. Sterba, "Simplified Catalytic Cracking Unit Meets Requirements of Smaller Refiners," *Pet. Ref.*, pp. 497–501, December, 1945.

Bland, W. F., "Cut Costs on New 3000 B/D Fluid 'Cracker' by Using Unified Reactor-Regenerator," *Pet. Proc.*, pp. 670–672, September, 1947.

Bozeman, H. C., "Tenneco Nearly Doubles Capacity of Its Cat Cracker," *Oil & Gas J.*, pp. 128–130, Dec. 4, 1961.

Brown, H. A., and M. J. Sterba, "Raw Crude Oil Is Charged Directly to 2600 B/D Catalytic Cracking Unit," *Pet. Proc.*, pp. 878–880, August, 1949.

Hyd. Proc., "U.O.P. Fluid Catalytic Cracking," p. 150, September, 1964.

Jacobs, R. W., "Panhandle's 'Kitten Cracker' Has 1,250-bbl Daily Charge," *Oil & Gas J.*, pp. 81–84, June 29, 1950.

Love, F. H., "Old Ocean Refinery Built Especially for Production of Aviation Gasoline," *Pet. Ref.*, pp. 51–58, September, 1945.

Pet. Proc., "Fluid Cracker Charges Heavy Distillate in Postwar Motor Gasoline Operation," pp. 46–48, January, 1947.

Pohlenz, J. B., "The Effect of Operational Variables in Fluid Catalytic Cracking of Petroleum," presented before Society of Chemical Engineers, Japan, Nov. 6–15, 1961.

Read, D., "Midget Fluid-catalyst Cracker Makes Progress Economical for the Smaller Refinery," *Oil & Gas J.* pp. 144–149, Apr. 20, 1946.

Read, D., "Process Systems Applicable to Military Aviation Fuel Manufacture," *Pet. Ref.*, pp. 130–131, March, 1951.

Shell Two-stage Fluid Catalytic Cracking

Two-stage fluid catalytic cracking was devised by Shell Development Co. to permit greater flexibility in shifting product distribution when dictated by demand.

As shown in the flow diagram of Fig. 3-5, virgin feed is first contacted with cracking catalyst in a riser reactor, i.e., a pipe in which fluidized catalyst and vaporized oil flow concurrently upward. Total contact time in this first stage is very short, on the order of seconds. High temperatures (875 to 1050°F) are employed to reduce undesirable coke laydown on catalyst without destruction of gasoline by secondary cracking. Other operating conditions in the first stage are a pressure of 16 psig and catalyst-oil ratio of 3 to 50/1. Volume conversion ranges between 20 and 70%.

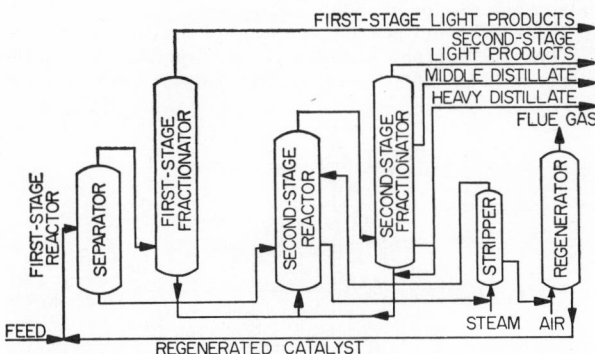

FIG. 3-5. Shell two-stage fluid catalytic cracking.

All or part of the unconverted or partially converted gas-oil product from the first stage is then cracked further in the second-stage fluid-bed reactor. Operating conditions herein are 900 to 1000°F and 16 psig. Catalyst-oil ratio is 2 to 15/1. Conversion in the second stage varies between 15 and 70%. The overall conversion range, therefore, is 50 to 80%.

The first commercial installation came on-stream in January, 1956, at the Shell Oil Co., Anacortes, Wash., refinery.

References include:

Heldman, J. D., et al., "Cat Cracking Now in Two Stages," *Pet. Ref.*, pp. 166–170, May, 1956.
Oil & Gas J., "Two Stage Cat Cracker Unveiled," p. 131, May 21, 1956.
Pet. Proc., "More Gasoline and Less Coke," pp. 54–57, June, 1956.
Pet. Ref., "Fluid Catalytic Cracking—2 Stage," p. 161, September, 1962.
Rehbein, C. A., et al., "The Economics of Two-stage Cat Cracking," *Oil & Gas J.*, pp. 108–119, June 15, 1959.

Moving-bed Catalytic Cracking

Airlift Thermofor Catalytic Cracking

Licensed by Mobil Oil Corp., Airlift TCC is a moving bed, reactor-over-regenerator continuous process for conversion of heavy gas oils into lighter high-quality gasoline and middle distillate fuel oils. Feed preparation may consist of flashing in a tar separator to get vapor feed. Tar separator bottoms are sent to a vacuum tower from which the liquid feed is produced. Mixed-phase feeds of prepared gas oils are also charged directly to the reactor.

Catalyst mechanical strength is good, and the metals-poisoning resistance of the bead catalyst has permitted the inclusion of residua as charge stock in some cases.

Typical feedstock might have a 25.0 to 30.0 API gravity and yield 35 to 55% debutanized gasoline (at 0 to 30% recycle) of 96 to 98 Research octane at 3.0 ml of TEL per gal, 25 to 50% light and heavy fuel oil, 10 to 18% butane-butylenes, and 3 to 5% coke, depending on the type of catalyst used.

Operating conditions in the reactor are about 840 to 920°F, 1.0 to 2.5/1.0 (v/hr/v) space velocity, 3.0 to 6.0/1.0 (v/v) catalyst-oil ratio, and 10 to 15 psig.

Gas-oil vapor-liquid flows downward through the reactor concurrent with the regenerated synthetic bead catalyst (Fig. 3-6). Catalyst is purged by steam at the base of the reactor and gravitates into the kiln for regeneration, accomplished by the use of air injected into the kiln. Approximately 70% of the carbon on the catalyst is burned in the upper kiln burning zone, the remainder in the bottom burning zone. Regenerated, cooled catalyst enters the lift pot, where low-pressure air transports it to the surge hopper above the reactor for reuse.

The present Airlift TCC unit is a simplification of the original TCC process introduced in 1943. The reactor is now mounted over a simplified kiln, and the catalyst-oil ratio has been increased by replacing the bucket elevators with the catalyst air lift.

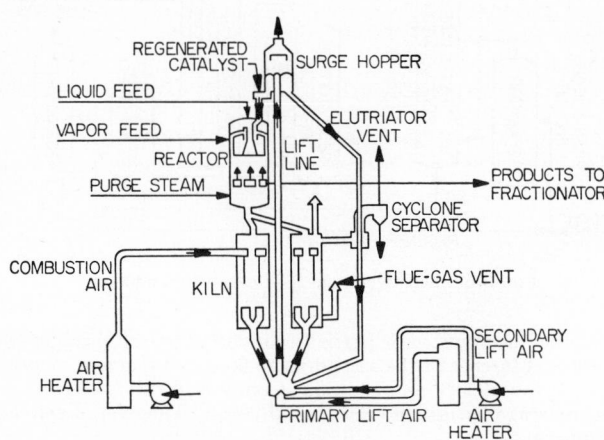

FIG. 3-6. Airlift Thermofor catalytic cracking (TCC).

The first commercial Airlift installation was put on-stream at Mobil Oil Co.'s Beaumont, Tex., refinery in October, 1950. Currently (1965), there are some 54 TCC units in commercial operation, having a capacity totaling about 900,000 B/D.

Airlift TCC references include:

Bland, W. F., "Prefabricated Catalytic Cracking Units Now Available in 'Packaged' Form," Pet. Proc., pp. 711–712, July, 1950.

Bowles, V. O., and H. D. Noll, "Economics of TCC Process in Plants of Moderate Refining Capacity," Oil & Gas J., pp. 83–89, Nov. 10, 1945.

Bourquet, J. M., et al., "The TCC Airlift," AIChE Meeting, Tulsa, Okla., September, 1960.

Chamberlain, J. P., and W. C. Meyer, "Delta's Cat Gets Twin Reactors and Common Lift System," Oil & Gas J., pp. 100–104, Aug. 12, 1963.

Curran, M. D., "Big Revamp and Some New Units Yield a Modern Refinery," Hyd. Proc., pp. 179–181, September, 1961.

Danner, A. V., "Socony-Vacuum TCC Units Move Catalyst by Gas Lift," Pet. Ref., pp. 179–182, September, 1950.

Eastwood, S. C., and T. E. Phalen, "Performance of New and Heavy Durabead Catalysts," WPRA Meeting, San Antonio, Tex., March, 1959.

Eastwood, S. C., et al., "What Mobil Learned about Durabead-5," Oil & Gas J., pp. 152–157, Oct. 29, 1962.

Elliott, K. M., and S. C. Eastwood, "Durabead-5, a New Cracking Catalyst," Oil & Gas J., pp. 142–144, June 4, 1962.

Evans, L. P., et al., "Durabead-5—How It Has Performed in Mobil Airlift TCC Units," Oil & Gas J., Sept. 9, 1963.

Hamilton, W. W., et al., "Wide Range of Feed Stocks Possible with Airlift TCC," Pet. Ref., pp. 71–78, August, 1952.

Noll, H. D., et al., "Commercial TCC Operations on Partially Vaporized Charge Stocks," NPA Meeting, Atlantic City, N.J., September, 1946.

Noll, H. D., and D. M. Luntz, "TCC Processing of Pennsylvania Grade Crude Oil Fractions," *Pet. Engr.*, p. 162, Jan. 15, 1948.
Oil & Gas J., "Cat Cracking at 100% Conversion," p. 92, Jan. 15, 1962.
Oil & Gas J., "Mobil's Secret New Cracking Catalyst Called Refining Breakthrough," pp. 96–97, Apr. 2, 1962.
Pet. Engr., "Twin Airlift TCC Units on Stream," pp. C11–16, April, 1951.
Pet. Ref., "Two New Air-lift Type TCC Units Put into Operation," pp. 95–96, March, 1951.
Thornton, Jr., D. P., "First 'Air-lift' TCC Unit," *Pet. Proc.*, pp. 146–149, February, 1951.
Thornton, Jr., D. P., "How Derby Gains Flexibility with Its New Air-lift TCC," *Pet. Proc.*, pp. 815–817, June, 1952.
Uhl, W. C., "TCC Unit Gives High Liquid Recovery by Operating on Vapor-Liquid Feed," *Pet. Proc.*, pp. 950–952, September, 1950.
Valentine, S., and T. B. Arnold, "Package TCC Capacity Up 50% at Vickers," Western Petroleum Refiners Assoc., San Antonio, Tex., June, 1957.
Withers, C. L., "TCC Flexible Refinery Operation," *Oil & Gas J.*, pp. 56–57, June 30, 1952.

Houdriflow Catalytic Cracking

Licensed by Houdry Process and Chemical Co., Houdriflow catalytic cracking is a continuous, moving-bed process employing an integrated single vessel for the reactor and regenerator kiln. Charge stock, sweet or sour, can be any fraction of the crude

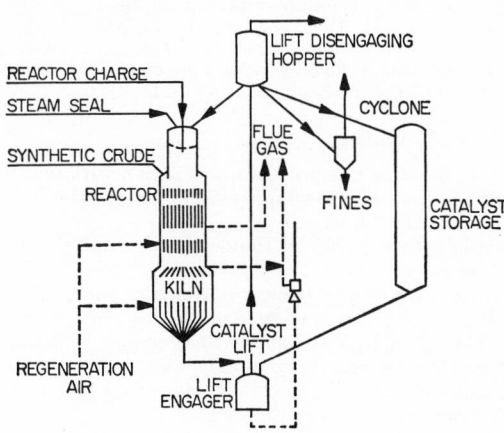

FIG. 3-7. Houdriflow catalytic cracking.

boiling between naphtha and penetration asphalt. Catalyst is transported from the bottom of the unit to the top in a gas lift employing compressed flue gas and steam as the motivating vapor, as shown in Fig. 3-7.

Houdriflow feed can be liquid, vapor, or a mixture of both. Normal operating conditions fall in the range of 850 to 950°F, 9 to 10 psig, 1.5 to 4.0/1.0 (v/hr/v) space velocity, and 3.0 to 7.0/1.0 (v/v) catalyst-oil ratio.

Recent designs employ a heat-balanced system, eliminating the need for cooling coils used in earlier models to remove regenerator heat from the kiln.

Reactor feed, emitting from a specially designed nozzle, and catalyst pass concurrently through the reactor zone to a disengager section, in which vapors are separated and then sent to a conventional fractionation system.

Spent catalyst, which has been steam-purged of residual oil, flows to the kiln for regeneration. Steam and flue gas are used to transport catalyst to the lift disengaging hopper, from which catalyst is fed to the reactor. Fines are separated continuously from a small portion of the circulating catalyst in the elutriator.

The gas-lift Houdriflow has replaced the earlier bucket-type Houdriflow moving-bed operation. The first commercial unit was placed on-stream at the Sun Oil Co.'s

Toledo, Ohio, refinery in May, 1950.   Today there are 22 units in the United States and abroad, ranging in size from 2,000 to 27,000 B/D.

Houdriflow references include:

Barton, P. D., "Sun's New Houdriflow Unit," *Oil & Gas J.*, pp. 232–234, 322–327, Mar. 29, 1951.
Boynton, F. B., "Producing 90-octane Gasoline with Houdriflow," *Pet. Ref.*, pp. 133–135, May, 1952.
Burtis, T. A., et al., "Economics of Moving Bed Catalytic Cracking Processes," *Chem. Engrg. Prog.*, pp. 97–101, February, 1949.
Hoge, A. W., et al., "Moving Bed Catalytic Cracking," *Petro/Chem Engr.*, pp. C32–44, December, 1960.
Hyd. Proc., "Houdriflow," p. 153, September, 1964.
Maerker, J. B., et al., "Moving Bed Recycle Catalytic Cracking Correlations," *Chem. Engrg. Prog.*, pp. 95–101, February, 1951.
Maerker, J. B., and J. W. Schall, "Correlation of Moving Bed Process Variables in Recycle Catalytic Cracking," *Houdry Pioneer*, pp. 1–8, August, 1950.
Mills, G. A., "Aging of Cracking Catalysts," *Ind. Engrg. Chem.*, p. 182, January, 1950.
Oil & Gas J., "Houdriflow Catalytic Cracking," p. 150, Apr. 5, 1965.
Peavy, C. C., et al., "Various Refinery Applications of Houdriflow Catalytic Cracking," WPRA Meeting, San Antonio, Tex., March, 1949, reviewed in *Pet. Proc.*, pp. 554–558, May, 1949.
Pet. Proc., "A Report on Houdriflow," pp. 137–138, February, 1949.
Schall, J. W., and J. C. Dart, "Catalytic Cracking of High Nitrogen Charge Stock," *Pet. Ref.*, pp. 101–103, March, 1952, and pp. 173–176, April, 1952.
Stackner, J., "Easier Control by Graphic Panel," *Pet. Proc.*, pp. 626–627, June, 1951.
Thornton, Jr., D. P., "New Houdriflow Cracking Unit Has High Catalyst-Oil Ratio, Low Preheat Requirements," *Pet. Proc.*, pp. 601–605, June, 1950.

Houdresid Catalytic Cracking

Houdresid catalytic cracking is a process which uses a variation of the continuous moving catalyst bed, designed to get high yields of high-octane gasoline and light distillate from reduced crude charge.

The process was originally licensed by Houdry Process and Chemical Co. but is no longer offered for license.

Residuum cuts ranging from crude tower bottoms to vacuum bottoms, including residua high in sulfur or nitrogen, can be employed as charge over a synthetic or natural catalyst.   Though equipment employed is similar in many respects to that used in Houdriflow units, novel process features modify or eliminate the adverse effects on catalyst and product selectivity usually resulting when heavy metals—iron, nickel, copper, and vanadium—are present in the fuel.

The Houdresid catalytic reactor and catalyst regenerating kiln (Fig. 3-8) are con-

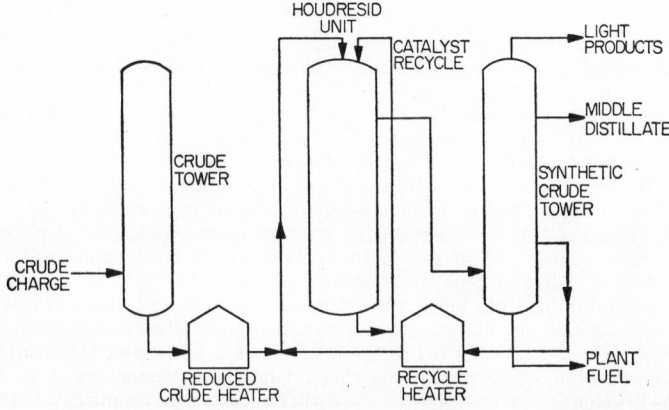

Fig. 3-8. Houdresid catalytic cracking.

tained in a single vessel. Fresh feed plus recycle gas oil are charged to the top of the unit in a partially vaporized state and mixed with steam. Hydrocarbon and catalyst flow patterns are similar to that employed in Houdriflow operation (Fig. 3-7).

Coke production (15.0 to 20.0 wt-%) is higher than that of the Houdriflow process when a 7.0 to 12.0 API heavy residuum with high contaminant concentration is charged. Debutanized gasoline yields of 30 to 37% were obtained having clear Research octane values between 89.0 and 93.0.

The first commercial installation of this process was placed on-stream in March, 1954, at the Sun Oil Co., Ltd., Sarnia, Ont., refinery.

Houdresid references include:

Bland, R. E., "How Houdresid Works on Heavy Stocks," *Pet. Ref.*, pp. 166–168, September, 1955.

Dart, J. C., et al., "Houdresid Process Cracks Residua," *Pet. Ref.*, pp. 153–156, June, 1955.

Oil & Gas J., "Houdresid Catalytic Cracking," p. 143, Mar. 19, 1956.

Oil & Gas J., "Houdresid Designed to Convert Entire Bottoms Fraction to Distillates without Significant Damage to Catalysts," p. 83, Mar. 28, 1955.

Pet. Ref., "Houdresid," pp. 238–239, September, 1956.

Pet. Ref., "Houdriflow and Houdresid," pp. 242–243, September, 1958.

World Pet., "Houdresid," pp. 47–49, July, 1955.

Suspensoid Catalytic Cracking

This process (Fig. 3-9) is once-through nonregenerative catalytic cracking in which cracking stock is mixed with a catalyst cycle oil slurry and passed through the coils of

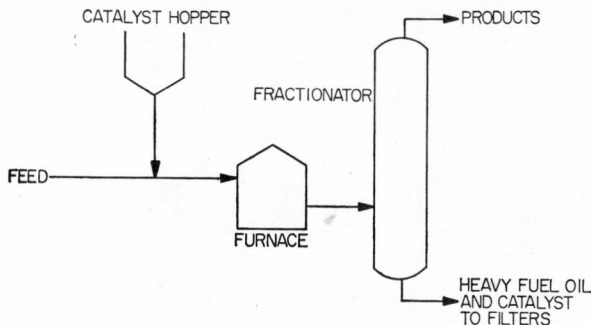

FIG. 3-9. Suspensoid catalytic cracking.

a fired heater. No reactor is employed, and catalyst is removed from the fractionator bottoms of heavy residual oil through use of rotary filters. Straight-run naphtha can be included in the charge for upgrading through the reforming mechanisms.

The process is actually a compromise between catalytic and thermal cracking. The main effect of the catalyst is to allow a higher cracking temperature and to assist mechanically in keeping coke from accumulating on the walls of the tubes. The normal catalyst employed is spent clay obtained from the contact filtration of lubricating oils (2 to 10 lb per barrel of feed). Regular fluid cracking catalyst can be employed and will give slightly higher gasoline yields of slightly lower octane number, but catalyst recovery and reuse would have to be practiced to justify the expense and would require more equipment investment. Where spent lube clay is available, use of new or fresh catalyst would be hard to justify economically.

Operating conditions at the coil outlet under "regular" suspensoid operation are 450 psi and 1050°F. Under "super" suspensoid operation, the conditions are 250 psi and 1090°F. The more severe operation necessitates use of special streamlined heater headers to minimize erosion.

A virgin charge of 35 to 43 API gravity and containing naphtha of 35 to 40% resulted in gasoline yields of 50 to 85% of 85 to 92 Research octane clear.

The first commercial installation of Suspensoid catalytic cracking was placed on-stream at the Imperial Oil Co., Ltd., of Canada refinery at Sarnia, Ont., in 1940. References include:

Burk, C. F., "New Developments in Suspensoid Catalytic Cracking," *Oil & Gas J.*, p. 100, Oct. 12, 1946.
Caesar, C. H., "Catalytic Effects in Suspensoid Cracking," *Pet. Proc.*, pp. 887–890, August, 1949.
Caesar, C. H., "Suspensoid Catalytic Cracking at Imperial Oil's Sarnia Refinery," *Oil & Gas J.*, p. 69, Aug. 27, 1947.
Caesar, C. H., "The Suspensoid Catalytic Cracking Process," Canadian Institute of Chemistry, June, 1947, Banff, Canada, reviewed in *Pet. Proc.*, pp. 929–930, December, 1947.
Foster, A. L., "Suspensoid Catalytic Cracking System Gives Nearly 60% Gasoline Yield," *Oil & Gas J.*, p. 43, May 4, 1944.
Purvin, R. L., "Suspensoid Catalytic Cracking," *Pet. Proc.*, pp. 328–331, May, 1947.
Sittig, M., "The Suspensoid Process," *Pet. Ref.*, pp. 125–126, November, 1950.

Fixed-bed Catalytic Cracking

Houdry Fixed-bed Catalytic Cracking

This is a fixed-bed cyclic regenerable catalytic cracking process for converting distillate charge stocks—virgin or coker gas oils—into yields of gasoline, light and heavy distillates, coke, butanes-butylenes, and fuel gas.

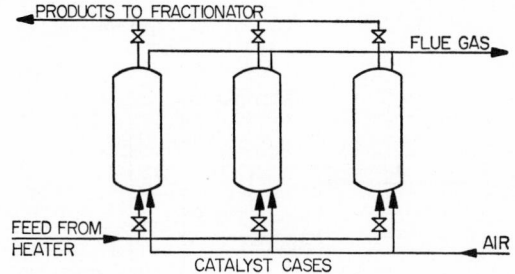

FIG. 3-10. Houdry fixed-bed catalytic cracking.

Synthetic and natural bead catalysts have been employed and have lasted more than 1½ years. Feedstocks having initial boiling points between 400 and 450°F and end points of 750 to 950°F have yielded 35 to 40% debutanized gasoline of 85.0 to 87.0 Research octane number clear. Light and heavy distillate yields ranged between 45 and 55%. Coke gas and mixed C_4 yields were 10 to 17%.

Operating conditions in the reactor for a typical feed have been in the range of 840 to 875°F, 7 to 30 psig, and 1.0 to 2.0 (v/hr/v) space velocity.

A typical cycle for Houdry Fixed-bed operation is (1) on-stream, 10 min; (2) purge, 5 min; (3) regeneration, 10 min; (4) steam purge, 5 min. A unit generally includes three catalyst cases (Fig. 3-10), with one being on-stream while the other two are being purged and regenerated.

Because of the elaborate instrumentation required and because of the economic attractiveness to the large and small refinery operations of the modern moving- and fluid-bed catalytic cracking processes, all the original Houdry Fixed-bed units have been replaced, and the process is no longer offered for license.

Historically, the Houdry Fixed-bed process was the first of the present-day catalytic cracking processes. It was preceded only by the McAfee batch process (employing a metallic halide catalyst) which has long since lost commercial significance. The first Houdry Fixed-bed commercial installation was placed on-stream by the Socony Vacuum Oil Co. (now Mobil Oil Co.) at Paulsboro, N.J., in June, 1936.

References include:

Bates, J. R., et al., "Composition of Catalytically Cracked Gasolines," *Ind. Engrg. Chem.*, pp. 147–152, February, 1942.
Carey, J. S., and H. W. Artendahl, "Catalytic Cracking Economics Complicate Comparisons," *Nat. Pet. News*, pp. R370–375, Oct. 16, 1940.
Evans, J. E., and R. C. Lassiat, "Combustion-gas Turbine in the Houdry Process," *Pet. Ref.*, pp. 461–466, November, 1945.
Hornberg, C. V., et al., "Bead Catalysts in Commercial Houdry Units," *Oil & Gas J.*, pp. 214–216, July 26, 1947.
Houdry, E. J., et al., "Catalytic Processing of Petroleum Hydrocarbons by the Houdry Process," *Proc. API*, vol. III, pp. 133–148, 1939.
Lassiat, R. C., and C. H. Thayer, "Improved Houdry Reactor Design," *Pet. Ref.*, pp. 453–456, September, 1946.
Pettibone, E. E., "Houdry Fixed Bed Cat Cracking Unit," *Pet. Engr.*, pp. 149–158, September, 1945.
Prickett, T. B., and R. H. Newton, "Developments in Houdry Fixed Bed Catalytic Processes," *Pet. Ref.*, p. 377, November, 1943.

Cracking Catalyst Treating

The latest technique developed by the refining industry to increase gasoline is treating of the catalysts from cracking units to remove metal poisons which accumulate on the catalyst. Thus, feedstocks which were once rejected because of high metals content can now be charged to catalytic cracking units.

Even though these processes might be considered auxiliary, they do have considerable influence on the existing catalytic cracking processes and, as such, are given process coverage in this section. Although these processes were developed (and are now commercially employed) for "fluid" cracking catalysts, they are claimed to have utility for moving-bed catalysts as well.

Nickel, vanadium, iron, and copper compounds contained in catalytic cracking feedstocks are deposited on the catalyst during the cracking operation. Both catalyst activity and selectivity are adversely affected if the metals are allowed to accumulate. Increased catalyst metal contents effect catalytic cracking yields by increasing coke, decreasing gasoline and butanes-butylenes, and increasing hydrogen "make." As metals accumulate, the cracking catalyst loses its relative activity, produces a lighter gas product, increases gas and coke factors, and maintains surface area.

Catalyst metal accumulation and attendant poisoning can be eliminated and/or controlled by (1) feedstock selection, (2) increasing catalyst addition rates, (3) removal of metals from feedstocks, and/or (4) treatment of catalyst after accumulating the metals.

Any of these approaches must be balanced against product value and operating costs to determine the most economic way of operating. However, the recent commercial development and adoption of cracking catalyst treating processes definitely improve refiners' overall catalytic cracking process economics.

Cracking catalyst treating references include:

Connor, J. E., et al., "Fluid Cracking Catalyst Contamination: Some Fundamental Aspects of Metal Contamination," *Ind. Engrg. Chem.*, pp. 272–276, February, 1957.
Donaldson, R. E., et al., "Metals Poisoning of Cracking Catalysts," *Ind. Engrg. Chem.*, p. 72, September, 1961.
Fowle, M. J., et al., "Controlled Catalysis in Catalytic Cracking," *Chem. Engrg. Prog.*, p. 66, December, 1961.
Gossett, E. C., "When Metals Poison Cracking Catalysts," *Pet. Ref.*, pp. 177–180, June, 1960.
Grane, H. R., et al., "Predict Poisoning Effects of Metals," *Pet. Ref.*, p. 168, May, 1961.
Grane, H. R., et al., "How to Predict Contaminant Coke Yield," *Oil & Gas J.*, p. 61, June 5, 1961.
Leum, L. N., and J. E. Connor, Jr., "Removal of Contaminants from Cracking Catalysts by Ion Exchange," ACS Meeting, Washington, D.C., March, 1962.
McEvoy, J. E., et al., "Distribution of Metal Contaminants on Cracking Catalysts," *Ind. Engrg. Chem.*, p. 865, May, 1957.
Voorhies, Jr., A., "Carbon Formation in Catalytic Cracking," *Ind. Engrg. Chem.*, pp. 318–322, April, 1945.

Demet

The Demet process for treating cracking catalysts was discovered and developed by Sinclair Refining Co. Licensing information can be obtained from Universal Oil Products Co.

Regenerated cracking catalyst is subjected to two pretreatment steps (pretreatment stage—Fig. 3-11): The first step effects vanadium removal; the second, nickel removal —to prepare the metals on the catalyst for chemical conversion to compounds (chemical treatment step) which can readily be removed through water washing (catalyst wash step). The treating steps include use of a sulfurous compound or sulfurizing vapor followed by chlorination with an anhydrous chlorinating agent (e.g., chlorine gas) and washing with an aqueous solution of a chelating agent (e.g., citric acid). The catalyst is then dried and further treated before returning to the cracking unit.

Commercial operation of a Demet unit on a fluid catalytic cracking catalyst (synthetic $SiO_2 \cdot Al_2O_3$) reduces nickel content 65 to 70%, vanadium 25 to 30%, and iron 15 to 20%. A typical catalyst feed before treatment might contain 260 ppm NiO,

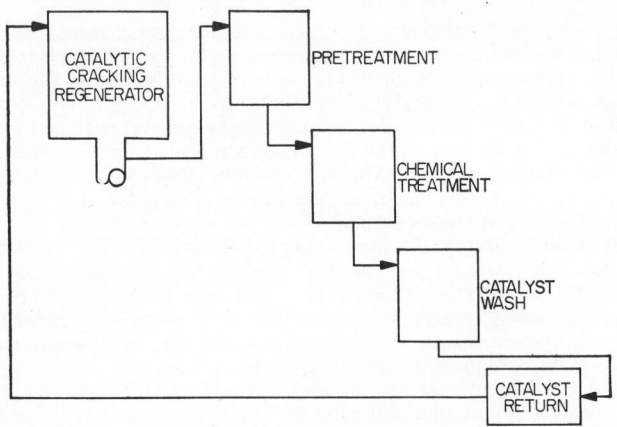

FIG. 3-11. Demet process for treating cracking catalysts.

3,960 ppm V_2O_5, and 2,460 ppm Fe; after treatment these metals would be reduced to 74, 1,629, and 2,010 ppm, respectively. Catalyst activity $(D + L)$ increased from 32.5 to 37.7, the gas factor decreased from 1.42 to 1.07, the coke factor decreased from 1.17 to 0.87, the weight per cent of Al_2O_3 decreased slightly from 26.8 to 24.8, and the hydrogen-producing factor decreased from 42 to 20.

Cited investment cost for a 5-ton per day Demet unit was $250,000. Operating costs (dollars per day), excluding royalty, for this unit would be $350—including chemicals and utilities ($90), operating maintenance and plant expense ($190), and depreciation of 10% per year ($70)—or $70 per ton of catalyst treated. This latter figure might be $38 per ton for a 20-ton day unit ($550,000 investment). Currently, fluid catalytic cracking catalyst costs $315 to $355 per ton.

The first Demet unit (10 tons per day) went on-stream at Sinclair's Wood River, Ill., refinery in 1961. It is used in conjunction with their 11,000-B/D FCC unit.

Demet references include:

Adams, N. R., and M. J. Sterba, "Economics of the Demet Process," AIChE Meeting, New Orleans, La., Mar. 10–14, 1963.
Chem. Week, "Removing Catalyst Poisons" (patents), p. 61, Nov. 3, 1962.
Gossett, E. C., "When Metals Poison Cracking Catalyst," *Pet. Ref.*, pp. 177–180, June, 1960.
Oil & Gas J., "New Demet Process Cleans Up Catalysts," p. 92, May 21, 1962.

Sanford, R. A., et al., "Better Yields after Treating Catalyst," *Pet. Ref.*, pp. 103–108, July, 1962.
Sanford, R. A., et al., "Demet Improves FCC Yields," *Oil & Gas J.*, pp. 92–96, Aug. 27, 1962.

Met-X

The Met-X cracking catalyst treating process was developed by Atlantic Refining Co. and is licensed by M. W. Kellogg Co. The process uses an ion-exchange resin to remove metal contaminants from the catalyst.

The process consists of cooling, mixing and ion exchange, separation, filtration, and resin regeneration. Moist catalyst from the filter is dispersed in oil and returned to the cracking reactor in a slurry.

Figure 3-12 shows a typical Met-X flow diagram. On a continuous basis, catalyst from a cracking unit is cooled and then transported to a stirred reactor and mixed with

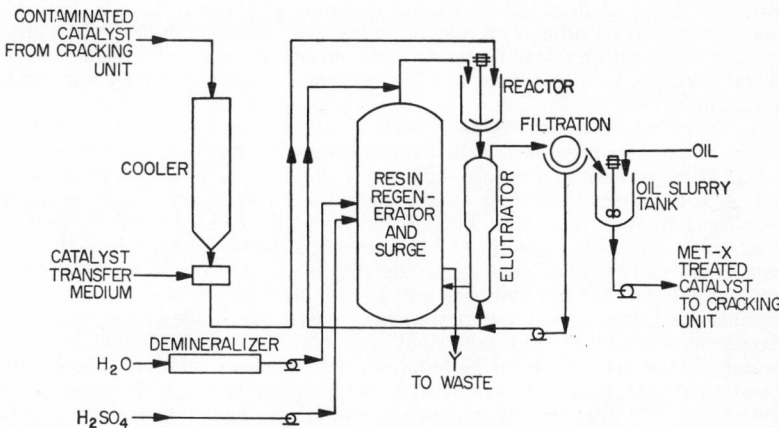

FIG. 3-12. Met-X process for treating cracking catalysts.

ion-exchange resin, which is introduced as a slurry. The catalyst and resin mixture is held at controlled conditions. After the reactor, the catalyst-resin slurry flows to an elutriator for separation. Catalyst slurry is taken overhead to a filter; the wet filter cake is slurried with oil and pumped into the catalytic cracker feed system. The resin leaves the bottom of the elutriator and is regenerated before returning to the reactor.

A regenerated catalyst from a Fluid cracking unit had the following properties before/after Met-X treatment: Carbon-producing factor, 2.0/1.0; ppm vanadium, 1,250/700; ppm nickel, 200/120; wt-% iron, 0.20/0.16.

Estimated investment cost for a 40-ton/day Met-X unit is $500,000. Operating costs (excluding royalty) for this size unit include $180 per day for chemicals and utilities; $105 per day for labor; $144 per day for maintenance and taxes (10% of investment), and $144 per day for depreciation (10-year life).

The first commercial Met-X unit went on-stream in November, 1961, at Atlantic's Philadelphia, Pa., refinery, with a capacity of 40 tons/day. It is operating in conjunction with a Fluid cracking unit of 48,000-B/D capacity.

Met-X references include:

Atteridg, P. T., "A Fresh Look at Solvent Decarbonizing," *Oil & Gas J.*, p. 72, Dec. 9, 1963.
Atteridg. P. T., and J. Humble, "Some Ways Met-X Can Pay for Itself," *Pet. Ref.*, p. 167, April, 1963.
Chem. Week, "Bonus in Catalyst Clean-up," Dec. 9, 1961.

Dilliplane, R. J., et al., "Met-X Boosts Gasoline Output 12%/Unit of Coke for Atlantic," *Oil & Gas J.*, p. 119, Aug. 5, 1963.
Fowle, M. J., et al., "Metals Removed from Catalyst," *Pet. Ref.*, p. 124, May, 1962.
Fowle, M. J., et al., "Metals Control—A New Dimension in Catalytic Cracking," API Refining Meeting, San Francisco, May 17, 1962.
Grane, H. R., et al., "The Behavior of Metal Contaminants in Catalytic Cracking," API Refining Meeting, Houston, Tex., May, 1961.
Oil & Gas J., "Met-X Unit Going on Stream," p. 110, Dec. 4, 1961.
Ozawa, J. K., et al., "Met-X Scores Success at United," *Oil & Gas J.*, p. 101, Feb. 10, 1964.
Stormont, D. H., "First Met-X Plant Is Now in Operation," *Oil & Gas J.*, pp. 74–77, Dec. 11, 1961.
Uhl, W. C., "Met-X Economics," *World Pet.*, p. 30, December, 1962.

HYDROCRACKING

Hydrocracking is perhaps the single most significant advance in petroleum refining technology in recent years. Acceptance by the industry has been cautious, but because of the great flexibility and other advantages, hydrocracking is destined to become an important adjunct to refining in the next decade. Basically it is an efficient, low-temperature, catalytic method of converting refractory middle-boiling or residual material to high-octane gasoline, reformer charge stock, jet fuel, and/or high-grade fuel oil.

Hydrocracking does not compete with catalytic cracking directly. Catalytic crackers charge virgin gas oils, while hydrocracker feed usually consists of refractive gas oils that are derived from cracking and coking operations. Currently hydrocracking will supplement rather than replace basic catalytic cracking in the new refinery construction patterns. Justification for hydrocracking currently is tied to the relative economics of gasoline and furnace-oil production. Because of the higher yields from hydrocracking, crude-oil availability is an important consideration. By adding a hydrocracker and hydrogen unit to existing facilities, a refiner can in some cases increase gasoline yield and at the same time reduce crude-oil throughput.

Hydrogen availability is an important economic factor when considering hydrocracking. Most refiners would be required to supplement reformer off-gas with a separate hydrogen facility to supply hydrocracking needs unless the operation was quite limited. (A discussion of processes for manufacturing hydrogen appears as Part 5 of this section, starting on page 3–.140)

Because of the nature of hydrogen and the high pressures involved, equipment costs are high. Units range from $350 to $600 per stream-day barrel. This is offset by the advantages listed by the licensors of the process:

1. Better balance of gasoline and distillate production
2. Improved gasoline pool octane quality and sensitivity
3. Greater gasoline yield
4. Reduction of fuel-oil make
5. Supplement to catalytic cracking to upgrade heavy cracked stocks, aromatic heavy cracked naphthas, cycle oils, coker oils

Chemically, hydrocracking can be regarded as a combination of cracking, hydrogenation, and isomerization. It is also a treating operation, since hydrogen combines with and practically eliminates contaminants in the feed such as sulfur, nitrogen, etc. (see Hydrogen Treating, page 3–38). At this writing, catalysts have not been identified with individual processes, but basic need would require a cracking catalyst such as silica-alumina in conjunction with a hydrogenating agent such as platinum, nickel, or tungsten oxide.

Commercial hydrocracking processes are operated at temperatures between 400 and 800°F and pressures from 100 to 2,000 psig. Operating severity and hydrogen consumption are dependent upon feedstock and the product distribution required, as well as the process itself and the catalyst used.

Process flow consists essentially of mixing hydrogen with feed, then heating and contacting with catalyst in a fixed-bed reactor at a specified hydrogen partial pressure.

Most feedstocks require pretreatment; in some processes this is the first step in the two-stage system.

Literature references include:

Canadian Chem. Process, "Hydrocracking; Combining Two in One," pp. 38–39, August, 1960.

Chem. Engr., "Refiners See Hydrocracking as Answer to Changing Demands," pp. 38–40, May 1, 1961.

Chem. Week, "Finding New Dollars in Distillation," pp. 33–36, Apr. 29, 1961.

Coonradt, H. L., et al., "Platinum-Acidic Oxide Catalysts for Hydrocracking," *Ind. Engrg. Chem.*, pp. 727–732, September, 1961.

Flinn, R. A., et al., "The Mechanism of Catalytic Cracking," *Ind. Engrg. Chem.*, pp. 153–156, February, 1960.

Giesler, John D., "Hydrocracking: A Means of Obtaining Higher Yields of Gasoline," WPRA Regional Meeting, Wichita, Kans., June 15–16, 1960.

Hyd. Proc., "Hydrocracking," p. 143, September, 1964.

Oil & Gas J., "Refiners Slow to Invest Dollars in Versatile Hydrocracking," p. 104, Sept. 18, 1961.

Sterba, M. J., et al., "Refinery Hydrogen Assets—Their Conservation and Effective Use," WPRA Annual Meeting, San Antonio, Tex., Apr. 10–12, 1961.

Stormont, D. H., "Hydrocracking . . . Its Big Impact Is Still Ahead," *Oil & Gas J.*, pp. 103–113, Feb. 12, 1962.

Uhl, W. C., "Processing Review," *World Pet.*, pp. 28–30, May, 1962.

Unzelman, G. H., and N. H. Gerber, "Hydrocracking—Today and Tomorrow," *Petro/Chem Engr.*, pp. 32–52, October, 1965.

Isomax

The Isomax process is a result of a cross-licensing arrangement between Chevron Research Co. (Isocracking) and Universal Oil Products Co. (Lomax). Each company offers hydrocracking catalysts. Heavy hydrocarbons are converted into high-grade gasoline and/or distillate fuels by a catalytic reaction in the presence of hydrogen.

A two-stage, fixed-bed catalyst system is used which operates under hydrogen pressures from 500 to 1,500 psig in a temperature range of 400 to 700°F (see Fig. 3-13

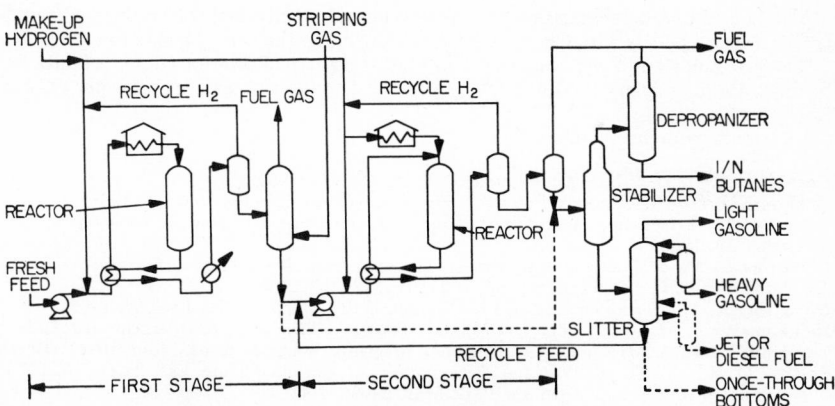

Fig. 3-13. Isomax hydrocracking—two-stage reactor system.

for an example of flow with middle distillate feed). Exact conditions depend upon the feedstock and product requirements. Hydrogen consumption is in the order of 1,000 to 1,600 scf per barrel of feed processed. Each stage has a separate hydrogen recycle system. Conversion may be balanced to provide products for variable requirements, and recycle can be taken to extinction if necessary. Fractionation can also be handled in a number of different ways to yield desired products.

Liquid yields are from 115 to 120 vol-% butanes and heavier; the ratio of iso to normal hydrocarbons is high. Fuel-gas make is low.

Light naphtha from Isomax is 100 Research octane or better with 3 ml of TEL per gal. Heavy naphtha is high in naphthenes and is excellent reformer charge stock. In some cases, heavy naphtha can be taken directly to the gasoline pool. Research octane will vary from 77 to 92 with 3 ml of TEL per gal. Both light and heavy naphthas have very little sensitivity (high Motor octane), and therefore road performance characteristics are excellent. Other products possible from Isomax are low-freeze-point jet fuels, low-pour high-cetane diesels, clean-burning furnace oil, etc.

Standard of California has installed a 62,500 B/D Isomax unit at its Richmond, Calif., refinery, the largest hydrocracker that has been constructed to date (mid-1966). Half of the feed is obtained by propane deasphalting heavy bottoms to produce feedstock with $1000°F+$ end point and is charged to first-section reactors. Heavy cycle oil from catalytic cracking joins first-section heavy fractions, along with a gas-oil stream, to feed second-section reactors. The ability to process these stocks to high-quality gasoline and middle distillates is largely a result of recent developments in catalyst technology.

Literature references (see also Isocracking and Lomax) include:

Eckhouse. J. G., et al., "Recent Commercial Isomax Operations," NPRA Gulf Coast Regional Meeting, Houston, Tex., Jan. 29–30, 1964.
Gould, G. D., et al., "Applications of Hydrocracking," API Refining Meeting, St. Louis, Mo., May 12, 1964.
Hyd. Proc., "Isomax," p. 147, September, 1964.
Read, D., et al., "Recent Isomax Developments for Maximum Gasoline or Middle Distillates," NPRA Annual Meeting, Houston, Tex., Apr. 2, 1965.
Scott, J. W., et al., "Isomax Gives Gasoline or Distillate," *Hyd. Proc.*, pp. 131–136, July, 1963.
Stormont, D. H., "Socal Sets New Process Route with Richmond Hydrocracking Complex," *Oil & Gas J.*, pp. 147–167, Apr. 25, 1966.
Uhl, W. C., "Processing Review," *World Pet.*, p. 28, May, 1962.

Lomax

The Lomax hydrocracking process was developed by Universal Oil Products Co. and is now licensed jointly with Chevron Research Co. under the name Isomax (see p. 3–17).

The first Lomax unit went on-stream in August, 1961, at Powerine Oil Co.'s Santa Fe Springs, Calif., refinery. The unit is 2,200 B/SD and uses catalyst-reformer off-gas for hydrogen supply.

Literature references include:

Chem. Engrg. News, "UOP Adds Lomax," p. 63, Apr. 11, 1960.
Eckhouse, J. G., "The Lomax Process," *Oil & Gas J.*, pp. 117–121, Nov. 6, 1961.
Oil & Gas J., "First Lomax Unit Going Up at Powerine's Santa Fe Springs Refinery," p. 68, Sept. 19, 1960.
Oil & Gas J., "New Hydrocracking Process Offers Refiners More Flexible Gasoline—Distillate Output, Lomax Process," pp. 102–104, May 23, 1960.
Oil & Gas J., "Powerine Operating First Lomax Unit," pp. 112–113, Feb. 12, 1962.
Reinkemeyer, L. R., et al., "The UOP Lomax Process—Two Years of Operation for Apco's Unit," NPRA Mid-continent Regional Meeting, Wichita, Kans., June 10–11, 1964.
Sterba, M. J., and C. H. Watkins, "Production of Gasoline From Distillates," WPRA Annual Meeting, San Antonio, Tex., Mar. 28–30, 1960.

Isocracking

The Isocracking process was developed by Chevron Research Co. and is now jointly licensed with Universal Oil Products Co. under the name Isomax (see p. 3–17).

The first Isocracker, considered a pilot operation and recently converted to other service, was a 1,000-B/D unit at the Richmond, Calif., refinery of Standard Oil Co. of California. It was started up in July, 1959. The first commercial unit, 7,500 B/SD, was started up in early 1962 by Standard Oil Co. (Ohio) at Toledo.

Literature references include:

Gould, G. D., and N. J. Patterson, "How Condensate Can Be Hydrocracked," *Hyd. Proc.*, pp. 167–170, September, 1961.
Robbers, J. A., et al., "Commercial Isocracking of Heavier Gas Oils," WPRA Annual Meeting, San Antonio, Tex., Apr. 10–12, 1961.
Scott, J. W., "How Isocracking Works," *Pet. Ref.*, pp. 115–160, April, 1960.
Scott, J. W., "How Much Can You Save with the Isocracking Process," *Pet. Ref.*, pp. 161–168, May, 1960.
Scott, J. W., "Isocracking: A Versatile New Refining Process," API Refining Meeting, Detroit, Mich., May 12, 1960.
Stormont, D. H., "Tidewater's Big Hydrogen Plant," *Oil & Gas J.*, p. 90, Jan. 9, 1961.
World Pet., "Isocracking—Cal Research's New Process," pp. 70–76, September, 1959.

H-Oil

Jointly developed and licensed by Cities Service Research and Development Co. and Hydrocarbon Research, Inc., H-Oil is basically a catalytic hydrogenation technique. During the reaction considerable hydrocracking takes place (see Fig. 3-14). The

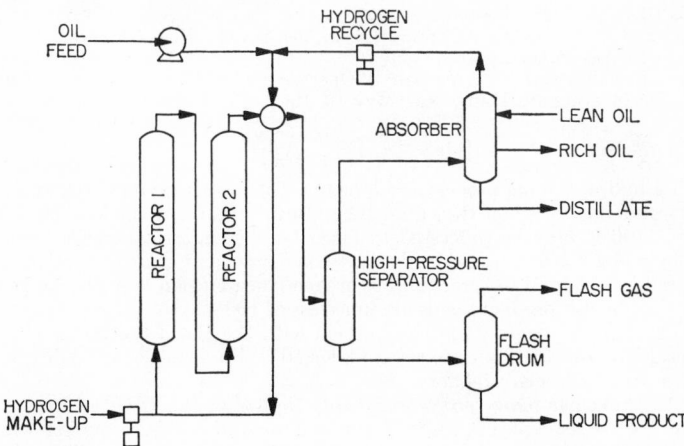

FIG. 3-14. H-Oil process.

process is used to upgrade heavy sour crudes and residual stocks to high-quality sweet distillates, thereby reducing fuel-oil yield. A modification of H-Oil called Hy-C cracking will convert heavy distillates to middle distillates and kerosine.

Oil and hydrogen are fed upward through the reactors as a liquid-gas mixture at a velocity such that catalyst is in continuous motion. Catalyst of small particle size can be used, giving efficient contact among gas, liquid, and solid with good mass and heat transfer. Part of the reactor effluent is recycled back through the reactors for temperature control and to maintain the requisite liquid velocity. The entire bed is held within a 20°F range. This provides essentially an isothermal operation with an exothermic process. Because of the movement of catalyst particles in the liquid-gas medium, deposition of tar and coke is minimized and fine solids entrained in the feed will not lead to reactor plugging. Catalyst can also be added and withdrawn from the reactor without upsetting operation.

Reactor effluent is cooled by exchange, and a vapor-liquid separation is made. After scrubbing in a lean-oil absorber, hydrogen is recycled. Liquid product is either stored directly or taken to fractionation prior to storage and blending.

Operating conditions range from 1,000 to 4,000 psig, 750 to 850°F, with hydrogen requirements from 300 to 2,000 scf per barrel of feed. Exact conditions depend upon feed and product requirements.   Investment costs range from $150 to $500 per B/SD, again depending on the type of operation.

The first commercial H-Oil installation, a 2,500-B/D unit, went on-stream in 1963 at Cities Service Refining Corp.'s refinery at Lake Charles, La.   This unit has operated predominantly on residual oil feed.

Literature references include:

Chem. Engrg. News, "Cities Service Details H-Oil Economics," p. 56, May 28, 1962.
Chervenak, M. C., et al., "H-Oil Process Treats Wide Range of Oils," Hyd. Proc., pp. 151–160, October, 1960.
Galbreath, R. B., and A. R. Johnson, "H-Oil Process Is Proven by First Commercial Unit," Hyd. Proc., pp. 121–124, September, 1963.
Griswold, C. R., and R. P. Van Driesen, "Commercial Experience with H-Oil," API Refining Meeting, St. Louis, Mo., May 12, 1964.
Hellwig, L. R., et al., "First H-Oil Unit Is Underway," Oil & Gas J., pp. 119–122, May 21, 1962.
Hyd. Proc., "H-Oil," p. 152, September, 1962.
Oil & Gas J., "H-Oil Process Ready," p. 71, Sept. 9, 1957.
Oil & Gas J., "H-Oil Process," p. 125, Apr. 3, 1961.
Pachler, H., et al., "This New Process Handles Residuals," Pet. Ref., pp. 201–204, September, 1959.
Pet. Ref., "H-Oil," p. 247, September, 1958.
Schuman, S. C., "Hydrogen Consumption in the H-Oil Process," Chem. Engrg. Prog., pp. 49–54, December, 1961.
Van Driesen, R. P., and N. C. Stewart, "Operation of a 2,500 BPD H-Oil Unit," API Refining Meeting, St. Louis, Mo., May 12, 1964.

Hy-C Cracking

This is a hydrocracking process developed by Cities Service Research and Development Co. and Hydrocarbon Research, Inc.   Both virgin and cracked gas oils to end points of 1100°F may be processed to lower-boiling, more valuable products, thus permitting wider control of middle distillate, kerosine, and gasoline production in the existing refinery.   In Hy-C cracking, hydrogen consumption is in the range of 1,000 to 2,000 scf/bbl and product yields are in excess of 100%.

Hy-C cracking is suggested in conjunction with the H-Oil process in a refinery in which residuum and distillate stocks are cracked in the presence of hydrogen to constitute the "all-hydrogen" refinery.

Hy-C cracking has many process elements similar to the H-Oil process.   Oil and hydrogen are passed upward through the reactor and contact catalyst which is in motion in an ebullating bed effecting an isothermal operation.   This also allows catalyst to be added and withdrawn from the reactor system on a continuous or intermittent basis.   Catalyst regeneration takes place exterior to the reactor, and downtime is not necessary for reactivation.

Operating pressure, space velocity, temperature, hydrogen flow rate, and catalyst replacement vary widely with the charge stock and the product requirement.   Catalyst type and composition may also be varied for specific applications.

Literature references include:

Chervenak, M. C., et al., "Hy-C Cracking," Chem. Engrg. Prog., pp. 53–59, February, 1963.
Hellwig, L. R., et al., "The Hy-C Cracking Process," API Refining Meeting, Philadelphia, Pa., May 16, 1963.
Johnson, A. R., and L. M. Rapp, "H-Oil and Hy-C Processing for Fuel Oil Elimination," NPRA Annual Meeting, San Antonio, Tex., Apr. 6–8, 1964.
Unzelman, G. H., and N. H. Gerber, "Hydrocracking—Today and Tomorrow," Petro/Chem Engr., pp. 32–52, October, 1965.

HDDV—Esso Research and Engineering

Hydrogen-Donor-Diluent-Visbreaking (HDDV), a possible substitute for conventional visbreaking, is not considered a commercially developed process.

The process is noncatalytic, and mild cracking takes place in the presence of a hydrocarbon fraction that donates hydrogen (see Fig. 3-15).   Tetrahydronaphthalene and partially hydrogenated refinery streams containing a high concentration of condensed aromatic ring compounds serve as thermal hydrogen transfer agents.

In a pilot-plant operation, residuum was mixed with a hydrogen donor diluent at ratios of 2 to 1 and 5 to 1 and the mix was cracked at 780 to 900°F and 400 psig. The hydrogen donor was a 700 to 900°F fraction of thermal tar partially hydrogenated with 400 scf of hydrogen per barrel. In recycle operation, the donor diluent is recovered from products and rehydrogenated. Residual fuel yield was decreased, gasoline quality improved, and middle distillate make increased over conventional visbreaking. Naphtha insolubles in the residual fuel were greatly reduced.

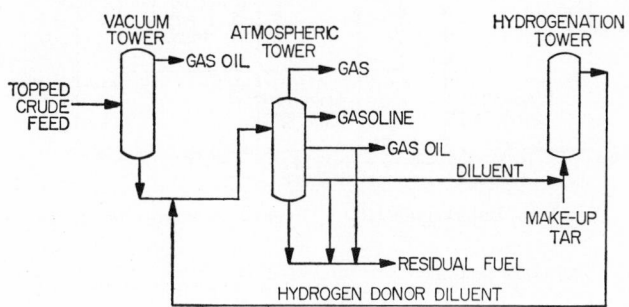

FIG. 3-15. Hydrogen-Donor-Diluent-Visbreaking (HDDV).

The process is suggested to be applicable whenever visbreaker operation is limited by fuel-oil qualities other than viscosity.

Literature references include:

Hyd. Proc., "Esso Reports on New HDDV Process," p. 45, February, 1962.
Uhl, W. C., "Process Review," *World Pet.*, p. 28, May, 1962.

Unicracking-JHC

This hydrocracking process is based on research programs of Union Oil Co. of California (Unicracking) and Esso Research and Engineering Co. (Esso Catalytic Hydrocracking). High-quality gasoline, jet fuel, and mid-barrel products are produced from catalytic cycle oil, coker, thermal, virgin gas oils, and heavy naphthas.

A fixed-bed catalytic process, Unicracking-JHC employs a high-activity catalyst with a high tolerance for sulfur and nitrogen compounds. Catalyst is regenerable. Design can be based upon a single-stage or a two-stage system with provisions to recycle to extinction.

A two-stage reactor system is shown in Fig. 3-16. Untreated feed, make-up hydrogen, and recycle gas are combined as charge to the first stage; gasoline conversion is as high as 60 vol-%. Reactor effluent is separated to recycle gas, liquid product, and unconverted oil. Second-stage operation may be either once-through or recycle cracking; conversion per pass is from 50 to 70 vol-%. Feed to the second stage is a mixture of unconverted first-stage oil and second-stage recycle.

Reactor temperatures range from 500 to 800°F, and operating pressures are from 1,000 to 2,000 psig. The type of feedstock and the product slate desired define specific operating conditions. Flexibility is such that feed may be converted to gasoline and lighter products or partially to middle distillates as required by end-product demand.

With complete conversion, C_5-180°F gasoline has Research and Motor octane from 99 to 101 with 3 ml of TEL per gal; the octane of the C_7-400°F naphtha varies with the aromatic content of the feed and can be as high as 90 Research octane at 3 ml of TEL per gal. Heavy naphtha is excellent reformer charge stock.

Investment costs range from $300 to $600 per barrel of feed and will vary with unit size.

Union Oil constructed the first unit, a 16,000-bbl, two-stage design at the Wilmington refinery. The unit was on-stream in late 1964.

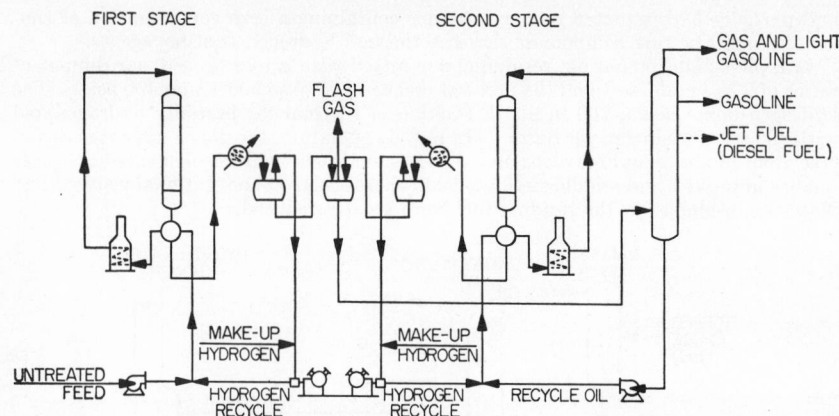

FIG. 3-16. Unicracking-JHC hydrocracking—two-stage design.

Literature references include:

Bradley, W. E., et al., "Unicracking-JHC Goes Commercial," NPRA Annual Meeting, Houston, Texas, Apr. 1–2, 1965.
Cheadle, G. D., et al., "Unicracking-JHC Process Broadens Commercial Application," API Refining Meeting, Houston, Texas, May 11, 1966.
Eilers, Robert D., "The New and Old Jointly Reduce Fuel Oils," California Oil World, pp. 128–139, January, 1965.
Huffman, H. C., et al., "A Team: Hydrocracking and Reforming," Hyd. Proc., pp. 181–186, June, 1964.
Huffman, H. C., et al., "Reforming of Unicracking-JHC Naphthas," API Refining Meeting, St. Louis, Mo., May 12, 1964.
Oil & Gas J., "Unicracking-JHC," p. 120, Apr. 5, 1965.

Unicracking

The unicracking process was developed by Union Oil Co. of California, and is now jointly licensed with Esso Research and Development Co. under the name Unicracking-JHC, described in detail above.

Literature references for Unicracking include:

Barnet, W. I., et al., "Extend Hydrocracking to Heavy Stocks," Pet. Ref., pp. 131–136, April, 1961.
Bradley, W. E., "Hydrogenation—Here Today and Here Tomorrow," Oil & Gas J., pp. 194–198, June 8, 1959.
Hansford, R. C., et al., "Unicracking—A Modern and Versatile Hydrocracking Process," WPRA Annual Meeting, San Antonio, Tex., Mar. 28–30, 1960.
Hansford, R. C., et al., "Unicracking Gives New Refining Step," Pet. Ref., pp. 169–176, June, 1960.
Hyd. Proc., "Unicracking," p. 154, September, 1962.
Peralta, B., et al., "New Developments in Unicracking," AIChE National Meeting, Los Angeles, Calif., Feb. 4–7, 1962.
Peralta, B., et al., "New Developments in Unicracking Technology," Chem. Engrg. Prog., pp. 41–46, April, 1962.
Stormont, D. H., "Hydrocracking—Its Big Impact Is Still Ahead," Oil & Gas J., p. 113, Feb. 12, 1962.

Esso Catalytic Hydrocracking

This is a hydrocracking process developed by Esso Research and Development Co. and now jointly licensed with Union Oil Co. of California under the name Unicracking-JHC.

Literature references include:

Stormont, D. H., "How Hydrogen Is Faring," *Oil & Gas J.*, p. 109, Feb. 11, 1963.
Uhl, W. C., "Processing Review," *World Pet.*, p. 28, May, 1962.

Gulf HDS

Developed by Gulf Research and Development Co., the Gulf HDS process is a regenerative fixed-bed process to upgrade petroleum residues by catalytic hydrogenation to refined heavy fuel oils or to high-quality catalytic charge stocks (see Fig. 3-17).

While desulfurization and quality improvement are the primary purposes of the process, if operating conditions and catalysts are varied, light distillates can be produced and viscosity of the heavy material can be lowered. Long on-stream cycles are maintained by reducing random hydrocracking reactions to a minimum. Whole crudes, virgin and cracked residuals may serve as feedstock.

Catalyst (70 to 85 cents per lb) is a metallic compound supported on pelleted alumina and may be regenerated *in situ* with air and steam or flue gas through a temperature cycle of 750 to 1200°F. On-stream cycles of 4 to 5 months can be obtained at desulfurization levels of 65 to 75%. Catalyst life is estimated at 2 years.

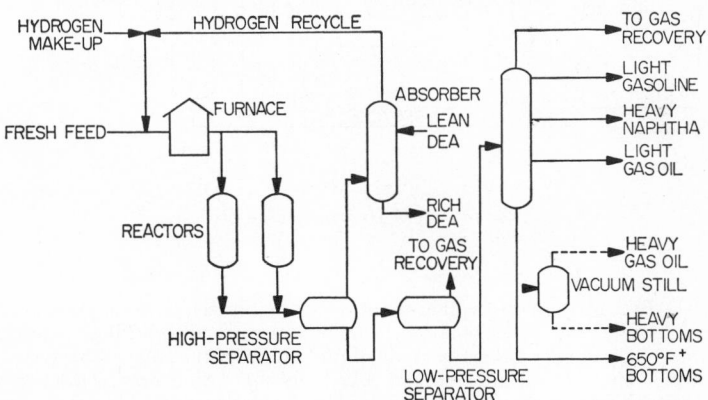

FIG. 3-17. Gulf HDS process.

Operating conditions are 750 to 850°F and 500 to 3,000 psig, with 400 to 1,000 scf per barrel of hydrogen required. Conditions vary with the feed and product requirements. Literature references include:

Beuther, H., "Recent Developments in the Technology of Residue Processing," AIChE National Meeting, Tulsa, Okla., Sept. 25–28, 1960.
Beuther, H., et al., "The Improved Gulf HDS Process," API Refining Meeting, Houston, Tex., May 11, 1961.
Beuther, H., and R. A. Flinn, "Gulf HDS Turns Resid into Cat Feed," *Pet. Ref.*, pp. 143–148, April, 1960.
Beuther, H., et al., "The Hydrocracking of Heavy Oils and Residuals," AIChE National Meeting, Houston, Tex., Feb. 7–11, 1965.
Davidson, R. L., "Hydrogen Processing," *Pet. Proc.*, pp. 115–138, November, 1956.
McAfee, J., et al., "The Gulf HDS Process for Upgrading Crudes and Residues," API Refining Meeting, St. Louis, Mo., May 11, 1955.
Oil & Gas J., "HDS Process," p. 126, Apr. 3, 1961.

H-G Hydrocracking

Developed by Gulf Research and Development Co. and now licensed by Houdry Process and Chemical Co., this process may be designed with either a single- or a two-stage reactor system for conversion of light and heavy gas oils to lower-boiling fractions. Design depends mainly on the feedstock characteristics.

The flow sheet in Fig. 3-17a shows a single-stage H-G Hydrocracker for processing heavy gas oil. Feed is mixed with recycle gas oil, make-up hydrogen, and hydrogen-rich recycle gas, and then heated and charged to the reactor. Reactor effluent is cooled and sent to a high-pressure separator. Hydrogen-rich gas is flashed off, scrubbed, and recycled to the reactor. Separator liquid then goes to a stabilizer for removal of butanes and lighter products. Stabilizer bottoms are taken to a fractionator for separation. Any unconverted material is recycled to the reactor. The process can be designed to maximize gasoline, jet fuels, heating oils, or LPG. Charge stock can vary from naphtha to deasphalted gas oils.

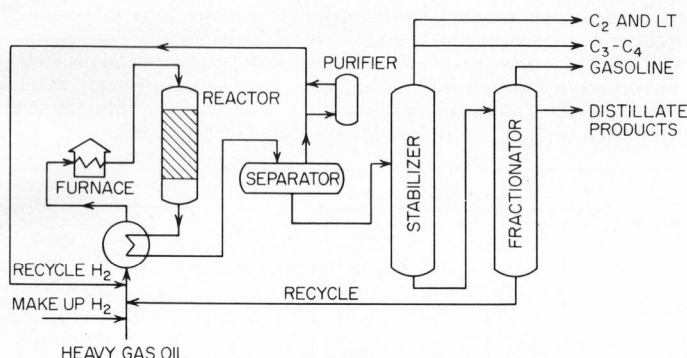

FIG. 3-17a.   H-G Hydrocracking process.

The H-G Process employs a dual-functioning highly selective catalyst providing optimum activity for hydrogenation, isomerization, and cracking. C_4+ yields (maximum gasoline) vary from 115 to 130 liquid vol-% of the feed. Catalyst may be regenerated in situ with air after normal coke build-up. An average run to regeneration is about 6 months; catalyst life is stated at 2 years plus.

The light gasoline cut from H-G Hydrocracking is composed mainly of pentane and hexane isomers and has leaded Research and Motor octane numbers exceeding 100. The naphtha cut is excellent reformer charge stock, especially when derived from refractory feeds. In some cases, heavy naphtha may be blended directly to gasoline. Kerosine, jet fuel, diesel fuel, and heating oils derived from the process are all of superior quality.

The first Gulf Hydrocracker was placed on-stream in 1962.

Literature references include:

Beuther, H., et al., "The Hydrocracking of Heavy Oils and Residuals," AIChE National Meeting, Houston, Tex., Feb. 7–11, 1965.
Beuther, H., and B. K. Schmid, "Reaction Mechanisms and Rates in Residue Hydrodesulfurization," Sixth World Petroleum Congress, June, 1963.
Gulf Research and Development Co., "Gulf Hydrocracking Process, Furnace Oil Distillates," Refinery Processes Division Manual.
Gulf Research and Development Co., "Gulf Hydrocracking Process, Heavy Gas Oil Distillate," Refinery Processes Division Manual.
Houdry Process and Chemical Co., "The H-G Hydrocracking Process," technical data sheet and brochure.
Hyd. Proc., "Gulf Hydrocracking," p. 144, September, 1964.
Oil & Gas J., "Hydrocracking," p. 121, Apr. 5, 1965.
Unzelman, G. H., and N. H. Gerber, "Hydrocracking—Today and Tomorrow," Petro/ Chem Engr., October, 1965.

Kellogg Hydrocracking

The process is licensed by M. W. Kellogg Co. and is unique in that a catalyst is not offered; instead, equipment is designed for any catalyst chosen by the refiner.

Advantages of the Kellogg design are said to include high isobutane recovery, economic heat removal (from reactors), and less hydrogen required per barrel processed. To effect a high isobutane recovery, Kellogg offers a specially designed product-separation section (see Fig. 3-17b). Reaction heat is removed by exchange with

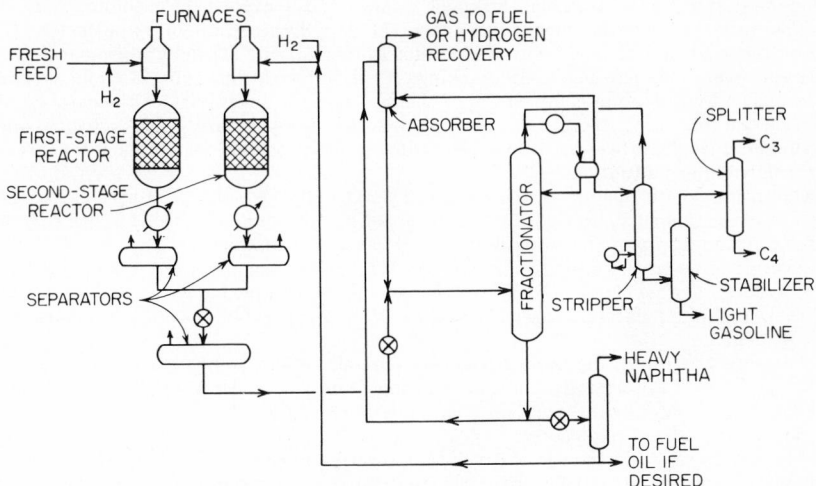

FIG. 3-17b.   Kellogg Hydrocracking Process.

recycle gas in the catalyst bed rather than by use of direct quench. Hydrogen requirements are lowered by introducing recycle as well as make-up hydrogen at the inlet of the hydrocracking zone. This results in better hydrogen-to-oil ratios.

Literature references include:

Oil & Gas J., "Hydrocracking Race Gains New Entry," pp. 57–58, Mar. 7, 1966.

BASF-IFB Hydrocracking

This process is jointly licensed by the Institut Français du Petrole and Badische Anilin- & Soda-Fabrik. It is designed specifically to satisfy the European distillate market by processing heavy sour feedstock. It also can be designed to maximize gasoline production. Catalyst is specially tailored to the type of feed material and end-products desired.

Literature references include:

Billion, A., et al., "What BASF-IFP Hydrocracking Will Do," *Hyd. Proc.*, pp. 129–134, March, 1966.

CATALYTIC REFORMING

Catalytic reforming is a continuous process to upgrade low-octane virgin, thermal, or heavy catalytically cracked naphthas into high-octane components for motor or aviation fuel blending or petrochemical usage. Volatility is increased, and sulfur content is reduced. Octane improvement of virgin naphtha is 20 to 50 Research octane numbers, depending upon charge quality and operating severity. Currently an average yield-octane relationship for reforming (excluding aromatic extraction) might be considered 82.0% yield (debutanized product based upon a 200 to 400°F virgin charge) of 96.0 to 99.0 Research octane number (at 3.0 ml of TEL per gal). The commercial processes available for use today can be broadly classified as of the moving-bed, fluid-bed, and fixed-bed types. The fluid- and moving-bed processes use mixed nonprecious-metal oxide catalysts in units equipped with separate regeneration facilities. Fixed-bed processes use predominantly platinum-containing catalysts in units equipped for cyclic, occasional, or no regeneration. Actually, 95% of the more than 2,000,000 B/D U.S. reforming capacity is of the fixed-bed type.

Basically, catalytic reforming is the rearranging of molecules in a gasoline boiling range material to give a higher antiknock quality at the expense of gasoline yield.

The primary reaction mechanisms are (1) dehydrogenation of naphthenes, (2) dehydrocyclization of paraffins, (3) paraffin isomerization, (4) dehydroisomerization of naphthenes, (5) paraffin hydrocracking, (6) desulfurization, and (7) olefin saturation. The hydrocarbon composition of the feed and the selectivity of the catalyst as well as the reforming operating severity—pressure, temperature, space velocity, and hydrogen recycle rate—determine the primary hydrocarbon reactions for a given reformer-refinery situation.

Operating conditions are in the ranges of 800 to 1000°F, 50 to 750 psig, 0.7 to 5.0 v/hr/v space velocity, and 3.0 to 10.0 moles of gas per mole of naphtha, gas recycle ratio, depending upon the process type and catalyst employed.

Average investment costs (1962) for catalytic reforming processes (including feed preparation) are reported at $350 to $650 per barrel of capacity with operating costs (direct, including catalyst costs) at 35 to 55 cents per bbl, excluding return on investment.

The catalytic reforming process was commercially nonexistent in the United States prior to 1940. Reforming operations during World War II were directed toward production of high-octane aviation components as well as chemicals to be employed in the production of explosives.

The rapid growth in the use of catalytic reforming in the refining industry was phenomenal from 1953 to 1959. By 1962, reforming capacity appeared to be leveling out at 19 to 20% of crude running capacity. However, current construction announcements indicate that refiners also intend to make full use of catalytic reforming for feedstocks other than virgin naphtha.

General catalytic reforming references include:

Brooks, B. T., et al., "The Chemistry of Petroleum Hydrocarbons," Reinhold Publishing Corporation, 1955, vol. 2.
Foster, A. L., "Catalytic Cracking and Reforming," *Pet. Engr.*, pp. C14–18, July, 1956.
Hettinger, Jr., W. P., et al., "Hydroforming Reactions; Effects of Certain Catalyst Properties and Poisons," *Ind. Engrg. Chem.*, pp. 719–730, April, 1955.
Kress, R. F., et al., "Reforming Heavy Catalytically Cracked Gasoline for High Octane Blend Stock," AIChE Meeting, Tulsa, Okla., September, 1960.
Nelson, W. L., "Operating Costs—Platinum Catalyst Reforming," *Oil & Gas J.*, p. 93, Sept. 14, 1959.
Nelson, W. L., "Feed Composition and Catalyst Costs in Catalytic Refining," *Oil & Gas J.*, pp. 137–138, Aug. 1, 1960.
Nelson, W. L., "In Octane Upgrading, What's the Value of Low-octane Gasoline?" *Oil & Gas J.*, p. 102, Dec. 19, 1960.
Nelson, W. L., "Reformate Sensitivity Varies Widely," *Oil & Gas J.*, p. 83, Jan. 9, 1961.
Nelson, W. L., "Is Reforming Feasible for 100-octane Production?" *Oil & Gas J.*, p. 81, May 22, 1961.
Oil & Gas J., "Catalysis in Petroleum Refining," pp. 148–162, Mar. 21, 1955.
Oil & Gas J., "100 Octane," pp. 144–158, Mar. 19, 1956.
Pet. Proc., "Catalytic Reforming," pp. 1157–1204, August, 1955.
Pet Proc., "Trends in Petroleum Refining," pp. 560–563, May, 1949.
Rushton, J. H., "The Evolution of Hydroforming," *Chem. in Canada*, pp. 33–36, November, 1954, and pp. 31–35, March, 1955.
Samuels, S. C., et al., "High Octane Components from Catalytic Reformates," *Ind. Engrg. Chem.*, pp. 73–76, January, 1959.
Sittig, M., "Catalytic Process Mechanics," *Pet. Proc.*, pp. 1048–1055, July, 1954.
Smith, R. B., "Kinetic Analysis of Naphtha Reforming," *Chem. Engrg. Prog.*, pp. 76–80, June, 1959.

Smith, R. B., and T. Dresser, "When to Discard Reformer Catalyst," *Pet. Ref.*, pp. 199–202, July, 1957.
Steel, R. A., et al., "Selecting a Catalytic Reforming Process," *Pet. Ref.*, pp. 167–171, May, 1954.
Thomas, R. W., and P. M. Arnold, "Petroleum Annual Review 1955," *Ind. Engrg. Chem.*, pp. 40–41A, January, 1956.
Vener, R. E., "Moving Bed Processes," *Chem. Engrg.*, pp. 173–206, July, 1955.
Wilson, J. L., and M. J. DenHerder, "Reforming Studies with Molybdena-Alumina Catalyst," ACS Meeting, Atlantic City, N.J., September, 1956.
Zielinski, R. M., "Relationship of the Composition of Reformer Feed to Reforming Yield," ACS Meeting, New York, September, 1957.

Fixed-bed No-swing Reactor

Fixed-bed, continuous catalytic reforming processes use an acid-type, pelleted catalyst containing platinum (0.01 to 1.0%) averaging approximately 0.5% on alumina or silica-alumina carrier. A pretreatment or guard case may be installed to ensure against catalyst contamination by arsenic, lead, copper, nitrogen, water, and sulfur.

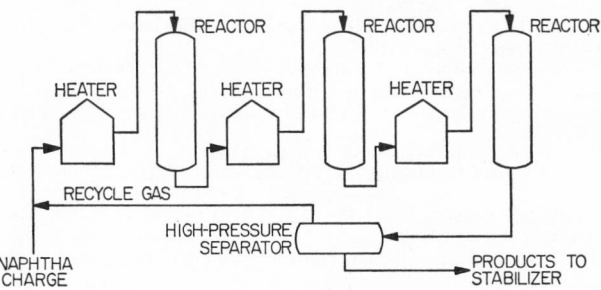

Fig. 3-18. Typical fixed-bed catalytic reforming process without swing reactor for catalyst regeneration.

Typical virgin naphtha feed has an initial boiling point of 175 to 250°F and a final boiling point of 350 to 400°F and is generally obtained from a prefractionator. The feed is added to a heater together with recycle gas (3.0 to 10.0 moles of H_2 per mole of liquid feed) containing hydrogen of 80 to 98% purity.

The temperature of 850 to 950°F is held by interreactor heaters (Fig. 3-18) to compensate for the endothermic reforming reaction. The pressure in the reactor is held at about 500 psig, though pressures in the range of 300 to 750 psig may be used, depending upon feed composition and boiling range and catalyst activity. Higher pressures are used to extend catalyst life when charging high-boiling stocks. Lower pressures favor dehydrogenation of naphthenes to aromatics and reduce hydrocracking. The space velocity (v/hr/v) is usually 1.0 to 5.0.

Except for the Platforming process, the processes described under this heading do have facilities for *in situ* regeneration of the catalyst under blocked-out operating conditions. Such regeneration is usually necessitated by prolonged high-severity operation or process upsets. Regeneration with air (plus steam in some instances) may take place at atmospheric pressure but normally is at 250 to 300 psig and 1000 to 1050°F. Regeneration in the case of Platforming is said to be possible but economically unjustifiable.

Catforming

Developed by Atlantic Refining Co. the main feature of this process which differentiates it from the other reforming processes in this category is the catalyst composition. The catalyst, produced by Engelhard Industries Inc., is a platinum, alumina,

silica-alumina composition which permits relatively high space velocities and results in very high hydrogen purity. Regeneration to prolong catalyst life is practiced on a blocked-out basis with a dilute air-in-steam mixture. McBride Oil and Gas Corp. at LaBlanca, Tex., placed the first Catforming unit on-stream in August, 1952.
Catforming references include:

Connor, Jr., J. E., et al., "Benzene Production over the Catforming Catalyst," *Ind. Engrg. Chem.*, pp. 152–156, January, 1955.
Eckel, J. G., and G. R. Worrell, "Why Dorchester Will Add Catalytic (Atlantic) Reforming," *Pet. Ref.*, pp. 146–147, April, 1957.
Faulkner, C. S., et al., "First Commercial 'Catforming' Unit on Stream at McBride Refinery," *Oil & Gas J.*, pp. 116–127, Nov. 24, 1952.
Foster, A. L., "Catforming Civilizes Naphthas," *Pet. Engr.*, pp. C33–36, February, 1955.
Fowle, M. J., et al., "Development of a Reforming Catalyst," *Advances in Chemistry Series*, no. 5, pp. 76–82, 1951.
Fowle, M. J., et al., "Tomorrow's Octanes," API Refining Meeting, San Francisco, Calif., May, 1952.
Fowle, M. J., et al., "In Reforming—It's the Catalyst That Counts," *Pet. Ref.*, pp. 156–159, April, 1952.
Grane, H. R., "A Look at Commercial Catforming," *Pet. Ref.*, pp. 112–115, September, 1953.
Grane, H. R., and J. A. Nevison, "Catforming for High Octane Numbers in Canada," *Ref. Engr.*, pp. C38–39, April, 1957.
Milner, B. E., "Atlantic Catforming," California Natural Gasoline Assoc., November, 1953.
Milner, B. E., "Catforming for Octanes," *World Pet.*, pp. 50–51, January, 1953.
Milner, B. E., et al., "Catforming at United Refining Company," *Pet. Ref.*, pp. 169–173, September, 1955.
Pet. Proc., "Catforming," pp. 1182–1183, August, 1955.
Pet. Proc., "39% Benzene Yield from a C_6 Straight-Run Fraction," pp. 249–250, March, 1951.
Thornton, Jr., D. P., "How a Small Plant Makes Octanes," *Pet. Proc.*, pp. 1159–1161, August, 1953.
Worrell, G. R., "Now 100 Octane from Natural Gasoline," *Pet. Ref.*, pp. 138–140, April, 1956.

Houdriforming

Houdriforming catalytic reforming is licensed by Houdry Process and Chemical Co. Catalyst may be regenerated if necessary on a block-out basis. A "guard case" catalytic hydrogenation pretreating stage—using the same Houdry catalyst as the Houdriformer reactors—is available for high-sulfur feedstocks. Lead and copper salts are also removed under the mild conditions of the guard case operation. The first commercial Houdriformer was placed on-stream in November, 1953, at Marcus Hook, Pa., by Sun Oil Co.
Houdriforming references include:

Beyler, D., et al., "Houdriforming for Aromatics," *Ind. Engrg. Chem.*, pp. 740–744, April, 1955.
Bland, R. E., "Houdriforming," *Pet. Engr.*, pp. C18–21, April, 1954.
Burtis, T. A., and H. D. Noll, "Houdriforming—Its Place in the Refining of Petroleum," WPRA Meeting, San Antonio, Tex., April, 1952.
Dart, J. C., et al., "Houdriforming at Low Pressure," API Refining Meeting, San Francisco, Calif., May, 1952, reviewed in *Pet. Proc.*, p. 842, June, 1952.
Heinemann, H., et al., "Houdriforming of Hydrocracked Naphthas: Composition of Naphthas from Gas Oil Hydrocracking," *Ind. Engrg. Chem.*, pp. 735–739, April, 1955.
Heinemann, H., et al., "Houdriforming Reactions—Studies with Pure Hydrocarbons," *Ind. Engrg. Chem.*, pp. 130–133, January, 1953.
Heinemann, H., et al., "Application of Houdriforming to Produce Aromatics and High-octane Motor Gasoline," AIChE Meeting, Galveston, Tex., October, 1951.
Heinemann, H., et al., "Houdriforming for Aromatics," *Pet. Engr.*, pp. C40–42, November, 1951.
Hyd. Proc., "Houdriforming," p. 160, September, 1964.
Kirkbride, C. G., "Houdriforming—A New Continuous Process for Reforming Petroleum Naphthas," *Pet. Proc.*, pp. 603–606, June, 1951.

Mills, G. A., et al., "Catalytic Mechanism (Houdriforming Reactions)," *Ind. Engrg. Chem.*, pp. 134–137, January, 1953.
Noll, H. D., "What to Reform and How," *Houdry Pioneer*, March, 1953.
Noll, H. D., et al., "Houdriforming Economics—Effect of Feed-stock Boiling Range and Reforming Severity," API Refining Meeting, New York, May, 1953.
Oil & Gas J., "Houdriforming," p. 145, Apr. 5, 1965.
Pet. Proc., "Houdriforming," pp. 1178–1179, August, 1955.
Stevenson, D. H., and G. A. Mills, "Hydropretreatment of Catalytic Reformer Feed," API Refining Meeting, St. Louis, Mo., May, 1955.

Platforming

The Platforming process is licensed by Universal Oil Products Co. The platinum-alumina catalyst includes a combined halogen—0.1 to 8.0% of fluorine or chlorine based on alumina. To guard against plant upsets, the feed is usually held at 375°F end point, although it may be higher. Regeneration may be practiced, although it is not usually resorted to.

The first unit went on-stream at the Old Dutch Refining Co. (now Marathon Oil Co.) refinery at Muskegon, Mich., in October, 1949. This was the first commercial reforming process using a platinum-containing catalyst.

Platforming references include:

Bland, W. F., "Platforming," *Pet. Proc.*, pp. 351–355, April, 1950.
Bogen, J. S., and V. Haensel, "Platformate—Properties and Performance," American Petroleum Institute, Cleveland, Ohio, May, 1950.
Cox, R. F., and R. H. Killgore, "Natural Gasoline Plant Platformer," *Pet. Proc.*, pp. 102–107, November, 1956.
Donaldson, G. R., and V. Haensel, "Platforming Reactions of Pure Hydrocarbons," *Ind. Engrg. Chem.*, pp. 102–104, September, 1951.
Donaldson, G. R., et al., "Dehydrocyclization in Platforming," *Ind. Engrg. Chem.*, pp. 731–735, April, 1955.
Egloff, G., "The Platforming Process," *J. Inst. of Pet.*, pp. 69–79, March, 1955.
Foster, A. L., "New Platformer Improves Octane Rating," *Pet. Engr.*, pp. C39–42, February, 1952.
Gerald, C. F., "Unifining—Platforming in One Unit," *Pet. Ref.*, pp. 216–217, May, 1956.
Haensel, V., "Platforming," Western Petroleum Refiners Assoc., San Antonio, Tex., March, 1950, reviewed in *Pet. Proc.*, pp. 356–360, April, 1950.
Haensel, V., and C. V. Berger, "Aromatics by Platforming," *Pet. Proc.*, pp. 264–267, March, 1951.
Haensel, V., and G. R. Donaldson, "How Paraffin Content Affects Quality of Motor Fuels from Platforming," *Pet. Proc.*, pp. 236–240, February, 1953.
Haensel, V., and D. H. Belden, "Platforming—A Progress Report," API Refining Meeting, Montreal, Quebec, May, 1956.
Kastens, M. L., and R. E. Sutherland, "Platinum Reforming of Gasoline," *Ind. Engrg. Chem.*, pp. 582–593, April, 1950.
Meerbott, W. K., et al., "Naphtha Processing over Platforming Catalyst," *Ind. Engrg. Chem.*, pp. 650–655, April, 1957.
Oil & Gas J., "Platforming," p. 141, Apr. 5, 1965.
Pet. Proc., "New Catalytic Reforming Process Shows High Octane Gain, Low Volume Loss," pp. 553–555, May, 1949.
Pet. Proc., "Penex-Platforming of Naphthas," pp. 66–67, August, 1956.
Pet. Proc., "Platforming," pp. 1176–1177, August, 1955.
Read, D., and P. C. Weinert, "100 Octane Gasoline by Once Through Platforming," *Oil & Gas J.*, pp. 105–108, Apr. 23, 1956.
Resen, L., "Segregated Feed Stocks Charge Two Catalytic Reforming Units," *Oil & Gas J.*, pp. 153–155, Nov. 10, 1958.
Sutherland, R. E., and D. D. Hanson, "Platforming Natural Gasoline," *Oil & Gas J.*, pp. 177–178, April, 1950.
Weinert, P. C., et al., "Three Years of Commercial Platforming," API Refining Meeting, San Francisco, Calif., May, 1952.

Sinclair-Baker Reforming

Arrangements for the use of the Sinclair-Baker RD-150 catalyst can be made through Engelhard Industries, Inc. The RD-150 catalyst (produced by Engelhard) contains about 0.6 wt-% platinum on alumina. This catalyst yields a high ratio of paraffin cyclization activity to hydrocracking activity and has the ability to be regenerated with dilute air. Normal Sinclair-Baker reforming consists of three in-series reactors with regeneration being practiced during infrequent intervals on an intermittent blocked-out basis (24 to 72 hr regeneration time). The Pure Oil Co. refinery at Heath, Ohio, and the Sinclair refinery at Marcus Hook, Pa., were the sites of the first commercial use of Sinclair-Baker reforming.

Sinclair-Baker references include:

Decker, W. H., "Sinclair-Baker Reforming," *Pet. Engr.*, pp. C30–32, April, 1954.
Decker, W. H., and D. Stewart, "Cat Reforming with In-place Regeneration," *Oil & Gas J.*, pp. 80–84, July 4, 1955.
Decker, W. H., and C. Rylander, "Cat Reforming with RD-150," *Oil & Gas J.*, pp. 88–91, Feb. 2, 1959.
Eisele, G. F., and R. Smith, "Reforming with Sinclair-Baker Catalyst at the Pure Oil Company's Heath Refinery," WPRA Meeting, Toledo, Ohio, May, 1956.
Pet. Proc., "Sinclair-Baker," pp. 1186–1187, August, 1955.
Pet. Proc., "Kellogg Sinclair-Baker," p. 1196, August, 1955.
Pet. Ref., "Catalytic Reforming," pp. 16–17, September, 1954.
Pet. Ref., "SBK Catalytic Reforming," pp. 214–215, September, 1956.
Reed, W. H., and R. R. Runge, "How to Regenerate Reformer Catalyst," *Oil & Gas J.*, pp. 226–229, Oct. 15, 1962.
Stanford, G. W., "First Kellogg-designed Cat Reformer," *Pet. Ref.*, pp. 190–192, September, 1955.
Teter, J. W., "New Catalytic Reforming Process," WPRA Regional Meeting, Casper, Wyo., September, 1953.
Teter, J. W., B. T. Borgerson, and L. H. Beckberger, "Sinclair's 'Cat' Reforming Process," *Pet. Proc.*, pp. 1519–1523, October, 1953.
Teter, J. W., and L. E. Olsen, "Research for Sinclair-Baker Catalytic Reforming," *World Pet.*, pp. 99–103, July 15, 1953.

Platinum Reforming (PR)

This is a process developed and used by the Socony Mobil Oil Co. and originally identified as "Sovaforming."

The process employs fixed-bed reactors with a platinum catalyst. It is a high-pressure (approximately 500 psig) process which generally operates with infrequent regeneration.

The first commercial installation was placed on-stream in November, 1954, at the Mobil Oil Co. refinery at Ferndale, Wash.

Sovaforming references include:

Backensto, E. B., et al., "High Temperature Hydrogen-sulfide Corrosion in Commercial Sovaforming Units," API Refining Meeting, Montreal, Quebec, May, 1956.
Ciapetta, F. G., "Special Report on Catalytic Reforming," *Petro/Chem Engr.*, pp. C19–31, May, 1961.
Oil & Gas J., "Socony Mobil Designs New Vapor-inlet Distributor," p. 79, June 22, 1959.
Oil & Gas J., "Socony Develops New Spherical Cat Reforming Reactor," p. 219, Aug. 17, 1959.
Pet. Proc., "Where Are We Going with Catalytic Reforming," pp. 1157–1204, August, 1955.
Pet. Ref., "Catalytic Reforming," pp. 164–165, September, 1962.
Sittig, M., and T. W. Warren, "How to Get Those Top Octanes," *Pet. Ref.*, pp. 230–279, September, 1955.

Fixed-bed Swing Reactor

Fixed-bed, continuous catalytic reforming can be classified by catalyst type: (1) cyclical regenerative with nonprecious-metal oxide catalysts and (2) cyclical regenerative with platinum-alumina catalysts. Both types use swing reactors (Fig. 3-19) to regenerate a portion of the catalyst while the remainder stays on-stream.

Cyclical regenerative, fixed-bed operation using platinum catalyst is basically a low-pressure operation—250 to 350 psig. The low-pressure process gives (1) higher gasoline yields because of fewer hydrocracking reactions, (2) higher product octane from a given naphtha charge, and (3) better hydrogen yields because of more dehydrogenation and less hydrocracking reactions.

Coke yield with attendant catalyst deactivation increases rapidly at low pressures. Cyclic regeneration affords operating flexibility to realize the advantages of low pressures while controlling the higher coke yield.

Investment costs are higher than the nonregenerative, fixed-bed processes, as an extra, or swing, reactor is needed.   An additional pretreat reactor for desulfurization or hydrogenation of a cracked feed might also be needed.

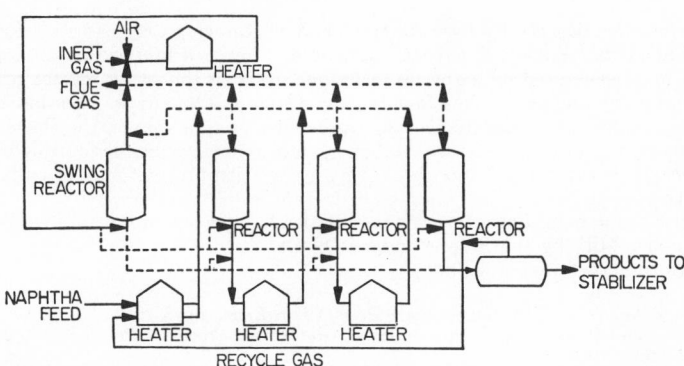

FIG. 3-19. Typical fixed-bed catalytic reforming process with swing reactor for catalyst regeneration.

The noticeable feature which differentiates these processes from the previously discussed nonregenerative processes is the swing reactor principle which permits regeneration with no interruption of on-stream time.

Fixed-bed Hydroforming

Fixed-bed Hydroforming was developed by Esso Research and Engrg. Co. in conjunction with Standard Oil Co. (Indiana) and M. W. Kellogg Co.   This process is licensed by M. W. Kellogg.   It was the first commercial process developed specifically for catalytic reforming.   The first unit went on-stream in November, 1940, at the Pan American Refining Co. (now American Oil Co.) refinery in Texas City, Tex.

The catalyst employed consists of 9.0% molybdenum oxide deposited on activated alumina granules or pellets.   The process is cyclic in that two of the four catalyst cases are regenerated while the remainder are on the process cycle.   The reaction takes place at about 1000°F, 50 to 150 psig, and 0.5 to 1.0 v/hr/v space velocity. Products include a hydrogen-rich off gas, 400°F end-point gasoline component, and a high-boiling aromatic polymer.   These units were used to produce toluene and aromatics for aviation gasoline during World War II, but none have been constructed since the war.   Some existing units have been converted to platinum catalyst usage.

Fixed-bed Hydroforming references include:

Armistead, Jr., G., "The Hydroforming Process," *Oil & Gas J.*, pp. 85–87, April, 1946.
Danziger, B. H., "Catalysts (Molybdenum)," *Ind. Engrg. Chem.*, pp. 1495–1500, August, 1955.
Gaylor, P. J., "Lengthen Hydroforming Process Cycle by Adding Reducible Sulfur to Feed," *Pet. Proc.*, p. 1479, October, 1952.
Greensfelder, B. S., et al., "Catalytic Reforming," *Chem. Engrg. Prog.*, pp. 561–568, October, 1947.
Hill, L. R., et al., "Hydroforming—A Catalytic Method for Naphtha Upgrading," *Nat. Pet. News*, p. R456, June 5, 1946.
McLaurin, N. H., et al., "Hydroforming and Thermal Reforming Operations on Sweet and Sour Heavy Straight-run Naphthas," *Pet. Ref.*, pp. 171–175, April, 1949.

Pet. Ref., "Hydroforming," pp. 112–113, September, 1952.
Saegebarth, E. O., "Catalytic Reforming Process of Standard Oil Company of California," *Pet. Engr.*, pp. 95–100, May, 1946.
Steiner, H., "Aromatics from Petroleum," *J. Inst. of Pet.*, pp. 410–435, July, 1947.
Swift, J. J., et al., "Aromatics Recovery," *Pet. Proc.*, pp. 81–87, January, 1953.
Webb, G. M., et al., "Chemical and Physical Properties of Alumina-Molybdena Oxide Catalyst for Hydroforming," *Pet. Proc.*, pp. 834–842, November, 1947.

Powerforming

Powerforming, licensed by Esso Research and Engineering Co., uses a platinum-on-alumina catalyst. Four or five reactors are used. Three or four are on-stream, while one is being regenerated by a special technique utilizing the swing reactor principle. The cycle on any one reactor in a four-reactor system is 3 to 5 days. For more favorable charge stocks or moderate octane-improvement requirements, the Powerformer may be operated at higher pressures and on a semiregenerative basis requiring blocked-out regeneration every 2 to 6 months. A hydrogenation pretreat may be employed if warranted.

The first commercial Powerforming installation was placed on-stream in July, 1955, at Baltimore, Md., by the Esso Standard Oil Co.

Powerforming references include:

Ciapetta, F. G., "Catalytic Reforming," *Petro/Chem Engr.*, pp. C19–31, May, 1961.
Holt, P. H., and R. R. Haig, "Data on Latest Reforming Process," *Pet. Ref.*, pp. 221–222, September, 1957.
Hyd. Proc., "Powerforming," p. 163, September, 1964.
Oil & Gas J., "Powerforming," p. 142, Apr. 5, 1965.
Oil & Gas J., "World's Largest Powerformer Will Be Ready in June," p. 74, Jan. 6, 1958.
Pet. Proc., "Powerforming Is No. 14," p. 69, April, 1956.
•*Pet. Proc.*, "Powerforming for Octanes," pp. 117–119, June, 1957.

Ultraforming

The Ultraforming process is licensed by Standard Oil Co. (Indiana). It is low-pressure reforming over a platinum catalyst (0.6%) on an alumina carrier. Regeneration of catalyst in the swing reactor restores catalyst activity without shutting down the unit.

Reformate of over 100 Research octane, clear, has been routinely obtained from Gulf Coast naphtha charge in the commercial units at C_5+ yields of about 75.0 vol-%. The first commercial installation of Ultraforming went on-stream in May, 1954, at the American Oil Co. refinery in El Dorado, Ark.

Ultraforming references include:

Birmingham, W. J., "Ultraforming," *Pet. Engr.*, pp. C35–38, April, 1954.
Brooks, J. A., "Ultraforming—8 Years Later," *Oil & Gas J.*, pp. 145–149, Sept. 10, 1962.
Forrester, J. H., et al., "Ultraforming, a Standard of Indiana Development," *Oil & Gas J.*, pp. 139–142, April, 1954.
Gumaer, R. R., and L. L. Raiford, "How First Ultraformer Has Performed," *Oil & Gas J.*, pp. 119–123, Aug. 8, 1955.
Hyd. Proc., "Ultraforming," p. 164, September, 1964.
Moore, T. M., "Ultraformer Tops 100 Octane," *Pet. Ref.*, pp. 174–178, September, 1955.
Nix, H. C., "Regenerative Platinum Catalyst Reforming," *Ref. Engr.*, pp. C13–16, June, 1957.
Oil & Gas J., "New Reforming Process," pp. 256–257, Sept. 14, 1953.
Oil & Gas J., "Ultraforming," p. 143, Apr. 5, 1965.
Pet. Proc., "Ultraforming," pp. 1184–1185, August, 1955.
Roberts, J. K., et al., "Regenerative Platinum-catalyst Reforming of Naphthas," Fourth World Petroleum Congress, Rome, Italy, June, 1955.
Steel, R. A., et al., "Selecting a Catalytic Reforming Process," *Pet. Ref.*, pp. 167–171, May, 1954.
White, P. C., et al., "Ultraform to Get 100 Octane, Clear," *Pet. Ref.*, pp. 171–177, May, 1956.

Moving-bed Reforming

Moving-bed catalytic reforming processes employ a single, moving-catalyst-bed reactor with continuous catalyst regeneration but are no longer considered commercially practical.

The catalyst is a nonprecious-metal oxide mixture in bead or pellet form which is transported between reactor and regenerator (Fig. 3-20) through use of mechanical or other means.

Depending upon catalyst type, typical feedstock for motor-gasoline usage has an initial boiling range of 150 to 175°F and an end-point range of 400 to 500°F. Feed-

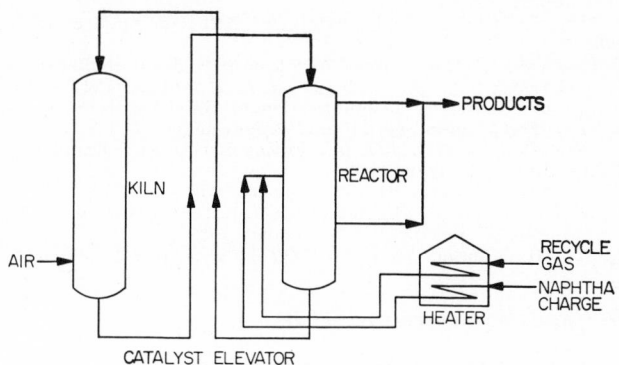

Fig. 3-20. Typical moving-bed catalytic reforming unit.

stock preparation is normally not a factor unless water has an adverse effect on the catalyst, and then drying is generally employed in the recycle gas stream.

Hyperforming

Hyperforming is a moving-bed reforming process developed by Union Oil Co. of California in the early 1950's but is now considered obsolete and no longer offered for licensing.

The process uses catalyst pellets of cobalt molybdate with a silica-stabilized alumina base. In operation, catalyst moves downward through the reactor with gravity flow and is returned to the top by means of a solids-conveying technique (Hyperflow), which moves the catalyst at low velocities and with minimum attrition loss. Feed naphtha vapor and recycle gas flow upward, countercurrent to catalyst. Regeneration of catalyst is accomplished in either an external vertical lift line or a separate vessel.

Hyperforming virgin naphtha and/or cracked naphtha feed (150 to 450°F) can result in 40 to 50 clear Research octane numbers improvement for the motor-fuel component. Sulfur and nitrogen removal is also accomplished. Light gas-oil stocks can also be charged to remove sulfur and nitrogen under mild hydrogenation conditions for the production of premium diesel fuels and middle distillates.

Operating conditions in the reactor are 400 psig and 800 to 900°F, the higher temperature being employed for straight-run naphtha feed. Hydrogen recycle is in the order of 3,000 scf per barrel of feedstock. Catalyst circulation is low—5 tons/hr (for a 10,000-B/D unit charging straight-run feed). Regeneration takes place at 950°F and 415 psig.

The first commercial Hyperforming unit was placed on-stream in May, 1955, at Calstate Refining Co.'s Signal Hill, Calif., refinery.

Hyperforming references include:

Berg, Clyde, "The Design of Hyperforming Units," *Pet. Ref.*, pp. 153–157, October, 1954.
Berg, Clyde, "Hyperforming," *Pet. Proc.*, pp. 1018–1023, July, 1953.
Berg, Clyde, "Hyperforming Makes Its Bow," *Oil & Gas J.*, pp. 286–293, Mar. 23, 1953.
Berg, Clyde, "Hyperforming—Newest Development in Reforming," *Pet. Ref.*, pp. 131–136, December, 1952.
Berg, Clyde, "Mechanical Features of the Hyperforming Process," *Mech. Engrg.*, pp. 19–22, January, 1955.
Berg, Clyde, "Signal Hill Plant Gets First Commercial Hyperformer," *Oil & Gas J.*, pp. 91–92, Aug. 2, 1954.
Berg, Clyde, "Union's Cobalt Molybdate Process for Refining High Sulfur Stocks," Oil Sands Project Conference, Canada, September, 1951, reviewed in *Pet. Proc.*, pp. 186–189, February, 1952.
Berg, Clyde, "The First Commercial Hyperformer," API Refining Meeting, Houston, Tex., May, 1954.
Bradley, W. E., et al., "Catalytic Desulfurization of High-sulfur Stocks by the Cobalt-molybdate Process," *Chem. Engrg. Prog.*, pp. 1–12, January, 1947.
Byrns, A. C., et al., "Catalytic Desulfurization of Gasoline by the Cobalt-molybdate Process," *Ind. Engrg. Chem.*, pp. 1160–1167, November, 1943.
Hendricks, G. W., et al., "Catalytic Desulfurization of Petroleum Distillates," ACS Meeting, Atlantic City, N.J., April, 1946.
Pet. Proc., "Hyperforming," pp. 1194–1195, August, 1955.
Pet. Proc., "New Lift Method for Moving Solids," pp. 1024–1025, July, 1953.
Pet. Ref., "Hyperforming," pp. 122–123, September, 1952.
Stormont, D. H., "Hyperformer Producing," *Oil & Gas J.*, p. 116, March, 1955.

Thermofor Catalytic Reforming (TCR)

This moving-bed reforming process was developed and licensed by Socony Mobil Oil Co. but is no longer in use.

The process used a synthetic-bead type of coprecipitated chromia and alumina catalyst.

Normal operating conditions in the reactor were 950 to 1000°F, 100 to 200 psig, and 0.7 space velocity (v/hr/v). The gas recycle unit ratio ranged from 3 to 9 moles gas per mole of naphtha. Catalyst-to-naphtha ratios had little effect on product yield or quality when varied over a wide range. Regeneration took place at essentially atmospheric pressure and 800 to 1050°F.

The catalyst flowed down through the reactor countercurrent and cocurrent with the naphtha-recycle gas feed, which entered the center of the reactor. The catalyst was transported from the base of the reactor to the top of the regenerator by bucket-type elevators. Reformate yields of C_5+ reformate (based on 170 to 400°F charge) had clear Research octane numbers of 85 to 95.

Reformate was normally rerun to remove a small volume of heavy aromatic polymer formed (about 2.0 vol-%).

TCR references include:

Chem. Engrg., "Thermofor Catalytic Reformer to Make High Octane Gasoline," pp. 242–244, August, 1952.
Dalton, S. D., and T. P. Simpson, "Catalytic Cracking and Reforming Processes for Increasing the Yield and Octane Number of Gasoline," *Proc. Third World Pet. Congr.*, The Hague, May–June, 1951.
Hughes, E. C., et al., "Co-gelled Chromia-Alumina Catalysts for Naphtha Reforming—II. Evaluation of Hydrocarbon Types and Naphthas," International Congress of Pure and Applied Chemistry, New York, September, 1951.
Oil & Gas J., "Thermofor Catalytic Reforming—An Answer to Demand for Higher Octane Numbers," pp. 357–368, Nov. 17, 1952.
Oil & Gas J., "Thermofor Catalytic Reforming Flow Diagram," pp. 142–143, Mar. 21, 1956.
Payne, J. W., et al., "Thermofor Catalytic Reforming," API Refining Meeting, San Francisco, Calif., May, 1952, reviewed in *Pet. Proc.*, pp. 966–969, July, 1952.
Pet. Proc., "Moving Bed Catalytic Reforming," pp. 601–602, June, 1951.
Pet. Proc., "Thermofor Catalytic Reforming," pp. 73–75, December, 1956.
Pet. Ref., "Thermofor Catalytic Reforming," pp. 216–217, September, 1956.

Fluid-bed Reforming

In catalytic reforming processes using a fluidized-solids-catalyst-bed (Fig. 3-21), continuous regeneration with a separate or integrated reactor is practiced to maintain catalyst activity by coke and sulfur removal. Cracked or virgin naphthas are charged with hydrogen-rich recycle gas to the reactor. A molybdena (10.0%) on alumina catalyst—not materially affected by normal amounts of arsenic, iron, nitrogen, or sulfur—is used.

Operating conditions in the reactor are about 200 to 300 psig, 900 to 950°F at a space velocity of 0.3 to 0.8 (w/hr/w). Recycle gas rates are 4,000 to 6,000 scf per barrel of feed with catalyst-oil weight ratios of 0.5 to 1.5. In the regenerator, 210 to 310 psig and 1000 to 1100°F conditions are used. Feedstock preparation, except for tailoring the boiling range for aromatic production, normally is not practiced.

Fluidized-bed operation with its attendant excellent temperature control prevents over- and under-reforming operations, resulting in more selectivity in the conditions needed for optimum yield of the desired product.

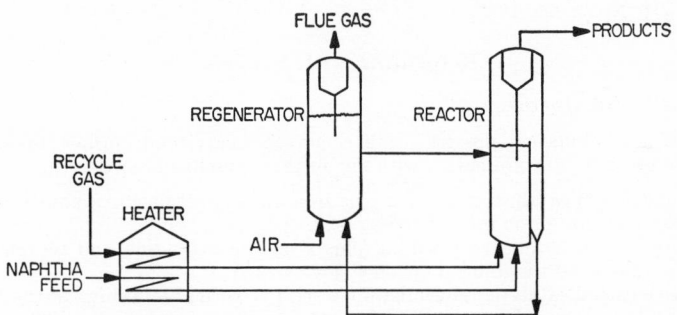

FIG. 3-21. Typical fluid-bed catalytic reforming unit.

Reforming severities resulting in yields of 70.0 to 80.0 vol-% reformate (C_5+) may have clear Research octane numbers of 98.0 to 93.0. Installation costs, depending on unit size, vary between $350 and $650 per barrel of charge. Operating costs, excluding return on investment, vary between 55 and 67 cents per barrel.

Fluid Hydroforming

The Fluid Hydroforming process is licensed by Esso Research and Engineering Co. and M. W. Kellogg Co. It employs the side-by-side arrangement of reactor and regenerator. The process used was essentially the Model I type. A Model II type process, which employs inert pellets along with the fluid catalyst to improve heat transfer, was announced but has not been built commercially as yet in the United States.

The first commercial Fluid Hydroforming unit was placed on-stream at the Destrehan, La., refinery of Pan Am Southern Corp. in June, 1953 (since dismantled). Only two or three units are in operation at present.

A modification of the basic Hydroforming process was the Orthoforming process, which employed the "in-line" stacked single-vessel design. The first commercial unit went on-stream in April, 1955, at the Whiting, Ind., refinery of American Oil Co.; it was destroyed by an explosion in August, 1955.

Fluid Hydroforming references include:

Christopher, R. G., "Orthoforming Unit Produced 100 Clear," *Pet. Ref.*, pp. 164–165, September, 1955.
Draeger, A. A., et al., "Production of High-octane Gasoline Component," *Pet. Ref.*, pp. 71–76, August, 1951.
Ferrell, R. D., "Destrehan Fluid Hydroformer Today," *Pet. Ref.*, pp. 121–124, April, 1955.

Gaylor, P. J., "Fluid Catalyst Technique Applied to Hydroforming," *Pet. Proc.*, p. 1218, December, 1948.
Harang, R. A., "Fluid Hydroforming Units," *Pet. Ref.*, pp. 135–138, October, 1953.
Murphree, E. V., "Fluid Hydroforming," *Advances in Chemistry Series*, no. 5, pp. 58–59, 1951.
Murphree, E. V., "Fundamentals of Fluid Hydroforming," *Pet. Ref.*, pp. 97–101, December, 1951.
Pet. Proc., "Economics in Catalytic Reforming," p. 1056, October, 1951.
Pet. Proc., "Orthoforming," pp. 1188–1189, August, 1955.
Pet. Proc., "Fluid Hydroforming," pp. 1190–1191, August, 1955.
Pet. Proc., "Moving Bed Catalytic Reforming," pp. 601–602, June, 1951.
Pet. Ref., "Fluid Hydroforming," p. 224, September, 1956.
Richards, J. C., "Orthoforming," AIChE Meeting, New York, December, 1954.
Shepardson, R. M., et al., "Fluid Hydroforming Pilot-plant; Results of 50 Barrel per Daily Operation," *Oil & Gas J.*, pp. 110, 136–141, May 19, 1952.
Siebold, J. E., et al., "Fluid Hydroforming; Full Scale Unit First to Use the Fluidized Catalyst Technique," *Oil & Gas J.*, pp. 111, 135–136, May 19, 1952.
Tyson, C. W., et al., "Model II Fluid Hydroformer Uses Shot-catalyst Principle," *Oil & Gas J.*, pp. 122–125, May 17, 1954.
Winslow, R. G., and F. C. Hanker, "Esso Fluid Hydroformer Performs," *Pet. Ref.*, pp. 214–216, November, 1955.

Reforming with Recycle

Iso-Plus Houdriforming

This is a combination process using a conventional Houdriformer, operated at moderate severity, in conjunction with one of three possible alternatives:

1. Conventional catalytic reforming plus aromatic extraction and separate catalytic reforming of the aromatic raffinate (Fig. 3-22A)
2. Conventional catalytic reforming plus aromatic extraction and recycle of the aromatic raffinate to the reforming stage (Fig. 3-22B)
3. Conventional catalytic reforming, followed by thermal reforming of the Houdriformate and catalytic polymerization of the C_3 and C_4 olefins from thermal reforming (Fig. 3-22C)

Typical feedstock to this type of unit consists of the conventional naphtha reforming charge. The use of a Houdry "guard case" permits charging stocks of relatively high sulfur content. Operating conditions are as described for the moderate fixed-bed reforming and aromatic extraction operations. Reformate yields of about 80.0% can result in antiknock quality of 100+ Research octane, clear.

Investment costs of $650 to $750 per barrel of daily charge capacity are typical for this process, with operating costs (excluding investment return) of 85 to 110 cents per barrel.

The first Iso-Plus unit went on-stream in Ravenna, Italy (Societa Azionaria Reffinazione Olii Minerali) in June, 1956. An Iso-Plus Houdriforming unit employing the world's largest Houdriforming unit (45,000 B/D) and an aromatics extraction unit has been successfully operating since 1957 at Tidewater Oil Co.'s refinery at Delaware City, Del.

Houdry Process and Chemical Co. is licensor for Iso-Plus.

Iso-Plus references include:

Heinemann, H., "Achieving 100 Octane Clear with Houdry Reforming Processes," *Oil in Canada*, pp. 110–122, August, 1956.
Heinemann, H., et al., "High Octanes from 'Iso-Plus' via the Thermal Reforming Route," *Pet. Proc.*, pp. 1570–1577, October, 1955.
Hyd. Proc., "Iso-Plus Houdriforming," p. 161, September, 1964.
Noll, H. D., "Iso-Plus Offers Three Routes to Higher Octane Gasoline," *Pet. Ref.*, pp. 135–137, September, 1955.
Noll, H. D., et al., "100 Clear Octane by the New Houdry 'Iso-Plus' Process," *Oil & Gas J.*, pp. 102–105, Mar. 28, 1955.
Oil & Gas J., "Iso-Plus Houdriforming," p. 150, Apr. 3, 1961.
Pet. Proc., "Iso-Plus," p. 1181, August, 1955.
World Pet., "Two New Reforming Processes Offered to Oil Refiners," pp. 30–56, April, 1955.

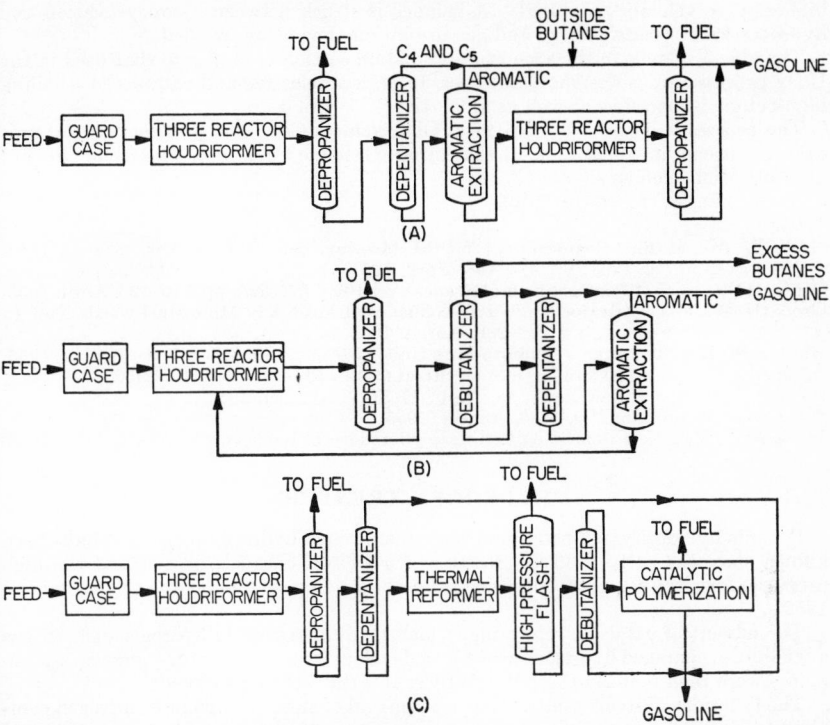

Fig. 3-22. Iso-Plus Houdriforming flow schemes: (A) reforming, plus aromatic extraction and separate reforming of the aromatic raffinate; (B) reforming, plus aromatic extraction and recycle of the aromatic raffinate; (C) reforming, followed by thermal reforming and catalytic polymerization.

Rexforming

Rexforming is a combination process using Platforming and aromatic extraction processes (Fig. 3-23) in which low-octane raffinate is recycled to the Platformer.

A 200 to 400°F virgin naphtha charge can yield typically (80.0 vol-%, approximate) a 98 to 100 Research octane reformate. Operating conditions in the reforming section, because of recycle, can be as much as 50°F lower than conventional Platforming, and

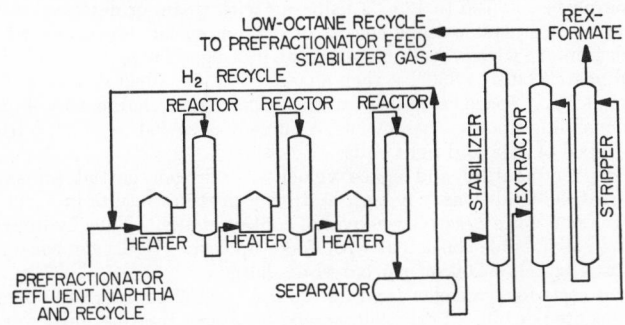

Fig. 3-23. Rexforming, combining Platforming and aromatic extraction.

higher space velocities are used. A balance is struck between hydrocyclization and hydrocracking, excessive coke and gas formation thus being avoided.

The glycol solvent in the aromatics extraction section is similar to that used in the Udex process. It is designed, however, to be less selective and extracts low-boiling high-octane isoparaffins as well as aromatics.

The process is licensed by Universal Oil Products Co. The first commercial unit came on-stream in April, 1956, at the Aurora Gasoline Co.'s (now Marathon Oil Co.) Detroit, Mich., refinery.

Rexforming references include:

Grote, H. W., et al., "Rexforming," WPRA Meeting, San Antonio, Tex., March, 1955, reviewed in *Pet. Proc.*, pp. 494–498, April, 1955.

Grote, H. W., et al., "Rexforming for Octanes over 100," *Pet. Ref.*, pp. 116–120, April, 1955.

Grote, H. W., et al., "Rexforming—Its Product Will Meet Any Motor-fuel Needs Now In Sight," *Oil & Gas J.*, pp. 233–236, Apr. 4, 1955.

Hunter, W. K., "Rexforming Looks to First Unit," *Pet. Ref.*, pp. 145–147, September, 1955.

Oil & Gas J., "First Rexformer Goes on Stream," pp. 106–107, Aug. 13, 1956.

Pet. Proc., "Cosden's New Rexformer," pp. 79–81, November, 1956.

Pet. Proc., "Rexforming," p. 1180, August, 1955.

World Pet., "Two New Reforming Processes Offered to Oil Refiners," pp. 30–56, April, 1955.

HYDROGEN TREATING

Benefits by "purifying" petroleum fractions through hydrogen processing have been known since the early 1930's. Because of a lack of cheap hydrogen and the high pressures formerly required, the process did not develop commercially until the middle 1950's.

The advent of catalytic reforming, which made inexpensive hydrogen-rich off-gas available, encouraged hydrogen-processing development. Operating-pressure requirements were then reduced through advances in catalyst technology.

The functions of hydrogen treating are removal of sulfur compounds, nitrogen compounds, and other impurities; hydrogen saturation of olefins and/or aromatics; and mild hydrocracking.

Hydrogen treating is used extensively to prepare reformer feedstock and to some extent for catalytic cracking feedstock preparation. Product upgrading of middle distillates, cracked fractions, lube oils, gasolines, and waxes by means of hydrogen treating is also widespread. Severity of treatment depends largely upon feedstock properties and the improvement required. Cracked stocks and heavy materials call for severe conditions.

Hydrogen treating is often justified for reasons other than the production of superior fuels. Yields are improved; waste-disposal problems caused by mercaptans, phenols, and thiophenols are almost eliminated; corrosion problems from sulfur, cyanides, and organic acids are reduced. Hydrogen treating also leads the way to sulfur recovery and subsequent reduction of air pollution by sulfur acid gases.

The most common catalyst is cobalt molybdate on an alumina carrier. Regeneration is accomplished at 700 to 1200°F using air with steam or flue gas. Catalyst life is 2 to 5 years. Catalyst poisons consist of carbon, sulfur, nitrogen, and polymers.

Feedstock is mixed with recycle and make-up hydrogen (50 to 90 mol-% H_2 reformer off-gas) and heated (400 to 850°F), then charged to a fixed-bed reactor at 50 to 1,500 psig (see Fig. 3-24). Space velocities are 0.5 to 20 (volume charge per volume catalyst per hour) depending upon process needs. Effluent is cooled, separated from recycle gas, and stripped of H_2S and light ends.

Along with temperature and space velocity, hydrogen partial pressure greatly influences treating conditions. A high hydrogen pressure results in a better rate of hydrogenation and more effective removal of contaminants. Thus hydrogen pressure is kept as high as possible consistent with make-up cost. Hydrogen consumption for treating is usually below 200 scf per barrel of charge.

Investment costs for hydrogen treating are $50 to $250 per bbl, and operating costs are 10 to 20 cents per bbl. Early commercial hydrogen treating units were installed by Esso Standard Oil Co. in the 1930's. World War II plants were pioneered by

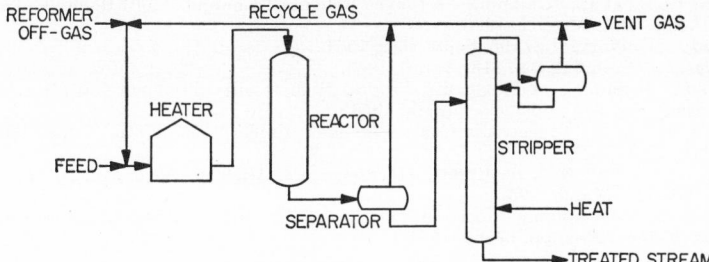

FIG. 3-24. Typical hydrogen treating flow scheme.

Humble Oil & Refining Co. and Shell Development Co. Heavy growth of hydrogen treating commenced in the middle 1950's and has spread through the entire petroleum refining industry.

General literature references include:

Beuther, H., et al., "For Better Hydrodesulfurization Activity of Promoted Molybdenum Oxide-Alumina Catalysts," *Ind. Engrg. Chem.*, pp. 1349–1350, November, 1959.

Bradley, W. E., et al., "Here Today and Here Tomorrow," *Oil & Gas J.*, pp. 194–198, June 8, 1959.

Connally, H. T., "Kerr-McGee's Experience with Catalytic Desulfurization of Cat Cracker Charge Stock," *Oil & Gas J.*, pp. 88–92, Sept. 14, 1959.

Coppoc, W. J., and C. E. Moser, "Use of Hydrogen in the Petroleum Industry," *Ind. Engrg. Chem.*, pp. 639–688, April, 1957.

Davidson, R. L., "Hydrogen Processing," *Pet. Proc.*, pp. 115–138, November, 1956.

Hyd. Proc., "Hydrotreating," pp. 185–196, September, 1964.

Kay, Herbert, "What Hydrogen Treating Can Do," *Pet. Ref.*, pp. 306–318, September, 1956.

Oil & Gas J., "Useful Tips on Refining Problems; Cat Reforming and Desulfurization," pp. 115–122, May 5, 1958.

Sittig, Marshall, and G. H. Unzelman, "Sulfur in Gasoline—An Economic Appraisal," *Pet. Proc.*, pp. 75–95, August, 1956.

Slyngstad, C. E., and F. L. Lempert, "Hydrogen Processing in Modern Refining," *Petro/Chem Engr.*, pp. C13–18, May, 1960.

Unifining

Licensed jointly by Union Oil Co. of California or Universal Oil Products Co., this is a regenerative, fixed-bed, catalytic process to desulfurize and hydrogenate refinery distillates of any boiling range. Contaminating metals, nitrogen compounds, and oxygen compounds are eliminated along with sulfur.

Catalyst is a cobalt-molybdenum-alumina type which may be regenerated *in situ* with steam and air.

Operating conditions are 500 to 800°F and 300 to 800 psig.

The first Unifiners went on-stream for United Refining Co. (Union 750 B/D) at Warren, Pa., and Standard Oil Co. of Ohio (UOP 13,000 B/D) at Lima, Ohio, in 1954.

Literature references include:

Davidson, R. L., "Hydrogen Processing," *Pet. Proc.*, pp. 115–138, November, 1956.

Day, H. W., "Tidewater's Delaware Refinery; Desulfurization Unit," *Oil & Gas J.*, vol. 55, pp. 183–184, May 27, 1957.

Eckhouse, J. G., "Unifining Upgrades Distillate Fuels," *Oil & Gas J.*, pp. 81–83, Aug. 30, 1954.

Gerald, C. F., "Unifining-Platforming in One Unit," *Pet. Ref.*, pp. 216–217, May, 1956.

Gerald, C. F., and L. O. Stines, "Commercial Unifining of Distillates and Gas Oil," *Pet. Ref.*, p. 176, May, 1958.

Grote, H. W., et al., "New Data on Unifining," *Pet. Ref.*, pp. 165–170, April, 1954.

Heinemann, H., "Don't Burn Your Excess Hydrogen," *Pet. Proc.*, pp. 1036–1040, July, 1953.

Hemmen, G. H., et al., "Unifining—A Proven Refining Technique," API Refining Meeting, St. Louis, Mo., May 11, 1955.
Hyd. Proc., "Unifining," p. 196, September, 1964.
Oil & Gas J., "Unifining," p. 113, Apr. 5, 1965.
Poll, H. F., "Unifining—Low Quality Crudes, High Quality Products," WPRA Annual Meeting, San Antonio, Tex., March, 1956.
Thornton, Jr., D. P., "Electronic Control for New Unifier," *Oil & Gas J.*, pp. 110–112, Apr. 9, 1956.
Watkins, C. H., and A. J. de Rosett, "Hydrogen Use High for Some Stocks," *Pet. Ref.*, pp. 201–204, March, 1957.
World Pet., "New Unifining Process a Treating Aid in Specialty Manufacturing Field," *Ann. Refinery Rev.*, pp. 63–65, 1954.

Hydrofining—Esso Research

Licensed by Esso Research & Engineering Co., this is a regenerative, fixed-bed catalytic process to desulfurize, hydrogenate, and improve the general quality of a wide range of refinery streams from gases through waxes.

Catalyst is a combination of cobalt and molybdenum oxides on an extruded alumina support and may be regenerated *in situ* using air mixed with steam or flue gas at 650 to 750°F. Regeneration requires 1 to 4 days. Catalyst life may exceed 2 years, depending upon service.

Operating conditions are 400 to 800°F and 50 to 800 psig. Hydrogen concentration to the unit is kept above 50%, and gas is free of carbon monoxide to maintain catalyst activity.

The first Esso Hydrofiner (2,500 B/SD) was placed in operation at Esso Standard Oil Co.'s Bayonne, N.J., refinery in 1950 to treat wax.

Literature references include:

Baeder, D. L., and C. W. Siegmund, "Hydrofining Is a Natural for Naphthas," *Oil & Gas J.*, pp. 122–124, 126, Feb. 21, 1955.
Carlsmith, L. E., and R. R. Haig, "Hydrogen Treating Helps Wide Range of Stocks," *Pet. Ref.*, pp. 233–235, September, 1957.
Davidson, R. L., "Hydrogen Processing," *Pet. Proc.*, pp. 115–138, November, 1956.
Jones, W. A., "Hydrofining Improves Low-cost-lube Quality," *Oil & Gas J.*, pp. 81–84, Nov. 1, 1954.
Hyd. Proc., "Hydrofining," p. 192, September, 1964.
Lewis, E. H., "Hydrofining Catalytic Cracking Feed Stocks," *Oil & Gas J.*, pp. 95–96, Apr. 18, 1955.
Morbeck, R. C., "Hydrofining of Middle Distillates," *Oil & Gas J.*, pp. 94–98, Jan. 3, 1955.
Oil & Gas J., "Hydrofining," p. 112, Apr. 5, 1965.
Pet. Engr., "Hydrofining," pp. C35–36, May, 1959.
Winslow, W. H., and J. Weikart, "Processing of Heavy High Sulfur Crude Oil in a Low-sulfur Refinery," API Refining Meeting, St. Louis, Mo., May 11, 1955.
Zimmerschied, W. J., "Improving Distillate Fuels by Hydrofining," API Refining Meeting, St. Louis, Mo., May 11, 1955.

Ultrafining

Licensed by Standard Oil Co. (Indiana) this is a regenerative, fixed-bed, catalytic process to desulfurize and hydrogenate refinery stocks from naphthas through lube stocks.

Catalyst is cobalt-molybdenum on alumina made to Standard Oil Co. (Indiana) specifications and may be regenerated *in situ* using an air-steam mixture. Regeneration requires 10 to 20 hr and may be repeated 50 to 100 times for a given batch of catalyst. Catalyst life is 2 to 5 years, depending on feedstock.

Usual operating conditions are about 600 to 800°F and 200 to 1,500 psig. The process uses reformer off-gas containing 70 mol-% hydrogen at a rate of 500 to 4,000 scf/bbl. Lower concentrations can be used for mild processing.

The process was first licensed to Great Northern Oil Co. for use at its Pine Bend refinery at St. Paul, Minn.

Literature references include:

Davidson, R. L., "Hydrogen Processing," *Pet. Proc.*, pp. 115–138, November, 1956.
Hyd. Proc., "Ultrafining," p. 195, September, 1964.
Weber, G., "Spotlight on Hydrogen," *Oil & Gas J.*, pp. 126–127, May 16, 1955.
Zimmerschied, W. J., et al., "Improving Distillates by Hydrofining," *Pet. Ref.*, pp. 153–155, May, 1955.

Vapor-phase Hydrodesulfurization

This is a regenerative, fixed-bed, catalytic process that desulfurizes and hydrogenates refinery naphthas boiling up to about 500°F. Olefins may be saturated without hydrogenating aromatics. The process is no longer in commercial use.

Catalyst is tungsten-nickel sulfide and may be regenerated with air and steam followed by sulfiding with hydrogen sulfide to restore activity completely.

Operating conditions are 450 to 700°F and 500 to 750 psig, with a gas rate of 6,000 scf of hydrogen per bbl.

The Vapor-phase Hydrodesulfurization process (11,150 B/D) was first installed at the Shell Oil Co. refinery at Dominguez, Calif., in 1944.

Literature references include:

Abbott, M. D., et al., "Vapor-phase Hydrodesulfurization," *Pet. Ref.*, pp. 118–122, June, 1955.
Abbott, M. D., et al., "Vapor-phase Hydrodesulfurization Applied to Cracked Naphthas and Light Straight-run Intermediates," API Refining Meeting, St. Louis, Mo., May 11, 1955.
Casagrande, R. M., et al., "Selective Hydrotreating over Tungsten Nickel Sulfide Catalyst," *Ind. Engrg. Chem.*, pp. 744–749, April, 1955.
Cole, R. M., and D. D. Davidson, "Hydrodesulfurization of Gasoline Fractions with Tungsten-Nickel Sulfide Catalyst," *Ind. Engrg. Chem.*, pp. 2711–2715, December, 1949.
Davidson, R. L., "Hydrogen Processing," *Pet. Proc.*, pp. 115–138, November, 1956.
Meerbott, W. K., and G. P. Hinds, Jr., "Selective Hydrotreating over Tungsten-Nickel Sulfide Catalyst—Pure Compounds," *Ind. Engrg. Chem.*, pp. 749–752, April, 1955.
Oil & Gas J., "Catalytic Hydrogenation of Octenes to Octanes," p. 50, May 26, 1938.

Trickle Hydrodesulfurization

Licensed by Shell Development Co., this is a fixed-bed, catalytic hydrogenation process for the removal of sulfur, nitrogen, and metal contaminants from middle distillates, catalytic cracker feeds, etc.

Catalysts are generally cobalt-molybdenum or nickle-molybdenum on alumina, both of which may be regenerated to original activity *in situ* with a mixture of steam and air.

Operating conditions are 600 to 800°F and pressures range from a few hundred to about 1,500 psig. Recycle-gas requirements are normally below 1,200 scf per barrel of charge; space velocity varies with feedstock and service.

Shell Refining and Marketing Co. Ltd. of Stanlow, England, put the first Trickle Hydrodesulfurization unit (7,000 B/SD) on-stream in April, 1955.

Literature references include:

Abbott, M. D., "Hydrogen Improves Cat Cracker Feed," *Pet. Ref.*, pp. 161–166, May, 1958.
Davidson, R. L., "Hydrogen Processing," *Pet. Proc.*, pp. 115–138, November, 1956.
Hoog, H., et al., "Shell Hydrodesulfurization Process," API Refining Meetings, New York, May 13, 1953.
Hoog, H., et al., "Desulphurization of Gas Oils by Hydrogenation and by Extraction," *Proc. Third World Petroleum Congr.*, sec. IV, The Hague, 1951.
Klinkert, H. G., and H. M. Penning, "Shell 'Trickle' Hydrodesulfurizer," *Pet. Ref.*, pp. 150–154, September, 1955.
Pet. Engr., "Trickle Hydrodesulfurization," pp. C27–28, January, 1959.

Hydrodesulfurization—Sinclair

Using a catalyst developed by Sinclair Research Laboratories, this is a regenerative, fixed-bed, catalytic process to remove sulfur, nitrogen, oxygen, and metal contaminants from a wide range of refinery streams.

Catalyst is extruded cobalt-molybdenum on alumina. Regeneration to original activity takes 24 to 72 hr with an air-steam mixture.　Numerous regenerations are possible with potential yields of 100 to 250 bbl/lb.

Operating conditions are 580 to 750°F and 200 to 800 psig with hydrogen recycle rates of 500 to 5,000 scf per barrel of feed.　Reformer off-gas with at least 65 mol-% hydrogen is used.

The first unit (10,000 B/D cycle oil) was placed on-stream Dec. 30, 1955, for the Sinclair Refining Co., Marcus Hook, Pa., refinery.

Literature references include:

Davidson, R. L., "Hydrogen Processing," *Pet. Proc.*, pp. 115–138, November, 1956.
Oil & Gas J., "Hydrodesulfurization—Sinclair," p. 136, Apr. 3, 1961.

Diesulforming

Diesulforming is a regenerative, fixed-bed, catalytic process to desulfurize and upgrade refinery naphtha, middle distillate, and gas-oil streams.

Catalyst ($1 per lb) is pelleted molybdenum and may be regenerated *in situ* with an air-steam mixture.　Activity is restored in 2 to 3 days.

Operating conditions are 650 to 800°F and 450 to 550 psig.　Minimum hydrogen recycle rate is 5,000 scf per barrel of feedstock.　Hydrogen concentration in the gas stream, when lowered, reduces desulfurization activity but is still effective at 50 mol-%.

The Diesulforming process was first placed in operation Aug. 12, 1953, at the Husky Oil Co.'s Cody, Wyo., refinery.

Literature references include:

Davidson, R. L., "Hydrogen Processing," *Pet. Proc.*, pp. 115–138, November, 1956.
Odasz, F. B., and J. V. Sheffield, "Diesulforming . . . Husky Oil Company's Desulfurization Process," *Oil & Gas J.*, pp. 203–204, Mar. 21, 1955.
Pet. Ref., "Diesulforming; Husky Oil Co.," p. 283, September, 1958.

Hydropretreating—Houdry

Licensed by Houdry Process and Chemical Co. (usually in conjunction with Houdriformers), this process is also called the "Houdry Guard Case."　It is a non-regenerative, fixed-bed, catalytic process to desulfurize and hydrogenate catalytic reformer charge stock.

Catalyst is either Houdriforming catalyst (containing platinum) or cobalt-molybdenum on alumina made by Houdry.

Operating conditions are 700 to 755°F with a high space velocity of 5 to 10 v/hr/v. Hydrogen produced in Houdriforming (600 to 1,000 cu ft/bbl) is ample to satisfy Hydropretreating of any Houdriformer feed.　Investment and operating costs are generally included with Houdriforming.

The first Hydropretreating units to purify Houdriformer feedstocks were installed at the Crown Central Petroleum Corp. refinery of Houston, Tex., 5,000 B/D; and at the Marathon Oil Co., Robinson, Ill., refinery, 13,000 B/D.

Literature references include:

Davidson, R. L., "Hydrogen Processing," *Pet. Proc.*, pp. 115–138, November, 1956.
Steveson, D. H., and G. A. Mills, "Hydropretreatment of Catalytic Reformer Feed," *Pet. Ref.*, pp. 117–210, August, 1955.
World Pet., "New Guard Case Houdriformer at Crown Central Refinery," pp. 44–46, February, 1955.

Autofining

Licensed by British Petroleum Co. Ltd., Autofining (Fig. 3-25) is a fixed-bed catalytic desulfurization process for a wide range of high-sulfur stocks from gasolines through gas oils.

Catalyst is pelleted cobalt-molybdate oxides on alumina and may be regenerated *in situ* at 800°F with steam-air or inert gas-air mixtures.　The length of a catalyst run varies with the feed material, from 200 to 1,000 hr.　Ultimate life has been reported as over 4 years in some instances.

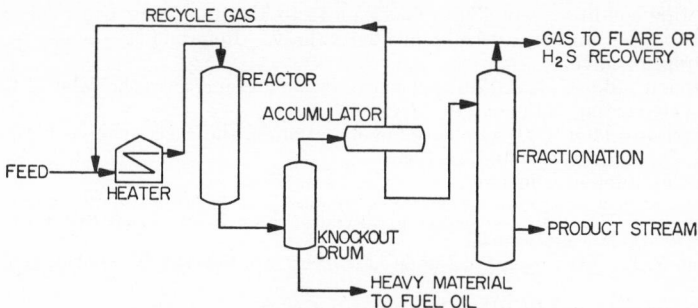

FIG. 3-25. Autofining catalytic desulfurization.

Operating temperatures (700 to 800°F) and pressures (50 to 200 psig) are controlled to give enough hydrogen by dehydrogenation of feedstock naphthenes for conversion of sulfur in the feedstock into H_2S.   No outside hydrogen is needed.   Excess hydrogen in the process gas (80% H_2) is recycled to the desulfurization zone.

The first commercial Autofining plant was installed at the Llandarcy refinery of National Oil Refineries Ltd., Wales, in July, 1952, and another in the BP refinery at Aden in 1954.

Literature references include:

Hyde, H. W., and F. W. B. Porter, "The Autofining Process: The Production of Low Sulfur
    Diesel Oils," *Proc. Fourth World Pet. Congress*, sec. III/C, New York, 1958.
Hyd. Proc., "Autofining," p. 185, September, 1964.
Jones, C. M., "Desulfurizing Process Has Wide Applications," *Pet. Engr.*, pp. C27–28,
    July, 1953.
Lomax, E. L., "Autofining Desulfurization Process," *World Pet.*, pp. 62–63, June, 1952.
Pet. Engr., "The Autofining Process," pp. C31–32, July, 1952.
Pet. Proc., "New Autofining Process," pp. 467–469, April, 1952.
Pet. Proc., "Three New Refining Processes," pp. 683–686, May, 1953.

Hydrofining—BP

Developed by British Petroleum Co. Ltd., this is a regenerative, fixed-bed hydrogen process designed to desulfurize virgin gas oil or blends of virgin gas oil and catalytic cracker cycle oil with little or no molecular breakdown.

Catalyst is pelleted cobalt-molybdenum oxide on alumina type and may be regenerated in 24 hr by a steam-air mixture at temperatures of 800 to 1100°F.

Operating conditions are 760 to 790°F and 500 to 1,000 psig with a gas recycle rate of up to 4,000 scf/bbl.

The process was first placed in operation early in 1955 at Kwinana, Western Australia, for the Australasian Petroleum Refinery Ltd. to produce low-sulfur diesel fuels (5,600 B/D).   M. W. Kellogg Co. constructed the unit using process data from British Petroleum Co.   An additional eight plants have been built or are in design.

Literature references include:

Sutherland, D. A., and F. W. Wheatley, "Desulfurizing by Hydrofining 'Down Under,'"
    Pet. Engr., pp. C37, 40, 42, 44, March, 1956.

Diolefin Hydrogenation

This is a regenerable, fixed-bed, catalytic process to hydrogenate diolefins to mono-olefins in alkylation feedstock with negligible olefin saturation.   Stocks processed are C_4 and C_5 fractions.

It is not now being actively licensed by Shell Development Co.

Catalyst is pelleted nickel sulfide on alumina and may be regenerated *in situ* at 750°F with a mixture of air and steam followed by sulfiding with hydrogen sulfide to restore activity completely.

Operating conditions are 480 to 660°F at 15 to 150 psig using 1-to-1 mol ratio of hydrogen to feed and space velocity of 2 to 8 v/hr/v.   Reformer off-gas may be used as the hydrogen source.

Treatment reduces alkylation acid consumption and increases the value of the feed-stock by converting butadiene to butylene.

The first unit (400 B/D) was placed in operation for the Shell Refining & Marketing Co. Ltd. at Stanlow, England, in 1949.

Literature references include:

Anderson, J. S. H., et al., "Diolefins in Alkylation Feed Stock," *Ind. Engrg. Chem.*, pp. 2295–2301, December, 1948.

Davidson, R. L., "Hydrogen Processing," *Pet. Proc.*, pp. 115–138, November, 1956.

Feed Preparation—Phillips Petroleum Co.

Licensed by Phillips Petroleum Co., the first unit of this hydrogen-treating process (10,000 B/D) was installed by Phillips at its Phillips, Tex., refinery for desulfurization, olefin saturation, and color stability of naphthas and kerosine.

Catalyst used is the molybdenum type.

Literature references include:

Davidson, R. L., "Hydrogen Processing," *Pet. Proc.*, pp. 115–138, November, 1956.

Platreating

Licensed by Universal Oil Products Co., this is a hydrogenation process to refine aromatic concentrates so that pure benzene and toluene may be recovered.   The process is operated in conjunction with Platforming and Udex.

The first Platreater was installed for the Roosevelt Oil & Refining Corp. at Mt. Pleasant, Mich., in January, 1953.

Literature references include:

Daudler, J. E., "New Processes Expand Market for Roosevelt Oil," *Pet. Proc.*, pp. 990–991, July, 1953.

Hydrodesulfurization—Kellogg

Furnished by M. W. Kellogg Co., this is a hydrogen treating process originally designed to desulfurize and purify catalytic reformer feedstock.   Charge stocks now include both virgin and cracked naphthas, kerosine, diesel stocks, furnace oils, and vacuum gas oils.   Sulfur, nitrogen, oxygen, and metal compounds are removed, and olefins are saturated to improve both color and stability.   The catalyst is cobalt molybdate on alumina.

Feedstock and hydrogen are charged to the reactor at from 650 to 800°F and 350 to 1,000 psig, depending upon the charge and the degree of improvement desired. Recycle gas rates are from 400 to 3,000 scf per barrel of charge.   Investment costs range from $75 to $150 per B/SD.

The first unit (3,000 B/D) went on-stream at Pure Oil Co.'s Newark, Ill., refinery to treat reformer feedstock.   Today the total capacity of units designed or constructed by Kellogg is over 465,000 B/SD.

Literature references include:

Davidson, R. L., "Hydrogen Processing," *Pet. Proc.*, pp. 115–138, November, 1956.

Hydrobon

Hydrobon was the name given to the Universal Oil Products Co.'s catalytic hydrogenation process before an arrangement was made with Union Oil Co. of California to license the process jointly under the name Unifining.

The only plant (13,000 B/D) licensed by UOP under the name Hydrobon was for Standard Oil Co. of Ohio at the Lima, Ohio, refinery.   On-stream date for this unit was 1954.

Literature references include:

Davidson, R. L., "Hydrogen Processing," *Pet. Proc.*, pp. 115–138, November, 1956.

Cobalt Molybdate Desulfurization

This process was developed by Union Oil Co. of California and became part of a patent arrangement with Universal Oil Products Co. to license Unifining jointly. The catalyst used by Union for Unifining differs in that it contains silica (Harshaw Chemical Co.).

The first hydrogen process under Union license was a 750-B/D unit for United Refining Co., Warren, Pa., and began operating in 1954.

Literature references include:

Berg, Clyde, "Catalytic Desulfurization of High Sulfur Stocks by the Cobalt Molybdate Process," *Chem. Engrg. Prog.*, pp. 1–12, January, 1947.
Berg, Clyde, "Union's Cobalt Molybdate Process for Refining High Sulfur Stocks," *Pet. Proc.*, pp. 186–189, February, 1952.
Byrns, A. C., et al., "Catalytic Desulfurization of Gasolines by Cobalt Molybdate Processes," *Ind. Engrg. Chem.*, pp. 1160–1167, November, 1943.
Davidson, R. L., "Hydrogen Processing," *Pet. Proc.*, pp. 115–138, November, 1956.

Catalytic Hydrodesulfurization (CHD)

This is a private hydrogen treating process developed and used by Mobil Oil Co. at its own refineries to the extent of over 350,000 B/D capacity. Utilization includes pretreatment of straight-run and thermal naphtha for reformer feed as well as desulfurization and stabilization of sour components of kerosine, diesel oil, and fuel oil.

It was previously identified as the 'Sovafining' process.

Literature references include:

Davidson, R. L., "Hydrogen Processing," *Pet. Proc.*, pp. 115–138, November, 1956.

Hydrotreating—Texaco Inc.

Developed by Texaco Inc., this process is installed at two company refineries (10,000 and 6,500 B/D) for desulfurization of catalytically cracked distillate and heavy cracked gasoline.

Literature references include:

Davidson, R. L., "Hydrogen Processing," *Pet. Proc.*, pp. 115–138, November, 1956.

Gulfining

This is a private hydrogen treating process developed by Gulf Research & Development Co. to upgrade cracked and straight-run distillates and fuel oils by reducing sulfur content and improving carbon residue, color, and general stability, along with slightly increasing the API gravity.

The process involves reacting hydrogen at 600 to 800°F with fuel-oil distillate streams in the presence of an unidentified catalyst at 600 psig.

The first two units, with a combined capacity of 45,000 B/D, were built at the Gulf Oil Co.'s Port Arthur, Tex., refinery and went on-stream January, 1957. Currently (the last half 1965), existing units have a combined capacity of 192,000 B/D and about five additional units are now planned.

Literature references include:

Gilmartin, R. P., "Hydrogen Treatment Improves Fuel Oils," *Oil & Gas J.*, pp. 85–88, Feb. 6, 1956.

Ferrofining

This mild hydrogen treating process was developed by British Petroleum Co. Ltd. to treat distilled and solvent refined lubricating oils. The process eliminates the need for acid and clay treatment (see Fig. 3-26).

Catalyst is a three-component material on an alumina base with life expectancy of 2 years or more. Hydrogen consumption is low and averages 25 scf/bbl. Reformer hydrogen is satisfactory, and recycle is usually unnecessary. Operating conditions may permit the reactor section to be constructed of relatively inexpensive carbon-$\frac{1}{2}$ molybdenum steel.

Operation includes heating the hydrogen-oil mixture and charging to a downflow catalyst-filled reactor. Separation of oil and gas is a two-stage operation, and gas is removed to the fuel system. Oil is then steam-stripped to control the flash point and dried in a vacuum column. A final filtering step removes catalyst fines prior to storage.

Investment costs are in the order of $140 per B/SD, based upon a 3,500-bbl unit. Treating costs are stated as less than clay costs for conventional finishing.

The first commercial unit, 3,100 B/SD went on-stream at SFBP Refinery, Dunkirk, France, November, 1961. Five units with a total capacity of 16,600 B/SD are in

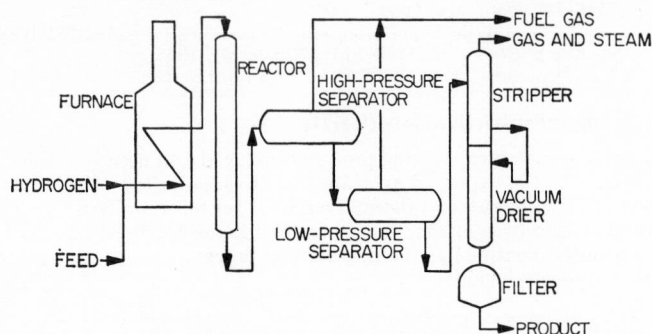

FIG. 3-26. Ferrofining hydrogen treating.

operation at BP refineries in France, Australia, Great Britain, and Germany, with a second unit under construction in BP's Hamburg refinery.

Literature references include:

Dare, H. F., and J. Demeester, "Ferrofining—First Commercial Unit on Stream," NPRA Annual Meeting, San Antonio, Tex., Apr. 2–4, 1962.
Hyd. Proc., "Ferrofining," p. 186, September, 1964.
Oil & Gas J., "Ferrofining," p. 141, Apr. 3, 1961.

ISOMERIZATION

Present isomerization applications in petroleum refining are to provide additional feedstock for alkylation units or high-octane fractions for gasoline blending. Straight chain paraffins (n-butane, n-pentane, n-hexane) are converted to respective isocompounds by continuous, catalytic (aluminum chloride, noble metals) processes. Natural gasoline, light straight-run gasoline, and LP gas can provide feed by first fractionating and/or splitting as a preparatory step. High volumetric yields (95% +) and 40 to 60% conversion per pass are characteristic of the isomerization reaction.

Nonregenerable aluminum chloride catalyst is employed with various carriers in a fixed-bed or liquid contactor. Platinum or other metal catalyst processes utilize fixed-bed operation and can be regenerable or nonregenerable (see Figs. 3-27 and 3-28 for general flow diagrams).

Reaction conditions vary widely depending upon the particular process and feedstock, 100 to 900°F and 150 to 1,000 psig. Residence time in the reactor is 10 to 40 min.

Investment costs are $300 to $500 per bbl, and operation costs range from 30 to 50 cents per bbl.

Isomerization found initial commercial application during World War II for making high-octane aviation-gasoline components and additional feed for alkylation units. Because of lowered alkylate demands after hostilities, all but 6 of an original 30 butane isomerization units (36,650 B/D) were converted or shut down. In recent years, because of greater demand for high-octane motor-fuel blend stock, new butane isomerization units have been installed to increase alkylation feed.

Attractive features for pentane and/or hexane isomerization are products with 93 to 107 Research octane at 3 ml of TEL per gal. Isomerized gasolines are considered excellent, since the road performance of isoparaffins is excellent in the present-day

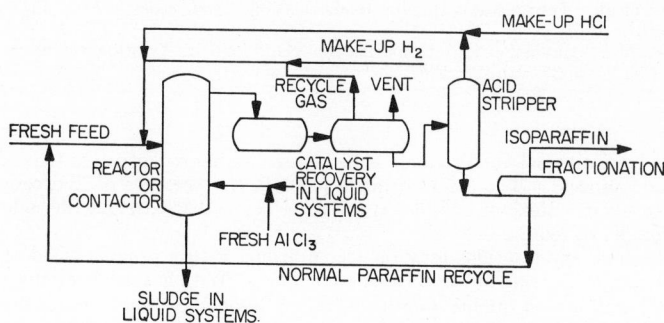

FIG. 3-27. Typical aluminum chloride isomerization flow.

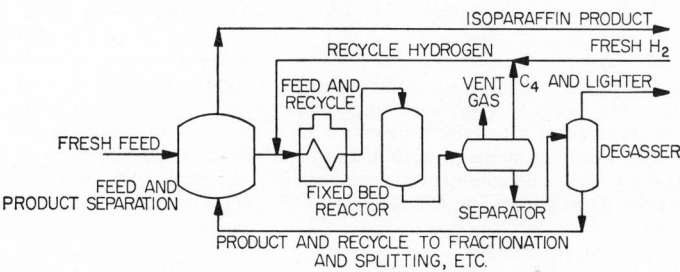

FIG. 3-28. Typical noble-metal isomerization flow.

high-compression automotive engine. Atlas Processing Co. of Shreveport, La., was the first to install a hexane isomerization process (Penex) for the production of a motor-fuel blending component.

General catalytic isomerization references include:

Armistead, Jr., G., "Isomerization and Isoforming—Pentane," *Oil & Gas J.*, p. 80, Oct. 5, 1946.

Chance, V. B., and G. F. Asselin, "Atlas Installs First Hexane, Penex Unit," *Oil & Gas J.*, p. 146, Feb. 16, 1949.

Coulthust, L. J., "Isomerization of Feed Stocks Charged to Alkylation Units," *Oil & Gas J.*, p. 37, Jan. 23, 1941.

Curry, S. W., "The Economics of Producing High Octane Pool Gasoline," WPRA Annual Meeting, San Antonio, Tex., Mar. 25–27, 1957.

Finger, J. S., "Fundamentals of Isomerization for Operators," *Pet. Ref.*, pp. 105–110, January, 1946.

Foster, A. L., "Facts and Possibilities of Alkylation and Isomerization Processes," *Oil & Gas J.*, p. 71, Apr. 15, 1943.

Foster, A. L., "Isomerization a Useful but As Yet Little Used Tool for the Refiner," *Oil & Gas J.*, Feb. 25, 1943, and Mar. 4, 1943.

Frost, A. V., "Improvement of Octane Rating of Gasoline by Isomerization," *Oil & Gas J.*, p. 165, July 29, 1944.

Hyd. Proc., "Isomerization Processes," pp. 171–180, September, 1964.
Lovell, R. G., "Distillation and Isomerization," *Pet. Engr.*, p. 118, September, 1944.
O'Donnell, J. P., "Shell's Wood River Plant . . . 100 Octane Manufacture," *Oil & Gas J.*, pp. 33–34, May 4, 1944.
Oil & Gas J., "Isomerization Processes," pp. 132–136, Apr. 5, 1965.
Parker, F. D., and E. G. Rogatz, "Refinery Processes for War Products," *Pet. Ref.*, pp. 105–115, December, 1943.
Perry, S. F., "Isomerization for the Past Two Years," *Ind. Engrg. Chem.*, pp. 2037–2039, September, 1952.
Perry, S. F., "Isomerization of Light Hydrocarbons," *Pet. Engr.*, pp. 185–188, July, 1947.
Pet. Ref., "Liquid Phase Isomerization," pp. 103–104, February, 1944.
Rabo, J. A., et al., "Pentane and Hexane Isomerization," *Ind. Engrg. Chem.*, pp. 733–736, September, 1961.
Simpson, F. M., "Alkylation and Isomerization Combined in Aviation Gasoline Plant," *Oil & Gas J.*, p. 37, Jan. 23, 1941.

Penex

Licensed by Universal Oil Products Co., Penex is a nonregenerative C_5 and/or C_6 isomerization process. The reaction takes place in the presence of hydrogen and a platinum catalyst. Reactor conditions are so selected that catalyst life is long and regeneration is not required.

Upgrading pentane fractions may be accomplished with a product yield of 99.5% and leaded quality increase of 7 to 8 Research octane. With hexane fractions, product yield is 97% and leaded quality increase about 9 Research octane. Reactor temperatures range from 500 to 900°F; pressures from 300 to 1,000 psi. Hydrogen requirements are low—49 scf/bbl for pentane isomerization and slightly higher for hexane isomerization.

Penex may be applied to many feedstocks by varying the fractionating system. Mixed feeds may be split into pentane and hexane fractions, and respective isofractions separated from each. Normal fractions would be processed in individual C_5 or C_6 Penex systems. Each stream could then be taken to storage or returned to fractionation for separation of isoproduct and recycle. The system can also be operated in conjunction with reforming of the C_7+ naphtha fraction.

The first C_5 Penex Isomerization unit was installed at the Phillips Petroleum Co. refinery at Rice, Tex., and commenced operation March, 1958. The first C_6 unit was installed at the Atlas Processing Co. refinery at Shreveport, La., and went on-stream in July, 1958.

Literature references include:

Belden, D. H., "Penex and Platforming Team Up," *Pet. Ref.*, pp. 149–152, October, 1956.
Belden, D. H., et al., "Upgrading the C_5 and C_6 Paraffins by the Penex Process," *Oil & Gas J.*, pp. 142–146, May 20, 1957.
Chance, V. B., and G. F. Asselin, "Atlas Installs First C_6 Penex Unit," *Oil & Gas J.*, p. 146, Feb. 16, 1959.
Chance, V. B., et al., "How C_6 Isomerization Works for Atlas; Penex Process," *Pet. Ref.*, pp. 216–218, June, 1960.
Grote, H. W., "Isomerize C_5 and C_6 with Penex," *Pet. Ref.*, p. 148, July, 1956.
Hyd. Proc., "Penex," p. 179, September, 1964.
Oil & Gas J., "High Octanes at Low Cost," p. 96, July 23, 1956.
Oil & Gas J., "Penex," p. 135, Apr. 5, 1965.
Pet. Engr., "Penex Catalytic Isomerization," pp. 19–20, February, 1960.
Pet. Proc., "Penex-platforming of Naphthas," pp. 66–67, August, 1956.

Isomerate

Licensed by Pure Oil Co., a division of Union Oil Co. of California, Isomerate is a continuous isomerization process designed to convert pentanes and hexanes into highly branched isomers. A rugged dual-function catalyst is used in a fixed-bed reactor system.

Operating conditions are mild, being less than 750 psig and 750°F. Outside hydrogen is added to the feed along with recycle gas. Usual operation would include fractionation facilities to allow the recycle of normal paraffins almost to extinction.

Example: 200°F end point, light deisopentanized gasoline feedstock of 64.4 clear Research octane is upgraded to 97.0 leaded (3 ml of TEL per gal) Research octane. With essentially pentane feed, the leaded isopentane Isomerate approaches 108 to 109 Research octane. The exact value depends upon feed preparation and degree of recycling.

Literature references include:

Folkins, H. O., and H. Hennig, "A Fresh Look at the Isomerate Process," *Oil & Gas J.*, pp. 120–123, June 6, 1960.
Hyd. Proc., "Isomerate," p. 175, September, 1964.
Oil & Gas J., "Isomerate Process," p. 136, Apr. 5, 1965.
Oil & Gas J., "New Process Expands High Octane Pool—Pure Oil Company's Isomerate Process," p. 86, Apr. 30, 1956.

Iso-Kel

Licensed by M. W. Kellogg Co., Iso-Kel is a fixed-bed, vapor-phase isomerization process using precious-metal catalyst and external hydrogen. A wide variety of feedstocks, including natural gasoline, pentane, and/or hexane cuts, can be processed. Fractionation design can vary. One approach is to yield separate isopentane and isohexane product streams for greater blending flexibility.

Operating conditions include reactor temperatures and pressures from 650 to 850°F and 350 to 600 psig, respectively.

A light naphtha charge showed a 97 vol-% combined stream yield having a 96 leaded (3 ml of TEL) Research octane. One natural-gasoline feed showed 99% yield and 99.8 leaded Research octane.

Payout time is estimated at 19 months for a 3,100-B/D unit incorporated in a 50,000-B/D refinery.

Literature references include:

Decker, W. H., "Get Octanes with This Isom Process; Iso-Kel Process," *Pet. Ref.*, pp. 201–204, April, 1960.
Oil & Gas J., "Kellogg Isomerization (Iso-Kel)," p. 155, Mar. 25, 1957.
Oil & Gas J., "Refiners Handed New Upgrading Tool," p. 76, Jan. 7, 1950.
Oil & Gas J., "Iso-Kel Process," p. 155, Apr. 3, 1961.
Pet. Proc., "The Iso-Kel Isomerization Process," p. 135, January, 1957.
Pet. Ref., "Iso-Kel, The M. W. Kellogg Company," p. 228, September, 1960.
Schwarzenbek, E. F., "Iso-Kel Process," *Pet. Ref.*, p. 182, May, 1957.
Schwarzenbek, E. F., "Isom Process Gives High Road Octanes," *Pet. Ref.*, pp. 215–220, September, 1957.

Isomate

Isomate is a nonregenerative C_5 and C_6 or C_6 naphtha isomerization process using aluminum chloride–hydrocarbon complex catalyst with anhydrous hydrochloric acid as a promotor. Hydrogen partial pressure is maintained to suppress undesirable reactions (cracking and disproportionation) and retain catalyst activity.

Isomerization proceeds at 240 to 250°F and 700 to 800 psig reactor conditions. Feed is saturated with anhydrous HCl in a prereactor absorber, then heated and combined with hydrogen, and charged to the reactor. Catalyst is added to the reactor separately, and the reaction takes place in the liquid phase.

Product is caustic- and water-washed, acid stripped, and stabilized before going to storage in "once-through operation." To provide recycle for a higher octane product, isomers are separated in a product splitter. C_6 naphthas and heavier material are removed from recycle in a rerun tower.

A Research octane of 97 (3 ml of TEL per gal) can be realized on mixed feed using once-through operation; 95 on hexane, once-through; and 106 on hexane (stabilized and rerun) with recycle. Yields of the above product are about 96, 96, and 82 vol-%, respectively. In the last case, an additional 10 to 12 vol-% of naphthenes is produced.

This process was developed by Standard Oil Co. (Indiana), and two commercial plants were operated during World War II to produce high-octane aviation blending stocks.

Literature references include:

Evering, B. L., et al., "Commercial Isomerization of Light Paraffins," *Nat. Pet. News*,
    p. R737, Nov. 1, 1944.
Krane, G., and E. W. Kane, "Isomate Process Adapted to Motor Fuel," *Pet. Ref.*, pp.
    177–182, May, 1957.
Murphy, G. B., "Catalytic Isomerization . . . Supply," *Nat. Pet. News*, pp. R401–402,
    Dec. 24, 1941.
Pet. Ref., "Isomate," p. 172, September, 1952.
Pet. Ref., "Isomate; Standard Oil Company (Indiana)," p. 255, September, 1956.
Oil & Gas J., "Isomate Process; Standard Oil Company (Indiana)," p. 159, Mar. 25, 1957.
Swearington, J. E., et al., "Isomate Process," *Pet. Proc.*, p. 140, October, 1946.
Swearington, J. E., et al., "The Isomate Process," *Amer. Inst. of Chem. Engrs. Trans.*,
    vol. 42, p. 573, August, 1946.

Pentafining

Developed by Atlantic Refining Co. and licensed by Engelhard Industries Inc.,
Pentafining is a regenerable pentane isomerization process using platinum catalyst on a
silica-alumina support and requiring outside hydrogen. A number of process combinations are possible.

For example, with natural gasoline as starting material, feed is depentanized and
heavy material goes to a low-pressure reformer. The pentane stream is split, and the
normal fraction is combined with recycle and make-up hydrogen, heated, and charged
to the reactor (300 to 700 psig and 800 to 900°F). Effluent has recycle hydrogen
removed, is degassed, then fractionated to separate normal and iso (95% purity) cuts.
Catalyst is regenerated with a 500 to 1000°F steam-air mixture.

Reforming may be tied in with the system to provide a convenient hydrogen source.
With natural gasoline feed (12 lb) to the two processes, a 93 to 94% yield and 100 leaded
octane product is realized.

Literature references include:

Grane, H. R., et al., "When to Isomerize Pentane and Hexane Fractions," *Pet. Ref.*, pp.
    172–176, May, 1957.
Hyd. Proc., "Pentafining," p. 180, September, 1964.
Oil & Gas J., "Pentafining," p. 154, Apr. 3, 1961.
Pet. Engr., "Pentafining," pp. C19–20, December, 1959.
Pet. Proc., "Isomerization of Normal Paraffins," p. 101, October, 1956.
Worrell, G. R., "Now 100 Octane from Natural Gasoline," *Pet. Ref.*, p. 138, April, 1956.

Butamer

Licensed by the Universal Oil Product Co., Butamer is designed to convert normal
butane to the isomer under mild operating conditions. Platinum catalyst on a rugged
support is used in a fixed-bed reactor system. A low hydrogen requirement can
readily be satisfied from reformer off-gas.

Operation can be designed for once-through or recycle operation and would normally
be tied in with alkylation unit deisobutanizer operations to provide additional feed
and savings in mechanical equipment costs.

Butane feed is mixed with hydrogen, heated, and charged to the reactor at moderate
pressure. Effluent is cooled prior to light gas separation and stabilization. The
resultant butane mixture is then charged to a deisobutanizer to separate a recycle
stream from the isobutane product. Volumetric yields are over 100% on recycle
operation, since light gas make is very low.

Investment costs for a 2,000-B/SD unit (isobutane product) are stated at $650 per
bbl, typical operating costs at 1.32 cents per gal.

Three units, ranging in capacity from 1,500 to 5,500 bbl, are now in operation.

Literature references include:

Grote, H. W., "Introducing Alkar and Butamer," *Oil & Gas J.*, pp. 73–76, Mar. 31, 1958.
Hyd. Proc., "Butamer," p. 172, September, 1964.
Oil & Gas J., "Butamer," p. 132, Apr. 5, 1965.

Butomerate

Butomerate is licensed by Pure Oil Co., a division of Union Oil Co. of California, and has been specially designed to isomerize normal butane to produce additional alkylation feedstock. Catalyst is stated as a "specially formulated and activated composition containing a small amount of non-noble hydrogenation metal on a high-surface-area support." The process operates with hydrogen recycle to eliminate catalyst coke deposition, but the isomerization reaction can go on for extended periods in the absence of hydrogen.

Feedstock should be dry and comparatively free of sulfur; water and sulfur constitute catalyst-sensitive materials. Pilot operations indicated that 100 ppm of water in the feed would depress the yield 2 to 3%.

Feed is dried, heated, mixed with hydrogen, and sent to the reactor. Operating conditions range from 300 to 500°F and 150 to 450 psig. Effluent is cooled and flashed, and liquid product is stripped of light material. Once-through operation yields product with 60% isobutane owing to low reactor temperature and other factors favoring equilibrium. With recycle operation to take n-butane to extinction, volumetric yield of isobutane is over 100%. Light hydrocarbons formed at conversions close to equilibrium are in the order of 1.5 wt-% or less; pentanes are close to 0.5%.

Literature references include:

Folkins, H. O., and E. L. Miller, "Butomerate Process—Something New in Upgrading Hydrocarbon Fractions," *Oil & Gas J.*, pp. 162–166, June 26, 1961.
Hyd. Proc., "Butomerate," p. 173, September, 1964.

Catalytic Isomerization

Licensed by Phillips Petroleum Co., this is a fixed-bed, vapor-phase butane isomerization process using aluminum chloride (deposited on bauxite) catalyst with HCl as a catalyst promotor.

Fresh catalyst is added to the system by sublimation into vaporized feed. In this manner it is deposited on a bauxite carrier in the pretreater and reactor chambers.

Dry n-butane feed is heated, mixed with recycle HCl and fed to the pretreater, then passed on to the reactor for additional catalyst contact. Reactor effluent is taken through a bauxite guard chamber to remove carry-over aluminum chloride, then liquefied in a series of coolers. Noncondensables are removed from the liquid stream in a vent gas absorber. Next, HCl is stripped and recycled to the feed. Mixed butanes are caustic-washed, sand-filtered, and separated in a splitter. High-purity isobutane goes to storage or is sent to alkylation feed. Normal butane from the splitter has "pentanes-plus" removed before being recycled to raw feed.

Literature references include:

Hyd. Proc., "Catalytic Isomerization," p. 174, September, 1964.

Liquid-phase Isomerization

This is a butane or pentane isomerization process licensed by Shell Development Co., in which liquid feed is contacted with a catalyst consisting of aluminum chloride dissolved in molten antimony trichloride. It can also handle hexanes.

Feed is dried, heated, and passed into a catalyst scrubber column where catalyst is continually purged of side-reaction-formed $AlCl_3$-hydrocarbon complex. A catalyst side stream from the reactor is continuously contacted with the feed, which extracts active material and carries it back to the reactor. The "complex" separates as sludge in the scrubber and is periodically drained. Water, sulfur, and olefins all increase catalyst consumption and are held to a minimum in the feed.

Next, recycle HCl is added to the feed and the mixture is passed to the reactor where it is contacted by the liquid $AlCl_3$-$SbCl_3$ catalyst. In the case of pentane and hexane isomerization, the reaction is effected in the presence of a small amount of hydrogen to suppress disproprotionation reactions. Reactor residence is short for high conversion—60% for butane, 75% for pentane, and about 45% for dibranched hexanes.

After the hydrocarbon stream separates from the catalyst by settling, it is passed to a distillation column where dissolved catalyst is removed. The stream is then cooled, stripped of HCl (HCl to recycle), and then caustic-washed as a final step before fractionation to separate the isomer product.

An important advantage claimed for the process is reduced product fractionation costs because of high conversion. High conversion is also suggested as an attraction to isomerize straight-run C_5-C_6 fractions in once-through operation.

The process was first used in 1944 to produce isopentane for aviation-gasoline blending. In recent years, the process has been adapted to motor-fuel blending stock manufacture. Currently there are eight plants in operation using Liquid-phase Isomerization.

Literature references include:

Evans, H. D., et al., "Isomerize Pentane with This New Reactor System," *Hyd. Proc.*, pp. 171–174, September, 1961.
Hyd. Proc., "Liquid Phase Isomerization—Shell Development Company," p. 178, September, 1964.
Oil & Gas J., "Liquid Phase Isomerization," p. 134, Apr. 5, 1965.
Schoofs, R. J., and E. B. Fountain, "Improved Pentane Isom Process Is Designed for Once-through Operation," *Oil & Gas J.*, pp. 149–152, Apr. 2, 1962.

ALKYLATION

Alkylation developments in petroleum processing in the late 1930's and during World War II were directed toward production of high-octane blending stock for aviation gasoline. The sulfuric acid process (Fig. 3-29) was introduced in the

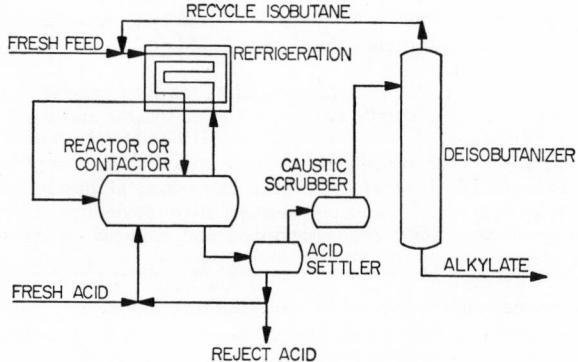

FIG. 3-29. Typical sulfuric acid alkylation flow scheme.

United States in 1938. HF alkylation (Fig. 3-30) was introduced by Phillips Petroleum Co. in December of 1942. Rapid commercialization took place during the war to supply military needs, but many of these plants were shut down upon cessation of hostilities.

In the middle 1950's, aviation-gasoline demand started to decline and motor-gasoline quality requirements rose sharply. Wherever practical, refiners shifted the use of alkylate to premium motor fuel. Alkylate end point was increased for this service, and total alkylate was often used without rerunning. To help improve the economics of the alkylation process and also the sensitivity of the premium gasoline pool, additional olefins were gradually added to alkylation feed. Licensers shifted emphasis on alkylation design from aviation gasoline to motor fuel. New plants were built to alkylate all the propylene and butylenes the refinery produced rather than the butane-butylene stream formerly used; some units have been built specifically to alkylate propylene. More recently *n*-butane isomerization has been utilized to produce additional isobutane for alkylation feed. Fractionation needs of the two operations suggest a close tie in current design. These advances have all combined to place alkylation in an important processing position as a means of obtaining octane numbers for today's premium motor fuels.

The alkylation reaction as practiced in petroleum refining is the union of an olefin (ethylene, propylene, butylenes, and amylenes) with isobutane to yield high-octane branched-chain hydrocarbons in the gasoline boiling range. Olefin feedstock is derived from the gas make of a catalytic cracker. Isobutane is recovered from refinery gases or produced by catalytic butane isomerization. Important in the reaction is a high isobutane-olefin ratio, and unit design is directed to this end.

Chemically alkylation can be accomplished as a thermal, thermal-catalytic, or catalytic reaction.

In thermal-catalytic alkylation ethylene or propylene is combined with isobutane at 125 to 450°F and 300 to 1,000 psig in the presence of metal halide catalysts such as aluminum chloride. Conditions are less stringent in catalytic alkylation. Olefins (C_3, C_4, and C_5) are combined with isobutane in the presence of an acid catalyst

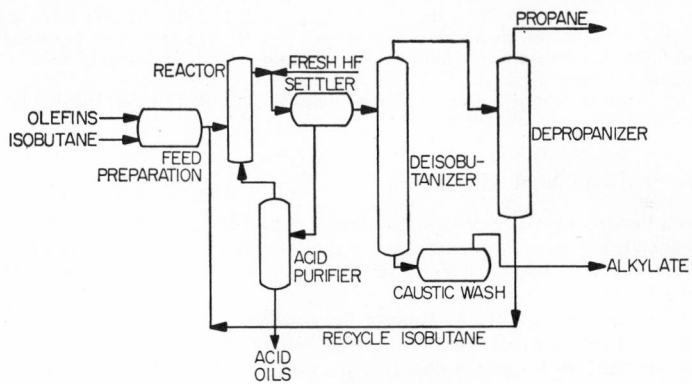

FIG. 3-30. Typical hydrofluoric acid alkylation flow scheme.

(sulfuric or hydrofluoric) at low temperatures and pressures (30 to 105°F and atmospheric to 150 psig). Alkylate yields, based upon olefin feed, range from 150 to 178% depending upon olefin molecular weight. Debutanized motor-fuel alkylates range from 90 to 95 Research octane clear and as high as 115 with 3 ml of TEL per gal. Sensitivity is practically nil and in the case of some leaded alkylates is negative. Road-performance characteristics are excellent with little or no depreciation of Research octane.

General literature references include:

Caesar, P. D., and A. W. Francis, "Mechanism of Low Temperature Alkylation of Isoparaffins," *Nat. Pet. News*, pp. R308 and R310, Oct. 1, 1941.
Foster, A. L., "Anhydrous HF Tomorrow's Refining Process Catalyst," *Oil & Gas J.*, p. 125, Jan. 7, 1943.
Foster, A. L., "Lost Factors in Post-war Alkylation," *Oil & Gas J.*, pp. 96–97, 119–120, 123–124, Oct. 27, 1945.
Foster, A. L., "Facts and Possibilities of Alkylation and Isomerization Processes," *Oil & Gas J.*, pp. 71, 81, 83, 85–86, 88, Apr. 15, 1943.
Gard, E. W., et al., "Alkylation and Its Influence on Utilization of Natural Gasoline," *Oil & Gas J.*, pp. 42–43, Dec. 14, 1939, and *Oil & Gas J.*, pp. 46–48, Dec. 21, 1939.
Garrison, A. H., "Alkylation Plant Takes Olefins Left by Selective Polymerization," *Nat. Pet. News*, pp. R212–213, R216, R219, June 12, 1946.
Goldsby, A. R., and D. K. Beavon, "Expansion of Sulfuric Acid Alkylation for Motor Fuel Production," API Refining Meeting, New York, May 28, 1959.
Hewson, J. E., "HF Alkylation," *Oil & Gas J.*, p. 189, Mar. 22, 1954.
Jones, E. K., "New Unit Gives 110 F-1 O.N. Alkylate," *Pet. Ref.*, pp. 157–158, July, 1959.
Knoble, W. S., and F. E. Hebert, "Key to Propylene Alkylation Found; Mobil Oil Company," *Pet. Ref.*, pp. 101–104, December, 1959.

Kunkel, J. H., "HF Alkylation Unit Placed on Stream in Record Time," *Pet. Engr.*, p. 80, September, 1944.
Mendious, W., "Alkylation Backbone of Aviation Gasoline," *Pet. Engr.*, p. 206, July 1, 1945.
Nat. Pet. News, "Anhydrous Hydrofluoric Acid; Properties as an Alkylation Catalyst," pp. R366–368, Oct; 28, 1942.
Nat. Pet. News, "Properties of Hydrogen Fluoride as Alkylation Catalyst Described," p. R174, May 27, 1942.
Oden, E. C., et al., "Propylene a Valuable Feed Stock for Alkylation," *Pet. Ref.*, pp. 103–108, April, 1950.
Park, M. S., "Revamp of an Alkylation Unit for Increasing Capacity," WPRA Regional Meeting, Casper, Wyo., Sept. 24–25, 1958.
Parker, F. D., and E. G. Ragatz, "Refinery Processes for War Products," *Pet. Ref.* pp. 105–115, December, 1943.
Pet. Proc., "Alkylation Plant Improvements," pp. 744–748, July, 1951.
Resen, F. L., "Four Ideas That Improve Cities Service Alkylation Unit Operation," *Oil & Gas J.*, pp. 178–180, Feb. 25, 1952.
Scott, J. A., and R. M. Cooper, "Economic and Successful Operation Obtained at HF Alkylation Plant," *Oil & Gas J.*, p. 204, Mar. 30, 1946.
Simmons, J. H., "Hydrogen Fluoride, the Catalyst," *Pet. Ref.*, pp. 83–87, July, 1943.
Stiles, S. R., "What's New in Alkylation Process," *Pet. Ref.*, pp. 103–106, February, 1955.
Vermillion, W. L., "Precision Control of Alkylation Unit Deisobutanizers," WPRA Regional Meeting, Houston, Tex., Feb. 15–16, 1961.
Wall, J. E., "How to Improve Alkylate Feed Stock for More and Better Avgas," *Pet. Proc.*, pp. 1438–1443, October, 1952.

Cascade Sulfuric Acid Alkylation

This is a low-temperature process employing concentrated sulfuric acid catalyst to react olefins (propylene, butylenes, and amylenes) with isobutane to produce high-octane aviation- or motor-fuel blend stock (104 to 112 Research octane at 3 ml of TEL per gal).

The process licensed by M. W. Kellogg Co. uses a Cascade reactor to give efficient contact of reactants, autorefrigeration to control temperature, series flow of acid and isobutane reactant with parallel-flow contact with olefins, and a settling zone in the reaction vessel.

Olefin feed is split into equal streams and charged to the individual reaction zones of the cascade reactor. Isobutane-rich recycle and refrigerant streams are introduced in the front of the reactor and pass in series flow through the reaction zones. The olefin is contacted with the isobutane and acid in the reaction zones, which operate at 35 to 45°F and 5 to 15 psig. Reaction temperatures are controlled by continuous refrigeration. Vapors are withdrawn from the top of the reactor, compressed, and condensed. Part of this refrigerant is sent to a depropanizer to control propane concentration in the unit. Depropanizer bottoms and the remainder of the refrigerant are combined and returned to the reactor. Spent acid is withdrawn from the bottom of the settling zone, while hydrocarbons spill over a baffle into a special withdrawal section. Effluent is hot-water-washed with caustic addition for pH control before being successively depropanized, deisobutanized, and debutanized. Alkylate can then be taken directly to motor-fuel blending or be rerun to produce aviation-grade blend stock.

Early installations utilized a jet-type reactor system requiring four reactors in parallel and almost three times as much isobutane recycle for the same alkylate make. The Cascade design has also reduced all auxiliary equipment needs. Acid consumption has been lowered to 0.4 to 0.45 lb of alkylate per gal for a yield of 172 vol-% of butylene feed.

Investment costs are stated at $550 to $950 per barrel of alkylate for units with a capacity of 1,000 to 5,000 B/SD of total alkylate.

Literature references include:

Bolles, W. L., "Alkylation Commands New Interest from Refinery," *Pet. Ref.*, pp. 150–151, September, 1951.
Borthick, G. D., et al., "Two Step Approach to Alkylation Efficiency," *Oil & Gas J.*, p. 88, June 4, 1956.

Chem. Engrg., "H₂SO₄ Alkylation Pictured Flow Sheet," pp. 212–215, September, 1951.
Hyd. Proc., "Cascade Sulfuric Acid Alkylation," p. 177, September, 1962.
Jewell, J. W., and C. C. King, "Sulfuric Acid Alkylation for Gasoline," *Oil & Gas J.*, p. 120, Aug. 16, 1951.
Oil & Gas J., "Cascade Sulfuric Acid Alkylation," p. 157, Apr. 3, 1961.
Pet. Ref., "Improvements in Sulfuric Acid Alkylation Process," p. 93, February, 1951.
Sager, F., "Sulfuric Acid Refining in the Production of Aviation and Motor Fuel," *Pet. Ref.*, pp. 127–130, June, 1946.

Phillips HF Alkylation

Licensed by Phillips Petroleum Co., this process uses regenerable hydrofluoric acid as a catalyst to unite olefins (propylene, butylenes, and amylenes) with isobutane to produce high-octane blend stock. Motor-fuel alkylate from butylene feed with 3 ml of TEL per gal will make 108 Research octane.

Dried charge is intimately contacted in the reactor with acid at 70 to 100°F and a high internal isobutane-to-olefin ratio. Some recent units are as high as 15 to 1. The mixture is separated in a settler, and acid is returned to the reactor. An acid side stream is continuously regenerated to 88% purity by fractionation to remove acid-soluble oils. HF consumption is about 0.20 lb per barrel of alkylate produced.

The hydrocarbon fraction from the settler is deisobutanized, and alkylate is run to motor-fuel storage. In aviation-fuel operation, the alkylate would also be debutanized and rerun before storage. Propane is separated from the isobutane fraction, acid-stripped, and caustic-washed before storage. Isobutane is recycled to the feed stream. Literature references include:

Bolles, W. L., "Alkylation Commands New Interest from Refiners," *Pet. Ref.*, pp. 150–151, September, 1951.
Findley, R. A., "Mechanical Operation of HF Alkylation Units," *Nat. Pet. News*, pp. R326–328, R386, and R388, May 2, 1944.
Frey, F. E., "Commercial Alkylation with Hydrogen Fluoride Catalyst," *Chem. & Met. Engrg.*, pp. 126–128, November, 1943.
Hyd. Proc., "HF Alkylation," p. 169, September, 1964.
Nat. Pet. News, "Special Catalyst Handling Equipment Provided in HF Alkylation Plant," pp. R243–244, June 2, 1943.
Oil & Gas J., "HF Alkylation," p. 139, Apr. 5, 1965.
Peters, W. D., and C. L. Rodgers, "New Alkylation Process," *Pet. Ref.*, p. 126, September, 1955.
Resen, F. L., "Revamping an HF Alkylation Unit," *Oil & Gas J.*, p. 84, July 7, 1952.
Thornton, Jr., D. P., "HF Alkylation Today," *Pet. Proc.*, pp. 488–491, May, 1951.
Vernon, W. F., and C. E. Baechler, "New Design Boosts Alkylate Quality," *Oil & Gas J.*, p. 120, Feb. 20, 1961.

UOP HF Alkylation

This process is licensed by Universal Oil Products Co. and employs regenerable hydrofluoric acid as a catalyst to unite olefins (propylene, butylenes, and amylenes) with isobutane. Recent modifications are tailored to motor-fuel product.

Dried charge is contacted intimately with acid at 60 to 100°F. Recycle isobutane is added to the feed and provides optimum isobutane-olefin ratio. The reaction mixture is separated in a settler. Slip-stream acid is purified by distillation and returned to the reaction zone. The hydrocarbon effluent is stabilized to a desired vapor pressure by means of an isostripper. Overhead from the stripper is isobutane, which is depropanized and recycled. Normal butane is removed from the alkylate as a side stream from the isostripper. The alkylate is then washed with dilute caustic and taken to storage.

A recent UOP innovation has incorporated a new reactor design similar to a counter-current liquid-liquid extractor. HF catalyst is fed to the top of the reactor, and olefin feed and isobutane are introduced at the mid-point. Contact is under countercurrent conditions such that high-molecular-weight products are immediately removed from the reaction zone while alkyl fluorides are returned to the reaction zone.

Leaded alkylate (3 ml of TEL per gal) ranges from 102 to 106+ Research octane,

depending upon feedstock; Motor octane from 104 to 108. Investment costs for a 1,000-B/SD unit (C_{5+} alkylate unit) are stated at \$900 to \$1,000 per bbl; operating costs from 1.75 to 2.35 cents per gallon of C_5+ alkylate depending upon feedstock. UOP units have also been designed using sulfuric acid catalyst.
Literature references include:

Bolles, W. L., "Alkylation Commands New Interest from Refiners," *Pet. Ref.*, pp. 150–151, September, 1951.
Fenske, E. F., "The Modern HF Alkylation Unit," AIChE National Meeting, Tulsa, Okla., Sept. 25–28, 1960.
Gerhold, C. G., et al., "Development of Hydrogen Fluoride Alkylation Process," *Pet. Engr.*, p. 256, July 1, 1944.
Hyd. Proc., "HF Alkylation," p. 170, September, 1964.
Linn, C. B., and A. V. Grosse, "Alkylation of Isoparaffins by Olefins in Presence of Hydrogen Fluoride," *Ind. Engrg. Chem.*, pp. 924–929, October, 1945.
McDonald, G. W. G., "HF Alky Incorporates New Design," *Hyd. Proc.*, pp. 137–140, March, 1962.
Nat. Pet. News, "AvGas Supply Enlarged through HF Alkylation," pp. R131–132, Apr. 29, 1942.
Oil & Gas J., "HF Alkylation," p. 140, Apr. 5, 1965.
Orr, A. R., "Why Cosden Chose HF Alkylation," *Oil & Gas J.*, pp. 102–104, Aug. 1, 1955.
Shanley, W. B., and H. J. Nebeck, "Numerous Improvements in HF Alkylation Will Cut Costs and Reduce Losses," *Oil & Gas J.*, pp. 94–96, 98, Dec. 1, 1945.
Thorton, Jr., D. P., "Three Unusual Features in New HF Alkylation Unit," *Pet. Proc.*, pp. 1570–1573, October, 1954.
Wheeler, Jr., H. K., and R. H. Judice, "HF Alkylation Unit Revamp Contributes to Scrap Drive," *Pet. Engr.*, pp. C5–8, C10, December, 1952.

Effluent Refrigeration Alkylation

Effluent refrigeration is a modification of sulfuric acid alkylation to maintain a high isobutane-olefin feed ratio and is licensed by Stratford Engineering Corp. Reactor effluent is used as a refrigerant to control reaction temperature (45 to 50°F) and at the same time to separate isobutane for recycle. The system can be used along with a conventional deisobutanizer to give up to 10-to-1 isobutane-to-olefin ratio in the reactor feed. Internal isobutane ratios are 700 to 1,000, with 68 to 75 vol-% isobutane provided from the hydrocarbon reactor effluent.

Fresh prepared olefin feed, deisobutanizer recycle, and refrigerant isobutane are contacted with acid catalyst in a liquid phase. Pressure is held at 50 psig or higher in the reactor and settler. Leaving the settler, the hydrocarbon effluent is discharged to a suction trap, and pressure is regulated to 3 to 5 psig. Self-cooling to 20 to 30°F is effected by vaporization of isobutane. Hydrocarbon effluent regulates the reaction temperature by passing through coils of the Stratco reactor, and a large amount of isobutane is vaporized. Isobutane vapors are separated from the liquid effluent, compressed and depropanized, and pass to the feed as recycle along with isobutane from conventional fractionation. Alkylate is treated in a conventional manner with caustic wash, water wash, and stabilization prior to storage.

Recent plants are designed to handle a mixture of propylene and butylene feed to yield debutanized alkylate of 103 to 107 Research octane with 3 ml of TEL per gal. Satisfactory operation has been reported on feedstock containing 97% propylene olefin. With butylene feed, Research octane is as high as 115 leaded. Investment costs range from \$280 to \$330 per B/SD for large units to \$650 to \$750 for small units.

The Texas Co. at Amarillo, Tex., built and operated the first effluent refrigeration unit early in 1954. Over 30 plants are either now operating or will be in the near future. Plants currently under construction have a total capacity of over 50,000 B/SD. Effluent refrigeration may be applied to existing plants as well as new construction.
Literature references include:

Goldsby, A. R., and D. H. Putney, "Texas Company's Amarillo Plant Uses Improved H_2SO_4 Alkylation Process," *Oil & Gas J.*, pp. 104–107, Sept. 19, 1955.
Goldsby, A. R., and D. K. Beavon, "Alkyl Units Revamped for More Output", *Pet. Ref.*, pp. 165–168, June, 1959.

Hyd. Proc., "Effluent Refrigeration Alkylation," p. 168, September, 1964.
Oil & Gas J., "Effluent Refrigeration Alkylation," p. 137, Apr. 5, 1965.
Putney, D. H., "Many Improvements Noted in Alkylation Plant Design," *Pet. Ref.*, pp. 89–94, June, 1951.
Putney, D. H., and O. J. Webb, "Sulfuric Acid Alkylation," *Pet. Ref.*, pp. 105–109, September, 1953.
Putney, D. H., and O. Webb, Jr., "Alkylation Now Goes after Propylene," *Pet. Ref.*, pp. 166–168, September, 1959.
Templeton, P. C., and B. H. King, "Sulfuric Acid Alkylation with Effluent Refrigeration and Its Effect on Refinery Operations," WPRA Regional Meeting, Wichita, Kans., June 21, 22, 1956.

Aluminum Chloride Alkylation (Diisopropyl Alkylation)

Licensed by Phillips Petroleum Co., this is a liquid-phase process to convert ethylene, propylene, and isobutane to high-octane blending stock in the presence of an aluminum chloride–hydrocarbon complex catalyst. Hydrochloric acid is used as a catalyst promotor.

Dried feedstock is passed through agitated reactors at about 120°F (see Fig. 3-31). Catalyst is added to the system by passing isobutane feed through a catalyst make-up

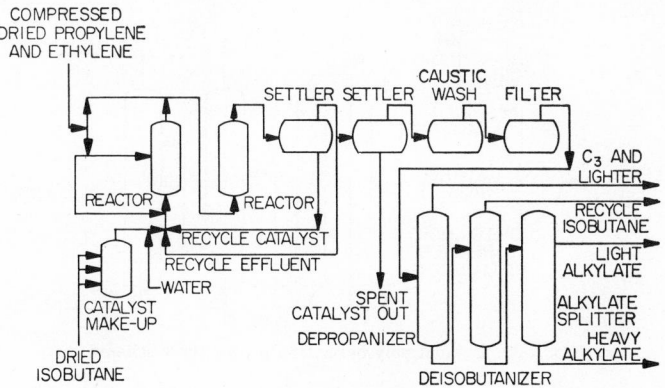

Fig. 3-31. Aluminum chloride alkylation (diisopropyl).

drum. A catalyst promotor (HCl) is used, and concentration is maintained by injecting a controlled amount of water into the reactor. Settlers separate the catalyst from the hydrocarbon stream. Catalyst is recycled to the reactors, as is also some of the hydrocarbon stream to control the reaction temperature. The main effluent stream is caustic-washed, filtered, and fractionated to remove propane and return isobutane to recycle. After the product stream is debutanized, heavy alkylate is removed in a splitter for end-point control of the light product.

Alkylate containing high percentages of diisopropyl is characterized by excellent blending properties and a high response to tetraethyl lead.

The process was first installed by Phillips and was operated during World War II to provide aviation-gasoline blending stock.

Literature references include:

Alden, R. C., et al., "The Story of Diisopropyl," *Oil & Gas J.*, pp. 70–73, Feb. 9, 1946.
Blunck, F. H., and D. R. Carmody, "Catalytic Alkylation of Isobutane with Gaseous Olefins," *Ind. Engrg. Chem.*, pp. 328–330, March, 1940.
Hyd. Proc., "Aluminum Chloride Alkylation," p. 166, September, 1964.
Ipatieff, V. N., "Catalytic Refining Trends," *Oil & Gas J.*, p. 86, Mar. 30, 1939.
Ipatieff, V. N., and A. V. Grosse, "Action of Aluminum Chloride on Paraffins," *Ind. Engrg Chem.*, p. 461, April, 1936.

POLYMERIZATION

Polymerization (Fig. 3-32) is a continuous, catalytic conversion of olefin gases to liquid condensation products.   Feed consists of propylenes and butylenes from cracking operations and may be a mixed feed for motor-gasoline production or selective olefins for codimer, tetramer, and other special products.   Feed is usually caustic- and water-washed to remove sulfur and nitrogen compounds.

The treated, olefin-rich feed stream is contacted with catalyst (sulfuric acid, copper pyrophosphate, phosphoric acid) at 300 to 425°F and 150 to 1,200 psig, depending upon feedstock and product requirement.   The reaction is exothermic, and temperature is usually controlled by heat exchange.   Stabilization and/or fractionation systems separate saturated and unreacted gases from the product.   Conversion of 90 to 97% of olefins is obtained for motor-gasoline operation.   Research octane is 93 to 99 clear.

Catalytic polymerization, originally developed by Universal Oil Products Co., came into use in the middle 1930's and became one of the first catalytic processes of the petroleum industry.   The first commercial plant started operations March, 1937, at the Shamrock Oil & Gas Corp. in Texas.   Development was rapid during World War II for special critical products.

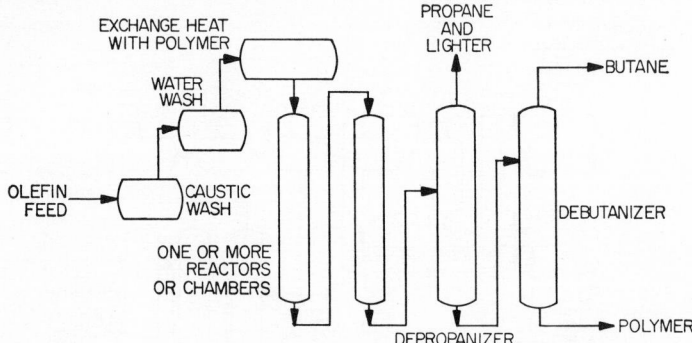

FIG. 3-32. Typical polymerization process flow scheme.

Recent trends are to use butylenes for alkylation rather than polymerization because of the superiority of isoparaffins over olefins in the automotive engine.   Most new construction will probably be paid out on the basis of special polymers from propylene.

General catalytic polymerization references include:

Beychock, M. R., "How to Control Poly Plant Recycle," *Pet. Proc.*, pp. 1403–1404, December, 1951.
Egloff, G., "Polymer Gasoline," *Ind. Engrg. Chem.*, pp. 1461–1467, December, 1936.
Ipatieff, V. N., "Aviation Gasoline by Polymerization and Alkylation of Cracked Gases," *Chem. Engrg. News*, pp. 1367–1368, Nov. 10, 1942.
McAllister, S. H., "Catalytic Polymerization of Butylenes by Sulfuric Acid," *Oil & Gas J.*, pp. 139–140, Nov. 12, 1937.
Norman, H. S., "Steady Increase in Production of Polymerized Motor Fuels," *Oil & Gas J.*, p. 29, Nov. 25, 1937.
Steffens, J. H., et al., "Correlation of Operating Variables in Catalytic Polymerization," *Chem. Engrg. Prog.*, pp. 269–278, April, 1949.
Sutherland, R. E., and J. L. Doegey, "Economics of Alkylation and Polymerization for Producing High Grade Gasoline," *Pet. Proc.*, pp. 679–680, June, 1949.

California Polymerization

Licensed by Chevron Research Co. and Hydrocarbon Research, Inc., this is a process to convert C_3 and/or C_4 olefins to high-octane motor fuel or propylene and benzene to cumene.

Catalyst is phosphoric acid on quartz chips and may be prepared in place in the reactor section. Regeneration consists of washing and steam-drying the quartz, contacting with acid in the reactor, and draining the excess acid.

Reaction temperatures and pressures for gasoline operation are 300 to 375°F and 150 to 600 psig. Heat of reaction is controlled by adding recycle saturates to the feed. Olefin conversions of 90 to 95 wt-% are obtained, yielding 93 (C_3 olefin) to 99 (C_4 olefins) F-1 clear octane debutanized polymer.

The first unit was built in 1937. California Polymerization references include:

Langlois, G. E., and J. E. Walkey, "Improved Process Polymerizes Olefins for High Quality Gasoline," *Pet. Ref.*, pp. 79–83, August, 1952.
Oil & Gas J., "Polymerization Process of California Research Corp.," p. 166, Mar. 25, 1957.

Solid Phosphoric Acid Condensation

Available with either chamber or tubular reactor, this process is licensed by Universal Oil Products Co. and catalytically converts propylenes and/or butylenes to high-octane gasoline or petrochemical polymers.

Pellet kieselguhr impregnated with phosphoric acid catalyst is used in either a chamber or tubular reactor. Exothermic reaction temperature is controlled in the chamber type by using saturates separated from the effluent as recycle to the feed and as quench between catalyst chamber beds. Tubular reactors are temperature-controlled by water or oil circulation around the catalyst tubes.

Reaction temperatures and pressures are 350 to 435°F and 400 to 1,200 psig. Conversion to motor gasoline (95 F-1 clear octane or better) is from 90 to 97 wt-% of olefin charge.

Olefins and aromatics may be united by alkylation for special applications at 400 to 600°F and 400 to 900 psig. A rerun column is required in addition to the usual fractionating.

Solid phosphoric acid condensation references include:

Cotton, E., "New Gulf Polymerization Unit in Port Arthur Texas Plant," *Oil & Gas J.*, p. 119, May 18, 1939.
DeLoach, D. C., "Catalytic Polymerization Unit Is a Feature of McMurrey Improvement Plan," *Oil & Gas J.*, pp. 74–75, Sept. 30, 1948.
Egloff, G., "Catalytic Production of Polymer Gasoline," *Oil & Gas J.*, pp. 140–142, Mar. 19, 1936.
Egloff, G., "Mechanisms of Olefin Polymerization," *Nat. Pet. News*, pp. 65–72, Nov. 20, 1935.
Egloff, G., "Polymerization a Tool of Great Economic Utility," *Nat. Pet. News*, p. 25, Nov. 20, 1935.
Egloff, G., et al., "Motor Fuels from Polymerization," *Oil & Gas J.*, p. 176, Nov. 12, 1937.
Faust, P. H., "What Takes Place in Polymerization," *Pet. Ref.*, pp. 99–104, July, 1944.
Hyd. Proc., "Solid Phosphoric Acid Condensation," p. 184, September, 1964.
Ipatieff, J. N., U.S. Patents 1,993,513 and 2,001,909, 1935.
Ipatieff, J. N., and B. B. Corson, U.S. Patent 2,113,654, 1938.
Ipatieff, J. N., and B. B. Corson, "Refining Gasoline with Solid Phosphoric Acid Catalyst," *Ind. Engrg. Chem.*, p. 1317, November, 1938.
Loehler, W. A., "Polymerization and Its Economics," *Chem. & Met. Engrg.*, p. 412, August, 1938.
Leenhouts, W. J., "Johnson Midget Polymerization Unit," *Oil & Gas J.*, pp. 65–66, Mar. 9, 1939.
O'Donnell, J. P., "Midget Poly Plant Uses Combined Feed," *Oil & Gas J.*, pp. 59–60, Aug. 24, 1939.
O'Donnell, J. P., "Exothermic Reactions Supply Entire Heat Requirements for Poly Units," *Oil & Gas J.*, p. 52, June 29, 1939.
Oil & Gas J., "Phosphoric Acid Condensation," p. 167, Mar. 25, 1957.
Oil & Gas J., "Sunray Makes a Codimer and Polymer Gasoline in Tandem Units," p. 34, Nov. 4, 1943.
Sanders, T. P., "Wilshire Completes First Polymerization Unit on West Coast," *Oil & Gas J.*, p. 48, Sept. 2, 1937.
Shanley, W. B., and G. Egloff, "Midget Polymerization," *Oil & Gas J.*, p. 119, May 18, 1939.

Thornton, D. P., "Cat Poly Unit; Tidewater Refinery at Delaware," *Pet. Proc.*, pp. 112–113, July, 1957.
Van Voorhees, M. G., "Cooperative Catalytic Poly Plant Operates at 1100 to 1200 Pounds," *Nat. Pet. News*, p. R270, Jan. 28, 1939.
Weinert, P. C., and G. Egloff, "Catalytic Polymerization and Its Commercial Applications," *Pet. Proc.*, pp. 585–586, June, 1948.

Bulk Acid Polymerization

Licensed by the Chevron Research Co., this is a process to produce high-octane polymer gasoline from all types of light olefin feed. Olefin concentration can be as high as 95%. Liquid phosphoric acid serves as catalyst.

Olefin feed is caustic- and water-washed and then contacted by liquid phosphoric acid very efficiently in a small reactor. Effluent and acid are separated in a settler, and acid is returned to the reactor through a cooler. Gasoline is first stabilized, then caustic-washed prior to being taken to storage. Acid carry-over is said to be slight, and conversion is in the order of 95 to 98%.

Heat of reaction is removed by circulation through an exchanger prior to contact with olefin feed. Catalyst activity is maintained by continuous addition of fresh acid and withdrawal of spent acid.

Literature references include:

Bethea, S. R., and J. H. Karchmer, "Propylene Polymerization in Packed Reactor—Liquid Phosphoric Acid Catalyst," *Ind. Engrg. Chem.*, pp. 370–377, March, 1956.
Kane, E. D., and G. E. Langlois, "Gasoline from a Liquid Poly Catalyst," *Pet. Ref.*, pp. 173–176, May, 1958.
Pet. Ref., "Bulk Acid Polymerization," p. 267, September, 1958.

DESULFURIZATION

Early desulfurization processes (Fig. 3-33) were similar to clay treating but utilized more severe operating conditions. Bauxite or fuller's earth-type catalysts desulfurize naphthas and light distillates, but middle and heavy distillates are treated with difficulty (cause catalyst coking) and an unstable product is yielded.

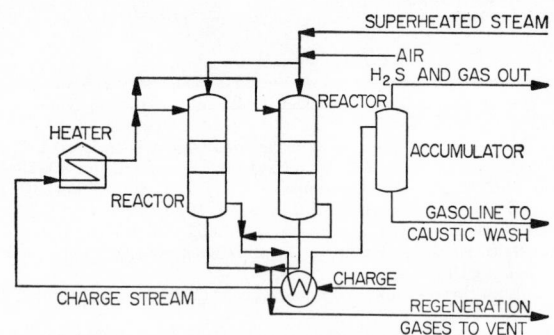

FIG. 3-33. Catalytic desulfurization—typical flow diagram.

The process is generally operated at low pressure and with temperatures in the 700°F range. No attempt is made to hydrogenate, although a small amount of naphthene reforming takes place which supplies some hydrogen for desulfurization. The main effort is to crack the carbon-sulfur bond, but some hydrocarbons are also cracked.

This type of process is being abandoned in favor of desulfurization employing the hydrogen treating principle. Because of the low cost of catalyst and simplicity, the process is still used to advantage in special isolated refining situations.

Cycloversion Desulfurization

Licensed by Phillips Petroleum Co., Cycloversion is a continuous, fixed-bed, catalytic process to desulfurize straight-run or cracked gasoline, kerosine, and other light fractions.

Catalyst may be optionally regenerated with air and steam at about 1200°F. Some units use single reactors and catalyst is discarded when spent. Bauxite catalyst has a life of 5,000 to 20,000 bbl/ton in nonregenerative service and as high as 100,000 bbl/ton when regenerated.

Feed is vaporized and passed over catalyst at 650 to 850°F and pressures up to 50 psig. Space velocity is in the order of 1 v/v/hr. All sulfur compounds except cyclic ones may be decomposed, forming hydrogen sulfide, which is eliminated mainly with vent gas. Remaining hydrogen sulfide is removed by caustic washing.

Product loss is 0.5 to 2.0%. Light straight-run products are almost completely desulfurized, while heavier cracked gasoline desulfurization is in the order of 40 to 60%.

The main application for Cycloversion was originally as a cracking process (980 to 1020°F and 70 to 90 psig) and reforming process (950 to 1050°F and 50 to 200 psig). The same equipment was used as for desulfurization, but with rerun and stabilization facilities. Most of these units have been converted to desulfurization service or are shut down.

The first Cycloversion unit was installed in 1940.

Literature references include:

Brooner, G. M., and M. W. Conn, "Recent Advances in Perco Catalytic Desulfurization," *Oil & Gas J.*, pp. 96–98, 115–121, Oct. 26, 1946.

Buell, A. E., and B. Skinner, "Catalytic Cracking of Texas Panhandle Gas Oil by Cycloversion Process," *Oil & Gas J.*, p. 87, May 5, 1945.

Buell, A. E., and P. M. Waddill, "Conversion of Thermal Cracking to Cycloversion," *Pet. Ref.*, pp. 83–87, October, 1944.

Daniels, V. W., and M. W. Conn, "Perco Cycloversion for the Small Refiner," *Pet. Proc.*, pp. 391–397, May, 1947.

Goldtrap, W. A., and Bradley Skinner, "Refinery Cycloversion for Premium Fuels," *Pet. Engr.*, pp. 174–179, 182, April, 1945.

Helmers, C. J., et al., "Cycloversion Process," *Oil & Gas J.*, p. 88, Dec. 13, 1947.

Oil & Gas J., "Cycloversion," p. 151, Mar. 22, 1947.

Pew, J. E., and A. E. Buell, "Catalytic Desulfurization Solves a Problem in High Sulfur Distillate," *Nat. Pet. News*, pp. R354–358, Oct. 2, 1940.

Schultze, W. A., and R. C. Alden, "Catalytic Desulfurization to Improve Aviation Blending Naphthas," *Ref. Nat. Gaso. Mfr.*, pp. 96–98, November, 1939.

Schultze, W. A., and C. J. Helmers, "Gas Oil Cracking by the Cycloversion Process," *Oil & Gas J.*, pp. 225–235, Apr. 13, 1944.

Gray Catalytic Desulfurization

Developed by the Gray Processes Corp., licensor until December, 1952, the present process patents are held by Pure Oil Co., a division of Union Oil Co. of California. This is a fixed-bed catalytic desulfurization process for gasoline and light distillates, using a solid absorbent-type catalyst such as fuller's earth. The process is continuous with two reactors in alternate regenerating and treating operation.

Gasoline vapors are passed through the catalyst bed at 700 to 750°F, at pressures depending on the system. Space velocity depends upon feedstock and desulfurization required. The product stream is cooled, and gas is separated. Final product treatment may include a caustic wash after stabilization to remove traces of hydrogen sulfide.

When cracked gasoline is being desulfurized, conventional clay pretreatment is needed for the most economical catalyst life, in the range of 1,200 bbl/ton.

Literature references include:

Amero, R. C., and W. H. Wood, "Catalytic Desulfurization of Cracked and Straightrun Gasolines," *Oil & Gas J.*, p. 82, May 24, 1947.

Martin, A. M., and L. Carson, "Catalytic Desulfurization by Use of the Gray Process," *Oil & Gas J.*, pp. 138–144, Mar. 26, 1942.

PART 2.   THERMAL PROCESSES

Thermal processes as described here are those which decompose, rearrange, or combine hydrocarbon molecules by the application of heat without the aid of catalysts. Commercial decomposition processes are designed to prevent coke formation as much as is practical (viscosity breaking or thermal cracking), or to control the process operation whereby coke can be recovered as a marketable product (coking). Molecular rearrangement to effect octane improvement of virgin naphtha feed can be accomplished through the thermal reforming process.

The major variables involved in thermal processing are type of feed, time, pressure, and temperature. Compounds of equal molecular weight can be listed in order of decreasing tendency to crack as follows: paraffins, olefins, diolefins, naphthenes, and aromatics. Heavier fractions generally are easier to crack than lighter fractions. Yield of light products increases with an increase in the time of reaction up to a point. Speed of reaction decreases with increase in reaction time. Pressure is a secondary variable. Its principal effect is to retain the heavier oil molecules in the zone of cracking at the temperature of decomposition as well as to control molecular density, thereby aiding heat transfer. The primary variable, temperature, ranges from 700 to 1100°F to accomplish decomposition at sufficiently high reaction rates for commercial processes.

Catalytic cracking and catalytic reforming for increasing gasoline production and upgrading virgin naphtha quality, respectively, have replaced their thermal counterparts in most of the present-day refineries. As a percentage of crude running capacity, thermal operations have decreased from 32 to 17% in the past decade. Coking and visbreaking (representing more than half of the thermal capacity now in use), however, still continue to play important roles in the modern refinery in that they permit more flexibility and control of the refiner's product-distribution picture.

THERMAL CRACKING

Thermal conversion processes are designed to increase the yield of gasoline obtainable from crude either directly—through production of thermal gasoline components from heavy charge—or indirectly—through production of light olefins suitable for polymerization charge. Thermal cracking processes can also be classified as to the physical state—mixed (or liquid) phase and vapor phase—in which the cracking mechanisms are carried out. Whether the cracking reactions are carried out in mixed or vapor phase depends upon the nature of the feed and the pressure and temperature conditions. Originally vessel construction imposed limits on these latter variables.

The advantages claimed for vapor-phase processes are (1) higher antiknock value of the motor fuel (more olefins), (2) relatively low installation cost, (3) elimination of pressure with consequently more safety appeal, and (4) adaptability to low-boiling-range distillates which cannot be cracked economically in the liquid phase.

The advantages for liquid-phase cracking can be summarized as follows: (1) lower fixed gas production, (2) easier treatment of product gasoline (more paraffinic fuel component), and (3) relatively lower fuel consumption.

Numerous thermal cracking processes—both mixed and vapor phase—have been patented and utilized by the industry since the first commercial cracking unit was placed on-stream at the Whiting Refinery of Standard Oil Co. (Indiana) in March, 1913 (Burton pressure distillation process). No attempt will be made to describe herein each and every thermal process, since the role of these processes in present-day refinery flow is of little or no significance.

The thermal cracking process was utilized initially to convert heavier charge oil into gasoline material. However, today it is almost impossible to justify economically the installation of thermal cracking facilities for gasoline production because of the current market demands for high-antiknock gasoline products. Existing thermal cracking units are being employed for cracking heavy catalytic cycle stocks that would usually go to fuel oil.

General references on thermal cracking include:

Appell, H. R., and C. V. Berger, "Decompositions of Hydrocarbons—Pyrolytic and Catalytic," *Ind. Engrg. Chem.*, pp. 1842–1848, September, 1955.

Armistead, Jr., G., "Thermal Cracking of Residual Feed Stocks to Liquid Residues," *Oil & Gas J.*, p. 94, Apr. 19, 1947.

Arnold, P. M., "Olefin Production by Thermal Cracking of Isobutane," *Oil & Gas J.*, pp. 87–99, July 8, 1945.

Burges, A. E., "Talco Refinery Designed Primarily for Producing Asphalt," *Pet. Engr.*, p. 53, December, 1937.

Clark, C. L., "Design of Cracking-still Tubes," *Oil & Gas J.*, pp. 33–34, Jan. 8, 1942.

Faust, P. H., "What Takes Place in Thermal Cracking," *Pet. Ref.*, pp. 66–68, January, 1943.

Ferris, S. W., "Limiting Compositions in Hydrocarbon Conversion," *Ind. Engrg. Chem.*, pp. 752–759, June, 1941.

Mithoff, R. C., and L. F. Schimansky, "Reaction Chambers on Thermal Cracking Units," API Annual Meeting, San Francisco, Calif., November, 1941.

Nelson, W. L., "Cracking Yields from Heavy Charge Stocks," *Oil & Gas J.*, July 22, 1948.

Nelson, W. L., "Pressure Drop in Pipestills—Effect of Distillation Range of Charge Stock," *Oil & Gas J.*, p. 323, June 24, 1948.

Nelson, W. L., "Cracking Temperature of Hydrocarbons," *Oil & Gas J.*, p. 120, Mar. 18, 1948.

Nelson, W. L., "Flashing Cracking Plant Residuum," *Oil & Gas J.*, Aug. 25, 1945.

Nelson, W. L., "Thermal Cracking Yields," *Oil & Gas J.*, p. 161, Feb. 16, 1959.

Nelson, W. L., "Operating Costs—Thermal Cracking," *Oil & Gas J.*, p. 122, May 4, 1959.

Rosen, R., "Kinetics and Chemistry of Cracking of Hydrocarbons," *Oil & Gas J.*, pp. 49–50, Feb. 13, 1941.

deRosset, A. J., and C. V. Berger, "Thermal and Catalytic Decomposition of Hydrocarbons," *Ind. Engrg. Chem.*, pp. 711–716, August, 1960.

Schwartz, F. G., and C. C. Ward, "Does Thermal Cracking Affect Stability," *Oil & Gas J.*, p. 177, Oct. 7, 1957.

Sung, J. C., et al., "Thermal Cracking of Petroleum," *Ind. Engrg. Chem.*, pp. 1153–1161, December, 1945.

Viscosity Breaking

Viscosity breaking is a comparatively mild thermal cracking process employed both for the reduction of viscosity and for the lowering of boiling range and pour point of heavy straight-run residuum stocks. Lighter distillate stocks can also be charged. Gasoline yield from viscosity breaking is affected primarily by the gravity of the feed, amount and characteristics of the fuel-oil product removed, and the gravity and end point of the gasoline product.

Operating conditions for a typical reduced crude feed are 925 to 975°F and 50 to 100 psig at the furnace outlet. Exit temperature of the reaction chamber is reduced to 830 to 850°F by injection of light gas oil.

Cracked products from the reaction chamber (Fig. 3-34) are fed to a flash distillation chamber. Vapor overhead is separated into a light distillate overhead product and light gas-oil bottoms in a fractionator. Liquid products from the flash chamber pass to a vacuum fractionator, which yields a heavy gas-oil distillate and a residual tar.

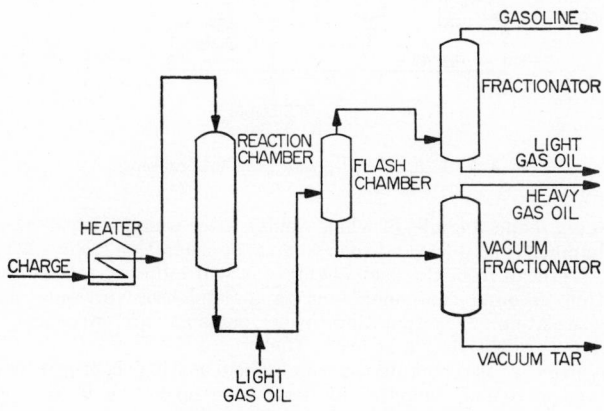

FIG. 3-34. Viscosity breaking.

In 1963 there were some 500,000 B/D of thermal capacity employed in visbreaking operations. This processing is primarily in use for feed preparation of lower-boiling gas oils for subsequent catalytic cracking. The amount of gasoline formed (8 to 10 vol-% of charge) is considered incidental.

Direct operating costs for these units range from 12 to 17 cents per bbl. The investment cost for the average size plant in "1962 dollars" is about $170 per bbl.

Viscosity-breaking references include:

Allen, J. G., et al., "Visbreaking High Vacuum Residua," AIChE Meeting, Kansas City, Mo., May, 1951, reviewed in *Pet. Proc.*, pp. 612–615, June, 1951.
Allinder, F. S., "Handling Reduced Crudes," *Pet. Ref.*, pp. 197–200, November, 1955.
Armistead, Jr., G., "Factors Affecting Yields in Thermal Cracking of Residue Oils to Liquid Residues," *Oil & Gas J.*, pp. 103–108, May 17, 1947.
Beuther, H., et al., "Thermal Visbreaking of Heavy Residues," *Oil & Gas J.*, pp. 151–157, Nov. 9, 1959.
Chem. Engrg., "Two New Petroleum Advances," pp. 196–198, July, 1951.
Egloff, G., and R. F. Davis, "Viscosity Breaking," *Modern Petroleum Technology*, The Institute of Petroleum, London, 1954.
Nelson, W. L., "Yields in Viscosity Breaking," *Oil & Gas J.*, p. 129, Mar. 10, 1962.
Nelson, W. L., "Viscosity of Cracked Fuel Oil," *Oil & Gas J.*, pp. 208–209, Mar. 24, 1952.
Nelson, W. L., "Viscosity Breaking—Unaccounted-for Losses," *Oil & Gas J.*, pp. 114–115, Mar. 31, 1952.
Nelson, W. L., "Operating Cost—Viscosity Breaking," *Oil & Gas J.*, p. 105, Mar. 7, 1960.
Wagner, C. R., "Viscosity-breaker Coil Operation," *Oil & Gas J.*, November, 1943.
Williams, R. W., and J. M. Naugle, "Vacuum Distillation and Vis-breaking," *Oil & Gas J.*, pp. 241–244, Mar. 23, 1950.

Mixed-phase Cracking

Mixed-phase cracking is a continuous thermal decomposition process for conversion of heavy products to gasoline-boiling components. In general, mixed-phase (also called liquid-phase) cracking processes employ rapid heating of the charging stock

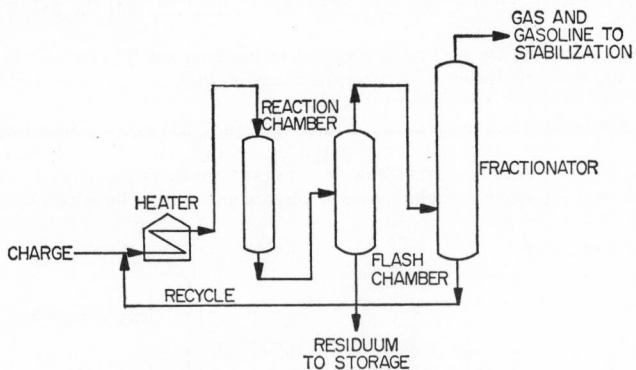

Fig. 3-35. Mixed-phase thermal cracking.

(kerosine, gas oil, reduced crude, or whole crude), after which it is passed to a digester or reaction chamber (Fig. 3-35) and then to a vapor separation tower where the vapors are cooled. Overhead from the flash chamber is sent either to a fractionating tower for specification gasoline components or to a condenser (pressure distillate) for eventual redistillation. Fractionating-tower bottoms are recycled, and flash-chamber bottoms are withdrawn as heavy fuel.

Pressure in mixed-phase cracking is normally held at 350 psig or greater to maintain a homogeneous phase and thereby eliminate coke formation in the tubes. Temperatures range between 750 and 900°F, depending on the nature of the charge.

A high ratio of liquid to vapor is maintained, and increasing temperature necessitates increase in pressure to prevent excessive vaporization.

Mixed-phase cracking references include:

Boelter, L. M., and R. H. Kepner, "Pressure Drop Accompanying Two-component Flow through Pipes," *Ind. Engrg. Chem.*, pp. 426–434, April, 1939.

Dittus, G., and J. Hildrebrand, "A Method of Determining the Pressure Drop for Oil-vapor Mixtures Flowing through Furnace Coils," *Trans. Amer. Soc. of Mech. Engrs.*, April, 1942.

McAdams, W. H., et al., "Vaporization inside Horizontal Tubes," *Trans. Amer. Soc. of Mech. Engrs.*, p. 193, April, 1942.

McReynolds, H., and J. M. Barron, "Thermal Cracking of Fluid Catalytic Cycle Gas Oils," *Pet. Ref.*, pp. 111–116, April, 1949.

Merryfield, G. E., and D. M. Little, "High Destruction Thermal Cracking of Catalytic Cycle Oil," *Oil & Gas J.*, pp. 242–249, Mar. 29, 1951.

Mithoff, R. C., and L. F. Schimansky, "Reaction Chambers on Thermal Cracking Units," *Ref. Nat. Gaso. Mfr.*, pp. 71–75, November, 1941.

Nelson, W. L., "Density of Vapors in Cracking Oils," *Oil & Gas J.*, p. 176, May 13, 1948.

Nelson, W. L., "Yields in Non-coking Thermal Cracking," *Oil & Gas J.*, p. 183, Feb. 25, 1952.

Nelson, W. L., "Thermal Cracking Yields and Rates," *Oil & Gas J.*, pp. 145–146, Oct. 18, 1951.

Nelson, W. L., "Relation of Cracking Yields to Viscosity of Cracked Fuel Oil," *Oil & Gas J.*, p. 88, June 29, 1950.

Nelson, W. L., "Thermal Cracking of Catalytic Cycle Stocks," *Oil & Gas J.*, pp. 75–76, Feb. 4, 1952.

Oil & Gas J., "UOP Reduces Royalty to 3 Cents a Barrell," p. 108, Oct. 21, 1944.

Oil & Gas J., "Thermal Cracking Processes," p. 181, Mar. 23, 1950.

Uren, L. C., et al., "Flow Resistance of Gas-Oil Mixtures through Vertical Pipes," *Amer. Inst. of Mech. Engrs. Tech. Publ.* 252, 1929.

Vapor-phase Cracking

Historically, vapor-phase cracking processes were installed for gasoline production, but even in the early days of thermal cracking, relative process economics favored use

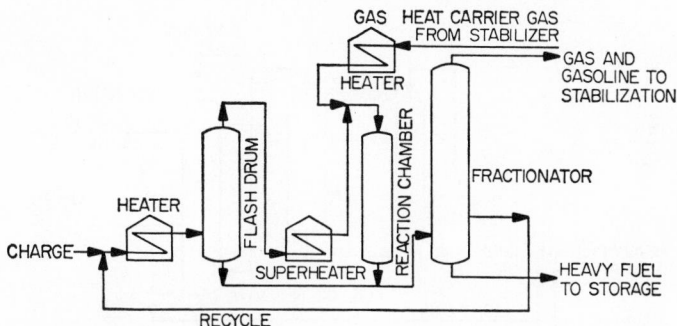

FIG. 3-36. Vapor-phase thermal cracking.

of mixed-phase units. Hard carbon (coke) was often deposited in vapor-phase unit heater tubes, causing failures. Compared with mixed-phase cracking, relatively larger reactors were required for these processes.

Vapor-phase cracking (Fig. 3-36) is a high-temperature, low-pressure thermal conversion process. Under these conditions, dehydrogenation reactions increase with the resultant higher yield of olefins and aromatics.

Operating conditions for vapor-phase reactions include pressures generally below 50 psig, temperatures between 1000 and 1100°F, and cracking times as low as 1 sec. Feedstocks can vary from gaseous hydrocarbons to gas oil.

Vapor-phase cracking is still successfully employed for the production of hydrocarbons—olefins and aromatics—for the petrochemical industry.
Vapor-phase cracking references include:

Berents, L. I., and V. E. Glushnev, "Vapor Phase Oxidizing Cracking of Oil Products," *Nat. Pet. News, Technical Section*, Nov. 7 and Dec. 5, 1945.
Gaylor, P. J., "Pyrolysis of Propane Gives Unsaturates Suitable for Synthetic Drying Oils," *Pet. Proc.*, p. 1009, September, 1951.
Heilman, H. H., "High Temperature Thermal Cracking of Petroleum Fractions," *World Pet.*, p. 86, November, 1947.
Kilpatrick, M. O., et al., "Phillips Pebble Heater," WPRA Annual Meeting, San Antonio, Tex., March, 1954.
Knaus, J. A., and J. Hanisian, "Olefin Production from Liquid Petroleum Stocks," API Refining Meeting, New York, May 27, 1959.
Linden, H. R., "Production of Oil Gas—Thermal Cracking," *Pet. Proc.*, pp. 1389–1396, December, 1951.
Linden, H. R., and R. E. Peck, "Gaseous Product Distribution in Hydrocarbon Pyrolysis," *Ind. Engrg. Chem.*, pp. 2470–2474, December, 1955.
Linden, H. R., et al., "Production of Natural Gas Substitutes by Thermal Cracking of Natural Gas Liquids," *Ind. Engrg. Chem.*, pp. 2475–2478, December, 1955.
Nat. Pet. News, Technical Section, "Acetylene and Other War Chemicals Made by Cracking in a Regenerative Furnace," pp. R270–276, Aug. 26, 1942.
Nelson, W. L., "Naphtha Cracking for Butadiene," *Oil & Gas J.*, p. 91, July 8, 1948.
Pet. Ref., "New Ethylene Process Uses Naphtha Pyrolysis," pp. 122–123, July, 1951.
Pet. Ref., "Catarole Process Produces High Yield of Olefin Gas and Aromatics," pp. 154–155, July, 1952.
Schultz, Jr., E. B., et al., "Pyrolysis of Crude Shale Oil," *Ind. Engrg. Chem.*, pp. 2479–2482, December, 1955.

Selective Cracking

This thermal conversion process scheme takes advantage of the fact that the optimum conditions for efficient cracking vary with the boiling range and/or hydrocarbon make-up of the feed.   Refractory stocks are cracked for a long period of time or at a higher temperature.   Less stable stocks are cracked at lower temperatures.

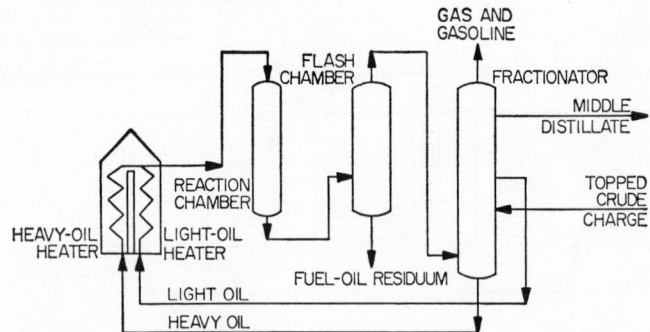

FIG. 3-37. Selective thermal cracking.

In a two-coil operation (Fig. 3-37), employing light and heavy oil from a topped crude charge, the operating conditions in the heavy oil heater might be 920 to 960°F and 300 to 500 psig outlet pressure.   For lighter gas-oil cracking, a temperature of 950 to 990°F and a pressure range of 500 to 700 psig might be employed.   When a wide-boiling-range material is employed, selective cracking eliminates the accumulation of the refractory low-boiling material in the recycle when the total charge is handled at low temperatures.   Likewise, this process eliminates the coke formation resulting from high-temperature cracking of the high-boiling material.

Relatively high yields of gasoline, middle distillate, and olefinic gases are obtainable from this process. The topped or reduced crude is fed to the bottom of a fractionator; end-point gasoline and middle distillate are removed as overhead and side streams, respectively, from the fractionator. From a lower side cut tray a higher-boiling distillate (including the low-boiling fractions of both recycle and charge stocks) is charged to the light oil heater. The heavy bottoms from the fractionating column are the charge to the heavy oil heater.

The heaters may be separate units or separate coils in one furnace. Though more expensive, the former is preferred for ease of control. The outlet of each coil or heater enters the top of a downflow reaction chamber operating without a liquid level. Fuel oil after stripping is withdrawn from the bottom of a flash chamber.

Gasoline yields can vary between 35 and 55% for a given charge (25.0 API Mid-continent) depending upon whether maximum furnace oil or maximum gasoline operations are being practiced in the refinery.

Selective cracking references include:

Holst, Jr., W. W., and A. B. Vought, "Two Stage Cracking Increases Capacity of Palembang Refinery," *Oil & Gas J.*, p. 305, Dec. 30, 1937.
Nat. Pet. News, Technical Section, "First 3-coil Cracking Unit in U.S. Goes on Stream at Sunray Refinery," pp. R675–678, Sept. 5, 1945.
Smoley, E. R., et al., "Theory of Combination Selective Cracking Demonstrated by Commercial Operations," *World Pet.*, p. 94, June, 1937.
Tuttle, R. B., "Addition of One Coil Increases Cracker's Efficiency by Two-thirds," *Oil & Gas J.*, pp. 83–85, Dec. 1, 1945.

Thermal Cracking of Naphtha

In this thermal conversion process, select low-octane fractions of catalytic naphtha are upgraded through thermal decomposition to higher quality material. The total combined catalytic and thermal-catalytic naphtha components yield a product of a quality comparable to that obtained from undercutting without the accompanying yield loss.

This process is designed to upgrade the heavy portion of catalytic naphtha which includes a portion of uncracked virgin charge stock resulting from overlapping at the crude tower fractionation.

Thermal cracking of catalytic naphtha removes naphthenes and paraffins, produces some heavy aromatics by condensation reaction, and produces a large quantity of olefins.

The use of spare thermal capacity as a remedy for reduced catalytic cracked naphtha yields or quality is an answer to those refiners confronted by the overlap problem.

This development in thermal processing is fairly recent, having been announced in September, 1954, by Sun Oil Co. Extensive pilot-plant work has been carried out, and limited plant scale tests made.

References on thermal cracking of naphtha include:

Appell, H. R., and C. V. Berger, "Decomposition of Hydrocarbons—Pyrolytic and Catalytic," *Ind. Engrg. Chem.*, pp. 1842–1848, September, 1955.
McReynolds, H., and J. M. Barron, "Thermal Cracking of Fluid Catalytic Cycle Gas Oils," *Pet. Ref.*, pp. 111–116, April, 1949.
Pollock, A. W., "Thermal Cracking of Cat Naphtha?" *Pet. Ref.*, pp. 127–128, February, 1955.

COKING

Coking is a thermal process for the continuous conversion of heavy, low-grade oils into lighter products. Feed can be such material as reduced crude, straight-run pitch, cracked tars, or shale oil. Depending on the existing market demand, the normal primary purpose of this process is to produce low-carbon gas oil for catalytic feedstock. Generally, gasoline, gas, and coke are secondary products. Calcination of petroleum coke can yield almost pure carbon or artificial graphite suitable for production of electrodes, motor brushes, dry cells, etc.

Basically, coking processes differ from thermal cracking processes in that reaction time of cracking is longer in the former. To accomplish this, drums, chambers, or reaction vessels are employed in the coking processes. Generally, the remaining equipment is similar in coking and thermal cracking operations. Two or more coke drums or chambers are normally provided in the majority of the coking operations in order that decoking can be accomplished in those vessels not on-stream without interrupting the semicontinuous nature of the process. Decoking operation has progressed from manual decoking to cable decoking and to the present-day drilling and hydraulic decoking methods.

Along with visbreaking, the present emphasis in thermal processing is on coking. Current coking capacity in the United States is 520,000 B/D, representing about 20,000 tons/day of coke production. Coking unit sizes range from 4,000 to 40,000 B/D of charge capacity. Depending on the size and type of process, investment costs range from $350 to $700 per bbl. Direct operating costs for coking in general average 32 cents per bbl for a 15,000-B/D unit.

The trend toward use of more heavy crude oils as well as the future prospect of employing shale oil and tar sands as sources of petroleum products should keep coking much in use in future refinery operations.

General references on coking include:

Adee, L. E., "Petroleum Coke Becomes Finished Product," *Oil & Gas J.*, pp. 115–116, Feb. 10, 1958.
Alexander, G. W., "Coke Removal by Steam," *Pet. Ref.*, pp. 301–303, June, 1948.
Armistead, Jr., G., "The Coking of Hydrocarbon Oils," *Oil & Gas J.*, p. 103, Mar. 16, 1946.
Colquette, R. T., and C. W. Peters, "Economics of Residual Petroleum Products," *Oil & Gas J.*, pp. 156–162, Apr. 14, 1952.
Court, W. F., "Hydraulic Decoking of Coke Chambers," *Oil & Gas J.*, pp. 179–184, Nov. 18, 1938.
Jones, D. G., "The Coking of Petroleum Residues," *Journal of Inst. of Pet. Technologists*, p. 895, July, 1935.
Jones, E. L., et al., "Realize a Profit from Refinery Coke," *Oil & Gas J.*, pp. 97–100, May 3, 1954.
Nelson, W. L., "Feasibility of Coking," *Oil & Gas J.*, p. 359, Mar. 23, 1953.
Nelson, W. L., "How to Compute Coking Yields," *Oil & Gas J.*, p. 132, Feb. 15, 1954.
Nelson, W. L., "Allowable Sulfur Content of Metallurgical Coke," *Oil & Gas J.*, p. 137, May 9, 1955.
Nelson, W. L., "Petroleum Coke in Steel Manufacture," *Oil & Gas J.*, p. 104, May 30, 1955.
Nelson, W. L., "Price of Petroleum Coke Will Be Competitive," *Oil & Gas J.*, p. 129, Oct. 31, 1955.
Nelson, W. L., "Properties of Petroleum Coke," *Oil & Gas J.*, Nov. 28, 1955.
Nelson, W. L., "Gas Production in Coking," *Oil & Gas J.*, p. 129, Sept. 3, 1956.
Nelson, W. L., "Operating Costs—Coking," *Oil & Gas J.*, p. 117, July 13, 1959.
Oil & Gas J., "Coke-cutting Tool Saves Time, Power," p. 151, May 13, 1957.
Oil & Gas J., "WPRA Question-Answer Session on Coking," pp. 178–183, Mar. 9, 1959.
Thomas, C. L., "Petroleum Coke and Coking; Progress in Petroleum Technology," *Trans. Amer. Chem. Soc.*, pp. 278–286, 1951.
Watson, K. M., "Cracking Fuel Oil to Coke," *Ref. Nat. Gaso. Mfr.*, p. 652, December, 1938.
Weber, G., "Petroleum Coke," *Oil & Gas J.*, pp. 151–154, Mar. 22, 1954.
Werstler, C. E., et al., "Inside a Coke Drum," *Oil & Gas J.*, pp. 98–100, Aug. 9, 1954.

Fluid Coking

This thermal process utilizes the fluidized-solids technique for continuous conversion of heavy low-grade oils into lighter products. Coking occurs in a thin, liquid film on circulating, fluidized, seed coke agitated by rising gaseous products in the reactor.

A typical feedstock of vacuum residuum yields a product of 50 to 60% gas oil (430 to 1015°F V. T.), naphtha of about 77 Research octane clear, and a coke yield (including about 5 wt-% on feed burned in the heater to heat the fluid coke to the reactor) of about 1.1 to 1.3 times the Conradson carbon in the feed.

Operating conditions are about 900 to 1050°F in the reactor at essentially atmospheric pressure. Fluid coke is preheated to about 1110 to 1200°F in the burner before it enters the reactor. Feed enters the reactor (Fig. 3-38) at 500 to 700°F. For better process control, recycle is practiced not only at the fractionator but also between the fractionator and reactor.   Normally, this latter stream (total feed to fractionator) is introduced to the reactor below the fresh feed point.

The fluid coking process is licensed by Esso Research and Engineering Co.   After it was announced in 1953, the first commercial unit went on-stream in 1954 at Carter Oil Co.'s refinery in Billings, Mont.   In 1960, the Model II Fluid Coker was announced. Basically, the original and Model II designs are the same; however, among other small modifications, the recycle stream is admixed with the fresh feed.   The resultant of these minor changes yields worthwhile savings in the areas of equipment investment (exchangers, pumps, piping, etc.) and instrumentation such that the capital investment and operating costs make this process even more attractive than the original design.

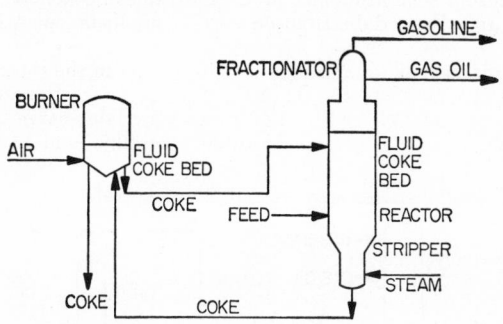

FIG. 3-38. Fluid coking.

The average investment cost for a 15,000-B/D unit is about $440 per bbl; average operating cost is 38 cents per bbl.   Claims for the Model II design are a savings in the investment cost of 35% and in the operating costs of 25%.

Fluid coking references include:

Barr, F. T., and C. E. Jahnig, "Fluid Coking and Fluid Coke," American Institute of Chemical Engineers, New York Section, October, 1954.
Cornforth, R. M., "Fluid Coking Shifts Product Distribution," *Pet. Ref.*, pp. 211–214, September, 1957.
Griffin, Jr., L. I., "Particle Size Control in Fluid Coking," American Institute of Chemical Engineers, Chicago Section, December, 1957.
Hyd. Proc., "Fluid Coking," p. 155, September, 1964.
Johnson, F. B., and R. E. Wood, "Fluid Coking—Yield Studies," WPRA Regional Meeting, Casper, Wyo., September, 1954.
Martin, H. Z., et al., "The Fluid Coking Process," WPRA Annual Meeting, San Antonio, Tex., March, 1954.
Molstedt, B. V., and J. F. Moser, Jr., "A Mechanically Fluidized Reactor," *Ind. Engrg. Chem.*, pp. 21–24, January, 1958.
Pet. Proc., "Coke and Lighter Products from Low Grade Oils by Fluid Coking," pp. 135–137, March, 1956.
Pet. Proc., "Fluid Coke Can Compete for Fuel and Metallurgical Coke Markets," p. 1664, November, 1954.
Pet. Proc., "New Fluid Coking Process," pp. 1316–1317, September, 1953.
Saxton, A. L., et al., "Lowered Investments Improve Economics of Fluid Coking," API Refining Meeting, Detroit, Mich., May 12, 1960.
Stormont, D. H., "Coker Runs a Record 578 Days," *Oil & Gas J.*, pp. 93–96, June 13, 1960.
Stracke, F. H., and F. H. Schiffer, "Fluid Coke—New Fuel for Steam Industry," *Oil & Gas J.*, pp. 75–78, Nov. 28, 1955.
Thornton, Jr., D. P., "Why Carter Likes Its Coker," *Pet. Proc.*, pp. 840–845, June, 1955.
Voorhies, Jr., A., and H. Z. Martin, "Fluid Coking of Residua," API Annual Meeting, Chicago, Ill., November, 1953.
Voorhies, Jr., A.. "The Fluid Coking Process: From Pilot Plant to Large Scale Application," California Natural Gasoline Assoc., Los Angeles, Calif., April, 1955.

Whitcombe, J. A., "The Fluid Coker at Tidewater's Delaware Refinery," *Oil & Gas J.*, p. 173, May 27, 1957.
Wright, R. O., et al., "Development of Model II Fluid Coker," AIChE National Meeting, Tulsa, Okla., September, 1960.

Delayed Coking

Delayed coking is used for converting any type of reduced crude to cracking feedstock.

Feed is heated through exchange or direct-fired heating and charged to a fractionating tower (Fig. 3-39). Tower bottoms are pumped through a pipe furnace and discharged to insulated coke drums. Vapors overhead from the drums are sent to the fractionating tower for separation into product streams of coker gas, gasoline, and gas oil. Depending upon desired distribution yield of products, operation can be recycle or single pass.

Furnace design is such as to minimize coke formation in the tubes. Steam generation is often practiced to recover some of the heat in the fractionation tower.

Typical yields for a 15 to 20 API gravity reduced crude charge having 6.0% Conradson carbon are 65 to 75% gas oil, 15 to 20% gasoline, and 12 to 17 wt-% coke.

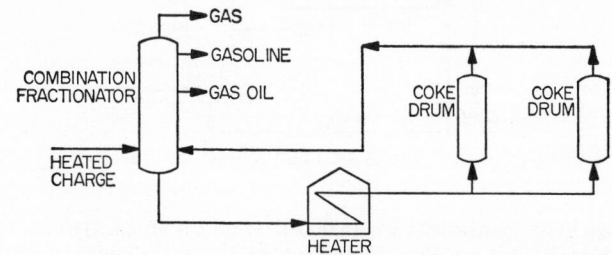

FIG. 3-39. Delayed coking.

The normal range of operating conditions for delayed coking units are 900 to 940°F furnace outlet temperature, 780 to 840°F coke drum temperature, and 10 to 70 psig coke drum pressure.

Numerous delayed coking unit designs are available, differing primarily in the equipment design, which in many instances is tailored to the exact demands of a given refining situation. Delayed coking units have been constructed and/or designed by Foster Wheeler Corp., M. W. Kellogg Co., Lummus Co., and Union Oil Co. of California. In the United States in 1962 there were 31 delayed coking units ranging in size from 3,500 to 35,000 B/D. Total capacity was about 400,000 B/D. Average operating cost for a 15,000-B/D unit is 30 cents per bbl.

References for Foster Wheeler-designed delayed coking units include:

Oil & Gas J., "Delayed Coker and Pour Reduction," p. 185, Mar. 23, 1950.
Pet. Ref., "Delayed Coking," pp. 148–149, September, 1952.
"Thermal Delayed Coking," *Foster Wheeler Bull.*, O-46-1, pp. 12–13.

References for Kellogg-designed delayed coking units include:

Armistead, Jr., G., "The Coking of Hydrocarbon Oils," *Oil & Gas J.*, pp. 103–111, Mar. 16, 1946.
Jewell, Jr., J. W., and J. P. Connor, "Economics of Petroleum Coking for 'Cat Cracking' Feed Preparation," AIChE Meeting, Tulsa, Okla., October, 1950, reviewed in *Pet. Proc.*, pp. 1199–1202, November, 1950.
Maass, Randal, and R. E. Lauterbach, "Delayed Coking Unit Processes Residuum into Gas Oil Charge for TCC Unit," *Pet. Proc.*, pp. 11–16, January, 1947.
Oil & Gas J., "Delayed Coking and Visbreaking," pp. 125–127, Mar. 22, 1954.
Pet. Ref., "The Kellogg Delayed Coking Process," pp. 108–109, July, 1953.

References for Lummus-designed delayed coking units include:

Diwoky, R. J., "Continuous Coking of Residuum by the Delayed Coking Process," API Annual Meeting, Chicago, Ill., Nov. 18, 1938.
Foster, A. L., "Coker Solves Heavy Fuel Problems," *Pet. Engr.*, pp. C53–62, April, 1951.
Fuchs, O. A., "New 420 Ton/Day Delayed Coking Unit," *Pet. Proc.*, pp. 1058–1062, October, 1950.
Kasch, J. E., and E. W. Thiele, "Delayed Coking," *Oil & Gas J.*, pp. 89–90, Jan. 2, 1956.
Mekler, V., and M. E. Brooks, "New Developments and Techniques in Delayed Coking," API Refining Meeting, New York, May 28, 1959.
Meyer, D. B., and J. C. Webb, "Coker Can Handle High Carbon Stock," *Pet. Ref.*, pp. 155–158, February, 1960.
Oil & Gas J., "Delayed Coking," p. 169, Mar. 19, 1956.
Pet. Ref., "Lummus Delayed Coking Process," pp. 102–103, July, 1953.

References for Union Oil-designed delayed coking units include:

Deal, J. M., "How Union Gets Top Octanes from Two California Plants," *Oil & Gas J.*, pp. 195–211, Mar. 19, 1956.
Stormont, D. H., "Union Whips Heavy Crude Bugaboo," *Oil & Gas J.*, pp. 76–77, Apr. 11, 1955.
Wilson, G. M., "New Refinery Begins with Coker," *Pet. Ref.*, pp. 149–150, July, 1955.

Decarbonizing

The decarbonizing thermal conversion process is designed to maximize coker gas-oil production and minimize coke and gasoline yields. Flow is essentially the same as shown in Fig. 3-39 for delayed coking.

Compared with conventional delayed coking, however, decarbonizing is operated at essentially lower temperatures and pressures. Operating pressures range from 10 to 25 psig. Heater outlet temperatures range from 900 to 910°F, and coke drum top temperatures between 775 and 785°F.

Yields for a 26 API gravity East Texas reduced crude (4.0% carbon residue) were 5.6 vol-% gasoline, 90.6 vol-% gas oil, and 5.0 wt-% coke. For a Mid-continent reduced crude charge (12.2% carbon residue, 14.4 API gravity), yields were 13.0 vol-% gasoline, 70.4 vol-% gas oil, and 16.5 wt-% coke.

Decarbonizing units have been designed by the Blaw Knox Co., Pittsburgh, Pa. The first commercial unit was installed for the McMurrey Refining Co., Tyler, Tex., in May, 1954.

Decarbonizing references include:

Eppard, J. H., et al., "Reducing Fuel Oil by Residuum Coking," *Pet. Ref.*, pp. 98–101, July, 1953.
Farrar, G. L., "Gasoline Production Rises to 60%—Decarbonizing Process Improves Flexibility," *Oil & Gas J.*, pp. 205–207, Mar. 21, 1955.
Gibson, C. E., "Three Years Experience Shows Low Pressure Coking Successful," *Ref. Engr.*, pp. C46–50, March, 1958.
Hood, R. J., "BP's Cat Crackers Get Clean Feed," *Oil & Gas J.*, pp. 107–115, Feb. 15, 1960.
Oil & Gas J., "Decarbonization Process," pp. 132–133, Mar. 22, 1954.
Pet. Ref., "The Blaw-Knox Decarbonization Process," pp. 106–107, July, 1953.

Low-pressure Coking

Low-pressure coking is a thermal conversion process designed for once-through, low-pressure operation. This process is similar to delayed coking operation except that no recycle is normally practiced and coke chamber operating conditions are 25 psig and 815°F.

Water is added to the coking heater feed to prevent furnace coking. Effluent from the coking chambers (Fig. 3-40) passes into a flash chamber. The bottoms therefrom are a small amount of heavy fuel; the overhead is the feed to the fractionator for separation into gas oil and gas and gasoline overhead.

Hydraulic decoking and a gantry crane coke-handling system are incorporated in the

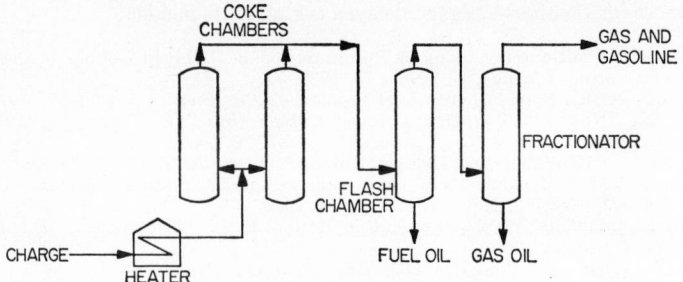

Fig. 3-40. Low-pressure coking.

process design of the unit installed in the McPherson, Kans., refinery of Nat. Co-op Refinery Assoc., placed on-stream in November, 1953.

Low-pressure coking references include:

Pet. Ref., "Low Pressure Coking," p. 157, February, 1954.

Pet. Ref., "Coking," p. 162, September, 1962.

Ward, J. W., and J. M. Holocek, "Chamber Coking at NCRA," *Pet. Ref.*, pp. 157–159, February, 1954.

Weber, G., "Petroleum Coke," *Oil & Gas J.*, pp. 151–154, Mar. 22, 1954.

Continuous Contact Coking

The Continuous Contact Coking process was developed in the late 1940's but is no longer available for licensing and is now considered obsolete. Only one unit was built—at the Sunray, Tex., refinery of Shamrock Oil and Gas Corp. in 1953.

The process employed the mass-flow lift principle for coke circulation and produced a pelleted coke.

Oil-wetted coke particles (seed coke) moved downward as dense bed in the reactor (Fig. 3-41), in which cracking, coking, and drying took place. Circulating coke was

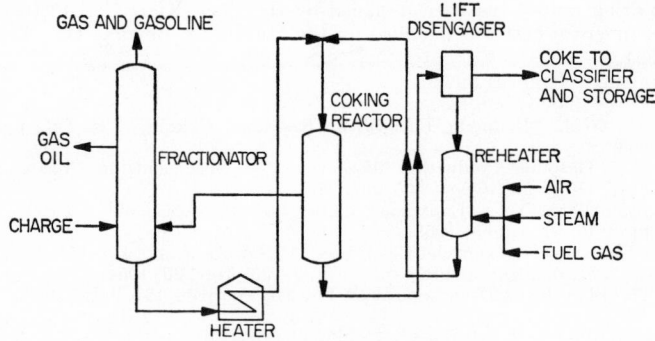

Fig. 3-41. Continuous contact coking.

raised in temperature in the reheater to supply the heat of reaction and also sensible heat to the feed. Reactor pressure powered the dense unagitated coke through the lift line to the disengager section. The ratio of oil to coke for normal operation was 1 lb of oil per 14 lb of coke circulated.

An average reactor temperature of 955°F and 35 psig pressure was employed for a 19.0 API gravity reduced crude charge (11.7% Conradson carbon). The yields obtained were 22.0 wt-% gasoline (76.5 Research octane, clear), 51.0 wt-% gas oil, and 12.5 wt-% coke.

Continuous Contact Coking references include:

Colquette, R. T., and C. W. Peters, "Residual Petroleum Products," *Oil & Gas J.*, pp. 156–174, Apr. 14, 1952.

Mekler, V., et al., "The Lummus Continuous Contact Coking Process," API Annual Meeting, Chicago, Ill., November, 1953, reviewed in *Pet. Proc.*, pp. 1882–1883, December, 1953.

Oil & Gas J., "Continuous Contact Coking," pp. 128–129, Mar. 22, 1954.

Pet. Ref., "Lummus Continuous Contact Coking," p. 104, July, 1953.

Schutte, A. H., "Application of Contact Coking to California Refining," *Oil & Gas J.*, pp. 70–72, Nov. 3, 1949.

Schutte, A. H., and W. C. Offutt, "Continuous Contact Coking," *Pet. Proc.*, pp. 769–775, July, 1949.

Smoley, E. R., and A. H. Schutte, "Continuous Contact Coking of Heavy Charge Material," Athabasca Oil Sands Conference, Canada, September, 1951, reviewed in *Pet. Proc.*, pp. 348–349, March, 1952.

High-temperature Oven Coking

High-temperature oven coking is a semicontinuous thermal conversion process designed to handle high-melting-point asphalt pitch and yield coke and coker gas oil as primary products. Treating the coke for sulfur removal results in a product acceptable as metallurgical grade (about 1.0 wt-% sulfur), even though the residuum feed may contain 4.0 to 5.0 wt-% sulfur.

A high-carbon residuum charge stock is sent to the pitch accumulator (Fig. 3-42) and thence to the heater—outlet conditions being 700°F and 30 psi. Three to four

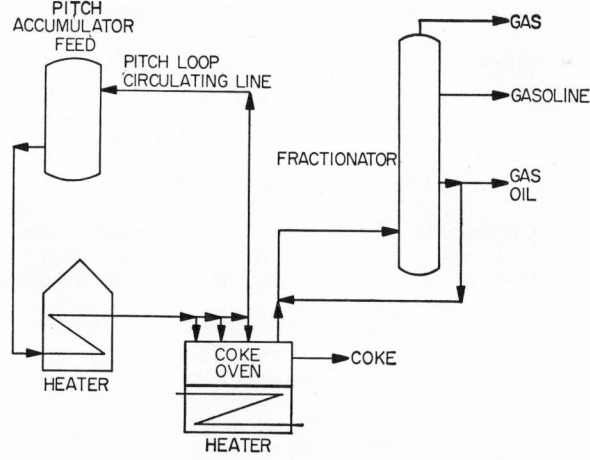

FIG. 3-42. High-temperature oven coking.

hours are required to charge the coke oven (about one-third of the heater effluent is circulated back to the accumulator). Coke ovens are heated via outside gas and maintained at 1800 to 2000°F. Vapors are collected in a main and sent to a fractionator for separation into the gas, gasoline, and gas-oil products (a portion of the gas oil is recycled to flush oil-control valves). After the cycle is complete, a pusher rod removes the coke for further sulfur removal and quenching prior to storage.

Yields for an asphalt-base pitch of 225°F melting point (35% Conradson carbon) were 45% coker gas oil (25 to 28 API); 35 to 37 wt-% coke, 12% coker gasoline (46 API, 66.0 Research octane, clear), and 4% absorption gasoline (70 API). The coke product after treating was 98.0% total carbon with 1.0 wt-% sulfur (pitch sulfur was 4.4 wt-%).

As in the past, the economics of coke production in any given situation depends on the existing market demand and product value. Production of low-sulfur coke from high-sulfur residue certainly contributes favorably to the economics of this refinery process.

High-temperature oven coking references include:

Curran, M. D., "Metallurgical Coke from Petroleum Pitches," WPRA Annual Meeting, San Antonio, Tex., March, 1954.
Curran, M. D., "Reducing Petroleum Residues to High Quality Solid Coke," *Oil & Gas J.*, pp. 100–102, Aug. 18, 1949.
Foster, A. L., "Knowles Ovens Solve Residuum Disposal Problem—Pay Satisfactory Profit," *Nat. Pet. News*, p. 29, Mar. 8, 1933.
Nelson, W. L., "Yields of Coke by Various Processes," *Oil & Gas J.*, p. 56, March, 1956.

OTHER THERMAL PROCESSES

Thermal Reforming

This continuous thermal process is used to convert, through molecular rearrangement, low-antiknock quality gasolines and naphthas into high-octane gasoline components. Secondary products of the process include olefin gases for polymerization

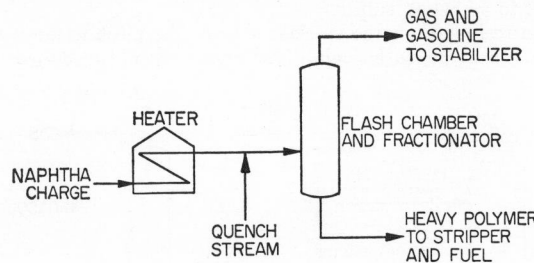

Fig. 3-43. Thermal reforming.

(not produced in catalytic reforming processes) and a heavy polymerized tar for heavy fuel usage. Thermal reforming equipment is similar to that employed in thermal cracking operation. With slight modification refiners have often employed the same equipment for these two processes.

Similar to the catalytic reforming feed, a typical charge is a virgin naphtha having an initial boiling point between 200 and 250°F and a final boiling point varying between 300 and 400°F. Natural gasoline and cracked material have also been employed as charge. A temperature of 950 to 1100°F at a pressure of 400 to 1,000 psig is reached at the exit of the heating section. A naphtha side stream from the fractionator is added to the heater effluent (Fig. 3-43) for quenching purposes to arrest too extensive decomposition reactions.

The thermal reforming process came into commercial use in the United States in 1930. In 1962, 100,000 B/D were listed as U.S. thermal reforming capacity. Except for that capacity employed as supplementary to catalytic reforming operations, the majority of the listed capacity is idle, having been replaced by catalytic reforming.

Process units have been designed by the Universal Oil Products Co., Des Plaines, Ill.

Thermal reforming references include:

Egloff, G., "Thermal Reforming for Olefin Production," *Oil & Gas J.*, pp. 157–163, July 29, 1944.
Feuchter, C. F., "Economics of Thermal Reforming," *Pet. Proc.*, pp. 682–686, April, 1949.
Gary, W. W., and N. R. Adams, "Economics of Reforming and Leading Mid-continent Gasoline," WPRA Meeting, Hot Springs, Ark., April, 1937.
Haensel, V., and M. J. Sterba, "Comparison of Platforming and Thermal Reforming," *Advances in Chemistry Series*, No. 5, pp. 60–75, 1951.

Mase, R. P., and N. C. Turner, "Thermal Reforming Plus Catalytic Polymerization," WPRA Meeting, Wichita, Kans., April, 1940.

McLaurin, N. H., et al., "Hydroforming and Thermal Reforming Operations on Sweet and Sour Heavy Straight Run Naphthas," WPRA Annual Meeting, San Antonio, Tex., March, 1949.

Nelson, W. L., "Octane Number of Reformed Naphthas," *Oil & Gas J.*, p. 118, Dec. 13, 1947.

Pet. Proc., "Improving Yields of Gasoline, Distillates from Pennsylvania Grade Crude Oil," pp. 329–334, April, 1948.

Pet. Ref., "Thermal Reforming Unit Improves Naphtha Octane Rating," p. 141, December, 1946.

Pet. Ref., "Thermal Reforming," pp. 144–145, September, 1952.

Read, D., "Cracking and Thermal Reforming," *Oil & Gas J.*, pp. 72–74, Mar. 4, 1948.

Tighe, H. F., "Method for Correlating Thermal Reforming Yield Data with Operating Conditions," *Pet. Proc.*, pp. 986–992, October, 1948.

Neohexane Alkylation

Neohexane alkylation is a noncatalytic alkylation process developed by Phillips Petroleum Co. It uses ethylene and isobutane as reactants to form neohexane. The process has not been used since World War II.

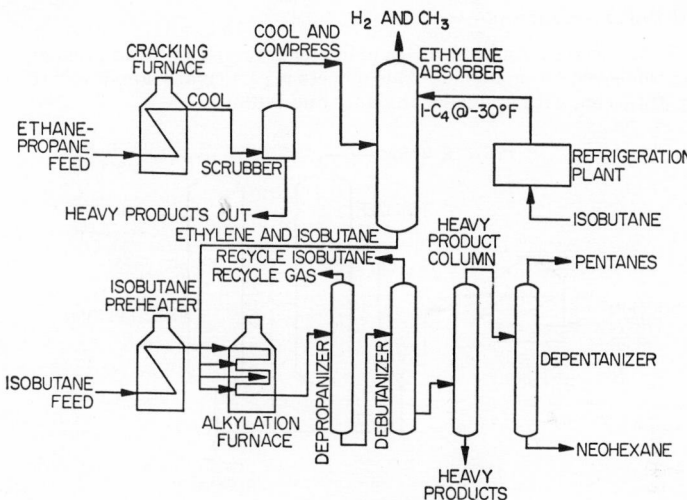

FIG. 3-44. Neohexane alkylation.

The process is not limited to neohexane production. By using different paraffin and/or olefin feeds, a variety of blending stocks may be produced.

A mixture of ethane and propane is cracked (Fig. 3-44) at about 1400°F and 6 to 8 psig for optimum ethylene formation. Gas products are freed of material heavier than C_2's by scrubbing, followed by compression and cooling. Ethylene is then absorbed by liquid isobutane at −30°F, and hydrogen and methane are removed from the system. This mixture is compressed to 4,000 to 5,000 psig and added, at ten points in the alkylation furnace, to a preheated (950°F) and compressed (4,000 to 5,000 psig) isobutane stream. Ratio of isobutane to ethylene is maintained at 9 or more to 1 in the reaction zone. Liquid yield is 70 wt-% based upon ethane-propane and isobutane consumed. Neohexane content of the liquid is 30 to 40%. Effluent is pressure-reduced, depropanized, debutanized, relieved of heavy products, and finally depentanized.

Neohexane is characterized by excellent aviation-fuel blending properties and a high response to tetraethyl lead. The material has an RVP of 9.5, a boiling point of 121°F, and an ASTM octane of 95.

The process was developed and installed commercially in early 1940. Operation continued until 1944.
Literature references include:

Alden, R. C., "Neohexane Plant Operating at Record High Pressures," *Nat. Pet. News*, pp. R234–240, June 26, 1940.
Foster, A. L., "Alkylation, Hydrogenation and Related Processes," *Pet. Engr.*, pp. 88–98, April, 1940.
Frey, F. E., "Motor Fuel Isoparaffins by Thermal Gas Polymerization," *Oil & Gas J.*, p. 60, Jan. 7, 1937.
Frey, F. E., and H. J. Hepp, "Non-catalytic Addition of Ethylene to Paraffin Hydrocarbons," *Ind. Engrg. Chem.*, p. 1439, December, 1936.
Iverson, J. O., and L. Schmerling, "Advances in Petroleum Chemistry and Refining," Interscience Publishers, Inc., New York, 1958, vol. 1, pp. 337–347.
Oberfell, G. G., and F. E. Frey, "Thermal Alkylation and Neohexane," *Oil & Gas J.*, p. 50, Nov. 23, 1939, and p. 70, Nov. 30, 1939.
Oil & Gas J., "Thermal Alkylation," p. 95, Mar. 26, 1942.
O'Kelly, A. A., and A. N. Sachanen, "Synthesis of Neohexane and Triptane," *Ind. Engrg. Chem.*, pp. 463–467, May, 1946.
Sachanen, A. N., and A. A. O'Kelly, "High Temperature Alkylation of Aromatic Hydrocarbons," *Ref. Nat. Gaso. Mfr.*, pp. 67–71, December, 1941.

Thermal Polymerization

Thermal polymerization converts C_4 and lighter gases from field and refinery sources into liquid condensation products. This process is particularly applicable to natural-gasoline plants, etc., with excessive propanes and butanes.

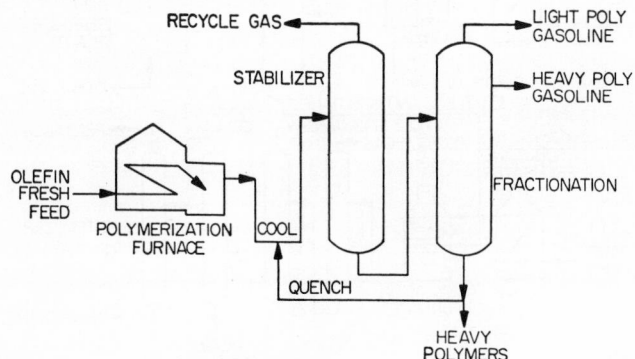

FIG. 3-45. Thermal polymerization.

Olefins are produced by thermal decomposition and polymerized by heat and pressure. Liquid feed, under a pressure of 1,200 to 2,000 psig, is pumped to a furnace (Fig. 3-45) and heated to 975 to 1100°F. Furnace effluent is cooled and stabilized, and polymer gasoline is separated by fractionation. Gas from the stabilizer is returned to vapor recovery to separate C_3's and C_4's for recycle.

Polymer yield is about 10 to 20% per pass of 87 to 88 F-1 clear octane gasoline. Based on C_3-C_4 feed, ultimate yields of poly gasoline are 50 to 72 wt-% and the heavy polymers or residue, 5 to 10%.

The thermal polymerization process developed by the Pure Oil Co. uses a single coil at high temperatures and pressure for both decomposition and polymerization. The Lummus Co. designed and installed the process.

The Unitary Thermal polymerization process was developed by Phillips Petroleum Co., Standard Oil Co. (Indiana), Standard Oil Development Co. (now Esso Research and Engineering Co.), Texaco Inc., and M. W. Kellogg and is owned by The Polymerization Process Corp. M. W. Kellogg licensed the process. The first commercial plant was installed at the Phillips Petroleum Co. refinery at Borger, Tex., in 1933.

Thermal polymerization as a process for manufacturing motor-gasoline components is practically nonexistent in the U.S. refining industry today.
Thermal polymerization references include:

Carey, J. S., "Commercial Aspects of the Unitary Thermal Polymerization Process," *Nat. Pet. News*, pp. 64–71, Nov. 4, 1936.
Chem. & Met. Engrg., "Polymer Gasoline," pp. 596, 607–608, November, 1935.
Cooke, M. B., et al., "Thermal Process for Polymerizing Olefin Bearing Gases," *Oil & Gas J.*, p. 57, Nov. 14, 1935.
Egloff, G., et al., "Motor Fuels from Polymerization," *Oil & Gas J.*, p. 176, Nov. 12, 1937.
Faust, P. H., "What Takes Place in Polymerization," *Pet. Ref.*, pp. 99–104, July, 1944.
Foster, A. L., "Poly Unit Cuts Crude Requirements 3–4 % at Atlantic Refinery," *Nat. Pet. News, Technical Section*, p. R259, June 1, 1938.
Foster, A. L., "Dehydrogenation and Polymerization of Light Hydrocarbons," *Pet. Engr.*, pp. 36–44, March, 1940.
Keith, Jr., P. C., and J. T. Ward, "Thermal Conversion of Hydrocarbon Gases to Gasoline," *Nat. Pet. News*, pp. 52–54, Nov. 20, 1935.
Koehler, W. A., "Polymerization and Its Economics," *Chem. & Met. Engrg.*, p. 412, August, 1938.
Lebedev, S. V., et al., "Kinetics of the Process of Thermal Polymerization of Butadiene," *Nat. Pet. News*, pp. R466–475, June 6, 1945.
Oil & Gas J., "Applications of Thermal Polymerization with Cost Data," p. 52, June 22, 1939.
Oil & Gas J., "Operation of New Atlas Thermal Polymerization Unit Described," p. 139, Mar. 19, 1936.
Oil & Gas J., "Thermal Polymerization," p. 96, Mar. 26, 1942.
Ridgway, C. M., and P. A. Maschivitz, "New Thermal Polymerization Unit Installed by Pure Oil," *Oil & Gas J.*, pp. 29–33, Oct. 30, 1941.
Smith, J. H., "Performance of Thermal Poly Transeconomizers," *Oil & Gas J.*, pp. 45–47, Aug. 3, 1950.

Polyforming and Gas Reversion

These continuous thermal processes convert straight-run naphthas and/or gas oils together with extraneous light hydrocarbon gases (predominantly C_3's and C_4's) into

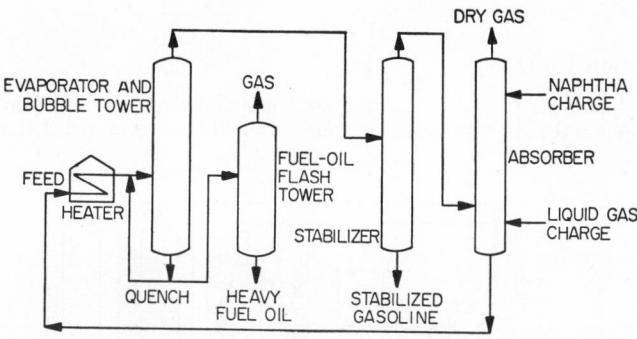

Fig. 3-46. Polyforming and gas reversion.

high-quality motor gasoline and fuel oil. Depending upon the charge, these processes are combined operations utilizing thermal polymerization, thermal reforming, thermal alkylation, and thermal cracking.

A typical operation (Fig. 3-46) involves the charging of a virgin naphtha to an absorber to pick up propane (80 to 90% propane recovery) and heavier gases. Transfer line pressure of the mixed liquid-gas–naphtha feed is 1,000 to 1,500 psig. Evaporator bottoms quench the transfer line (temperature 1020 to 1120°F) to 650 to 700°F. The evaporator pressure is about 400 psig. Net evaporator bottoms are flashed in a flash tower yielding fuel oil as a product and gas. Evaporator overhead is routed to a stabilizer, from which condensable gases are separated from the polyform product for reuse in the absorber along with the outside liquid gas charge.

Polyforming yields for a comparable naphtha feed and quality product are about 5% higher than conventional thermal reforming.

If the gas charge is sent through a heater coil prior to mixing with the naphtha charge in the heater, the process is known as the Gas Reversion process.

Both Gas Reversion and Polyforming processes have been essentially replaced by the modern catalytic reforming processes.

The Gas Reversion process was developed by Phillips Petroleum Co. The Polyforming process was developed by Gulf Oil Corp.

Polyforming and Gas Reversion references include:

Armistead, Jr., G., "The Gas Reversion and Polyform Processes," *Oil & Gas J.*, pp. 189–195, July 27, 1946.
Bogk, J. E., et al., "The Naphtha Polyform and Gas Reversion Processes," API Annual Meeting, Chicago, November, 1940.
Cameron, D. F., and R. T. Weaver, "Still Polyforming in Your Plant?" *Pet. Ref.*, pp. 161–164, July, 1955.
Goldtrap, W. A., and E. L. Jones, "Self-contained Naphtha Polyforming Correlation," *Pet. Engr.*, pp. 269–272, October, 1948.
Hirsch, J. H., et al., "Correlation of Operating Variables in the Naphtha Polyform Process," API Annual Meeting, Chicago, November, 1946, reviewed in *Pet. Proc.*, pp. 61–68, January, 1947.
Offutt, W. C., et al., "Naphtha Polyforming," API Annual Meeting, Chicago, November, 1946, reviewed in *Pet. Proc.*, pp. 267–276, December, 1946.
Offutt, W. C., et al., "Polyforming with Outside Gas," API Annual Meeting, Chicago, November, 1946, reviewed in *Pet. Proc.*, pp. 278–288, December, 1946.
Offutt, W. C., et al., "Gas Oil Polyforming," *Pet. Proc.*, pp. 753–768, October, 1947.
Offutt, W. C., et al., "Utilization of Refinery Gases by the Polyform Process," NPA Annual Meeting, Atlantic City, N.J., September, 1948, reviewed in *Pet. Proc.*, pp. 1083–1090, November, 1948.
Smoley, E. R., and V. O. Bowles, "Polyform and Gas Reversion Processes," *Oil & Gas J.*, pp. 143–146, Nov. 6, 1941.
Teitsworth, R. C., "Improve Polyform Operations with These Correlations," *Pet. Ref.*, p. 91, February, 1953.
Wall, J. D., "How to Predict Polyform Gasoline Octane," *Pet. Proc.*, pp. 1032–1035, July, 1953.

Combination Units

Combination units (Fig. 3-47) incorporate a group of thermal conversion processes predicated upon the selective cracking principle applied to segregated charge stocks to

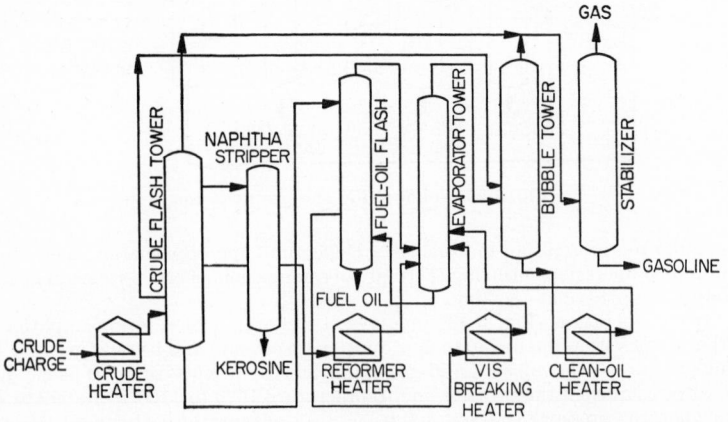

FIG. 3-47. Combination thermal conversion scheme.

obtain optimum product yields and quality with maximum operating flexibility. Crude running is combined with visbreaking, reforming, and thermal cracking. Other processes that might also be included in combination cracking operation are coking, vacuum flashing, selective cracking of several boiling-range stocks, and polymerization.

Each individual process involved in combination cracking has been described elsewhere in this glossary. Yield and operating conditions will not be repeated here. The main advantages to combining and integrating their operations include (1) large amounts of heat—characteristic of cracking processes—can be recovered, (2) selective cracking of individual stocks at optimum operating conditions can be carried out, and (3) many thermal processes, because of their general similarity, lend themselves to dual or multiple use of equipment, thereby eliminating duplication which inherently exists when processes are considered singularly.

Basically, the design of a combination unit varies with the specific need of the refiner. However, each unit is designed for the segregation of cracking stocks and their independent selective cracking under optimum conditions. Segregation is predicated upon charge stock boiling range and refractivity. Cracking operations are carried out in specially designed heaters to obtain maximum conversion, larger yields, and longer runs.

Process design and construction of the combination cracking process is handled by Lummus Co.

Foster Wheeler Corp. has designed and constructed a combination unit involving the operations of distillation, visbreaking, and vacuum flashing.

Combination processing references include:

Al-Pachachi, N., et al., "Combination Unit Lowers Cost at New Baghdad Refinery," *Pet. Proc.*, pp. 58–61, January, 1956.
Armistead, Jr., G., "Combination Units Facilitate Refining," *World Pet.*, p. 369, November, 1933.
Egloff, G., and W. L. Nelson, "Modern Cracking Process," *Oil & Gas J.*, p. 50, Jan. 25, 1936.
Nelson, W. L., "Combination Topping Plants," *Oil & Gas J.*, p. 101, Dec. 23, 1944.
Smoley, E. R., et al., "Theory of Combination Selective Cracking Demonstrated by Commercial Operations," *World Pet.*, p. 94, June, 1937.
Smoley, E. R., "Combination Cracking Economics," *Ref. Nat. Gaso. Mfr.*, p. 184, May, 1938.
Williams, N., "Atlantic Refinery on Gulf Coast," *Oil & Gas J.*, p. 70, Feb. 23, 1939.

PART 3.　SOLVENT PROCESSES

Solvent refining processes are of a physical nature only, and both the desirable and undesirable constituents are recovered in their original state. Feeds to these processes are either high- or low-boiling materials. The higher-boiling feedstocks include lube oil or potential lube oil blending stocks, kerosine, diesel oil and other middle distillates, and gas oil or catalytical cracking charge stocks.

Basically the processes in use today for middle distillate and heavier feeds can be classified as:

1. Treating or deasphalting—applied to full-boiling materials such as gas oils and stocks from which lube cuts are made. Solvents are precipitants which divide the petroleum charge into constituents based on molecular size.

2. Treating or refining—applied to lube boiling-range charges to remove by selective extraction aromatics, naphthenes, and other materials that adversely affect viscosity index. Feed to these processes has usually been deasphalted. The effluent is normally fed to dewaxing processes.

3. Dewaxing operations—applied to lube components or full-boiling material prior to final clay treating. Wax in a finished lube would result in a product of high pour point, which reduces fluidity of the oil at low temperatures because of crystallization. Dewaxing operations are often combined with wax production, producing simultaneously dewaxed low-pour-point lube oil and high-melting-point, low-oil-content wax.

Solvent processes are also widely used in the recovery of C_6-C_{10} aromatics from gasoline boiling-range streams, particularly catalytic reformate. These aromatics are primarily used as building blocks in the petrochemical industry, although market conditions often dictate the use of one or more of the aromatic streams as gasoline blending components.

Historically, liquid SO_2 treatment of kerosine can be considered the forerunner of present-day solvent refining methods. However, the impetus for modern lube treating came in the late 1920's and early 1930's, when the increased demand for high-quality lubricating oils made it necessary to use a wider variety of crude oils for lubricating stocks, which in turn required solvent treating methods. Solvent treating for lighter aromatics recovery on a large-scale commercial basis was brought about by the current demands of the petrochemical industry and the needs of our country during World War II.

General solvent refining references include:

Armistead, Jr., G., "Better Technology Equals Better Lube Oils," *Oil & Gas J.*, p. 1154, May 4, 1946.
Kalichevsky, V. A., "Modern Methods of Refining Lubricating Oils," *ACS Monograph Series*, No. 76, Reinhold Publishing Corporation, 1938.
Kalichevsky, V. A., "Modern Solvent Refining of Lubricating Oils," *Pet. Proc.*, pp. 1168–1172, December, 1948; pp. 32–36, January, 1949; pp. 145–148, February, 1949; pp. 254–257, March, 1949; pp. 415–421, April, 1949.
Kalichevsky, V. A., and K. A. Kobe, "Petroleum Refining with Chemicals," Elsevier Publishing Company, Amsterdam, 1956.
Lupfer, G. L., "Newest Solvent Lube Plant," *Pet. Proc.*, pp. 34–38, January, 1951.
Nelson, W. L., "100 Years of Processing Advances," *Oil & Gas J.*, pp. 204–206, July 25, 1960.
Nolting, H. F., and R. B. Selund, "Modernized Motor Oil Facilities," *Pet. Proc.*, pp. 76–80, January, 1953.
Pet. Proc., "World's Largest Solvent Lube Units Installed in Cit-Con Refinery," pp. 939–940, October, 1948.
Reeves, E. J., "Optimum Temperature Gradient in Selective Solvent Extraction," *Ind. Engrg. Chem.*, pp. 1490–1492, July, 1949.
Reeves, E. J., and E. P. Hardin, "Temperature Gradient in Solvent Extraction of Lubricating Oils," *Pet. Ref.*, pp. 89–90, January, 1950.
Reman, G. H., and J. G. Van der Vusse, "Applying RDC to Lube Extraction," *Pet. Ref.*, pp. 129–134, September, 1955.
Reman, G. H., and R. B. Olney, "The Rotating Disc Contactor, A New Tool for Liquid-Liquid Extraction," AIChE Meeting, New York, December, 1954, reviewed in *Pet. Proc.*, pp. 230–232, February, 1955.
Scheibel, E. G., "Find Best Solvent with This Chart," *Pet. Ref.*, pp. 189–194, May, 1961.
Smith, A. S., and J. E. Funk, "Extraction of Aromatics from Hydrocarbon Mixtures," *Pet. Engr.*, Reference Annual, pp. 239–245, 1944.
Smoley, E. R., "The Outlook for Lube Oils," *Pet. Proc.*, pp. 735–738, August, 1948.
Smoley, E. R., and D. Fulton, "Modern Manufacture of Lubricating Oils," *Pet. Proc.*, pp. 594–608, August, 1947.
Thornton, Jr., D. P., "Pure Oil's First Solvent Lube Plant Now in Operation at Smiths' Bluff," *Pet. Proc.*, pp. 245–246, March, 1950.
Treybal, R. E., "Solvent Extraction," *Ind. Engrg. Chem.*, pp. 53–63, January, 1952.
Von Berg, R. L., and H. F. Wiegandt, "Liquid-Liquid Extraction," *Chem. Engrg.*, pp. 189–200, June, 1952.

DEASPHALTING AND TREATING (GAS OIL, LUBE STOCKS)

Carbon residue and coke-forming tendencies of lube-oil stocks, gas oils, or middle distillates are reduced through the extraction or precipitant action of solvents on the asphaltic and resinous materials which are present in either solution or colloidal form. Sulfur and heavy metals are removed, and color is improved.

Deasphalting solvents may be divided into two major groups: low-molecular-weight hydrocarbons, particularly propane, and oxygenated compounds, such as alcohols or esters. Deasphalting and dewaxing or extraction operations may be combined.

Many processes capable of removing asphalt from hydrocarbons were employed by the petroleum industry before solvent methods were developed. Distillation, refining with clay or sulfuric acid, and treatment with metal chlorides are the best known.

Solvent refining, when employed for preparation of feed to catalytic cracking units, can be considered in competition with vacuum distillation, coking, and visbreaking.

Solvent refined cracking feed approaches virgin feed in relation to its ease of processing and the characteristics of the resulting products.

General deasphalting and treating references include:

Atteridg, P. T., and R. P. Cox, "Better Cat Feed by Solvent Extraction," *Pet. Ref.*, pp. 145–150, May, 1960.
Berg, C., "Refining Capacity Up," *Petro/Chem Engr.*, pp. 45–48, January, 1962.
Bozeman, H. C., "Cat Cracking Begins Second 20 Years," *Oil & Gas J.*, pp. 67–78, July 23, 1962.
Bray, U. B., and W. H. Bahlke, "Refining with Liquid Propane," *Science of Petroleum*, 1938, vol. III, p. 1966.
Dunning, H. N., and J. W. Moore, "Propane Removes Asphalts from Crudes," *Pet. Ref.*, pp. 247–250, May, 1957.
Hood, R. J., "Preparation of Feed Stock for Catalytic Cracking," AIChE Meeting, San Francisco, Calif., December, 1959.
McMillan, K. K., "Processing of Feed Stocks for Catalytic Cracking," AIChE Meeting, San Francisco, Calif., December, 1959.
Nelson, W. L., "Completeness of Processing—a Lube-oil Refinery," *Oil & Gas J.*, p. 141, May 12, 1958.
Nelson, W. L., "Operating Costs—Propane Decarbonizing or Deasphalting," *Oil & Gas J.*, p. 145, Mar. 9, 1960.
Nelson, W. L., "Cost of Asphalt Manufacture," *Oil & Gas J.*, p. 170, Oct. 29, 1962.
Oil & Gas J., "Lube-capacity Cushion Grows," pp. 128–129, Apr. 2, 1962.
Stormont, D. H., "Lube Capacity Edging up Slightly," *Oil & Gas J.*, pp. 66–68, Aug. 28, 1961.

Furfural Extraction—Gas Oil

Furfural extraction is a continuous process for the removal of aromatics, nitrogen, sulfur, and organometallic, acidic, and unstable compounds that affect burning quality, engine cleanliness, and cetane number. Charge stocks include diesel fuel, catalytic cracking gas oil, and middle distillate burner oils. This is an outgrowth of the Furfural Refining process employed in lube-oil treating.

The extracting temperatures and solvent ratios are generally lower for gas oil than for lubricating-oil charge stocks. Solvent ratios in the range of 0.25 to 1.0, 0.40 to 1.0, and 0.80 to 1.0 have been employed in treating cycle oil for catalytic cracker charge, cycle oil for diesel fuel, and virgin gas oil for middle distillate usage, respectively.

Yields of product raffinate are dependent upon the type and characteristics of the charge stock and the degree of quality desired. An 82.0% yield was obtained from a West Texas virgin gas oil with sulfur reduction from 1.12 to 0.49% and cetane number improvement from 53.2 to 62.9.

Following azeotropic distillations in the raffinate and extract strippers (Fig. 3-48), the overhead vapors (azeotropic mixtures) from these columns are condensed and sent to a vessel for decanting. Three liquid phases separate in the decanter. The bottom phase, predominantly furfural, is returned to the solvent system. The intermediate

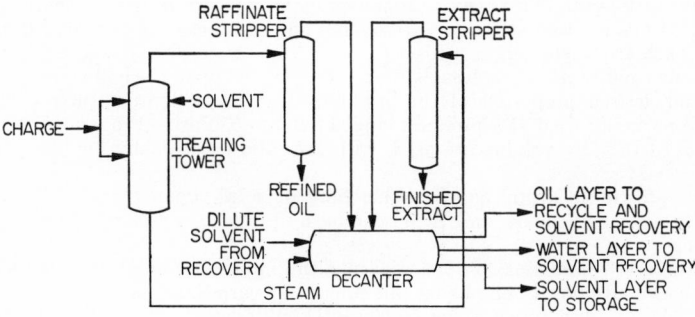

FIG. 3-48. Furfural extraction—gas oil.

phase, predominantly water, is charged to a tower for water removal.   The light-oil phase is sent to a tower for extraction of the furfural with water.

This process is licensed by the Texaco Development Corp.   In 1964, there were eight commercial installations.

Furfural extraction references include:

Carter, R. T., "Furfural Refining Scores on Four Counts," *Oil & Gas J.*, p. 157, Mar. 22, 1954.

Dickinson, R. R., and C. B. Longsworth, "Use of Furfural Refining to Improve Catalytic Cracking Feed Stocks," AIChE Meeting, San Francisco, Calif., December, 1959.

Gross, H. H., et al., "Furfural Refines Gas Oil," *Oil & Gas J.*, pp. 211–213, Mar. 23, 1950.

Hyd. Proc., "Furfural Extraction of Gas Oils," p. 216, September, 1964.

Kemp, Jr., L. C., et al., "Furfural as a Selective Solvent in Petroleum Refining," *Ind. Engrg. Chem.*, pp. 220–227, February, 1948.

Oil & Gas J., "Furfural Refining of Cycle Stocks," p. 168, Mar. 19, 1956.

Tears, C. F., "Catalytic Cycle-stock Treating Unit," *Oil & Gas J.*, pp. 119–124, June 15, 1953.

Texaco Development Co., "Furfural Process Adapted to Operate on Wide Variety of Charge Stocks," *Pet. Ref.*, pp. 196–198, September, 1950.

Propane Deasphalting

Propane deasphalting is an extraction process in which the desirable oil from the charge is dissolved in the propane solvent and asphaltic materials are precipitated as asphalt.   Normally, vacuum-reduced crude of various boiling ranges is charged for finishing into cylinder stock, bright stock, or other lubricating stock.

The solubility of oil in propane decreases with increase in temperature and is significantly influenced by pressure.

Liquid propane is contacted countercurrently (Fig. 3-49) with descending heavy oil in the deasphalting tower.   Deasphalted oil is separated from propane by evaporation

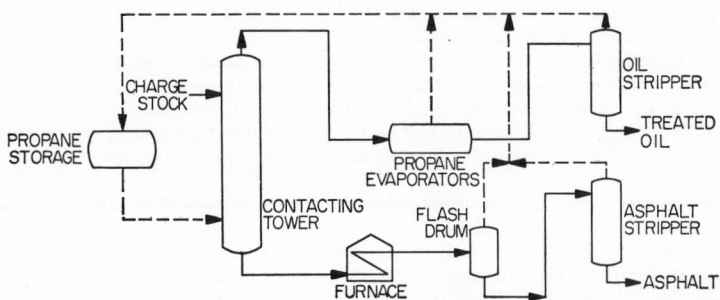

Fig. 3-49. Propane deasphalting.

and steam stripping.   The heavy asphalt-propane mixture is heated, flashed, and stripped.   Propane from strippers is recovered and compressed for reuse.   Evaporation and flash are at condenser pressure.

Operating conditions in the deasphalting tower depend upon the boiling range of the charge and desired properties of the product.   Top tower temperatures of 130 to 180°F are normally used at a pressure range of 400 to 550 psi.   Propane-oil ratios of 6 to 1 to 10 to 1 by volume are used, with the ratio occasionally being as high as 13 to 1.

Asphalt present in an oil can be either hard or soft.   Pennsylvania oils contain practically no hard asphalt, and their refining by propane is referred to as Propane Deresining.

Since the first commercial propane deasphalting unit went on-stream in 1934, the industry has continuously employed this process or variations thereof for lube stock treating.   Direct operating cost for a 5,000-B/D unit averages about 25 cents per bbl.

This process is licensed by the M. W. Kellogg Co.

Propane deasphalting references include:

Bahlke, W. H., et al., "Deasphalting Acid Treating of Lubricants with Propane," *Oil & Gas J.*, p. 44, July 22, 1937.
Bray, U. B., et al., "Use of Propane in Lubricating Oil Refining," *Pet. Ref.*, pp. 333–336, 353–359, September, 1934.
Dickinson, J. T., and N. R. Adams, "Lubricant Stocks Improved by Propane Deasphalting," *Oil & Gas J.*, pp. 185–200, Mar. 30, 1946.
Ditman, J. G., and F. T. Mertens, "Propane Deasphalting of Residua," *Pet. Proc.*, pp. 162–168, July, 1952.
Graff, P. T., and H. O. Forrest, "Propane Precipitation of Petroleum Resins," *Ind. Engrg. Chem.*, pp. 294–298, March, 1940.
Hoiberg, A. J., "History, Source and General Nature of Asphaltic Bitumen," *Pet. Ref.*, pp. 1–7, January, 1947.
Kalichevsky, V. A., "Technology—Deasphalting Lubes," *Pet. Engr.*, pp. C11–14, February, 1956.
Livingston, M. J., and J. T. Dickinson, "Packed Towers Useful in Solvent Refining; Special Uses of Propane," *Nat. Pet. News*, pp. 25–29, July, 1935.
McCluer, W. B., et al., "Application of Propane to Manufacture of Pennsylvania Lubricating Oils," *Oil & Gas J.*, pp. 209–210, Nov. 18, 1938.
Oil & Gas J., "Operating Costs—Propane Deasphalting," p. 145, May 9, 1960.
Pet. Ref., "Propane Deasphalting and Fractionation," pp. 227–228, September, 1962.
Reinkemeyer, L. R., "Making High Quality Asphalt with a Propane Deasphalting Unit," *Oil & Gas J.*, p. 166, Sept. 7, 1958.
Thegze, V. B., et al., "Rotating Disk Contactors Perform Well in Propane Deasphalting of Lube Oil," *Oil & Gas J.*, pp. 90–94, May 8, 1961.
Wilson, R. E., et al., "Liquid Propane, Use in Dewaxing, Deasphalting, and Refining Heavy Oils," *Ind. Engrg. Chem.*, pp. 1065–1078, September, 1936.

Propane Decarbonizing

Propane decarbonizing is a solvent extraction process used to recover catalytic cracking feed from heavy residues. Since butane, alone or with propane, can also be used as the solvent, this process is often referred to as solvent decarbonizing. Decarbonized and demetalized oil is recovered from topped or vacuum-reduced crude charge.

Process flow and equipment are essentially the same as that employed for the deasphalting of lubricating-oil stocks. Extraction temperatures generally range from 150 to 250°F with pressures of 400 to 600 psi. The solvent-oil ratio and amount of temperature gradient employed are determined by feed, solvent, and product characteristics and requirements. Yields of 40 to 75% decarbonized oil have been obtained from a reduced crude charge. Feedstocks with Conradson carbon in the range of 12.0 to 22.0 wt-% have been reduced to a range of 2.0 to 5.5.

Investment costs of $150 to $200 per bbl of charge and operating costs of about 10 to 13 cents per bbl of charge, excluding return on investment, have been claimed for units of 10,000 to 20,000 B/D capacity.

This process is licensed by M. W. Kellogg Co.

Propane decarbonizing references include:

Dickinson, J. T., and N. R. Adams, "Lubricant Stocks Improved by Propane Deasphalting," *Oil & Gas J.*, pp. 185–200, Mar. 30, 1946.
Dimmig, H., and N. L. Dickinson, "Propane Deasphalting of 'Cat Cracker' Feed Stock," *Nat. Pet. News*, pp. 48–49, Oct. 4, 1944.
Ditman, J. G., and J. C. Dunmyer, Jr., "New Cat Feed Made by Deasphalting," *Pet. Ref.*, pp. 187–192, May, 1960.
Ditman, J. G., "Propane Deasphalting for the Production of Catalytic Cracking Feed," AIChE Meeting, San Francisco, Calif., December, 1959.
Durland, L. V., and C. Nysewander, "Propane Deasphalting at Utah Oil Refining Co.," *Oil & Gas J.*, pp. 216–218, Mar. 23, 1950.
Farrar, G. L., "Decarbonizing Process Improves Flexibility," *Oil & Gas J.*, pp. 205–207, Mar. 21, 1955.
Kraemer, C. W., "Propane Deasphalting Process," *Oil & Gas J.*, pp. 228–233, Mar. 30, 1946.
Oden, F. C., and E. L. Foret, "Deasphalting Crude Residuum for Catalytic Cracker," *Ind. Engrg. Chem.*, pp. 2088–2095, October, 1950.

Oil & Gas J., "Solvent Decarbonizing," p. 130, Apr. 5, 1965.
Oil & Gas J., "Operating Costs—Propane Decarbonizing," p. 145, May 9, 1960.
Pet. Ref., "Solvent Decarbonizing," p. 222, September, 1964.
Reinkemeyer, L. R., "Use of Propane Deasphalting Unit for Making Asphalt," WPRA Regional Meeting, Wichita, Kans., June, 1959.
Weber, G., "Reduced Crude Decarbonization," *Oil & Gas J.*, pp. 54–76, June 8, 1950.

Propane Fractionation

Propane Fractionation is a continuous extraction process for the segregation of long vacuum residuum into two or more grades of lubricating-oil stock. This process, employing liquid propane as the solvent, is a development of the Propane Deasphalting process. Products of this process are stocks corresponding to heavy neutral distillate and bright stock with better color, carbon-residue, and viscosity-index properties than corresponding fractions from vacuum distillation. Asphalt is also produced.

Residuum feed is charged to the primary tower (Fig. 3-50), where it is countercurrently extracted with liquid propane. Temperature control is such that only the

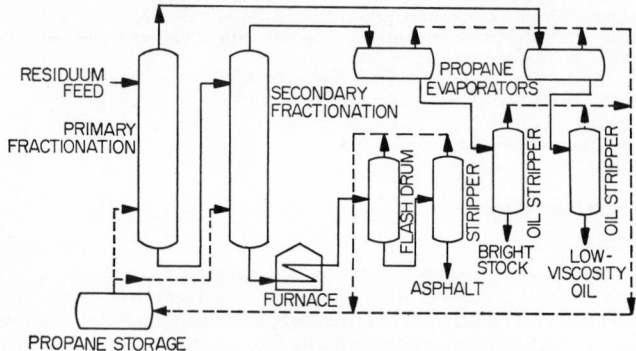

Fig. 3-50. Propane fractionation for lube-oil stocks.

heavy stock corresponding to heavy neutral distillate dissolves in the propane for overhead withdrawal. Propane is removed in evaporators and a steam stripper. The primary tower bottoms are countercurrently extracted with propane in the secondary tower for production of bright stock at lower temperature and pressure conditions. Bright stock is separated from propane in a manner similar to that employed for the primary tower overhead. The asphalt and propane mixture is separated after heating by flashing and steam stripping.

There are three commercial installations of this process on-stream. Investment costs for Propane Fractionation units vary from $350 to $720 per bbl of charge for units of 5,000 to 1,000 B/D capacity, respectively. The process is licensed by M. W. Kellogg Co.

Propane Fractionation references include:

Ditman, J. G., and L. Nilssen, "The Separation of High-molecular Weight Petroleum Fractions By Propane Fractionation," API Refining Meeting, San Francisco, Calif., May, 1962.
Graff, P. T., and H. O. Forrest, "Propane Precipitation of Petroleum Resins," *Ind. Engrg. Chem.*, pp. 294–298, March, 1940.
Hyd. Proc., "Propane Deasphalting and Fractionation," p. 219, September, 1964.
Johnson, P. H., et al., "Recovery of Catalytic Cracking Stock by Solvent Fractionation," *Ind. Engrg. Chem.*, pp. 1578–1585, August, 1955.
Pet. Ref., "Propane Fractionation," p. 269, September, 1956.
Weber, G., "Pennzoil Carries Propane Fractionation to Advanced New Stage," *Oil & Gas J.*, pp. 78–79, 109, 110, Jan. 21, 1952.

HF Extraction

HF extraction is a liquid-liquid extraction process for removing sulfur and coke formers from virgin and cracked light naphthas, middle distillates, and gas oils.

The charge stock, after passing through an absorber (Fig. 3-51), is contacted counter-currently with liquid hydrofluoric acid in an extraction tower. The overhead product raffinate is sent to a tower for HF removal. The solvent is recovered from the extract through use of evaporators and a stripper.

This process is relatively insensitive to variations in temperature and pressure over practical ranges. Normally, temperatures between 100 and 125°F and pressures below 100 psi are employed. Solvent-to-oil ratios are low, being in the range of 0.15 to 0.30 to 1.0.

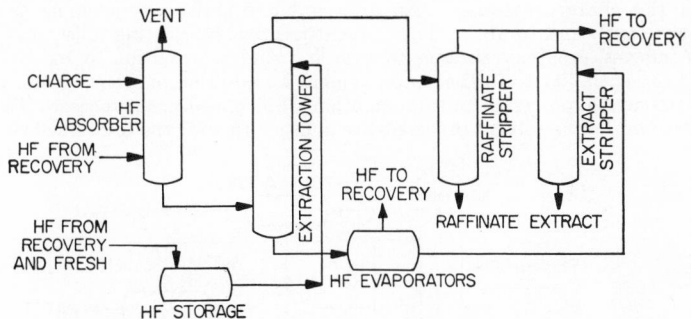

Fig. 3-51. HF extraction for removing sulfur and coke formers.

HF treatment of kerosine, gas oil, and cycle oil has resulted in 85 to 95% raffinate product yields at 60 to 90% desulfurization.

No commercial processes have been installed to date.

HF extraction references include:

Hughes, E. C., et al., "Separation of Sulfur and Aromatics from Petroleum with BF₃ and HF," *Ind. Engrg. Chem.*, pp. 750–753, March, 1951.
Lien, A. P., and B. L. Evering, "Hydrofluoric Acid Extraction of High-sulfur Virgin Petroleum Stocks," *Ind. Engrg. Chem.*, pp. 874–879, April, 1952.
Lien, A. P., et al., "Extraction of Sulfur Compounds with Hydrogen Fluoride," *Ind. Engrg. Chem.*, pp. 2698–2702, December, 1949.
Pet. Ref., "HF Extraction," pp. 188–189, September, 1952.
Scafe, E. T., "Sulfur in Light Petroleum Products," *Pet. Ref.*, pp. 87–92, September, 1946.

SO₂ Extraction

This continuous liquid-liquid extraction process is used for removing aromatic and sulfur-bearing compounds from paraffinic and naphthenic charge materials. Feed-stocks include light naphtha, kerosine, gas oil, catalytic cycle oil, and high and medium lubricating oil. Other applications of this process will be described in the Lube Refining Section (Edeleanu SO₂ Processing).

When this process is used for gas-oil treatment, a temperature of about +60°F is employed. Yields of 65 to 90% resulted in 10 to 70% sulfur reduction and an improvement in diesel index at solvent-oil ratios (weight) varying between 0.25 to 1.0 to 1.0. Treatment of catalytic cycle stock may reduce coke laydown in subsequent cracking by 50% or more.

Commercial units of this process totaling approximately 115,000 B/D capacity have been designed and built by Stone and Webster Engineering Corp.

SO₂ extraction references include:

Dickey, S. W., "Diesel Fuel of 50-cetane Value Produced in New Sulfur Dioxide Extraction Plant," *Pet. Proc.*, pp. 538–542, June, 1948.

Dryer, C. G., et al., "Solvent Extraction of Diesel Fuels," *Ind. Engrg. Chem.*, pp. 813–821, July, 1938.

Obergfell, P., "Treatment of Light Distillates by the SO₂ Extraction Process," *Pet. Proc.*, pp. 660–664, June, 1949.

Pet. Ref., "SO₂ Extraction," p. 230, September, 1962.

Steffen, E., and E. Saegebarth, "The Improvement of Diesel Fuels by Extraction with Sulfur Dioxide," *Ref. Nat. Gaso. Mfr.*, pp. 12–14, January, 1938.

Stone and Webster Engineering Corp., "Modified Edeleanu Process," *Pet. Ref.*, p. 237, September, 1951.

LUBES—REFINING

Solvent treating is the most widely used method of refining lubricating oils. These processes yield products that meet the requirements of a modern lubricant by removing from the charge material those undesirable constituents such as aromatics, naphthenes, and unsaturates. The solvent-treated lubricating oils have higher viscosity indexes, greater resistance to gum and sludge formation by oxidation, and increased susceptibility to further improvement by addition of selective additives.

Most commercial processes for this operation are single-solvent processes (Fig. 3-52). A few mixed or double-solvent processes are in use. In 1963 the total finished lube-oil

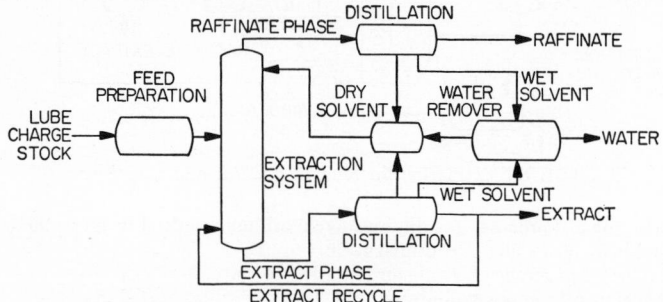

FIG. 3-52. Typical single-solvent extraction process for refining lubricating oils.

capacity of U.S. refineries was 191,215 B/D. Almost 250,000 B/D of solvent extraction capacity existed for treating these lubes, of which only 75,000 B/D were double-solvent processes.

Recent pilot-plant work on a thermal-diffusion-based method of removing undesirable constituents from lube base stocks indicates that a thermal-diffusion process might soon reach commercial reality.

Al-Naqib, and J. W. Mitchell, "World's Newest Complete Lube Plant," *Oil & Gas J.*, pp. 101–105, Aug. 11, 1958.

Berg, C., "Refining Capacity Up—Thermal Diffusion Refining," *Petro/Chem Engr.*, p. 48, January, 1962.

Fitzgerald, G. F., "Established Processes Make Aviation Lubricants," *Pet. Ref.*, pp. 179–184, October, 1942.

Fons, J. L., "Standard's New Heavy-oils Refinery at Casper," WPRA Regional Meeting, Casper, Wyo., September, 1956.

Foster, A. L., "Refining Methods for Lubricating Oils," *Pet. Engr.*, pp. 25–28, July, 1940.

Kalichevsky, V. A., "Lubricating Oils—Properties & Applications," *Ref. Engr.*, pp. C14–17, January, 1957.

Kalichevsky, V. A., "Processing of Lubricating Oils," *Pet. Ref.*, pp. 75–82, February, 1948.

Kalichevsky, V. A., "Manufacture of Lubricating Oils," *Pet. Ref.*, pp. 112–116, September, 1945.

Kalil, P., "Commonly Used Lubricant Additives," *SAE J.*, pp. 76–77, December, 1961.

King, E. P., "Lubricating Oil Manufacture," *Oil & Gas J.*, pp. 122–134, Mar. 16, 1953.

Nelson, W. L., "Lube Additives—Major Cost Item," *Oil & Gas J.*, p. 116, Aug. 27, 1962.

Oil & Gas J., "Thermal-diffusion Process Gives Better Lubes," p. 61, Mar. 27, 1961.

Pet. Engr., "Manufacture of Lubricating Oils," pp. 98–100, November, 1942.

Furfural Refining

Furfural Refining is a single-solvent process for removing aromatic, naphthenic, olefinic, and unstable hydrocarbons from a lubricating-oil base stock, thereby improving its viscosity index and stability characteristics. The solvent furfural is an aldehyde, manufactured from agricultural wastes such as oat hulls, corncobs, bagasse, etc. The treating is conducted in counterflow towers (packed or employing the Rotating Disc Contactor design) or multistage units, normally operating at a treating temperature range of 100 to 250°F. A high-temperature gradient in the treating section permits high yield and quality of refined oil for a given solvent dosage. Furfural has high stability and can be recovered using distillation temperatures ranging from 400 to 550°F. Solvent losses are only 0.03 to 0.06 vol-% of solvent circulated.

Direct operating costs in 1963 for a 2,000-B/D furfural extraction unit for lube treating (2.0 to 1.0 solvent-to-oil ratio) ranged from 35 to 45 cents per bbl.

This process has been developed and is licensed by the Texaco Development Corp. It was first commercially applied in 1933 at the Indian Refining Co. plant at Lawrenceville, Ill. In 1964, approximately 65,000 B/D furfural refining capacity existed for handling lube stocks in U.S. refineries.

Furfural Refining references include:

Brown, L. C., "Pilot Plant Aids Operation of Furfural Extraction Unit," *Oil & Gas J.*, pp. 112–114, June 14, 1947.

Brown, L. C., and C. F. Tears, "New Furfural Solvent Lube Plant," *Pet. Proc.*, pp. 492–496, 557, July, 1947.

Bryant, G. R., et al., "The Solvent Refining of Lubricating Oils with Furfural," *Oil & Gas J.*, p. 53, Apr. 16, 1935.

Evans, W. V., and M. B. Aylesworth, "Some Critical Constants of Furfural," *Ind. Engrg. Chem.*, pp. 24–27, January, 1926.

Garwin, L., and E. C. Barber, "Mass Transfer Data for Furfural Extraction," *Pet. Ref.*, pp. 144–148, January, 1953.

Kemp, Jr., L. C., et al., "Furfural as a Selective Solvent in Petroleum Refining," *Ind. Engrg. Chem.*, pp. 220–227, February, 1948.

Livingston, M. J., and J. T. Dickinson, "Packed Towers Useful in Solvent Refining," *Nat. Pet. News*, pp. 25–29, July, 1935.

Manley, R. E., et al., "Use Furfural as the Extraction Solvent to Make Motor Oils Have a Lower Viscosity," *Pet. Ref.*, pp. 420–31, December, 1933.

Mertens, F. T., "Refining Pennsylvania Oils with Furfural," *Oil & Gas J.*, pp. 47–48, Nov. 7, 1940.

Mertens, F. T., "Baghdad's New Lube Plant Uses Furfural Refining," *Pet. Ref.*, pp. 239–240, September, 1957.

Oil & Gas J., "Furfural Refining," p. 128, Apr. 3, 1961.

Pet. Ref., "Furfural Refining," p. 217, September, 1964.

Smoley, E. R., et al., "Cost and Extent of Solvent Refining and Dewaxing of Lubricating Oils," *World Pet.*, pp. 76–79, November, 1938.

Weber, G., "Gulf Treating Lubricating Stock from Various Crude Oils in New Solvent Unit," *Oil & Gas J.*, pp. 30–32, July 2, 1936.

Woelfel, W. C., et al., "Determination of Furfural in Lubricating Oils," *Pet. Engr.*, pp. C42–44, July, 1952.

Ziegenhain, W. T., "Simple, Compact and Unusually Flexible Vacuum Asphalt Unit Completed at Sunray Refinery," *Oil & Gas J.*, p. 44, May 30, 1933.

Phenol Extraction

This is a liquid-liquid extraction process for the removal of aromatic constituents of lubrication oils, thereby improving the viscosity index, oxidation stability, and resistance to sludging, carbon formation, and varnish deposition. All grades of oils, including gas oils and catalytic stocks, can be treated. The solvent phenol is employed in either anhydrous or aqueous form.

Modern plants employ centrifugal extractors or Rotating Disc Contactors; other plants use towers, generally with plates. Feed is preheated, passed through an absorption tower to pick up phenol from an aqueous mixture (obtained in the phenol recovery step), and added to the centrifugal extractor or to the treating tower for anhydrous phenol treatment.

Treating temperatures vary with the charge material. For lubricating oils the treating tower is maintained at 130 to 200°F.

One of the features of the process is the use of water to reduce the solubility of oil in phenol at the bottom of the tower (in a centrifugal extractor system phenolic water from the phenol water drum is injected into the feed prior to entering the extractor). Water induces oil solubility in the extract solution, thereby generating reflux by rejecting the least soluble components from the extract phase before it is removed from the system.

Treatment of an SAE 40 lube stock (28.3 API, 77.4 SSU viscosity at 210°F, and 0.16 Conradson carbon) with a 2.1-to-1.0 solvent-to-oil ratio yielded an 88 vol-% raffinate (30.2 API, 72.3 SSU viscosity at 210°F, 0.04 Conradson carbon, and a dewaxed viscosity index of 93).

Total U.S. refinery capacity for Phenol Extraction for lube stocks was about 95,000 B/D in 1963. Direct operating costs for a 2,000-B/D-capacity unit at that time ranged from 30 to 40 cents per bbl.

M. W. Kellogg Co. is the licensor for this process under the patents and patent rights of Esso Standard Oil Co., Union Oil Co. of California, Standard Oil Co. (Indiana), and M. W. Kellogg Co.

Phenol Extraction references include:

Freitas, J. A. D., "See What Brazil Added to Its Bahia Refinery," *Pet. Ref.*, pp. 185–188, September, 1961.
Hyd. Proc., "Phenol Extraction," p. 218, September, 1964.
Kalichevsky, V. A., "Estimation of Properties of Solvent-refined Products," *Pet. Ref.*, pp. 17–22, January, 1947.
Kenny, D. W., and W. B. McCluer, "Refining Pennsylvania Lube Oils by Phenol Extraction," *Oil & Gas J.*, pp. 48–49, 56, Jan. 16, 1941.
Nelson, W. L., "Phenol Solvent Treating," *Oil & Gas J.*, p. 78, Mar. 10, 1945.
Stines, D. E., "Some Recent Trends in Phenol Plant Design and Operation," *Oil & Gas J.*, pp. 75–76, Mar. 19, 1936.
Stratford, R. K., and J. L. Huggett, "Phenol Used in Production of High-grade Lubricants at Port Jerome Refinery," *Oil & Gas J.*, pp. 44–46, Dec. 27, 1934.
Stratford, R. K., et al., "Successfully Use Phenol as Selective Solvent in Manufacturing Lubricants and Special Oils,",*Proc. World Pet. Congr.*, p. 362, 1933.
Stratford, R. K., et al., "Use of Phenol as Selective Solvent in the Production of High-grade Lubricating Oils," *Ref. Nat. Gaso. Mfr.*, pp. 458–462, December, 1933.

Edeleanu SO₂ Processing

Edeleanu SO_2 treating, although still used to extract aromatics from feedstocks in the kerosine and gas-oil boiling range, has been largely replaced by other processes in the refining of lubricating oils. It had been used to improve oxidation stability of transformer, turbine, and very light lubricating-oil blending stocks. Because of the low solvent power of liquid SO_2, in the absence of auxiliary solvents, it was of little usefulness in the preparation of high-viscosity-index motor oils. Use of liquid SO_2 (liquid SO_2-benzene) as a heavy lubricating-oil-treating process is discussed later.

Liquid SO_2 plants have been built by Stone and Webster Engineering Corp.

Operating temperatures range from $-20°F$ for naphthas and kerosines to $+60°F$ for high-pour-point stocks. Solvent ratios range from 0.5 to 3.0 to 1.0, the heavier charge material requiring the higher solvent-oil ratio.

Edeleanu SO_2 process references include:

Cottrell, O. P., "Edeleanu Refining Process May Be Applied to Treat Practically All Petroleum Fractions," *Oil & Gas J.*, p. 64, Nov. 30, 1933.
Edeleanu, L., "Refining with Liquid SO₂," *J. Inst. Pet. Technology*, pp. 900–920, December, 1932.
Edeleanu, L., "New Method of Refining Petroleum by Liquid SO₂," *Nat. Pet. News*, pp. 41–43, August, 1926.
Kain, W., "Improvement in Solvent Refining of Lubricating Oils by Edeleanu Process," *Ref. Nat. Gaso. Mfr.*, pp. 553–557, November, 1932.
Oil & Gas J., "Edeleanu Process," pp. 214–215, Mar. 29, 1951.
Pet. Ref., "Modified Edeleanu Process," pp. 237–238, September, 1951.

Pet. Ref., "SO₂ Extraction," p. 230, September, 1962.
Plank, R., "Edeleanu SO₂ Process Used in Treating Lubricating Oils," *Nat. Pet. News*, pp. 63–71, November, 1928.

Chlorex Extraction

Chlorex Extraction is a single-solvent process which exhibits good selectivity and high solvent power at low extraction temperatures. Typical feedstocks are Mid-continent stocks for viscosity-index improvement or highly paraffinic oils (derived from Pennsylvania crudes) for improved viscosity indexes and increased yields from percolation filters.

Stability of chlorex obtained from the commercial product is improved by removal of the small amount of hydrochloric acid present.

Pennsylvania residual stocks contain resins but are almost free of asphaltenes. This permits refining with single solvents without deasphalting. Extraction temperatures are 100 to 125°F. Mid-continent distillates or deasphalted residue can be extracted without wax removal at 30 to 80°F. High chlorex specific gravity facilitates separation of the two liquid phases. However, even though extraction efficiency increases directly with increased solvent-oil ratio and inversely with temperature, high wax-containing charge stocks would require dewaxing if high recoveries and low corrosion rates are to be maintained at the efficient subatmospheric-temperature treating conditions.

Countercurrent multistage extractors (four to seven stages) or extraction towers are employed for chlorex extraction. The raffinate phase contains 15 to 25% solvent. Chlorex is recovered at 300 to 325°F at a vacuum of 26 to 28 in. of Hg using steam.

This process was developed by Standard Oil Co. (Indiana) in the early 1930's and was used extensively through the 1940's and early 1950's for the production of high-quality lube stocks. It has now been largely replaced by more modern processes.

Chlorex Extraction references include:

Adams, G. F., et al., "Route to Waxes and Lubes Modernized at DX's Refinery," *Pet. Ref.*, pp. 189–194, September, 1961.
Bahlke, W. H., et al., "Commercial Application of the 'Chlorex' Process in the Manufacture of High Grade Motor Oils," *Oil & Gas J.*, p. 60, Oct. 26, 1933.
Barton, P. D., "Modern Lubricating-oil Plant at Marcus Hook," *Pet. Eng.*, pp. 206–218, January, 1944.
Kalichevsky, V. A., "Estimation of Properties of Solvent-refined Products," *Pet. Ref.*, pp. 93–98, January, 1947.
Nelson, W. L., "Comparison of Solvent Treating Processes," *Oil & Gas J.*, p. 113, Feb. 10, 1945.
Page, J. M., et al., "Production of Lubricating Oils by Extraction with Dichloroethyl Ether," *Ind. Engrg. Chem.*, pp. 418–423, April, 1933.
Williams, D. B., "Application of the Chlorex Process to Pennsylvania Lubricating Oils," *Nat. Pet. News*, pp. 26–32, May, 1935.

Nitrobenzene Extraction

This single-solvent process of high solvent powers is applicable to lubricating oils or residua from any crude. Charge materials varying between light distillates (150 SSU at 100°F) and heavy residua (950 SSU at 210°F) have been successfully processed. The presence of wax in the charge does not interfere with treatment, nor is deasphalting a prerequisite operation for the charge material. Clay treatment is normally employed for color and carbon-residue improvement.

High solvency power of nitrobenzene enables use of low extraction temperatures (30 to 100°F) and solvent-oil ratios of 0.5 to 2.5 to 1.0. Refrigeration equipment may be necessary. Nitrobenzene recovery from the oil takes place at about 350°F by application of vacuum and steam.

A "cascade" countercurrent five-stage treating system is employed to contact oil with the solvent. For a typical waxy Mid-continent SAE 30 lube distillate charge (25.5 API, 75 SSU at 210°F viscosity) treated with a 1 to 1 solvent-oil ratio at 50°F, the raffinate product yield was 75% (30.2 API, 63 SSU at 210°F viscosity).

This process was developed and is commercially employed by Atlantic Refining Co. Currently (1963) about 2,500 B/D of Nitrobenzene Extraction capacity for lube oils is in use in U.S. refineries.

Nitrobenzene Extraction references include:

Ferris, Jr., S. W., and W. F. Houghton, "The Nitrobenzene Process for Lubricating Oils," *Ref. Nat. Gaso. Mfr.*, pp. 560–567, 581–583, November, 1932.

Ferris, Jr., S. W., et al., "The Nitrobenzene Process—Installation and Operating Costs," *Ref. Nat. Gaso. Mfr.*, pp. 435–444, December, 1933.

Foster, A. L., "Refining Lubricants by Solvent Extraction," *Pet. Engr.*, pp. 40–46, September, 1940.

Myers, W. A., "Superior Lubricants via the Nitrobenzene Process," *Oil & Gas J.*, pp. 81–91, Mar. 19, 1936.

Oil & Gas J., "Lube-capacity Cushion Grows," pp. 128–129, Apr. 2, 1962.

Pet. Ref., "Nitrobenzene Extraction," pp. 56–57, September, 1952.

Duo-Sol Extraction

Duo-Sol is a double-solvent lube-treating process for the simultaneous deasphalting and solvent treating of a lube or any residual or distillate base stock. Propane is employed as the paraffinic solvent for the deasphalting operation (dissolves the oil as

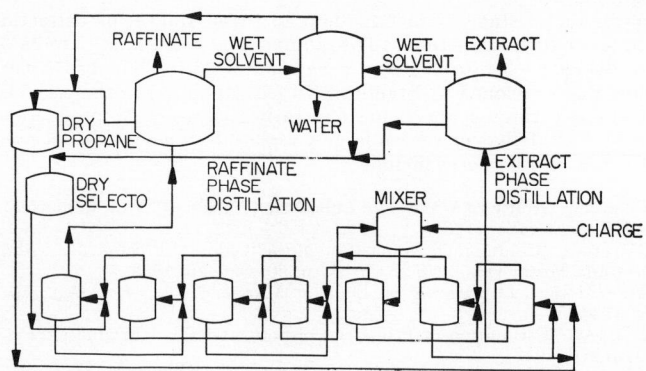

FIG. 3-53. Duo-Sol extraction.

the raffinate and precipitates the asphalt). Cresylic acid, usually containing from 20 to 40% phenol, is the naphthenic solvent (dissolves the precipitated asphalt and the undesirable aromatics, naphthenes, color bodies, and low-VI constituents). (Pure phenol may be used.) The cresylic acid–phenol mixture is referred to as "selecto." Because of the dilution of oil with propane, the viscosity of the raffinate phase is low and its separation from the extract phase is relatively easy. Under normal treating temperatures the solubility of wax in propane is sufficiently high to permit solvent extraction of most oils in the undewaxed state.

Commercial units are multistage (seven to nine) counterflow type. Fresh oil is introduced into the extraction system (Fig. 3-53) two or three stages from the extract exit end. Propane and selecto are introduced at the extract and raffinate outgoing stages, respectively.

Operating conditions include 120 to 150°F extraction temperature, although certain charge stocks have been treated at as low as 90°F. Solvent-to-oil ratios are 4.0 to 1.0 (or higher) by weight and by volume for propane and selecto, respectively. The phenol content of selecto is 35 to 40% by volume or higher. Water content of selecto is held below 0.15%.

The number of variables in Duo-Sol Extraction is considerable. Normally speaking the following generalities apply: (1) At constant propane-to-selecto ratio, an increase in total solvents-to-oil ratio improves viscosity index and carbon residue of the raffinate; (2) at constant selecto-to-oil ratio, an increase in propane lowers viscosity index and may lower or increase carbon residue; (3) at constant propane-to-oil ratio, an increase in selecto improves viscosity index and carbon residue.

This process was developed by Max B. Miller Co. and is licensed by the Milwhite Co., Inc. In 1963 there was approximately 53,000 B/D of Duo-Sol Extraction capacity in U.S. refineries for treating lubricating-oil stocks.

Duo-Sol Extraction references include:

Albright, J. C., "Union Oil Two-solvent Lube Unit Operating," *Pet. Engr.*, pp. C11–14, February, 1950.
Barton, P. D., "Modern Lubricating-oil Plant at Marcus Hook," *Pet. Engr.*, pp. 206–218, January, 1944.
Cobb, M. L., "Double Solvent Extraction of Residual Stocks," *Pet. Engr.*, pp. C47–56, May, 1949.
Hightower, J. V., "Refining Lubricating Oil by Solvent Extraction," *Chem. Engrg.*, pp. 82–85, April, 1935.
Nat. Pet. News, "New Solvent Lubricating Oil Plant Designed for Varying Blends of Crudes," pp. R7–10, Jan. 3, 1945.
Oil & Gas J., "Duo-Sol Extraction," pp. 210–211, Mar. 29, 1951.
Pet. Ref., "Duo-Sol," p. 223, September, 1962.
Tuttle, M. H., and M. B. Miller, "Duo-Sol Process in Manufacture of Lubricating Oils," *Ref. Nat. Gaso. Mfr.*, p. 453, December, 1933.
Tuttle, M. H., "Performance, Flexibility of the Duo-Sol Process," *Oil & Gas J.*, pp. 13–14, Jan. 7, 1935.

Liquid SO$_2$–Benzene Process

This is a mixed-solvent process for the treating of lubricating oils to effect viscosity-index improvement (and dewaxing, simultaneously, if conditions are so chosen). The liquid SO$_2$ is a highly selective solvent for aromatic and other nonparaffinic hydrocarbons but has low solvent capacity. Mixing benzene with the SO$_2$ increases the solvent capacity and retains the selectivity.

Varying the percentage of benzene in the solvent mixture makes it possible to select the most advantageous treating conditions for any charge stock in order that the desired specification product can be obtained. At a given extraction temperature, an increase in the percentage of benzene increases the solvent power of the mixture.

The design of the plant is similar to that of a single-solvent process, but the solvent recovery system is more elaborate to permit separate recovery of the solvents.

Extraction temperature is about 25°F. The mixed solvent-to-oil ratio is about 2.0 to 1.0. Finished lubricating oils have 90 to 100 viscosity index, with improved carbon residue and oxidation stability. Treating of a Gulf Coast lube stock (20.0 API, 66 SSU viscosity at 210°F, 23 VI, 0.24 wt-% carbon residue) with 25% (volume) benzene and 75% SO$_2$ yielded a 74% raffinate product (25.3 API, 60.5 SSU viscosity at 210°F, 63 VI, 0.06 wt-% carbon residue).

In 1963 there was approximately 23,000 B/D of SO$_2$-benzene processing in use in U.S. refineries for treating lubricating-oil stocks.

Liquid SO$_2$–benzene process references include:

Albright, J. C., "Process Description—Liquid SO$_2$–Benzene," *Nat. Pet. News*, p. 25, October, 1935.
Armistead, Jr., G., "Improved Technology Equals Better Lube Oils," *Oil & Gas J.*, pp. 126–131, Apr. 27, 1946.
Cottrell, O. P., "Scope and Flexibility of the Edeleanu Process," *Ref. Nat. Gaso. Mfr.*, pp. 432–434, December, 1933.
Foster, A. L., "Refining Lubricants by Solvent Extraction," *Pet. Engr.*, pp. 40–46, September, 1940.
Gard, E. W., and E. G. Ragatz, "Fundamentals of the Solvent Refining Operations as Applied to the Petroleum Industry," *Oil & Gas J.*, pp. 49–52, June 6, 1940.
Kain, W., "Improvement in Solvent Refining of Lubricating Oils by Edeleanu Process," *Ref. Nat. Gaso. Mfr.*, pp. 553–557, November, 1932.
Stockman, L. P., "Liquid SO$_2$–Benzol Process," *Oil & Gas J.*, p. 6, Aug. 16, 1934.

LUBES—DEWAXING

Dewaxing processes are designed to remove wax from lubricating oils in order that the product will exhibit good fluidity characteristics at low temperatures (low pour point). The mechanism of solvent dewaxing can be either the separation of wax as a solid which has been crystallized from the oil solution at low temperature or the separation of wax as a liquid which has been extracted at temperatures above the melting point of the wax through preferential selectivity of the solvent. This latter method has not been developed to the commercial stage. The former mechanism is the basis upon which all existing commercial solvent dewaxing processes of significant importance operate.

Solvents for dewaxing should (1) have substantially complete solvent action on the wax-bearing oil between 100°F and minimum chilling temperature of the oil-solvent mixture; (2) have complete solvent action on the normally liquid components present in the waxy oil but substantially no solvent power for the wax precipitated at the minimum chilling temperature; (3) be readily available at low cost; (4) be chemically stable, noncorrosive, and not highly toxic; (5) have a boiling point that permits recovery from oil and wax by distillation; (6) be unaffected and easily separated from water; and (7) have latent-heat and specific-heat characteristics conducive to economical refrigeration.

All the processes have the following steps in common: Feedstock is contacted with solvent; mixture is chilled and precipitated wax is separated; solvent is recovered from wax and dewaxed oil for reuse. Depending upon the process type, outside or autorefrigeration is employed; wax separation is accomplished through filtration or centrifuging.

Solvent dewaxing processes are often carried out in conjunction with wax deoiling processes. Modern solvent methods, which began to be employed in significant amounts commercially in the middle 1930's, have superseded the old methods, which included cold settling, filter pressing, and centrifuging, using naphtha. However, the same basic principles exist today that governed the older nonsolvent processes. In 1963, the total solvent dewaxing capacity in U.S. refineries was slightly higher than 160,000 B/D.

General lube dewaxing references include:

Deen, H. E., and G. R. Williges, "Tests Show Additives Can Up Dewaxing Thruput," *Pet. Ref.*, pp. 143–146, September, 1963.
Kalichevsky, V. A., "Solvent Dewaxing Processes," *Pet. Proc.*, pp. 145–148, February, 1949.
King, E. P., "Some Aspects of Lubricating Oil Manufacture," WPRA Regional Meeting, Beaumont, Tex., February, 1953.
Mallow, J. E., "Better Lubrication at Lower Cost," *Oil & Gas J.*, p. 101, Apr. 28, 1958.
Nelson, W. L., "Cost of Lube Rerun Distillation Operations," *Oil & Gas J.*, p. 96, May 1, 1961.
Nelson, W. L., "Cost of Lube Protection," *Oil & Gas J.*, pp. 107–108, Jan. 15, 1962.
Nelson, W. L., "Summarizing Lube Manufacture—Naphthene Base," *Oil & Gas J.*, p. 43, Dec. 24, 1962.
Nelson, W. L., "Summarizing Lube Manufacture—Paraffin Base," *Oil & Gas J.*, p. 137, Feb. 11, 1963.
Rautschka, R., et al., "Examine Solvent Dewaxing," *Pet. Ref.*, pp. 165–168, March, 1957.
Stormont, D. H., "World's Most Modern Lube-oil Terminal," *Oil & Gas J.*, pp. 84–87, July 16, 1962.
Zurcher, P., "Notes on Dewaxing-Removal of Petroleum," *Pet. Ref.*, pp. 121–126, November, 1951.

Solvent Dewaxing

This process, sometimes erroneously referred to as the benzol-acetone process, employs any single or mixed solvent, except hydrocarbon solvents having less than five carbon atoms. Mixtures of methyl ethyl ketone (MEK) and toluol are now in general use. Other ketones are also used, either alone or mixed with an aromatic solvent. Examples of single solvents employed by one or more licensees of the solvent dewaxing process are methyl isobutyl ketone and methyl n-butyl ketone.

The composition of the solvent mixture depends largely upon the type of stock being dewaxed. The more paraffinic stocks are dewaxed with a solvent containing less ketone than is generally used for dewaxing naphthenic or asphaltic stocks. The use of a solvent mixture of this type is an example of the dual opposing solvent dewaxing theory—both oil and wax are relatively insoluble in one solvent (MEK), while both oil and wax are soluble in all proportions in the other solvents (benzol and toluol). Hence, proper proportioning of the two opposing solvent actions permits control of wax solubility.

The solvent and charge oil are mixed (Fig. 3-54) with a solvent-oil ratio varying between 1.0 to 1.0 and 4.5 to 1.0, depending upon the nature and viscosity of the charge. The mixture is then chilled through heat exchange and outside refrigeration to a temperature a few degrees below the desired pour point of the dewaxed oil (e.g., −12°F for cylinder stock charge). Low pressures are employed throughout the process. The chilled oil and precipitated wax are then filtered, normally using a continuous rotary vacuum filter. Wash solvent is used on the wax filter cake before it is removed. The solvent-oil mixture is sent to the solvent recovery system for distillation in a multiple-effect evaporating unit. The solvent is removed from the wax slurry mix in a single-effect evaporating unit.

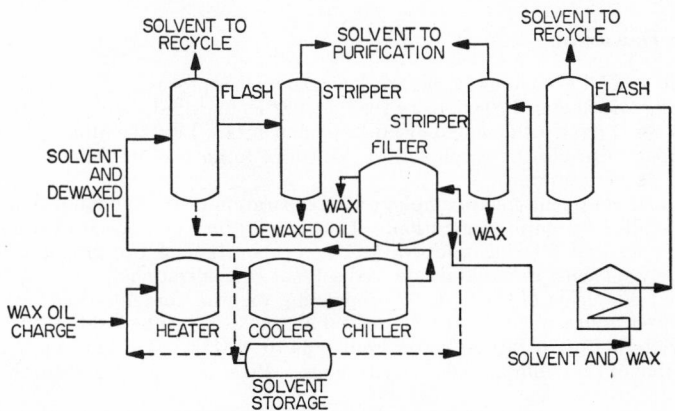

Fig. 3-54. Typical lube dewaxing flow diagram.

Total direct operating costs for MEK dewaxing for a 4,000-B/D unit range from 37.0 to 65.0 cents per bbl, based on 1962 values.

This process is the most widely used solvent dewaxing process in the petroleum industry today. In 1963, over 80% of U.S. refinery solvent dewaxing processing was of this type, utilizing MEK or mixtures of MEK and toluol, MEK and benzol, etc. The first commercial unit was installed in 1927. It is licensed by the Texaco Development Corp.

MEK Dewaxing references include:

Chenault, W. B., "Sinclair's Solvent Dewaxing Unit," *Oil & Gas J.*, p. 49, Aug. 3, 1939.
Ebner, E. E., and F. T. Mertens, "Solvent Dewaxing and Recrystallizing Units Included in New Lubricating Oil Refinery," *Nat. Pet. News*, pp. R265–270, May 3, 1944.
Gee, W. P., et al., "Two-solvent System Offers Advantages in Solvent Dewaxing," *Nat. Pet. News*, pp. 29–33, May, 1936.
Gover, F. X., and G. R. Bryant, "Solvent Dewaxing of Oil with Benzene and Acetone," *Ref. Nat. Gaso. Mfr.*, pp. 222–228, December, 1933.
Hall, F., and B. Y. McCarty, "Miscibility Relations in Solvent Dewaxing," *Nat. Pet. News*, pp. R15–25, July 14, 1937.
Hinman, J. M., and R. R. Maddocks, " 'Reverse Sequence' Dewaxing Cuts Solvent, Refrigeration Needs," *Pet. Proc.*, pp. 1215–1218, November, 1949.
Hyd. Proc., "Solvent Dewaxing," p. 223, September, 1964.
Mitchell, R. S., "MEK Drying Discussed at Rochester A. I. Ch. E. Meet.," *Oil & Gas J.*, p. 337, Oct. 4, 1951.

Mueller, A. J., "Mixture of Ketones Used to Dewax Motor Lubricants," *Oil & Gas J.*, pp. 54–56, Apr. 18, 1940.
Myers, W. A., "Atlantic Refining Co. Uses Mixture of Acetone-Benzene for Dewaxing," *Oil & Gas J.*, pp. 78–80, Mar. 18, 1935.
Oil & Gas J., "Operating Cost—MEK Dewaxing," p. 157, Oct. 30, 1961.
Oil & Gas J., "Solvent Dewaxing," p. 127, Apr. 3, 1961.
Pet. Proc., "First Use of Centrifugal Compressors for Plant Refrigeration," pp. 149–150, February, 1949.
Skelton, W. E., "Solvent Refining and Dewaxing of Lubricating Oil," *Oil & Gas J.*, pp. 137–144, Feb. 22, 1947.
Smoley, E. R., et al., "Solvent Refining of Lubricating Oils," *Ref. Nat. Gaso. Mfr.*, pp. 535–540, November, 1938.
Smoley, E. R., and D. Fulton, "Modern Manufacture of Lubricating Oils," *Pet. Proc.*, pp. 594–608, August, 1947.
Thornton, Jr., D. P., "Pure Oil's First Solvent Lube Plant Now in Operation at Smiths Bluff," *Pet. Proc.*, pp. 245–247, March, 1950.
Tiedje, J. L., and D. M. MacLeod, "Higher Ketones as Dewaxing Solvent," *Pet. Ref.*, pp. 150–154, February, 1955.
Weber, G., "Ketone-Benzene Dewaxing Process Is Employed by Gulf at Port Arthur," *Oil & Gas J.*, pp. 46–48, July 16, 1936.

Propane Dewaxing

Lube oils and lube blending stocks are dewaxed by this process. Feedstocks ranging from paraffin distillates of 70 viscosity (SSU at 100°F) to cylinder stock of 250 viscosity (SSU at 210°F) have yielded zero or lower pour points. Like the other solvent processes, propane can also be employed to deoil wax to an acceptable market product (Propane Deoiling process).

Continuous rotary filters are employed for separation of the wax-oil mixture which has been chilled by autorefrigeration. The temperature is reduced at a controlled rate to the desired filtering temperature by evaporation of the propane, which is continuously replaced to maintain desired solvent concentrations.

Dewaxed oil yields of 60 to 85% (depending on wax content of oil) have been obtained from a deasphalted solvent-treated charge at solvent-oil ratios of 2.5 to 3.0 to 1.0. Filtering temperatures range from −34 to −41°F. The entire process, with the exception of the chilling cycle, is continuous. Pressure in most parts of the system is 180 to 200 psi.

This process is simple and economical. One fluid—propane—serves as solvent, diluent, refrigerant, blanketing gas, and filter blow back. In dewaxing oils of varying viscosities, the only solvent adjustment required is the ratio of propane to oil. Reduction of pressure on the oil-solvent mixture results in evaporation of the solvent with attendant refrigerating effect.

Total direct operating costs for a 5,000-B/D Propane Dewaxing unit range from 30.0 to 47 cents per bbl (1962 basis).

This process is licensed by M. W. Kellogg Co. It is employed commercially in some 13 refineries in the United States. In 1963, the total Propane Dewaxing capacity in these refineries was about 50,000 B/D.

Propane Dewaxing references include:

Anderson, A. P., et al., "Propane Deasphalting and Dewaxing of Mid-continent Residuum," *Oil & Gas J.*, pp. 103–106, June 4, 1936.
Bahlke, W. H., et al., "Dewaxing Oils in Propane Solution with Self-refrigeration," *Ref. Nat. Gaso. Mfr.*, pp. 229–234, December, 1933.
Berne-Allen, Jr., A., and L. T. Work, "Solubility of Refined Paraffin Waxes in Petroleum Fractions," *Ind. Engrg. Chem.*, pp. 806–812, July, 1938.
Bray, U. B., et al., "Results Obtained from Use of Propane During Dewaxing and Treating Lubrication Oil," *Oil & Gas J.*, pp. 14–16, Nov. 2, 1933.
Brown, E. E., "Continuous Rotating Filters in Propane Dewaxing," *Oil & Gas J.*, p. 107, June 4, 1937.
Conine, R. C., "Pennsylvania Refiners Learn of Advantages of Propane Dewaxing," *Oil & Gas J.*, pp. 17, 33, Dec. 20, 1934.
Cooke, M. B., et al., "Continuous Deasphalting, Deresining and Dewaxing by Sun-Alco Process," *Nat. Pet. News*, p. R186, April, 1938.

Griswold, J., "Mechanism of Paraffin-wax Filtration; Continuous Drum Filter Dewaxing of Cylinder Stocks in Propane Solution," *Trans. Amer. Inst. of Chem. Engrs.*, pp. 505–515, 1939.
Hyd. Proc., "Propane Dewaxing," p. 220, September, 1964.
Lacey, W. N., and B. H. Sage, "Properties of Pure Propane," *Oil & Gas J.*, pp. 49–52, Feb. 14, 1935.
Moulton, R. W., and V. K. Loop, "A Study of the Composition of Petroleum Waxes from Western Crude Oils," *Pet. Ref.*, pp. 161–164, April, 1945.
Nelson, W. L., "Operating Costs for Propane Dewaxing and the Older Dewaxing Processes," *Oil & Gas J.*, p. 101, Nov. 27, 1961.
Oil & Gas J., "Propane Dewaxing," pp. 208–209, Mar. 29, 1951.
Pet. Engr., "Manufacture of Lubricating Oils," pp. 72–78, December, 1942.

Liquid SO₂–Benzene Dewaxing

This process has been described under the section Lubes—Refining (Liquid SO_2–Benzene) from the standpoint of removing low-viscosity-index constituents from lubricating oil charge. The ratio of liquid SO_2 to benzene is considerably lower in dewaxing than in the lube refining process. The quantity of liquid SO_2 varies between 15 and 30% as compared with the greater than 50% volume employed in solvent refining.

As in Propane Dewaxing, SO_2 evaporation furnishes internal refrigeration for the precipitation of wax prior to removal via closed continuous rotary filters. This process can advantageously be employed to follow Liquid SO_2–Benzene extraction for viscosity-index improvement through adjustment of the solvent composition. Deoiling of slack wax to high-melting-point wax can also be accomplished with proper adjustment of the necessary process flow and variables.

Liquid SO_2–Benzene Dewaxing references include:

Bray, U. B., et al., "Results Obtained from Use of Propane During Dewaxing and Treating Lubricating Oil," *Oil & Gas J.*, pp. 14–16, Nov. 2, 1933.
Bray, U. B., "Solvent Treatment of California Lubricating Stocks, Particularly with SO₂, SO₂–Benzene," *The Science of Petroleum*, Oxford University Press, 1938, vol. III, pp. 1893–1903.
Cottrell, O. P., "Edeleanu Refining Process May Be Applied to Treat Practically All Petroleum Fractions," *Oil & Gas J.*, p. 64, Nov. 30, 1933.
Kalichevsky, V. A., and K. A. Kobe, "Benzene–Liquid SO₂ Dewaxing Process," *Petroleum Refining with Chemicals*, Elsevier Publishing Company, Amsterdam, 1956.
Oil & Gas J., "Dewaxing with Liquid Sulfur Dioxide–Benzene," p. 131, Dec. 29, 1938.
Stephens, M. M., and O. F. Spencer, "Liquid SO₂ Benzene," *Petroleum Refining Processes*, The Pennsylvania State University Press, University Park, Pa., 1956.

Separator-Nobel Dewaxing

This process is referred to as the S-N Dewaxing process, or the trichloroethylene process. The solvent is a chlorinated hydrocarbon and process flow is essentially as described in the solvent dewaxing flow diagram (Fig. 3-54). One difference, however, does exist. Most chlorinated hydrocarbon solvents have high specific gravities (greater than 1.0), and for this reason centrifuging rather than filtering methods are used for separating the wax.

In practice the solvent-oil ratio ranges from 0.67 to 1.5 to 1.0. Charge and solvent are contacted at 110 to 120°F, and the mixture is then chilled to a temperature of 5 to 20°F below the desired pour point of the product, depending upon the properties of the charge stock. Cooling rate varies between 8 and 15°F/hr. The chilled solution is pumped to vaportight centrifuges, from which dewaxed oil is discharged (from the center of a centrifugal bowl) under relatively low centrifuge speeds—8,000 to 9,000 rpm.

Trichloroethylene is removed from the oil and wax phases by distillation at about 230°F, followed by steam stripping.

Separator-Nobel Dewaxing process references include:

Backlund, N. O.. "Dewaxing and Acid-refining Mineral Oils." *J. Inst. Pet. Technology*, pp. 1–34, 1933.

Foster, A. L., "The Technology of Refining Processes," *Pet. Engr.*, pp. 33–40, August, 1940.
Jones, L. D., "De Laval Separator-Nobel Process," *The Science of Petroleum*, Oxford University Press, London, 1938.
Kalichevsky, V. A., "Separator-Nobel Dewaxing Process," *Modern Methods of Refining Lubricating Oils*, Reinhold Publishing Corporation, New York, 1938.

Bari-Sol Process

This dewaxing process, like the S-N process, employs a solvent which is heavier than either the oil to be treated or the wax to be separated.   Like the S-N process also, the heavy chlorinated solvent-wax-containing hydrocarbon mixture is amenable to separation by centrifugal dewaxing operation.

The solvent employed commercially is a mixture of ethylene dichloride (78%) and benzol (22%); the composition will vary with the type of stock to be dewaxed.   Benzol is required, as ethylene dichloride has insufficient solvent power toward the oil constituents at the lower temperatures.

Solvent-oil ratios of about 3.0 to 1.0 are employed.   The mixture is heated to about 110°F and then chilled by ammonia refrigeration.   A chilling temperature of −8° to −12°F has been employed in dewaxing Mid-continent stocks to a 0°F pour point.

After centrifuging, the wax phase is diluted with further quantities of solvent (solvent-oil ratio 8.0 to 1.0) and recentrifuged at a lower temperature than the first operation, −15° to −20°F.   Solvent is recovered from the oil and wax phases at a temperature of 290°F.   This process has been developed by the Sharples Specialty Co. and Max B. Miller and Co., Inc.   Currently (1963) about 2,000 B/D of Bari-Sol solvent dewaxing capacity exists in U.S. refineries.

Bari-Sol process references include:

Albright, J. C., "Solvent Dewaxing Process at Ponca City Refinery," *Ref. Nat. Gaso. Mfr.*, pp. 287–292, March, 1936.
Armistead, Jr., G., "Better Lube Oils," *Oil & Gas J.*, pp. 126–131, Apr. 27, 1946.
Backlund, N. O., "Dewaxing and Acid-refining Mineral Oils," *J. Inst. Pet. Technology*, pp. 1–24, March, 1933.

Urea Dewaxing

The Urea Continuous Dewaxing process is highly selective and, in contrast to other dewaxing techniques, is carried out without the use of refrigeration.   Evaluation of the time, temperature, and separation conditions on a commercial-process basis indicates that Urea Dewaxing could not usually compete economically with other solvent dewaxing processes for processing the regular motor lubricating oil stocks.   However, when applied to lighter lubricating materials (viscosity 50 to 220 SSU at 100°F, pour point 0 to 15°F, gravity 26 to 30 API) that have already been dewaxed using conventional methods, Urea Dewaxing yields products (viscosity 64 to 246 SSU at 100°F, pour point −25 to −55°F, gravity 25 to 29 API) that are particularly useful for refrigerator oils, transformer oils, automatic-transmission fluids, hydraulic fluids, and Arctic lubricants.   Product yields from this process range from 85 to 95 vol-%.

Process flow for Urea Dewaxing is almost the same as that shown for typical dewaxing (Fig. 3-54).   Charge and solvent urea are not subjected to refrigeration but are mixed continuously in a stirred contactor.   Urea and an activator (water, esters, ketones, or alcohols such as methanol) form an adduct with the paraffinic waxes of the charge.   The activator is used to complete adduct formation in a reasonable time.   Effluent from the contactor is filtered, the dewaxed oil is separated from the solvents by evaporation, and the urea and wax complex is decomposed in the urea recovery system.

Urea Dewaxing references include:

Brooks, B. T., et al., *The Chemistry of Petroleum Hydrocarbons*, Reinhold Publishing Corporation, New York, 1954, vol. 1, pp. 241–274.
Domask, W. G., and K. A. Kobe, "Urea Complexes—Fundamentally Speaking," *Pet. Ref.* pp. 128–133, April, 1955.
Fetterly, L. C., "Urea Complexes—How They're Applied," *Pet. Ref.*, pp. 134–137, April, 1955.

Hoppe, A., and H. Franz, "Low Pour Oils Made by Urea Process," *Pet. Ref.*, pp. 221–224, May, 1957.
Oil & Gas J., "Urea Dewaxing," p. 168, Mar. 25, 1957.
Oil & Gas J., "Urea Helps Two-step Process Make Easy Pouring Lubes," p. 74, Apr. 1, 1957.
Rogers, T. H., et al., "Light, Low-pour Lubes from Waxy Crudes," *Oil & Gas J.*, pp. 107–110, July 15, 1957.
Rogers, T. H., et al., "Urea-Dewaxing Gets More Emphasis," *Pet. Ref.*, pp. 217–220, May, 1957.

Benzene-Acetone Dewaxing

The Benzene-Acetone Continuous Dewaxing process is one of the forerunners of the benzol-ketone-type processes—MEK and benzol, MEK and toluol—now in wide use in the refining industry. Process flow is generally that shown in Fig. 3-54. Process description is similar to that presented for the MEK Dewaxing process.

Actually acetone and methyl ethyl ketone (antisolvents) are pretty much interchangeable when used with benzene (miscible solvent) as a mixed solvent. The quantity of acetone required to suppress the solubility of wax in the mixed dewaxing solvent and to obtain satisfactory filtering rates is less than that of MEK. However, MEK is less likely to cause oil separation than acetone. Also, the boiling point of MEK is somewhat higher than acetone, which would tend to increase, somewhat, evaporation losses of the solvent acetone.

Circumstances that have favored use of benzene-acetone as a commercial lube dewaxing process include:

1. Lack of necessity for low-pour-point lubricants. Consistently mild climates often do not require lube pour points much below 10°F.

2. Processing economics in a refinery operating in a mild climate. Such a plant can recover the larger part of the benzene and acetone (because of their low boiling points) with low-pressure exhaust steam; the remainder of the solvent recovery can be made with 220-psig steam.

3. Supply and availability of benzene and acetone.

A 200-B/D benzene-acetone commercial lube dewaxing plant has been successfully operating at the Elbyn refinery in Moshaton, Greece. It was put on-stream in 1950.

Benzene-Acetone Dewaxing references include:

Kalichevsky, V. A., "Refining of Lubricating Oils—Dewaxing," *Pet. Engr.*, pp. C70–74, March, 1956.
Kalichevsky, V. A., and K. A. Kobe, *Petroleum Refining with Chemicals*, Elsevier Publishing Company, Amsterdam, 1956.
Konstas, A. S., "They Use Benzene-Acetone to Dewax," *Pet. Ref.*, pp. 241–242, September, 1957.

WAXES

Processes employed for the production of waxes are aimed at deoiling the petroleum wax concentrate (slack wax) which is a product of the dewaxing processes.

The wax deoiling operations, which are physical in nature, are similar to those used in dewaxing oils and often vary only in the choice of treating conditions. Oil as well as low-melting-point waxes are separated from the waxy concentrate charge to obtain a product of desired specifications. The deoiling of wax is sometimes combined with the dewaxing of oil operation.

Slack wax may contain 10 to 50% oil, which is removed by sweating (not discussed herein) or by refining with solvents.

Dewaxing of residual stocks yields wax concentrates containing microcrystalline waxes (petrolatum). Deoiling is accomplished only through solvent processes to yield 140 to 200°F melting-point products.

Solvents employed in the crystallization and production of wax are similar to those employed in the dewaxing of oils: naphtha, benzene-acetone, methyl butyl ketone, liquid sulfur dioxide and kerosine, liquid propane, and hexane and butyl acetate. Naphtha is usually not employed in the deoiling of microcrystalline wax.

MEK deoiling is now the most widely used process because it can be conducted in the same plant used for MEK dewaxing (the highest capacity dewaxing process in use in U.S. refineries). Direct operating costs for a 500-B/D MEK deoiling plant thus operated in 1963 might range from 50 to 85 cents per barrel of product wax. If MEK deoiling were not conducted in conjunction with the MEK dewaxing operation—thereby losing the combined plant labor burden advantage—operating costs would be 1.5 to 1.8 times higher than those cited.

General references on wax processing include:

Blaylock, J. E., "Determination of Oil Content of Paraffin and Microcrystalline Waxes," *Pet. Ref.*, pp. 133–135, November, 1952.

Brooks, K. W., "Wax Technology Faces Some Changes," *Oil & Gas J.*, pp. 89–90, Aug. 29, 1960.

Clarke, E. W., "Crystal Types of Pure Hydrocarbons in the Paraffin Wax Range," *Ind. Engrg. Chem.*, pp. 2526–2535, November, 1951.

Clary, B. H., "Petroleum Waxes Dominate Field," *Oil & Gas J.*, p. 103, Jan. 12, 1953.

Dean, J. C., "Processing Materials from Petroleum," *Chem. Engr. News*, pp. 1164–1167, July 10, 1945.

Edwards, R. T., "Composition and Tensile Strength of Paraffin Waxes," *Ind. Engrg. Chem.*, pp. 2555–2558, December, 1955.

Edwards, R. T., "A New Look at Paraffin Waxes," *Pet. Ref.*, pp. 180–186, January, 1957.

Fischer, W., "Try This New Way to Get Oil-free Wax," *Pet. Ref.*, pp. 236–238, September, 1957.

Ghublikian, J. R., et al., "Atlantic's New $10,000,000 Wax Plant," *Oil & Gas J.*, p. 83, Aug. 31, 1959.

Gibson, J. W., et al., "Isomerization of High Molecular Weight n-Paraffins," *Ind. Engrg. Chem.* pp. 113–116, February, 1960.

Lund, H. A., "Meeting Quality in Wax Crystallization," *Pet. Proc.*, pp. 326–331, March, 1952.

Nelson, W. L., "Wax and Lube Content of Crude Oil—An Approximation," *Oil & Gas J.*, pp. 122–124, Feb. 5, 1962.

Nelson, W. L., "Specifications of Microcrystalline Waxes," *Oil & Gas J.*, p. 105, Feb. 19, 1948.

Nelson, W. L., "References to Questions on Wax," *Oil & Gas J.*, p. 93, Jan. 4, 1960.

Nelson, W. L., "Lubes, Waxes Have High Packaging and Shipping Costs," *Oil & Gas J.*, p. 157, July 30, 1962.

Nelson, W. L., "Cost of Wax Deoiling," *Oil & Gas J.*, p. 116, May 7, 1962.

Phillips, J., "Refine Waxes for Suitable Properties," *Pet. Ref.*, pp. 193–198, September, 1959.

Rugar, G. F., "Chlorinated Paraffin Wax in Flame Retardant Thermoplastic," *Mod. Plastics*, pp. 148–149, January, 1953.

Sayre, J. E., and C. J. Marsel, "The $100 Million Market for Waxes," *Chem. Week*, pp. 29–50, Sept. 27, 1952.

TAPPI, "Tensile Strength of Paraffin Wax," pp. 154A–156A, October, 1954.

Turner, W. R., et al., "Properties of Paraffin Waxes," *Ind. Engrg. Chem.*, pp. 1219–1226, June, 1955.

Wax Fractionation

Wax Fractionation is a physical process operated for the production of wax of low oil content. The product wax can be crystalline (paraffin) or microcrystalline with various melting points, depending upon the nature of the charge. Crystalline waxes having oil contents less than 0.5% can be produced.

A selected portion of wax from a concentrate is crystallized in the presence of a large amount of solvent, which is added to the concentrate (Fig. 3-55) before crystallization is commenced. The solvent-concentrate mixture is first pumped to a chiller, where crystallization starts, and then to a filter. The wax cake is washed with cold solvent. The cake is continuously discharged from the primary filters, mixed with additional solvent, and the wax again continuously separated by means of a second filter after a final solvent wash. The discharge from the second filter is fed to the product wax and solvent recovery operation. Second-stage filtrate is recirculated to the feed of the first-stage filter.

Filtrate from the primary filters is charged to soft wax and solvent recovery opera-

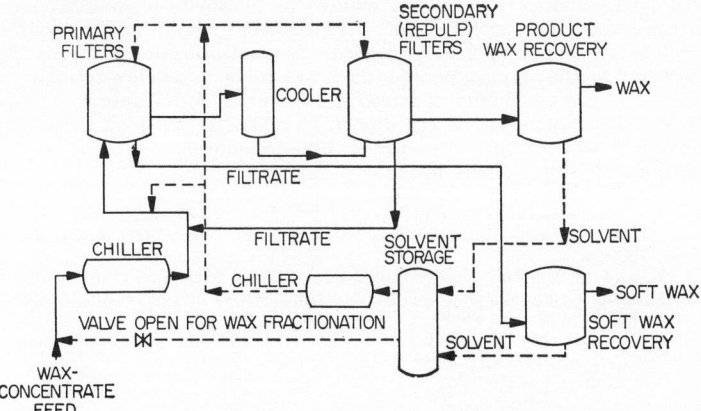

FIG. 3-55. Wax fractionation and manufacturing.

tion.   The solid wax is separated from the liquid phase (in both filtration stages) at temperatures suitable for the desired degree of fractionation, with the practical range being 5 to 60°F.

This process is often combined with oil dewaxing with overall solvent dosages varying from 3.5 to 9.0 to 1.0, depending upon the nature of the charge material.

This process is licensed by Texaco Development Corp.

Wax Fractionation references include:

Chem. Engrg. Prog., "New Paraffin Wax Plant—Design and Equipment," pp. 92–99, September, 1959.

David, R. A., and P. M. Huemmer, "Production of Waxes from California Crudes," *Chem. Engrg. Prog.*, pp. 174–176, April, 1947.

Hyd. Proc., "Wax Fractionation," p. 225, September, 1964.

Oil & Gas J., "Wax Fractionation," pp. 212–213, Mar. 29, 1951.

Schutte, A. H., "The Emulsion Deoiling Process," *Ref. Nat. Gaso. Mfr.*, p. 83, November, 1940.

Wax Manufacturing

This physical-type process is used on high-oil-content wax-containing charge stocks for the production of oil-free wax.   Depending upon the nature of the charge concentrate, the various melting-point crystalline (paraffin) products can have oil contents below 0.1 to 0.3%.   Microcrystalline waxes of low oil content (1% or less) can also be produced.

Originally, the process was a development of Texaco Development Corp. and Union Oil Co. and was licensed by Texaco.

Solvents such as MEK-benzol are employed at solvent-to-oil ratios of 2.5 to 1 (low-viscosity distillate charge source) to 7.0 to 1 (viscous residual charge source containing microcrystalline waxes).   The crystallization of the total wax or of a desired selected portion of the concentrate feed is effected by first cooling in the absence of solvent or in the presence of a limited and carefully controlled amount of solvent.   Solvent in small increments is then added until all the desired wax types have crystallized.   Filtration is accomplished after sufficient cold solvent is added to permit separation of the liquid and solid phases.   Operation and flow of the process after the "chilling" operation are essentially the same as that shown in Fig. 3-55 for wax fractionation.

In 1954, Union Oil started operating their wax-manufacturing facilities at their Oleum Refinery with water-saturated methyl isobutyl ketone (wet MIBK) as the deoiling solvent.   The solvent is used in the water-saturated condition and is added to the cooled melted slack wax feed as required for optimum crystallization.   Flow is as described above for the MEK-benzol-type solvent (shown in Fig. 3-55), except that no solvent-drying equipment is needed.   Solvent-to-fresh-feed ratios are very low—1 to 1 to 2 to 1 for typical paraffin slack wax feeds.   Such low ratios are due to full utilization of the techniques of incremental dilution and filtrate recirculation in this process.   The properties of the solvent-wet MIBK—good oil miscibility and wax antisolvency—permit maximum use of these techniques.

Wax manufacturing references include:

Albright, J. C., "High-melting Paraffin Waxes," *Pet. Ref.*, p. 121, July, 1947.

Brown, T., "Principles and Methods of Refining Paraffin Wax," *Oil & Gas J.*, p. 103, Dec. 28, 1939.

David, R. A., and P. M. Huemmer, "Production of High Melting Point Paraffin Waxes from California Crude Waxes," AIChE Meeting, San Francisco, Calif., August, 1946, reviewed in *Pet. Proc.*, pp. 86–87, October, 1946.

Ebner, E. E., and F. T. Mertens, "Oil Dewaxed and Wax Deoiled in Simultaneous Operation," *Pet. Ref.*, pp. 118–122, June, 1944.

Jenkins, V. N., "Refining of High Melting Point Waxes," *Oil & Gas J.*, p. 98, Mar. 25, 1943.

Oil & Gas J., "Wax Manufacturing," pp. 212–213, Mar. 29, 1951.

Pet. Ref., "Wax Manufacturing, MIBK," p. 235, September, 1962.

Pullen, E. A., et al., "Improve Wax Manufacture with MIBK," WPRA Annual Meeting, San Antonio, Tex., March, 1959.

Thornton, Jr., D. P., "New Method of Slabbing Wax," *Pet. Proc.*, pp. 963–967, September, 1951.

Warnecke, J. G., and P. S. Backlund, "Try MIBK in Your Wax De-oiling Unit," *Pet. Ref.*, pp. 189–192, August, 1958.

Continuous Wax Molding

This is a continuous automatic materials-handling operation for processing liquid wax into solid slabs of specified weight.

Microcrystalline or paraffin wax in liquid form is cooled or heated to a predetermined molding temperature and delivered to a jacketed steel hopper (Fig. 3-56),

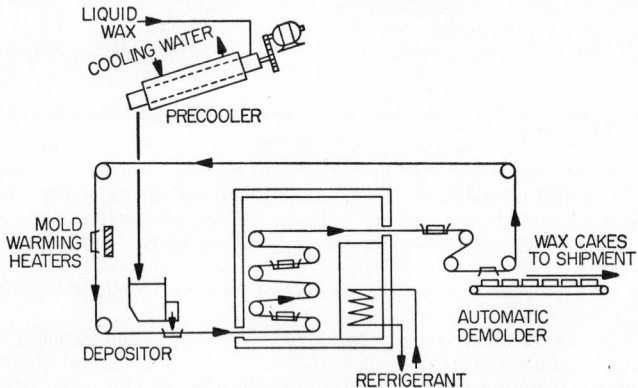

FIG. 3-56. Continuous wax molding.

from which it is removed by piston-in-cylinder metering into mold pans.   Metering is accurate to ±0.5 to 1.0 wt-% of a given stroke setting.   The conveyor carries the wax through a refrigeration unit in which the residence time and amount of cooling are determined by the characteristics and quantity of the wax being processed. Solidified wax slabs are then conveyed for packaging and for shipment.

The first unit was installed in December, 1950, at the Beaumont, Tex., refinery of Magnolia Petroleum (now Mobil Oil Co.).

Continuous Wax Molding references include:

Below, L. H., et al., "Wax is End Product, Not By-product," *Oil & Gas J.*, pp. 124–125, Apr. 13, 1959.
Chem. Engrg. News, "New Continuous Process Permits Wax to Be Molded at One-fifth Labor Cost," p. 3200, Aug. 6, 1951.
Jennings, L. H., et al., "Continuous Molding of Paraffin Wax Is Effected," *Pet. Ref.*, pp. 97–100, August, 1951.
Kalichevsky, V. A., "Continuous Wax Molding Lowers Cost," *Oil Forum*, pp. 288–291, August, 1951.
Kandra, S. F., and W. D. Manz, Jr., "Less Moisture Means Faster Wax-cake Molding," *Oil & Gas J.*, pp. 113–115, Jan. 5, 1959.
Keyes, B., "Continuous Molding of Petroleum Waxes," *Oil & Gas J.*, pp. 128, 131, 171–172, Apr. 14, 1952.
Thornton, Jr., D. P., "Lube Oil Manufacture," *Pet. Proc.*, pp. 245–247, May, 1950.

EXTRACTION OF AROMATICS

Solvent extraction processes for aromatics are continuous processes employing the unit operations of distillation, extraction, and absorption—alone or in combination—to separate aromatics from hydrocarbon mixtures.

The desired product from all these processes is normally the aromatics, per se, either for petrochemical usage or for blending into the gasoline stream as a high-octane component. However, the dearomatized raffinate—kerosine or light odorless paraffins—can also be the primary product for use as jet fuel or specialty nonaromatic solvents.

Feedstock can be almost any refinery stream—straight-run, cracked, or reformed naphtha—having a wide or narrow boiling range anywhere between 150 to 700°F. The feedstock preparation and operating conditions are dictated by the nature of the feed as well as the type of process with its attendant vapor-liquid, liquid-liquid, or liquid-solid phase relationships.

Investment costs for aromatics recovery processes are relatively high when compared with the cost of the aromatic producing processes—predominantly catalytic reformers. Investment costs on some small units might be as high as $2,000 per barrel of feed, with direct unit operating costs up to $1.50 per bbl.

Historically, commercial processes for the manufacture and subsequent separation of aromatics from petroleum came into extensive use at the outset of World War II. Coal-based chemicals (which were the chief commercial source of aromatics at that time) were insufficient to satisfy the high demand for nitration-grade toluene.

Along with the liquid-liquid extraction, selective adsorption, and simple and extractive distillation mechanisms employed to separate aromatics, azeotropic distillation is sometimes used commercially in the solvent recovery step of these processes. However, it will not be treated in this section.

General references for extraction of aromatics include:

Adams, G., and B. G. Casteel, "Petroleum Aromatics—What's Ahead?", *Oil & Gas J.*, pp. 171–175, Sept. 4, 1961.
Bramston-Cook, H. E., "Petroleum Plays a Growing Role in Manufacturing Aromatics," *Pet. Proc.*, pp. 1145–1148, August, 1952.
Cubbage, T. L., "Petrochemicals—Tops in Value and Getting Bigger," *Oil & Gas J.*, pp. 114–120, Mar. 27, 1961.
Derr, E. L., *Solvent Extraction*, ASTM D-2 Research Division, New Orleans, La., February, 1957.
Fedor, W. S., "Major Aromatics in Transition, 1961–1965," *Chem. Engrg. News*, pp. 116–134, Mar. 20, 1961; pp. 130–144, Mar. 27, 1961.
Haines, Jr., H. W., et al., "P-Xylene from Petroleum," *Ind. Engrg. Chem.*, pp. 1096–1103, June, 1955.
Maisel, D. S., "Aromatics from Petroleum," *Pet. Proc.*, pp. 1185–1192, August, 1953.
Mark, D., "BTX Aromatics—What Happened to the Gravy," NPRA Annual Meeting, San Antonio, Tex., Apr. 1, 1963.
Medcalf, E. C., et al., "Aromatics Recovery by Solvent Refining," *Pet. Ref.*, pp. 97–100, July, 1951.

Meerbott, W. K., et al., "Aromatics Manufacture over Platforming Catalyst," *Ind. Engrg. Chem.*, pp. 2026–2032, October, 1954.
Oil & Gas J., "Refiners Moving in on Aromatics," pp. 44–46, Aug. 25, 1958.
Oil & Gas J., "Special Benzene Properties Make Aromatics an Important Branch of Hydrocarbon Family," p. 106, Oct. 24, 1960.
Oil & Gas J., "Aromatics Pace Petrochemicals at Suntide," pp. 80–81, June 5, 1961.
Oil & Gas J., "Oil Industry Moves Further into Aromatics," pp. 116–117, July 3, 1961.
Oil & Gas J., "Joint Aromatics Unit Goes on Stream," pp. 102–103, Mar. 12, 1962.
Oil & Gas J., "Separating Aromatic Hydrocarbons," p. 105, Apr. 8, 1963.
Oil & Gas J., "Petrochemicals Growing 8–10% Yearly," pp. 46–48, Jan. 13, 1964.
Oostermeyer, J., "Petroleum Steps into Breach as Supply Source for Critical Materials," *Pet. Proc.*, pp. 651–654, September, 1947.
Pet. Proc., "Quick Freeze Helps Solve Problem of Recovering Para-Xylene," pp. 731–732, June, 1952.
Pet. Proc., "How to Recover Aromatics," p. 1197, August, 1955.
Pet. Proc., "Modified Fractional Crystallization Promises Better Xylene Separation," pp. 307–308, March, 1955.
Scheibel, E. G., "Find Best Solvent with This Chart," *Pet. Ref.*, pp. 189–194, May, 1961.
Scheibel, E. G., "Figure Multicomponent Extraction," *Pet. Ref.*, pp. 227–246, September, 1959.
Sherwood, P. W., "Whither Petrochemicals," *Oil & Gas J.*, pp. 73–79, Dec. 16, 1963.
Somekh, G. S., "How to Improve Aromatics Extraction," *Pet. Ref.*, pp. 157–163, October, 1963.
Sweeney, W. J., and E. H. McArdle, "Highly Aromatic Petroleum Solvent Naphthas," *Ind. Engrg. Chem.*, pp. 787–791, June, 1941.
Treybal, R. E., "Solvent Extraction," *Ind. Engrg. Chem.*, pp. 53–62, January, 1952.
Von Berg, R. L., and H. F. Wiegandt, "Liquid-Liquid Extraction," *Chem. Engrg.*, pp. 189–200, June, 1952.

Extractive Distillation

Extractive Distillation is a vapor-liquid phase process for continuous recovery of individual nitration-grade aromatics (benzene, toluene, or xylene) from appropriate petroleum fractions (usually catalytic reformate). A solvent is used to increase vapor-pressure differences among components.

Chemical feed pretreat, sulfuric and/or phosphoric acid, or clay feed pretreat might be necessary if olefins are present. Prefractionation (Fig. 3-57) concentrates a single aromatic into a close boiling cut. Aromatic concentrate is then extractively distilled with a solvent—usually phenol for benzene or toluene recovery and mixed cresylic

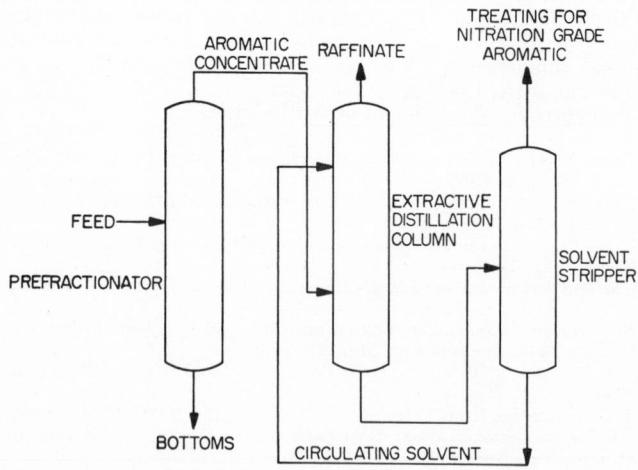

Fig. 3-57. Extractive distillation for aromatics recovery.

acids for xylene recovery—and removed as bottoms to a stripper for the solvent recovery. To meet nitration-grade specifications, acid treating and redistillation may be necessary.

Yields of 97 to 99% aromatic recovery of 99 + % purity are obtained. The prefractionation and extractive distillation column both have 60 to 70 trays. The solvent stripper contains about 30 trays.

The process has been licensed by Shell Development Co. but is rapidly being supplanted by its Sulfolane Extraction process. The first commercial unit came on-stream in 1940.

Extractive Distillation references include:

Dunn, C. L., and G. E. Liedholm, "Shell Extractive Distillation Process for Recovery of High Purity Aromatics," API Refining Meeting, San Francisco, May, 1952, reviewed in *Pet. Proc.*, pp. 842–843, June, 1952.
Gerster, J. A., "Advances in Distillation Separations," *Ind. Engrg. Chem.*, pp. 253–257, February, 1955.
Maisel, D. S., "Aromatics from Petroleum," *Pet. Proc.*, pp. 1185–1192, August, 1953.
Oil & Gas J., "Benzene Manufacture," pp. 204–205, Mar. 29, 1951.
Pet. Proc., "Aromatics—Nitration Grade," pp. 1224–1226, August, 1954.
Pet. Proc., "Extractive Distillation," p. 1198, August, 1955.
Pet. Ref., "Extractive Distillation for Aromatics Recovery," pp. 242–243, September, 1952.
Pet. Ref., "Extractive Distillation," p. 198, December, 1955.
Pet. Ref., "Shell Benzene Recovery Process," pp. 233–234, September, 1951.
Smith, A. S., and J. E. Funk, "Extraction of Aromatics from Hydrocarbon Mixtures," *Pet. Engr.*, Reference Annual, pp. 239–245, 1944.

Udex Extraction

Udex Extraction is a liquid-liquid phase recovery process for the selective extraction of high-purity aromatics from hydrocarbon mixtures through use of a solvent mixture of diethylene glycol and water (8 to 10%).

An outstanding feature of this process is its ability to handle wide-boiling-range feedstocks to yield a high-purity mixture of aromatics which can be clay treated and fractionated into nitration grades of benzene and toluene and a high-purity concentrate of xylenes and ethylbenzene. A mixture of C_9, C_{10}, and C_{11} aromatics as xylene fractionator bottoms is also a product if the boiling range of the charge is wide enough.

The process employs an efficient continuous multistage countercurrent extraction column. The feed, which may be pretreated to eliminate high conjugated dienes or alkenyl-aromatics, rises countercurrent to the descending solvent, which extracts all aromatics present (Fig. 3-58). The stripper recovers solvent (returned as recycle) from the rich-oil extractor bottoms. The extract overhead (from the stripper) is settled and sent to clay treating and fractionation for aromatics separation. The extractor overhead is essentially pure paraffins and leaves the system as raffinate from the water wash column overhead.

Adjustment of the solvent-hydrocarbon feed ratio allows control of extraction efficiency. Higher-boiling aromatic fractions require a greater proportion of solvent.

Pressures are just high enough to maintain liquid-phase conditions. Temperatures are relatively low.

This process is licensed by Universal Oil Products. The first commercial unit was placed on-stream at the Houston refinery of the Eastern States Petroleum Co. in 1952.

Udex Extraction references include:

Block, H. S., and R. C. Wackher, "Pure Aromatics from Cracked Naphthas," *Pet. Ref.*, pp. 145–149, February, 1955.
Broughton, D. B., and E. R. Fenske, "Fractionation of Mixed Aromatics from Udex Extraction," *Ind. Engrg. Chem.*, pp. 714–717, April, 1955.
Broughton, D. B., and H. W. Grote, "Use of the Udex Process for Refinery Octane Balancing," API Refining Meeting, New York, May 28, 1959.
Grote, H. W., "The Udex Process," *Chem. Engrg. Prog.*, pp. 43–48, August, 1958.
Jackson, W. K., et al., "Making High Purity Aromatics," *Pet. Proc.*, pp. 233–237, February, 1954.

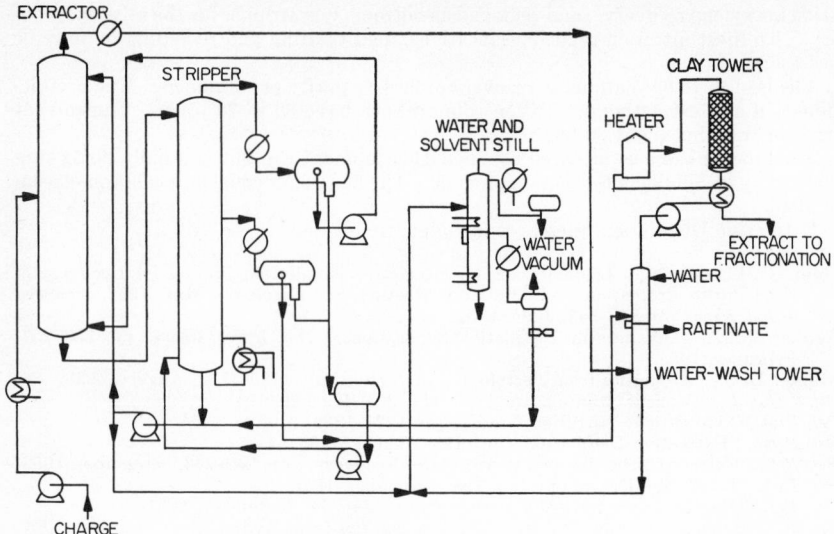

Fig. 3-58. Udex solvent extraction for aromatics.

Kerns, G. D., "Operating Cost Lowest in the Industry—Udex Unit," *Oil & Gas J.*, pp. 91–93, Feb. 29, 1960.
Maisel, D. S., "Aromatics from Petroleum," *Pet. Proc.*, pp. 1185–1192, August, 1953.
Oil & Gas J., "Udex," p. 128, Apr. 5, 1965.
Pet. Proc., "Udex," p. 1199, August, 1955.
Pet. Ref., "Udex Extraction," p. 304, November, 1957.
Read, D., "Platforming and Udex Extraction," API Refining Meeting, San Francisco, May, 1952, reviewed in *Pet. Proc.*, pp. 839–842, June, 1952.
Read, D., "Production of High-purity Aromatics for Chemicals," *Pet. Ref.*, pp. 97–103, May, 1952.
Reidel, J. C., "53 Million Gallons of BTX Annually—Capacity of Sun's $15,000,000 Marcus Hook Refinery," *Oil & Gas J.*, pp. 94–96, Mar. 29, 1954.
Resen, F. L., "New Process Unveiled," *Oil & Gas J.*, pp. 55–56, Mar. 3, 1952.
Stewart, L. D., et al., "Maybe Aromatics Production Is Your Dish," *Oil & Gas J.*, pp. 110–113, Feb. 6, 1961.
Thornton, Jr., D. P., "First Complete Udex 'BTX' Plant Makes Nitration-grade Products," *Pet. Proc.*, pp. 384–387, March, 1953.
Thornton, Jr., D. P., "First Udex Unit Recovers Benzene," *Pet. Proc.*, pp. 498–501, April, 1952.

Modified SO₂ Extraction

Modified SO_2 Extraction is a liquid-liquid phase process for the selective extraction of aromatics from hydrocarbon mixtures in virgin naphthas, catalytic reformates, or heavier petroleum distillates. The mixture is solvent-extracted (Fig. 3-59) with liquid sulfur dioxide, and the aromatic portion is concentrated in the extract phase, which is in turn contacted with a wash oil (a kerosine fraction in the case of a catalytic reformate charge) for removing the lower-boiling nonaromatics from the extract to the raffinate phases. The raffinate and extract streams are sent to respective stripper and fractionator systems to recover SO_2 and wash oil for recycle to the extraction tower.

Feed pretreatment for total catalytic reformate charge normally consists of addition of inhibitor to prevent resin formation in storage or under low-temperature operation of the extractor tower. Water is also removed from feed through use of alumina driers.

Unique aspects of this process for handling reformate feeds include (1) high recoveries (98.0 to 99.0%) of benzene and toluene-xylene fractions, which have been extracted simultaneously; (2) one-stage separation can be used to remove SO_2 from the raffinate; and (3) methanol has been utilized for the azeotropic separation of non-aromatics from benzene.

Primary extraction of aromatics and extract stripping are carried out at a temperature of $-20°F$. Investment costs can be $1,000 to $1,100 per barrel of charge with total operating cost, excluding return on investment, being around $1 per barrel of feed.

Historically, this process is a modification of that suggested by Edeleanu in 1907 for the removal of aromatics from kerosine by SO_2 extraction. The first commercial

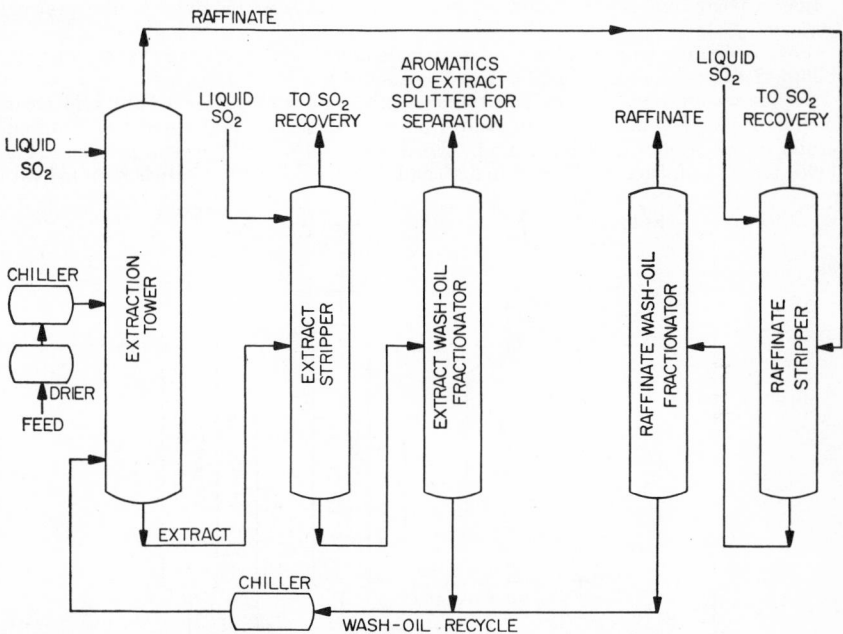

FIG. 3-59. Modified SO_2 extraction for aromatics.

installation of the process as it is used today was made by the Baytown Ordinance Works (now owned by Humble Oil & Refining Co.), Baytown, Tex., during World War II.

Modified SO_2 Extraction references include:

Arnold, R. C., and A. P. Lien, "Sulfur Dioxide Extraction of Sulfur Compounds and Aromatics," *Ind. Engrg. Chem.*, pp. 234–240, February, 1955.

Daly, J. B., and G. Weyermuller, "Aromatics by SO_2 Extraction," *Chem. Proc.*, pp. 10–12, June, 1955.

Maisel, D. S., "Aromatics from Petroleum," *Pet. Proc.*, pp. 1185–1192, August, 1953.

Marshall, C. H., "Production of Aromatics," *Chem. Engrg. Prog.*, pp. 313–318, June, 1950.

Oil & Gas J., "Edeleanu Process," pp. 214–215, Mar. 29, 1951.

Pet. Proc., "Modified SO_2 Extraction," p. 1200, August, 1955.

Pet. Ref., "Modified SO_2 Extraction for Aromatics Recovery," pp. 242–243, September, 1952.

Pet. Ref., "Modified Edeleanu Process for Recovery of Aromatics," pp. 237–238, September, 1951.

Pet. Ref., "Modified SO_2 Extraction," p. 202, December, 1955.

Ratliff, R. A., and W. B. Strobel, "New Look on How to Produce Aromatics by SO_2 Process," *Pet. Ref.*, pp. 151–155, May, 1954.

Resen, F. L., "New SO_2 Aromatic Extraction Unit Being Installed by Conoco at Lake Charles," *Oil & Gas J.*, pp. 313–316, June 16, 1952.

Wilkinson, W. F., et al., "The Sulfur Dioxide Extraction Process for Aromatic Hydrocarbon Recovery," *Chem. Engrg. Prog.*, pp. 257–262, May, 1953.

Arosorb

The Arosorb process separates aromatics (primarily) from various refinery streams (boiling anywhere between 150 and 700°F) through liquid-solid phase relationships. The process involves cyclic selective adsorption-desorption in fixed silica-gel beds. Olefins and nonhydrocarbon liquids can also be removed.

Feedstock is pretreated over activated alumina driers to reduce water (below 20 ppm) and remove poisons such as traces of nitrogen, sulfur, and oxygen compounds. Organic polar and readily polymerizable olefins are also removed in the pretreat section.

For low-boiling feedstocks the gel desorbent boils higher than the charge; for high-boiling feedstocks a lower-boiling desorbent is used.

In the recovery of benzene and toluene—nitration grade after a light acid treatment—of 99.0+% purity at high yields from a catalytic reformate charge, a desorbent crude xylene stream (65.0%) is used. Feed passes into one of several gel cases (Fig. 3-60) for 30 min, then is diverted to a second gel case. The first case is then fed with

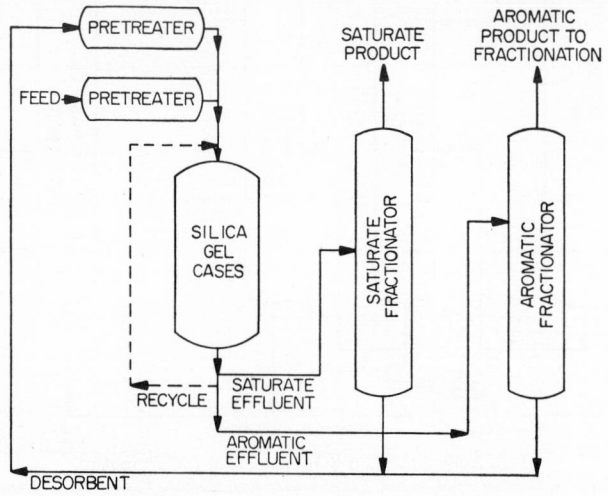

FIG. 3-60. Arosorb process for aromatics recovery.

the desorbent, which pushes the saturates out of the silica-gel bed and then displaces the adsorbed aromatics.

Approximate operating conditions are 130°F and 150 psi. Depending upon the size of the unit, investment costs are $700 to $2,200 per barrel of charge capacity and total operating costs, excluding return on investment, are $0.75 to $1.80 per barrel of feed.

For a typical 250 to 400°F catalytic reformate charge stock (63% aromatics), the yield of the aromatic concentrate was 59.0 vol-% (93% aromatics). The saturate concentrate contained 19% (volume) aromatics. For this particular operation, benzene (60 vol-%) and a saturate fraction (160 to 190°F) were the desorbent stream. This stream was the overhead from the two fractionators shown in Fig. 3-60.

Although the Arosorb process was originally developed by Sun Oil Co., Universal Oil Products Co. now has exclusive rights to offer this process to others as a part of its Unisorb process.

The first commercial unit was put on-stream by Petrocarbon Chemicals, Inc., Irving, Tex., in 1951.

Arosorb references include:

Chem. Engrg., "Adsorbing Aromatics," pp. 234–236, September, 1951.
Chem. Engrg., "Silica Gel Removes Aromatics," p. 374, February, 1953.

Chem. Engrg. News, "Silica Gel Employed for Separation of Aromatics," pp. 3015–3016, July 23, 1951.
Davis, W. H., et al., "The Arosorb Process, A New Refining Tool," *Oil & Gas J.*, pp. 112–120, May 19, 1952.
Davis, W. H., et al., "The Arosorb Process in Refinery Operations," API Refining Meeting, San Francisco, May, 1952, reviewed in *Pet. Proc.*, pp. 843–844, June, 1952.
Gaylor, P. J., "Continuous Selective Adsorption on Silica Gel," *Pet. Proc.*, pp. 527–529, April, 1952.
Guthrie, V. B., "Aromatics by Adsorption," *Pet. Proc.*, pp. 833–835, August, 1951.
Harper, J. I., et al., "The Arosorb Process," *Chem. Engrg. Prog.*, pp. 276–280, June, 1952.
Hirschler, A. E., and T. S. Mertens, "Liquid-phase Adsorption Studies Related to the Arosorb Process," *Ind. Engrg. Chem.*, pp. 193–202, February, 1955.
Maisel, D. S., "Aromatics from Petroleum," *Pet. Proc.*, pp. 1185–1192, August, 1953.
Pet. Engr., "New Adsorption Process Recovers Aromatics," pp. C21–24, August, 1951.
Pet. Proc., "Arosorb," p. 1201, August, 1955.
Pet. Ref., "Arosorb," pp. 238–239, September, 1952.
Pet. Ref., "Arosorb," p. 193, December, 1955.
Shuman, F. R., and D. G. Brace, "Versatile Process for Aromatics Recovery," *Pet. Engr.*, pp. C9–14, April, 1953.
Weber, G., "New Arosorb Process," *Oil & Gas J.*, pp. 58–59, July 19, 1951.

Cyclic Adsorption

Cyclic adsorption is used for the separation of aromatics from petroleum hydrocarbons by a cyclic selective adsorption-desorption operation in fixed silica-gel beds. Like the Arosorb process, this liquid-solid phase process appears to have its greatest utility in the recovery and concentration of nitration-grade aromatics with 99.0% yields of 99.0+% purity from catalytic reformate. The process can also be employed for dearomatization of stocks for the production of purified raffinates such as odorless paint thinners or high-quality kerosines.

An alumina drier and guard cases serve to remove water and sulfur and diolefins, respectively. Steps in the cyclic process (Fig. 3-61) are (1) extraction of the adsorb-

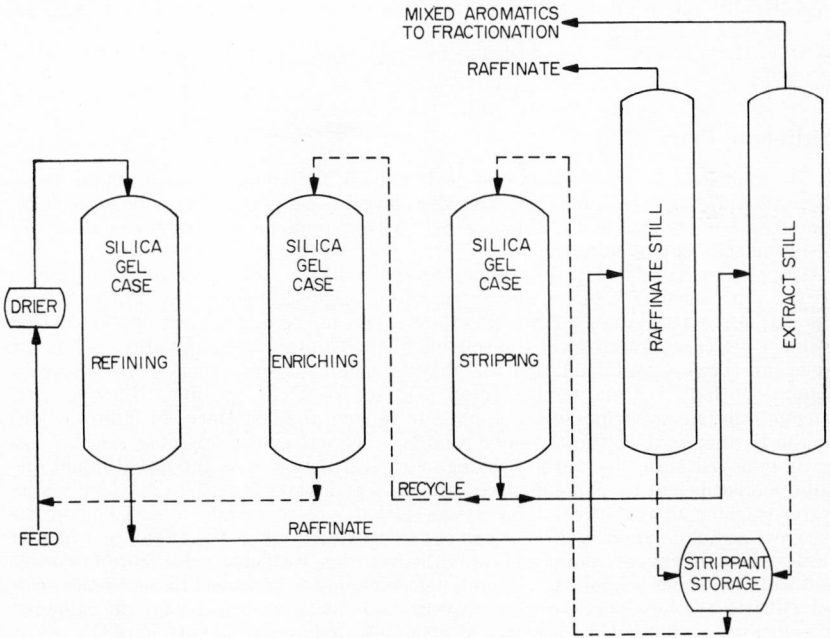

Fig. 3-61. Cyclic adsorption for aromatics recovery.

able material from the feed (refining), (2) concentration of the adsorbed phase (enriching), and (3) stripping for recovery of the extract and regeneration of the gel (stripping).

The refining period is of 30 min duration when catalytic reformate is the charge material. Kerosine charge requires a 2-hr period. The strippant employed for reformate is kerosine or pentane fraction. For kerosine treatment, light straight-run gasoline may be used. Separate fractionating columns handle the separation of the aromatics into pure products.

Employing a kerosine charge stock, refining enriching adsorption steps operate at about 80 to 100°F. The stripping cycle employs around 350°F. The highest pressure is that required to pump feed through the adsorption towers, generally not over 100 psig.

Proper choice of adsorbent for use in both the guard chambers and main adsorption columns is important because the cost of adsorption refining is markedly dependent upon selectivity, diffusivity, size, and life of the adsorbent. The first three properties are important in determining the size and investment required for a projected plant, and the rate of decline in activity of the adsorbent is an important factor in the operating cost of the plant. Investment costs of $525 and $1,000 per barrel of charge capacity have been estimated for units of 8,000 and 2,000 B/SD, respectively. However, as with most fixed-bed cyclic operations with attendant complex instrumentation, high operating costs preclude their use today.

The process is licensed by Chevron Research Co. No commercial plant has been installed to date.

Cyclic adsorption references include:

Eagle, S., and J. W. Scott, "Refining by Adsorption," *Pet. Proc.*, pp. 881–884, August, 1949.
Guthrie, V. B., "Aromatics by Adsorption," *Pet. Proc.*, pp. 833–835, August, 1951.
Humphreys, R. L., "Cyclic Adsorption Refining," *Pet. Ref.*, pp. 99–101, September, 1953.
Pet. Proc., "Cyclic Adsorption," p. 1202, August, 1955.
Pet. Ref., "Cyclic Adsorption Refining," p. 85, September, 1954.
Seybert, E. K., "Silica Gel Adsorption for Separation of Liquid Hydrocarbons," *Pet. Proc.*, pp. 1150–1153, August, 1952.
Spengler, G., and K. Krenkler, "Selective Adsorption of Hydrocarbons," *Pet. Ref.*, pp. 111–114, July, 1952.
Weil, B. H., "Adsorption Combined with Desulfurization," *Pet. Proc.*, p. 1181, November, 1949.

Sulfolane Extraction

The Sulfolane Extraction process is based on the combination mechanisms of liquid-liquid extraction and extractive distillation. The extractive solvent is sulfolane—$(CH_2)_4SO_2$—made by reacting SO_2 with butadiene and hydrogenating the resulting sulfolene to sulfolane.

The process was developed by Shell Development Co., primarily for the manufacture of pure benzene, toluene, and xylenes from catalytic reformate. Bottoms from the extractive distillation column (Fig. 3-62) are a mixture of solvent and aromatics. The extract is separated from the solvent in the recovery column. Because of the excellent thermal and oxidation stability of sulfolane, operation of the recovery column at high bottom temperatures (345 to 355°F) is possible. Furthermore, vacuum and steam stripping are applied to remove the last traces of hydrocarbons in the bottom part of the recovery column. Solvent in the outgoing streams has to be removed either by water washing or by distillation. As the liquid-liquid distribution coefficient for sulfolane over aromatics and water is close to 1, which makes water washing unattractive, a rectifying section for the extract is installed in the recovery column, yielding a near-solvent-free extract. For the raffinate which is saturated with solvent, the liquid-liquid distribution coefficients are more favorable and water washing is applied. Overall heat economy is improved by using the near-solvent-free overhead water of the recovery columns as wash water for the raffinate. The spent wash water is then reused as stripping medium in the bottom of the recovery column.

Various units of this type are now in operation.

Units currently using diethylene glycol as the extractive solvent can be converted, with slight modifications, to sulfolane use. Advantages claimed for these units are (1) greater selectivity for aromatics, (2) reduced heat requirements (about 50% per barrel of feed), (3) increased unit capacity (about 50%), if required. One converted diethylene glycol unit in Europe is in commercial use employing sulfolane.

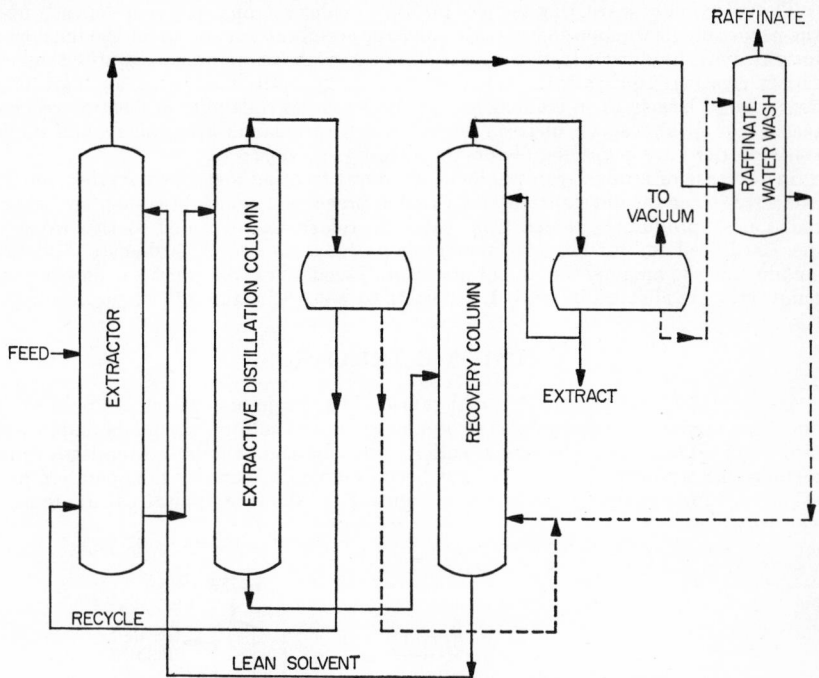

Fig. 3-62. Sulfolane extraction.

The Sulfolane Extraction process is licensed by Universal Oil Products Co. Sulfolane solvent is made by Shell on a commercial basis.

Sulfolane Extraction references include:

Beardmore, F. S., and W. C. Kosters, "Application of Sulpholane in Udex Units," *J. Inst. of Pet.*, p. 1, January, 1963.

Chem. Engrg. News, "Use of Sulfolane as a Selective Solvent for Aromatics," p. 50, May 18, 1964.

Deal, Jr., C. H., et al., "A Better Way to Extract Aromatics," *Pet. Ref.*, pp. 185–192, September, 1959.

Hyd. Proc., "Sulfolane Extraction," p. 224, September, 1964.

Oil & Gas J., "Sulfane Process Bids for Aromatics Extraction Jobs," pp. 262–264, Sept. 23, 1963.

PART 4.   TREATING PROCESSES

Contaminates are present in crude oils in varying quantities and combinations. They may include organic compounds containing sulfur, nitrogen, and oxygen; dissolved metals and inorganic salts; soluble salts dissolved in emulsified water carried by the oil; and in many oils, resins, and asphalt materials.

These contaminating materials, being undesirable, are usually removed or con-

verted to a harmless form.   This is accomplished at some intermediate refining stage
or just prior to sending the finished product to storage.   The purposes are to prevent
(1) corrosion of equipment, (2) catalyst destruction, and (3) inferior finished products
(i.e., poor color, instability, corrosion, bad odor, etc.).

Numerous treating chemicals and combinations are used, and most fall into one or
more of the following classifications: (1) acid, (2) alkali, (3) solvent, (4) oxidizing
agent, or (5) adsorption agent.

The selection of a treating process for a particular refining situation depends upon
the nature of the fraction to be treated and the specifications stated for the finished or
intermediate product.   In most cases, there is a choice of process routes that can
satisfy treating requirements.   Final selection is generally made after an engineering
study based largely upon economics.   In some cases, availability of treating reagent,
market for spent reagent, disposal problems to the stream or atmosphere, and similar
considerations are governing factors in process selections.

In the case of straight-run products, chemical treating methods have become less
attractive since the development of hydrogen processing, which became practical as a
result of byproduct hydrogen from catalytic reforming.   Parallel to this were the
increased need and further development of processes to remove acid gases (hydrogen
sulfide) for the manufacture of sulfur.   Catalytically cracked fractions, on the other
hand, still are chemically treated, generally by some variation of caustic washing.

CAUSTIC TREATING

Caustic treating of petroleum products is as old as the industry itself.   Product odor
and color are improved by removing organic acids and sulfur compounds (naphthenic
acids, mercaptan sulfur, hydrogen sulfide, phenolic compounds).   Countless equipment modifications and processes have been designed to perform the operation most
effectively for specific requirements.   Figure 3-63 shows a typical caustic-treating
flow diagram.

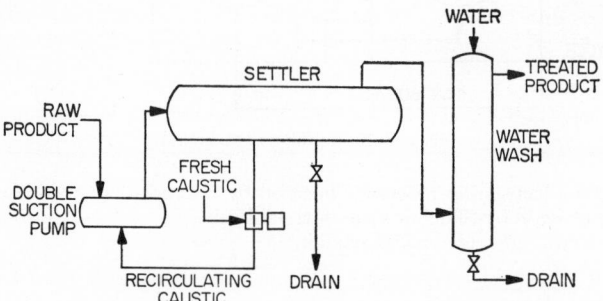

Fig. 3-63. Typical caustic treating flow diagram.

Stage treating, or countercurrent washing, provides more complete and effective
use of reagents and removes a greater percentage of the less reactive contaminants
(i.e., mercaptans).   Water washing may be used to remove carry-over caustic.

Caustic solutions ranging from 5 to 20 wt-% are used at 70 to 110°F and 5 to
40 psig.   Usually this closely represents run-down stream conditions.   High temperatures and strong caustic are usually avoided because of the danger of color body
formation and stability loss with some distillates.   Caustic-to-product treating ratios
vary from 1 to 1 to 1 to 10.

Processes incorporating caustic regeneration are popular.   Very often caustic
washing is used as a preliminary treatment ahead of some other treating or refining
step to effect savings of costly chemicals or to protect catalyst.   It may be used as a
final wash in clean-up service to neutralize carry-over acids.

United States patents for caustic treatment of oils were issued as early as 1863, and improvements have been made to the present day.
General literature references include:

Borgstrom, P., "Effect of Caustic Soda on Mercaptan Sulfur in Naphtha," *Ind. Engrg. Chem.*, p. 250, March, 1936.
Crary, R. W., and M. M. Holm, "Recovery of Mercaptans from Cracked Naphtha Fractions," *Ind. Engrg. Chem.*, pp. 1389–1395, December, 1937.
Happel, J., and D. W. Robertson, "Removal of Mercaptans from Naphtha by Caustic," *Ind. Engrg. Chem.*, pp. 941–943, August, 1935.
Henderson, L. M., et al., "Tetraethyllead Susceptibilities of Gasoline—Doctor Treatment vs. Caustic Washing," *Ind. Engrg. Chem.*, p. 27, January, 1939.
Kalichevsky, V. A., "Sweetening and Desulfurization of Light Petroleum Products—Part VI, Alkaline Reagents," *Pet. Ref.*, p. 111, April, 1951.
Ref. Nat. Gaso. Mfr., "Lime Wash Used for Hydrogen Sulfide Removal," pp. 121–122, April, 1940.
Ref. Nat. Gaso. Mfr., "Propane Treating with Caustic Solution," pp. 387–388, October, 1940.
Ridgway, C. M., "Improving Gasoline by Caustic Washing," *Oil & Gas J.*, p. 83, Mar. 31, 1938.
Windle, G. S., "Thiophenol in Cracked Gasoline," *Pet. Ref.*, pp. 41–45, February, 1944.

Nonregenerative Treating

Simple Caustic Treating

Nonregenerative caustic treating is generally economically applied when contaminating materials are in low concentration and waste disposal is not a problem.

Traces of H_2S, etc., are removed from liquefied gas and light gasoline streams in numerous ways with calcium, ammonium, or sodium hydroxide solutions. Common practice makes use of sodium hydroxide in a single-stage system. With good contact H_2S can be reduced to less than 0.0001 wt-% with accompanying partial mercaptan removal.

Cracked gasoline streams are often treated to remove phenols and mercaptans in a continuous two- or three-stage caustic scrubber system. Ring-packed extraction towers with two or three caustic layers are also used. Phenols which accumulate in the spent solutions facilitate extractions of mercaptans. Recovery of spent caustic solutions for sale is often a partial incentive for nonregenerative operation.

Use of nonregenerative systems is on the decline because of the waste-disposal problems and availability of numerous other processes which effect more complete removal of contaminating materials.
Literature references include:

Albright, J. C., "Mercaptan & Hydrogen Sulfide Removal by Caustic Scrubbing," *Ref. Nat. Gaso. Mfr.*, p. 437, September, 1938.
Kalichevsky, V. A., "Sweetening and Desulfurization of Light Petroleum Products— Part VI, Alkaline Reagents," *Pet. Ref.*, p. 111, April, 1951.
Morrell, J. C., and D. J. Bergmann, "Designing Equipment for Chemical Treatment of Oil Distillates," *Chem. & Met. Engrg.*, pp. 291–295, May, 1928.
Pet. Ref., "Mercaptans Removed Prior to Doctor Treating," pp. 103–106, April, 1942.
Ref. Nat. Gaso. Mfr., "Propane Treating with Caustic Solutions," pp. 387–388, October, 1940.
Ridgway, C. M., "Improving Gasoline by Caustic Washing," *Oil & Gas J.*, pp. 83–84, Mar. 31, 1938.
Sierra, A. V., "Better Caustic Utilization within a Modern Refinery Treating Plant," *Oil & Gas J.*, p. 252, Jan. 22, 1953.
Walker, N. E., and E. B. Kenney, "Removing & Converting," *Pet. Proc.*, April, 1956.

Polysulfide Treating

This is a nonregenerative, chemical treating process used to remove elemental sulfur from refinery liquids.

Caustic-polysulfide solution is prepared by dissolving 1 lb of commercial sodium sulfide (50 to 60%) and 0.1 lb elemental sulfur to each gallon of 15° Baumé caustic. Sulfur must be completely dissolved to avoid contamination of stock being treated. In the refinery, sodium sulfide may also be prepared by passing hydrogen sulfide through caustic solution.

Ordinary caustic washing equipment is used, and temperature is held at about 120°F for polysulfide treating.   Good contacting equipment is necessary (i.e., a double-suction-centrifugal mixing pump).   A ratio of 60% hydrocarbon and 40% polysulfide at the pump discharge yields maximum sulfur removal efficiency.

The solution is most active at a polysulfide concentration of about $Na_2S_{2.5}$.   Activity decreases rapidly when concentration reaches $Na_2S_{3.5-4.0}$.   When the solution is discarded, at least 20% of the old solution is retained and mixed with fresh caustic-sulfide solution.   This eliminates the need to add free sulfur.   If the hydrocarbon being treated also contains hydrogen sulfide in addition to free sulfur, it may be possible simply to add fresh caustic.

Literature references include:

Brooner, G. M., "The Selection of Treating Processes for Gasoline Plants," *Pet. Ref.*, pp. 95–99, April, 1941.
Vesseloysky, V., and V. A. Kalichevsky, "Action of Alkali Hydroxides on Elementary Sulfur and Mercaptans Dissolved in Naphtha," *Ind. Engrg. Chem.*, pp. 23, 181–184, 1931.

Regenerative Treating

Steam-regenerative caustic treating (Fig. 3-64) is essentially directed toward removal of mercaptans from light straight-run gasoline.

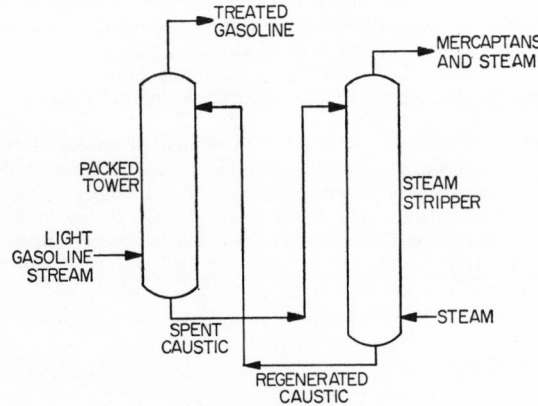

FIG. 3-64. Typical regenerative caustics treating.

Caustic is regenerated by steam blowing in a stripping tower.   The nature and concentration of the mercaptans to be removed dictate the quantity and temperature of stream.

Caustic gradually deteriorates because of the accumulation of material that cannot be removed by stripping.   Caustic quality is maintained by continuous or intermittent discard and replacement of a minimum amount of operating solution.

Literature references include:

Henderson, L. M., et al., "Caustic Washing of Gasoline and Caustic Regeneration," *Oil & Gas J.*, p. 114, Mar. 28, 1940.
Kalichevsky, V. A., "Sweetening and Desulfurization of Light Petroleum Products—Part VI, Alkaline Reagents," *Pet. Ref.*, p. 111, April, 1951.
Oil & Gas J., "Low Cost H₂S Extraction and Mercaptan Reduction," p. 80, May 25, 1944.
Pet. Ref., "Effective Treating for a Small Plant," pp. 118–119, April, 1943.
Phenix, Jr., B. C., "Regeneration of Alkaline Solutions," *Oil & Gas J.*, p. 89, Sept. 22, 1958.

Dualayer Distillate Process

Licensed by Mobil Oil Corp., this process (Fig. 3-65) extracts organic acidic substances (including mercaptans) from cracked or virgin distillate fuels, using Dualayer reagent. Stability, carbon residue, and storage compatibility of the distillate fuels are improved.

Reagent is continuously prepared in a settler, separating into a mixture of essentially caustic (sodium or potassium hydroxide) and cresylic acid under controlled conditions. The lower phase, 42 to 50% caustic saturated with cresylate salt, is used for distillate treating. The settler also serves to separate treated distillate (top layer) and remove excess cresylate salt and impurities (middle layer).

In a typical operation distillate is mixed with reagent at about 130°F and passed to the settler, where three layers are separated with the aid of electrical coagulation.

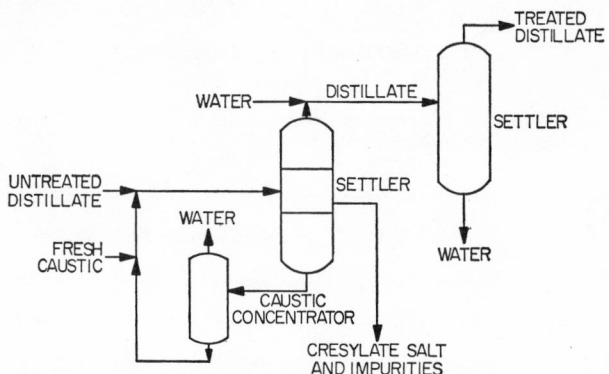

FIG. 3-65. Dualayer distillate process.

Product is withdrawn from the top layer, water-washed, and taken to storage. Dualayer reagent is withdrawn from the bottom layer, relieved of excess water of reaction, fortified with fresh caustic, and recycled.

Literature references include:

Duval, C. A., et al., "First Dualayer Treater for Fuel Oil," *Pet. Ref.*, pp. 142–144, September, 1955.
Scheumann, W. W., "Chemical and Hydrogen Treating," *Pet. Proc.*, pp. 53–57, April, 1956.

Dualayer Gasoline Process

Developed and licensed by Mobil Oil Corp., this Dualayer process (Fig. 3-66) is regenerative. It is used to extract mercaptan sulfur from LP gas, gasoline, and naphtha, using Dualayer reagent. Stability, odor, and TEL susceptibility of the fuels are improved.

Reagent is prepared in a settler and under controlled conditions separates into two phases: a mixture of caustic and cresylic acid salt. The upper phase, potassium or sodium cresylate, is used in gasoline treating.

Gasoline, free of H_2S, is contacted with Dualayer solution (4 to 6 vol-%) at 120°F in at least two stages, water-washed, and taken to storage. The treating solution is diluted with water (60 to 70% of the solution volume) and taken to a stripper where mercaptans, gasoline, and excess water are removed for further processing.

Water is reused to dilute the reagent, gasoline is recovered and returned to the contactor, and mercaptans are processed for sale or disposal. The stripped treating solution is returned to the settler along with a proper amount of fresh caustic to establish the equilibrium necessary to obtain a regenerated reagent.

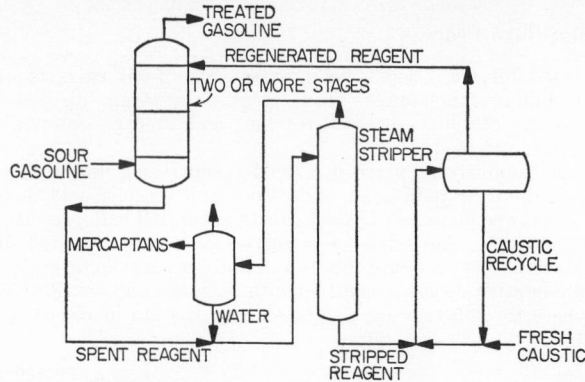

FIG. 3-66. Dualayer gasoline process.

The process requires in the order of 10 lb of steam per barrel of treated gasoline. Literature references include:

Chem. Engrg. News, "Two Layer Extraction of Mercaptans from Petroleum Products," p. 1508, Apr. 12, 1954.

Duval, G. A., and V. A. Kalichevsky, "Dualayer Gasoline Treating Process," *Pet. Ref.*, p. 161, April, 1954.

Duval, G. A., and V. A. Kalichevsky, "Treating Gasoline by Dualayer Process," *Oil & Gas J.*, p. 122, Apr. 12, 1954.

Duval, G. A., and V. A. Kalichevsky, "Twist a Problem into a Profit," *Pet. Proc.*, pp. 691–693, May, 1954.

Greek, B. F., et al., "Dualayer Gasoline Process to Remove Mercaptans," *Ind. Engrg. Chem.*, pp. 1938–1944, December, 1957.

Electrolytic Mercaptan

Developed by the American Development Corp., aqueous caustic solutions (generally sodium or potassium hydroxide) are used, as shown in Fig. 3-67, to extract mercaptans from refinery streams. An electrolytic process is used to regenerate the solution. Disposal problems are minimized, while reagent utility is maximized.

Charge stock is precaustic-washed to remove H_2S and contacted countercurrently with the treating solution in a mercaptan extraction tower. The treated gasoline is taken to storage. Spent solution is mixed in a surge drum with some regenerated

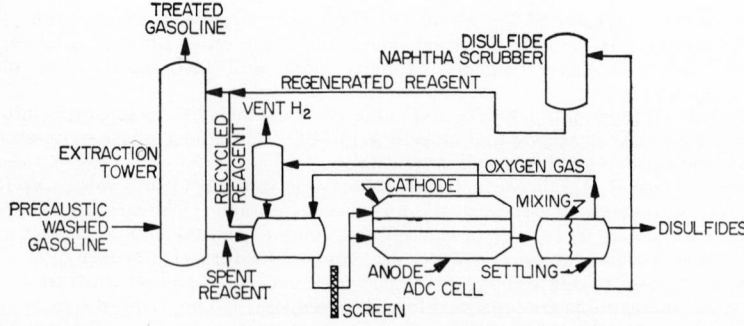

FIG. 3-67. Electrolytic mercaptan extraction.

solution along with byproduct oxygen from the ADC cell. The mixture is screened and pumped to the ADC cell in two streams—the major portion through the anode sections, where mercaptans are converted to disulfides and oxygen is formed, and the remaining portion through the cathode section, where hydrogen is formed (H_2 vented).

Catholyte is recycled to spent caustic, and anolyte is agitated to effect additional conversion of mercaptans and then settled. Oxygen and disulfides are separated from the regenerated solution, which is then naphtha-scrubbed for further disulfide removal. Washed treating solution (about 18° Baumé) is recycled to the extractors.

Literature references include:

Duff, D. M., "Mercaptan Removal Ups Octane," *Oil & Gas J.*, pp. 115–117, July 19, 1954.
Fiske, C. E., and R. Miller, "Economics of ADC Electrolytic Mercaptan Process," *Pet. Engr.*, p. C48, November, 1954.
Pet. Ref., "Gaylor Caustic Recovery Process," p. 119, June, 1950.
Zandona, O. J., and C. W. Rippie, "Low Cost Removal of Sulfur Compounds from Motor Fuels," *Oil & Gas J.*, p. 80, June 2, 1952.
Zandona, O. J., and C. W. Rippie, "New Caustic Regeneration Process Improves Mercaptan Sulfur Removal," *Pet. Proc.*, pp. 136–137, February, 1951.

Ferrocyanide Process

The Ferrocyanide process (Fig. 3-68) is a regenerative chemical treatment for removing mercaptans from straight-run naphthas, LP gas, natural and recycle gasolines,

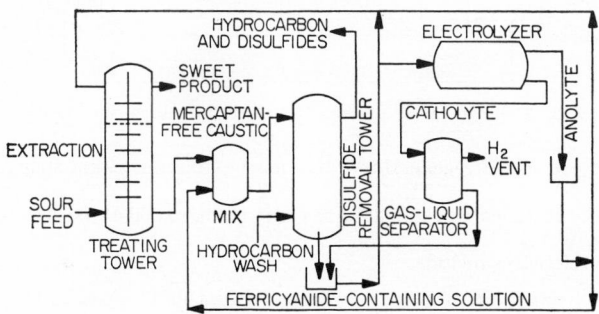

Fig. 3-68. Ferrocyanide process for mercaptan removal.

etc., using caustic-sodium ferrocyanide reagent. Sodium ferrocyanide is added to caustic in small concentrations and, when converted to ferricyanide, functions as an oxidation agent in both the extraction and regeneration steps.

Gasoline is prewashed with caustic to remove H_2S and acid oils, then washed countercurrently at stream temperature in a tower with treating agent. The tower may be designed to incorporate internal mixing. It also can include extraction and sweetening sections. Operation is arranged so that gasoline contacts caustic with ferrocyanide during extraction and caustic with ferricyanide during sweetening.

Spent solution (containing mercaptans and ferrocyanide) is mixed with fresh solution (containing ferricyanide). Mercaptans are converted to caustic insoluble disulfides and are removed by a countercurrent hydrocarbon wash. Sulfur-free solution (containing ferrocyanide) is then divided into two streams, one of which returns to the extraction tower. The remaining stream is taken to an electrolyzer to convert ferrocyanide to ferricyanide. Part of the cell effluent goes to sweetening in the extraction tower and the remainder to spent-solution mercaptan-oxidation.

Literature references include:

Salmon, J. H., and R. Miller, "New Process for Mercaptan Removal," *Pet. Ref.*, pp. 155–157, September, 1955.

Mercapsol

Developed and licensed by Pure Oil Co., a division of Union Oil Co. of California, Mercapsol (Fig. 3-69) is a regenerative chemical treating process for extracting mercaptan sulfur (93 to 99%) utilizing aqueous sodium (or potassium) hydroxide containing mixed cresols, naphthenic acids, and phenol as solubility promoters. Stability, odor, and TEL susceptibility are improved.

Gasoline is precaustic-washed, coalesced, and contacted countercurrently with Mercapsol solution at the stream temperature. Treated gasoline is removed from the top of the tower, settled, passed through a coalescer, inhibited, and taken to storage. Spent Mercapsol solution is first settled and lightly stripped to remove gasoline, then

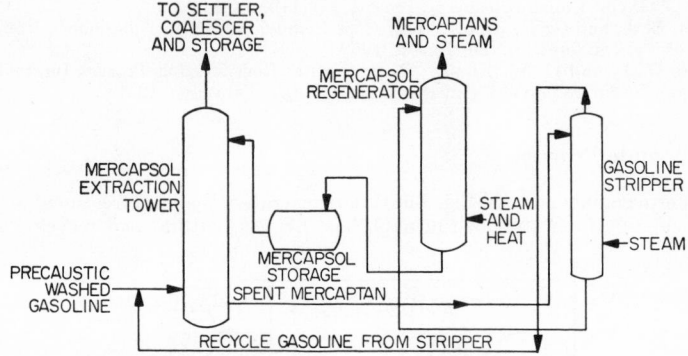

Fig. 3-69. Mercapsol process for extracting mercaptan sulfur.

taken to the Mercapsol regenerator where mercaptans are steam-stripped from the reagent.

Pure Oil Co. installed and operated the first plant (5,000 B/D) in April, 1941, at Cabin Creek, W.Va.

Literature references include:

Bond, D. C., "Regeneration of Caustic Solutions for Gasoline Treating by Catalytic Oxidation," *Oil & Gas J.*, pp. 83–84, Dec. 8, 1945.

Kalichevsky, V. A., "Sweetening and Desulfurization of Light Petroleum Products—Part VI, Alkaline Reagents," *Pet. Ref.*, p. 111, April, 1951.

MacKusiek, B. L., and H. A. Alves, "Mercapsol Process for Gasoline Treating," *Oil & Gas J.*, pp. 126–127, Apr. 13, 1944.

Solutizer—Steam Regenerative

Developed by Shell Development Co., this is a chemical treating process (Fig. 3-70) for mercaptan extraction of gasoline or naphtha to odor-sweet specification

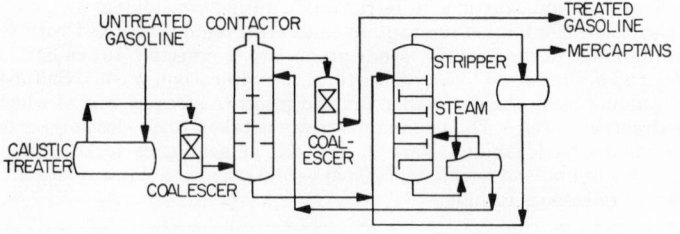

Fig. 3-70. Solutizer—steam regenerative.

using "solutizers" (potassium isobutyrate, potassium alkyl phenolate, etc.) in strong potassium hydroxide solution.

The raw stream is prewashed with caustic, then relieved of mercaptans by extracting with solutizer solution at 100°F. The ratio of treating solution to gasoline is 0.1 to 0.2. Entrained and dissolved gasoline is separated from mercaptan-rich solutizer solution after dilution with regenerator condensate.

Regeneration is then effected by heating and steam blowing at 270°F in a stripping column. Steam and mercaptans are then condensed and separated. Regenerated solutizer solution is returned to the extraction operation. Stream consumption is 10 to 30 lb per barrel of gasoline treated.

The product is essentially free of TEL antagonistic mercaptans and thereby effects savings when blended to octane specifications.

The process was originally developed in 1939, with the first plant (15,000 B/D) being placed in operation at Shell Oil Co.'s Wood River, Ill., refinery May 17, 1940.

Literature references include:

Border, L. E., "Shell Operating First Solutizer Treating Plant at Wood River," *Oil & Gas J.*, pp. 55–56, Nov. 7, 1940.
Border, L. E., "Solutizer, A New Principle Applied to Gasoline Sweetening," *Chem. & Met. Engrg.*, p. 776, November, 1940.
Border, L. E., "Solutizer Process for Sweetening Gasoline," *Oil & Gas J.*, pp. 36–38, July 18, 1940.
La Croix, H. N., "Solutizers Improve Treating of Four Types of Gasoline," *Nat. Pet. News*, pp. R72–80, Mar. 5, 1941.
Walker, H. E., and E. B. Kenny, "Removing and Converting Mercaptans," *Pet. Proc.*, pp. 58–66, April, 1956.
Yabroff, D. L., "Action of Solutizers in Mercaptan Extraction," *Ind. Engrg. Chem.*, pp. 950–953, July, 1940.
Yabroff, D. L., "Extraction of Mercaptans with Alkaline Solutions," *Ind. Engrg. Chem.*, pp. 257–262, February, 1940.
Yabroff, D. L., et al., "Solutizer Sweetening," *Oil & Gas J.*, pp. 35–36, Dec. 19, 1940.

Solutizer—Air Regenerative

Developed and licensed by Shell Development Co., this process (Fig. 3-71) is used for mercaptan removal from heavy gasolines. It is identical with the Solutizer—Steam Regenerative process, except for the regeneration step.

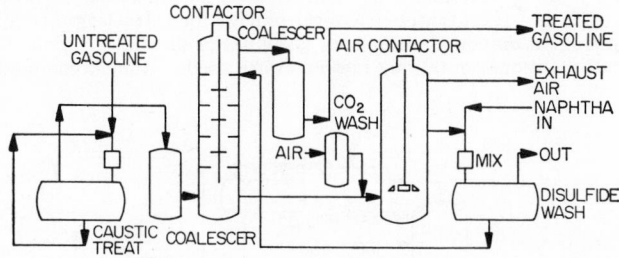

FIG. 3-71. Solutizer—air regenerative.

Gasoline-free, spent solutizer solution is contacted with air (CO_2 free) in a column or turbo-aerator at about 160°F and 40 to 100 psig for 10 to 20 min. Disulfides, formed by oxidation of mercaptans, are extracted with a wash naphtha. Part of the regenerated solutizer solution is passed through a filter to remove traces of oxidation products before being recycled to the extraction tower.

This process is being substituted for the Tannin Solutizer process because of fewer side reactions leading to reagent contamination. Newer units use air regeneration without tannin catalyst, and some old units have been converted.

Literature references include:

Walker, H. E., and E. B. Kenny, "Removing and Converting Mercaptans," *Pet. Proc.*, pp. 58–66, April, 1956.

Solutizer—Tannin

The Solutizer—Tannin is an early variation of the Solutizer—Air Regenerative process for extracting mercaptans from gasoline using a tannin-catalyzed oxidation regeneration step. The process was jointly developed by the Mobil Oil Corp. and Shell Development Co. Shell's solutizer and Socony's tannin-catalyzed air regeneration techniques were combined.

Treating solution from the extraction step is heated to 110 to 130°F and taken to a regenerating column. Air blowing, in the presence of tannin, catalytically oxidizes mercaptans to corresponding disulfides. Presence of tannin in the solutizer reagent also aids in converting some mercaptans in the gasoline to disulfides.

New units eliminate tannin and use higher regeneration temperatures. This reduces undesirable side reactions and lowers chemical costs.

Literature references include:

Happel, J., and S. P. Cauley, "Tannins and Allied Chemicals in Mercaptan Removal Process," *Ind. Engrg. Chem.*, pp. 1655–1659, December, 1947.
La Croix, H. N., "Improved Refining of Sour Crudes by New Tannin Solutizer Process," *Oil & Gas J.*, p. 50, Feb. 10, 1944.
O'Donnell, J. P., "Tannin Solutizer Process Practically Automatic; Saves 6.5 Cents per Barrell," *Oil & Gas J.*, pp. 45–47, July 1, 1944.
Pet. Proc., "New Solutizer Unit Has Unique Features for Disulfide Removal and Regeneration," pp. 965–966, December, 1947.
Pet. Proc., "Tannin Solutizer Process Continues to Show Possibilities," pp. 7–8, September, 1946.
Pet. Proc., "Use of Tannins as Oxidation Catalyst in Regenerating Treating Solutions," pp. 146–147, October, 1946.
Tuttle, R. B., "Cities Service Tannin Solutizer Unit," *Oil & Gas J.*, pp. 84–85, Sept. 13, 1947.
Walker, H. E., and E. B. Kenney, "Removing and Converting Mercaptans," *Pet. Proc.*, pp. 58–66, April, 1956.

Unisol Mercaptan Extraction

Developed by Atlantic Refining Co. and licensed by Universal Oil Products Co., Unisol (Fig. 3-72) is a regenerative chemical process for extracting mercaptan sulfur and certain nitrogen compounds from sour gasolines or distillates. Sodium or potassium hydroxide solutions containing methanol are used. Almost complete (99% +)

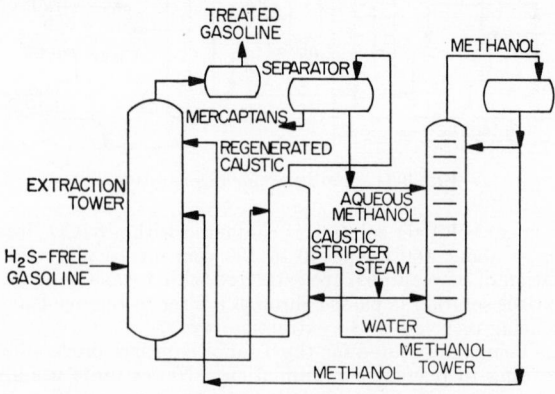

Fig. 3-72. Unisol mercaptan extraction.

mercaptan removal from gasoline is effected, accompanied by improved TEL susceptibility and product stability.   Distillates (heating oil, etc.) are improved in odor and stability, with 95%+ mercaptan removal.

Gasoline, free of H_2S, is contacted countercurrently with aqueous-caustic-methanol solution at about 100°F.   Caustic enters the top of the extraction tower, while methanol is injected at the center.   Upon leaving the tower, gasoline is settled, inhibited, and taken to storage.

Spent caustic is regenerated in a stripping tower (290 to 300°F) where methanol, water, and mercaptans are removed.   Mercaptans are separated from methanol, which is then fractionated to about 95% purity before being recycled.   Caustic strength varies from 45 (regenerated) to 50° Baumé (fresh).

The process was developed by Atlantic Refining Co. on a pilot-plant basis in 1941 at Philadelphia, Pa.   One of the first commercial units was installed at the Big West Oil Co. at Kevin, Mont., and operation started in the fall of 1942.

Literature references include:

Bent, R. D., and J. H. McCullough, "Atlantic Refining Is First Year with Unisol Process,"
    Oil & Gas J., p. 95, Sept. 9, 1948.
Brandt, P. L., and J. O. Hougen, "Treat Cracked Distillates with Caustic Methanol Solu-
    tion," *Oil & Gas J.*, pp. 98–103, Mar. 30, 1939.
Brown, H. A., and J. R. Strong, "Frontier Refining Co. Installs Unisol Treating Plant at
    Cheyenne," *Oil & Gas J.*, pp. 236–237, Mar. 24, 1949.
Field, H. W., "Malodorus Mercaptans—Caustic Methanol Process," *Oil & Gas J.*, pp.
    40–41, Sept. 25, 1941.
Lowry, C. D., and F. C. Moriarty, "Unisol Process Improves Octane Number and TEL
    Susceptibility," *Oil & Gas J.*, pp. 105–106, Nov. 3, 1945.
Lyles, H. R., "Better Gasoline Treating," *Pet. Proc.*, pp. 655–657, May, 1955.
Mason, C. F., et al., "Caustic Methanol Process," *Oil & Gas J.*, p. 116, Nov. 6, 1941.

ACID TREATING

Sulfuric Acid Treating

Sulfuric acid treating is a continuous or batch chemical treating process that is used to remove sulfur, precipitate asphaltic material, and improve stability, color, and odor of a wide variety of refinery stocks.

A standard method of treating pressure distillate for many years was acid treatment, then distillation with steam, followed by doctor sweetening.

Acid strength varies from over 100 (fuming) to 80%, with 93% being common. The weakest suitable acid is used for each particular treating situation to reduce sludge losses due to chemical action upon aromatic and/or olefinic hydrocarbons. Temperature is kept low with strong acid (25 to 50°F), but higher temperatures (70 to 130°F) may be practical and economical if material is to be rerun.   In continuous systems contact time may be as little as 1 min or, in batch treating of some kerosines, as high as 40 min.

Disadvantages of the process are cost, high processing losses, and the necessity for redistillation in many cases.   Present application is in the field of kerosine and special solvents.

Sulfuric acid has been used since the early days of the industry but has been replaced to a large degree by regenerative caustic washing employing promoters and, more recently, hydrogenation.

Literature references include:

Albright, J. C., "Treating Cracked Gasoline Containing Sulfur," *Pet. Engr.*, p. 108, May,
    1937.
Altshuler, J. A., et al., "Advantages of Stratcold Process Demonstrated Commercially,"
    Ref. Nat. Gaso. Mfr., p. 181, April, 1937.
Mapstone, G. E., "Chemistry of Acid Treatment of Gasoline," *Pet. Ref.*, pp. 142–150,
    November, 1950.
Nelson, W. L., "Finishing Pressure Still Distillate into Gasoline," *Ref. Nat. Gaso. Mfr.*,
    pp. 247–250, July, 1936.

Oil & Gas J., "Sulfuric Acid for Removing the Different Forms of Sulfur," p. 45, Nov. 10, 1938.
Sager, F., "Modern Commercial Acid Treating Reproduced in Laboratory," *Ref. Nat. Gaso. Mfr.*, pp. 57–61, February, 1938.
Scheumann, W. W., "Chemical & Hydrogen Treating," *Pet. Proc.*, pp. 53–57, April, 1956.
Walton, J. S., "Continuous Acid Treatment of Cracked Distillates," *Ref. Nat. Gaso. Mfr.*, pp. 94–97, March, 1936, and pp. 160–163, April, 1936.
Wood, A. E., et al., "Action of Petroleum Refining Agents on Naphtha," *Ind. Engrg. Chem.*, pp. 169–171, February, 1926.

Nalfining

This is a continuous chemical treating process, employing acetic anhydride followed by a concentrated caustic rinse, primarily for light naphthas and distillates. It chemically converts contaminants into harmless oil-soluble compounds.

Anhydride is injected by a proportioning pump into the product stream (Fig. 3-73), where it reacts with oxygen to form the ester, with sulfur to form the thioester, with nitrogen to form substituted amides, and with complex organic impurities to form

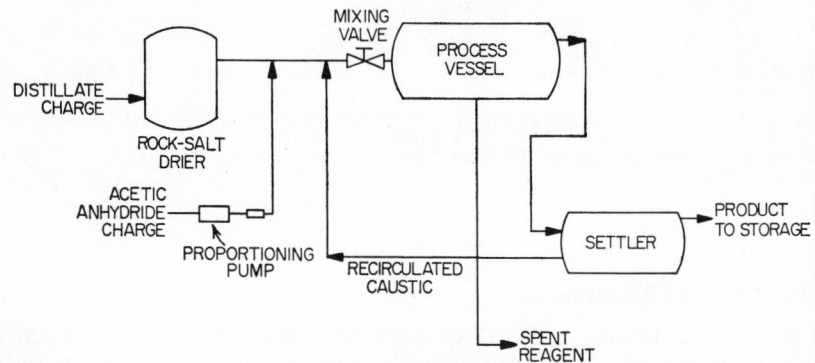

FIG. 3-73. Nalfining.

harmless reaction products. Color, stability, and odor are all improved. Water haze resulting from processing is eliminated. The caustic-wash step neutralizes resultant corrosive acetic acid, and the whole operation involves no appreciable product loss.

Initial capital outlay for equipment is low, and existing treating equipment can generally be adapted. Aluminum storage tanks, pumps, and lines are required to handle the anhydride. Investment is in the order of 50 to 75 cents per barrel of processing capacity. Chemical costs range from a few mills a barrel for odor control to 4 to 5 cents a barrel for fuel oil and sour gasoline streams.

Literature references include:

Chem. Engrg. News, "Nalfining Upgrades Distillates," pp. 42–43, June 8, 1959.
Inter. Oilman, "Unique New Product-upgrading Process," pp. 296–297, September, 1959.
Pet. Ref., "Nalfining: Unique New Upgrading Method for Low-cost Processing of Distillates," pp. 56–57, October, 1959.

CLAY TREATING

Natural clay adsorbents and synthetic adsorbents are used to treat refinery products to improve color and odor; to eliminate water, suspended impurities, resinous or asphaltic substances, nitrogen compounds, oxygenated compounds, and some sulfur compounds; and to absorb preferentially some hydrocarbons (olefins).

In general, three approaches have been taken to clay treating: percolation through coarse clay (lube oils, waxes, specialties), contact with powdered clay (light and heavy distillates), and vapor-phase contact through a loosely packed bed (gasoline and light distillates).

Percolation filters are 15 to 30 ft high and may hold up to 100 tons of clay. Oil is either gravitated or pumped through the bed. Spent clay is washed with solvent and steamed *in situ*, then removed and regenerated by burning.

The contact process mixes finely divided clay with the refinery stream, then separates the two by filtration. Regeneration is not practical. Any type of refinery fraction may be treated by this process.

Vapor-phase clay treating is used for light petroleum fractions. The stream is passed through a loose clay bed slightly above the vaporization point of the fraction. Reactive olefinic material is polymerized and absorbed as gum.

Clay treating was widely accepted in the 1930's and early 1940's. Since then, however, advances in treating methods, gum inhibitors, and cracking technology have all contributed to the decline of clay treating of gasoline and distillates. Adsorbent clays are still used to improve color of lubes, but solvent extraction processes are now standard practice. Clay is frequently used as a polishing process after solvent extraction.

Literature references include:

Chenault, W. B., and A. E. Miller, "New Lubricating-oil Decolorizing and Clay Reactivating Process," *Ref. Nat. Gaso. Mfr.*, p. 79, November, 1941.

Cooke, M. B., and A. W. Hayford, "Distillate Treating Now Highly Developed and Simplified," *Ref. Nat. Gaso. Mfr.*, pp. 84–87, 115–122, March, 1934.

Davis, L. L., "Underlying Principles of Contact Filtration," *Ref. Nat. Gaso. Mfr.*, p. 90, March, 1928.

Funsten, S. R., "Percolation Filtration Process Development," *Ref. Nat. Gaso. Mfr.*, p. 201, June, 1934.

Funsten, S. R., "Recent Developments in the Use of Contact Clays," *Ref. Nat. Gaso. Mfr.*, pp. 237–249, June, 1937.

Holland, W. W., "Osterstrom Vapor Phase Treating System," *Ref. Nat. Gaso. Mfr.*, pp. 26–27, January, 1932.

Johnson, W. A., "Decolorization of Petroleum Waxes by Absorbent Percolation," *Pet. Proc.*, p. 673, September, 1947.

Kalichevsky, V. A., and J. W. Ramsey, "Lubricating Oils in Contact with Clays," *Ind. Engrg. Chem.*, pp. 941–943, August, 1933.

Kauffman, H. L., "Effect of Size and Shape of Filter on Percolation Filtration," *Ref. Nat. Gaso. Mfr.*, pp. 74–78, April, 1928.

Laughlin, C. D., "New Way to Stability; Continuous Percolation," *Pet. Ref.*, p. 30, September, 1957.

Myers, H. C., and R. M. Owen, "Economic Survey of Thermofor Installations in Socony-Vacuum Lubricating Oil Refineries," *Oil & Gas J.*, p. 93, Aug. 4, 1945.

Nelson, W. L., "Finishing Pressure Still Distillate into Gasoline," *Ref. Nat. Gaso. Mfr.*, pp. 247–250, July, 1936.

Ridgway, C. M., et al., "Study of Recovery of Oil from Spent Percolation Filters," *Oil & Gas J.*, p. 154, Nov. 18, 1938.

Simpson, T. P., and J. W. Payne, "Thermofor Kiln & Clay Regeneration Processes for Maintaining Decolorizing Clays at High Efficiencies," *Oil & Gas J.*, p. 147, Nov. 17, 1939.

Smoley, E. R., and D. Fulton, "Modern Manufacture of Lubricating Oils," *Pet. Proc.*, p. 594, August, 1947.

Continuous Contact Filtration

Developed by the Filtrol Corp., this is a continuous clay treating process (Fig. 3-74) to finish lubricants, waxes, or special oils after acid treating, solvent extraction, or distillation.

Finely divided adsorbent is mixed with the charge stock and heated to 200 to 350°F. The slurry is then fed to a steam stripping tower. Following withdrawal from the tower, the slurry is cooled and vacuum-filtered, then vacuum-stripped for additional product specification control (flash, odor, etc.). Oil is then cooled, blotter-pressed, and taken to storage. Yields normally are over 98% of the charge material.

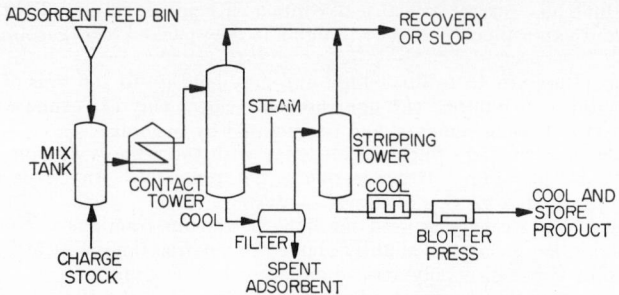

FIG. 3-74. Continuous contact filtration.

Literature references include:

Davidson, R. C., "Contact Filtration," *Oil & Gas J.*, pp. 116–128, Mar. 25, 1943.
Moore, M. M., and H. M. Gwyn, Jr., "Filtrol Fractionation Plant," *Ref. Nat. Gaso. Mfr.*, pp. 45–48, February, 1939.
Brant, V. L., "Bauxite Treaters Save for D-X Sunray," *Oil & Gas J.*, p. 103, Sept. 9, 1957.

Thermofor Continuous Percolation (TCP)

Developed and licensed by Mobil Oil Corp., this is a continuous, regenerative, clay treating process (Fig. 3-75) to stabilize and decolorize lubricants or waxes that have been distilled, solvent-refined, or acid-treated. The process also has application outside the petroleum industry.

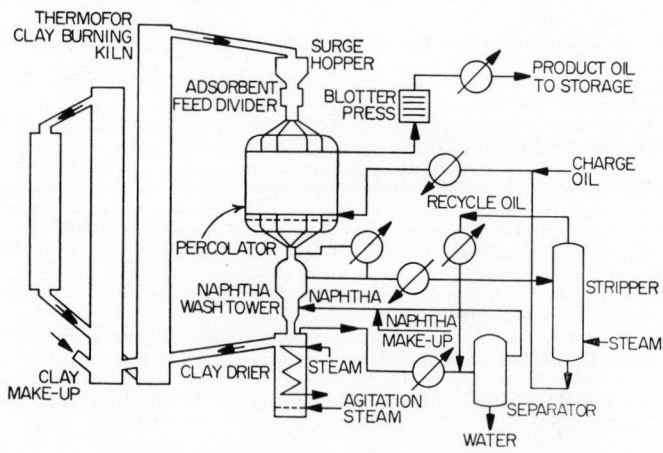

FIG. 3-75. Thermofor continuous percolation (TCP).

Charge stock is heated (125 to 350°F) and evenly charged into the bottom of a clay-filled tower by means of a multinozzle distributor. After percolating in countercurrent flow through the clay bed, stock is passed through a blotter press and taken to storage.

Spent clay is withdrawn continuously from the bottom of the tower, or percolator, and regenerated clay is added to the top of the bed to maintain a constant level. A multipipe draw-off and feed-leg system ensures uniform clay distribution and movement down the percolator. Clay feed rate may be set to control product color.

Spent clay flows countercurrent to a wash naphtha, is dried and regenerated in a Thermofor clay burning kiln.

Wash naphtha is separated from oil in a steam stripper and returned to the wash system.   Oil is recycled to the charge stock.

The first TCP unit (2,100 B/D) was installed in 1953 at the Mobil Oil Co. Ltd. refinery at Coryton, England.

Literature references include:

Chem. Week, "Continuous Way to Whiten Waxes," Feb. 6, 1960.
Evans, L. P., et al., "Thermofor Continuous Percolation," *Oil & Gas J.*, pp. 116–120, June 8, 1953.
Myers, H. C., and R. M. Owen, "Economic Survey of Thermofor Installations in Socony-Vacuum Lubricating Oil Refineries," *Oil & Gas J.*, pp. 93–97, Aug. 4, 1945.
Pet. Proc., "Lube Oil Improvement by Thermofor Continuous Percolation," pp. 137–139, July, 1956.

Percolation Filtration

Developed by Mineral & Chemicals Philipp Corp., this is a continuous-flow, cyclic-regenerative, liquid-phase, clay treating process (Fig. 3-76) to improve color, odor, and stability of lube oils and waxes.

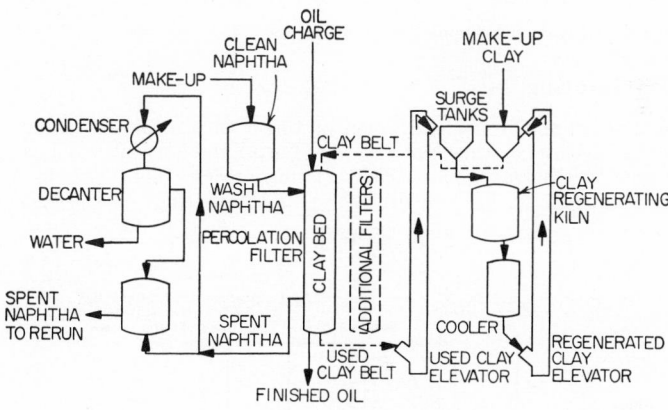

Fɪɢ. 3-76. Percolation filtration.

Oil is filtered through a bed containing 10 to 50 tons of fuller's earth or activated bauxite (20/60 or 30/60 mesh) and taken to storage.   Two or more beds are used alternately on an operating and regenerating cycle.

Spent clay is washed free of oil with naphtha, then steamed to remove naphtha. Rerunning separates the oil-naphtha mix for recycle.   After steaming, clay is carried to a kiln and carbonaceous material is removed by burning.

Literature references include:

Pet. Ref., "Percolation Filtration," p. 272, September, 1956.

Alkylation Effluent Treating

Jointly developed by DX Sunray Oil Co. and M. W. Kellogg Co., this is a continuous, regenerative, liquid-percolation process (Fig. 3-77) to remove 99% of acidic and neutral esters and carry-over acid from sulfuric acid alkylation effluent.

Reactor effluent is coalesced in a vessel containing glass wool and steel mesh, then charged alternately through two bauxite (medium mesh) towers.   Bauxite preferentially adsorbs impurities.   After contact has reached about $\frac{3}{4}$ bbl per pound of

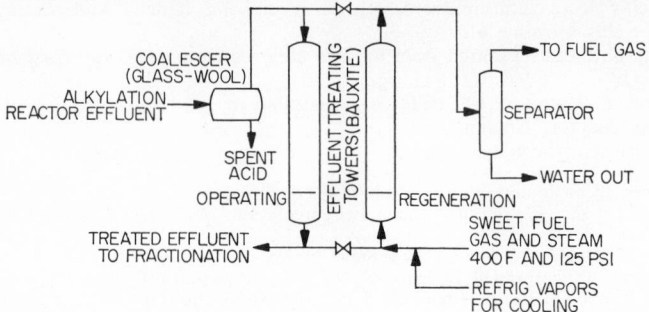

FIG. 3-77. Alkylation effluent treating.

adsorbent, percolators are switched and regeneration is effected with a mixture of
steam and gas.   Ultimate catalyst life is about 150 bbl/lb.

The process is claimed to be much more effective than conventional caustic and
water washing, but with comparable economic payout.

Literature references include:

Oil & Gas J., "Alkylation Effluent Treating," p. 180, Mar. 19, 1956.

Gray Clay Treating

Licensed by Pure Oil Co., a division of Union Oil Co. of California (originally
developed and licensed by the Gray Processes Corp.), this is a continuous, regenerative
(or nonregenerative), vapor-phase clay treating process (Fig. 3-78) to polymerize

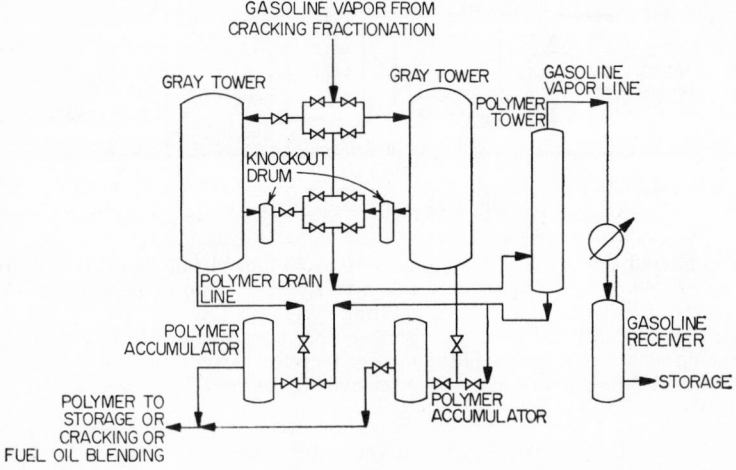

FIG. 3-78. Gray clay treating.

diolefins selectively and remove other gum-forming constituents found in thermal
gasolines.

For continuous operation two or more towers (10 ft diameter and up to 30 ft high)
are used in parallel.   Spent fuller's earth (30/60 mesh) is either discarded or regenerated in outside kilns.

Hydrocarbon vapors are passed through the bed at temperatures (250 to 475°F) slightly above the condensation point.   Contact time is 20 to 400 sec.   Diolefins polymerize and drain from the base of the towers to polymer accumulators.   Remaining polymers are separated from gasoline by a fractionating step before the product is condensed and taken to storage.

Hydrocarbon vapors go directly to Gray towers from cracking units after removal of noncondensable gases.

Gray Clay Treating (8-ton tower) was first installed commercially for motor-fuel treating in the fall of 1924 at the Barnsdall Refining Co. at Barnsdall, Okla.   The process was widely accepted in the 1930's and 1940's.   Since its utility was tied closely with thermal processing, Gray Clay Treating has gradually become obsolete.

Literature references include:

Albright, J. C., "12,000 Barrels through One Ton of Clay," *Ref. Nat. Gaso. Mfr.*, pp. 71–73, May, 1931.
Cobb, A. W., and L. Carlson, "Pilot Plant Study of the Gray Clay Treating Process," *Oil & Gas J.*, pp. 33–36, Nov. 19, 1942.
Goode, R. E., "Clay Volume and Vapor Velocity Govern Color and Gums," *Ref. Nat. Gaso. Mfr.*, p. 79, December, 1931.
Mandelbaum, M. R., and P. F. Swanson, "Clay Treating of Motor Fuel," *Oil & Gas J.*, p. 226, May 14, 1936.
Morrell, J. C., and Gustav Egloff, "Chemical Processes in Refining of Petroleum Distillates," *Oil & Gas J.*, p. 57, Mar. 26, 1936.
Price, R. H., and P. L. Brandt, "Pan American Treats over Million Barrels Gasoline with Same Earth," *Oil & Gas J.*, pp. 54–58, Mar. 29, 1935.

OXIDATION SWEETENING

A group of sweetening processes has been developed to convert foul-smelling alkyl mercaptans to less objectionable disulfides by oxidation.   Choice of a process depends upon the particular refining situation.

A general equation may be stated as follows:

$$4RSH + O_2 = 2RSSR + 2H_2O$$

Because disulfides reduce TEL susceptibility of gasoline streams, these processes are gradually being abandoned.   Recent trends are to processes which completely remove mercaptans.

Doctor Sweetening

The Doctor Sweetening process (Fig. 3-79), sometimes called Sodium Plumbite treating, is a continuous or batch chemical treating process to convert mercaptans in sour refinery stock to disulfides utilizing sodium plumbite solution and elementary sulfur.

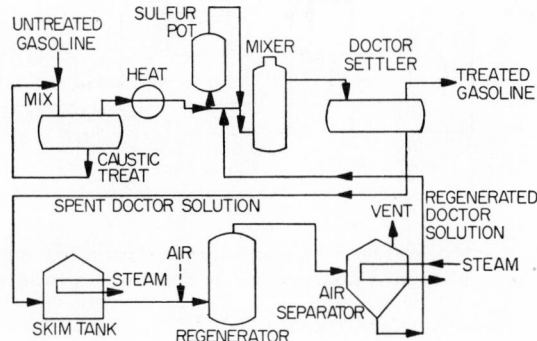

Fig. 3-79. Doctor sweetening (sodium plumbite treating).

Reagent is prepared by dissolving litharge (lead oxide) in 5 to 20 wt-% sodium hydroxide solution. The amount of litharge that will go into solution (1 to 3%) depends upon temperature and caustic strength.

In continuous treating, sour stock is precaustic-washed to remove H_2S and other acidic components. A slip stream picks up a set amount of sulfur and rejoins the main stream, which is then intimately mixed with regenerated doctor solution at 85 to 120°F. The mixture is settled, and gasoline is taken to storage. Spent reagent is heated and skimmed in a settler and air-blown at 150 to 175°F to accomplish regeneration. Fresh reagent must be added periodically.

Doctor treating is still practiced even though it is one of the oldest sweetening processes of the refining industry. Recent advances in hydrogen treating have reduced the need for doctor treating of distillates. Gasoline treating trends are away from doctor treating because of the deleterious effects of disulfides on the octane of leaded fuels.

Literature references include:

Berger, C. W., "Economics of Doctor Treating," Ref. Nat. Gaso. Mfr., pp. 255–259, July, 1936.

Berger, C. W., "Naphtha & Lead Recovery System Proves Economical," Oil & Gas J., pp. 81–82, Mar. 11, 1937.

Blair, Jr., C. M., "Selection and Application of Break-inducing Reagents in Doctor Treating," Pet. Ref., p. 379, August, 1946.

Bottomly, H., "Economics in Hypochlorite Treating," Ref. Nat. Gaso. Mfr., pp. 359–362, September, 1936.

Ellis, E. W., "Determination of Lead in Doctor Solution," Ref. Nat. Gaso. Mfr., pp. 335–336, August, 1936.

Harwick, C. D., "Use of Oxygen in Gasoline Treating," Oil & Gas J., p. 62, Sept. 1, 1949.

Henderson, L. M., et al., "Tetraethyl Lead Susceptibilities of Gasoline—Doctor Treatment vs. Caustic Washing," Ind. Engrg. Chem., p. 27, January, 1939.

Kalichevsky, V. A., "Sweetening & Desulfurization of Light Petroleum Products—Part III, Sodium Plumbite Treatment," Pet. Ref., p. 129, January, 1951.

Lachman, A., "Chemistry of the Doctor Sweetening Process," Ind. Engrg. Chem., pp. 354–357, April, 1931.

Lowry, Jr., C. D., et al., "Factors in Doctor Sweetening," Ind. Engrg. Chem., pp. 1275–1279, November, 1938.

Rampino, L. D., and M. J. Gorham, "Add Doctor Injection for Better Inhibitor Sweetening," Pet. Proc., pp. 46–49, January, 1957.

Inhibitor Sweetening

This is a continuous chemical treating process (Fig. 3-80) to sweeten low-mercaptan-content gasoline using a phenylene-diamine-type inhibitor, air, and caustic.

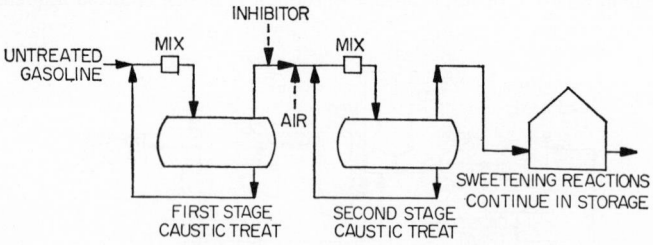

FIG. 3-80. Inhibitor sweetening.

Inhibitor and air are injected between stages during caustic washing. Mercaptan disappearance may be attributed first to reaction with caustic, then to oxidation to disulfides both during washing and in storage. Traces of caustic must be carried with gasoline to effect rapid storage sweetening.

Cracked gasoline of low mercaptan sulfur content may be inhibitor sweetened in absence of caustic, but the reaction is accompanied by excessive peroxide formation.

Inhibitor sweetening has been used extensively with numerous variations since its discovery in 1946 but generally in connection with caustic washing to minimize disulfide and peroxide formation. Peroxides contribute to deterioration of stability, which leads to engine uncleanliness.

Phenylene-diamine-type inhibitors are marketed by Universal Oil Products Co., Tennessee Eastman Corp., E. I. du Pont de Nemours & Company, Ethyl Corporation, and others.

Literature references include:

Barringer, C. M., "Antioxidant Sweetening of Gasolines," *Ind. Engrg. Chem.*, pp. 1022–1027, May, 1955.
Rampino, L. D., and M. J. Gorham, "Chemistry of Inhibitor Sweetening," *Pet. Proc.*, pp. 1146–1149, August, 1955.
Rampino, L. D., and M. J. Gorham, "Add Doctor Injection for Better Inhibitor Sweetening," *Pet. Proc.*, pp. 46–49, January, 1957.
Rosenwald, R. H., "Chemistry of Inhibitor Sweetening," *Pet. Proc.*, September, 1951.
Rosenwald, R. H., "Chemistry of Inhibitor Sweetening," *Pet. Proc.*, October, 1956.
Walker, H. E., and E. B. Kenney, "Removing and Converting Mercaptans," *Pet. Proc.*, April, 1956.

Hypochlorite Sweetening

This is a chemical treating process to convert mercaptans in straight-run or natural gasoline to less offensive sulfur compounds. The principal reaction produces disulfides with limited formation of sulfones and sulfonic acids, depending upon reagent strength. Either alkaline sodium or calcium hypochlorite may be used.

Reagent may be prepared in the refinery by passing chlorine gas through a 10% alkaline solution at 95°F. Prepared bleaching powder (calcium hypochlorite) which contains about 65% chlorine may also be purchased.

When more than a trace of hydrogen sulfide is present, an alkaline prewash reduces chemical costs and prevents the formation of elemental sulfur. An alkaline afterwash is frequently necessary to remove undesirable chlorinated byproducts. Treating temperatures are 95 to 110°F.

The process became well established in the 1930's but is now limited to the treating of natural gasoline and special solvents. Simplicity of the process remains an advantage, even with unfavorable economics because of the high cost of chlorine.

Literature references include:

Birch, S. F., "Hypochlorite Process in Refining," *Oil & Gas J.*, pp. 190–194, May 23, 1929.
Birch, S. F., "Application of Hypochlorite Process," *Oil & Gas J.*, p. 38, May 30, 1929.
Bottomly, H., "Economics in Hypochlorite Treating," *Ref. Nat. Gaso. Mfr.*, pp. 359–362, September, 1936.
Dunstan, A. E., and B. T. Brooks, "Refining of Gasoline and Kerosene by Hypochlorites," *Ind. Engrg. Chem.*, pp. 1112–1115, December, 1922.
Kalichevsky, V. A., "Sweetening and Desulfurization of Light Petroleum Products—Hypochlorite Treatment," *Pet. Ref.*, p. 95, February, 1951.
Kirkwood, G. M., "Treating Pennsylvania Gasoline and Kerosene with Hypochlorite," *Pet. Engr.*, p. 108, February, 1932.
Ref. Nat. Gaso. Mfr., "Treating Natural Gasoline with Various Types of Hypochlorite Solutions," p. 278, July, 1933.
Reid, G. W., "Modern Treating Methods in Refining," *Oil & Gas J.*, p. 119, June 11, 1925.
Ziegenhain, W. T., "Simplify Gasoline Treating Problems," *Oil & Gas J.*, p. 29, July 9, 1931.

Bender Process

Developed by Sinclair Refining Co. at Marcus Hook, Pa., in 1940 and licensed by the Petreco Division of Petrolite Corp., this is a continuous, fixed-bed catalytic treating process (Fig. 3-81) used to sweeten kerosine, jet fuel, or No. 2 fuel, employing lead sulfide catalyst. Sweetening is effected by converting mercaptans in disulfides.

Controlled amounts of sulfur, alkali, and air are added to the product stream and passed through lead sulfide catalyst beds. Proper control leads to a sweet, noncorrosive product. The catalyst bed is regenerated continuously.

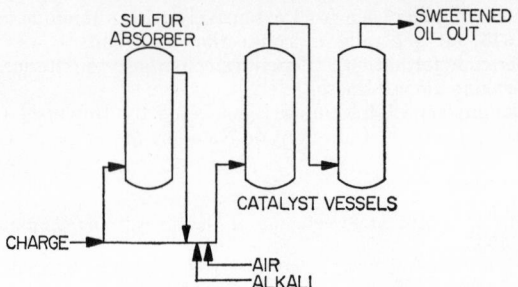

FIG. 3-81. Bender sweetening.

Literature references include:

Altshuler and Graves, "Refinements in Sweetening Technique," *Ref. Nat. Gaso. Mfr.*, p. 272, June, 1937.
Bender, R. O., U.S. Patents 2,272,594-6, July 21, 1942.
Ealey, L., "Bender Lead Sulfide Process Proves Economical," *Oil & Gas J.*, pp. 25–27, Jan. 22, 1942.
Fritz, I. T., "Bender Lead Sulfide Treating Process," *Oil & Gas J.*, p. 180, Feb. 23, 1950.
Happel and Robertson, "Lead Sulfide . . . Dry Sweetening Agent," *Oil & Gas J.*, p. 125, Mar. 31, 1938.
Waterman, L. C., and R. A. Wiley, "Bender Process Sweetens at No Loss," *Pet. Ref.*, p. 182, September, 1955.

Merox Process

Licensed and developed by Universal Oil Products Co., Merox is a combination process for mercaptan extraction and sweetening of gasoline and lighter-boiling-range materials. The process can also operate separately as a mercaptan extractor or mercaptan sweetener, depending upon treating requirements and product economics. When used for sweetening alone the process is most applicable to jet fuels, kerosines, and other middle distillates.

Merox catalyst, basically a cobalt salt, is insoluble in oil, noncorrosive, and may be used dissolved in caustic solution or on a suitable solid support. Catalyst costs are about 0.05 to 0.1 cents per bbl for the soluble catalyst. Catalyst costs are lower when the catalyst is used in the solid form on a support.

In a typical commercial Merox combination process (Fig. 3-82), gasoline at run-down stream temperature is prewashed with caustic before being contacted by catalyst and caustic in the extractor. Gasoline is then settled, and air is injected before it is passed

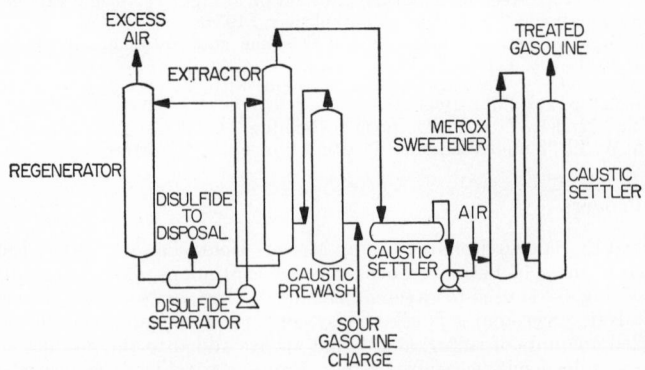

FIG. 3-82. Merox sweetening.

into the Merox sweetener.   The treated product is then settled again, inhibited, and taken to storage.

In the regeneration step, caustic is pumped from the bottom of the extractor and mixed with air in the oxidizer.   Disulfides and excess air are separated from the reagent in the disulfide separator.   Regenerated caustic is then recirculated to the top of the extractor.

The main advantage of the Merox process is that it can perform the dual function of extracting easily removed mercaptans and converting the remaining mercaptans to disulfides.   More effective desulfurization is claimed by this process, since more complete regeneration is made possible by the presence of the catalyst in caustic.

Literature references include:

Brown, K. M., et al., "Low-cost Way to Treat High-mercaptan Gasoline," *Oil & Gas J.,*
    pp. 73–78, Oct. 26, 1959.

COPPER SWEETENING

Copper sweetening of hydrocarbon streams consists of converting mercaptans to disulfides by contact with oxygen in the presence of copper chloride.   Cupric chloride

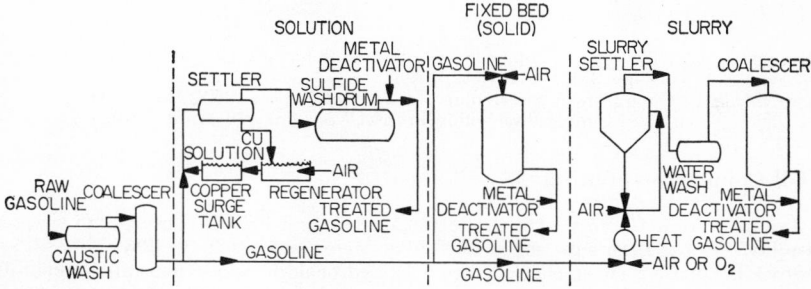

FIG. 3-83. Three typical copper oxidation sweetening systems: solution, fixed bed, and slurry.

is reduced to cuprous chloride during the mercaptan conversion.   Cuprous chloride is regenerated to cupric chloride with oxygen.

Copper sweetening is applied commercially as a continuous treating operation in three variations (Fig. 3-83): solid, slurry, and solution.   Generally the solid process is applied to light straight-run gasoline while the wet process can handle many stocks, including cracked distillates.

The refinery stream is treated to remove reactive nitrogen and sulfur compounds (H_2S and free S).   Oxygen or air is injected into the stream, and it is contacted with copper chloride catalyst at 80 to 120°F.   The treated stream is settled to recover catalyst, washed with water or sulfide solution, and protected with a metal deactivator before going to storage.   In the solution or liquid process, catalyst is regenerated by air blowing in a separate tank.   Catalyst make-up requirements are slight.

Copper sweetening is simple, but the catalyst is corrosive, and there is no reduction of total sulfur.

Copper sweetening is still practiced but is being replaced by processes which remove sulfur.

The first installation was made in early 1931.   It was offered as an alternate to the then-popular doctor treating process.

Literature references include:

Happel, J., et al., "Critical Analysis of Sweetening Processes and Mercaptan Removal,"
    Oil & Gas J., p. 136, Nov. 12, 1942.
Oil & Gas J., "Copper Chloride Treater Handles Straight-run or Cracked Gasoline," p. 37,
    Apr. 10, 1941.

Pet. Ref., "Skelly Blends Gasoline before Sweetening," pp. 220–221, June, 1945.
Robinson, P. M., "Total Removal of Sulfur Compounds Is Goal of Sweetening Techniques," *Nat. Pet. News*, pp. R298–300, Aug. 21, 1940.
Schulze, W. A., and A. E. Buell, "Sweetening Gasoline More Effective, Less Expensive with Cupric Chloride," *Nat. Pet. News*, pp. 24c, 24f, 24h, Oct. 23, 1935.
Schulze, W. A., and F. E. Frey, U.S. Patent 1,980,555, Nov. 13, 1934.
Schulze, W. A., and L. S. Gregory, "Solution and Gasoline Losses Low in Copper Sweetening Processes," *Nat. Pet. News*, pp. 34, Oct. 7, 1936.

Phillips Copper Sweetening

Developed by Phillips Petroleum Co. and licensed until 1954, the liquid, solid, and slurry variations of this process are still in commercial use.

Gasoline stable to color and gum formation in the presence of air is mixed with air and filtered through a solid bed of absorbent material impregnated with copper reagent.

Air-sensitive stocks are contacted with copper solution in the absence of air, and the copper solution is regenerated in a separate tank. Gasoline is sodium sulfide-washed to remove traces of copper.

Reagent is made up with copper sulfate and sodium chloride in the liquid process.

Literature references include:

Albright, J. C., "Two-in-one Copper Sweetening Unit Lowers Treating Inhibitor Costs," *Nat. Pet. News*, pp. R211–212, Oct. 6, 1937.
Conn, M. W., "Perco Copper Sweetening," *J. Inst. of Pet.*, p. 196A, May, 1941.
Pet. Ref., "Perco Solid Copper Sweetening Process," pp. 113–116, April, 1940.

UOP Copper Sweetening

Available from Universal Oil Products Co., this is a fixed-bed process to sweeten gasoline. It has been superseded by UOP's Merox process (p. 3–128).

Stock is caustic-washed, then washed with hydrochloric acid to neutralize alkalinity and remove organic base compounds. Contact is then made with copper reagent in a bed of Italian pumice at about 100°F. Temperature is dependent upon the molecular weight of mercaptans. Trace copper is removed by contact with zinc sulfide or pumice.

Reagent consists of a compound of ammonium chloride and copper sulfate.

Literature references include:

Ref. Nat. Gaso. Mfr., "Cushing Plant Has New Copper Treating Unit," pp. 117–119, April, 1940.

Linde Copper Sweetening

In commercial use since 1935, this process is licensed by the Linde Division of Union Carbide Corp. and is a slurry process to treat straight-run or cracked gasolines and distillates.

Reagent is made up with 200-mesh clay and cupric chloride.

Charge stock is heated, oxygen is added, and the stock is then contacted with a clay–cupric chloride slurry at 80 to 100°F. The mixture is settled, and the treated stock is drawn off and washed with sodium sulfide to remove traces of copper.

Literature references include:

Atkinson, R. G., "Slurry-type Copper Treater Stops Octane Number Losses," *Nat. Pet. News*, pp. R450–452, Dec. 11, 1940.
Deering, F. A., "Slurry-type Copper Chloride Treater," *Oil & Gas J.*, pp. 43–44, Sept. 5, 1940.
Mann, G. L., "Technology, Operation, and Results from Linde Copper Sweetening Process," *Oil & Gas J.*, pp. 195–202, Mar. 22, 1947.
Schieman, C. T., "Linde Copper Sweetening Process," *Pet. Engr.*, pp. 184–190, May, 1947.

HYDROGEN SULFIDE (AND ACID GAS) REMOVAL

The elimination of hydrogen sulfide and other acidic material from natural and refinery gases has become increasingly important in recent years.

Popular processes are those which use a regenerable reagent and recover the hydrogen sulfide. In operation and principle the processes are similar to the general scheme shown in Fig. 3-84. Gases are contacted in towers countercurrently with alkaline solutions. Spent solutions are blown with steam and heated to remove hydrogen sulfide and regenerate the solution.

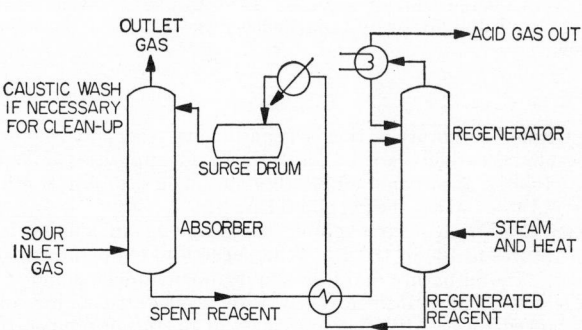

Fig. 3-84. Typical hydrogen sulfide removal scheme.

Whether other material, such as carbon dioxide, hydrogen cyanide, mercaptans, carbonyl sulfide, carbon disulfide, and water vapor, is present will influence the choice of process.

The use of caustic, except for the removal of trace amounts of acid gases, has been largely abandoned because of excessive cost and disposal problems.

Processes employing air blowing to regenerate reagent contribute to air contamination whether spent to the atmosphere directly or fed to a furnace.

Removal processes are as follows:

Name	Reaction	Regeneration
Caustic soda....	$2NaOH + H_2S \rightarrow Na_2S + 2H_2O$	None
Lime..........	$Ca(OH)_2 + H_2S \rightarrow CaS + 2H_2O$	None
Iron oxide......	$FeO + H_2S \rightarrow FeS + H_2O$	Partly by air
Seaboard.......	$Na_2CO_3 + H_2S \leftrightarrows NaHCO_3 + NaHS$	Air blowing
Thylox.........	$Na_4As_2S_5O_2 + H_2S \rightarrow Na_4As_2S_6O + H_2O$	Air blowing
	$Na_4As_2S_6O + \frac{1}{2}O_2 \rightarrow Na_4As_2S_5O_2 + S$	
Girbotol........	$2RNH_2 + H_2S \leftrightarrows (RNH_3)_2S$	Steaming
Phosphate......	$K_3PO_4 + H_2S \leftrightarrows KHS + K_2HPO_4$	Steaming
Phenolate......	$NaOC_6H_6 + H_2S \leftrightarrows NaHS + C_6H_5OH$	Steaming
Carbonate......	$Na_2CO_3 + H_2S \leftrightarrows NaHCO_3 + NaHS$	Steaming

Newer processes for removing hydrogen sulfide from refinery streams also accomplish oxidation of the sulfide to free sulfur as part of the regeneration step. Sulfur is precipitated as a finely divided solid and subsequently recovered by settling or filtration. Liquid product may be recovered where it is possible to heat the suspension above the melting point of sulfur.

General literature references include:

Baehr, Hans, "Gas Purification and Sulfur Recovery," *Ref. Nat. Gaso. Mfr.*, pp. 237–244, June, 1938.

Buck, B. O., and Angus R. S. Leitch, "Try Gas Treating with Hot Carbonate," *Pet. Ref.*, pp. 241–246, November, 1958.

Carvlin, G. M., "The Use of Sodium Phenolate for Hydrogen Sulfide Removal," *Ref. Nat. Gaso. Mfr.*, pp. 225–233, June, 1938.

Graff, R. A., "Corrosion in Amine Type Gas Processing Units," *Ref. Engr.*, pp. C12–14, March, 1959.

Gregory, L. B., and W. G. Scharmann, "Carbon Dioxide Scrubbing by Amine Solutions," *Ind. Engrg. Chem.*, pp. 514–520, May, 1937.

Powell, A. R., "Recovery of Sulfur from Fuel Gases," *Ind. Engrg. Chem.*, pp. 789–796, July, 1939.

Riesenfeld, F. C., and J. F. Mullowney, "Giammarco-Vetrocoke Processes for Acid Gas Removal," NGAA Annual Meeting, Dallas, Tex., Apr. 22–24, 1959.

Rieve, R. W., et al., "Sulfur Recovery from Refinery Gases," *Oil & Gas J.*, pp. 180–184, June 8, 1959.

Girbotol

Licensed by Girdler Corporation, this is a continuous, regenerative process to separate hydrogen sulfide, carbon dioxide, and other acid impurities from natural and refinery gases. It uses an organic amine reagent (mono-, di-, or triethanolamine), frequently referred to as MEA, DEA, and TEA.

Ethanolamines are strong, water-soluble bases having an affinity for hydrogen sulfide at temperatures to 160 to 180°F. When heated to temperatures above 200°F, they become less basic in nature and lose affinity for hydrogen sulfide.

Impure gas is contacted countercurrently with regenerated amine solution in an absorber (i.e., packed tower with Raschig rings) at operating temperatures (100 to 150°F) and pressures. Hydrogen sulfide-rich amine solution is heated to 200°F or above by heat exchange with regenerated reagent, then steam-stripped to remove acid gases. Amine solution is cooled to about 100°F and returned to the absorber through a surge drum.

Selection of reagent depends upon the particular refining situation and the product requirements. Liquid hydrocarbons may also be treated using a similar procedure.

The first natural-gas application for hydrogen sulfide removal was at the Shell Oil Co. plant at Hobbs, N.M. It was not until 1938, at the Atlantic Refining Co. refinery near Port Arthur, Tex., that hydrogen sulfide was removed from refinery gas using the Girbotol process.

Literature references include:

Bottoms, R. R., and W. R. Wood, "The Girbotol Purification Process," *Ref. Nat. Gaso. Mfr.*, pp. 105–107, March, 1935.

Gordon, J. D., "Liquid Hydrocarbon Treating with Aqueous Amine Solutions," *Oil & Gas J.*, p. 202, Apr. 13, 1944.

Gregory, L. B., and W. G. Scharmann, "Carbon Dioxide Scrubbing by Amine Solutions," *Ind. Engrg. Chem.*, pp. 514–520, May, 1937.

Norris, W. E., and F. R. Clegg, "Investigation of Girbotol Unit Charging Cracked Refinery Gases Containing Organic Acids," *Pet. Ref.*, pp. 739–741, November, 1947.

Reed, R. M., "Improved Design, Operating Techniques for Girbotol Absorption Processes," *Pet. Proc.*, p. 907, December, 1947.

Storrs, B. D., and R. M. Reed, "Application of the Girbotol Process to Industry," *Nat. Pet. News*, p. R250, Oct. 29, 1941.

Wood, W. R., and B. D. Storrs, "The Girbotol Purification Process," *Ref. Nat. Gaso. Mfr.*, pp. 234–236, June, 1938.

Glycol-amine Gas Treating

Developed and licensed by Fluor Corporation, this is a continuous, regenerative process to dehydrate and remove acid gases simultaneously from natural or refinery gases. A mixture of an aqueous amine (generally monoethanolamine) and di- or triethylene glycol make up the treating reagent. A typical solution might contain 20% amine, 70% glycol, and 10% water.

Sour gas is contacted countercurrently in a bubble-tray tower with regenerated solution at approximately 100°F and operating line pressure. Purified gas is removed at the tower top. Spent solution is heated by exchange with regenerated solution, charged to a regenerator at about 300°F, and steam-stripped. Excess water and acid gases are removed; reagent is cooled and returned to the top of the bubble-tray tower. The reagent boiling point is raised by the presence of glycol so that acid gases may be stripped at a low regenerator pressure.

Literature references include:

Chapin, W. F., "High Purity Natural Gas by the Glycol-amine Process," *Pet. Ref.*, p. 109, June, 1947.

Phosphate Desulfurization

Developed by Shell Development Co., this is a continuous, regenerative process to separate hydrogen sulfide from natural gas, refinery gas, or liquid hydrocarbons by means of tripotassium phosphate solution. A typical solution contains about 30% tripotassium phosphate in water and will selectively absorb hydrogen sulfide in the presence of carbon dioxide.

Sour gas is contacted countercurrently with regenerated reagent in a bubble-plate extractor at about 100°F (temperatures as high as 200°F are possible) and operating gas pressure. Purified gas is taken overhead from the extractor. Spent reagent, carrying hydrogen sulfide, is removed from the extractor base, heated by exchange with regenerated reagent, charged to the regenerator, and steam-stripped at about 240°F. Hydrogen sulfide and water vapor are taken from the top of the regenerator. Steam can be used directly in the tower for stripping, in which case a reconcentrator operates in conjunction with the regenerator to regulate the water content of the reagent. Regenerated reagent is then cooled and returned to the extractor.

In treating some gases, the reagent is returned to the extractor from two points of the regenerator. The two streams consist of a lean reagent and a very lean reagent. The very lean reagent is charged to the top of the extractor, while the lean reagent is charged farther down to contact the more sour gas. A large steam saving is effected by the practice.

For liquid hydrocarbon treating and some gas treating, reagent is returned to the extractor top as a single stream. Packed towers are generally used to extract hydrogen sulfide from liquids.

Literature references include:

LaCroix, H. N., and L. J. Coulthurst, "Hydrogen Sulfide Removal," *Ref. Nat. Gaso. Mfr.*, pp. 90–94, August, 1939.
Mullen, J. M., "Hydrogen Sulfide Removal by Using Tripotassium," *Oil & Gas J.*, pp. 37–38, Apr. 13, 1939.
Rosebaugh, T. W., "The Shell Phosphate Process for the Removal of Hydrogen Sulfide," *Ref. Nat. Gaso. Mfr.*, pp. 245–247, June, 1938.

Alkazid Process

This is a process for removing H_2S and CO_2 from natural or refinery gases using concentrated aqueous solutions of amino acids. Badische Anilin- & Soda-Fabrik AG of Ludwigshafen am Rhein introduced the process on an industrial scale, and currently it is in use in more than 50 plants in Europe. Some plants have been in service for more than 20 years.

Operation consists of a continuous recycle process, and sour gases or liquids are treated with aqueous alkazid solutions in an absorber at line pressure. Spent solution is regenerated by means of heat and is then returned to the absorber via a heat exchanger and cooler.

Alkazid salts (Alkazid M and Alkazid DIK) are nonvolatile, and their solutions are stable alkaline-reacting liquids. Absorption capacity for H_2S and CO_2 is very high, but very low for hydrocarbons. Alkazid DIK gives excellent results for selective removal of H_2S from gases containing CO_2 and for purification of liquid hydrocarbons. Alkazid DIK solution may also be used for H_2S storage (about 60 volumes of H_2S to the volume of solution). Alkazid M solution is used to remove both H_2S and CO_2 from gases and accomplishes very high scrubbing efficiency (98.6%+).

Corrosion does not represent a problem in the Alkazid process when aluminum is substituted for stainless steel in reboilers, heat exchangers, etc.

Literature references include:

Leuhddemann, Rolf, et al., "The Alkazid Process for Removing H_2S and CO_2 from Natural or Plant Gas," *Oil & Gas J.*, pp. 100–104, Aug. 3, 1959.

Hot Potassium Carbonate Process

The most practical application of this process is for treating natural gas at pressures above 250 psi having acid gas content of from 5 to 50%.   Acid gas may be reduced to as low as 0.5%.

The system is similar to an amine unit except that absorption and regeneration steps both occur at the same temperature and no liquid heat exchangers are required.   In a natural-gasoline plant the usual location for the process is on the compressor discharge before cooling and dehydration.

In operation, acid-bearing gases enter the base of the contactor, either hot or cold, and are contacted countercurrently by carbonate reagent.   Treated gas is removed from the top of the contactor.   Rich carbonate solution is taken from the bottom of the contactor to the regenerator where heat of decomposition is added.   Carbon dioxide is removed from the top of the regenerator.   In some cases, water is recovered in a condenser; in others, it is removed with acid gas.

A number of other variations of the Carbonate process have been designed by Petrocon Engineering Co. to fit various schemes of operation.
1. Split-stream carbonate system
2. Water carbonate process
3. Partially cooled carbonate process
4. Carbonate-amine process

The greatest advantage of a carbonate system is stated as lower operating costs. Units are flexible and may be easily converted to amine systems.

Literature references include:

Buck, B. O., and Angus R. S. Leitch, "Try Gas Treating with Hot Carbonate," *Pet. Ref.*, pp. 241–246, November, 1958.
Eickmeyer, A. G., "The Economics of Acid Gas Removal by Hot Carbonate," *Oil & Gas J.*, p. 106, Sept. 22, 1958.

Giammarco-Vetrocoke Processes

These gas treating processes were developed by G. Giammarco of SPA Vetrocoke, Italy, for the removal of hydrogen sulfide and of carbon dioxide.   Licensing is by Fluor Corp.   A number of plants are in commercial operation in Europe, and one plant is currently under construction in the United States for acid gas removal from high-pressure natural gas.

One process (Fig. 3-85) is for removal of carbon dioxide from gases essentially free of

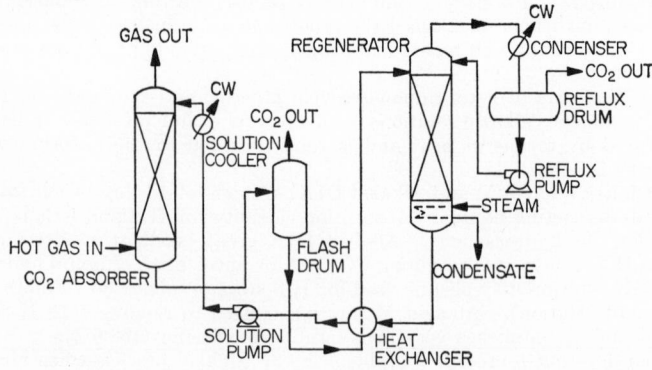

FIG. 3-85. Giammarco-Vetrocoke carbon dioxide removal process with steam regeneration.

hydrogen sulfide. The other (Fig. 3-86) is for selective hydrogen sulfide removal and conversion to free sulfur. The two processes may also be combined, as shown in Fig. 3-87. Both are based upon absorption of acid gases in an alkaline solution containing additives to speed the absorption and desorption rate. The treating solution is non-corrosive, and operating temperatures range from 75 to 300°F.

In the hydrogen sulfide removal process, the reagent consists of sodium or potassium carbonates containing a mixture of arsenites and arsenates. Treated gas from a well-designed unit will contain less than 1 ppm H_2S.

Hydrogen sulfide is first absorbed by reaction with arsenite; the resulting compound is then converted to monothioarsenate by reaction with arsenate. Decomposition

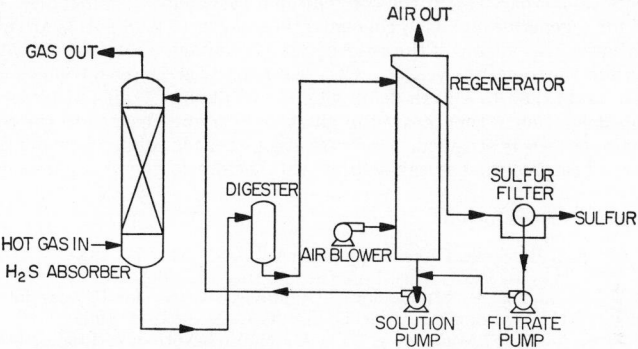

Fig. 3-86. Giammarco-Vetrocoke hydrogen sulfide removal process with air regeneration.

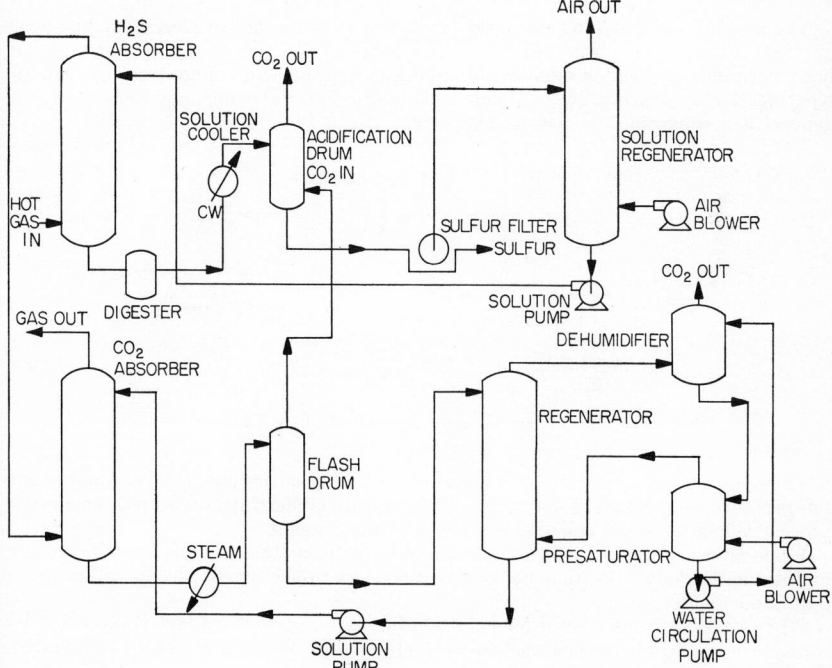

Fig. 3-87. Giammarco-Vetrocoke combination carbon dioxide and hydrogen sulfide removal.

to elemental sulfur and the conversion of trivalent arsenic to the pentavalent form may be accomplished by air blowing or a combination of acidification with CO_2 and air blowing.   The choice depends upon process requirements.

Sulfur may be separated by filtration or flotation.   Water washing yields relatively pure sulfur.   Further purification may be accomplished by recrystallization.

When CO_2 is to be recovered for further processing from feedstock containing both H_2S and CO_2, a two-stage system is used and H_2S is removed first.

The carbon dioxide removal process utilizes hot aqueous alkali carbonate solution which is activated by arsenic trioxide, selenous or tellurous acid.   Commercially, the arsenic compound is most commonly used.   Higher absorption and desorption rates are possible as a result of the additive, and this, in turn, reduces stripper and absorber requirements as compared with the conventional carbonate systems; also, less steam is required for regeneration. CO_2 concentrations down to 0.05 vol-% are obtainable.

Feed gas enters the bottom of the absorber and is contacted by lean reagent, purified, and removed at the top of the vessel. CO_2-rich reagent is removed from the bottom of the absorber and taken to a flash drum where part of the CO_2 is released.   The partially stripped solution is then heated in an exchanger and charged to the regenerator where remaining CO_2 is stripped.   Indirect steam heat is supplied to the base of the regenerator.   Lean solution is taken from the bottom of the stripper to repeat the cycle.

Literature references include:

Oil & Gas J., "Two Processes Unveiled at NGAA," p. 92, Apr. 27, 1959.
Pet. Week, "New Way to Purify Gas," pp. 64, 66, May 29, 1959.
Riesenfeld, F. C., and J. F. Mullowney, "Giammarco-Vetrocoke Process for Acid Gas Removal," NGAA Annual Meeting, Dallas, Tex., Apr. 22–24, 1959.
Riesenfeld, F. C., and J. F. Mullowney, "The Giammarco-Vetrocoke Process for Acid Gas Removal," *Oil & Gas J.*, pp. 86–91, May 11, 1959.

ELECTRIC TREATING

The principle of breaking oil-water emulsions with the aid of electricity has been applied to the field of desalting since the middle 1930's.   Recently, the process has been improved to include emulsion breaking in acid or acid sludge treating, caustic treating, doctor treating, etc.   A wide variety of refinery streams may be treated.   A general flow diagram is shown in Fig. 3-88.

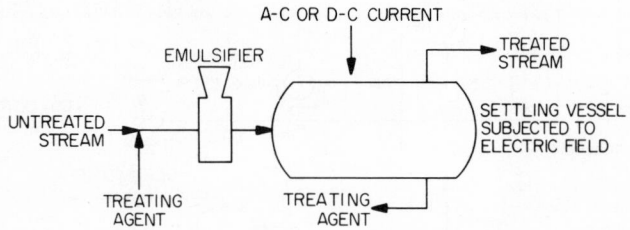

Fig. 3-88. Electric treating—general flow diagram.

The hydrocarbon stream is intimately mixed with the treating agent at appropriate temperature and pressure conditions.   The mixture is then introduced into an electric process vessel for rapid coagulation of the treating agent.

When alternating current is used, treating-agent particles are caused to vibrate in a high potential field.   Particle contact is increased to the extent that small droplets unite and rapidly settle.

When direct current is used for emulsion breaking, particle contact is increased by controlled migration toward the positive electrode.   When sufficiently coagulated, the treating agent settles.

General claims for electric treating are that much better contact is possible with very rapid settling. At the same time, chemical consumption is reduced and carry-over is lowered to a minimum.

Electrofining

Licensed by the Petreco Division of Petrolite Corp., this is a continuous electric process used with a wide variety of chemical treating processes to provide rapid separation of hydrocarbon-reagent dispersions. The process may be applied to treating of gasoline, kerosine, diesel fuel, lube distillates, and other refinery products.

Reagent is dispersed into the refinery stream under the conditions required for the particular treating process being applied, and the mixture is subjected to the action of an electrostatic field in a vessel. Rapid separation of reagent facilitates its continuous automatic withdrawal, accompanied by recirculation of reagent if required.

Normal electrical requirements are a 460-volt 60-cycle primary circuit, and power consumption is less than 1 kw.

The first commercial electric precipitating unit for desludging acid-treated, thermally cracked gasoline (16,000 B/D) was installed at the Torrence, Calif., refinery of General Petroleum Corp. in December, 1951.

Literature references include:

Gandsey, L. J., and R. L. Pettefer, "Richfield Treaters Have Flexibility," *Pet. Ref.*, September, 1959.

Pet. Ref., "Three Electric Processes—Electric Distillate Treating," pp. 125–129, September, 1953.

Stenzel, R. W., and L. C. Waterman, "The Role of Electrofining in the Treatment of Refinery Distillates," WPRA Meeting, Houston, Tex., Feb. 7–8, 1957.

Stenzel, R. W., "Clean-up of Jet Fuels—Electrically," *Ref. Engr.*, January, 1959.

Electrical Precipitation

Licensed by Howe-Baker Engineers, Inc., this is a continuous electric process to improve separation of hydrocarbon-reagent dispersions. Application may be made to most refinery chemical treating processes and a wide variety of refinery stocks.

Feedstock is mixed with treating reagent in a mixing valve and discharged into the precipitator vessel. The mixture is separated as a result of passage through an electric field, then the product is taken to storage, water- and reagent-free. Reagent is continuously drawn from the bottom of the vessel and recirculated until spent. Temperature control is not critical, but temperatures will generally be 100 to 150°F (stream conditions), while pressure is set by the stock being treated.

Equipment cost is stated at about $50,000 for electrifying an existing vessel and includes power supply, electrode assemblies, electrical controls, etc.

The first commercial installation was applied to caustic and water washing of catalytic heating oil in a Gulf Coast refinery in 1955.

Literature references include:

Phillips, R. J., "Cut Costs in Distillate Treating," *Pet. Ref.*, pp. 194–202, September, 1955.

Ref. Engr., "Electrical Distillate Treating," p. 37, October, 1959.

Sulfining

Licensed by the Howe-Baker Engineers, Inc., this is a treating process (Fig. 3-89) employing sulfuric acid in conjunction with electrostatic mixing to remove sulfur compounds from petroleum distillates. It is particularly applicable to the pretreatment of reformer feedstocks.

Acid varying in strength from 88% spent alkylation acid to 98% fresh commercial acid may be used. Operating temperature and pressure are usually that of the stream being treated but will vary with treating design requirements.

Untreated hydrocarbons are first intimately contacted with acid in a mixing vessel with a specially designed, multistaged turbine blade agitator. The hydrocarbon-acid dispersion is then taken to the reaction stage vessel and subjected to electrostatic mixing to oxidize contaminants completely and dissolve them into the acid phase.   Acid is continuously separated and recycled from this vessel.

Hydrocarbon is then mixed with 5 to 10% strength caustic solution to react with carry-over acid (less than 50 ppm).   Separation from caustic takes place in the neutralization vessel, and caustic is recycled.

Hydrocarbon is then water-washed and once again electrically coalesced in a separation vessel.   When reformer feedstock is treated, the stream is passed through driers as a final stage.

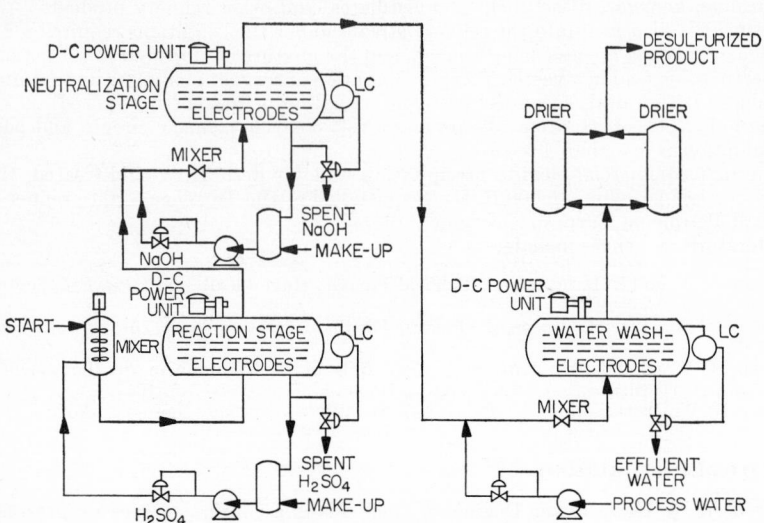

FIG. 3-89. Sulfining electrostatic desulfurization.

The first commercial sulfiner unit went into operation at the East Chicago, Ind., refinery of the Cities Service Oil Co. in October, 1958.   Design charge to this unit was 8,000 B/SD of Mid-continent straight-run naphtha, and erected cost of the unit was reported at about $350,000.

Literature references include:

Phillips, R. J., and H. G. Napier, "Electrostatic Desulfurization of Reformer Feedstocks," API Refining Meeting, New York, May 28, 1959.
Phillips, R. J., and H. G. Napier, "New Desulfurization Process Onstream," *Ref. Engr.*, pp. C18–22, March, 1959.
Ref. Engr., "Electrostatic Desulfurization," p. 35, September, 1959.
Youngblut, K. C., "How the First Electrostatic Desulfurizer Has Performed for Cities Service," *Oil & Gas J.*, pp. 78–80, Aug. 24, 1959.
Youngblut, K. C., "Electrostatic Desulfurization of Reformer Feedstock at the East Chicago Refinery, Cities Service Oil Co.," WPRA Regional Meeting, El Paso, Tex., Apr. 29–30, 1959.

DESALTING

Brine containing sulfates and chlorides is often associated with crude oil in the form of droplets in suspension or emulsion.   Inorganic salts, particularly chlorides, break down during processing and cause serious corrosion and fouling of equipment.   This, in turn, causes reduced throughput and other undesirable side effects.   Removal of these salts is often essential to maintain an economical operating cycle.   A salt content of 20 lb per 1,000 bbl is considered a maximum that can be tolerated in crude oil, but desalting operations are generally aimed at a much lower figure.

Three general approaches have been taken to the desalting of crude oil: mechanical, chemical, and electrical, all shown in Fig. 3-90. Numerous variations of each have been devised, and the selection of a particular process is dependent upon the type of salt dispersion, viscosity of crude, etc.

Simple brine suspensions often may be removed from crude oil by heating under whatever pressure is necessary to prevent vapor loss (200 to 300°F and 50 to 250 psig), then settling in a large vessel. Coalescence is aided by passage through packed towers (sand, gravel, excelsior, fiber glass, etc.)

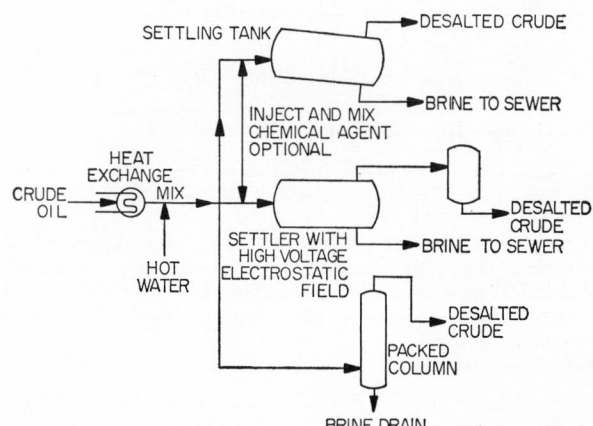

FIG. 3-90. General desalting scheme showing three approaches: mechanical, chemical, and electrical.

The use of chemicals is often necessary to break the more stable emulsions of brine. Oil is generally heated, mixed with wash water and chemicals, then settled for from a few minutes to 2 hr. Treating agents available include various alkaline brines, soaps, fatty acids, sulfonates, and long-chain alcohols.

A high potential field across the settling vessel will also aid coalescence and is used both with and without the aid of chemicals.

General literature references include:

Allison, J. J., "Electric Dehydrating and Desalting," *Ref. Nat. Gaso. Mfg.*, p. 267, June, 1937.
Armistead, Jr., G., "The Desalting of Petroleum Oils," *Oil & Gas J.*, pp. 81–89, Mar. 23, 1946.
Blair, Jr., C. M., "Removal of Inorganic Salts from Petroleum," *Oil & Gas J.*, p. 52, Apr. 4, 1940.
Egloff, G., et al., "Improper Acidizing Presents Desalting Problems to Refining Industry," *Oil & Gas J.*, pp. 57–58, Oct. 14, 1937; p. 66, Oct. 21, 1937.
Foster, A. L., "Desalting Crude Oil," *Pet. Engr.*, pp. C34–38, August, 1951.
Fugua, F. D., "Crude Oil Desalting in Refining Operations," *Pet. Ref.*, pp. 140–146, April, 1946.
Jones, E. R., "Salt Removal from Crude Oil," *Ref. Nat. Gaso. Mfr.*, pp. 208–210, May, 1937.
Van Dedem, G. W., "Chemical Methods for Separating Petroleum Emulsions," *Oil & Gas J.*, p. 65, Aug. 12, 1937.

Electric Desalting—Petreco

Licensed by the Petreco Division of Petrolite Corp., this is a continuous process to eliminate inorganic salts, solids, and other impurities from refinery charge stock by means of electric precipitation.

From 2 to 15% water is emulsified into the untreated crude oil, which has been heated to 180 to 300°F. Operating pressure is that required to maintain the oil and water in a liquid state. The emulsion is broken upon being subjected to a high-potential electrostatic field in a suitable vessel. Dissolved salts and impurities are then removed with the water.

Literature references include:

Albright, J. C., "Salt Removal from Refinery Charge Stock," *Pet. Engr.*, p. 94, November, 1941.
Bonawitz, W. A., et al., "Desalting Crude Oil," *Pet. Engr.*, p. 98, November, 1947.
Fisher, L. E., et al., "Crude Oil Desalting to Reduce Refinery Corrosion Problems," *Materials Protection*, pp. 8–11, 14–17, May, 1962.
Pet. Ref., "Electric Desalting," p. 299, September, 1958.
Waterman, L. C., and J. R. Moechel, "Crude Oil Desalting a Must for Modern Refineries," *Ref. Engr.*, October, 1957.
Waterman, L. C., "Crude Desalting: Why and How," *Hyd. Proc.*, February, 1965.

Electrical Desalting—Howe-Baker

Licensed by Howe-Baker Engineers, Inc., this is a continuous process to eliminate inorganic salts and other impurities from raw crude oil based, on the principle of settling in an electrostatic field.

From 4 to 8% water is intimately mixed with crude oil and heated to 125 to 250°F (operating pressure of the crude system may be maintained). Injection of an emulsion-breaking chemical may be made when necessary. The stable emulsion is then pumped to a settler where the aqueous phase is dropped out by the application of a high-voltage electrostatic field.

Literature references include:

Foster, A. L., "Desalting Crude Oil," *Pet. Engr.*, pp. C34–38, August, 1950.
Pet. Ref., "Electrical Desalting," p. 300, September, 1958.
Ref. Engr., "Electrical Crude Desalting," *Process Notebook*, October, 1959.

Chemical Desalting

Chemical emulsion breakers are compounded and marketed specifically for chemical crude-oil desalting as described in the general discussion above.

Design of chemical desalting equipment varies widely, and no specific process is licensed. Service and assistance may generally be obtained from the chemical supplier.

When a chemical is used for emulsion breaking during desalting, it may be added at one or more of three points in the system:

1. To oil before it is emulsified with fresh water
2. To fresh water before mixing with oil
3. To mixture of oil and water

Chemical emulsion breakers are available from Tetrolite Division, Petrolite Corp., and Visco Products Co.

Literature references include:

Blair, Jr., C. M., "Removal of Inorganic Salts from Petroleum," *Oil & Gas J.*, p. 52, Apr. 4, 1940.
Kirkpatrick, W. H., "Chemical Treating of Crude Oil Emulsions," *Pet. Ref.*, pp. 622–624, November, 1948.
Van Dedem, G. W., "Chemical Methods for Separating Petroleum Emulsions," *Oil & Gas J.*, p. 65, Aug. 12, 1937.

PART 5.   HYDROGEN MANUFACTURE

In recent years the trend has been to increase hydrogen treating of refinery streams, not only for product upgrading but also for feedstock preparation (see Hydrogen Treating, p. 3–38). Much of the hydrogen for such processing is recovered from catalytic reformer off-gas, currently the largest single source of hydrogen in petroleum refining. When hydrogen from naphtha reforming is insufficient, processes to produce hydrogen independent of the refining operation have been used.

The introduction of hydrocracking as a commercial process focused even greater attention on hydrogen from external sources. Hydrocracking requires from 1,000 to 3,000 cu ft of hydrogen per barrel of daily capacity and cannot be designed for maximum usefulness in the average refining situation without supplemental hydrogen. At this writing, the real impact on hydrogen demand by the hydrocracking process has yet to be felt.

Most external hydrogen is manufactured by either the steam methane reforming or the partial oxidation process. These methods also appear to be the favored routes to future hydrogen, although old processes are now being reexamined and improved.

Steam methane reforming is used extensively in the United States because of the readily available methane as a starting material. Partial oxidation of hydrocarbons is used outside the United States, since economics favor using liquid hydrocarbon as feedstock. Recently steam reforming has also been adapted to liquid fuels. Both methods require a treating operation to remove CO_2.

Recent developments also enable the refiner to treat end gases and recycle gases to recover and purify valuable hydrogen previously burned in the furnace or at the flare.

Hydrogen-purification methods play an important part in the cost of refinery hydrogen, now stated at about 30 cents per Mcf. Plants are generally designed for 95% hydrogen purity. This may be accomplished by running partially synthesized gases through shift converters and carbon dioxide scrubbers. Remaining traces of CO and CO_2 may be removed in a methanator, by copper liquor scrubbing, or by the use of molecular sieves. Earlier purification methods included scrubbing with amine solutions, aqueous ammonia, or caustic solutions. Low-temperature purification and absorption in petroleum oils are also under study as possible routes to hydrogen clean-up.

New catalyst development has also improved the economics of hydrogen manufacture and extended the type of hydrocarbon feedstocks that can be used as starting materials.

Less important methods for the manufacture of hydrogen are as follows:

Where small amounts are required with low initial investment, the steam-methanol or ammonia dissociation processes are applicable.

Steam-methanol involves catalytic cracking of methanol at 300 psi and about 500°F with the formation of hydrogen and carbon dioxide. CO_2 is removed with an amine solution. Low-cost methanol is an essential requirement.

Ammonia dissociation also requires low-cost starting material. Gas produced contains 75% hydrogen, 25% nitrogen.

Electrolysis of water produces an extremely pure hydrogen (99.95% +), but power cost must be low ($\frac{1}{2}$ cents per kwhr or less).

The steam-iron process is one of the oldest commercial processes for producing hydrogen. Steam is passed over reduced iron oxide at temperatures of 1400 to 1900°F. Catalyst is regenerated by reversing the reaction with producer or water gas at from 1400 to 1900°F. Total hydrogen produced in the United States from this process is relatively low.

The steam-water gas process has been utilized on a large scale for many years to produce hydrogen for ammonia and methanol syntheses. Water gas (40% CO, 50% H_2) is reacted catalytically (iron oxide activated with chromium) with steam at 700 to 900°F to produce CO_2 and H_2. Relatively pure hydrogen is obtained by scrubbing out CO_2 with amine solutions.

Literature references include:

Arnold, J. H., and W. T. Dixon, "From By-product Hydrogen to Anhydrous Ammonia," *Pet. Proc.*, pp. 62–66, January, 1956.
Baker, D. F., "Low Temperature Processes Purify Industrial Gases," *Chem. Engrg. Prog.*, pp. 399–402, September, 1955.
Brownlie, David, "Bulk Production of Hydrogen," *Ind. Engrg. Chem.*, pp. 1139–1146, October, 1938.
Byrne, P. J., et al., "Recent Progress in Hydrogenation of Petroleum," *Ind. Engrg. Chem.*, pp. 1129–1135, October, 1932.
Chem. Engrg. News, "Catalyst Cuts Hydrogen Plant Cost," pp. 46–48, Feb. 11, 1963.
Chem. Week, "Help for Hydrogen Hungry CIP," p. 61, Oct. 21, 1961.

Davidson, R. L., "Hydrogen Processing," *Pet. Proc.*, pp. 115–138, November, 1956.
Heinemann, Heinz, "Don't Burn Your Excess Hydrogen," *Pet. Proc.*, pp. 1036–1040, July, 1953.
James, G. R., "Which Process Best for Producing Hydrogen?" *Chem. Engrg.*, p. 161, Dec. 12, 1960.
James, R. L., "Hydrogen—Its Markets and Potential Uses," API Refining Meeting, St. Louis, Mo., May 11, 1955.
Johanson, E. S., et al., "When Small Hydrogen Plants Pay," *Hyd. Proc.*, pp. 119–124, January, 1964.
Murphree, E. V., et al., "Hydrogenation of Petroleum," *Ind. Engrg. Chem.*, pp. 1203–1212, September, 1940.
Oil & Gas J., "Cheaper Hydrogen Object of Research," pp. 146–147, Oct. 7, 1963.
Oil & Gas J., "New Catalyst Lowers Hydrogen Costs," p. 106, Feb. 25, 1963.
Oil & Gas J., "New Source of Hydrogen; Refinery Off-gases," pp. 78–79, July 8, 1963.
Oriolo, D. J., and D. R. McIlvain, "More Hydrogen for Petrochemicals via Purification Processes," *Oil & Gas J.*, pp. 102–107, Feb. 4, 1963.
Palazzo, D. F., "Low Temperature Recovery of Hydrogen from Refinery Gases," ACS Meeting, Atlantic City, N.J., Sept. 17–21, 1956.
Pet. Proc., "CO-free Hydrogen Can Be Made by Using Waste Thermal Cracking Gas," p. 549, April, 1955.
Pet. Proc., "How Sohio Makes Ammonia," pp. 47–50, February, 1956.
Pet. Week, "How to Make Hydrogen Cheaper," p. 53, Mar. 10, 1961.
Pfeiffer, C., and H. J. Sandler, "Hydrogen from Catalytic Reformer Off-gas," API Refining Meeting, St. Louis, Mo., May 11, 1955.
Stormont, D. H., "Special Report on Hydrogen," *Oil & Gas J.*, pp. 110–123, Mar. 19, 1962.
Unzelman, G. H., and N. H. Gerber, "Hydrogen," *Petro/Chem Engr.*, pp. 26–37, October, 1966; pp. 28–48, December, 1966.
Updegraff, N. C., "Need Hydrogen? Examine This Process," *Pet. Ref.*, pp. 175–178, September, 1959.
Voogd, J., and Jack Tielrooy, "Make Hydrogen by Naphtha Reforming," *Hyd. Proc.*, pp. 144–148, March, 1963.

Steam-methane Reforming

Steam-methane reforming, a continuous catalytic process, has been in use for several decades as a means of producing hydrogen.

The original U.S. patent was issued to Mittasch and Schneider in 1915 and assigned to Badische Anilin- & Soda-Fabrik for a "process of producing hydrogen." The first commercial continuous steam-methane reformer was installed in 1930–1931 at the Bayway, N.J., refinery of Standard Oil Company of New Jersey.

Process and catalyst improvements through the years have extended the feedstocks to include natural gas, refinery gases, propane, butane, and (more recently) liquid hydrocarbons (see separate discussion below).

The reaction in steam-methane reforming is as follows:

$$CH_4 + H_2O \rightarrow CO + 3H_2$$

Heavier gaseous hydrocarbons react as follows:

$$C_nH_m + nH_2O \rightarrow nCO + (0.5m + n)H_2$$

A flow diagram for a typical process is shown in Fig. 3-91. Feed is first desulfurized (to protect the reforming catalyst) by passage through activated carbon, sometimes preceded by caustic and water washes. Desulfurized feed is then mixed with steam and passed over nickel-based catalyst in a reforming furnace (vertical tubes). Operating temperatures range between 1350 and 1550°F; outlet pressure is generally in excess of 400 psig but can be varied according to hydrogen processing needs. Effluent gases are then cooled by the addition of steam or condensate to a point suitable for CO shift conversion, about 700°F. In the shift converter CO reacts with steam in the presence of iron oxide catalyst to form CO_2 and H_2; the reaction may be accomplished in several stages. Converter effluent gas is cooled, and CO_2 is removed by amine washing—or some similar type of absorbing agent. Remaining CO_2 can be removed by use of additional converters and amine systems or by methanation of the residual CO_2. Typical H_2 by the process is 99+% pure.

Recent developments in metallurgy, catalysts, and furnace design have reduced the cost of hydrogen by steam-methane reforming in recent years. Furnace tubes were formerly rolled welded stainless steel, and operation was at low pressure. Incoloy tubes have improved life and raised pressures to 30 to 40 psig. Centrifugally cast tubes, used widely today, made reforming possible at pressures of 400 psig and above.

Catalysts have improved process economics through better manufacturing techniques, providing uniformity, improved carriers, and longer catalyst life.

In 1963, an improved CO converter catalyst was offered (a mixture of zinc, chromium, and copper oxides) allowing 400°F converter temperature and more favorable chemical equilibrium. Plant cost is reduced by elimination of one or more shift converters and associated CO_2 removal equipment.

Numerous furnaces have been developed for steam-methane reforming; roof-fired, floor-fired, and side-fired are used with vertical tubes. Various flame sizes, tube spacings, burner positions, etc., provide optimum temperature gradient for the reaction.

Some of the companies offering hydrogen manufacturing equipment include Foster Wheeler Corp., Alcorn Combustion Co., and Selas Corporation of America. Some

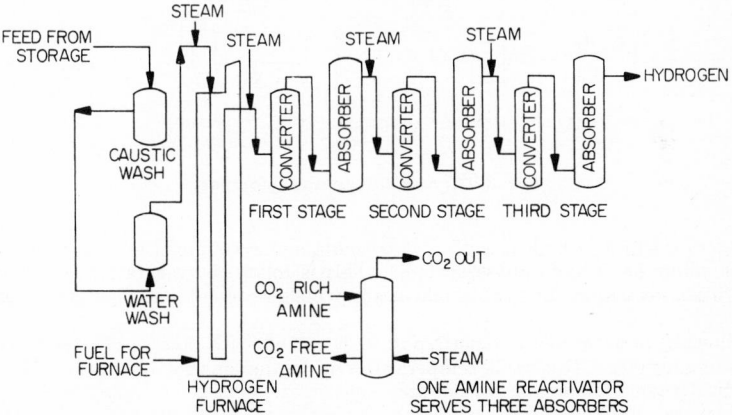

FIG. 3-91. Steam-methane process for making hydrogen.

of the companies offering catalysts for hydrogen manufacture are Chemetron Chemicals, Division of the Chemetron Corp.; Catalyst and Chemicals, Inc.; and Nalco Chemical Co.

Literature references include:

Chem. Engrg. News, "Catalyst Cuts Hydrogen Plant Cost," pp. 46–48, Feb. 11, 1963.

Fox, J. M., and J. C. Yarze, "Catalytic Control of Carbon Deposition in Steam Hydrocarbon Reforming," ACS Division of Petroleum Chemistry Meeting, New York, Sept. 8–13, 1963.

Kenard, Jr., R. J., "Steam Methane Reforming for Hydrogen Production," *World Pet.*, March and April, 1962.

Kityen, Maurice R., and Jack Tielrooy, "What's New in Steam Methane Reformers," *Pet. Ref.*, April, 1961.

Oil & Gas J., "New Catalyst Lowers Hydrogen Ammonia Costs," p. 106, Feb. 25, 1963.

Oil & Gas J., "Cheaper Hydrogen Object of Research," pp. 146–147, Oct. 7, 1963.

Pfeiffer, Carl, and H. J. Sandler, "Ammonia from Cat Reformer Off-gas—Six Ammonia Syntheses Gas Preparation Processes," *Pet. Ref.*, pp. 145–152, May, 1955.

Selas Corporation of America, "Steam Reforming," Bulletin SR-1, May, 1963.

Speed, D. W., and J. H. Noble, "Construction Materials for Steam-Methane Reforming Equipment," ACS Division of Petroleum Chemistry Meeting, New York, Sept. 8–13, 1963.

Uhl, W. C., "Processing Review," *World Pet.*, p. 28, October, 1963.

Unzelman, G. H., and N. H. Gerber, "Hydrogen," *Petro/Chem Engr.*, pp. 26–37, October, 1966; pp. 28–48, December, 1966.
Updegraff, Norman C., "Need Hydrogen? Examine This Process," *Pet. Ref.*, pp. 175–178, September, 1959.

Steam-naphtha Reforming

Steam-naphtha reforming, a continuous catalytic process for the production of hydrogen from liquid hydrocarbons, is similar to steam-methane reforming. A wide variety of naphthas in the gasoline boiling range has performed successfully, including feeds containing as high as 35% aromatics.

The basic problem in attempts to steam-reform liquid feedstocks has been carbon formation in the furnace tubes and resultant pressure increase due to tube blockage. New catalyst developments have provided the solution to the problem, and these are available from a number of sources.

A general flow diagram for the process is shown in Fig. 3-92.

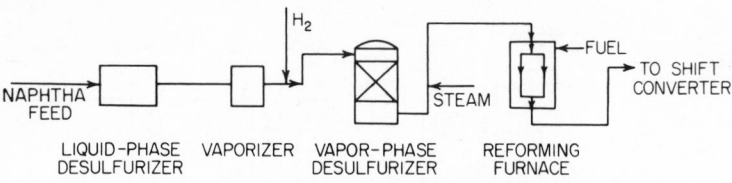

FIG. 3-92.   Steam-naphtha reforming.

When feedstock is high in sulfur, it is sometimes economical to remove the bulk of the sulfur in a liquid-phase system. This is followed by catalytic vapor-phase desulfurization with hydrogen for final cleanup; olefins present in the feed are saturated during this step.

Following pretreatment, vaporized feed is mixed with steam and taken to the reforming furnace. Operating temperatures are in the range of 1250 to 1500°F; units may be designed for pressures from atmospheric to 300 psig.

In general, differences between steam-naphtha and steam-methane reforming are associated with sulfur removal, feed vaporization, and catalyst. In the purification train, additional CO_2 removal capacity is required with the steam-naphtha process because of the lower H/C ratio of the feedstock.

Application of the process has been mainly in Europe, where liquid feeds are in surplus. Some plants have been designed to operate with natural gas, naphtha, or a combination feed.

Literature references include:

Fox, J. M., and J. C. Yarze, "Catalytic Control of Carbon Deposition in Steam-Hydrocarbon Reforming," ACS Division of Petroleum Chemistry Meeting, New York, Sept. 8–13, 1963.
Gignier, J. P., and J. H. Quible, "Cat Reform Naphtha for Ammonia," *Hyd. Proc.*, pp. 153–156, March, 1965.
Imperial Chemical Industries, Ltd., "The ICI Steam Naphtha Process," brochure.
Unzelman, G. H., and N. H. Gerber, "Hydrogen," *Petro/Chem. Engr.*, pp. 26–37, October, 1966; pp. 28–48, December, 1966.
Voogd, J., and J. Tielrooy, "Make Hydrogen by Naphtha Reforming," *Hyd. Proc.*, pp. 144–146, March, 1963.

Synthesis Gas Generation Process—Texaco

This is a continuous, noncatalytic process for the production of hydrogen by partial oxidation of gas or liquid hydrocarbons (i.e., fuel oil, natural gas, etc.) Operation is essentially a noncatalytic flame reaction of oxygen with gas or liquid fuel.

A carefully controlled mixture of preheated fuel and oxygen is fed to the top of the reactor (Fig. 3-93). Free oxygen first reacts with fuel to produce CO_2, water vapor, and heat. A secondary exothermic reaction between the gases and fuel forms hydrogen and CO. In the case of liquid fuel feed, steam is also injected to the reactor.

Effluent from the reactor is then charged to a shift converter with high-pressure steam, which is generated at no cost by utilizing the heat content of the generator-effluent gas in a direct-quenching system or a waste-heat boiler. In the shift converter, CO is converted to CO_2 with accompanying production of H_2 (mole H_2 for mole CO_2).

A major economic improvement in the process has resulted from raising the generator pressure to 1,500 psig or higher. By this means, purified gas can be fed directly without compression to ammonia synthesis, hydrocracking, or other reaction systems.

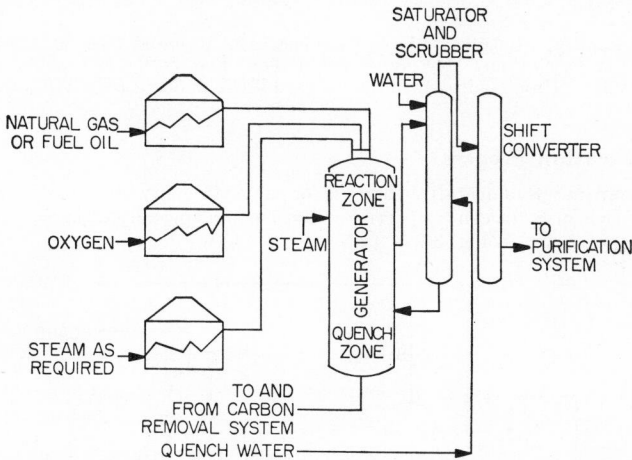

FIG. 3-93. Texas synthesis-gas generation process.

Substantial improvements in steam supply to shift conversion, surplus steam for external use, and energy requirement for CO_2 removal from the product gas are also obtained.

Another recent improvement to the process is a carbon-extraction system prior to shift conversion. With natural-gas feed, carbon yield is negligible; with residual oils, the carbon yield, on a single-pass basis, is controlled at the most economic level, about 2% of charge. Gases from the generator are water scrubbed to remove the last trace quantities of carbon particles. The carbon is then extracted with naphtha and can be transferred to fuel oil. The oil-carbon mixtures are quantitatively consumed by burning in a boiler or by recycling to the generator so that no net carbon is produced.

Reactor temperatures range from 2000 to 2700°F; pressures from atmospheric to over 1,500 psig. Preheat temperatures are in the range of 1200°F but depend upon feed composition. Oxygen requirements are from 250 to 275 cu ft per Mcf of synthesis gas.

Hydrogen yield in the crude gas from the reactor with natural-gas feed is about 60%; from heavy fuel oil, 45%. Hydrogen content of synthesis gas is raised to 92 to 98% by shift conversion.

The gas cleanup system depends upon final hydrogen purity needs. For CO_2 removal, scrubbing with cold methanol, hot potash, amine solution, caustic, and various other solvents has been used. To remove CO and traces of other gases, washing with liquid nitrogen can be used to produce extremely high-purity hydrogen. Copper-liquor solutions are also effective for CO removal, especially when nitrogen is not desired in the product.

Literature references include:

Duff, Barrett S., "Ammonia—Cost of Manufacturing from Five Different Raw Materials," *Pet. Proc.*, pp. 223–228, February, 1955.

Eastman, duBois, "Production of Synthesis Gas by Partial Oxidation," Div. Gas & Fuel Chem. Symposium, Minneapolis, September, 1956, *Synthetic Fuels & Chemicals*, No. 35-5, pp. 71–80.

Eastman, duBois, "Synthesis Gas by Partial Oxidation," *Ind. Engrg. Chem.*, pp. 1118–1122, July, 1956.

Marion, C. P., and W. L. Slater, "Manufacture of Tonnage Hydrogen by Partial Combustion—The Texaco Process," *World Petroleum Congress*, Frankfurt, Germany.

Oil & Gas J., "Four Improvements Feature Texaco Partial-oxidation Process," pp. 127–128, 131, Oct. 7, 1963.

Olson, H. N., and Schneider, P. E., "Synthesis Gas by Texaco Partial Oxidation Process," WPRA Regional Meeting, Corpus Christi, Tex., Jan. 27–28, 1960.

Reidel, J. C., "A First: Hydrogen by Partial Oxidation," *Oil & Gas J.*, pp. 86–91, June 7, 1954.

Slater, W. L., and R. M. Dille, "Partial Combustion of Residual Fuels at Elevated Pressures," AIChE National Meeting, Houston, Tex., Feb. 7–10, 1965.

Stormont, D. H., "How Hydrogen Is Synthesized," *Oil & Gas J.*, pp. 119–123, Mar. 19, 1960.

Shell Gasification Process

This is a continuous noncatalytic process for converting any hydrocarbon feed into a gas mixture rich in hydrogen by partial oxidation in a specially designed reactor. A flow diagram is shown in Fig. 3-94.

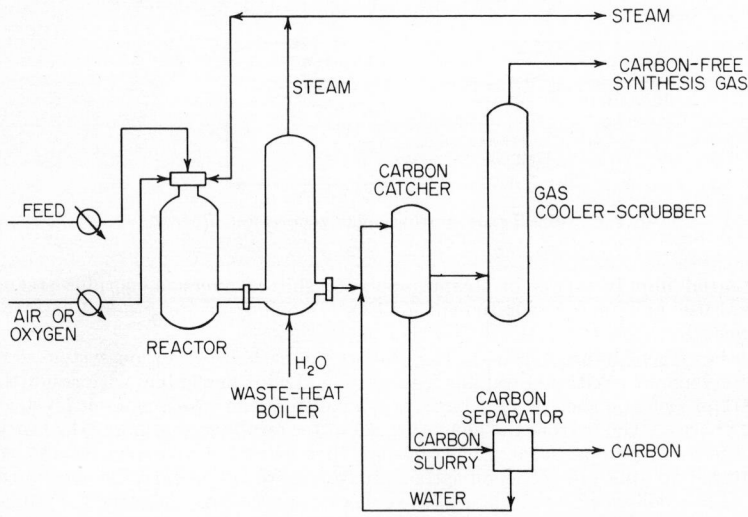

FIG. 3-94.   Shell gasification process.

Hydrocarbon feed and oxidant (either oxygen or air) are preheated separately and mixed in a combustion chamber, where part of the feed is burned and the remainder is thermally cracked by the heat of combustion. Cracked products then combine with the combustion products to form the final gas mixture.

A temperature moderator is included in the reaction mixture. When air is the oxidant, nitrogen serves as a moderator. When oxygen is used, either steam or carbon dioxide is used, depending upon final product requirements. Reactor temperature ranges from 2000 to 2700°F.

Since gas leaving the reactor is high in sensible heat, an important feature of the process is a patented waste-heat boiler. High heat-transfer rates are effected even though the product contains residual carbon. Exchange is from slightly below gasification temperature to close to the temperature of steam produced.

Residual-carbon make varies with the feedstock, but in any case the gas product retains less than 1 ppm, most of which is removed in a carbon catcher. A water wash then removes all but traces of carbon from the gas.

Carbon is removed as a slurry, processed in a carbon separator, and pelleted; water is suitable for recycle to the process. Carbon pellets may be used either as fuel or as raw material for a range of carbon-based products. When liquid feedstocks are used, a carbon production of 2 to 3 wt-% of the feed is typical. With natural gas feed, there is negligible residual carbon production.

Although the H_2/CO ratio is altered when shifting from a heavy to a light feedstock, the thermal efficiency of the process remains about the same. One set of test results using 95 vol-% oxygen with natural gas, straight-run naphtha (99 to 246°F), and heavy fuel oil (1 lb of feedstock to the reactor) showed crude-gas production varying from 49.9 scf for heavy fuel oil to 62.3 scf for natural gas. The composition of the crude gas (dry) varied from 93 vol-% $(CO + H_2)$ when charging heavy fuel oil to 95.4 vol-% $(CO + H_2)$ when charging natural gas. The H_2/CO ratio varied from 0.98 for heavy fuel oil to 1.76 for natural gas.

Literature references include:

Hyd. Proc., "Synthesis Gas—Shell Development Co.," p. 273, November, 1965.
Shell Development Co., "The Shell Gasification Process," brochure.
Singer, S. C., Jr., and L. W. Ter Haar, "Reducing Gases by Partial Oxidation of Hydrocarbons," *Chem. Engrg. Prog.*, pp. 68–74, July, 1961.
Ter Haar, L. W., "Developments in Gasification of Hydrocarbons for Chemical Synthesis," Institute of Gas Engineers and Institute of Fuel Meeting, Hastings, England, Sept. 10–14, 1962.

Hypro

Hypro is a continuous, catalytic process (Fig. 3-95) for the production of hydrogen from refinery off-gas or natural gas. It is licensed by Universal Oil Products Co.

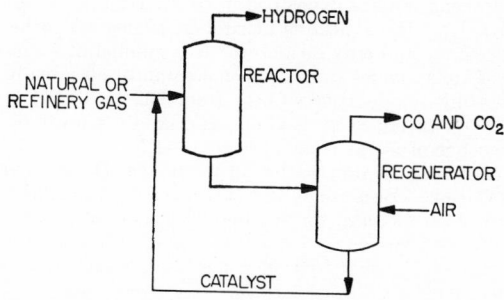

Fig. 3-95. Hypro process for hydrogen manufacture.

The process is designed both to recover hydrogen in off-gas and to convert natural gas to hydrogen according to the following decomposition reaction:

$$CH_4 \rightleftharpoons C + 2H_2$$

Hydrogen is disengaged from carbon (solid) by phase separation at about 93+ vol-% purity. The principal contaminant is methane.

Thermal requirements for the process are low, generally close to the theoretical amount needed for conversion; burning carbon produced in the reaction often supplies the total energy input necessary, since steam is not required for reaction control. Any excess heat can be made available as carbon monoxide and added to the refinery fuel system.

Catalyst identity has not been disclosed but is stated to be a stable, rugged, non-corrosive solid with long life.

Operating costs are said to be low because of low heat requirements for the process; they will vary with the composition of the feed to the unit and thus with each refining situation, since free hydrogen in off-gas is recovered at very low cost.

Estimated costs have been published for a Hypro unit to produce 24.6 million scfd hydrogen in a 50,000-bbl refinery processing Mid-continent crude. Hydrogen make is derived from both free hydrogen and hydrocarbons in the off-gas from hydrocracking. Investment costs were stated at $125 per Mscfd design capacity, operating cost at about 20 cents per Mscf of hydrogen.

Literature references include:

Hyd. Proc., "Hypro," p. 232, September, 1964.

Oil & Gas J., "Hypro Can Make Cheaper Hydrogen," p. 90, Apr. 9, 1962.

Pohlenz, J. B., and L. O. Stine, "New Process Makes Hydrogen from Fuel Gases," *Hyd. Proc.*, pp. 191–194, May, 1962.

Pohlenz, J. B., and L. O. Stine, "Retention of Hydrogen Values in Petroleum Refining," NPRA Meeting, San Antonio, Tex., Apr. 2–4, 1962.

Pohlenz, J. B., and L. O. Stine, "New Process Promises Low Cost Hydrogen," *Oil & Gas J.*, pp. 82–85, Apr. 23, 1962.

PART 6.   MISCELLANEOUS PROCESSES

DEHYDROGENATION

Dehydrogenation is a process for splitting hydrogen from a hydrocarbon or mixture of hydrocarbons, either catalytically or thermally. In the catalytic process, either mono- or diolefins may be produced.

Thermal dehydrogenation has been used to some extent with gas feeds such as ethane, isobutane, etc., for special refining situations. The process is less selective than the catalytic operation and tends to form lighter molecules in preference to the desired olefin. Operating conditions range from 1400 to 2200°F and 5 to 50 psia.

Catalytic dehydrogenation has developed in recent years in the production of basic materials (butadiene) for the synthesis of rubber, resins, and other petrochemicals. The process is selective and may be adapted to a number of feedstocks. Catalysts in general consist of metal oxides supported on ceramic or alumina bases (Cr_2O_3, NiO, TiO_2, etc.). Operating temperatures range from 932 to 1380°F with pressures of about 1 atm. The operating cycle is short, and catalyst must be regenerated frequently in the presence of air at from 1100 to 1300°F.

The main application for the process during World War II was to produce additional olefin feed for alkylation. Because of renewed interest in alkylation for motor fuel, attention has again been directed toward dehydrogenation as a means of providing olefin feed from saturated gases.

Another significant fact is that hydrogen may be obtained as a byproduct in the range of about 80% purity.

General dehydrogenation references include:

Burgin, J., et al., "Dehydrogenation of the Lower Paraffins over Activated Alumina Catalysts," *Oil & Gas J.*, pp. 48–55, Sept. 8, 1938.

Donnell, C. K., "Upgrading Saturates with Dehydro-reforming," *Pet. Ref.*, pp. 186–188, May, 1957.

Frey, F. E., and W. F. Hyppke, "Equilibrium Dehydrogenation of Ethane, Propane & Butanes," *Ind. Engrg. Chem.*, pp. 54–59, January, 1933.

Heinemann, Heinz, "Digest of U.S. Patents (29) on Dehydrogenation and Dehydrodehalogenation of Hydrocarbons," *Pet. Ref.*, pp. 142–154, April, 1944.

Karzhev, V. I., et al., "High Octane Gasoline Obtained by Catalytic Dehydrogenation," *Oil & Gas J.*, pp. 50–53, June 9, 1938.

Komarewsky, V. I., and C. H. Riesz, "Production of Butadiene by Single Stage Catalytic Dehydrogenation of Butane," *Oil & Gas J.*, pp. 33–39, Sept. 17, 1942.

Riesz, C. H., "Catalytic Dehydrogenation of Natural Gas Hydrocarbons," *Oil & Gas J.*, pp. 67–69, 96, July 15, 1944.

Catadiene

This is a continuous, catalytic process, licensed by Houdry Process Corp., Philadelphia, Pa., to dehydrogenate light hydrocarbons to their corresponding mono- or diolefins. Mixtures of hydrocarbons may be processed for the manufacture of chemicals, chemical intermediates, and refinery feedstocks (see Fig. 3-96).

Dehydrogenation is accomplished by passing preheated hydrocarbon vapors, under controlled conditions, over catalysts of the chromic oxide-alumina type $Cr_2O_3 \cdot Al_2O_3$. The process is adiabatic, and heat exchange is eliminated within the reactors. Heat capacity of the catalyst bed is controlled by using an inert granular material along with the active catalyst.

Regeneration of catalyst is accomplished with air at elevated temperature. Reactors are arranged in banks of three to eight units. One reactor is on the process cycle, while the second is on regeneration receiving air, and the third on purge.

For the production of monoolefins, approximately atmospheric pressure and temperatures in the range of 1050 to 1100°F are used. Diolefins are produced at higher temperatures with subatmospheric pressures.

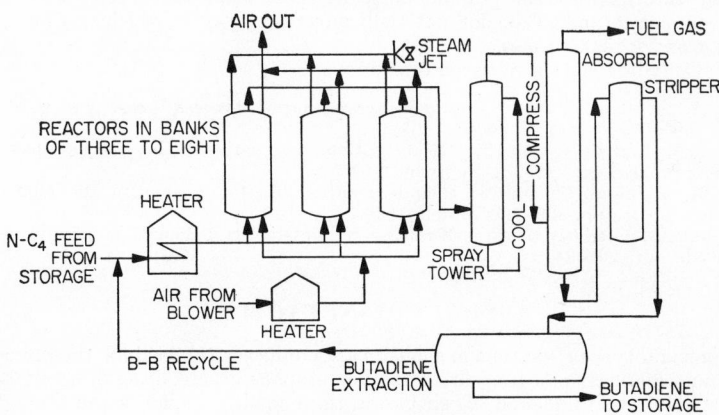

Fig. 3-96. Catadiene dehydrogenation.

When the process is operated for butadiene production, butadiene is extracted from the reactor effluent and the remaining B-B cut is recycled.

The first commercial plants were operated during World War II by Sun Oil Co. at Toledo and Standard Oil Co. of California at El Segundo.

Houdry Butane Dehydrogenation references include:

Donnell, C. K., "Upgrading Saturates with Dehydro-reforming," *Pet. Ref.*, pp. 186–188, May, 1957.
Hyd. Proc., "Butane Dehydrogenation," p. 227, September, 1964.
Oil & Gas J., "Catadiene," p. 125, Apr. 5, 1965.
Pet. Ref., "Dehydrogenation," pp. 208–209, September, 1949.

Dehydrogenation—UOP

A continuous, catalytic dehydrogenation process (Fig. 3-97) capable of converting C_2 to C_4 hydrocarbons to corresponding olefins.

Charge material (butane) is heated to from 932 to 1380°F and passed through a catalyst (chromium oxide on alumina) chamber of the tubular heat-exchanger type under pressures from 10 to 50 psi. By recycling and using optimum conditions, 90 to 95% conversion to olefins is obtained. The reaction is exothermic, and catalyst temperature is carefully controlled. Two or more reactors are used on alternate process and regeneration cycles. Reactor effluent is cooled and compressed before light gases

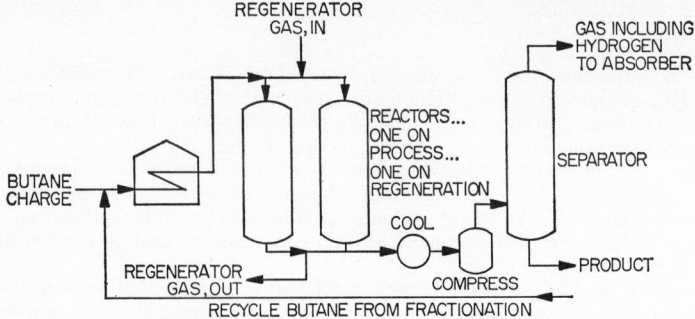

FIG. 3-97. Dehydrogenation—UOP.

(H_2, C_1, C_2's, and C_3's) are separated from butane and butenes.   Fractionation is used to separate the butane for recycle.

The operating cycle is short (about 1 hr), and catalyst contamination is mainly coke. Sulfur compounds and CO do not materially affect the catalyst, which may be regenerated as many as 2,000 times.

UOP Dehydrogenation references include:

Grosse, A. V., et al., "The Catalytic Dehydrogenation Process," *Ref. Nat. Gaso. Mfr.*, pp. 100–107, November, 1939.

Grosse, A. V., and V. N. Ipatieff, "Catalytic Dehydrogenation of Gaseous Paraffins," *Ind. Engrg. Chem.*, pp. 268–272, February, 1940.

Grosse, A. V., et al., "Catalytic Dehydrogenation of Mono-olefins to Di-olefins," *Ind. Engrg. Chem.*, pp. 309–311, March, 1940.

Van Winkle, M., *Aviation Gasoline Manufacture*, McGraw-Hill Book Company, New York, 1st ed., pp. 120–124.

ASPHALT MANUFACTURE

Two general types of petroleum asphalts are produced: (1) residual, those produced by vacuum distillation, steam distillation, or propane deasphalting of a suitable cut from asphalt base crude, and (2) air blown, those made by oxidation of asphalt-base reduced crude.

For the manufacture of paving asphalt, two grades of residue can be produced (one soft and one hard) by distillation.   Blending the two grades, along with lighter components, will meet a whole series of paving asphalt specifications.

Air blowing may be either a continuous or a batch process.   Oxidation catalysts such as ferric chloride and phosphorus pentoxide are used to impart special characteristics and reduce blowing time.

In the continuous process (Fig. 3-98), asphaltic residue from a vacuum tower is fed

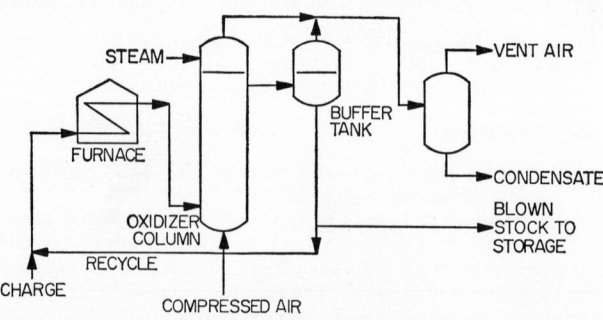

FIG. 3-98. Asphalt manufacture.

to the bottom of a column.   Air is introduced at the tower bottom at 400 to 600°F at a rate of 30 to 50 cfm per ton of charge.   The reaction is exothermic, and steam is often introduced to control the temperature and slow down the oxidation rate.   The amount of air contact and recycle is adjusted to produce the desired hardness.   Oxidized asphalt overflows the top of the column to a tank from which withdrawal is made for both storage and recycle.   Most roofing asphalts are produced by air blowing.

Whether an asphalt is prepared by distillation or oxidation depends upon specifications desired.

Literature references include:

Abraham, H., *Asphalts and Allied Substances*, D. Van Nostrand Company, Inc., Princeton, N.J., 1945, 5th ed.
Alexander, S. H., and A. J. Hoiberg, "Progress in Asphalt Uses and Specifications," *Pet. Engr.*, pp. C11–14, March, 1958.
Barth, E. J., "Know Your Asphalts," *Pet. Ref.*, September, 1957 to August, 1958 (series).
Barth, E. J., "Tailor-made Asphalts," *Pet. Engr.*, pp. C22–25, March, 1958.
Egloff, G., and J. C. Morrell, "Asphalt from the Cracking Process," *Ind. Engrg. Chem.*, pp. 679–680, June, 1931.
Fan, Yun-Nan, "Why Is Continuous Asphalt Best?" *Pet. Proc.*, pp. 862–864, June, 1955.
Griffith, John M., "Progress in Asphalt Technology," WPRA Meeting, San Antonio, Tex., Mar. 16–18, 1959.
Guthrie, V. B., and W. A. Bussard, "Surge Ahead for Asphalt," *Pet. Proc.*, pp. 39–44, January, 1957.
Kastens, M. L., "Paving Asphalt from California Crude Oil," *Ind. Engrg. Chem.*, pp. 548–557, April, 1948.
Kirschbraun, Lester, "Asphalt Emulsion Has Many Uses in the Chemical Plant," *Chem. & Met. Engrg.*, pp. 477–479, August, 1929.
Lomax, E. L., "Qaeyarah Asphalt Plant Begins Operation in Iraq," *World Pet.*, pp. 41–43, August, 1955.
Milburn, H. M., and J. T. Pauls, "Types of Road Oils and Their Use in Road Building," *Nat. Pet. News*, pp. 28–36, July 11, 1934.
Mullins, G. M., "Road Oils," *Nat. Pet. News*, pp. 25–30, Aug. 10, 1932.
Oil & Gas J., "Asphalt Changes During Processing," p. 123, Apr. 27, 1959.
Pet. Proc., "Modern Asphalt Refinery Can Be Neat and Trim," p. 332, March, 1952.
Pruess, D. B., "Cities Service Places Modernized Asphalt Refinery Onstream," *Pet. Engr.*, pp. C6–8, October, 1958.
Ref. Nat. Gaso. Mfr., "Vacuum Recovery of Penetration Asphalt," p. 78, April, 1942.
Reinkemeyer, L. R., "How Anderson-Prichard Uses 3000-bbl per Day Propane . . . Unit for Making High-quality Asphalt," *Oil & Gas J.*, p. 166, Sept. 7, 1959.
Shearon, W. H., "Catalytic Asphalt," *Ind. Engrg. Chem.*, pp. 2122–2132, October, 1953.

GAS AND LIQUID DRYING

Removal of water vapor from gases is achieved by compression, cooling, and refrigeration; by contact with hygroscopic chemicals such as flake caustic or calcium chloride, etc; or by absorption of water into diethylene glycol (see H_2S removal).

Water is removed from refinery liquids to reduce corrosion, prevent loss of water-soluble additives, etc., and may be accomplished by contact with chemicals such as rock salt, calcium sulfate, etc.

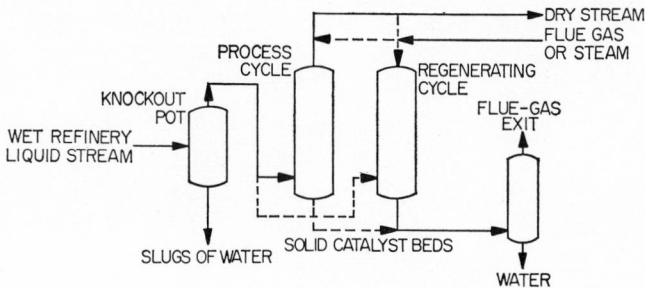

FIG. 3-99. Gas and liquid drying.

The use of anhydrous catalysts in refinery processing often requires drying of both gases and liquids, and methods common to both include the use of clays (fuller's earth), activated alumina, and others. Solid adsorbents may be regenerated (Fig. 3-99) by blowing the bed with steam or hot flue gas at 300 to 400°F.

Generally two or more catalyst beds are used alternately in a drying and regeneration cycle. With liquid refinery streams, a settler is provided to separate slugs of water prior to passage through the drying chamber.   Gas streams containing alkaline material or cracked residue will cause plugging, and if hydrogen sulfide is present, free sulfur may be deposited.

Very recently, molecular sieves (crystals of dehydrated aluminum silicate) have been used in natural-gas drying.   They are said to have twice the dehydration capacity of conventional desiccants.

Literature references include:

Capell, R. G., and R. C. Amero, "Dehydration Liquids and Gases with Granular Absorbents," *Oil & Gas J.*, p. 37, June 18, 1942.

Chem. Engrg. News, "Linde Unmasks Zeolite Process," p. 45, May 11, 1959.

Clark, E. L., "How to Use Molecular Sieves for Natural-gas Treating," *Oil & Gas J.*, pp. 120–123, Apr. 27, 1959.

Daugherty, R. A., "A New Way to Dehydrate Gas," *Pet. Engr.*, pp. E13–16, September, 1956.

Moore, E. R., and W. G. Cutler, "Dehydrate with Calcium Chloride," *Oil & Gas J.*, pp. 166–169, July 27, 1959.

Nelson, W. L., "Drying Liquids with Absorbents," *Oil & Gas J.*, p. 72, June 1, 1946.

Pet. Proc., "Temperature Required for Dryer Regeneration," p. 747, July, 1950.

Senatoroff, N. K., "Application of Diethylene Glycol-Water Solution for Dehydration of Natural Gas," *Oil & Gas J.*, p. 98, Dec. 15, 1945.

Section 4

EQUIPMENT

By HAROLD S. BELL*

Consulting Engineer, Summitt, N.J.

and

LAWRENCE LOWY

Consulting Engineer, New York, N.Y.

HEATERS AND FURNACES

Direct-fired Heaters

The direct-fired heater is used in all distillation units, in many cracking systems, and in several processes such as alkylation, polymerization, and reforming.

Fundamentally, the heater consists of a combustion chamber for heat release, surrounded by tubes through which the oil flows to absorb heat by both radiation and convection.

Nelson identifies them by types as listed below and shown in Fig. 4-1[1,†]

a. Large box-type
b. Separate convection
c. Down convection
d. Straight-up
e. A-frame
f. Circular

g. Large isoflow
h. Small isoflow
i. Equiflux
j. Double upfired
k. Radiant wall

The box types, a and b, are most prevalent. They are best suited for large capacities. Attention must be paid to the hot-tube problem in all types for those tubes subject to both radiation and convection. The tubes most susceptible to overheating are shown in Fig. 4-1 in black.

One solution to overheating is the introduction of cold feed into the first tubes of the convection bank, a so-called screen bank. A late trend is to eliminate oil flow through the convection bank and substitute steam superheaters or other economizing sections. Direct-fired box-type heaters have been built for heat releases of up to 400,000 Btu/hr.

Stills i and k are best adapted to close control of the rate of heat input and for temperatures in the range of 1000 to 1500°F.

* Deceased.
† Superior numbers indicate references at end of section.

4-1

Low rates of heat absorption are best attained in the vertical circular furnaces f, g, and h.

Photographs of typical units in service are shown in Figs. 4-2 to 4-4. Theories of combustion and of radiant and convection heat transfer are given in the literature.[2,3] Two criteria are applied when rating a furnace.

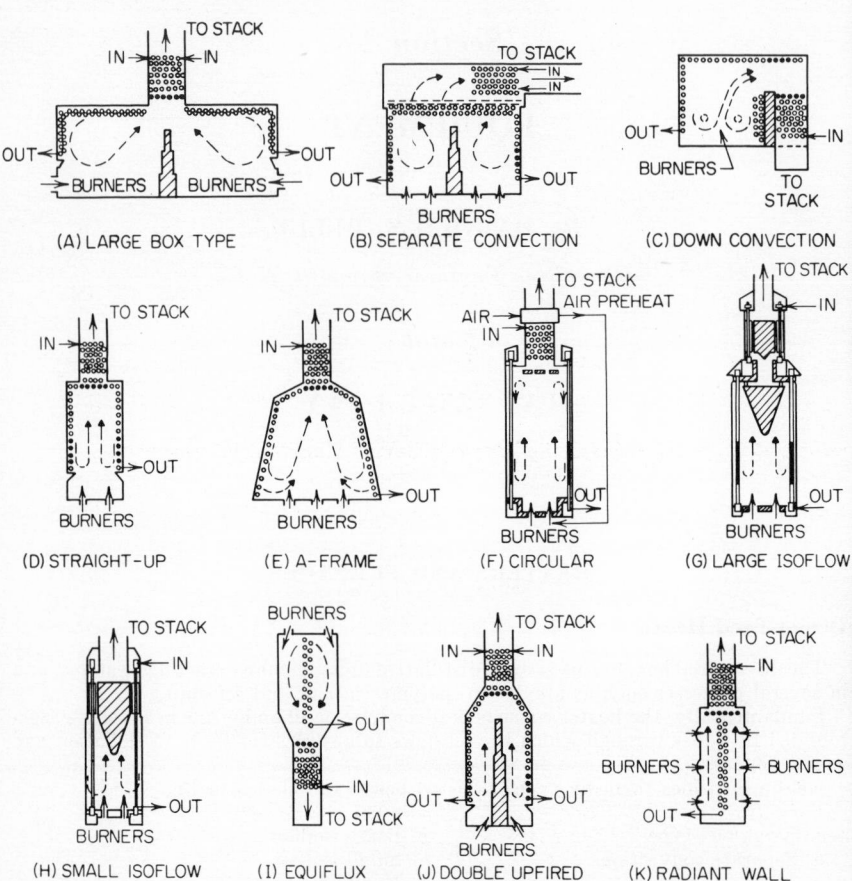

Fig. 4-1. Basic types of pipestill heaters.[1]

1. The possible heat release. For oil and gas firing, furnace volumes are based on a heat release of 20,000 to 30,000 Btu/cu ft/hr. Only the highest-grade refractories are suitable for the higher figure.

2. The amount and arrangement of the heat-absorbing surface.

It has been demonstrated that 50 to 55 per cent of the net heat release in the furnace is generally absorbed in the radiant section, i.e., the part of the furnace in which the tubes are exposed to the radiant effect of flame, hot gases, and hot refractory walls. The allowable rate of absorption depends upon the service. The rate is expressed in Btu per square foot of projected area and not upon the actual surface of the tube. For normal tube velocities of 5 to 7 fps (figured as oil at 60°F) 20,000 to 30,000 Btu/sq ft of projected area is a normal figure for distillation and cracking service. For lubricants or heavy oils subject to carbonization at relatively low temperatures, a value of 15,000 to 20,000 is recommended.

In the convection section of the heater, transfer rates depend upon the temperature

FIG. 4-2. Horizontal box-type direct-fired heater.   (*Courtesy of Foster Wheeler Corp.*)

FIG. 4-3. Double upfired heater.   (*Courtesy of Universal Oil Products Co.*)

Fig. 4-4. Radiant wall heater.   (*Courtesy of Selas Corp. of America.*)

difference between the oil and the flue gas, and the velocity of the flue gas through the
tube bank.   A good design calls for flue-gas velocities through the tube bank of 14 to
16 fps.   Under such conditions transfer rates of 7 to 8 Btu/sq ft of tube surface/hr/
mean log temperature difference may be expected.

Losses from the total heat release include radiation to the atmosphere and heat
remaining in the exit flue gases.   In a well-designed and operated furnace, these will
approximate 5 and 15 per cent, respectively.   For a total heat release of 50 million
Btu/hr, the distribution will approximate

	Btu/hr	%
Heat adsorbed:		
Radiant section....................	27,500,000	55
Convection section.................	12,500,000	25
Total.........................	40,000,000	80
Heat lost:		
Radiation (through setting)..........	2,500,000	5
To stack (atmosphere)..............	7,500,000	15
Total.........................	10,000,000	20

The efficiency may be expressed as 80 per cent.

The above figures presuppose close control of excess air in the furnace.   Some
excess air is essential to ensure complete combustion, but above 25 per cent stack
losses increase and the efficiency is reduced.   However, in some cases where lower

flame temperatures are advisable with sensitive stocks, a higher percentage of excess air is permissible at the sacrifice of some efficiency.

Burners

Burners for direct-fired heaters are classified as follows, with additional information for each type as to its operating cost for burning 100 gal of fuel oil:[4]

Oil-pressure atomizing (5 cents/100 gal fuel oil)
Low-pressure air-atomizing (7.5 cents/100 gal fuel oil)
Intermediate air-atomizing (10 to 15 cents/100 gal fuel oil)
Steam-atomizing (20 to 30 cents/100 gal fuel oil)
High-pressure air-atomizing (60 cents/100 gal fuel oil)

While steam-atomizing types may be expensive to operate, they receive wide usage because of their ability to handle most fuels, including the liquid fuels encountered in today's refinery operations which are little more than waste products. In addition, steam is usually available in the plant.

Burners of this type use either an outside or an inside mix arrangement. In the former, oil and steam are delivered from separate openings. The inside mix utilizes a steam-oil mix inside the chamber.

Atomizing or foaming properties of a fuel will affect the efficiencies of these burners. Residual fuels atomize well, while gas oils or distillates must be mixed with residual fuels to facilitate atomization.

Materials such as asphaltic residues, acid sludge, mixes, and dirty residues are best handled with air-atomizing burners which use low-pressure compressed air.

Viscous fuels are best preheated to hold viscosities down to 130 to 250 SSU. Care must be taken to avoid vaporizing the fuel because vaporization will result in erratic burner operation.

In addition to heating the high-viscosity fuels, the almost solid waste products used, such as tars, have to be recirculated continuously past the burners to prevent solidification in the line. Such lines for feeding and circulating these very heavy fuels require strainers, insulation, and even tracing to reduce plugging.

Heater Tubes

Carbon-steel tubes are used in furnaces to a very limited extent, and only where temperatures are low and corrosive conditions are not severe. Various alloys of steel with chromium, molybdenum, and nickel are more commonly used.

Chromium resists corrosion that may be caused by hydrogen sulfide, free sulfur, and organic sulfur compounds at pipe-heater temperatures. It also resists oxidation at high temperatures. Five per cent chromium in a steel will reduce oxidation at 1200°F to 12 or 13 per cent of that encountered with carbon-steel tubes.

Molybdenum increases the resistance to creep or metal flow at high temperatures.

Nickel of itself does not add to the high-temperature strength of alloys, nor does it contribute to corrosion or oxidation resistance. However, high-chromium alloys tend to become brittle, and nickel acts as a toughening agent to give them excellent corrosion and oxidation resistance and a high creep-stress value.

When metals are subjected to high temperature and stress over extended periods of time, a slow deformation termed "creep" will take place. The "creep stress" is defined as the stress in pounds per square inch for 1 per cent elongation over a given period of time. The determinations are made over a period of 1,000 hr, extrapolated and reported for 10,000 and 100,000 hr. Table 4-1 gives results of creep tests for several alloys used in tube manufacture. The coefficients of expansion of alloy tubes are presented in Table 4-2.

Tubes are connected at the ends by return bends located outside the furnace proper. The tube ends are expanded into the fitting. Removable plugs are placed opposite each tube end for inspection and cleaning. The plugs are held in position by a locking bar between wings on the fitting proper and a heavy set screw, as shown in Fig. 4-5.

Table 4-1. Creep-stress Data for Carbon, Carbon Molybdenum, and Chrome-alloy Steels*

Creep stress in psi for 1 % elongation in 10,000 and 100,000 hr

Material	Analysis	1 %, 10,000 hr (0.1 %, 1,000 hr)				
		800°F	900°F	1000°F	1100°F	1200°F
Carbon steel (killed)..	0.10–0.20 % C, 0.25 % max Si	19,500	11,000	7,200		
Carbon-molybdenum steel...............	0.10–0.20 % C, 0.44–0.65 % Mo	33,500	24,500	12,000		
Chrome steel.........	1.65–2.25 % Cr, 0.44–0.65 % Mo	......	......	11,000	6,700	3,400

Material	Analysis	1 %, 100,000 hr (0.01 %, 1,000 hr)				
Carbon steel (killed)..	0.10–0.20 % C, 0.25 % max Si	13,500	8,400	, 2,100		
Carbon-molybdenum steel...............	0.10–0.20 % C, 0.44–0.65 % Mo	22,000	17,200	6,500		
Chrome steel....... ...	1.65–2.25 % Cr, 0.44–0.65 % Mo	......	......	5,600	2,300	1,000

* Courtesy of Babcock & Wilcox Co.

Another type comprises a removable U bend with sectional bodies which are keyed together, shown in Fig. 4-6. The U-bend design offers smooth flow through the fitting, and the sectional body permits the removal of one tube at a time from the furnace.

Fittings are available in a wide range of alloys and in designs for service to 2000 psi at 800°F and to 1,500 psi at 1000°F. Service at temperatures above 1100°F is not recommended. Welded fittings have been used for more severe conditions, but their use entails the cutting and rewelding of a tube replacement.

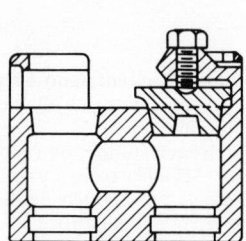

Fig. 4-5. Cast-steel junction box with removable plugs.[3]

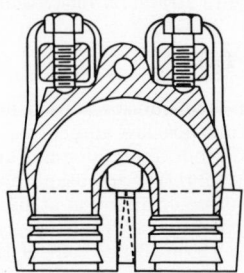

Fig. 4-6. U-bend junction box with sectional body.[3]

The pressure drop through the furnace depends on several unknown factors. Two examples are the ratio of vapor to liquid throughout the coil, and the amount of carbon deposit within the tubes. Several formulas have been suggested, but they are of limited value. A very wide range of pressure drops through furnace tubes for different services may be expressed as follows:

Crude-oil distillation, 150 to 200 psi
Lube oil in solution, 100 to 175 psi
Cracking coils, 250 to 600 psi

Designers of furnace coils rely upon their available data from the actual operation of similar installations.

Table 4-2. Properties of Tubing for High-temperature and High-pressure Service*
Linear expansion, mean coefficients, in./in./°F $\times$ 10^{-6}

	Temp.	Exp. coeff.		Temp.	Exp. coeff.
			Croloy 5 Si		
Carbon steel			5 Cr, 0.50% Mo,		
0.10–0.20% C......	70– 400	6.90	1.50 Si..........	70– 800	7.08
	70– 600	7.40		70–1000	7.29
	70– 800	7.80		70–1200	7.38
	70–1000	8.10		70–1300	7.44
	70–1200	8.30			
			Croloy 7		
Carbon-molybdenum			7% Cr, 0.50% Mo,		
0.50% Mo..........	70– 400	6.63	0.75% Si........	70– 400	6.28
	70– 800	7.68		70– 800	6.74
	70–1000	7.94		70–1000	7.02
	70–1200	8.28		70–1200	7.20
Croloy 2			Croloy 9		
2% Cr, 0.50% Mo...	70– 450	7.45	9% Cr, 1.25% Mo..	70– 300	6.29
	70– 750	7.78		70– 600	6.67
	70–1050	8.45		70– 900	7.00
	70-1150	8.50		70–1200	7.30
			Croloy 16–13–3		
Croloy 2¼			16% Cr, 13% Ni,		
2.25% Cr, 1.00% Mo	73– 402	6.90	3% Mo..........	75– 400	9.60
	73– 609	7.21		75– 800	10.20
	73– 809	7.54		75–1000	10.50
	73–1015	7.86		75–1210	10.80
	73–1207	8.09			
			Croloy 18–8		
Croloy 3M			18% Cr, 9.00% Ni,		
3% Cr, 0.90% Mo...	70– 410	7.51	0.07% C max.....	68– 600	9.88
	70– 802	7.75		68– 800	10.08
	70–1005	7.81		68–1000	10.20
	70–1209	8.06		68–1200	10.41
				68–1292	10.50
Croloy 5			Croloy 25–20		
5% Cr, 0.50% Mo...	70– 400	6.40	25% Cr, 20% Ni....	105–1320	9.20
	70– 600	6.80		70–1832	10.60
	70– 800	7.10			
	70–1000	7.20			
	70–1200	7.30			

* Courtesy of Babcock & Wilcox Co.

Kilns

Kilns are used in refining principally for the regeneration of filtering media, generally fuller's earth. The operation is one of burning from the pores of the material the coloring matter absorbed by the clay during the filtration period. Temperatures must be sufficient to accomplish this result, but not high enough to disintegrate the clay. The maximum temperature is between 1000 and 1100°F.

In older installations, rotary kilns similar to those of the cement industry are still in use. The clay travels down the inclined refractory-lined barrel of the kiln toward the burner end. Combustion products travel upward countercurrent to the clay particles, which fall through the hot gases as the barrel rotates. A smaller rotating cooler is placed below the kiln outlet to reduce the temperature of the discharged material.

Figure 4-7 illustrates a rotary kiln made by the Bonnot Co. The standard sizes with ratings are given in Table 4-3. Few refiners operate at such ratings, 30 to 50 per cent of the figures given in the table being more common.

The multiple-hearth furnace is in general usage. The wedge furnace of the Bethlehem Foundry and Machine Co. and the Nichols-Herreshoff furnace of the Nichols

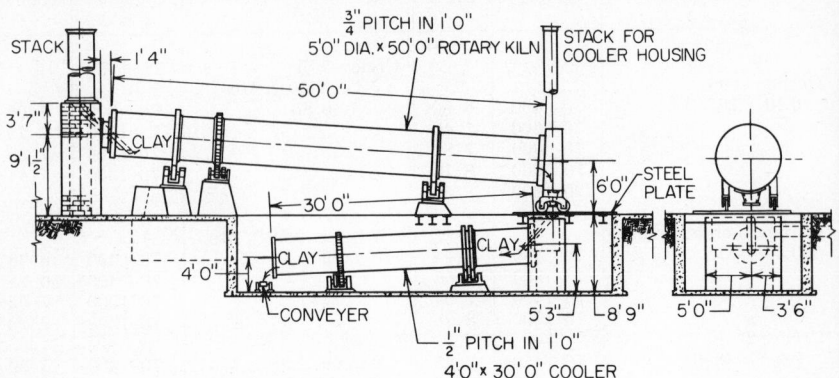

FIG. 4-7. Arrangement of rotary kiln and cooler.[2]

Engineering & Research Corp. are representative. A cylindrical steel shell with refractory lining supports a number of hearths. Rabble arms extending from a revolving center shaft propel the fuller's earth over the hearths as it falls from the top to the bottom of the kiln. Figure 4-8 is illustrative of the Nichols-Herreshoff design.

The principal difference between the two multiple-hearth furnaces is in the method of cooling. The wedge design employs water cooling for the shaft rabble arms and the final cooling hearth, whereas the Herreshoff furnace utilizes air. The two furnaces are available in diameters from 10 to 22 ft, representing hearth areas of 311 to 2,776 sq ft, respectively. Heights accommodate six to eight hearths. Rates of feed for regeneration of fuller's earth are reported as $2\frac{1}{2}$ to 3 psf of hearth area.

Table 4-3. Capacities of Rotary Burners

Diam, ft	Length, ft	Capacity, tons/hr
4	40	1
5	40	2
6	40	3
6	60	4

Kilns, usually termed regenerators, are extensively used in catalytic cracking units but are integral parts of process design. In this case, the free carbon accumulated as the catalyst flows through the reactor furnishes the fuel to maintain the necessary combustion. The kilns of the Thermofor, Fluid, UOP, and Houdriflow units are all of this type. The differences are in the internal arrangements and methods of directing the catalyst flow. In these processes, the catalyst emerges from the kilns at temperatures sufficiently high (850 to 1000°F) to furnish the heat for the reactors.

FRACTIONATING TOWERS

The separation of distillate streams from a complex mixture of hydrocarbons is an essential feature of many oil-refining processes. Such separations are usually achieved by fractionation.

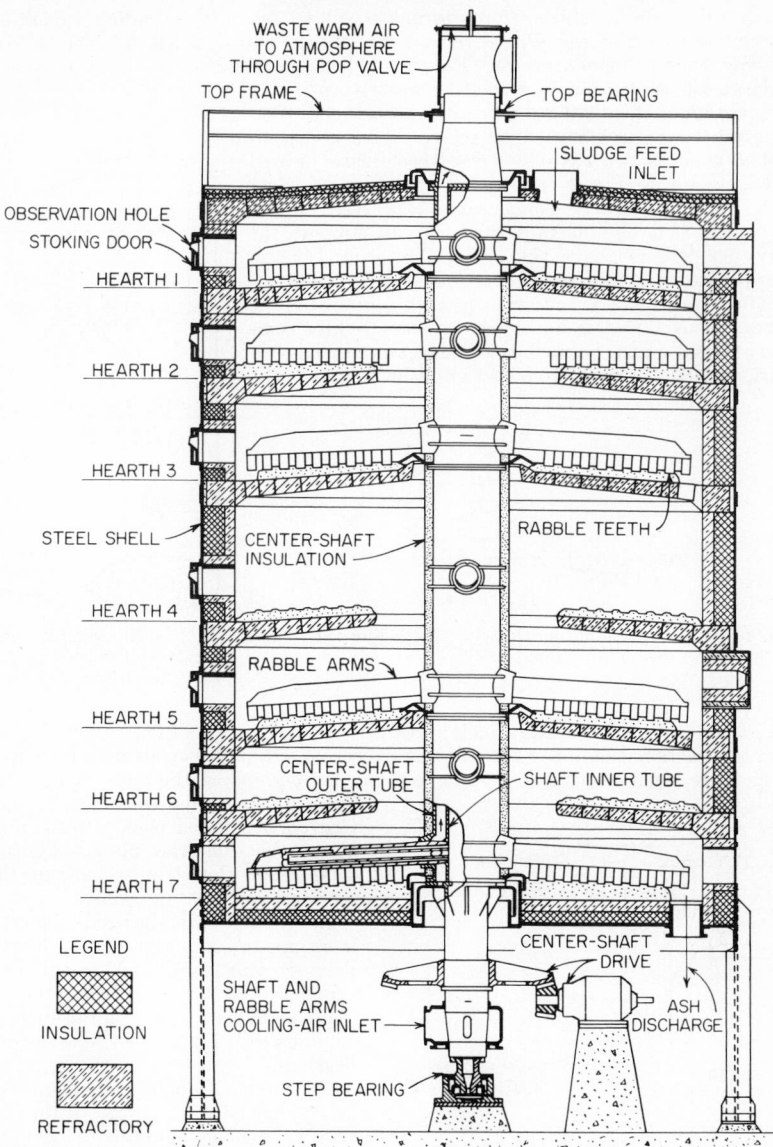

FIG. 4-8. Nichols-Herreshoff multiple-hearth furnace. (*Courtesy of Nichols Engineering & Research Corp.*)

Figure 4-9 illustrates the basic theory of fractionation, which is well described in the literature.[5,6,7] Vapors ascending through a series of vessels will partially condense in the liquid descending from the vessel above. The rectified vapor leaves the last vessel to be condensed for the desired product. Residual is withdrawn from the lowest vessel.

In practice, the separate vessels are incorporated into one piece of equipment—a fractionating tower, which affords intimate contact between ascending vapors and descending condensates. Plates replace the vessels, and liquid is retained on each plate by overflow pipes or weirs. Figure 4-10 is diagrammatic of a section of a simple perforated-plate type of fractionating tower.

A portion of the distillate taken overhead from a fractionating tower is returned as reflux. This is accomplished in one of two ways: by a partial condenser located above the tower (Fig. 4-11) or by a reflux pump feeding back a portion of the condensates (Fig. 4-12). The second method eliminates a structure to support the heavy condensers above the top of the tower. It also affords better and more flexible temperature control.

Side streams are withdrawn from the liquid on the plates. This liquid is never free of some lighter components and must be stripped to meet product specifications. Such stripping is accomplished outside the

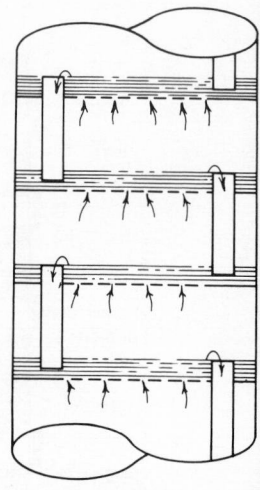

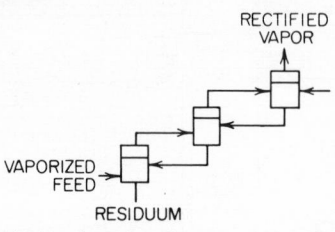

FIG. 4-9. Basic fractionation theory. Ascending vapors are condensed by descending liquid.

FIG. 4-10. Simple perforated-plate type of fractionating tower.

tower in a short plate column such as shown in Fig. 4-13. As the liquid withdrawn from the fractionating tower flows downward through the stripper, steam is introduced at the base, thus reducing the vapor tension and permitting the more volatile constituents to revaporize. The revaporized portion is fed back into the tower above the side-draw plate. The bottoms from the stripper column then meet product specifications.

In the same way, the residual product is stripped of any light ends. For this purpose four to six additional plates are placed in the tower below the feed plate, and steam acts as in the stripping columns to bring the bottoms to specifications.

If a supply of hot oil is available, such as residuals from a primary tower, it may be used as a source of heat in a reboiler, causing reevaporation of a side stream and thus acting as a stripper.

Figure 4-14 illustrates the application of a reboiler to a two-tower assembly. Side streams are not shown. It should be noted that the secondary tower may be considered the top section of a single tower. In the case shown, the heat from the reboiling coil causes reevaporation of the light ends of the residual from the second tower. If the entire complex is in a single tower, the second-tower residual will correspond to a side stream.

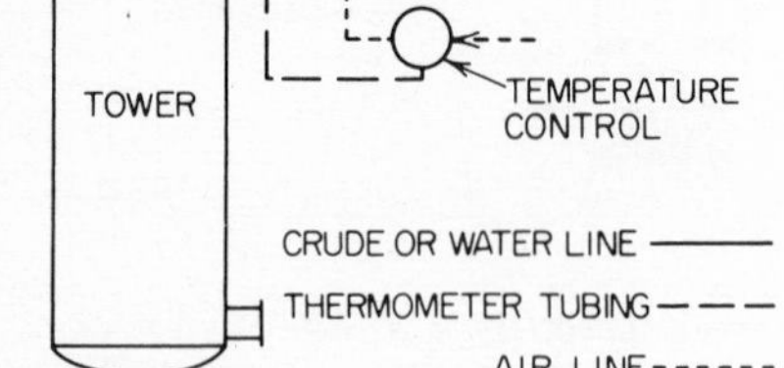

FIG. 4-11. Tower with reflux condenser located overhead.

CRUDE OR WATER LINE ———
THERMOMETER TUBING — — —
AIR LINE - - - - - -

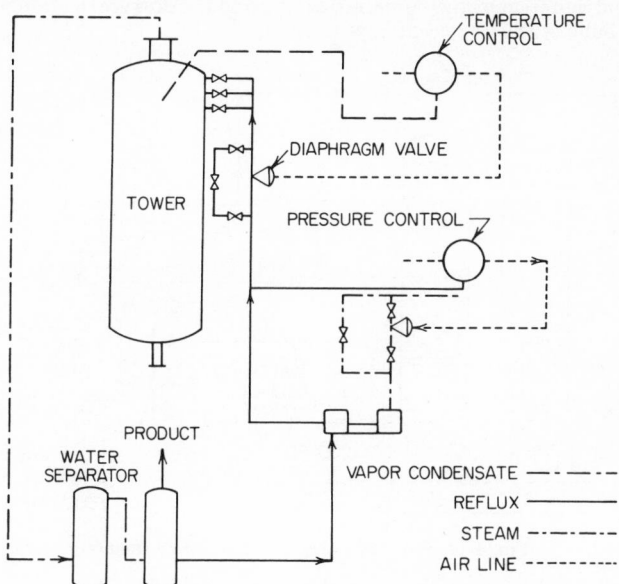

Fig. 4-12. Tower with pump for returning reflux.

Packed Columns

An early approach to scientific fractionation was the packed column. For constant vapor velocities and small volumes, good results are attained. Stone crushed to uniform size, pipe nipples, and Raschig rings are among the materials used as packing. The glass beads used in modern laboratories are illustrative.

But there are two inherent difficulties. The columns, if of any size, will channel, and there is no way to provide side-stream draw-offs. Practically, such towers serve only as dephlegmators and are little used in commercial installations.

Plate Columns

The bubble-tray column is most widely used by petroleum refiners. Riser pipes allow the ascending vapors to pass through the plate, as shown in Fig. 4-15. The combined area of the risers is 7 to 11 per cent of the tray area. Over each riser is placed a bubble cap with a serrated or slotted skirt. The slot areas per tray vary from 12 to 15 per cent of the column cross section. The vapors ascending through the vapor nipples are deflected downward by the cap to exit through the slots below the liquid level on the tray. The object of the cap is to obtain even diffusion of vapors through the liquid. Deformation of the

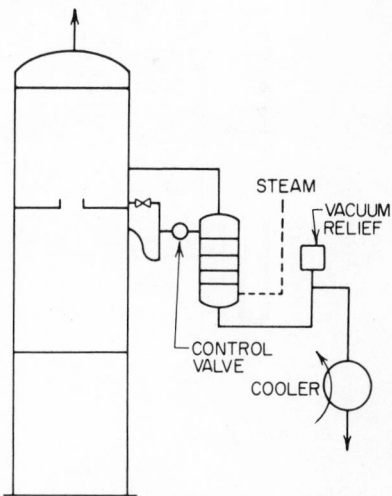

Fig. 4-13. Side-stream stripper for distillation tower.

bubbles and agitation of the liquid is essential and the caps are so arranged that collision with bubbles from adjacent caps is attained.

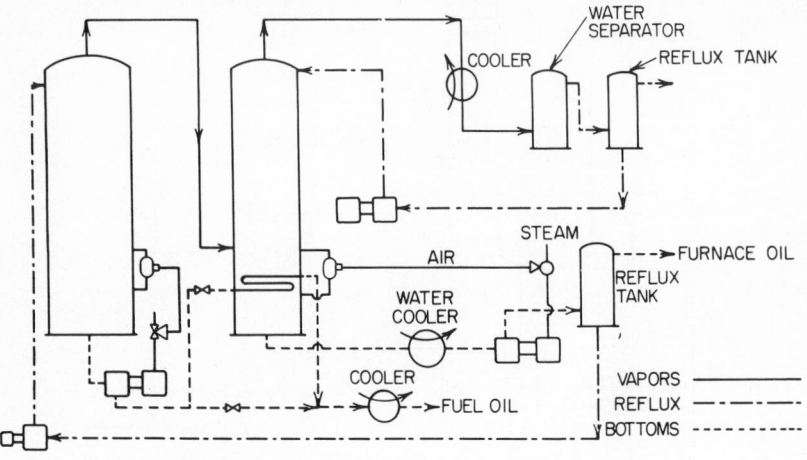

FIG. 4-14. Reboiler applied to two-tower assembly.

Liquid level on the tray is maintained by downspouts or weirs sealed at the bottom by the liquid on the tray below.

Shell Design

Fractionating towers rarely exceed bottom-liquid temperature in excess of 800 to 900°F, and follow the ASME *Code for Unfired Pressure Vessels.*

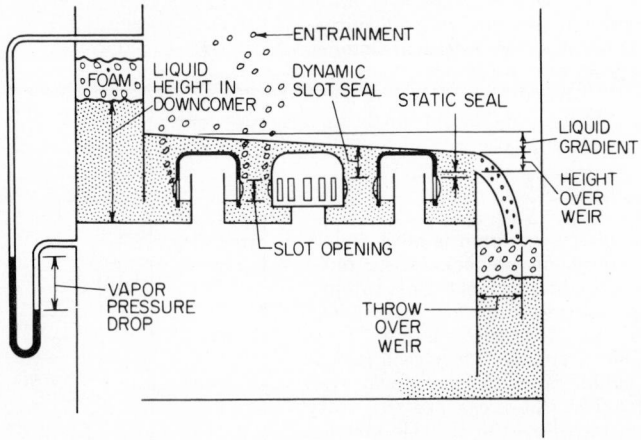

FIG. 4-15. Section of conventional bubble tray, showing risers, bubble caps, weir.[8]

Means of entering the tower for inspection of trays and caps is provided by side manholes. Practice indicates tray spacing of 18 in. for tower diameters under 5 ft, and 24 in. for larger towers. It is customary to place manheads at 90° between successive trays.

The diameter of the tower is proportional to the amount of vapor ascending. W. L. Nelson[9] offers the formula

$$C = 220D^2R$$

in which C = capacity, feed, B/D
D = diameter, ft
R = per cent residuum, expressed as a decimal
The relationship is shown by Fig. 4-16.

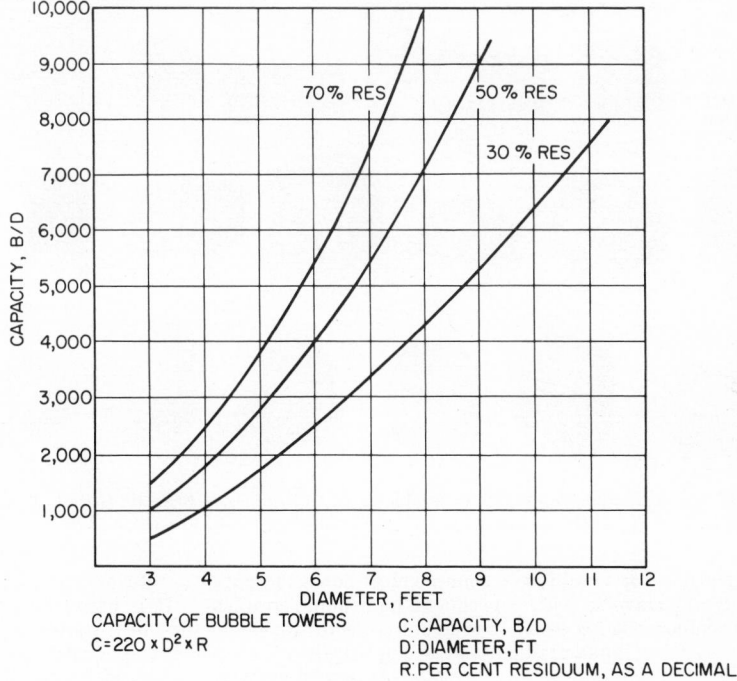

Fɪɢ. 4-16. Relationship between fractionating tower diameter and capacity.[2]

For the fractionation of streams under vacuum, towers have been built up to 33 ft in diameter. The height of the tower depends on the number of trays which, in turn, depend on the degree of fractionation desired and the ratio of vapor to liquid in the feed. As a guide, Table 4-4 is offered. The tallest tower of present record is 227 ft.

Table 4-4. Typical Towers, Classified According to Degree of Rectification, Typical Numbers of Trays, and Ratio of Vapor Rate to Feed Rate

Degree of rectification	No. of trays	Ratio of vapor to feed
1—Stripping still..............	10– 20	Vapor = 20 % of feed
2—Primary fractionator........	20– 40	Vapor = feed
3—Secondary fractionator.......	40– 50	Feed = 50 % of vapor
4—Splitter....................	50– 70	Feed = 25 % of vapor
5—Superfractionator...........	70–100	Feed = 10 % of vapor

The heat balance within a tower must be protected, and hence shells are well insulated under a weatherproof coating or jacket.

Trays

As diameters increase, attention must be given to the maintenance of even depths of liquid over the tray areas. Otherwise vapors will channel to the area of lowest

FIG. 4-17. Typical shapes and sizes of bubble caps.   (*Courtesy of Fritz W. Glitsch & Sons, Inc.*)

liquid depth.   For small towers, simple cross flow is acceptable.   Reverse flow offers a long liquid traverse with consequent high liquid gradient.   It is used when the liquid-to-vapor ratio is low.   The double-pass arrangement provides shorter liquid traverse with low liquid gradient and is adopted when the liquid-to-vapor ratio is high. Intermediate weirs which control the liquid level are also used when the liquid vapor ratio is high.   Termed the "cascade type," it is essential for towers of large diameters.

Pressed-steel bubble caps have superseded the earlier cast-iron caps.   Available in many shapes and sizes, they are illustrated in Fig. 4-17.

Several proprietary trays are in general use in refineries.   The Flexitray (Fig. 4-18) utilizes a design similar to a clapper valve over circular openings in the plate for vapor passage.   The lift of graded weights of caps with increased vapor loads ensures distribution.   There is no obstruction to liquid flow, hence no gradient across the tray.

As fractionation has become better understood, there has been some return to the economy and efficiency of the perforated tray.   Of this design the ripple tray (Fig. 4-19) is an example.   No weirs or downspouts are necessary as descending liquid

FIG. 4-18. Flexitray design.   (*Courtesy of Koch Engineering Co., Inc.*)

flows to the tray below through the perforations in the valleys of the corrugations. Alternate trays are placed at 90° to assure good distribution.

The Turbogrid tray utilizes the perforated-tray principle by means of parallel slots covering the entire cross section of the tower. Both liquid and vapor pass through the slots.

FIG. 4-19. Ripple tray. (*Courtesy of Stone & Webster Engineering Corp.*)

The Uniflex tray embraces a series of pressed-steel S-shaped members. The sections interlock and one edge is slotted in the usual bubble-cap method. The other edge forms the weir for descending liquid to drop to the tray below.

It is often desirable to install mist extractors in the top of a tower to guard against entrainment or "carry-over" of minute liquid globules in the overhead stream. To be

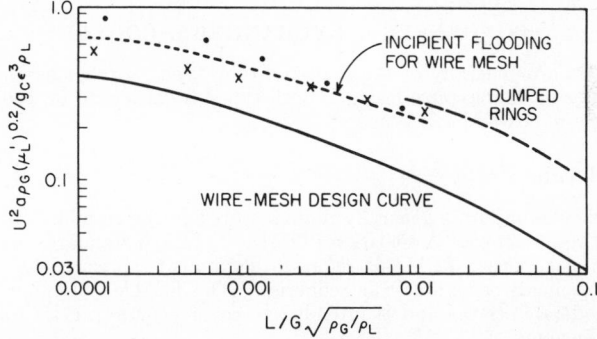

FIG. 4-20. Effect of liquid entrainment load on allowable gas velocity through Demisters.® (*Courtesy of Otto H. York Co., Inc.*)

effective, the material used must have ample extended surface in relation to volume. Wire mesh or metal cloth is used in layers to a thickness of 4 to 8 in. The tortuous vapor channel through the Demister forces the minute droplets into contact with the metal surface. Here, with other droplets, they form larger globules which finally fall to the fractionating tray below. One manufacturer rates the operating limits of mist extractors as shown by Fig. 4-20.

A well-designed bubble column must have some flexibility in capacity to meet varying load factors and liquid-vapor ratios. Bolles,[8] from a study of the operation of many bubble towers, offers Fig. 4-21, which illustrates the large area of satisfactory operation for a typical example.

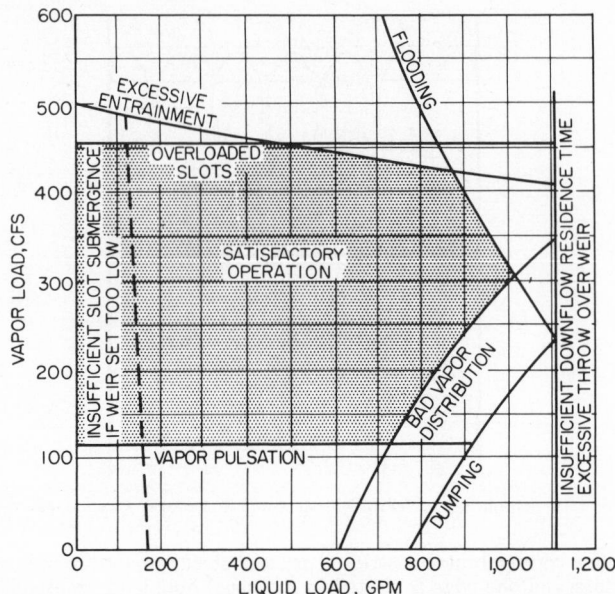

FIG. 4-21. Satisfactory operating range of bubble towers. (*Courtesy of Petroleum Processing magazine.*[8])

CONDENSERS, EXCHANGERS, COOLERS

Condensers are generally of the shell-and-tube type. Exchangers may be of the shell-and-tube or double-pipe design. Both types are also used for coolers, with the inclusion of the box-and-coil type.

Shell-and-tube Units

Shell-and-tube units are generally manufactured to the standards of the Tubular Exchanger Manufacturers Association (TEMA). TEMA standards are divided into two classes of service. TEMA-R refers to equipment fabricated for the generally severe requirements encountered in refineries. The TEMA-C standard is applicable to more moderate service and is intended to cover general service conditions with maximum economy.

The commonest exchanger or condenser encountered in refineries is of the shell-and-tube design, shown at the top of Fig. 4-22. Tubes are rolled into a stationary tube sheet at one end. A free-floating head at the other end of the tubes allows for expansion. Liquid flow is directed by baffles in the fixed and floating heads. Counterflow of liquids or vapors is directed by baffles in the shell. Shells and heads generally conform to the ASME *Code for Unfired Pressure Vessels.*

Shell-and-tube equipment may be subdivided according to design:

1. Fixed tube sheets at both ends: one or two passes in both shell and tubes, shells cross-baffled. Recommended only when the differential between tube and shell expansion is low. See bottom of Fig. 4-22.

2. A floating head at one end: single shell pass, cross baffles in shell.   Floating head must be removed before tube bundle can be pulled.

3. A floating head at one end, which may be removed with the tube bundle: two or

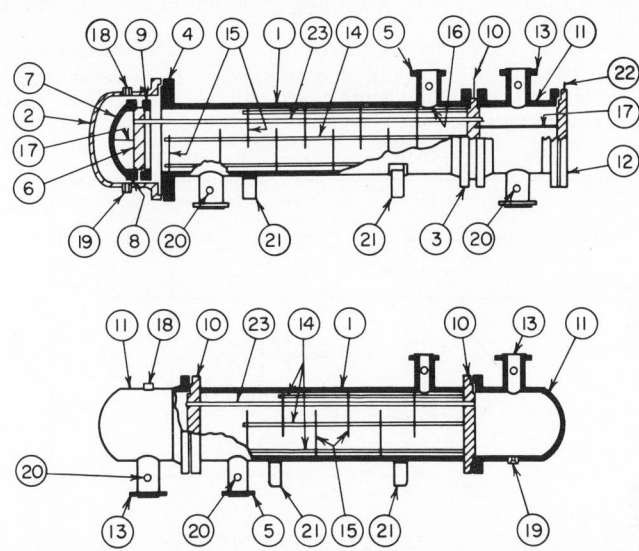

Fig. 4-22. Nomenclature of heat-exchanger components, showing floating-head type and fixed-tube-sheet exchangers. Typical parts and connections, for illustrative purposes only, are numbered for identification. Standards of Tubular Exchanger Manufacturers Association, 1962.[10]   (1) Shell; (2) shell cover; (3) shell flange channel end; (4) shell flange cover end; (5) shell nozzle; (6) floating tube sheet; (7) floating-head cover; (8) floating-head flange; (9) floating-head backing device; (10) stationary tube sheet; (11) channel or stationary head; (12) channel cover; (13) channel nozzle; (14) tie rods and spacers; (15) transverse baffles or support plates; (16) impingement baffle; (17) pass partition; (18) vent connection; (19) drain connection; (20) instrument connection; (21) support saddles; (22) lifting lugs; (23) tubes.

more passes in both shell and tubes, cross or longitudinal baffles in shell.   This design is more frequently used in refineries.

A design which eliminates the second tube sheet employs U tubes, thus permitting free tube expansion, shown in Fig. 4-23.   There are two tube passes and with a longitudinal baffle also two shell passes.   While the tube bundle may be pulled and tubes cleaned externally, internal cleaning is difficult.   Economy of first cost is offset by higher maintenance.

The different types of construction may be further classified by the service conditions:

1. Liquid-to-liquid exchangers
2. Vapor-to-liquid exchangers
3. Condensers, vapor-to-water
4. Reboilers
5. Coolers

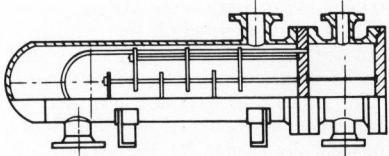

Fig. 4-23. The U-tube heat-exchanger design eliminates the second tube sheet, thus permits free tube expansion.

The selection of which medium to pass through the tubes and which to pass through the shell of an exchanger is important.   In condensers it is natural to pass the vapors through the shell and the water through the tubes.   Water may deposit scale which a tube cleaner will easily remove but which would present a problem if on the external

surface.   Also, the vapors will pass through the shell without excessive pressure drop.
Any medium likely to leave deposits should pass through the tubes.

In general, the medium of higher pressure and greater corrosiveness should pass
through the tubes.   The more viscous medium or the one of less volume is generally
passed through the shell.

One economic factor should be considered.   Increments of surface do not yield
proportionate heat recovery.   A study of a group of residual exchangers is shown
graphically by Fig. 4-24.   The maximum practical recovery is taken arbitrarily at
100 per cent.   The curve shows the percentage of optimum heat recovery for lesser

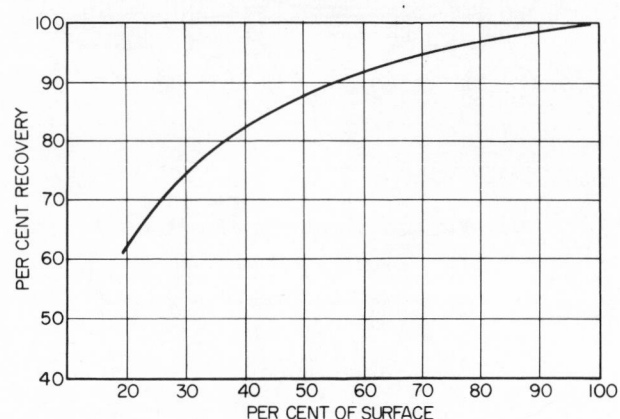

FIG. 4-24.   Percentage of optimum heat recovered for lesser percentages of surface.

percentages of surface.   The curve shows that 90 per cent of the optimum recovery
may be attained on about 55 per cent of the required surface.

Materials

Cast-iron shells, heads, channels, tube sheets, and cover plates are considered satisfactory for relatively low temperatures, to 400°F, and pressures under 150 psi.   For
higher temperatures and pressures, shells and all the above parts should be of steel.
With the advent of resistance and fusion welding, steel shells are generally fabricated
from pipe of API specifications.

Table 4-5 presents the specifications of tubing.   Tubes under $\frac{3}{4}$ in. OD are not
easily cleaned and are rarely used.   Carbon-steel tubes are in general use for exchanger
service.   For condensers, Admiralty metal (Cu 70 per cent, Sn 1 per cent, Zn 29 per
cent) is recommended for temperatures up to 500°F.   The introduction of ammonia as
a neutralizing agent into the vapor stream must be avoided.   For severe sulfur corrosion, 4–6 chrome–$\frac{1}{2}$ molybdenum and also 18–8 stainless-steel tubes are used.

For tubes 1 in. OD and less, a thickness of 12 to 18 BWG is recommended, with
preference for 16 BWG.   For tubes of greater than 1 in. OD, the BWG in use is 10 to
16, with preference for 14 BWG.

Exchanger-tube lengths are standardized by TEMA specifications at 8, 10, 12, 16,
and 20 ft.   Lengths of 16 and 20 ft are preferred in the interests of economy, and most
manufacturers standardize their exchangers for these two lengths.

The specification sheet of TEMA is shown in Fig. 4-25.   An inquiry for shell-and-
tube equipment should include all the pertinent service data, items 10 to 25 inclusive,
and the fouling factor that should be applied, item 29.   Other data should include
full information on the two mediums as to corrosive characteristics.   Other items
should be left to the manufacturer.

Table 4-5. Characteristics of Exchanger Tubing

OD of tubing	BWG gage	Thickness, in.	Internal area, sq in.	Sq ft external surface per ft length	Sq ft internal surface per ft length	Weight per ft length brass, lb	Weight per ft length copper, lb	Weight per ft length steel, lb	ID tubing, in.	Constant C^*	OD/ID	Metal area (transverse metal area)
9/16	21	0.032	0.1952	0.1473	0.1304	0.196	0.206	0.183	0.4985	305	1.128	0.053
5/8	10	.134	.100	.1636	.0937	.76	.800	.703	.357	155	1.750	.207
5/8	11	.12	.1170	.1636	.101	.70	.735	.647	.385	182	1.62	.190
5/8	12	.109	.130	.1636	.107	.649	.683	.605	.407	202	1.53	.177
5/8	13	.095	.149	.1636	.1142	.581	.61	.54	.435	232	1.44	.158
5/8	14	.083	.165	.1636	.1205	.520	.548	.481	.458	257	1.36	.142
5/8	15	.072	.1817	.1636	.126	.460	.482	.43	.481	283	1.30	.125
5/8	16	.065	.1924	.1636	.1295	.420	.44	.39	.495	300	1.26	.114
5/8	18	.049	.2181	.1636	.1379	.326	.342	.301	.527	340	1.19	.0887
5/8	20	.035	.2419	.1636	.146	.238	.25	.22	.555	377	1.13	.0649
3/4	10	.134	.1822	.1963	.1265	.95	1.04	.882	.482	284	1.56	.260
3/4	11	.120	.2043	.1963	.1335	.87	.918	.81	.510	319	1.47	.238
3/4	12	.109	.223	.1963	.14	.807	.845	.75	.532	348	1.41	.219
3/4	13	.095	.247	.1963	.147	.718	.752	.67	.56	385	1.34	.195
3/4	14	.083	.268	.1963	.153	.640	.67	.591	.584	418	1.28	.174
3/4	15	.072	.289	.1963	.159	.563	.592	.522	.606	451	1.24	.153
3/4	16	.065	.302	.1963	.163	.514	.54	.48	.620	471	1.21	.140
3/4	17	.058	.314	.1963	.166	.463	.494	.429	.634	490	1.18	.128
3/4	18	.049	.334	.1963	.1707	.396	.417	.367	.652	521	1.15	.108
3/4	20	.035	.3632	.1963	.179	.289	.306	.267	.680	567	1.10	.0786
7/8	8	.165	.2333	.2297	.143	1.348	1.42	1.251	.545	364	1.60	.3681
7/8	10	.134	.2890	.2297	.158	1.134	1.24	1.06	.607	451	1.44	.312
7/8	11	.12	.317	.2297	.167	1.04	1.096	.968	.635	494	1.38	.284
7/8	14	.083	.394	.2297	.186	.759	.798	.702	.709	615	1.23	.207
7/8	16	.065	.436	.2297	.1952	.608	.636	.562	.745	680	1.17	.165
7/8	18	.049	.474	.2297	.2034	.467	.49	.432	.777	740	1.12	.127
7/8	20	.035	.508	.2297	.210	.340	.357	.314	.805	793	1.09	.0927
1	8	.165	.355	.2618	.175	1.578	1.67	1.471	.670	554	1.49	.4304
1	10	.134	.421	.2618	.192	1.34	1.41	1.24	.732	657	1.37	.364
1	11	.12	.455	.2618	.200	1.22	1.28	1.13	.76	710	1.32	.3317
1	12	.109	.479	.2618	.205	1.12	1.18	1.04	.782	748	1.28	.306
1	13	.095	.515	.2618	.213	.99	1.042	.92	.810	804	1.24	.270
1	14	.083	.5463	.2618	.2183	.88	.923	.813	.834	852	1.20	.239
1	15	.072	.576	.2618	.225	.77	.813	.714	.856	899	1.17	.209
1	16	.065	.594	.2618	.228	.70	.737	.65	.870	927	1.15	.191
1 crimped	16	.065	.50	.2618	.228	.70	.737	.65		780		.191
1	18	.049	.639	.2618	.236	.54	.568	.50	.902	997	1.11	.146
1	20	.035	.679	.2618	.243	.391	.409	.361	.930	1060	1.07	.106
1¼	7	.18	.622	.3272	.232	2.213	2.36	2.057	.890	970	1.40	.6052
1¼	8	.165	.662	.3272	.240	2.069	2.18	1.912	.920	1033	1.36	.565
1¼	10	.134	.757	.3272	.258	1.73	1.83	1.597	.982	1180	1.27	.470
1¼	11	.12	.801	.3272	.2644	1.56	1.633	1.45	1.01	1250	1.24	.426
1¼	12	.109	.838	.3272	.271	1.44	1.52	1.328	1.032	1308	1.21	.389
1¼	13	.095	.882	.3272	.279	1.27	1.33	1.172	1.06	1376	1.18	.345
1¼	14	.083	.923	.3272	.284	1.12	1.173	1.04	1.084	1440	1.15	.304
1¼	16	.065	.985	.3272	.293	.89	.934	.823	1.12	1537	1.12	.242
1¼	17	.058	1.012	.3272	.297	.80	.83	.738	1.134	1578	1.10	.215
1¼	18	.049	1.042	.3272	.301	.682	.714	.629	1.152	1625	1.08	.185
1¼	20	.035	1.092	.3272	.308	.498	.521	.454	1.180	1704	1.06	.135
1½	16	.065	1.474	.3927	.359	1.08	1.132	.996	1.37	2300	1.10	.293
1½	14	.083	1.397	.3927	.349	1.36	1.43	1.256	1.334	2179	1.123	.368
2	13	.095	2.573	.5236	.474	2.09	2.20	1.933	1.81	4015	1.10	.569

$*$ Liquid velocity $= \dfrac{\text{lb per tube per hr}}{C \times \text{sp gr of liquid}}$ in fps (sp gr of water at 60°F $= 1.0$).

NAME OF MANUFACTURER
EXCHANGER SPECIFICATION SHEET

#				
1			JOB NO.	
2	CUSTOMER		REFERENCE NO.	
3	ADDRESS		INQUIRY NO.	
4	PLANT LOCATION		DATE	
5	SERVICE OF UNIT		ITEM NO.	
6	SIZE	TYPE	CONNECTED IN	
7	GROSS SURFACE PER UNIT	SHELLS PER UNIT	SURFACE PER SHELL	
8	PERFORMANCE OF ONE UNIT			
9		SHELL SIDE		TUBE SIDE
10	FLUID CIRCULATED			
11	TOTAL FLUID ENTERING			
12	VAPOR			
13	LIQUID			
14	STEAM			
15	NONCONDENSABLES			
16	FLUID VAPORIZED OR CONDENSED			
17	STEAM CONDENSED			
18	GRAVITY-LIQUID			
19	VISCOSITY-LIQUID			
20	MOLECULAR WEIGHT-VAPORS			
21	SPECIFIC HEAT-LIQUIDS	BTU/LB/°F		BTU/LB/°F
22	LATENT HEAT-VAPORS	BTU/LB/°F		BTU/LB/°F
23	TEMPERATURE IN	°F		°F
24	TEMPERATURE OUT	°F		°F
25	OPERATING PRESSURE	PSI		PSI
26	NUMBER OF PASSES PER SHELL			
27	VELOCITY	FPS		FPS
28	PRESSURE DROP	PSI		PSI
29	FOULING RESISTANCE			
30				
31	HEAT EXCHANGED-BTU/HR		MTD (Corrected)	
32	TRANSFER RATE-SERVICE	EFF. SURFACE PER UNIT		
33	CONSTRUCTION			
34	DESIGN PRESSURE	PSI		PSI
35	TEST PRESSURE	PSI		PSI
36	DESIGN TEMPERATURE	°F		°F
37	TUBES NO. OD BWG	LENGTH	PITCH	
38	SHELL ID OD			
39	SHELL COVER	FLOATING HEAD COVER		
40	CHANNEL	CHANNEL COVER		
41	TUBE SHEETS-STATIONARY	FLOATING		
42	BAFFLES-CROSS	TYPE		
43	BAFFLE-LONG	TYPE		
44	TUBE SUPPORTS			
45	GASKETS			
46	CONNECTIONS-SHELL-IN	OUT	RATING	PSI
47	CHANNEL-IN	OUT	RATING	PSI
48	CORROSION ALLOWANCE-SHELL SIDE		TUBE SIDE	
49	CODE REQUIREMENTS		TEMA CLASS	
50	WEIGHTS-EACH SHELL	BUNDLE	FULL OF WATER	
51	NOTE: INDICATE AFTER EACH PART WHETHER STRESS RELIEVED (SR) AND WHETHER RADIOGRAPHED (X-R)			
52	REMARKS:			

FIG. 4-25. Exchanger specification sheet of Tubular Exchanger Manufacturers Association.

Coil-in-box Units

Coil-in-box condensers have been largely superseded by shell-and-tube equipment. An exception is bent-tube sections (Fig. 4-26). In these, the tubes are installed with a slight curvature between headers. The flow of the hot medium through the tubes increases this curvature, which causes outside scale accumulation to drop off the tubes. The bent-tube sections are also used as atmospheric coolers (Fig. 4-27).

Coils submerged in a tank of water are often used as coolers of residuals from distillation towers. Because of high temperatures, coils are fabricated of steel pipe. Coils may be installed with runs in series, or in parallel between headers. For products such as Bunker C fuel oil or asphalt, the continuous coil is preferable as it may be blown free when desired.

FIG. 4-26. Bent-tube exchanger section. (*Courtesy Baldwin-Lima-Hamilton Corp., Industrial Equipment Div.*)

FIG. 4-27. Submerged bent-tube sections in an atmospheric cooler. (*Courtesy of Baldwin-Lima-Hamilton Corp., Industrial Equipment Div.*)

Double-pipe Exchangers

Double-pipe exchangers are extensively used for liquid-to-liquid service; true counterflow is attained. Figure 4-28 is a typical assembly. The sections of a double-pipe exchanger may be assembled in stacks for either series or parallel flow, and for a wide range of capacity and temperature differentials.

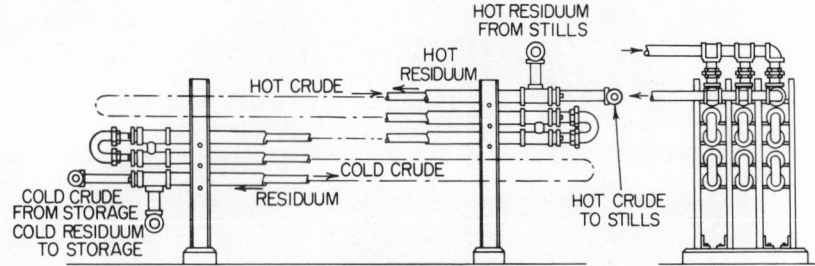

FIG. 4-28. Typical double-pipe exchanger assembly.

Fin tubes (Fig. 4-29) are largely used in double-pipe exchangers. The longitudinal fins are fusion-welded to commercial tubing. The number of fins and their OD are variable in accordance with the duty and conditions to be met. A double-pipe exchanger may be constructed with a removable head at the rear end and straight-through tubes for cleaning, or with U tubes which may be withdrawn through the removable rear head. The effective transfer surface of the external or finned side of

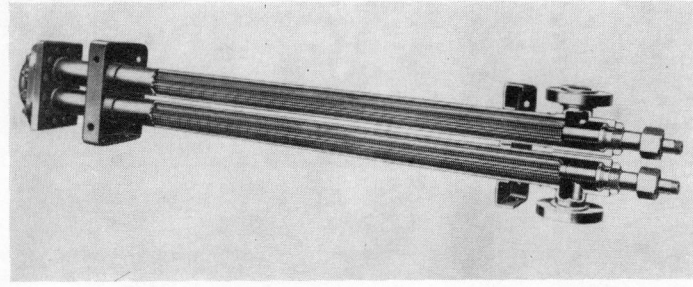

FIG. 4-29. Double-pipe exchanger elements, finned inner tubes. (*Courtesy of Baldwin-Lima-Hamilton Corp., Industrial Equipment Div.*)

the tube is dependent upon the number and OD of the fins, and ratios up to 9 to 1 are available. Finned tubes are not adapted for use in conventional shell-and-tube design, as the large diameter of the fins does not permit withdrawal of the tubes or bundles through the tube sheets.

Air-cooled Condensers

Air-cooled condensers (Fig. 4-30) are in use in some locations where water supply is a problem. The condenser consists of a bank of tubes, preferably finned, with header connections through which vapors and/or liquids pass. Air at high velocity is passed over the tubes by either forced or induced draft, depending on whether the fan is below or above the tube bank. Four blade fans 12 ft in diameter with tip speeds to

12,000 fpm are used. Variable-speed controllers and variable pitch of fan blades effect the volume control of air.

Fig. 4-30. Air-cooled heat exchanger. (*Courtesy of The Trane Co.*)

Combination Air-Water Units

Units are available which combine a water-cooling tower with a condenser or cooler. The air is cooled in the tower before passing over the finned condensing sections, as shown in Fig. 4-31. Temperatures are controlled by louvers and by the speed of the fans. The combination is effective where ambient air temperatures might be too high during hot summer months.

VESSELS

Exclusive of storage tanks, refinery vessels cover a wide range of designs. Such items as reaction chambers, the shells of fractionating towers, regenerative kilns, conventional filters, and similar equipment are classified as vessels.

Most refinery vessels fall within the scope of the ASME *Code for Unfired Pressure Vessels*. This code has been adopted by the political subdivisions of the United States listed in Table 4-6, as well as by all the provinces of Canada.

Certain vessels are not subject to manufacture under the ASME *Code for Unfired Pressure Vessels*. Such exemptions include those having internal or external pressures of 15 psi or less, those subject to Federal control, and those containing water (either air-cushioned or not) intended for domestic use.

Riveted construction of pressure vessels is obsolete for refinery equipment, and only welded vessels will be considered. Vessels are predominately fabricated from carbon and low-carbon alloy steels (ferritic). Under certain conditions, austenitic steels such as chrome-nickel alloys of the stainless group are used. With the exception of

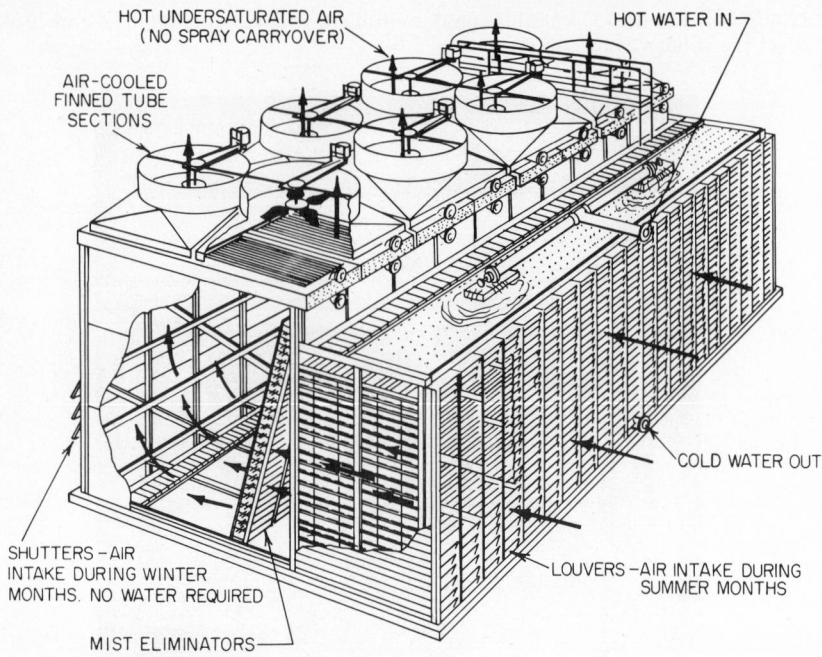

HOT UNDERSATURATED AIR
(NO SPRAY CARRYOVER)

HOT WATER IN

AIR-COOLED
FINNED TUBE
SECTIONS

COLD WATER OUT

SHUTTERS –AIR
INTAKE DURING WINTER
MONTHS. NO WATER REQUIRED

LOUVERS –AIR INTAKE DURING
SUMMER MONTHS

MIST ELIMINATORS

FIG. 4-31. The Combin-Aire ©, a combination water- and air-cooling unit, is covered by patents owned by the Hudson Engineering Corp.

Table 4-6. U.S. States, Territories, and Cities Requiring ASME Code Compliance

States

Alaska	Kansas	N. Carolina	Vermont
Arkansas	Massachusetts*	Ohio	Washington
California	Minnesota	Oregon	Wisconsin
Idaho	Mississippi	Pennsylvania	
Indiana	Nebraska†	Tennessee	
Iowa	Nevada	Utah	

Territories

District of Columbia   Panama Canal Zone   Puerto Rico

Cities and Counties

Buffalo, N.Y.	Los Angeles, Calif.	Seattle, Wash.
Chicago, Ill.	Miami, Fla.	Spokane, Wash.
Dade County, Fla.	Nashville, Tenn.	St. Louis, Mo.
Dearborn, Mich.	New Orleans, La.	St. Louis County, Mo.
Detroit, Mich.	Omaha, Nebr.	Tacoma, Wash.
E. St. Louis, Ill.	Phoenix, Ariz.	Tampa, Fla.
Jefferson Parish, La.	Richmond, Va.	Tulsa, Okla.
Kansas City, Mo.	San Francisco, Calif.	

* Only compressed-air vessels must be ASME.
† Only steam vessels over 150 psi must be ASME.

the corrosion problem, the controlling factor in the selection of the steel to be used is the tensile strength.

The ASME code expresses the relationship of pressure, diameter, and shell thickness for cylindrical vessels by the formula

$$t = \frac{PR}{SE - 0.6P}$$

in which t = thickness, in.
$\quad\quad P$ = allowable pressure, psi
$\quad\quad S$ = allowable stress, psi
$\quad\quad E$ = joint efficiency, %
$\quad\quad R$ = inside radius, in.
For spherical shells the formula is

$$t = \frac{PR}{2SE - 0.2P}$$

Welded-joint efficiencies allowed by the ASME code are given in Table 4-7.[11]

Table 4-7. Welded-joint Efficiencies Allowed by ASME Code*,[11]

Type of joint and radiography	Efficiency allowed, %	Code reference
Double-welded butt joints:		
Fully radiographed................................	100	
Spot-radiographed................................	85	Par. UW–12
No radiograph....................................	70	Table UW–12
Single-welded butt joints (backing strip left in place):		Par. UW–52
Fully radiographed................................	90	Par. UCS–25
Spot-radiographed................................	80	
No radiograph....................................	65	

* From Robert Chuse, *Unfired Pressure Vessels*, McGraw-Hill Book Company, New York, 4th ed., p. 53. Copyright © 1960. Used by permission.

The shell thicknesses for various welded-joint efficiencies and vessel diameters are presented in the charts of Figs. 4-32 through 4-35. The charts are based on an allowable stress of 13,750 psi. The thickness given by the charts may be corrected for other allowable stress values by application of the following constants:

Stress, psi	Constant
11,250	1.222
12,500	1.100
15,000	0.917
16,250	0.846
17,500	0.786
18,750	0.733

It is customary to add a corrosion allowance of from $\frac{1}{16}$ to $\frac{1}{8}$ in. to the design thickness of shells. Clad steels are available with stainless-steel, monel-metal, and nickel-steel linings which are rolled integral with a carbon-steel plate.

Two classifications of heads are generally used, the ellipsoidal of Fig. 4-36 and the torispherical of Fig. 4-37. The thickness of the ellipsoidal head of Fig. 4-36 is expressed

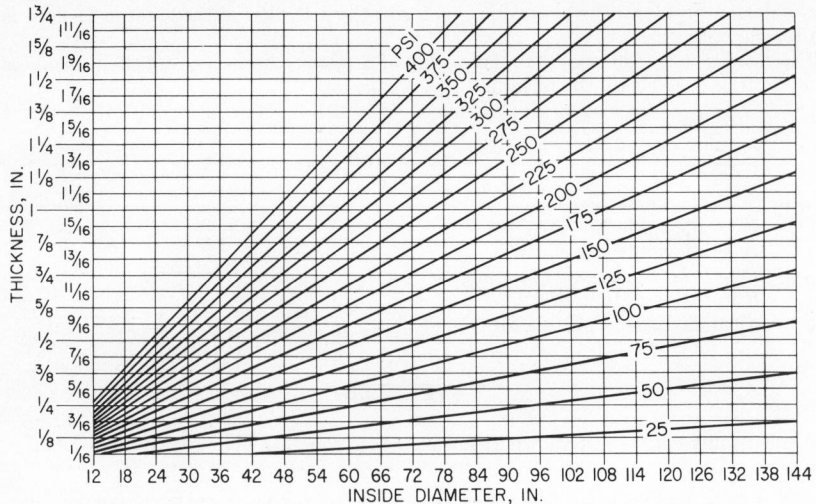

FIG. 4-32. Cylindrical-shell-thickness chart for 70 per cent joint efficiency.[11] Indicated thicknesses do not include corrosion allowance. (*From Robert Chuse, Unfired Pressure Vessels, McGraw-Hill Book Company, New York, 4th ed, p. 27. Copyright © 1960. Used by permission.*)

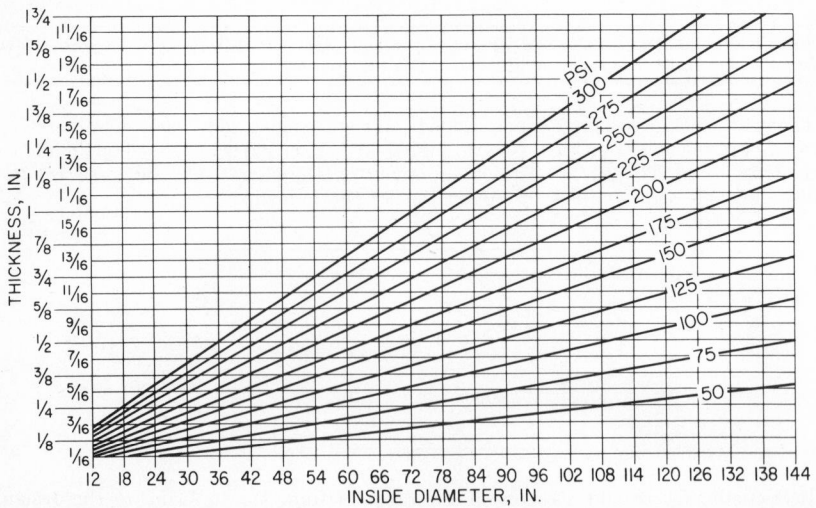

FIG. 4-33. Cylindrical-shell-thickness chart for 80 per cent joint efficiency.[11] Indicated thicknesses do not include corrosion allowance. (*From Robert Chuse, Unfired Pressure Vessels, McGraw-Hill Book Company, New York, 4th ed, p. 27. Copyright © 1960. Used by permission.*)

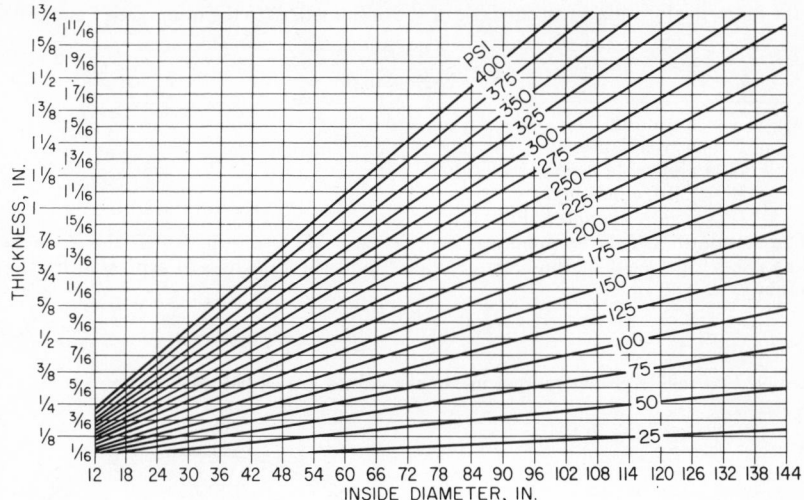

FIG. 4-34. Cylindrical-shell-thickness chart for 85 per cent joint efficiency.[11] Indicated thicknesses do not include corrosion allowance. (*From Robert Chuse, Unfired Pressure Vessels, McGraw-Hill Book Company, New York, 4th ed, p. 28. Copyright © 1960. Used by permission.*)

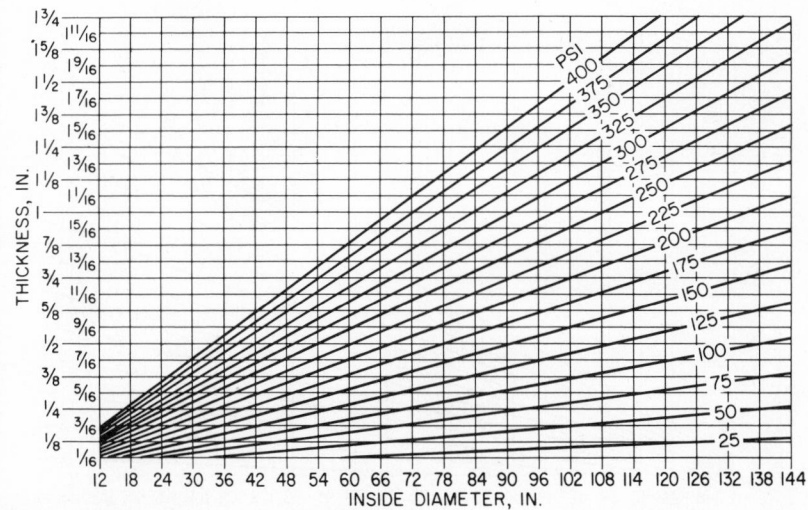

FIG. 4-35. Cylindrical-shell-thickness chart for 100 per cent joint efficiency.[11] Indicated thicknesses do not include corrosion allowance. (*From Robert Chuse, Unfired Pressure Vessels, McGraw-Hill Book Company, New York, 4th ed, p. 28. Copyright © 1960. Used by permission.*)

by the following formula when the ratio of the major axis D to the minor axis h is 2 to 1:

$$t = \frac{PD}{2SE - 0.2P}$$

in which t = minimum thickness, in.
P = internal pressure, psi
S = allowable stress, psi
E = joint efficiency, %
D = inside diameter, in.

The standard ASME torispherical head (Fig. 4-37) is one in which the knuckle radius r is 6.0 per cent of the crown radius L and the thickness is expressed by the formula

$$t = \frac{0.885PL}{SE - 0.1P}$$

in which L = inside crown radius, in.; other values are as before

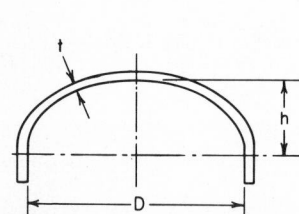

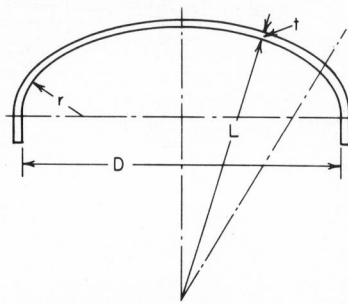

FIG. 4-36. Ellipsoidal head (D = major axis; h = minor axis).

FIG. 4-37. Torispherical head (L = inside crown radius; r = knuckle radius).

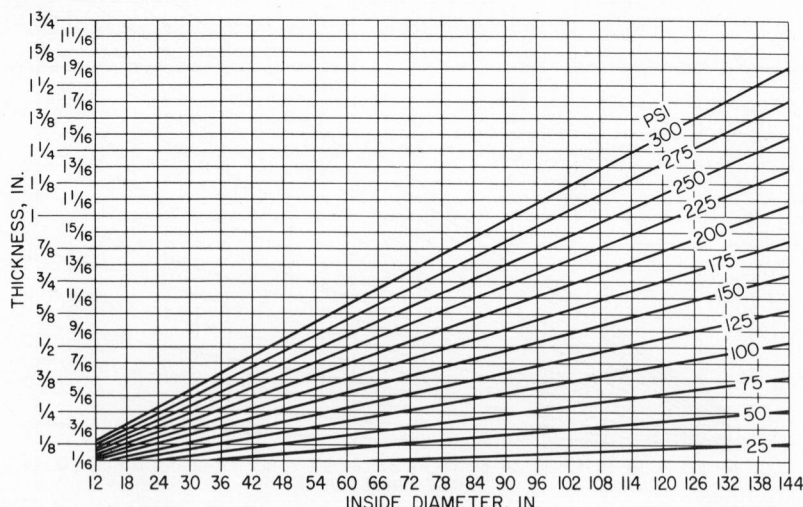

FIG. 4-38. Internal-pressure thickness chart for ellipsoidal one-piece heads.[11] (*From Robert Chuse, Unfired Pressure Vessels, McGraw-Hill Book Company, New York, 4th ed, p. 35. Copyright © 1960. Used by permission.*)

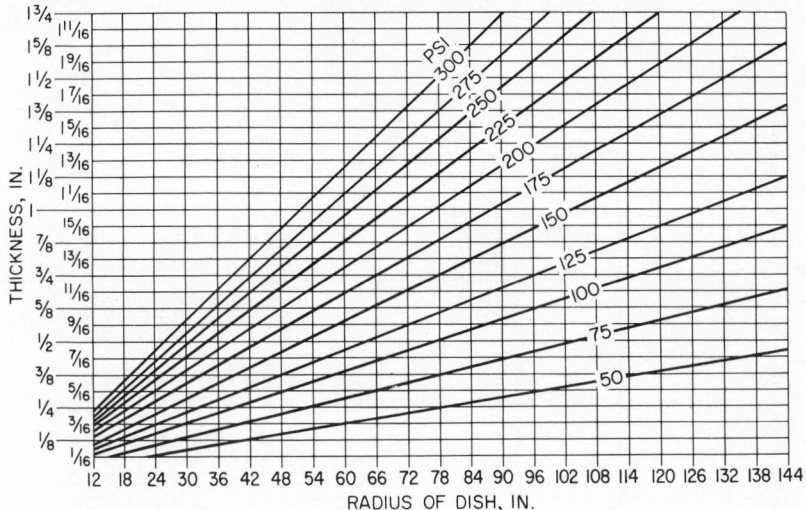

Fig. 4-39. Internal-pressure thickness chart for torispherical one-piece heads (flanged and dished).[11] *(From Robert Chuse, Unfired Pressure Vessels, McGraw-Hill Book Company, New York, 4th ed, p. 35. Copyright © 1960. Used by permission.)*

Head thickness may be determined from Figs. 4-38 and 4-39. The charts are based upon a stress value of 13,750 psi. The preceding constants for other stress values are applicable.

The ASME suggested method of welding joints is shown by Table 4-8.

The ASME code establishes tolerances for cylindrical pressure vessels. Allowable out of roundness when under internal pressure (the maximum diameter less the mini-

Table 4-8. Suggested Joint-welding Sketch and Procedure*,11

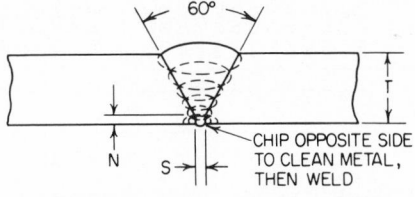

Plate thickness T, in.	No. of passes	Electrode diam, in.	Avg voltage	Avg amperage	Angle of groove, deg	Nose at root N, in.	Plate clearance S, in.
$\frac{3}{16}$	3	$\frac{1}{8}$	22	120	60	$\frac{1}{16}$	0
$\frac{1}{4}$	3	$\frac{1}{8}$	22	120	60	$\frac{1}{16}$	0
$\frac{3}{8}$	4	$\frac{5}{32}$	24	170	60	$\frac{1}{16}$	$\frac{1}{16}$
$\frac{7}{16}$	2	$\frac{5}{32}$	24	170	60	$\frac{1}{8}$	$\frac{1}{8}$
	3	$\frac{3}{16}$	26	210			
$\frac{1}{2}$	2	$\frac{5}{32}$	24	170	60	$\frac{1}{8}$	$\frac{1}{8}$
	4	$\frac{3}{16}$	26	210			

* From Robert Chuse, *Unfired Pressure Vessels*, McGraw-Hill Book Company, New York, 4th ed., p. 99. Copyright © 1960. Used by permission.

Table 4-9. Maximum Allowable Stress Values in Tension for Carbon and Low-alloy Steel, in Psi, Class UCS*.[7]

Material and specification No.	Grade	Nominal composition	P No.	Spec. min tensile	Notes	For metal temperatures not exceeding °F											
						−20 to 650	700	750	800	850	900	950	1000	1050	1100	1150	1200
Plate steels;																	
Carbon steels																	
SA 7			1	60,000	a, b	12,650											
SA 113	C		1	48,000	a	11,050											
SA 201	A	C, Si	1	55,000	...	13,750	13,250	12,050	10,200	8,350	6,500	4,500	2,500				
SA 201	B	C, Si	1	60,000	...	15,000	14,350	12,950	10,800	8,650	6,500	4,500	2,500				
SA 212	A	C, Si	1	65,000	...	16,250	15,500	13,850	11,400	8,950	6,500	4,500	2,500				
SA 212	B	C, Si	1	70,000	...	17,500	16,600	14,750	12,000	9,250	6,500	4,500	2,500				
SA 283	A		1	45,000	a	10,350											
SA 283	B		1	50,000	a	11,500											
SA 283	C		1	55,000	a	12,650											
SA 283	D		1	60,000	a	12,650											
SA 285	A		1	45,000	c, d	11,250	11,000	10,250	9,000	7,750	6,500						
SA 285	B		1	50,000	c, d	12,500	12,100	11,150	9,600	8,050	6,500						
SA 285	C		1	55,000	c, d	13,750	13,250	12,050	10,200	8,350	6,500						
SA 300				......	e												
Low-alloy steels																	
SA 202	A	Cr, Mn, Si	4	75,000	...	18,750	17,700	15,650	12,600	9,550	6,500	4,500	2,500				
SA 202	B	Cr, Mn, Si	4	85,000	...	21,250	19,800	17,700	12,800	9,900	6,500	4,500	2,500				
SA 203	A, D	2½ and 3½ Ni	4	65,000	...	16,250	15,500	13,850	11,400	8,950	6,500	4,500	2,500				
SA 203	B, E	2½ and 3½ Ni	4	70,000	...	17,500	16,600	14,750	12,000	9,250	6,500	4,500	2,500				
SA 204	A	C, ½ Mo	3	65,000	...	16,250	16,250	16,250	15,650	14,400	12,500	10,000	6,250				
SA 204	B	C, ½ Mo	3	70,000	...	17,500	17,500	17,500	16,900	15,000	12,750	10,000	6,250				
SA 204	C	C, ½ Mo	3	75,000	...	18,750	18,750	18,750	18,000	15,900	13,000	10,000	6,250				
SA 302	A	Mn, ½ Mo	3	75,000	...	18,750	18,750	18,750	18,000	15,900	13,000	10,000	6,250				
SA 302	B	Mn, ½ Mo	3	80,000	...	20,000	20,000	20,000	19,100	16,800	13,250	10,000	6,250				
SA 357		5 Cr, ½ Mo	5	60,000	f	......	13,400	13,100	12,800	12,400	11,500	10,000	7,300	5,200	3,300	2,200	1,500
SA 387	A	½ Cr, ½ Mo	3	65,000	...	16,250	16,250	16,250	14,750	14,200	12,500	11,000	7,500				
SA 387	B	1 Cr, ½ Mo	4	60,000	...	15,000	15,000	15,000	15,000	14,400	13,100	11,000	7,800	5,000	2,800	1,550	1,000
SA 387	C	1¼ Cr, ½ Mo-Si	4	60,000	...	15,000	15,000	15,000	15,000	14,400	13,100	11,000	7,800	5,500	4,000	2,500	1,200
SA 387	D	2¼ Cr, 1 Mo	5	60,000	...	15,000	15,000	15,000	15,000	14,400	13,100	11,000	7,800	5,800	4,200	3,000	2,000
SA 353	A	9 Ni	10	90,000	...	22,500											
SA 353	B	9 Ni	10	95,000	...	23,750											

4-30

* Extracted from the 1962 edition of the ASME *Boiler and Pressure Vessel Code, Unfired Pressure Vessels*, with permission of the publisher, the American Society of Mechanical Engineers, New York 17, N.Y.

The stress values in this table may be interpolated to determine values for intermediate temperature.

Stress values in restricted shear such as dowel bolts, rivets, or similar construction in which the shearing member is so restricted that the section under consideration would fail without reduction of area shall be 0.80 times the values in the table.

Stress values in bearing shall be 1.60 times the values in the above table.

a These stress values are one-fourth the specified minimum tensile strength multiplied by a quality factor of 0.92, except for SA 283, Grade D, and SA 7.

b These allowable stress values apply also to structural shapes.

c Flange quality in this specification not permitted over 850°F.

d For service temperatures above 850°F, it is recommended that killed steels containing not less than 0.10 per cent residual silicon be used. Killed steels which have been deoxidized with large amounts of aluminum and rimmed steels may have creep and stress-rupture properties in the temperature range above 850°F, which are somewhat less than those on which the values in the above table are based.

e The stress values to be used for temperatures below −20°F when steels are made to conform with Specification SA 300 shall be those that are given in the column for −20 to 650°F.

f Maximum allowable stress values for temperatures below 700°F are given below.

Spec.	Grade	No.	−20 to 400	500	600	650
SA 357	...	5	15,000	14,500	14,000	13,700

4–31

mum diameter) must not exceed 1.0 per cent of the nominal diameter.   No plate shall have an undertolerance exceeding 0.01 in., or 6.0 per cent of the ordered thickness, whichever is less.

The effect of temperature on the maximum allowable stress tensions of steels used in vessel fabrication must be considered in design.   Table 4-9 illustrates this effect for the plate of usual specifications.

Another effect of high temperatures and stresses over long periods of time is the slow deformation termed "creep."   The creep stress is defined as the stress in psi for 1.0 per cent elongation over a given period of time.   The tests are generally made over a period of 1,000 hr and the results extrapolated for 10,000 and 100,000 hr.   Creep-stress data for steels in customary use were given earlier in Table 4-1.

Fatigue of the metal should receive consideration in design.   For instance, alternate heating and cooling of a vessel with ellipsoidal or torispherical heads will cause flexure on the head knuckles.   Forecast of this effect is difficult.   Longer radii of knuckles and complete freedom of expansion will help alleviate the danger of fatigue.

In all welded seams, the grain structure of the plate becomes coarse just beyond the weld fusion line and loses its ductility.   The weld-metal zone is usually stronger than the plate.   To correct the loss of ductility, the code requires stress relieving for certain design and service conditions:

1. All carbon-steel welded joints if their thicknesses exceed 1.25 in. or $(D - 50)/120$, in which D equals the inside diameter or 20 in., whichever is greater
2. Vessels containing lethal substances
3. Carbon- and low-alloy-steel vessels for services at temperatures below 20°F
4. Other specified applicable service conditions

Stress relieving consists of heating the entire vessel in a furnace to a temperature of about 1200°F.   The closed furnace is then allowed to cool gradually to atmospheric temperature, a procedure taking considerable time, depending on the tonnage of the vessel.   Major fabricators of refinery vessels have large furnaces capable of containing vessels of maximum shipping diameter.   In the case of very tall towers, for example, the heat may be applied in two or more steps with an overlap between the section.

Testing of Vessels

The methods of testing welds and allowable joint efficiencies are given in Table 4-10.   Full radiography is required for all joints in plates over 1½ in. thick, and for vessels containing lethal substances.   Radiographic inspection may be by X ray or gamma ray, using a radioactive-isotope capsule.   The latter is valuable for heavy plate thickness.   Other methods of weld inspection of more limited application are magnetic-particle and ultrasonic.

The code permits the examination of welded joints by sectioning or trepanning, wherein a specimen of the weld metal is removed for study by an etching process.   The method is considered a substitute for radiography.

Specifications for shop-fabricated vessels may require that a hydrostatic-pressure test be applied to the finished piece of equipment.   The code calls for a pressure at least 1½ times the designed working pressure.   Under certain conditions a pneumatic test may be used at a pressure of 1¼ times the maximum allowable.

Refractory Linings

Refractory linings are essential for many vessels used in petroleum refining, such as reaction chambers, regulators, and kilns.   They must be capable of withstanding service temperatures of 2000 to 3000°F, and must possess low coefficients of thermal expansion.   In many locations, resistance to abrasion is necessary.   When a platinum catalyst is used, refractories should be selected with a minimum iron content.

The thermal conductivity of refractories approximates those given in Table 4-11.   The greater the density, the higher the thermal conductivity.   The lighter, more porous materials have better insulating values.

Table 4-10. Methods of Testing Welds and Allowable Joint Efficiencies in Unfired Pressure Vessels*,11
Radiographic Requirements

Remarks	Code reference
Full radiography	
X Ray all joints over 1¼ in. (see Code Par. UCS–57, UW–2, and U–1(e) for required radiographing on lesser thickness)	Par. UW–11(a)2
Circumferential butt joints of nozzles and sumps not exceeding 10 in. in diameter of 1⅜ in. in thickness do not require X Ray	Par. UW–11(a)5
Vessels containing lethal substances must be fully radiographed	Par. UW–11(a)1
	Par. UW–2
Efficiency of fully radiographed joints	Par. UW–12
	Table UW–12
Radiographic techniques	Par. UW–51
Carbon and low-alloy steel vessels	Par. UCS–57
High-alloy steel vessels	Par. UHA–33
Clad-plate vessels	Par. UCL–35
Spot radiography	
General recommendations	Par. UW–11(b)
Spot radiographing required for larger weld-joint efficiencies	Par. UW–12
	Par. UW–52
	Table UW–12
Minimum length of spot radiograph, 6 in.	Par. UW–52(c)
If cladding is included in computing required thickness, spot radiograph is mandatory	Par. UCL–23(c)
Vessels having 50 ft or less of main seams require one spot examination; larger vessels require one spot for each 50 ft of welding	Par. UW–52(b)1
Additional spots may be selected to examine welding of each welder or welding operator	Par. UW–52(b)2
Hydrostatic and Pneumatic Tests	
Hydrostatic tests	
Test must be at least 1½ times the maximum design pressure multiplied by the lowest ratio of the stress value for the test temperature to that for the design temperature	Par. UG–99
	Par. UG–21
	Par. UA–60(e)
Combination units must be so tested that hydrostatic pressure is on one chamber without pressure in the other parts	
Include corrosion allowance in calculating test pressure	
Inspection must be made at a pressure not less than ⅔ of test pressure	
Test cast-iron vessels at 2 times the design working pressure; for design pressure under 30 psi, test at 2½ times design pressure	Par. UCI–99
Vessels in service under internal pressure	Par. UA–814
Test for clad-plate vessels	Par. UCL–52
Pneumatic tests	
Pneumatic test may be used instead of hydrostatic test when:	Par. UG–100
1. Vessels are so designed and supported that they cannot safely be filled with water	
2. Vessels for service in which traces of testing liquid cannot be tolerated are not easily dried	
Test pressure must not be less than 1¼ times maximum allowable pressure	
All attachment welds of a throat thickness greater than ¼ in. must be given a magnetic-particle or penetrating-oil test before pneumatic test	Par. UW–50
Special precautions should be taken when using air or gas for testing	Par. UG–100, footnote 2

* From Robert Chuse, *Unfired Pressure Vessels*, McGraw-Hill Book Company, New York, 4th ed, pp. 53–55. Copyright © 1960. Used by permission.

Table 4-11. Approximate Thermal Conductivity of Refractory Brick and Insulating Materials*,12

Btu/hr/sq ft/°F/in. thickness

Type of brick	0	200	400	600	800	1000	1200	1400	1600	1800	2000	2200	2400	2600	2800
Chrome	9.0	9.5	10.0	10.5	11.0	11.5	12.0	12.5	13.0	13.5	14.0	14.5	15.0	15.5	16.0
Fire-clay†	6.4	6.8	7.2	7.6	8.0	8.4	8.8	9.2	9.6	10.0	10.4	10.8	11.2	11.6	12.0
Forsterite	14.2	13.5	12.8	12.3	11.8	11.4	11.0	10.7	10.5	10.4	10.3	10.3	10.4	10.6	10.8
70% alumina	12.8	12.3	11.9	11.6	11.2	10.9	10.7	10.4	10.3	10.1	10.0	9.9	9.9	9.9	10.0
90% alumina	22.4	20.4	18.7	17.3	16.1	15.2	14.3	13.9	13.5	13.3	13.2	13.2	13.3	13.5	13.8
Magnesite, fired	39.6	36.4	33.5	30.7	28.4	26.3	24.5	23.1	21.9	21.0	20.4	20.1	20.0	20.2	20.5
Silica	6.0	6.8	7.5	8.3	9.0	9.8	10.5	11.3	12.0	12.8	13.5	14.3	15.0	15.8	16.5
Lightweight silica	1.9	2.4	2.9	3.4	3.9	4.3	4.8	5.3	5.8	6.3	6.8	7.2	7.7	8.2	8.7
Insulating fire:															
Group 16	0.62	0.74	0.87	0.99	1.12	1.24	1.37	1.49	1.62						
Group 20	0.80	0.93	1.07	1.20	1.34	1.47	1.61	1.74	1.88	2.01	2.15				
Group 23	1.28	1.44	1.59	1.75	1.91	2.06	2.22	2.38	2.53	2.69	2.85	3.00			
Group 26	1.45	1.62	1.78	1.95	2.11	2.28	2.44	2.61	2.77	2.94	3.10	3.27	3.43	3.60	
Group 28	1.93	2.09	2.25	2.41	2.57	2.73	2.89	3.05	3.21	3.37	3.53	3.69	3.85	4.01	
Group 30	2.20	2.36	2.52	2.68	2.84	3.00	3.16	3.32	3.48	3.64	3.80	3.96	4.12	4.28	

* From C. L. Mantell, *Engineering Materials Handbook*, McGraw-Hill Book Company, New York, Copyright © 1958. Used by permission.
† The conductivities given for fireclay brick are approximate mean values for brick having a density of 130 lb/cu ft.
NOTE: For brick of different makes but of the same type, class, and composition, the conductivities are approximately proportional to their densities (weights per cubic foot). The figures given above are approximate mean values and are regarded as sufficiently accurate for most engineering purposes. Thermal-conductivity data reported by different observers are frequently not in close agreement.

Castable refractory material is generally used for vessel linings, replacing the block forms previously used. The installation procedure generally followed is to weld steel studs to the shell spaced at 10 to 12 in. Castable refractory of good insulating value is then applied by air gun to a thickness equal to the length of the studs. Lightweight expanded metal is then spot-welded to the tips of the studs. This is followed by an application of a layer of denser castable refractory which is more resistant to abrasion. The first coat is 3 to 4 in. thick, the final coat being $1\frac{1}{2}$ to 2 in.

One-component castable linings are used for vessels where service conditions are moderate. The Bigelow-Liptak Corp. recommends V-shaped anchors welded to the

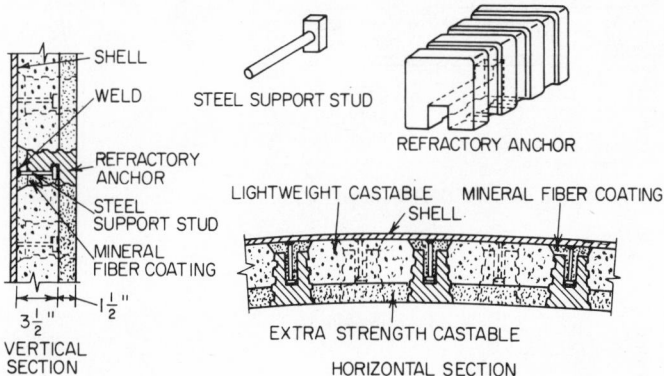

FIG. 4-40. Refractory anchor and typical arrangement for monolithic linings in catalytic cracking units. (*Courtesy of Harbison-Walker Refractories Co.*)

shell at 10-in. centers. The anchors are alternately long and short and are turned at various angles. The castable lining is applied with an air gun to a thickness of 2 to 3 in.

When vessels are within shipping clearances they may be factory-lined. A method used by the Harbison-Walker Refractories Co. is illustrated by Fig. 4-40. Studs with square heads are welded to the shell. Over these is placed a precast refractory anchor with internal face in the plane of the final surface. The anchor is grooved to afford bonding strength to the castables. The first coat, lightweight material with insulating value, is applied to a thickness sufficient to be flush with the square heads of the studs. Note the refractory precast anchors project beyond to furnish a bond for the second application of desired thickness.

PUMPS

Reciprocating Steam-driven Pumps

When steam is available and the exhaust can be used for heating or in processes, steam-driven direct-acting reciprocating pumps are often economical. They may be classified as simplex and duplex, descriptive of one or two liquid cylinders. They may be further classified as single-acting or double-acting, piston or plunger types, and inside- or outside-packed.

A cross section of a direct-acting steam pump is shown in Fig. 4-41, a simplex pump in Fig. 4-42, and a duplex pump in Fig. 4-43. Either type may be single- or double-acting. A single stroke of the piston fills the liquid cylinder which is discharged on the reverse stroke for the single-acting pump, while in the double-acting design liquid is drawn in and discharged on each stroke of the piston. Simplex and single-acting pumps should be equipped with air chambers on the discharge sides to dampen flow pulsations.

Reciprocating steam pumps are rated upon the cylinder displacement, the rpm or piston speed, and the characteristics of the liquid. This rating is usually on the basis

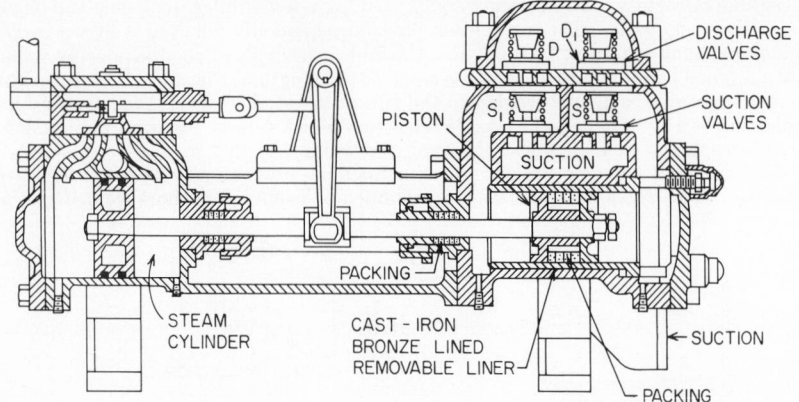

FIG. 4-41. Cross section through the steam and water ends of a direct-acting steam pump.[13]

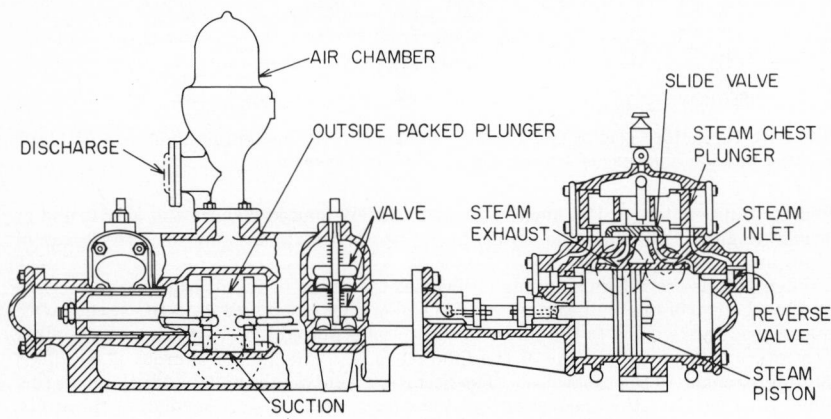

FIG. 4-42. A simplex outside-packed plunger pump.[14]

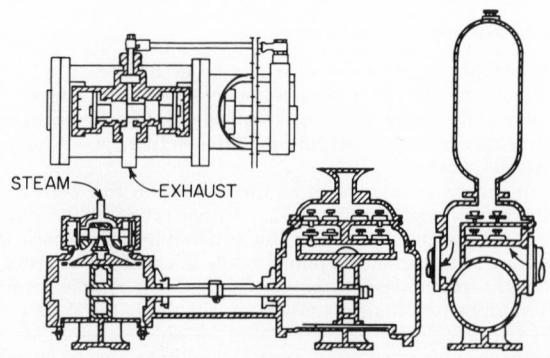

FIG. 4-43. Water end of a double-acting steam-driven reciprocating pump.[7]

Table 4-12. Maximum Recommended Piston Speed for Reciprocating Pumps[15]

Pump size*	Max recommended piston speed, fpm	Capacity of duplex pump, gpm of H_2O†	Pump size*	Max recommended, piston speed, fpm	Capacity of duplex pump, gpm of H_2O†
2⅛ × 1⅜ × 2	30	3.93	12 × 4½ × 12	90	126.4
2½ × 1⅜ × 2	30	3.93	12 × 4½ × 18	110	154.5
			12 × 5½ × 12	90	188.8
3 × 2 × 3	40	11.1	12 × 6 × 12	90	224.8
3 × 2¼ × 4	50	17.6	12 × 7 × 13	95	322.8
3½ × 2¾ × 4	50	26.2	12 × 7 × 18	110	373.8
3½ × 3½ × 4	50	42.5	12 × 7½ × 18	110	429.2
			12 × 8 × 12	90	399.5
4½ × 2¾ × 4	50	26.2	12 × 8 × 18	110	488.3
4½ × 3 × 4	50	31.2	12 × 10 × 18	110	763.0
4½ × 3 × 6	65	40.6	12 × 12 × 20	115	1,148.0
4½ × 5 × 6	65	112.7	12 × 12 × 24	120	1,198.5
5¼ × 3½ × 5	60	51.0	14 × 8 × 20	115	510.5
5½ × 4 × 7	70	77.6	14 × 8½ × 20	115	576.3
			14 × 20 × 18	110	3,060.0
6 × 2 × 6	65	18.0			
6 × 4 × 6	65	72.1	15 × 7 × 18	110	373.8
6 × 5½ × 12	90	188.8	15 × 8 × 18	110	488.3
6 × 5¾ × 6	65	149.1	15 × 10½ × 20	115	879.0
6⅛ × 6 × 12	90	224.8	15 × 10½ × 24	120	917.6
6½ × 4 × 10	80	88.8			
6½ × 4⅛ × 8	75	88.5	16 × 3½ × 12	90	76.5
			16 × 8 × 18	110	488.3
7 × ¾ × 10	80	3.1	16 × 10½ × 18	110	841.1
7½ × 2½ × 10	80	34.7	16 × 12 × 24	120	1,198.5
7½ × 4½ × 7	70	98.3	16 × 14 × 24	120	1,631.4
7½ × 4½ × 10	80	112.4			
7½ × 5 × 6	65	112.7	18 × 6 × 18	110	274.7
7½ × 5 × 10	80	138.7	18 × 8 × 16	105	466.1
7½ × 6 × 10	80	199.8	18 × 8 × 18	110	488.3
7½ × 7¼ × 10	80	291.5	18 × 8¼ × 24	120	566.5
7½ × 7½ × 6	65	253.6	18 × 9 × 16	105	589.9
			18 × 9 × 18	110	618.1
8 × 6 × 12	90	224.8	18 × 9½ × 18	110	688.5
8 × 6½ × 12	90	263.8	18 × 14 × 24	120	1,631.4
8 × 7½ × 12	90	351.1			
8 × 8 × 12	90	399.5	20 × 5 × 20	115	199.4
8 × 9½ × 12	90	563.1	20 × 6 × 12	90	224.6
			20 × 9 × 15	100	561.9
10 × 4½ × 10	80	112.4	20 × 12 × 12	90	898.9
10 × 6 × 10	80	199.8	20 × 12 × 24	120	1,198.5
10 × 6 × 12	90	224.8			
10 × 6½ × 12	90	263.8	21 × 7½ × 24	120	468.2
10 × 7 × 12	90	305.8			
10 × 7½ × 12	90	351.1	22 × 10 × 24	120	832.3
10 × 8 × 10	80	355.1	22 × 12 × 24	120	1,198.5
10 × 8 × 12	90	399.5	22 × 14 × 20	115	1,560.0
10 × 8 × 18	110	488.3			
10 × 10 × 18	110	763.0	24 × 3½ × 24	120	102.0
			24 × 36 × 24	120	10,780.0
12 × 1¼ × 12	90	9.8			
12 × 2½ × 18	110	47.7	36 × 8½ × 24	120	601.4

* Pump size is given as (stm. cyl. diam.) × (liq. cyl. diam.) × (stroke).

† GPM is for max recommended speed with 15 per cent slippage and neglecting the size of the piston rod. The normal design speed is about 75 per cent of this figure.

MECHANICAL EFFICIENCY			
		DIFFERENTIAL PRESSURE	
		TO 300 PSI	ABOVE 300 PSI
STROKE, IN.	PISTON PATTERN	PACKED PLUNGER	PACKED PLUNGER
3	55	50	41
4	60	55	44
6	65	60	51
8	68	65	54
10	72	68	57
12	74	70	59
16	76	73	63
18	78	75	65
20	79	77	66
24	80	78	67
USE 90% OF ABOVE IF VISCOSITY EXCEEDS 4,000 SSU			

1. SELECT PISTON SPEED, STROKE, AND MECHANICAL EFFICIENCY FROM TABLES.

2. DETERMINE LIQUID CYLINDER DIAMETER FROM GRAPH.

3. ON RUN AROUND CHART, SELECT PROPER STEAM CYLINDER DIAMETER.

	BASIC PISTON SPEED, FPM SIMPLEX STEAM PUMP	
STROKE, IN.	WATER OR COLD LIGHT OIL	OIL OVER 300°F VOLATILE LIQUIDS
3	30	16
4	33	21
6	42	28
8	50	33
10	58	40
12	65	45
16	73	52
18	80	58
20	85	62
24	90	68

SPEED CORRECTION FACTORS	
	FACTOR
DUPLEX STEAM	0.90
SIMPLEX POWER	1.20
DUPLEX POWER	1.40
VERTICAL TRIPLEX	1.40

VISCOSITY	
500 TO 1,000 SSU	0.90
1,000 TO 2,000	0.80
2,000 TO 3,000	0.75
3,000 TO 4,000	0.70
4,000 TO 5,000	0.66
5,000 TO 6,000	0.63
6,000 TO 7,000	0.59
7,000 TO 8,000	0.55
8,000 TO 9,000	0.52
9,000 TO 10,000	0.50

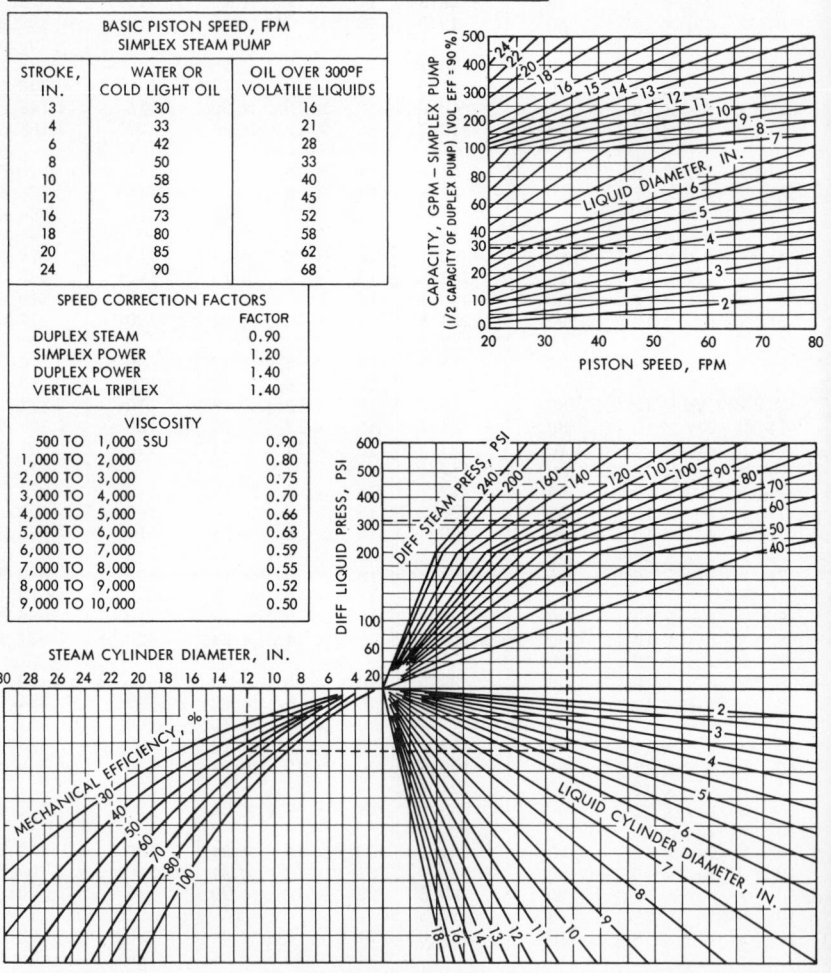

FIG. 4-44. Reciprocating-pump capacities as related to mechanical efficiency, piston speed, differential liquid pressure, and differential steam pressure.[15]

of water as the fluid. The rpm is a function of the piston speed. Piston speed is furthermore related to the length of the stroke; the shorter the stroke the slower the piston speed. Table 4-12 rates many combinations of steam and liquid ends with different strokes. The rating is for water. The available liquid pressure derived from the pump is dependent upon the ratio of the areas of the liquid and steam cylinders. The horsepower developed by any pump is the weight of liquid raised a certain distance against a total lift (head) or the foot-pounds of work done. The theoretical horsepower is

$$hp = \frac{WH}{33,000}$$

in which W = weight of fluid pumped/min
H = total head, suction lift plus discharge head, ft, head corrected for the specific gravity of the liquid

To the theoretical horsepower must be applied a factor expressing the efficiency of the pump. For steam pumps, the rating of the efficiency is usually expressed in

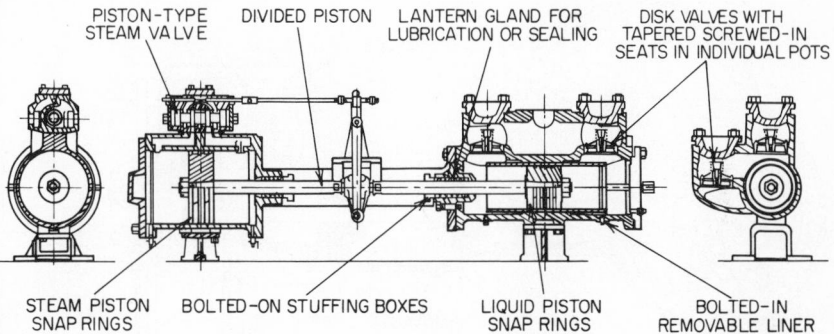

FIG. 4-45. Duplex outside-packed steam pump with valves located in separate pots above the liquid cylinders. (*Courtesy of Worthington Corp.*)

pounds of steam per horsepower-hour. The steam consumption depends on the rpm (or piston speed) and the net effective steam pressure (live minus exhaust). Figure 4-44 indicates the differential liquid pressure of a range of pumps determined from the formula

$$\Delta P_f(\text{fluid}) = \frac{D_s{}^2}{D_F{}^2} \times \Delta P_s(\text{steam}) \times E$$

in which ΔP_f = increment of pressure added to fluid, psi
D_s = diameter of steam cylinder
D_F = diameter of fluid cylinder
ΔP_s = differential steam pressure (live minus exhaust), psi
E = mechanical efficiency

When pumping viscous or hot oils, it is necessary to decrease the piston speed to assure complete filling of liquid cylinders and smooth flow through the valves. Figure 4-44 relates the various factors entering into the selection of reciprocating pumps.

Two types of liquid ends for pumps are available: the piston type as previously discussed, and the plunger type. The latter is outside-packed (see Fig. 4-42). Plunger pumps, with boxes outside the pump where they may be observed and easily repacked, are recommended for corrosive conditions for hot oils and for light pressures.

For temperatures to 500°F, cast-iron liquid ends are standard. Cast steel may be used to temperatures of 650 or 700°F. Above this temperature, forged steel should be used. Pumps of outside-packed design and with forged-steel liquid ends have been used for temperatures to 1000°F.

Spring-loaded disk valves are standard for water and light-oil service. Wing-guided, clapper, and ball valves are used for viscous and hot oils. Figure 4-45 illustrates a duplex outside-packed steam pump with valves located in separate pots above the liquid cylinders.

Steam pumps have excellent suction characteristics; they may be operated at any capacity below rating, an advantage when starting with viscous oils. They need no relief valves as they will automatically stall when maximum pressure is reached. Ease of maintenance is a feature. But they are economical only where steam at suitable pressure is available and where there is use for the exhaust steam.

Reciprocating Power-driven Pumps

When steam is not available, the advantage of the piston and plunger types of pumps may be realized by substitution of mechanical drive. In this way two, three, or five cylinders may be grouped on one mechanized drive in either horizontal (Fig. 4-46) or vertical (Fig. 4-47) position. The liquid ends may be single- or double-acting, and either of piston or plunger type. It is necessary to use reducing gears to adapt

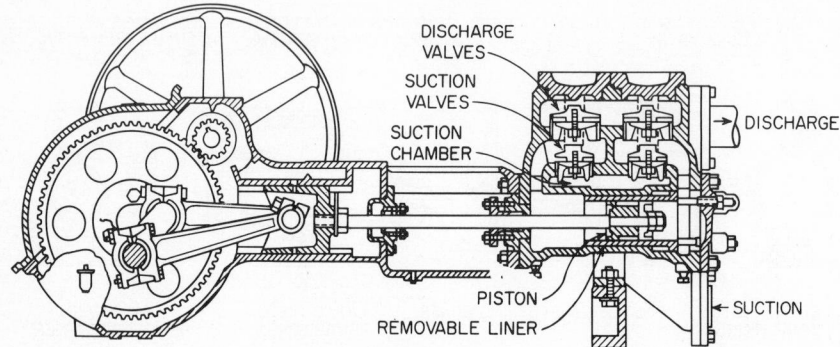

FIG. 4-46. Cross section through power-type piston pump with horizontal drive.[13]

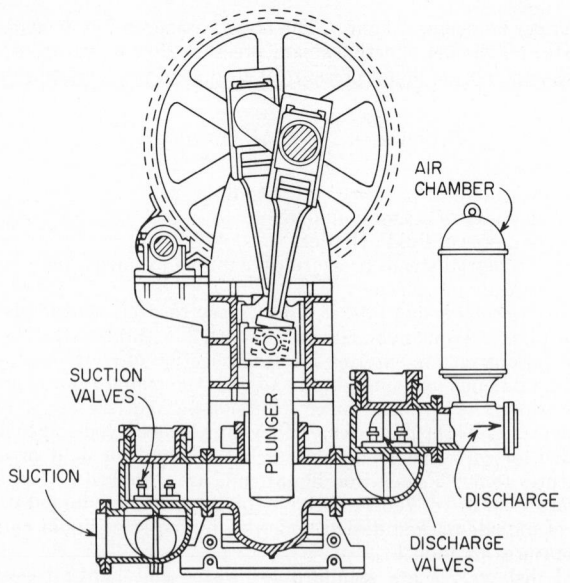

FIG. 4-47. Cross section through vertical-plunger, power-driven pump.[13]

the speed of the driver to the speed of the relatively slow piston or plunger speed.

To protect the motor or turbine prime mover against excessive pressure or shock, such as closure of the discharge line, a preset relief should be placed upon the pump discharge relieving into the pump suction.

Centrifugal Pumps

Basically a centrifugal pump consists of a set of vanes rotating within a fixed casing. Energy is imparted to the fluid by means of centrifugal force. It is estimated that 60 per cent of all pumps used by industry are of this type and that 75 per cent of these pumps are within a range of service requirements that can be met by standardization. Centrifugal pumps can be classified as follows:

1. Number of stages
   a. Single
   b. Multiple
2. Casing type
   a. Volute
   b. Double volute
   c. Circular
   d. With diffuser vanes
3. Shaft position
   a. Horizontal
   b. Vertical
4. Suction
   a. Single
   b. Double

Turbine, axial-flow, and mixed-flow pumps are sometimes erroneously classified as centrifugal pumps. But as they generally are low-head and large-volume types, they rarely find use in refineries.

Classification 1 is self-explanatory. The single-stage pump (Fig. 4-48) accomplishes

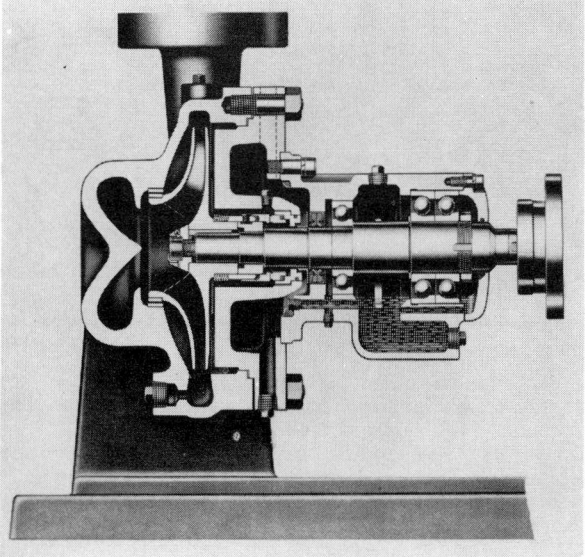

Fig. 4-48. Single-stage centrifugal pump. (*Courtesy of Ingersoll-Rand Co.*)

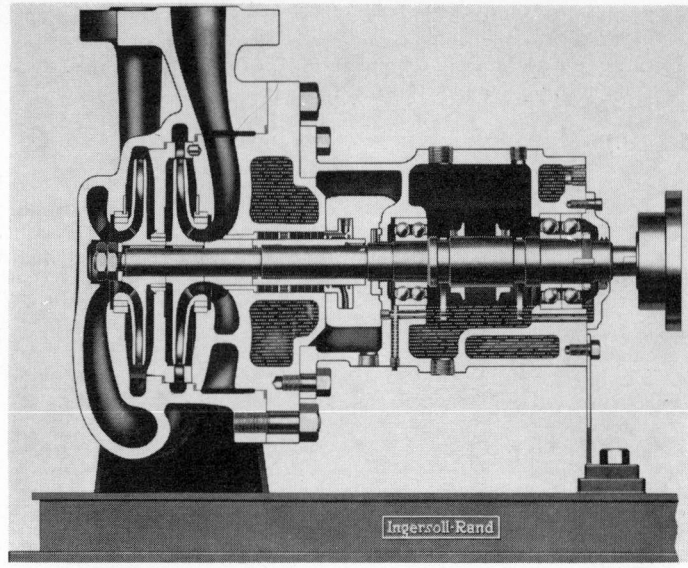

Fig. 4-49. Multistage centrifugal pump.   (*Courtesy of Ingersoll-Rand Co.*)

its duty with one impeller.   The multistage discharges from one impeller to the next to build up pressure (Fig. 4-49).

Under classification 2, the volute type of casing with or without diffuser vanes and with one or more stages is the most extensively used (Fig. 4-50).

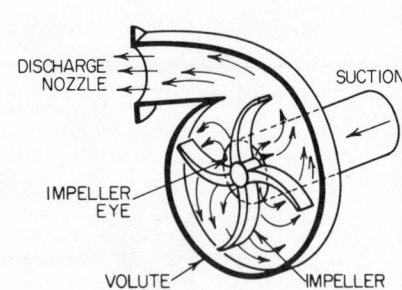

Fig. 4-50. Typical volute-type casing cen-trifugal pump.

Fig. 4-51. Operation of a single-suction volute-type centrifugal pump.

Whether the shaft is horizontal or vertical (classification 3) depends upon the service. Most refinery pumps have shafts in the horizontal position.   In pumping from sub-merged tanks, the vertical mounting has advantages.

The single or double suction in classification 4 establishes the method of introduction of the fluid into the eye of the impeller.   Single suction indicates this entry from one side, while double suction is from both sides.

Figure 4-51 is the elemental diagram of a single-suction volute-type centrifugal pump, and Fig. 4-52 presents the construction of a single-stage unit.

The double-volute design introduces a second volute cast integral with the casing (Fig. 4-53). In the single-volute pump, unequal pressure is exerted between casing and impeller with resultant side thrust and possible deflection of the rotating element. The introduction of the second volute equalizes this pressure.

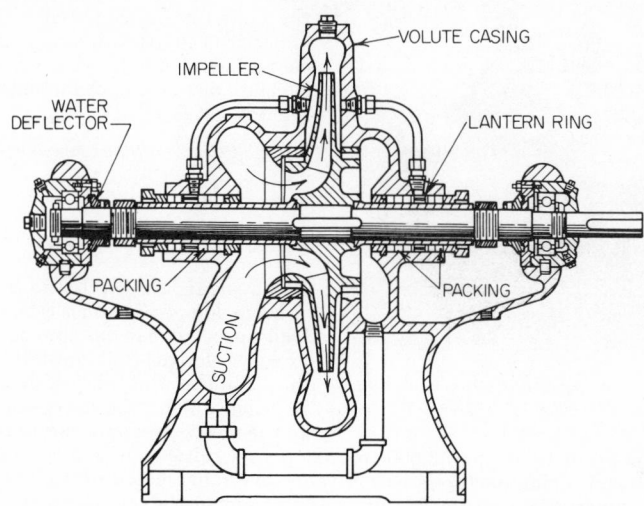

FIG. 4-52. Single-suction, single-stage, volute-type centrifugal pump.[13]

Diffuser vanes (classification 2, Fig. 4-54) are used for high heads and efficiencies. Independently of the above classification, centrifugal pumps may be further divided into two types: (1) radially split casings, and (2) horizontally or axially split casings (Fig. 4-55).

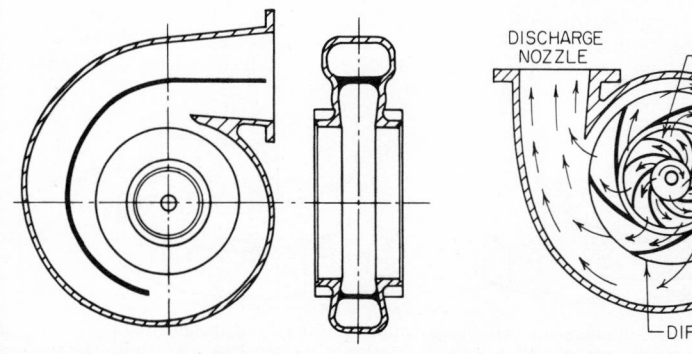

FIG. 4-53. Cross section of double-volute pump casing, showing the curved vane or splitter.

FIG. 4-54. Flow through a diffuser-type centrifugal pump.

The radially split casing is advantageous in first cost. For the end-suction type, the suction piping must be dismantled for inspection or repair of the impeller. The pump with horizontally split casing may be opened for such purposes without disturbing the piping.

Impellers for centrifugal pumps can be open (Fig. 4-56, right), enclosed with side walls or shrouds (Fig. 4-56, left), or semiopen with a single shroud. Open-type impellers are generally used for small inexpensive pumps or for pumps handling abrasive liquids. The closed impeller is almost universally used for refinery service.

The head generated by a pump impeller varies both with the square of its diameter and with its rotative speed. It is obvious that several combinations will produce the same results (Fig. 4-57). The great majority of centrifugal pumps are directly connected to 60-cycle a-c motors. The synchronous speeds of motors in general refinery use are

Poles	Synchronous speed
2	3600
4	1800
6	1200
8	900

Pump speeds are selected to conform.

From Fig. 4-57, which is based on full-load speed, it can be seen that several impeller designs will meet the specified duty and that the pump with the smallest impeller will be the lowest in first cost.

Fig. 4-55. Axially split casing centrifugal pump. (*Courtesy of Worthington Corp.*)

In the most commonly encountered range of capacities up to 1,000 gpm, 3,600-rpm pumps are suitable, with most standardized lines of the pump manufacturers within the above range based on 3,600 rpm.

Pump designers use the term "specific speed" to define the rpm at which a geometrically similar impeller would run if it were of such a size as to discharge 1 gpm against 1 ft of liquid head. It has no direct relationship to the actual speed of the pump. Specific speed is indicative of the shape of the impeller. Expressed as a formula,

$$N_s = \frac{\text{gpm} \times \text{rpm}}{H^{3/4}}$$

in which N_s = specific speed
H = head per stage, ft

Radial-vane centrifugal pumps for refinery service, both single and double suction, approximate specific speeds from 1,000 to 4,000 rpm. Mixed- and axial-flow pumps are classified for higher specific speeds.

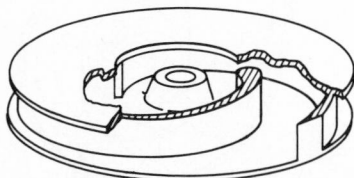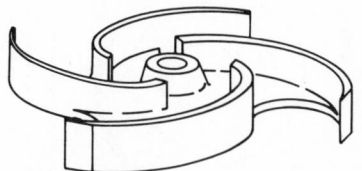

Fig. 4-56. Typical impellers for centrifugal pumps. At left is the enclosed or shrouded impeller. It has two vanes and is called a trash-type impeller. At right is the open impeller with four vanes.

The two factors governing the performance of a centrifugal pump are the peripheral speed of the impeller and the capacity of the pump. The former establishes the head pressure. The latter depends on pump size and flow area through casings and vanes. As the pressure increases, the flow decreases to a point of maximum pressure and zero flow, a point termed the "shut-off" pressure of the pump. The pump will continue to operate at rated speed with no flow.

The power input depends on head and volume pumped. At shut-off pressure and no flow, the power required is a minimum, increasing as the pressure drops and the volume increases. The pump designer has established the peak efficiency within the range of the anticipated conditions of service. The different conditions of operation are brought together graphically by characteristic curves, of which Figs. 4-58 and 4-59 are typical. Pump manufacturers supply these curves as a guide to selection. The information for standardized pumps is often presented in tabular form.

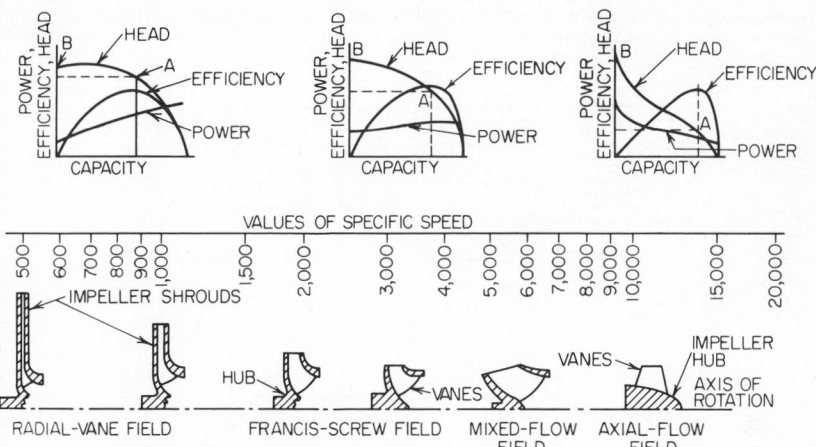

FIG. 4-57. Some typical impeller outlines, with their specific speeds and performance characteristics. (*Courtesy of Worthington Corp.*)

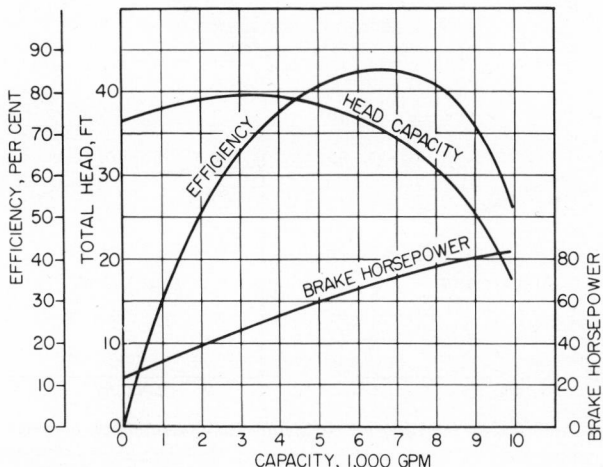

FIG. 4-58. Typical characteristic curve for a pump with overloading power curve with reduction in head.

A pump for a well-determined set of operating conditions may be selected for maximum efficiency at a closely defined point on the curve as shown in Fig. 4-60, left side. If conditions are not so well known or a range of pressures may be foreseen, a pump with a flat performance curve is desirable, such as in Fig. 4-60, right side.

Any pump casing will accommodate several different impeller diameters, which has led to the use of composite curves (Fig. 4-61). From these curves it is possible to select any one of the group of pumps that will fulfill the required conditions of service.

Centrifugal pumps range from the giant low-head irrigation pumps with 144-in. suction inlet and 600,000-gpm capacity to the very small laboratory type with a few gallons capacity. Multistage pumps of standard design are capable of operating at pressure up to 5,500 psig.

Single-stage process pumps are available for pressures to 600 psig and temperatures up to 750°F. Table 4-13 gives data and dimensions for one manufacturer's standard line of pumps. For higher pressures, multistage pumps are used. Higher temperatures require special design. Hot-oil pumps are furnished for service up to 1000°F and 1,500 psig.

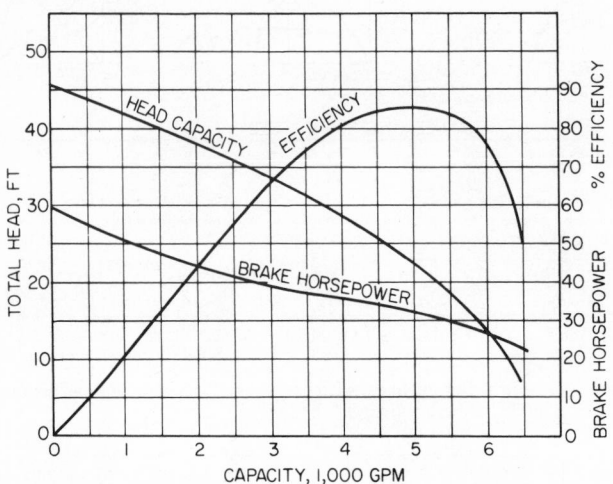

Fig. 4-59. Typical characteristic curve for a pump with overloading power curve with increase in head.

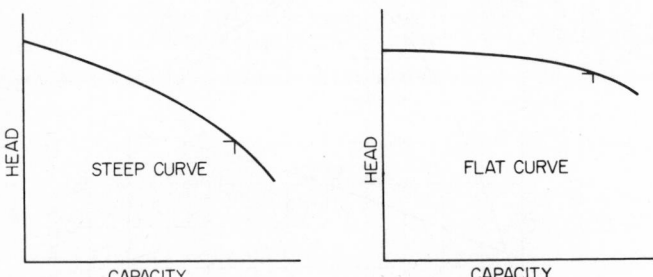

Fig. 4-60. Pump selection for steep and flat characteristic curves.

The suction lift for a centrifugal pump requires careful consideration. The term "net positive suction head" (NPSH) is the total head at the suction flange less the vapor pressure of the liquid at pumping temperature. With centrifugal pumps, it is essential to have a positive suction head so that the fluid may flow into the eye of the impeller.

The capacities of centrifugal pumps are rated in units of water. The viscosity of the fluid affects this rating, and a correction factor must be applied. Brake horsepower is increased, there is a reduction in available discharge head, and capacity is slightly affected adversely.

Centrifugal pumps normally operate with clockwise rotation, as seen from the motor or other prime mover. The position of the suction flange on the vertically split, single-suction, horizontal pump is opposite the eye of the impeller.   For double-suction pumps, it may be located at the side or bottom.   Discharge nozzles may be adapted to the purchaser's desires.

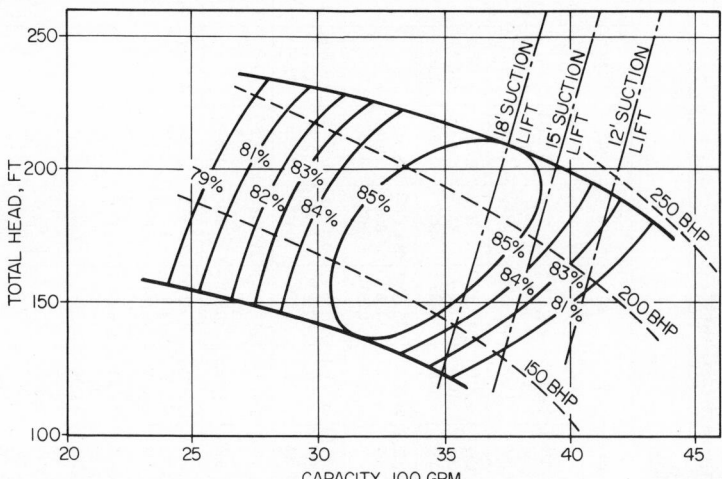

Fig. 4-61. Rating curve of a 10-in. double-suction single-stage pump.   (*Courtesy of Worthington Corp.*)

If complete information about the fluid to be handled is furnished to the manufacturer, his recommendations on materials should be followed.   The material standards generally adhered to are

Casing....................	Cast iron
Impeller...................	Bronze
Impeller ring...............	Bronze
Casing ring................	Bronze
Diffuser...................	Cast iron or bronze
Shaft (sleeve used)..........	Steel
Shaft (no sleeve)............	Stainless steel or steel
Shaft sleeve................	Bronze
Glands....................	Bronze

All-iron pumps are available for certain corrosive liquids, a .d all-bronze for sea water.

Cast-iron casings are seldom used for pressures above 1,500 psi or temperatures above 350°F.   For higher pressures and temperatures, steel is used.   Cast iron also loses strength and becomes brittle at very low temperatures, and alloy cast iron or steel is recommended.

Inquiries for quotations on centrifugal pumps should give complete information about the nature of the fluid to be pumped and the service anticipated.   But the detailed design should be left to the manufacturer.   A typical inquiry sheet is illustrated by Table 4-14.

Rotary Pumps

This type of pump is not to be confused with the centrifugal pump.   The basic design is that of two spur gears, one driven and one idling (Fig. 4-62).   As the gears rotate, fluid is trapped between the teeth and the casing and is propelled around to the discharge.   A rotary pump combines the constant-discharge characteristic of the centrifugal pump with the positive-discharge feature of a reciprocating pump.   The spur-gear pump is designed for relatively slow speed operation, up to about 600 rpm.

Table 4-13. Portion of Modern Pump-rating Chart*

Each cell lists: Size, type / hp / rpm

gpm	pump	20	25	30	35	40	45	50	55	60	70	80	90	100	110	120	130
200	a	2½, CF-1 / 1.6 / 1,010		2½, CF-1 / 2.4 / 1,170	2½, CF-1 / 2.8 / 1,240	2½, CF-1 / 3.25 / 1,310	2½, CF-1 / 3.75 / 1,380	2½, CF-1 / 4.2 / 1,440	2½, CF-1 / 4.65 / 1,500	2½, CF-1 / 5.05 / 1,550	2½, CF-1 / 6.1 / 1,660	2½, CF-1 / 6.85 / 1,750	2½, CF-1 / 7.9 / 1,845	2½, CF-1 / 9.0 / 1,930			
200	b		3, CF-1 / 2.25 / 830	3, CF-1 / 2.6 / 890	3, CF-1 / 3.0 / 945	3, CF-1 / 3.3 / 1,000	3, CF-1 / 3.6 / 1,048	3, CF-1 / 4.1 / 1,095	3, CF-1 / 4.6 / 1,140	3, CF-1 / 5.0 / 1,185	3, CF-1 / 6.2 / 1,275	3, CF-1 / 7.3 / 1,355	3, CF-1 / 8.3 / 1,430	3, CF-1 / 9.4 / 1,500	3, CF-1 / 10.5 / 1,568	3, CF-1 / 11.9 / 1,635	3, CF-1 / 13.1 / 1,700
225	a	2½, CF-1 / 1.9 / 1,055	2½, CF-1 / 2.3 / 1,130	2½, CF-1 / 2.7 / 1,205	2½, CF-1 / 3.2 / 1,280	2½, CF-1 / 3.65 / 1,345	2½, CF-1 / 4.15 / 1,410	2½, CF-1 / 4.7 / 1,470	2½, CF-1 / 5.1 / 1,530	2½, CF-1 / 5.55 / 1,585	2½, CF-1 / 6.5 / 1,685	2½, CF-1 / 7.6 / 1,785	2½, CF-1 / 8.6 / 1,875	2½, CF-1 / 9.7 / 1,960			
225	b		3, CF-1 / 2.6 / 850	3, CF-1 / 2.9 / 907	3, CF-1 / 3.3 / 962	3, CF-1 / 3.6 / 1,015	3, CF-1 / 4.0 / 1,065	3, CF-1 / 4.6 / 1,113	3, CF-1 / 5.0 / 1,155	3, CF-1 / 5.5 / 1,205	3, CF-1 / 6.6 / 1,290	3, CF-1 / 7.8 / 1,370	3, CF-1 / 8.9 / 1,445	3, CF-1 / 9.8 / 1,510	3, CF-1 / 11.1 / 1,580	3, CF-1 / 12.5 / 1,647	3, CF-1 / 13.8 / 1,710
250	a	2½, CF-1 / 2.2 / 1,095	2½, CF-1 / 2.6 / 1,170	2½, CF-1 / 3.1 / 1,245	2½, CF-1 / 3.6 / 1,315	2½, CF-1 / 4.1 / 1,385	2½, CF-1 / 4.6 / 1,450	2½, CF-1 / 5.1 / 1,510	2½, CF-1 / 5.6 / 1,565	2½, CF-1 / 6.2 / 1,620	2½, CF-1 / 7.1 / 1,720	2½, CF-1 / 8.2 / 1,815	2½, CF-1 / 9.2 / 1,900	2½, CF-1 / 10.5 / 1,990			
250	b		3, CF-1 / 2.9 / 870	3, CF-1 / 3.2 / 927	3, CF-1 / 3.6 / 980	3, CF-1 / 4.0 / 1,035	3, CF-1 / 4.4 / 1,085	3, CF-1 / 4.9 / 1,130	3, CF-1 / 5.5 / 1,175	3, CF-1 / 5.9 / 1,220	3, CF-1 / 7.2 / 1,305	3, CF-1 / 8.3 / 1,385	3, CF-1 / 9.5 / 1,460	3, CF-1 / 10.4 / 1,525	3, CF-1 / 11.7 / 1,593	3, CF-1 / 13.2 / 1,660	3, CF-1 / 14.6 / 1,720
300	a	2½, CF-1 / 3.0 / 1,190	2½, CF-1 / 3.5 / 1,265	2½, CF-1 / 4.05 / 1,335	2½, CF-1 / 4.55 / 1,400	2½, CF-1 / 5.1 / 1,465	2½, CF-1 / 5.65 / 1,530	2½, CF-1 / 6.25 / 1,590	2½, CF-1 / 6.75 / 1,645	2½, CF-1 / 7.3 / 1,700	2½, CF-1 / 8.5 / 1,800	2½, CF-1 / 9.65 / 1,885	2½, CF-1 / 10.9 / 1,970				
300	b		3, CF-1 / 3.5 / 920	3, CF-1 / 3.8 / 972	3, CF-1 / 4.3 / 1,025	3, CF-1 / 4.8 / 1,075	3, CF-1 / 5.3 / 1,125	3, CF-1 / 5.8 / 1,173	3, CF-1 / 6.5 / 1,220	3, CF-1 / 7.0 / 1,260	3, CF-1 / 8.2 / 1,340	3, CF-1 / 9.6 / 1,420	3, CF-1 / 10.8 / 1,490	3, CF-1 / 12.2 / 1,555	3, CF-1 / 13.1 / 1,620	3, CF-1 / 14.9 / 1,690	3, CF-1 / 16.4 / 1,750

Total head, ft

Table 4-13. Portion of Modern Pump-rating Chart* (Continued)

Total head, ft

gpm		20	25	30	35	40	45	50	55	60	70	80	90	100	110	120	130
350	Size, type		3, CF-1	3, CF-1	3, CF-1	3, CF-1	3, CF-1	3, CF-1	3, CF-1	3, CF-1	3, CF-1	3, CF-1	3, CF-1	3, CF-1	3, CF-1	3, CF-1	
	hp		4.1	4.6	5.1	5.5	6.3	6.8	7.6	8.2	9.4	10.8	12.2	13.5	15.0	16.5	
	rpm		970	1,025	1,075	1,120	1,175	1,220	1,265	1,305	1,380	1,455	1,525	1,595	1,660	1,720	
	Size, type				4, CF-1	4, CF-1	4, CF-1	4, CF-1	4, CF-1	4, CF-1	4, CF-1	4, CF-1	4, CF-1	4, CF-1	4, CF-1	4, CF-1	4, CF-1
	hp				4.2	4.8	5.6	6.2	7.0	7.7	8.4	10.3	11.8	13.6	15.3	17.6	20.1	21.8
	rpm				905	965	1,020	1,070	1,125	1,170	1,215	1,305	1,387	1,467	1,537	1,610	1,677	1,745
400	Size, type		3, CF-1	3, CF-1	3, CF-1	3, CF-1	3, CF-1	3, CF-1	3, CF-1	3, CF-1	3, CF-1	3, CF-1	3, CF-1	3, CF-1	3, CF-1		
	hp		4.9	5.4	6.0	6.5	7.2	8.0	8.6	9.4	10.7	12.4	13.9	15.2	16.9		
	rpm		1,025	1,080	1,130	1,170	1,225	1,270	1,310	1,350	1,425	1,500	1,567	1,635	1,700		
	Size, type				4, CF-1	4, CF-1	4, CF-1	4, CF-1	4, CF-1	4, CF-1	4, CF-1	4, CF-1	4, CF-1	4, CF-1	4, CF-1	4, CF-1	4, CF-1	
	hp				4.8	5.6	6.2	6.8	7.5	8.3	9.1	11.0	12.6	14.6	16.5	18.6	20.9	
	rpm				930	990	1,040	1,090	1,140	1,185	1,230	1,320	1,400	1,480	1,550	1,620	1,690	

* From Igor T. Karassik and Roy Carter, *Centrifugal Pumps*, McGraw-Hill Book Company, New York, Table 17.4, p. 220. Copyright © 1960. Used by permission.

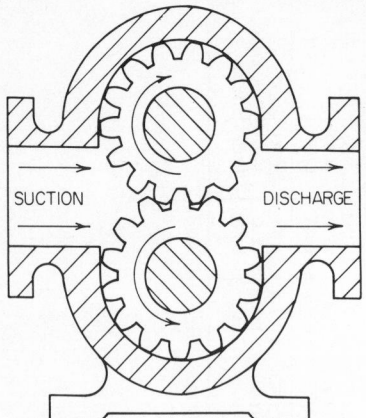

FIG. 4-62. Gear-type pump. The spaces between the gear teeth fill on suction, squeeze out on discharge.

For motor or turbine drive, gears must be interposed. The spur-gear pump is well adapted to handle heavy oils which provide lubricating value, but they are not suitable for light oils. The suction lift attainable is comparable with that of the positive-displacement pump.

Modifications of the rotary spur-gear pump include

Herringbone gears (Fig. 4-63). These gears permit smooth operation at speeds to 1,800 rpm and are suitable for direct-motor connection. Though designed to pump oils with some lubricating value, they may be adapted to water and light-oil service.

Helical or spiral gears (Fig. 4-64). These fall within the service classification of the herringbone type.

Screw-type rotary pumps have value where viscous oils are to be handled and where a high suction lift is desired. They offer the advantages of the reciprocating pump as a

Table 4-14. Summary of Essential Data Required in Selection of Centrifugal Pumps*

1. Number of units required
2. Nature of the liquid to be pumped
   Is the liquid:
   a. Fresh or salt water, acid or alkali, oil, gasoline, slurry, or paper stock?
   b. Cold or hot and if hot, at what temperature? What is the vapor pressure of the liquid at the pumping temperature?
   c. What is its specific gravity?
   d. Is it viscous or nonviscous?
   e. Clear and free from suspended foreign matter or dirty and gritty? If the latter, what is the size and nature of the solids, and are they abrasive? If the liquid is of a pulpy nature, what is the consistency expressed either in percentage or in lb/cu. ft of liquid? What is the suspended material?
   f. What is the chemical analysis, pH value, etc.? What are the expected variations of this analysis? If corrosive, what has been the past experience, both with successful materials and with unsatisfactory materials?
3. Capacity
   What is the required capacity as well as the minimum and maximum amount of liquid the pump will ever be called upon to deliver?
4. Suction conditions
   Is there:
   a. A suction lift?

   b. Or a suction head?
   c. What is the length and diameter of the suction pipe?
5. Discharge conditions
   a. What is the static head? Is it constant or variable?
   b. What is the friction head?
   c. What is the maximum discharge pressure against which the pump must deliver the liquid?
6. Total head
   Variations in items 4 and 5 will cause variations in the total head?
7. Is the service continuous or intermittent?
8. Is the pump to be installed in a horizontal or vertical position? If the latter:
   a. In a wet pit?
   b. In a dry pit?
9. What type of power is available to drive the pump and what are the characteristics of this power?
10. What space, weight, or transportation limitations are involved?
11. Location of installation
    a. Geographical location
    b. Elevation above sea level
    c. Indoor or outdoor installation
    d. Range of ambient temperatures
12. Are there any special requirements or marked preferences with respect to the design, construction, or performance of the pump?

* Courtesy of Worthington Corp., "Basic Factors in Preparing a Centrifugal Pump Inquiry," Worthington Reprint 21RP-477, Table I, p. 36.

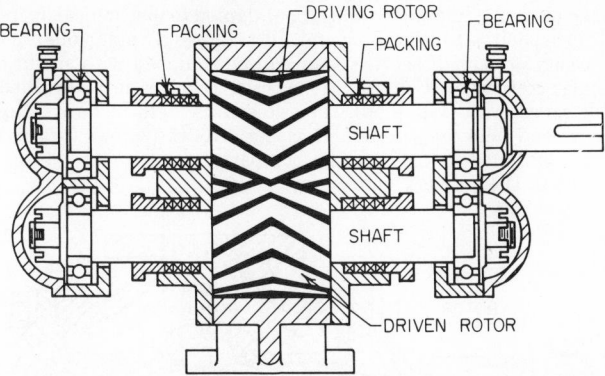

Fig. 4-63. Herringbone pump. Permits smooth operation at speeds up to 1,800 rpm. This design has bearings separate from the liquid being pumped to protect the bearings from abrasives and corrosives.

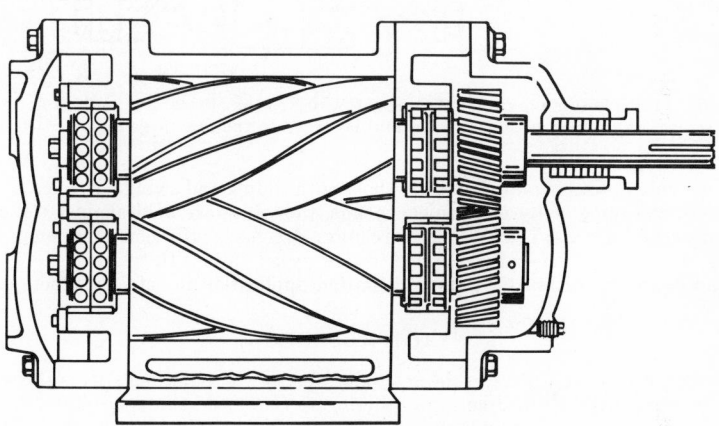

Fig. 4-64. Typical helical- or spiral-gear pump.[13]

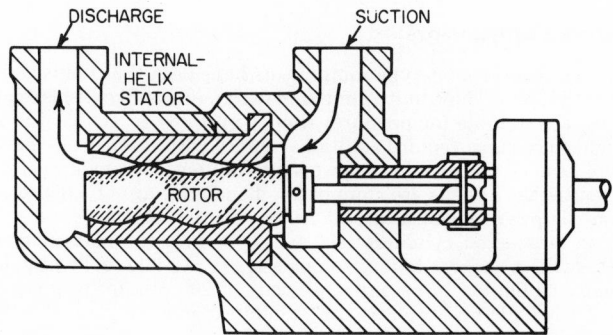

Fig. 4-65. Single-screw pump. The rotor revolves and oscillates in a two-threaded liner and jumps the liner threads to force the liquid to the discharge outlet.

positive-acting piece of equipment and can be adapted to mechanical operation.   They are available in capacities to 1,000 gpm and will successfully pump oils of high viscosity. For high-viscosity or asphalt service, they can be equipped with steam jacketing.

In the single-screw type (Fig. 4-65), the screw revolves in a threaded liner, while in the double-screw type (Fig. 4-66), the two screws mesh to form a liquidtight seal between screws and between screws and casing.   As the screws turn in normal rotation, the fluid contained in the separate compartments between the screws and the casing is forced to the delivery end.

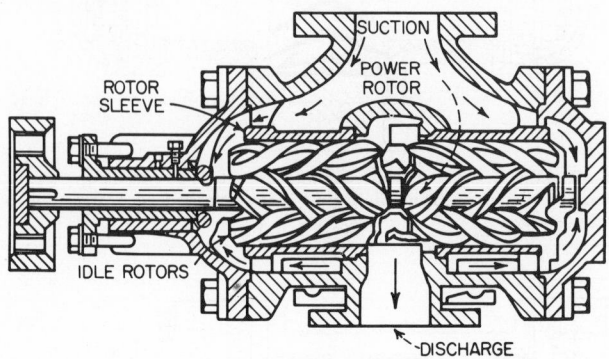

Fig. 4-66. Multiple-screw pump.   As one screw drives the others, fluid enters at the ends of the screws, is trapped in the threads, and is forced to the discharge end.

Other types of rotary pumps include those with sliding and swinging vanes.   Either by centrifugal force or by mechanical means, the vanes are held against the casing. Hinged buckets are also used.   There are many designs for this classification of rotary pumps, but they are generally confined to slow-speed units with relatively low capacities and discharge pressures, thus find limited application in refinery practice.

COMPRESSORS

Refinery use of compressors is generally related to handling air.   Those units developing less than 5 psi, from atmospheric, are classified as fans or blowers.

Compression equipment is classified in three broad categories: reciprocating displacement type, rotary displacement type, and centrifugal type (radial or axial flow). Any one of these types, or combinations, may be used for single-, double-, or multiple stage compression.

Reciprocating Compressors

Reciprocating displacement-type compressors have been the standard in the industry until recent years.   These units, similar in design to positive-displacement reciprocating pumps, are suitable for pressures from 1 up to 80 to 100 psig in single-stage design.   Higher pressures require additional stages with intercoolers between stages to absorb the heat of compression.   Two-stage units will provide ratings up to 500 psig, while multistage designs are equipped to handle ratings of 5,000 psig or higher.

The piston compressors will utilize any type of prime mover.

Generally in large-sized cylinders (300 to 5,000 cfm), the limit of pressure to an individual stage is set at three to four compressions.   An inlet of 50 psig, for instance, would normally be limited to an outlet pressure of 200 psig for a single compression stage.   Smaller cylinders (50 to 300 cfm) would expand the value to seven or eight compressions.

Cylinder construction follows the same designs used in reciprocating pumps; single- and double-acting cylinders are used (Figs. 4-67 and 4-68).   Double-acting units are limited to total pressures of 500 to 600 psi, whereas the single-acting plunger provides better service at pressures above these values.

Plungers are ring-packed or packed with soft or semipliable metallic packing using gland and stuffing-box design. Jacketing of cylinders and heads is important

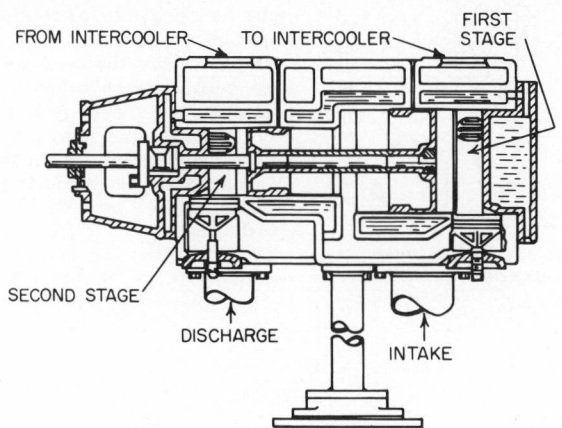

FROM INTERCOOLER   TO INTERCOOLER   FIRST STAGE

SECOND STAGE

DISCHARGE   INTAKE

FIG. 4-67. Two-stage, single-acting opposed piston in a single-step-type cylinder.[7]

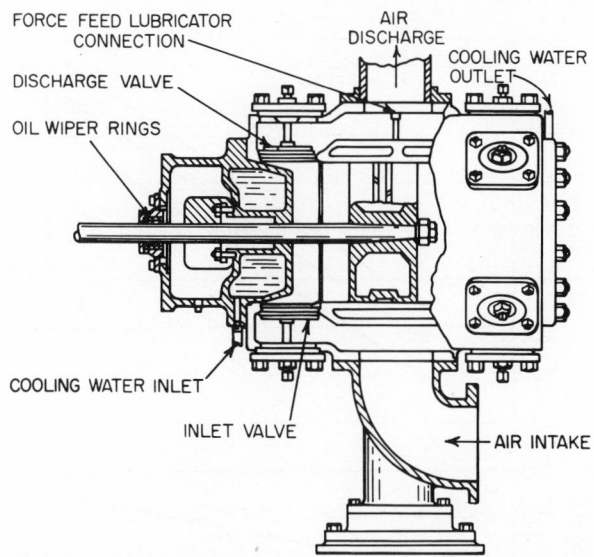

FORCE FEED LUBRICATOR CONNECTION

AIR DISCHARGE

COOLING WATER OUTLET

DISCHARGE VALVE

OIL WIPER RINGS

COOLING WATER INLET

INLET VALVE

AIR INTAKE

FIG. 4-68. Typical double-acting piston and compressor cylinder.[7]

for good heat removal. Piston speeds are limited to 650 fpm on smaller units and 750 fpm on the larger sizes, although some designs have exceeded 800 fpm. This has been made possible by the development of lightweight plate and strip valves.

Rotary Compressors

Rotary-displacement-type compressor designs are classified as blowers where developed pressures vary from 6 oz to 12 psig. In one unit of this type (Fig. 4-69), two

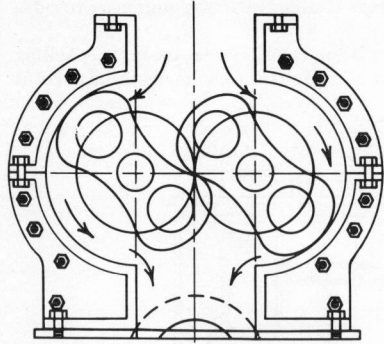

FIG. 4-69. Cross-section schematic of typical rotary positive-displacement-type gas pump.

revolving impellers within a close-fitting case actually trap and move definite volumes of gas from inlet to outlet in successive movements. The action occurs twice for each revolution of each impeller or four times for the combined revolutions of both impellers.

Impellers are designed for close clearance between each other and the casing to reduce the slip or leakage that would occur through such passages. Units such as this have capacities up to 10,000 cfm at a discharge pressure of 10 psi with inlet pressure at 14.7 psia, 68°F, and a rotary speed of 600 rpm. Figure 4-70 indicates a typical characteristic curve for a general-use blower. The formula generally used for determining horsepower is expressed as follows:

$$hp = \frac{Q(p_2 - p_1)}{33,000}$$

in which Q = cu ft of air compressed/min
  p_1 = initial pressure, psf
  p_2 = final pressure, psf

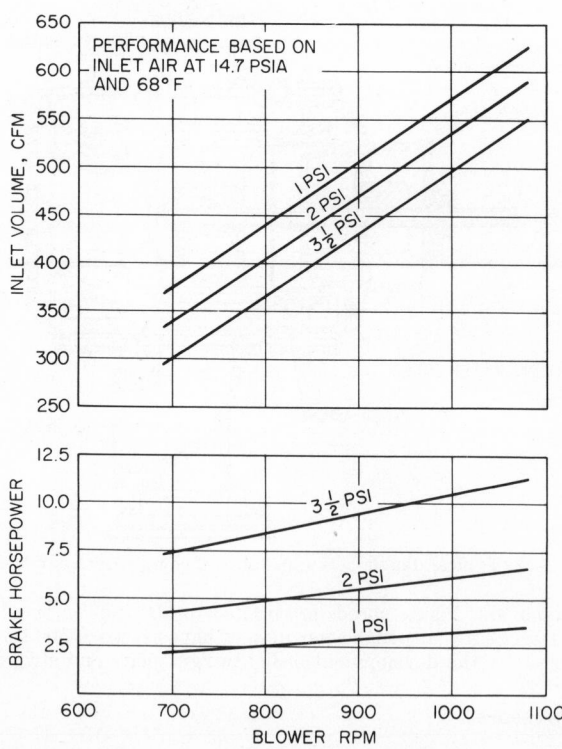

FIG. 4-70. Typical capacity-power curve of rotary displacement-type blower. (*Courtesy of Roots Product Dept., Dresser Clark, Div. of Dresser Industries, Inc.*)

Another type uses a water seal between suction and discharge chambers. This liquid-packing-ring type of blower is also used as a vacuum pump (Fig. 4-71).

The rotor, revolving in an elliptical case filled with water, causes the water to act as a liquid piston, thereby displacing air in the spaces between vanes of the impeller. Such units will handle discharge pressures to 75 psi and vacuum to approximately 27 in. of mercury. They are particularly well suited for processes requiring oil-free air or gas.

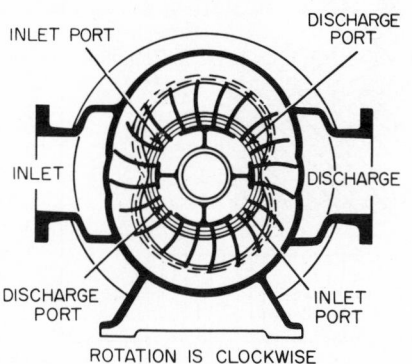

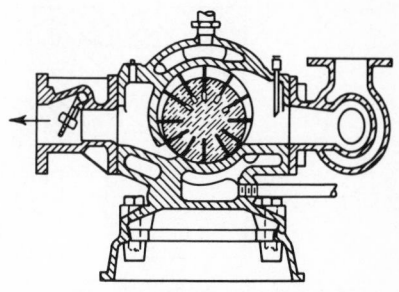

Fig. 4-71. The liquid-packing-ring-type blower can also be used as a vacuum pump.

Fig. 4-72. Cross section of typical sliding-vane-type rotary compressor.

Still another rotary design uses sliding vanes on a rotor revolving within an eccentric housing, as in Fig. 4-72. Gas is trapped within each pair of centrifugally driven vanes and compressed as the rotation carries the section into the smaller portion of the eccentric pattern. These units are capable of 125-psi discharge and capacities up to 2,000 cfm.

Centrifugal Compressors

Within recent years, the marked increase in volumes of gases to be handled, the limitations on reciprocating equipment to meet these needs economically, and the development and improvements in the design of centrifugal compressors have led to increased importance and use of this design in lieu of the positive-displacement types. This is not to be construed as meaning that the centrifugal compressor will replace the other types. Each has its place. The reciprocating compressor lends itself to applications of low volume and high pressure.

High volume and low pressure are the practical application for the centrifugal design. It offers high speed and direct drive from turbine or motor; it is relatively light, free from vibration, and lacking in close clearances. Gas is delivered in a steady stream, free from pulsations. In addition, it provides flexible and easy regulation of output based on volume, suction, or discharge pressure as the control medium.

Some idea of how these various units fit the overall field of compressor requirements can be gained by referring to Fig. 4-73, where units are grouped on a basis of discharge pressure plotted against capacities.

Another method of comparison is by reference to developed adiabatic head, shown in Fig. 4-74. To understand this, it is well to define some of the terms associated with gas compression. In these discussions, it is important to note that centrifugal compressor performance is materially affected by the density and characteristics of the gas being handled.

Compression occurring without heat externally added or removed during the compression cycle and also developing an increase of temperature of the gas itself is defined as adiabatic compression. It is expressed as

$$pv^k = \text{constant}$$

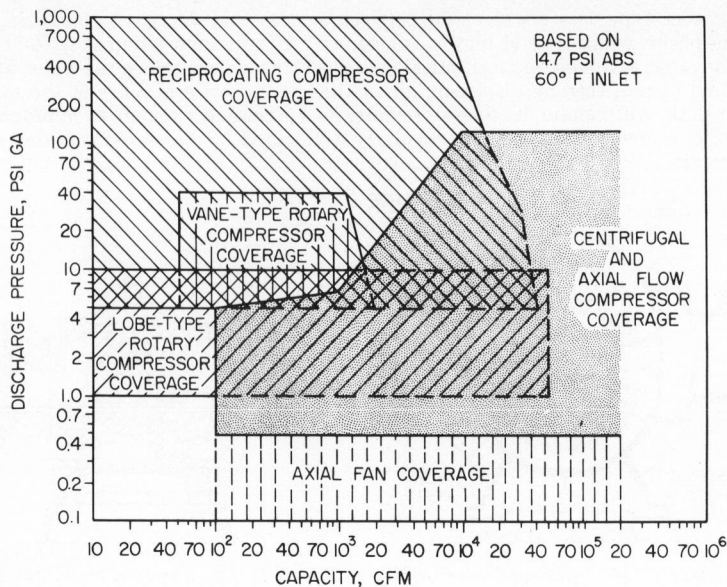

FIG. 4-73. Chart of approximate field for compressor types based on discharge pressure.[16]

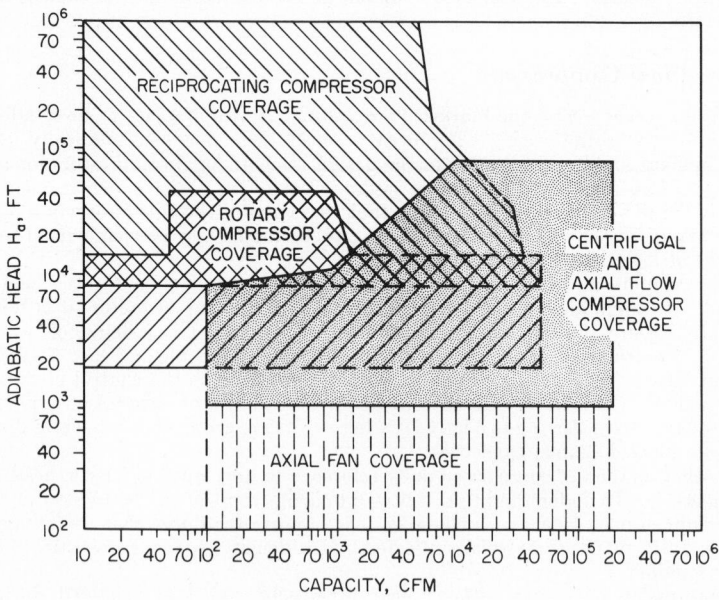

FIG. 4-74. Chart of approximate field for compressor types based on adiabatic heat.[16]

in which p = absolute pressure, psi
  v = specific volume, cu ft/lb
  k = ratio of specific heat of gas at constant pressure to that at constant volume ($k = c_p/c_v$)

When the heat of compression is removed so that gas temperature remains constant, it is known as isothermal compression, pv^1 = constant.

Both forms are expressed as polytropic compression where pv^n = constant. When $n = 1$, polytropic compression is isothermal; when $n = k$, it is adiabatic. The pressure-volume diagram in Fig. 4-75 indicates various slopes for gas-compression cycles. This slope is a function of the exponent n. Since centrifugal compressors are not cooled, a compression cycle of 100 per cent efficiency would produce an adiabatic character. However, all compressors develop some heat, because of inefficiencies, and actual compression becomes polytropic of a type where n is greater than k.

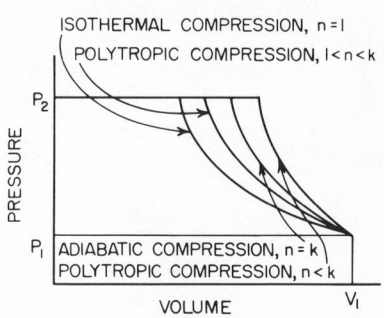

FIG. 4-75. Pressure-volume diagram of polytropic compressions.[16]

In order now to get a measure of centrifugal compressor capacity, we follow the same procedure used with centrifugal pumps; i.e., we use terms corresponding to "total developed head" and "total required head." The former refers to the energy in ft-lb/lb of fluid or gas, transmitted to the fluid by the compressor, based on its speed, dimensions, and configuration. The latter term refers to the work required to raise 1 lb of fluid from one energy level to another. When the fluid is a compressible gas and is transported by an uncooled centrifugal compressor, this difference between pressure levels is termed adiabatic head H_a:[16]

$$H_a = RT_1 \frac{k}{k-1} (r^{k-1/k} - 1)$$

in which T_1 = initial gas temperature, abs °R
  R = gas constant (1.544/molecular weight)

With this term, it becomes possible to rate and analyze compressors on the basis of head-capacity curves, as in Fig. 4-74, just as is the case with centrifugal pumps.

Single-stage centrifugal compressors will generally cover discharge pressure ranges of 1 to 15 psig. One design is designated as overhung type in which the impeller wheel is overhung on the extended shaft. A second type is referred to as pedestal type wherein the impeller wheel is supported in an overhung position on its own shaft with its bearings contained in a bearing pedestal; the pedestal may also support the blower casing. The shaft is flexibly coupled to the driver. The third classification for single-stage centrifugal units is the double-inlet type. In this instance a double-sided impeller is mounted on a shaft supported by bearings on each side.

Multistage centrifugal compressors are generally air-cooled but for large volume capacities may be provided with water jackets or intercoolers. Ranges covered will start at as low as 6 psig.

Classification covers two arrangements. A single-inlet type receives gas at one end, passing it successively through the various stages and discharging it at the other end. A double-inlet unit uses double the number of impellers, arranged in opposed groups. Gas enters at both ends and is discharged at the center.

Many of the details and features of centrifugal compressors for refinery service are covered in API Standard 617.[17] Of particular interest is a consideration of shaft sealing. Seals should be suited to operating speeds, variations in suction conditions, and other special considerations. Recommendations cover four types, as shown in Fig. 4-76.

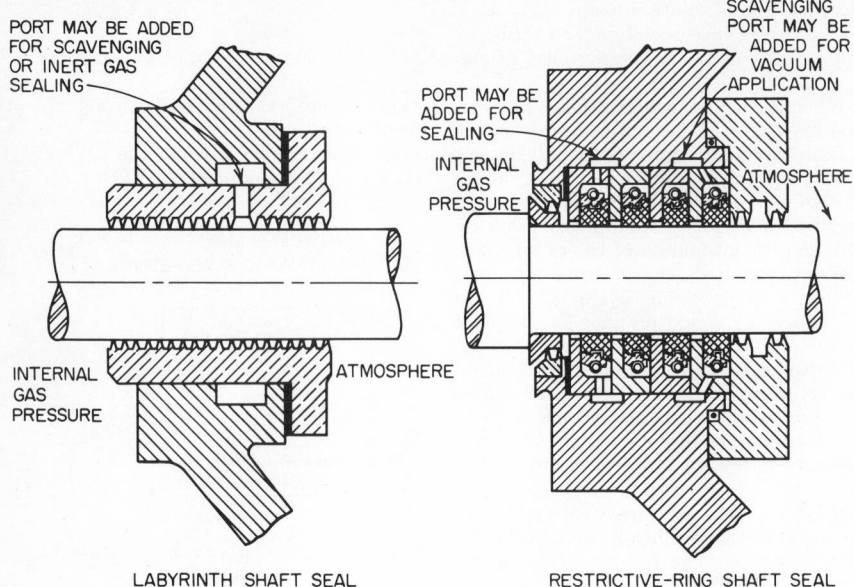

LABYRINTH SHAFT SEAL   RESTRICTIVE-RING SHAFT SEAL

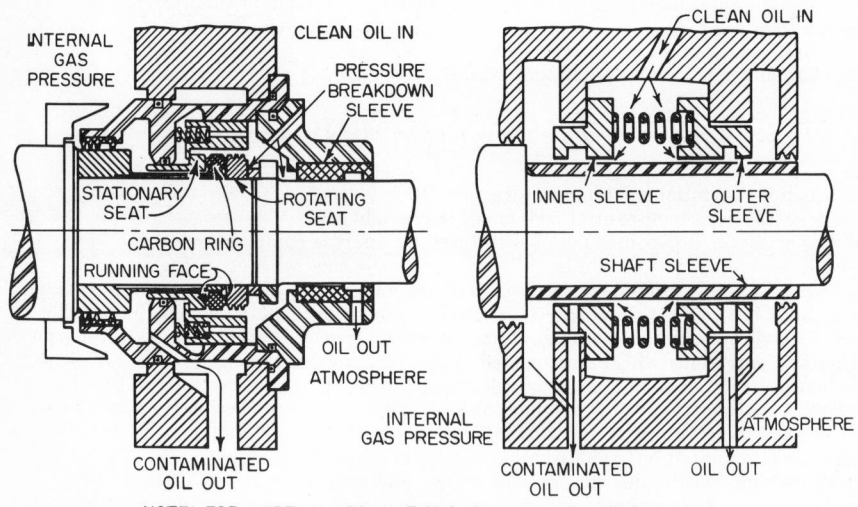

NOTE: FOR CERTAIN APPLICATIONS, SEAL OIL IS NOT REQUIRED

MECHANICAL (CONTACT) SHAFT SEAL   LIQUID-FILM SHAFT SEAL

FIG. 4-76. Types of shaft seals for centrifugal compressors.[17]

PRIME MOVERS

Steam Turbines

The steam turbine offers many advantages as a prime mover. It is compact, requires a minimum of floor space, is capable of considerable overload, and its speed is adaptable to rotary compressor and centrifugal pump drives, its initial cost is moderate, its efficiency is good, and it is available in capacities beyond those of the reciprocating steam engine.

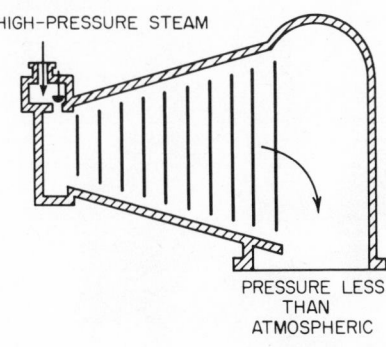

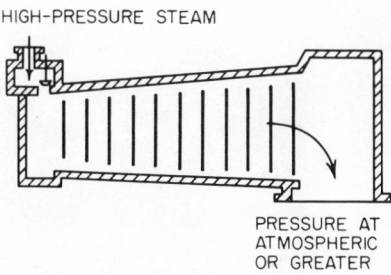

FIG. 4-77. Straight compressing turbines are useful where steam consumption must be kept to a minimum.

FIG. 4-78. Straight noncondensing turbines are useful where exhaust steam has other uses and when very low-cost steam is available.

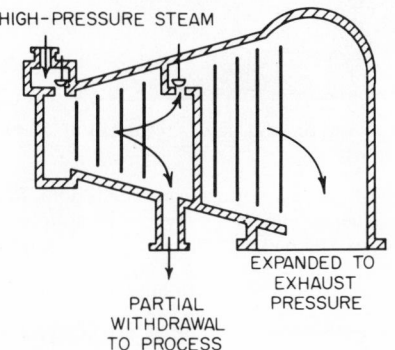

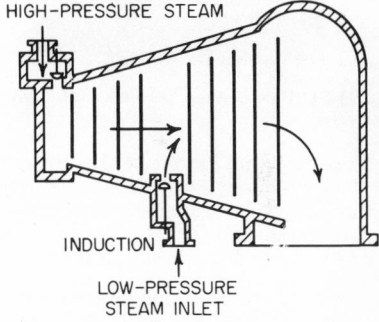

FIG. 4-79. Extraction-type turbines allow partial withdrawal of intermediate-pressure steam for process use.

FIG. 4-80. Induction-type turbines are used when there is sufficient low-pressure process steam to furnish a portion of the power requirements.

The combinations of steam pressures, speed, and horsepower are almost infinite. The largest mechanical-drive turbine of record is rated at 24,000 hp. Laboratory units of 10 hp are in use. Speeds from 1,000 to 12,000 rpm are available. Any combination of steam conditions may be met.

Steam turbines may be classified as impulse turbines or reaction turbines. In the impulse design, steam expansion occurs only in the stationary blades or nozzles. Reaction turbines utilize revolving nozzles.

Steam turbines may be further classified as straight condensing (Fig. 4-77), noncondensing (Fig. 4-78), and extraction (Fig. 4-79). A turbine may be further designed for induction operation (Fig. 4-80), wherein available low-pressure steam from a process source is inducted into the turbine to furnish a portion of the power requirements. A combination of both extraction and induction design is also available.

In view of the usual demand for process steam, either the noncondensing (operating against a back pressure) or the extraction type (also termed "bleeder") is generally favored in refinery practice.

Double extraction is possible in the case of Fig. 4-79 by adding another takeoff. For instance, it may be desirable to bleed 100-psi steam for process work and 15-psi steam for general heating, with any steam not utilized passed to a condenser.

A single-stage turbine with two rows of rotating buckets is illustrated by Fig. 4-81.

FIG. 4-81. Single-stage turbine with two rows of rotating buckets. (*Courtesy of Terry Turbines.*)

The American Petroleum Institute Standard 615 presents recommendations for mechanical-drive steam-turbine selection for general refinery service. The standard is applicable where steam characteristics do not exceed 850 psig and 750°F, and speeds do not exceed 12,000 rpm.

Integrally forged wheel and shaft construction is preferred with speeds limited as follows:[18]

Pitch diam of wheel, in.	Speed, rpm
32	6,000–7,000
25	7,600–9,500
20	9,500 plus

Steam-turbine casings are generally made of cast iron for steam temperatures to 500°F. Above this figure they should be of steel. Blades and shrouds are preferably of 11-13 chrome steel or monel metal.

Steam turbines are available for variable-speed operation or for governor control at fixed speeds. In general, the higher the speed the lower the cost, because the diameter of the rotating elements decreases. Gear reduction, either built-in or separate, is extensively used with high-speed turbines.

The efficiency of the steam turbine compares favorably with that of the steam engine. Theoretical steam consumption of condensing turbines is given in Table 4-15.

Table 4-15. Theoretical Steam Rates, lb/kwhr*,1ᵛ

Pressure, psig	Saturated		500		600		700		800		900		1000	
	— Exhaust pressure, in. Hg abs													
	1	2	1	2	1	2	1	2	1	2	1	2	1	2
100	10.20	11.32	9.72	10.24	8.72	9.60	8.18	8.98	7.67	8.38	7.19	7.83	6.74	7.31
150	9.58	10.52	8.82	9.68	8.30	9.08	7.81	8.51	7.34	7.97	6.89	7.46	6.47	6.99
200	9.17	10.01	8.54	9.32	8.04	8.75	7.57	8.21	7.12	7.70	6.69	7.22	6.30	6.77
250	8.88	9.66	8.35	9.07	7.85	8.52	7.39	7.99	6.96	7.50	6.55	7.05	6.17	6.62
300	8.67	9.40	8.21	8.90	7.71	8.34	7.26	7.83	6.83	7.36	6.44	6.91	6.07	6.49
350	8.49	9.19	8.10	8.76	7.60	8.20	7.15	7.70	6.73	7.23	6.35	6.80	5.98	6.39
400	8.35	9.01	8.03	8.66	7.51	8.09	7.06	7.60	6.65	7.14	6.27	6.71	5.91	6.31
450	....	....	7.97	8.59	7.44	8.01	6.99	7.51	6.59	7.05	6.21	6.63	5.86	6.24

Pressure, psig	550		600		650		700		800		900		1000	
500	7.64	8.22	7.39	7.94	7.15	7.68	6.94	7.43	6.53	6.99	6.15	6.57	5.81	6.18
600	....	....	7.31	7.83	7.06	7.56	6.84	7.32	6.44	6.87	6.07	6.46	5.72	6.08
1,000	....	....	....	....	....	....	6.67	7.09	6.23	6.62	5.86	6.21	5.53	5.86
1,400	....	....	....	....	....	....	....	....	6.16	6.53	5.77	6.10	5.43	5.73
1,600	....	....	....	....	....	....	....	....	....	....	5.75	6.07	5.40	5.70
2,000	....	....	....	....	....	....	....	....	....	....	5.73	6.04	5.37	5.65
3,000	....	....	....	....	....	....	....	....	....	....	5.80	6.10	5.36	5.63

* From T. Baumeister and L. S. Marks, *Mechanical Engineers' Handbook*, McGraw-Hill Book Company, New York, 6th ed., Table 2, p. 9-81. Copyright © 1958.

The efficiencies of large units, 5,000 to 7,500 kw rating, approximate 72 to 75 per cent; small units of 1,000 to 3,000-kw, 65 to 67 per cent.

Steam Engines

Steam engines find little application in refinery engineering. The steam and gas turbine and the electric motor are better suited as prime movers for process equipment.

Gas Turbines

In its simplest form, a gas turbine consists of a centrifugal compressor supplying air for combustion, a combustion chamber with a fuel-injection system, and a turbine

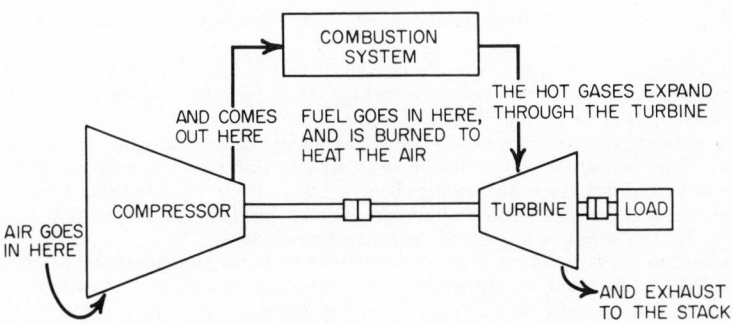

Fig. 4-82. Diagram of a simple-cycle gas turbine.

wheel through which the expansion of hot gases provides rotary power to the turbines. Figure 4-82 illustrates the simple-cycle gas turbine. The heated gases leave the combustion chamber at about 1500°F and expand through the turbine in much the same way that steam expands through a steam turbine. The power developed is sufficient to drive the air compressor mounted on the same shaft and meet the load on the turbine. Figure 4-83 illustrates the cross section of a 6,750-hp simple-cycle, single-shaft gas turbine.

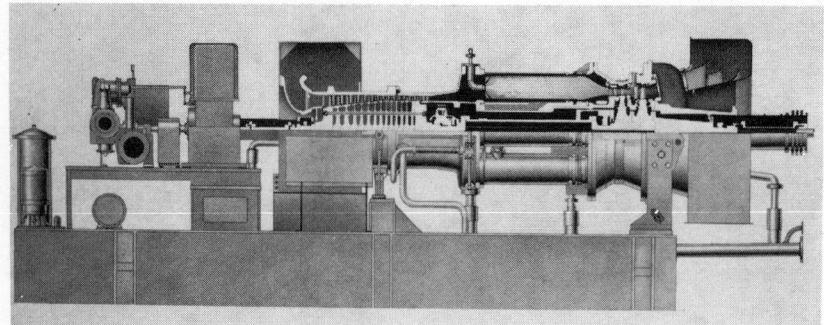

FIG. 4-83. Cross section of a 5,300-kw simple-cycle, single-shaft, base-mounted gas turbine. (*Courtesy of General Electric Co.*)

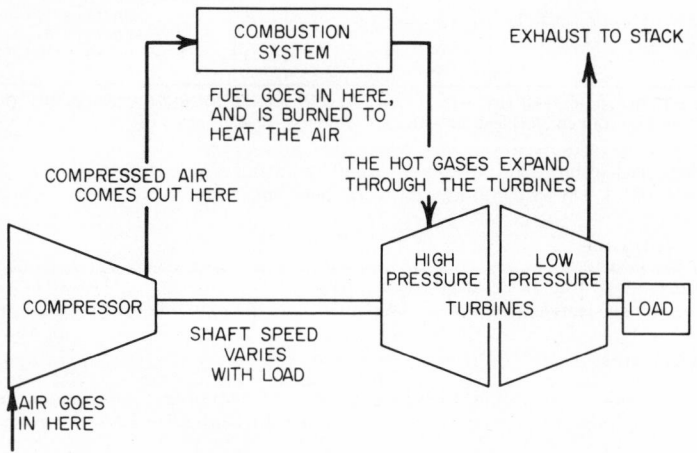

FIG. 4-84. Diagram of a simple-cycle two-shaft gas turbine.

Where variable speed is desired, a two-shaft turbine may be used. The hot gases first expand through a high-pressure turbine, then pass to a low-pressure unit which carries the power load (Fig. 4-84).

The exhaust temperature of the spent gases exiting from the gas turbine is about 850°F. The contained heat may be recovered in different ways when operation is continuous or when the best possible efficiency is desired. Waste-heat boilers and air preheaters for the combustion chamber may be used. When operated with such auxiliaries, the complex is termed "regenerative cycle."

Natural or waste refinery gases are excellent fuels for the gas turbine. Distillate fuel oils may be used by direct injection into the combustion chamber. Heavy residual fuel oils present difficulties. Dual-fuel units are available, capable of burning distillate oil or gas or any combination of the two under automatic control.

Two characteristics of the gas turbine must be given consideration:
1. At elevations from sea level to 1,000 ft, the turbine rating is satisfactory.   At 2,000 ft, the duration factor becomes 0.963; at 3,000 ft, 0.927; and at 5,000 ft, 0.853.
2. At low ambient temperatures, a fundamental characteristic of the gas turbine is its ability to develop more horsepower.   A. L. Vaughan[10] reports that a 5,700-hp turbine operated at 126 per cent of rated capacity when the ambient temperature was 30°F, and at 88 per cent at 75°F ambient temperature.

Depending on size and type of auxiliary regenerative equipment, the fuel consumption of the gas turbine is reported at 10,000 to 11,000 Btu/bhp-hr.   Efficiencies of 20 to 24 per cent are reported.

Gas turbines are available in sizes from 1,000 to 30,000 bhp.   Costs range from $55 per installed horsepower for small units to $25 per installed horsepower for large units of the simple-cycle type.   Regenerative units will cost 10 to 12 per cent more.

For refinery applications, gas turbines find their principal uses in the mechanical drive of large rotary compressors and occasionally for high-volume centrifugal pumps. They provide an excellent means of providing emergency electrical generating capacity.

Direct-current Motors

The direct-current (d-c) motor, because of danger from sparking brushes and poorer economy than possible from alternating-current (a-c) distribution, finds little application in the oil refinery and will not be discussed.

Alternating-current Motors

Standard polyphase induction motors have two windings.   One is on the stator or stationary part, and one is on the revolving part or rotor.   There is no electrical connection from the rotor winding to any source of electrical energy, the necessary voltage and current in the rotor section being produced by induction from the stator winding. Two types of rotor windings are used.   The "squirrel cage" is used extensively in refineries and the "wound rotor" is used occasionally for special conditions.

Modifications of the basic squirrel-cage motor include
1. Enclosed explosion-proof type in class I, division 1 locations.
2. Pressurized motors for use in class I, division 1 areas.   The motor housing is designed to be airtight, and nitrogen or other inert gas is forced through the casing.
3. Fan-cooled motors for use in areas of high ambient temperatures.
4. Motors with built-in electric space heaters for locations of extreme cold.
5. Back-geared motors, wherein built-in gearing is used to reduce the normal motor speed.
6. Variable-speed motors.   The induction motor for a given number of poles is a constant-speed machine.   The speed may be modified by changing the number of poles or by the use of slip rings.

The slip of an induction motor is the ratio of the difference between rpm of the rotating magnetic field and the actual rotor speed to the rotating-magnetic-field speed. The slip may vary from 4 to 8 per cent at full load and for motors to 75 hp.

API Bulletin RP-540 classifies refinery voltage systems as follows:[21]

Low: 600 volts and less
Medium: 601 to 15,000 volts
High: 15,000 volts, plus

Low-voltage distribution is restricted, supplying utilization equipment directly. Medium voltage is used for distribution and for utilization voltage for large motors. High voltage is used by utilities for transmission and is usually stepped down for refinery distribution and utilization.

For standard three-phase 60-cycle current, 440-volt motors are preferable.   For service in excess of 600 volts, 2,300-volt motors are acceptable and higher voltages are permissible under carefully analyzed conditions.

The economic breakpoint between the installation of low-voltage motors (440-volt range) and high-voltage motors (2,300 volts upward) is between motors within the range of 100 to 150 hp.

The speed of the a-c motor is established by the number of poles in the motor and the frequency of the supplied current, and may be expressed as

$$\text{rpm} = \frac{120f}{p}$$

in which f = frequency of current, cycles
p = number of poles

The efficiency of the squirrel-cage and wound-rotor a-c induction motor increases with size and loading (Table 4-16). Full-load current in amperes is given in Table 4-17.

Table 4-16. Approximate Performance Data for Induction Motors*

Hp	Weight, lb	Amp	Power factor, %			Efficiency, %		
			½ load	¾ load	Full load	½ load	¾ load	Full load
3-phase, 220-volt, 60-cycle, 1,750-rpm, squirrel-cage type								
1	40	3.4	48	61.5	71	70	75.5	77.5
2	45	5.8	52	66	75	77.8	80.5	80.5
5	80	14.0	64	75	68.1	78	81	82
10	145	28	63	74	80	81	83	84
20	240	54	68	76	81	84	86	86.5
40	515	98	79	85	87	90.0	90.0	90.4
100	975	242	80.5	86	88	90	90.5	91.2
3-phase, 220-volt, 60-cycle, 1,750-rpm, wound-rotor type								
25	480	32	76.5	84.3	87.5	87	88.7	88.9
50	850	121	78.8	86.5	89.7	88	89	89
100	2618	233	87.2	90.5	89.5	86	88	88

* Courtesy of Westinghouse Corp.

One phenomenon to be considered in all a-c calculations is the ratio of the true power or watts to the apparent power or volt-amperes. The ratio is termed the power factor and is expressed as a decimal or percentage (Table 4-16). The power factor is best at full load and increases with the size of the motor. All effort should be made to have a-c motor drives loaded to top capacity.

The overall power factor of a system may be increased by the application of a separately excited synchronous motor or by the capacitor. Manufacturers should be consulted.

The standard a-c motors are rated for a maximum of 115 per cent overload above nameplate when the ambient temperature does not exceed 40°C (104°F). They will carry a short overload of 50 per cent in excess of rating.

Most motors used in the refining industry drive centrifugal pumps, blowers and compressors, and types of equipment that present no particular problems relative to starting torque. For these, motor characteristics should be specified as not less than the "normal" torque in accordance with motor manufacturers' standards. It may be noted that in the case of centrifugal pumps and compressors, more torque is required

Table 4-17. Approximate Full-load Currents of Motors[*,22]

Hp	Three-phase a-c motors, squirrel-cage and wound-rotor induction type, amp			Single-phase a-c motors, amp		D-c motors, amp	
	220 volts†	440 volts	550 volts	115 volts	230 volts	115 volts	230 volts
½	2	1	0.8	9.8	4.9	5.4	2.7
¾	2.8	1.4	1.1	13.8	6.9	7.4	3.7
1	3.5	1.8	1.4	16	8	9.6	4.8
1½	5	2.5	2.0	20	10	13.2	6.6
2	6.5	3.3	2.6	24	12	17	8.5
3	9	4.5	4	34	17	25	12.5
5	15	7.5	6	56	28	40	20
7½	22	11	9	80	40	58	29
10	27	14	11	100	50	76	38
15	40	20	16	.....	....	112	56
20	52	26	21	.....	....	148	74
25	64	32	26	.....	....	184	92
30	78	39	31	.....	....	220	110
40	104	52	41	.....	....	292	146
50	125	63	50	.....	....	360	180
60	150	75	60	.....	....	430	215
75	185	93	74	.....	....	536	268
100	246	123	98	.....	....	.....	355
125	310	155	124	.....	....	.....	443
150	360	180	144				
200	480	240	192				

These values of full-load currents are average for all speeds and frequencies of continuous-duty motors.

* From T. Baumeister and L. S. Marks, *Mechanical Engineers' Handbook*, McGraw-Hill Book Company, New York, 6th ed., Table 22, p. 15-100. Copyright © 1958. Used by permission.

† For full-load currents of 208- and 200-volt motors, increase the corresponding 220-volt full-load currents by 6 and 10 per cent, respectively.

Table 4-18. Capacities of Percolation-type Filters[2]

Diam, ft	Depth, ft	Charging capacity, tons of clay	Rate of flow, gpm			Avg throughput, B/D (averaged over the year)		
			Cylinder stocks	Paraffin oils	Neutral oils	Cylinder stocks	Paraffin oils	Neutral oils
6	15	7	0.50	0.58	0.66	10	14	13
8	20	17	0.88	1.03	1.18	19	26	25
8	25	21	0.88	1.03	1.18	20	27	26
10	20	25	1.38	1.61	1.84	27	38	36
10	25	32	1.38	1.61	1.84	29	41	38
10	30	38	1.38	1.61	1.84	31	43	40

to bring the motor to rated speed with the pump discharge valve open than when it is closed.

FILTERS

Percolation and Plate

Percolation through beds of adsorptive material has long been practiced by the oil industry to improve the color of lubricating oils.   Other methods are gradually replacing the system.   The equipment is quite simple.   The filters are upright cylindrical

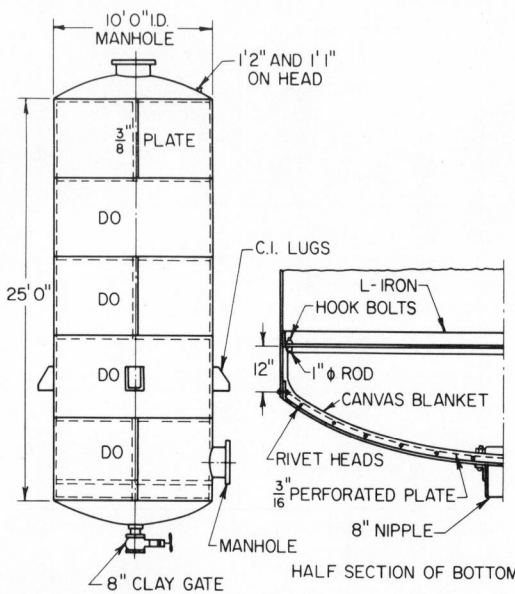

FIG. 4-85. Typical dimensions and bottom half section of a 32-ton filter.

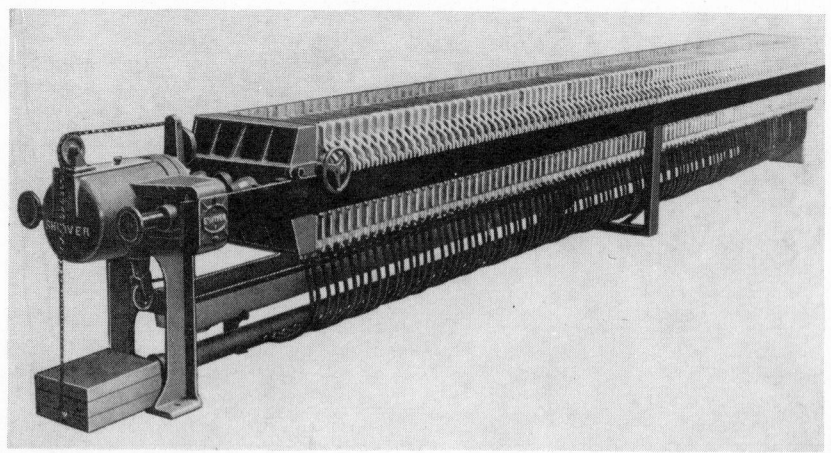

FIG. 4-86. Plate-and-frame-type filter.   (*Courtesy of T. Shriver & Co.*)

shells (Fig. 4-85). A canvas blanket is placed over a perforated drainage plate to retain the filter medium, fuller's earth or bauxite. Fifteen feet of depth is considered essential. Above 30 ft in depth, channeling may occur. Table 4-18 presents performance figures for the usual sizes of percolating filters.

Plate-and-frame filters have been used for the separation of paraffin wax from distillates. A plate-and-frame filter press is illustrated by Fig. 4-86, and the flow through the plates by Fig. 4-87.

Contact Filtration

The advent of "contact" filtration and solvent-wax extraction has made the above two types of equipment obsolete, particularly for new refineries.

The contact system of decolorization consists, briefly, of agitating fine clay or other adsorptive medium together with the oil, mixing them thoroughly, and then bringing them to the desired temperature. The mixture is then passed to filters of cell or rotary type.

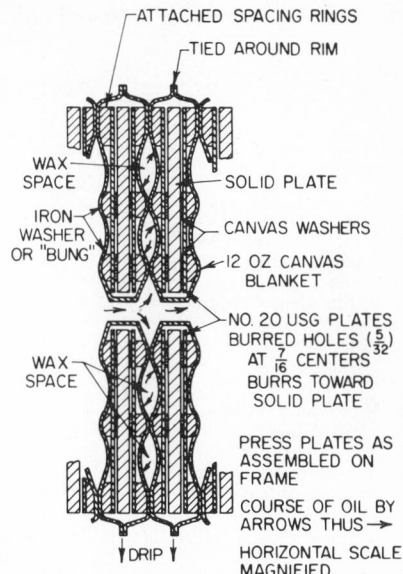

FIG. 4-87. Cross-section diagram to show course of oil through filter-press blankets.

The Sweetland press, extensively used for contact filtration, shown in Fig. 4-88, is of the cell type. The leaves or cells consist of circular leaves of heavy metal screen covered by a canvas-filter-cloth bag. The oil and clay slurry is pumped through the shell at 40 to 50 psi; the oil passes through the filter cloth and the clay is retained. The clarified oil passes to sight feed glasses, one to each filter leaf.

FIG. 4-88. Sweetland automatic filter for contact filtration in oil processing. (*Courtesy of Dorr-Oliver, Inc.*)

Table 4-19. Capacity of Sweetland Filter Presses*

Filter numbers	1	2A	2	3	5	7	10	12
Inside diam, in	10	16	16	25	25	25	31	37
Inside length, in	20½	19	36½	37	61	82	109	145
2-in. filter leaf spacing:								
Number of leaves	9	9	18	18	30	41	54	72
Filter area, sq ft:								
Smooth rim	8	23	46	111	185	252	523	1,004
Grooved rim	7.5	21	42	105	175	239	500	965
Max cake capacity, cu ft, with ½-in. cake	0.3	1	2	4.8	8	11	22	42
3-in. filter leaf spacing:								
Number of leaves	7	.....	12	12	20	27	36	48
Filter area, sq ft:								
Smooth rim	6	.....	31	74	123	166	349	670
Grooved rim	5.7	.....	28	70	117	158	333	642
Max cake capacity, cu ft, with 1-in. cake	0.5	.....	2.5	6.4	10	14	28	56
4-in. filter leaf spacing:								
Number of leaves	5	.....	9	9	15	20	27	36
Filter area, sq ft:								
Smooth rim	4.5	.....	23	55	92	123	262	502
Grooved rim	4	.....	21	53	88	117	250	483
Max cake capacity, cu ft, with 1½-in. cake	0.6	.....	3	7.2	12	15	33	63
Total weight of filter filled with water, lb	550	1,500	2,300	5,500	7,300	9,350	16,500	29,600
Shipping weight, lb:								
Domestic	500	1,300	2,100	4,800	6,200	7,900	13,500	24,000
Export	600	1,400	2,200	5,400	7,600	9,300	16,500	29,000

* Courtesy of Dorr-Oliver, Inc.

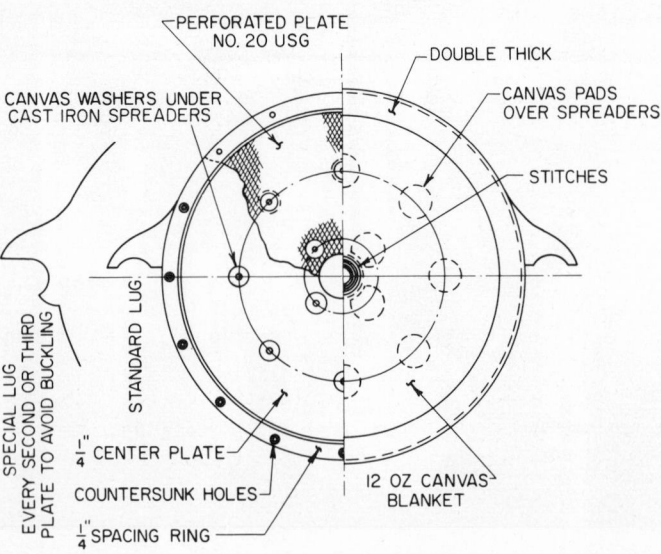

Fig. 4-89. Cross section of assembled filter-press plate with spacing rings attached.

The filter may be backwashed with naphtha or other suitable solvent   The bottom half of the shell is hinged and may then be dropped for discharge of the retained spent clay.   The capacities of Sweetland presses manufactured by Dorr-Oliver are given in Table 4-19.   In contact filtration, removal of the last trace of clay may be desired and a frame-and-plate press is then used through which the filtrate from the primary filters is passed.   The plates may be covered with canvas or closely woven metallic cloth. Blotting paper is often used.   Figure 4-89 illustrates a typical filter-press plate.

Rotary Filtration

Rotary filters are used extensively in solvent-dewaxing plants.   Basically the rotary filter consists of a drum revolving partially submerged in a tank of the unfiltered oil/wax mixture.   The drum is covered with the filtering medium of woven metal

FIG. 4-90. Rotary-drum pressure solvent-oil dewaxing filters for propane dewaxing.   (*Courtesy of Goslin-Birmingham Mfg. Co., Inc.*)

cloth which in turn may be covered with a canvas blanket.   A vacuum is applied within the drum.   As the drum revolves, it picks up a layer of solid materials from the charge mixture, to be later scraped from the drum.   Figure 4-90 illustrates a rotary-drum filter used in propane dewaxing.   Filtration rate is governed by the speed of drum rotation.   Speeds are adjustable from 1 to 8 rpm, as required for the character of the feedstock.

A modification of the rotary filter is the Oliver Precoat Filter.   The construction is similar to a conventional rotary but allows for precoating the drum with a layer of filter aid, usually diatomaceous earth.   The coat is built up by operating the filter a sufficient length of time with a slurry of the filter aid in the underneath tank.   The customary thickness of the precoat is 3 in.   The knife blade is adjusted to take the wax from the revolving drum plus a very thin cut of the precoat (Fig. 4-91).   The precoat is shoved away when the drum is again coated for the next cycle.

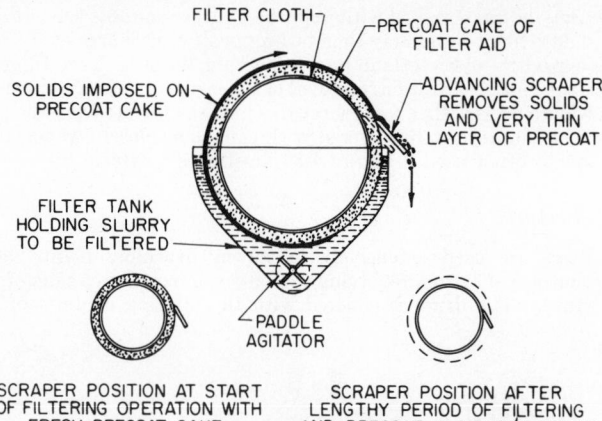

FILTER CLOTH ¬

┌ PRECOAT CAKE OF
   FILTER AID

SOLIDS IMPOSED ON ¬
PRECOAT CAKE

┌ ADVANCING SCRAPER
  REMOVES SOLIDS
  AND VERY THIN
  LAYER OF PRECOAT

FILTER TANK
HOLDING SLURRY
TO BE FILTERED

└ PADDLE
  AGITATOR

SCRAPER POSITION AT START
OF FILTERING OPERATION WITH
FRESH PRECOAT CAKE

SCRAPER POSITION AFTER
LENGTHY PERIOD OF FILTERING
AND PRECOAT IS NEARLY SPENT

Fig. 4-91. Operating principle of a continuous precoat-type filter. (*Courtesy of Dorr-Oliver, Inc.*)

REFRIGERATION

Refrigeration has several uses in an oil refinery. With the exception of self-refrigeration, as in propane dewaxing, two systems are in use. They are compression and absorption.

Mechanical refrigeration is expressed in tons. A ton of refrigeration is based upon the melting effect of 2,000 lb of ice in 24 hr. Since the latent heat of fusion of ice is 144 Btu/lb, the unit of measurement is 288,000 Btu/24 hr, or 12,000 Btu/hr, or 200 Btu/min. The amount of refrigeration necessary to chill a substance from one temperature to another in 1 day is expressed by the following formula:

$$T = \frac{WS(t_1 - t_2)}{288,000}$$

in which T = tons of refrigeration
W = weight of substance chilled in 24 hr
S = specific heat of substance
t_1 = initial temperature, °F
t_2 = final temperature, °F

If there is any change of state, the value of the latent heat must be added. For example, the latent heat of fusion of the paraffin-wax content in a distillate is approximately 45 Btu/lb.

In some cases calcium chloride brine is used to transfer refrigeration to the substance being chilled. The "gallon degree" is then useful in checking the amount of refrigeration being transferred. This unit is the heat required to raise the temperature of 1 gal of brine by 1°F/min. This quantity may be expressed by

$$\text{Btu} = 8.35gs$$

in which g = specific gravity of the brine
s = specific heat of the brine

The gallon degrees per ton is

$$G_t = \frac{34,491}{gs}$$

Knowing the quantity of brine circulated per minute and the drop in temperature through the brine cooler, the tonnage rating of the equipment at that moment is given

by the formula
$$T = 0.042gsV(t_1 - t_2)$$
in which T = rating, tons
g = specific gravity of brine
s = specific heat of brine
V = gpm of brine circulated
t_1 = initial brine temperature, °F
t_2 = final brine temperature, °F

Compression Systems

The compression system is generally favored. The operation of this system is based upon the use of a volatile substance, usually ammonia, which exists only as a gas at atmospheric pressures and normal temperatures. When compressed to a sufficiently high pressure and cooled, the gas is reduced to a liquid. The reevaporation of

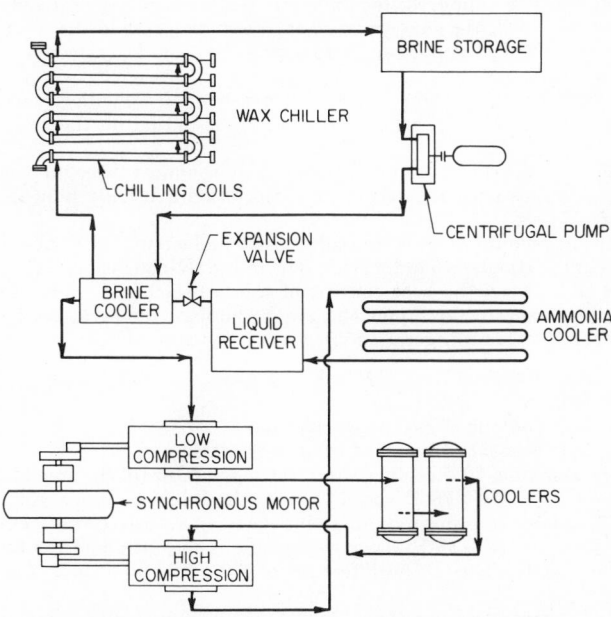

FIG. 4-92. Schematic diagram of a typical compression-refrigeration system.

the liquid when the pressure is reduced absorbs heat and constitutes the refrigerating effect.

The compression-refrigeration system is a closed cycle consisting fundamentally of

The compressor, which draws the vapor from the evaporating coils and compresses it, delivering it to

The condenser, in which the latent heat and the heat from compression are removed, usually by cooling water. The liquefied material then passes through a needle valve and is expanded in

The evaporating coils, where the liquid is again vaporized, absorbing the heat from the material to be cooled and producing the refrigeration.

Auxiliary equipment includes receivers, intercoolers, and traps.

Figure 4-92 is diagrammatic of a compression machine in which brine is cooled and circulated as the refrigerating medium. The brine circuit is often eliminated, especially for low-temperature work, and the liquid ammonia is expanded directly into the chilling equipment. The properties of ammonia may be obtained from standard tables.

The pressure on the condenser, together with friction loss, must be supplied by the compressor. This figure is usually termed head pressure. It will depend entirely upon the temperature of the water available for condensing purposes. Referring to the standard tables and assuming the ammonia is cooled to within 8° of the water temperature, it is apparent that with water at 60°F and ammonia at 68°F, the condensing gage pressure will be 109.6 lb. When the water is 80°F and the ammonia 88°F, the pressure will increase to 160.1 lb. The figures will be increased slightly at the compressor to allow for friction losses.

Ammonia exists as a liquid at a normal atmospheric temperature of 60°F only under about 93 lb pressure. To reach low temperatures in the evaporator the pressures must be reduced. The desired temperature of the material being cooled is the criterion that establishes the pressure in the evaporating coils or chamber of the chilling machine. The ammonia must be 8 to 10°F below the material being chilled for efficient interchange of heat. If 0°F is desired, the ammonia temperature should be about −8°F. With ammonia at −8°F, the gage pressure on the suction side of the compressor will be 10.3 lb. For −20°F, the pressure is zero psig; for −30°F a vacuum of 7.4 in. of mercury exists.

As the pressure decreases, there is a large increase in the volume of vapors as they return to the compressor. Reference to the tables shows that for these examples, the relation is

At a −30°F chill, the compressor will have to handle almost twice the volume of gas as with 0°F. For this reason, single-stage compression is usually limited to suction pressures above 5 psig.

To establish the amount of ammonia compressed and circulated by the compressor, refer to tables in the standard handbooks. The refrigerating effect of 1 lb of ammonia is the difference between the heat content of the liquefied gas before and after its expansion. To evaporate the liquid and reach the final temperature desired, heat must be absorbed. This refrigerating effect may be expressed as

$$R = h - H$$

in which R = Btu/lb
     h = heat content of gas after expansion
     H = heat content of liquid at condenser pressure

For example, assuming 70°F cooling water, the temperature of the liquefied ammonia would be about 78°F. At 78°F and 133 psig, the liquid contains 129.7 Btu. If −20°F chill temperature is desired, an ammonia temperature in the cooler of about −28°F is necessary. At −28°F, the gage pressure is zero psig and the heat content in 1 lb of the gas is 602.1 Btu. The difference, or 472.4 Btu, is the heat absorbed from the oil.

Since 1 ton of refrigeration per day is equivalent to 200 Btu/min, it follows that the amount of ammonia circulated is expressed by

$$Q = \frac{200}{R}$$

in which Q = lb ammonia
     R = refrigerating effect for 1 lb

Multiplying Q by the specific volume, V at intake pressure will give the volume of gas to be handled.

The conventional drive for the compression machine is the electric motor. In large units, synchronous motors are recommended. Built in as part of the compressor, they eliminate the flywheel, save space, and improve the plant's overall power factor. Figure 4-93 illustrates the horsepower per ton required for compression refrigeration for different head and suction pressures.

Ammonia condensers may be of conventional atmospheric double-pipe or shell-and-tube type. The heat to be removed by the cooling water is theoretically equivalent to the heat of vaporization of the ammonia at the temperature of the condenser. The amount of water depends on the inlet and outlet temperatures. For example, if the surface of the condenser is so proportioned that water entering at 60°F will leave at 75°F, then each pound of water will remove 15 Btu. If the condenser pressure is 133 psig, corresponding to a temperature of 78°F, the latent heat per pound of ammonia is 500.7 Btu. The water required amounts to 500.7/15, or 33 lb of water per lb of ammonia circulated. The amount of ammonia per ton of refrigeration is determined

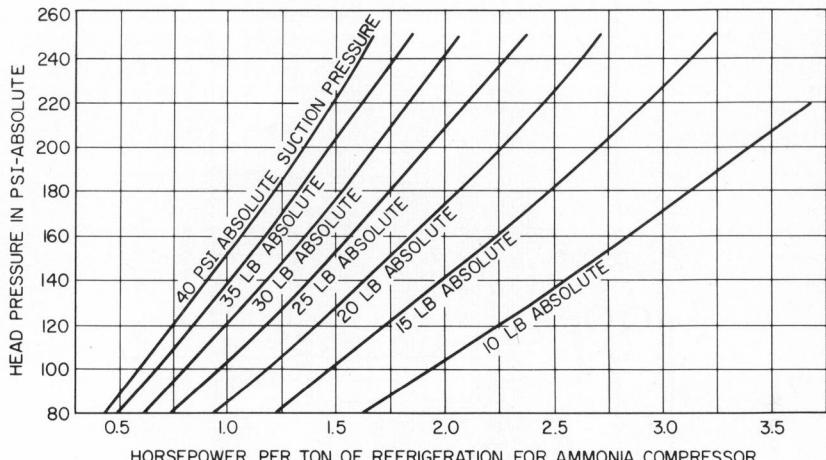

FIG. 4-93. Chart of brake horsepower for ammonia compression. Theoretical 1 hp 10 per cent; no liquid cooling; no superheat at beginning of compression.[2]

by previous formula $Q = 200/R$, from which the amount of water required for the plant may be calculated.

Absorption Systems

When exhaust steam is available and when chill temperatures not lower than 0°F will suffice, the absorption system of refrigeration has advantages. Figure 4-94 illustrates a large installation. The system relies upon the fact that ammonia is readily absorbed in water.

Starting with the generator, the concentrated solution of ammonia in water, termed strong liquor, is heated by steam coils. The vapors evolved are ammonia gas and some steam due to the existing partial pressure. This "wet gas" enters the rectifier where the temperature is reduced sufficiently to condense out some of the entrained water vapor. The so-called "dry gas" then enters the condenser and is liquefied. The head pressure on the generator serves the purpose of the pressure maintained by the compressor in the compression system. The liquid ammonia, under condenser pressure (this pressure is a function of the temperature desired), is then expanded through the cooling coils. The gas then passes to the absorber, where it is absorbed into the weak liquor drawn from the generator. Since the concentration of ammonia in water depends on the temperature, the absorber must operate at low temperatures. Since the generator is at a high temperature, the hot weak liquor must first be cooled. This cooling is accomplished by the weak-liquor cooler. The cold strong liquor leaving the absorber is collected in a receiver and pumped to the generator. To preheat it and to assist in cooling the weak liquor, a heat exchanger is used. Sometimes the strong-liquor feed is sprayed over an analyzer mounted over the generator, consisting of a plate or tray tower. The hot gases ascend and exchange heat, and dephlegmation of some of the water vapors ensues, thus conserving steam and relieving some of the load on the rectifier.

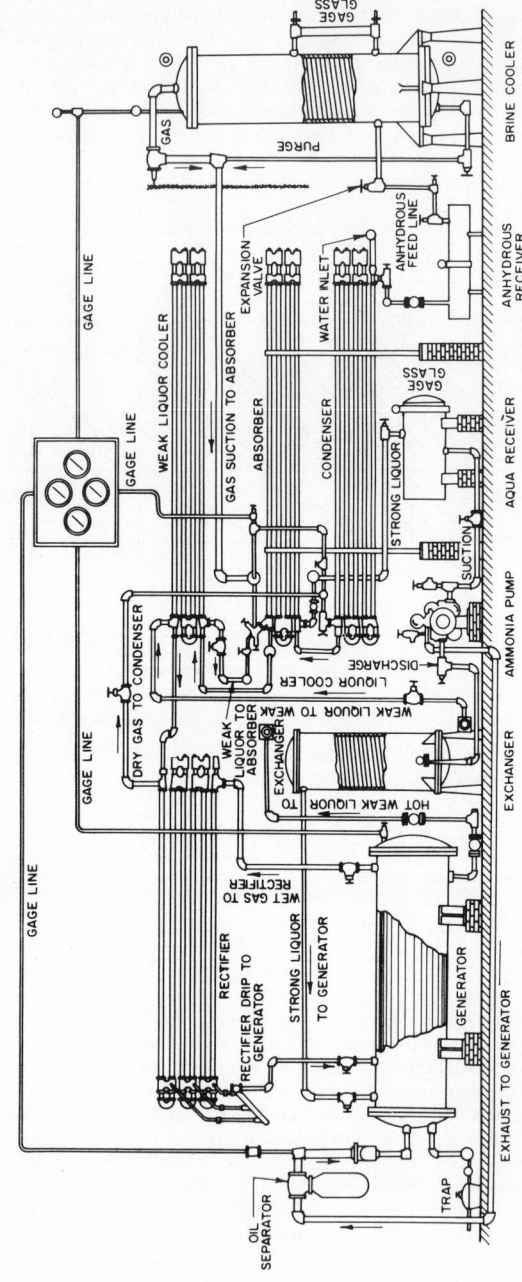

FIG. 4-94. Absorption-refrigeration machine with double-pipe rectifier, absorber, and condenser. (*Courtesy of Worthington Corp.*)

4-74

In Fig. 4-94, the diagram shows a double-pipe absorber. Some prefer a shell type wherein internal water-cooling coils are placed. The brine cooler may be eliminated if direct expansion of ammonia into the chillers is advisable. The rectifier, condenser, and coolers and exchangers are all pieces of equipment designed to abstract heat from liquids or gases. Atmospheric and double-pipe coils and shell-and-tube units are in use, the choice being governed by the conditions outlined for similar equipment for the compression system.

Comparisons

The steam consumption of the absorption-refrigerating plant depends on the head pressure to be developed, which in turn depends on the temperature of the condensing water. Figure 4-95 is indicative.

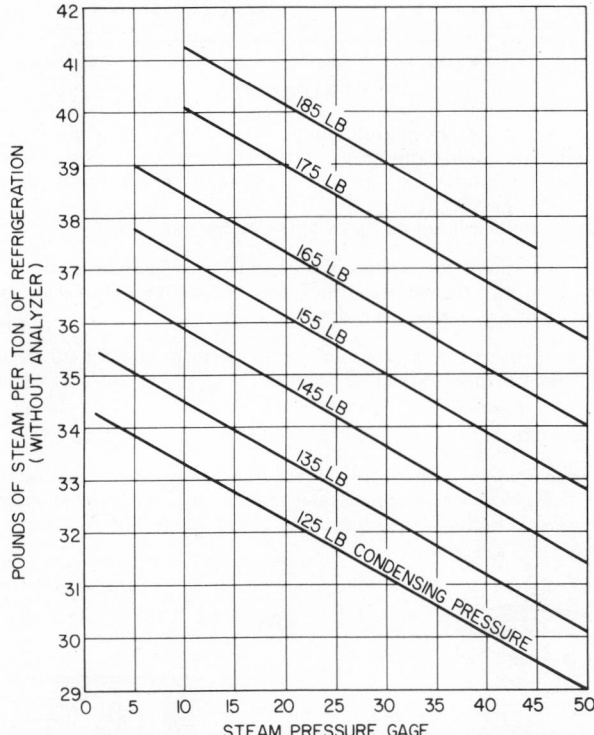

FIG. 4-95. Pounds of steam per ton of refrigeration produced by absorption-refrigerating machines.[2]

The cooling-water requirements are slightly greater than those for the compression machine. Table 4-20 gives the water consumption per ton for various temperatures of the water to attain a chilling temperature of 0°F. The values are typical but influenced somewhat by the condition of the cooling surfaces.

In comparing the two types of systems, the compression system is easy to take out and return to service (1 or 2 hr). Control, by compressor clearance pockets, is limited to about 50 per cent of full load. It has lower water consumption, and is adapted to chill temperatures below 0°F.

Absorption systems are uneconomical unless exhaust steam (15 to 25 psig) is available. Other features are long periods of operation without shutdown, only one moving part (the ammonia pump), higher water consumption, greater flexibility, and the fact that may be operated at any load factor up to full capacity.

**Table 4-20. Water Consumption of Absorption-refrigeration Machine
When Cooling Brine to 0°F**

Temp of cooling water, °F	Water/ton of refrigeration, gpm
60	1.75
65	2.20
70	2.85
75	3.60
80	4.58
85	5.90
90	7.50

INSTRUMENTS

Process instruments for the petroleum process industry are primarily concerned with measurements of liquids and gases. It is essential to measure those conditions which will provide the information needed to make decisions: temperature, pressure, flow, level, viscosity, gravity, etc.

Each measuring device is made up of three or four sections:

1. The primary element for sensing the condition to be measured
2. The conversion and transmission means to convey the representative signal
3. Suitable indicating or recording devices

A final step is to use these measurements to perform control functions or return operating conditions to preestablished values.

Primary Elements

The primary element for temperature measurement is a heat-sensing device which will, via mechanical, electrical, or pneumatic means, produce a signal analogous to the temperature sensed. Perhaps the oldest and simplest method of temperature meas-

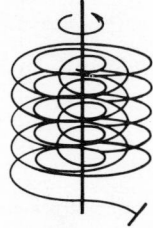

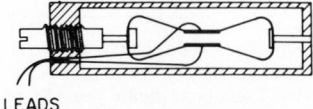

LEADS

Fig. 4-96. Bimetallic-type element for temperature measurement.

Fig. 4-97. Cross section of typical bimetallic-element temperature-measuring device.

urement is the movement produced when two metals of different thermal-expansion coefficients are bonded. As the bimetallic element is heated, it will bend, thereby producing a change in a switch, shaft, or other device for transmitting the signal (Fig. 4-96).

Figure 4-97 illustrates a contact device utilizing low-expansion struts which carry contacts in a high-expansion tube. A screw regulates temperature at which contacts break. Ranges of instrument sensing elements of this type fall within the limits of −40 to 1400°F with accuracies of ±0.5°F.

Another design utilizes the volumetric expansion characteristic of liquids and gases. These meet standards established by ISA and are designated:

Class I (Fig. 4-98): Liquid-filled thermal thermometer, bulb filled with organic liquid and connected by capillary to pressure spiral or helix. (Class V uses mercury instead.) The distortion of a mechanical linkage due to bulb expansion or contraction causes a change in scale indication. It is usually necessary to compensate for ambient effects with a second thermal system (Fig. 4-99). Ranges available are from −200 to 1000°F with accuracies of ±½ to 1 per cent.

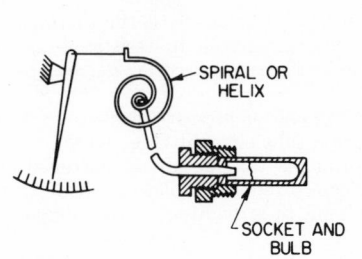

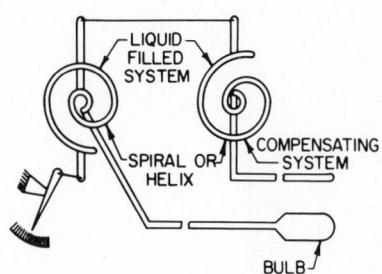

FIG. 4-98. Operation of liquid-filled thermal thermometer.

FIG. 4-99. Liquid-filled thermal-thermometer system with second thermal system to compensate for ambient-temperature effects.

Class II (Fig. 4-100): Vapor-pressure thermal thermometer, volatile liquid in bulb produces vapor pressure to effect signal. Interface must be in bulb with liquid in capillary. To assure interface's remaining in bulb, a nonvolatile liquid is used for transmission (Fig. 4-101). Ranges may go from −300 to 600°F.

Class III: Gas-filled thermal thermometer in which the gas fill, under pressure, provides gas distortion of a pressure element. Ranges are −460 to 1000°F with accuracies of ±½ to 1 per cent.

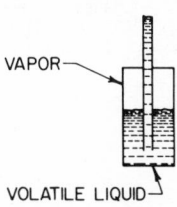

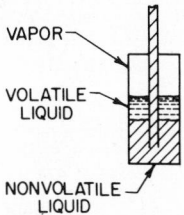

FIG. 4-100. Cross section of vapor-pressure thermal thermometer.

FIG. 4-101. Cross section of vapor-pressure thermal thermometer with nonvolatile liquid to maintain interface in bulb.

A well-accepted element, because of its simplicity and reliability, is the thermoelectric pyrometer or thermocouple. The current flow produced by two wires of dissimilar metal joined at the end with junctions maintained at different temperatures is a result of absorption of heat at the hot junction. The voltage produced is proportional to the difference of hot- and cold-junction temperatures and is measured by a millivoltmeter or potentiometer. Metal combinations for wires include

Platinum and platinum–10 % rhodium or platinum–13 % rhodium.........................	To 2700°F, accuracy of ±0.25 %
Copper and constantan.......................	−200 to 700°F, ±0.3 to 0.8 % per span
Iron and constantan..........................	−200 to 1400°F, accuracy as above
Chromel and alumel..........................	−200 to 2300°F, accuracy as above

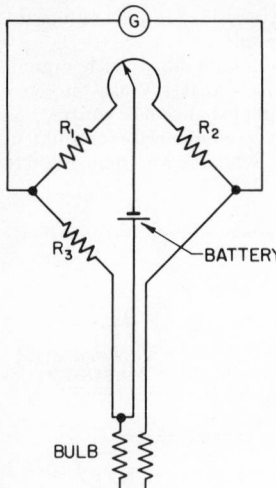

FIG. 4-102. Wheatstone-bridge measurement with resistance thermometer.

Another type of temperature measurement, used where quick response and high precision are desired, is the resistance thermometer. This utilizes the principle that the change in resistance of an element varies with change in temperature. It usually makes the measurement with a Wheatstone-bridge circuit. Figure 4-102 illustrates one in which a coil of fine wire such as platinum, nickel, or copper is wound on an insulated frame. Ranges will go from -325 to $1200°F$, with accuracies of $\pm$ 0.25 to 1.0 per cent.

A recent addition to this family is the thermistor in which semiconductors increase in resistance as temperature increases. These are used for narrow spans of measurement.

Conventional devices for pressure measurement consist of the Bourdon-tube element (Fig. 4-103). It operates on the principle that a tube of flattened cross section, in the form of an incomplete circle or a spiral or a helix, will tend to straighten as the pressure increases (A).

Variations include the multiple-diaphragm unit (B), in which the pressure acts on the metallic bellows but is balanced against the calibrated spring. C illustrates a force-balance gage where line pressure acts against the underside of the diaphragm and is balanced by pressure from the air supply imposed on top. The air condition duplicates the underside condition (to achieve balance) by the combination of a restriction in the supply line, and the position of the bleed valve.

A relatively recent method for pressure measurement which has proved effective is the strain-gage load cell (D). Small resistance elements are bonded to a unit which

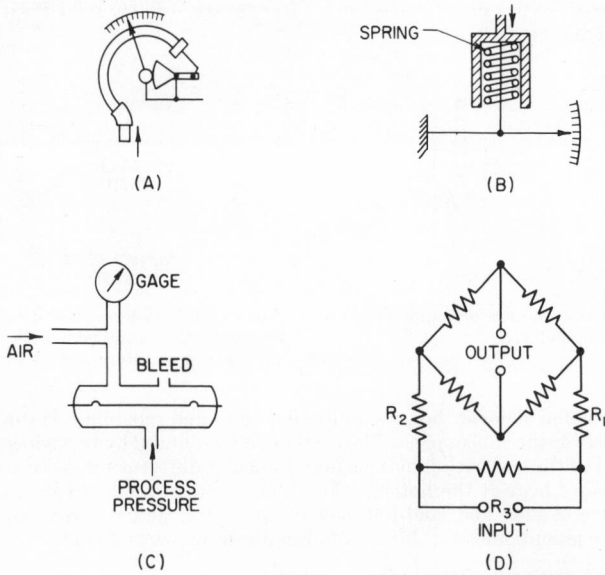

FIG. 4-103. Conventional pressure-sensing elements. (A) Flattened Bourdon-tube element tries to open out when pressure is exerted internally. (B) Multiple-diaphragm variation. (C) Force-balance gage where line pressure acts against underside of diaphragm, balanced by air pressure imposed on top. (D) Strain-gage load cell.

expands under internal pressure. An increase of pressure is recognized as a physical stretching action which increases the resistance of the elements. The measure of changes in resistance provides accurate pressure values.

Complete details concerning recommended practices on temperature and pressure measurement will be found in the ISA and API Standards.[23]

Inferential Elements

The flow measurement of liquids and gases can be handled in a number of ways, depending on accuracy desired as opposed to cost of equipment and variables to be controlled or accounted for. Perhaps the oldest and most commonly used method for process-measurement work is the orifice plate. The orifice meter is classed with a

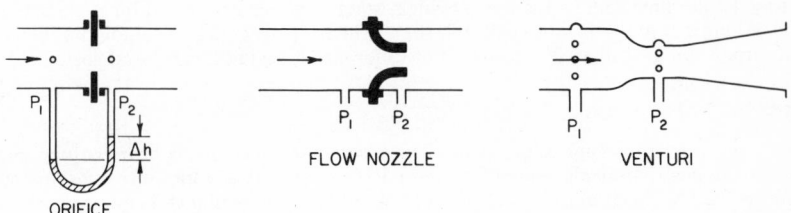

FIG. 4-104. Flow measurement by controlled pressure drop.

group of meters designated head type, inferential. They generally take three forms (Fig. 4-104):

1. Thin-plate orifice
2. Flow nozzle
3 Venturi tube

These primary elements, when placed in a flowing stream, cause a definite flow configuration to take place. The resulting pressure and velocity from this shaping of the flow can be used as an indirect measurement, from which the flow can be inferred by use of known physical laws. In each of the three forms, the static pressure is taken off at two points, between which the cross section and the linear velocity of the flowing stream change.

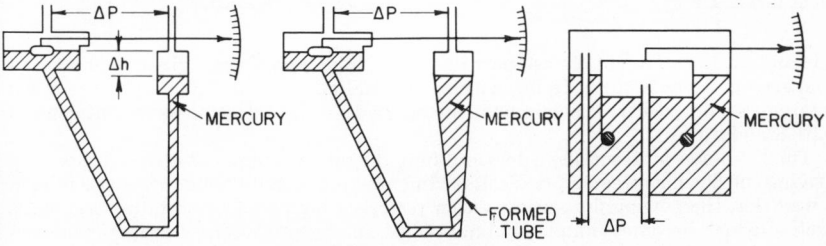

FIG. 4-105. Types of differential-pressure gages to read liquid-flow rates.

The secondary element includes a differential-pressure gage, from which the difference in pressure between the two connections can be, read or converted for transmission purposes. It can be sensed as a float position and transmitted as a square-root function or as a linear function (a range tube chamber in which pressure fluctuation measured is formed to a variable parabolic cross section to extract the square root) or as an opposed-bellows-type movement to position a torque tube. Figure 4-105 illustrates these various secondary devices.

The inferential meter is influenced by installation conditions. The most important consideration is to provide a uniform flow pattern free of swirls or other flow distortions. This is usually accomplished by straight pipe of recommended lengths upstream and downstream of the metering section. There should be no appreciable pressure pulsations. Foreign solids in the stream can deposit in front of the meter and change the stream cross section or may deposit on the inner bore of a Venturi or nozzle, or on the leading edge of the orifice plate. Changes in the roughness of the pipe wall upstream of the meter can affect performance. Elbows, valves, or fittings upstream, and in some cases downstream, are objectionable because they distort the flow pattern and make the meter readings subject to a variable error.

A complete summary of inferential meters can be found in the publication of the Fluid Research Committee of the ASME.[24] Such meters, in addition to those above, include eccentric and segmented orifices, the ASME nozzle, Venturi tube, and flow tubes. Each offers some advantages which, when a meter is selected for a particular application, will make it more suitable than others. Advantages include differences in accuracy, rangeability, life, recovery of differential, laying length, and cost.

Inventory Metering

Another important type of meter used extensively for accounting or inventory purposes is the positive-displacement meter, or PD meter. It is a high-precision instrument in which an accuracy of ± 0.1 per cent over a flow range of 5 to 1 can be expected (viscosity, temperature, and gravity conditions taken into consideration), provided

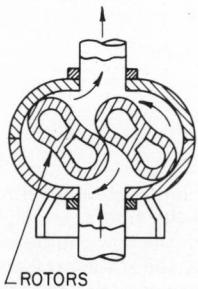

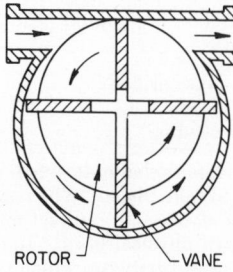

FIG. 4-106. Cross section showing interlocking rotating segments of a positive-displacement flowmeter.

FIG. 4-107. Cross section showing rotating eccentric cylinder of one type of positive-displacement flowmeter.

adequate calibration checks are part of the routine operations. Sizes for meters of this type go to as large as 16 in., with corresponding rates to 15,000 bph. Detailed information on positive-displacement meters and associated facilities is contained in API Standard 1101.[25]

The PD meter is essentially a device akin to the rotary pump, but driven rather than driving the flowing stream. It is compartmented into equal-volume segments in such a way that the flowing fluid drives either reciprocating pistons or rotating segments, each of which becomes completely liquid-filled and then is dumped in its progression from start to finish position. In essence it is much like using a dipper to transfer liquid from one tank to another. The count of each compartment filling and emptying is done through a rotary output device driving a counter. Figures 4-106 and 4-107 illustrate two types.

Relatively new developments in the field of metering have lead to a departure from larger and larger PD and inferential meters. Such developments include the turbine or velocity meter.

The term "velocity meter" is defined as a metering device in which the reaction of the primary element to the fluid flow is nominally proportional to the velocity of the stream. A meter of this type, the turbine meter, utilizes a turbine-type rotor in an axial-flow position and is driven by the fluid stream (Fig. 4-108).

Mounted within the body of the meter rotor is a small permanent magnet polarized at right angles to the axis of rotation.   As the rotor spins, the magnet sets up an alternating current in a coil of wire held within its field.   The total number of pulses generated by a sensing element is directly proportional to the total quantity of liquid passed through the unit.   The frequency of the alternating current produced is directly proportional to the rate of flow of the liquid through the sensing element.

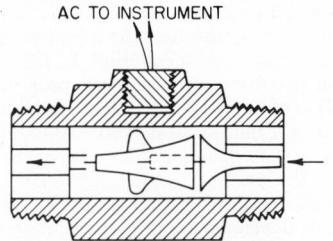

FIG. 4-108. A typical velocity-meter type of flowmetering device.

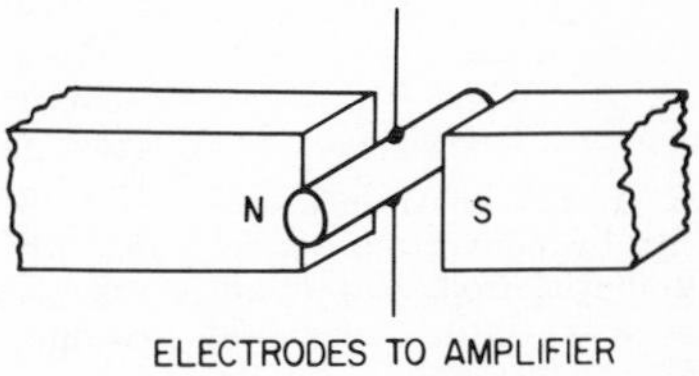

FIG. 4-109. Electromagnetic meter for flow measurement.

Another recent development is the electromagnetic meter (Fig. 4-109) where a voltage is developed across a conductor when the conductor is moved across a magnetic field.   The voltage is directly proportional to the velocity of the conductor.   In the electromagnetic meter, the fluid stream is the conductor and the stream is passed through an adequate field created by two poles of an electromagnet.   It offers no obstruction to flow, but the fluid must be a suitable conductor.

Level Measurement

Closely allied with measurement of flow for inventory and accounting purposes is that of level.   This is handled in a number of ways, depending on accuracy desired as opposed to cost of equipment and variables to be controlled or accounted for.   Perhaps the one most commonly used method for measuring liquid level is the float instrument, though it is gradually giving way to more accurate and trouble-free devices.

The simplest form of level indication is the conventional gage glass mounted in a vertical position, having its top and bottom connected to the vessel well above and below the points of level desired.   Special designs provide for high-pressure, low- and high-temperature conditions, armoring, etc.   Float instruments consist of a hollow ball or float riding the liquid level and mechanically connected to a linkage to transmit its position (Fig. 4-110).   Special provisions are made to overcome wave action, temperature effects, linkage friction, etc.   Properly weighted floats can be used to indicate interface levels between immiscible liquids.

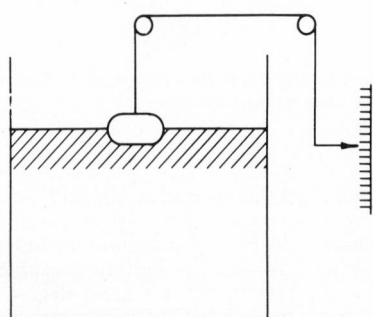

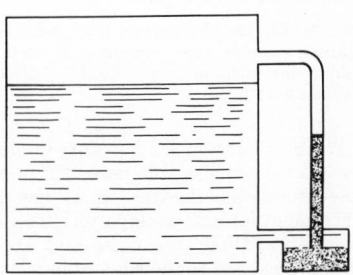

FIG. 4-110. Measurement of liquid level with a float gage.

FIG. 4-111. Hydrostatic gage with differential arrangement for liquid-level measurement.

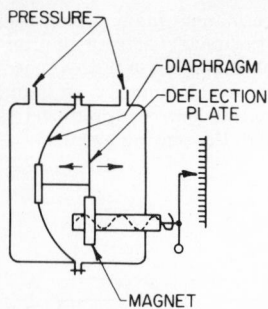

FIG. 4-112. Level measurement by converting sensed differential from diaphragm movement into a magnetic action to rotate the indicator device.

Hydrostatic gages (Fig. 4-111) utilizing a differential arrangement to compensate for pressure over the liquid are suitable for small- and medium-sized tanks and vessels. Many variations of this are available. Figure 4-112 indicates a method for converting the sensed differential from diaphragm movement into a magnetic action to rotate the indicator device. Another simple device is the air-bubbler system, wherein the air pressure necessary to bubble the air through the liquid provides a direct analogous measure of level (Fig. 4-113).

A recently developed method coming into prominence utilizes the principle of capacitance. A shielded probe (two capacitors) or a Teflon-coated probe immersed in the liquid measures the capacitance developed by the wetted portion of the probe, the dielectric remaining essentially constant. Applying this to a Wheatstone bridge, the circuit develops a signal proportional to level.

A still more recent development is the application of gamma-ray absorption. A source of gamma rays in the vessel gives off radiations detected by a Geiger-counter tube external to the tank. Geiger output decreases as the level rises and can be calibrated in terms of level (Fig. 4-114).

Quality Measurement

All the foregoing elements of measurement have been quantitative. For years, these have provided the basic information necessary to make decisions. However, as processes have become more involved, conditions more critical and end products more demanding of closer tolerances, considerable attention has been given to developing reliable and accurate instruments to measure the quality of materials: viscosity, specific gravity, pH, oxygen content, thermal conductivity, composition.

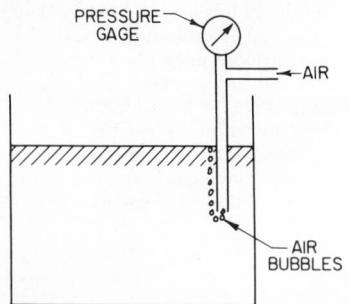

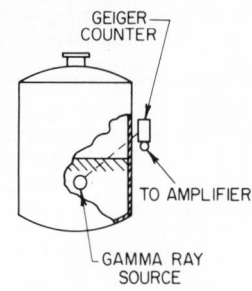

FIG. 4-113. Level measurement with air-bubbler system. Air pressure needed to bubble air through the liquid is directly analogous to level.

FIG. 4-114. Level measurement by gamma-ray absorption.

While laboratory instruments to measure these variables are well known and widely used, it is impractical to use them in present-day process operations without considerable production delays. It is also a difficult feat to modify a laboratory-proved instrument for day-to-day process application where the needs are continuous, subject to field handling and abuse and absence of interpretation. Yet a great degree of success has been achieved, and much more will result in applying qualitative measurement to continuous processing work because the need is constantly increasing.

The measurement devices for specific properties such as gravity, viscosity, color,

and dielectric are older and well established.   Viscosity instruments must generally
provide information on the basis of kinematic viscosity, usually in terms of centipoises
(cp).   In addition, the values must be corrected to standard-temperature conditions.
In one design, a bypass flow provides a fixed, restricted flow path through which the
liquid is pumped at a constant rate.   The pressure drop across this length gives a
viscosity reading.   By immersing the flow path in a bath maintained within close
temperature limits, the readings are obtained at standard conditions.

Another arrangement (Fig. 4-115) causes a bypass flow through a rotameter to
position a viscosity-sensitive bob or float
after manual adjustment has brought the
nonsensitive float to a predetermined bypass
flow rate.

A more recent design measures the damping effect of the measured liquid on an oscil-
lating reed.   The rate of damping of blade
vibration, fed through a computer, gives a
continuous readout of viscosity.

Specific-gravity measurement has long
been a well-accepted requirement in processing.   It is now commonplace to effect
accuracies of at least 0.1, and generally
0.05 API.   The gravity-sensing devices
operate by weighing small samples of liquid
continuously.

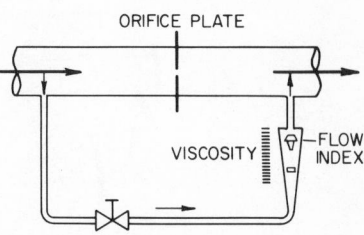

Fig. 4-115. Viscosity measurement with
rotameter in bypass, liquid flow positioning the bob or float to give a viscosity
reading.

One type of gravitometer weighs a small known volume of the flowing stream that is
bypassed through a pressuretight container (Fig. 4-116).   The container is supported
on a knife-edge balance with a recording pen linked directly to the balance arm.   A
bimetallic coil in contact with the fluid acts on the balance arm to provide temperature
compensation.   Another approach uses a submerged float in a small chamber through
which the liquid flows.   Float position, partly restrained by a chain, is electrically
detected and translated into pen movement to directly record API gravity (Fig. 4-117).
Again, temperature compensation allows recording at 60°F.

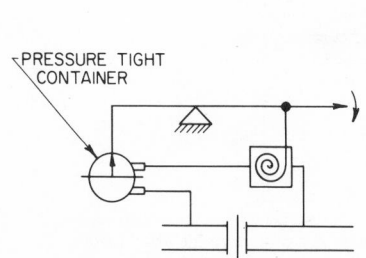

Fig. 4-116. System to measure specific gravity. Small known volume of flowing stream
is bypassed through a pressuretight container, and weighed.

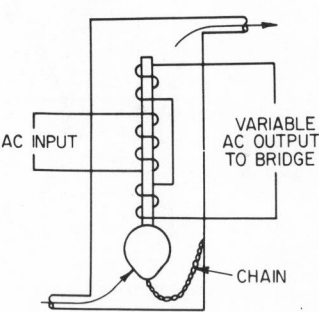

Fig. 4-117. Specific-gravity measurement
using submerged float in small chamber
through which liquid flows.   Float position,
partially restrained by a chain, is electrically detected and transmitted.

A third approach uses a radioactive gamma-ray source and detector to measure
gravity.   The intensity of radiation reaching the detector depends on the density of
the material between the source and detector.   The reading of product under the
sensor is a direct function of gravity.

As process operations have become more complex, it has become almost essential
to provide analyzers to determine the course or progress of reactions.   This is true
where composition of materials must be checked.

One type of analyzer for this purpose uses the absorbing powers of infrared waves to analyze single components in a process stream continuously.   It is set to measure any two infrared wavelengths simultaneously.   One is a region in which the component of interest absorbs strongly and the other where it does not.   The two measurements are ratioed and reported as percentages of component.   The source radiation is chopped mechanically into two beams 180° out of phase, and of slightly different path.

The beams alternately pass through sample cells, a reflecting system and slit (Fig. 4-118).   Both beams pass through a prism and are reflected by separate mirrors along

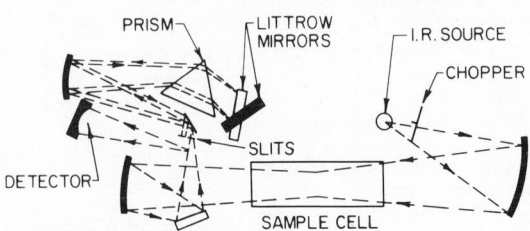

Fig. 4-118. On-stream analysis of composition by infrared absorption.

different paths back through the prism and reflecting system to a detector thermocouple.   If beams are alike, there is no signal; if different because of different absorption, the a-c results are amplified to drive a servo to attenuate and null-reference the beam.   The attenuator position indicates the sample composition.   Reproducibility for such systems is usually 1 per cent of full scale.

Another analyzer uses ultraviolet light.   In this design, the light of desired wavelength (range 0.22 to 1.2 microns) is selected by the optical filter F in Fig. 4-119 and

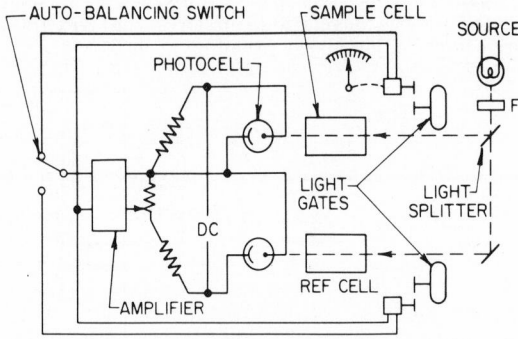

Fig. 4-119. On-stream analysis of composition by ultraviolet absorption.

split into two beams, one passing through the sample cell and the other through the reference cell.   Unabsorbed light passed through cells falls on opposed photocells in the bridge circuit; the unbalance drives a servomotor to move the light gate in the sample path to restore balance.   The ultraviolet analyzer is less selective than the infrared units but is more sensitive, giving readings ranging from 0 to 100 per cent and, in some cases, down to 0 to 1 ppm.

The mass spectrometer is yet another analyzer of the spectrum category (Fig. 4-120). Specimen gas (used for materials that can be vaporized) at low pressure enters the tube where high-velocity electrons are produced between filament and plate by electron-accelerating voltage A; these strike molecules and produce ionized molecules and molecular fragments which are accelerated into the tube by the high voltage.   Ions enter the field of a magnet which curves their paths proportional to mass.   Individual masses can be focused by varying the ion-accelerating field, to enter the slit and discharge at the ion collector. The ion current is amplified and recorded vs. accelerating voltage in terms of mass number. Different materials give characteristic mass peaks which are mathematically analyzed for composition.

Gas chromatography is one of the newer laboratory developments which has been adapted to process measurement. Means are provided to analyze gases and low-boiling liquids continuously (Fig. 4-121). A controlled quantity of gas is swept into

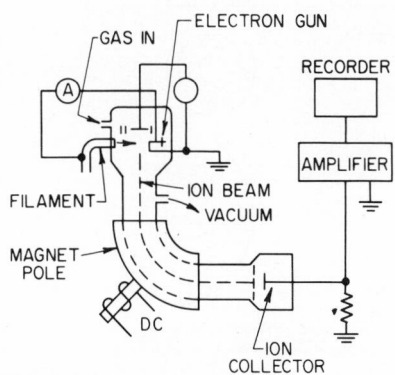

Fig. 4-120. On-stream analysis of composition by mass spectrometer.

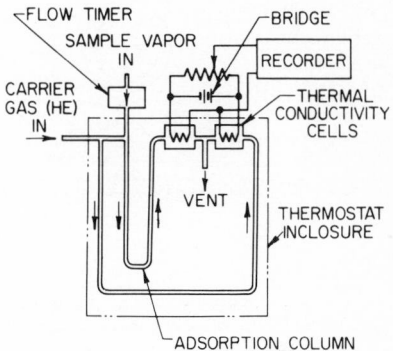

Fig. 4-121. On-stream analysis of composition by gas chromatography.

the column (packed with an adsorbent solid or an inert solid coated with adsorbent liquid) by means of a carrier gas, usually helium. Sample flow is controlled and intermittent, but carrier flow continues, first sweeping out the least firmly held component, followed in turn by others more strongly held. Sample carrier flow passes through one cell of a thermal-conductivity bridge with pure carrier as reference passing through the other cell.

BLENDING AND COMPOUNDING

Blending is the term applied to the bulk batch or in-line mixing of two or more crude or process stocks to form a composite intermediate or finished product.

Batch Blending

Batch blending is accomplished in tanks by circulation or in-tank mixers.

For circulation blending, a high-capacity low-head centrifugal pump takes suction from the tank and returns the liquid through one or more nozzles at high velocity. The nozzles enter the tank at an angle of 7 to 12° from the tank diameter to produce a swirling motion throughout the liquid.

For in-tank mixing, three-blade propellers are placed within the tank with motors and controls mounted outside the shell. The propellers are placed near the bottom of the tank, at an angle of 7 to 12° with the tank diameter, and in a position to induce a clockwise rotation of the liquid in the tank (see Figs. 4-122 and 4-123). The propeller blades not only direct the flow axially but also set up a rotary or spiral motion within the stream.

The ratio of liquid depth to tank diameter affects the results obtained with internal propeller-type mixers. The higher the ratio the more efficient the blending operation. In smaller tanks, 1 to 1 is an advisable ratio. For larger tanks built for blending service, the height should be the maximum for the allowable soil load. The horsepowers and blade diameters of in-tank propeller-type mixers and their application for various sizes of tanks are presented in Table 4-21.

The time required to effect a complete mixture of components will depend on the size and shape of the tank, the capacity and number of mixers, and the gravity differential between the components. As an example, the blending of two or three stocks (Fig. 4-124) in a 50,000-bbl tank of standard dimensions for premium-grade gasoline

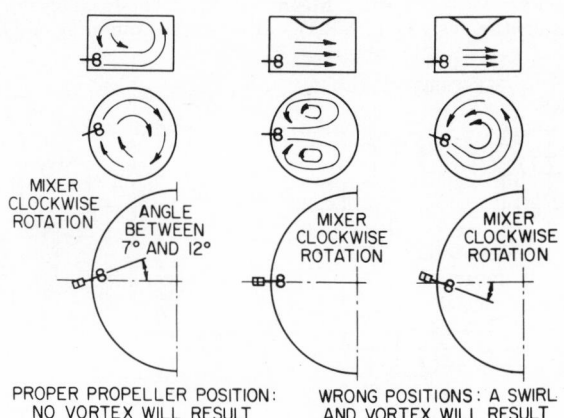

FIG. 4-122. Mixing actions possible from selection of positions for side-entering propeller mixers.

FIG. 4-123. In-tank mixer sets up swirling action in liquid. (*Courtesy of Mixing Equipment Co., Inc.*)

requires about 1 hr with two 25-hp propellers. If tetraethyl lead is added, the time will be approximately doubled. For military-grade aviation fuels, $2\frac{1}{2}$ to 3 hr would be required. A single 25-hp propeller in the same tank would take six to seven times as long in each case to effect the blend. Shell Pipe Line Co. reports results for the blending of stratified crude oils, shown graphically by Fig. 4-125.

With the mixing propellers in operation, it is possible to introduce an additional liquid component behind the propeller, as shown in Fig. 4-126.

When blending light products by the "in-tank" method, the question of evaporation losses must be considered.   It is preferable to conduct the operation in tanks where provisions are made to reduce such losses.

Table 4-21. Operating Data for In-tank Propeller-type Mixers

Tank capacity, bbl	No. of mixers	Blade diam, in.	Total hp
1,000	1	14	2
2,500	1	18	5
5,000	1	20	7½
10,000	1	26	10
15,000	1	26	15
20,000	1	26	15
25,000	1	26	20
30,000	1	26	25
35,000	2	26	30
40,000	2	26	40
45,000	2	26	40
50,000	2	26	50
55,000	3	26	60
85,000	3	26	75
100,000	4	26	100

Often it is necessary to add small percentages of special chemicals to impart additional qualities to the product.   An important one is tetraethyl lead.   Because of TEL's toxicity, cost, and personnel hazard, it is added with a separate equipment

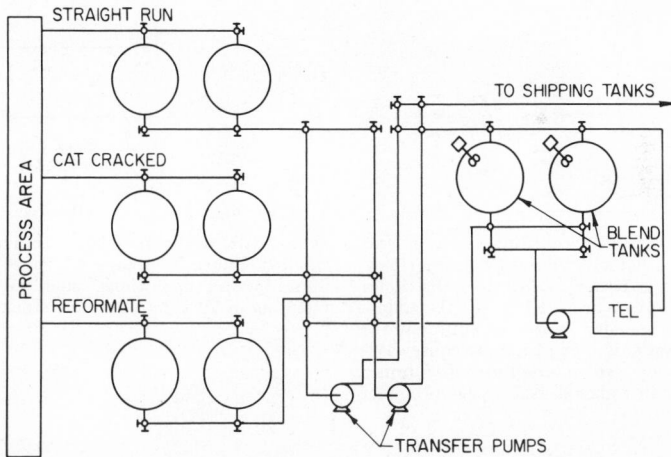

Fig. 4-124. Blending system for three feedstocks to produce a premium gasoline stock.

system.   The proportions of such chemicals, only a few cubic centimeters per gallon, require precise control.   The arrangement generally used to add TEL utilizes an eductor system as shown in Fig. 4-127.   In operation, the TEL is first introduced into a solution with the gasoline in the auxiliary circulating tank.   This solution may then be introduced to the product in the large blending tank with better control.

In-line Blending

In recent years, batch or in-tank blending has given way to the constant demand for more efficient methods to produce high-octane gasolines, resulting in the design of new equipment to increase the number of components in finished gasolines. The refiner is faced with the need to prepare and blend carefully seven, eight, or even a dozen

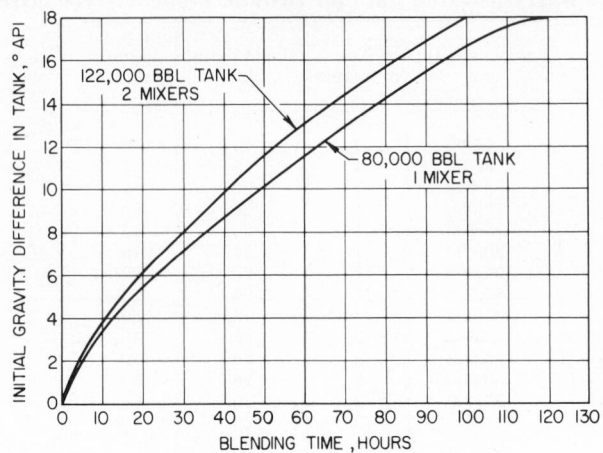

Fig. 4-125. Blending time for stratified crude oils in a 122,000-bbl tank. (*Courtesy of Shell Pipe Line Co.*)

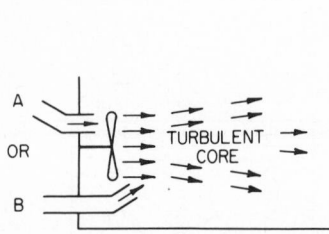

Fig. 4-126. Method for introduction and mixing of a light liquid with a heavier-base liquid. It is better to introduce the liquid above the mixer propeller (point *A*), unless the liquid density is greatly different or vapor pressure is high enough to cause cavitation, under which circumstances liquid should be introduced below the propeller (point *B*).

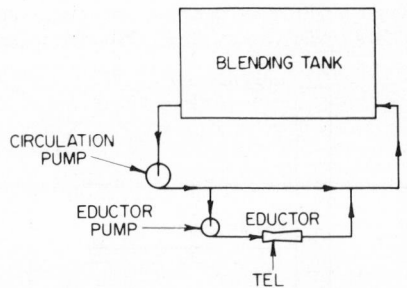

Fig. 4-127. Eductor system to introduce small volumes of liquid additives during liquid-mixing operations, such as during blending of TEL into gasoline stocks.

components in proper proportions and in large volumes. In addition, these same problems became apparent for jet and diesel fuels, kerosine, and distillate fuels. These updated needs have been met through a developing trend to in-line blending of products.

In-line blending (Fig. 4-128) provides for continuous proportioning and mixing of components. Considering a gasoline blend, this means base stock, dyes, additives, and TEL. The most advanced systems provide proper safeguards in the event of upsets. By applying the theory of the closed control loop (continuously metering each stream, comparing the signal with the established condition, and adjusting the

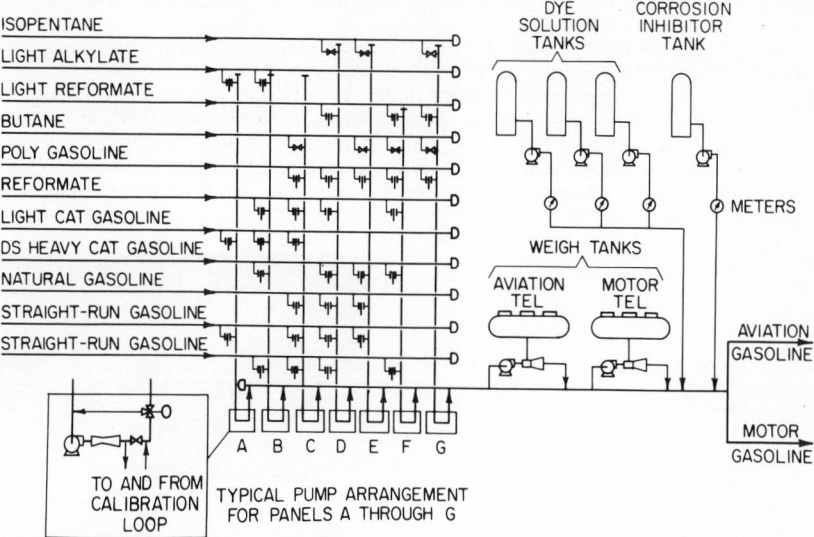

FIG. 4-128. How Tidewater handled in-line blending at Delaware City, Del. (*Petroleum Processing, July, 1957.*)

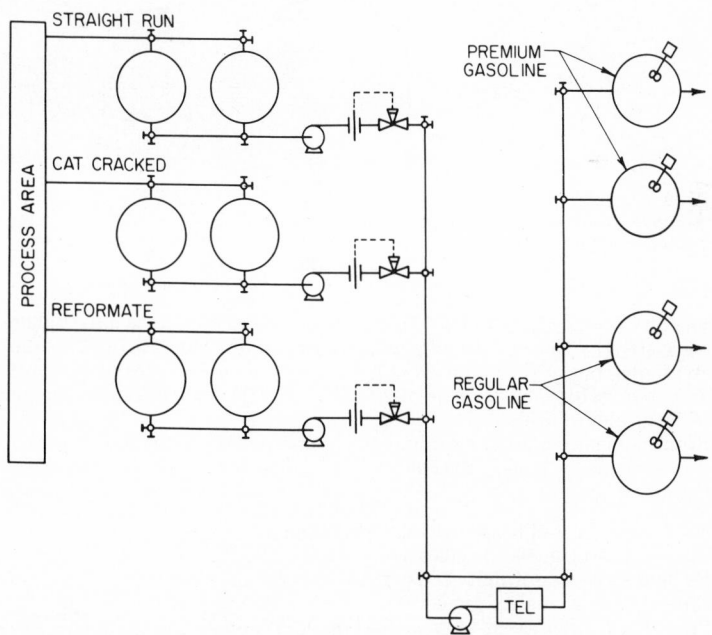

FIG. 4-129. Schematic diagram showing how a batch-blending system (as shown in Fig. 4-124) can be converted into on-stream blending.

rate if it does not match), all streams will stay in proper ratios to each other and with end-product demand.

If conditions prohibit balancing the system, automatic controls can cut off all component feed to the blend manifold, recycle each component, and thereby prevent further blending. Or, it can start a memory device to total the short or overfeed of the particular component so that when the difficulty is corrected, the shortage or excess can be made up. Metering of the individual streams may be of the positive-displacement, velocity, or differential types. Transmission can be direct mechanical, pneumatic, or electrical. Blend rates can be set manually, by process variables or by output demand.

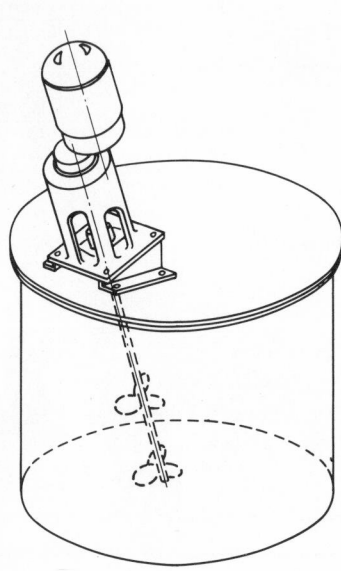

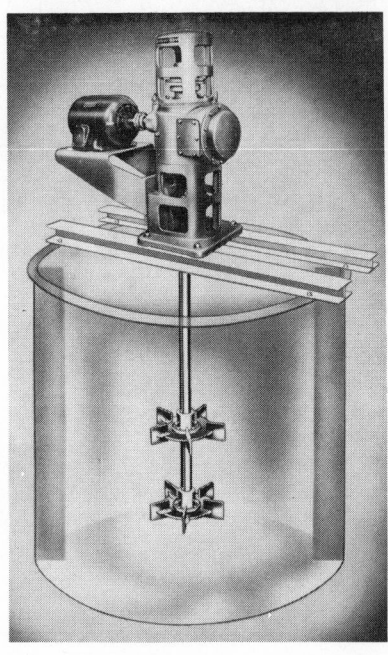

FIG. 4-130. Angular mount of propeller mixer to give top-to-bottom turnover of liquid contents during mixing.

FIG. 4-131. Turbine mixers are used for compounding large batches of lubricating materials. (*Courtesy of Mixing Equipment Co., Inc.*)

Almost any existing batch-blending system can be converted to a partial on-stream blending system by the addition of a few pieces of equipment (Fig. 4-129; pumps, control valves, meters), or the adoption of basic operating procedures such as installation of separate pumps for each component, or the installation of instruments for controlling rates of flow, plus ·continuous recorders.

Continuous blending provides initial and long-range advantages over tank blending or partial in-line systems. Specifically the advantages possible from continuous blending are

1. Elimination of duplicate tankage for shipping
2. Blending to ultimate specification
3. Reduction in loss of light ends
4. Elimination of quality giveaway
5. Provision for increased tank utilization
6. Elimination of operator attendance and error, reduction in power costs
7. Increased safety

Compounding Lubricants

Lubricants are compounded to meet specifications for gravity, viscosity, viscosity index, pour, color, and other qualities. Additives to control these qualities are often used. Animal and vegetable oils are often used in compounding. The problem is quite different from that of blending light products.

The older methods embrace in-tank mixing. Small propeller-type mixers are shell-mounted horizontally, as in blending tanks, or mounted on the tank roof in an angular off-center position to achieve top to bottom turnover as in Fig. 4-130. This type is

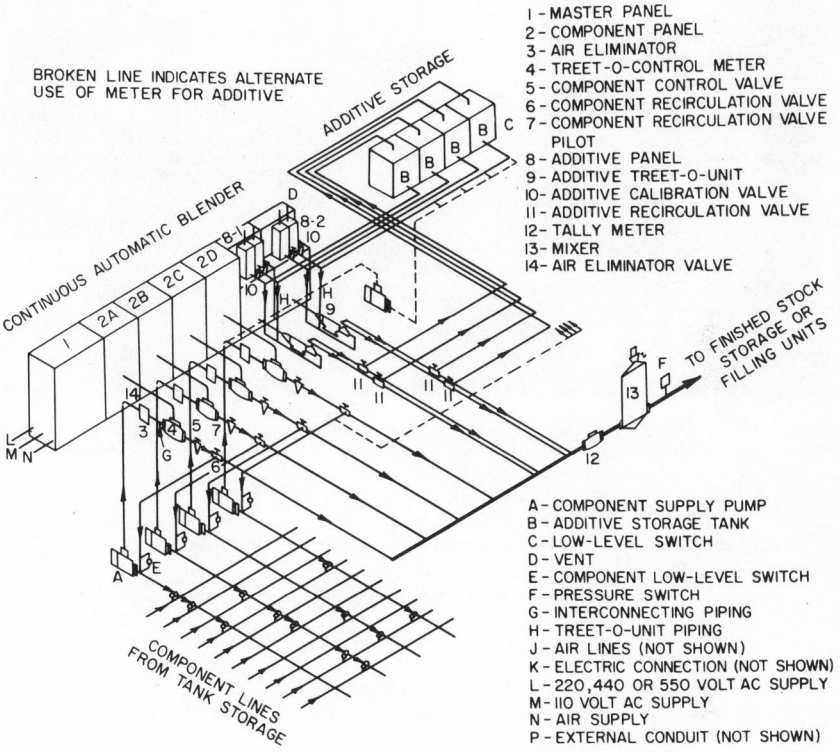

```
1 - MASTER PANEL
2 - COMPONENT PANEL
3 - AIR ELIMINATOR
4 - TREET-O-CONTROL METER
5 - COMPONENT CONTROL VALVE
6 - COMPONENT RECIRCULATION VALVE
7 - COMPONENT RECIRCULATION VALVE
    PILOT
8 - ADDITIVE PANEL
9 - ADDITIVE TREET-O-UNIT
10- ADDITIVE CALIBRATION VALVE
11- ADDITIVE RECIRCULATION VALVE
12- TALLY METER
13- MIXER
14- AIR ELIMINATOR VALVE
```

```
A - COMPONENT SUPPLY PUMP
B - ADDITIVE STORAGE TANK
C - LOW-LEVEL SWITCH
D - VENT
E - COMPONENT LOW-LEVEL SWITCH
F - PRESSURE SWITCH
G - INTERCONNECTING PIPING
H - TREET-O-UNIT PIPING
J - AIR LINES (NOT SHOWN)
K - ELECTRIC CONNECTION (NOT SHOWN)
L - 220,440 OR 550 VOLT AC SUPPLY
M - 110 VOLT AC SUPPLY
N - AIR SUPPLY
P - EXTERNAL CONDUIT (NOT SHOWN)
```

BROKEN LINE INDICATES ALTERNATE USE OF METER FOR ADDITIVE

Fig. 4-132. Typical automatic lube-oil blending system.[2]

available in sizes to 3 hp. For larger batches, the turbine-type mixer is used, shown in Fig. 4-131. Baffles on the shell provide a top-to-bottom flow pattern.

The advent of broader product needs and critical additives has led to continuous or in-line blending of lubricants. Here, as in the case of in-line blending of gasoline and fuel-oil products, continuous compounding has many advantages. The volumes of the components are proportioned in an overall integrated pattern so that each increment of flow contains the proper proportions of each component (Fig. 4-132). Control is effected for both overall rate and individual-component proportions.

An important step in compounding of lubricants is the dehydrating of the product to assure low moisture content. One system uses a homogenizer (Fig. 4-133). The combined streams are spun into an extremely thin film onto a rapidly rotating centrifugal disk, then caught by a collector ring to fall into a receiver from which a discharge pump passes the finished product to storage or drums for shipment.

When compounding lubricants, temperature control is essential. If tanks are used, they should be well insulated to conserve the heat of the components as they reach the

FIG. 4-133. Lube-oil proportioning panel and homogenizer unit to proportion up to four base stocks with up to four additives simultaneously. (*Courtesy of Cornell Machine Co.*)

tank. Steam exchangers are recommended with automatic temperature control. The continuous systems have exchangers embodied in the apparatus.

COOLING TOWERS

In locations where water is scarce, it is essential to reduce water consumption. The earlier spray ponds often used by public utilities served this purpose but had inherent disadvantages. Large areas were required, losses from wind drift were high, and the total loss of water reached 6 to 7 per cent of the volume sprayed. Refineries today utilize cooling towers.

In any type of spray pond or tower, the heat loss is made up of both direct transfer and the latent heat absorbed from that portion of the water because of evaporation.

The temperature of the cooled water depends more on the temperature and humidity of the atmosphere and fineness of the spray than on the initial temperature of the water. True heat transfer takes place only when the atmosphere is fully saturated.

At high humidity, 80 to 90 per cent, the temperature of the effluent water may be brought to within 12 to 14°F of the ambient-air temperature with a total temperature drop of 35 to 40°F. When the humidity is low, 20 to 30 per cent, the effluent may be cooled to about 8°F below the air temperature with a total effluent drop of 40 to 45°F. The loss of water by evaporation is approximately 0.15 lb/deg of water effluent drop of temperature/100 lb of water discharged.

The modern cooling tower consists of a structure of some height, over and through which the warm water to be cooled is sprayed. Forced circulation of air countercurrent to the descending water droplets is provided (see Fig. 4-134). The type of draft may be forced, with fans at the bottom of the tower, or induced, with fans located at the top. The ventilating slots at the sides and the baffles within the tower are of wood which will resist the effects of the descending water.

To permit accurate and variable control of the air volume, variable-pitch fans have

FIG. 4-134. Large 20-cell redwood counterflow cooling tower. (*Courtesy of Foster Wheeler Corp.*)

FIG. 4-135. Air-cooled condensers. (*Courtesy of Baldwin-Lima-Hamilton Corp., Ind. Equip. Div.*)

been developed with marked power savings. The pitch of the fans may be varied automatically by thermostatic control.

Where climatic conditions are favorable, towers entirely air-cooled are in use to cool condenser water for recirculation, and for condensing and subcooling distillate streams to temperatures within 20 to 25°F of the ambient-air temperature (Fig. 4-135). With normal summer temperatures, there is insufficient subcooling for light hydrocarbons. A combination of an atmospheric tower and the air condenser should be considered. The atmospheric section may be shut off when ambient-air temperatures are low during winter months, and brought into service during hot humid weather.

STORAGE TANKS

Refinery tankage may be classified as follows:

1. Standard
2. Conservation
   Floating roofs
   Lifter roofs
   Roofs with expanding membranes
   Spheroids
3. Special
   Spheres
   Miscellaneous

Standard Tanks

Data on standard tanks with welded steel shell and cone or globe roof are given in Tables 4-22 through 4-24.[26] Standard tanks with steel-cone roofs are suitable for

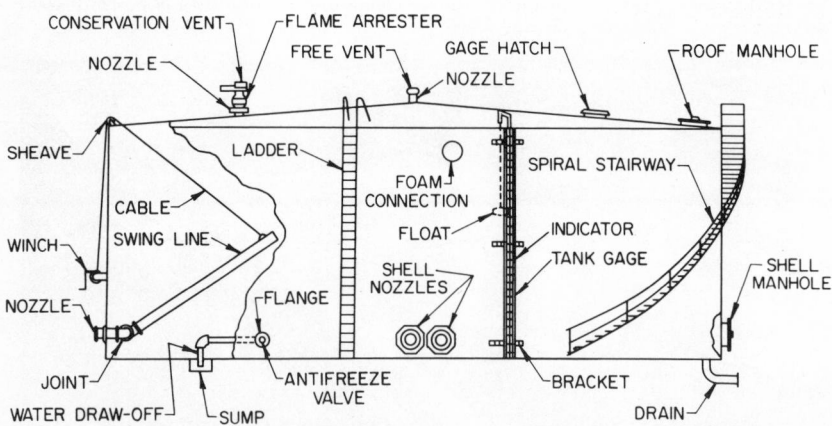

FIG. 4-136. Typical arrangement of fittings for a fixed-roof tank. (*Petroleum Processing, July, 1956.*)

storage of heavier crude oils, residual and distillate fuel oil, lubricating oil, diesel oil, and kerosine. A section of a cone-roof tank with normal complement of fittings is shown by Fig. 4-136.

Tank Appurtenances

Shell manholes are placed on the bottom ring of plates. The minimum inside diameter is 20 in., and perferably this dimension should be 24 or 30 in. For tanks exceeding 75 ft in diameter, two manholes placed diametrically opposite are recommended for ventilation and convenience when cleaning. Attention must be given to

Table 4-22. Typical Sizes and Corresponding Nominal Capacities for Tanks with 72-in. Butt-welded Courses[26]

Tank diam, ft	Approx capacity per ft of height, bbl	Tank height, ft								
		12	18	24	30	36	42	48	54	60
		No. of courses in completed tank								
		2	3	4	5	6	7	8	9	10
10	14.0	170	250	335	420	505				
15	31.5	380	565	755	945	1,130				
20	56.0	670	1,010	1,340	1,680	2,010	2,350	2,690		
25	87.4	1,050	1,570	2,100	2,620	3,150	3,670	4,200	4,720	5,250
30	126	1,510	2,270	3,020	3,780	4,530	5,290	6,040	6,800	7,550
35	171	2,060	3,080	4,110	5,140	6,170	7,200	8,230	9,250	10,280
40	224	2.690	4,030	5,370	6,710	8,060	9,400	10,740	12,090	13,430
45	283	3,400	5,100	6,800	8,500	10,200	11,900	13,600	15,300	17,000
50	350	4,200	6,290	8,390	10,490	12,590	14,690	16,790	18,880	20,980
60	504	6,040	9,060	12,090	15,110	18,130	21,150	24,170	27,190	30,220
70	685	8,230	12,340	16,450	20,560	24,680	28,790	32,900	37,010	41,130
80	895	10,740	16,120	21,490	26,860	32,230	37,600	42,970	48,350	53,720
90	1,133	13,600	20,390	27,190	33,990	40,790	47,590	54,390	61,180	67,980
100	1,399	16,790	25,180	33,570	41,970	50,360	58,750	67,140	75,540	83,930
120	2,014	......	36,260	48,340	60,430	72,510	84,600	96,690	108,800	120,900
140	2,742	......	49,350	65,800	82,250	98,700	115,100	131,600	148,000	164,500
160	3,581	......	......	......	107,400	128,900	150,400	171,900	193,400	214,900
180	4,532	......	......	......	136,000	163,200	190,400	217,500	244,800	254,300
200	5,595	......	......	......	167,900	201,400	235,000	268,600	**284.500**	D = 174
220	6,770	......	......	......	203,100	243,700	284,400	322,300	D = 194	
								D = 219		

The nominal capacities given in the table are based on the formula

$$\text{Capacity (42-gal bbl)} = 0.14D^2H$$

where D = center-line shell diameter

H = listed tank height

Capacities and diameters shown in boldface type in the last three columns are maximum for the tank heights shown, based on the $1\frac{1}{2}$-in. maximum permissible thickness of shell plates and maximum allowable design stresses.

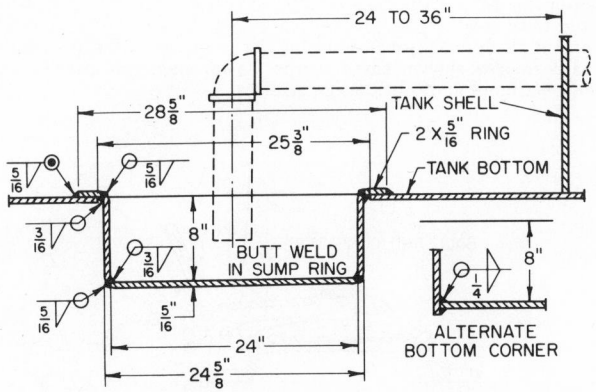

Fig. 4-137. Tank sump and draw-off through which water is siphoned out through the tank shell.[26] Erection-procedure notes: 1. Cut hole in bottom plate. 2. Make neat excavation to conform to shape of draw-off sump. 3. Place and weld sump.

Table 4-23. Typical Sizes and Corresponding Nominal Capacities for Tanks with 96-in. Butt-welded Courses[26]

Tank diam, ft	Approx capacity per ft of height, bbl	Tank height, ft						
		16	24	32	40	48	56	64
		No. of courses in completed tank						
		2	3	4	5	6	7	8
10	14.0	225	335	450				
15	31.5	505	755	1,010	1,260			
20	56.0	900	1,340	1,790	2,240	2,690		
25	87.4	1,400	2,100	2,800	3,500	4,200	4,900	5,600
30	126	2,020	3,020	4,030	5,040	6,040	7,050	8,060
35	171	2,740	4,110	5,480	6,850	8,230	9,600	10,960
40	224	3,580	5,370	7,160	8,950	10,740	12,530	14,320
45	283	4,530	6,800	9,060	11,330	13,600	15,860	18,130
50	350	5,600	8,390	11,190	13,990	16,790	19,580	22,380
60	504	8,060	12,090	16,120	20,140	24,170	28,200	32,230
70	685	10,960	16,450	21,930	27,420	32,900	38,380	43,870
80	895	14,320	21,490	28,650	35,810	42,970	50,130	57,300
90	1,133	18,130	27,190	36,260	45,320	54,390	63,450	72,520
100	1,399	22,380	33,570	44,760	55,950	67,140	78,340	89,530
120	2,014	......	48,340	64,460	80,580	96,690	112,800	128,900
140	2,742	......	65,800	87,740	109,700	131,600	153,500	175,500
160	3,581	......	......	114,600	143,200	171,900	200,500	229,200
180	4,532	......	......	145,000	181,300	217,500	253,800	**238,100**
200	5,595	......	......	179,100	223,800	268,600	**274,200**	**D = 163**
220	6,770	......	......	216,700	270,800	**322,300**	**D = 187**	
						D = 219		

The nominal capacities given in the table are based on the formula

$$\text{Capacity (42-gal bbl)} = 0.14 D^2 H$$

where D = center-line shell diameter
H = listed tank height

Capacities and diameters shown in bold face type in the last three columns are maximum for the tank heights shown, based on the 1½-in. maximum permissible thickness of shell plates and maximum allowable design stresses.

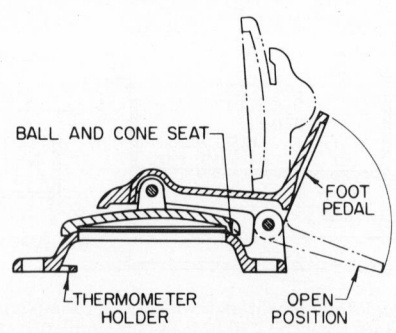

BALL AND CONE SEAT

FOOT PEDAL

THERMOMETER HOLDER

OPEN POSITION

Fig. 4-138. Gastight nonsparking gage and thief-hole cover for petroleum storage tanks.

Table 4-24. Shell-plate Thicknesses for Typical Sizes of Tanks[26]
With 72-in. Butt-welded Courses

Tank diam, ft	6	12	18	24	30	36	42	48	54	60	Max allowable height* for diameters listed, ft
	1	2	3	4	5	6	7	8	9	10	
	Shell-plate thickness, in.										
10	3/16	3/16	3/16	3/16	3/16	3/16					
15	3/16	3/16	3/16	3/16	3/16	3/16					
20	3/16	3/16	3/16	3/16	3/16	3/16	3/16	3/16			
25	3/16	3/16	3/16	3/16	3/16	3/16	3/16	0.19	0.20	0.22	
30	3/16	3/16	3/16	3/16	3/16	3/16	0.19	0.21	0.24	0.26	
35	3/16	3/16	3/16	3/16	3/16	0.19	0.21	0.24	0.27	0.30	
40	3/16	3/16	3/16	3/16	0.19	0.21	0.24	0.28	0.31	0.35	
45	3/16	3/16	3/16	3/16	0.19	0.23	0.27	0.31	0.35	0.39	
50	1/4	1/4	1/4	1/4	1/4	0.26	0.30	0.35	0.39	0.43	
60	1/4	1/4	1/4	1/4	0.26	0.31	0.36	0.41	0.47	0.52	
70	1/4	1/4	1/4	0.25	0.30	0.36	0.42	0.48	0.54	0.61	
80	1/4	1/4	1/4	0.27	0.34	0.41	0.48	0.55	0.62	0.69	
90	1/4	1/4	1/4	0.31	0.38	0.46	0.54	0.62	0.70	0.78	
100	1/4	1/4	0.25	0.34	0.43	0.51	0.60	0.69	0.78	0.86	
120	5/16	5/16	0.31	0.41	0.51	0.62	0.72	0.83	0.93	1.03	
140	5/16	5/16	0.35	0.47	0.60	0.72	0.84	0.96	1.08	1.21	
160	5/16	5/16	0.40	0.54	0.68	0.82	0.96	1.10	1.24	1.38	65.3
180	5/16	5/16	0.45	0.61	0.76	0.92	1.08	1.24	1.39	....	58.2
200	5/16	0.32	0.50	0.67	0.85	1.02	1.20	1.37	....	....	52.5
220	3/8	3/8	0.55	0.74	0.94	1.13	1.32	....	....	....	47.8

the area of metal cut from the shell plate at a point near the maximum stress. A reinforcing plate is essential. Details of manholes, covers, and reinforcement are presented in the API Standard 650, *Welded Steel Tanks for Oil Storage.*[26]

Roof manholes are of lighter construction. They should be located approximately above the shell manholes and 6 to 8 ft in from the roof circumference.

Nozzles for connection of piping to the shell consist of short lengths of pipe with weld-neck or cast-steel pipe flanges at outer ends. The assembly is welded to the shell. When the pipe diameter is 3 in. or larger, a shell reinforcing plate is used. For pipe smaller than 3 in., a coupling may be used, welded to the shell, and with internal standard pipe threads for the outside connection. Figure 4-136 shows the location of many common tank appurtenances.

Water will accumulate in the bottom of any tank because of condensation of the moisture in the air drawn in through the vents. A weld ell is welded to the bottom plate with the neck extended beyond the circumference of the shell. Cast steel may be used. An alternative (Fig. 4-137) is the installation of a sump from which the water is siphoned out through the shell. With such construction, the water draw-off valve may be placed on the pipe inside the tank to prevent freezing. An extension shaft through the shell operates the valve.

Unless automatic gaging is used, openings in the roof must be installed for introduction of the gage pole or tape for sampling contents. Such fittings are made of alumi-

Table 4-24. Shell-Plate Thicknesses for Typical Sizes of Tanks[26] (Continued)
With 96-in. Butt-welded Courses

Tank diam, ft	Tank height, ft								Max allowable height* for diameters listed, ft
	8	16	24	32	40	48	56	64	
	No. of courses in completed tank								
	1	2	3	4	5	6	7	8	
	Shell-plate thickness, in.								
10	3/16	3/16	3/16	3/16					
15	3/16	3/16	3/16	3/16	3/16				
20	3/16	3/16	3/16	3/16	3/16	3/16			
25	3/16	3/16	3/16	3/16	3/16	0.19	0.20	0.23	
30	3/16	3/16	3/16	3/16	0.19	0.21	0.24	0.28	
35	3/16	3/16	3/16	0.19	0.20	0.24	0.28	0.33	
40	3/16	3/16	3/16	0.19	0.23	0.28	0.32	0.37	
45	3/16	3/16	0.19	0.21	0.26	0.31	0.36	0.42	
50	1/4	1/4	1/4	0.25	0.29	0.35	0.40	0.46	
60	1/4	1/4	1/4	0.27	0.34	0.41	0.48	0.55	
70	1/4	1/4	0.25	0.32	0.40	0.48	0.56	0.65	
80	1/4	1/4	0.27	0.37	0.46	0.55	0.64	0.74	
90	1/4	1/4	0.31	0.41	0.52	0.62	0.72	0.83	
100	1/4	0.25	0.34	0.46	0.57	0.69	0.80	0.92	
120	5/16	5/16	0.41	0.55	0.69	0.83	0.97	1.10	
140	5/16	0.31	0.47	0.64	0.80	0.96	1.13	1.29	
160	5/16	0.35	0.54	0.73	0.91	1.10	1.29	1.47	65.3
180	5/16	0.40	0.61	0.82	1.03	1.24	1.45	....	58.2
200	5/16	0.44	0.67	0.91	1.14	1.37	....	....	52.5
220	3/8	0.48	0.74	1.00	1.25	....	....	....	47.8

* Based on the 1½-in. maximum permissible thickness of shell plates and the maximum allowable design stresses.

num or steel for bolting or welding to the roof. A minimum diameter or opening of 6 in. is recommended, preferably 8 in. One gage seal for tanks up to 30 ft in diameter will suffice. For tanks 30 to 50 ft in diameter, two are recommended; 50 to 80 ft, three; four for larger tanks. For oils with flash points above 100°F and tanks operating at atmospheric pressure, the self-closing type with nonsparking seats, as in Fig. 4-138, is in general use. As conservation vents impose a slight pressure within the tank, a toggle arrangement may be added to ensure tightness of the closed gage hatch.

The cone roof of an oil tank is usually of 3/16-in. plate (see Table 4-25). It will resist very little internal pressure and must be vented to the atmosphere. For less volatile products with flash points above 100°F, an open vent is satisfactory. If of cast metal or made from pipe and fittings, it is in the shape of an inverted U to shed storm water, as in Fig. 4-139. Low-flash products, under 100°F, when stored in cone-roof tanks, are customarily vented through pressure-vacuum valves, termed conservation vents because they have a favorable influence on evaporation losses.

The conservation vent imposes a slight pressure on the tank contents and opens to break any vacuum in reverse flow (see Fig. 4-140). Such valves are set to open at

¾ to 1 in. of water pressure or vacuum.   They reduce evaporation losses to some
extent.   Generally, 1 oz (1½ in. of water) pressure will reduce the losses due to breathing by about 7½ per cent.

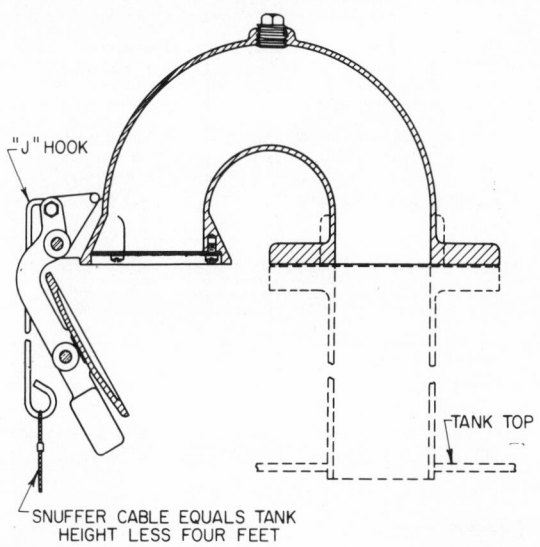

FIG. 4-139. Free vent for cone-roof tank.   (*Courtesy of Shand & Jurs Co.*)

The API offers as a guide Table 4-26 (abbreviated)[27] for the required vent capacity
for standard cone-roof tanks.   While the volumes given in the tables are sufficient for
normal breathing, filling, and evaporation, they may not suffice in an emergency, and
pressure-relief vents are installed.   Such vents operate at higher pressures than the
settings for conservation vents, usually at
about 3 in. of water.   Several types are
available, among them the mushroom design shown in Fig. 4-141.

Flame arrestors, as shown in Fig. 4-142,
are an adjunct of tanks containing volatile
products and are designed to prevent the
propagation of flame through flammable
mixtures escaping from the tank.   The
general principle follows that of the Davy
safety lamp used in mines.   The issuing
stream of gas is broken up and the heat
generated by the flame is cooled or dissipated.   Authorities differ upon their efficiency.   If used, flame arrestors are placed
next to the tank between the vent and the
tank.   Note that there is a pressure drop
across the arrestor bank which should be
considered in the selection of any vent size.

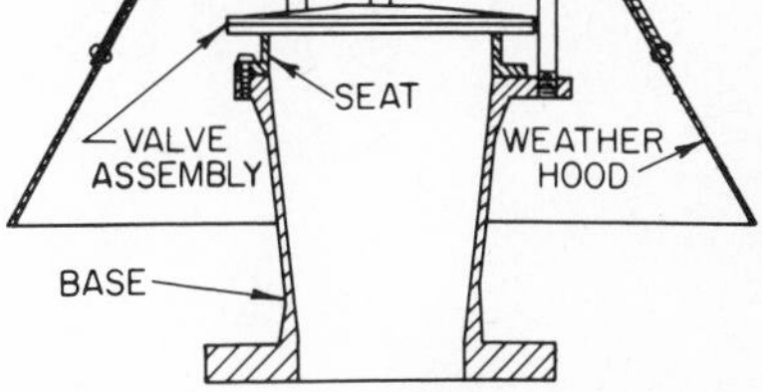

FIG. 4-140. Conservation-type vent for
cone-roof tanks.   (*Courtesy of Shand &
Jurs Co.*)

All vents should be provided with a flame
snuffer at the outlet.   This is in effect a
clapper valve, as shown in Fig. 4-139, which
is weighted to hang open normally and which may be closed by chain action from
the ground.

Swing lines within the tank are often used and serve two purposes.   They permit withdrawal of tank contents from any level and when in elevated positions are added safety factors against valve or piping leaks, as shown in Fig. 4-136.

Table 4-25. Characteristics of Steel-tank Roofing*

Thickness, in.†	Weight, psf	Operating pressure, oz/sq in.
$\frac{1}{16}$ (16 gage)	2.553	0.284
$\frac{5}{64}$ (14 gage)	3.187	0.354
$\frac{7}{64}$ (12 gage)	4.473	0.497
$\frac{1}{8}$ (11 gage)	5.107	0.568
$\frac{9}{64}$ (10 gage)	5.740	0.638
$\frac{5}{32}$ (9 gage)	6.374	0.708
$\frac{11}{64}$ (8 gage)	7.00	0.778
$\frac{3}{16}$† (7 gage)	7.65	0.850
$\frac{1}{4}$ (3 gage)	10.20	1.133

* *Pet. Proc.* p. 112, July, 1956.
† Minimum thickness specified by API Standard 650.

Other tank appurtenances include thermometer wells, float gages, and high- and low-level alarms. Gage and temperature readings may be transmitted electrically to a central console.

Evaporation Losses

Evaporation losses are substantial when volatile products are stored in standard tanks with gastight steel roofs. Each cycle of temperature change, diurnal or seasonal, creates breathing of the gas volume above the liquid surface. As the temperature

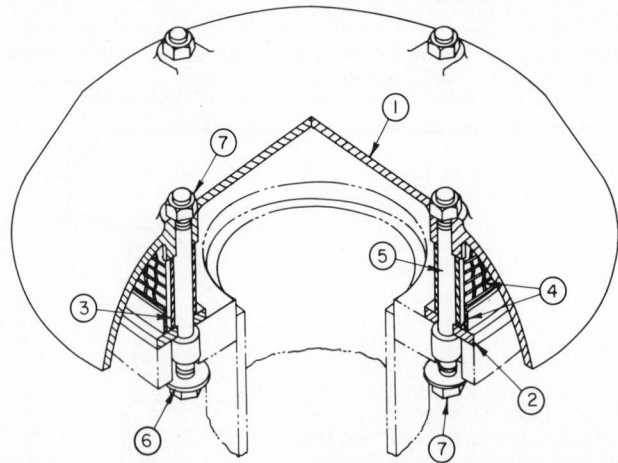

FIG. 4-141. Mushroom-type free vent for cone-roof tanks. (1) Hood; (2) base plate; (3) spacer; (4) screen and frame; (5) stud; (6) washer; (7) hexagonal nuts. (*Courtesy of Shand & Jurs Co.*)

decreases the gas volume contracts and air is drawn into the vessel. When the temperature rises, the volume expands, forcing the vapor-saturated air out through the tank vents. During this cycle, air and vapor reach equilibrium, and as the tank exhales, some of the light ends of the stored product are carried away. The develop-

Table 4-26. Thermal Venting Capacity for Cone-roof Tanks, Cu ft/hr of Air[27]

Tank capacity,* bbl	In-breathing (vacuum) all stocks	Out-breathing (pressure)	
		Flash point 100°F and above	Flash point below 100°F
1,000	1,000	600	1,000
2,000	2,000	1,200	2,000
3,000	3,000	1,800	3,000
4,000	4,000	2,400	4,000
5,000	5,000	3,000	5,000
10,000	10,000	6,000	10,000
15,000	15,000	9,000	15,000
20,000	20,000	12,000	20,000
25,000	24,000	15,000	24,000
30,000	28,000	17,000	28,000
35,000	31,000	19,000	31,000
40,000	34,000	21,000	34,000
45,000	37,000	23,000	37,000
50,000	40,000	24,000	40,000
60,000	44,000	27,000	44,000
70,000	48,000	29,000	48,000
80,000	52,000	31,000	52,000
90 000	56,000	34,000	56,000
100,000	60,000	36,000	60,000
120,000	68,000	41,000	68,000
140,000	75,000	45,000	75,000
160,000	82,000	50,000	82,000
180,000	90,000	54,000	90,000

For tanks with more than 20,000-bbl capacity, the requirements for the vacuum condition are very close to the theoretically computed value of 2 cu ft of air/hr/sq ft of total shell and roof area.   For tanks of less than 20,000-bbl capacity, the thermal in-breathing requirement for the vacuum condition has been limited to 1 cu ft of air/hr for each barrel of tank capacity.   This is substantially equivalent to a mean rate of vapor-space-temperature change of 100°F per hour.

The required out-breathing capacity for stocks with a flash point at or above 100°F has been assumed as 60 per cent of the required in-breathing capacity, as the tank-roof and shell temperatures cannot rise as rapidly under any conditions as they can drop during a sudden cold rain.

For stocks with a flash point below 100°F the thermal pressure-venting requirements have been made equal to the vacuum requirements, in order to allow for vaporization at the liquid surface and the higher specific gravity of the tank vapors.

* Interpolate for intermediate sizes.

ment of the theory on breathing losses is set forth in references 28 through 31.   Figure 4-143 summarizes these losses for cone-roof tanks.

When the contents of a tank are withdrawn, air enters through the vents.   When again filled, the air is forced out, again carrying light vapors with it.   Such filling losses depend on the number of turnovers in a given length of time.   Figure 4-144 summarizes the theory.   Filling losses are not so significant as breathing losses but do present an appreciable amount.

The theory of evaporation losses from standard cone-roof tanks has been substantiated by actual tests by 37 companies as reported to the API Symposium.[32]   Tests conducted by eight companies to determine filling losses are likewise summarized in Table 4-27.

Conservation Tanks

Evaporation losses may be materially reduced by

1. Allowing no vapor space above the liquid level: Floating roofs

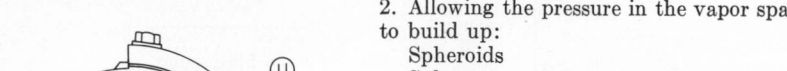
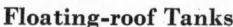
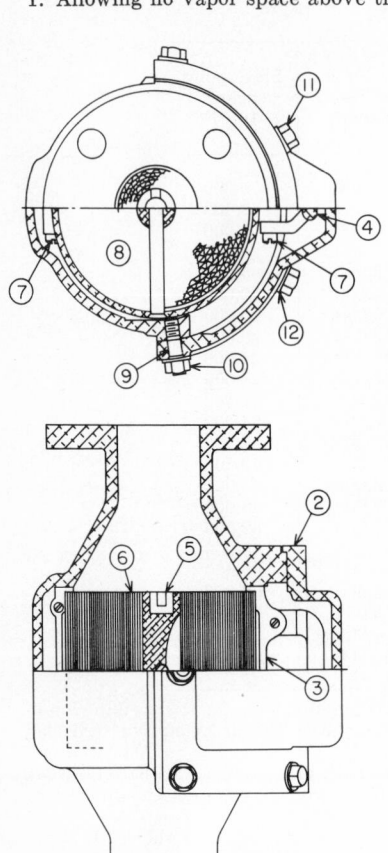

2. Allowing the pressure in the vapor space to build up:
   Spheroids
   Spheres
   Other types of pressure storage
3. Allowing the air-vapor mixture to change volume at constant or variable pressure:
   Lifter roof
   Common gas holder

Floating-roof Tanks

These embrace the original pan-type tank roof which has been largely superseded by the pontoon and double-pontoon types. A cutaway view showing roof drainage and auxiliaries is shown in Fig. 4-145. Typical seals about the rim are shown in Fig. 4-146. In the double-deck type, the pontoons cover the entire area of liquid surface. With no air-vapor space above the liquid, evaporation losses are limited to those caused by the small peripheral exposure at the seals. Such a tank eliminates 90 to 95 per cent of the evaporation losses.

Insofar as filling losses are concerned, there are practically none. Hence floating roofs are most advantageous for the storage of volatile products at locations where considerable in-and-out movement is anticipated.

The pan type is the least expensive but also is the least stable and should not be considered at locations where excessive rainfall and snow loads may be expected. There is considerable heat transfer from the roof deck to the tank contents, with possible resultant boiling of volatile products.

The pontoon type is more stable and reduces the possibility of underdeck boiling. The double-deck pontoon type practically eliminates boiling from solar heat on the deck. It is also the most stable in design.

All floating roofs offer advantages in addition to the control of evaporation and filling losses:

1. The elimination of moisture, such as encountered in the vapor space of other types, greatly reduces the corrosion problem.

Fig. 4-142. Typical flame arrestor for use on tanks storing volatile products. (1) Body; (2) cover; (3) tube bank shell; (4) tube bank handle; (5) tube bank core; (6) tube bank; (7) filister-head screw; (8) tube bank pin; (9) stud; (10) hexagonal nut; (11) hexagonal-head cap screw; (12) washer. (*Courtesy of Shand & Jurs Co.*)

This is particularly true when sour crude oil is stored.

2. When evaporation takes place, ordinarily it is the lighter hydrocarbons that escape. These fractions are responsible for easy starting of internal-combustion engines and have a favorable influence on the antiknock properties of motor fuels. One authority reports that a volumetric loss of 1.0 per cent from a typical gasoline

results in a rise of 5°F in the 10 per cent point of the distillation curve, a decrease of 0.5 psi in Reid vapor pressure, and a 1-point loss in the octane rating of the fuel.

3. The elimination of the vapor space above the oil surface results in the safest storage insofar as fire risk is concerned. This is reflected in the low insurance ratings for this type of storage.

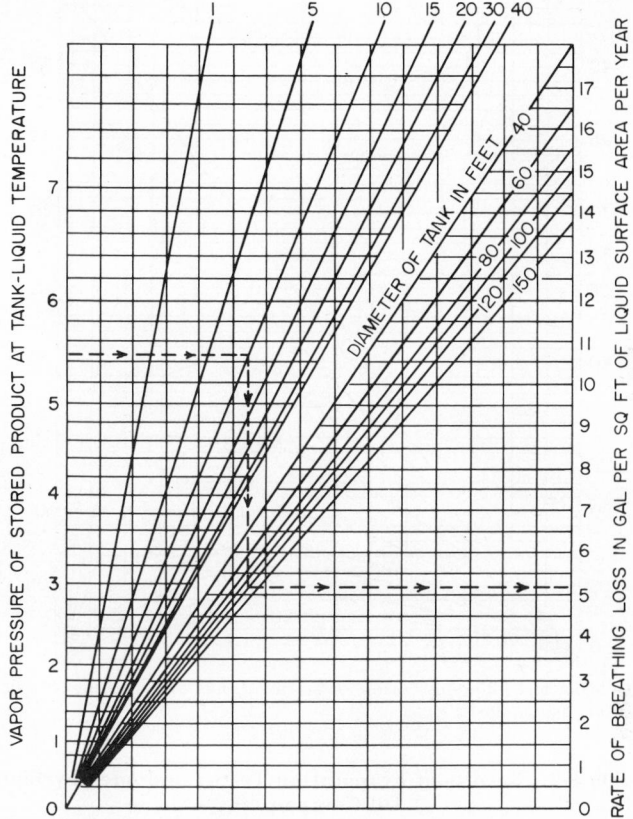

FIG. 4-143. Breathing losses from cone-roof tanks with conservation vents. (*Trans. Am. Inst. Chem. Engrs.* 38-1.)

All types of floating roof are adaptable to all diameters of vertical cylindrical storage tanks.

Limited advantages, similar to those of floating roofs, may be attained in cone-roof tanks with addition of

1. A blanket of microballoons which consist of minute plastic spheres filled with inert gas and floated over the oil surface. A 1-in. layer is reported to reduce evaporation losses of motor gasoline by 50 per cent.

2. A segmental floating plastic cover, as in Fig. 4-147, where relatively narrow segments are introduced through a shell manhole and reassembled to form a floating blanket on the oil surface.

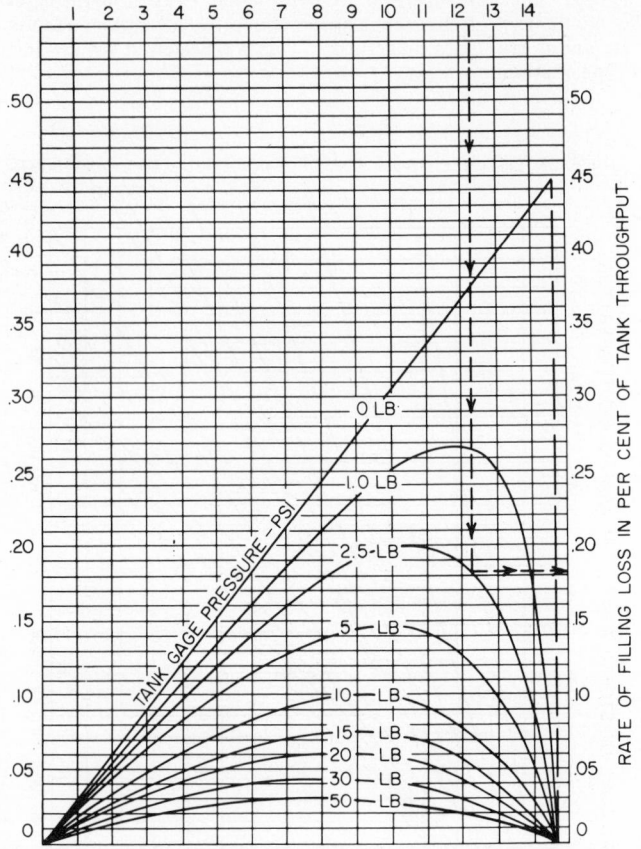

Fig. 4-144. Theoretical filling-loss rates for storage tanks and pressure vessels. (*Trans. Am. Inst. Chem. Engrs.*, 38-1.)

Table 4-27. Results of Evaporation Tests Conducted by Eight Oil Companies[2]

Tank size			Avg loss bbl year 8 companies
Nominal capacity, bbl	Diam, ft	Shell height, ft	
5,000	30	40	154
5,000	35	30	169
10,000	42½	40	297
20,000	60	40	570
55,000	100	40	1,382
67,000	100	48	1,536
80,000	120	40	1,895
125,000	150	40	2,753
150,000	150	48	3,022

Pressure Storage

To prevent breathing losses effectively, a container must be capable of withstanding the maximum pressure normally attained without venting. Vapor pressures of light refinery products and distillates will rarely exceed 10 to 12 psi absolute by the Reid method at 100°F. At the liquid surface for temperatures encountered in storage, the pressure required to prevent evaporation will range from 2 to 5 psig.

The Reid vapor pressure of natural gasolines ranges from 10 to 30; for LP gas, from 30 to 40. Such products require higher storage pressures to eliminate evaporation losses. Pressure storage may therefore be divided into two classifications:

1. Low-pressure, normally designed for 2.5 to 5 psig. Spheroids and hemispheroids come within this classification, although spheroids have been constructed for a pressure of 15 psig. Such tankage is suitable for aviation, motor, and jet fuels.

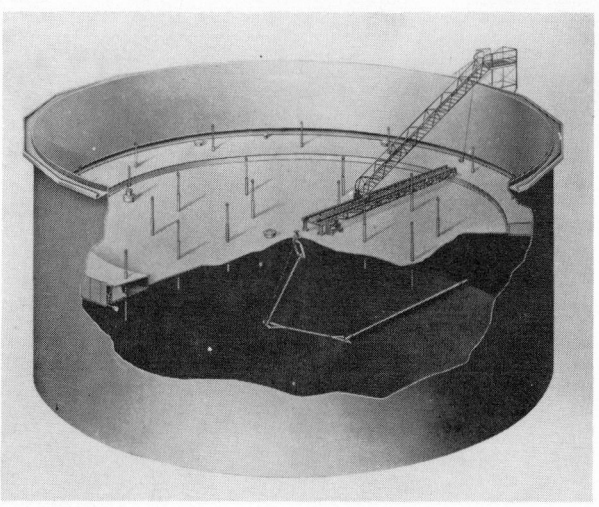

Fig. 4-145. Cutaway of swing-jointed roof drain and auxiliaries for floating-roof tank. (*Courtesy of Chicago Bridge & Iron Co.*)

2. High-pressure for storage at 30 to 200 psig. Spheres and horizontal tanks with bumped heads are examples.

The plain spheroid (Fig. 4-148) is built in the shape of a toy rubber balloon containing air and water, and resting on a flat surface. The stress in the shell is equal in all directions. The plain spheroid is available in capacities to 40,000 bbl and for pressures to 30 psig. For larger capacities one or more nodes are introduced, as shown in Fig. 4-149. Multiple-noded spheroids are built with capacities of up to 120,000 bbl and for pressures of up to 15 psig.

Spheres (Fig. 4-150) are built for storage under higher pressures than possible with spheroids. They are constructed with capacities of up to 30,000 bbl. The working pressure to stay within the API-ASME code decreases with increasing diameter. For example, a 1,000-bbl sphere may be constructed for a 200-psig working pressure, a 5,000-bbl sphere for 120 psig, and a 30,000-bbl sphere (60 ft diameter) for 50 psig.

Variable-vapor-space Storage

The third classification of conservation storage containers is the type that permits vapor-volume changes without venting to the atmosphere.

The lifter roof shown in Fig. 4-151 is in effect a gas holder mounted on a standard storage tank. As the air-vapor mixture above the oil surface expands or contracts,

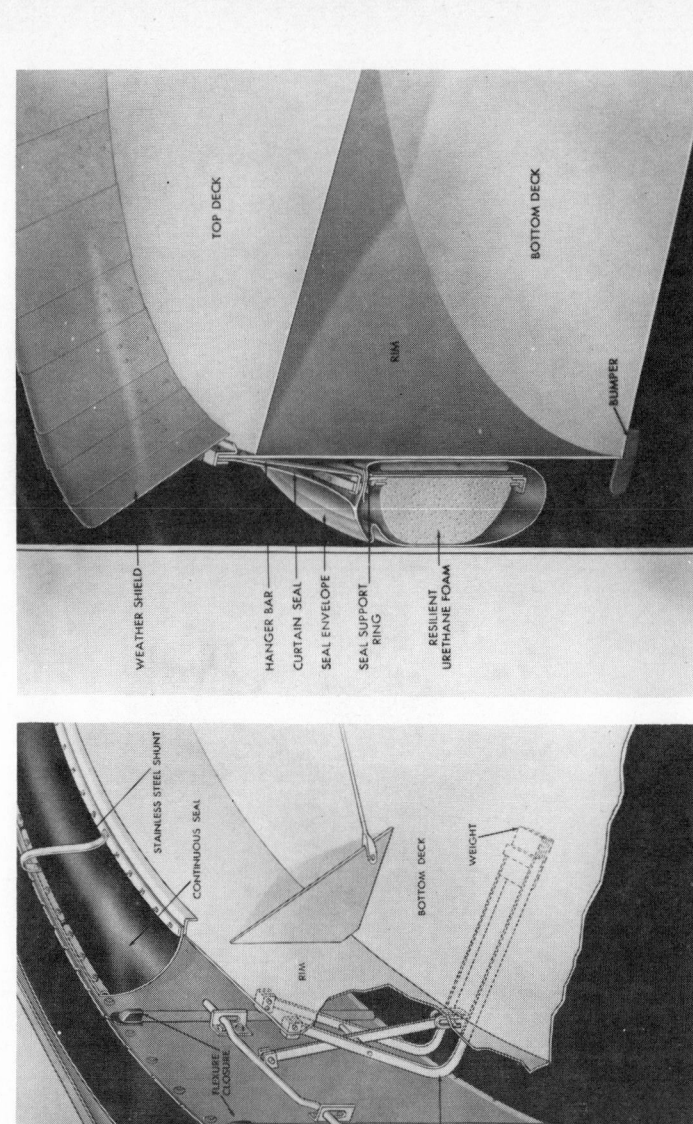

TOP DECK

BOTTOM DECK

RIM

BUMPER

WEATHER SHIELD

HANGER BAR

CURTAIN SEAL

SEAL ENVELOPE

SEAL SUPPORT RING

RESILIENT URETHANE FOAM

STAINLESS STEEL SHUNT

CONTINUOUS SEAL

BOTTOM DECK

WEIGHT

RIM

FLEXURE CLOSURE

WIND GIRDER

TANK SHELL

SEALING RING

PANTAGRAPH HANGER

FLEXURE

Fig. 4-146. Sealing floating-roof tanks. At left, flexure sealing ring; at right, resilient foam-fabric seal. (*Courtesy of Chicago Bridge & Iron Co.*)

4-106

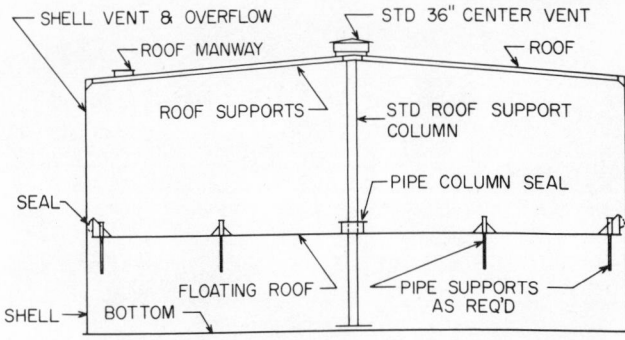

TYPICAL SCHEMATIC OF A COVERED FLOATING ROOF TANK

FIG. 4-147. Typical schematic of an internal floating roof installed in a conventional cone-roof tank.[33]

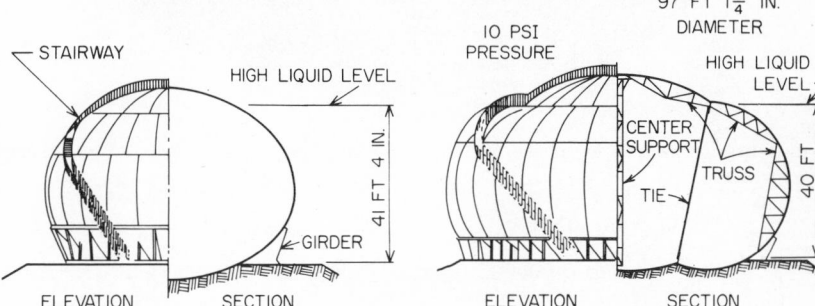

FIG. 4-148. Cutaway diagram of a 20,000-bbl plain spheroid. (*Courtesy of Chicago Bridge & Iron Co.*)

FIG. 4-149. Cutaway diagram of a 40,000-bbl spheroid. (*Courtesy of Chicago Bridge & Iron Co.*)

FIG. 4-150. Forty-three-foot-diameter sphere for storage of products under high pressure. (*Courtesy of Chicago Bridge & Iron Co.*)

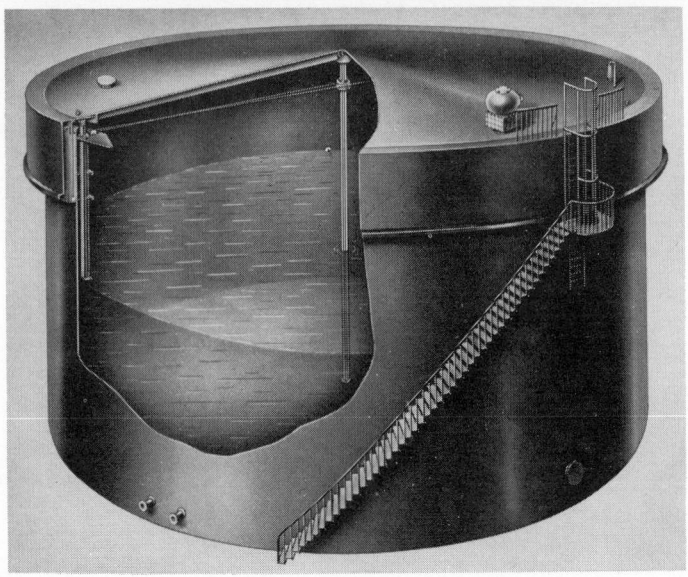

FIG. 4-151. Cutaway of expansion (or lifter) roof tank.   (*Courtesy of Graver Tank & Mfg. Co.*)

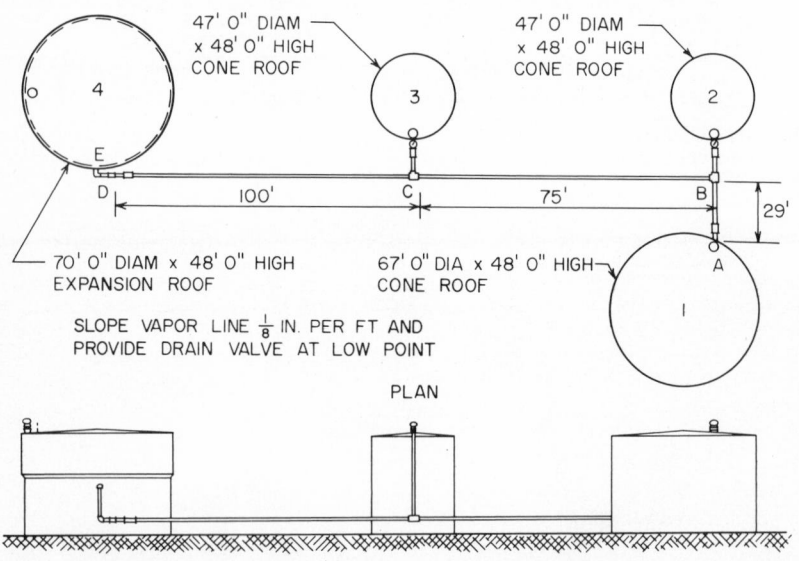

FIG. 4-152. Application of expansion roof to a group of cone-roof tanks.   (*Courtesy of Graver Tank & Mfg. Co.*)

the lifter roof rises or falls.   There is no breathing loss, provided the expansion of the air-vapor space does not exceed the capacity of the lifter roof.   If the capacity of the lifter is ample, one or more additional cone-roof tanks may be accommodated.

The vapor space of the additional tanks is connected to the vapor space of the tank equipped with a lifter, as shown in Fig. 4-152.   This type of roof operates at low pressures of ½ to 1 oz/sq in.   When excess pressure develops, the air-vapor mixture is vented to the atmosphere.   Lifter roofs follow gas-holder design with liquid seals. The rise may be up to 15 ft, depending on the volume of the anticipated expansion of the air-vapor mixture.   They are available for tanks with capacities up to 120,000 bbl.

Another type of construction which allows vapor-volume changes but restrains them from escape to the atmosphere is the vapor dome mounted on a standard gastight cone roof (Fig. 4-153).   A flexible impervious membrane of nylon impregnated with synthetic rubber, and hemispherical in shape, is attached at the equator of the dome and

FIG. 4-153. Vapor dome allows vapor-volume changes but prevents vapors from escaping to atmosphere.   (*Courtesy of Pittsburgh-Des Moines.*)

allowed to hang downward freely.   As it is inflated, it reaches the restraint of the steel shell of the dome and hence gives a vapor-expansion volume equal to a sphere of the dome's diameter.   Vapor domes are available in diameters to 66 ft, and volumes of 150,000 cu ft.   They may be placed upon the roofs of existing tanks.   If properly selected for size, air-vapor mixtures may be accommodated from other tanks.

A modification of the vapor dome is a separate cylindrical shell supporting the dome from ground level.   Spheres operating on the same principle are also used.   Both the two last named are usually installed where multiple-tank connection is being considered.

A recapitulation of the most advantageous applications of various types of tanks is

Pan-type floating roofs for
   Sour crude oils
Pontoon and double-deck floating roofs for
   Light crude oils
   Motor and aviation gasoline and jet fuels
   (Note: Recommended where annual throughput is high)
Spheroids and hemispheroids for
   Motor and aviation gasolines and jet fuels
   Natural gasoline
   (Note: Recommended when annual throughput is low)
Spheres for
   Natural gasoline
   LP gas

Lifter roofs, vapor domes, and vapor spheres for
All volatile products where throughput is not high; recommended only when it is desirable to apply conservation methods to existing tankage

Cost of Storage Tanks

The cost of API welded-steel tanks with and without cone or floating roofs erected in 1957 is shown by Fig. 4-154.   Nelson or ENR indices may be used for subsequent years.   The costs are exclusive of foundations and painting.   Sandblasting, priming, and two coats of paint cost $16 to $18 per square (100 sq ft) in the same year.

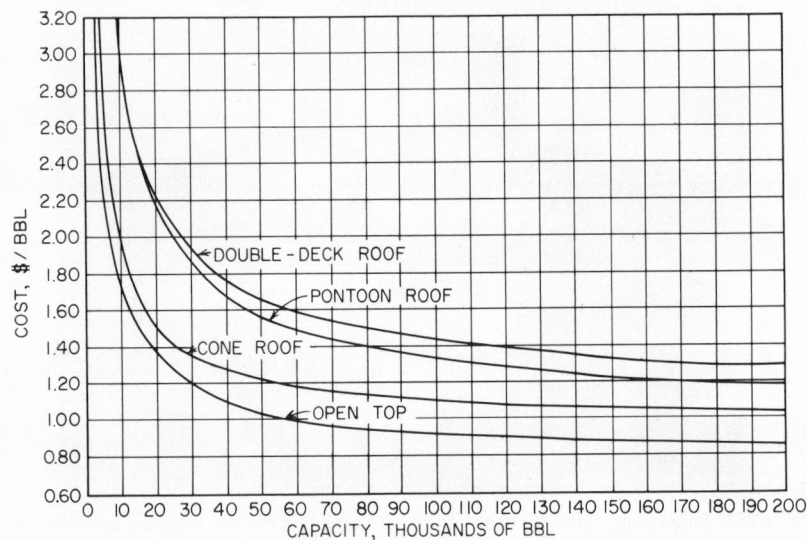

FIG. 4-154. Cost of API welded-steel tanks for oil storage, 1957.[2]

The cost of spheroids varies too widely for close figures.   In general, capacity for capacity, they will cost $2\frac{1}{2}$ to 4 times standard cone-roof tanks.
The cost of spheres depends upon capacity and working pressure and may be approximated by Fig. 4-155.
Relative costs of various types of storage tanks are reported by *Petroleum Processing*.[29]

Conventional gastight cone-roof tanks......................	1.0
Floating roofs, pan-type...................................	1.10–1.20
Floating roofs, pontoon and double-deck.....................	1.25–1.50
Hemispheroids..	1.50–2.00
Lifter roofs...	1.10–1.20
Spheroids..	3.00–5.00
Spheres..	5.00–10.00

Painting

The preceding costs do not allow for painting.   An outline specification includes sandblasting or wire brushing (power equipment to remove all mill scale) followed by a prime coat of red lead and oil and followed by two finish coats.   Costs per square (100 sq ft), in 1959 approximated: sandblasting, $6/$8; prime and two finish coats, $8/$11.   A single repaint coat averages $5 to $6 per 100 sq ft.

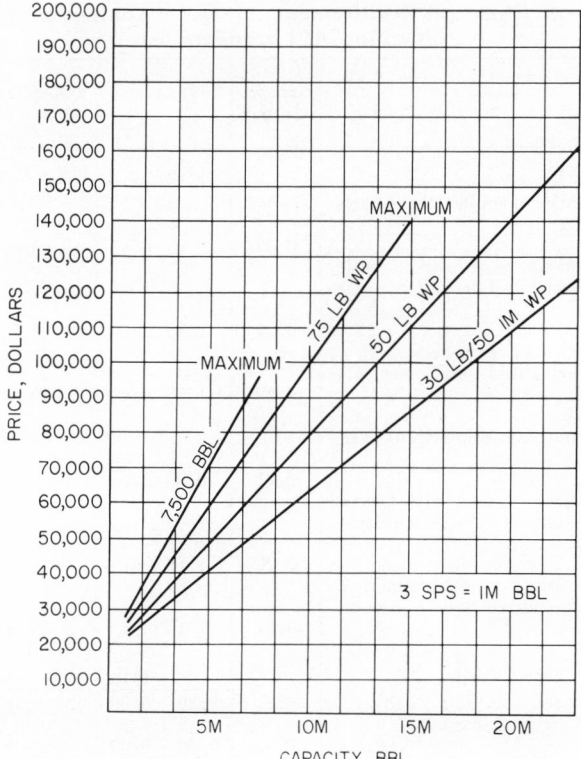

FIG. 4-155. Cost of spheres for pressure storage, 1957.[2]

PIPING AND FITTINGS

Piping

The latest edition of the *ASA Standard Code for Refinery Piping*, B31.3, applies. API Standard 5-L covers standard piping, and 5-LX covers high-test line pipe.

Specification 5-L offers the chemical requirements of the steel for ordinary line pipe. Tables 4-28 and 4-29 from this specification describe the process for manufacture of the steel. The tensile requirements are shown in Table 4-30.

High-test line pipe, 5-LX, conforms to more rigorous standards with greater tensile and bursting strength than for pipe manufactured under API Standard 5-L. Three grades, X-42, X-46, and X-52, are specified as shown in Table 4-31, which gives the chemical requirements. Table 4-32 lists the physical properties.

The use of the terms *standard*, *extra strong*, and *double extra strong* have been superseded in the oil industry. Pipe is now ordered by wall thickness or by schedule number. The schedule number for each size of pipe takes into consideration the wall thickness and is approximately equal to 1,000 P/S, in which P is design pressure and S is allowable stress.

There are many factors affecting the value of P. Joint efficiency, external pressure, if any, and temperature must be taken into consideration. A corrosion factor may be advisable.

Table 4-28. Chemical Requirements for Ladle Analyses of Standard
Line Pipe (API Standard 5-L)

	Carbon, %, max	Manganese, %		Phosphorus, %		Sulfur, %, max
Seamless: Electric-furnace, open-hearth, or basic-oxygen:						
Grade A	0.22	0.90 max		0.04 max		0.05
Electric-furnace, open-hearth, basic-oxygen, or killed deoxidized basic-bessemer:						
Grade B	0.27	1.15 max		0.04 max		0.05
Killed deoxidized acid-bessemer, or killed deoxidized basic-bessemer:						
Grade B	0.22	1.15 max		0.10 max		0.05
Electric-welded: Electric-furnace, open-hearth or basic-oxygen:						
Grade A	0.21	0.90 max		0.04 max		0.05
Electric-furnace, open-hearth, basic-oxygen, or killed deoxidized basic-bessemer:						
Grade B	0.26	1.15 max		0.04 max		0.05
Killed deoxidized acid-bessemer, or killed deoxidized basic-bessemer:						
Grade B	0.21	1.15 max		0.10 max		0.05
		min	max	min	max	
Lap-welded or butt-welded:						
Electric-furnace	....	0.30	0.60	.....	0.045	0.060
Open-hearth or basic-oxygen:						
Class I	....	0.30	0.60	.....	0.045	0.060
Class II	....	0.30	0.60	0.045	0.080	0.060
Bessemer	....	0.30	0.60	.....	0.110	0.065

Table 4-29. Process of Manufacture of Standard Line Pipe
(API Standard 5-L)

Process	Grade or class designation
Steel pipe:	
Seamless, electric-flash, or electric-resistance welded:	
Electric-furnace, open-hearth, or basic-oxygen	A, B
Bessemer*	B
Electric induction-welded† (6⅝-in. and smaller, 0.280-in. wall thickness and less):	
Electric-furnace, open-hearth, basic-oxygen, or bessemer*	B
Lap-welded or butt-welded:	
Electric-furnace	
Open-hearth or basic oxygen	I, II
Bessemer	

* Killed deoxidized bessemer steel. These requirements apply to both acid-bessemer steel and basic-bessemer steel.
† The induction-welding process is tentative, and the application of the API monogram to pipe made by this process is not permitted.

Table 4-30. Tensile Requirements for Standard Line Pipe
(API Standard 5-L)

	Yield strength, min, psi	Tensile strength, min, psi	Elongation, min, in.	
			2 in.	8 in.
Steel pipe				
Seamless, electric-flash or electric-resistance welded:				
Electric-furnace, open-hearth, or basic-oxygen, grade A	30,000	48,000	†	
Electric-furnace, open-hearth, basic-oxygen, or bessemer, grade B	35,000	60,000	†	
Electric induction-welded:*				
Electric-furnace, open-hearth, basic-oxygen, or bessemer, grade B	35,000	60,000	†	
Lap-welded or butt-welded:				
Electric-furnace	25,000	45,000	...	22‡
Class I open-hearth or basic-oxygen	25,000	45,000	...	22‡
Class II open-hearth or basic-oxygen	28,000	48,000	...	20‡
Bessemer	30,000	50,000	...	18‡

* Tentative.

† Elongation requirements for grades A and B seamless and electric-welded steel line pipe shall be as follows:

Tabulated wall thickness, in.	Elongation in 2 in., min, %		Tabulated wall thickness, in.	Elongation in 2 in., min, %	
	Grade A	Grade B		Grade A	Grade B
0.312 and larger	35.00	30.00	0.156	26.25	22.50
0.281	33.25	28.50	0.125	24.50	21.00
0.250	31.50	27.00	0.094	22.75	19.50
0.219	29.75	25.50	0.062	21.00	18.00
0.188	28.00	24.00			

The above table gives the computed minimum elongations value for each $\frac{1}{32}$-in. decrease in wall thickness. If the tabulated wall thickness is less than 0.312 in. and other than shown above, the minimum elongation requirement shall be determined by the following formulas.

Grade	Formula
A	$E = 56t + 17.50$
B	$E = 48t + 15.00$

wherein E = per cent elongation and t = tabulated wall thickness of specimen, in.

‡ The elongation and gage-length requirements for butt-welded pipe, 1.050 in. outside diameter ($\frac{3}{4}$ in. nominal) and smaller, shall be as follows:

Outside diam, in.	Nominal size, in.	Gage length, in.	Elongation, min, %
1.050, 0.840	$\frac{3}{4}$, $\frac{1}{2}$	6	18
0.675, 0.540	$\frac{3}{8}$, $\frac{1}{4}$	4	18
0.405	$\frac{1}{8}$	2	18

Table 4-31. Chemical Requirements for Ladle Analyses of High-test Line Pipe (API Standard 5-LX)

Process of manufacture	Grade	Carbon, % max	Manganese, % max	Phosphorus, % max	Sulfur, % max
Seamless:					
Electric-furnace, open-hearth, basic-oxygen or killed deoxidized basic-bessemer:*					
Nonexpanded.............	X42	0.29	1.25	0.04	0.05
Nonexpanded.............	X46, X52	0.32	1.35	0.04	0.05
Cold-expanded†...........	X42, X46, X52	0.29	1.25	0.04	0.05
Killed deoxidized acid-bessemer or killed deoxidized basic-bessemer:*					
Nonexpanded.............	X42	0.24	1.25	0.10	0.05
Nonexpanded.............	X46, X52	0.27	1.35	0.10	0.05
Cold-expanded‡...........	X42, X46, X52	0.24	1.25	0.10	0.05
Welded:					
Electric-furnace, open-hearth, basic-oxygen or killed deoxidized basic-bessemer:*					
Nonexpanded.............	X42	0.28	1.25	0.04	0.05
Nonexpanded.............	X46, X52	0.31	1.35	0.04	0.05
Cold-expanded	X42, X46, X52	0.28	1.25	0.04	0.05
Killed deoxidized acid-bessemer or killed deoxidized basic-bessemer:*					
Nonexpanded.............	X42	0.23	1.25	0.10	0.05
Nonexpanded.............	X46, X52	0.26	1.35	0.10	0.05
Cold-expanded............	X42, X46	0.23	1.25	0.10	0.05
Cold-expanded............	X52	0.24	1.25	0.10	0.05

* Pipe made from killed deoxidized basic-bessemer steel shall conform to one of the analyses shown above, as specified on the purchase order.
† For cold-expanded seamless pipe in sizes 20 in. and larger, the maximum carbon content shall be 0.28 per cent.
‡ For cold-expanded grades X42 and X46 seamless pipe in sizes 20 in. and larger, the maximum carbon content shall be 0.23 per cent.

Screwed couplings for refinery piping are seldom used, except for small sizes. Couplings should be of *line-pipe* type. These differ from the couplings furnished with merchant pipe. The barrel is longer and recessed at the ends, and the threads are different. Table 4-33, from API Standard 5-L, conveys information for standard-weight threaded-line-pipe trade designation "Pipe with T&C" in sizes through 4 in.

Refinery piping is welded in place by the manual shielded metal-arc process wherever possible. Where joints are necessary, weld-neck flanges are customarily used. Pipe is ordered with plain ends beveled for welding. The usual angle of bevel is 37½° from a line at 90° to the axis of the pipe, with a tolerance of +2½°.

Pipe is classified under three methods of manufacture:

1. Seamless, in which a white-hot billet is mechanically pierced by a mandrel guided by rolls. The hot hollow cylinder thus formed is drawn through finishing rolls to proper diameter and wall thickness. Commercial seamless pipe is available in diameters up to 14 in.

2. Electric-resistance-welded pipe is made by forming a flat plate of desired thickness into tubular shape of desired diameter. The edges of the plate are then welded automatically between pressure rolls. Electrodes are placed on both sides of the

**Table 4-32. Tensile Requirements for High-test Line Pipe
(API Standard 5-LX)**

Grade	Yield strength, min, psi	Tensile strength, min, psi	Elongation in 2 in., min, %				
			Tabulated wall thickness, in.*				
			0.500 to 0.312, incl.	0.281	0.250	0.219	0.188
X42	42,000	60,000	25.00	23.75	22.50	21.25	20.00
X46	46,000	63,000	23.00	21.50	20.00	18.50	17.00
X52	52,000	66,000†	22.00	20.00	18.00	16.00	14.00

* If the tabulated wall thickness is less than 0.312 in. and other than shown above, the minimum-elongation requirement shall be determined by the following formula:

Grade	Formula
X42	$E = 40t + 12.50$
X46	$E = 48t + 8.00$
X52	$E = 64t + 2.00$

wherein E = per cent elongation and t = tabulated wall thickness of specimen, in.

† For grade X-52 pipe in sizes 20 in. and larger, with wall thickness 0.375 in. and smaller, the minimum tensile strength shall be 72,000 psi.

seam, and current is passed across the abutting edges. The resistance to the current produces welding temperature, and the pressure from the rolls squeezes the edges together. Electric-resistance-welded pipe is available with outside diameters of from 4½ to 18 in.

3. Electric-fusion-welded pipe is confined to larger sizes. The plate is rolled and formed as in case 2. The pipe is then welded in two passes, first inside with the welding head mounted on a retractable boom while the plate is mechanically held in proper position. Mounted on rolls, the semiwelded pipe passes a stationary welding fixture for the outside seam. This order is reversed by some mills and the outside weld is made first. It is customary to apply high hydraulic pressure to expand the pipe while it is held in restraining dies. This step not only beneficially cold-works the metal but brings the pipe to an accurate outside diameter. Fusion-welded pipe is available in outside diameters from 18 in. up to any reasonable size.

The test pressures for pipe are specified by the API Standard 5-L for line pipe and 5-LX for high-test pipe. In general, test values are predicated on 60 per cent of the minimum yield strength for line pipe. For high-test line pipe, these values are 75 per cent for 8 in. and smaller, 85 per cent for 10 to 18 in., inclusive, and 90 per cent for 20 in. and larger. Test pressures are given for line pipe in Table 4-34.

All pipe may be cut to exact length in the mill, but for general purposes it is customary to order in 25- or 50-ft random lengths.

When ordering pipe, it is necessary to specify the applicable API or equivalent standard which will take care of the chemical and physical qualities of the steel and the proper tests to be applied. Further information necessary includes

Grade: A or B or X-42, X-46, or X-52
Seamless or electric weld
Outside diameter
Schedule number or wall thickness (Note: If minimum wall thickness is desired, it should be so specified)
Plain or beveled ends, or T&C
Lengths, exact or random

Table 4-33. Standard-weight Threaded-line-pipe Dimensions, Weights, and Test Pressures (API Standard 5-L)

Size, nominal, in.	Outside diam, in. D	Nominal weight threads and coupling, lb/ft	Wall thickness, in. t	Inside diam, in. d	Calculated weight, lb/ft — Plain end w_{pe}	Calculated weight, lb/ft — Threads and coupling w	Calculated coupling weight, lb	Test pressure, psi, min — Steel — Butt-welded	Test pressure, psi, min — Steel — Lap-welded and grade A	Test pressure, psi, min — Steel — Grade B	Test pressure, psi, min — Iron — Open-hearth and wrought
⅛	0.405	0.25	0.068	0.269	0.24	0.25	0.04	700	700	700	700
¼	0.540	0.43	0.088	0.364	0.42	0.43	0.09	700	700	700	700
⅜	0.675	0.57	0.091	0.493	0.57	0.57	0.13	700	700	700	700
½	0.840	0.86	0.109	0.622	0.85	0.86	0.24	700	700	700	700
¾	1.050	1.14	0.113	0.824	1.13	1.14	0.34	700	700	700	700
1	1.315	1.70	0.133	1.049	1.68	1.69	0.54	700	700	700	700
1¼	1.660	2.30	0.140	1.380	2.27	2.30	1.03	800	1,000	1,100	800
1½	1.900	2.75	0.145	1.610	2.72	2.74	0.90	800	1,000	1,100	800
2	2.375	3.75	0.154	2.067	3.65	3.71	1.86	800	1,000	1,100	800
2½	2.875	5.90	0.203	2.469	5.79	5.88	3.27	800	1,000	1,100	800
3	3.500	7.70	0.216	3.068	7.58	7.67	4.09	800	1,000	1,100	800
3½	4.000	9.25	0.226	3.548	9.11	9.27	5.92	1,200	1,200	1,300	950
4	4.500	11.00	0.237	4.026	10.79	11.01	7.59	1,200	1,200	1,300	950

Table 4-34. Regular-weight Plain-end Line-pipe Dimensions, Weights, and Test Pressures (API Standard 5-L)

Size, outside diam, in. D	Plain-end weight, lb/ft	Wall thickness* in. t	Inside diam,* in. d	Test pressure, psi, min		
				Steel		Iron
				Lap-welded and grade A	Grade B	Open-hearth and wrought†
3½	6.63	0.188	3.124	1,900	2,200	1,500
3½	7.58	0.216	3.068	2,200	2,500	1,800
3½	8.68	0.250	3.000	2,500	2,500	2,100
3½	9.67	0.281	2.938	2,500	2,500	2,300
4½	8.64	0.188	4.124	1,500	1,800	1,200
4½	10.00	0.219	4.062	1,700	2,000	1,400
4½	10.79	0.237	4.026	1,900	2,200	1,500
4½	11.35	0.250	4.000	2,000	2,300	1,600
4½	12.67	0.281	3.938	2,200	2,500	1,800
4½	13.98	0.312	3.876	2,500	2,500	2,000
6⅝	12.89	0.188	6.249	1,000	1,200	800
6⅝	14.97	0.219	6.187	1,200	1,400	950
6⅝	17.02	0.250	6.125	1,400	1,600	1,100
6⅝	18.97	0.280	6.065	1,500	1,800	1,200
6⅝	21.07	0.312	6.001	1,700	2,000	1,400
6⅝	23.06	0.344	5.937	1,900	2,200	1,500
6⅝	25.03	0.375	5.875	2,000	2,400	1,600
8⅝	16.90	0.188	8.249	800	900	650
8⅝	19.64	0.219	8.187	900	1,100	750
8⅝	22.36	0.250	8.125	1,000	1,200	850
8⅝	24.70	0.277	8.071	1,200	1,300	950
8⅝	27.74	0.312	8.001	1,300	1,500	1,000
8⅝	28.55	0.322	7.981	1,300	1,600	1,000
8⅝	30.40	0.344	7.937	1,400	1,700	1,100
8⅝	33.04	0.375	7.875	1,600	1,800	1,300
8⅝	38.26	0.438	7.749	1,800	2,100	1,500
10¾	24.60	0.219	10.312	750	850	600
10¾	28.04	0.250	10.250	850	1,000	700
10¾	31.20	0.279	10.192	1,000	1,200	800
10¾	34.24	0.307	10.136	1,000	1,200	800
10¾	38.20	0.344	10.062	1,100	1,300	900
10¾	40.48	0.365	10.020	1,200	1,400	1,000
10¾	48.19	0.438	9.874	1,500	1,700	1,200
12¾	33.38	0.250	12.250	700	800	600
12¾	37.45	0.281	12.188	800	950	650
12¾	41.51	0.312	12.126	900	1,000	700
12¾	43.77	0.330	12.090	1,000	1,200	800
12¾	45.55	0.344	12.062	1,000	1,200	800
12¾	49.56	0.375	12.000	1,100	1,200	850
12¾	57.53	0.438	11.874	1,200	1,400	1,000

Table 4-34. Regular-weight Plain-end Line-pipe Dimensions, Weights, and Test Pressures (API Standard 5-L) (Continued)

Size, outside diam, in. D	Plain-end weight, lb/ft	Wall thickness,* in. t	Inside diam,* in. d	Test pressure, psi, min		
				Steel		Iron
				Lap-welded and grade A	Grade B	Open-hearth and wrought†
14	45.68	0.312	13.376	800	950	650
14	50.14	0.344	13.312	900	1,000	700
14	54.57	0.375	13.250	950	1,100	750
14	63.37	0.438	13.124	1,100	1,300	900
14	72.09	0.500	13.000	1,300	1,500	1,000
16	52.36	0.312	15.376	700	800	550
16	57.48	0.344	15.312	750	900	600
16	62.58	0.375	15.250	850	1,000	700
16	72.72	0.438	15.124	1,000	1,100	800
16	82.77	0.500	15.000	1,100	1,300	900
18	59.03	0.312	17.376	600	750	500
18	64.82	0.344	17.312	700	800	550
18	70.59	0.375	17.250	750	900	600
18	82.06	0.438	17.124	900	1,000	700
18	93.45	0.500	17.000	1,000	1,200	800
20	65.71	0.312	19.376	550	650	450
20	72.16	0.344	19.312	600	700	500
20	78.60	0.375	19.250	700	800	550
20	91.41	0.438	19.124	800	900	650
20	104.13	0.500	19.000	900	1,000	700

* The weight per cubic inch is slightly less for wrought iron than for steel. Since wrought-iron pipe is made with the same weight per foot as steel pipe, the wall thickness of wrought-iron pipe is 2 to 3 per cent greater than the tabulated values above, and the inside diameter is correspondingly smaller.

† Test pressures shown apply to seamless, electric-welded, and lap-welded open-hearth-iron pipe, and to lap-welded wrought-iron pipe.

NOTE: Lap-welded pipe is not produced currently in outside diameters larger than 16 in.

For pipelines, pipe may be obtained mill-coated against corrosion. For refineries, this procedure is rarely followed.

The application of a plastic coating; extruded and bonded to the pipe, has recently been developed for pipes of various sizes. This coating is composed of a priming adhesive of a modified rubber blend applied hot to a thickness of 5 mils. Over this, a coating of high-density polyethylene is extruded onto the pipe. The thickness of the extruded polyethylene plastic coating ranges from 0.25 in. for 1½-in. pipe, up to 0.040 in. for 8-in. pipe. The cost is approximately 10 to 18 per cent above that for the bare pipe.

The cost of line pipe over several[34] years is illustrated by Table 4-35. The costs are for carload lots, f.o.b. mill. Certain mill extras may follow, including cutting to exact length, minimum wall specification, special end preparation, and quantity differentials.

Cast-steel Flanged Fittings

This class of fitting has a broad use in the refinery field. They are available in test pressures to 9,000 psi. Ratings are specified by series numbers. The series number is an indication of the allowable working pressures in psi.   The effect of temperature must be considered.   The series numbers include 150, 300, 400, 900, 1500, and 2500. Series 150 is rated 150 psi at 500°F.   All others are in accordance with the series number at some definite temperature.   Table 4-36 gives information on the test pressures and the pressure-temperature relationship for all series.

As a general guide, carbon steel is used for temperatures to about 750°F.   Above this temperature, alloy steels are essential.   Reference may be made to ASTM Standard 217 for carbon molybdenum and low-chromium molybdenum steels which are suitable for temperatures to 1100°F, and to ASTM Standard A-157 for other alloys used for special conditions.

Table 4-35. Approximate Cost of Welded and Seamless Pipe
(Out of-stock Prices; Carload Lots Are 20 % Less),[34] Dollar per Foot, 1946 and 1956*

Pipe size, in.	Standard (Schedule 40 and 30)		Extra heavy (Schedule 80 and 60)		Double extra heavy, seamless
	Welded	Seamless	Welded	Seamless	
½	0.04(0.09)		0.06(0.12)		
1	0.08(0.17)		0.10(0.22)		
1½	0.12(0.27)		0.17(0.37)		
2	0.16(0.36)	0.20(0.44)	0.23(0.50)	0.30(0.65)	0.59(1.30)
3	0.34(0.74)	0.39(0.84)	0.47(1.02)	0.56(1.23)	1.15(2.53)
4	0.53(1.17)	0.52(1.14)	0.73(1.60)	0.77(1.69)	1.68(3.70)
6	1.05(2.20)	0.90(1.97)	1.35(2.95)	1.44(3.15)	3.27(7.18)
8		1.20(2.62)		2.10(4.59)	4.45(9.78)
10		1.56(3.41)		2.64(5.79)	
12		2.16(4.73)		3.20(7.00)	
14		2.22		3.53	
16		2.52		4.06	
18		2.84		4.58	
20		3.14		5.10	
24		3.78		6.15	

* February, 1956, prices in parentheses (Tulsa).

The dimensions of cast-steel fittings are standardized and obtainable from manufacturer's literature.

Companion flanges for refinery service are customarily forged from steel to the same dimensions as the flanges on valves and fittings.   They carry the same series number and pressure-temperature ratings.   The weld-neck flange is in general use within the industry for all conditions of service and especially for high pressures and temperatures.

Slip-on flanges are sometimes used in close quarters, or where great accuracy of length of pipe runs is essential.   Slipped over the pipe, the hub is then welded to the pipe.   The strength of slip-on flanges under internal pressure is about two-thirds that of weld-neck flanges, and they are limited to a 2½ in. pipe diameter in the 1500 series. They are not made for series 2500.   Slip-on flanges should be limited to series 150 and 300, and restricted from any severe conditions.

The lapped or Van Stone joint consists of lips formed on both ends of the pipe held together by a pair of free slip-on flanges.   The strength of this joint is comparable with that of the slip-on flange joint.   Its principal advantage is in the ease of dismantling and reassembling a complex piping system.   Often of value for steam piping in a powerhouse, it is rarely used in the refinery.

Flanges screwed onto the pipe are not recommended for refineries, except in small sizes and moderate conditions of service.   Screwed fittings and unions are preferable.

Flanges for Fittings

Flanges for fittings, cast steel, forged, and weld-neck, are all made to standard dimensions, and drilling templates are available from manufacturers' catalogs.

The facing of the flanges and the type of gasket used are most important. The commonest design is the raised face. For series 150 and 300, $\frac{1}{16}$ in. is standard, with $\frac{1}{4}$ in. for the heavier series. Plain raised faces with composition gaskets are suitable for relatively low temperatures of about 250°F and for pressures in accordance with

Table 4-36. Pressure-temperature Ratings for Carbon-steel Pipe Flanges and Flanged Fittings According to ASA B16.5-1961*

Primary service-pressure ratings, psig†	150	300	400	600	900	1,500	2,500
Hydrostatic shell test pressure, psig	425	1,100	1,450	2,175	3,250	5,400	9,000
Service temperatures, °F	Max. allowable nonshock service pressure, psig						
100	275	720	960	1,440	2,160	3,600	6,000
150	255	710	945	1,420	2,130	3,550	5,915
200	240	700	930	1,400	2,100	3,500	5,830
250	225	690	920	1,380	2,070	3,450	5,750
300	210	680	910	1,365	2,050	3,415	5,690
350	195	675	900	1,350	2,025	3,375	5,625
400	180	665	890	1,330	2,000	3,330	5,550
450	165	660	875	1,320	1,975	3,255	5,430
500	150‡	625	835	1,250	1,875	3,125	5,210
550	140	590	790	1,180	1,775	2,955	4,925
600	130	555	740	1,110	1,660	2,770	4,620
650	120	515	690	1,030	1,550	2,580	4,300
700	110	470	635	940	1,410	2,350	3,920
750	100	425	575	850	1,275	2,125	3,550
800	92	365	490	· 730	1,100	1,830	3,050
850	82	300‡	400‡	600‡	900‡	1,500‡	2,500‡
900	70	225	295	445	670	1,115	1,855
950	55	155	205	310	465	770	1,285
1000	40	85	115	170	255	430	715

* Applies to ASTM A105-58T Grades I and II and A181-58T Grades I and II (150- and 300-lb classes only) carbon-steel forgings; ASTM A216-58T WCB carbon-steel castings; ASTM A350-55T Grade LF1 forged and ASTM A352-58T Grade LCB cast low-temperature carbon steels.

† Ratings apply also to flanged-end and butt-welding-end valves.

‡ Primary service pressure ratings are in boldface type.

the rating of the fitting. If the raised face is serrated, temperatures to 650°F may be safely handled with ring gaskets of suitable material, such as composition asbestos or metal-clad asbestos.

For more severe service, male-and-female and tongue-and-groove facings are furnished. Flat-ring gaskets are used, either composition or metallic, wherein a soft center core is jacketed in a thin metal covering. These types of joints will develop the full pressure rating of the flanges to a temperature of 750°F.

The ring-type design embraces a groove cut in each of the pairs of matching flanges. Into the groove is inserted a metal ring of soft carbon steel or iron. The cross section of the ring may be oval or octagonal in shape. Such a joint will meet all conditions of pressure and temperature for which the fitting or flange is suitable. If exceptionally high temperatures or unusual corrosive conditions are anticipated, alloy-steel ring gaskets are available. Alloy-steel rings should be heat-treated after fabrication to render the steel as soft as possible.

Figure 4-156 illustrates various types of facings. Table 4-37 presents dimensions and data on bolting.

Pipe Fittings

Fittings in common use in refineries may be classified as

1. Small screwed fittings
   a. Cast iron
   b. Malleable iron
   c. Forged steel
2. Series flanged fittings
   a. Cast iron
   b. Cast steel
3. Weld fittings

Small screwed fittings are used for instrument and control pipe and similar nonprocess lines. Cast- and malleable-iron screwed fittings are available in two classifications, 125 and 250 psi safe working pressure. For nonshock cold-water service, malleable unions are manufactured to service ratings of 600 psi. Cast-iron and malleable fittings should not be used at temperatures above 350°F.

Forged-steel fittings are adaptable to high temperatures and are available with ratings up to 6,000 psi. They are recommended for all temperatures in excess of 350°F.

Flanged fittings are available in two series, 125 and 250.

Class 125 is described by

Size, in	1–12	14–24
Saturated steam, psi	125	100
Nonshock cold service	175	150

Flanges for 125-lb fittings have plain faces with smooth tool surfaces. Flange diameter and drilling correspond to series 150 cast steel.

Class 250 is described by

Sizes, in	1–12
Saturated steam, psi	250
Nonshock cold service, psi	400

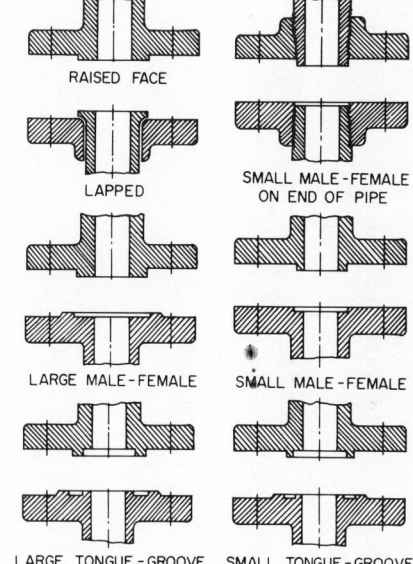

RAISED FACE

LAPPED

SMALL MALE-FEMALE ON END OF PIPE

LARGE MALE-FEMALE

SMALL MALE-FEMALE

LARGE TONGUE-GROOVE

SMALL TONGUE-GROOVE

FIG. 4-156. Typical flange facings.[35]

Flanges for 250-lb fittings have $\frac{1}{16}$-in. raised faces. Bolt circle and drilling for 125 and 250 series in cast iron correspond to those of series 150 and 300 cast-steel fittings, respectively.

The use of cast-iron flanged fittings is limited to relatively low pressures and cold working temperatures, such as water and drainage lines. They are not considered suitable for process or transfer lines where the shock of a cold fire stream on a heated line might occur.

Weld fittings are fabricated from pipe and include 45, 90, and 180° bends, tees, reducers, weld-neck flanges, and all varieties of fittings necessary to comprise a piping complex. Ends are beveled to agree with the ends of beveled pipe. The composition of steel corresponds to that of the pipe used, and the thickness corresponds to the allowable stress-temperature relationship. The allowable working pressures are shown in

Table 4-37. Dimensions of Steel Flanges for Primary Service-pressure Rating of 150 Psig—Table 14, ASA B16.5-1961
All dimensions in inches

SCREWED · SLIP-ON-WELDING · SOCKET WELDING ($\frac{1}{2}$ TO 3 ONLY) · LAPPED · BLIND · WELDING NECK

Nominal pipe size	Outside diameter of flange, O	Thickness of flange, min, C	Diameter of hub, X	Hub diameter beginning of chamfer welding neck, A	Length through hub — Screwed slip-on socket welding, Y	Length through hub — Lapped, Y	Length through hub — Welding neck, Y	Thread length screwed, T	Bore — Slip-on socket welding, min, B	Bore — Lapped, min, B	Bore — Welding-neck socket welding, B	Corner radius of bore of lapped flange and pipe, r	Depth of socket, D
$\frac{1}{2}$	$3\frac{1}{2}$	$\frac{7}{16}$	$1\frac{3}{16}$	0.84	$\frac{5}{8}$	$\frac{5}{8}$	$1\frac{7}{8}$	$\frac{5}{8}$	0.88	0.90	0.62	$\frac{1}{8}$	$\frac{3}{8}$
$\frac{3}{4}$	$3\frac{7}{8}$	$\frac{1}{2}$	$1\frac{1}{2}$	1.05	$\frac{5}{8}$	$\frac{5}{8}$	$2\frac{1}{16}$	$\frac{5}{8}$	1.09	1.11	0.82	$\frac{1}{8}$	$\frac{7}{16}$
1	$4\frac{1}{4}$	$\frac{9}{16}$	$1\frac{15}{16}$	1.32	$1\frac{3}{16}$	$1\frac{3}{16}$	$2\frac{3}{16}$	$1\frac{1}{16}$	1.36	1.38	1.05	$\frac{1}{8}$	$\frac{1}{2}$
$1\frac{1}{4}$	$4\frac{5}{8}$	$\frac{5}{8}$	$2\frac{9}{16}$	1.66	$1\frac{3}{16}$	$1\frac{3}{16}$	$2\frac{1}{4}$	$1\frac{3}{16}$	1.70	1.72	1.38	$\frac{3}{16}$	$\frac{9}{16}$
$1\frac{1}{2}$	5	$1\frac{1}{16}$	$2\frac{9}{16}$	1.90	$\frac{7}{8}$	$\frac{7}{8}$	$2\frac{7}{16}$	$\frac{7}{8}$	1.95	1.97	1.61	$\frac{1}{4}$	$\frac{5}{8}$

2	6	3/4	3 1/16	2.38	1	1	2 1/2	1	2.44	2.46	2.07	5/16	11/16
2 1/2	7	7/8	3 9/16	2.88	1 1/8	1 1/8	2 3/4	1 1/8	2.94	2.97	2.47	5/16	3/4
3	7 1/2	15/16	4 1/4	3.50	1 3/16	1 3/16	2 3/4	1 3/16	3.57	3.60	3.07	3/8	13/16
3 1/2	8 1/2	15/16	4 13/16	4.00	1 1/4	1 1/4	2 13/16	1 1/4	4.07	4.10	3.55	3/8	
4	9	15/16	5 5/16	4.50	1 5/16	1 5/16	3	1 5/16	4.57	4.60	4.03	7/16	
5	10	15/16	6 7/16	5.56	1 7/16	1 7/16	3 1/2	1 7/16	5.66	5.69	5.05	7/16	
6	11	1	7 9/16	6.63	1 9/16	1 9/16	3 1/2	1 9/16	6.72	6.75	6.07	1/2	
8	13 1/2	1 1/8	9 11/16	8.63	1 3/4	1 3/4	4	1 3/4	8.72	8.75	7.98	1/2	
10	16	1 3/16	12	10.75	1 15/16	1 15/16	4	1 15/16	10.88	10.92	10.02	1/2	
12	19	1 1/4	14 3/8	12.75	2 3/16	2 3/16	4 1/2	2 3/16	12.88	12.92	12.00	1/2	
14	21	1 3/8	15 3/4	14.00	2 1/4	3 1/8	5	2 1/4	14.14	14.18	To be	1/2	
16	23 1/2	1 7/16	18	16.00	2 1/2	3 7/16	5	2 1/2	16.16	16.19	specified	1/2	
18	25	1 9/16	19 7/8	18.00	2 11/16	3 13/16	5 1/2	2 11/16	18.18	18.20	by	1/2	
20	27 1/2	1 11/16	22	20.00	2 7/8	4 1/16	5 11/16	2 7/8	20.20	20.25	pur-	1/2	
24	32	1 7/8	26 1/8	24.00	3 1/4	4 3/8	6	3 1/4	24.25	24.25	chaser	1/2	

Table 4-38. Allowable Working Pressures for Refinery Piping—Grade A Carbon Steel

(Values apply also to ASTM A106, Grade A seamless pipe)

Nominal pipe size	Weight or schedule No.	Wall thick-ness	Allowable working pressures For temperatures, °F, not to exceed				
			−20 to 100	300	500	650	750
½	ST 40	0.109	1,810	1,640	1,480	1,360	1,210
¾	ST 40	0.113	1,550	1,410	1,270	1,170	1,040
	XS 80	0.154	2,770	2,510	2,270	2,080	1,850
1	ST 40	0.133	1,690	1,530	1,380	1,270	1,130
	XS 80	0.179	2,780	2,520	2,280	2,090	1,860
	160	0.250	4,580	4,150	3,750	3,440	3,070
	XX	0.358	7,630	6,920	6,250	5,730	5,110
1¼	ST 40	0.140	1,450	1,320	1,190	1,090	970
	XS 80	0.191	2,400	2,170	1,960	1,800	1,610
	160	0.250	3,550	3,210	2,900	2,660	2,370
	XX	0.382	6,350	5,760	5,200	4,770	4,250
1½	ST 40	0.145	1,340	1,220	1,100	1,010	900
	XS 80	0.200	2,230	2,020	1,820	1,670	1,490
	160	0.281	3,600	3,260	2,950	2,700	2,410
	XX	0.400	5,790	5,250	4,740	4,340	3,870
2	ST 40	0.154	1,180	1,070	970	890	790
	XS 80	0.218	2,000	1,810	1,630	1,500	1,340
	160	0.343	3,680	3,340	3,020	2,760	2,470
	XX	0.436	5,030	4,560	4,120	3,780	3,370
2½	ST 40	0.203	1,480	1,340	1,210	1,110	990
	XS 80	0.276	2,260	2,050	1,850	1,690	1,510
	160	0.375	3,360	3,050	2,750	2,520	2,250
	XX	0.552	5,480	4,970	4,490	4,110	3,670
3	LW	0.120	510	462	417	382	341
	ST 40	0.216	1,320	1,190	1,080	990	880
	XS 80	0.300	2,050	1,860	1,680	1,540	1,370
	160	0.438	3,300	2,990	2,700	2,480	2,210
	XX	0.600	4,880	4,420	3,990	3,660	3,260
3½	LW	0.120	445	404	365	334	298
	ST 40	0.226	1,220	1,110	1,000	920	820
	XS 80	0.318	1,920	1,740	1,570	1,440	1,280
		0.636	4,510	4,090	3,700	3,390	3,020
4	LW	0.120	396	359	324	297	265
	ST 40	0.237	1,160	1,050	950	870	780
	XS 80	0.337	1,830	1,660	1,500	1,370	1,220
	160	0.531	3,190	2,890	2,610	2,390	2,130
	XX	0.674	4,250	3,850	3,480	3,190	2,840
5	LW	0.134	391	354	320	293	262
	ST 40	0.258	1,040	940	850	780	700
	XS 80	0.375	1,670	1,520	1,370	1,260	1,120
	160	0.625	3,080	2,790	2,530	2,310	2,060
	XX	0.750	3,830	3,470	3,130	2,870	2,560

Table 4-38. Allowable Working Pressures for Refinery Piping—Grade A Carbon Steel (Continued)
(Values apply also to ASTM A106, Grade A seamless pipe)

Nominal pipe size	Weight or schedule No.	Wall thick-ness	Allowable working pressures For temperatures, °F, not to exceed				
			−20 to 100	300	500	650	750
6	LW	0.134	328	298	269	246	220
	ST 40	0.280	970	880	790	730	650
	XS 80	0.432	1,650	1,500	1,360	1,240	1,110
	160	0.718	3,010	2,730	2,460	2,260	2,010
	XX	0.864	3,730	3,380	3,060	2,800	2,500
8	LW	0.148	298	270	244	224	200
	ST 40	0.322	880	800	720	660	590
	XS 80	0.500	1,500	1,360	1,230	1,120	1,000
	XX	0.875	2,850	2,580	2,330	2,140	1,910
	160	0.906	2,960	2,690	2,430	2,220	1,980
10	LW	0.165	284	257	232	213	190
	ST 40	0.365	820	750	670	620	550
	XS 60	0.500	1,190	1,080	980	900	800
	160	1.125	2,990	2,710	2,450	2,250	2,000
12	LW	0.180	272	247	223	204	182
	ST	0.375	720	650	590	540	476
	XS	0.500	1,000	910	820	750	670
	160	1.312	2,960	2,690	2,430	2,220	1,980
14	10	0.250	389	353	319	292	261
	ST 30	0.375	650	590	530	485	433
	XS	0.500	910	830	750	680	610
16	10	0.250	341	309	280	256	228
	ST 30	0.375	570	520	463	424	378
	XS 40	0.500	800	720	650	600	530
18	10	0.250	303	275	248	227	203
	ST	0.375	510	454	411	376	335
	XS	0.500	710	640	580	530	469
20	10	0.250	272	247	223	204	182
	ST 20	0.375	450	408	369	338	301
	XS 30	0.500	640	580	520	473	422
22		0.250	247	224	202	185	165
	ST	0.375	408	370	335	306	273
	XS	0.500	580	520	469	430	384
24	10	0.250	226	205	185	170	151
	ST 20	0.375	375	340	307	281	251
	XS	0.500	530	475	429	393	350
26	ST	0.375	346	313	283	260	232
	XS	0.500	484	438	396	363	324
30	ST	0.375	300	272	245	225	201
	XS 20	0.500	418	379	342	314	280
34	ST	0.375	264	240	216	198	177
	XS	0.500	368	334	302	276	247
36	ST	0.375	248	225	204	186	166
	XS	0.500	348	315	285	261	233
42	ST	0.375	213	193	174	160	142
	XS	0.500	298	270	244	223	200

Table 4-39. Allowable Working Pressures for Refinery Piping—Grade B Carbon Steel

(Values apply also to ASTM A106, Grade B seamless pipe)

Nominal pipe size	Weight or schedule No.		Wall thick-ness	Allowable working pressures for temperatures, °F, not to exceed				
				−20 to 100	300	500	650	750
½	ST	40	0.109	2,260	2,050	1,850	1,700	1,470
¾	ST	40	0.113	1,940	1,760	1,590	1,460	1,260
	XS	80	0.154	3,460	3,140	2,830	2,590	2,240
1	ST	40	0.133	2,110	1,910	1,730	1,580	1,370
	XS	80	0.179	3,470	3,150	2,840	2,610	2,250
		160	0.250	5,720	5,200	4,680	4,290	3,710
	XX		0.358	9,540	8,660	7,800	7,160	6,180
1¼	ST	40	0.140	1,810	1,650	1,480	1,360	1,180
	XS	80	0.191	3,000	2,720	2,450	2,250	1,940
		160	0.250	4,430	4,020	3,620	3,320	2,870
	XX		0.382	7,940	7,210	6,490	5,960	5,140
1½	ST	40	0.145	1,680	1,520	1,370	1,260	1,090
	XS	80	0.200	2,780	2,530	2,280	2,090	1,800
		160	0.281	4,500	4,080	3,680	3,380	2,910
	XX		0.400	7,230	6,560	5,910	5,430	4,690
2	ST	40	0.154	1,470	1,340	1,210	1,110	960
	XS	80	0.218	2,490	2,260	2,040	1,870	1,620
		160	0.343	4,600	4,180	3,770	3,450	2,980
	XX		0.436	6,290	5,710	5,140	4,720	4,070
2½	ST	40	0.203	1,850	1,680	1,510	1,390	1,200
	XS	80	0.276	2,820	2,560	2,310	2,120	1,830
		160	0.375	4,200	3,810	3,430	3,150	2,720
	XX		0.552	6,850	6,220	5,600	5,140	4,440
3	LW		0.120	640	580	520	477	412
	ST	40	0.216	1,640	1,490	1,350	1,230	1,070
	XS	80	0.300	2,560	2,320	2,090	1,920	1,660
		160	0.438	4,130	3,750	3,370	3,100	2,670
	XX		0.600	6,090	5,530	4,980	4,570	3,950
3½	LW		0.120	560	510	455	417	361
	ST	40	0.226	1,530	1,390	1,250	1,150	990
	XS	80	0.318	2,400	2,180	1,960	1,800	1,550
			0.636	5,640	5,120	4,610	4,230	3,650
4	LW		0.120	494	449	404	371	320
	ST	40	0.237	1,440	1,310	1,180	1,080	940
	XS	80	0.337	2,280	2,070	1,870	1,710	1,480
		160	0.531	3,980	3,620	3,260	2,990	2,580
	XX		0.674	5,310	4,820	4,340	3,990	3,440
5	LW		0.134	488	443	399	366	316
	ST	40	0.258	1,300	1,180	1,060	980	840
	XS	80	0.375	2,090	1,900	1,710	1,570	1,350
		160	0.625	3,850	3,500	3,150	2,890	2,500
	XX		0.750	4,780	4,340	3,910	3,590	3,100
6	LW		0.134	410	373	336	308	266
	ST	40	0.280	1,210	1,100	990	910	790
	XS	80	0.432	2,070	1,880	1,690	1,550	1,340
		160	0.718	3,760	3,410	3,070	2,820	2,440
	XX		0.864	4,660	4,230	3,810	3,500	3,020

Table 4-39. Allowable Working Pressures for Refinery Piping—Grade B
Carbon Steel (Continued)

Nominal pipe size	Weight or schedule No.	Wall thickness	Allowable working pressures for temperatures, °F, not to exceed				
			−20 to 100	300	500	650	750
8	LW	0.148	372	338	305	279	241
	ST 40	0.322	1,100	1,000	900	830	720
	XS 80	0.500	1,870	1,700	1,530	1,400	1,210
	XX	0.875	3,560	3,230	2,910	2,670	2,310
	160	0.906	3,700	3,360	3,030	2,780	2,400
10	LW	0.165	354	322	290	266	230
	ST 40	0.365	1,030	930	840	770	670
	XS 60	0.500	1,490	1,350	1,220	1,120	970
	160	1.125	3,740	3,400	3,060	2,810	2,420
12	LW	0.180	340	309	278	255	221
	ST	0.375	890	810	730	670	580
	XS	0.500	1,250	1,140	1,020	940	810
	160	1.312	3,700	3,360	3,030	2,780	2,400
14	10	0.250	486	442	398	365	315
	ST 30	0.375	810	740	670	610	530
	XS	0.500	1,140	1,030	930	850	740
16	10	0.250	426	387	349	320	276
	ST 30	0.375	710	650	580	530	458
	XS 40	0.500	990	900	810	750	640
18	10	0.250	378	344	310	284	245
	ST	0.375	630	570	520	470	406
	XS	0.500	880	800	720	660	570
20	10	0.250	340	309	278	255	221
	ST 20	0.375	570	520	460	422	364
	XS 30	0.500	790	720	650	600	520
22		0.250	308	280	252	231	200
	ST	0.375	510	463	417	383	331
	XS	0.500	720	650	590	540	464
24	10	0.250	282	256	231	212	183
	ST 20	0.375	468	425	383	351	304
	XS	0.500	660	600	540	491	424
26	ST	0.375	432	393	354	324	280
	XS	0.500	610	550	494	453	392
30	ST	0.375	374	340	306	281	243
	XS 20	0.500	530	474	427	392	338
34	ST	0.375	330	300	270	248	214
	XS	0.500	460	418	377	345	298
36	ST	0.375	310	282	254	233	201
	XS	0.500	434	395	355	326	282
42	ST	0.375	266	242	218	200	173
	XS	0.500	372	338	305	279	241

Table 4-40. Thermal Expansion of Pipe in Inches per 100 Ft*,[35]

Saturated steam, vacuum in. Hg below 212°F, pressure, psig above 212°F	Temp, °F	Cast iron	Carbon and carbon-molybdenum steel	Wrought iron	4-6% Cr alloy steel	12% Cr stainless steel	18 Cr-8 Ni stainless steel	Copper	Brass
	−200	−1.058	−1.282	−1.289	−1.250	−1.170	−2.030	−1.955	−2.065
	−180	−0.982	−1.176	−1.183	−1.150	−1.070	−1.850	−1.782	−1.890
	−160	−0.891	−1.066	−1.073	−1.030	−0.970	−1.670	−1.612	−1.705
	−140	−0.797	−0.948	−0.955	−0.970	−0.870	−1.480	−1.428	−1.508
	−120	−0.697	−0.826	−0.833	−0.800	−0.750	−1.300	−1.235	−1.308
	−100	−0.593	−0.698	−0.705	−0.700	−0.630	−1.090	−1.040	−1.098
	−80	−0.481	−0.563	−0.570	−0.550	−0.520	−0.880	−0.835	−0.888
	−60	−0.368	−0.428	−0.435	−0.430	−0.400	−0.670	−0.630	−0.673
	−40	−0.248	−0.288	−0.295	−0.290	−0.270	−0.450	−0.421	−0.452
	−20	−0.127	−0.145	−0.152	−0.145	−0.130	−0.225	−0.210	−0.227
	0	0	0	0	0	0	0	0	0
	20	0.128	0.148	0.154	0.140	0.140	0.223	0.238	0.233
	32	0.209	0.230	0.249	0.234	0.234	0.356	0.366	0.373
	40	0.263	0.285	0.313	0.280	0.280	0.446	0.451	0.466
Vacuum, in. Hg 29.39	60	0.391	0.448	0.468	0.430	0.430	0.669	0.684	0.690
28.89	80	0.522	0.580	0.628	0.600	0.600	0.892	0.896	0.920
27.99	100	0.660	0.753	0.787	0.750	0.750	1.115	1.134	1.150
26.48	120	0.799	0.910	0.958	0.900	0.900	1.338	1.366	1.390
24.04	140	0.924	1.064	1.113	1.050	1.050	1.545	1.590	1.625
20.27	160	1.073	1.223	1.275	1.220	1.220	1.784	1.804	1.865
14.63	180	1.218	1.383	1.445	1.370	1.370	2.000	2.051	2.100
6.45	200	1.368	1.546	1.626	1.520	1.520	2.230	2.296	2.340
0	212	1.451	1.643	1.721	1.600	1.600	2.361	2.428	2.467
Pressure, psig 2.5	220	1.507	1.707	1.784	1.675	1.675	2.460	2.516	2.580
10.3	240	1.653	1.875	1.958	1.825	1.825	2.680	2.756	2.830
20.7	260	1.804	2.038	2.127	2.000	2.000	2.920	2.985	3.070
34.5	280	1.958	2.205	2.313	2.150	2.150	3.130	3.218	3.315
52.3	300	2.106	2.374	2.478	2.320	2.320	3.375	3.461	3.565
74.9	320	2.268	2.545	2.648	2.470	2.470	3.615	3.696	3.820
103.3	340	2.416	2.717	2.836	2.625	2.625	3.840	3.941	4.065
138.3	360	2.573	2.884	3.023	2.820	2.780	4.075	4.176	4.320
180.9	380	2.732	3.066	3.198	2.980	2.980	4.346	4.424	4.560
232.4	400	2.881	3.230	3.369	3.140	3.130	4.560	4.666	4.825
293.7	420	3.055	3.421	3.568	3.300	3.300	4.800	4.914	5.080
366.1	440	3.218	3.595	3.748	3.470	3.470	5.045	5.154	5.340
451.3	460	3.384	3.784	3.944	3.650	3.650	5.335	5.408	5.600
550.3	480	3.556	3.955	4.128	3.800	3.800	5.540	5.651	5.925
664.3	500	3.720	4.151	4.325	4.000	4.000	5.800	5.906	6.120
795.3	520	3.893	4.342	4.525	4.150	4.150	6.050	6.148	6.380
945.3	540	4.063	4.525	4.714	4.350	4.340	6.320	6.410	6.650
1,115	560	4.238	4.715	4.905	4.540	4.500	6.572	6.640	6.920
1,308	580	4.414	4.906	5.116	4.740	4.640	6.835	6.919	7.170
1,525	600	4.598	5.102	5.303	4.920	4.850	7.100	7.184	7.440
1,768	620	4.769	5.292	5.508	5.100	5.020	7.370	7.432	7.715
2,041	640	4.955	5.482	5.698	5.280	5.180	7.630	7.689	7.980
2,346	660	5.133	5.686	5.915	5.470	5.350	7.900	7.949	8.240
2,705	680	5.315	5.875	6.108	5.670	5.550	8.170	8.196	8.515
3,080	700	5.502	6.084	6.329	5.850	5.700	8.425	8.472	8.780
	720	5.681	6.280	6.521	6.050	5.900	8.670	8.708	9.050
	740	5.879	6.490	6.747	6.220	6.040	8.932	8.999	9.324
	760	6.073	6.688	6.948	6.430	6.280	9.220	9.256	9.600
	780	6.262	6.901	7.162	6.600	6.480	9.480	9.532	9.870
	800	6.460	7.105	7.356	6.800	6.680	9.750	9.788	10.150
	820	6.652	7.319	7.605	7.000	6.890	10.020	10.068	10.425
	840	6.843	7.517	7.800	7.200	7.090	10.270	10.308	10.690
	860	7.049	7.743	8.043	7.400	7.300	10.540	10.610	10.975
	880	7.248	7.953	8.248	7.580	7.500	10.820	10.971	11.250
	900	7.452	8.168	8.487	7.770	7.720	11.075	11.156	11.545
	920	7.668	8.400	8.715	7.970	7.950	11.350	11.421	11.815
	940	7.862	8.610	8.937	8.170	8.140	11.620	11.707	12.120
	960	8.073	8.830	9.148	8.360	8.350	11.900	11.976	12.420
	980	8.279	9.051	9.395	8.560	8.550	12.150	12.269	12.720
	1000	8.490	9.276	9.624	8.760	8.750	12.432	12.543	13.080

* From Crocker and King, *Piping Handbook*, 5th ed., McGraw-Hill Book Company, New York. Copyright © 1967. Used by permission.

Tables 4-38 and 4-39 for grades A and B carbon steel. The tables are abstracted from ASTM Specification A-106.

When joints are indicated, weld-neck flanges are used. Flange dimensions and bolting correspond to Table 4-37.

Weld fittings are available in 45, 90, and 180° elbows, both short and long radius and also for crosses, tees, etc. Ends are beveled for welding, $36\frac{1}{2}° \pm 2\frac{1}{2}°$, as for pipe. Manufacture includes fittings made from all grades of pipe.

Bolting

Bolts connecting two flanges must be selected to exert sufficient pressure to maintain a tight gasket but must not exert strains in flanges or fittings. For theory on the relationship of flange, gasket, and bolts, reference is directed to the *Piping Handbook*.[35]

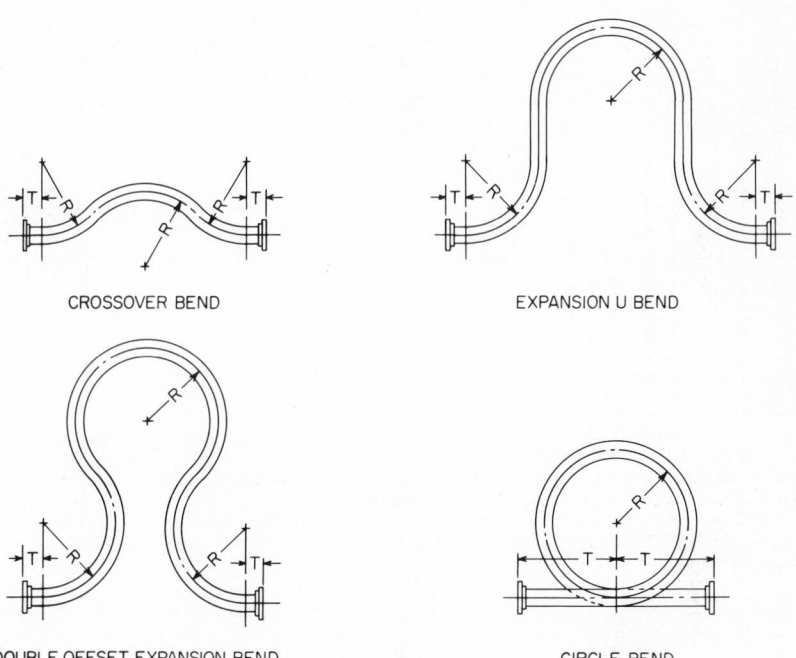

CROSSOVER BEND   EXPANSION U BEND

DOUBLE OFFSET EXPANSION BEND   CIRCLE BEND

Fig. 4-157. Configuration of expansion bends for refinery piping. See Table 4-41 for dimensions.

For cast-iron flanges, carbon-steel machine bolts with square head and hexagon nuts may be used at temperatures from −20 to 500°F when operating pressures do not exceed 500 psig. For temperatures below −20°F, alloy steel must be used.

For temperatures above 500°F, stud bolts with nuts on each end are used. Chrome-molybdenum alloy steels are suitable for temperatures to 1100°F. Stainless-steel bolts are used at higher temperature.

Expansion

Table 4-40 gives the thermal expansion of commonly used piping materials over a wide temperature range. Such expansion is compensated for by expansion joints, either a bend or loop fabricated from pipe, or a mechanical joint.

The bends fabricated from straight lengths of pipe are illustrated by Fig. 4-157.

Table 4-41. Dimensions of Pipe Bends Shown in Fig. 4-157*,[35]

Pipe size	Minimum tangent ends			Radii					
				Recommended		Minimum		†Extreme minimum	
	Sc'd	Weld	Lap	Rad.	Arc.	Rad.	Arc.	Rad.	Arc.
	T	T	T	R	90	R	90	R	90
1	2¼	2¼		7	0′11″	2¼	0′ 3 9/16	2″	0′ 3⅛
1¼	2½	2½		8	1′ 0 9/16	2¾	0′ 4⅜	2½	0′ 3 15/16
1½	2¾	2¾		10	1′ 3 11/16	3¼	0′ 5⅛	3	0′ 4¾
2	3	3		1′0	1′ 6⅞	4½	0′ 7 1/16	4″	0′ 6¼
2½	4	4		1′2	1′10″	6	0′ 9 7/16	5″	0′ 7⅞
3	4	5	6	1′6	2′ 4¼	7	0′11	6	0′ 9 7/16
3½	5	5	6	1′8	2′ 7⅜	9	1′ 2⅛	8	1′ 0 9/16
4	5	5	6	2′0	3′ 1 11/16	1′0	1′ 6⅞	10	1′ 3 11/16
5	6	5	7	2′6	3′11⅛	1′6	2′ 4¼	1′3	1′11 9/16
6	7	6	7	3′0	4′ 8½	2′0	3′ 1 11/16	1′6	2′ 4¼
8	9	6	8	4′6	7′ 0 13/16	2′8	4′ 2¼	2′0	3′ 1 11/16
10	12	7	10	5′6	8′ 7 11/16	4′0	6′ 3⅜	2′6	3′11⅛
12	14	7	10	7′0	11′ 0	5′0	7′10¼	3′0	4′ 8½
14	16	7	14	8′0	12′ 6⅞	6′0	9′ 5⅛	4′0	6′ 3⅜
16	18	8	16	10′0	15′ 8½	7′0	11′ 0	5′6	8′ 7 11/16
18	18	8	18	11′0	17′ 3⅜	8′0	12′ 6⅞	6′6	10′ 2½
20	18	8	18	12′0	18′10¼	9′0	15′ 1⅝	8′0	12′ 6¾
24	18	9	20	15′0	23′ 6¾				

Full dimension sketch or blueprint should accompany all inquiries or orders for bends. Drawings submitted should include dimensions R and T and center-to-end dimensions.

* From Crocker and King, *Piping Handbook*, 5th ed, McGraw-Hill Book Company, New York. Copyright © 1967. Used by permission.

† When "Extreme minimum" radius bends are required, pipe should be Schedule 80 (extra strong) or heavier. "Recommended" and "Minimum" radii bends can be made from Schedule 40 (standard-weight) pipe.

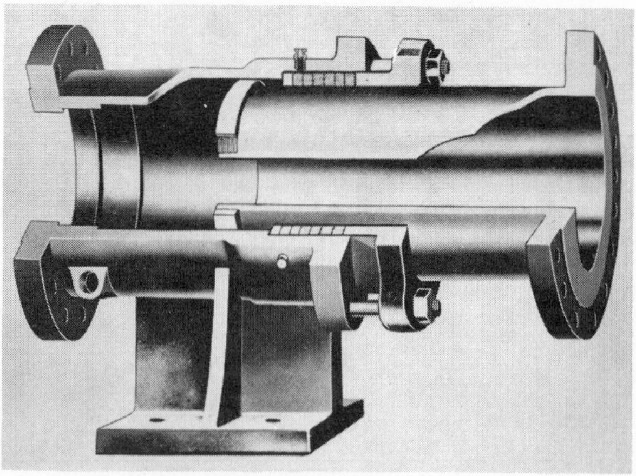

FIG. 4-158. Single-traverse-type, packed, slip expansion joint. (*Courtesy of Adsco.*)

The allowable lengths of straight pipe permissible for a U bend are given in Table 4-41.

With the advent of welded piping systems, expansion bends assembled from 90° weld ells and straight runs of pipe have largely superseded the fabricated bends of Fig. 4-157. Bends of both types are subject to a combination of stresses.[35]

The ability of both types of fabricated pipe bends to accommodate expansion may be enhanced by *cold springing* the bend into position. The method consists of mechanically expanding the bend in the opposite direction to that of the thrust developed as the line expands with temperature.

Mechanical expansion joints are available with packed slip joints or bellows joints.'

Packed-spring expansion joints are available in all normal sizes with screwed and flanged faces when of cast iron, or flanged faces and weld end when of steel. Figure 4-158 illustrates a packed, single-traverse slip joint. Double-traverse joints are also available.

Depending on the size of pipe, the traverse per end varies from 2 to 12 in. Packed slip joints are not recommended for high pressures and temperatures.

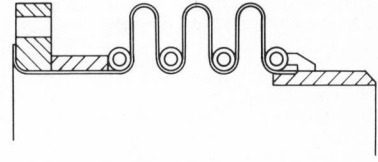

The bellows type is illustrated schematically in Fig. 4-159. The number of waves in the bellows determines the allowable axial traverse of the joint. Copper bellows may be used for moderate pressures and temperatures up to 400°F, and for liquids noncorrosive to copper. Reinforcing rings may be used to increase the pressure rating of the bellows. Stainless steel is used for more rigorous service. This type of joint is available with 125 or 250 psi cast-iron bodies with screwed or flanged faces, and series 150 and 300 with flanged or weld ends. Available service ratings range up to 350°F in cast iron and up to 800°F for steel.

Fig. 4-159. Bellows-type pipe-expansion joint with reinforcing rings.

VALVES

Gate Valves

The temperature and pressure ratings of gate valves correspond to those of fittings. In cast iron, they are available in cold-working nonshock pressures for water, oil, and gas at 125 and 250 psi and 24 in., and at 800 psi up to 12 in. The drilling of flanges for 125 psi valves agrees with that for series 150 steel fittings and valves; that for 250 psi agrees with series 300. The 800 psi (hydraulic) flanges agree with series 600.

Cast-iron valves in refineries are limited to water, steam, and similar services. They are not used for petroleum products within process limits, and rarely for yard transfer lines.

Gate valves, whether of iron or of steel, fall within two definite classifications:

1. The inside-screw or nonrising stem (NRS)
2. The rising-stem, or outside-screw-and-yoke (OS&Y)

The NRS type is less costly and is adapted for use where clearance is limited. It should not be used where corrosive liquids may come in contact with the threads on the stem.

Where clearance is adequate, the OS&Y valve is recommended for oil service. The rising stem indicates the position of the valve at a glance. The threads of the stem are not in contact with the liquid.

Gate valves are further classified by the type of closure, which includes

1. The solid single-wedge gate (Fig. 4-160).
2. The split-wedge or double-disk valve in which the two disks are forced against the seats by the wedging action of the stem.
3. The parallel-seat or slide valve of Fig. 4-161, which is sometimes preferred.

Tightness depends upon the fluid pressure exerted upon the disk.   It offers an advantage that the disk cannot be jammed into the body and hence the valve may be more easily opened.   For this reason it may be favored for motor-operated valves.

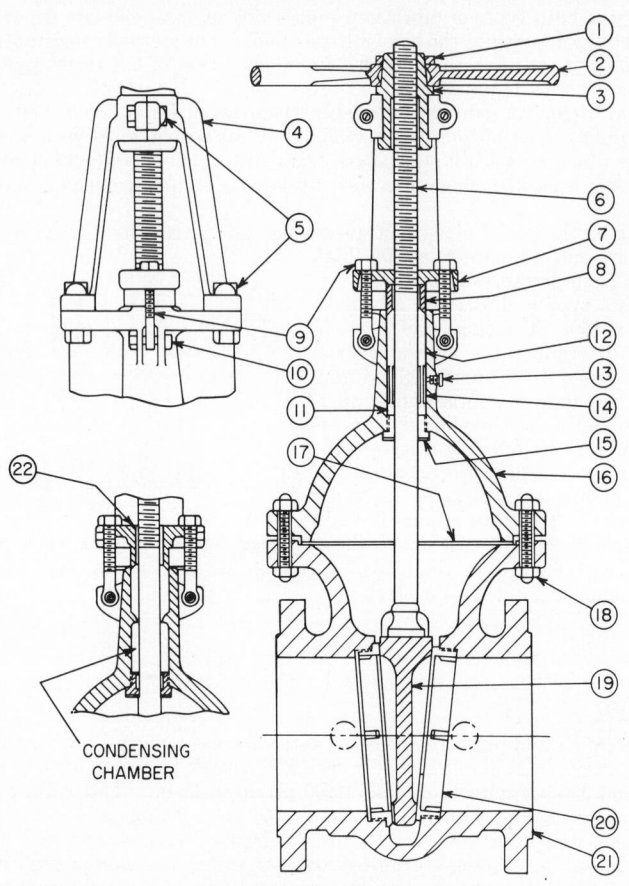

CONDENSING CHAMBER

Fig. 4-160. Wedge gate valve, outside-screw and yoke (OS&Y), rising stem.   (1) Handwheel nut; (2) handwheel; (3) yoke nut; (4) yoke; (5) yoke bolting; (6) stem; (7) gland flange; (8) gland; (9) gland bolts or gland eyebolts and nuts; (10) gland lug bolts and nuts; (11) stem wiper packing; (12) stem packing; (13) drain plug; (14) lantern; (15) bonnet back-seat bushing; (16) bonnet; (17) bonnet gasket; (18) bonnet bolts and nuts; (19) wedge-type gate; (20) seat ring; (21) body; (22) one-piece gland (alternate).   (*API Standard 600.*)

Cast-iron valves are available in all sizes with either screwed or flanged ends.   Cast-steel valves are furnished with flanged or weld ends and are normally available in the following sizes:

Series	Max size, in.
150	24
300	24
400	20
600	18
900	14
1500	12

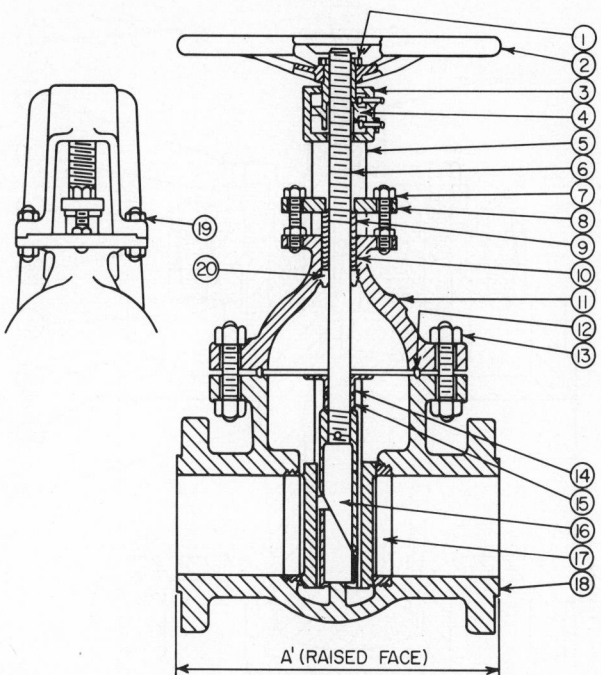

FIG. 4-161. Parallel-seat gate valve, outside-screw and yoke (OS&Y), rising stem. (1) Handwheel nut; (2) handwheel; (3) yoke-nut retaining ring; (4) yoke nut; (5) yoke; (6) stem; (7) gland studs and nuts; (8) gland flange; (9) gland; (10) stem packing; (11) bonnet; (12) bonnet gasket; (13) bonnet studs and nuts; (14) disk guide; (15) back-seat collar; (16) double-disk-type gate; (17) seat ring; (18) body; (19) yoke studs and nuts; (20) bonnet back-seat bushing. (*API Standard 600.*)

Plug Valves

API Standard 600 for refinery valves classifies plug valves under four general design groups:

1. Short pattern in series 150 and 300. This pattern has the same face-to-face dimensions as gate valves of the corresponding series.

2. Regular pattern in which the plug port is larger. Face-to-face dimensions exceed those of corresponding series gate valves.

3. The Venturi pattern in which plug port and body are designed to approximate a Venturi throat. Pressure loss is reduced.

4. The round-port full-bore pattern. The circular bore corresponds to that of attached pipe or fitting. The plugs are necessarily large, and overall dimensions therefore exceed those of types 1, 2, and 3.

Plug valves for refinery service are of the lubricated type, as shown in Fig. 4-162. The screw applying pressure to the lubricant passes through the plug stem. Grooves are cut on the periphery of the plug. Upon application of lubricant pressure, the taper plug is lifted slightly from the valve body, permitting easier rotation of the plug. The lubricant also serves to seal the plug against line pressure.

Rockwell's Hypreseal valve is a modification of the standard design. Figure 4-163 illustrates the valve. The taper plug is inverted and the design permits packing under full line pressure.

Plug valves are available in all the metals and alloys used in gate-valve manufacture,

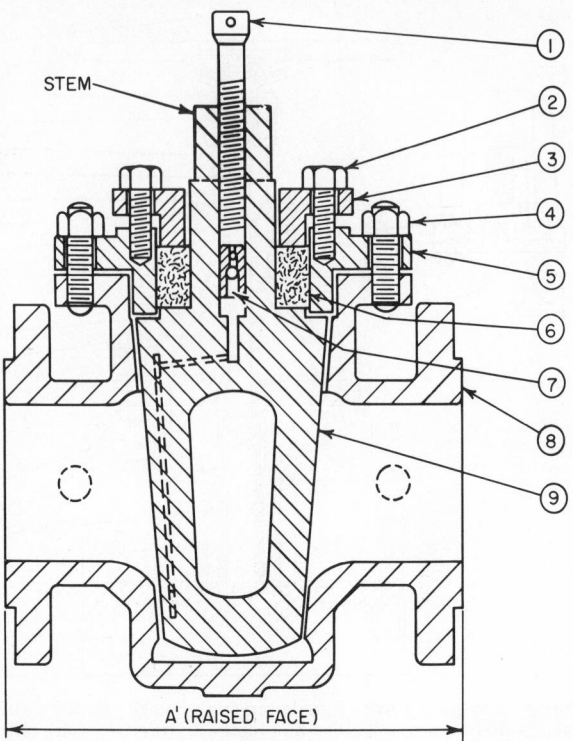

STEM

① ② ③ ④ ⑤ ⑥ ⑦ ⑧ ⑨

A' (RAISED FACE)

FIG. 4-162. Plug valve.   (1) Lubricant screw; (2) gland stud and nut; (3) gland; (4) cover studs and nuts; (5) cover; (6) stem packing; (7) lubricant check valve; (8) body; (9) plug.

and they carry the same series ratings for pressures and temperatures.   In the iron-body design, they are manufactured to test pressures of 1,600 psi in a range of sizes to 24 in., and with steel bodies for all series to 2500.

Series 150 and 300 steel plug valves are available in sizes to 12 in.   Higher series are manufactured in sizes to 10 in.

FIG. 4-163. Hypreseal plug valve with inverted plug taper. (*Courtesy of Rockwell Mfg. Co.*)

Three-way plug valves are illustrated in Fig. 4-164 and four-way in Fig. 4-165.

When pressures and sizes indicate, plug valves should be equipped for spur- or worm-gear operation. Table 4-42 gives the recommendations of API Standard 600. Two types of gear operators are in general use. The spur gear is illustrated by Fig. 4-166.

While lubricated plug valves carry the same pressure-temperature ratings as the corresponding series for gate valves, the type of lubricant influences the operation of these valves. A lubricant must allow the plug to turn easily, yet must have sufficient body to seal the plug tightly against line pressure. Elevated temperatures and the nature of the fluid passing through the valve affect the lubricant.

Inquiries to manufacturers should specify the nature of the fluid and the maximum temperature anticipated.

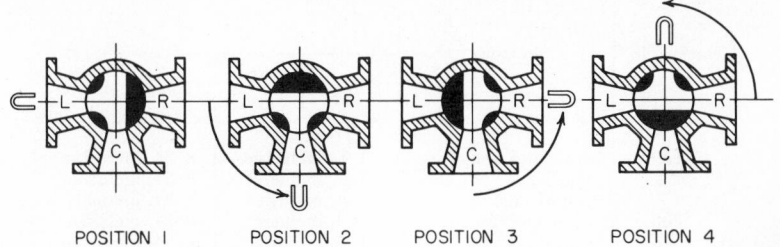

POSITION 1   POSITION 2   POSITION 3   POSITION 4

FIG. 4-164. Four basic settings for a three-way, three-port, 270° turn multiport plug valve. (*Courtesy of Rockwell Mfg. Co.*)

Globe Valves

Globe valves have limited use in refineries. They are available in both nonrising stem and outside-screw-and-yoke designs, as shown in Fig. 4-167. Valves follow the pressure-temperature ratings of gate valves: in cast iron, series 125 and 250 psi; in steel, series 150 to 1500. In cast iron, ends may be screwed or flanged; in steel, screwed, flanged, or weld type. The face-to-face dimension flange diameter and bolting specifications correspond to those for gate valves of the same rating. Sizes available in the standard series ratings range from ¼ to 2 in.

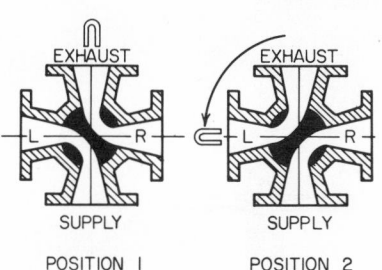

EXHAUST   EXHAUST

SUPPLY   SUPPLY

POSITION 1   POSITION 2

FIG. 4-165. The two basic settings for a four-way, four-port, 90° turn multiport plug valve. (*Courtesy of Rockwell Mfg. Co.*)

FIG. 4-166. Plug valve equipped with spur gearing and handwheel. (*Courtesy of Rockwell Mfg. Co.*)

Forged-steel Valves

Forged-steel valves, machined from solid forgings, find their greatest use in lines at elevated pressures and temperatures and for smaller diameters. Ends may be screwed flanged, socket weld, or weld neck. Gate, globe, check-angle, and needle valves are available. A section of a wedge OS&Y gate valve of series 1500 with ring-joint faces is shown by Fig. 4-168.

Heavy-duty forged-steel valves are also available to 6,000 psi cold-working pressures, and in a variety of standard dimensions. Manufacturers should be consulted for details.

Table 4-42. Plug-valve Stem Operation: Direct vs. Gear
(API Standard 600)

Rating, lb	Pattern	Size, in.	
		Direct operation	Gear operation
150	Short and venturi	1-8, inclusive	8-24, inclusive
	Regular	1-6, inclusive	6-24, inclusive
	Round port	1-4, inclusive	4-24, inclusive
300	Short and venturi	1-8, inclusive	8-24, inclusive
	Regular	1-6, inclusive	6-24, inclusive
	Round port	1-4, inclusive	4-24, inclusive
400	Venturi	4-6, inclusive	6-24, inclusive
	Regular	4	4-24, inclusive
	Round port	4	4-24, inclusive
600	Venturi and regular	1-4, inclusive	4-12, inclusive*
	Round port	1-3, inclusive	3-12, inclusive*
900	Venturi and regular	3	3-12, inclusive*
	Round port	. . . .	3-12, inclusive*
1,500	Venturi and regular	1-3, inclusive	3-12, inclusive*
	Round port	1-2, inclusive	2-12, inclusive*
2,500	Venturi and regular	1-3, inclusive	3-8, inclusive

* Requirements for sizes above 12 in. will be developed if and when a demand is indicated.

Bronze Valves

For small sizes up to 3 in., bronze is used to manufacture various types of gate, globe, and check valves. Pressure ratings are as high as 2,500 psi for cold-water and oil service. Bronze should not be considered at working temperatures above 550°F.

Small bronze valves are generally furnished with screwed ends and are used extensively for water and relatively low-pressure steam service.

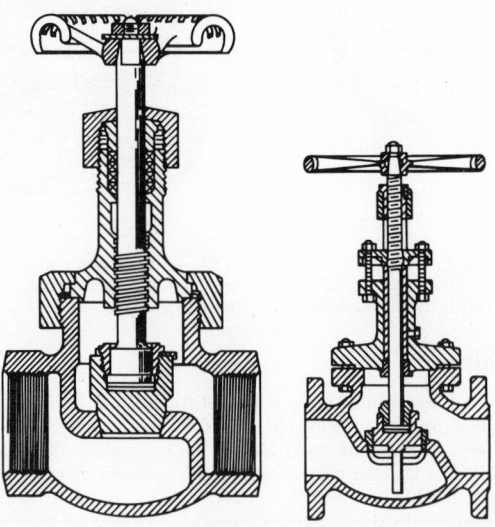

Fig. 4-167. Globe valves. At left, inside-screw rising-steam type; at right, outside-screw and yoke (OS&Y) type.

Bar-stock Valves

This type of valve is available in sizes from ⅛ to 1 in. Bodies are machined from carbon-steel bars and are available only with screwed ends, limited to globe, angle, and needle types.

Pressure ratings range to 3,000 psi, cold nonshock service, and temperatures up to 750°F.

INSULATION

In an age where processing temperatures move further from ambient conditions, where better and more accurate process control is needed, and where economies in equipment and fuel become of paramount importance, the efficient use of heat-containing methods is of significant value. Prior to 1920, little attention was given to this subject. Temperatures for processes were not extremely high, nor were fuel costs a matter of economy. Since then, the need for efficient and economic heat conservation has grown more important as processes go to higher and lower temperature extremes.

The property most significant in terms of heat insulation is thermal conductivity. This is a measure of the rate of heat transfer through a material in one direction per unit thickness per unit time (Btu/sq ft/°F /in./hr as an example). Table 4-43 indicates thermal-conductivity values for various materials. It will vary with temperature, increasing as temperature rises. Air is extremely low in thermal conductivity, and therefore a good insulation barrier. In fact, many insulating materials owe their good properties to the air trapped in the pores of the insulation.

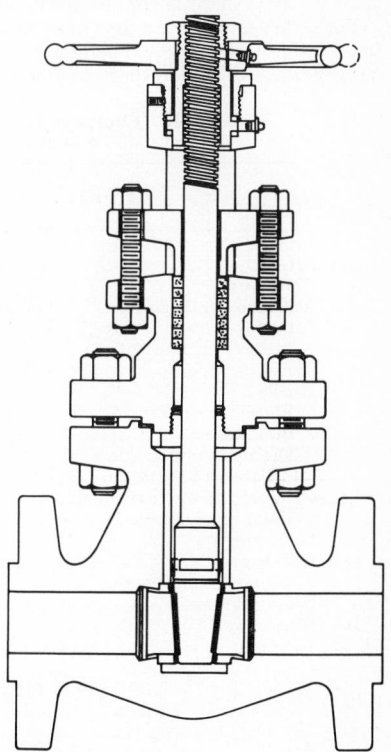

Fig. 4-168. Cross section of forged OS&Y wedge gate valve. (*Courtesy of Henry Vogt Machine Co.*)

Types of Heat Loss

Heat-transfer losses take place in any one of, or a combination of, three ways:

1. *Conduction.* Transmission of heat from one part of a body to another or one body to another in direct contact with it by molecular vibration.

2. *Convection.* Transfer of heat between a fluid and a surface by the circulation or movement of the fluid. It may be free when fluid motion is due to the difference in density between hotter and cooler parts, and forced when circulation is induced by external force such as pump or fan.

3. *Radiation.* Transmission of heat energy through space from one object to another by means of electromagnetic waves of very long wavelengths. The space between objects is not warmed. Radiant energy may be absorbed, becoming thermal energy, and cause an increase in the temperature of the absorbing body.

In most applications of industrial insulation, conduction loss through the insulation is the significant heat-transfer value. It can be expressed, for flat surfaces, by

$$Q = \frac{t_1 - t_2}{x/k}$$

in which Q = heat transferred, Btu/sq ft/hr
t_1 = temperature of warmer surface, °F
t_2 = temperature of colder surface, °F
x = insulation thickness, in.
k = thermal conductivity

For a curved body or a pipe in particular, the heat path increases as heat flow moves from inside to outside.   Therefore, increasing the insulation thickness increases the outside surface with respect to inner and thereby adds resistance to flow of heat.

Table 4-43. Thermal Conductivities k of Various Materials
k = Btu/(hr)(sq ft)(°F)(in.)

Material	Mean temp, °F	Approx k
Air.................................	32	0.16
Aluminum...........................	100	1,416.0
Asbestos, 36 lb/cu ft..................	200	1.32
Asbestos-cement sheets................	70	5.2
Asbestos felt, laminated...............	100	0.40
Asbestos paper, corrugated, 4 plies per in..	100	0.55
Bricks:		
Refractory (average).................	1000	9.0
Building...........................	70	5.0
Diatomaceous earth, across strata......	400	0.61
Parallel to strata..................	400	1.0
Calcium silicate, hydrous..............	100	0.38
Cellular glass, blocks.................	100	0.42
Concrete............................	....	7.2
Corkboard..........................	85	0.30
Fiber insulating board.................	70	0.34
Glass (average)......................	....	6.0
Glass wool, 4 lb/cu ft.................	100	0.28
Hair felt...........................	85	0.26
Ice................................	32	15.6
Magnesia, 85% carbonate of Mg.........	100	0.41
Mineral wool, 4 lb/cu ft...............	100	0.28
Sawdust, 12 lb/cu ft..................	70	0.40
Snow...............................	32	3.2
Steel...............................	70	312.0
Water..............................	80	4.20
Wood (avg)..........................	....	1.2

Figure 4-169 indicates variations in heat-transfer values for changing pipe sizes with a given insulation thickness.   All insulation manufacturers provide full tables and curves for determining characteristics and applicable values.

A material used to insulate heated surfaces of 200°F or greater must have a number of desirable properties.   It needs to have low thermal conductivity as well as ability to maintain its physical condition under temperatures to be encountered.   In addition, it must withstand external effects such as walkway support, abrasion, moisture, corrosive atmosphere, and impact.

Three requirements of insulating materials can be defined for use in elevated-temperature ranges:

1. *Composition*

    a. Fibrous type—consists of materials matted or felted in sheet form (paper, blankets, or batts); blocks, boards or molded shapes that provide varying degrees of rigidity; or loose, fluffy pellets to be poured.   The pellets, when mixed with clay, become insulating cement.

    b. Molded type—consists of materials formed in blocks or segments used primarily on tanks and equipment.

2. *Shape or use*

    a. Pipe insulation
    b. Sheet, block, or board
    c. Cements, finishes, fillers

3. *Temperature limits*

Approx max
temp, °F

    a. Corrugated asbestos paper............................. 300
    b. Fibrous glass, molded with resin binder.................. 450
    c. Carbonate of magnesia, 85 % magnesia.................. 600
    d. Laminated sponge felts............................... 700
    e. Amosite fibers, molded............................... 750
    f. Cellular glass....................................... 700–800
    g. Felted glass fibers, no binder......................... 1000
    h. Felted rock or slag fibers............................ 1200
    i. Hydrous calcium silicate.............................. 1200
    j. High-temperature calcium silicate...................... 1800
    k. Diatomaceous earth, with asbestos..................... 1500–1900
    l. Diatomaceous silica brick, calcined brick................ 1600–2000
    m. Cements and fillers.................................. Up to 1900

In recent years, increasing high-temperature applications for processing have led to newer high-temperature fibers as indicated in Table 4-44.

A new development is designated reflective insulation. It consists of multiple concentric mirror sheets of metal, each of which acts as a thermal barrier by reflecting heat back to its source.

Augmenting Insulation

Generally, elevated process temperatures require that insulation be augmented with some external heat to offset insulation losses. This may take the form of small-diameter tubing or pipe fastened to the process line (tracing), or inserted inside it

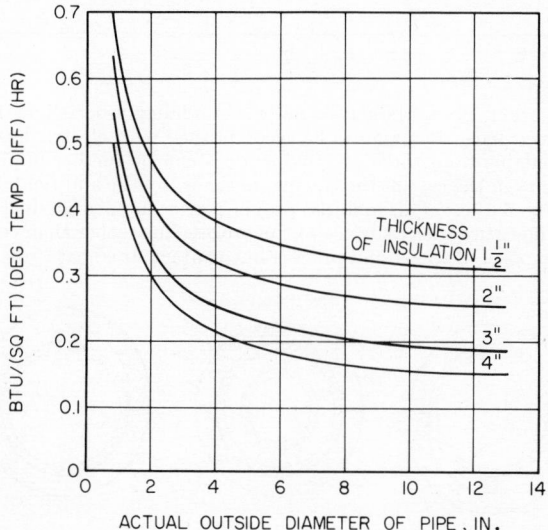

Fig. 4-169. Variation with pipe size of rate of heat transfer through a given thickness of insulation.[36]

Table 4-44. High-temperature Fibers for Process Use As Insulation, Gaskets, Packing, and Filters*

Fiber and manufacturer	Composition	Form	Temp range, °F
Fiberfrax (Carborundum Co., Niagara Falls, N.Y.)	Aluminum silicate plus sodium borate or zirconium dioxide	Bulk, blankets, felt, paper, tubing, coating cement, chopped, board, vacuum-cast shapes, block, roving, rope, yarn, cord, tape, cloth, square braid	1000–2300
Cerafiber, Cerafelt (Johns-Manville, New York)	Aluminum silicate plus titanium dioxide	Felt, board, molded shapes, bulk	1000–2300
Kaowool (Babcock & Wilcox, New York)	High-purity Georgia kaolin	Blanket, strip, bulk	1000–2300
Micro Quartz (Johns-Manville)	High silica (over 96 %)	Felt, web, bulk	Up to 3000
Refrasil (H. I. Thompson, Los Angeles, Calif.)	High silica (over 96 %)	Fabric, felt, sleeves, roving, yarn, bulk	Up to 2200
Sil-Temp (Haveg Industries, Wilmington, Del.)	High silica (98–99 %)	Fabric, mats, yarn, tubing, roving	Up to 3000
Graphite (National Carbon Co. Division of Union Carbide Corp., New York)	99.96 % graphitic carbon	Cloth, felt	Up to 4500
Carbon (National Carbon Co.)	91 % carbon	Cloth, felt	Up to 4500
Hitco-C (H. I. Thompson).	Carbon (usually mixed with phenolics for improved performance)	Cloth, felt	Up to 4000
Pluton B (Minnesota Mining & Mfg. Co., St. Paul, Minn.)	Carbon-containing	Cloth	Not available
Tipersul (Du Pont, Wilmington, Del.)	Potassium titanate	Bulk, sheet, paper, pulp, block	1300–2100

* *Chemical Week*, June 9, 1962, p. 61.

(gut line), or a jacket (pipe of larger diameter surrounding process line) through which heating medium is fed. It may also be accomplished with electric heating cables.

The heat available from these external sources can be further improved, even as much as ten times in the case of the tracing methods, by the addition of heat-transfer cement, as in Fig. 4-170. This material provides a continuous conducting section for the heat from the entire tracer surface to the process line rather than conducting only that heat at the point of tracer and process-line contact; the heat from the remainder

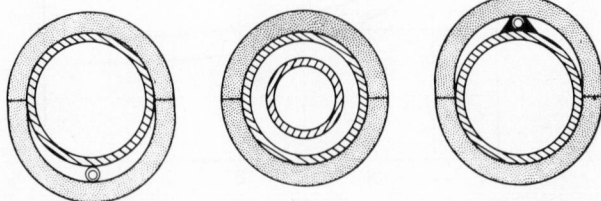

FIG. 4-170. The efficiency of steam tracing with insulation (left) is greater than that of insulation alone (center) but can be materially improved by adhering the tracing to the pipe with a heat-transfer cement (right).

of the tracer surface must transfer by radiation and convection through the very effective insulation of the air gap with no heat-transfer cement to eliminate this insulating effect.

Insulation Finishes

Finishes for outer surfaces of insulating materials are cotton duck or canvas for indoor application, and weatherproof roofing felt for outdoor use. Metal jackets of aluminum, stainless, or galvanized metal are commonly used now as additional protection against fire, to improve appearances, and to provide greater durability.

Vapor Barriers

While the discussion has been concerned with insulation in terms of containing heat, the same principles apply for low-temperature applications. Here the need is to keep heat out or maintain minimum *heat gain*. This generally results in considerably

Table 4-45. Permeability of Various Materials to Water Vapor*,36

Material	Permeability, perms
Aluminum foil, 0.0025 in	0.005– 0.01
Aluminum foil, 0.001 in	0.01 – 0.05
Asphalt-saturated 15-lb felt, coated with 25 lb asphalt/100 sq ft	0.05
Cork, 1 in	6.19
Duplex laminated kraft (30-30-30)	0.20
Duplex laminated kraft, reinforced	0.70
Fiberboard, ¾ in	12.5
Foil-faced kraft paper	0.01
Foil-faced reflective insulation, double-faced	0.08 – 0.13
Gypsum lath, metallic aluminum backing	0.09 – 0.39
Laminated paper and foil	0.01
Mineral wool, unprotected, 4 in	29.07
Plaster, 2 coats aluminum paint	1.15
Plaster, fiberboard, or gypsum lath	19.7 –20.6
Plywood, 2 coats aluminum paint	1.29
Plywood, 2 coats asphalt paint	0.43
Polyethylene, 0.004 in	0.10
Roll roofing, smooth, 40–65 lb/108 sq ft	0.13 – 0.17
Sheathing paper, asphalt-impregnated, glossy	0.17 – 2.05
Vinyl membrane, 0.004 in	0.8 – 2.0

* From A. C. Wilson, *Industrial Thermal Insulation*, McGraw-Hill Book Company, New York, Table A-15, p. 273. Copyright © 1959. Used by permission.

thicker insulation than the corresponding thickness for high-temperature application. In addition, the potential cooling of moisture-laden air and the resultant deposit of water can easily reduce insulation effects, and even destroy the insulating material. The method employed to solve this problem consists of erecting a vapor barrier on the warm side of the insulation, the object being to prevent moisture movement to the cold side. Table 4-45 lists the resistances of various materials suited for such application.

Selecting Insulants

The selection of a proper insulating material or elimination of such a need must be evaluated from a number of considerations. Aside from the standpoint of economics, it must also be considered as a safety item to protect personnel from burns and to provide fire protection.

Table 4-46. Industrial Insulation Application Guide*,[36]

Material	Attachment method	Job requirements or conditions	Finishing methods
		Insulated Pipe	
Molded-pipe insulation *without jacket* (used where special or separately applied finishes are required)	Wire or band in place with 3 fasteners per 3-ft section and no fewer than 2 fasteners per piece	For finished *indoor* work requiring good moisture resistance or washability	Finish with 6- or 8-oz canvas dipped in resin-base lagging adhesive. Additional brush coat of some adhesive may be applied for better appearance

Wire or strap designation	Insulation	For finished *indoor* work with good moisture, rot, and flame resistance	Finish with Fiberglas fabric or asbestos cloth dipped in resin-base adhesive
16-gage annealed wire for hot lines or copper-clad steel for cold or outdoor lines	Up to 12 in. OD	*Indoor work* requiring extra-smooth, finished appearance	Cover with red rosin paper, canvas with 6- or 8-oz canvas pasted with resin-base adhesive. Paint
14 gage (same as above)	12–17 in. OD	Widely used *outdoor* finish	Rag or asbestos roofing felts, 35–45–55 lb wired on 6-in. centers or banded on 8–12-in. centers. Use unsaturated asbestos felts where fire resistance is required
$\frac{3}{8}$ × 0.015 in. galvanized	17–26 in. OD		
$\frac{1}{2}$ × 0.015 in. galvanized	26–36 in. OD		
		Outdoor finish. Durability varies with the selected mastics. Often more durable than felts	Spray, brush, or trowel (according to OD of insulation). Emulsified or cutback asphalt, or vinyl mastic with open-mesh glass fabric between 2 wet coats
		For insulation subject to *heavy abuse or outdoors*	Aluminum or rolled galvanized sheet metal held in place with sheet-metal screws or bands. Thickness of metal 0.015–0.018 in.
		Inexpensive all-purpose finish. Hot work	Light corrugated aluminum, banded or stapled
		All-purpose flexible finish. Available in colors	Vinyl jacket. Zip on over insulation and seal zipper

Molded-pipe insulation *with* factory-applied canvas (used on pipes up to 12 in. where factory-applied canvas jacket is required)	Seal laps with paste. Paste mix should contain mold inhibitor	Concealed hot lines not to be painted	None required
	Seal laps with resin-base lagging adhesive or suitable moisture-proof adhesive	Exposed *indoor* hot lines—moisture-resistant work (painting can be eliminated)	Brush with 1 or 2 coats of resin-base adhesive depending on appearance desired
		Minimum exposed *indoor* treatment	Painting done by others
		For finished *indoor* work with good moisture, rot, and flame resistance	Finish with Fiberglas or asbestos cloth dipped in resin-base adhesive

Table 4-46. Industrial Insulation Application Guide*,36 (Continued)

Material	Attachment method	Job requirements or conditions	Finishing methods
		Insulated Pipe	
		Indoor work requiring extra-smooth, finished appearance	Cover with red rosin paper, canvas with 6- or 8-oz canvas pasted with resin-base adhesive. Paint
		Outdoor finish. Durability varies with the selected mastics. Often more durable than felts	Spray, brush, or trowel (according to OD of insulation), emulsified or cutback asphalt or vinyl mastic with open-mesh glass fabric between 2 wet coats
		For insulation subject to *heavy abuse or outdoors*	Aluminum or rolled galvanized sheet metal held in place with sheet-metal screws or bands. Thickness of metal 0.015–0.018 in.
		All-purpose finish. Hot work	Light corrugated aluminum, banded or stapled
		All-purpose flexible finish. Available in colors	Vinyl jacket. Zip on over insulation and seal zipper
Molded-pipe insulation with *factory*-applied waterproof jackets (35-, 45-. and 55-lb roofing felts). (Used on horizontal runs only. Vertical runs should have the jacket field applied)	On horizontal runs seal all circumferential joints with 4-in. factory-supplied seal strips. Refer to Wire or strap designation above for description of fastener. Lap joint to be at the 4 or 8 o'clock position to shed water	Inexpensive *outdoor* finish	None required
Blanket-type pipe insulation	Band or wire selvage edges using galvanized bands or annealed wires on 8-in. centers. Lace mesh at all circumferential joints on vertical runs with 18-gage annealed wire	Most accepted all-purpose finish for *blanket-type* pipe insulation	Jacket of aluminum sheet or rolled galvanized sheet metal 0.015–0.018-in. thick. Apply ½-in. wide compatible metal bands on 12-in. centers. Use sheet-metal screws for vertical lines
		Insulated Pipe Fittings	
Molded fitting insulation (screwed ells and tees to 8-in. pipe)	Apply with a minimum of 3 staples or 2 wires per fitting. When necessary to "build up" the thickness of the fitting to conform to the rest of the insulated line coat with a hydraulic setting cement	Minimum for *indoor work* (useful on *concealed* work or where painting is omitted)	Apply a brush coat of insulation coating or vinyl mastic

Table 4-46. Industrial Insulation Application Guide*,[36] (Continued)

Material	Attachment method	Job requirements or conditions	Finishing methods
		Insulated Pipe Fittings	
Mitered segments from molded-pipe insulation (not suitable on screwed ells and tees)	Fittings can be fabricated with adhesive or each pair of mitered pieces can be wired in place with annealed steel or soft copper wire. Point up where necessary with an insulating cement to obtain a smooth base for finish	*Indoor* work or lines to be painted	Apply 6-oz canvas or lagging tape with resin-base adhesive. (Include 0.002-in. aluminum foil under fabric for lines below 32°F)
Hydraulic-setting insulating cement (ideal for small pipe sizes)	Apply in standard manner recommended by manufacturer	*Indoor work*, longer service life, and where moisture resistance is important	Apply knitted glass lagging tape with resin-base adhesive. (Include 0.002-in. aluminum foil under tape for lines operating below 32°F)
Flexible wrapped insulation (small fittings, concealed fittings, for all screwed ells and tees)	Wrap around fitting enough layers to obtain required thickness. Hold in place with tying cord or glass tape. Make a smooth finish by troweling a thin coat of a hydraulic setting cement to fill over the voids	*Outdoor* weather-resistant finish. Durability varies with mastic selected and thickness applied	Tack coat of asphalt or vinyl mastic to hold glass-fabric reinforcement. Trowel on asphalt or vinyl mastic finish. (Include 0.002-in. aluminum foil before applying tack coat of asphalt mastic for lines operating below 32°F)
		Insulation of Equipment—Round Surfaces	
		Minimum indoor finish or base for further finish	Apply suitable metal mesh,† lacing edges. Apply $\frac{1}{2}$-in. coat of an insulating cement.
Metal-mesh blankets. Round equipment over 36 in. OD. NOTE: Blankets are available with $\frac{3}{8}$-in. metal rib which provides an air space between the insulation and hot surface, improving overall thermal performance	Best for round or small vessels. Apply with $\frac{3}{4}$-in. bands on 12–18-in. centers. Lace metal edges that butt together	Fair *indoor* finish	To minimum indoor finish (above) finish with a finishing cement and apply 6- or 8-oz canvas pasted. Painting by others
	Good for horizontal tank work: Impale over welded pins on 12-in. centers. Use washers to hold blanket in place. Lace metal edges	Good *indoor* finish (does not require painting)	To minimum indoor finish (above) apply a finishing cement and a resin-base adhesive, 6- or 8-oz canvas, and a final brush coat of adhesive. (On work below 32°F use 0.002-in. aluminum foil under the canvas finish. Adhere foil by brush coating back of foil with rubber-base adhesive)

Table 4-46. Industrial Insulation Application Guide*,[36] (Continued)

Material	Attachment method	Job requirements or conditions	Finishing methods
		Insulation of Equipment—Round Surfaces	
Block. For round equipment over 36 in. OD	Band in place with ½-in. galvanized-steel bands on work up to 48 in. diam. Bands spaced on 12–18-in. centers. (The temp limit where bands will not break due to expansion of equipment can be established by suppliers of metal bands)	*Best indoor* finish with added puncture and fire resistance (does not require painting) *Minimum outdoor* finish	Same as above but use a Fiberglas fabric in place of 6- or 8-oz canvas To minimum indoor finish (above) apply glass fabric or poultry netting and trowel or spray one coat of asphalt or vinyl mastic. (For work below 32°F use 0.002-in. aluminum foil under the glass fabric or poultry netting)
		Most economical for work on *large outdoor storage tanks*	Install large sheets of corrugated metal, asbestos cement boards, or 20-gage flat sheet. Fasten sheets with stud-welded pins or sheet-metal screws to vertical channels or with batten strips held by welded studs. Sheets at each course held with stainless-steel S clips and strapping
	Hold in place with cross wires from welded rods or studs. Point up		
	Miter to fit diameter of vessel. Impale over pins on 12-in. centers. Point up. Use washers to hold insulation		
Blocks and boards	Impale over welded pins on 12-in. centers. Point up. Use washers to hold board in place	*Minimum indoor* finish or base for further finish	Apply suitable metal-mesh† lacing edges. Apply ½-in. coat of an insulating cement
	For side-wall and bottom surfaces with deep stiffeners. Apply over 6 in. × 6 in. × 6 gage road mesh. Use patented metal clips to hold insulation in place	Fair *indoor* finish	To minimum indoor finish (above) apply a finishing cement and 6- or 8-oz canvas pasted. Painting by others
	Impale boards over finger strapping on 12-in. centers. Recommended for multiple-layer construction only. Hold in place with cross wires from welded rods or studs. Point up	Good *indoor finish* (does not require painting)	To minimum indoor finish (above) apply finishing cement and a resin-base adhesive, 6- or 8-oz canvas, and a final brush coat of adhesive. (On work below 32°F, use 0.002-in. aluminum foil with rubber-base adhesive)

Table 4-46. Industrial Insulation Application Guide*,36 (Continued)

Material	Attachment method	Job requirements or conditions	Finishing methods
Insulation of Equipment—Round Surfaces			
Metal-mesh blankets (select suitable metal facing from literature). NOTE: Blankets are available with ⅜-in. metal rib which provides an air space between insulation and hot surface. This improves overall thermal performance of blanket	Most suitable for round or small vessels. Apply with ¾-in. bands on 12–18-in. centers. Lace all metal edges that butt together Best all-around recommendation. Impale over welded pins on 12-in. centers. Use washers to hold blanket in place. Lace metal edges Good for horizontal work. Apply with wire ties to the surface of the vessel by No. 9 wire or cable. Lace metal edges	*Best indoor* finish with added puncture and fire resistance (does not require painting) *Minimum outdoor* finish. Best for low-cost work on large outdoor surfaces	Same as above but use a Fiberglas fabric in place of 6- or 8-oz canvas To minimum indoor finish (above) apply glass netting and trowel or spray one coat of asphalt or vinyl mastic. (For work below 32°F, use 0.002-in. aluminum foil under the glass fabric or poultry netting) Install large sheets of galvanized corrugated metal, asbestos-cement boards, or 20-gage flat sheet. Fasten sheets with stud-welded pins or sheet-metal screws to vertical channels or with batten strips held by welded studs. Hold sheets at each course with stainless-steel S clips and strapping

* From A. C. Wilson, *Industrial Thermal Insulation*, McGraw-Hill Book Company, New York Table 5-1, pp. 82–89.   Copyright ⓒ 1959.   Used by permission.
† Metal mesh—poultry netting, expanded metal lath, etc.

The determination of the proper material will involve such factors as operating temperatures to be encountered, upper and lower temperature limits, space limitation, durability of material under external conditions, and appearance. Attention must also be given to the optimum thickness to provide the desired results with the minimum cost. A considerable body of opinion exists which designs insulation to maintain a maximum of 200°F on the outside skin of the insulating jacket (stainless steel) and 180°F when no jacket is used.

One important caution in the case of insulating austenitic stainless steels concerns the development of stress corrosion in the presence of soluble chlorides. The type of insulation best suited for such installation calls for a maximum of sodium silicate (inhibits stress corrosion) and a minimum of soluble chlorides.

The application of insulation material, in addition to selection of proper material and thickness, must also provide for methods of securing, treatment of fittings, values, etc., and finishing. In addition, consideration must be given to supports. Table 4-46 indicates prevailing practice in the method of applying various types of insulating materials.

Figure 4-171 illustrates a typical metal-covered insulated hot pipe. Insulation is wired with 16-gage galvanized wire on 9-in. centers, and the ends twisted and bent back into the insulation. Joints are pointed with insulating cement.

Long vertical lines require supports spaced at 12-ft intervals, as in Fig. 4-172. Resilient insulation at these sections acts as expansion joints.

Insulation at pipe bends can be fabricated from short radial sections. Pieces should butt tightly against adjacent sections, and joints and seams should be filled with insulating cement, as in Fig. 4-173.

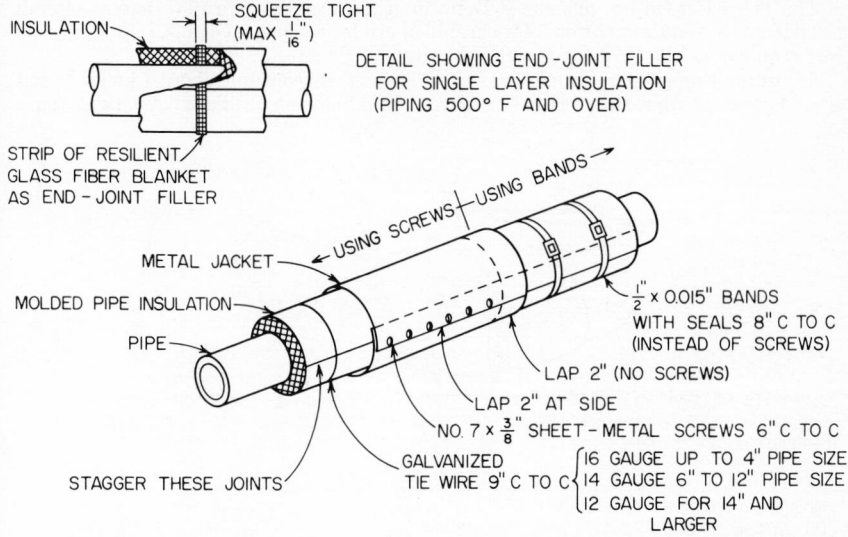

FIG. 4-171. Typical metal-covered insulation showing use of bands and screws.[36]

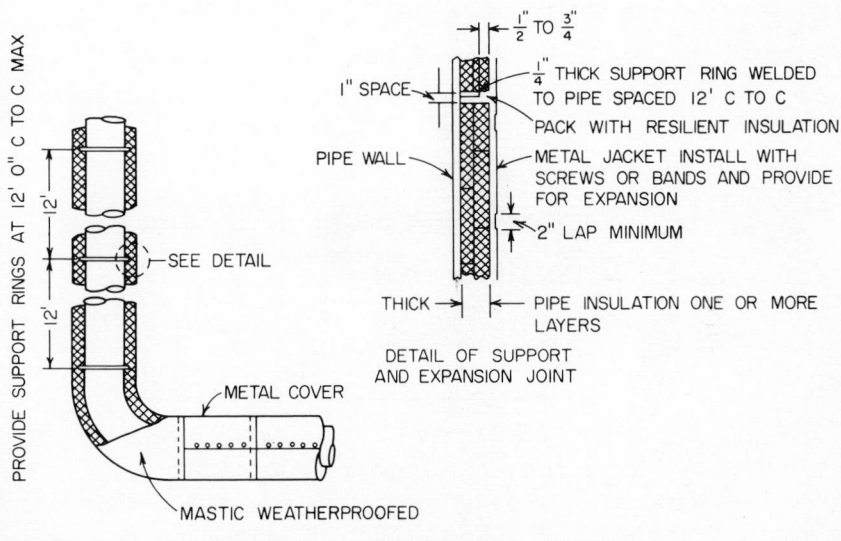

FIG. 4-172. Method for insulation of vertical lines.[36]

Typical sections for valves and flanges, both insulated and uninsulated, are shown in Figs. 4-174 and 4-175.

ELECTRICAL

Electrical requirements for the modern refinery of today are complex, broad, and growing. A typical load spread among various refinery departments is given in Table 4-47.

The Tidewater Oil Co. refinery at Delaware City, Del., built in 1957, had an overall electric-power requirement of 7.4 kwhr/bbl of crude charged.   Of this, 0.31 kwhr/bbl was required as the load for the crude units.

For estimating purposes, main-area yard lighting will consume about 1 kw/acre and secondary areas approximately 0.5 kw/acre.   Buildings can be figured at 0.8 watt/sq ft

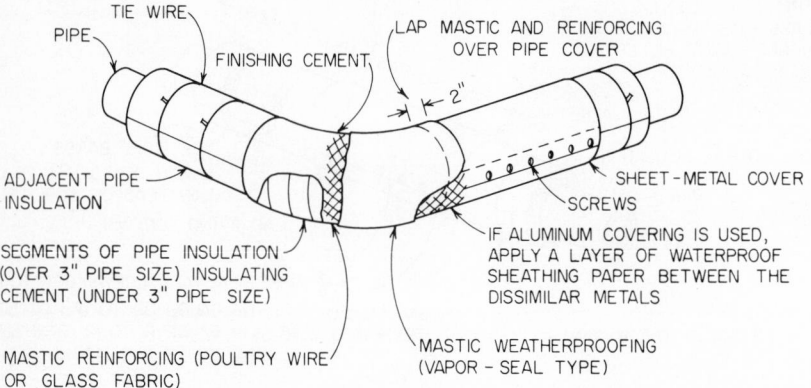

Fig. 4-173. Method for installing insulation at an elbow fitting.[36]

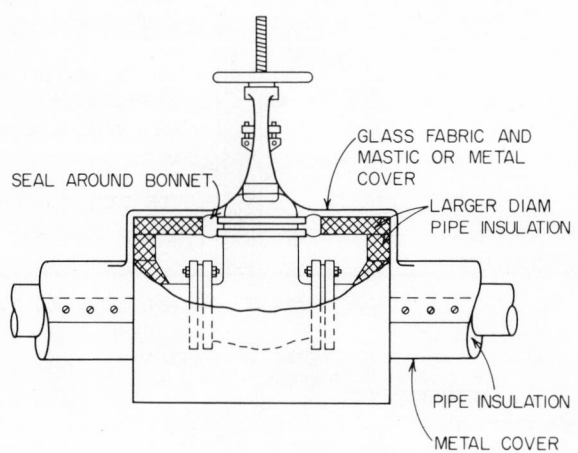

Fig. 4-174. Typical insulation of process valves.[36]

of floor area.   Illumination for distillation units and cracking plants will vary from 1.0 to 1.5 kw/1,000 bbl of charging capacity.   The larger the unit, the lower the figure.

Whatever the electrical-distribution system for the refinery, it should be coordinated with the process-unit design.   In this way the needs for reliability can be kept within the bounds of economy and yet fulfill the overall policy of plant design and operation.

The choice of the system for power distribution will be predicated on the type of power available and the nature of the load.   Many technical publications dealing with such system design are available for a review of the possibilities.   In a situation where power is supplied to a large complex, improved voltage regulation is obtained through the use of distribution centers.   Power from the generating source is distributed via feeders to the center. From this location it goes to the individual use points through mains where transformers step down voltage to required ratings.

Allowable current-carrying capacities for conductors and the number of conductors for various conduit sizes are shown in Tables 4-48 and 4-49.

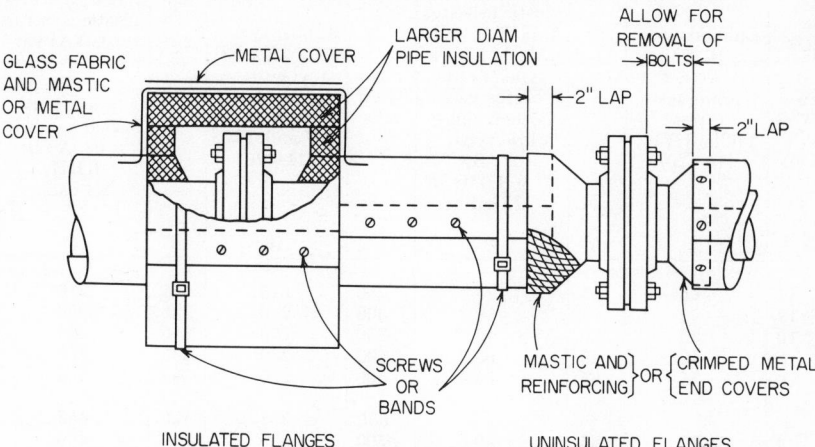

FIG. 4-175. Flange insulation. Pipe covering is aluminum and reinforcing is galvanized poultry mesh. A tar-felt separator should be placed between the dissimilar metals at the point of contact.[36]

Table 4-47. Typical Electric-power Requirements[2]

Service	Unit	kw
Distillation.............................	1,000 bbl	12–18
Cracking...............................	1,000 bbl	25–35
Treating...............................	1,000 bbl	3–5
Wax pressing...........................	100 bbl	10–12
Wax centrifuging.......................	100 bbl	48–60
Solvent plants.........................	100 bbl	8–12
Can manufacturing......................	1,000 cans	6–8
Building illumination..................	1,000 sq ft	0.8–1.0
Process illumination...................	1,000 bbl	0.5–1.0
Yard-area illumination.................	1 acre	0.3–0.4

Transformers

Transformers perform the function of reducing high-voltage transmissions, sent in this fashion for reasons of economy, to lower values suitable for operating the electrical machinery and equipment. One classification rates them as single- or three-phase. A three-phase design incorporates the magnetic circuits of three single-phase transformers, resulting in lower weight and size. However, the three-phase possibility of a single-phase failing makes it somewhat less flexible. Such a failure shuts down the complete transformer, whereas a failure of a single-phase unit in a group of three requires only a single-phase spare to replace it.

Transformers may also be classified according to function. They may be used for power, distribution, instrument, or network application. All generate heat in pro-

Table 4-48. Allowable Current-carrying Capacities of Copper Conductors, Amp*,[22]

Not more than three conductors in raceway or cable.† Based on room temperature of 86°F

Size, AWG	Rubber types: ‡ R, RW, [RU-(14-2)]; thermoplastic types: ‡ T and TW, (14-4/0)	Type RH	Paper; thermoplastic asbestos type TA; varnished cambric type, V; asbestos varnished cambric type, AVB; § RHH; MI cable	Size, M cir mils	Rubber types: ‡ R, RW, RU (14-6); thermoplastic types: ‡ T and TW, (14-4/0)	Type RH	Paper; thermoplastic asbestos type, TA; varnished cambric type, V; asbestos varnished cambric type, AVB; RHH; MI cable
	A	B	C		A	B	C
14	15	15	25	250	215	255	270
12	20	20	30	300	240	285	300
10	30	30	40	350	260	310	325
8	40	45	50	400	280	335	360
6	55	65	70	500	320	380	405
				600	355	420	455
4	70	85	90	700	385	460	490
3	80	100	105	750	400	475	500
2	95	115	120	800	410	490	515
1	110	130	140	900	435	520	555
0	125	150	155	1,000	455	545	585
00	145	175	185	1,250	495	590	645
000	165	200	210	1,500	520	625	700
0000	195	230	235	1,750	545	650	735
				2,000	560	665	775

Correction Factors for Room Temperatures over 86°F

Temp, °F	104	113	122	131	140	158	167	176
Column A	0.82	0.71	0.58	0.41	0.00			
Column B	0.88	0.82	0.75	0.67	0.58	0.35	0.00	
Column C	0.90	0.85	0.80	0.74	0.67	0.52	0.43	0.30

*From T. Baumeister and L. S. Marks, *Mechanical Engineers' Handbook*, McGraw-Hil Book Company, New York, 6th ed., Table 20, p. 15-96. Copyright © 1958. Used by permission.

† For four to six conductors the current is 80 per cent. For seven to nine conductors the current is 70 per cent.

‡ The code rubber insulations in the above National Electrical Code table include those made from natural and synthetic rubber, neoprene, and other vulcanizable materials. Because thermoplastic insulation, as distinct from thermosetting, may stiffen at temperatures below −14°F, care should be used in its installation at such temperatures.

§ The current-carrying capacities for type RHH conductors for sizes 14, 12, and 10 are the same as for type RH conductors in this table; column C shows the capacities of the remaining RHH sizes.

For aluminum wire the allowable carrying capacity of No. 12 is 15 amp (75 per cent of Cu); that of No. 10 is 25 amp (83 per cent of Cu). In general the allowable carrying capacity is between 75 and 80 per cent of those given in the table for the respective sizes of copper wire with the same kind of covering for sizes from No. 12 to 300 MCM. Above 300 MCM the average carrying capacity is 81.5 per cent that of copper, increasing to 84 per cent for 2,000 MCM.

Table 4-49. Minimum Size of Conduit or Tubing, In.*,[22]

Types RF-32, R, RW, RU, RH, RH-RW, TF, T, and TW, 600 volts

Size of wire, AWG or cir mils	No. of wires in conduit								
	1	2	3	4	5	6	7	8	9
No. 18	½	½	½	½	½	½	½	¾	¾
16	½	½	½	½	½	½	¾	¾	¾
14	½	½	½	½	¾	¾	1	1	1
12	½	½	½	¾	¾	1	1	1	1¼
10	½	¾	¾	¾	1	1	1	1¼	1¼
8	½	¾	¾	1	1¼	1¼	1¼	1½	1½
6	½	1	1	1¼	1½	1½	2	2	2
4	½	1¼	1¼	1½	1½	2	2	2	2½
3	¾	1¼	1¼	1½	2	2	2	2½	2½
2	¾	1¼	1¼	2	2	2	2½	2½	2½
1	¾	1½	1½	2	2½	2½	2½	3	3
0	1	1½	2	2	2½	2½	3	3	3
00	1	2	2	2½	2½	3	3	3	3½
000	1	2	2	2½	3	3	3	3½	3½
0000	1¼	2	2½	3	3	3	3½	3½	4
250,000	1¼	2½	2½	3	3	3½	4	4	5
300,000	1¼	2½	2½	3	3½	4	4	5	5
350,000	1¼	3	3	3½	3½	4	5	5	5
400,000	1½	3	3	3½	4	4	5	5	5
500,000	1½	3	3	3½	4	5	5	5	6
600,000	2	3½	3½	4	5	5	6	6	6
700,000	2	3½	3½	5	5	5	6	6	
750,000	2	3½	3½	5	5	6	6	6	
800,000	2	3½	4	5	5	6	6		
900,000	2	4	4	5	6	6	6		
1,000,000	2	4	4	5	6	6			
1,250,000	2½	5	5	6	6				
1,500,000	3	5	5	6					
1,750,000	3	5	6	6					
2,000,000	3	6	6						

* From T. Baumeister and L. S. Marks, *Mechanical Engineers' Handbook*, McGraw-Hill Book Company, New York, 6th ed., Table 21, p. 15-99. Copyright © 1958. Used by permission.

viding the energy conversion, and cooling means must be provided. Air-cooled units will utilize either free or forced ventilation from electric fans. Liquid-cooled systems use water or oil and depend on thermal movement or forced agitation to circulate coolant, which may in turn be either air- or water-cooled. Other transformer accessories include tap-changing facilities to provide changes in ratio of primary to secondary windings, and relief devices to relieve pressure built up in liquid cooling systems due to abnormal operating conditions. Figure 4-176 illustrates a typical design.

Distribution centers or subcenters may be of the multicircuit classification in which a large power transformer is connected to a section of metal-clad switchgear, the latter consisting of switches, circuit breakers, bus bars, instruments, relays, etc. It is used where more than one feeder circuit is to be taken from the transformer. Where only one circuit is taken from the station, it is defined as a single-circuit unit. A load center is another form of substation or distribution center suited to low-voltage distribution and usually located at the center of the load to reduce long heavy bus runs and cable.

Switchgear

Switchgear elements can be further defined:

1. *Bus bars.* Flat copper-strip conductors sized for a current density of approximately 1,000 amp/sq in. (Fig. 4-177).

2. *Switches.* Circuit-closure devices designed to carry 1,000/amp/sq in. (at contact surfaces, current density of 50 amp/sq in. desirable). Types include *knife* (for low-

FIG. 4-176. 5,000-kva transformer, oil-filled, self-cooled; high voltage of 46 kv, low voltage of 15 kv.   (*Courtesy of General Electric Co.*)

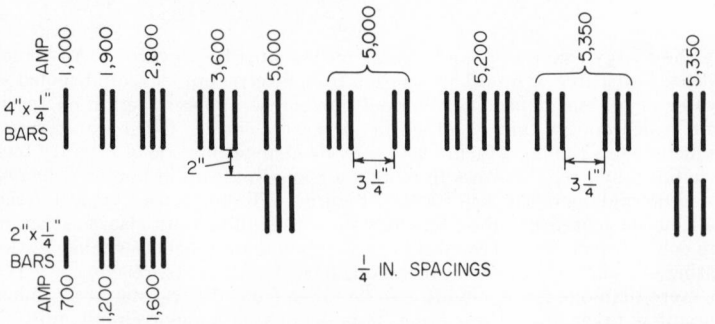

FIG. 4-177. Load-carrying capacity of bus bars.[22]

tension circuits, throw vertically with blade side of switch disconnected from power source when open to avoid accidental contact), *plug* (for high-voltage circuits when current is small), *copper-brush* (used on circuit breakers, uses auxiliary break between carbon blocks to prevent burning copper as result of arcing), *oil-break* (for high-voltage work where oil is needed to extinguish the arc).

3. *Circuit breakers.* Switches equipped with tripping device, usually solenoid actuation, to open circuit when (a) current exceeds set-value overload, (b) current decreases beyond set-value underload, or (c) voltage decreases beyond set-value undervoltage.

Purchased Power

Often, in terms of convenience and flexibility, it is more economical to purchase electric power rather than produce it for internal consumption. In this event, best utilization is made of it, if the following points are carefully considered:[37]

1. Location of all loads, present and future.
2. Energy requirements of each load, present and future.
3. Electric-power source, phase, voltage, capacity.
4. Best utilization voltage for the installation—common ones are 440, 762, and 880.
5. Design and cost of necessary distribution lines, transformer banks, switches, etc.
6. Proper types of electric motors and load evaluation.
7. Rate selection and evaluation.
8. Overall costs of using purchased electric energy for life of project.

Emergency Power

Even with fully reliable electric-power facilities there is always the possibility of failure and the need to maintain essential service. Power reliability in refinery operations is of paramount importance where long on-stream periods and high productivity must be maintained; emergency shutdowns must be avoided. A power failure of even a few seconds' duration may cause production losses of hours or even days. Consideration of, and provision for, insurance against such occurrences through standby facilities are therefore called for. These may consist of a number of sources of emergency power. Selection may be made from dual electrical-distribution system, alternate feeders and loop arrangements, a switchover to steam service on critical electrical-driven service, banks of batteries for communication and lighting, and gasoline-, diesel- or gas-engine drives for generating units.

A prime factor in determining the advisability and extent of emergency facilities is the total cost in relation to the accumulated savings afforded by maintaining full or partial production and reducing time to return on-stream during a power failure. Infrequent failures of short duration and only partial shutdown will usually not justify emergency systems of any consequence.

REFERENCES

1. Nelson, W. L., *Petroleum Refinery Engineering*, McGraw-Hill Book Company, New York, 1958, 4th ed.
2. Bell, H. S., *American Petroleum Refining*, D. Van Nostrand Company, Inc., Princeton, N.J., 1959, 4th ed.
3. *Guide for Inspection of Refinery Equipment*, American Petroleum Institute, New York, 1958, chap. IX.
4. Griswold, J., *Fuels, Combustion, and Furnaces*, McGraw-Hill Book Company, New York, 1946.
5. Robinson, C. S., and E. R. Gilliland, *Elements of Fractional Distillation*, McGraw-Hill Book Company, New York, 1950, 4th ed.
6. Taylor, H. S., and S. Glasstone, *Treatise on Physical Chemistry*, D. Van Nostrand Company, Inc., Princeton, N.J., 1942, 3d ed.
7. Perry, J. H., *Chemical Engineers' Handbook*, McGraw-Hill Book Company, New York, 1963, 4th ed.
8. Bolles, W. L., "Optimum Bubble-cap Tray Design," *Pet. Proc.*, p. 95, March, 1956.

9. Nelson, W. L., "Approximate Capacity of Topping Tower," *Oil & Gas J.*, p. 72, Apr. 22, 1943.
10. Davis, R. S., and P. H. Spitz, "HPI Equipment Selection Handbook," *Petro/Chem Engr.*, p. 41, May, 1965.
11. Chuse, R., *Unfired Pressure Vessels*, McGraw-Hill Book Company, New York, 1960, 4th ed.
12. Mantell, C. L., *Engineering Materials Handbook*, McGraw-Hill Book Company, New York, 1958.
13. Kristal, F. A., and F. A. Annett, *Pumps*, McGraw-Hill Book Company, New York, 1953, 2d ed.
14. Staniar, W., *Plant Engineering Handbook*, McGraw-Hill Book Company, New York, 1959, 2d ed.
15. Jacks, R. L., "Pumps, Fans and Blowers," Pump-Selection 2, 3, *Oil & Gas J.* reprint.
16. Karassik, Igor J., "Process Engineers Guide to Centrifugal Pumps," *Chem. Engr.* reprint.
17. API Standard 617, June, 1958.
18. *Mechanical-drive Steam Turbines for General Refinery Services*, API Standard 615, 1961, 2d ed.
19. Keenan and Keyes, "Theoretical Steam Rate Tables," American Society of Mechanical Engineers, 1938.
20. Vaughn, A. L., ASME Petroleum Conference, Paper 56-PET-13, September, 1956.
21. *Recommended Practice for Electrical Installation in Petroleum Refineries*, API Bull. RP-540, 1959.
22. Baumeister, T., and L. S. Marks, *Mechanical Engineers' Handbook*, McGraw-Hill Book Company, New York, 1958, 6th ed.
23. *Recommended Practices*, RP 1.1–1.7, Instrument Society of America, July, 1959.
24. *Report of the ASME Special Research Committee on Fluid Meters*, American Society of Mechanical Engineers, 1959.
25. *Measurement of Petroleum Liquid Hydrocarbons by Positive Displacement Meters*, API Standard 1101, 1960.
26. *Welded Steel Tanks for Oil Storage*, API Standard 650, 1961.
27. *Guide for Tank Venting*, API Bull. RP-2000, 1952.
28. *Technical Bulletin 20*, Chicago Bridge & Iron Co., 1949.
29. Bussard, W. A., "Evaporation Losses and Their Control in Storage," *Pet. Proc.*, pp. 103–126, July, 1956.
30. Boberg-Water, "Oil Storage Tank Foundations," *CB&I Tower*, March, 1951.
31. Guthrie, V. B., *Petroleum Products Handbook*, McGraw-Hill Book Company, New York, 1960.
32. "Evaporation Losses of Petroleum from Storage Tanks," Part II, *API Proc.*, vol. 32 (I), November, 1952.
33. Martin, Richard, "Covered Floaters," *Petro/Chem Engr.*, p. 23, August, 1965.
34. Nelson, W. L., "Prices of Pipe, Tubing, and Bends," *Oil & Gas J.*, p. 20, Oct. 15, 1956.
35. Crocker, S., *Piping Handbook*, McGraw-Hill Book Company, New York, 1945, 4th ed.
36. Wilson, A. C., *Industrial Thermal Insulation*, McGraw-Hill Book Company, New York, 1959.
37. Stevenson, J. M., "Lowest Purchased Electric Power," *Pet. Engr.*, p. B19-25, February, 1960.

Section 5

MATERIALS OF CONSTRUCTION

By ROBERT B. NORDEN

Senior Editor
Chemical Engineering
New York, N. Y.

FLUID CORROSION

When selecting construction materials for a particular fluid system, it is important to take into consideration first the characteristics of the solution, giving special attention to extraneous factors that may influence corrosion. Since these factors would be peculiar to a particular system, it would be impractical to attempt to offer a set of hard-and-fast rules to cover all situations.

The material or materials from which the system is to be fabricated is the second essential consideration; therefore, a knowledge concerning the characteristics and also the general behavior of materials when exposed to certain environments is essential.

In the absence of actual corrosion information for a particular set of fluid conditions, a reasonably good selection would be possible from data based on resistance of materials to a very similar environment. These data, however, should be used with some reservations. Good practice calls for applying such data for preliminary screening from which materials would be selected for further study in the fluid system under consideration.

General Corrosion

Metallic Materials. Pure metals and their alloys tend to enter into chemical union with the elements of a corrosive medium to form stable compounds similar to those found in nature. When metal loss occurs in this way, the compound formed is referred to as the corrosion product and the metal surface is spoken of as being corroded. The general corrosion-resisting properties of a number of metals and alloys are given in Tables 5-1 and 5-2.

Corrosion is a complex phenomenon that may take any one of several forms. It is usually confined to the metal surface and is called *general* corrosion. But it sometimes occurs along grain boundaries or other lines of weakness because of a difference in resistance to attack or local electrolytic action.

According to the electrochemical concept, the complete corrosion reaction is divided into an anodic portion and a cathodic portion, occurring simultaneously at discrete points on metallic surfaces. Flow of electricity from the anodic to the cathodic areas may be generated by local cells set up either on a single metallic surface (due to local point-to-point differences on the surface) or between dissimilar metals.

Table 5-1. General Corrosion Properties of Some Metals and Alloys*

Ratings: 0 unsuitable. Not available in form required or not suitable for fabrication requirements or not suitable for corrosion conditions.
1 poor to fair.
2 fair. For mild conditions or where periodic replacement is possible. Restricted use.
3 fair to good.
4 good. Suitable when superior alternatives are uneconomic.
5 good to excellent.
6 normally excellent.
Small variations in service conditions may appreciably affect corrosion resistance. Choice of materials is therefore guided wherever possible by a combination of experience and laboratory and site tests.

Material	Liquids — Nonoxidizing or reducing media — Acid solutions, excluding hydrochloric e.g., phosphoric, sulfuric, most conditions, many organics	Neutral solutions, e.g., many nonoxidizing salt solutions, chlorides, sulfates	Alkaline solutions, e.g. — Caustic and mild alkalies, excluding ammonium hydroxide	Ammonium hydroxide and amines	Oxidizing media — Acid solutions, e.g., nitric	Neutral or alkaline solutions, e.g., persulfates, peroxides, chromates	Pitting media,† acid ferric chloride solutions	Natural waters — Fresh-water supplies — Static or slow-moving	Turbulent	Sea water — Static or slow-moving	Turbulent	Gases — Common industrial media — Steam — Moist, condensate	Dry at high temp; promoting slight dissociation	Furnace gases with incidental sulfur content — Reducing, e.g., heat-treatment furnace gases	Oxidizing, e.g., flue gases	Ambient air, city or industrial
Cast iron, flake graphite, plain or low alloy	1	3	4	5	0	4	0	4	3	4	2	4	4	1	1	3
Ductile iron (higher strength and hardness may be attained by composition and heat-treatment or both)	1	3	4	5	0	4	0	4	4	4	3	4	4	1	1	3
Ni-Resist corrosion-resistant cast iron, type 1 (14 Ni; 7 Cu; 2 Cr; bal. Fe)	4	5	5	5	0	5	0	5	5	5	5	5	5	3	2	4
Ni-Resist corrosion-resistant cast iron, type 2 Cu free (20-30 Ni; 2-3 Cr; bal. Fe)	4	5	5	6	0	5	0	5	5	5	5	5	5	3	2	4
Ni-Resist corrosion-resistant cast iron, ductile (24 Ni; bal. Fe)	4	5	5	6	0	5	0	5	5	5	5	5	5	3	2	4
14% silicon iron	6	6	2	5	6	6	3	5	5	5	5	6	4	4	3	6

Material															
Mild steel, also low-alloy irons and steels	1	3	4	4	5	4	0	5	4	0	4	0	4	2	4	4
Stainless steel, ferritic 17% Cr type	2	4	4	4	6	4	5	6	4	0	6	5	4	4	1	6
Stainless steel, austenitic 18 Cr; 8 Ni type	3	4	5	5	6	5	6	6	5	0	6	6	5	5	2	6
Stainless steel, austenitic 18 Cr; 12 Ni; 2.5 Mo type	4	5	5	5	6	5	5	6	5	1	6	5	5	6	3	6
Stainless steel, austenitic 20 Cr; 29 Ni; 2.5 Mo; 3.5 Cu type	5	6	6	5	6	6	5	6	6	2	6	5	6	6	4	6
Ni-o-nel nickel-iron-chromium alloy (40 Ni; 21 Cr; 3 Mo; 1.5 Cu; bal. Fe)	6	6	5	5	6	6	5	6	6	2	6	5	6	6	4	6
Hastelloy alloy Cᵃ (55 Ni; 17 Mo; 16 Cr; 6 Fe; 4 W)	5	6	5	4	6	6	4	6	6	5	6	4	6	6	5	6
Hastelloy alloy Bᵇ (61 Ni; 28 Mo; 6 Fe)	6	5	4	3	6	4	0	6	6	0	3	0	4	3	4	5
Hastelloy alloy D (82 Ni; 10 Si; 4 Cu)	6	6	3	5	6	6	6	6	6	1	5	2	4	6	3	6
Inconel nickel-chromium alloy (78 Ni; 15 Cr; 7 Fe)	6	6	6	6	6	6	4	6	6	1	6	3	6	6	6	6
Copper-nickel alloys up to 30% nickel	4	5	5	5	6	6	6	6	6	1	4	0	5	5	5	5
Monel nickel-copper alloy (68 Ni; 30 Cu; 2 Fe)	5	6	6	6	6	6	4	6	6	1	5	1	6	6	6	6
S Monel nickel-copper cast alloy (66 Ni; 30 Cu; 4 Si)	5	6	6	5	6	6	4	6	6	1	5	1	5	5	6	6
K Monel age hardenable Ni-Cu alloy (67 Ni; 30 Cu; 3 Al)	5	6	6	5	6	6	3	6	6	0	5	1	4	5	5	4
A nickel—commercial (99.4 Ni)	4	5	4	2	6	5	4	6	6	0	4	0	3	4	4	4
Copper and silicon bronze	4	4	2	2	5	5	4	5	6	0	3	0	3	2	2	6
Aluminum brass (76 Cu; 22 Zn; 2 Al)	3	4	2	2	5	5	4	5	6	0	3	0	3	2	2	6
Nickel-aluminum-bronze (80 Cu; 10 Al; 5 Ni; 5 Fe)	4	4	4	5	5	5	4	6	6	0	4	0	4	4	2	6
Bronze, type A (88 Cu; 5 Sn; 5 Ni; 2 Zn)	4	5	0	2	5	5	0–5	6	6	0	0–4	0–5	0	0	4	5
Aluminum and its alloys	1	3	2	6	2	5	5	5	4	0	2	0	6	2	0	5
Lead, chemical or antimonial	5	5	6	2	0	5	5	5	6	0	2	0	2	6	6	4
Silver	4	6	2	0	0	2	6	6	6	6	6	6	0	2	6	3
Titanium	3	6	6	5	5	6	6	6	6	2	6	6	6	2	6	3
Zirconium	2	6	6	5	6	6	6	6	6	2	6	6	6	2	6	3

5–3

ᵃ Also Chlorimet 3. ᵇ Also Chlorimet 2.

ᵃ Also Chlorimet 3.

Table 5-1. General Corrosion Properties of Some Metals and Alloys* (Continued)

Material	Gases (*Cont'd.*) Halogens and derivatives — Halogens — Moist, e.g, chlorine below dew point	Halogens — Dry, e.g, fluorine above dew point	Halide acids, moist, e.g, hydrochloric hydrolysis products of organic halides	Hydrogen halides, dry‡ e.g, dry hydrogen chloride, °F.	Available forms	Cold formability in wrought and clad form	Weldability	Max. strength annealed condition × 1,000 psi	Remarks¶
Cast iron, flake graphite, plain or low alloy	0	2	0	2 < 400 1 < 750	Cast	No	Fair§	45	ASTM-A48-48
Ductile iron (higher strength and hardness may be attained by composition and heat-treatment or both)	0	2	0	2 < 400 1 < 750	Cast	No	Good§	67	ASTM A339, A396-55 MIL-I-1466 (ORD), MIL-I-17166 (Ships), AMS 5313, 5316
Ni-Resist corrosion-resistant cast iron, type 1 (14 Ni; 7 Cu; 2 Cr; bal. Fe)	0	2	3	3 < 400 2 < 750	Cast	No	Good§	22-31	AMS-5392 MIL-G-858A Hyd. Inst. No. 115
Ni-Resist corrosion-resistant cast iron, type 2 Cu free (20-30 Ni; 2-3 Cr; bal. Fe)	0	2	3	3 < 400 2 < 750	Cast	No	Good§	22-31	MIL-G-858A Hyd. Inst. No. 115 Types 3 Ni-Resist has same corrosion resistance
Ni-Resist corrosion-resistant cast iron, ductile (24 Ni; bal. Fe)	0	2	3	3 < 400 2 < 750	Cast	No	Good§	56	AMS-5394 MIL-I-18397 (Ships)
14% silicon iron..........	0	0	4	1 < 400	Cast	No	No	22	Very brittle, susceptible to cracking by mechanical and thermal shock
Mild steel, also low-alloy irons and steels	0	3	0	3 < 400 1 < 750	Wrought, cast	Good	Good	67	High strengths obtainable by alloying, also improved atmospheric corrosion resistance. See ASTM specifications for particular grade
Stainless steel, ferritic 17% Cr type....	0	2	0	2 < 400	Wrought, cast, clad	Good	Good§	78	AISI Type 430 ASTM corrosion- and heat-resisting steels
Stainless steel, austenitic 18 Cr; 8 Ni types	0	2	0	3 < 400	Wrought, cast, clad	Good	Good	90	AISI Type 304 ASTM corrosion- and heat-resisting steels. Stabilized or ELC types used for welding
Stainless steel, austenitic 18 Cr; 12 Ni; 2.5 Mo type	0	3	2	4 < 400 3 < 750	Wrought, cast, clad	Good	Good	90	AISI Type 316 ASTM corrosion- and heat-resisting steel. ELC type used for welding
Stainless steel, austenitic 20 Cr; 29 Ni; 2.5 Mo; 3.5 Cu type	1	3	3	4 < 400 3 < 750	Wrought, cast	Good	Good	90	ACI Cn-7M. Good resistance to sulfuric, phosphoric, and fatty acids at elevated temperatures
Ni-o-nel nickel-iron-chromium alloy (40 Ni; 21 Cr; 3 Mo; 1.5 Cu; bal. Fe)	2	3	3	4 < 400 3 < 750	Wrought, cast, clad	Good	Good	100	Special alloy with good resistance to sulfuric, phosphoric, and fatty acids. Resistant to chlorides in some environments

Corrosion-resistant materials table (data courtesy of International Nickel Co.)

Material	Rating 1	Rating 2	Rating 3	Dry-corrosion rating / max temp, °F	Forms available			Strength, 1000 lb/in²	Remarks
Hastelloy alloy C[a] (55 Ni; 17 Mo; 16 Cr; 6 Fe; 4 W)	5	4	4	4/750, 3/900	Wrought, cast, clad	Fair	Good	145	Excellent resistance to wet chlorine gas and sodium hypochlorite solutions
Hastelloy alloy B[b] (61 Ni; 28 Mo; 6 Fe)	1	3	5	4/750, 3/900	Wrought, cast, clad	Fair	Good	135	Resistant to solutions of hydrochloric and sulfuric acids
Hastelloy alloy D (82 Ni; 10 Si; 4 Cu)	1	1	2	3/900, 3/400	Cast	No	§	90–110	Greatest application in hot concentrated solutions of sulfuric acid
Inconel nickel-chromium alloy (78 Ni; 15 Cr; 7 Fe)	2	5	3	5/400, 4/900, 3/750, 2/400	Wrought, cast, clad	Good	Good	90	Wide application in food and pharmaceutical industries
Copper-nickel alloys up to 30% nickel	1	5	2	4/400, 4/900, 3/750, 2/400	Wrought, cast, clad	Good	Good	38–62	High-iron types excellent for resisting high-velocity effects in condenser tubes
Monel nickel-copper alloy (68 Ni; 30 Cu; 2 Fe)	2	6	3	6/400, 3/750, 2/900	Wrought, cast, clad	Good	Good	77	Widely used for sulfuric acid pickling equipment. Also for propeller shafts in motor boats. Take precautions to avoid sulfur attack during fabrication
S Monel nickel-copper cast alloy (66 Ni; 30 Cu; 4 Si)	2	4	3	6/400, 3/900	Cast	No	No	100	Nongalling characteristics. Excellent for bearings or bushings. High strength developed by heat-treatment
K Monel age hardenable Ni-Cu alloy (67 Ni; 30 Cu; 3 Al)	2	6	3	6/400, 3/900	Wrought, cast	Fair	Good	99–155	High strength obtainable by heat-treatment. Take precautions to avoid sulfur attack during fabrication
A nickel—commercial (99.4 Ni)	2	6	2	6/400, 5/750, 3/900, 2/400	Wrought, cast, clad	Good	Good	54	Widely used for hot concentrated caustic solutions. Take precautions to avoid sulfur attack during fabrication
Copper and silicon bronze	0	5	2	4/400, 3/900	Wrought, cast, clad	Excellent	Fair	29	Unsuitable for hot concentrated mineral acids or for high-velocity HF
Aluminum brass (76 Cu; 22 Zn; 2 Al)	0	4	2	2/400, 3/750	Wrought, cast	Good	Fair	60	May develop localized corrosion in sea water
Nickel-aluminum-bronze (80 Cu; 10 Al; 5 Ni; 5 Fe)	0	4	3	3/400	Wrought, cast	Good	Fair	60–80	Ship propellers an excellent application
Bronze, type A (88 Cu; 5 Sn; 5 Ni; 2 Zn)	0	4	3	3/400, 1/750	Cast	No	§	45	High strengths obtainable by heat-treatment. Not susceptible to dezincification
Aluminum and its alloys	0	6	0	3/400	Wrought, cast, clad	Good	Good	9–90	Extent of corrosion dependent upon type and concentration of acidic ions. Wide range of mechanical properties obtainable by alloying and heat-treatment
Lead, chemical or antimonial	0	1	3	0, 400	Wrought, cast, clad	Excellent	Good	2	High-purity "chemical lead" preferred for most applications
Silver	5	5	3	4/400, 2/750	Wrought, cast, clad	Excellent	Good	21	Used as a lining
Titanium	6	0	1	0	Wrought, cast	Fair	Good§	6–90	Red fuming HNO₃ may initiate explosions. Good resistance to solutions containing chlorides
Zirconium	6	1	6	0	Wrought, cast	Fair	Good§		

* Data courtesy of International Nickel Co.
† On unsuitable materials these media may promote potentially dangerous pitting.
‡ Temperatures are approximate.
§ Special precautions required.
‖ Many of these materials are suitable for resisting dry corrosion at elevated temperatures.
a Also Chlorimet 3. b Also Chlorimet 2.

5–5

Table 5-2. Corrosion Resistance of Selected Materials of Construction

Material of construction	Cast iron	Duriron	Low-alloy steel	5130 steel	AISI 416	AISI 431	AISI 304	AISI 321-347	AISI 316	Brass	Aluminum bronze	Silicon bronze	Phosphor bronze	Cupronickel	Nickel	Monel	Tantalum	Inconel	Hastelloy C	Aluminum	Carbon
Acetic acid	D	...	D	D	C	C	B	B	A	C	B	C	C	C	B	B	A	A	A	B	A
Acetone	A	A	A	A	A	A	A	A	A	A	A	A	A	A	A	A	A	A	A	C	A
Alumina	A	A	A	A	A	A	A	A	A	A	A	A	A	A	A	A	A	A	A	B	B
Aluminum chloride	D	...	D	D	D	D	D	D	C	D	B	B	B	B	B	A	A	A	A	D	A
Ammonia, anhydrous	A	A	A	A	A	A	A	A	A	A	A	A	A	A	A	A	D	A	A	...	A
Ammonia, moist	A	A	A	A	A	A	A	A	A	D	D	D	D	D	C	C	D	A	A	...	A
Asphalt	A	A	A	A	A	A	A	A	A	A	A	A	A	A	A	A	A	A	A	A	A
Barium hydroxide	A	A	A	A	A	A	A	A	A	B	B	A	A	A	A	A	A	A	C	A	
Benzene	A	A	A	A	A	A	A	A	A	A	A	A	...	A	A	A	A	A	A	A	A
Brines	C	B	C	...	D	D	C	C	B	C	B	B	B	B	A	A	A	B	A	C	A
Butane	A	A	A	A	A	A	A	A	A	A	A	A	A	A	A	A	A	A	A	A	D
Calcium chloride	C	B	C	C	D	D	D	D	B	C	B	B	B	B	A	A	A	A	A	A	A
Calcium hydroxide	A	A	A	A	A	A	A	A	A	B	A	A	A	A	A	...	A	A	A	A	A
Calcium hypochlorite	D	C	D	D	D	D	D	D	C	D	D	D	D	D	D	D	A	D	A	B	A
Carbon tetrachloride, dry	A	A	A	A	A	A	A	A	A	A	A	A	A	A	A	A	A	A	A	A	A
Carbon tetrachloride, moist	C	A	C	C	C	C	C	C	C	C	B	B	B	B	A	A	A	A	A	D	A
Chlorine, dry	A	A	A	A	B	B	B	B	B	A	A	A	A	A	A	A	A	A	A	C	A
Chlorine, moist	D	B	D	D	D	D	D	D	D	D	D	D	D	D	D	D	A	D	A	D	A
Copper chloride	D	C	D	D	D	D	D	D	D	D	D	D	D	D	D	D	A	D	C	D	A
Crude oil	A	A	A	A	A	A	A	A	A	B	C	C	C	A	A	A	A	A	A	A	A
Ethers	A	A	A	A	A	A	A	A	A	A	A	A	A	A	A	A	A	A	A	A	A
Ethylene glycol	A	A	A	A	A	A	A	A	A	A	A	A	...	A	A	A	A	A	A	A	
Formaldehyde	C	A	C	C	B	A	A	A	A	A	B	A	A	A	...	A	A	A	A	B	A
Freon	A	A	A	A	A	B	A	A	A	A	A	A	A	A	A	A	A	A	A	A	A
Fuel oil	A	A	A	A	A	A	A	A	A	A	A	A	A	A	A	A	A	A	A	A	A
Furfural	...	A	B	B	A	A	A	A	A	A	C	A	A	A	...	A	...	A	...	A	A
Gasoline	A	A	A	A	A	A	A	A	A	B	A	A	A	A	A	A	A	A	A	A	A
Hydrocarbons, pure	A	A	A	A	A	A	A	A	A	A	A	A	A	A	A	A	A	A	A	A	A
Hydrochloric acid	D	C	D	D	D	D	D	D	D	D	D	C	C	C	C	B	B	A	B	A	D
Hydrofluoric acid	D	D	D	D	D	D	D	D	D	D	C	C	C	C	B	B	D	B	A	D	B
Hydrogen	A	A	A	A	A	A	A	A	A	A	A	A	A	A	A	A	A	A	A	A	D
Hydrogen sulfide, dry	A	A	A	A	A	A	A	A	A	A	A	A	A	A	A	A	A	A	A	A	A
Hydrogen sulfide, moist	C	B	C	C	C	B	B	B	B	C	D	D	D	D	B	B	A	B	B	A	A
Kerosine	A	A	A	A	A	A	A	A	A	A	A	A	A	A	A	...	A	A	A	C	A
Lime	A	A	A	A	A	A	A	A	A	A	A	A	A	A	A	A	A	A	A	...	A
Methyl alcohol	B	A	B	B	A	A	A	A	A	A	A	A	A	A	A	A	A	A	A	A	A
Natural gas	A	A	A	A	A	A	A	A	A	A	A	A	A	A	A	A	A	A	A	A	A
Nitrogen	A	A	A	A	A	A	A	A	A	A	A	A	A	A	A	A	A	A	A	A	D
Oxygen	A	A	A	A	A	A	A	A	A	A	A	A	A	A	A	A	A	A	A	A	D
Paraffin	A	A	A	A	A	A	A	A	A	A	A	A	A	A	A	A	A	A	A	A	A
Phosphoric acid	D	B	D	D	D	D	B	B	A	D	C	C	C	C	C	B	A	B	A	D	A
Potassium hydroxide	B	B	B	B	B	A	A	A	A	A	C	B	B	B	A	A	A	D	A	D	A
Propane	A	A	A	A	A	A	A	A	A	A	A	A	A	A	A	A	A	A	A	A	A
Sea water	C	A	C	C	D	C	C	C	B	B	B	B	B	A	B	A	A	A	A	B	A
Sewage	B	A	B	B	A	A	A	A	B	B	B	B	B	A	A	A	A	A	A		
Sodium carbonate	A	A	A	A	A	A	A	A	A	C	B	B	B	A	A	A	A	A	D	A	
Sodium hydroxide	B	B	B	B	B	A	A	A	A	C	B	B	B	A	A	A	D	A	D	A	
Sodium hypochlorite	D	B	D	D	D	D	D	D	C	D	C	C	C	C	D	D	A	A	A	D	D
Steam	B	B	B	B	B	A	A	A	A	A	C	B	B	B	C	C	A	B	A	A	A
Sulfur, dry	A	A	A	A	A	A	A	A	A	A	B	B	B	A	A	A	A	A	A	A	A
Sulfur, molten	A	A	A	A	A	A	A	A	A	A	D	D	D	D	D	D	D	A	A	A	
Sulfur chloride, dry	A	A	A	A	A	A	A	A	A	A	A	A	A	A	A	A	A	A	A	A	A

Ratings:

A: *Excellent.* Suitable for equipment which requires the maintenance of close tolerances for maximum operating efficiency. Corrosion rate better than 0.005 in./year.

B: *Good.* Suitable where maintenance of close fits not critical, also for exposure conditions where contamination of the solution with small percentages of corrosion product not objectionable. Corrosion rates are 0.020 to 0.005 in./year.

C: *Fair.* Suitable for piping and tanks where corrosion allowance will result in satisfactory performance. Not suitable for maintenance of close tolerances or where the corrosion product will contaminate the process solution. Corrosion rates are usually 0.10 to 0.20 in./year.

D: *Unsatisfactory.* Corrosion losses too high to be suitable for any type of equipment.

Nonmetallic Materials. As stated, corrosion of metals applies specifically to chemical or electrochemical attack. The deterioration of plastics and other nonmetallic materials which are susceptible to swelling, crazing, cracking, softening, etc., is essentially physiochemical rather than electrochemical in nature. The general chemical resistances of important plastics are given in Table 5-3.

Nonmetallic materials either can be rapidly deteriorated when exposed to a particular environment or, at the other extreme, can be practically unaffected. Under some conditions, a nonmetallic may show evidence of gradual deterioration; however, it is seldom possible to evaluate its chemical resistance by measurements of weight loss alone, as is most generally true for metals.

Localized Corrosion

Intergranular Corrosion. Selective corrosion in the grain boundaries of a metal or alloy without appreciable attack on the grains or crystals themselves is called intergranular corrosion. When severe, this attack causes a loss of strength and ductility out of proportion to the amount of metal actually destroyed by corrosion.

Alloys such as the austenitic stainless steels and the aluminum-copper alloys of Duralumin, when improperly heated, become susceptible to intergranular corrosion because of the precipitation of intergranular compounds.

The austenitic stainless steels that are not stabilized or that are not of the extra-low-carbon types when heated in the temperature range of 850 to 1550°F have chromium-rich compounds (chromium carbides) precipitated in the grain boundaries. This causes grain-boundary impoverishment of chromium and makes the affected metal susceptible to intergranular corrosion in many environments.

Austenitic stainless steels stabilized with columbium or titanium to decrease carbide formation or containing less than 0.03 per cent carbon are normally not susceptible to grain-boundary deterioration when heated in the above temperature range. Unstabilized austenitic stainless steels or types with normal carbon content should be given a solution anneal to be immune to intergranular corrosion. This consists of heating to 2000°F, holding at this temperature for a minimum of 1 hr per 1 in. of thickness, followed by rapidly quenching in water (or, if impractical because of size, rapidly cooling with an air-water spray).

When improperly heat-treated, aluminum-copper alloys of Duralumin became susceptible to selective grain-boundary attack. This form of attack is attributed to the precipitation of relatively large particles of the $CuAl_2$ constituent at the grain boundaries, which results in depletion of copper from the grain boundaries of adjacent aluminum-copper solid-solution material. Depletion of copper in the grain-boundary material causes the affected metal to become anodic to both the $CuAl_2$ precipitate and the Al-Cu solid solution, and intergranular corrosion will progress in some environments by galvanic behavior. When properly annealed, the copper will be in solid solution, or when properly age-hardened, the $CuAl_2$ constituent formed will be in a finely divided state, so that in either case, performance will be substantially that of a single-phase alloy.

Pitting Corrosion. Pitting is a form of corrosion that develops in highly localized areas on a metal surface. This results in the development of cavities called pits. They may be deep cavities with small diameters or relatively shallow depressions. Pitting examples include aluminum and stainless alloys in aqueous solutions containing chloride. Inhibitors are sometimes helpful in preventing pitting.

Stress-corrosion Cracking. Corrosion can be accelerated by stress which is either residual internal stress in the metal or externally applied stress. Internal stresses are produced by nonuniform deformation during cold working (bending, shearing, punching, etc.), by unequal cooling from high temperature, or by internal structural rearrangements involving volume change.

Stresses induced in rivets and bolts and by press and shrink fits can also be classified as internal stresses. Tensile stresses at the surface usually of a magnitude equal to the yield stress are necessary to produce stress-corrosion cracking. However, failures of this kind have been known to occur at lower stresses.

Virtually every alloy system has specific environment conditions which will produce

Table 5-3. Chemical Resistance of Important Plastics

	Polypropylene polyethylene	CAB*	ABS†	PVC‡	Saran§	Polyester glass¶	Epoxy glass	Phenolic asbestos	Fluorocarbons	Chlorinated polyether (Penton)	Polycarbonate
10% H_2SO_4	Excel.	Good	Excel.	Excel.	Excel.	Excel.	Excel.	Excel.	Excel.	Excel.	Excel.
50% H_2SO_4	Excel.	Poor	Excel.	Excel.	Excel.	Good	Excel.	Excel.	Excel.	Excel.	Excel.
10% HCl	Excel.	Excel.	Excel.	Excel.	Excel.	Excel.	Excel.	Excel.	Excel.	Excel.	Excel.
10% HNO_3	Excel.	Poor	Good	Excel.	Excel.	Good	Good	Fair	Excel.	Excel.	Excel.
10% Acetic	Excel.	Good	Excel.	Excel.	Excel.	Excel.	Excel.	Excel.	Excel.	Excel.	Excel.
10% NaOH	Excel.	Fair	Excel.	Good	Fair	Fair	Excel.	Poor	Excel.	Excel.	Excel.
50% NaOH	Excel.	Poor	Excel.	Excel.	Fair	Poor	Good	Poor	Excel.	Excel.	Excel.
NH_4OH	Excel.	Poor	Excel.	Excel.	Poor	Fair	Excel.	Poor	Excel.	Excel.	Excel.
NaCl	Excel.	Excel.	Excel.	Excel.	Excel.	Excel.	Excel.	Excel.	Excel.	Excel.	Excel.
$FeCl_3$	Excel.	Excel.	Excel.	Excel.	Excel.	Excel.	Excel.	Excel.	Excel.	Excel.	Excel.
$CuSO_4$	Excel.	Excel.	Excel.	Excel.	Excel.	Excel.	Excel.	Excel.	Excel.	Excel.	Excel.
NH_4NO_3	Excel.	Excel.	Excel.	Excel.	Excel.	Excel.	Excel.	Good	Excel.	Excel.	Excel.
Wet H_2S	Excel.	Excel.	Excel.	Excel.	Excel.	Excel.	Excel.	Excel.	Excel.	Excel.	
Wet Cl_2	Poor	Poor	Excel.	Good	Poor	Poor	Poor	Excel.	Excel.	Excel.	
Wet SO_2	Excel.	Poor	Excel.	Excel.	Good	Excel.	Excel.	Excel.	Excel.	Excel.	
Gasoline	Poor	Excel.	Excel.	Excel.	Excel.	Excel.	Excel.	Excel.	Excel.	Fair	Excel.
Benzene	Poor	Poor	Poor	Poor	Fair	Good	Excel.	Excel.	Excel.	Fair	Fair
CCl_4	Poor	Poor	Poor	Fair	Fair	Excel.	Good	Excel.	Excel.	Fair	Poor
Acetone	Poor	Poor	Poor	Poor	Fair	Poor	Good	Poor	Excel.	Good	Good
Alcohol	Poor	Poor	Excel.	Excel.	Excel.	Excel.	Excel.	Excel.	Excel.	Excel.	Excel.

Ratings are for long-term exposures at ambient temperatures (less than 100°F).
* Cellulose acetate butyrate.
† Acrylonitrile butadiene styrene polymer.
‡ Polyvinyl chloride, Type I.
§ Chemical resistance of Saran-lined pipe is superior to extruded Saran in some environments.
¶ Refers to general-purpose polyesters. Special polyesters have superior resistance, particularly in alkalies.

stress-corrosion cracking. The time of exposure required for failure will vary from minutes to years. Typical examples include cracking of cold-formed brass in ammonia environments, cracking of austenitic stainless steels in the presence of chlorides, cracking of Monel in hydrofluosilicic acid, and the caustic-embrittlement cracking of steel in caustic solutions.

This form of corrosion can be prevented in some instances by eliminating the high stresses. Stresses developed during fabrication, particularly during welding, are frequent sources of trouble. Stress relieving or annealing should always be considered where metals will be exposed to known environments that cause stress-corrosion cracking. Of course, temperature and solution concentration also are important factors in this type of attack.

The presence of chlorides does not generally cause cracking of austenitic stainless steels where temperatures are below about 120°F. However, where temperatures are high enough to concentrate chloride on the stainless surface, cracking may occur where the chloride concentration in the surrounding medium is a few parts per million.

Typical examples are cracking of heat-exchanger tubes at the crevices in rolled joints and under scale formed in the vapor space below the top tube sheet in vertical heat exchangers. The cracking of stainless steel under insulation is caused when moisture leaches chlorides from the insulation and the chlorides are then concentrated on the hot surfaces.

In handling caustic, welded steel can be used without developing caustic embrittlement cracking if the temperature is below 120°F. If the temperature is higher and particularly if the concentration is above about 30 per cent, cracking at and adjacent to non-stress-relieved welds frequently occurs.

Galvanic Corrosion. Galvanic corrosion is the corrosion rate above normal that is associated with the flow of current to a less active metal (cathode) in contact with a more active metal (anode) in the same solution.

Table 5-4 shows the anodic-cathodic grouping of various metals. This table should be used with caution, since exceptions to these groupings in actual use are possible.

However, as a general rule when dissimilar metals are used in contact with each other and are exposed to an electrically conducting solution, combinations of metals should be chosen that are as close as possible in the galvanic series. Coupling two metals widely separated in this series generally will produce accelerated attack on the more active metal. Often, however, protective oxide films and other effects will tend to reduce galvanic corrosion. Galvanic corrosion can, of course, be prevented by insulating the metals from each other. For example, where plates are bolted together, specially designed plastic washers can be used.

Area effects in cathodic corrosion are very important. Corrosion is caused by a flow of current from the anode, or more active metal, to the cathode, or less active metal.

An unfavorable area ratio is a large cathode and a small anode. Corrosion of the anode may be 100 to 1,000 times more than if the two areas were the same. This is the reason stainless steels are susceptible to rapid pitting in some environments. Steel rivets in a copper plate will corrode much more severely than a steel plate with copper rivets.

Crevice Corrosion. Crevice corrosion occurs within or adjacent to a crevice formed by contact with another piece of the same or another metal or with a non-metallic material. When this occurs, the intensity of attack is usually more severe than on surrounding areas of the same surface.

This form of corrosion can result because of a deficiency of oxygen in the crevice.

Oxygen Concentration Cell. The oxygen concentration cell is an electrolytic cell where the driving force to cause corrosion results from a difference in the amount of oxygen in solution at one point as compared with another. Corrosion is accelerated where the oxygen concentration is least, for example in a stuffing box or under gaskets, etc.

This form of corrosion will occur under solid substances that may be deposited on a metal surface and thus shield it from ready access to oxygen. Redesign or changes in mechanical conditions are methods that must be used to overcome this situation.

Metal-ion Concentration Cell. The metal-ion concentration cell is a cell established on a metal surface due to different concentrations of its ions in the electrolyte

Table 5-4. Galvanic Series of Metals and Alloys
Corroded End (Anodic, or Least Noble)

Magnesium
Magnesium alloys
Zinc
Aluminum 2S
Cadmium
Aluminum 17ST
Steel or iron
Cast iron
Chromium-iron (active)—stainless Type 410
Ni-Resist cast iron
18-8 chromium-nickel-iron (active) stainless Type 304
18-8-3 chromium-nickel-molybdenum-iron (active)—stainless Type 316
Lead-tin solders
Lead
Tin
Nickel (active)
Inconel nickel-chromium alloy (active)
Hastelloy alloy C (active)
Brasses
Copper
Bronzes
Copper-nickel alloys
Monel nickel-copper alloy
Silver solder
Nickel (passive)
Inconel nickel-chromium alloy (passive)
Chromium-iron (passive)—stainless Type 410
Titanium
18-8 chromium-nickel-iron (passive)—stainless Type 304
18-8-3 chromium-nickel-molybdenum-iron (passive)—stainless Type 316
Hastelloy alloy C (passive)
Silver
Graphite
Gold
Platinum

Protected End (Cathodic, or Most Noble)

where it is in contact with the metal surface. These variations in concentration result in local differences in potential, thus allowing the establishment of a local cell. The anodic or active area is in contact with the solution of lower metal-ion concentration. A cell of this type can be established, for instance, where a high-copper alloy passes through water-soaked wood into water.

Erosion. Erosion is the destruction of a metal by the combined action of corrosion and abrasion or attrition by a liquid or gas with or without suspended matter. When the liquid or gas contains particles harder than the metal surface affected, erosion will be by the combined action of corrosion and abrasion. When the liquid or gas does not contain suspended matter or contains matter which is softer than the metal, then erosion will be by corrosion and attrition.

Harder materials, changes in velocity or environment, and the addition of inhibitors are methods used to prevent erosion attack.

Corrosion Fatigue. Corrosion fatigue is a reduction by corrosion of the ability of a metal to withstand cyclic or repeated stresses. The surface of the metal plays an important role in this form of damage, as it will be the most highly stressed and at the same time subject to attack by the corrosive media. Corrosion of the metal surface will lower fatigue resistance, and stressing of the surface will tend to accelerate corrosion.

Under cyclic or repeated stress conditions, protective oxide films are ruptured faster than new protective films can be formed. Such a situation frequently results in formation of anodic areas at the points of rupture. These produce pits which serve as stress concentration points for the origin of cracks which cause ultimate failure.

Cavitation Erosion. The formation of transient voids or vacuum bubbles in a liquid stream passing over a surface is called cavitation. This is often encountered around propellers, rudders, and struts and in pumps. Cavitation corrosion is often blamed for the damage resulting from the collapse of such vacuum bubbles, but more correctly, this should be called cavitation erosion. When this collapse occurs on a metal surface, there is an explosive effect which can destroy protective films, greatly accelerating corrosion. Redesign or a more resistant metal is generally required to avoid this problem.

Impingement Attack. Impingement attack is corrosion associated with turbulent flow of a liquid, as at the entrance of a condenser tube or around bends in a pipeline.

Hydrogen Embrittlement. Loss of ductility of a metal caused by the entrance or absorption of hydrogen into the metal is called hydrogen embrittlement. Under some conditions, at elevated temperatures, decarburization, fissuring, and cracking of carbon steels will result when hydrogen combines with carbon in the steel to form methane. Alloys which form stable carbides, such as the low-chromium alloys, provide a solution to this form of attack.

Structural Corrosion

Graphitic Corrosion. Graphitic corrosion usually involves gray cast iron in which metallic iron is converted into corrosion products, leaving a residue of intact graphite mixed with iron corrosion products and other insoluble constituents of cast iron.

When the layer of graphite and corrosion products is impervious to the solution, corrosion will cease or slow down. If the layer is porous, corrosion will progress by galvanic behavior between graphite and iron. The rate of this attack will be approximately the same as that for the maximum penetration of steel by pitting. The layer of graphite formed may also be effective in reducing the galvanic action between cast iron and more noble alloys, such as bronze used for valve trim and impellers in pumps.

Low-alloy cast irons frequently demonstrate a superior resistance to graphitic corrosion, apparently because of their denser structure and the development of more compact and more protective graphitic coatings. Highly alloyed austenitic cast irons show considerable superiority over gray cast irons to graphitic corrosion because of the more noble potential of the austenitic matrix plus more protective graphitic coatings.

Carbon steels heated for prolonged periods at temperatures above 850°F may be subject to the segregation of carbon which is transformed into graphite. When this occurs, the structural strength of the steel will be affected. Killed steels or low-alloyed steels of chromium and molybdenum or chromium and nickel should be considered for elevated-temperature services.

Dezincification. Dezincification is corrosion of a brass alloy containing zinc in which the principal product of corrosion is metallic copper. This may occur as plugs filling pits (plug type) or as continuous layers surrounding an unattacked core of brass (general type).

The mechanism may involve overall corrosion of the alloy followed by redeposition of the copper from the corrosion products or selective corrosion of zinc or a high-zinc phase to leave copper residue. This form of corrosion is commonly encountered in brasses that contain more than 15 per cent zinc and can be either eliminated or reduced by the addition of small amounts of arsenic, antimony, or phosphorus to the alloy.

Other Forms of Deterioration

Biological Corrosion. The metabolic activity of microorganisms can either directly or indirectly cause deterioration of a metal by corrosion processes. This activity can (1) cause a corrosive environment, (2) create electrolytic concentration cells on the metal surface, (3) alter the resistance of surface films, (4) have an influence on the rate of anodic or cathodic reaction, and (5) alter the environment composition.

Microorganisms associated with corrosion are of two types, aerobic and anaerobic.

Aerobic microorganisms readily grow in environments containing oxygen, while the anaerobic species thrive in environments virtually devoid of atmospheric oxygen. The manner in which many of these bacteria carry on their chemical processes is quite complicated and in some cases not fully understood. The role of sulfate-reducing bacteria (anaerobic) in promoting corrosion has been extensively investigated. The sulfates in slightly acid to alkaline (pH 6 to 9) soils are reduced by these bacteria to form calcium sulfide and hydrogen sulfide. When these compounds come in contact with underground iron pipes, conversion of the iron sulfide occurs. As these bacteria will thrive under these conditions, they will continue to promote this reaction until failure of the pipe occurs.

Rusting. Rusting is the corrosion of iron or iron-base alloys. It results in the formation of corrosion products on the surface consisting mainly of hydrated ferric oxide.

Season Cracking. Season cracking is an older term used for the cracking of a metal or alloy as a result of combined corrosion and internal stress. It is usually applied to the stress cracking of brass.

Water-line Attack. Water-line attack is the corrosion that occurs at the air-liquid interface when a metal is partially immersed in a liquid. The susceptibility to this form of corrosion varies among different materials.

Crazing. Crazing is the development of hairline cracks in a plastic material. These cracks may be in the surface or through the mass of the plastic.

Checking. Checking is the deterioration of an organic coating characterized as breaks in the coating that do not penetrate it completely and usually has a pattern similar to that on an alligator's skin. These breaks are usually visible.

Chalking. Chalking is the deterioration of an organic coating or metal that, upon exposure, results in a powdery, chalky residue on the surface.

Factors Influencing Corrosion

Solution Acidity. Since the discharge of hydrogen ions takes place in most corrosion reactions, acidity of a solution as represented by the concentration of hydrogen ions (pH) is a most important factor. Also, acid (low pH) solutions are, as a general rule, more corrosive than neutral (pH 7) or alkaline (high pH) solutions.

In the case of ordinary iron or steel, the dividing line between rapid corrosion in acid solutions and moderate or slow corrosion in nearly neutral or alkaline solutions occurs at about pH 4.5. With the amphoteric metals, such as aluminum or zinc, highly alkaline (high pH) solutions may be even more corrosive than acid solutions.

The tendency for metals to corrode by displacing hydrogen ions from solution is indicated in a general way by their position in the electromotive series shown in Table 5-5. It should be noted that the potential values given in this table apply only to the conditions where the metal is in contact with a solution in which the activity of the ion indicated is 1 mole per 1,000 g of water. In any other solution, different values for the potentials would be developed.

Metals above hydrogen in this series displace hydrogen more readily than do those below hydrogen. A decrease in hydrogen-ion concentration (acidity) tends to move hydrogen up relative to the metals, while an increase in the metal-ion concentration tends to move the metals down relative to hydrogen.

Whether or not hydrogen evolution will occur in any case is determined by several other factors in addition to the concentrations of hydrogen and metallic ions. These other factors include the phenomenon of hydrogen overvoltage, which does not come within the scope of this discussion. The reader is referred to standard works on physical chemistry, electrochemistry, and corrosion for a detailed discussion of this subject.

Oxidizing Agents. In some corrosion processes, such as the solution of zinc in hydrochloric acid, hydrogen may be evolved as a gas. In others, such as the relatively slow solution of copper in sodium chloride, the removal of hydrogen, which must occur in order that corrosion may proceed, is effected by a reaction between hydrogen and some oxidizing chemical, such as oxygen, to form water.

Because of the high rates of corrosion which usually accompany hydrogen evolu-

Table 5-5. Electromotive Series

| Metal | Molal electrode potential at 77°F (25°C) | |
	Ion	Volt
Magnesium........	Mg++	−2.34
Aluminum.........	Al+++	−1.67
Zinc.............	Zn++	−0.76
Chromium........	Cr+++	−0.71
Iron.............	Fe++	−0.44
Cadmium.........	Cd++	−0.40
Nickel...........	Ni++	−0.25
Tin.............	Sn++	−0.14
Lead.............	Pb++	−0.13
Hydrogen.........	H+	Arbitrary zero point
Copper...........	Cu++	+0.34
Silver...........	Ag+	+0.80
Palladium........	Pd++	+0.83
Mercury..........	Hg++	+0.85
Platinum.........	Pt++	+1.2
Gold............	Au+++	+1.42

tion, metals are rarely used in solutions from which they evolve hydrogen at an appreciable rate. As a result, most of the corrosion observed in practice occurs under conditions where the oxidation of hydrogen to form water is a necessary part of the corrosion process. For this reason, oxidizing agents are often powerful accelerators of corrosion, and in many cases, the oxidizing power of a solution is its most important single property as far as corrosion is concerned.

Oxidizing agents that accelerate the corrosion of some materials may also retard corrosion of others through the formation on their surface of oxides or layers of adsorbed oxygen which make them more resistant to chemical attack. This is true of metals like aluminum, magnesium, and chromium and, to a lesser extent, iron and nickel. These metals are able to contribute the inhibiting effect to other metals with with they may be alloyed. This property of chromium is responsible for the principal corrosion-resisting characteristics of the stainless steels.

It follows, then, that oxidizing substances, such as dissolved air, may accelerate the corrosion of one class of materials and retard the corrosion of another class. In the latter case, the behavior of the material usually represents a balance between the power of oxidizing compounds to preserve a protective film and their tendency to accelerate corrosion when the agencies responsible for protective film breakdown are able to destroy the films.

Temperature. The rate of corrosion tends to increase with rising temperature. Temperature also has a secondary effect through its influence on the solubility of air (oxygen), which is the most common oxidizing substance influencing corrosion.

If all the air is boiled out of a dilute sulfuric acid solution, the rate of corrosion of copper-base alloys, which are not dependent upon maintenance of an oxide film for protection, will decrease considerably. On the other hand, the corrosion rate of stainless steels will increase considerably through the loss of the oxidizing substance (dissolved oxygen) needed to maintain their protective film.

Velocity. An increase in the velocity of relative movement between a corrosive solution and a metallic surface tends to accelerate corrosion. This effect is due to the higher rate at which the corrosive chemicals, including oxidizing substances (air), are brought to the corroding surface and to the higher rate at which corrosion products, which might otherwise accumulate and stifle corrosion, are carried away. The higher the velocity, the thinner will be the films through which corroding substances must penetrate and through which soluble corrosion products must diffuse.

Whenever corrosion resistance results from the accumulation of layers of insoluble corrosion products on the metallic surface, the effect of high velocity may be either to prevent their normal formation or to remove them after they are formed. Either effect allows corrosion to proceed unhindered. This occurs frequently in small-diameter tubes or pipes through which corrosive liquids may be circulated at high velocities (e.g., condenser and evaporator tubes), in the vicinity of bends in pipe lines, and on propellers, agitators, and centrifugal pumps. Similar effects are associated with cavitation erosion.

Films. Once corrosion has been started, its further progress very often is controlled by the nature of films, such as the passive films, that may form or accumulate on the metallic surface. The classic example is the thin oxide film that forms on stainless steels. Insoluble corrosion products may be completely impervious to the corroding liquid and, therefore, completely protective, or they may be quite permeable and allow local or general corrosion to proceed unhindered.

Films that are nonuniform, or discontinuous, may tend to localize corrosion in particular areas or to induce accelerated corrosion at certain points by initiating electrolytic effects of the concentration-cell type. Films may tend to retain or absorb moisture and thus, by delaying the time of drying, increase the extent of corrosion resulting from exposure to the atmosphere or to corrosive vapors.

It is generally agreed that the characteristics of the rust films that form on steels determine their resistance to atmospheric corrosion. The rust films that form on low-alloyed steels are more protective than those that form on unalloyed steel.

In addition to films that originate at least in part in the corroding metal, there are others that originate in the corrosive solution. These include various salts, such as carbonates and sulfates, which may be precipitated from heated solutions and insoluble compounds, such as "beer stone," which form on metal surfaces in contact with certain specific products. In addition, there are films of oil and grease that may protect a material from direct contact with corrosive substances. Such oil films may be applied intentionally or may occur naturally, as in the case of metals submerged in sewage or equipment used for the processing of oily substances.

Other Effects. Stream concentration can have important effects on corrosion rates. Unfortunately, corrosion rates are seldom linear with concentration over wide ranges. In equipment such as distillation columns, reactors, and evaporators, concentration can change continuously, making prediction of corrosion rates rather difficult. As an example of the complexity of this effect, Type 304 stainless can show a maximum corrosion rate at 40 per cent acid concentration; at higher concentrations the rate can decrease to a constant value at around 90 per cent. Concentration is important during plant shutdown. The presence of moisture that collects during cooling can turn innocuous chemicals into dangerous corrosives.

As to the effect of time, no universal law governs the reaction for all metals. Some corrosion rates can remain constant with time over wide ranges; others can slow down with time; some alloys will have increased corrosion rates with respect with time. Situations can develop where the corrosion rate can follow a combination of these paths. Therefore, extrapolation of corrosion data and corrosion rates should be done with utmost caution.

High-temperature Attack

Physical Properties. The suitability of an alloy for high-temperature service (800 to 2000°F) will depend on properties inherent in the alloy composition and on the conditions of use. Crystal structure, density, thermal conductivity, electrical resistivity, thermal expansivity, structural stability, melting range, and vapor pressure are all physical properties basic to and inherent in individual alloy compositions.

Of usually greater relative importance in this group of properties is expansivity. A surprisingly large number of metal failures at elevated temperatures is the result of excessive thermal stresses originating from constraint of the metal during heating or cooling. Such constraint in the case of hindered contraction can cause rupturing of the metal.

Another important physical property is alloy structural stability. This means freedom from formation of new phases or drastic rearrangement of those originally present within the metal structure as a result of thermal experience. Such changes may have a detrimental effect upon strength or corrosion resistance or both.

Mechanical Properties. Mechanical properties which are of wide interest involve creep, rupture, and short-time strengths; various forms of ductility; and resistances to impact and fatigue stresses. Creep strength and stress rupture are usually of greatest interest to designers of stationary equipment such as vessels and furnaces.

Corrosion Resistance. Possibly of greatest importance then either physical or mechanical properties is the ability of an alloy's chemical composition to resist the corrosive action of various hot environments. The forms of high-temperature corrosion which have received the greatest attention are oxidation and scaling.

Chromium is an essential constituent in alloys to be used above 1000°F. It diffuses rapidly and accumulates at the surface of the steel, providing a tightly adherent oxide film that materially retards the oxidation process. Silicon is a useful element in imparting oxidation resistance to steel. It will enhance the beneficial effects of chromium. Also, for a given level of chromium, experience has shown that oxidation resistance improves as the nickel content increases.

Aluminum cannot be used as an alloying element in steel to improve oxidation resistance; the amount required interferes with both workability and high-temperature strength properties. However, the development of high-aluminum surface layers by various methods (including spraying, cementation, or dipping) is a feasible means of improving the heat resistance of low-alloy steels.

Special Problems in Refineries

While much of the equipment in refineries is subject to the usual problems of temperatures, pressures, and corrosive and erosive environments, there are some special problems that all too often cause premature equipment failures.

Temper Embrittlement. Low-alloy chromium steels are particularly susceptible to temper embrittlement. After exposure to 850 to 1000°F followed by slow cooling, these steels very often exhibit brittleness at room temperatures. In some cases, this can be so severe that the steel fractures like glass when struck with a hammer.

A simple laboratory test for temper embrittlement is to normalize a specimen, then stress-relieve the steel at 1200°F. If the ratios of impact values of rapidly cooled specimens to those of slowly cooled specimens are greater than 1, the specimens are temper brittle. Fortunately, the amount and rate of this embrittlement are decreased by the addition of at least 0.5 per cent Mo, although the tendency may not be eliminated entirely.

Graphitization. Experience has shown that steels with less than 0.7 per cent Cr tend to graphitize in weld-heat-affected zones when held at 750 to 1050°F for long times. This phenomenon involves separation of carbon from carbides. Segregated graphite particles can seriously affect the mechanical properties of steel so that it will crack along a plane of segregated graphite.

Many refinery engineers follow the recommendations of an API report (Wilson, *Amer. Pet. Inst.*, vol. 35, pp. 209–248, 1955) and limit use of low-alloy steels with less than 0.5 per cent Mo to 850°F. There is some evidence that 0.5% Cr-0.5% Mo still can graphitize at 750 to 1050°F, and many refineries use 1% Cr-0.5%Mo steels.

Hydrogen Attack. Hydrogen can dissociate thermally or catalytically at the metal surface and diffuse into steel in atomic form, then recombine into molecular hydrogen by using some internal defect of the metal as nuclei.

At atmospheric temperatures, the hydrogen can cause external or internal blisters on the steel. In some cases, these blisters can be chipped out and the metal repaired by welding. But usually the blisters are too numerous to make repair practical. While the use of fully deoxidized steels might be helpful, the best remedy is to change process conditions to decrease or eliminate the formation of atomic hydrogen.

In a number of cases, hydrogen at atmospheric temperatures has caused cracking and embrittlement of steels and alloys. Nonferrous alloys and austenitic stainless steels are resistant to this attack.   If the temperature is around 1000 to 1300°F, hydrogen will react with carbon on the surface of the steel, removing it and causing decarbonization of the surface.   This will reduce the strength of the steel.

Between about 1300°F and room temperature, hydrogen can dissociate and diffuse into the metal structure, combine with carbon, and form methane.   The pressure

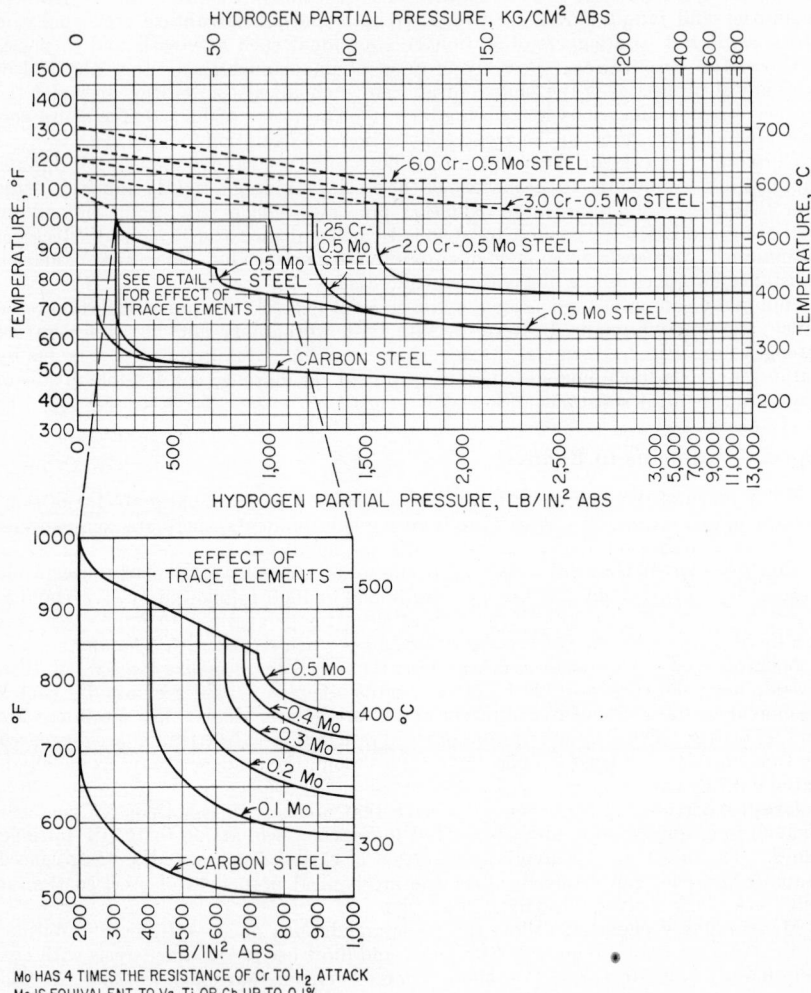

Fɪɢ. 5-1. Operating limits for carbon and alloy steels in contact with hydrogen at high temperatures.   (*Hyd. Proc., May, 1965, p. 186.*)

from this gas formation can build up until internal fissures form and permanently weaken the steel.   The attack also depletes the steel of carbon, further decreasing strength properties.

This high-temperature hydrogen attack can be prevented by using alloying elements such as molybdenum and chromium to stabilize the carbides and block the hydrogen reaction (see Fig. 5-1).   When the operating conditions and partial pressure of hydrogen are known, a steel composition can be selected from Fig. 5-1 which is resistant to hydrogen attack.

Sulfide Attack. Sulfur compounds in crude oils can cause severe corrosion problems in crude distillation as well as in thermal and catalytic cracking units. The corrosion becomes significant usually at above 500°F. This type of corrosion can be reduced by using steels containing 2, 5, or 9 per cent chromium (Fig. 5-2).

In the presence of hydrogen sulfide or hydrogen–hydrogen sulfide mixtures, at about 800 to 1200°F, steels with 9 per cent chromium have limited value in controlling corrosion. It should be noted that organic sulfur compounds found in petroleum streams tend to decompose on clean metal surfaces, frequently releasing hydrogen sulfide and creating the condition just discussed. In this situation, austenitic stainless steels (even though they have disadvantages, as discussed later in this section) or steels with aluminized surfaces are used

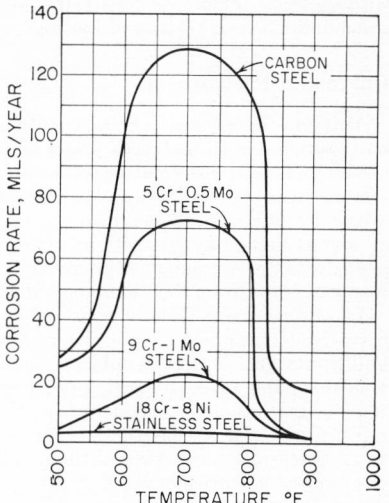

Fig. 5-2. Corrosion of steels in crude oil containing 1.5 per cent sulfur.

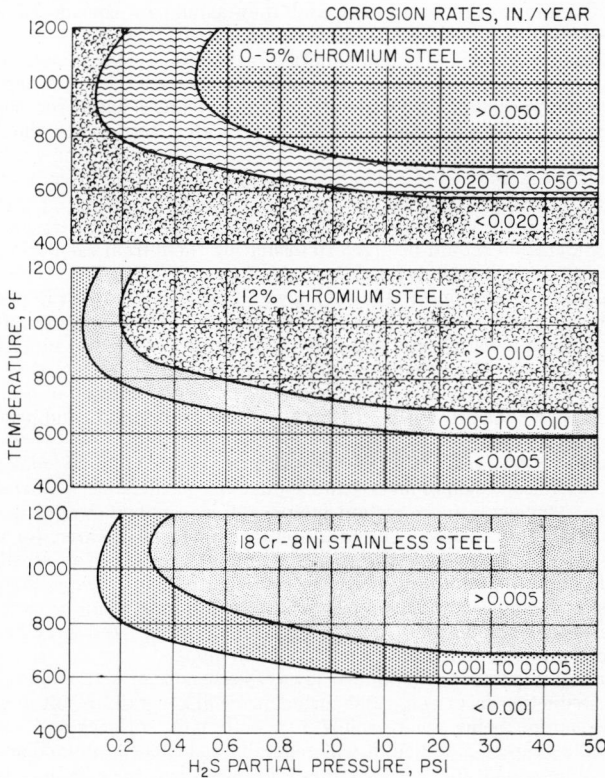

Fig. 5-3. Corrosion resistance of steels to hydrogen sulfide in the presence of hydrogen.

with success.　Figure 5-3 illustrates the corrosion resistance of various steels to hydrogen sulfide in the presence of hydrogen.

Corrosion Prevention

Material Selection.　Selection of materials for process equipment should be based on experience with materials under similar conditions.　Corrosion tests may be necessary to develop information to permit selection of adequate materials.　Where such tests are made, procedures and conditions should be as realistic as practical to simulate operating and exposure conditions with respect to temperature, aeration, and velocity as well as chemical composition.　Hot or cold wall effects should be considered where heat-exchange equipment is involved.　The possibility of the occurrence of one of the forms of corrosion other than general corrosion should be considered.

In selecting materials of construction for equipment which would be exposed to corrosive environments, careful consideration should be given to the practical aspects and limitations of design, fabrication, installation, and maintenance.　For instance, if field welding or cold working must be done on equipment for installation or maintenance reasons and subsequent heat-treatment is not practical, then it is necessary to select a material of construction not susceptible to attack by intergranular corrosion or stress-corrosion cracking in the non-heat-treated condition.

Permissible corrosion rates are an important factor and will vary with equipment.　Appreciable corrosion can be permitted for tanks and lines if anticipated and allowed for in design thickness, but essentially no corrosion can be permitted in fine-mesh wire screens, orifices, and other items where small changes in dimensions are critical.

In many instances use of nonmetallic materials will prove to be attractive from an economic and performance standpoint and they should be considered where their strength, temperature, and design limitations are satisfactory.

Design Considerations.　The installation of baffles, stiffeners, and drain nozzles and the location of valves and pumps should be made so that free drainage will occur and washing can be accomplished without holdup.　Means of access for inspection and maintenance should be provided wherever practical.　Butt joints should be used wherever possible.　If lap joints employing fillet welds are used, the welds should be continuous on the process side.　The use of dissimilar metals in contact with each other should generally be avoided, particularly if they are widely separated in their nominal positions in the electromotive series (see Table 5-5)　If they are to be used together, consideration should be given to insulating them from each other or making anodic material area as large as possible.

Equipment should be supported in such a way that it will not rest in pools of liquid or on damp insulating material.　Porous insulation should be weatherproofed or otherwise protected from moisture and spills to avoid contact of the wet material with the equipment.　Specifications should be sufficiently complete to ensure that the desired composition or type of material will be used and the right condition of heat-treatment and surface finish will be provided.　Inspection during fabrication and prior to acceptance is desirable.

Environment Changes.　Simple changes in environment may make an appreciable difference in corrosion of metals and should be considered as a means of combating corrosion.　Oxygen is an important factor, and its removal or addition may cause marked changes in corrosion.　The treatment of boiler feed water, for instance, to remove oxygen greatly reduces the corrosiveness of the water on steel.　Inert-gas purging and blanketing of many solutions, particularly acidic media, generally minimizes corrosion of copper and nickel-base alloys by minimizing air or oxygen content.　Corrosiveness of acid media to stainless alloys, on the other hand, may be reduced by aeration because of the formation of passive oxide films.

Reduction in temperature will almost always be beneficial with respect to reducing corrosion.　Reduction in velocity and turbulence will generally result in reduced corrosion, an exception being where solids may collect on surfaces and cause pitting.　Where pH values can be modified, it will generally be beneficial to hold acid level to a minimum.　Where acid additions are made in batch processes, it may be beneficial to add them last so as to obtain maximum dilution and minimum acid concentration and exposure time. Alkaline pH values are less critical than acid values with respect to controlling corrosion. Elimination of moisture can and frequently does minimize, if not prevent, corrosion of metals, and this possibility of environmental alteration should always be considered.

Inhibitor Addition. The use of various substances or inhibitors as additives to corrosive environments to decrease corrosion of metals in the environment is an important means of combating corrosion. This is generally most attractive in closed or recirculating systems where the replacement cost of inhibitor is low. However, it has also proved to be economically attractive for many once-through systems. Inhibitors are effective as the result of their controlling influence on the cathode- or anode-area reactions.

Typical examples of inhibitors used for minimizing corrosion of iron and steel in aqueous solutions are the chromates, phosphates, and silicates. These minimize corrosion by increasing anodic polarization and are called anodic inhibitors. Organic sulfide and amine materials are frequently effective in minimizing corrosion of iron and steel in acid solution. In this instance they control cathodic polarization and are called cathodic inhibitors.

The use of inhibitors is not limited to controlling corrosion of iron and steel. They frequently are effective with stainless steel and other alloy materials. The addition of copper sulfate to dilute sulfuric acid will, for example, essentially stop the corrosion of stainless steels in hot dilute solutions of this acid, whereas the uninhibited acid causes rapid corrosion.

The effectiveness of a given inhibitor generally increases with increase in concentration, but those considered practical and economically attractive are used in quantities of less than 0.1 per cent by weight.

In some instances the amount of inhibitor present is critical in that a deficiency may result in localized or pitting attack with the overall results being more destructive than where none of the inhibitor is present. Consideration for use of inhibitors should therefore include review of experience in similar systems or investigation of requirements and limitations in new systems.

Cathodic Protection. Two methods of providing cathodic protection for minimizing corrosion of metals are in use today. These are the sacrificial-anode method and the impressed-emf method. Both depend upon making the metal to be protected the cathode in the electrolyte involved. Examples of the sacrificial-anode method include the use of zinc, magnesium, or aluminum as anodes in electrical contact with the metal to be protected. These may be anodes buried in the ground for protection of underground pipelines or as attachments to surfaces of equipment such as condenser water boxes or on ships' hulls. The current required is generated in this method by corrosion of the sacrificial-anode material.

In the case of the impressed emf, the direct current is provided by external sources and is passed through the system by use of essentially nonsacrificial anodes such as carbon, noncorrodible alloys, or platinum buried in the ground or suspended in the electrolyte in the case of aqueous systems. The requirements with respect to current distribution and anode placement vary with resistivity of soils or electrolyte involved.

The practical problem increases as the resistivity of the electrolyte decreases. It is generally impractical, for instance, to use cathodic protection where dilute acids are involved as the corroding solution or electrolyte. It is also impractical where the dimensions or configuration of equipment to be protected are such that adequate placement of anodes cannot be provided, e.g., heat-exchanger tube bundles.

Coatings and Linings. The use of nonmetallic coatings and lining materials in combination with steel or other materials has been and will continue to be an important type of construction for combating corrosion (see Table 5-6).

Organic coatings of many kinds are used as linings in equipment such as tanks, piping, pumping lines, and shipping containers, and they are often an economical means of controlling corrosion, particularly where freedom from metal contamination is the principal objective. One principle that is now generally accepted is that thin brush- or spray-applied nonreinforced paintlike coatings of less than 10 mils thickness should not be used in services where full protection is required in order to prevent rapid attack of the substrate metal. This is because most thin coatings as commercially

Table 5-6. Properties and Chemical Resistance of Organic Coatings[a]

	Alkyd						Acrylic	Bitumi-nous	Cellulose		
	Alkyd	Alkyd-amine	Alkyd-phenolic	Alkyd-silicone	Alkyd-urea	Styren-ated alkyd	Acrylic	Bitumi-nous	Nitro-cellulose	Buty-rate	Ethyl cellulose
Chemical resistance:											
Exterior durability	E	E	E	E	E	F	E	F	E	E	E
Salt spray	E	VG	E	E	G	G	E	E	E	E	…
Solvents, alcohols	F	G	E	G	G	G	P	P	G	G	P
Solvents, gasoline	G	E	E	G	E	F	G	P	G	G	P
Solvents, hydrocarbons	G	E	E	G	E	E	F	P	F	F	F
Solvents, esters, ketones	P	P	F	P	F	P	P	P	P	P	P
Solvents, chlorinated	P	P	P	P	P	P	P	P	P	P	P
Beverages, food	F	G	VG	G	G	VG	VG	E	G	G	P
Salts	VG	E	E	VG	E	E	VG	G	G	VG	E
Ammonia	P	P	P	P	P	P	P	…	P, P	P, P	G
Alkalies[b]	F, P, P	VG, G	VG, G, F	VG, G	G, G	G, VG	G, F	E, −, −	E, G, F	G, F, P	G
Acids, mineral[c]	F, P, P	VG, F, P	G, F, P	G, P, P	F, P, P	G, P, P	G, F, P	G, −, −	P, P, P	P, P, P	G, G
Acids, oxidizing[c]	P, P, P	F, P, P	F, P, P	P, P, P	F, P, P	F, P, P	F, P, P	E, P, −	P, P, P	P, P, P	G, F, P, P
Acids, organic (acetic, formic, etc.)[e]	F	P, P, P	P	P, G	P, P, P	P, P, P	P, F, P	…	P, F	P, F	F, P, P
Acids, organic (oleic, stearic, etc.)[c]	P	P	P	G	P	P	P	E, P	P	P	G, −, −
Acid, phosphoric	P	P	P	G	F	F	F	E	P	P	…
Water (salt, fresh)	F	G	G	G	G	G	E	E	G	E	E
Physical properties:											
Sward rocker hard. (8th day)	24	30	34	16–30	28	28	24	…	30	26	25–30
Flexibility	E	VG	G	VG	VG	G	E	E	E	E	E
Abrasion resistance, cycles[d]	3500	>5000	>5000	4000	>5000	>5000	2500	…	2500	2500	…
Max. svc. temp., °F	200	250	250	450	225	200	180	200	180	180	300
Toxicity	None	Slight	None	None	Slight	Slight	None	…	None	None	None
Dielec. properties	VG	E	G	G	G	G	VG	E	E	E	E
Adhesion to:											
Ferrous metals	E	E	E	VG	E	F	VG	E	VG	VG	E
Nonferrous metals	F	E	E	VG	VG	F	VG	E	G	G	P
Old paints	VG	G	G	E	G	VG	P	…	P	P	P

Table 5-6. Properties and Chemical Resistance of Organic Coatings[a] (Continued)

	Epoxy						Chlorinated polyether	Chlorinated polypropylene	Fluorocarbon (air-dried)	Furane	Phenolic
	Epoxy-amine	Epoxy-ester	Epoxy-furane	Epoxy-melamine	Epoxy-phenolic	Epoxy-urea					
Chemical resistance:											
Exterior durability	G	E	E	E	E	VG	E	E	E	G	E
Salt spray	VG	E	E	E	E	E	E	E	E	E	E
Solvents, alcohols	G	F	E	E	E	E	E	E	E	E	E
Solvents, gasoline	E	VG	E	E	E	E	E	G	E	E	E
Solvents, hydrocarbons	VG	F	G	VG	E	E	E	P	E	E	F
Solvents, esters, ketones	G	F	E	E	E	VG	E	P	P	E	F
Solvents, chlorinated	E	VG	E	E	E	E	E	P	E	E	E
Beverages, food	E	E	E	E	E	E	E	G	E	E	E
Salts	E	E	E	E	E	P	E	G	E	E	P
Ammonia	E, E	P	G	E, E	F	E, E	E, E	VG, VG	E, E	E	P, P
Alkalies[b]	E, VG, G	E, E	G	E, VG, G	E, E	E, VG, F	E, E	E, G, G	E, E, E	E	G, F, P
Acids, mineral[c]	G, P, P	G, F, P	F	G, F, P	E, E	F, P, P	E, E, E	G, P, P	E, E, E	G	G, F, P
Acids, oxidizing[c]	P, F, P	F, P, P	F, G	G, F, P	E, VG, P	F, P, P	E, E, E	G, P, P	E, E, E	F, G	G, F, P
Acids, organic (acetic, formic, etc.)[e]	F, F	F, E	F	G	E, E, VG	E	E	G	E	E	E
Acids, organic (oleic, stearic, etc.)	F	P	E	G	E	G		G		E	F
Acid, phosphoric	G	E	E	G	E	G		G		E	E
Water (salt, fresh)	G	VG	E	G	E	G	E	E	E	E	E
Physical properties:											
Sward rocker hard. (8th day)	36	30	24	36	44	34		24	<10	28	38
Flexibility	F	E	E	VG	VG, E	VG	F	VG	G	F	G
Abrasion resistance, cycles[d]	>5000	>5000		>5000	>5000	>5000	>5000	5000	1000		>5000
Max. svc. temp., °F	400	300	350	400	400	400	300	250	550	300	350
Toxicity	None	None	None	None	None	None	None	Slight	Slight	None	None
Impact resistance	G	E	G	VG	VG	G	F	G	E	F	G
Dielec. properties	VG	VG	G	VG	VG	VG	E	E	E	F	E
Adhesion to:											
Ferrous metals	E	E	E	E	E	E	VG	G	VG	F	E
Nonferrous metals	E	E	E	E	E	E	G	VG	VG	F	E
Old paints	G	VG	E	P	P	P	P	E	P	E	G

5–21

Table 5-6. Properties and Chemical Resistance of Organic Coatings[a] (Continued)

	Polyamide (nylon)	Polyester	Polyethylene	Rubber						Vinyl	Vinyl-alkyd (approx. 1:1)
				Chlorinated rubber	Neoprene	Hypalon	Viton	Silicone	Urethane		
Chemical resistance:											
Exterior durability	P	E	P	E	E	E	E	E	E	E	E
Salt spray	F	E	VG	E	E	E	E	E	E	E	E
Solvents, alcohols	G	G	E	E	E		E	F	F–E	F	G
Solvents, gasoline	G	E	P	G	F	P	E	VG	F–G	E	E
Solvents, hydrocarbons		E	P	P	P		E	P	F–E	G	P
Solvents, esters, ketones		P	G	P	G	P	P	P	F	P	P
Solvents, chlorinated		P	P	G	P	P	E	F	P–E	P	P
Beverages, food		G	E		F	F	E	G	VG	E	F
Salts		G	E	G	E	E	E	E, F	E	E	G
Ammonia	G, G	P	E		E	P	E	G, G		E	P
Alkalies[b]	P, P, P	E	P, G	E, E, G	E, E	E, F	E, E	P, P, P	P	E, E	G, P
Acids, mineral[c]		P	VG, VG, VG	E, E, F	E, G, P	E, G, F	E, E, E	P, P, P	VG, F	E, E, G	VG, G, P
Acids, oxidizing[c]		P	VG, F, P	G, F, P	F, P, P	G, F, P	E, E, E	P, P, G	G, F, P	E, VG, G	P, P, P
Acids, organic (acetic, formic, etc.)[c]	P, P, P	P, P, P		F	P, P, P	F, P, P	E, E, E	E, P, P	G, F, P	E, P, P	G, F, P
Acids, organic (oleic, stearic, etc.)	VG	F	G	E	P	E	E	E	G, E, P	E	E
Acid, phosphoric		F	VG		VG	E	E	E	E	E	E
Water (salt, fresh)	F	G	E	E	E	E	E	E	E	E	E
Physical properties:											
Sward rocker hard. (8th day)		30		24	<10	<10	<10	16	35–65	20	26
Flexibility	G	G	E	VG	E	E	G	F	E	E	E
Abrasion resistance, cycles[d]	300	3500	225	>5000	5000	5000	1000	2500	>5000	>5000	2500
Max. svc. temp., °F		200		200	200	250	550	550	300	150	180
Toxicity	VG	None	None	Slight	None		Slight	None	Slight	None	None
Impact resistance	G	F	E	E	E	E	E	E	E	E	E
Dielec. properties		G	E		F	VG		E	E	E	G
Adhesion to:											
Ferrous metals	VG	F–P	E	F	VG	VG	VG	F	E	G	VG
Nonferrous metals	VG	F–P	E	VG	VG	VG	VG	E	E	VG	G
Old paints		P	P	F		F	P	E	G	F	P

[a] These data are intended only as a preliminary selection guide. Final selections should be made after consulting with coating formulator. Source: *Materials* in *Design Engineering*, Mid. October, 1962, issues, pp. 364–367.

Key: E = excellent; VG = very good; G = good; F = fair; P = poor.

[b] Two ratings are for dilute (20%) and concentrated, respectively.
[c] Three ratings are for dilute (10%), medium (10–30%), and concentrated, respectively.
[d] Taber GS-10 wheel.
[e] Not recommended with nitric acid.
[f] Not recommended with strong acetic acid solutions.

applied contain defects or holidays and these lead to early failures due to corrosion of the substrate metal, even though the coating material is resistant. Spark testing of coating-type linings is always desirable for immersion-service applications in order to detect holiday-type defects in the coating.

The most dependable linings for corrosive services are those which are bonded directly to the substrate and are built up in multiple-layer or laminated effects to thicknesses greater than 100 mils. These include the glass-fiber-reinforced resin systems and the elastomeric and plasticized plastic systems. Good surface preparation and thorough inspections of completed lining, including spark testing, should be considered as minimum requirements for any lining applications.

Ceramic or carbon-brick linings are frequently used as a facing lining over plastic or membrane linings where the service temperatures exceed those which can be handled by the unprotected materials. This type of construction permits processing of materials that are too corrosive for handling in metal constructions.

Metal- and glass-coated or -lined steel has been available for some time as a method of avoiding the high cost of a solid alloy. Nickel, for instance, can be deposited by electroplating or by chemical processes. Such deposits, however, tend to be slightly imperfect or porous if less than 5 mils thick. Glassed steel has all the advantages and disadvantages of glass, except that it has high tensile strength.

Steel clad with an alloy is another approach to this problem. There are a number of metal-cladding methods in general use today. In one, a sandwich is made of the corrosion-resistant metal and carbon steel. This is sealed around the edges by welding and hot rolling to produce a pressure weld between the plates. In a second technique, the metal is brazed to a steel backing plate. These sheets, usually available in small sizes, must be welded together to make larger plates.

In a third process, a loose liner is fastened to a carbon-steel shell by electric resistance welds spaced about $1\frac{1}{2}$ to 2 in. apart. A fourth process is the strip-liner method where a liner is installed in a vessel using sheets or strips of various sizes by welding.

A new process involves explosive forming. The corrosion-resistant metal is bonded to a steel backing metal by the force generated by properly positioned explosive charges. Relatively thick sections of metal can be bonded into plates by this technique.

A compound of a metal in the form of a film can be produced on a metal surface by immersing the metal or spraying it with a selected chemical solution. This provides improved corrosion resistance to the metal or increases the adhesion of organic coatings to be applied later. Examples are anodizing of aluminum, phosphate coatings on steel or zinc, etc.

Zinc coatings applied by various means have good corrosion resistance to many atmospheres. They have been used extensively as coatings on steel. Zinc has the advantage of being anodic to steel and therefore will protect exposed areas of steel by electrochemical action.

Cadmium, being anodic to steel, behaves quite similarly to zinc in providing corrosion protection when applied as a coating on steel. Tests should be conducted on zinc and cadmium coatings when it becomes necessary to determine the most economical selection for a particular environment.

Lead has good general resistance to various atmospheres. As a coating, it has had its greatest application in the production of terneplate, which is used as a roofing, cornicing, and spouting material.

Aluminum coatings on steel will perform in a manner similar to zinc coatings. Aluminum has good resistance to many atmospheres and, in addition, is anodic to steel and will galvanically protect exposed areas. Aluminum-coated steel products are quite serviceable under high-temperature conditions where good oxidation resistance is required.

PROPERTIES OF MATERIALS

Carbon Steel

Carbon steel is the most common, cheapest, and most versatile metal used in petroleum refinery applications. It has excellent ductility, permitting many cold-forming

Table 5-7. Properties of Metals and Alloys*

Material	Nominal composition (essential elements), %	Form and condition	Typical mechanical properties				Typical physical constants								
			Yield strength (0.2% offset), 1,000 psi	Tensile strength, 1,000 psi	Elongation in 2 in., %	Hardness, Brinell	Density, lb/cu in.	Specific gravity	Melting point, °F	Specific heat (32–212°F) Btu/(lb)(°F)	Thermal expansion coefficient (32–212°F) $\times 10^{-6}$ (in.)/(in.)/(°F)	Thermal conductivity (32–212°F) Btu/(sq ft)(hr)(°F/in.)	Electrical resistivity (68°F), ohms/cir mil ft	Tensile modulus of elasticity $\times 10^6$ psi
FERROUS ALLOYS														
Carbon steels[a] AISI-SAE 1020	Fe bal, Mn 0.45, Si 0.25, C 0.20	Annealed / Hot-rolled / Hardened (water quench 1000°F temper)	38 / 42 / 62	65 / 68 / 90	30 / 32 / 25	130 / 135 / 179	0.284	7.86	2760	0.107	6.7	360	60	30
300-M[b]	C 0.43, Mn 0.80, Si 1.60, Ni 1.85, Cr 0.85, Mo 0.38, V 0.08	Hardened (oil quench, 600°F temper)	240	290	10	535	0.283	7.84	2740	0.107	6.5	400	70	30
Wrought iron	Fe bal, Slag 2.5	Hot-rolled	30	48	30 (in 8 in.)	100	0.278	7.70	2750	0.11	6.35	418	70	29
Ingot iron	Fe 99.9 plus	Hot-rolled / Annealed	29 / 19	45 / 38	26 / 45	90 / 67	0.284	7.86	2795	0.108	6.8	490	57	30.1
Cast-gray iron (ASTM A48-48 Class 25)	C 3.4, Si 1.8, Mn 0.5, Fe bal.	Cast (as cast)	......	25 min.	0.5 max.	180	0.260	7.20	2150	......	6.7	310	400	13 ± 1.5
Malleable iron	C 2.5, Si 1, Mn 0.55 max	Cast (annealed)	33	52	12	130	0.264	7.32	2250	0.122	6.6	......	180	25
Ductile iron (Mg-containing)	C 3.4, Si 2.5, Mn 0.40, P 0.1 max., Ni 0–1, Mg 0.06, Fe bal.	Cast / Cast (as cast) / Cast (quench temper)	53 / 68 / 108	70 / 90 / 135	18 / 7 / 5	170 / 235 / 310	0.26	7.2	2100	......	7.5	228	360	25
Ductile iron (Mg-containing) (heat-resistant)	C 3.3, Si 4.3, Mn 0.4, P 0.1 max., Ni 0–1, Mg 0.06, Fe bal.	Cast (annealed, as cast)	60	80	10	220	0.26	7.2	2100	......	......	200	437	

Ni-Resist ductile iron (Mg-containing)	C 2.8, Si 2.5, Mn 1, P 0.2 max., Ni 20, Cr 2, Mg 0.1, Fe bal.	Cast (as cast)	35	60	10	175	0.268	7.4	2250	....	10.4	....	610	18.5
Ni-Resist Type D-3	C 2.60 max., Si 2.5, Mn 0.6, P 0.2 max., Ni 30.0, Cr 3.5	Cast	35	61	12	150	0.27	7.45	2250	....	7.0	....	....	14.0 ± 1.5
Ni-Resist Type D-4	C 2.60 max., Si 5.5, Mn 0.6, P 0.2 max., Ni 30.0, Cr 5.5	Cast	41	60	2.5	190	0.27	7.45	2200	....	8.0	....	....	13.0 ± 1.5
Ni-Hard cast iron Type 2	C 2.7, Si 0.6, Mn 0.5, Ni 4.5, Cr 2.0, Fe bal.	Sand-cast / Chill-cast (temper)	.....	55 / 75	..... /	550 / 625	0.275	7.70	2150	....	4.8	99	....	25
Ni-Hard Type 1	C 3.5, Si 0.6, Mn 0.5, Ni 4.5, Cr 2.0, Fe bal.	Sand-cast / Chill-cast (temper)	..... /	40 / 50	..... /	600 / 700	0.275	7.70	2150	....	4.8	99	....	25
Ni-Resist cast iron Type 1	C 2.8, Si 2.0, Mn 1.2, Ni 15.5, Cr 2.5, Cu 6.5, Fe bal.	Cast (as cast)	.....	27	2	150	0.264	7.3	2250	0.110	10.3	276	842	13.0 ± 1.5
Ni-Resist Type 2	C 2.8, Si 2.0, Mn 1.0, Ni 20.0, Cr 2.5, Fe bal.	Cast (as cast)	.....	27	2	140	0.264	7.3	2250	0.116	9.6	276	1023	15.6 ± 1.5
Ni-Resist Type 4	C 2.6 max., Si 5.5, Mn 0.6, Ni 30.0, Cr 5.5, Fe bal.	Cast		30	2	180	0.268	7.4	2200	0.120	7.3	261	962	15.0 ± 1.5
Ni-Resist ductile iron Type D-2	C 2.90 max., Si 2.5, Mn 1.0, P 0.2 max., Ni 20.0, Cr 2.5	Cast	34	62	14	160	0.268	7.41	2250	....	10.4	93	614	17.5 ± 1.5

[a] Carbon-steel mechanical properties are strongly influenced by the carbon level; the physical properties will not be appreciably changed.

[b] Type of ultra-high-strength steel, attaining tensile strengths of 220,000 to 300,000 psi.

Table 5-7. Properties of Metals and Alloys* (Continued)

Material	Nominal composition (essential elements), %	Form and condition	Typical mechanical properties				Typical physical constants							
			Yield strength (0.2% offset), 1,000 psi	Tensile strength, 1,000 psi	Elongation in 2 in., %	Hardness, Brinell	Density, lb/cu in.	Specific gravity	Melting point, °F	Specific heat (32–212°F) Btu/(lb)(°F)	Thermal expansion coefficient (32–212°F) $\times 10^{-6}$ (in./in.)/(°F)	Thermal conductivity (32–212°F) Btu/(sq ft)(hr)/(°F/in.)	Electrical resistivity (68°F), ohms/cir mil ft	Tensile modulus of elasticity $\times 10^6$ psi
Stainless Steels														
Stainless steel Type 201	C 0.15 max., Mn 5.5–7.5, Cr 16.0–18.0, Ni 3.5–5.5, N 0.25 max.	Mill-annealed strip	50	115	60	194	0.28	7.7	2550–2650	0.12	8.8	113	414	28.6
Stainless steel Type 202	C 0.15 max., Mn 7.5–10.0, Cr 17.0–19.0, Ni 4.0–6.0, N 0.25 max.	Mill-annealed strip	50	100	60	184	0.28	7.7	2550–2650	0.12	....	113	414	28.6
Stainless steel Type 301	Fe bal., Cr 17, Ni 7, C 0.08–0.20	Annealed / Cold-rolled[c]	30 to 165	100 to 200	72 / 15[d]	160 / 385	0.29	8.02	2550–2590	0.12	9.4	112.8	435	28
Stainless steel Type 302	Fe bal., Cr 18, Ni 8, C 0.08–0.20	Annealed / Cold-rolled[c]	30 to 165	90 to 190	60 / 8[d]	160 to 400	0.29	8.02	2550–2590	0.12	9.6	112.8	435	28
Stainless steel Type 304	Fe bal., Cr 19, Ni 9.0, C 0.08 max.	Annealed / Cold-rolled[d]	30 to 160	85 to 185	62 / 8[d]	160 to 400	0.29	8.02	2550–2650	0.12	9.6	113	435	28
Stainless steel Type 304L	Fe bal., Cr 19, Ni 10, C 0.03 max.	Annealed / Cold-drawn	30 / 95	80 / 125	60 / 25	150 / 277	0.29	8.02	2550–2650	0.12	9.6	113	435	28
Stainless steel Type 309	Fe bal., Cr 23, Ni 13, C 0.20 max.	Annealed / Cold-rolled[c]	30 to 120	82 to 150	50 / 4[d]	165 / 275	0.29	8.02	2550–2650	0.12	8.3	96	470	29
Stainless steel Type 310	Fe bal., Cr 25, Ni 20, C 0.25 max.	Annealed	40	100	50	165	0.29	8.02	2550–2650	0.12	8.0	96	470	29

Table of stainless steel properties (values shown as "annealed value / cold-rolled (or heat-treated/cold-drawn/hardened) value" where two are given):

Material	Composition	Condition												
Stainless steel Type 316	Fe bal., Cr 18, Ni 11, Mo 2.5, C 0.10 max.	Annealed / Cold-rolled[e]	30 to 120	90 to 150	50 / 8[d]	165 / 275	0.29	8.02	2500-2550	0.12	8.9	113	445	28
Stainless steel Type 316L	Fe bal., Cr 17, Ni 12, C 0.03 max., Mo 2	Annealed / Cold-drawn	30 / 60	80 / 90	60 / 45	150 / 190	0.29	8.02	2500-2550	0.12	8.9	113	445	28
Stainless steel Types 321 and 347 (321 has Ti) (347 has Cb)	Fe bal., Cr 18, Ni 1.0, C 0.10 max., Ti 4 × carbon min. or Cb 8 × carbon min.	Annealed / Cold-rolled	30 to 120	85 to 150	50 / 5[d]	160 / 300	0.286	7.92	2550-2600	0.12	9.3	110	435	28
Stainless steel Type 330	Fe bal., Ni 36, Cr 16	Hot-rolled / Cold-drawn (anneal) / Cold-drawn (heat-treated)	55 / .... /	100 / 80 / 150	35 /	200 /	0.284	7.86	2555	0.11	6.3, 8.8 (68-932°F)	90	600	
Stainless steel Type 410	Fe bal., Cr 12.5, C 0.15 max.	Annealed / Heat-treated	40 / 115	75 / 150	30 / 15	150 / 300	0.28	7.75	2700-2790	0.11	5.5	173	340	29
Stainless steel Type 414	Fe bal., Cr 12.5, Ni 2.5	Annealed / Heat-treated	80 / 150	100 / 200	22 / 17	217 / 387	0.28	7.75	2600-2700	0.11	6.1	173	420	29
Stainless steel Type 420	Fe bal., Cr 13, C 0.35	Annealed / Heat-treated	60 / 200	98 / 250	28 / 8	180 / 480	0.28	7.75	2650-2750	0.11	5.7	173	330	29
Stainless steel Type 430	Fe bal., Cr 16, C 0.12 max.	Annealed / Cold-rolled	40 / 95	70 / 110	35 / 10	165 / 225	0.28	7.75	2600-2750	0.11	6.0	180	360	29
Stainless steel Type 431	Fe bal., Cr 16, Ni 2	Annealed / Heat-treated	85 / 150	120 / 195	25 / 20	250 / 400	0.28	7.75	2600-2700	0.11	6.5	140	430	29
Stainless steel Type 446	Fe bal., Cr 25, C 0.35 max.	Annealed	50	80	30	165	0.27	7.45	2600-2750	0.12	5.8	145	402	29
Stainless steel 17-4 PH	Fe bal., Cr 17, Ni 4, Cu 4, Co 0.35, C 0.07	Annealed / Hardened	110 / 180	150 / 195	12 / 13	363 / 404	0.28	......	....	....	6	124	....	28.5

c The cold-roller properties depend upon composition; Types 302 and 304 are not rolled often in excess of 175,000 psi tensile strength.

d The values for elongation (per cent in 2 in.) are obtainable in the steel cold-rolled to the *maximum stated* yield strength and tensile strength. For lower values of tensile strength, elongation will be correspondingly higher.

e Types 316, 321, and 374 are used chiefly in the annealed condition.

5–27

Table 5-7. Properties of Metals and Alloys* (Continued)

Material	Nominal composition (essential elements), %	Form and condition	Typical mechanical properties				Typical physical constants								
			Yield strength (0.2% offset), 1,000 psi	Tensile strength, 1,000 psi	Elongation in 2 in., %	Hardness, Brinell	Density, lb/cu in.	Specific gravity	Melting point, °F	Specific heat (32-212°F) Btu/(lb)(°F)	Thermal expansion coefficient (32-212°F) ×10⁻⁶ (in./in.)/(°F)	Thermal conductivity (32-212°F) Btu/(sq ft)(hr)(°F/in.)	Electrical resistivity (58°F), ohms/cir mil ft	Tensile modulus of elasticity ×10⁶ psi
Stainless steel 17-7 PH	Fe bal., Cr 17, Ni 7, Al 1, C 0.09	Annealed / Hardened	40 / 185	130 / 200	30 / 9	165 / 404	0.282	....	....	....	8.5	117	....	29
Stainless steel HNM	Fe bal., Cr 18, Mn 3.5, Ni 9.5, C 0.30	Annealed / Hardened	56 / 124	116 / 168	57 / 19	192 / 352	0.284	....	....	....	8.4	....	....	29
Stainless steel, stainless W	Fe bal., Cr 17, Ni 7, Ti 0.7, Al 0.2, C 0.07	Annealed / Hardened	75-115 / 150-185	120-150 / 170-210	8-15 / 8-16	255 / 365	0.28	....	....	....	6	130	600	28
Carpenter stainless No. 20ℓ	C 0.07 max., Mn 0.75, Si 1.00, Cr 20.00, Ni 29.00, Mo 2.00 min., Cu 3.00 min.	Annealed	35	85	50	160	0.289	8.02	....	0.12	9.4	145.2 (212°F)	451	28
		Cast Stainless Steels												
Cast 12 Cr alloy (CA-15) (C means corrosion-resistant casting, Alloy Casting Institute designations)	C 0.15 max., Mn 1.00 max., Si 1.50 max., Cr 11.5-14, Ni 1.00 max., Fe bal.	Air-cooled from 1800°F. Tempered at 600°F / Air-cooled from 1800°F. Tempered at 1400°F	150 / 75	200 / 100	7 / 30	390 / 185	0.275	7.61	2750	0.11	6.4 (70-1000°F)	174 (212°F)	468	29
Cast 12 Cr alloy (CA-40)	C 0.20-0.40, Mn 1.00 max., Si 1.50 max., Cr 11.5-14, Ni 1.00 max., Fe bal.	Air-cooled from 1800°F. Tempered at 600°F / Air-cooled from 1800°F. Tempered at 1400°F	165 / 67	220 / 110	1 / 18	470 / 212	0.275	7.61	2725	0.11	6.4 (70-1000°F)	174 (212°F)	456	29

Alloy	Composition	Condition												
Cast 20 Cr alloy (CB-30)	C 0.30 max., Mn 1.00 max., Si 1.00 max., Cr 18–22, Ni 2 max.	Annealed	60	95	15	195	0.272	7.53	2725	0.11	6.5 (70–1000°F)	153.6 (212°F)	456	29
Cast 28-4 alloy (CC-50)	C 0.50 max., Mn 1.00 max., Si 1.00 max., Cr 26–30, Ni 4.00 max., Fe bal.	As cast	65	70	2	212	0.272	7.53	2725	0.12	6.4 (70–1000°F)	151.2 (212°F)	462	29
		Air-cooled from 1900°F	65	97	18	210								
Cast 29-9 alloy (CE-30)	C 0.30 max., Mn 1.50 max., Si 2.00 max., Cr 26–30, Ni 8–11, Fe bal.	As cast	60	95	15	170	0.277	7.67	2650	0.14	9.6 (70–1000°F)	....	510	25
Cast 20-10 alloy (CF-8)	C 0.08 max., Mn 1.50 max., Si 2.00 max., Cr 18–21, Ni 8–11, Fe bal.	Water-quenched (1950–2050°F)	37	77	55	140	0.280	7.75	2600	0.12	10.0 (70–1000°F)	110.4 (212°F)	457.2	28
Cast 20-10 alloy (CF-20)	C 0.20 max., Mn 1.50 max., Si 2.00 max., Cr 18–21, Ni 8–11, Fe bal.	Water-quenched (above 2000°F)	36	77	50	163	0.280	7.75	2575	0.12	10.4 (70–1000°F)	110.4 (212°F)	467.4	28
Cast 20-10-2,5 alloy (CF-8M)	C 0.08 max., Mn 1.50 max., Si 1.50 max., Cr 18–21, Ni 9–12, Mo 2.5, Fe bal.	Water-quenched (1950–2050°F)	42	80	50	156–170	0.280	7.75	2550	0.12	9.7 (70–1000°F)	112.8 (212°F)	492	28
Cast 18-38 alloy (HU) (H means heat-resistant)	C 0.35–0.75, Mn 2.00 max., Si 2.50 max., Cr 17–21, Ni 37–41, Fe bal.	As cast	40	70	9	170	0.290	8.03	2450	0.11	9.7 (70–2000°F)	106.8 (70–1000°F)	630	27
		Aged	43	73	5	190								

1 Carpenter stainless No. 20-Cb same composition except for columbium + tantalum, eight times the carbon minimum.

Table 5-7. Properties of Metals and Alloys* (Continued)

Material	Nominal composition (essential elements), %	Form and condition	Yield strength (0.2% offset), 1,000 psi	Tensile strength, 1,000 psi	Elongation in 2 in., %	Hardness, Brinell	Density, lb/cu in.	Specific gravity	Melting point, °F	Specific heat (32–212°F) Btu/(lb)(°F)	Thermal expansion coefficient (32–212°F) ×10⁻⁶ (in./in.)/(°F)	Thermal conductivity (32–212°F) Btu/(sq ft)(hr)(°F/in.)	Electrical resistivity (68°F), ohms/cir mil ft	Tensile modulus of elasticity ×10⁶ psi
					Typical mechanical properties						Typical physical constants			
Cast 12-60 alloy (HW)	C 0.35–0.75, Mn 2.00 max., Si 2.50 max., Cr 10–14, Ni 58–62, Fe bal.	As cast	36	68	4	185	0.294	8.14	2350	0.11	9.2 (70–2000°F)	92.4 (70–212°F)	672	25
		Aged	52	84	4	205								
Cast 15-65 alloy (HX)	C 0.35–0.75, Mn 2.00 max., Si 2.50 max., Cr 15–19, Ni 64–68, Fe bal.	As cast	36	65	9	176	0.294	8.14	2350	0.11	9.5 (70–2000°F)	....	....	25
		Aged	44	73	9	185								
Cast 9 Cr alloy (HA)	C 0.20 max., Mn 0.35–0.65 max., Si 1.00 max., Cr 8–10, Fe bal.	Annealed	65	95	23	180	0.279	7.72	2750	0.11	7.5 (70–1200°F)	204 (70–1000°F)	420	29
		Normalized	81	107	21	220								
Cast 28-4 alloy (HC)	C 0.50 max., Mn 1.00 max., Si 2.00 max., Cr 26–30, Ni 4 max., Fe bal.	As cast	65	70	2	190	0.272	7.53	2725	0.12	7.7 (70–2000°F)	214.8 (70–1000°F)	462	29
		Aged	80	115	18									
Cast 28-7 alloy (HD)	C 0.50 max., Mn 1.50 max., Si 2.00 max., Cr 26–30, Ni 4–7, Fe bal.	As cast	48	85	16	190	0.274	7.58	2700	0.12	9.2 (70–2000°F)	214.8 (70–1000°F)	486	27
Cast 29-9 alloy (HE)	C 0.20–0.50, Mn 2.00 max., Si 2.00 max., Cr 26–30, Ni 8–11, Fe bal.	As cast	45	95	20	200	0.277	7.68	2650	0.14	11.1 (70–2000°F)	120 (70–1500°F)	510	25
		Aged	55	90	10	270								

Alloy	Composition	Condition												
Cast 21-10 alloy (HF)	C 0.20-0.40, Mn 2.00 max., Si 2.00 max., Cr 19-23, Ni 9-12, Fe bal.	As cast	45	85	35	165	0.280	7.75	2550	0.12	10.9 (70-2000°F)	160.8 (70-1000°F)	480	28
		Aged	50	100	25	190								
Cast 25-12 alloy (HH)	C 0.20-0.50, Mn 2.00 max., Si 2.00 max., Cr 24-28, Ni 11-14, Fe bal.	As cast—Type 1	50	80	25	185	0.279	7.72	2500	0.12	10.8 (70-2000°F)	130.8 (70-1000°F)	450-510	27
		Type 2	40	85	15	180								
		Aged—Type 1	55	86	11	200								
		Type 2	45	92	8	200								
Cast 28-15 alloy (HI)	C 0.20-0.50, Mn 2.00 max., Si 2.00 max., Cr 26-30, Ni 14-18, Fe bal.	As cast	45	80	12	180	0.279	7.72	2550	0.12	10.8 (70-2000°F)	98.4 (70-1000°F)	. . . . . .	27
		Aged	65	90	6	200								
Cast 25-20 alloy (HK)	C 0.20-0.60, Mn 2.00 max., Si 2.00 max., Cr 24-28, Ni 18-22, Fe bal.	As cast	50	75	17	170	0.280	7.75	2550	0.12	10.1 (70-2000°F)	130.8 (70-1000°F)	540	29
		Aged	50	85	10	190								
Cast 30-20 alloy (HL)	C 0.20-0.60, Mn 2.00 max., Si 2.00 max., Cr 28-32, Ni 18-22, Fe bal.	As cast	52	82	19	192	0.279	7.72	2600	0.12	10.1 (70-2000°F)	150.8 (70-1000°F)	574	29
Cast 20-25 alloy (HN)	C 0.20-0.50, Mn 2.00 max., Si 2.00 max., Cr 19-23, Ni 23-27	As cast	38	68	17	160	0.283	7.83	2500	0.11	. . . . . .	. . . . . .	. . . . . .	27
Cast 15-35 alloy (HT)	C 0.35-0.75, Mn 2.00 max., Si 2.50 max., Cr 13-17, Ni 33-37, Fe bal.	As cast	40	70	10	180	0.286	7.96	2425	0.11	9.8 (70-1000°F)	136.8 (70-1000°F)	600	27
		Aged	45	75	5	200								
Cast 20-10-2.5 alloy (CF-12M)	C 0.12 max., Mn 1.50 max., Si 1.50 max., Cr 18-21, Ni 9-12, Mo 2.5, Fe bal.	Water-quenched (from above 2000°F)	42	80	50	156-170	0.280	7.75	2550	0.12	9.7 (70-1000°F)	112.8 (212°F)	492	28

Table 5-7. Properties of Metals and Alloys* (Continued)

Material	Nominal composition (essential elements), %	Form and condition	Typical mechanical properties				Typical physical constants							
			Yield strength (0.2% offset), 1,000 psi	Tensile strength, 1,000 psi	Elongation in 2 in., %	Hardness, Brinell	Density, lb/cu in.	Specific gravity	Melting point, °F	Specific heat (32–212°F) Btu/(lb)(°F)	Thermal expansion coefficient (32–212°F) $\times 10^{-6}$ (in./in.)/(°F)	Thermal conductivity (32–212°F) Btu/(sq ft)(hr) (°F/in.)	Electrical resistivity (68°F), ohms/cir mil ft	Tensile modulus of elasticity $\times 10^6$ psi
Cast 20-10 Cb alloy (CF-8C)	C 0.08 max., Mn 1.50 max., Si 2.00 max., Cr 18–21, Ni 9–12, Cb or Cb-Ta, Fe bal.	Water-quenched (1950–2050°F)	38	77	39	149	0.280	7.75	2600	0.12	10.3 (70–1000°F)	111.6 (212°F)	426	28
Cast 20-10 alloy (CF-16F)	C 0.16 max., Mn 1.50 max., Si 2.00 max., Cr 18–21, Ni 9–12, Mo 1.5 max., Fe bal.	Water-quenched (from above 2000°F)	40	77	52	150	0.280	7.75	2550	0.12	9.9 (70–1000°F)	112.8 (212°F)	432	28
Cast 25-12 alloy (CH-20)	C 0.20 max., Mn 1.50 max., Si 2.00 max., Cr 22–26, Ni 12–15, Fe bal.	Water-quenched (from above 2000°F)	50	88	38	190	0.279	7.72	2600	0.12	9.6 (70–1000°F)	98.4 (212°F)	504	28
Cast 25-20 alloy (CK-20)	C 0.20 max., Mn 1.50 max., Si 2.00 max., Cr 23–27, Ni 19–22, Fe bal.	Water-quenched (from 2100°F)	38	76	37	144	0.280	7.75	2600	0.12	9.2 (70–1000°F)	98.4 (212°F)	540	29
Cast 25-20 alloy (CN-7M)	C 0.07 max., Mn 1.50 max., Si 1.00, Cr 18–22, Ni 21–31, Fe bal.	As cast	30	65	30–45	130–150	0.287	8.02	2650	0.12	9.7 (70–1000°F)	145.2 (212°F)	537.6	24

Material	Composition	Condition												
Durimet 20 (Stainless steel alloy)	C 0.07, Mn 1.5, Si 1.5, Cr 20.0, Ni 29.0, Mo 2.0, Cu 3.0	Cast, annealed	30	65	35 min.	130	0.286	8.02	2650	0.12	8.6	145 (212°F)	......	
Iron-silicon alloy (Duriron)	Si 14.50, C 0.85, Mn 0.65, Mo nil, Fe bal.	Cast only	......	16	Nil	520	0.255	7.0	2300	0.13	7.4 (68–392°F)	......	......	23
Fe-Si-Mo alloy (Durichlor)	Si 14.50, C 0.85, Mn 0.65, Mo 3.00, Fe bal.	Cast only	......	16	Nil	520	0.255	7.0	2300	0.13	7.2	......	......	23

NONFERROUS ALLOYS
Aluminum and Alloys

Material	Composition	Condition												
Aluminum alloy No. 1100	Al 99 plus	Annealed-0	5	13	45	23	0.098	2.71	1190–1215	0.23	13.1 (68°F)	1540	18	10
		Cold-rolled-H14	17	18	20	32						......	19	
		Cold-rolled-H18	22	24	15	44						1510 (68°F)		
3003	Al bal., Mn 1.2	Annealed-0	6	16	40	28	0.099	2.73	1190–1210	0.23	12.9 (68°F)	1340	21	10
		Cold-rolled-H14	21	22	16	40						1100	25	
		Cold-rolled-H18	27	29	10	55						1075 (68°F)	26	
5052	Al bal., Mg 2.5, Cr 0.25	Annealed-0	13	28	30	47	0.097	2.68	1125–1200	0.23	13.2 (68°F)	960 (68°F)	30	10.2
		Cold-rolled and stabilized-H–34	31	38	14	68						......	......	
		Cold-rolled and stabilized-H–38	37	42	8	77						960 (68°F)	30	
6061	Al bal., Si 0.6, Cr 0.15, Mg 1.0, Cu 0.25	Annealed-0	8	18	25	......	0.09	2.70	1100–1205	......	13.0	1390	20	10
		Cold-rolled and stabilized-T4	21	35	22									
		Cold-rolled and stabilized-T6	40	45	12									

Table 5-7. Properties of Metals and Alloys* (Continued)

Material	Nominal composition (essential elements), %	Form and condition	Yield strength (0.2% offset), 1,000 psi	Tensile strength, 1,000 psi	Elongation in 2 in., %	Hardness, Brinell	Density, lb/cu in.	Specific gravity	Melting point, °F	Specific heat (32–212°F) Btu/(lb)(°F)	Thermal expansion coefficient (32–212°F) ×10⁻⁶ (in./in.)/(°F)	Thermal conductivity (32–212°F) Btu/(sq ft)(hr)(°F/in.)	Electrical resistivity (68°F), ohms/cir mil ft	Tensile modulus of elasticity ×10⁶ psi
6063	Al bal., Si 0.4, Mg 0.7	Annealed-0	7	13	30	25	0.098	2.70	1140–1205	....	13.0 (68°F)	1390 (68°F)	20	10.0
		Artificially aged–T5	21	27	12	60								
		Heat-treated and artificially aged–T6	31	35	12	73								
7075	Al bal., Zn 5.6, Cu 1.6, Mg 2.5, Cr 0.3	Annealed-0	15	33	17	60	0.101	2.80	890–1180	0.23	12.9 (68°F)	840 (68°F)	34	10.4
		Heat-treated and artificially aged–T6	73	83	11	150								
43	Al bal., Si 5.0	Sand-cast-F	8	19	8	40	0.097	2.69	1065–1170	0.23	12.3 (68°F)	990 (68°F)	27	10.3
		Permanent mold-cast-F	9	23	10	45								
		Die-cast-F	16	30	7									
195	Al bal., Cu 4.5, Si 0.8	Sand-cast; heat-treated-4	16	32	8.5	60	0.102	2.81	970–1190	0.23	12.7 (68°F)	960 (68°F)	30	10.3
		Sand-cast; heat-treated and artificially aged–T6	24	36	5	75		.....	....	....	....	960 (68°F)		
214	Al bal., Mg 3.8	Sand-cast-F	12	25	9	50	0.096	2.65	1110–1185	0.23	13.4 (68°F)	960 (68°F)	29	10.3
356	Al bal., Mg 0.3, Si 7.0	Sand-cast; heat-treated-T6	24	33	3.5	.....	0.097	2.68	1035–1135				....	10.3

Material	Composition	Condition												
Nickel silver 18% (wrought) 752	Cu 65, Zn 17, Ni 18	Annealed	25	58	40	70	0.316	8.75	2030	0.09	9.0	230	173	18
		Cold-rolled (HT)[a]	70	85	4	170								
		Cold-drawn wire (HT)	......	105										
Nickel silver 12% (cast)	Cu bal., Zn 20, Ni 12.5, Pb 10.5, Sn 2	Cast (HT, HT)	18	35	15	55	0.320	8.9	1830	0.12	10.0	190	204	16
Nickel silver 10% 745	Cu 65, Zn 25, Ni 10	Annealed	20	55	45	60	0.313	8.67	1870	0.09	9.0	320	125	17.5
		Cold-rolled (HT)	70	88	5	180								
		Cold-drawn wire (HT)	......	110										
Cupronickel 10% 706	Cu 88.35, Ni 10, Fe 1.25, Mn 0.4	Annealed	22	44	45	......	0.323	8.94	2090	0.09	9.3	310	113	18
		Cold-drawn tube	57	60	15									
Cupronickel 30% 715	Cu 68.90, Ni 30, Mn 0.60, Fe 0.50	Annealed	22	55	45	70	0.323	8.94	2240	0.09	8.5	200	220	22
		Cold-drawn	60	75	20	150								
		Cold-rolled	70	77	5	155								
Cupronickel 55–45 (Constantan)	Cu 55, Ni 45	Annealed	30	60	45	......	0.321	8.89	2300	......	8.1	155	290	24
		Cold-drawn	50	65	30									
		Cold-rolled	65	85	20									
Copper	Cu 99.9 plus	Annealed	10	32	45	42	0.322	8.91	1980	0.092	9.3	2700	10.3	17
		Cold-drawn	40	45	15	90								
		Cold-rolled (HT)	40	46	5	100								
Red brass (wrought) 230	Cu 85, Zn 15	Annealed	15	40	50	50	0.316	8.75	1875	0.09	9.8	1100	28	17
		Cold-drawn	55	70	15	120								
		Cold-rolled	60	75	7	135								
Red brass (cast)	Cu 85, Zn 5, Pb 5, Sn 5	Cast (as cast)	17	35	25	60	0.317	8.75	1810–1840	......	10.2	500	63	13
Gilding metal 210	Cu 95.0, Zn 5.0	Cold-rolled	50	56	5	114	0.320	......	1950	0.09	10.0	1600	......	17
Commercial bronze 220	Cu 90.0, Zn 10.0	Cold-rolled	54	61	5	125	0.318	......	1910	0.09	10.2	1300	......	17

[a] Hard temper (HT). Hard temper and heat-treated (strip) (HT, HT).

Table 5-7. Properties of Metals and Alloys* (Continued)

Material	Nominal composition (essential elements), %	Form and condition	Yield strength (0.2% offset), 1,000 psi	Tensile strength, 1,000 psi	Elongation in 2 in., %	Hardness, Brinell	Density, lb/cu in.	Specific gravity	Melting point, °F	Specific heat (32-212°F) Btu/(lb)(°F)	Thermal expansion coefficient (32-212°F) ×10⁻⁶ (in.)/(in.)/(°F)	Thermal conductivity (32-212°F) Btu/(sq ft)(hr)(°F/in.)	Electrical resistivity (68°F), ohms/cir mil ft	Tensile modulus of elasticity ×10⁶ psi
Yellow brass (high brass) 268	Cu 65, Zn 35	Annealed Cold-drawn Cold-rolled (HT)	18 55 60	48 70 74	60 15 10	55 115 180	0.306	8.47	1710	0.09	10.5	830	40	15
Muntz metal 280	Cu 60, Zn 40	Annealed	20	54	45	80	0.303	8.39	1660	0.09	10.8	870	37	15
Cartridge 70-30 brass 260	Cu 70.0, Zn 30.0	Cold-rolled	63	76	8	155	0.308	....	1750	0.09	11.1	840	....	16
Architectural bronze 385	Cu 57.0, Zn 40.0, Pb 3.0	Annealed	20	60	30	95	0.306	....	1630	....	11.6	850	....	14
Phosphor bronze 10% 524	Cu 90, Sn 10, P 0.25	Spring temper	....	122	4	241	0.317	....	1830	....	10.2	350	....	16
Phosphor bronze 5% 510	Cu 94.75, Sn 5, P 0.25	Annealed Cold-drawn wire (HT) Cold-rolled (HT)	20 65	50 130 80	50 2 8	60 160	0.320	8.86	1920	0.09	9.4	480	69	16
Aluminum brass 687	Cu 76.0, Zn 22.0, Al 2.0, As trace	Annealed	27	60	55	82	0.301	....	1780	....	10.3	700	....	16
Naval brass 464	Cu 60, Zn 39.25, Sn 0.75	Annealed Cold-drawn	22 40	56 65	40 35	90 150	0.304	8.41	1625	0.09	11.0	810	40	15
Admiralty brass 443 (inhibited)	Cu 71, Zn 28, Sn 1, As, Sb, or P present	Annealed	20	53	65	60	0.308	8.53	1720	0.09	10.2	770	42	16
Manganese bronze 675	Cu 58.5, Zn 39.2, Fe 1, Sn 1, Mn 0.3	Annealed Cold-drawn	30 50	60 80	30 20	95 180	0.302	8.36	1645	0.09	11.2	700	45	15

Material	Composition	Condition												
High-silicon bronze A 655	Cu 96, Si 3, Mn, Zn, or Fe	Annealed	22	58	60	70	0.308	8.53	1865	0.09	9.5	225	160	15
		Cold-drawn	60	90	20	180								
		Cold-rolled	60	95	7	190								
Low-silicon bronze B 651	Cu 96, Si 0.8–2.0, Mn 0.7 max., Fe 0.8 max.	Annealed	15	40	50	F55	0.316	......	1890–1940	0.09	9.9	360	......	......
		Hardened	55	70	15	B80 (Rockwell)								
Aluminum bronze 612	Cu 92, Al 8	Annealed	25	70	60	80	0.281	7.78	1900	0.09	9.2	490	70	17
		Hard	65	105	7	210								
Ni-Vee bronze Type A	Cu 88, Ni 5, Sn 5, Zn 2	As cast	22	50	40	85	0.32	8.8	1950	......	11	......	......	15
		Tempered	40	65	10	130	0.32	8.8	1950	......	11	......	......	15
		Heat-treated	55	85	10	180	0.32	8.8	1950	......	11	......	......	15
Ni-Vee bronze Type B	Cu 87, Ni 5, Sn 5, Pb 1, Zn 2	As cast	20	45	30	80	0.32	8.8	1925	......	10	......	......	13
		Tempered	30	60	8	120								14
Copper beryllium 172	Be 1.9, Co 0.25, Cu bal.	Annealed (SA)[i]	......	70	45	B60 (Rockwell)	0.298	8.26	1600–1800	......	9.3	750–900	......	17
		(SA, HT)	......	175	6	C38								19
		Cold-rolled (HT)	......	110	5	B99								17
		(HT, HT)	......	200	2	C42 (Rockwell)								19

Lead and Alloys

Material	Composition	Condition												
Chemical lead	Pb 99.9, Cu 0.06, Bi 0.005 max.	Rolled	1.9	2.6	50	5	0.410	11.35	621	0.030	16.4	240	124	2
Antimonial lead	Pb 94, Sb 6	Cast	......	6.8	22	12	0.393	10.90	554	0.032	15.1	200	140	3
		Rolled	......	4.1	47	9								
Tellurium lead	Pb 99.85, Te 0.04, Cu 0.06	Rolled	2.2	3	45	6	0.410	11.35	621	0.030	16.4	240	124	2

[i] Solution annealed (SA). Solution annealed, heat-treated (SA, HT)

5–37

Table 5-7. Properties of Metals and Alloys* (Continued)

Material	Nominal composition (essential elements), %	Form and condition	Typical mechanical properties				Typical physical constants							
			Yield strength (0.2% offset), 1,000 psi	Tensile strength, 1,000 psi	Elongation in 2 in., %	Hardness, Brinell	Density, lb/cu in.	Specific gravity	Melting point, °F	Specific heat (32–212°F) Btu/(lb)(°F)	Thermal expansion coefficient (32–212°F) $\times 10^{-6}$ (in./in.)/(°F)	Thermal conductivity (32–212°F) Btu/(sq ft)(hr)(°F/in.)	Electrical resistivity (68°F), ohms/cir mil ft	Tensile modulus of elasticity $\times 10^6$ psi
Soft solder 50–50	Sn 50, Pb 50	Cast	....	6.8	50	14	0.321	8.89	421	0.051	13.1	310	93	
Soft solder 60–40	Sn 60, Pb 40	Cast	....	7.1	45	15	0.306	8.5	374	0.055	12.2	330	90	
Magnesium Alloys														
Magnesium alloy AZ31B	Mg bal., Al 3.0, Zn 1.0, Mn 0.20 min.	Rolled-plate (strain-hardened then partially annealed)	24	37	18	......	0.064	1.77	1160	0.245	14.5	540	55	6.5
		Rolled-sheet (strain-hardened then partially annealed)	32	42	15	73	......	......	......	......	......	540	55	
		Annealed	22	37	21	56	......	......	......	......	......	540	55	
		Extruded	28	38	14	......						540	55	
Magnesium alloy AZ80A	Mg bal., Al 8.5, Zn 0.2, Mn 0.15 min.	Extruded	36	49	11	60	0.065	1.80	1130	0.245	14.5	350	87	6.5
		Extruded (age-hardened)	39	53	6	82								
		Forged (age-hardened)	34	50	6	72								
Magnesium alloy AZ91A and AZ91B	Mg bal., Al 9.0, Zn 0.6, Mn 0.13 min.	Die-cast (as cast)	22	33	3	63	0.065	1.81	1105	0.245	14.5	390	102	6.5
Magnesium alloy AZ92A	Mg bal., Al 9.0, Zn 2.0, Mn 0.10 min.	Sand-cast (as cast)	14	24	2	65	0.066	1.83	1100	0.245	14.5	360	84	6.5
		Sand-cast (solution heat-treated)	14	40	12	63	......	......	......	......	......	310	101	

		Sand-cast (solution heat-treated and aged)	21	40	2	83	......	.....	....	....	....	410	74	
		Sand-cast (age-hardened)	14	24	2	66	......	.....	....	....	....	410	74	
Magnesium alloy AZ91C	Mg bal., Al 8.7, Zn 0.7, Mn 0.13 min.	Sand-cast (as cast)	14	24	2	52	......	.....	....	....	....	370	82	6.5
		Sand-cast (solution heat-treated)	14	40	14	55	0.065	1.81	1105	0.245	14.5	320	97	
		Sand-cast (solution heat-treated and aged)	19	40	5	66	......	.....	....	....	....	390	78	
Magnesium alloy HK31A	Mg bal., Th 3.0, Zr 0.7	Sand-cast (solution heat-treated and aged)	15	30	8	55	0.065	1.79	1200	0.245	14.5	690	46	6.5
		Rolled-sheet (strain-hardened then partially annealed)	29	37	8	57	......	.....	....	....	....	780	37	
Magnesium alloy HZ32A	Mg bal., Th 3.0, Zn 2.1, Zr 0.7	Sand-cast (age-hardened)	14	29	7	57	0.066	1.83	1198	0.245	14.5	740	39	6.5
Magnesium alloy ZK61A	Mg bal., Zn 5.7, Zr 0.80	Extruded (age-hardened)	28	45	8	70	0.066	1.83	1175	0.245	14.5			

Nickel and Alloys

Nickel (pure)	Ni 99.99	Annealed	8.5	46	30	.....	0.322	8.91	2650	0.11	7.4	543	41	30
Nickel 210 (cast)	Ni 95.6, Cu 0.5, Fe 0.5, Mn 0.8, Si 1.5, C 0.8	As-cast	25	57	22	110	0.301	8.34	2450–2600	0.13	8.85 (70–1400°F)	410	125	21.5
Nickel 200	Ni(+Co) 99.40, C 0.06, Mn 0.25, Fe 0.15, S 0.005, Si 0.05, Cu 0.05	Annealed	20	70	40	100	0.321	8.89	2615–2635	0.13	6.6	420	53	30
		Hot-rolled	25	75	40	110	......	.....	....	....	....	....	57	
		Cold-drawn	70	95	25	170								
		Cold-rolled	95	105	5	210								
Nickel 201	Ni(+Co) 99.50, C 0.02, Mn 0.20, Fe 0.15, S 0.005, Si 0.05, Cu 0.05	Annealed	15	60	50	90	0.321	8.89	2615–2635	0.11	7.2	420	50	30
		Hot-rolled	25	60	45	105								
		Cold-drawn	65	95	15	150								
		Cold-rolled												

Table 5-7. Properties of Metals and Alloys* (Continued)

Material	Nominal composition (essential elements), %	Form and condition	Yield strength (0.2% offset), 1,000 psi	Tensile strength, 1,000 psi	Elongation in 2 in., %	Hardness, Brinell	Density, lb/cu in.	Specific gravity	Melting point, °F	Specific heat (32–212°F) Btu/(lb)(°F)	Thermal expansion coefficient (32–212°F) × 10⁻⁶ (in./in.)/(°F)	Thermal conductivity (32–212°F) Btu/(sq ft)(hr)(°F/in.)	Electrical resistivity (68°F), ohms/cir mil ft	Tensile modulus of elasticity × 10⁶ psi
Nickel 211	Ni(+Co) 95.00, C 0.10, Mn 4.75, Fe 0.05, S 0.05, Si 0.05, Cu 0.02	Annealed Hot-rolled Cold-drawn	35 50 80	75 90 100	40 35 25	140 150 190	0.315	8.78	2600	0.11	7.4	335	110	30
Nickel 212	Ni(+Co) 97.85, C 0.05, Mn 1.95, Fe 0.05, S 0.005, Si 0.04, Cu 0.03	Annealed Hot-rolled Cold-drawn	35 50 80	75 90 100	40 35 25	140 150 190	0.319	8.86	2600	0.11	7.4	335	85	30
Duranickel 301	Ni(+Co) 93.90, C 0.15, Mn 0.25, Fe 0.15, S 0.005, Si 0.55, Cu 0.05, Al 4.50, Ti 0.45	Annealed Annealed age-hardened Spring Spring, age-hardened	45 125	100 170 175 205	40 25 5 10	160 330 320 370	0.298	8.26	2550–2620	0.104	7.2	128 137	280 260	30
Permanickel	Ni(+Co) 98.65, C 0.25, Mn 0.10, Fe 0.10, S 0.005, Si 0.06, Cu 0.02, Ti 0.45, Mg 0.35	Annealed Annealed, age-hardened Spring Spring, age-hardened	45 125 195	105 175 180 210	45 25 5 10	160 325	0.316	8.75	2550–2620	0.106 (70–750°F)	7.2	400	94.5	30
Nickel 220	Ni(+Co) 99.65, C 0.06, Mn 0.10, Fe 0.05, S 0.005, Si 0.05, Cu 0.02, Mg 0.04	Annealed	20	70	40	100	0.0321	8.89	2615–2635	0.11	7.2	420	57	30
Nickel 225	Ni(+Co) 99.50, C 0.07, Mn 0.10, S 0.005, Si 0.20, Cu 0.02, Fe 0.05	Annealed	20	70	40	100	0.321	8.89	2615–2635	0.11	7.2	420	57	30

Alloy	Nominal composition, %	Condition	20	70	40	100	0.321	8.85	2615–2635	0.110	7.2	420	57	31
Nickel 233	Ni(+Co) 99.55, C 0.09, Mn 0.20, Fe 0.05, S 0.005, Si 0.05, Cu 0.02	Annealed	20	70	40	100	0.321	8.85	2615–2635	0.110	7.2	420	57	31
Incoloy alloy 800	Ni(+Co) 31.90, C 0.04, Mn 0.75, Fe 46.05, S 0.007, Si 0.35, Cu 0.27, Cr 20.60	Annealed	40	90	40	150	0.290	8.01	2540–2600	0.120	8.2	85	600 (76°F)	28.5
Incoloy alloy 801	Ni(+Co) 32.20, C 0.04, Mn 0.85, Fe 44.60, S 0.007, Si 0.35, Cu 0.15, Cr 20.80, Ti 1.00	Annealed	50	90	40	160	0.287	7.99				....	....	28.5
Incoloy alloy 901	Ni(+Co) 42.65, C 0.05, Mn 0.45, Fe 33.90, S 0.010, Si 0.40, Cu 0.10, Cr 13.45, Al 0.25, Ti 2.50, Mo 6.20	Annealed	45	110	45	160	0.297	8.23	....	....	7.75	93	.... 662 (76°F)	30.0
		Annealed, aged	105	165	26	310	....	....	....	....	....	....	....	
Monel alloy 400	Ni(+Co) 66.15, C 0.12, Mn 0.90, Fe 1.35, S 0.005, Si 0.15, Cu 31.30	Annealed	35	75	40	125	0.319	8.83	2370–2460	0.127	7.5	174 (212°F)	290	26
		Hot-rolled	50	90	35	150								
		Cold-drawn	80	110	25	190								
		Cold-rolled	110	110	5	240								
Monel alloy 410 (cast)	Ni(+Co) 64.0, Cu 31.5, Fe 1.0, Mn 0.8, Si 1.5, C 0.20	As cast	....	75	35	140	0.312	8.63	2400–2450	0.13	9.17	186	320	23
Monel alloy R-405	Ni(+Co) 66.35, C 0.18, Mn 0.90, Fe 1.35, S 0.50, Si 0.15, Cu 31.00	Hot-rolled	45	85	35	145	0.319	8.84	2370–2460	0.127	7.8	180	290	26
		Cold-drawn	75	100	25	200								
Monel alloy 402	Ni(+Co) 58.10, C 0.12, Mn 0.90, Fe 1.20, S 0.005, Si 0.10, Cu 39.55	Hot-rolled	65	90	35	175	0.320	8.85	....	....	....	....	305	26
		Cold-drawn	85	95	25	215								

Table 5-7. Properties of Metals and Alloys* (Continued)

Material	Nominal composition (essential elements), %	Form and condition	Yield strength (0.2% offset), 1,000 psi	Tensile strength, 1,000 psi	Elongation in 2 in., %	Hardness, Brinell	Density, lb/cu in.	Specific gravity	Melting point, °F	Specific heat (32–212°F) Btu/(lb)(°F)	Thermal expansion coefficient (32–212°F) ×10⁻⁶ (in.)/(in.)/(°F)	Thermal conductivity (32–212°F) Btu/(sq ft)(hr)/(°F/in.)	Electrical resistivity (68°F), ohms/cir mil ft	Tensile modulus of elasticity ×10⁶ psi
Monel alloy 403 Rods, bars, forgings, plate sheet, strip, wire	Ni(+Co) 57.85, C 0.12, Mn 1.90, Fe 0.15, S 0.006, Si 0.30, Cu 39.65	Hot-rolled	45	85	40	145	0.320	......	......	......	7.7 (80–200°F)	172	319	25.2
		Cold-drawn	85	90	30	200	......	......	......	......				
Monel alloy K-500 Rods, bars, forgings, plate, sheet, strip, wire	Ni(+Co) 65.25, C 0.15, Mn 0.60, Fe 1.00, S 0.005, Si 0.15, Cu 29.60, Al 2.75, Ti 0.45	Annealed	45	100	40	155	0.305	8.47	2400–2460	0.127	7.4	130	350	26
		Annealed, age-hardened	100	155	25	270								
		Spring	140	150	5	300								
		Spring, age-hardened	160	185	10	335								
Monel alloy 501 Rods, bars, forgings, wire	Ni(+Co) 64.75, C 0.25, Mn 0.60, Fe 1.00, S 0.005, Si 0.15, Cu 29.85, Al 2.85, Ti 0.45	Hot-finished, annealed	45	95	35	160	0.305	8.44	2400–2460	0.127	7.0	130	350	26.0
		Hot-finished, annealed, age-hardened	90	145	25	275								
Monel weldable alloy 411 centrifugal castings only	Ni(+Co) 62.0, Cu 31.5, Fe 2.0, Mn 0.8, Si 1.5, C 0.20, Cb added	As cast	35	75	35	140	0.312	8.63	2400–2450	0.13	9.17 (70–1100°F)	186	....	23
Monel alloy 505	Ni(+Co) 63.0, Cu 29.5, Fe 2.0, Mn 0.8, Si 4.0, C 0.08	Annealed	75	110	8	220	0.302	8.36	2250–2350	0.13	8.87 (70–1000°F)	136	380	24
		As cast or annealed and aged	110	135	2	340	0.302	8.36	2250–2350	0.13	8.87	136	380	24

Alloy	Composition	Condition													
Inconel alloy 600 (wrought)	Ni(+Co) 76.40, C 0.04, Mn 0.20, Fe 7.20, S 0.007, Si 0.20, Cu 0.10, Cr 15.85	Annealed Cold-drawn	35 100	90 130	45 20	150 200	0.304	8.43	2540–2600	0.109	7.0	104 113 (212°F)	623 (76°F)	31.0
Inconel alloy 610 (cast)	Ni(+Co) 71.5, Cu 0.5, Fe 8.0, Mn 1.0, Si 2.0, Cr 16.0, C 0.20	As cast	38	80	15	175	0.300	8.302	540–2600	0.11	8.92 (70–1400°F)	104	700	23
Inconel alloy 705 (weldable)	Ni(+Co) 68.5, Cu 0.5, Fe 9.0, Mn 1.0, Si 5.6, Cr 15.5, C 0.20	As cast	90	95	2	340	0.292	8.06	2540–260	.....	9.20 (70–1400°F)	.....	760	25
Inconel alloy 705	Ni(+Co) 6.80, Cu 0.5, Fe 8.0, Mn 1.0, Si 5.5, Cr 15.5, C 0.20	As cast Annealed and aged	90 95	95 100	2 2	340 340	0.292	8.06	2540–2600	.....	9.20 (70–1100°F)	.....	760	25
Inconel alloy X-750	Ni(+Co) 72.85, C 0.04, Mn 0.65, Fe 6.80, S 0.007, Si 0.30, Cu 0.05, Cr 15.15, Al 0.75, Ti 2.50, Cb(+Ta) 0.85	Annealed Annealed, age-hardened	50 115	115 175	50 25	150 300	0.298	8.25	2540–2600	0.105	6.7	102	735 (76°F)	31.0
Inconel alloy 751	Ni(+Co) 72.50, C 0.05, Mn 0.55, Fe 6.85, S 0.007, Si 0.35, Cu 0.05, Cr 15.15, Al 1.15, Ti 2.30, Cb(+Ta) 0.95	Annealed Annealed, age-hardened	75 110	125 175	50 20	176 337	0.298	8.25	.....	.....	.....	.....	743 (76°F)	30.0
Inconel alloy 721	Ni(+Co) 71.40, C 0.03, Mn 2.20, Fe 6.70, S 0.007, Si 0.10, Cu 0.04, Cr 16.40, Al 0.05, Ti 3.05	Annealed Annealed, age-hardened	45 100	110 145	50 7	160 300	0.298	8.25	.....	.....	.....	.....	.....	31.0

Table 5-7. Properties of Metals and Alloys* (Continued)

Material	Nominal composition (essential elements), %	Typical mechanical properties					Typical physical constants								
		Form and condition	Yield strength (0.2% offset), 1,000 psi	Tensile strength, 1,000 psi	Elongation in 2 in., %	Hardness, Brinell	Density, lb/cu in.	Specific gravity	Melting point, °F	Specific heat (32–212°F) Btu/(lb)(°F)	Thermal expansion coefficient (32–212°F) $\times 10^{-6}$ (in./in.)/(°F)	Thermal conductivity (32–212°F) Btu/(sq ft)(hr)(°F/in.)	Electrical resistivity (68°F), ohms/cir mil ft	Tensile modulus of elasticity $\times 10^{6}$ psi
Inconel alloy 722	Ni(+Co) 74.35, C 0.05, Mn 0.60, Fe 6.50, S 0.007, Si 0.20, Cu 0.05, Cr 15.20, Al 0.60, Ti 2.40	Annealed Annealed, age-hardened	45 95	110 155	50 30	160 275	0.298	8.31	2540–2600	0.105	7.6	102	750 728 (76°F)	31.0
Inconel alloy 604	Ni(+Co) 73.55, C 0.04, Mn 0.20, Fe 7.55, S 0.005, Si 0.30, Cu 0.03, Cr 16.00, Cb(+Ta) 2.30	Annealed	.....	125	.....	.....	0.305	8.44	.....				620	31.0
Inconel alloy 700	Ni 45.0, C 0.16, Mn 0.10, Fe 7.0, S 0.008, Si 0.25, Cr 15.0, Al 3.0, Ti 2.20, Mo 3.0, Co 28.0	Hot-rolled Heat-treated	106	168	.....	321	0.295	8.17	2450–2600	0.11	6.8 9.3 (70–1600°F)	86	777	32
Inconel alloy 702	Ni(+Co) 78.00, C 0.02, Mn 0.05, Fe 0.30, S 0.007, Si 0.15, Cu 0.05, Cr 15.85, Al 3.00, Ti 0.60	Annealed	45	105	55	150	0.302	8.37	.....	.....	6.95	.....	.....	31.5
Inconel alloy 713	Ni(+Co) bal., Cr 13.0, C 0.13, Mo 4.5, Cb 2.0, Al 6.0, Ti 06	Investment-cast	102	120	6	.....	0.286	7.91	2300–2350	.....	6.1			

Material	Composition	Condition												
Inconel alloy 804	Ni(+Co) 42.10, C 0.06, Mn 0.85, Fe 25.60, S 0.007, Si 0.60, Cu 0.40, Cr 29.70, Al 0.25, Ti 0.40	Annealed	45	100	40	175	0.286	7.92	....	....	8.0	....	635	32.0
Nimonic 75	Ni(+Co) 77.40, C 0.10, Mn 0.45, Fe 0.50, S 0.007, Si 0.45, Cu 0.05, Cr 20.50, Al 0.15, Ti 0.35	Annealed	55	115	40	168	0.302	8.35	2530–2590	0.11	6.8	86.5	655	27
Nimonic 80A	Ni(+Co) 74.45, C 0.05, Mn 0.55, Fe 0.55, S 0.007, Si 0.20, Cu 0.05, Cr 20.45, Al 1.25, Ti 2.40	Annealed	60	115	60	185	0.296	8.25	2530–2590	0.10	6.5	84.5	745	31
Nimonic 90	Ni(+Co) 57.00, C 0.05, Mn 0.50, Fe 0.45, S 0.007, Si 0.20, Cu 0.05, Cr 20.55, Al 1.65, Ti 2.60, Co 16.90	Annealed	90	155	......	260	0.298	8.25	2470–2530	....	6.5	86.5	690	31
Ni-o-nel alloy 825	Ni 40, Cr 21, Fe 31, Mo 3.0, Cu 1.75, Mn 0.60, Si 0.40, C 0.05	Annealed / Cold-drawn	45 /	95 / 155	40 / 10	185 / 255	0.294	8.14	....	....	....	....	....	28.5
Ni-Mo alloy (Chlorimet 2)	Ni 62.00, Mo 32.00, Fe 3.00 max., Si 1.00, C 0.10	Cast	55	80	5	230	0.333	9.24	2460	....	4.7	....	....	27
Ni-Cr-Mo alloy (Chlorimet 3)	Ni 60.00, Cr 18.00, Mo 18.00, Fe 3.00 max., Si 1.00, C 0.07	Cast only	50	75	10	220	0.325	8.94	2380	0.092	7.0 (68–392°F)	....	....	24.5

Table 5-7. Properties of Metals and Alloys* (Continued)

Material	Nominal composition (essential elements), %	Form and condition	Yield strength (0.2% offset), 1,000 psi	Tensile strength, 1,000 psi	Elongation in 2 in., %	Hardness, Brinell	Density, lb/cu in.	Specific gravity	Melting point, °F	Specific heat (32–212°F) Btu/(lb)(°F)	Thermal expansion coefficient (32–212°F) ×10⁻⁶ (in./in.)/(°F)	Thermal conductivity (32–212°F) Btu/(sq ft)(hr)(°F/in.)	Electrical resistivity (68°F), ohms/cir mil ft	Tensile modulus of elasticity ×10⁶ psi
		Typical mechanical properties					Typical physical constants							
Illium G	Ni bal., Cr 22.5, Fe 6.5, Mo 6.4, Cu 6.5, Mn, Si	As cast	50	68	32	200	0.310	8.31	2375	0.105	7.5	84	735	24
Illium 98	Ni 55.00, Cr 28.00, Mo 8.5, Cu 5.5, Mn 1.25, Fe 1.00	Cast	......	54	18	155								
Hastelloys														
Hastelloy alloy B	Ni bal., Mo 28, Fe 5, Mn, Si	Sand-cast (anneal)	50	80	8	199	....	....	2410–2460	0.091	5.6	72	812	26.5
		Rolled (anneal)	56	120	50	215	0.334	9.24						30.8
		Investment cast	54	85	14	209								28.5
Hastelloy alloy C	Ni bal., Mo 16, Cr 16, Fe 5, W 4, Mn, Si	Sand-cast (anneal)	50	78	5	199	....	....	2320–2380	0.092	6.3	61	834	26
		Rolled (anneal)	71	130	45	204	0.323	8.94						30.9
		Investment cast	50	80	10	215								24.5
Hastelloy alloy D	Ni bal., Si 10, Cu 3, Mn	Sand-cast (anneal)	118	118	0–2	321	0.282	7.80	2030–2050	0.108	6.1	145 (at 72°F)	680	28.9
Hastelloy alloy F	Ni 46, Cr 22, Fe 21, Mo 6.5, Cb + Ta 2, Mn, Si	Hot-rolled (anneal)	50	110	50	255	0.296	8.2	2350	0.1025	8.7 (70–600°F)	....	676	29

Titanium and Alloys / Other Nonferrous Alloys

Material	Composition	Condition												
Hastelloy alloy R-235	Cr 15.5, Fe 10, Mo 5.5, Ni bal., Al, Ti	Hot-rolled, mill-annealed	75	140	35	244	0.296	8.22	2350–2400	0.1096	6.7 (76–212°F)	63 (70–600°F)	800	29.4
Hastelloy alloy X	Co 1.5 max., Fe 18.5, Cr 22.0, Mo 9.0, W 0.6, C 0.15 max. (wrought), C 0.20 max. (cast), Ni bal.	Wrought sheet Mill-annealed / As investment cast	52 / ...... / 46.5	113.2 / 67.0	41.0 / 17.0	194 / 172	0.297	8.23	2350	0.1046	7.90 (70–600°F)	62.8	712	28.6

Titanium and Alloys

Material	Composition	Condition												
Titanium (commercially pure)	Ti bal., Fe 0.2 max., N_2 0.05 max., C 0.08 max., H_2 0.015 max. O_2 0.4 max.	Annealed	75	85	23	200	0.163	4.54	3040	0.125	5.0 (70–1000°F)	114	370	15
Titanium-chromium-iron-molybdenum alloy (Ti-5 Al-2.5 Sn)	Ti bal., Al 5, Sn 2.5, Fe 0.5 max., C 0.05 max., H_2 0.015 max., N_2 0.05 max.	Annealed	120	127	15	34 (Rockwell C)	0.162	4.50	3000	0.125	5.3 (70–1000°F)	54	1015	16
Titanium-aluminum-vanadium alloy (Ti-6 Al-4V)	Ti bal., Al 6.0, V 4.0, Fe 0.25 max., C 0.08 max., H_2 0.0125 max., N_2 0.05 max.	Annealed / Heat-treated	130 / 145	140 / 155	14 / 10	32 (Rockwell C) / 37 (Rockwell C)	0.161	4.46	3000	0.135	5.3 (70–1000°F)	50	1020	16.5

Other Nonferrous Alloys

Material	Composition	Condition												
Antimony	Sb 100	As cast	......	1.56	......	42	0.249	6.62	1166	0.049	5.5	131	234	11.3
Columbium			35	40	......	......	0.31	......	4474	0.06	3.8	375		
Columbium 1Zr	Zr 0.8–1.2, Cb bal.		135	48	......	......	0.31	......	4350					

5–47

Table 5-7. Properties of Metals and Alloys* (Continued)

Material	Nominal composition (essential elements), %	Form and condition	Typical mechanical properties				Typical physical constants							
			Yield strength (0.2% offset), 1,000 psi	Tensile strength, 1,000 psi	Elongation in 2 in., %	Hardness, Brinell	Density, lb/cu in.	Specific gravity	Melting point, °F	Specific heat (32–212°F) Btu/(lb)(°F)	Thermal expansion coefficient (32–212°F) $\times 10^{-6}$ (in./in.)/°F	Thermal conductivity (32–212°F) Btu/(sq ft)(hr)(°F/in.)	Electrical resistivity (68°F), ohms/cir mil ft	Tensile modulus of elasticity $\times 10^6$ psi
Columbium F48	W 13.5–16.5, Mo 4.5–5.5, Zr 0.85–1.15, Cb bal.	......	110	120	......	......	0.34	......	4500					
Columbium FS82	Ta 33, Zr 0.8, Cb bal.	......	50	68	......	......	0.37	......	4550					
Columbium D31	Ti 10, Mo 10, Cb bal.	......	90	100	......	......	0.29	......	4100					
Gold	Au 99.99	Hard / Annealed	......	30 / 17.5	2 / 40	48 / 28	0.692	19.3	1945	0.056	7.9	2060	14.7	10.8
Haynes Stellite alloy 21	C 0.25, Cr 28, Ni 2.5, Mo 5.5, Co bal.	As investment cast	82.0	103	8.0	313 max.	0.299	8.30	2465	0.1006	7.83 (70–600°F)	100.6 (at 392°F)	527	36.0
Haynes Stellite alloy 31 (X-40 Cast)	C 0.50, Cr 25.5, Ni 11, W 7.5, Co bal.	As investment cast	80.0	113	8.0	313 max.	0.311	8.61	2500	0.0981	7.84 (70–600°F)	102.7 (at 392°F)	....	36.0
Haynes Stellite alloy 25	C 0.15 max., Cr 20.0, Ni 10.0, W 15.0, Mn 1.5, Co bal.	Wrought sheet Mill-annealed	63	140	60.0	244	0.330	9.15	2425–2570	0.0924	7.61 (70–600°F)	64.9	532	34.2
Haynes Stellite alloy 36	C 0.40, Cr 19, Ni 10, W 15.0, Mn 1.5, Co bal.	As investment cast	90	103	5.0	298 max.	0.326	9.04	2535	0.092	7.65 (70–600°F)	65	532	33.8
Iridium	Ir 100	Annealed	....	36	......	175	0.808	22.4	4449	....	3.7	407	29.4	75

Material	Composition	Condition												
Molybdenum	Mo 99.9 plus	As rolled	75	100	30	250	0.369	10.22	4730	0.061	2.67	900	31.3	46.0
		Stress-relieved	75	100	30	240								
		Recrystallized	50	70	45	190								
Mo Alloy	Mo bal., Ti 0.5	As rolled	90	120	30	290	0.369	10.22	4730	0.061	3.06	816	31.3	46.0
		Stress-relieved	90	120	30	280								
		Recrystallized	60	80	40	200								
Multimet	Ni 19.0–21.0, Co 18.5–21.0, Cr 20.0–22.5, Mo 2.5–3.5, W 2.0–3.0, Fe bal., C 0.08–0.16, N 0.10–0.20, Cb + Ta 0.75–1.25	Mill-annealed sheet	58	118	49	194	0.296	8.20	2340–2400	0.104		101 (at 392°F)	560	29
		Mill-annealed bar	54	111	55	189								
		Sand-cast	54	98	23	180								
		As investment cast	58	101	31	180 max.								
Platinum	Pt (commercial)	Hard		65	2	101	0.773	21.4	3215	0.057	5.0	465	65	22.0
		Annealed		27	28	65								
Platinum-iridium	Pt 90, Ir 10	Hard		80	2	169	0.776	21.5	3299		5.0		146	25.0
		Annealed	34	53	23	104								
Platinum-rhodium	Pt 90, Rh 10	Hard		93	3	169	0.720	19.93	3353				117	21.2
		Annealed	18.3	50	36	79								
Platinum-ruthenium	Pt 90, Ru 10	Hard		145	2	210	0.173	19.8	3344				255	31.5
		Annealed	47.6	91	28	156								
Palladium	Pd (commercial)	Hard		55		91	0.433	12.0	2829		6.5	488	63.5	16.3
		Annealed	7.6	30	30	47								
Palladium-ruthenium	Pd 95.5, Ru 4.5	Hard		132	3	184	0.433	12.0	2425				145.5	20.4
		Annealed	51	85	26	120								
Palladium-silver	Pd 60, Ag 40	Hard	94	100		176	0.415	11.5	2530°F liquidus 2425°F solidus				264.2	22.2
		Annealed	15	47	40	87								
Palladium alloy 934	Pd 35, Pt 10, Au 10, Ag 30, Cu 15	Annealed	61	96	24	180	0.430	11.9	1985				214.8	
		Heat-treated	125	146	10	Aged 280								
Rhodium	Rh 100	Annealed		80		119	0.448	12.44	3571		4.6	611	27	50

Table 5-7. Properties of Metals and Alloys* (Continued)

Material	Nominal composition (essential elements), %	Form and condition	Yield strength (0.2% offset), 1,000 psi	Tensile strength, 1,000 psi	Elongation in 2 in., %	Hardness, Brinell	Density, lb/cu in.	Specific gravity	Melting point, °F	Specific heat (32–212°F) Btu/(lb)(°F)	Thermal expansion coefficient (32–212°F) ×10⁻⁶ (in./in.)/(°F)	Thermal conductivity (32–212°F) Btu/(sq ft)(hr)(°F/in.)	Electrical resistivity (68°F), ohms/cir mil ft	Tensile modulus of elasticity ×10⁶ psi
					Typical mechanical properties			Typical physical constants						
Silver (pure)	Ag 99.9 plus	Annealed / Cold-rolled	12 / 38	23 / 43	45 / 6	30 / 90	0.379	10.50	1760	0.056	10.6	2900	9.8	10.5
Sterling silver	Ag 92.5, Cu bal.	Hard / Annealed	50 / 20	64 / 41	4 / 26	125 / 65	0.376	......	1635	....	10.5	2510	12.08	10.5
Silver, coin	Ag 90, Cu bal.	Hard / Annealed	53 / 23	65 / 42	4 / 26	125 / 70	0.374	......	1615	....	10.5	2490	12.13	11
Tantalum	Ta 99.9 plus	Annealed sheet / Unannealed sheet	45 / 100	60 / 110	37 / 3	55 / 123	0.60	16.6	5425	0.036	3.6	377	74.6	
Tantalum 10W	W 10, Ta bal.	Annealed	158	160	......	......	0.61	......	5516					
Tin	Sn 100	As cast	......	2.1	70	3.9	0.263	7.29	449	0.0954	12.8	428	66	6
Tungsten	W 100	Hard (sheet) / Annealed / Hard (wire)	360 / ... / 540	400 / ... / 600	... / 0–8 / ...	... / 290 / ...	0.697	19.3	6092 / 6092	0.034 / 0.034	2.4 / 2.4	1390 / 1390	33.08 / 33.08	53.0 / 53.0
Waspaloy	C 0.10 max., Mn 0.50, S 0.030 max., Si 0.75 max., Cr 18.00–21.00, Co 12.00–15.00, Mo 3.50–5.00, Ti 2.75–3.25, Al 1.00–1.50, Zr 0.05–0.12, B 0.008 max., Fe 2.00 max., Cu 0.10, Ni remainder	Vacuum melted forgings	115	185	25	298–346	0.295	8.18	2400	......?	8.92 (70–1500°F)	80–160 (70–1500°F)	....	31.4

5–50

Material	Composition	Condition												
Udimet 500	C 0.15 max., Mn 0.75 max., Si 0.75 max., S 0.015 max., Cr 15.0–20.0, Ti 2.50–3.25, Al 2.50–3.25, Co 13.0–20.0, Mo 3.0–5.0, Fe 4.0 max., Cu 0.15 max., Ni bal.	Forged and heat-treated	125	195	12–18 (in 1 in.)	346–363	0.290	8.03	2550–2600	......	6.8–8.8 (32–1800°F)	75.6 (32–1700°F) 177.6	......	31.05
Zinc	Zn bal., Pb 0.08	Hot-rolled (long.)	......	19.5	65	38	0.258	7.14	786	0.094	18	746	36.56	
		Hot-rolled (transv.)	......	23	50	......					12.8			
		Cold-rolled (long.)	......	21	50									
		Cold-rolled (transv.)	......	27	40									
Zilloy-15	Zn bal., Mg 0.010, Cu 1.00	Hot-rolled (long.)	......	29	20	61	0.259	7.18	792	0.0957	19.3	725	38	
		Hot-rolled (transv.)		40	10	......					11.7			
		Cold-rolled (long.)		36	25	80								
		Cold-rolled (transv.)		46	10									
Zilloy-40	Zn bal., Cu 1.00	Hot-rolled (long.)	......	24	50	52	0.259	7.18	792	0.0957	16.6	......	37.5	
		Hot-rolled (transv.)		30	35									
		Cold-rolled (long.)		31	40	60								
		Cold-rolled (transv.)		40	30									
Zirconium commercial	O_2 0.07, C 0.15, Hf 1.90, Zr bal.	Annealed	53	65	24	B80 (Rockwell)	0.237	......	3380	0.118	2.9	95	......	14
Zircaloy 2	Sn 1.46, Fe 0.12, Ni 0.005, Zr bal., other 0.25	Annealed	50	60	37	B90 (Rockwell)	0.237	......	3300	......	3.6	95		

operations. Carbon steel also is the most weldable of all commercial metals. It is two-thirds the weight of lead but three times heavier than aluminum. In refinery applications, it is generally used at temperatures of less than 1000°F.

For annealed steel with low carbon (0.2 per cent C), tensile strength can be 55,000 psi. For annealed high-carbon (0.4 per cent C) steel, tensile strength will be 100,000 psi. Cold working can boost tensile strength up to 180,000 psi or higher. Table 5-8 gives typical mechanical and physical properties of representative grades of carbon steel.

Over the years, various types of carbon steels have been developed, such as the structural and pressure-vessel steels which are divided into flange and firebox grades. There are only minor metallurgical variations between these types. The important differences are in quality or tighter specifications. Some steels may be fully deoxidized ("killed" with silicon or aluminum added) or partially deoxidized (semikilled).

There are a number of standards and specifications for carbon steel as bar, pipe, plate, etc. The American Society for Testing Materials publishes numerous material specifications on many materials of construction. Table 5-8 gives some data on ASTM standards. The American Iron and Steel Institute also issues specifications on a variety of carbon and alloy steels (Tables 5-9 through 5-13). The American Society of Mechanical Engineers (ASME), the American Standards Association (ASA), and the American Petroleum Institute (API) are also active in this area.

The corrosion resistance of steel depends upon the formation of an oxide surface film. However, resistance to corrosion is somewhat limited. Carbon steel should not be used in contact with dilute acids. Though not recommended with sulfuric acid concentrations below 90 per cent, at concentrations between 90 and 98 per cent steel can be used up to the boiling point of the acid, and between 80 and 90 per cent it is serviceable at room temperature. For sulfuric acid concentrations greater than 102

Table 5-8. Typical ASTM Specifications for Carbon-steel Plate

ASTM Specification	Tensile strength, psi	Yield point min., psi	Elon- gation* in 8 in., min. %	Car- bon* max. %, plate to 2 in.	Mn* max., %	Ni, %	Mo, %	Cr, %
A-285 firebox:								
Grade A......	45,000-55,000	24,000	29	0.17	0.80			
Grade B......	50,000-60,000	27,000	27	0.22	0.80			
Grade C......	55,000-65,000	30,000	25	0.30	0.80			
A-285 flange:								
Grade C......	55,000-65,000	30,000	24	....	0.80			
A-201 firebox:								
Grade A......	55,000-65,000	30,000	25	0.24	0.80			
Grade B......	60,000-72,000	32,000	22	0.27	0.80			
A-212 firebox:								
Grade A......	65,000-77,000	35,000	21	0.31	0.90			
Grade B......	70,000-85,000	38,000	19	0.33	0.90			
A-203 firebox:								
Grade A......	65,000-77,000	37,000	21	0.17	0.80	2.00-2.75		
Grade B......	70,000-85,000	40,000	19	0.20	0.80	2.00-2.75		
Grade C......	65,000-90,000	43,000	18	0.25	0.80	2.00-2.75		
A-204 firebox:								
Grade A......	65,000-77,000	37,000	21	0.21	0.90	....	0.40-0.60	
Grade B......	70,000-85,000	40,000	19	0.23	0.90	....	0.40-0.60	
Grade C......	75,000-90,000	43,000	18	0.26	0.90	....	0.40-0.60	
A-387 firebox:								
Grade A......	65,000-82,000	40,000	20	0.21	0.84	....	0.40-0.70	0.46-0.79
Grade B......	60,000-82,000	35,000	21	0.17	0.69	....	0.40-0.70	0.75-1.31
Grade C......	60,000-85,000	35,000	21	0.17	0.69	....	0.40-0.70	0.94-1.56
A-302 firebox:								
Grade A......	75,000-95,000	45,000	17	0.23	1.35	....	0.41-0.64	
Grade B......	80,000-100,000	50,000	17	0.23	1.55	....	0.41-0.64	

* These values can vary with gage; refer to specification for full details.

Table 5-9. American Iron and Steel Institute Standard Steels: Ladle
Chemical Ranges and Limits for Basic Open-hearth and
Acid Bessemer Carbon Steels

AISI No.	Chemical composition limits, %			
	C	Mn	P, max.	S, max.
C 1008	0.10 max.	0.25/0.50	0.040	0.050
C 1010	0.08/0.13	0.30/0.60	0.040	0.050
C 1011	0.08/0.13	0.60/0.90	0.040	0.050
C 1012	0.10/0.15	0.30/0.60	0.040	0.050
C 1015	0.13/0.18	0.30/0.60	0.040	0.050
C 1016	0.13/0.18	0.60/0.90	0.040	0.050
C 1017	0.15/0.20	0.30/0.60	0.040	0.050
C 1018	0.15/0.20	0.60/0.90	0.040	0.050
C 1019	0.15/0.20	0.70/1.00	0.040	0.050
C 1020	0.18/0.23	0.30/0.60	0.040	0.050
C 1021	0.18/0.23	0.60/0.90	0.040	0.050
C 1022	0.18/0.23	0.70/1.00	0.040	^.050
C 1023	0.20/0.25	0.30/0.60	0.040	0.050
C 1024	0.19/0.25	1.35/1.65	0.040	0.050
C 1025	0.22/0.28	0.30/0.60	0.040	0.050
C 1026	0.22/0.28	0.60/0.90	0.040	0.050
C 1027	0.22/0.29	1.20/1.50	0.040	0.050
C 1029	0.25/0.31	0.60/0.90	0.040	0.050
C 1030	0.28/0.34	0.60/0.90	0.040	0.050
C 1031	0.28/0.34	0.30/0.60	0.040	0.050
C 1033	0.30/0.36	0.70/1.00	0.040	0.050
C 1035	0.32/0.38	0.60/0.90	0.040	0.050
C 1036	0.30/0.37	1.20/1.50	0.040	0.050
C 1037	0.32/0.38	0.70/1.00	0.040	0.050
C 1038	0.35/0.42	0.60/0.90	0.040	0.050
C 1039	0.37/0.44	0.70/1.00	0.040	0.050
C 1040	0.37/0.44	0.60/0.90	0.040	0.050
C 1041	0.36/0.44	1.35/1.65	0.040	0.050
C 1042	0.40/0.47	0.60/0.90	0.040	0.050
C 1043	0.40/0.47	0.70/1.00	0.040	0.050
C 1045	0.43/0.50	0.60/0.90	0.040	0.050
C 1046	0.43/0.50	0.70/1.00	0.040	0.050
C 1049	0.46/0.53	0.60/0.90	0.040	0.050
C 1050	0.48/0.55	0.60/0.90	0.040	0.050
C 1051	0.45/0.56	0.85/1.15	0.040	0.050
C 1052	0.47/0.55	1.20/1.50	0.040	0.050
C 1053	0.48/0.55	0.70/1.00	0.040	0.050
C 1055	0.50/0.60	0.60/0.90	0.040	0.050
C 1060	0.55/0.65	0.60/0.90	0.040	0.050
C 1070	0.65/0.75	0.60/0.90	0.040	0.050
C 1078	0.72/0.85	0.30/0.60	0.040	0.050
C 1080	0.75/0.88	0.60/0.90	0.040	0.050
C 1084	0.80/0.93	0.60/0.90	0.040	0.050
C 1085	0.80/0.93	0.70/1.00	0.040	0.050
C 1086	0.80/0.93	0.30/0.50	0.040	0.050
C 1090	0.85/0.98	0.60/0.90	0.040	0.050
C 1095	0.90/1.03	0.30/0.50	0.040	0.050

per cent, steel is good up to temperatures of 140°F. Usually, steel is not used with hydrochloric, phosphoric, or nitric acid.

When iron contamination is permissible, steel can be used to handle caustic soda at concentrations as high as 75 per cent and at temperatures of 212°F. Stress relieving is sometimes necessary to reduce caustic embrittlement.

Table 5-10. American Iron and Steel Institute Standard Steels: Ladle Chemical Ranges and Limits for Basic Open-hearth Resulfurized Carbon Steels*

AISI No.	Chemical composition limits, %			
	C	Mn	P, max.	S
C 1108	0.08/0.13	0.50/0.80	0.040	0.08/0.13
C 1109	0.08/0.13	0.60/0.90	0.040	0.08/0.13
C 1110	0.08/0.13	0.30/0.60	0.040	0.08/0.13
C 1113	0.10/0.16	1.00/1.30	0.040	0.24/0.33
C 1115	0.13/0.18	0.60/0.90	0.040	0.08/0.13
C 1116	0.14/0.20	1.10/1.40	0.040	0.16/0.23
C 1117	0.14/0.20	1.00/1.30	0.040	0.08/0.13
C 1118	0.14/0.20	1.30/1.60	0.040	0.08/0.13
C 1119	0.14/0.20	1.00/1.30	0.040	0.08/0.13
C 1120	0.18/0.23	0.70/1.00	0.040	0.08/0.13
C 1125	0.22/0.28	0.60/0.90	0.040	0.08/0.13
C 1126	0.23/0.29	0.70/1.00	0.040	0.08/0.13
C 1132	0.27/0.34	1.35/1.65	0.040	0.08/0.13
C 1137	0.32/0.39	1.35/1.65	0.040	0.08/0.13
C 1138	0.34/0.40	0.70/1.00	0.040	0.08/0.13
C 1139	0.35/0.43	1.35/1.65	0.040	0.12/0.20
C 1140	0.37/0.44	0.70/1.00	0.040	0.08/0.13
C 1141	0.37/0.45	1.35/1.65	0.040	0.08/0.13
C 1144	0.40/0.48	1.35/1.65	0.040	0.24/0.33
C 1145	0.42/0.49	0.70/1.00	0.040	0.04/0.07
C 1146	0.42/0.49	0.70/1.00	0.040	0.08/0.13
C 1151	0.48/0.55	0.70/1.00	0.040	0.08/0.13

* For easy machining.

Brines and sea water corrode steel at slow rates, and the metal can be used where iron contamination is not objectionable. Also, carbon steel is little affected by neutral water or organic chemicals. Many large water tanks are fabricated from carbon steel, as well as storage tanks for organic solvents, naphthalene, etc.

Wrought Iron

Ferrous metal which includes a minutely and evenly distributed quantity of iron silicate (slag) is called wrought iron. The slag exists as a discrete and separate phase in the form of silicate fibers. The composition of standard wrought iron has about 0.02 per cent C, 0.12 per cent Si, 0.03 per cent Mn, 0.12 per cent P, and 0.02 per cent S, with the balance as iron and 3 per cent slag. A modification of this is called 4D, a composition which contains 0.06 per cent Mn. A nickel grade contains 3.25 per cent Ni, and a manganese variation has 1 per cent Mn.

Wrought iron is tough and ductile, with excellent fatigue and shock resistance because of the silicate fibers. It can be worked easily by hot- or cold-forming techniques. Machinability is about equivalent to free machining steels. It can be welded by all common processes. Tables 5-7 and 5-14 give chemical, mechanical, and physical properties of wrought iron.

The corrosion resistance of wrought iron is somewhat better than that of carbon steel. Wrought iron has excellent resistance to atmospheric corrosion (structural applications) and to soil corrosion (water lines). One theory is that the silicate fibers tend to hinder corrosive attack and prevent pitting. Wrought iron can handle alkalies and alkaline solutions without difficulty, but dilute acids will cause rapid failure.

With mechanical loading, a safe operating temperature for wrought iron is around

Table 5-11. American Iron and Steel Institute Standard Alloy Steels

Open-hearth and electric-furnace alloy steels, bars, billets, blooms, and slabs
(See Notes below)
The ranges and limits in this table apply to steel not exceeding 200 sq in.
in cross-sectional area

AISI No.	Chemical composition ranges and limits, %							
	C	Mn	P, max.	S, max.	Si	Ni	Cr	Mo
1330	0.28/0.33	1.60/1.90	0.040	0.040	0.20/0.35			
1335	0.33/0.38	1.60/1.90	0.040	0.040	0.20/0.35			
1340	0.38/0.43	1.60/1.90	0.040	0.040	0.20/0.35			
1345	0.43/0.48	1.60/1.90	0.040	0.040	0.20/0.35			
3140	0.38/0.43	0.70/0.90	0.040	0.040	0.20/0.35	1.10/1.40	0.55/0.75	
E3310	0.08/0.13	0.45/0.60	0.025	0.025	0.20/0.35	3.25/3.75	1.40/1.75	
4012	0.90/0.14	0.75/1.00	0.040	0.040	0.20/0.35	. . . .	. . . .	0.15/0.25
4023	0.20/0.25	0.70/0.90	0.040	0.040	0.20/0.35	. . . .	. . . .	0.20/0.30
4024	0.20/0.25	0.70/0.90	0.040	0.035/ 0.050	0.20/0.35	. . . .	. . . .	0.20/0.30
4027	0.25/0.30	0.70/0.90	0.040	0.040	0.20/0.35	. . . .	. . . .	0.20/0.30
4028	0.25/0.30	0.70/0.90	0.040	0.035/ 0.050	0.20/0.35	. . . .	. . . .	0.20/0.30
4037	0.35/0.40	0.70/0.90	0.040	0.040	0.20/0.35	. . . .	. . . .	0.20/0.30
4042	0.40/0.45	0.70/0.90	0.040	0.040	0.20/0.35	. . . .	. . . .	0.20/0.30
4047	0.45/0.50	0.70/0.90	0.040	0.040	0.20/0.35	. . . .	. . . .	0.20/0.30
4063	0.60/0.67	0.75/1.00	0.040	0.040	0.20/0.35	. . . .	. . . .	0.20/0.30
4118	0.18/0.23	0.70/0.90	0.040	0.040	0.20/0.35	. . . .	0.40/0.60	0.08/0.15
4130	0.28/0.33	0.40/0.60	0.040	0.040	0.20/0.35	. . . .	0.80/1.10	0.15/0.25
4135	0.33/0.38	0.70/0.90	0.040	0.040	0.20/0.35	. . . .	0.80/1.10	0.15/0.25
4137	0.35/0.40	0.70/0.90	0.040	0.040	0.20/0.35	. . . .	0.80/1.10	0.15/0.25
4140	0.38/0.43	0.75/1.00	0.040	0.040	0.20/0.35	. . . .	0.80/1.10	0.15/0.25
4142	0.40/0.45	0.75/1.00	0.040	0.040	0.20/0.35	. . . .	0.80/1.10	0.15/0.25
4145	0.43/0.48	0.75/1.00	0.040	0.040	0.20/0.35	. . . .	0.80/1.10	0.15/0.25
4147	0.45/0.50	0.75/1.00	0.040	0.040	0.20/0.35	. . . .	0.80/1.10	0.15/0.25
4150	0.48/0.53	0.75/1.00	0.040	0.040	0.20/0.35	. . . .	0.80/1.10	0.15/0.25
4320	0.17/0.22	0.45/0.65	0.040	0.040	0.20/0.35	1.65/2.00	0.40/0.60	0.20/0.30
4337	0.35/0.40	0.60/0.80	0.040	0.040	0.20/0.35	1.65/2.00	0.70/0.90	0.20/0.30
E4337	0.35/0.40	0.65/0.85	0.025	0.025	0.20/0.35	1.65/2.00	0.70/0.90	0.20/0.30
4340	0.38/0.43	0.60/0.80	0.040	0.040	0.20/0.35	1.65/2.00	0.70/0.90	0.20/0.30
E4340	0.38/0.43	0.65/0.85	0.025	0.025	0.20/0.35	1.65/2.00	0.70/0.90	0.20/0.30
4422	0.20/0.25	0.70/0.90	0.040	0.040	0.20/0.35	. . . .	. . . .	0.35/0.45
4427	0.24/0.29	0.70/0.90	0.040	0.040	0.20/0.35	. . . .	. . . .	0.35/0.45
4520	0.18/0.23	0.45/0.65	0.040	0.040	0.20/0.35	. . . .	. . . .	0.45/0.60
4615	0.13/0.18	0.45/0.65	0.040	0.040	0.20/0.35	1.65/2.00	. . . .	0.20/0.30
4617	0.15/0.20	0.45/0.65	0.040	0.040	0.20/0.35	1.65/2.00	. . . .	0.20/0.30
4620	0.17/0.22	0.45/0.65	0.040	0.040	0.20/0.35	1.65/2.00	. . . .	0.20/0.30
4621	0.18/0.23	0.70/0.90	0.040	0.040	0.20/0.35	1.65/2.00	. . . .	0.20/0.30
4718	0.16/0.21	0.70/0.90	0.040	0.040	0.20/0.35	0.90/1.20	0.35/0.55	0.30/0.40
4720	0.17/0.22	0.50/0.70	0.040	0.040	0.20/0.35	0.90/1.20	0.35/0.55	0.15/0.25

Table 5-11. American Iron and Steel Institute Standard Alloy Steels
(Continued)

AISI No.	Chemical composition ranges and limits, %							
	C	Mn	P, max.	S, max.	Si	Ni	Cr	Mo
4815	0.13/0.18	0.40/0.60	0.040	0.040	0.20/0.35	3.25/3.75	....	0.20/0.30
4817	0.15/0.20	0.40/0.60	0.040	0.040	0.20/0.35	3.25/3.75	....	0.20/0.30
4820	0.18/0.23	0.50/0.70	0.040	0.040	0.20/0.35	3.25/3.75	....	0.20/0.30
5015	0.12/0.17	0.30/0.50	0.040	0.040	0.20/0.35	....	0.30/0.50	
5046	0.43/0.50	0.75/1.00	0.040	0.040	0.20/0.35	....	0.20/0.35	
5115	0.13/0.18	0.70/0.90	0.040	0.040	0.20/0.35	....	0.70/0.90	
5120	0.17/0.22	0.70/0.90	0.040	0.040	0.20/0.35	....	0.70/0.90	
5130	0.28/0.33	0.70/0.90	0.040	0.040	0.20/0.35	....	0.80/1.10	
5132	0.30/0.35	0.60/0.80	0.040	0.040	0.20/0.35	....	0.75/1.00	
5135	0.33/0.38	0.60/0.80	0.040	0.040	0.20/0.35	....	0.80/1.05	
5140	0.38/0.43	0.70/0.90	0.040	0.040	0.20/0.35	....	0.70/0.90	
5145	0.43/0.48	0.70/0.90	0.040	0.040	0.20/0.35	....	0.70/0.90	
5147	0.45/0.52	0.70/0.95	0.040	0.040	0.20/0.35	....	0.85/1.15	
5150	0.48/0.53	0.70/0.90	0.040	0.040	0.20/0.35	....	0.70/0.90	
5155	0.50/0.60	0.70/0.90	0.040	0.040	0.20/0.35	....	0.70/0.90	
5160	0.55/0.65	0.75/1.00	0.040	0.040	0.20/0.35	....	0.70/0.90	
E50100	0.95/1.10	0.25/0.45	0.025	0.025	0.20/0.35	....	0.40/0.60	
E51100	0.95/1.10	0.25/0.45	0.025	0.025	0.20/0.35	....	0.90/1.15	
E52100	0.95/1.10	0.25/0.45	0.025	0.025	0.20/0.35	....	1.30/1.60	
								V
6118	0.16/0.21	0.50/0.70	0.040	0.040	0.20/0.35	....	0.50/0.70	0.10/0.15
6120	0.17/0.22	0.70/0.90	0.040	0.040	0.20/0.35	....	0.70/0.90	0.10 min.
6150	0.48/0.53	0.70/0.90	0.040	0.040	0.20/0.35	....	0.80/1.10	0.15 min.
								Mo
8115	0.13/0.18	0.70/0.90	0.040	0.040	0.20/0.35	0.20/0.40	0.30/0.50	0.08/0.15
8615	0.13/0.18	0.70/0.90	0.040	0.040	0.20/0.35	0.40/0.70	0.40/0.60	0.15/0.25
8617	0.15/0.20	0.70/0.90	0.040	0.040	0.20/0.35	0.40/0.70	0.40/0.60	0.15/0.25
8620	0.18/0.23	0.70/0.90	0.040	0.040	0.20/0.35	0.40/0.70	0.40/0.60	0.15/0.25
8622	0.20/0.25	0.70/0.90	0.040	0.040	0.20/0.35	0.40/0.70	0.40/0.60	0.15/0.25
8625	0.23/0.28	0.70/0.90	0.040	0.040	0.20/0.35	0.40/0.70	0.40/0.60	0.15/0.25
8627	0.25/0.30	0.70/0.90	0.040	0.040	0.20/0.35	0.40/0.70	0.40/0.60	0.15/0.25
8630	0.28/0.33	0.70/0.90	0.040	0.040	0.20/0.35	0.40/0.70	0.40/0.60	0.15/0.25
8637	0.35/0.40	0.75/1.00	0.040	0.040	0.20/0.35	0.40/0.70	0.40/0.60	0.15/0.25
8640	0.38/0.43	0.75/1.00	0.040	0.040	0.20/0.35	0.40/0.70	0.40/0.60	0.15/0.25
8642	0.40/0.45	0.75/1.00	0.040	0.040	0.20/0.35	0.40/0.70	0.40/0.60	0.15/0.25
8645	0.43/0.48	0.75/1.00	0.040	0.040	0.20/0.35	0.40/0.70	0.40/0.60	0.15/0.25
8650	0.48/0.53	0.75/1.00	0.040	0.040	0.20/0.35	0.40/0.70	0.40/0.60	0.15/0.25
8655	0.50/0.60	0.75/1.00	0.040	0.040	0.20/0.35	0.40/0.70	0.40/0.60	0.15/0.25
8660	0.55/0.65	0.75/1.00	0.040	0.040	0.20/0.35	0.40/0.70	0.40/0.60	0.15/0.25
8720	0.18/0.23	0.70/0.90	0.040	0.040	0.20/0.35	0.40/0.70	0.40/0.60	0.20/0.30
8735	0.33/0.38	0.75/1.00	0.040	0.040	0.20/0.35	0.40/0.70	0.40/0.60	0.20/0.30
8740	0.38/0.43	0.75/1.00	0.040	0.040	0.20/0.35	0.40/0.70	0.40/0.60	0.20/0.30
8742	0.40/0.45	0.75/1.00	0.040	0.040	0.20/0.35	0.40/0.70	0.40/0.60	0.20/0.30
8822	0.20/0.25	0.75/1.00	0.040	0.040	0.20/0.35	0.40/0.70	0.40/0.60	0.30/0.40

Table 5-11. American Iron and Steel Institute Standard Alloy Steels
(Continued)

AISI No.	Chemical composition ranges and limits, %							
	C	Mn	P, max.	S, max.	Si	Ni	Cr	Mo
9255	0.05/0.60	0.70/0.95	0.040	0.040	1.80/2.20			
9260	0.55/0.65	0.70/1.00	0.040	0.040	1.80/2.20			
9262	0.55/0.65	0.75/1.00	0.040	0.040	1.80/2.20	....	0.25/0.40	
E9310	0.08/0.13	0.45/0.65	0.025	0.025	0.20/0.35	3.00/3.50	1.00/1.40	0.08/0.15
9840	0.38/0.43	0.70/0.90	0.040	0.040	0.20/0.35	0.85/1.15	0.70/0.90	0.20/0.30
9850	0.48/0.53	0.70/0.90	0.040	0.040	0.20/0.35	0.85/1.15	0.70/0.90	0.20/0.30

Standard Boron Steels
These steels can be expected to have 0.0005 % minimum boron content

AISI No.	C	Mn	P, max.	S, max.	Si	Ni	Cr	Mo
50B40	0.38/0.43	0.75/1.00	0.040	0.040	0.20/0.35	....	0.40/0.60	
50B44	0.43/0.48	0.75/1.00	0.040	0.040	0.20/0.35	....	0.40/0.60	
50B46	0.43/0.50	0.75/1.00	0.040	0.040	0.20/0.35	....	0.20/0.35	
50B50	0.48/0.53	0.75/1.00	0.040	0.040	0.20/0.35	....	0.40/0.60	
50B60	0.55/0.65	0.75/1.00	0.040	0.040	0.20/0.35	....	0.40/0.60	
51B60	0.55/0.65	0.75/1.00	0.040	0.040	0.20/0.35	....	0.70/0.90	
81B45	0.43/0.48	0.75/1.00	0.040	0.040	0.20/0.35	0.20/0.40	0.35/0.55	0.08/0.15
86B45	0.43/0.48	0.75/1.00	0.040	0.040	0.20/0.35	0.40/0.70	0.40/0.60	0.15/0.25
94B15	0.13/0.18	0.75/1.00	0.040	0.040	0.20/0.35	0.30/0.60	0.30/0.50	0.08/0.15
94B17	0.15/0.20	0.75/1.00	0.040	0.040	0.20/0.35	0.30/0.60	0.30/0.50	0.08/0.15
94B30	0.28/0.33	0.75/1.00	0.040	0.040	0.20/0.35	0.30/0.60	0.30/0.50	0.08/0.15
94B40	0.38/0.43	0.75/1.00	0.040	0.040	0.20/0.35	0.30/0.60	0.30/0.50	0.08/0.15

Note 1: Grades shown in the above list with prefix letter E generally are manufactured by the basic electric-furnace process. All others are normally manufactured by the basic open-hearth process but may be manufactured by the basic electric-furnace process with adjustments in phosphorus and sulfur.

Note 2: The phosphorus and sulfur limitations for each process are as follows:

Basic electric furnace........... 0.025 max. %
Basic open hearth.............. 0.040 max. %
Acid electric furnace............ 0.050 max. %
Acid open hearth............... 0.050 max. %

Note 3: Minimum silicon limit for acid open-hearth or acid electric-furnace alloy steel is 0.15 per cent.

Note 4: Small quantities of certain elements are present in alloy steels which are not specified or required. These elements are considered as incidental and may be present to the following maximum amounts: copper, 0.35 per cent; nickel, 0.25 per cent; chromium, 0.20 per cent; and molybdenum, 0.06 per cent.

Note 5: Where minimum and maximum sulfur content is shown it is indicative of resulfurized steels.

Table 5-12. Typical Physical Properties[a] of Low-alloy AISI Steels

AISI type	Melting temperature, °F	Thermal conductivity Btu/(hr) (sq ft)(°F /ft), 212°F	Coefficient of thermal expansion (0–1200°F) per °F	Specific heat (68–212°F), Btu/(lb) (°F)
13XX	....	27	7.9×10^{-6b}	0.10–0.11
23XX	2600–2620	38.3[c]	8.0×10^{-6}	0.11–0.12
25XX	2610–2620	34.5–38.5[c]	7.8×10^{-6}	0.11–0.12
40XX	....	27	8.3×10^{-6b}	0.10–0.11
41XX	....	24.7[d]	....	0.11
43XX	2740–2750	21.7[c]	8.1×10^{-6}	0.107
46XX	....	27[d]	6.3×10^{-6e}	0.10–0.11
48XX	2750	26[f]	8.6×10^{-6}	
51XX	2720–2760	27–34[g]	7.4×10^{-6h}	0.10–0.11
61XX	....	27	8.1×10^{-6b}	0.10–0.11
86, 87XX	2745–2755	21.7[c]	8.2×10^{-6}	0.107
92, 94XX	....	27	8.1×10^{-6b}	0.10–0.12

[a] Density for all low-alloy steels is about 0.28 lb/cu in.
[b] 68 to 1200°F.
[c] 120°F.
[d] 68°F.
[e] 0 to 200°F.
[f] 75°F.
[g] 32 to 212°F.
[h] 100 to 518°F.

650°F. Heating coils, exhaust lines, and smokestacks are some typical process applications of wrought iron.

Low-alloy Steels

Alloy steels contain one or more alloying agents to improve mechanical and corrosion-resistant properties over those of carbon steel (Tables 5-7, 5-9 to 5-11, 5-14, and 5-15).

A typical low-alloy grade steel used in refinery service contains 0.40 per cent C, 1.25 per cent Cr, 0.5 per cent Mo, and 0.75 per cent Si. Many other alloying agents are used to produce a large number of standard and proprietary grades. Usually, carbon-moly and the low and intermediate chromium-moly steels (up to 9 per cent Cr, 1 per cent Mo) can be used at temperatures up to 1200°F. Chromium and silicon improve hardness, abrasion resistance, corrosion resistance, and resistance to oxidation. Molybdenum provides strength at elevated temperatures. Chromium steels are used to handle sulfur compounds at high temperatures.

The tensile strengths of these alloys are in the 75,000- to 125,000-psi range, but hardening heat-treatments can produce alloys with tensile strengths as high as 225,000 psi. Working, machining, and welding properties of low-alloy steels are similar to those of carbon steel when allowance is made for the increased strength.

The addition of small amounts of alloying materials greatly improves corrosion resistance to atmospheric environments but does not have much effect against liquid corrosives. The alloying elements produce a tight, dense, adherent rust film. But in acid or alkaline solutions, corrosion is about equivalent to that of carbon steel. However, the greater strength of low-alloy steel permits thinner walls in process equipment. Some of the new high-strength steels are shown in Table 5-16.

Table 5-13. Typical Mechanical Properties[a] of Low-alloy AISI Steels

AISI type	Tensile strength, 1,000 psi	Yield strength (0.2% offset), 1000 psi	Elongation (in 2 in.), %	Reduction of area, %	Hardness, Brinell	Impact strength (Izod), ft-lb
1330[b]	122	100	19	52	248	
1335[c]	126	105	20	59	262	
1340[c]	137	118	19	55	285	
2317[c]	107	72	27	71	222	84
2515[c]	113	94	25	69	233	85
E2517[c]	120	100	22	66	244	80
4023[d]	120	85	20	53	255	
4032[e]	210	182	11	49	415	
4042[f]	235	210	10	42	461	
4053[g]	250	223	12	40	495	
4063[h]	269	231	8	15	534	
4130[i]	200	170	16	49	375	25
4140[j]	200	170	15	48	385	16
4150[k]	230	215	10	40	444	12
4320[d]	180	154	15	50	360	32
4337[k]	210	140	14	50	435	18
4340[k]	220	200	12	48	445	16
4615[d]	100	75	18	52	. . .	42
4620[d]	130	95	21	65	. . .	68
4640[l]	185	160	14	52	390	25
4815[d]	150	125	18	58	325	44
4817[d]	. . . .	. . .	15	52	355	36
4820[j]	. . . .	. . .	13	47	380	28
5120[d]	143	114	13	45	302	6
5130[m]	189	175	13	51	380	
5140[m]	190	170	13	43	375	16
5150[m]	224	208	10	40	444	
6120[n]	125	94	21	56	. . .	28
6145[o]	176	169	16	52	429	20
6150[o]	187	179	13	42	444	13
8620[p]	122	98	21	63	245	76
8630[p]	162	142	14	54	325	42
8640[p]	208	183	13	43	420	18
8650[p]	214	194	12	41	423	
8720[p]	122	98	21	63	245	76
8740[p]	208	183	13	43	420	18
8750[p]	214	194	12	41	423	
9255[q]	232[q]	215	9	21	477	6
9261[p]	258[r]	226	10	30	514	12

[a] Properties are for materials hardened and tempered as follows: [b] water-quenched from 1525°F, tempered at 1000°F; [c] oil-quenched from 1525°F, tempered at 1000°F; [d] pseudocarburized 8 hr. at 1700°F, oil-quenched, tempered 1 hr. at 300°F; [e] water-quenched from 1525°F, tempered at 600°F; [f] oil quenched from 1500°F, tempered at 600°F; [g] oil-quenched from 1475°F, tempered at 600°F, [h] oil-quenched from 1450°F, tempered at 600°F; [i] water-quenched from 1550–1600°F, tempered at 800°F; [j] oil-quenched from 1550°F tempered at 800°F; [k] oil-quenched from 1525°F, tempered at 800°F; [l] normalized at 1650°F, reheated to 1475°F, oil-quenched, tempered at 800°F; [m] normalized at 1625°F, reheated to 1550°F, water-quenched, tempered at 800°F; [n] carburized 10 hr. at 1680°F, pot-cooled, oil-quenched from 1525°F tempered at 300°F; [o] normalized at 1600°F, oil-quenched from 1575°F, tempered at 1000°F; [p] oil-quenched, tempered at 800°F; [q] normalized at 1650°F, reheated to 1625°F, quenched in agitated oil, tempered at 800°F; [r] normalized at 1600°F, reheated to 1575°F, quenched in agitated oil, tempered at 800°F.

Table 5-14. Standard Designations and Chemical Composition Ranges for Heat- and Corrosion-resistant Castings

Cast-alloy designation	Wrought-alloy type (see Note A)	Composition, % (balance Fe)							
		C	Mn max.	Si max.	P max.	S max.	Cr	Ni	Other elements
CA-15	410	0.15 max.	1.00	1.50	0.04	0.04	11.5-14	1 max.	Mo 0.05 max.†
CA-40	420	0.20-0.40	1.00	1.50	0.04	0.04	11.5-14	1 max.	Mo 0.05 max.†
CB-30	431	0.30 max.	1.00	1.50	0.04	0.04	18-22	2 max.	
CB-7Cu	17-4PH	0.07 max.	1.00	1.00	0.04	0.04	15.5-17	3.6-4.6	Cu 2.3-3.3
CC-50	446	0.50 max.	1.00	1.50	0.04	0.04	26-30	4 max.	
CD-4MCu	· · ·	0.040 max.	1.00	1.00	0.04	0.04	25-27	4.75-6.00	Mo 1.75-2.25, Cu 2.75-3.25
CE-30	· · ·	0.30 max.	1.50	2.00	0.04	0.04	26-30	8-11	
CF-3	304L	0.03 max.	1.50	2.00	0.04	0.04	17-21	8-12	
CF-8	304	0.08 max.	1.50	2.00	0.04	0.04	18-21	8-11	
CF-20	302	0.20 max.	1.50	2.00	0.04	0.04	18-21	8-11	
CF-3M	· · ·	0.03 max.	1.50	1.50	0.04	0.04	17-21	9-13	Mo 2.0-3.0
CF-8M	D319(316)	0.08 max.	1.50	2.00	0.04	0.04	18-21	9-12	Mo 2.0-3.0
CF-8C	347	0.08 max.	1.50	2.00	0.04	0.04	18-21	9-12	Cb 8 × C min., 1.0 max.
CF-16F	303	0.16 max.	1.50	2.00	0.17	0.04	18-21	9-12	Mo 1.5 max., Se 0.20-0.35
CG-8M	317	0.08 max.	1.50	1.50	0.04	0.04	18-21	9-13	Mo 3.0-4.0
CH-20	309	0.20 max.	1.50	2.00	0.04	0.04	22-26	12-15	
CK-20	310	0.20 max.	1.50	2.00	0.04	0.04	23-27	19-22	
CN-7M	· · ·	0.07 max.	1.50	*	0.04	0.04	18-22	21-31	Mo-Cu*
CY-40	· · ·	0.40 max.	1.50	3.00	0.015	0.015	14-17	Bal.	Fe 11.0 max.
CZ-100	· · ·	1.00 max.	1.50	2.00	0.015	0.015	· · ·	95 min.	Fe 1.50 max.
M-35	· · ·	0.35 max.	1.50	2.00	0.015	0.015	· · ·	Bal.	Cu 26-33, Fe 3.50 max.
HA	· · ·	0.20 max.	0.35-0.65	1.00	0.04	0.04	8-10		Mo 0.90-1.20
HC	446	0.50 max.	1.00	2.00	0.04	0.04	26-30	4 max.	Mo 0.5 max.†
HD	327	0.50 max.	1.50	2.00	0.04	0.04	26-30	4-7	Mo 0.5 max.†
HE	· · ·	0.20-0.50	2.00	2.00	0.04	0.04	26-30	8-11	Mo 0.5 max.†
HF	302B	0.20-0.40	2.00	2.00	0.04	0.04	19-23	9-12	Mo 0.5 max.†
HH	309	0.20-0.50	2.00	2.00	0.04	0.04	24-28	11-14	Mo 0.5 max.† N 0.2 max.
HI	· · ·	0.20-0.50	2.00	2.00	0.04	0.04	26-30	14-18	Mo 0.5 max.†
HK	310	0.20-0.60	2.00	2.00	0.04	0.04	24-28	18-22	Mo 0.5 max.†

Table 5-14. Standard Designations and Chemical Composition Ranges for Heat- and Corrosion-resistant Castings
(Continued)

Cast-alloy designation	Wrought-alloy type (see Note A)	Composition, % (balance Fe)							
		C	Mn max.	Si max.	P max.	S max.	Cr	Ni	Other elements
HL	....	0.20-0.60	2.00	2.00	0.04	0.04	28-32	18-22	Mo 0.5 max.†
HN	....	0.20-0.50	2.00	2.00	0.04	0.04	19-23	23-27	Mo 0.5 max.†
HT	330	0.35-0.75	2.00	2.50	0.04	0.04	13-17	33-37	Mo 0.5 max.†
HU	....	0.35-0.75	2.00	2.50	0.04	0.04	17-21	37-41	Mo 0.5 max.†
HW	....	0.35-0.75	2.00	2.50	0.04	0.04	10-14	58-62	Mo 0.5 max.†
HX	....	0.35-0.75	2.00	2.50	0.04	0.04	15-19	64-68	Mo 0.5 max.†

* There are several proprietary alloy compositions falling within the stated chromium and nickel ranges, and containing varying amounts of silicon, molybdenum, and copper.

† Molybdenum not intentionally added.

Designations with the initial letter "C" indicate alloys generally used to resist corrosive attack at temperatures less than 1200°F. Designations with the initial letter "H" indicate alloys generally used under conditions where the metal temperature is in excess of 1200°F. The second letter represents the nominal chromium-nickel type; the nickel content increasing in amount from "A" to "Z." For example, "F" stands for the 19 per cent Cr-9 per cent Ni, "K" for the 25 per cent Cr-20 per cent Ni, and "W" for the 12 per cent Cr-60 per cent Ni alloy types. Numerals following the letters indicate the *maximum* carbon content of the corrosion-resistant alloys; carbon content may also be designated in the heat-resistant grades by following the letters with a numeral to indicate the *midpoint* of a ±0.05 per cent carbon range. If special elements are included in the composition, they are indicated by the addition of a letter to the symbol. Thus, "CF-8M" is an alloy for corrosion-resistant service, of the molybdenum-containing 19 per cent Cr-9 per cent Ni type with a maximum carbon content of 0.08 per cent.

Note A: Wrought-alloy type numbers are listed only for the convenience of those who want to determine corresponding wrought and cast grades. Because the cast-alloy chemical composition ranges *are not the same* as the wrought composition ranges, buyers should use cast-alloy designations for proper identification of castings.

Note B: Most of the standard grades listed are covered for general applications by ASTM Specifications A 296 and A 297. ASTM Specifications A 217, A 351, A 362, A 447, A 448, A 451, and A 452 also apply to some of the grades.

Table 5-15. Properties of Alloy Steels for Heat-resistant Tubular Products

Alloy grade	Nominal composition, %								Room-temperature mechanical properties*					
	Carbon	Manganese	Sulfur, max.	Phosphorus, max.	Silicon	Chromium	Molybdenum	Titanium	Tensile strength, psi	Yield strength, psi	Elongation, % in 2 in.	Brinell hardness	Charpy impact, ft-lb	Recorded service temperature, °F
Carbon steel, killed	0.10–0.20	0.30–0.60	0.045	0.04	0.25	......	......	......	60,000	42,500	46	117	46	1050
½ Mo	0.10–0.20	0.30–0.80	0.045	0.045	0.10–0.50	......	0.45–0.66	......	64,000	49,000	47	141	53	1050
½ Cr–½ Mo	0.10–0.20	0.30–0.61	0.045	0.045	0.10–0.30	0.50–0.81	0.45–0.66	......	62,000	43,000	50	140	52	1075
1 Cr–½ Mo	0.15 max.	0.30–0.61	0.045	0.045	0.50 max.	0.80–1.25	0.45–0.66	......	69,000	34,500	40	130	46	
1¼ Cr–½ Mo	0.15 max.	0.30–0.60	0.03	0.03	0.50–1.00	1.00–1.50	0.45–0.66	......	74,000	55,000	42	153	53	1150
2 Cr–½ Mo	0.15 max.	0.30–0.60	0.03	0.03	0.50 max.	1.65–2.35	0.45–0.66	......	69,500	46,000	44	140	52	1175
2¼ Cr–1 Mo	0.15 max.	0.30–0.60	0.03	0.03	0.50 max.	1.90–2.60	0.87–1.13	......	70,000	45,000	49	147	54	1175
3 Cr–1 Mo	0.15 max.	0.30–0.60	0.03	0.03	0.50 max.	2.65–3.35	0.80–1.06	......	69,000	40,000	53	146	53	1200
5 Cr–½ Mo	0.15 max.	0.30–0.60	0.03	0.03	0.50 max.	4.00–6.00	0.45–0.65	......	70,000	38,500	47	139	46	1300
5 Cr–½ Mo–Si	0.15 max.	0.30–0.60	0.03	0.03	1.00–2.00	4.00–6.00	0.45–0.65	......	88,000	59,000	40	172	60	1250
5 Cr–½ Mo–Ti	0.12 max.	0.30–0.60	0.03	0.03	0.50 max.	4.00–6.00	0.45–0.65	4 × carbon (0.70 max.)	62,500	35,000	52	123	58	1250
7 Cr–½ Mo	0.15 max.	0.30–0.60	0.03	0.03	0.50–1.00	6.00–8.00	0.45–0.65	......	72,500	37,500	45	148	40	1250
9 Cr–1 Mo	0.15 max.	0.30–0.60	0.03	0.03	0.25–1.00	8.00–10.00	0.90–1.10	......	73,500	43,500	41	145	46	1350

* Fully annealed.

Cast Irons

Generally, cast iron is not a particularly strong or tough structural material, but it is one of the most economical. For this reason various grades and types of cast iron are used widely in the process industries. Representative mechanical and physical properties of various types of cast irons are given in Table 5-7.

Gray Cast Iron. This material is low in cost and easy to cast into intricate shapes. It contains carbon, silicon, manganese, and iron. The carbon (1.7 to 4.5 per cent) is present as combined carbon and graphite. Combined carbon is distributed in the matrix as iron carbide (cementite), while free graphite occurs as thin flakes dispersed throughout the body of the metal. Various strengths of gray iron are produced by varying the size, amount, and distribution of the graphite.

Gray iron has outstanding damping properties (ability to absorb vibration) as well as wear resistance. However, gray iron is brittle and has poor resistance to impact or shock. Machinability is excellent, and gray iron castings can be welded with proper techniques and preheating.

Gray iron castings are not usually considered corrosion resistant, although they do resist atmospheric corrosion as well as attack by natural or neutral waters and neutral soils. Dilute acids, however, and acid-salt solutions will attack gray iron. Gray iron is resistant against concentrated acids (nitric, sulfuric, phosphoric), as well as most alkaline and caustic solutions.

White Cast Iron. This material is brittle and difficult to machine. It is made by controlling composition and rate of solidification of the molten iron so that all carbon present is in the combined form. White cast iron is abrasive and wear resistant and is used as liners, grinding balls, dies, and pump impellers.

Malleable Iron. This material is made from white cast iron. In composition, it is cast iron with free carbon as dispersed nodules. This arrangement produces a tough, relatively ductile material. Total carbon content is about 2.5 per cent.

Two types are produced, standard and pearlitic (combined carbon plus nodules). Standard malleable iron is easily machined; pearlitic less so. Both types will withstand bending and cold working without cracking. Large welded areas are not recommended with fusion welding because welds are brittle. Corrosion resistance is about the same as gray cast iron.

Ductile Cast Iron. This category includes a group of materials with good strength, toughness, wear resistance, and machinability. These irons contain combined carbon and dispersed nodules of carbon. The compositions are about the same as for gray iron, but with more carbon (3.7 per cent) than for malleable iron. The spheroidal graphite reduces the notch effect produced by graphite flakes, making these materials more ductile.

There are, as mentioned earlier, a number of grades of ductile iron. Some have maximum toughness and machinability; others have maximum resistance to oxidation. Generally their corrosion resistances are similar to gray iron. Ductile iron, however, can be used at higher temperatures than gray iron—up to 1100°F and higher.

Alloy Cast Irons

Cast irons are used because of their mechanical properties but are not usually considered corrosion resistant. To improve corrosion resistance, cast alloys have been developed. There are a number of such materials on the market. Tables 5-7 and 5-14 list chemical, mechanical, and physical properties of typical alloy cast irons.

High-silicon Cast Irons. These irons contain 13 to 16 per cent silicon and have excellent corrosion resistance. These materials are known as Duriron and Corrosiron. Adding 3 per cent Mo yields a product called Durichlor.

While the high silicon content improves corrosion resistance, it lowers some of the mechanical properties compared with gray iron. Silicon irons are hard and brittle and do not stand up well to shock and impact. Also these irons are difficult to machine. Threads must be formed by direct casting. Welding is possible, although Durichlor is more difficult to weld than Duriron. Special welding rods and heat-treatments must be used.

Table 5-16. Compositions of High-strength Construction Steels

Representative High-strength Low-alloy Steels (Approx. 50,000 psi yield strength)[d]

Designation[a]	Composition, wt %[b]								Producer
	C	Mn	P	S	Si	Cu	V	Others	
Tri-Ten[c]	0.18	1.14	0.023	0.034	....	0.28	0.045	....	U. S. Steel Corp.
Kaisaloy 1 (regular)	0.10	0.72	0.021	0.031	0.48	0.26	0.042	0.27 Ni, 0.18 Cr, 0.06 Mo, 0.08 Ti	Kaiser Steel Corp.
Kaisaloy 2 (cold forming)	0.11	0.63	....	....	0.36	0.27	0.08	0.25 Ni, 0.10 Cr, 0.04 Mo, 0.01 Ti	Kaiser Steel Corp.
Kaisaloy 3 (wear resistance)	0.09	0.38	....	....	0.25	0.14	0.06	0.03 Ni, 0.10 Cr, 0.03 Mo, 0.01 Ti	Kaiser Steel Corp.
Mayari-R	0.12 max.	0.50/1.00	0.08/0.12	0.05 max	0.20/0.90	0.50 max.	....	1.00 max. Ni, 0.40/1.00 Cr, 0.10 max. Zr	Bethlehem Steel Co.
Armco High Strength No. 5	0.22 max.	1.25 max.	0.04 max.	0.05 max.	0.30 max.	0.20 min.	....	....	Armco Steel Corp.
Armco High Strength No. 1 (corrosion)	0.15 max.	0.40/0.95	0.04/0.07	0.04 max.	0.15 max.	0.50/0.75	0.02 min.	0.60/0.90 Ni	Armco Steel Corp.
Armco High Strength No. 3 (cold drawing)	0.10 max.	0.30/0.90	0.04 max.	0.05 max.	0.10 max.	0.20 max.	....	....	Armco Steel Corp.
N-A-X High-tensile (corrosion)	0.18 max.	0.50/0.90	0.04 max.	0.04 max.	0.60/0.90	....	0.02 min.	0.40/0.70 Cr, 0.03/0.12 Zr	Great Lakes Steel Corp.
N-A-X Finegrain (cold forming)	0.20 max.	1.00 max.	0.04 max.	0.04 max.	0.40/0.90	0.20 max.	....	0.03/0.12 Zr	Great Lakes Steel Corp.
Dynalloy	0.15 max.	0.60/1.00	0.05/0.10	0.05 max.	0.30 max.	0.30/0.60	....	0.40/0.70 Ni, 0.05/0.15 Mo	Allan Wood Steel Co.
Maxeloy	0.15 max.	1.00/1.25	0.04/0.07	0.04 max.	0.60/0.90	0.30/0.50	0.05/0.10	0.25 max. Ni, 0.20/0.30 Cr	Crucible Steel Co. of America
Hi-Steel	0.12 max.	0.50/0.90	0.05/0.12	0.05 max.	0.15 max.	0.95/1.30	....	0.45/0.75 Ni, 0.08/0.18 Mo, 0.12/0.27 Al	Inland Steel Co.
Republic "50"	0.15 max.	0.50/1.00	0.04 max.	0.05 max.	....	0.30/1.00	....	0.40/1.10 Ni, 0.30 max. Cr, 0.10 min. Mo	Republic Steel Corp.

Designation							Others	Producer
Stelcoloy No. 1	0.10/0.14	0.90/1.10	0.04 max.	0.05 max.	0.18/0.25	0.30/0.40	0.45/0.75 Ni, 0.40/0.60 Cr	Steel Co. of Canada
Stelcoloy No. 2	0.08/0.12	0.60 max.	0.08/0.12	0.035 max.	0.30 max.	0.40/0.60	0.65/1.00 Ni, 0.40/0.60 Cr	Steel Co. of Canada
U.S.S. Cor-Ten	0.12 max.	0.20/0.50	0.07/0.15	0.05 max.	0.25/0.75	0.25/0.55	0.65 max. Ni, 0.30/1.25 Cr	U. S. Steel Corp.
Yoloy "E" HS	0.18 max.	0.90 max.	0.10 max.	0.05 max.	......	0.20/0.50	0.40/1.00 Ni, 0.20/0.35 Cr, 0.40 max. Mo	Youngstown Sheet & Tube Co.
Yoloy "E" HSC	0.18 max.	0.90 max.	0.08 max.	0.05 max.	......	0.20/0.50	0.40/1.00 Ni, 0.20/0.35 Cr	Youngstown Sheet & Tube Co.
Yoloy "E" HSR	0.18 max.	0.90 max.	0.10 max.	0.05 max.	0.35 max.	0.20/0.50	0.40/1.00 Ni, 0.20/0.35 Cr	Youngstown Sheet & Tube Co.
Yoloy "E" ACR	0.10 max.	0.60 max.	0.05 max.	0.05 max.	......	0.25/0.50	0.60 max. Ni, 0.35 max. Cr	Youngstown Sheet & Tube Co.

a For each steel listed, the designation used is a trade name.

b When ranges are given, the producer's specification is being cited; otherwise, the compositions listed are considered to be typical.

c The composition higher in C and Mn is for heavy products; the composition lower in C and Mn is for sheet, strip, and light plate.

d To ASTM Specifications A 440, A 441, and A 242.

Representative Extra-high-strength Steels (80,000 to 115,000 psi yield strength)

Designation[a]	Composition, wt %[b]									Producer
	C	Mn	Si	Cr	Ni	Mo	V	Cu	Others	
HY-80:										
Thicknesses to 1⅛ in., incl.	0.22 max.	0.10/0.40	0.15/0.35	0.90/1.40	2.00/2.75	0.23/0.35	......	......	......	Various
Thicknesses over 1¼ in.	0.23 max.	0.10/0.40	0.15/0.35	1.35/1.85	2.50/3.25	0.30/0.60				
T-1	0.10/0.20	0.60/1.00	0.15/0.35	0.40/.80	0.70/1.00	0.40/0.60	0.03/0.10	0.15/0.50	0.002/0.006 B	U. S. Steel Corp.
T-1 Type A	0.12/0.21	0.70/1.00	0.20/0.35	0.40/.65	......	0.15/0.25	0.03/0.08	......	0.005/0.005 B	U. S. Steel Corp.
Jalloy-S	0.10/0.20	1.10/1.50	0.15/0.30	......	......	0.20/0.30	......	......	0.01/0.03 Ti	Jones & Laughlin Steel Corp.
N-A-XTRA	0.15	0.80	0.70	0.60	......	0.15	......	......	0.072 Zr	Great Lakes Steel Corp.

a For each steel listed, the designation used is a trade name. HY-80 designates a steel with 80,000 psi minimum yield strength.

b When ranges are given, the producer's specification is being cited; otherwise, the compositions listed are considered to be typical.

Table 5-16. Compositions of High-strength Construction Steels (Continued)

Representative Low-alloy Steels (150,000 to 300,000 psi yield strength)

Designation	Composition, wt %							
	C	Mn	Si	Cr	Ni	Mo	V	Others
Chromium-Molybdenum Types								
AISI 4130[a]	0.28/0.33	0.40/0.60	0.20/0.35	0.80/1.10	......	0.15-0.25	......	......
AISI 4140[a]	0.38/0.43	0.75/1.0	0.20/0.35	0.80/1.10	......	0.15/0.25	......	......
MBMC No. 1[b]	0.42/0.46	0.70/0.90	1.50/1.70	0.60/0.90	......	......	0.10 min	......
Airsteel X-200[b]	0.41/0.46	0.75/1.0	1.40/1.75	1.90/2.25	......	0.45/0.60	0.03/0.08	......
Chromium-Nickel-Molybdenum Types								
AISI 4340[a]	0.38/0.43	0.60/0.80	0.20/0.35	0.70/0.90	1.65/2.00	0.20/0.30	......	
AMS 6434[c]	0.31/0.38	0.60/0.80	0.20/0.35	0.65/0.90	1.65/2.00	0.30/0.40	0.17/0.23	
Ladish D-6-AC[b]	0.42/0.48	0.60/0.90	0.15-0.30	0.90/1.20	0.40/0.70	0.90/1.10	0.05/0.10	
300-M (Trident)[b]	0.41/0.46	0.65/0.90	1.45/1.80	0.70/0.95	1.65/2.00	0.30/0.45	0.05 min	
HY-Tuf[b]	0.25	1.30	1.5	......	1.80	0.40	......	
USS Strux[b]	0.40/0.47	0.75/1.00	0.50/0.80	0.80/1.05	0.60/0.90	0.45/0.60	0.01 min	0.0005 B min

[a] Designation of the American Iron and Steel Institute.
[b] Trade name.
[c] Designation for Aeronautical Material Specification.

Some Martensitic Stainless Steels

Designation	Composition, wt %[a]								Producer
	C	Mn	Si	Cr	Ni	Mo	V	Others	
AISI 420[b]	Over 0.15	1.0 max.	1.0 max.	12/14	......	...	...	......	Various
Rex 448[c]	0.15	......	......	11.5	......	0.45	0.15	0.45 Cb	A British development
Crucible 422[c]	0.20	......	......	13.0	0.75	0.95	0.25	1.0 W	Crucible Steel Co.
12 MoV[c]	0.25	0.50	0.50	12.0	0.50	1.0	0.30	......	U. S. Steel Corp.
Stainless W[c]	0.10	1.0	1.0	17.0	7.0	...	...	1.0 Ti	U. S. Steel Corp.
17-4PH[c]	0.07	1.0	1.0	16.5	4.0	...	...	4.0 Cu, 0.35 Cb	Armco Steel Corp.
Greek Ascoloy[c,d]	0.18	......	......	13.0	2.0	0.5	...	3.0 W	Various
AISI 431[b]	0.20 max.	1.0 max.	1.0 max.	15/17	1.25/2.50	...	...	......	Various

[a] Nominal composition, unless ranges are given.
[b] Designation of the American Iron and Steel Institute.
[c] Trade name.
[d] Also known as Lescalloy 5616, Unitemp 1415 W, 418 Special, and tungsten stainless.

Typical Semiaustenitic Stainless Steels

Designation[a]	Composition, wt %								Producer
	C	Mn	Si	Cr	Ni	Mo	Al	N	
17-7PH	0.09 max.	1.0 max.	1.0 max.	16.0/18.0	6.5/7.75	......	0.75/1.5	......	Armco Steel Corp.
PH15-7Mo	0.09 max.	1.0 max.	1.0 max.	14.0/16.0	6.5/7.75	2.0/3.0	0.75/1.5	......	Armco Steel Corp.
AM 350	0.12 max.	0.90	0.5 max.	16.0/17.0	4.0/5.0	2.5/3.25	......	0.07/0.13	Allegheny Ludlum
AM 355	0.15 max.	0.95	0.5 max.	15.0/16.0	4.0/5.0	2.5/3.25	......	0.05/0.13	Allegheny Ludlum

[a] For each steel listed, the designation used is a trade name.

Several Low-carbon High-nickel Maraging Steels

Designation[a]	Composition, wt %						
	C, max.	Ni	Ti	Al	Co	Mo	Cb
18 Ni (200)	0.03	17/19	0.15/0.25	0.05/0.15	8/9	3.0/3.5	
18 Ni (250)	0.03	17/19	0.3/0.5	0.05/0.15	7/8.5	4.6/5.1	
18 Ni (300)	0.03	18/19	0.5/0.7	0.05/0.15	8.5/9.5	4.7/5.2	
20 Ni	0.03	18/20	1.3/1.6	0.15/0.35	......	......	0.3/0.5
25 Ni	0.03	25/26	1.3/1.6	0.15/0.35	......	......	0.3/0.5

[a] The numbers in parentheses refer to the nominal yield strength, in thousands of pounds per square inch, to which it is possible to heat-treat the steel.

High-strength Cold-rolled Austenitic Stainless Steels

Type	Designation	Composition, %									
		C	Mn	Si	P	S	Cr	Ni	N	Mo	V
17 Cr-7 Ni	AISI 301[a]	0.80/0.20	2.00 max.	1.00 max.	0.03 max.	0.03 max.	16.0/18.0	6.0/8.0	0.10/0.15		
17 Cr-7 Ni	MicroMach (301 N)[b]	0.08/0.12	2.00 max.	1.00 max.	0.03 max.	0.03 max.	17.0/17.6	6.0/6.5	0.25 max.		
Cr-Ni-Mn	AISI 201[a]	0.15 max.	5.5/7.5	1.00 max.	0.06 max.	0.03 max.	16.0/18.0	3.5/5.5	0.25 max.		
High Mn	USS Tenelon[c]	0.10	14.5	......	......	......	17	......	0.40		
High Mn	USS 17-5 MnV[c]	0.10	15.0	0.05	......	......	17.0	5.0	0.35	2.0	0.75

[a] Designation of the American Iron and Steel Institute.
[b] Trade name. The steel is produced by the Washington Steel Corp. and by Jones and Laughlin Steel Corp.
[c] Trade name. The steel is produced by the United States Steel Corp.

Silicon irons are very resistant to oxidizing and reducing environments, this resistance depending on the formation of a passive film. Silicon irons are used widely in sulfuric-acid service, since they are unaffected by sulfuric at all acid strengths and up to the acid boiling point.

The Mo-containing iron Durichlor is especially recommended for hydrochloric acid, although the presence of oxidized chlorides, such as ferric chloride, will cause pitting.

Because they are very hard, silicon irons are recommended for combined corrosion-erosion service. To overcome some of the mechanical disadvantages of silicon irons, a number of other cast alloys have been developed.

High-nickel Irons. One of this group, Hastelloy D, contains 82 per cent Ni, 3 per cent Cu, 9 per cent Si, and 2 per cent Fe. This alloy is superior to silicon irons in its resistance to mechanical and thermal shock. Machining of Hastelloy D is easier than with silicon irons.

Hastelloy D is recommended for sulfuric acid service, since it is practically unaffected by all concentrations of sulfuric acid up to and including the boiling point. However, it is not recommended for nitric or chromic acid services.

Another group of cast-iron alloys is called Ni-Resist. These materials are related to gray cast iron, having high carbon contents (3 per cent) and fine graphite flakes distributed throughout the structure. Nickel contents vary from 13.5 to 36 per cent, and some have 6.5 per cent Cu.

Generally, nickel-alloy castings have superior toughness and impact resistance when compared with gray irons. The nickel-alloy castings can be welded and machined.

The corrosion resistance of these cast nickel alloys is superior to that of the cast irons but less than that of pure nickel. There is little attack from neutral or alkaline solutions, but oxidizing acids such as nitric acid are highly detrimental. Cold, concentrated sulfuric can be handled.

The Ni-Resist alloys have excellent heat resistance. Some grades are serviceable at temperatures as high as 1500°F. Also, a ductile variety of Ni-Resist is available, as well as a hard variety for abrasive service.

Stainless Steels

There are more than 70 standard types of stainless steels and many special alloys. These steels are produced in the wrought form (AISI types), and as cast alloys (ACI types, Table 5-14). Generally, they all are iron alloys containing 12 to 30 per cent chromium, 0 to 22 per cent nickel, and minor amounts of carbon, columbium, copper, molybdenum, selenium, tantalum, or titanium. They are heat- and corrosion-resistant, noncontaminating, and easily fabricated into complex shapes. The physical and mechanical properties of a number of these alloys are given in Table 5-7.

There are three groups of stainless alloys: martensitic, ferritic, and austenitic.

Martinsitic Alloys. These contain 12 to 20 per cent chromium with controlled amounts of carbon and other additives. Type 410 is a typical member of this group. These alloys can be hardened by heat-treatment, which can increase tensile strength from 80,000 to 200,000 psi.

Corrosion resistance is inferior to austenitic stainless steels; therefore martensitic stainless steels are generally used in mildly corrosive environments (atmospheric, fresh water, organics).

Ferritic Stainless. These steels contain 15 to 30 per cent chromium, with low carbon contents (0.1 per cent). The higher chromium content improves corrosion resistance. Type 430 (17 per cent Cr) is a typical example.

These steels lose strength rapidly as the temperature rises. The strength of ferritic stainless can be increased by cold working but not by heat treatment. Fairly ductile ferritic grades can be fabricated by all standard methods. They are fairly easy to machine with high-speed equipment. Welding is no problem, although it requires skilled operators.

Corrosion resistance of ferritic stainless is rated good, but not against reducing such as hydrochloric acid. Mildly corrosive solutions and oxidizing media are handled without harm.

Austenitic Stainless.   These are the most corrosion-resistant of the three groups. These steels contain 16 to 26 per cent chromium and 6 to 22 per cent nickel.   Carbon is kept low to minimize carbide precipitation.

These alloys can be work-hardened, but heat-treatment will not cause hardening. Tensile strength in the annealed condition is about 85,000 psi, and work hardening can increase this to 300,000 psi.

Austenitic stainless steels are tough and ductile.   They can be fabricated by all standard methods, but they are not easy to machine, as they harden and gall.   Rigid machines, heavy cuts, and high speeds are essential.   Welding is readily performed, but welding heat (750 to 1650°F) may cause chromium carbide precipitation, thus depleting the alloy of some chromium and lowering its resistance to liquid corrosives. High-temperature properties are not affected.

For mild service, this is not serious, but for severe corrosive service, the carbides must be put back into solution by heat-treatment.   This is not possible in field welds. To avoid precipitation, special stainless steels stabilized with titanium, columbium, or tantalum have been developed (Types 321, 347, and 348).   Another approach to the problem is the low-carbon steels, such as Types 304L and 316L, with a maximum carbon content of 0.03 per cent.

Type 302 is the basic alloy of the austenitic stainless group.   Types 304 and 304L are low-carbon versions of Type 302.   Types 316, 316L, and 317, with 2.5 to 3.5 per cent molybdenum, are the most corrosion resistant.

In this group, nickel greatly improves corrosion resistance over straight chromium stainless.   Even so, the chromium-nickel steels, particularly the 18-8 alloys, perform best under oxidizing conditions, since their resistances depend on oxide films on the surfaces.   Reducing conditions and chloride ions destroy this film.   Chloride ions, combined with high tensile stresses, will cause stress-corrosion cracking.

Austenitic stainless steels have excellent resistances to naphthenic acids and sulfidic compounds at high temperatures, as well as to liquid corrosion by oxidizing acids. However, in sulfuric acid without inhibitors, stainless steel (Type 316) can be used only below the boiling point at concentrations less than 5 per cent or more than 85 per cent.

Cast Stainless Alloys.   These are used widely in pumps, valves, and fittings. They are designated under the Alloy Casting Institute system (Table 5-14).   All corrosion-resistant alloys have the letter C plus a second letter (A to N) to denote increasing nickel content.   Numerals indicate maximum carbon.

Though a rough comparison can be made between ACI and AISI types, compositions are not identical and analysis cannot be used interchangeably.   Foundry techniques require a rebalancing of the wrought chemical compositions.   However, corrosion resistance is not affected by these composition changes.   Typical numbers of this group are CF-8 (similar to Type 304 stainless) and CF-8M (similar to Type 316 stainless).

In addition to the C grades, a series of heat-resistant grades of ACI cast alloys is produced.

Precipitation-hardening Steels.   These were developed as stainless steels which could be hardened by heat treatment.   They are very strong and hard and have moderate corrosion resistances.

A typical PH steel (containing 17 per cent Cr, 7 per cent Ni, 1.1 per cent Al) has high strength, good fatigue properties, and good resistance to wear and cavitation corrosion.   A large number of these steels are available commerically.   Essentially, they contain chromium and nickel with small amounts of one or more alloying agents, such as copper, aluminum, beryllium, molybdenum, nitrogen, phosphorus, or titanium.

Medium Alloys

One group of alloys with somewhat better corrosion resistance than stainless steels includes the medium alloys.   A popular member of this group is the "20 alloy," developed to fill the need for a material with better sulfuric acid resistance than that of the stainless steels.

A number of companies make this alloy under various trade names.   Durimet 20 is the well-known cast version, containing 0.07 per cent C, 29 per cent Ni, 20 per cent Cr, 2 per cent Mo, and 3 per cent Cu. The ACI designation of this alloy is CN-7M. A wrought form of this alloy is known as Carpenter 20. Another proprietary 20 alloy contains about 24 per cent Ni and 20 per cent Cr. The physical and mechanical properties of a number of medium alloys are given in Table 5-7.

The strength of a 20 alloy can be increased by cold working but not by heat treatment. Welding characteristics are good—similar to those of cast austenitic stainless steels.

In sulfuric acid service, the 20 alloy is almost supreme. It provides satisfactory resistance to all concentrations below 175°F. Also the alloy is not attacked by nitric acid below 205°F and 60 per cent concentration. Cold hydrofluoric acid also can be handled below 20 per cent, but hydrochloric will corrode 20 alloy.

Equipment made from 20 alloy includes pipe, tanks, pumps, valves, and agitators.

Ni-o-nel, another medium alloy, has 42 per cent Ni, 21 per cent Cr, 3 per cent Mo, and 2.25 per cent Cu. It is resistant to reducing and oxidizing environments. Its low carbon with some titanium prevents carbide precipitation during welding. It can be machined and welded and is fully resistant to all concentrations of sulfuric acid at room temperatures. At 175°F, it resists concentrations up to 60 per cent.

In nitric acid it is fully resistant to all concentrations at room temperature, even nitric acid containing small amounts of chlorides. However, Ni-o-nel is not recommended for hydrochloric acid, and it is not so resistant against caustic solutions as is nickel.

Hastelloy F, a third type of medium alloy (44 per cent Ni, 22 per cent Cr, 6.5 per cent Mo), is resistant to oxidizing and reducing media. It tends to work-harden more than austentic stainless, so fabricating procedures call for frequent annealing. Welding characteristics are similar to those of Type 316 stainless. Its primary use has been in sulfite pulp digesters. Hastelloy F has excellent resistance to stress-corrosion cracking.

High Alloys

The materials called high alloys contain relatively large percentages of nickel. The physical and mechanical properties of typical high alloys are given in Table 5-7.

Hastelloy B contains 61 per cent Ni, 28 per cent Mo, 5.5 per cent Fe, and 1 per cent Cr. It is available in wrought and cast forms. Work-hardening presents some fabrication difficulties, and machining is somewhat more difficult than for Type 316 stainless. Conventional welding methods can be used on Hastelloy B.

The alloy, with unusually high resistance to all concentrations of hydrochloric acid at all temperatures, has proved useful in butane-isomerization units. Sulfuric acid attack is low for all concentrations at 150°F, but the corrosion rate increases with temperature. Oxidizing acids and salts rapidly corrode Hastelloy B, but alkalies and alkaline solutions cause little damage. This alloy is also known for its high-temperature properties. It is good up to 1600°F in reducing environments and up to 1500°F in oxidizing media.

The Chlorimets also belong to the high alloys. Chlorimet 2 has 63 per cent Ni, 32 per cent Mo, and 3 per cent Fe. It is available only in cast form, mainly in valves and pumps. This is a tough alloy, very resistant to mechanical and thermal shock. It can be machined with carbide-tipped tools and welded with metal-arc techniques.

The alloy has excellent resistance to reducing or neutral solutions. Resistance to hydrochloric acid at all concentrations to the boiling point is excellent.

Chlorimet 3 is the chromium variation of Chlorimet 2. It is available only in the cast form and contains 18 per cent chromium. This alloy fills the need for a material able to withstand strong oxidizing and reducing situations. It does not depend on a thin film for resistance.

Hastelloy C is very similar to Chlorimet 3, with 54 per cent Ni, 16 per cent Mo, 5.5 per cent Fe, and 15.5 per cent Cr. It is available in wrought or cast forms. Fabrication, machining, and welding present no unusual problems, except that Hastelloy C tends to work-harden. Inert-gas-shielded or metal-arc welding processes are usually recommended.

Hastelloy C resists all concentrations of hydrochloric acid at room temperature. The alloy resists the effects of wet and dry chlorine, as well as hypochlorite and chlorine dioxide solutions. As a high-temperature alloy, it is resistant to oxidizing and reducing environments to 2100°F.

Inconel (80 per cent Ni, 13 per cent Cr, 7 per cent Fe) should also be mentioned as a high alloy. It contains no molybdenum. The corrosion-resistant grade is recommended for reducing and oxidizing environments, particularly at high temperatures. When heated in air, the alloy resists oxidation up to 2000°F. The alloy is outstanding in resisting corrosion by gases if they are essentially sulfur-free.

Another high-alloy group includes the Illium alloys, available in cast and wrought forms. Illium R, the wrought alloy, contains 64 per cent Ni, 22 per cent Cr, and 0.5 per cent Mo. Excellent in fresh and sea water, these alloys are well suited for oxidizing acids such as sulfuric, nitric, and phosphoric.

Nickel and Nickel Alloys

The commercial metal known as nickel is available in practically any mill form as well as in castings. Cold-worked nickel is one of the most ductile materials known. It can be machined easily and joined by welding.

Generally, oxidizing conditions favor corrosion while reducing conditions retard attack. Neutral and alkaline solutions, sea water, and mild atmospheric conditions do not affect nickel. The metal is widely used for handling alkalies, particularly in concentrating, storing, and shipping of high-purity caustic soda. Chlorinated solvents and phenol are often refined and stored in nickel to prevent product discoloration and contamination. Table 5-7 gives physical and mechanical properties of pure nickel and a number of its alloys.

A large number of nickel-based alloys are on the market. Many have been mentioned under the discussion of alloy castings and high alloys. But one other nickel alloy should be mentioned. This is Monel, with 67% Ni and 30% Cu. It is available in all standard forms, is ductile and tough, and can be fabricated and joined readily.

Corrosion resistance of Monel is generally superior to that of its components. It is more resistant than nickel in reducing environments, more resistant than copper in oxidizing environments. It can be used for relatively dilute sulfuric acid (below 80 per cent) service, although aeration will increase the corrosion rate. Monel will handle hydrofluoric acid at concentrations up to 92 per cent at temperatures up to 235°F. Alkalies have little effect on this alloy, but it will not stand up against very highly oxidizing or reducing environments.

Aluminum and Alloys

Aluminum and its alloys are made in practically all the forms in which metals are produced, including castings. With a thermal conductivity of 60 per cent of that of pure copper, unalloyed aluminum is used in many heat-transfer applications. Its high electrical conductivity makes aluminum popular in electrical applications. Aluminum is one of the most workable of metals, and it is usually joined by inert-gas-shielded arc-welding techniques. Table 5-7 gives physical and mechanical properties of aluminum and a number of its alloys.

Commercially pure aluminum has a tensile strength of 10,000 psi, but it can be strengthened by cold working. One limitation of aluminum is that strength declines greatly above 300°F, and 400°F is usually considered the highest permissible safe temperature for aluminum when strength is important. However, aluminum has excellent low-temperature properties and can be used at −450°F.

Aluminum has high resistance to atmospheric conditions, as well as industrial fumes and vapors and fresh, brackish, or salt waters. Many mineral acids attack aluminum, although the metal can be used with concentrated nitric acid (above 82 per cent) and sulfuric acid (99 per cent).

Aluminum cannot be used with strong caustic solutions. Mild basic solutions, when inhibited will not attack aluminum. Salts of strong acids and weak bases, except salts of halogens, have little effect. Most organic acids will be corrode aluminum.

A number of aluminum alloys are available, many with improved mechanical properties over pure aluminum. The wrought, heat-treatment aluminum alloys have tensile strengths of 13,000 to 33,000 psi, as annealed. When these alloys are fully hardened, strengths can go as high as 83,000 psi. However, aluminum alloys usually have lower corrosion resistance than the pure metal. The Alclad alloys were developed to overcome this shortcoming. Alclad consists of a sacrificial aluminum-alloy layer metallurgically bonded to a core alloy.

When alloy steels do not give adequate corrosion protection, particularly from sulfidic attack, steel with an aluminized suface coating can be used. A spray coating of aluminum on a steel is not likely to spall or flake, but the coating is usually not continuous and may leave some areas of the steel unprotected.

Hot-dipped "aluminized steel" gives a continuous coating and has proved satisfactory in refinery applications, particularly where sulfur or hydrogen sulfide is present. It is also used to protect insulating lagging and weather shields for equipment. The coated steel resists fires better than solid aluminum.

Copper and Alloys

Copper and its alloys are widely used in petroleum refinery applications, particularly where heat and electrical conductivity are important factors. The thermal conductivity of copper is twice that of aluminum and 90 per cent of that of silver. Table 5-7 gives physical and mechanical properties of copper and its alloys, while Table 5-17 gives the standard CABRA number designations for wrought copper and its alloys.

A large number of copper alloys is available, including brasses (Cu-Zn), bronzes (Cu-Sn), and cupronickels. Pure copper has excellent low-temperature properties and is used at $-300°F$. Brazing and soldering are common joining methods for copper, although welding, while difficult, is possible.

Generally, copper has high resistance to industrial and marine atmospheres, sea water, alkalies, and solvents. Oxidizing acids rapidly corrode copper. However, the alloys have somewhat different properties from commercial copper (usually tough-pitch copper).

Brasses with up to 15 per cent zinc are ductile but difficult to machine. Machinability improves with increasing zinc content up to 36 per cent. Brasses with less than 20 per cent zinc have corrosion resistance equivalent to copper but with better tensile strengths. Brasses with 20 to 40 per cent zinc have lower corrosion resistance and when ammonia is present are subject to dezincification and stress-corrosion cracking.

Bronzes, commonly called phosphor bronzes, are similar to brasses in mechanical properties and to high-zinc brasses in corrosion resistance (except that bronzes are generally not affected by stress cracking). Aluminum and silicon bronzes combine good strength and corrosion resistance.

Cupronickels (10 to 30 per cent nickel) are becoming very important as copper alloys. They have the highest corrosion resistance of all copper alloys and find application as heat-exchanger tubing. Resistance to sea water is particularly outstanding.

Lead and Alloys

Chemical lead, acid lead, and copper lead are the usual grades used in industry. The small amounts of silver and copper in these leads add to the corrosion resistance and improve creep and fatigue resistance. Table 5-7 gives physical and mechanical properties of lead and its alloys.

Antimony hardens lead and improves its physical characteristics up to the boiling point of water. However, it lowers the melting point, and while it may not detract from the corrosion resistance of lead, seldom does it improve it.

Tellurium in extruded lead products retards grain growth and increases fatigue resistance. Corrosion resistance is comparable to chemical lead.

In lead-clad products, the corrosion resistance of lead is combined with the strength of steel or the high heat transfer of copper.

Table 5-17. Standard CABRA Number Designations for Wrought Copper and Copper Alloys

Previous commonly accepted name	Number	Previous commonly accepted name	Number
Coppers		**Copper alloys** (*cont.*)	
Oxygen-free certified	101	Extra high leaded brass	356
Oxygen-free	102	Free cutting brass	360
Oxygen-free with silver	104	Leaded Muntz metal, uninhibited	365
Oxygen-free with silver	105	Leaded Muntz metal, arsenical	366
Electrolytic tough pitch	110	Leaded Muntz metal, antimonial	367
Electrolytic tough pitch, anneal resist	111	Leaded Muntz metal, phosphorized	368
Tough pitch with silver	113	Free-cutting Muntz metal	370
Tough pitch with silver	114	Forging brass	377
Tough pitch with silver	116	Architectural bronze	385
Phosphorus-deoxidized, low residual phosphorus	120	Admiralty, uninhibited	442
		Admiralty, arsenical	443
Phosphorus-deoxidized, high residual phosphorus	122	Admiralty, antimonial	444
		Admirality, phosphorized	445
Fire-refined tough pitch	125	Naval brass, 63.5 %	462
Fire-refined tough pitch with silver	127	Naval brass	464
Fire-refined tough pitch with silver	128	Naval brass, arsenical	465
Fire-refined tough pitch with silver	130	Naval brass, antimonial	466
Arsenical tough pitch	141	Naval brass, phosphorized	467
Phosphorus-deoxidized arsenical	142	Naval brass welding and brazing rod	470
Phosphorus-deoxidized tellurium-bearing	145	Brazing alloy	472
Sulfur-bearing	147	Naval brass, medium-leaded	482
Zirconium copper	150	Naval brass, high-leaded	485
		Phosphor bronze E	502
		Phosphor bronze A	510
Copper alloys		Phosphor bronze	518
		Phosphor bronze C	521
		Phosphor bronze D	524
Cadmium copper	162	Phosphor bronze B	532
Beryllium copper	172	Phosphor bronze B-1	534
Chromium copper	182	Phosphor bronze B-2	544
Chromium copper	184	Phsophor bronze B-2 (P 0.50 max.)	546
Chromium copper	185	Aluminum bronze D	614
Gilding, 95 %	210	Low-silicon bronze B	651
Commercial bronze, 90 %	220	High-silicon bronze A	655
Jewelry bronze, 87.5 %	226	Manganese brass	667
Red brass, 85 %	230	Manganese bronze B	670
Low brass, 80 %	240	Manganese bronze A	675
Cartridge brass, 70 %	260	Bronze, low fuming (nickel)	680
Yellow brass, 66 % (sheet)	268	Bronze, low fuming	681
Yellow brass, 65 % (rod and wire)	270	Aluminum brass, arsenical	687
Yellow brass, 63 %	274	Silicon brass	692
Muntz metal, 60 %	280	Silicon red brass	294
Brazing alloy	298	Copper nickel, 5 %	704
Leaded commercial bronze (low lead)	310	Copper nickel, 7 %	705
Leaded commercial bronze	314	Copper nickel, 10 %	706
Leaded commercial bronze (nickel-bearing)	316	Copper nickel, 11 %	708
Leaded red brass	320	Copper nickel, 20 %	710
Low leaded brass (tube)	330	Copper nickel, 30 %	715
High leaded brass (tube)	332	Copper nickel, 40 %	720
Low leaded brass	335	Nickel silver 65-10 (sheet)	745
Medium leaded brass, 64.5 %	340	Nickel silver 65-18	752
High leaded brass, 64.5 %	342	Nickel silver 65-15	754
Medium leaded brass, 62 %	350	Nickel silver 65-12	757
High leaded brass, 62 %	353	Nickel silver 55-18	770

SOURCE: Copper and Brass Research Association, New York. In addition to alloys above, other alloys with no commonly accepted trade names have been given numbers.

Vessels designed for operation at high temperatures, fluctuating temperatures, or vacuum are usually made of steel with a lead lining bonded directly to the steel. Heating coils of copper with an external lead cladding are in common use, as are steel pipes and valves with internal lead cladding.

Among the many variables determining the suitability of lead or any other material may be temperature, concentration, grade of lead or type of alloy, presence of oxygen, degree of abrasion-erosion, and presence of impurities.

Lead relies for its high resistance to corrosion upon a thin protective coating that forms on its surface.

When the coating is one of the highly insoluble lead salts, such as the sulfate, carbonate, or phosphate, resistance to corrosion is high and the environment generally promotes self-healing upon mechanical injury to the film.

On the other hand, if a soluble film forms, such as the nitrate, acetate, or chloride, little protection is afforded and the lead may corrode further. Likewise, if insoluble protective coatings are removed mechanically (as by abrasion or erosion) or dissolved chemically (as in the case of some mixed corrosives), corrosion resistance is similarly reduced.

Dilute nitric acid is an example of a corrosive which reacts with lead to form a soluble salt. It is also capable of reacting with lead sulfate to reduce its function as a protective coating, depending on the acid concentration. Obviously, where lead linings are protected by a brick lining, the protective coating or film will not be disturbed.

Titanium

Titanium has become important as a construction material only since 1946. It is strong and of medium weight (see Table 5-7). Corrosion resistance is superior in oxidizing and mild reducing media. A recent development is a Ti-Pd alloy with superior resistance in reducing environments. Titanium is usually not bothered by impingement attack, crevice corrosion, and pitting attack in sea water. Its general resistance to sea water is superior to stainless. Titanium is resistant to nitric acid at all concentrations, except red fuming nitric. The metal also resists ferric chloride, cupric chloride, and other hot chloride solutions.

However, there are a number of disadvantages to titanium which limit its use. One is cost: On a per pound basis titanium is fifteen times more expensive than Type 316 stainless steel. And titanium is not easy to form: It has a high spring back and tends to gall. Welding must be carried out in an inert atmosphere. However, prices are dropping and fabrication techniques are becoming "standard," promising increasing titanium applications in the process industries.

Zirconium

This is another relatively new material, originally considered as a construction material for atomic reactors. Reactor-grade zirconium contains very little hafnium, which would alter the neutron-absorbing properties of zirconium. Commercial-grade zirconium for process applications, however, contains 2.5 per cent hafnium (see Table 5-7). Zirconium resembles titanium from a fabrication standpoint. All welding must be done under an inert atmosphere.

Zirconium also resembles titanium in corrosion resistance; however, in hydrochloric acid, zirconium is more resistant. It also resists all chlorides except ferric and cupric. There are a number of alloys of titanium and zirconium with mechanical properties superior to those of the pure metals. The zirconium alloys are referred to as Zircaloys.

Tantalum

The physical properties of tantalum are similar to mild steel, except that tantalum has a higher melting point (see Table 5-7). Tantalum is ductile and malleable and can be worked into intricate forms. It can be welded by a number of techniques. The metal is practically inert and to many oxidizing and reducing acids (except fuming sulfuric). It is attacked by hot alkalies and hydrofluoric acid. Its cost limits its use to heating coils, bayonet heaters, coolers, and condensers operating under severe conditions.

Table 5-18. Properties of Glass and Silica

	Pyroceram	96 % silica	Borosilicate	Glass lining
Specific gravity, 77°F............	2.60	2.18	2.23	2.56
Water absorption, %.............	0.00	0.00	0.00	
Gas permeability...............	Gastight	Gastight	Gastight	
Softening temp., °F.............	2282	2732	1508	
Specific heat, 77°F.............	0.185	0.178	0.186	
Mean specific heat (77–752°F)....	0.230	0.224	0.233	
Thermal conductivity, mean temp. 77°F Btu/(sq ft)(hr)(°F/in.)....	25.2	....	7.5	
Linear thermal expansion, per °F, (77–572°F)....................	32×10^{-7}	4.4×10^{-7}	18×17^{-7}	
Modulus of elasticity, psi $\times 10^{-6}$..	17.3	9.6	9.5	
Poisson's ratio..................	0.245	0.17	0.20	
Modulus of rupture, psi $\times 10^{-3}$...	20	5–9	6–10	
Knoop hardness, 100 g...........	698	532	481	480
Knoop hardness, 500 g...........	619	477	442	
Adhesion strength, psi............	....	....	....	5–10,000
Max. operating temp., °F........	....	....	....	500*
Thermal shock resistance, temp. diff., °F......................	....	....	....	305(higher for special grades)

* Special glasses are available for high temperatures. A new one is polycrystalline glass capable of withstanding 1000°F (Nucerite).

Glass and Glassed Steel

Glass has excellent resistance to all acids except hydrofluoric. It is also subject to attack by hot alkaline solutions. Glass is particularly suitable for piping where transparency is desirable.

The chief drawback, of course, is brittleness, and glass is also subject to damage by thermal shock. However, glass, when armored with polyester fiberglass, can readily be protected against breakage. On the other hand, glassed steel combines the corrosion resistance of glass with the working strength of steel (see Table 5-18).

Accordingly, glass linings are resistant to all concentrations of hydrochloric acid to 300°F, to dilute concentrations of sulfuric to the boiling point, to concentrated sulfuric to 450°F, and to all concentrations of nitric acid to the boiling point. Acid-resistant glass with improved alkali resistance (up to pH 12) is available, as well as thermal-shock-resistant glassed steel capable of withstanding a 260°F temperature difference at a vessel temperature of 250°F.

Nucerite is a new ceramic-metal composite made in a similar manner to glassed steel. Controlled high-temperature firings chemically and physically bond the ceramic to steel, nickel-base alloys, and refractory metals. Nucerite resists corrosive hydrogen chloride gas, chlorine, or sulfur dioxide at 1200°F. It resists all acids, except HF, up to 350°F. Impact strength is eighteen times that of safety glass; abrasion resistance is superior to porcelain enamel. It has three to four times the thermal-shock resistance of glassed steel.

Plastic Materials

In comparison with metallic materials, the use of plastics is limited to relatively moderate temperatures and pressures (450°F is considered high for plastics). Plastics are also less resistant to mechanical abuse and have high expansion rates, low strengths (thermoplastics), and only fair resistance to solvents. However, they are lightweight, are good thermal and electrical insulators, are easy to fabricate and install, and have low friction factors. Table 5-19 gives typical physical and mechanical properties of plastics.

Table 5-19. Typical Physical and Mechanical Properties of Plastics (73.4°F)

Material	Specific gravity	Thermal conductivity, Btu/(hr)(sq ft)(°F/ft)	Coefficient of thermal expansion, 10^{-5}/°F	Specific heat, Btu/(lb)(°F)	Flammability, in./min	Modulus of elasticity in tension, 10^5 psi	Tensile strength, 1,000 psi	Elongation (in 2 in.), %	Hardness, Rockwell	Impact strength (Izod notched), ft-lb/in. notch	Modulus of elasticity in flex, 10^5 psi	Flexural strength, 1,000 psi	Compression strength (0.1% offset), 1,000 psi	Maximum recommended service temperature, °F	Heat distortion temperature, °F (264 psi)
Acetal polymer	1.425	0.16	4.5	0.35	1.1	4.1	10.0	15 (total)	M94, R120	1.4	4.1	14.1	5.2	185	255
Acrylics, cast:															
General-purpose, Type 1	1.17–1.19	0.12	4.5	0.35	0.5–2.2	3.5–4.5	6–9	2–7	M80–90	0.4	3.5–4.5	12–14	12–14	140–160	150–180
General-purpose, Type 2	1.18–1.20	0.12	4.5	0.35	0.5–1.8	4.0–5.0	8–10	2–7	M96–102	0.4	4.0–5.0	15–17	14–18	150–200	190–225
Acrylonitrile butadiene styrene (ABS):															
High-impact	1.04	0.08–0.12	4.7	0.35–0.38	1.3	2.6–2.9	4.5–8.5	20–50	R85–118	3–6		7.5–11		150	185–215
Extra-high impact	1.01–1.06	0.08–0.12	4.7–5.6	0.35–0.38	1.3	2.1–3.1	5–8	20–50	R85–100	5–7		6.8–8.9			180–218
Low-temperature impact	1.02	0.08–0.12	4.7–5.6	0.35–0.38	1.3	2–3	3–5	30–200	R75–95	6–10		5–8			185–224
Alkyds, molded	2.22–2.24	0.35–0.60	1–3		Self-ext.		3–4			0.30–0.35		7–10	16–20	350	350–400
Cellulose acetate:															
Medium, Type 1	1.23–1.34	0.10–0.19	4.4–9.0	0.3–0.4	0.5–2.0		2.7–6.5	18–54	R68–115	1.1–4.0	1.1–3.5		14–25		196
Hard, Type 2	1.29–1.34	0.10–0.19	4.4–9.0	0.3–0.4	0.5–2.0		6–8.5	6–31	R112–123	0.4–1.9	2.6–4.0		25–36		158–210
Soft, Type 3	1.27–1.34	0.10–0.19	4.4–9.0	0.3–0.4	Self-ext.		4.6–7.5	47–40	R106–121	0.6–2.3	1.9–3.4		22–33		121–137
Cellulose acetate butyrate (CAB):															
Medium, Type 1	1.16–1.24	0.10–0.19	6–9	0.3–0.4	0.5–1.5		4.9–5.7	47–66	R79–112	4.4–6.9	0.93–1.7				
Hard, Type 2	1.19–1.25	0.10–0.19	6–9	0.3–0.4	0.5–1.5		5.0–6.8	38–54	R108–114	0.6–2.4	1.5–2.0				
Soft, Type 3	1.15–1.22	0.10–0.19	6–9	0.3–0.4	0.5–1.5		2.9–3.8	60–74	R23–42	7.5–10	0.74–1.3			255	
Chlorinated polyether (Penton)	1.4		6.6		Self-ext.		6	130 (total)	R100	33	1.3	5			185
Epoxies, cast:															
General-purpose	1.12–2.4	0.1–0.8	1.7–5.0		0.3 to self-ext.	1.9–3.0	2–12	2–6	M75–110	0.2–0.7	0.4–1.5	8–20	20–40	175	250
Heat-resistant	1.15–3.2	0.1–0.8	2.8–3.3			0.38–0.65	5–14	2–5	M90–110	0.2–1.5	0.4–1.5	8–20	25–40	400	500
Fluorocarbons:															
Polytrifluorochloroethylene	2.15	0.145	3.88	0.22	Non	0.5–0.7	4.6–5.7	125–175	R110–115	3.5–3.6	2–2.5	3.5	2.0	380	150–178
Polytetrafluoroethylene	2.1–2.3	0.14	5.5	0.25	Non		2.5–6.5	250–350	J75–95	2.5–4.0	0.6	1.6	0.7–1.8	500	
Fluorinated ethylene-propylene	2.14–2.17	0.11	8.3–10.5	0.28	Non		2.5–3.5	300–900	D55	No break	0.8			400	
Melamines, unfilled	1.48		2–4		Self-ext.	7–15	4.5–7.5		M105–120	0.2–0.6	13	11–14	40–45	210	295
Phenolics, molded (no filler)	1.24–1.90				Self-ext.						7–15	7–12	18–32	300–425	300–350
Phenolics, cast:															
Type 1, mechanical and chemical	1.31		3.3–4.4		Self-ext.	4–5	6–9		M93–120	0.30–0.45	3–5	11–17	14–18		170–195
Polyamides, molded:															
Nylon 66 (0.2% water)	1.14	0.14	5.5	0.3–0.5	Self-ext.	4.1	11.8	60	M79, R118	0.9	4.1	13.8	7–9.7	275–300	145
Nylon 6	1.14	0.10–0.14	4.6–5.4	0.4	Self-ext.	2.5–3.4	10.2–12	300	R105–118	1.2–3.0				225–250	
Nylon …	1.1														

Material															
Polycarbonates	1.20	0.11	3.9		Self-ext.	3.2	9–10.5	60-000 (total)	M70, R118	12–16	3.8	11–13	11	250–300	280–290
Polyester, cast: Allyl type	1.30–1.45	0.12	2.8–5.6	0.26–0.55		2–3	4.5–7	500–725	M92–118	0.18–0.32	3–8	6–14	20–26	300	120–320
Styrene type, rigid	1.12–1.46	0.10–0.12	3.9–5.6	0.30–0.55		1.5–6.5	4–10	<5	M65–115	0.18–0.40	3–9	7–19	12–37	250–300	120–420
Polyethylene: Type I, low density	0.91–0.925	0.19	8.9–11.0	0.55	1.0	0.21–0.27	1.4–2.5	500–725	C73 (Shore)		13–27			212	175–200 (soft. pt.) 215
Type II, medium density	0.926–0.940	0.19	8.3–16.7	0.55	1.0		2.0	200	D55 (Shore)		43			250	250 (soft pt.)
Type III, high density	0.942–0.960	0.19	8.3–16.7	0.46–0.55	1.0		2.9–4.0	25–400	D60 (Shore)	0.4–6.0	90–125			250	250 (soft pt.)
Polypropylene	0.89–0.91	1.3	3.5	0.45		1.4–1.7	5.0	500–700 (total)	R85–95 (Shore)	1.02	1.4–1.7	8.1		250	130–140
Polystyrenes: General-purpose	1.04–1.07	0.058–0.09	3.3–4.8	0.30	1–1.5	4–5	5–8	1.5–2.5	M68–80	0.25–0.35	4–5	8–15	11.5–16	140–160	165–190
Heat, chemical resistant	1.05–1.11	0.046–0.09	3.6–3.8	0.30	0.4–1.0	4–6	10–11	1–4	M78–88	0.25–0.50	4–6	11–17	12–17	175–190	200–220
Polyvinyl butyral: Rigid	1.08–1.12		4.4–12.7	0.4	Slow	3.5–4.0	4–8.5	5–60	L95	1.2		10		115	61.5
Flexible	1.05						0.5–3	150–450	10–100 (Shore)						
Polyvinyl chloride: Type I, rigid	1.32–1.44	0.07–0.10	2.8–3.3		Self-ext.	3.5–4.0	5.5–9.0	5–25	R117	0.25–1.2	3.8–5.4	12–16	11–12	150–165	140–170
Type II, flexible	1.20–1.55	0.07–0.10			Self-ext.	0.004–0.03	1–3.5	200–450	R117	Variable		14–17		150–220	
Polyvinyl dichloride	1.5					0.40	7.5–9.0	4.5		0.8–5.0				210	
Polyvinylidene chloride (Saran)	1.68–1.75	0.053	8.78	0.32	Self-ext.	0.7–2.0	3–5	15–25	M50–25	2–8	10–13	4–7	16–20	150–212	130–150
Silicones, molded: Mineral filler	1.8–2.0	0.09–0.97	2.78–3.23		Self-ext.		4–4.3		M89	0.25–0.30	10–13	6.8–7.5	16–20	>700	>900
Ureas, molded: Cellulose-filled	1.52		4.4–12.7		Self-ext.	13–16	5–10	1.0	E94	0.24–0.35	18–32	10–18	25–38		266–280

Reinforced Plastics

Material															
Epoxies: Woven-glass filled	1.6–1.85				Self-ext.		40–85		M100	12–18	30–46	65–120	45–52	250–400	350–450
Filament wound	1.7–2.2	20–60					80–250		M98–120	40–60	50–70	100–270	45–70	500	
Phenolic: Glass fabric filled	1.80–1.95	0.15	5.0–6.0	0.23	Self-ext.	34	58			15	48	87	47.5	525	500–600
Asbestos fiber	1.908	0.17		0.30	Self-ext.	55	46.1				49.8	52.5	20.8	>350	500–600
Polyester: Glass fiber reinforced	1.6–2.0	0.15	1.0–1.4	0.3	Self-ext.		25–55		M100	13–18	20–38	40–75	25–45	250–400	390–550
Silicones: Glass fabric reinforced	1.6–1.93						20–40					23–47	9–24	450–500	

ASTM test results referring to thermoplastics: sp gr, D 792; thermal conductivity, C 177; coefficient of thermal expansion, D 696; flammability, D 635; modulus of elasticity, D 638, D 790; tensile strength, D 638, D 651; elongation, D 638, D 651; impact strength, D 785; hardness, D 638; flexibility strength, D 790; compression strength, D 695; heat distortion temperature, D 648.

Generally, plastics have excellent resistance to weak mineral acids and are unaffected by inorganic salt solutions—areas where metals are not entirely suitable. Since plastics do not corrode in the electrochemical sense, they offer another advantage over metals: Most metals are affected by slight changes in pH, minor impurities, or oxygen content, while plastics will remain resistant to these changes.

The important thermoplastics used commercially are polyethylene, acrylonitrile butadiene styrene (ABS), polyvinyl chloride (PVC), cellulose acetate butyrate (CAB), vinylidene chloride (Saran), fluorocarbons (Teflon, Kel-F), chlorinated polyether (Penton), polycarbonates, polypropylene, nylons, and acetals (Delrin).

Important thermosetting plastics are general-purpose polyester glass reinforced, bisphenol-based polyester glass, epoxy glass, and phenolic asbestos.

The most chemical-resistant plastic commercially available today is tetrafluoroethylene or TFE (Teflon). This thermoplastic is practically unaffected by all alkalies and acids except fluorine and chlorine gas at elevated temperatures and molten metals. It retains its properties up to 500°F. Chlorotrifluoroethylene or CFE (Kel-F) also possesses excellent corrosion resistance to almost all acids and alkalies up to 350°F.

A new Teflon has been developed from the copolymerization of tetrafluoroethylene and hexafluoropropylene. This resin, FEP, has similar properties to TFE except that it is not recommended for continuous exposures at temperatures above 400°F. Also, FEP can be extruded on conventional extrusion equipment, while TFE parts must be made by complicated "powdered-metallurgy" techniques.

Thermoplastic Materials.   Polyethylene is the lowest cost plastic commercially available. Mechanical properties are generally poor, particularly above 120°F, and pipe must be fully supported. Carbon-filled grades are resistant to sunlight and weathering.

Unplasticized polyvinyl chlorides (Type I) have excellent resistance to oxidizing acids (other than concentrated) and to most nonoxidizing acids. Resistance is good to weak and strong alkaline materials, but resistance to chlorinated hydrocarbons is not good.

Acrylonitrile butadiene styrene polymers (ABS) have good resistance to nonoxidizing and weak acids but are not satisfactory with oxidizing acids. The upper temperature limit is about 150°F. Resistance to weak alkaline solutions is excellent. They are not satisfactory with aromatic or chlorinated hydrocarbons but have good resistance to aliphatic hydrocarbons.

Chlorinated polyether (Penton) is a new thermoplastic. This material can be used continuously up to 255°F, intermittently up to 300°F. Chemical resistance is between polyvinyl chloride and the fluorocarbons. Dilute acids, alkalies, and salts have no effect. Hydrochloric, hydrofluoric, and phosphoric acids can be handled at all concentrations up to 225°F. Sulfuric acid at concentrations over 60 per cent and nitric acid over 25 per cent cause degradation, as do aromatics and ketones.

Acetals have excellent resistance to most organic solvents but are not satisfactory for use with strong acids and alkalies.

Cellulose acetate butyrate is not affected by dilute acids and alkalies or gasoline, but chlorinated solvents cause some swelling. Nylons resist many organic solvents but are attacked by phenols, strong oxidizing agents, and mineral acids.

Among other newer materials are polypropylene and the polycarbonates. The chemical resistance of polypropylene is about the same as polyethylene, but it can be used at 250°F. Polycarbonate, a relatively high-temperature plastic, can be used up to 300°F. Resistance to mineral acids is good. Strong alkalies slowly decompose it, but mild alkalies do not. It is partially soluble in aromatic solvents and soluble in chlorinated hydrocarbons.

Thermosetting Materials.   Among the thermosetting materials are phenolic plastics filled with asbestos, carbon, graphite, or silica. Relative low costs, good mechanical properties, and chemical resistances (except against strong alkalies) make phenolic plastics popular for chemical equipment.

Furane plastics, filled with asbestos, have much better alkali resistance than phenolic asbestos. They are more expensive than the phenolics but also offer somewhat higher strengths.

General-purpose polyester resins, reinforced with fiberglass, have good strengths and good chemical resistances except to alkalies. Some special materials in this class, based on bisphenol, are more alkali resistant. The temperature limit for polyesters is about 200°F.

Epoxies reinforced with fiberglass have very high strengths and resist heat. Chemical resistance of the epoxy resin is excellent in nonoxidizing and weak acids, but not good against strong acids. Alkaline resistance is excellent in weak solutions. Chemical resistance of epoxy-glass laminates may be affected by any exposed glass in the laminate.

Phenolic asbestos, general-purpose polyester glass, Saran, and CAB are adversely affected by alkalies, and thermoplastics generally show poor resistance in organic materials.

Rubber and Elastomers

To meet the demands of the petroleum industry, rubber processors are continually improving their products. A number of synthetic rubbers have been developed, and while none has all the properties of natural rubber, each is superior in one or more ways. Table 5-20 indicates typical applications of various elastomers, while Table 5-21 gives physical properties for natural and synthetic rubbers.

Natural rubber is resistant to dilute mineral acids, alkalies, and salts, but oxidizing media, oils, benzene, and ketones will attack it. Hard rubber is made by adding 25 per cent or more of sulfur to natural or synthetic rubber and, as such, is both hard and strong. Chloroprene or Neoprene rubber is resistant to attack by ozone, sunlight, oils, gasoline, and aromatic or halogenated solvents.

Styrene rubber has chemical resistance similar to natural. Nitrile rubber is known for resistance to oils and solvents. The resistance of butyl rubber to dilute mineral acids and alkalies is exceptional; its resistance to concentrated acids, except nitric and sulfuric, is good. Silicone rubbers, also known as polysiloxanes, have outstanding resistances to high and low temperatures as well as to aliphatic solvents, oils, and greases.

Table 5-20. Where Various Rubbers Are Used

Type of rubber	Features
Butadiene-styrene	General purpose; poor resistance to hydrocarbons, oils, and oxidizing agents
Butyl	General purpose; relatively impermeable to air; poor resistance to hydrocarbons and oils
Chloroprene	Good resistance to aliphatic solvents; poor resistance to aromatic hydrocarbons and many fuels
Chlorosulfonated polyethylene	Excellent resistance to oxidation, chemicals, and heat; poor resistance to aromatic oils and most fuels
cis-Polybutadiene	General purpose; poor resistance to hydrocarbons, oils, and oxidizing agents
cis-Polyisoprene	General purpose; poor resistance to hydrocarbons, oils, and oxidizing agents
Ethylene-propylene	Excellent resistance to heat and oxidation
Fluorinated	Excellent resistance to high temperature, oxidizing acids, and oxidation; good resistance to fuels containing up to 30 % aromatics
Natural	General purpose; poor resistance to hydrocarbons, oils, and oxidizing agents
Nitrile (butadiene-acrylonitrile)	Excellent resistance to oil, but not resistant to strong oxidizing agents; resistance to oils is proportional to the acrylonitrile content
Polysulfide	Good resistance to aromatic solvents; unusually high impermeability to gases; poor compression set and poor resistance to oxidizing acids
Silicone	Excellent resistance over unusually wide temperature range (−150 to +500°F); fair oil resistance; poor resistance to aromatic oils, fuels, high-pressure steam, and abrasion
Styrene	Synonymous with butadiene-styrene

Table 5-21. Physical Properties of Natural and Synthetic Rubbers

	Specific gravity	Tensile strength, psi	Hardness, Shore	Max. temp., continuous use, °F*	Abrasion resistance	Effect of sunlight†
IIR(Butyl, GR-I)	0.91	2,300–3,000	40–70	250–300	Excellent	None
NR(natural rubber)	0.93	3,000–4,500	20–100	175	Excellent	Deteriorates
Hard rubber	1.20–1.95	4,000–10,000	65–95	220		
CR(Neoprene, GR-M)	1.25	2,000–3,500	30–90	250	Excellent	None
SBR(Styrene, Buna S)	0.94	1,600–3,700	35–90	175	Excellent	Deteriorates
NBR(Nitrile, Buna N)	0.99	500–4,000	40–90	250	Excellent	Slight
Polysulfide (Thiokol ST)	1.35	700–1,250	50–80	212	Good	None
Silicone (high temp.)		700–800	45–65	600	Good	None
Polyvinyl chloride (Koroseal)	1.32	2,400–3,000	80–90	160	Good	None
Chlorosulfonated polyethylene (Hypalon)	1.2	500–3,000	55–95	250	Excellent	None
Fluoroelastomers (Viton A)	1.85	2,000–3,000	60–95	500	Excellent	None
(Kel-F 3700)	1.85	3,500	60	400	Excellent	None
Vistanex	0.9	200–550			Excellent	None
Polyisoprene	0.93	2,000–3,000	40–80	175	Excellent	Deteriorates
Polybutadiene	0.94	2,500	45–80	175	Excellent	Deteriorates
Ethylene-propylene	0.93	3,600–4,000	60–65	175	Excellent	Excellent

* Maximum temperature suitable for service depends greatly upon the exact service conditions. Maximum temperature for use as a packing can be much higher than the maximum temperature suitable for tank lining. Individual cases should be referred to the supplier for recommendations.
† Effect of exposure to sunlight under tension.

Chlorosulfonated polyethylene, known as Hypalon, has outstanding resistance to ozone and oxidizing agents, except fuming nitric and sulfuric acids. Oil resistance is good. Fluoroelastomers (Viton A, Kel-F) combine excellent chemical and high-temperature resistance. Polyvinyl chloride elastomer (Koroseal) was developed to overcome some of the limitations of natural and synthetic rubbers. It has excellent resistance to mineral acids and petroleum oils. Rubber and elastomers are widely used as lining materials.

The cis-polybutadiene and cis-polyisoprene rubbers are close duplicates of natural rubber. The new ethylene-propylene rubbers have excellent resistances to heat and oxidation.

Carbon and Graphite

The chemical resistance of impervious carbon or graphite depends somewhat on the type of resin binder used to make the material impervious. Generally, impervious graphite is completely inert to all but the most severe oxidizing conditions. This property, combined with excellent heat transfer, has made impervious carbon and graphite very popular in heat exchangers, as brick linings, and in piping and pumps. One limitation of these materials is low tensile strength. Table 5-22 gives average properties for graphite and carbon.

Two types of resin impregnates are employed in manufacturing impervious graphite. The standard impregnant is a phenolic resin suitable for service in most acids, salt solutions, and organic compounds. A modified phenolic impregnant is recommended for service in alkalies and oxidizing chemicals. However, neither type of impervious graphite is recommended for use with over 60 per cent hydrofluoric, over 20 per cent nitric, over 96 per cent sulfuric acids, or 100 per cent bromine, fluorine, or iodine.

Porcelain and Stoneware

Porcelain and stoneware materials are about as resistant to acids and chemicals as glass, but with the advantage of greater strength. This is offset somewhat by poor thermal conductivity, and the materials can be damaged by thermal shock fairly easily.   The properties of stoneware and porcelain are given in Table 5-23.

Porcelain enamels are used to coat steel, but the enamel has slightly inferior chemical resistance.   Some refractory coatings, capable of taking very high temperatures, are also available.

Wood

Though fairly inert chemically, wood is readily dehydrated by concentrated solutions and hence shrinks badly when subjected to the action of such solutions.   It is also slowly hydrolyzed by acids and alkalies, especially when hot.   In tank construction, if sufficient shrinkage once takes place to allow crystals to form between the staves, it becomes very difficult to make the tank tight again.

A number of manufacturers now offer wood impregnated to resist acids or alkalies or the effects of high temperatures.

Gaskets, Packing, and Seals

Table 5-24 gives data on the important properties of gasket, packing, and sealing materials.   Note that nonmetallic and metallic materials are included, as well as those from both natural and synthetic sources.

Brick Construction

Brick-lined construction can be used for many severely corrosive conditions where high alloys would fail.   Common bricks are made from carbon, red shale, or acidproof refractory materials.   Red-shale brick is not used above 350°F because of spalling. Acidproof refractories can be used up to 1600°F.

A number of cement materials are used with brick.   Standard are phenolic and furane resins, polyesters, sulfur, silicate, and epoxy-based materials.   Carbon-filled polyesters and furanes are good against nonoxidizing acids, salts, and solvents.   Silica-filled resins should not be used against hydrofluoric or fluosilicic acids.   Sulfur-based cements are limited to 200°F, while resins can be used to about 350°F.   The sodium silicate based cements are good against acids to 750°F.

Fire-clay Refractories

Fire clays can be divided into plastic clays and hard flint clays.   They may also be classified as to alumina content.   Table 5-25 gives data to help in the selection of high-temperature refractory materials.

Firebricks.   These are usually made of a blended mixture of flint clays and plastic clays which is then formed, after mixing with water, to the required shape.   Some or all of the flint clay may be replaced by highly burned or calcined clay, called grog.   A large proportion of modern bricks is molded by the dry-press or power-press process where the forming is carried out under high pressure and with a low water content.

Fire-clay bricks are used for boiler settings, kilns, incinerators, and many portions of steel and nonferrous metal furnaces.   They are resistant to spalling but are not generally suitable for use under severe load conditions.

High-alumina Brick.   These are manufactured from raw materials rich in alumina, such as diaspore.   They are graded into groups with 50, 60, 70, 80, and 90 per cent alumina content.   When well fired, these bricks contain a large amount of mullite and less of the glassy phase than is present in the firebricks.   Corundum is also present in many of these bricks.

High-alumina brick are generally used for unusually severe temperature or load conditions, but they cost more than firebrick.

Silica Brick.   These are manufactured from crushed ganister rock containing about 97 to 98 per cent silica.   A bond consisting of 2 per cent lime is used, and the bricks are fired in periodic kilns at temperatures of 2700 to 2800°F for several days until a stable volume is obtained.   They are especially valuable where good strength

Table 5-22. Typical Average Properties of Graphite and Carbon*

(Measured with grain and at room temperature)

Type of product (Dimensions in inches)	Bulk density, g/cc	Porosity, %	Strength, psi†		Elastic modulus, 10⁶ psi	Specific resistance, 10⁻⁵ ohm-in.	Thermal conductivity, Btu/(hr)(sq ft)(°F/ft)‡	(α) Coef. of thermal expansion, 10⁻⁷/°F§
			Compressive	Flexural				
Graphite electrodes, anodes, cylinders, and plates:								
to 2¾ diam. and to ¾ thick..........	1.58	30	5,200	2,600	1.5	33	88	6
3 to 5¾ diam. and ¾ to 5¾ thick......	1.58	30	4,400	2,200	1.4	34	85	8
6 to 12 diam. and 6 to 12 thick.......	1.57	30	3,400	1,700	1.2	35	83	10
6 to 12 diam. and 12 to 17¾ thick.....	1.55	31	3,200	1,600	1.1	38	76	8
14 to 35 diam. and 20 to 24 thick......	1.54	32	1,700	850	0.5	38	76	6
40 and 45 diam......................	1.65	27	2,500	750	0.6	41	71	11
Medium-grain dense graphite, cylinders and plates:								
to 2¾ diam. and to ¾ thick..........	1.68	25	5,000	2,800	1.8	32	91	8
3 to 11 diam. and 2 to 12 thick.......	1.72	24	5,000	2,400	1.7	34	85	12
12 to 18 diam. and 16⅜ to 17⅜ thick...	1.72	24	5,000	2,400	1.5	34	85	12
20 to 24 diam. and 20 × 20 cross section...	1.70	24	5,500	2,200	1.2	35	83	15
20 to 24 diam. and 24 × 24 and 30 cross section.	1.70	24	5,500	2,200	1.2	35	83	15
30 to 50 diam. and 20 × 47 cross section...	1.78	21	5,100	2,200	1.3	43	67	13
Nuclear graphite......................	1.70	25	6,000	2,400	1.5	29	100	12
Fine-grain premium graphite...........	1.73	23	8,300	4,000	1.5	43	68	13
High-density premium graphite.........	1.84	19	8,400	3,700	1.7	47	63	11
Recrystallized graphite...............	1.95	13	7,200	5,400	2.7	28	104	3
Graphite brick.......................	1.56	31	3,100	1,650	1.4	34	86	10
Graphite pipe........................	1.67	26	5,000	2,800	1.7	34	86	10
KARBATE impervious graphite brick....	1.91	0.7	9,000	4,700	2.2	34	86	24
KARBATE impervious graphite pipe.....	1.87	0.7	9,000	4,700	2.2	34	86	24
Porous graphite, Grade 60............	1.05	52	600	400	0.3	120	50	11
Porous graphite, Grade 45............	1.04	53	500	300	0.3	130	45	11
Porous graphite, Grade 25............	1.03	53	400	200	0.2	150	40	11
Carbon electrodes:								
8 diam. and equivalent rectangular....	1.57	21	2,400	1,100	1.2	110	9	13
10 to 14 diam. and equivalent rectangular....	1.57	21	2,000	800	1.0	140	9	13
17 to 45 diam. and equivalent rectangular....	1.57	21	1,700	400	0.7	200	9	13
50 to 55 diam.......................	1.63	18	1,700	350	0.7	200	7	13
Carbon furnace lining stock (24 × 30 cross section)	1.57	21	2,000	550	0.8	200	9	13

Carbon refractory brick	1.65	17	3,600	1,200	16	350	1.2	18
Carbon chemical brick	1.56	22	8,800	2,600	4	160	1.9	13
Carbon pipe	1.55	22	9,000	2,600	3	150	1.9	13
KARBATE impervious carbon brick	1.76	2	10,000	4,400	3	160	2.8	20
KARBATE impervious carbon pipe	1.77	2	10,000	4,400	3	160	2.8	20
Porous carbon, Grade 60	1.05	48	1,000	600	1	700	0.5	16
Porous carbon, Grade 45	1.04	47	900	500	1	700	0.4	16
Porous carbon, Grade 25	1.03	47	800	300	1	700	0.3	16
Carbon—graphite mechanical materials:								
General purpose	1.83	(b)	33,000	8,800	12	...	4.0	13
Grease, oil, jet fuel service	1.81	(b)	40,800	8,800	7	...	4.1	18
Chemical service	1.75	(b)	16,000	4,800	18	...	1.9	20
High-temperature service (1200°F +)	1.77	(b)	36,300	8,800	8	...	3.4	15

* The physical properties given in this table are averages for the most commonly used stock in each type of product. Materials having other properties are available.

† No loss in strength for carbon up to 1700°C and for graphite up to 2500°C.

‡ To convert to gram calories-cm/(sec)(cm²)(°C) multiply by 0.00413.

§ To calculate lineal expansion to any temperature, see below.

(b) Porosity is controlled to fit specific applications. Range 0 to 8 per cent.

Thermal expansion to any temperature: The lineal expansion in *per cent* for a temperature rise from 70°F to a final temperature t (°F)

$$= \frac{(a + K)(t - 70)}{100,000}$$

The lineal expansion in *per cent* for a temperature rise from 21°C to a temperature t (°C)

$$= \frac{1.8(a + K)(t - 21)}{100,000}$$

NOTE: In arithmetic determination of lineal expansion in per cent, do not use 10^{-7} as a multiplier for a and K. Values of a for different sizes and grades of carbon and graphite are in the table. Values of the constant K are given in the table below and are dependent on the final temperature.

Temp, °F	K	Temp, °F	K	Temp, °F	K
212	0	1400	6.1	3000	10.2
400	1.1	1600	6.7	3400	11.2
600	2.4	1800	7.2	3800	12.2
800	3.6	2000	7.7	4200	13.2
1000	4.6	2200	8.2	4600	14.2
1200	5.5	2600	9.2		

Table 5-23. Properties of Stoneware and Porcelain

	Stoneware	Porcelain
Specific gravity	2.2–2.7	2.4–2.9
Hardness, Mohs' scale	6.5	7.5
Modulus of rupture, psi	3–7,000	8–15,000
Modulus of elasticity, psi	5–10×10^6	10–$15 = 10^6$
Compressive strength, psi	40–60,000	60–90,000
Pore volume, %	1.5	0.2–0.5
Water absorption, %	0.5–4.0	0–0.5
Linear thermal expansion, per °F	2.4×10^{-6}	2.5×10^{-6}
Thermal conductivity, Btu/(sq ft)(hr)(°F/in.)	8–22	8–10

is required at high temperatures. They have lowered alumina content and often lowered porosity.

Although silica bricks spall (crack during temperature changes) readily below red heat, they are very stable if the temperature is kept above this range, and for this reason they stand up well in regenerative furnaces. Any structure of silica brick should be heated slowly. A large structure often requires 2 weeks or more to bring up to the working temperature.

Magnesite Bricks. These are made from crushed magnesium oxide which is produced by calcining raw magnesite rock to high temperatures. A rock containing several per cent of iron oxide is preferable, as this permits the rock to be fired at a lower temperature.

Magnesite bricks are generally fired at a comparatively high temperature, though large tonnages of unburned brick are now produced. The latter are made with special grain sizing and a bond such as an oxychloride.

Magnesite bricks are not so resistant to spalling as are fire-clay bricks.

Chrome Bricks. These are manufactured in much the same way as magnesite bricks but are made from natural chromite ore. Commercial ores always contain magnesia and alumina. Unburned hydraulically pressed chrome brick are also made.

Basic bricks combining various properties of magnesite and chromite are now made in large quantities and have advantages over either material alone for some purposes.

Insulating Firebricks. This is a class of brick which consists of a highly porous fire clay or kaolin. They are lightweight (about one-half to one-sixth the weight of fireclay) with low thermal conductivity, yet are sufficiently temperature resistant to be used successfully on the hot side of the furnace wall. This permits thin walls of low thermal conductivity and low heat content. The low heat content is particularly valuable in saving fuel and time on heating, allows rapid changes in temperature to be made, and permits rapid cooling.

These bricks are made in a variety of ways, such as mixing organic matter with the clay and later burning it out to form pores. Or a bubble structure can be incorporated in the clay-water mixture, which is later preserved in the fired brick. The insulating firebricks are classified into several groups according to the maximum use limit; the ranges are up to 1600, 2000, 2300, and 2600 and above 2800°F.

Insulating Refractories. These are used extensively in chemical-process furnaces and in oil stills or heaters. They usually have a life equal to the heavy brick that they replace. They are particularly suitable for constructing experimental or laboratory furnaces because they can be cut or machined readily to any shape.

There are a number of types of special brick. High-burned kaolin refractories are particularly valuable under conditions of severe temperature and heavy load or for severe spalling conditions, as in the case of high-temperature oil-fired boiler settings. Another brick for the same uses is a high-fired brick of Missouri aluminous clay.

There is a number of bricks on the market made from electrically fused materials, such as fused mullite, fused alumina, and fused magnesite. These bricks, although high in cost, are particularly suitable for certain severe conditions.

Table 5-24. Important Properties of Gasket Materials*

Material	Max. service temp., °F	Important properties
Rubber (straight):		
Natural.................	225	Good mechanical properties. Impervious to water. Fair to good resistance to acids, alkalies. Poor resistance to oils, gasoline. Poor weathering, aging properties
Styrene-butadiene (SBR)....	250	Better water resistance than natural rubber. Fair to good resistance to acids, alkalies. Unsuitable with gasoline, oils and solvents
Butyl....................	300	Very good resistance to water, alkalies, many acids. Poor resistance to oils, gasoline, most solvents (except oxygenated)
Nitrile..................	300	Very good water resistance. Excellent resistance to oils, gasoline. Fair to good resistance to acids, alkalies
Polysulfide................	150	Excellent resistance to oils, gasoline, aliphatic and aromatic hydrocarbon solvents. Very good water resistance, good alkali resistance, fair acid resistance. Poor mechanical properties
Neoprene.................	250	Excellent mechanical properties. Good resistance to nonaromatic petroleum, fatty oils, solvents (except aromatic, chlorinated, or ketone types). Good water and alkali resistance. Fair acid resistance
Silicone..................	600	Excellent heat resistance. Fair water resistance; poor resistance to steam at high pressures. Fair to good acid, alkali resistance. Poor (except fluorosilicone rubber) resistance to oils, solvents
Acrylic..................	450	Good heat resistance but poor cold resistance. Good resistance to oils, aliphatic and aromatic hydrocarbons. Poor resistance to water, alkalies, some acids
Chlorosulfonated polyethylene (Hypalon)	250	Excellent resistance to oxidizing chemicals, ozone, weathering. Relatively good resistance to oils, grease. Poor resistance to aromatic or chlorinated hydrocarbons. Good mechanical properties
Fluoroelastomer (Viton, Fluorel 2141, Kel-F)	450	Can be used at high temperatures with many fuels, lubricants, hydraulic fluids, solvents. Highly resistant to ozone, weathering. Good mechanical properties
Asbestos:		
Compressed asbestos-rubber sheet	To 700	Large number of combinations available; properties vary widely depending on materials used
Asbestos-rubber woven sheet	To 250	Same as above
Asbestos-rubber (beater addition process).........	400	Same as above
Asbestos composites........	To 1000	Same as above
Asbestos-TFE..............	500	Combines heat resistance and sealing properties of asbestos with chemical resistance of TFE
Cork compositions............	250	Low cost. Truly compressible materials which permit substantial deflections with negligible side flow. Conform well to irregular surfaces. High resistance to oils; good resistance to water, many chemicals. Should not be used with inorganic acids, alkalies, oxidizing solutions, live steam

Table 5-24. Important Properties of Gasket Materials* (Continued)

Material	Max. service temp., °F	Important properties
Cork rubber...............	300	Controlled compressibility properties. Good conformability, fatigue resistance. Chemical resistance depends on kind of rubber used
Plastics:		
TFE (solid)............... (tetrafluoroethylene, Teflon)	500	Excellent resistance to almost all chemicals and solvents. Good heat resistance; exceptionally good low-temperature properties. Relatively low compressibility and resilience
TFE (filled)...............	To 500	Selectively improved mechanical and physical properties. However, fillers may lower resistance to specific chemicals
TFE composites...........	To 500	Chemical and heat resistance comparable to solid TFE. Inner gasket material provides better resiliency and deformability
CFE..................... (chlorotrifluoroethylene, Kel-F)	350	Higher cost than TFE. Better chemical resistance than most other gasket materials, although not quite so good as TFE
Vinyl....................	212	Good compressibility, resiliency. Resistant to water, oils, gasoline, and many acids and alkalies. Relatively narrow temperature range
Polyethylene..............	150	Resists most solvents. Poor heat resistance
Plant fiber:		
Neoprene-impregnated wood fiber...................	175	Nonporous; recommended for glycol, oil, and gasoline to 175°F
GRS-bonded cotton........	230	Good water resistance
Nitrile rubber-cellulose fiber.	....	Resists oil at high temperatures
Vegetable fiber, glue binder..	212	Resists oil and water to 212°F
Vulcanized fiber..............	....	Low cost, good mechanical properties. Resists gasoline, oils, greases, waxes, many solvents
Inorganic fibers..............	To 2200°F	Excellent heat resistance, poor mechanical properties
Felt:		
Pure felt....................	....	Resilient, compressible and strong, but not impermeable. Resists medium-strength mineral acids and dilute mineral solutions if not intermittently dried. Resists oils, greases, waxes, most solvents. Damaged by alkalies
TFE-impregnated.........	300	Good chemical and heat resistance
Petrolatum or paraffin-impregnated..................	....	High water repellency
Rubber-impregnated.......	....	Many combinations available; properties vary widely depending on materials used
Metal:		
Lead.....................	500	Good chemical resistance. Best conformability of metal gaskets
Tin......................	....	Good resistance to neutral solutions. Attacked by acids, alkalies
Aluminum................	800	High corrosion resistance. Slightly attacked by strong acids, alkalies
Copper, brass..............	....	Good corrosion resistance at moderate temperatures
Nickel....................	1400	High corrosion resistance
Monel....................	1500	High corrosion resistance. Good against most acids and alkalies, but attacked by strong hydrochloric and strong oxidizing acids
Inconel...................	2000	Excellent heat, oxidation resistance
Stainless steel..............	....	High corrosion resistance. Properties depend on type used

Table 5-24. Important Properties of Gasket Materials* (Continued)

Material	Max. service temp., °F	Important properties
Metal composites		Many combinations available; properties vary widely depending on materials used
Leather	220	Low cost. Limited chemical and heat resistance. Not recommended against pressurized steam, acid or alkali solutions
Glass fabric		High strength and heat resistance. Can be impregnated with TFE for high chemical resistance
Packing and Sealing Materials		
Rubber (straight)	To 600	See Gasket Materials for properties. Mainly used for ring-type seals, although some types are available as spiral packings
Rubber composites: Cotton-reinforced	350	High strength. Chemical resistance depends on type of rubber used; however, most types are noted for high resistance to water, aqueous solutions
Asbestos-reinforced	450	High strength combined with good heat resistance
Asbestos: Plain, braided asbestos	500	Heat resistance combined with resistance to water, brine, oil, many chemicals. Can be reinforced with wire
Impregnated asbestos	To 750	Environmental properties vary widely depending on type of asbestos and impregnant used. Neoprene-cemented type resists hot oils, gasoline, and solvents. Oil and wax-impregnated type resists caustics. Wax-impregnated blue asbestos type has high acid resistance. TFE-impregnated type has good all-around chemical
Asbestos composites	To 1200	End properties vary widely depending on secondary material used
Metals: Copper	To 1500	Properties depend on other construction materials and form of copper used. Packing made of copper foil over asbestos core resists steam and alkalies to 1000°F. Packing of braided copper tinsel resists water, steam, and gases to 1500°F
Aluminum	To 1000	Resists hot petroleum derivatives, gases, foodstuffs, many organic acids
Lead	550	Many types are available
Organic fiber: Flax	300	Good water resistance
Jute	300	Good water resistance
Ramie	300	Good resistance to water, brine, cold oil
Cotton	300	Good resistance to water, alcohol, dilute aqueous solutions
Rayon	300	Good resistance to water, dilute aqueous solutions
Felt	300	See Gasket Materials
Leather	To 210	Good mechanical properties for sealing. Resistant to alcohol, gasoline, many oils and solvents, synthetic hydraulic fluids, water
TFE	To 500	Available in many forms, all of which have high chemical resistance
Carbon-graphite	700	Good bearing and self-lubricating properties. Good resistance to chemicals, heat

* From *Materials in Design Engineering*, pp. 111–126, Reinhold Publishing Corporation, New York, December, 1959.

Table 5-25. Data to Help Select High-temperature Nonmetallic Materials of Construction

	Magnesia	Zircon $ZrO_2 \cdot SiO_2$	Bonded 99% Al_2O_3	Bubble alumina	Fused cast alumina	Zirconia stabilized	SiC	Silicon-nitride-bonded SiC	Magnesite	Chrome $FeO \cdot Cr_2O_3$	Forsterite $MgO \cdot SiO_2$
Melting point, °F	5072	4390	3660	3330-3550	3640	4710	Diss. at 4082	Diss. at 3452	5070	3540-3990	3461
Bulk density, lb/cu ft	175	205	165-175	80-85	175-219	275	153-162	168-178	160-175	170-190	150-155
Thermal conductivity, Btu/(hr)(sq ft)(°F/in.)	18 at 2000°F	22 at 2192°F	24 at 2200°F	7.7 at 2400°F	24-31 at 200°F	7 at 2400°F	109 at 2200°F	113.5 at 2200°F	20 at 2200°F	12 at 2400°F	10.3 at 2200°F
Coefficient of thermal expansion, in./°C	13.5×10^{-6}, 25°-800°C	5.5×10^{-6}, 20°-1200°C	7.4×10^{-6}, 25°-1400°C	8.6×10^{-6}, 25°-1400°C	$4.9\text{-}8.6 \times 10^{-6}$, 25-1400°C	5.5×10^{-6}, 20°-1200°C	4.4×10^{-6}, 25°-1400°C	4.4×10^{-6}, 25°-1400°C	14.7×10^{-6}, 0°-1400°C	10.4×10^{-6}, 0°-1540°C	11.0×10^{-6}, 0°-1425°C
Mean specific heat, Btu/(lb)(°F)	0.283, 25°-800°C	0.15, 75°-210°F	0.33	0.32	0.28-0.32	0.20, 20°-1200°C	0.288, 25°-1400°C	0.288	0.27, 0°-1400°C	0.21	0.25
Emissivity at 1500°F	0.51, 86-2012°F	0.53	0.47, 32-2552°F	0.55, 32-2552°F	0.51, 32-2552°F	0.53, 0-1200°F	0.93, 0-2552°F	0.93, 0-2500°F	0.48, 60-1200°F	0.97, 60-1200°F	0.95, 0-1200°F
Permeability, cu ft/(hr)(sq ft)(in.)	3-61	10-78	3-51	>100	0.5-2	4-32	1-6	1-4	7-20	6-14	10-25
Apparent porosity, %	19	11-26	22	54-67	2.5-7.5	19-50	13	8	20-28	20-28	24-27
Mohs' hardness	6.0	7.5	9.0	2.5-5.0	9.2	6.5-7.5	9.6	9.6	5.0-5.5	5.5	7.5
Modulus of rupture, psi		157 at 2732°F	200-900 at 2462°F	63 at 2462°F	703->1,500 at 2732°F	100-980 at 2732°F	800-315 at 2462°F	5,640 at 2462°F	750-3000 at 68°F	25-125 at 2462°F	450-800 at 68°F
Electrical resistivity, ohm-cm	4×10^{6} at 1500°F; 10,500 at 2500°F	High at 2732°F	$1.2 \times 10^{13}\text{-}1 \times 10^{6}$, 580-2000°F	$1.2 \times 10^{13}\text{-}1 \times 10^{6}$, 580-2000°F	175-55, 1832-2552°F	12.90-1.6, 2300-3090°F	7420-745, 1832-2732°F			650-40, 1500-2500°F	
Wt., lb/9-in. brick	10	11-12.5	9.7-10.4	4.8	10-13	14-15	9-9.5	9.9-10.4	10.0-11.5	11.0-12.0	9.1-9.8
Thermal shock resistance	Poor	Good <3150°F	Fair	Fair	Poor	Fair	Excellent	Excellent	Fair	Fair	Fair
Abrasion resistance	Poor	Fair	Good	Poor	Excellent	Fair	Excellent	Excellent	Fair	Fair	Good
Use limit, °F	4172, oxid.; 3092, red.	3100	3300	2800-3200	3300	4600	2800-3000	2900-3000	2900-3200	2800-3200	3000
Cost, dollars/9-in. brick	2.81	1.39	3.12	1.52-3.34	4.65-6.30	10.80	2.14	4.28	0.68-0.70	0.55-0.58	0.57-0.60

Bricks of silicon carbide, either recrystallized or clay-bonded, have high thermal conductivities and are used in muffle walls and as slag-resisting materials.

Other types of refractory which find certain limited use are forsterite, zirconia, and zircon. Acid-resisting brick consisting of a dense body (like stoneware) are used to line tanks and conduits in the chemical industry.

Low-temperature Materials

The low-temperature properties of metals have created some unusual problems in fabricating cryogenic equipment. Most metals lose their ductility and impact strength at low temperatures, although in many cases yield and tensile strengths increase as the temperature goes down.

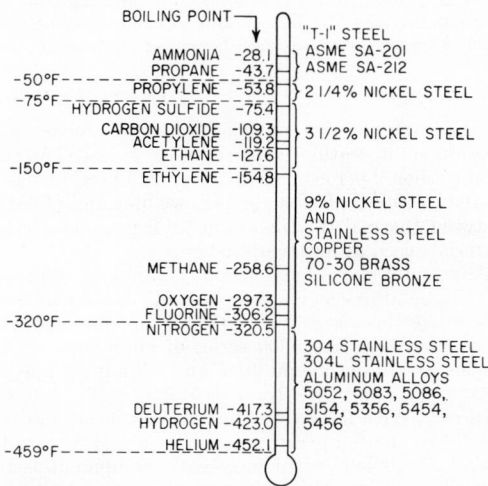

FIG. 5-4. What metals to use as the temperature drops. This chart shows schematically the various metals used for low-temperature application in environments defined by the normal boiling points of certain liquefied gases.

It is important to choose materials resistant to shock. Usually a minimum **Charpy** value of 15 ft-lb (keyhole notch) is specified at the operating temperature. For severe loading, a value of 20 ft-lb is recommended. Ductility tests are performed on notched specimens, since smooth specimens usually show amazing ductility.

Steels. The ASME and ASTM have set up an A 300 classification which covers carbon and alloy steels suitable to −150°F. In addition, the ASME Code adds certain features on fabrication, such as stress relieving, to ensure suitability of these alloys in low-temperature service. A large number of steels are made to the A 300 classification. Steels A 201 and A 212 are specially treated carbon steels for −50°F service. Among the alloy steels are A 203 Grade A or B for −75°F and A 203 Grade D or E for service to −150°F (see Fig. 5-4).

Stainless Steels. Chromium-nickel steels are inherently suitable for service from −300 to −425°F. Table 5-26 shows some AISI grades used at low temperatures. Type 304 is the most popular. The original cost of stainless may be higher than another metal, but ease of fabrication (no heat-treatment) and welding combined with high strength offsets the higher initial cost.

Sensitization or formation of chromium carbides can occur in several stainless steels during welding, and this will affect impact strength. However, tests have shown that impact properties of Types 304 and 304L are not greatly affected by sensitization but the properties of 302 are impaired at −300°F.

Table 5-26. Metals and Alloys for Low-temperature Service*

ASTM Specification and Grade	Recommended lowest service temp., °F
Carbon and alloy steels:	
T-1	− 50
A 201, A 212, flange or firebox quality	− 50
A 203, Grades A and B (2¼ Ni)	− 75
A 203, Grades D and E (3½ Ni)	−150
A 353 (9% Ni)	−320
Stainless steels Types 302, 304L, 304, 310, 347	−425
Aluminum alloys 5052, 5083, 5086, 5154, 5356, 5454, 5456	−425
Copper alloys, silicon bronze, 70-30 brass, copper	−320

* Based on a series of articles on low-temperature metals appearing in *Chem. Eng.* from May 20 to Oct. 31, 1960. See also McClintock and Gibbons, *Mechanical Properties of Structural Materials at Low Temperatures*, National Bureau of Standards, June, 1960.

Nickel Steel.   Low-carbon 9 per cent nickel steel is a ferritic alloy developed for use in cryogenic equipment operating as low as −320°F.   ASTM Specifications A 300 and A 353 cover low-carbon 9 per cent nickel steel (A 300 is the basic specification for low-temperature ferritic steels).   Refinements in welding and (ASME Code approved) elimination of postweld thermal treatments make 9 per cent steel competitive with many low-cost materials used at low temperatures.

Aluminum.   Aluminum alloys have an unusual ability to maintain strength and shock resistance at temperatures as low as −425°F.   Good corrosion resistances and relatively low costs make these alloys very popular for low-temperature equipment. For most welded construction, the 5000 series of aluminum alloys is widely used. These alloys are the aluminum-magnesium and aluminum-magnesium-manganese types.

Copper and Alloys.   With few exceptions, the tensile strengths of copper and its alloys increase rapidly as the temperature goes down.   However, the low structural strength of copper is a problem when large-scale equipment is being constructed. Therefore, alloys must be used.   One of the most successful alloys for low temperatures is silicon bronze, which can be used to −320°F with safety.

Insulation.   Perlite, balsa wood, cellular glass, and mineral wood can all be used at temperatures down to about −260°F.   But still better insulations are needed to cope with liquid hydrogen, oxygen, and helium.   Randomly dispersed reflective flakes are not good enough.   Two very promising insulations have been developed by Linde.

One is a medium-quality variety consisting of 10 to 20 layers of aluminum foil per inch, separated by a 3-micron glass-fiber mat.   It is a relatively low-density (2 lb/cu ft), low-cost insulation with a thermal conductivity of 0.11×10^{-3} Btu/(hr) (ft)(°F) between 80 and −297°F.   A higher quality variety consists of 40 to 80 layers per inch of aluminum foil and submicron glass-fiber paper.   This is somewhat heavier than the other insulation (4.7 lb/cu ft) and more costly, but it has the very low thermal conductivity of 0.025×10^{-3} Btu/(hr)(ft)(°F) over the same temperature range.

High-temperature Materials

Successful applications of metals in high-temperature process service depend on an appreciation of certain engineering factors.   The important alloys for service up to 2000°F are shown in Table 5-27.

Among the most important properties are creep, rupture, and short-time strengths (see Figs. 5-5 and 5-6).   Creep relates initially applied stress to rate of plastic flow. Stress rupture is another important consideration at high temperatures, since it relates stress and time to produce rupture.   As the figures show, ferritic alloys are weaker than austenitic compositions, and in both groups molybdenum increases strength.

Austenitic castings are much stronger than their wrought counterparts.   And higher

Table 5-27. Important Commercial Alloys for High-temperature Process Service

	Nominal composition, %			
	Cr	Ni	Fe	Other
Ferritic steels:				
Carbon steel..............	....	....	bal.	
2¼ Chrome.............	2¼	....	bal.	Mo
Type 502................	5	....	bal.	Mo
Type 410................	12	....	bal.	
Type 430................	16	....	bal.	
Type 446................	27	....	bal.	
Austenitic steels:				
Type 304................	18	8	bal.	
Type 321................	18	10	bal.	Ti
Type 347................	18	11	bal.	Cb
Type 316................	18	12	bal.	Mo
Type 309................	24	12	bal.	
Type 310................	25	20	bal.	
Type 330................	15	35	bal.	
Nickel-base alloys:				
Nickel...................	....	bal.		
Incoloy..................	21	32	bal.	
Hastelloy B..............	....	bal.	6	Mo
Hastelloy C..............	16	bal.	6	W, Mo
60/15...................	15	bal.	25	
Inconel..................	15	bal.	7	
80/20...................	20	bal.		
Hastelloy X..............	22	bal.	19	Co, Mo
Multimet................	21	20	bal.	Co
Rene 41.................	19	bal.	5	Co, Mo, Ti
Cast irons:				
Ductile iron.............	....	....	bal.	C, Si, Mg
Ni-Resist, D-2...........	2	20	bal.	Si, C
Ni-Resist, D-4...........	5	30	bal.	Si, C
Cast stainless (ACI types):				
HC.....................	28	4	bal.	
HF.....................	21	11	bal.	
HH.....................	26	12	bal.	
HK.....................	26	20	bal.	
HT.....................	15	35	bal.	
HW.....................	12	bal.	28	
Super alloys:				
Inconel X...............	15	bal.	7	Ti, Al, Cb
A 286..................	15	25	bal.	Mo, Ti
Stellite 25.............	20	10	Co-base	W
Stellite 21 (cast)...........	27.3	2.8	Co-base	Mo
Stellite 31 (cast)...........	25.2	10.5	Co-base	W

strengths are available in Inconel, cobalt-based Stellite 25, and iron-based A 286. Other properties which become important at high temperatures include thermal conductivity, thermal expansion, ductility, alloy composition, and stability.

In many cases, strength and mechanical properties become of secondary importance in process applications compared with resistance to the corrosive surroundings. All common heat-resistant alloys form oxides when exposed to hot oxidizing environments. Whether the alloy is resistant depends upon whether the oxide is stable and forms a protective film. Thus mild steel is seldom used above 950°F because of excessive

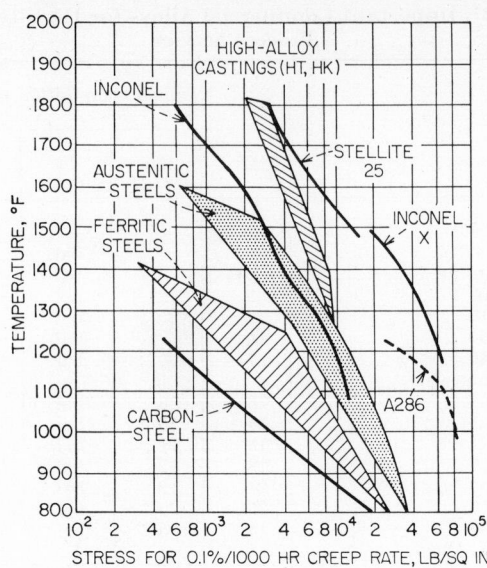

Fɪɢ. 5-5. Effect of creep on metals for high-temperature use.   (*Chem. Eng., Dec. 15, 1958, p. 139.*)

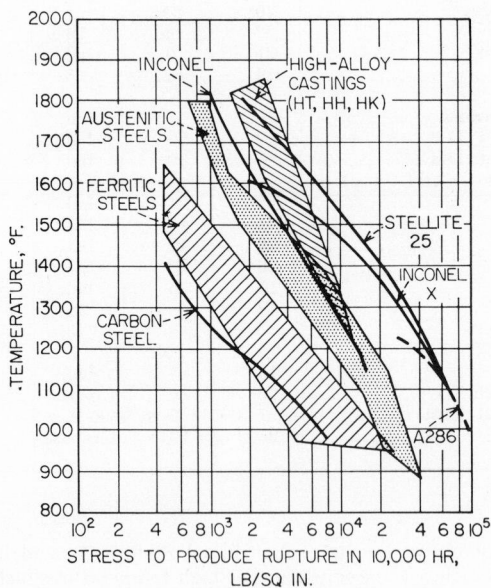

Fɪɢ. 5-6. Rupture properties of metals as a function of temperature. (*Chem. Eng., Dec. 15, 1958, p. 139.*)

scaling rates. Higher temperatures require chromium (see Fig. 5-7). Thus Type 502 steel, with 4 to 6 per cent Cr, is acceptable to 1150°F; 9 to 12 per cent Cr steel will handle 1300 to 1400°F; 14 to 18 per cent Cr extends the limit to 1500°F; and 27 per cent Cr can be used up to 2000°F.

The well-known austenitic stainless steels have excellent oxidation resistance: up to 1650°F for 18-8 and up to 2000°F for 25-12 (and Inconel and Incoloy). The cobalt-based alloys, of which Stellite 25 is an example, show excellent strengths up to 2000°F.

Another useful element in imparting oxidation resistance to steel is silicon (complementing the effects of chromium). In the lower chromium ranges, 0.75 to 2 per cent silicon is more effective than chromium on a weight-percentage basis. The influence of 1 per cent silicon in improving the oxidation rate of steels with varying chromium contents is shown in Fig. 5-8.

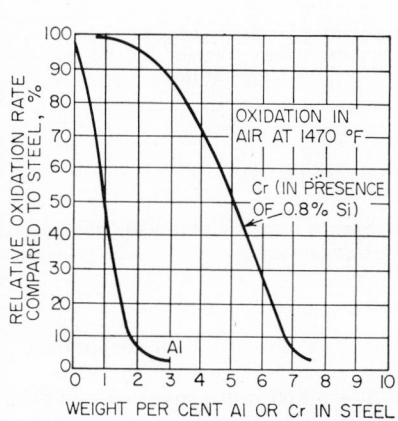

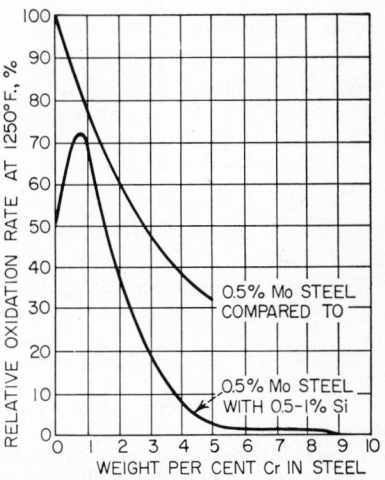

Fig. 5-7. How Cr and Al reduce steel oxidation.

Fig. 5-8. Effect of silicon on oxidation resistance.

Aluminum also improves the resistance of iron to oxidation as well as sulfidation, but its use as an alloying agent is limited because the amount required interferes with workability and high-temperature strength properties of the steel. However, development of high-aluminum surface layers by spraying, dipping, and cementation is a feasible means of improving the heat resistance of low-alloy steels.

One of the most important applications for high-temperature materials is in furnace tubes and furnace-tube supports. The alloys often used for furnace tubes where temperatures exceed 1500°F are the 25 per cent Cr, 20 per cent Ni alloys (Type 310 stainless steel), 78 per cent Ni, 25 per cent Cr alloys (Inconel), and 35 per cent Ni, 21 per cent Cr, 34 per cent Fe alloys (Incoloy). Two common heat-resisting castings used in petroleum-refinery furnaces are 25 per cent Cr, 12 per cent Ni and 28 per cent Cr, 8 per cent Ni (ACI designations HH and HE).

For temperatures much above 2000°F, only the refractory metals are suitable. Much work has been done on these metals in connection with atomic and rocket programs. All the refractory metals have melting points above 3000°F, but each has distinct advantages and disadvantages (see Table 5-28).

Properties of important nonmetallics for high-temperature applications are summarized in Table 5-25 and discussed on page 5-88.

Table 5-28. Comparison of Properties of Refractory Metals

Melting point, °F	Element	Advantages	Disadvantages
6180	Tungsten	Highest melting point; non-volatile oxide to at least 2500°F	Highest density; oxidizes rapidly; brittle at low temperatures
5425	Tantalum	Very high melting point; non-volatile oxide; ductile	High density; oxidizes rapidly; least abundant
4730	Molybdenum	High melting point; less dense than tungsten or tantalum; moderately ductile at room temperature	Extremely high oxidation rate (volatile oxide)
4380	Columbium	High melting point; nonvolatile oxide; ductile; moderate density	Oxidizes rapidly
3435	Chromium	Extremely oxidation resistant; lightest of refractory metals	Lowest melting point of refractory metals; brittle at low temperatures

CORROSION-TESTING METHODS

Numerous techniques are available for corrosion testing in petroleum refineries. The method which best suits the purpose of the test must be selected. The principal types of tests are:
1. Laboratory tests, including inspection tests
2. Plant tests
3. Field tests, exposing samples to the atmosphere, to soils, or to waters (fresh, brackish, saline)
4. Service tests, the use of a material for part of some operating equipment, such as an evaporator tube, or for small-scale pilot-plant equipment

Laboratory Tests

Laboratory corrosion tests are useful to study the chemistry and mechanism of corrosion and to indicate the environments in which a particular metal or alloy may be used satisfactorily. Such tests help determine the possible effects of metals and alloys on the characteristics of an environment to which they may be exposed, e.g., contamination by corrosion products of materials in process, transportation, or storage. They serve as control tests in making corrosion-resisting metals or alloys and help determine the value of changes in composition or treatment in developing such alloys. Also, laboratory tests can be used to determine whether a metal, alloy, or protective coating conforms to a specification requiring a certain performance in a specified corrosion test.

Metals and alloys do not respond alike to all the influences of the many factors that control corrosion. Consequently, it is impractical to establish standard laboratory procedures for corrosion testing, except for inspection purposes. Even so, some details of laboratory testing require careful attention to assure useful results. There are a number of excellent references on this subject.[1-6]

Laboratory corrosion tests are often the best means of arriving at a preliminary selection of the most suitable materials to use in corrosive service. Unfortunately, no existing laboratory test will predict accurately the behavior of the selected material in the plant. The problem is that any standardized laboratory test reports only the chemical resistance of the proposed material to the corrosive agent but fails to consider numerous other factors entering into the behavior of the materials in the plant.

Total Immersion Test

This is an unaccelerated test method which gives results which agree approximately with those obtained on the large scale when the other factors are considered. This test helps to eliminate the materials which obviously cannot be used. Further narrowing of the selection from among those not eliminated in the total immersion test can be made on the basis of a knowledge of the properties of the materials concerned and the working conditions or by constructing larger scale equipment of the proposed material in which the operating conditions can be simulated.

For a discussion of immersion test methods, see *International Nickel Co. Publication* F10 and *Bulletin,* Corrosion.

Test Piece. The shape of the test piece does not affect results within a reasonable range of ratio of length of edge to surface, except for the accelerating effect of the internal segregations exposed on the edge sections. In general the proportion of edge section should be kept low.

Fine polishing is unnecessary, although tool marks and oxide coatings should be removed. Many corrosion determinations have been made on roughened and smoothed pieces.

Composition of Solution. The composition of the solution should be controlled to the fullest extent possible. When results are reported, the solution should be described as completely as possible.

Use 250 cc of the corroding solution per test piece of given area (4.6 sq in.) for a fairly rapid corrosion rate (0.01-in. penetration per month). The volume should be increased in proportion for pieces of greater area.

Temperature of Corroding Solution. For theoretical work in determining the effect of temperature, it is necessary to use an accurately regulated thermostat capable of $\pm 2°F$ control. When results are applied to plant practice, similarly careful consideration must be given to temperature control in all specific tests.

In comparison of different metals, care must be taken that the temperature is the same for all tests. A convenient way to control temperature is to use a constant-boiling liquid.

Time of Exposure. Results of corrosion tests vary enormously with the time of exposure of the material to the corroding solution. This variation is due to initial electrochemical surface actions such as overvoltage and to the period of time required for the formation of a protective coat. Initial effects of this kind must be neglected if a corrosion factor for use over a long period of time is to be obtained. One way is to measure the rate over an interval of time after this initial high corrosion rate has decreased and become constant.

The experimental data available at present indicate that corrosion should be determined at the end of 48, 96, and 240 hr. Sufficient accuracy for comparative work may usually be obtained by neglecting the corrosion over the first 48 hr and averaging the rates over the second 48-hr period.

Preparing Test Pieces. Coatings are removed from test pieces by dissolving them or removing them mechanically. The error due to a particular method has been determined. The first 48 hr of immersion in the corroding solution may be considered a preparation of the test piece for the final test.

A metal test piece should be 2 by 1 by 0.1 in. (area 4.6 sq in.). The dimensions should be accurate to 0.01 in. to save the time of measurement and area calculation. Other shapes may be used within a range of ratio of length of edge to total area of test piece between 4 and 8. Tool marks should be removed by successive use of file and emery. Exceedingly fine finishes are unnecessary, but the surface should be clean and reasonably smooth as, for example, finished with 120-grain emery.

Testing Apparatus. An apparatus for testing has been described by Fraser, Ackerman, and Sands.[7] Jars containing the testing solution are surrounded by an electrically heated water bath, the temperature of which is controlled thermostatically. For tests at temperatures near atmospheric, a suitable temperature is 86°F (30°C). This is usually slightly above room temperature and, therefore, easy to control.

The stirrups (glass rods or hooks or Teflon cord) by which the specimens are supported are attached to an arm actuated by a crank mechanism by which the specimens are moved in a vertical, circular path. This ensures that all parts of the specimen move at exactly the same velocity. Also, it is easy to make connections for electrical measurements without the use of such devices as mercury cups or commutators.

The testing solution may be saturated with air or other gases introduced through an Alundum thimble. The Alundum aerator forms very small gas bubbles which favor rapid solution of the gases. Merely bubbling gases through a glass tube drawn to a fine tip is not an adequate method of aeration. A glass chimney is placed over the aerator to prevent impingement of gas bubbles on the specimens.

Support of Specimens. This will vary with the apparatus used but should be designed to insulate specimens from one another and from any metallic container or supporting device used with the apparatus. The supporting device and container should not be affected by the corroding agent to an extent that might cause contamination of the testing solution so as to change its corrosiveness. The shape and form of the specimen support should be such as to avoid, as much as possible, any interference with free contact of the specimen with the corroding solution.

Where it is desired to set up conditions favoring contact corrosion, "deposit attack," or other forms of concentration-cell action, the means by which these types of attack are favored should be such as to ensure exact reproducibility from specimen to specimen and test to test.

Duration of Test. This will be determined by the nature and purpose of the test. In some cases it will be desirable to expose a number of specimens so that certain of them can be removed at definite time intervals in order to provide a measure of change of corrosion rates with time. With some materials it may be important to repeat tests using the same specimens to eliminate any effects due solely to the original condition of the metallic surface. However, a procedure which requires removal of solid corrosion products between periods of exposure will not measure accurately normal changes of corrosion with time.

Calculating Test Results. If W is the loss in weight (in grams) of the test piece during the second 48-hr immersion, A the area of the test piece in square inches, S the density of the metal in grams per cubic centimeter, and t the time of exposure in hours ($t = 48$), then C is the rate of chemical corrosion expressed as inches penetrated per month. Another popular way to express corrosion rates is in milligrams per square decimeter per day (mdd).

The higher the rate of corrosion, the shorter may be the testing period. The duration of a laboratory test need not be longer than the number of hours calculated by dividing the corrosion rate, in mdd into 10,000. For example, where the rate of corrosion is 500 mdd, the test need not be run for more than 20 hr, but for a rate of 50 mdd, the duration should be 200 hr. This method of estimating the proper duration of a test is applicable only where corrosion is uniform and is useful only as an aid in deciding, after a test has been made, whether or not it is desirable to repeat the test for a longer period. The duration of any test should be reported.

If the metal develops pitting, this factor must be included in the results, since failure in any case will occur when a pit has entirely penetrated the metal. The magnitude of this effect can be determined by grinding on a metallographic grinding set until all the pits have just disappeared and solid metal is reached. The loss in weight during this grinding is determined, and the pitting calculated as follows:

$$C = \frac{24 \times 30 \times W}{(2.54)^3 A S t} \text{ or } \frac{43.9 W}{A S t}$$

To calculate the pitting corrosion, let p be the loss in weight (in grams) from grinding out pits and d the rate of penetration of the metal both by normal corrosion over the entire surface and by local action due to pitting:

$$d = \frac{43.9(W + p)}{A S t}$$

Reliability of Results. As in other branches of engineering, it is necessary to apply a factor of safety to the results obtained. The factor varies with the degree of confidence in the applicability of the results. Ordinarily, a factor of 3 to 10 is considered normal.

Effect of Oxygen. Oxygen must be taken into account for each problem. In general, oxygen increases the rate of corrosion markedly.

Effect of Electrolysis. This is a frequent source of trouble on a large scale. Not only is the use of different metals in the same piece of equipment dangerous, but the effect of cold working may be sufficient to establish potential differences of objectionable magnitude between different parts of the same piece of metal. Riveting, even when extreme precautions are taken to use rivets of identical composition with the sheet, is likely to establish a potential difference. Even such slight working as threading without subsequent annealing has led to rapid failure from electrolysis. The mass of metal in chemical apparatus is ordinarily so great and the electrical resistance consequently so low that a very small voltage can cause a very high current.

Improper heat-treatment of a solid-solution-type alloy may convert it from the solid-solution type to the nonhomogeneous type. The results is that it will fail more rapidly than even its poorest constituent. Welding also, if not properly done, may leave the weld of a different physical or chemical composition from the body of the sheet and cause the development of stray currents. A simple test for weld homogeneity is to treat a sample of the welded sheet containing a portion of both weld and sheet with a suitable solvent such as nitric acid or aqua regia until about half the thickness of the metal has been dissolved. If heterogeneity exists, evidence of differential attack should be seen.

Effect of Liquid Velocity. Increased velocity of the corrosive medium removes corrosion products faster and increases the corrosion rate. This is especially true if the flow is turbulent or if the liquid carries suspended matter. Frequently, slight alterations in equipment design can be made to reduce this factor. Several methods are available for laboratory testing of high-velocity effects involving rotating disks.[3]

Effect of Local Concentration. Both local variations in temperature and crevices that permit the accumulation of corrosion products can allow the formation of concentration cells. The result will be accelerated local corrosion.

Effect of Temperature. In the laboratory, the temperature of the test specimen is that of the liquid in which it is immersed and the measured temperature is actually that at which the reaction is taking place. In the plant, however, where the heat is supplied through the metal to the liquid in many cases, the temperature of the film of (corrosive) liquid on the inside of the vessel may be a number of degrees higher than that registered by the thermometer. As the relation between temperature and corrosion is logarithmic, the rate of corrosion increase is very rapid as temperature rises.

As with other chemical reactions, the corrosion rate ordinarily increases two to threefold for each 10°F temperature rise, the actual relation being that of the equation $\log K = A + B/T$, where K represents the rate of corrosion and T absolute temperature. This relationship, although expressed mathematically, is more qualitative than quantitative.

When a metal is used as a heating coil or other heat-transfer application, it must be evaluated under heat-transfer conditions. Groves[8] has developed a good test method to measure corrosion under heat-flux conditions.

Effect of Impurities. The effect of impurities in either structural material or corrosive material is marked and may be either accelerating or decelerating. For reliable results, the actual materials to be used should be tested, not mere types of these materials. It is better to test the actual plant solution and the actual metal or nonmetal than to rely upon a duplication of either. As little as 0.01 per cent of certain organic compounds can reduce the rate of solution of steel in sulfuric acid 99.5 per cent, and 0.05 per cent bismuth in lead will increase the rate of corrosion over 1,000 per cent under certain conditions.

Plant Corrosion Tests

It is not always practical or convenient to investigate corrosion problems in the laboratory. In many instances it is difficult to discover just what are the conditions of service and to reproduce them exactly. This is especially the case with processes involving changes in the composition and other characteristics of the solutions as the process is carried out, as, for example, in evaporation, distillation, polymerization, sulfonation, or synthesis.

With many natural substances also, the exact nature of the corrosive is uncertain and is subject to changes not readily controlled in the laboratory. Or the corrosiveness of the solution may be influenced greatly by or even may be due principally to a constituent present in minute proportions. In such cases, the mass available in the limited volume of a laboratory setup would be exhausted by the corrosion reaction early in the test, and consequently the results over a longer period of time would be misleading.

Another difficulty sometimes encountered in laboratory tests is that the contamination of the testing solution by corrosion products may change its corrosive nature to an appreciable extent. In such cases it is often best to conduct the corrosion-testing program by exposing specimens in operating equipment under the actual conditions of service. Other advantages of this procedure are that it is possible to test a large number of specimens at the same time, little technical supervision is required, and it is unnecessary for the investigator to know the exact nature of the corrosive solution.

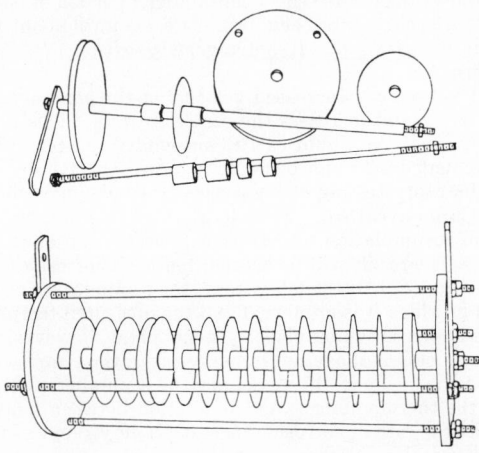

Fig. 5-9. Specimen holder for testing spools of materials.

Sometimes it is necessary to choose materials for equipment to be used in a process developed in the laboratory and not yet in plant-scale operation. A good procedure in such cases is to construct a pilot plant, using either the cheapest materials available or some other materials selected on the basis of past experience or from laboratory tests. While the pilot plant is being operated to check the process itself, specimens can be exposed in the operating equipment as a guide to the choice of materials for the large-scale plant or as a means of confirming the suitability of the materials chosen for the pilot plant.

In carrying out plant tests it is necessary to install the test specimens so that they will not come into contact with other metals and alloys; this avoids having their normal behavior disturbed by galvanic effects. It is also desirable to protect the specimens from possible mechanical damage.

Test Specimen. The method discussed here is in accord with ASTM A 224-46. The specimen holder is shown in Fig. 5-9. Specimens are machined to 2.233 in. with a $2\frac{1}{64}$-in.-diameter hole in the center. It is convenient to make cast specimens $\frac{3}{16}$-in. thick. With this thickness, it is common practice to make the diameter 2.083 in. and again make the exposed area, when installed on the specimen holder, exactly 0.5 sq cm.

Specimens are prepared for the test in the same manner as described for laboratory corrosion tests. These specimens can then be weighed readily on an analytical balance. Because of their uniform size, the time involved in changing weights on the balance pan is minimized. For certain purposes, special specimens may be prepared, such as specimens made by welding or soldering pieces of metal together.

Specimen Holder. When specimens are being assembled for a test, a supporting bracket is slid over the threaded end of the center rod and two nuts are screwed on to lock it in place. Next, an insulating tube is slid over the rod and then a large-end disk which bears against the bracket. Then an insulating spacer is slid over the insulating tube, then a specimen, then another spacer, and another specimen, and so on until all the available space has been utilized.

It is customary to mount at least two specimens of each metal or alloy for checking purposes. Then the second end disk is put on, followed by the other supporting bracket and the second set of nuts, which are drawn up tightly and locked. Four additional rods are inserted in the holes located around the peripheries of the end disks, and are drawn up tightly with nuts. The four rods serve two purposes: They afford protection to the specimens against mechanical abuse, and they make the assembly very strong. One of the rods passes through holes in the supporting brackets to hold them firmly.

The spacing of specimens is regulated by the length of the spacer; usually it is ⅝ in. Short spacers may be used to provide for any variations in thickness of specimens.

To investigate galvanic effects, the insulating spacer may be replaced by one of metal (of the normal, or shorter, length) made from one of the materials of the galvanic couple. Ordinarily, the metal spacer is exposed to the corrosive attack and is considered part of the metal specimen comprising one half of the couple. But if desired, the exposed surface of the metal spacer can be coated with an insulating lacquer.

Similarly, corrosion of the specimens forming the couple may be confined to adjacent faces by lacquering the backs. When a galvanic arrangement of this kind is used, it is common practice to use a compression spring, made of a corrosion-resisting material, between a spacer and the end disk. The spring takes up any slack that might result from thinning of specimens by corrosion.

The choice of materials from which to make the holder is important. These do not need to be durable enough to last indefinitely, but they must last long enough to ensure satisfactory completion of the test. Metallic parts are generally made of nickel or Monel for exposure to the majority of corrosives and of some type of chromium-nickel stainless steel for tests in oxidizing acids, such as nitric. Insulating materials used are Bakelite, procelain, neoprene, Teflon, and glass. Bakelite answers most purposes. Its principal limitations are unsuitability for use at temperatures over 300°F and lack of adequate resistance to concentrated alkalies and certain organic compounds, such as coal-tar products.

Specimen Positioning. The method of supporting the specimen holder during the test period is important. The preferred position is with the long axis of the holder horizontal; thus dripping of corrosion products from one specimen to another is avoided. The holder must be located so as to cover the conditions of exposure to be studied. It may have to be submerged or exposed only to the vapors or located at liquid level, or holders may be called for at all three locations.

Exposure at the liquid level may be accomplished most readily by building a suitable float, usually of wood, around the holder so that it will float on the surface. Various means have been utilized for supporting the holders in liquids or in vapors. The simplest is to suspend the holder by means of a heavy wire or light metal chain. Holders have been strung between heating coils, clamped to agitator shafts, welded to evaporator tube sheets, etc.

In a few special cases, the standard spool-type specimen holder is not applicable and a suitable special test method must be devised to apply to the corrosion conditions being studied. An example is shown in Fig. 5-10. Here the specimen is a strip, ½ in. wide by 3 in. long, bent at each end. Another method is an API test, shown in Fig. 5-11. Note that there is no provision for electrical insulation of the test specimens in the API test.

For conducting tests in pipelines of 3-in. diameter or larger, a spool holder, as shown in Fig. 5-12, has been used which employs the same disk-type specimens as used on the standard spool holder. This frame is so designed that it may be placed in a pipeline in any position without permitting the disk specimens to touch the wall of the pipe.

As with the strip-type holder, this assembly does not materially interfere with the fluid flow through the pipe and permits the study of corrosion effects prevailing in the pipeline.

Another way to study corrosion in pipelines is to install in the line short sections of pipe of the materials to be tested. These test sections should be insulated from each other and from the rest of the piping system by nonmetallic couplings. It is also good practice to provide insulating gaskets between the ends of the pipe specimens

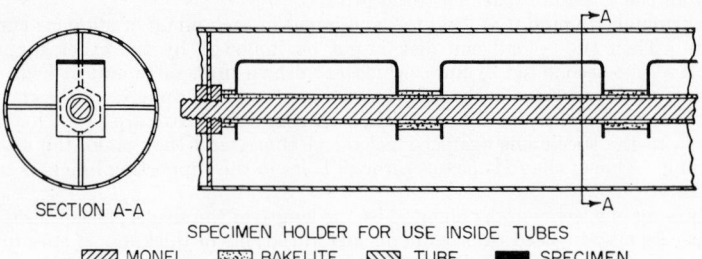

SECTION A-A

SPECIMEN HOLDER FOR USE INSIDE TUBES

▨ MONEL　　▥ BAKELITE　　◩ TUBE　　■ SPECIMEN

Fig. 5-10. Specimen holder for testing strips inside tubes.

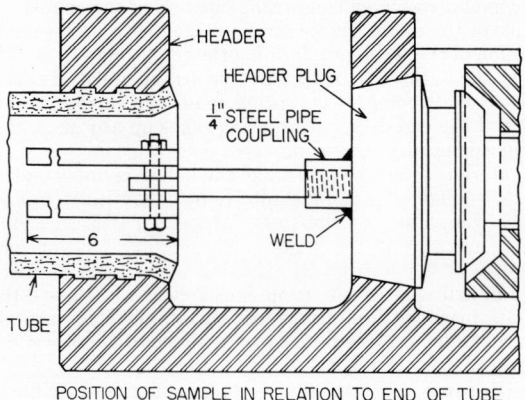

POSITION OF SAMPLE IN RELATION TO END OF TUBE

Fig. 5-11. API method for testing specimens inside pipe.

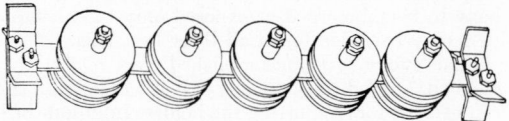

Fig. 5-12. Spool specimen holder for large pipe.

where they meet inside the couplings. Such joints may be sealed with various types of dope or cement. It is desirable in such cases to paint the outside of the specimens so as to confine corrosion to the inner surfaces.

It is occasionally desirable to expose corrosion-test specimens in operating equipment without the use of specimen holders of the type described previously. This can be done by attaching specimens directly to some part of the operating equipment and by providing the necessary insulation against galvanic effects as shown in Fig. 5-13. The suggested method of attaching specimens to racks is very suitable for the exposure of specimens to corrosion in sea water, where metal racks should be used to avoid destruction by teredos (shipworms).

Specimen Cleaning. The methods of cleaning specimens after a test are the same as those described in connection with laboratory corrosion testing.

Specimen Evaluation. Although loss of weight is used as the principal measure of corrosion, this should be supplemented always by an examination of each specimen before it is cleaned. Note should be made of tarnishing, pitting, intergranular deterioration, and the nature of corrosion products.

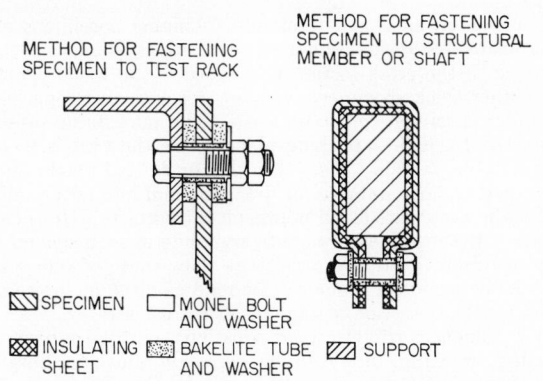

METHOD FOR FASTENING SPECIMEN TO TEST RACK

METHOD FOR FASTENING SPECIMEN TO STRUCTURAL MEMBER OR SHAFT

▧ SPECIMEN   ☐ MONEL BOLT AND WASHER

▨ INSULATING SHEET   ▨ BAKELITE TUBE AND WASHER   ▨ SUPPORT

Fig. 5-13. Method of attaching specimen to operating equipment.

Corrosion Probes

In the corrosion probe method, the electrical resistance of a test specimen is used as a measure of its cross-sectional area. As corrosion reduces the cross section, the electrical resistance increases.[9,10]

The test specimen on the end of the probe is a wire, strip, or thin-wall tube 0.001 to 0.080 in. thick and attached to lead wires. The specimen end of the probe can be placed in any type of refinery equipment which is provided with a suitable entrance. The wires pass through a pressure-tight insulating seal mounted in a fitting and

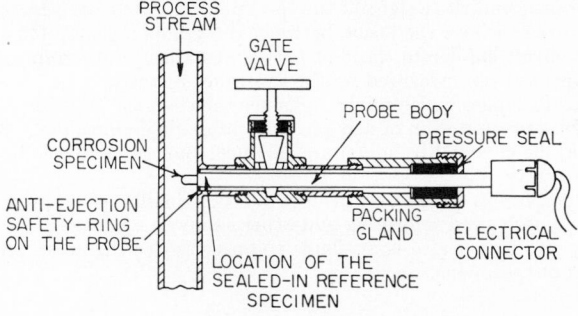

PROCESS STREAM

GATE VALVE

PROBE BODY

PRESSURE SEAL

CORROSION SPECIMEN

ANTI-EJECTION SAFETY-RING ON THE PROBE

PACKING GLAND

ELECTRICAL CONNECTOR

LOCATION OF THE SEALED-IN REFERENCE SPECIMEN

Fig. 5-14. Typical retractable corrosion probe.

are connected to a calibrated electronic instrument which measures the electrical resistance.

When modified by a factor which depends on the shape and the initial thickness of the specimen, the measuring instrument reads directly in microinches of corrosion. Because changes in temperature influence the electrical resistance, probes are equipped with one or more reference specimens made of the same material as the test specimen but protected against corrosion by a coating or by being embedded in cement or plastic (Fig. 5-14).

The electrical conductivity of corrosion products, except some sulfides of high purity formed in clean sulfidic systems, is negligible compared with the conductivity of the metal. Therefore, corrosion products usually need not be removed, and measurements can be made while the probe is in the operating vessel. High resolution can be achieved in comparing electrical resistances, so small amounts of corrosion can be detected and measured. Detection of 1 to 2 μin. of corrosion is commonplace, and higher resolutions can be reached under favorable circumstances. With such sensitivity, a corrosion rate of 0.005 in./year may be detected from measurements taken only a few hours apart.

Resolution depends on specimen thickness. Thinner specimens give higher sensitivity per microinch of corrosion, but they corrode away sooner. Measurements are dependable only until the cross section is reduced by 20 to 30 per cent. For long corrosion tests, either thick specimens with low resolution or replaceable retractable probes with thinner specimens must be used. The retractable probe diagramed in Fig. 5-14 is inserted through a packing gland while the unit is on-stream. If the probe is not retractable, its body has a threaded or flanged portion for mounting.

The probe method is most suitable for detecting and measuring uniform corrosion, that occurring at the same rate at all points on the surface. However, the rate may change with time. Because measurements are made in sequence on the same probe specimen, they are more mutually consistent than those of one set of weight-loss specimens. Thus, because the probe can be read as often as desired, the method is especially suited for the detection of changes in corrosion rates.

Anything that influences the electrical resistivity may be interpreted as uniform corrosion. Pitting, cracking, intergranular corrosion, and metallurgical effects may change the electrical resistance and must be considered. The probe specimen should be examined visually and metallographically after the test. If the probe specimen cannot be removed for examination before it is consumed, a coupon of the same material should be exposed during the probe readings and examined later.

Pitting complicates interpretation of the readings. Scattered pitting does not influence the electrical resistance much more than it would the weight loss of a coupon. When pitting severely decreases the cross section, the increase in resistance from small additional pitting may be large and may be mistaken for rapid uniform corrosion. Hence, only the first third of the corrosion-versus-time curve is reliable, and any rise indicating fast corrosion beyond this limit must be interpreted with caution.

An occasional deep pit or crack should not upset the results of probe measurements. A hole with a diameter as large as one-half of the cross section changes the resistance by only 2 per cent and the weight of the specimen by a little over 1 per cent. Intergranular corrosion, where the space between the grains is converted to a corrosion product, has effects similar to those of partial cracking. Intergranular corrosion of stainless steels has been measured by the resistance method.

Metallurgical changes which might influence the electrical resistance include phase transformations, precipitation of new phases, and gradual annealing. Sometimes the resistance may be changed by metallurgical effects not recognizable by usual metallographical means.

Corrosion probes have been used widely in crude distillation units, water systems, vapor recovery units, and reforming and other catalytic operations.

Laboratory equipment also is available to measure electrical resistance and corrosion potential of specimens.

REFERENCES

1. *ASTM Spec. Tech. Pub.* 93, 1952.
2. Uhlig, H. H., *The Corrosion Handbook*, John Wiley & Sons, Inc., New York, 1948.
3. La Que, F. L., *Proc. ASTM*, vol. 51, p. 495, 1951.
4. Champion, F. A., *Corrosion Testing Procedures*, Chapman & Hall, Ltd., London, 1952.
5. Schreir, L. L., ed., *Metallic Corrosion Handbook*, George Newnes, London, 1963.
6. La Que, F. L., and H. R. Copson, *Corrosion Resistance of Metals and Alloys*, p. 107, Reinhold Publishing Corporation, New York, 1963.
7. Fraser et al., *Ind. Eng. Chem.*, vol. 19, pp. 332–338, 1927.
8. Groves, N. D., et al., *Corrosion*, vol. 17, p. 173t, 1961.
9. Troscinski, E. S., et al., *Mech. Eng.*, July, 1961, p. 47.
10. Dravnieks, A., *Use of Corrosion Probes in Petroleum Refineries*, paper presented at 25th Mid-year Meeting, API Division of Refining, May 9, 1960.

Section 6

CHEMICALS AND CATALYSTS

By VITTORIO DE NORA

and

JOHN A. KING

Oronzio de Nora Impianti Elettrochimico
Milano, Italy

This section, a guide to the principal chemicals and catalysts used in petroleum processing, is divided into Part 1, Chemicals, and Part 2, Catalysts. Since many of the chemicals are also used as catalysts, cross references are used where appropriate to avoid duplication.

Chemicals used in petroleum processing can be classified into the following general groups: treating agents, solvents, laboratory reagents, and catalysts. Only the major chemicals used are covered here. For quick reference, they are listed below:

acetic acid	hydrogen
acetone	hydrogen fluoride
aluminum compounds	methyl ethyl ketone
barium compounds	methyl isobutyl ketone
benzene	natural oils
bone char (see also clay)	nitrobenzene
cadmium compounds	phenol
calcium compounds	phosphorous compounds
caustic soda (see sodium hydroxide)	potassium compounds
chlorine	propane
Chlorex (see dichloroethyl ether)	sodium carbonate
clay	sodium hydroxide
copper chloride	sodium hypochlorite
cresol	sodium phenolate
dichloroethyl ether	sodium plumbite
ethanolamines	sulfur chlorides
ethylene dichloride	sulfur dioxide
ethylene glycol	sulfuric acid
formaldehyde	toluene
furfural	trichlorethylene

Catalysts used in petroleum processing can be readily classified in three areas: by catalyst function, by process, and by catalyst material. Part 2 presents listings of catalysts by process and by catalyst material. Catalysts are not listed by function, since this area has not been accurately determined for many processes. Processes covered including cracking, reforming, hydrotreating, isomerization, alkylation, and polymerization.

PART 1. CHEMICALS

Acetic Acid

Acetic acid is used (as is hydrochloric acid) to break emulsions. It is sometimes added to sulfuric acid in small quantities to increase treating capabilities for distillate color improvement, and is used as a treating agent itself (in the presence of zinc) for reduction of sulfur content. Another use is in extraction of polymers from cracked distillates using an acetic acid–ether mixture, or in separation of wax using an acetic acid–methanol mixture.

It is irritating and can cause burns, because it easily penetrates the skin. Inhalation can cause irritation of the mucous membranes, and the fumes can cause conjunctivitis. In case of skin or eye contact, flush immediately with water. For internal contact, swallow magnesia or chalk in water. The upper tolerable limit is 10 ppm in air. Because of its flammability and explosive limit of 4 per cent in air, it should be stored in a sealed vessel and in a cool place. Goggles, rubber gloves, and apron are recommended for handling.

Acetone

Mixtures of a nonpolar solvent and acetone are used in regeneration of treating clays. It is also used in azeotropic distillation for benzene isolation. Mixtures of acetone with propyl alcohol or isopropyl alcohol are used as a solvent in determining the oil content of waxes.

Acetone-benzene mixtures are used in lube-oil dewaxing (particularly where MEK availability is low) and in deoiling paraffin waxes (mixtures as high as 1 to 1).

Acetone is volatile and highly flammable. It should be stored in a cool place, away from constant sources of ignition. Personnel exposed to high concentrations of acetone should wear airline respirators or twin-cartridge chemical gasmasks of a type approved for this material. Good ventilation should be provided, and chemical goggles should be worn if it causes irritation to the eyes in the concentration being used. One suffering from narcotic or toxic effects of acetone should immediately be removed to fresh air. If the symptoms are severe, call a physician.

Aluminum Compounds

Aluminum chloride (first used in refining as a cracking catalyst in 1915), and also used as an alkylation catalyst, is now used as an isomerization catalyst for isomerization of butane, pentane, and hexane feeds. (See also Isomerization in Part 2, Catalysts.) Aluminum oxide (bauxite) is an important cracking-catalyst constituent.

Aluminum naphthenates are used as a detergent additive for lubricating oils, as are aluminum phenates, aluminum soaps (in general), and aluminum stearate. Aluminum stearate is also extensively used as a corrosion-inhibition additive for product oils.

Except for aluminum chloride and aluminum oxide (bauxite), these compounds of aluminum are not hazardous to handle either as solids or in solution. Because aluminum chloride may cause some allergic symptoms, protective clothing may be appropriate for its use. Bauxite dust over 50 million particles per cubic foot has been known to cause pulmonary diseases similar to silicosis; therefore, respiratory equipment is called for when it is handled.

Barium Compounds

Barium hydroxide has several important uses in refining petroleum: to treat spent caustic solutions from caustic treating, to neutralize acid-treated oils, to precipitate naphthenic acids, to prevent possible foaming before caustic soda treating for mercaptan removal, and to remove inorganic salts from furfural prior to refining.

Barium salts of alkyl phenol sulfide, methallyl phenol, or derivatives of hydrogenated wax are used as an additive for oxidation inhibition in lube oils. Barium sulfonates and esters of aromatic sulfonic acids are used as detergent additives for lube oils.

Barium hydroxide has a caustic action on eyes, skin, and mucous membranes, and is a poison which causes injury to the heart. Protective goggles and clothing are required and exposed areas must be washed immediately to prevent burns. Non-alkaline barium salts must be handled with care to prevent inhalation or other form of intake because of their toxic effect.

Benzene

Benzene is used in mixed-solvent extraction processes with liquid sulfur dioxide for treating lube oils to improve viscosity index and simultaneously remove waxes. Extraction temperature is about 25°F and the mixed solvent/oil ratio is about 2.0 to 1.0. Increased benzene proportion in mixed solvent increases solvent action.

It is sometimes used in methyl ethyl ketone–benzene mixture in the Solvent Dewaxing Process (Texaco Development Corp.) for removal of wax from lubricating oils. Mixed solvent ratios range from 1.0 to 1.0 to 4.5 to 1.0, depending on wax content of feedstock.

Benzene is both toxic and flammable, with explosive limits of 1.5 to 8.0 per cent by volume in air. Acute poisoning results from the inhalation of vapors above 3,000 ppm for even short periods of time. The effects are rapid: symptoms are a tightening of the leg muscles, dizziness, excitement, breathlessness, and the sensation of pressure on the lower central part of the forehead. In advanced cases of acute poisoning, the patient becomes confused and hysterical, laughs, shouts, curses, sings, becomes very obstinate. Chronic poisoning results from daily exposure to low but unsafe concentrations over prolonged periods, such as weeks or months. Its use can cause an irritation of the skin, because of the defatting effect. The liquid is highly volatile and flammable and introduces a potential fire hazard wherever it is stored, handled, or used.

If benzene is spilled on the person, clothing should be immediately removed to avoid prolonged skin contact. If it enters the eyes, they should be flushed thoroughly for at least 15 min with clean warm water. Medical attention must be secured at once. In case of unusual exposure to the vapor, for first aid the patient should be removed to fresh air and if possible collapse should be prevented by having him lie down without a pillow, quiet and warm, but not hot. Keep benzene away from heat and open flame and store in a cool place. Containers should be kept sealed, plainly labeled, and should be handled only with adequate ventilation.

Bone Char

Bone char is used, as are several natural adsorbents, to decolorize oil. Carbonaceous adsorbents have generally been superseded by clays (see Clay).

Bone char in the usual dust form must be handled to avoid exposure to the eyes, because it causes conjunctivitis and blepharitis. Goggles should be worn by the operator.

Cadmium Compounds

For distillate desulfurizing, cadmium–ammonium chloride or cadmium hydroxide are used in solution; cadmium chloride or cadmium sulfide (mixed with steel wool) in solid form have also been used.

Cadmium oleate and cadmium naphthenate have been used as lube-oil additives to prevent oxidation. Cadmium dithiocarbamate and cadmium sulfonate have been used as a detergent additive.

Cadmium and its compounds are highly poisonous (maximum concentration allowed is 0.1 mg/cu m of air). The compounds are dangerous in varying degrees, but the effect is common: lung damage on inhalation with possible kidney and bone damage as well as blood change. Symptoms of exposure are diarrhea, nausea, and abdominal pain. Exposure to the eyes causes conjunctivitis. Personnel handling compounds of cadmium must wear goggles and protective clothing, and ventilation must be adequate.

Calcium Compounds

Slaked lime (calcium hydroxide) and quicklime (calcium oxide) are sometimes used as a substitute for sodium hydroxide to neutralize acid-treated oils, and to remove hydrogen sulfide and organic acids from oils. Because lime is only slightly soluble in water, and because calcium soaps or esters formed by reaction with acids may be soluble in oil with emulsion formation, long retention times for the reaction (sometimes days), and then percolation through charcoal for emulsion breaking may be required.

Calcium carbonate may be substituted for lime in the previous application with the advantage of almost complete removal of mercaptans and sulfides. Addition of 1.0 per cent lithium carbonate results in naphthenic-acid decomposition in the oil treat, when treating at 570 to 890°F.

Calcium chloride is used as a desiccant and to remove traces of moisture from oil when required.

Calcium hypochlorite action on oils is similar to that of sodium hypochlorite, whereby complete desulfurization results from oxidation of sulfides and mercaptans. It is applied either before or after sulfuric acid treatment but usually before final distillation. Saturated chlorination of lime, followed by use of the supernatant liquor, produces the required hypochlorite.

Calcium hydroxide has a typical caustic reaction; is irritating to the skin, eyes, and mucous membranes. Safety goggles, protective clothing, and a respirator are required for handling. Calcium carbonate causes allergic symptoms; so sensitive personnel should wear protective clothing. Calcium chloride causes inflammation of the eyes; so safety goggles must be worn and eyes must be irrigated thoroughly when exposed. Solid calcium hypochlorite causes eye inflammation and is very irritating to mucous membranes. Extensive irrigation of the eyes is required if they are exposed.

Chlorine

Where complete removal of disulfides is required, chlorine gas may be used as the oxidizing agent to convert them to sulfonyl halides in hot sodium hydroxide solutions. Such treatment also removes mercaptans by converting them first to disulfides and then to the final halide form, which is soluble in the aqueous phase.

Bentonite (treating clay) may be regenerated by chlorine gas.

Spent "doctor solution" (see Sodium Plumbite) may be regenerated by chlorine oxidation followed by caustic treatment, although air oxidation has largely superseded this technique.

Chlorine is used to prepare calcium and sodium hypochlorite, which are used as treating agents (see Calcium Compounds, Sodium Compounds).

Chlorine may be added to fuels to improve the cetane number.

Chlorine is very dangerous to the eyes, can cause inflammation and varying degrees of damage. High concentrations can cause pneumonitis and edema of the lungs. In concentrations of 1,000 ppm it is invariably and rapidly fatal. The least concentration with detectable odor is 3.5 ppm. The least concentration necessary to cause throat irritation is about 15 ppm and to cause coughing is about 30 ppm. Thus the odor of chlorine is a warning.

Symptoms from inhalation of chlorine are respiratory reflexes, causing coughing, smarting of the eyes, and a feeling of chest discomfort, nausea, and vomiting.

A patient who has been exposed must be promptly removed from the toxic area. All constricting clothing about the neck should be loosened. Oxygen should be administered in all cases. Olive oil may be effective for the eyes.

Containers of chlorine under pressure must be stored in a cool place away from acute fire hazards, and should be plainly labeled. They should never be dropped, because they may break and allow large volumes of chlorine to escape. This gas is heavy and will cling to the ground, and therefore may cause a great deal of damage to personnel near its point of release. Personnel who must be exposed to chlorine must have protective equipment. The eyes should be protected, as must be the respiratory tract; an approved respirator is recommended. If the concentration is quite high, protective clothing should be worn to avoid contact with the skin.

Clays

A wide variety of clays is used as adsorbents to improve color, odor, and stability of waxes and lube oils; and as catalysts in cracking operations. Table 6-1 lists char-

Table 6-1. Characteristics of Adsorbent Clays*

Adsorbent clay	Activation temp, °F	Bulk density, g/cc	% voids	% pores	Surface area, sq m/g
Alumina (activated):					
Grade F-1	400	0.88	44	51	200
Grade H-151	400	....	...	...	300-350
Bauxite:					
Arkansas	450	1.33	44	20	27
	700	0.97	43	64	236
	900	0.95	42	49	178
	1300	0.94	42	52	151
South American	900	0.96	36	54	187
Silica gel	400	0.78	41	38	602
Fuller's earth	400	0.56	44	60	126
	600	0.55	...	...	124
	900	0.54	43	64	118
	1100	0.56	42	64	119
	700	0.68	41	57	260
Bentonite	900	0.67	41	57	236
	1300	0.67	41	56	208
Diatomaceous earth	400	0.32	49	44	4
Activated carbon:					
Columbia carbon	....	0.42	44	60	1,397
Darco carbon	....	0.36	51	66	560
Bone char	....	0.68	47	46	110

* W. L. Nelson, *Petroleum Refinery Engineering*, McGraw-Hill Book Company, New York, 1958, 4th ed.

acteristic properties of many of these materials. Table 6-2 gives some of their typical uses. Their use as catalysts is discussed in Part 2 of this section.

Copper Chloride (Cupric chloride)

This compound is used to convert mercaptans to insoluble disulfides in the copper-sweetening process, in which the copper chloride promotes oxidation of mercaptans to disulfides and is itself reduced to cuprous chloride. This process is being largely replaced by the Merox process, which reduces total sulfur content. Copper chloride specifications for various sweetening processes are given in Table 6-3.

Copper chloride may cause serious irritation of the skin although it is not considered very dangerous. Intake of very large quantities may cause cirrhosis of the liver, preceded by violent vomiting, purging, and intense abdominal pain. Normal protective clothing and goggles are recommended for personnel.

Table 6-2. Adsorbent Clays and Their Uses in Petroleum Processing

Alumina...............	Laboratory-scale separation of polynuclear aromatic hydrocarbons according to the number of rings per molecule
Bauxite...............	Catalysis in Gray, Stratford, and Osterstrom processes for polymerization of reactive materials that impart color or form gums in stored cracked gasoline
	Catalysis of virgin-gasoline desulfurization processes: Gray, Cycloversion, and Houdry Treating
	Adsorption treating of lubricating-oil stocks, petrolatum, waxes, kerosine, and gasoline (Perco process)
Silica gel...............	Analytical separation of aromatics, olefins, and paraffins
Fuller's earth...........	Catalysis of Gray process for gasoline stabilization; in this role it competes with bauxite
Bentonite...............	Widely used for clarification of petroleum oils

Table 6-3. Uses of Copper Chloride

Process	Licenser	Copper chloride form
Perco gasoline . . .	Phillips Petroleum Co.	Solid bed of $CuCl_2$ impregnated bed of adsorbent
Perco lube oil....	Phillips Petroleum Co.	$CuCl_2$ solution
U.O.P...........	Universal Oil Products Co.	NH_4Cl—$CuSO_4$ mixture dissolved in Italian pumice
Linde...........	Linde Div. Union Carbide Corp.	$CuCl_2$ saturated solution impregnated on 200-mesh clay
Airco-Hoover....	Air Reduction	$CuCl_2$ saturated solution impregnated on diatomaceous earth

Cresol

Cresol is used in the Duo-Sol extraction process for extraction of high-viscosity-index, light-color, low-carbon-residue lubricants from any residual or distillate base stock. Cresol is mixed with phenol to prepare Selecto solvent which has an affinity for asphalts, aromatics, naphthenes, and color bodies. The solvent usually consists of 20 to 40 per cent phenol and 60 to 80 per cent cresol.

Cresol is very dangerous in contact with the eyes and is a moderate fire hazard. The toxic effects resemble those of phenol poisoning, but ortho- and para-cresol are considered more poisonous than phenol. If large areas of the skin have been wet with concentrated cresols, serious and possibly fatal poisoning may result. Following intake, phenol collapse may occur, consisting of exhaustion, followed by coma and death.

Removal of exposed personnel from the area will cause the symptoms to decline rapidly. However, exposure to high concentrations may be serious and possibly fatal; therefore, even though the symptoms may seem mild, a physician should be consulted. Personnel who must be exposed to this material must wear protective equipment and take care to avoid spillage on the clothing because it is readily absorbed into the body through the skin. Rubber gloves, safety shoes with rubbers, or rubber shoes and rubber aprons are recommended. If the concentration of the vapor of this material in the air is high, respirators of an approved design are required.

Dichloroethyl Ether

This material is used as Chlorex solvent in the Chlorex extraction process to improve viscosity index and yields from percolation filters of Mid-Continent stocks or highly paraffinic oils. It is colorless and chloroform-like with free acidity as HCl not more than 0.005 per cent; ethylene dichloride content should be not more than 1.00 per cent.

This chemical is a moderate fire hazard, and definitely toxic. It can cause permanent damage to the eyes. Concentrations of 550 and 1,000 ppm are very irritating to nasal passages and eyes upon brief exposure. Deep inhalations at this concentration causes nausea. On the other hand, 260 ppm has a similar effect, but to a lesser degree, and this concentration, although objectionable, is not intolerable. Concentrations of 100 ppm are objectionable because they are slightly nauseating and irritating. A concentration of 35 ppm is easily noticeable by an odor which is slightly offensive but nonirritating. The chief hazard comes from contact with low concentrations for long periods of time, which can cause great irritation of the respiratory system. It is regarded as harmful to the liver, and its effects can be fatal.

Personnel who must be exposed to considerable quantities of dichloroethyl ether should wear adequate personal protection. Recommended are chemical safety goggles for the eyes and an approved type of respirator for the lungs and respiratory tract.

Ethanolamines

Several ethanolamines, including monoethanolamine (MEA), diethanolamine (DEA), and triethanolamine (TEA), are used for the removal or recovery of water, hydrogen sulfide, or carbon dioxide from gaseous streams. They are used in aqueous solution in the Girbotol process (Girdler Corp.) to remove H_2S or CO_2 selectively, and are used in the Glycol-Amine Gas Treating process (The Fluor Corp., Ltd.) in an aqueous mixture with diethylene glycol or triethylene glycol to remove water, H_2S, or CO_2. The bases for selection of the various ethanolamines are summarized in Table 6-4.

Table 6-4. Characteristics of Various Ethanolamines

Characteristic	MEA	DEA	TEA	MDEA*
Molecular weight	61.1	105.1	149.2	119.2
Concentration of 2.5N solution, %	15.2	25.5	36.0	28.0
Approx cost, $ per gal.	0.35–0.40	0.50–0.55	0.75–0.80	1.80–1.85
Approx vapor pressure (pure) 100°F	0.03	0.0061	0.0001	6.004
Advantages	(1) Ease of reclammation (2) Low cost	Unreactive with COS and CS$_2$		
Disadvantages	Reactivity with COS and CS$_2$	Too expensive	Too expensive	Too expensive
Usual application	Most low-temperature streams with S	Streams with COS or CS$_2$	Selective absorption of H$_2$S in the presence of CO$_2$	
Usual concentration, %	15–20	10–25	30	30

* Methyldiethanolamine.

Use of the Glycol-Amine Gas Treating process is usually for removal of H_2S or CO_2 from high-pressure natural gas. For this purpose mixtures of 10 to 30 per cent MEA, plus 45 to 85 per cent DEA or TEA and 5 to 25 per cent water, are usually used. The advantages are simultaneous gas dehydration, and less steam consumption than with ethanolamine-water systems. The disadvantages are loss of amine because of vaporization and difficult regeneration of solvent.

Ethanolamines must be handled with regard for their low flash points (200°F for MEA). They must be stored in a cool, well-ventilated area away from open flame, and acute fire hazard, and also strong oxidizing agents.

Ethylene Dichloride

Ethylene dichloride is used in the Bari-Sol process as a constituent of the extraction solvent for increasing viscosity index of lube oils by removing waxes. The solvent used usually consists of 78 per cent ethylene dichloride and 22 per cent benzol. Solvent requirements are dictated by the solvent/oil ratio of 3.0 to 1.0. The operating temperature is about 110°F. The ethylene dichloride must be free of contaminants which remain dissolved below −20°F, which is the second-chilling-stage temperature.

Ethylene dichloride is flammable, and toxic by inhalation, by contact with the skin or mucous membranes, and by ingestion. Headache, depression, mental confusion, fatigue, or nausea may result from contact. Diarrhea with bloody stools, suppression of urine, and blood in the urine may also result. Prolonged or repeated contact with the skin causes dermatitis. Ethylene dichloride can cause serious eye damage. For first aid, quick removal from exposure is important. If breathing has ceased, start artificial respiration. Call a physician. If it gets into the eyes, they should be washed promptly with copious amounts of warm water for at least 15 min. If it has been ingested, the patient should be induced to vomit by drinking a glass full of mustard water, lukewarm salt water, or warm soapy water.

Because of the fire hazard, store ethylene dichloride in a cool, well-ventilated place away from open flames. When traces of moisture are present, it can corrode steel, iron, and certain other metals, and it should therefore be stored in as dry a location as possible. Personnel who must use this material should do so only with adequate ventilation and should avoid skin contact. Chemical safety goggles should be worn. Head protection in the form of a soft hat and protective clothing should be provided for those who work with this material.

Ethylene Glycol

This material is used in the selective recovery of benzene, toluene, and xylenes from petroleum stocks. It is the main constituent of the extraction solvent in the U.O.P. Udex process.

Ethylene glycol is a flammable liquid and a moderate fire hazard. Ingestion causes pupils to become dilated and nonreactive. Contact causes eye irritation. Although about as toxic as methyl alcohol, it is considered less dangerous because it is not absorbed so readily.

Formaldehyde

Formaldehyde is used as a laboratory reagent and solvent, and has been proposed for use as a chemical-treatment agent for degumming cracked gasolines.

Concentrated formaldehyde imposes a serious hazard to person and property, because of its flammability, volatility, and irritating and in some instances damaging effect on human tissues. The explosive limits are 7.0 to 73.0 per cent by volume in air. The vapors have a suffocating effect and irritate mucous membranes, affecting the upper respiratory tract and the eyes. High vapor concentrations cause coughing, constriction in the chest, a sense of pressure in the head, weakness, heart palpitations, and sleeplessness. Ingestion may result in fatal effects, and prolonged or frequent exposure may cause hypersensitivity and dermatitis. Adequate ventilation must be provided to prevent concentration in air from exceeding 10 ppm. Personnel handling the liquid must be equipped with waterproof gloves, vaportight goggles, gasmasks, and rubber aprons. The liquid should be stored in a cool place away from flames and other fire hazards.

Furfural

Furfural is used in the Texaco Development Corp. Furfural Extraction process for extraction of diesel fuel oils, burning oils, cracking stocks, and cycle oils. It removes materials with a low cetane number, in addition to undesirable materials with low stability or high acidities and sulfur, organometallic, and nitrogen compounds. For this application the water and oil content of make-up and regenerated furfural should be minimized, although small percentages are tolerable with process adjustments. Solvent ratios of 0.25 to 1.0, 0.40 to 1.0, and 0.80 to 1.0 may be used for extraction of diesel fuel, virgin gas oil, and middle distillates, respectively.

Furfural is also used in Texaco's Furfural Refining process for extractive removal of aromatic, naphthenic, olefinic, and unstable hydrocarbons from lube oils to improve viscosity index and stability. Furfural solvent regeneration must not be carried out over 450 to 550°F; above this temperature it breaks down.

Furfural is harmful to the eyes and is a moderate fire hazard. The vapor irritates mucous membranes and poisons the central nervous system, but because of its low volatility its toxic effect is small. Danger of prolonged exposure to it is overlooked because of its low volatility. It should be stored in a cool well-ventilated place away from areas of acute fire hazard.

Hydrogen

Hydrogen is used in the pretreatment of catalytic reforming feed to reduce the content of catalyst poisons, such as nitrogen compounds. It is also used in the treatment of fuel stocks to reduce sulfur content; in the case of gasoline, susceptibility to octane improvement by agents such as tetraethyl lead is thus increased. Hydrogen treating also improves gasoline quality by reducing the concentration of unsaturates without affecting the aromatics content.

Table 6-5 lists some of the hydrogen uses in petroleum processing.

Hydrogen has a very high calorific value (61,060 Btu/lb) and has an ignition temperature between 982 and 1000°F. When not produced at the plant site, it is shipped in steel cylinders at a pressure of 120 to 150 atm. Explosive limits are 4.1 to 74 per cent by volume in air. Hydrogen is not toxic, although it can cause asphyxiation by replacing the oxygen required to sustain life. It should be stored in a cool place away from areas of fire hazard, particularly open flames. Containers must be labeled and personnel cautioned. It is an odorless, colorless gas, slightly soluble in water, and its presence is not apparent.

Hydrogen Fluoride (Hydrofluoric Acid)

Hydrogen fluoride is used mainly as an alkylation catalyst (see Alkylation in Part 2). It has been proposed by Standard Oil Co. (Indiana) for the removal of sulfur and coke-forming compounds from virgin and cracked light naphthas, middle distillates, and gas oils in a liquid-liquid extraction process.

Hydrogen fluoride is an extremely corrosive and poisonous liquid. Large amounts in the eye cause immediate blindness, and vapor or liquid is very dangerous when in contact with the skin or any part of the body, or if ingested. It causes medically serious and very painful skin burns. Its sharp, penetrating odor warns againt inhalation of toxic quantities; 50 ppm exposure or more can be fatal in an exposure of from 30 to 60 min. Burns caused by 20 per cent or weaker solutions do not become noticeable for some hours, whereas burns from solutions of 20 to 60 per cent are detected almost at once, and 60 to 100 per cent acid is felt immediately. If skin is contacted, use a drenching shower immediately. Personnel should understand operation of all safety equipment when working with hydrofluoric acid. Compresses of ice, alcohol, or magnesium sulfate should be applied to contacted skin or, if possible, it should be dipped into magnesium sulfate solution for at least 30 min.

Required safety equipment includes bubbler fountains for the eyes and drenching showers to flush acid from the person quickly. Clothing which has been exposed to this material should be washed before reuse and clothing which is used daily by workers with this material should be washed each day before reissue. Long gauntlet-type gloves made of a resistant material, such as neoprene or rubber, are required, as are rubber shoes or boots and chemical safety goggles with transparent plastic face shields to protect the eyes and face. Airline respirators should be used to avoid inhalation.

Table 6-5. Hydrogen Uses in Petroleum Processing

Processes	Purpose	Hydrogen consumption	Process charge	Operating temp, °F	Operating pressure, psig
Hydrotreating:					
Autofining (British Petroleum Co.)	Desulfinization of distillate feeds	None	Straight-run distillates up to 700°F end point	750	100–200
Ferrofining (British Petroleum Co.)	Color and stability improvement of lube oils	0.15 lb/bbl	Solvent-refined or straight distillate lube oils	n.a.	n.a.
Gulf HDS (Gulf Research and Development Co.)	Reduces sulfur content ᵉf high-sulfur stocks	Only make-up gas	High-sulfur residual stocks	750–850	500–3,000
Gulfining (Gulf Research and Development Co.)	Reduce sulfur, improve color and stability, increase API gravity	n.a.	Straight-run distillates and fuel oil	600–800	600
Hydrodesulfurization (M.W. Kellogg Co.)	Remove sulfur, nitrogen, oxygen, metal compounds from distillates	300 scf/bbl	Virgin or cracked distillate	650–800	350–1,000
Hydrofining (Esso Research and Engineering Co.)	Remove sulfur or other impurities from any one of several stocks	n.a. (hydrogen source above 50 mole %)	Cracked or virgin naphtha, solvents, jet fuel, heating oils, lube oil, wax	550–800	50–800
Selective hydrogenation (British Petroleum Co.)	Refine byproducts from gasoline steam cracking	100–230 scf/bbl	Byproduct from gasoline steam cracking	To 350	300–600
Trickle hydrosulfurization (Shell Development Co.)	(1) Remove sulfur or nitrogen from distillates. (2) Hydrogenate aromatics (polycyclic). (3) Remove heavy metals from catalytic-cracking feed	For treated West Texas vacuum gas oil, 530 scf/bbl For treated catalytic cycle stock, 500 scf/bbl	Distillates from gasoline to heavy gas oils	600–800	300–1,500
Ultrafining [Standard Oil Co. (Indiana)]	Desulfurize and remove nitrogen from stock, and saturate olefins	170 scf/bbl (dilute H_2 addition 500–4,000 scf/bbl at 70 mole %)	Variety of stocks	600–800	200–1,500

Table 6-5. Hydrogen Uses in Petroleum Processing (Continued)

Processes	Purpose	Hydrogen consumption	Process charge	Operating temp, °F	Operating pressure, psig
Unifining (Union Oil Co. of California— Universal Oil Products)	Remove sulfur, nitrogen, oxygen from stocks	n.a.	Straight-run or cracked fractions	500–800	300–800
Hydrocracking:					
H-Oil (Hydrocarbon Research, Inc.)	Upgrade and desulfurize heavy sour crude and residue	800–1,200 scf/bbl	Heavy sour crude tars, residues	n.a.	n.a.
Isomax (California Research Corp. Universal Oil Products Co.)	Upgrade distillates to gasoline or middle distillates	n.a.	Distillates	n.a.	n.a.
Unicracking (Union Oil Co. of California)	To produce premium gasoline, jet fuel, midbarrel products	n.a.	Catalytic cycle oil, gas oils and heavy naphthas	500–800	500–2,000
Lomax (Universal Oil Products)	Reduce heavy stocks to gasoline	n.a.	Heavy stocks	Up to 850	Up to 2,000
Hydrodealkylation processes (Hydeal, Unidac, Detol, Humble, Sun Oil, HOA, Texaco)	To produce naphthalene, benzene, and toluene	n.a.	Heavy reformate	Over 1,000	100–1,000

Hydrofluoric acid should be stored in lead carboys, or in containers made of Teflon, Kel-F, oleoresin wax, or gutta-percha, in a cool place away from acute fire hazard, and in a ventilated and isolated area out of the direct rays of the sun.

Methyl Ethyl Ketone (MEK)

Methyl ethyl ketone (MEK) is used in the Solvent Dewaxing process (Texaco Development Corp.) to remove wax from lube oils. Mixtures of MEK plus toluene are generally used, although benzene sometimes takes the place of toluene. Mixed solvent ratios range from 1.0 to 1.0 to 4.5 to 1.0, depending on the wax content of the feedstock. It is also used as a laboratory solvent.

Because explosive limits are 1.8 to 11.5 per cent by volume in air, MEK must be stored in a cool, dry place away from acute fire hazards. Safety goggles and respirator are recommended for operators working with MEK, because of its irritating effect on eyes and nose.

Methyl Isobutyl Ketone (MIBK)

This solvent is used for deoiling of high-quality refined waxes in Union Oil Co. of California's Wax Manufacturing process, which is applicable for both high-melting "hard" and low-melting "soft" waxes. Methyl isobutyl ketone of commercial quality is suitable for this application.

Because it is relatively flammable, MIBK must be stored in a cool dry area away from acute fire hazards.

Natural Oils

Several natural oils are used in the production of lubes and greases. These may be categorized as follows: vegetable oils, animal fats, fish and marine mammal oils, and secondary products from the treatment of natural products. Table 6-6 lists the

Table 6-6. Properties of Principal Natural Oils

Oils	S.G. (15/15°C)	Solidifying point, °C	Iodine no.	Acid value	Unsaponifiable matter	Saponification value
Vegetable oils:						
Cottonseed	0.92 (25°/25°C)	+12 to −13	103–111	0.6–0.9	1.1	194–196
Soybean	0.93	−10 to −16	122–134	0.3–1.8	......	189–194
Rapeseed	0.91–0.92	−10	96–105	0.4–1.0	1.5	168–179
Linseed	0.93–0.94	−19 to −27	175–202	1–3.5	0.4–1.2	188–195
Castor	0.96–0.97	−12	84	0.1–0.8	0.6	175–183
Animal fats and oil:						
Lard (fatty tissue)	0.93–94	27–30	47–67	0.5–0.8	......	195–203
Tallow (bovine)	0.91–0.92	2–75	56–61	0.2–0.3	......	194–199
Neat's-foot oil	0.91–0.92	+10 to −2	58–75	0.1–0.6	0.1–0.7	193–199
Fish and marine oil:						
Whale	0.92	−2 to 0	90–146	2	1–4	160–202
Sperm	0.88	16	80–84	13.2	39–42	120–137
Shark	0.92	....	115–139	......	....	157–164

principal natural oils used in the petroleum processing industry, and their useful properties. Most of these oils and fats, although only slightly flammable and in general nontoxic, must be handled with some regard for their properties.

Nitrobenzene

This compound is used in the Nitrobenzene Extraction process (Atlantic Refining Co.) to increase viscosity index and extract carbon and sludge-forming compounds from lube oils. Extraction takes place at 30 to 100°F and requires low nitrobenzene/oil ratios from 0.5 to 1.0 to 0.3 to 1.0. The nitrobenzene should be anhydrous.

Nitrobenzene is flammable, presents a moderate fire hazard, and is highly toxic when taken internally or absorbed through the skin. Continued external exposure to nitrobenzene vapors may result in scaly dermatitis. The lower limit of flammability in air is 1.8 per cent by volume at 200°F. The maximum tolerable concentration in air for safe conditions is 5 ppm. Persons exposed to this material must wear chemical goggles, rubber gloves, and protective clothing, and must work in well-ventilated areas.

Phenol

Phenol is used in the Duo-Sol Extraction process as a constituent of Selecto solvent along with cresol (see Cresol). Pure phenol must be used in this application. It is also used in the Phenol Extraction process (M. W. Kellogg Co.) to improve viscosity index, color, and oxidation resistance, and to reduce carbon and sludge-forming tendencies of lube oils.   The solvent, either in anhydrous form or in aqueous form, is dehydrated prior to use in the extraction system.

Phenol is corrosive to body tissue and highly toxic, affecting the central nervous system and causing edema of the lungs, kidneys, liver, pancreas, and spleen.   Skin or eyes contacted by phenol must be washed freely with soap and water and, if possible, with alcohol or glycerin to dissolve any residue.   Phenol can cause death if ingested; liquid such as salt water, sodium bicarbonate solution, or milk should be swallowed, followed by a demulcent such as cornstarch paste.   Gastight goggles and positive-pressure hose masks such as airline respirators are suggested.   Rubber gloves, aprons, and boots are recommended.   Spilled phenol must be immediately attended to. Areas in which phenol is handled must be freely ventilated and away from acute fire hazards.

Phosphorous Compounds

1. Four polymerization processes utilize phosphorous compounds as catalysts (see Polymerization in Part 2).   These are phosphoric acid on Kieselguhr, copper pyrophosphate on charcoal, liquid phosphoric acid, and phosphoric acid on quartz.   2. Phosphorous pentoxide also serves as a catalyst for air blowing of distilled asphalt to obtain relatively high consistency/pliability ratios.

Phosphoric acid is a toxic and corrosive liquid, and is dangerous to the eyes.   The acid anhydride (phosphorous pentoxide) reacts violently with water, releasing much heat, and is therefore harmful to the eyes and skin as well as internal tissues if ingested. Rubber gloves and apron, and boots are suggested for personnel handling either the acid or the anhydride.   The latter must be stored in a dry place away from sprinkler systems and all other moisture sources.

Potassium Compounds

Potassium hydroxide (caustic potash) is used in some instances as a treating agent to remove acids from petroleum (see Sodium Hydroxide).

Potassium phosphate is used in gas treating to remove hydrogen sulfide in phosphate desulfurization process (Shell Development Co.). K_3PO_4 solution selectively removes H_2S without removing CO_2; the solution contains approximately 30 per cent K_3PO_4.

Potassium hydroxide is harmful to the eyes (it can cause dissolution of the cornea and consequent blindness) and to internal and external body tissues.   Its high exothermic heat of reaction is another source of hazard.   Gas- and dust-tight goggles with full face shield, rubber gloves, apron, and boots are suggested.   Safety showers must be readily available.

Propane

Propane is used primarily in the refinery in solvent-extraction processes—deasphalting, decarbonizing, and dewaxing.

Two propane deasphalting processes (one developed by Foster Wheeler Corp. and the other by M. W. Kellogg Co.) are available for separation of asphalt and two grades of lubricating oils from heavy residues by solvent extraction.   In the Kellogg process, propane/oil ratios of 2.5 to 1.0 to 5.0 to 1.0 are used for treating heavy- and low-viscosity residue, respectively.   Top tower temperatures of 130 to 180°F at 400 to 550 psig dictate propane quality.   For multistage extraction such as in the Foster Wheeler process lower temperatures are required and propane/oil ratio may be 9.0 to 1.0 to 11.0 to 1.0 or higher.

The Propane Decarbonizing process (M. W. Kellogg Co.) recovers catalytic-cracking feed from heavy fuel residues.   The process is also called Solvent Decarbonizing because butane-propane mixtures are sometimes the solvent used.   Top temperature ranges between 160 and 220°F and the pressure from 450 to 600 psig.   Propane used is commercial LPG.

The Propane Dewaxing process (M.W. Kellogg Co.) extracts wax from lube oils and lube blending stocks.   Propane quality is dictated by filtering temperature of

−34 to 41° at 180 to 200 psig.   Solvent requirements are dicated by solvent/oil ratios of 2.5 to 1.0 to 3.0 to 1.0.

Propane is also used in the Duo-Sol process for extraction of high-viscosity-index, light-color, low-carbon-residue lubricants from any residual or distillate base stock. (The other solvent used in this process is Selecto solvent, a blend of cresol and phenol). In the Duo-Sol process, propane serves as extractant for raffinate while the Selecto solvent preferentially dissolves the asphalts, aromatics, naphthenes, and color bodies. The extraction is carried out at 120 to 150°F.

Propane gas is explosive in air mixtures containing 2.4 to 9.5 per cent by volume. It must be stored in a cool, isolated, and ventilated area free of acute fire hazard.   In high concentration the gas has an anesthetic effect but is not toxic.

Sodium Carbonate (Soda Ash)

Sodium carbonate may be used in place of sodium hydroxide as an agent to neutralize acids in petroleum processing streams.   (See Sodium Hydroxide for applications.)

Sodium carbonate dust or solution irritates the respiratory tract.   Therefore, it must be handled with care to avoid such contact; gastight goggles with a respirator should be used to avoid excessive dust inhalation.

Sodium Hydroxide (Caustic Soda)

Sodium hydroxide (caustic soda) is used in solution as a treating agent to remove acidic substances remaining in petroleum products, particularly those which cause undesirable color or odor, and which inhibit lead susceptibility.   Among them are naphthenic acids, mercaptans, hydrogen sulfide, and phenolic compounds.   The solutions used range from 5 to 50 per cent (by weight) and are used at 70 to 110°F and 5 to 40 psig, depending on the process.   For mercaptan removal, a small volume of more concentrated caustic solution is more effective then a larger volume of less concentrated caustic solution; higher temperatures effect an increase in the transfer of mercaptans to caustic solution by 50 per cent for every 20°F temperature increase. Sodium hydroxide solutions are also used to wash oil before or after clay treatment to stabilize it, to neutralize acid-treated products, to manufacture grease, and in water-deionization units.   Table 6-7 indicates the concentration and method of use for each of the caustic uses.

This material has an extremely corrosive action on all body tissues and must therefore be handled with great care to avoid contact.   Even mists, vapors, and dusts of this compound present a hazardous condition.   Ingestion in any form results in damage to the mucous membranes and can cause perforation and/or scarring of internal tissues.   In cases of accidental contact with caustic removal must be effected immediately and the areas must be rinsed with floods of water.   Caustic soda dissolves the cornea of the eye and can cause immediate blindness if allowed to contact the eye for short periods of time even in weak concentrations.   Always call a physician in the event of an accident with caustic soda.

In handling caustic soda, spillage must be immediately neutralized with acetic acid or any available weak acid and completely washed away.   Gastight safety goggles, rubber apron and gloves, rubbertight shoes, and tight cotton coveralls are required for all personnel in the area, whether or not they handle caustic.   As a solid, this compound must be stored in a cool dry place, because of its deliquescent property and extremely high heat of solution.   Drums must be kept closed when not is use.

Solutions normally used in the petroleum industry may be stored or transferred in vessels or piping made of steel, cast iron, nickel, copper alloys, monel 400, Saran chlorinated polyether, rubber, epoxy resins, Haveg 60 or 61, polyethylene, or PVC. Selection of a material of construction should be based on relative cost, fabricability requirements, and availability.

Sodium Hypochlorite

Sodium hypochlorite is used in hypochlorite sweetening to convert mercaptans in straight-run or natural gasoline to disulfides.   It is prepared by passing chlorine gas

Table 6-7. Caustic Soda Use and Critical Specifications

Treating processes	Conversion	Caustic concentration	Comments
Nonregenerative...	Removal of H_2S and mercaptan traces	Dilute caustic (alternately dilute calcium or ammonium hydroxide)	
Polysulfide........	Removal of elemental sulfur from refinery liquids	1 lb Na_2S (50–60 %) plus 1 lb sulfur to each gallon of 15° Be caustic (11 %)	Solution most active at polysulfide concentration Na_2S_{25}
Regenerative......	Removal of mercaptans from light straight-run gasolines	Same as nonregenerative process	Caustic regeneration is effected by blowing steam through solution to convert mercaptans to insoluble disulfides
Dualayer distillate (Socony Mobil Oil Co., Inc.)	Extract organic acidic material such as mercaptans from cracked or virgin distilled fuels	43 % caustic solution saturated with cresylate salt used for distillate treating (5 % of oil volume used)	(1) Contact with distillate layer is at 130°F. (2) Reagent is prepared in settler tank
Dualayer gasoline (Socony Mobil Oil Co., Inc.)	Extract mercaptans from LPG, gasoline, and naphtha	Same Dualayer reagent used (2–3 % of gasoline volume used)	Contact takes place at 120°F in at least two stages
Electrical distillate treating (Howe-Baker Engineers, Inc.)	To remove salts and other impurities from most distillates	15°Be caustic (23 %)	High-voltage direct current used to break oil-caustic emulsion
Electrolytic Mercaptan (American Development Corp.)	To remove mercaptans from hydrocarbon mixtures	18°Be caustic (23 %)	
Ferrocyanide (American Development Corp.)	To remove mercaptans from straight-run naphtha, LPG, natural, and recycle gasolines	Caustic–sodium ferrocyanide solution	Ferricyanide functions as an oxidizing agent when formed in caustic solution
Mercapsol (Pure Oil Co.)	To remove mercaptan sulfur from gasoline	Caustic soda (or caustic potash) solution containing mixed cresols (concentration undisclosed)	
Merox (Universal Oil Products Co.) (this process replacing doctor and copper sweetening)	To remove mercaptans from gasoline and low-boiling fractions and to convert mercaptans in heavy stocks to disulfides	Caustic solution of undisclosed concentration plus Merox catalyst	Caustic solution is regenerated by air blowing through solution (converts sodium mercaptide to disulfides which may be separated)
Solutizer-regenerative (Shell Development Co.)	To remove mercaptans from sour gasoline	Caustic solution concentration undisclosed	Caustic solution used for prewash
Unisol mercaptan extraction (Universal Oil Products Co.)	To extract mercaptans and some nitrogen compounds from sour gasoline or distillates	Caustic solution 45°Be (42 %) regenerated 50°Be (50 %) fresh	Caustic is mixed with methanol and used at 100°F. Regeneration at 290–300°F removes methanol, water, and mercaptans

Table 6-7. Caustic Soda Use and Critical Specifications (Continued)

Treating processes	Conversion	Caustic concentration	Comments
Chelate sweetening (Mobil Oil Co.)	In first step of chelate-sweetening process, extraction of mercaptan from gasoline by caustic treating	Caustic solution	Oxygen-bearing chelate oxidizes mercaptan to mercaptide in second step
Copper sweetening (Linde Co.)	Prewash with caustic to reduce mercaptan content of stream	5–10 % caustic solution	
Inhibitor sweetening	To remove mercaptans from liquids	Caustic soda solution 10–30 % (14–36°Be)	

Other uses	Caustic specifications
Neutralization of acid-treated products...	Dilute caustic solution (about 10 %) used to neutralize gasoline and lube oils after acid treatment
Grease manufacture....................	Caustic soda for soap ingredient of grease is generally about 75 % solution, slightly more than stoichiometric quantity used to assume reaction completion
Water deionization.....................	Caustic soda solution to regenerated anion deionizer section. Must be chemically-pure-grade solution, supplied by mercury cells
Product stabilization..................	Dilute caustic solution (5–10 %) used to neutralize products. Treated with acid clays

through 10 per cent caustic soda solution at 95°F to obtain the correct concentration of sodium hypochlorite in aqueous solution.

Sodium Phenolate

Sodium phenolate is used in gas treating to remove hydrogen sulfide, and also as a color stabilizer for gasoline.

Sodium Plumbite

Sodium plumbite is used in the doctor sweetening process to convert mercaptans in hydrocarbon streams to disulfides. Doctor reagent consists of 5 to 20 per cent (weight) sodium hydroxide solution with 1 to 3 per cent (weight) lead oxide. This process is being supplanted by the Merox and other processes which remove mercaptans rather than merely changing them to less objectionable compounds. Some doctor reagent is sometimes used in the Bender Sweetening process.

Sodium plumbite, as do other compounds of lead, can cause serious diseases of the optical system. (Extent of lead-poisoning hazard is proportional to solubility in body fluids.) Ingestion of lead compounds affects the gastrointestinal tract and nervous system and may cause varying degrees of disturbance in them.

Sulfur Chlorides

Sulfur chloride and sulfur dichloride are solvents used in laboratory and plant washing and extractives.

These solvents react with water to release noxious sulfur dioxide and hydrogen chloride, and must be kept away from moisture sources (see Sulfur Dioxide).

Sulfur Dioxide

This chemical is used in the Sulfur Dioxide Extraction process (Stone and Webster Engineering Corp.) to extract aromatic hydrocarbons and sulfur-bearing compounds from paraffins and naphthenic hydrocarbons. The present process is a modification of the Edeleanu process. Operating temperature range is from $-20°F$ for naphtha treating to $+60°F$ for more viscous stocks. Solvent ratios range from 0.5 to 1.0 to 1.7 to 1.0 as stock heaviness increases.

It is also used in the Sulfur Dioxide–Benzene Extraction process which is used to simultaneously improve viscosity index and remove waxes from lube oils. Extraction temperature is about 25°F and the mixed solvent ratio is about 2.0 to 1.0. Decreased sulfur dioxide proportion in mixed solvent increases solvent action.

Sulfur dioxide is irritating, corrosive, and toxic. Allowed concentration in air is 10 ppm. It must be stored in a cool isolated area away from acute fire hazard. In areas where concentrations in air exceed 100 ppm, personnel must wear respirators and gastight goggles.

Sulfuric Acid

This acid is used as a chemical-treating agent to remove aromatics that cause smoky flames in kerosine, and to remove or dissolve resinous materials, asphaltic substances, and sulfur as mercaptans. Acid concentrations of 93 per cent (66°Be) are generally used except for lubricating oil, for which 98 per cent acid is sometimes used. Acid (93 to 103 per cent) used at 25 to 30°F removes sulfur from gasoline, and aromatics from burning oils. Conditions for the treatment of several petroleum products are shown in Table 6-8.

Table 6-8. Conditions for Sulfuric Acid Treatment

Temp, °F	Product	Acid concentration, %	Acid amount, lb/bbl
70–90	Natural gasoline	80	2
	Straight-run gasoline	93	5
	Pressure distillate	93	0–8 (depends on sulfur content)
90–130	Solvents	93	0–5
110–180	Kerosine	93	1–15
	Lube oils	93	0–60

Sulfuric acid is highly corrosive to metals, is a powerful oxidizing agent, and dissolves in water or reacts with organic compounds liberating considerable heat. Body tissue is destroyed rapidly when contacted by liquid or mists. Spillage and waste must be neutralized (do not add water to a container filled with this acid). Store containers away from daylight in a cool, ventilated area. Personnel should wear protective hat and rubber gloves, boots, apron, and emergency showers must be close to working areas.

Toluene

Toluene is used with methyl ethyl ketone in Solvent Dewaxing process (Texaco Development Corp.) to remove waxes from lube oils. Mixed solvent/oil ratios range from 1.0 to 1.0 to 4.5 to 1.0, depending on the wax content of the process feed.

Toluene is a highly flammable liquid (explosive limit in air is 1.3 to 6.8 per cent by volume) and is toxic, causing corneal damage, intoxication, and (in severe exposures) death. It must be stored in a cool, ventilated area away from acute fire hazard and must be handled by personnel protected by safety goggles and protective clothing.

Trichlorethylene

Trichlorethylene has been used in the S-N or Separator-Nobel Dewaxing process, to extract carbon- and sludge-forming constituents of lube oils and to increase their viscosity index.   Solvent/oil ratios used in practice range from 0.67 to 1.0 to 1.5 to 1.0. Solvent is charged at 110 to 120°F and stripped at 230°F.   It is sometimes used in lube-oil dewaxing processes such as the Bari-Sol process in place of ethylene dichloride.

This chemical is nonflammable at ambient temperature but moderately flammable at higher temperatures.   It forms explosive mixtures with strong alkalies (dichloroacetylene).   Vapors are irritating and intoxicating, but the liquid is not toxic unless ingested, in which case it stops kidney functions.   Suggested protective equipment includes gastight rubber gloves and apron, safety goggles, and oxygen supply for application in the event of personnel intoxication.

PART 2.   CATALYSTS

The petroleum processing industry is without doubt the largest user of catalysts in the chemical industry.   The catalytic materials include both solids and liquids and range all the way from common clay to the precious metals.   Table 6-9 lists the various catalytic processes commonly used in petroleum processing and the materials employed.

Table 6-9. Key to Principal Applications for Catalyst Materials

Catalyst material	Processing application					
	Cracking	Reforming	Hydro-treating	Isomerization	Alkylation	Polymerization
Alumina....................	X	X	X			
Aluminum chloride*..........	...	...	...	X	X	
Antimony trichloride..........	...	...	...	X		
Bauxite....................	X	...	...	X		
Bentonite clay...............	X					
Clay*.....................	X					
Cobalt-molybdena............	...	...	X			
Cobalt molybdate.............	...	X	X	X		
Cobalt oxide.................	...	X				
Copper.....................	X					
Copper pyrophosphate........	...	...	...	...	...	X
Hydrochloric acid.............	...	...	...	X	X	
Hydrofluoric acid.............	...	...	...	...	X	
Iron oxide...................	X					
Kaolin clay..................	X					
Magnesia...................	X					
Molybdena..................	...	X				
Molybdenum.................	...	X				
Nickel sulfide................	...	...	X			
Phosphoric acid..............	...	...	...	...	...	X
Platinum....................	...	X	X			
Potassium...................	X					
Silica-alumina...............	X	X	X	X		
Sulfuric acid*................	...	...	...	...	X	
Tungsten nickel sulfide........	...	...	X			

* See also Part 1 in this section.

Cracking

The principal difference between the commerical catalytic-cracking processes is in the method of handling the catalyst rather than the catalyst choice. The choice of catalyst-handling method dictates the catalyst characteristics: (1) fluid-bed process, (2) moving-bed process, and (3) fixed-bed process. For fluid-bed processing the catalyst is in powder form, 5 to 100 mesh, or in the form of microspheres with stability to withstand 875 to 975°F and pressures of 8 to 10 psig.

The catalyst/oil volume ratios range from 5 to 1 to 30 to 1 for the different processes, although most processes are operated at 10 to 1.

For moving-bed processes the catalyst is in the form of ⅛-in. beads with catalyst/oil volume ratios in the range from 4 to 1 to 7 to 1.

For fixed-bed operation (no longer used in the United States) the catalyst is in pellet form.

Table 6-10. Performance Factors for Typical Cracking Catalysts

Catalyst	Performance	Typical yield from 900°F gas-oil cracking			CO_2/CO ratio in flue gas
		Gasoline vol. %	Coke	Octane of gasoline	
Synthetic:					
Silica-alumina....	Gives highest octane gasoline.	43	4.2	95.8	1.0–1.2
Silica-magnesia...	Gives highest gasoline yields. Sometimes a poor rate of burning. Sometimes poor regeneration due to pore jamming	49.0	4.1	92.8	1.6–2.0
Natural:					
Bentonite clays...	High stability	46.5	4.4	94.7	1.7–2.2
Kaolin clays......	Similar to silica-alumina except higher stability	n.a.	n.a.	n.a.	n.a.

All commercially used cracking catalysts are "insulator catalysts" possessing strong protonic acid properties. They function as catalyst by altering the cracking-process mechanism through an alternate mechanism involving (1) chemisorption by proton donation at a high pK value and (2) desorption resulting in cracked oil and theoretically restored catalyst. (Catalyst usage is actually an appreciable cost item, ranging from 0.1 to 1.0 lb catalyst per lb of charge.)

A basic understanding of this mechanism leads to an awareness of the limitation of cracking-catalyst use. For example, all cracking catalysts are poisoned by proton-accepting vanadium. This basis for catalytic cracking applies to use of both natural and synthetic catalysts. Natural catalyst is used (sometimes mixed with synthetic) to start up a cracking unit because fresh synthetic catalyst regenerates too rapidly, thus closing its pores and trapping internal carbon deposits.

Common semisynthetic catalysts have the following dry analysis: Al_2O_3 (33 to 34.5 per cent), Na_2O (0.01 to 0.04 per cent), Fe (0.1 per cent), SO_4 (0.4 to 0.5 per cent), with a pore volume of 0.7 cc/g, surface area of 250 to 300 sq m/g, and over 90 per cent particle-size distribution less than 100 microns.

Table 6-10 lists some of the commoner cracking catalysts and their performance factors.

Reforming

The composition of a reforming catalyst is dictated by the compositions of the reformer charge and the desired reformate. Reforming consists of two types of chemical reactions which are catalyzed by two different types of catalysts: (1) isomerization of straight-chain paraffins and isomerization (simultaneous with hydrogenation) of olefins to produce branched-chain paraffins, and (2) dehydrogenation/hydrogenation of paraffins to produce aromatics, and olefins to produce paraffins.

Table 6-11. Reforming Catalysts

Process	Isomerization component	Hydrogenation dehydrogenation component	Form	Licensor	Regeneration	Bed type
Catforming (Atlantic Refining Co.)	Pt (0.3–0.7 %)	SiO₂/Al₂O₃	Pellets	Davison Chemical (catalyst)	Occasional	Fixed
Fluid hydroforming..	MoO₃ (10 %)	Al₂O₃	Powder (+inert pellets)	Esso + M.W. Kellogg	Separate	Fluid
Houdriforming.......	Pt (Houdry type 3D)	Al₂O₃	Pellets	Houdry Process Chemical Co.	Occasional	Fixed
Hyperforming.......	CoMoO₄	SiO₂/Al₂O₃	Pellets	Union Oil Co. California	Separate	Moving
Iso-Plus Houdriforming	Pt	Acid support	Pellets	Houdry Process Chemical Co.	Occasional	Fixed
Orthoforming (in-line fluid reforming)	MoO₃	Al₂O₃	Powder	M.W. Kellogg Co.	Separate	Fluid
Platforming.........	Pt (0.1–1.0 %)	Al₂O₃ + halogen (0.1–0.8 %)	⅛-in. pellets	Universal Oil Products Co.	None	Fixed
Powerforming (Esso Research Development Co.)	CoMoO₄	Al₂O₃	Pellets	(Catalyst by Davison Chemical Co.)	Separate	Fixed
Rexforming.........	Pt	Al₂O₃	Pellets	Universal Oil Products Co.	None	Fixed
Sinclair-Baker.......	Pt (0.6 %) (RD − 150)	Al₂O₃	Pellets	(Catalyst by Baker Co.)	Cyclic	Fixed
Sinclair-Baker-Kellogg	Pt (0.6 %) (RD − 160)	Al₂O₃	Pellets	(Catalyst by Baker Co.)	Cyclic	Fixed
Sovaforming.........	Pt (0.6 %) (RD − 150)	Al₂O₃	Pellets	(Catalyst by Baker Co.)	Occasional	Fixed
Thermofor Catalytic Reforming (Socony Mobil Oil Co.)	C₂O₃	Al₂O₃	Beads	(Catalyst by Socony Mobil Co.)	Separate	Moving
Ultraforming........	Pt (0.6 %)	Al₂O₃	Pellets	(Catalyst by American Cyanamid Co.)	Cyclic	Fixed

Usually the isomerization catalyst (molybdena, platinum, chromia, or cobalt molybdate) is supported on an acid support (alumina on silica) which catalyzes hydrogenation/dehydrogenation.

Other acid constituents such as combined halogens (chloride or fluoride) promote the second reaction. Specificity of reaction products results from proper selection of catalyst and composition; platinum is the most selective catalyst for reforming.

Table 6-11 indicates catalysts used for the presently commercial reforming processes.

Hydrotreating

The principal catalyst for hydrotreating (which accounts for the bulk of the hydrogenation operations in petroleum processing) is a form of cobalt molybdena supported on alumina. Other catalysts such as platinum, molybdena, nickel-tungsten sulfide, and nickel are in limited use in the United States.

Special catalysts such as nickel-tungsten sulfide (for desulfurizing cracked naphthas) and nickel sulfide (for selective hydrogenation of diolefins) also have some application in the United States.

In general, catalyst composition must be balanced to meet particular processing needs. Table 6-12 lists the catalysts used in commercially available processes.

Isomerization

During World War II aluminum chloride was the catalyst used to isomerize butene, pentane, and hexane. Since then, supported metal catalysts have been developed for use in high-temperature processes which operate in the range 700 to 900°F and 300 to 750 psi, while aluminum chloride plus hydrogen chloride are universally used for the low-temperature processes.

Table 6-12. Characteristics of Some Hydrotreating Catalysts

Process	Licensor	Catalyst type	Support or form	Characteristics
Autofining	British Petroleum	Cobalt-molybdena	Pelleted on alumina	2-year life 200- to 2,000-day regeneration cycle 0.005–0.02 lb/bbl of feed consumption
Diolefin hydrogenation	Shell Development	Nickel sulfide	Pelleted on alumina	
Ferrofining	British Petroleum	....	....	0.03 lb/bbl of feed consumption
Diesulforming	Husky Oil	Molybdenum	Pellet	24- to 36-day cycle
Distillate hydrogenation	Gulf Oil	Tungsten-nickel sulfide	Pellet	
Gulf HDS	Gulf Oil	Unidentified metal	Alumina	2-year life 0.0725 lb/bbl of feed consumption
Gulfining	Gulf Oil	Cobalt molybdenum	Pellet	
Gulfinishing	Gulf Oil	Undisclosed		
Hydrodesulfurization	M. W. Kellogg	Cobalt molybdenum	Pellet	0.008 lb/bbl of feed consumption
Hydrofining	Esso Research	Cobalt and molybdenum oxides	Alumina pellets	0.0002–0.04 lb/bbl of feed consumption
Hydropretreating	Houdry Process	Type 3 platinum catalyst or cobalt molybdenum on alumina		
Pentafining	Atlantic Refining	Platinum	Silica-alumina	Fixed-bed regeneration
Residuum hydroconversion	Humble Oil	Cobalt molybdate	Alumina	600–1,000 hr regeneration cycle
Sinclair hydrotreating	Sinclair	National Aluminate Corp. type RD-154		
Sovafining	Socony Mobil	Not disclosed		
Trickle hydrodesulfurization	Shell Oil	Cobalt molybdenum	Alumina	Less than 0.01 lb/bbl of feed consumption
Ultrafining	Standard Oil (Indiana)	Not disclosed		Less than 0.003 lb/bbl of feed consumption
Unifining	Universal Oil Products	Not disclosed		

Because aluminum chloride is volatile at commercial reaction temperatures and is somewhat soluble in hydrocarbons, techniques are employed to prevent its migration from the reactor. This catalyst is nonregenerable and is utilized in either a fixed-bed or liquid contactor.

Table 6-13 lists the catalysts used in some commercial processes.

Table 6-13. Isomerization Catalysts

Low-temperature processes	Catalyst	HQ cons. %	Selectivity
Vapor phase:			
Anglo-Jersey.................	Impregnated bauxite	4	95
Phillips....................	Impregnated bauxite		
Shell.......................	Impregnated bauxite	2–14	95
Liquid-phase:			
UOP.......................	Complex on quartz chips	5	97
Standard Oil (Ind.)...........	Liquid complex	4	97
Shell......................	Dissolved in SbCl$_3$	5	97

High-temperature processes	Catalyst	Regeneration
Butamer.......................	Platinum	None required
Iso Kel.......................	Precious metal	Regenerable
Isomerate.....................	Nonnoble metal	Infrequent
Penex.........................	Platinum	None required
Pentafining...................	Platinum	Steam-air

Alkylation

Sulfuric acid, hydrogen fluoride, and aluminum chloride are the only commercially utilized catalysts for this process.

Sulfuric acid is used with propylene and higher-boiling feeds, but not with ethylene because it reacts to form ethyl hydrogen sulfate. Suitable catalyst contains 85 per cent titratable acidity (minimum). Acid catalyst is pumped through the reactor and forms an air emulsion with reactants; emulsion is held at 50 per cent acid. Rate of deactivation varies with feed type and i-butane charge rate. Butene feeds cause less acid consumption than the propylene feeds; when butane feed ratio to olefin feed is 5 to 1 and 12 to 1, acid consumption is 0.9 and 0.3 bbl/gal, respectively.

Aluminum chloride is not widely used as an alkylation catalyst. When employed, hydrogen chloride is used as a promoter and water is injected to activate Lewis acid catalyst. The form of catalyst is an AlCl$_3$-hydrocarbon complex, and AlCl$_3$ concentration is 63 to 84 per cent. AlCl$_3$ consumption is 1.4 to 3.0 lb/bbl of product.

Hydrogen fluoride is used for alkylation of higher-boiling olefins with i-butene and propylene. The advantage of HF is that it is more readily separated and recovered from the resulting product. Usual concentration is 85 to 92 per cent titratable acid, with about 1.5 per cent H$_2$O. Consumption is about 0.2 to 0.8 lb/bbl of product.

Polymerization

Phosphates are the principal catalyst for polymerization. The commercially used catalysts are liquid phosphoric acid (California Research), phosphoric acid on kieselguhr (UOP), copper pyrophosphate pellet (M. W. Kellogg), and phosphoric acid film on quartz (California Research—Hydrocarbon Research). The latter is the least active, but the most used and easiest one to regenerate simply by washing and recoating; the serious disadvantage is that tar must occasionally be burned off the support. The California Research process using liquid H$_3$PO$_4$ catalyst is far more responsive to attempts to raise production by increasing temperature than the other processes. Catalyst life is 100 to 200 gal of product per lb H$_3$PO$_4$.

Section 7

MAINTENANCE AND CONSTRUCTION

By EVERETTE KERNS*

Mechanical Engineer
Amanda, Ohio

Maintenance of an oil refinery means keeping the plant in good operating condition for making quality products in an efficient manner. As safe operation of equipment is uppermost in the minds of refinery management, the equipment must be kept in good repair through the use of sound materials of construction and skilled workmanship. Management will also establish a degree or level for maintenance from an appearance standpoint which has a bearing on employee and public opinion and acceptance of products in general.

Construction as discussed here means the adding of any new installation within a refinery. It may be the revamping of or an addition to an existing process unit, facility, or building. It can mean the building of an entirely new unit in a plant or of a whole new refinery. Part 2 of this section, dealing with construction, starts on page 7–44.

PART 1. MAINTENANCE

The proper maintenance of a modern oil refinery today is a major undertaking, requiring well-qualified men who have become specialists in their fields.

Such maintenance has become a necessity because of the many advances in refining techniques over the years. We are no longer living in an age where a few reciprocating pumps are connected by small-size piping to a few atmospheric vessels. In the modern plant are found much rotating equipment operating at high speeds, dozens of pressure vessels carrying a wide range of pressures, and miles of piping making up the piping systems.

Many experienced refiners well remember when the only pieces of centrifugal equipment found within a refinery were a few low-pressure steam turbines. There were no centrifugal pumps or compressors and practically no automatic instruments and equipment. Maintenance of a plant in those days was quite simple, and almost any individual who was mechanically inclined could do a fairly good job. The tools required could almost be carried to the job site by the workmen.

Today, centrifugal pumps and compressors have replaced practically all the reciprocating type, with the latter used only for special services. Speeds of centrifugal compressors in the 10,000-rpm range are common, and pumping speeds and capacities have increased greatly.

* Mr. Kerns was formerly Head, Engineering Service Section, Standard Oil Co. (Ohio), Cleveland, Ohio.

The average mechanic is no longer qualified to maintain equipment of this type. He must be schooled and trained by experts, frequently at the manufacturer's plant. Oftentimes he will spend days with the manufacturer watching a piece of equipment being assembled, noting the proper steps taken, and learning of tolerances required. While the mechanic is learning all about the parts and the method of assembly and disassembly, the instrument man is being schooled on all instruments and automatic equipment that accompany the machine. Many of these installations are as automatic as they can possibly be made. Knowing the principles of operation and how to keep these many devices operating satisfactorily requires much skill and know-how.

Today, vessels 30 and 40 ft in diameter are common, and many fractionating towers stand more than 200 ft above grade. Piping systems have lines large enough for a man to walk through, where at one time a 6-in. line was considered large. To maintain vessels and piping systems in these large sizes requires skilled trades and a great deal of specialized tools and equipment.

Great sums of money are expended each year for the many tools and specially designed equipment to maintain a modern plant. Management can no longer rely largely on manpower alone in this age of mechanization. Much of this equipment calls for a large initial investment as well as highly trained operators. A careless operator can cost the company thousands of dollars, whereas an efficient one can help to raise the dividends.

Organization

For a maintenance department to operate efficiently, it must be well organized. Management, starting right at the top, must be sincerely in accord with the program and then back it. Cooperation between maintenance and operating personnel on all levels is absolutely essential to success. Misguided and sloppy operating procedures can result in many man-hours of excessive maintenance work. Furthermore, if all jobs become emergencies, the priority system will fail. Operating and maintenance personnel must get together on a common basis. Plant operating personnel must be included within the area of decision on maintenance problems. It is they who decide which work is to be done first.

Maintenance personnel decide how to do the job, who is to do it, and the number of people required. It should always be recognized that many of these decisions cannot be made from the front office. Field supervisors should be given the responsibility and the authority to make and carry out their decisions.

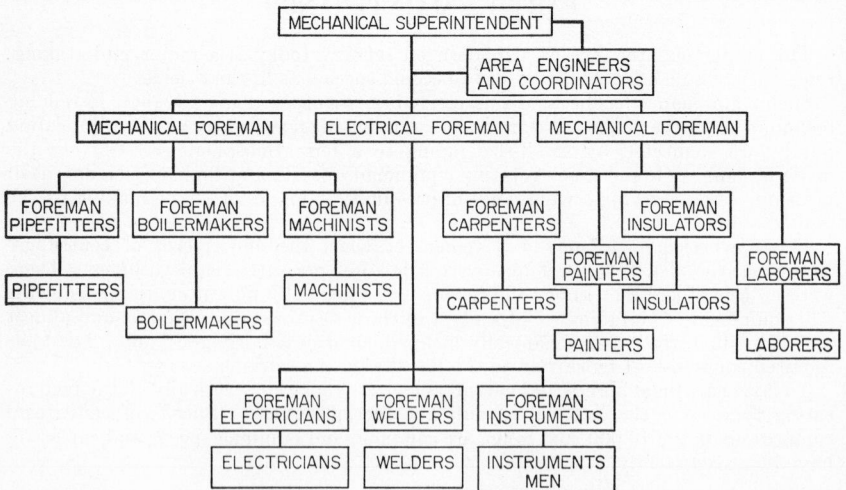

Fig. 7-1. Typical mechanical department organization chart.

Good organization alone may not achieve good human relations but will provide the structure on which good relations can be built. By a good organization we mean one in which duties and functions are clearly drawn with a minimum of overlapping, one in which line and staff functions are properly segregated, and one in which responsibility and the necessary authority go hand in hand (Fig. 7-1).

Every refinery today has a chief engineer and in most cases a staff. One of their activities is to prepare estimates for new work within the plant as well as for replacements. A great deal of information that goes into the making of these estimates comes from the mechanical department. This is true even though the department may not actually do the work; it does supply the man-hour requirements for the trades involved. For this reason there must be a close relationship between the engineering and maintenance groups.

As the greater part of all work originates in the operating department, the mechanical department should be advised of any contemplated work as soon as possible. Where materials or equipment are involved, it may require weeks or even months for procurement. Usually, the first step to take here is to submit a work order; if the work involved is complicated, a meeting should be arranged. This meeting should include the mechanical superintendent, the operating superintendent, the chief engineer, and the plant manager or his assistant. Other personnel should be included as necessary. The meeting should establish:

1. The real need for the repair or installation
2. The urgency to get it done
3. When it should be done
4. Whether or not it can be worked into a regularly scheduled shutdown (assuming a shutdown is required to do the work)
5. If safety of operation is involved (see Sec. 9)
6. In an emergency installation calling for an unscheduled shutdown, how it will affect the operating schedule
7. The manpower involved (can the plant men handle the work or will an outside contractor be needed?)
8. What the job will cost
9. What the material or equipment delivery schedules are

These are some of the vital points that must be considered and which clearly indicate that there must be a close relationship between the mechanical department and the operating personnel, the engineers, the operating budget and scheduling group, management, inspection, and the safety department.

In order to keep maintenance costs at a reasonable figure the department should operate with a minimum basic crew, which means operating with a backlog of maintenance work. Right here the question arises: How large can the backlog afford to be? If the list continues to grow, some job-completion dates get pushed further and further away, which could lead to one or more undesirable situations: Unsafe operating conditions could develop, the cost of doing the job could increase, or morale could be affected by employees feeling that "we can't get anything fixed around here." This is where the mechanical superintendent is really on the spot. He cannot afford to have too many men on his payroll, and still he is expected to take care of all mechanical work, including emergencies, without letting the plant "go to pot." This is not an easy assignment, to which every mechanical superintendent will agree.

To operate at maximum mechanical efficiency all manpower should be scheduled. But no sooner is this done than an unscheduled emergency shutdown occurs, bad weather conditions come along, or some unexpected breakdown develops. With a tight schedule any of these events can completely disrupt a schedule if they require many man-hours. Actually, there is no real solution to the problem.

Past records are a great asset to a mechanical group and should be made available. Records that should be maintained accurately and up-to-date at all times include:

1. Inspection records, showing corrosion rates, retiring limits, and anticipated replacement periods
2. Manpower available, according to crafts

3. Overtime records
4. Cost records for performing various operations
5. Preventive-maintenance records
6. Master unit shutdown schedule
7. Backlog list
8. List of work orders, with dates
9. Monthly cost of operating the department, broken down to show labor, materials, and overhead
10. Maintenance costs, by units
11. Quarterly and annual cost reports for the department

All the above are tools that serve as guides in determining and in improving the effectiveness of the department.

Communications

Little does it matter how well maintenance is organized or how good the relationship is with other refinery groups or how well scheduling of manpower is carried out or how good the records might be.   Without communications the whole system breaks down.

Meetings are a good means of communicating, provided meeting memorandums are written immediately after a meeting and copies sent to everyone present and others who might be interested.   It is far better to have too many copies in circulation rather than too few.   The objection to this idea, of course, is that not everyone has time (or will take time) to read all that is circulated, but he at least cannot say, "I was not properly informed."

Meetings can be and often are carried to extremes, to the point that manpower is not wisely utilized.   There are other ways of communicating, which will be covered later.   The main point to remember here is that there must be a good communication system to make certain that the right people are being informed promptly, accurately, and in some detail.

It can be truthfully said that maintenance costs play an important part in refinery operating costs.   Both the maintenance and operating departments must continually review their operating procedures to hold costs to a minimum if the cost of making a gallon of gasoline is to be competitive.   This, naturally, is the prime objective of every plant manager.

SHOPS AND EQUIPMENT

Location

Selecting the best location for refinery maintenance shops within a plant requires a great deal of thought.   The items to consider include:

1. Proximity to operating units
2. Location with respect to plant access
3. Accessibility to railroad sidings
4. Free area around the shops
5. Location on primary or secondary roads
6. Distance from petroleum storage facilities

Although there may be others, these points must be considered for a safe and efficient operation.

1. Proximity to Operating Units.   There is always traffic between operating units and maintenance shops, and especially so at shutdown periods.   To reduce travel time it is advantageous to have the shops located near the units.   This may at first appear as a minor item, but a few minutes added to each trip will result in many man-hours at the end of the year.   In looking at this point alone, perhaps the best location would be to have the units surrounding the shops.   But such a location may be difficult as we consider other points mentioned above.

2. Location with Respect to Plant Access. Since many outside deliveries are made directly to the shops, it is advisable to have the shops close to the plant entrance. For safety reasons outside trucks should be kept away from plant operations if at all possible. This would include operating units, storage facilities, and rail and truck loading and unloading racks. Plant personnel are trained and are familiar with the operations; they know what to do in an emergency. This is not true with outsiders, and when abnormal conditions develop, only experienced personnel should be involved. Furthermore, truck traffic in case of a fire should be limited to plant equipment.

3. Accessibility to Railroad Sidings. Shops should be located on a good railroad siding for convenience in handling materials and equipment which come to the plant by rail. The siding and shops location should be such as not to interfere with truck traffic and personnel safety. Thought should be given to means used in switching cars—whether plant or outside locomotive power is used and its routing through the plant. If switch engines must pass plant operations, they must do so at safe distances. Even under these conditions safety rules must be established and strictly adhered to.

4. Free Area around the Shops. As all types of automotive and power equipment move in and around the shops buildings, ample parking and maneuvering space must be provided. A roadway, wide enough for passing, should encircle the building or buildings, with parking areas specified. As much moving-type equipment becomes larger year by year, it is wise to have a good safety factor for this outside working area. A common mistake is the failure to look far enough into the future and predict what conditions may be 3, 5, or 10 years hence.

5. Location on Primary or Secondary Roads. As long as there is a choice of locating shops on a primary or a secondary road, it is far better to have them on a secondary road. Since traffic is usually heavy on primary roads, the location of shops on a secondary will tend to relieve traffic conditions during rush periods.

6. Distance from Petroleum Storage Facilities. Maintenance shops are loaded with sources of ignition, and for this reason they should be located at good safety distances from all petroleum products and/or vapors. The subject of what constitutes "good safety distances" is discussed in Sec. 9. But as a word of caution here, it might be well to review some refinery fire reports and note the great distances that petroleum vapors sometimes travel to pick up a source of ignition.

Layout

The layout of the interior of the shops buildings must likewise be given much consideration.[6] The storage of pipe should be near the pipe-threading or cutting operations. Steel plate and structural members should be near fabrication and welding. In other words, the sequence of operations should move along a well-planned pattern. Nothing will disrupt operations within a shop more than the unnecessary movement of materials (Fig. 7-2).

Ample working space should be provided around machines for free movement of materials and personnel. This is particularly true for pumps, tubular units, compressors, and small-size vessels. This equipment is often removed from foundations and taken to the shop for repairs and/or maintenance. Adequate spacing makes it possible for mobile equipment to enter the area and remove these items. The unit layout should provide roadways for the mobile equipment.

The alignment of pumps as shown in Fig. 7-3 and locating them along a roadway provide for easy maintenance of pumping units. Tubular units should likewise be lined up along a roadway so that tube bundles can be handled by traveling cranes or mobile units. An example of this layout is shown in Fig. 7-4. All working areas must be well-lighted. Both these mean not only better working conditions but also greater safety of operation.

As refineries were built years ago, the mechanical crafts moved into buildings that might have been located anywhere within the refinery limits. Some buildings might have already existed which were revamped and were used for the various crafts. Then as refineries increased in capacity over the years (and most of them have), these facilities housing the crafts were merely added to in order to meet the needs. This resulted in a very inefficient operation in many cases. Materials-handling facilities

FIG. 7-2. The sequence of operations in a modern machine shop moves along a well-planned pattern without unnecessary movement of materials. (*Courtesy of Tidewater Oil Co., Delaware City, Del.*)

FIG. 7-3. Aligning pumps along a roadway provides for easy maintenance. (*Courtesy of Tidewater Oil Co., Delaware City, Del.*)

were inadequate, maintenance work was done in different locations, and much time was lost in chasing work and materials from one location to another.

Management has seen the great need for centralization, preferably under one roof, and during the past 10 years or so a number of modern shops buildings have been constructed. Many of these are air conditioned. Quite a number have eliminated partitions between the operations of the different crafts, which permits both uninterrupted travel of overhead cranes and better observation of the entire shop from any location. This design is also better suited for over-all shop lighting.

FIG. 7-4. Tubular units may be located along a roadway so that bundles can be handled by traveling cranes or mobile units. (*Courtesy of Tidewater Oil Co., Delaware City, Del.*)

In addition to having overhead cranes, it is quite practical to have a number of small jib cranes, pivoted from the crane columns, for moving lighter loads. Fork-lift trucks are advantageous in moving parts and materials that are palletized. Several of these motorized trucks may be useful in a large shops building.

It may be concluded that a well-laid-out central shops building, with plenty of working space, good lighting, comfortable working conditions, and an easy flow of materials, will result in higher efficiency. It is only reasonable to expect better workmanship, which is desirable when dealing with modern equipment demanding precision work. The greater efficiency could very well lead to a lower manpower requirement or a shorter turnaround period, either of which is most desirable.

Shop Expediter

Much of the work that passes through a refinery shop will require work done by more than one craft, meaning that work must be moved from one location within the shop to another. A shop expediter is usually assigned to coordinate the flow of such work. In a common shop, this operation becomes relatively simple. Furthermore, if work is being followed by an expediter, it becomes much easier for him to follow the job than to have to chase from one part of the refinery to another.

Effective coordination, which is necessary to keep the work moving through the central shop at all times, requires cooperation with the shop expediter, the individual shop foremen, zone supervisors, and field foremen. The shop expediter should be notified of all special or unusual requirements which may arise so that the most efficient control of the work may be had at all times.

The shop expediter should have the following over-all responsibilities:

1. Upon receipt of a job in a shop he expedites and controls the flow of the job among shop areas. He keeps informed at all times as to the status of every job moving through more than one shop.

2. He orders necessary materials from the warehouse and requests their delivery at the proper time.

3. He receives all materials coming into the shop and dispatches them to the proper craft areas.

4. Upon shop completion, he arranges for and expedites the movement of the job to the proper field site. Automotive equipment should be made available to him for greater efficiency.

Qualified Mechanics

Working with equipment in a modern refinery requires well-trained and qualified mechanics who possess interest, accuracy, patience, and skill. It takes years of experience and training along with these qualities to make a good mechanic.

Oftentimes the building of a new unit involving a different process will bring in new and different types of equipment that call for special training. The men must be of a type to be able to take this training and to make effective use of it.

Many plants are today conducting training or refresher courses for their mechanics, which is an excellent way to keep the older men on their toes and to train younger personnel for the future. Companies supplying equipment have been extremely cooperative in furnishing well-qualified men to assist with these courses. They put valuable information and data into the hands of those who need it rather than into the files in the head office, where it often gathers dust and does little good. Manufacturers are willing to offer this service, feeling that it is a direct means of communicating with those in intimate contact with their equipment. If they can get only one point across that will prevent future trouble or an emergency shutdown, the time spent is well worth while. More and more companies are realizing this fact and are resorting to plant training courses for mechanics.

Mobile Maintenance Equipment

An amazing amount of mobile equipment is used in maintaining a modern refinery. All this equipment, regardless of cost, must be justified by management before it is purchased. The justification is usually found in the claim that its use will result in greater safety for personnel, reduce unit downtime, or save manpower. In many cases, all three of these items are true.

Today you will find in a plant all kinds of standard and special trucks, standard and specially designed cranes (Fig. 7-5), lowboys, and other special motorized equipment. One reason for all this equipment is the need for moving so much material and equipment within a plant from one place to another. There are hundreds of heat exchangers in an average-sized refinery that are moved from place to place, pumps are often taken to the shops for repairs, heater tubes are taken from storage for heater installations, vessels are moved from one location to another, and so on.

Truck cranes and crane cars have been considerably improved in recent years. Today's equipment has far more maneuverability, greater capacity, and much longer reach than in years past. Thus, the trend in a refinery is to provide more of the mobile handling units and cut down on built-in facilities for materials handling. Before laying out a unit or a new plant, engineers should determine what mobile equipment will be needed for maintenance and then make the necessary provisions. These include clearances around permanent installations as well as overhead clearances. Ample working spaces around equipment for handling and storage should not

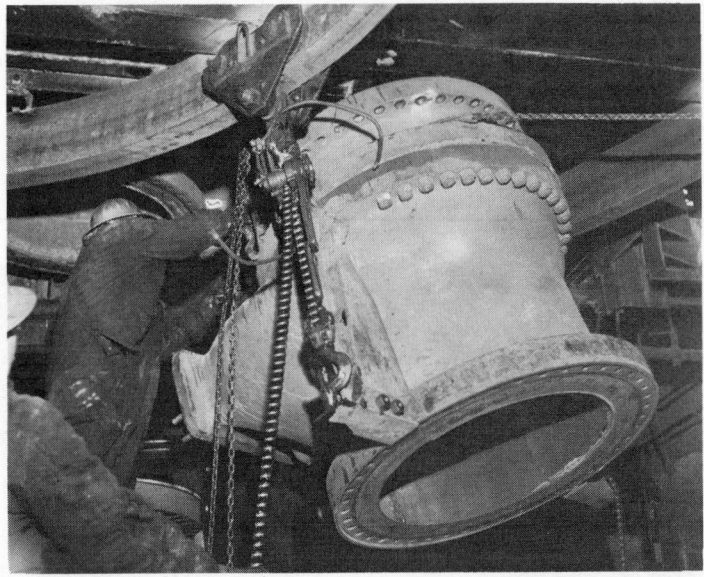

FIG. 7-5. Specially designed handling equipment, such as this trolley for handling a 34-in. catalyst slide valve, can give greater safety and reduce unit downtime. (*Petroleum Processing, August, 1947.*)

be overlooked. A wise layout engineer will anticipate future designs for mobile equipment and take these into consideration. This is not always done, with the result that maintenance suffers in the future.

In an average plant there are tons of refuse or waste materials that must be gathered and disposed of daily. In the past trash and other wastes were collected in cans that were picked up daily by trucks. These cans were handled by manpower, requiring much hard labor and resulting in a costly operation. Today, refuse is usually gathered in large metal containers having capacities up to 15 cu yd. These are picked up by special trucks requiring only one man for operation and no lifting whatsoever on his part. An empty container is taken to a location and left in position while the full one is taken away for disposal. By this means tons and tons of waste material can be moved by one man in the course of a day. Although this type of truck and container was initially used for the moving of waste materials, they are today used for moving

FIG. 7-6. Refinery refuse can be collected in large metal containers having capacities up to 15 cu yd. Special trucks require only one man to remove and empty the containers. (*Courtesy of Dempster Dumpster.*)

materials of all kinds. The portable containers are designed for dumping with a tipping action, while others have a drop-bottom design. This type of equipment is shown in Fig. 7-6. Either of these designs may be used for loading all kinds of materials, new or used, into trucks or gondola cars.

Within the course of a year an average-size refinery will move hundreds of tubular bundles, pumps, and drivers and thousands of pieces of lumber, pipe, and tubing from one place to another. Equipment manufacturers have designed straddle-type mobile units that do an excellent job in moving such items. Being designed with rather high undercarriage clearances, they can be driven right over the material to be moved until they straddle it. The material is then lifted from its supports and transported to

Fig. 7-7. Straddle carriers can be used to move heavy castings and other parts of refinery equipment. (*Courtesy of Clark Equipment Co.*)

another location. As these units are built in different sizes and capacities, they have become a very useful piece of equipment in the materials-handling field. Many materials that are palletized can be easily moved with a one-man operation. Heavy castings and all parts of equipment can be moved with straddle trucks as shown in Fig. 7-7. Most refinery pipe and tube sizes can be moved easily from storage to the point of usage or from one location to another also by straddle trucks. This eliminates much fixed materials-handling equipment and saves a vast amount of manpower. These types of mobile units are used extensively in refinery maintenance and construction work.

STORES

Inventory and Records

The amount of inventory that should be kept in stores has always been a big question, and it probably will always be a problem.

It is most important to have maintenance spare parts and materials available when needed and at the same time to keep the inventory investment within a reasonable figure and not have the shelves loaded with a surplus. This is not an easy task, and any variation in either direction from the ideal position can be costly. A stores manager does, however, have some rules and pointers that can help:

1. Review usage periodically and develop minimum and maximum levels for each item.

2. Keep the storeroom safeguarded by assigning attendants and/or by locked enclosures with a strict control of parts and materials withdrawn.

3. Segregate and correctly identify parts and materials purchased for special jobs.

4. Maintain complete and up-to-date stores records.

5. Make a study of past records and try to establish a working pattern.

6. Compare the size of inventory with that of other plants of like size, taking into account the type of equipment in each case.

7. Develop a good communications system between stores and other departments that place orders.

8. Have a good follow-up system on orders and establish delivery dates.

9. Post stores withdrawals daily.

It is most essential that stores be operated on an efficient basis. Stores should be staffed with well-trained and qualified personnel and with a large enough staff to give prompt service when service is needed. Having workmen standing in line waiting for service can become a costly item and should be avoided.

Much thought should be given in laying out a storehouse. Parts and materials should be classified, correctly identified in a legible manner, and kept systematically. Looking for and sorting out materials when needed can be time-consuming and expensive. It is not a simple matter to develop an efficient stores system, and guidance and council should be obtained from various sources.

When parts and materials are issued by stores, the costs are charged to a work-order number covering the job and the total costs posted on the work order. If tabulating equipment is available, and it should be in an average plant, inventory controls and material job-cost information can be readily determined with a minimum of clerical help.

Some work has been done by a few companies to determine scientifically the size of an inventory that should be maintained. These results are based on certain facts that can be developed by any plant, such as time between placing an order and delivery date and the demand for the item in question.[13] With this information a scientific approach is possible by the use of high-speed calculators in determining the quantity to order and the time for reordering.

Storage

The storage of materials, both inside and outside the warehouse, should follow a systematic order. Inside storage should be so designed as to minimize walking, particularly on fast-moving items. Service can be greatly hampered if the stores attendants have to walk great distances to obtain popular items.

Likewise, the storage of materials outside should follow a well-thought-out plan. These items include pipe, heater tubes, steel plate, structural steel, and the like. The storage of these should be adjacent to a railroad siding and close to the stores office. Racks should be built above grade for this storage and arranged so as to be accessible for crane servicing and truck loading, without congestion of traffic. This latter point is important, especially at a turnaround period. Roads in the area should be readily passable the year around.

UNIT TURNAROUNDS

Scheduling

There are two general types of refinery schedules: operating schedules, which cover the length of runs of operating units, and maintenance schedules, which cover the work of the mechanical crafts. Both types must be carefully coordinated for an efficient refinery operation.

It is common practice to have operating schedules developed for 1 to 2 years in advance. These indicate the shutdown periods of each of the operating units. Such schedules are based on marketing demands as well as seasonal conditions. As weather greatly affects maintenance work, no major shutdown is scheduled for the winter months (in the case of a northern refinery) or when there is likely to be inclement weather. Maintenance crews cannot operate efficiently under bad weather conditions, which consequently prolongs the shutdown time. This results in higher maintenance costs and loss of valuable production. Both situations should be avoided.

Manpower should be scheduled for all maintenance, including capital improvements, maintenance and repair work required throughout the plant, and work needed for scheduled unit shutdowns. The requirements are based on:

1. Work orders received from the operating group
2. Work as recommended by the inspection engineers
3. Capital changes and improvements
4. Regular cleaning of the units
5. Work that might be required to make a change or tie-in at some future date
6. A backlog that might have occurred at a previous shutdown

In studying the above, the mechanical superintendent can readily visualize a great deal of work that can be done prior to the shutdown date. All work that can be done during this pre-shutdown period, such as building scaffolds and getting maintenance equipment lined up and in place, will usually tend to reduce the downtime. Here, we should toss out a warning that this early work, although most important, must be done with full knowledge that the unit is still operating and that hazards often exist. Extreme care must be exercised for safety reasons (see Sec. 9).

After the manpower requirements of the various crafts have been determined for the entire shutdown, there is a good possibility that some crafts may be found to be under-manned while others are overstaffed. This condition could be resolved by shifting men from one assignment to another if this can be done wisely, or perhaps outside help will have to be obtained. This can be accomplished by hiring additional men on a temporary basis or, better still, by contracting out some phase of the work.

Such contracting is being done more and more on turnaround work. The main advantage of this procedure is that it does not disrupt an existing organization. But to carry out this plan, early scheduling of work is a "must" to get the contractor lined up to fit into the schedule of the plant. Note that mention was made of contracting "some phase of the work." This detail is most important, so that the contractor's work is entirely separated from the refinery men's work to remove any possibility of confusion. Of course, all contracted work should be spelled out in great detail in the "scope of work" description. Contract maintenance is discussed in detail beginning on page 7-32.

The scheduling of manpower in a refinery is not an easy task, but an attempt should be made, and after wide experience a fairly good job can be done. The better the job, the better the payout.

Preplanning Shutdowns

The timing on preplanning a scheduled shutdown of an operating unit is most important and varies with the unit size. But detailed planning of the mechanical work ahead of time is essential for an efficient operation. This planning should begin from a few weeks to perhaps a year or longer in advance of the shutdown date.

Lubricating the threads of large-size bolts of flanges, such as tubular units (Fig. 7-8), several days prior to a shutdown saves time in removing nuts. Penetrating oil ensures faster work, thus reducing downtime to a minimum. Cumulative effects of thorough planning and preliminary preparation save thousands of dollars of potential profits.

Lubrication of threads is only one of the preparatory steps. Others include erection of scaffolding, tagging valves for removal or repacking, attaching blanks to lines where they will be used, moving equipment and tools to the job site, prefabricating replacements, and so forth (Fig. 7-9).

If it is a normal turnaround where only cleaning and minor repairs are needed, the planning period may be very short. Factors that call for a much longer period include:

1. A general inspection
2. Preventive maintenance
3. Mechanical maintenance work inside vessels, such as installing liners or replacing fractionating trays
4. Replacing equipment
5. Installing additional equipment

While a unit is operating, both the mechanical and operating departments are preparing lists of work to be done at the next shutdown. At the same time, the preventive-maintenance group is listing work that can be done only at a shutdown.

FIG. 7-8. Lubricating the threads of large-size bolts of flanges several days prior to a shutdown will save time in removing the nuts. (*Petroleum Processing, August, 1947.*)

The list developed by the operating department should be supplemented by individual work orders, and these in turn should be discussed with the mechanical departments perhaps 2 months ahead of the shutdown. The reason for this long lead factor is to allow sufficient time for planning the work. In many cases materials or parts must be ordered, and the job might involve some prefabrication. Each job must be reviewed in detail to determine what is involved and the approximate manpower needed to complete it. This in itself is sometimes a real assignment that requires a great deal of experience. But it is much better to know what is needed ahead of time rather than after the unit is off-stream.

After the complete list has been compiled, the exact manpower requirements by crafts are established, which will determine the number of shifts and/or overtime required. Then a detailed work schedule is prepared listing each job separately and the number of men required in each craft (Fig. 7-10). From this, the length of the shutdown can be determined.

When any of the five factors listed above as calling for more advance planning are involved in a shutdown, the preplanning becomes much more involved and the lead-time period increases.   Where a general inspection is scheduled, the chief metal inspector enters the picture in a big way.   He will spend days reviewing his records to determine the degree or extent of inspection necessary for each piece of equipment, the vessels, and the piping system.   For a sizable unit this list is quite lengthy and represents many man-hours of review and preparation of recommendations which should be discussed with management.   All this work should be done well in advance of the scheduled shutdown to allow time for following through on recommendations.

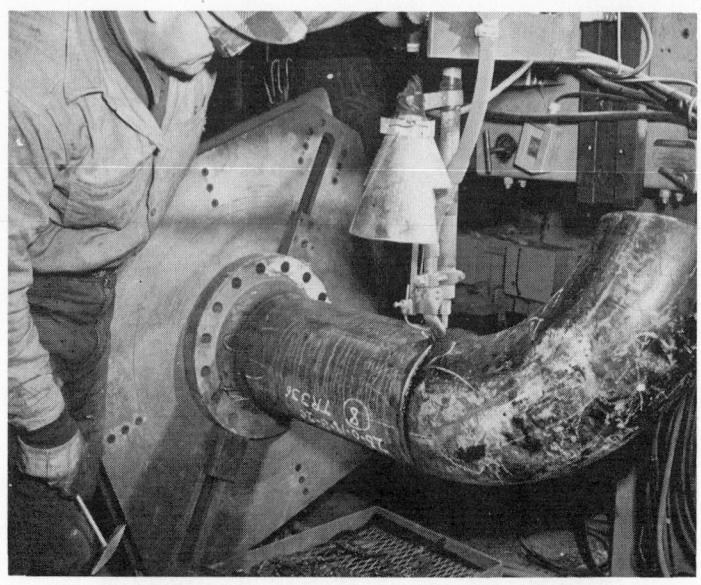

Fig. 7-9. Prefabricated replacements, such as piping sections, save time in returning shut-down units to operation.   (*Courtesy Tidewater Oil Co., Delaware City, Del.*)

Much preventive-maintenance work must be done during the shutdown period.   A review of the records by the preventive-maintenance group should be made well in advance of the shutdown to give the mechanical superintendent an indication of what work is required and to schedule the crafts accordingly.

During this preplanning period there should be meetings scheduled for all foremen to review the scope of the work.   One individual who should attend all these pre-shutdown meetings is the safety engineer.   Too often this man is forgotten by the operating and mechanical departments, while at the same time management fails to include him as a part of plant management.

Most plants have operating procedures which have been prepared with safety in mind.   In areas where procedures have not been prepared, it is all the more important that the safety engineer be included to review the steps to be taken from a safety standpoint.   Safety records clearly show that the failure to follow a definite procedure or the taking of a short cut often results in accidents.

It is considered good practice to review operating procedures prior to a unit shutdown.   As the length of unit runs has increased materially in recent years (to as much as 2 years or longer on some units), it is quite possible that unit operators might forget some of the shutdown and startup procedures from one shutdown to the next.   Furthermore, with these long runs it is quite possible that one or more new operators may appear and to them the shutdown may be their first one.   This point must be recognized by the foreman so that procedures can be reviewed with all concerned.

Crafts and work description	Men req'd	Hours req'd	Total man-hours	Man-hours avail.	Add. MHs req'd
Pipefitters:					
Normal routine maintenance...............	60	40	2,400		
Connecting new accumulator..............	8	20	160		
Running 100' 6'' pipe....................	6	8	48		
Total................................	...	...	2,608	2,500	108
Boilermakers:					
Pull and clean 10 exchanger bundles........	6	20	120		
Retube two bundles......................	4	40	160		
Repair one head.........................	2	16	32		
Total................................	...	...	312	260	52
Machinists:					
Replace one pump impeller................	3	8	24		
Install new pump bearings................	2	6	12		
Overhaul furnace fan.....................	4	20	80		
Total................................	...	...	116	96	20
Carpenters:					
Build 3 scaffolds........................	4	14	56		
Build concrete form......................	2	8	16		
Total................................	...	...	72	72	0
Insulators:					
Normal insulation repair..................	4	30	120		
Insulate 100' 6'' line.....................	2	8	16		
Total................................	...	...	136	140	0
Electricians:					
Normal light and circuit checking..........	4	20	80		
Install 2 new lights......................	2	16	32		
Total................................	...	...	112	100	12
Instrument men:					
Usual instrument maintenance.............	4	18	72		
Install new flowmeter.....................	2	8	16		
Replace flow recorder....................	2	9	18		
Total................................	...	...	106	110	0
Welders:					
Weld 100' 6'' pipe.......................	4	6	24		
Misc. welding...........................	3	10	30		
Total................................	...	...	54	60	0

FIG. 7-10. Typical work schedule, crude unit.

Also, closer supervision must be exercised in all such cases. Quite frequently accidents can be traced to the inexperienced or improperly supervised operators.

Chemical Cleaning

The cleaning of refinery equipment in the sizes we have today is no small item. However, chemical cleaning is used to some extent in certain equipment and where the deposits are affected by the cleaning agents. Under these conditions, the cost of cleaning with chemicals is usually less than by other methods, as is the time required.

Table 7-1. Typical Deposits Encountered in Refinery Heat Exchangers in Which Chemical Cleaning Can Be Effective

Type of unit	Deposit	Remarks
Crude distillation, bottoms exchangers, shell side (feed)	$\frac{1}{16}$-in. black, oily material, composed mainly of iron oxide, ferric oxide, organics, a few phosphates and silicates, and much iron sulfide	This is typical of residual deposit encountered in crude distillation and topping processes or where crude has not yet been subjected to cracking temperatures
Crude distillation, gasoline, naphtha, and kerosene overhead condensers, shell side	Flake deposit, consisting of iron oxide and sulfide, with little organic	Typical fractionating-column overhead products formed in crude distillation, cracking, and vapor-recovery operations. Sometimes cupric sulfide and iron disulfide will be found in these.
Overhead and side-product condensers and coolers, tube side (water)	Dead algae from cooling tower, iron oxide, oil from contamination, carbonates, silica or silicates, phosphates, etc. from plant cooling-water-treatment chemicals	Just about any combination of these can be expected on the water side

Table 7-2. Typical Chemical Cleaning Jobs—Condensers, Coolers, Exchangers

Type of unit	Purpose of cleaning	Method	Results and remarks
Catalytic cracker, gasoline condenser	To remove water deposits from tube side, composed of 22 % organic, 50 % iron oxide, 10 % phosphates, 12 % silicates, 6 % calcium sulfate	1. Hot alkaline-detergent solvent to remove organic and disintegrate silicate and sulfate 2. Heated inhibited HCl to dissolve oxides and phosphates	*Before After* Gasoline temperature to storage, °F.......... 125 90
Debutanizer overhead, condensers in vapor-recovery plant	To remove water deposits from tube side while units are in operation; deposit composed of 15 % organic, 10 % silica, 3 % sulfates, heavy iron oxides	1. Alkaline-detergent solvent injected to remove algae and mud 2. Concentrated inhibited HCl injected into water inlet header	*Before After* Transfer coefficient... 40.8 58.5 Water rate, gpm...... 229 343 Reflux temperature, °F 152 101 Reflux drum pressure, psi................. 92 70
Bottoms to crude feed exchangers and pancake coils in crude distillation unit	To remove deposits from tube side, composed of 48 % organic (oil), 35 % crude salts, remainder iron oxide and iron sulfide	1. Degrease with hot kerosine emulsion 2. Heated alkaline-oxidation solvent to disintegrate partially carbonized fraction 3. Heated inhibited HCl to remove iron oxide and sulfide	Because of design features of pancake coils, no other method was possible. Effect of treatment was an increased throughput of 5,000 B/D
Griscolm-Russell sections in heavy naphtha cooling service	To remove product deposits from tube side, composed of 10 % organic, 20 % iron sulfide, heavy cupric sulfide	1. Hot alkaline-oxidizing solvent to dissolve copper compound 2. Heated inhibited HCl with reducing agent to remove oxidized compounds and iron sulfide	Very satisfactory results. Previous approach had been to pass superheated steam through tubes and then shock with water spray on outside. This was severe on units, causing distortion and warping*

*Loose material also caused operating troubles later. These troubles were avoided by successful chemical cleaning methods.

Some refineries have been using this cleaning method for years and have concluded that it definitely has a place in refinery maintenance.

One rather wide application is in cleaning deposits from shell-and-tube-type equipment.[19] On the water side of this type of equipment will be found carbonates, silicates, oxides, phosphates, and organic matter (algae and bacteria). On the product side will be found oil, partially carbonized oil, crude salts, sulfides, oxides, and residues from special processes. These latter may include:

1. Hydrofluoric acid sludge (HF alkylation)
2. Antimony trichloride and aluminum chlorides (butane isomerization)
3. Ferro-ferricyanide (vapor recovery systems)
4. Sulfuric acid sludge (H_2SO_4 alkylation)
5. Catalyst fines (catalytic cracking)

Through the years experience has shown that chemical cleaning has been effective in certain heat exchangers, as shown in Table 7-1. Table 7-2 shows typical cleaning jobs on condensers, coolers, and exchangers, while Table 7-3 indicates typical cleaning,

Table 7-3. Typical Chemical Cleaning Jobs—Columns

Type of service	Purpose of cleaning	Method	Results and remarks
Gasoline tower, 4 × 51 ft	One tray plugged. Thin black flakes 1 mm thick, composed of 23 % oil and organic, 55 % iron oxide, 22 % iron sulfide	1. Oil and organic fractions removed with a hot alkaline-oxidizing solution 2. Heated inhibited HCl solvent	Trays, caps, and downcomers clean
Propane fractionating tower, 3 × 86 ft, 50 bubble cap trays	Brown, crumbly deposit up to ¼ in. thick, composed of iron oxide and sulfide	1. Cascade-heated, inhibited HCl over trays, while draining to waste 2. Close bottom drain valve and fill; soak, intermittent agitation with gas	Inspection showed unit to be clean
Absorber column, 3 × 60 ft	Paraffin deposits around caps causing high velocity and carry-over	Alkaline solvent with wetting agent, heated by injecting steam	Paraffin removed, resulting in normal operating conditions
Propane rectifying column, 5 × 100 ft	Cyanide deposits up to 3 in. thick on trays	Alkaline solvent with wetting agent, heated by injecting steam	Deposits completely removed; before treatment tower was removing very little propane
Vapor recovery plant consisting of 8 × 52 ft still, 5 × 56 ft absorber, 4 × 29 ft stripper	Oil, iron oxide and sulfide, and free sulfur	1. Alkaline boilout with wetting agent 2. Heated inhibited HCl soak	Stripper and absorber columns clean. Patches of carbonized oil remaining on top tray of still
Treater scrubber column packed with Raschig rings	Iron sulfide and caustic residues	Heated inhibited HCl	Pressure drop and capacity restored to normal

jobs on columns. Table 7-4 lists the typical refinery processes that use chemical cleaning.

From these four tables it can readily be seen that chemical cleaning has a rather wide application in refinery maintenance. But here is a good place to offer a word of caution. The use of many common solvents may be restricted when aluminum- and zinc-coated tubes are being dealt with. Also, there is evidence of chlorides causing intergranular corrosion when hydrochloric acid is used. However, it is important to note that the use of the acid per se causes the corrosion. When hydrochloric acid is used on stainless steel, the equipment must be thoroughly flushed.

Inhibited dilute sulfuric, sulfamic, and phosphoric acids are used by one service company on stainless-steel-equipped units whose design is such that they cannot be completely flushed. In some cases where the entire system, including piping, is stainless (no carbon steel), this company has used 20 to 40 per cent nitric because of the passivating effect of the nitric acid on stainless steel.

Aluminum-coated tubes can be cleaned with sulfuric, phosphoric, and chromic acids. Zinc-coated tubes are confined to the use of chromic acid. Special care must be taken to preserve the coating. Some progress has been made in the use of weak alkaline solvents on aluminum and zinc.

Table 7-4. Typical Refinery Process Equipment Subject to Chemical Cleaning

Process or operation	Equipment
Reforming	Waste-heat boilers, gasoline condensers, compressors
Alkylation	Fractionating towers, refrigeration equipment
Isomerization	HCl stripper, overhead condensers, accumulators, vent-gas scrubbers, catalyst removal column, deisobutanizer columns, deethanizer columns
Polymerization	Catalyst warmer banks, product condensers
Catalytic cracking	Surface condensers, gasoline condensers, waste-heat boilers, slurry exchangers, hot-oil pump gland system, gas-oil exchangers
Crude distillation	Naphtha coolers, pancake coils, crude feed exchangers
Thermal cracking	Gasoline condensers, crude exchangers
Stabilization and vapor recovery	Stabilizers, compressors, absorbers, flare and gas-supply lines to heater burners
Alkymer units	Tanks, lines
Sulfuric acid plant	Acid contact towers, waste-heat boilers
Toluene unit	Rundown lines, towers, process lines, exchangers
Heavy oils	Wax machines, open-box condensers, ammonia generators, sweating ovens, contact clay filters, jet condensers, caustic scrubbers
Utility department	Economizers, water lines, boilers, boiler feed lines, filters
Gas treating	Caustic tower
Miscellaneous	Propane dewaxing stripper, propane stripper condensers, jet condensers

Free-machining (SAE 1115) bolts or similar material are not resistant to inhibited acids. Such bolts, running 0.16 to 0.23 per cent sulfur, are attacked by inhibited acids. But steel bolts with low sulfur content, about 0.02 to 0.06 per cent, are resistant. There is also some evidence that 4140 alloys used in pressure-vessel bolts are subject to failure, although these are low in sulfur.

Consideration must be given to the kind of gasket materials present before deciding to use chemical cleaning.

Flushing after cleaning is also a very important consideration for satisfactory results.

Sandblast Cleaning

There are a number of instances where sandblasting is used extensively as a means of cleaning. This is particularly true in the case of tubular-type bundles, tower trays, pump casings, and many other items.

Although chemical cleaning has a definite place in cleaning tubular-type units in position, the wet sandblast method is used where chemicals fail to do the job satisfactorily. One company that specializes in making such equipment is the Hydro-Blast Corporation of Chicago. The equipment consists of a blast room made of structural and steel plate and large enough to accommodate the equipment to be cleaned. At each end of the room there are swinging doors with an access door provided in the wall on the side opposite the operator. Each door has a warning light mounted over it which is automatically switched on when the high-pressure water pumps are put in operation. Observation ports are provided along the walls to give the operator good vision of the work. A typical unit is shown in Fig. 7-11.

The interior of the blast-room floor is made of subway-type grating supported by structural members strong enough to support safely a fork-lift truck and its load. Illumination for the room is supplied by strong marine-type floodlights equipped with impact- and heat-resistant lenses.

Inside the blast room are located high-pressure water nozzles mounted in a mechanism that is hydraulically operated for nozzle movement. This affords an easy means for changing the angle of the nozzle. The gun, containing the nozzle, is mounted on rails located along the inside of the room on the operator's side. Horizontal and vertical movements of the gun are effected by electric motors. All movements of the nozzle and carriage are controlled by the operator from a movable control box mounted on rails along the outside of the wall.

Special sand hoses lead from the sand hopper to the guns, and the sand is sucked in a water slurry through these hoses into the stream by venturi action of the nozzle.

Fig. 7-11. Wet sandblast methods are used to clean where chemicals fail to do the job. (*Courtesy of Hydro-Blast Corp., Chicago.*)

After striking the equipment being cleaned, the sand and water mixture, together with removed rust and scale, falls through the floor grating and is washed to a basket screen, which removes the larger particles. The sand, water, and entrained fines are washed through the screen to a sump, from which they are pumped back to the sand hopper to complete the cycle.

In many cases the outsides of tube bundles are cleaned on the site after being removed from their shells. If the deposits are fairly soft, a high-pressure water stream or a combination of steam and water will often do a good cleaning job. Many hydrocarbon deposits may be softened by soaking the entire bundle in vats containing hot chemical solutions. Solutions containing Oakite or other chemical cleaning agents are often quite effective in softening deposits to where high-pressure water streams produce excellent results.

Where chemical cleaning solutions are not used, and where there are open passages between the rows of tubes, a high-pressure water stream containing some sand will often clean satisfactorily. This method is far more effective if the tubes are placed on square pitch. If the bundles are rotated, the high-pressure streams can be directed through the passageways, resulting in a good cleaning action (Fig. 7-12).

For effective cleaning with high-pressure streams inside the tubes, there must be an opening. This is especially true if sand is used in the stream. Without an opening, the wet sand will merely add to the plugging and the effort is worthless. An opening is often created by drilling through the tube first and then following with the high-pressure hose method.

Sand must keep on the move for any cleaning action to result. For this reason it is much easier to clean the outside of a tube bundle than the inside. With the small openings inside the tubes the tendency to plugging is much greater.

Where tubular-type bundles are being cleaned, it is much better to have the tubes located on square pitch rather than on triangular pitch. The square pitch provides for better passageways for the sand to pass through. Although tubes on square pitch will sometimes require larger bundles and consequently will cost more initially, they may be less expensive in the long run when cleaning is considered.

FIG. 7-12. High-pressure water streams containing some sand often will give good cleaning action. (*Petroleum Processing, February, 1951.*)

It should be added here that cleaning should always be considered in the original design. Engineers should not be governed by the initial cost figure of a tubular unit but should consider all the design features to determine if some of the cost-reducing features might lead to more costly maintenance. It is always wise to have the maintenance engineers in the plant review the initial design from a maintenance standpoint.

Table 7-5 illustrates the type of materials found on tube bundles and the results obtained by cleaning with wet sandblasting.

Table 7-5. Results Obtained by Cleaning Tube Bundles with High-pressure Wet Sandblasting

Type of deposit	Shell-side results	Tube-side results
1. Carbonaceous:		
a. Soft, buttery..........	Complete removal	Complete removal
b. Oily, granular..........	Complete removal	Complete removal
c. Tarry.................	Ineffective*	Ineffective†
d. Thin, powdery..........	Complete removal	
2. Water deposits:		
a. Silt..................	Complete removal	Complete removal
b. Algae.................	Complete removal	Complete removal
c. Iron oxide.............	Complete removal	Ineffective‡

* Alternate method: chemical soaking plus wet sandblasting.
† Alternate method: chemical soaking plus drilling.
‡ Alternate method: chemical soaking.

Inspection

Soon after operating pressures and temperatures rose above atmospheric, refiners recognized the need for inspection of equipment. This need was further verified by an increase in the number and severity of fires. As we all know, fires can be extremely costly. But above all, where lives are involved, fires must be kept to a bare minimum.

One who has lived through the early days of refinery inspection many times has heard the claim that "inspections are costly." True, perhaps, but lives cannot be replaced, equipment repair costs money, and loss of production can be a really sizable item.   Management soon became convinced of these facts, and today every refinery has an inspection group or is covered by inspection in one way or another.

What type of individual makes a good inspector? He should be curious, mechanically inclined, and full of ambition and, above all, have no fear of getting dirty.   An inspector's job is the dirtiest job in a refinery if he does it well.   Although inspection is dirty work, it does have its rewards.   It is one of the best ways to become familiar with all refinery equipment, and it affords an excellent opportunity to learn something about operation.

An inspector should be found in his office only during the report-making period. In other words, one trait of a good inspector is that he will be out with the equipment to become familiar with it and to understand its operation.   Seldom should a day pass without his spending a portion of his time in the plant.   One good reason for this is that all piping systems within the operating areas should be identified, preferably by a line sketch.   This means that each piece of pipe, every valve and fitting should carry some identifying number.   Piping changes are constantly being made, and records will soon become obsolete and inaccurate if the inspector does not stay on top of these changes.

Also worth mentioning is the fact that many alloys are used today in piping systems, and the ideal way to identify them is by color coding.

To point up the need for an inspector keeping up with piping changes and for having the piping clearly identified, let us assume that in a certain line the valves and fittings are of an alloy material.   Let us further assume that for some reason one of the valves was not readily identified as being alloy and that this valve developed a leaking condition.   The operator writes a work order to have it replaced, and the replacement valve happens to be carbon steel.   The inspector is unaware of this change.   His records still show the valve as being alloy, and he schedules inspections accordingly. But the carbon steel, being less resistant to corrosion, develops a high corrosion rate and an eventual failure, resulting in a fire, an emergency shutdown, and loss of production.   If the loss of production cannot be recovered, the net effect is a loss in earnings. All this points up the need for the inspector to be "on his toes" at all times and to have a well-developed system worked out.

Each year more and more inspection work is done by engineers.   A technically trained inspector is in a much better position to exercise sound judgment in determining the extent and type of inspection required for each unit and each piece of equipment.   He is better able to plan a shutdown and organize his work and carry it out in an able and efficient manner.   His training is better for calculating metal thicknesses and retiring limits.   Much of his work is related to codes of various kinds, so his position is contingent on his understanding and interpretation of them.   For unfired pressure vessels we have the *ASME Code*, as well as some individual state codes.   The piping code most generally used throughout the industry is ASA B 31.3, *Petroleum Refinery Piping*, published by the American Society of Mechanical Engineers. In addition, the American Petroleum Institute has published a number of *Recommended Practices* for use and guidance over the years, and more are added each year.   In the case of inspection, the API *Guide for Inspection of Refinery Equipment* includes:

Chapter I.   Introduction
Chapter II.   Conditions Causing Deterioration or Failures
Chapter III.   General Preliminary and Preparatory Work
Chapter IV.   Inspection Tools
Chapter V.   Preparation of Equipment for Safe Entry and Work

It is a known fact that corrosion rates increase as velocities become higher. The higher velocities tend to keep the surfaces clean of any deposits, thereby permitting corrosive materials to attack the surfaces. This condition shows up in such places as 90° bends and in return bends on heaters. More and more corrosion-resistant materials are being used in these areas and in many cases result in a good payout. It costs money to break flanges for inspection purposes only, and in many cases an alloy material will eliminate this expense. Of course, sonic-type instruments are quite extensively used for determining metal thickness.

One all-important job of an inspection group is the keeping of records—accurate and up-to-date. Plant management is not interested in wading through a lengthy and detailed inspection report on a unit 3 months after the startup. But management is vitally interested in a brief and well-planned résumé of the findings as soon as the facts are assembled. Inspectors should keep a running record of their measurements and findings at each inspection period. The longer the record, the more valuable it becomes, assuming that operating conditions have not drastically changed.

An inspection record in which plant management is always interested is for pressure vessels. Such vessels represent a large part of an operating unit, and the failure of any one may readily shut down the entire unit. If a sudden replacement is necessary, it could mean a lengthy period of no production that could run into thousands of dollars. As vessels, and especially pressure vessels, play such an important part in refinery operations, it is most essential that thorough and detailed inspections be made and that complete, accurate, and up-to-date records be maintained. A typical pressure-vessel record may be found in Fig. 7-13 in which all pertinent data relating to the vessel are tabulated. The following Fig. 7-14 indicates the general dimensions of the vessel as well as a running record of the shell thicknesses.

A mere glance at these tabulations will show the corrosion rate, if any, and indicates a direct comparison of the present thicknesses with the original dimensions. These data are of immense value to management.

Another typical record sheet showing pertinent vessel data for all vessels in a unit is covered in Fig. 7-15, which includes valuable data, not only to the inspection group, but to operating management as well.

Reviewing and analyzing these continuous records prior to a scheduled shutdown are "musts." These records are invaluable for preparing recommendations for the mechanical department. The review should be made far enough in advance of the shutdown so that plans can be made by the mechanical group for disassembly or replacement. These records also establish a means for determining corrosion rates and hence dates for retirement. Determining the rate of corrosion will set the basis for frequency of inspections and the extent of inspection necessary. This is particularly true today, as operating units have become so large that piping systems represent a real mass of piping measuring miles in length. Every flange that need not be broken results in a saving of time and money.

PRESSURE VESSEL RECORD

Name of Unit _____ Vessel Name _____

Battery No. _____ Original Item No._____

_____ Works Date _____

HISTORY

Estimate No. _____ Date Received _____
Order No. _____ Date Installed _____
Mfgd. By _____ Company No. _____
Mfgrs. Serial No. _____ Company Inspector _____
Mfgrs. Inspector _____
Mfgrs. Test Press._____

DESCRIPTION

Drawing No.
  Fabricators _____
  Contractors _____
  Company _____
Position (Vertical or Horizontal) _____
Code Constructed_____
  Code_____ Year _____
  Code Stamp _____
Material Specified and Grade or Type
  Base_____
  Lining_____
  Thickness_____
Stress Relieved (Original)_____
Radiographed (Original)_____
  Complete _____
  Weld Intersections _____
Size—Nominal ID_____
  Length Base Line to Base Line_____
Design—Pressure, psi_____
  Temperature, F_____
  Stress, psi _____
Maximum Allowable Operating Pressure, psi_____
  Temperature, F_____
  Limited by_____
Shell
  Type of Construction _____
  Joint Efficiency_____
  Type of Support _____
  Interior or Exterior Stiffeners_____
  Original Thickness_____
  Corrosion Allowance_____
Manways
  No. _____
  Size _____
  Flange Rating _____
  Reinforcement
    Factory or Field _____

Top Head
  Type: Elliptical_____Hemispherical_____
    Dished—Cr. R. _____Kn. R. _____
    Conical—Angle _____
    Flat _____
  Joint Efficiency _____
  Original Thickness_____
  Corrosion Allowance _____
  Manways
    No. _____
    Size _____
    Flange Rating _____
    Reinforcement
      Factory or Field _____
Bottom Head
  Type: Elliptical_____Hemispherical_____
    Dished—Cr. R. _____Kn. R. _____
    Conical—Angle _____
    Flat _____
  Joint Efficiency _____
  Original Thickness_____
  Corrosion Allowance _____
  Manways
    No. _____
    Size _____
    Flange Rating _____
    Reinforcement
      Factory or Field _____
Nozzles
  Minimum Flange Rating_____
  Type Facing _____
  Openings Reinforced _____
Remarks _____
_____
_____
_____

NOTE: A copy of this sheet shall be prepared for each individual vessel in a unit. If new vessels are installed or any changes are made to present vessels affecting "Description" items, a new or revised copy of this sheet shall be submitted with the current inspection report.

FIG. 7-13. Typical form for a permanent pressure-vessel record. (*Courtesy of American Petroleum Institute.*)

Inspection Manual

Every refinery inspector should prepare an inspection outline or procedure. Some call it an inspection guide or an inspection manual. The reference in this writing will be to a manual.

It is not easy to prepare such a tool for the inspectors. To do the job well requires much experience as well as a great deal of time and effort, not only to prepare it but to keep it up to date. But with the amount of equipment found in an average refinery and the inspectors needed to do an efficient job, a good, complete manual is a necessity.

VESSEL INSPECTION SHEET

SHELL DWGS.	LINING DWGS.	INTERNAL DWGS.	EST. NO.
FABR.			
CONTR.			
CO.			

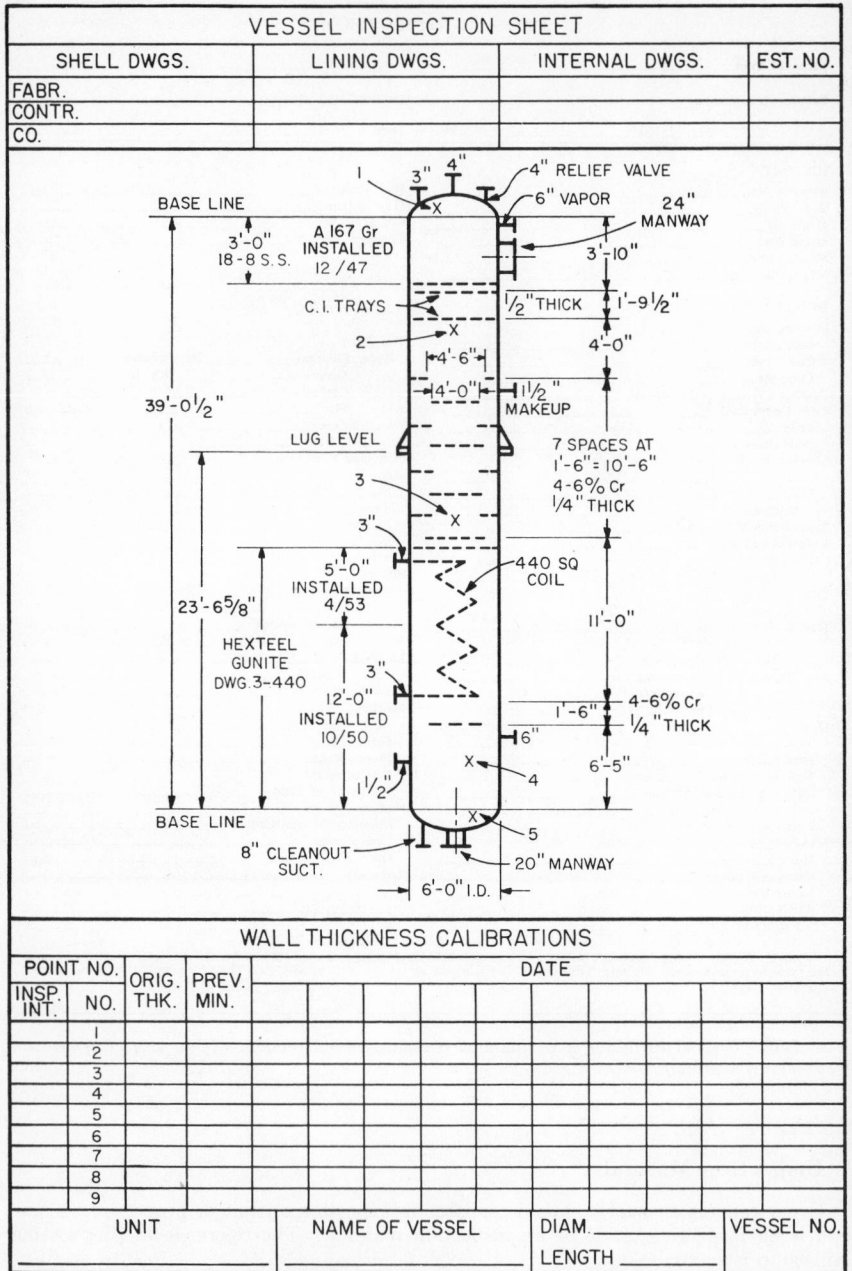

WALL THICKNESS CALIBRATIONS

POINT NO.		ORIG. THK.	PREV. MIN.	DATE							
INSP. INT.	NO.										
	1										
	2										
	3										
	4										
	5										
	6										
	7										
	8										
	9										

UNIT	NAME OF VESSEL	DIAM. _____ LENGTH _____	VESSEL NO.

FIG. 7-14. Typical vessel-inspection sheet for recording thickness measurements in order to calculate corrosion rates. (*Courtesy of American Petroleum Institute.*)

PRESSURE VESSEL RECORD

Name of Unit_____   Battery No._____

_____Works   Inspection and Test No._____   Date_____

NAME OF VESSEL				
COMPANY VESSEL AND SKETCH NO.				
OPERATING DATA				
Hours Under Pressure to Date............				
SERVICE DATA				
Average Maximum Operating Pressure, psi..				
Average Maximum Operating Temperature, F				
Top				
Bottom				
No. Trays				
No. Baffles				
No. Coils				
INSPECTION AND TEST DATA				
Inspector				
Nominal ID				
Minimum Thickness				
Shell				
Location				
Top Head				
Bottom Head				
Joint Efficiency—Shell				
Heads				
Head Factor—Top				
Bottom				
Last Inspection				
Inside Welds				
Outside Welds				
Under Insulation				
Extent Inspection				
Test Pressure				
Vessel, psi				
Coils, psi				
Test Medium				
Time Pressure Held—(1)+(2)..........				
Maximum Allowable Operating Pressure, psi.				
Maximum Allowable Operating Temperature,				
F				
Limited by				
Working Stress, psi—Operating Temperature				
Approved Operating Pressure, psi.........				
Approved Operating Temperature, F......				
Safety Valve Setting, psi...............				
PROTECTIVE LINING DATA				
Drawing No.				
Date Installed				
Material and Type....................				
Section Lined				
Date Repaired				
Extent of Repairs..................				
Date Previous Lining Removed.........				
Cause of Removal.........				
REMARKS				

Fig. 7-15. Suggested form for recording the physical condition and allowable operating conditions of all pressure vessels on an operating unit. (*Courtesy of American Petroleum Institute.*)

A well-written manual serves as a "bible" for the inspector. It points out the need for inspection, the equipment to be covered by inspection, the extent of inspection, the data expected, and forms for recording the data. Some forms may be used interchangeably, while others have to be special and apply only to a particular unit. This leads to stacks of forms that must be prepared and maintained in a current status.

A complete manual enables an inspector to do a better and more thorough job of planning the inspection work. It assists him in determining the locations which may require special preparation, tools, or inspection equipment. He can also better predict the areas wherein the work load is heaviest so he can concentrate the inspection efforts on these potential trouble spots. All of this tends to reduce the shutdown period.

The success of any inspection program is dependent not only on the extent of the preliminary and preparatory work but on how the actual inspection is conducted, the inspection data obtained, and the use made of the collected data. The manual should cover this all-important last point. Data as gathered have no value unless they are put to some beneficial use, with a follow-up system. Like many other situations, details can easily get buried in the course of busy weeks and months, and some important item may be completely forgotten. It is the responsibility of the inspector to keep the ball rolling and make certain that his procedure will not permit such to occur.

On-stream Inspection

Every effort should be made to do as much planning and inspection work as possible prior to the shutdown. Some inspection work can always be accomplished while the units are on-stream or during a short process turnaround, and this is an area where an inspector can usually accomplish a great deal to reduce work after the unit comes off-stream. A thorough review of the process flow and the operation may reveal that the total inspection of a unit is unnecessary at any one shutdown.

A successful inspection program hinges on the work done in planning inspections, knowledge of the deterioration of the equipment and the conditions causing the change, and the ability to work closely with others in the organization. The inspector must have a good understanding of the process operations and a comprehensive knowledge of the equipment. He must be able to locate and identify the form of deterioration within the equipment. Oftentimes samples of products and deposits may be obtained and an analysis made while the unit is still operating.

The preparation of new inspection forms or revision of old forms should be done during operation. With constant piping changes, piping inspection sketches are continually being revised. A revised sketch will often affect the continuous records, which likewise should be changed. Here is a listing of some things that an inspection can do while the units are operating:

1. Prepare report from previous inspection.
2. Get forms prepared or revised.
3. Keep up with changes made by the mechanical department.
4. Study flow diagrams and process operations.
5. Have samples taken and analyzed for possible corrosive conditions.
6. Keep management informed on probable replacements.
7. Review continuous inspection records.
8. Review current publications.
9. Check inspection tools for workability and accuracy.
10. Take thickness measurements on piping systems, vessels, and other equipment where such can be done safely. (This item alone is a great timesaver.)
11. Handle all correspondence.
12. Get all necessary forms lined up for the next inspection.

Tools for Inspection

To carry out inspection work as it should be done requires a good assortment of inspection tools. Some of these tools are on the market and can be purchased, while others are special home-made gadgets used for one purpose or another. An ingenious inspector will have a good assortment of these special tools at his disposal. The same type of inspector will dream up easier and better ways of carrying out his work effectively. As an example, the inspection of the inside of an all-steel heater stack required the building of a scaffold inside, from bottom to top. This alone was time-consuming and costly. Today, many plants have mobile cranes with tall booms to which can be attached a cable with a bosun's chair (Fig. 7-16). An inspector can ride up and down inside (or outside) a stack or a vessel in the chair and conduct whatever inspection is necessary. To communicate between the inspector and the crane operator, earphones and chest microphones are used quite satisfactorily.

One of the most modern and up-to-date techniques used in refinery inspection today is the ultrasonics method. It is being used quite extensively in obtaining metal thicknesses in equipment, vessels, and piping systems. The portable resonance type is most generally used for convenience and practicability. Most refinery inspection groups use this method for determining thicknesses of metals, which is essential in establishing corrosion rates and retirement dates and in gathering data for developing

FIG. 7-16. Many plants have mobile cranes with tall booms which can be attached to a cable with a bosun's chair for inspections. (*Petroleum Processing, August, 1947.*)

shutdown schedules. This method of inspection is much preferred to the older scheme where test holes were drilled and piping systems had to be dismantled for actual measurement purposes. With this advanced technique many more readings can be taken in less time, with the result that the inspector is able to do a more thorough job of inspection.

In order to use this ultrasonic method of inspection effectively, special procedures are required as well as a thorough knowledge of the principles involved. The operator must have a background of experience to get accurate and reliable results.

Inspection tools are covered in detail in Chap. IV of the API *Guide for Inspection of Refinery Equipment.*[24]

PREVENTIVE MAINTENANCE

The term "preventive maintenance" is common in most oil refineries today. It may be defined as maintaining a plant in such condition as to prevent an emergency situation in the case of operating equipment or to avoid excessive expenditures at one time in the area of nonoperating equipment. This method of increased assurance of continuous and better operation began in most plants several years ago. Since then, the degree of coverage has greatly increased, and in many cases, complete plants have adopted this program of stopping trouble before it goes too far.

In order to have a successful preventive-maintenance program, there should be:[18]

1. A mutual understanding between maintenance and operation
2. Good maintenance records
3. Good workmanship on the part of maintenance mechanics
4. An adequate inspection program
5. A good corrective maintenance program
6. Good administration of the over-all program

The mechanism of operating a preventive-maintenance program is vital to its success and requires a man above average ability to do the job well.

In our discussion here we shall think of this program applying only to equipment on operating units, and not to buildings, tankage, and the like. So unit operators play an important part in the program by making regular and periodic inspections of their equipment. They should be constantly looking for unusual or abnormal conditions. The preventive-maintenance engineer is responsible for follow-up, analysis, and record keeping. On large units having critical equipment, maintenance might have its own men making these checkups. The line organization is expected to contribute heavily to the program to make it worth while.

When a new plant is started up or new installations are dealt with, weaknesses usually develop early and the outstanding trouble spots are easily recognized. As these are corrected, a routine approach designed to identify the less obvious weaknesses becomes increasingly important. The routine approach to corrective maintenance involves identifying the critical component in each mechanical repair and carefully recording it. When probable trouble spots are identified, a more detailed investigation should be made, and in many cases some change will bring the equipment into better balance.

Evaluating Preventive Maintenance

No single denominator has been devised for evaluating maintenance within a refinery. One reason for this lies in the fact that there are so many interrelations between individual pieces of equipment and major groupings of equipment. There are, however, a number of indicators used for controlling and evaluating the preventive-maintenance function. Statistics used for measurement purposes include:

1. Ratio of emergency (or breakdown) maintenance to total maintenance
2. Ratio of overtime hours to total hours
3. Per cent compliance with daily work schedule
4. Ratio between actual production and scheduled capacity
5. Maintenance man-hours per unit of production
6. Frequency of maintenance jobs, broken down to units
7. Ratio of preventive maintenance to total maintenance work

There is a linear relationship among most of these factors. When one variable goes up or down, expect its associated variables to follow the same pattern. When one or more of the variables do not follow this pattern over an appreciable length of time, take it as an indication of a change—either good or bad. Most of these factors are relevant to the problem of control when considered individually. When each of the factors is recorded in a systematic manner and analyzed regularly, it is a simple matter to spot trends. Then examine these trends critically, but do not place too much weight on them unless they are continued over a reasonable length of time or reflected in associated variables.

Maintenance management's ability to interpret control statistics should continue to improve as the program continues. This in turn brings about more reliable forecasting and allows the transfer of a larger portion of mechanical work to preventive maintenance.

Rotating Equipment

One class of equipment that nets the highest return on preventive maintenance is rotating equipment, particularly as the speeds of such equipment have greatly increased over the years. This includes such equipment as centrifugal pumps, centrifugal compressors, and turbines. In this class of equipment alone, there are many factors upon which satisfactory operation depends, including lubrication, packing, vibration, and associated piping.

Lubrication

All moving parts must have adequate lubrication at all times. However, over-lubrication can result in difficulties as well as underlubrication. Likewise, lubrication pressure is an important item, as well as lubrication temperature. A high pressure on the system could mean too small lubricating lines, the build-up of carbon deposits within the piping, or the deposition of foreign matter in the piping. Any of these could lead to the starvation of oil to the bearings and eventually bearing failure.

A low oil pressure could mean an oil leak, a loose bearing, or an inaccurate pressure gage. Such items as these would, of course, be listed on an inspection card and would be checked periodically by either an operator or an individual from the mechanical department.

An abnormally high temperature could be caused by improper cooling, resulting in high bearing temperatures that could lead to failures. Reading and recording bearing oil temperatures periodically and at regular intervals are most important. If a temperature begins to rise, the thermometer or temperature instrument should be checked at once for accuracy.

In all lubricating-oil systems, the oil should be changed at regular intervals for successful operation. The equipment manufacturer will cover this point in his instruction manual, and it is very unwise to lengthen these change periods over his recommendations.

In gas engines, the combustion gases may contain moisture and sulfur compounds which, when combined, can cause corrosion of the internal parts by the formation of an acid. In like manner, corrosive compounds may get into the lubrication system through the air breather vent and contaminate the lubricant. Furthermore, particles of dust, dirt, oxides, ash, acid fumes, and the like are also contaminants that can cause trouble to the cylinder walls, pistons, rings, and valves. Any of these abrasive materials will cause excessive wear to moving parts. It is a good idea to test the lubricant in a system periodically for acidity and contaminants. If atmospheric conditions are bad, testing should be done as often as experience indicates is necessary.

If operating temperatures are above normal, there is always the possibility of a lube-oil breakdown. If oxidation takes place to any great extent, a sticky, varnish-like material is formed which combines with other foreign materials and really causes trouble. Any sludge build-up in a system can impair circulation, causing lube oil to fail as a coolant. Also, failure of the lube-oil pump to circulate properly can result in bearing failures. Therefore, lube-oil operating temperatures must be maintained near normal.

Packing

The manufacturer of equipment requiring packing will recommend the correct kind and type of packing in his manual. It is always considered good practice to follow his recommendations for the best results. The mechanical department and particularly those in preventive maintenance should see that the correct kind, type, and size of packing is available in stores. The substitution of some other kind or type is seldom satisfactory.

Proper tension on the packing should be checked periodically. A packing that is too tight can cause a high bearing temperature, and one that is to loose can cause leakage. Any leakage will result in a loss of product and may also create a safety problem. Both these are important points to remember. A continuous loss of product over a period of time can amount to some rather astounding figures. On the other hand, as safety is high on most managers' lists, any leakage should be corrected at once.

Vibration

As operating speeds have increased greatly over the years, the need for good, well-designed foundations has become apparent. There is very little comparison in the design of a foundation for a piece of rotating equipment operating at 1,750 rpm and that operating at 10,000 rpm. A well-designed and -constructed foundation for these higher speeds is a must. Most essential, over and above a good foundation, is the tightness of the anchor bolts. This important item should be given constant attention in the maintenance program.

If vibration becomes apparent on the permanently installed tachometer or by periodically checking with a portable instrument, the tachometer should be checked for accuracy. One that is used continuously on a permanently mounted installation often loses its accuracy.

Any excessive vibration created within a piece of rotating equipment can carry through into the piping system and perhaps into other equipment.

Piping Supports

Too much emphasis cannot be placed on the need for properly supported and correctly braced piping systems. Often piping that is not adequately supported will exert excessive strains on equipment. If this equipment happens to be a pump, compressor, or turbine, these abnormal stresses may be great enough to distort the piece of equipment, causing misalignment and eventual breakdown.

Once a piping system is correctly designed and installed, it does not necessarily follow that the system will remain in that same state. Many times settlement of supports or shifting of supporting members will result in a bad condition. Any piping system connected to rotating equipment must be given constant attention to maintain proper supports and bracing with a minimum amount of vibration.

The *Petroleum Refinery Piping Code*, ASA B 31.3, 1959, Part 5, covers expansion, flexibility, structural attachments, supports, and restraints.[26] This code can be of real assistance in avoiding potential trouble.

Turbine Safety

Some very serious and fatal accidents have occurred in recent years because of improper maintenance practices on steam turbines. Most steam turbines have overspeed shutdown safety devices to prevent excessive speeds that could lead to turbine failures. These safety devices are well designed and will function properly if correctly maintained, but many companies overlook this maintenance item until accidents occur because of malfunctioning.

The atmosphere in an oil refinery is filled with varying degrees of vapors and foreign matter, some of which are corrosive in nature. Any slight corrosion or deposition of matter on moving parts can readily make the parts inoperative. This is what often happens in the working parts of the safety shutdown mechanism. As a precaution, most preventive maintenance as practiced in plants today will cover this safety item in their program. All that is required is a regular and faithful manual operation of the safety devices to make certain they will operate properly if and when called on. These checking intervals may have to be as often as daily, as actual experience dictates. Some plant managers were alarmed to learn that these overspeed trips would not function a month after being checked. But here again, as previously mentioned, atmospheric conditions can make a great difference. Each plant must work out its own frequency schedule.

Procedures

There are several basic procedures that should be followed in setting up a workable preventive-maintenance program in any plant. These include:

1. All machinery and equipment should be numbered for identification.
2. Equipment record cards should be compiled.
3. Routine inspection and oiling routes should be determined and clearly defined on master cards, as:

    a. Equipment to be oiled, type of lubricant, and number and type of fittings.

    b. Inspection requirements—intense, medium, or minor.

    c. Amount of minor repairs performed at the time.

    d. Periodic trips—daily, weekly, monthy, etc.

    e. Time required for each routine trip.

4. Equipment record cards for certain types of equipment should be kept in an open file where attention is required. Each card should indicate the date for attention.
5. Maintenance employees should be supplied with master work sheets on which completed jobs should be indicated. These individuals should also itemize requests for repair work as noted during the inspection. Work orders should then be issued.
6. The control plan reveals repair costs of all jobs. Actual costs should be posted on the equipment cards so that repairs by type of equipment are known, thus enabling the engineer to determine if routine inspection and oiling are necessary, if major repairs are required, or if it would be more economical to replace the equipment.

At the beginning of a preventive-maintenance program the details and clerical work should not be excessive and the groups should be well qualified to perform the necessary functions. Only the most critical equipment should be included in the program in the beginning, with additional equipment added as conditions warrant. Critical equipment would include rotating equipment and especially that operating at relatively high speeds. Other equipment falling in this category would be equipment where high maintenance costs exist, such as a reciprocating pump or compressor.

Records

No preventive-maintenance records are of value unless they are properly maintained and correctly analyzed by qualified personnel and remedial actions are taken promptly where necessary. The records should indicate the established schedules, and these schedules must be adhered to in every detail (Fig. 7-17). These accurate and detailed records should be studied, and oftentimes by so doing, better and more economical schedules can be attained. Perhaps daily schedules may be lengthened to weekly schedules. Conversely, it may be found that monthly schedules are too long and must be shortened. Inspections should not be made and records maintained unless necessary. Making an inspection just for the sake of making one adds nothing except to the cost of operation.

With accurate and up-to-date records many major troubles can be predicted ahead of time. This leads to a possible scheduled shutdown at a suitable time, thus avoiding conflicts with other major work. Furthermore, indications of troubles discovered in their early stages can in many cases prevent their becoming major problems. Too much cannot be said about the importance of keeping records and having them studied constantly by trained people capable of detecting troubles before they occur.

Plant Practices

Before a preventive-maintenance program is developed, it must be realized that any successful plan requires qualified people who will take an interest in the work. It must also be recognized that clerical help is needed to maintain neat and up-to-the-minute records. The records must be studied constantly, and unusual findings reported promptly to the right people. Then comes the necessary action that should have management's backing. At times flaws are found in one of these steps with the result that the program has lost its effectiveness and the return on the investment has been decreased.

An excellent payout practice is the reporting of abnormal situations to the plant management.   With detailed and accurate costs indicated on the cards it is easy to detect where high maintenance costs exist, and if these costs continue at a high level over a period of time, management should know about it.   It could be that the piece of equipment in question is not suitable for the service, it may not be operated correctly, or it may have exceeded its service life and should be replaced.   It could be that the wrong equipment was purchased initially.   Many factors can enter into the picture, but these can be singled out once it is known where excessive costs exist.

EQUIPMENT	GAS COMPRESSOR		REFINERY			
MAKE-TYPE			LOCATION			
PLANT RECORD NO.			SERIAL NO.			
SIZE	CAPACITY		SERVICE			
SPEED	HP		NSPH			
STAGES	SUCTION PRES.		DISCHARGE PRES.			
DRIVER			GEAR REDUCER PR NO.			
DATE INSTALLED			ESTIMATE NO.			

MAINTENANCE RECORD

DATE	JOB NO.	DESCRIPTION		COSTS			
			MHs	LABOR	MATL.	TOTAL	
	1310	INSPECTED INTERIOR AND REPLACED					
		ONE BEARING					
	1480	CHANGED LUBRICATING OIL					
	1502	CHECKED VIBRATION					

FIG. 7-17. Typical record card for preventive maintenance.

Another weakness in plant practices is the failure to carry through on the established schedule.   Operating and preventive-maintenance schedules should be worked out together so that the necessary preventive-maintenance work can be done on a regularly scheduled unit shutdown.   However, product demand conditions may make it more advisable to continue operating the unit for a longer period or a quick shutdown may be scheduled, in which case time will not permit the necessary maintenance work. In either case, the plan suffers.   The result could then be an emergency shutdown, which is usually more costly.

Preventive maintenance is just a part, although a vital part, of the over-all maintenance program.   A good preventive-maintenance program goes right along with good planning and scheduling, an adequate work-order system, and, of course, good management.

CONTRACT MAINTENANCE

As refinery units have increased in size over the years, it has become more and more difficult to maintain a mechanical force large enough the year around to service the units properly at turnaround periods.   The peak manpower requirement at the shutdown of a large unit is tremendous.   Building up the various crafts to handle this peak condition, however, then results in the problem of finding work for them when all units are operating.   As the lengths of runs increase, which they have, it has become more of a problem.

One approach to the problem is to hire from the outside during these peak periods. This is not a wise procedure, and very few companies consider it. It is not only expensive from the standpoint of hiring, but the company becomes involved in the area of benefit plans, which alone creates another problem.

A very sound solution, which many companies are using, is contract maintenance. Stated very simply, contract maintenance is the practice of contracting with an outside firm to provide craft labor to handle part or all of the maintenance manpower needs of a plant. These maintenance men are a part of the outside organization, the contractor, similar to workers on a construction job. They are subject to plant supervisory and administrative control.

Kinds of Contract Maintenance

Generally speaking, four kinds of contract maintenance services will be considered here:[14] (1) labor supply, (2) periodic turnaround job, (3) permanent or continuing supplemental maintenance force, and (4) entire maintenance management. There are advantages and disadvantages in each.

1. Labor Supply. In this case, the contractor merely serves as a labor broker, supplying needed manpower at specified times. This manpower may include engineers, craft labor, watchmen, or any other kind.

With this type of contract maintenance, however, there is no contract supervision, as the company merely adds the supplemental force to the existing plant staff. Although the required amount of manpower may be on the job, the working relationship among the existing force, the new men, and supervision is not always the best, with the result that the desired results may not be attained. The contract broker feels no responsibility for any labor trouble or difficulty that may develop, even to the point of having a picket line placed around the plant. Experience indicates that this arrangement does not work well.

2. Periodic Turnaround Job. Here the contractor comes into a plant with a crew, including supervisers, at turnaround or peak periods of labor demand. This type is used extensively and has worked satisfactorily where the outside force and the company's crafts are separated and work in different and well-defined areas.

Here the responsibility can be delegated and at the same time tied down. The supervisors have direct control over their own men, and the men know that they must produce or find themselves without a job. Also, if men are not qualified to do the necessary work, they can easily be replaced. The contractor, at the same time, knows that he is being watched and judged by others surrounding him, so it behooves him to do his best.

Sometimes the question is asked if the highly skilled outside mechanics do not cost the company more than its own men. While the actual hourly rates for a maintenance contractor in accordance with construction trades practices are higher, the total man-hour costs for contract labor are lower because of the fringe benefits which must be provided for every man on the plant payroll. The contractor requires only the minimum necessary facilities for the use of his forces and will have no employee benefits except those required by law and by the terms of the construction contracts. There are also savings by virtue of having a flexible force.

3. Permanent or Continuing Supplemental Maintenance Force. A number of plants today are going to this form of contract maintenance. Among the **advantages** are these:[11]

The contractor's permanent working force becomes familiar with the plant's facilities, organization, and working rules.

The supervision is able to work out a good relationship with the company's personnel.

Although contracts are drawn for a 1-year period, they may be terminated sooner if unsatisfactory situations develop.

The contractor's men develop more of a personal interest in the job and a feeling of being a part of the plant.

By having a permanent working nucleus of trades in the plant at all times, they are in a much better position to take the lead and help the supplemental manpower called in at peak periods.

This supplemental manpower is more likely to work harder to become permanent employees.

In many instances, the efficiency of the men is higher than that of the company's personnel. This often shows up at the "knock-off to wash up" periods before noon and quitting time. Some good examples have been set in this area.

The company need not buy and maintain as much maintenance equipment. The contractor may either furnish it himself or rent it on a prearranged basis.

As a contractor has only service to sell, he knows that he must maintain this service at a high level to retain his contract.

A contractor operating over a large area of the country can shift men from one area to another to meet peak demands.

When skilled trades work in different plants, they become familiar with different kinds of equipment, whereas a company's employees are limited to the equipment within its own plant.

A contractor does not keep an unqualified or a physically incapable worker.

If a company has only an average maintenance program, a good contract maintenance force coming in and working on a continuing basis may very well "show up" this average maintenance.

Maintenance costs may be obtained in a shorter time.

A maintenance superintendent may shift work to the contractor and thereby keep his work schedule up to date.

Contract men may be used in starting up a new unit when many problems usually arise and additional manpower is needed.

On the other hand, there are many **disadvantages** of this type of contract maintenance:

The plant must provide separate shop facilities for certain classes of work, such as welding.

Permanent change-house facilities must be provided.

A separate working area within the plant should be assigned.

Permanent parking facilities should be maintained.

Contractor's men may not be so safety-conscious as company personnel, although better contractors have in general developed some very good safety records.

Administrative costs would be prohibitive in a plant with small maintenance needs.

There may be a question of qualification of men in skilled trades.

Facilities must be provided for keeping and storage of tools.

There is potential difficulty with plant union members.

4. Entire Maintenance Management. When a refinery is built in a new area, thought and consideration should be given to a complete contract maintenance program. Here the contractor would be responsible for planning, coordination, and supervision of all maintenance work. Under such an arrangement the company requires only enough maintenance staff to coordinate the requirements of management and the operating staff with the contractor's representatives. All the advantages of direct management of maintenance by the company's personnel can be preserved under a complete plant maintenance contract. Other advantages include:

Experienced and capable maintenance engineering is assured.

A planning and coordination group is affiliated with the mechanical force.

There is a flexible supply of experienced maintenance specialists under a field supervisory staff familiar with plant equipment.

There is a complete staff familiar with the plant management, operating, and service personnel.

Mechanics with a high degree of special skills are available.

There is complete flexibility in the staff as well as in the work force.

Management is relieved of the responsibility of hiring and training men for each of the crafts.

Management is saved time in developing and negotiating labor contracts.

It must not be concluded that contract maintenance is the answer to all refinery maintenance problems. There are, however, applications where one type or another should be given consideration.

CONTROLLING MAINTENANCE COSTS

It is estimated that the cost of plant maintenance in all industry exceeds $14 billion annually and that approximately 5 cents of every sales dollar is now spent in this area.[2] As more and more plant processes are mechanized and higher speeds, temperatures, and pressures are introduced, maintenance expenditures will increase still more.

Management is meeting the challenge by giving increased attention to better methods of maintenance cost control. These efforts are especially significant in the refining industry, where maintenance costs represent a great portion of the total plant costs.

It is most difficult to estimate a figure, as maintenance costs vary greatly, depending on the age and condition of the operating equipment. Maintenance costs on a new unit are nominal at first but increase with time. This is a factor in deciding when replacements are necessary.

To control and reduce maintenance costs effectively, we must apply control methods that parallel those used in production operations. Since many maintenance tasks are generally nonrepetitive in nature, however, they have received less attention from industrial engineers than have the more highly repetitive direct labor tasks.

The very nature of maintenance work requires an approach different from that to direct labor. It might even be said that a different philosophy is required.

There are four fundamentals that must be incorporated in any maintenance cost-control plan:

1. You must develop a system of formalized work requests that includes the reporting of time and materials used.

2. Activities of maintenance personnel must be scheduled and planned.

3. You must develop a means for measuring the over-all effectiveness of all maintenance crews.

4. You must have a preventive-maintenance program to maintain plant and equipment at the lowest cost over a period of time.

In addition to these fundamentals, there must also be a sound and functioning maintenance organization and program, and management and maintenance supervision must fully understand the program and endorse its principles. A good program reduces maintenance costs by providing a means for management to plan work to a better advantage. It also provides a method for evaluating the working efficiency of the various crafts.

Work-order Requests

The basic requisite of a good plan is that a formal work request must be prepared by authorized personnel and approved by management. A written work-order system provides a record of requests for maintenance jobs, a record of the approval by qualified executives, a basis for accumulating time and material charges, and a means for developing cost consciousness on the part of maintenance foremen.

The work-order form should be of a convenient size and should provide for such pertinent data as:

Dates of issuance, start, and completion
Department or charge cost center
Machine or equipment number
Charge account number
Authorized approvals
Complete work description
A sample form is shown in Fig. 7-18.

Work orders may be originated by the operating department and then forwarded to the maintenance department after approval. Or requests may be telephoned to the maintenance dispatcher, who prepares and schedules individual jobs. There is no reason, of course, why some orders cannot come directly from the maintenance men, including such items as painting, housekeeping, safety, and the like.

But whatever the source may be, no work should be done before first contacting the safety man. The foreman in direct charge of the work is responsible for making this contact. The best procedure is, of course, to have the safety man's approval in

NO.	DATE		REFINERY		
ISSUED BY		ISSUED TO			
APPROVED BY		DATE			
SAFETY CHECKED BY		DATE		TIME	
WORK STARTED		TO BE COMPLETED BY			
TYPE JOB (CHECK ONE)	MAINTENANCE ___	AFE CAPITAL ___	AFE EXPENSE ___		
PRIORITY (CHECK ONE)	EMERGENCY ___	REGULAR ___	SHUTDOWN ___		
MATERIAL CODE	LABOR CODE		AFE NO.		

LOCATION OF JOB

DESCRIPTION OF WORK

SKETCH	MECHANICAL OFFICE USE		
	CRAFTS	MHs	COST
	PIPEFITTERS		
	BOILERMAKERS		
	CARPENTERS, ETC		
	COSTS		
	MAINTENANCE		
	AFE CAPITAL		
	AFE EXPENSE		
JOB COMPLETED	SIGNED		

FIG. 7-18. Typical work-order request form.

writing before any work is started. This plan not only may avoid confusion later but may prevent a serious accident. See also Sec. 9 for a further discussion of safety procedures during maintenance work.

One of the advantages of the work-order system is that it may reveal situations where uneconomical use is being made of personnel. There are cases where maintenance men were making or fabricating items that could be purchased more economically.

Work Schedules

Each day, work orders should be selected in order of priority and the following day's work should be planned for each craft. All major items should be projected on a planning sheet that shows the sequence of operations, the number of employees required, and the sequence by crafts. Job progress should be followed daily and per cent completion noted on the chart.

Jobs may be scheduled far in advance when they fit in with shutdowns or with available dates predicted on production schedules. Priorities of smaller jobs will change from day to day, but the work load should be reviewed periodically so that corrective action can be taken before employees are overloaded or idle.

By referring to planning charts, the maintenance superintendent can:

1. Keep informed on over-all progress
2. Direct his attention to jobs that are behind schedule
3. See clearly what effect rush jobs have on those in progress and which jobs can be conveniently delayed if necessary
4. Keep operating personnel informed as to job status without requiring special field checks that are normally time-consuming and often inaccurate
5. Minimize overtime and week-end assignments

Periodic meetings of maintenance department supervisors are necessary for the proper planning and scheduling of work. The listing of the major jobs on a planning board placed in the maintenance office, so that status can be quickly noted, is a valuable aid in such meetings. The maintenance superintendent should attend these meetings to observe progress and at the same time determine if jobs are being completed in a standard required time or some jobs are being dragged out. This latter case can result in excess maintenance costs. The superintendent should be in a position to observe these conditions.

True vs. Project Maintenance

Maintenance work in a refinery may be divided into two basic categories: true maintenance and project maintenance.

True maintenance is work done to keep existing facilities in good condition. It includes repairs, routine servicing, emergency jobs, housekeeping, painting, and replacements of worn-out equipment.

Project maintenance is new work done in connection with existing facilities which can be justified on a return-on-investment basis. This includes additions, modifications, rearrangements, removals and demolitions, and so forth.

Cost reductions may be possible in each of these two classifications. Three ways of accomplishing this end in the case of true maintenance are:

1. Reduce the quantity of work to attain the same objective.
2. Reduce material costs by using alternate materials without sacrificing safety. It may be that the alternate materials selected do not have the same service life but replacements may be made with a monetary saving in the long run. All such replacements must be justified on costs, taking into account replacement cost, money invested, downtime, and possible loss of production.
3. By contract maintenance, where such work can be contracted to some outside company on an agreeable basis. (See Contract Maintenance, page 7-32.)

In the case of project maintenance there are at least four good ways of reducing costs:

1. Eliminate work that cannot be economically justified.
2. Use alternate and less costly materials without reducing the safety level.
3. Make the installation in a less expensive manner. In most installations there are different ways of approach to accomplish the same end.
4. Contract the work to an outside company. There are many instances where contractors can make an installation for less money than can an oil company. This is particularly true where an appreciable difference in overhead costs exists. The size of the contracting company makes a big difference in this area.

Each maintenance job is actually an individual investment that should be justified on the basis of return. Postponing maintenance work to some future date may reduce present costs, but in the long run the over-all costs may be far greater. Painting at the right time is a good example. Another would be the installation of protective linings, where corrosion exists, before the corrosion allowance is entirely gone.

On an average, the need for true maintenance will accrue at a constant rate or combination of rates which may be related to calendar time or operating throughput. Over short periods of time, such as a month, actual expenditures will seldom parallel the pattern of average need. One reason for this is that the need for true maintenance tends to be cyclic rather than uniform, with peaks occurring when cycles coincide. Another reason is that the actual performance of required work can be and often is delayed or advanced in accordance with the availability of manpower and equipment downtime. Still another reason is the occurrence of emergencies at irregular intervals.

Cost Reports

In most cost systems today, reports are issued at the end of comparatively short "cost periods" that vary from 1 week to 90 days. The most frequently used period is 1 month. In all cases, the length of one or two periods is much too short to develop a meaningful pattern of true maintenance costs. It is considered a good idea to plot these monthly costs on a chart, and any variation will be readily called to the superintendent's attention for an investigation.

Designing for Lower Costs

An excellent place to look at reducing maintenance costs is back in the design stage. It is always easier and simpler to accept an earlier design of a unit, even to the extent of specifying the same materials of construction. This plan may be a means of saving time and money at the time but perhaps not in the long run. Each operating unit or even an individual piece of equipment should be considered by the design engineer. Some questions he should ask are:

1. Is the most modern design being considered from a maintenance standpoint?
2. Is the layout such that working crews have sufficient space for performing their work?
3. Are clearances between various pieces of equipment great enough for personnel to get around safely?
4. Have suitable roadways been provided for moving in maintenance equipment?
5. Are tubular units, pumps, compressors, and the like accessible so that parts can be removed and taken to the shops for maintenance and/or repairs?
6. Have overhead clearances been considered from a working-space and safety standpoint?
7. Has sufficient space been provided for easy access of fire-fighting equipment?
8. In the case of an operating unit, have safe distances been used between the new unit and existing units or facilities (including buildings, roadways, railroads, etc.)?
9. Have the best metals been considered for the service? Note that we say "considered," not "selected." In many instances the use of the best materials may be prohibitive from a cost standpoint, and periodic replacements of a less expensive material may be more economical in the long run. However, all practical materials and metals should be considered and their usage based on economics.

GENERAL MAINTENANCE

Painting

As labor and material costs have greatly increased over the years, the cost of surface protection has become a sizable item in refinery maintenance. In addition to labor and material, other factors which influence painting costs include surface preparation, paint quality, and film thickness.

Surface Preparation

In the past, probably too much blame for paint failures was placed on the coating material and not enough blame on surface preparation and coating application.

There are several methods of preparing a metal surface but it appears that sandblasting is still the best and may not be the most costly. There have been cases where sandblasting was well in line with other surface-preparation methods. But it goes without saying that metal surfaces must be properly prepared before any coating is applied if a normal life is expected. This means that, when competitive bids are being obtained, the job must be covered with a good set of job specifications followed by close field inspection. It is always wise to remove mill scale, particularly for outdoor exposure. This may be done by sandblasting, pickling the material before erection, or letting it weather for some 6 months during the winter (assuming that your plant is located in an area where the temperature drops below freezing). If the last method if used, the sandblasting in the spring or summer will be a much easier job. If the scale is not removed before painting, it is almost certain to pop off later, causing coating failure.

Paint Quality

As the material to be applied costs about the same as the labor for application, it is advisable to use a good time-tested material that will spread well and cover evenly. Also, the material should have good coverage properties so that when applied it will have sufficient film thickness. Some coatings fail to have this property, and hence their life is much shorter. As a basis for coating selection, refer to Table 7-6.

Table 7-6. Basis for Protective Coating Selection

Exposure	Corrosion rate	Type of protection	Film thickness, mils	Average life, years	Cost, cents per sq ft per year
Rural atmosphere....	Low	Oleoresinous coatings	4–6	6	3–4
Fume-contaminated atmosphere.......	Low	Oleoresinous coatings	6–8	4	5–7
Heavy fumes........	Low	Asphalt mastics	125	5	5–6
Heavy fumes........	Low	Oleoresinous coatings	6–8	$2\frac{1}{2}$	8–11
Splash or spillage....	Low	Asphalt mastics	125	*	*
Splash or spillage....	Low	Plastics	8–10	*	*
Splash or spillage....	Low	Neoprene	6–20	*	*
Immersion..........	High (.05 in./year)	Linings	60–250	5–10	30–60

* Varies considerably.

Film Thickness

The job specifications should indicate the film thickness required. Table 7-6 also includes some suggested thicknesses for typical installations. For oil refineries which would be classified as having fume-contaminated atmospheres, the film thickness should have a minimum of 6 mils, meaning that bare metal would probably require four coats. Of these, two should be primer coats, followed by two finish coats. Some might question the second primer coat, but it is recommended to assure a minimum of voids under the finish coats. There is too much danger of bleeding through small pinholes and causing a film failure.

Reducing Painting Costs

Much effort and research have been directed toward getting a good first-coat paint job. Thoughts have been given to designs, methods of applying, and the best kinds of coatings for the service.

A cost breakdown shows that surface preparation will vary from 10 to 15 cents per sq ft, material from 4 to 10 cents per sq ft, and labor for application from 5 to 10 cents per sq ft.[15] As material and application labor account for roughly half the total cost, some savings can result by (1) using designs to reduce the surface areas to be painted, (2) applying coatings by the hot-spray method, and (3) using coatings with a high solids content.

Reduced Surface Area. All those associated with protecting metals from corrosion are familiar with the advantages of designing to eliminate or reduce sharp edges, sharp corners, crevices, and areas practically inaccessible for painting. Another approach is to reduce the surface area to be painted. Structural engineers consider it standard practice to use tees and wide flange beams instead of back-to-back members. The surface area per lineal foot of a 4 by 5-in. tee is 1.5 sq ft, while two 3 by 3-in. angles back to back have 2 sq ft of area.

Hot-spray Paints. Another development in the field of maintenance painting is the application of coatings by the hot-spray method. Here the result is a better flow-out and consequently fewer pin holes and a better film build-up per pass. Also, better results are obtained in cooler weather.

Time studies made on an application of mastics to several storage tanks by the regular method vs. the hot-spray method showed that the application labor was reduced by 50 per cent. This was accomplished because of the higher build-up per pass and because the painter did not tire so quickly, as the hose used from the heater was only ½ in. as compared with 1 in. for normal spray application. Tests have indicated that a much higher film build-up can be obtained with the hot-spray method when spraying plastisols, phenolic-tung oil, and certain alkyd–linseed-oil-base coatings.

Another practical advantage of the hot-spray method is that surfaces may be sandblasted, hot-spray primed, and hot-spray finish-coated in a shorter time period, the chances for intercoat contamination by dust, moisture, and chemicals thus being reduced. Furthermore, with a greater film thickness there will be greater durability.

High-solids Paints. Numerous studies made on protective coatings have indicated that adequate film thickness is necessary to achieve a mechanical barrier. With most low-solids paints, this requires, by conventional application methods, four or five coats, which are not only costly but often impractical. This fact has prompted paint manufacturers to develop paints with high solids contents. Products are now available which can deposit in one coat a dry film thickness up to 10 mils. Adequate film thickness can be obtained, therefore, with only a two-coat application, thereby saving the other two coats.

Lubrication

One of the main reasons for high maintenance costs in refinery operations is improper lubrication. It accounts for many unscheduled shutdowns. Equipment bearings must be adequately lubricated if normal runs are to be expected, and the low-cost operation of machinery depends more on lubrication than on any other factor.

Lubrication in practice means reduction of friction and wear to prolong the life of equipment and at the same time to reduce the power required to operate it. All rubbing surfaces must be coated with a good lubricating film for satisfactory operation, regardless of the kind or type of bearing or moving parts.[7]

It is not sufficient merely to have lubrication; it is just as important to have the correct lubricant for the service. Often a lubricant fails under operating conditions, resulting in a bearing failure and consequently an emergency shutdown.

Most lubricating systems have filters for removing dirt, rust, and carbon particles. These filters must be cleaned and replaced periodically to assure continuous and satisfactory operation. Such filters should always be listed on the preventive-maintenance list.

On large rotating equipment the lubricating system is often an auxiliary system consisting of a reservoir for oil storage, circulating pumps, filters, etc. All this equipment requires constant attention for maximum operating service. The oil reservoir is considered a safety item and must be blanketed with an inert gas. It is deemed good practice to check the lubricating oil occasionally for foreign matter and moisture and to keep records of the findings.

Another good policy is to strain lubricant drawn from a system at the change period to detect the presence of any foreign matter or materials. Often small pieces of bearings may be found, indicating potential trouble from within. Lubricants must be changed at regular periodic intervals as covered by the manufacturer's manual on operating procedures.

Moisture

A little moisture in a lubricating system can often spell trouble. How does moisture get into a system? There are several possibilities.

1. Water may have been trapped at a low point in the piping system as a result of a hydrostatic test.

2. In winter weather the cooling water is often so cold that condensation occurs on the walls of suction passages and the cylinders of gas and air compressors. To combat this condition an oil should be specified that will properly lubricate despite condensation. Another solution is to recirculate part of the cooling water to maintain an inlet cooling temperature safely above the dew point in the suction line under the coldest condition. Throttling the cooling-water outlet should be avoided, as it is difficult to maintain the proper temperature and flow conditions.

3. Tubes in the lube cooler might leak.

4. In the case of air compressors, the location of the inlet suction line is important. Excessive moisture may be taken in from some outside source, such as an exhaust steam discharge.

Compressor and Packing Lubrication

Carbon and gum deposits are the enemies of most machinery. During break-in periods lubrication should be applied to compressor cylinders and packing in quantities to encourage both the flushing away of minute wear particles and the formation of a good surface glaze. The correct way to accomplish this is to follow the manufacturer's recommendations and use the minimum number of drops of oil per minute. To determine when this minimum has been reached, pull a valve or two and examine the cylinder walls, pistons, and rings for a uniform oil film. There should be no dry spots. Metallic packing should be examined in the same manner.

The manufacturer's instruction book usually specifies the number of drops of oil required per minute. These drops are usually listed for gravity sight feed lubricators. If sight glasses are liquid-filled, drop volume is approximately three times as great. This must be taken into account when adjusting force-feed lubricators.

Oil Specification for Compressors[7]

Viscosity characteristics also must meet the manufacturer's recommendations, which frequently differ for crankcase and cylinder. The oil used in the crankcase may be subject to severe splashing, as in a single horizontal machine where dippers and counterweights pick up oil and distribute it around the machine. Oil for this service must therefore be foam resistant. A force-pressure-feed crankcase with an oil cooler uses a lower viscosity oil than one without an oil cooler. Normally, the recommendation falls in the range of 60 to 80 SSU at 210°F but may vary with the type of unit and the installation.

For cylinders, packing, and inlet passages, the usual recommendation is a straight-run oil with a viscosity of 50 to 60 SSU at 210°F. The flash point of the oil should be between 375 and 400°F. If condensate conditions are encountered (as in winter), a 3 to 5 per cent moisture-absorbent compound may be added. Oils should be as carbon-resistant as possible. This compound cylinder oil should not be used in the crankcase because it may be harmful to babbitted bearings.

Housekeeping

Housekeeping within a refinery means an orderly and efficient arrangement of materials, operations, tools, equipment, supplies, storage, and facilities in general.[1] In a limited sense we normally think of housekeeping as "keeping in order a facility that already exists." Good housekeeping plays such an important part in the safety of a plant that there is sometimes a tendency to pigeonhole it as a part of the safety program. But, like safety, it is everybody's business; again, like safety, it is primarily management's responsibility to assure the existence of a clean and orderly work environment.

Plans for new structures or alterations in existing buildings should be studied with a view to good housekeeping arrangements. It is frequently noticed that new installations introduce serious hazards which result in future problems. Maintenance work can be greatly curtailed without an orderly and systematic arrangement of parts, tools, supplies, and equipment. This is particularly true at busy turnaround periods when valuable time can be wasted if tools and parts must be searched out.

When a repair job has been completed, the tools, leftover materials, and parts should at once be assembled and returned to their proper places. A job is not finished until all protective devices have been returned to their keeping place, supplies stored, and debris disposed of. Fire ladders borrowed for the job should be returned to their respective stations.

It is not uncommon to find oily wastes and greasy rags, empty pop bottles, burned-out light bulbs, pieces of lumber, nuts and bolts, pipes, valves, and fittings lying around long after the job is finished. None of these leads to good housekeeping, and they should not be tolerated.

A good housekeeping program requires teamwork, and there must be full support from all levels of management and employees. If there is a key person in such a program, it is probably the front-line supervisor, since it is his responsibility and duty to see that clean-up is one of the first orders of business.

All employees should have a part in the program. Operating personnel should have certain routine duties assigned in keeping their areas clean and orderly, while the mechanical group should be expected to clean up after completing each job. Maintaining a high standard of cleanliness is a continuing job. To permit an accumulation over a period of time and then resort to a "crash cleanup program" to bring it back to standard are not satisfactory. It results in excessive expense, and employees often lose interest in the program. Employees prefer a clean and orderly place in which to spend their working hours and are generally receptive in helping to maintain a high standard of housekeeping. After becoming accustomed to working in such an atmosphere, they point with pride to their place of employment. The feeling becomes contagious and will be reflected throughout the community and in improved company public relations.

There are other things that can be done in a refinery to improve housekeeping.[8] Early refineries located much of their piping systems underground. Leaks often developed, saturated the area, and created unsightly conditions. Today, most piping is located above grade, on either stanchions or sleepers (Fig. 7-19). In either case, leaks can be readily detected and quickly repaired. Also, it is much easier and cheaper to maintain the piping and keep it painted. Excavations are far less frequent, and grade areas can be kept more tidy.

A good refinery road system, constructed for all weather conditions and periodically swept, is a substantial asset in the program. Removal of dust and dirt from the roads results in a cleaner sewer system, which often eliminates possible plugging and need for cleaning. Clean roads reduce the possibility of dirt blowing into rotating equipment, where it could cause bearing failures. Likewise, it decreases the chances of foreign matter being blown into the eyes of employees, first-aid treatment being required and time possibly lost.

Painting is done in a plant to preserve surfaces and for appearance. Here some thought should be given to frequent painting. It is often more economical to paint when only one coat is required rather than allowing the surface to deteriorate to a point where cleaning, priming, and possibly two additional coats are needed. Furthermore, the one-coat policy keeps the plant looking better at all times.

Many plants use a multicolor painting scheme on piping systems. It makes a good appearance, and it serves as a means for service identification. A color scheme tied into service makes it much easier for all personnel to identify lines carrying different products. This results in greater safety, as the color of the line will indicate if the contents require special precautions, such as goggles and gloves. Light and heavy

FIG. 7-19. Most modern piping installations in refineries are located above grade, on either stanchions or sleepers, so that leaks can be detected and repaired readily. (*Courtesy of Mobil Oil Co., Paulsboro, N.J.*)

products can also be differentiated by color. An American standard, known as *A Scheme for the Identification of Piping Systems*, ASA A13.1, serves as a guide for identifying piping systems.

Methods that can be used in a plant to instill good housekeeping practices include:

1. Follow a satisfactory preventive-maintenance program for all equipment.
2. Paint equipment and other facilities regularly to maintain good appearance at all times and reduce deterioration.
3. Repair leaks in flanges, pump glands, and valves quickly.
4. Provide modern housekeeping tools, equipment, and materials.
5. Mow tank farms and grass-covered areas frequently.
6. Provide suitable receptacles for discarded paper, rags, food scraps, and other waste materials. Empty these receptacles before they reach the overflow stage.
7. Avoid open ditches and bell holes for any length of time.
8. Keep paved areas flushed down.
9. Make good use of crushed stone in unpaved areas where it will benefit personnel or add to appearance.

By this time someone is ready to ask "how can we afford all this and still remain competitive?" It would be difficult to develop figures, but there are many intangible advantages which ultimately result in lower over-all operating expenses. Some of these—and each is important—are higher employee morale, fewer accidents, fewer fire losses, less spillage and waste of products, lower maintenance costs, a better place to work, and improved public relations.

IDLE EQUIPMENT

Protection

There is almost always a certain amount of equipment in a plant that is termed "idle equipment." It acquires this status for one of several reasons:

1. It may be equipment that has been replaced because of size or capacity.
2. As processes change, equipment replacements are often called for, thereby idling existing equipment.
3. Corrosion may require the replacement of a pressure vessel, although it may still be satisfactory for further service under lower operating conditions.
4. Building new operating units often replaces old or obsolete units, but some parts of the old units may be good for further service.
5. Lack of product demand may call for shutting down certain units or equipment for a temporary period.

Whenever good equipment which might be used at some future date is removed from service, it must be given special treatment to make sure its usable condition is maintained. It is a known fact that spare or idle equipment will deteriorate faster than that in service. Operating equipment usually carries some temperature which prevents condensation of moisture which is so destructive from a corrosion standpoint. Refinery fumes when combined with a moist surface will almost always result in attack of the surface.

When equipment becomes idle, an attempt should be made to determine the length of this idle period. As protection costs money, an appreciable expenditure cannot be justified for a short idle period. In any case, an approximate estimate should be made to determine the degree of protection, which may be a compromise between no protection and all-out protection.

The equipment involved may be almost any kind that is found within a refinery: pumps, compressors, pressure vessels, tubular equipment, heaters, storage tanks, cooling towers, instruments, nonpressure vessels, turbines, piping systems, electrical equipment, etc. The kind and type of protection for this variation of equipment differ greatly and involve many details. Recommended practices for such protection are covered in detail in the American Petroleum Institute publication *Guide for Inspection of Refinery Equipment.*[24]

Returning to Service

Returning idle equipment to active service should be accompanied by a thorough and detailed inspection. No matter how many precautions were taken when it was put in "moth balls," there may have been some deterioration in one form or another. One common occurrence is the breakdown of the weatherproofing of insulated vessels. Water-soaked insulation adjacent to a steel surface can result in rather high corrosion rates at atmospheric temperatures. Where appropriate, inspection of idle equipment being returned to service should include calculations for pressure and temperature conditions in the new service.

PART 2.   CONSTRUCTION

There was a time when the general manager or vice-president or maybe even the president of a refinery would telephone a contractor and order a new operating unit having a certain rated capacity that would produce products meeting given specifications.

Today, most refineries have engineers who play an all-important part in the construction of new facilities or the revamping of existing ones. They enter the picture as soon as the idea is conceived and remain active until after the unit or piece of equipment goes into service.   The same is true with plant buildings, tankage, piping systems, or any plant facility.

ORGANIZATION

General Engineering

Most oil companies having more than one refinery have a general engineering division or department.   This group handles the type of engineering which is mechanical in nature, as differentiated from process engineering, and usually consists of mechanical, civil, industrial, metallurgical, electrical, and other engineers.   The specialized consultants are still another group that has become active in recent years.

The chief engineer is generally head of this division or department and is usually given authority and responsibility for such phases of construction projects as the budget, planning, design, construction, quality of work, and cost control.

Project Engineer

Under the direct supervision of the chief engineer are the project engineers, one of whom is assigned to a construction job in its very early stages and works in that assignment until after operation of the new unit begins.

Much of the chief engineer's authority is assigned to the project engineer.   It is therefore most important that the project engineer's authority be clearly defined and understood so that all groups will know the relationship that exists.   This is an area where confusion often develops unless there is a clear and concise understanding, particularly since his duties usually take him across organizational lines.

The project engineer's assignment is not an easy one to carry out because of the many contacts that he must make with a job of any great size.   He must rely upon the services and advice of many individuals and groups.   He should be a supersalesman to get all these services to fall in line with an established schedule which carries a definite completion date.   Slippage in the schedule is often difficult to recover.   On the other hand, it is practically impossible to adhere strictly to a construction schedule during all phases of the operation.   But in view of all this, he is supposed to shoot for this goal.   To do this successfully, he must:

1. Have complete knowledge of the job
2. Have an interest in his work
3. Possess keen personal and professional integrity
4. Have common sense and good judgment
5. Get along well with people
6. Have physical ability to handle the job
7. Be able to organize men
8. Be able to make decisions readily and accurately and stand back of them
9. Respect the abilities of the individual specialists and groups
10. Organize the work for maximum efficiency
11. Get others to carry out their duties on schedule
12. Keep costs within the budget
13. Prepare meaningful reports for management
14. Complete the project on schedule

It takes time to develop a qualified project engineer.   The training is not derived from reading a book but comes from years of actual experience, with effective guidance and supervision from those who have already qualified and with a variety of rotational assignments included.   The payout to the company is tremendous with a good project engineer on the job, while a poor selection can be costly.   Some of the advantages of having a top-notch engineer on a project are:

1. Coordination of effort between the various engineering groups
2. Coordination of work between the general engineering group and the plant engineers
3. Better engineering design
4. More efficient utilization of manpower
5. Better control of costs
6. More assurance of meeting the construction schedule

Project Planning

One all-important activity of the project engineer is the planning of the project, which starts with the selection of the best qualified subordinates. After taking this first step, the engineers should carefully outline the detailed steps to carry out the project successfully. Establishing a sequence for these steps is the next consideration. There is a normal sequence of events that should clearly be spelled out so that a firm understanding exists among the different engineers or groups of engineers. Without this procedure there is a good chance of confusion developing later as the project gets under way.

Of course, projects vary widely because of their many sizes and the types of equipment involved. But the work should be divided among the various engineers as evenly as possible for the best utilization of manpower. Also, it is better, if possible, to concentrate an engineer's efforts on one project rather than to divide his attentions among two or more. However, the size of some projects might not always make this practical.

Nontechnical work can often be delegated to trainees, junior engineers, or technicians. Any switch of manpower in this direction will lead to better use of the technically trained and more experienced men. It helps to maintain a higher morale and a lower turnover, resulting in higher productivity and better job satisfaction. When the planning or work schedule is prepared, make sure that the men who will be working on the project have a part in developing the schedule. This always leads to a better relationship. It gives them the desired sense of identity with management's objectives and the means of attaining those objectives.

In the planning of a project, it should not be overstaffed so that the men are not kept busy. Nothing can contribute more to dissatisfaction than idleness. The ideal situation is to have the man's time scheduled 100 per cent at all times.

A well-planned schedule for a project will lead to greater assurance of materials and equipment arriving at the site as needed. Nothing can delay a construction schedule more than a poor schedule of deliveries. The lack of some particular material or piece of equipment can delay the entire job. Conversely, equipment arriving far in advance of the installation date may have to be protected from the weather. This can be costly but may be essential if the equipment is to be in good operating condition when operations begin. The project engineer must have a good knowledge of the lead factors needed for procurement of materials and equipment.

Where a project is handled entirely by the refinery, there is a distinct advantage in having a close relationship between the project engineer and the design group. This relationship should not necessarily be limited to the project engineer but should include all engineers working on the job. Too often the designers will carry out their work in line with their thinking only to find later that a wrong concept existed of some phase of the project, which can lead to extra design work and added expense. It is therefore an excellent idea to have the chief design engineer attend the initial meetings of the project planning group to obtain a clear understanding of the complete scope of the work as well as a thorough knowledge of the materials of construction.

Design Models

Within the past few years more and more companies have resorted to the use of design models as an aid in the engineering and construction of new facilities. Properly used, such models can:

1. Reduce design costs
2. Reduce maintenance costs
3. Indicate where clearances are inadequate
4. Show unsafe spacing of equipment and vessels
5. Assist in the location of fire-water hydrants
6. Aid in training for both the mechanical and process departments

Design models are made to scale with all the important vessels and equipment included and the piping connected. As much detail is usually shown as is feasible.[2] Often the positioning of heaters, towers, vessels, and pumps by themselves will indicate adequate spacing, but when the piping system is completed, clearances may look entirely different. If enough clear space is not provided, it not only may create a hardship on the maintenance people but may result in an unsafe condition for operating personnel. Means for fire protection must always be considered, and sufficient clearances provided. Getting all details worked out in the beginning can save much time and money later. This is one great advantage of models—to detect design errors early while they can still be corrected on paper without costly construction changes later.

It is not easy to form a mental picture of steel structures, platforms, ladders, piping arrangements, access areas, trucking lanes, and the like from two-dimensional plans. To show all these and many other details requires a great number of sketches and drawings for even a small unit. The model does away with much of this drafting work and at the same time provides a means for mutual understanding between the contractor and the company. Decisions are made faster and with greater conviction.

A maintenance group review of a model can pay big dividends. Its ideas and opinions are of great assistance in evaluating, for example, refinery traffic both for routine maintenance and for a typical unit turnaround. Questions can be raised and discussed involving mobile equipment clearances and dimensions, complicated lifts in congested areas, location of stockpiles, temporary toolrooms and workshops, and movements of personnel in and around the unit.

The location of permanent handling facilities such as booms and trolleys can be examined to assure maximum use and minimum interference, with a corresponding saving of future field time. Overhead cranes and gantry cranes at grade can be evaluated against mobile equipment for the specific installation. Once in reviewing a model, it was found that provision had been made for lowering fractionating trays from a tower into an area that was completely inaccessible for a truck. Merely moving the boom to another location around the tower solved the problem.

Elevated platforms are almost always a subject for discussion.[10] As a rule, the location, size, type, etc., are spelled out in the general specifications, but often this is not sufficient for a specific installation. Much time is spent interpreting specifications because the particular situation is not clearly understood or the detail is not completely covered. Maintenance must present its case during the formative design stage so that platforms are adequate for personnel, materials, and tools for a major turnaround. The design model makes it practical for both the design and maintenance people to review each platform quickly for specific conditions of expected use.

Tall structures in particular, such as those required for fluid coking and catalytic cracking units, can be more intelligently planned with the aid of a model. These units usually undergo major internal inspection and repair at shutdown periods, and ample space on elevated platforms should be provided for temporary storage of materials without too much congestion of the working area for personnel. Although maintenance can furnish much foresight, it is effective only if presented before the demands of schedules freeze the design. A scale model, showing each step of the design from its inception, can be used very effectively to supplement argument and to demonstrate, if necessary, the most feasible design.

As for the costs of models, they vary considerably, depending on the size and complexity of the layout. Even though many hours of tedious manpower go into their making, design changes later on in the drafting stage are even more costly and field changes are often prohibitive. Design models should be given due consideration in any complex construction project. A start-up date for the unit of only a few days sooner might well cover the cost of the model.

Design for Lower-cost Maintenance

Most companies have developed engineering specifications and standards for designers to follow to assure an adequate and safe operating facility. Frequently, these specifications and standards are concerned with only the broad general features, clearances, and dimensions and are not always aimed at reducing maintenance costs. As maintenance is a very important part of refinery costs, the specifications and standards should include details on maintenance based on present-day practices.

Alloy metals. The use of alloys in equipment, vessels, and piping systems increases the initial cost but makes longer runs possible and reduces maintenance. The field lining of an existing vessel such as a fractionating tower, for example, is a costly and time-consuming operation. Furthermore, such a liner is seldom as satisfactory as a shop-installed liner. The job specifications should specify alloy shop-installed lining materials if there is any possibility of their being needed later on. The same is true with any other vessel, large-size piping at high elevation, or any part of a unit where future replacement would be impractical or relatively costly. If nonalloy materials will give a reasonable life and can be replaced, if necessary, in the future at reasonable cost, this should be considered. It is difficult to justify the use of alloys initially based on the life of a unit. But here again it is a matter of economics that should be worked out during the design stage.

Access. Practically every process unit must be opened up, cleaned, inspected, and repaired at more or less regular intervals. Design can cut maintenance costs by planning for this repetitive work, by providing ample access when laying out an operating area, and by seeing that manufacturers provide aids to maintenance in their equipment designs.[3] Ample working space should be provided for maintenance. The design should call for built-in facilities where needed. These design provisions for maintenance should be based on up-to-date experience and should be economically sound. A separate specification clearly spelling out the maintenance aids to be built into new equipment, a new unit, or a new refinery is the best way to assure sound, economic design.

Designing for maintenance usually increases the first cost of a unit, even though each item is checked and found economically justified. Surprisingly, though, this incremental investment often saves so many expense dollars that it has a better payout than the over-all investment in the unit.

Equipment Layout. As to the design layout of equipment in a process area, like pieces of equipment should be near each other or in groups and located so they can be easily reached for maintenance. All pumps handling products not subject to freezing should be located outdoors and not inside buildings. In the open, they can be located adjacent to a roadway and out from beneath low-clearance structures. Located in this manner and with ample clear space provided around each pump, they can be easily serviced. Replacing an impeller becomes a simple job, and renewing a bearing becomes a much easier chore. Complete pump and pump-driving units can readily be picked up and taken to the shops for repair work.

Tubular-type Units. For ease of maintenance of tubular-type units, such as exchangers, coolers, and condensers, it is preferable to have a removable bundle of the pull-through, floating-head type. This design has several distinct advantages: The bundle can be easily removed without unbolting and removing the shell cover or the floating head; after a bundle has been cleaned or repaired, it can be tested for leaks and then reinstalled in the shell, ready to operate—all without removing the floating head.

Although tubular bundles may be designed for steam or chemical cleaning, mechanical cleaning is often necessary. So if heavy fouling is expected, the tube layout and pitch should provide cleaning lanes (square pitch) for rodding and jetting through the bundle.

To facilitate the handling of parts, every exchanger should have lifting lugs permanently attached to any shell cover or channel cover that weighs more than about

75 lb. Tapped holes or pulling studs already in the channel tube sheet make it easy to attach a pulling plate to the tube bundle. A tapped hole in the floating head makes it easy to handle the head when it is removed.

Piping connected to tubular-type units should be spooled so that piping connections can be broken loose without handling long lengths of piping and heavy valves. All pressure bolting such as end flanges should be through-bolted, using full-length threaded studs. Perhaps an exception to this would be the use of stud bolts to keep high-pressure units more economical.

Fractionating Columns. On fractionating columns, ready and easy access to the internals should be provided by external manholes, properly located and adequately sized. The trays should have removable sections or manway openings in the larger sizes. A manhole should be located at each tray level on a tower in heavy-fouling service. If the service is clean, manholes can be spaced several trays apart with a maximum of no more than four or five tray spacings. This makes inspection and cleaning, if necessary, more practical. For tray manways, covers that are removable from either above or below the tray allow flexibility in the work schedule. Manways should be aligned vertically to simplify the hoisting of parts and tools inside the column. Fractionating caps removable by one man working on one side of the tray will speed column maintenance.

External manholes that are 20 in. or more in diameter (never less than 18 in.) provide easy access. Hinging manhole covers, instead of swinging them on a bolt or on a makeshift davit, reduces the size of the maintenance crew and provides greater safety. Steel platforms for working should be located at each manhole, or all manholes must be accessible from a platform.

A well-designed master davit is often needed atop a column to handle heavy relief valves, boxes of bubble caps, tray parts or sections, and valves or fittings. But with or without a davit, the designer should orient the manholes and platforms to provide a clear or open space for lowering column parts to grade. Also, enough clear space at grade should be allowed for access and movement of mobile equipment for handling the parts.

Fired Heaters. In the design of fired heaters, provisions should be made for the cleaning and pulling of tubes. Doors on return bend housings should be in sections that can be easily handled or operated. When open, these doors should give clear access to all return bends or headers and yet not interfere with the work platforms. Permanent platforms or platform supports for temporary staging should be provided to give ready access to all clean-out type return bends. Racks near by for temporarily storing the plugs or return bends as they are removed help to ensure that each one will be returned to its original seat. Heaters located near the plot edge make access easier and afford greater safety. The open area allocated for pulling horizontal tubes should be on the plot so that roads are not blocked. A heater designed with maintenance in mind will have trolley beams to handle heavy loads.

Utility Stations. One maintenance aid often overlooked by designers is the provision of utility stations in the operating areas. These stations can be located on the structure columns and should have proper-sized connections for the required utilities, such as steam, air, water, and electricity. Stations should be so located that working areas, whether at grade, on structures, or in buildings, can be reached with no more than 50 ft of hose or extension cord. It is also advisable to have stations for plugging in electric welding machines in these areas.

Clearances. Clearances must be carefully considered by the designer in laying out a process area. After clearances have been established, it is necessary to check during the detailing stage to assure that these clear areas are not obstructed by motor-operated valves with their bypasses, piping supports, structural bracing, piping drains or vents, and the like. Some clearances for roadways and equipment are shown in Table 7-7.

Monorails, Cranes. Tubular units (exchangers, coolers, condensers) that have their center lines more than 40 ft above grade are usually beyond the capacity of mobile equipment. Therefore, each unit located above this level should have a monorail for servicing. With a properly designed monorail, workmen can easily handle the channel cover, channel, tube bundle, and shell cover. At the channel end, the monorail should extend out over a lowering well. Provision of walkways on the monorail structure, so that slings can be easily placed around the bundle, is often overlooked.

Tubular units with horizontal center lines less than 40 ft above grade usually are more economically handled by mobile equipment. For maintenance economy, a group of tubular units at grade should be lined up in a row side by side, with a bundle-pulling area provided adjacent to a roadway or accessway. Any large number in a row calls for consideration of a traveling gantry crane (Fig. 7-4), especially if the units are large and difficult to handle. A gantry can pay for itself because an expensive mobile crane is not tied up for long periods.

Table 7-7. Minimum Clearances for Roadways and Equipment

Minimum vertical distances:
Over main roadways.................................... 16 ft
Over secondary roadways............................... 12 ft
Above floors in buildings............................. 7 ft 6 in.
Above elevated platforms.............................. 7 ft 6 in.
At grade in operating areas........................... 9 ft
Minimum horizontal distances:
Exchangers at grade, shell cover end*..................... 5 ft
Exchangers at grade, channel end*...................... 5 ft plus tube bundle length
Exchangers elevated on platforms, either end to handrail.... 4 ft, 6 in.
Aisle between piping around exchangers and vessels......... 3 ft
Between shells of vessels, horizontal and vertical........... 4 ft
Aisle between piping around pumping and compressor units.. 3 ft
Width of elevated platforms............................ 3 ft
Width of elevated walkways............................ 2 ft, 6 in.
* Measured from shell flange.

Several large compressors in line, found in almost every refinery, require good built-in handling facilities for efficient maintenance. Often the only facilities available consist of two parallel monorails, one over each bank of cylinders, and frequently the travel along these monorails is blocked by overhead piping, suspended lights, or some other obstruction. It then becomes necessary for someone to design a bridge crane that will properly service the compressors and engine cylinders, as well as the inter- and aftercoolers.

Some types of units may justify more or less special built-in features for handling. For example, the turnaround of a large catalytic cracking unit may warrant the installation of a pipe chute for trash, a vacuum-cleaning system, and a passenger-freight elevator of adequate size and speed. Catalyst-handling facilities are another item for consideration.

It can be readily seen that the designer must give a great deal of thought to materials and equipment handling in the design of new facilities. He should work closely with the refinery's mechanical department so that all avenues are completely covered.

Plant Engineering

Most refineries today have an engineering group to handle local engineering problems. It works closely with the mechanical department and assists in the day-to-day work. In many cases the plant engineering group handles complete projects up to a certain size. It may do the design work, purchasing, project engineering, and field engineering. This is the most economical arrangement, providing the job is within the scope of its activities from a size standpoint. Plant management, however, should not approve such an arrangement unless it is sure that the local group can satisfactorily handle the work. This decision should be made before the project has ever begun; switching to another group after the project has been started is often costly, time-consuming, and confusing. Either of these last two factors can easily lead to a delayed completion date and an over-all loss to the company.

Coordination of Contractor's Work

A vast amount of work in a refinery construction project is done by outside contractors, and some by materials and equipment manufacturers. Whatever the circumstances there must be a direct working relationship between them and the company or plant.

On a major contract job such a building a new process unit there is much engineering work that precedes actual construction. Once the contractor has been selected and a project engineer assigned, the contractor's engineers and the company's engineering groups get together to review and discuss each others' mechanical engineering specifications and standards. Here the contractor is inclined to use his own as they are written. On the other hand, the company has spent much time and money in developing its ideas and naturally wants to use them. Actually, there is a difference between the two sets; those of the contractor are more general while the company's include details covering its own likes developed with years of experience. After a complete review of the two sets the one used is generally a combination of the two. This makes it a little more difficult for the contractor, as he must familiarize himself with all the additions. But as this is more or less a standard procedure, he is used to this practice.

Once the mechanical specifications have been decided, the drafting work follows in accordance with these approved specifications and standards. As drawings are produced, they are usually checked by the company's engineers to make certain the right interpretations have been made and correctly included. As a general rule, the company will station some engineers in the contractor's office to work with the latter's engineers. The company's engineers are available daily to answer the thousands of questions that arise and also to keep information and data flowing back to their home base. These contacts are most important from a communications viewpoint. Then, as the need arises, general meetings are held either at the contractor's office or at the company's location. Other engineers are invited to these meetings, including some from the plant. Items are discussed, decisions reached, and minutes written which serve as a basis for further progress.

A very important step in this coordination work is the development of the process flow diagrams and their review. These will include many mechanical features which concern the materials and mechanical equipment group as well as the process engineers. At these reviews the kinds, types, and locations of equipment are discussed, as well as pressures, temperatures, valving, and the like. These meetings often run from 1 to 3 days, resulting in the release of much valuable information and many decisions needed to progress with the contractor's work.

If a model is in the picture, work on it begins as soon as preliminary drawings are released and continues until the model has been completed. The company's resident engineers follow the progress of the model building until a time when a general model review meeting is called to review the unit visually. Here again, many details are covered and decisions reached. It is at this meeting that the plant maintenance men play an important part. They observe the unit for the first time from a maintenance standpoint. They note roadways, accessways, location of equipment, and clearances between vessels, structures, and equipment. They will discuss maintenance equipment with the contractor and then jointly determine if sufficient clearances have been provided for this equipment. For anyone interested in maintenance work, the value of a scale model and the importance of a maintenance group review at this stage of engineering are readily apparent. As models are constructed from preliminary drawings, it is the opportune time to make whatever changes might be deemed necessary.

Field Office

As engineering work progresses in the contractor's office, the field construction crew begins to move to the refinery site to begin field operations. Here coordination of ideas, decisions, and work continues at a high pitch. Previously, there were only two segments involved: the company's office and the contractor's office. Now we have the third, the field office. This is the real test for communications, as all previous decisions and conclusions must carry through to the plant site or confusion will soon develop. It should not be concluded that confusion can be avoided entirely, but it can be minimized. This should always be the goal of engineers: to reduce confusion as much as possible.

As the contractor's engineering work reaches its peak and begins to taper off, field work takes on momentum. Materials and equipment are placed on order, and subcontractors enter the picture. All this increases the work load on the company's engineers, as more avenues have been opened.

Inspection and Expediting

Oil companies, being vitally interested in safety and quality construction, consider it wise to spend considerable money to make inspections. Even though contractors may assume this responsibility, companies often assist or supplement the inspection work. Inspection today covers most equipment, vessels, piping, tubing, valves, special fittings, and other items too numerous to mention. It opens a wide field and involves much manpower requiring a great deal of coordination.

It is good practice to have the company's and the contractor's inspectors visit the supplier's plant at the same time. This affords a better opportunity to discuss problems and come to decisions without delaying the work.

In addition to inspection in subcontractor's shops, there is always a certain amount of expediting that is necessary. This work is usually done by the contractor, as he is responsible for keeping construction work on schedule. The company has no direct interest in this phase of the work.

Subcontractors

As the field construction work gets under way, the contractor will often subcontract portions of the work to other contractors, usually those located in the area. Even though these subcontractors are responsible to the prime contractor, the company has an interest in seeing that they operate under the company's safety rules and other plant rules and regulations.

Field Engineering

As outside contractors and manufacturers' representatives come to a plant to perform work, they need help and guidance from many different angles. They may be coming to the plant for the first time and be unfamiliar with the company's safety rules, plant security policies, and change-house facilities and may not be known to the plant's personnel.

The plant management will usually assign an engineer or a group of engineers to the job of assisting them under the heading of "field engineering." This activity is actually twofold. In addition to providing service to these outsiders, the field engineers work along with them, checking details, drawings, specifications, and the like. In the case of a major process unit, this work becomes quite a project and a most important one. There are many arrangements and contacts that must be made, first to get the job started and then to keep it going on a prearranged work schedule. As work progresses, it calls for more and more field service, mostly engineering and specialists' assistance.

Keeping the job moving and in accordance with previously prepared drawings, engineering specifications, and standards requires constant checking. One portion demanding a great deal of attention is that of welding. As much field construction today involves electric welding, this phase becomes a major item.

The first problem in welding is to get the required number of qualified welders on the job. Sometimes this is quite difficult, especially in an area where the demand for welders is great. They often have to be brought in from other parts of the country.

In the testing and qualifying of welders the percentage of failures is sometimes extremely high, requiring that a greater number of prospects must be tested. This can develop into an expensive item.

As the welders become qualified and begin welding on the structures and piping systems, thorough inspection is required to assure high-quality welds. The welding inspector is not expected to check the mechanical properties or the chemical composition of the materials, as this is the job for the engineer. The inspector does cover the procedure qualification tests and checks these in great detail. The inspector should be quite familiar with the welding process, including the characteristics of the welding equipment. Some of the things that a welding inspector should look for, in addition to those listed later in this section under Field Welding (see page 7-65), are:[20]

1. Intermittent heavy electric loads cutting into the power line that feeds the resistance welders

2. Heavy intermittent electric loads on a feeder line of rectifier or transformer welders

3. The use of welding electrodes that are not the same as the ones used in the procedure

4. Laminations in the joints of the materials being joined

5. A change in the joint design, which may occur because of an error in the preparation of a bevel or for some other reason

6. Improper or incomplete removal of defects that have been found and have been repaired

Some of the field engineers on a major project should be civil engineers with training in soils, structures, and the like. Even though the nature of the soils at the project site must be known before foundations are designed, variations may be discovered during excavation. This calls for experienced and well-qualified civil engineers in the building of foundations of any kind. Mistakes here can be extremely costly.

Another all-important activity of field engineers is the checking of equipment installations, which must be correct for trouble-free operation. Highly trained equipment specialists usually assist with this work. Although the contractor's men may be experienced, it still takes the experience and services of the company's engineers and specialists to assist in checking all the minute details. This may appear as a duplication of effort and a waste of manpower, but actually it is money well spent. Most companies today follow this practice and consider it well worth while.

When field engineers are on the job at all times, they may readily observe the progress of all phases of the project. It is important that work progresses in line with the schedule, or there may be a delay in startup operations. If delays occur, and they often do, it may be necessary to shift manpower, work overtime, or perhaps hire additional manpower.

A construction completion date is a very important date. Much depends on it. Production schedules and market requirements are often geared to it, and any appreciable delay can mean real loss to the company. Every official in the company who is interested in the project is watching this startup date with a keen eye. This throws a great deal of responsibility on the field engineers and especially the chief field engineer, who is directly responsible. The point that he must not overlook is the fact that the finishing touches of construction work are extremely slow. Progress can be readily observed until it gets above the 90 or 95 per cent completion mark. A great deal of manpower is spent on the last few per cent without showing much progress. This becomes more critical as the 100 per cent is approached. Experienced field engineers are well aware of this and make allowances in the preparation of the work schedule.

One very essential point that should be mentioned is the qualifications of the chief field engineer. On a major project he should be "tops." Lack of field experience or the inability to coordinate work can mean a real loss to the company. He must delegate work; he must be a good organizer and get along well with people; he must be well liked and respected by all his associates. He is not an average engineer but one who has many qualifications vital to a good construction organization. But having a number of good qualities is not enough.   He must have had a lot of field experience.

Often a field engineer is assigned to projects long before he should be, which is unfair to him.   Any engineer who wants field engineering work and has the qualifications should work under the guidance and close supervision of an experienced man. After he has acquired the actual field experience, he will then qualify for a responsible position in the construction field.

COMMUNICATIONS

Communications is one of the leading problems today.   As companies become larger and construction projects more complex, more and more letters and reports are necessary to keep all interested parties informed.   Management looks forward to reports from many different sources to keep informed at all times.   For this reason reporting must be prompt and accurate.   Nothing is more disconcerting to an engineer than to have his boss find out indirectly about some important development that should have been reported directly by him.

Means of Communications

Communications can be made in several ways: in person, by telephone, by memorandum, or by a written report.[17]   Regardless of the means used, the reporting must be prompt, accurate, and complete.   For a better understanding, the written form is much preferred, but oral messages are faster.

In Person.   This excellent way to report allows you to show and explain with your voice, gestures, and facial expressions and to use a wide variety of visual aids. It gives you the opportunity to express yourself in your own words.   You learn more about people this way, and the meeting of the minds is a test of quick thinking, but it is too time-consuming for busy executives.

By Telephone.   This is the quickest for a short oral report, but words and figures can be misunderstood or perhaps even forgotten if not recorded at the time.   Both parties should keep a dated notation of the conversation.

By Memorandum.   This is the quickest written report (except the informal note) and is recorded with at least two dated copies.   One disadvantage is that these memos are easily lost along with other mail, which may reduce the effectiveness of memos.   A little notation of "Personal" or "Confidential" may solve this problem.

The Written Report.   This is the most formal and complete but the slowest because it requires study and consequently more time.   It is the most accurate and serves as a future reference and record.   Many copies can be made, which is generally the case.

There are as many different kinds of reports as there are bosses and jobs.   They may be engineering in nature, or they may be compilations of facts or figures.   In refinery work, they may be financial, progress, or process reports.   There are also periodic reports on maintenance and construction covering costs, man-hours, progress, and the like.   There are many reports covering materials and details of construction of equipment used within a refinery.   For best results, these reports should assume a rather definite form, as indicated in Table 7-8.

Table 7-8. Sequence and Importance of the Elements in a Written Report

Subject or topic	Importance
1. Cover	Secondary
2. Title page	Secondary
3. Table of contents	Optional
4. Letter of authorization, letter of transmittal, foreward, or preface	Optional
5. General summary and recommendations	Most important
6. Suggestions and recommendations	Very important
7. Body—step-by-step discussion of the subject	Very important
8. Appendix—including charts, illustrations, photographs, diagrams, figures, graphs, maps, biography, etc	Important

The "Summary and Recommendations" are listed at the beginning of the report, not buried in the body or at the end. The boss is interested in knowing the results of your findings and your recommendations without having to wade through pages of detailed material; therefore, your "Summary and Recommendations" should be clear and brief. If he needs further information, he can then turn to the remainder of the report for the details. This might happen and often does; therefore the facts must be there in abundance to convince him that there was a sound basis for your conclusions.

As written reports today take on rather broad coverage, we must be certain that they can be readily understood by all recipients. This means the use of words that are simple, clear, and accurate.

Some general suggestions in report writing include:

1. Follow the order of news reported in the daily newspaper: "Summary and Recommendations" in boldface type (underlined if typewritten) and the body in standard type.

2. Keep one specific reader in mind—the man or men who will get the report.

3. Explain clearly and briefly what the report covers.

4. Proceed from the general to the specific. First give the frame, followed by the full picture.

5. Do not bury important points under an avalanche of details.

6. Be consistent in spelling, abbreviations, graph scales, and form.

7. Use plenty of headings, subheads, and underlining to emphasize the main points.

In writing and submitting a formal written report you are actually doing three things: communicating, selling an idea, and selling yourself. To accomplish these points successfully your should be somewhat of a salesman, and in order to be a good salesman, you must believe in your product—your idea or ideas. Of course, all these ideas and facts must be well grounded as derived through studies, research, and an accumulation of data that lead to the final sound conclusion. Then with all these data in the background you are in a position to sum them up in a convincing manner to make the sale. Keep the main points before the reader so that he cannot possibly escape them, and he is more likely to come to the same conclusion as you have reached.

CONSTRUCTION SITE

Plant Layout

Most engineers have come to believe that the layout of a new unit or a new plant is one of the most important engineering steps.[4] Layout vitally affects payout. It affects every major cost of a process unit, including initial cost, operation, maintenance, safety, and expansion. A quick glance at some of the older units proves that in many cases very little time and study went into their planning. Experience has shown that making changes or additions to many of these units is extremely costly.

In developing a method of approach for an over-all layout of a new unit or plant, the engineers should:

1. Establish the objectives to be attained.

2. Evaluate the advantages and disadvantages of the site.

3. Develop requirements and guides for the master layout.

4. Develop principles for layout of additional process areas.

5. Study effect on costs, particularly maintenance.

6. Consider the use of a design model.

In looking at the history of process units and complete refineries it is practically impossible to find one that has not experienced growth over the years. So there should be two primary objectives in the initial layout. One is to create a plant that is low in initial cost and will have low maintenance and operating costs throughout its operation. The other is to provide adequately for growth and change.

In the selection of a suitable site there are many things to consider, such as size of the available area, relationship of the area to other equipment or operating units, proximity to service facilities (roadways, utilities, railroads, highways, waterways, water supply), cost of clearing the site, soil conditions, and so on. Each of these points must be considered separately and carefully weighed. It is always good to have an alternate site in mind so that one can be weighed against the other from the standpoint of advantages, including costs. In locating a completely new plant there may be several alternates that are used for comparison.

When the master layout for a new plant is developed, numerous requirements must be kept in mind, especially those pertaining to safety, utilities, local regulations, and codes. Usually these requirements are quite definite, and comparatively little leeway is available, particularly as to their effect on operation and maintenance costs. There are, however, other factors to consider that are not affected by the above, such as:

1. Over-all shape of the plant
2. Size and shape of the process blocks
3. Location and layout of the plant service areas
4. Transportation services—trucking, rail, water
5. Provision for growth and change

Shape of Plant and Process Areas

The over-all shape of a plant should be such as to facilitate operation, to produce short runs of major piping and utilities, and to reduce travel time to areas of major maintenance work. These objectives are best attained by using a rectangular or nearly square shape for the process section of the plant. These sections should then be divided into rectangular blocks of nearly uniform size. A block size of about 300 by 500 ft has worked out quite well for many types of process plants. Sometimes a large process unit may require an entire block, but generally a block will accommodate two or more units that are closely related process-wise.

Road System

Each of these process blocks should have roads along all four of its sides to provide access for maintenance. These roads should be straight, paved, and two lanes wide. Roads bounding the process blocks thus establish the pattern for a rectangular grid system of roads for the entire plant. In addition, a paved accessway should run crosswise through the block from one road to the parallel road opposite. Usually this accessway can be positioned to be adjacent to each of the process units within the block.

Travel of handling equipment from roads or accessways into the operating areas should not be blocked by pipeways, curbs, and the like. Sharp curves and dead ends should be avoided. Any part of a plant should be accessible by road from at least two directions, especially for personnel protection and fire fighting.

Plant Service Areas

Today's process plant usually needs supporting facilities such as administrative offices, clerical offices, control laboratory, locker building, cafeteria, maintenance shops, warehouse, and garage. Generally these facilities should be grouped in a plant service area, preferably located near the main entrance. Grouping will reduce the number of separate buildings, as more than one service can sometimes be included in a building (such as warehouse and shops). Locating the facilities near the main plant entrance eliminates considerable pedestrian and vehicular travel through process areas. Maintenance buildings are a good example of these advantages. When the maintenance building is near the main entrance, delivery trucks and large vans need not drive through restricted areas to reach the warehouse.

Transportation Services

Considerable attention should be given to location of transportation services, particularly railroad trackage. Often a plant will need permanent trackage for receiving materials and shipping product.   When not properly planned, trackage can be an obstacle to operation and maintenance.   Facilities requiring rail service should be located so as to minimize the number of spur tracks and road crossings.   A track should not be placed astride a roadway, nor should a track be strung alongside several process units to reach the one unit that needs rail service.

Growth and Change

A good layout will provide for growth in two ways.   First, blocks for additional process and storage areas will be earmarked and clearly indicated on the plot plan showing long-range development.   Second, space for adding bays to buildings or for adding process equipment at a unit will be indicated on the plot plans.   This earmarked space can be put to profitable use in the interim period—for outside storage, for maintenance work areas, or for certain facilities that can be easily relocated at reasonable cost.

Maintenance Factors

The over-all objective in laying out a process block is to arrange the equipment to satisfy in the most practical manner requirements of operation, maintenance, appearance, and cost.   From the standpoint of maintenance, items deserving attention include the grouping of similar equipment together and the mounting of equipment at grade.

Grouping of Equipment.   Generally, major equipment is first laid out to agree with the sequence of process flow.   Then, where economical, like items of equipment are grouped—first within the operating sections, next within the unit, and sometimes within adjacent units.   For example, heat exchangers go in one bank, pumps in one row, elevated condensers on one structure.   This grouping makes it much easier and less costly to provide good access, adequate handling facilities, and work space around equipment.

Mounting at Grade.   Where possible, grade mounting of equipment should be used instead of elevated locations on structures.   For example, condensers can sometimes be located at grade and thus eliminate completely the need for a structure and a tube bundle-pulling frame.   Grade mounting may increase the plot area needed, but it reduces the investment in structures and the cost of maintenance handling.

FOUNDATIONS

Soil Classification

Soils can be classified into two large general groups—noncohesive and cohesive.[21]

The **noncohesive** materials generally have a single grain structure in which the individual soil grains are in direct contact with each other.   Typical of this class are sands and gravels having the general structural arrangement where voids appear between the particles.

The **cohesive** soils usually have complex structures.   These either may be of sedimentary origin or result from long weathering of the present material.   When of sedimentary origin their structures show flocculation of the soil particles into loose honeycombs during deposition.   This flocculent or honeycomb structure is typical of that found in many clays.

Illustrating the differences in these structures is the range in porosity found in different soils.   The percentage of pore volume in sands usually ranges from about 30 per cent for very dense to about 45 per cent for very loose sands.   Pore space in cohesive soils may range from below 20 to as much as 95 per cent, i.e., soils in which only 5 per cent of the entire mass is solids.

Single-grain-structure soils may range from very dense soils, in which the individual grains of sand or gravel are tightly packed together, to quite loose structures with relatively large voids formed by arching of the grains.

Soil Settlement

Settlement of a load imposed upon a dense sand structure results almost entirely from elastic distortion of the soil grains. This is small, and consequently settlements of foundations on dense sand are small. With loose sand, the shearing forces and distortions caused by a load destroy some of the arches, permitting volume change in the soil mass; consequently, the settlements of foundations on loose sands are several times those on dense sands.

Vibration is even more effective than static loads in breaking down the soil arches of loose sand, and loose sands subjected to vibrations, such as those created by reciprocating pumps and compressors, can settle materially.

Actually, the designer studying a new site must evaluate both the possibility of troublesome settlement from compaction of sand by vibration and what corrective or preventive measures are feasible.

Sands in nature generally range from medium loose to medium dense, and almost any natural sand will be compacted by vibration. Furthermore, since in most formations there are variations in density, this settlement will not be uniform. The magnitude of settlement and, to some extent, its rate will be determined by the initial density of the sands. Dense sands will settle much less and much more slowly than loose sands.

Another factor in the rate of settlement of sand under vibration is the relationship between the foundation and the ground-water level. Sands far above ground-water level generally have films of moisture surrounding the individual grains, and these films develop surface tension at points of contact between the grains. The forces developed by surface tension are small, but the weight of any individual sand grain also is small, and these forces usually are capable of holding the grains in position, which greatly reduces movement under vibration. Sands in which surface tension is absent may be subject to rapid volume change with vibration. Such a condition is found in sand masses near or below ground-water level, where the pore spaces are filled completely with water, and in sands that are almost completely dry, as, for example, the blown sands of desert regions.

Compacting

Loose deposits of granular soil can be compacted prior to construction. The two methods most commonly used are Vibroflotation and pile driving. Vibroflotation is a patented method. A large vibrator, rather similar in basic principles to a concrete vibrator, is inserted in the soil mass. At the same time, water jets located at the top and bottom of the vibrator flood the mass of soil being compacted. With this method, it is generally necessary to add additional sand as the work progresses to compensate for the reduction in volume of the soil mass.

Piles of wood, concrete, or sand have been used successfully both in this country and abroad. In the construction of sand piles, a casing is driven into the ground, filled with sand, and then withdrawn. The vibration from driving the pile and the displacement from the added mass of sand effectively compact the soil.

This method was used to compact the soil underlying a second compressor plant installed in a refinery in California where an earlier installation was located on sand that had not been compacted. In the case of the second plant sand piles having a 16 in. diameter were driven on a 4 by 5-ft spacing over the area of the proposed compressor building and to a distance of about 20 ft on either side. These piles varied in length from 25 to 40 ft, because at these depths a denser structure was encountered. The total volume of material added within the area of compaction was the equivalent of nearly 2 ft of soil if spread over the surface. Yet, despite this, the surface of the ground settled as much as 6 in.

A comparison of the soil-settlement conditions of the two compressor plants is shown in Fig. 7-20. This graph shows the settlement that took place over a period of 8 years on loose sand while on compacted sand there was no significant settlement after 18 months of operation.

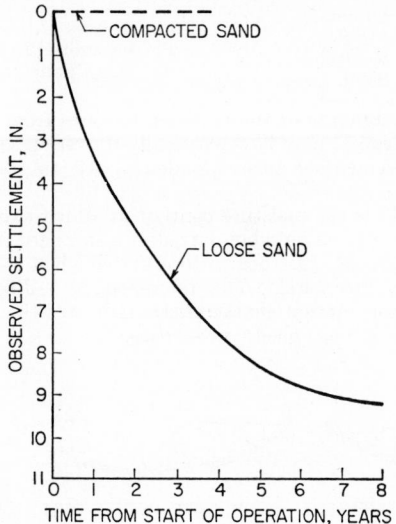

FIG. 7-20. Settlements of compressor plants built on loose and compacted sands.

Clay Soils

Vibration has little effect on clay soils, but trouble with these soils may originiate from other causes. As previously mentioned, clay soils generally have a loose, honeycomb, or flocculent structure. Individual particles of a true clay soil are extremely small, as are the pore spaces between them. When the surface of such a soil is loaded, the structure of the soil will readjust to a denser state, squeezing and distorting the individual pore spaces. This distortion, however, can occur only as fast as the water contained within the pores is squeezed out. Because of their extremely fine-grained character, clays are of low permeability, and this process is slow. Actually, many clays are less permeable than good concrete. Consequently, structures located on clay are subject to long-term settlements. The magnitude of settlement is dependent primarily on the load imposed and the previous stress history of the material.

Preconsolidation of clays to values materially in excess of the weight of the existing overburden is common. At some locations overlying deposits, which have existed once, may have been eroded. In the northern portion of the United States the ice of the continential glaciers may have overridden the area. More commonly, preconsolidation is caused by drying. The hardening and stiffening of clay as it dries, which is a familiar phenomenon, can preconsolidate clay soils effectively throughout the depth to which drying extends.

The standard penetration test may be used for judging whether a clay soil is so soft that excessive settlement should be anticipated. This penetration test consists of driving a standard sampling spoon or sampler, using a weight of 140 lb falling freely from a height of 30 in. The first 6 or 7 in. of penetration are not considered, but the blows required to drive the sampler the ensuing foot into the ground are counted. For the average compressor plant, soils requiring less than six blows per foot of penetration should be considered questionable.[23] Additional information may be derived by comparing the natural moisture contents of the soil with the Atterberg limits.

A clay suspension flows as a liquid and has practically no shearing strength. The latter is gradually built up as the suspension dries and the clay mass passes through the following stages, which are known as the Atterberg limits or limits of consistency:[33]

Stage of the mass	Limits between stages
Liquid.............	
	Liquid limit (LL)
Plastic.............	
	Plastic limit (PL)
Semisolid..........	
	Shrinkage limit
Solid..............	

These are moisture contents of the mass as it passes from stage to stage. When clay is in the plastic stage, it may flow plastically if overloaded. Plastic flow may be defined as a slow movement of an overloaded plastic substance without change in volume.

The liquid limit (LL) is the moisture content at which two sections of a standard clay cake barely touch each other when jarred in a standard manner (Fig. 7-21). At the plastic limit (PL), a clay just begins to crumble when rolled out into threads of about $\frac{1}{8}$ in. diameter (Fig. 7-21). The difference, $LL - PL$, is the plasticity index (PI or I_P). This important soil characteristic indicates the range of moisture contents within which the soil has plastic properties.

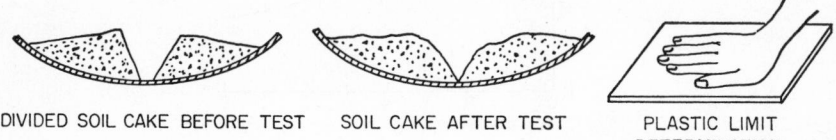

DIVIDED SOIL CAKE BEFORE TEST   SOIL CAKE AFTER TEST   PLASTIC LIMIT DETERMINATION

FIG. 7-21. Determination of Atterberg limits, liquid and plastic, for clay suspensions.

PI, as it is simply termed in current practice, is of importance in specifications for the use of soils as engineering materials and in soil classification. The shrinkage-limit concept finds considerable application in testing water-sensitive, expandable clays. It is a stage when a saturated clay exposed to drying starts to lose its water content; then air gradually invades the pores, and the soil changes its color from dark to light. It is obvious that the shrinkage limit is the moisture content, w_o, at saturation and may be determined by the equation

$$w_o = \frac{62.4}{D} - 0.381$$

where D is the weight of 1 cu ft of perfectly dry material.

To depths of 50 ft or more, soils that have been consolidated only by existing overburden generally will have moisture contents slightly below the liquid limit. Soils that have been preconsolidated will have moisture contents significantly lower than the liquid limit and frequently approaching the plastic limit. Considerable settlement usually will occur if foundations are placed on soils with moisture contents above or near the liquid limit.

Pile Foundations

The most common method of avoiding settlements under adverse soil conditions is to carry foundations through the compressible strata to more stable underlying soils. Piles are most commonly used for this purpose (Fig. 7-22). The use of so-called friction piles, which are merely driven into a clay bed of uniform consistency, can hardly be recommended. When, however, piles can reach appreciably stiffer soils or when they can reach firm sands, they form an excellent foundation. The basic principles of the pile foundations are well known, and a number of good references, for example Chellis,[22] are available.

Tops of wood or steel piles should be carried at least 18 in. into the mat. Concrete piles should be reinforced, and the reinforcement extended into the mat a sufficient distance to provide anchorage.

In designing foundations it should be noted that vertical piles, although not subject to excessive bending stresses, will be loaded additionally by the consolidating clay soil in which they will be embedded, since this soil will tend to hang up on the piles. This problem of negative friction must be given consideration in establishing pile loads and the total number of piles to be installed.

Fig. 7-22.   Steel piles are often used where soils will not withstand the loads of foundations without settlements.   (*Courtesy of Tidewater Oil Co., Delaware City, Del.*)

DIKES AND RETAINING WALLS

Dikes and retaining walls as found around operating units and in tank fields should be well designed and constructed to serve their purposes adequately.   In operating areas there is usually one of two kinds of diking: either a depressed area surrounded by a curb or a grade area enclosed by low curbing.   The curbing in either case is most generally poured concrete, but concrete-block construction is sometimes used.

Where mobile equipment is needed within the area, an advantage of having the area depressed is that a ramp inside the closure will serve for easy entry.   Where the retaining wall is built at grade, two ramps are required and more free area is needed. At times, fabricated steel closures are used in accessways instead of ramps, but these are not satisfactory.   They are seldom made to fit tightly, and often they are removed for entry and not replaced.

All walled enclosures must have drains large enough to handle all oil or water (or both) that might collect in the enclosed area.   Many times drain openings are insufficient in size, and when partial plugging takes place, overflow conditions result.

This could be particularly bad in case of a fire.   It is therefore much safer to have these drain or sewer connections oversized to take care of partial clogging.   This is a "must" for the covers over the drainage connections.

In tank fields all storage tanks should be surrounded by retaining dikes.   The only exception would be isolated tanks, where a spill or major shell failure would not involve other tankage, equipment, or property.   But as a general rule, tanks are surrounded by earthen dikes having a total capacity required by code or law.   Some states have regulations and many cities have ordinances that govern these capacities.

The real problem is not the building of dikes around tankage but the maintaining of the dikes.   To prevent erosion, they must be covered and in most cases the cover is grass with the usual weeds.   To control growth, mowing must be done periodically. No matter how well the dikes are designed and built or what kind of mower is used, there is always need for some hand mowing, which is time-consuming and costly. But there is no other solution, as unattended dikes will soon become great fire hazards, to say nothing of the poor appearance.   They are often sprayed for weed control, but there is still grass requiring attention.

Some refiners have figured that, after new dikes are correctly built, it is more economical in the long run to stabilize the soil.   This is done by spraying the surface with a cutback asphalt material.   Experience shows that this method has been very satisfactory.   Although the original cost may be somewhat higher, the payout is good, as there is little maintenance.   Where plants have cutback material and apply it themselves, this form of protecting dike should be considered.

Tank dikes must be cut at times to install new piping connections or replace old lines or for some other reason.   Whatever the reason, the dike should be repaired and put back in its original condition.   This not only falls in the category of poor housekeeping if not done, but it creates an unsafe condition if the tank is returned to service with an open dike.   In case of a tank failure or fire other tankage or plant facilities could be involved and the loss could increase many fold.

An item that is found in almost every fire insurance company's report covering a plant is the poor maintenance of dikes.   Unfortunately, dikes do not remain in their original condition because of uneven settling, erosion, traffic over the top, piping movement, and the like.   This means that dikes need constant attention and especially so after the winter months have passed.   Often this repair work never gets on the maintenance department's list, and if it does make the list, it is near the bottom. It is true, however, that this work can best be done during the summer and fall months.   Every plant should have a policy of having all dikes repaired during this period where such repairs are needed.   (Additional information on dikes for storage areas appears in Sec. 8, Offsite Facilities, pages 8–27 to 8–28.)

FIREPROOFING

Of the thousands of tons of structural steel used in a refinery much of it is used to support loads, including fractionating towers and vessels of all kinds.   For fire reasons, all structural steel around units should be fireproofed up to the load-bearing line.   This is a common practice that has been used for years.   Although the vertical and horizontal steel members have always been protected, there have been differences of opinion as to whether the bracing members should be so protected.   The answer seems quite simple: If they are needed for supporting the load, then the bracing structurals should be fireproofed.

Concrete is still considered the best material for fireproofing steel.   Other lightweight materials have been used with fair success, but when durability and low maintenance costs are taken into account, concrete is the preferred material.   The initial cost may be a little higher, but once applied, the fireproofing can be forgotten. Actual fires have proved that concrete will stand up under the most severe conditions and provide good protection for the minimum period of 4 hr used for design in figuring fire protection.   It is convincing that concrete has the necessary characteristics. Conversely, many pictures after fires have shown great damage and losses to vessels and equipment because of no fireproofing.   Failure to protect load-bearing steel structures properly is no way to cut costs.   Even though refineries have a good fire-loss record, there have been many devastating fires over the years. It is wise to assume that fires and losses can occur in any refinery and to design installations to withstand such conditions.

Fireproofing must be weathertight so that rain and condensation do not enter between it and the steel, resulting in corrosion of the steel. This may be avoided on vertical columns by welding flashing just above the protected steel. The concrete itself should be given a coat of waterproofing to reduce its porosity and seal the surface. One good coating will last indefinitely, so that little maintenance is needed.

A mistake sometimes made is the failure to fireproof properly the skirts of towers. If the outside of a tower is insulated, the insulation is often extended on to the base of the skirt, that being considered sufficient fireproofing. Insulation is no substitute for fireproofing, particularly where fire-water streams are likely to be playing on the surface. Insulating materials will not take that abuse. A very satisfactory means of fireproofing tower skirts on the outside is to lay brickwork from the foundation up to the flashing, which is welded just above the point where the head and shell join. This provides a good weatherproof design. If a tower skirt is larger than 4 ft in diameter, then the inside of the skirt should likewise be protected from fire conditions. Here a Gunnite material is most generally used.

Steel pipe supports in and around unit operating areas, as well as the horizontal supporting members, should be fireproofed up to the load line. This is usually done either with formed concrete or by Gunniting. The latter method can create something of a mess in the area and may have a damaging effect on equipment having moving parts (pumps, compressors, motors, valves, and the like). All such parts must be protected before any Gunniting is done. Lubricating-oil systems must be given special protective care.

Fireproofing must be subjected to periodic inspections. Water damage seems to be the greatest problem. Inspection will reveal any evidence of moisture getting to the steel, and if such a condition is found, removal of some protection is recommended to note the condition of the steel beneath. Lack of attention can cause excessive corrosion of the steel members, resulting in a very unsafe condition.

INSULATION

Insulating materials are used in refineries to conserve heat, keep excessive heat from materials, or protect personnel. Refiners usually include some figure, such as 150°F, in their specifications as the maximum temperature for unprotected hot surfaces. This limit applies regardless of the economics for heat conservation.

The engineer who selects insulating material for the job must also consider its maintenance. As process units require cleaning and inspection periodically, a certain amount of insulation is almost always removed for one reason or another and must then be replaced. In addition to having the necessary heat-resisting qualities, the specified material should be durable enough to be removed, salvaged, and reapplied if necessary.

Protective Coverings

The life and efficiency of insulation applied to vessels, piping systems, and equipment depend, to a great extent, on how well it is protected against the weather, moisture, chemicals, fire, and mechanical abuse. All insulating materials must have some form of protection.

Weather Protection

For years the usual form of weather protection on insulated vessels and equipment has been a material made from asphalt emulsion, containing asbestos fibers and fillers. On insulated piping, the material has been a heavy paper impregnated with an asphaltic-base material. The main objection to these materials is that rain and moisture will enter at the joints and connections, resulting in a wet and inefficient insulation. The age of the mastic materials seems to affect their bonding and elastic properties, causing cracking.   Improvements have been made in materials, but where high temperatures exist, the materials still fail at points of contact.

The use of flashing around piping connections on vessels has been a great improvement in waterproofing.   This gets away from the temperature objection in many cases and improves the tightness of the joint.

Many piping systems use **galvanized sheeting** for weather protection, which also provides fairly good protection from mechanical abuse.   Some installations of this type, especially on steam lines, have been in service for years.   Today, however, the life of galvanized material seems to be much shorter.   The coating fails, rusting begins, and painting must follow.   This condition has caused many refiners to resort to aluminum, which is now used quite extensively on vessels and piping.

Aluminum sheeting, being light in weight, is easy to handle, bend, cut, and form. There is practically no maintenance required and no painting.   If atmospheric conditions are on the corrosive side, the material may be alloyed to resist attack.   Being corrosion resistant, it may be used in light gage, such as 0.020 in.   With these advantages, aluminum has become popular as a weather-protective material over insulation and quite competitive with other materials.

Many fractionating towers today are covered with aluminum sheets, usually of the corrugated type.   With a one and one-half corrugation lap on the vertical seams and about 3 in. on the end seams, the result is a tight job.   If the attachments are flashed and corrugated aluminum sheets used on the sides, the finished job, with good application, should be among the best.

Another weatherproofing material that does an excellent job is a **fiber-glass cloth,** applied and then coated with a mastic material.   The use of the cloth reduces cracking.   A fiber-glass job, correctly applied, will give years of satisfactory service.

Moisture

If moisture, other than from weather conditions, is a problem, then a moisture-resistant material should be used.   Most insulating materials are absorbent by nature.   If the material is to be applied in a tunnel, underground, or in any location with moisture present, it should by all means be resistant to moisture.   One material with this characteristic is known as Thermobestos.   It is made in both block and pipe forms.   It may be soaked in water and it still retains its mechanical strength, or it may be dried out and its conductivity is the same as originally.

Mechanical Abuse

If insulation is applied where it will receive much mechanical abuse, the proper protection should be used.   Insulation near a materials-lowering well is a good example.   Under this condition, only a metal covering should be considered, and even then it might be wise to install a steel guard or shield to protect the insulation further.

Mobile cranes often cause damage to insulation, and extra care on the part of the crane operator is about the only answer to this problem.   Insulated lines near grade are often damaged by truck traffic, and here guardrails or posts can be helpful.

Installation of Insulation

Where piping insulation must be removed periodically for inspection or maintenance, a metal covering is the best kind.   As aluminum is light in weight and easy to handle, it is preferred in most cases.   If the removal is to be frequent, an insulation with the aluminum attached should be considered.

In new piping systems, all joints and seams that had not previously been pressure-tested should be insulated only after the field pressure test has been satisfactorily completed.

Insulation in a refinery is a maintenance item, and it can be a sizable one if good designs and practices are not used.   Heat losses can be appreciable, so the best insulation obtainable for the service should be used.   But the best insulation fails to give good results if correct installation techniques are not employed. This means experienced and well-qualified applicators. There are too many short cuts in application to accept the lowest bid without a most thorough analysis. The best materials, installed in a workmanlike manner, will still require some maintenance after the first turnaround. This maintenance should be prompt and of the same quality of materials and workmanship as the original.

FIELD WELDING

Electric welding is today the accepted way of "joining" in the refinery. Not too many years ago welding was limited mostly to gas welding in limited applications. Today, gas welding is used only in special cases, while electric welding has come into being and has largely replaced couplings and flanges in piping and riveted joints in tanks and vessels.

The use of electric welding did not come about all of a sudden but developed gradually over the years. The improvement of electrodes has been a big factor. The coating of rods advanced the application of electric welding and added much to the safety and dependability of welded joints. Another factor that greatly influenced the use of welding was the advancement of inspection techniques. Plant management had to be sold that it had welded joints that were free from cracks, porosity, slag inclusions, voids, and other flaws. Where lives and millions of dollars worth of equipment are involved, there had to be some assurance that failures would not occur. One means of determining the soundness of weld metal is by nondestructive testing, which is quite dependable when done by qualified personnel and is discussed under Weld Inspection Methods, page 7-67.

Code Requirements

All refinery pressure vessels and pressure piping are welded in accordance with recognized codes. Somes states have their own published codes, covering welding procedures and welders qualifications. In Ohio, for example, the standard qualifications are specified in Ohio Section IX, *Standard for Welding Requirements Covering the Ohio Boiler Inspection Laws and Rules for Construction of Unfired Pressure Vessels.*[32]

Where there is no state requirement, the qualification of welding procedures, welders, and welding operators is in accordance with Section IX, ASME *Boiler and Pressure Vessel Code,*[27] or ASA B 31.3, *Petroleum Refinery Piping Code.*[26]

The construction of pressure vessels should be in accordance with the ASME *Boiler and Pressure Vessel Code,*[27] the API-ASME *Unfired Pressure Vessel Code,*[25] or the ASME *Unfired Pressure Vessel Code,*[28] except in Ohio, where the *Ohio Boiler Code* or the *Ohio Rules for Construction of Unfired Pressure Vessels*[29] prevail.

Pressure piping should be constructed in accordance with ASA B 31.3, *Petroleum Refinery Piping Code,*[26] or in Ohio, the *Ohio Specific Safety Requirements Covering the Installation of Pressure Piping Systems,*[30] or for refrigeration piping, the *Ohio Specific Safety Requirements Covering the Installation of Mechanical Refrigeration Systems and Equipment.*[31]

States other than Ohio may have their own specific requirements for pressure vessels and pressure piping also and should be checked before proceeding.

Whatever code is used, all welders within a refinery doing strength welding must have the procedures qualified and they themselves must have their work tested and approved before they begin work.

In recent years, electric welders have been in great demand and schools have been established to train men. As with anyone completing formal training, the principles and procedures may be perfectly clear but the actual experience is lacking. This lack explains why so many schooled welders have difficulty passing the requirements tests or in consistently producing good welds in the field. It is unfortunate that a good prospect must be turned away because of his lack of actual experience. He must then turn to nonpressure type of work to gain this experience.

As refinery welding calls for a certain amount of noncode work, it does afford a good training field for new welders, and each plant has some of these men. They may acquire experience in tack welding and in welding nonpressure parts. Much can be done in giving training and guidance by the more experienced men, and in many cases good welders have been made.

Welders' Problems

In field welding, there are a number of factors or problems that face an electric welder. He must be able to cope with all of them to qualify as a first-class welder. Some of these problems are:[16]

1. It is important to have a sound stringer or first pass bead. This bead must penetrate to the root of the weld, and the weld metal be fused or bonded to the base metal. The bead must be continuous with no voids. There should be no appreciable undercutting of the welding groove.

2. There should be no foreign materials on the welded surfaces, such as rust, grease, or scale.

3. The root face dimensions of two pipe ends being joined should be the same.

4. An improper gap between ends being welded presents difficulties.

5. Misalignment of two ends is bad.

6. The joining of two ends having different wall thicknesses adds a problem.

7. Two lines of the same size may have wide variations of tolerances. One end may have a tolerance all on the plus side, while the other end may have a minus diameter tolerance. It is possible that the outside-diameter variations and the wall-thickness variations could be additive to increase the offset to approximately $\frac{5}{32}$ in. for a 3.0-in.-OD by 0.375-in. wall pipe. This is appreciable and would present quite a problem, especially on the first pass. The more a line-up deviates from uniformity, the greater is the skill demanded of the welder. Also, his speed of welding is reduced and the possibilities of unsound welds are greater.

8. Too high a current setting of the welding machine for the rods used will raise the electrode temperature and cause the coating to fail. As the coating breakdown becomes more severe, the welding process progresses from a shielded-arc process to an unshielded-arc process. This results in a lower tensile strength weld with less ductility.

9. A change in current setting should be made for the second weld pass, which is sometimes called the "hot pass." This higher current ensures good penetration into the stringer bead, burning out the stringer bead undercut and at the same time filling in voids.

10. The stringer bead should be thoroughly cleaned before the second pass is applied in order to remove the possibility of slag inclusions.

11. The correct size electrode is most important to penetrate into the groove properly.

12. As weld metal cools, certain gases are less soluble in the metal and are released, forming voids. The ones that do not escape to the surface remain in the weld as porosity. Excessive heat, defective electrode coatings, and improper electrode manipulation may contribute to porosity.

13. In the final pass, pinholes and undercutting may occur. These are usually caused by poor technique or defective coatings. If a welder holds too long an arc or moves the electrode away too quickly from the sides of the pass, undercutting will result.

14. The welder must be certain of the material being welded. Knowing if a material is aluminum, brass, or stainless steel is not enough. There is a wide variation of analysis in each of these classes, and the welder must be sure of the kind and type of material.

Metal Identification

To get an accurate metal identification would require a spectrographic or a quantitative chemical analysis.[12] Such analyses are expensive and time-consuming. There are a few tests, however, that can provide an idea of the metal composition. It is not too difficult to differentiate between aluminum, brass, and steel. If the material is steel, a distinction can be made between castings, forgings, pipe, plate, etc. Some simple tests that may be used would include:

1. **Spark testing** involves comparing the sparks of the unknown material with those of a known material, using information generally available in the literature or from steel suppliers' catalogues. Different analyses of steels will produce sparks of varying characteristics when the material is held against a grinding wheel. Nonferrous metals fail to produce sparks.

2. **Nitric acid (dilute or concentrated) spot testing** is another method of checking metal composition. A drop will not affect stainless steel or nickel iron but will produce a green or blue-green color with monel or copper-nickel and a brown color with low-carbon steel.

3. **Chisel testing** is frequently used to determine if a material is gray cast iron, malleable iron, or steel. This identification can readily be made with a sharp-edged chisel and a hammer. By trying to cut a continuous chip (not over $\frac{1}{16}$ in. thick), one can tell which of the metals is involved. With cast iron a continuous chip is impossible, while malleable iron will produce a short, slightly curved chip. A long, continuous, and curled chip can be obtained from steel.

4. **Magnet testing** helps in identification, by separating materials into three groups:

Group 1, **strongly magnetic:** carbon and low-alloy steels, wrought iron, cast iron, pure nickel, stainless irons (under 17 per cent chrome)

Group 2, **slightly magnetic:** stainless steel (18-8 types, when cold worked), monel, high-nickel alloys

Group 3, **non-magnetic:** stainless steel (18-8 type, annealed), copper-base alloys, aluminum-base alloys, zinc-base alloys

Magnet testing has special significance with stainless-steel identification, since the nonmagnetic or slightly magnetic materials require no preheat or stress relieving while those that are magnetic do require it.

Weld-inspection Methods

Fortunately, there are several methods for determining good, sound weld metal over and above visual inspection. All nondestructive testing methods, other than visual, require a probing medium to indicate the soundness of the weld material beyond the visual surface. These probing media may be classified into two categories:[5] motion of matter and transmission of energy. Some methods use a combination of these two.

Motion-of-matter Tests

Generally, nondestructive testing methods employing the motion of matter require the defect coming to the surface. Leak testing was one of the first nondestructive methods used on welded joints. There are several types of leak tests, which involve the motion of water, air, or helium passing through a defective weld seam.

Where a highly competitive, relatively inexpensive product is involved, the simplest way of leak testing is to put 3 to 5 lb of air pressure inside the container and swab the outside with a soap solution. Another form of leak test employs high-pressure air in tubing or pipe submerged in water to indicate leaks.

The next step in upgrading of leak tests is what is called a proof test. This is a hydraulic test in which the pipe or container is filled with water and pressure applied so that the stress is near the yield stress point of the material and can result in the fracturing of defective welds.

Probably the most sensitive of all leak tests is the helium mass spectrometer test. Helium can pass through crevices which are too small for air or water to permeate. Minute traces of helium leakage are detectable by the mass spectrometer test. This test is frequently used in complicated heat-exchanger and pressure-vessel testing.

Another matter-in-motion test is the liquid penetrant test. In its original form this test, known as the machine shop oil and whiting test, is as old as the metal industry.

The surface to be inspected is covered with an oil or dye which penetrates into any invisible crack or seam. The surface is then wiped off so that no visible trace of the oil or dye remains. The area is coated with an opaque, chalky liquid. When dry, the opaque coating acts as a blotter to draw out the penetrant from any cracks that might exist. This is a very effective method for detecting a crack of practically any magnitude which comes to the surface.

Transmission-of-energy Tests

Nondestructive testing methods employing the transmission of energy can locate defects entirely within the interior of the weld.

The oldest of the probing methods using transmission of energy is radiography. The source of radiant energy is either an X-ray tube or a radioactive isotope such as cobalt-60. Maximum capacity of X-ray radiography is generally considered to be 16 in. of steel for the highest voltage machines. For checking welds, X-ray machines of from 100 to 400 kv are used for steel up to 4 in. thick. The maximum thickness of steel that can be radiographed with a cobalt-60 source is about $7\frac{1}{2}$ in., regardless of the size of the capsule.

The ultrasonic method generally used for weld inspection is the pulse echo system. In this equipment, sound waves above 250 kc are generated in a crystal from a high-frequency electrical generator. The crystal, enclosed in plastic, can be moved by hand along the welded seam. When the pulse encounters a flaw within the weld, it sends back an echo which is in turn picked up by the crystal and indicated on a cathode-ray tube as a peak or blip. As the search crystal is moved back and forth along the weld, an entire weld section can be scanned for flaws. There is practically no limitation on the maximum thickness of materials that can be examined by this method.

Magnetic-particle Inspection Method

The magnetic-particle inspection method, used widely in industry during the past 30 years, uses both the principle of matter in motion and the principle of transmission of energy. It is applicable only to testing of magnetic materials.

The method requires that the piece to be tested be magnetized, which usually is accomplished by wrapping a coil around the section and passing a large current through it. Any defects or flaws at or near the surface of the part create magnetic leakage fields. These leakage fields attract magnetic particles when they are applied to the surface of the part, indicating the extent of the defect. This method of inspection has been found very sensitive for locating surface cracks. Any cracks beneath the surface will not be detected.

REFERENCES

1. Burroughs, Jr., E. O., "Safe Housekeeping in Chemical Plants," *Safety Maintenance*, August, 1960.
2. Chandler, Harold J., "Control Maintenance by Work Measurement," *Chem. Engrg.*, p. 120, Feb. 6, 1961.
3. Evans, Henry P., and Bromley, Ditmas, "How Good Design Reduces Maintenance," *Pet. Ref.*, p. 124, January, 1958.
4. Evans, Henry P., "Plant Layout," *Pet. Engr.*, p. C-14, March, 1960.
5. Gibson, Glenn J., "Quality Control," *Welding J.*, p. 225, March, 1961.
6. Haldane, Robert, "Refinery Shop—Layout and Operation," *Pet. Ref.*, p. 79, May, 1950.
7. Henderson, H. P., "Proper Lubrication—Longer Compressor Life," *Pipe Line Engr.*, p. D-26, April, 1958.
8. Hull, C. D., "Good Housekeeping Can Pay for Itself," *Pet. Ref.*, p. 156, March, 1959.
9. Innis, T. J., and J. L. Nanney, "High Pressure Wet Sandblasting Cleans Esso Refinery Equipment," *Oil & Gas J.*, p. 86, Sept. 15, 1952.
10. Kershaw, H., and A. F. Hollowell, "Models, a New Maintenance Tool," *Pet. Ref.*, p. 124, January, 1958.
11. Knight, Alan T., "Maintaining Refinery Efficiency through Contract Maintenance," paper presented before National Petroleum Association, Apr. 21, 1960.

12. Kolb, Ray M., "Successful Welding for Maintenance," *Pet. Ref.*, p. 207, May, 1958.
13. Mayer, John W., and Paul Taylor, "Scientific Inventory Control," *Chem. Engrg.*, p. 140, June 27, 1960.
14. McCone, Alan, "Maintenance by Contract," *Factory*, July, 1959.
15. Monack, M. L., H. W. Shockley, and J. R. Allen, "New Ways to Reduce Painting Costs," *Pet. Ref.*, p. 125, January, 1957.
16. Morel, Richard D., "Pipeline Welding," *Pipe Line Engr.*, p. D-51, May, 1958.
17. Murphy, Dennis, "Communication," *Pet. Engr.*, p. E-19, February, 1957.
18. Price, Ralph N., "Program for Prevention," *Chem. Engrg.*, p. 196, Apr. 20, 1959.
19. Reidel, John C., "Chemical Cleaning Is Cheaper, too," *Oil & Gas J.*, p. 78, Dec. 19, 1955.
20. Ronay, Bela, "What Is the Welding Inspector's Job?" *Welding Engr.*, p. 47, January, 1961.
21. Swiger, W. F., "What You Should Know to Design a Compressor Foundation," *Pipe Line Engr.*, p. 624, August, 1958.
22. Chellis, R. D., *Pile Foundations*, McGraw-Hill Book Company, New York, 1951, Chaps. 1 and 2.
23. Terzaghi, Karl, and R. B. Peck, *Soil Mechanics in Engineering Practice*, John Wiley & Sons, Inc., New York, 1948, pp. 11, 32, and 473.
24. *Guide for Inspection of Refinery Equipment*, American Petroleum Institute.
25. *Unfired Pressure Vessel Code*, API-ASME.
26. *Petroleum Refinery Piping*, ASA B 31.3, American Society of Mechanical Engineers, 1959.
27. *ASME Boiler and Pressure Vessel Code*, American Society of Mechanical Engineers, Sec. IX covering "Qualification of Welding Procedures, Welders, and Welding Operators."
28. *ASME Unfired Pressure Vessel Code*, American Society of Mechanical Engineers.
29. *Ohio Rules for Construction of Unfired Pressure Vessels*, Sec. VIII, The Industrial Commission of Ohio, Columbus, Ohio.
30. *Ohio Specific Safety Requirements Covering the Installation of Pressure Piping Systems*, The Industrial Commission of Ohio, Columbus, Ohio.
31. *Ohio Specific Safety Requirements Covering the Installation of Mechanical Refrigeration Systems and Equipment*. The Industrial Commission of Ohio, Columbus, Ohio.
32. *Ohio Section IX, Standard for Welding Requirements Covering the Ohio Boiler Inspection Laws and Rules for Construction of Unfired Pressure Vessels*, The Industrial Commission of Ohio, Columbus, Ohio.
33. Krynine, Dimitri P., and William R. Judd, *Principles of Engineering Geology and Geotechnics*, McGraw-Hill Book Company, New York, 1957, chap. 4.

Section 8

OFFSITE FACILITIES AND UTILITIES

By GEORGE C. PATTERSON

The M. W. Kellogg Co.
Division of Pullman Incorporated
New York, N.Y.

BASIS FOR DESIGN OF OFFSITE FACILITIES

Offsite facilities include all oil, utility, auxiliary, waste-disposal, and service installations located outside process unit battery limits. Functionally, offsite facilities serve the following purposes:

1. Storage, blending, and shipping of crude, intermediate, and final products
2. Preparation and distribution of utilities and auxiliary materials
3. Collection and disposal of liquid and gaseous waste materials
4. Storage of supplies and spare parts and housing of maintenance equipment and of administrative and process control offices

In order to initiate and proceed with the design of offsite facilities, there are fundamental data that should be known or determined. Three major categories of data affecting basic offsite facility design are as follows:

1. **Economic factors and basic philosophy,** including data pertaining to the payout time required to justify incremental improvements, the philosophy concerning future provisions, and preference for maintenance and operating methods. Also, general policy concerning the permanence and appearance of the major offsite structures should be known.

2. **Engineering and technical factors,** including information available from meteorological records, land and water boring investigations, existing maps (topographic, hydrographic, and geologic), and reports from detailed site investigations concerning sources of water and analyses, waste disposal, and other similar basic design and operating data.

3. **Legal and fiscal obligations,** including Federal, state, and local statutes, ordinances, and codes, the requirements of which affect the design.

Economic Factors and Basic Philosophy

A philosophy as to required life of offsite structures must be developed, since the criteria applicable to process equipment are not suitable for offsites. In general, economic evaluations and material selection for offsite installations are based on longer periods of time than are assumed for process units. Exceptions are facilities such as roads and drainage systems, which can be planned for short service periods and improved or upgraded as a maintenance operation.

8-1

To determine payout for incremental improvement and to facilitate evaluation and choice of alternates, cost data should be available on fuels, electrical power, general utilities such as water and steam, auxiliaries such as chemicals, basic construction materials, labor, and various direct and indirect factors such as freight, taxes, and insurance.

A philosophy must be developed as to preference for type and materials of construction for major items, including buildings, harbor and river structures (piers wharves, and water intake stations), oil separators and lagoons, and roads, parking areas, gates, and fencing. Most of these are not covered by generally accepted specifications or standards such as ASA, NEMA, and the like.

Future requirements for plant expansion must be established in principle. Allocation of space for future offsite facilities and future process units is a basic factor to be considered in plant layout. In addition, future needs may influence size or capacity of many facilities. A general policy as to built-in oversizing of initial installations must be formulated.

Design of offsite facilities will be influenced by decisions regarding degree of integration of process and offsite operations; shelters and enclosures to protect equipment and personnel; extent, type, and mobility of equipment-handling facilities; use of contract maintenance and/or cleaning; and extent of shop equipment and facilities.

Engineering and Technical Factors

Crude and Product Systems

Storage-tank capacities for crude and product services must be determined. Common criteria for such capacities include marketing requirements, seasonal storage needs, tanker loading schedules, etc. A combination of operating and cost considerations will be involved in the selection of storage-vessel types and their materials of construction.

Blending facilities, batch or in-line, will be selected to suit the storage and shipping facilities, which, in turn, will exercise control over the selection of tank mixers (propeller, jet).

Loading and unloading facilities will involve decisions on the number of berths or spots for ships, barges, rail cars, or tank trucks. The sizes and capacities, as well as locations, of crude and product pipelines associated with loading and unloading will also influence refinery design and operation.

Other factors requiring study and decisions in the design of crude and product systems include:

1. Location of loading, transfer, and blending pumps (centralized or decentralized).

2. Accessories for storage vessels (manholes, stairways, ladders, gage hatches, vents, water drains, fire-protection fittings, piping connections, heating coils, and/or suction heaters) to suit type of vessel and material stored.

3. Gaging, metering, sampling, and blending facilities needed to ship products or receive charge stock.

4. Storage of additives, modifiers, dyes, and similar chemicals. Facilities for each depend on method of handling.

5. Slop tanks or process charge tanks for rerouting of off-specification product. These depend on process sequence, product loading facilities, pumping and transfer arrangements.

6. Pipe routing and support methods for hot and cold lines inside and outside dikes; needs to provide for expansion, anchoring, pipe guides, road or ditch crossings.

7. Access walkways, stairs, ladders, handrails, etc., for dikes, tanks, valve manifolds, loading facilities.

8. Special installations at pump stations, storage tanks, pipelines, loading facilities, etc. These include thermal relief systems, meter calibration loops, proving tanks, special instruments, scraper traps, emergency evacuation systems, soakaways, blanketing facilities, purging or equalizing systems, vents for pressurized storage.

9. Corrosion protection of tankage, piping, and other metal equipment or structures. Consider both internal and external conditions.

10. Shelter and protection for personnel and equipment. Consider ambient conditions, equipment design, plant practices, and special obligations such as are imposed by TEL suppliers for lead blending plants.

11. Lighting of facilities and areas. This must comply with safety and operating procedures.

12. Ground systems, required for storage tanks, pumps, and loading facilities (railroad, truck, ship).

Water Systems

Services and groupings of water systems must be established to suit plant requirements and water availability. Refinery systems generally include fire water, cooling water, service water, sanitary or drinking water, process water, and boiler feed water. Various combinations of source and distribution system can be utilized.

Flowing quantities and pressures necessary for users in process units, auxiliary units, and offsite should be determined. Both normal and emergency conditions should be developed.

Other aspects of water systems requiring decisions include:

1. Type of cooling-water system (once-through or recirculated, cooling tower, spray pond, air cooling), materials of construction, and supplementary details

2. Source of water for each service or grouping of services with alternates (municipal supply, sea, river, lake, wells), possible reuse of waste water

3. Type of collection facilities, such as intake structures, flumes, dams, wells, etc., for each water system

4. Design of distribution systems and equipment; pumps and pumping stations, piping-system arrangement, routing, supports, other details; specifications for materials and equipment

5. Possible need for fish screens, settling ponds, filters, chemical softening or clarifying facilities; shelters, tanks, mixers, pumps; control of algae, bacteria, or pH

6. Design of storage structures (tanks, ponds, reservoirs) to provide surge capacity in the water systems

7. Type of instruments to control or record flow of water in refinery water systems; manual, automatic, or remotely operated valves

Steam Systems

The first step in designing a refinery steam system is to determine pressure and temperature levels required in process units and offsites for steam, exhaust steam, and condensate. An over-all steam balance is usually needed to ascertain most economical steam usage. Flowing quantities must be established for users and makers of steam, and the quantity of steam available from waste heat must be determined.

Equipment capacities, specifications, and materials must be determined for steam generators, feed-water pumps, deaerators, water-treating facilities, and other related equipment. The type of fuel for boilers and related fuel efficiency need to be known. Other aspects of steam systems involved in design are:

1. Deaerators and water-treating specifications, chemical treatment for boiler water

2. Steam-distribution systems (make-up and pressure-reducing stations to control system pressure levels, desuper-heating if necessary, material specifications for piping and accessories)

3. Boiler blowdown systems (blowdown drums or pits and disposal facilities)

Refinery Fuel Systems

Possible sources of refinery fuel include process tail gas, purchased natural gas, heavy fuel oil, petroleum coke, unsalable distillates, and carbon monoxide. The required quantity, temperature, and pressure of each type of fuel must be determined.

Provisions must be made for mixing gaseous fuels, for controlling gas pressures, for pumping and heating fuel oil, and for storing oil and gas.

Refinery Air Systems

In the design of refinery air systems, the required quantities and pressures for plant, instrument, and other air systems must be determined, and standby and crossover provisions should be provided. Specifications and capacities should be established for compressors, inter- and aftercoolers, receiver vessels, and air driers when required.

Air-distribution systems between compressor locations and process and offsite users must be provided. These air facilities are frequently included within refinery utility plant limits to simplify air-distribution systems.

Blowdown, Pump-out, Relief, and Flare Systems

Suitable blowdown, pump-out, relief, and flare services must be established to meet process requirements. While certain of these systems are located completely within process-unit limits, others are classified as both process and offsite systems. Major aspects in establishing these services are:

1. Flow quantity and material composition.
2. Possible combinations of process upsets and emergencies for relief and flare systems.
3. Location and type of facility needed to suit flow conditions. The height of a flare stack is established by quantity and material and by location of other facilities. The effect on adjoining properties should be investigated.
4. Auxiliary equipment needed, such as water sprays, smokeless burners, knockout drums, ignition devices, and pump-out pumps.

Fire Protection

Fire protection, to be effective and economical, must be an early part of refinery design. Essential considerations are the extent and type of fire prevention needed, applicable local practices, codes, and insurance underwriters' requirements.

Other fire safety items essential in early planning are:

1. Fire-water system arrangement, capacity, and pressure level
2. Size of hazards, use of foam, and extent of mobile equipment
3. Foam protection system for fixed-roof storage tanks, spill areas, and other risks, including foam fire trucks, fixed installations, portable first-aid stations, and shelters for equipment and stores
4. Local standards and reciprocal fire-protection arrangements and minimum desirable interchangeability of equipment
5. First-aid fire-fighting equipment
6. Special fire-fighting systems, such as carbon dioxide protection and water sprays
7. Other requirements, including warning and communication systems, protective clothing, special gear and equipment, and buildings and shelters

Miscellaneous Utility Systems

Chemical preparation and distribution systems must be developed to suit plant requirements. These include caustic, acid, phenol, MEK, etc.

Solid-handling systems for catalyst and product should be established, and truck and rail facilities and warehouse arrangements and equipment adapted to suit.

Occasionally, refrigeration for low-temperature storage is required, as is inert gas or dry air for special storage facilities.

Electric Power and Communications

The source of electrical energy for the refinery must be determined. If purchased power is to be utilized, the voltage level and quantity of power available must be known. When power is generated in the refinery, the size of generators should be established in accordance with the philosophy on spare generation, with generator voltage usually established at one of the NEMA standard values. Following this, the plant power-distribution method (overhead or underground) should be established.

Once the over-all aspects of the refinery power system are established, decisions are needed for:

1. Voltage levels for various users.
2. Motor sizes.
3. Distribution voltages in the plant.
4. Substation capacity and location to suit loads. One main incoming switch and transformation station is usually provided, with other substations located to serve major power users.
5. Hazardous and nonhazardous area designations. (These are discussed in detail in Sec. 9 under Hazardous Electrical Areas, pages 9-59 to 9-63.)
6. Switch, switchgear, motor starter, and circuit-breaker specifications.
7. Shelters and enclosures for electrical gear or weatherproof or explosionproof gear in place of enclosures.
8. Area- and street-lighting requirements.
9. Telephone, signal, and fire-warning systems; type of equipment and installation details.

Sewers and Waste-disposal Systems

Types and groupings of sewer and waste-disposal systems must be established to suit process requirements and local conditions. Refinery disposal systems include clean, oily, storm, chemical, ballast, and sanitary wastes in various combinations. Flowing quantities and approximate composition of various wastes for each system must be determined.

Final waste-disposal outlets must be selected, as well as the separating and treating facilities needed for each type or group of wastes. Methods for disposing of separated oil and sludges must be developed.

Refinery economics require careful study of the possible reuse of waste water in refinery service, both to conserve water and reduce contamination of refinery effluent.

Design decisions associated with waste-disposal systems are:

1. Material and equipment specifications for sewers and waste-disposal facilities
2. Need for pumps or ejectors in the systems
3. Arrangement of sewer and ditch systems
4. Provisions for gaging flows and temperatures for record and tax purposes

Site Preparation and Investigation

Prior to any site preparation or construction, maps of the property and region must be obtained or prepared, both topographic and hydrographic. In addition, detail drawings of all existing installations and facilities are essential.

The nature of the terrain upon which construction is planned should be studied in detail. Important terrain factors include surface growth, topsoil, subsoil, ground water, rock location and outcropping, frost, stream bed or alluvial deposit locations, ease of excavation, surface drainage pattern, and permeability (as per cent runoff).

Subsurface soil characteristics are important. These include soil bearing, settlements, density, shrinkage, corrosiveness, grounding characteristics, vibration resistance, permeability, and angle of repose. In addition, pile-driving test data may be required.

The weather history of an area can have significant effect on design and construction of facilities. Designers should have access to meteorological data for the area, including temperature, humidity, rainfall intensity and duration, wind intensity and direction, snowfall, earthquake intensity, barometric pressure, tide or river stage, and wave heights.

Other basic aspects of site investigations include:

1. Transportation facilities available at the site, locations of railroads, highways, and marine facilities, as well as their capacities and limitations
2. Building materials available at the site; suitability of local sand, gravel, rock, and select earthen material for use in concrete, foundations, or construction
3. Specifications for materials for and installation of roads, fences, culverts, drainage ditches, lagoons, parking areas, etc., with fullest possible use of local and site materials

Refinery Buildings

Certain refinery services are more efficient when housed and possibly grouped. These services include operation offices, control and research laboratories, locker and wash facilities, workshops, equipment and supply storage, chemical storage, administration offices, fire fighting, and lunch facilities.

The sizes of buildings and shelters for grouping of services must be determined, based on the number of people using the facilities, equipment housed, and quantity of material stored. Also, the type of construction, material specifications, and building furnishings and equipment must be developed. Prefabricated buildings should be considered.

Heating, ventilating, and air-conditioning requirements for these buildings must be determined, and demand for water, steam, and electrical power. Special systems and equipment often are needed, such as building sprinkler systems, water-pressuring equipment, water heaters, cooling towers, etc.

Legal and Fiscal Obligations

Design of offsite facilities is affected by numerous statutes, codes, and ordinances. Those applicable at a particular site may vary widely from those in force at an adjoining locality. In some instances approvals or permits are necessary prior to construction. Compliance with other requirements may be dependent solely on inspection or similar police powers.

A general but not necessarily inclusive listing of legal and fiscal obligations which should be checked into includes:

1. Fiscalization of crude or product
2. Customs or bonded storage
3. Weights and measures
4. Public health or sanitation
5. Pure stream and harbor pollution
6. Air pollution or smoke abatement
7. Navigational regulations, aids, and lights
8. Pilotage rules
9. Aeronautic—heights and lighting
10. Railroad clearances and right of way
11. Atmospheric storage tanks
12. Pressure storage tanks
13. Unfired pressure vessels
14. Fired boilers
15. Air receivers
16. Buildings
17. Plumbing and sewerage
18. Electrical
19. Concrete structures
20. Steel structures
21. Pressure piping
22. Safety
23. Noise
24. Vibration
25. Labor

REFINERY LAYOUT CONSIDERATIONS

The over-all arrangement of a refinery will exert a major influence on plant operation, safety, personnel relations, and investment cost. A good plot arrangement must be based on the primary function of producing refined products, but consideration also should be given to supplemental facilities such as utility preparation, refinery service buildings, and product shipping. As a result of the growing scarcity of suitable industrial property many refineries are now being built on marshy or hilly terrain. Plant arrangements must be so developed as to obtain maximum use of such sites at reasonable cost.

General factors which affect the plant layout and the interrelation of various facilities include (1) physical features of site, (2) refinery oil systems, (3) utility and auxiliary systems, (4) service facilities, and (5) future expansion.

Physical Features of Site

Physical features of a plant site have a major effect on all facilities included in the plant. These features, together with local data as to transportation facilities, labor, repair-shop facilities, climatic conditions, availability of water and power, etc., must be considered in the course of site selection. Generally, however, a site is selected as the most acceptable of various alternates in a given locality, and the plant plot plan must be adapted to suit conditions.

Topographic and Property Survey

Topographic and property-line survey maps of a site are necessary in preparing a plant layout. When convenient, one drawing can be used for all necessary information. Topographic data should consist of surface contours drawn at intervals of not more than 2 ft locating high and low points, ditches, etc. When the site is relatively flat, contours should be at 1-ft intervals.

Location of streams, swamps, ponds, tree groups, and other unusual features should be shown on the survey maps. Details as to location, sizes, and elevations of existing roads, railroads, structures, equipment, and pipelines are required. Roads, railroads, and pipelines immediately adjoining the plant should also be located on the site map.

Information should be accumulated regarding the uses of adjoining property, which can affect location of facilities for many reasons. In some instances noise, odor, and light could be objectionable to adjoining property owners. In other instances neighboring operations could constitute a potential fire hazard. In general, it is advisable to allow buffer zones ranging in width from 150 to 250 ft from plant fence to the nearest building or equipment.

Utility and Service Entry

The location or probable point of entry of utilities and services must be determined. Piped systems such as natural gas and established water supply should be located. Point of disposal of waste water should be known. If purchased electrical power is available, locations of both temporary and permanent power lines are needed.

Aspects of the site related to receipt of crude oil and shipment of product must be developed. Crude pipeline routes should be located. Highway and rail access for product shipping as well as for other services is necessary. When water shipment of crude or product is contemplated, hydrographic data for the navigable stream are needed. Information as to sizes and types of vessels must also be developed to permit proper orientation of facilities for loading or unloading.

Geological and Geographic Features

Use of portions of the site for specific installations should be based on suitability for support of structures and tanks. Where foundation conditions vary widely, the location of process units should favor areas having high soil-bearing values. Prefer-

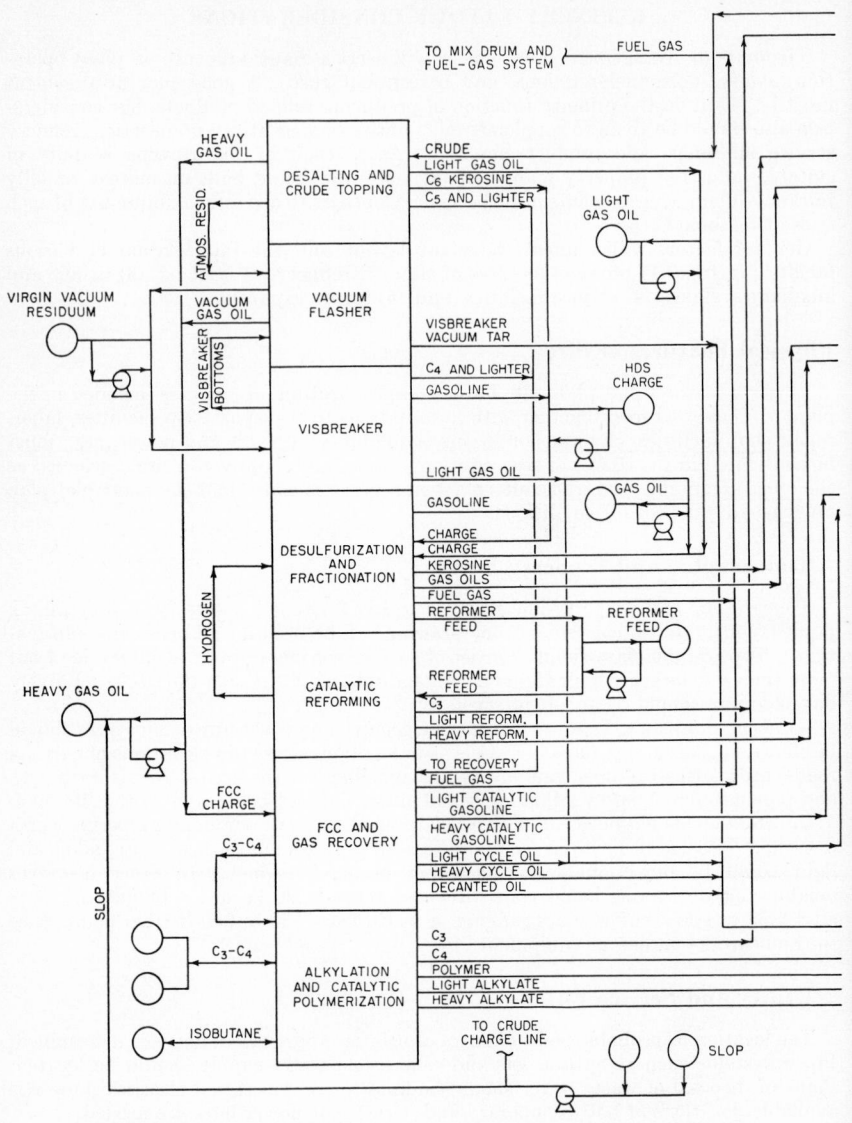

Fig. 8-1. Block flow diagram

ably, subsurface investigations should be made to determine allowable soil-bearing values or underground rock contours. Subsurface investigations should also determine ground-water levels.

The direction of the prevailing winds at the site should be established to permit proper orientation of furnaces, flares, dusty operations (such as sulfur handling), and possibly malodorous equipment. For winter operations in windy regions and in high-snowfall localities, it may be desirable to locate certain portions of the refinery to minimize interference from drifting. The relative rainfall rate should also be known to determine if site drainage of certain areas will be critical.

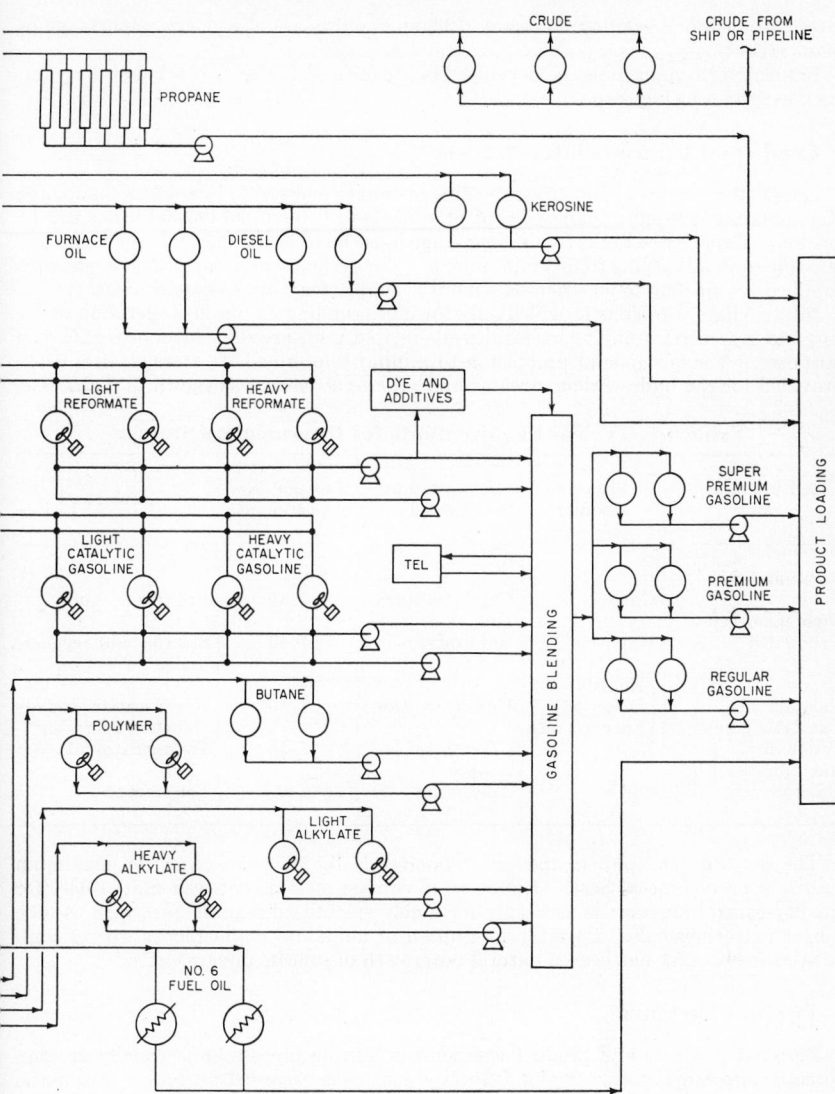

of typical simple refinery.

Information should be obtained as to trees and vegetable growths. Wooded areas must be grubbed clear. In some cases, removal of peat or marshy soil will be necessary. In general, topsoil must be removed where process units and major structures will be located. Areas should be reserved for soil of peat and other high-organic-content soil.

Refinery Oil Systems

Included under the general subject of refinery oil systems are facilities related to storage of crude, intermediate and final products, blending and shipping, and the interrelationship of refining processes.   These facilities are of primary importance in plant layout.

Refinery oil systems must be developed in the form of a refinery block flow diagram, such as shown in Fig. 8-1.

Crude and Intermediate Systems

Crude oil is received from crude pipeline or tanker and stored in not less than three storage tanks.   If more than one type of crude is to be used, additional tanks will be needed.   Capacity of tanks for crude storage must be based on method and frequency of crude receipt, varying from a minimum of 7 days crude charge capacity for refineries supplied by pipeline to as much as 4 months supply for winter-isolated areas.

Intermediate storage is provided only for units running on blocked operation or for surge to keep certain units on stream while related units are shut down or for startup purposes.   Though general practice is to minimize intermediate storage, it is often provided for the hydrocarbon streams and process sequences as shown in Table 8-1.

Table 8-1. Typical Requirements for Intermediate Storage

Feed preparation unit	Flowing medium	Process unit charged	Storage days at charge rate	Remarks
Vacuum dist. Coker Visbreaker	Gas oil	Catalytic cracking	7–15	For turnaround
Crude dist.	Naphtha	Reformer	7–10	Startup and regeneration
Crude dist. Vacuum dist. Catalytic crack	Kerosine Diesel oil Furnace oil	Desulfurization	7–40	Regeneration and/or blocked operation
Crude dist. Vacuum dist.	Residuum	Visbreaker Coker	7–15	For turnaround
Gas recovery	C₃-C₄	Alkylation	1–2	For startup

The general tendency in modern refineries is to integrate units for maximum reutilization of process heat.   Rundown of process streams between units solely for quality-control purposes is used only for highly specialized manufacture such as lube oils or petrochemicals.   Physical integration of units into major blocks with a common control center has been a natural outgrowth of process integration.

Product Systems

Finished products and product components leaving process units usually are run directly into large storage tanks.   Quality control is exercised by proper automatic control of process variables in units.   Products can be blended directly by combining streams leaving units or by blending components which are stored separately.   There are various alternate methods of implementing the above basic schemes.   The capacity, number, and size of product storage tanks will vary with the blending arrangement.   The refinery oil block flow diagram must be developed in sufficient detail to establish blending method and tankage.

In general, refineries include tankage for 30 to 60 days' storage of product make.   Fortanker shipping of product it is usually necessary to provide one or two tanks for each service having a total capacity equal to the tanker capacity to allow for sampling and gaging.

In a large plant this storage quantity fits into the normal storage pattern, but in a small refinery extra tanks may be necessary.   If tanker shipping is a seasonal operation, one set of tanks can be used for combined services such as gasoline and middle distillates.

Layout of storage tanks and related facilities will be affected by general pump arrangement.   Pumps can be arranged in groups, or individual pumps can be provided for a tank or pair of tanks.   The arrangement of pumps in groups (the number of groups will be a function of refinery size) results in centralized operation but requires long runs of large-size suction piping.   Use of individual pumps will result in a low-cost installation through savings in piping costs.

For purposes of plant layout, space requirements must be developed for various facilities.   The largest components of refinery oil systems are oil-storage tanks. General principles involved in tankage installations are discussed later in this section (pages 8-17 to 8-33).   Generous allotments of space should be assumed for related offsite items such as pump groups, TEL plants, blending facilities, etc.   The assumed space requirements can be verified in later phases of design.   Some approximations of over-all space requirements are listed in Table 8-2.

Table 8-2. Over-all Space Requirements

Item	Space per item, sq ft
Pumps	150
TEL plants	4,000
Blenders	3,000
Truck loading per truck	1,000
Rail loading per car	1,000

Space requirements for process units should be developed from the actual units, but it is frequently necessary to establish a general refinery arrangement initially.   For approximate space requirements, major groupings of units can be assumed to require a space about 250 ft wide by 500 ft long.   The usual new refinery will include from two to four such spaces plus allowance for future groupings.

Utility and Auxiliary Systems

Refinery utility and auxiliary systems include all facilities related to preparation and distribution of steam, air, water, fuel, electrical power, chemicals, and catalysts, as well as facilities related to disposal of solid, liquid, and gaseous waste materials.

Utility Plants

Facilities for generation of steam and electrical power, when required, are usually located within definite battery limits of a utility plant.   In certain refineries using CO boilers, the utility plant is located adjacent to or contiguous with a process unit. This results in minimum length of flue gas line.

In addition to steam boilers and electrical generators, utility plants include boiler feed-water treating facilities, deaerators, boiler feed pumps, fuel-oil heating and pumping equipment, fuel-gas mixing systems, and facilities for preparation of plant and instrument air.

When electrical power for all or part of the refinery load is generated in the refinery utility plant, the main switchgear and transformers are located in this plant.   It is also common in refineries using purchased electrical energy exclusively to locate the main substation in or adjacent to the utility plant.

Cooling towers can be located within the utility plant limits with circulating cooling-water pumps and fire-water pumps which take suction from the cooling-tower basin adjacent to the tower.   When make-up cooling water must be treated to remove sediment, decrease hardness, etc., it is frequently economical to combine the initial phases of boiler feed make-up treatment with cooling-water treating.   In such instances, water-treating facilities could be situated in the utility plant without reference to location of cooling towers.

The space required for a refinery utility plant depends on the equipment included and its location in reference to process units.   When the plant is incorporated in a major process block, less space is needed than if a separate block is established. Usually a space 225 ft long by 300 ft wide is ample for most utility plants.

Water Systems

Water pumping stations for once-through cooling water or for make-up water must be located at points where reliable quantities of suitable water are obtainable. Though this station may be located outside the refinery, it should be as close as possible to major water users in order to minimize distribution.

As noted before, water-treating plants and cooling towers are sometimes located outside utility plant battery limits. This is preferable for a number of reasons. If extensive sedimentation basins are required, ample space must be provided. Silt disposal also must be considered; the simplest disposal is to return silt to the stream downstream of the water intake. Wind direction is an important factor in determining location of cooling towers. Spray effect on adjoining equipment and fire hazard must both be considered.

Chemicals and Catalysts

Chemicals and catalysts are received in both liquid and solid forms. When possible, these materials are stored in drums or bags in a warehouse-type building. Storage tanks are provided for acid, caustic, phenol, MEK, etc. In remote areas liquid bulk shipment of materials such as caustic and phenol are not available and melt facilities must be provided.

Chemical and catalyst warehouses are usually located near a road and rail siding to facilitate handling. Bulk catalyst, acid, caustic, phenol, and similar materials which do not constitute a fire hazard may be located adjacent to or in process areas. Flare stacks or ground flares must be located at a point remote from process units or other hazardous areas. Site topography, wind direction, and the nature of adjoining property will influence placing of this equipment.

Gaseous and Liquid Wastes

Facilities for liquid blowdown, when required, are often included within process-unit battery limits. Whether in or near process areas, it is important that blowdown stacks be remote from furnaces and that suitable segregated sewers are provided.

Waste-disposal and -treating plants usually consist of oil separators and holding basins or lagoons plus, when necessary, chemical and biological treating facilities. These facilities require a large space which can be of marginal caliber. Normally a low portion of the plant is preferred. A 3- to 4-acre site is the minimum necessary for this purpose.

Service Facilities

Service facilities included in refineries comprise all equipment, buildings, and space devoted to refinery administration, process control, maintenance, and accounting. Administrative and similar facilities may be provided for nonrefining departments to a limited extent, but garages, warehouses, shops, and offices for marketing, transportation, and pipeline operations are not included in refinery service facilities.

Most refinery service facilities are included in or associated with buildings and are usually grouped in one or two refinery blocks. Separate or combined buildings are normally provided for administration, operations, laboratory, change, gate control, vehicle garage, firehouse, workshop, and warehouse.

A common grouping of service facilities is one including administration, laboratory, operations, and gatehouse. The gatehouse portion frequently combines plant police, timekeeping, personnel office, and first-aid rooms. Adminstration and operations offices are often combined and located adjacent to the laboratory. For some plants, it is necessary to locate the gatehouse, plant police, and time-keeping in a building adjacent to the change house. In large refineries, it is preferable to have the operations office and laboratory closer to various process units than is desirable for general offices. The administration building should be so situated as to permit access to this building independently of the refinery in general.

Table 8-3. Service Facility Space Requirements

Name, function, and factors causing variations	Area, sq ft	
	Minimum	Maximum
Operations		
Office space for supervisory process and maintenance personnel. Size of building varies with complexity of refining operations and number of process units more than with refinery throughput.	Included in administration building as 25 % of building	8,500
Laboratory		
Process control and offices with some research for large refineries. Size of building varies with complexity of refining operation and number of products and product components.	3,500	12,600
Change house		
Locker and wash facilities. Varies directly with plant manpower, which is dependent on number of process units, size of utility plant, extent of offsites, and maintenance provisions.	2,200	4,100
Shop		
Maintenance and repairs with some fabrication. Varies with plant location and availability of local shop facilities and with type of maintenance, i.e., refiner's or contract.	6,200	22,500
Warehouse		
Storage of spare parts, maintenance material, and supplies. Varies with plant location, number of process units, extent of offsites, and refinery throughput.	Warehouse 7,200, chemical stores 800	16,100
Chemical stores		
Storage of chemicals, catalyst, paints, and oil. Varies with plant location and refinery throughput.	Included in general warehouse as 10 % of building	4,000
Administration		
Office space for management, engineering, accounting, and general clerical. Size of building and type of facilities vary with refiner's organization and plant location.	Administration 7,500, operations 2,500, total 10,000	22,000
Firehouse		
Storage of fire-fighting equipment and supplies. Varies with size of storage tanks with cone roofs.	1,000	4,000
Canteen		
Lunchroom for plant personnel. Includes kitchen for large refineries. Varies with plant location and refiner's practices.	1,200	7,000
Storage yard		
Outside storage of bulky material and equipment varies with shop facilities, maintenance practices, location, and refinery throughput.	20,000	60,000

Buildings housing shops, warehouses, garages, and fire equipment are frequently grouped. Shops and warehouses should be arranged to allow for convenient servicing of process units. Space must be provided adjacent to shops and warehouses for bulk material storage and for storage of large equipment. Road and rail access to storage areas and buildings is essential.

Change facilities can be located in various places, sometimes at the gatehouse, sometimes centralized at a point close to shops and process plants. In some refineries, locker and wash facilities are located both in shop buildings and in or near process areas, thus providing separate facilities for maintenance and operating personnel. The use of contract maintenance may require separate facilities. For contract maintenance, toilet rooms and washrooms without lockers and benches are usually satisfactory.

Service facilities also include canteens or cafeterias. In some plants, lunchrooms with facilities for beverages only are adequate. Canteen facilities are for direct use by nonoperating personnel only. Facilities for operating personnel are provided within process units.

Usual space requirement for various service functions are given in Table 8-3.

Future Expansion

Future expansion for a refinery is of two definite types: (1) known expansion planned for a definite time period and (2) unknown expansion. Known expansion can usually be expressed in terms of increasing refinery throughput or upgrading product. The implicit time period is generally conditioned to market requirements.

Provision for known expansion is usually made by apportioning definite space for specific purposes. The required space for additional processing equipment and offsite tankage can be determined with reasonable accuracy. Some known future expansion is restricted solely to offsite facilities such as tankage for additional flexibility in handling crude and intermediate and final product and does not necessarily involve additional pumping facilities.

Expansions of a known type involving reservation of space for predetermined purposes may result in portions of a refinery remaining unoccupied for a period and result in increased cost. Such unoccupied spaces in the center of an established plant are costly. They are served by roads, fire-protection systems, and drainage, and oil and utility piping runs must be lengthened in order to allow space to remain unoccupied.

Expansion of an unknown type involves apportioning of areas to increase refinery throughput, upgrade product, or diversify manufacture. Logical and orderly future expansion can be effected only by allowing space adjacent to each type of facility for expansion. As a result, new process units will adjoin existing units and new tankage will adjoin existing tankage. Expansion of the refinery utility plant will normally be allowed for in the original layout.

When diversification of manufacture is introduced in a refinery, it is frequently advisable to expand by installing segregated facilities. Diversification will generally involve adding units for the manufacture of petrochemicals, asphalt, or lube oils. With each of these products a different type of product-handling scheme is introduced. Storage and shipping facilities are sufficiently different from normal petroleum product practices to justify installation of segregated facilities. The refinery is then considered only a source of feedstock and utilities for such future plants.

BASIC LAYOUT SCHEMES

For a given set of conditions as defined by the preceding factors, a wide variety of plant layouts can be developed. The most acceptable layouts are normally variations of a few basic schemes.

As the basic function of a refinery is the production of refined hydrocarbons, the best layout will favor this operation. Ideally a refinery should be considered as a box into one end of which crude oil is pumped and from the opposite end of which various refined products flow. Obviously the best plant arrangement follows the general flow pattern of crude to process units to shipping, with tankage interposed only as necessary. Storage tanks are merely surges or bumps in pipelines in accordance with this concept.

Following this basic philosophy, a refinery can be considered as a group of general areas arranged to include crude storage, process units, intermediate storage, product storage, shipping, utilities, administration and service, and waste disposal.

Some of these areas may be subdivided into two or more portions. Utilities can be separated into a utility plant for steam, air, etc., and a cooling-water area. Shipping facilities are sometimes subdivided by mode of transportation. All the above areas should be expandable for future purposes.

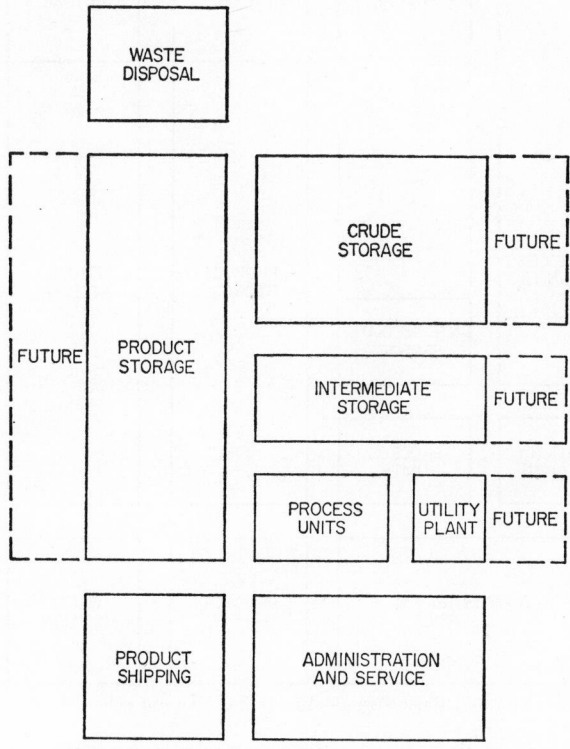

Fig. 8-2. Basic L-shaped refinery layout scheme.

In addition to arranging areas to follow the oil-flow pattern, other interrelations should be applied. Utility-plant and cooling-water facilities should be located close to process units to minimize runs of utility piping. Service facilities such as shops and warehouse should be close to process units for convenience. Administration-type buildings should be isolated from odors or noise but convenient to process facilities for supervisory purposes. Waste-disposal facilities should be located to minimize odors and hazard from or to adjoining equipment.

Two basic refinery layout schemes are shown in Figs. 8-2 and 8-3. These schemes show L-shaped and straight-line patterns of process facilities with both layouts adaptable to expansion. Unoccupied areas can be left for future expansion between adjoining areas if so desired.

All areas of the refinery should be subdivided into blocks. The maximum size of blocks should be as established by storage-tank layout. When possible the maximum size measured at roads should be 900 ft. Blocks for process units and for small tankage should preferably be limited to a 600-ft maximum dimension.

Blocks of facilities should be subdivided and served by refinery roads. Whenever possible roads should be straight and extend from end to end of the plant. Every point in the plant except minor service facilities should be served by alternate routes so as to be accessible in the event of blockage of a single route by fire or construction. Roads or future roads should parallel the refinery fence around the complete periphery of the plant.

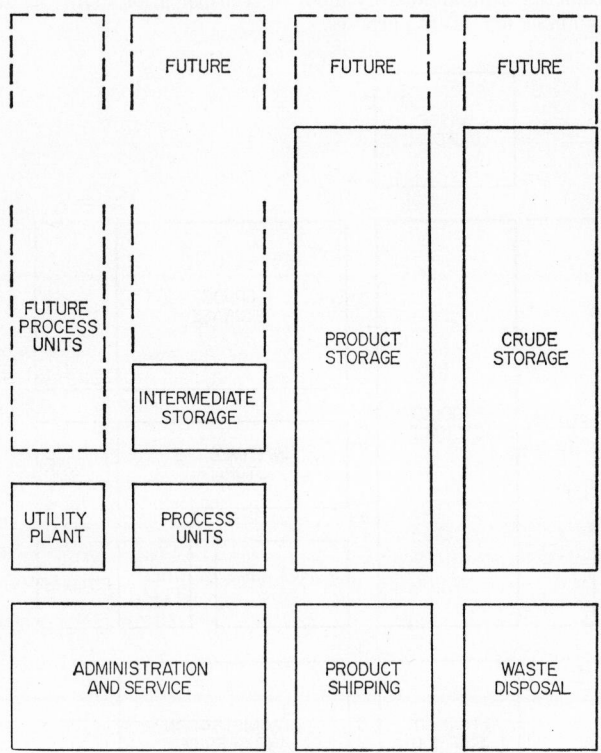

Fig. 8-3. Basic straight-line refinery layout scheme.

Refinery roads should be of two general types: main and secondary. Main roads should lead from the main entrance to administration and service facilities and along at least one side of all process units. The refinery utility plant should be located on a main road. Main roads should also serve the refinery truck-loading station.

Minimum width of main roads should be 20 ft of pavement with at least 3-ft-wide improved shoulders. A main road should be centered in a right of way not less than 150 ft wide. If a major pipe bank parallels a road, a minimum clear space of 100 ft should be provided from the road center line to the edge of right of way on the pipe bank side of the road. Greater widths are preferred. Roads for access to loading areas and service buildings are frequently provided with greater widths of paving and right of way.

Minimum width of secondary roads should be 16 ft of pavement with at least 2-ft-wide improved shoulders. A secondary road should be centered in a right of way not less than 100 ft wide. The width of the right of way should be measured to the toe of the slope of dikes, to a building line, or to the nearest point on any piece of equipment such as a pump or tank.

EVALUATION OF PLOT PLANS

Once various plant layouts have been developed based on the schemes and factors described above, each must be evaluated on the basis of the following factors: (1) safety, (2) operation, (3) maintenance, (4) expansion, (5) construction, (6) initial cost. Table 8-4 lists various factors to be considered under each of these areas and gives a correlation value for each one.   To evaluate a group of plot plans, each factor should be rated on each plot plan on the basis of 10.   Each rating is then multiplied by the correlation value for that item to obtain an adjusted rating and a summation of points.

TANKAGE SELECTION AND INSTALLATION

Purpose of Storage Vessels

In the processing of petroleum, sizable inventories of crude, semifinished, and finished hydrocarbons are required.   Both atmospheric and pressure storage vessels are used.   A major refinery offsite cost is represented by storage facilities and related piping, access roads, dikes, and fire and safety equipment.

Atmospheric Storage

The major portion of oil, water, and other liquids stored in refineries is contained in atmospheric storage tanks.   These are normally steel vessels which operate at or only slightly above atmospheric pressure.   The capacity of various storage tanks is set by processing, blending, shipping, and marketing requirements.

Types of Tanks

The most frequently used storage tanks are welded-steel tanks fabricated in accordance with the API 650 Specification.   Small tanks generally conform to API 12F *Specification for Small Welded Production Tanks*.   Other API tank specifications are for tanks which are considered portable and are employed in producing fields.   These include bolted tanks (API 12B) and large prefabricated welded production tanks (API 12D).   Tanks for water storage should be fabricated in accordance with the AWWA specification which takes account of the greater weight of water and the need for a corrosion allowance.

The main distinguishing feature of atmospheric oil-storage tanks is the type of roof employed.   The two basic types of roofs are **fixed** and **floating**.   The choice between types of roof should be predicated on (1) evaporation loss, (2) fire risk, (3) product contamination from atmosphere, and (4) maintenance cost resulting from corrosion.

Evaporation losses vary with the type of material stored and the tank operating cycle.   The two causes of evaporation losses are tank filling and breathing.   Filling losses are influenced by the throughput of the plant and methods and frequency of shipping.   Breathing losses are caused by variations of ambient atmospheric conditions and depend on the vapor pressure of the material and the volume of vapor space in the tank.

When the stored material is subject to ready ignition, a floating roof is desirable to reduce the risk of fire.   Such materials, for which the tank vapor space is usually in the explosive range, include crude petroleum, gasoline components, jet fuels, heavy naphtha, and kerosines with flash points below about 80°F.

Some water-soluble solvents, lubricating oils, and materials adversely affected by air are occasionally stored in floating-roof tanks.   Alternatively, fixed-roof tanks with inert gas blankets are also used to protect air-sensitive products.

Materials which can evolve corrosive vapors, such as crude petroleum and some gasoline components, often require special types of floating-roof tanks to reduce the effects of such corrosive vapors.

Table 8-4. Plot Plan Evaluation

Factor	Arrangement A Plot plan drawing No.				Arrangement B Plot plan drawing No.			
	Remarks	Rating	Correl. value	Adjusted rating	Remarks	Rating	Correl. value	Adjusted rating
A. *Safety:*								
1. Does relative layout of facilities minimize hazard of fire or explosion?			10				10	
2. Is access to various facilities satisfactory for fire fighting or escape?			5				5	
3. Is fire control and extinguishment affected by wind, grades, fire-water, system, foam system, dike volumes, etc.			10				10	
B. *Operation:*								
1. Does plot result in low-cost operations as related to following:								
a. Oil transfer			5				5	
b. Utilities			5				5	
c. Product loading			5				5	
d. Waste disposal			5				5	
2. Is convenience of operation and supervision affected by following:								
a. Terraces and grades			5				5	
b. Relative locations of services			5				5	
c. Access by road and rail			5				5	
C. *Maintenance:*								
1. Will layout result in reasonable cost of maintenance of offsite facilities including earthen dikes and other structures, roads, and piping?			10				10	
2. Is arrangement of equipment, access roadways, and extent of paving convenient for maintenance?			5				5	

Notes: Rate each factor of each arrangement on basis of value 10 for best ranging downward to 1 for worst. Multiply by correlation value to obtain adjusted rating. Summation of adjusted ratings is indication of plot plan preference.

Table 8-4. Plot Plan Evaluation (Continued)

Factor	Arrangement A Plot plan drawing No.				Arrangement B Plot plan drawing No.			
	Remarks	Rating	Correl. value	Adjusted rating	Remarks	Rating	Correl. value	Adjusted rating
D. *Expansion:*								
1. Is amount of space for expansion adequate?			2				1	
2. Is location of space for expansion in proper relationship with regard to present and future?			1				1	
3. Is additional property available?			1				1	
4. Can crude and product systems be expanded?			3				3	
5. Can utility generating and distribution facilities be expanded?			3				3	
E. *Construction:*								
1. Does layout minimize problems related to physical factors such as foundation conditions, ground and surface water, sloping terrain, etc.?			3				3	
2. Is working and storage space available for present and future?			2				2	
F. *Initial cost:*								
1. Does layout minimize initial cost of following:								
a. Earth moving, roads, and railroads			2				2	
b. Oil piping			2				2	
c. Utility piping			1				1	
d. Electrical distribution			1				1	
e. Pumps			1				1	
2. Is initial cost affected by following:								
a. Need for special structures such as retaining walls			1				1	
b. Amount of waste space			1				1	
c. Relocation of existing facilities			1				1	

Fluids which are vapors at ambient temperatures can also be stored in atmospheric tanks as liquids at low temperatures. These tanks normally operate at low pressures (measured in inches of water) and are therefore constructed in accordance with API Standard 620. Proper insulation of low-temperature atmospheric storage tanks is important.

Size and Capacities

Tank capacities are initially dictated by plant process and operating requirements· Generally, the principles involved have been discussed under Refinery Layout, pages 8-7 to 8-14. The process and operating requirements must be tailored to obtain an economical design consistent with site limitations and safety from fire or explosion.

While atmospheric storage tanks are usually referred to as "standard-size" tanks, it would be more proper to refer to typical or suggested sizes. The so-called standard tank sizes are of three general types:

1. API specifications list typical sizes. The sizes listed in API Standard 650 are based on usual rolling-mill practices. Basically two lists of sizes are included in this specification utilizing multiple courses of 6- and 8-ft-wide plates.

2. Historical standard sizes or those tank sizes which were commonly used in production fields and refineries. An example of this is the well-known 55,000-bbl 100-ft-diameter by 40-ft-high tank.

3. Manufacturers' standards which have been developed by each manufacturer from tanks previously constructed by him. The sizes will usually reflect proprietary designs, sources of material, and shop practices.

Table 8-5. Approximate Maximum Tank Diameters

Height, ft	Diameter, ft	Capacity, bbl
36	294	435,600
40	264	390,300
42	251	370,400
48	219	322,300
54	194	284,500
56	187	274,200
60	174	254,300
64	163	238,100

Maximum diameter permitted by API Standard 650 for liquids of specific gravity 1.00 or lighter. Diameters are limited by maximum plate thickness of $1\frac{1}{2}$ in. Allowable stress 21,000 psi, joint efficiency 85 per cent. Thickness calculated by the formula

$$t = 0.0001456 \ D(H - 1)S$$

in which t = minimum thickness, in.
D = nominal inside diameter of tank, ft
H = height from bottom of course under consideration to top of top angle or to bottom of any overflow which limits tank filling height, ft
S = specific gravity of liquid to be stored but in no case less than 1.0

The sizes of storage tanks used in a refinery should be the most economical ones which satisfy site conditions. Tanks frequently range in height from 15 to 56 ft, although higher tanks have been constructed. Diameters of tanks are limited by API standard requirements of material, plate thickness, and height. Table 8-5 lists limiting diameters for various heights for tanks designed in accordance with API 650. Large-capacity tanks can be built as noncode tanks with diameters and heights greater than those permitted by the requirements of API 650.

When soil-bearing conditions permit, the tall tanks are usually most economical as to both tank and foundation cost. Tall tanks also conserve space in some plant arrangements. Tall floating-roof tanks are definitely less costly. Small-capacity tanks ranging in diameter from 12 to 25 ft are frequently limited to 18 ft high to permit roof access by means of vertical ladder. When tank heights exceed 18 ft, it is customary to provide stairways leading to the roof. Tank heights usually do not exceed one and one-half diameters. Various typical sizes and capacities are listed in Table 8-6.

Table 8-6. Typical Storage-tank Sizes

Diameter, ft	Height, ft	Capacity, bbl	Diameter, ft	Height, ft	Capacity, bbl
15	18	500	70	40	27,400
20	18	1,000	70	48	32,880
20	24	1,345	73⅓	40	30,000
25	18	1,500	75	48	37,300
25	24	2,000	80	40	35,000
25	30	2,500	80	45	40,000
25	36	3,000	80	48	42,500
27	24	2,400	90	40	45,000
30	24	3,000	90	45	50,500
30	30	3,750	90	48	54,384
30	36	4,536	100	40	55,000
30	40	5,000	100	45	62,500
34	30	4,800	100	48	67,000
35	30	5,100	110	40	76,100
35	36	6,100	110	48	80,000
36⅔	40	7,500	120	40	80,000
40	30	6,715	120	45	90,000
40	32	7,100	120	48	96,000
40	36	8,000	130	48	113,000
40	40	8,900	134	40	100,000
40	42	9,400	134	48	120,000
41	36	8,400	140	40	109,500
42½	40	10,000	140	48	131,500
44	36	9,500	150	40	125,000
45	36	10,000	150	48	150,000
45	42	11,895	160	40	142,500
45	45	12,600	160	45	160,000
48	40	12,500	160	48	170,000
48	48	15,000	170	48	192,000
49	36	12,000	180	40	180,000
50	40	14,000	180	48	216,000
50	45	15,600	200	40	224,000
52	40	15,000	200	48	268,000
60	40	20,000	220	40	267,000
60	48	24,000	220	45	301,000
67	40	25,000	220	47	325,000
67	48	30,000	292	28	334,000
70	36½	25,000			

As tank diameters increase, the cost of fixed-roof structures (columns and beams) approaches the cost of floating roofs. While the break-even cost is greater than the cost of the largest tanks now in use, the advantages derived from use of floating-roof tanks may offset the cost saved by a cone roof. Large-diameter tanks with heights appreciably greater than the limitations of API 650 Standard would probably be the most economical type if built with floating roofs. Most of the extremely large diameter tanks which have been built are quite low to suit poor soil-bearing conditions.

Fixed-roof Tanks

Fixed-roof atmospheric storage tanks are normally freely vented to the atmosphere. The most commonly used type of fixed roof is the cone roof, which is the most inexpensive storage tank available.

Cone roofs can be either column supported or self-supported. Self-supported roofs are actually supported at the vessel shell, and trusses are used to carry normal roof beams. This type of cone roof may be required for tanks located on unevenly yielding supports to avoid uneven settlement of the roof and shell. Self-supported roofs are frequently used for tanks up to 20 ft in diameter at no incremental cost. As the tank diameter increases beyond 20 ft, the self-supporting fixed roof is more costly than column-supported roofs. This incremental cost ranges from about 5 per cent at a 25 ft diameter up to slightly over 20 per cent at a 150 ft diameter.

The self-supporting roofs for small tanks may be built with a steeper slope than column-supported roofs and are called dome or umbrella roofs. This type of roof has a stronger roof to shell joint and can operate at a small pressure (inches of water). In the event of an explosion, greater damage to shell plates may result because of this joint.

The usual column-supported cone roof is built with columns and supporting beams carried by the columns. The roof plates are normally laid loosely on supporting beams and lap-welded to the adjoining plate. Plates are not attached to supporting beams. It is common to assume the maximum working pressure in the tank to be equal to the unit weight of roof plates so that the plates barely lift off the supporting beams. When $\frac{3}{16}$-in.-thick roof plates are used, this pressure is 7.65 psf or $1\frac{1}{2}$ in. of water.

Fixed-roof tanks incur filling losses proportional to the product inage and outage. Breathing losses also occur in accordance with vapor pressure of stored material and normal liquid and vapor volume. This type of tank must be vented to atmosphere with the use of free or controlled vents. Vents are described in greater detail in a later section.

Column-supported cone roofs are preferred for all installations of tanks over 20 ft diameter. When soil conditions control, the more costly self-supporting roof may be required. When a group of small tanks is arranged for access by connecting walkways instead of individual stairways or ladders, self-supporting roofs may be used to simplify walkway support.

Floating-roof Tanks

Floating-roof tanks consist of a bottom and cylindrical shell to contain liquid. The roof of the tank floats in the liquid and rises and falls within the shell. The upper edge of the tank shell is supported by a wind girder, which can also serve as a walkway. The floating roof is fitted with a seal to close the space between the roof and shell.

Floating roofs are made in three general types, namely, (1) pan, (2) pontoon, and (3) double deck. The various types with a few major variations are shown in Fig. 8-4.

Pan roofs are basically open dishes which float on the liquid surface. They are unstable and may overturn and sink. Because of this and only a slight cost advantage over the minimum pontoon roof, pan roofs are seldom used.

Pontoon roofs consist of a deck plate supported by one or more pontoons. When pontoons are arranged to keep the deck completely out of the liquid, the roof is referred to as a high- or clear-deck pontoon roof. If the deck is below the liquid level, the roof is called a low deck. The minimum pontoon roof has pontoons under approximately 30 per cent of the deck area. The volume of pontoons necessary must be determined by snow or rain loads anticipated.

A double-deck roof has two complete plate decks with suitable structural stiffness between the decks. The bottom deck is continuously submerged in the stored liquid. The empty space between the two decks serves as insulation for the product. Double-deck roofs are most costly in larger sizes. In sizes under 50 ft diameter, double-deck roofs are as cheap as or cheaper than pontoon roofs.

Floating-roof tanks are used in preference to fixed-roof atmospheric storage tanks for volatile materials and liquids which evolve corrosive vapors. The major aims are to minimize product losses and to control fires. The type of floating roof used can also be influenced by the material stored.

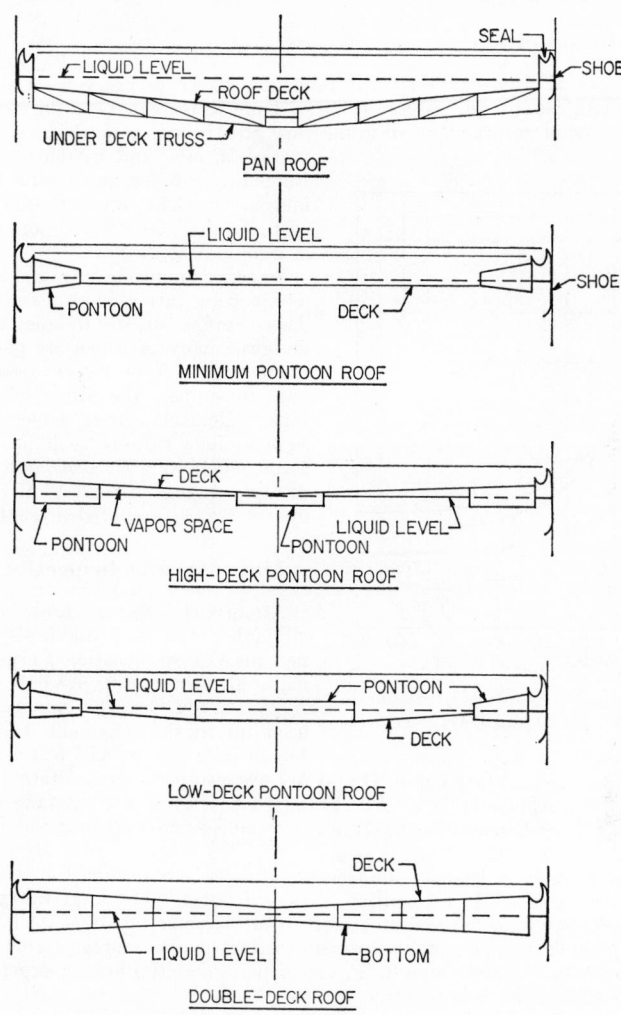

FIG. 8-4. The types of floating roofs for storage tanks.

Volatile fluids which evolve large amounts of vapor when heated should be insulated from the effects of sun temperature. Materials such as gasoline should be stored in high-deck pontoon-roof tanks to allow for an insulating vapor space. When the vapor collecting under a floating roof can be corrosive, the roof plates can be damaged rapidly. In such cases low-deck roofs in which the deck is submerged in the stored fluid are preferred. Low-deck pontoon roofs are used for crude petroleum. Double-deck roofs provide insulation plus a submerged deck and are, therefore, suitable for use with any material. Double-deck roofs are preferred for use with highly volatile corrosive fluids.

Floating roofs are used to minimize fire hazard in the storage of volatile liquids by avoiding explosive mixtures in the vapor spaces of tanks. Actually the incidence of fires in floating-roof tanks is as high or higher than for fixed-roof tanks, but these fires (usually seal fires) are easily extinguished and cause little damage. For fire prevention and extinguishment of floating tank seal fires see Sec. 9.

Storage-tank Costs

Relative costs of atmospheric storage tanks are shown in Fig. 8-5. Costs are in terms of dollars per barrel. While costs of large tanks can be estimated with a reasonable degree of accuracy, small tanks vary greatly with manufacturer, locality, materials, etc., and therefore, estimates from Fig. 8-5 for these tank costs are inaccurate. The relative difference in cost for double-deck vs. pontoon roofs is also indicated.

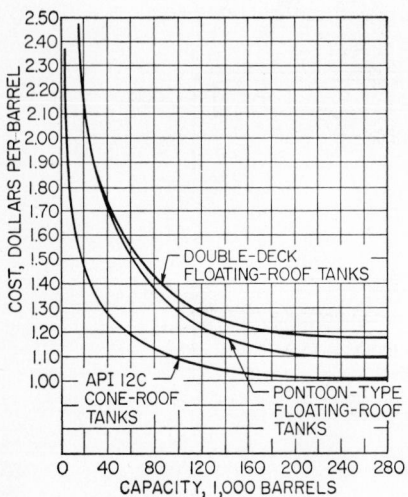

FIG. 8-5. Approximate cost of atmospheric storage tanks. Based on 1964–1965 prices.

So-called noncode tanks have been designed by various 'tank manufacturers. These tanks utilize unique structural designs, many of which are patented to reduce cost. One design uses tension bars to support the upper edge of the tank. Noncode tanks have been designed for various volumes up to 1,000,000 bbls, with associated costs indicating savings of $0.25 to $0.30 per bbl over generally accepted present costs.

Materials and Inspection

Atmospheric storage tanks are generally fabricated from materials as specified in API specifications, usually plate material that meets ASTM Standard A 283 Grade C. This material can be used up to the maximum thickness of $1\frac{1}{2}$ in. allowed by API 650. For plate less than $\frac{3}{4}$ in. thick, A 283 Grade D and A-7 are also permitted. Alternate designs are permitted by Appendix D of API Standard 650 using A 283 Grade C, A 131 Grades A, B, C, and C-normalized. In most instances a more economical design can be achieved by use of alternates.

Tank failures resulting from brittle fracture of steel have occurred in recent years in large-diameter tanks. In recognition of these failures, many large tanks are now constructed of plate material manufactured to meet requirements of ASTM A 131. When it is possible to perform hydrostatic testing of new storage tanks with the ambient temperature approaching 32°F, use of plate material having superior notch toughness characteristics is imperative.

In the design of tanks which do not comply with API specifications, many tank manufacturers have taken advantage of plate materials with greater strength. API 650 specifies an allowable working stress of 21,000 psi with an 85 per cent joint efficiency. When stronger materials are used, larger tanks can be built without using plate thickness greater than $1\frac{1}{2}$ in. Naturally when tanks are not designed to meet the API Code, other requirements such as plate thickness can also be exceeded if proper welding techniques are used.

The API specifications permit inspection of welds by random trepanning or random radiography. As a further outgrowth of brittle fracture failures, use of trepanning has been largely replaced by nondestructive testing methods. While it is common practice to require more rigorous inspection of pressure vessels, such as 100 per cent radiographing, the relative size of atmospheric storage tanks makes the cost of such procedures prohibitive.

Tank Details

Details are included in API Standard 650 for various items as follows: (1) shell manholes, (2) shell nozzles, (3) large, rectangular flush-type cleanouts, (4) roof manholes, (5) roof nozzles, (6) water drawoff elbows, (7) drawoff sumps, (8) scaffold cable supports.

Shell manholes usually have 20- or 24-in. diameters except for special purposes. Suction heaters and propeller mixers are mounted in manholes of suitable size. In tanks smaller than 50 ft diameter, one shell manhole is used. Larger tanks are provided with two shell manholes. When tank diameters exceed 120 ft, it is often desirable to add a third shell manhole.

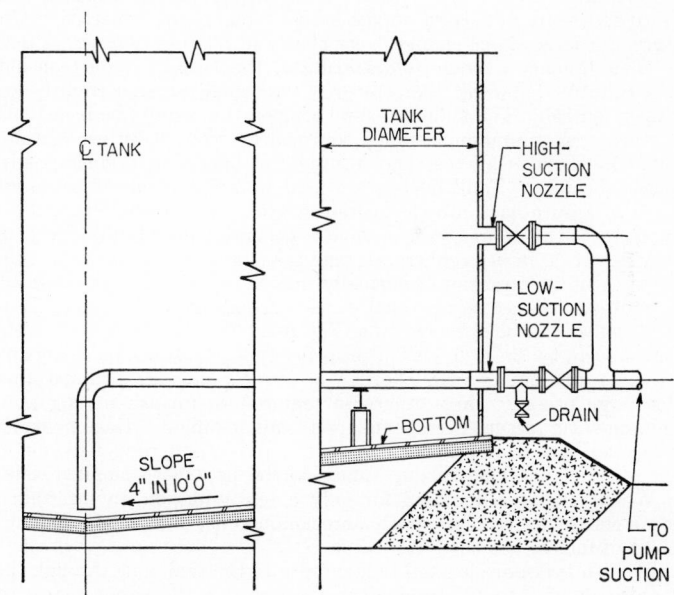

FIG. 8-6. Details of cone-down tank bottom and related connections.

Shell nozzles are spaced vertically and circumferentially to provide clear space between reinforcing pads. To simplify piping arrangement shell nozzles may be located vertically so that the bottom of all piping will rest on common supports. Shell nozzles are oriented either radially or parallel to a major tank axis to simplify piping.

Rectangular cleanouts are intended for use in dirty service such as crude tanks. With the increase in use of sludge mixers, recirculation, and other methods of passing crude bottom sediment into process equipment, the use of cleanout doors is decreasing.

Roof manholes serve to provide light and ventilation during tank repair. The manholes are also located adjacent to level gages to facilitate inspection and repair of this equipment. Manholes usually have a 20-in. diameter. Similar to shell manholes, one roof manhole is provided on small tanks, tanks larger than 75 ft diameter have two roof manholes, and tanks in excess of 120 ft in diameter have three roof manholes.

API drawoff sumps are provided at tank drains. Drain connections are usually 3 in. minimum for tanks up to 50 ft diameter. Four-inch drains are used for larger tanks. Cone-down bottoms are frequently used for crude tanks to permit drawoff of bottom sediment. Details of cone-down bottoms and related connections are shown in Fig. 8-6.

Accessories

Most commonly used accessories for storage tanks are heaters, mixers, venting devices, and gages. These items can be fabricated from common materials or purchased as proprietary items.

Fluids can be heated in tanks by immersion-type heaters or in heaters fitted in the suction piping. Immersion-type heaters can be made up of coils of pipe fitted at the bottom of the tanks. Greater efficiency, generally at lower installed cost, can be realized by use of proprietary heaters fabricated from finned pipe. Suction heaters are generally finned-pipe heat exchangers which can be fitted through a tank manhole. In some installations external suction heaters are used, located close to the tank outlet nozzle. External heaters require more extensive piping and separate foundations.

Tank heat losses are not based on the lowest temperature expected. Usually the temperature of a large vessel changes very slowly in relation to external temperature changes. The January average temperature of the locality under consideration is generally a suitable design air temperature. For small tanks a slightly lower temperature may control. The maintained oil temperature should be based on the tank operating cycle. A minimum installed cost will be realized by use of tank heaters exclusively. Lower over-all cost may result from operating savings resulting from use of suction heaters. Tank heaters are used to maintain an oil temperature just high enough to ensure flow into the suction heater, and the suction heater increases the temperature to the desired value. As it is generally not advisable to permit entry or maintenance of oil in atmospheric storage tanks at temperatures exceeding 250°F, insulation for heat conservation is normally uneconomical. Guards or insulation for operators protection must be provided in the normal manner. Approximate values of tank heat losses can be obtained from Fig. 8-7.

Tank mixers can be "in-tank jet" or propeller type. In-tank jet mixers use normal transfer pumps and piping with addition of jets for mixing. Pumps must have a higher discharge pressure than otherwise required to furnish mixing power. One pump may, however, serve several tanks with minor piping fittings provided in each tank. Propeller mixers require one or more drivers and propellers for each tank dependent on size, but long mixing times can be used with small mixers to effect mixing. When mixing is required for only a few tanks on an intermittent basis, propeller mixers are preferable. With more common use of pipeline product blending, employment of mixers has been decreased.

Mixers of both types are located tangentially to the tank axis through the point of entry at an angle of 8 to 10° to avoid static arcing in the vapor space; the mixing stream in operation should never break the surface of the liquid of the tank. When mixers are used in tanks containing water or sludge bottoms, a mixing shield may be installed to minimize contamination. Shields are usually ¼-in. plate located 12 in. below the bottom of the jet or propeller. Plates are usually 3 ft wide and 6 ft long.

Storage-tank vents serve to permit inflow and outflow of vapor during pumping operations or as a result of thermal changes. The design of venting devices is covered in the API *Venting Guide* RP-2000.

Two general types of vents are (1) open vents, which are permitted for use with fluids having flash points in excess of 100°F, and (2) pressure and vacuum vents used for fluids having low flash points. Open vents can be simple U bends fitted with bird screens. Chain-operated snuffing flaps are often fitted to open vents for fire extinguishment. Pressure and vacuum vents usually consist of two weight-loaded pallets, one opening for in-breathing, the other for out-breathing.

Vapor-conservation systems can be used in conjunction with any type of vent. In general, these systems consist of interconnections of various tanks to balance inflow and outflow of vapor. Adjustable vapor spaces are usually provided to allow for surges in vapor volume.

Proprietary venting equipment can be selected from manufacturers' catalogues to suit conditions and from the API guide. Vents for floating-roof tanks, to protect the tank when the roof is at the low position, and for the tank rim are provided by the tank manufacturer.   Flame arrestors are sometimes installed in venting apparatus. These devices can clog with dirt or ice and create a hazard far greater than their questionable value in fire prevention.

Gaging liquid levels in storage tanks is done with automatic float-type indicators or through gage hatches.   Float-type indicators can be read from external gage boards or from tapes at local indicators.   The tape-type indicator can also be provided with a transmitter for remote reading.   Gage hatches combined with thief hatches

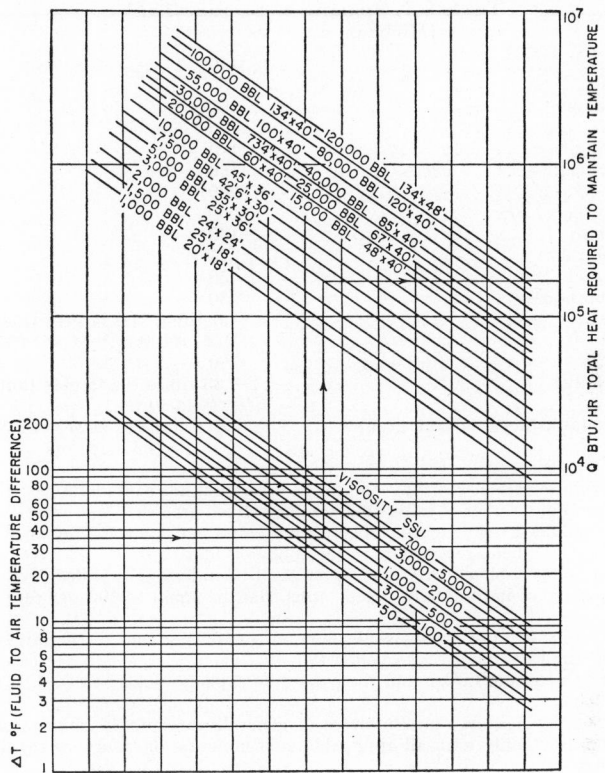

FIG. 8-7. Approximate heat loss for storage tanks.

for withdrawal of samples are provided on tank roofs.   Greater reproducible accuracy can be obtained by placing the gage hatch on a manhole instead of an unstiffened tank roof plate.

On floating-roof tanks, a slotted gage well can be used for gaging and sampling. These are accessible from the access stairway without going down onto the roof. Slotted gage wells also serve as antirotational devices.

Location and Spacing

Tanks containing flammable liquids should be located in diked areas, except those containing material having a high flash point (asphalt, lubricating oil) at storage temperature.   The volume of the diked area should be equal to the volume of the largest tank in a group plus the volume of the remaining tanks below the height of the dike.   Tanks larger than 70,000 bbl should be individually diked.   Tanks larger than 65 ft diameter should be accessible from roads on two sides of the tank.

Dikes are formed of masonry or earth. Masonry dikes are frequently used for heights up to 2½ ft. Earthen dikes are usually not less than 3 ft high. Dikes may be equal in height to one-half the height of the lowest tank enclosed. A desirable maximum height for safe access is 6 ft, although this is exceeded for large tanks. Earth dikes are constructed with side slopes of about 1½ or 2 horizontal on 1 vertical, as necessary to suit the soil.

General tank spacing criteria are listed in Table 8-7.

Table 8-7. Spacing of Storage Tanks
(Minimum distances in feet)

Process units	150
Power plants	50
Fire-water pumps	150
Cooling tower	100
API separator	100
Blowdown drum	150
Flare stack	200
Furnaces	200
Ethyl blending	100
Truck loading racks	25
Rail loading racks	50
Transfer pump house	50
Property line	1.0 times the largest tank dimension $\leqq$ 120 ft
Buildings	100
Tank shell to dike	0.33 times the largest tank dimension (min 10 ft)
Two adjacent tanks in same tank group	0.5 times the largest tank dimension of the smaller tank

Notes:

1. The maximum capacity of any one diked area shall not exceed 150,000 bbl.

2. The maximum number of tanks in a common diked area shall not exceed six when the dike capacity is 30,000 bbl or more. There shall be no limit on the number of tanks when the capacity of the diked area does not exceed 30,000 bbl.

3. Tanks of 70,000 bbl capacity and larger shall be individually diked.

4. Dike capacity of individually diked tanks shall be equal to 100 per cent of the tank capacity.

5. Dike capacity of tank groups shall be equal to 100 per cent of the largest tank plus the volume of the remaining tanks below the dike height, except in the case of boilover stocks stored in cone-roof tanks, and then the dike capacity shall be equal to 100 per cent of all the tankage.

6. Minimum spacings are representative of generally acceptable refinery practices but do not necessarily comply with all applicable governmental or insurance regulations for a specific locality.

Piping Arrangements

Piping at storage tanks is provided for hydrocarbons, utilities such as steam and air, and foam for fire extinguishment. Three major design factors affecting piping at storage tanks are tank settlement, solar expansion, and fire.

The dikes surrounding tanks are intended to impound oil spills, therefore, piping passing through dikes must be sealed in the dike to contain oil. Insulated lines must be provided with special seals. Bare lines are usually sealed by the surrounding earth, but in some cases anchor plates should be provided to ensure sealing. As a result of these seals, pipes must be considered anchored in dikes for flexibility purposes.

The vertical and horizontal movements of tanks and piping resulting from settlement and thermal expansions are preferably provided for by using piping arrangements which include bends. In some cases expansion loops are necessary. When extreme settlements are possible, flexible joints are necessary. To allow for forces and moments and for shocks due to explosions, steel valves at all tank nozzles are considered mandatory.

Foam piping should be supported at the foam chamber but at no other point in the vertical rise to allow for explosion. An ample horizontal leg should be provided before a support is installed from the tank shell or from grade.

Foundations

In the design of tank foundations, the soil-bearing capacity must be known together with soil consolidation and settlement. If anticipated, large settlements of tanks (1 ft or more) can be accepted provided these are uniform. Slight differential settlements are extremely troublesome with floating-roof tanks. Floating roofs can bind owing to out-of-round resulting from uneven settlement.

General principles are included in API Standard 650 Appendix B, *Recommended Practice for Construction of Foundations for API Vertical Cylindrical Oil-storage-tanks.* Typical details of storage-tank foundations are shown in Fig. 8-8.

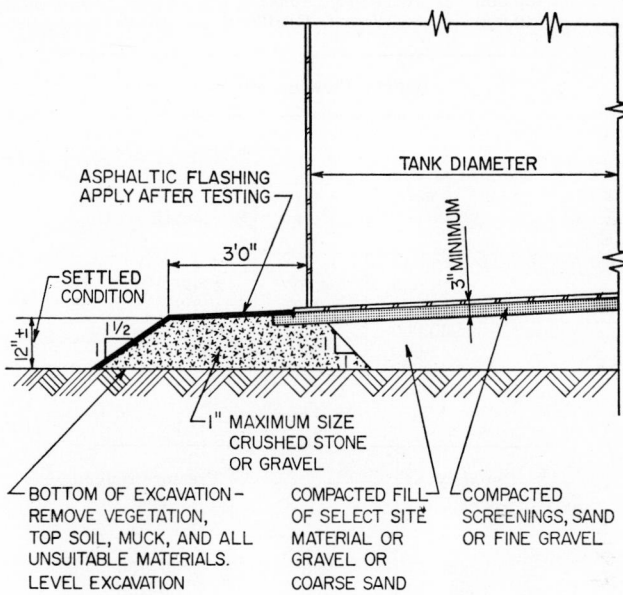

FIG. 8-8. Details of typical storage-tank foundation.

Pressure Storage

Pressure storage used in refineries is generally limited to LP gas and to high-vapor-pressure gasoline components such as pentane. LP gas is usually composed of propane, propylene, butane and butylenes, alone or in combinations. These materials are usually costly to produce and to store. Because of the high value of these products and the expense of storage (frequently twenty times as costly as atmospheric storage) pressure storage facilities should be designed carefully.

Types of Vessels

Pressure storage vessels used for LP gas and high-vapor-pressure hydrocarbons are cylinders, spheres, and spheroids. The type and working pressure of the vessel should correspond to the highest vapor pressure of the individual material or mixture at the design operating temperature.

Table 8-8. Capacities and Sizes of Pressure Storage Vessels*

Spheroids (Plain)

Capacity, bbl	Size		Pressure range, psig
	Diam	Height	
2,500	31'4"	23'2"	10–75
5,000	41'2"	27'2"	10–75
7,500	47'5"	30'8"	10–60
10,000	52'4"	33'5"	10–60
15,000	57'10"	40'0"	10–50
20,000	66'9"	41'4"	10–50
25,000	69'2"	47'7"	10–40
30,000	76'11"	46'8"	5–40
40,000	85'1"	50'10"	5–40

Bullets, Pressure 250 Psig

Capacity, gal	Size	
	Diam	Str. length
18,000	8'10"	39'0"
20,000	8'10"	41'0"
30,000	8'10"	67'0"
42,000	12'0"	42'0"

Spheres

Capacity, bbl	Diam	Pressure range, psig
1,000	22'3"	30–150
1,500	25'6"	30–150
2,000	28'0"	30–150
2,500	30'3"	30–150
3,000	32'6"	30–150
4,000	35'3"	30–150
5,000	38'0"	30–100
6,000	40'6"	30–100
7,500	43'6"	30–100
10,000	48'0"	30–75
12,000	51'0"	30–75
15,000	59'4"	30–75
20,000	60'6"	30–50
25,000	65'0"	30–50
30,000	69'0"	30–50

* Larger sizes available to special design.

Sizes, Capacities, and Pressure Ratings

In general, propane is stored in cylindrical vessels designed for a working pressure of about 250 psig. Three general sizes are commonly sold as "off-the-shelf" designs. Propane-butane mixtures and butanes are generally stored in spheres with working pressures ranging up to 85 psig. Because of cost, storage spheres are not stress relieved, and therefore, the vessel size is limited by maximum thickness permitted by the ASME Code. Spheroids are used for storage of butanes and pentanes for working pressures ranging up to 50 psig.

Typical sizes, capacities, and pressure ratings of pressure storage vessels are listed in Table 8-8.

Storage-vessel Costs

Costs of pressure storage vessels vary with individual conditions to a greater extent than atmospheric storage vessels. This is to some extent a result of standardization in the higher pressure range. A guide to vessel cost can be obtained from Fig. 8-9.

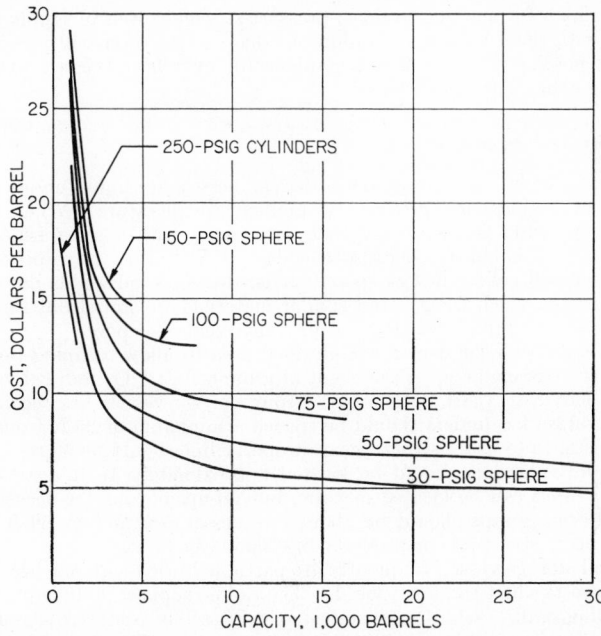

FIG. 8-9. Approximate cost of pressure storage vessels. Based on 1964–1965 prices.

Materials and Procurement

Pressure storage vessels having working pressures not exceeding 15 psig are designed and tested in accordance with API Standard 620. Vessels having working pressures in excess of 15 psig are designed and tested in accordance with ASME *Code for Unfired Pressure Vessels*.

The material most commonly employed for various pressure storage vessels is ASTM A-285 Grade C flange-quality steel. This type of vessel is wholly or partially shop fabricated, and usually testing procedures are far more stringent than those applied to atmospheric vessels.

The majority of the vessels used for pressure storage is produced as either proprietary or off-the-shelf items. Normally it is more economical to use a larger or higher specification off-the-shelf cylinder for a specific service than to order a custom-made vessel.

Details and Accessories

Cylindrical storage vessels are usually supported on two saddles in a horizontal position. Various appurtenances and piping connections are located at one end of the vessel to simplify access and permit partial burial of the vessel. Platforms and access ladders or stairways are usually provided for groups of vessels. Accessories include relief valves, pressure gages, thermometers, and liquid-level gages. Two full-sized relief valves are usually provided on each vessel. A three-way cock is installed to permit servicing of one valve, while full protection is furnished by the other valve.

Spheres are supported on steel pipe section legs. Spheroids on the other hand are normally supported on a sand pad at grade. These vessels are provided with access stairways and platforms as necessary for inspection and servicing of accessories. These accessories also include relief valves, pressure gages, thermometers, and liquid-level gages. Relief valves usually consist of two pilot-actuated pressure-relief valves and one vacuum relief valve.

It is customary for manufacturers of pressure storage vessels to supply all necessary accessories with these vessels. In addition, the manufacturer will provide all necessary piping nozzles plus one 20-in. manhole for cylindrical vessels and two 20-in. manholes on spheres and spheroids.

Location and Spacing

The hydrocarbons stored in pressure storage vessels are liquid under pressure but are vapors at atmospheric pressure at ambient temperature. When such liquids escape from pressure storage, they flash into vapor with a self-refrigerating effect that subcools the balance. Because the flashed hydrocarbons vapors are heavier than air, they will collect in low spots. It is, therefore, advisable not to use dikes around vessels in which LP gas and similar materials are stored but rather to allow for free flow of air around vessels to scour away escaped vapors.

Vessels should be located in a well-drained area to allow escaping liquids to flow away from the storage areas in the event of ground fires. Cylindrical vessels should be located 3 to 5 ft apart in groups. Groups of six vessels are recommended by NFPA. Groups of cylinders should be spaced a minimum of 25 ft from each other. The clear distance to other structures or property line should be 50 ft.

Spheres and spheroids should be located approximately 10 ft apart at the tank equators. Vessels can be located in pairs, but groups of four are occasionally used. Spacing between groups should be about one vessel diameter or 50 ft. The clear distance to other structures or property line should be 50 ft.

Cylindrical storage vessels frequently are partially buried with one head and part of the straight to the first saddle exposed. The saddle support is designed to act as an earth-retaining wall. When so installed, the vessel is protected against fire and maintained at a reasonably constant soil temperature. When special vessels of long length are so installed, the soil temperature can be used as design temperature, resulting in a lower design pressure.

Piping Arrangements

As a result of the greater hazard associated with pressure storage, the provisions made to ensure flexibility in piping systems must be even more carefully developed than are those for atmospheric storage. Normally, however, piping is smaller and no dikes are present in which piping is anchored. When groups of vessels are used, it is customary to consider the group to be operated as a unit.

At pressure storage vessels, all valves must be steel valves of suitable rating. Connections which can be opened to atmosphere (vents, drains, etc.) are usually double blocked or valved and plugged. For butanes and lighter hydrocarbons, piping connections for level and pressure gages are restricted to an opening equivalent to a

No. 40 drill to minimize escape of fluid in the event of a failure.   Other connections in these services except relief valves are usually fitted with excess flow-check valves or quick-closing safety valves which can be actuated automatically or manually.

Suction outlets on pressure storage vessels are usually provided with internal standpipes to prevent entry of water.   Bottom manholes on such vessels must be fitted with drains.

Foundations

Foundations for cylinders and spheres are similar to those used for process-type equipment.   Spheroids are usually supported on sand pads as shown in Fig. 8-10.

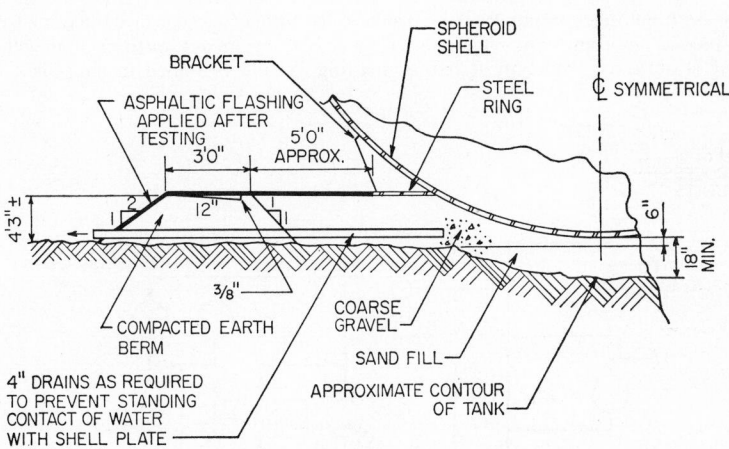

Fig. 8-10. Typical sand-pad foundation for spheroid.

Low-temperature Storage

Study of per barrel costs for pressure storage cylinders, spheres, and spheroids and atmospheric storage tanks shows that, as the storage pressure decreases, the cost of vessel also decreases.   Naturally as the storage pressure is considered decreased, the storage temperature must also be reduced.   To maintain material at low temperatures, insulation and refrigeration must be provided.   Comparative initial cost of storage must be adjusted to account for the cost of insulation and refrigeration equipment plus the larger operating and maintenance costs.

In the development of the costs of refrigeration equipment, it is found that the duty necessary to cool product rundown from processing temperature to storage temperature is frequently a major percentage of the total duty.   In most refineries, products are shipped or sold at ambient temperatures and corresponding pressures. If the storage temperature is extremely low, additional problems are introduced.

Various studies of low-temperature storage of hydrocarbons and related fluids show this method to be economically attractive.   A great deal of development is still under way in this field, for example, the shipment of liquefied methane by tanker.

PRODUCT BLENDING AND SHIPPING

Methods of Blending and Shipping

Process units produce various product components and base stocks which must be combined or blended with suitable additives to manufacture finished products. These finished products are generally grouped into the broad categories of gasolines, middle distillates, and fuel oils.

Different methods of product blending are used to suit variations in type of product, available components, operating procedures, shipping and marketing requirements, and storage facilities. Blending methods which are normally employed include (1) batch blending, (2) partial in-line blending, and (3) continuous in-line blending.

Petroleum products are shipped in bulk and in containers. The procedure of packaging products into containers (drums and various size cans) involves special techniques and equipment. Bulk shipments utilize pipeline, marine, road, and rail facilities. This section is limited to description of bulk shipping by marine, road, and rail equipment only.

Batch Blending

In batch blending, components of a product are added together in a tank one by one or in partial combination. The materials are then mixed until a homogeneous product is obtained. A typical batch-blending system is shown in simplified form in Fig. 8-11.

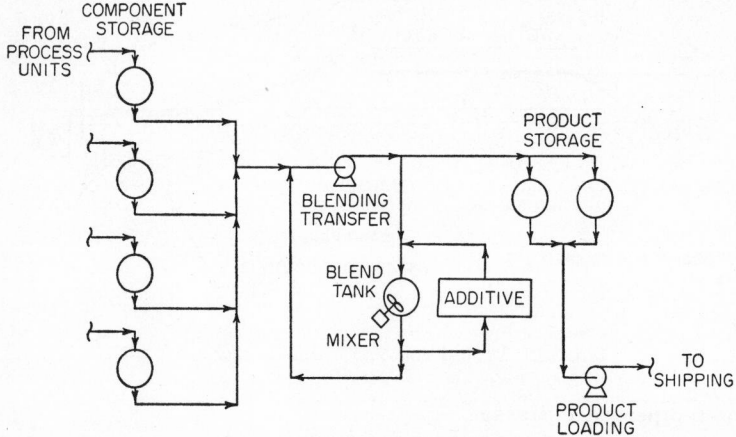

FIG. 8-11. A typical batch blending system.

Components are run from process units to component storage tanks. If the process unit produces a material which has a constant composition, it is not necessary to prepare an analysis of the component prior to blending. When different grades of material are produced dependent on operating conditions (e.g., reformate varying with severity of operation), samples must be analyzed prior to each blend. Each component stream is pumped separately into the blending tank, and the tank gaged after each addition. Additives, such as TEL and gasoline dye, are then added in batches. The contents of the tank are then mixed thoroughly. After laboratory analysis, the blended product is pumped to storage or shipping tanks.

When butane is blended into gasoline, the batch procedure is varied to bring butane into the tank after line-blending it with another component. In this manner the butane is dissolved in the heavier material with less butane waste and without the fire hazard introduced by static electricity generated by in-tank blending of the butane.

Equipment

Storage tanks are required for components, for mixing, and for finished products and shipping. Required capacity can be concentrated in the form of components or of finished product dependent on shipping requirements in terms of the size of plant. The capacity of mixing or blending tanks is not included in total storage. These tanks are usually of sufficient capacity to allow for blending 2 to 3 days' product make per batch.

Components are transferred to the mixing tank by one or more pumps. A pump of relatively high capacity is required because each component must be transferred and gaged separately. In general, each component transfer pump should be of sufficient size to fill the blending tank in 3 hr or a similar reasonable operating period. The same pumps are normally used to transfer product to final storage.

The blended material must be mixed thoroughly by means of propeller mixers, in-tank jet mixers, or recirculation pump and piping. Propeller-type mixers are mounted directly to the tank shell and are usually motor-driven, requiring electrical wiring and switchgear. For large tanks, multiple mixers are required, for example three 25-hp mixers would be required for an 80,000-bbl gasoline tank.

For mixing blends by in-tank jet mixers, a pump is necessary. The pump takes suction from the blend tank and discharges, through suitable piping, back to the tank. If the discharge into the tank is through a distribution spider or a swing line, a large-capacity pump is necessary, and mixing is considered complete when the entire tank contents have been pumped at least once through the pump.

A jet mixer consists of a nozzle instead of a spider or swing line. This nozzle is directed upward from the bottom of the tank at an angle. The high velocity of the jet stream induces circulation of the entire contents of the tank. The pump used with the in-tank jet mixer can be a low-capacity high-head pump such as TEL eductor booster pump in gasoline service.

Application of Batch Blending

Batch blending is most adaptable to use in small refineries in which a limited variety of blends is to be produced. In a small plant the cost of extra blending tanks, pumps, and related equipment may not be so large as the cost of instrumentation and equipment needed for in-line blending. The unavailability of a sufficient number of trained operators may influence selection of the simpler batch-blending system.

Partial In-line Blending

Partial in-line blending is accomplished by adding together product components simultaneously in a pipeline at approximately the desired ratio without necessarily obtaining a finished specification product. Final adjustments and additions are required, based on test, to obtain specification product. In partial in-line blending, mixing is required only for final adjustments. Additives such as TEL and dye for gasoline are usually added as a batch into the blending header during the final stages of the blend or during the final adjustment stage.

A typical partial-in-line blending system is shown in simplified form in Fig. 8-12.

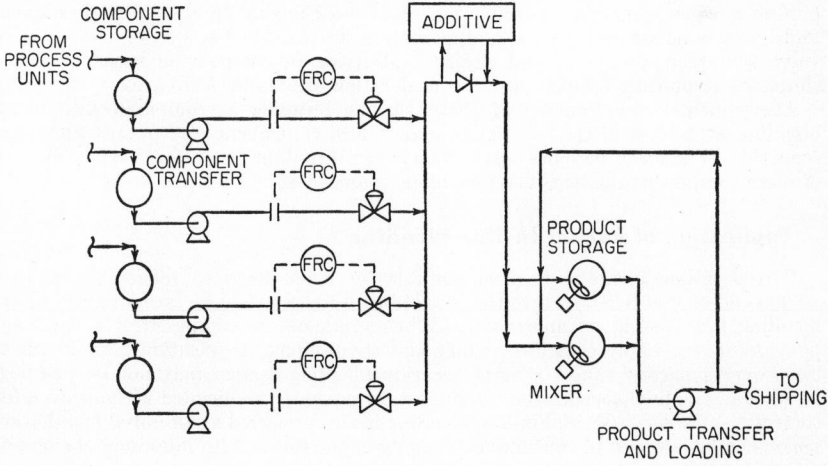

FIG. 8-12. Typical partial in-line blending system.

Product streams are run down to component storage as in the batch-blending system. The required components are then pumped simultaneously from each component tank through appropriate flow controllers into a blending header. Mixing of the components is accomplished by turbulence in the header as the combined components flow to the finished-product storage tank.

Additives are introduced into the blend by use of a bypass stream with a suitable booster pump to an eductor or by use of a proportioning pump delivering a premix. Although batch addition of these materials is commonly used, some partial in-line blenders employ continuous additive systems. A further refinement adapted to partial in-line blending of gasoline is use of a Reid vapor-pressure controller to adjust flow of butane to the blending header.

Equipment

Storage tanks are required for each component and for final-product storage. A large proportion of total storage is frequently concentrated in component storage when this type of blending is used. No mixing or blending tank is required.

An individual pump is required for each component. The capacity of pumps must be established to permit simultaneous pumping and delivery of one day's blend to product tanks within a reasonable time. Usual practice is to complete a blending operation in about a 6-hr period.

The quantity of each component of a blend must be proportioned by use of a flowmeter and control valve. Flow controllers are set for the predetermined flow rate, and flow is recorded. Deviations from the required total can be determined and corrected. Flowmeters used for partial in-line blending need not be extremely accurate for satisfactory operation. Accuracy ranges of approximately 5 per cent as attained with orifice meters are suitable. Velocity meters, flow nozzles, venturis, and positive-displacement meters are also used when greater accuracy is desired.

Mixers are necessary in final storage tanks for final correction of blends by addition of components. The use of mixers in component tanks is dependent on material stored. When components are of constant quality and composition, a single component run-down and storage tank can be used and mixers are unnecessary. When components vary in quality and composition at frequent intervals, multiple run-down and storage tanks are preferable and mixers are used to obtain a uniform component.

Advantages and Disadvantages

A number of advantages are realized through use of partial in-line blending. Blending time is substantially reduced because of (1) simultaneous instead of consecutive pumping of components, (2) reduction of over-all mixing time, (3) elimination of multiple gaging operations. Fewer operators are needed because of reduction in valve switching, pumping, and gaging. Metering of components such as butane simplifies accounting by eliminating consideration of butane shrinkage.

The principal disadvantage of partial in-line blending as compared with batch blending is the cost of the additional meters, flow controllers, and pumps which are required. These can be more costly than a blend tank in a small refinery. The use of more complex equipment increases maintenance cost.

Application of Partial In-line Blending

Partial in-line blending is most suitable for moderate-sized refineries (between 10,000 and 40,000 B/SD) where the cost of blend tanks would be excessive and where blending time should be minimized. When products are composed of a few components of reasonably constant quality and the product specification permits variations over a narrow range, a more accurate blending system may not be justified. In such a case, final specification product can frequently be blended without need for correction of blend. Partial in-line blending is also preferred as an initial installation which can be adapted to continuous blending in the future with minimum changes of pumps, piping, and tankage.

Continuous In-line Blending

In continuous in-line blending, all components of a product and all additives are blended in a pipeline simultaneously with such accuracy that at any given moment finished specification product may be obtained directly from the line.   As a result of accuracy and safeguards included in the system, no provisions for reblending or correction of blends is necessary.

A typical continuous in-line blending system is shown in simplified form in Fig. 8-13.

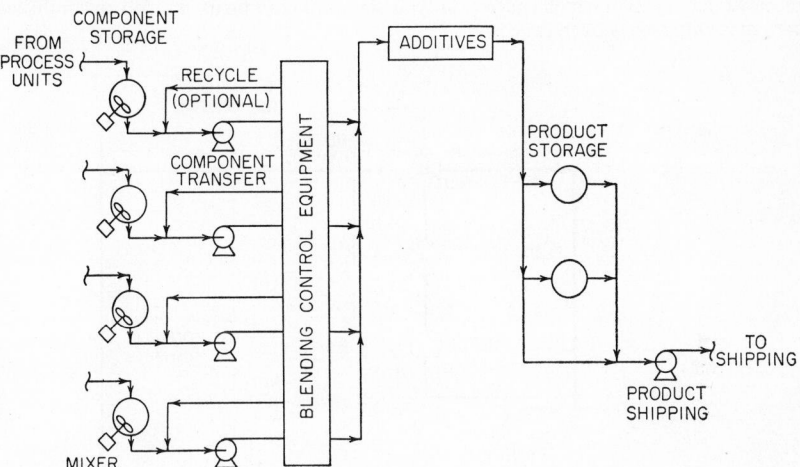

Fig. 8-13. Typical continuous in-line blending system.

Product streams are run down to component storage with at least two tanks provided for each component.   Samples of components are test blended in the plant laboratory, and blends are analyzed to determine the most suitable proportions of components and additives for a desired finished specification product.   The required components and additives are then pumped simultaneously at controlled rates into a blending header.   Various methods of controlling individual flow rates have been developed with interlock provisions to ensure delivery of only specification material.   Products can be sent to final storage, can be delivered to product pipeline for transmission to remote terminals, or can be loaded directly into bulk transports such as tankers or barges.

Equipment

Storage tanks are required for components and for most additives.   Storage of blended product will be only as required to suit shipping methods.   In general, the greatest proportion of storage can be in component form with a minimum of finished-product storage.

An individual pump is required for each component.   Pumps are also required in conjunction with dye and additive preparation and delivery.   Dyes and certain additives are stored in solution form and added by proportioning-type pumps.   Booster pumps are provided for TEL service to develop motive power for eduction.   Pumps are also required for unloading and preparation of additive solutions.

The quantity of each component of a blend must be proportioned by accurate methods to ensure that specification product is available from the blending header at any instant.   Recording flowmeters and flow-control valves used to proportion components are simular to those used for partial in-line blending, but a greater degree of accuracy is necessary.   After calibration of product meters with actual product at actual flowing temperature, an accuracy of $\frac{1}{4}$ to $\frac{1}{2}$ per cent should be attained. Orifice meters are unsuitable for use in continuous in-line blenders. Most installations utilize positive-displacement meters or venturis. Velocity meters are also considered sufficiently accurate for blending service.

To ensure continued accuracy under varying conditions, blending equipment should be designed to provide for adjustment of individual component flow in proportion to total flow. Failure of the system to readjust should result in complete shutdown of blending operations by means of flow stoppage or recycle. Accuracy of this nature is necessary when product is delivered to pipelines or bulk transports. For some installations a simpler system which controls only at set points can be used. When conditions vary, manual reset is necessary.

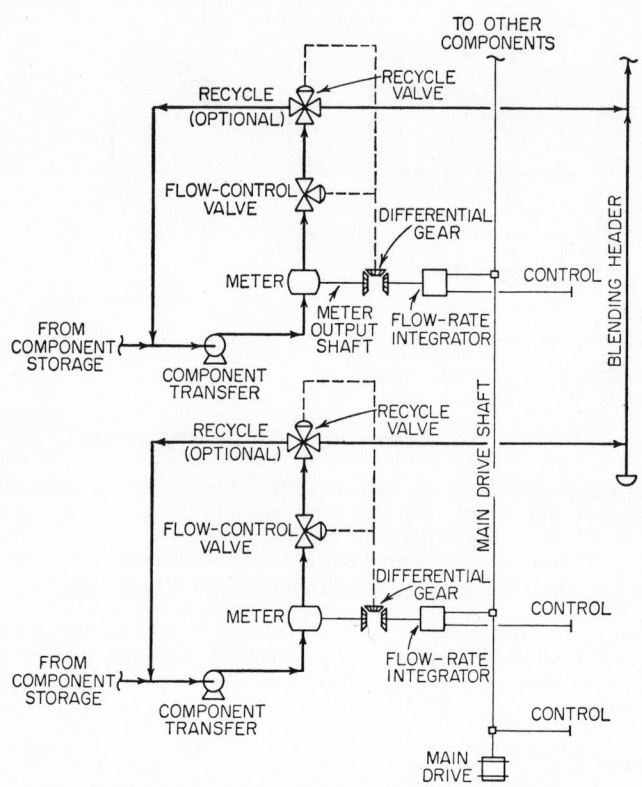

Fig. 8-14. Typical arrangement of mechanical blending control equipment.

Two types of blending controls are used to adjust component flows to desired rates: mechanical and electronic. In the mechanical system, rotary motion generated by the component meter is matched through a differential gear device against a preset rotary motion. When the metered rate differs from the preset rate, pneumatic or electrical controls are actuated to adjust the flow-control valve to change the flow rate to the desired quantity. Figure 8-14 shows a typical arrangement of mechanical blending equipment. In the electronic system, electronic pulses generated by the component meter are matched against preset pulses generated by an electronic device. Differences in pulse rates are detected by a digital totalizer which feeds back a signal to adjust the flow-control valve to change the flow rate to the desired quantity. Figure 8-15 shows a typical arrangement of electronic blending equipment.

Reid vapor-pressure control of butane addition for gasoline blending is frequently employed with continuous in-line blending.

To ensure accuracy of blends, it is necessary to calibrate meters frequently. One method of meter calibration utilized is to remove the meter from the system and replace it with a calibrated spare meter. This requires use of a prover tank or similar device, which in some plants is also used for checking product shipping meters.

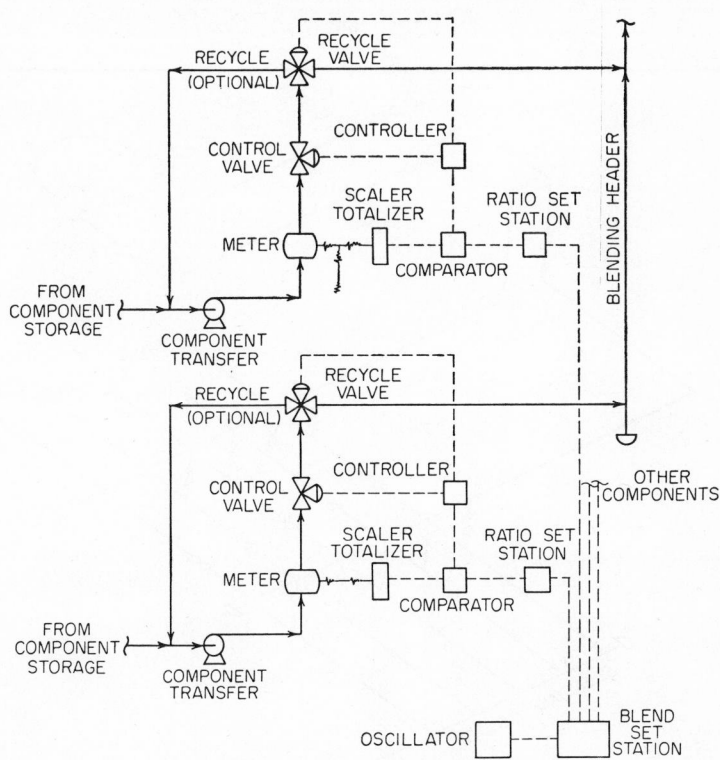

FIG. 8-15. Typical arrangement of electronic blending control equipment.

As a result of the work involved in replacing a large number of meters frequently, most refineries which use continuous in-line blending have incorporated a metering loop in the blending system. Such a loop is basically a pipe of known length and, therefore, known volume. The flowing fluid forces a "pig" through the loop passing control points from which signals are sent for comparison. A typical arrangement is shown in Fig. 8-16.

Most loops are made up of 600 to 1,000 ft of 8-in. pipe to ensure sufficient accuracy. With electronic controls and digital totalizers, great accuracy can be achieved in short lengths.

Advantages and Disadvantages

Through the use of continuous in-line blending the following advantages can be realized: (1) reduced blending time; (2) minimum finished-product storage, since components are stored and blended as required; (3) minimum operating personnel; (4) increased accuracy, eliminating need to "give away" additional quality; (5) maximum flexibility in blending various product grades; and (6) reduction in loss through weathering of finished products.

Certain disadvantages are associated with the continuous in-line blending system. When products are transferred directly to pipeline or bulk transport, a complete blender is required for each product which must be loaded simultaneously. For example, if a tanker must be loaded with two grades of gasoline simultaneously, two blenders are necessary. Alternately, the advantage of reduced product tankage cannot be realized.

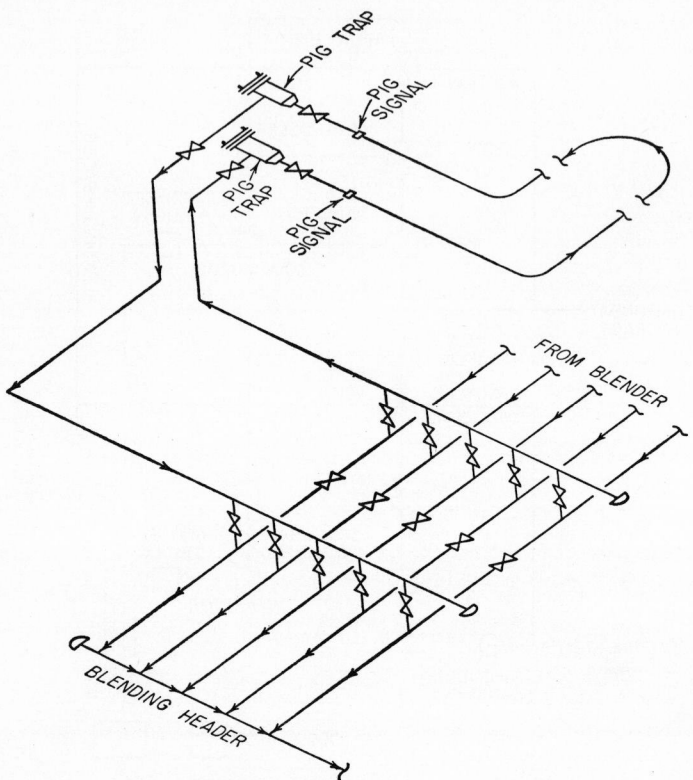

Fig. 8-16. Typical arrangement of calibration loop for an in-line blending system.

Another disadvantage of this type of blending is the cost. It is possible that the cost of even a single system cannot be justified for a small refinery. A further disadvantage is the extreme difficulty in correcting errors if they do occur. As a result of the safeguards built into the equipment, the only errors possible are the simple human blunders of opening the wrong valve or missetting the control.

Application of Continuous In-line Blending

Continuous in-line blending is best for use in larger refineries which make several grades of product. When multiple components are used, the greater accuracy inherent in this system is justified. When product can be transferred directly to pipeline or bulk transport without use of final-product storage, continuous in-line blending is the only satisfactory system.

Modifications to In-line Blending

Various adaptations of in-line blending have been used or proposed. Many of these methods merit consideration for specific installations.

Off-unit Blending

Many refineries are made up of a limited number of closely integrated process units each producing streams of uniform and consistent quality materials. When this type of refinery ships a limited number of products with few grade variations, off-unit blending is applicable.

As the name implies, components are blended into products directly as they come off the unit, usually under control of process operators. A small volume of some streams must be bypassed to storage while certain blends are being made. These components must be returned for inclusion in other product blends.

Storage tanks are required for the total make of finished products. In addition, some component tanks are required to suit grade variations. Final storage tanks should be provided with propeller or jet mixers to permit corrections of blends after testing. When a single grade of a product is produced, component storage is not required and such product tanks are not fitted with mixers.

This type of blending system does not justify use of extremely accurate meters. When product specifications are stringent, blends can be corrected in final storage tanks if necessary. In general, the degree of accuracy used for partial in-line blending is satisfactory.

Pumps are required for blending additives and dye and for return of surplus component for use in blends. The process-unit run-down pumps will be required to have additional discharge pressure in order to overcome losses in blending meters and control valves. This will involve a minor incremental pump cost.

Off-unit blending offers various advantages when adaptable. These include (1) minimum operators, (2) minimum storage tanks, and (3) fewer pumps. The disadvantages of this system are the lack of flexibility of refinery and process-unit operations with possible product "give-away."

This system is applicable only to moderate-sized refineries which have closely integrated process units. The refinery must also produce a limited number of uniform products, preferably from a consistent type of crude.

Base-stock Blend System

The base-stock blend system is a further adaptation of "off-unit" blending. Instead of finished products being produced directly from the process unit, two base stocks are blended. These base stocks are not produced simultaneously; therefore, some components must be bypassed to storage and returned when needed. The base stocks are subsequently blended to produce finished products.

In general, the base stocks should meet specifications which straddle the desired product grade range. In this manner, various quantities of each base stock will be necessary to produce any grade of product. As the composition of each base stock is determined by a laboratory analysis before final product blending, the accuracy of the off-unit blending is unimportant. Similarly, greater flexibility in process-unit operation and charge-stock composition is acceptable.

The blending of base stocks into final products can be performed by two-component blenders which are both accurate and inexpensive. This makes it possible to eliminate final-product storage when this blending system is used.

Tanks are needed for base stocks and for bypassed components. The quantity of components should be quite small. At least two tanks must be provided for each base stock to allow for analysis and blending from one tank while filling the other tank. Tanks should be provided with mixers to obtain uniform base stocks.

Pumps are needed for return of components and for final-product blending. When blends are transferred directly to pipeline or transport, no final storage tank and shipping pumps are necessary.

The base-stock blending system incorporates most of the advantages of the off-unit blending system with some of the advantages of continuous in-line blending. The flexibility of this system is greater than that obtainable from the off-unit system.

The base-stock blending system is applicable only to refineries with integrated process units. The spread of product grades permissible with this system is not limited, however.

Two-component Blenders

Two-component blenders have been developed for use in blending of fuel oils and asphalts at loading stations. In these blenders, two components are metered and blended continuously in combinations to produce any desired product. Some models of these blenders are portable for use at various points within the plant. These devices use standard instruments and valves and are relatively inexpensive. A diagram of one type of two-component blender is shown in Fig. 8-17.

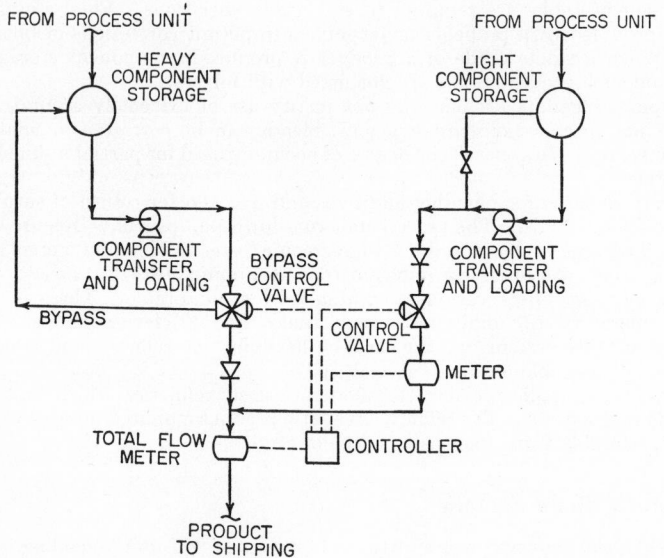

FIG. 8-17. Diagram of a two-component blending system.

Dye, Additive, and TEL Blending

In addition to various hydrocarbon components, most petroleum products are also blended with dyes and additives, the most common of which is tetraethyl lead. Blending methods used in conjunction with these materials vary with the basic product blending system.

As previously stated, for both batch and partial in-line blending, additives, dye, and TEL are introduced into the blend in batches. Liquids such as inhibitors or top lubes are pumped into the blend by small proportioning-type pumps. The quantity blended is determined by gaging a small tank, by weighting, or by metering. Dye is usually added dry by vacuum eduction out of a drum.

TEL is added to batch or partial in-line blending systems by educting the TEL mix out of drums or weigh tanks, depending on the size of the refinery. The blended product is used as motive fluid in the eductor. In order to avoid leaking or spilling of TEL, this material is never handled under pressure.

TEL mix is shipped to refineries in drums, trucks, rail cars, or marine shipping cylinders. TEL is also educted out of these containers into the weigh tank. Sizes of weigh tanks in common use are 4,250, 8,500, and 15,000 gal.

For continuous in-line blending, additives, dye, and TEL must be added continuously. Liquid additives are pumped using meters and the previously described control equipment either to adjust a control valve or change the pumping rate of a proportioning-type pump. Dye is added in the same manner using a dye solution prepared by educting dry dye into light naphtha. TEL is added by educting mix out of a weigh tank. The flow rate is controlled by signals transmitted by a totalizing meter in the blending header. The signals adjust the movement of a traveling poise on the scale. The relative balance of the scale controls the flow of TEL mix to the eductor. The blended product is used as driving fluid. A separate eductor with an independent motive fluid system is generally used to transfer TEL from shipping container to weigh tank. Simplified flow diagrams of TEL blending systems are shown in Fig. 8-18.

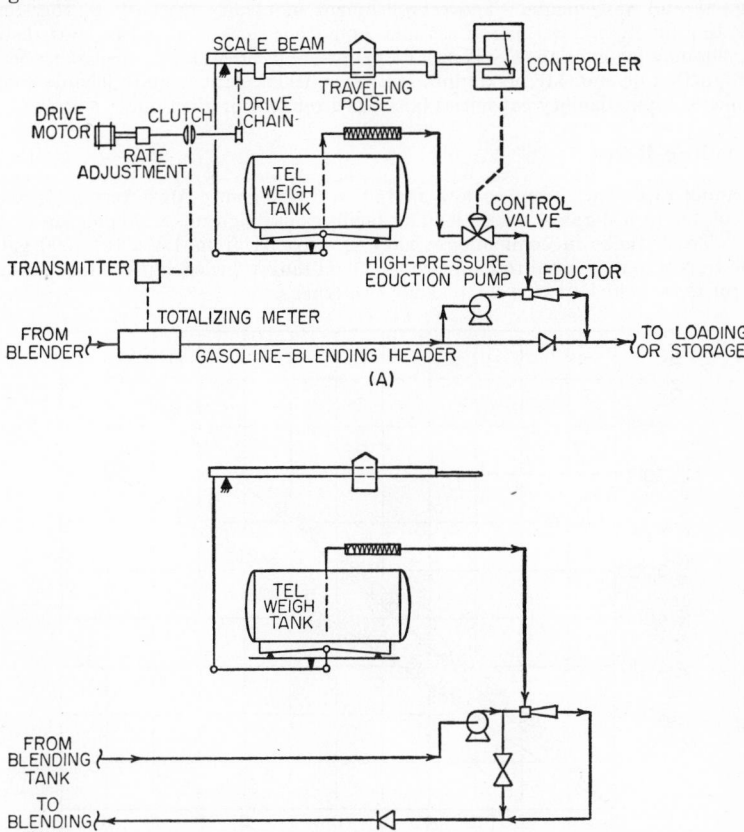

Fig. 8-18. Simplified flow diagram of a TEL blending system. (A) Loss in weight TEL blending system. (B) Batch TEL blending system. Batch blending shown; partial in-line blending similar.

Cost Comparison

A number of cost comparisons have been made for specific blending systems and are cited in published papers and articles on product blending. These comparisons are normally acceptable for purposes of choice among alternates, but may be weighted by some factor such as capacity of product storage tankage needed to suit loading requirements. Specific studies should be made to determine the optimum blending system for a particular application.

When detailed cost-comparison studies cannot be developed, the following criteria may be utilized to aid in system selection: (1) Batch blending has the highest investment and operating cost.   (2) With equal volumes of product tankage, partial in-line and continuous in-line blending investment costs are comparable for larger plants. (3) Continuous in-line blending has lowest annual operating cost.   (4) Major reductions in investment cost for continuous in-line blending are possible.

Truck-loading Facilities

The product shipping facilities required in a refinery vary with the size of the plant, the local market, the location of other refineries, and the details of various sales contracts and agreements.   The establishment of facility capacity is, therefore, a marketing function.   Because of seasonal and other variations and product distribution, shipping facility capacity is usually expressed in broad terms such as percentage of production or monthly maximum, etc.   The usual plant is quite flexible, and the installed shipping facility capacities far exceed refinery production.

Loading Rates

Loading rates vary from as low as 150 to 1,000 gpm.   Most terminals load at rates of 300 to 550 gpm.   Large loading facilities use high rates, ranging up to 1,000 gpm.   Truck tanks in common use vary in capacity from 1,300 to 6,500 gal.   A single tractor may be used to haul two 6,500-gal tank vehicles, but the loading operation for these vehicles is the same as for two trucks.

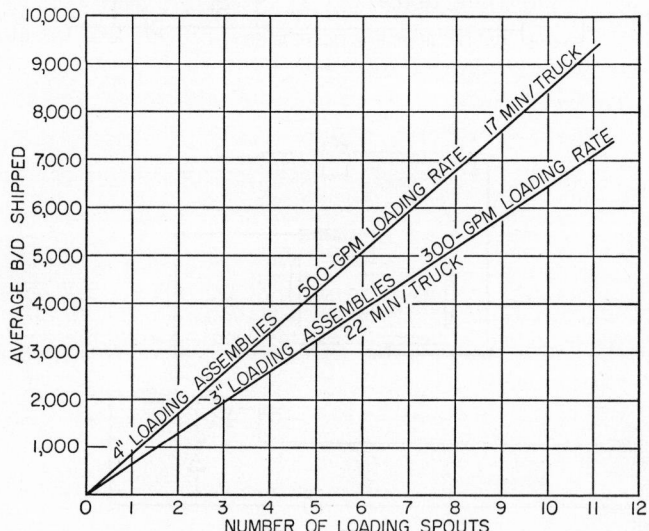

FIG. 8-19. Number of spouts required for loading 3,500-gal tank trucks.   Truck size, 3,500 gal.   Loading any one product in 4 hr of an 8-hr day.   Loading 5 days per week.   Total loading time equals filling time plus 10 min.

The number of loading spouts required for each product varies with (1) truck size, (2) number of loading hours per day, (3) number of loading days per week, and (4) time required for positioning, hookup, and depositioning of truck.   Figure 8-19 shows the number of spouts required for assumed conditions for a 3,500-gal truck.   Figure 8-20 shows loading pump rates for the same conditions.   Similar diagrams can be developed for each special set of conditions.

Equipment

Usual equipment required for a truck-loading operation is shown in Fig. 8-21. Pumps are usually located adjacent to tanks.   These should have flat head capacity

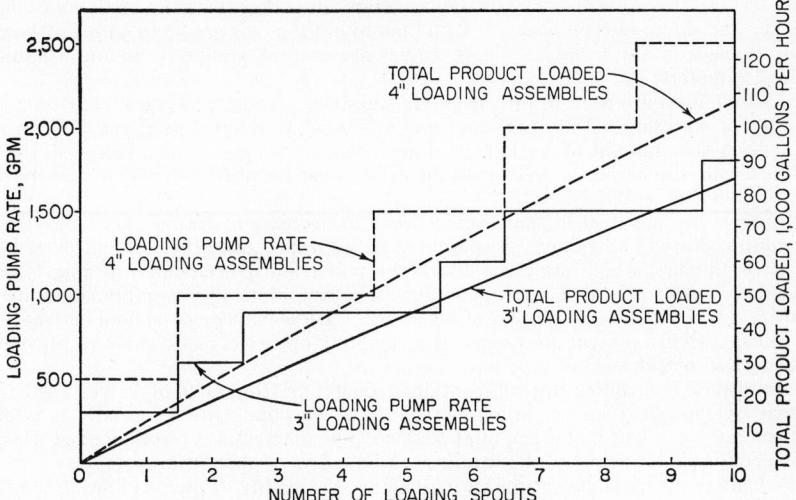

Fig. 8-20. Loading pump capacity needed to load 3,500-gal tank trucks. Total filling time per 3,500-gal truck: 3-in. assemblies, 22 min; 4-in. assemblies, 17 min.

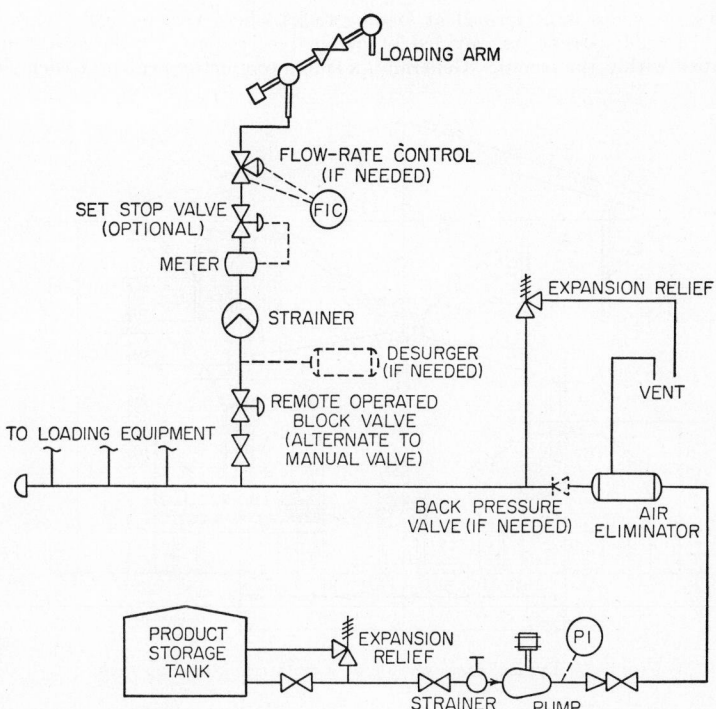

Fig. 8-21. Schematic diagram of usual equipment needed for tank-truck loading.

characteristics to provide a reasonably constant discharge pressure under varying capacity and discharge conditions.   Usual pump differentials are 35 to 45 psi without major changes in static head.   Check valves are used at pumps to maintain liquid in the flowmeters.

Air eliminators are used to disengage air and other vapors which would affect the accuracy of metering.   Disengaging of vapor is done at about 3 psig, and if there is not at least this amount of static head difference between the air eliminator and the loading spout discharge, a back pressure valve must be provided.   This may be a swing-type check valve.

Desurgers are installed in some installations to decrease hydraulic shock resulting from quick shutoff.   Strainers are provided to keep dirt and other foreign particles out of the meters, which are normally of the positive-displacement recording type.

Set stop valves are used to stop product flow automatically at a predetermined quantity set on the set-stop counter of the meter.   These valves can be used in remote-controlled systems and can also serve as a remotely operated block valve to prevent unauthorized withdrawal of product.

Rate-of-flow controllers are self-contained flow-indicating control valves used to prevent overspeedi::g and wear of meters.   The flow indicator is usually a pitot venturi, and a straight meter run of at least six pipe diameters is recommended when the controller is downstream of a strainer, globe valve, or short-radius elbow.

The loading arm is a weight- or spring-balanced assembly of pipe and swing joints which will reach various points on trucks of a range of heights.   A controlled closing loading valve is included in the assembly.   This decreases the flow rate rapidly to a small percentage of capacity, after which shutoff is slow to prevent shock.

Details and Arrangements

Loading of trucks is performed at loading racks where various loading arms are grouped.   Usually islands are provided for loading both sides.   The piping and meters are grouped within the island.   Generally, a single product is loaded at each side of

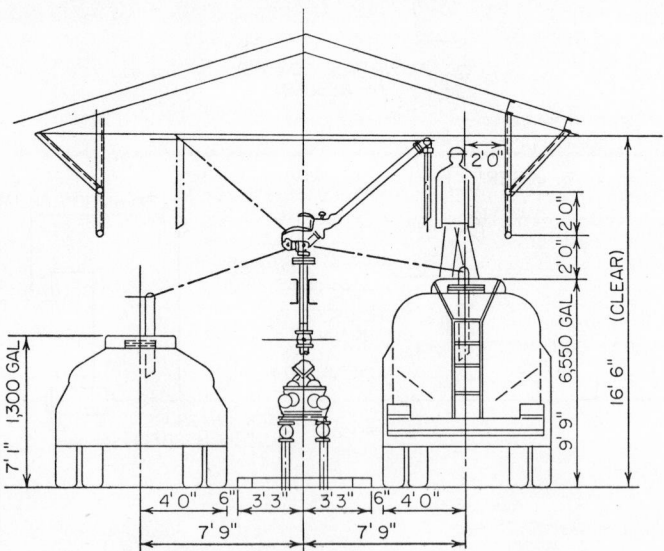

FIG. 8-22. Typical cross section at a tank-truck loading rack.

each island and loading arms serve only one side.   It is possible to group as many as six products on a single rack with loading arms serving either side.

A roof is usually provided to shelter the truck and operators.   At some racks, platforms are provided for access to the loading trucks. Trucks vary widely in size; therefore, in many plants access platforms have been found to be useless and all access is by truck catwalk only. Grab rails are provided suspended from the loading rack roof.

A typical section through truck-loading rack is shown in Fig. 8-22.

Rail-loading Facilities

As discussed under Truck-loading Facilities, the loading requirements must be established to suit marketing conditions. The usual installation is based on anticipated maximums and is, therefore, frequently oversize.

Loading Rates

Rail tank car sizes are reasonably uniform. Most cars have a capacity of 10,000 gal, and very few cars are smaller than 9,000-gal size. Most facilities are designed to load 10,000-gal cars. It is usual to assume that cars are loaded twice during an 8-hr-day shift, 5 days per week.

When the quantity of a product which must be shipped by rail is expressed in barrels per stream day, the number of 10,000-gal cars loaded per 24-hr day is

$$\text{No. of cars/day} = \frac{\text{B/SD} \times 42 \text{ gal/bbl} \times 7 \text{ days/week} \times 24 \text{ hr/day}}{10,000 \text{ gal/car} \times 40 \text{ hr/week}}$$

$$= 0.01765 \times \text{B/SD}$$

The number of loading spouts for each product is equal to one-half of the number of cars per day based on two loadings per 8-hr-day shift.

Pumping capacity should allow for filling a string of cars in 3 hr, reserving the balance of the time for spotting and shifting. Pumping capacity is

$$\text{gpm} = \frac{10,000 \text{ gal/car} \times \text{No. of cars}}{3 \text{ hr} \times 60 \text{ min/hr}}$$

Equipment

Rail shipments are usually gaged and weighed to determine the quantity. No metering facilities are required at rail loading installations. Equipment is, therefore, limited to pumps and loading arms.

Pumps for railroad loading may be used for other services such as truck loading if both are operated infrequently. In general, loading pumps should discharge a minimum quantity of 500 gpm at a discharge pressure of 35 psig.

Loading arms for railroad loading are generally quite simple as a result of the fixed track position and uniform car size. Cars are usually filled through drop tubes (long or short) which can be fastened to the loading arm with a quick coupler. With this arrangement only one simple swing joint is needed. No valving is required at the end of the loading arm. Valves are usually located in the vertical risers operable from the loading platform.

Rail loading racks usually consist of an elevated walkway serving parallel railroad tracks. Drop platforms are provided for access to the car filling dome. These platforms are spaced about 41 ft apart to allow for uncoupling of even the largest cars. One product is usually loaded at each drop platform. A general arrangement of a loading rack is shown in Fig. 8-23.

Tanker and Barge Loading Facilities

Tankers and barges are loaded and unloaded at piers or quays generally called docks or wharves. Facilities for handling petroleum are preferably separated from facilities used for general cargo. In some instances, loading and unloading are done through submarine pipelines at deep-water anchorages.

Loading Rates

Tankers range in size from small coastal vessels of 10,000-bbl capacity up to super-tankers of 250,000-bbl capacity. A few special supertankers have been built holding up to 800,000 bbl. The larger tankers are used in crude service, while the T-2 tanker is used in product service. The T-2 tanker has a capacity of 141,000 bbl and can carry three different petroleum products. Some T-2 tankers have been modified to carry an additional number of products.

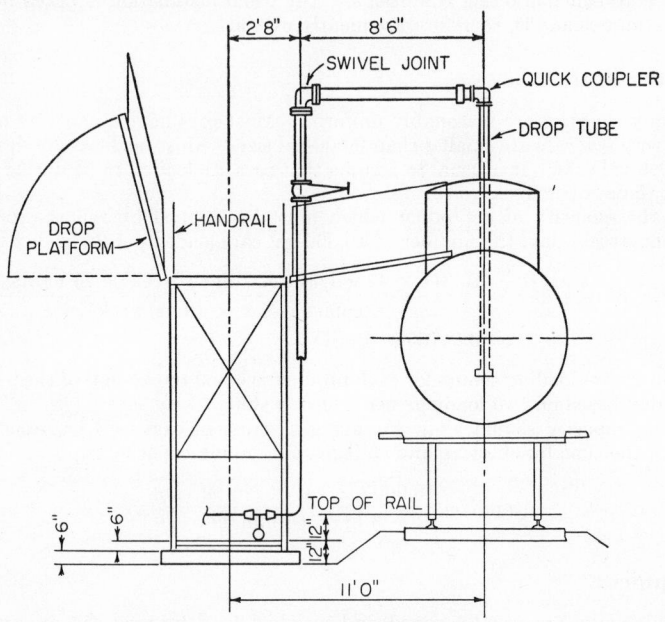

Fig. 8-23. General arrangement of a typical railroad tank-car loading rack.

Tanker piping and pumping systems are designed for relatively high rates. The usual T-2 tanker pumps can handle 8,000 bph. Supertankers can pump as high as 22,000 bph. A desirable tanker product-loading rate for each product is 8,000 to 10,000 bph. It is general practice to load two products into a tanker simultaneously.

Barges are flat-bottomed, shallow-draft vessels used to transport petroleum products short distances in canals, harbors, or inland waterways. While there are exceptions, barges have capacities of 600 to 2,000 bbl. Loading rates for barges are usually limited to about 2,000 bph.

Equipment

Equipment required for tanker loading includes pumps, hose or flexible loading pipe, and handling cranes or structures. Pumps are usually centrifugal pumps having discharge pressures of about 100 psig. Pressure drops in hoses or flexible pipes range from 15 to 25 psi. Frequently there is a considerable static head loss to the deck manifold of an empty tanker. Runs of piping from pumps to pier heads may be sizable.

When rubber hose is used, it is supported by a dockside derrick plus the tanker boom. Recent installations have employed combinations of hose and pipe or flexible assemblies of pipe and swivel joints supported by structures. Some systems use the pipe as structure. Automatic adjustment for tide and tanker draft is incorporated in this equipment.   Hose and various assemblies are available in sizes from 2 to 12 in. diameter with 8- and 10-in. sizes used most frequently for products.

Quantities loaded aboard tankers are measured by gaging storage tanks and tanker compartments.   Only ships' bunker fuel is metered before loading.

Barges are usually loaded through hoses supported by dockside derricks and in some cases by derricks on the barges.   To conserve space, barges are frequently moored two or three abreast and the hose is manhandled across the inboard barges to the outboard barge.   The hose size is accordingly limited to a 6-in. size weighing about 8 lb/ft.

Products loaded into barges are measured by metering.   Meters are located on the dock near the hose connections.   The metering facilities should include strainers and, when necessary, flow-rate controllers.

Details and Arrangements

Docks are specialized structures, and a wide variety of arrangements are possible to suit specific conditions.   The general requirements at a dock are (1) sufficient depth of water for the tanker or barge size expected; (2) mooring facilities for the required number of tankers; (3) a suitable working and storage platform for oil transfer hose, pipe, and related equipment; (4) slop- and spill-collection facilities; (5) ballast unloading piping; (6) fire-fighting facilities; (7) access pipeway, above or below water, from the working platform to shore; (8) means for transporting personnel, equipment, and dry cargo from shore to the working platform, such as a roadway; and (9) lighting and communication facilities.

COOLING SYSTEMS

Refining operations are frequently conducted at elevated temperatures.   Heat is added to material being processed by heating in fired heaters, by exchange with hot process streams, by exchange with steam, and by direct contact with stripping steam. Other heat is accumulated as waste heat from coke-burning operations.   Cooling systems are provided to remove heat as necessary in process sequences and before storage of final products.

In a rough over-all sense, refineries must be in heat balance.   All heat added in the form of fuel burned, steam consumed, or coke burned must be removed by the cooling system.   Exchange of heat among various hydrocarbon streams is an internal operation having no effect on an over-all balance.   Losses of heat from equipment or piping are usually extremely small in comparison with total heat.   Similarly heat remaining in products can be neglected in a rough balance.   An accurate heat balance, considering all incoming and outgoing stream temperatures and all losses, could be made but would serve no useful purpose.   The rough balance described above is, however, a handy checking tool.

Methods of Cooling

Refinery cooling can be effected by various methods including (1) exchange with other streams, (2) air cooling, (3) water cooling, and (4) refrigeration.   The method, or combination of methods adaptable to a specific application is often influenced by limiting process conditions.   Climatic and other geographic considerations also influence the cooling method.

Heat Exchange

Heat exchangers transferring heat from one process fluid to another are a specialized application of cooling.   Either shell-and-tube exchangers or double pipe exchangers are employed in this service.   The fluid temperatures and temperature differentials must be within suitable ranges for economical utilization of exchange.   In general when the inlet temperature of the fluid being cooled is 200°F or higher, heat exchange will be economical.   Proper use of heat exchange for cooling results in reduction in combined cost of process and offsite equipment with coincident operating and maintenance savings.

Air Cooling

Air coolers are adaptable to a wide range of cooling and condensing services. Maximum economy is achieved when these coolers are used to cool fluids to an outlet temperature approximately 40°F higher than the design air temperature. When necessary, air coolers can be used in the temperature ranges normally considered most suitable for heat exchange and to within 20°F of the design air temperature.

Air coolers usually consist of an arrangement of finned pipe. Air is circulated over the fins by a fan. The coolers occupy a sizable amount of space but are frequently located in structures over yard pipe banks, thus minimizing the effect on plot size. The use of air coolers, which are completely within process units, results in increased process-unit investment costs in most cases but effects savings in offsite cooling-system costs.

The cost of maintenance for air coolers is considered to be one-half the cost of maintenance of a water exchanger. In addition, maintenance costs of cooling-water systems and piping are reduced when air coolers are used. Operating costs associated with air coolers are usually lower than operating costs of water coolers and related equipment.

Water Cooling

Water is the traditional material for cooling and condensing services in refineries. Water cooling is applicable for temperatures over a wide range. The maximum inlet temperatures of fluid being cooled is limited by water quality to prevent excessive deposits or corrosion of exchanger tubes. The minimum temperatures to which fluid can be cooled are approximately 15°F above water inlet temperature.

Water cooling, unlike air cooling, is not normally self-contained within a process unit. Water-cooling systems may include heat exchangers, pumping equipment, distribution piping, water intake stations, and cooling towers. Facilities such as water intake stations and cooling towers are usually offsite facilities. Heat exchangers can be shell-and-tube units or finned-pipe or box coolers. These are usually located within process units either in groups or spaced between other types of equipment.

Refrigeration

When colder temperatures than those obtained by use of water are required, some form of refrigeration is used. Normally these installations are situated within process units but occasionally are employed in conjunction with offsite installations such as dry-air blanketing systems for products.

Virtually all refrigeration installations use air or water cooling of refrigerant. For temperatures ranging down to approximately 50°F, cold-water evaporative cooling systems are used because of low equipment cost. These systems are, however, relatively inefficient in terms of steam and condensing water. When colder temperatures are required for process cooling, refrigeration cycles are necessary. Package refrigeration units can be used for small duties; larger systems are composed of individual assemblies of compressors, condensers, and receivers employing suitable refrigerants.

Cooling-water Systems

Cooling-water systems in refineries are either once-through type or recirculated type. In a **once-through system,** pumps take suction from a source of supply such as a river, lake, or other body of water and deliver the water to process units or to water users at other locations. After passing through cooling equipment, the hot cooling water is conducted to a point of disposal through a pressure system of piping or through a gravity-flow system. Gravity-flow return systems are frequently a combination of closed and open conduits.

In the **recirculated type** of system, pumps take suction from a cooling-tower basin and deliver water to cooling equipment. After passing through water users, the hot cooling water is discharged through a pressure-return system to the top of the cooling tower.

Choice of System

The choice between a once-through and recirculated cooling-water system must be based on availability of sufficient water of satisfactory quality, on process temperatures, on atmospheric conditions, and on equipment maintenance and operating costs.

As various industrial processes utilize larger quantities of water, available supplies remain limited. In certain arid areas, such as Los Alamos, domestic sewage has been treated and used for cooling-tower make-up. The initial treatment of domestic sewage required before industrial use is the same as that necessary before any other method of sewage disposal. The cost of supplemental treatment, in addition to standard sewage disposal, of cooling-water make-up is not materially greater for sewage treatment plant effluent than for moderately hard or corrosive waters.

Refinery waste waters are also reusable as cooling-tower make-up. When they are used in this manner, reduction in phenols and mercaptans is effected in the tower. Sea water is also suitable for make-up purposes when other waters are not available. When corrosive waters are used for cooling water, the cost of cooling equipment including pumps and piping is greater because of use of more corrosion-resistant materials. This must be considered when sewage, refinery-waste waters, or sea water is used in cooling-water systems. Process temperatures in conjunction with air temperatures and humidity may limit use of cooling towers. In hot, humid areas sufficiently low temperatures cannot be obtained by cooling towers for effective process use.

Cost Comparison

The selection of a cooling-water system is often predetermined by the availability of water. In arid areas where water is scarce, once-through systems are obviously eliminated from consideration and recirculated cooling water systems are chosen automatically. Similarly where fresh-water supplies are limited and salt water is available, a once-through system is selected. At plants located on rivers or lakes or where abundant supplies of both salt and fresh water are present, a choice must be made between a recirculated or once-through system.

The choice of cooling-water system must be based on comparisons of initial and operating costs. When the distance from cooling-water pumps to the water users is equal for systems being compared, the operating cost of a recirculated system is approximately 40 per cent greater than the operating cost of a once-through system. Relative initial costs of system vary with climatic conditions, distance from pumps to users, complexity of inlet structures, and corrosive nature of water, In general, recirculated cooling-water systems having a capacity smaller than 17,000 gpm have a lower initial cost than once-through systems.

Relative initial costs of systems, excluding corrosion effects, can be estimated from Fig. 8-24. In this figure, the effect of corrosive water on the cost of heat exchangers and piping is excluded.

Establishing Cooling Requirements

The initial step in the design of a cooling-water system is to determine design temperatures and system capacity. The system capacity varies with design temperatures as limited by process conditions. The usual cooling ranges are between 25 and 30°F. Inlet temperatures to process coolers vary in general from 75 to 85°F, and outlet temperatures range between 100 and 115°F. Inlet temperatures to cooling equipment are established by ambient conditions, while outlet conditions are set by type and quality of water. Maximum temperatures in heat-exchange equipment must be limited in order to prevent corrosion or deposition of solids.

Cooling duties of process plants are determined on the basis of average fractionation data or pilot-plant results or from other process plants. Duties are frequently adjusted to allow for extra recycle or other safety factors. When heat exchangers are used in alternate operational cases, the design duty of the heat-exchange equipment is the largest of the various duties. In any process plant there may be a large number of coolers designed for alternate operations having noncoincidental maximum duties.

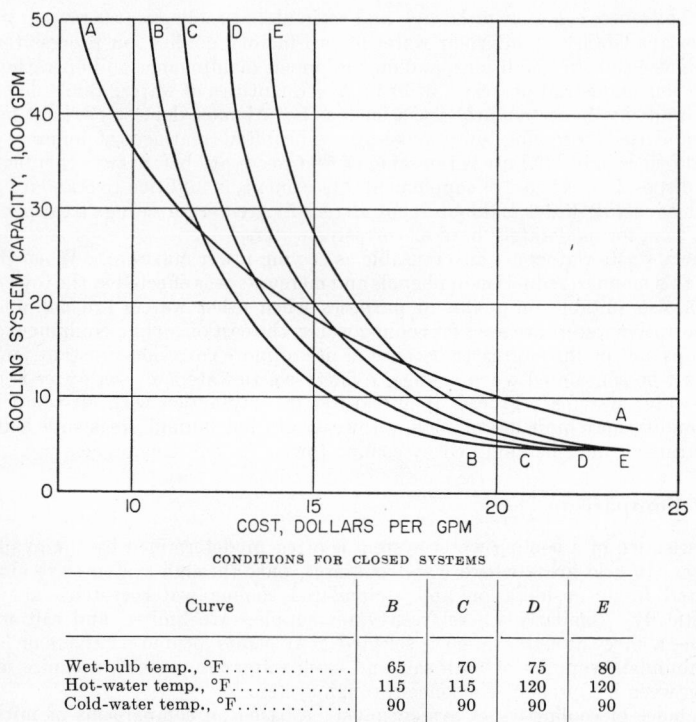

CONDITIONS FOR CLOSED SYSTEMS

Curve	B	C	D	E
Wet-bulb temp., °F...............	65	70	75	80
Hot-water temp., °F..............	115	115	120	120
Cold-water temp., °F.............	90	90	90	90

Fig. 8-24. Relative initial costs of cooling water systems. Costs are based on U.S. Gulf Coast direct material and labor. Systems include cooling towers or water-intake station, pumps, piping, electrical gear and wiring, miscellaneous equipment, and foundations. *A*: Once-through system. *B, C, D*, and *E*: Closed systems.

The process cooling-water requirement of a unit can be considered to be the sum of the water requirements of the various coolers in the plant without the addition of additional safety factors. This will result in a design water quantity which is 5 to 10 per cent greater than the normal requirement.

In a similar manner the cooling-water requirements for utility services, such as condensers for steam-turbine drivers, lubricating-oil coolers, and pump glands, can be considered to be the summation of the individual design requirements. Definite safety factors are included in these design quantities. Condenser duties are based on steam-turbine design flow quantities, which include pump or compressor safety factors and turbine safety factors. While it may be possible to use the full design water requirement for a single condenser serving an overloaded turbine, it is unrealistic to assume all turbines overloaded simultaneously. Because of the cumulative safety factor associated with utility equipment, the design water requirement for these purposes may be 10 to 15 per cent greater than the normal water requirement.

When the total plant requirement for cooling water for process and utility users is established as the summation of the maximum design duties of individual items of equipment, certain other safety factors are introduced.   As a result of the limiting process design temperatures and steam condenser design temperatures, full utilization of the potential cooling range is not realized.   This has no effect in a once-through system but can result in excess capacity in a recirculated system.   The total actual duty of a cooling tower in btu is less than the apparent duty obtained by multiplying theoretical temperature differential by flowing quantity.   This unused cooling-tower duty may in some cases amount to as much as 10 per cent of design duty.

Safety factors are also introduced into plant cooling-water requirements by design ambient conditions which are generally based on short-term periods.   For example, the usual design wet-bulb temperature for cooling towers is the temperature which is exceeded 5 per cent of the time or for 150 hr during a 4-month summer period.   Actual wet-bulb temperatures are lower than design wet-bulb temperatures 95 per cent of the time during the 4 summer months, resulting in potentially colder water at most times.   Similarly, the basis for establishing the inlet temperature of once-through cooling-water systems is usually conservative.   These temperatures are usually based on the average of maximum observed water temperatures during summer periods.

System Components

The usual cooling-water-system design will not incorporate standby apparatus other than pumps.   Because of the general reliability of pumping equipment it is only necessary to install one full- or part-capacity standby pump with driver.   For a selected type of system, water pump capacities should be established to minimize initial cost of the installation including equipment, foundation, electrical gear if required, and lead piping.   Operating costs need be considered only when choosing between alternative drivers.

Table 8-9. Economical Sizes of Vertical Motor-driven Pumps

System capacity, gpm	No. of pumps	Pump capacity, gpm
5,000	2	5,000
10,000	3	5,000
15,000	3	7,500
20,000	5	5,000
25,000	5	6,250
30,000	6	6,000
35,000	5	8,750
40,000	6	8,000
45,000	6	9,000
50,000	6	10,000

For cooling-water systems it is economical to use motor- or steam-turbine-driven horizontal pumps or motor-driven vertical pumps in sizes up to 10,000 or 12,000 gpm. Vertical steam-turbine-driven pumps are limited in application because of the high cost of the angle drives required for large pumps.   When condensing steam turbines are used, pumps larger than 12,000 gpm may be economical.   For installations using motor-driven pumps, the cost of motor and switchgear must be considered for each case.   Table 8-9 can be used as a general guide to help establish the minimum number of motor-driven vertical pumps needed.

Pressure Drops in Piping

Cooling-water pump differential pressure is established to include pressure losses in process units, supply and return headers, and pump suction and discharge leads, as well as static head losses.   For a once-through system in which water is discharged to sewers after passing through cooling equipment, the pressure in the supply header at process-unit battery limit may be established at approximately 25 psig.   This pressure corresponds to the total pressure drop in piping and equipment in the unit.   When a return pressure system of the once-through type or a recirculation system is used, the pressure drop in piping and equipment in the unit between the inlet and outlet points will be approximately 30 psi.

Supply and return headers are usually sized on a moderately conservative basis. A balance can be established between incremental pipe cost and incremental cost of pumps, drivers, and electrical or steam supply.   In most systems the optimum-size pipe for use in offsite cooling-water headers will have a unit pressure drop equal to or less than 0.5 psi per 100 ft.   When header lengths are extremely long, appreciably lower pressure drops may be desirable.

Pressure drops in pump suction and discharge leads are usually appreciably large in relation to pump differential.   The high cost of valves and pipe fittings used at pumps justifies reductions in line sizes.   Good, low-cost piping installations are obtained when lead sizes are the same as pump nozzle sizes, the cost of reducing fittings thus being eliminated.   Because of the cost of valves in discharge leads, it may be more economical to use smaller size piping provided the total drop in discharge lead is limited to 2 or 3 psi.   Suction leads can be established on the basis of a unit pressure drop not exceeding 0.5 psi per 100 ft.

Static head losses include all differences in elevation between cooling-water pumps and point of discharge of hot water.   In a once-through system the point of discharge will be either grade at the hot-water sewer inlet or elevation of the discharge end of a pressure return.   In a recirculation system, the static head loss is the difference in elevation between the cooling-water pumps and the top of cooling tower.   Most cooling-tower manufacturers express head loss as measured above a given datum including a small loss in distribution boxes or spray nozzles.

Cooling Towers

Cooling towers function by direct removal of heat from water by air flow and by vaporizing a portion of the water.   Both forms of cooling are accomplished by a counterflow of air and water.   Towers are constructed of wood, metal, or concrete with wood or plastic packing for distribution of water flow.   Towers can have gravity air flow, or fans can be used for forced- or induced-draft air flow.   In the United States, wood towers with induced-draft fans are in common use.   In Europe concrete venturi-type natural-draft towers are commonly favored.   The natural-draft tower has a much lower operating cost than the induced- or forced-draft type.

Tower Losses

As stated above, a portion of the water passing over a cooling tower is vaporized. Any solids which this water contains are left behind and increase the concentration of solids in the water.   In order to limit the concentration of solids and to prevent their deposition on cooling surfaces, it is necessary to blow down an additional amount of water.   A further water loss occurs when water drifts off the tower in the wind. This is windage or drift loss.   Other liquid losses are called miscellaneous losses.

Cooling-tower losses are usually evaluated as follows: (1) Evaporation loss, the vaporized water, is approximately equal to 1 per cent of tower throughput for each 10°F of cooling-tower temperature differential; (2) drift loss is limited in the design of the tower to 0.2 per cent of cooling-tower throughput; (3) miscellaneous liquid loss is assumed to be equal to one-seventh of the evaporation loss minus the drift loss; and (4) blowdown is determined on the basis of reducing the concentration of solids to 4 or 5 cycles of concentration.   For 5 cycles of concentration, blowdown in gallons per minute, $B = 10.7(\Delta T)\text{gpm}_T 10^{-5}$, in which ΔT is differential temperature and gpm_T is cooling-tower throughput in gallons per minute.

Design Details

In each cooling-water system there are certain nonrecoverable uses of water such as pump gland cooling, wash water, etc. The water loss by these uses may be equal to or greater than the desired system blowdown, and no other water need be withdrawn from the system.

Block valves or similar means of isolation are provided in the return piping to each cooling-tower cell to allow for cleaning or for temperature control during cold weather. When the cooling tower can be partially shut down without resulting in a complete shutdown of related refining facilities, it is preferred that the cooling-tower basin be segregated into separate compartments for each cooling-tower cell or group of cells.

Cooling-tower basins are usually made sufficiently large to impound 10-min throughput of water. This allows for a supply of cold water in the event of fan failure. With fans out of service, cooling water can be circulated for a period of about 30 to 40 min before becoming too hot for effective cooling. This is sufficient time to permit restoration of fans or shutdown of process equipment in a reasonably orderly manner.

Pumping Stations and Water Intake

A wide variety of arrangements can be applied at water intakes and pumping stations. Pumping stations are similar whether used at cooling towers or in rivers, lakes, or other bodies of water for once-through systems. Intake pumping stations may, of course, serve purposes other than cooling, such as boiler feed make-up, drinking water, or process water. Fire-water pumps are frequently located at water pumping stations for cooling towers or once-through cooling systems.

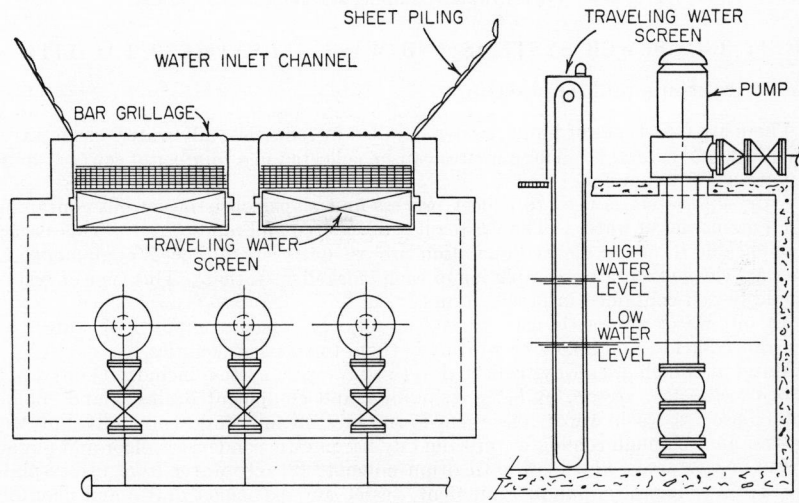

FIG. 8-25. Typical water-pumping station, submerged pumps.

Factors which should be considered in design of a water pumping station include (1) geophysical characteristics of the site, (2) materials of construction available, (3) available utilities together with relative cost, (4) character and temperature of water in various portions of body of water, (5) remoteness of location for access by operators, and (6) necessary auxiliaries such as fish screens, grillages, traveling water screens, chlorinators, and handling facilities.

Typical pumping stations are shown in Figs. 8-25 and 8-26.

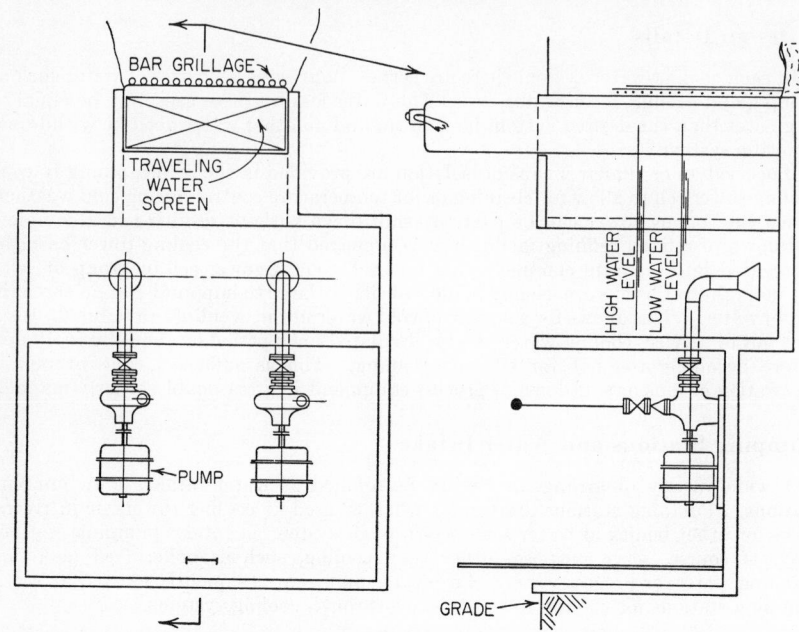

FIG. 8-26. Typical water-pumping station, external pump.

REFINERY SEWER SYSTEMS AND WASTE-TREATMENT FACILITIES

Sewer Systems and Application

There are four types of liquid wastes handled in refineries: oily water, clean water, chemical, and sanitary. These wastes can be collected in a number of sewer systems in various combinations.

A dry-oil system is used to collect process-unit slops consisting of oils containing small quantities of water. The waste oil is dewatered and reprocessed. This system is applicable to use in blocked-operation process units where process equipment and lines are cleaned out after completion of each blocked operation. This type of system should be self-contained in process units.

An oily-water system is used to collect refinery wastes composed of water, oil, neutralized acids, and alkalies and convey them to treating facilities.

Materials which are always collected in the oily-water system include (1) oily water from process-unit vessels, exchangers, pumps, and equipment drains; gland cooling water; process wash water; oily blowdown; oily steam condensate; oily hot well outlets; once-through cooling water from oily barometric condensers, floor and paving drains in oily areas, and blowdown drum effluent; (2) oily water from utility-plant oily-steam condensate, pump, equipment, vessel, and exchanger drains and floor and paving drains in oily areas; (3) oily water from offsite-area blowdown drum effluent.

Materials which are frequently collected in the oily-water system but could be diverted to another system include (1) waste from process-unit neutralized acids and alkalies, once-through cooling water from cooler boxes, rainfall from oil-free areas, roof drainage, cooling-tower blowdown, chemical treating unit effluent, sanitary drains, and septic-tank effluent; (2) waste from utility-plant water-treating sludge, boiler blowdown, rainfall from oil-free areas, sanitary drains, and septic-tank effluent; and (3) waste from offsite neutralized-chemical effluent.

Drainage from offsite areas which is usually collected in the oily-water system but can be provided with special facilities includes pump group drains, truck and rail loading area drainage, and oily-steam condensate.

A clean-water system collects waste water which has little or no contact with oil and conveys it to a treating plant. This system includes closed conduits and open ditches and is used for once-through cooling water, boiler blowdown, feed-water treating sludge, rainfall from clean areas, neutralized acids and alkalies, gutter and roof drains, cooling-tower blowdown, lavatory sink and shower drains, diked area drains, fire water, and steam condensate. In special cases septic-tank effluent can be included in this system.

Sanitary sewer systems collect wastes from toilet facilities and convey these to a treating system or municipal sewer system. Treating-system effluent may be discharged to the oily-water system, or to a closed clean-water system, or to a separate sewer leading to a disposal point.

Chemical wastes are separated from the petroleum during processing or are obtained from chemicals used in processing. The effect of these wastes should be studied, and segregation and treatment formulated accordingly. In general, if chemical wastes affect only oxygen demand and solids content of the plant wastes, they can be discharged to the oily- or clean-water sewer. When other characteristics of waste water are affected (pH, taste and odor, toxicity, color, or turbidity), special treating facilities should be included at the source of wastes.

Design of Collection Systems

Based on the types of wastes, the various systems required in a plant can be selected. The quantities flowing in each system can then be determined, and design details established. Refinery waste-collection systems are similar in many respects to systems used in municipalities and other industrial plants.

Table 8-10. Time of Concentration

1. The time of concentration is the time in minutes it takes a drop of water to enter a ditch and to move to the end of that particular ditch. It is equal to the distance between these two points in feet divided by sixty times the design velocity. Each subsequent ditch would then have a higher time of "concentration."

2. Recommended maximum velocities of flow in ditches are tabulated for cohesive and noncohesive materials as established by site survey.

3. Velocities are for ditches 36 in. deep. For ditches 12 in. deep multiply by 0.8. For ditches 24 in. deep multiply by 0.9.

4. For lined ditches use a design velocity of 6 fps.

Maximum Mean Ditch Velocity

Cohesive material	Velocity, fps		
	Loosely compacted	Fairly compacted	Compact
Sandy clay (sand content is less than 50%)...	1.48	2.95	4.26
Heavy clayey soils........................	1.31	2.79	4.10
Clays.......................................	1.15	2.62	3.94
Lean clayey soils..........................	1.05	2.30	3.44
Sandy silt................................	0.87	1.90	2.84

Noncohesive material	Velocity, fps
Coarse sand..........	1.00
Fine gravel...........	2.50
Medium gravel........	5.00

Table 8-11. Rainfall Rate

Rainfall rates can be determined from the following equations for a locality as listed. In the formula, t = time of concentration in minutes as determined from Table 8-11 but not greater than 120 min. These equations are based on a storm frequency of once in 5 years.

No.	Equation	No.	Equation
1	$R = \dfrac{247}{t + 29}$	5	$R = \dfrac{81}{t + 13}$
2	$R = \dfrac{190}{t + 25}$	6	$R = \dfrac{75}{t + 12}$
3	$R = \dfrac{131}{t + 19}$	7	$R = \dfrac{48}{t + 12}$
4	$R = \dfrac{97}{t + 16}$		

Rainfall Data

State	Locality	Equation No.
Arkansas	El Dorado	2
	Little Rock	2
	Magnolia	2
	Stephens	2
California	Bakersfield	7
	Dominguez	7
	El Segundo	7
	Fillmore	7
	Los Angeles	7
	Martinez	7
	Oleum	6
	Richmond	7
	San Francisco	7
	Santa Barbara	7
	Torrance	7
	Ventura	7
	Wilmington	6
Colorado	Denver	5
Georgia	Atlanta	2
	Brunswick	2
Illinois	Chicago	3
	Lawrenceville	3
	Lemont	3
	Lockport	3
	Robinson	3
	Wood River	3
Indiana	East Chicago	3
	Hammond	3
	Indianapolis	3
	Whiting	3
Kansas	Eldorado	2
	Garden City	3
	Kansas City	2
	Ulysses	3
	Wichita	2
Kentucky	Ashland	3
	Louisville	3
Louisiana	Baton Rouge	1
	Destrehan	1
	Lake Charles	1
	Shreveport	1
	Sterlington	1
Maryland	Baltimore	2
Massachusetts	Everett	2
Michigan	Lansing	4

Table 8-11. Rainfall Rate (Continued)

State	Locality	Equation No.
Missouri...............	Kansas City	2
	Sugar Creek	2
Montana...............	Billings	7
	Great Falls	7
New Jersey..............	Barber	2
	Bayonne	2
	Elizabeth	2
	Jersey City	2
	Linden	2
	Newark	2
	Paulsboro	2
	Perth Amboy	2
	Port Monmouth	2
	Westville	2
New York...............	New York	2
	Olean	3
North Dakota...........	Mandan	3
Ohio....................	Cincinnati	3
	Cleveland	3
	Dayton	3
	Heath	3
	Lima	3
	Toledo	3
Oklahoma..............	Barnsdall	2
	Cushing	2
	Drumwright	2
	Duncan	2
	Enid	2
	Oklahoma City	2
	Ponca City	2
	Tulsa	2
Oregon.................	Portland	7
Pennsylvania...........	Bedford	3
	Bradford	3
	Marcus Hook	3
	Philadelphia	3
	Rouseveille	3
	Titusville	3
Tennessee..............	Memphis	3
Texas..................	Amarillo	3
	Baytown	1
	Beaumont	1
	Big Spring	2
	Corpus Christi	2
	Dallas	2
	Deer Park	1
	El Paso	6
	Fort Worth	2
	Houston	1
	Nederland	1
	Port Arthur	1
	Texas City	1
	Wichita Falls	2
Utah...................	Salt Lake City	7
Virginia................	Norfolk	2
	Richmond	3
Washington.............	Anacortes	7
	Ferndale	7
West Virginia...........	Parkersburg	3
Wyoming...............	Casper	5
	Greybull	7
	Parco	7

Surface Runoff

Information about the site soil must be known to establish ditch velocities and slopes, runoff rates, etc. Based on this information surface runoff can be calculated using the rational formula $Q = AIR$, in which Q is runoff in cubic feet per second, A is the area drained in acres (43,560 sq ft), and R is the rainfall rate based on time of concentration. Time of concentration is shown in Table 8-10, and rainfall rate in Table 8-11. I, the coefficient of runoff, can be derived from Fig. 8-27.

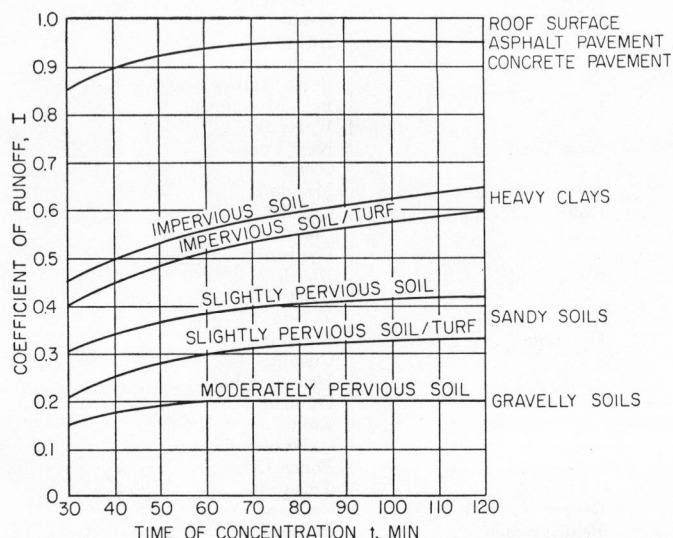

FIG. 8-27. Coefficient of rain runoff I for use with Table 8-10. Note: Applicable to flat sites.

Rainfall collected within diked areas is not considered in determining surface runoff. These areas are considered ponding areas. Water collected in diked areas is discharged to the clean-water system after the bulk of rain water has been drained from the refinery.

Design of Ditches

Ditches of various sizes must be designed and located throughout the refinery to drain areas of surface runoff and other clean water. Ditches should preferably have side slopes of $1\frac{1}{2}$ horizontal and 1 vertical. For some site materials flatter side slopes are necessary.

An over-all hydraulic gradient for ditches must be established. Ditches need not be sloped; the bottom elevation of a particular ditch will remain constant. Hydraulic gradient is developed at culverts crossing refinery roads. The hydraulic gradient for a ditch will be equal to the difference in elevation of inlet and outlet of the culvert serving the ditch divided by the length of ditch plus culvert.

In most cases, allowable velocity in the ditch will establish ditch dimensions. Velocities are those listed in Table 8-10, Time of Concentration. The minimum ditch depth should be 12 in. Depth of larger ditches should be set at even increments of 2 or 3 in. When necessary, ditch size can be determined by using Fig. 8-28.

Design of Clean-water Sewers

Gravity-flow sewers may be used in each of the refinery collection systems, including the clean-water system. Quantities flowing in the clean-water sewer will include rainfall as described under Surface Runoff above, cooling water in a once-through system, and any other oil-free water.

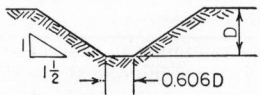

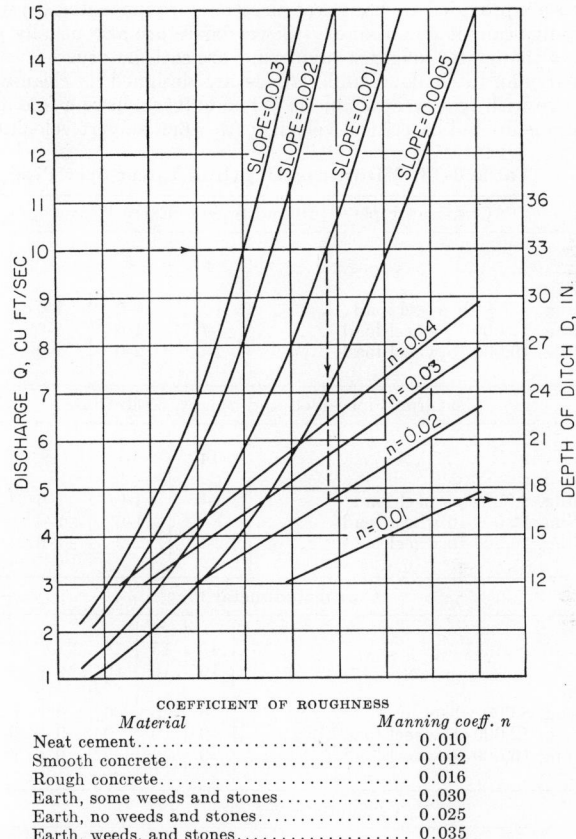

COEFFICIENT OF ROUGHNESS	
Material	Manning coeff. n
Neat cement............................	0.010
Smooth concrete.........................	0.012
Rough concrete..........................	0.016
Earth, some weeds and stones..............	0.030
Earth, no weeds and stones................	0.025
Earth, weeds, and stones..................	0.035
Stiff clay................................	0.025
Prefabricated asphaltic lining..............	0.015

FIG. 8-28. Determination of flow capacities of ditches. Curves based on Manning formula. $Q = 1.486R^{2/3}S^{1/2}n^{-1}$. Velocity $= 68.4Q/D^2$.

Gravity-flow sewers in refineries are usually designed to flow at three-fourths of full depth; therefore design flow should be increased by a factor of 1.10 to obtain equivalent full flow of sewer. It is customary to select sewer pipe size by the Hazen-Williams formula. Suitable C values are listed in Table 8-12.

Table 8-12. Hazen-Williams Coefficient

Pipe material	C
Corrugated metal...............	40
Cast iron and steel..............	100
Clay tile pipe...................	110
Concrete pipe...................	120
Asbestos cement pipe............	140

Sewer velocities are usually maintained at 2 fps minimum to prevent settlement of solids. Unless otherwise established by velocity, the minimum sewer slope used in refineries is 1 ft per 1,000 ft. The minimum sewer size recommended is 4 in. Concrete pipe is available as 6-in. size minimum, and corrugated steel as 8-in. size minimum.

Sewer boxes are provided at changes of direction for pipe materials, such as concrete, for which fittings cannot be obtained. Sewer boxes are also usually provided about every 500 ft, although far longer straight runs are satisfactory.

Culverts carrying ditch flows under roads are designed as clean-water sewers as described above with two exceptions: (1) When culverts have a free discharge at the outlet, the size is limited by critical velocity; (2) when culvert velocities exceed twice

Table 8-13. Minimum Depth of Cover over Pipe

Cast-iron Pipe—Hub and Spigot, Extra Heavy

Diameter, in...............................	4	6	8	10	12
Min. cover, ft-in.:					
H-10 Loading, 8,000-lb wheel load.........	1-0	1-0	1-0	1-0	1-6
H-15 Loading, 12,000-lb wheel load........	1-0	1-6	1-6	1-6	2-0
H-20 Loading, 16,000-lb wheel load........	1-6	2-0	2-0	2-0	2-0

Cast-iron Pipe—Bell and Spigot, 50-lb Class

Diameter, in...............................	14	16	18	20	24
Min. cover, ft-in.:					
H-10 Loading, 8,000-lb wheel load.........	1-0	1-0	1-6	1-6	2-0
H-15 Loading, 12,000-lb wheel load........	1-6	1-6	2-0	2-0	2-6
H-20 Loading, 16,000-lb wheel load........	2-0	2-0	2-6	2-6	3-0

Corrugated-metal Pipe

Diameter, in...............................	12	15	18	21	24	30	36	42
Gage.......................................	16	16	16	16	14	14	12	12
Min. cover, in.:								
H-10 Loading, 8,000 wheel load.............	6	6	6	6	6	6	6	6
H-15 Loading, 12,000-lb wheel load..........	9	9	9	9	9	9	9	9
H-20 Loading, 16,000-lb wheel load..........	9	9	9	9	9	9	9	9

Concrete Sewer Pipe (C14)

Diameter, in...............................	6	8	10	12	15	18	24
Min. cover, ft-in.:							
H-20 Loading, 16,000-lb wheel load.......	1-6	1-6	2-0	2-6	2-6	3:0	3-0

Reinforced-concrete Pipe (C76)

Diameter, in...............................	12	15	18	24	30	36	42
Min. cover, ft-in.:							
H-20 Loading, 16,000-lb wheel load.......	1-6	1-6	2-0	2-0	2-0	2-0	2-0

the design ditch velocity, both the inlet and outlet areas around the culvert should be protected against erosion by riprap or lining material. If the culvert discharges against the side of a ditch, this area should be protected.

Suggested minimum cover over the top of pipe to road grade for various loading and piping materials is listed in Table 8-13.

Oily-water Flows

Oily water is primarily collected in process units. Sewers serving units should be designed for the greater of two possible combinations. Both combinations include normally flowing process waste as previously defined, including equipment drains, gland sealing water, process wash, etc. The design combinations are as follows: (1) rainfall plus process waste with the sewer flowing at three-fourth of full depth (Equivalent full capacity is obtained by multiplying design flow by 1.10. Because distances are relatively short in process units, time of concentration is short and rainfall rates are high. Time of concentration includes time to flow across roofs and pavement and should never be less than 15 min at the process-unit outlet.) and (2) process waste plus expected fire-water runoff with sewer flowing full. For fire-water runoff, hoses are estimated to discharge 250 gpm each, but in no case is the fire-water runoff entering the oily water sewer expected to exceed 750 gpm.

Design of Oily-water Sewers

In general the criteria outlined for design of gravity sewers for clean-water systems apply to oily-water systems. The main variations concerning sewer boxes are as follows: (1) Sewer lines leaving hazardous areas such as process units, pumping stations, etc., must be sealed at the first sewer box which they enter to minimize spread of fire. This is done by causing the end of the incoming pipe to be completely submerged in liquid. (2) Sewer boxes are provided at all points where seals are needed and at intervals of 300 ft for 15-in. lines and smaller or 500 ft for 18-in. lines and larger.

Sanitary Effluents

Frequently, local public-health codes and regulations control the design of these facilities and may differ from procedures described herein. Collection systems are provided from buildings containing sanitary facilities to public sanitary sewers, to septic tanks, or to central sewage-treating plants.

The simplest disposal system utilized is to discharge sanitary sewage to a septic tank and to dispose of the septic-tank effluent in a leaching field or a closed sewer system such as the oily-water sewer. When used, the septic tank should be as close as possible to the building with long runs to the disposal field or sewer. Design flows for sewers and capacities of septic tanks are given for varying numbers of single-shift people in Table 8-14.

The general design principles for gravity-sewer systems as outlined for clean-water sewers apply here. The velocity in sanitary sewers should not be less than 2 fps. Portions of these sewers which carry solids (upstream of the septic tank or central

Table 8-14. Design Flows and Septic-tank Capacities

No. of people served	Design flows, gpm	Septic-tank capacity, gal
1–11	50	450
12–19	75	800
20–30	100	1,200

treating plant and complete run to the public sewer) should slope a minimum of 10 ft per 1,000 ft.   In some plants, it may be necessary to pump sanitary sewage.

Piping material used for sanitary sewage must be selected to withstand corrosive conditions when ambient temperatures are relatively high.   Asbestos cement and clay tile are frequently specified for this service.

Physical Details of Collection Systems

Waste-collection systems include various sewer boxes, neutralizing sumps, ditches, roads, etc., which can be standardized within a particular plant.   Some of the more important details are described below.

Sewer Boxes

Most sewer boxes are box type consisting of a bottom slab and cast-in-place concrete walls.   Details of this type are shown in Fig. 8-29.   When concrete pipe is available, this material can be adapted for use as sewer boxes.   Where fittings for concrete sewer pipe are available, a tee can be used in lieu of a sewer box.   All boxes must be provided with suitable covers.

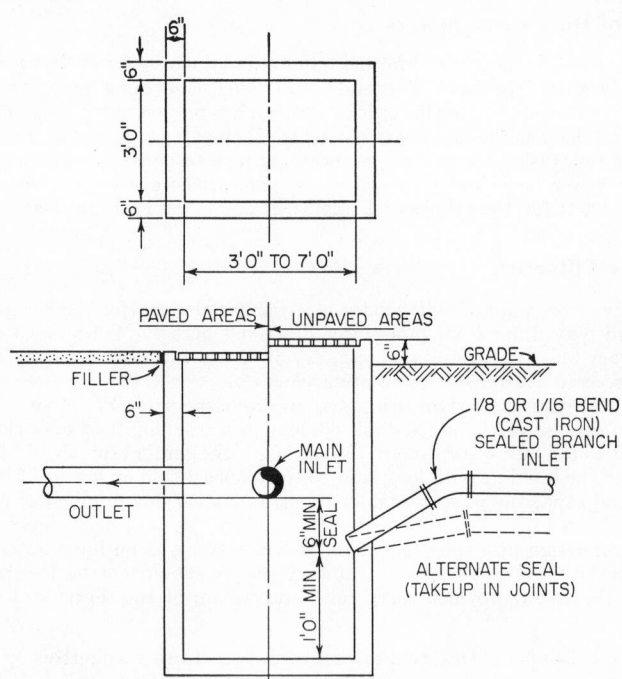

FIG. 8-29.   Details of rectangular sewer box with bottom slab and cast-in-place concrete walls.

Sumps

Sumps are used in collection systems for oily-water wastes in remote areas and for chemicals.   Oil sumps can be considered miniature oil separators and can be so designed.   Basically an oily-water sump is a large box in which the velocity of flow is low.   The outlet of the box must be baffled to retain oil.   Some type of skimming device is required to permit withdrawal of oil to a storage pit or tank.   Portable equipment is generally used to empty the storage pit.

Chemical sumps are of two types: isolation and neutralization.   Isolation sumps normally have a closed outlet and are discharged only when they contain nonhazardous or nontoxic materials such as storm water.   When chemicals enter the sump, they must be pumped out.

Neutralization sumps usually have an open outlet and discharge neutralized material constantly, although provisions are made to plug the outlet.   Neutralization is effected by passing the chemical waste through a neutralizing bed or by adding a neutralizing agent under manual or automatic control.

Drainage Ditches and Roads

Refinery roads provide access to various portions of the plant for movement of product, operations, maintenance of equipment, and fire protection.   Roads are paralleled by a network of product and utility piping and drainage facilities (sewers and ditches).   Layout and construction make it necessary to consider roads and ditches as a common structure.

Roads vary in width from about 10 to about 30 ft pavement width.   In addition, shoulders having a minimum width of 3 ft are provided.   Main roads around administration areas are often curbed and provided with storm-water drains.   These drains discharge to ditches in the more functional areas of the plant.   Road intersections should have a radius of curvature at the pavement edges to suit the vehicles using the roads.   A 20-ft radius is considered satisfactory for most roads.   Details of a typical road are shown in Fig. 8-30.

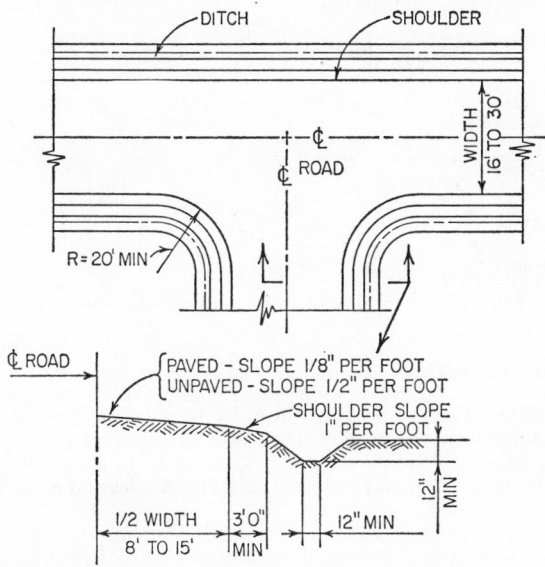

FIG. 8-30. Details of typical refinery road.

Typical ditch sections are shown in Fig. 8-31.   A lined ditch is recommended for high-velocity flows.   Ditches should not be shallower than 12 in., and to simplify construction, the bottom width should be 12 in.   Depths should vary in even increments to suit flows and culvert elevations.

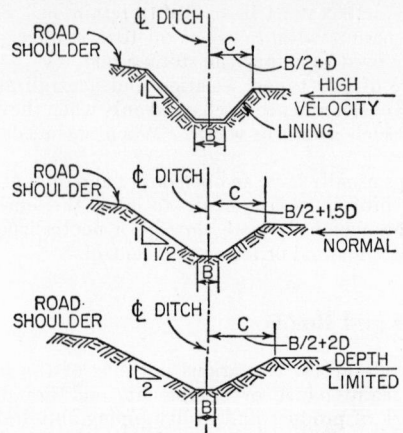

Fig. 8-31. Details of sections of typical refinery ditches.

Dike Drainage

Dike area drains are normally closed and are opened only to empty collected water. If products are spilled, they must be pumped out. Typical drain details are shown in Fig. 8-32.

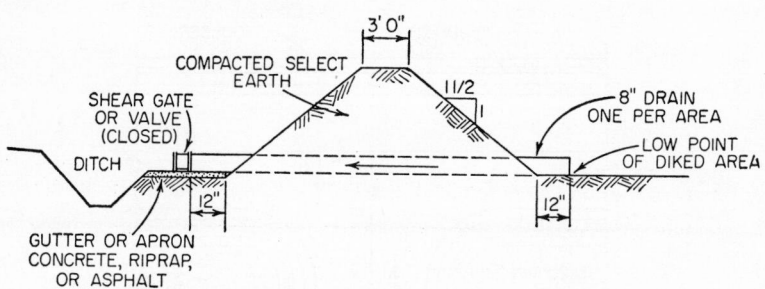

Fig. 8-32. Typical drain details through a diked area.

Waste-treating Facilities

Waste-treating facilities are provided in order to obtain a refinery effluent which will meet legal requirements as to oil content, toxicity, odor, color, taste, pH, biological oxygen demand (BOD), chemical oxygen demand (COD), and turbidity. Requirements affecting toxicity, odor, and taste will limit the sulfides and phenols which can be discharged into the effluent.

Treating facilities for refinery wastes can be subdivided into three general categories: (1) facilities directly related to petroleum processing such as H_2S stripping, mercaptan extraction, and caustic neutralization (these facilities should be incorporated in process-unit design to minimize waste-disposal problems); (2) facilities for separation of nonemulsified oil and oil-bearing sludges from waste water, including oil-water separators, effluent holding basins or lagoons, and ballast water separators; and (3) special treating facilities to remove residual oil in all forms, to remove phenols and remaining sulfides, to remove all sediment, to satisfy BOD and COD, and generally to improve odor, color, and taste.

Design Data

Design data necessary in planning waste-treating facilities must include effluent requirements of the legal agency which controls the body of water into which ultimate disposal is made. These requirements are of two general natures, the performance type and the specific type. The **performance type of requirement** may simply

Table 8-15. Sample Refinery Effluent Requirements

Item	Requirement
pH	6.5–8.5
Total oil	15 ppm
Phenol	0.2 ppm
Sulfides	0.5 ppm
Mercaptans	0.5 ppm

specify that the quality of the plant effluent be equal to the quality of the present stream flow, or it may state that objectionable odors and tastes are not permitted.

Specific requirements for waste-treating facilities give definite limits of contaminants in terms of parts per million and specify color and turbidity by some general standard. These requirements may be in terms of the effluent as it leaves the plant or of the stream after mixing with the effluent. A sample requirement for refinery effluent quality is shown in Table 8-15.

Necessary design data also include complete information as to quantity and composition of each of the waste streams which will be included in the effluent. These normally can be broken down into the various systems previously described. Specific

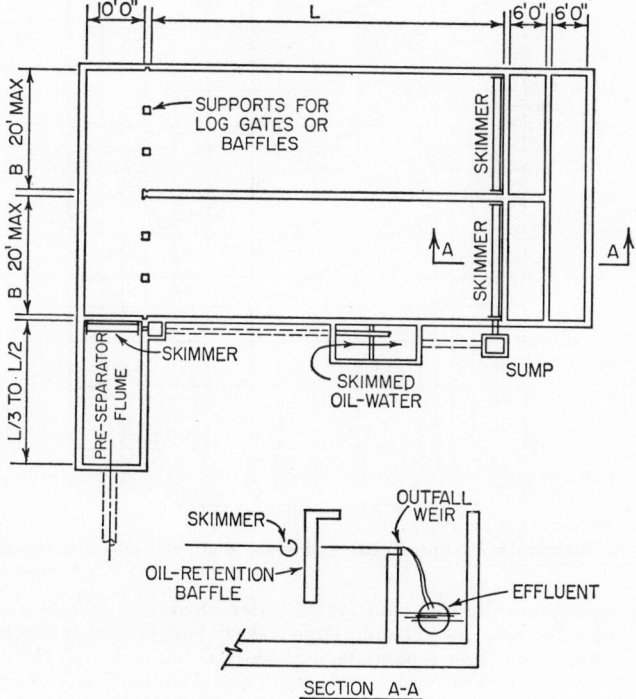

FIG. 8-33. Schematic arrangement of typical API-type oil-water separator.

data required include quantity, temperature, specific gravity of oil, pH, phenol and sulfide content, ammonia content, turbidity, BOD, COD, and other chemical compositions which will require treatment or affect the treating processes. Certain of this information can be ascertained only after sampling typical waste flows and must be estimated for design purposes.

Oil-water Separator

Oil-water separators are designed in accordance with the API *Manual on Disposal of Refinery Wastes*, Vol. I. The API separator design is based on Stokes' law, in which it is assumed that the controlling oil particle has a diameter of 0.015 cm. Experience and experimentation show that these separators are most effective if width and flowing depth are limited to 20 and 8 ft, respectively, in order to minimize any short-circuit effect. Oil-separator design is completely described in Chap. 2 of the API *Manual*, which also includes charts to facilitate design. A schematic arrangement of an API type of oil-water separator is shown on Fig. 8-33.

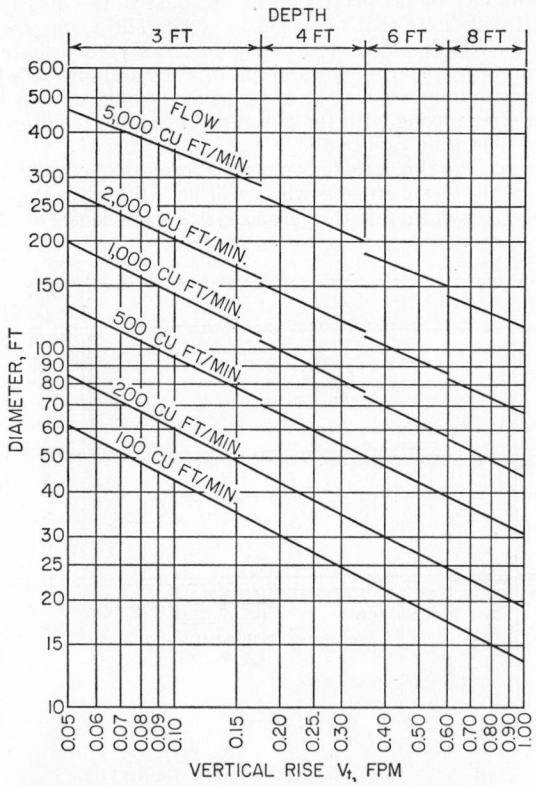

FIG. 8-34. Estimation of approximate diameters of circular oil-water separators.

Circular separators are also used for oil-water separation in refineries. These can be economical installations to handle large flows but are subject to short circuiting and effects of wind. In these separators, oily water is brought in at the center and flows radially outward. The vertical velocity V_t of an oil particle is determined by Stokes' law in accordance with the API *Manual*. The horizontal velocity of waste water is determined by assuming that water flows out radially in all directions, and the cross-sectional flow area is the area of a cylinder at the point in question.   Separation is assumed to begin when the horizontal velocity equals 3 to 6 fpm.   Separator diameter is made 40 to 50 per cent greater than the calculated diameter at which an oil particle reaches the surface.   The diameter of a circular separator can be estimated from Fig. 8-34.

Effluent Holding Basins

These basins are used to retain and dilute effluent waters.   They serve to make the effluent uniform, prevent large accidental discharge of contaminants, and also improve quality of the effluent.   Some oil separation is effected, and some reduction in BOD and increase in dissolved oxygen can be achieved.

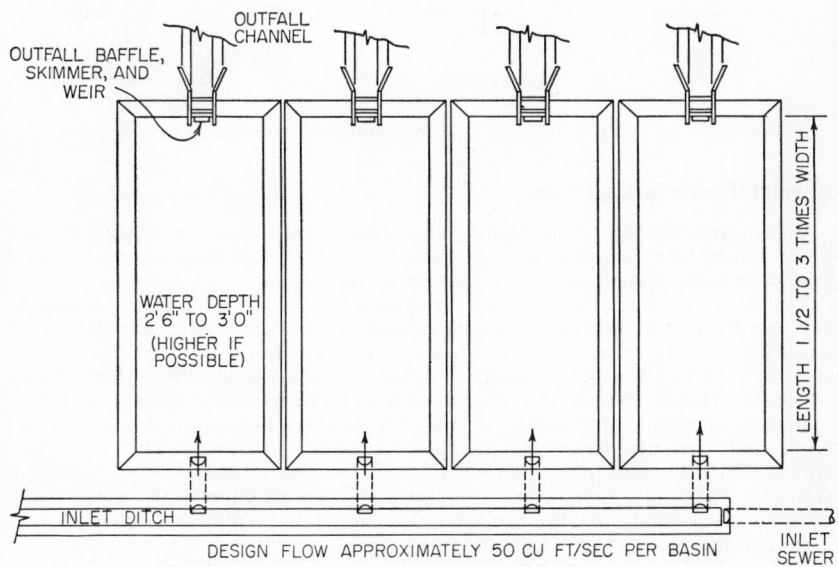

Fig. 8-35. Schematic diagram of typical effluent holding basin.

When possible, natural lagoons should be used for these basins.   In general a series of parallel basins is more effective than a single basin, whether basins are natural or man-made.   Basins should have a capacity equivalent to 45-min flow at peak conditions.   Basins should have a minimum depth of about $2\frac{1}{2}$ ft and should have a horizontal velocity ranging from 3 fpm at normal flow to 10 fpm at peak flow.   Each basin should be designed for 25 to 50 cfs.   A schematic diagram of holding basins is shown in Fig. 8-35.

Ballast-water Systems

Ballast water from product tankers cannot be pumped directly into harbor water without treatment.   The amount of treatment necessary will vary with the locality. The simplest arrangement is to pump all ballast into an open-top steel tank, hold the water in this tank as long as possible, skim the separated oil, and discharge the clarified water directly to the harbor or through the refinery holding basin.   In some cases, all the water is passed through an API separator before discharge.   In other cases, special treatment and emulsion breaking are necessary.

For a T-2 product tanker, a ballast discharge of 20,000 bbl is considered usual. This can be handled in a 24,000-bbl tank. A typical ballast-water tank is shown on Fig. 8-36.

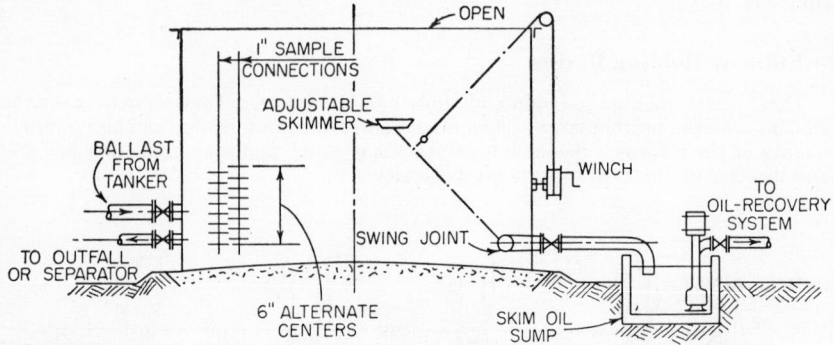

FIG. 8-36. Typical ballast water tank to separate oil from tanker ballast water.

Special Treating Facilities

While special waste-treating facilities are becoming more commonplace in refineries, the design of this equipment is specialized. Therefore, this section is limited to a brief description of some applicable treating arrangements.

To obtain greater oil removal, **air floatation** is employed. Air is mixed intimately with waste water, improving the gravity separation of oil and water. If still more oil must be removed from waste water, **chemical flocculation** using lime, alum, etc., is necessary. This also reduces suspended solids and improves turbidity. Flocculation results in virtually 100 per cent removal of oil in all forms, including emulsions.

Removal of phenols and sulfides is effected by **bacteriological methods.** Trickle filters can be used with stone, cinders, or plastic packings. The bacteria which develop on the packing oxidize phenol and sulfides. The other commonly used method is activated sludge in which air and recycled sludge are mixed with the waste water. Bacteria in the sludge oxidize the unwanted materials. These two processes have been used in series with either one first.

Color and odor improvement and general enhancement of effluent quality can be effected by **filtration.** Filter media used include activated carbon and proprietary precoat materials on vacuum filters.

Sterilization of effluent plus general odor and taste improvement can be effected by **ozonization** or **chlorination.** Chlorine is not considered desirable by some authorities because of possible reaction with phenol.

Incidental to various phases of special waste-treating processes are pH and temperature adjustments. These improve flocculation, bacteriological action, etc.

PRESSURE RELIEF AND BLOWDOWN

Selection of Systems

Pressure-relief and blowdown systems are used for disposal of vapors and liquids discharged by various pressure-relieving devices such as safety and relief valves, rupture disks, pressure-control valves, and furnace emergency blowdown valves. Facilities included in these systems are contained within process units and in offsite.

Relief and safety valves discharge to two types of systems: (1) **open systems,** in which the vapors or liquids are permitted to discharge directly to atmosphere either individually or collectively, and (2) **closed systems,** in which the vapors or liquids are conducted through a gathering header to a safe disposal facility such as flare, burning pit, or blowdown drum.

Most open systems are wholly contained within process units; therefore, the relief and blowdown facilities included in offsite facilities are those related to closed systems.

Vapors

The method of disposal of vapors varies with the hazardous nature of the vapor and its molecular weight, meteorological conditions, equipment spacing, and discharge frequency. Nonhazardous chemicals, ammonia, steam, and air are usually discharged through open systems directly to atmosphere. Hydrocarbons with a molecular weight greater than 80 are usually conducted through a closed system to a water-quench blowdown drum.

Hydrocarbons having a molecular weight less than 80 which are discharged on a continuous basis (for example, a bypass stream during startup) are usually discharged through a closed system to a flare or burning pit. Hydrocarbons having a molecular weight less than 80 which are discharged intermittently from relief valves or rupture disks may be discharged through open systems directly to atmosphere unless otherwise restricted by the following: (1) local ordinances or plant practices requiring closed systems, (2) concentrations of the contaminants at ground or adjacent platform levels exceeding explosibility limits, and (3) meteorological considerations such as severe temperature inversions of long duration.

Acid, toxic, and other hazardous gases which are discharged on a continuous basis are usually conducted through closed systems to suitable disposal facilities. Carbon monoxide, for example, is usually disposed of at a flare or furnace burner, while H_2S is generally brought through segregated piping to a separate burner. When these materials are discharged intermittently from safety or relief valves, open systems are sometimes employed when restrictions similar to those described for hydrocarbons do not apply.

Liquids

The method of disposal of liquids varies with the hazardous nature of the fluid, the vapor pressure, and the temperature of the material. Water and nonhazardous chemicals are discharged by open systems to a sewer. Hydrocarbon liquids which vaporize at atmospheric pressure are generally discharged through closed systems to a flare or burning pit.

Heavy hydrocarbons which are not expected to vaporize at atmospheric pressures and flowing temperature may be discharged through a closed system to a blowdown drum and vent. Heavy hydrocarbons which, because of temperature, might be expected to evolve large amounts of vapor should be discharged to a water-quench blowdown drum and stack.

Equipment

Discharge of liquids or vapors from equipment such as pumps, compressors, and furnaces must also be considered. Discharges from safety valves located at the discharge of pumps or compressors are usually returned to the suction or tank or drum from which suction is taken.

Coils of furnaces in hydrocarbon and hazardous chemical service are usually provided with emergency blowdown facilities. The blowdown discharges to closed systems which terminate at facilities as described above for liquids. Blowdown valves at furnaces are remotely operated by motors or manual devices.

Types of Facilities

Relief valves, pressure-control valves, piping headers, etc., related to pressure-relief and blowdown systems are usually completely within process units and are, therefore, not considered in this section. The facilities normally but not exclusively located offsite include water-quench blowdown drums, blowdown drum and vent, and flare stacks.

Water-quench Blowdown Drum

A water-quench blowdown drum is a vertical drum with an outlet stack in which hot liquids can be cooled prior to ultimate disposal. The hot hydrocarbon liquid enters the drum above the liquid level. The entering hot liquid automatically actuates the water flow. Baffles are provided to implement cooling. Additional water can be introduced manually. Steam for dispersal of vapors can also be introduced into the stack manually. A typical water-quench blowdown drum is shown in Fig. 8-37.

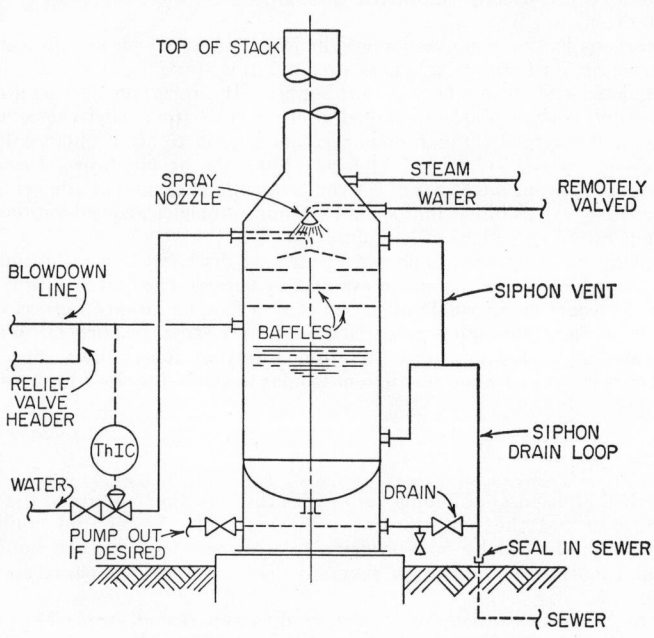

FIG. 8-37. Typical water-quench blowdown drum for disposal of hot liquids.

One drum is normally provided to serve two adjoining process units, dependent on distance between units. Drums must always be isolated from sources of ignition as should sewers into which drums discharge.

Water is generally obtained from a source independent of cooling water, such as fire water or city water. The quantity of water required is determined on the basis of the maximum flow of vapors that can be quenched at atmospheric pressure. It is assumed that 40 to 50 per cent of the water will vaporize. The drum size is based on the quantity of blowdown vapors plus water.

Stacks are sized to have an exit velocity not exceeding 300 fps. Stacks normally extend to a height 10 ft above the highest working platform within a radius of 70 ft. In most offsite locations this requirement does not apply and the height is controlled by maximum permissible ground concentration of contaminant.

Blowdown Drum and Vent

Blowdown of liquid hydrocarbons which are not expected to vaporize appreciably is usually collected in a horizontal drum fitted with a vent. The drum is sized to hold the total amount of liquid contained in the coils of the largest furnace with allowance for vapor space. The vent is made large enough to release the steam used for blowdown.

Drums are provided with an emergency overflow to the sewer. The drum drain is

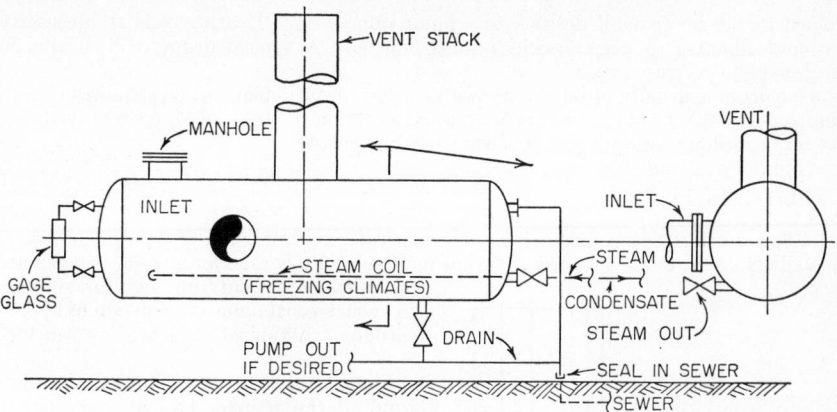

FIG. 8-38. Typical blowdown drum and vent for liquid-hydrocarbon collection.

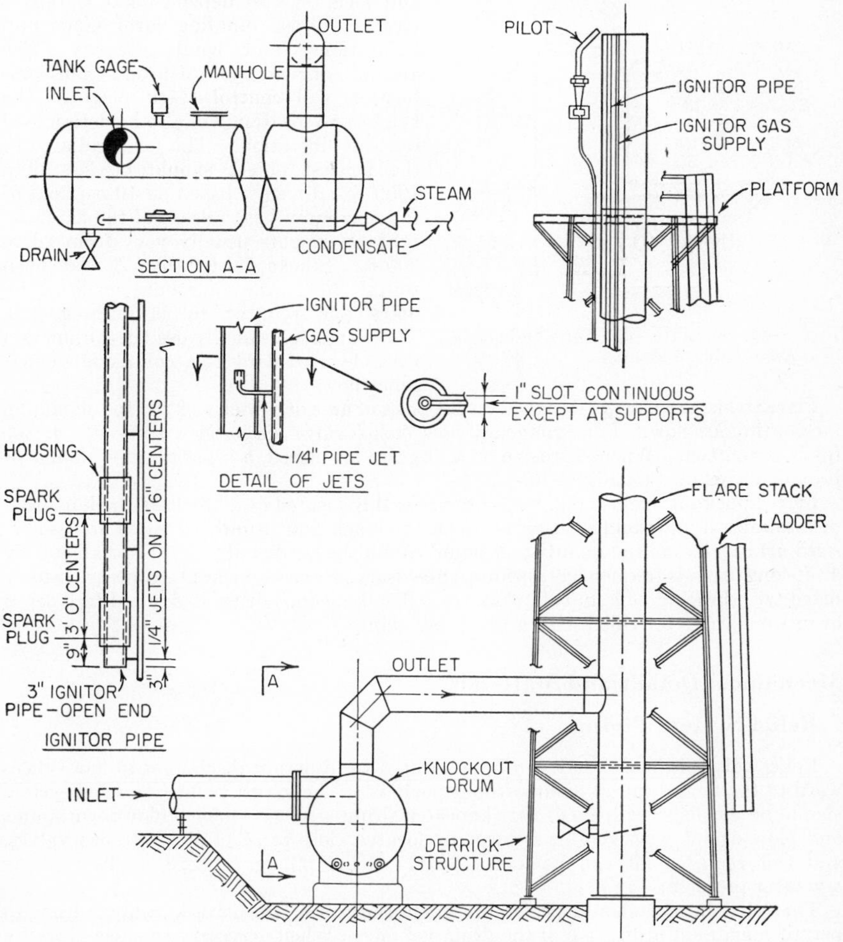

FIG. 8-39. Typical flare stack for burning refinery waste gases.

piped to the sewer or, if desired, to a pump-out pump. Heating coils are necessary in cold climates to keep viscous liquids flowing. A typical drum is illustrated in Fig. 8-38.

One drum is usually provided for two adjacent units, dependent on spacing between units. The height of the vent is established solely on the basis of location of facilities, as no hazardous concentrations of vapors are expected.

Flare Stacks

A flare stack is a burning stack for safe disposal of hydrocarbon vapors. The vapors usually pass through a knockout drum in which liquids are eliminated. Stacks are provided with pilots and ignitor systems to ensure continuous combustion of hydrocarbons. A typical stack is shown in Fig. 8-39.

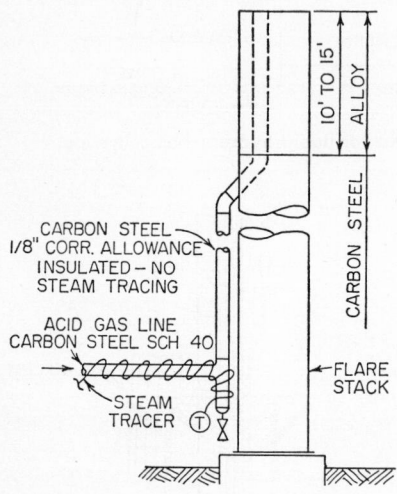

FIG. 8-40. Separate pipe for conducting acid gases to flare stack.

Flare stacks are usually located downwind of the refinery and at least 300 ft from any process unit or equipment handling low-flash-point materials. The height and location also depend on topography, location of surrounding farm lands and habitations, and wind intensity. The ground concentration of products of combustion will control stack height. The height and location will also be determined by heat liberation. The heat radiated to the nearest object should be less than 1000 Btu/hr/sq ft based on 10 per cent of the maximum heat release of the flare.

Stack heights usually vary from 20 to 300 ft. Stacks shorter than 20 ft or horizontal pipes are sometimes enclosed by dikes and referred to as burning pits. These are occasionally used in conjunction with elevated flares to accommodate emergency flows.

Flare stacks are usually designed on the basis of an exit velocity of 200 fps maximum for continuous flow. For emergency flow (relief valve discharge), a velocity of 300 fps is permitted. When corrosive or acid gases are flared, a separate pipe should be provided as shown on Fig. 8-40.

Flare knockout drums (Fig. 8-39) are preferably located near the flare stack in order to minimize liquid condensation between the stack and drums. Drums are usually sized arbitrarily, as the quantity of liquid cannot be predicted. A 10-ft-diameter by 40-ft-long drum is frequently considered necessary. Flare knockout drums are usually fitted with heating coils in cold climates. The knockout drum is drained to a sewer or in some cases provided with a pump-out pump.

Mechanical Details and Materials

Relief System Piping

Relief valves in most cases can be so located as to permit discharges to flow downward to the final disposal point without pockets. The lowest point in a relief system should be the disposal drum or flare knockout drum. Water-quench blowdown drums and blow drums with vents are normally located close to relief or blowdown valves, and this requirement can be met readily. Relief piping to flare stacks presents a greater problem.

The relief header should preferably enter the top of the flare knockout drum to permit maximum utilization of the drum volume. When necessary to effect this, the drum can be buried in earth. In some instances it is preferable to use elevated supports from the process unit to the drum location.

In some cases pockets in flare lines are unavoidable. Pockets may drain to small knockout drums fitted with heating coils or pump-out pumps. In cold climates, protective heating is necessary to prevent condensation of H_2S.

Similarly the separate acid gas line should be run from the source of discharge to a drip leg or knockout pot at the base of the flare stack. This run should not be pocketed. The entire run should be provided with protective heating.

Flare Stack and Supports

Flare stacks can be self-supporting, guy supported, or supported by a derrick-type structure. Self-supporting stacks are limited by general economy to height-diameter ratios of 30 to 1. When stacks are located in remote areas, guys can be used effectively. Guys are generally located at 120° intervals and run to one or two levels of the stack. In congested areas, guy locations may interfere with structures and equipment.

Derrick structures are usually economically justifiable for heights greater than 100 ft up to heights approaching 200 ft. This type of flare stack support is frequently required because of space limitations. Access is required to the flare pilot burners and to other equipment which can be located at the top of the flare. Vertical ladders are usually provided for access. When derrick structures are used, the ladders are located in the structure instead of being supported from the stack.

Ignition Systems

Various types of ignition systems are used to ensure continuous burning of flares and pilots. The most commonly used types are the flame-propagation type and the flame-front type. The flame-propagation type is shown in Fig. 8-39. In this ignition system a slotted pipe is run from the bottom to the top of stack. A series of gas jets are located within the slotted pipe on 30-in. centers. The flowing gas is ignited by two spark plugs near the base of the stack and ignites successive jets as it travels up the stack. The pilot is lit by the flame. The gas shutoff valve and ignition control station are usually remotely located.

In the flame-front system, gas and air are admitted into the ignition pipe in a combustible ratio. After enough gas has been admitted to fill the line, the gas is ignited by a spark plug. The flame travels through the pipe until it reaches the pilot. Gas and air valves and the ignition control station are also remotely located for this type of ignitor.

One or more pilots are provided on the flare stack to ensure continuous burning. At low flows the flare and pilots may be extinguished by high winds. To minimize extinguishment of the flare and to prevent flash backs of flame into the knockout drum and relief-valve piping, it is desirable to use a fuel gas bleed into the relief system. The use of a bleed is superior to installation of flame arrestors or seal pots.

Smokeless Flares

Flare stacks generally burn with a lazy reddish smoky flame. In some localities smokeless flares are required to comply with ordinances related to air contamination and smoke abatement. The main cause of smoky burning of refinery flares is insufficient combustion air. It is also believed that polymerization may occur in the flame, resulting in heavy hydrocarbons which burn with a smoky flame.

One of the accepted smokeless flares is a proprietary system using a series of steam jets at the top of the flare pipe to induce greater air flow and to cool the flame. This type of smokeless flare is extremely effective, resulting in an invisible flame in the daytime. Steam flow is regulated by a flow-indicating controller set to operate over the continuously flared range. During emergency flaring operations, the flame is smoky. This type of flare uses a large volume of steam, varying from 0.2 to 0.4 lb of steam per pound of hydrocarbon.

Extremely effective smokeless burning has been obtained by using ground flares which have a series of concentric rings of large diameter in which the waste gases are burned. Earlier designs used steam and water sprays, but recent installations have been quite successful without sprays. This type of smokeless flare depends solely on inducing a large flow of combustion and secondary air.

Materials

Furnace blowdown valves are necessarily of the same rating and material as the furnace transfer line. Downstream of the blowdown valve to the drum or stack, carbon steel is generally used having the same rating as the furnace transfer line but not higher than 300 lb ASA. The use of carbon steel can be justified by the short-term intermittent nature of the operation.

Relief-valve discharge piping offsite from process units to flare stacks is usually extremely low pressure, permitting use of minimum-thickness carbon-steel piping. Materials as light as 10-gage spiral weld pipe have been employed in this service. The flare stack design is determined by structural requirements rather than pressure, resulting in a more usual pipe wall thickness.

The upper 10 to 15 ft of flare stacks (including accessories, platforms, etc., within this range) are fabricated from alloy steel of 18 per cent Cr–8 per cent Ni or higher. The upper portion of acid gas lines are also made of alloy material.

The horizontal runs of acid gas lines are usually installed as Schedule 40 pipe. The vertical run of this line is generally required to have a $\frac{1}{8}$-in. corrosion allowance, as this run is not provided with protective heating.

REFINERY UTILITY PLANTS

General Description

The centralized grouping of equipment and facilities for generation and preparation of refinery utilities is referred to as the "utility plant." This plant might include facilities related to production and distribution of steam, exhaust steam, boiler feed water, plant fuel, plant and instrument air, electrical power, cooling water, fire water, drinking water, and various chemicals. Facilities for many of the utilities listed above are frequently not centrally located in the utility plant but are situated in process areas or elsewhere in the refinery. When an abundant supply of electrical power is available inexpensively from a reliable source, electrical generators are not installed in refinery utility plants, although the main substation may be there.

The major functions of the refinery utility plant are those directly related to steam production for process purposes and the production and/or distribution of steam and electricity for plant drivers. This section describes systems related to these major functions including specifically steam, exhaust steam, boiler feed water, and electrical power. Facilities related to other systems are described in other parts of this section.

Design Conditions

The following design conditions must be determined for utility-plant facilities and systems: (1) the basis for establishing equipment design capacity and design pressure and temperature, (2) the method of establishing flowing quantities in piping systems with corresponding temperature and pressure for normal, emergency, and design conditions.

Steam and Exhaust-steam Systems

Steam and exhaust steam are used in refineries for a number of purposes. Certain uses are essential to the process and operations of the plant, while others are employed to effect costs savings or for convenience. Uses of steam and exhaust steam include (1) process uses such as stripping, fuel atomizing, etc.; (2) process heating in exchangers, reboilers, and tracers; (3) offsite and utility-plant heating; (4) drivers in services which require maximum dependability; (5) large steam drivers which have a lower initial equipment cost than an electric driver; (6) large drivers in general when use of steam will result in substantial operating-cost savings over electricity; and (7) drivers in general when electricity is not available or is available only in limited quantity.

Steam for refinery use can be obtained from one or more of the following: fired steam generators, unfired steam generators, turbine exhaust, turbine extraction, or pressure-reduction valve throttling higher pressure steam.

Steam and exhaust steam systems in refinery utility plants are integral portions of the entire refinery system. Steam and exhaust steam can be generated and used at various levels to satisfy numerous conditions and requirements. The basic refinery system must be established in order to determine the capacity and operating levels of equipment in the utility plant. The general principles for establishing refinery steam and exhaust-steam systems are shown in Fig. 8-41.

Refinery steam systems are established and equipment capacities are determined through a series of approximations. These approximations are as follows:

Approximation 1. Use data from curves and calculations for process units and for offsites in general to establish a total steam requirement. Select pressure levels tentatively by comparison with other plants.

Approximation 2. Estimate quantities from process flow diagrams for units and from block flow diagrams for offsites. Estimate pump and compressor horsepowers from basic flowing data without selecting equipment. Follow steps outlined below to determine flow rates and pressure and temperature levels.

Approximation 3. Tabulate quantities from process flow diagrams, equipment data sheets for selected equipment, offsite block flow diagrams, etc. Follow steps outlined below to determine flow rates and pressure and temperature levels.

1. Determine the exhaust-steam demand for process heating, reboilers, offsite heating, building heating, etc. Estimate the quantity of exhaust required for the deaerator.

2. Determine the medium-pressure steam demand for process stripping, heating, reboilers, offsite heating, jets, fuel atomizing, etc.

3. Determine the quantity of medium-pressure steam generated in process units.

4. Tabulate major equipment which must be steam driven.

5. Tabulate minor equipment which must be steam driven.

6. Tabulate large equipment which shall preferably have condensing-extraction steam drive.

7. Tabulate minor equipment which may have steam drivers or steam-driven standbys.

8. Assume a pressure condition for high-pressure steam.

9. Balance the total exhaust demand (Step 1) with exhaust supply from Step 5 using Group B drivers and Step 4 using Group C drivers. Small differences can be brought into balance using Group B drivers or standbys listed in Step 7.

10. Determine the total demand for medium-pressure steam as the sum of item 2 and the quantity used by Group B drivers per Step 9.

11. Balance the medium-pressure steam demand (Step 10) against the supply from process-unit generation (Step 3), extraction steam (Step 6), and exhaust from Group A drivers. Trim the steam supply by changing extraction steam from Group D drivers.

12. Supply additional steam drivers as fully condensing Group D drivers.

13. Determine the total normal demand for high-pressure steam, and verify the pressure assumed in Step 8. Check the balance, if necessary, by recalculating on the basis of an alternate steam pressure.

14. Adjust the complete balance for emergency conditions, eliminating services which would be inoperative and adding emergency loads.

In a plant having a number of process units which can be operated independently of other units, it is desirable to balance the steam at each unit independently of other units. Similarly it is preferable to have the refinery utility plant self-sufficient in relation to exhaust-steam requirements.

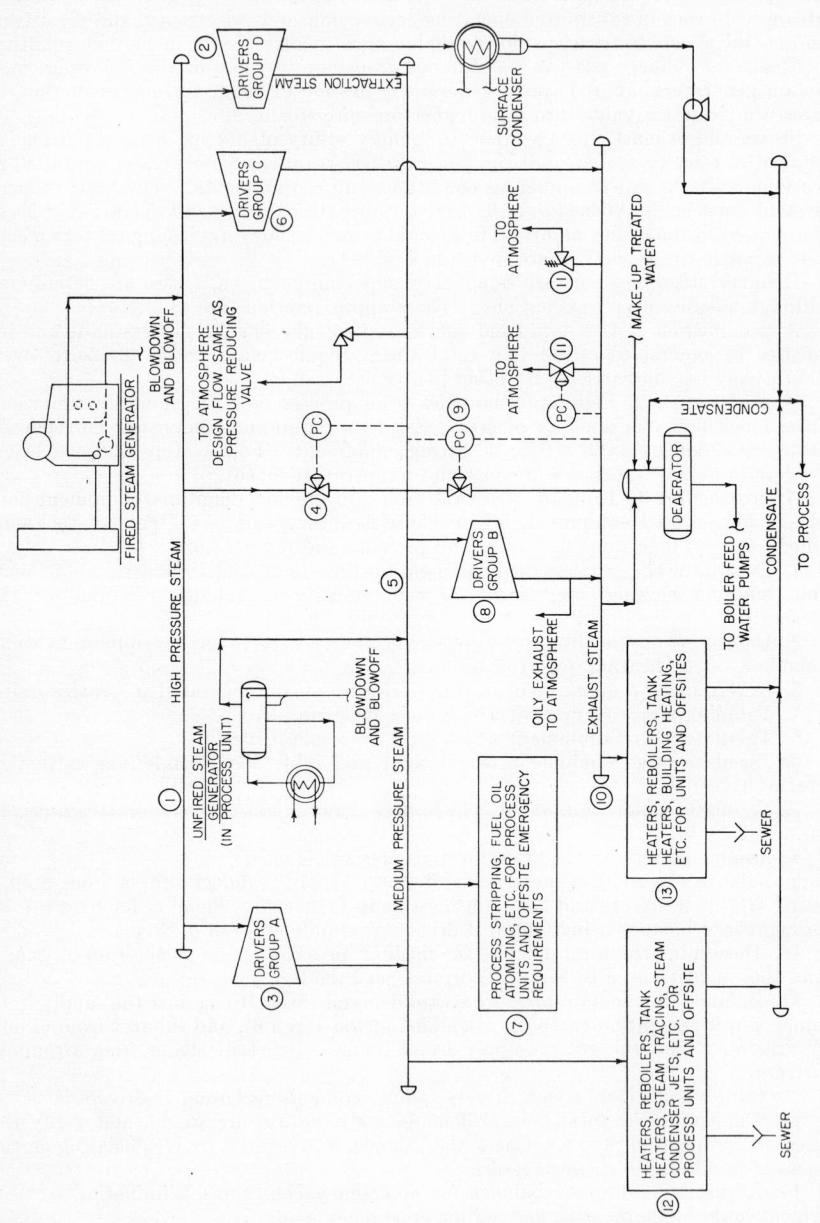

1.

Preferred conditions	Alt. 1	Alt. 2
Nominal pressure, psig........	625	400
Max. at source, psig.........	650	425
Min. at user, psig..........	600	400
Max. temperature, °F.........	750	750
Pipeline flange rating, lb...	600	400

2. Large equipment such as process compressors, electrical generators, or offsite water pumps shall be driven by Group D drivers when exhaust requirements of the plant have been met by other drivers in Group C. These services require reliability and economy, which are most easily achieved through use of steam drives. Condensing extraction turbines shall be used for drivers which meet a base load such as process unit electrical power but also must intermittently supply extra demand such as product loading pump power. These drivers can extract medium-pressure steam on a steady basis and meet fluctuations through increases in the amount of steam condensed. Drivers operating at constant load may be fully condensing turbines.

3. Group A drivers shall be used to make up deficiencies between demand for medium-pressure steam and the amount produced by steam generators. Extraction steam from Group D drivers shall be preferred as a source of medium-pressure steam. Large drivers shall preferably be Group C.

4. Normal flow—none. Design flow—this shall be the largest of following: (a) process unit emergency; (b) extraction quantity of largest Group D driver; (c) flow to largest Group A driver; (d) flow to largest Group C driver.

5. Nominal range 140 to 200 psig.

Preferred conditions

Nominal pressure.............	160 psig
Max. at source..............	175 psig
Min. at user...............	150 psig
Temp. at source max.........	470°F
Temp. at user nom..........	377°F

6. Group C drivers shall be used to produce all required exhaust steam. This is the preferred driver for large equipment in process units. Group C drivers can also be used for boiler feedwater pumps and for boiler fans.

7. Pressure—about 100 psig. Quantities—Process furnace atomizing 1 lb of steam per pound of fuel oil. Steam boilers—1 per cent of steam make.

8. Drivers in Group B are relatively small and are generally used in services requiring maximum reliability. These drivers are also used to "trim" deficiencies between exhaust demand and the supply from Group C drivers. Most standby drivers are in Group B resulting in greater overall flexibility.

9. Normal flow—none. Design flow—This shall be the largest of following: (a) sum of offsite exhaust users plus 25 per cent of deaerator design exhaust flow (b) flow from largest Group C driver; (c) flow from largest Group A driver.

10. Nominal range—10 to 50 psig.

Preferred conditions

Nominal pressure.............	40 psig
Max. at source..............	50 psig
Min. at user...............	40 psig
Temperature max.............	350°F
Pipeline flange rating.......	125 lb

11. Design flow shall be largest of following: (a) total flow from all nonoperating standby drivers; (b) flow from largest Group C driver.

12. Medium-pressure steam is used for heating purposes only when required for high temperature or where pressure drops are large. For steam tracers quantity required is 10 lb/hr per assembly averaging 75 ft of traced pipe in process units or 180 ft of traced pipe offsite.

13. Exhaust steam is preferred for these services. Condensate from isolated equipment shall be discharged to sewer.

FIG. 8-41. General design principles and conditions for establishing refinery steam and steam-exhaust systems.

Steam Generation

Steam in refinery systems is generated in either fired or nonfired steam generators. Nonfired generators are usually located in process units and utilize waste heat or exchange with hydrocarbons being processed. Fired steam generators or boilers used in utility plants are either package or built-in-place installations. **Package installations** are factory-assembled boilers supplied to the job site in sizes up to about 80,000 lb/hr for pressures up to 600 psig. Fire tube construction is used in sizes up to about 10,000 lb/hr, and water tube construction for larger sizes. **Built-in-place installations,** in which the various components are assembled and erected on the job site, are available in a full range of sizes and pressures. These can be power boilers using fuel oil, fuel gas, or coal or CO boilers using the CO gas formed in the regeneration section of a fluid cat cracker with fuel oil or fuel gas as supplemental fuel.

Equipment Capacity

In refinery utility plants, steam should be generated in two or more steam generators having a total installed rated capacity not less than 133 per cent of normal flow. All boilers shall be considered to be operating at partial load to supply normal refinery requirements.

The total installed capacity shall also be at least equal to the emergency steam requirement of the refinery, assuming that built-in-place equipment can be operated at 110 per cent of rated capacity for a short term. An additional factor to be considered is shutdown of boilers. When one boiler is out of service, the remaining boiler or boilers should be capable of supplying process and offsite stream requirements with either of two major process units also shut down. At this time boilers must operate at or below rated capacity.

The design condition for an individual boiler and associated piping and auxiliaries shall be equal to the rated capacity of the boiler. The emergency design condition for each boiler shall be equal to 110 per cent of the rated capacity of the boiler with pressure and temperature assumed equal to those at design condition.

The maximum operating pressure of boilers is equal to the feed-water inlet pressure and exceeds the boiler outlet pressure by the sum of the internal losses. The outlet pressure of a boiler is expressed as the pressure at steam drum or superheater outlet not including valve or piping pressure drops. Internal losses in boilers include losses in economizers, superheaters, and attemperators. The design pressure of boilers is 10 per cent greater than the maximum operating pressure but not less than 25 psi in excess of this pressure.

Installation Details

In warm and moderate climates, boilers can be installed outdoors with the firing aisles located completely outdoors. In extremely cold climates, semioutdoor installations should be used with the fronts of the boilers and firing aisles located inside buildings while the rest of the boilers are located outdoors. Most boilers can be obtained either right or left hand. Use of opposite-hand boilers may permit common use of platforms for firing, soot blowers, etc., and common sets of pipe supports.

CO boilers should be located reasonably close to the regenerator producing CO gas in order to minimize large-diameter piping runs. The minimum distance from CO boilers to process equipment in process units should be 40 ft. On CO boiler installations, the depressuring chamber should be located about 25 ft from the main and bypass seal pots.

Boilers are usually equipped with fuel burners, forced- and/or induced-draft fans, flue gas ducts, stacks, soot blowers, controls and instruments, platforms and ladders, and other related auxiliaries and accessories. Boiler installations include suitable instrumentation for combustion and boiler-feed-water control. Combustion controls usually provide for adjustment of fuel flow to maintain steam pressure and for adjustment of air flow to suit required air-to-fuel ratio.   Controls are normally suitable for combination (oil-gas) firing of steam generators, including flow indication control of atomizing steam.   Feed-water controls provide for regulation of feed-water flow in proportion to steam flow modified by steam drum water level.   Feed-water regulators should have a minimum pressure drop of 25 psig across the valve at design flow rate to ensure suitable range of control.   In high-pressure steam-generator service, required pressure drops are 50 to 75 psi.

Boiler Feed Water and Condensate Systems

Boiler feed water is used as make-up to fired and nonfired steam generators located in utility plants and in process units.   Boiler feed water is also used for process purposes when other available waters are of inferior quality.   This water in refinery usage is a deaerated mixture of condensate and treated water.   The equipment used for treating must be selected to suit the available make-up water, due account being taken of the quantity of water required.

Boiler feed water is normally heated and deaerated in a deaerating feed heater. Common practice is to have this equipment float on the pressure of the refinery exhaust system.   It is preferable, however, to utilize pressure control to insure steady operating conditions in the deaerator.   Feed-water exchangers usually are not economically justifiable in refineries.

Pressure and Quantity

Boiler feed water must be supplied to steam generators at sufficient pressure to provide for design flow concurrent with relief-valve discharge.   The boiler-feed-water pump discharge pressure must accordingly be equal to the sum of the following: (1) relief valve relieving pressure of $1.06 \times$ boiler design pressure (Note: boiler design pressure is established by the manufacturer as maximum operating pressure plus 5 to 10 per cent), (2) static head to liquid level, (3) feed-water control-valve loss (25 to 75 psi), (4) pipe friction losses at design flow, and (5) preheater loss, if any.

The installed capacity of steam generators includes a definite safety factor (33 to 50 per cent recommended above).   It would, therefore, be illogical to include additional compound safety factors in establishing the capacity of boiler-feed-water pumps and of deaerators.   The maximum steam-generating capacity will be either rated capacity or, for built-in-place boilers, 110 per cent of rated capacity.   Equipment should be capable of meeting this capacity plus boiler blowdown plus other requirements such as for process water.   To ensure ability to meet normal steam requirements at all times, two or three boiler-feed-water pumps should be installed.   Normally, a single full sized deaerator is considered adequate.   To provide for surge the deaerator should have a storage compartment with a design volume equivalent to approximately 10-min capacity.

Feed-water Pumps

The temperature and pressure conditions at boiler-feed-water pump suction should be based on the design equilibrium conditions of the deaerator.   The deaerator must be installed at a sufficiently high elevation to allow for pipe friction losses and for the required pump net positive suction head (NPSH).   The required NPSH can be assumed to range between 12 and 16 ft for initial design purposes.

Boiler-feed-water pumps in some plants can operate at as low as 10 per cent of pump capacity during periods of partial refinery operation or turnaround.   In such cases the pumps should be equipped with low-flow recirculation lines discharging back to the deaerator.   These lines can be fitted with restriction orifices sized to pass a minimum flow or with flow-indication actuated control valves.

High-pressure multistage pumps having discharge pressure of about 800 psi frequently must be provided with a balance line for leakage from labyrinth packing glands to the deaerator.

Deaerators

Deaerators are used primarily to prevent corrosion in boilers, feed-water pumps, and piping by removing dissolved oxygen and carbon dioxide from boiler feed water. In addition, deaerators heat boiler feed water to a uniform temperature, conserving heat in excess exhaust steam and minimizing temperature shock in boiler drums.

Deaerators are vertical pressure vessels fitted with sprays or trays arranged to disengage gas by the scrubbing action of steam. Cold make-up water usually passes through a vent condenser which condenses steam mixed with vent gases. Vertical or horizontal storage tanks are provided under the deaerating section of the vessel. While a 10-min storage capacity is recommended, some storage compartments allow for 2-min capacity.

Make-up water supply to the deaerator must be at a pressure sufficient to overcome the following losses: (1) pipe friction, (2) static head to the top of the deaerator, (3) vessel operating pressure ranging up to maximum plant exhaust-steam pressure unless reduced by a control valve, (4) make-up water control valve loss of about 5 psi, and (5) nozzle loss for a spray-type deaerator of 20 to 25 psi.

The pressure of condensate must be established on a similar basis as for make-up water except that no control valve should be provided and fluid must be considered to be a flashing mixture in most instances.

Deaerators which float on the exhaust system are protected against overpressuring by the system relief valves. Equipment suppliers normally provide a nominal relief valve, which should be considered a sentinel valve, and for large vessels a vacuum breaker. When the deaerator operates at a pressure lower than that in the exhaust system, protection against overpressuring must be provided by a relief valve or loop seal.

Water-treating Equipment

Water-treating equipment should be capable of supplying all necessary make-up water. This should be at least equal to the total capacity of boiler-feed-water pumps minus the normal amount of condensate return. To allow for startup of the utility plant, this treating capacity should be not less than 120 per cent of the rated capacity of a single boiler. Storage of boiler feed water should be provided for about an 8-hr period at normal rate of consumption. Two storage tanks should be provided for this purpose.

Treating of boiler feed make-up water is performed primarily to effect removal of turbidity (suspended material) and reduction of hardness. As a result of having relatively large quantities of steam used without being returned as condensate, the make-up water requirements for the average refinery are appreciable. Consequently, it is not advisable to establish extremely high standards for make-up feed water.

Turbidity is removed by filtration, sedimentation, and coagulation. Hardness is reduced by chemical reaction in cold-lime or hot-process softeners or by cation and anion exchange as in zeolite treaters or demineralizers. Cold-lime or hot-process softeners can serve dual functions in reducing both turbidity and hardness. Filtration is always used in conjunction with other treatments as well as alone. Ion exchange is also frequently used as a final treatment following chemical softening.

Filtration and Sedimentation

Removal of turbidity is most frequently effected by **filtration.** The commonly used type of filter is the pressure type. This consists of a cylindrical steel shell containing a filter bed of sand and gravel or anthracite. Water enters at the top and percolates down through the bed to a collector system at the bottom. Periodically the flow is reversed to wash out accumulated material.

Multiple units are provided to permit backwash of one unit while remaining units are in operation. Pressure drops range from 10 to 15 psi for a clean filter to 20 psi prior to backwash. Various operating arrangements are available for manual, semi-automatic, or fully automatic backwash and rinsing.

When turbidity is extremely high, additional facilities are necessary prior to filtering. The simplest type of facility for reduction of turbidity is a **sedimentation basin** using gravity separation. Such basins require a relatively large area and are rather ineffective in removing fine particles. In general, when water is extremely turbid, it is preferable to combine turbidity reduction with softening, using chemicals for coagulation and softening.

Softening

So-called **cold-lime softening** is effected by reacting lime and possibly soda ash with water and separating clear water from the sludge. Simultaneously chemicals such as alum or ferric sulfate can be added to reduce turbidity. In this process turbidity can be reduced to 10 ppm while hardness is reduced to 50 ppm as $CaCO_3$.

Separate mixing chambers and coagulation basins with sludge collectors are occasionally used. Most cold-lime softeners utilize a combined vessel such as the Spaulding precipitator. In this equipment chemicals and water are mixed in a central cone by agitation and flow downward under the edge of the cone into the outer sludge-filtering zone. The sludge and clear water separate, and the clear water passes over an overflow weir. Sludge is withdrawn from the sludge concentration section. It is necessary to provide mixing tanks and pumps for chemical feed. A clear-water surge tank of about 10-min capacity should be installed downstream of the softener. Cold-lime softeners operate at atmospheric pressure, usually in open vessels. The pressure of incoming raw water need only be sufficient to overcome static head losses, water losses, pipe friction losses, and control-valve losses. Whenever possible softeners should be located above or partially above grade to simplify subsequent pumping operations.

Hot-process softeners operate by reacting lime, soda ash, or phosphate with raw water and separating clear water from the sludge. This is similar to the cold-lime softening except that the reaction is more rapid and complete at high temperatures. This equipment will also reduce the silica content of water. With the use of lime, the hardness of water can be reduced to 15 ppm as $CaCO_3$. Greater reduction can be achieved using phosphate.

The hot-process softener employs the principle of the Spaulding precipitator. The incoming raw water is usually sprayed into the vessel and mixed with steam and chemicals. In this process separate deaerators can be eliminated either by treating total make-up or by building a deaerator compartment into the softener. Sludge formation and clarification are similar to the cold-lime softener. Sludge recirculation is often employed to provide a better sludge blanket.

Chemical mix tanks and pumps are required for hot-process softeners. Clear-water storage for about a 10-min period is usually provided in the deaerator compartment or in a clear-water storage section. Hot-process softeners operate at 5 psig minimum or float on the exhaust system. Usually incoming water must be provided at pressure sufficient to overcome the following losses: (1) pipe friction, (2) static head to the top of the softener, (3) vent condenser, (4) spray nozzle, (5) water flowmeter, (6) water-level control valve, and (7) vessel operating pressure (exhaust steam pressure).

Zeolite exchangers soften water by exchanging calcium and magnesium cations in the zeolite. The zeolite can be regenerated by treatment with a salt solution which restores the sodium cation to the zeolite and removes calcium and magnesium in soluble forms. Zeolites are insoluble granular materials which possess the cation-exchange property. Certain zeolites are suitable for hot waters, while others are used only for cold water. Cold zeolite exchange can be effected in gravity-type exchangers, although in refinery utility plants it is most usual to employ pressure-type exchangers for both cold and hot zeolite softening. Brine storage tanks or pits and measuring tanks are required for regenerating systems.

A pump is usually provided for pressuring salt solution into exchangers. With cold zeolite it is possible to use an eductor taking suction on a brine solution. Multiple zeolite units are installed to permit backwash regeneration and rinsing of one unit while remaining units are in operation. The pressure drop through zeolite exchangers is greatest when one unit is out of service and usually ranges between 20 and 25 psi.

The normal pressure drop is in the order of 15 to 20 psi.   Zeolite exchangers can be equipped for various manual and automatic operating cycles.

Demineralization

The demineralizing process of water treating consists of a two-stage exchange of cations and anions.   In the first stage hydrogen ions are exchanged for cations of calcium, magnesium, and sodium in water, forming acids with the anions of bicarbonate, sulfate, and chloride.   Carbonic acid formed from bicarbonates breaks down into carbon dioxide and water.   The other acids are ionized in the second-stage anion exchanger.   In the anion exchanger insoluble compounds of sulfate and chloride are formed with the exchange material.   The first-stage hydrogen-ion exchanger is regenerated with dilute sulfuric acid.   The second-stage anion exchanger is regenerated with a solution of soda ash.   Carbon dioxide formed in the first-stage exchanger is removed in a degasifier if the quantity of gas is appreciable or can be removed in the plant deaerator.   Chemical tanks and pumps are required for regenerating solutions. Exchangers are installed as multiple units to permit backwash regeneration and rinsing of a single unit while the remaining units are in operation.   Pressure drops through these units are similar to those for zeolite exchangers, that is, 20 to 25 psi during backwash and 15 to 20 psi normally.

The requirements for water treatment vary widely with the quality of water and the purposes for which the water will be used.   Better quality water is required for high-pressure superheated steam than for low-pressure saturated steam.   The selection of systems and equipment should be based on design and recommendations of water-treating equipment suppliers or qualified water-treating consultants.

Electrical Power

The use of electrical power and the degree of dependence thereon vary for each refinery.   For large equipment the general trend is toward installing extremely reliable machines without standby.   Many of these large machines are driven by electrical motors for general reasons of economy.   Following this philosophy to a logical conclusion, a completely electrically operated refinery is workable if the source of electrical power is absolutely reliable.

Electrical power is used for drivers for all types of equipment in process units and offsite.   Electrical power is also used for process purposes such as desalting, precipitation, and heating.   Refinery lighting and communication systems are powered by electricity.

Refinery electrical power can be purchased from an outside power source or generated in the refinery utility plant.   Purchased power is available at various voltage levels and normally must be transformed prior to use to either, 2,300 or 4,160 volts, the general levels used in refineries.   Standard voltages for electrical generators in the size range up to and including, 18,750 kva are 2,400, 4,160, 6,900, or 13,800.   Within this size range, power may be generated at 2,400 or 4,160 volts except when supplemental to purchased power at either of the other levels.   When the size of electrical generator exceeds 18,750 kva the standard voltage is 13,800 and generated power must be transformed to a lower level.

Electrical substations are usually based on distribution at voltages as listed in Table 8-16.

Table 8-16. Refinery Voltage Levels

Motors larger than 100 or 125 hp	4,160 volts, three phase, 60 cycles
Alternate to above	2,300 volts, three phase, 60 cycles
Motors 1 to 100 or 125 hp	440 volts, three phase, 60 cycles
Motors ½ to ¾ hp for process service	440 volts, three phase, 60 cycles
Motors ½ to ¾ hp for nonprocess service	120 volts, single phase, 60 cycles
Lightning, instruments, and motors smaller than ½ hp	120 volts, single phase, 60 cycles*

* Derived from a 208/120-volt three-phase four-wire 60-cycle grounded-neutral system.

Power Generation

Electrical power can be generated by steam-turbine or gas-turbine generators. Steam-turbine drivers are also usually most economical for large pumps and compressors from a standpoint of initial and operating cost. In a refinery, the logical first choice for electrical generators will, therefore, be steam-turbine-driven machines. These should be noncondensing or condensing extraction machines in preference to fully condensing. Turbines should use high-pressure steam and furnish medium- or low-pressure steam to other users.

When low-cost fuel (gas or diesel fuel) is available, it is sometimes possible to eliminate use of high-pressure steam and to utilize gas-turbine drivers for process equipment and electrical generators. This is attractive in conjunction with steam generation using gas-turbine exhaust and producing medium-pressure steam for process use.

Diesel-engine-driven electrical generators are least preferable for refinery use because of the greater salability of fuel as marine diesel or No. 6 fuel oil. These engines can be adapted to heat recovery producing hot water or low-pressure steam to improve over-all efficiency, but the capital investment required is relatively large in terms of Btu recovery. The initial cost of a simple diesel generating plant is, however, low, and the flexibility for expansion by adding relatively small units can be attractive in some plants.

When the utility plant is the sole source of electrical power, the installation should consist of two or more generators having an installed rated capacity of not less than 133 per cent of normal load. It is usually preferable to operate all generators at partial load to supply normal requirements. The total installed capacity should at all times be sufficient to permit startup of equipment in accordance with normal or prescribed emergency operating procedures. It is usual to assume the power factor to be 80 per cent in establishing the required generating capacity.

Steam-turbine generators are usually installed in fully enclosed buildings. When condensers are used with turbines, the condensers are located at grade with generators supported on an elevated structure above. Only the elevated structure is necessarily fully enclosed. Gas-turbine generators are usually installed in open-sided shelters. Waste-heat equipment can be supported from the deck which forms the shelter roof. Diesel generators are usually housed in fully enclosed single-story structures. Handling equipment, consisting of trolley beams or cranes, is necessary for maintenance purposes.

High-voltage switchgear and generator controls are located in fully enclosed buildings, usually at the generator level when elevated. In moist, hot climates, control and switchgear rooms are air conditioned. Air intakes for gas turbines and for diesel engines should be equipped with filters. Simple manually cleaned filters are adequate except in dusty areas, where some form of washed filter is desirable.

Electrical Distribution

Electrical distribution originates at the incoming power station or at the generating equipment in the utility plant. When required, power is transformed to 4,160 or 2,300 volts at these points. Power for drivers in major process areas is distributed from a substation in or near the process areas. When distances are short, underground distribution from the main distribution switchgear to process areas is desirable. Distribution to substations serving all other users such as offsite oil services, water pumps, and buildings should be by overhead cable.

All substations should include transformers for reducing voltage to distribution system levels. Primary and secondary protective and switching equipment should also be provided. Small transformers for low-voltage power, lighting, and street lighting can be located on distribution poles or structures. A common pole system can be used for power distribution and street lighting.

REFINERY AIR SYSTEMS

The purpose of refinery air systems is to supply dry air for instruments and plant air for air-powered tools and various other purposes. Dry air is also used for air motors which are intermittently operated and thus subject to freezing in cold climates. Plant air is employed in steam-air decoking of furnaces, catalyst lift, catalyst transfer, and reformer regeneration.   High-pressure air is used for engine starting for refinery drivers.   For certain products adversely affected by moisture (for example, lube oils discolor and water-soluble solvents are diluted), dry-air blankets can be substituted for inert gas for protection in storage tanks.

Design Conditions—Plant Air

The capacity of a plant air system should be established to supply air for catalyst regeneration, for plant maintenance during turnaround, for fluid catalytic cracking (FCC) unit startup, or for steam-air decoking.   As all these operations can be scheduled, the system should not be designed for coincident operation.   When other design criteria are not available, an estimate of probable requirements can be made based on the crude charge rate of the refinery.   A minimum of 500 scfm should be provided for plants up to 10,000 B/SD crude charge rate.   A minimum of 1,300 scfm should be provided for plants having a crude capacity of 70,000 B/SD or more.   Plants of intermediate size should fall within the above range.   Air requirements for regeneration of catalytic reforming and desulfurizing units varies with charge rate and type of process.   The usual range of required air is 300 to 600 scfm.

Table 8-17. Air Requirements for Tools

Tool	Size	Flow, scfm
Tube expander...............	Up to 3″ tube	60
Hand riveters................	8″ to 10″ stroke	40
Chipping hammer...........	1″ to 4″ stroke	25
Impact wrench..............	1¾″	50
Motor......................	15 hp	240
Rotary steel drill............	2″	90
Nut runner.................	1″	45
Sandblast..................	3,000 lb/hr sand	375
Tube cleaners..............	....	200

Aeration air for startup of fluid catalytic cracking units is dependent on unit charge rate and type of unit.   A 6,000-B/SD fresh feed unit may require 300 scfm for startup. This ranges up to as high as 2,000 scfm for units charging 35,000 B/SD of fresh feed.

When plant air is used to supply air motors, the piping must be designed to pass the required flow (200 to 400 scfm per motor).   The capacity of other associated equipment usually is not affected by this quantity, as it is intermittent.

Air for steam-air decoking is used during the burning operation.   Air required varies with furnace tube size and is usually 10 per cent of steam flow or 0.6 lb/sq ft/sec.

Air consumption for various commonly used pneumatic tools based on 90 psig at tool is tabulated in Table 8-17.   Air should be delivered to distribution headers in refinery utility plant at 100 psig and not more than 100°F.

Design Conditions—Dry Air

The capacity of dry-air systems will usually be established by requirements of the instrument system.   It is customary to allow 0.5 scfm of dry air for each instrument pilot in process units.   The total requirements vary with types of process facilities and instruments installed.   These total requirements range between 200 and 900 scfm for reasonably complex refineries with crude capacities between 20,000 and 70,000 B/SD.   In cold climates dry air may be used for air motors to avoid freezing of intermittently operated equipment.

The usual design conditions are for delivery of dry air to an instrument air header in the utility plant at a minimum pressure of 40 psig.   Motor air is required at a pressure of 100 psig in the distribution header to allow for 90 psig minimum at motor. Usually this requires separation into two dry-air systems.

All dry air should be at a temperature not exceeding 100°F and a winter dew point compatible with ambient conditions. Dry-air quantity for blanketing storage tanks should be based on requirements of API venting code and pumping inage and outage rates. As the air is used at low pressure (1½ to 2 in. of water), it is frequently preferable to provide a separate air system for blanket air.

High-pressure Air

High-pressure air systems for engine starting (about 250 psig) are normally furnished with the engine. These are single-purpose systems located adjacent to the equipment which they serve.

Equipment

Equipment used in plant air systems includes compressors, aftercoolers, and air receivers. Dry-air systems include air driers in addition to the above equipment. The arrangements commonly used for equipment are as follows:

1. Two compressors, each with a separate aftercooler, but both discharging to a single air receiver. The single air receiver supplies the plant air system and the air drier. Air from the air drier passes through a pressure-reducing valve before entering the instrument air system.

2. Two compressors, each with a separate aftercooler and each discharging to a separate air receiver. Instrument air is dried by passing through an air drier. Standby service is provided for instrument air only, by a pressure-reducing valve from the plant air system into the instrument air system upstream of the drier.

3. Two compressors plus one stand-by compressor, each with separate aftercooler and discharging to two air receivers. Air is taken from either or both receivers to the plant air system and to the instrument air system through an air drier and pressure-reducing valve.

A typical arrangement of equipment and facilities for air systems is shown in Fig. 8-42.

Compressors

Centrifugal, rotary, or reciprocating compressors may be used in air services. These are available with steam-turbine, motor, steam-engine, or gas-engine drives. The usual refinery plant and instrument air services can be classified as low-capacity, high-pressure installations adaptable to use of reciprocating compressors. Plant air service is normally intermittent; therefore, constant-speed, electric-motor-driven compressors operating with off-on pressure control are suitable. With this control, single-stage unloaders are used.

Small compressors, with drivers of about 15 bhp or less, can be operated continuously with on-off unloading to maintain pressure in the system. Instrument air service is fairly constant as to flowing quantity. Usually the normal discharge is, however, considerably less than compressor design discharge. To match the required discharge it is desirable to vary driver speed with the use of a steam-engine or steam-turbine driver. When minimum initial cost is desired, a steam-engine-driven compressor exhausting to atmosphere should be installed. The air side of instrument-air compressors should be of the nonlubricating or nonoily type to eliminate carry-over of oil.

Coolers

When required, intercoolers for air compressors are furnished integrally mounted. The equipment supplier can furnish a suitable aftercooler, or these can be selected from manufacturers' standard coolers based on compressor discharge temperature. The in-line-type cooler with integral separator is preferred. When an integral sep-

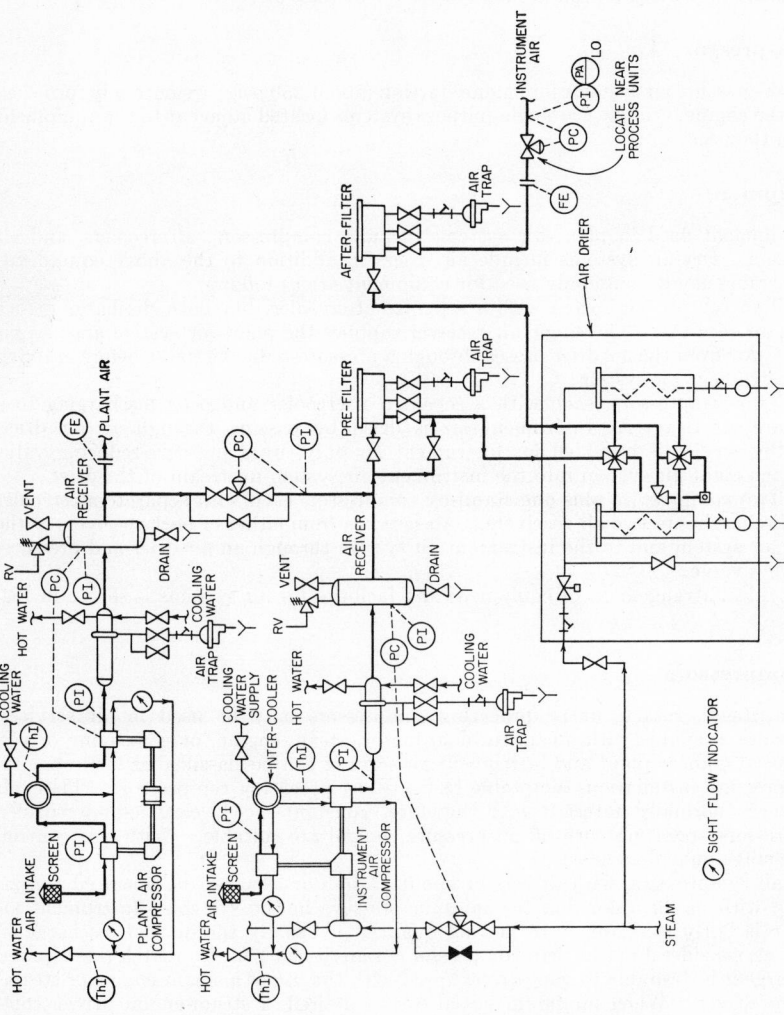

FIG. 8-42. Typical arrangement of equipment and facilities for a refinery air system.

arator is not provided, an independently mounted unit with air trap must be installed in the outlet piping.

On small machines the aftercooler is frequently furnished mounted on the frame and connected to a common compressor cylinder jacket and intercooler water system. On large machines, the aftercooler should be provided with a separate water system. The intercooler and cylinder jackets on large machines are usually connected in series with a single inlet and visible outlet.

Receivers

Air receivers are used to provide surge capacity, to act as pulsation dampeners, to collect water and grease held in suspension in compressed air, to reduce friction of air in pipe systems, and to cool air before entering transmission systems. Air receivers should be large enough to furnish 50 per cent of full demand for a period of 8 min with final system pressure equal to 50 per cent of initial pressure. For a 100-psig system this is equal to 7.67 times actual cubic feet per minute or one times the standard cubic feet per minute of the system. When the plant air system provides standby service for instrument air, a nominally sized instrument air receiver can be installed and the plant air receiver should be sized for the combined capacity of the systems.

Driers

Air driers are provided for removal of moisture in dry-air systems. Wet, cooled, compressed air is passed through a desiccant-filled drying tower and emerges as a dry gas. Two drying towers are provided, one of which is regenerated while the other is used for drying. A prefilter should be furnished upstream of the air drier to protect the desiccant bed from contamination by entrained oil or water. An afterfilter of a similar type should be located downstream of the drier to trap out any desiccant that may be carried over with the dried gas.

Air driers are usually purchased as complete packages mounted on common base plates equipped for manual, semiautomatic, or automatic regeneration. Steam is commonly used for regeneration, although electrical regeneration can be obtained.

Piping

Suction and discharge piping for air compressors should be sized so that pulsation dampeners are not required. As a minimum, suction and discharge lines should be compressor-intake or discharge-nozzle size. Suction and discharge lines should have a minimum number of changes in direction. Where changes are necessary, long-radius bends should be employed and lines should be anchored securely. Piping from the compressor to the aftercooler and the air receiver should be not less than discharge-nozzle size. Compressor suction inlets should be provided with dry-type air-intake screens with weather screens. Suction inlet screens installed in cold climates are usually provided with steam or electric heaters.

Air-distribution piping is sized on a rather liberal basis. The maximum allowance for pressure drop from air receiver to user is 10 psi. Preferably the loss in a distribution system should be considerably less than this amount. There are no precise criteria for sizing plant air lines. The main header for plant air should not be less than 3 in. in size and should not exceed 6 in. Instrument air lines should be sized for a maximum total pressure drop from receiver to process units of 10 psi.

REFINERY FUEL SYSTEMS

Refinery fuel is used for production of heat for process applications and for generation of steam in refinery utility plants. Fuel is produced and consumed in various portions of the refinery. Fuel systems include facilities for collection, preparation, and distribution of fuel to users. The commonly used refinery fuels are oil and gas.

Fuel Selection

The selection of a refinery fuel must be based on the cost of the fuel. Material which cannot be sold readily in commercial channels has the least monetary value and is normally used as plant fuel. Liquid materials commonly diverted to refinery fuel include visbreaker tar, FCC decanted oil, vacuum tower bottoms, lube extracts, and waxes. The majority of these materials would be difficult to blend to a commercial fuel of acceptable specification because of viscosity, sulfur content, or presence of foreign matter. Gaseous materials diverted to refinery fuel are those which cannot be processed to salable products economically and frequently include H_2, CH_4, H_2S, and C_2H_6. The relative values of all the above materials as salable products or as fuel will also be affected by availability of natural gas for plant fuel.

Most refineries do not produce a sufficient quantity of fuel gas and unsalable liquids to meet heat requirements of process furnaces and steam generators. Additional fuel is required, consisting of either natural gas or bunker fuel products.

In most refineries both liquid and gaseous fuels are used simultaneously. It is common practice to burn one fuel, gas or oil, as a base and to use the other material to trim to meet the heat requirement of the furnace or boiler. Many furnaces and boilers are equipped with combination burners suitable for gas or oil. Exceptions to this are reforming or hydrogen-treating processes in which frequently only gas-fired furnaces are installed. For some reforming processes gaseous fuel is considered essential for process control. Furnaces which are operated when no gas fuel is available, as during refinery startup, must be equipped for oil burning. Oil is commonly used for supplementary fuel for CO boilers, as this tends to simplify operations without CO burning.

Fuel-oil Systems

The purpose of fuel-oil systems is to ensure a constant regulated supply of oil to burners of steam boilers and process furnaces. The system includes facilities for storage, pumping, heating, and distribution of oil at suitable pressures and viscosities so that atomization and burning are possible. A typical plant fuel-oil system is shown in Fig. 8-43.

In general, oil is drawn from the storage tank and pumped through a heater to raise the temperature, thus lowering the viscosity to the desired value. The oil is distributed to all users, and to ensure proper temperature, viscosity, and pressure at the burners, an excess of oil is supplied to the header from which the burner leads originate. This excess is recirculated through a return fuel-oil line back to storage. The ultimate aim in fuel-oil-system design should be to ensure that the supply of suitable fuel to each furnace will not fluctuate with operational changes in other portions of the refinery.

Storage

One or more fuel-oil storage tanks should be provided. These tanks should have a capacity of not less than 5 days' supply at normal firing rates of the furnaces and the boilers which are usually oil fired. If refinery fuel is obtained from a number of sources, it may be necessary to blend the stored material, in which case at least two tanks are required. It is often possible to line blend the refinery fuel as it leaves the various units without precise control of the blend.

Tanks should be cone roof, fitted with tank heaters and/or suction heaters. In most plants heaters are seldom operated because the oil leaves the process units at a high enough temperature to ensure pump suction. This temperature should be limited, however, to a maximum of 240°F to minimize possibility of boil-over due to vaporization of water in the tanks. Storage tanks are insulated in cold climates, but this is not otherwise necessary because heat losses are compensated for by the heat of the recirculated material.

Pumping and Heating Equipment

Equipment for heating fuel oil to the temperature corresponding to the desired viscosity and for pumping is usually combined in a set mounted on a common base or frame. This fuel-oil set is commonly installed in the refinery utility plant with the fuel-oil storage tanks immediately adjoining.

Oil should be delivered to burners at a pressure of approximately 100 psig for control and atomization. To provide this pressure and to allow for pipeline losses, oil should be discharged from the fuel-oil set at a pressure ranging between 125 and 150 psig.

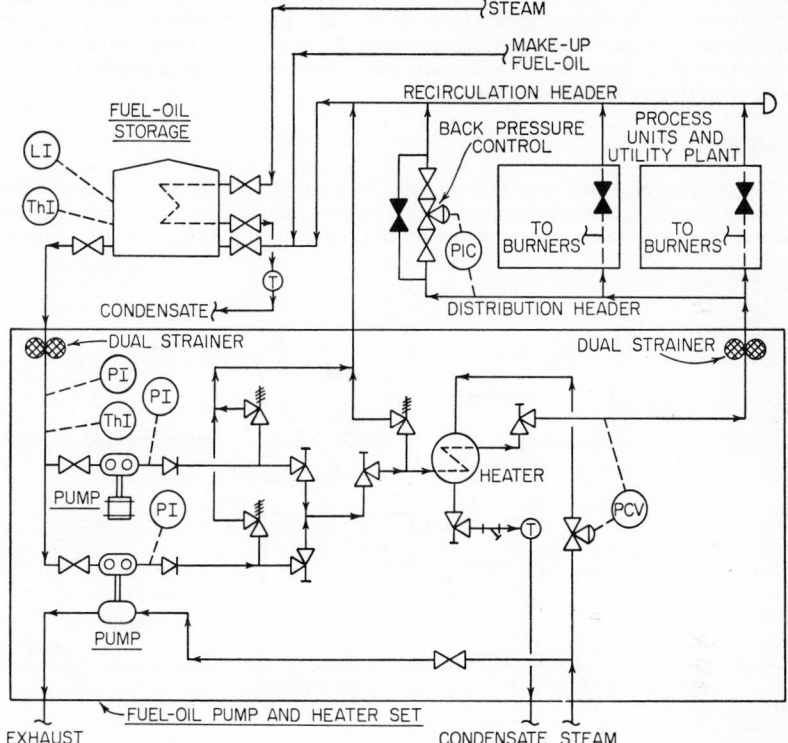

Fig. 8-43. Typical refinery fuel-oil system to ensure constant supply of oil to boiler and furnace burners.

If separate components are assembled for pumping and heating, the pump discharge pressure must also include losses in heater, strainer, and connecting piping.

The temperature at which oil is discharged from the fuel-oil set should vary with material being pumped. For proper atomization, the temperature at the fuel-oil set outlet should be sufficient to lower the viscosity of the oil to 30 to 40 cs (140 to 180 SSU). The design capacity of the fuel-oil pumping and heating equipment should be equal to 125 per cent of design requirements of the plant assuming simultaneous firing at design rate of all fuel-oil-burning furnaces and boilers. This allows for 25 per cent recirculation of oil.

In a fuel-oil set, the pumps, heaters, and controls are mounted on a common base plate with all interconnecting piping provided. These include two full-size pumps preferably rotary type with one motor and one turbine driver. Heaters are steam heaters of the fixed tube sheet type in which single tubes are rolled in. On duplex sets, two heaters are provided. Fuel-oil temperature is regulated by controlling steam

flow to the heater.   Relief valves are located on the discharge side of the pumps and
on the oil heater.   Relief-valve discharges should be piped back to the oil storage
tank.   Fuel-oil sets are equipped with duplex strainers with $\frac{1}{16}$-in. mesh at the inlet
to intercept dirt and foreign material.   Duplex strainers with $\frac{1}{32}$-in. mesh are
provided at the outlet to strain out precipitated carbon particles.

Piping

Suction lines to fuel-oil pumping equipment should be sized for a unit pressure drop
not exceeding 0.3 psi per 100 ft.   The line should be at least as large as the inlet
nozzle of fuel-oil set or pump.   Discharge piping supplying burners and recirculation
piping returning to storage should be sized for a unit pressure with protective heating.
Separate nozzles should be provided on storage tanks for make-up of fuel oil, recirculation, and withdrawal of oil.   The arrangement of nozzles should, if possible, minimize
short circuiting of recirculated oil.

Fuel-gas Systems

The purpose of a fuel system is to provide a supply of fuel gas to steam boilers,
process furnaces, gas engines, and gas turbines at a regulated pressure and reasonably
constant heating value.   The system includes collecting piping, mixing drum controls,

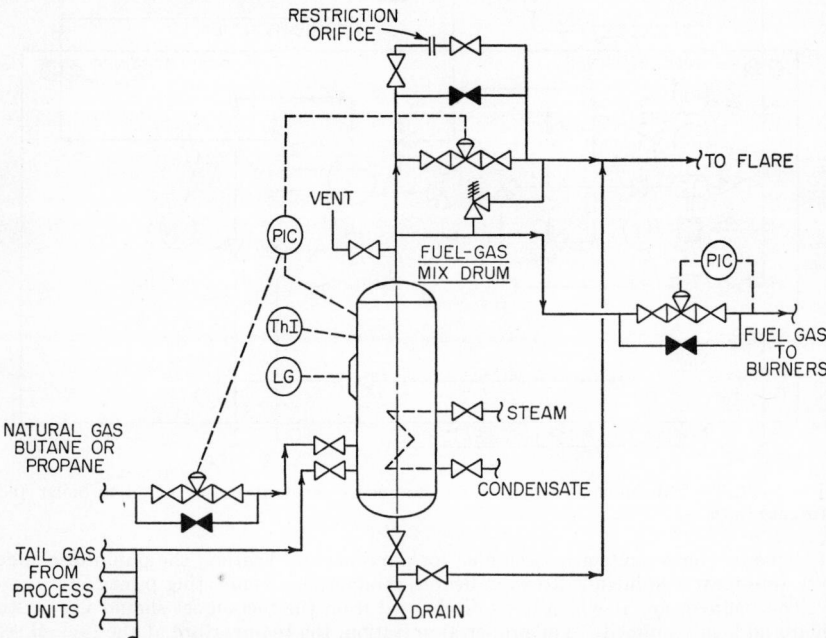

Fig. 8-44. Typical refinery fuel-gas system to ensure constant supply of gas to boilers,
furnaces, engines, and turbines.

and distribution piping.   When necessary, standby storage of LP gas may be included.
A typical fuel-gas system is shown in Fig. 8-44.   Tail gas from various process units
is collected and brought to a mix drum.   Natural gas or LP gas from storage is added
to the mix drum controlled by a flow-indicating controller preset to bring the anticipated make to a predetermined heating value.   The mixed gas is distributed at controlled pressure.   Excess quantity of gas is dumped to flare by a pressure controller.

Mix Drum

Gas is usually furnished to headers at furnace or boiler burners at a pressure of 15 to 20 psig. To allow for piping-line losses and control-valve losses, the operating pressure of the fuel-gas mix drum should be 30 to 40 psig. When a gas turbine is installed in the refinery, the gas pressure at the turbine must be 125 to 120 psig, varying with manufacturer. This equipment should be supplied by a separate system (usually operated directly from natural gas), or a gas booster compressor should be provided. The fuel-gas mix drum should be designed for a flow equal to the total simultaneous requirements of all normally gas-fired furnaces and boilers in the plant. Gas will be properly mixed if the mixing drum is sufficiently large to separate all liquid droplets from the gas; thus the general criteria used for sizing knockout drums is applicable. The length of the vessel should be equal to approximately two diameters. The vessel should be fitted with mixing baffles and a steam coil for vaporizing liquid carry-over or LP gas.

Piping

Gas-distribution piping between the mix drum and headers at burners should be sized on the basis of a unit pressure drop between 0.2 and 0.5 psi per 100 ft. Lines to isolated low-capacity burners, such as flare pilots, may require protective heating in extremely cold climates. For other users, such as furnaces, the normal knockout pots are adequate to ensure elimination of liquid.

REFERENCES

1. American Petroleum Institute, *Manual on Disposal of Refinery Wastes*, 1959, 6th ed., Vol. I.
2. Armistead, Jr., George, *Safety in Petroleum Refining and Related Industries*, John G. Simmonds & Co., Inc., 1959, 2d ed.
3. Bland, W. F., "Products Blending," *Pet. Proc.*, pp. 142–143, July, 1957.
4. Chesler S., and B. W. Jesser, "Some Aspects of the Design and Economic Problems Involved in the Safe Disposal of Inflammable Vapors from Safety Relief Valves," *Trans. ASME*, February, 1952.
5. Combes, C. L., et al., *Car Builders' Cyclopedia*, Simmons-Boardman Publishing Corporation, 1957, 20th ed.
6. Cowell, W. D., "Calculating Tank Heat Losses," *Ref. Eng.*, September, 1959.
7. Crane, B. G., "Which Blending Method Is Best?" *Pet. Proc.*, pp. 82–90, April, 1956.
8. Elkin, H. F., "How Sun Oil Licked Its Water Problem; Pollution Control and Conservation by Biological Oxidation and Reuse of Refinery Waste Water," *Oil & Gas J.*, pp. 118–120, May 28, 1956.
9. Jacobs, R. B., F. H. Blank, and F. W. Scheineman, "Beware Process Explosion Hazards," *Pet. Ref.*, pp. 361–374, September, 1959.
10. Lamb, I., *Oil Tanker Cargoes, Their Safe and Efficient Handling*, Charles Griffin & Company, Ltd., London, 1954.
11. Lange, K. L., "How to Specify Tanks," *Pet. Ref.*, pp. 99–104, August, 1959.
12. Mayo, H. C., and R. E. Daze, "Turbine or Motor Drives," *Pet. Ref.*, pp. 110–115, August, 1957.
13. McRae, A. D., et al., "Biological Oxidation of Phenolic Waste Water," *Oil & Gas J.*, pp. 223–227, August 20, 1956.
14. Neumann, E. D., G. J. Reno, and L. C. Burroughs, "Modern Waste-disposal Facilities at Shell's Anacortes Refinery," *Proc. API*, vol. 38, sec. III, Refining, pp. 293–303, 1958.
15. Prater, N. H., and D. W. Antonacci, "How to Estimate Cooling Tower Costs," *Pet. Ref.*, pp. 119–124, April, 1959.
16. Reed, Paul, and R. C. Deering, "Products Automatically Blended In-Line," *Oil & Gas J.*, Oct. 4, 1954.
17. Sax, N. I., *Handbook of Dangerous Materials*, Reinhold Publishing Corporation, New York, 1951.
18. Stewart, Oswald, "Plant Site Selection Guide," *Factory*, May, 1957.
19. Thornton, Jr., D. P., "How to Get What You Need When You Order Cooling Equipment," *Pet. Proc.*, Nov., 1948.
20. Wooler, R. G., *Tankerman's Handbook*, Edward W. Sweetman, 1946.

Section 9

PERSONNEL AND PLANT SAFETY

By JAMES H. HERBERT*

Consultant
Accident and Fire Prevention
Christmas Cove, Maine

A sound accident- and fire-protection program can help **more than any other single factor** in the development of an efficient operation with high employee morale.

Keeping personnel at a high degree of interest and enthusiasm while working efficiently is the greatest problem of supervision. Explaining **how** work procedures are safe interests a man. The supervisor, without forfeiting his supervisory rights, can ask his men for their safety ideas while at the same time giving his own whenever he lays out a job. Actually, such a practice builds morale because it gives his men a feeling of participation.

The **extra bonus** to the safety approach is that the safest way is, in the last analysis, invariably the most efficient way.

SAFETY PROGRAM CHECK LIST

To examine an accident- and fire-protection program to see if it is doing everything it should, the following questions should be answered. More details concerning each individual one are given later in this section on the pages indicated.

1. Are necessary statistics available to determine where the problems exist? (See pages 9-2 to 9-6.)

2. Is there a company or plant safety policy? (See pages 9-6 to 9-7.)

3. In checking a job applicant's references, is information obtained as to his work safety record? (See page 9-6.)

4. Is there a **formal** safety indoctrination program for new employees? (See pages 9-6 to 9-7.)

5. Is there a **formal** posted program for the use of personnel protective equipment and clothing? (See page 9-7.)

6. Is there a **formal** recorded safety inspection for equipment, such as hand tools, elevators, hoists, rope, and cable? (See pages 9-7 to 9-11.)

7. Is there an off-the-job safety program? (See pages 9-11 to 9-12.)

8. Do all supervisors point out in detail hazards of job assignments at the time employees are put on a job, and do they check back to see if their warnings are understood? (See page 9-12.)

9. Is there a **formal** prompt investigation by immediate supervision of all injuries? (See page 9-12.)

* Mr. Herbert was formerly Safety Director, Socony Mobil Oil Co., New York, N.Y.

10. Is safety performance part of the "performance reviews" for all employees, including supervisors and department heads?   (See page 9–13.)

11. Are written approvals required for repetitive activities, such as:
   a. Maintenance department or outside contract personnel working on operating equipment
   b. Possible exposure to toxic materials, including radioactive matter
   c. Entering vessels
   d. Burning and welding in questionable areas
   e. Blinding and plugging
   f. Starting furnace fires

(See pages 9–13 to 9–16.)

12. During the course of the design, selection, and placement of new equipment, is a **formal** safety study made to determine such matters as:
   a. Safe location in relation to other equipment
   b. Safety guards for moving parts, stairways, runways, ladders, electrical equipment, hot lines, and so forth
   c. Fire protection

(See pages 9–16 to 9–18.)

13. Is there **formal** safety consideration of toxicity, flammability, and radioactivity when it is decided to use new chemicals or materials?   (See pages 9–18 to 9–19.)

14. When new processes are developed, is there **formal** safety consideration of their operation and maintenance to determine possible hazards to personnel?   (See pages 9–16 to 9–17.)

15. When new products are developed, is there **formal** safety study of possible hazards to users?   (See page 9–18.)

16. Is there **formal** first-air-to-the-injured training for plant personnel?   (See page 9–24.)

17. Is there **formal** driver training for plant truck drivers, fork-lift truck operators, and plant passenger-car drivers?   (See page 9–24.)

18. Is there a **formal** fire-fighting training program for plant personnel?   (See pages 9–27 to 9–29.)

19. Is there a **formal** disaster-control program?   (See page 9–29.)

20. Is there a mutual-aid program with other petroleum processing plants in the area?   (See page 9–30.)

21. Are there **formal** instructions for "outside" contractors?   (See page 9–18.)

Economics of Safety

Accident- and fire-protection problems are solved most easily if we understand their magnitudes thoroughly.   The humane aspects are incontrovertible, but we need statistics to appreciate the dollars-and-cents value of safety—or cost if safety is lacking.   If more were known about the total costs of accidents and fires, there would be fewer of them.

The following National Safety Council report came from a thorough investigation of accident costs:

"A beverage concern reported numerous cuts from broken bottles, but the compensation and medical cost were not considered serious by the employer.   Investigation disclosed that the total loss because of bottle breakage was more than **30 times** the injury costs.   Correction of the accident causes, therefore, led not only to reduction of injuries but also to operating savings far greater than the injuries would have indicated."

Some companies have found that when the **total** costs of injuries, fires, and property damage claims have been determined, they amount to as much as 10 to 15 cents and more in some years for **each share** of stock.

On-the-job Accident Reports

The details of an accident to an employee should be reported on a regular form by supervisors.   Such a form is shown in Fig. 9-1.   Reports should be made for non-disabling as well as disabling injuries.

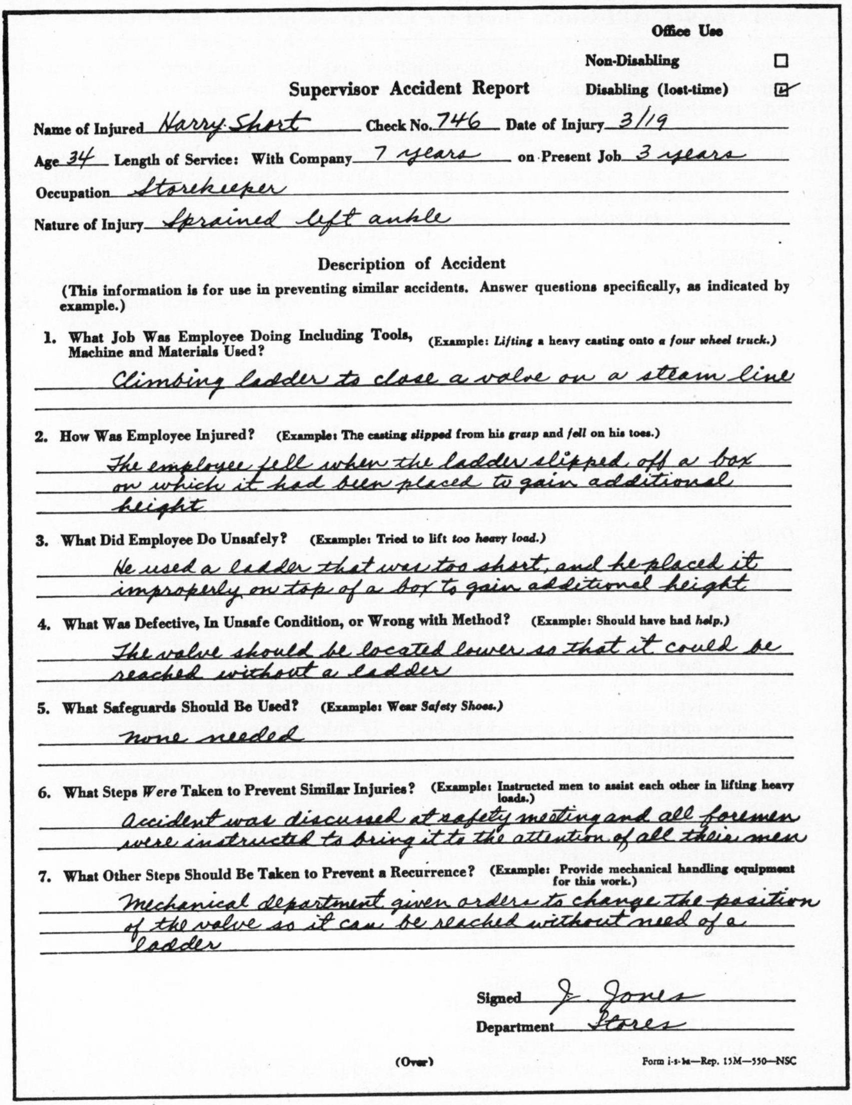

Office Use

Non-Disabling ☐

Supervisor Accident Report Disabling (lost-time) ☑

Name of Injured *Harry Short* Check No. *746* Date of Injury *3/19*

Age *34* Length of Service: With Company *7 years* on Present Job *3 years*

Occupation *Storekeeper*

Nature of Injury *Sprained left ankle*

Description of Accident

(This information is for use in preventing similar accidents. Answer questions specifically, as indicated by example.)

1. What Job Was Employee Doing Including Tools, Machine and Materials Used? (Example: *Lifting a heavy casting onto a four wheel truck.*)

   Climbing ladder to close a valve on a steam line

2. How Was Employee Injured? (Example: *The casting slipped from his grasp and fell on his toes.*)

   The employee fell when the ladder slipped off a box on which it had been placed to gain additional height

3. What Did Employee Do Unsafely? (Example: *Tried to lift too heavy load.*)

   He used a ladder that was too short, and he placed it improperly on top of a box to gain additional height

4. What Was Defective, In Unsafe Condition, or Wrong with Method? (Example: *Should have had help.*)

   The valve should be located lower so that it could be reached without a ladder

5. What Safeguards Should Be Used? (Example: *Wear Safety Shoes.*)

   none needed

6. What Steps *Were* Taken to Prevent Similar Injuries? (Example: *Instructed men to assist each other in lifting heavy loads.*)

   Accident was discussed at safety meeting and all foremen were instructed to bring it to the attention of all their men

7. What Other Steps Should Be Taken to Prevent a Recurrence? (Example: *Provide mechanical handling equipment for this work.*)

   Mechanical department given orders to change the position of the valve so it can be reached without need of a ladder

Signed *J. Jones*

Department *Stores*

(Over) Form i-s-le—Rep. 15M—550—NSC

FIG. 9-1. Report form which can be used by supervisors to record detailed information on an accident. (*Courtesy of National Safety Council.*)

Many companies now pay the difference between workmen's compensation and full pay for varying periods of time based on years of service. This payment can amount to much more than the compensation figure. In addition, the following costs should be considered:

1. Wage cost of employees not injured but who were made idle due to accident
2. Premium pay rates for employees who replace injured employees on unscheduled hours
3. Cost of supervisors' time required in connection with the accident
4. Upon the injured employee's return to work, if he is placed on "light" work, the difference between the pay which he receives and the value of the light work
5. Property damaged by the accidents

Table 9-1. API Guide Sheet for Fire Investigations and Reports

The lessons that can be learned from some fires may be of much help to fire-protection engineers in preventing injuries and property loss in their companies' operations. One of the difficulties in reporting this information is knowing what is desired. The following outline may serve as a guide for such reporting. As fires differ widely in nature, the reporter should select those items that are directly applicable to the fire in question and prepare his report accordingly. It is suggested that the following outline be read completely before drafting the report.

I. *Confidential information for API only:*
   1. Name of company and location of plant or property involved
   2. Date of fire
   3. May API staff distribute copies of the following information anonymously to members of the Central Committee on Safety and Fire Protection and its reporting committees? Please answer in letter of transmittal

II. *Summary of the fire:*
   1. A brief description of the incident. (This permits reader to obtain an over-all picture of the case prior to reading the full report.)
   2. Lessons learned: These factors are of particular importance:
      a. Lessons in operations, both as to fire prevention and fire control
      b. Lessons in design and layout of operating equipment involved
      c. Lessons in fire fighting and supervision of fire fighting
      d. Miscellaneous, such as first aid to injured, mutual-aid plans, use of public fire-fighting services, public relations, etc.

III. *Detail information on the fire:*
   1. Time fire started and when extinguished
   2. Weather and wind conditions, temperature, and humidity
   3. Applicable laboratory characteristics of the fuels involved, etc.:
      a. Name of product (butane, gasoline, crude oil, etc.)
      b. Flash or vapor pressure (volatility factors) and actual temperature of product at time of ignition
      c. What was the source of fuel that started the fire if more than one fuel was involved
   4. Source of ignition that started the fire. (If unknown, evaluate likely causes.)
   5. Operations that led up to the start of the fire:
      a. If tanks: the type, size, capacity, amount of oil involved, vent sizes, etc.
      b. If operating equipment: temperatures, pressures and brief description of equipment involved
      c. A brief description of operations leading up to start of fire
   6. A narrative account of the fire itself.
      a. Description of the initial fire and its early development
      b. Spread or lack of spread. Were protection devices (dikes, etc.) involved?
      c. Describe explosion or rupture and effect thereof
      d. How did automatic controls function?
   7. Details of fire fighting:
      a. Narrative account including:
         (1) How soon was fire reported?
         (2) Description of fire fighting equipment used
         (3) How soon fire fighting started
         (4) What were shutdown operations, as applied to control of fire?
         (5) Important steps in fire fighting activities.
         (6) When was fire under control and when extinguished?
         (7) What water and foam rates used, etc.
   8. Extent of damage:
      a. Nature of injuries
      b. Nature and damage to equipment
      c. Monetary loss (if available)

IV. *Supplemental photographs, maps, etc.,* may be helpful and may reduce amount of description needed.

The majority of fires may be described fairly easily and understandably under the applicable headings. The supplemental detail may be supplied in narrative form also under these headings.

A single copy of such report, mailed to American Petroleum Institute, Division of Science and Technology, 1271 Avenue of the Americas, New York 20, N.Y., with supplemental instructions as to disposition of the report, is desired.

Off-the-job Accident Reports

As outlined on page 9-11, Off-the-job Safety Program, absenteeism from off-the-job accidents can be from five to twenty times as much as absenteeism from on-the-job accidents. Many companies pay full wages for such absences.

Most payroll departments show a combined cost for absenteeism due to off-the-job accidents and illnesses. The payroll department can separate the figures, but the information as to the cause of absence will have to come from the supervisors.

Fires and Explosions

Table 9-1 shows the "API Guide Sheet For Fire Investigations and Reports." It is excellent. Reporting fires also to the API helps acquaint the entire industry with problems met and lessons learned.

It is helpful to secure a written statement from each person closely involved in a fire—his opinion of the cause, how it spread, and how it was extinguished.

Table 9-2. Detailed Cost of a "$25,000 Fire"

Listed below are detailed cost items which could exist in a so-called $25,000 fire which shut down a 20,000-bbl cracking unit for 7 days:

1. Replacement cost of pumps, valves, lines, fittings, instrumentation, insulation, wiring, etc., damaged or destroyed by the fire (this is the amount usually used as the "horseback" cost figure for a fire) $25,000
2. Foam used in fire fighting 4,000
3. Unit operators idled 7 days (they work, but on repairs necessitated by the fire) 2,000
4. Wages for fire crew and men on "standby" 400
5. Wages and doctor bills for injured fire fighters 210
6. Recharging small fire extinguishers 105
7. Product loss 400
8. Labor for maintenance men cleaning up unit 600
9. Labor for removing damaged equipment and installation of new equipment, including overtime 7,600
10. Damage to employee's glasses 25
11. Damage to fire hose 350
12. Fire investigation work 800
13. Maintenance crew idled by fire 75

Partial total $40,565

Plus the profit which could have accrued from the sale of approximately 70,000 bbl of gasoline which could have been run from that unit had it not been for the fire (worth on today's market about $308,700).

Complete cost figures should be tabulated for each fire. Some companies base figures on "insured" or "book" values only. Table 9-2 shows what can really be involved in a so-called $25,000 fire. The actual cost is a far cry from the cost based on the insured or book value of the property destroyed.

Consider the following in determining true costs of fires:

1. Actual replacement cost of damaged equipment
2. Downtime of damaged equipment
3. Downtime of equipment not damaged but which can operate only in connection with damaged equipment
4. Overtime paid to fill orders delayed by fire
5. Loss of time of employees idled by fire but paid
6. Damage to raw and finished products
7. Installation cost of replacement equipment
8. Labor for cleaning up the location
9. Fire-fighting agents such as foam
10. Damage to fire-fighting equipment
11. Fees paid to outside fire-fighting companies
12. Claims paid as a result of injuries to nonemployees or their property

The following should also be considered in determing the costs of accidents and fires, even though they are not "additive" figures. Regardless, they could be the **most expensive items.**

1. Inconvenience to customers
2. Loss of customers
3. Bad employee relations
4. Unduly strict community ordinances resulting from employee or public injuries or property damage
5. Inefficiency of untrained employees who replaced injured employees

Plant Safety Policy

Most companies have safety policies. Too often, however, safety policies are highlighted for the first year or two and then fade into oblivion. A policy should be kept before all employees and be referred to regularly by **all** levels of management.

Safety References for Job Applicants

It would not be good judgment to hire someone whose safety record in prior employment or off the job was bad. If a person has a series of injuries in certain work, either he has not been properly trained or he is not suited to that type of work. The time taken to check safety records is a good investment.

New Employees Safety Program

How many injuries and how many explosions happen because someone "didn't know"? There is nothing more fundamental than a formal safety indoctrination program. The dividends are high.

The word **formal** means that the program has a definite outline. It should not be "hit or miss" depending upon the day-to-day mood of the instructor. Essential subject matter should be given so that basic safety principles are learned immediately.

Safety indoctrination programs should be kept up to date. Following is a list of subjects to be considered:

1. Company's safety policy.
2. How a petroleum processing plant differs from others.
3. Hazards in petroleum processing plants: fires, gases, heat, pressure, vacuum, hydrogen sulfide, steam, and chemicals.
4. Danger of fire, sparks, and smoking in certain areas.
5. Accidents can be prevented.
6. Mechanical safeguarding.
7. Description of supervisor's and employee's roles in safety.
8. Report of unsafe condition to supervisor.
9. Explanation of plant's policy for wearing safety goggles, shoes, hats, gloves, and other personal protective equipment, illustrated by showing broken goggles, dented hats, dented safety shoes, etc.
10. Distribute rule books, if available, but do not forget that few people will read them.
11. Instruction on how to lift safely.
12. Demonstrate how to use the right tools for the job.
13. Instruct in the proper use of ladders.
14. Emphasize do not enter any tank, tank car, or vessel without **formal** authorization because of possibility of gases or lack of oxygen.
15. Give a general explanation of **formal** permits for open fires, welding, entering vessels, etc.
16. Instruct in care in riding trucks; hazards in jumping on and off moving vehicles.
17. Warn of the dangers in horseplay.
18. Emphasize the danger in using equipment without safety guards.
19. Warn against undue haste.

20. Emphasize the need and reason for good housekeeping.
21. Give suggestions for off-the-job safety at home during recreation or travel.

Two meetings to indoctrinate employees in safety are recommended, each about an hour long.

Safety films are recommended. Excellent ones are available from the National Safety Council (see page 9-74).

The second meeting is advised because new employees miss much the first time. This training should be of a general nature and not related to a specific job. The unit supervisor in question should point out specific hazards on the job.

New technical employees should be included in safety indoctrination programs. Few technical schools include thorough safety training as a part of their instruction. They expect the employers to furnish safety instruction.

Protective Equipment Program

In some cases, employees learn of the availability of safety equipment (glasses, hats, shoes, and other clothing) only by word of mouth and are reluctant to ask for it because they think that their foremen or fellow employees may call them sissies.

No manager or supervisor wants to lose employees as a result of injuries. But how many times we see managers and supervisors "look the other way" when they pass men who are not wearing prescribed protective equipment!

Many employees do not know whether an area is or is not one for hard hats or safety glasses. Conspicuous signs will clarify the question.

What is the employer's position in regard to paying for all or part of prescription safety glasses or for safety shoes? Posted copies of company policies clarify the policy and, furthermore, help to sell the program. The following policies are typical:

Hard Hats. The company designates, with conspicuous signs, all areas where hard hats are to be worn. The company furnishes hard hats without charge.

Safety Shoes. The company recommends that all employees wear safety shoes at all times on the job. They can be purchased at the stockroom at cost (or x per cent less than cost). Employees may purchase a reasonable number of shoes each year for their own use, including shoes to be worn off the job.

Eye Protection. The company designates, with conspicuous signs or verbal instructions of supervisors, all areas where safety goggles or other suitable eye protection must be worn. When an employee must wear a corrected eyeglass lens to perform his work, the company assumes full cost (or x per cent of the full cost) of safety glasses made from prescriptions furnished by a qualified eye specialist. The company may, however, provide spectacle shields to be worn over an employee's own prescription glasses. When prescription safety glasses are broken on or off the job, they will be replaced by the company (without charge or at x per cent of cost).

Protective Clothing. The company provides protective clothing without charge where the supervisor determines that such clothing should be worn.

Disciplinary Action. When the company requires that protective equipment be worn, employees who do not do so will be subject to disciplinary action.

Equipment Inspection Program

There should be a formal, recorded safety-inspection program for such equipment as hand tools, elevators, hoists, ropes, and cables. Too often the man using the equipment simply **looks at it** before he uses it. But visual inspections by themselves are not sufficient. Nothing should be left to the casual "look."

With the use of tool cribs, warehouses, and storerooms, there is ample opportunity to have **formal** checks on equipment. When tools are "tagged" and the tag carries the initials of the person saying that the equipment is safe, you can be confident that it is safe. Most people will double-check themselves before signing that the equipment is safe.

Unsafe ropes and cables have caused many injuries, but they have also caused many "near-misses," which do not cause injuries but which do result in considerable property damage.   This damage is rarely, if ever, recorded, but it contributes greatly to the economic waste factor.

Rope and cable manufacturers are most cooperative in helping to develop safety checks for their products.   The following suggestions for the care and inspection of fiber rope, wire rope, and chain hoists are condensed from the National Safety Council's *Accident Prevention Manual:*

Fiber Rope

Inspection.   New fiber rope should be inspected thoroughly before being placed in service.   Rope in service should, under ordinary circumstances, be inspected at least every 30 days, oftener if it is used to support scaffolding on which men work. If it is exposed to acids or caustics, it should be inspected daily.

Inspection should mean examination of the entire length of the rope for wear, abrasions, broken or cut fibers, displacement of yarns or strands, variation in size or roundness of strands, discoloration, and rotting.   To inspect the inner fibers for breakage from overload, the rope should be untwisted in several places to see if the inner yarns are bright, clear, and unspotted.

A rope that has lost its feel of pliability or stretch or in which the fibers have lost their luster and appear dry and brittle should be replaced.

Care of Ropes.   Ropes should not be kinked.   Ropes should be cleaned before being stored by washing with a moderate hose stream.   A wet rope should be hung up or laid in a loose coil in a dry place until it is thoroughly dry.   A wet rope should never be left where it may freeze.

Rope should not be stored or used where it will be exposed to acids or acid fumes. Strong alkali, drying oil, and paint are also injurious.   Rope should not be dragged on the ground or against rough or sharp objects.   Sharp bends over an unyielding surface cause extreme tension on the fibers.

Wire Rope

Causes of Deterioration.   Wire-rope deterioration varies considerably with the conditions of service.   Causes of wire-rope deterioration include wear, corrosion, kinks, fatigue, drying out of lubrication, overloading and overwinding, and mechanical abuse.

The safety and efficiency of hoisting-rope installations can be greatly increased by the use of sheaves and drums of suitable size and design, proper lubrication, and good maintenance of the rope and hoisting equipment.

Lubrication.   Regular application of a suitable lubricant to wire rope designed for hoisting prevents corrosion, wear from friction, and drying out of the core.

Periodic cleaning eliminates dirt, abrasive particles, and corrosion-producing moisture.   Cleaning fluids should not be used because of their detrimental effect on the core lubricant.   A compressed-air or steam jet or other mechanical method cleans a rope effectively and thoroughly.   Hand methods use a cloth and a brush.

Inspection and Replacement.   The frequency of inspections and replacement of a wire rope depends on service conditions.   At regular intervals, a specially trained inspector should examine the ropes on which human life depends.   The inspector checks especially for wear of the crown wires, broken wires, kinking, high strands, loose wires, nicking, lubrication, and changes in the cross section of rope at various locations.   In some cases, elongation is measured as an indication of fatigue.

Since there is no accurate method of determining the remaining strength of a rope in service, experience and judgment of all factors, combined with the length of time in service and the tonnage hoisted or other unit for judging the work done by the rope, determine when it should be discarded.

Chains

Inspection.   Most chain failures can be detected in advance if the proper inspection procedure is followed.

A good inspection plan provides for daily inspection by the personnel using the chain to detect those links which are visibly unsafe. At least twice a year or oftener if needed, a thorough inspection should be made by a trained man.

Wear. The maximum wear which can be permitted depends somewhat on the conditions under which the chain is used. It is important that all the links be calipered because, under certain conditions of usage, the links may not wear uniformly. Table 9-3 gives factors for the reduction of working load limits of chains due to wear.[2]

Table 9-3. Correction Table for Reduction of Working Load Limits of Chain Due to Wear[2]

Nominal size of original chain, in.		Reduce working-load limits of chain by the following per cent or remove from service when diameter of stock at worn section is as follows		
Fraction	Decimal	5%	10%	Remove
1/4	0.250	0.244	0.237	0.233
3/8	0.375	0.366	0.356	0.335
1/2	0.500	0.487	0.474	0.448
5/8	0.625	0.609	0.593	0.559
3/4	0.750	0.731	0.711	0.671
7/8	0.875	0.853	0.830	0.783
1	1.000	0.975	0.949	0.895
1 1/8	1.125	1.100	1.070	1.010
1 1/4	1.250	1.220	1.190	1.120
1 3/8	1.375	1.340	1.310	1.230
1 1/2	1.500	1.460	1.430	1.340
1 5/8	1.625	1.590	1.540	1.450
1 3/4	1.750	1.710	1.660	1.570
1 7/8	1.875	1.830	1.780	1.680
2	2.000	1.950	1.900	1.790

Elongation. Elongation of chain, usually caused by overloading, is the warning that a fracture is likely to occur.

Safe Practices. Following inspection, the chain should be proof-tested, if possible, to the proof-test load recommended by the manufacturer for the chain under test. After testing, a link-by-link inspection for damage done in testing, if any, should follow.

Unless full and adequate facilities for repair are available, chain showing faults by inspection should be returned to the manufacturer for reconditioning.

Alloy-steel chains and heat-treated carbon-steel sling chains should never be annealed or normalized because these processes will reduce the hardness.

Wrought-iron crane chain should be annealed periodically. Annealing or normalizing should be done only by the chain manufacturer or by persons who have received training in the methods recommended by the manufacturer and who have the necessary equipment.

In the use of chains, a number of recognized safe practices will do much to prevent failures. Some of these are:

1. Never splice a chain by inserting a bolt between two links.
2. Never put a strain on a kinked chain.
3. Do not use a hammer to force a hook over a chain link.
4. Tag each chain to indicate the maximum load which it can be used to lift.
5. Remember that increasing the angle between the legs of a chain sling and the vertical increases the strain on the sling.
6. Use chain attachments of the same material as the chain.
7. Use hooks of forged steel.

8. See that the load is always properly set in the bowl of the hook.
9. Where practicable, use hooks with safety latches and handles.

Hand Tools

The American Petroleum Institute's Central Committee on Accident Prevention and its Central Committee on Fire Protection have reviewed research data on the ability of hand-tool sparks to ignite hydrocarbons, as well as fire records of its member companies. The conclusion reached was that sparks from hand tools made of either steel or nonferrous metal have not contributed significantly as a source of ignition for accidental fires.

Mail One Copy To:

Industrial Dept.
National Safety Council
425 N. Michigan Ave.
Chicago 11, Ill.

QUARTERLY SUMMARY
OFF-THE-JOB INJURIES

COMPANY NAME AND ADDRESS _____

QUARTER ENDING _____ YEAR _____

AVERAGE NUMBER OF EMPLOYEES DURING QUARTER _____

INJURIES (Indicate Fatalities by Superscripts)

     1. Transportation . . . .   _____

     2. Home . . . .   _____

     3. Public . . . .   _____

     4. Total . . . .   _____

TOTAL MAN-DAYS LOST . . . .   _____

DESCRIBE FATALITIES BRIEFLY: _____

_____

_____

REMARKS:

Date _____ Name of Individual Reporting _____

Fig. 9-2a. Report form for off-the-job injuries. (*Courtesy of National Safety Council.*)

"Therefore, the Institute's position is that the use of special nonferrous hand tools, sometimes referred to as 'nonsparking,' is not warranted as a fire-prevention measure applicable to oil and gas operations."*

Off-the-job Safety Program

There is commonly at least five and often twenty times as much absenteeism from off-the-job injuries as from on-the-job injuries.   Many companies pay benefits equivalent to full wages for varying periods of time for absenteeism due to off-the-job injuries. The expense can run considerably higher than for injuries on the job.

It must not be forgotten that a considerable amount of such absenteeism occurs among supervisory and executive groups, where wages are not only high but where it is difficult to make temporary replacements.

A safety program aimed at off-the-job injuries will pay extra dividends because an employee will be more likely to think of safety on an "around-the-clock" basis.   This will, in turn, make him more conscious of safety on the job.

Figure 9-2a illustrates a reporting form for off-the-job injuries, and Fig. 9-2b gives instructions for determining frequency rates for such injuries.

* Statement approved for publication by the API Committee on Safety of the Board of Directors, Feb. 3, 1956.

Off-the-job injuries.   All injuries, exclusive of those on the job while in the employ of the company reporting, which result in employee's (1) loss of one or more full working days or (2) inability to work on one or more nonscheduled work days.   No medical opinion is needed.   Fatal injuries are to be included.

1. *Transportation:* injuries caused by or resulting from automobile, truck, bicycle, bus, streetcar, motorcycle, railroad, boat, pedestrian activity involving traffic, etc.

2. *Home:* injuries, in the home or home yard area, caused by firearm, machinery, tools, fire, explosion, exposure (heat or cold), electricity, toxic material, fall, slip, improper lifting, hot object or material, sharp object, object striking, overexertion, animal, insect, etc.

3. *Public:* injuries, other than transportation and home, caused by firearm, fire, explosion, fall, slip, object striking, animal, insect, fight, assault, exposure (heat or cold), sports, toxic material (poison ivy, for instance), electricity, etc.

Man-days Lost.   Total calendar days lost from work plus all intervening holidays, vacations, and other nonscheduled time during which employee is unable to work because of injury to himself.   Injury may have occurred in previous quarter, but count days lost in current quarter if employee has not returned to work.   Do not include scheduled charges for permanent partial disabilities or fatalities, but use only actual days lost.   There would be no day lost in a fatality if accident and death occurred on same day.   Otherwise include only the days lost up to time of death.

Fatalities.   If employee dies from injury previously reported, indicate type of accident or activity and date of original occurrence in "Remarks."

Frequency.   Use 312 hr per employee for exposure hours during a month.   This represents "active" off-the-job time: 8 hr/day for 5 days a week, plus 16 hr/day for 2 days of rest per week; the sum multiplied by 4⅓ weeks per month.

Example: A plant working 10,000 employees experiences five off-the-job injuries—each causing at least a 1-day loss of time from work during March.   The off-the-job injury frequency rate for the month will be

$$\frac{5 \times 1,000,000}{312 \times 10,000} = 1.60$$

The same plant with one on-the-job injury during the month would have an occupational frequency rate of

$$\frac{1 \times 1,000,000}{1,730,000} = 0.58$$

This plant would have an off-the-job frequency rate about three times greater than its on-the-job frequency rate.

FIG. 9-2b. Instructions for filling out form shown in Fig. 9-2a.

National Safety Council suggestions for development of an off-the-job program, abstracted from its *Accident Prevention Manual*,[1] include:

1. Encourage planning and support of off-the-job safety by employees. Help publicize their efforts.
2. Allow discussion time for off-the-job safety in special meetings or as part of regular safety work.
3. Collect and analyze off-the-job accident data.
4. Allocate a reasonable amount in the budget to promote and publicize off-the-job safety for employees.

Employee interest in off-the-job safety programs can be stimulated by many means, including:

1. An off-the-job safety calendar, with a promotional "theme" for the month publicized in plant publications, discussed in safety meetings, and emphasized in special, seasonal publications
2. Planned use of films, special booklets, posters, home-inspection check lists, and similar materials available through insurance companies, local councils, the NSC, etc.
3. Prize contests based on departmental off-the-job records for the year, results of home inspection (self-imposed), safety essays, poster design by children of employees, feature articles, etc.
4. Publicity and awards for the best home, traffic, vacation, and other off-the-job safety suggestions
5. Letters from department heads to employees to emphasize seasonal and holiday hazards or sent just before the employee's vacation, etc.

Some employers have given strong support to community safety efforts made by local police and fire departments, service clubs, schools, women's organizations, and civic groups to extend their own safety education to their employees' leisure hours.

In addition, the National Safety Council has prepared a packet of safety materials to help promote an off-the-job safety program.

Employee "Hazard" Education

Keeping personnel at a high degree of enthusiasm and interest, while working efficiently, is the greatest problem of supervision. To meet this problem, there is no better aid to management than the development of safe practices and then directing personnel to adhere to them closely.

All personnel can be asked by their supervisors, preferably on an individual basis, to suggest ideas which will bring about safer working practices. Work procedures can always be made safer.

Not too many people will work hard to develop ideas which will bring efficiency to their jobs for fear of reducing the number of jobs, but they will enthusiastically do so to make their jobs safer.

Work in refineries is not generally of the repetitive assembly-line type. A man may repair the same pump several times or the same type of pump, but the conditions are not always the same, with either the pump or the surrounding area. This gives the supervisor many opportunities to discuss possible work hazards with his men. He can give them his safety ideas and ask for theirs each time he lays out a job.

Injury Investigations

In considering that most injuries are the result of failures in job planning or in failure to follow well-developed plans, a prompt investigation by a supervisor as to the cause of an injury is the best way to determine whether or not the job was planned properly, if the employee followed the plan, if he understood the plan, and if he was well trained. The minor injury or fire can also be used as an excellent medium for spotting impending trouble.

An investigation form or check list is shown in Fig. 9–1 (page 9–3).

Personnel Safety Ratings

One of the best-known firms in the United States places "safety" as the number one item on its performance-review check lists.

Supervisors who did not direct their personnel safely are not considered for further advancement, and if their unsatisfactory safety record continues, they are relieved of supervisory responsibilities.

Hazardous Activity Work Approvals

Many serious refinery accidents and fires have been caused by misunderstood **verbal** approvals and by opening or entering process equipment without a prior formal check with operating people to start work **at a given time.**

Formal **written** approvals are now required in some plants before maintenance can start on equipment which is in operation. Verbal approvals are not acceptable because operators may forget to pass the word along or note it in the log.

A signed approval program is no guarantee that work will not start without proper approval, because people can still forget. It does, however, provide another check point.

Work Approval Check List

Work permit forms should be numbered serially and made up in pads with at least two copies for each number. Some items to consider in preparing a work approval check list are:

1. Spaces should be allowed and designated for the following:

   a. Name of employee to make the repair
   b. Leader of the group if more than one repair man
   c. Shop in which the repair man is employed
   d. Name of the unit where the work is to be done
   e. Item of equipment on the unit where the work is to be done
   f. Type of repair work to be done
   g. Time and date when the work is to start

2. A limit should be stated for the work permit, that is, the hour and date. A work permit for a unit in operation should not be continued from one day or shift to another.

3. If pumps, lines, valves, or other fittings are to be repaired, the operating department personnel approving the work permit should write on the form pertinent safety factors, such as type and temperature of product handled, normal pressure or vacuum on line, and any possibility of toxic vapors.

4. A space should be provided on the work-permit form for the operating personnel to write any unusual conditions surrounding the work area and to designate need for protective clothing.

5. Spaces should be provided to show the time the work is completed and for signatures by the maintenance leader who is to perform the work and the stillman or chief operator in charge of the unit.

6. One copy of the work-permit form should be given to the maintenance man; a second copy is to be kept by the operator and posted on the control-room bulletin board.

7. When the work is completed, the maintenance department copy should be filed for a designated period of time in the maintenance department files. The operating department form should be filed at the operating unit. It probably would not be necessary to file these copies for more than a week.

8. Work-permit form pads should be kept in the control room of every operating unit.

9. Maintenance men should be forbidden to proceed with any work on operating units while they are operating unless the maintenance man in charge has a copy of a

signed work permit on his person or unless the operating department supervisor's written authorization is granted.

10. A space should be provided for the operator to advise as to the need for fire-equipment readiness.

When necessary to perform "hot" work or when men are required to enter tanks or other vessels, the following items are recommended for work permit forms:

1. Each permit is to be numbered with at least two copies: one copy for the employee who is to perform the "hot" work or enter a vessel, a second copy retained by the operator where the vessel is situated.

2. Provide spaces for the following:

   a. Tank or vessel number
   b. Type of work to be performed
   c. Time when the work is to be started
   d. Time and date of permit

3. The permit form is to be made out originally by the operating department representative in charge. Initials should be provided opposite each of the following statements (conditions not necessary should be crossed out but should also be initialed):

   a. Equipment has been gas freed.
   b. All lines leading to the vessel have been blinded.
   c. All sewer or drain openings in the area are covered.
   d. Fire equipment is ready on a standby basis.
   e. Life lines and belts are provided.
   f. Fresh air mask is to be worn.
   g. Area has been checked for H_2S, combustible vapors, CO, hydrogen, and arsine.
   h. Radioactive materials have been removed or shielded.

4. Space should be provided to show date and time when work is completed and for both maintenance department and operating department representatives to sign prior to and following the work.

5. The person conducting the tests for hydrocarbon gases, H_2S, CO, hydrogen, or arsine should sign the form.

Line Blinding

During unit turnarounds on all types of petroleum processing equipment, decisions are made far in advance as to just exactly which lines or vessels are to be sealed off. Lists are made designating the location. The person who installs a blind or plug initials the item on the list to certify to the fact that he did it, as does the person who later removes it. When units are ready to go back on-stream, it is easy to check all forms to see whether or not installed blanks have been removed.

Figure 9-3 shows a suggested blinding form quoted from correspondence in connection with preparation of the API manual *Safe Maintenance Practices in Refineries*.[3]

Starting Furnace Fires

Serious explosions occur because well-known fundamentals are not adhered to in lighting refinery heaters and furnaces. It is usually a question of someone failing to do something which he knows perfectly well he should do.

A check list will act as a reminder for a man lighting unit heaters. A separate check list should be developed for each unit and posted at a conspicuous location near the heater. A check-list pad, to be carried by the person lighting the fire, will also help him to remember.

The following items should be kept in mind when lighting instructions are prepared for each heater:

1. List all dampers which are to be opened.
2. Check fuel supply and pressure.

Battery Combination Unit Blinding Record

To be made out in triplicate
Third copy to maintenance headquarters after blinds installed
Second copy to maintenance headquarters after blinds removed
Original to stay at unit

Date _____
Hour to start shutting down _____

Sheet No. 1 of 9 sheets

No.	Description	Location	Size, in.	Approximate hour blind can be installed	Foreman to initial blinds to be installed	Installed by			Removed by		
						Date	Operator	Mechanic	Date	Operator	Mechanic
1	Gas to unit at main	Northside atmospheric furnace	6	7							
2	Gas to atmospheric furnace	Northside atmospheric furnace	6	7							
3	Pilot gas roadside atmospheric furnace	Overhead above burner	2	7							
4	Pilot gas tower side atmospheric furnace	Pipe trench north furnace	2	7							
5	Gas to vacuum furnace	On walk between vacuum and railroad furnace	3	7							
6	Pilot gas to vacuum furnace	Overhead roadside vacuum furnace	2	7							
7	Waste gas to recovery plant	Upper platform water separator and gasoline stripper	6	7							
8	Waste gas to recovery plant	Alongside C-8 kerosine cooler	6	7							
9	Kontol disconnect and plug union at discharge of pump		½	12							
10	Kontol (carrying agent) disconnect and plug union at discharge Kontol pump	-	¾	12							
11	Ammonia disconnect and plug union downstream of orifice		½	3							
12	Gasoline feedback	Suction of J-34 pump	3	3							
13	Lube feedback	Vacuum pump room near J-26 pump	2	3							
14	Caustic to J-1 disconnect and plug	Atmospheric pump room	1	3							
15	Caustic to J-1A disconnect and plug	Atmospheric pump room	1	3							

Fig. 9-3. Suggested blinding form to be used to record installation and removal of line blinds.[3]

3. Drain the condensate from the gas fuel lines.
4. Purge the furnace thoroughly for a designated time.
5. The operator should stand aside to avoid flash back.
6. The operator should wear a full shirt, sleeves rolled down.
7. The operator should wear a face shield.
8. A special torch should be built and used to light heaters.
9. Kerosine or diesel oil—not gasoline or light naphtha—should be used for a torch.
10. Light the torch and place it at the burner before the gas is turned on.
11. Should fires go out while lighting, turn the gas off and purge the furnace thoroughly before attempting to relight burners.

Safety and Equipment Design

Location of New Equipment

An engineering design group may take every possible precaution to provide injury and fire protection for a given unit, but what about the relationship of the new unit

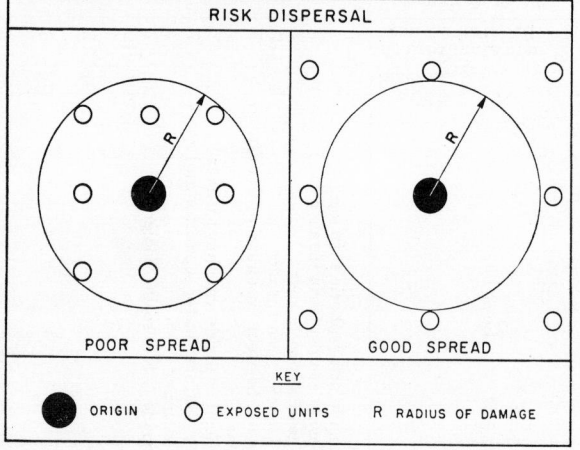

FIG. 9-4. Principle of radius of destruction as applied to dispersal of risk.[4]

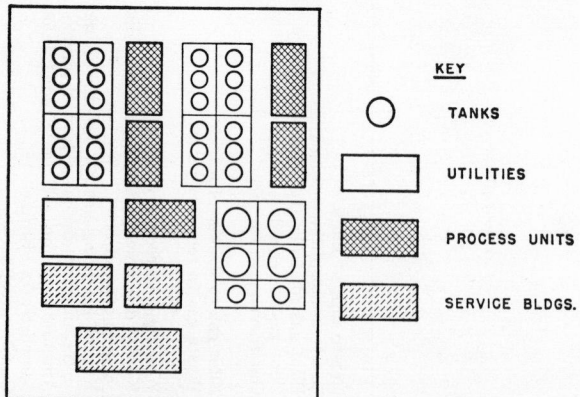

FIG. 9-5. Poor spacing in plant layout. Tanks are located in close proximity to refining units, which in turn are located close to one another.[4]

to other equipment already in place? Potentially dangerous equipment sometimes is placed directly next to a ready source of fire or a heavy concentration of personnel. The same condition may also apply to a pressure vessel or a particular type of storage tank.

Figure 9-4 illustrates the principle of radius of destruction as applied to dispersal of equipment. In the good spread only the unit of origin will suffer damage, whereas with the poor spread all the exposed units will be damaged. Figure 9-5 shows poor equipment spacing, while Fig. 9-6 shows good equipment spacing.[4]

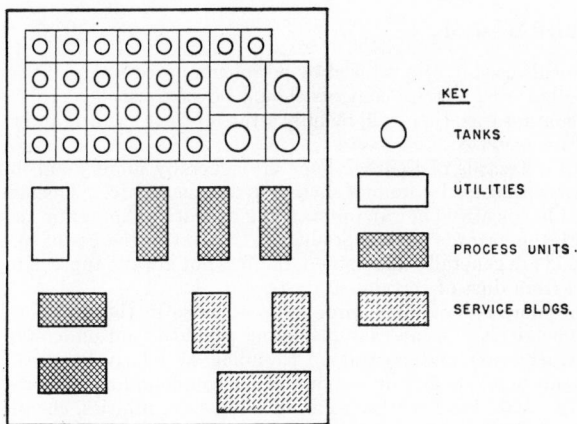

FIG. 9-6. Good spacing in plant layout. The various elements of the plant are well separated.[4]

It is urged that the paper, *Occurrence and Nature of Hydrocarbon Detonations*, by R. B. Jacobs, Standard Oil Co. (Indiana), presented at the 39th Annual Meeting of the American Petroleum Institute, be read.

Consideration should also be given to flame propogation through pipe, electrical conduit, and sewers. The U.S. Bureau of Mines has an outstanding lecture and demonstration on this subject.[5]

Safety Guards

Various industry codes and state regulations now define basic safety guard protection required. Though design engineers follow such rules to the letter, operating unit designs and conditions change rapidly.

Design engineers without field operating experience are sometimes not aware of operating hazards. Consequently, a helpful step at the design stage is to have an operating man and a safety man check the design engineers at various stages, bringing their attention to practical safety problems. To be effective, these men should not be under the jurisdiction of the engineering department. Bringing in operating and safety people, especially from the area where the equipment is to be used, lends an up-to-the-minute approach. Many of the safety aids are major items, such as location of motors, pumps, exchangers, or relief valves. Repair and maintenance men should be able to get at them easily. Differences of opinion should be settled by top operating line management.

Questions may arise concerning the mechanical guarding of machinery. Much helpful information has been published on this subject. Two recommended sources are "The Principals and Techniques of Mechanical Guarding," *U.S. Department of Labor Bulletin* 197, and *Accident Prevention Manual for Industrial Operations*, 4th ed. Chaps. 23 to 27, National Safety Council.

Fire Protection

Precautions concerning the prevention of injuries at the design stage apply also to fire protection. This matter is frequently left for consideration after the new equipment is in the process of being installed, but this is too late for some of the major steps which may be required for the best protection. A study and consultation at the design stage with the people who are going to have to fight any fires makes good sense.

New Product Safety Studies

Toxic and Fire Hazards

Many chemicals used in or produced by petroleum processing plants are life hazards as well as fire hazards. Strict controls should be established so that everyone who may come in contact with them will be fully aware of their hazards and the precautions which should be exercised.

Toxic and fire hazards of hydrocarbons are generally understood, but such is not always the case with some byproduct such as hydrogen sulfide. The manufacturers of tetraethyl lead have carried out an outstanding educational program to acquaint their customers with the hazards of that product. Dangers in the use of hydrofluoric acid and sulfuric acid are generally understood, but it is not always appreciated that trouble can be caused from dust of certain catalysts.

Some of the more serious toxic problems can arise in the use of relatively small quantities of chemicals. As an example, many people do not understand the hazards of carbon tetrachloride and mercury. Handling of relatively small quantities of alkalies and acids in laboratory operations, in the compounding of lubricants, or in the manufacture of grease has been the cause of serious eye injuries, chemical burns, and sickness from inhalation of toxic vapors.

In the use of new chemicals it is advisable to check possible use hazards with *Chemical Safety Data Sheets*, available from the Manufacturing Chemists Association, 1825 Connecticut Ave., N.W., Washington, D.C. 20009. Further checking can be done in determining the "maximum allowable concentration" of a particular chemical by reviewing reports of the American Conference of Governmental Hygienist, available from the Superintendent of Documents, Washington, D.C.

Publication 49M of the National Fire Protection Association gives data on hazardous chemicals. The fire as well as life hazards of these materials are shown, along with recommended storage procedures and fire-fighting phases. (This publication is also included in Vol. II of NFPA's *National Fire Codes*.)

Equipment can be purchased readily to determine if hydrocarbon atmospheres are in the explosive range and if hazardous atmospheres exist.

Excess Noise

Noise is now recognized as a cause of impaired hearing as well as lowered work efficiency. Instruments can be secured to determine the intensity level of sound.

Noise problems in petroleum processing plants are frequently found around furnaces, can-filling operations, compressors, turbines, repair work, and tube cleaning operations. Exhausts are also the source of noise problems. A noise problem can be best eliminated or controlled at the design stage. Acoustical consultants can be called upon to help analyze noise problems.

Regular audiometric examinations by plant medical personnel are advisable for those persons subject to noise.

Overheating

Heat can be enervating; air conditioning for shops and offices is proving to be an economical step as well as one which provides greater comfort. Great strides are being made in the comfort of personnel exposed to heat in the repairing of furnaces, towers, tanks, or other equipment in confined spaces by the use of blowers and cool air. Plans for repair work should include consideration of properly conditioned air supply for personnel working in confined areas.

Physical Examinations

In all toxic problems, including those where there is a question with new processing units or products, plant doctors and nurses can be of inestimable value by early detection of toxic symptoms during regular physical examinations or during first-aid treatment.

Radioactive Hazards

Radioactive material purchases should be cleared in advance with safety department people. The plant purchasing department should not purchase any chemicals or radioactive material unless the order form is okayed by the plant safety director, laboratory head, or some other person designated by the plant manager.

A program for guarding against injury from radioactive materials is described on page 9–55.

Safety Checks during Operations or Repair

Research people will exercise their normal safety checks in the development of a process, but a neutral group of operating and safety personnel should also study the process to see where hazards can exist during the course of its operation or during periods when it is being repaired. The group would look especially for toxic gases as well as regular hydrocarbon gases and the presence of hydrogen, CO, H_2S, arsine, and so on.

Inerting Gas Atmospheres

Some companies maintain "inert gas baths" in the vapor spaces of closed containers (such as storage tanks and tanker and barge compartments) during periods when there is vapor space in such containers in order to prevent possible ignition of petroleum vapors which are in the explosive range with ambient temperatures.

Figure 9-7 shows the percentages of CO_2, exhaust gas, or nitrogen required to provide an inert atmosphere for various percentages of gasoline vapors.

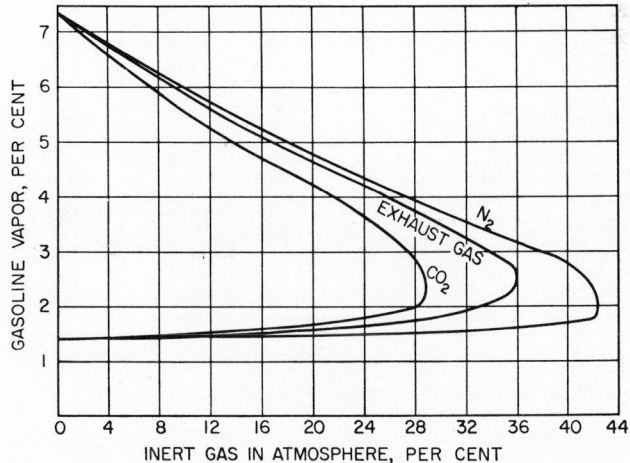

Fig. 9-7. Limits of flammability of gasoline vapor in various air–inert-gas atmospheres.[6]

Table 9-4. Flash Points and Limits of Flammability in Air for Common
Hydrocarbons and Petrochemicals[7]

Material	Flash point, °F	Flammable limit in air, vol %	
		Lower	Upper
Acetaldehyde	−36	4.1	55
Acetic acid (glacial)	109	5.4	16.0*
Acetic anhydride	129	2.7	10
Acetone	0	2.6	12.8
		2.40*	
Acetylene	Gas	2.5	81†
Ammonia (anhydrous)	Gas	16‡	25‡
Benzene	12	1.4*	7.1
Butadiene	Gas	2.0	11.5
n-Butane	Gas	1.9	8.5
iso-Butane	Gas	1.8	8.4
Butene-1	Gas	1.6	9.3
Butene-2 (trans)	Gas	1.8	9.7
Butyl alcohol	84	1.4	11.2
Carbon disulfide	−22	1.3	44
Carbon monoxide	Gas	12.5	74
Cumene	111		
Cyclohexane	−4	1.3	8
Diethanolamine	305		
Ethane	Gas	3.0	12.5
Ethyl alcohol	55	4.3	19
Ethyl benzene	59	1.0	
Ethyl chloride	−58	3.8	15.4
Ethylene	Gas	3.1	32
Ethylene dichloride	56	6.2	16
Ethylene glycol	232	3.2	
Ethylene oxide	<0	3	100
n-Heptane	25	1.2	6.7
iso-Heptane (mixture)	<0	1.0	6.0
n-Hexane	−7	1.2	7.5
iso-Hexane (mixture)	<−20	1.0	7.0
Hydrogen	Gas	4.0	75
Hydrogen sulfide	Gas	4.3	45
Isoprene	−65		
Methane	Gas	5.3	14.0
Methyl alcohol	52	7.3	36
Methyl chloride	Gas	10.7	17.4
Methyl ethyl ketone	21	1.8	10
Naphthalene	174	0.9	5.9
n-Pentane	<−40	1.5	7.8
iso-Pentane	<−60	1.4	7.6
Phenol	175		
Propane	Gas	2.2	9.5
n-Propyl alcohol	59	2.1	13.5
iso-Propyl alcohol	53	2.0	12
Propylene	Gas	2.4	10.3
Propylene glycol	210	2.6	12.5
Styrene	90	1.1	6.1
Toluene	40	1.4	6.7
Triethanolamine	355		
o-Xylene	63	1.0	6.0
m-Xylene	77	1.1	7.0
p-Xylene	77	1.1	7.0

* At 212°F.
† 100 per cent at <5.8 psig, normal atmospheric temperature, gas dry; 100 per cent at
<16.1 psig and 157°F, gas saturated with water vapor.
‡ Nonflammable in air except in comparatively high concentrations.

Table 9-4 shows the flash point and the lower and upper limits of flammability in air for certain common hydrocarbons. Such information will assist in establishing too lean, too rich, or inert atmospheres in the handling of these products.

SAFETY ORGANIZATION IN A PLANT

Safety Director

The person responsible for safety in a plant should have good working knowledge of refinery operating units as well as of general hazards in connection with handling hydrocarbons and general plant maintenance work problems. It is an excellent spot to be in to become well acquainted with all activities and personnel of a plant.

Duties of Safety Director

1. Establish procedures for securing the following data concerning plant injuries and fires:

   a. Causes of accidents and fires, minor and major
   b. Statistics showing disabling injury frequency and severity rates for plant on a monthly basis

2. Recommend management steps which can be taken to prevent recurrences of accidents and fires.
3. Prepare a safety indoctrination program for new employees.
4. Recommend to plant manager safety topics for use at regular meetings with top management.
5. Give supervisors suggestions of topics to discuss at safety meetings with employee groups.
6. Develop for approval by plant management **formal** permit procedures and forms for repetitive hazardous activities.
7. Suggest safety contests among departments.
8. Study available safety services such as booklets, posters, films, and so on, and recommend which material to use to the plant manager.
9. Suggest safe procedures for unit turnarounds.
10. Recommend a disaster-control program.
11. Organize first-aid training courses for the entire plant personnel.
12. Suggest protective clothing **and** equipment where indicated.
13. Work with safety people in other plants and companies to learn of new safety approaches.
14. Establish a procedure with the purchasing and storeroom departments for use when hazardous chemicals and radioactive materials are purchased and distributed.
15. Keep advised as to toxicity problems.
16. Suggest the style and amount of fire-fighting equipment, and suggest automatic test and maintenance procedures.
17. When changes are made in equipment, determine from engineering and operating personnel any conditions wherein hazards to personnel may exist or where fires or explosions may be caused. As indicated, recommend emergency escapes for personnel, fire-fighting equipment, and safe operating procedures.
18. Develop training programs for plant employees to teach the best methods of fighting fires. Hold drills regularly.
19. For new equipment, see that plans are made in advance to guard employees from moving parts.
20. When new equipment is planned, work with design and engineering department personnel regarding the safest way to operate and repair such equipment.
21. Work with other petroleum processing plants in the area to establish mutual-aid programs. Work closely with the municipal fire chief.
22. Keep advised as to state and local requirements regarding injury and fire prevention.

Position of Safety in the Organization

Safety is most effectively directed when the staff person in charge of the activity reports directly to the plant manager or general refinery superintendent.

The position of the safety man should be one of recommending or suggesting rather than "doing." The "doing" has to be done by top management and line supervisors.

Though safety is a staff responsibility, it loses its stature when it reports to staff people because it is thereby insulated from top management. This is especially true when safety people have to report to staff people who have not had line plant supervisory experience.

Training might be considered an activity associated with safety, but most training people today are experts in the *techniques* of training, not in the subject matter itself.

The safety activity has been placed under the employee relations activity in some operations, but there is little in common between the safety activity and the many responsibilities now falling under the heading of employee relations. The employee relations manager generally represents the management at the collective bargaining table, and it does not help the safety effort in the minds of the employees to have it directed by the same person who bargains with their representatives.

Accident and Fire Protection

It is difficult to separate the accident- and fire-protection responsibilities. If a fire or explosion occurs and people are hurt, who has the staff responsibility—the fire-protection man or the accident-prevention man? There cannot help but be conflict where these activities are under two different heads.

The engineering department should have responsibility for such matters as establishing final specifications for tank diking, venting, and fixed fire-fighting equipment, and it should consider safety in the selection of each individual item of equipment: valve, tank, pressure vessel, fitting, and so on. The safety man, however, should have the responsibility of recommending to the engineering and design departments any new ideas which he may have learned as a result of working with safety people in other plants or as a result of experiences in his own particular plant. If the safety man does not believe that the engineering department is sound in its accident- or fire-protection specifications, he should be able to refer any differences to the plant manager for settlement.

Safety Committees

Safety committees can be of great help, but it should be kept in mind that safety cannot be turned over to a committee. The final responsibility for safe operation of a plant and the authority to do something about it rest with the plant manager and the various supervisors. A plant manager or supervisor should use a committee to assist him in making his decisions, but the decisions and actions should be his and **not** the committee's.

A safety committee is usually composed of the top supervisory group, with the safety man acting as secretary. The manager runs the meetings, and the subject matter of such meetings is generally concerned with plant safety policy matters rather than safety details in each department. Such meetings do provide an opportunity, however, to review general safety procedures which are being carried out in each department.

The plant manager sets the pace for the safety activities throughout the entire plant. The degree of sincere, intense interest on the part of the plant manager tells the story as to whether the accident- and fire-protection record in the plant is excellent, mediocre, or bad.

Individual department supervisors often select three of four men on a "roving" basis to serve as a department safety committee. Each man may serve for 6 months with a new man coming on each month. This method of rotation keeps continuity of thinking, and over a period of time all men in the department become members of the committee.   Just as in the case of a plant manager's committee, a supervisor should use a committee to assist him, but the decisions and actions should be his.   A half hour or hour meeting once a month with a well-planned agenda is generally sufficient.

Employee Safety Meetings

If employee meetings are to be held, it is recommended that they be held on a department or small-group basis rather than bringing all available men in a plant to one big meeting.   The latter step can be followed, of course, for a motion picture which would be of common interest to everyone.

In any one department, employee group meetings are most effective when a supervisor, assistant supervisor, or lead man, depending upon the number in each group, brings his people together for a matter of 10 or 15 min.   Sometimes 5 min is sufficient. The supervisor should have a definite message thought out in advance.

During big maintenance jobs, some supervisors bring their men together each morning before they start on the job and talk to them for perhaps 5 min, citing some of the "near misses" of the day before.   These meetings are very effective.

Safety Meeting Subject Material

Subject matter for supervisor's safety meetings should cover not only the general causes of injuries and how to prevent them but reviews of recent plant accidents and preparation for upcoming work.   In a review of recent accidents, major or minor, occurring within the last 30 days, it is sometimes helpful to have an injured employee discuss how his injury occurred and measures that have been taken to prevent recurrences.

Typical safety meeting topics include falls, head injuries, toe injuries, eye injuries, finger injuries, lifting injuries, stumbling and slipping injuries, horseplay, burns, electrical hazards, toxic hazards, explosive hazards of hydrocarbon gases, hazards of removing machinery guards, danger of entering closed vessels, welding-arc rays, loose clothing around moving machinery, and danger of hot oil vapors.

Of course, no safety programs is complete without repeated explanation of plant formal work permits and the reasons for them and discussion of types and causes of off-the-job injuries.

Promoting the Safety Effort

Proper advertising and promotion work well with safety just as they do in any field of endeavor.   There is a wealth of helpful, promotional information available through the National Safety Council: posters, films, safety kits, pamphlets, booklets, banners, talks for supervisors, and complete safety programs.   A copy of the current edition of *Catalogue of Occupational Materials* should be secured from the Council.

Safety contests among departments in a plant as well as among plants in a company help to bring about group interest through competition.   Standards for classifying and measuring work-injury experience are discussed on pages 9–69 to 9–74.

Token awards to employees in winning departments or plants sometimes help. A list of suggested awards can be found in the National Safety Council *Catalogue* mentioned above.   A surprise steak dinner to a winning department in a plant safety contest arouses interest.   There are many ways to create enthusiasm, but the pressure must constantly be kept on the "spring."   As an example, a new safety program can be launched before an old one "dies out."

Another activity which helps any plant safety program is the **Wise Owl Club.** Membership in this group is granted an employee in any plant operation whose sight is saved from injury or loss through the use of spectacles with safety lenses, either with or without correction, or other recognized or approved eye-protection devices such as goggles or face shields.   To enroll, a letter should be addressed to the Wise Owl Club of America, c/o The National Society for the Prevention of Blindness, 16 East 40th St., New York, N.Y. 10016.

First-aid Training

Training all employees in first aid is a help to any plant safety program; it teaches the results of carelessness. First-aid training can be secured from the American Red Cross, the U.S. Bureau of Mines, and (in Canada) the St. John's Ambulance Association. The American Petroleum Institute has published an excellent *First-aid Training Guide*,[8] approved by both its Central Committee on Accident Prevention and its Medical Advisory Committee. The guide provides the fundamentals of training in the six basic emergency measures: artificial respiration, control of bleeding, treatment for shock, care of open wounds and burns, care of dislocations and fractures, and transportation of the injured. This guide also outlines the petroleum industry's program for the safety of the individual employee, as well as the personnel protective equipment available and its function.

Safe-driving Program

Any plant-wide safety program should concern itself with safe driving of motor vehicles. The National Safety Council has developed a safe driving program, termed its "Defensive Driving" formula, and has published the booklet *A Professional Code for Defensive Driving*.[9]

It is recommended that copies of this code be given to all drivers, but it must be remembered that merely passing out safety booklets is not sufficient. Supervisors should see that all drivers are thoroughly versed in the excellent suggestions which appear in the code. The six basic points of defensive driving in the code can be used as the basis for a **formal** driver-training program even before the booklets are distributed. They are:

1. Drive without having a preventable accident.
2. Know and strictly observe all applicable traffic rules and regulations.
3. Be constantly alert for illegal acts and driving errors of other drivers, and be willing to adjust your driving to prevent accidents from such acts or errors.
4. Know the special hazards presented by abnormal, unusual, or changing conditions in the mechanical functioning of your vehicle, type of road surface, weather, amount of light, kind of traffic, your personal health and physical condition, and your state of mind, and intelligently adjust your driving accordingly.
5. Know the rules of right of way, but be willing to yield the right of way whenever necessary to avoid an accident.
6. Adopt an attitude of confidence that one can drive without ever having a preventable accident.

Plant Watchmen and Guards

The information following is abstracted from the National Fire Protection Association's booklet *The Watchman*[10] and can be used as instruction and guidance for plant watchmen or guards charged with responsibility for protecting property.

It is not a matter of luck that many operations, large and small, have enjoyed immunity from serious losses; it is the result of careful planning, which indicates that the plant has trained personnel to protect its property.

Watchmen should be concerned with identification of all persons entering and leaving the plant; fire, theft, and accident prevention; air, water, oil, gas, and steam leaks; power and light; waste; property damage; housekeeping; violation of plant rules; control of traffic; and other matters peculiar to the property being protected.

Selecting the Watchman

Each plant should draw up a set of qualifications and requirements to be used in selecting watchmen personnel. Following is a list of suggested points to be considered in hiring watchmen:

1. Very young or extremely old men should not be employed in such positions. Young men, generally, lack judgment and sense of responsibility; old men are not likely to be alert or to have enough physical strength and endurance.

2. He should be of medium height and weight and in good health and physical condition.

3. He should have a high school education or equivalent.

4. He should be a citizen of the country.

5. He should have no police record and should be of good character.

6. A new watchman should be placed on probation for about 6 months and then pass a written examination before he is given a final appointment.

The watchman should be given an annual physical examination and an annual written examination of pertinent facts and information.

Character, habits, and reliability of the watchman must be unquestioned. He should never smoke while making a round. Good appearance is a most desirable quality. Courtesy and self-control are absolute requirements. Whatever the assignment may be, he should do his work as instructed and protect the employees' and the plant's properties.

Number of Watchmen

Adequate protection requires complete coverage of all plant property. Most watchmen will be on duty for an 8-hr shift and must then be relieved. Enough watchmen should be provided so that each man can cover his assigned route in 40 min, allowing ample time to check valves, windows, doors, lights, etc. It is not good judgment to increase the area covered by a watchman beyond that which he can patrol reasonably in 40 min by cutting down his rest period or by omitting any building or room from his route.

Where two or more watchmen are employed, one should be appointed as supervisor. Each route should be patrolled continuously by the same man, who will be held responsibile for anything that occurs on his assigned route during his shift. However, it is desirable to alternate the men every other night so they are familiar with all parts of the plant.

It is a good practice to adopt a plan whereby the watchmen check with each other or with an attendant on duty on every round so that in case of sickness or accident an investigation can be made and help obtained.

Duties

The watchman should patrol the plant property to protect it from fires, vandalism, or other damage and to prevent or detect pilferage and other irregularities or infractions of plant rules and regulations. He should observe carefully general conditions of buildings and equipment, detecting and reporting any irregularities or defects in plumbing, piping, lighting, power, and other equipment and all apparent or potential fire and accident hazards. He should make certain that fire-fighting equipment, aisles, passageways, valve pits, fire escapes, and fire doors are secured properly and have not been rendered ineffective. He should know how to turn in fire alarms and assist in extinguishing fires.

The watchman should be required to make a daily report, preferably on a printed form. This report should list all items and hazards pertaining to watchman service and also will serve as a constant reminder of important things to look for in making rounds. All dangerous and unsatisfactory conditions should be brought to the attention of responsible parties.

Gate Watchmen

The duties of a gate watchman include the guarding of an assigned entrance to the plant area, opening and closing of gates, observing the movement of employees and visitors entering or leaving, and determining that all are identified properly. When badges or passes are issued for the admission of employees to the premises, the watchman should actually see them and not pass anyone merely because he recognizes him. He should require passes for all bundles and packages being removed from the premises.

Watchmen should be thoroughly acquainted with the entire property and the location of fire-alarm boxes and should be able to direct the municipal fire department and other municipal agencies in emergencies.

Routes

An important part of good property protection is carefully laid-out routes. A plan or other record of the route should be prepared and should be available to the watchman. The route should be so laid out that the watchman is required to pass through the entire area. It should not force him to retrace his steps any more than is absolutely necessary to cover all parts of the property. It should be laid out so as to prevent short cuts by stairways, elevator, or bridges.

Check Points

Some of the important things to check, particularly on the first round, are:

1. Outside doors and gates closed and locked; windows, skylights, fire doors, and fire shutters closed.

2. Oily waste, rags, paint residue, rubbish, etc., removed from buildings or in approved containers.

3. Fire apparatus in place and not obstructed.

4. Aisles clear.

5. Motors or machines turned off.

6. All offices, conference rooms, and smoking areas checked for carelessly discarded smoking materials.

7. All gas and electric heaters, coal and oil stoves, and other heating devices on the premises checked.

8. All hazardous manufacturing processes left in a safe condition. The temperature of driers, annealing furnaces, etc., which continue to operate during the night and on holidays and week ends noted on all rounds.

9. Hazardous materials, such as gasoline, rubber cement, and other flammable and highly volatile combustibles kept in proper containers or removed from buildings.

10. All sprinkler valves open with gages indicating proper pressures. If not open, report immediately.

11. All rooms checked during cold weather to determine if heated properly.

12. All leaking water faucets and air valves closed. If unable to stop leaks, report the condition.

13. Give particular attention to new construction or alterations underway.

The watchman should leave his route only when he is properly relieved, never merely because his time has expired. If the watchman's duties include the starting of equipment such as lighting furnaces or ovens or starting machinery on the last round, he should be instructed thoroughly. Heating equipment, especially if automatically controlled, should be checked.

Recording Systems

A clock record of the watchman's movements should be maintained. Several methods and systems are available for this purpose. Central-station watchman's service is the best supervisory service obtainable. It has been found to be the most reliable and trustworthy, and the largest credits in fire insurance rates are given for this type of service.

Where the plant is large and a number of watchmen are employed, the private proprietary system can be installed. By using this method, the operator located at the control center can supervise the movements of the watchmen on their routes. At operations where only a few men are required, watchmen's portable clocks may be used.

The watchman should study and have a working knowledge of the National Fire Protection Association standards for protective signaling systems[11,12] if they apply.

Training

Instruction and training of the watchman should be the duty of a responsible person, preferably the chief of plant protection or other official in charge of the property. Advantage should be taken of training courses for watchmen which are available under the fire training programs in a number of states. When the watchman is being instructed and trained, all possible emergencies that may confront him should be anticipated.

All watchmen should be furnished a copy of the NFPA *Inspection Manual*[13] and should study and have a working knowledge of it.

FIRE PROTECTION IN REFINERIES

The American Petroleum Institute has published *Fire Protection in Refineries* to "provide a better understanding of fire protection problems and of the steps necessary to insure the safe storing, handling, processing, and shipping of petroleum and petroleum products in refineries." Much of the following, particularly the parts on organization, training, and equipment, is abstracted from that publication.[14]

Organization

Personnel Selection

In selecting personnel for fire fighting, two possibilities are suggested:

1. Regular company employees on shift who can be spared from regular duties for fire fighting, supplemented by employees who reside near the refinery
2. Company employees on shift, supplemented by the company fire department, a municipal or county fire department, or by industrial-disaster or mutual-aid groups

Instead of the assignment of specific individuals within the plant, some refineries select company shift personnel for fire-fighting duties on the basis of job assignment or classifications. Thus the desired number of men are available at all times, regardless of such consideration as changes in shift, days off duty, and vacations. If the fire-fighting personnel includes shift employees only, each member should be trained as to his duties both during normal plant operations and in the event of a fire. Consideration should also be given to avoiding the presence of too many fire fighters, who may not only cause confusion but might expose personnel unnecessarily.

Arrangements with volunteer or other supplemental groups should be definite and specific so that no confusion will result as to who is in charge of the various phases of the fire-fighting activities.

An individual should be designated to head up the fire-fighting organization. In some refineries the fire chief or the supervisor of the fire-protection staff will take charge of fire fighting. In others the fire chief or the staff supervisor may serve as an assistant to a line supervisor. In the case of large fires, fire-fighting duties may include shutting down of equipment and other related emergency activities as well as fighting the fire itself. The division of responsibilities for various coordinated emergency activities should be decided in advance.

In some areas issuance of company identification to key personnel and members of fire-fighting squads has proved to be desirable to facilitate their passing through road blocks when proceeding to the plant during an emergency. Rapid and reliable means of communication, supplemented by mobile portable radio, should be placed at the disposal of the fire-fighting supervisor.

Training for Average Fires

Those charged with handling fire-fighting equipment should be trained periodically. In every case, instructors should be thoroughly familiar with characteristics and limitations of equipment used and should be prepared for various contingencies to assure that there will be a minimum of hazards to employees. This training should be done on drill grounds, using various types and sizes of live fires.

Drill Grounds

Simulated fires at drill grounds should be adequate to provide proper training. Training should not be limited to pit, tank, and other two-dimensional fires. Adequate and realistic props can be constructed from scrap equipment at low cost. These can be used to simulate the more common jet, spray, or other three-dimensional fires usually encountered on the job.

Drill grounds should be built to accommodate fairly large-scale fires, i.e., adequate for the use of the large portable equipment normally employed by assigned fire crews. Men trained on large fires do not become overly confident. They lose their fear of fire but develop a wholesome respect for it. They also learn the limitations of their equipment and, in an emergency, are better prepared to act with intelligence and good judgment and generally conduct themselves in such a way as to avert danger to themselves.

Drill-ground training may be by skilled instructors from the plant fire department or staff, or the skilled instructors may act as teachers for the foremen or others who, in turn, may teach the men who will serve under their leadership when the emergency occurs.

Types of Training

Classroom Instruction. This basic training includes table-top demonstrations, fire triangle, and theory regarding the burning characteristics of various fuels and the types of fire-fighting mediums used.

Training on Drill Grounds. This includes four types of training:

1. Primary training in the use of one-man equipment, such as hand extinguishers, small hose lines, fixed monitors, or deluge sets.

2. Secondary training in the handling of equipment, such as large hose lines, by groups. Included in this training should be a study of fire mains and foam systems.

3. Brigade training in the handling of large foam equipment, towers, and related equipment, which involves team work by various groups.

4. Training of outside help, such as the personnel of municipal or other outside fire departments. An opportunity for such personnel to learn and understand the refinery's fire-fighting problems will ensure proper coordination and maximum benefit from their assistance.

Hypothetical Fire Drills. On-the-job hypothetical fires should relate the drill-ground work to potential emergency problems in operating areas. This can be accomplished if the refinery fire chief and an operating unit supervisor jointly plan a simulated emergency. The alarm is sounded for the hypothetical drill. The fire fighters bring up the appropriate fire-fighting equipment. Meanwhile, the operating personnel simulate the emergency shutdown operations by hanging tags or small fiber gaskets, marked "open" or "closed," on valves, pumps, etc. At the conclusion of the drill the foreman and head operator should go over the unit, checking for accuracy of the steps taken to shut down the equipment. The lessons learned from the drill are then discussed with the operators, and the drill may be repeated to correct errors.

Such drills should be conducted periodically in all sections of the refinery. If feasible, some of these drills may be extended to include turning on spray systems, fixed monitors, use of large hose lays, and sometimes the use of assigned fire crews.

Training for Major Fires

Because large fires are rare, perhaps too little attention has been given to their special problems. Top-level supervisors, fully capable of handling difficult day-to-day operating problems, may fail completely if suddenly confronted with the necessity of making quick decisions for the organization of personnel and equipment to fight a fire of great magnitude.

If supervisors have not been trained adequately in the techniques of fighting even moderate-scale fires, or if they are not familiar with the limitations of their fire-fighting equipment, they may have a false sense of security when they are faced with an emergency and may endanger themselves and their men or they may cause the fire to get out of control.

A breakdown in communications or liaison between operations—particularly if fire fighting must be done in the dark, in smoke-filled areas, or over a large and complex fire-fighting front—may result in overlapping or conflicting supervision.

Advance Planning for Major Fires

In the handling of large, complex fires, fire fighting is only one part of the job. Related emergency activities must be smoothly coordinated with the fire fighting, including:

1. First aid for the injured
2. Shutting down of equipment and rerouting of fuel from the fire area
3. Special emergency maintenance work
4. Control of utilities
5. Auxiliary traffic control
6. Shutting off extraneous use of water from fire mains when the fire whistle is blown
7. Transportation and marshaling of reserve manpower and fire-fighting equipment
8. Liaison among the foregoing activities, particularly if multiple fire fronts are involved
9. Public relations

Emergency Headquarters Plan

To coordinate essential activities, some refineries have an emergency headquarters plan. The plan outlined below is for a major refinery and may have to be scaled down for smaller organizations. Experience has shown that the best way to set up such a plan is to have it follow the existing normal departmental organization so that the normal structure of the refinery organization will be disturbed to the least extent.

A suggested plan is to have the ranking supervisor in the refinery set up his emergency headquarters in a nearby but safe location where telephone and other communications are readily available. The main gates and key offices of the refinery should be notified as to the headquarters location, and assigned group leaders should report as early as possible to the headquarters and establish proper liaison with the fire-fighting front.

The departmental supervisor in the unit or area where the fire has broken out (or his alternate in the event the supervisor is not at his post at the time) should be in charge of emergency shutdown operations. Supervision of the actual fire fighting should be the responsibility of the refinery's acting fire-fighting supervisor or his alternate on duty. Other supervisors, specifically assigned, should be in charge of transportation, reserve manpower, utilities, first aid, and other major activities. All these supervisors should report immediately to emergency headquarters and maintain continuous liaison by radio, runners, and other means of communication until the emergency is over.

Obviously, every effort should be made to avoid overlapping supervision and direction or to lose communication with headquarters. Advance planning and occasional outdoor drills can establish the proper coordination among sectional commands.

Permanent Fire Committee

Those who compose the emergency headquarters organization may constitute a permanent management fire committee. This committee should meet periodically to study fire-prevention and fire-fighting problems. It should also meet promptly after any important fire and potentially serious emergency so that a policy for correction may be established.   It can also train subordinates to act for the committee. In some refineries the fire alarm is sounded for serious close calls.   This plan has several advantages.   The fire-fighting equipment and organization are available if the emergency gets out of hand.   Answering such calls provides unplanned fire drills which may be studied in post-mortem sessions as to over-all policies, response, and coordination.

Study of past fires is an important and often little-appreciated phase of the fire-training program and should be on the agenda for the committee.   When possible, reports of other fires should be studied to determine "how would we handle a fire of that type?"

In any planning for major disasters every plant should take into consideration the booklet *Radioactive Fallout and What It Means to the Petroleum and Gas Industries.*[15] Copies can be secured from the U.S. Department of the Interior, Washington 25, D.C.

Mutual-aid Plans

Mutual-aid plans among nearby oil company plants are helpful.   Oil company representatives in an area should get together regularly on a voluntary basis to determine what they can do to help one another.   The steps agreed on should be spelled out in writing, and test drills held regularly.   It is always helpful to have the local municipal fire department chief serve as a regular member of the group.

Mutual-aid programs can be most effective if voluntary joint safety inspections are made quarterly, during which the group takes turns inspecting the properties involved for accident and fire hazards.

The members of the National Petroleum Refiners Association in certain areas have been parties to outstanding mutual-aid disaster programs.   For those plants which are interested in developing such an organization, it is recommended that an inquiry be directed to the National Petroleum Refiners Association, 1725 Desales St., NW., Washington, D.C. 20006.

Alarm and Signal Systems

Some plants use central-station as well as proprietary signaling systems.   Under the central-station system, an alarm or signaling service is provided by an outside qualified agency.   Alarms indicating fire, smoke, flow of water into sprinkler systems, trespassing, or any other conditions which can be supervised by an electrical circuit can be sent to and recorded at a central station.   Preplanned action, such as calling the fire or police departments, can then be carried out by the outside agency.

Under a proprietary system, electrically supervised circuits carry alarms to a plant supervisory station such as the head watchman's office.   Such alarms can be recorded as to time and location.   Proprietary and central-station systems are sometimes tied together.

There have been many recent developments in both types of equipment.   NFPA Codes 71 and 72[11,12] will assist plant operators when such systems are being considered.

Fire-control Equipment

The first objective of fire fighting is to extinguish a fire before it becomes a large one or to hold a large fire in check until adequate help arrives.   Often the use of small water-hose lines equipped with combination fog and straight-stream nozzles or the use of a correct hand-type extinguisher, brought immediately to the scene of a small fire and used with intelligence, has proved most effective while awaiting assigned fire crews.

Primary fire-fighting equipment, ready for quick use by the men on the job, is of two types: portable small hose lines and extinguishers and fixed systems which are either automatic or manual.

Detailed information on installation and operation of equipment may be obtained from the manufacturer's published data or from NFPA 10, *Standard for the Installation, Maintenance, and Use of Portable Fire Extinguishers.*

Water Systems

Water is particularly valuable in fighting oil, gas, and Class A fires. It is usually safe to use on petroleum fires, except injudiciously on hot viscous oils.

In order to obtain the most practicable protection in a refinery, the quantity of water, the number of fire hydrants, the amount of fire hose, special fire-extinguishing equipment, and the number of men for fire fighting should be based upon sound fire-protection engineering. NFPA 24, *Standard for Outside Protection*,[16] describes many details for designing water systems. Other recommendations specific to refineries follow.

Water Supply

The required water supply for fire fighting may be obtained from a combination of high-pressure mains, low-pressure processing mains, and other sources. Each important unit of the processing area may be looped with high-pressure fire mains of such a size that, at the desired pressure, they can supply the critical areas with water at the desired rate and pressure.

Usually the water-main grid system is designed so that probable plant expansion will be accommodated. Gate valves should be installed in such an arrangement that only part of the system will be out of service during failures or repairs. These block valves or, in the case of buried systems, post indicator valves should be marked for easy identification.

The recommended high-pressure water supply for a refinery unit or for a processing area may depend on the quantity, availability, and dependability of water from other sources such as the low-pressure process mains, the municipal water supply, and natural water sources. Also, other available sources of process water, such as cooling towers and drainage sewers, should be considered. The high-pressure water supply should equal the minimum requirement under adverse conditions, adequate to handle one major fire and resultant exposure.

This high-pressure water should be supplied from a dependable source such as a body of water. If the normal source is not dependable, an emergency 4- to 6-hr supply such as a full storage tank or reservoir should be provided.

Fire lines, valves, hydrants, pumps, storage tanks if used, and portable water-type extinguishers should be protected from freezing. Lines should be designed and installed to avoid overstressing by earthquake, severe shock from mechanical impact, and damage by fire exposure. Diagrams of lines and valves should be posted at key points throughout the plant, and any change or alteration to the system should be recorded immediately.

Pumping Equipment

Centrifugal pumps, preferably those having a relatively flat characteristic curve, are generally provided for supplying fire systems because they provide a steady non-pulsating flow of water at uniform pressure and they can idle against closed valves for a period without damage to the pump or connected equipment. In some instances, relief valves or governors are provided to prevent overpressuring the system.

Fire system pumps are usually classed in one of three categories: primary, supplementary, and standby.

The primary and supplementary pumps frequently are electrically or steam-turbine driven. The standby pumps are usually driven by either gasoline or diesel engines and may be remotely located from the other pumps, sometimes utilizing a different source of supply.

Fire Hydrants and Reels

The type of hydrant used will depend upon the climatic conditions. Standard American Water Works Association self-draining hydrants are generally used in freezing climates.

Special hydrants with one or two $1\frac{1}{2}$-in. outlet connections, as well as standard hydrants drilled and tapped for $1\frac{1}{2}$-in. outlets or standpipes protected against freezing, are satisfactory for inside locations or for structures where manpower is limited.

Among the suitable types of fire hydrants available are the following which comply with the standard specification of the American Water Works Assn:

1. Four-inch main valve opening with two $2\frac{1}{2}$-in. hose outlets
2. Five-inch main valve opening with three $2\frac{1}{2}$-in. hose outlets
3. Six-inch main valve opening with four $2\frac{1}{2}$-in. hose outlets

Any of these hydrants can be provided with pumper outlets having capacity equal to that of the $2\frac{1}{2}$-in. outlets. It is understood that a pumper outlet cannot be used simultaneously with the hose outlets.

Additional information is provided in NFPA 29C, *Standard Specifications for Fire Hydrants for Private Fire Service.*[17] The 4-in. fire hydrant with two $2\frac{1}{2}$-in. hose outlets, one or both fitted with a valve, is suitable for use with conventional straight-stream nozzle-tip sizes from $\frac{3}{4}$ to $1\frac{1}{8}$ in. For a greater number of such tips or larger tips, 5- or 6-in. hydrants are more suitable. When quick-acting hose reels are used instead of hydrants, they should be equipped with a separate hose outlet.

Hose connections should be identical with those on the hoses of local fire department, or adequate adapters should be provided so that delay will be minimized when outside or mutual-aid assistance is required.

For a unit or a process area, enough hydrants or quick-acting reels should be placed strategically around the area so that an adequate supply of water will be available where required to handle a major fire regardless of wind direction.

The normal distance between fire hydrants is 150 to 300 ft, depending on layout of the area, water requirements, and number of outlets on the hydrants. Whenever practicable, the distance between a fire hydrant and a building or structure which is to be protected should be at least 50 ft.

The number of hydrants or quick-acting hose reels in an area is usually determined by the design and type of process, although one hydrant or reel for each 250 gpm of the high-pressure water allocated is usually considered sufficient.

For ease of operation some refineries use hydrants equipped with double-disk-type outlet valves. For proper drainage of the hydrants in cold climates, a $\frac{1}{4}$-in. hole is drilled at the bottom of the disk on the pressure side. If a self-draining hydrant with bottom shutoff is equipped with only one valve, the first hose should be connected to the outlet not provided with the valve so that the second hose can be connected or disconnected without having to shut off the hydrant. The second outlet on a quick-acting hose reel should be equipped with a gate valve.

Fire Hose

The use of $2\frac{1}{2}$- and $1\frac{1}{2}$-in. fire hose is standard practice in most refineries. At a given refinery, all fire-hose couplings should be standardized.

The $2\frac{1}{2}$-in. hose is used primarily for heavy cooling streams inasmuch as three or more men are required to handle each $2\frac{1}{2}$-in. line depending upon the system pressure. It is general practice to carry this hose on mobile equipment, which permits reduction of total hose required as compared with numerous hand-drawn reels throughout the refinery.

For rapid availability of water in process areas, permanently connected hose reels have been used extensively. These reels are provided usually with $1\frac{1}{2}$-in. hose, which is the maximum size of line that can be handled normally by one man. In some refineries 1-in. hose is preferred for this application because one man can protect himself by holding the nozzle with one hand while using the other hand to turn a valve or do other emergency operations. Because of the high-pressure drop through this small hose, hose lengths on these reels generally are limited to 100 ft to assure adequate nozzle pressures. Also, some $1\frac{1}{2}$-in. hose usually is carried on the mobile equipment for use in tank fields or other outlying areas.

Reliability of fire hose is an important consideration. Hose with hydrostatic-test pressures of at least 300 psi for $1\frac{1}{2}$-in. and 400 psi for $2\frac{1}{2}$-in. sizes normally is used.

All hose should be subjected to annual hydrostatic retests, usually at some pressure below that applied to new hose as determined by fire-line pressure and service conditions. Neoprene-covered and neoprene-lined hose is widely used in refineries because of its resistance to oil and chemical deterioration as well as to mildew and rot. Cotton-jacketed hose must be carefully cleaned and dried after use. For additional information, refer to NFPA 198, *Care of Fire Hose*.[18]

Hose Nozzles

Fire-hose nozzles of both the straight-stream and water-spray (or water-fog) types are widely used. Many nozzles of the combination type are available which can deliver either a water spray or straight stream.

Straight water streams are utilized primarily for cooling of equipment, although they have been used effectively for sweeping burning hydrocarbon spills away from exposed equipment. They usually provide longer range than do water-spray nozzles and will still provide reasonably good streams at nozzle pressure as low as 45 to 50 psig.

For $2\frac{1}{2}$-in. nozzles, straight-stream tips of $\frac{3}{4}$ to $1\frac{1}{8}$ in. are utilized, whereas on the $1\frac{1}{2}$-in. nozzle, $\frac{1}{2}$-in. tips are conventional. Play pipes or hose straps should be used with the $2\frac{1}{2}$-in. straight-stream nozzles to facilitate holding the hose. It is general practice to use shutoff-type nozzles to provide for control of the hose line by the nozzle man.

Most water-spray nozzle manufacturers recommend nozzle pressures greater than 70 psig for best water dispersion. When handled by trained men, water spray is particularly effective for extinguishing fires involving high-flash products, chilling or dispersing hot vapors and preventing them from igniting, cooling equipment exposed to fire, cooling gasoline and LP gas fires so that they can be more readily extinguished, and frothing out heavy viscous oils fires.

Reference should be made to the NFPA *Handbook of Fire Protection*[19] for a thorough discussion of nozzle types and water-spray patterns, as well as water capacities, range, and so forth.

Monitors and Deluge Sets

Monitors operated by one man can be used to supply as much water as fire hoses connected to standard hydrants and handled by three to six men.

For greater flexibility the monitor may be equipped with stacked tips of various sizes. This is desirable to provide flexibility where variable water pressures or demands may be expected. Adjustable fog nozzles may be used where shorter range spray cooling or extinguishment is the problem.

Deluge sets for use with two or more hose streams are preferred by some. In other cases, fixed monitors may be desirable for use at spot locations. Because of the limited area of coverage from fixed monitors, careful consideration should be given to their location to provide for maximum effectiveness. These locations should be reviewed during both the design and final construction stages of a job in order to avoid undue obstructions from pipelines, support columns, etc. In some cases, fixed monitors are provided with shields to protect the operator from radiant heat.

The installation of fixed monitors in an area does not eliminate the need for fire hydrants for hand line use inasmuch as it is impractical to provide monitors to cover the entire area completely.

Water-spray and Water-fog Systems

Water-spray Systems

Water-spray cooling systems usually are provided to minimize fire-exposure damage by keeping a water film on exposed surfaces. The effect on fire intensity is a secondary consideration. Water spray is particularly adapted for cooling uninsulated steel structures, elevated pipe lanes or ways, vessels, spheres, manifold pits, and similar installations with a minimum quantity of water. At the same time, water spray provides excellent protection by which men can get closer to a fire.

It is necessary that the water actually gets on the surface to be protected and that the coverage is complete and adequate to avoid excessive metal temperatures. Areas not actually wetted may be overheated, which might result in thermal stresses or weakening of the metal. To obtain the desired cooling, 0.10 to 0.25 gpm of water per square foot may be required. An allowance should be made for windage loss, evaporation of water other than at the surface, and the efficiency of the means of application. Low-velocity water application may be the most effective; however, coarse drops of water are less susceptible to winds and drafts. Reference should be made to NFPA 15, *Standard for Water Spray Systems for Fire Protection.*[20]

Water-fog Systems

Water-fog systems are intended to reduce fire intensity by the mixing of water with fuel vapors or by the contact of fine drops or a very fine mist of water with the oil surface. Water fog on contact can cool the surface of a high-flash oil to a temperature below its flash point. Some of the water will be converted into steam, which, in turn, tends to exclude oxygen from the surface and reduce fire intensity of even low-flash oils.

Water fog is effective on viscous oils or high-flash oils where areas are within the range of the spray. Except under ideal conditions, it is seldom effective for extinguishment of fires in gasoline or other low-flash products (less than 150°F).

Fog nozzles utilize high-velocity turbulence or impingement to break the water into a fine, well-dispersed discharge pattern and require a water pressure ranging from 50 psi up. Reference should be made to *NBFU Research Report* 10, "The Mechanism of Extinguishment of Fire by Finely Divided Water."[21]

Strainers

Permanently installed water-spray or water-fog systems should be well-protected with suitable strainers to prevent dirt or trash from reaching or clogging the nozzles. It is advisable to install master strainers where the water supply enters the spray systems and also individual strainers at each separate spray head.

Carbon Dioxide Systems

An inert gas, such as carbon dioxide (CO_2), discharged into a closed room or into enclosed spaces may be an effective extinguishing agent. A carbon dioxide system is one method of extinguishing fires in petroleum pump rooms, electrical installations, and special machinery or apparatus such as is used in laboratories.

The minimum carbon dioxide requirement should be calculated separately for each individual fire hazard, as prescribed by NFPA 12, *Standard for Carbon Dioxide Extinguishing Systems.*[22]

Dry Chemical Systems

The application of dry chemicals is effective for the control and extinguishment of fires which may occur during the processing and handling of flammable liquids, solids, and gases. This extinguishing agent is composed of specially treated sodium bicarbonate in dry powder form with components for producing free flow and water repellency. Being nonconductive, it is suitable for fires which involve energized electrical equipment. Rapid fire control and flame reduction also may be obtained when dry chemical is used on combustible materials, such as wood and paper, but in addition, a quenching agent such as water often must be used to extinguish the remaining embers. The dry chemical may be used simultaneously with water fog without practical damage to the powder. The water will not only quench embers and cool hot surfaces but will also reduce flame size and thereby make the fire easier to extinguish with the dry chemical.

NFPA 17, *Standard for Dry Chemical Extinguishing Systems*, provides information concerning such installations.[23]
(Note: Dry chemical which is compatible with foam or vice versa should be considered when the two agents may be used simultaneously.)

Foam Systems

During and shortly after World War I, foam became widely accepted as the basic fire-extinguishing medium for oil fires. Today, the uses of water spray and fog for unit and equipment fire fighting and of other types of portable extinguishers for primary use have increased to the extent that the use of foam largely is restricted to tank and major ground fires.

This reduction in scope of foam usage results from limitations in its effectiveness: Foam cannot be depended upon to extinguish three-dimensional fires of any magnitude, but where a blanketing effect is required over a moderately large area, there is no suitable substitute for foam.

The effectiveness of foam lies in its ability to form a stable, fire-resistive blanket over the burning surface. Partial destruction of the foam releases water which, in the case of heavier oils, serves to cool the oil surface and minimize vapor evolution. This effect, however, is not noticeable on light crude oils, gasoline, and similar commodities where a full blanket vapor seal is required. Obviously, foam is ineffective on gas and LP gas fires, because the vapors can pass up through the foam and burn above it.

General data relating to foam installations are given in NFPA 11[24] and NBFU 11,[25] both titled *Standard for Foam Extinguishing Systems*. Manufacturers' catalogs and engineering data sheets also are fruitful sources of information.

Air Agitation

This is a method of fighting fires in oil storage tanks by injecting air or an inert gas near the bottom of the tank to induce an upward flow of cold subsurface oil to cool the surface of the burning liquid. Extinguishment may be gained in a short time in the case of a fire in a tank containing a cold high-flash product.

Steam Smothering

The use of steam for smothering fires, such as one which might occur as a result of a tube leak in a furnace or header box, is general practice today. Steam smothering lines which lead to the furnaces should be equipped with remote-control valves. The use of steam is usually most effective in relatively small, confined areas.

NATIONAL FIRE CODES

Many of the National Fire Protection Association codes are of priceless help to the petroleum industry. Copies of relevant codes should be available in the technical and safety offices of every processing plant.

The following material on flammable liquids and foam systems is abstracted from NFPA 30, *Flammable Liquids Code*,[26] and NFPA Code 11, *Foam Extinguishing Systems*.[24] Additional codes are listed on page 9–42.

Flammable Liquids—Tank Storage*

Flammable liquids are those having a flash point below 200°F and a vapor pressure not exceeding 40 psia at 100°F.

The three classes of flammable liquids are:

Class I, those with flash points at or below 20°F
Class II, those with flash points above 20°F but at or below 70°F
Class III, those with flash points above 70°F but below 200°F

* Abstracted from NFPA 30, *Flammable Liquids Code*.[26]

Classes II and III liquids heated to or above their flash points should be subject to the requirements for Class I or II liquids. This code may also be applied to high-flash-point liquids heated to or above their flash points even though, when not heated, they are outside its scope.

Tank Construction

Aboveground, field-erected vertical tanks for atmospheric storage should be built to American Petroleum Institute Standards; atmospheric tanks in accord with API 12A, *Specification for Oil Storage Tanks with Riveted Shells*, 7th ed., September, 1951, or Number 12C, *Specification for Welded Oil Storage Tanks*, 15th ed., March, 1958, and low-pressure tanks in accord with API 620, *Recommended Rules for the Design and Construction of Large, Welded, Low-pressure Storage Tanks*, 1st ed., February, 1956, and Addenda, February, 1958.

Spacing between Tanks

The distance between two flammable-liquid storage tanks should not be less than 3 ft.

For tanks above 50,000-gal individual capacity storing any flammable liquid, except crude petroleum in producing areas, the distance should not be less than half the diameter of the smaller tank.

The minimum separation between a liquefied petroleum-gas container and a flammable liquid tank should be 20 ft. Suitable means should be taken to prevent the accumulation of flammable liquids under adjacent liquefied petroleum-gas containers such as by diking, diversion curbs, or grading. When flammable-liquid tanks are diked, the liquefied petroleum-gas containers should be outside the diked area and at least 10 ft away from the center line of the dike. These provisions need not apply when liquefied petroleum-gas containers of 125-gal or less capacity are installed adjacent to Class III flammable liquid tanks of 275-gal or less capacity.

Vents

Normal Breathing

For normal tank breathing, tanks should have normal venting capacity sufficient to permit filling and emptying, plus breathing from temperature changes, without distortion of tank shell or roof.

Tanks storing Classes I and II flammable liquids need to be equipped, where practical, either with venting devices which are normally closed when not under pressure or vacuum or with approved flame arresters, except that tanks under 2,500-gal capacity for Class I liquids and tanks under 3,000-bbl capacity for crude petroleum in producing areas may have open vents. (Note: Further guidance can be found in API RP 2000, *Guide for Tank Venting*.)

Emergency Relief

Every aboveground storage tank needs some form of construction or device that will relieve excessive internal pressure, caused by exposure fires, that might rupture the tank shell or bottom.

In a vertical tank, this construction may take the form of a weakened roof seam. The joint between the roof and shell of a tank 36 ft or more in diameter, if designed and built as an atmospheric storage tank according to API standards, can be the weakened seam for this purpose.

Where the entire dependence for such additional relief is on some device other than a weak roof seam or joint, the total venting capacity of both normal and emergency vents should be enough to prevent rupture of the shell or bottom of the tank if vertical or of the shell or heads if horizontal. Such a device may be a self-closing manhole cover or one using long bolts that permit the cover to lift under internal pressure or an additional or larger relief valve or valves.

The outlet of all vents and vent drains on tanks designed for 0.5 psi or greater pressure should be arranged to discharge so as to prevent localized overheating of any part of the tank in the event vapors from such vents are ignited.

Dikes and Walls

Tanks or groups of tanks containing crude petroleum should be diked or other suitable means taken to prevent discharge of liquid from endangering adjoining property or reaching waterways. The diked enclosure should have a capacity not less than that of the tank or tanks served by the enclosure.

For flammable liquids other than crude petroleum, individual tanks or groups of tanks, because of proximity to waterways, character of topography, or nearness to structures of high value or to places of habitation or assembly, should be diked or the yard should be provided with a curb or other suitable means taken to prevent the spread of liquid onto other property or waterways. Where a diked enclosure is needed, the volumetric capacity of the diked area should not be less than the capacity of the largest tank within the diked area.

Dike Construction

Dikes or retaining walls should be of earth, steel, concrete, or solid masonry designed to be liquid tight and to withstand a full hydraulic head. Earthen dikes 3 ft or more in height require a flat section at the top not less than 2 ft wide. The slope should be consistent with the angle of repose of the material of which the dikes are constructed. Dikes should be restricted to an average height of not more than 6 ft above the exterior grade. Unless means are available for extinguishing a fire in any tank containing crude petroleum, dikes and walls enclosing such tanks should be provided at the top with a flareback section designed to turn back a boil-over wave. This flareback is not required when approved floating-roof tanks are enclosed.

Drainage

Drains to remove rain water from diked areas should be kept closed normally and be designed so that when in use they will not permit flammable liquids to enter natural water courses, public sewers, or public drains if such would be a hazard. Where pumps control drainage from the diked area, they should not be self-starting.

Other Requirements

Tanks should rest directly on the ground or on foundations or supports of concrete, masonry, piling, or steel. Exposed piling or steel supports need protection by fire-resistive materials with a fire-resistance rating of at least 2 hr.

Each connection to an aboveground tank storing flammable liquids, located below normal liquid level, should have an internal or external control valve located as close as practicable to the shell of the tank. Except for flammable liquids whose chemical characteristics are incompatible with steel, such valves, when external, and their connections to the tank should be steel.

Tanks for storage of Classes I and II flammable liquids should not be installed inside buildings without special provisions. NFPA Code 30 should be carefully studied before any such installation is made. (Note: It is highly recommended that such liquids not be stored inside buildings, even in underground tanks.)

Fixed Foam Protection of Outdoor Tanks*

A fixed foam discharge outlet is a device permanently attached to a tank by means of which foam is introduced into the tank. A Type I outlet delivers foam directly onto the surface of the burning liquid without undue submergence or undue agitation of the

---

* The material in this and the following two sections on foam protection is abstracted from NFPA 11, *Foam Extinguishing Systems*.[24]

surface of the liquid.   A Type II outlet is not supplemented with means for delivering foam without undue submergence or undue agitation of the surface of liquid.

Fixed installations are piped from a central foam house to the tanks, discharging through fixed delivery outlets on the tanks.   Any required pumps should be permanently installed.

In semiportable installations, tanks can be equipped with fixed discharge outlets and piping which terminates at a safe distance from the tanks.   The fixed piping installation may or may not include a foam maker.   Necessary foam-producing materials, foam-making apparatus, hose, and portable foam towers are transported to the scene after the fire starts and are connected to the piping.

Rate of Application

The minimum rates of discharge to foam discharge outlets protecting an individual tank containing liquid hydrocarbons are:

1. For chemical foam systems with stored solutions, 0.5 gpm of A solution and 0.5 gpm of B solution for each 10 sq ft of liquid surface area of the tank protected.

2. For dry-powder foam-generator systems the water rate to the generators shall be at least 1 gpm for each 10 sq ft of liquid surface area of the tank to be protected.

3. For air-foam systems the delivery rate to the foam makers shall be at least 1 gpm of water (including stabilizer) for each 10 sq ft of liquid surface area of the tank to be protected.

4. For highly volatile materials such as casinghead gasoline (25 to 40 lb Reid vapor pressure), rates possibly as high as double normal rates may be needed.

Foam-producing Materials

Sufficient form-producing materials provided should be available to permit operation of the apparatus at the needed delivery rates for the minimum periods of time shown in Table 9-5 plus some residual quantity after the emergency and until a complete reorder of supplies can be obtained.

Table 9-5. Minimum Operating Time in Minutes to Be Provided for with Fixed Foam Outlets and Portable Foam Towers[24]

Liquid hydrocarbon in storage	Fixed foam installations		Portable foam towers	
	Type I outlet	Type II outlet	Type I outlet	Type II outlet
Lubricating oils; dry viscous residuum (more than 50 sec Saybolt-Furol at 122°F); dry fuel oils, etc., with flash point above 200°F..............	15	25	25	35
Kerosine; light furnace oils; diesel fuels, etc., with flash point over 110 to 200°F.................	20	30	30	50
Gasoline, naphtha, benzol, and similar liquids with flash point below 110°F..................	30	55	55	65
All crude petroleums...........................	30	55	55	65

A quantity of foam-producing materials should also be provided sufficient to produce foam or foam solutions to fill the feed lines actually installed between the source and the most remote tank.

Dry-powder Generators

It is assumed that dry-powder generators (dual- or single-powder type) consume 1.25 lb of powder per gallon of water.   Where "listings" of dry-powder generators and powder by nationally recognized testing laboratories show powder consumption less than 1.25 lb per gallon of water, such lower figure may be used when the generator is used in the manner on which the listing was based.

Foam Hose Streams

Approved foam hose stream equipment should be provided in the quantities shown in Table 9-6 in addition to tank foam installations as supplementary protection for

Table 9-6. Minimum Foam Hose Stream Requirements[24]

Largest tank diameter, ft	Min number of hose streams required	Min operating time, min*
Up to 35................	1	10
Over 35, to 65...........	1	20
Over 65, to 95...........	2	20
Over 95, to 117.5.........	2	30
Over 117.5..............	3	30

* Based on simultaneous operation of the minimum number of hose streams required. Adjustment may be made where streams of greater capacity are provided.

ground fire.   The equipment for producing foam hose streams should have a water rate (or solution rate) of at least 50 gpm in addition to quantities required for tank areas.

Additional foam-producing materials should be provided to permit operation of the hose stream equipment simultaneously with tank foam installations for the periods of time set forth in Table 9-6.

Fixed Discharge Outlets

Tanks should be provided with approved discharge outlets as given in Table 9-7.

Fixed discharge outlets should be securely attached to the tank shell, so located and connected as to preclude the possibility of the tank contents overflowing into the foam system lines.   They should be designed to permit movement or distortion from fire and securely attached so that displacement of the roof is not likely to subject them to serious injury.

Table 9-7. Minimum Number of Fixed Foam Discharge Outlets and Portable Foam Towers Required for Various Sizes of Tanks[24]

Largest tank diameter, ft (or equivalent area)	Min number of fixed foam discharge outlets	Min number of portable foam towers
Up to 80.................	1	1
Over 80, to 117.5...........	2	2
Over 117.5, to 140..........	3	3
Over 140, to 160*..........	4	4
Over 160, to 180*..........	5	5
Over 180, to 200*..........	6	6

* Since there has been no experience with foam application to fires in oil tanks over 140 ft diameter, requirements for foam protection on tanks above this size are based on extrapolation of data from successful extinguishments in smaller tanks.

In tanks containing liquids subject to evaporation during storage, fixed outlets should be provided with an effective and durable vapor seal, frangible under low pressure, to prevent entrance of vapors into outlets and pipelines.

It is desirable that at least one portable tower be provided as supplementary protection in the event that a fixed discharge outlet is damaged by an explosion within the tank.

Portable Foam Protection of Outdoor Tanks

This section relates to those systems in which the foam is applied to the burning surface of a tank through approved portable towers, which are placed in operating position after the fire starts, together with powder, generators, hose connections, etc. (in the case of dry chemical systems), or liquid foam stabilizer, proportioning devices, etc. (in the case of air-foam systems). Towers may also be used with wet storage systems.

Generally, portable towers are to be regarded as limited in scope and effectiveness and as not affording the same degree of protection as fixed systems. Portable-tower systems require an adequate number of men to place and maintain the apparatus in operation and in some cases special truck units for the ready transportation of the equipment to the location of the fire.

Rate of Application

The minimum rate of discharge to portable foam towers should be the same as for fixed foam towers, as should be the total foam supplies to be maintained.

Foam-producing Materials

The foam-producing materials provided should be sufficient to permit operation of the apparatus at the required delivery rates for the minimum periods of time shown in Table 9-5.

It is assumed that dry-powder generators consume 1.25 lb of powder per gallon of water.

The requirements for hose streams and filling pipelines are the same as for fixed foam towers (see Table 9-6).

Approved foam towers should be available for tanks as given in Table 9-7.

Foam Hose Protection of Outdoor Tanks

Foam hose streams are usually recommended as auxiliary protection in conjunction with fixed piping systems and portable towers. In some cases, however, they are suitable when used alone, as in protection of horizontal cylindrical tanks, floating-roof tanks, and vertical tanks not over 30 ft diameter or over 20 ft high.

Rate of Application

The minimum rate of discharge from foam hose streams protecting vertical tanks containing liquid hydrocarbons should be:

1. For chemical foam systems with stored solutions, 0.8 gpm of A solution and 0.8 gpm of B solution for each 10 sq ft of liquid surface area to be protected.

2. For dry-powder foam-generator systems, at least 1.6 gpm of water (including stabilizer) to the generator for each 10 sq ft of liquid surface area of the tanks to be protected.

3. For air-foam systems the delivery rate shall be at least 1.6 gpm of water (including stabilizer) to the foam makers for each 10 sq ft of liquid surface area of the tanks to be protected.

4. For highly volatile materials such as casinghead gasoline (25 to 40 lb Reid vapor pressure), higher rates of application are required, possibly as high as double those specified above.

Foam-producing Materials

The quantity of foam-producing materials provided should be sufficient to permit operation of the apparatus at the required delivery rates for the minimum periods of time shown for portable foam towers, Type II outlet, in Table 9-5.

It is assumed that dry-powder generators have the same requirements as fixed foam towers.

In plants where only horizontal cylindrical, floating-roof, or pressure tanks (e.g., spheroids) are in service, the following minimum number of foam hose streams, with water or solution rates of at least 50 gpm, are needed: for tanks less than 65 ft diameter, one hose stream; for tanks 65 ft in diameter and larger, two hose streams; where more than one horizontal tank is enclosed by a single dike and the aggregate capacity of the tanks exceeds 35,000 gal, at least two foam hose streams are needed.

Unlined fabric hose should not be used with foam equipment.

Other Safety Codes

Table 9-8 lists those National Fire Protection Association codes, including those previously discussed in this section, which should be available for current reference in any petroleum processing plant.

SAFE PRACTICE PROCEDURES

Over the years many safe practice procedures have been developed for various repetitive activities. Many have been reduced to writing by API accident- and fire-protection committees, while others which have not been so formalized are generally accepted industry practices.

Safe Maintenance Practices

The following comments regarding safe maintenance practices are taken from API committee correspondence in connection with the preparation of the API manual *Safe Maintenance Practices in Refineries*.[3]

Mechanical and Physical Hazards

Potential hazards exist in most refinery equipment, some of which are outlined below. These hazards should be expected and guarded against on every maintenance job. A thorough knowledge of each job to be performed will reveal mechanical and physical conditions that could be hazardous.

Preshutdown Period

When a unit is coming down and the operating department is readying the equipment for shutdown activities, maintenance personnel should be alert for:

1. Hot lines and equipment
2. Rotating and reciprocating equipment
3. Furnace gases and vapors
4. Scaffolds and rigging erected for a specific job

Shutdown Period, Primary

When a unit is being opened, maintenance personnel may encounter:

1. Lines under pressure (steam, liquid, hydrocarbon, gas)
2. Hot lines and equipment
3. Toxic gases or oxygen deficiency when lines or vessels are open

Table 9-8. National Fire Protection Association Standards and Codes of Particular Interest to the Petroleum Processing Industry

Note: The NFPA publications listed below can be purchased from the National Fire Protection Association, 60 Batterymarch St., Boston 10, Mass. All of them are included in a volume of the National Fire Code as indicated (e.g., NFC Vol. II), and most of them can be purchased separately.

NFPA No.	Title	NFC vol.
	Extinguishing Appliances	
11	*Foam Extinguishing Systems*, 1960, 76 pp.	IV
12	*Carbon Dioxide Extinguishing Systems*, 1962, 84 pp.	IV
13	*Sprinkler Systems, Installation of*, 1961, 164 pp.	IV
15	*Water Spray Systems*, 1962, 56 pp.	IV
17	*Dry Chemical Extinguishing Systems*, 1958, 48 pp.	IV
198	*Fire Hose, Care, Maintenance, Use*, 1958, 48 pp.	VII
	Extinguishing Auxiliaries	
20	*Centrifugal Fire Pumps, Installation of*, 1962, 116 pp.	IV
21A	*Steam Fire Pumps, Operation and Maintenance*, 1937	IV
22	*Water Tanks for Private Fire Protection*, 1962, 148 pp.	IV
23	*Fire Department Hose Connection for Sprinkler and Standpipe Systems*, 1931, 8 pp.	IV
24	*Outside Protection (Yard Mains for Sprinklers, Standpipes, etc.)*, 1962, 52 pp.	IV
27	*Private Fire Brigades, Organization, Training, and Equipment*, 1955, 32 pp.	VII
29A	*Gate and Check Valves, Specifications*, 1933	IV
29B	*Indicator Posts*, 1961, 12 pp.	IV
29C	*Fire Hydrants, Private Fire Service*, 1955, 16 pp.	IV
	Flammable Liquids	
30	*Flammable Liquids Code*, 1962, 76 pp.	I
325	*Flammable Liquids, Gases, and Volatile Solids, Fire Hazard Properties of*, 1960, 132 pp.	I
325A	*Flash Point Index of Trade Name Liquids*, 1959, 160 pp.	I
	Electrical	
70	*National Electrical Code*, 1962, 529 pp.	V
71	*Central Station Protective Signalling Systems, Installation, Maintenance, and Use of*, 1962, 32 pp.	V
72	*Proprietary, Auxiliary, and Local Protective Signalling Systems, Installation, Maintenance, and Use of*, 1962, 44 pp.	V
77M	*Static Electricity*, 1961, 64 pp.	V
78	*Protection against Lightning*, 1959, 52 pp.	V
	Marine	
304L	*Petroleum Wharves, Ordinance for*, 1938, 12 pp.	VI
307	*Operation of Marine Terminals*, 1961, 32 pp.	VI
	Miscellaneous	
49M	*Hazardous Chemicals Data*, 1962, 84 pp.	II
58	*Liquefied Petroleum Gases, Storage and Handling of*, 1961, 108 pp.	I
68	*Explosion Venting*, 1954, 60 pp.	II
69	*Inerting for Fire and Explosion Protection*, 1956, 52 pp.	II
214	*Water Cooling Towers*, 1961, 16 pp.	III
601	*The Watchman*, 1956, 20 pp.	VII
801	*Laboratories Handling Radioactive Materials, Safe Practice for*, 1955, 46 pp.	II

Shutdown Period, Secondary

When the plant is down and isolated and full maintenance crews begin work, check the area in general for:

1. Electric motors or other power units coupled to pumps (tag out, when pump is being worked on)
2. Opening of lines, exchangers, or vessels when welding is in process
3. Oil spills
4. Access to areas where work is in progress overhead
5. Disconnected piping where units are interconnected
6. Open sewers and trenches
7. Congested working areas
8. Adequacy of staging or scaffold to be used
9. Electric welding arcs
10. Facilities for safe handling of equipment
11. Housekeeping practices

Opening and Repairing Furnaces

Guard against:

1. Use of defective or inadequate tools or equipment
2. Blowing dust particles
3. Loose brickwork and tube hangers
4. Congested passageways
5. Improper use of tools or safety equipment, including electric extension cords
6. Entry without permit or failure to adhere to requirements of entry permits
7. Inadequate staging
8. Working directly above or below another employee
9. Permitting passage in areas where tools or other objects might fall
10. Failure to use safety equipment

Tower Repairs

Check for:

1. Entry permits or gas tests
2. Hanging particles (coke, steel)
3. Sharp objects
4. Proper staging
5. Improper blinding of pipelines connected to the vessel
6. Presence of oxygen-deficient atmosphere, toxic vapors, or corrosive liquids (acids, caustics)
7. Inadequate ventilation
8. Trapped pockets of liquid hydrocarbons
9. Excessive heat
10. Improper use of tools, including electric extension cords
11. Congested passageways
12. Communication equipment
13. Presence of iron or hydrogen sulfide in some vessels
14. Slippery or oily surfaces on interior of vessel
15. Employees working directly above or below another employee

Protective Clothing and Equipment

Protective clothing and equipment should be available to employees where needed. Workmen should know what is available to them and when and how best to use it. A well-balanced educational program is the best way to instill this knowledge in the minds of supervisors and workers.

Coordinated Safety Meetings

Before work on any maintenance project is begun, a safety meeting should be held to discuss the job in detail and to coordinate activities of the various groups concerned. The schedule of the shutdown operations should be thoroughly understood, including shutdown procedures and timing. A review of vessel or line blinding requirements, as well as the schedule for the startup operation after turnover, should also be made.

Operating personnel responsible for the shutdown procedure should be present at this safety meeting and should thoroughly understand the maintenance steps to be taken. The engineering personnel should be present to discuss design of new or altered equipment involved in the shutdown. Representatives of the fire and safety departments should be included.

Administration of the maintenance project, including the work lists, turnovers permits required, and procedures to be followed in case of emergency, should be outlined. It is especially important that the general safety aspects and potential hazards of the project be discussed fully by operating, fire, safety, and maintenance personnel. Outlined procedures should be followed during the entire hutdown period .,

Electrical Work

General rules for the safe performance of electrical work should be established. The maintenance supervisor should assure himself that the workmen understand these general rules before proceeding.

Proper permits should be obtained for necessary hot work or excavation work. Specific instructions should be issued to electrical maintenance personnel for work on equipment or lines having various voltages.

Written authorization to proceed with work in operating areas should be obtained from the operating supervisor. When the work has been completed, the maintenance supervisor should so notify the operating supervisor. Electrical circuits should be deenergized before work begins, if possible.

Whenever circuits have been deenergized for repairs, an appropriate tag should be placed at the disconnecting device of the circuit being worked on to assure that no one will energize that circuit while work is in progress. If work has to be performed on energized lines, appropriate precautions should be taken, including the use of rubber gloves, rubber blankets, and insulated tools.

When working in an elevated location the workman should wear a safety belt or other protective device. Periodic inspection of safety belts, ropes, and falls is important to be sure they are in good working order. (See Equipment Inspection Procedures, page 9-7.)

Maintenance work on high-tension lines should be performed by at least two men working together. In such elevated line work there should be one man on the ground to serve the workmen aloft. Electrical workers should avoid using metallic measuring tapes.

Pneumatic Hammers and Chisels

Pneumatic hammers and chisels used in maintenance present the possibility of eye injuries from flying scale or chips and from tools and bits flying from the gun. Observance of the following recommendations may prevent a serious accident:

1. Always wear chipping goggles when working with portable power tools.
2. Never aim the pneumatic gun at another worker.
3. When the gun is not in use, even momentarily, set the safety catch on the trigger and remove the tool.
4. Point the tool toward the ground or floor when not in use.
5. Should the tool become detached from the air hose, do not attempt to grab the hose; shut off the air supply.
6. Use only the proper couplings to connect the hose. The use of short pipe nipples is dangerous.

7. Do not disconnect the hose from the gun or attempt to repair the gun until the air supply has been closed down and the air bled down through the tool.

8. Do not attempt to hold a backup wrench when using a rotary air-driven wrench. Hands should be kept in the clear while placing the wrench.

Hot Tapping

To avoid shutting down the unit, it is more expedient in some instances to tie into a system with a new line or connection while the line is in service. With careful planning, hot tapping can be done safely. Specific approval of the operating supervisor and appropriate permits must be obtained before hot-tapping operations begin. The line to be hot-tapped must be positively identified, and the operating supervisor should mark the specific point on the line where the tap is to be made. He should also ascertain that the metal thickness of the line is sufficient to prevent burning through during welding and that the connection (nipple and flange) to be installed is the correct size and length for the hot-tapping equipment.

During hot-tapping operations the operating supervisor should coordinate activities with the operating department to ensure a continuous flow of stock through the line. Electric arc welding is preferable for attaching the stub. The maintenance supervisor should approve any alternate method of welding proposed. After the stub is welded in place and before the line is drilled, the nipple and weld on the hot tap should be tested for leaks by bolting a test pump to the hot tap flange and applying a pressure test. When hot tapping is done on small lines, it is very important that the drill be continually observed to avoid drilling through the opposite side of the pipe.

Scaffolds

It is important that scaffolds be adequate for the intended use and that the workmen engaged in the building and use of scaffolds be thoroughly briefed concerning potential hazards within the area in which the scaffold is to be used. They should avoid clamping staging supports to pipelines or using pipelines as base supports for staging.

Hoisting equipment or hand lines and buckets or canvas bags should be used to raise and lower material for staging. It may be necessary to use rope falls for hoisting heavy sections or staging boards to elevated scaffold sections. Workmen should not stand on pipelines while erecting staging. The ground men should never stand under material being hoisted from or lowered to the ground.

The scaffold should be constructed of first-class material only. It should be lashed or securely fastened by other means to the structure being served. Access to each working level of the scaffold may be maintained by ladders fastened to the scaffold. Under no circumstances should a workman climb up or down the scaffold using cross bracing or other structural members for access.

After the construction of a scaffold, surplus blocking, clamps, and other staging materials may be removed. The scaffold should be used only for the specific purpose for which it was designed. Overloading it with maintenance materials and tools or other unwarranted equipment should be avoided. Workmen should avoid jumping onto the scaffold or dropping heavy objects to the platforms or runways.

Protection of the structure from moving objects such as cranes and hoists should receive special attention. Any damage to the scaffolding should be repaired immediately. Men found to be unsteady when working above ground level, because they are not physically or mentally suited to such work, should not be permitted on scaffolding.

Hot-work Precautions*

When welding or cutting in an area which is not specifically designated as a safe area, a written work permit should be obtained from a designated authority before the job is started. The work location should be gas-free and away from flammable

* Abstracted from API *Accident Prevention Manual* 3, 2d ed, 1953, "Gas and Electric Cutting and Welding."

materials or products. The direction of the wind should be considered when hot work is to be performed in the vicinity of operating equipment so that flammable vapors will be prevented from entering the work area. Sampling or gaging in the area should not be permitted unless approved by the individual who signed the hot-work permit. Hot work on tanks or vessels which contain flammable liquids should not be permitted while adjacent equipment is being steamed, ventilated, or flushed of sediment.

When welding or cutting is to be done above oily ground, the ground in the area should be flushed with water or covered with clean dirt or sand or other precautions should be taken. All sewer manholes or catch basins nearby should be covered, or if the basin is equipped with a trap, the basin may be continuously flushed with water.

Fire-extinguishing Equipment

The fire chief or designated individual who issues the work permit should specify on the permit the necessary standby fire-protection equipment or personnel.

Work on Vessels, Exchangers, Drums, and Tanks

Before any work is done inside a vessel, the vessel should be thoroughly cleaned, gas-freed, and gas-tested. Periodic gas-indicator retests may be desirable while welding or cutting proceeds. If the indicator registers more than a trace of vapor, work should be stopped immediately and the source of vapor should be located and removed. Indicator tests should be repeated in a vessel which previously was pronounced gas-free if work therein has been delayed or suspended for more than a few hours. Indicator tests should not be conducted during steaming or ventilating operations.

When the vessel contains internal equipment, an inspection should be made to ensure that all oil or flammable material has been removed where sparks may fall on bubble-cap trays, weirs, or other internal pans.

If welding is to be done on the outside surface of the vessel, and if the area is otherwise safe for the use of an open flame, the vessel need not be gas-freed if one of the following procedures is used:

1. If the vessel is partly filled with water and the welding is done not less than 1 ft below the level of the water surface
2. If the vessel is partly filled with liquid product, provided welding is done not less than 1 ft below the level of the liquid and conditions are such that there is no chance of burning through the vessel wall (hot tapping is also permissible under these conditions)
3. If the vessel is known from chemical analysis or other reliable evidence to contain an atmosphere incapable of being ignited because it is "too rich" or because of deficiency of oxygen

Note that, if steam is used as a means of displacing oxygen, it is necessary that every part of the vessel be heated by the steam to a temperature of at least 170°F so that the oxygen concentration will be reduced to a point where no ignitible mixture can be formed. Visible discharge of steam from an opening is not sufficient evidence that the atmosphere within is not explosive.

If carbon dioxide is used as the displacing medium, it should be introduced in such a way as to displace completely the gas in **all parts** of the vessel, taking into account that carbon dioxide is heavier than air, and the effectiveness of the purging must be determined by chemical analysis.

When the oxygen content has been reduced to 10 per cent by volume, it is safe to assume that the vessel cannot contain an ignitible mixture (except where hydrogen or carbon monoxide is the sole combustible involved).

The ordinary combustible-gas indicator **cannot** be used to distinguish between an explosive mixture and one rendered nonexplosive by the addition of carbon dioxide or other inert gas. For more precise figures of safe oxygen concentration, see *U.S. Bureau of Mines Bulletin* 503.[6]

Hot work should not be performed on a vessel or piece of equipment in a complete plant unit while other parts of the unit are operating unless precautions are taken to prevent the release of flammable liquids and vapors into the area.   Sometimes, when adjacent equipment is operating, it is permissible to perform work inside a vessel if a positive gas-free air pressure is maintained and there is no possibility of vapors drifting into the vessel.

Work in Buildings

Caution should be exercised when welding or cutting in buildings such as compressor rooms, receiving houses, and can- or drum-filling rooms while any equipment within these buildings is in operation.   The equipment to be worked on should be freed of gas or oil, and if feasible, other equipment should be depressured.   The area should be ventilated until all traces of gas or vapor are eliminated.

Work on Pipelines

Whenever feasible, pipelines which are to be cut or welded should be drained of product and, if necessary, gas-freed.   Periodic tests with a combustible-gas indicator, should be made as the work proceeds.   The possibility of a flash, which may result from oil on the inner surface of the pipe, may be minimized by having steam, carbon dioxide, or other inert gas in the line under slight pressure and noticeable at a vent.

The most important safety procedure when welding or cutting on a pipeline is to disconnect or blind the line.

A closed gate valve is not a positive block.   The actual determination that a line is safe is a step-by-step procedure which must be followed closely.

When it is necessary to weld on or "hot tap" a pipeline while the line is in operation, the work can be done safely if the area is safe for use of an open flame and if the pipe wall thickness, as well as the method, has been checked by a qualified person.

Protection of Personnel

Steps should be taken during welding or cutting operations to guard against the following hazards to personal health:

1. Ultraviolet light rays from electric welding
2. Poisonous substances, such as lead, cadmium, and zinc fumes
3. Poisonous gases, such as carbon monoxide
4. Flying particles during chipping and slag removal
5. Oxygen deficiency of the air in small, poorly ventilated spaces

Further details on this subject will be found in ASA Z 49.1, Sec. 7, "Protection of Personnel," and Sec. 8, "Ventilation and Health Protection."[27]

Hot-work Permits

Before any work is started, a permit signed by the plant fire chief or designated individual should be obtained.   Each permit for hot work should state all precautions necessary for the specific job.   It should be in writing and should be issued for one working shift only.   When it is necessary to continue the work on the second or third shift, the designated supervisor on each succeeding shift should countersign the permit.

After he has made a careful analysis and has determined that oil and gas lines to the equipment have been disconnected or blanked and that the equipment has been cleaned and gas-freed, the area supervisor may issue a permit for the mechanical supervisor to proceed with the work.   If the work continues beyond one day, the foregoing procedure should be repeated each day the work continues.   The fact that a permit has been issued in no way relieves the welder or burner of his responsibility for the safe execution of the assigned task.   Any hazardous condition should be reported promptly.

Safety Instructions to Contractors

Many serious injuries as well as fires and explosions have occurred in otherwise safely operated plants when contractors' employees have failed to exercise the same safety precautions as are followed by plant personnel. The following items specifically relating to the procedures of a plant should be reviewed by a contractor with his employees.

1. The reason for the plant wishing to maintain a safe operation.

2. An outline of hazards common to petroleum processing equipment, storage tanks, pumps, and lines.

3. Explanation as to why contractors' men should leave the scene of a fire. Plant fire crews are trained to handle such problems.

4. Designations of smoking areas, if any, and the hazards of "sneaking" a smoke in the wrong place.

5. A review of special plant **formal** written permit procedures.

6. Need for written approval of plant management before blocking any roads or passageways on account of possible blocking of fire trucks.

7. Reason for prohibition against walking around operating equipment or starting work on it unless written permission is granted by the operator in charge.

8. Explanation of policy to secure written approval of operator in charge before shutting off any electric lines or before removing them in any demolition job to determine if such lines are "dead."

9. Plant speed limit for contractors' trucks and cars.

10. Advice to the effect that contractors' men must not **assume** anything around a refinery. In cases of doubt, the operator in charge should always be consulted.

Cleaning Storage Tanks

Precautionary measures should be followed in every stage of tank cleaning, and each plant should develop its own list of safe practices. Items shown in the check list, Table 9-9, later in this section (page 9–49) should be considered. The following material, derived from *API Bulletin* 2016, "Cleaning Tanks Which Have Contained Gasoline and Similar Low-flash Products," should be reviewed:

1. What product has been stored in the tank: flammable, high flash point, toxic?

2. For crude-oil tanks:

   a. If hydrogen sulfide is present in amounts above 20 ppm, protective respiratory equipment should be worn by all personnel. The area **in the vicinity** of the tank, as well as **in the tank,** should be tested frequently to observe any changes in hydrogen sulfide concentration. After an area has become safe, it can soon become hazardous because sediment has been disturbed. A man should never be permitted to work alone in an area suspected of being contaminated by hydrogen sulfide. General precautions as outlined earlier in this section should be reviewed.

   b. A determination should be made as to whether or not the crude tank at some time contained leaded gasoline and was or was not thoroughly cleaned prior to being placed in crude service. If the latter is the case, it should be considered lead-hazardous and the precautions outlined earlier for cleaning tanks having contained leaded gasoline should be followed. The sludge and scale in such a tank may contain lead for an indefinite period. If there is any doubt, the tank should be considered lead-hazardous.

   c. If the flash point of the crude is low, the same precautions should be exercised as for gasoline and light naphthas.

3. For high-flash-point product tanks: Tanks which have contained high-flash-point products, such as kerosine and the various fuel oils, are generally not hazardous, but care must nevertheless be exercised. The tank may have contained leaded gasoline in prior service and may not have been thoroughly cleaned since that service. There is always the possibility of contamination with light products. Also, vapors may enter the tank through product and foam lines. Tests should be made to determine if an explosive mixture exists and if the atmosphere is safe for breathing. If "sour"

Table 9-9. Tank Cleaning Check List

A form listing the following questions and providing appropriate spaces after each for answering "yes," "no," or "not applicable" and the initials of the person answering each question should be completed before cleaning starts on any storage tank.

1. Does tank contain low-flash product?
2. Does tank contain leaded gasoline?
3. Did tank ever contain leaded gasoline?
4. If answer to No. 3 is "yes," was it made lead-hazard-free?
5. Does tank bottom leak?
6. Has vapor-testing equipment been checked?
7. Is hydrogen sulfide present?
8. Is procedure established for regular testing for petroleum vapors outside as well as inside tank during entire cleaning process?
9. Is procedure established for regular testing for hydrogen sulfide?
10. Has personnel protective equipment been checked to see if clean and in good repair?
11. Do personnel have knowledge of artificial respiration?
12. Have plans been made for safe disposal of sludge and product washed from tank?
13. Are any personnel carrying mechanical lighters or matches?
14. After tank has been vapor-freed, is it deficient in oxygen?
15. Have all lines to tank been blanked, including foam lines?
16. If water hose line is to be used, is nozzle bonded to tank?
17. Can vapors from other tanks, lines, or sewers drift into tank being cleaned?
18. When vapors are being expelled from tank being cleaned, can they drift to source of ignition?
19. If personnel have to go on tank roof, are roof plates safe?
20. If roof plates are unsafe, have planks been put in place firmly?
21. Is it necessary to barricade roads to keep vehicles from area?
22. Are electrical fittings and portable lights of explosionproof type where they may be exposed to vapors?
23. Are all personnel unnecessary to the cleaning operation to be kept away?
24. If gasoline-driven or any nonvaporproof equipment must be used in vicinity of tank, is procedure established to observe wind direction and regular checking of atmosphere in area of such equipment?
25. When tank is to be entered, has inspection been made for hazards such as loose rafters, angle irons, columns, or swing lines?
26. In case of dust in tank during repair work, are personnel directed to wear masks and goggles in case of rust?
27. Are fire extinguishers and hose with water spray and foam laid out and ready for immediate use?

products have been stored in a tank, a check should be made for hydrogen sulfide, especially if any sludge is present.

Leaded-gasoline Tanks

A tank which has been used for mixing or storing gasoline to which lead antiknock compounds have been added is a potential source of lead hazard throughout the cleaning process. Although the tank may be vapor-free and safe as far as fire, explosion, and toxic hazards are concerned, it is not necessarily safe as far as the lead hazard is concerned.

Any tank which has at some time contained leaded gasoline, even though used to store other products but without prior cleaning, should always be considered lead-hazardous. The sludge and scale in such a tank may contain lead for an indefinite period. Intense heat, such as in torch cutting and welding, may evolve dangerous vapors from lead compounds. Scaling and scraping dry internal surfaces may contaminate the atmosphere with lead-containing dust.

Entering

A workman who enters a tank which has contained leaded gasoline should wear respiratory protective equipment through which fresh air is continuously supplied under positive pressure to the facepiece.   He should continue to wear this equipment until material which may release significant concentrations of lead vapors has been removed.

Hand-operated blower-type equipment can be used under a wide range of circumstances to furnish fresh air under positive pressure to the facepiece.   Experience thus far with other types of equipment which furnish fresh air to the facepiece, however, would seem to indicate that such equipment can be used safely only under the supervision of trained and competent men.

A workman who enters a tank which has contained leaded gasoline should dress in specially provided clothing.   Light-colored outer clothing is recommended so that men working inside the tank can be more readily seen and contamination of clothes can be immediately recognized.   At the end of the day, such clothing should be removed and laundered before reuse.   Each worker should bathe at the end of each day's work.

Workmen should not carry tobacco or foodstuffs into a tank which has contained leaded gasoline.   Hose mask sets, boots, gloves, and tools should be cleaned at the end of each day's work and at the end of the job.

Sludge from tanks which have contained leaded gasoline is dangerous to handle and should be buried or otherwise disposed of as recommended by the lead suppliers.

Repairs

The same personnel precautions should also be followed during repair work unless the tank has been well cleaned.   Usually cold work produces no additional hazard. However, if perceptible dust results from scaling or scraping of the sides or bottom of the tank or from any other cause, a dust respirator should be worn.

It has been industry practice to consider a tank free of lead hazard when:
1. Sludge has been removed.
2. Adherent material has been scraped from the surfaces which have been in direct contact with the sludge and has been removed from the tank.
3. The tank has been swept free of liquids.
4. The tank has been thoroughly ventilated after the foregoing operations.

Concrete tanks, tanks with leaky bottoms, or tanks having contained absorbent material or having interior surfaces coated with absorbent material cannot be rendered lead-free.

The foregoing applies to cold work only.   Surfaces of a tank which have been in contact with leaded gasoline should be cleaned down to bare metal over any area which might be made excessively hot by welding or by other operations which require the application of heat.   As an alternative to cleaning down to bare metal, welders (after an open-fire permit has been approved) may use fresh-air respiratory equipment or a welder's facepiece designed for use with a hose mask through which fresh air is supplied.

Manufacturers of antiknock compounds which contain lead provide advisory services in connection with the entry or cleaning of storage tanks which have contained leaded gasoline.

Fuel and Crude Tanks

The following comments apply to the cleaning of tanks which have contained low-flash products, including crude, naphthas, gasoline, and gasoline-type jet fuels (JP-4).

For safety, from the standpoint of toxicity, the maximum concentration of gasoline vapor in air should not exceed 500 ppm, or approximately 4.0 per cent of the lower flammable limit for gasoline vapors for exposure for an 8-hr day.   Concentrations up to 20 per cent of the lower flammable limit may be tolerated for short periods.

Anyone entering a tank before it has been ventilated which has been closed for an extended period should wear air-supplied respiratory protective equipment.

Before cleaning operations are started, the supervisor should conduct a briefing session to ensure that cleaning personnel know potential hazards as well as the correct order and methods of procedure. Tank-cleaning personnel should know potential sources of ignition, including static electricity, and methods for their exclusion or control. Instructions should include the safe use of potentially hazardous equipment, such as electric motors and gasoline engines. Personnel should be provided with and trained in the proper use of personal protective equipment and necessary mechanical equipment.

A survey should be made of the surrounding area to avoid hazards of drifting vapors from other tanks, lines, or sewers. If tank roof plates are thin, planks should be placed on the roof to distribute the added weight of workmen.

To ensure control of ignition sources during tank cleaning:

1. Roads leading into the area should be posted or barricaded to prevent entry of vehicles.

2. Welding, burning, or other maintenance work in the area which might create a source of ignition should be stopped.

3. Electrical equipment or portable lights used at ground level outside the tank should be of the explosion- or vaporproof type.

4. Personnel should be warned against smoking or carrying matches and other potential sources of ignition.

Removing Product

Before any shell or roof manway is opened, the following should be done:

1. Residual oil should be pumped or drained to the lowest possible level.

2. Temperatures permitting, water should be added through an existing piping connection (not through a roof opening) to float any remaining product out of low spots.

3. Valves and lines to the tank should be closed, and the lines between the valves and the tank should be drained or flushed. Blinds should be installed adjacent to the tank, or the lines should be disconnected and the ends closed with blind flanges or plugs.

Vapor-freeing

It is best that the tank be freed completely of flammable vapors before other steps are undertaken. Vapor freeing implies the complete replacement, with fresh air, of the hydrocarbon vapors in the tank. In the initial stage of vapor freeing, work in the area should be kept to a minimum.

Vapor freeing by natural ventilation requires only the removal of roof and shell manway covers, preferably in that order, allowing natural air circulation to displace the tank vapors. These vapors may drift for a considerable distance along the ground, so the area surrounding the tank should be evacuated except for the person making necessary vapor tests.

Mechanical ventilation requires some device to educe vapor from a tank or displace vapors by blowing air into the tank. A ventilation rate of several air changes per hour is required to bring the tank vapor concentration down quickly below the flammable range. A slower ventilation rate of one air change per hour or even less will, however, produce satisfactory results. Frequent vapor testing is the only safe way to determine the progress of tank-ventilating procedure.

An explosionproof exhaust fan or steam- or air-powered eductor should be started at a low delivery rate before the shell manway is opened in order to establish a pressure differential so that on removal of the shell manway cover there is no vapor released at ground level.

An electric-driven blower or fan at ground level and discharging into the shell manway of a tank should be explosionproof unless it is located well away from the possible areas of vapor travel.

Portable power equipment, such as a gasoline-engine-driven air compressor or blower, used in isolated areas presents an additional source of ignition. Gasoline-engine-driven equipment should be located out of range of possible vapor travel from the open shell manway and on the upwind side of the tank if possible and preferably outside diked enclosures. A canvas duct may be used to convey air from the blower outlet to the tank.

When steam is used to ventilate a tank, it should be introduced through some connection at or near the bottom of the tank. To avoid the possibility of excessive pressure or of drawing a vacuum in the tank, a roof manway or gage hatch must remain open during the steaming period and until the atmosphere in the tank has cooled.

To displace vapors effectively, steam should be supplied at a rate sufficiently high to raise the temperature of the tank atmosphere to at least 170°F. Steam ventilation is not usually considered the best practice for vapor-freeing large tanks, especially during cold weather. There is seldom adequate steam to vapor-free them, and besides the steam will condense as rapidly as it is introduced and displacement of flammable vapors will cease.

Steam may generate static electricity. The nozzle of the steam hose should be bonded to drain off static charges generated by the flow of steam. However, a charge is generated in the steam exhaust after it leaves the nozzle and can charge metallic objects which are insulated from the ground inside a tank. Sparks may occur between the charged object and a grounded object.

Vapor Testing

Samples of vapor should be taken at the exhaust outlet and surrounding area and checked. Care must be taken when sampling to avoid drawing moisture into the indicator. Such moisture may result in erroneous readings. When vapor concentration has been reduced to 50 per cent of the lower flammable limit and air is entering the shell manways, the presence of men around the tank no longer need be restricted. However, introduction of potential ignition sources within the area should still be subject to rigorous control.

When the vapor concentration in the mixture leaving the tank is reduced to approximately 20 per cent of the lower flammable limit, the first objective has been essentially accomplished. However, this condition is not necessarily permanent, and ventilation should be continued. The exact vapor concentration which will be considered safe before proceeding with the next step in the work will depend on the program set up for sludge removal. This, in turn, will depend on the size of the tank, the facilities available, the amount of sludge, and other factors.

If the tank has not contained leaded gasoline and the indicator registers a reading below 4.0 per cent of the lower flammable limit, the tester may enter the tank without respiratory equipment to make further tests inside the tank. During this testing, ventilation should be continued.

Cleaning from Outside

When the tank is temporarily vapor-free, cleaning may be started after removal of remaining manway covers, riveted door sheets, or bolted cleanout cover plates. Initial cleaning should be performed from outside the tank. Water-hose streams directed through open manways, rotating nozzles pointing inward from the tank shell, and numerous similar devices have been successfully used to dislodge sludge and float it to a water-draw or pump-out connection. Ventilation should be continued, stirring of a sludge will release vapors and increase vapor concentration.

Discharging sludge-laden water into a closed sewer or into a guarded sump which can be protected from accidental ignition is preferable. If sludge-laden water is discharged into an open pit, sewer, or ditch, sources of ignition must be avoided. Discharge into a public sewer should be avoided.

Pumping equipment used for the removal of sludge and excess water from tanks should, preferably, be air or steam driven. If it is necessary to use electric-power- or gasoline-engine-driven equipment, the following precautions are recommended:

1. An adequate flow of fresh air should be circulated through the tank and exhausted from the roof manway.

2. Gasoline-engine-driven equipment should be on the windward side of the tank.

3. The area around the tank should be tested for flammable vapors.

4. If a gasoline-engine-driven pump is used to remove tank residues, it should be attended and properly maintained for continuous operation during the period of tank cleaning.

5. If at any time the flow of fresh air into the tank is stopped, the gasoline-engine-driven pump should be stopped immediately.

Entering

A tank which has not previously contained leaded gasoline may be regarded safe for entry without respiratory equipment provided flammable-vapor tests show that the tank atmosphere is below the allowable concentration, 4.0 per cent of the lower flammable limit or 500 ppm.

When the tank is entered, its interior should be inspected for any materials which might fall. It is desirable to continue forced ventilation, regardless of test results for flammable vapors, until sludge has been removed. Tests for flammable vapors should be repeated at frequent intervals throughout the entire cleaning period.

While workmen are inside a tank completing the cleaning process, one man should be available outside the tank to assist those within in the event of an emergency. Life lines attached to the D ring of a workman's harness may be used for added protection. The outside observer should also have adequate respiratory equipment available. Someone qualified to administer artificial respiration and simple first aid should be available. Lighting should be provided by means of explosionproof lamps.

The simplest method for sludge removal is to wash, brush, or sweep the sludge into piles and shovel it into buckets or wheelbarrows to remove it from the tank, sweep and wash down the tank with a water-hose stream, and remove remaining moisture from the tank with an absorbent such as sawdust, spent clay, or rags.

If the tank has floor-level cleanout openings, much of the sludge can be removed by flushing it out with a high-pressure water stream. Without such openings, pumps of the self-priming type or steam- or water-operated ejectors may be used.

Vacuum-tank trucks provide a fast and efficient method for removing and hauling sludge from tanks. The area of operation must be vapor-free, and the truck should be located upwind from the tank.

Repairs

Should cold-work repairs raise perceptible dust, personal protective equipment should be worn. If repairs involve hot work, surfaces to be heated, including roofs and structurals, should be free of liquid hydrocarbons and ignitible scale deposits. Continued ventilation will minimize accumulation of flammable vapors in the tank.

Repairs to original or false tank bottoms may be accomplished by drilling and tapping the bottom to provide connections by which carbon dioxide, water, or other purging agents may be introduced.

Cleaning Tanks without Vapor Freeing

Cleaning tanks which are not vapor-free should be avoided whenever possible. They may contain a flammable atmosphere.

If such cleaning is essential, however, the area surrounding the tank should be kept free of ignition sources from the time the tank cleaning starts until the tank is closed and ready for return to service. Under no circumstances should electric-power-operated tools or lights or any equipment operated from an extension cord be permitted in the tank. Only dry-cell-powered flashlights, safety lanterns, or cap lamps should be used.

The tank should be pumped out and drained of gasoline to the lowest possible level. In some cases, it may be possible to remove additional product through the water-draw connections. Such operations should be completed before the manway covers are removed.

Workmen removing manway covers or working near open manways should keep to the **upwind side** and should avoid inhaling vapors from the tank. Under certain conditions, respiratory equipment should be used.

Every practical effort should be exerted to perform the necessary cleaning without entering the tank. Sometimes sediment or scale may be removed from the tank by directing a stream of water from a hose through the manway while pumping out or draining. The hose nozzle should be bonded to the tank.

Workmen entering a tank should be provided with protective gloves and boots and with protective respiratory equipment. If the tank is entered through the roof manway, a ladder should be inserted through the manway and secured and a life line should be used. Workmen within the tank should be under the constant observation of a responsible individual outside the tank.

Contracting Tank Cleaning

Because of possible dangers to plant personnel and equipment, tank cleaning contractors should be asked to follow the preceding precautionary measures.

Storing JP-4 and Similar Products

Because the vapor space in cone-roof tanks storing JP-4 and products with similar Reid vapor pressures (2 to 3 lb) is in the explosive range in temperate climates, it is recommended that all such products in those areas be stored in floating-roof tanks. If this is not practical and storage in cone-roof tanks is necessary, precautions should be taken.

When pumping JP-4 into an air-empty tank, some plants pump a small amount of light product into the tank first so that the atmosphere quickly goes into the "too rich" condition.

The following suggestions are recommended for cone-roof tanks:

1. There should be no overshot or "splash" filling.
2. Automatic gages or line sampling should be used to avoid the necessity of personnel going on the roof to open hatches.
3. If manual sampling and gaging are necessary, such should not be done until 1 hr after pumping into the tank.
4. When first pumping into a cone-roof tank, the pumping rate should be limited to not more than 3 fps linear velocity until 6 ft of product is in the tank. Maximum barrels-per-hour rates for various outlet sizes which are within this velocity limit follow:

Outlet size, in..	2	2½	3	3½	4	5	6	8	10	12
Rate, bbl/hr...	43	64	93	125	165	250	360	670	1,040	1,465

5. Pressure and vacuum vent valves should be in good working order.
6. No object which can contain a charge of static electricity should float on the surface of the liquid.
7. Care should be exercised to prevent build-up of water bottoms which can develop a static charge.

In filling tank cars and tank trucks:

1. The fill pipe should be extended to the bottom and pumping rate should be low until the fill pipe outlet is covered with product.
2. Filling should not be carried on during a lightning storm.
3. Bonding wires should be in place between the tank shell and the loading lines before the manhole cover is opened; no sampling or gaging should be done or temperature taken until the tank is full or until pumping has ceased for half an hour.

Hot-oil Tanks

Many serious fires have started from storing oils in cone-roof tanks at temperatures in excess of 212°F. The principal danger arises from the chance of water being introduced into the hot oil.   The resulting steam forces the hot oil up to the roof of the tank, ruptures the roof, and the oil then flows over the side of the tank.

If products must be stored at temperatures above 212°F, the following precautions should be taken:

1. Drain all water from products to be stored.
2. Circulate products continuously in hot oil tanks so that small quantities of water cannot settle and collect in the bottom of the tank.
3. Prohibit steam coils in the tank for heating the oil.   Desired temperatures can be maintained by circulating to a nearby furnace.   Such circulation will accomplish the objective outlined in item 2 above.
4. Keep inert gas in the vapor space so that the atmosphere is below the explosive range, or maintain the vapor space in a "too rich" condition.
5. Construct diversion dikes to keep "boil-over" from reaching a source of ignition.
6. Use antifoaming agents.[28]

Radioactive Materials Safety

One person in a plant should be given responsibility for directing the safe handling and storage of radioactive materials.   He should be known as the plant radiological officer and can be under the direction of the safety director.

Table 9-10. Helpful Booklets When Planning a Safety Program for the Use of Radioactive Materials*

41. *Medical X-ray Protection up to Two Million Volts*
42. *Safe Handling of Radioactive Isotopes*
48. *Control and Removal of Contaminants in Laboratories*
50. *X-ray Protection Design*
51. *Radiological Monitoring Methods and Instruments*
52. *Maximum Permissible Amounts of Radioisotopes in the Human Body and Maximum Permissible Concentration in Air and Water*
53. *Recommendations for the Disposal of Carbon-14 Wastes*
54. *Protection against Radiation from Radium, Cobalt-60, and Cesium-137.*
59. *Permissible Dose from External Sources of Ionizing Radiation*
62. *Report of the International Commission on Radiological Units and Measurement*

* All these can be ordered from the Superintendent of Documents, Government Printing Office, Washington 25, D.C.

The radiological officer should be advised by the department ordering radioactive materials as to the expected uses of the materials or equipment.   The department ordering should also advise the purchasing department and the storeroom where the product is to be received.   The purchasing department should confirm the order to the radiological officer, and the storeroom should advise the radiological officer when the material or equipment is received.

All personnel who are expected to use radioactive materials or equipment should be trained in:

1. General hazards of radiation exposure
2. Hazards of specific materials or equipment
3. Proper handling and storing methods
4. Monitoring of personnel and maintenance of monitoring reports and records
5. Radiation warning signs
6. Transportation of materials in the plant
7. Surveys and checks of materials, equipment, storage areas, and work areas
8. Emergency procedures for sudden exposure or loss of materials

These steps should be in writing in a **formal** plant manual, and the radiological officer should see that it is kept up to date.   A list of government booklets which will be of help in the preparation of such a plant manual appears in Table 9-10.

Floating-roof Tank Hazards

When a floating-roof tank is first filled with a product having a Reid vapor pressure similar to JP-4, it is advisable first to pump in a "light" product slowly so that the vapor space below the roof will go quickly to a "too rich" mixture.

When products in a floating-roof tank are gaged or sampled manually, the gager or sampler should arrange that someone else in the plant will know of his whereabouts and investigate should he not report back within a certain period of time. If hydrogen sulfide is present, protective equipment should be worn and another employee should stand by.

Floating Plastic Blankets

Evaporation losses in cone-roof tanks can be reduced by plastic blankets, but for safety's sake it should be assumed that JP-4 or gasoline vapors in the space above the blanket are always in the explosive range when floating blankets are used unless tests show otherwise.

The following precautions in the use of such blankets should be exercised:

1. Use automatic gages, or sample from product lines; avoid open gage hatches.

2. Do not gage or sample from the roof until at least 1 hr after completion of pumping. This time should allow any static charge at the surface of the liquid, at the blanket edges, or at the blanket gage hole to dissipate.

3. The gager should remove his gloves and touch the tank top with his bare hand to equalize any electrical differential between his body and the tank before opening the gage hatch; the tape or gaging chain should be kept in contact with the rim of the hatch while being lowered into the tank.

4. No hand gaging or sampling should be done during an electrical storm.

5. Liquids should not be poured back into a tank.

6. A gager should not carry a cigarette lighter.

7. If the floating blanket is metallic, a bonding wire should be run from the blanket to the tank roof.

When products with a Reid vapor pressure similar to JP-4 are stored, a possible value in the use of floating blankets may be to keep the vapor space below the explosive range. However, in very hot climates it might be better not to use the blanket because the vapor space would probably have a better chance of remaining in a "too rich" condition without the blanket and therefore be less hazardous.

Tanker Loading and Unloading

Serious hazards can exist in the loading and unloading of petroleum products at marine docks. Tanker or barge captains, as well as dock foremen, should have procedures to follow during loading and unloading operations, such as the following:

1. Do not clean ship's product compartments while docked.

2. Loading hoses are to be drained on the dockside, not blown clear by air or compressed gas into a ship's compartment.

3. Keep flame arrester screens at ullage hatches in place during the entire loading or unloading period, except when gages, temperatures, or samples are taken.

4. Keep hatch covers closed while the ship is docked.

5. Keep all doors and portholes in the tanker midship house and on the forward side of the afterpart of the ship closed during loading operations. Port and starboard doorways on the bridge deck may be open. Personnel entering the midship house during loading should use passageway doors on the captain's deck. Only port, starboard, and galley doors should be used to enter the afterpart of the ship. The doors should be closed except when in use.

6. Fire-water hose lines should be laid on deck, ready to use.

Special precautions to be observed by personnel on the dock are:

1. Watch for movement of the ship during loading or unloading which may strain the hoses; advise the ship's loading officer of movement, stop pumping if indicated.

2. Watch for compartment overflowing.

3. Have shore fire-water hoses ready for immediate use.

4. Prohibit smoking on the dock.

5. Drain loading and unloading hoses on the dockside, do not blow by air or other compressed gas into a ship's tank.

6. JP-4, kerosine, Nos. 1 and 2 fuel oil, or any other similar product should be pumped slowly when first discharging into a tanker compartment until the tanker loading officer advises that the "bellmouth" has been covered.

7. Determine that bonding wires for "stray" currents are in place.

8. Observe to determine if tanker personnel take safety precautions requested of them. If not, advise the tanker's executive officer.

Sandblasting Tanks in Service

There has been considerable debate over the possible hazards of sandblasting storage tanks while in service. The following is abstracted from a discussion of this subject by Holly P. Bradley.[29]

Static electricity build-up is common in this type of work. If the sandblasting equipment, the operator, and the tank are maintained at the same potential by bonding attachments, however, static build-up and consequently dangerous arcs will not occur.

Sandblasting does create sparks, but they are not hot enough to ignite flammable vapors. There is some heat build-up from the abrasive action of sand on the tank steel, but this is no problem, since the temperature build-up is well below the ignition temperature of petroleum vapor—approximately 450°F.

To sandblast a tank roof in service, the tank is isolated by closing the valves on all incoming and outgoing lines.

The "no smoking" rule must be observed.

When the roof of a cone-roof tank is being sandblasted, all fire screens must be in place, all snuffers closed, and all gage hatches closed except one which must be removed. All gage hatch covers and other tank appurtenances should be checked to see that they are not leaking. If necessary, water-soaked burlap may be placed over these openings to provide additional protection. Also, sandblasting must not be done on any tank which has corroded to the extent that leakage occurs.

At the opening from which the hatch is removed, a specially designed vent should be installed to permit "breathing" of the tank caused by atmospheric changes.

The tank and the generator must be grounded (not more than 25 ohms resistance), and the hose must be the conductive type. This requirement applies to all tanks— either cone- or floating-type roofs.

The generator, hose, nozzle, and the operator must be bonded and grounded to the tank roof and shell. On floating-roof tanks, the tank and the roof must be bonded and grounded at the top and bottom of the ladder and from the roof to the sealing ring shoe. Also, the generator, nozzle, hose, and the operator must be bonded into the same circuit with the tank and roof.

When a floating-roof tank is being sandblasted, the tank should be virtually full in order to reduce the dead air space.

All openings in sight-gage appurtenances from which vapor may escape into the atmosphere should be sealed off. All vents, gage hatches, etc., must be closed tightly to prevent vapors from escaping. Sandblasting must not be done on any floating roof of which the secondary seal is defective. Adequate fire-fighting equipment must be available on the roof and manned, ready for use.

A sandblast helmet and gloves should be worn by the operator when sandblasting.

Painting Tanks in Service*

This procedure was prepared to establish a standard practice for preparing and painting the exterior of tanks containing products with a flash point of 125°F or lower.

* Abstracted from "Safe Practice Pamphlet 29," *National Petroleum Association Report* F&S 60-32.

A written permit good for one day only should be issued to the maintenance group before preparation and painting are begun.

Regardless of the flash point, tanks containing material bearing hydrogen sulfide should be checked. When concentrations are above 0.04 per cent H_2S (400 ppm) the tank should first be emptied and the gas concentration reduced below 0.002 per cent H_2S (20 ppm).

When the roof, roof fittings, side walls, and bottom do not leak, the tank should be filled so that the vapor space is reduced to a minimum without the possibility of overflowing due to expansion.

Air-driven tools of nonsparking materials may be used on the side walls up to within 2 ft of the roof; from there on up, the sides and roof must be cleaned with nonsparking hand motive tools.

In circumstances where a tank cannot be held full for the complete preparation and painting operation, the tank may be painted when partially filled, provided the preparation has been completed and the priming coat has been applied. Jobs of this nature should be kept at a minimum.

Under no conditions should pumping in or out of the tank be permitted while the job is actually in progress.

Hydrogen Sulfide Hazards

Hydrogen sulfide is five or six times as deadly as carbon monoxide. The maximum allowable concentration for H_2S is 20 parts of H_2S per million parts of air (0.002 per cent H_2S). When exposure to H_2S is greater than 20 ppm but in concentrations of less than 2.5 per cent by volume in open areas, an approved gas mask with a yellow canister should be worn. Self-contained breathing apparatus or fresh-air breathing equipment should be worn for higher concentration or in enclosed areas. Protective goggles also should be worn.

Some of the more hazardous areas of H_2S exposure are tank gaging, sample taking, water draw-off, pump houses, and low manifolds and in cleaning pipes, pumps, sewers, tanks, towers, drains, and other vessels.

Hydrogen sulfide can be found in scale and sludge. Tank roof corrosion is high in the presence of H_2S.

Hydrogen sulfide is an explosive gas. The lower explosive limit is 4.3 per cent, and the higher is 45 per cent. In the presence of H_2S, men should work in pairs.

A note of caution: In high concentrations, the odor of H_2S is noticed for only a brief moment because the sensory nerves are quickly immobilized by the gas.

Engine-starting–Air-line Explosions

Safety committees of the Natural Gas Processors Assn. and of the American Petroleum Institute have been working together to secure broad experience as to explosions and their possible causes when rolling and starting gas-fueled engines with air. In a recent survey by those two organizations, 63 companies returned helpful information.

No final conclusions have been reached, but the following information from the "Joint Report Covering Summary and Answers to Questionnaire on Engine Starting-Air Line Explosions (FP/141) Obtained by Natural Gas Processors Assn. and American Petroleum Institute Fire Protection Engineering Subcommittee," dated May, 1960, may be of help to plants faced with this problem.

During the last 10 years, companies reporting indicated that there have been a total of 63 starting–air-line explosions which resulted in 13 people receiving minor injuries and one person being killed.

"In most of the cases, the starting-air line explosions have ruptured lines and fittings adjacent and away from the engines. Only two companies reported damage was done to lines and fittings on the air receivers. A few companies have experienced paint being burned on lines and air storage tanks, indicating an internal fire. **All of the companies that have experienced air line explosions indicated that explosions always occurred when air was being used to start an engine.** In most of the explosions that were reported most of the companies indicated that there was evidence of the power cylinder check valve being stuck open and in most cases they were not sure that the checks were operating freely and properly. . . . "Only one company indicated that rupture discs in the air line header has prevented the air line itself from rupturing when an explosion occurred. In cases where companies have had starting air line explosions, eight companies felt the source of fuel came from the lubricating oil in starting air compressor. Twelve companies thought that gas was backing from power cylinder into air line. Four thought it was from lubricating oil in starting air housing on engine. One company felt the explosions appeared to be exothermic and are not dependent upon fuel similar to TNT, but in this case oxygen complexes of carbon and oil. Two companies thought the source of ignition for the fuel in starting air line explosions was from carbon deposits in the line. Thirteen companies thought the source of ignition was the power cylinder. None of the companies thought it was caused by static electricity and only four companies thought that heat of compression in air compressors was the source of ignition.

"Fifty companies reported that they have a second block valve between starting air valve and engine with a bleeder between these two valves to keep gas from coming back from power cylinder while engine is operating.

"Companies have various criteria as to when the compressors should be reconditioned and not too many of them use oil consumption as a gauge for reconditioning compressors. Most of the companies follow a definite schedule for blowing the condensate from drips and air receivers and the frequency ranges from once every eight hours to once every week. The majority of the companies employ a definite schedule for checking the mechanical condition of their air compressors. Air compressors are checked by either operators or mechanics by most companies at least once a day and, of course, the length of time between complete overhauls depends upon the amount of time the machine is operated, with the make and type of compressor being a contributing factor. . . ."

Blowing Hydrocarbon Lines with Air

Dangers in blowing hydrocarbon lines with air are threefold:
1. When lines are blown into a storage tank, tanker, barge, or any container, the air bubbling up to the surface of the liquid can develop a charge of static electricity on the surface. If the hydrocarbon vapors at the surface are in an explosive range, they can be ignited should the static discharge.
2. When lines are blown into any container and the vapors from the product are in the flammable range, vapors may be blown from the container to a source of ignition. The ignited vapors can then flash back to the storage tank or container, thereby setting the contents on fire.
3. When air is used to clear lines of products, it can sometimes be "trapped" in operating units where later exposure to hydrocarbon vapors can help to form an explosive mixture.

Hazardous Electrical Areas

The Committee on Refinery Equipment, Division of Refining, American Petroleum Institute, has prepared a publication entitled *The API Recommended Practice for Classification of Areas for Electrical Installations in Petroleum Refineries*.[30] Various sections of this publication are abstracted in this section.

The following discussion is only a guide and should be used with sound engineering judgment.

National Electrical Code

The National Electrical Code[31] is used widely as a guide to good electrical practices and has been adopted as law at various state and local levels. Most installations, regardless of legal considerations, conform as a matter of good practice. The API's recommended practice is not an attempt to rewrite or otherwise supersede the NEC or other applicable codes or ordinances but to supplement them.

Once an area has been classified as to hazard, there is little difficulty in understanding the NEC requirements for equipment. The problem is to determine the existence, degree, and extent of a classified hazardous area. While the NEC has established certain criteria for this determination, the API publication considers the application of these criteria to petroleum refineries.

Article 500 of the NEC recognizes two types of hazardous areas directly and a third by indirection:

Division 1 Areas: Those likely to be hazardous under normal conditions

Division 2 Areas: Those likely to be hazardous under abnormal conditions only, such as equipment failure

Nonhazardous Areas: Locations which cannot be classified as Division 1 or 2

Installations for Division 1 areas use "explosionproof" equipment. Division 2 installations use equipment that does not provide ignition sources under normal conditions of full operation.

Degree of Hazard

Once a hazardous area has been identified, its degree of hazard must be determined. As stated previously, the criterion for Division 1 is whether the location is likely to be hazardous under normal conditions. "Normal" conditions for Division 1 does not necessarily mean that everything is working properly. An operation might be so sensitive to control that frequent operation of the relief valves would be considered "normal." If the valves released flammable liquid or vapor to the atmosphere, the adjacent area would be classed as Division 1. If, however, the relief valves operate infrequently under unusual conditions, it would not be considered "normal."

Similarly, in areas where frequent maintenance and repair are viewed as "normal," the accompanying release of significant quantities of flammable liquid or vapor would classify the location as Division 1.

Extent of Hazardous Areas

The final step is to determine the extent of a hazardous area, the most difficult step of all. A good beginning is to remember that hydrocarbons are generally heavier than air, therefore:

1. With no walls, enclosures or other barriers, air currents, or other disturbing forces, assume that a vapor will disperse in all directions and the horizontal area covered by the vapor will be a circle.

2. Concentrations of heavier-than-air vapors released at or near grade level are most likely below grade, are next most likely at grade, and decrease as height above grade increases. In open locations away from the immediate point of release, freely drifting heavy vapors from a source near grade seldom reach ignition sources more than 6 or 8 ft above grade. For lighter-than-air gases, there is little or no potential hazard at and below grade, greater potential hazard being above grade.

3. Elevated or depressed sources of release will alter the zones of potential hazard.

Air currents may substantially alter the outline of the limits of potential hazard. Thus, area limits recommended for Division 1 or Division 2 locations must be recognized from experience rather than from any theoretical diffusion of vapors.

There is usually a twilight zone between a Division 1 location and a nonhazardous location, a normally nonhazardous zone which could be hazardous under abnormal conditions, such as unfavorable air currents, an abnormally large release of flammable material, etc. For cases in which an unpierced barrier, such as a blank wall, serves to prevent vapor spread, there would be no twilight zone.

Consider a process vessel containing hydrocarbons and which normally does not release flammable material. There is no Division 1 location. However, the vessel might leak under abnormal conditions, and therefore surrounding the vessel is a Division 2 zone.

The significance of Division 2 is that it is the twilight zone which normally exists between a source of hazard and a nonhazardous area. In these areas the law of probability comes into play.

Process equipment does not fail very often. Furthermore, the NEC requirements for electrical installations in Division 2 locations are such that a spark can occur in a flammable vapor-air mixture only in the event of a breakdown of electrical equipment. Such breakdowns do not happen frequently either.

Correctly evaluated, a petroleum installation will be found to be a multiplicity of Division 1 locations of extremely limited extent. Probably the most numerous offenders are packing glands.

Even heavy vapor is rapidly dispersed in a freely ventilated area. For that reason, areas out of doors or those having ventilation equivalent to normal outdoor conditions are generally classed as Division 2. However, wherever ventilation is restricted, vapor concentrations can develop and the situation may well justify the entire area of restricted ventilation being classed as Division 1.

The following method for determining the degree and extent of classified areas has been developed by survey and analysis of the practices followed in the petroleum refining industry:

1. For refinery process areas with unrestricted ventilation where flammable liquids or vapors are handled only in closed systems, refer to Figs. 9-8 and 9-9.

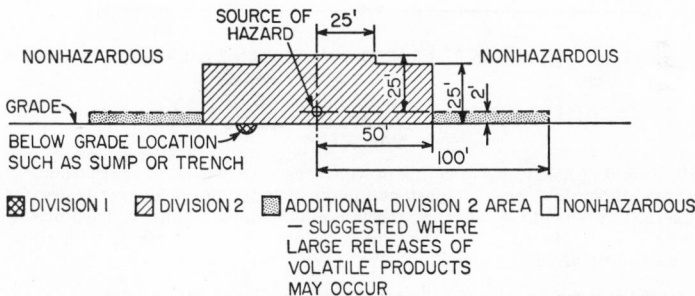

Fig. 9-8. Extent of hazardous area in a freely ventilated process area (source of hazard located near grade).[30] Note: Distances given are for average refinery installations; they must be used with judgment, with consideration given to all factors discussed in RP 500.

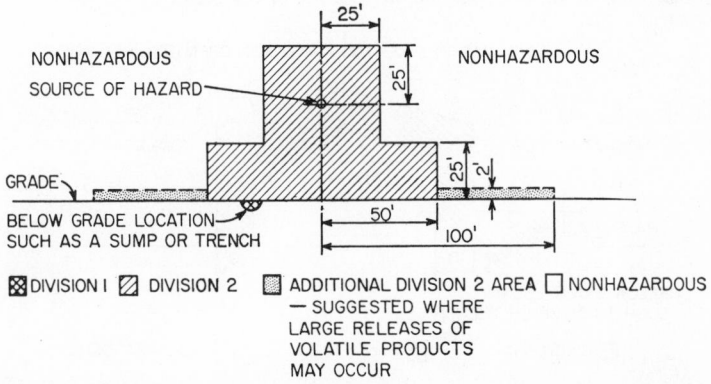

Fig. 9-9. Extent of hazardous area in a freely ventilated process area (source of hazard located above grade).[30] Note: Distances given are for average refinery installations; they must be used with judgment, with consideration given to all factors discussed in RP 500.

a. *Extent of Division* 1: Negligible for above-grade locations. Entire area of open, below-grade locations such as pits, sump, and trenches. This should not be interpreted to mean that the earth itself is hazardous. However, buried conduits may collect hazardous liquids or vapors which can then be communicated to otherwise safe locations unless properly sealed.

b. *Extent of Division* 2: Use clearances given in Figs. 9-8 and 9-9.

2. For refinery process area with restricted ventilation, refer to Fig. 9-10.

a. *Extent of Division* 1: Use clearances given in Fig. 9-10.

b. *Extent of Division* 2: Use clearances given in Fig. 9-10.

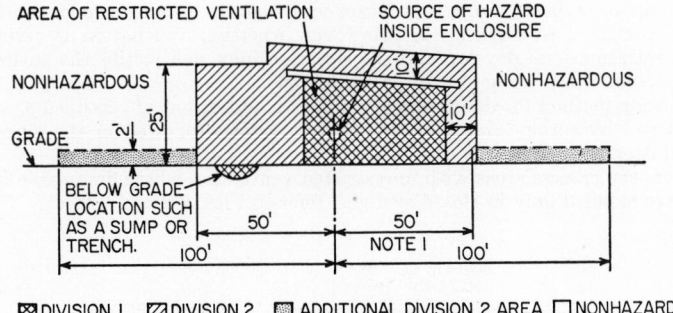

Fig. 9-10. Extent of hazardous area in a process area with restricted ventilation.[30]   Note 1: Apply above dimensions or 10 ft past perimeter of building, whichever is greater; except past unpierced vaportight walls, the area is nonhazardous.   Note 2: Distances given are for average refinery installations; they must be used with judgment, with consideration given to all factors discussed in RP 500.[30]

3. For tanks in freely ventilated areas, refer to Fig. 9-11.

a. *Extent of Division* 1: Vapor space within tanks and immediate vicinity of vents; otherwise negligible.

b. *Extent of Division* 2: Use clearances given in Fig. 9-11.

4. For loading racks, use Art. 515, Sec. 515-2, of the NEC.[31]

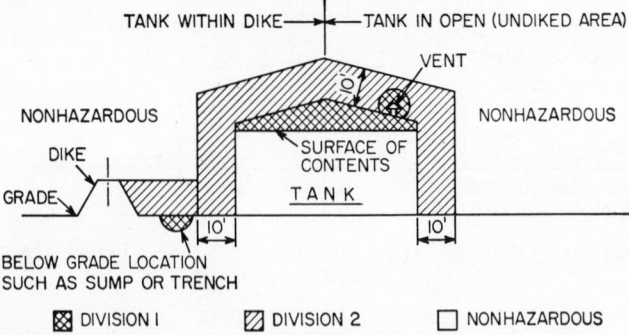

Fig. 9-11. Extent of hazardous area in and around a refinery tank.[30]   Note 1: For floating-roof tanks, the area above the tank roof and within the shell is classified Division 1.   Note 2: High filling rates or blending operations involving Class 1 liquids may require extending the boundaries of classified areas.   Note 3: Distances given are for average refinery installations; they must be used with judgment, with consideration given to all factors discussed in RP 500.[30]

Method to Determine Hazardous Area Extent

With the preceding background material and the principles set forth, the API publication recommends that the existence, degree, and extent of classified areas be determined by considering a series of questions. Each room, section, or area is to be considered individually to determine its classification.

Step 1—Need for Classification

The need for classification is indicated by an affirmative answer to one of the following questions:
1. Is the presence likely of flammable liquids or vapors having flash points less than 70°F?
2. Are any flammable liquids or vapors having flash points under 200°F likely to be handled, processed, or stored at temperatures above their flash points?

Step 2—Assignment of Classification

Assuming an affirmative answer from Step 1, use the following questions to determine the assignment of classification.

Division 1 locations may be distinguished by an affirmative answer to any one of the following:
1. Is a hazardous gas or vapor concentration likely to exist in the air under normal operating conditions?
2. Is a flammable atmospheric concentration likely to occur frequently because of maintenance, repairs, or leakage?
3. Would a failure of process, storage, or other equipment be likely to cause as electrical-system failure simultaneously with the release of flammable gas or liquid?
4. Is the location an area not freely ventilated and one in which flammable liquids or vapors are handled or processed in other than a suitable, well-maintained piping system without valves, screwed or flanged fittings, and meters or are stored in other than suitable closed containers?
5. Is the area below the surrounding elevation or grade such that flammable liquids or vapors may accumulate in it?

Division 2 locations may be distinguished by an affirmative answer to any one of the following:
1. Is the location a freely ventilated one in which flammable liquids or vapors are processed or handled in closed systems (other than suitable, well-maintained piping systems consisting only of pipe, valves, fittings, and meters) from which they can escape only during abnormal conditions, such as accidental blowing of a gasket or rupture of a pipe, or are stored in other than suitable closed containers?
2. Is the location adjacent to a Division 1 location, or can vapor be communicated to the location as through trenches, pipe, or ducts?
3. If positive mechanical ventilation is used, could failure or abnormal operation of the ventilating equipment permit atmospheric vapor mixtures to build up to hazardous concentrations?

Step 3—Extent of Classified Areas

The extent of a classified location is determined by applying the recommended distances given in Figs. 9-8 through 9-11 or, for locations not covered in these figures, the recommended distances specified in Art. 515, Sec. 515-2 of the NEC.[31]

Step 4—NEC Grouping of Atmospheric Mixtures

Equipment must be selected, tested, and approved for the specific flammable material involved, inasmuch as maximum explosive pressures and safe operating temperatures vary widely with composition of flammable material.

Electricity: Static, Lightning, Stray

The American Petroleum Institute Committee on Static Electricity and Stray Currents, under the direction of the Central Committee on Fire Protection, has prepared the manual *Protection against Ignitions Arising out of Static, Lightning, and Stray Currents*.[32]

This publication, API RP 2003, describes some of the conditions which have resulted in oil fires ignited by electrical sparks and arcs from so-called natural causes, as well as the methods which the petroleum industry currently is applying for preventing such ignitions. The material which follows has been abstracted from API RP 2003.

Static Electricity

Products handled at temperatures at or above their flash points and products delivered into tanks which contain flammable vapor-air mixtures are considered here.

Tank Trucks

When tank trucks are loaded through open domes, one end of a bond wire is electrically connected to a fill stem, piping, or inherently connected steel loading rack and the other end to the tank truck. This latter connection is made before the dome is opened and remains in place until the dome has been closed. The bond prevents the build-up of a difference in electrical potential between the fill stem and the tank truck.

Bond wires are not needed around flexible metallic joints or swivel joints, because high resistance (as high as several million ohms) can be tolerated for static dissipation.

Downspouts which extend into the body of the truck tank, used principally to reduce evaporation losses, should be long enough to extend to the bottom of the tank.

Static protection is not required when tank trucks are loaded or unloaded with conductive or nonconductive rubber hose, flexible metallic tubing, or pipe connection through or from tight top or bottom outlets (not open domes), as no gap exists over which a spark can occur.

Unloading with suction unloaders through the open dome requires the same protection as the loading of a tank truck, because flow through the piping can produce potentials between pipe and truck tanks.

If bonding is to the rack, it is essential that piping and rack be electrically interconnected. In such cases the grounding of the rack provides no additional protection against static ignition.

Tank Cars

The resistance of tank cars to ground through the rails may be assumed adequately low to prevent the accumulation of a hazardous static charge **on the tank body.**

Stray currents on spur tracks are discussed in more detail on page 9-69. Loading lines must be bonded to the rails to prevent a difference in electrical potential between the fill pipe and the tank car which might result from stray currents. "Splash" filling should be avoided, and downspouts should extend to the bottom of the tank car.

The unloading of tank cars through open domes using suction pipes requires no protection against static sparks between the cars and the suction pipes in the domes, since the cars are adequately bonded to the rails and they, in turn, are electrically connected to the unloading piping. Neither does the unloading of tank cars from fixed top or bottom outlets by means of conductive or nonconductive rubber hose, flexible metallic tubing, or pipe connections require any additional protection against static sparks inasmuch as the flammable material is in a closed system.

Drums and Cans

Filling of drums and cans on conveyors or on other metallic foundations inherently bonded to the filling arm requires no additional protection against possible static accumulation and discharge. Nor is further bonding needed at the possible spark gap between vessel and spout required for cans and drums filled through flexible metallic spouts which can be maintained in contact with the vessel throughout the filling operation. The current produced by such filling operations is always small, and an electrical path is sufficiently conductive to prevent incendiary sparks.

Service-station Deliveries

No bonding is required during the delivery of gasoline from service-station tanks to fuel tanks of motor vehicles, irrespective of whether the hose and nozzle are conductive or nonconductive. Experience over many years in the fueling of millions of motor vehicles indicates no static ignition hazard.

Aircraft

A bonded hose is not necessary for the safe fueling of aircraft if prior to and during the fueling the hose nozzle is bonded to the plane with a short bond wire and clip.

Tank Ships

There is no static hazard connected with the loading and unloading of steel tank ships and barges, except for the static charges which may accumulate on the surface of the liquid from turbulence in loading. The hull of the ship will not accumulate a static charge because the ship is in contact with the water and thus is inherently grounded. (Bonding cables which are often used between ship and shore are for protection against stray currents only.)

Storage Tanks

The presence of water in oil is one of the important factors in the generation of static electricity on oil surfaces by turbulence. Large charges can build up on the surface of refined oils (normally quite "dry") when water settles downward from the oil surface. A sufficient distribution of water droplets retained throughout the oil body, as in oil-with-water emulsions, appears to reduce static hazard.

The method of introducing oil into a tank should avoid settling of water through an oil body and avoid surface agitation, as by "splash filling" or by filling at high velocity while the oil level is low.

Outlets of downcomers on overshot fill lines should be arranged so that the incoming oil is discharged horizontally rather than straight downward. Siphon breakers which permit air or vapor to enter the downcomer should not be used.

During the initial stages of tank filling, there is more opportunity for the incoming stream to produce agitation or turbulence. Some companies, therefore, limit the velocity of the incoming liquid stream to 3 fps until 6 ft of depth has been reached or, in the case of a floating-roof tank, until the floating roof is buoyant. In the light of present knowledge, it is not possible to recommend any maximum velocity of flow.

Agitators

Air-blown batch agitators are prolific generators of static electricity. The static charges accumulate on the surface of the oil following agitation and during settling of the treating solution and the water wash. Some charges can be detected during agitation and during the water (rain) wash. Charges increase immediately after agitation has stopped and also after the rain wash has been turned off. The charges are quite intensive to begin with, resulting in sparks darting across the surface of the oil, and diminish over a period of 5 to 10 min.

A steam blanket is the protection usually applied to air-blown agitators for the treatment of stocks which contain ignitible vapor-air mixtures at the oil surface. Other methods are inert gas blanketing and the elimination of a free oil surface by use of a closed circulating system (commonly called the continuous treating system). Some companies have tried wire screens and similar devices to obtain a short-circuiting effect on the free oil surface.   The use of such devices has not been successful, and they are not recommended.

Jet and Propeller Mixing

A test was made from static-electricity generation during low-velocity propeller mixing; no charges were detected.   Such propeller mixers have been in use for many years without evidence of trouble from static generation.   In-tank jet mixing and high-velocity propeller mixing are more recent, and their use is increasing.   Some adverse experience has been reported relative to static generation during high-velocity mixing.   There may be additional hazard if a water bottom is present.

Wearing Apparel

A great many fabrics may generate static electricity when they are brought into contact with other materials and then separated or when rubbed on various substances. Most synthetic fabrics (nylon, orlon, dacron, rayon, etc.) are somewhat more active generators than natural fabrics.

Both rubber- and leather-soled shoes generate static when dragged across dry carpeting or other nonconductive surfaces during periods of low humidity.

Lightning

The most accurate and concise discussion of the subject of lightning is *National Bureau of Standards Handbook* 46: *Code for Protection against Lightning*.   It is sponsored by the Bureau, the American Institute of Electrical Engineers and the National Fire Protection Association (NFPA 78).   In 1952 it was approved by the American Standards Association (ASA C 5).   Of especial interest to fire-protection engineers in the petroleum industry is Part III: "Protection of Structures Containing Flammable Liquids and Gases" (ASA C 5.3-1959).

A direct lightning stroke may be destructive to nonmetallic structures when the latter become part of the path of the stroke between the earth and the charged cloud. Flammable vapor-air mixtures in contact with such a direct stroke will be ignited. In years past most of the damage to oil properties from lightning resulted from the ignition of flammable vapor-air mixtures in oil-storage tanks and reservoirs.   The principal product affected has been crude oil.   In almost all these fires wooden-roofed steel tanks and concrete reservoirs with wooden roofs have been involved.   Occasionally the contents of a steel tank with a steel cone roof have been ignited by lightning, but apparently this has been because the tank roof was not vaportight.   Also a few floating-roof seal fires have occurred on gasoline and crude-oil storage tanks as a result of lightning.

In addition to the hazard from a direct lightning stroke, there may be secondary effects in the form of smaller sparks between objects, also "corona" sparks at high points, all within the induced electrostatic field (area) which is "neutralized" by the main stroke.

Aside from the foregoing, damage from lightning has been infrequent—particularly with respect to tall steel structures found in refineries and natural-gasoline plants. Trucks, ships, and tank cars have been practically free of damage caused by lightning.

Ground Rods

Artificial grounding with driven ground rods neither decreases nor increases the chance of a tank being struck, nor does it reduce the possibility of ignition of the tank contents.

Wooden-roof Tanks

The roof boards of wooden-roof tanks may become ignited by direct lightning strokes, and even high-flash-point oil in the tank may be ignited when the burning roof boards fall into it. If the flash point of the oil is below its storage temperature, direct lightning strokes may also ignite the vapors which escape through the wooden roofs.

Occasionally, protection has been provided by steel, direct-hit lightning towers, approximately 125 ft high with or without additional 25-ft spikes above the latticework. Current treatises on lightning protection recommend designs based on a radius of protection equal to the height of the tower in important cases or up to twice the height in less important cases. Tower layouts based on a cone assumption of $R = 2H$ have thus far proved effective for large-area concrete reservoirs with wooden roofs. The towers are placed around the tank or tanks in such a pattern as to bring the wooden roofs within the overlapping "cones of protection." The towers are electrically grounded and sometimes are interconnected at their tops.

Cone-roof Tanks

The usual means of protection for steel tanks with cone roofs is to maintain the roof structure and its appurtenances in a vaportight condition. This has prevented the ignition of vapors which might otherwise occur through perforated roof sheets or other roof openings. Pressure-vacuum relief valves, without commercial flame arresters, have proved to be satisfactory closures for vent openings. To reduce the ignition hazard, some companies forbid the opening of gage hatches during lightning storms.

Floating-roof Tanks

The chance of igniting vapors at the vapor seal on a floating-roof tank by direct lightning stroke to the tank rim is reduced if the roof seal is tight.

Steel Equipment

Much of the steel equipment in a refinery is immune to the effect of direct lightning strokes; however, concrete and masonry foundations are sometimes provided with artificial grounds to prevent damage to the foundation and to prevent side flashes from a direct lightning stroke. Tests and field experience indicate that elevated steel structures which are adequately grounded shield nearby structures of lesser height against direct lightning strokes.

Steel Process Equipment

With respect to process equipment which consists principally of steel vessels resting on steel or concrete structures, experience indicates that there have been no fires caused by lightning other than the occasional ignition of flammable vapor at vents.

Other Lightning Hazards

A lightning cloud sets up (induces) an electrostatic field over a large area of the earth's surface below the cloud. Within this field some equipment may become highly charged. A lightning stroke will cause the field to collapse, and the release of bound charges on equipment may cause sparks or corona discharge. Sparks may discharge either into the atmosphere at elevated points on the equipment or between insulated and grounded objects. These sparks can ignite flammable vapor-air mixtures.

Vapors which escape from pressure-relief valves at elevated discharge points sometimes have been ignited during lightning storms, as have vapor-air mixtures in storage tanks and reservoirs equipped with nonvaportight steel or wooden roofs. Explosions, followed by fires, have occurred in the space between floating roofs and the sealing ring which presses against the shell of the tank. Seal fires of this type are readily extinguished by appliances such as hand dry-chemical extinguishers and by 1.5-in. (or smaller) foam streams.

One form of protection in areas of high lightning incidence for wooden-roof steel tanks storing oils at or above their flash points consists of a system of wires suspended above the tank roof and grounded at the outer ring. However, modern design tends toward elimination of wooden roofs for atmospheric steel tanks.

Floating roofs, steel cone roofs, and similar vaportight all-steel equipment are virtually standard today. If this equipment is maintained vaportight, it is substantially free of any ignition hazard from lightning sparks.

Tank Ships

Damage to ships from lightning has been infrequent since the advent of the steel hull and steel masts, and no special protective design is required. No adverse experience from lightning has been reported in the case of tank ships whose hatches were closed at the time of a strike.

Current regulations of the U.S. Coast Guard permit pressure-vacuum valves in lieu of flame arresters on mast vent outlets. Regulations of the U.S. Coast Guard require that the loading and unloading of tankers be suspended during lightning storms. Some companies have rules that loading and unloading of tankers be suspended and all openings into tanks be closed when there are lightning storms in the immediate vicinity.

Stray Currents

A stray current is any electrical current, not deliberately applied, which may flow through piping and connected vessels which are normally located in more or less intimate contact with the ground.

Stray currents are of two types, depending on the source: those from power-line leakage and those generated by galvanic action associated with soil contacts.

Stray currents from either source seldom have sufficient potential to cause sparks, but the arcs which result from contact breaking, such as opening a pipe run, may be hot enough to ignite petroleum vapors.

Stray-current Hazards

The differences in electrical potential from stray currents have caused arcs and ignited petroleum vapors when pipelines were cut or separated, as at flanges. Therefore, all buried piping, especially in densely built-up industrial areas, may be assumed to carry stray power currents. Even in the absence of power sources, buried piping may carry currents of galvanic origin. Tanks and other metallic vessels, especially those buried in or resting on the ground, must be assumed to be part of the electrical circuit of their connecting pipe systems unless such circuits have been artificially interrupted by insulating flanges.

Hazard Control in Pipelines

Gas or light oil pipelines carrying heavy stray currents should, where stray currents are known or suspected, be connected by a short heavy-gage bond wire or "jumper" across the point where the line is to be cut or separated. Such a bond must have a reasonably low electrical resistance.

Additional precautions in opening pipelines are:

1. A buried cable should not be cut unless it has first been established that it is safe to do so.

2. Unless someone familiar with the situation is on hand to identify underground wiring, excavation should not be permitted in the vicinity of a cathodic-protection system.

3. Fire-control equipment should be provided in advance, especially for the protection of personnel.

4. The ground return cables of welding machines should be attached directly to the work. The practice of attaching such cables to pipes or steel structures remote from the welding operations may cause arcs or sparks at unexpected locations where flammable vapors are present.

Cathodic-protection Systems

Sometimes the insulating devices and special circuits (e.g., buried cables) used by the corrosion engineer present an additional fire-control problem associated with stray currents. The cathodic-protection system proper (including power sources, cables, etc.) requires its own specific precautionary measures. These are problems for individual engineering study.

Spur Tracks

Pipelines which serve tank-car loading or unloading spots on spur tracks may be, as a result of stray currents, at a different potential from that of the rails. Stray currents may flow in the pipelines or in the rails. The usual protection against stray-current arcs is to bond both rails to the pipelines which serve the loading or unloading facilities. This bond is a permanent electrical connection of one No. 4 or not less than two No. 6 American Wire Gage wires plus an adequate ground. Insulated pipe joints between the loading or unloading facilities and the connecting yard pipelines give additional protection against stray currents.

Sometimes spur tracks connect with electrified main lines, cross electric railway tracks, or are equipped with tail-circuit signal systems. In all such circumstances insulating flanges are placed in the rail joints of the spur track so that the latter will be completely insulated from the source of any return rail currents. Most spurs on main-line rail systems are built and equipped for protection against electrical sparks in accordance with the applicable provisions of *Circular* 17-D, issued by the Association of American Railroads.

Wharf Lines

If stray currents are present in wharf piping, connecting and disconnecting the ship's hose may produce arcs because the resistance of the ship's hull to ground (water) is exceedingly low. However, when piping is grounded as well as bonded to the ship, such arcs are seldom dangerous.

Some companies bond wharf pipelines to the ship with bonding cables both before and during the time hoses are connected. This bond may not eliminate the chance for arcing when hose flanges are made up or separated, as a stray current will follow all available paths.

Some companies insulate the piping of the wharf from the pipeline system with insulating flanges at the points where the pipelines enter the wharf structure and also ground all wharf piping to water, or the insulating flanges may be installed in the pipe risers to the hose connections. Where combinations of these practices are followed, bonding cables are frequently omitted.

Special study is required when steel wharves or piers are cathodically protected. Bonding cables can be eliminated and hose risers can be provided with insulating flanges so that dissipation of cathodic currents from the wharf to the ship's structure will be prevented.

RECORDING AND MEASURING WORK-INJURY EXPERIENCE

Statistics on work-injury experience are useful in evaluating the relative need for accident-prevention activities, the seriousness of the accident problem, the effectiveness of safety activities, and the progress made in accident prevention. Figure 9-12 is a typical chart based on injury statistics.

Uniform methods of recording and measuring work-injury experience are necessary so that valid comparisons can be made among various departments or units in a refinery, among different plants or companies in the industry, and even among the refining and other industries, as in Fig. 9-12.

The Z 16 Standard of the American Standards Association is widely used for this purpose, and the following definitions and rules for measuring work injuries are quoted from the Z 16 Standard.[34] Although the Standard is not quoted in its entirety, the items given here can be of general help in measuring work injury experience.

The methods outlined in the Z 16 Standard for classifying work injuries are independent of workmen's compensation laws and rulings of workmen's compensation agencies. Also, the fact that an employee or employer did not have control over

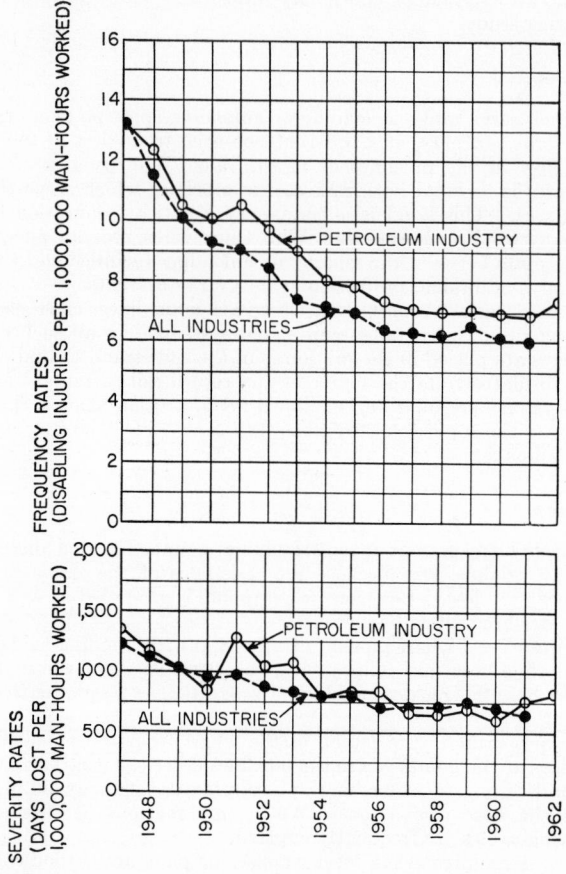

FIG. 9-12. Frequency and severity rates of disabling injuries in all industries vs. the petroleum industry.[33]

the cause of a work injury is not a criterion for excluding the injury from application of the Z 16 Standard.

Definitions

Work injury is any injury suffered by a person which arises out of and in the course of his employment. (Wherever the word "injury" is used in this standard, it shall be construed to include also occupational disease and work-connected disability.)

Occupational disease is a disease caused by environmental factors, the exposure to which is peculiar to a particular process, trade, or occupation and to which an employee is not ordinarily subjected or exposed outside or away from such employment.

Death is any fatality resulting from a work injury, regardless of the time intervening between injury and death.

Permanent total disability is any injury other than death which permanently and totally incapacitates an employee from following any gainful occupation or which results in the loss of or the complete loss of use of any of the following in one accident:

1. Both eyes
2. One eye and one hand or arm or leg or foot
3. Any two of the following not on the same limb: hand, arm, foot, or leg

Permanent partial disability is any injury other than death or permanent total disability which results in the complete loss or loss of use of any member or part of a member of the body or any permanent impairment of functions of the body or part thereof, regardless of any preexisting disability of the injured member or impaired body function.

The following injuries are **not** classified as permanent partial disability:

1. Inguinal hernia, if it is repaired (for unrepaired hernia, see Table 9-11, page 9–72)
2. Loss of fingernails or toenails
3. Loss of tip of finger without bone involvement
4. Loss of teeth
5. Disfigurement
6. Strains or sprains which do not cause permanent limitation of motion
7. Simple fractures to the fingers and toes, also such other fractures as do not result in permanent impairment or the restriction of normal function of the injured member

Temporary total disability is any injury which does not result in death or permanent impairment but which renders the injured person **unable to perform a regularly established job** which is open and available to him **during the entire time interval** corresponding to the hours of his regular shift **on any one or more days** (including Sundays, days off, or plant shutdown) subsequent to the date of the injury.

Medical treatment injury is an injury which does not result in death, permanent impairment, or temporary total disability, but which requires medical treatment (including first aid).

Disabling injury is a work injury which results in death, permanent total disability, permanent partial disability, or temporary total disability as defined above. **These are the injuries used in calculating the standard injury frequency and severity rates.**

Lost time injury is the same as disabling injury as defined above. Use of the term "disabling injury" is preferred.

Regularly established job is one which has not been established especially to accommodate an injured employee, either for therapeutic reasons or to avoid counting the case as a temporary total disability.

Total days charged is the combined total, for all injuries, of:

1. All days of disability resulting from temporary total injuries
2. All scheduled charges assigned to fatal, permanent total, and permanent partial injuries

Days of disability is the total of full calendar days on which the injured person was unable to work as a result of a temporary total injury. The total does not include the day the injury occurred or the day the injured person returned to work, but it does include all intervening **calendar** days (including Sundays, days off, or plant shutdown). It also includes any other full days of inability to work because of the specific injury, subsequent to the injured person's return to work.

Scheduled charge is the specific charge (in days) assigned to a permanent partial, permanent total, or fatal injury. For a list of these charges see Evaluation of Severity below and Table 9-11.

Exposure is the total number of employee-hours worked by all employees including those in operating, production, maintenance, transportation, clerical, administrative, sales, and other activities.

Disabling injury frequency rate is the number of disabling injuries per 1,000,000 employee-hours of exposure.

Table 9-11. Scheduled Charges for Work Injuries[34]

A. For Loss of Member—Traumatic or Surgical

Fingers, thumb, and hand:

Amputation involving all or part of bone*	Thumb	Fingers				Hand
		Index	Middle	Ring	Little	
Distal phalange.....	300	100	75	60	50	
Middle phalange....	...	200	150	120	100	
Proximal phalange..	600	400	300	240	200	
Metacarpal........	900	600	500	450	400	
Hand at wrist........	...	...	...	...	...	3,000

Toe, foot, and ankle:

Amputation involving all or part of bone*	Great toe	Each of other toes	Foot
Distal phalange..........	150	35	
Middle phalange.........	...	75	
Proximal phalange.......	300	150	
Metatarsal..............	600	350	
Foot at ankle.............	...	...	2,400

Arm:

Any point above† elbow, including shoulder joint............................ 4,500
Any point above wrist and at or below elbow............................... 3,600

Leg:

Any point above† knee... 4,500
Any point above ankle and at or below knee................................. 3,000

B. Impairment of Function

One eye (loss of sight), whether or not there is sight in the other eye............. 1,800
Both eyes (loss of sight) in one accident...................................... 6,000
One ear (complete industrial loss of hearing), whether or not there is hearing in the
    other ear... 600
Both ears (complete industrial loss of hearing) in one accident.................. 3,000
Unrepaired hernia.. 50

* If the bone is not involved, use actual days lost and classify as temporary total disability.
† The term "above" when applied to the arm means toward the shoulder and when applied to the leg means toward the hip.

Disabling injury severity rate is the total days charged per 1,000,000 employee-hours of exposure. (It is recommended that this rate be rounded to the nearest whole number.)

Evaluation and Severity

Death resulting from work injuries shall be assigned a time charge of 6,000 days each.

Permanent total disability resulting from work injuries shall be assigned a time charge of 6,000 days each.

Permanent partial disability, either traumatic or surgical, resulting from work injuries shall be assigned charges as provided in Table 9-11. These charges shall be used whether the actual number of days lost is greater or less than the scheduled charges or even if no days are lost at all.

Measures of Injury Experience

The **disabling injury frequency rate** is based on the total number of deaths and permanent total, permanent partial, and temporary total disabilities which **occur** during the period covered by the rate. The rate relates these injuries to the hours worked during the period and expresses them in terms of a million-hour unit by use of the following formula:

$$\text{Disabling injury frequency rate} = \frac{\text{number of disabling injuries} \times 1,000,000}{\text{employee-hours of exposure}}$$

The **disabling injury severity rate** is based on the total scheduled charges for all deaths, permanent total and permanent partial disabilities, and the total days of disability from all temporary total injuries which occur during the period covered by the rate. The rate relates these days charged to the hours worked during the period and expresses the loss in terms of a million-hour unit by use of the following formula:

$$\text{Disabling injury severity rate} = \frac{\text{total days charged} \times 1,000,000}{\text{employee-hours of exposure}}$$

Injuries Charged to Date of Occurrence. An injury and all days lost or charged because of it shall be charged to the date on which the injury occurred, except as noted in the following paragraph.

Nonaccident Injuries. For injuries such as bursitis, tendosynovitis, silicosis, etc., which do not arise out of specific accidents, the date of the injury shall be the date when the injury is first reported.

Classification of Special Cases

Inguinal hernia shall be considered a work injury only if it is precipitated by an impact, sudden effort, or severe strain and meets **all** the following conditions:

1. There is a clear record of an accident or an incident, such as a slip, trip, fall, sudden effort, or overexertion.
2. There is actual pain in the hernial region at the time of the accident or incident.
3. The immediate pain was so acute that the injured employee was forced to stop work long enough to draw the attention of his foreman or fellow employee, or the attention of a physician was secured within 12 hr.

Hernia other than inguinal shall not be considered under the above conditions but shall be considered in the same way as any other injury.

Back injury or strain shall be considered a work injury only if it meets all the following conditions:

1. There is a clear record of an accident or an incident such as a slip, trip or fall, sudden effort, overexertion, or blow on the back.
2. The physician authorized to treat the case is satisfied, after a complete review of the circumstances of the accident or incident, that the injury could have arisen out of the accident or incident.

A back condition which is revealed while an employee is performing his normal regular duties but which neither results from nor is caused by an accident or incident shall not be considered a work injury.

Aggravation of Preexisting Condition.   If aggravation of a preexisting physical deficiency arises out of and in the course of employment, the resulting disability shall be considered a work injury and shall be classified according to the ultimate extent of the injury, except that, if the injury is an inguinal hernia or a back injury, the conditions cited above apply.

Horseplay.   An injury inflicted by or arising out of horseplay during employment shall be considered a work injury.

Animal and insect bites are work injuries if they arise out of and in the course of employment.

Skin irritations and infections, such as dermatitis, poison ivy, etc., are work injuries if they arise out of and in the course of employment.

Muscular disabilities, such as bursitis, tendosynovitis, etc., are work injuries if they arise out of and in the course of employment.

Exposure to Temperature Extremes.   An injury which results from exposure to temperature extremes (heat or cold) is a work injury if it arises out of and in the course of employment.

Athletic Activities.   An injury to an employee resulting from participation in athletic activities, whether or not they are company-sponsored, shall be considered a work injury only if the participant was paid by the company for these activities.

External Events.   An injury which results directly from an external event of such proportions and character as to be beyond the control of the employer, such as a tornado, twister, hurricane, earthquake, flood, conflagration, or explosion originating outside of employment or from an immediate secondary event, such as a fire, boiler explosion, or falling electric wire, shall be classified as a work injury only if the victim was a policeman, fireman, member of a disaster or emergency squad, utility lineman, or other employee who is assigned duties in connection with such events.

Materials for Plant Accident- and Fire-protection Programs

In addition to the films and pamphlets listed below, which represent only a small percentage of the large number of such aids available to safety people, all persons interested in accident and fire protection should make full use of the many services provided by such organizations as:

American Petroleum Institute
National Petroleum Refiners Association
Western Oil and Gas Association
National Safety Council
National Fire Protection Association
U.S. Bureau of Mines

Addresses for these groups are included in Section 13.

Special attention is called to the assistance which can be secured from the Bureau of Mines and from the Director of the Flammable Liquids Field Service of the National Fire Protection Association.   Special reference is made to the help which can be secured from the Naval Research Laboratory, Washington, D.C., in regard to the use of foams in extinguishing tank fires.

Films

1. *Fire at Whiting, Ind., Refinery of American Oil Co.* (American Oil Co., 910 S. Michigan Ave., Chicago, Ill. 60680)

2. *Fire at Signal Hill* (Ethyl Corp., 100 Park Ave., New York, N.Y. 10017)

3. *Kansas City Bulk Plant Fire, August 18, 1959* (Television Station KMBC, Kansas City, Mo.)*

* Before showing this film it is recommended that a critique of it be secured from the Technical Service Department of the American Petroleum Institute.

4. *Tank Wagon Fires* (Socony Mobil Oil Co., 150 East 42d St., New York, N.Y. 10017)

5. *Hazards of Water* (Ethyl Corp.)

6. *Danger! Air!* (Ethyl Corp.)

7. *Pipe Line on Wheels* (Petroleum Chemicals Division, E. I. du Pont de Nemours & Company, Area Sales Offices)

8. *Static Electricity* (U.S. Bureau of Mines)

9. *Flame Propagation* (U.S. Bureau of Mines)

10. *Fire under Control* (Socony Mobil Oil Co.)

11. *Extinguishment of Bulk Plant Fires* (National Fire Protection Association, 60 Batterymarch St., Boston, Mass. 02110)

12. *Detonations* (*American Oil Co.*)

13. *Safe Handling of Light Ends* (Du Pont Company)

14. *Foreman Training in Safety* (a series of four films of the National Safety Council, 425 N. Michigan Ave., Chicago, Ill. 60611):
Let's Talk about Safety (NSC film 147.01)
Take a Talky Break (NSC film 147.02)
Setting 'Em Straight (NSC film 147.03)
Let Everybody Help (NSC film 147.04)

15. *Everywhere, All the Time* (National Safety Council, film 171.04)

16. *Fire Control Film List* (National Fire Protection Association, 60 Batterymarch St., Boston, Mass. 02110)

Pamphlets

1. *Hazards of Water* (American Oil Co.)

2. *Hazards of Air* (American Oil Co.)

3. *How to Light a Furnace* (American Oil Co.)

4. *Safe "Ups and Downs" for Refinery Units* (American Oil Co.)

5. *Hazards of Electricity* (American Oil Co.)

6. *Safe Handling of Light Ends* (American Oil Co.)

7. Papers on operating safety presented before meetings of the Division of Refining, American Petroleum Institute (copies can be secured from the American Petroleum Institute):
Safety in Process Operations by John J. Coates
Development of Safe Procedures by R. C. Steinhoff, Jr.
Are Standard Procedures the Whole Answer? by James R. Howard
Protective Facilities for Refinery Process Units by William C. Bluhm
Risk Analysis and Safety by C. Henry Austin
A Pattern for Process Safety by J. C. Ducommun
Startup-Shutdown Procedures for a Large Crude Oil Distillation Unit by W. S. Bonnell and J. A. Burns
Safe "Ups and Downs" for a Refinery Unit by A. H. Hayes and R. M. Melaven
Detonation—Old Processes Are Not Immune by O. A. Pipkin
Occurrence and Nature of Hydrocarbon Detonations by R. B. Jacobs
Towers Are Touchy by R. W. Ballmer
Safe Operation of Refinery Flare Systems by W. C. Bluhm
Safe Operation of Spheres by H. T. Fuller and R. E. Bistline
Safer Operations through Committee Action by J. N. Kayser and R. W. Williams
The Human Factor in Operating Practices by M. A. Pappas
Education of Employees in the Safe Handling of HF in the Alkylation Process by J. A. Price
What Status Process Safety? by J. C. Ducommun

8. Pamphlets on process safety available from the American Petroleum Institute:
Cleaning Petroleum Storage Tanks—Section A, Crude-Oil and Unfinished-Products Tanks, *API Accident Prevention Manual* 1, 1955
"Cleaning Tanks Used for Gasoline or Similar Low-flash Products," *API Bulletin* 2016, 1961
Fire Protection in Natural Gasoline Plants, API RP 2002, Secs. A and B, 1956

"Gas and Electric Cutting and Welding," *API Accident Prevention Manual* 3, 1953
"Cleaning Mobile Tanks Used for Transportation of Flammable Liquids," *API Accident Prevention Manual* 13 (Sec. A, "Tank Vehicles," 1959, Sec. B, "Tank Cars," 1958)

Guide for Tank Venting, API RP 2000, 1952
Fire Protection in Refineries, API RP 2001, 1959
Sparks from Hand Tools, API study, 1956
First-aid Training Guide and Safety Information, 1958

9. Available from the United States Bureau of Mines, 4800 Forbes St., Pittsburgh, Pa. 15213:

"Explosion of Dephlegmator at Cities Service Oil Co. Refinery, Ponca City, Okla.," *Bureau of Mines Report of Investigations* by M. G. Zabetakis, 1959
"Safety at Gas-processing Plants," *Bureau of Mines Bulletin* 588

Reference should also be made to Table 9-8, NFPA Standards and Codes (page 9-42), and Table 9-10, Booklets on Radioactive Materials (page 9-55), as well as to the list of references at the end of this chapter.

REFERENCES

1. *Accident Prevention Manual*, National Safety Council, 4th ed.
2. *Safety Code for Cranes, Derricks, and Hoists*, American Standards Association B 30.2, 1943.
3. *Safe Maintenance Practices in Refineries*, American Petroleum Institute Manual, in preparation.
4. Austin, C. Henry, "Risk Analysis and Safety," paper presented at 40th Annual Meeting, American Petroleum Institute, Chicago, Nov. 14, 1960.
5. *U.S. Bur. Mines Inform. Circ.* IC 7980, 4800 Forbes St., Pittsburgh, Pa. 15213.
6. Coward, H. F., and G. W. Jones, "Limits of Flammability of Gases and Vapors," *U.S. Bur. Mines Bull.* 503, 1952.
7. *National Fire Codes*, vol. I, "Flammable Liquids and Gases," National Fire Protection Association.
8. *First-aid Training Guide*, American Petroleum Institute No. 2017B, 1962.
9. *A Professional Code for Defensive Driving*, National Safety Council, 1959.
10. *The Watchman*, National Fire Protective Association, NFPA 601, 1956.
11. *Central Station Protective Signalling Systems; Installation, Maintenance, and Use*, National Fire Protective Assn., NFPA 71, 1962.
12. *Proprietary, Auxiliary, and Local Protective Signaling Systems; Installation, Maintenance, and Use*, National Fire Protection Association, NFPA 72, 1962.
13. *NFPA Inspection Manual*, National Fire Protection Association, 1959.
14. *Fire Protection in Refineries*, American Petroleum Institute, RP 2001, 4th ed., 1959.
15. *Radioactive Fallout and What It Means to the Petroleum and Gas Industries*, U.S. Department of the Interior, Office of Oil and Gas, June, 1960.
16. *Outside Protection (Yard Mains for Sprinklers, Standpipes, etc.)*, National Fire Protection Association, NFPA 24, 1962.
17. *Fire Hydrants, Private Fire Service*, National Fire Protection Association, NFPA 29C, 1955.
18. *Fire Hose, Care, Maintenance, Use*, National Fire Protection Association, NFPA 198, 1958.
19. *Fire Protection Handbook*, National Fire Protection Association, 12th ed., 1962.
20. *Water Spray Systems*, National Fire Protection Association, NFPA 15, 1962.
21. "Mechanism of Extinguishment of Fire by Finely Divided Water," *Natl. Board Fire Underwriters, Res. Rept.* 10, 1955.
22. *Carbon Dioxide Extinguishing Systems*, National Fire Protection Association, NFPA 12, 1962.
23. *Dry Chemical Extinguishing Systems*, National Fire Protection Association, NFPA 17, 1958.
24. *Foam Extinguishing Systems*, National Fire Protection Association, NFPA 11, 1960.
25. *Foam Extinguishing Systems*, National Board of Fire Underwriters, NBFU 11, 1960.
26. *Flammable Liquids Code*, National Fire Protection Association, NFPA 30, 1962.
27. *Safety in Welding and Cutting*, American Standards Association, ASA Z 49.1–1958.
28. Waddell, P. M., et al., "Silicones Combat Foaming in Residuum Storage," *Oil & Gas J.*, Feb. 27, 1961.
29. Bradley, Hollis P., "Sandblasting Petroleum Storage Tanks in Service," *Natl. Safety Congr. Trans.*, vol. 19, Petroleum Industry, 1960.

30. *The API Recommended Practice for Classification of Areas for Electrical Installations in Refineries,* American Petroleum Institute, API RP 500, 2d ed., 1957.
31. *National Electrical Code,* National Fire Protection Association, NFPA 70, 1962.
32. *Protection against Ignitions Arising out of Static, Lightning, and Stray Currents,* American Petroleum Institute, API RP 2003, 1956.
33. *Annual Summary of Injuries in the Petroleum Industry for* 1962, American Petroleum Institute.
34. *Method of Recording and Measuring Work Injury Experiences,* American Standards Association, ASA Z 16.1-1954.
35. Jacobs, R. B., "Occurrence and Nature of Hydrocarbon Detonations," Paper presented at the 39th Annual Meeting of the American Petroleum Institute.

Section 10

PROCESS CONTROL AND INSTRUMENTATION

By MATTHEW J. DE PASQUALE*

Senior Instrument Engineer
The M. W. Kellogg Co.
New York, N.Y.

and

RUNNE C. OHRBERG†

Instrument Analytical Section Engineer
The M. W. Kellogg Co.
New York, N.Y.

Process control and instrumentation have grown to a relative level of maturity in the petroleum processing field. Without these elements present-day plants would be virtually nonexistent. Instrumentation generally and process control specifically are directly responsible for processes which are complex, integrated, and large, both physically and in terms of throughput. Faster, more critical and more hazardous operations can be carried out because of suitable control equipment. Processes hitherto unfeasible economically are successfully managed with appropriate instrument equipment.

Instrumentation, in the broad sense, serves two functions—measurement and control.

Measurement is the acquisition of pertinent process knowledge which is useful and, indeed, even essential to the operation of a process. Such knowledge usually relates to a significant physical property of the process material or equipment; to temperature or pressure; to flow rates in a continuous process or to total quantity in batch processes; to level; to special properties, such as viscosity, specific gravity, acidity, and vapor pressure; and (most recently) to direct composition measurement.

Control, the other important function of instrumentation, is the maintaining of a process variable at some predetermined value. This function is by far the predominant influence of instrumentation on petroleum processing. It makes use of the first function (measurement) as a necessary part of its own operation.

These functions are represented in simple block-diagram form in Fig. 10-1. Measurement is characterized by a single path (an **open loop**) originating at the process and terminating at some measuring instrument. Control is characterized by an encircling path (a **closed loop**) originating at the process, moving through to the measurement,

* Mr. DePasquale is now with Mobil Oil Co., New York, N. Y.
† Mr. Ohrberg is now with Buck Sales Inc., Buckingham, Pa.

control, and process valve and finally back to the process itself. By means of a self-contained reference (the desired value of the process variable) the control device can continuously compare the measured variable with a desired value, manipulate the valve, and thus modify the process as required.

Since the usual methods of process control involve the feedback or measurement path, the control apparatus recognizes not the actual process variable, rather, it feels only the output of the measuring apparatus. Clearly, then, any errors in the measuring apparatus will result in a corresponding inability of the control system to maintain the controlled variable accurately.

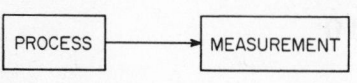

(A) MEASUREMENT (OPEN LOOP)

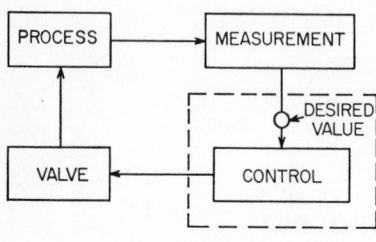

(B) CONTROL (CLOSED LOOP)

Fig. 10-1. Block-diagram representation of measurement (*A*) and control (*B*).

Though far less common, open-loop control systems can also be used wherein the control device acts upon a process by manipulating a valve in response to a related measurement. The outstanding difference between this and the closed loop is that no feedback path exists. Expressed simply, a measured process variable, acting through a controller, produces a valve action to cause another process variable to change in accordance with some predetermined relationship between the two variables. Such an open-loop system is shown in block-diagram form in Fig. 10-2. In addition to control errors due to an inaccurate measuring system, the open-loop control suffers additionally from the fact that the control effect is not "fed back" to compare continuously with a reference. Rather, it relies on the quality of the original system design to accomplish its purpose.

Sensing the desired process variable and comparing it with a predetermined desired value are obvious advantages over merely sensing a related variable and acting on the process in some predetermined way. Thus, in petroleum process control at least, closed-loop control involving measurement of the vital process variable is emphasized.

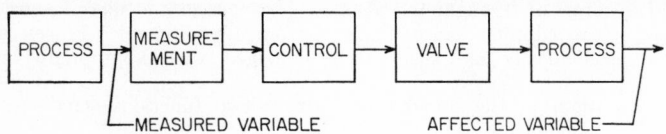

Fig. 10-2. Block-diagram representation of open-loop control.

In a discussion of instrumentation and control for petroleum processes, one cannot ignore the importance of manual control, a special case of the closed-loop control system described above. The distinction between manual and automatic control is one of components rather than principle. If we replace the word *control* of the closed-loop control system in Fig. 10-1 with the word *operator*, we can appreciate the substitution of the human senses for measurement evaluation and human decision making followed by manipulation for control.

Clearly then, where continuous measurement evaluation and valve manipulation are not necessary, manual control is sufficient. The operator applies human discretion in evaluating the measurement and moving the valve, relying heavily upon training and experience and recognizing the peculiarities of his unit. While equipment for an automatic closed-loop control system cannot duplicate those human characteristics, it is capable of handling most process control applications. In fact, the functions it does perform far surpass the same functions performed by a human being.

Since the control mechanism initiates the corrective action of the control loop, it is important that its basic nature suit the problem at hand. Three types, or modes, of control action are generally available for petroleum process applications. Each action performs some operation upon the error signal (the difference between the desired value of the process variable and the measurement of that variable) in a predetermined manner. Consequently the input to the control block is the error signal. The controller acts upon this input to produce an output (or valve manipulating) signal in accordance with its control mode. Thus a controller with **proportional action** creates an output signal proportional to the error signal. **Reset action** (mathematically, integral action) creates an output signal which varies with time in proportion to the error signal. **Rate action** (mathematically, derivative action) creates an output signal proportional to the rate of change of the error signal.

In petroleum processes the most commonly used control mode is a combination of proportional and reset. This combination creates the desired continuous throttling action and eliminates the offset or static error encountered in proportional control alone. In less critical services proportional control suffices. Rate action is occasionally used in combination with proportional or with proportional and reset to minimize peak deviations or overshoots in processes containing large thermal capacities or physical volumes. This subject is treated extensively in the available literature and need not be detailed here.

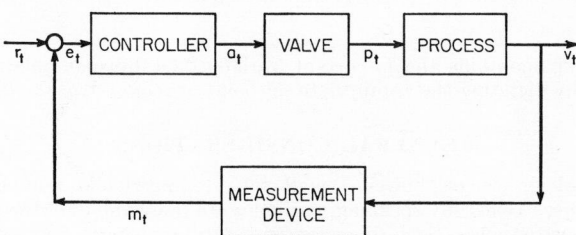

FIG. 10-3. Closed-loop temperature-control system.

The block-diagram representation of Fig. 10-3 illustrates the closed-loop process control system. A certain process variable, the controlled variable v_t, is measured by a temperature-sensing device whose output signal m_t is compared with a control point or reference r_t. Any difference between the measured and desired temperature manifests itself as an error signal e_t acting upon the controller. The controller produces an output signal a_t which acts upon the control valve. The valve position p_t influences the process in a manner tending to overcome disturbing influences upon the controlled temperature.

In some processes, disturbing influences upon the controlled variable are difficult to counteract because they must travel through the entire process and sensing system, often large and slow acting, before they can be evaluated and corrected. This difficulty can be circumvented by the use of a **cascaded control** system wherein certain disturbances or load changes can be sensed, evaluated, and counteracted before they affect the significant controlled variable. An additional controller (the inside loop of the cascade system) accomplishes this.

The classic example of a cascade system in practical use involves the firing controls of a large furnace. Fuel is throttled to maintain an appropriate outlet temperature. If, however, the fuel header pressure changed and caused a new fuel rate, it would not be until the process stream felt the effect of the modified firing rate that a simple temperature-control loop could correct for the load change. The process block of Fig. 10-3 can be redrawn as shown in Fig. 10-4a. If a new controller is added to maintain fuel rate, shown in Fig. 10-4b, a fluctuation in the fuel header pressure would be felt immediately as a change in fuel rate. This could be corrected by the valve without its having to manifest itself upon the outlet temperature through a large, slow process path, which is the furnace. The outside loop of the cascade system (the temperature

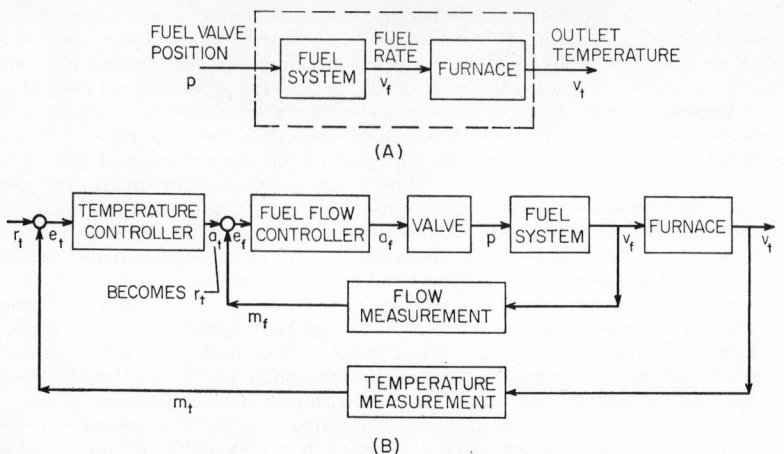

(A)

(B)

Fig. 10-4. Fuel-rate controls for large furnaces: (*A*) simple temperature-control loop; (*B*) cascade control system with temperature resetting flow.

controller) still maintains the important variable, not by manipulating the valve directly, but by dictating the appropriate set point or reference of the flow controller.

GENERAL CONSIDERATIONS

Every petroleum process, however small or large, consists of a number of variables and parameters.   Generally speaking, variables are those values of the process which are subject to variations.   Such things as temperature, pressure, flow rates, and levels are all process variables.   Their values may change with changing conditions, or they may be deliberately changed by the operator.

Process parameters, on the other hand, are those values generally not subject to variation.   Such things as line and vessel sizes, reactor bed volumes, physical and chemical properties of materials, and similar values are parameters.   For purposes of this discussion they are assumed to be distinct in value, an inherent and unchanging part of any process.   By their very nature, parameters generally need not be measured.   Since they are "built in" to any process, they necessarily dictate how the process behaves.

Measurement, then, is confined to process variables.   Control is applied to process variables and also manipulates other process variables to achieve control.   By way of distinction we refer to these as **measured variables, controlled variables,** and **manipulated variables.**   In petroleum processes, as in most other continuous hydraulic processes, the manipulated variables are most often control-valve settings, though they may occasionally be variable-speed rotating machinery or variable-stroke reciprocating machinery.   At any rate, manipulated variables always relate to a device in the system capable of establishing an appropriate energy balance.

Naturally, a process may have numerous independent control loops.   These would obviously not exceed the number of manipulated variables available.   When every variable capable of being manipulated has a controller associated with it, the system has been completely constrained and operates at one particular steady-state condition.   In practice, however, it is not necessary to control all the process variables but only a group of the most important ones.   The system then operates close to a particular steady state, minor fluctuations permitted by virtue of a group of uncontrolled, relatively weak degrees of freedom.

The control engineer selects those variables to be controlled as best will fulfill the process objectives.   The choice of manipulated variables or control valves is likewise important from the standpoint of control-system sensitivity, pressure drop and valve design, and system stability. These subjects and others related to control theory form the basis of many detailed reference works and will not be dealt with here.

The typical present-day petroleum process is, relatively speaking, slow. Because of this its control is static or steady-state control. The performance of the system is determined by maintaining certain quantities fixed relative to certain frames of reference. Occasionally the operator alters the references as the situation demands, usually rather slowly, and the process seeks its new steady state. Such a system can be defined in terms of the algebraic relationships between process variables and parameters. The parameters form the constants of the relationships between the independent process variables (cooling-water rates, heating-steam rates, valve positions, etc.) and the dependent process variables (process temperatures, pressures, flows, etc.). True, dynamic situations occur as the system journeys from one steady state to another, but these phenomena, whose behavior is described by differential equations, form a more specialized consideration of petroleum process behavior and control-system design. When warranted, the dynamic behavior must be carefully studied.

Material Balance

The most basic concept related to process control is that of material balance. A single flow stream splitting into two streams serves as an example of this principle. The algebraic relationship governing the flow behavior of this system, shown in Fig. 10-5, is $W_1 = W_2 + W_3$. For an uncontrolled system, this is the only relationship, one involving all three process variables, W_1, W_2, and W_3.

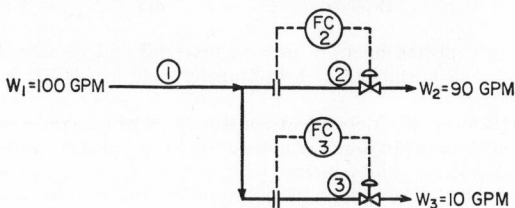

FIG. 10-5. Fully controlled material-balance system for split process stream.

Since the single relationship describes how any one variable depends upon the other two, the system is said to have two degrees of freedom. Two variables can vary independently. In fact, the number of degrees of freedom is merely the difference between the total number of process variables of the system and the number of independent relationships associating them.

The maximum number of independent control loops which could be applied to this system are two, corresponding to the number of degrees of freedom. Each controller can be considered as adding one more equation. With two control systems the total number of independent equations would become three, corresponding to the number of variables. Such a system of equations has a unique solution; that is, the process would operate at a particular steady state. Let us suppose that streams 2 and 3 were flow controlled as seen in Fig. 10-5. We would have the case of a system completely constrained; it would have no degrees of freedom, since W_3 is dictated by controller FC-3, W_2 is dictated by controller FC-2, and W_1 follows the system material-balance equation previously stated: $W_1 = W_2 + W_3$.

Conceivably, the system could be adequately operated with one control loop only. If W_2 were the critical flow then FC-2 would be retained and FC-3 omitted. In this case the three-variable system is defined by only two equations. Hence it retains one degree of freedom. In planning such a system, one must appreciate that changes can occur. If, for example, stream 3 were subject to 10 per cent flow changes in either direction, then stream 1 would consequently feel flow changes of 1 per cent in either direction. If such a situation is tolerable in operation, then FC-3 is not necessary and the system performs satisfactorily with its one relatively weak degree of freedom.

Heat Balance

A second important concept related to the control of petroleum processes is heat balance. A simple heat exchanger, shown in Fig. 10-6 serves to illustrate. The process variables seem to be

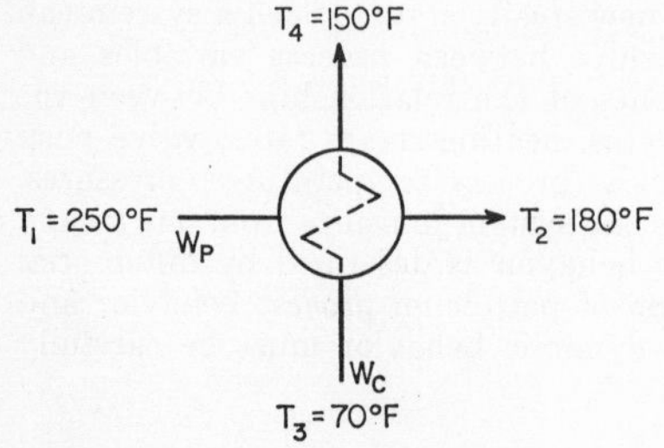

$T_4 = 150°F$

$T_1 = 250°F$ W_p

$T_2 = 180°F$

W_c
$T_3 = 70°F$

FIG. 10-6. Simple heat-exchanger system for heat-balance calculations.

W_p, flow quantity of process material
W_c, flow quantity of cooling medium
T_1, inlet temperature of process material
T_2, outlet temperature of process material
T_3, inlet temperature of cooling medium
T_4, outlet temperature of cooling medium

However, the inlet temperatures T_1 and T_3 are imposed upon the system from other systems external to the one considered, so they should not be included as variables of this system. A further condition is that the heat-exchanger area A is constant and that neither the process material nor the cooling material changes phase. Thus the system contains only four process variables: W_p, W_c, T_2, and T_4; it is described by two system equations

$$W_p C_p (T_1 - T_2) = UA(mtd)$$
$$W_c C_c (T_4 - T_3) = UA(mtd)$$

where C_p and C_c are specific heats of process material and cooling medium, respectively, U is the over-all coefficient of heat transfer, and mtd is the logarithmic mean temperature difference.

The maximum number of independent control loops which could be applied to this system are two, since there are four variables and two equations, giving two degrees of freedom. Normally one is chiefly interested in the process outlet temperature, so it is the controlled variable. The system obviously offers two manipulated variables, W_p and W_c. These can be manipulated directly or indirectly as shown in Fig. 10-7.

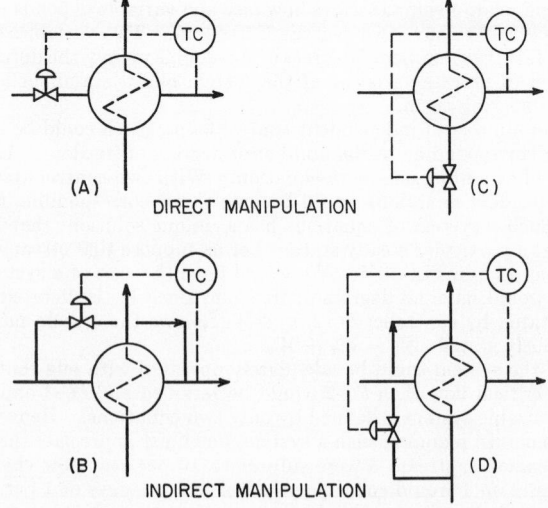

(A)

(C)

DIRECT MANIPULATION

(B)

(D)

INDIRECT MANIPULATION

FIG. 10-7. Methods for controlling process outlet temperatures.

The process flow quantity cannot usually be manipulated per se, as indicated in Fig. 10-7a, since other considerations usually govern its rate. In such a case the quantity of process material entering the heat exchanger can be manipulated by indirect means, that is, by utilizing a bypass control valve on the process side of the exchanger (see Fig. 10-7b). A more direct and more common way for controlling is by manipulation of the cooling-medium flow as shown in Fig. 10-7c.

Occasionally, heat exchange between two process streams is required; furthermore, both streams have their flow rates set by other considerations. Clearly, then, any manipulation must be by indirect means, by utilizing a bypass control valve on the process (hot) side of the exchanger, as in Fig. 10-7b, or on the process (cold) side of the exchanger, as in Fig. 10-7d.

Both the concepts of material and heat balance, while they apply to the simple systems shown, are equally applicable to larger, more complex systems. The simple junction in the material-balance system shown in Fig. 10-5 could be replaced with any process configuration, and the principles would still apply. Such a system would be a process with a single feed and two products. Likewise, heat balance can be approached from a broader view than a single heat-exchange system.

OPERATION OF PROCESS EQUIPMENT

Fractionating Columns

The single most important piece of equipment in a petroleum processing plant is the fractionating column. As such, it merits careful consideration when it comes to control instrumentation requirements for satisfactory operation. While there are many varying schemes for controlling fractionating columns that have come about by long years of operating experience, a simple column serves to illustrate some basic principles of tower instrumentation (see Fig. 10-8).

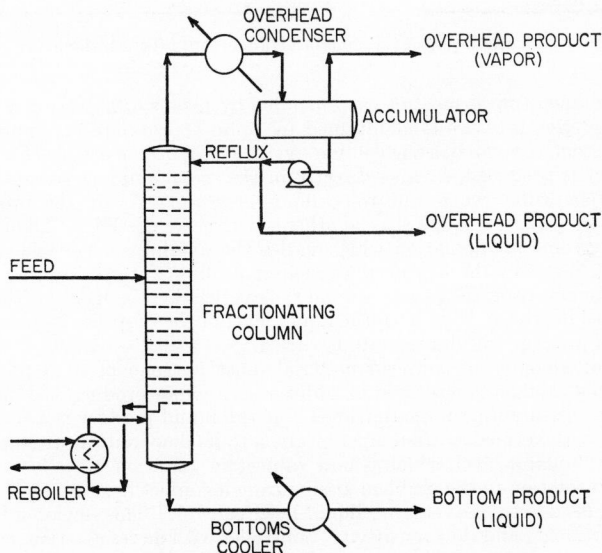

FIG. 10-8. Fractionating column and related equipment.

Like the over-all process, it is desirable to have the column operating at some steady set of conditions. Sudden changes in any of the system variables cause deviations from the desired steady state which diminish the operating efficiency of the column. Basically the column is made up of a material-balance system consisting of a feed stream and two or more product streams and two heat-exchange systems—one for overhead heat removal and the other for bottom heat addition.

The simplest and most effective means for establishing stable column operation is direct flow control of the feed stream. Next, and as a direct result of feed flow control, is direct flow control of the reflux stream. The rate of total product withdrawal must correspond, in a system in balance, to the feed rate. The specific quantity of overhead or bottom product depends upon the desired product specification, either overhead or bottom. If, for a given feed, a certain bottom-product specification is desired, the bottom-product rate is thereby determined, there existing only a given quantity of that product in the feed. Usually the column is designed with some liquid holding time, making level controlled tower bottom a practical way of withdrawing that product. The balance of the material then moves toward the opposite end of the column, thus setting the overhead product rate.

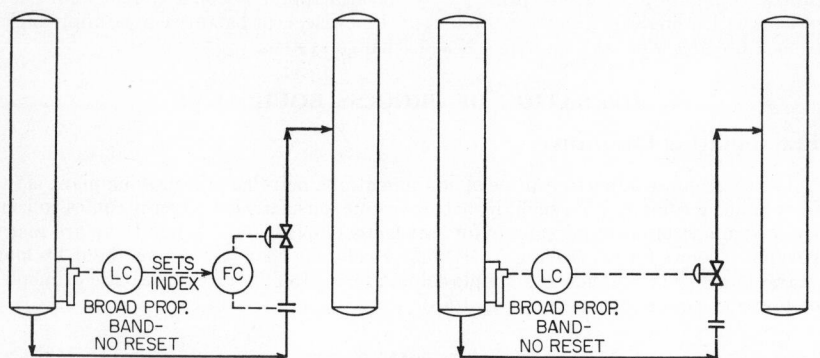

Fig. 10-9. Averaging liquid-level control arrangements for fractionating columns.

Column temperature is most often controlled by means of adjusting reboiler duty. If this is the case, pressure is maintained by some appropriate manipulation in the overhead system. A valve in the column overhead vapor line or, where a vapor overhead product is produced, a valve directly in the vapor product stream leaving the column system will serve to control column pressure. Where the total overhead product is withdrawn as liquid, several other schemes are possible. All of them, however, involve some manipulation which varies the overhead condensing rate. One, for example, throttles the overhead condenser cooling water. Another bypasses a portion of the hot overhead vapors around the overhead exchanger. A third operates with a flooded overhead drum with the liquid level in the overhead condenser varying as the liquid product withdrawal rate is varied.

A sufficient amount of overhead material must be condensed to provide reflux. Where a liquid product is produced in addition to a vapor product, the liquid may be withdrawn to maintain accumulator level. If the liquid product is a feed stream to another tower, then consideration must be given to its flow control, with careful supervision of accumulator level, which is now subject to variation.

A common solution to the problem arising from a conflict between level control and flow control is the use of averaging liquid-level control. This device utilizes a very broad proportional band (low sensitivity) controller with no reset action, either setting the index (control point) of a flow controller in the liquid effluent stream or manipulating the valve directly (see Fig. 10-9). Changes in level result in very gradual and tolerable changes in the flow rate. While level is not controlled precisely, it is permitted to make unharmful swings between upper and lower limits. In most cases these upper and lower level limits should be monitored to avoid undesirable operating conditions.

Reactors

In petroleum processes, reactor operation is usually based upon predetermined temperature and pressure conditions. Where a reaction is exothermic, heat must be removed in some manner to maintain proper reaction temperature. In such cases, the process design includes internal sprays, cooling coils, built-in steam-generating equipment, or some other device suitable for removing the heat generated by the reaction. For endothermic reactions, heat must be added in some manner to maintain proper reaction temperature. It may be added directly to the feed stream prior to its entry into the reactor, or as in the case of fluidized catalytic cracking reactors, it may be transported to the reactor by means of a circulating stream of heated catalyst. In either case, the reaction temperature is determined by the appropriate addition of heat to the reactor system.

Pressure is maintained, as it is in most hydraulic systems, by establishing a proper balance between influx and effluent. Since many reactor charge rates are flow controlled, the most likely location for reactor pressure control is in the effluent stream as seen in Fig. 10-10a. An alternate method, where direct throttling of the effluent stream is not feasible, is one in which an excess gas make stream is throttled (see Fig. 10-10b). In fixed catalyst bed reactors it is convenient to furnish inlet and outlet

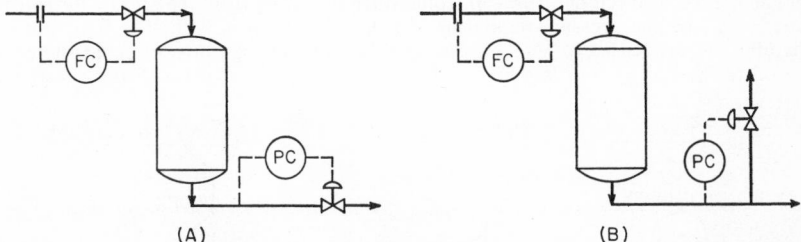

(A)   (B)

FIG. 10-10. Reactor pressure control by (A) throttling of effluent stream and (B) throttling of excess gas make.

pressure-indicating devices for the determination of differential across the reactor bed. These not only furnish an indication of the catalyst bed condition but offer a safeguard against abnormal internal forces on the catalyst bed structure.

Fired Heaters

Process fired heaters (or furnaces), while often complex in design, have relatively simple control systems. As in most process reactors, the process stream is flow controlled, it representing the major charge stream to the process equipment following. In addition to maintaining its allied process steady, flow-controlled furnace charge materially simplifies the firing operation by holding the most significant system variable constant.

Often, because of their large physical size, furnaces are designed as multipass units. While there is no rigorous way to determine if it is necessary to flow-control each pass or merely the total flow, a few special instances can be pointed out as typical of those where individual flow control would be required. The first is where coking of the furnace tubes can occur. In this situation, as coking progresses, the hydraulic resistance of the multipass unit becomes progressively unbalanced and the manually set flows would likewise become unbalanced. The second is where partial flashing or vaporization is to occur in the furnace tubes. Once again, slight imbalances in flow cause corresponding imbalances in the degree of vaporization. This leads to further resistance change, and the imbalance can be continuously aggravated. Where a relatively clean tube condition can be expected, and where no vaporization occurs, a total flow controller with hand-set individual streams can assure a relatively reasonable degree of operating stability. In either case, however, it is wise to include sufficient pressure- and temperature-indication points to describe adequately the behavior of each furnace pass.

Relative to firing control, this is usually limited to direct manipulation of the fuel stream. In petroleum process furnaces the combustion air is usually not manipulated jointly with fuel as is normally done on steam-generation equipment. The major operating variable is, in almost every instance, furnace outlet temperature. Occasionally, where coking can become severe, a point somewhere ahead (or upstream) of the outlet location is selected to minimize the detrimental effect of coking on the temperature element. Thus, regardless of whether the function of the furnace is to crack, vaporize a hydrocarbon mixture, or simply to add sensible heat, temperature control is used.

As noted previously, cascade control is widely used to overcome upsets in the fuel system without waiting for the ensuing process upsets to occur. The example then given was one of temperature resetting flow (see Fig. 10-4b). However, pressure control can be equivalent to flow control when used in conjunction with a fixed restriction such as a burner tip. This is usually preferred in most furnace-firing control problems, since pressure is generally easier to measure. Additionally, flow measurement in fuel gas can be further complicated by header pressure variations. Specifically, if fuel gas is fired, the temperature-controller output is used to set the index (or control point) of a fuel-gas pressure-regulating valve. If fuel oil is fired, the temperature-controller output is used to set the index (or control point) of a pressure-control instrument in the fuel-oil system (see Fig. 10-11). Furthermore, in fuel-oil firing, where it is

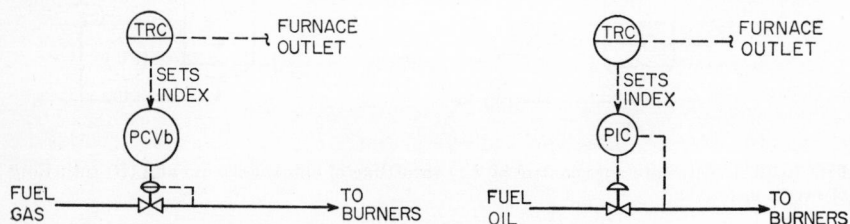

Fig. 10-11. Furnace firing controls for gas and oil fuels.

necessary to use steam for atomizing the fuel, a scheme for maintaining a proper pressure ratio between the steam and fuel-oil pressures is shown in Fig. 10-12.

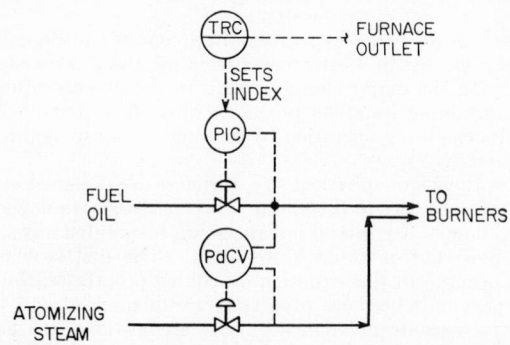

Fig. 10-12. Control for steam atomization of fuel oil.

Draft gages are required to measure the pressure on the hot side of the furnace. The appropriate pressure (a slightly negative pressure just below the furnace arch) is usually achieved by manually setting the damper in the furnace stack.

Heat Exchangers

As previously pointed out in the material under Heat Balance, temperature control (or other variable control) at heat exchangers is accomplished either by direct throttling of the streams through the exchanger or by means of bypass manipulation. Where a phase change occurs, heat-exchanger control requires more than simple consideration. Take the case of a steam heater as an example. Usually, saturated or nearly saturated steam is used. Rather than sensible heat, the steam gives up its latent heat and changes to a liquid in doing so.

The most straightforward method for controlling a steam heat exchanger is shown in Fig. 10-13a. It offers the best control, the manipulation being made directly upon

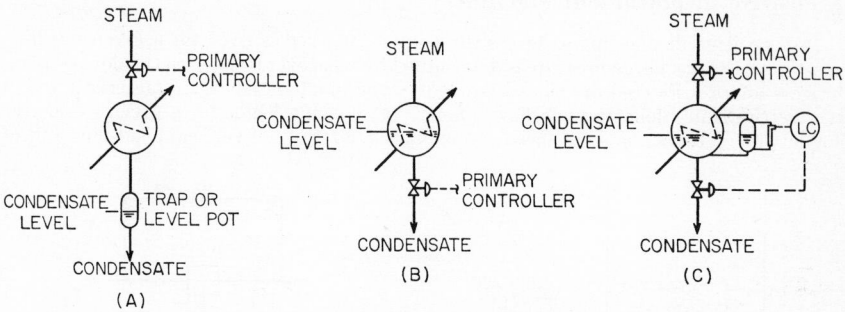

Fig. 10-13. Controlling steam heat exchangers: (A) control on steam inlet, (B) control on condensate outlet, (C) controls on both steam inlet and condensate outlet.

the significant value—the steam flow. Since condensate is collected by a trap, the most efficient use is made of the exchanger surface. Depending upon the quantity of condensate produced, removal is accomplished by means of the trap (as shown) or by means of a level-controlled pot. In either case, the exchanger tubes are completely exposed.

Herein lies the chief disadvantage of this otherwise ideal system. Often, under conditions of low load, the minimum condensing pressure can fall below the condensate system pressure, making condensate removal impossible unless a pump or vacuum trap is added to the system.

Two alternatives are suggested. A simple variation to the ideal system is one in which the control valve is located in the condensate outlet instead of in the steam inlet (see Fig. 10-13b). Liquid is permitted to build up, covering the tubes to vary exchanger duty. A steady state is reached when the level has established itself to produce the required effective exchanger area, with the steam condensing at essentially header pressure. Control by this method is coarse, since manipulation of the condensate stream often involves undesirable lags associated with liquid-level build-up and suppression.

Should fine control be required where low condensing pressures prevent the use of the simple ideal system, an additional element of control can be added to yield the needed action. The primary controller is arranged to manipulate a valve in the steam inlet. A level-controlled condensate pot, so located that the condensate level in the exchanger can be varied, is added to the system as shown in Fig. 10-13c. This enables external manipulation of effective exchanger area, thereby condensing steam at a pressure suitable for discharge into the condensate system while retaining the advantages of a primary control valve in the steam inlet.

Phase change can occur in the other direction, too. Process coolers evaporating a refrigerant are frequently found in low-temperature petroleum processes. The same fundamental principles apply as do in condensing systems. Fine control is still achieved by manipulation of the vapor stream; level control can be applied to change the effective exchanger area.

Pumps and Compressors

From an energy standpoint, pumps and compressors achieve the opposite effect of control valves in that stream energy is increased by pumps and compressors and is dissipated by control valves. From a control standpoint each device is a suitable final control element depending, of course, on whether energy must be added or removed to achieve a particular process energy level. Often both elements are present in one system, the pump or compressor adding a discrete quantity of energy and the valve trimming out that variable portion not required by the process. The arrangement of any system depends largely upon the nature of the pump or compressor.

Positive-displacement Machinery

In a positive-displacement device the volume handled is fixed for a given machine speed and the discharge pressure is determined by whatever resistance the down stream process offers. Plotted on a head-flow curve the machine has a characteristic which is a vertical line, denoting one flow quantity at varying heads for a given speed (see Fig. 10-14). If the machine speed can be varied, a family of vertical head-flow curves

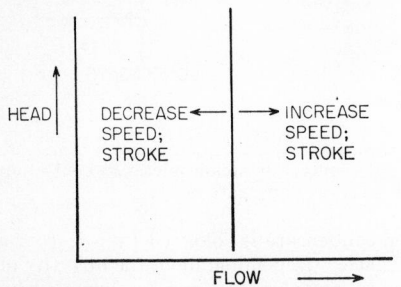

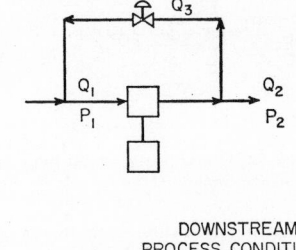

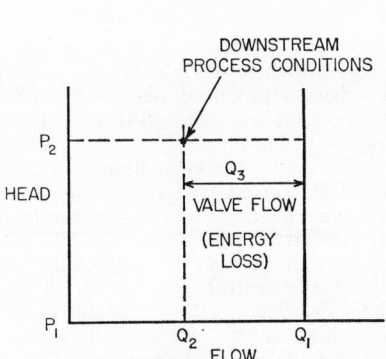

FIG. 10-14 (above). Flow characteristics of a typical positive-displacement pump.

FIG. 10-15 (right). Kickback in positive-displacement pump to control pump output.

exist and some degree of flow variation can be achieved. Another method of varying the flow through a reciprocating device is by furnishing means for adjusting the reciprocating stroke length. Externally, the same effect as speed or stroke control is accomplished by permitting excess material to flow back to the suction side, as illustrated both physically and graphically in Fig. 10-15. If the kickback valve arrangement is used, there may even be a direct throttling valve downstream in stream Q_2.

Internally, a control means similar to kickback may be accomplished stepwise by unloaders.

Rotating Machinery (Centrifugal)

In a centrifugal device, the volume handled varies with head for a given machine speed. Plotted on a head-flow curve, the centrifugal machine may have a typical characteristic as shown in Fig. 10-16. If the machine speed can be varied, a family of head-flow curves exists and some degree of flow variation can be achieved. Unlike a positive-displacement machine, a control valve can be placed directly in the discharge stream to set a flow rate, without the necessity for a controlled kickback stream.   Then any difference between the head produced by the machine *at that flow* and the pressure established by the downstream process is manifested as pressure drop across the valve. This is illustrated both physically and graphically in Fig. 10-17.

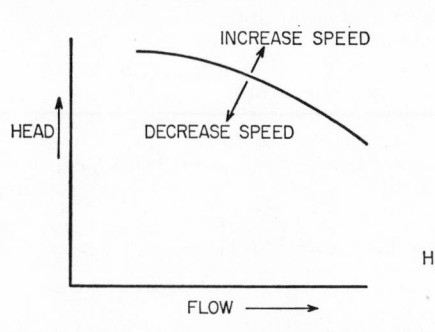
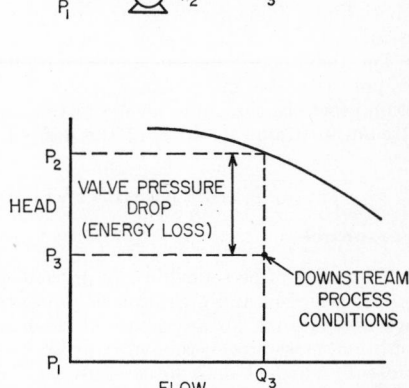

FIG. 10-16 (above). Flow characteristics of a typical centrifugal pump.

FIG. 10-17 (right). Flow control of centrifugal pump by throttling the pump discharge.

On large centrifugal compressors handling major process gas streams it is desirable, from an economical standpoint, to find the appropriate operating point by speed control in order to avoid the energy dissipation of a control valve in the process stream. Occasionally these large machines are electric motor driven rather than turbine driven and do not lend themselves readily to speed control.   In this case, operation is at one speed as in the case of most centrifugal pumps handling liquid, except that suction throttling is preferred to discharge throttling from an energy-consumption standpoint. This energy saving is related to the compressible nature of the fluid handled, and as such, this system of control is not utilized on centrifugal pumps handling liquid.   In fact, suction throttling in liquid pumping service is deliberately avoided to prevent undesirable suction conditions.

In special liquid applications where severe erosion or corrosion can occur, a valve in the centrifugal pump discharge may be avoided and speed control used instead.

Maintaining Proper Liquid Levels

As material progresses through a chain of process equipment, phase change forms a vital part of the operation.   To pin down the location of the phase change requires appropriate instrumentation.

Take the case of an accumulator vessel following a condenser.   Figure 10-18 shows

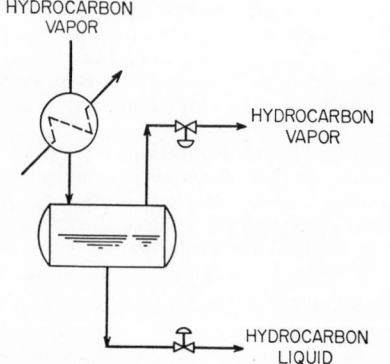
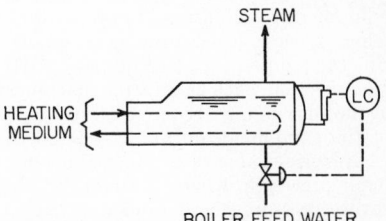

FIG. 10-18 (left). Condenser and liquid-vapor separator system.

FIG. 10-19 (above). Liquid-level control in an unfired boiler.

such a system in which a hydrocarbon vapor is partially condensed, the total exchanger effluent entering a drum for liquid-vapor separation. The vapor leaving forms part of the pressure system. It may be throttled to maintain system pressure. The liquid leaving usually represents a charge stream for another element of process equipment or a product stream. The former may be flow controlled for stable operation of the equipment it feeds. In such a case a level record of the drum is essential, it usually being carried as a second pen on the flow-controlling instrument. In the latter case, where liquid effluent represents a product stream, a level-controlled drum to remove as much liquid as the system produces is usually preferred.

The above considerations apply, in principle, to systems in which process material is evaporated. An unfired boiler serves to illustrate (see Fig. 10-19). As steam generation proceeds, the liquid level tends to drop. Level control on the boiler-feed-water stream maintains the level at the desired value.

INSTRUMENTATION FOR PROTECTION

Personnel

Protection of personnel in any process operation is related, to a large extent, to the proper selection and operation of process equipment and to the careful handling of process material, for any abuse of these represents a potential personal hazard. In addition, most processes have areas where fire, explosion, poison, or other hazards are present and which must be carefully monitored. Analytical instruments are available to detect the presence of dangerous concentrations of toxic gases and vapors, and devices are available to detect flame failure. These not only detect the existence of the hazardous condition but initiate visual and audible alarms and can be arranged to actuate other devices automatically for eliminating or minimizing the dangerous condition (see also Sec. 9).

Process Material

Adverse temperature conditions can cause deterioration of process materials or the formation of undesirable material within the process. The transfer lines from furnaces to process equipment are likely locations for the formation of coke or the undesirable cracking of the process material. Reactor catalyst can deteriorate, and its useful life is shortened when it is exposed to abnormally high temperatures for long periods. Monitoring and alarming are often necessary at such locations.

The physical condition of some process material, such as the configuration of catalyst pellets in a reactor bed, is subject to change (packing, for example) if the pressure rises across the catalyst bed. Such a condition should be easily detectable to assure good reactor performance.

Process Equipment

Another important aspect of temperature monitoring is the one of protecting process equipment from adverse temperature conditions. This need is especially significant where process vessels may be internally lined to protect the vessel shell from excessive process temperatures. Equipment which generates large quantities of heat energy from friction, such as rotating machinery bearings, and from electrical resistance, such as motor windings, requires careful temperature monitoring to assure proper operation of the equipment and its associated cooling system.

Pressure is always a potential destroyer of equipment; this is especially so in petroleum processes, where the highly volatile confined process materials are subject to severe pressure fluctuations as energy levels change. Relief valves and other emergency pressure-relieving devices are essential on all equipment subject to excessive pressure, particularly those to which heat energy is added during the normal course of the process operation. Occasionally, on low-pressure or vacuum vessels, vacuum breaking is required to prevent equipment collapse, particularly where sudden abnormal cooling can occur.

Rotating machinery requires protection against overspeeding. On centrifugal gas compressors, additional protection is required to prevent abnormally low flows through the machine. The lack of flow causes an inability of the machine to meet the downstream process pressure, resulting in a flow reversal. The mechanical damage which could occur from such a condition can be avoided by the use of a flow-actuated kickback stream around the centrifugal machine.

Level monitoring on drums preceding certain types of process equipment assures that the equipment receives the appropriate material. For example, a furnace feed drum with the proper level is assurance that adequate material is available to absorb the large quantities of thermal energy in the furnace, thus avoiding tube burnout. A sufficiently low compressor suction drum level will assure that only gases reach the machine, preventing the damage which would result from liquids reaching the rotating parts.

ACCOUNTING OF UTILITIES

Since economic considerations are most vital in any process operation, it is important to account for the utilities consumed. Steam, a major source of power and process heating (in addition to its use as a stripping medium, for fire fighting, and for general process service), is normally metered at its entry to each process unit. Where various steam pressure levels exist, each is metered. Water, a major source of process cooling (in addition to its use for fire fighting, personnel protection, drinking, and general process service), is metered at least at its prime source and often at each process unit. Air for general plant service and the more expensive clean and dry air for pneumatic instruments are often metered at their source.

Electrical systems are furnished with kilowatt-hour meters for a total block and often in individual feeders for partial blocks covering each process unit.

Both fuel oil and fuel gas are metered at least at entry to each process unit; in the case of heavy users such as large process furnaces and fired boilers, metering for accounting purposes may be done at individual user equipment.

BASIC TYPES OF PROCESS MEASUREMENT

Temperature

Measurement of temperature plays an important role in petroleum processes, since, in addition to the obvious purpose of determining sensible heat, it offers an excellent indication of material state and composition.

In fractionating columns, temperature measurement is found at feed points and overhead, bottom, and intermediate drawoff points. Where intermediate refluxes exist, temperature is measured at the reflux entry point.

In reactors, temperatures are measured in the reaction space as well as in the inlet and outlet streams; these serve useful purposes both for the normal reaction cycle and for a regeneration cycle as well. In high-temperature processes, particularly where internally insulated reactors are used, reactor metal temperature is an important consideration, and provision is made for its measurement.

Fired heaters are furnished with temperature measurement throughout. On the process side, such measurements are made at the inlet and outlet of convection tubes, the outlet of radiant tubes, and (where a tube bank is extremely long) at intermediate points. Each pass naturally requires its own set of measurements, except that inlet temperature is common to all passes. On the hot side, measurements are made in the radiant section, in the gases leaving the radiant section and entering the convection section, and in the outlet flue gases.

Heat exchangers may have temperature measurement on the inlet and outlet of both the streams.

Gas compressors require temperature measurement on suction and discharge sides to determine inter- and after-cooler requirements.

Temperature measurement is additionally useful in determining the effect of combining process streams, in splitting flow through multishell exchanger units, in determining process energy loss to the atmosphere, and in indicating the extent of flashing when volatile materials flow through high-resistance systems.

Most temperature measurements in petroleum processes are made by means of thermocouples to facilitate bringing the measurements to a centralized location. Locally at the equipment, bimetallic or filled system thermometers are used. To a lesser extent, usually where a high measurement accuracy is required, resistance thermometers are used. For some transmitting applications, filled system types are used.

Temperatures in petroleum processes will vary from lows of about −150°F on low-temperature refrigerants for light ends recovery processes to highs of about 1800°F in process furnace fireboxes.

Pressure

Like temperature, pressure is a valuable indication of material state and composition. In fact, these two measurements considered together are the primary evaluating devices of petroleum materials.

In fractionating columns, pressure measurement is of primary importance. Occasionally, the pressure difference between top and bottom may be made to indicate tower loading.

In reactors, pressure is measured in the reaction space and, as previously suggested, at inlet and outlet to indicate the condition of the reactor internals.

Fired heaters require pressure measurement for process considerations and to serve as a guide to the condition of tubes. This is particularly important if the heat addition represents severe cracking of the charge. Pressure measurements also serve to maintain a rough split through multipass furnaces without individual flow control.

Pumps, compressors, and other process equipment associated with pressure changes in the process material are furnished with pressure-measuring devices. Thus, pressure measurement becomes an indication of energy increase or decrease, degree of agitation across mixing valves, operation of the equipment, etc.

Most pressure measurements in petroleum processes are by elastic element devices, either direct connected for local use or transmission type to a centralized location.

Pressures in petroleum processes will generally vary from lows of about 75 mm of mercury absolute in vacuum distillation units to highs of at least 1,000 psi in thermal reforming units.

Flow

The measurement most representative of continuous process operation and economics is flow. In petroleum processing, flow control predominates over all other types, hence the importance of proper and sufficient flow measurement.

At fractionating columns, flow is of primary importance on feed and reflux streams. Occasionally, where temperature is not indicative of reboiler duty, flow measurement of the heating medium is used. Reactor and furnace charge flows, either over-all or through individual passes, are almost always controlled. All manually set streams require some flow indication or some easy means for occasional sample measurement.

For accounting purposes, feed and product streams are metered. In addition, utilities to individual and grouped equipment are measured to determine operating costs and equipment performance.

Most flow measurements in petroleum processes are by variable head devices. To a lesser extent, variable area and positive-displacement types are used, as are the many other available types as special metering situations arise.

Flows in petroleum processes may vary from lows of a few cubic centimeters per minute for inhibitor injection to highs of 150,000 B/D on large process units.

Level

While most petroleum processes are classified as continuous, this is not literally so with regard to liquid streams. In fact, many process vessels, notably columns and drums, actually serve to interrupt the passage of liquid in some manner, depending upon the nature of the manipulation made upon the flow stream to the following item of process equipment.

At the inlet to a process unit, a feed surge drum receives material from storage or from the product stream of a preceding unit. The rate at which material is received does not instantaneously coincide with the feed rate to the process equipment following the drum, usually a furnace, a tower, or a reactor. From storage, material may be level-controlled into the drum; when material enters from a preceding unit (where all the product from that unit must be received), a continuous drum-level record must be provided.

Column overhead reflux drums likewise fall into two main categories relative to operating level considerations. Since column load can be expected to vary somewhat during operation, it is reasonable to expect a varying amount of condensed overhead to enter the drum. Except for a few special fractionating operations, overhead reflux (representing a portion of the reflux drum effluent) is nearly always flow-controlled. The question of level recording vs. level controlling depends largely upon the manner in which liquid product is withdrawn. As mentioned previously in the discussion on fractionating-column instrumentation, reflux drums will be level-controlled if liquid product is removed to storage; a continuous drum-level record usually is provided if the liquid product is a feed stream to another tower or if that stream is otherwise flow-controlled.

Fractionating-column bottoms serve as temporary inventory for liquid bottoms product and require the same considerations as drums.

Level measurement is of extreme importance on equipment whose specific function is to separate liquid and vapor. This equipment includes, in addition to the reflux drums mentioned above, drums following equipment in which material is partially condensed or vaporized, surge drums in which undesirable vapors are permitted to separate, and knockout drums in which undesirable liquids are permitted to collect without passing on to equipment which could be damaged by that liquid. In addition, level measurement and control assure the proper liquid flow into and out of equipment in which material is completely evaporated or condensed.

Level measurement also plays an important role in monitoring interfaces in the separation of immiscible phases in settling drums, such as interfaces between hydrocarbons and water, between gasoline and caustic solution in treating systems, between sulfuric acid and reactor effluent in the alkylation reaction, and between hydrocarbons and diethanolamine in hydrogen sulfide removal equipment.

Finally, level is significant in petroleum processing in so far as it serves to account for material on hand such as catalyst inventory in the reactor, stripper, and regenerator of a fluid catalytic cracking unit, process material in feed and product storage drums; water in boiler feed drums; and an assortment of process treating, inhibiting, washing, and purging materials.

Most level measurements in petroleum processes are by external devices such as gage glasses, level displacers, and ball floats. Less frequently and depending upon special circumstances, internal displacers and ball floats are used.

Levels may vary from several inches in liquid knockout equipment to several feet in most columns and drums to 30 or 40 ft in large storage containers.

Composition

Direct analysis of a process stream in petroleum refining is the ultimate use of process variable measurement to monitor plant operation. While temperatures and pressures have long been used to infer process material compositions, the last 10 years have seen the development of basic physical-chemical principles applied to practical instruments for measuring compositions of process materials. Additionally, conditions which cannot be inferred and which are vital to efficient plant operation are capable of direct measurements. Finally, continuous measurements of stream compositions place the knowledge in the hands of the operator without the long delays associated with laboratory quality-control tests and put composition measurements on a par with other process variables which can be controlled for operating efficiency.

In catalytic cracking, oxygen analysis of regenerator flue gases is important in establishing air requirements for suitable catalyst regeneration and can be useful in determining the quantity of coke burned. Carbon monoxide and carbon dioxide analyses of the same stream further supplement the knowledge of the combustion operation taking place and may be additionally useful in determining the onset of "afterburning" in the regenerator exit passages.

In catalytic reforming, continuous analysis of the recycle gas indicates the hydrogen content of that stream during the normal process operation. During regeneration of the same unit, oxygen analysis assures carefully controlled catalyst regeneration with the proper proportions of air and inert gas. Combustibles analysis may be used in conjunction with the generation of the inert gas.

The majority of process stream analyzers in petroleum refining operations are or will ultimately be associated with equipment whose major function is to change the composition of the process material. These include the outlet of cracking furnaces, reactors, and fractionating columns. Commonly used are analyzers for detecting C_1, C_2, and C_3 components in fuel-gas and tail-gas streams, for detecting C_3 and C_4 components in splitter product streams, and for detecting C_4 in debutanized gasoline.

Although there are numerous types of analytical instruments in use for petroleum process measurement, those generally encountered utilize thermal-conductivity, paramagnetic, and energy-absorbtion properties of materials to make the measurement; they may be designed for either single-component or multicomponent analysis.

Miscellaneous

Many other measurements play important roles in petroleum process instrumentation, though not to the same extent as those already mentioned.

Viscosity is used to verify specification on certain marketable petroleum materials such as automotive lubricants. It provides means for identifying material in a pipeline flowing a variety of materials in sequence and an indication of mixture qualities, solution concentrations, treating material strength, etc.

Specific gravity, liquid, supplements flow measurement, especially where major product streams are metered for process-unit accounting and where products are loaded for transportation or for marketing; provides a rough indication of crude oil grade or composition; is used to blend gas oils for catalytic cracking unit feed; can be correlated to composition or heating value; provides an indication of mixture qualities, solution concentrations, treating material strength, etc.; and determines acid concentration in the sulfuric acid alkylation process.

Specific gravity, gas, supplements flow measurement, especially where major product streams are metered for process-unit accounting and where products are distributed or marketed; provides an indication of heating value; and can be correlated to composition where a light material such as hydrogen is an important component (tail gas, fuel gas, recycle gas in catalytic reforming, etc.).

Color indicates the amount of additives and the degree of refinement, detects thermal decomposition of lube-oil stocks, and indicates the presence of darkly colored contaminants.

Vapor pressure helps monitor processing of straight-run naphtha or gasoline, checks in-line blending of motor fuels, and safeguards transportation and storage of volatile materials.

Flash point establishes ignition specifications for fuels, determines safety limitations, detects contamination by light ends, and helps monitor stripping operations.

Distillation properties, though usually limited to sample-type measurement (including initial, intermediate, and end-point determination), are important for indicating the quantities of low- and high-boiling components in petroleum products and are useful in evaluating the operating quality, purity, and economy of motor fuels, heating oils, and solvents.

APPLICATIONS OF INSTRUMENTATION TO PROCESSES

Utilizing the principles set forth in the previous discussion, measurement and control flow sheets have been developed for a typical group of the processes covered in Sec. 3. These are shown in Figs. 10-20 through 10-32. The instrumentation indicated reflects broad requirements and does not attempt to include every item which

(*Text continues on page 10–32*)

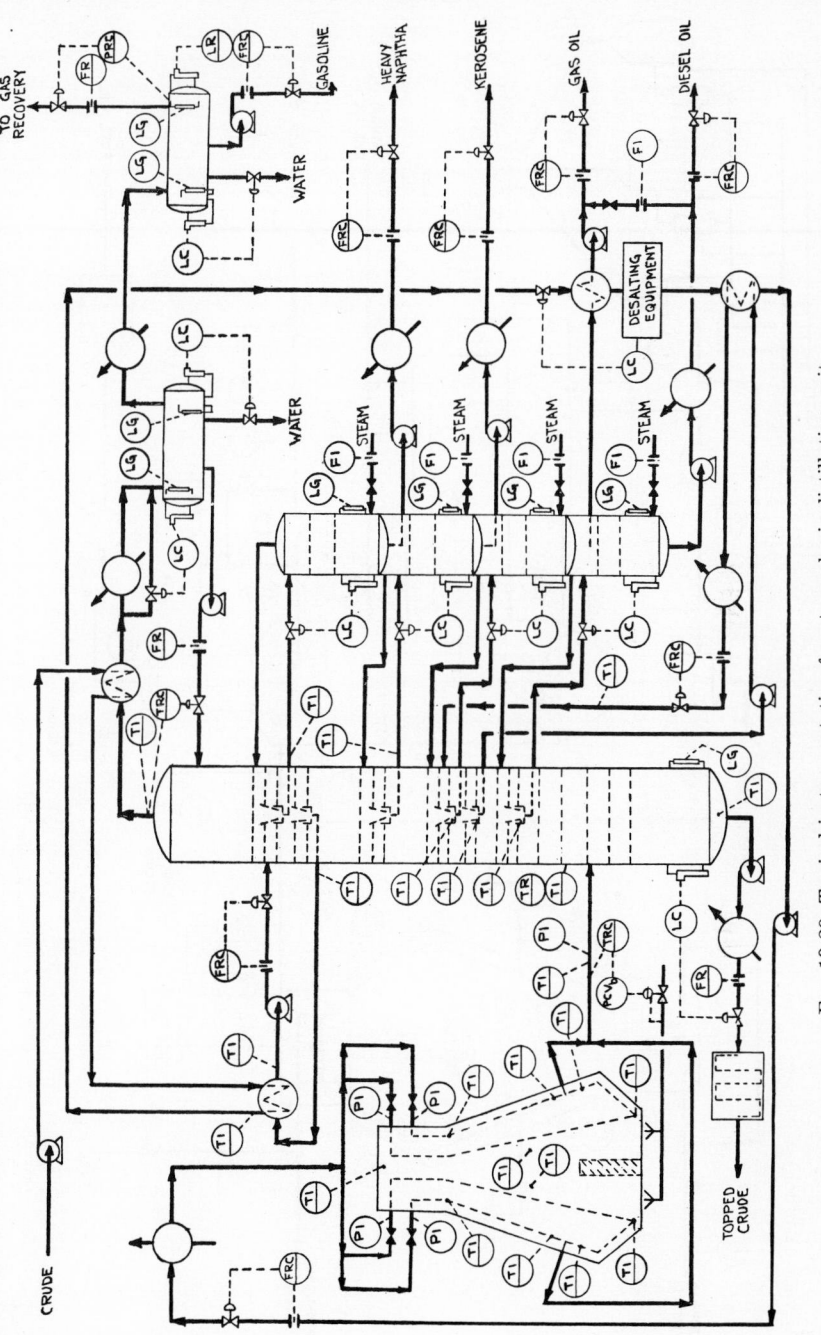

FIG. 10-20. Typical instrumentation for atmospheric distillation unit.

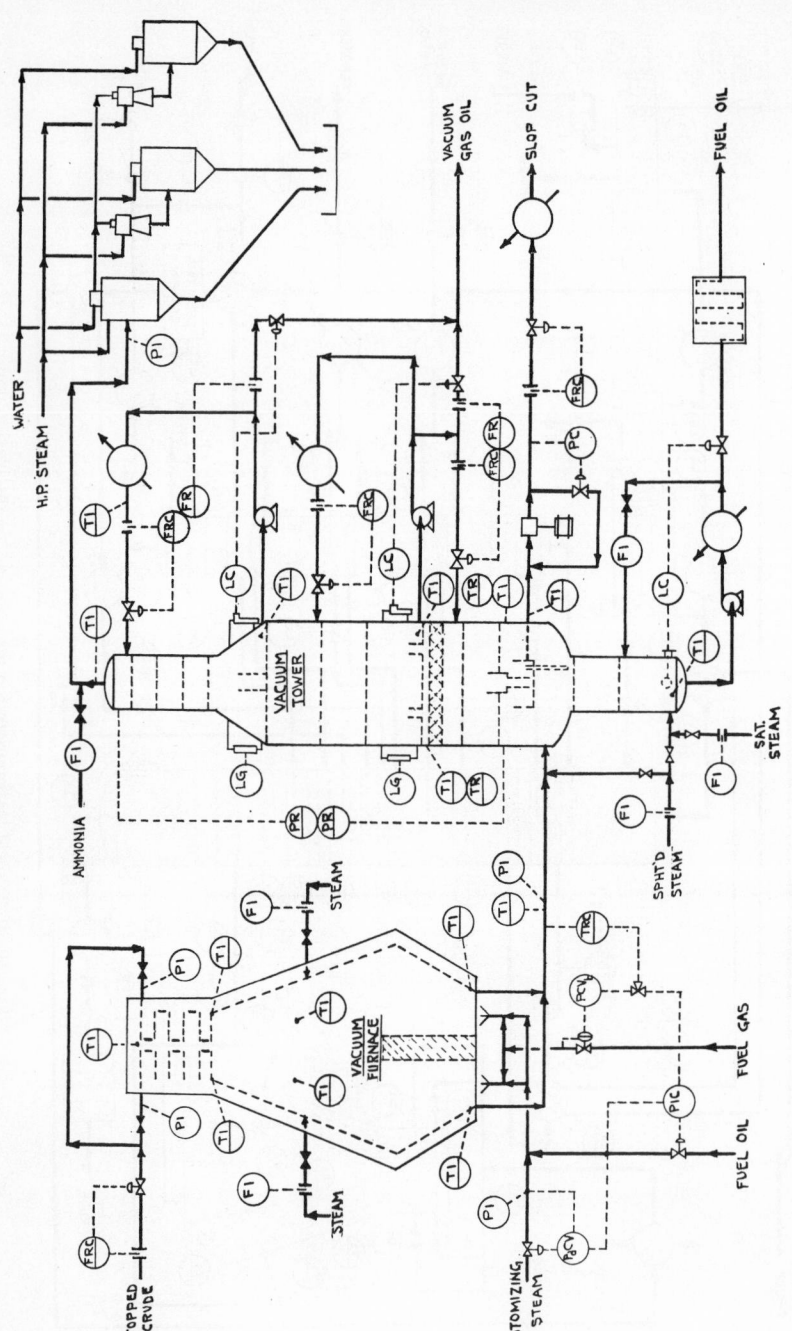

FIG. 10-21. Typical instrumentation for vacuum distillation unit.

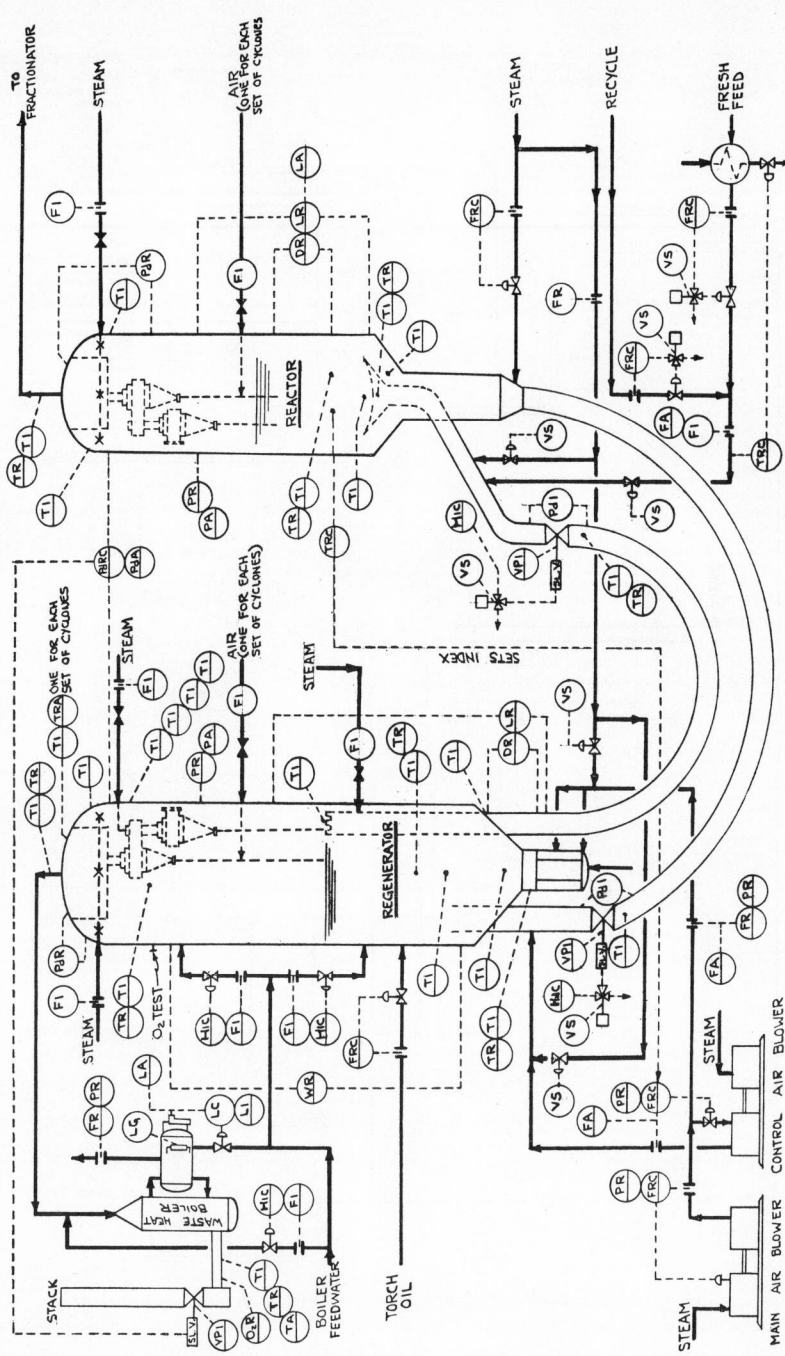

Fig. 10-22. Typical instrumentation for Model IV catalytic cracker.

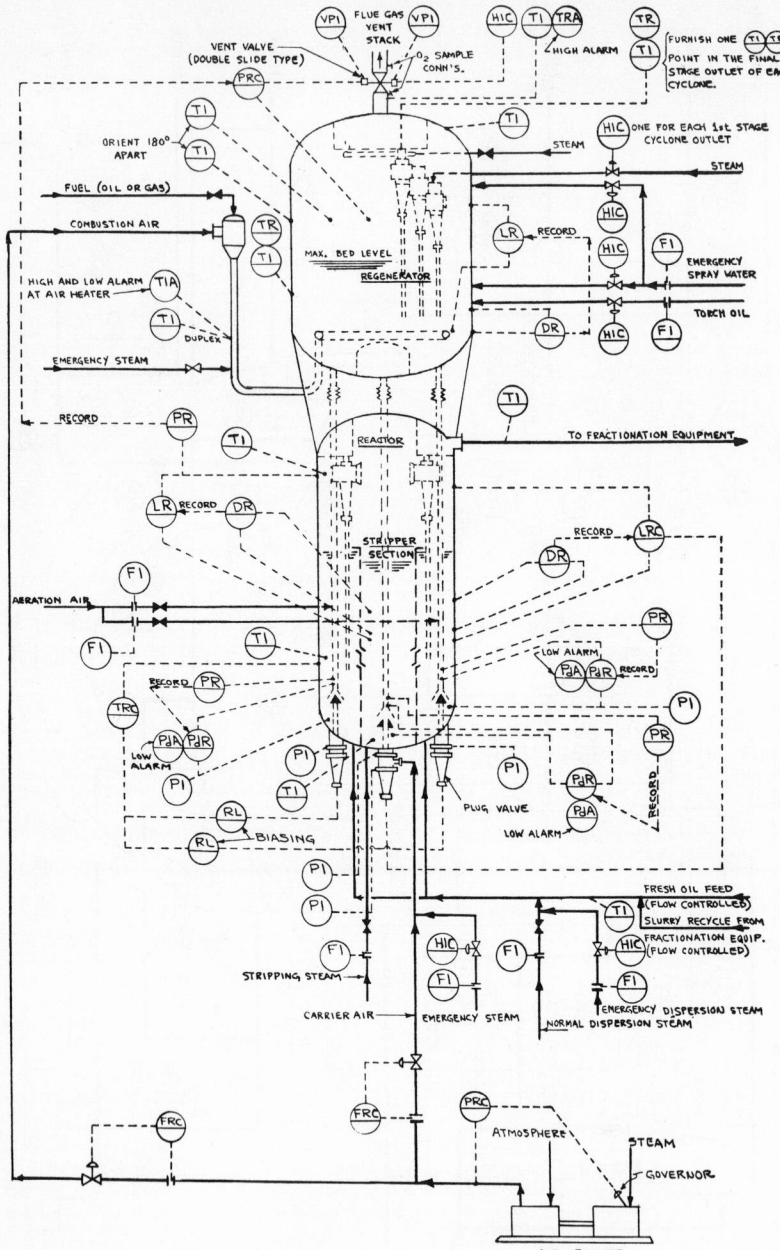

FIG. 10-23. Typical instrumentation for Model B Orthoflow converter.

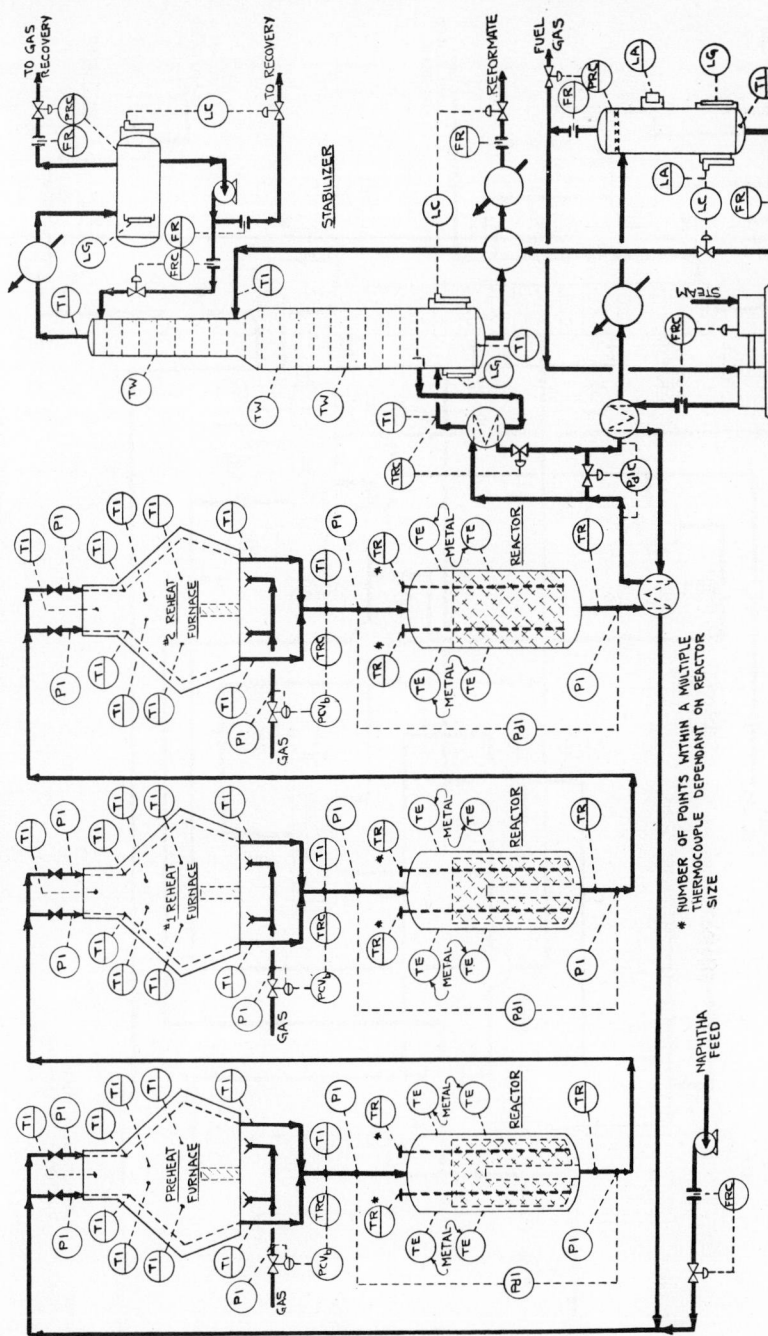

FIG. 10-24. Typical instrumentation for catalytic reformer (SBK) without regeneration facilities.

10–23

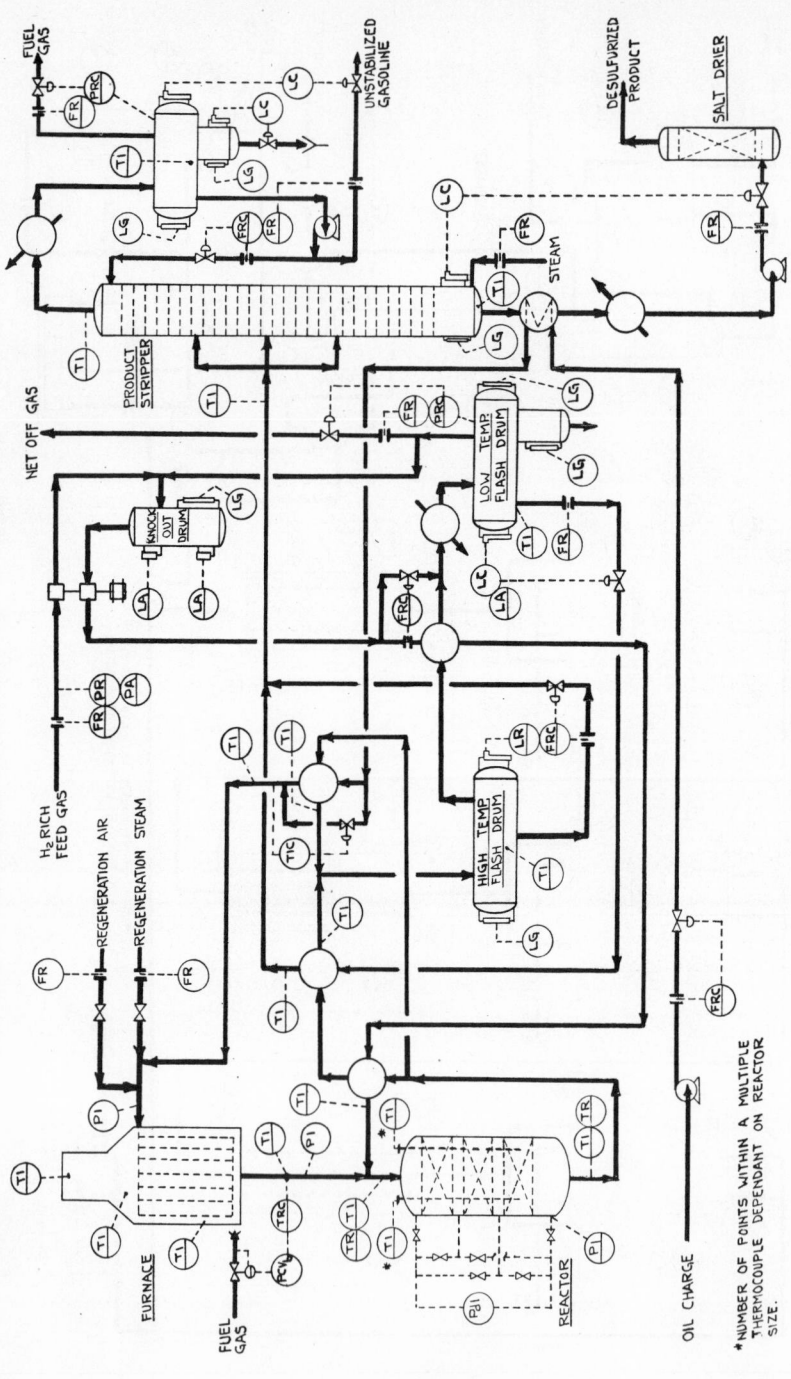

FIG. 10-25. Typical instrumentation for desulfurization unit.

* NUMBER OF POINTS WITHIN A MULTIPLE THERMOCOUPLE DEPENDANT ON REACTOR SIZE.

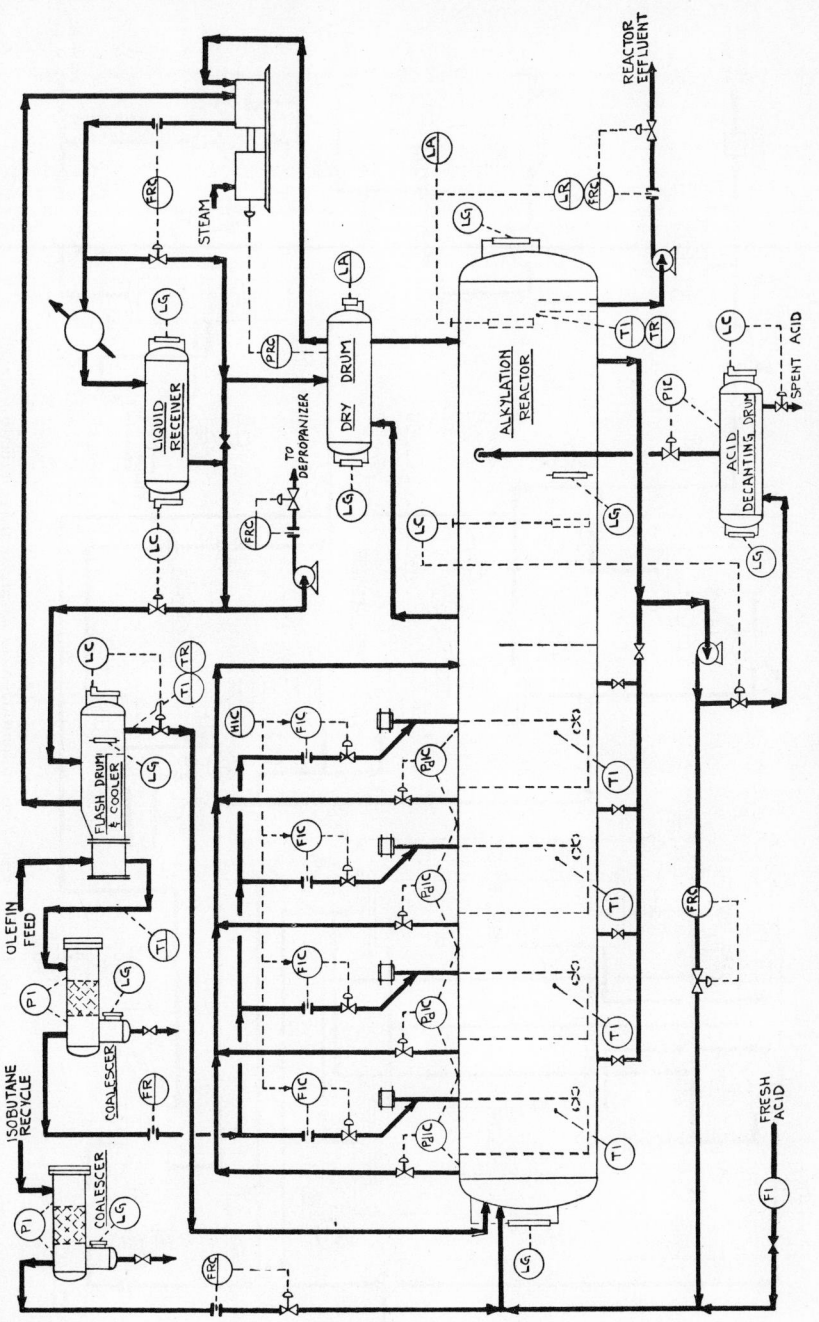

FIG. 10-26. Typical instrumentation for sulfuric acid alkylation reactor.

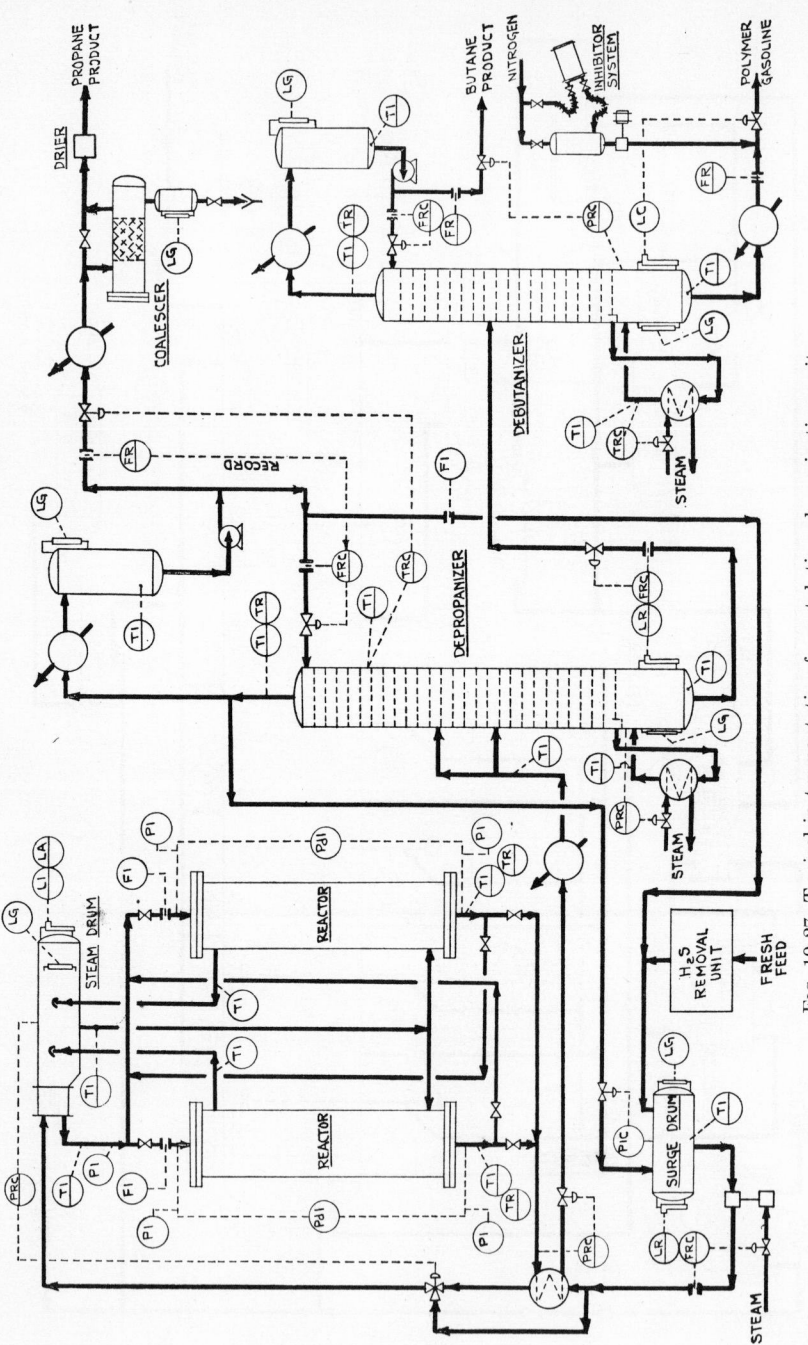

Fig. 10-27. Typical instrumentation for catalytic polymerization unit.

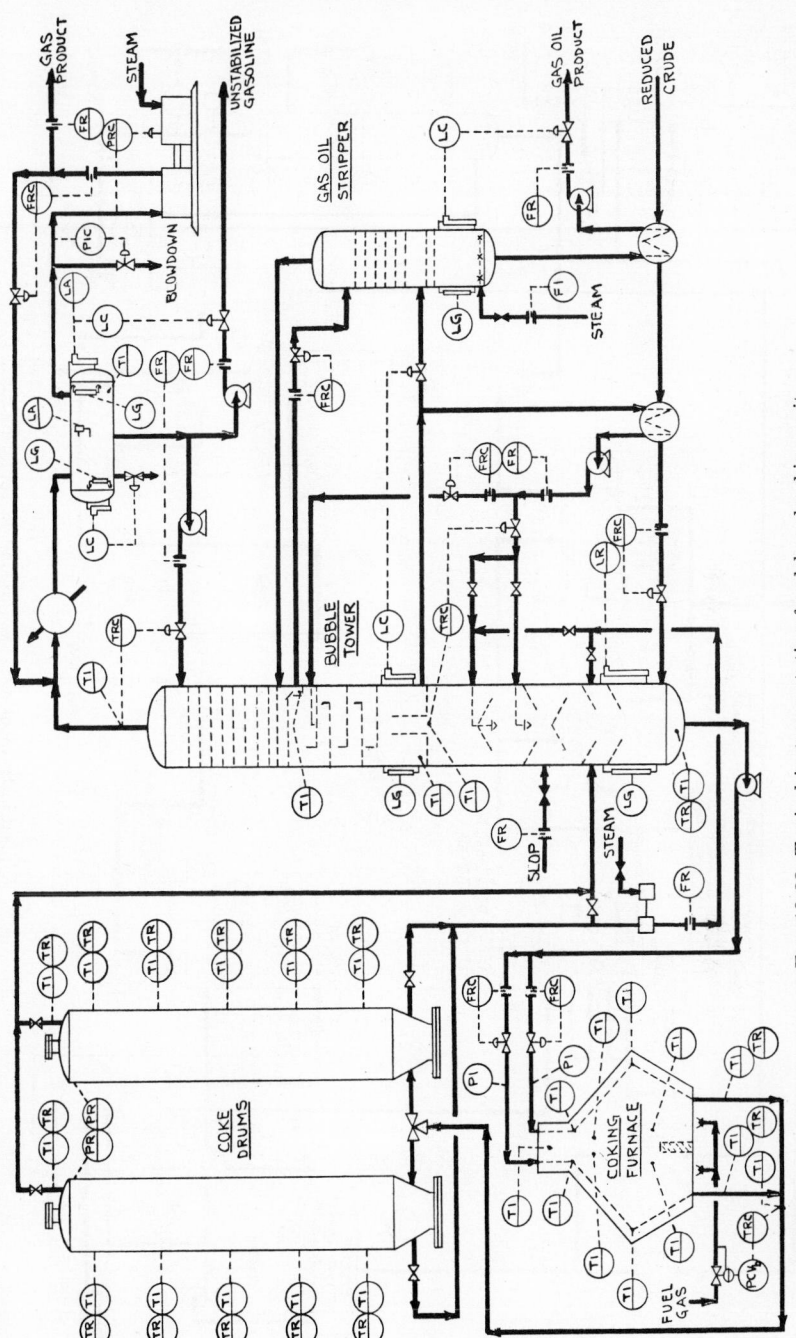

FIG. 10-28. Typical instrumentation for delayed coking unit.

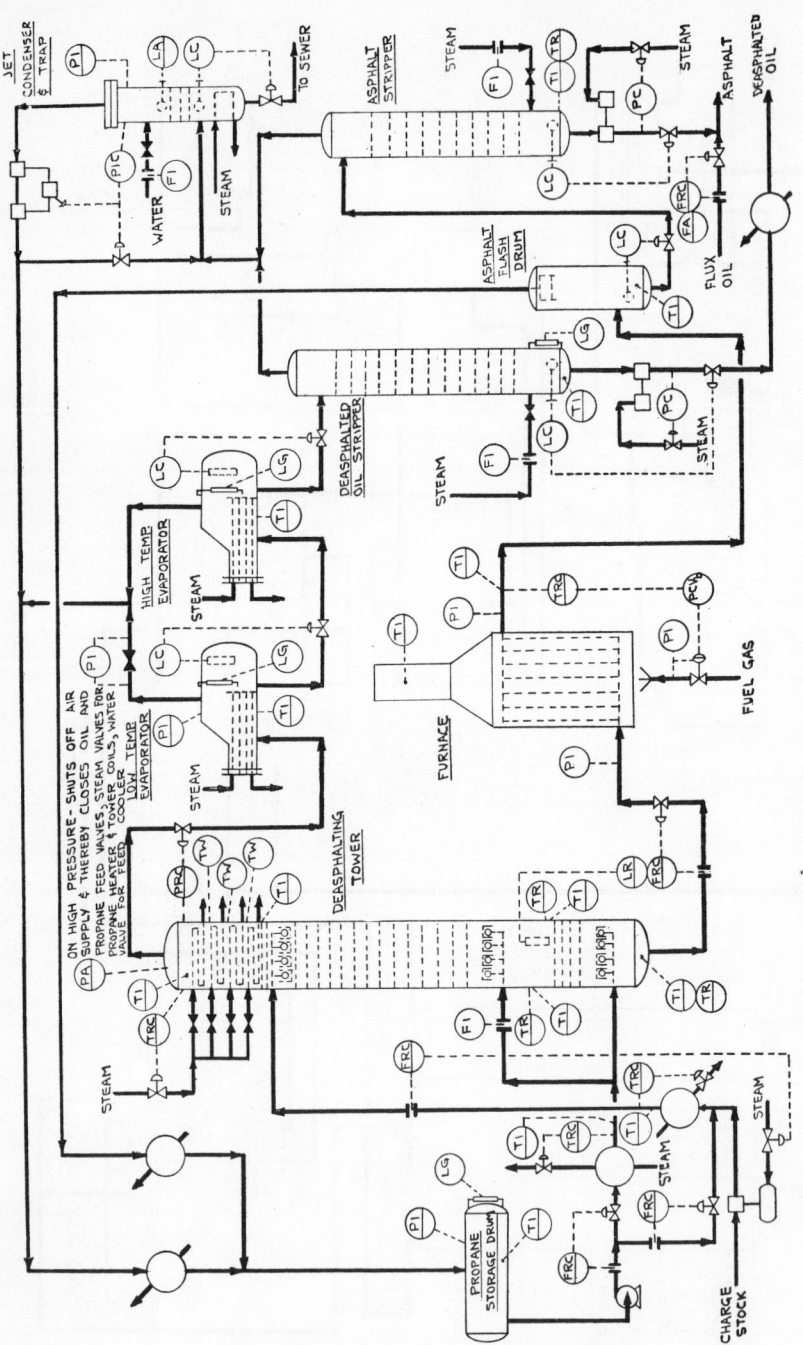

FIG. 10-29. Typical instrumentation for propane deasphalting unit.

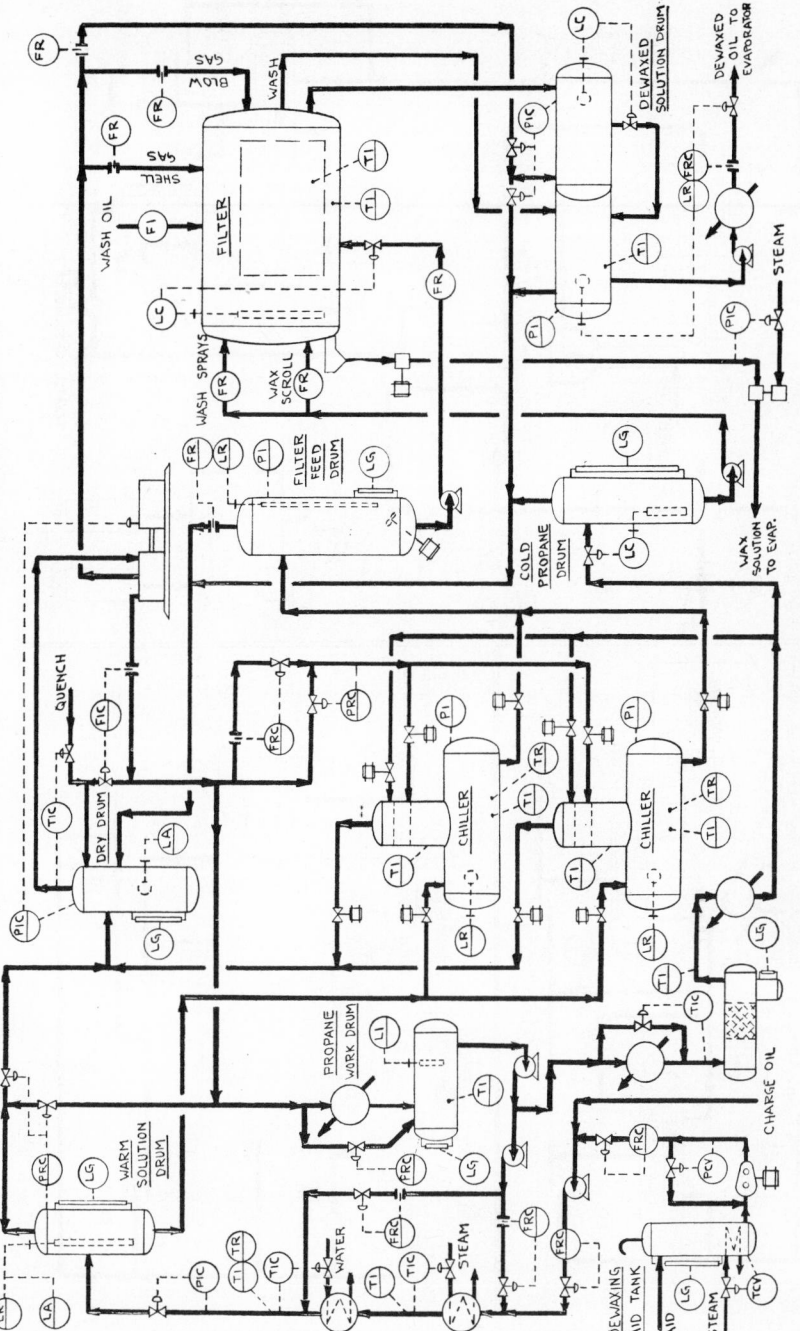

Fig. 10-30. Typical instrumentation for propane dewaxing unit.

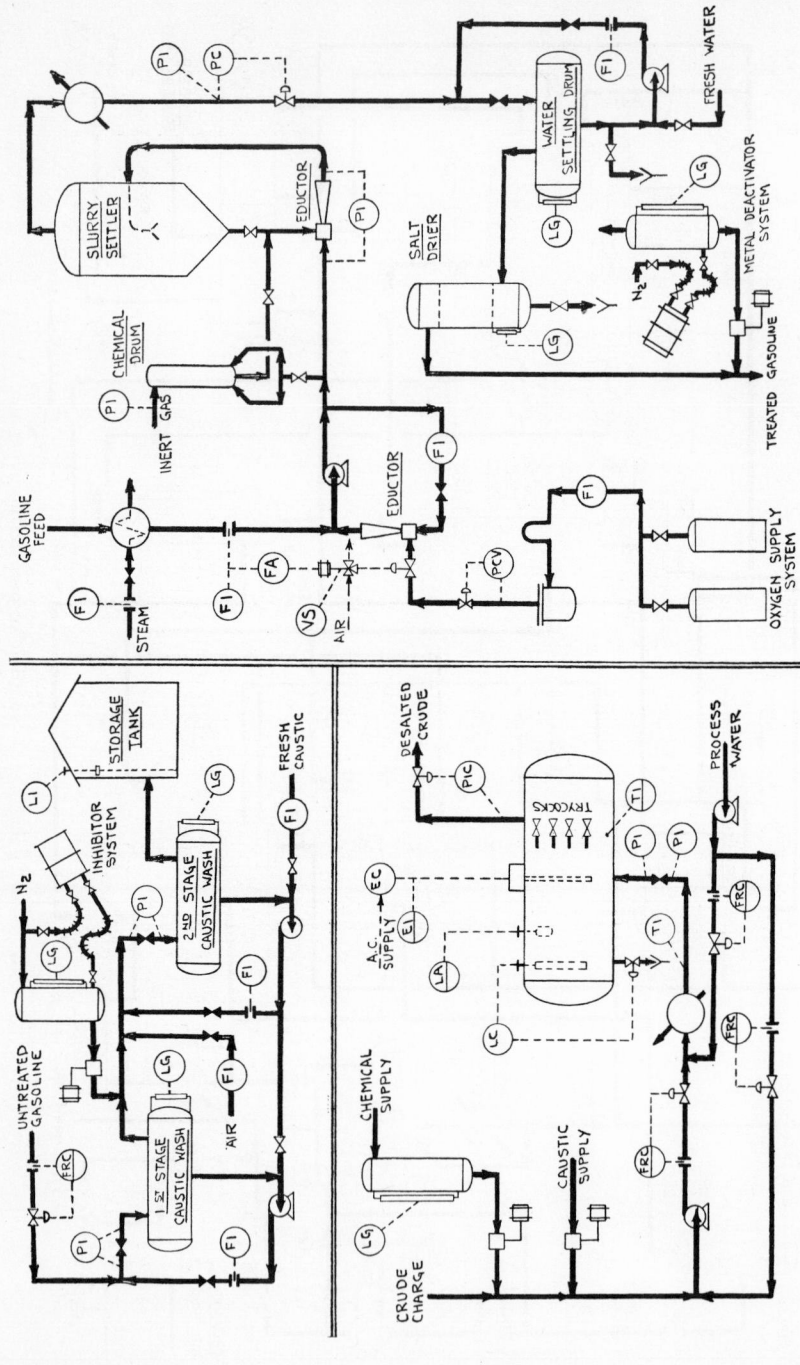

Fig. 10-31. Typical instrumentation for refinery treating processes: (*a, upper left*) inhibitor sweetening, (*b, lower left*) Howe-Baker electric desalter, (*c, right*) slurry copper sweetener.

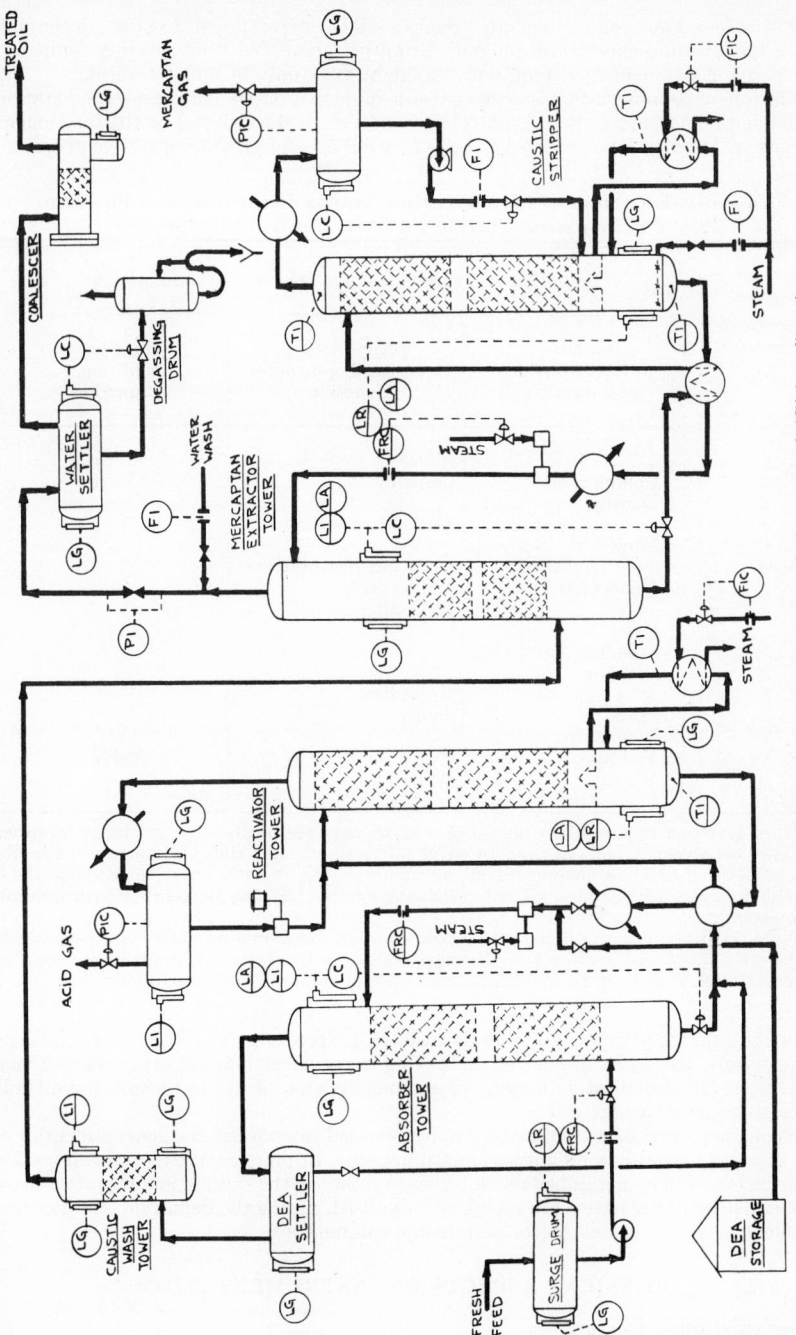

FIG. 10-32. Typical instrumentation for hydrogen sulfide removal (Girbotol) unit.

would be found in a complete plant. Such detail varies from plant to plant, depending upon the personal experience, judgment, and preferences of the operating company; the major instrumentation requirements might vary only in minor respects.

The following information is presented as a guide in interpreting the instrumentation noted on the measurement and control flow sheets. It is suggested by the Instrument Society of America *Recommended Practice* ISA RP 5.1 and provides a system of symbol

Table 10-1. Letters of Identification Noting Definition and Position of Instrumentation on Flow Control Sheets

| Upper-case letter | Definition and permissible positions in any combination | | |
	First letter, process variable or actuation	Second letter, type reading or other function	Third letter, additional function
A	....	Alarm	Alarm
C	Conductivity	Control	Control
D	Density		
E	....	Element (primary)	
F	Flow		
G	....	Glass (no measurement)	
H	Hand (actuated)		
I	....	Indicating	
L	Level		
M	Moisture		
P	Pressure		
R	....	Recording	
S	Speed	Safety	
T	Temperature		
V	Viscosity	....	Valve
W	Weight	Well	

Note 1: When required, the following may be used optionally as a first letter for other process variables: A may be used to cover all types of analyzing instruments. Readily recognized self-defining chemical symbols such as CO_2, O_2, etc., may be used for these specific analysis instruments. The self-defining symbol pH may be used for hydrogen-ion concentration.

Note 2: Although not a preferred procedure, when considered necessary it is permissible to insert a lower-case r after F to distinguish flow ratio. Likewise, lower-case d may be inserted after T or P to distinguish temperature difference or pressure difference.

identification for petroleum process flow sheets; for recording in specifications, listings, requisitions, and order sheets; for indicating items on piping and other construction drawings; for identification tagging of equipment; and for use in technical and trade literature and drawings.

Symbols consist of a combination of letters used to establish the general identity of an item with regard to its purpose and function. In practice, the letter combination may be followed by a number which serves to establish the specific identity of an item. General identifying letters are shown in Table 10-1, noting the definition and position. Table 10-2 shows the complete permissible combinations.

PHYSICAL ASPECTS OF INSTRUMENTATION

Control-house Design

Because even a small petroleum process consists of many vital process variables, operation of most units is carried out in a central location from which an operator can monitor the over-all process. This requires reasonably rapid and accurate data

Table 10-2. Complete General Identifications (All Permissible Combinations of Letters Given in Table 10-1)

Second and third letters (type of device)

First letter (process variable or actuation)	Controlling devices — Separate controllers Recording	Controlling devices — Separate controllers Indicating	Controlling devices — Blind	Controlling devices — Self-actuated (integral regulating valves)	Controlling devices — Safety (relief) valves	Measuring devices — Recording	Measuring devices — Indicating	Measuring devices — Glass devices for observation only (no measurement)	Alarm devices — Recording	Alarm devices — Indicating	Alarm devices — Blind	Primary element	Wells
Temperature..... T	TRC	TIC	TC	TCV	TSV	TR	TI		TRA	TIA	TA	TE	TW
Flow..... F	FRC	FIC				FR	FI	FG	FRA	FIA	FA	FE	
Level..... L	LRC	LIC	LC	LCV		LR	LI	LG	LRA	LIA	LA		
Pressure..... P	PRC	PIC	PC	PCV	PSV	PR	PI		PRA	PIA	PA	PE	
Density..... D	DRC	DIC	DC			DR	DI		DRA	DIA			
Hand..... H		HIC	HC	HCV									
Moisture..... M	MRC	MIC	MC			MR	MI		MRA	MIA	MA	ME	
Conductivity..... C	CRC	CIC				CR	CI		CRA	CIA	CA	CE	
Speed..... S	SRC	SIC	SC	SCV	SSV	SR	SI		SRA	SIA	SA		
Viscosity..... V	VRC	VIC				VR	VI	VG	VRA	VIA			
Weight..... W	WRC	WIC				WR	WI		WRA	WIA		WE	

accumulation to enable the operator to evaluate process performance and initiate changes in plant conditions. Hence, the chief functions of a well-planned control house are:

1. The display, in some convenient pattern, of essential process variables such as temperature, pressure, flow, level, composition, etc.
2. The availability of control instrumentation with which operating values of important variables are selected and maintained

In a large, integrated petroleum refinery, especially where individual process units are connected with a minimum of intermediate storage, one large control house with complete data display, but only generalized or over-all control instrumentation, may be provided. This may be supplemented by smaller control houses for individual process units in which indicating and recording devices together with control instrumentation for the specific unit are contained.

Instrumentation selected for display in a control house may include:

1. Most controlled variables, particularly furnace and heat-exchanger outlet temperatures, important fractionating-column and reactor-zone temperatures, all major process equipment pressures, most of the vital process stream flows, and levels necessary for the satisfactory operation of process equipment
2. Most recorded variables, particularly multipass furnace temperatures, important fractionating-column and reactor-zone temperatures, process equipment pressures requiring close monitoring, process-unit feed and product streams, vessel levels (particularly where outlet flow rates are determined by independent considerations), and stream compositions
3. Indication of general process temperatures (sampling rather than continuous, usually)
4. Indication of utility conditions, including pressure of steam, air, water, and fuel headers
5. Audible and visible alarm devices for signaling off-normal conditions
6. Special manual stations for starting or stopping, opening or closing, advancing, switching, or otherwise modifying the position or operating condition of process equipment

Instrument display varies with regard to (1) the type of instruments selected, (2) the manner in which they are displayed, and (3) their general location in the control house (that is, on vertical panels or on desk-type consoles). With the atmospheric distillation unit of Fig. 10-20 serving as an example, several control panel arrangements are developed to illustrate some possible approaches to vertical panel design.

Figure 10-33a shows an arrangement of large case instruments on nonpictorial panels. The majority of instruments are placed two per panel with the larger potentiometric temperature recorders and recorder controllers in the lower rows. From left to right the instruments are located relative to their function in the process, with feed flow control upper left to product flow controllers extreme right. A general temperature indicator has been located on the first panel, at an elevation for easy reading consistent with its frequent use. Although not indicated on the instrument flow sheet in Fig. 10-20, it has been assumed, by way of example, that auxiliary board-mounted devices include four utility pressure gages and six alarms associated with the utilities and major process variables. These have been located across the top of the panels and are shown in a variety of configurations per panel. Such a control board is typical of petroleum refinery panels from the early 1930's to about the mid-1950's.

Figure 10-33b shows an arrangement of miniature instruments on a pictorial background: a full graphic panel. The graphic panel displays each instrument in its exact role in the process, the background including a pictorial representation of all the major process equipment and streams. Instrument density has been substantially increased, it being about one and a half instruments per linear foot of panel compared with about one instrument per foot of panel for the arrangement with large instruments. The multipoint temperature indicator and recorder, as well as gages and alarms, have been located on a nongraphic panel to the left.

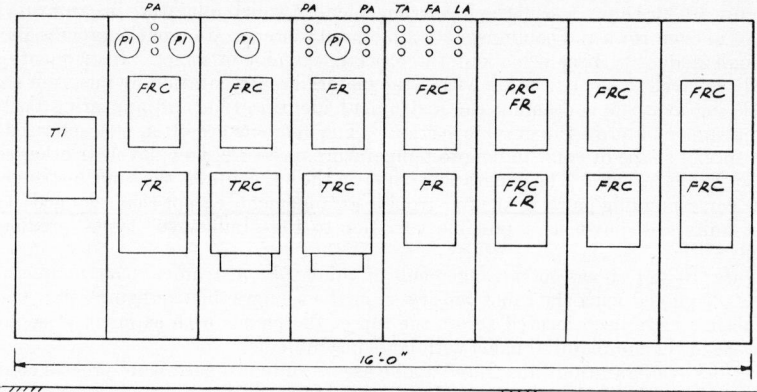

(a) Large Case Conventional

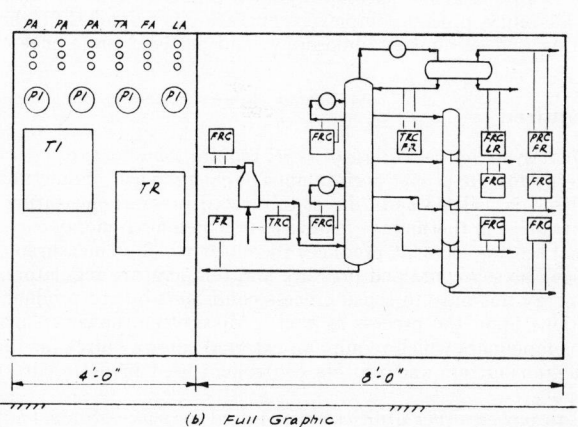

(b) Full Graphic

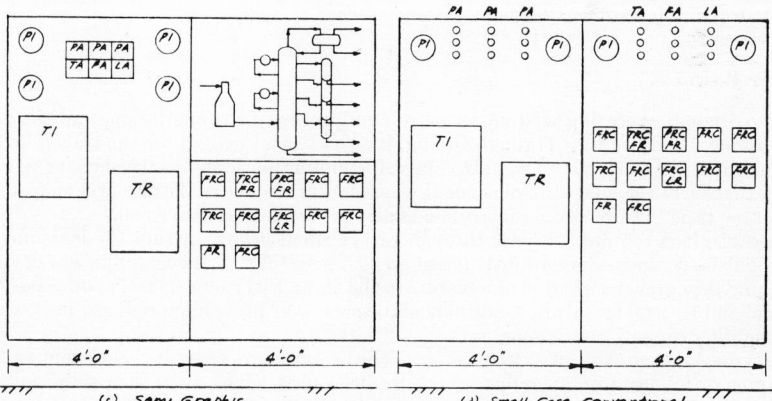

(c) Semi Graphic   (d) Small Case Conventional

Fig. 10-33. Control-house instrument-board arrangements.

Figure 10-33c shows a semigraphic arrangement in which miniature instruments are located in even rows and columns, with a pictorial representation of the process above. Instrument density becomes about three per linear foot of panel. Instruments are usually referred to some location within the pictorial representation by means of small identifying symbols containing the instrument identification and appearing both in the instrument and in the graphic section. The symbols are often characterized by some special shape or color to denote temperature, pressure, flow, level, or other suitable service category. The nongraphic panel on the left contains the same instruments as the corresponding panel in the full graphic arrangement, except that "ganged" type alarm units are shown as a possible variation to the "bull's-eye" types previously shown.

Figure 10-33d shows an arrangement of miniature instruments on nonpictorial panels, displayed with the same density as in the semigraphic configuration. Gages and alarms have been located across the top of the panels with as much freedom of choice as in the nonpictorial panel with large instruments.

The four configurations illustrated above are all suited to petroleum process control board design. Variations to the basic schemes are made as conditions require. The user generally is guided in his selection by such considerations as initial investment cost, space availability, process complexity, operating personnel skill, choice of instruments, operating and accounting philosophy, and personal preference.

Utilities Required

All instrument systems, regardless of type, require some form of operating medium to provide energy for the proper performance of each device. Sometimes the process energy level itself is utilized as in direct-connected pressure-measuring devices with elastic elements or as in thermometric devices in which heat-energy exchange between the process and sensing element produces the effect which is measurement. Control devices too, such as self-contained pressure and temperature regulators, make use of the process energy not only to sense process conditions but to produce the required corrective actions upon the process as well. More often, however, instrument systems include components which require an external energy source for their operation. Measurement transmitters and process controllers used in petroleum processes fall into this category.

In addition to power, other utilities are required for successful instrument installations. These may be materials continuously flowing into the system or statically located in the instrument leads as sealing fluids.

Air Power

The original operating medium in petroleum process instrumentation, and the one still widely used, is air. Through tubing, it serves as a medium for the transmission of measurement and control signals. In valve motor operators it furnishes the power for actuation of the final control element whose motor may be of the diaphragm, piston, or rotary type. In recording instruments air may be used to drive charts.

Because it is required to pass through rather small orifices within the instrument mechanism, it is necessary for instrument air to be carefully cleaned. Sufficient drying is required to prevent freezing of moisture in the air on two counts: first, upon exposure to low ambient-temperature conditions and, second, as pressure is reduced in flowing through instrument restrictions.

A typical instrument air system to serve one or more process units consists of an air compressor followed by a combination filter-drier unit. The air is then subjected to one stage of regulated pressure reduction into a distribution system which transports it throughout the unit. At individual user devices such as local transmitters, controllers, final control elements, and the control house itself, the air may be further filtered and its pressure regulator-reduced before entering the instrument devices. For operating reliability, a standby air supply is made available and can be cut in automatically when the original supply fails.

Electric Power

More recently, process instruments have become available which use electrical signals for the transmission of measurement and control signals. Electrical energy is also used to power final control elements containing electric motors or solenoids, to drive recorder charts, to energize audible and visible alarm units, to illuminate instruments, and to furnish energy for heating instrument enclosures.

Though originating at the main electric power supply for the plant, instrument power supply is served by automatic standby equipment in the event of power failure.

For energizing process instrumentation that creates measurement and control signals, voltage regulation may be necessary. This can be accomplished on an over-all process-unit basis or, as in the case of several scattered devices, within each instrument component as required.

In special instances, frequency regulation of alternating-current power supply is necessary to assure proper performance of instrument components.

Hydraulic Fluid Power

Although not used too frequently, hydraulic systems may be found in conjunction with petroleum process instrumentation, for piloting signals for measurement and control, and for use as power to drive final control elements. The last is common where electric control instruments are required to manipulate a valve without using air as a power source.

The general requirements for adequate maintenance of hydraulic components involves assurance of a leakproof system to avoid material and pressure loss and to avoid contamination of the fluid, prevention of excessive temperatures or other detrimental environmental conditions, and a regular check on fluid inventory to assure reliable operation of the system.

Steam Power

Steam is commonly used for heating instrument installations subject to adverse temperature conditions. This may be accomplished by containing instruments in steam-heated enclosures or by carefully and selectively wrapping instrument components and connecting piping with steam leads. The combined effect of the steam conditions and the actual installation configuration should be sufficient to eliminate the adverse temperature effects upon the process material and instrument without creating other undesirable conditions such as vaporization or deterioration of the process material or damage to instrument parts as a result of excessive heating.

Purge Materials

Many instrument installations require the continuous addition of a medium which prevents process material from entering the instrument and its leads. This purge material must be readily available and at a sufficiently high pressure to be introduced into the process without difficulty; it should not be detrimental to the process or instrument with which it comes into contact; it must be immune to the prevailing ambient conditions.

For making pressure, differential pressure, and level measurements in process vessels containing fluidized catalyst beds, instrument leads are purged to prevent plugging with catalyst. In fluid catalytic cracking units, for example, such instrument leads may be purged with air, steam, or a hydrocarbon fluid, depending upon which portion of the process equipment is involved. Generally, one may use the bed fluidizing medium itself as a suitable purge material.

Often, process streams contain entrained solids which can cause instrument malfunctioning if the leads plug. Usually, a similar material, devoid of solids, is available which can then be used to purge such installations, making pressure, level, and flow measurement in these applications possible.

Sealing Materials

Whereas purge materials completely prevent process material from entering any part of the instrument system external to the process, sealing materials occupy a limited part of the system by filling only a portion of the connecting leads to the instrument. These materials are used to protect the instrument from harmful process fluids. In addition, they confine the process material to regions where it cannot be adversely affected by ambient conditions. Finally, seals may be used to eliminate or at least minimize steam or electric heating of instrument installations.

In view of the above-mentioned objectives, sealing materials must not be detrimental to the process fluid or instrumentation; must be immune to both ambient and process conditions in so far as they must not evaporate, freeze, or become viscous; and must be immiscible with and otherwise unaffected by the process fluid.

Some typical materials for sealing instrument installations are ethylene glycol for hydrocarbon fluids, kerosine for caustic or sulfuric acid solutions, and dibutyl phthalate for steam, water, and wet air.

NEW AREAS OF ACTIVITY

Data-handling Systems

Traditionally, data presentation in petroleum processing has been pretty much confined to those methods illustrated above in the discussion on Control-house Design. Such designs have been, and will continue to be, successful in achieving the operating objectives of most process units. But with the trend toward integrated refineries, operation of the entire facility from a central location can become burdened by:

1. Excessive data required for critical operation
2. A growing demand for additional data relative to subsequent plant design and other scientific activity
3. The increased interest in accurate cost-accounting information
4. The large number of operating personnel required in the control room

While the development of miniature and electronic process instrumentation should help substantially in overcoming the problems inherent in measurement and control of large, integrated plants, the use of specially designed data-handling systems offers alternate methods for monitoring process operations and for acquiring data.

A typical data-handling system for a petroleum processing installation may perform the following functions:

Gathering Data. Information fed into a petroleum process data-handling system originates with conventional instrument components measuring the required process variables. Where electronic instrumentation is involved, having been selected, perhaps, in anticipation of its subsequent use with either data-handling or computing equipment, no special modification of the measurement signals is required. Where pneumatic components furnish measurement signals, these must be converted to proportional electrical signals for use by the data-handling equipment.

Converting Data. Most process measurements are made by analog, or smoothly varying, means. The data-handling system, with a view toward its ultimate objectives, converts the analog inputs to digital form. This conversion affords a distinct advantage for both accounting and some aspects of plant operation.

Under certain operating circumstances, however, the trend information one can get from a continuous chart record is of more value than tabulated digital data. So while the recording type of receiving instrument is not due for immediate elimination as a result of more advanced data presentation, the number of recorders as well as the more conventional alarm devices used will undoubtedly decline.

Computing Data. Data, as measured, are often not in their final usable form. Data-handling equipment is capable of performing simple mathematical computations on input variables. Such sensing devices as thermocouples and some filled-system thermometers, as well as many special measurement devices, do not have linear signal outputs. The data-handling equipment can perform the necessary linearization.

Since the majority of flow measurements in a petroleum process are of the variable-head square-root type, square-root extraction circuitry is frequently required. In addition, since process operating conditions are subject to frequent variation, automatic temperature and pressure compensation of uncorrected flow measurement can be performed before final presentation of the corrected flow data.

Scanning Process Variables. This function involves the automatic and sequential monitoring of process variables to determine if any of these have deviated from some predetermined desired value. This is an especially valuable function of data-handling equipment in petroleum processing in view of the many operating conditions requiring monitoring, particularly the numerous important temperature-measurement points common to petroleum equipment.

Logging Process Data. The ultimate function of a data-handling system is to present the information which has been gathered, converted, modified by computation, and scanned for abnormal conditions in some usable form. For refinery operation this takes the form of a standard log sheet on which final data are automatically typed at regular intervals. This takes the place of the usual hand logging of control board instruments done, perhaps, every hour. In addition, a log sheet may be automatically activated when abnormal conditions occur or on demand of the operator.

While the development of data-handling equipment stemmed largely from efforts in military, aeronautical, and other scientific pursuits requiring high-speed data acquisition, a system for a typical petroleum application may be designed for, say, several hundred input measurements which it can process at a rate of about one hundred per minute.

Modern data-handling techniques can enhance the petroleum process in so far as they are capable of relieving the operating personnel of tedious and monotonous logging, performing numerous measurements in a short period of time and in a reasonable amount of space, overcoming inherent inaccuracies resulting from human error, and continuing to operate during plant upsets when operating personnel are distracted from their usual monitoring functions. In this regard, these techniques offer a desirable supplement to conventional plant instrumentation.

A more far-reaching potential for data-handling systems in petroleum processing relates to the monitoring of complex processes, to pilot plant investigations, and to their use in compiling operating data for the creation of mathematical models and for updating existing mathematical models during computer-controlled operation.

Computer Control

The typical process-control system utilizing the feedback principle is an analog computer. In this instance the process consists of a single process variable and one manipulating value (the control valve) related to that variable. The objective, or performance criterion, is to maintain the process variable at some desired value. The system senses the variable and determines any deviation, its direction, and its magnitude. The control device then manipulates the control valve in an effort to reduce to zero or at least minimize the deviation from the desired value. In a sense, it attempts to optimize performance by holding a particular process variable at a value selected by the plant operator.

Extending this concept of optimization of individual variables to an entire process, we might conceive of:

1. A broader process objective, or plant performance criterion, compared with desired values for individual variables
2. An entire family of manipulated variables related to that broader objective, compared with individual valve positions responding only to their respective controlled variables

A cascade control system can be seen as a fundamental extension of the single-variable control loop. In an example mentioned earlier, the value of fuel flow to a large furnace was not held at a value selected by the plant operator. The limitation of such a fixed control-point controller lay in the fact that it would have represented, at best, a compromise performance criterion. There was clearly a broader operating objective, that is, the maintaining of an appropriate furnace outlet temperature. Consequently any fuel flow necessary to achieve this end was desirable. The system developed, one of an outlet temperature controller dictating an appropriate set point for the fuel flow controller, provided the necessary coupling of two related process variables. Furthermore, it established one of the variables as directly related to the main operating objective and the other to the subordinate role of finding the one best fuel valve position.

Similarly, in large processes with many and interrelated items of equipment and functions and valves which can be manipulated, it is evident that for any given set of operating conditions there is **one best set** of valve positions to achieve a particular operating objective. Since valve positions in a petroleum process are dictated by the demands of each process controller, another way of visualizing plant optimization is to consider a means for establishing the **one best set** of control-point settings to achieve that same operating objective. One obvious advantage of such an approach is that the original scheme of control remains essentially in its original form, the overriding or optimizing system serving as an additional refinement to the existing instrumentation. It then can be removed from the process without upsetting the original mode of operation.

For optimizing a process there are usually three broad operating objectives that may be pursued. These are:

1. Producing the greatest quantity of a given product
2. Producing an end product of the highest purity
3. Producing a given end product with a minimum of operating cost

Normally the operator makes his control-point settings and establishes operating conditions in such a manner as to achieve one of these objectives or, at least, approach the objective. Petroleum processes are often so complicated as to defy such an approach. The number of variables involved is too great for the operator to inter-relate logically. In optimization, the task of choosing the correct set of control points is approached in a somewhat more rigorous manner than is found in human operation.

As individual process controllers may be either closed loop or open loop, so may plant optimization be either direct (with feedback) or predetermined (without feedback). The direct optimization system maintains a particular process value at a minimum or maximum. It accomplishes this by sensing the value or computing it from other measurable values and manipulating the process to maintain the minimum or maximum peak value of the variable. Initially, the method would be applicable to systems containing few variables.

The simplest case is a two-dimensional system, or one containing one optimized variable and one manipulated variable. Graphically such a system is represented by a curve as shown in Fig. 10-34. The single curve represents the relationship between the manipulated and optimized variables for only one set of operating conditions. As load changes occur, the relationship changes and, as is likely the case in most processes, the peak value of the important (or optimized) variable occurs at different values of the manipulated variable. In fact, it is equally likely that the magnitude of the peak also changes, producing a family of curves representing many operating conditions. If, for example, the optimized variable represents product quality, a unique value of the manipulated variable would exist at any operating condition to produce optimum quality product, independent of any artificially controlled variable such as temperature, pressure, or flow. That is, the system would adjust itself to achieve the desired end directly.

A three-dimensional system, or one containing one optimized variable and two manipulated variables, becomes significantly more involved, since it is necessary to find the correct combination of manipulated variables to optimize. Physically speaking, it is comparable to finding, by some trial-and-error method, the highest point on a surface; this, opposed to the relatively simple task of finding the high point on a plane curve for the two-dimensional system.

A predetermined optimizing system makes use of a close mathematical model of the actual process. The equations describing the process behavior, together with the equation for some performance criterion, form the working parts of a predetermined optimizing system. As data are gathered from the operating plant, the system equations are solved and the process manipulated in accordance with achieving the desired optimization. In effect, the equipment evaluates the important uncontrolled or uncontrollable process variables and determines the correct control points for the controlled variables.

A further sophistication of optimization by model is a system which, through regular investigation of the process, can continuously modify the equations to match the process more closely.

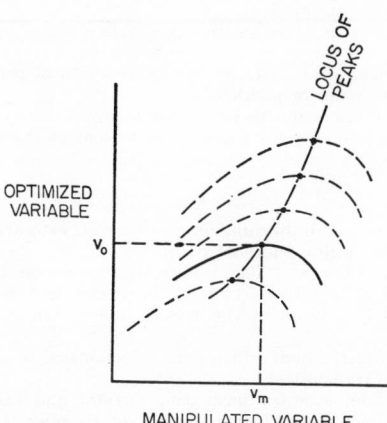

FIG. 10-34. Two-dimensional system illustrating maximum peak optimization.

One important aspect of process optimization is the establishment of constraints. Clearly the mathematical representation of a process behavior must be subject to defined limits. Process manipulation cannot be tolerated to the point where personal safety is jeopardized, nor can equipment operate beyond its limitations, nor can process material be exposed to abnormal temperatures and pressures. Likewise, limitations are encountered naturally as a result of exhaustible energy levels and material inventories.

GLOSSARY OF INSTRUMENTATION AND CONTROL TERMS

Following is a glossary of terms pertaining to instrumentation and control. The selection is limited to those expressions ordinarily associated with and used within the petroleum industry. Many terms, however, are common to instrumentation in a broader sense, since it is necessary to include those terms in the language pertaining to measurement, control, and instrument components and to the design features of both the instrument components and the process plant.

A

Absolute Humidity. Water-vapor content expressed directly as composition.

Absolute Measurement. A measurement in terms of the fundamental standards of length, mass, and time.

Absolute Viscosity. The tangential force per unit area of either of two horizontal planes at unit distance apart, one of which is fixed while the other moves with unit velocity, the space between the planes being filled with the substance.

Absorption. The process whereby some or all of the energy of sound waves or electromagnetic radiations is transferred to the substance through which they travel.

Accuracy. A number or quantity indicating how close an instrument is to the actual value of a variable. It may be expressed as a percentage of the full-scale reading of the instrument.

Ammeter. An instrument for measuring the magnitude of an electric current and provided with a scale usually graduated in amperes, milliamperes, microamperes, or kiloamperes.

Amplifier. A device whose output is an increased function of an input signal and which draws power therefore from a source other than the input signal.

Amplitude. The maximum departure of a value from the average or mean value, such as the extent of the vibratory movement of a swinging pendulum.

Amplitude Ratio. The ratio of the recorded amplitude of any process variable to that of a preceding amplitude or wave.

Analog. As applied to a computer or other instrumental device, this defines one which handles variables as continuous values that can be expressed in graphic curves.

Analysis. The ascertainment of the identity and/or the concentration of the constituents or components of a material.

Annunciator. A device which gives audible or visible warning when a value differs from a predetermined value.

Attenuation. The decrease with distance in the direction of propagation of a quantity associated with a traveling wave or particle.

Automatic Control. The operation in which the value of a controlled condition is compared with a desired value and corrective action dependent on the difference is taken without human intervention.

B

Balance. To cause to be in equilibrium; in an electrical network, an adjustment in one branch to cause no current to flow in another branch.

Bimetallic Thermometer Element. The temperature-sensitive bimetal of a bimetallic thermometer. The bimetal comprises two or more metals mechanically associated in such a way that relative expansion of the metals due to temperature change produces bending of the bimetal.

Blind Transmitter. An instrument which transmits a signal to another location without indicating its value at the transmitter.

Bolometer. A primary detector for measuring current and radiant power utilizing a temperature-sensitive resistor and which depends for its operation on the temperature difference maintained between the primary detector and its surroundings.

Bridge. An electrical network used in many instruments. One branch of the bridge connects two points of equal potential and hence carries no current when the circuit is properly adjusted or balanced.

C

Calibration. The operation of making an adjustment or marking a scale so that the readings of an instrument conform to an accepted standard, the checking of readings by comparison with an accepted standard, the errors or corrections revealed by this comparison.

Calibration Curve. The plot of a calibration.

Calorimetry. The measurement of heat content.

Capacitance. The change in quantity contained per unit change in a reference variable. It is measured in units of quantity divided by the reference variable. In a process vessel it may be cubic feet of liquid per foot of liquid height or liquid surface area.

Capacity. A measure of the maximum quantity of energy or material which can be stored within the confines of a stated piece of equipment. It is measured in units of quantity. In a process vessel it may be cubic feet of liquid.

Cascade. The use of one controller to alter the set point of another controller, with both controllers associated with related variables of one process.

Closed Loop (or Feedback Control System). A control system in which the controlled quantity is measured and compared with a standard representing the desired performance. Any deviation from the standard is fed back into the control system in such a sense that it will reduce the deviation of the controlled quantity from the standard.

Cold Junction. The free ends of a thermocouple which are connected to an instrument.

Continuous Process. A process in which for extended periods uninterrupted flows of proportioned components enter and products leave a system.

Control Point (or Set Point). The value of controlled variable which, under any fixed set of conditions, the automatic controller operates to maintain.

Controlled Medium. That process material in which a variable is controlled.

Controlled Variable. That quantity or condition of the controlled system which is directly measured and controlled.

Correction. That quantity which is added to a calculated or observed value to obtain the correct value. It is the negative of the error.

Correction Time (or Stabilizing Time). The time required for the controlled variable to reach and stay within a predetermined band about the control point following any change of the independent variable or operating conditions.

Corrective Action. The variation of the manipulated variable produced by the controlling means.

Cycle Control. The automatic changing of the control point of a variable as a function of time. It is often accomplished by a rotating cam on which rides a cam follower that resets the control point.

Cycling. A periodic change of the controlled variable.

D

Damping. That property of a system which causes dissipation of energy and hence causes decay in amplitude of oscillations. Critical damping is the minimum amount required to change motion from oscillatory to aperiodic. A smaller degree is called underdamping; a larger degree overdamping.

Dead Time (or Transportation Delay). A pure delay between two related actions measured in units of time.

Dead Zone (or Neutral Zone). The largest range of values of the measured variable to which the instrument will not effectively respond.

Derivative Action (or Rate Action). Action in which there is a predetermined relation between the instantaneous rate of change of the controlled variable and the position of the final control element.

Derivative Time. In derivative action, the value relating the magnitude of correction of the final control element to the rate of change of the controlled variable.

Deviation. The difference at any instant between the value of the controlled variable and the desired value.

Diaphragm Motor Valve. A control valve which responds to the signal from a controller and uses air pressure as the activating force.

Differential Gap. In two-position action, the smallest range of values through which the controlled variable must pass in order to move the final control element from one to the other of its final positions.

Digital. As applied to a computer or other instrumental device, this defines one which handles variables in actual number form.

Distillation. The separation of the components of a liquid mixture by partial vaporization of the mixture and separate recovery of vapor and residue.

Disturbance. See Load Change.

Drift. The deviation of the instrument indication with time from an initial value when the measured variable and ambient conditions are constant.

Droop (or Offset or Load Error). The steady-state difference between the control point and the value of the controlled variable corresponding to the set point.

E

Elastic Element. The chamber within which a pressure medium is confined. It may be a Bourdon tube, a single flat or corrugated diaphragm, a bellows, or a combination of any of these.

Electronic Controller. An electronic device which operates to correct or limit deviation of a measured value from a selected reference by electric means.

EMF. Electromotive force, the electric potential difference between the terminals of any device used as a source of electrical energy.

Error. The difference between the measured value and the desired value of a controlled variable. It is the negative of the correction.

F

Fail-safe. The ability of a control system to indicate its inoperability and to revert to a safe static condition in case of a component failure or a power failure.

Feedback. In a transmission system or a section thereof, the returning of a fraction of the output to the input.

Feedback Control System. See Closed Loop.

Fidelity. The degree with which a system or a portion of a system accurately reproduces at its output the essential characteristics of the signal which is impressed upon its output.

Final Control Element. The part of the control system which receives impulses from the controller and directly changes the value of the manipulated variable.

Floating Action. That in which there is a predetermined relation between the deviation and the rate of motion of a final control element.

Fluidity. The property of a substance expressing its ability to flow, as contrasted to viscosity or resistance to flow. It is the mathematical reciprocal of viscosity.

Fractional Distillation. A distillation in which distillates boiling in different temperature ranges are separately collected.

Fractionating Column. In fractional distillation, an apparatus designed to bring into intimate contact the rising vapor and falling liquid.

Frequency Response. Two relations between sets of sinusoidal inputs and the resulting outputs. One relates frequencies to the output-input amplitude ratios; the other to phase difference between output and input.

G

Gage. A device applied to the point of measurement, containing the primary measuring element.

Gain. The increase in signal power in transmission from one point to another.

Galvanometer. An instrument for indicating or measuring a small electric current or a function of the current by means of a mechanical motion derived from electromagnetic or electrodynamic forces which are set up as a result of the current.

Gamma Ray. An electromagnetic radiation from disintegrating isotopes, used sometimes to measure process levels.

Gas Chromatography. An analytical method of separating similar vapors by selective adsorption through the flow of gas through a special column.

H

Half-life. The time required for disintegration of one-half the atoms of a radioactive sample.

Head. Pressure resulting from gravitational forces on liquids, measured in terms of the depth below a free surface of the liquid which is the zero reference head.

Holdup. The total amount of process material tied up in a plant or in some component thereof.

Hot Junction. The process end of a thermocouple system; this applies, in a manner of speaking, even though the process temperature is lower than the instrument, or cold junction, temperature.

Hot-wire Instrument. An electrothermic instrument which depends for its operation on the heating effect of a wire carrying a current.

Hydraulic Gage. A gage specifically constructed for service at extremely high pressures and where water or a noncorrosive liquid is the pressure medium.

Hydrostatic-head Gage. A gage employing one or more elastic chambers and differing from the ordinary types of pressure gage only in the graduation of the scale. The scale is usually graduated to show the head of water or other liquid in feet.

Hysteresis. When the relation between magnetizing field strength and magnetization is not single valued but depends on the past history, the material is said to exhibit magnetic hysteresis. By analogy, elastic hysteresis is that between strain and stress; instrument hysteresis is that between input and response. The phenomenon commonly appears as a closed loop in the plot relating the two variables during a cyclic process. The form of this loop may be affected by the rate of cycling, by the number of previous cycles, and by the time elapsed since the previous cycle. Flexibility and loose fits in linkage and backlash in gearing cause similar effects.

I

Immersion Length. Of a thermometer, the length from the free end of the well, bulb, or element, if unprotected, to the point of immersion in the medium the temperature of which is being measured.

Impedance. The complex ratio of a forcelike quantity (force, pressure, voltage, temperature, or electric field strength) to a related velocity-like quantity (velocity, volume velocity, current, heat flow, or magnetic field strength).

Independent Variable. That independent quantity or condition which, through the action of the control system, directs the change in the controlled variable according to a predetermined relationship.

Indicating Instrument. Any measuring device which is read by observing the position of a pointer on a scale.

Inductance. The property of a system element which tends to oppose change of a velocity-like quantity (velocity, volume velocity, current, etc.).

Inertia. The property of any material to resist change in its state of motion.

Infrared. The region of the electromagnetic spectrum extending from 0.78 to 300 microns.

Instrument. A device for measuring the value of a quantity under observation; a physical device for replacing the perceptual, communicative, manipulative, or intellectual abilities of man; a physical device which senses a physical quantity or deals with the information resulting thereby.

Integral Action (or Reset Action). That in which the final control element is positioned in accordance with a time integral function of the controlled variable.

Integral Time. In integral action, the value relating to the magnitude of the deviation of the controlled variable to the rate of change of the final control element.

Integrator. As applied to a flowmeter, a device that computes the total flow over a period of time, based on the rate of flow recorded by the flowmeter.

Interlock. A device actuated by the operation of some other device with which it is directly associated to govern succeeding operations of the same or allied devices.

Input. The signal fed into a control system, computer, or other device.

K

Kinematic Viscosity. The ratio of absolute viscosity to the density of a fluid.

L

Lag. Any deviation from instantaneously complete response to an input signal.

Limit Switch. A switch which is operated by some part or motion of a power-driven machine or equipment to alter the electrical circuit associated with the machine or equipment.

Linear. A relationship existing between two quantities such that a change in one quantity is exactly proportional to a change in the other quantity.

Liquid-level Manometer. A differential manometer employing a liquid as the movable partition and providing means for observing the change in level of one or both of the free surfaces.

Load Cell. A weighing transducer which converts force to a hydraulic, electric, or pneumatic signal to actuate an instrument.

Load Change (or Disturbance or Upset). A change in operating conditions which requires a change in the flow of the manipulated medium to maintain the desired control point.

Load Error. See Droop.

Logger, Data. An automatic data-handling system that receives instrument signals and converts them into typed figures.

M

Manometer. A device for measuring pressure or differential pressure.

Manual Controller. A controller having all its basic functions performed by devices which are operated by hand.

Measurement Device. An assembly of one or more basic elements with other components and necessary parts to form a separate, self-contained unit for performing one or more measurement operations.

Manipulated Variable. A quantity or condition that is varied by an automatic controller so as to affect the value of the controlled variable.

Memory. Any device into which information can be introduced and then extracted at a later time.

Miniaturization. The design and production of a miniature that will perform the same functions as the larger sized original.

Motor Operator. A portion of the controlling means which applies power for operating the final control element.

N

Natural Frequency. The frequency of free oscillation of a body or system.

Negative Feedback. Feedback which results in decreasing the amplification.

Neutral Zone. See Dead Zone.

Noise. Any undesired sound; by extension, any unwanted disturbance of the same nature as the signal, which tends to interfere with the reception of that signal.

Nonlinearity. The deviation of any functional relationship from direct proportionality.

Null Balance. A method of measuring that directly compares an unknown quantity with a like known value, usually after balancing an electrical circuit to zero (null).

O

Objective Variable. That quantity or condition which is controlled by virtue of its relation to the controlled variable and which is not directly measured for control.

Offset. See Droop.

Ohmmeter. A direct-reading instrument for measuring electric resistance and provided with a scale, usually graduated in either ohms or megohms.

Optical Pyrometer. A temperature-measuring device comprising a standardized comparison source of illumination and some convenient arrangement for matching this source, either in brightness or in color, against the source whose temperature is to be measured. The comparison is usually made by the eye.

Optimize. To create the most advantageous conditions for processing, through automatic control.

Orifice Meter. A constriction flowmeter in which the constriction takes the form of an orifice in a thin plate.

Oscillation. The variation, usually with time, of the magnitude of a quantity with respect to a specified reference when the magnitude is alternately greater and smaller than the reference.

Output. The signal emerging from a control system or other instrumental device.

Overshoot. The amount of overtravel of an indicator or recorder beyond its final steady deflection when a new constant value of the measured quantity is suddenly applied to an instrument.

P

Permanent Pressure Drop. The unrecoverable loss in pressure which occurs when a fluid passes through a throttling device.

Pilot-operated. A device in which the energy transmitted through the primary element is either supplemented or amplified by energy from another source.

Pirani Gage. A bolometric vacuum gage which depends for its operation on the thermal conduction of the gas present. Pressure is measured as a function of the resistance of the heated filament.

Pitot Tube. A slender flow-sensing element with an open-ended tube facing upstream to measure impact pressure and another whose open end is parallel to the flow path to measure static pressure.

Pitot-Venturi Tube. A combination of a pitot and a venturi tube where the venturi encloses the pitot tube.

Pneumatic Controller. A mechanism which measures the value of a variable quantity or condition and operates to correct or limit deviation of this measured value from a selected reference by pneumatic means.

Positioning Action. Action in which there is a predetermined relation between the value of the controlled variable and the position of the final control element.

Positive-displacement Meter. A generic term for instruments in which metering is effected by alternately filling and emptying containers of known and fixed capacities so that fluid passes through the instrument in a series of successive units of volume.

Positive Feedback. Feedback which results in increasing the amplification.

Potentiometer. An instrument for measuring an unknown electromotive force or potential difference by balancing it, wholly or in part, with a known potential difference produced by the flow of known currents in a network of known electrical resistances.

Pressure. Force per unit area.

Primary Element. That portion of the measuring means which first either utilizes or transforms energy from the controlled medium to produce an effect which is a function of change in the value of the controlled variable.

Process. The collective functions performed in and by the equipment in which a variable is to be controlled.

Proportional Action. That in which the controller adjusts the final control element in proportion to the deviation of the controlled variable.

Proportional Band. The range of measured value needed to cause maximum possible change in the final control-element setting, i.e., the amount of pen movement necessary to give full valve movement. It is usually expressed as a percentage of the full-scale range.

Pyrometer. A thermometer of any kind usable at relatively high temperature.

R

Radiation. Energy propagated through space or a material medium as waves. Used in many forms such as infrared, visible light, ultraviolet, and gamma rays for analysis and other measurement.

Radiation Pyrometer. A pyrometer in which the radiant power from the object or source to be measured is utilized in the measurement of its temperature. The radiant power, within wide- or narrow-wavelength bands filling a definite solid angle, impinges upon a suitable detector. The detector is usually a thermocouple or thermopile, a bolometer responsive to the heating effect of the radiant power, or a photosensitive device connected to a sensitive electrical instrument.

Rangeability. A ratio of the maximum to minimum flow of a control valve, within which flow characteristics are maintained to stated limits.

Rate Action.   See Derivative Action.

Ratio Control.   A type of control which uses a measurement of one variable to adjust the control point for another variable.   Often used for proportioning two flows such as column feed and reflux.

Readout.   The visual indication and printed record of any process variable.

Receiver.   That part of an instrument system which receives a signal from a transmitter and converts it into some usable form.

Recording Instrument.   An instrument whose primary function is to make a continuous record of a process condition.

Regulator.   A device which measures a variable and limits the deviation of this value from a selected reference.

Relay.   A measuring or controlling accessory which uses additional energy for supplementing or amplifying its ultimate action.

Repeatability (or *Reproducibility*).   A measure of the ability of an instrument to reproduce the same results when subjected to the same test conditions.

Reproducibility.   See Repeatability.

Reset Action.   See Integral Action.

Resistance.   Opposition to flow, measured in units of potential change required to produce unit change in flow.

Resistance Thermometer.   An electric thermometer which operates by measuring the electrical resistance of a material whose resistance is known to be a certain function of temperature.

Response.   The quantitative expression of the output of a system as a function of its input under conditions which must be explicitly stated.

S

Scale.   A fixed part of an instrument on which appear the graduations that show the value of the quantity being measured.

Scanning.   An electronic means of automatically and sequentially monitoring process variables.

Self-contained Instrument.   An instrument which has all the necessary equipment built into the case or made a corporate part thereof.

Self-regulation.   An inherent characteristic of the process which aids in limiting deviation of some variable.

Set Point.   See Control Point.

Signal.   An action (visible, audible, electrical, hydraulic, pneumatic) used to convey information in an instrument system.

Simulation.   The representation of a physical system, such as a process and its instrumentation, by a computer and related accessory equipment.

Speed of Response.   The time required for an instrument to react to a change in one of the process variables.

Stability.   That property of a process and its related equipment and appurtenances which permits it to approach a steady-state condition when all disturbances have ceased.

Standard Cell.   In a potentiometer, a cell which serves as a standard of electromotive force.

Static Error.   The deviation of an instrument reading from that of the absolute value of a static variable.

Straightening Vanes.   Vanes placed between a flow-measuring device and a flow-disturbing fitting to establish normal, long-pipe flow conditions.

Supply Pressure.   The pressure of air supplied to a pneumatically operated device.

T

Telemetering.   The measurement and transmission, over a considerable distance, of variable quantities.

Temperature.   The relative hotness or coldness of a body as determined by its ability to transfer heat to its surroundings.   There is a temperature difference between two bodies if, when they are placed in thermal contact, heat is transferred from one body to the other. The body which loses heat is said to be at the higher temperature.

Temperature Compensation.   Construction of a device in a manner which makes it relatively unaffected by temperature change.

Thermocouple.   A pair of dissimilar conductors so joined that an electromotive force is developed by the thermoelectric effects when the two junctions are at different temperatures.

Thermopile.   A closed circuit of several thermocouples connected in series to multiply their effect, used for detecting and measuring small temperature differences.

Transducer.   A device which receives energy (or a signal) from one system and transmits it to another.   The nature of the input and output signals may be similar or different.

10-48 PETROLEUM PROCESSING HANDBOOK

Transmitter. A device that modulates a separate power source in accordance with changes in a measured variable and transmits a corresponding signal over some distance.

Transportation Delay. See Dead Time.

Turndown. The ratio of normal maximum flow to minimum controllable flow through a final control element.

Two-position Action. That in which a final control element is moved from one of two fixed positions to the other.

U

Ultrasonic. Relating to sound in the frequency range above about 15 kc/sec.

Ultraviolet. Electromagnetic radiation extending from the visible spectrum at the violet end up to the region of low-frequency X rays, with wavelengths from about 136 to 4,000 angstroms.

Upset. See Load Change.

V

Vacuum. Any gaseous space at pressures below atmospheric.

Valve Positioner. An auxiliary system on control valves which is capable of applying full energy to or removing all energy from a valve motor until the stem is positioned as required by the controller output.

Variable Area Meter. A flow-measuring device in which a float contained in a vertical tapered tube, large end up, is lifted to a position of equilibrium between the downward weight of the float and the upward force of the fluid moving through an annular space around the float.

Venturi. A tube of varying cross section formed by an approach cone, a throat section, and an exit cone. In a flow stream it causes a reduction in pressure which may be related to the flow rate.

W

Wheatstone Bridge. A simple bridge network for measuring electrical resistance, with four resistances (one the unknown quantity) arranged in a diamond configuration. A battery is connected across one pair of opposite corners, and a galvanometer across the other pair. When one of the resistances is so adjusted as to cause no deflection of the galvanometer, the unknown resistance can be evaluated.

REFERENCES

1. Considine, Douglas M., ed., *Process Instruments and Controls Handbook*, McGraw-Hill Book Company, New York, 1957.
2. Rhodes, Thomas J., *Industrial Instruments for Measurement and Control*, McGraw-Hill Book Company, New York, 1941.
3. Freilich, Arthur H., "Automatic Data Handling Systems," presented at Western Petroleum Refiners Association, Wichita, Kans., June, 1955, Regional Meeting.
4. Stern, Robert K., Automatic Intelligence Gathering Systems," *ISA J.*, May, 1955.
5. Anonymous, "Instrument Terminology," *Petr. Proc.*, April, May, and June, 1957.
6. Stern, Joshua, "Glossary of Terms," sec. 13 in reference 1, above.
7. *Instrumentation Flow Plan Symbols*, Tentative Recommended Practice of the Instrument Society of America, RP 5.1, May 12, 1949.
8. "Seal Fluids," *Foxboro Bull.* 458, p. 146.
9. Yanak, J. D., "Instrument Panels Made for Easier Operation," *Pet. Ref.*, March, 1958.
10. Nelson, W. L., *Petroleum Refinery Engineering*, McGraw-Hill Book Company, New York, 4th ed., 1958.
11. Eckman, Donald P., *Automatic Process Control*, John Wiley & Sons, Inc., New York, 1958.
12. Young, A. J., *An Introduction to Process Control System Design*, Instruments Publishing Company, Pittsburgh, 1955.
13. Tucker, G. K., and D. M. Wills, *A Simplified Technique of Control System Engineering*, published by Minneapolis-Honeywell Regulator Company, Brown Instruments Division, Philadelphia, 1958.
14. Yanak, J. D., and A. M. Calabrese, "Continuous Testers," Parts I and II, *Control Engineering*, McGraw-Hill Book Company, New York, April and June, 1959.

Section 11

PETROLEUM PRODUCTS

By VIRGIL B. GUTHRIE

*Editor, Petroleum Products Handbook**
Danbury, Conn.

A constantly growing number of commercial products is being manufactured from petroleum, with a continually widening range of consumer applications.

A survey conducted by the American Petroleum Institute a few years ago listed 2,347 products made to individual formulas by petroleum refineries and allied petrochemical plants in this country. They range from gases to solids. Table 11-1 shows the number in 17 classes. The volume of consumption of several individual products runs into hundreds of millions of barrels annually, as shown in Table 11-2.[1,†]

Table 11-1. Products Made by the U.S. Petroleum Industry‡

Class of product	Number of products
Fuel gas	1
Liquefied gases	13
Gasolines:	40
Motor	19
Aviation	9
Other (tractor, marine, etc.)	12
Gas turbine (jet) fuels	5
Kerosines	10
Distillates (diesel fuels and light heating oils)	27
Residual fuel oils	16
Lubricating oils (light and heavy, compounded and uncompounded)	1,156
White oils:	100
U.S. Pharmacopoeia	78
Technical	22
Rust preventives	65
Transformer and cable oils	12
Greases	271
Waxes (crystalline and microcrystalline)	113
Asphalts (including road oils, tars, etc.)	209
Cokes	4
Carbon blacks	5
Chemicals, solvents, miscellaneous	300
	2,347

‡ From *Am. Petrol. Inst. Inform. Bull.* 11, based on a survey of oil-supplying companies made by the API.

* Mr. Guthrie was formerly Chief Editor of *Petroleum Processing* and Managing Editor of *National Petroleum News.*

† Superior numbers refer to references at end of section.

The limit in number and diversity of petroleum products and in the extension of their application has by no means been reached. Through advanced technology and practical research by oil companies, new products will continue to be added to the long list of those recovered or made from petroleum.

Table 11-2. United States Demand for Petroleum Products, Yearly Totals*

(In thousands of barrels)

Year	Motor fuel	Kero-sine	Dist. fuel oil	Jet fuel	Residual fuel oil	Lube oils	Wax	Coke	Asphalt and road oil	Total U.S. demand
1963	1,634,835	73,090†	747,155‡	191,625†	541,295‡	43,800	4,015	32,485	124,100	3,924,480‡
1962	1,584,691	97,901†	732,405	178,667†	545,813	43,615	3,965	31,624	121,087	3,796,029
1961	1,533,173	97,248†	694,356	151,623†	548,678	41,534	4,390	30,480	113,555	3,641,280
1960	1,511,670	99,340†	685,268	135,962†	559,439	42,676	4,438	26,057	110,576	3,585,820
1959	1,479,907	109,979	659,406	104,241	558,800	42,717	4,556	35,550	108,546	3,439,078
1958	1,435,897	113,279	653,426	94,177	531,067	39,472	4,300	31,119	102,295	3,315,213
1957	1,392,953	107,701	616,090	72,961	548,801	41,215	4,430	27,026	96,096	3,218,619
1956	1,373,079	117,324	615,856	72,155	562,813	43,933	4,340	24,877	99,433	3,213,187
1955	1,334,205	116,808	581,128	56,286	557,057	42,477	4,056	24,403	92,642	3,087,775
1954	1,230,595	118,311	526,347	45,852	522,317	38,537	3,925	19,776	83,793	2,832,424
1953	1,205,775	114,467	488,075	34,483	560,474	40,497	3,889	17,599	78,818	2,775,321
1952	1,142,987	121,253	476,986	20,126	555,165	38,165	3,443	13,924	77,954	2,664,407
1951	1,089,566	123,241	447,278	....	564,397	42,292	3,246	14,481	72,274	2,569,827
1950	994,290	117,844	394,885	....	553,793	38,853	3,238	15,021	65,574	2,375,057
1949	913,713	102,672	329,278	....	496,021	33,101	2,255	14,427	57,188	2,118,250
1948	871,270	112,220	340,576	....	500,543	35,983	2,348	11,670	57,989	2,113,678
1947	795,015	102,519	298,273	....	518,510	36,481	2,478	10,082	54,090	1,989,803
1946	735,417	89,088	242,894	....	480,029	34,891	2,271	9,029	49,192	1,792,786
1945	696,333	75,573	226,084	....	523,423	35,334	2,403	9,214	40,855	1,772,685
1944	632,482	71,812	209,320	....	512,020	32,363	2,261	8,327	39,689	1,671,263
1943	568,238	68,598	208,110	....	467,008	31,459	2,092	5,250	38,854	1,521,426
1942	589,110	69,767	185,661	....	405,697	29,057	1,951	5,036	44,217	1,449,908
1941	667,505	69,469	172,824	....	383,422	30,255	1,871	8,143	44,465	1,485,779
1940	589,940	68,776	160,851	....	340,163	24,690	1,275	7,034	36,031	1,326,620
1939	555,509	60,503	134,973	....	323,488	23,713	1,162	7,108	34,939	1,231,076
1938	523,003	56,360	117,449	....	291,883	21,233	995	5,589	32,002	1,137,123
1937	519,352	54,972	116,841	....	325,514	23,323	1,062	5,765	29,830	1,169,682
1936	481,606	51,428	102,757	....	307,884	22,323	1,077	6,266	27,874	1,092,754
1935	434,810	47,645	86,028	....	280,695	19,661	933	6,703	21,614	983,686
1934	410,339	44,234	74,824	....	265,547	18,484	857	7,540	20,302	920,164
1933	377,003	38,493	64,748	....	258,957	17,152	1,263	9,962	17,074	868,488
1932	373,900	33,221	65,101	....	243,056	16,614	945	9,592	19,300	835,482

Authority: U.S. Bureau of Mines Annual Petroleum Statements.

* This total includes liquefied gases, miscellaneous products, still gases, and processing losses.

† Starting in 1960, kerosine-type commercial jet fuels are reported separately from kerosine and are included in jet fuels.

‡ Demand for gasoline, kerosine, distillate and residual fuel oils in 1963 is not strictly comparable to earlier years because of product reclassification. Total demand includes, in addition to products listed separately, liquefied petroleum gases, miscellaneous products, still gases, and product losses.

Spotlight on Minor Products. Many oil companies today are giving greater attention to the so-called minor products from processing crude oil, both as a means for widening their markets and as source material for new products. These lesser products include the small-volume materials and byproducts which remain after the gasolines, naphthas, kerosine, and distillate and residual fuel oils are taken from the crude. In many instances markets are being found for them where they do not have to be sold under the intense competition which the mass-volume products face. Hence they return higher values to the oil companies manufacturing them.

The minor petroleum products include liquefied petroleum gas, refinery still gases, jet fuels, lubricants, wax, asphalt, and coke. The quantity marketed of liquefied petroleum gas, made from both natural-gas and refinery-gas sources, has already reached volume proportions. A third of the current output goes into the manufacture of chemicals and synthetic rubber. Refinery still gases, formerly burned as fuel, now supply liquefied petroleum gas and also olefins for petrochemicals. Jet-fuel demand is reaching volume proportions. New uses for wax, coke, and asphalt are expanding their markets at a more rapid rate than for the conventional petroleum products.

Two factors, one economic and the other technological, will give impetus to the oil companies' efforts to find markets which will return higher values for petroleum than when it is sold for some of its present uses:

1. Refiners will pay more in the future for the crude oil they process in their plants. More intensive exploration, deeper drilling, and other costs of finding new fields and bringing the oil to the surface will increase the cost of crude, at least of that produced in this country. With the higher product costs that will follow advances in crude-oil postings, petroleum may feel even more intense competition in the heating and power-generation field from coal and natural gas. Even electricity may be competitive in some areas.

2. Improved processing technology is constantly providing better tools for refiners to manufacture new and improved products from their crude and working stocks. Five decades ago thermal cracking first applied commercially the principle of changing the size of the molecules in petroleum hydrocarbons and rearranging their molecular structure as a means for increasing the yield of gasoline from crude oil. More recently, the commercial development of catalytic cracking, reforming, and hydrocracking, to mention only a few of the new processes, has greatly extended the possibilities in transforming the hydrocarbon compounds in petroleum into more valuable working materials.

The declining rate of refinery production of residual fuel oil represents a significant achievement by oil companies in converting crude petroleum into products of higher value. In years past residual fuel oil was considered a byproduct of refining and frequently sold for less than the price of the crude from which it was produced. From 1940 to 1965 the yield of residual has been reduced from 24.4 to 8.1 per cent of crude oil processed, as refiners found the means to convert this material into more valuable forms of liquid fuels and other products.

The current trend toward wider demand for the so-called minor petroleum products is shown in Fig. 11-1. While the requirements for the mass volume products (gasoline, kerosine, and distillate and residual fuel oils) increased 26 per cent from 1953 through 1963, demand for the other products as a group more than doubled, with a

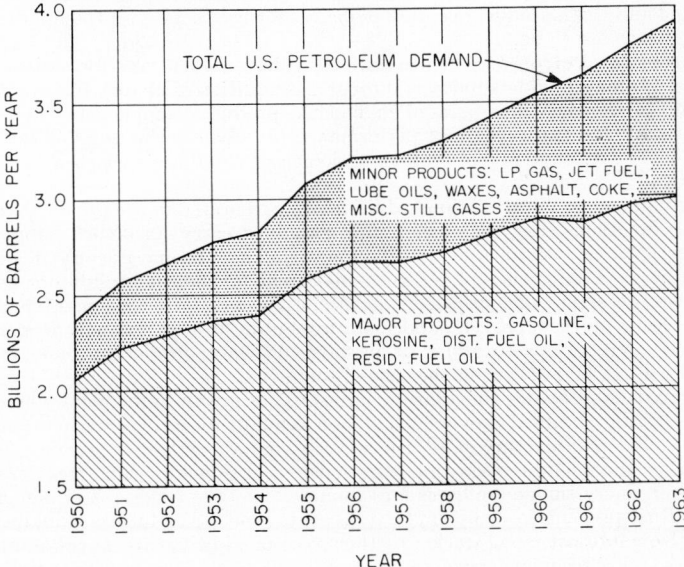

FIG. 11-1. Relative growth in U.S. demand for major petroleum products, 1950–1963.

growth of 131 per cent. This was in part owing to the rapid expansion in the use of jet fuels. The market for liquefied petroleum gases grew widely in this decade. Total U.S. demand for all petroleum products increased 41 per cent in the 1953–1963 decade.

The Amazing Petrochemicals. The manufacture of chemicals and chemical intermediates from petroleum and natural-gas hydrocarbons is a dramatic example of the conversion of such materials into higher-value products. Within two and a half decades the manufacture of petrochemicals has grown to be a recognized industry, with about 600 plants in the United States and Canada with a plant investment of about $8 billion. Some of the plants are owned and operated by oil companies, some by chemical concerns, and in a few cases they are joint ventures.[2]

The individual petrochemicals made from petroleum and natural gas number into the hundreds. They include industrial and household chemicals, fertilizers, paints and other coatings, intermediates for the manufacture of such large-volume and diversified products as synthetic rubber and plastics, and many other items. Several individual petrochemicals sell in volumes of over 1 billion pounds a year, which is a large-volume product in the chemical field. Total production of petrochemicals in the United States in 1965 was estimated at 92.3 billion pounds with a market value of over $8.7 billion.

Values for the petrochemicals are not directly comparable to those for conventional petroleum products. Intermediate processing, handling, high degrees of product purity, and different methods of marketing are involved with the former. Some indication of the higher values realized for petrochemicals, however, can be seen from the fact that the wholesale value of all gasoline, distillate fuel oil, and gas oil sold in the United States in 1962 was $10.65 billion, and about 68.0 per cent of all crude processed went into these products. By comparison, the value of petrochemicals produced in that year was about $6.78 billion, although less than 5 per cent of the production of petroleum and natural gas in this country went into petrochemicals.

A New Role for Kerosine. Of the conventional petroleum products, demand for which since 1932 is shown in Table 11-2, kerosine will likely show the greatest rate of growth in the next few years. This increase is expected because of its use as an aircraft gas turbine and jet fuel by both the commercial airlines and the military services. This demand in 1963 was over 500,000 bbl daily and by 1970 may reach 700,000 bbl.[3] The latter figure equals the total current consumption of kerosine for all other uses. The coming needs may require refiners further to adjust their yields of kerosine and distillate fuel oils if all demands for products from the middle-of-the-barrel fractions are to be met.

Additives for Petroleum Products. A widely used means for expanding the usefulness of many products today is through the addition of quality-improver agents. These additives, when incorporated in finished products, supplement their natural characteristics and improve their performance or broaden the areas of their suitability.[4] They are found in motor and aviation gasolines, diesel fuels, heating oils, and lubricants of all kinds.

In many instances more than one additive is incorporated in the product, each with a specific function. For example, motor gasolines carry antiknock suppressants, color dyes, antioxidants, antirust additives, antistall agents, and preignition additives, among others. Lubricating-oil additives may include oxidation inhibitors, corrosion preventives, detergent and dispersant agents, oiliness improvers, film-strength additives, viscosity-index improvers, and foam inhibitors, to mention only in part those that can be used. They are added to the finished refinery product in amounts from a small per cent to 30 per cent in the case of some oils.

In the past two decades the additive industry has grown to the point where it now supplies substances valued at over $600 million annually for incorporation in petroleum products. The manufacture of additives has added a new line of products for some refiners, since many additives are made from petroleum stocks. For example, sulfonic acids from refinery sludge are bases for saponification, emulsifying, and demulsifying agents. Rust-prevention compounds, detergents, and other lubricant additives are prepared from lubricating-oil stocks. Other types of additives are manufactured from naphthenic acids recovered from some crudes. Polyethylene improves the natural qualities of some petroleum waxes.

Future Growth. Still further expansion in the use of petroleum was discussed in 1959 by W. J. Sweeney, vice-president, Esso Research and Engineering Co.[5] He said in part:

"Establishing new outlets for oil is like developing a whole series of new businesses. It takes imagination, sound economic thinking, painstaking experiment and risk capital. In our work in its present stage, we are giving imagination and paper studies free rein, while holding tightly on the halter of experimental expense. We have had some promising minor developments and we hope for bigger ones. The industry (petroleum) as a whole appreciates that growth and finding new outlets are synonymous, and when a concerted effort gets rolling it cannot help but result in an accelerated growth in demand for petroleum products.

"As a summary, I'd like to deal not with the studies we have made but with the conclusions these studies suggest.

"First, there has always been competition between various forms of energy and there always will be. This competition is in the field of old products as well as new. So the competition and what it can do must continually be watched.

"Second, more than we realize our markets in the past have been handed to us by the inventors of oil-consuming devices. We must help improve these devices to make them as efficient as possible, so they can stand up against the competition of devices that use other forms of energy to do the same job for the consumer.

"Third, we should aid and abet the invention and development of new oil-consuming devices, both those conceived by others and those invented by ourselves.

"Fourth, we should depend more on ourselves than we have in the past to find new uses for oil. We should recognize that this is no easy task and will require both joint and individual effort and expense to be effective.

"Finally, other industries like the chemical, electrical, pharmaceutical, aviation are doing this all the time—and successfully, too. There is no reason why we can't do it better."

Mr. Sweeney's remarks were pointed up later by the announcement[6] that Esso Research and Engineering had developed a method for using heavy fuel oil in blast furnaces in place of coke in ironmaking. When widely used, this development could create a new large-volume use for what is still considered a petroleum byproduct.

Knowledge of the physical and chemical properties of petroleum products that determine their performance capabilities is essential to their efficient use. Oil companies and consumers alike are fortunate in that widely approved and accepted test methods for determining performance standards are available for a great many products.[7] The following pages review the test methods, performance standards, and the fields of use to which they apply for many individual petroleum products.

LIQUEFIED PETROLEUM GAS

Liquefied petroleum gas, also termed LP gas, LPG, and "bottled gas," differs in its characteristics and in its marketing procedures from other volume petroleum products.

Since its chief components are propane and/or butane, it is transported, stored, and handled to the point of consumption under moderate pressure to maintain it as a liquid. This feature requires equipment differing from that used for conventional products.

Before a market can be established for its principal use as a domestic fuel, the stoves, heaters, and other appliances that will burn this fuel have to be sold and installed on the consumers' premises.

From an early date, this latter factor brought into the marketing of LP gas hardware stores, appliance and farm supply firms, and other dealers who, having sold the equipment, took on the distribution of the fuel itself. Many established oil companies also market LP gas. However, even in the case of oil companies, the special handling and product-selling features create a marketing operation that is often organized as a business or division separate from that handling conventional products.

Despite these unusual conditions, the LP gas market in the United States has expanded at a phenomenal rate during the past three decades. Total sales in 1963 were 11,370,000,000 gal (271,000,000 bbl). LP gas today ranks fourth in volume among petroleum products marketed, exceeded only by gasoline, distillate fuel oil, and residual fuel oil. Figure 11-2 shows the growth trend of the principal uses from 1951 through 1963.

Transportation facilities have been expanded and improved. Today at least 10 pipeline systems transport finished LP gas from the Southwest producers to the Middle West, North Central, and Southeast areas of the country.

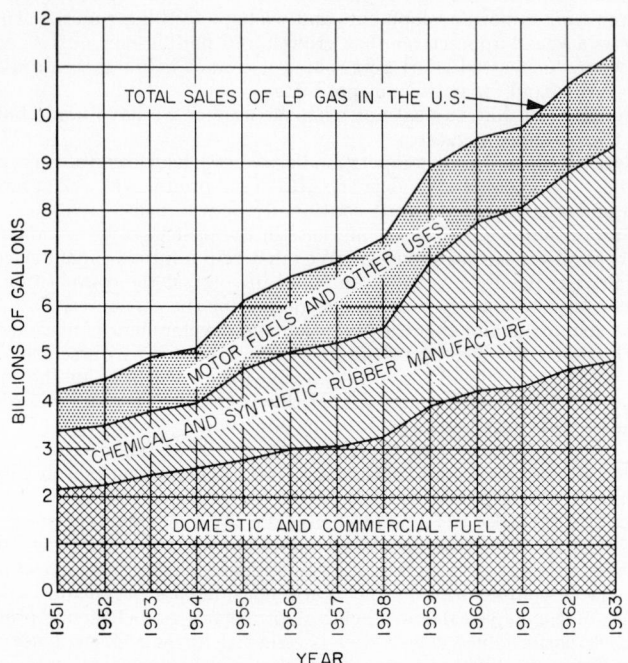

FIG. 11-2. Growth trend for the large-volume uses of LP gas.

Developments in the principal uses of LP gas are discussed in the annual report on this subject made by Phillips Petroleum Company.[8]

Uses for LP Gas

Uses for LP gas include house heating and other household, farm, and agricultural uses. Sales in this category in 1963 made up nearly 45 per cent of the total LP gas market. Use in this field is expected to increase as LP gas pipelines open new territories and more efficient transportation methods in general make this product more competitive with natural gas and other fuels.

The fastest growing market for LP gas is in the manufacture of chemicals and chemical intermediates, including synthetic-rubber components. This usage has increased nearly threefold in the past decade, as shown in Fig 11-2. Total demand for petrochemicals of 4,371,000,000 gal in 1963 was 37.8 per cent of the total LP gas market.

The market for LP gas in chemicals manufacture is expected to increase during the next few years owing to the large number of petrochemical projects, either new plants or extensions, now under way.

Sales of LP gas as motor fuel have had a large growth in the last decade, but this use is small compared with the total market. LP gas motor fuels are used in industrial trucks, farm tractors, irrigation engines, and local trucking operations.

Table 11-3. Marketed Production of LP Gas in the United States, 1951 to 1964
(Millions of gallons)

Year	Domestic and commercial[a]	Motor fuel	Industrial and misc.	Gas manu-facturing	Chemical mfg.	Rubber components	Total
1951	2,167	290	269	282	845	375	4,227
1952	2,266	371	339	260	871	371	4,477
1953	2,479	498	374	222	967	391	4,932
1954	2,627	547	402	192	1,050	308	5,126
1955	2,801	652	556[b]	214	1,493[c]	406	6,123
1956	3,001	773	630	212	1,601	418	6,636
1957	3,067	805	685[d]	231	1,732	418	6,939
1958	3,294	852	806	239	1,899	372	7,462
1959	3,935	890	871	183	2,526[e]	514	8,919
1960	4,225	898	707	157	3,031	550	9,860
1961	4,318	880	671	169	3,759[f]	...	9,798
1962	4,713	932	753	174	4,159	...	10,729
1963	5,053	999	930	217	4,371	...	11,570
1964	5,397	1,069	1,042	228	4,678	...	12,414

Authority: All years except 1964 compiled from U.S. Bureau of Mines reports; 1964 data estimated by Phillips Petroleum Company.

[a] House heating and other household and farm uses. Included also are sales by domestic distributors but used for commercial or industrial purposes.

[b] Includes more complete coverage of refinery fuel.

[c] Includes more complete coverage of LP-gas mixture streams containing ethane and methane.

[d] Includes volume used in secondary recovery of petroleum.

[e] Includes more complete coverage of ethane and/or ethylene and mixtures containing more than 50 per cent of either or both.

[f] Includes rubber components.

Table 11-3[8] presents details on various uses for LP gas, and the diversity of these applications in the individual fields is given in Table 11-4.[9]

Supplies

About 75 per cent of the LP gas produced in this country is extracted from natural gas and contains only paraffin hydrocarbons, principally propane and butane. Refinery gases supply nearly 25 per cent of the production, and olefin hydrocarbons are found in these. Table 11-5 lists LP gas production as reported by the U.S. Bureau of Mines. Not all LP gas produced is marketed as LP gas fuel.

To differentiate between LP gas from natural gas and that from refinery gases, the latter is commonly termed liquefied refinery gas, LR gas. Very little of the refinery butanes-butylenes reach the LP-gas fuel market because their value is higher for use as petrochemical feedstocks and motor-fuel blending. Propane containing only a minor amount of propylene is the product chiefly sold to the LP-gas industry for fuel uses. Table 11-6 lists the average properties of commercial propane and butane contained in marketed LP gas.[9]

Odorization

With a few exceptions in the case of refinery gases, there is not sufficient odor in LP gas produced to warn of its presence in dilute concentrations with air. Most states, however, require that a distinct warning of its presence in air be provided before marketing, which makes it necessary to add a malodorant to the fuel. This is done at the point of shipment. Forty-eight of the fifty states have an odorization requirement based on standards of the National Fire Protection Association.[10] This

Table 11-4. Uses of Liquefied Petroleum Gas[9]

Domestic and Farm

House heating	Stock tank heating
Cooking	Forage dehydration
Water heating	Sterilization
Refrigeration	Pasteurization
Clothes drying	Scalding
Incineration	Stock branding
Weed burning	Irrigation engines
Tobacco curing	Stationary engine fuel
Tractor, truck motor fuel	Air conditioning
Brooder fuel	Swimming-pool heating
	Crop drying

Commercial

Baking ovens	Wallpaper steamers
Deep-fat fryers	Salamanders
Hot plates	Lead pots
Space heating	Tar kettles
Steam boilers	

Industrial

Heat treating	Soldering and brazing
Carburizing	Galvanizing
Die casting	Malleablizing
Core baking	Engine-block testing
Mold drying	Glass-plant operations
Flame cutting	Textile processing
Lithographing ovens	Steam generation
Vitreous enamel baking	Space heating
Industrial tractors and lift trucks	Forging
Torches	Motor fuel
Brick and lumber kilns	Cannery cookers
Calendering	Foundry work
Pottery kilns	Singeing

Petrochemicals (Raw Materials for)

Alcohols	Synthetic-rubber components
Aldehydes	Acetates
Ethylene and ethylene oxide	Nitroparaffins
Organic acids	Plastics
Detergents	Ketones
Ethylene glycol	

Chemical Use (Direct Application)

Dewaxing and deasphaltizing
Degreasing of food products
Solvent extraction of vegetable oils
Color control of soaps

Utility

Enrichment (to raise or maintain constant Btu content)
Peak shaving
Standby
Distribution as such (undiluted or LP gas and air)

Transportation

Busses	Industrial tractors and lift trucks
Trucks	Taxicabs
Tractors (farm)	Automobiles

requirement is 1.0 lb of ethyl mercaptan, 1.0 lb of thiophane, or 1.4 lb of amyl mercaptan per 10,000 gal of LP gas.

Specifications

Most LP gas is marketed under definitions and specifications of the Natural Gas Processors Association (NGPA).[11] Definitions of the four commercial grades of LP gas and their specifications are as follows:

Table 11-5. LP-gas Production*
(Millions of gallons)

Year	Total	From natural gas sources		From refinery sources	
		MM gal	% of total	MM gal	% of total
1950	4,214	2,993	71.0	1,221	29.0
1951	4,950	3,562	72.0	1,388	28.0
1952	5,490	4,189	76.3	1,301	23.7
1953	6,153	4,754	77.3	1,399	22.7
1954	6,639	5,204	78.4	1,435	21.6
1955	7,805	5,973	76.5	1,832	23.5
1956	8,482	6,300	74.3	2,182	25.7
1957	8,900	6,655	74.8	2,244	25.2
1958	9,151	6,731	73.6	2,420	26.4
1959	10,717	7,875	73.6	2,842	26.4
1960	10,662	7,763	72.7	2,898	27.3
1961	11,242	8,296	73.8	2,946	26.2
1962	11,350	8,539	75.2	2,811	24.8
Totals	105,755	78,834	74.2 (avg)	31,919	25.41 (avg)

* Source: U.S. Bureau of Mines.
Note: Change in method of reporting accounts for increase in percentage of refinery production starting in 1954. Prior to that, LP gas produced in refineries and distributed by pipelines for chemical processing and plant fuel was not reported.

Commercial Propane

Commercial propane shall be a hydrocarbon product composed predominantly of propane and/or propylene and shall conform to the following specifications:

Vapor Pressure. The vapor pressure at 100°F as determined by NGPA LPG Vapor Pressure Test shall not be more than 210 psig pressure.

95 Per Cent Boiling Point. The temperature at which 95 per cent of volume of the product has been evaporated shall be −37°F or lower when corrected to a barometric pressure of 760 mm Hg, as determined by the NGPA Weathering Test for Liquefied Petroleum Gases.

Residue. The product shall pass the nonvolatile residue test and shall pass the oil-ring test—each as determined by the NGPA Method for Determining Residues in Liquefied Petroleum Gases.

Volatile Sulfur. The unstenched product shall not contain volatile sulfur in excess of 15 grains per 100 cu ft as determined by NGPA Volatile Sulfur Test.

Corrosive Compounds. The product shall cause no more discoloration to a polished copper test strip when such product is subjected to the NGPA LPG Corrosion Test than the discoloration of Standard copper strip Classification 1, as described in ASTM Method D 130-56, Table I, Copper Strip Corrosion by Petroleum Products.

Dryness. The product shall be dry as determined by the NGPA Propane Dryness Test (Cobalt Bromide Test).

Commercial Butane

Commercial butane shall be a hydrocarbon product composed predominantly of butanes and/or butylenes and shall conform to the following specifications:

Vapor Pressure. The vapor pressure at 100°F as determined by NGPA LPG Vapor Pressure Test shall not be more than 70 psig.

95 Per Cent Boiling Point. The temperature at which 95 per cent of volume of the product has evaporated shall be 36°F or lower when corrected to a barometric

Table 11-6. Average Properties of Commercial Propane and Commercial Butane*

Properties	Commercial propane	Commercial butane
Vapor pressure, psig:		
At 70°F	124	31
At 100°F	192	59
At 105°F	206	65
At 130°F	286	97
Specific gravity of liquid at 60/60°F	0.509	0.582
Initial boiling point at 14.7 psia, °F	−51	15
Weight per gal of liquid at 60°F, lb	4.24	4.84
Dew point at 14.7 psia, °F	−46	24
Specific heat of liquid at 60°F, Btu/(lb)(°F)	0.588	0.549
Cu ft gas at 60°F, 30 in. Hg, per gal liq at 60°F	36.28	31.46
Specific volume of gas, at 60°F, 30 in. Hg, cu ft/lb	8.55	6.50
Specific heat of gas at 60°F (C_p), Btu/(lb)(°F)	0.404	0.382
Specific gravity of gas (air = 1) at 60°F, 30 in. Hg	1.52	2.01
Ignition temperature in air, °F	920–1020	900–1000
Max flame temperature in air, °F	3595	3615
Per cent gas in air for max flame temperature	4.2–4.5	3.3–3.4
Max rate of flame propagation in 25-mm tube:		
Cm/sec	84.9	87.1
In./sec	33.4	34.3
Limits of flammability, % gas in air:		
At lower limit	2.4	1.9
At max rate of flame propagation	4.7–5.0	3.7–3.9
At upper limit	9.6	8.6
Required for complete combustion:		
Cu ft O_2/cu ft gas	4.9	6.3
Cu ft air/cu ft gas	23.4	30.0
Lb O_2/lb gas	3.60	3.54
Lb air/lb gas	15.58	15.3
Products of complete combustion:		
Cu ft CO_2/cu ft gas	3.0	3.9
Cu ft H_2O/cu ft gas	3.8	4.6
Cu ft N_2/cu ft gas	18.5	23.7
Lb CO_2/lb gas	3.0	3.1
Lb H_2O/lb gas	1.6	1.5
Lb N_2/lb gas	12.0	11.8
Ultimate CO_2, % by volume	13.9	14.1
Latent heat of vaporization at boiling point:		
Btu/lb	185	167
Btu/gal	785	808
Total heating values (after vaporization):		
Btu/cu ft	2522	3261
Btu/lb	21,560	21,180
Btu/gal	91,500	102,600

* Courtesy of Phillips Petroleum Co.

pressure of 760 mm Hg, as determined by the NGPA Weathering Test for Liquefied Petroleum Gases.

Volatile Sulfur. The unstenched product shall not contain volatile sulfur in excess of 15 grains per 100 cu ft as determined by NGPA Volatile Sulfur Test.

Corrosive Compounds. The product shall cause no more discoloration to a polished copper test strip when such product is subjected to the NGPA LPG Corrosion Test than the discoloration of Standard copper strip Classification 1, as described in ASTM Method D 130-56, Table I, Copper Strip Corrosion by Petroleum Products.

Dryness. The product shall not contain free, entrained water.

Butane-Propane Mixtures

Butane-propane mixtures shall be hydrocarbon products composed predominantly of mixtures of butanes and/or butylenes with propane and/or propylene and shall conform to the following specifications:

Vapor Pressures. The vapor pressure at 100°F as determined by NGPA LPG Vapor Pressure Test shall not be more than 200 psig pressure.

95 Per Cent Boiling Point. The temperature at which 95 per cent of volume of the product has evaporated shall be 36°F or lower when corrected to a barometric pressure of 760 mm Hg, as determined by the NGPA Weathering Test for Liquefied Petroleum Gases.

Volatile Sulfur. The unstenched product shall not contain volatile sulfur in excess of 15 grains per 100 cu ft as determined by NGPA Volatile Sulfur Test.

Corrosive Compounds. The product shall cause no more discoloration to a polished copper test strip when such product is subjected to the NGPA LPG Corrosion Test than the discoloration of Standard copper strip Classification 1, as described in ASTM Method D 130-56, Table I, Copper Strip Corrosion by Petroleum Products.

Dryness. The product shall not contain free, entrained water.

Product Designation. Butane-propane mixtures shall be designated by the vapor pressure at 100°F in pounds per square inch gage. To comply with the designation the vapor pressure of mixtures shall be within +0 lb −5 lb of the vapor pressure specified. For example: A product specified as 95 lb LP gas shall have a vapor pressure of at least 90 lb but not more than 95 lb, at 100°F.

Propane HD 5

Propane HD 5 shall be a special grade of propane for motor fuel and other uses requiring more restrictive specifications than commercial propane and shall conform to the following specifications:

Vapor Pressure. The vapor pressure at 100°F as determined by NGPA LPG Vapor Pressure Test shall not be more than 200 psig pressure.

95 Per Cent Boiling Point. The temperature at which 95 per cent of volume of the product has evaporated shall be −37°F or lower when corrected to a barometric pressure of 760 mm Hg, as determined by NGPA Weathering Test for Liquefied Petroleum Gases.

Residue. The product shall pass the nonvolatile residue test and shall pass the oil-ring test—each as determined by the NGPA Method for Determining Residues in Liquefied Petroleum Gases.

Volatile Sulfur. The unstenched product shall not contain volatile sulfur in excess of 10 grains per 100 cu ft as determined by the NGPA Volatile Sulfur Test.

Corrosive Compounds. The product shall cause no more discoloration to a polished copper test strip when such product is subjected to the NGPA LPG Corrosion Test than the discoloration of Standard copper strip Classification 1, as described in ASTM Method D 130-56, Table I, Copper Strip Corrosion by Petroleum Products.

Dryness. The product shall be dry as determined by the NGPA Propane Dryness Test (Cobalt Bromide Test).

Composition. The propylene content of the product shall not exceed five liquid volume per cent, and the product shall contain a minimum of 90 liquid volume per cent of propane.

Test Methods

In the markets where LP gas is used for heat or power, the NGPA specifications are adequate. These standards place no limits on the propylene or butylene contents, and in the majority of cases none is needed. In certain solvent and controlled-atmosphere applications, however, the presence of the unsaturated hydrocarbons is detrimental, and the purchaser avoids this condition by specifying that the LP gas must be obtained from a natural-gas source.[9]

Most LP gas sold as a petrochemical feedstock is on the basis of NGPA specifications. A petrochemical manufacturer may, however, have to treat the LP gas feedstock for his particular process. Sulfur compounds may have to be further removed if the gas is to be used with catalysts that are susceptible to poisoning from such compounds. Some petrochemical processes require removal of the unsaturated compounds.

Where the LP gas is to be used in the solvent extraction of vegetable oils and the color control of soaps, it must be odorless, and this may require special treating.

Laboratory tests for implementing LP gas specifications have been developed by the American Society for Testing and Materials (ASTM), NGPA, and the California Natural Gasoline Association (CNGA). Among the tests are:

Sampling Liquefied Petroleum Gases, ASTM D 1265-55.[12] NGPA Publication 2140[11] carries the same test procedure. Sulfur in Petroleum Products, Including LP Gas, by Lamp Combustion, ASTM D 1266-62T.[13] Vapor Pressure of Liquefied Petroleum Gases, ASTM D 1267-55, also NGPA 2140. Unsaturated Light Hydrocarbons, Silver-Mercuric Nitrate Method, ASTM D 1268-55.

Other LP-gas test methods given in NGPA 2140 include specific gravity, corrosion (copper-strip method), volatile sulfur, propane dryness, propane residue, and weathering tests for butane-propane mixtures. Test methods given in CNGA Bulletin TS 441, Part III, are vapor pressure, specific gravity, weathering test analysis, determination of isobutane in propane, and determination of water content of a liquefied petroleum gas.

LP-gas Handling

LP gas is stored and shipped under moderate pressure as a liquid and is used under ambient temperatures and pressures as a gas. Special storage and handling equipment is required from the point of manufacture to the consumer. Unless butane with its lower vapor pressure is to be handled exclusively, tanks with a design pressure of 250 psig are used. Such tanks are suitable for commercial propane or butane and all propane-butane mixtures. Large-scale storage of LP gas is underground in dissolved salt cavities or mined caverns or above ground in spheres or horizontal or vertical tanks. Bulk-plant piping systems are designed to withstand the maximum pressures exerted by LP gas.

Some LP gas is transported in pipelines from the point of manufacture as a finished product or in raw streams or mixed with crude oil or natural gas. In the last three cases, further processing is necessary to separate the LP-gas hydrocarbons from the other materials in the mixtures. LP gas is also transported in tankers, barges, tank cars, tank trucks, and portable tank containers and by cylinders as small as the 20-lb so-called self-service containers.

Various authorities, Federal, state, and industry, have promulgated regulations to ensure the safe storage and handling of LP gas in all steps from the point of manufacture to the point of use. The NGPA has developed tentative standards for the underground storage of LP gas.[14] The NFPA Pamphlet 58[10] sets forth standards for the storage and handling of LP gas. The American Petroleum Institute publishes standards[15] covering the design and construction of LP-gas installations at marine and pipeline terminals, natural-gasoline plants, and refineries. The California Natural Gasoline Association also published tentative specifications and test methods[16] for LP gas. The Association of American Railroads issues recommended practices relating to the unloading of tank cars, grounding of tanks, and prevention of sparks.

The Interstate Commerce Commission classifies LP gas as a "flammable condensed gas," and any interstate movement in any medium is subject to the general regulations for this classification. Individual states, in many instances, have promulgated regulations governing the transportation, storage, handling, and utilization of LP gas.

GASOLINE

Gasoline is more than the product which refiners make and sell in the largest volume. Its output is the largest of any of the basic industries in the United States. The total production of 408 billion lb of gasoline in 1962 exceeded the output of steel, of lumber, of corn, and of other materials the public consumes in volume.   The per capita output of gasoline was 2,380 lb in 1962.

For the past 50 years, refiners in this country have operated their plants primarily for the production of gasoline.   First thermal cracking, then thermal re-forming, and later catalytic cracking and other processes were developed to provide the maximum output of gasoline from the crude oil run to refinery stills.   By 1965, the U.S. average yield of gasoline was 44.9 per cent of total crude run to stills.   The demand for gasoline in this country in 1965 was 73,499,000,000 gal (1,749,988,000 bbl).   For comparison, the total demand for kerosine, distillate fuel oil, and residual fuel oil combined was less than 61 billion gal.

Gasoline is defined as a petroleum fuel designed for use in reciprocating, spark-ignition, internal-combustion engines; this definition outlines the basic field of utilization of this product.   Other applications for gasoline are of small volume.   In the primary field of application, however, there is a variety of applications.   These include fuel for automotive ground vehicles of all types, for reciprocating aircraft engines, for inboard and outboard marine engines, and for other engines ranging in unit volume consumption down to the power plants in lawn mowers for city homes and for country estates.   Other small-scale uses include fuel in pressure appliances such as field stoves, heating and lighting units, and blowtorches.   The principal uses of gasoline as fuel for internal-combustion engines are given in Table 11-7.[17]   Basically,

Table 11-7. The Distribution of Gasoline Consumption in the United States in 1961

Type of use	Millions of gallons
Motor vehicles:	
Passenger cars	42,033
Trucks	16,443
Busses	830
Total	59,306
Aviation gasoline	2,336
Boating, recreational	650
Other uses*	1,686
Total domestic demand	63,978

Authority: American Petroleum Institute, *Petroleum Facts and Figures*, 1963.
* Includes farm tractors, industrial and miscellaneous uses.

the demand for this refinery product will grow in proportion to the growth in the number of automotive vehicles in operation in the United States.

General Characteristics

Gasoline is a mixture of hydrocarbons of four basic types differing in their properties according to the number of carbon and hydrogen atoms in the molecule and in the arrangement of the atoms.   There are several hundred different hydrocarbon compounds in various proportions in a single commercial gasoline.   The four basic types of hydrocarbons are paraffins, olefins, naphthenes, and aromatics.   With the processes available today the refiner can use hydrocarbon compounds of these four types to tailor his gasoline to commercial requirements.   In addition to hydrocarbon compounds, there are minute quantities of other compounds in gasoline containing sulfur, oxygen, or nitrogen.

In order to do its work in the internal-combustion engine, gasoline is mixed with air and the mixture burned in the engine cylinder at the proper time in the engine cycle.   Theoretically, the proper ratio of air to gasoline for perfect combustion is 15 parts by weight of air to 1 part of gasoline.   However, when burned in an engine this mixture generally does not give either maximum power or maximum economy. S. Bennett Hill and John G. Moxey, Jr.,[18] point out that to function satisfactorily in an internal-combustion engine, the gasoline must:

1. Burn smoothly and quietly in the cylinder without detonation or knocking.
2. Evaporate readily enough to supply a combustible mixture of fuel vapor and air to the cylinder when the engine is started cold.

3. Not be so volatile that it boils in the fuel pump or fuel lines when the engine is hot, resulting in vapor lock.

4. Under normal running, with the engine warm, it must be sufficiently volatile so that a considerable portion is vaporized in the intake manifold and the disturbing effects of liquid in the manifold are minimized.

5. It must not contain components of such low volatility that they are not vaporized and burned in the hot cylinder.

6. It must evaporate completely and cleanly without leaving solid or gummy deposits in either the fuel or the induction system.

Motor-fuel Gasoline

About 90 per cent of the total gasoline consumption in the United States is as fuel for civilian automotive vehicles. Table 11-8 gives the annual consumption for various

Table 11-8. Consumption of Gasoline by U.S. Motor Vehicles, 1948 to 1961
(Millions of gallons)

Year	Passenger cars	Motor trucks	Busses	Total
1961	42,033	16,443	830	59,306
1960	41,169	15,882	827	57,878
1959	40,056	15,453	823	56,332
1958	38,095	14,514	809	53,418
1957	36,769	14,271	825	51,865
1956	35,326	13,978	802	50,106
1955	33,548	13,308	771	47,627
1954	30,915	12,541	755	44,211
1953	30,249	11,418	937*	42,604
1952	28,729	10,844	890*	40,463
1951	26,990	10,188	836*	38,014
1950	25,238	9,526	782*	35,546
1949	22,957	8,666	711*	32,334
1948	21,369	8,189	709*	30,338

Authority: Bureau of Public Roads, American Petroleum Institute.
* School and nonrevenue busses included with commercial busses.

types of vehicles. Most refiners produce and market more than one grade of motor gasoline, differing principally in antiknock quality. Refiners also change the volatility properties of their motor fuels, depending on the atmospheric temperatures at which the vehicle is to operate. The changes are by geographical areas. Winter gasoline is the most volatile, as reflected in the ASTM distillation temperatures, and summer gasoline the least volatile. Table 11-9[18] gives average distillation temperatures for four representative sections of the country.

The product transportation systems of the oil companies in the United States are based principally on their requirements to move motor-fuel gasoline. In 1962 it would have required over 100,000 tank cars each day, each with 15,000 gal capacity, to move the gasoline shipped from U.S. refineries. Actually, almost half of that volume moves in product pipelines, and large volumes also move by motor carriers and by inland water and coastwise vessel transportation. The greater part of motor-fuel gasoline reaches the consumer through the gasoline stations of the oil companies. Garages, parking lots, and country stores also are outlets. Recently, gasoline pumps have been installed at supermarkets in many localities.

Performance Requirements

Antiknock Quality. This is one of the most important performance properties of motor gasoline, imparting resistance to knock, or "ping," in the engine. This

Table 11-9. Average Distillation Temperatures for Premium-
and Regular-grade Motor Gasoline, °F

Region	10%		50%		90%	
	Prem.	Reg.	Prem.	Reg.	Prem.	Reg.
Middle East:						
Summer*	123	122	218	212	320	345
Winter	110	110	210	202	317	339
Northern Middle West:						
Summer	127	127	214	211	325	334
Winter	111	113	204	206	321	335
Southwest:						
Summer	128	126	216	208	322	338
Winter	113	110	205	195	320	328
West Coast:						
Summer	126	126	219	206	334	332
Winter	117	117	216	199	332	323

Source: *Surveys of Motor Gasolines*, Bureau of Mines.
* Summer of 1963, winter of 1962–1963:

property may be secured in part through the refiner's choice of crude oil and processing
techniques, in part by blending the knock-suppressant additive tetraethyl lead into
the gasoline, and to some degree by improved engine design. The higher the antiknock quality of the fuel, the more power and efficiency the engine designer can build
into his engine.

Figure 11-3[18] shows the trend in Research octane numbers for the average regular and
premium-grade gasolines sold in the United States from 1930 through 1957. Figure

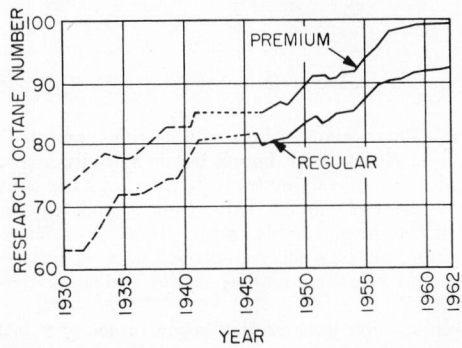

FIG. 11-3. Trend in antiknock quality for average U.S. gasolines, 1930–1962.

11-4[18] shows the corresponding trend in engine performance for the average passenger
car in the United States during the same period. With motor fuels of today's quality,
engine knock is not a problem for most motorists.

The antiknock quality of motor gasolines is normally expressed in octane numbers.
Two laboratory methods have been standardized for obtaining octane numbers of
motor fuels by comparing their knock tendencies in standard test engines with reference fuels blended from normal heptane and isooctane. The proportion of isooctane
in the reference fuel which matches the knock tendency of the fuel under test is
termed its octane number.

The Research method, ASTM D 908,[19] and the Motor method, ASTM D 357,[19] use the same basic test engine but operate under different conditions. The Research method is accepted as a better guide of antiknock quality of fuels when vehicles are operated under mild conditions associated with low engine speeds. When operation is at high engine speed or under heavy load conditions, the Motor octane number may become of equal or greater importance. The difference between the Research and the Motor octane numbers of a given fuel is known as its sensitivity. A high difference is taken to indicate greater sensitivity of the fuel to changes in severity of operating conditions of the engine.

Since some commercial gasolines now have antiknock quality above 100 octane number, the American Society for Testing and Materials has adopted a scale of octane numbers above 100 to correspond to varying amounts of tetraethyl lead added to 100 per cent isooctane, as shown in Table 11-10.[19] For example, a test fuel matching a blend of isooctane containing 1.0 ml of tetraethyl lead would have an octane number on this scale of 108.6.

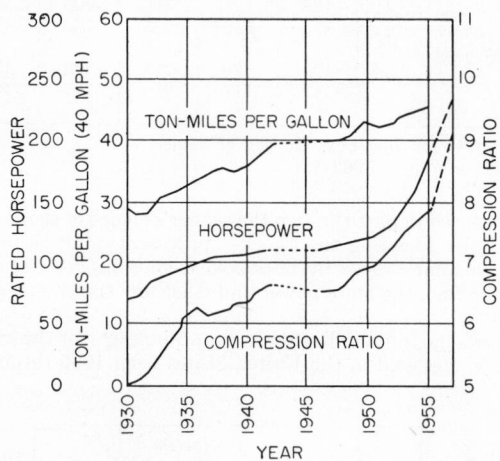

Fig. 11-4. Trends in engine performance for average U.S. passenger cars.

Surface Ignition.   This is another form of abnormal combustion in an automobile engine that is associated closely with knock but is actually a separate phenomenon. It is defined as the initiation of a flame front by any hot surface other than the spark discharge (such as overheated exhaust valves, spark-plug electrodes, or porcelains) prior to the arrival of the normal flame front. Principal offenders are combustion-chamber deposits. Gasolines have been improved to remedy this condition by more careful control of tail-end volatility and by use of additives that alter the chemical composition of deposits.

Volatility.   Gasolines differ in their distillation range, or volatility, since they are a mixture of a large number of hydrocarbons with different boiling points. Volatility is measured by a standardized laboratory distillation test, ASTM D 86.[20] The most commonly used boiling points in the distillation test are the temperatures at which 10, 50, and 90 per cent of the gasoline is evaporated, indicating front-end, middle, and tail-end volatilities, respectively. Gasoline volatility controls its performance for starting, vapor lock, warm-up, and crankcase dilution.

Starting.   Approximately 10 per cent of the gasoline in the fuel-air stream must evaporate and reach the cylinder as a vapor in order to give prompt starting. The controlling volatility requirement for starting is shown by the 10 per cent evaporated temperature from the distillation test. One of the principal differences between summer and winter gasolines is the amount of the lowest boiling hydrocarbons they contain. In states where below-freezing weather is common, the winter gasoline may contain 10 to 14 per cent butane; in warm-weather gasoline the proportion may be about 5 to 8 per cent.

Starting requirements in a motor fuel are of significance when the problem of vapor lock is considered. If too much of the light hydrocarbons is included to give satisfactory starting characteristics, the motorist may encounter vapor lock on warmer days. On the other hand, if the proportion of the light hydrocarbons is kept low to protect against vapor lock, the motorist may have difficulty in starting.

Vapor Lock. This is a complex problem resulting from the fact that gasoline boils and forms vapor somewhere in the fuel system between the fuel tank and the carburetor discharge nozzle. Boiling may take place in the line to the pump, in the pump, in the line from the pump to the carburetor, in the carburetor bowl, or even in the carburetor fuel passages. The vapor formed interrupts the normal flow of liquid fuel, and the engine will lose power and may misfire or stall completely.

Table 11-10. Scale for Determination of Octane Numbers above 100

TEL*	Octane number	TEL*	Octane number
0.0	100.0	2.1	113.1
0.1	101.3	2.2	113.4
0.2	102.5	2.3	113.7
0.3	103.5	2.4	114.0
0.4	104.4	2.5	114.3
0.5	105.3	2.6	114.5
0.6	106.0	2.7	114.8
0.7	106.7	2.8	115.0
0.8	107.4	2.9	115.3
0.9	108.0	3.0	115.5
1.0	108.6	3.2	116.0
1.1	109.1	3.4	116.4
1.2	109.6	3.6	116.8
1.3	110.1	3.8	117.2
1.4	110.5	4.0	117.5
1.5	111.0	4.5	118.3
1.6	111.4	5.0	119.1
1.7	111.7	5.5	119.7
1.8	112.1	6.0	120.3
1.9	112.5		
2.0	112.8		

* Milliliters of TEL per gallon in isooctane.

The vapor pressure of the gasoline at 100°F is considered the key to vapor lock. The Reid vapor pressure test, ASTM D 323,[20] is considered by the ASTM as "the best single criterion of freedom from vapor lock in automotive equipment. When used in conjunction with the initial portion of the distillation curve, the vapor-lock relationships become more precise."

Carburetor Icing. Before the engine is warm, if the temperature of the air entering the carburetor is near freezing and the humidity high, ice may form on the throttle plate and adjacent parts. When the car stops at a traffic light and the throttle is closed, the space around the throttle, which normally lets the engine idle, is blocked with ice. The air flow is restricted, and the engine stalls. Carburetor icing can be remedied by control of the volatility of the gasoline, by adding antistall agents to the fuel, and in engine design by providing additional heat around the critical part of the carburetor throttle body.

Other factors influenced by the volatility of motor gasoline are crankcase dilution and, to some extent, combustion deposits. To correct these conditions, ASTM Specification D 439[20] sets the following volatility limits for motor gasoline for normal driving: 140°F temperature maximum for 10 per cent evaporated for winter, 149°F for fall and spring, and 158°F for summer; 284°F maximum for 50 per cent evaporated temperature; 392°F maximum for 90 per cent evaporated temperature; and Reid vapor pressures of 15 lb for winter, 11.5 lb for fall and spring, and 10 lb for summer.

Oxidation Stability. In certain gasolines, particularly those produced by cracking, easily oxidizable hydrocarbons may be present which have a tendency to form gummy materials in the gasoline in storage and handling. Before refiners learned to control these gummy oxidation products, gum deposits sometimes formed at the carburetor float valve, in the venturi area of the carburetor nozzle, around the throttle, or on valve stems. Today the refiner removes the small amounts of undesirable, oxidizable hydrocarbons present in the raw gasoline. Most retail gasolines also contain an oxidation-inhibitor additive.

Two standard laboratory tests[20] are provided for determination of gum in gasoline. ASTM D 381, Existent Gum by Air Jet Evaporation, shows the gum actually present in the gasoline at the time of the test. ASTM D 525, Oxidation Stability of Gasoline, is related to the tendency of a gasoline to form gum during storage.

Odor. While this property has no effect on the performance of motor gasoline, it is objectionable to the motorist. Refiners remove or neutralize the sulfur compounds known as mercaptans, which are the principal sources 'of offensive odor in gasolines. Standard tests are available to determine mercaptans in gasoline.

Color. Most motor gasolines are artificially colored with gasoline-soluble dyes. The U.S. Department of Health, Education and Welfare requires that any gasoline containing tetraethyl lead be colored to warn the public that it contains lead and should not be used other than as a motor fuel.

Gasoline Additives

Nonfuel extraneous substances, known as additives, are incorporated into gasoline by refiners to improve specific performance characteristics of the motor fuel. While some of these additives have been the subject of controversy, most of them are effective to a certain degree, and the value of many has been proved. The gasoline additives have been grouped into eight classes, and their uses described by Hill and Moxey as follows:[18]

Antiknock Compounds. First introduced in 1922, tetraethyl lead (TEL) is today in common use throughout the world. Its effect in improving the octane number of motor gasoline is well established, the response varying with the hydrocarbon composition of the gasoline. In spite of much research work, the exact mechanism by which TEL works to suppress knock is not known. In simple terms, it is visualized that the compound is decomposed by the heat in the combustion chamber, giving rise to particles which then influence the chemical reactions involved in the combustion of the fuel in such a way as to promote smooth combustion to the exclusion of knock.

Tetraethyl lead has certain well-recognized disadvantages such as tending to increase deposits in the combustion chamber, tending to increase exhaust valve burning, and tending to foul spark plugs. These disadvantages have been minimized through improved engine design, improved gasolines, and the use of other additives which modify the nature of the deposits.

Tetraethyl lead is used in concentrations up to 3 ml per gal, which is equivalent to 0.08 per cent by volume of the fuel into which it is blended. In the low concentrations in which it is present in gasoline, it is nonhazardous when handled with the precautions and special equipment for blending insisted upon by the TEL manufacturers.

Dyes. Those used in gasoline are the oil-soluble type. They are present to the extent of about 5 ppm. No engine-operation problems are reported in connection with their use.

Antioxidants. These are used widely to inhibit the oxidation reaction which forms gums and thereby tend to improve the stability of gasoline in use or in storage. There are two major classes of antioxidant compounds, aromatic amines and alkyl-substituted phenols. They are used in concentrations of about 50 ppm in gasolines. They do not give rise to engine-operation troubles.

Metal Deactivators. These are used in addition to antioxidants to destroy the catalytic effect of copper in promoting the gum-forming oxidation reaction. The concentration used is about 4 ppm. No engine-operation difficulties are reported from their use.

Antirust Agents. These additives are added to gasoline at the refinery to protect pipelines and storage tanks from the corrosive condition caused by the minute amounts of water usually present in gasoline. This corrosive condition, however, is seldom a problem in automobiles.

Antistall Agents. One type of antistall agent, such as isopropanol or methanol, acts as an antifreeze agent to prevent carburetor icing. Another type, a surface-active agent, forms a thin film on the metal surface so that ice will not stick to it.

Antipreignition Agents. This class of additives, also known as deposit modifiers, acts to change the character of the combustion-chamber deposits to give less tendency to induce preignition, thus reducing the tendency of the engine to knock as the car builds up mileage. Phosphorus compounds are increasingly used as antipreignition agents in concentrations of around 200 ppm.

Upper-cylinder Lubricants. Many refiners incorporate from 0.2 to 0.5 per cent of a light lubricating oil or similar material into their gasolines to provide extra lubrication for the engine intake valves and the top ring belt area. This light oil serves to prevent the deposition of gummy deposits in the intake system but may contribute to combustion-chamber deposits.

Specifications and Standards

Most refiners market more than one grade of gasoline. The difference in these grades, usually a minimum of a regular and a premium grade, is based on the octane rating of the fuels, which ranges from around 83 to over 100. The refiners also change the volatility properties of their gasolines seasonally and also for the geographical areas where the fuel is distributed.

Standard specifications for three grades of motor gasoline, covering both regular and premium grades, are issued by the ASTM. Type A is for use under normal conditions, Type B is for use where a gasoline of greater overall volatility than Type A is desired, and Type C is for use where a relatively nonvolatile fuel is desired. Table 11-11 shows the detailed requirements in the ASTM Specification D 439-60T.[20] These are minimum standards and revised frequently as the overall quality of motor fuels marketed by the oil companies is improved.

Some states have adopted specifications for motor gasolines marketed within their boundaries. Some of these state specifications are based on the ASTM Standards. For tax-collection purposes, many states have adopted a definition for motor fuel. In these definitions, volatility requirements are higher than in standard specifications in order to cover all products sold as motor fuel.

The government specifications for automotive motor gasolines, VV-G-76,[18] cover three volatility classifications of both regular and premium grades of commercial gasolines " . . . for use in automotive engines under all climatic conditions." In general, the chemical and physical properties prescribed are the same as or similar to those in the ASTM Specifications.

Both the ASTM and the government specifications recognize the differing seasonal and geographical volatility requirements for motor fuel in the United States and provide for an automatic variation by the refiner in the 10 per cent distillation points and in the vapor pressures to meet seasonal changes in temperature. Both sets of specifications classify the states into the same four groups or classifications and name the class or type of fuel for each grouping of states for each month. Table 11-12 shows the seasonal requirements in the ASTM Specifications.

Alcohol Blends

Blends of ethanol (ethyl alcohol) with gasoline are not used as motor fuels to any extent in the United States.[18] Their use is, however, quite common in some countries —particularly those which have not been self-sufficient in their gasoline supply but

Table 11-11. Detailed Requirements for Gasoline (ASTM 439-60T)

Gasoline	Minimum percentages to be evaporated at temperatures, °F shown below					Distillation residue, max, %	Vapor pressure, max, lb[h]			Research method octane number,[b] min	Copper strip corrosion, max	Gum, max, mg per 100 ml	Sulfur
	10 per cent			50 %	90 %		W^a	F^a	S^a				
	W^a	F^a	S^a										
Type A.......	140	149	158	284	392	2	15.0^g	11.5	10^i	87 or 96^c	No. 1	5^e	f
Type B.......	140	149	158	257	356	2	15.0^g	11.5	10^i	87 or 96^c	No. 1	5^e	f
Type C.......	167	167	167	284	392	2	15.0^g	11.5	10^i	d	No. 1	5^e	f

^{a} W, F, and S denote the seasonal variations indicated in Table 11-12.

^{b} In all cases the octane number shall be agreed upon between the purchaser and the seller.

^{c} The numerical values shown are minimum values currently encountered in service stations. The lower value pertains to regular-price gasolines, and the higher value to premium-price gasolines. For more detailed information on current levels for both Research and Motor octane numbers, as well as for other characteristics of motor gasoline, reference is made to the series of semiannual reports, issued as Information Circulars (I.C.) by the U.S. Bureau of Mines and entitled National Motor Gasoline Survey.

^{d} The information available does not permit designation of a minimum Research octane number value for Type C gasoline.

^{e} In the case of gasoline containing added nonvolatile material, the gum requirement shall apply to the base stock.

^{f} The technical data available do not afford an adequate basis for specifying maximum sulfur content. At the time of this report, gasolines containing up to 0.25 per cent sulfur (ASTM Methods D 90 and D 1266) were distributed within the United States.

^{g} The maximum vapor pressure shall be 13.5 lb in Section 3 and in March and November in Section 2, (see Table 11-12).

^{h} Lower maximum vapor pressures may be required for operations at high altitudes or where abnormally high fuel system temperatures are encountered as in some heavy-duty equipment [(see Section 1 (b))] and in some heavy-duty operations.

^{i} These values shall be 9.5 lb, max, in Arizona, California, Colorado, Nevada, New Mexico, and Utah.

can supply agricultural alcohol. There has been agitation in this country to promote the use of blends of gasoline and alcohol as motor fuels by legislation providing for tax advantage or subsidy to the alcohol producers.

Where ethanol blends are used, the percentage of alcohol varies from as low as 10 to as high as 40 per cent. The common 95 per cent alcohol, containing 5 per cent water, is not miscible with gasoline and will not form a homogeneous mixture. Therefore, the alcohol used must be anhydrous so that it will mix in all proportions with gasoline.

Ethanol has high antiknock value. Its blending octane number in the range of 20 per cent added to an 85-octane-number gasoline is about 125. From this standpoint, it is a desirable component of motor fuel. Its volatility also is good. But there is about one-third less energy in a gallon of ethanol than in a gallon of gasoline. An ethanol blend, therefore, gives appreciably less mileage than gasoline. Ethanol-gasoline blends are sensitive to water, which, if present beyond a critical amount, will cause the blend to separate into two layers. This intolerance to water is more of a problem for the refiner or wholesale marketer than for the motorist.

A number of technical studies have been made of ethanol-gasoline blending.[21]

Aviation Gasoline

Aviation-gasoline production by U.S. refineries reached a peak of 192,440,000 bbl in 1944, the period of maximum wartime effort by the United States and its allies. A large portion of the fuel supplied was less than 100 octane, and the demand was almost entirely from the military services. By 1946 aviation-gasoline requirements from U.S. refineries had dropped to 14,480,000 bbl.

The growth of civil aviation in the past decade, together with the continuing

Table 11-12. Schedule for Geographical Seasonal Variation in ASTM Gasoline Specification, D 439-60T

Territory	January	February	March	April	May	June	July	August	September	October	November	December
									Month			
Section 1	W	W	W or F	F	F or S	S	S	S	S or F	F	F or W	W
Section 2	W	W or F	W or F	F or S	F or S	S	S	S	S or F	F	F or W	F or W
Section 3	W	W or F	F or S	F or S	S	S	S	S	S	S or F	F	F or W
Section 4	F	F or S	F or S	S	S	S	S	S	S	S or F	S or F	F

Section 1	Section 2		Section 3	Section 4
Idaho	Colorado	Nebraska	Alabama	Arizona[b]
Iowa	Connecticut	Nevada	Arizona[a]	Florida
Maine	Delaware	New Jersey	Arkansas	Louisiana
Michigan	District of	New York	California	New Mexico
Minnesota	Columbia	Ohio	Georgia	Texas[b]
Montana	Illinois	Oregon	Mississippi	
New Hampshire	Indiana	Pennsylvania	New Mexico[a]	
North Dakota	Kansas	Rhode Island	North Carolina	
South Dakota	Kentucky	Utah	Oklahoma	
Vermont	Maryland	Virginia	South Carolina	
Wisconsin	Massachusetts	Washington	Tennessee	
Wyoming	Missouri	West Virginia	Texas[a]	

[a] North of 33 deg latitude.
[b] South of 33 deg latitude.

military needs, again increased demand for aviation gasoline to the point where production at U.S. refineries was 95,774,000 bbl in 1958. Of this volume almost 89,000,000 bbl was 100 octane and higher. The output of all other grades was only 7,685,000 bbl. Since 1958, production has declined. Table 11-13[22] shows U.S. production of aviation gasoline since 1944.

The U.S. refiners have demonstrated their ability to supply aviation gasoline in the volume and of the high antiknock values required. The future demand for this product in civil aviation depends on the extent to which the airlines turn to the use of gas-turbine- and jet-engine-propelled aircraft in their scheduled operations. This trend is now under way. Table 11-14[22] shows the decline in civil aviation gasoline consumption starting in 1959 and in military requirements starting in 1960.

The Bureau of Mines reports aviation-gasoline production at U.S. refineries in four octane range groups: 91/98, 100/130, 108/135, and 115/145. The first figure in the grade refers to the cruising or lean-mixture rating of the fuel; the second to the take-off or rich-mixture rating. In addition to these four grades, refiners market an 80/87 grade.

Tentative specifications for these five grades for civilian use are given in ASTM Standard D 910-63T.[20] Principal requirements are given in Table 11-15. The specification prescribes that the product " . . . shall consist of blends of refined hydrocarbons derived from crude petroleum, natural gasoline, or blends thereof with synthetic hydrocarbons or aromatic hydrocarbons, or both." United States Military Specification MIL-F-5572A covers four grades of aviation gasoline: 80, 91/96, 100/130, and 115/145.

Many of the gasoline problems of automotive engines are also common to aviation gasoline when used for spark-ignition reciprocating engines.[18] The airplane engine, however, has many additional special requirements for its fuel, some arising from the

Table 11-13. U.S. Production of Aviation Gasoline, 1944 to 1964
(Thousands of barrels)

Year	100 octane and above	Other grades	Net production*
1964	43,111	3,774	52,963
1963	46,750	4,088	53,309
1962	49,795	4,316	55,937
1961	57,179	5,275	62,341
1960	62,183	6,225	70,898
1959	81,542	7,182	88,969
1958	88,813	7,685	95,774
1957	85,638	8,390	95,696
1956	86,230	9,442	97,807
1955	80,583	8,996	89,602
1954	75,736	9,241	84,298
1953	75,972	13,841	87,485
1952	66,773	13,577	78,000
1951	57,210	14,958	66,274
1950	39,235	12,036	46,349
1949	35,215	11,738	44,285
1948	33,421	12,825	42,961
1947	17,867	17,429	28,190
1946	5,342	20,070	14,480
1945	124,215	28,180	141,233
1944	136,130	60,253	192,440

Authority: American Petroleum Institute, *Petroleum Facts and Figures*, 1965.
* Includes adjustments for production of alkylates and transfers to automotive fuel.

Table 11-14. Consumption of Aviation Gasoline by U.S. Civil Aircraft and Military Procurement, 1951 to 1961.[22]
(Thousands of barrels)

Year	U.S. civil consumption			Military procurement*	Total consumption
	Domestic	Internat. and terr.	Total civil		
1961	18,960	3,675	22,365	36,469	59,604
1960	23,005	5,310	28,316	37,247	65,563
1959	28,280	8,570	5036,8	47,123	83,983
1958	29,442	8,669	38,111	33,011	71,122
1957	28,955	8,115	37,100	41,381	78,481
1956	24,885	7,359	32,244	43,125	75,369
1955	22,190	6,259	28,449	42,413	70,862
1954	18,921	5,314	24,235	43,513	67,748
1953	16,853	4,731	21,589	41,741	63,330
1952	14,473	4,513	18,986	36,069	54,995
1951	12,194	4,071	16,265	35,000	51,205

Authority: American Petroleum Institute, *Petroleum Facts and Figures*, 1963.
* Includes purchases in both the United States and foreign countries.

fact that many engines today use direct fuel injection into the cylinder in place of the conventional carburetor. Another important difference is the use of superchargers in airplane engines.

Knock in an airplane engine is very damaging and may, in a short time, burn a hole in the piston head. The pilot must be sure of the octane rating of the gasoline he is using and the amount of power he can get out of his engine without developing detonation or knock. The antiknock quality of aviation gasoline is controlled carefully by specifications of the different grades for different planes and types of service.

The need for rating aviation gasolines at over the 100 top limit of the octane scale brought about the development of two methods for making the higher ratings. In one, the octane scale was extended upward by equating antiknock quality to isooctane plus so many milliliters of tetraethyl lead per gallon of fuel. The other scale above 100 octane is the performance-number scale. Performance number is the percentage of power that can be taken out, without knocking, of an average engine on the given fuel compared with that which can be obtained with a 100-octane-number fuel, or isooctane. For example, a fuel with a performance number of 130 is capable of giving, without knocking, 130 per cent of the power that would be maximum for isooctane.

Antiknock values for aviation gasoline are usually given in octane numbers up to and including 100. Above 100, the figures used are either performance numbers or isooctane plus so much TEL. Performance numbers have been related to octane numbers both above and below 100, as shown in Table 11-16.[18]

The method for antiknock evaluation of aviation gasolines for crusing conditions, also known as the lean-mixture rating, is given in ASTM Method D 614-63T, Knock Characteristics of Aviation Fuels by the Aviation Method. The test to simulate take-off conditions, which gives the so-called rich-mixture rating, is given in ASTM Method D 909-63T, Knock Characteristics of Aviation Fuel by the Supercharge Method.[20]

Performance Requirements

Heat of Combustion. In aviation gasoline, the heat of combustion is more important than in motor gasoline, since the more energy per pound of fuel, the less the fuel load required for a specified trip. Specifications provide for a minimum net heat of combustion expressed in Btu's per pound. Heat of combustion is determined by ASTM D 240.[20] Since net heat of combustion correlates with the product of the aniline point and the API gravity, specifications may provide that the heat-of-combustion determination may be waived if the aniline-gravity constant exceeds a certain figure. The aniline point is determined by ASTM D 611.[20]

Volatility. Much the same volatility problems are encountered in airplane engines equipped with carburetors as in automotive engines. The problems are more serious, however, in the airplane. To safeguard from vapor lock, Reid vapor pressures are held to 7 lb maximum. Starting is controlled by the maximum 10 and 50 per cent evaporated temperatures. Warm-up is controlled by the maximum 50 per cent point. Final evaporation is controlled by the maximum 90 per cent point. The 90 per cent evaporation point for civil use usually is set at 257°F maximum to ensure complete vaporization. Volatility requirements of airplane engines equipped with fuel injection are less severe, but the same volatility specifications are used for both types of engines. Volatility of aviation gasolines is determined by the same methods as for motor gasolines.

Oxidation Stability. These requirements are more severe than for motor gasoline, and different test methods are commonly used. Gum present in the sample is determined by ASTM Aviation Gasoline Specification D 910, Sec. (j).[20] The potential gum test for aviation gasolines is described in ASTM D 873.[20]

Freezing Point. Since aviation gasolines may be subjected to very low temperatures in flight at high altitudes, it is important that, as the gasoline cools, there be no freezing or separation of crystals which might impede the free flow of the fuel. Specifications usually include the provision for a maximum freezing point of minus 76°F, which limits the amount of benzene that may be present in the fuel.

Table 11-15. Detailed Requirements[a] for Aviation Gasolines (ASTM 910-63T)

	Grade 80-87	Grade 91-98	Grade 100-130	Grade 108-135	Grade 115-145	Test Method[b]
Knock value, min, octane number, lean rating...	80	91	100	Isooctane plus 0.22 ml of tetraethyl lead per gallon	Isooctane plus 0.47 ml of tetraethyl lead per gallon	ASTM D 614
Knock value, min, octane number, rich rating...	87	98	Isooctane plus 1.28 ml of tetraethyl lead per gallon	Isooctane plus 1.68 ml of tetraethyl lead per gallon	Isooctane plus 2.8 ml of tetraethyl lead per gallon	ASTM D 909
Color[m]...	Red[c]	Blue	Green	Brown	Purple	Section 9 (c)
Dye content:						
Permissible blue dye,[d] max, mg per gal...	0.5	5.7	4.7	3.1	4.7	
Permissible yellow dye,[e] max, mg per gal...	None	None	7.0	None	None	
Permissible red dye,[f] max, mg per gal...	8.65	None	None	2.7	3.27	
Permissible orange dye,[g] max, mg per gal...	None	None	None	6.0	None	
Tetraethyl lead,[h] max, ml per gal...	0.5[e]	2.0	3.0	3.0	4.6	ASTM D 526
Net heat of combustion,[i] min, Btu per lb...	18,700	18,700	18,700	18,800	18,800	ASTM D 240

Requirements for All Grades

		Test Method
Distillation temperature, °F:		
10 % evaporated, max...	158	ASTM D 86
50 % evaporated, max...	221	
90 % evaporated, min...	212	
90 % evaporated, max...	257	
Final boiling point, max... °F...	338	
Sum of 10 and 50 % evaporated temperatures, min, °F...	307	
Distillation recovery, min, %...	97	
Distillation residue, max, %...	1.5	
Distillation loss, max, %...	1.5	
Acidity of distillation residue...	Shall not be acid	ASTM D 1093
Vapor pressure, max, lb...	7.0	ASTM D 323
Copper strip corrosion, max...	No. 1	ASTM D 130
Gum by copper dish, max, mg per 100 ml...	5	Section 9 (j)
Potential gum (5 hr aging gum),[j] max, ng per 100 ml...	6	ASTM D 873

Visible lead precipitate,[k] max, mg per 100 ml	3	ASTM D 873
Sulfur, max, %	0.05	ASTM D 90
Freezing point, max, °F	Minus 76	Section 9 (m)
Water tolerance	Volume change not to exceed ±2 ml	ASTM D 1094
Permissible gum inhibitors[l] max, lb per 1,000 bbl (42 gal)	4.2	

[a] Requirements contained herein are absolute and are not subject to correction for tolerance of the test methods. If multiple determinations are made, average results shall be used.

[b] The test methods indicated in this table are described or referred to in Section 9, ASTM 910–63T.

[c] If mutually agreed upon between the vendor and the purchaser, Grade 80–87 may be required to be free from tetraethyl lead. In such a case, the fuel shall not contain any dye and the color as determined in accordance with Method of Test for Saybolt Color of Refined Petroleum Products (Saybolt Chromometer Method) (ASTM Designation D 156) shall not be darker than +20.

[d] The only blue dye which shall be present in the finished gasoline shall be essentially 1,4-dialkylamino-anthraquinone.

[e] The only yellow dye which shall be present in the finished gasoline shall be essentially p-dimethylaminoazobenzene (Color Index No. 19).

[f] The only red dye which shall be present in the finished gasoline shall be essentially methyl derivatives of azobenzene-4-azo-2-naphthol (methyl derivatives of Color Index No. 248).

[g] The only orange dye which shall be present in the finished gasoline shall be essentially benzene-azo-2-naphthol (Color Index No. 24).

[h] The tetraethyl lead shall be added in the form of an antiknock mixture containing not less than 61 per cent by weight of tetraethyl lead and sufficient ethylene dibromide to provide two bromine atoms per atom of lead. The balance shall contain no added ingredients other than kerosine, and approved inhibitor, and blue dye, as specified, herein.

[i] The net heat of combustion determination may be waived if the product of the gravity of the fuel as determined in accordance with the Method of Test for API Gravity of Petroleum and Its Products (Hydrometer Method) (ASTM Designation D 287) in degrees API and the aniline point of the fuel as determined in accordance with Method of Test for Aniline Point and Mixed Aniline Point of Petroleum Products and Hydrocarbon Solvents (ASTM Designation D 611) in degrees Fahrenheit is equal to or above 7400 in the case of Grades 80–87, 91–98, and 100–130, and equal to or above 8200 in the case of Grades 108–135, 115–145. The product of the aniline point and the API gravity shall be known as the aniline-gravity constant.

[j] If mutually agreed upon between the vendor and the purchaser, aviation gasoline may be required to meet a 16-hr aging gum test (ASTM Method D 873) instead of the 5-hr aging gum test. In such a case, the gum content shall not exceed 10 mg per 100 ml and the visible lead precipitate shall not exceed 4 mg per 100 ml. In such fuel the permissible gum inhibitors shall not exceed 8.4 lb per 1,000 bbl (42 gal).

[k] The visible lead precipitate requirement applies only to leaded fuels.

[l] Permissible gum inhibitors are as follows:
N,N'di-secondary-butyl-para-phenylenediamine
2,4, dimethyl-6-tertiary-butylphenol
2,6-ditertiary butyl, 4-methylphenol.
These inhibitors may be added to the gasoline separately or in combination, in total concentration not to exceed 4.2 lb of inhibitor (not including weight of solvent) per 1,000 bbl (42 gal).

[m] These colors have been approved by the Medical Director Chief, Division of Occupational Health, U.S. Department of Health, Education and Welfare.

11–25

Table 11-16. Performance-number Scale for Aviation Gasolines

Below 100		Above 100		
PN*	ON†	PN*	TEL‡	ON†
60	81.3	100	....	100.0
62	82.8	104	0.10	101.3
64	84.2	108	0.22	102.7
66	85.6	112	0.35	104.0
68	86.8	116	0.51	105.3
70	88.0	120	0.69	106.7
72	89.1	124	0.90	108.0
74	90.2	128	1.14	109.3
76	91.2	132	1.43	110.7
78	92.1	136	1.77	112.0
80	93.0	140	2.17	113.3
82	93.8	144	2.65	114.7
84	94.7	148	3.22	116.0
86	95.4	152	3.90	117.3
88	96.2	156	4.72	118.7
90	96.9	160	5.71	120.0
92	97.6			
94	98.2			
96	98.8			
98	99.4			

* Performance number.
† Octane number.
‡ Milliliters of TEL per gallon in isooctane.

Water Tolerance. Water must be low in aviation gasoline for handling and operational requirements. Alcohols and some other substances increase the solubility of water in the fuel. Their presence can be determined by shaking a measured volume (80 ml) of the gasoline with 20 ml of water and noting the change in the volume of water by ASTM D 1094.[20] Specifications usually provide that this change shall not be more than plus or minus 2 ml.

Color. The color of aviation gasoline is fixed by standard colors designated in the specifications as an additional safeguard to ensure that the correct grade of gasoline is loaded into the plane.

Additives. These compounds are used in aviation gasolines only as agreed upon in the specifications by the supplier and the purchaser. ASTM Specifications provide for only three types of additives: dye, tetraethyl lead, and gum inhibitor. Individual contracts may provide for the use of other additives.

Marine Gasoline

The sale of gasoline for use in marine engines represents a small but fast-growing market in areas adjacent to rivers, lakes, and coastal waterways. Consumption in both the inboard and outboard types of engines is estimated at around 650,000,000 gal in 1959, less than 2 per cent of the consumption of gasoline in passenger automobiles. It is estimated there are currently about 7 million gasoline-engine boats in operation in the United States, mostly for recreational purposes.

Two types of gasoline are used in marine engines. Conventional regular and premium grades of motor gasoline supply about 80 per cent of this market; the so-called marine white gasolines supply the other 20 per cent.[18] The current trend is toward the greater use of the conventional motor gasolines in both inboard and outboard engines, since marine-engine manufacturers are designing their equipment to use these fuels.

The marine white gasolines are unleaded, undyed products usually protected with antioxidant and metal deactivator additives. They generally are of relatively low Reid vapor pressure (around 7 lb) for greater safety.

In the past the octane requirements of marine engines have been lower than with automotive engines. At present, however, horsepower and compression ratios of marine engines are being increased to take full advantage of the antiknock qualities of the available motor gasolines. Motorboat engines operate most of the time under the severe conditions of nearly full throttle and at high engine speeds. Hence, Motor octane is more significant than Research number as an indication of antiknock capabilities of the fuel. In the modern high-compression marine engines, a Motor octane of 80 or higher is required usually. Under the severe operating conditions that marine engines generally encounter, the need for adequate octane quality in the fuel is more important than in automobile engines, where operating severities vary and the problem of knock is intermittent.

High resistance to oxidation and gum formation formerly was important in marine gasolines because copper was used in the fuel systems, particularly the tanks. Now copper is being eliminated, and galvanized steel is used in inboard fuel tanks and aluminum or terne plate in outboard engine tanks. In the special marine gasolines sold today, additional antioxidant and metal deactivator additives are sometimes used to provide extra protection under prolonged storage conditions. In general, the same ASTM tests as for motor gasolines are used to measure oxidation stability and gum content of marine gasolines.

Gasoline Tractor Fuels

Conventional regular-grade motor gasoline constitutes nine-tenths of the fuel consumed in farm tractors today. The remainder is diesel fuel, LP gas, and kerosine. The use of diesel fuel and LP gas is growing, while the consumption of the kerosine-type fuels is declining. The use of gasoline has been made more attractive with the repeal, in 1956, of the 4-cent Federal gasoline tax for such use.

Tractor engines using conventional motor gasolines are designed specifically for it, and many of the performance characteristics required for automotive gasoline are applicable to its use in tractors. Since tractor engines are lower in compression ratios than passenger-car engines (6 to 1 in 1956), octane requirements are met with regular-grade fuels.

The total consumption of all fuels by farm tractors in 1959 was 3,678,470,000 gal, about 6 per cent of the fuel consumed in motor vehicles. The total number of tractors on U.S. farms in 1959 was 4,688,500, a growth of 37 per cent in the 1948–1958 decade.

Stove Gasoline

Some refiners market an unleaded product known as stove gasoline. Such gasoline is specified by the Federal government as " . . . suitable for use in stationary internal-combustion, spark-ignition engines, and as fuel in gasoline pressure appliances such as field stoves, heating units, blow torches, and so on." The amount of tetraethyl lead present shall not exceed a trace, such as might result from contamination.

A trace of TEL is defined as 0.65 ml or less of tetraethyl lead per gallon of gasoline. Additives, when used, must conform to the list and concentrations approved by the Department of the Army. Physical and chemical requirements in the Federal specifications for gasoline, unleaded, VV-G-109, are shown in Table 11-17.[18]

AIRCRAFT GAS-TURBINE FUELS

Jet-fuel requirements for gas-turbine-powered aircraft are small compared with the total supply of petroleum products. However, the proportion will increase as the demand from both military services and commercial airlines increases for jet fuels.

In 1957 military demand for jet fuels in this country had reached nearly 200,000 bbl daily, while that of the commercial airlines was less than 16,000 bbl daily. By 1963 military demand for jet fuel had grown to 317,000 bbl daily, and that of the

Table 11-17. Physical and Chemical Requirements for Gasoline, Unleaded: Federal Specification VV-G-109

Property	Test limit
Distillation:	
10% evaporation, °F max	158
50% evaporation, °F max	266
90% evaporation, °F max	365
Residue, % max	2
Vapor pressure (Reid), lb max	10
Gum, ml per 100 ml gasoline, max	4
ASTM Motor octane number, min	62
Sulfur, % max	0.25
Corrosion	None
Oxidation stability, minutes, min	480
Color, min	15

commercial airlines had expanded to over 200,000 bbl. In the same period total aviation-gasoline needs dropped from 201,000 bbl daily to 137,000 as shown in Fig. 11-5. One authority has forecast[23] that military requirements under peacetime conditions might reach 350,000 bbl daily by 1965.

The military services, necessarily planning ahead for a huge emergency demand of liquid fuels of all types, look upon the petroleum industry's supply of gasoline as the logical source of the major part of their jet-fuel requirements. Their overall requirements in 1962 were made up of around 74 per cent gasoline, 14 per cent kerosine, and about 11 per cent distillate fuels. The commercial airlines, for most of their fuel requirements, use a kerosine-type fuel.

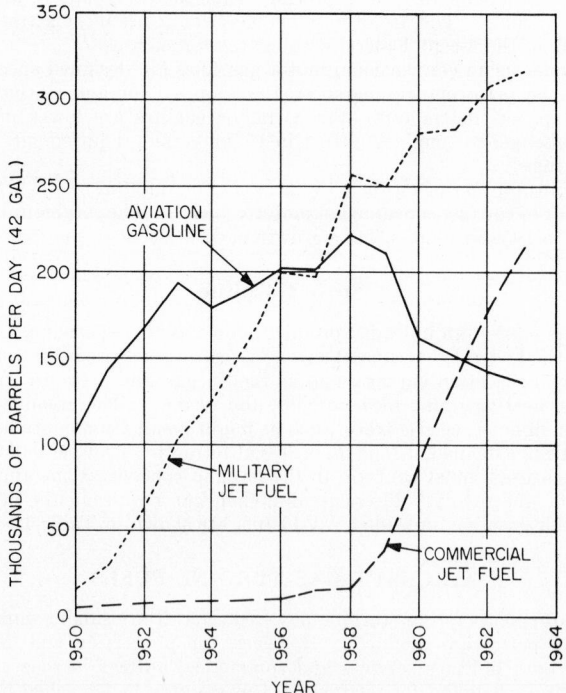

FIG. 11-5. Consumption of commercial and military jet fuels and aviation gasoline in the United States, 1950–1963 (Bureau of Mines data).

From both military and civilian sources, requirements for jet fuels or their components from refiners' kerosine and distillate fuel-oil fractions in a few years may reach 300,000 bbl daily.   Civilian needs for distillate-type products, such as kerosine, diesel fuel, light heating oils, and others, now amount to over 2,000,000 bbl daily.   Flexibility in overall refinery operations will be required to meet these additional needs without an undue increase in the volume of crude oil charged to stills.   Supplies and requirements for jet fuels are shown in Table 11-18.

Table 11-18. Supply and Requirements for Jet Fuels in the United States, 1952 to 1962
(Thousands of barrels)

Year	Production*	Imports	Total supply	Military requirements†	Commercial requirements
1962	169,246	10,897	180,143	112,287	65,707
1961	143,110	10,045	153,155	104,357	49,441
1960	121,407	12,731	133,788	102,674	33,159
1959	92,933	13,682	106,600	94,269	14,432
1958	74,739	21,169	95,908	95,150	6,063
1957	63,322	9,185	72,507	95,479	1,930
1956	66,443	7,763	74,206	79,559	800
1955	56,648	4,297	60,945	65,425	
1954	46,550	929	47,479	41,782	
1953	35,747	.....	35,747	27,756	
1952	20,929	.....	20,929	12,292	

Authority: U.S. Bureau of Mines and Military Petroleum Supply Agency.
* Beginning with 1959 includes commercial kerosine type fuel formerly included in kerosine production.
† Prior to 1959 for fiscal year ending June 30.   For calendar year from 1959 on.

The first civil operations of gas-turbine-powered aircraft in North America were with a British-manufactured turboprop engine.   The Canadian airline operating these engines uses the kerosine-type JP-4 fuel specified by the U.S. military services, while the American airline uses the D. Eng. R.D. 2482 type of kerosine (see Table 11-19). Recently the ASTM has tentatively adopted specifications (D 1655-63T) for the three fuels shown in Table 11-20.   Types A and A-1 are kerosine-type fuels, while Type B is a wide-cut material similar to the military JP-4 fuel.   Approved oxidation, corrosion, and metal deactivator additives may be used in both the military and civil aircraft gas-turbine fuels.

Types of Aircraft Turbine Engines

Because the earliest applications of gas-turbine engines in aircraft used the jet thrust principle, the popular tendency is to refer to all gas-turbine-powered aircraft as "jet" aircraft and the fuels for all types are known as "jet" fuels.   This designation is incorrect for aircraft powered by gas turbines driving conventional propellers. Such engines should be referred to as "turboprop" or "prop-jet" aircraft.

In all types of aircraft gas turbines,[24] atmospheric air is compressed by the compressor of the engine, heated in the combustion section by burning fuel in it, expanded through a turbine, and exhausted to the atmosphere through a jet or tail pipe.   In the turbojet gas-turbine engine, the greater portion of the energy is extracted from the hot gases as they expand through the turbine and is used to drive the air compressor and engine accessories, such as fuel and oil pumps, electrical generators, etc.   The remaining portion is jet thrust to propel the aircraft.

In the turboprop engine, the energy is extracted from the hot gases as they pass through the turbine and used to drive a conventional-type propeller as well as the air

Table 11-19. Principal Characteristics of Selected Military Aircraft Turbine Fuels

	United States					Great Britain
	JP-1 (MIL-F-5616 Amend. 1)	JP-3 (MIL-F-5624D)	JP-4 (MIL-F-5624D)	JP-5 (MIL-F-5624D)	JP-6 (MIL-F-25656 Amend. 1)	D. Eng. R.D. 2482 (Issue 3)
Gravity, °API, max	35	50–60	45–57	36–48	37–50	40.0–51.0
Flash point, °F, min	110			140		100
Viscosity, centistokes, at:						
0°F, max						
−30°F, max				16.5	15	6
−40°F, max	10					
Freezing point, °F, max	−76	−76	−76	−40	−65	−40
Color, 18 in. Lovibond, max	+12					4
Distillation, °F:						
IBP, min					250	
10% evap., max					350	
20% evap., max		240	290			
50% evap., max	410	350	370	400	425	392
90% evap., max	490	470	470		500	
EP, max	572			550	600	572
Sum of EP + 50%, min						
Residue, %, max	1.5	1.5	1.5	1.5	1.5	2
Loss, %, max	1.5	1.5	1.5	1.5	1.5	1.5
Reid vapor pressure, lb		5–7	2–3			
Total sulfur, %, max	0.2	0.4	0.4	0.4	0.4	0.2
Mercaptan sulfur, %, max	0.005	0.005	0.005	0.005	0.005	0.005
Total aromatics, %, max	20	25	25	25	25	20
Olefins, %, max		5	5	5	5	5
Bromine number, max	3.0					5
Net heat of combustion, Btu/lb, min		18,400	18,400	18,300	18,400	18,300
Aniline-gravity constant, min		5,250	5,250	4,500	5,250	4,500
Smoke point, mm, min				20	20	
Smoke volatility index, min		54	54			
Copper strip corrosion, max	*	*	*	*	*	*
Water tolerance, ml, max	2	1	1	1	1	1
Existent gum, mg/100 ml, max	5	7	7	7	5	3
Total potential residue, mg/100 ml, max	8	14	14	14	10	6
Inorganic acidity, max						Nil

* See detailed specifications.

compressor and accessories. Because of the additional energy extracted to drive the propeller, there is some, but relatively little, energy in the exhaust of the turboprop engine to produce jet thrust.

The bypass engine, also called the turbofan or ducted-fan engine, is a variation of the turbojet engine. The energy extracted from the hot gases is used to drive not only the combustion air compressor and accessories but also a fan which supplies air to the annular duct between the internal and external casings of the engine. The

Table 11-20. Principal Characteristics of Proposed ASTM Aircraft Turbine Fuel Specifications (D 1655-63T)*

Property	Type A	Type B	Type A-1
Gravity, °API:			
Max.	51	57	51
Min.	39	45	39
Distillation temp, °F:			
10% evap., max.	400	....	400
20% evap., max.	....	290	
50% evap., max.	450	370	450
90% evap., max.	....	470	
Final boiling point, °F, max.	550	....	550
Distillation residue, %, max.	1.5	1.5	1.5
Distillation loss, %, max.	1.5	1.5	1.5
Vapor pressure, lb, max.	....	3	
Flash point, °F:			
Min.	110 or legal	....	110 or legal
Max.	150	....	150
Pour point, °F, max.	Minus 40	....	
Freezing point, °F, max.	Minus 40	Minus 60	Minus 58
Viscosity at −30°F, centistokes, max.	15	....	15
Net heat of combustion, Btu/lb:			
Min.	18,400	18,400	18,400
Copper strip corrosion:			
3 hr at 122°F, max.	No. 1	....	No. 1
2 hr at 212°F, max	....	No. 1	
Total acidity, mg KOH/g, max.	0.1	....	0.1
Sulfur, %, max.	0.3	0.3	0.3
Mercaptan sulfur, %, max.	0.003	0.003	0.003
Water tolerance, vol. change not to exceed, ml	±1	±1	±1
Existent gum, mg/100 ml, max.	7	7	7
Total potential residue 16 hr, mg/100 ml, max.	14	14	14
Aromatics vol., %, max.	20	20	20
Olefins vol., %, max.	....	5	

* Source: *ASTM Standards, Part 17, Petroleum Products*, 1964.

purpose of the fan is to increase propulsive efficiency and thus reduce fuel consumption through mixing relatively low-velocity bypass air with high-velocity gas leaving the turbine to produce a bypass air-exhaust gas mixture having a velocity closer to that of the aircraft.

The optimum applicability lies below about 500 to 550 mph for turboprop engines (limited by propeller efficiency) and above this speed range for turbojet and bypass engines. The bypass engine is limited to a somewhat lower maximum speed range than the turbojet because of aerodynamic considerations involving efficiency of the fan. The specific fuel consumption of the turboprop engine is superior to that of the turbojet. That of the bypass engine is somewhere between those of the turboprop and the turbojet.

Basic Fuel Systems

Turbine-engine fuel systems are composed of a fuel pump, fuel metering control, and injection nozzles, as shown in Fig. 11-6.[24]   Fuel pumps, usually of the gear or wobble plate piston type, provide the necessary pressure for use in the fuel-control unit and nozzles.

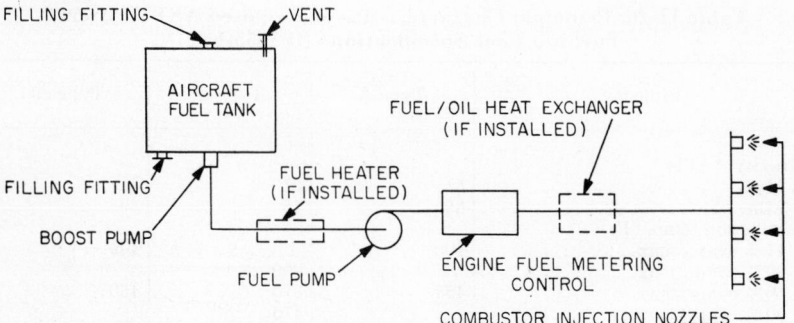

FIG. 11-6. Aircraft turbine-engine fuel system.[24]

The fuel metering control must accurately regulate flow of the fuel over the entire operating range from starting to full power and from sea level to high altitudes.   The injection nozzles spray the metered fuel into the combustion area in the proper pattern to produce optimum combustion under all conditions of operation.

Combustion Systems

Aircraft gas-turbine-engine combustion systems are of three main types.   In the "can" arrangement, each flame tube is surrounded by its own outer liner.   The cans are arranged in a ring around a main engine shaft.   Gas flow in each can is separated from the others except for interconnecting tubes which serve to equalize the pressures and aid in the initial ignition of the engine.

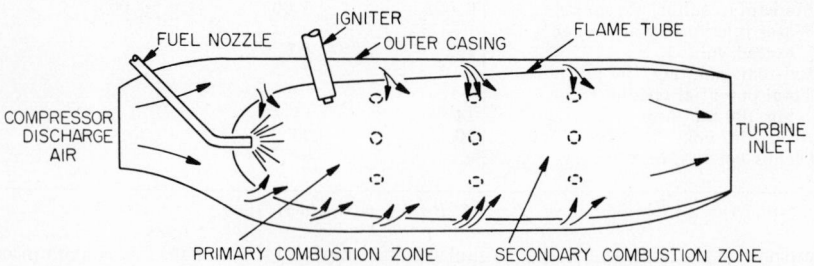

FIG. 11-7. Aircraft gas-turbine combustion chamber.[24]

In the "annular" arrangement, the inner flame tube and the outer casing are annular members concentric around the main shaft.   The "cannular" arrangement is a combination of the other two systems in that the flame tubes are contained in a single outer casing which is annular in cross section.

The combustion process takes place in a combustion chamber, a generalized sketch of which is shown in Fig. 11-7.[24]   Air received from the engine compressor enters the outer casing and flows through holes and baffles into the flame tube.   A portion, called primary air, is mixed with the fuel spray in approximately the stoichiometric ratio to support combustion. The remainder, or secondary air, is used to cool the surface of the flame tube, to stabilize combustion in the primary zone by recirculation and turbulence, and to cool the combustion gases in the secondary zone to temperatures which can be tolerated by the turbine inlet nozzles, blading, and other downstream parts.

Combustion is initiated by an ignition source, such as a spark plug, which is shut off as soon as combustion is started, since, once initiated, combustion is self-sustaining and continuous. Afterburning (reheating) is used frequently in military turbojet engines to increase thrust temporarily for take-off or combat purposes. This is done by injecting fuel into the exhaust gas after it leaves the turbine. While it can produce large increases in thrust, fuel consumption increases greatly with afterburning.

Supplementary fluid injection, either water or a water-alcohol mixture, can be used to increase the power output of both turbojet and turboprop engines. The fluid may be injected into the inlet of the compressor or directly into the combustion area. Supplementary fluid injection is for intermediate use, primarily take-off.

Aircraft Fuel Properties

Samples of 44 commercial jet fuels sold in the United States in 1963 were studied and analyzed by the Bureau of Mines' Petroleum Experimental Station at Bartlesville, Okla., under a cooperative agreement with the American Petroleum Institute. Under this agreement, surveys of aviation fuels, both gasoline and jet, are made annually. Average properties of the commercial jet fuels sampled and analyzed in 1963 are given in Table 11-21.[25]

Following is a listing of some of the more important fuel requirements and properties.[24]

Table 11-21. Average Properties of 44 Commercial Jet Fuels Sold in the United States in 1963[25]

Property	Average values		
	Type A	Type B	Type A-1
Gravity, °API	43.9	52.0	43.3
Distillation temperatures, °F:			
Initial boiling point	338	137	333
10% point	369	222	360
20% point	382	254	372
50% point	410	315	401
90% point	464	423	464
Final boiling point	500	480	501
Evaporation, % at 400°F	37.9	82.6	48.4
Freezing point, °F	−54	−76	−64
Viscosity, kinematic, at −30°F	8.85	2.94	8.04
Water tolerance, ml	0.2	0.1	0.2
Aniline point, °F	145.8	132.9	139.1
Aniline-gravity constant	6.401	6.911	6.058
Sulfur, wt%:			
Total	0.055	0.044	0.071
Mercaptan	0.0002	0.0006	0.0004
Aromatic content, vol %	14.1	12.3	15.3
Olefin content, vol %	1.2	0.9	1.0
Smoke point, mm	24.3	26.4	23.2
Gum, mg/100 ml, steam jet at 450°F:			
Existent	0.8	0.7	0.7
16-hr accelerated	1.8	1.3	1.6
Net heat of combustion, Btu/lb	18,600	18,703	18,571

Volatility. The tendency of a fuel to vaporize is one of its most important physical properties. Volatility affects ground and aircraft tank vaporization losses, vapor lock, engine starting and operation, and general safety. These conditions are emphasized because aircraft fuel tanks are usually vented to the atmosphere to avoid the complexity and weight penalties of pressurized or conservation storage.

No single test method can be used to describe completely the volatility of a turbine fuel. It is normally measured and controlled by a combination of ASTM distillation and Reid vapor-pressure or flash-point tests.

Volume-Weight Relationships. Fuel is usually metered into aircraft tanks and into the combustion zone of the engine on a volumetric basis. Knowledge of the weight of the fuel aboard, however, is essential to assure that take-off and landing weight limitations are not exceeded. In addition, aircraft performance and range calculations are based on fuel volume-weight relationships when used in conjunction with heat of combustion, since fuel used in the combustion process is noted on a weight basis. Established standards are available to determine these relationships.

Combustion. Ideally, the combustion of fuel in a turbine engine should transfer the maximum possible energy from the fuel to the air by complete fuel combustion over a wide range of operating conditions without adversely affecting the life of the engine parts or producing adverse effects external to the engine. In practice, this is difficult if not impossible to achieve. Deviations from the ideal occur in combustion efficiency and stability, deposit formation, engine life, and exhaust smoke and odor. In the final analysis, fuel performance in the engine must strike a balance among factors of fuel cost, fuel performance, and parts life.

Viscosity. Fuel must remain sufficiently fluid at low temperatures to flow through openings in fuel tank baffles and into the inlets of pumps through screens. Wax must not form in fuel systems to plug screens or filters. If the fuel is too viscous, excessive pump power will be needed or the injection nozzle spray patterns will be altered and ignition will be difficult.

Kinematic viscosity is commonly used in fuel specifications to describe the required fuel-flow properties down to the temperature at which the fuel ceases to act as a Newtonian fluid (one where viscosity is independent of rate of shear). Below this point, available tests are not entirely satisfactory to express the flow properties of turbine fuels in design calculations for aircraft fuel systems. Maximum freezing points and sometimes pour points are incorporated in specifications to assure freedom from wax problems and sufficient fuel mobility.

Contamination. Safe, dependable aircraft operation demands fuels free of contaminants. Contamination can be from solids, such as finely divided rust and pump wear materials, dust, sand, or soft particles. It also can be from an extensive variety of other liquid materials, including other types of fuels. Solid contaminants should be removed by settling and adequate filtration prior to placing the fuel in aircraft tanks. Contamination from other fuels can be avoided by using separate storage and handling facilities for the aircraft fuels.

Water may be present in dissolved or free (undissolved) forms or a combination of the two. The presence of dissolved water is prevented by adequate fuel settling and fuel dehydration in ground handling equipment to remove free water. Most turbine-fuel specifications include a water-tolerance test to assure that the fuel will not dissolve excessive quantities of water and that it contains no appreciable amounts of water-soluble materials. One way of attacking the problem of fuel-system icing caused by free water is to use anti-icing additives.

Stability. Resinous, insoluble materials in fuel can form gummy, varnishlike deposits in fuel controls or on other fuel-system components and result in engine malfunctions. Turbine fuel stocks should contain no more than a trace of gum, which should not increase appreciably in quantity during extended storage. For this reason, specifications generally limit both existing gum and total potential residue content. Approved types of gum inhibitors are frequently listed in specifications.

Odor. The odor of unburned fuel is important in public and labor relations. This problem is frequently aggravated with the heavier fuels, such as kerosine, which are relatively slow to evaporate. Most fuel specifications limit the maximum mercaptan sulfur content as such or limit it by providing that the fuel be negative (sweet) in the doctor test.

Flammability. As with other hydrocarbon fuels, turbine fuels are a hazard when subject to ignition in locations other than in engine combustion areas. The safety hazards involved depend not only upon the particular type of fuel but also upon ambient conditions in the adjacent atmosphere, the geometry and materials of construction of the containing equipment, and the types of available ignition sources.

While the relative hazards vary with different fuels, the known precautions should be followed meticulously at all stages in the handling and use of aircraft gas-turbine fuels.

DIESEL FUELS

Sales of diesel fuel in the United States were 254,040,000 bbl in 1964.[26] This is a volume market exceeded only by gasoline, distillate-type heating oils, residual fuel oils for all uses, and liquefied petroleum gas. Table 11-22 shows sales of diesel fuel in the United States for all uses.

Table 11-22. Sales of Diesel Fuel Oil in the United States by Uses, 1940 to 1962*
(Thousands of barrels)

Year	Railroads	Vessels	Gas and electric plants	Industrial including oil co. fuel	Military	Misc., including automotive fuel†	Total
1964	86,274	15,793	2,318	26,294	5,827	117,534	254,040
1963	86,005	14,995	2,537	24,987	6,924	106,341	241,789
1962‡	84,456	15,472	2,521	22,655	7,578	89,729	222,411
1961‡	81,485	14,247	2,321	20,543	6,727	77,825	203,178
1960‡	82,600	18,377	2,823	21,246	6,315	74,562	205,923
1959‡	85,672	18,535	3,420	19,936	7.302	68,725	204,590
1958	81,601	18,204	3,169	17,481	9,321	65,186	194,962
1957	83,569	19,533	3,055	18,663	8,934	49,684	183,438
1956	84,449	17,586	3,147	19,129	7,653	48,870	180,834
1955	80,064	15,869	3,430	17,865	7,794	44,215	169,237
1954	72,907	14,539	3,738	17,146	6,141	38,809	153,280
1953	70,669	15,849	4,257	16,973	6,942	34,980	149,670
1952	63,306	16,158	4,939	17,095	7,081	30,551	139,130
1951	55,159	13,026	5,743	16,296	5,913	25,485	121,622
1950	45,084	12,119	6,564	14,972	4,650	21,333	104,722
1945	12,036	13,374	3,067	6,254	2,950	8,371	66,412
1940	1,838	12,979	1,631	2,180	1,115	4,926	24,669

* Authority: American Petroleum Institute, *Petroleum Facts and Figures*, 1965.
† Includes fuel used in trucks, busses, tractors, and heavy moving equipment.
‡ Includes Alaska beginning in 1959 and Hawaii in 1960.

Grades of Diesel Fuels

Distillate fuel oils from the refinery in general have the properties needed for high-speed diesel engine fuels, such as volatility, ignition qualities, and viscosity. Medium- and low-speed engines use the heavier distillates and blends of these with selected residual fuels. The fractions distilled from crude petroleum which are suitable for diesel fuels are shown in Fig. 11-8.[27]

The ASTM, in D 975,[28] has classified diesel fuels for commercial use into three grades, the properties of which are shown in Table 11-23.

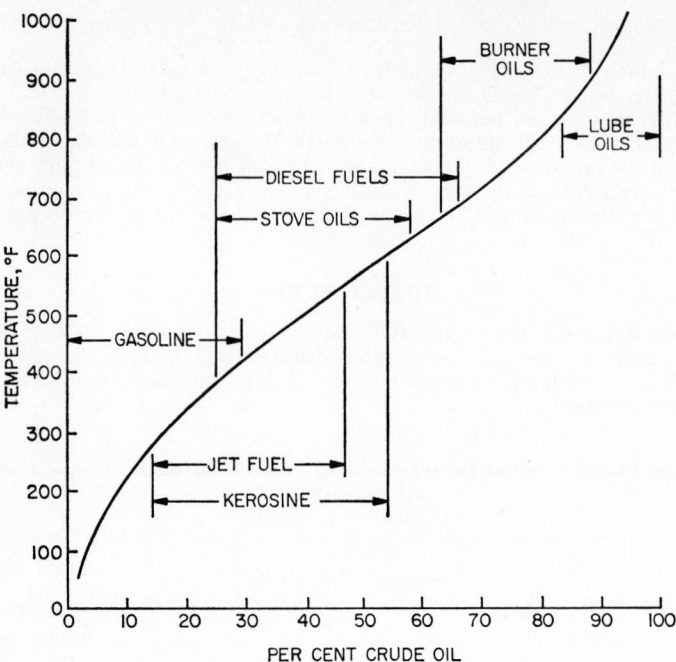

FIG. 11-8. Proportion of diesel fuel and other petroleum products processed from crude petroleum.

Table 11-23. Limiting Requirements for Diesel Fuel Oils[28]

Grade of diesel fuel oil	Flash point, °F	Pour point, °F	Water and sediment, % by volume	Carbon residue on 10% residuum, %	Ash, % by weight	Distillation temp, °F 90% point	Distillation temp, °F End point	Viscosity at 100°F, kinematic, centistokes or SSU	Viscosity at 100°F, kinematic, centistokes or SSU	Sulfur, % by weight	Copper strip corrosion	Cetane number
	Min	Max	Max	Max	Max	Max	Max	Min	Max	Max	Max	Min
No. 1-D {A volatile distillate fuel oil for engines in service requiring frequent speed and load changes}	100 or legal	...	Trace	0.15	0.01	...	625	1.4	2.5 (34.4)	0.50	No. 3	40
No. 2-D {A distillate fuel oil of low volatility for engines in industrial and heavy mobile service}	125 or legal	...	0.10	0.35	0.02	675	...	1.8 (32.0)	5.8 (45)	1.0	No. 3	40
No. 4-D {A fuel oil for low- and medium-speed engines}	130 or legal	...	0.50	....	0.10	...	...	5.8 (45)	26.4 (125)	2.0	...	30

From Tentative Classification of Diesel Fuel Oils, ASTM D 975-60T. Footnotes are omitted which deal with special conditions pertaining to pour point and cetane-number ratings.

Grade 1-D comprises the class of volatile fuel oils from kerosine to the intermediate distillates. It is used in high-speed engines operating at frequent and relatively wide variations in loads and speeds and for applications where fuel temperatures are abnormally low.

Grade 2-D includes distillate gas oils of lower volatility used in high-speed engines at relatively high loads and uniform speeds or in engines not requiring the higher volatility or other properties of Grade 1-D.

Grade 4-D covers the more viscous distillates and blends of these distillates with residual fuel oils. This grade is used in low- and medium-speed engines operating with sustained loads at substantially constant speed.

Federal Specification VV-F-800 classifies diesel fuel oils into the following four grades, having the properties given in Table 11-24:[27]

Table 11-24. Property Requirements of Diesel Fuels in Federal Specification VV-F-800[27]

Property	Grade DF-A (arctic)	Grade DF-1 (winter)	Grade DF-2 (regular)	Grade DF-4 (heavy)
Flash point, °F, min	100, or legal	100, or legal	125, or legal	130, or legal
Cloud point, °F, max	Minus 50			
Pour point, °F, max	Minus 70			
Kinematic viscosity at 100°F:				
Centistokes, min	1.4	1.4	1.8	5.8
Centistokes, max	4.0	4.0	6.0	20.6
Water and sediment, % by volume, max	0.03	0.05	0.05	0.50
Sulfur, %, max	0.25	0.5	1.0	1.5
Carbon residue on 10% residuum, %, max	0.12	0.15	0.25	0.5
Ash, %, max	0.01	0.01	0.02	0.10
Corrosion, copper strip, 3 hr at 122°F, ASTM number, max	2	3	3	
Ignition quality, cetane number min:	40	40	40	35
Distillation				
50% point, °F	Record	Record	Record	Record
90% point, °F, max	550	575	675	
End point, °F, max	575	625	725	
Gravity, °API	Record	Record	Record	Record

Footnotes are omitted dealing with special conditions pertaining to cloud point, pour point, and carbon residue.

Grade DF-A (arctic grade) is for high-speed, automotive-type diesel engines and pot-type burner space heaters where mean ambient temperatures lower than minus 25°F occur and where it is impractical to obtain or store diesel and burner fuels. This grade of diesel fuel should not be used for slow-speed stationary engine applications.

Grade DF-1 (winter grade) is for high-speed automotive service where ambient temperatures as low as minus 25°F occur. This grade of diesel fuel may be used for medium-speed stationary engine applications where fuel heating facilities are not available.

Grade DF-2 (regular grade) is for all automotive, high-speed diesel engines and for medium-speed engines where ambient temperatures are above 20°F.

Grade DF-4 (heavy grade) is for low- and medium-speed diesel engines.

In general, the government's Grade DF-2 conforms in properties to the ASTM 2-D and its Grade DF-4 to the ASTM 4-D.

Diesel fuel oils sold in the United States are graded in four classes in the 1966 survey of country-wide sales of this product conducted by the Bureau of Mines.[29] The classes are:

1. Diesel fuel oils for city bus and similar operations
2. Fuels for diesel engines in trucks, tractors, and similar operations
3. Fuels for railroad diesel engines
4. Heavy-distillate and residual fuels for large stationary and marine diesel engines

The Bureau's annual surveys are based on the study and analysis of samples sold in five geographical regions of the United States. The data on the characteristics of these fuels are presented separately for each region. Table 11-25 shows the average

Table 11-25. Average of Selected Properties of Four Classes of Diesel Fuel Oils Sold in the Central Region* of the United States in 1966[29]

Property	Class 1, Type C-B	Class 2, Type T-T	Class 3, Type R-R	Class 4, Type S-M
Gravity, °API	41.9	37.3	34.8	34.0
Viscosity at 100°F:				
Kinematic, cs	1.84	2.54	2.74	2.79
Saybolt Universal, sec	32.1	34.6	35.2	35.4
Sulfur content, wt %	0.142	0.223	0.287	0.543
Aniline point, °F	148.6	146.2	140.2	139.3
Ramsbottom carbon residue on 10% residuum, %	0.057	0.088	0.117	0.163
Ash, %	0.0005	0.0009	0.0010	0.0023
Cetane number	51.1	50.0	47.0	46.7
Distillation test:				
IBP, °F	356	380	388	397
10%, °F	393	430	440	448
50%, °F	440	490	502	509
90%	501	557	574	582
EP, °F	542	600	618	622

* Central region: Minnesota, Iowa, Wisconsin, Illinois, Indiana, Missouri, Kansas; parts of Oklahoma, Michigan, Kentucky, Arkansas, Texas, Nebraska, and the Dakotas.

of selected properties for Classes 1, 2, 3, and 4 in the central region (Minnesota, Iowa, Wisconsin, Illinois, Indiana, Missouri, Kansas, and parts of Oklahoma, Michigan, Kentucky, Arkansas, Texas, Nebraska, and the Dakotas).

Performance Properties of Diesel Fuels

The ignition quality of a diesel fuel is usually expressed in cetane value as determined by a mixture of cetane, $C_{16}H_{34}$, which has a high ignition quality, and alpha methyl naphthalene, $C_{11}H_{10}$, which has a low ignition quality. The percentage of cetane in the reference fuel by volume is termed the cetane number. The cetane value of a fuel is determined by a standard ASTM test method, D 613-62T.[30]

A method for direct estimation of the ASTM cetane number of distillate fuels from their API gravity and mid-boiling points is also published by the ASTM.[28] The addition of ignition-improver additives to diesel fuels does not change the API gravity or the distillation characteristics of the fuel. Hence the calculated cetane index does not give the correct cetane number for a fuel containing ignition-improver additives.

Diesel fuels of high cetane number differ from those of lower cetane numbers by having shorter ignition lag when the fuel is injected into the engine cylinder. Fuels of high cetane values also ignite at lower compressed-air temperatures than do the low-cetane fuels. In general, the higher the cetane number, the lower the temperature at which the engine can be started. After starting at low temperatures, engines can be brought to steady running more quickly on the high-cetane fuel than on the low-cetane fuel.

Combustion knock in a diesel engine and shock loading of pistons, bearings, and other engine parts result when a fuel having too low a cetane number is used for the size and type of engine and conditions under which the engine is being operated. Low-cetane fuels may cause more rapid accumulation of varnish and carbonaceous deposits when the engine is idling at light-load operation. High-cetane fuel will help reduce the production of acrid odor and fumes (cold smoke) during light-load, cool-running conditions, but ignition quality has only a minor effect on black (hot) smoke. In general, ignition quality has a negligible effect on output and fuel economy.

Influence of Sulfur. The effect of sulfur content on diesel-engine wear and deposits varies and depends largely on operating conditions. Much evidence supports the fact that increasing precentages of sulfur in the fuel are responsible for higher rates of cylinder wear. The standard method of testing for sulfur is ASTM D 129.[28]

Viscosity. This is an important characteristic of diesel fuels. It affects pump leakage and the power required to operate the pump. With lighter viscosity fuels, the pump clearances must be closer or special seals must be provided to prevent undue leakage rates past the plunger and to lengthen the injection period. The higher viscosity fuels create pressure loadings on the plunger and cams, which must, therefore, be provided with adequate bearing surface to permit the injection of fuel at the rates prescribed.

Viscosity also has an influence on the size of fuel particles sprayed into the cylinder through the injection nozzle. The standard method of testing for kinematic viscosity is ASTM D 445; for conversion to Saybolt viscosity, ASTM D 2161.[28]

Volatility. This is an important property of diesel fuels, depending on engine size. In higher speed engines, a too-volatile fuel may cause detonation and vapor-lock problems in the fuel injection system.

Ash-forming Materials. These materials are abrasive and contribute to engine wear and to engine deposits. Soluble metallic soaps, another form of ash, have little effect on wear but contribute to deposits. Carbon residue is a factor in deposit control. The standard method of testing for ash is ASTM D 482; for carbon residue, ASTM D 524.[28]

Additives. As little as 0.1 to 0.3 per cent by volume of amyl nitrate as an ignition-improver agent will upgrade conventional cracked stable distillate fuels into diesel fuels of acceptable cetane number. Engine tests have shown that cetane numbers gained with amyl nitrate additives are equal to *natural* cetane numbers.

Auxiliary starting-aid fluids help overcome starting difficulties for trucks, busses, and construction equipment powered by diesel engines and which often are stored outdoors and exposed to low temperatures. These fluids are composed of hydrocarbon blends with ether or ether and heptane. A priming mechanism attached to the engine injects the fluid while the cold engine is being cranked with a starter.

Utilization of Diesel Fuels

Railroad diesel engines have created the largest single market for diesel fuel in this country. In 1964, their requirements were 86,274,000 bbl, 34 per cent of all diesel-fuel sales (Table 11-22). Thus the railroads have been forced to seek fuels other than straight-run distillates having cetane numbers of 50 or better and volatilities ranging up to 625°F end point. Railroad operators also seek lower grade fuels to reduce operating expenses but must balance the reduced cost of low-grade fuels against the resulting higher maintenance costs that arise from these heavier fuels.

The trend has been toward the adoption of heavy distillates and blends of distillates and residual fuels where geographically practicable. Experimental work has been carried out by some roads with a dual-fuel system, one a distillate fuel used for starting and idling and the other a blend of distillate and residual fuel oils for high-engine-speed, high-load operations.

In a classification of diesel fuels based on service requirements, the Bureau of Mines has included one grade, Class 3, for railroad diesel engines. The fuel[27] has an indicated end point of 650 to 700°F and cetane numbers within the range of 40 to 45,

Table 11-26. Physical Properties of Four Types of Railroad Diesel Fuels*

	ASTM Method	Type of fuel			
		A	B	C	D
Flash point, °F, min or legal	D 93	150	150	150	150
Viscosity:					
SSU at 100°F	D 88	33–45	33–50	320 max	
SSF at 122°F	D 88	....	....	....	210 max
Cetane No., min	D 613	43	34		
Sulfur, % max	{ D 90	0.75	1.00		
	{ D 129	....	....	2.50	2.50
Sediment and water, % max	D 96	0.05	0.05	0.50	2.0
Pour point, °F, max	D 97				
Summer	....	+15	+25	+20	+45
Winter	....	0	−5	+20	+45
Carbon residue, % max (total fuel)	D 189	....	....	9.5	
Ash, % max	D 482	....	....	0.10	0.20
Corrosion test 3 hr	D 130	Not darker than No. 3	Not darker than No. 3		
Thermal-stability test (FS 346.1)	....	....	....	Pass No. 2 tube	

* Southern Pacific Railroad Co. Footnotes are omitted dealing with special conditions pertaining to carbon residue and pour point.

though in some cases with cracked stocks the cetane numbers were as low as 35. Four types of railroad diesel fuels which have been under test by western railroads are listed in the following, and their properties are given in Table 11-26.[27]

Type A: Straight-run distillates for use in small- and medium-bore diesel engines other than locomotives and for diesel locomotives where fuel is delivered by truck

Type B: A blend of straight-run and cracked distillates or all cracked distillates for general use in locomotive diesel engines

Type C: A blend of residual stocks, either straight-run or cracked, and distillate stocks, either straight-run or cracked, for use in diesel engines on locomotives and other specified uses

Type D: Residual fuel cut back with blending stock as required to meet this specification, for use in steam locomotives, power-plant boilers, etc.

Automotive Diesel Fuels

The fastest growing use of diesel oils is as fuel for automotive equipment (Table 11-22). Cetane number is more important for this service than for other uses. Specifications in many instances call for cetane numbers over 40 for city bus and similar operations where relatively small high-speed diesel engines are in service.[31]

The higher cetane number is important for low-temperature starting, fast warm-up, combustion smoothness, reduced carbonaceous deposits, and less exhaust odor and smoke.

The more volatile fuels in general will provide the best performance, particularly with regard to smoke and odor, for engines operating with rapidly fluctuating loads and speeds, as in busses and trucks. Flash point is not related to engine performance, but it is important for safety in fuel handling and storage. Flash point is normally specified to meet insurance and fire regulations.

Diesel fuels for city bus and similar operations are covered in No. 1-D of the ASTM Specifications shown in Table 11-23[28] for "engines in service requiring frequent speed and load changes." A fuel of 40 minimum cetane number and 625°F maximum final boiling point is required. Flash point of 100°F minimum is specified. The ASTM No. 2-D fuel is also listed for trucks and tractors in industrial and heavy mobile service.

Stationary Diesel Fuels

Heavy-duty diesel engines used in power plants and for heavy industrial purposes require a balance between the blending of distillate and heavier fuels to control overall maintenance costs.[27] This is of even greater concern where the diesel power plant must compete with other forms of prime movers, such as steam turbines and gas turbines. Economy trends are toward the use of heavy distillates and residual fuels, as classified in ASTM No. 4-D. This grade covers the class of more viscous distillates and blends of these distillates with residual fuels. They are applicable for use in low- and medium-speed engines in services involving sustained loads at substantially constant speed.

Where residual fuels having low ignition qualities are used, a preliminary charge of light distillate with suitable ignition-quality characteristics may be injected into the engine. The dual-fuel engine either uses a fuel oil for initial ignition by compression and ignites a fuel with lower cetane number or uses an initial charge of fuel readily ignitable by compression ignition, causing the burning of a gaseous fuel which will not ignite by compression temperatures alone.

Marine Diesel Fuels

Marine diesel fuels can be placed in the following four categories:[27]
Type I: A light distillate fuel similar to ASTM No. 1-D classification.
Type II: A distillate of about the characteristics of ASTM No. 2-D.
Type III: A heavy distillate or blend of distillate with some residual material.
Type IV: Various blends of residual with distillate material generally classified as light bunker fuel oils. Also included in this group are the heavy bunker fuel oils commonly called Bunker C or boiler oil.
Table 11-27[27] shows these types of fuels as related to marine applications. As with diesel operators in other fields, shipowners are conscious of operating costs and the trend is toward the use of lower cost fuels for the main propulsion engine.

HEATING OILS

The heating oils include kerosine and distillate and residual fuel oils. The consumption of petroleum products for space heating ranks second only to motor gasoline.

Table 11-27. Diesel Fuels in Marine Service*

Engine group	Application of engine	Predominant fuel in use	Secondary fuel in use	Exploratory fuel†
Group A (1,200 rpm and up)	"Automotive type," high-speed for main propulsion and auxiliaries	Type II	Type I	
Group B (700–1,200 rpm)	Intermediate horsepower for main propulsion and auxiliaries	Type II	Type III	
Group C (365–700 rpm)	Intermediate horsepower for main propulsion and large auxiliaries	Types II and III	. . . .	Type IV
Group D (365 rpm and lower)	Main propulsion and some large auxiliaries	Type III	Type II	Type IV

* *ASTM Spec. Tech. Publ.* 167.
† A significant amount of service experience to consider as another fuel type.

Total sales classified as heating oils in 1962 were 658,104,000 bbl.   Sales of gasoline as motor fuel in the same year were 1,585,000,000 bbl.

Various types of refinery products serve for space heating as well as for other uses where the oil is consumed in a burner.   Kerosine (range oil or No. 1 grade distillate heating oil, a product of 43°API gravity or above) has wide use in pot-type burners for space heating, as well as in cook stoves, water heaters, and other appliances.   Four grades of distillate fuel oils, with gravities ranging from 43 to 20°API gravity, are

Table 11-28. Sales in the United States of Heating Oils by Grades*
(Thousands of barrels)

Year	Range oil†	Distillate heating oils					Residual oils		Total, all heating and range oils
		No. 1	No. 2	No. 3‡	No. 4	Total dist.§	No. 5	No. 6	
1963	96,045	35,600	390,403	.....	23,156	465,315	28,431	96,817	670,452
1962	99,243	37,137	392,519	.....	20,375	466,830	28,872	96,292	674,438
1961	96,684	38,217	365,061	19,151	18,146	440,575	26,472	94,625	658,356
1960	94,167	42,496	341,651	22,491	16,217	422,855	28,962	96,126	642,110
1959	100,644	39,708	327,559	22,628	12,715	402,610	28,931	82,919	615,104
1958	102,809	42,617	324,957	18,743	12,836	399,153	30,443	75,196	607,601
1957	92,482	50,193	276,631	25,285	8,103	360,212	28,704	52,708	534,106
1956	103,485	50,321	275,816	25,681	8,009	359,827	29,505	58,096	550,913
1955	101,705	48,977	258,093	24,898	7,247	339,215	30,211	56,071	527,202
1954	100,801	43,540	233,021	21,973	6,006	304,540	28,084	50,761	484,186
1953	98,273	41,266	200,109	21,434	4,689	267,498	29,705	52,119	447,595
1952	102,813	40,487	196,255	23,491	3,146	263,379	30,393	48,758	445,343
1951	102,847	41,929	180,166	24,811	2,852	249,758	29,551	46,613	428,769
1950	94,662	43,958	149,526	25,295	2,168	220,947	28,403	44,313	388,325
1949	78,523	35,936	128,088	24,784	1,579	190,387	24,510	35,904	329,324
1948	84,163	30,630	127,637	39,986	1,771	200,024	24,418	34,221	342,826
1947	74,114	26,562	113,462	37,233	1,102	178,359	23,686	32,716	308,875
1946	60,564	20,275	86,172	31,886	1,304	139,637	19,795	29,939	249,935
1945	51,021	15,554	75,824	27,810	2,154	121,342	17,404	26,470	216,237

* Where use is primarily for heating residences, industrial, commercial, and other buildings.
† Includes kerosine and No. 1 distillate heating oil sold as range oil.
‡ Not reported separately after 1961.
§ Total includes kerosine sold as No. 1 distillate heating oil.

tailored by refiners to meet the needs of automatic central burners, largely in homes. Two grades of residual fuel oils, with gravities ranging down to 12°API, are used for space heating in burners with preheating facilities.   Sales of these oils, as classified for space heating, are shown in Table 11-28.[32]   The trend in sales for distillate and residual heating oils is shown in Fig. 11-9.

Until recently, about 80 per cent of refiners' production of kerosine has been used for space heating and allied domestic uses.   However, supplying jet aircraft fuels will now expand the proportion of kerosine going into other uses.   Of the refinery output of distillate fuel oils, over 65 per cent now goes into heating oils and a large proportion into diesel fuels.   Of residual fuel-oil production and imports into the United States, only about 20 per cent is classified as used in space heating.   Other large uses for residual fuel oil, in marine installations and in burners for heat and power generation, are discussed briefly in later portions of this section.

To a greater extent than other volume petroleum products, heating oils face outside competition for their markets.   This comes from natural gas, coal, and electricity, which are all accepted as space-heating fuels.   Electricity is likely to become a more important competitor, particularly in areas of low-cost power.   Other means of space heating, such as solar energy, atomic energy, and the fuel cell, are all possible future competition with oil for space-heating markets.

Relative cost is an important factor in the selection of any fuel for space or other heating purposes. Cost not only includes the price at the source of supply, as for natural gas in the producing fields, but also transportation and handling costs. The convenience of oil as fuel for heating, including storage and automatic deliveries through the degree-day system, is an important advantage which the oil companies have not used to the fullest advantage in meeting the competition of other fuels.

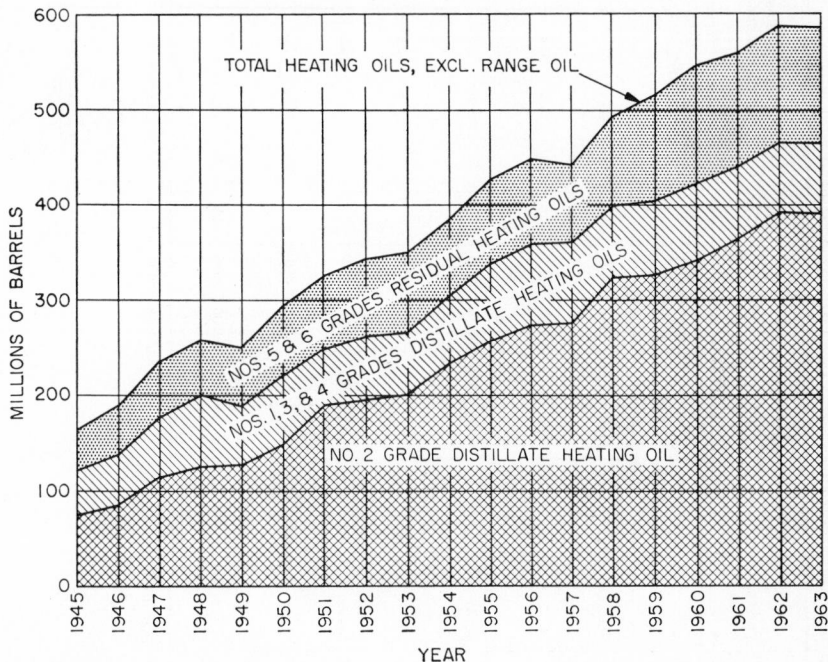

FIG. 11-9. Sales of No. 2 grade distillate heating oil compared with other grades of distillate and residual oils. See Table 11-28.

Proper consideration is not given, in many cases, to the cleanliness and convenience of using fuel oil. The liquid fuels offer distinct advantages in delivery and storage once suitable facilities are provided. Where portability is concerned, as in marine use, oil is the preferred fuel.

Heating-oil Specifications

Heating-oil specifications have been formalized by several agencies. The ASTM, in D 396,[33] has prepared tentative standards for six grades of fuel oils, the principal requirements for which are given in Table 11-29. Wider applications than for space heating are covered by these specifications. The ASTM Specification defines fuel oils as "hydrocarbon oils free from acid, grit and fibrous or other foreign matter likely to clog or injure the burner or valves." The ASTM Standards have been promulgated as commercial standards by the U.S. Department of Commerce.

The Federal government has issued specifications for five similar grades of fuel oils in its Fuel Oil, Burner Standard, VV-F-815, principal requirements for which are shown in Table 11-30.[34] Burner fuel is defined as a "homogeneous product consisting of hydrocarbon oil free from grit, acid, and fibrous or other foreign material likely to clog or injure burner or valves. Blending of various compatible grades of

Table 11-29. Detailed Requirements for Fuel Oils (D 396-63T)*

Grade of fuel oil	Flash point, °F Min	Pour point, °F Max	Water and sediment, % by volume Max	Carbon residue on 10% bottoms, % Max	Ash, % by weight Max	Distillation temperatures, °F 10% point Max	90% point Max	90% point Min	Saybolt viscosity, sec Universal at 100°F Max	Universal at 100°F Min	Furol at 122°F Max	Furol at 122°F Min	Kinematic viscosity, centistokes At 100°F Max	At 100°F Min	At 122°F Max	At 122°F Min	Gravity, °API Min	Copper strip corrosion Max
No. 1 {A distillate oil intended for vaporizing pot-type burners and other burners requiring this grade of fuel}	100 or legal	0	trace	0.15	....	420	550	....	37.93	32.6	....	....	2.2	1.4	....	....	35	No. 3
No. 2 {A distillate oil for general-purpose domestic heating for use in burners not requiring No. 1 fuel oil}	100 or legal	20	0.10	0.35	....	....	640	540	125	45	....	....	(3.6)	(2.0)	....	....	30	....
No. 4 {Preheating not usually required for handling or burning}	130 or legal	20	0.50	....	0.10	....	....	....	300	150	....	....	(26.4)	(5.8)	....	....		
No. 5 (Light) {Preheating may be required depending on climate and equipment}	130 or legal	....	1.00	....	0.10	....	....	....	750	350	....	....	(65)	(32)				
No. 5 (Heavy) {Preheating may be required for burning and, in cold climates, may be required for handling}	130 or legal	....	1.00	....	0.10	....	....	....	(9000)	(900)	(40)	(23)	(162)	(75)	(81)	(42)		
No. 6 {Preheating required for burning and handling}	150	....	2.00	....	....	....	....	....	....	....	300	45	....		(638)	(92)		

* Recognizing the necessity for low-sulfur fuel oils used in connection with heat-treatment, nonferrous metal, glass, and ceramic furnaces and other special uses, a sulfur requirement may be specified in accordance with the following table:

Grade of fuel oil	Sulfur, max, %
No. 1	0.5
No. 2	1.0
No. 4	No limit
No. 5	No limit
No. 6	No limit

Other sulfur limits may be specified only by mutual agreement between the purchaser and the seller.

Table 11-30. Physical and Chemical Requirements for Burner Oils in Government Specification, Fuel Oil, Burner, VV-F-815

F.S. grade No.	Description	Flash point, °F	Pour point, °F	Water and sediment, % Max	Carbon residue on 10% residuum, % Max	Ash, % Max	Distillation 10% point Max	90% point Max	End point Max	Saybolt Universal at 100°F Max	Min	Furol at 122°F Max	Min	Kinematic centistokes At 100°F Max	Min	At 122°F Max	Min	Gravity, °API Min
1	Distillate oil intended for vaporizing pot-type burners and other burners requiring this grade	100 or legal	0	Trace	0.08	.....	420	.....	625	.....	.....	.....	.....	2.2	1.4	.....	.....	35
2	Distillate oil for general purpose domestic heating for use in burners not requiring No. 1	100 or legal	5	0.10	0.33	.....	.....	675	.....	40	.....	.....	.....	4.3	.....	.....	.....	26
4	Oil for burner installations not equipped with preheating facilities	130 or legal	20	0.50	.....	0.10	.....	.....	.....	125	45	.....	.....	26.4	5.8	.....	.....	
5	Residual-type oil for burner installations equipped with preheating facilities	130 or legal	.....	1.00	.....	0.10	.....	.....	.....	.....	150	40	.....	.....	32.1	81	.....	
6	Oil for use in burners equipped with preheaters permitting a high-viscosity fuel	150 or legal	.....	2.00	.....	.....	.....	.....	.....	.....	.....	300	45	638	.....	638	92	

the material to produce an intermediate grade is permitted; however, such blending shall be accomplished by mechanical agitation or mixing in a tank prior to loading into transport equipment."

On the Pacific Coast, the accepted standards for burner oils are the so-called Pacific Specifications. These include two distillate oils, P.S. 100 and 200, which correspond in general to the ASTM Grades 1 and 2, and two grades of residual oils, P.S. 300 and 400. Grade 300 is for use without preheating.[35]

Kerosine and Distillate Fuels

Kerosine, better known as range oil in the space-heating field, is a broad term for the virgin refinery product of 43 to 45°API gravity and in the 350 to 550°F distillation range. It is also classified as Grade 1 distillate heating oil. The volume of sales for it are reported in Table 11-28.

In heating uses, kerosine, or range oil, has an advantage over other fuels when used in pot burners and range burners. Its name of range oil comes from its early use in perforated-sleeve or range burners in cooking ranges. In this type of burner, the oil is evaporated from a metal pot or base. It is the preferred fuel for this equipment, since it has a lower tendency to form carbon deposits than other grades of heating oils. As Grade 1 distillate heating oil, kerosine is also used in automatic central-heating installations.

Sales of four types of distillate heating oils are shown in Table 11-28. Sales of Grade 3 are relatively small and declining. Both the ASTM and the Federal government specifications for the Grade 3 have been dropped. It is a medium-light oil used primarily for domestic central heating in the western and some middle-western states. Grade 4 may be a blend of distillate and residual fuel oils.

Consumption of the Grade 2 is the largest of any of the heating oils, either distillate or residual. In fact, sales of Grade 2 in 1963, over 390 million barrels, were more than half the total sales of all types of heating oils. The growth in the use of Grade 2 heating oil has been rapid in the past decade. This trend, as shown in Fig. 11-9, is likely to continue and to replace virgin kerosine and Grade 1 because of the growing demands on kerosine-type fuels for jet aircraft.

Types of Burners and Controls

The high-pressure gun burner is used in about 80 per cent of the oil-fired heating equipment now in the field.[34] Sales of low-pressure burners were 11 per cent in 1957 and of vertical rotary burners 5 per cent. Sales of vaporizing burners of the pot and range type were about 1 per cent of the total in 1957.

A typical high-pressure gun-burner design is shown in Fig. 11-10. The burner pumps oil at 100 psig to a nozzle from which it is fed into a combustion chamber as a conical pattern of fine drops. Figure 11-11 illustrates a typical high-pressure atomiz-

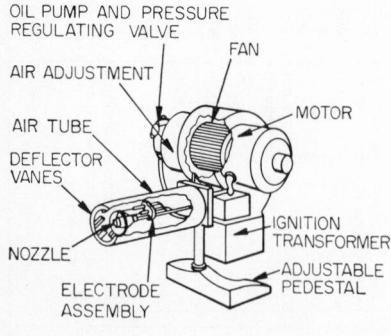

FIG. 11-10. High-pressure atomizing oil burner.

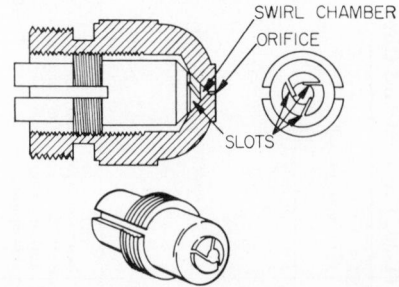

FIG. 11-11. Typical high-pressure atomizing nozzle.

ing nozzle design.[34] The nozzle is mounted in the center of a blast tube, which carries combustion air supplied by an integral squirrel-cage fan into the burning zone. Ignition is provided by a spark between two electrodes generated by a built-in ignition transformer. A shutter is provided on the fan inlet to control the supply of combustion air.

Three important efficiency concepts for oil burners are (1) seasonal efficiency, (2) thermal efficiency, and (3) running or absorption efficiency. For home heating units, the seasonal efficiency is essentially the ratio of the heat needed to the heat energy of the fuel supplied.[34]

For field use, the best index of the performance of a central-heating unit is the absorption efficiency or running efficiency. This test provides a measure of the heat loss from a system in the flue gas during burner operations. This is the main heat loss affecting normal operation, and it is thus a useful performance criterion.

Running efficiency is also called "CO_2-stack temperature efficiency." Two measurements are required to determine absorption efficiency: an analysis of the flue gases for CO_2 and the net stack temperature, that is, the temperature difference between the flue gases and the room air. Analysis for smoke from a central oil heating installation is also an essential of a burner performance test. Practical kits are available to the oil company's service men for determining the CO_2 in flue gases, thermometers for determining stack temperatures, and for the analysis of smoke in flue gases.

Smoke Density. A device for evaluating the smoke density in flue gases of equipment burning distillate fuels has been developed. It may be used in the field as well as the laboratory. A test smoke spot is obtained by pulling a fixed volume of flue gas through a fixed area of standard filter paper. The color or shade of the spot thus produced is visually matched with a standard scale, and the smoke density is expressed as smoke spot number.

Properties Affecting Burning Qualities

The aromatic, paraffinic, and naphthenic contents of heating oils influence their performances. These contents are particularly important because they affect smoking tendency for critical equipment and heat of combustion.[34] Tests to determine percentages of the three hydrocarbon classes are difficult and expensive and are seldom used as control tests. But there is a relationship between gravity and paraffin content. Highly paraffinic stocks usually have high API gravities. However, gravity is only an indication of burning quality and its main value is to check product uniformity.

Diesel Index. This is a useful laboratory method to predict the burning qualities of distillate fuels. This relationship is determined by testing for both aniline point and API gravity. The aniline point, ASTM D 611,[33] is often used as a relative measure of the aromatic content of the sample. The smoke-point test of the Institute of Petroleum (British)[36] is used widely for measuring smoking tendency in kerosine. Heat-of-combustion values, although seldom used in specifications, are necessary when the efficiency of a burner unit is determined. Except where extreme accuracy is required, gross and net heating values can be determined from available tables.[37]

Viscosity. This property regulates the rate of fuel flow to the burner and therefore controls the heat output. The reaction to viscosity changes differs in different types of burners. In atomizing burners, viscosity affects the quality of the oil spray. Viscous oils tend to reduce the oil-spray angle and also to form large droplets.

The viscosity of No. 1 heating oil is usually determined by the kinematic viscosity test, ASTM D 445; for No. 2 oil, the Saybolt test is used, ASTM D 88.[33] The use of the kinematic viscosity test for this fuel is growing.

Carbon Deposits. Some distillate fuel oils will not burn completely without forming tarry, cokelike deposits which may collect on the internal parts of burners. Vaporizing burners are more critical of carbon formation than atomizing-gun burners. Two tests are used to measure the carbon-forming tendency of light burner oils: the wick char and the carbon residual tests. The wick char test method, developed by the British Institute of Petroleum,[36] is the most commonly used for kerosine.

There are two recognized ASTM carbon residue tests, the Ramsbottom (ASTM D 524) and the Conradson (ASTM D 189). Since No. 1 and No. 2 oils are commonly low in carbon residue, both of these tests are usually made on the 10 per cent residue remaining after 90 per cent of the sample has been distilled off.[33]

Volatility. Distillate heating oils volatility must be kept uniform to avoid the necessity of changing burner settings. A good distillate heating oil must contain enough light or volatile material to be ignited easily but not heavy nonvolatile material which would be difficult to burn.

The ASTM standard distillation test, ASTM D 86,[33] is used to control the volatility of light burning oils. The usual criterion for kerosines is the 10 per cent boiling point and the end-point temperature. For No. 2 fuel oil, the 90 per cent point is the more critical temperature.

Sulfur Content. Sulfur in heating oils can cause corrosion in boilers and flues under certain temperature conditions in the stack. Typical heating oils contain less then 0.5 per cent sulfur (Table 11-31), and field experience shows that there are no significant corrosion problems with such oils when stack temperatures are over 400°F, typical of home heating units in the United States.

Table 11-31. Characteristics of Three Grades of Distillate Heating Oils Sold in Eastern United States in 1960*

Property	Grade 1	Grade 2	Grade 4
Gravity, °API......................	42.6	34.9	21.2
Viscosity at 100°F, cs...............	1.79	2.61	15.41
Sulfur, wt %......................	0.071	0.249	0.77
Ramsbottom carbon residue, wt %....	0.052	0.116	3.30
Distillation, °F:			
Initial boiling point...............	349	370	422
10% point......................	390	432	496
50% point......................	437	499	674
Final boiling point...............	533	629	754

* Authority: O. C. Blade, *Burner Fuel Oils*, Bureau of Mines, Bartlesville, Okla., 1960.

Low-sulfur distillate fuels are used in connection with heat treatment and other special industrial purposes. Where such uses are intended, a maximum sulfur content in No. 1 fuel of 0.5 weight per cent may be specified and for No. 2 fuel a 1.0 weight per cent. Usually, sulfur content runs lower than this.

Storage Stability. Distillate fuel oils may be subject to long periods of storage in refiners' or distributors' tanks before they are burned. Under such conditions the oils, if unstable and chemically reactive, may form degradation products which deposit in storage tanks, distribution systems, or burners. The color of the product may be affected. Instability became a problem some 15 years ago when cracked stocks in the middle distillate range came into use to meet the fast-growing demand for heating oils.

Modern refining and treating techniques have gone far to make the cracked distillate oils stable in storage. It is now common practice to use additives to help control distillate fuel-oil stability. The composition of these additives varies widely. They include sulfonates, polymers, amines, metal phenolates, and combinations of these.[34]

Two types of additives can be used, particularly to control sediment stability. The dispersing type does not prevent the formation of sediment or sludge in the oil but rather keeps the sediment well dispersed in the oil and reduces sediment size so it can pass through filter and burner parts without plugging. The sediment inhibitors, on the other hand, eliminate or reduce greatly the formation of organic sediment in fuels. In actual practice, many heating-oil additives consist of both a dispersant and a sediment inhibitor.

There is no standard method of test for determining the stability of heating oils in storage. Several tests which have been developed by oil companies are in use today. These are of two types: long-term storage tests and accelerated aging tests.[34]

Residual Fuel Oils

The consumption of residual fuel oils for space heating was 23 per cent of this product in 1963. These sales include a large amount of imports as well as domestic refinery output. Table 11-32 shows the sales of residual fuel oil by principal uses. The use of residual fuel oil in steam-powered vessels is larger in volume than that used for space heating. Consumption in industrial applications, including fuel for oil company use, is more than twice as large as the volume used in space heating. In addition to firing steam boilers, residual fuel oils are used in other industrial applications requiring the generation of heat, as in open-hearth furnaces and soaking pits in the steel mills, in kilns in the cement and related industries, and in furnaces n petroleum refining and chemical manufacturing.

Table 11-32. U.S. Sales of Residual Fuel Oils by Principal Uses, 1945 to 1963*
(Thousands of barrels)

Year	Vessels	Railroads	Industrial uses†	Heating oils	Gas and electric power plants	Miscellaneous‡	Total§
1963	76,502	5,342	196,245	125,248	99,615	43,570	538,522
1962	84,415	5,501	201,463	125,164	88,261	42,893	547,697
1961	87,308	5,347	196,765	121,097	87,881	43,188	541,586
1960	94,084	5,610	200,667	125,088	85,408	37,565	548,872
1959	102,049	5,613	213,878	111,850	82,208	38,754	554,352
1958	106,269	5,772	202,697	93,118	76,424	47,087	531,367
1957	123,651	6,953	217,038	81,412	75,950	38,946	543,950
1956	117,445	10,575	231,078	87,601	73,962	40,877	561,538
1955	115,128	15,018	226,417	86,282	75,966	38,172	556,983
1954	108,790	16,122	212,286	78,845	70,749	33,922	520,714
1953	114,324	28,477	217,991	81,824	85,352	36,761	564,729
1952	110,412	40,489	212,794	79,151	70,497	42,930	556,273
1951	107,007	54,998	211,335	76,164	70,550	43,334	563,388
1950	92,947	60,878	201,374	72,716	93,062	33,231	554,208
1949	89,362	63,467	174,300	60,414	80,092	27,298	494,933
1948	95,763	89,588	174,417	58,639	56,812	31,278	506,497
1947	101,900	97,500	177,757	56,402	60,964	26,006	520,529
1946	88,185	100,305	157,065	49,734	50,921	40,910	487,060
1945	100,365	112,297	148,512	43,874	34,532	102,685	542,265

* Source: American Petroleum Institute, *Petroleum Facts and Figures*, 1963.
† Includes fuel for oil company use.
‡ Includes military use.
§ Includes heating oil. See Table 11-28.

In 1945, railroad locomotives used a larger volume of residual fuel oil to generate steam than the total required today for space heating. But the dieselization of the railroads has reduced this market for heavy fuel oils to a few million barrels per year. Another wide use of residual fuel oils is in electric power plants and, to some extent, in the generation of gas to distribute for household use or to enrich manufactured gas.

In most applications, the fuel oil is burned to produce heat. The necessary appliances include the preheater, the burner itself, and suitable control mechanisms. The quality of the oil must be such that the burner may fulfill five basic requirements:[38]

1. Deliver the oil into the combustion zone in finely divided particles
2. Introduce the air required for combustion
3. Mix the air and finely divided fuel oil in the right proportions
4. Form a flame of the desired size and shape
5. Complete combustion in the available time and space

The two accepted commercial grades of residual fuel oil marketed in the United States (except on the Pacific Coast) are Grades 5 and 6 of the ASTM and the Federal government. Grade 5 is described as "a residual type oil for burner installations equipped with preheating facilities" and the Grade 6 as "an oil for use in burners equipped with preheaters permitting a high viscosity fuel." The Pacific Specification Grade 300 calls for "a residual fuel oil for use without preheating in furnaces and burners requiring a low-viscosity oil." Grade 400 requires "a residual fuel oil for use in furnaces and burners equipped with preheaters permitting a high viscosity oil."[33]

For space heating, Grade 5 is described as "a medium-heavy oil used primarily for heating industrial and larger commercial buildings and usually requiring preheating." Grade 6 is described as "a heavy oil used primarily for heating large commercial and industrial installations, usually requiring preheating."[32] Table 11-33 shows the characteristics of Grades 5 and 6 of residual fuel oil sold in 1962.

Table 11-33. Characteristics of Two Residual-type Heating Oils Sold in Eastern United States in 1962*

Property	Grade 5	Grade 6
Gravity, °API	17.1	12.3
Viscosity:		
Kinematic at 100°F, cs	60.2	
Furol at 122°F, sec	25.8	170.2
Sulfur content, %	1.07	1.33
Ramsbottom carbon residue on 100% sample, %	6.7	10.7
Ash, %	0.035	0.41
Water and sediment, vol %	0.16	0.15

* Authority: O. C. Blade, *Burner Fuel Oils*, Bureau of Mines, Bartlesville, Okla., 1962.

Residual-fuel-oil Specifications

The properties that influence the performances of the residual fuel oils when burned to produce heat, either for space heating or most other uses, are:

Heat of Combustion. The heating value (heat of combustion) is often the primary consideration in the choice of oil over competitive fuels—coal or natural gas. This can be determined in the laboratory by means of a bomb calorimeter test, ASTM D 240.[33] Heat of combustion at constant volume of fuel oils and pure hydrocarbons can be estimated from their specific gravities within an accuracy of about 1 per cent. Tables for calculating air requirements and fuel-gas analysis are available.[38]

Specific Heat. This property is sometimes required in estimating the amount of heat necessary to preheat the fuel. Specific heat can be estimated with reasonable accuracy from the specific gravity of the fuel oil by established relationships.[37]

Volatility. The fact that residual fuel oils contain both high- and low-boiling components may present a problem unless the fuel selected is suited to the particular furnace. The volatile portions may vaporize and burn without leaving sufficient time for the complete combustion of the heavy portions. When the fuels are blends of vacuum-still residuum and distillate fuel oil, the amount of distillate may be approximated by the percentage of the sample distilled at 650°F, which is the approximate end point of the distillate cutter stock.[38]

Sulfur. High sulfur contents in residual oils may present problems through atmospheric pollution and through corrosion in furnaces and boilers. Operation of boiler installations at temperatures above the dew point of the stack gases, while minimizing corrosion, may result in reduced efficiency of the installation. Magnesium and calcium compounds, such as magnesia, lime dolomite, and calcium or magnesium soaps, are effective if used in adequate concentrations.

Contaminants in Ash. High vanadium and sodium contents in residual fuel oils present problems in some installations through their concentration in the ash.

Sodium is largely removed in the refining steps. No economical means for removing vanadium from residual fuel oil is yet available. Most fuel oils prepared from crude oils produced in the United States are low in vanadium.

The selection of a suitable corrosion-resistant alloy for critical boiler parts is expensive, and the material chosen must give optimum resistance to ash of the particular composition expected in the fuel oil to be burned. Corrosion and fouling can be largely eliminated by the use of magnesium-bearing additives in the fuel oil.

Instability and Incompatibility. Various types of instability occur in residual fuel oils, developing particularly during periods of storage. Incompatibility appears most marked when fuels containing a large proportion of thermally cracked tars are blended with fuels made from highly paraffinic long residues.

ASTM Test Procedure D 96[33] is a centrifuge method to determine the water and sediment in fuel oils, as well as in crude oils. Incompatibility may be indicated if the fuel blend has a higher water and sediment content than either of its components.

Other special properties are required in residual fuel oils supplied to the Navy or other military services. Two grades, Navy Special and Heavy, are covered in the Navy Residual Fuel Oil Specifications, MIL-F-859D.[38] The Special Grade is intended for all steam-powered vessels of the Navy; the Heavy Grade is for use in other steam-powered vessels in government service, for shore plants, and for emergency use in naval vessels. Both grades are lower in viscosity than Grade 6 in the ASTM Specifications.

A fluidity test is included to give a better indication of pumpability than the pour test alone. A thermal stability test is provided to indicate fuels which are likely to cause preheater fouling. An explosiveness test is included to determine the explosiveness of atmospheres in storage facilities on shipboard.

MISCELLANEOUS LIGHT OILS

Kerosine

Kerosine was first processed from crude oil in the early Pennsylvania oil fields, before the Drake well was drilled in 1859. Then known as coal oil because a similar product was being distilled from coal, it took the place of the dwindling supplies of whale oil as an illuminant.

Kerosine has found many new uses over the years, and its output has grown steadily. Today it is over 157,000,000 bbl annually. The largest volume uses now are as heating oils and as fuel for gas-turbine-powered aircraft. Both these uses were discussed earlier in this section.

Kerosine is still consumed in considerable volume as an illuminating oil, although its sales for this purpose are not classified separately from other uses. In this field, it serves many essential uses—in lighthouse beacons, miners' lamps, railroad signal systems, and field camps. Closely allied uses are in cooking ranges, water heaters, and other home appliances.

Specifications

Commercial kerosine is defined by the ASTM as a "refined petroleum distillate suitable for use as an illuminant when burned in a wick lamp."[39] The ASTM does not publish a specification for kerosine. The Federal government provides specifications for three grades for use as illuminants. One grade, VV-K-211c,[40] requires that the product be a "petroleum fraction, free from water, additives, foreign and/or suspended matter and suitable for use as an illuminating oil." Specification VV-O-391,[41] mineral seal oil, requires a "petroleum fraction, free from water, glue and suspended matter, suitable for use where a high-flash illuminating oil is required." Specification VV-O-381 is for a long-time burning oil. Physical and chemical properties of the three grades for the Federal specifications are shown in Table 11-34.

Performance Properties

The nature of kerosine and its wide use as an illuminating oil require not only that the product have the proper burning qualities, as determined by scientific test methods,

Table 11-34. Physical and Chemical Requirements Specified by the Federal Government for Three Grades of Illuminating Oil and Designated Test Methods

Property	Kerosine		300 mineral seal		Long-time burning	
	Limit	ASTM Test Method	Limit	ASTM Test Method	Limit	ASTM Test Method
Flash point, °F min........	115	D 56	250	D 92	115	D 56
Burning test, hr, min.......	16	D 187	20	D 239	120–156	D 219
Sulfur, % max.............	0.13	D 129	.....	......	0.10	D 129
Color, Saybolt chrom, no darker than.............	+21	D 156	+16*	D 156	+21*	D 156
Color, Saybolt chrom, after heating 16 hr, no darker than...................	+16	D 156				
End point, °F max........	572	D 86	.....	......	599	D 86
Cloud point, °F...........	5	D 97	32†	D 97	0†	D 97

* Minimum.
† Maximum.

but also that it be safe to transport, handle, and use in lamps and burners in homes and other buildings. Specifications to ensure both these requirements have been in effect for many years in many states and municipalities. Some states still inspect kerosine for safety purposes. Flash point and other properties are legally prescribed and must be met by the product sold there, regardless of the use to which the kerosine is put.

Distillation range, particularly final boiling point, and sulfur content are important properties of kerosine when used as an illuminant, and they are included in the requirements of some states. Color is sometimes included, for it is the first property that comes to the attention of the consumer, even though it has little significance in the performance of the product. Color also can indicate a long storage period for the product. The property requirements in state specifications for kerosine are summarized in Table 11-35.[40]

The important performance properties for kerosine and the tests by which they are determined are:

Flash Point. The lowest temperature at which vapors arising from the oil sample will ignite momentarily, or flash, on application of a flame under the prescribed test conditions in ASTM D 56[39] is an important safety as well as performance property. The minimum allowable flash temperature is generally placed above the prevailing ambient temperature.

The inclusion of the flash-point requirement in kerosine specifications goes back to the early days of the petroleum industry when this oil was the principal product made at the refinery and gasoline was a byproduct. The natural tendency then was to include in what was sold as kerosine some of the more volatile refinery fractions which today go into gasoline. Explosions resulting from the transportation and use of kerosine in lamps, cook stoves, and heaters led to the adoption of means to reduce this hazard. This was done by setting up a minimum flash-point requirement which was, in effect, the lower limit of flammability of the product. In the government property standards and in those of many states, the stipulated minimum flash point is 115°F.

Burning Test. This procedure determines the ability of a kerosine to burn steadily and cleanly over a stipulated number of hours. The average rate of burning, change in the shape of the flame, density and color of the chimney deposit, and condition of the wick are the factors by which the burning quality of the oil is determined. The standard test for kerosine (ASTM D 187)[39] provides that the oil be burned in a wick lamp for a specified period in a room free from drafts.   The test period in the government and most state specifications is 16 hr.   More severe test methods (ASTM D 219) are used for determining the burning quality of the special grade of kerosine used in semaphore signal lamps, with a test period of 120 to 144 hr.

Where standard burning test equipment is not available, it is pointed out,[40] or if there is not enough time to make extended burning tests, a reasonably reliable idea of the quality of an illuminating oil for a particular purpose can be obtained from a judicious consideration of its gravity, flash point, viscosity, pour point, sulfur content, color, distillation range, and cloud point.

Table 11-35. Properties for Kerosine Specified in State Inspection Laws[40]

State	Flash point, °F, min	End point, °F, max	Sulfur %, max	Color, Saybolt, no darker than	Burning test hr, min	Cloud point, °F
Alabama...............	115	572	0.125	+21		
Arkansas..............	140*					
Colorado.............	115	625	0.20	+16	16	5
Florida...............	115	572	0.13	+16		
Georgia...............	115	572	0.125	+16	16	
Idaho.................	115	626	0.13	+16	16	5
Illinois...............	115	550	.....	+16		
Indiana...............	120	550	.....	+16		
Kansas................	115					
Louisiana.............	115	625	0.125	+16	16	5
Minnesota............	120	600				
Mississippi...........	115	575	0.25	+16	16	5
Montana..............	110					
Nebraska.............	115	625	0.125	+16	16	5
New Hampshire........	110					
New Mexico...........	115	550	0.30	....	16	
New York.............	100					
North Carolina........	115	572	0.13			
North Dakota.........	115	572	0.13	+16	16	5
Oklahoma.............	115					
Pennsylvania..........	110*					
South Carolina........	110†	570				
South Dakota.........	115	572	0.13	+16		
Tennessee.............	112	600	0.13	+16	16	5
Wisconsin.............	110	572	.....	+16		
Wyoming.............	115	625	0.125	+16	16	5

* Fire point.
† Elliot closed-cup method.

Smoke Point.   The most widely used test for measuring the smoking tendency of kerosine is the smoke test of the Institute of Petroleum.[42]   The results provide some indication of the hydrocarbon composition of the fuel.

Distillation Test.   The boiling range of kerosine is an indication of the viscosity of the product.   The Federal government and most state specifications are limited to the final boiling point as determined by ASTM D 86.

Viscosity.   There is no viscosity requirement in the specifications of the Federal government or states.   The ASTM states on this subject:[43]

"For illuminating oils, the value of a viscosity test is mainly to determine whether they may be of abnormally high viscosity, which may not allow them to flow properly through the wicks of lamps and burners; or may be of very low viscosity, which may cause unsteady flames.   The performance of burning oils is influenced less by viscosity than by the tendency to encrust the wick."

11-54 PETROLEUM PROCESSING HANDBOOK

Sulfur. The government standards provide a maximum limitation of 0.13 per cent, as determined by ASTM D 129. State requirements vary from 0.125 to 0.3 per cent maximum. In general, the sulfur content in kerosine is important only when the oil is to be burned under conditions where the sulfur oxides that may be produced during combustion must be limited, as in lamps and in heating equipment not connected by a flue to the outside atmosphere.

Color. This property has little significance in indicating performance quality, but it is included in most specifications because it comes to the attention of the consumer and it is desirable to market a kerosine of uniform color. Color is determined by ASTM D 156.

Cloud Point. This test of an illuminating oil or a heating oil to be used in a wick appliance indicates the temperature at which the wick may become coated with wax particles which may lower the burning qualities. Most specifications call for a 5°F cloud point, as determined by ASTM D 97.

Industrial Naphthas

The refiners' ability to manufacture a large number of liquid hydrocarbon compounds in a narrow range of boiling points and with high degrees of purity from natural gas and petroleum has developed a wide and growing market for petroleum oils as industrial naphthas. The term includes solvents, thinners (for example of the type used in the paints and varnish industries), and diluents (as used in the manufacture of pharmaceuticals and insecticides).

The original industrial naphtha from petroleum was a selected naphtha from the straight-run processing of certain crude oils. This classification of petroleum products today includes also pentane, hexane, and heptane produced largely from natural gas and the aromatics—benzene, toluene, and xylene—now recovered from petroleum by modern refining methods. The industrial naphthas have a wide variety of properties and serve many industrial and domestic uses. The total volume used has grown to the point that one authority[44] states that in the overall volume employed these products now rank second to the universal solvent water. The wide applications for the industrial naphthas are given in Table 11-36.

Table 11-36. Typical Applications for Industrial Naphthas[44]

Oil extraction	Dry cleaning	Textile manufacturing
Adhesives	Printing inks	Leather goods
Asphalt compounds	Rubber industry	Degreasing
Papermaking	Metal cleaning	Alcohol denaturant
Floor coverings	Rosin extraction	Silicone compounds
Wax compounds	Type cleaning	Rustproofing compounds
Metallic dryers	Textile printing	Polishes
Chlorinated rubber	Machine cleaning	Coatings
Brake linings	Resin solutions	Chemical intermediates
Pharmaceuticals	Desludging solvents	Pesticides

Scientific achievements in many solvent-consuming industries have contributed to the rapid growth in the volume of these products. The development of the entire synthetic-resin industry and the wide use of these resins would have been greatly curtailed had not the industrial naphthas essential to these operations been available. The developments in the consuming industries and by the oil companies have complemented one another. The industrial naphthas are consistently lower in cost than the organic solvents, such as the alcohols, ketones, and esters.

Industrial naphthas are usually highly complex mixtures of hydrocarbons, except for the relatively pure compounds such as hexane or toluene, and in many instances do not lend themselves to scientific classification based on their properties. They vary from paraffin-type hydrocarbons to the aromatic, and in some cases, they may contain naphthenes and olefins. Through usage, a classification has evolved which is generally used today. These classes are aliphatics, aromatics, intermediates, and odorless. It is common practice to divide the aliphatics into paraffins and naphthenes.

In general,[44] when the dominating hydrocarbons comprising an industrial naphtha are paraffinic and/or naphthenic, the naphtha is classified as an aliphatic. More specific, however, is the practice in industry of classifying the industrial naphthas based on their solvency, as determined by the Kari-Butanol laboratory test.[39] Naphthas with a Kari-Butanol number of 45 or lower are classified as aliphatics without regard to their chemical compositions. The aromatics, the industrial grades of benzene, toluene, and xylene, have Kari-Butanol solvency numbers of 98 or higher. This classification includes also a few special naphthas, such as the alkyl naphthalenes. The intermediate classification covers the straight-run and blended naphthas with Kari-Butanol solvency numbers between 45 and 98. A fourth classification, the odorless solvents, has been set up to cover a relatively new grade of industrial naphthas. They are characterized not by their solvency but by being, for all practical purposes, free from odor.

Specifications

To date, 15 industrial solvents have been standardized in their specifications listed by the ASTM. These are listed in Table 11-37, and the properties of three grades

Table 11-37. List by Title of the Industrial Solvents and Related Materials for Which Specifications Have Been Prepared by the American Society for Testing and Materials

Name	Designation
Prepared by ASTM Committee D-2 on Petroleum Products and Lubricants	
Petroleum spirits (mineral spirits)	D 235
Heavy petroleum spirits (heavy mineral spirits)	D 965*
Stoddard solvent	D 484
Prepared by ASTM Committee D-16 on Industrial Aromatic Hydrocarbons and Related Materials	
Nitration-grade benzene	D 835
Industrial-grade benzene	D 836
Industrial 90 benzene	D 837*
Refined solvent naphtha	D 838
Crude light solvent naphtha	D 839
Crude heavy solvent naphtha	D 840
Nitration-grade toluene	D 841
Industrial-grade toluene	D 842
Nitration-grade xylene	D 843
Industrial-grade xylene	D 844
Five-degree xylene	D 845
Ten-degree xylene	D 846

* Tentative.

prepared by the ASTM Committee D-2 on Petroleum Products and Lubricants are given in Table 11-38.[39] The Federal government publishes specifications for over 20 industrial solvents and related products.[41] Figure 11-12 illustrates the distillation ranges of a representative group of the widely used aliphatic industrial naphthas.

Performance word Tests

Solvency. This is one of the most important characteristics required of the industrial naphthas, since much of the use of these products is as a solvent. The wide range in the nature of the products themselves, together with the large number of materials for which industries require solvents, has made it impossible to set up a single standard test method for evaluating this characteristic in a naphtha. In many instances, individual consuming companies have developed methods for determining solvency for their own requirements.

In general, ASTM standard test methods are used to indicate the qualitative and

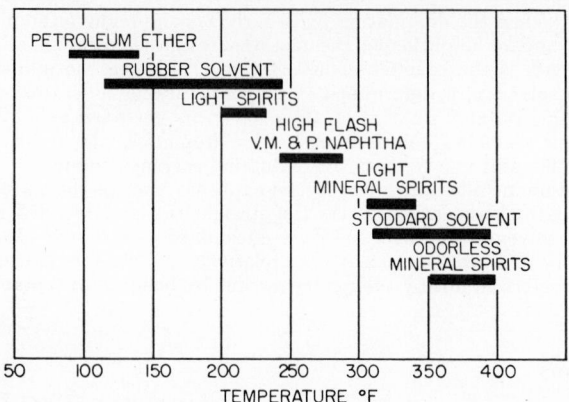

FIG. 11-12. Distillation ranges of a group of aliphatic industrial naphthas.

relative abilities of naphthas to hold in solution the more generally employed solutes, particularly those used in the protective coating industry.   These methods are:

1. Aniline Point and Mixed Aniline Point of Hydrocarbon Solvents, ASTM D 1012[39]
2. Kari-Butanol Value of Hydrocarbon Solvents, ASTM D 1133
3. Nitrocellulose Diluting Power of Hydrocarbon Solvents, ASTM D 1134
4. Heptane Number of Hydrocarbon Solvents, ASTM D 1132
5. Viscosity Reduction Power of Hydrocarbon Solvents, ASTM D 1311

In general, the higher the aromatic content of an industrial naphtha, the greater its solvency.   Hence, an analysis of the naphtha to determine its "hydrocarbon breakdown" is desirable in determining the suitability of the material for some applications.   Several standard ASTM procedures are available to indicate the nature of hydrocarbons present.

Distillation Range.   This test procedure is used to supply a relative indication of the evaporation rate of one naphtha as compared with another.   The distillation range of the aliphatic naphthas is usually determined by ASTM Test Method D 86; that of the aromatics, benzene, toluene, and xylene, by ASTM D 850.   ASTM D 1078 is a special distillation procedure to determine the distillation range of lacquer solvents. Several of the naphthas derived from natural gasoline are so volatile that a special procedure for determining their boiling range has been standardized, ASTM D 216.

Evaporation Rate.   This property is related to volatility and is of significance to many consuming industries, since it provides information on the time required for any specific naphtha to dry completely.   Many variable factors are involved in the determination of this property of the solvent, such as the ambient temperature, atmospheric pressure, relative humidity, vapor pressure of the solvent at its boiling point, and heat of vaporization.   Several test procedures for determining evaporation rate have been developed, but because of the complexities involved, no method has yet been standardized.   Among these are the Amsco method,[44] the modified Jolly balance,[44] and the method using the thin-film evaporometer developed by Shell Oil Co.

Gravity and Density.   The primary use of either the specific-gravity or the API-gravity tests is to establish volume-temperature and volume-weight relationships which are essential in commercial transactions.   These relationships are advantageous for those consuming industries whose formulas are established on a pound basis rather than a gallon basis.   The accepted methods for determining the necessary volume-temperature and volume-weight relationships are the ASTM-IP Petroleum Measurement Tables, ASTM D 1250.[39]   Standard ASTM methods are available for establishing the specific gravity and the API gravities of naphthas.

Volatility.   This property is of major importance in considering the application of the industrial naphtha for a specific purpose.   It is also the basis of municipal

Table 11-38. Properties of Three Industrial Solvents for Which Specifications Have Been Prepared by ASTM Committee D-2 on Petroleum Products and Lubricants

Petroleum Spirits (Mineral Spirits), D 235

Scope. These specifications apply only to petroleum distillates.
Properties. The material shall conform to the following requirements:

Appearance.............................	Clear and free of suspended matter and water
Color................................	Water white, not darker than No. 21 Saybolt Chromometer
Flash point, min.......................	100°F (38°C)
Blackening............................	Shall not blacken or corrode clean metallic copper in 30 min. at boiling point of spirits

Distillation:
% recovered at 350°F (177°C), min.....	50
End point...........................	410°F (210°C)
Acidity of distillation residue.............	Neutral

Heavy Petroleum Spirits (Heavy Mineral Spirits), D 965*

Scope. These specifications apply only to petroleum distillates used as thinners for slow-setting paints and varnishes.
Properties. The materials shall conform to the following requirements:

Appearance...........................	Clear and free of suspended matter and water
Color................................	Water white, not darker than No. 21 Saybolt chromometer
Flash point, min.......................	125°F (51.5°C)
Blackening............................	Shall not blacken or corrode clean metallic copper in 30 min at boiling point of spirits

Distillation:
Initial boiling point, min..............	340°F (171°C)
95% (by volume)....................	460°F (238°C)
End point, max.....................	485°F (251.5°C)
Acidity of distillation residue.............	Neutral

* Tentative.

Stoddard Solvent, D 484

Scope. These specifications cover a grade of petroleum distillate of low flammability used in dry cleaning.
General Requirements. Stoddard solvent shall be a petroleum distillate, clear and free from suspended matter and undissolved water, and free from rancid and objectionable odor. The odor shall be typical of a "sweet" refined naphtha.
Detailed Requirements. Stoddard solvent shall conform to the following detailed requirements:

Color.......................................	Water white or not darker than 21
Corrosion at 100°C..........................	Not more than extremely slight discoloration of the copper test strip, or shows no greater corrosion than a mutually approved reference strip
Doctor test..................................	Negative
Sulfuric-acid absorption, max, %................	5
Flash point, min, °F..........................	100

Distillation:
Percentage recovered at 350°F (176°C), min....	50
Percentage recovered at 375°F (190°C), min....	90
End point, max, °F..........................	410
Residue, max, %............................	1.5
Acidity......................................	No acid reaction to methyl orange shown by residue from distillation

and other regulations to minimize fire and explosion hazards. The test procedure generally used to determine volatility is the flash point. The three standard procedures for determining flash point are:

1. The ASTM Closed Cup (Tag Closed Cup), ASTM D 56, normally used for naphthas flashing below 175°F.
2. The Pensky-Martens Closed Cup, ASTM D 93, to determine the flash point of heavy fuel oils. Its application to industrial naphthas is limited to unusually high-boiling-point products.
3. Cleveland Open Cup, ASTM D 92, used to determine the flash point, as well as the fire point, of petroleum products except fuel oils and those products having an open-cup flash point below 175°F.

Color. The color of an industrial naphtha is important to many consumers as well as producers of the product. This is particularly true in the paint, varnish, lacquer, and dry-cleaning industries. For naphthas having a "water-white" to "straw" color, color is determined by the ASTM Tentative Method of Test for Saybolt Color of Refined Petroleum Products, ASTM D 156-53T. For naphthas darker in color than pale straw, the color determination is made by the ASTM Tentative Method of Test for Color of Lubricating Oils and Petrolatum, ASTM D 1500-58T.

Distribution and Handling

Originally the industrial naphthas were marketed by the oil companies manufacturing them through the same channels as for conventional petroleum products. In the past 25 years, however, the marketing of the solvents has become more specialized. Some oil companies have separate divisions or subsidiary companies to handle them. The industrial naphthas occupy the same position as chemical raw materials used by consuming industries, and their distribution in some cases is through the same channels as the chemicals.

The nature of the industrial naphthas and the conditions under which they are used provide many problems which are not encountered with the conventional petroleum products. They must be held to uniform specifications from the refiner to the consumer. Storage tanks, pumps, transfer lines, and all transporting equipment must be segregated from that used for other products. Because of the flammable and sometimes toxic characteristics of the industrial naphthas, municipalities and other agencies have established regulations which must be observed by both the distributor and the consumer of the products. Liaison research between manufacturer and consumer is sometimes required to meet new applications of the naphthas, and technical sales service often supplements other sales methods. Fortunately, the special conditions encountered in handling and utilization of the industrial naphthas are widely recognized by various technical agencies, and test methods, product specifications, and handling procedures have been set up on sound bases. The ASTM and the Federation of Paint and Varnish Production Clubs have standardized test procedures. Data on toxic properties is available through the API,[45] the U.S. Public Health Service, and the American Conference of Governmental Hygienists. Regulations for transportation and handling are supplied by the Interstate Commerce Commission and municipal and other public agencies.

Petroleum Pesticides

The insecticidal properties of petroleum have been known since the eighteenth century. However, the expansion of the so-called pesticide industry for control of the losses in crops, livestock, industrial supplies, and household furnishings has come in the years since World War II. Through military research came DDT, the aerosol bomb, and practical knowledge of the use of petroleum solvents and other products in this field. The pesticide business is estimated to have today an annual sales volume of over $500 million annually.

The term pesticide is usually applied to those products which are used for killing insects and other orthropods and weeds and fungi.[46] Individually, the control products are known as insecticides, herbicides, and fungicides. In a broad sense the

Table 11-39. Petroleum Products Used as Pesticides

INSECTICIDES

Emulsified spray oils for fruit trees:
Lubricating-type oils, 70–120 viscosity, SSU at 100°F

Geneva classification	API gravity	SSU at 100°F	Unsulfonated residue, %
Dormant...............	28	90–120	75–85
Delayed dormant........	32	90–120	90–92
Summer................	32	90–120	95–96

California classification, summer oil	% Dist. at 636°F	Unsulfonated residue, min, %
Light...........................	64–79	90
Light medium...................	52–61	92
Medium.........................	40–44	92
Heavy medium..................	28–37	92
Heavy..........................	10–25	92

Mosquito Larvicides: Usually No. 2 heating oil type or more aromatic; often used with added DDT
Household Sprays: Deodorized kerosine-type oil often with aromatic distillate, always with insecticide toxicants; both spray and aerosol

FUNGICIDAL OILS
Lubricating-type oil used as mist on bananas to control Sigatoka disease

HERBICIDES
General: Aromatic distillates
Preemergence: Highly aromatic, quite volatile
Selective: Stoddard solvent

TOBACCO DESUCKERING
Emulsions of white oils

SOLVENTS
Xylene and low-boiling aromatics
Middle-boiling aromatics, kerosine extracts
High-boiling aromatics (from catalytically cracked gas oils)
Nonaromatics: isoparaffinic distillates, white oils

term herbicide is also used for products which inhibit sprouting and defoliate plants. Pesticides are also classified in four categories, depending on their utilization: (1) household applications; (2) agricultural, for use on crops; (3) nonagricultural, such as mosquito control; and (4) industrial, for preventing loss and spoilage in warehouses and storage plants.

Various types of petroleum products in themselves possess toxic properties for insects and weeds. Another important role is for petroleum solvents and other oils as carriers for active pesticide toxicants. The insecticides form the largest group of pesticide products. Table 11-39 lists the principal petroleum products used as pesticides.

Insecticides

Kerosine was early employed as a spray oil to combat insects in citrus plants. Emulsions of kerosine, soap, and water were found less damaging to fruit trees, and there is today wide use of petroleum-oil emulsions for the control of scale insects and mites on fruit trees. The oils are supplied in the form of concentrates from pesticide manufacturers, and the water is added at the point of application. The viscosity of the petroleum oil is a factor in the uniform spreading of the spray. The aromatic components have been found to contribute most to the phytotoxicity of the spray oil, its tendency to harm the plant or tree. The unsulfonated residue (USR) is an adverse indication of aromaticity and, therefore, is an important feature of the specifications of horticultural oils. In general, the higher the USR, the safer the oil to use.

Dormant oils, since they are applied before foliage begins to grow in the spring, usually have a lower USR than summer or verdant oils, which must be carefully selected to avoid phytotoxicity. Other materials may be added to oil spray emulsions, such as cryolite, rotenone, or certain fungicides. Federal law requires that spray oils, for either horticultural or household use, be labeled to show their flammability and toxic characters.

A wide variety of petroleum oils has been used for control of mosquito larvae, including kerosene, diesel fuel, heating oils, Stoddard solvent, aromatic fractions, crankcase drainings, and black oil. One important requirement of a larvicide oil is its ability to spread rapidly on the surface of the water. Emulsifying agents are used to secure this property. With the advent of DDT and other synthetic organic insecticides, the volume of mosquito oil used has dropped.

The first commercial household insecticides, fly sprays, were a kerosine solution of the active portion of pyrethrum flowers, known as pyrethrins. Today, most pesticide manufacturers purchase concentrates from pyrethrum processors. Among other chemicals used in household insecticides are chlordane, DDT, lindane, malathion, and perthane.

The advent of DDT (dichloro-diphenyl-trichloroethane) into commercial use after World War II brought about a new type of household insecticides, the residual spray.[46] This consisted of a solution of a stable active toxic chemical in a petroleum-base oil. When the oil evaporates, a film of insecticide remains which retains its toxic properties for a period of time, depending on its volatility, stability, atmospheric factors, and cleaning. A widely used insecticide of this type is composed of 5 per cent of DDT, 15 per cent of an aromatic petroleum solvent, and 80 per cent of a deodorized kerosene. More recently, refined bottoms from the production of aviation- and motor-fuel alkylates have been used as base oils in household sprays. The bottoms are redistilled and treated so they are practically odorless.

Sprays for livestock are of two general types, oil-base sprays and emulsion concentrates. The selection of the base oil, as well as the toxic agent, is most important in these sprays, since they must not injure the animal. Pale or white oils are used, rather than deodorized kerosene. The emulsion concentrates are formulated with the toxicants, emulsifiers, and aromatic solvents and then diluted with water to the desired concentration.

Another important development of World War II in the insecticide field was the aerosol bomb. The original bombs, which were widely used in malaria regions, consisted of the insecticide solution together with the high-pressure liquefied-gas refrigerant Freon 12 in a heavy steel container. It was later found that low-pressure propellents could be used and the package could be made much lighter. The aerosol bomb is not only easier and more convenient to use than conventional spraying, but it also possesses certain advantages over liquid spray. The insecticide is emitted in the form of very minute particles, which float like a cloud in the air for a considerable time. Sales of aerosol and pressurized insecticides have grown rapidly and amounted to over 43,500,000 units in 1963.

Herbicides

This weed-killer type of pesticide is of two general kinds, selective and nonselective.[46] The former destroys the weed but does not affect the crop; the latter kills all types of plants. Millions of gallons of the nonselective herbicides are used annually in treating railroad rights of way and along roadside fences and ditches where weeds are a fire hazard in the summer. A wide variety of petroleum oils are used. Light aromatic oils cause acute toxicity; heavy oils, such as diesel and fuel oils, cause chronic toxicity. Often the herbicidal oils are fortified with chemical toxicants to increase their activity.

Preemergence sprays are used with many crops after the seeds have been planted but before they have started to grow and the ground surface is covered with small weeds, which are vulnerable to light petroleum distillates ranging from mineral spirit to diesel fuel. Stoddard solvent, kerosine, and other light distillates are used as selective herbicides with certain crops where the weeds can be killed without leaving a flavor in the crop. Crops treated in this manner include beets, berries, carrots, celery, cotton, flax, grapes, onions, parsnips, and soya beans. Some orchard crops are also treated with selective herbicides. One petroleum oil widely used in this service has a distillation range of around 300 to 400°F, API gravity of 40 to 44°, flash point above 100°F, and aromatic content of 14 to 15 per cent.

Fungicides

It is only recently that the value of petroleum oils as fungicides has been realized. The oil is applied to banana and other crops in the form of a fine mist by portable sprayers and by aircraft. White mineral oil has been found effective in preventing the growth of suckers on tobacco plants.

Pesticide Solvents

The trend toward the use of emulsion concentrates and their application by spraying has emphasized the importance of petroleum solvents in agricultural pesticides. The emulsion concentrates consist of the insecticide dissolved in a suitable solvent, together with an emulsifier. The most important attribute of the solvent is to dissolve a sufficient quantity of the pesticide for effective use.

Three types of aromatic solvents are employed for the manufacture of pesticidal emulsion concentrates: (1) low-boiling solvents or solvents in the xylene boiling range (275 to 365°F); middle-boiling aromatics, such as kerosine extracts (350 to 525°F); and high-boiling solvents (450 to 675°F) usually obtained from catalytically cracked gas oils.

For agricultural applications where the solvent used must be low in aromatics, isoparaffin fractions obtained from alkylate bottoms are used, although they are relatively poor solvents.

Because of the flammable and generally toxic properties of pesticides, the handling and use of all types is closely regulated by Federal and state laws.[46] The *API Toxicological Reviews* include petroleum oils used in the pesticide field.[45]

PETROLEUM LUBRICANTS

Lubricants are unique among the petroleum products marketed, indeed among most merchandised items, in that, as their quality has improved, their period of use has been extended and the relative volume of their sales has diminished.

The most conspicuous example of this trend is with the automotive oils. The better performance which the refiners have built into their crankcase oils has encouraged the car manufacturers, service agencies, and some oil companies to advise longer intervals between drain periods. One authority has stated[47] that motorists today drive nearly twice as many miles on a barrel of lubricant as they did at the end of World War II.

The result of this doctrine of longer periods of use has been to slow down significantly the rate of growth in the consumption of automotive lubricants. Similarly, with industrial oils and greases, the improvement in the product itself, together with better design of gear housings, use of plastics, and other new materials for surfaces encountering friction, improvement in oil circulating and reclamation systems, among other means, has reduced the rate of growth in the use of these lubricants. In the last 10 years, the demand in the United States for petroleum lubricants has increased 20 per cent while that for all petroleum products has grown nearly 40 per cent. Figure 11-13 compares the growth in demand for lubricants in the United States since the end of World War II with that for general industrial activity. Table 11-40 shows the consumption of automotive and industrial lubricants in the United States from 1954 through 1960.

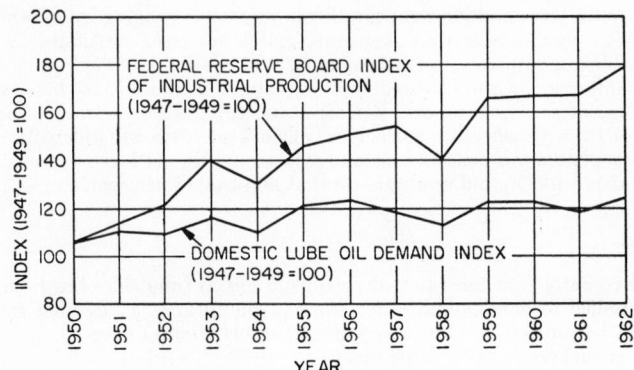

Fig. 11-13. Index of lubricating-oil demand in the United States 1950–1962, as related to the index of industrial production. United States Lube Index is based on the average of 34,985,000 bbl for the 1947–1949 period.

The lag in the demand for lubricants is receiving attention from the oil companies and their associations. The National Petroleum Association holds annual clinics on the subject. Special attention is being directed to the future trend in the use of these products. It is recognized that some of the factors which now hold down the growth in the use of lubricants will continue to make themselves felt in future years. Without doubt further improvements will be made in the performance value of lubricants by

Table 11-40. Consumption of Lubricants, 1954 to 1962*,†
(Thousands of barrels)

Year	Exports	Total domestic	Automotive lubricants‡	Industrial lubricants	Motor-oil consumption
1962	17,167	44,640	20,160	16,780	20,170
1961	16,606	41,640	22,070	16,700	20,085
1960	15,319	43,612	22,735	16,800	20,845
1959	13,739	42,717	22,500	17,000	20,610
1958	13,036	39,439	20,969	15,300	19,300
1957	13,826	41,215	21,679	16,890	20,115
1956	13,859	43,933	23,372	16,620	21,990
1955	14,298	42,477	23,151	19,328	21,255
1954	15,075	38,537	21,195	17,343	19,460

* Source: *National Petroleum News Factbook*, 1963.
† Includes greases.
‡ Includes motor oils, chassis lubricants, automotive transmission fluids, etc.

the refiners who manufacture them. New types of additives will enhance the qualities of the petroleum oils and widen the range of operating conditions where they can be used. Synthetic oils will replace petroleum lubricants in certain applications where operating conditions are most severe. Improvements in oiling systems, both automotive and industrial, will add to the period of usefulness of the lubricant itself. The effect on oil consumption of the smaller cars with their smaller crankcase capacities is not yet known.

Offsetting these various factors which will tend to hold down the overall consumption of lubricants, the oil companies plan educational programs to convince motorists that more frequent attention to lubrication is true economy in the life of their cars. They believe some car manufacturers have gone too far in advising oil changes every 3,000 or 6,000 miles. The American Petroleum Institute has revised its drain-interval policy and now recommends that the oil companies advise an oil change every 30 days in winter and 60 days in summer, with a maximum service period of every 2,000 miles. Future developments may include a faster means for draining and refilling the crankcase at the service station, thus making it less of a chore for the motorist. Another possibility is that entirely new markets may be found for the refiners' lubricating oils and oil stocks as has happened with other established petroleum products. Intensive research by the oil companies may expedite the discovery of new markets in nonlubrication uses.

Lubricant Types and Grades

Hundreds of different lubricant products are made by many refiners and compounders. A survey made by the API a few years ago reported 1,156 lubricating oils, light and heavy, compounded and uncompounded, of individual formulas and 271 grease products made by U.S. companies. This does not include such specialty products as white oils, transformer oils, rust preventives, and other products made from refiners' lubricating stocks.

The two broad classes of lubricants are automotive and industrial. Of the former the largest volume is motor oils, as shown in Table 11-40. There are far more kinds of industrial than of automotive lubricants. A third class of lubricants, much smaller in volume, includes oils and greases used for industrial nonlubricating purposes. Oils in this class, as reported by the Chemicals Branch, Industry Division of the Bureau of Census,[48] are mainly oils derived from petroleum, used for processing, testing, and other nonlubricating purposes. They do not include fuels, solvents, asphalts, or petroleum chemicals. Their principal uses are classified in Table 11-41.

Table 11-41. Nonlubricating Uses for Petroleum Products*

Apron dressings	Heat-transfer oils	Petroleum sulfonates
Brick oils	Ink oils	Quenching oils
Cable oils	White oils	Tempering oils
Coal spray oils (if from finished or unfinished lubricating-stock oils)	Belt dressings	Rub-roll oils
	Defoaments	Rust-preventive oils and compounds
Cordage oils	Polishing oils	Spray oils
Flotation oils	Launching grease; base and slip coats	Tanners' products
Form oils and compounds	Paint and putty oils	Textile-processing oils
Fruit and vegetable preservatives	Paper-processing oils	Transformer oils

* Source: *Facts for Industry*, Series M29-C-06, Chemicals Branch, Industry Division, Bureau of the Census, Washington, D.C., Jan. 31, 1958.

The great number of types of lubricants manufactured and the variety in their applications and the service in which they are used make impossible the kind of standard specifications which have been set up by the ASTM and other scientific agencies for some petroleum products. The Society of Automotive Engineers has adopted uniform viscosity classifications for crankcase oils and transmission and axle lubricants. These are for guidance in selecting lubricants of suitable viscosity for seasonal temperatures and engine service conditions. Associations of manufacturers of some industrial equipment, such as air compressors and gears, provide performance standards for lubricants used in their equipment. The Federal government and the armed services have adopted specifications for hundreds of oils and greases covering wide ranges of service and operating conditions.

Lubricant Additives

Additives of many types are widely used today in lubricating oils and greases. They include petroleum and nonpetroleum compounds. In most cases the additive is only a small proportion of the volume of the straight oil, but it may run as high as 30 per cent as in the case of marine diesel-engine oil. Additives for lubricants, as for all petroleum products, are more than nostrums to put into the product in prescribed dosages. They are incorporated into the finished oils by the refiner, and as much care is taken in their selection as in the steps of processing the oil itself.

In addition to the specific properties which the additives impart to the oil, they must possess certain general properties[49] to allow them to be effectively incorporated in the petroleum product. They must be soluble in the base petroleum product over the temperature range encountered in service. In the case of motor oils, they should be insoluble in water. Even though effective, they must not discolor the oil in which they are incorporated and thus give it the appearance of a used oil. The additives must be of low volatility, and not vaporize when exposed to high temperatures. They should retain stability during blending and storage of the product of which they are a component. They must be compatible with other additives which may be added to the same product and avoid visible evidence of such reactions as a change in color of the product or a throwdown of insolubles.

Additives for crankcase oils may be singly functional or multifunctional, and they may enhance or add to the oil one or more of the qualities listed below:

1. Decrease oil oxidation and reduce the corrosion of certain types of sensitive bearing materials
2. Provide the oil with greater resistance to wear of main and connecting rod bearings, particularly when full-film lubrication is lacking, and of piston rings and other engine parts
3. Improve the dispersant-detergent qualities of the oil and thus promote engine cleanliness
4. Improve the viscosity-temperature relationship of an oil over the range of temperatures at which it is to be used
5. Reduce the pour point, the lowest temperature at which the oil will pour or flow under specified test conditions
6. Prevent foaming of the oil in the crankcase
7. Prevent rusting when water enters oil in the crankcase

Some of the additives for engine oils are used to improve the performance of transmission fluids. Additives are used today to give higher load-carrying capacity to gear lubricants.

Industrial oils for many uses are fortified with the same type of additives as are incorporated in crankcase oils, such as viscosity-index improvers, pour-point depressants, antioxidant and antifoam agents, and detergent inhibitors.[49] Other types of additives may be used with industrial oils to meet the specific application, such as to extend the load-carrying capacity of the oil. Oiliness agents may be incorporated to increase film strength. Tackiness agents give greater cohesion to the oil and prevent dripping or spattering from a bearing. Fatty oils are used as additives to increase the resistance of the straight oil to displacement by water on the surface to be lubricated.

Additives used in lubricating greases include oxidation inhibitors, corrosion inhibitors, color stabilizers, dyes, film-strength agents, metal deactivators, rust inhibitors, additives to impart stringiness to the grease, and structure modifiers for soap-oil systems.

Automotive Lubricants

Three types of automotive lubricants have been defined and classified by the Society of Automotive Engineers.[50] These are (1) crankcase oils, (2) transmission and axle lubricants, and (3) fluids for hydraulic torque converters and fluid couplings. The last named group is a special type of transmission oil used in automatic transmissions.

The classification in each group is in terms of viscosity, based on the temperatures at which the oils are to be used. The property of viscosity is recognized as the most important characteristic of lubricating oils for automotive use. There are seven viscosity classifications for crankcase oils, five for the conventional transmissions (manually shifted) and rear-axle oils and four tentative groupings for hydraulic torque converter and fluid coupling oils. Table 11-42 shows the SAE classification of crankcase oils and of transmission and axle lubricants.

Table 11-42. SAE Classifications for Crankcase Oil, Transmission, and Axle Lubricant Viscosities
Crankcase Oils

SAE viscosity number	Viscosity range, SSU			
	At 0°F		At 210°F	
	Min	Max	Min	Max
5W	....	4,000		
10W	6,000*	Less than 12,000		
20W	12,000†	48,000		
20	....	....	45	Less than 58
30	....	....	58	Less than 70
40	....	....	70	Less than 85
50	....	....	85	110

* Minimum viscosity at 0°F can be waived provided viscosity at 210°F is not below 40 SSU.
† Minimum viscosity at 0°F can be waived provided viscosity at 210°F is not below 45 SSU.

Transmission and Axle Lubricants

SAE viscosity number	0°F		210°F	
	Min	Max	Min	Max
75	....	15,000		
80	15,000*	100,000		
90	....	....	75	120†
140	....	....	120	200
250	....	....	200	

* The minimum viscosity at 0°F can be waived if the viscosity is not less than 48 SSU at 210°F.
† The maximum viscosity at 210°F can be waived if the viscosity is not greater than 750,000 SSU at 0°F (extrapolated).

The viscosity values are given in terms of Saybolt seconds Universal (SSU) by ASTM Method of Test D 88. Kinematic viscosities can be converted into Saybolt seconds by the use of tables given in ASTM D 567.

In order to assist the public, oil refiners, and automotive-equipment manufacturers in identifying crankcase oils suitable for particular operating conditions, the American Petroleum Institute has defined six types of service in which these oils are most frequently used.[51] Each service class is identified by letters, which provides a convenient means for the engine manufacturer to indicate the service requirements of his various designs and hence their lubrication requirements. Oil companies use the letter designations to indicate the class of service for which each of their brands is suitable. Classifications MS, MM, and ML apply to gasoline automotive-type engines and DS, DM, and DG to diesel automotive-type engines. The API service classifications for crankcase oils are:

Service MS. Service typical of gasoline or other spark-ignition engines used under unfavorable or severe types of operating conditions and where there are special lubrication requirements, for deposit, wear, or bearing corrosion control, due to operating conditions, engine design, or fuel characteristics.

Service MM. Service typical of gasoline and other spark-ignition engines used under moderate to severe operating conditions but presenting problems of deposit or bearing corrosion control when crankcase-oil temperatures are high.

Service ML. Service typical of gasoline and other spark-ignition engines having no special lubrication requirements and having no design characteristics sensitive to deposit formation.

Service DS. Service typical of diesel engines operating under very severe conditions or having design characteristics or using fuel tending to produce excessive wear or deposit.

Service DM. Service typical of diesel engines operating under severe conditions or using fuel of a type normally tending to promote deposits and wear but where there are design characteristics or operating conditions which may make the engine either less sensitive to fuel effects or more sensitive to residues from lubricating oil.

Service DG. Service typical of diesel engines in any operation where there is no severe requirement for wear or deposit control due to fuel, lubricating-oil, or engine-design characteristics.

Performance Evaluation

Because of the great variety in the types of automotive engines, transmissions, and axles and in the operating conditions they must meet, their lubricants are evaluated by actual use in the full-scale units for which they are intended. The large oil companies, as well as the automobile companies and other agencies, maintain engine stands in their laboratories, where lubricants and fuels are continually under test. A standard test engine and standard procedures for testing crankcase oils and other automotive lubricants under actual service have been developed under the joint auspices of the oil companies and automobile companies.

The Federal government has adopted six engine tests and two rear-axle tests to assist in identifying the type of lubricants which it specifies. Three diesel engines and one gasoline engine are used in these tests. The Federal government has also set up three test methods using full-scale rear axles to assist in identifying rear-axle lubricants. A full-scale test method for automatic transmission fluids developed by General Motors Corp. is in general use.

The Fuels and Lubricants Committee of the SAE has worked up tentative specifications for automotive central systems fluids, both petroleum-base and synthetic-base oils.[50] These specifications are offered to assist in the development of central hydraulic systems for highway vehicles and cover fluids designed to operate in the circulating system from -45 to $275°F$.

A wide number of standard ASTM laboratory tests are used to determine the individual properties of automotive and industrial lubricants. Viscosity is generally the most important controlling property of lubricating oils. It is used for identifying grades of oils and for following the performance of oils in service. Other widely used laboratory tests include API gravity, carbon residue, cloud and pour points, color determination, corrosive tendencies, foaming characteristics, sulfur content, saponification, and neutralization number.[52]

Industrial Lubricating Oils

Industrial lubricating oils fall into two general classifications:[53] those required for the power plant of the shop or factory and those for the production equipment of the factory. The lubricants in the second category vary widely in their properties, depending on the types of equipment required for the specific product being manufactured.

Oils required to meet the needs of the power plant are used in the following areas:

1. External lubrication of engines and general machinery
2. Steam-engine cylinders
3. Diesel-engine crankcases and cylinders
4. Steam-turbine bearings and reduction gears
5. Speed-reduction gears where pressure- or bath-lubricated
6. Air, gas, or refrigeration compressors

In the production side of the plant, the following classes of lubricating oils are used:

Oils for motor bearings	Hydraulic fluids
Circulating oils	Wire-rope lubricants
Industrial gear oils	Spindle oils
Oil spray lubricants	Pneumatic tool oils
Instrument oils	Insulating oils

Metalworking and cutting oils, soluble oils, grinding oils

As with automotive oils, standard specifications for the general use of industrial lubricants are not available. Technical groups of manufacturers' associations sometimes set up methods for evaluating lubricants used in their equipment. The Federal government and the armed services issue specifications for grades of industrial lubricants for classified uses.

The ASTM Standard Methods of Test are used for laboratory determinations of the properties of industrial lubricants. The ASTM lists such methods of test for general industrial oils, cutting oils, turbine oils, and electrical insulating oils.

Important properties of industrial lubricants for specific uses are discussed in the following:[53]

Engine and Machine Oils. For lubricating the elements of reciprocating engines, pumps, compressors and crosshead guides, pins, and rocker-arm bearings. Also for rotating parts elsewhere in the plant, including pillow blocks, line-shaft bearings, and sleeve-type bearings found in assembly lines. Operating conditions include temperature, speed, and pressure. Two principal properties required in the oils are viscosity and demulsibility.

Steam Cylinder Oils. Operating conditions affecting the selection of the oil are steam temperature, pressure, moisture content of the steam, the superheat present, and the use of the exhaust steam. The cylinder oil must be refined to function at temperatures normally well above 250°F and in some classes of service temperatures up to 1000°F. The types of refinery stocks used in cylinder oils under varying steam pressure limits are given in Table 11-43. Where even small percentages of moisture are present in the steam, compounded oils are used for emulsifying purposes.

Diesel-engine Oils. The operation of diesel engines in many types of industrial service differs from that of automotive-type diesels. Since the same oil often must serve the entire engine, its characteristics must be such that it will meet the prevailing load and temperature conditions. It must be closely fractionated to avoid undue volatilization at high temperatures. The emulsification tendency must be low to avoid permanent emulsions in the presence of water. The oils are generally fortified with additives to impart oxidation resistance and dispersion-detergent ability.

Steam-turbine Oils. Operating problems affecting lubrication are possibility of water contamination, chance of elevated temperatures, and contamination by dust, dirt, rust, or other solid foreign materials. The viscosity of the oil is determined by the type of turbine, whether direct drive, reduction geared, or hydraulic, and the service in which it is operated. Neutralization number by ASTM D 664[52] is an indication of the oxidation tendency of the oil. Demulsibility is another indication of performance value.

Speed-reduction Gear Oils. Basic types of lubricants range from a straight residual oil for exposed or bath or hand-lubricated gears, operating at average speeds

Table 11-43. Types of Stocks Used in Manufacturing Cylinder Oils[53]

Type of cylinder stock	Intended service	Steam pressure limits, lb	Average % compound
Steam-refined straight mineral	Where condensate is returned to boiler	Below 150	
Steam-refined compounded	For saturated steam conditions	Below 150	3–10
Bright stocks vacuum-distilled and solvent-refined	Where rapid atomization is required		
	Where rapid separation from steam is desired	200 max	4–6
	Where steam is used for process heating or condensate is used for ice manufacture		
Fire stocks vacuum-distilled and dewaxed	For heavy-duty service, super-heat, or high-temperature saturated steam or where a low pour test is desired as in worm gear lubrication	250 max	4–8
Fire stocks (partially dewaxed) after vacuum distillation	For high-pressure high-temperature steam which is comparatively dry	Above 175	May or may not be compounded up to 5%

and low loads, to extreme-pressure oils with chemically active additives for splash or pressure lubrication of gears where high speed and low torque or low speed and high torque or shock loading prevails.

Compressor Oils. Selection of the oil depends upon the type of gas to be handled and whether the compressor is reciprocating, rotary, or centrifugal. In an air compressor, the flash point and rate of evaporation of the lubricant indicate its carbon-forming tendency. In gas compressors, straight mineral oils are generally suitable for most types of gas. In refrigeration compressors, an important property of the oil is its ability to flow at low temperatures and, if congealing should occur, to develop a minimum of wax.

Motor Bearing Oils. The type of bearing in the electric motor, whether sleeve or ball or roller, and the wide variations in motor sizes and operating temperatures influence the selection of the oil as to viscosity. Pour point is an important indication of the ability of the oil to flow at low temperatures.

Hydraulic Fluids. Resistance to flammability may be an important requirement of the fluid, whether petroleum or synthetic base. Since lubrication is an auxiliary factor, viscosity and resistance to oxidation, rusting, and excessive foaming are important considerations.

Industrial Circulating Oils. Basic requirements of the lubricant are oxidation resistance, load-carrying capacity, a high viscosity index, good demulsibility, and resistance to foaming. Volatility is a factor in foaming, since lower viscosity oils free themselves from foam more readily than heavy oils.

Wire-rope Lubricants. Depending on the type of service, the lubricant must resist dripping at high temperatures and cracking in cold weather. It must be tacky enough to stick well, be free from water, and resist emulsification. The viscosity must be such that it will penetrate the assembly of strands and wires in the rope.

Industrial Gear Oils. In pressure circulating systems, the lubricant, while protecting the gears, must also lubricate the bearings. An extreme-pressure additive is incorporated in some of these oils. Turbine-grade oils of the proper viscosity are fortified with antioxidants, rust retarders, and foam dispersers to protect against a breakdown in service.

Spindle Oils. These are light-viscosity oils used in textile spinning processes and applied to the high-speed grinding machine spindle in metalworking and to the cutting spindle in woodworking machinery. They must maintain positive bearing lubrication to reduce frictional wear, and they must serve as a cushion to prevent vibration or wobble of the spindle. Control of the viscosity range is important. Foaming may lead to loss of oil. Additives used are rust and corrosion inhibitors and antioxidant and detergent-dispersant agents.

Instrument Lubricants. Effective lubrication of the intricate mechanisms in any instrument or control device is provided when the oil (1) stays on the parts to be lubricated and does not creep or spread to other parts, (2) provides maximum stability and resistance to oxidation and gumming, (3) is compatible with electrical insulating materials, (4) possesses maximum oiliness since boundary lubrication conditions may prevail, and (5) prevents tarnish or oxidation of certain metallic contact surfaces.

Pneumatic Tool Oils. The oil must be resistant to oxidation and prevent rusting of tool parts. It must have ample film strength and be able to stick to steel surfaces in the presence of moisture. To meet this condition the oil is compounded with a fatty oil, such as rapeseed oil, to develop emulsification properties.

Insulation and Transformer Oils. These are low-viscosity oils used in electric switches and transformers and meeting two basic requirements, adequate fluidity and freedom from moisture. Viscosity and pour points are important in determining fluidity.

Metalworking and Cutting Oils. Oils in such applications must remove heat while at the same time providing lubrication at the face of the cutting tool and the metal being cut. Severity of the service and the hardness and chemical composition of the metal being cut are primary factors in the selection of this industrial lubricant.

Lubricating Greases

A lubricating grease is defined[54] as a solid or semisolid lubricant consisting of a thickening agent in a fluid lubricant. The liquid is almost always the major component and generally is a petroleum oil. In the past conventionally refined oils of moderate viscosity index were principally used. The tendency now in manufacturing lubricating greases is to use an oil of higher VI in order to improve the performances of greases over wider operating ranges. For special applications such as in aircraft, where wide temperature variations are encountered, synthetic lubricating fluids are used. These may include diesters, silicones, and some of the polyalkylene glycols and derivatives.

The major thickening agents used in manufacturing lubricating greases are soaps of aluminum, barium, calcium, lithium, sodium, and strontium. Lesser amounts of nonsoap types of thickeners consist of compounds of inorganic origin, such as modified clays or fine silicas, or compounds of organic origin, such as arylureas or phthalocyanine pigments.

Solids, or fillers, are sometimes added to lubricating greases for specific types of applications in concentrations varying from a fraction to several per cent by volume. These fillers are in most cases inorganic materials such as asbestos, graphite, metal oxides, metal powders or flakes, or metal sulfides, among others.

Additives are now incorporated in lubricating greases as in many petroleum products. Types of additives used include antioxidants, corrosion and rust inhibitors, color stabilizers, dyes, extreme-pressure or film-strength agents, metal deactivators, additives to impart stringiness, and structure modifiers for soap-oil systems.

Selecting a Lubricating Grease

In the selection of a lubricant, many applications will be found where a grease has advantages over a lubricating oil. Some of these are:
1. Less frequent application is necessary where a grease is used.
2. Grease has suitable physical characteristics for the method of application.
3. Dripping and spattering of the lubricant are largely eliminated where a grease is used.
4. Less expensive seals are required for a grease-lubricated bearing.
5. Lubricating grease will ensure some degree of lubrication even if a new application is neglected for some time.

6. The proper grease will cling to metal better than an oil.

7. Lubricating grease is preferable under extreme operating conditions such as high temperatures, extreme pressures, low speeds, shock loading, and where bearings operate intermittently or in reverse.

8. Grease provides better lubrication where machine parts are badly worn.

9. In certain cases, the use of lubricating grease simplifies mechanical design, as in the case of vertical bearings.

Among the disadvantages to be found in the use of grease are (1) greases are not as good coolants as oil; (2) oil will flush contaminants from the bearing or other part more readily than grease; (3) when the lubricant is changed, less labor is required with oil than with grease; (4) more problems are encountered in the storage of grease than with oil; (5) a gallon of grease will cost more than the equivalent amount of oil, but it will be used, in general, over a much longer period of time.

The general considerations which will determine what lubricating grease is to be selected include:[54]

1. Speeds likely to be encountered
2. Loads encountered and whether continuous or shock loading
3. Temperatures, both ambient and operating
4. Moisture conditions and whether continuous or intermittent
5. Other contaminants likely to be present, dust, moisture, etc.
6. Method and frequency of application
7. Life expectancy of the lubricant
8. Possibility of contamination by the lubricant of the material being processed

As with lubricating oils, grease falls into two main categories of use, automotive and industrial. A survey of the sales of lubricating oils and greases in the United States in 1960,[48] made by the Chemicals Branch, Industry Division, of the Census Bureau, showed total grease sales of 3,931,204 bbl (315 lb to the barrel). Of this total, 2,630,015 bbl were automotive greases and 1,301,189 were industrial greases. A small volume of sales was for nonlubricating purposes.

Automotive Greases

Growing use is being made of the so-called multipurpose gear lubricant corresponding to the API Service GL-4 classification.[50] This term designates grease of the character, structure, and consistency meeting the individual requirements for chassis lubricant, wheel bearing lubricant, universal joint lubricant, water-pump lubricant, and cup grease. These greases have high load-carrying capacity and high rust-preventive qualities and are resistant to water and high temperatures. They have great stability and are relatively easier to pump and handle at low temperatures.

These superior qualities are imparted to the grease by various additives, which must be compatible with one another as well as with the lubricant. The multipurpose grease must also be compatible with other greases with which it may come in contact. The grease is applied simultaneously to the various parts either automatically or by hand pressure. Wide use is still made of the standard automotive lubricating greases, which include chassis and wheel-bearing lubricants and greases for water-pumps, universal joints, and steering gears.

Many laboratory methods for determining the performance properties of lubricating grease have been developed under the auspices of Committee D-2 on Petroleum Products and Lubricants of the ASTM and have been standardized by that agency. Some well-recognized but unstandardized service tests have been developed by the National Lubricating Grease Institute and other agencies. Table 11-44 lists the ASTM standard methods of test for lubricating greases. A few of the most important of these test methods and their significance are discussed briefly:

Consistency. This property is measured by ASTM Test Method D 217, which provides an empirical estimation of the consistency of lubricating greases by measurement of the extent of penetration of a standard cone. Normally, tests are reported in worked penetration, which is the penetration of the sample after it has been brought to 77°F, and then subjected to 60 strokes in a standard grease worker.

Table 11-44. ASTM Standard Methods of Tests for Lubricating Greases*

Method number	Title
D 128-61†	Analysis of Lubricating Grease
D 1092-62†	Apparent Viscosity of Lubricating Greases
D 217-60T†	Cone Penetration of Lubricating Grease
D 566-42†	Dropping Point of Lubricating Grease
D 1261-55†	Effect of Grease on Copper
D 972-56†	Evaporation Loss of Lubricating Greases and Oils
D 1262-55†	Lead in New and Used Greases
D 1263-61†	Leakage Tendencies of Automotive Wheel Bearing Greases
D 664-58†	Neutralization Value (Acid and Base Numbers) by Potentiometric Titration
D 974-58T	Neutralization Value (Acid and Base Numbers) by Color-indicator Titration
D 942-50†	Oxidation Stability of Lubricating Greases by the Oxygen Bomb Method
D 270-61	Sampling Petroleum and Petroleum Products
D 1264-63†	Water Washout Characteristics of Lubricating Greases
D 1403-62†	Cone Penetration of Lubricating Grease Using One-quarter Scale Cone Equipment (One-quarter Scale of Cone and Worker Specified in D 217)
D 1404-56T	Estimation of Deleterious Particles in Lubricating Grease
D 1402-58†	Effect of Copper on Oxidation Rate of Grease
D 1741-60T	Functional Life of Ball Bearing Greases
D 1478-63	Low-temperature Torque of Ball Bearing Greases
D 1742-60T	Oil Separation from Lubricating Grease During Storage
D 1831-61T	Roll Stability of Lubricating Grease
D 1743-60T	Rust Preventive Properties of Lubricating Greases

* *ASTM Standards, Part 17, Petroleum Products,* 1964.

† Also approved as American Standard by the American Standards Association.

Dropping Point.   This is the temperature at which, under ASTM D 566, the grease passes from a semisolid to a liquid state under conditions of the test.   While this value has no relation to service performance, it is pointed out that a grease cannot be expected to provide satisfactory performance at temperatures higher than the dropping point.

Evaporation Loss.   ASTM D 972 provides for the determination of evaporation loss of grease in applications where evaporation loss is a factor.

Apparent Viscosity.   ASTM D 1092 is a procedure for measuring the apparent viscosity of lubricating greases in the temperature range of -65 to $100°F$.   The results may be related to the ease of handling and dispensing and to starting and running torques of grease-lubricated mechanisms.

Leakage Tendencies.   This test is intended to evaluate the leakage tendencies of wheel bearing greases under prescribed laboratory test conditions.   The method, ASTM D 1263, is not the equivalent of long-time service tests.

Water Washout Characteristics.   ASTM D 1264 is intended to evaluate the resistance of a lubricating grease to washout by water from a bearing when tested at 100 and 175°F under prescribed laboratory conditions.   It is not considered the equivalent of service tests.

Oxidation Stability.   A method for determining the resistance of greases to oxidation where the grease is renewed only at long intervals, as in coatings on anti-friction bearings and on motor parts, for example, is provided in ASTM D 942.   This test is not intended for predicting the stability of greases under dynamic service conditions or when stored in commercial containers.

Multifunctional Greases

These greases containing 3 to 10 per cent of graphite, lead powder, or molybdenum disulfide are recommended for certain severe applications on cars and trucks.   A small proportion of an extreme-pressure lubricant may be included in the composition. They are sometimes used on ball-joint suspensions and found superior to a chassis lubricant.   They are also used on tractor fifth wheels and in certain applications on trailers.

Industrial Greases

The listing given in Table 11-45 is typical of the wide number and variety of industries and of applications other than automotive where lubricating greases are used. While general types and grades of grease may be recommended for the needs of a specific industry, conditions of operation in the individual plant may require the selection of greases of an entirely different character.

Table 11-45. Typical Industries and Applications Where Lubricating Greases of Specific Properties Are Required[54]

Cement plants	Paper mills	Instrument lubrication
Coal mining	Public buildings	Logging and sawmill machinery
Construction equipment	Railroads and rolling stock	Mining and quarrying
Foodstuff processing	Steel mills	Oil-field equipment
Household appliances	Clay-products machinery	Printing machinery
Laundry machinery	Communications apparatus	Rubber plants
Machine-tool operations	Farm machinery	Ships and vessels
Nuclear power plants	Glassmaking machinery	Textile mills

It is well to keep in mind, states one authority,[54] that the motion of parts to be lubricated consists of either rolling or sliding and that, under such conditions, the lubricating grease should prevent metal-to-metal contact and thus reduce friction. The magnitude of the rolling or sliding forces, as well as the temperature of operation and the possibility of contamination, determine the grade and type of lubricating grease to use.

PETROLEUM WAXES

New markets continue to expand the demand for the petroleum waxes. Several relatively new uses individually dwarf in volume the consumption of wax in candle manufacturing. This was the original use for the byproduct of the refiners in the early Pennsylvania oil fields, which resulted from the necessity for separating the naturally occurring wax and petrolatum from the crude oil they were processing. Total demand for the wax produced in refineries in the United States in 1964 was 5,330,000 bbl, almost three times the requirement for this product in 1930. See Table 11-46.[55]

The varied and growing uses for the petroleum waxes, including paraffin and microcrystalline wax and petrolatum, have lifted this material from the byproduct class in the operations of some refiners. One authority states:[56]

"Our view is that wax consumption will continue to show a net increase both in present uses, where it offers many advantages over competitive materials, and in new uses yet to be developed. Our way of looking at wax is colored by a basic processing philosophy. Wax is a primary product, rather than a by-product of our refinery operations. Special crude oils are run for the wax they contain and the oil recovered is a by-product of this operation."

The largest single use of petroleum waxes is in the paper and paperboard industry, both in the manufacture of paper and in the coating and impregnating of paper and paperboard for protective wrappings for foods. Over 80 per cent of the petroleum waxes used in this country is for these purposes. It is estimated that wax consumption for the total paper and paperboard industry will be a billion pounds by 1968.

By 1910 wax bread wrappers were not uncommon, and by 1920 the use of wax papers for the moisture-loss protection of candy and baked goods amounted to many thousand tons annually. During the 1920's the use of waxed paper as a heat sealing carton was developed and waxed paper for the protection of frozen meat and soap followed shortly. In the 1930's the waxed-paper milk bottle made its appearance and quickly came into general use. The use of wax-impregnated wrappings and containers continues to grow. Both paraffin and microcrystalline waxes serve this field of application.

Table 11-46. U.S. Production and Demand, Petroleum Wax, 1930 to 1964*
(Thousands of barrels, 280 lb per bbl)

Year	Production	Demand			
		Exports	Domestic	Total (1,000 bbl)	Total (1,000 lb)
1964	5,352	1,734	3,596	5,330	1,492,400
1963	5,126	1,455	3,809	5,264	1,473,920
1962	5,353	1,429	3,965	5,394	1,510,320
1961	5,781	1,237	4,390	5,627	1,575,560
1960	5,896	1,333	4,438	5,771	1,615,880
1959	5,630	1,033	4,556	5,589	1,564,920
1958	5,252	911	4,300	5,211	1,459,080
1957	5,461	1,023	4,430	5,453	1,526,840
1956	5,367	920	4,430	5,350	1,498,000
1955	5,293	1,248	4,056	5,304	1,485,120
1950	4,462	1,193	3,238	4,431	1,260,680
1945	2,921	566	2,403	2,969	831,320
1940	1,833	678	1,274	1,952	546,560
1935	1,608	821	933	1,754	491,120
1930	1,956	1,046	864	1,910	534,800

* American Petroleum Institute, *Petroleum Facts and Figures*, 1965.

Refinements in manufacture have aided this and other uses for wax. Before 1948[57] the principal material for waxing paper was unmodified paraffin wax. Although this coating fulfilled the primary requirement of moistureproofing, there were deficiencies in this material. These have been overcome and other properties of the wax improved through the use of polyethylene additives and the production of higher quality waxes. It is estimated that about 20 million pounds of polyethylene are used annually for the purpose of wax modification.

Candlemaking continues to be one of the large uses of paraffin wax.[58] The utilitarian use of candles, together with religious and decorative needs, consumes between 60 and 70 million pounds annually. Pharmaceuticals and cosmetics constitute the major demand for highly refined petrolatums. Less-refined grades supply varied industrial requirements. Paraffin wax is used for chlorination to the extent of around 23 million pounds annually. Table 11-47 lists the wax consumption in the United States by type of application.

Table 11-47. Typical Annual Wax Consumption in the United States[58]
(Based on 1956–1957 Bureau of Mines reports)

Application	Volume, lb (000)
Sanitary containers	240,340
Waxed wrappers	251,320
Candles	63,440
Drugs, cosmetics, chemicals	48,800
Chlorination	23,000
Rubber	6,588
Electrical goods	4,880
Textiles	4,800
Matches	5,368
Miscellaneous*	197,900
Total domestic	846,436
Export	378,300
U.S. production represented	1,220,736

* Pyrotechnics, adhesives, crayons, polishes, paint removers, metal castings, etc.

Types of Petroleum Waxes

Paraffin wax and petrolatum were among the earliest petroleum products manufactured. They were separated from the wax distillate, the higher-boiling fractions taken from crude oil by the early Pennsylvania refiners. Basically, petroleum wax and petrolatum are made today by separation but with many refinements in techniques, including the use of solvents and centrifuging. The paraffin waxes have melting points in the range of 120 to 140°F. They are hard and brittle and have a tendency to harden into large crystals.

When the need developed, principally in motor oils, for making oils which were more fluid at low temperatures, more waxy materials were removed from the refiners' heavier lubricating-oil stocks and residues. From these materials, including tank bottoms, the microcrystalline waxes were developed. They are higher in melting point than the paraffin waxes and harden into small, needlelike crystals. Some types are ductile. Table 11-48 shows the properties of paraffin wax and of three grades of

Table 11-48. Comparison of Major Wax Types Produced in the United States[58]

Wax	Characteristic	No. of carbon atoms	Melting point, °F	Viscosity at 210°F, SSU	Crystals
Paraffin.............	Brittle	18–56	122–140	40	Plates
Motor oil...........	Brittle	26–42	145–170	50	Needles
Residual...........	Flexible	36–70	145–175	65–100	Small needles
Tank bottom........	Hard	40–70	180–200	......	Very small needles

microcrystalline waxes. The microcrystalline waxes are produced in much smaller volumes than the paraffin waxes. They have wide application in coatings.

As first produced, petrolatums were derived from tank bottoms and from the waxy substance that plugged casings and collected on pump rods in the wells in the Pennsylvania oil fields. Petrolatum is solid at ordinary temperatures but contains both solid and liquid hydrocarbons. It is described[58] as a "colloidal system in which the solid hydrocarbon components are the external phase and those that are liquid are the internal phase. The wax has taken up the liquid phase, resulting in an amorphous jelly-like mass that gave the material the original name of petroleum jelly."

Petrolatum is distinguished by its semisolid state at room temperatures and by easy deformation under slight pressure at this temperature. Its melting point ranges from 110 to 175°F, and its consistency from 40 to 300 penetration at 77°F. As a chemically inert, salvelike emollient base, it has wide cosmetic and pharmaceutical uses, and less refined grades have many industrial applications. The United States Pharmacopoeia has set up the property characteristics of petrolatums for pharmaceutical uses, including solubility, specific gravity, consistency, alkalinity, residue on ignition, organic acids, fixed oils, fats or rosin, and color.

Tests and Properties

Two degrees of paraffin wax refinement are generally recognized in commerce,[58] crude-scale and fully refined. Crude-scale paraffin has a melting-point range of 118 to 135°F, is a color darker than +21, and contains more than 0.5 per cent oil. The fully refined grade has a melting point of 118 to 149°F and color of +21 or lighter and contains not over 0.5 per cent oil.

The analytical procedures for testing petroleum waxes are of two types: those which measure its physical properties and those which measure its functional or performance properties. Many of these tests have been worked out jointly by Committee D-2 on Petroleum Products and Lubricants of the ASTM and the Technical

Association of the Pulp and Paper Industry, representing the largest industrial users of petroleum waxes. The following are among the analytical test procedures:

Melting Point (ASTM D 87, D 127).[59] This property has both direct and indirect significance in most wax utilization. Since wax is a mixture of hydrocarbons of different melting points, melting point is usually expressed as a range of temperatures for both paraffin and microcrystalline wax.

Color (ASTM D 156). Color of paraffin wax is an index of the degree of its refinement, the wax while molten being compared with the Saybolt chromometer. For determining the color of petrolatum darker than −16, the test procedure in D 155 is used.

Oil Content (ASTM D 721). This method uses methyl ethyl ketone as a solvent and is applicable to petroleum waxes having a melting point of 105°F or higher and containing not more than 15 per cent oil. It measures the differential solubility of oil and wax under prescribed conditions.

Needle Penetration (ASTM D 1321). This tentative standard test procedure is used in microcrystalline wax specifications as an empirical means for estimating their consistency by measurement of the extent of penetration by a standard needle. The consistency of petrolatum is determined by Method D 937.

Carbonizable Substances (ASTM D 612). This test is applicable to paraffin wax to determine whether it conforms to the standard of quality required for pharmaceutical use.

Odor and Taste. These properties, while important in many wax applications, have been difficult to standardize in test procedures. Wax producers have experienced personnel who follow a prescribed procedure in routine production testing.

Other Physical Tests. There are other tests that can be used with the petroleum waxes. These include API gravity (ASTM D 287), flash point (ASTM D 92), and viscosity (ASTM D 88 or D 445). They are not usually given by the producer unless especially requested.

Functional Tests

Test methods have been developed for the petroleum waxes to evaluate their performance in specific applications. Some of these methods have been worked out jointly by the Technical Association of the Pulp and Paper Industry in cooperation with Committee D-2 of the ASTM. Some of these functional tests are described below:

Blocking Point (ASTM D 1465). This tentative test method is a procedure for determining the blocking point and the picking point of paraffin wax. The blocking point is the lowest temperature at which waxed papers will stick together sufficiently to injure the surface films and performance properties. The picking point is the temperature at which the first film disrupture occurs on the waxed paper.

Sealing-strength Test. This test designates the amount of force required to separate two sealed samples of wax-coated paper of specified size prepared under certain conditions. Adequate sealing strength is a requisite of self-sealing wax wraps for food.

Water-vapor Transmission Test. The efficiency of a wax-paper barrier in preventing the passage of moisture vapor is the sum of the individual efficiencies of the paper and the wax to resist the transfer. Moisture-vapor resistance is necessary with food wrappings to keep moisture from leaving the package or, conversely, to keep moisture out of the container.

Flexibility Test. This is used with both paraffin- and microcrystalline-wax papers to indicate their resistance to creasing.

Gloss Test. The gloss of a paraffin film is measured by the amount of light reflected from a waxed surface illuminated with a light of known intensity. This property is important for its "eye appeal."

Chemical Stability. This performance test is primarily that of measuring resistance to oxidation, which, when it occurs, produces odor and/or taste and degradation of other functional characteristics of the wax. Paraffin wax is more susceptible to air oxidation than microcrystalline wax.

Synthetic paraffin waxes have been produced commercially in the past decade by the Fischer-Tropsch reaction technique. Their unique character and high purities make them suitable for replacing certain natural vegetable waxes and for modifiers for conventional petroleum waxes. Melting points of the synthetic waxes are uniformly higher than for the conventional petroleum grades, ranging from 214 to 220°F. They are capable of taking a high gloss.

Oxidized microcrystalline waxes are now available for applications which require an appreciable saponifiable content.[58] Here petroleum wax comes the closest to replacing the more expensive products such as Carnuba wax. After oxidation, the wax has a melting point of 180 to 200°F, is moderately hard, and has a good color. Tank-bottom-type microcrystalline wax is the preferred material for oxidation.

PETROCHEMICALS

The oil and natural gas companies, with the chemical manufacturing concerns, have spawned a large-scale industry which produces chemical products from petroleum. Some 30 per cent of the chemicals manufactured in the United States today are petrochemicals or from petrochemical materials.

Hardly two and a half decades old, the petrochemical industry has grown until today over 300 companies are engaged in this field in the United States and Canada. There are about 700 operating petrochemical plants. In 1952, when data on petrochemical operations first became classified separately, 105 companies operated 176 plants. Many oil companies have set up separate petrochemical subsidiaries; some are joint ventures with chemical companies.

Capital investment in petrochemical plants in the United States in 1963 was estimated at $12.9 billion. Investment in petroleum refining facilities was put at around $16.5 billion.

A widely accepted definition of a petrochemical reads: "A chemical compound or element recovered from petroleum or natural gas, or derived in whole or in part from petroleum or natural gas hydrocarbons and intended for chemical markets."[60] The hundreds of individual petrochemical items include synthetic rubbers, plastics, synthetic fibers, detergents, solvents, sulfur, ammonia and ammonia fertilizers, and carbon black.

Petrochemicals are produced from the entire gamut of hydrocarbons, ranging from those found in natural and refinery gases to refinery sludges and bottoms from lubricating-oil stocks, sometimes from crude oils. Their entire output consumes only about 5.5 per cent of the natural gas and crude oil produced in the United States today.

Some petrochemical plants are units of petroleum refineries; some are adjacent to refineries and draw their feedstocks by pipeline from the refinery. Some plants are situated along natural-gas transmission lines near their markets and take gas from the line for processing. A few plants are located in market areas and transport their raw materials long distances.

Scope of Operations

Total output in 1964 of the varied groups of petrochemicals produced in the United States was 83.3 billion pounds. Estimated output for 1965 was 92.3 billion pounds. This latter figure approaches five times the 21.4-billion-pound output of petrochemicals in the United States in 1952. Figure 11-14 shows output for the period 1952 to 1964.

Marketed value of petrochemicals in 1964 was $7.91 billion, and the estimated value for 1965 was $8.73 billion. The value of the 1952 production was $2.7 billion. Table 11-49 shows volumes and values of U.S. petrochemicals by types.

The rate of growth of petrochemicals marketed exceeds the growth of the entire chemical manufacturing industry. Their output in 1965 of 92.3 billion pounds was 32.6 per cent of total chemicals production of 282.8 billion pounds. In 1953 the volume of petrochemicals was 25.3 billion pounds, 22.0 per cent of all chemicals. The market values of petrochemicals in 1965, $8.73 billion, was 60.1 per cent of the value of all chemicals. In 1953 the value of the petrochemicals, $3.22 billion, was

52.7 per cent of the value of all chemicals. The higher relative value of the petro-chemicals to all chemicals, as compared with their volume ratios, is due to the large production of the high-value aliphatic petrochemicals.

Of the total 1965 output of petrochemicals of 92.3 billion pounds, the largest production, 53.5 billion pounds, was in the aliphatic group, which includes butadiene, the alcohols, and other large-volume chemicals. The volume of aliphatic petro-chemicals produced has nearly tripled since 1953 and now supplies about 90 per cent of the volume of all aliphatic chemicals. The value of the aliphatic output was 58 per cent of the total value of petrochemicals output in 1965.

The volume of the aromatic petrochemicals in 1965 was 14.9 billion pounds, about 16 per cent of total petrochemical output. This group has had the largest growth in production of the petrochemicals. The output in 1965 was nearly five times that in 1955 and reflects the increasing use of petroleum benzene, toluene, and xylenes to

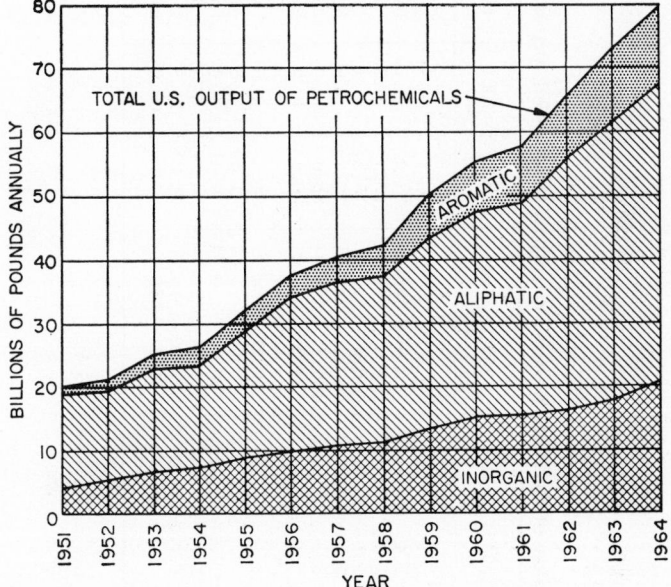

FIG. 11-14. Growth pattern of U.S. petrochemical production since 1951.

replace the coal aromatic chemicals. The value of the aromatic petrochemicals in 1965 was over 18 per cent of the total petrochemicals.

The output of inorganic petrochemicals in 1965, 23.9 billion pounds, was 24 per cent of total petrochemicals output. The gain in this group of nearly 170 per cent from 1955 was largely in ammonia and ammonia chemicals.

The figures on volume and value of petrochemicals appearing in Table 11-49 were prepared from data compiled in accordance with a system established by Dr. Robert L. Bateman, formerly with Union Carbide Corp. His system handles these data to avoid counting hydrocarbon molecules more than once during their conversion into finished chemical products. The totals therefore are frequently below figures issued by the government, in which the total is pyramided by counting a molecule each time it passes from one processor to another.

Future Growth

Forecasts of the expansion in petrochemicals in coming years are based on one premise which, it seems, will continue to hold good. The raw materials from which

Table 11-49. Volume and Value of Petrochemicals Output in the United States, 1930 to 1965

	1930	1940	1945	1950	1955	1959	1960	1961	1962	1963	1964	1965*
Volume, billions of lb:												
Aliphatic	0.1	1.3	5.4	11.77	20.0	30.1	32.2	33.7	39.6	43.4	48.7	53.5
Aromatic		0.05	1.4	0.95	3.2	6.5	7.9	8.7	9.8	11.3	13.4	14.9
Inorganic	0.38	1.04	1.15	3.42	9.0	13.6	15.2	15.4	16.4	18.2	21.2	23.9
All petrochemicals	0.48	2.39	7.95	16.14	32.2	50.2	55.3	57.8	65.8	72.9	83.3	92.3
All chemicals	....	....	....	....	134.8	184.3	185.0	192.6	225.7	244.8	267.4	282.8
% petrochemicals	....	....	....	....	24.0	27.2	30.0	30.0	29.1	29.8	31.2	32.6
Value, billions of $:												
Aliphatic	....	....	....	....	2.77	4.21	4.83	5.04	4.75	5.44	5.57	6.12
Aromatic	....	....	....	....	0.58	1.04	1.30	1.35	1.37	1.36	1.47	1.63
Inorganic	....	....	....	....	0.36	0.54	0.61	0.62	0.66	0.73	0.87	0.98
All petrochemicals	....	....	....	....	3.71	5.79	6.74	7.01	6.78	7.53	7.91	8.73
All chemicals	....	....	....	....	7.22	10.25	11.23	11.5	11.89	12.88	13.53	14.49
% petrochemicals	....	....	....	....	51.4	56.5	60.2	60.9	57.0	58.4	58.5	60.1

* Estimated.

petrochemicals are manufactured, natural gas and petroleum, will continue to be available in abundant supply and at relatively stable prices.

More than 130 new plants or additions were reported under way at the end of 1964. This compares with 47 such projects in 1958.

Petrochemical output in 1965 was estimated at 92.3 billion pounds, compared with 83.3 billion in 1964. Various authorities estimate that output will be well over 100 billion pounds by 1968.

Plastics apparently will be, for the next few years, the fastest growing outlets for some petrochemicals. Total production of plastics in 1964 was estimated at 9.75 billion pounds, with 1965 production estimated at 10.481 billion pounds.[62]

Types and Sources of Petrochemicals

There are three general groups of petrochemicals, based on their chemical composition and structure: (1) aliphatic, (2) aromatic, and (3) inorganic.[60]

An aliphatic petrochemical is an organic compound (a compound containing carbon) which has an open chain of carbon atoms, whether normal or branched, saturated or unsaturated. Important petrochemicals in this group include acetic acid, acetic anhydride, acetone, butadiene, ethyl alcohol, ethyl chloride, ethylene dichloride, ethylene glycol, formaldehyde, isopropyl alcohol, and methyl alcohol. Most of the aliphatic petrochemicals are made from the C_1 through C_4 hydrocarbons (methane, ethane, propane, butane).

An aromatic petrochemical is also an organic compound but one that contains or is derived from the basic benzene ring (six carbon atoms). Important in this group are benzene, toluene, and xylene, commonly known as the B-T-X group, plus phenol and styrene.

Figure 11-15 gives the structural chemical formulas for some typical petrochemicals in each of these groups and illustrates the difference between the open-chain aliphatics and the benzene-ring aromatics.

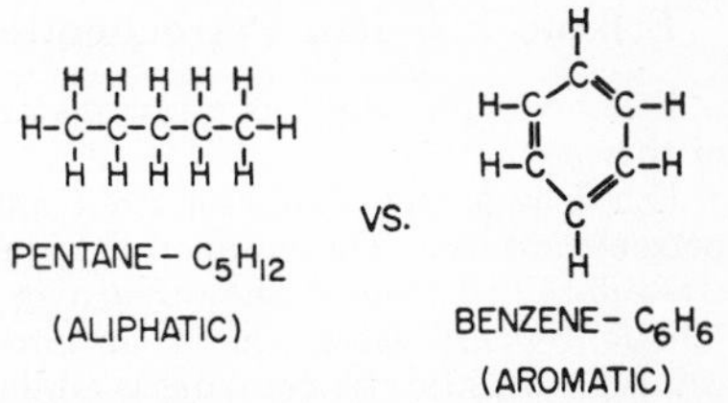

PENTANE– C_5H_{12}

(ALIPHATIC)

VS.

BENZENE– C_6H_6

(AROMATIC)

FIG. 11-15. Typical structural chemical formulas, showing difference between aliphatic (open-chain) and aromatic (closed-ring) petrochemicals.

An inorganic petrochemical is one which does not contain carbon atoms. Typical here are sulfur, ammonia and its derivatives (nitric acid, ammonium nitrate, ammonium sulfate, urea), and carbon black. Carbon black is included in this group because it is a basic element carbon, not a compound containing carbon (see later discussion of carbon black).

The aliphatics represent by far the bulk of all petrochemicals, over 58 per cent. They are also the most important group dollarwise. Although the aromatics are the smallest in volume, the inorganics are the least important in terms of value, as shown in Table 11-49.

The bulk of petrochemicals use the light C_1-C_4 hydrocarbons as raw materials. These include methane, ethane, propane, and butane (both iso- and normal butanes). All these gases are found in natural gas, which is their principal source for petrochemical manufacture, but they are also recovered from refinery gases. The refinery gases are especially valuable because they contain quantities of the more reactive olefins.

Also important as raw materials for petrochemicals are the aromatic hydrocarbons—benzene, toluene, and xylene. They are not naturally present in crude oil, except in rare instances and in small amounts. By means of catalytic re-forming, the non-aromatic hydrocarbons can be converted into aromatics through dehydrogenation and cyclization.

It must be borne in mind that many petrochemicals can be made from more than one hydrocarbon raw material. For example, acetaldehyde can be made from methane (via acetylene), from ethane (via dehydrogenation or oxidation of ethyl alcohol), and from propane and butane (via oxidation). The raw material which a particular plant operator might want to use will depend on what is economically available and on what other products he also wants to make.

Methane Petrochemicals

Methane, the major component of natural gas, is the source of liquid anhydrous ammonia, which is used as a fertilizer as such or converted into other petrochemicals. It can be oxidized to produce nitric acid, which is reacted with more ammonia to yield ammonium nitrate, which is used both for fertilizer and explosives. Ammonia is reacted with sulfuric acid to yield ammonium sulfate, another fertilizer. Urea, used as a fertilizer and chemical intermediate and to make plastics, adhesives, and fire-resistant paints, is made from ammonia by reacting it with carbon dioxide.

Methyl alcohol, or methanol, is an important methane-derived petrochemical, largest use of which is in making formaldehyde. Large quantities go into nonpermanent antifreezes. Isobutyl alcohol is a byproduct of methanol synthesis.

Carbon black is made from methane by controlled combustion of natural gas. At present this use of methane is declining, while petroleum distillates and residual oils are being used to a greater extent to produce carbon blacks. (The manufacture and uses of carbon black are discussed later in this section.)

Ethane-Ethylene Petrochemicals

Ethane is important as a raw material primarily as a source of ethylene by means of thermal cracking.

Ethylene is the most important and versatile raw material for the production of petrochemicals. The number of primary products, secondary products, and derivatives obtained from it runs well over a hundred. While ethylene is obtained from ethane, refinery gases are also an important source. The increased use of synthetic catalysts in catalytic cracking is adding to the supply of refinery ethylene.

Polyethylene, the "squeeze-bottle plastic," made by the polymerization of ethylene, is the most popular of the ethylene petrochemicals, and its uses are expanding rapidly.

Ethylene oxide is a major ethylene derivative, made by direct oxidation and by the chlorohydrin route. The oxide, in turn, is converted to ethylene glycol (permanent antifreeze), the various ethanolamines, acrylonitrile, and other derivatives.

Ethyl alcohol, or ethanol, is another important ethylene derivative. It is made by the direct catalytic hydration of ethylene or by absorbing ethylene in sulfuric acid to give a mixture of ethyl sulfates, which in turn is hydrolized to crude ethyl alcohol. Ethyl alcohol is used as an antifreeze and as a solvent and extraction agent. It is a source for acetaldehyde and other chemicals.

Ethyl benzene, made by the reaction of ethylene and benzene, is used in the production of styrene, which is important in the manufacture of synthetic rubber and polystyrene. Ethylene chloride and dichloride are other important ethylene petrochemicals.

Propane-Propylene Petrochemicals

Propane is the largest single hydrocarbon raw material for petrochemicals. Propane can be oxidized directly to give methyl alcohol, formaldehyde, acetaldehyde, acetone, other alcohols, and oxygenated products. The bulk of the propane, however, is used to make the more reactive olefins, ethylene, and propylene.

Isopropyl alcohol is one of the most important petrochemicals derived from propylene. This alcohol has a number of important uses, a major one being the conversion to acetone by dehydrogenation. Acetone finds application as a solvent and extractant and in the manufacture of plastics. Propylene trimer and propylene tetramer and propylene oxide are important among the petrochemical products.

Chlorination of propylene yields allyl chloride, which is a source material for allyl alcohol, synthetic glycerin, and epoxy resins.

A relatively new propylene petrochemical is polypropylene, a high-molecular-weight material which gives promise of being competitive in some applications with polyethylene and styrene resins.

Butane-Butylene Petrochemicals

Butane is a major component of liquefied petroleum gas (LP gas). Isobutane and the butylenes are feedstocks for poly gasoline and alkylates. Some quantities of the butanes and butylenes go into the manufacture of petrochemicals. The oxidation of butane yields acetic acid, acetaldehyde, methyl alcohol, propionic acid, butyric acid, and propyl and butyl alcohols. Through catalytic dehydrogenation, butane is converted into normal and isobutylenes. Most of the n-butylenes are used to make butadiene, an important component in GR-S synthetic rubber. Also made from n-butylene is secondary butyl alcohol, which is used to make methyl ethyl ketone, the solvent. Isobutylene is used largely to make butyl rubber in combination with isoprene.

Aromatic Hydrocarbon Petrochemicals

Most important as raw material for petrochemicals are the B-T-X group of the aromatics, benzene, toluene, and xylene. These products traditionally in the past have been recovered as byproducts of the coal-tar industry. They are now being recovered in increasing volumes from petroleum by the catalytic re-forming process.

Benzene is widely used in reactions with other petrochemicals. With ethylene it gives ethyl benzene, which is converted to styrene, an important synthetic-rubber component. Reacted with propylene, benzene gives cumene, an intermediate for phenol and acetone. As a raw material, benzene is used to make phenol. A growing use is in the manufacture of adipic acid for nylon.

Toluene is largely used as a solvent and in the manufacture of trinitrotoluene for explosives.

Xylene is recovered as three separate isomers—paraxylene, metaxylene, and orthoxylene—which in turn are source materials for polyester fibers, isophthalic acid, and phthalic anhydride, among other petrochemicals.

Miscellaneous Petrochemical Sources

Aside from the major hydrocarbon sources described in the preceding, other miscellaneous sources for petrochemical manufacture include:[60]

Hydrogen sulfide, recovered from natural and refinery gases, is an important source of elemental sulfur.

Hydrogen, recovered as a byproduct of the catalytic re-forming process, is used for ammonia synthesis in a few refineries.

Pentane, one of the light hydrocarbons recovered from the distillation of crude and from cracking and refining operations, can be converted to amyl alcohols.

n-Pentanes are a source of amyl phenol.

Cyclohexane can be oxidized to adipic acid for nylon production.

Heptene is converted to isooctyl alcohol by the Oxo process.

Cresols (cresylic acids—ortho-, meta-, and paracresol isomers) produced during certain refining operations are recovered from waste streams as marketable products.

Naphthenic acids occurring in some crude oils and removed during topping unit operations can be recovered as sodium naphthenates.

Naphthalene is recovered by distillation from crude and cracked petroleum.

Sulfonates, in the past removed as byproducts from the sulfuric acid treatment employed in refining white oils, today are recovered as primary petrochemicals.

Isopentane and isopentene are dehydrogenated to give isoprene for butyl synthetic rubber.

Paraffin waxes are halogenated to give chlorinated paraffins.

Gas oils and even heavier fractions are cracked to yield light hydrocarbons (particularly ethylene) and in a few cases are used to make the synthesis gas for ammonia. The major commercial petrochemicals and their important properties are listed in Table 11-50.

Table 11–50. Important Properties of Major Commercial Petrochemicals

Petrochemical	Chemical formula	Physical appearance	Melts at		Boils at		Specific gravity	Density, lb/gal	Hydrocarbon raw material
			°F	°C	°F	°C			
Acetaldehyde	CH_3CHO	Colorless liquid	−190.3	−123.5	68.4	20.2	0.783 (18/4°C)	6.50 (20°C)	Methane, ethane, propane, butane
Acetic acid	CH_3COOH	Colorless liquid	61.9	16.6	244.6	118.1	1.051 (20/20°C)	8.64 (20°C)	Methane, ethane, butane
Acetic anhydride	$(CH_3CO)_2O$	Colorless liquid	−90.4	−68	283.8	139.9	1.0830 (20/20°C)	9.0 (20°C)	Ethane
Acetone	CH_3COCH_3	Colorless liquid	−137.7	−94.3	133	56.1	0.7972 (15°C)	6.64 (15°C)	Propane
Acetylene	C_2H_2	Colorless gas	−115.0	−81.8	−119.2	−84	0.91 (air = 1)		Methane
Acrolein	CH_2CHCHO	Colorless to yellowish liquid	−125.9	−87.7	126.5	52.5	0.8427 (20/20°C)	7.03 (20°C)	Propylene
Acrylonitrile	CH_2CHCN	Colorless liquid	−117.4	−83	171.3	77.4	0.8004 (25°C)	6.7 (20°C)	Methane, ethylene
Adipic acid	$COOH(CH_2)_4COOH$	Colorless to yellowish crystals	305.6	152	509	265 (at 100 mm)	1.360 (20/4°C)		Cyclohexane
Adiponitrile	$CN(CH_2)_2CN$	Colorless liquid			582.8	306	0.962 (20/4°C)		Methane
Allyl alcohol	CH_2CHCH_2OH	Colorless liquid	−200.2	−129	206.4	96.9	0.8520 (20/4°C)	7.11	Propylene
Allyl chloride	CH_2CHCH_2Cl	Colorless liquid	−210.1	−134.5	113	45	0.9382 (20/4°C)	7.83	Propylene
Ammonia	NH_3	Colorless gas or liquid	−107.9	−77.7	−28.1	−33.4	0.5971 (gas) 0.817 (liquid, 79°C)		Methane
Ammonium nitrate	NH_4NO_3	Colorless or white solid	337.3	169.6	410 (decomposes)	210 (decomposes)	1.725		Methane
Ammonium sulfate	$(NH_4)_2SO_4$	White to brownish-gray crystals	955.4 (decomposes)	513 (decomposes)			1.77		Methane
Amyl alcohols	$C_5H_{11}OH$	Colorless liquid	−110.0	−78.9	280.4	138	0.8240 (20/20°C)	6.9 (20°C)	Pentane
Amyl chlorides	$C_5H_{11}Cl$	Colorless liquid	−76	−60			0.88 (20°C)	7.33	Pentane
Amyl naphthalenes	$C_{10}H_7 \cdot C_5H_{11}$	Yellow to brown liquid					0.965 (20°C)		n-Pentenes
Amyl phenol	$C_5H_{11}C_6H_4OH$	Straw-colored solid						8.0	n-Pentenes
Benzene	C_6H_6	Colorless liquid	42	5.56	176.2	80.1	0.879 (20/4°C)	7.32	Benzene
Benzene hexachloride	$C_6H_6Cl_6$	White to yellowish solid	234	112	(gamma isomer-lindane)				Butane
Butadiene	$CH_2CHCHCH_2$	Colorless gas	−164	−108.9	24.1	−4.41	0.6211 (20°C)		Butane
n-Butyl alcohol	$CH_3CH_2CH_2CH_2OH$	Colorless liquid	−111.8	−79.9	243.9	117.7	0.810 (20/4°C)	6.75 (20°C)	Butane
sec-Butyl alcohol	$CH_3CH_2CHOHCH_3$	Colorless liquid	−174.5	−114.7	211.1	99.5	0.808 (20/4°C)	6.74	Butylenes
Butylene-1	$CH_3CHCHCH_2$	Colorless gas	−202	−130	20.7	−6.3			Butane
Butylene-2	$CH_3CHCHCH_3$	Colorless gas	−196.6 (both forms)	−127	38.7 (cis), 33.6	3.7 (cis form), 0.88 (trans form)			Butane
Butyric acid	$CH_3CH_2CH_2COOH$	Colorless liquid	−17.8	−7.9	326.7	163.7	0.9587 (20/4°C)		Butane
Carbon black	C	Black amorphous solid	6332	3500	7592	4200	1.8–2.1		Methane, oils
Carbon dioxide	CO_2	White solid or colorless liquid or gas	−68.8 (at 5 atm.)	−56	−108.8	−78.2	1.057 (liquid) 1.56 (solid)		Methane

Name	Formula	Appearance					Sp gr		Combustion products
Carbon disulfide	CS_2	Colorless liquid	−163.5	−108.6	115.3	46.3	1.263 (20/4°C)	10.48	Methane
Carbon tetrachloride	CCl_4	Colorless liquid	−9.4	−23	170.2	76.8	1.595 (20/4°C)	13.22	Methane
Chloroform	$CHCl_3$	Colorless liquid	−82.3	−63.5	142.2	61.2	1.489 (20°C)	12.4	Methane
Cumene	$C_6H_4CH(CH_3)_2$	Colorless liquid	−140.8	−96	306.9	152.7	0.8620	7.19	Propylene, benzene
Cyclohexane	C_6H_{12}	Colorless liquid	43.3	6.3	177.3	80.7	0.779		
Diethanolamine	$NH(CH_2CH_2OH)_2$	Faintly colored liquid	82.4	28	516.4	269.1	1.092 (30/20°C)	9.1	Methane, ethylene
Diethylene glycol	$CH_2OHCH_2OCH_2CH_2OH$	Colorless liquid	17.6	−8	473	245	1.1184	9.35	Ethylene
Dipropylene glycol	$(CH_3CHOHCH_2)_2O$	Colorless liquid			449.2	231.8	1.0252 (20/20°C)	8.4 (20°C)	Propylene
Epichlorohydrin	CH_2OCHCH_2Cl	Colorless liquid	−72.6	−58.1	239.4	115.2	1.179–1.184	9.78	Propylene
Ethyl alcohol	CH_3CH_2OH	Colorless liquid	−169.6	−112	172.9	78.3	0.7905	6.578	Ethylene
Ethyl benzene	$C_6H_5C_2H_5$	Colorless liquid	−139.0	−94.98	277.2	136.2	0.867	7.21	Ethylene, benzene
Ethyl chloride	CH_3CH_2Cl	Colorless gas	−216.9	−138.3	54.1	12.27	0.917 (6/6°C)		Ethylene
Ethylene	CH_2CH_2	Colorless gas	−272.2	−169	−152.5	−102.5	0.978		Ethane, propane
Ethylene diamine	$NH_2CH_2CH_2NH_2$	Colorless liquid		9.1	242.6	117	0.892		Ethylene
Ethylene dibromide	CH_2BrCH_2Br	Colorless liquid	48.4		268	131	2.17–2.18	18.1	Ethylene
Ethylene dichloride	CH_2ClCH_2Cl	Colorless liquid	−31.9	−35.5	182.3	83.5	1.2554 (20/40°C)	10.4	Ethylene
Ethylene glycol	CH_2OHCH_2OH	Colorless liquid	1.4	−17	387	197.2	1.1155	9.31	Ethylene
Ethylene oxide	CH_2CH_2O	Colorless gas	−167.8	−111	51.3	10.7	0.882	7.25 (20°C)	Ethylene
Formaldehyde	$HCHO$	Colorless gas	−180.4	−118	−2.2	−19	0.815 (−20°C)		Methane, propane
Glycerin	$CH_2OHCHOHCH_2OH$	Colorless to yellowish liquid	64.2	17.9	554	290	1.260	10.5	Propylene
Hydrazine	H_2NNH_2	Colorless liquid	35.6	2.0	234.5	113.5	1.011		Methane
Hydrogen	H_2	Colorless gas	−434.4	−259.1	−423	−252.8	0.06948		Methane
Hydrogen cyanide	HCN	Colorless liquid	8.1	−13.3	78.1	25.6	0.6970		Methane
Hydrogen sulfide	H_2S	Colorless gas	−118.8	−83.8	−76.4	−60.2	0.1895		
Isobutyl alcohol	$(CH_3)_2CHCH_2OH$	Colorless liquid	−162.4	−108	226.2	107.9	0.798 (25/4°C)	6.65	Methane, butane
Isobutylene	$(CH_3)_2CCH_2$	Colorless gas	−218.2	−139	19.6	−6.9			Butane
Isophthalic acid	$C_6H_4(COOH)_2$	White crystalline solid	662	350					Metaxylene
Isoprene	$CH_2C(CH_3)CHCH_2$	Colorless liquid	−230.8	−146	93.4	34.08	0.6808 (20/4°C)	6.55	Isopentane, isopentene
Isopropyl alcohol	$(CH_3)_2CHOH$	Colorless liquid	−126.2	−97.9	180.3	82.4	0.7863		Propylene
Maleic anhydride	$(CHCO)_2O$	White needles or flakes	127.4	53	391.5	199.7	0.934 (20/4°C)		Benzene
Metaxylene	$C_6H_4(CH_3)_2$	Colorless liquid	−53.3	−47.4	281.8	138.8	0.8684 (15°C)	6.59	Methane, propane, butane
Methyl alcohol	CH_3OH	Colorless liquid	−144	−97.9	148.5	64.7	0.792 (20/4°C)	7.68	Methane
Methyl chloride	CH_3Cl	Colorless gas	−143.7	−97.6	−10.7	−23.7	0.92	6.71	Methane
Methyl ethyl ketone	$CH_3COCH_2CH_3$	Colorless liquid	−123.3	−86.4	175.3	79.6	0.805 (20/4°C)		n-Butylene
Methyl mercaptan	CH_3SH	Colorless gas	−189.4	−123.1	45.7	7.6			
Methyl styrene	$C_6H_5C(CH_3)CH_2$	Colorless liquid	−9.8	−23.2	329.7	165.4	0.9062 (25°C)	11.07	Propylene, benzene
Methylene chloride	$CHCl_2$	Colorless liquid	−142.6	−97	107.6	42	1.335 (15/4°C)	8.4	Methane
Monoethanolamine	$NH_2CH_2CH_2OH$	Colorless liquid	176.4	80.2	338.9	170.5	1.0179		Ethylene
Naphthalene	$C_{10}H_8$	White flakes or powder	−42.3	−41.3	424.2	217.9	1.152		
Nitric acid	HNO_3	Colorless to yellowish liquid			186.8	86	1.530		Methane
Nonyl phenol	$C_9H_{19}C_6H_4OH$	Straw-colored liquid					0.968		
Orthoxylene	$C_6H_4(CH_3)_2$	Colorless liquid	−13.9	−25.5	291.9	144.4	0.880 (20/4°C)	7.36	Propylene

Table 11-50. Important Properties of Major Commercial Petrochemicals (Continued)

Petrochemical	Chemical formula	Physical appearance	Melts at °F	Melts at °C	Boils at °F	Boils at °C	Specific gravity	Density, lb/gal	Hydrocarbon raw material
Paraxylene	$C_6H_4(CH_3)_2$	Colorless liquid or crystals	55.8	13.2	281.3	138.5	0.861 (20/4°C)	13.46 (26°C)	Methane
Perchloroethylene	CCl_2CCl_2	Colorless liquid	−8.3	−22.4	250	131	1.60–1.615		Propylene, benzene
Phenol	C_6H_5OH	White crystals			358.5	181.4	1.071 (25/4°C)		Orthoxylene, naphthalene
Phthalic anhydride	$C_6H_4(CO)_2O$	White flakes or needles	267.4	130.8	544.1	284.5	1.527 (4°C)		Ethylene
Polyethylene	$(C_2H_4)_x$	Colorless liquid to white solid							
Polystyrene	$(C_6H_5CHCH_2)_n$	Colorless solid							Ethylene, benzene
Propionic acid	CH_3CH_2COOH	Colorless liquid	−7.6	−22	285.3	140.7	0.987		Butane
Propyl alcohol	$CH_3CH_2CH_2OH$	Colorless liquid	−197	−127	208	97.8	0.804 (20/4°C)	6.7	Butane
Propylene	CH_3CHCH_2	Colorless gas	−301.4	−185.1	−52.6	−47	1.46 (0°C)		Propane
Propylene dichloride	$CH_3CHClCH_2Cl$	Colorless liquid	−112	−80	205.3	96.3	1.1583	9.6	Propylene
Propylene glycol	$CH_3CHOHCH_2OH$	Colorless liquid			370.8	188.2	1.0381	8.6	Propylene
Propylene oxide	$CH_3(CHCH_2)O$	Colorless liquid	−155.9	−104.4	93	33.9	0.8304	6.9	Propylene
Propylene tetramer	$CH_3(CH_2)_2CH_3$	Colorless liquid	−28.5	−33.6	415.4	213	0.7600		Propylene
Propylene trimer	$CH_3(CH_2)_2CHCH_2$	Colorless liquid			302	150	0.7433		Propylene
Styrene	$C_6H_5CHCH_2$	Colorless to yellowish liquid	−23.1	−30.6	293.4	145.2	0.903 (20/4°C)	7.55	Ethylene, benzene
Sulfur	S	Amorphous or solid yellow element	248 (approx)	120 (approx)	832.3	444.6	1.9556 (amorphous), 2.046 (solid)		
Terephthalic acid	$C_8H_6O_4$	White crystals or powder							Paraxylene
Thiophene	C_4H_4S	Colorless liquid	−37.3	−38.5	183	84	1.0644 (20/4°C)		Butane
Toluene	$C_6H_5CH_3$	Colorless liquid	−139.2	−95.1	231.4	110.8	0.866 (20/4°C)	7.21	
Trichloroethylene	$ClCHCCl_2$	Colorless liquid	−99.4	−73	189	87.2	1.466	12.2	Methane
Triethanolamine	$N(CH_2CH_2OH)_3$	Pale-yellow liquid	70.2	21.2	680	360	1.12	9.4	Ethylene
Triethylene glycol	$HO(C_2H_4O)_2C_2H_4OH$	Colorless liquid	19	−7.2	549.3	287.4	1.1254	9.4	Ethylene
Urea	NH_2CONH_2	White crystals or powder	270.9	132.7	decomposes	decomposes	1.335 (20/4°C)		Methane
Vinyl chloride	CH_2CHCl	Colorless gas	−255.5	−159.7	7	−13.9	0.9121	7.6	Methylene, ethylene
Xylenes	$C_6H_4(CH_3)_2$	Colorless liquid	(see data for isomers—meta-, ortho-, paraxylene)						

CARBON BLACK

While this product comes within the definition of a petrochemical, being an element recovered from petroleum and natural gas, its data are generally reported separately.[66]
Carbon black is a semigraphitic form of carbon prepared in a fine state of subdivision by the partial combustion of hydrocarbons. It is the most finely divided and blackest pigment available to industry. Its remarkable jetness and high tinctorial power give it preference over other black pigments such as lampblack and bone black. It has remarkable ability to reinforce and strengthen rubber, and the use of this material by the rubber-tire industry accounts for 90 per cent of the production of carbon black. An early use was as a pigment in the printing-ink and paint and lacquer industries. While first used to color plastic articles, it has been recently

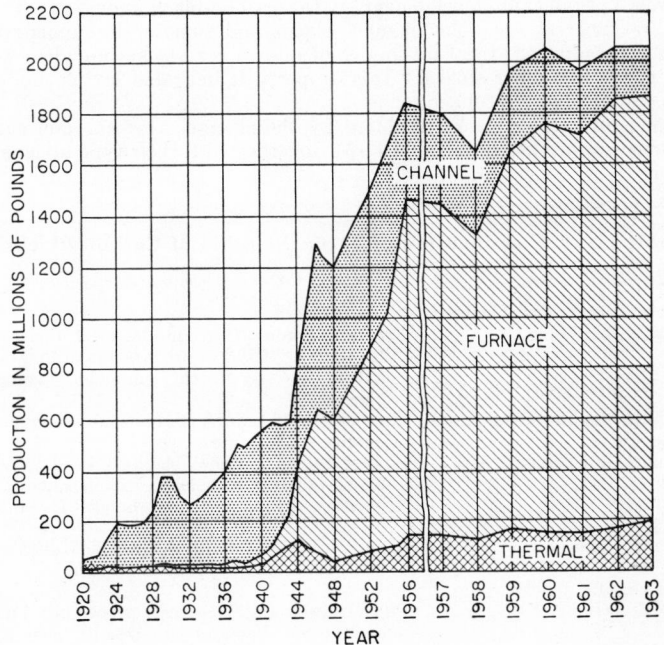

Fig. 11-16. United States production of carbon black by processes, 1920–1963.[66]

found to be a strengthening agent for some plastic applications. Among other small-scale uses is the manufacture of dry cells and carbon paper.

Total production of carbon black in the United States in 1963 was 2,056,205,000 lb, its growth in the past two decades paralleling the expansion in the tire-manufacturing industry. Of the 1962 production, 1,683,796,000 lb were furnace blacks, 178,321,000 lb channel blacks, and 194,088,000 lb thermal blacks. The growth rate since 1920 is shown in Fig. 11-16.

Manufacture and Types

The properties of carbon black are largely determined by the process by which it is manufactured. The channel process involves burning natural gas in a deficiency of air.[67] When the process is modified, particle size of the black can be altered through a range of 10 to 30 millimicrons. These fine-particle-size blacks are used in the paint and lacquer industry and in high-grade printing inks.

Coarser blacks, of 200 to 500 millimicrons diameter, are produced by the thermal process, in which natural gas is passed on a "make" cycle through checker-brick furnaces preheated to 2500 to 3000°F. These blacks, which produce softer rubber stocks, are used in tire carcasses.

The desire to improve recovery and produce blacks with reinforcing ability intermediate between channel and thermal blacks resulted in the development of the gas-furnace process. Here special slotted or drilled burners are used, through which natural gas and air in alternating layers are fired into large insulated furnaces. The blacks produced by this process have a particle size of 70 millimicrons. Although they do not provide sufficient reinforcement for use in tire treads, they are superior to thermal blacks and have also found application in a wide variety of rubber goods in which abrasive wear is not a factor.

The substitution of liquid hydrocarbons or oils for natural gas in the furnace process was a significant development in the carbon-black industry. The blacks produced are finer in size than channel blacks and superior in rubber-reinforcing properties. In addition, the finer grades of oil-furnace blacks provide a high level of conductivity in rubber stocks. This property is desirable in the production of antistatic and conducting rubber goods.

The grades of carbon blacks produced by the channel, thermal, and gas- or oil-furnace processes are listed in Table 11-51, together with their type designation and applications.

Table 11-51. Classification of Typical Grades of Carbon Blacks[67]

Grade and type	General properties and uses
A. Color and ink channel blacks:	
Channel, high color..........	Premium grades of carbon black. Impart maximum blackness. Used in preparation of highest quality automotive synthetic enamels, lacquers, and plastics
Channel, medium color......	Used chiefly in paints, enamels, and lacquers of good quality
Channel, low color..........	Lowest priced channel blacks. Used in news inks, black paper, inexpensive enamels, coatings, and plastics. General-utility blacks for inexpensive color
Channel, long flow..........	Lithographic and halftone inks. Multiple-use carbon papers and typewriter ribbons
B. Rubber channel blacks:	
Easy processing, EPC.......	Fully reinforcing. Easiest processing and lowest hysteresis. Used in tire treads, heels and soles, mechanical goods
Medium processing, MPC....	Fully reinforcing. Medium in processing and in properties. Used in tire treads, heels and soles, mechanical goods
Hard processing, HPC.......	Fully reinforcing. Hard processing. Hard stock. Maximum hysteresis. Used in tire treads
Conductive, CC.............	Fully reinforcing. Hard processing. High hysteresis. Good electrical conductivity. Used in sand-blast hose, conductive soles and heels
C. Rubber gas-furnace blacks:	
Semireinforcing, SRF........	Semireinforcing. High loading capacity. Easy processing. Good resilience and flex resistance. Used in tire carcass and bead insulation, mechanical goods, footwear and soling, wire jackets, belts, hose, packings
High modulus, HMF........	Moderate reinforcement and resilience. Easy processing. Used in tire carcass and side walls, footwear, mechanical goods
Fine furnace, FF............	Good reinforcement. Moderate heat build-up. Used in truck-tire carcass, breaker, cushion

Table 11-51. Classification of Typical Grades of Carbon Blacks[67] (Continued)

Grade and type	*General properties and uses*
D. Rubber oil-furnace blacks:	
General purpose, GPF.......	Moderate reinforcement, high resilience, low compression set, good processing smooth-out. Used in tire carcass, tread base, side walls, sealing rings, cable jackets, hose, soling, extruded stripping
Fast extruding, FEF.........	Good reinforcement and resilience, excellent processing, low compression set. Used in tire carcass, tread base and side wall, butyl inner tubes, hose, extruded stripping
High abrasion, HAF.........	Fully reinforcing. Excellent processing. Moderate heat build-up. Good flex resistance. Used in tire treads and camel back, heels and soles, mechanical goods
Intermediate abrasion, ISAF.	Fully reinforcing. Good processing. Medium heat build-up. Good flex resistance. Used in tire treads and camel back, heels and soles, mechanical goods
Superabrasion, SAF.........	Maximum reinforcement. Hard processing. High heat build-up. Used in tire treads and camel back, heels and soles, mechanical goods
Conductive, CF.............	Excellent electrical conductivity. Good retention after overmilling or flexing. Used in antistatic and conductive rubber goods, belts, hose, flooring, soling, heels, static straps
E. Rubber thermal blacks:	
Medium thermal, MT........	Lowest reinforcement. High loading capacity. Soft stocks. High resilience. Low stress-strain properties. Used in wire insulation and jackets, mechanical goods, footwear, belts, hose, packings, stripping mats, proofed goods
Fine thermal, FT............	Low reinforcement, low hardness, high elongation, good tear resistance, high resilience, and good flex resistance. Used in natural-rubber inner tubes, inflations, footwear uppers, mechanical goods

Properties and Evaluation

Since carbon black retains its identity and remains as a discrete phase in all applications, its performance can be predicted to a large degree on the basis of particle size.[67] The ultimate particles of even the coarsest blacks are too small to be seen under the optical microscope. However, under the electron microscope, the particles are clearly resolved. The structure of carbon blacks has been established by X-ray diffraction studies, and these studies have substantiated the theory that it is the overall dimension of the particle which determines the level of properties for the individual grades of blacks.

Measurement of particle size or the related property of surface area constitutes one of the principal procedures for evaluating and controlling quality of carbon blacks. Determination of particle size by the electron microscope provides precise data for standardization but is hardly suited for plant control. Measurement of surface area is a more common practice. The most reliable procedure for this is by determination of the nitrogen-adsorption isotherm at 195°C.[68]

Of the chemical properties of carbon black, volatile content is important in many applications. The blacks contain varying amounts of combined oxygen and hydrogen. The combined oxygen and hydrogen are released from the black as CO, CO_2, and H_2 at 1000°C in an accepted test procedure. The loss in weight on heating in the absence of air is termed the volatile content and is reported as percentage by weight of the original sample of the black. High-color and lithographic-ink blacks may contain 12 to 18 per cent of volatile matter, the rubber grades of channel blacks around 5 per cent, and furnace blacks usually contain less than 1.5 per cent.

Table 11-52. Sales of Carbon Black to Key Industries, 1956 to 1963[69]

(Thousands of pounds)

Industry	1956	1957	1958	1959	1960	1961	1962	1963
Rubber	1,244,651	1,271,562	1,192,162	1,463,239	1,362,912	1,382,893	1,551,204	1,629,905
Ink	42,047	43,153	40,645	47,366	47,980	42,987	41,162	46,471
Paint	13,231	11,951	10,997	13,828	12,270	15,267	15,766	13,008
Other	3,100	4,700	7,133	7,816	6,456	18,858	31,765	38,036
Total domestic	1,303,029	1,331,366	1,250,937	1,532,249	1,429,618	1,460,005	1,639,897	1,727,420
Export	425,328	459,671	440,542	513,143	543,032	522,331	442,437	370,928
Total sales	1,728,357	1,791,037	1,691,479	2,045,392	1,972,650	1,982,336	2,082,334	2,098,348

Other quality evaluations of blacks, such as ash and grit, are checked regularly and held within specification limits.

Applications of Carbon Blacks

The rubber industry consumes over 90 per cent of all carbon black produced, the largest use being in the manufacture of tires. The reinforcing ability of carbon black is of great importance here. The tread of a passenger-car tire contains about 30 per cent by weight of reinforcing black and will provide some 40,000 miles of road wear, many times the wear possible to obtain with an inert rubber filler, such as clay or whiting. The toughness and wear properties imparted by carbon black apply to both natural and synthetic rubber.

High carbon-black loadings are also employed in tire carcasses, tread bases, side walls, and inner tubes, providing such specific properties as resistance to abrasive wear and to cracking and lowering the generation of heat. More than a dozen separate grades of carbon black are used by the tire-manufacturing industry, the major difference among the grades being particle size.

In extruded rubber goods, blacks are used to provide fast extrusion rates and conformity to die dimensions. The coarser grades of black are compounded to high loadings in the manufacture of wire insulation, matting, and mechanical goods. Various blacks are compounded into rubber to provide stocks with either high electrical resistance or good conducting properties.

The use of carbon black as a pigment for printing inks is the second largest application, although much smaller than the volume used in rubber manufacture. News inks generally contain 5 per cent by weight of black in a mineral oil. Lithographic inks may contain as high as 18 per cent. Carbon black is the favored black pigment in the paint and lacquer industry. It has recently been found that the loading of certain plastics with carbon black improved their properties. The manufacture of dry cells and of carbon paper consumes small volumes of black. The sales of carbon black to U.S. industries are shown in Table 11-52.[69]

Lampblacks[67] are manufactured by slowly burning selected oils and tars in a restricted supply of air. These blacks are of large particle size, possess little reinforcing ability in rubber, and are lower in jetness and coloring power. They are of value as tinting pigments in certain paints and lacquers. In most applications they have been replaced by carbon blacks.

Acetylene blacks, produced by the thermal decomposition of acetylene, possess a high degree of structure or chaining tendency. Their particle size is about 40 millimicrons. They provide high elastic modulus and high conductivity in rubber stocks.

PETROLEUM ASPHALT

Future growth in the requirements for petroleum asphalt in the United States is directly related (1) to the highway building program of the states and the Federal government and (2) to the extent to which asphalt is used in road construction instead of other materials, principally concrete.

In 1963, a total of 16,947,389 short tons (93,210,639 bbl) of petroleum asphalt, including road oil, was used in paving. This was 74.8 per cent of total sales. Sales for roofing materials were 3,821,281 tons, 16.9 per cent. Miscellaneous sales were 1,879,151 tons, 8.3 per cent of the total, 22,647,821 tons. Our imports of natural asphalt are negligible (Table 11-53, Fig. 11-17).[70]

The extent of highway construction from year to year is a growth factor for asphalt which is beyond any direct control by the oil companies. Developing an improved product and its more efficient use in road building and maintenance are matters of continuing research study and can lead to a greater proportional use of asphalt for this purpose. Such studies are carried on by The Asphalt Institute and some of the 50 or more oil companies in this country which manufacture asphalt from crude petroleum. J. E. Buchanan, president of The Asphalt Institute, stated that the research program of this agency ranges from the molecular structure of asphalt hydrocarbons to new and improved applications for this refinery material.

Asphalt is expected to be improved for highway use, stated W. J. Sweeney, vice-president, Esso Research and Engineering Co.[71] "By the mid-1960's," he said, "we will most likely see the introduction of asphalt binders which are extremely resistant to air oxidation and weather attack. The result will be a marked reduction of cracking and pot holes. A little later on, we will probably be riding on roads and highways made of chemically-modified asphalt pavements which have several times the load bearing strength of present compounds."

Table 11-53. Consumption of Petroleum Asphalt in the United States, 1930 to 1963*

(Short tons; basis for conversion is 5.5 bbl to the ton)

Year	Paving	Roofing	Miscellaneous	Total
1930	2,617,016	1,042,653	494,325	4,153,994
1931	2,888,023	837,936	341,681	4,067,639
1932	2,726,128	689,105	371,056	3,786,289
1933	2,470,271	703,317	291,028	3,464,616
1934	2,936,076	750,063	324,893	4,011,033
1935	2,869,033	910,737	495,987	4,275,757
1936	3,793,799	1,073,819	583,243	5,450,861
1937	3,907,143	1,092,025	488,398	5,487,561
1938	4,215,629	1,124,552	426,754	5,766,935
1939	4,233,759	1,227,610	526,975	5,988,344
1940	4,686,550	1,220,920	451,485	6,358,955
1941	5,428,394	1,679,122	559,707	7,667,223
1942	5,563,830	1,846,581	618,203	8,028,614
1943	4,966,733	1,657,991	615,412	7,240,136
1944	4,852,130	1,739,977	480,969	7,073,076
1945	4,811,298	2,045,348	422,171	7,278,817
1946	5,087,111	2,568,827	797,429	8,453,367
1947	6,184,718	2,835,114	875,321	9,895,153
1948	6,712,460	2,611,092	873,736	10,197,288
1949	6,923,850	2,351,471	773,773	10,049,094
1950	7,879,804	2,846,623	1,023,039	11,749,466
1951	8,834,014	2,962,444	986,677	12,783,135
1952	10,082,336	2,599,740	992,648	13,674,724
1953	10,488,156	3,459,355	1,417,524	15,365,035
1954	11,314,373	3,250,132	1,462,777	16,027,282
1955	12,072,673	3,514,909	1,411,642	16,999,224
1956	13,701,253	3,410,814	1,638,712	18,750,779
1957	13,240,382	2,818,626	1,619,981	17,678,989
1958	14,548,826	3,101,187	1,694,062	19,344,075
1959	14,580,735	3,298,938	1,894,915	19,774,588
1960	14,673,505	3,525,024	1,855,345	20,053,874
1961	15,317,737	3,635,542	1,755,093	20,708,372
1962	17,322,046	3,841,991	1,932,221	22,096,258
1963	16,947,389	3,821,281	1,879,151	22,647,821

* Based on U.S. Bureau of Mines data.

A new technique for the treatment of railroad roadbeds may open a large-volume new market for asphalt. The roadbed is sprayed with hot asphalt, and then stone aggregate is sprinkled on the surface to seal the ballast.[72]

Asphalt is naturally present in most crude petroleum in varying degrees.[70] The removal of light distillates and a substantial portion of the heavier distillates provides asphaltic products of semisolid to solid consistency, depending on the amount of distillate remaining. These are termed asphalt cements. By permitting a large portion of the oily fraction to remain or by blending oily distillates with asphalt cement, a group of products normally termed road oils is obtained.

Blending asphalt cements with lighter petroleum distillates, such as naphthas, gasoline, or kerosine, provides a group of materials known as cutback asphalts. The emulsification of asphalt cements with chemically treated water provides emulsified asphalts. Road oils, cutback asphalts, and emulsified asphalts are classified as liquid asphalts.

Asphaltic residuals, such as road oils, can be subjected to a blowing process to provide various types of materials for use in special or industrial applications. Some asphalt cements for paving purposes are also produced by partial blowing of softer

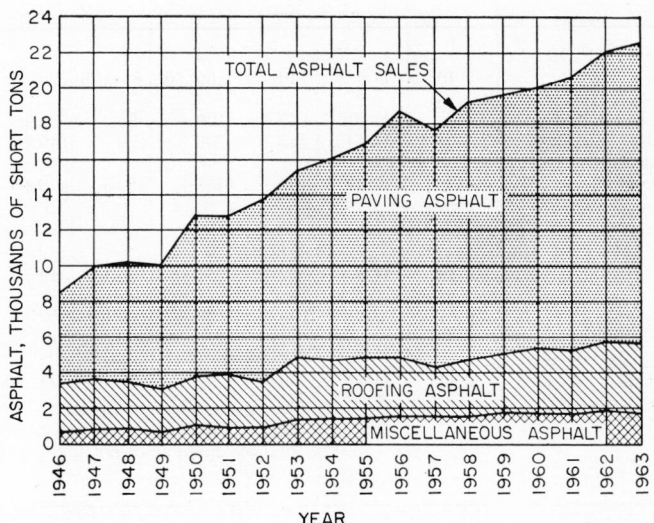

FIG. 11-17. Sales of petroleum asphalt, including road oil, by uses in the United States, 1946–1963 (see Table 11-53). Conversion is 5.5 bbl to a short ton. Source: Bureau of Mines reports on sales of asphalt.

grades of asphalt cement. In some cases, the blowing is done in the presence of catalysts.

Types and Grades

Asphalt cement is a thermoplastic material varying in consistency from firm to hard at normal temperatures. For application or for mixing with other materials it must be heated to fluid consistency. Four grades of paving asphalt and one grade for industrial and special uses are classified by the Asphalt Institute.[73] The basis of the grade classification is the penetration test. These specifications are given in Table 11-54. The American Society for Testing and Materials[74] includes four grades of asphalt cement in addition to those of The Asphalt Institute, with penetrations respectively of 50 to 60, 70 to 85, 100 to 120, and 150 to 200.

The liquid asphalts of the road-oil and cutback groups are of three standard types: (1) slow curing, (2) medium curing, and (3) rapid curing. The slow-curing products, often called road oils, are usually a residual material produced by the fractional distillation of certain crudes. They can also be prepared by blending an asphalt cement with an oily petroleum fraction. The medium-curing and rapid-curing liquid asphaltic materials, known as cutback asphalts, are a blend of asphalt cement with naphtha, gasoline, or kerosine.

Each of the three types is produced in six grades which have been standardized by The Asphalt Institute. The Furol viscosity and the distillation range are the principal varying characteristics for each of the six grades. The prefix for each grade denotes the type: SC for slow curing, MC for medium curing, and RC for rapid curing. The

Table 11-54. Specifications for Asphalt Cement of The Asphalt Institute[73]

Characteristics	AASHO* Test Method	ASTM Test Method	Industrial and special	Paving			
Penetration, 77°F, 100 g., 5 sec	T 49	D 5	40–50	60–70	85–100	120–150	200–300
Viscosity at 275°F:							
Saybolt Furol, SSF	......	E 102	120+	100+	85+	70+	50+
Kinematic, cs	......	D 445	240+	200+	170+	140+	100+
Flash point (Cleveland open cup), °F	T 48	D 92	450+	450+	450+	425+	350+
Thin-film oven test	T 179	......	....	....	....	....	....
Penetration after test, 77°F, 100 g., 5 sec, % of original	T 49	D 5	52+	50+	45+	42+	37+
Ductility:							
At 77°F, cms	T 51	D113	100+	100+	100+	60+	....
At 60°F, cms	......	......	....	....	....	....	60+
Solubility in carbon tetrachloride, %	T 44†	D 4†	99.5+	99.5+	99.5+	99.5+	99.5+
General requirements	......	......	The asphalt shall be prepared by the refining of petroleum. It shall be uniform in character and shall not foam when heated to 350°F				

* American Association of State Highway Officials.
† Except that carbon tetrachloride is used instead of carbon disulfide as solvent, Method No. 1 in AASHO Method T 44 or Procedure No. 1 in ASTM Method D 4.

suffix, or grade number, from 0 to 5, denotes the consistency range. In each type Grade 0 is the most liquid, Grade 5 the most viscous, with the intervening numbers in an orderly progression in consistency. Thus the designations for slow curing are SC-0 through SC-5, for medium curing MC-0 through MC-5, and for rapid curing RC-0 through RC-5.

The emulsified asphalt group of liquid asphaltic materials consists of minute globules of asphalt suspended in chemically treated water. The globules are generally in the colloidal-size range. Three standard grades of emulsified asphalt have been designated by The Asphalt Institute: rapid setting (RS), medium setting (MS), and slow setting (SS). The names indicate the relative rates at which the coalescence of the asphalt particles can be expected.

Tests for Asphalt Cement

The penetration test is a measure of the consistency or relative hardness of the asphalt and is the basis for its classification by grades. Under prescribed test conditions (ASTM D 5),[74] the penetration of a needle into the asphalt cement in units of $\frac{1}{10}$ mm is termed the penetration of the material. For a given set of conditions, the needle will penetrate farther into a soft asphalt than into harder material. Thus the soft cements have the higher penetration number.

Flash Point (ASTM D 92).[75] This tells the user how much the material can be safely heated without danger of an instantaneous flash in the presence of an open flame. It is seldom included in asphalt specifications.

Loss on Heating Test (ASTM D 6).[74] Since the material must be heated for application or mixing, usually to temperatures not exceeding 325°F, this test was devised to prevent the inclusion in the asphalt of volatile materials to the point that its original properties might be changed.

Ductility. In many applications this property is an important characteristic of asphalt cements. It is measured by a standard "extension" type of test described in ASTM D 113.

Solubility. This test procedure (ASTM D 4)[74] dissolves the cement in a suitable solvent, carbon tetrachloride being preferred, to determine the presence of minerals and other foreign matter.

Tests for Liquid Asphalt

Flash Point. This test is performed on all cutback asphalts and road oils, with the desired limits included in most specifications. The Cleveland open-cup method (ASTM D 92) is used for SC products. Because of the more volatile nature of the RC and MC types, test procedure for determining their flash points is prescribed in ASTM D 1310.[75] Flash-point determination is not made for emulsified asphalts because volatility is not a problem at the temperatures at which these materials are used.

Viscosity. The consistency of the liquid asphalts is measured by the Saybolt-Furol viscosity test prescribed in ASTM D 88.[75] Additional requirements for emulsified asphalts are given in ASTM D 244.[74]

Distillation. Since road oils and cutback asphalts are combinations of asphalt cement and petroleum distillates, the properties and relative amounts of both constituents are of importance in the application and performance of these materials. ASTM Test Procedure D 402[74] prescribes the standard method for separating by distillation the two constituents in order to determine the relative proportions of each in the material. The amounts and properties of the asphalt cement contained in an emulsified asphalt are determined by ASTM D 244.

Residue from Distillation. The asphalt residue from liquid asphalts is subjected to further tests to determine its specific characteristics in conformance with specification requirements. The penetration, ductility, and solubility tests previously described are made on residues from RC, MC, and emulsified types of liquid asphalts. The residue from the SC type is tested by the "float-test" method described in ASTM D 139,[74] which is a modified viscosity or consistency test.

Settlement, Sieve Tests. For emulsified asphalt ASTM D 244 is used to determine the tendency of asphalt globules to settle out during storage. The sieve test, also described in D 244, complements the settlement test and is used to determine the percentage of asphalt present in emulsified asphalt in the form of large globules. The demulsibility test, also prescribed in D 244, indicates the rate at which the colloidal asphalt globules will coalesce in emulsified asphalts when spread in thin films on soil or stone.

The Blown Asphalts

The type known as blown or oxidized asphalt is produced by blowing air through a residual oil at temperatures usually in the range of 400 to 600°F.[70] Depending on the characteristics of the residual oil and the length of the blowing time, asphaltic materials of varying characteristics are produced. These asphalts are used in a variety of industrial and special applications, including the construction of built-up roof coverings and the coating of underground transmission pipes.

For this material, the softening-point test (ASTM D 36)[74] is a measure of the softness or hardness of the material. It is used in conjunction with the penetration test to furnish a general indication of the relative temperature susceptibility of two or more asphalts of the same penetration measurement.

PETROLEUM COKE

Coke is a byproduct of the thermal cracking of reduced crudes and residuums, but it has become sufficiently important as a marketable commodity to be a factor in some refiners' selection of a thermal cracking process.   It is produced in about 40 refineries in the United States.

The output of marketable coke in 1962 was 6,325,000 short tons (31,624,000 bbl), nearly three times the output a decade earlier (Table 11-55).   The Bureau of Mines reported total coke production of 78,724,000 bbl at U.S. refineries in 1962.   Its data include nonmarketable catalyst coke, which forms on catalysts in cracking operations and is burned off at the plant.

Table 11-55. Production of Marketable Petroleum Coke in the United States, 1945 to 1962*

Year	Short tons (1,000's)	Barrels (1,000's)†
1945	1,300	6,500
1946	1,400	7,000
1947	1,560	7,800
1948	1,750	8,750
1949	1,800	9,000
1950	1,900	9,500
1951	2,070	10,350
1952	2,215	11,075
1953	2,500	12,500
1954	2,750	13,750
1955	2,900	14,500
1956	3,400	17,000
1957	3,850	19,250
1958	4,520	22,600
1959	4,694	23,470
1960	5,211	26,057
1961	6,090	30,480
1962	6,325	31,624

* Data from F. L. Shea, Jr., Great Lakes Carbon Corp.
† Bureau of Mines conversion basis of five 42-gal barrels to 1 short ton.
Note: The above figures were obtained by estimating and deducting carbon deposited on cracking catalyst, which is included in the Bureau of Mines figures on supply and demand for coke in U.S. refineries.

Raw marketable coke is an economic domestic and industrial fuel in areas convenient to a refinery source of supply.   In refined form it is made into aluminum anodes, furnace electrodes and liners, carbonaceous pastes and cements, and a variety of carbon and graphite products and is marketed over wide areas.   Recent developments indicate a large potential use in the coming atomic age of highly purified graphite from petroleum coke as construction material in nuclear plants.   An important property here is good neutron-moderating ability.[76]

Starting with atmospheric pressure-still coking in 1928, several coking processes have been developed, charging reduced crude or residuums, primarily for the production of distillates for the refiner to convert into gasoline.

The present major source of production of marketable coke in the United States is the delayed coking process developed by the Standard Oil Co. (Indiana).   In this process the bottoms from the delayed coker fractionator are charged to coke drums arranged in pairs.   One is on stream while the other is off stream and cooling. Hydraulic equipment removes the coke from the drums.   The size distribution of the coke is largely dependent on the manner of hydraulic decoking.

The continuous Fluid coking process of the Esso Research and Engineering Co. utilizes the fluidized solids technique developed for Fluid catalytic cracking, except that no catalyst is used.   Coke formed in the thermal cracking operation is laid down on fluidized seed coke in the reactor and is continuously withdrawn into a coke burner. A portion of the coke is returned to the reactor to maintain the cracking temperature. The remainder, after cooling, is removed as a stream of fine particles.

Factors in Coke Quality

The interrelated factors influencing the nature and quality of petroleum coke are (1) feedstock variables, (2) processing variables, and (3) engineering variables.[76]

Composition and character of the feedstock are basic to the quality of the coke and hence its potential utilization. Sulfur content and metallic constituents in the feedstock have an important effect on the quality of the coke. Numerous efforts have been made to correlate sulfur content in the feedstock with the sulfur content of the coke produced but with limited success even with high-sulfur residual oils. In view of the importance of the sulfur content of coke to its marketable value, considerable research work is being carried on to develop commercial methods for desulfurizing petroleum coke.[77] The metallic constituents in coke are almost as important as sulfur in determining its quality. Usually, high and undesirable metal contaminants are found in high-sulfur cokes, while cokes from low-sulfur feedstocks are low in metal contaminants.

Among the processing variables inherent in coking operations which influence coke quality, the time-temperature-pressure relationship is important. The reduction of recycling to yield gas oils of different quality reduces coke formation and, consequently, coke quality. Coke-removal procedures in any process definitely affect the nature and quality of the coke produced.

Among engineering variables, whether the process used is batch, semicontinuous, or continuous affects the characteristics of the coke produced. Design factors and size of the installation influence the nature of the coke. The refiner's handling and storage facilities contribute to some extent to the physical and size qualities of the coke produced.

Test Methods

Standard methods of the ASTM for testing the composition and properties of petroleum coke are not available. ASTM Committee D-5 on Coal and Coke has prepared standard methods for the laboratory sampling and analysis of coal and coke which, though not directly applicable, are used for the analysis of petroleum coke (Table 11-56).[78]

Table 11-56. ASTM Methods of Test for Coke[78]

Number	Title
D 346	Sampling Coke for Analysis
D 271	Laboratory Sampling and Analysis of Coal and Coke*
D 167	Test for Volume of Cell Space of Lump Coke
D 141	Drop Shatter Test for Coke
D 294	Tumbler Test for Coke
D 293	Test for Sieve Analysis of Coke
D 292	Test for Cubic Foot Weight of Coke
D 547	Test for Index of Dustiness of Coal and Coke

* Published in *Book of ASTM Standards*, part 8, 1962.

The Great Lakes Carbon Corp. has established test procedures especially for petroleum coke as a result of 15 years of cooperative work in its refineries. The GLC analytical methods include procedures for both a gross sample and an analytical sample, for removal of moisture without removal of volatile constituents, for removal of organic volatile matter without appreciable oxidation of the residual portion of the sample, and for the accurate determination of the sulfur content of a sample in about 10 min. These test procedures include methods for the separation into two groups of the six elements silicon, iron, calcium, vanadium, nickel, and titanium for further colorimetric analysis as constituents of the ash from petroleum coke. Water-soluble constituents in the coke are extracted with boiling distilled water. A 1-cu-ft volume of coke having particle sizes measuring 1 in. or less is weighed to obtain the weight per cubic foot of the coke.

Calcination

To be suitable for special uses, such as a conductor of electricity, petroleum coke as produced in a refinery is calcined. This is a high-temperature treatment in which the carbon-hydrogen ratios of the material are increased from about 20 to 1,000 and higher.[76] There is a decrease in volatile material with rising calcination temperatures, and for most purposes devolatilization and dehydrogenation are complete at about 2100°F. A number of commercial processes have been developed to accomplish calcination. Calcination at high temperatures makes available to the electroprocessing industries a type of amorphous carbon of high purity.

Utilization of Raw Coke

Figure 11-18[76] shows diagrammatically the fuel and nonfuel uses for raw petroleum coke. About half the total refinery output finds its way into the low-ash industrial and domestic fuel markets. In industrial heating outlets, petroleum coke made by

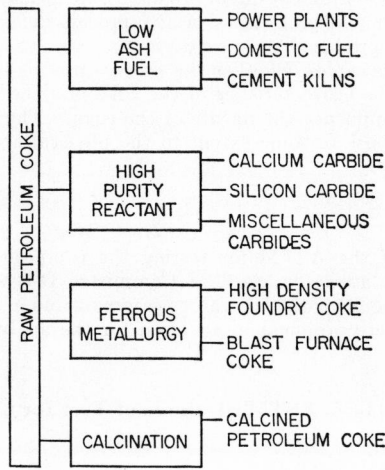

Fig. 11-18. Fields of utilization of raw petroleum coke.

the delayed coking process is burned in lump sizes mixed with coal in automatic stokers. In pulverized form it is used by itself or with powdered coal.

Coke made by the Fluid coking process is a fuel for the steam and power industries, sometimes with supplementary fuel to sustain ignition. A percentage of run-of-pile delayed coke is available in lump form, which is marketed as domestic fuel under favorable economic conditions. Fines from delayed coke are often briquetted into domestic fuel. Under suitable economic conditions cement kilns use pulverized petroleum coke because of its low ash and high Btu content.

High-quality and high-purity petroleum coke is used in the production of commercial carbides. Commercial calcium carbide is produced by the fusion of lime with coke in an electric furnace at about 2000°C. Silicon carbide is produced by heating a charge of carbon and silica in a resistance furnace. The carbon is a low-sulfur, low-ash anthracite coke or petroleum coke. Boron carbide is made by heating a mixture of boric acid, petroleum coke, and kerosine in a gastight resistance furnace to about 2500°C.

The use of coke in blast furnaces and in foundries represents the largest markets for metallurgical coke, approaching 75 million tons annually in volume. At present the cokes used are carbonized from coal. As reserves of high-quality coking coals in locations convenient to steel mills become depleted, more attention will be given to the use of petroleum coke as supplementary material. Under present conditions, however, this market appears to offer only limited possibilities for petroleum coke.

Utilization of Calcined Coke

A large portion of the raw petroleum coke destined for use in industrial fields is calcined. A schematic diagram of the principal outlets for calcined coke is given in Fig. 11-19.[76] The largest market for this form of coke is in the form of anodes for the

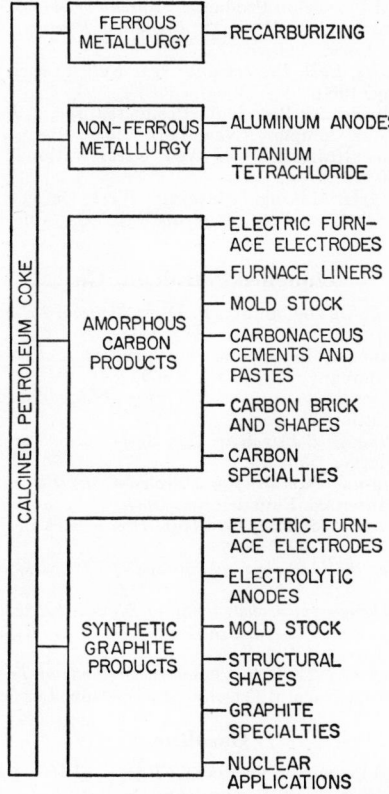

FIG. 11-19. Principal outlets for calcined petroleum coke.

electrolytic reduction of aluminum ore to make high-purity metal. The consumption of calcined coke in aluminum production in the United States in 1959 is estimated at around 1,115,000 short tons on the basis of an average of 0.42 lb of coke for 1 lb of aluminum. Advancing aluminum requirements and improved technology in the electrolytic reduction of alumina will further develop the production, processing, and utilization of petroleum coke in this field. A minor but growing use for low-volatile, low-ash petroleum coke is in the production of titanium metal. Another minor market for calcined coke of particular properties is as an additive for molten ferrous metals.

Substantial tonnages of calcined coke are now required for the manufacture of synthetic graphite products. These include electric-furnace electrodes, electrolytic anodes, structural shapes, and graphite specialty products.

Over recent years, synthetic graphite produced from petroleum coke has created an assured position for itself in nuclear applications.   Purity specifications for reactor-grade graphites for both industrial and military applications are most stringent.   The advantages inherent in petroleum coke graphites for nuclear applications are good neutron-moderating ability, desirable physical properties, and low cost.   Other advantages are ease of machinability and availability in a variety of types, shapes, and sizes.

REFERENCES

General

1. "Crude Petroleum and Petroleum Products," *Bureau of Mines Minerals Yearbook*, 1963.
2. Guthrie, Virgil, "Petrochemicals," sec. 12; *Petroleum Products Handbook*, McGraw-Hill Book Company, New York, 1960.
3. Mount, W. S., *Jet Fuels, Past, Present and Near Future*, Society of Automotive Engineers, Annual Meeting, 1960.
4. Guthrie, Virgil, "Additives for Petroleum Products," sec. 2, *Petroleum Products Handbook*, McGraw-Hill Book Company, New York, 1960.
5. Sweeney, W. J., *Better Utilization and New Uses*, National Petroleum Association, Annual Meeting, 1959.
6. "Fuel Oil Can Aid in Iron Making," *Petroleum Week*, Oct. 14, 1960.
7. *ASTM Standards, Parts 17 and 18, Petroleum Products*, American Society for Testing and Materials, 1964.

Liquefied Petroleum Gas

8. Benz, George R., W. F. DeVoe, and A. F. Dyer, *Annual LP-gas Sales Review*, Phillips Petroleum Co., 1964.
9. Guthrie, Virgil, "Liquefied Petroleum Gas," sec. 3, *Petroleum Products Handbook*, McGraw-Hill Book Company, New York, 1960.
10. *Bulletin 58, Standards for the Storage and Handling of Liquefied Petroleum Gases*, National Fire Protection Association, Boston.
11. *Publication 2140-62, Liquefied Petroleum Gas Specifications and Test Methods*, Natural Gas Processors Association, Tulsa.
12. *Manual on Measuring and Sampling of Petroleum and Petroleum Products*, American Society for Testing Materials, Philadelphia, 1957.
13. *ASTM Standards, Parts 17 and 18, Petroleum Products*, American Society for Testing and Materials, 1964.
14. *Tentative Standards for the Underground Storage of LP-gases*, National Gas Processors Association, Tulsa.
15. *API Standard 2150, Design and Construction of Liquefied Petroleum Gas Installations at Marine and Pipeline Terminals, Natural Gasoline Plants, Refineries and Tank Farms*, American Petroleum Institute.
16. *Tentative Specifications and Tentative Standard Methods of Test for Liquefied Petroleum Gases, TS 441*, California Natural Gasoline Association, Los Angeles.

Gasoline

17. "Crude Petroleum and Petroleum Products," *Bureau of Mines Minerals Yearbook*, 1958.
18. Guthrie, Virgil, "Gasoline," sec. 4, *Petroleum Products Handbook*, McGraw-Hill Book Company, New York, 1960.
19. *ASTM Manual for Rating Motor Fuels by Motor and Research Methods*, American Society for Testing Materials, 1960.
20. *ASTM Standards, Parts 17 and 18, Petroleum Products*, American Society for Testing and Materials, 1964.
21. Bridgeman, O. C., "Utilization of Ethanol-Gasoline Blends as Motor Fuels," *Ind. Eng. Chem., Ind. Ed.*, vol. 28, 1936.
22. *Petroleum Facts and Figures*, American Petroleum Institute, New York, 1965.

Aviation Gas-Turbine Fuels

23. Mount, W. W., *Jet Fuels, Past, Present and Near Future*, Annual Meeting, Society of Automotive Engineers, 1960.
24. Guthrie, Virgil, "Aircraft Gas Turbine Fuels and Lubricants," sec. 5, *Petroleum Products Handbook*, McGraw-Hill Book Company, New York, 1960.
25. Blade, O. C., *Survey of Aviation Fuels*, Petroleum Research Center, Bureau of Mines, Bartlesville, Okla., 1963.

Diesel Fuels

26. *Petroleum Facts and Figures*, American Petroleum Institute, New York, 1965.
27. Guthrie, Virgil, "Diesel Fuel Oils," sec. 6, *Petroleum Products Handbook*, McGraw-Hill Book Company, New York, 1960.
28. *ASTM Standards, Parts 17 and 18, Petroleum Products*, American Society for Testing and Materials, 1964.
29. Blade, O. C., *Diesel Fuel Oils*, Bureau of Mines Petroleum Experiment Station, Bartlesville, Okla., 1966.
30. *ASTM Manual for Rating Diesel Fuels by the Cetane Method*, American Society for Testing Materials, 1959.
31. *Symposium on Diesel Fuels, STP 167*, American Society for Testing Materials.

Heating Oils

32. *Petroleum Facts and Figures*, American Petroleum Institute, New York, 1963.
33. *ASTM Standards, Parts 17 and 18, Petroleum Products*, American Society for Testing and Materials, 1964.
34. Guthrie, Virgil, "Distillate Heating Oils," sec. 7, *Petroleum Products Handbook*, McGraw-Hill Book Company, New York, 1960.
35. *Handbook of Oil Burning*, Oil Heat Institute.
36. *Standard Methods for Testing Petroleum and Its Products, IP, 57/55*, The Institute of Petroleum, London.
37. *Natl. Bur. Std. Misc. Publ.* 97.
38. Guthrie, Virgil, "Residual Fuel Oils," sec. 8, *Petroleum Products Handbook*, McGraw-Hill Book Company, New York, 1960.

Miscellaneous Light Oils

39. *ASTM Standards, Parts 17 and 18, Petroleum Products*, American Society for Testing and Materials, 1964.
40. Guthrie, Virgil, "Kerosine, Miscellaneous Light Oils," sec. 11, part 1, *Petroleum Products Handbook*, McGraw-Hill Book Company, New York, 1960.
41. *Reference List of Specifications for Petroleum and Allied Products*, Military Petroleum Supply Agency, Washington.
42. *Standard Methods for Testing Petroleum and Its Products*, The Institute of Petroleum, London.
43. *Significance of ASTM Tests for Petroleum Products*, American Society for Testing Materials.
44. Guthrie, Virgil, "Industrial Naphthas, Miscellaneous Light Oils," sec. 11, part 2, *Petroleum Products Handbook*, McGraw-Hill Book Company, New York, 1960.
45. *Toxicological Reviews*, American Petroleum Institute.
46. Guthrie, Virgil, "Petroleum Pesticides, Miscellaneous Light Oils," sec. 11, part 3, *Petroleum Products Handbook*, McGraw-Hill Book Company, New York, 1960.

Lubricants

47. Coble, J. A., *What's Ahead for Lubricants?* National Petroleum Association, Cleveland, 1960.
48. *Facts for Industry*, Series M-29-C-06, Chemical Branch, Industry Division, Bureau of the Census.
49. Guthrie, Virgil, "Additives for Petroleum Products," sec. 2, *Petroleum Products Handbook*, McGraw-Hill Book Company, New York, 1960.
50. *SAE Handbook*, Society of Automotive Engineers, New York, 1961.
51. *API Classification of Internal Combustion Engine Service to Guide the Choice of Crankcase Oils*, American Petroleum Institute.
52. *ASTM Standards, Parts 17 and 18, Petroleum Products*, American Society for Testing and Materials, 1964.
53. Guthrie, Virgil, "Industrial Lubrication," sec. 9, part 2, *Petroleum Products Handbook*, McGraw-Hill Book Company, New York, 1960.
54. Guthrie, Virgil, "Lubricating Greases," sec. 9, part 3, *Petroleum Products Handbook*, McGraw-Hill Book Company, New York, 1960.

Petroleum Wax

55. *Petroleum Facts and Figures*, American Petroleum Institute, New York, 1965.
56. Canfield, K. S., and A. C. Smith, Jr., *The Refiner Looks at Wax*, National Petroleum Association, Mid-year Meeting, 1960.

57. Dannenbrink, R. A., and G. E. Unmuth, *Response of Waxes to Polyethylene Additives*, National Petroleum Assn., Mid-year Meeting, 1960.
58. Guthrie, Virgil, "The Petroleum Waxes," sec. 10, *Petroleum Products Handbook*, McGraw-Hill Book Company, New York, 1960.
59. *ASTM Standards, Parts 17 and 18, Petroleum Products*, American Society for Testing and Materials, 1964.

Petrochemicals

60. Guthrie, Virgil, "Petrochemicals," sec. 12, *Petroleum Products Handbook*, McGraw-Hill Book Company, New York, 1960.
61. Benz, George R., W. F. DeVoe, and A. F. Dyer, *LP-gas Sales Approach 10 Billion Gallons in 1960*, Phillips Petroleum Co.
62. Stamp, D. E., *Plastics in Construction in 1959*, Monsanto Chemical Co.
63. "Petrochemicals, Their Volume Grows and Their Applicants Multiply," *Pet. Week*, Aug. 19, 1960.
64. Sweeney, W. J., *A Look Ahead*, National Highway Users Conference, 1960.
65. Groppe, Henry, "Recent Developments in Chemicals and Raw Materials from Petroleum," American Chemical Society Symposium, San Francisco, 1958.

Carbon Black

66. *Carbon Black Report*, U.S. Bureau of Mines, 1963.
67. Guthrie, Virgil, "Carbon Black," sec. 15, *Petroleum Products Handbook*, McGraw-Hill Book Company, New York, 1960.
68. Emmet, P. H., and T. DeWitt, *Ind. Eng. Chem. Anal. Ed.*, vol. 13, p. 28, 1941.
69. *Petroleum Facts and Figures*, American Petroleum Institute, New York, 1963.

Petroleum Asphalt

70. Guthrie, Virgil, "Petroleum Asphalt," sec. 13, *Petroleum Products Handbook*, McGraw-Hill Book Company, New York, 1960.
71. Sweeney, W. J., *A Look Ahead*, National Highway Users Conference, Washington, 1960.
72. "Railroad Roadbeds, New Asphalt Market," *Pet. Week*, Dec. 9, 1960.
73. *Specifications for Asphalt Cements, Series 2*, The Asphalt Institute, 1963.
74. *ASTM Standards on Bituminous Materials for Highway Construction, Waterproofing and Roofing*, American Society for Testing Materials, 1962.
75. *ASTM Standards, Parts 17 and 18, Petroleum Products*, American Society for Testing and Materials, 1964.

Petroleum Coke

76. Guthrie, Virgil, "Petroleum Coke," sec. 14, *Petroleum Products Handbook*, McGraw-Hill Book Company, New York, 1960.
77. Correspondence with F. L. Shea, Jr., Director of Research, Great Lakes Carbon Corp.
78. *ASTM Standards on Coal and Coke*, part 8, American Society for Testing and Materials, 1962.

Section 12

PHYSICAL PROPERTIES OF HYDROCARBONS

By DR. A. S. BRUNJES

Former Manager, Technical Information Group
The Lummus Company
Newark, N.J.

The charts and tabulations of data presented in this section have been selected after careful evaluation of the available published information. Many sources were consulted and not without some resulting conflicting values. Publications of the API and the ASTM have been drawn upon freely, along with the recent literature, in the selection of the final values.

Twenty years of experience in collecting, collating, and disseminating technical information to process and project engineers and ten years in actual process design have helped in the selection of what is considered the essential information needed by practical engineers for refinery design and operation.

In order to keep the length of the chapter reasonable, many items have been omitted that some might feel should have been included. This unfortunately cannot be avoided.

Table 12-1. Physical Properties of Hydrocarbons[a] and Refinery Gases (Part I)*

No.	Compound	Boiling point at 1 atmos, °F	Vapor pressure at 100°F, psia	Melting point in air at 1 atmos, °F	Critical constants Pressure, psia	Critical constants Temp, °F	Critical constants Volume, cu ft/lb	Specific gravity	API gravity at 60°F, °API[d]	Density of the liquid at 60°F[b,c] Lb/cu ft	Density of the liquid at 60°F[b,c] Lb/gal	Coefficient of expansion at 60°F[b]	Refractive index of liquid, ND 77°F	Kinematic viscosity of the liquid, centistokes At 100°F	Kinematic viscosity At 210°F	No.
1	Hydrogen, H_2, MW 2.016	−422.99	......	−434.5	188	−399.82	0.515	......	......	4.43[g]	......	......	......	......	......	1
2	Carbon monoxide, CO, MW 28.01	−312.6	......	−337.1	508	−218.2	0.0532	......	......	50.6[g]	......	......	......	......	......	2
3	Carbon dioxide, CO_2, MW 44.01	−109.3[a]	......	−69.86	1073	+87.98	0.0343	......	......	94.4[g]	......	......	......	......	......	3
4	Hydrogen sulfide, H_2S, MW 34.08	−75.4	396.9	−122.0	1306	212.7	0.0465	0.79	47.6	60.0[g]	6.58	......	1.374	0.1746	0.1440	4

Paraffins

No.	Compound	Boiling point at 1 atmos, °F	Vapor pressure at 100°F, psia	Melting point in air at 1 atmos, °F	Critical constants Pressure, psia	Critical constants Temp, °F	Critical constants Volume, cu ft/lb	Specific gravity	API gravity at 60°F, °API[d]	Density of the liquid at 60°F[b,c] Lb/cu ft	Density of the liquid at 60°F[b,c] Lb/gal	Coefficient of expansion at 60°F[b]	Refractive index of liquid, ND 77°F	Kinematic viscosity of the liquid, centistokes At 100°F	Kinematic viscosity At 210°F	No.
5	Methane, CH_4, MW 16.042	−258.68	......	−296.46	673.1	−115.78	0.0991	0.3[i]	340[i]	18.70[i]	2.5[i]	......	......	......	......	5
6	Ethane, C_2H_6, MW 30.068	−127.53	......	−297.89	709.8	+90.32	0.0788	0.3771[e]	243.7[e]	23.52[e]	3.144[e]	......	......	......	......	6
7	Propane, C_3H_8, MW 44.094	−43.73	190	−305.84	617.4	206.26	0.0278	0.5077[e]	147.2[e]	31.57[e]	4.220[e]	0.00152[e]	......	......	......	7
8	n-Butane, C_4H_{10}, MW 58.120	+31.10	51.6	−217.03	550.1	305.62	0.0702	0.5844[e]	110.6[e]	36.39[e]	4.865[e]	0.00117[e]	1.3292[e]	......	......	8
9	2-Methyl propane (isobutane)	10.89	72.2	−255.28	529.1	274.96	0.0724	0.5631[e]	119.8[e]	35.05[e]	4.686[e]	......	......	......	......	9
10	n-Pentane, C_5H_{12}, MW 72.146	96.93	15.570	−201.50	489.5	385.5	0.0690	0.6312	92.7	39.29	5.253	0.00087	1.35472	0.330[e]	......	10
11	2-Methyl butane (isopentane)	82.13	20.44	−255.3	483	369	0.0685	0.6248	95.0	38.89	5.199	0.00090	1.35088	......	......	11
12	2,2-Dimethyl propane (neopentane)	49.10	35.9	+2.21	464.0	321.08	0.0674	0.5967	105.6	37.14[e]	4.965[e]	0.00104[e]	1.399[e]	......	......	12
13	n-Hexane, C_6H_{14}, MW 86.172	155.73	4.956	−139.63	440.0	454.1	0.0685	0.6640	81.6	41.34	5.526	0.00075	1.37226	0.4137	......	13
14	2-Methyl pentane (isohexane)	140.49	6.767	−244.61	440.1	435.7	0.0681	0.6579	83.6	40.96	5.746	0.00078	1.36873	......	......	14
15	2,2-Dimethyl butane (neohexane)	121.53	9.856	−147.77	450.7	420.1	0.0667	0.6540	84.9	40.72	5.443	0.00078	1.36595	......	......	15
16	n-Heptane, C_7H_{16}, MW 100.198	209.17	1.620	−131.10	396.8	512.62	0.0682	0.6882	74.1	42.85	5.728	0.00069	1.38511	0.5214	0.3425[e]	16
17	2-Methyl hexane (isoheptane)	194.09	2.271	−180.90	400	495.0	0.0685	0.6830	75.7	42.53	5.685	0.00068	1.38227	......	......	17
18	2,2,3-Trimethyl butane (triptane)	177.59	3.374	−12.84	437.2	497.0	0.0631	0.6945	72.2	43.25	5.782	0.00069	1.38692	......	......	18
19	n-Octane, C_8H_{18}, MW 114.224	258.2	0.537	−70.23	362.1	563.7	0.0682	0.7068	68.7	44.01	5.883	0.00062	1.39505	0.6476	0.4039	19
20	2-Methyl heptane (isooctane)	243.76	0.768	−164.27	364	547.5	0.0685	0.7021	70.0	43.72	5.845	0.00061	1.39257	......	......	20
21	2,5-Dimethylhexane (diisobutyl)	228.39	1.101	−132.16	362	530	0.0676	0.6980	71.2	43.46	5.810	0.00065	1.39004	......	......	21
22	2,2,4-Trimethyl pentane (isooctane)	210.63	1.708	−161.28	374.7	520.07	0.0676	0.6963	71.7	43.35	5.795	0.00065	1.38898	......	......	22

Table 12-1. Physical Properties of Hydrocarbons and Refinery Gases (Part I Continued)

No.	Compound	Boiling point at 1 atmos, °F	Vapor pressure at 100°F, psia	Melting point in air at 1 atmos, °F	Critical constants Pressure, psia	Critical constants Temp, °F	Critical constants Volume, cu ft/lb	Specific gravity	API gravity at 60°F, °API[d]	Density of the liquid at 60°F[b],[c] Lb/cu ft	Density Lb/gal	Coefficient of expansion at 60°F[b]	Refractive index of liquid, ND 77°F[b]	Kinematic viscosity At 100°F	At 210°F	No.
23	n-Nonane, C_9H_{20}, MW 128.250	303.44	0.179	−64.33	332	610.5	0.0679	0.7217	64.6	44.94	6.008	0.00063	1.40311	0.8087	0.4766	23
24	2-Methyl octane (isononane)	289.87	0.26	−112.72	(343)	(604)	(0.069)	0.7175	65.7	44.68	5.973	0.00058	1.4008	......	......	24
25	n-Decane, $C_{10}H_{22}$, MW 142.276	345.42	0.073	−21.39	304	651.9	0.0679	0.7341	61.3	45.72	6.112	0.00055	1.40967	1.004	0.5591	25
26	2-Methyl nonane (isodecane)	332.60	......	−102.37	318	(650)	(0.066)	0.7306	62.2	45.50	6.082	0.00064	1.4075	......	......	26
27	2,7-Dimethyl octane (diisoamyl)	319.77	1.51	−65	(317)	(643)	(0.066)	0.7285	62.7	45.36	6.064	0.00059	1.4062	......	......	27

Mono- and Diolefins

No.	Compound	Boiling point at 1 atmos, °F	Vapor pressure at 100°F, psia	Melting point in air at 1 atmos, °F	Critical constants Pressure, psia	Critical constants Temp, °F	Critical constants Volume, cu ft/lb	Specific gravity	API gravity at 60°F, °API[d]	Density Lb/cu ft	Density Lb/gal	Coefficient of expansion at 60°F[b]	Refractive index of liquid, ND 77°F[b]	Kinematic viscosity At 100°F	At 210°F	No.
28	Ethene, C_2H_4, MW 28.052	−154.68	......	−272.47[f]	742.1	49.82	0.0706	......	......	......	......	......	......	......	......	28
29	Propene, C_3H_6, MW 42.078	−53.86	226.4	−301.45[f]	667	197.4	0.0689	0.5220	139.6[e]	32.49[e]	4.343[e]	0.00189[e]	......	......	......	29
30	1-Butene, C_4H_8, MW 56.104	+20.73	63.05	−301.63[f]	583	295.6	0.0689	0.6013[e]	103.8[e]	37.43[e]	5.004[e]	0.00116[e]	......	......	......	30
31	cis-2-Butene	38.70	45.54	−218.04	600	324.3	0.0503	0.6271[e]	91.4[e]	39.04[e]	5.219[e]	0.00098[e]	......	......	......	31
32	trans-2-Butene	33.58	49.80	−157.09	600	311.9	0.0503	0.6100[e]	100.5[e]	37.97[e]	5.076[e]	0.00107[e]	......	......	......	32
33	2-Methyl propene (isobutene)	19.58	63.40	−220.63	579.8	292.51	0.0513	0.6004[e]	104.2[e]	37.37[e]	4.996[e]	0.00120[e]	......	......	......	33
34	1-Pentene, C_5H_{10}, MW 70.130	85.94	19.115	−265.40	586	376.9	(0.0672)	0.6457	87.6	40.20	5.374	0.00089	1.36835	......	......	34
35	2-Methyl-1-butene	88.09	18.399	−215.61	(514.4)	(390)	(0.0672)	0.6557	84.3	40.83	5.458	0.00090	1.3746	......	......	35
36	Propadiene (allene) C_3H_4, MW 40.062	−30.1	(168)	−213.3	(793)	248	(0.0649)	0.657[e]	83.9[e]	40.90[e]	5.468[e]	......	......	......	......	36
37	1,2-Butadiene, C_4H_6, MW 54.088	+51.53	(20)	−213.14	(653)	(339)	(0.0649)	0.658[e]	83.5[e]	40.90[e]	5.47[e]	0.00098[e]	......	......	......	37
38	1,3-Butadiene	24.06	58.7	−164.05	628	306	0.0654	0.6272[e]	91.4[e]	39.05[e]	5.220[e]	0.00113[e]	......	......	......	38
39	1,2-Pentadiene, C_5H_8, MW 68.114	112.74	11.479	−215.07	(590.8)	(446)	(0.0649)	0.6977	71.3	43.44	5.807	0.00083	1.41773	......	......	39
40	2-Methyl-1,3-butadiene (isoprene)	93.32	16.672	−230.71	(558.4)	(412)	(0.0650)	0.6861	74.7	42.72	5.711	0.00086	1.41852	......	......	40

Table 12-1. Physical Properties of Hydrocarbons and Refinery Gases (Part I Continued)

No.	Compound	Boiling point at 1 atmos, °F	Vapor pressure at 100°F, psia	Melting point in air at 1 atmos, °F	Critical constants			Specific gravity	API gravity at 60°F, °API[d]	Density of the liquid at 60°F[b,c]		Coefficient of expansion at 60°F[b]	Refractive[b] index of liquid, ND 77°F	Kinematic viscosity of the liquid, centistokes		No.
					Pressure, psia	Temp, °F	Volume, cu ft/lb			Lb/cu ft	Lb/gal			At 100°F	At 210°F	
	Acetylenes															
41	Acetylene, C_2H_2, MW 26.036	-119.0	....	-114[f]	905	97.4	0.0695	0.615[p]	98.6[p]	....	....	....	....	....	....	41
42	Methyl acetylene, C_3H_4, MW 40.062	-9.8	....	-152.9	(750)	251.0	(0.0656)	(0.63)	(93.1)	(39.22)	(5.243)	....	....	....	....	42
43	Ethyl acetylene C_4H_6, MW 54.088	+46.53	(39)	-194.30	(683.4)	375.0	(0.0654)	0.65[e]	86.1[e]	40.9[e]	5.47[e]	....	....	....	....	43
44	Dimethyl acetylene, C_4H_6, MW 54.08	80.58	21.5	-26.07	(737.4)	419.0	0.0654	0.6965	71.6	43.37	5.798	0.00087	1.3893	....	....	44
45	Phenyl acetylene, C_8H_6 MW 102.128	287.06	(0.34)	-40	611.4	720	0.0526	(0.933)	(20.2)	(58.1)	(7.77)	0.00048	1.5485	....	....	45
	Naphthenes and Aromatics															
46	Cyclopropane, C_3H_6, MW 42.078	-27.04	....	-197.36	797	256	....	0.699[e]	70.7[e]	35.08	4.690	0.00087[e]	1.362[e]	....	....	46
47	Cyclobutane, C_4H_8, MW 56.104	54.52	....	-131.31[f]	....	....	....	....	....	43.57[e]	5.824[e]	0.00070	1.40363	....	....	47
48	Cyclopentane, C_5H_{10}, MW 70.130	120.67	9.914	-136.96	654.7	461.48	0.0593	0.7505	57.0	46.73	6.247	0.00071	1.40700	0.499	....	48
49	Methyl cyclopentane, C_6H_{12}, MW 84.156	161.26	4.503	-224.42	549.0	499.30	0.0607	0.7535	56.2	46.93	6.274	0.00068	1.42354	0.569	....	49
50	Cyclohexane, C_6H_{12}, MW 84.156	177.33	3.264	+43.80	588.0	536.5	0.0592	0.7834	49.1	48.79	6.522	0.00066	1.49792	0.953	....	50
51	Benzene, C_6H_6, MW 78.108	176.18	3.224	+41.96	714	552.0	0.0535	0.8845	28.4	55.096	7.3653	0.00060	1.49413	0.5870	....	51
52	Toluene, C_7H_8, MW 92.134	231.12	1.032	-138.98	590	605.5	0.0564	0.8719	30.8	54.309	7.2601	0.00054	1.49320	0.5584	0.341	52
53	Ethylbenzene, C_8H_{10}, MW 106.160	277.13	0.371	-138.96	540	651.2	0.0556	0.8717	30.8	54.299	7.2589	0.00054	1.50295	0.6428	0.390	53
54	o-Xylene, C_8H_{10}	291.94	0.264	-13.33	530	674.8	0.0584	0.8848	28.4	55.111	7.3673	0.00055	1.49464	0.740	0.428	54
55	m-Xylene, C_8H_{10}	282.39	0.326	-54.17	510	649.9	0.0584	0.8687	31.4	54.114	7.2339	0.00054	1.49325	0.591	0.366	55
56	p-Xylene, C_8H_{10}	281.03	0.342	+55.87	500	649.4	0.0556	0.8657	31.9	53.924	7.2086	0.00054		0.631	0.372	56

* Notes for this table follow Table 12-2.

Table 12-1. Physical Properties of Hydrocarbons and Refinery Gases (Part II)*

No.	Compound	Aniline point, °F	Heat capacity 60F and constant pressure, Btu/lb, °F — Gas, ideal state	Heat capacity — Liquid at 1 atm	Heat of vaporization at the normal boiling point 1 atm, Btu/lb	Heat of combustion — Gross, to form H_2O (liquid) + CO_2 (gas) Btu/lb	Gross Btu/gal	Net, to form H_2O (vapor) + CO_2 (gas) Btu/lb	Net Btu/gal	Flammability Lower	Flammability Higher	ASTM octane — Motor method D357, clear	ASTM octane — Research method D908, clear	No.
1	Hydrogen, H_2, MW 2.016		3.408	2.232g	192.4	61,070g		51,600g		4.1	74.2			1
2	Carbon monoxide, CO, MW 28.01		0.2456	0.516g	92.7	4,350g		4,350g		12.5	74.2			2
3	Carbon dioxide, CO_2, MW 44.01		0.1991		244.6g									3
4	Hydrogen sulfide, H_2S, MW 34.08		0.2385	0.480g	238.0	7,100g		6,545		4.3	45.5			4
	Paraffins													
5	Methane, CH_4, MW 16.042		0.5271		219.22					5	15			5
6	Ethane, C_2H_6, MW 30.068		0.4097	0.9256	210.41	22,163	69,680	20,270	63,728	2.9	13	+0.05[k]	+1.6[k,l]	6
7	Propane, C_3H_8, MW 44.094		0.3885	0.5920	183.05	21,470	90,624	19,766	83,432	2.1	9.5	97.1	1.8[k,l]	7
8	n-Butane, C_4H_{10}, MW 58.120	181.6	0.3908	0.5636	165.65	21,108	102,692	19,493	94,833	1.8	8.4	89.6[i]	93.8[l]	8
9	2-Methyl propane (isobutane)	225.7	0.3872	0.5695	157.51	21,059	98,681	19,443	91,110	1.8	8.4	97.6	+0.10[k,l]	9
10	n-Pentane, C_5H_{12}, MW 72.146	159.3	0.3883	0.542	153.59	20,905	109,795	19,339	101,568	1.4	8.3	62.6[i]	61.7[l]	10
11	2-Methyl butane (isopentane)	170.6	0.3827	0.5353	147.13	20,861	108,477	19,300	100,360	1.4	8.3	90.3	92.3	11
12	2,2-Dimethyl propane (neopentane)	216[h]	0.3914	0.554	135.59	20,809	103,339	19,248	95,585	1.4	8.3	80.2	85.5	12
13	n-Hexane, C_6H_{14}, MW 86.172	155.5	0.3864	0.5333	143.95	20,758	114,727	19,232	106,298	1.2	7.7	26.0	24.8	13
14	2-Methyl pentane (isohexane)	164.8	0.389	0.5264	138.67	20,730	113,520	19,206	105,169	1.2	(7.7)	73.5	73.4	14
15	2,2-Dimethyl butane (neohexane)	178.2[h]	0.382	0.5165	131.24	20,684	112,606	19,160	104,307	1.2	(7.7)	93.4	91.8	15
16	n-Heptane, C_7H_{16}, MW 100.198	157.5	0.3853	0.5276	135.99	20,657	118,348	19,157	109,748	1.0	7.0	0.0	0.0	16
17	2-Methyl hexane (isoheptane)	165.2	0.390	0.522	131.58	20,632	117,316	19,133	108,790	(1.0)	(7.0)	46.4	42.4	17
18	2,2,3-Trimethyl butane (triptane)	162.0	0.3812	0.498	124.20	20,601	119,137	19,103	110,475	(1.0)	(7.0)	40.1[r]	+1.8[k]	18
19	n-Octane, C_8H_{18}, MW 114.224	159.1	0.3845	0.5230	129.51	20,579	121,085	19,099	112,381	0.96				19
20	2-Methyl heptane (isooctane)	165[h]	(0.393)	0.5173	127.2	20,559	120,167	19,080	111,521	0.98		23.0[l]	20.6[l]	20

Table 12-1. Physical Properties of Hydrocarbons and Refinery Gases (Part II Continued)

No.	Compound	Aniline point, °F	Heat capacity 60F and constant pressure, Btu/lb,°F — Gas, ideal state	Heat capacity — Liquid at 1 atm	Heat of vaporization at the normal boiling point 1 atm, Btu/lb	Gross, to form H_2O (liquid) + CO_2 (gas) Btu/lb	Gross Btu/gal	Net, to form H_2O (vapor) + CO_2 (gas) Btu/lb	Net Btu/gal	Flammability limits, volume % in air mixture — Lower	Flammability — Higher	ASTM octane nos. — Motor method D357, clear	ASTM — Research method D908, clear	No.
21	2,5-Dimethylhexane (diisobutyl)	172.4[h]	(0.373)	0.5114	122.8	20,539	119,350	19,060	110,756	(0.98)	......	55.7	55.2	21
22	2,2,4-Trimethyl pentane isooctane	175.1	(0.380)	0.4892	116.69	20,543	119,067	19,065	110,499	1.0	......	100.0	100.0	22
23	n-Nonane, C_9H_{20}, MW 128.250	164.7	(0.3840)	0.5220	126.65	20,519	123,298	19,055	114,500	0.87[m]	2.9	......	......	23
24	2-Methyl octane (isononane)	171.5[h]	(0.396)	0.516	122.9	20,500	122,466	19,036	113,721	(0.85)	......	......	......	24
25	n-Decane, $C_{10}H_{22}$, MW 142.276	170.6	(0.3835)	0.5207	118.68	20,470	125,136	19,019	116,262	0.78[m]	2.6	......	......	25
26	2-Methyl nonane (isodecane)	176.5[h]	(0.398)	0.517	......	20,453	124,418	19,002	115,588	(0.75)	......	......	......	26
27	2,7-Dimethyl octane (diisoamyl)	174.2	(0.386)	0.510	......	20,436	123,943	18,985	115,142	(0.75)	......	......	......	27
	Mono- and Diolefins													
28	Ethene, C_2H_4, MW 28.052	......	0.3622	......	207.57	......	......	......	......	2.7	34	75.6	+0.03[k]	28
29	Propene, C_3H_6, MW 42.078	......	0.3541	0.585	188.18	......	......	......	......	2.0	10	84.9	+0.2[k]	29
30	1-Butene, C_4H_8, MW 56.104	......	0.3548	0.535	167.94	20,677	103,469	19,316	96,656	1.6	9.3	80.8[l]	97.4	30
31	cis-2-Butene	......	0.3269	0.5271	178.91	20,611	107,589	19,250	100,483	(1.6)	......	83.5	100	31
32	trans-2-Butene	......	0.3654	0.5351	174.39	20,584	104,503	19,222	97,592	(1.6)	......	......	......	32
33	2-Methyl propene (isobutene)	58.8	0.3701	0.549	169.48	20,546	102,650	19,186	96,851	(1.6)	......	......	......	33
34	1-Pentene, C_5H_{10}, MW 70.130	66.2	0.3635	0.5196	154.46	20,550	110,457	19,189	103,140	1.4	8.7	77.1	90.9	34
35	2-Methyl-1-butene	......	0.3705	0.5266	156.31	20,453	111,655	19,092	104,224	(1.4)	......	81.9	+0.2[k]	35
36	Propadiene (allene) C_3H_4, MW 40.062	......	0.3439	......	(218)	(20,710)	(113,240)	(19,755)	108,022	(2.0)	(12)	......	......	36
37	1,2-Butadiene, C_4H_6, MW 54.088	......	0.3458	0.5408	(181)	(20,461)	(112,106)	(19,463)	(106,308)	2.0	11.5	......	......	37
38	1,3-Butadiene	......	0.3412	0.5079	(174)	20,026	104,716	18,982	99,257	2.0	11.5	......	......	38

Table 12-1. Physical Properties of Hydrocarbons and Refinery Gases (Part II Continued)

No.	Compound	Aniline point, °F	Heat capacity 60F and constant pressure, Btu/lb,°F — Gas, ideal state	Heat capacity — Liquid at 1 atm	Heat of vaporization at the normal boiling point 1 atm, Btu/lb	Gross, to form H_2O (liquid) + CO_2 (gas) Btu/lb	Gross Btu/gal	Net, to form H_2O (vapor) + CO_2 (gas) Btu/lb	Net Btu/gal	Flammability Lower	Flammability Higher	Motor method D357, clear	Research method D908, clear	No.
39	1,2-Pentadiene, C_5H_8, MW 68.114	……	0.360	(0.59)	(160)	(20,383)	118,548	(19,264)	(112,039)	(1.5)	……	……	……	39
40	2-Methyl-1,3-butadiene (isoprene)	……	0.357	0.5245	(153)	19,939	114,051	18,834	107,730	(1.5)	……	8.10	99.1	40

Acetylenes

No.	Compound	Aniline point, °F	Gas, ideal state	Liquid at 1 atm	Heat of vaporization Btu/lb	Gross Btu/lb	Gross Btu/gal	Net Btu/lb	Net Btu/gal	Flammability Lower	Flammability Higher	Motor method D357, clear	Research method D908, clear	No.
41	Acetylene, C_2H_2, MW 26.036	……	0.3966	……	(175)	(20,633)	(108,366)	(19,680)	(103,359)	2.5	80	……	……	41
42	Methyl acetylene, C_3H_4, MW 40.062	……	0.3545	……	(179)	(20,469)	(112,149)	(19,411)	(106,351)	……	……	……	……	42
43	Ethyl acetylene C_4H_6, MW 54.088	……	0.3513	0.5946	(197)	(20,300)	(117,883)	(19,242)	(111,738)	……	……	……	……	43
44	Dimethyl acetylene, C_4H_6, MW 54.08	……	0.3372	0.5474	……	……	……	……	……	……	……	70.2	85.9	44
45	Phenyl acetylene, C_8H_6, MW 102.128	……	(0.35)	(0.58)	(151)	……	……	……	……	……	……	……	……	45

Naphthenes and Aromatics

No.	Compound	Aniline point, °F	Gas, ideal state	Liquid at 1 atm	Heat of vaporization Btu/lb	Gross Btu/lb	Gross Btu/gal	Net Btu/lb	Net Btu/gal	Flammability Lower	Flammability Higher	Motor method D357, clear	Research method D908, clear	No.
46	Cyclopropane, C_3H_6, MW 42.078	……	……	……	……	……	……	……	……	2.4	10.4	……	……	46
47	Cyclobutane, C_4H_8, MW 56.104	……	……	0.413	……	……	……	……	……	1.8	……	……	……	47
48	Cyclopentane, C_5H_{10}, MW 70.130	62.2	0.2712	0.4216	167.34	20,187	126,128	18,826	117,623	1.4	……	89.9[j]	+0.10[k]	48
49	Methyl cyclopentane, C_6H_{12}, MW 84.156	91.4	0.3010	0.4407	147.83	20,129	126,308	18,768	117,766	1.2	8.35	80.0	91.3	49
50	Cyclohexane, C_6H_{12}, MW 84.156	87.8	0.2900	0.4322	153.7	20,038	130,707	18,677	121,827	1.3	7.8	77.2	83.0	50
51	Benzene, C_6H_6, MW 78.108	<−22	0.2404	0.4098	169.34	17,991	132,519	17,257	127,115	1.3	7.9	+2.8[k]	+2.8[k]	51
52	Toluene, C_7H_8, MW 92.134	<−22	0.2599	0.4017	156.2	18,251	132,522	17,422	126,501	1.2	7.1	+0.3[k]	+5.8[k]	52
53	Ethyl benzene, C_8H_{10} MW 106.160	<−22	0.2795	0.4114	145.7	18,494	134,264	17,594	127,735	0.99	6.7	97.9	+0.8[k]	53
54	o-Xylene, C_8H_{10}	<−4	0.2914	0.4418	149.1	18,445	136,058	17,556	129,426	1.1	6.4	100.0	……	54
55	m-Xylene, C_8H_{10}	−22	0.2782	0.4045	147.4	18,441	133,421	17,542	126,916	1.1	6.4	+2.8[k]	+4.0[k]	55
56	p-Xylene, C_8H_{10}	−22	0.2769	0.4083	146.1	18,445	132,990	17,546	126,507	1.1	6.6	+1.2[k]	+3.4[k]	56

* Notes for this table follow Table 12-2.

Table 12-2. Physical Properties of Gaseous Hydrocarbons C₁ to C₅

No.[r]	Compound	Compressibility factor of the gas $Z = PV/RT$		Specific gravity of real gas at 60F, 1 atm referred to air as unity	Specific volume of gas at 60F 1 atm		Heat capacity of gas at constant pressure[q] at 60F 1 atm Btu/lb/°F	Heat of combustion of real gas at 60F and constant pressure, Btu/cu ft		Air required for combustion of real gas at 60F 1 atm	
		At the critical point Z_c	At 60F 1 atm Z		cu ft gas /lb gas	cu ft gas/gal liquid		Gross to form H_2O (liquid) + CO_2 (gas)	Net to form H_2O (gas) + CO_2 (gas)	Cu ft air/cu ft gas	Lb air /lb gas
5	Methane, CH_4, MW 16.042	0.289	0.9981	0.55491	23.6113	......	0.5271	1011.6	910.77	9.563	17.233
6	Ethane, C_2H_6, MW 30.068	0.285	0.9916	1.0469	12.5151	......	0.4097	1783.7	1631.5	16.845	16.090
7	Propane, C_3H_8, MW 44.094	0.277	0.9820	1.5503	8.4515	35.775[e]	0.3885	2653.3	2358.3	24.300	15.674
8	n-Butane, C_4H_{10}, MW 58.120	0.274	0.9667	2.0757	6.3120	30.752[e]	0.3908	3374.4	3114.2	32.089	15.459
9	2-Methyl propane (isobutane)	0.283	0.9669	2.06805	6.3355	29.745[e]	0.3872	3352.15	3092.9	31.970	15.459
10	n-Pentane, C_5H_{12}, MW 72.146	0.269	0.9435	2.6400	4.9629	26.115	0.3883	4249.1	3929.3	40.466	15.328
11	2-Methyl butane (isopentane)	0.268	0.9482	2.6269	4.9876	25.980	0.3827	4218.7	3900.25	40.265	15.328
12	2,2-Dimethyl propane (neopentane)	0.269	(0.95)	2.622	4.997	24.86[e]	0.3914	4197	3879	40.19	15.33
28	Ethene, C_2H_4, MW 28.052	0.269	0.9940	0.9740	13.4524	......	0.3662	1608.5	1507.3	14.398	14.783
29	Propene, C_3H_6, MW 42.078	0.274	0.9839	1.4765	8.8736	38.618[e]	0.3541	2371.7	2218.3	21.827	14.783
30	1-Butene, C_6H_8, MW 56.104	0.277	0.9694	1.9982	6.5571	32.871[e]	0.3548	3177.9	2970.3	29.538	14.782
31	cis-2-Butene	0.276	(0.97)	1.997	6.561	34.30[e]	0.3269	3168	2960.5	29.52	14.78
32	trans-2-Butene	0.276	(0.97)	......	......	33.37[e]	0.3654	3163	2957		
33	2-Methyl propene (isobutene)	0.276	(0.97)	1.997	6.561	32.84[e]	0.3701	3156	2949	29.52	14.78
34	1-Pentene, C_5H_{10}, MW 70.130	......	(0.95)	2.549	5.141	27.68	0.3635	4028.5	3763.5	37.68	14.78
35	2-Methyl-1-butene	......	(0.95)	2.549	5.141	28.10	0.3705	4010	3745	37.68	14.78

Table 12-2. Physical Properties of Gaseous Hydrocarbons C_1 to C_5 (Continued)

No.ʳ	Compound	Compressibility factor of the gas $Z = PV/RT$		Specific gravity of real gas at 60F, 1 atm referred to air as unity	Specific volume of gas at 60F 1 atm		Heat capacity of gas at constant pressureᵍ at 60F 1 atm Btu/lb/°F	Heat of combustion of real gas at 60F and constant pressure, Btu/cu ft		Air required for combustion of real gas at 60F 1 atm	
		At the critical point Z_c	At 60F 1 atm Z		cu ft gas /lb gas	cu ft gas/gal liquid		Gross to form H_2O (liquid) + CO_2 (gas)	Net to form H_2O (gas) + CO_2 (gas)	Cu ft air/cu ft gas	Lb air /lb gas
36	Propadiene (allene), C_3H_4, MW 40.062....	.....	(0.98)	1.411	9.283	.....	0.3439	2249	2146	19.48	13.80
37	1,2-Butadiene, C_4H_6, MW 54.088....	.....	(0.97)	1.925	6.806	37.36ᵉ	0.3458	3031	2875	27.06	14.06
38	1,3-Butadiene.............	0.271	0.975	1.9153	6.841	35.771ᵉ	0.3412	2954.8	2800	26.92	14.06
39	1,2-Pentadiene, C_5H_8, MW 68.114......	.....	(0.95)	2.475	5.293	30.78	0.360	3885	3674	35.165	14.21
40	2-Methyl-1,3-butadiene (isoprene).......	.....	(0.96)	2.450	5.349	30.59	0.357	3762	3553	34.80	14.21
41	Acetylene, C_2H_2, MW 26.036........	0.274	0.9925	0.9057	14.4664	.....	0.3966	1483.8	14.331	12.021	12.273
42	Methyl acetylene, C_3H_4, MW 40.062......	.....	(0.98)	1.411	9.283	.....	0.3545	2241	2138	19.48	13.80
43	Ethyl acetylene, C_4H_6, MW 54.088.....	.....	(0.95)	1.966	6.665	35.99ᵉ	0.3513	3098	2939	27.63	14.06
44	Dimethyl acetylene, C_4H_6, MW 54.088..	.....	(0.96)	1.945	6.736	39.11	0.3372	3044	2886	27.34	14.06
46	Cyclopropane, C_3H_6, MW 42.078.........	.....	(0.967)ᵍ	1.805ᵍ	7.825ᵍ	.....	0.2712	.....	.....	35.800ᵍ	14.786
48	Cyclopentane, C_5H_{10}, MW 70.130......	0.276	(0.968)ᵍ	2.240ᵍ	5.848ᵍ	34.997ᵍ		.....	.....		
1	Hydrogen, H_2, MW 2.016............	0.304	0.4962	0.0695	187.9	.....	3.408	325.1	274.8	2.382	34.226
2	Carbon monoxide, CO, MW 28.01........	0.294	0.9995	0.967	13.50	.....	0.2484	323.4	323.4	2.382	3.464
3	Carbon dioxide, CO_2, MW 44.01.........	0.275	0.9947	1.529	8.54	.....	0.1991	.....	.....		
4	Hydrogen sulfide, H_2S, MW 34.08........	0.284	1.0010	1.190	10.98	.....	0.2373	637	586	7.146	6.074

Notes on Tables 12-1 and 12-2

Values in parentheses are estimated by the methods outlined in ASTM-STP 109A, p. 62.

Values in Table 12-2 for hydrogen, carbon monoxide, carbon dioxide, and hydrogen sulfide from *The Gas Engineers Handbook*, McGraw-Hill Book Company, New York, 1934, and values calculated by the author.

[a] Condensed from *ASTM Spec. Tech. Publ.* 109A, The American Society for Testing and Materials, Philadelphia, 1963.

[b] For the air-saturated hydrocarbon at 1 atm.

[c] Apparent values from weights in air.

[d] Absolute values from weights in vacuum.

[e] At saturation pressure.

[f] At saturation pressure (triple point).

[g] At boiling point.

[h] Critical solution temperature instead of aniline point.

[i] Apparent values for methane at 60°F (15.56°C).

[k] The + sign and the number following signify that the octane number of the compound corresponds to that of 2,2,4-trimethyl pentane with the indicated number of milliliters of tetraethyl lead added.

[l] Average value from octane numbers of more than one sample.

[m] Extrapolated to room temperature from higher temperatures.

[n] For heats of combustion of gaseous hydrocarbons, C_1 to C_5, see Table 12-2.

[o] At sublimation point.

[p] Specific gravity at 119°F/60°F (sublimation point).

[q] Value for the ideal gas.

[r] Numbers in Table 12-2 correspond to compounds as listed in Table 12-1.

[s] Sublimation point.

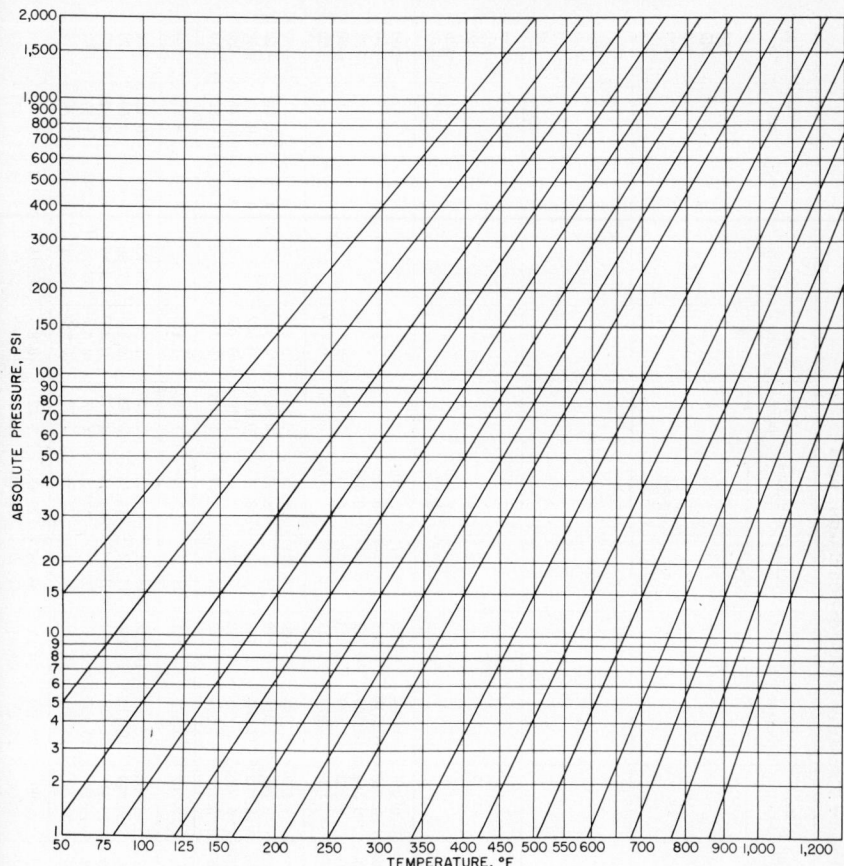

Fig. 12-1. Vaporization chart for hydrocarbons. (*From O. A. Hougen and K. M. Watson, Industrial Chemical Calculations, 2d ed., John Wiley & Sons, Inc., New York, 1936, p. 73.*)

Table 12-3. Vapor Pressure of Hydrocarbons, psia

Temp, °F	Methane	Ethylene	Ethane	Propylene	Propane	i-Butene	1-Butene	trans-2-Butene	cis-2-Butene	1,3-Butadiene	i-Butane	n-Butane	i-Pentane	n-Pentane	Neopentane	Cyclopentane	n-Hexane	Temp, °F
-250	21.47																	-250
-240	31.9	0.253																-240
-230	46.4	0.488	0.269															-230
-220	64.5	0.885	0.489															-220
-210	87.6	1.519																-210
-200	115.7	2.484	0.842															-200
-190	150.0	3.899	1.385															-190
-180	191.5	5.899	2.189															-180
-170	240.0	8.64	3.339	0.191														-170
-160	297.0	12.31	4.938	0.325														-160
-150	364	17.08	7.099	0.531	0.380													-150
-140	440	23.17	9.952	0.837	0.605													-140
-130	527	30.8	13.64	1.276	0.932													-130
-120	627	40.2	18.32	1.892	1.394						0.229	0.109						-120
-110			24.15	2.732	2.030	0.250	0.240			0.210	0.359	0.176						-110
-100		65.555	31.31	3.853	2.887	0.388	0.374	0.246	0.195	0.329	0.547	0.276						-100
-90			39.98	5.32	4.017	0.585	0.565	0.378	0.303	0.501	0.810	0.422						-90
-80		100.98	50.34	7.21	5.481	0.860	0.832	0.564	0.460	0.742	1.171	0.625						-80
-70			62.63	9.59	7.344	1.236	1.196	0.823	0.679	1.073	1.657	0.906	0.197	0.12				-70
-60		148.45	77.02	12.55	9.680	1.739	1.684	1.173	0.979	1.518	2.295	1.285	0.296	0.179				-60
-50		206.29	93.76	16.19	12.57	2.308	2.325	1.639	1.382	2.106	3.121	1.787	0.434	0.268				-50
-40			113.1	20.59	16.09	3.249	3.153	2.249	1.914	2.869	4.17	2.439	0.622	0.392		0.20	0.06	-40
-30		289.91	135.0	25.87	20.33	4.330	4.207	3.032	2.604	3.483	5.49	3.273	0.874	0.562		0.29	0.10	-30
-20			159.9	32.13	25.4	5.684	5.527	4.025	3.485	5.069	7.12	4.326	1.034	0.789		0.42	0.15	-20
-10			188.1	39.43	30.95	7.395	7.16	5.266	4.594	6.593	9.12	5.64	1.205	1.088		0.59	0.217	-10
0		388.03	219.7	47.90	37.81	9.404	9.16	6.797	5.972	8.461	11.53	7.25	1.892	1.475		0.81	0.311	0
+10			254.9	57.69	45.85	11.876	11.57	8.665	7.662	10.728	14.41	9.20	2.870	1.97	6.23	1.10	0.438	+10
20		507.88	294.0	68.92	55.00	14.831	14.46	10.919	9.713	13.449	17.82	11.56	3.726	2.594	7.88	1.48	0.606	20
30			337.1	81.75	65.70	18.33	17.89	13.61	12.176	16.684	21.8	14.36	4.777	3.370	9.85	1.95	0.825	30
40		654.23	385.0	96.30	77.80	22.44	21.92	16.80	15.107	20.49	26.5	17.66	6.05	4.325	12.20	2.546	1.107	40
50			437.5	112.7	91.50	27.22	26.60	20.55	18.56	24.95	31.91	21.53	7.59	5.489	14.96	3.278	1.464	50
60			494.2	131.1	106.9	32.74	32.02	24.90	22.60	30.10	38.08	26.03	9.42	6.887	18.18	4.172	1.911	60
70			558.3	151.6	124.3	39.05	38.23	29.94	27.29	35.61	45.09	31.20	11.57	8.559	21.9	5.254	2.464	70

80	3.143	6.552	26.2	10.538	14.10	37.13	53.07	42.33	32.40	35.59	45.5	46.25	143.6	174.3	630.7
90	3.966	8.095	31.52	12.862	17.04	43.89	62.07	49.99	38.53	42.23	53.60	54.42	165.0	199.6	715.9
100	4.956	9.914	37.02	15.570	20.44	51.57	72.22	58.7	45.54	49.80	63.27	63.64	188.7	227.2	
110	6.136	12.043	43.19	18.7	24.3	60.22	83.54	68.53	53.53	58.38	72.62	73.99	214.8	257.7	
120	7.531	14.517	50.14	22.3	28.8	69.95	96.18	79.56	69.55	68.07	85.13	85.58	243.4	291.0	
130	9.168	17.37	57.91	26.4	33.26	80.79	110.2	91.91	72.71	78.96	96.71	98.48	274.5	327.5	
140	11.075	20.65	66.60	31.11	38.80	92.87	125.7	105.7	84.10	91.13	112.07	112.8	304.8	367.2	
150	13.283	24.4	76.24	36.4	45.04	106.2	142.7	121.0	96.81	104.7	129.08	128.6	345.8	410.5	
160	15.823	28.6	86.92	42.4	52.03	120.9	161.3	137.9	110.9	119.7	141.44	146.0	385.0	457.2	
170	18.73	33.85	98.70	49.0	59.83	137.0	181.7	156.5	126.6	136.3	164.96	165.1	426.0	508.0	
180	22.03	39.28	111.7	56.5	68.49	154.5	203.9	176.8	143.8	154.6	183.31	186.0	473.2	562.8	
190	25.8	44.34	125.8	64.76	78.09	173.6	227.9	199.2	162.8	174.7	207.27	208.7	523.4	622.9	
200	30.0	52.04	141.4	73.89	88.68	194.3	254.0	223.5	183.6	196.7	228.58	233.4	575.0		
210	33.83	59.48	158.3	83.94	100.3	216.8	282.3	249.9	206.3	220.6	256.51	260.1			
220	38.91	67.71	176.7	94.99	113.1	241.2	312.8	278.5	231.0	246.7	281.21	289.0			
230	44.57	76.83	196.6	107.1	127.0	267.6	345.7	309.5	257.7	274.8	312.96	320.3			
240	50.86	86.89	218.2	120.3	142.2	296.3	381.2	342.8	286.8	305.3	347.06	354.1			
250	57.84	97.74	241.5	134.7	158.7	327.4	419.4	378.7	318.1	338.1	377.05	390.4			
260	65.54	108.65	266.5	150.3	176.6	361.0	460.5	417.1	351.8	373.4	415.27	429.6			
270	74.03	122.5	293.4	167.3	195.9	397.1	505.6	458.2	388.1	411.3	448.94	471.8			
280	83.35	136.6	322.2	185.6	216.8	435.9		502.1	426.9	451.8	491.56	517.4			
290	93.54	151.8	353.0	204.9	239.1	477.8		548.8	468.4	495.2	528.76	567.0			
300	104.7	167.1	385.9	226.7	263.2	523.5		598.6	512.7	541.2					
310	116.8	185.8	421.0	249.7	288.9				559.9	590.2					
320	130.1	204.6	458.3	274.4	316.4				610.0						
330	144.4	225.1		300.8	345.7										
340	159.9	246.8		329.1	374										
350	176.5	269.7		359.4	405										
360	194.5	290.4		391.9	442										
370	213.9	320.2		426.8											
380	234.6	348.2		464.7											
390	256.9	377.9													
400	280.6	409.6													
410	306.0	443.1													
420	333.0	478.7													
430	361.3	516.3													
440	392.1	557.7													
450	424.5	600.8													
460		646.7													

Table 12-3. Vapor Pressure of Hydrocarbons, psia (Continued)

Temp, °F	Cyclohexane	Methyl cyclopentane	n-Heptane	Triptane	n-Octane	Diisobutyl	2,2,4-Trimethyl pentane	n-Nonane	n-Decane	Benzene	Toluene	Ethyl benzene	o-Xylene	m-Xylene	p-Xylene	n-Propyl benzene	iso-Propyl benzene	Temp, °F
-10		0.20		0.206														-10
0		0.28		0.291														0
+10		0.40		0.404						0.183s								+10
20		0.55		0.551						0.284s								20
30		0.75		0.741			0.235			0.432s								30
40		1.008	0.288	0.983		0.184	0.325			0.643s								40
50	0.918	1.332	0.398			0.257	0.443			0.88l	0.240							50
60	1.212	1.738	0.541	1.287		0.354	0.596			1.170	0.331							60
70	1.579	2.242	0.726	1.664	0.216	0.479	0.789			1.534	0.448							70
80	2.033	2.856	0.960	2.128	0.297	0.641	1.033			1.987	0.599	0.202			0.186			80
90	2.590	3.604	1.254	2.693	0.402	0.845	1.336			2.544	0.791	0.276	0.194	0.242	0.254			90
100	3.264	4.503	1.620	3.374	0.537	1.101	1.708	0.179	0.073	3.224	1.032	0.371	0.264	0.326	0.342		0.188	100
110	4.075	5.575	2.069	4.188	0.709	1.419	2.163	0.245		4.045	1.331	0.492	0.353	0.434	0.454	0.192	0.254	110
120	5.042	6.842	2.616	5.153	0.925	1.808	2.711	0.330		5.028	1.698	0.645	0.467	0.571	0.596	0.259	0.340	120
130	6.185	8.330	3.276	6.288	1.192	2.282	3.367	0.438		6.195	2.146	0.836	0.610	0.742	0.774	0.344	0.449	130
140	7.526	10.064	4.065	7.615	1.521	2.852	4.147	0.576	0.219	7.570	2.687	1.073	0.789	0.955	0.994	0.452	0.586	140
150	9.089	12.072	5.002	9.154	1.922	3.534	5.066	0.748	0.293	9.178	3.335	1.363	1.010	1.217	1.265	0.588	0.756	150
160	10.899	14.382	6.106	10.929	2.406	4.343	6.142	0.961	0.387	11.047	4.104	1.715	1.280	1.536	1.594	0.756	0.966	160
170	12.981	17.02	7.398	12.964	2.985	5.294	7.393	1.222	0.505	13.205	5.013	2.139	1.608	1.922	1.990	0.962	1.223	170
180	15.363	20.03	8.899	15.283	3.674	6.407	8.839	1.541	0.652	15.681	6.077	2.646	2.003	2.385	2.466	1.214	1.534	180
190	18.07	23.43	10.633	17.91	4.487	7.699	10.500	1.924	0.834	18.508	7.316	3.248	2.475	2.935	3.030	1.518	1.908	190
200	21.14	27.3	12.623	20.88	5.440	9.190	12.396	2.384	1.057	21.715	8.750	3.957	3.034	3.586	3.696	1.882	2.353	200
210	24.6	31.53	14.897	24.21	6.549	10.902	14.551	2.930	1.326	25.34	10.400	4.787	3.694	4.350	4.478	2.316	2.881	210
220	28.5	36.29	18.00	27.93	7.833	12.855	16.99	3.574	1.651	29.41	12.288	5.753	4.465	5.242	5.388	2.829	3.502	220
230	32.3	41.59	21.3	32.24	9.311	15.073	19.73	4.329	2.039	33.0	14.437	6.870	5.363	6.275	6.442	3.432	4.228	230
240	36.96	47.42	24.8	37.04	11.002	17.58	22.80	5.208	2.499	39.0	16.82	8.155	6.401	7.468	7.657	4.135	5.071	240
250	42.14	53.91	28.6	42.36	12.928	20.40	26.23	6.227	3.041	44.0	19.62	9.625	7.596	8.837	9.049	4.950	6.046	250
260	47.89	61.02	32.7	48.24	15.110	23.56	30.03	7.399	3.676	50.0	22.70	11.299	8.964	10.399	10.636	5.892	7.165	260
270	54.23	68.82	37.4	54.72	17.57	27.10	34.21	8.742	4.414	57	26.14	13.195	10.522	12.174	12.437	6.973	8.445	270
280	61.19	77.03	42.4	61.82	20.33	31.04	38.84	10.272	5.268	65	29.98	15.335	12.288	14.181	14.472	8.207	9.901	280
290	68.82	86.74	48.0	69.59	23.42	35.42	43.96	12.008	6.252	72	33.6	17.738	14.281	16.441	16.761	9.610	11.549	290
300	77.17	96.94	52.0	72.02	26.9	40.26	49.56	13.968	7.377	82	38	20.43	16.522	18.975	19.32	11.197	13.408	300

Temp	15.495	12.986	22.19	21.81	19.030	23.42	43	91	8.660	16.170	55.71	45.59	30.7	87.28	60.9	107.9	86.28
310	15.495	12.986	22.19	21.81	19.030	23.42	43	91	8.660	16.170	55.71	45.59	30.7	87.28	60.9	107.9	86.28
320	17.83	14.993	25.37	24.96	21.83	26.75	48.5	100	10.114	18.64	62.42	51.45	34.5	97.25	68.8	119.9	96.19
330	20.43	17.236	28.89	28.45	24.94	30.43	55	111	11.757	21.39	69.71	57.87	39.1	108.0	77.5	132.8	107.0
340	23.31	19.73	31.94	32.33	28.38	34	62	124	13.603	24.44	77.66	64.89	44.2	119.7	86.5	146.6	118.6
350	26.50	22.50	36.04	36.60	32.18	38	69	136	15.672	27.83	86.25	72.53	50	132.2	96.0	161.6	131.1
360	30.02	25.57	40.53	41.31	36.38	43.8	77	150	17.98	31.13	95.52	80.84	55.5	145.6	106.5	177.5	145.9
370	33.60	28.95	45.46	46.47	40.98	49.4	85	166	20.54	35.17	105.45	89.84	62	159.9	117.5	194.7	159.2
380	37.79	32.5	50.85	52.11	46.02	55	94	182	23.39	39.61	116.21	99.57	68.6	175.3	130.0	213.0	174.9
390	42.36	36.4	56.72	58.27	51.52	61	105	201	26.11	44.48	127.8	110.1	76.5	191.6	143.0	232.5	191.6
400	47.34	41	63.17	64.97	57.71	68.5	116	221	29.46	49.87	140.1	121.4	85	209.0	156.5	253.3	209.5
410	52.78	46	70.08	72.25	64.01	75	127	242	33.16	55.62	153.4	133.5	95	227.4	171.0	275.5	223.4
420	58.66	51	77.61	80.12	71.06	83	140	262	37.22	61.96	167.4	146.5	105	246.6	186.5	299.2	248.9
430	65.03	56	85.76	88.64	78.68	91	153	294	41.66	68.83	182.6	160.4	116	267.6	203.5	324.4	270.6
440	71.92	62.5	94.76	97.82	86.89	100	168	307	46.53	76.31	199.3	175.3	127	289.5	223.0	351.2	293.6
450	79.35	69	104.0	107.7	95.74	110	183	332	51.84	84.39	216.0	191.1	140	312.5	243.0	379.5	317.9
460	87.33	76	114.2	118.3	105.2	121	200	362	57.62	93.12	234.4	207.9	153	336.7	264.0	409.5	343.7
470	95.91	83.5	125.1	129.7	115.4	132	218	394	63.88	102.5	253.9	225.8	168	362.1	286.0	441.7	371.0
480	105.1	91.5	136.8	141.3	126.3	144	237	421	70.70	112.6	274.7	244.7	184	388.9	310.0	476.0	399.7
490	114.9	100	149.3	154.8	137.9	158	256	455	78.08	123.5	297.1	264.8	201	416.8	336.0	512.7	430.1
500	125.4	110	162.8	168.6	150.3	173	277	490	86.02	135.1	320.9	286.0	220		362.5		461.9
510	136.6	121	177.2	183.4	163.5	188	299	525	94.60	147.6	345.8	308.4	239		392.0		495.5
520	148.5	130	192.2	199.0	177.5	203	320	567	103.8	160.9		331.9	260				530.7
530	161.1	142	208.4	215.6	192.3	221	346	610	113.8	175.1		336.7	283				
540	174.6	155	225.5	233.1	208.1	240	372	655	124.5	190.2			310				
550	188.8	169	243.7	251.7	224.8	260	402	701	135.9	206.3			327				
560	203.8	183	262.9	271.3	236.8	281	432		148.1	223.4			350				
570	219.6	198	283.3	292.0	260.9	303	462		161.2	241.5							
580	236.4	214	304.8	213.8	280.4	327	499		175.1	260.7							
590	254.0	233	327.4	336.7	300.9	354	537		189.9	280.9							
600	272.5	253	351.3	360.8	322.4	380	578		205.6	302.4							
610	291.9	273	376.4	386.4	345.0	409			222.7	325.0							
620	312.4	294	402.8	412.5	368.7	439			240.1								
630	333.7	316	430.5	440.2	393.5	473			258.9								
640	356.2	340	459.6	471.4	419.4	508			278.8								
650	379.6	365	490.1	499.5	446.5	544			299.9								
660	404.1	388			474.7												
670	429.6	415			504.7												
680	456.3	450															

s = solid; l = liquid.

References for Table 12-3, Vapor Pressure of Hydrocarbons

1. F. D. Rossini, ed., "Selected Values of Physical and Thermodynamic Properties of Hydrocarbons and Related Compounds," API Project 44, Carnegie Press, Pittsburgh, 1953, and looseleaf tables extant December, 1963. Vapor pressure for all hydrocarbons in the table from 0.2 to 30 psia were taken from this reference.
2. Griswold, J., and W. B. Brooks, *Refiner*, vol. 26, 818, December, 1947. Contains data from 60°F to the critical point for propylene, propane, isobutane, isobutene, 1-butene, 1,3-butadiene, *n*-butane, trans-2-butene, cis-2-butene, isopentane, and *n*-pentane.
3. Glazer, F., and H. Ruland, *Chem. Ing. Tech.*, vol. 29, p. 772, 1957, contains data for 33 technically important substances.
4. Sage, B. H., and W. N. Lacey, "Thermodynamic Properties of the Lighter Paraffin Hydrocarbons and Nitrogen," API Research Project 37 Monograph, American Petroleum Institute, 1950. Includes data on methane, propane, *n*-butane, isobutane, and *n*-pentane.
5. Sage, B. H., and W. N. Lacey, "Some Properties of the Light Hydrocarbons, Hydrogen Sulfide and Carbon Dioxide," API Research Project 37 Monograph, American Petroleum Institute, 1955. Includes data on propylene, 1-butene, and *n*-nonane.
6. Hachmuth, K., G. Hanson, and M. L. Smith, *Trans. Am. Inst. Chem. Engrs.*, vol. 42, p. 975, 1946, contains data for isobutane, isobutylene, 1-butene, 1,3-butadiene, *n*-butane, trans-2-butene, cis-2-butene, vinyl acetylene, ethyl acetylene, diacetylene, 1,2-butadiene, and dimethyl acetylene.
7. Brown, G. G., G. G. Oberfell, and R. C. Alden, "Natural Gasoline and the Volatile Hydrocarbons, Section I", NGAA, 1948, contains a tabulation for each 10°F from 40°F to the critical temperature of vapor-pressure data for ethylene, acetylene, ethane, propylene, propane, cyclopropane, isobutane, isobutylene, 1-butene, 1,3-butadiene, *n*-butane, cyclobutane, isopentane, *n*-pentane, cyclopentane, *n*-hexane, benzene, cyclohexane, methylcyclohexane, *n*-heptane, toluene, *n*-octane, ethyl benzene, *n*-nonane, and *n*-propyl benzene.
8. Young, S., *Sci. Proc. Roy. Dublin Soc.*, vol. 12, p. 374, 1910, includes data for isopentane, *n*-pentane, *n*-hexane, diisopropyl, *n*-heptane, diisobutyl, *n*-octane, cyclohexane, and benzene.
9. Thompson, G. W., "The Antoine Equation for Vapor Pressure Data," *Chem. Rev.*, vol. 38, no. 1, pp. 1–39, 1946.
10. Thodos, G., *Ind. Engrg. Chem.*, vol. 42, pp. 1514–1526, 1950, includes reference selections and mathematical methods for vapor-pressure computation for saturated hydrocarbons methane through decane.

Specific References for Particular Hydrocarbons

METHANE
11. Mathews, C. S., and C. O. Hurd, *Trans. Am. Inst. Chem. Engrs.*, vol. 42, p. 55, 1946.
ETHYLENE
12. York, R., Jr., and E. F. White, *Trans. Am. Inst. Chem. Engrs.*, vol. 40, p. 227, 1944.
ETHANE
13. Barkelew, C. H., J. L. Valentine, and C. O. Hurd, *Chem. Engrg. Progr.*, vol. 43, p. 25, 1947.
PROPYLENE
14. Hanson, G. H., *Trans. Am. Inst. Chem. Engrs.*, vol. 42, p. 959, 1946. This article also contains thermodynamic data for saturated propane, isobutane, isobutylene, and *n*-butane.
15. Canjar, L. N., M. Goldman, and H. Marchman, *Ind. Engrg. Chem.*, vol. 43, p. 1186, 1951.
PROPANE
16. Stearns, W. V., and E. J. George, *Ind. Engrg. Chem.*, vol. 35, p. 602, 1943, and references 4 and 14.
ISOBUTENE
References 2, 6, and 13.
1-BUTENE
17. Beattie, J. A., and S. Marples, Jr., *J. Am. Chem. Soc.*, vol. 72, p. 1449, 1950, and references 2, 5, and 6.
TRANS-2-BUTENE
References 2 and 6.

Cis-2-butene
References 2 and 6.

1,3-Butadiene
18. Meyers, C. H., C. S. Cragoe, and E. F. Mueller, *J. Res. Natl. Bur. Std.*, vol. 39, p. 507, 1947.
19. Meyers, C. H., R. A. Scott, F. G. Brickwedde, and R. P. Rand, Jr., *J. Res. Natl. Bur. Std.*, vol. 35, p. 39, 1945, and references 2 and 6.

Isobutane
References 4, 6, and 14.

n-Butane
20. Prengle, H. W., Jr., L. R. Greenhaus, and R. York, Jr., *Chem. Engrg. Progr.*, vol. 44, p. 863, 1948, and references 2, 4, 6, and 14.

Isopentane
21. Isaac, R., Kun Li, and L. N. Canjar, *Ind. Engrg. Chem.*, vol. 46, p. 199, 1954, and references 2 and 8.

n-Pentane
22. Bryden, J. W., N. Walen, and L. N. Canjar, *Chem. Engrg. Progr.*, Symposium Series 49, no. 7. p. 151, 1953, and references 2, 4, and 8.

2,2-Dimethyl Propane (Neopentane)
23. Beattie, J. A., D. R. Douslin, and S. V. Levine, *J. Chem. Phys*, vol. 19, p. 948, 1959.

Cyclopentane
24. Kay, W. B., *J. Am. Chem. Soc.*, vol. 69, p. 1273, 1947.   This reference includes data for methyl cyclopentane, ethyl cyclopentane, and methyl cyclohexane.

n-Hexane
25. Kay, W. B., *J. Am. Chem. Soc.*, vol. 68, p. 1336, 1946.
26. Thomas, G. L., and S. Young, *J. Chem. Soc.*, 1071, 1895, corrected as noted in reference 27, 1910, p. 68.   Also reference 8.

Cyclohexane
27. Young, S., and E. C. Fortey, *J. Chem. Soc.*, vol. 75, p. 873, 1895, corrected in "Annual Tables of Chemistry, Physics and Technology, Constants and Numerical Values," vol. 1, p. 68, 1910, Schimmel, Leipzig, and references 3 and 7.

Methyl Cyclopentane
Reference 24.

n-Heptane
28. Frost, A. A., and P. R. Kalkwarf, *J. Chem. Phys.*, vol. 21, p. 264, 1953.
29. Kobe, K., H. R. Crawford, and R. W. Stephenson, *Ind. Engrg. Chem.*, vol. 47, p. 1767, 1955.
30. Young, S., *J. Chem. Soc.*, vol. 73, p. 675, 1898.

2,2,3-Trimethyl Butane (Triptane)
Calculated above 30 psia by Antoine equation (reference 9).

n-Octane
31. Young, S., *J. Chem. Soc.*, vol. 77, 1145, 1900, reference 8 and tables cited in reference 27, and reference 8.

2,5-Dimethyl Hexane (Diisobutyl)
32. Young, S., and E. C. Fortey, *J. Chem. Soc.*, vol. 27, p. 1136, 1900, reference 8 and calculations using reference 9.

2,2,4-Trimethyl Pentane (isooctane)
33. Kay, W. B., and F. M. Warzel, *Ind. Engrg. Chem.*, vol. 43, p. 1150, 1951.

n-Nonane
Calculated above 30 psia from references 9 and 10.

n-Decane
Calculated above 30 psia from references 9 and 10.

Benzene
34. Organic, E. I., and W. R. Studhalter, *Chem. Engrg. Progr.*, vol. 44, p. 847, 1948.
35. Bender, P., G. T. Furukawa, and J. R. Hyndman, *Ind. Engrg. Chem.*, vol. 44, p. 387, 1952, and references 7 and 8.

Toluene (methyl benzene)
36. Krase, N. W., and J. B. Goodman, *Ind. Engrg. Chem.*, vol. 22, p. 13, 1930.

Ethylbenzene
37. Reference 7.

o-, m-, p-Xylene
Reference 3 and unpublished calculations by the author.

n-Propyl benzene
Reference 7.

Isopropyl benzene
Reference 3 and unpublished calculations by the author.

Table 12-4. Relation between Pounds per Gallon, Specific Gravity, and Degrees API*

°API	Tenths of degrees									
	0	1	2	3	4	5	6	7	8	9
0	8.962	8.956	8.949	8.942	8.935	8.928	8.922	8.915	8.908	8.901
	1.0760	1.0752	1.0744	1.0736	1.0728	1.0720	1.0712	1.0703	1.0695	1.0687
1	8.895	8.888	8.881	8.875	8.868	8.861	8.855	8.848	8.841	8.835
	1.0679	1.0671	1.0663	1.0655	1.0647	1.0639	1.0631	1.0623	1.0615	1.0607
2	8.828	8.821	8.815	8.808	8.802	8.795	8.788	8.782	8.775	8.769
	1.0599	1.0591	1.0583	1.0575	1.0568	1.0560	1.0552	1.0544	1.0536	1.0528
3	8.762	8.756	8.749	8.743	8.736	8.730	8.723	8.717	8.710	8.704
	1.0520	1.0513	1.0505	1.0497	1.0489	1.0481	1.0474	1.0466	1.0458	1.0451
4	8.698	8.691	8.685	8.678	8.672	8.666	8.659	8.653	8.646	8.640
	1.0443	1.0435	1.0427	1.0420	1.0412	1.0404	1.0397	1.0389	1.0382	1.0374
5	8.634	8.627	8.621	8.615	8.608	8.602	8.596	8.590	8.583	8.577
	1.0366	1.0359	1.0351	1.0344	1.0336	1.0328	1.0321	1.0313	1.0306	1.0298
6	8.571	8.565	8.558	8.552	8.546	8.540	8.534	8.527	8.521	8.515
	1.0291	1.0283	1.0276	1.0269	1.0261	1.0254	1.0246	1.0239	1.0231	1.0224
7	8.509	8.503	8.497	8.490	8.484	8.478	8.472	8.466	8.460	8.454
	1.0217	1.0209	1.0202	1.0195	1.0187	1.0180	1.0173	1.0165	1.0158	1.0151
8	8.448	8.442	8.436	8.430	8.424	8.418	8.412	8.406	8.400	8.394
	1.0143	1.0136	1.0129	1.0122	1.0114	1.0107	1.0100	1.0093	1.0086	1.0078
9	8.388	8.382	8.376	8.370	8.364	8.358	8.352	8.346	8.340	8.334
	1.0071	1.0064	1.0057	1.0050	1.0043	1.0035	1.0028	1.0021	1.0014	1.0007
10	8.328	8.322	8.317	8.311	8.305	8.299	8.293	8.287	8.282	8.276
	1.0000	0.9993	0.9986	0.9979	0.9972	0.9965	0.9958	0.9951	0.9944	0.9937
11	8.270	8.264	8.258	8.252	8.246	8.241	8.235	8.229	8.223	8.218
	0.9930	0.9923	0.9916	0.9909	0.9902	0.9895	0.9888	0.9881	0.9874	0.9868
12	8.212	8.206	8.201	8.195	8.189	8.183	8.178	8.172	8.166	8.161
	0.9861	0.9854	0.9847	0.9840	0.9833	0.9826	0.9820	0.9813	0.9806	0.9799
13	8.155	8.150	8.144	8.138	8.132	8.127	8.122	8.116	8.110	8.105
	0.9792	0.9786	0.9779	0.9772	0.9765	0.9759	0.9752	0.9745	0.9738	0.9732
14	8.099	8.093	8.088	8.082	8.076	8.071	8.066	8.061	8.055	8.049
	0.9725	0.9718	0.9712	0.9705	0.9698	0.9692	0.9685	0.9679	0.9672	0.9665
15	8.044	8.038	8.033	8.027	8.021	8.016	8.011	8.006	8.000	7.995
	0.9659	0.9652	0.9646	0.9639	0.9632	0.9626	0.9619	0.9613	0.9606	0.9600
16	7.989	7.984	7.978	7.973	7.967	7.962	7.956	7.951	7.946	7.940
	0.9593	0.9587	0.9580	0.9574	0.9567	0.9561	0.9554	0.9548	0.9541	0.9535
17	7.935	7.930	7.925	7.919	7.914	7.909	7.903	7.898	7.893	7.887
	0.9529	0.9522	0.9516	0.9509	0.9503	0.9497	0.9490	0.9484	0.9478	0.9471
18	7.882	7.877	7.871	7.866	7.861	7.856	7.851	7.846	7.841	7.835
	0.9465	0.9459	0.9452	0.9446	0.9440	0.9433	0.9427	0.9421	0.9415	0.9408
19	7.830	7.825	7.820	7.814	7.809	7.804	7.799	7.793	7.788	7.783
	0.9402	0.9396	0.9390	0.9383	0.9377	0.9371	0.9365	0.9358	0.9352	0.9346
20	7.778	7.773	7.768	7.762	7.757	7.752	7.747	7.742	7.737	7.732
	0.9340	0.9334	0.9328	0.9321	0.9315	0.9309	0.9303	0.9297	0.9291	0.9285
21	7.727	7.722	7.717	7.711	7.706	7.701	7.696	7.691	7.686	7.681
	0.9279	0.9273	0.9267	0.9260	0.9254	0.9248	0.9242	0.9236	0.9230	0.9224
22	7.676	7.671	7.666	7.661	7.656	7.651	7.646	7.641	7.636	7.632
	0.9218	0.9212	0.9206	0.9200	0.9194	0.9188	0.9182	0.9176	0.9170	0.9165
23	7.627	7.622	7.617	7.612	7.607	7.602	7.597	7.592	7.587	7.583
	0.9159	0.9153	0.9147	0.9141	0.9135	0.9129	0.9123	0.9117	0.9111	0.9106
24	7.578	7.573	7.568	7.563	7.558	7.554	7.549	7.544	7.539	7.534
	0.9100	0.9094	0.9088	0.9082	0.9076	0.9071	0.9065	0.9059	0.9053	0.9047
25	7.529	7.524	7.519	7.514	7.509	7.505	7.500	7.495	7.491	7.486
	0.9042	0.9036	0.9030	0.9024	0.9018	0.9013	0.9007	0.901	0.8996	0.8990

Table 12-4. Relation between Pounds per Gallon, Specific Gravity, and Degrees API* (Continued)

°API	Tenths of degrees									
	0	1	2	3	4	5	6	7	8	9
26	7.481	7.476	7.472	7.467	7.462	7.458	7.453	7.448	7.443	7.438
	0.8984	0.8978	0.8973	0.8967	0.8961	0.8956	0.8950	0.8944	0.8939	0.8933
27	7.434	7.429	7.424	7.420	7.415	7.410	7.406	7.401	7.397	7.392
	0.8927	0.8922	0.8916	0.8911	0.8905	0.8899	0.8894	0.8888	0.8883	0.8877
28	7.387	7.383	7.378	7.373	7.368	7.364	7.360	7.355	7.351	7.346
	0.8871	0.8866	0.8860	0.8855	0.8849	0.8844	0.8838	0.8833	0.8827	0.8822
29	7.341	7.337	7.332	7.328	7.323	7.318	7.314	7.309	7.305	7.300
	0.8816	0.8811	0.8805	0.8800	0.8794	0.8789	0.8783	0.8778	0.8772	0.8767
30	7.296	7.291	7.287	7.282	7.278	7.273	7.268	7.264	7.259	7.255
	0.8762	0.8756	0.8751	0.8745	0.8740	0.8735	0.8729	0.8724	0.8718	0.8713
31	7.251	7.246	7.242	7.238	7.233	7.228	7.224	7.219	7.215	7.211
	0.8708	0.8702	0.8697	0.8692	0.8686	0.8681	0.8676	0.8670	0.8665	0.8660
32	7.206	7.202	7.198	7.193	7.188	7.184	7.180	7.176	7.171	7.167
	0.8654	0.8649	0.8644	0.8639	0.8633	0.8628	0.8623	0.8618	0.8612	0.8607
33	7.163	7.158	7.153	7.149	7.145	7.141	7.137	7.132	7.128	7.123
	0.8602	0.8597	0.8591	0.8586	0.8581	0.8576	0.8571	0.8565	0.8560	0.8555
34	7.119	7.115	7.111	7.106	7.102	7.098	7.093	7.089	7.085	7.081
	0.8550	0.8545	0.8540	0.8534	0.8529	0.8524	0.8519	0.8514	0.8509	0.8504
35	7.076	7.072	7.068	7.063	7.059	7.055	7.051	7.047	7.042	7.038
	0.8498	0.8493	0.8488	0.8483	0.8478	0.8473	0.8468	0.8463	0.8458	0.8453
36	7.034	7.030	7.026	7.022	7.018	7.013	7.009	7.005	7.001	6.997
	0.8448	0.8443	0.8438	0.8433	0.8428	0.8423	0.8418	0.8413	0.8408	0.8403
37	6.993	6.989	6.985	6.980	6.976	6.972	6.968	6.964	6.960	6.955
	0.8398	0.8393	0.8388	0.8383	0.8378	0.8373	0.8368	0.8363	0.8358	0.8353
38	6.951	6.947	6.943	6.939	6.935	6.930	6.926	6.922	6.918	6.914
	0.8348	0.8343	0.8338	0.8333	0.8328	0.8324	0.8319	0.8314	0.8309	0.8304
39	6.910	6.906	6.902	6.898	6.894	6.890	6.886	6.882	6.878	6.874
	0.8299	0.8294	0.8289	0.8285	0.8280	0.8275	0.8270	0.8265	0.8260	0.8256
40	6.870	6.866	6.862	6.858	6.854	6.850	6.846	6.842	6.838	6.834
	0.8251	0.8246	0.8241	0.8236	0.8232	0.8227	0.8222	0.8217	0.8212	0.8208
41	6.830	6.826	6.822	6.818	6.814	6.810	6.806	6.802	6.798	6.794
	0.8203	0.8198	0.8193	0.8189	0.8184	0.8179	0.8174	0.8170	0.8165	0.8160
42	6.790	6.768	6.782	6.779	6.775	6.771	6.767	6.763	6.759	6.756
	0.8155	0.8151	0.8146	0.8142	0.8137	0.8132	0.8128	0.8123	0.8118	0.8114
43	6.752	6.748	6.744	6.740	6.736	6.732	6.728	6.724	6.720	6.716
	0.8109	0.8104	0.8100	0.8095	0.8090	0.8086	0.8081	0.8076	0.8072	0.8067
44	6.713	6.709	6.705	6.701	6.697	6.694	6.690	6.686	6.682	6.679
	0.8063	8.8058	0.8054	0.8049	0.8044	0.8040	0.8035	0.8031	0.8026	0.8022
45	6.675	6.671	6.667	6.663	6.660	6.656	6.652	6.648	6.645	6.641
	0.8017	0.8012	0.8008	0.8003	0.7999	0.7994	0.7990	0.7985	0.7981	0.7976
46	6.637	6.633	6.630	6.626	6.622	6.618	6.615	7.611	6.607	6.604
	0.7972	0.7967	0.7963	0.7958	0.7954	0.7949	0.7945	0.7941	0.7936	0.7932
47	6.600	6.596	6.592	6.589	6.585	6.582	6.578	6.574	6.571	6.567
	0.7927	0.7923	0.7918	0.7914	0.7909	0.7905	0.7901	0.7896	0.7892	0.7887
48	6.563	6.560	6.556	6.552	6.548	6.545	6.541	6.537	6.534	6.530
	0.7883	0.7879	0.7874	0.7870	0.7865	0.7861	0.7857	0.7852	0.7848	0.7844
49	6.526	6.523	6.520	6.516	6.512	6.509	6.505	6.501	6.498	6.494
	0.7839	0.7835	0.7831	0.7826	0.7822	0.7818	0.7813	0.7809	0.7805	0.7800
50	6.490	6.487	6.484	6.480	6.476	6.473	6.469	6.466	6.462	6.459
	0.7796	0.7792	0.7788	0.7783	0.7779	0.7775	0.7770	0.7766	0.7762	0.7758

Table 12-4. Relation between Pounds per Gallon, Specific Gravity,
and Degrees API* (Continued)

°API	Tenths of degrees									
	0	1	2	3	4	5	6	7	8	9
51	6.455	6.451	6.448	6.445	6.441	6.437	6.434	6.430	6.427	6.423
	0.7753	0.7749	0.7745	0.7741	0.7736	0.7732	0.7728	0.7724	0.7720	0.7715
52	6.420	6.416	6.413	6.410	6.406	6.402	6.399	6.396	6.392	6.389
	0.7711	0.7707	0.7703	0.7699	0.7694	0.7690	0.7686	0.7682	0.7678	0.7674
53	6.385	6.381	6.378	6.375	6.371	6.368	6.365	6.361	6.357	6.354
	0.7669	0.7665	0.7661	0.7657	0.7653	0.7649	0.7645	0.7640	0.7636	0.7632
54	6.350	6.347	6.344	6.340	6.337	6.334	6.330	6.326	6.323	6.320
	0.7628	0.7624	0.7620	0.7616	0.7612	0.7608	0.7603	0.7599	0.7595	0.7591
55	6.316	6.313	6.310	6.306	6.303	6.300	6.296	6.293	6.290	6.287
	0.7587	0.7583	0.7579	0.7575	0.7571	0.7567	0.7563	0.7559	0.7555	0.7551
56	6.283	6.280	6.276	6.273	6.270	6.266	6.263	6.259	6.256	6.253
	0.7547	0.7543	0.7539	0.7535	0.7531	0.7527	0.7523	0.7519	0.7515	0.7511
57	6.249	6.246	6.243	6.240	6.236	6.233	6.229	6.226	6.223	6.219
	0.7507	0.7503	0.7499	0.7495	0.7491	0.7487	0.7483	0.7479	0.7475	0.7471
58	6.216	6.213	6.209	6.206	6.203	6.199	6.196	6.193	6.190	6.187
	0.7467	0.7463	0.7459	0.7455	0.7451	0.7447	0.7443	0.7440	0.7436	0.7432
59	6.184	6.180	6.177	6.174	6.170	6.167	6.164	6.161	6.158	6.154
	0.7428	0.7424	0.7420	0.7416	0.7412	0.7408	0.7405	0.7401	0.7397	0.7393
60	6.151	6.148	6.144	6.141	6.138	6.135	6.132	6.129	6.125	6.122
	0.7389	0.7385	0.7381	0.7377	0.7374	0.7370	0.7366	0.7362	0.7358	0.7354
61	6.119	6.116	6.113	6.109	6.106	6.103	6.100	6.097	6.094	6.090
	0.7351	0.7347	0.7343	0.7339	0.7335	0.7332	0.7328	0.7324	0.7320	0.7316
62	6.087	6.084	6.081	6.078	6.075	6.072	6.068	6.065	6.062	6.059
	0.7313	0.7309	0.7305	0.7301	0.7298	0.7294	0.7290	0.7286	0.7283	0.7279
63	6.056	6.053	6.050	6.047	6.044	6.040	6.037	6.034	6.031	6.028
	0.7275	0.7271	0.7268	0.7264	0.7260	0.7256	0.7253	0.7249	0.7245	0.7242
64	6.025	6.022	6.019	6.016	6.013	6.010	6.007	6.004	6.000	5.997
	0.7238	0.7234	0.7230	0.7227	0.7223	0.7219	0.7216	0.7212	0.7208	0.7205
65	5.994	5.991	5.988	5.985	5.982	5.979	5.976	6.973	5.970	5.967
	0.7201	0.7197	0.7194	0.7190	0.7186	0.7183	0.7179	0.7175	0.7172	0.7168
66	5.964	5.961	5.958	5.955	5.952	5.949	5.946	5.943	5.940	5.937
	0.7165	0.7161	0.7157	0.7154	0.7150	0.7146	0.7143	0.7139	0.7136	0.7132
67	5.934	5.931	5.928	5.925	5.922	5.919	5.916	5.913	5.910	5.907
	0.7128	0.7125	0.7121	0.7118	0.7114	0.7111	0.7107	0.7103	0.7100	0.7096
68	5.904	5.901	5.898	5.895	5.892	5.889	5.886	5.883	5.880	5.877
	0.7093	0.7089	0.7086	0.7082	0.7079	0.7075	0.7071	0.7068	0.7064	0.7061
69	5.874	5.871	5.868	5.866	5.863	5.860	5.857	5.854	5.851	5.848
	0.7057	0.7054	0.7050	0.7047	0.7043	0.7040	0.7036	0.7033	0.7029	0.7026
70	5.845	5.842	5.839	5.836	5.833	5.831	5.828	5.825	5.823	5.820
	0.7022	0.7019	0.7015	0.6012	0.7008	0.7005	0.7001	0.6998	0.6995	0.6991
71	5.817	5.814	5.811	5.808	5.805	5.802	5.799	5.796	5.793	5.791
	0.6988	0.6984	0.6981	0.6977	0.6974	0.6970	0.6967	0.6964	0.6960	0.6957
72	5.788	5.785	5.782	5.779	5.776	5.773	5.771	5.768	5.765	5.762
	0.6953	0.6950	0.6946	0.6943	0.6940	0.6936	0.6933	0.6929	0.6926	0.6923
73	5.759	5.757	5.754	5.751	5.748	5.745	5.743	5.740	5.737	5.734
	0.6919	0.6916	0.6913	0.6909	0.6906	0.6902	0.6899	0.6895	0.6892	0.6889
74	5.731	5.728	5.726	5.723	5.720	5.718	5.715	5.712	5.709	5.706
	0.6886	0.6882	0.6879	0.6876	0.6872	0.6869	0.6866	0.6862	0.6859	0.6856
75	5.703	5.701	5.698	5.695	5.693	5.690	5.687	5.685	5.682	5.679
	0.6852	0.6849	0.6846	0.6842	0.6839	0.6836	0.6832	0.6829	0.6826	0.6823

Table 12-4. Relation between Pounds per Gallon, Specific Gravity, and Degrees API* (Continued)

| °API | Tenths of degrees | | | | | | | | | |
	0	1	2	3	4	5	6	7	8	9
76	5.676	5.673	5.671	5.668	5.665	5.662	5.660	5.657	5.654	5.652
	0.6819	0.6816	0.6813	0.6809	0.6806	0.6803	0.6800	0.6796	0.6793	0.6790
77	5.649	5.646	5.643	5.641	5.638	5.635	5.632	5.630	5.627	5.624
	0.6787	0.6783	0.6780	0.6777	0.6774	0.6770	0.6767	0.6764	0.6761	0.6757
78	5.622	5.619	5.617	5.614	5.611	5.608	5.606	5.603	5.600	5.598
	0.6754	0.6751	0.6748	0.6745	0.6741	0.6738	0.6735	0.6732	0.6728	0.6725
79	5.595	5.592	5.590	5.587	5.584	5.582	5.579	5.577	5.574	5.571
	0.6722	0.6719	0.6716	0.6713	0.6709	0.6706	0.6703	0.6700	0.6697	0.6693
80	5.568	5.566	5.563	5.561	5.558	5.556	5.553	5.550	5.548	5.545
	0.6690	0.6687	0.6684	0.6681	0.6678	0.6675	0.6671	0.6668	0.6665	0.6662
81	5.542	5.540	5.537	5.534	5.532	5.529	5.526	5.524	5.522	5.519
	0.6659	0.6656	0.6653	0.6649	0.6646	0.6643	0.6640	0.6637	0.6634	0.6631
82	5.516	5.514	5.511	5.508	5.506	5.503	5.501	5.498	5.496	5.493
	0.6628	0.6625	0.6621	0.6618	0.6615	0.6612	0.6609	0.6606	0.6603	0.6600
83	5.491	5.489	5.486	5.483	5.480	5.477	5.475	5.472	5.470	5.467
	0.6597	0.6594	0.6591	0.6588	0.6584	0.6581	0.6578	0.6575	0.6572	0.6569
84	5.465	5.462	5.460	5.458	5.455	5.453	5.450	5.448	5.445	5.443
	0.6566	0.6563	0.6560	0.6557	0.6554	0.6551	0.6548	0.6545	0.6542	0.6539
85	5.440	5.437	5.435	5.432	5.430	5.427	5.425	5.422	5.420	5.417
	0.6536	0.6533	0.6530	0.6527	0.6524	0.6521	0.6518	0.6515	0.6512	0.6509
86	5.415	5.412	5.410	5.407	5.405	5.402	5.400	5.397	5.395	5.392
	0.6506	0.6503	0.6500	0.6497	0.6494	0.6491	0.6488	0.6485	0.6482	0.6479
87	5.390	5.387	5.385	5.382	5.380	5.377	5.375	5.372	5.370	5.367
	0.6476	0.6473	0.6470	0.6467	0.6464	0.6461	0.6458	0.6455	0.6452	0.6449
88	5.365	5.363	5.361	5.358	5.356	5.353	5.351	5.348	5.346	5.343
	0.6446	0.6444	0.6441	0.6438	0.6435	0.6432	0.6429	0.6426	0.6423	0.6420
89	5.341	5.338	5.336	5.334	5.331	5.329	5.326	5.324	5.321	5.319
	0.6417	0.6414	0.6411	0.6409	0.6406	0.6403	0.6400	0.6397	0.6394	0.6391
90	5.316	5.314	5.312	5.310	5.307	5.305	5.302	5.300	5.297	5.295
	0.6388	0.6385	0.6382	0.6380	0.6377	0.6374	0.6371	0.6368	0.6365	0.6362
91	5.293	5.291	5.288	5.286	5.283	5.281	5.278	5.276	5.274	5.271
	0.6360	0.6357	0.6354	0.6351	0.6348	0.6345	0.6342	0.6340	0.6337	0.6334
92	5.269	5.266	5.264	5.262	5.260	5.257	5.254	5.252	5.250	5.248
	0.6331	0.6328	0.6325	0.6323	0.6320	0.6317	0.6314	0.6311	0.6309	0.6306
93	5.246	5.243	5.240	5.238	5.236	5.234	5.232	5.229	5.227	5.225
	0.6303	0.6300	0.6297	0.6294	0.6292	0.6289	0.6286	0.6283	0.6281	0.6278
94	5.222	5.220	5.217	5.215	5.213	5.210	5.208	5.206	5.204	5.201
	0.6275	0.6272	0.6269	0.6267	0.6264	0.6261	0.6258	0.6256	0.6253	0.6350
95	5.199	5.196	5.194	5.192	5.190	5.187	5.185	5.183	5.180	5.179
	0.6247	0.6244	0.6242	0.6239	0.6236	0.6233	0.6231	0.6228	0.6225	0.6223
96	5.176	5.174	5.172	5.170	5.167	5.164	5.162	5.160	5.158	5.156
	0.6220	0.6217	0.6214	0.6212	0.6209	0.6206	0.6203	0.6201	0.6198	0.6195
97	5.154	5.151	5.149	5.146	5.144	5.142	5.140	5.138	5.136	5.133
	0.6193	0.6190	0.6187	0.6184	0.6182	0.6179	0.6176	0.6174	0.6171	0.6168
98	5.131	5.129	5.126	5.124	5.122	5.120	5.118	5.116	5.113	5.111
	0.6166	0.6163	0.6160	0.6158	0.6155	0.6152	0.6150	0.6147	0.6144	0.6141
99	5.109	5.107	5.105	5.102	5.100	5.098	5.096	5.092	5.091	5.089
	0.6139	0.6136	0.6134	0.1631	0.6128	0.6126	0.6123	0.6120	0.6118	0.6115

Table 12-4. Relation between Pounds per Gallon, Specific Gravity, and Degrees API* (Continued)

°API	Degrees									
	0	1	2	3	4	5	6	7	8	9
100	5.086	5.065	5.043	5.021	5.000	4.979	4.957	4.936	4.915	4.895
	0.6112	0.6086	0.6060	0.6034	0.6008	0.5983	0.5957	0.5932	0.5907	0.5883
110	4.876	4.855	4.835	4.815	4.795	4.776	4.757	4.738	4.719	4.699
	0.5659	0.5835	0.5811	0.5787	0.5763	0.5740	0.5717	0.5694	0.5671	0.5648
120	4.680	4.662	4.644	4.625	4.608	4.589	4.572	4.554	4.536	4.518
	0.5325	0.5603	0.5581	0.5559	0.5538	0.5516	0.5495	0.5473	0.5452	0.5431
130	4.502	4.484	4.468	4.450	4.434	4.417	4.400	4.384	4.368	4.352
	0.5411	0.5390	0.5370	0.5349	0.5329	0.5309	0.5289	0.5270	0.5250	0.5231
140	4.335	4.319	4.303	4.288	4.273	4.257	4.242	4.226	4.211	4.196
	0.5211	0.5192	0.5173	0.5154	0.5136	0.5117	0.5099	0.5080	0.5062	0.5044
150	4.181	4.166	4.152	4.137	4.122	4.108	4.093	4.079	4.065	4.051
	0.5026	0.5008	0.4991	0.4973	0.4956	0.4938	0.4921	0.4904	0.4887	0.4870
160	4.037									
	0.4854									

* Values from 0 to 100°API, American Petroleum Institute Tables. Values from 100 to 160°API calculated by The Lummus Co.

Table 12-5. Viscosity Conversion Factors

Viscosity of water at 68°F (20°C) = 1.002 centipoises
$$= 1.0038 \text{ centistokes}$$
Viscosity in centipoises $\div$ 100 = viscosity in poises, gm/(sec) (cm)
Viscosity in centipoises $\times$ 10 = viscosity in millipoises
Viscosity in centipoises $\times$ 10^4 = viscosity in micropoises
Viscosity in centipoises $\times$ 0.000672 = viscosity in lb/(sec) (ft)
Viscosity in centipoises $\times$ 0.000209 = viscosity in (lb force) (sec)/sq ft
Viscosity in centipoises $\times$ 2.42 = viscosity in lb/(hr) (ft)
Viscosity in centipoises $\div$ density at same temperature = viscosity in centistokes
Viscosity in centipoises $\times$ 69 $\times$ 10^5 = viscosity in reyns (lb) (sec)/sq in.

References for Density Charts—Figures 12-2 and 12-3

1. Densities for liquid hydrocarbons in their normal liquid state were taken from reference 1 under References for Table 12-3 on vapor pressure. Values from −50 to 140°F for the seven C_4 hydrocarbons, propylene and propane, and iso and normal pentane were compared with API44 from "Liquid Densities of Hydrocarbons Found in Commercial C_4 Mixtures," Natl. Bur. Std. (U.S.) *Letter Circ.* LC-736 (Nov. 23, 1945).
2. Data for methane, ethylene, ethane, propylene, and propane were derived from references 11, 12, 13, 14, 15, and 16, Table 12-3.
3. Data for isobutane, isobutylene, and *n*-butane were taken from reference 14, Table 12-3.
4. Data for 1,3-butadiene were taken from reference 19, Table 12-3.
5. Data for 1-butene and *n*-nonane were taken from reference 5, Table 12-3.
6. Data for cis-2, and trans-2-butene were taken from reference 1, Table 12-3.
7. Data for isopentane were taken from references 21 and 8, Table 12-3.
8. Data for *n*-pentane, *n*-hexane, diisopropyl, *n*-heptane, diisobutyl, *n*-octane, and cyclohexane were taken from reference 8, Table 12-3.
9. Data for benzene were taken from references 8 and 34, Table 12-3.
10. Data for toluene were taken from M. Altschul, *Z. Phys. Chem.*, vol. 2, p. 574, 1895; and Von Hirsch, *Ann. Phys.*, vol. 69, p. 456, 1899.
11. Data for *o*-, *m*-, and *p*-xylene were taken from Von Hirsch, *Ann. Phys.*, vol. 69, p. 456, 1899.

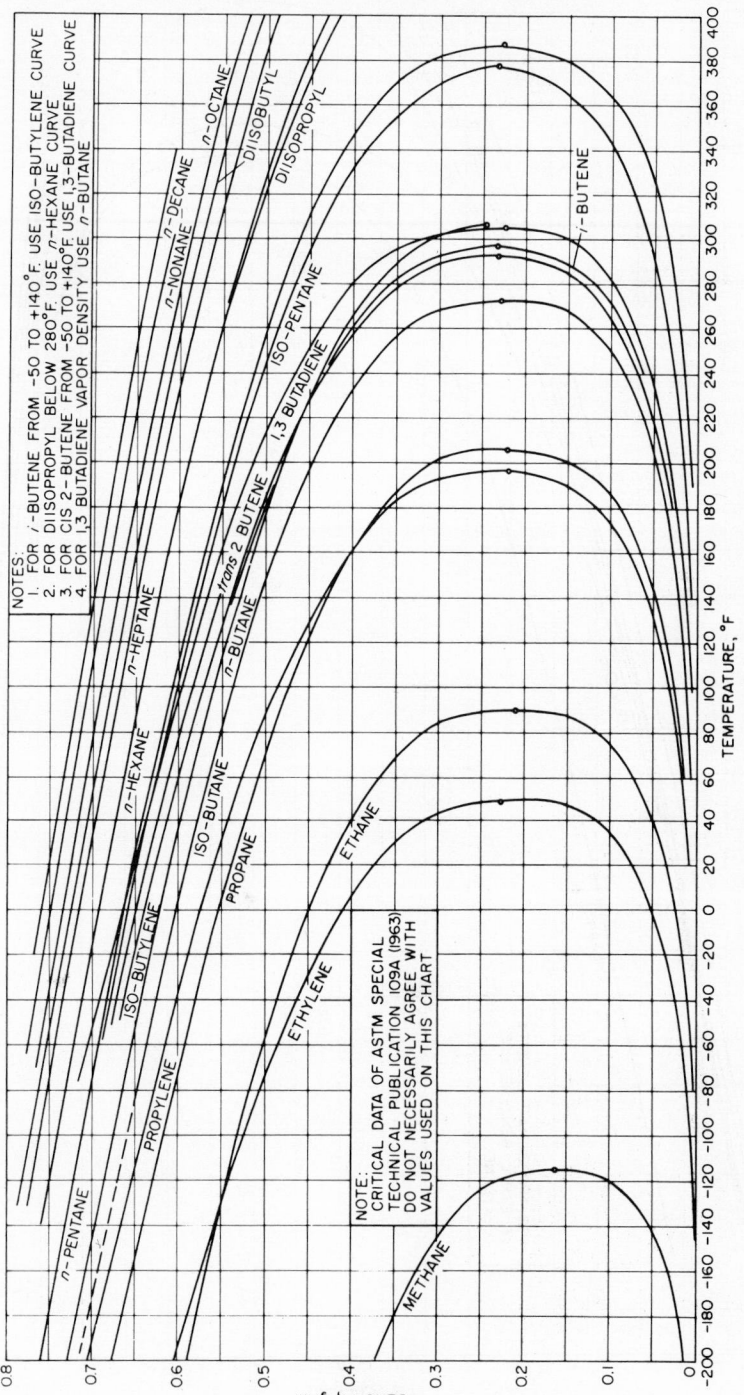

Fig. 12-2. Densities of pure hydrocarbons, saturated liquid and vapor.

12–23

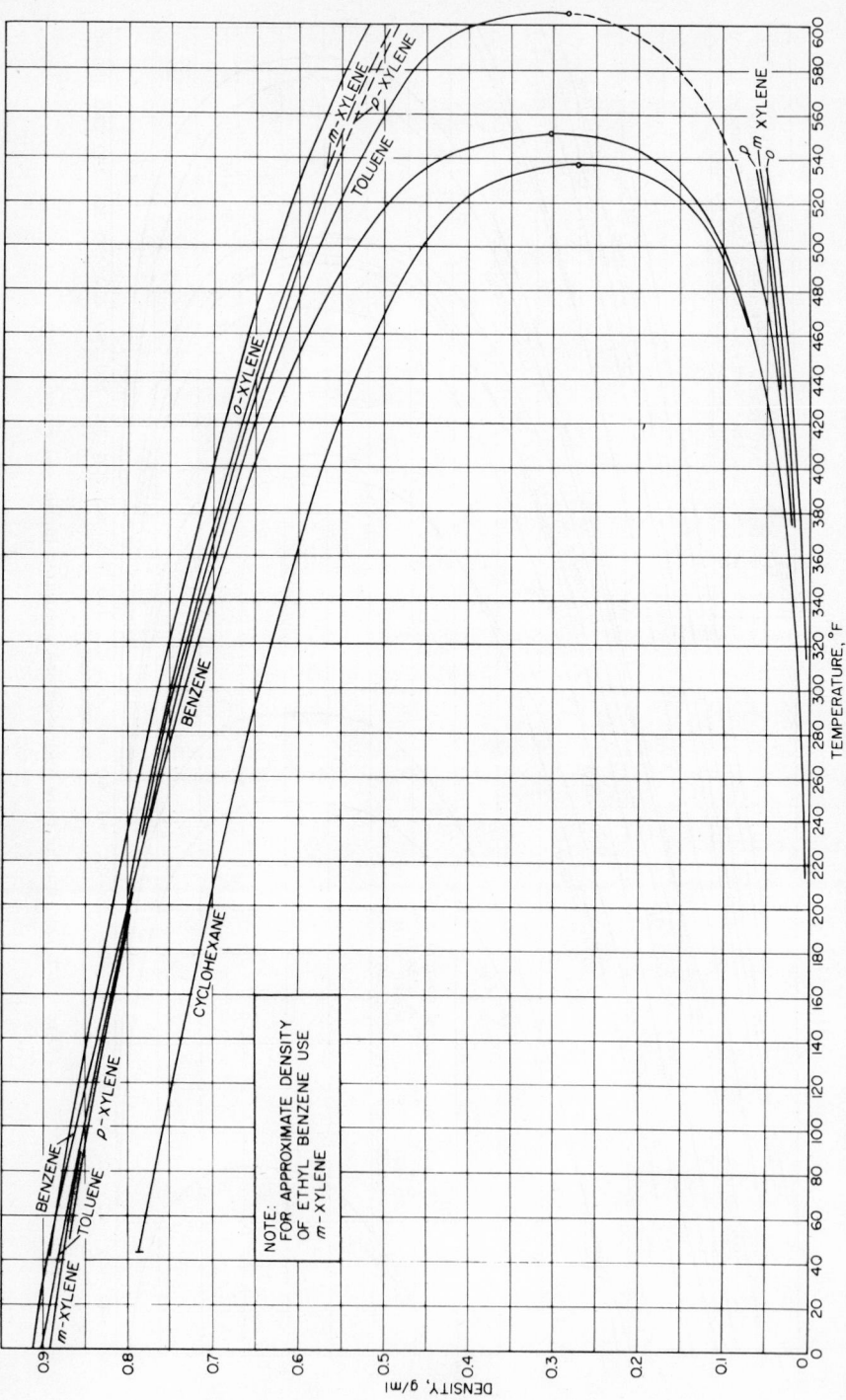

FIG. 12-3. Densities of aromatic hydrocarbons, saturated liquid and vapor.

NOTE:
FOR APPROXIMATE DENSITY
OF ETHYL BENZENE USE
m-XYLENE

DENSITY, g/ml

TEMPERATURE, °F

Average Boiling Points, Critical Properties, and Characterization Factor of Petroleum Fractions

The various average boiling points are defined as follows:

1. *Volumetric Average Boiling Point* (VABP):

$$t_B = \frac{t_{10} + t_{30} + t_{50} + t_{70} + t_{90}}{5}, \text{ where } t_{10}, \text{ etc., are the temperatures in °F of the 10 per cent,}$$

etc., points (by volume) of the ASTM or Engler distillation curve of the mixture.

2. *Weight Average Boiling Point* (WABP):

$t_B = f_{w1}(T_{B1}) + f_{w2}(T_{B2}) + \cdots$, where f_{w1} is the weight fraction of component 1 and T_{B1} is the boiling point in °R for pure component 1.

3. *True Molal Average Boiling Point* (TMABP):

$T_B = f_{m1}(T_{B1}) + f_{m2}(T_{B2}) + \cdots$, where f_{m1} is the mol fraction of component 1 in the mixture.

4. *Cubic Average Boiling Point* (CABP):

$T_B = [f_{v1}(T_{B1})^{1/3} + f_{v2}(T_{B2})^{1/3} + \cdots]^3$, where f_{v1} is the volume fraction of component 1 in the mixture.

5. *Mean Average Boiling Point* (MABP):

$$T_B = \frac{T_B \text{ (cubic avg)} + T_B \text{ (molal avg)}}{2}$$

Recent developments of correlations now permit the use of the volumetric average boiling point for the determination of most properties.

In the case of high-boiling stocks where the vapor temperature exceeds 700°F, it is preferable to use the data from a distillation under reduced pressure (about 10 mm of mercury) in order that the liquid temperature does not enter the range of cracking. The data from the low-pressure distillation are then converted to atmospheric pressure by means of the vaporization chart (Fig. 12-1). If no low-pressure data are available on high-boiling stocks, it is best to take the 50 per cent temperature of the ASTM or Engler distillation as the volumetric average boiling point.

The various properties of petroleum fractions are determined as follows:

1. The **true critical temperature** of the mixture is read from Fig. 12-60 with the ASTM volumetric average boiling point (VABP) and API gravity as arguments.

2. Similarly, the **true critical pressure** of the mixture is read from Fig. 12-61 with the same parameters and the ASTM 10 to 90 per cent slope.

3. The **pseudo-critical pressure** is read from Fig. 12-62 with VABP and pseudo-critical temperature as arguments.

4. Figure 12-59 gives corrections to be added or subtracted from VABP to obtain the other average boiling points. The mean average boiling point is useful with Fig. 12-58 for determining hydrogen content of petroleum fractions.

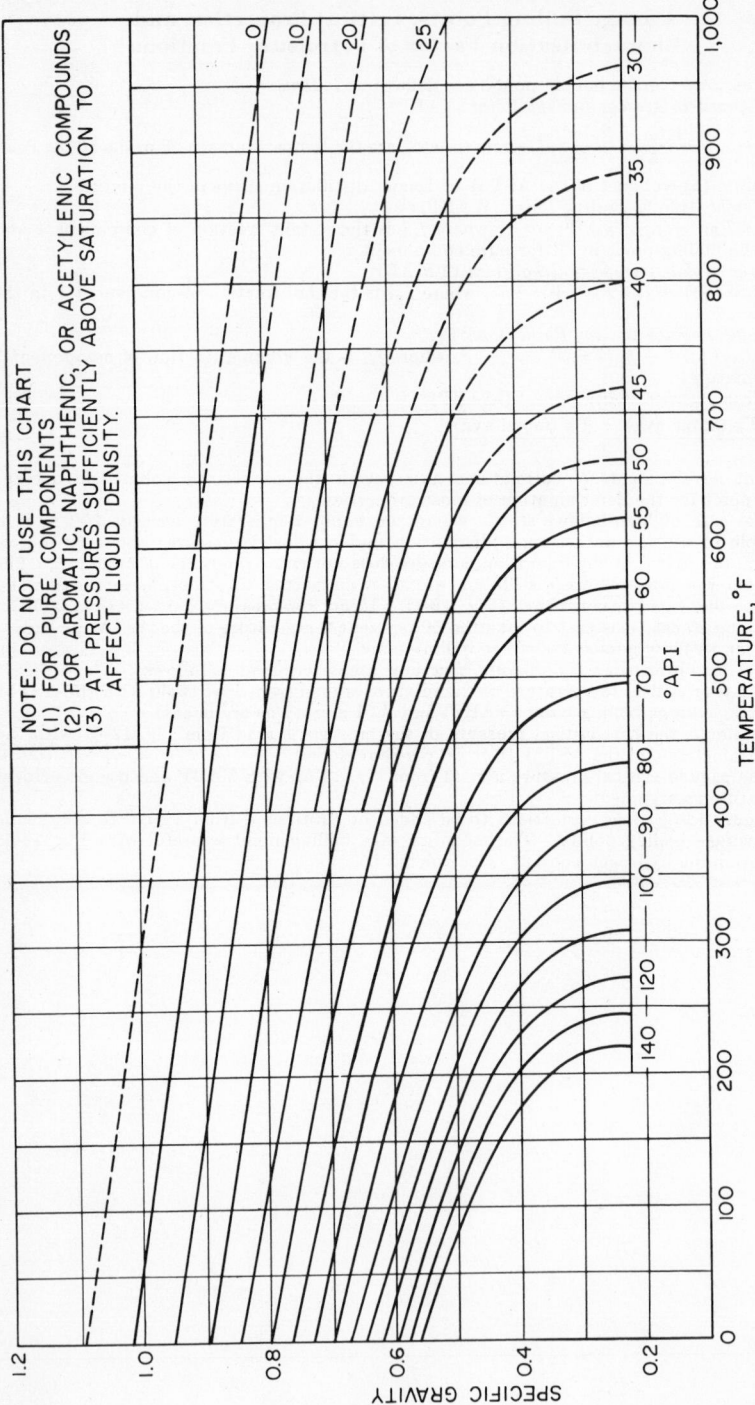

NOTE: DO NOT USE THIS CHART
(1) FOR PURE COMPONENTS
(2) FOR AROMATIC, NAPHTHENIC, OR ACETYLENIC COMPOUNDS
(3) AT PRESSURES SUFFICIENTLY ABOVE SATURATION TO AFFECT LIQUID DENSITY.

TEMPERATURE, °F

SPECIFIC GRAVITY

°API

Fig. 12-4. Approximate specific gravities of saturated liquid-petroleum fractions. (*Based on data from W. L. Nelson, Oil & Gas J., Jan. 27, 1938, p. 184; NGAA Gravity Conversion Tables, 1942.*)

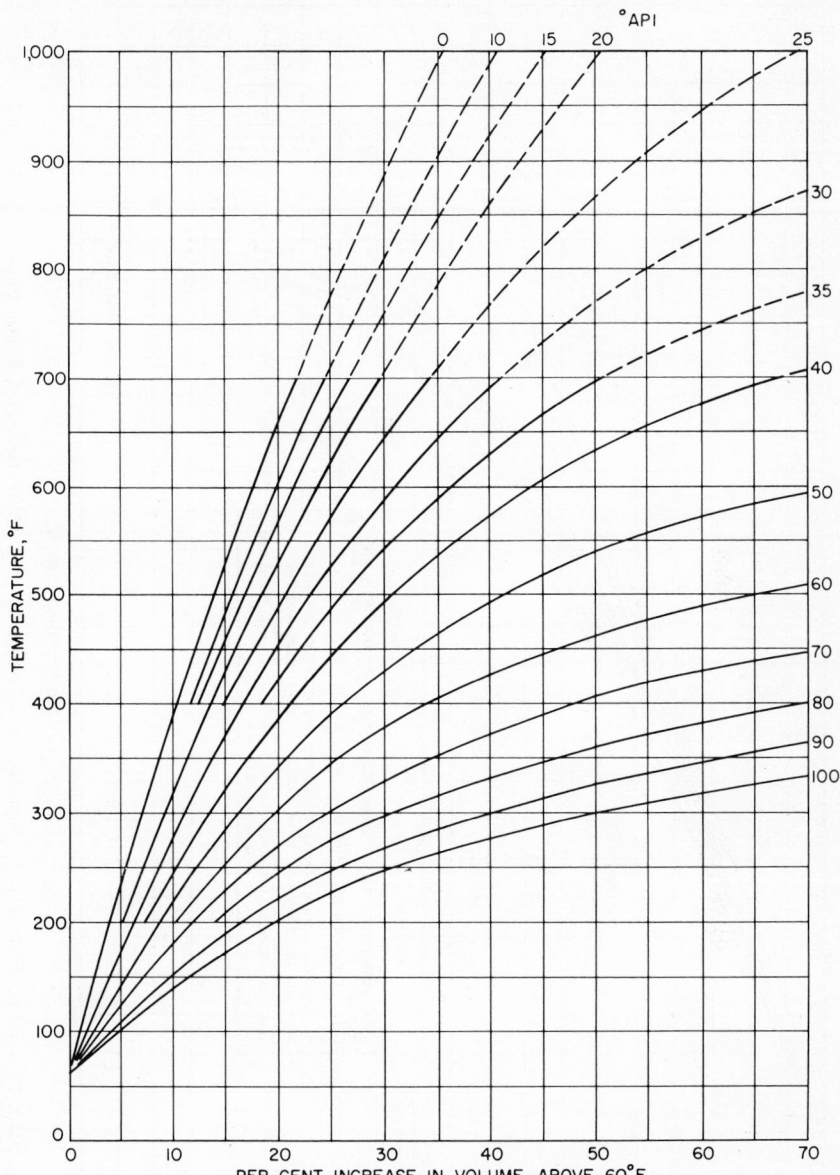

FIG. 12-5. Approximate expansion of saturated liquid-petroleum fractions. (*Based on data from W. L. Nelson, Oil & Gas J., Jan. 27, 1938, p. 184.*)

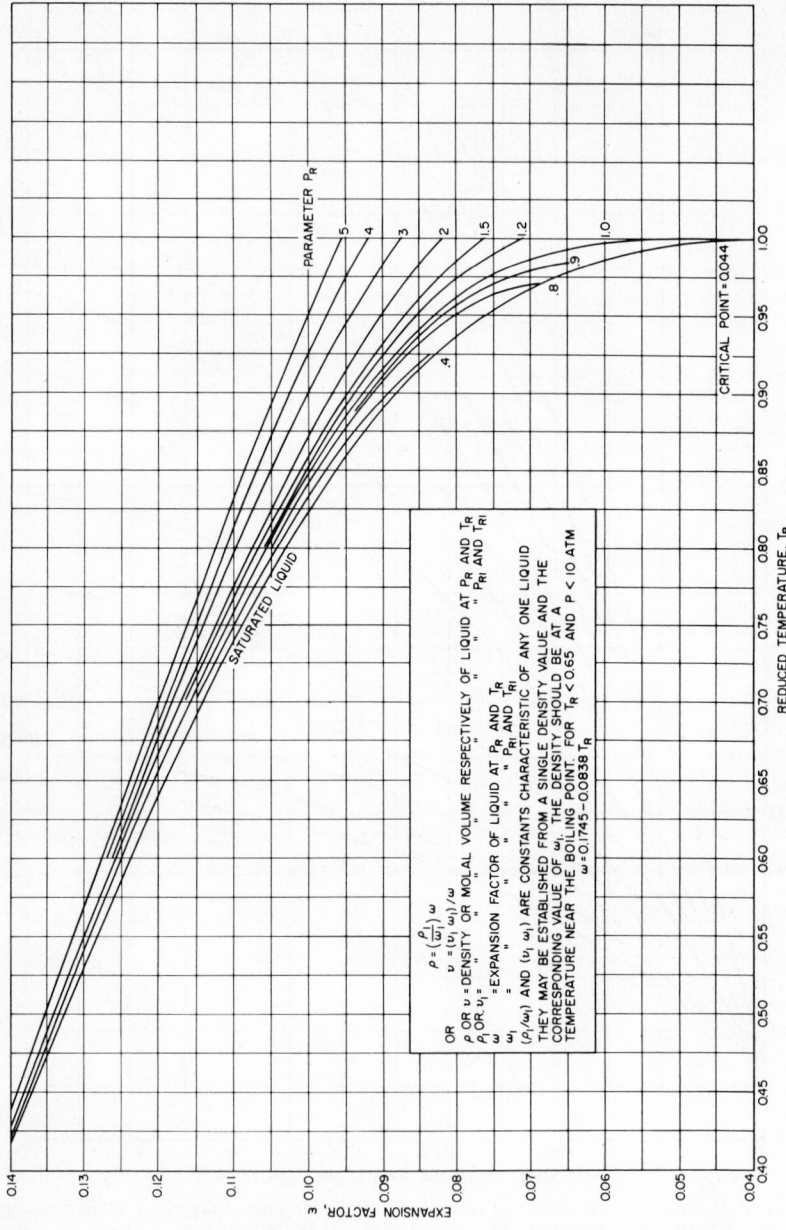

Fig. 12-6. Thermal expansions and compressibilities of liquids. (K. M. Watson, Ind. Engrg. Chem., vol. 35, p. 889, 1943.)

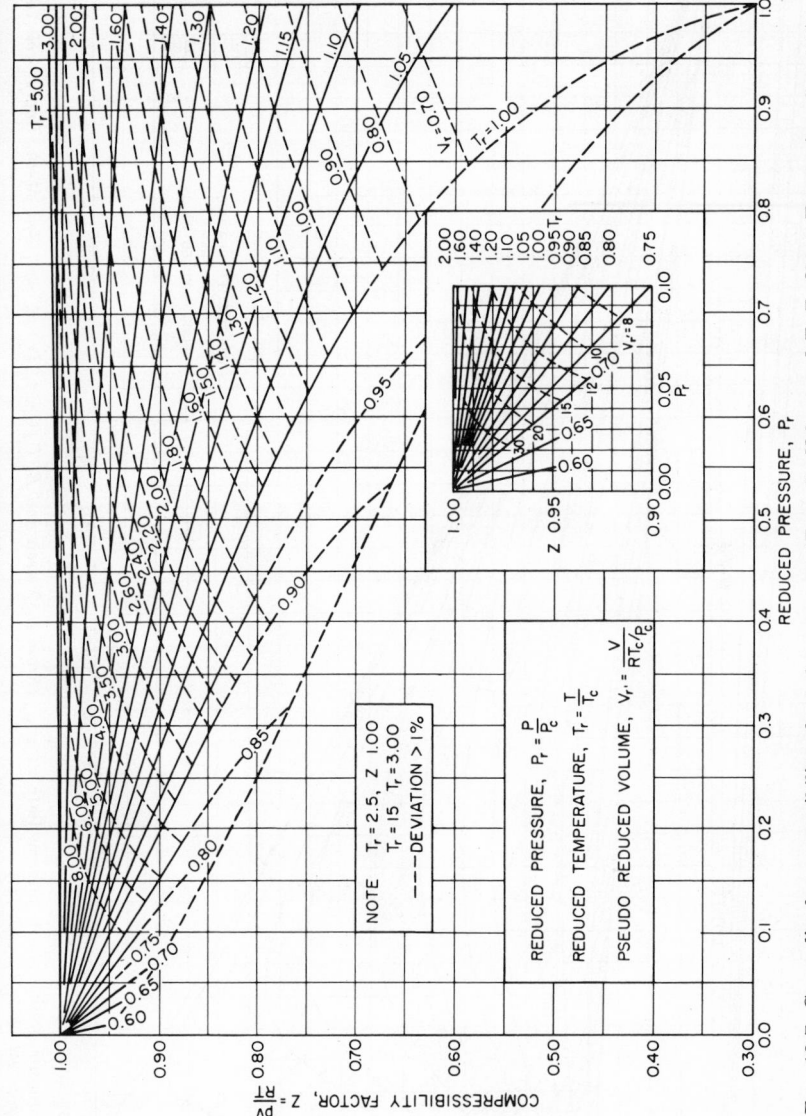

FIG. 12-7. Generalized compressibility chart, low range. (*From L. Nelson and E. F. Obert, Trans. Am. Soc. Mech. Engrs., vol. 76, p. 1057, 1954, Chem. Engrg., p. 203, July, 1954.*)

NOTE $T_r = 2.5$, Z 1.00
$T_r = 15$ $T_r = 3.00$
---DEVIATION $> 1\%$

REDUCED PRESSURE, $P_r = \dfrac{P}{P_c}$

REDUCED TEMPERATURE, $T_r = \dfrac{T}{T_c}$

PSEUDO REDUCED VOLUME, $V_r = \dfrac{V}{RT_c/P_c}$

COMPRESSIBILITY FACTOR, $Z = \dfrac{PV}{RT}$

REDUCED PRESSURE, P_r

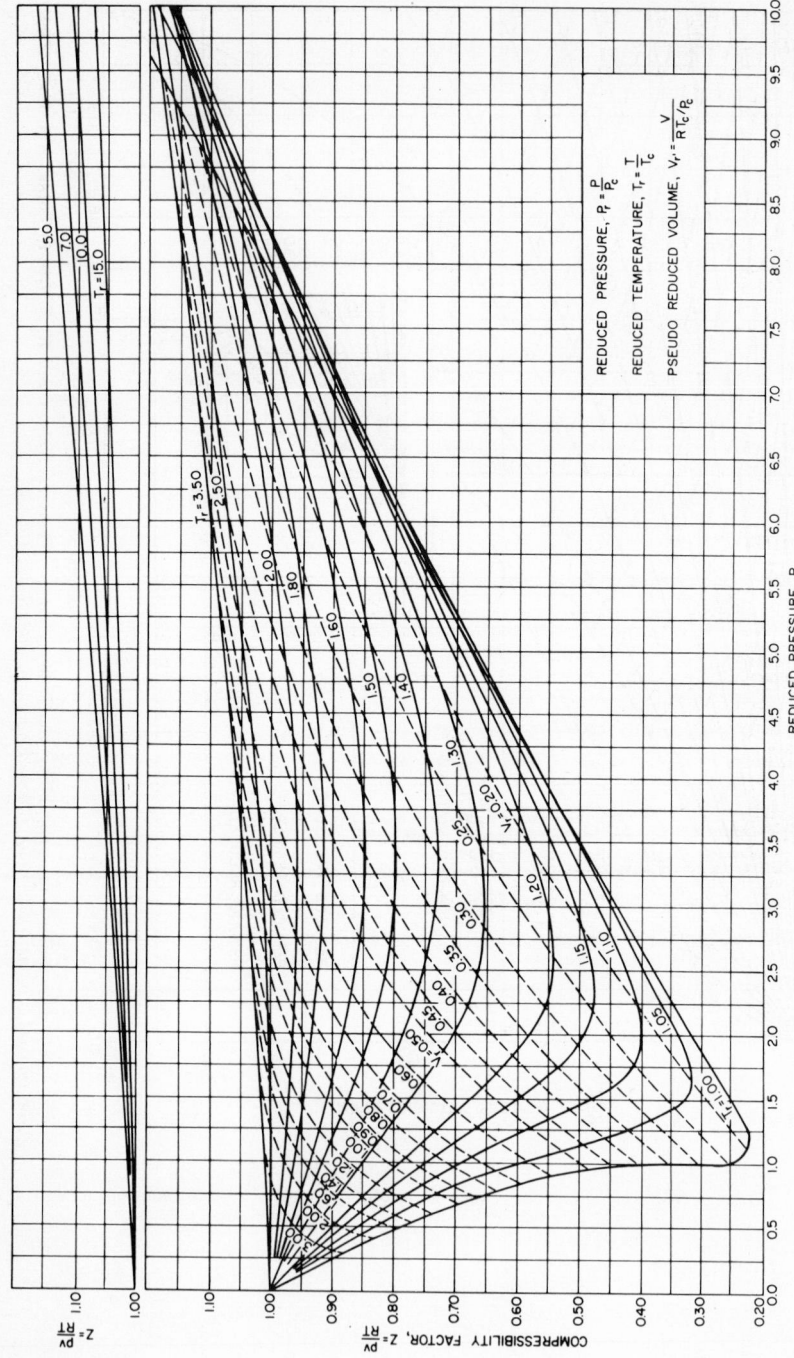

Fig. 12-8. Generalized compressibility chart, intermediate range. *(From L. Nelson and E. F. Obert, Trans. Am. Soc. Mech. Engrs., vol. 76, p. 1057, 1954, Chem. Engrg., vol. 203, July, 1954.)*

REDUCED PRESSURE, $P_r = \dfrac{P}{P_c}$

REDUCED TEMPERATURE, $T_r = \dfrac{T}{T_c}$

PSEUDO REDUCED VOLUME, $V_r = \dfrac{V}{RT_c/P_c}$

COMPRESSIBILITY FACTOR, $Z = \dfrac{pv}{RT}$

REDUCED PRESSURE, P_r

$Z = \dfrac{pv}{RT}$

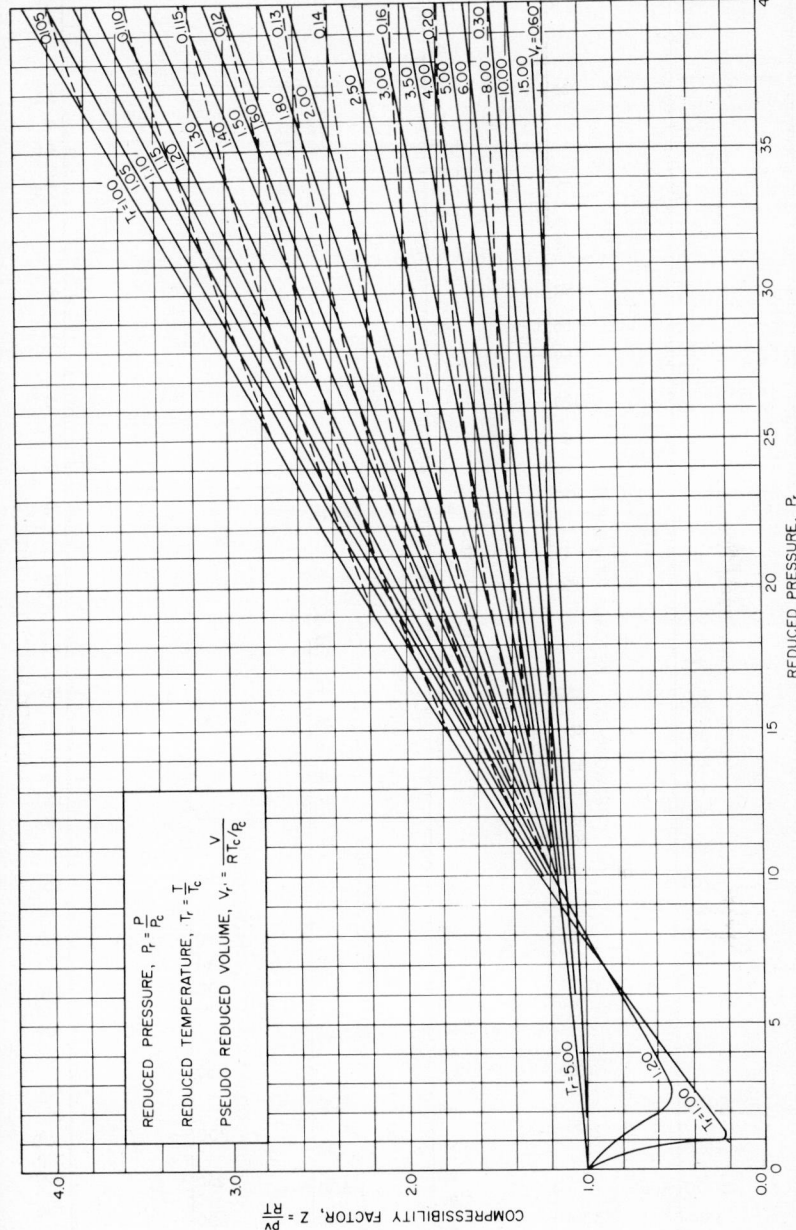

Fig. 12-9. Generalized compressibility chart, high range. (*From L. Nelson and E. F. Obert, Trans. Am. Soc. Mech. Engrs., vol. 76, p. 1057, 1954, Chem. Engrg., p. 203, July, 1954.*)

COMPRESSIBILITY FACTOR, $Z = \dfrac{Pv}{RT}$

REDUCED PRESSURE, P_r

REDUCED PRESSURE, $P_r = \dfrac{P}{P_c}$

REDUCED TEMPERATURE, $T_r = \dfrac{T}{T_c}$

PSEUDO REDUCED VOLUME, $V_r = \dfrac{V}{RT_c/P_c}$

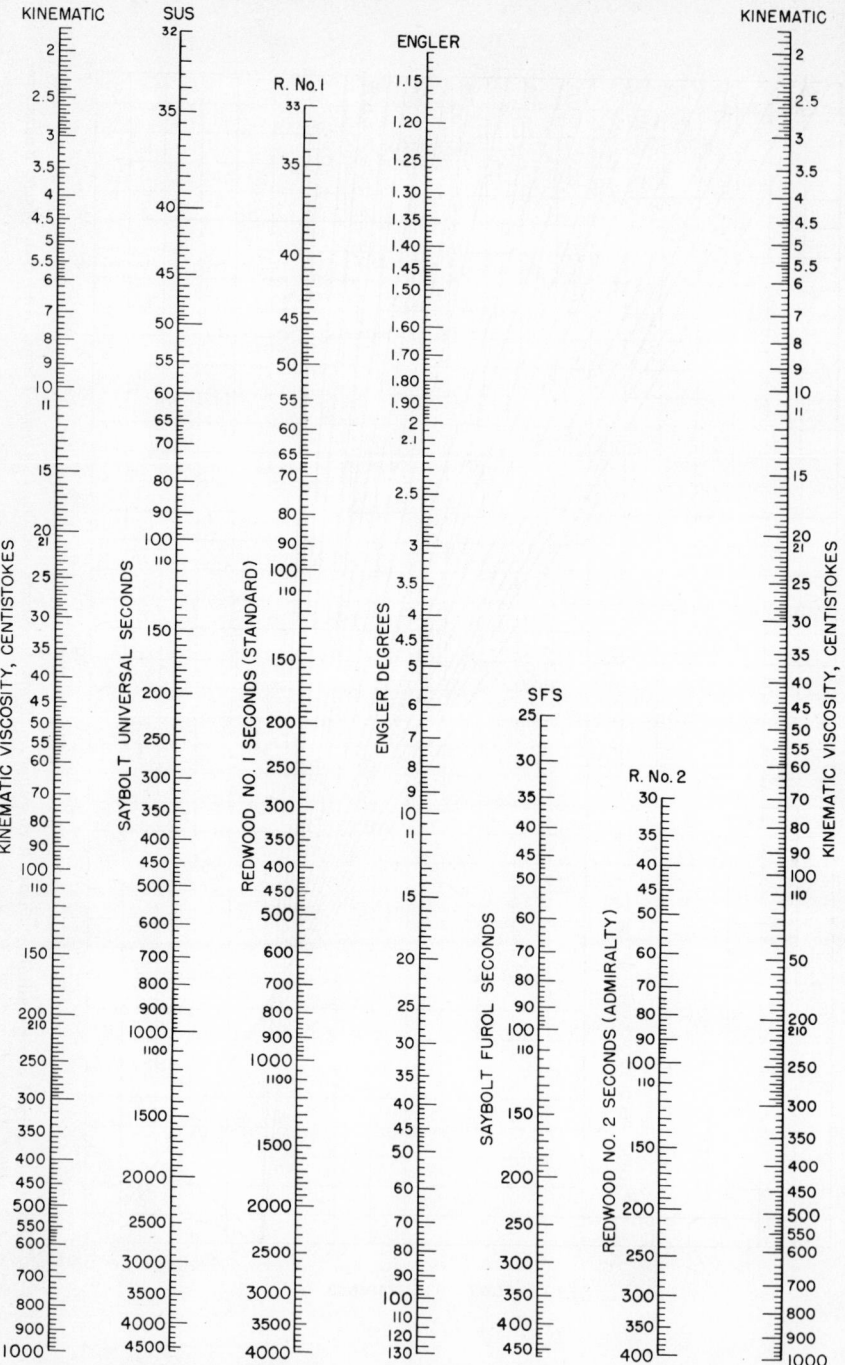

Fig. 12-10. Viscosity conversion chart. (*Courtesy of Texaco, Inc.*) Line up straight edge so centistoke value on both kinematic scales is the same. Viscosities at the *same temperature* on all scales are then equivalent. To extend range of only the kinematic, Saybolt Universal, redwood No. 1, and Engler scales: Multiply by 10 the viscosities on these scales between 100 and 1,000 centistokes on the kinematic scale and the corresponding viscosities on the other three scales. For further extension, multiply these scales as above by 100 or a higher power of 10. (Example: 1,500 centistokes = 150 × 10 CS − 695 × 10 SUS = 6,950 SUS.)

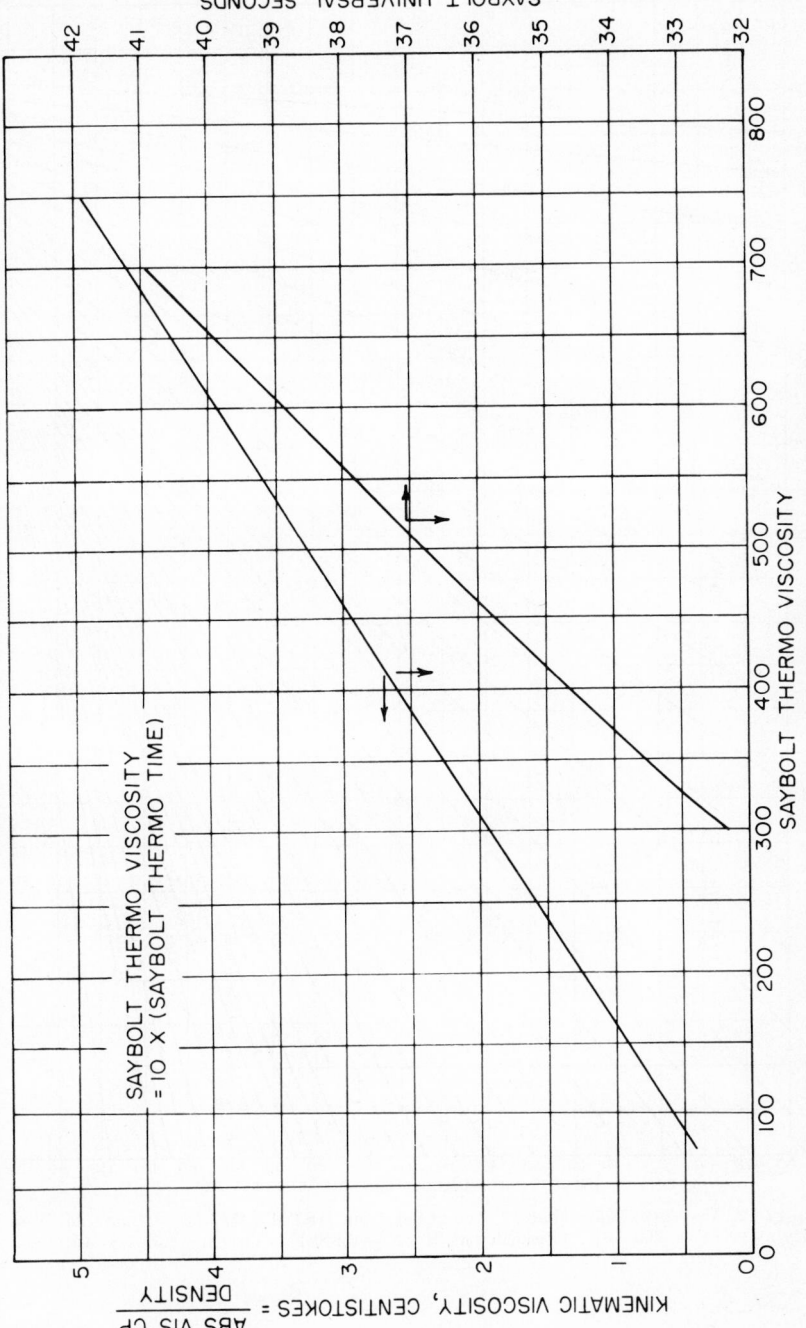

SAYBOLT UNIVERSAL SECONDS

SAYBOLT THERMO VISCOSITY

SAYBOLT THERMO VISCOSITY
= 10 × (SAYBOLT THERMO TIME)

KINEMATIC VISCOSITY, CENTISTOKES = $\dfrac{\text{ABS VIS CP}}{\text{DENSITY}}$

Fig. 12-11. Thermoviscosity, kinematic viscosity conversion chart. (*Nelson, Oil & Gas J., p. 151, June 14, 1954.*)

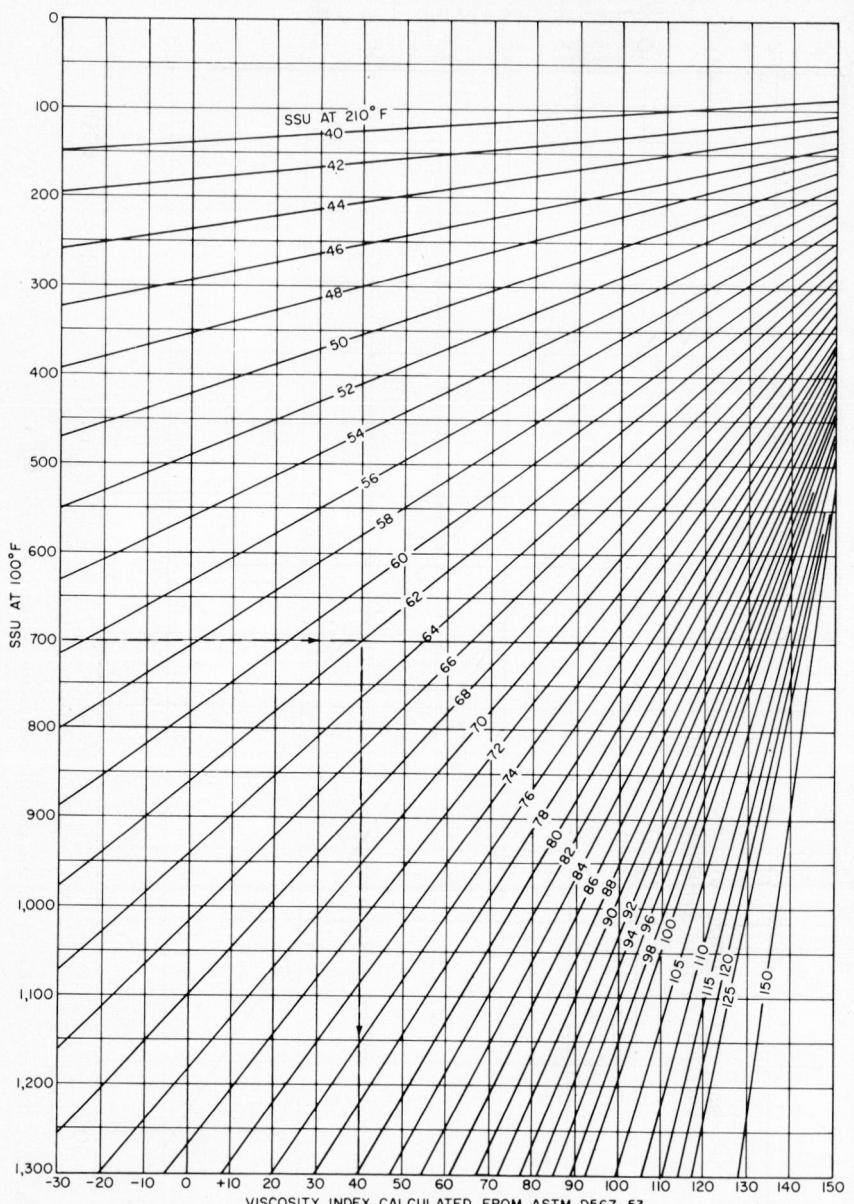

FIG. 12-12. Viscosity index chart. Calculated from ASTM D567-53. (Example: Viscosity at 100°F = 700 SSU. Viscosity at 210°F = 62 SSU. Viscosity index = 40.)

$L = \left[\text{LOG LOG} (\nu + 0.8) \right] \times 1{,}000$

$H = 0.87L + 154$

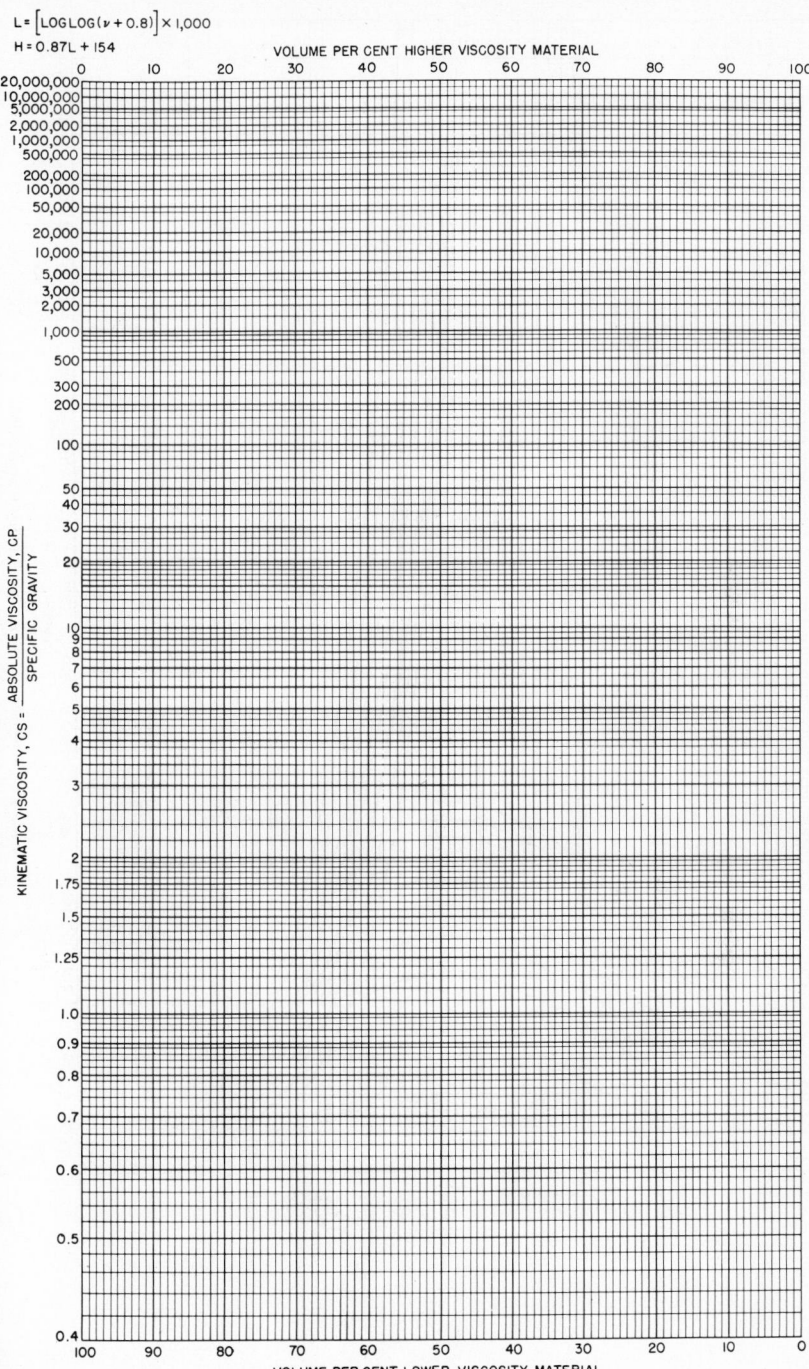

FIG. 12-13. Viscosity blending chart.

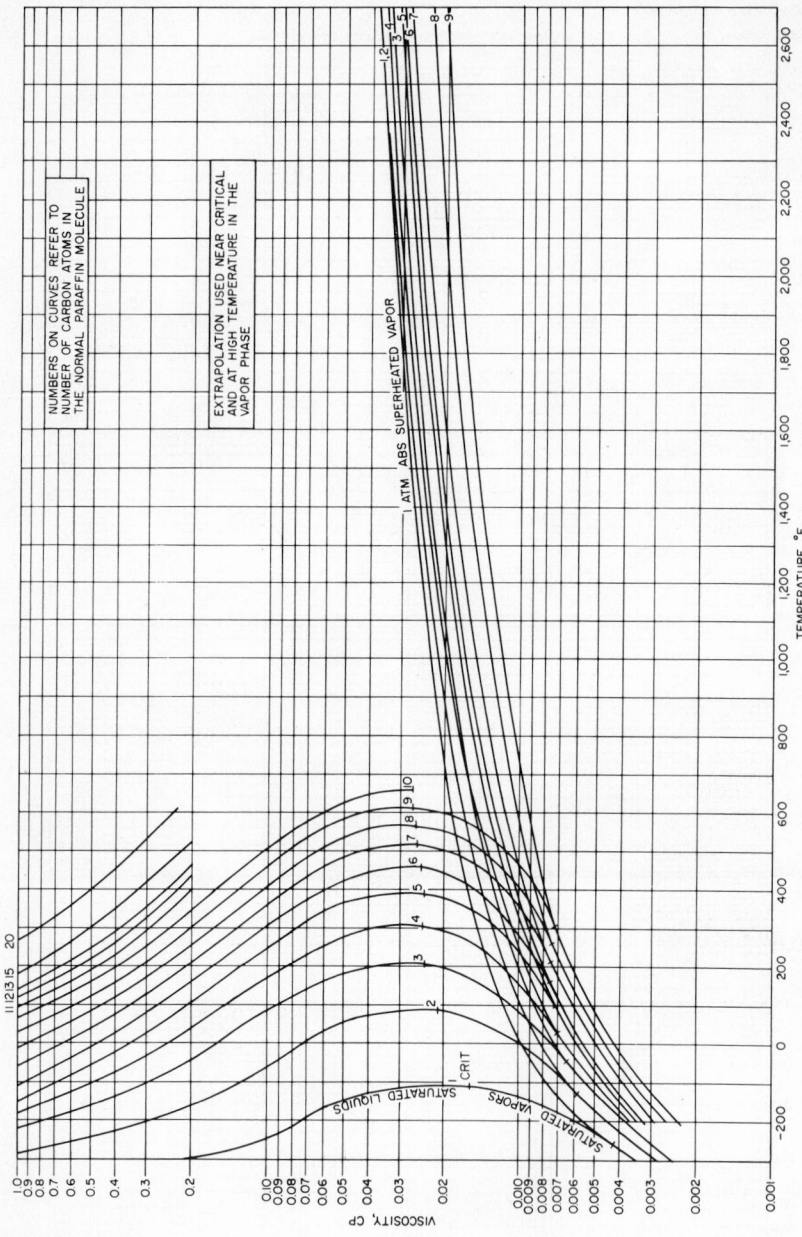

FIG. 12-14. Viscosities of paraffin hydrocarbons.

NUMBERS ON CURVES REFER TO
NUMBER OF CARBON ATOMS IN
THE NORMAL PARAFFIN MOLECULE

EXTRAPOLATION USED NEAR CRITICAL
AND AT HIGH TEMPERATURE IN THE
VAPOR PHASE

1 ATM ABS SUPERHEATED VAPOR

SATURATED LIQUIDS

SATURATED VAPORS

CRIT

VISCOSITY, CP

TEMPERATURE, °F

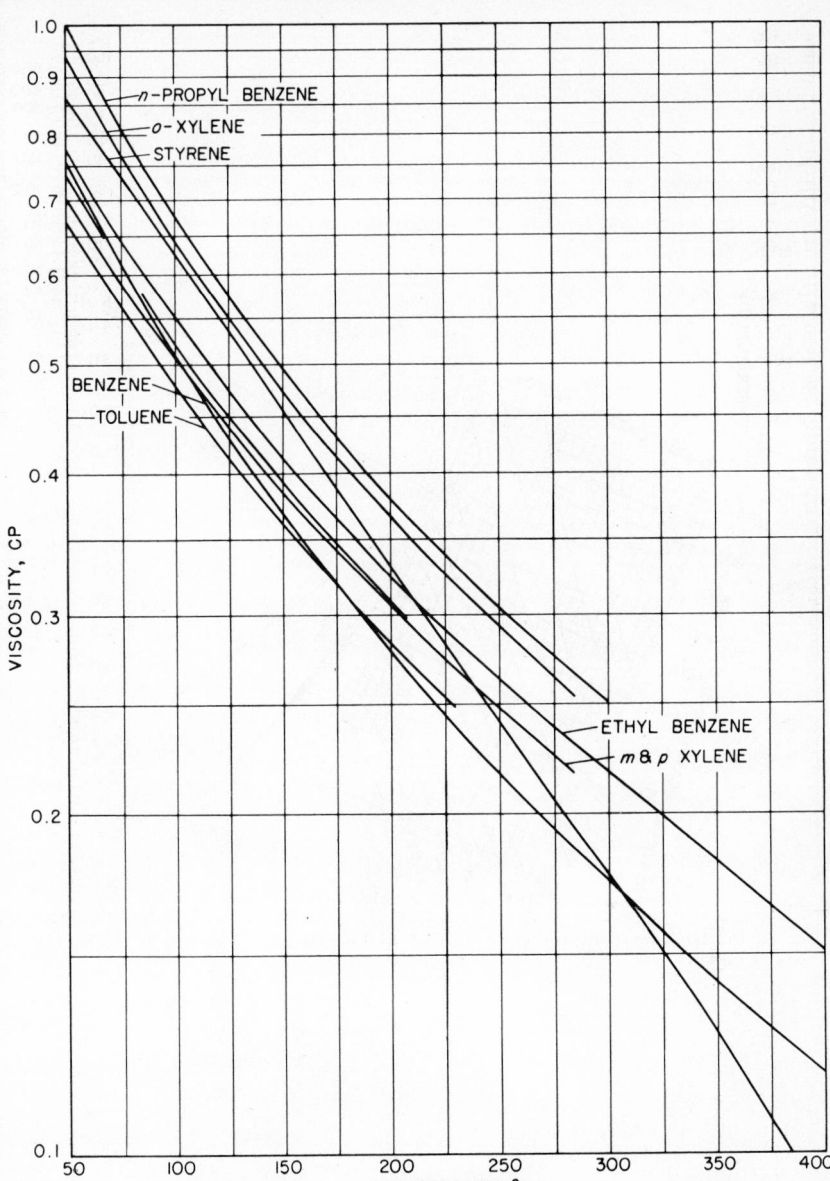

Fig. 12-15. Viscosities of liquid aromatic hydrocarbons.   (API 44.)

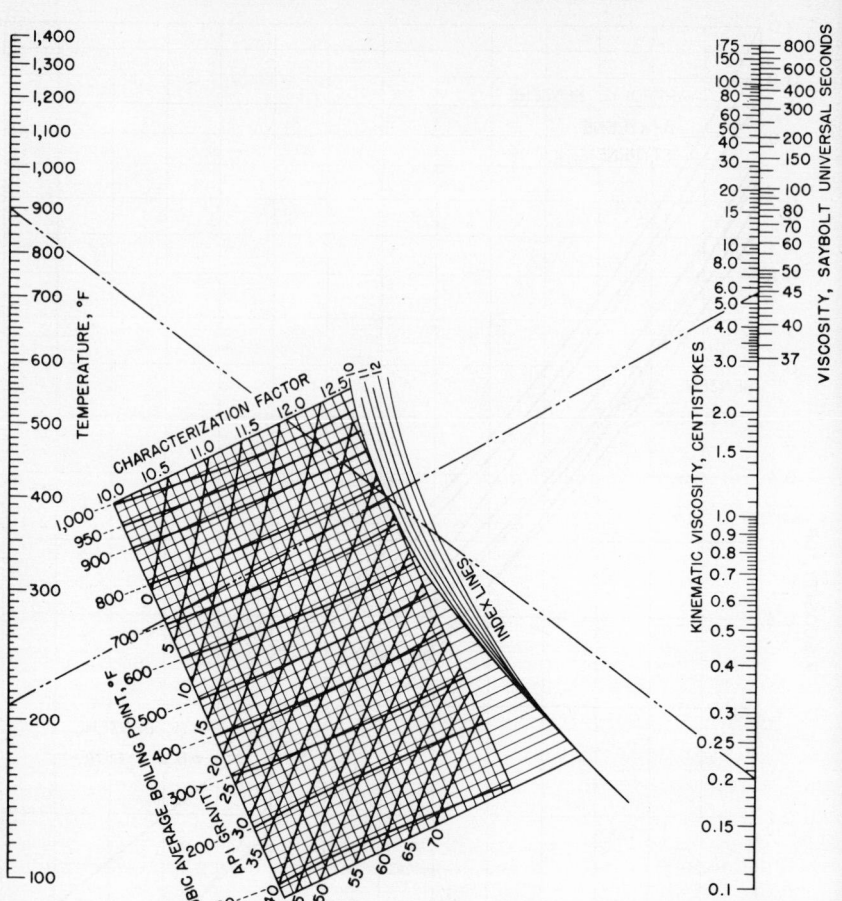

FIG. 12-16. High-temperature viscosities of hydrocarbons. (*Courtesy of Universal Oil Products Co.*)

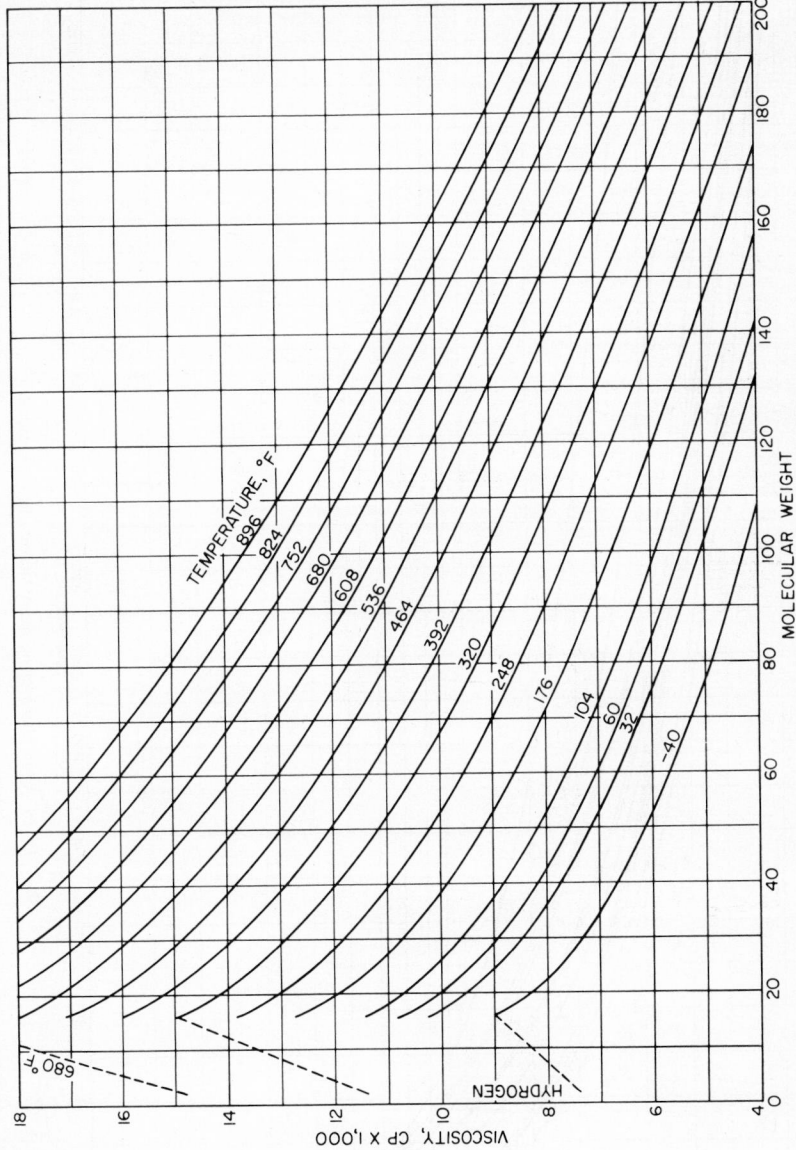

FIG. 12-17. Viscosities of gaseous paraffins at 1 atm pressure. (*Bicher and Katz, Trans. AIMME, vol. 155, p. 251, 1944.*)

12-39

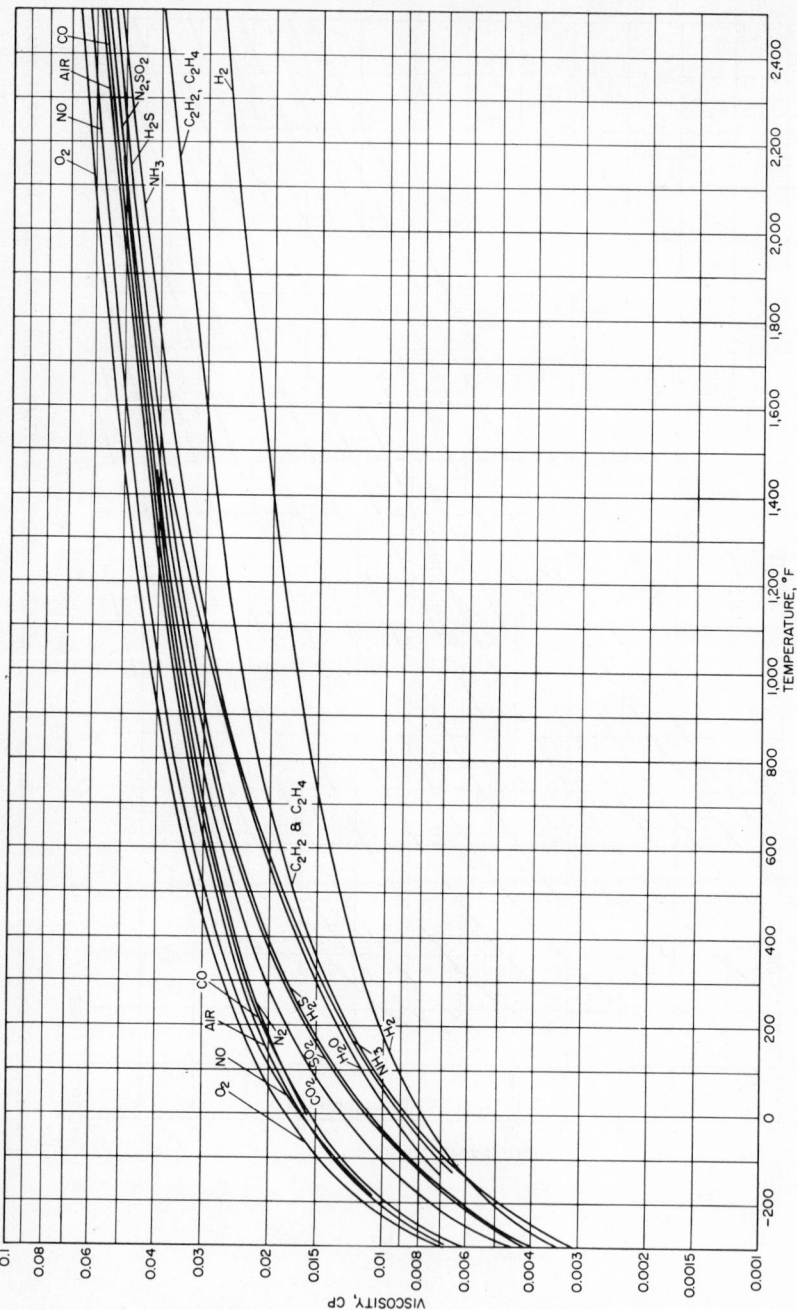

FIG. 12-18. Viscosities of common gases, low pressure.

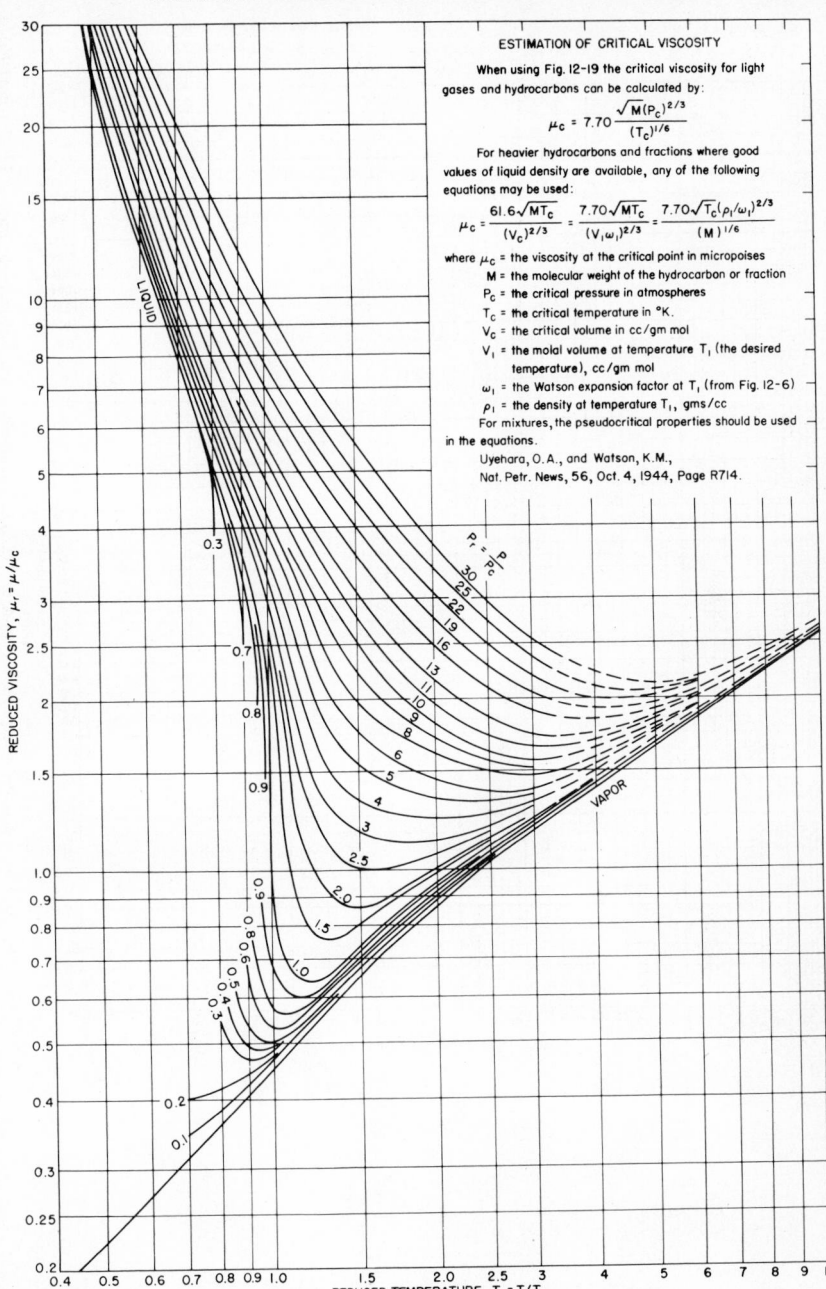

ESTIMATION OF CRITICAL VISCOSITY

When using Fig. 12-19 the critical viscosity for light gases and hydrocarbons can be calculated by:

$$\mu_c = 7.70 \frac{\sqrt{M}(P_c)^{2/3}}{(T_c)^{1/6}}$$

For heavier hydrocarbons and fractions where good values of liquid density are available, any of the following equations may be used:

$$\mu_c = \frac{61.6\sqrt{MT_c}}{(V_c)^{2/3}} = \frac{7.70\sqrt{MT_c}}{(V_1\omega_1)^{2/3}} = \frac{7.70\sqrt{T_c}(\rho_1/\omega_1)^{2/3}}{(M)^{1/6}}$$

where μ_c = the viscosity at the critical point in micropoises
M = the molecular weight of the hydrocarbon or fraction
P_c = the critical pressure in atmospheres
T_c = the critical temperature in °K.
V_c = the critical volume in cc/gm mol
V_1 = the molal volume at temperature T_1 (the desired temperature), cc/gm mol
ω_1 = the Watson expansion factor at T_1 (from Fig. 12-6)
ρ_1 = the density at temperature T_1, gms/cc
For mixtures, the pseudocritical properties should be used in the equations.

Uyehara, O.A., and Watson, K.M.,
Nat. Petr. News, 56, Oct. 4, 1944, Page R714.

Fig. 12-19. Generalized reduced viscosities.   Estimation of critical viscosity.   (*Chemical Process Principle Charts, 1st ed., O. A. Hougen and K. M. Watson, John Wiley & Sons, Inc., New York, 1946.*)

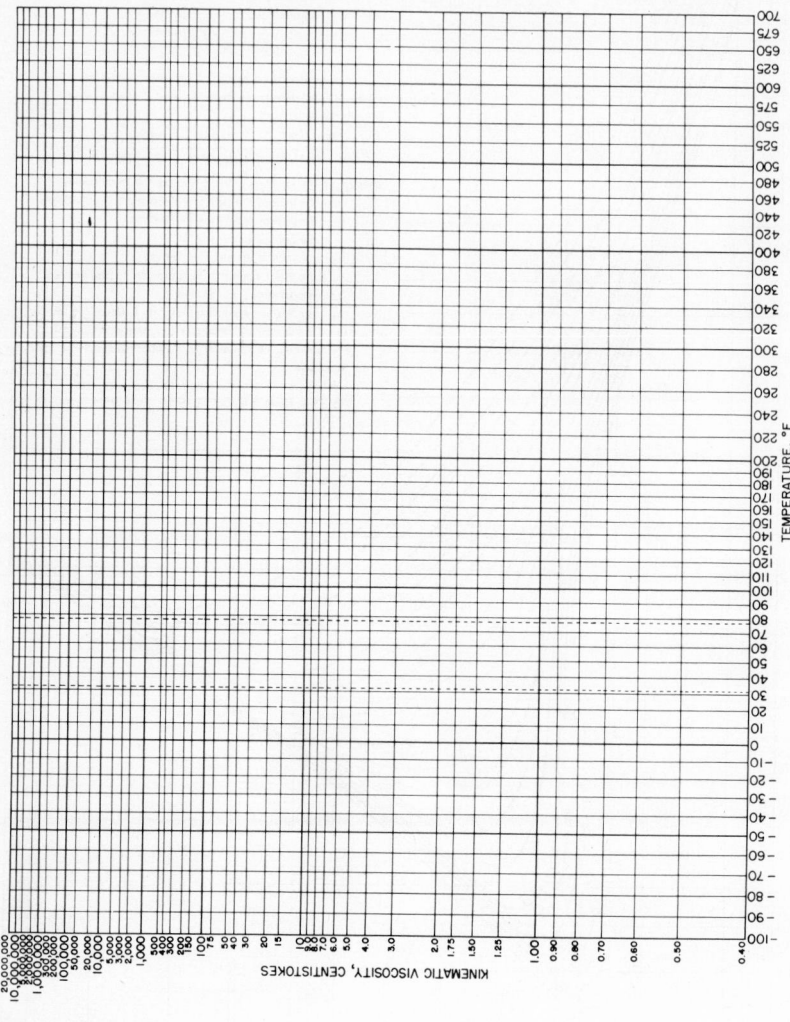

Fig. 12-20. Modified ASTM viscosity-temperature chart. (*U.S. Naval Research Laboratory; now ASTM Chart F.*)

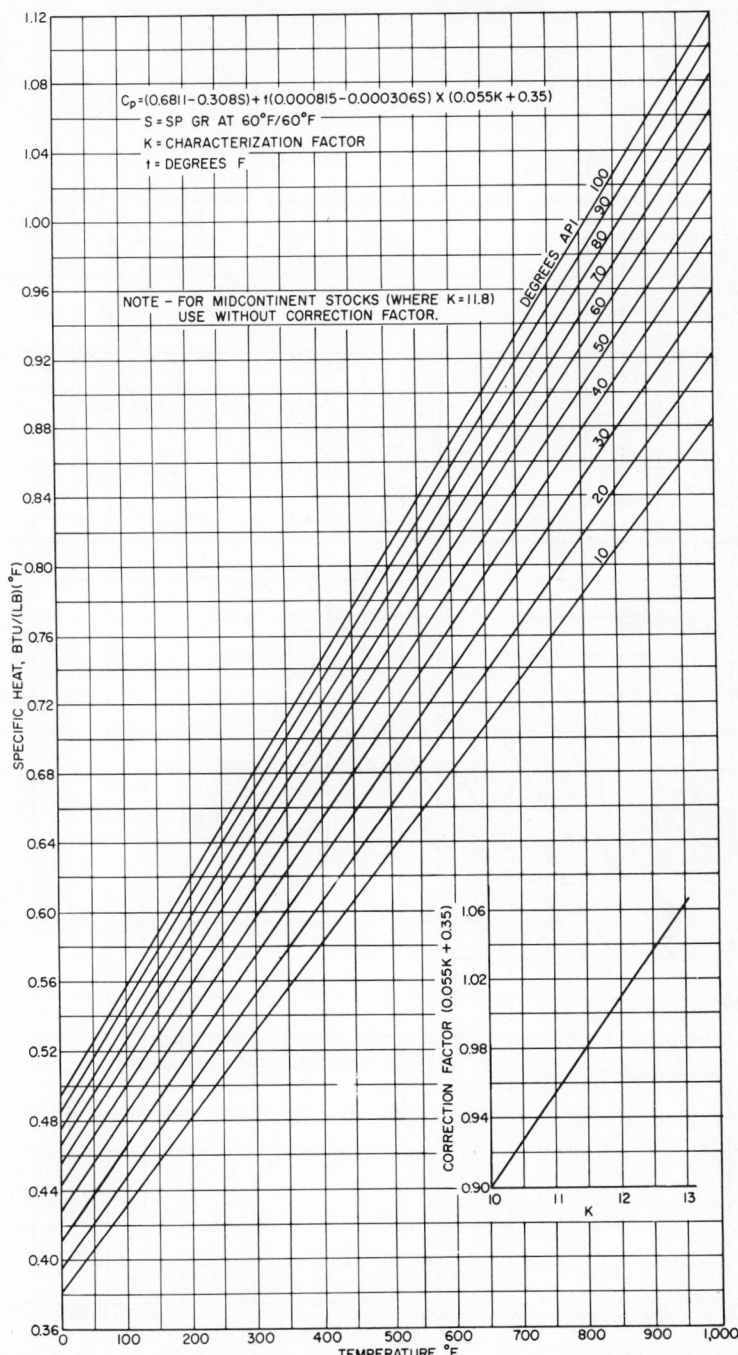

Fig. 12-21. Specific heats of petroleum fractions, liquid state. (*Standards of TEMA, 108, 1952.*)

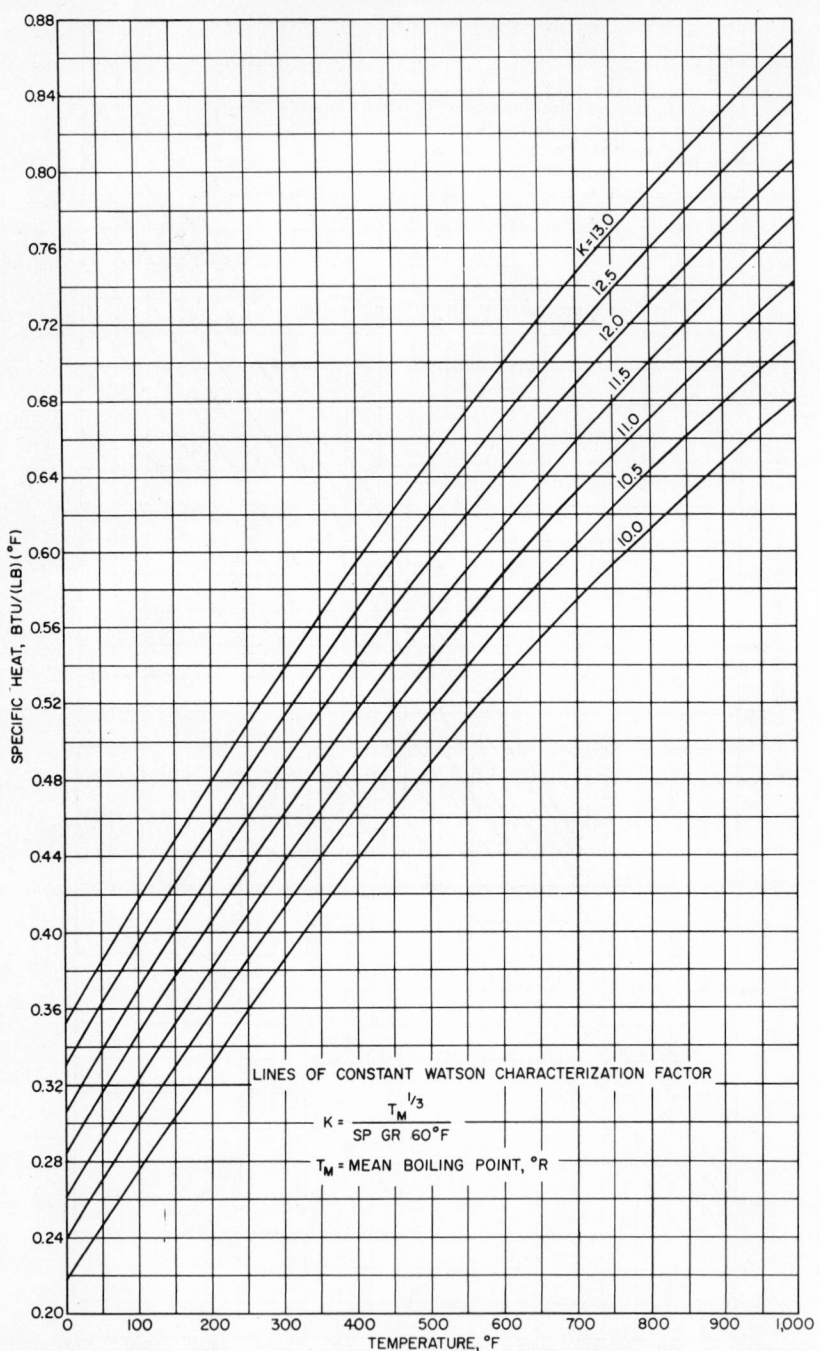

FIG. 12-22. Specific heats of petroleum fractions, vapor phase. (*Fallon and Watson, National Petroleum News, pp. R372, R375, June 7, 1944.*)

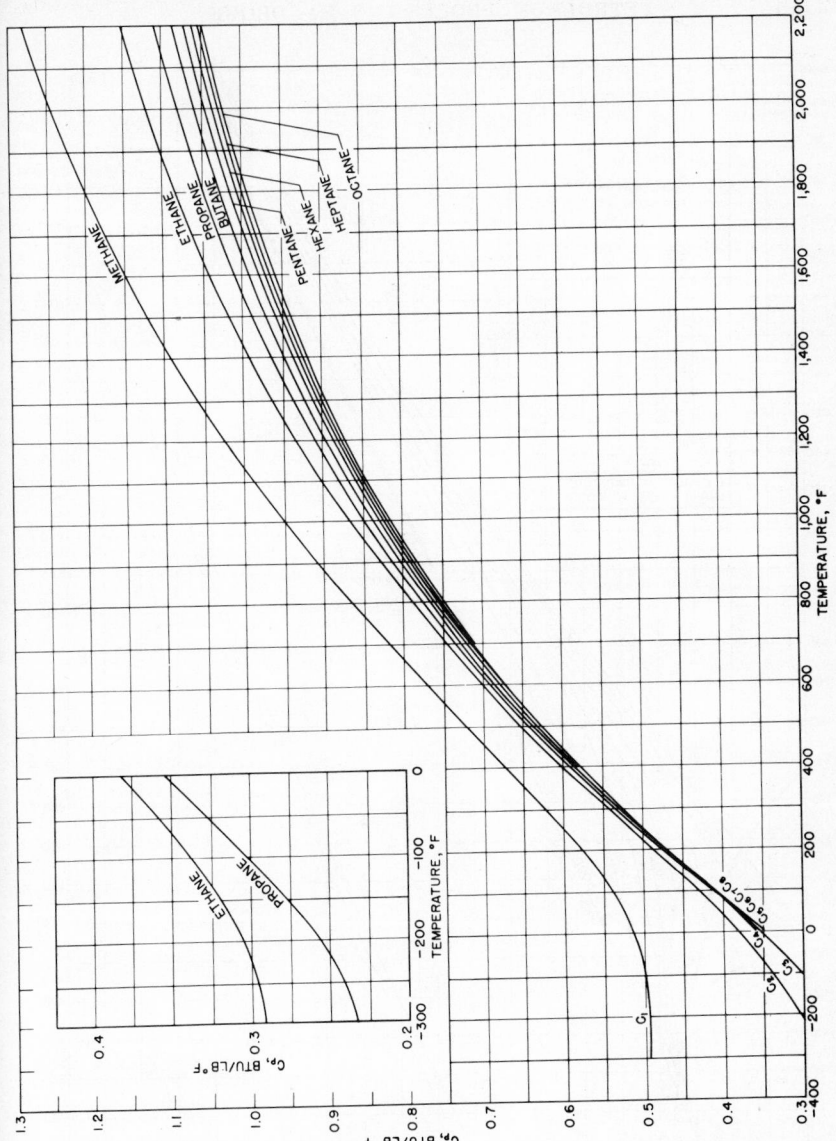

FIG. 12-23. Specific heats of normal paraffin vapors at 1 atm pressure. (*API 44, pp. 655–660, 1952.*)

12-45

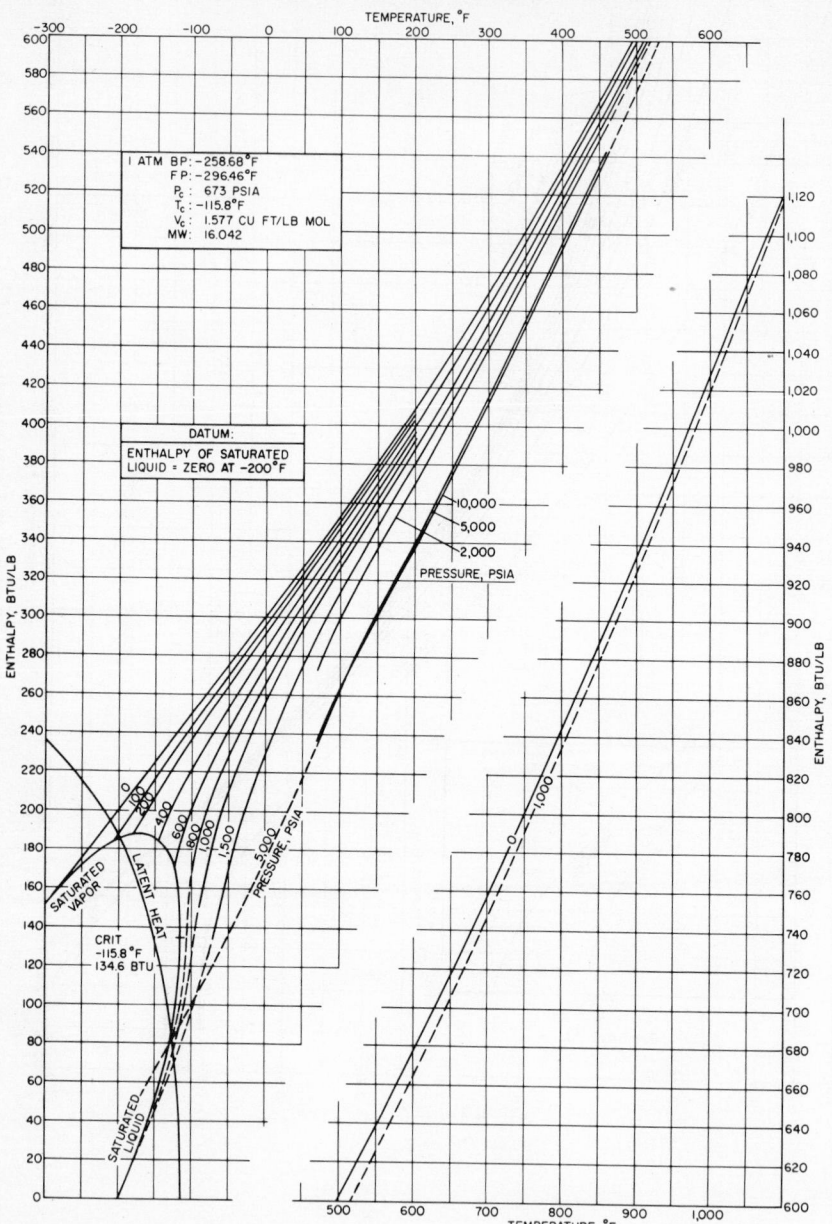

FIG. 12-24. Temperature-enthalpy chart for methane.   Based on data from the following sources:

  −300 to +100°F, zero pressure line: API 44 Tables, 1950.

  −200 to +500°F, 100 to 1,500 psia lines: C. S. Matthews and C. O. Hurd, *Trans. Am. Inst. of Chem. Engrs.*, vol. 42, p. 55, 1946.

  +100 to +460°F, 2,000 to 10,000 psia lines: B. H. Sage and W. N. Lacey, API 37 Monograph, 1950.

  +460 to 1100°F, O. A. Hougen and K. M. Watson: *Chemical Process Principles*, Part II, John Wiley & Sons, Inc., New York, 1947, p. 495.

  Below +100°F, extrapolation of compressed liquid lines by analogy to higher paraffins and by law of corresponding states.

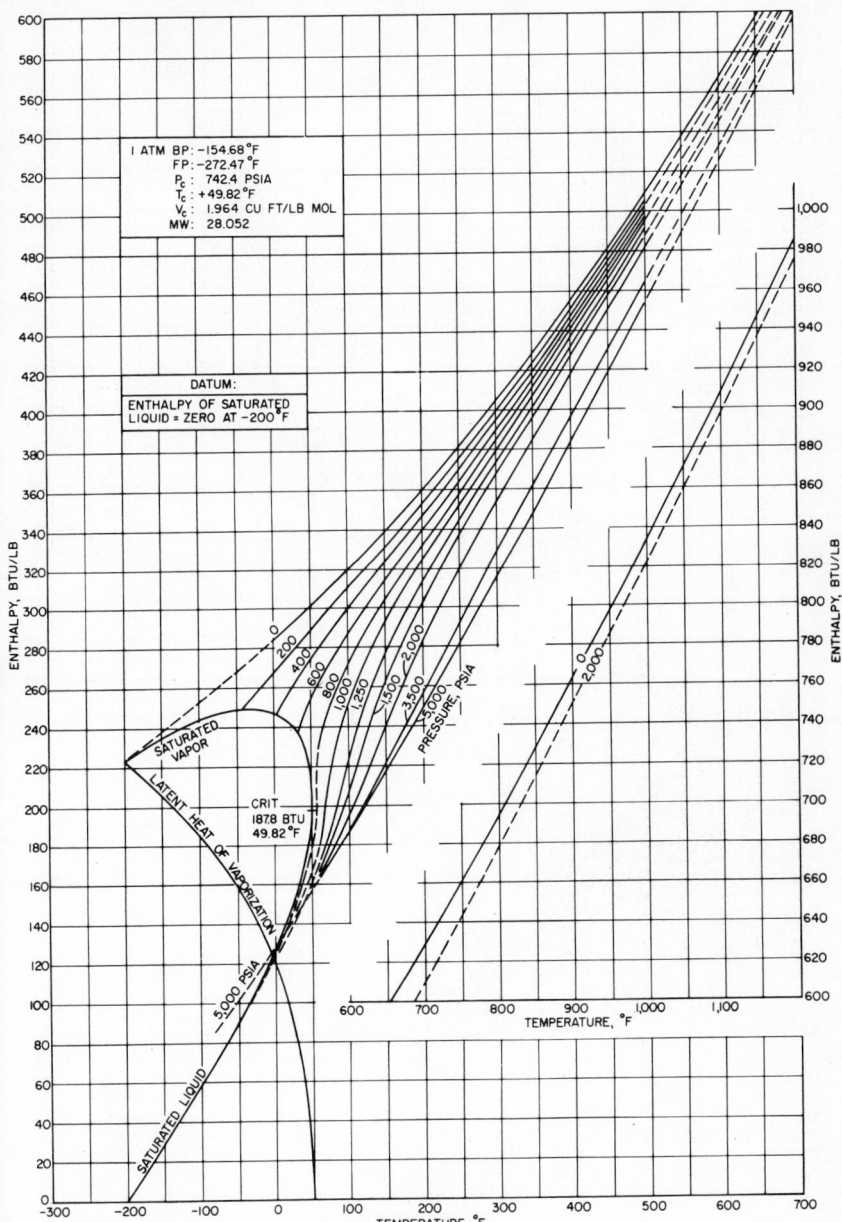

FIG. 12-25. Temperature-enthalpy chart for ethylene.   Based on data from the following sources:
—272 to +500°F, R. York, Jr., and E. F. White, Jr., *Trans. Am. Inst. of Chem. Engrs.*, vol. 40, p. 227, 1944.
0 to 1200°F, zero pressure line: API 44 Tables, 1950.
+500 to 1200°F, O. A. Hougen and K. M. Watson, *Chemical Process Principles*, Part II, John Wiley & Sons, Inc., New York, 1947, p. 495.
Below 60°F, extrapolations by analogy to higher hydrocarbons.

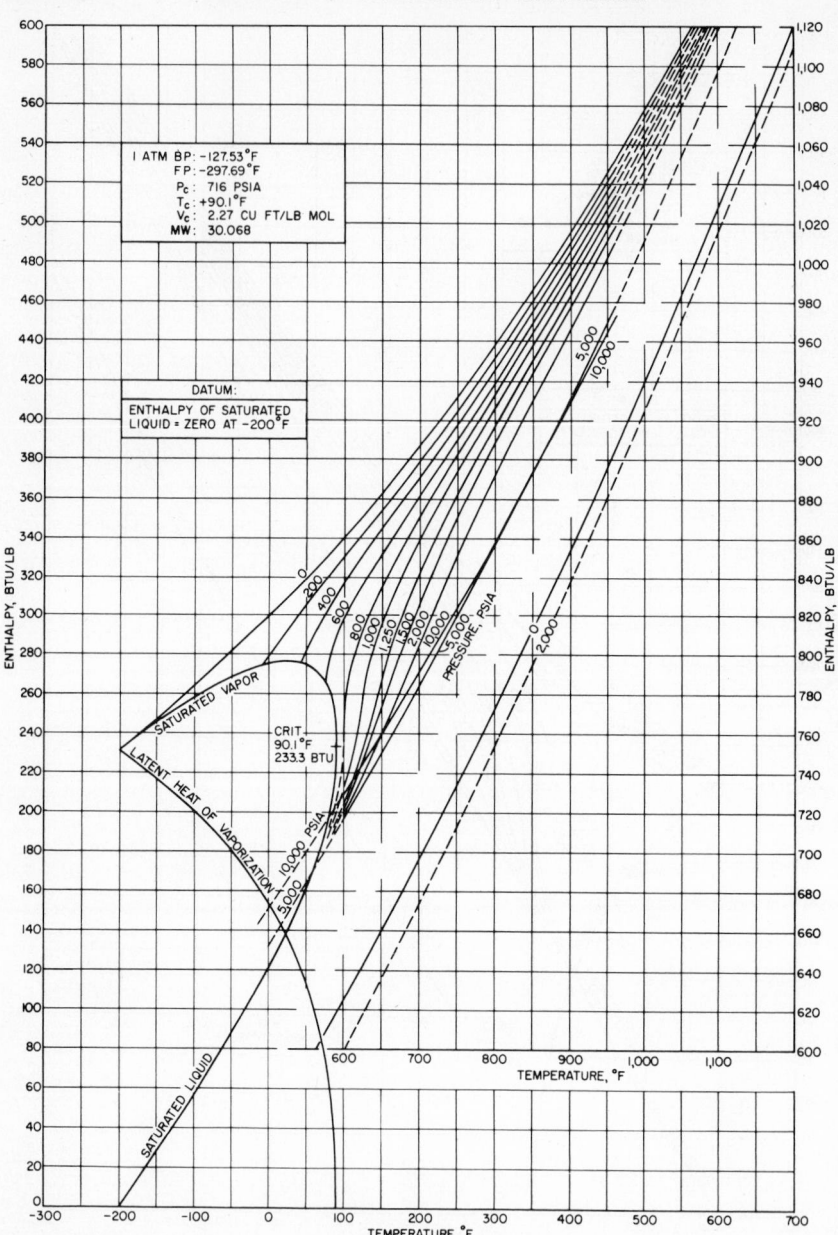

I ATM BP: -127.53°F
FP: -297.69°F
P_c: 716 PSIA
T_c: +90.1°F
V_c: 2.27 CU FT/LB MOL
MW: 30.068

DATUM:
ENTHALPY OF SATURATED
LIQUID = ZERO AT -200°F

CRIT
90.1°F
233.3 BTU

FIG. 12-26. Temperature-enthalpy chart for ethane. Based on data from the following sources:

−200 to 1200°F, zero pressure line: API 44 Tables, 1950.

−200 to +100°F, C. H. Barkelew, J. L. Valentine, and C. O. Hurd, *Chem. Engrg. Progr.*, vol. 43, p. 25, 1947.

+100 to +460°F, B. H. Sage and W. N. Lacey, API 37 Monograph, 1950.

+460 to 1200°F, O. A. Hougen and K. M. Watson, *Chemical Process Principles*, John Wiley & Sons, Inc., New York, 1947, p. 495.

Below +100°F, extrapolation of compressed liquid lines by analogy to higher paraffins.

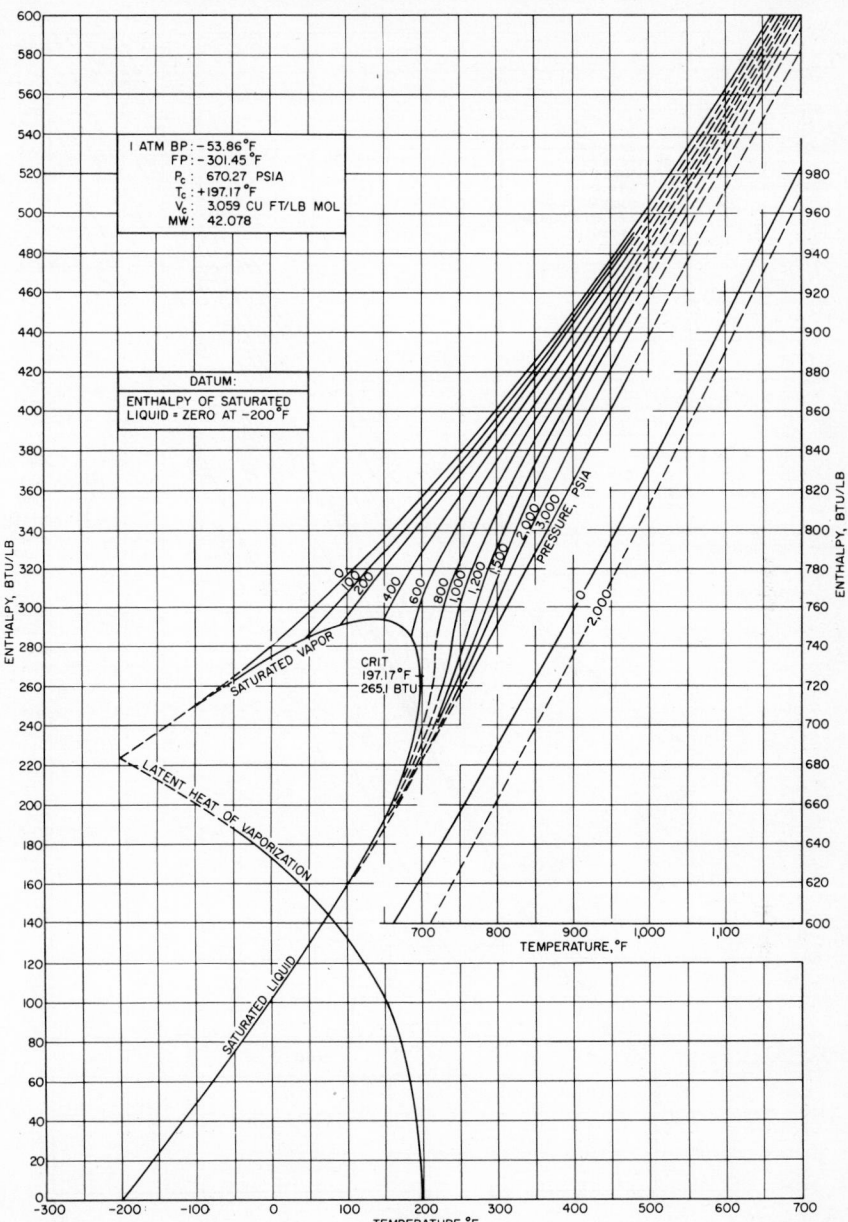

FIG. 12-27. Temperature-enthalpy chart for propylene. Based on data from the following sources:

−200 to 1200°F, zero pressure line: API 44 Tables, 1950.

−200 to −50°F, saturated liquid line: T. M. Powell and W. F. Giauque, *J. Am. Chem. Soc.*, vol. 61, p. 2366, 1939.

−50 to +480°F, L. N. Canjar, M. Goldman, and H. Marchman, *Ind. Engrg. Chem.*, 43, p. 1186, 1951.

+480 to 1200°F, O. A. Hougen and K. M. Watson, *Chemical Process Principles*, Part II, John Wiley & Sons, Inc., New York, 1947, p. 495.

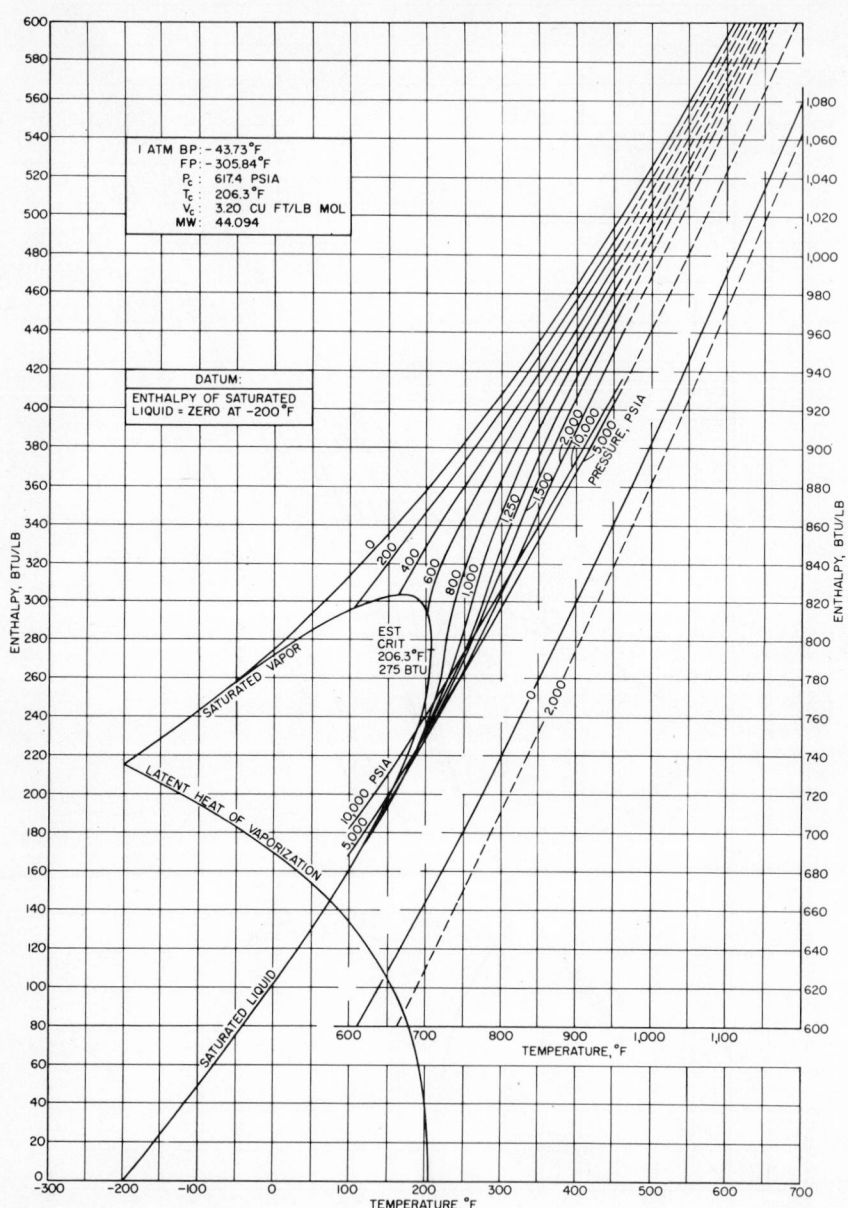

FIG. 12-28. Temperature-enthalpy chart for propane.   Based on data from the following sources:
   −300 to 1200°F, zero pressure line: API 44 Tables, 1950.
   −200 to −80°F, saturated liquid line: J. D. Kemp and C. J. Egan, *J. Am. Chem. Soc.*, vol. 60, p. 521, 1938.
   −80 to +100°F, W. V. Stearns and E. J. George, *Ind. Engrg. Chem.*, vol. 35, p. 602, 1943.
   +460 to 1200°F, O. A. Hougen and K. M. Watson, *Chemical Process Principles*, Part II, John Wiley & Sons, Inc., New York, 1947, p. 495.

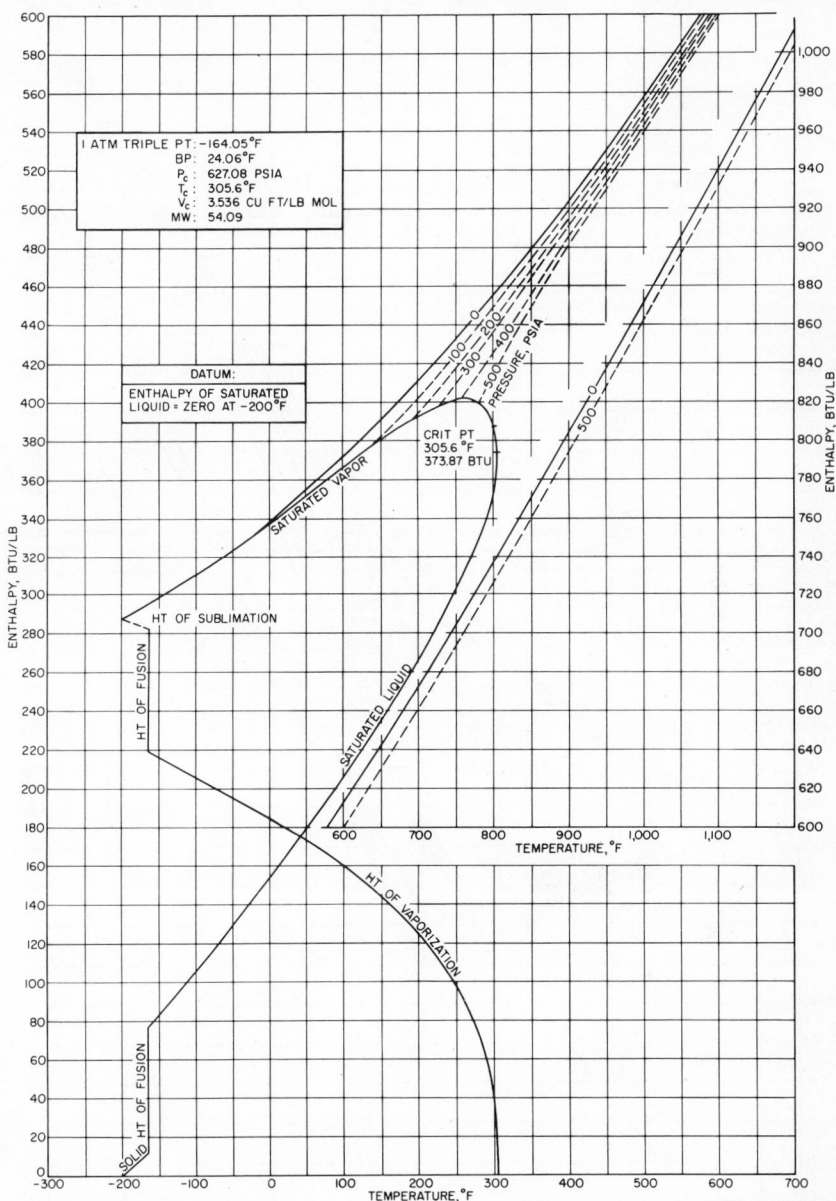

FIG. 12-29. Temperature-enthalpy chart for 1,3-butadiene. Based on data from the following sources:

Solid curves, O. H. Meyers, C. S. Cragoe, and E. F. Mueller, Table and Mollier chart of thermodynamic properties of 1,3-butadiene, *J. Res. Natl. Bur. Std.*, vol. 39, p. 507, 1947.
R. B. Scott, C. H. Meyers, R. D. Rands, F. G. Brickwedde, and S. N. Bekkedahl, *J. Res. Natl. Bur. Std.*, vol. 35, p. 39, 1945.

305 to 600°F, zero pressure curve: API 44 Tables, 1950.

100 to 500 psia, effect of pressure on enthalpy: O. A. Hougen and N. M. Watson, *Chemical Process Principles*, Part II, John Wiley & Sons, Inc., New York, 1947.

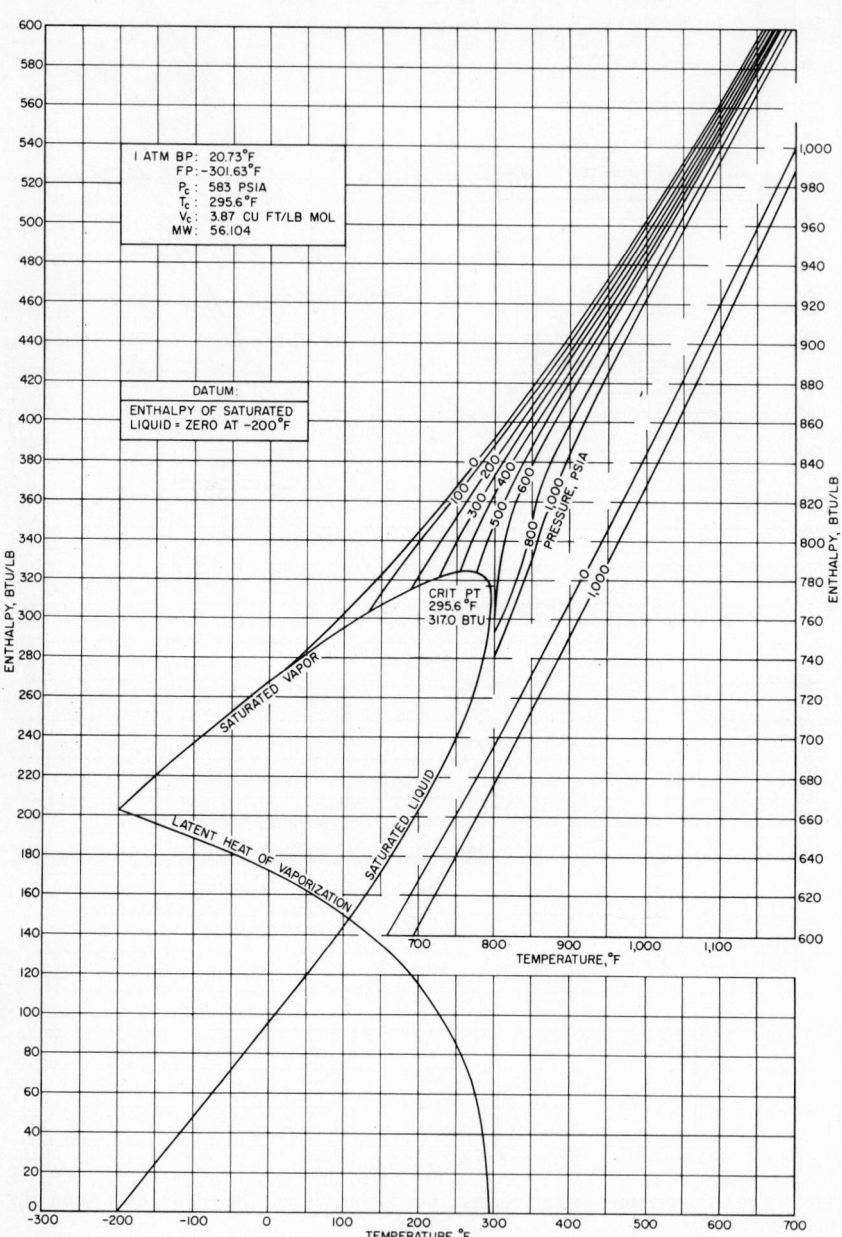

FIG. 12-30. Temperature-enthalpy chart for 1-butene. Based on data from the following sources:

Saturated liquid curve: −200 to 0°F, S. S. Todd and G. S. Parks, *J. Am. Chem. Soc.*, vol. 58, 134, 1936; 0 to 295.6°F, saturated vapor curve minus latent heat.

Saturated vapor curve: −200 to 0°F, saturated liquid curve plus latent heat; 0 to 295.6°F, Hougen and Watson correlation.

Latent heat curve: API 44, Table 8M; D. H. Gordon, *I & E Chem.*, vol. 35, p. 851, 1943.

Pressure curves: 0 psia, API 44, Table 8U-E; 100 to 1000 psia, Hougen and Watson correlation.

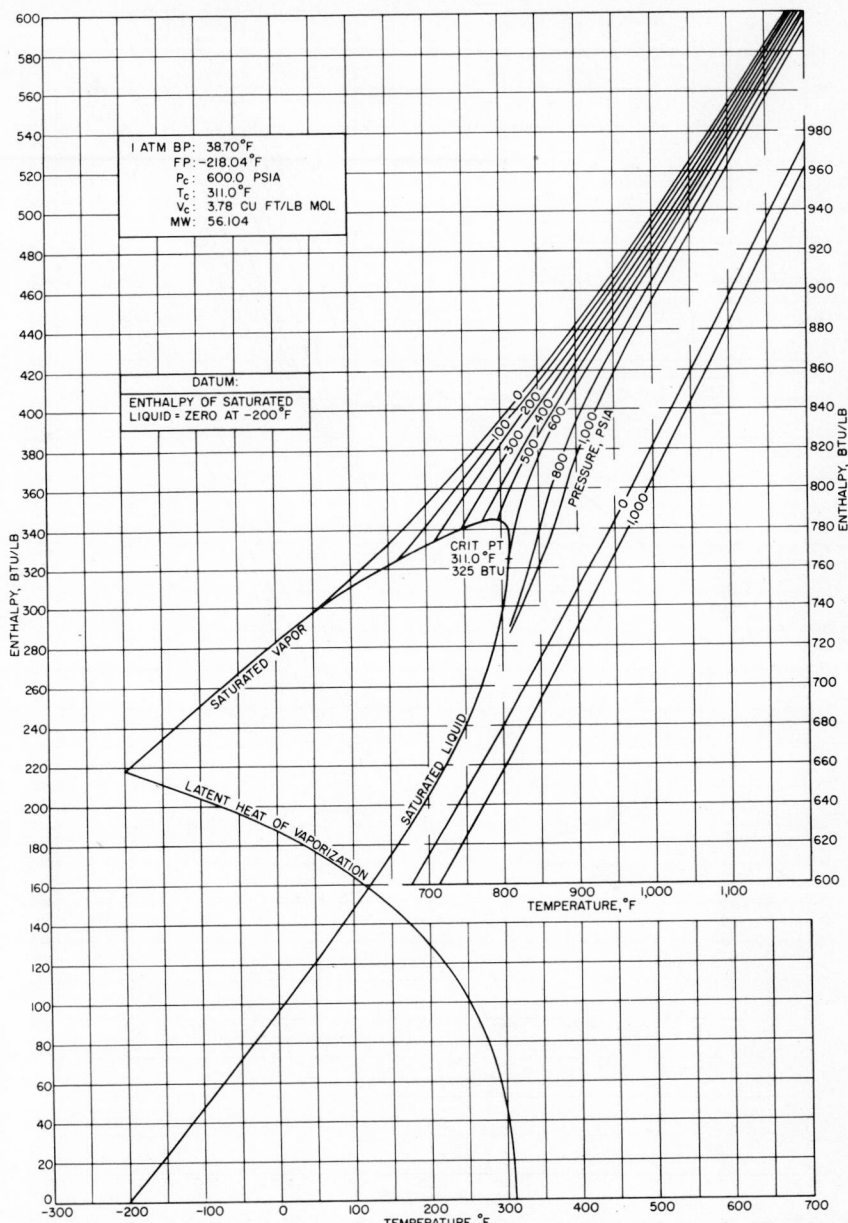

FIG. 12-31. Temperature-enthalpy chart for cis-2-butene.   Based on data from the following sources:
Saturated liquid curve: −200 to 80°F, *J. Res. Natl. Bur. Std.*, vol. 33, p. 1, 1944; 80 to 311°F, saturated vapor curve minus latent heat.
Saturated vapor curve: −200 to −18°F, saturated liquid curve plus latent heat; −18 to 80°F, *J. Res. Natl. Bur. Std.*, vol. 33, p. 1, 1944; 80 to 311°F, Hougen and Watson correlation.
Latent heat curve: −200 to −18°F and 80 to 311°F, D. H. Gordon, *I & E Chem.*, vol. 35, p. 651, 1943; −18 to 80°F, *J. Res. Natl. Bur. Std.*, vol. 35, p. 1, 1944.
Pressure curves: 0 psia, API 44, Table 8U-E; 100 to 1000 psia, Hougen and Watson correlation.

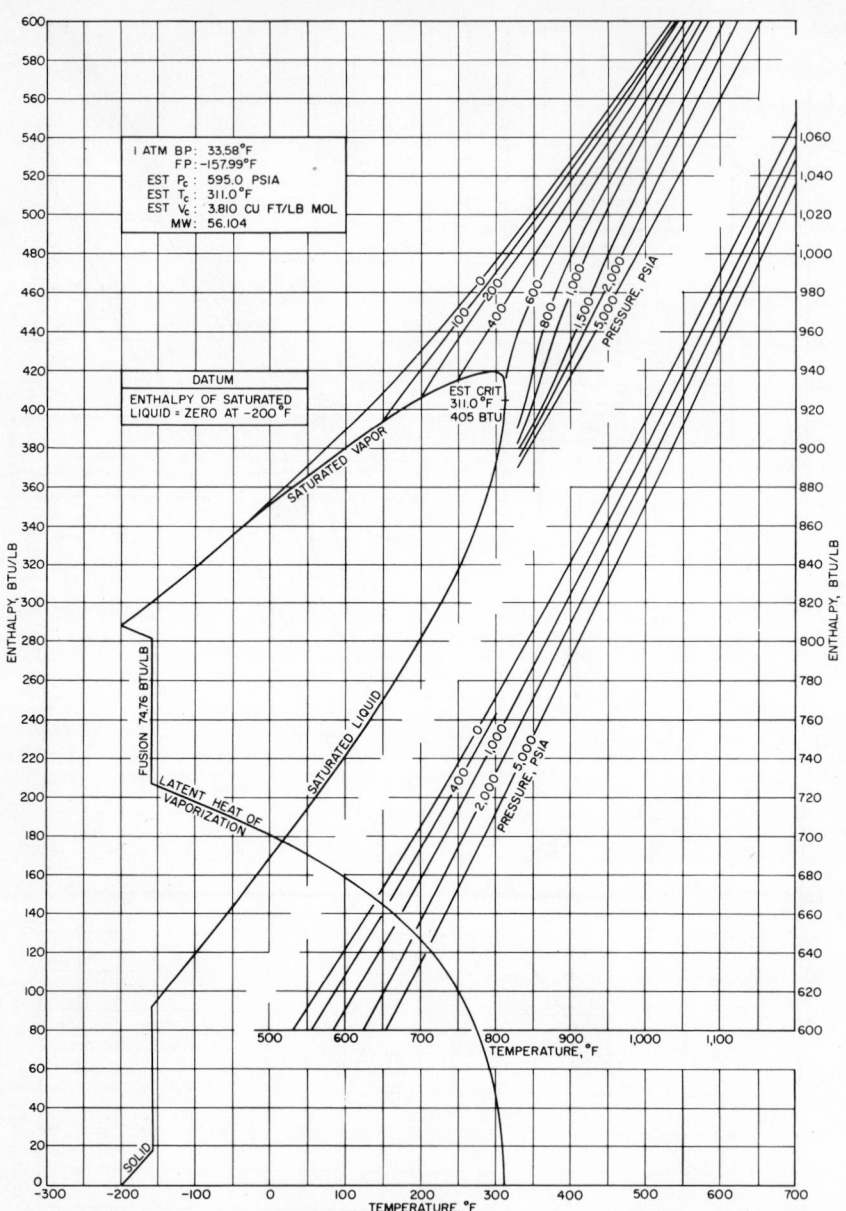

I ATM BP: 33.58°F
F P: -157.99°F
EST P_c: 595.0 PSIA
EST T_c: 311.0°F
EST V_c: 3.810 CU FT/LB MOL
MW: 56.104

DATUM
ENTHALPY OF SATURATED
LIQUID = ZERO AT -200°F

EST CRIT
311.0°F
405 BTU

SATURATED VAPOR

FUSION 74.76 BTU/LB

SATURATED LIQUID

LATENT HEAT OF VAPORIZATION

PRESSURE, PSIA

SOLID

ENTHALPY, BTU/LB

TEMPERATURE, °F

Fig. 12-32. Temperature-enthalpy chart for trans-2-butene. Based on data from the following sources:

−200 to +33.58°F, saturated solid and liquid lines: L. Guttmann and K. S. Pitzer, *J. Am. Chem. Soc.*, vol. 67, p. 324, 1943.

−200 to +311.0°F, latent heats: Guttmann and Pitzer, *J. Am. Chem. Soc.*, vol. 67, p. 324, 1943. Gordon method.

−200 to +1200°F, zero pressure line: API 44 Tables, 1951.

−200 to +1200°F, pressure correction to enthalpy: Hougen and Watson method.

Critical data: Estimate by Cragoe, *Natl. Bur. Std. Letter Circ.* 736, 1943.

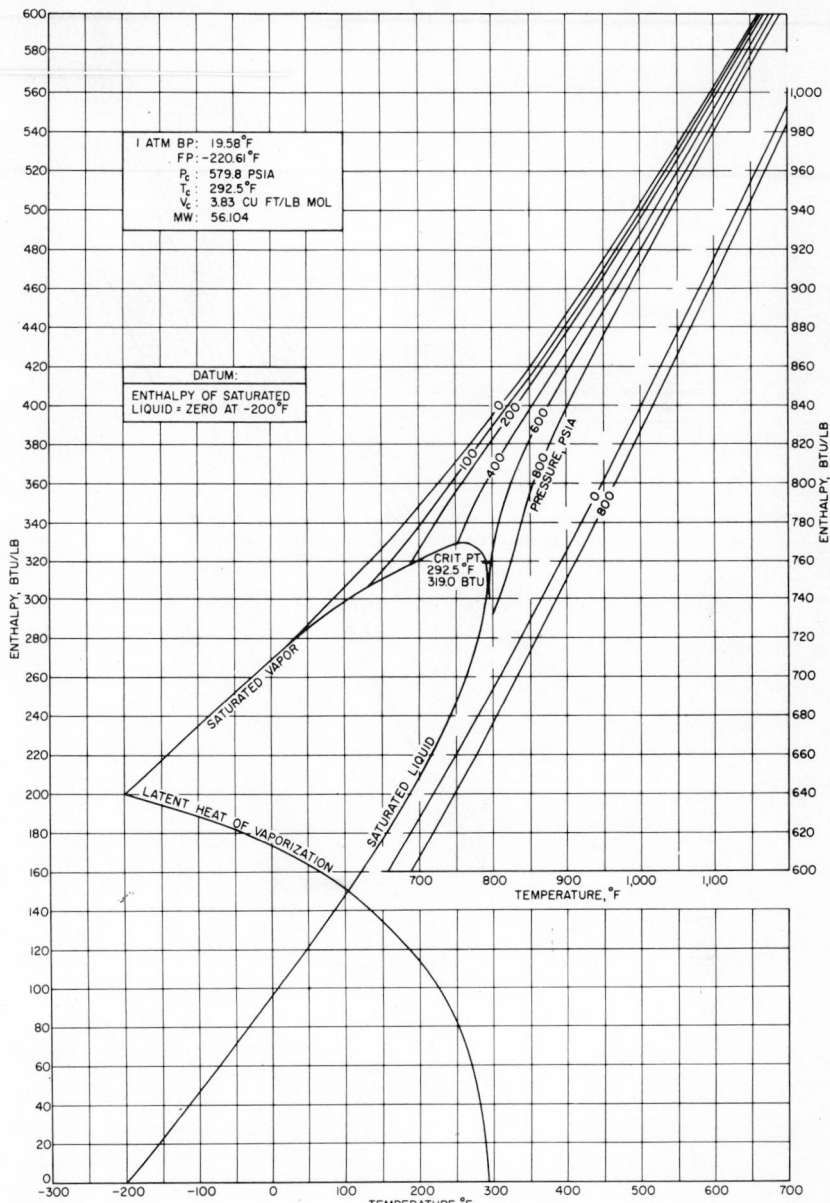

FIG. 12-33. Temperature-enthalpy chart for isobutene.   Based on data from the following
sources:
Saturated liquid curve: −200 to 0°F, S. S. Todd and G. S. Parks, *J. Am. Chem. Soc.*,
vol. 58, p. 134, 1936; 0 to 292.5°F, saturated vapor curve minus latent heat.
Saturated vapor curve: −200 to 0°F, saturated liquid curve plus latent heat; 0 to 292.5°F,
Hougen and Watson correlation.
Latent heat curve: API 44, Table 8M; G. H. Hanson, *Trans. Am. Inst. of Chem. Engrs.*,
vol. 42, p. 959, 1946.
Pressure curves: 0 psia, API 44, Table 8U-E; 100 to 800 psia, Hougen and Watson
correlation.

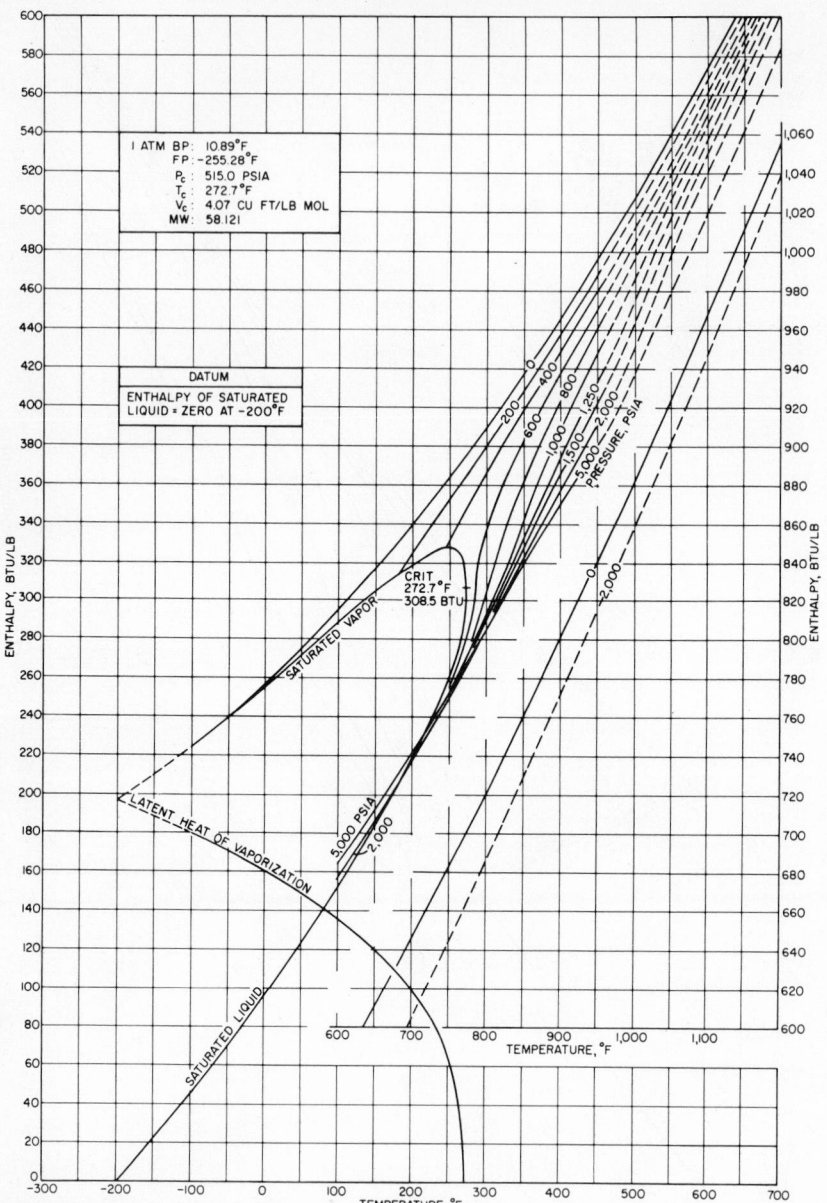

Fig. 12-34. Temperature-enthalpy chart for isobutane. Based on data from the following sources:

−200 to 0°F, saturated liquid line: J. G. Aston, R. M. Kennedy, and S. C. Schumann, *J. Am. Chem. Soc.*, vol. 62, p. 2059, 1940.

−100 to 1200°F, zero pressure line: API 44 Tables, 1950.

−100°F, latent heat: Clapeyron equation, with vapor-pressure data from API 44 Tables, 1950.

+100 to +460°F, B. H. Sage and W. N. Lacey, API 37 Monograph, 1950.

+460 to 1200°F, O. A. Hougen and K. M. Watson, *Chemical Process Principles*, Part II, John Wiley & Sons, Inc., New York, 1947, p. 495.

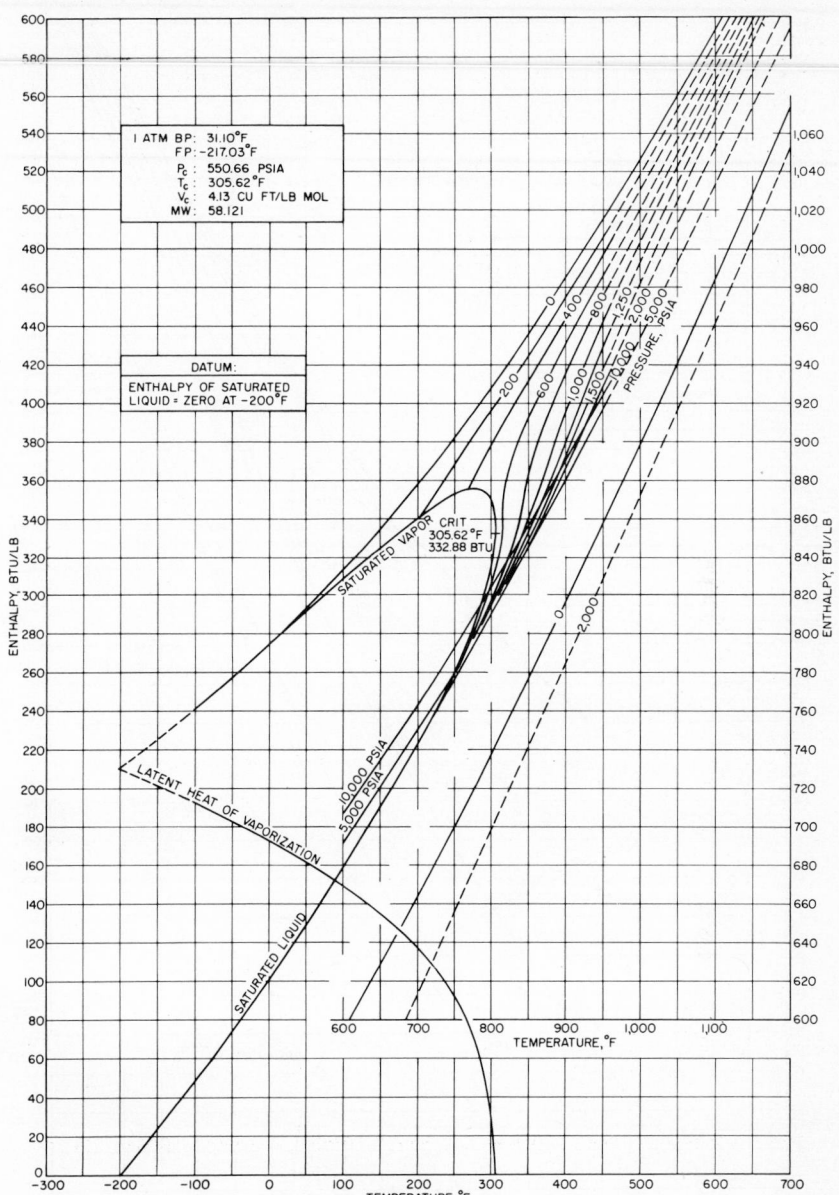

I ATM BP: 31.10°F
FP: -217.03°F
P_c : 550.66 PSIA
T_c : 305.62°F
V_c : 4.13 CU FT/LB MOL
MW: 58.121

DATUM:
ENTHALPY OF SATURATED
LIQUID = ZERO AT -200°F

CRIT
305.62°F
332.88 BTU

Fig. 12-35. Temperature-enthalpy chart for n-butane.   Based on data from the following sources:

−200 to +311°F, saturated liquid line: J. G. Aston and G. H. Messerly, *J. Am. Chem. Soc.*, vol. 62, p. 1917, 1940.

−100 to 1200°F, zero pressure line: API 44 Tables, 1950.

+31.1 to +100°F, M. W. Prengle, Jr., L. R. Greenhaus, and R. York, Jr., *Trans. Am. Inst. of Chem. Engrs.*, vol. 44, p. 863, 1948.

+100 to +460°F, B. H. Sage and W. N. Lacey, API 37 Monograph, 1950.

+460 to 1200°F, O. A. Hougen and K. M. Watson, *Chemical Process Principles*, Part II, John Wiley & Sons, Inc., New York, 1947, p. 495.

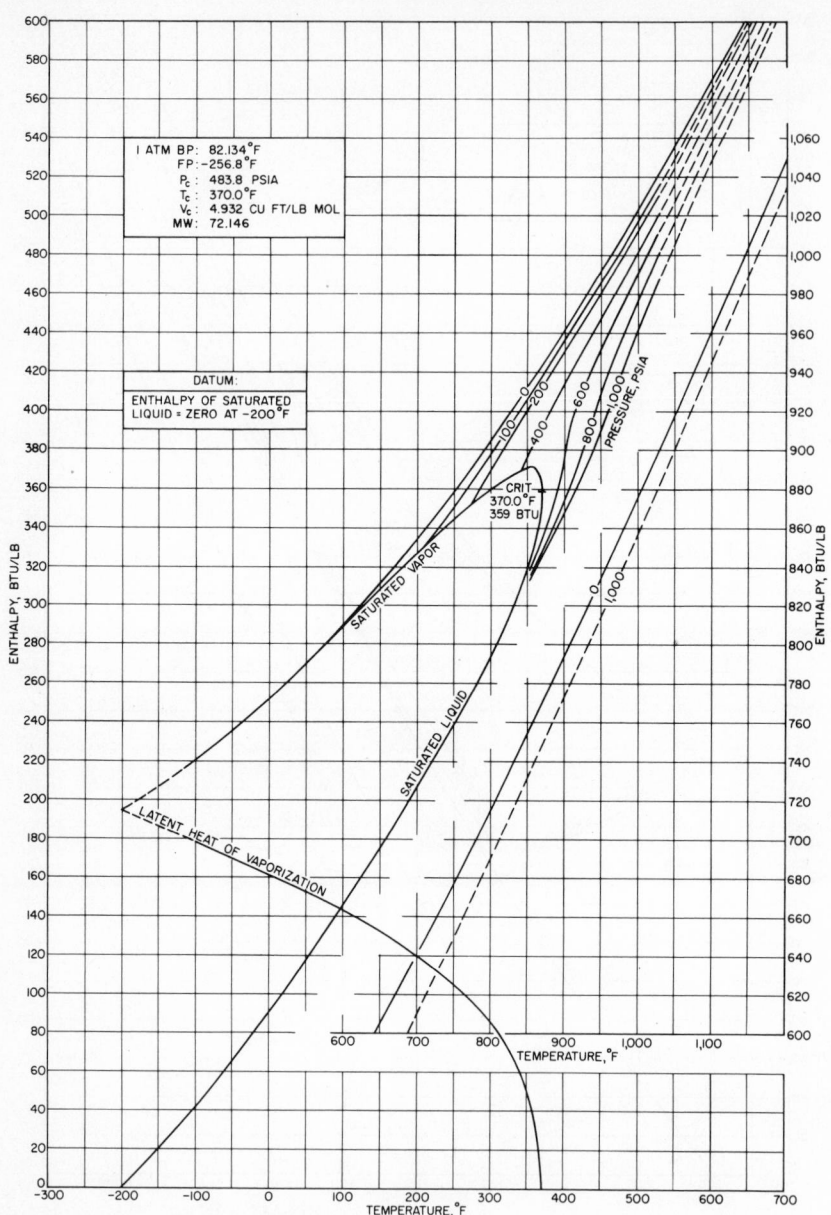

Fig. 12-36. Temperature-enthalpy chart for isopentane.   Based on data from the following sources:

−200 to +59°F, saturated liquid line: G. B. Guthrie, Jr., and H. M. Huffman, *J. Am. Chem. Soc.*, vol. 65, p. 1139, 1943.

−100 to 1200°F, zero pressure line: API 44 Tables, 1950.

+59 to +536°F, *PVT* data of S. Young, *Proc. Phys. Soc. (London)*, vol. 13, pp. 602–663, 1895.

+536 to 1200°F, O. A. Hougen and K. M. Watson, *Chemical Process Principles*, Part II, John Wiley & Sons, Inc., New York, 1947, p. 495.

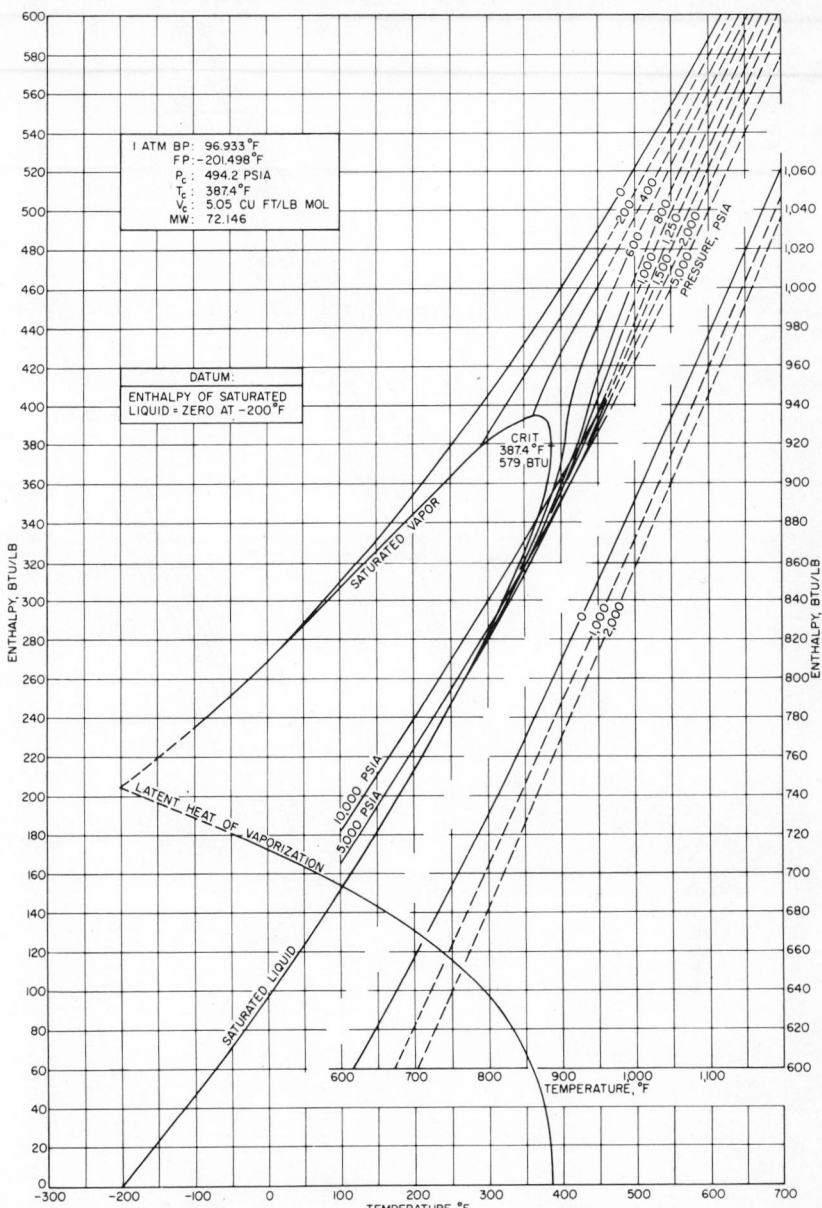

FIG. 12-37. Temperature-enthalpy chart for *n*-pentane.   Based on data from the following sources:

—200 to 0°F, saturated liquid line: G. H. Messerly and R. M. Kennedy, *J. Am. Chem. Soc.*, vol. 62, p. 2988, 1940.

—100 to 1200°F, zero pressure line: API 44 Tables, 1950.

0°F, latent heat: Clapeyron equation with vapor-pressure data from API 44 Tables, 1950.

+100 to +460°F, B. H. Sage and W. N. Lacey, API 37 Monograph, 1950.

+460 to 1200°F, O. A. Hougen and K. M. Watson, *Chemical Process Principles*, Part II, John Wiley & Sons, Inc., New York, 1947, p. 495.

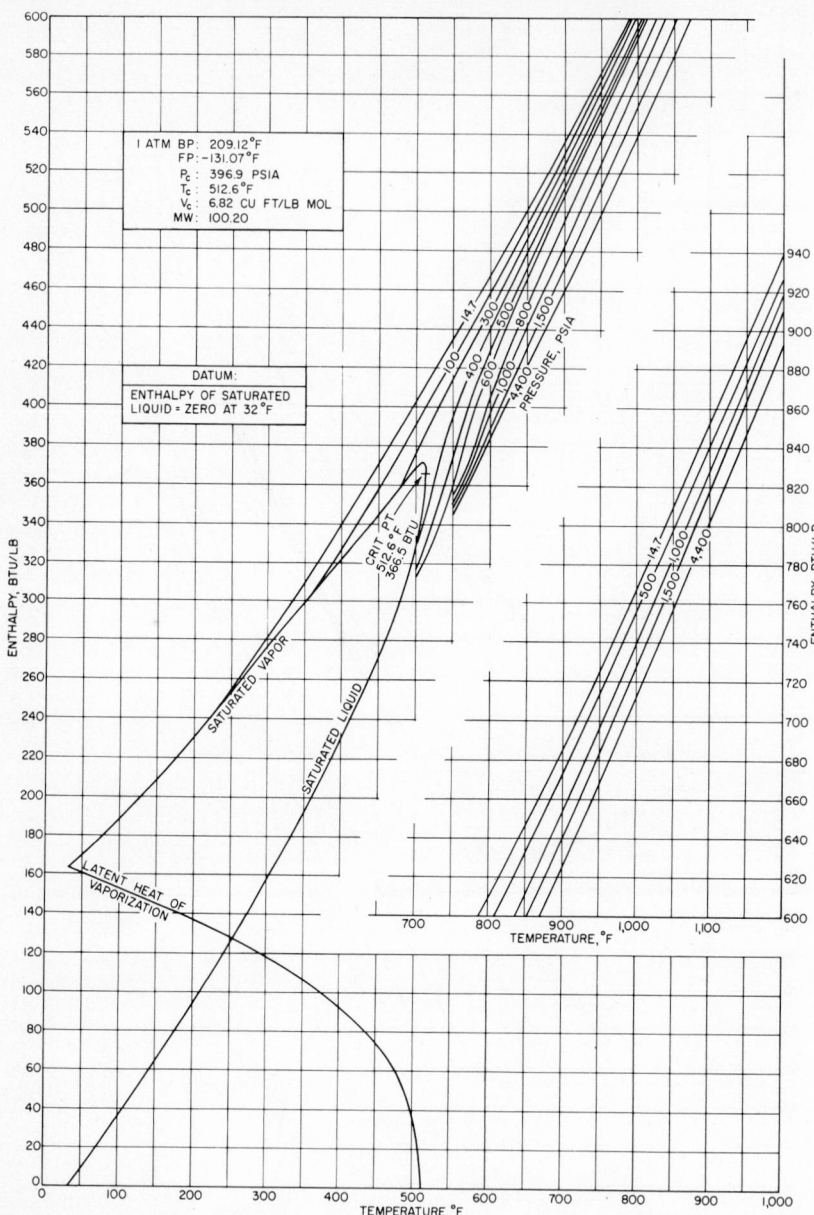

FIG. 12-38. Temperature-enthalpy chart for n-heptane.   Based on data from the following sources:

Saturated liquid curve: 32 to 70°F, G. S. Parks et al., *J. Am. Chem. Soc.*, vol. 52, p. 1032, 1930; 77 to 209°F, extrapolation; 209 to 512°F, E. B. Stuart et al., *Chem. Engrg. Progr.*, vol. 46, p. 311, 1950, E. R. Gilliland and M. D. Parekh, *Ind. Engr. Chem.*, vol. 34, p. 360, 1942.

Latent heat curve: S. Young, *Sci. Proc. Soc. Dublin*, N.S., XII, p. 374, 1910.

Pressure curves: E. B. Stuart et al., *Chem. Engrg. Progr.*, vol. 46, p. 311, 1950; below 600°F (400 to 4400 psia), Hougen and Watson correlation.

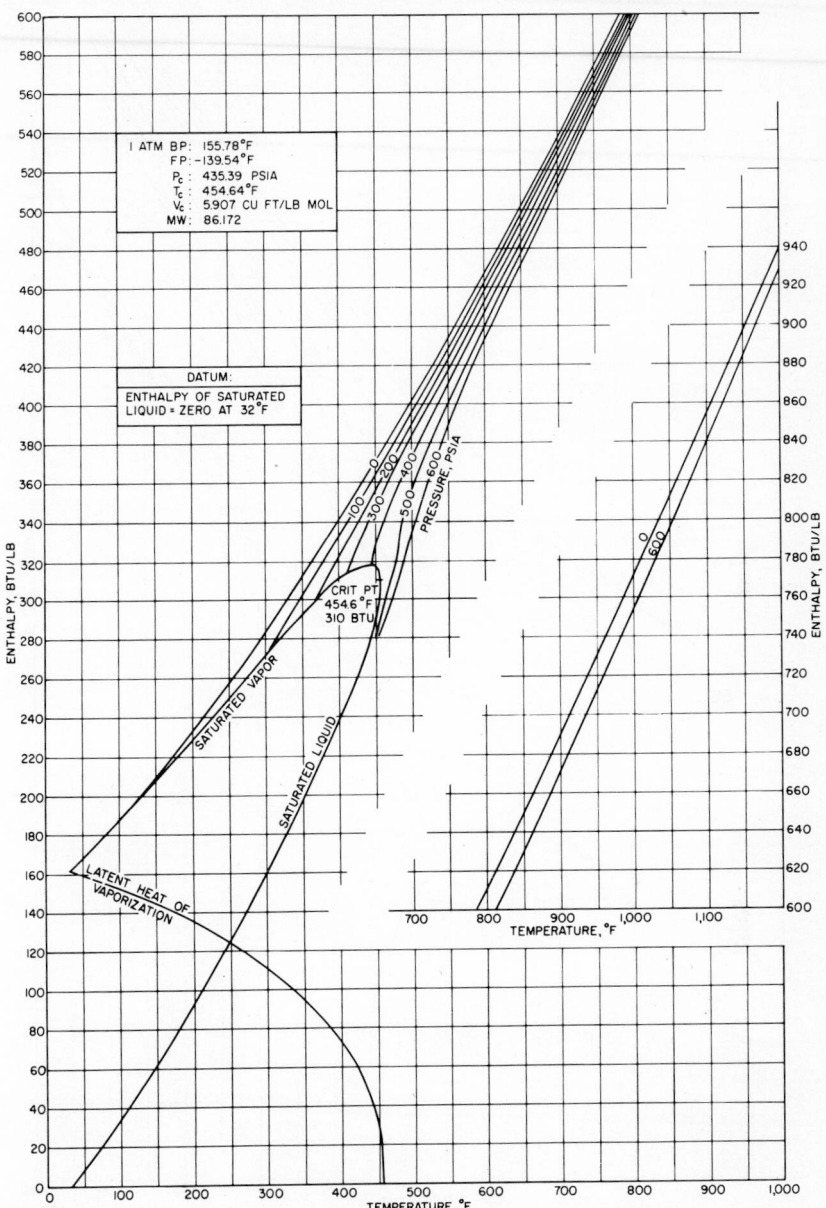

FIG. 12-39. Temperature-enthalpy chart for *n*-hexane.   Based on data from the following sources:
    Saturated curves: vapor, O. A. Hougen and K. M. Watson, *Chemical Process Principles*, Part II, p. 495; liquid, vapor minus latent heat.
    Latent heat curve: S. Young, *Sci. Proc. Soc. Dublin, N.S.*, XII, p. 374, 1910.
    Pressure curves: 0 psia, API 44, Table 20U-E; 100 to 600 psia, O. A. Hougen and K. M. Watson, *Chemical Process Principles*, Part II, p. 495.

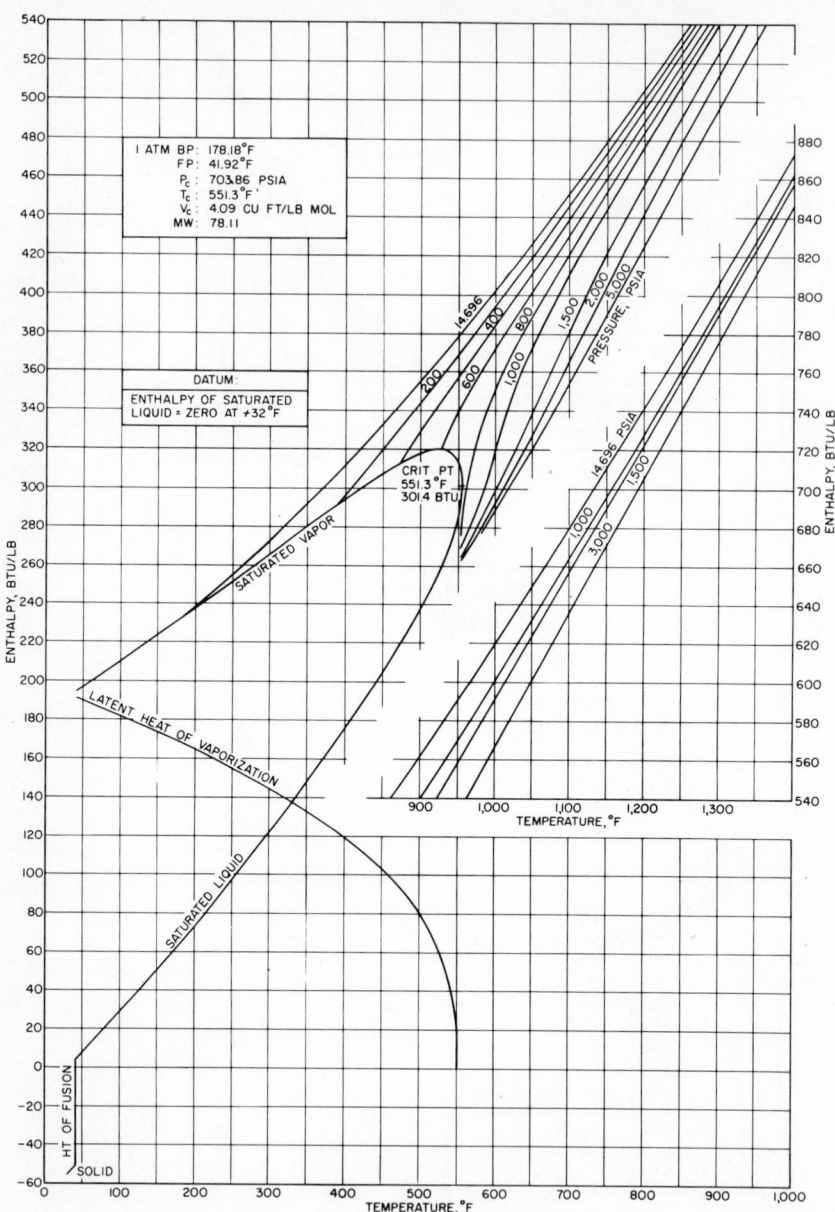

FIG. 12-40. Temperature-enthalpy chart for benzene. Based on data from the following sources:

E. Flock et al., *J. Res. Natl. Bur. Std.* 6, p. 893, 1931.

E. I. Organick and W. R. Studhalter, *Chem. Engrg. Progr.*, vol. 44, p. 847, 1948.

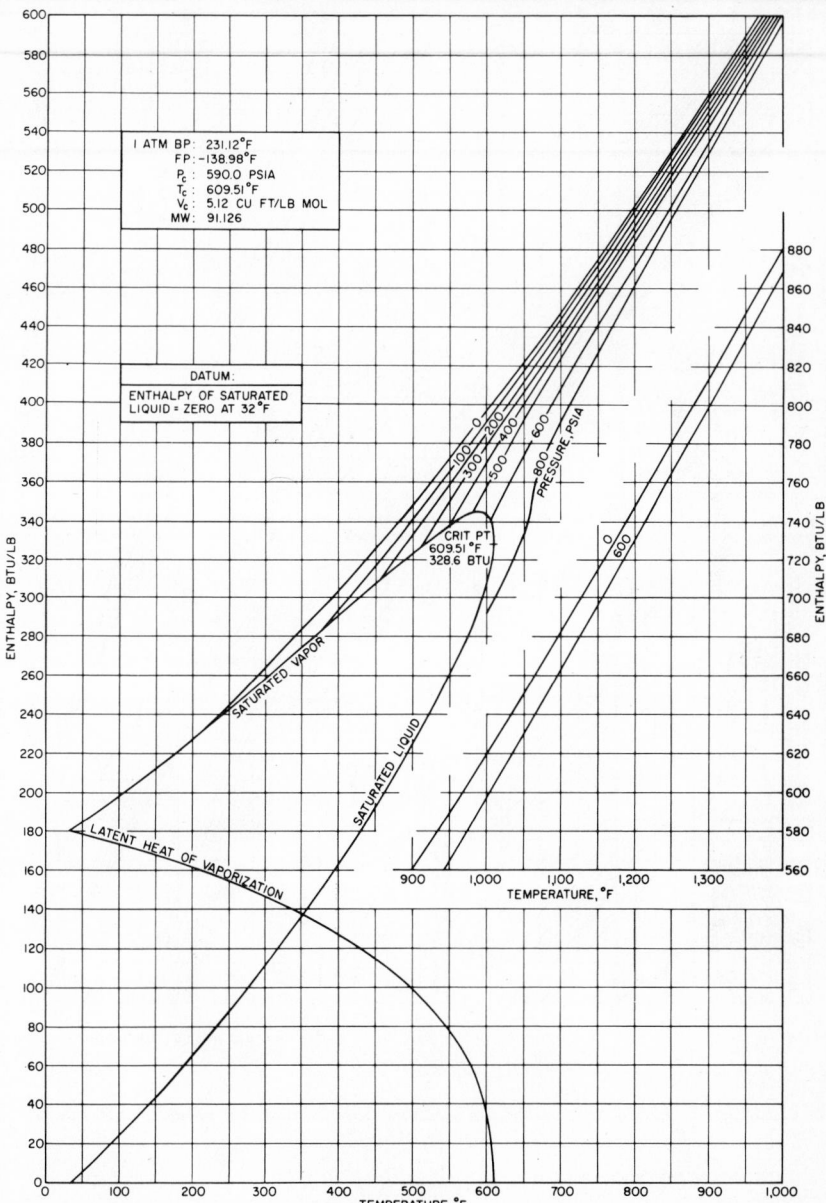

FIG. 12-41. Temperature-enthalpy chart for toluene. Based on data from the following sources:
Saturated curves: vapor, Hougen and Watson correlation; liquid, saturated vapor minus latent heat.
Latent heat curve: API 44, Table 8M; D. H. Gordon, *Ind. Engrg. Chem.*, vol. 35, p. 851, 1943.
Pressure curves: 0 psia, API-44, Table 5U-E; 100 to 800 psia, Hougen and Watson correlation.

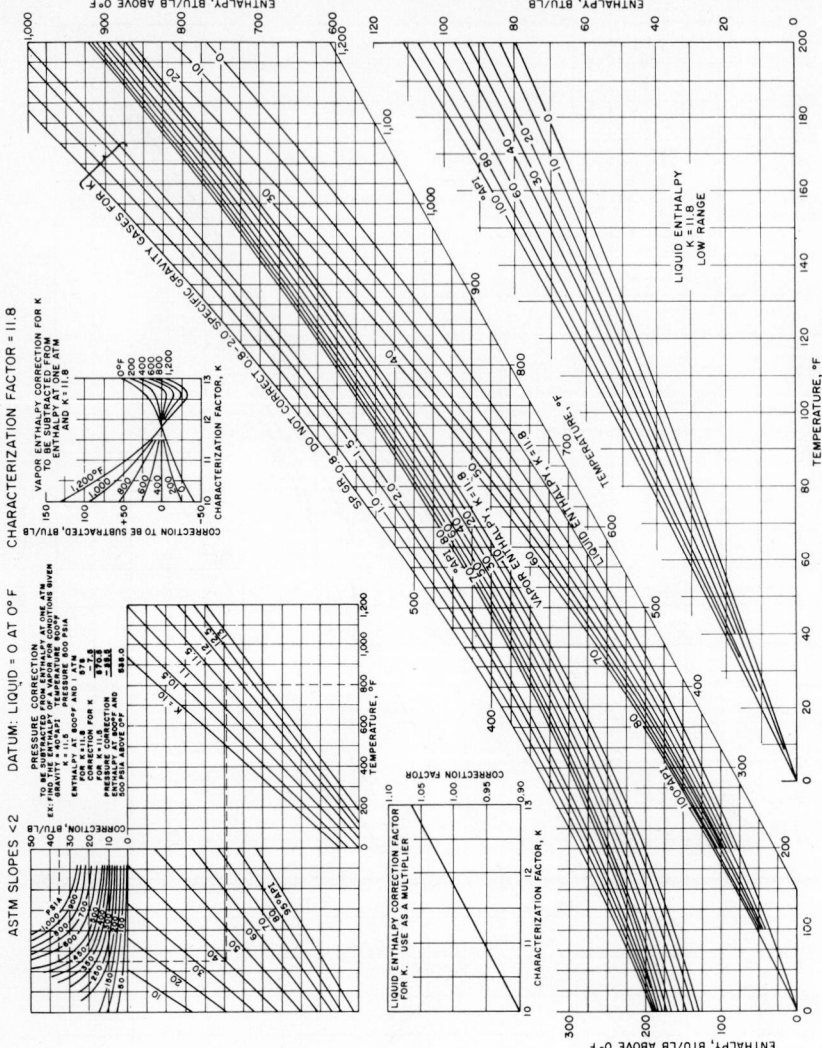

Fig. 12-42. Temperature-enthalpy chart, petroleum fractions.

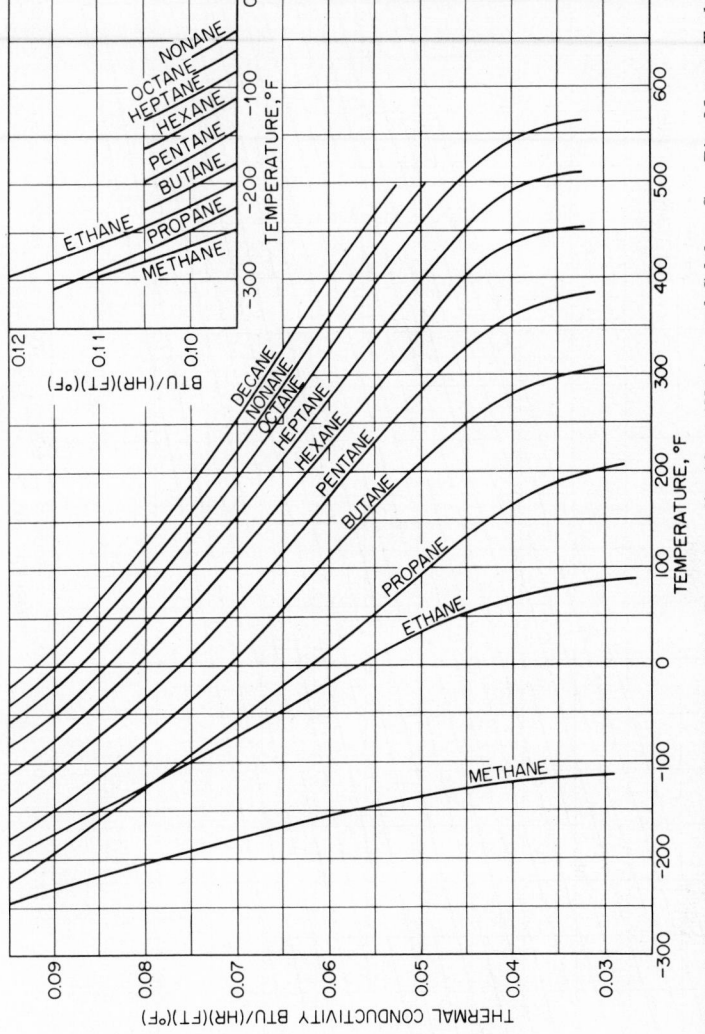

FIG. 12-43. Thermal conductivities of hydrocarbon liquids. (*Naziev and Goluben, Ser. Fiz.-Mat. 1 Teckn. Nauk., No. 6, 113–117, 1962; Naziev, Izv. Vysshikh Uchebn. Zavedenii, Neft. 1 Gas, No. 4, 65–68, 1966.*)

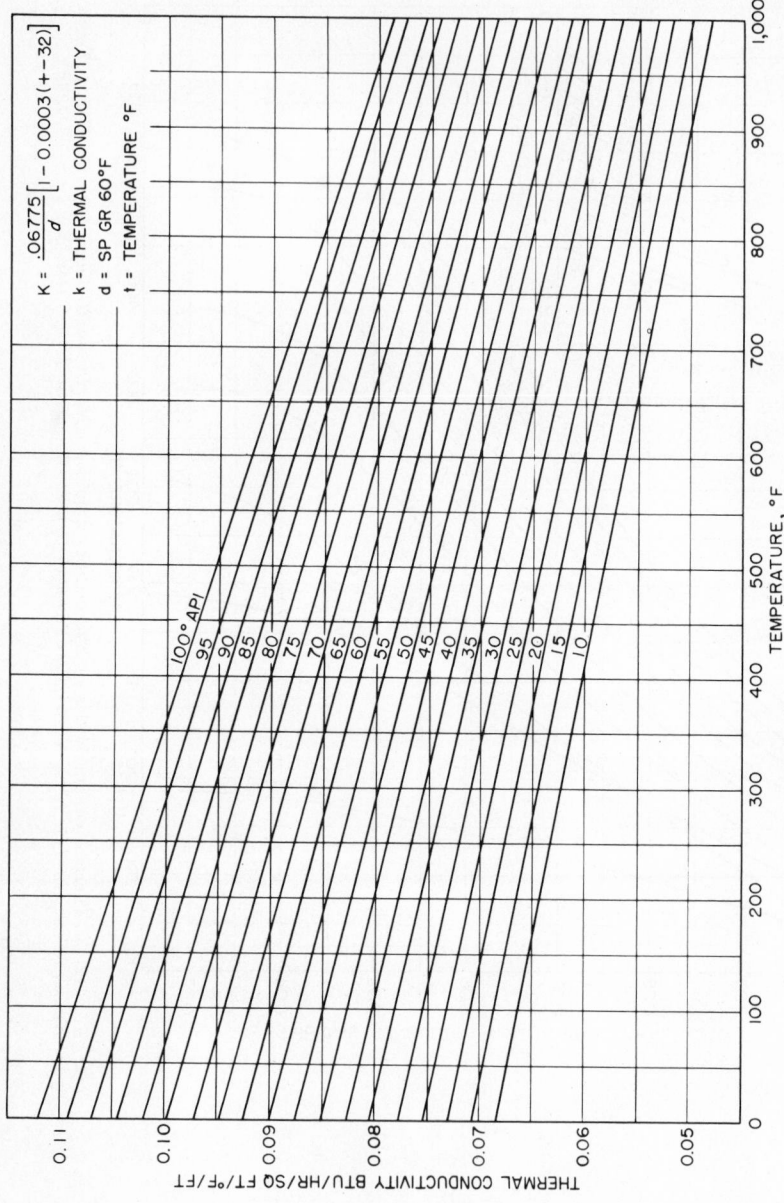

FIG. 12-44. Thermal conductivities of liquid petroleum products. (*U.S. Bur. Std. Misc. Publ. 97.*)

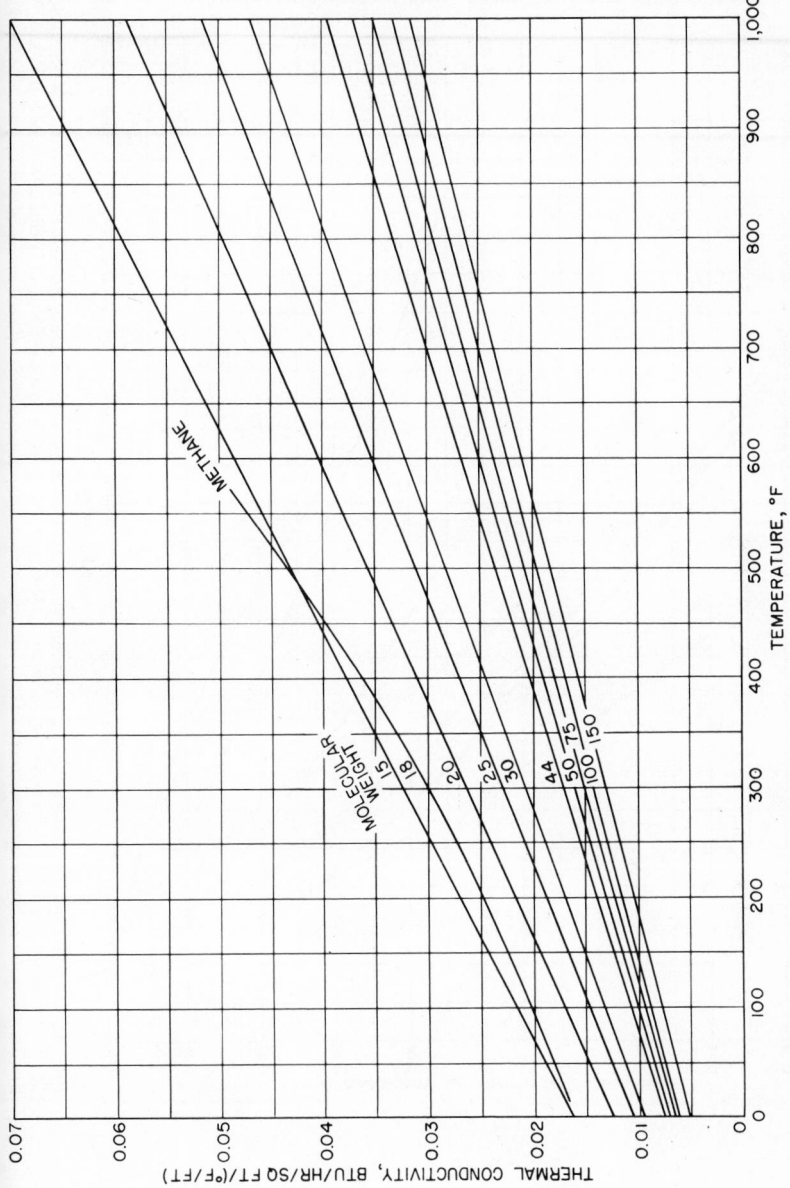

FIG. 12-45. Thermal conductivities of paraffinic hydrocarbon gases. (From Maxwell's Data Book on Hydrocarbons. Copyright 1950, D. Van Nostrand Company, Inc., Princeton, N.J.)

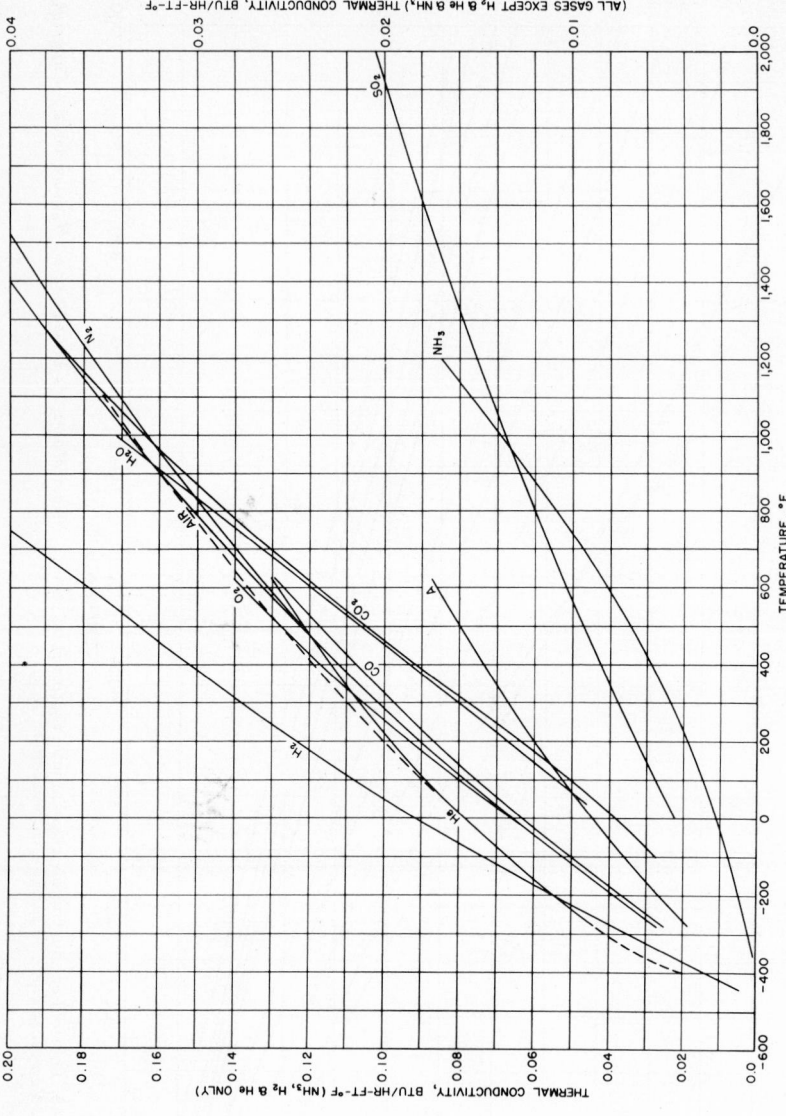

FIG. 12-46. Thermal conductivities of common gases at 1 atm pressure. Data sources: (1) NACA-NBS tables. (2) J. Hilsenrath and Y. S. Touloukian, *Trans. ASME*, vol. 76, p. 967, 1954. Data of Akin, *Trans. ASME*, vol. 72, p. 751, 1950, on helium, same source as (2).

12–68

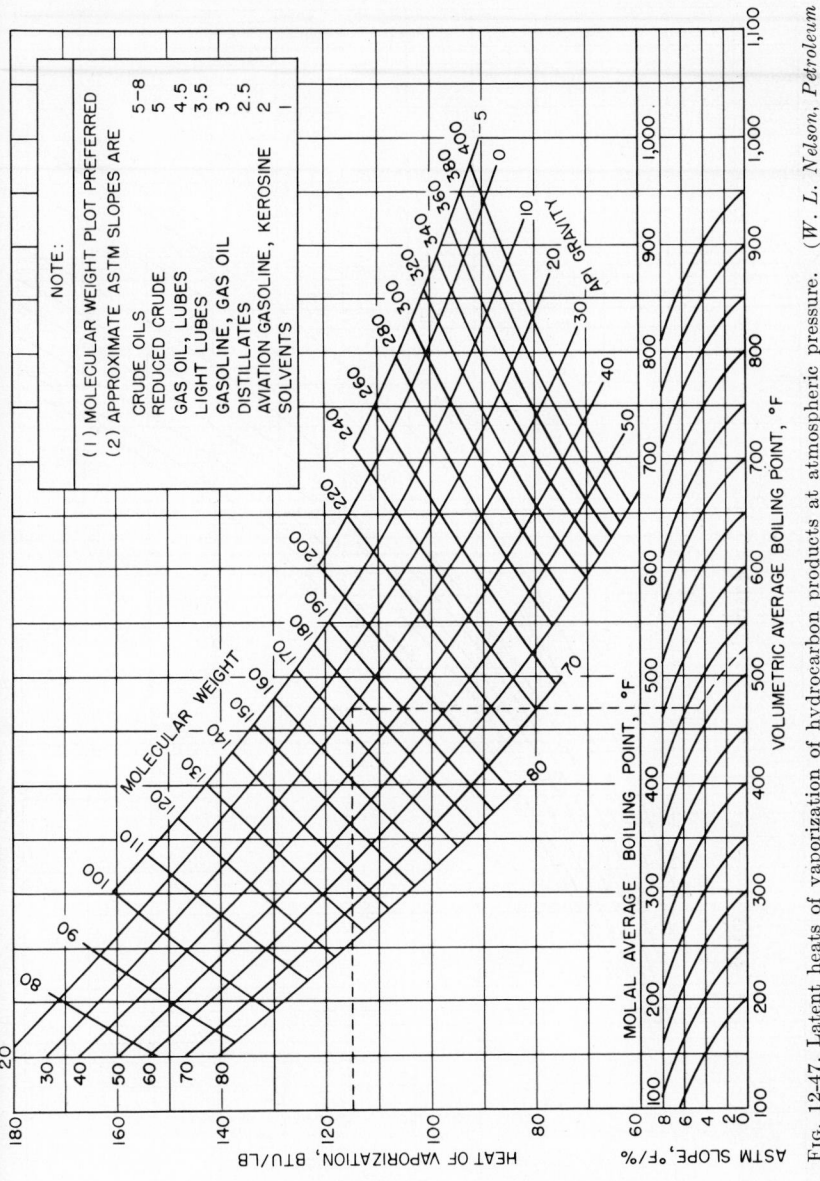

Fig. 12-47. Latent heats of vaporization of hydrocarbon products at atmospheric pressure. (*W. L. Nelson, Petroleum Refinery Engineering, McGraw-Hill Book Company, New York, 1949, 3d ed., p. 142.*)

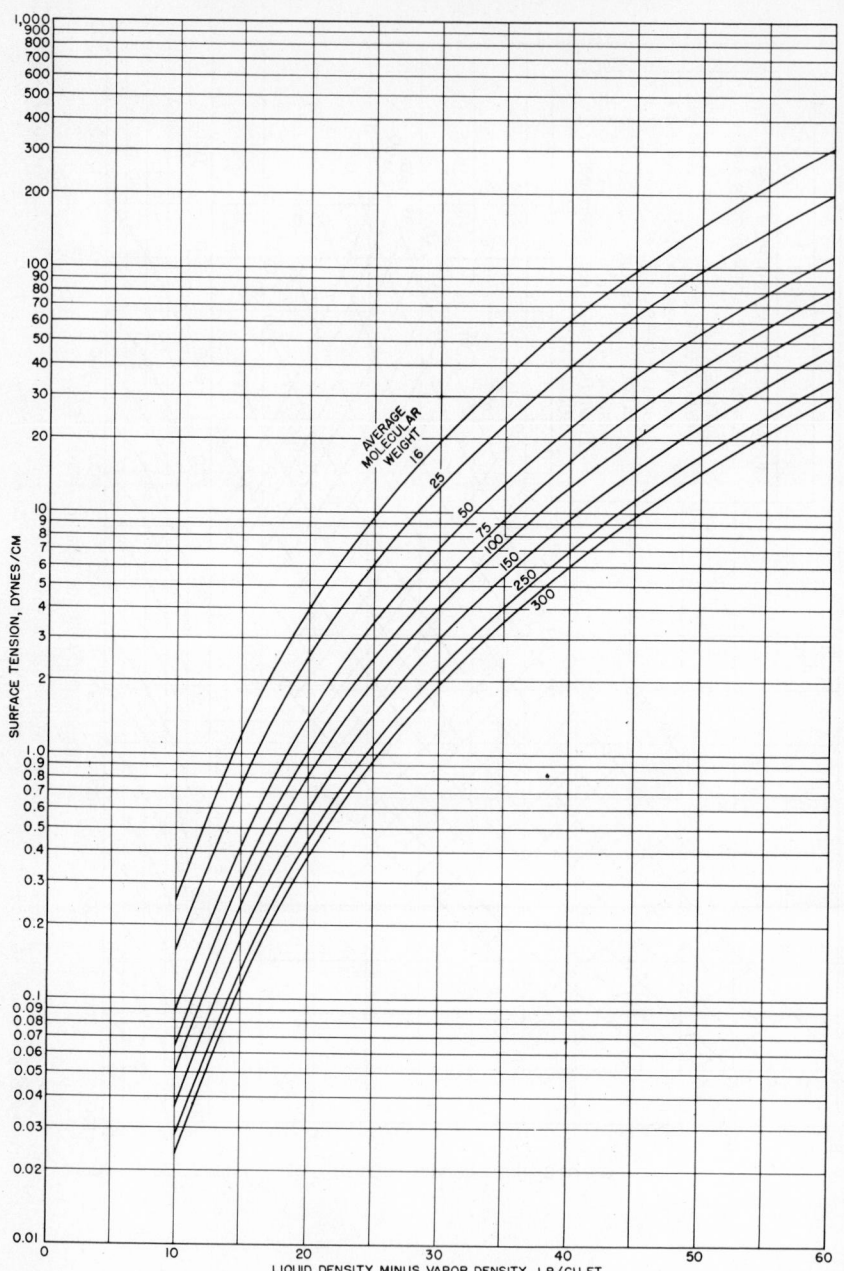

FIG. 12-48. Surface tensions of hydrocarbon mixtures vs. liquid-vapor density. (*Baker and Swerdloff, Oil & Gas J., p. 141, Dec. 5, 1955.*)

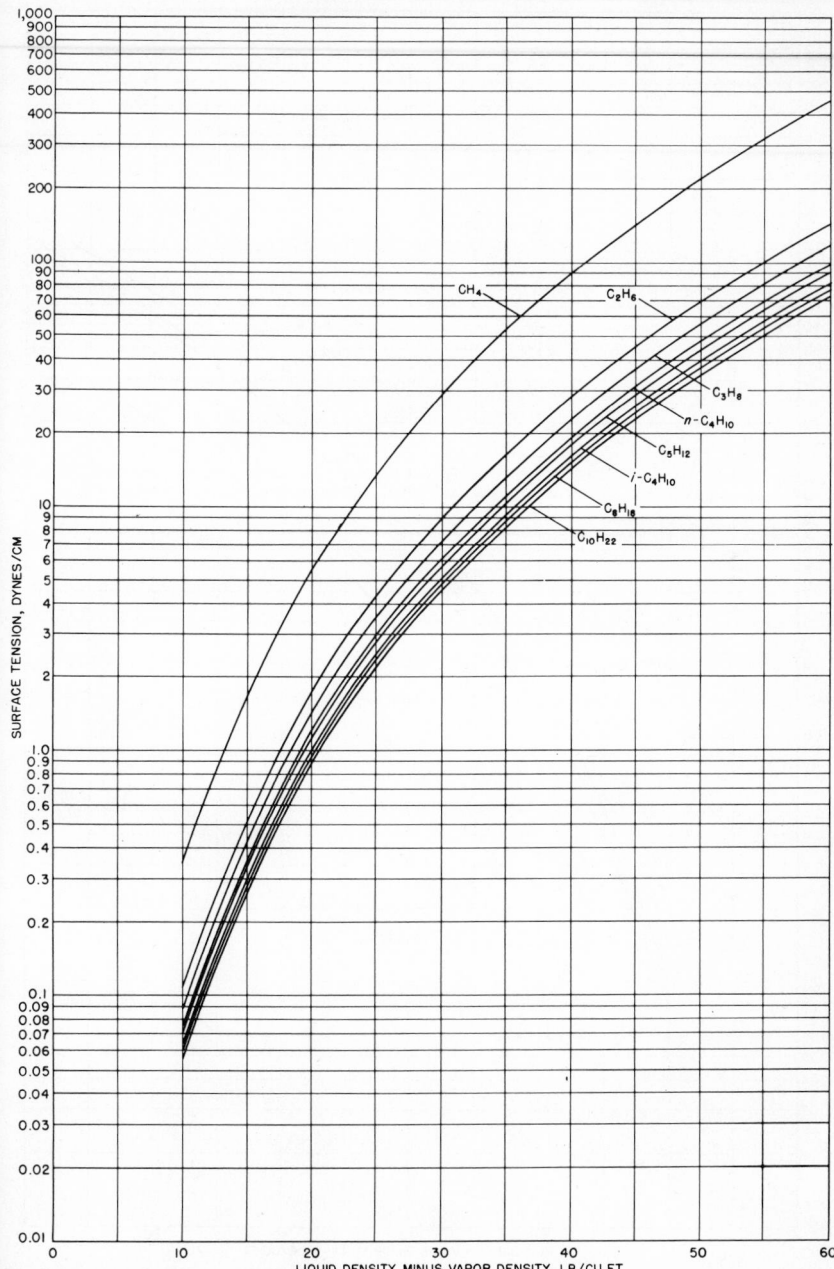

FIG. 12-49. Surface tensions of paraffins vs. liquid-vapor density. (*Baker and Swerdloff*, *Oil & Gas J., p. 141, Dec. 5, 1955.*)

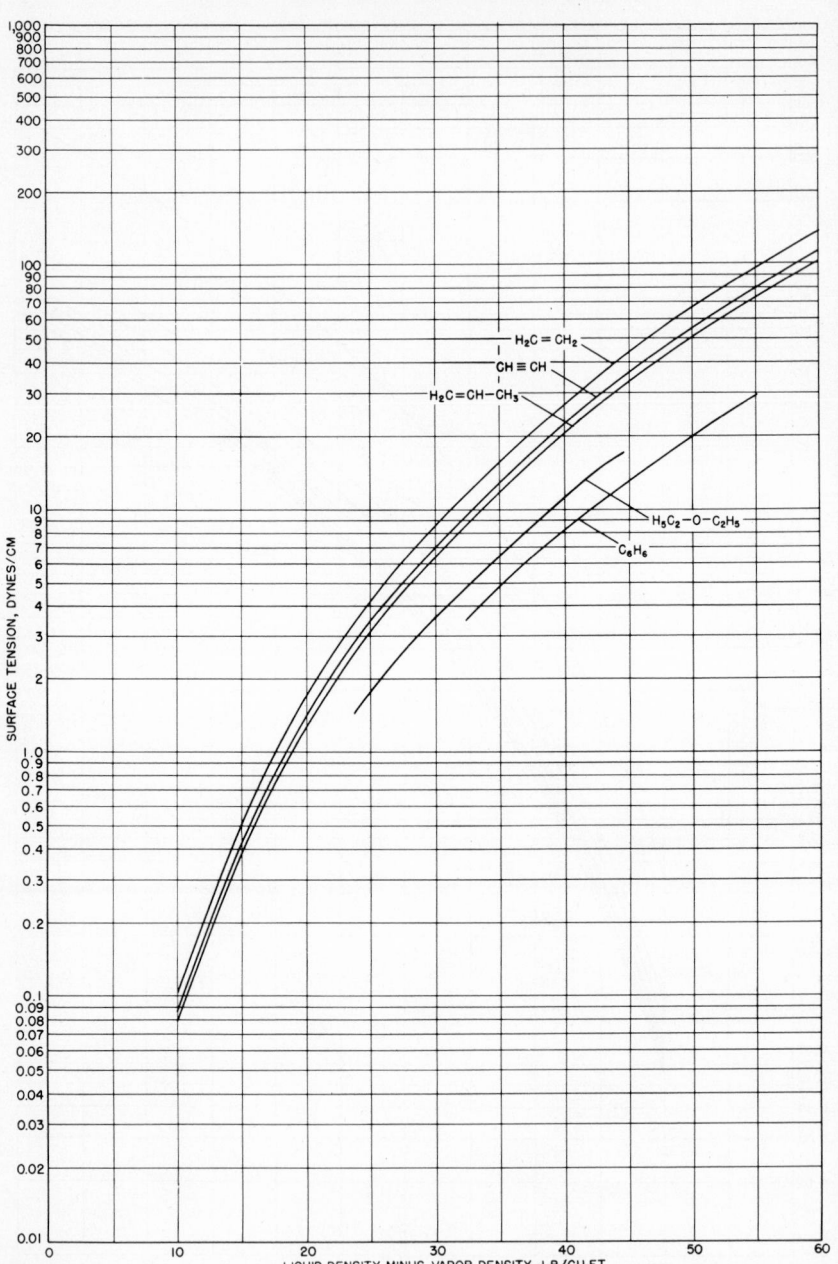

Fig. 12-50. Surface tensions of olefins and miscellaneous compounds vs. liquid-vapor density. (*Partington, An Advanced Treatise on Physical Chemistry, vol. 11, p. 142. Baker and Swerdloff, Oil & Gas J., p. 141, Dec. 5, 1955.*)

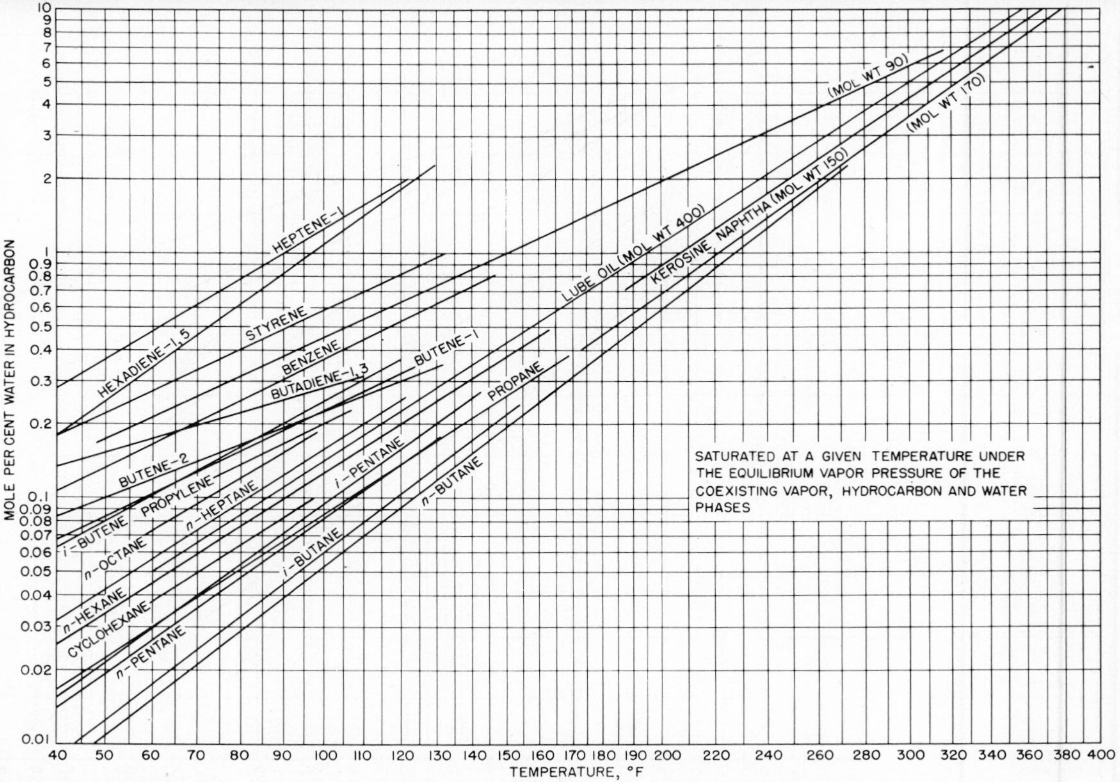
MOLE PER CENT WATER IN HYDROCARBON
TEMPERATURE, °F
SATURATED AT A GIVEN TEMPERATURE UNDER THE EQUILIBRIUM VAPOR PRESSURE OF THE COEXISTING VAPOR, HYDROCARBON AND WATER PHASES
(MOL WT 90)
(MOL WT 170)
KEROSINE NAPHTHA (MOL WT 150)
LUBE OIL (MOL WT 400)
HEPTENE-1
STYRENE
HEXADIENE-1,5
BENZENE
BUTADIENE-1,3
BUTENE-1
PROPANE
BUTENE-2
PROPYLENE
i-PENTANE
n-BUTANE
n-HEPTANE
i-BUTENE
n-OCTANE
i-BUTANE
n-HEXANE
CYCLOHEXANE
n-PENTANE

FIG. 12-51. Solubilities of water in hydrocarbons and petroleum fractions. [*W. F. Hoot et al., Pet. Ref., vol. 36, no. 5, p. 255, 1957. Gentry and Gunther, Oil & Gas J., Feb. 28, 1955, p. 131 (propylene). Natural Gasoline Supply Men's Association, p. 126, Apr. 15, 1951. (mol. wt. 90 petroleum).*]

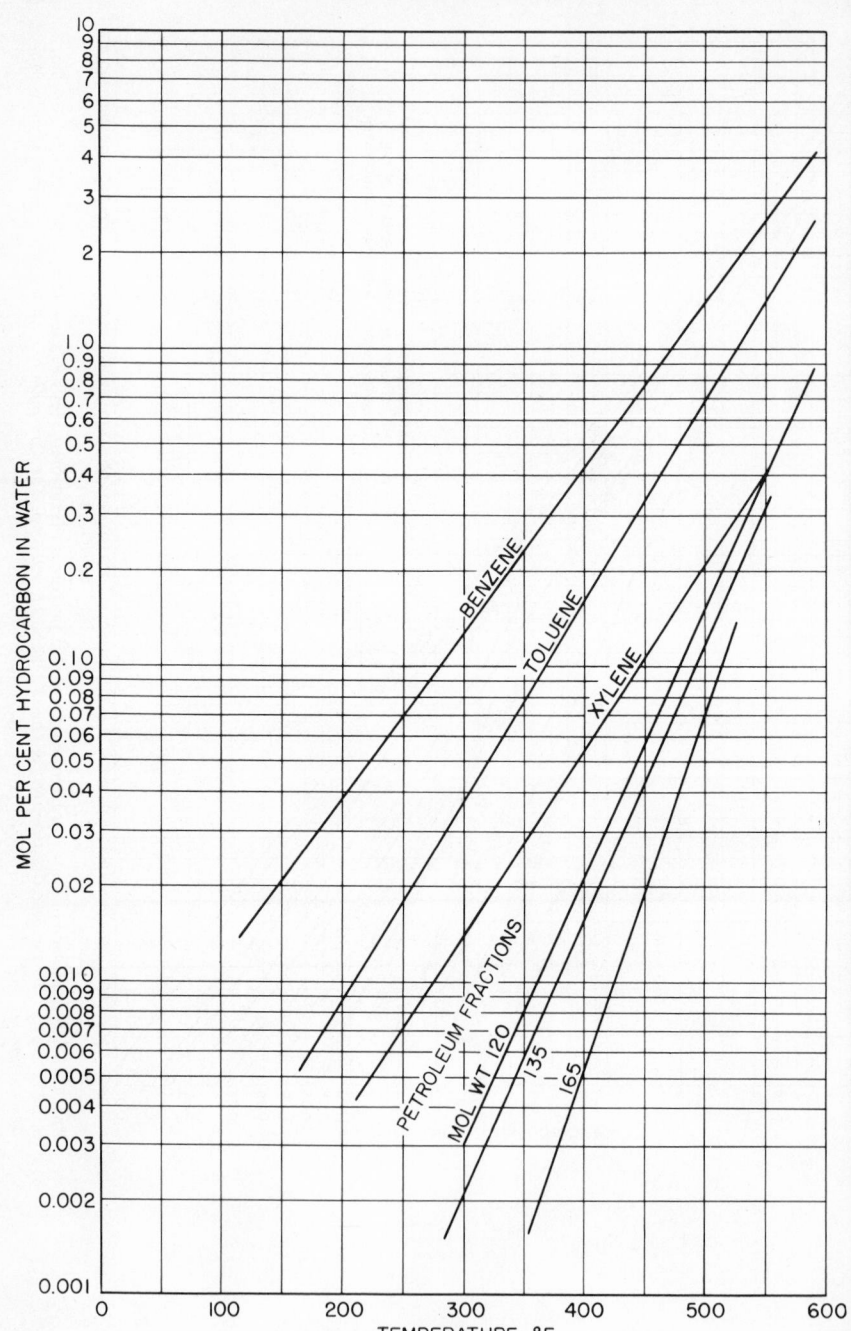

Fig. 12-52. Solubilities of hydrocarbons and petroleum fractions in water at total system pressure. (*Griswold, Ind. Engrg. Chem., vol. 34, p. 808, 1942.*)

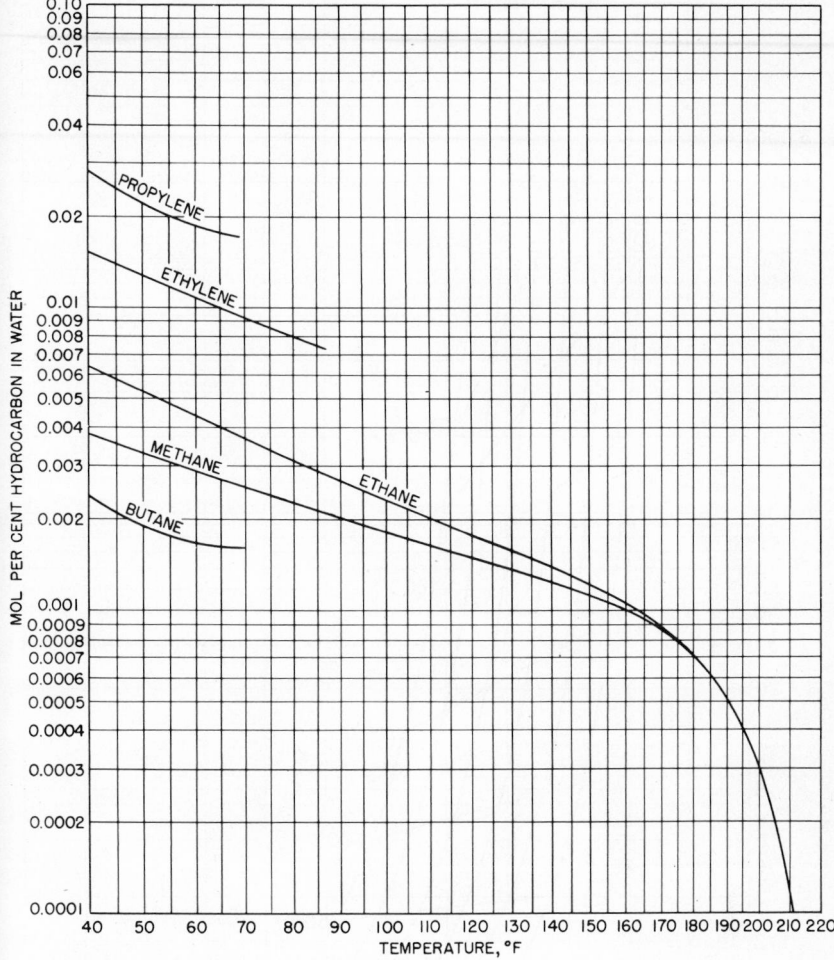

FIG. 12-53. Solubilities of hydrocarbons and petroleum fractions in water under total pressure of 1 atm. (*A. Seidell, Solubility of Inorganic and Organic Compounds, vol. 2, and supplement to 3d ed., D. Van Nostrand Company, Inc., Princeton, N.J., 1952.*)

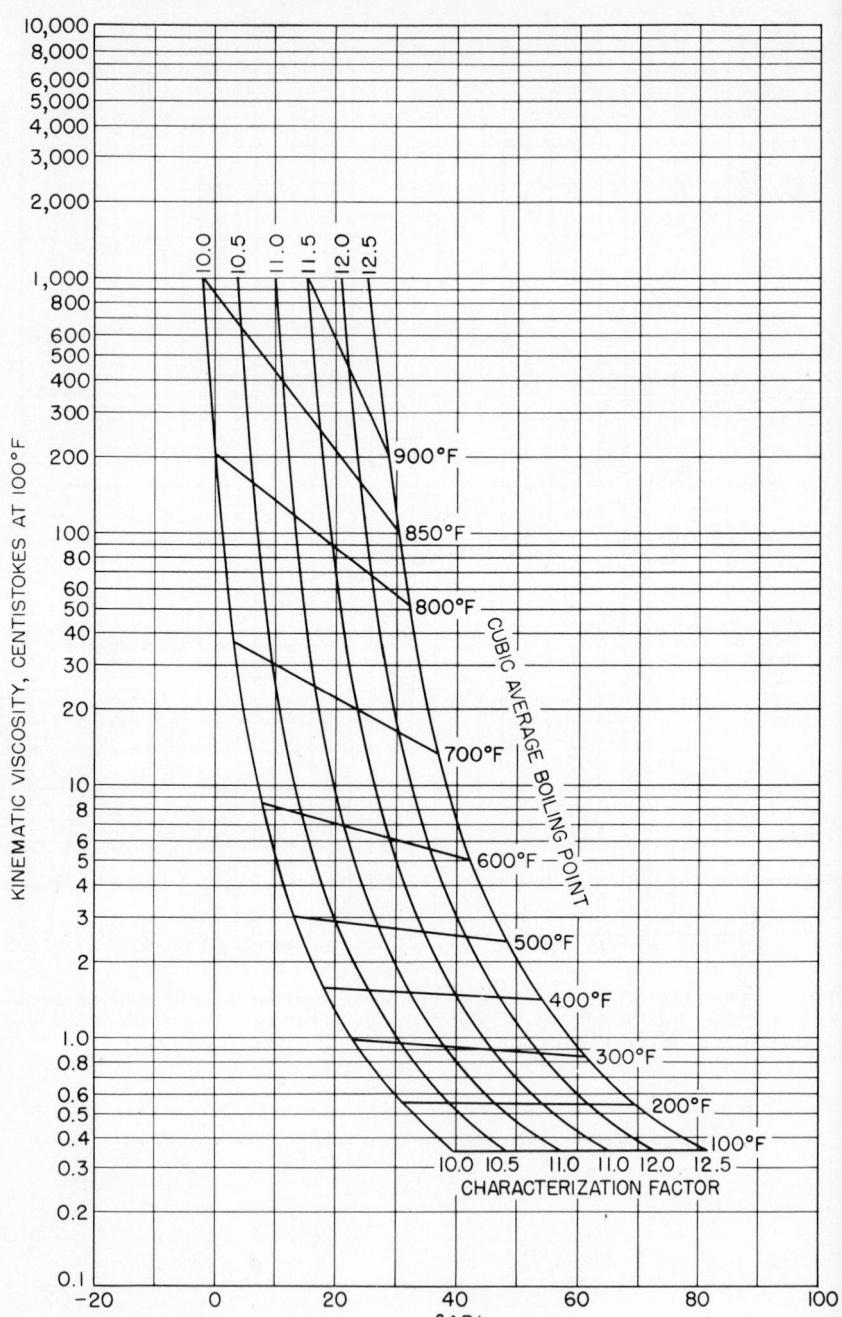

Fig. 12-54. Relation between viscosities at 100°F and characterization factors. (*Watson, Nelson and Murphy, Ind. Engrg. Chem., vol. 27, p. 1470, 1935.*)

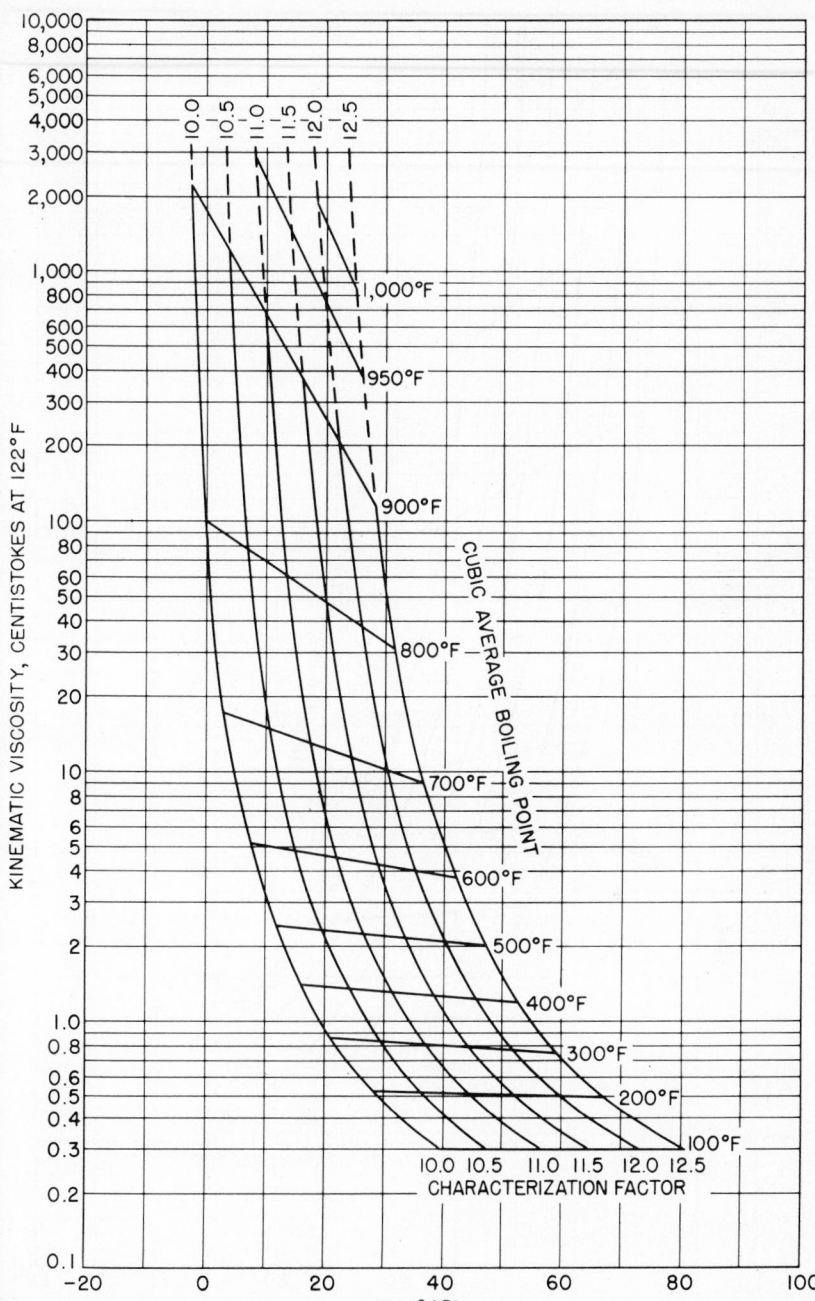

FIG. 12-55. Relation between viscosities at 122°F and characterization factors. (*Chemical Process Principle Charts, 1st ed., O. A. Hougen and K. M. Watson, John Wiley & Sons, Inc., New York, 1946.*)

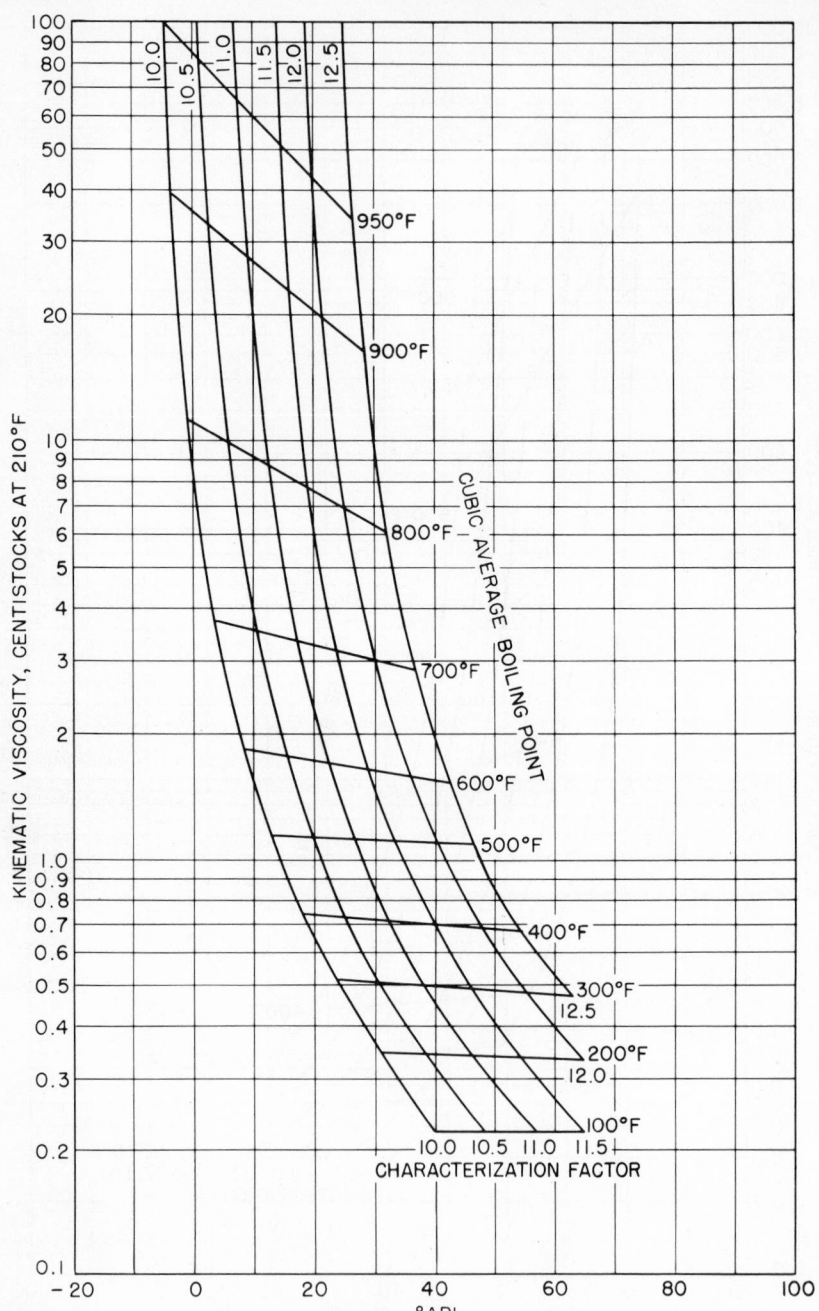

Fig. 12-56. Relation between viscosities at 210°F and characterization factors. (*Oil & Gas Journal.*)

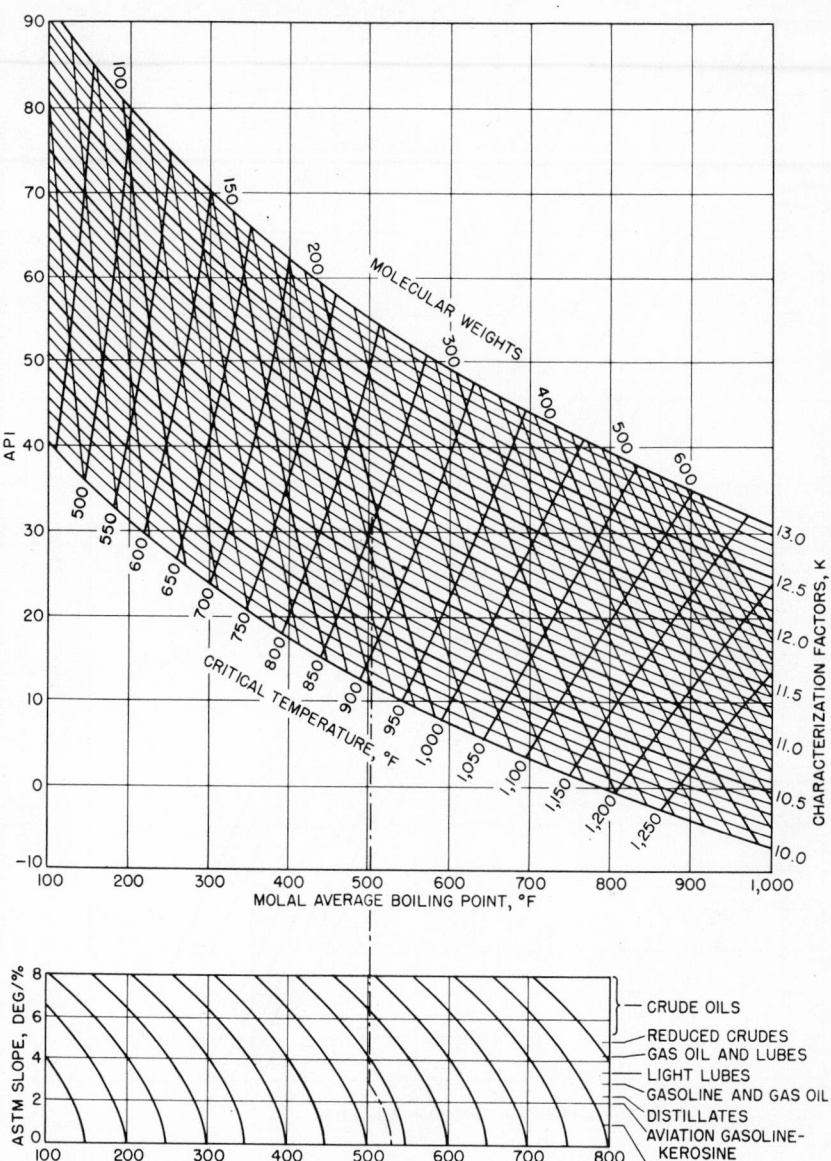

Fig. 12-57. Relation between molecular weights, pseudo-critical temperatures, characterization factors, and gravities of petroleum fractions. (*Chemical Process Principles, vol. 1, O. A. Hougen and K. M. Watson, John Wiley & Sons, Inc., New York, 1947.*)

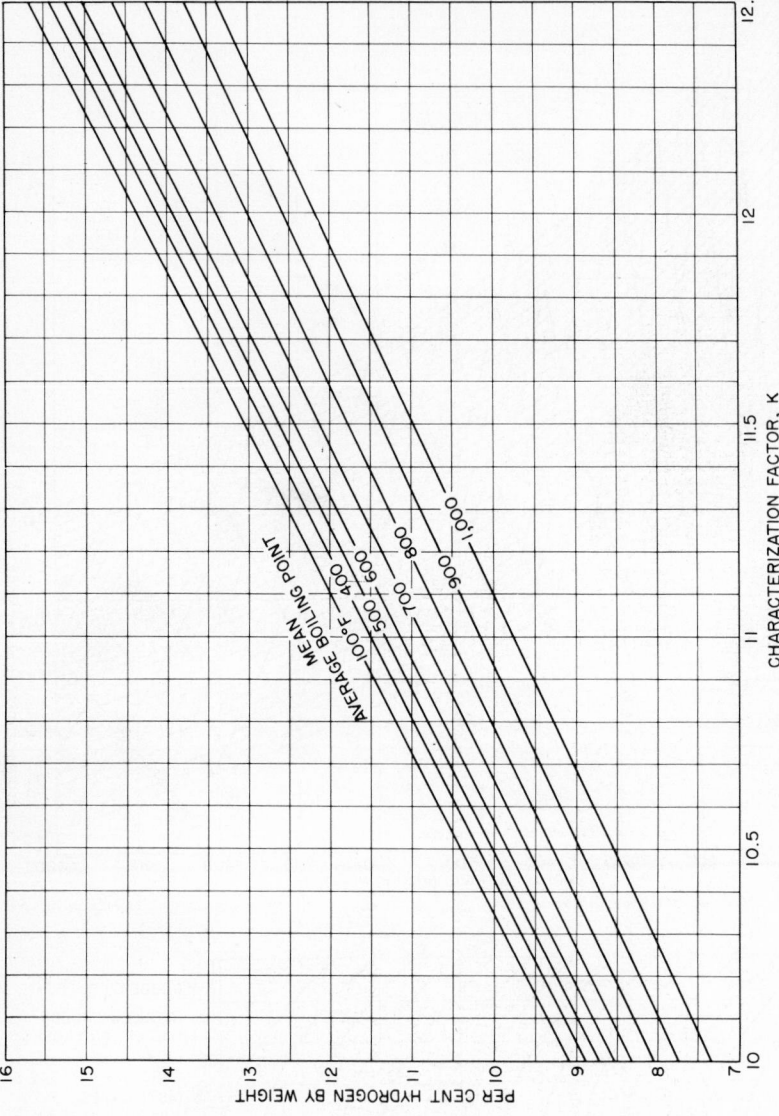

FIG. 12-58. Hydrogen contents of petroleum fractions from characterization factors. Note: Use with caution on highly aromatic low-boiling materials. For California and Mexican stocks specific data should be obtained. (Chemi-

PER CENT HYDROGEN BY WEIGHT

CHARACTERIZATION FACTOR, K

MEAN BOILING POINT
AVERAGE BOILING POINT—100°F 400
500 600
700
800
900
1,000

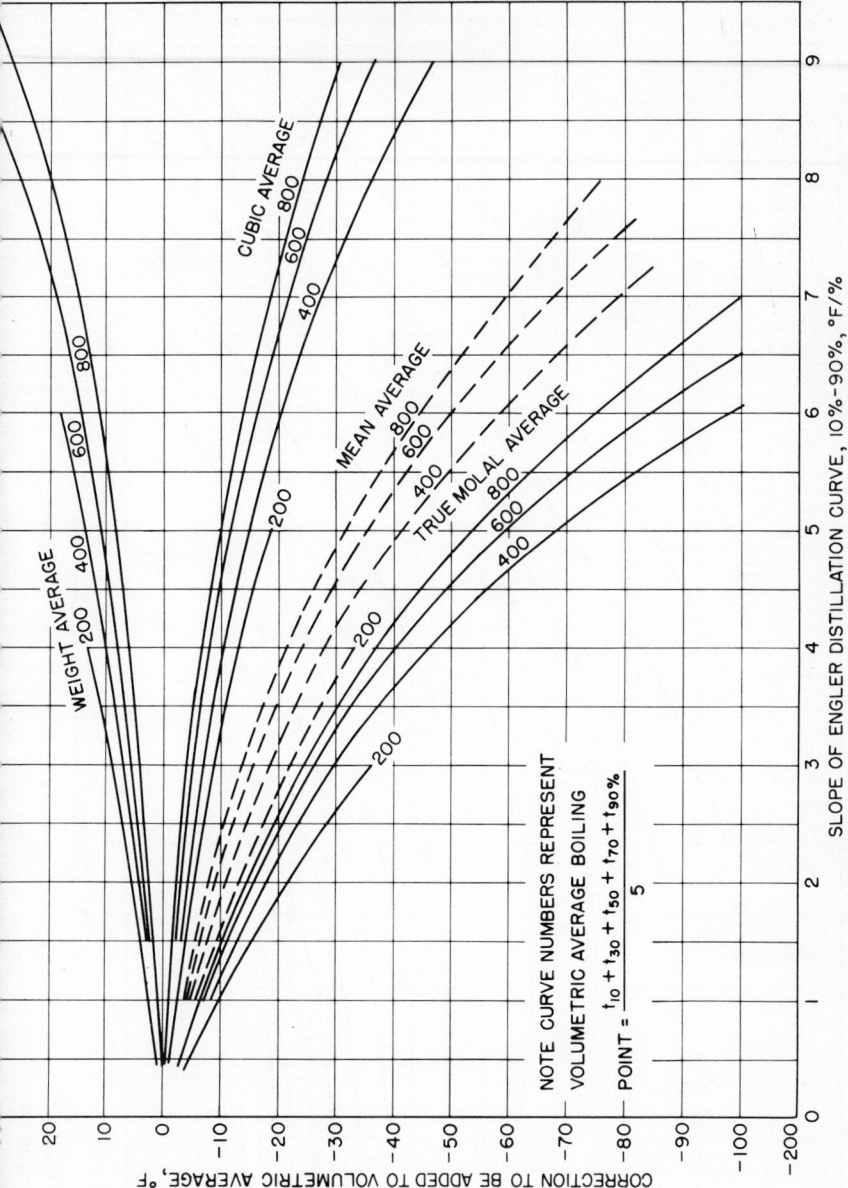

FIG. 12-59. Average boiling-point corrections. (*Ind. Engrg. Chem.*)

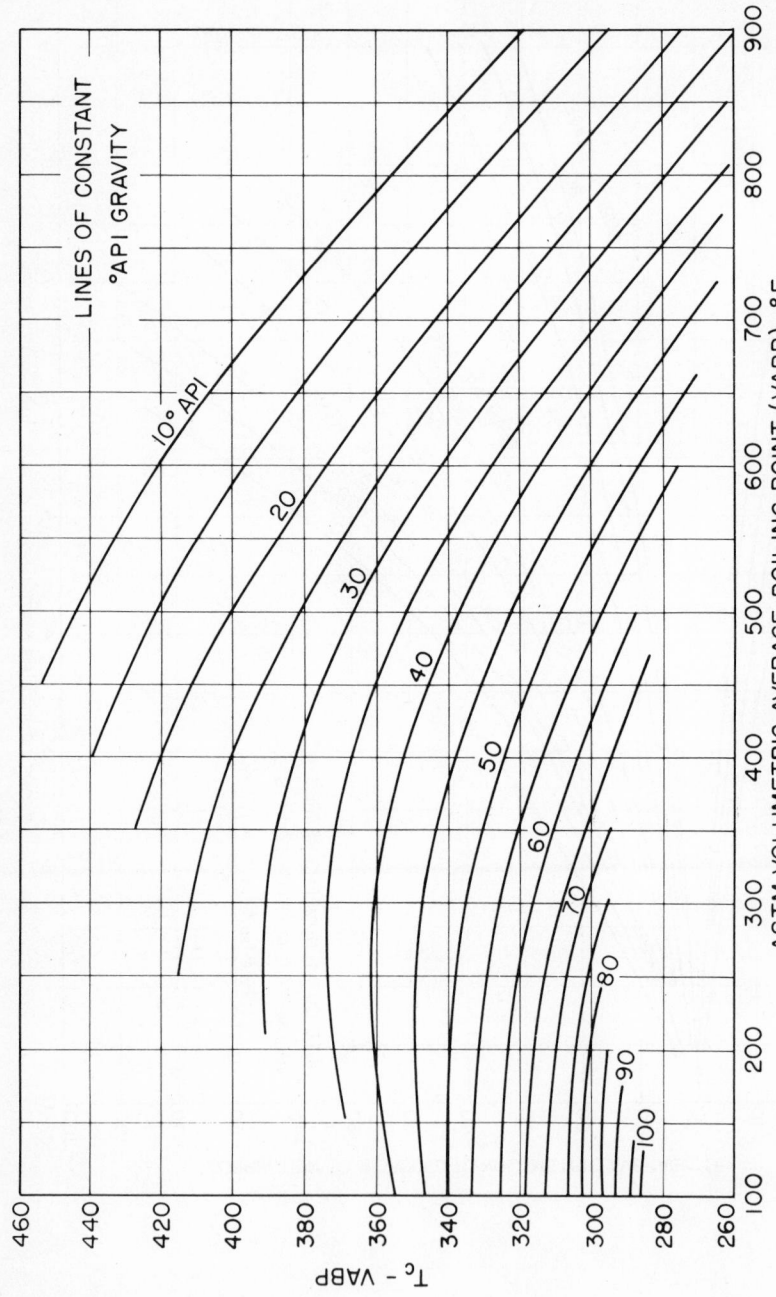

FIG. 12-60. True critical temperatures of petroleum fractions. (*Applied Hydrocarbon Thermodynamics, Wayne C. Edmister, Gulf Publishing Co., Houston, 1961.*)

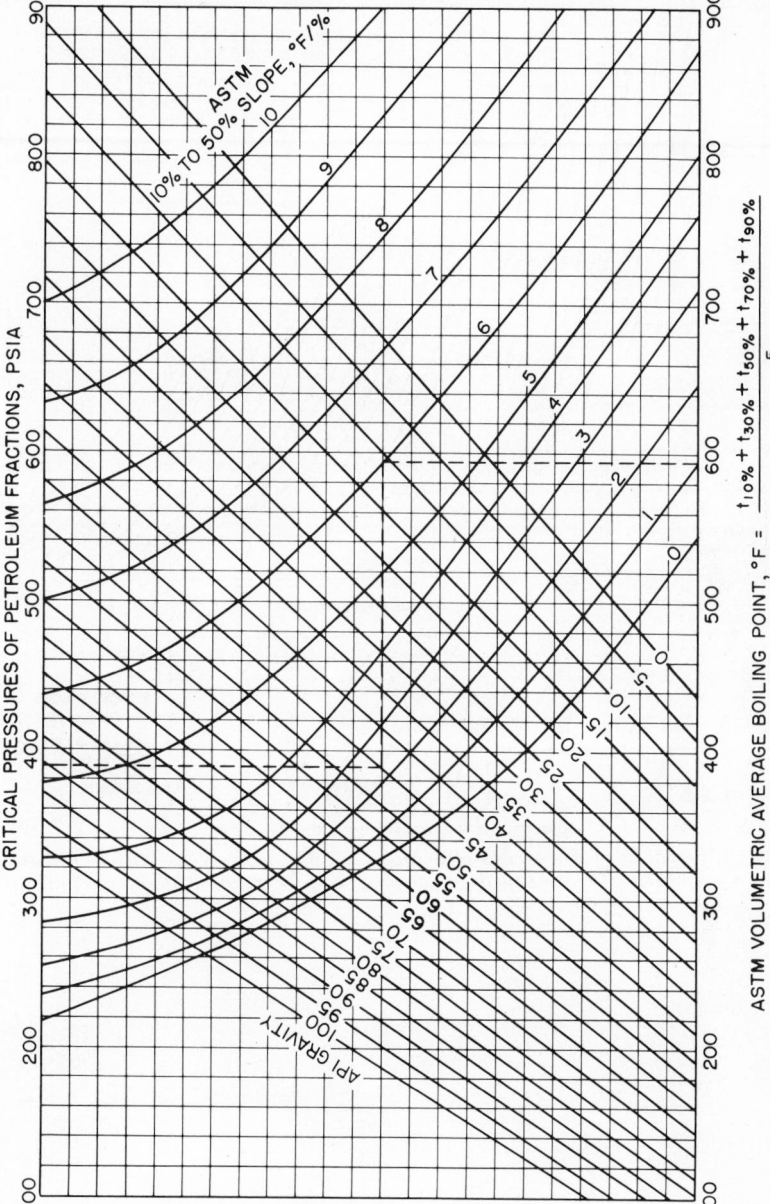

CRITICAL PRESSURES OF PETROLEUM FRACTIONS, PSIA

ASTM VOLUMETRIC AVERAGE BOILING POINT, °F = $\dfrac{t_{10\%} + t_{30\%} + t_{50\%} + t_{70\%} + t_{90\%}}{5}$

FIG. 12-61. True critical pressures of petroleum fractions. (*Applied Hydrocarbon Thermodynamics, Wayne C. Edmister, Gulf Publishing Co., Houston, 1961.*)

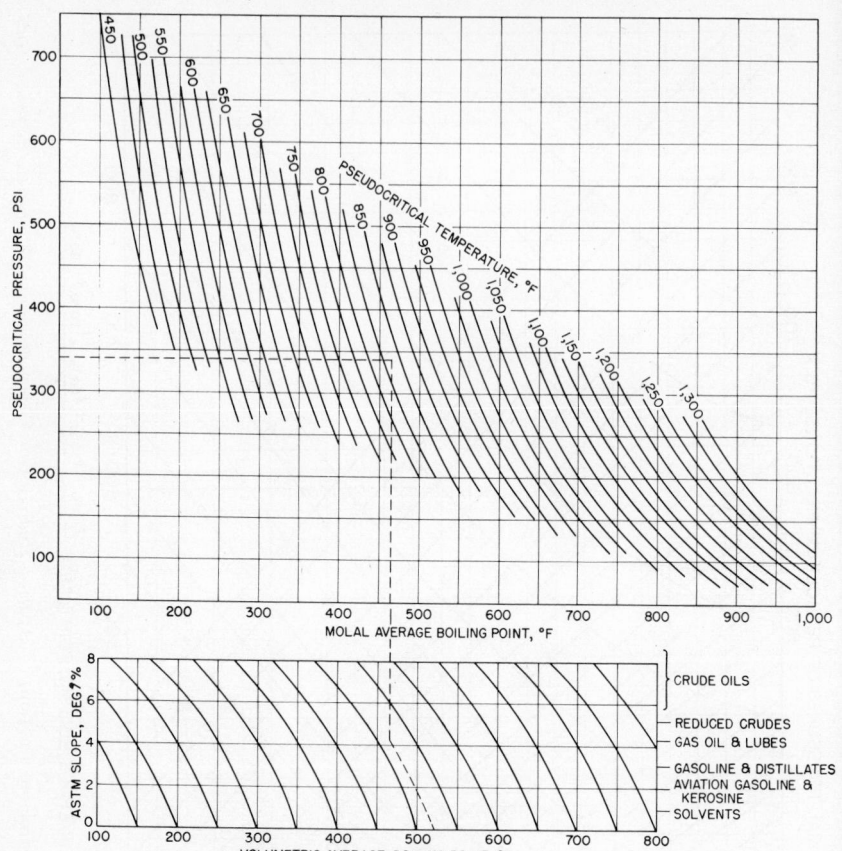

FIG. 12-62. Pseudo-critical pressures of hydrocarbon mixtures. (*W. L. Nelson, Petroleum Refinery Engineering, McGraw-Hill Book Company, New York, 1958, 4th ed., p. 178.*)

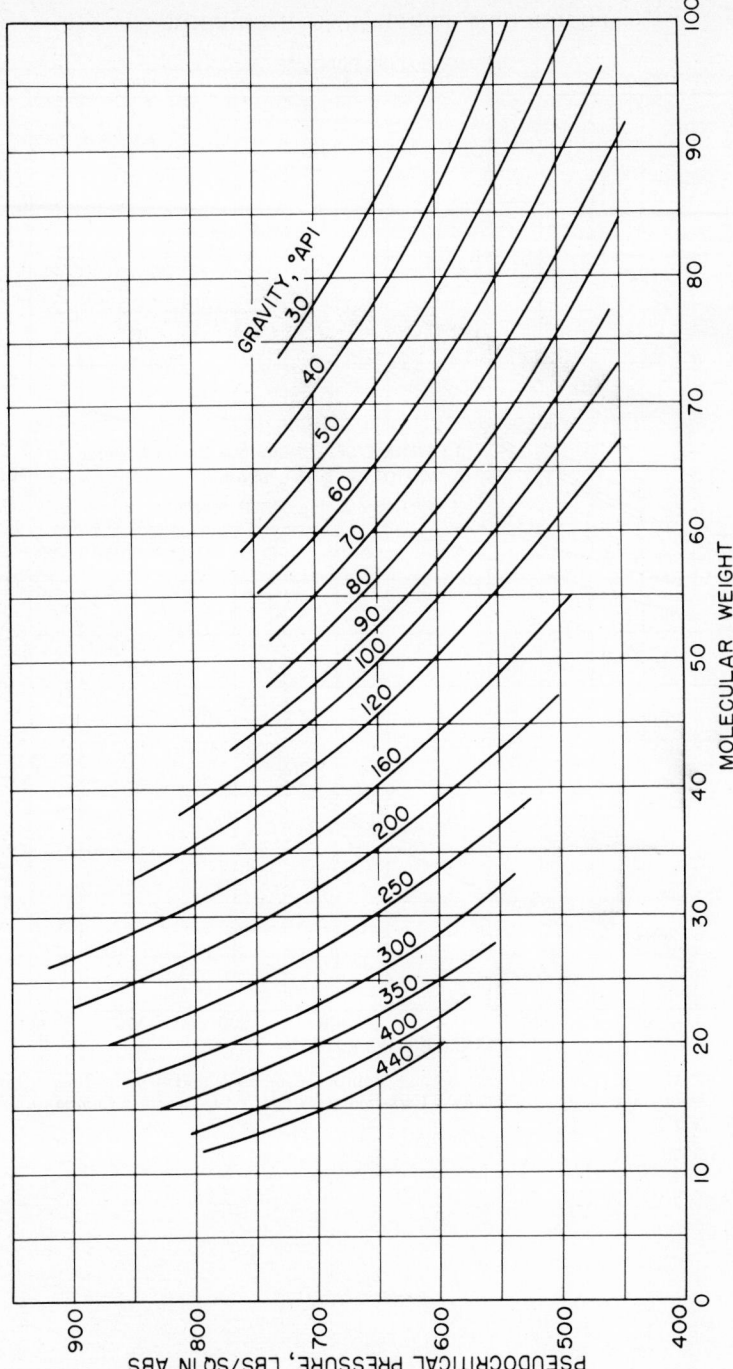

Fig. 12-63. Pseudo-critical pressures of hydrocarbon gases. (*UOP Booklet 222, Ind. Engrg. Chem., vol. 29, p. 1408, 1937.*)

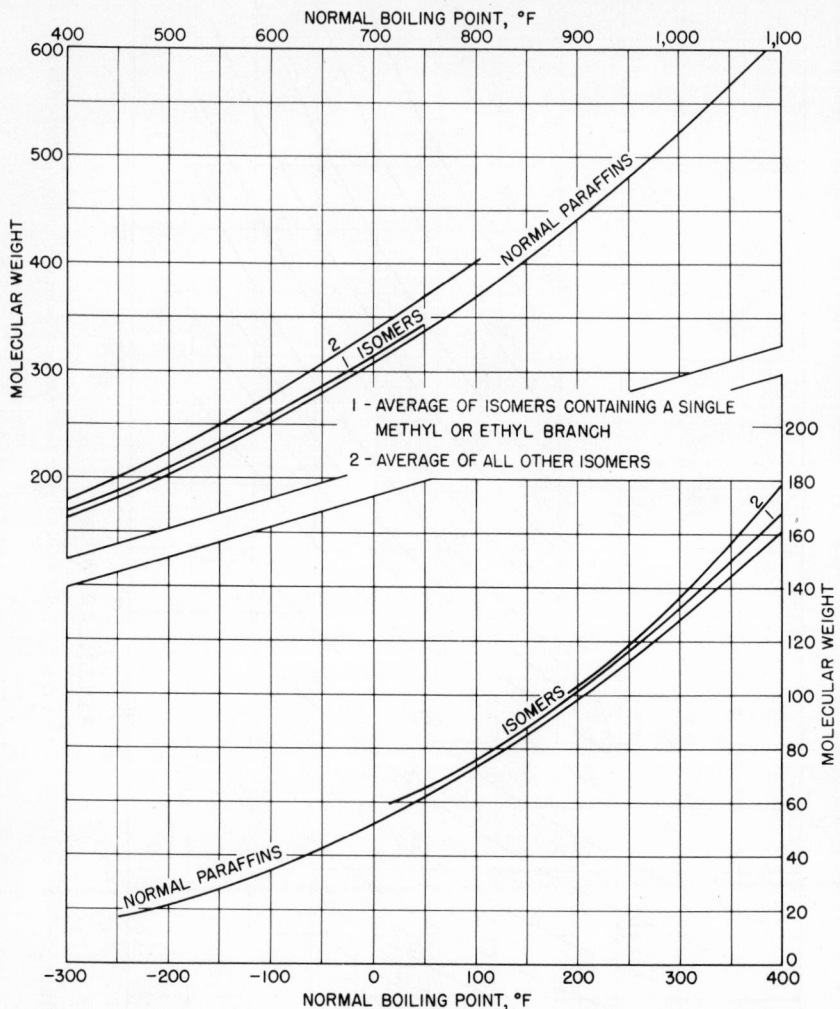

FIG. 12-64. Molecular weights vs. boiling points of normal and isoparaffins. (*From Maxwell's Data Book on Hydrocarbons. Copyright 1950, D. Van Nostrand Company, Inc., Princeton, N.J.*)

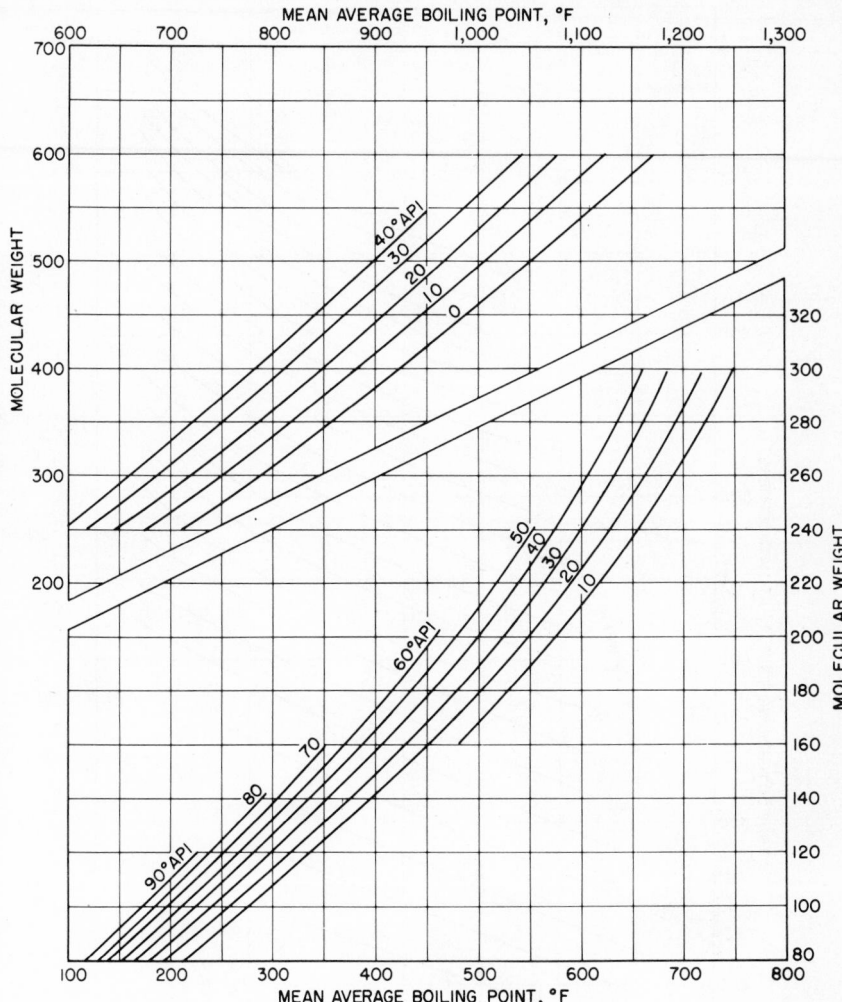

FIG. 12-65. Molecular weights vs. boiling points of petroleum fractions. (*From Maxwell's Data Book on Hydrocarbons. Copyright 1950, D. Van Nostrand Company, Inc., Princeton, N.J.*)

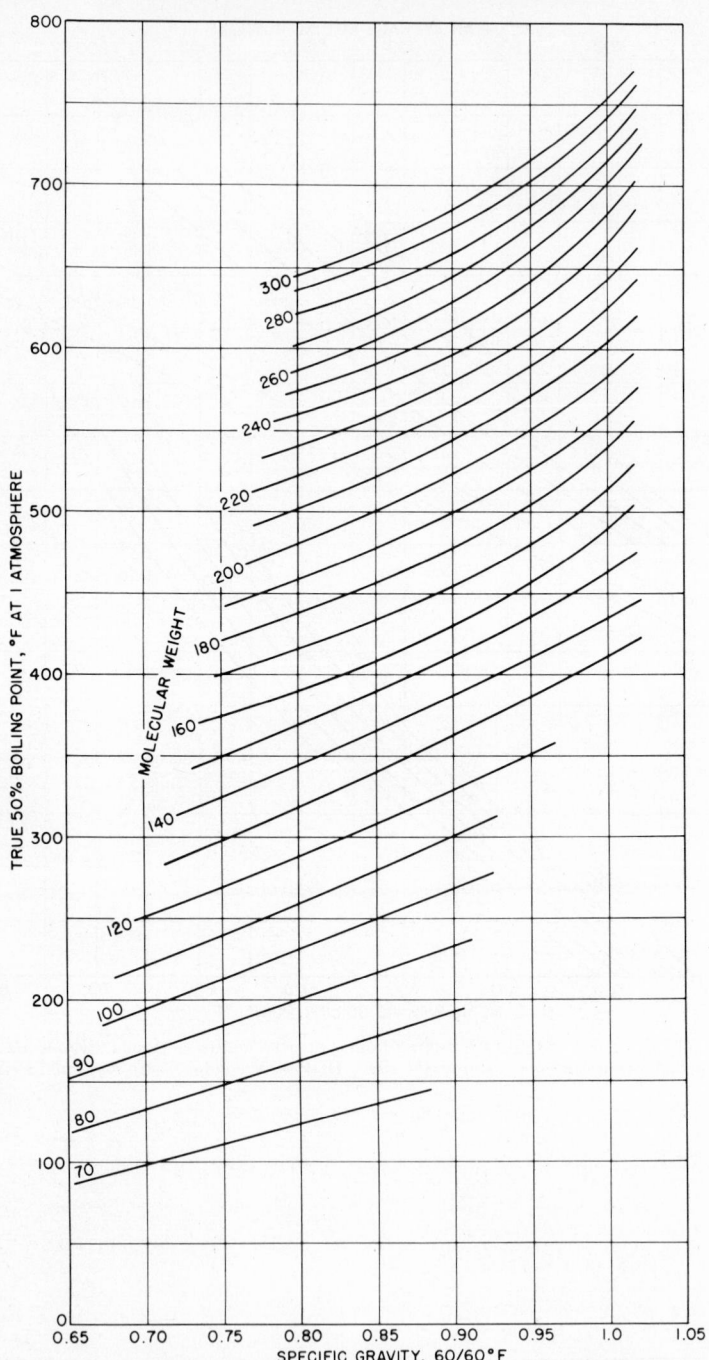

Fig. 12-66. Molecular weights on true 50 per cent boiling point vs. gravities of petroleum fractions. (*Mills et al., Ind. Engrg. Chem., vol. 38, p. 443, 1946.*)

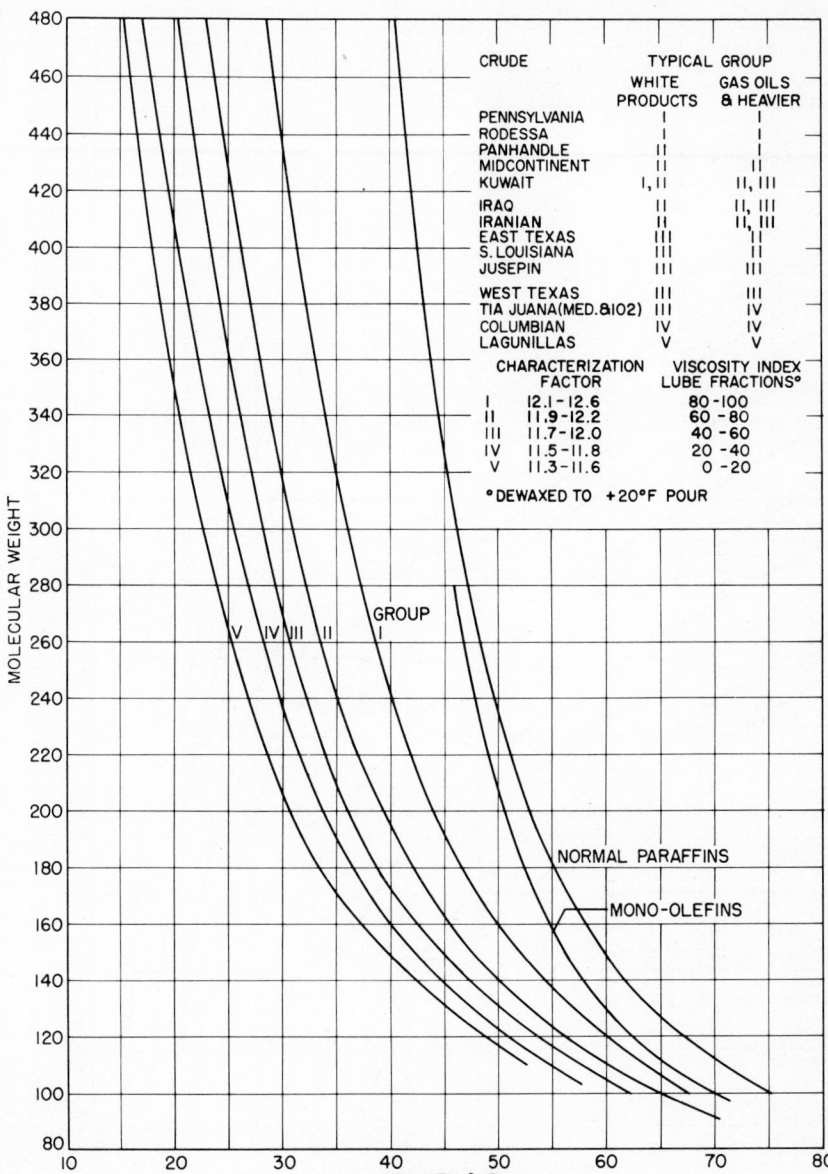

CRUDE	WHITE PRODUCTS	TYPICAL GROUP GAS OILS & HEAVIER
PENNSYLVANIA	I	I
RODESSA	I	I
PANHANDLE	II	I
MIDCONTINENT	II	II
KUWAIT	I, II	II, III
IRAQ	II	II, III
IRANIAN	II	II, III
EAST TEXAS	III	II
S. LOUISIANA	III	II
JUSEPIN	III	III
WEST TEXAS	III	III
TIA JUANA(MED.&I02)	III	IV
COLUMBIAN	IV	IV
LAGUNILLAS	V	V

	CHARACTERIZATION FACTOR	VISCOSITY INDEX LUBE FRACTIONS°
I	12.1 – 12.6	80 –100
II	11.9 –12.2	60 – 80
III	11.7 –12.0	40 – 60
IV	11.5 –11.8	20 – 40
V	11.3 –11.6	0 – 20

°DEWAXED TO +20°F POUR

FIG. 12-67. Approximate molecular weights vs. API gravities of typical crude fractions. (*From Maxwell's Data Book on Hydrocarbons. Copyright 1950, D. Van Nostrand Company, Inc., Princeton, N.J.*)

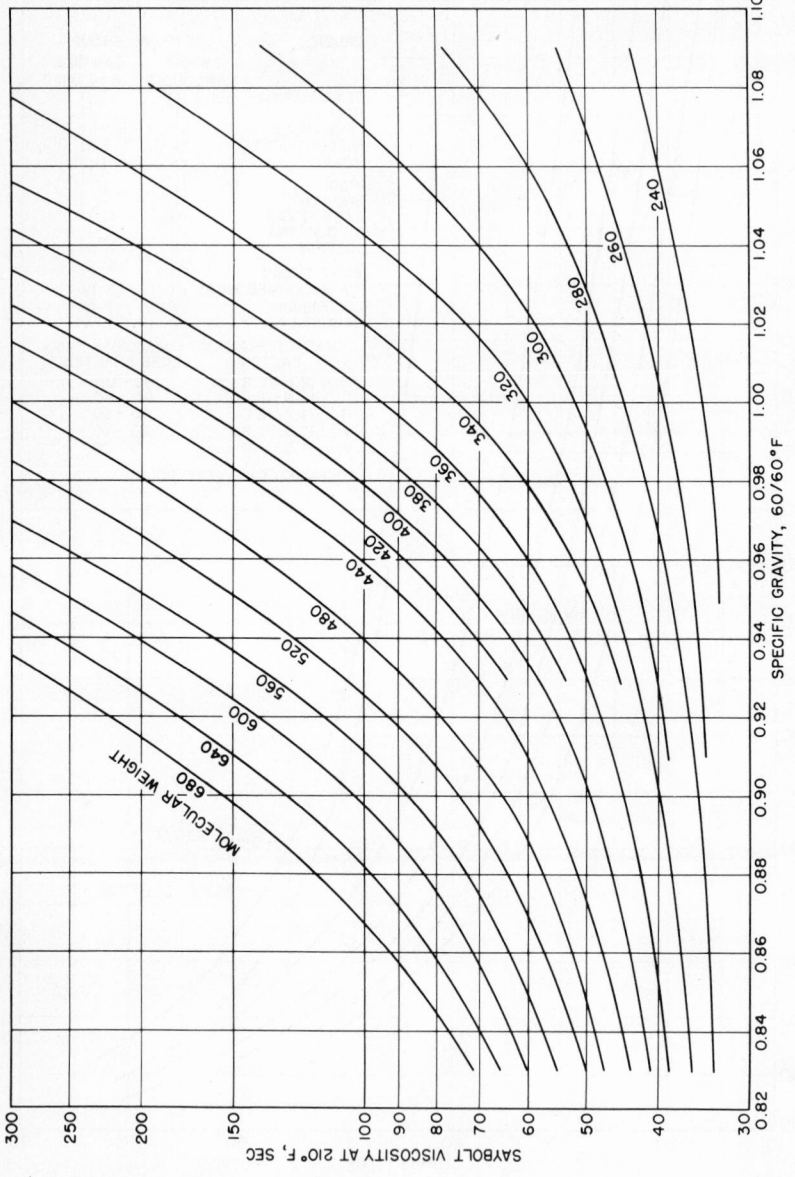

FIG. 12-68. Molecular weights of lube oils vs. Saybolt Universal viscosities at 210°F and specific gravities. (*Mills et al. Ind. Engrg. Chem., vol. 38, p. 442, 1946.*)

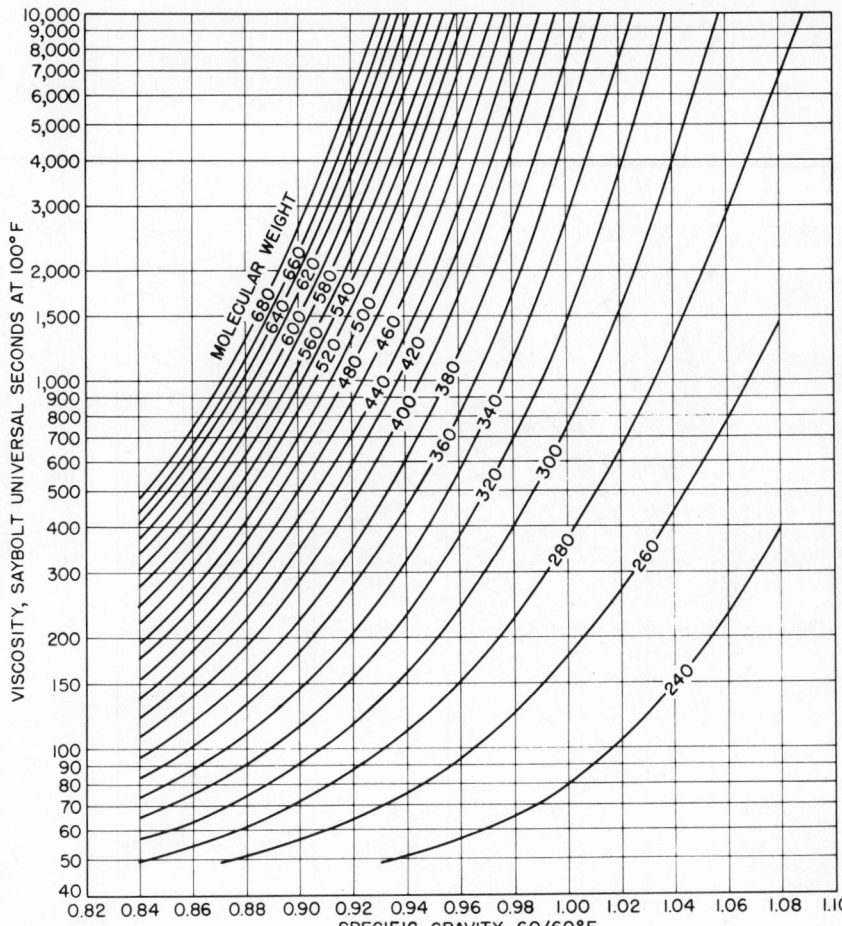

Fig. 12-69. Molecular weights of lube oils vs. Saybolt Universal viscosities at 100°F and specific gravities. (*Mills et al., Ind. Engrg. Chem., vol. 38, p. 442, 1946.*)

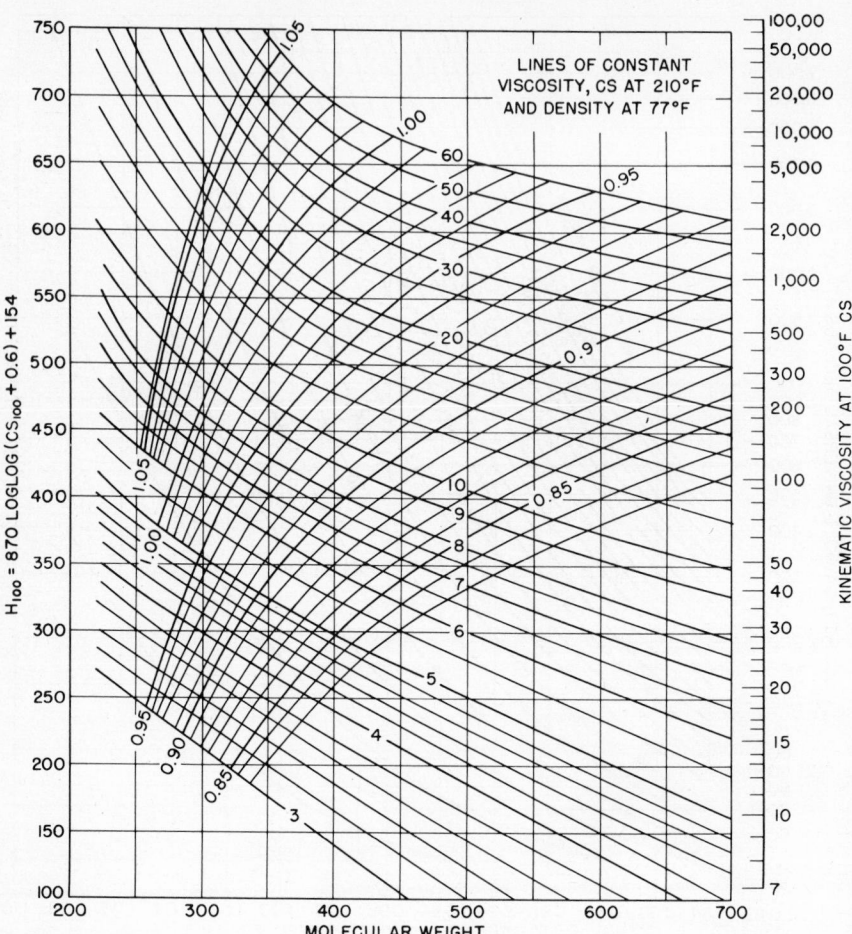

FIG. 12-70. Molecular weights of lube oils vs. kinematic viscosity at 100°F, density, and H_{100} blending-factor correlation. (*Hirschler, J. Inst. Pet., vol. 32, p. 148, 1946.*)

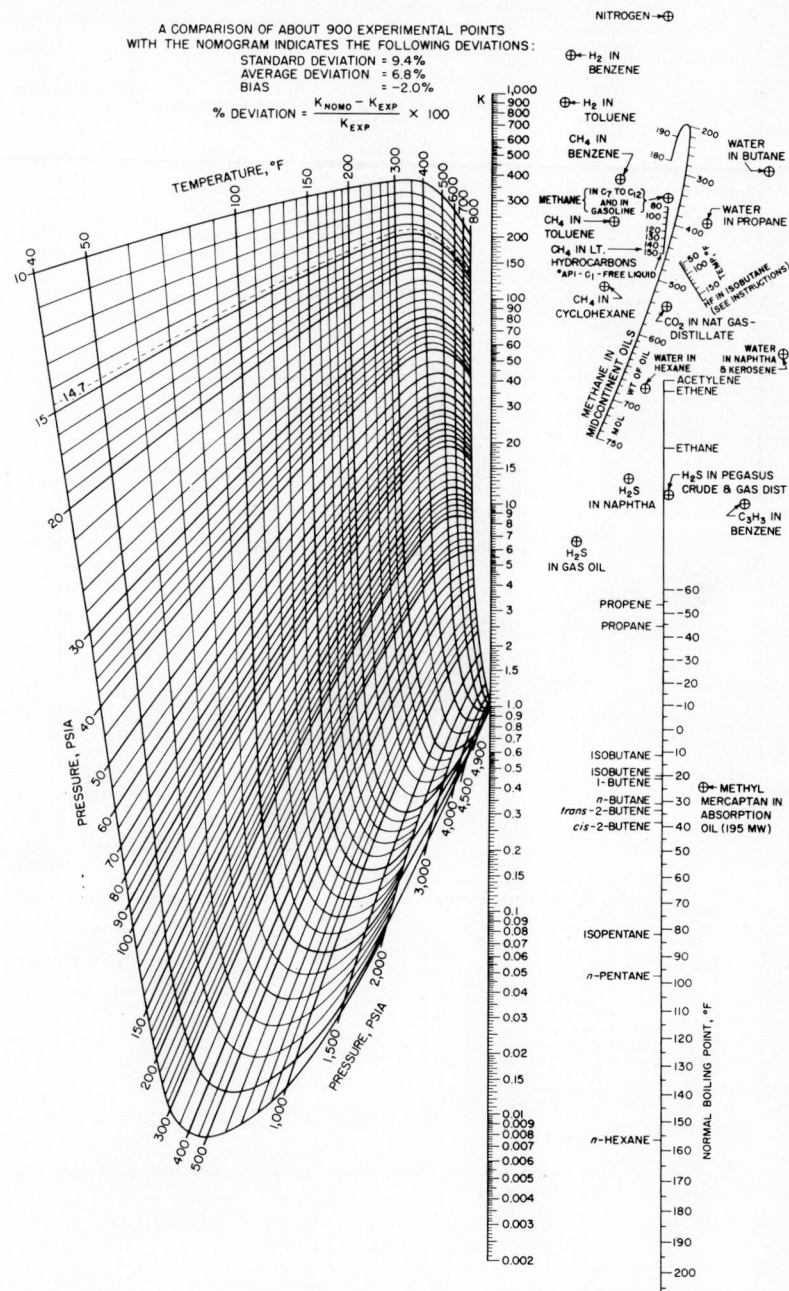

Fig. 12-71. Vapor-liquid equilibria for hydrocarbons and gases, high-temperature (40 to 800°F). (*Courtesy of Mobil Oil Co.*)

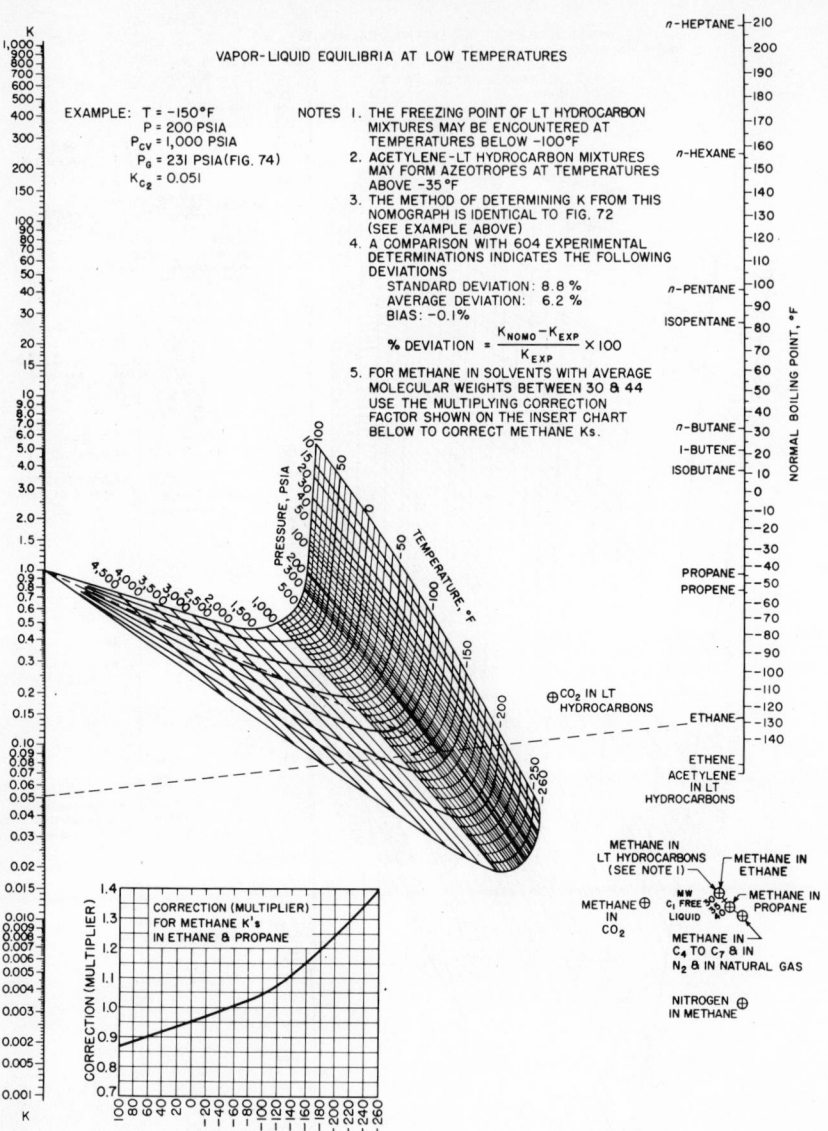

VAPOR-LIQUID EQUILIBRIA AT LOW TEMPERATURES

EXAMPLE: T = –150°F
          P = 200 PSIA
          P_{cv} = 1,000 PSIA
          P_G = 231 PSIA (FIG. 74)
          K_{C_2} = 0.051

NOTES 1. THE FREEZING POINT OF LT HYDROCARBON
         MIXTURES MAY BE ENCOUNTERED AT
         TEMPERATURES BELOW –100°F
      2. ACETYLENE-LT HYDROCARBON MIXTURES
         MAY FORM AZEOTROPES AT TEMPERATURES
         ABOVE –35°F
      3. THE METHOD OF DETERMINING K FROM THIS
         NOMOGRAPH IS IDENTICAL TO FIG. 72
         (SEE EXAMPLE ABOVE)
      4. A COMPARISON WITH 604 EXPERIMENTAL
         DETERMINATIONS INDICATES THE FOLLOWING
         DEVIATIONS
                 STANDARD DEVIATION: 8.8 %
                 AVERAGE DEVIATION: 6.2 %
                 BIAS: –0.1%

         % DEVIATION = $\dfrac{K_{NOMO}-K_{EXP}}{K_{EXP}} \times 100$

      5. FOR METHANE IN SOLVENTS WITH AVERAGE
         MOLECULAR WEIGHTS BETWEEN 30 & 44
         USE THE MULTIPLYING CORRECTION
         FACTOR SHOWN ON THE INSERT CHART
         BELOW TO CORRECT METHANE Ks.

CORRECTION (MULTIPLIER)
FOR METHANE K's
IN ETHANE & PROPANE

Fig. 12-72. Vapor-liquid equilibria for hydrocarbons and gases, low-temperature (–260 to 100°F). (*Courtesy of Mobil Oil Co.*)

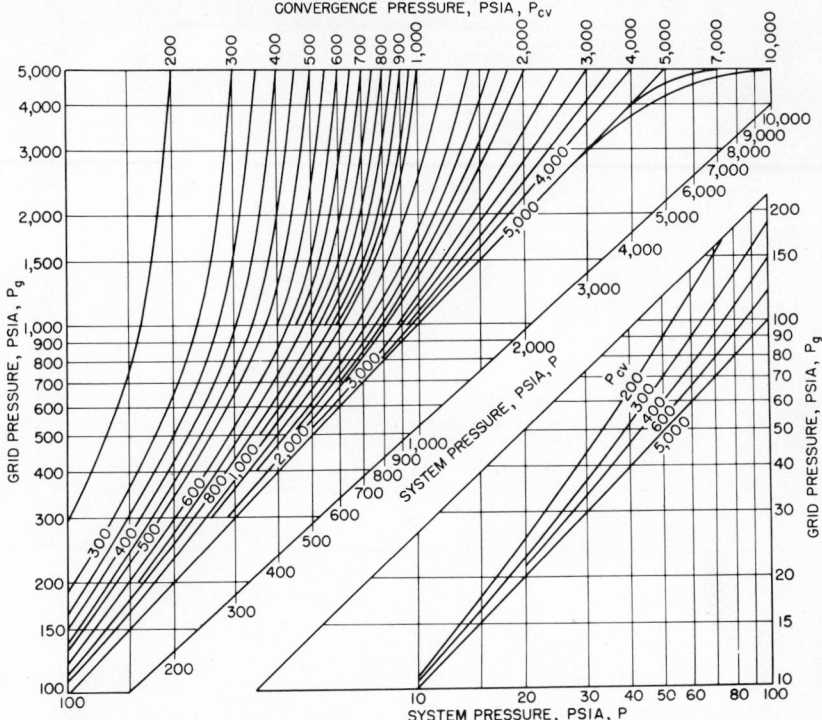

Fɪɢ. 12-73. Grid pressures vs. system pressures.   (*Courtesy of Mobil Oil Co.*)

Section 13

SOURCES OF INFORMATION

By MORRIS D. SCHOENGOLD

Technical Information Division
Esso Research and Engineering Company

The printed literature related to any field of endeavor serves a twofold purpose—that of reporting what is currently new in the field and that of recording what has previously been accomplished. In other words, the purposes are "current awareness" and "retrospective searching."

The basic tools for current awareness are the scientific journals which publish results of original research and the news magazines which report these results. The important tools for retrospective searching are the secondary journals which abstract and index what has been previously reported as new in the primary journals or what has already been patented.

This section discusses the literature of the petroleum processing industry from these two points of view. It considers the primary journals which report new research and developments, then describes the technical societies which hold meetings and publish proceedings of interest to the refiner. This is followed by a description of the secondary sources. An annotated listing of the more important books and monographs published since 1950 is also included. The section concludes with a discussion of the information available from the government, including technical reports, statistics, market data, prices, etc.

PRIMARY SOURCES OF INFORMATION

There are three basic methods of keeping up with current developments in any field: (1) reading the scientific journals which are published at regular or irregular intervals by scientific societies and private publishers, (2) reading the news magazines which report new developments, and (3) attending meetings sponsored by the scientific societies.

Part 1 which follows contains a list of the scientific and news journals which are considered to be of value to the refiner. The list gives the title of the publication, the publisher, frequency of publication, price, and the date publication started.

Part 2 contains a listing of the scientific and trade societies which hold meetings of potential interest to the refiner. The list gives the name of the organization, headquarters city, its aims, frequency of meetings, approximate date of meetings, and publications.

A generally neglected source of primary information is the patent literature. In many cases patents are the best source for the latest developments, particularly in the field of technology. Although U.S. patents issue approximately 3 years after they have been filed, they may still be the most current announcement of new developments.

It is thus possible to keep up with current technology by keeping informed about newly issued U.S. patents. These are issued each Tuesday and are announced in the *U.S. Official Gazette*. This publication, issued by the U.S. Patent Office, lists each new patent, giving its title, assignor, assignee, filing date, class number, and one claim, generally the broadest. In addition, the *Official Gazette* also contains information about new trademarks, an index by title, patentee, and assignee.

Searching through the U.S. patents is not a job for the amateur because the system of classification is rather complicated. It is strongly suggested that the assistance of an expert be sought.

SECONDARY SOURCES OF INFORMATION

The functions of the secondary sources are to report, summarize, abstract, and index the information originally published in the primary sources. This function is of great importance in searching out what has previously been reported in the technical literature. There are two types of secondary sources; the abstract and indexing services, and books and monographs.

Part 3 of this section contains a list of the major abstracting and indexing journals of interest to the refiner. In Part 4 there is a list of the more important books published, mainly since 1950. This listing covers various aspects of petroleum refining and petroleum products.

One area which causes a great deal of confusion is that of statistical information. It is frequently impossible to obtain much-needed statistics because the sources are obscure or unknown. Part 5 has been compiled in order to assist the refiner in this area. The list contains information as to the sources of statistical information from government agencies and from private sources outside the government.

The publications issued by various API fundamental research projects are important sources of information and data for the petroleum processing industry. These projects are the following:

Project 6—Hydrocarbons in Petroleum. Dr. B. J. Mair, Carnegie Institute of Technology, Pittsburgh, Pa.

Project 37—The Fundamentals of Hydrocarbon Behavior. Dr. B. H. Sage, California Institute of Technology, Pasadena, Calif.

Project 42—Synthesis and Properties of High Molecular Weight Hydrocarbons. Dr. J. A. Dixon, Pennsylvania State University, University Park, Pa.

Project 44—Data on Hydrocarbons and Related Compounds. Dr. B. J. Zwolinski, Texas A. & M. Research Foundation, College Station, Tex.

Project 48—Synthesis, Properties and Identification of Sulfur Compounds in Petroleum. Dr. John S. Ball, H. T. Rall, and Dr. John P. McCullough, U.S. Bureau of Mines, Bartlesville, Okla., and Laramie, Wyo., and F. G. Bordwell, Northwestern University, Evanston, Ill.

Project 52—Nitrogen Constituents in Petroleum. Dr. C. A. VanderWerf, University of Kansas, Lawrence, Kans., Dr. John S. Ball, U.S. Bureau of Mines, Laramie, Wyo., and Dr. John P. McCullough, U.S. Bureau of Mines, Bartlesville, Okla.

Project 54—New Spectroscopic Techniques in Hydrocarbon Thermodynamics. Dr. G. C. Pimentel, University of California, Berkeley, Calif.

Project 55—Hydrothermal and Structural Studies of Minerals. Dr. R. Roy and Dr. G. W. Brindley, Pennsylvania State University, University Park, Pa.

Project 56—Isolation, Identification and Characterization of Metal-bearing Organic Compounds Contained in Petroleum and Natural Asphalts. Dr. J. M. Sugihara, University of Utah, Salt Lake City, Utah.

Project 57—Characterization of Porous Media. Dr. R. Ullman, Polytechnic Institute of Brooklyn, Brooklyn, N.Y.

Project 58—Synthesis and Purification of Hydrocarbon Standard Samples. Dr. K. W. Greenlee, Ohio State University, Columbus, Ohio, and Dr. A. J. Streiff, Carnegie Institute of Technology, Pittsburgh, Pa.

PART 1. JOURNALS

Title	Publisher	Frequency	Price	Year started
A.I.Ch.E.Journal	American Institute of Chemical Engineers, 345 E. 47th Street, New York, N.Y. 10017	Quarterly	$25/year to nonmembers	1908
Air Pollution Control Association Journal	Air Pollution Control Association, 4400 Fifth Avenue, Pittsburgh, Pa. 15213	Monthly	$15/year to members $50/year to nonmembers	1951
Air & Water News	McGraw-Hill Publications, 330 W. 42nd St., New York, N.Y. 10036	Weekly	$90/year	1966
American Gas Association Monthly	American Gas Association, 420 Lexington Avenue, New York, N.Y. 10017	Monthly	$5/year	1912
American Gas Journal	Petroleum Engineer Publ. Co., Box 1589, Dallas, Tex. 75221	Monthly	$3/year	1859
American Institute of Chemical Engineers Journal (called *A.I.Ch.E.Journal*, which see)				
American Oil Chemists' Society Journal	American Oil Chemists' Society, 35 E. Wacker Dr., Chicago, Ill. 60601	Monthly	$8/year	1917
American Petroleum Institute Proceedings	American Petroleum Institute, 1271 Avenue of the Americas, New York, N.Y. 10020	Annually	50¢/section	1921
American Petroleum Institute Quarterly (discontinued publication in 1959)	American Petroleum Institute, 1271 Avenue of the Americas, New York, N.Y. 10020			
American Petroleum Institute Statistical Bulletin	American Petroleum Institute, 1271 Avenue of the Americas, New York, N.Y. 10020	Weekly	$12.50/year	1920
Angewandte Chemie	Verlag Chemie, G.m.b.H. Hauptstrasse 694, Weinheim/ Bergstrasse, Germany	Semi-monthly	14.50 marks/ quarter	1888
Angewandte Chemie—International English Edition	Verlag Chemie, G.m.b.H. Hauptstrasse 694, Weinheim/ Bergstrasse, Germany	Monthly	$15/year	1962
ASLE Transactions American Society of Lubrication Engineers	Academic Press, 111 5th Avenue, New York, N.Y. 10003	Semi-annually	$18/volume	1958
Asphalt Institute Quarterly	Asphalt Institute, University of Maryland, College Park, Md.			1949
Association Française des Techniciens du Petrole—Bulletin	Association Française des Techniciens du Petrole, 14 Ave. De La Grande Arme, Paris 17, France	Bimonthly		
ASTM Bulletin (name changed to *Materials Research and Standards*, which see)				
Automotive Industries Annual Statistical Number	Chilton Company, Chestnut & 56th Sts., Philadelphia, Pa. 19139	Semi-monthly	$3/year	1899
Bitumen, Teere, Asphalte, Peche und verwandte Stoffe	Strassenbau, Chemie und Technik Verlagsgesellschaft m.b.h., 17(a) Heidelberg Germany	Monthly	48 DM/year	1950

Title	Publisher	Frequency	Price	Year started
Brennstoff-Chemie	Verlag W. Girardet, Essen, Germany	Monthly	60 Dm/year	1920
Brennstoff-Wärme-Kraft; Zeitschrift für Energiewirtschaft und technische Uberwachung	Verein Deutscher Ingenieure G.m.b.H., Prinz-Georg Strasse 77, Dusseldorf, Germany	Monthly	18 DM (3 months)	1949
British Chemical Engineering	Heywood & Co., Ltd., Drury House Russell Street, Drury Lane, London, WC 2	Monthly	40 shilling/year	1956
Butane-Propane News	Chilton Company, Chestnut & 56 Sts., Philadelphia, Pa. 19139	Monthly	$3/year	1939
Canadian Chemical Processing	Southam-McLean Publications, Ltd., 1450 Don Mills Road, Don Mills, Ontario, Canada	Monthly	$6/year in Canada, $10/year in U.S.A.	1917
Canadian Journal of Chemical Engineering (Formerly: *Canadian Journal of Technology*)	Chemical Institute of Canada, 48 Rideau Street, Ottawa 2, Canada	Bimonthly	$6/year in Canada, $7/year in U.S.A.	1929
Chemical & Processing Engineering	Lomond Technical Press Ltd., London	Monthly	1 pound, 10 shilling/year	1920
Chemical Engineering	McGraw-Hill Publications, 330 W. 42nd Street, New York, N.Y. 10036	Biweekly	$3/year in U.S.A., $5/year in Canada	1902
Chemical & Engineering News	American Chemical Society, Washington, D.C. 20036	Weekly	$6/year to nonmembers	1923
Chemical Engineering Progress	American Institute of Chemical Engineers, 345 E. 47th Street, New York, N.Y. 10017	Monthly	$25/year to nonmembers	1947
Chemical Engineering Science	Pergamon Press Ltd., 122 E. 55th Street, New York, N.Y. 10022	Monthly	$60/year	1951
Chemical Processing	Putnam Publ. Co., 111 E. Delaware Place, Chicago, Ill. 60611	Biweekly	$15/year or gratis	1938
Chemical Week	McGraw-Hill Publications, 330 W. 42nd Street, New York, N.Y. 10036	Weekly	$3/year	1914
Chemie-Ingenieur-Technik	Verlag Chemie G.m.b.H., Hauptstrasse 127, Weinheim/ Bergstrasse (17a), Germany	Monthly	96 marks/year	1928
Chemistry and Industry	Society of Chemical Industry, 14 Belgrave Square, London, SW 1, England	Weekly	5 pounds/year	1881
Chimie et Industrie	Chimie et Industrie, 28 rue St. Dominique, Paris, France	Monthly	$37/year	1917
Combustion	Combustion Publ. Co., 200 Madison Avenue, New York, N.Y. 10016	Monthly	$4/year	1929
Combustion and Flame (Quarterly Journal of the Combustion Institute)	Butterworth, Inc., Washington, D.C.	Quarterly	$16/year	1957
Control Engineering	R. H. Donnelley Corp., 466 Lexington Avenue, New York, N.Y. 10017	Monthly	$3/year	1954
Corrosion	National Association of Corrosion Engineers, 1061 M & M Building, Houston, Tex.	Monthly	$10/year to nonmembers	1945
Corrosion Science	Pergamon Press Ltd., 122 E. 55th Street, New York, N.Y. 10022	Quarterly	$30/year	1961

Title	Publisher	Frequency	Price	Year started
Corrosion Technology	Technology Publ., The Tower, Brookgreen Road, Hammersmith, London, W 6	Monthly	$8/year	1954
Electrical Engineering	American Institute of Electrical Engineers, 345 E. 47th Street, New York, N.Y. 10017	Monthly	$12/year to nonmembers $6/year to members	1887
Electrochemical Society Journal	Electrochemical Society, 30 East 42nd St., New York, N.Y. 10017	Monthly	$24/year to nonmembers	1902
Electrochemical Technology	Electrochemical Society, 30 E. 42nd Street, New York, N.Y. 10017	Bimonthly		1963
Engineer	Morgan Bros. Ltd., 20 Essex Street, London, WC2	Weekly	5 pounds 10.0/year	1856
Engineering	36 Bedford Street, London, WC2	Weekly	5 pounds, 10 s/year	1866
Engineer's Digest	E.D. Publ. Inc., 366 Madison Avenue, New York, N.Y. 10010	Monthly	$7.50/year	1940
Erdöl und Kohle Erdgas-Petrochemie (contains abstracts)	Industrieverlag von Hernhausser Kg., Hamburg, Germany	Monthly	$18/year	1948
Fluid Handling	Fuel & Metallurgical Journals, Ltd., 17/19 John Adams St., London, WC2	Monthly	$4/year	1950
Foreign Petroleum Technology (vols 1–10, 1933–1942) (ceased publication)				
Fuel. A Journal of Fuel Science	Butterworth's Scientific Publ., Washington, D.C.	Bimonthly	$16/year	1922
Fueloil & Oil Heat	Heating Publishers, Inc., 2 W. 45th Street, New York, N.Y. 10036	Monthly	$3/year	1942
Gas	Chilton Co., Inc., 198 S. Alvarado St., Los Angeles, Calif. 90057	Monthly	$3/year	1925
Gas Age	Moore Publishing Co., 25 West 45th Street, New York, N.Y. 10036	Biweekly	$5/year	1883
Das Gas und Wasserfach	R. Oldenbourg Verlag, Munich 8, Germany	Semimonthly	27 DM/6 months	1858
Highway Research Board Proceedings	National Academy of Sciences, National Research Council, Washington, D.C.	Annual		1920
Hydrocarbon Processing	Gulf Publ. Co., P. O. Box 2608, Houston, Tex. 77001	Monthly	$3/year	1922
Industrial and Engineering Chemistry	American Chemical Society, Washington, D.C. 20036	Monthly	$2.50–$5.00/year to members (depending on number of quarterlies)	1923
Industrial and Engineering Chemistry, Fundamentals	American Chemical Society, Washington, D.C. 20036	Quarterly	For subscription price see under Industrial and Engineering Chemistry	1962
Industrial and Engineering Chemistry, Process Design and Development	American Chemical Society, Washington, D. C. 20036	Quarterly	For subscription price see under Industrial and Engineering Chemistry	1962

Title	Publisher	Frequency	Price	Year started
Industrial and Engineering Chemistry, Product Research and Development	American Chemical Society, Washington, D.C. 20036	Quarterly	For subscription price see under Industrial and Engineering Chemistry	1962
Industrial Chemist (ceased publication, 1964)				1925
L'Industrie du Petrole	Oliver Lesourd, Paris 8, France	Monthly		1933
Institut Français du Petroles, Revue et Annales des Combustibles Liquids	Societé des Editions Tecnip, 7 Rue Nelation, Paris, XV	10 times/ year	$20.25/year	1946
Institute of Fuel Journal	18 Devonshire Street, Portland Place, London, W 1	Monthly	12s, 6/number	1926
Institute of Petroleum Journal	Institute of Petroleum, 61 New Cavendish St., London, W 1	Monthly	6 pounds, 6 shillings/ year	1914
Institute of Petroleum Review	Institute of Petroleum, 61 New Cavendish St., London, W 1	Monthly	25 shillings/ year	1947
Institute Spokesman (name changed to NLGI Spokesman, which see)				
Institution of Chemical Engineers (London) Transactions	Institution of Chemical Engineers, 16 Belgrave Square, London, SW 1	Bimonthly	4 pounds, 10 shillings/year	1923
International Chemical and Process Industries (now called Chemical & Processing Engineering, which see)				
International Journal of Heat and Mass Transfer	Pergamon Press, 122 E. 55th St., New York, N.Y. 10022	Monthly	$40/year	1960
International Oilman	Oil Forum, Inc., 258 Majestic Bldg., Ft. Worth, Tex.	Monthly	$2/year	1947
ISA Journal	Instrument Society of America, 530 William Penn Pl., Pittsburgh, Pa. 19119	Monthly	$4/year (nonmembers)	1954
ISA Transactions	Instrument Society of America, 530 William Penn Pl. Pittsburgh, Pa. 19119	Quarterly	$15/year	1962
Journal of Applied Chemistry	Society of Chemical Industry, 14 Belgrave Square, London	Monthly	15 pounds/year	1881
Journal of Applied Mechanics (Series E of Transactions of the ASME)	American Society of Mechanical Engineers, 345 E. 47th St., New York, N.Y. 10017	Quarterly	$10/year (nonmembers)	1935
Journal of Basic Engineering (Series D of Transactions of the ASME)	American Society of Mechanical Engineers, 345 E. 47th St., New York, N.Y. 10017	Quarterly	$10/year (nonmembers)	1957
Journal of Chemical and Engineering Data (Formerly: Ind. & Eng. Chem., Chemical & Engineering Data Series)	American Chemical Society, 1155 16th St. N.W., Washington, D.C. 20006	Quarterly	$18/year (nonmembers)	1959
Journal of Engineering for Industry (Series B of Transactions of the ASME)	American Society of Mechanical Engineers, 345 E. 47th St., New York, N.Y. 10017	Quarterly	$10 to nonmembers	1959

Title	Publisher	Frequency	Price	Year started
Journal of Engineering for Power (Series A of Transactions of the ASME)	American Society of Mechanical Engineers, 345 E. 47th St., New York, N.Y. 10017	Quarterly	$10 to non-members	1959
Journal of Heat Transfer (Series C of Transactions of the ASME)	American Society of Mechanical Engineers, 345 E. 47th St., New York, N.Y. 10017	Quarterly	$10/year (non-members)	1959
Journal of Research of the National Bureau of Standards—Section A, Physics and Chemistry	Superintendent of Documents, Washington, D.C. 20402	Bimonthly	$4/year	1959
Journal of Research of the National Bureau of Standards—Section C, Engineering and Instrumentation	Superintendent of Documents, Washington, D.C. 20402	Quarterly	$2.25/year	1959
Journal of Scientific Instruments	Institute of Physics, 47 Belgrave Square, London, SW 1	Monthly	6 pounds/year	1923
LP Gas	Moore Publishing Co., Inc., Ojibway Bldg., Duluth, Minn. 55802	Monthly	$3/year	1941
Lubrication Engineering	American Society of Lubrication Engineers, 5 North Wabash Ave., Chicago, Ill. 60602	Monthly	$6/year	1945
Materials Protection	National Association of Corrosion Engineers, Houston, Tex.	Monthly	$5/year	1962
Materials Research & Standards (formerly ASTM Bulletin)	American Society for Testing and Materials, Philadelphia, Pa.	Monthly	$5/year	1921
Mechanical Engineering	American Society of Mechanical Engineers, 345 E. 47th St., New York, N.Y. 10017	Monthly	$3.50 members/year, $7 non-members/year	1906
Natural Gas Processors Association Proceedings	Natural Gas Processors Association, 429 Kennedy Bldg., Tulsa, Okla. 74103	Annual		1922
NLGI Spokesman	National Lubricating Grease Institute, 4638 J. C. Nichols Parkway, Kansas City, Mo. 64112	Monthly	$5/year	1937
NPN (National Petroleum News)	McGraw-Hill Publications, 330 W. 42nd St., New York, N.Y. 10036	Monthly	$6/year	1909
NPN Bulletin	McGraw-Hill Publications, 330 W. 42nd St., New York, N.Y. 10036	Weekly	$60/year	1961
Oil and Gas International	Petroleum Publ. Co., 211 S. Cheyenne Ave., Tulsa, Okla.	Monthly	$10/year	1961
Oil and Gas Journal	Petroleum Publ. Co., 211 S. Cheyenne Ave., Tulsa, Okla.	Weekly	$6/year	1902
Oil Engines and Gas Turbine	Temple Press, Ltd., Bowling Green Lane, London, EC 1	Monthly	$5.50/year	1933
Oil Forum (name changed to International Oilman, which see)				
Oil in Canada	Stovel-Advocate Publishing Co., 365 Bannatyne Ave., Winnipeg 2, Manitoba, Canada	Weekly	$7/year Canada, $10 year U.S.A.	1937
Oil, Paint and Drug Reporter	Schnell Publ. Co., 100 Church St., New York, N.Y. 10007	Weekly	$5/year	1871
Petro/Chem Engineer	Petroleum Engineer Publishing Co., P.O. Box 1589, Dallas Tex. 75221	Monthly	$3/year	1929

Title	Publisher	Frequency	Price	Year started
PetroChemical News	Wm. F. Bland Co., 54 W. 40th St., New York, N.Y. 10018	Weekly	$85/year	1963
Petroleum	Technology Publishing Ltd., 229–243 Shepherds Bush Road, London, W 6	Monthly	$12/year	1939
Petroleum Chemistry (English translation of *Nefte Khimiia*)	Pergamon Press, 122 E. 55th St., New York, N.Y. 10022	Quarterly	$85/year	1962
Petroleum Engineer (now *Petroleum Management*, which see)				
Petroleum Management	Petroleum Engineer Publishing Co., 800 Davis Bldg., Dallas, Tex. 75221	Monthly	$6/year	1929
Petroleum Press Service	24, Ludgate Hill, London, EC 4	Monthly	1 pound, 10 shillings/year	1934
Petroleum Processing (discontinued with the September, 1957, issue)				
Petroleum Refiner (now *Hydrocarbon Processing* which see)				
Petroleum Refining Developments	Ethyl Corp., Detroit, Mich.	Monthly		1954
Petroleum Times	Temple Press Ltd., Bowling Green Lane, London, EC 1	Biweekly	78 shillings/year	1899
Petroleum Week (ceased publication 1955–1961)				
Petroleum World Oil (retitled *Western Oil and Refining*)	A. W. Boulton, Co., 1700 W. 8th St., Los Angeles, Calif. 90017	Monthly	$3/year	1910
Petroleum Zeitschrift (now *Erdöl und Kohle*, which see)				
Pipe Line Industry	Gulf Publ. Co., Box 2608, Houston, Tex. 77001	Monthly	$2/year	1954
Platt's Oilgram News Service	McGraw-Hill Publications, 330 W. 42nd St., New York, N.Y. 10036	Daily	$300/year	
Power	McGraw-Hill Publications, 330 W. 42nd St., New York, N.Y. 10036	Monthly	$3/year	1882
Power Engineering	Technical Publ. Co., 308 E. James St., Barrington, Ill.	Monthly	$10/year	1896
Reports on the Progress of Applied Chemistry	Society of Chemical Industry, 56 Victoria St., London, SW 1	Annual		1916
Review of Scientific Instruments	American Institute of Physics, 335 E. 45th St., New York, N.Y. 10017	Monthly	$11/year (nonmembers	1930
Revue Petroliere	Cie Française d'Editions, 40, Rue du Colisee, Paris	Monthly	$14/year	1922
Rivista dei Combustible		Monthly		1947
SAE Journal	Society of Automotive Engineers, Inc., 485 Lexington Ave., New York, N.Y. 10017	Monthly	$12/year	1917
Schmiertechnik	Karl Marklein Verlag, Engerstrasse 21a, Dusseldorf, Germany	Bimonthly	DM 21	1954

Title	Publisher	Frequency	Price	Year started
Scientific Lubrication	Scientific Publ., Broseley, Shropshire, England	Monthly	40s/year	1948
Society of Chemical Industry (London) Journal (name changed to Journal of Applied Chemistry, which see)				
Techniques du Petrole	11, Rue de Magdebourg, Paris	Monthly	$15.50/year	
VDI Zeitschrift (Verein Deutsche Ingenieure)	V.D.I. Verlag, Bongard Strasse 3, Dusseldorf, Germany	3 times/ month	$20/year	1857
Water Pollution Control Federation Journal	Water Pollution Control Federation, 4435 Wisconsin Ave., NW, Washington, D.C. 20016	Monthly	$10/year (non-members)	1928
Wear: International Journal on Fundamentals of Friction, Lubrication, Wear and Their Control in Industry (contains an abstract section)	Elsevier Publ. Co., New York, N.Y.	Irregular	$15/volume	1957
Western Oil and Refining (formerly Petroleum World Oil, which see)				
World Petroleum	Mona Palmer, 25 W. 45th St., New York, N.Y. 10036	Monthly	$12/year	1930

PART 2. TECHNICAL SOCIETIES

Air Pollution Control Association—4400 Fifth Ave., Pittsburgh, Pa. 15200
Publishes *Journal* and *APCA Abstracts*. Meets annually, generally in June.
American Chemical Society—1155 Sixteenth St. NW, Washington, D.C. 20036
Professional society of chemists and chemical engineers. Has a large number of divisions based on subject lines such as organic chemistry, physical chemistry, rubber, and petroleum. Publications: *Chemical & Engineering News*, weekly; *Chemical Abstracts*, semimonthly; *Journal*, semimonthly; *Journal of Physical Chemistry*, monthly; *Journal of Organic Chemistry*, monthly; *Journal of Agriculture and Food Chemistry*, biweekly; *Analytical Chemistry*, monthly; *Industrial and Engineering Chemistry*, in sections; *Chemical Reviews*, bimonthly; *Journal of Chemical Education*, monthly; *Rubber Chemistry and Technology*, quarterly; *Journal of Chemical Documentation*. Holds semiannual meetings, generally in the spring and fall.
American Gas Association—420 Lexington Ave., New York, N.Y. 10017
Provides information on all phases of gas industry; keeps staff of experts in nearly every field of the gas industry; develops operating and performance standards, specifications, good operating practices, and standards on construction and performance of appliances and equipment; assembles wide variety of statistical, economic, financial, and market studies; conducts cooperative research and development programs. Publications: *A. G. A. Monthly*; *A. G. A. Information Service*, biweekly; *A. G. A. Index*, monthly; *A. G. A. Proceedings*; *Gas Facts*, annual; *Gas Data Book*, annual; *A. G. A. Rate Service*. Maintains a library which answers questions, loans books, and compiles bibliographies and reports. Holds annual meetings, generally in October.
American Institute of Chemical Engineers—345 E. 47th St., New York, N.Y. 10017
Conducts research projects; establishes standards for chemical engineering curricula. Publications: *Chemical Engineering Progress*, monthly; *A.I.Ch.E. Journal*, quarterly; *Symposium Series*, irregular; *Standard Testing Procedures*, irregular. Holds annual meeting, generally in December.

American Institute of Mining, Metallurgical and Petroleum Engineers—11 Broadway, New York, N.Y. 10011

Publications: *Mining Engineering*, monthly; *Journal of Metals*, monthly; *Journal of Petroleum Technology*, monthly; *Transactions of the Metallurgical Society of AIME*, bimonthly. Holds annual meeting, generally in February or March.

American Oil Chemists' Society—35 E. Wacker Dr., Chicago, Ill. 60601

Publishes laboratory manual on methods of analysis, sponsors short courses. Publications: *Journal of the AOCS*, monthly; *Book of Official and Tentative Methods*, annual. Holds semiannual meetings, generally in May and October.

American Petroleum Institute—1271 Ave. of the Americas, New York, N.Y. 10020

Provides extensive publication and information services; conducts fundamental research on petroleum; maintains extensive petroleum library. Publications: *Petroleum Today*, quarterly; *Proceedings*, annual; also publishes manuals, booklets, and other materials on production, refining, marketing, transportation, safety, fire protection, standards, codes, and statistics. Holds annual meeting, generally in November.

American Society for Metals—Metals Park, Novelty, Ohio

Disseminates educational information about the making, using, and treatment of metals and related engineering materials; conducts conferences, séminars, and lectures on metals; sponsors National Metal Exposition; biennial Western Metal Exposition; annual National & Western Metal Congresses. Publications: *Metal Progress*, monthly; *Metals Review*, monthly; *Review of Metal Literature*, monthly; *Metals Engineering Quarterly*; *Transactions*, annual; *Metals Handbook*. Holds annual meeting, generally in October or November.

American Society for Testing and Materials—1916 Race St., Philadelphia, Pa. 19103

Aims to promote the knowledge of materials of engineering and the standardization of specifications and testing methods. Sponsors research projects. Publications: *Materials Research and Standards*, monthly; *Proceedings*, annual; *Yearbook*; *Index to Standards*, annual; *Book of Standards*, triennial. Holds annual meetings, generally in June.

American Society of Lubrication Engineers—5 No. Wabash Ave, Chicago, Ill. 60602

Sponsors joint committees with industry to advance the science of lubrication; sponsors annual short courses in lubrication engineering. Publications: *Lubrication Engineering*, monthly; *Transactions*, semiannual. Holds annual meeting, generally in May.

American Society of Mechanical Engineers—345 E. 47th St., New York, N.Y. 10017

Conducts research, develops boiler and pressure vessel, and power test codes; serves as sponsor for the American Standards Association in developing safety codes and standards for equipment. Publications: *Mechanical Engineering*, monthly; *Transactions*, quarterly; *Applied Mechanics Reviews*, monthly; *Mechanical Catalog*, annual; *Smog News*, semimonthly; *Journal of Applied Mathematics and Mechanics*, bimonthly. Holds annual meeting, generally in November.

American Society of Safety Engineers—5 No. Wabash Ave., Chicago, Ill. 60602

Professional society of safety engineers, safety directors, etc. Publication: *Engineering for Safety*, monthly. Holds annual meeting, generally in the fall.

Asphalt Institute—Asphalt Institute Bldg., University of Maryland, College Park, Md. 20740

Conducts extensive program of education, research, and engineering relating to asphaltic products. Publication: *Asphalt Institute Quarterly*; also specifications, manuals, booklets, etc., on asphalt construction. Holds annual meeting, generally in December.

Association of Asphalt Paving Technologists—1224 E. Engineering Bldg., Ann Arbor, Mich. 48104

Professional society of engineers and chemists engaged in asphalt paving or related fields. Publication: *Proceedings*, annual. Holds annual meeting, generally in January or February.

Association of Petroleum Re-Refiners—1500 N. Quincy St., Arlington, Va. 22207

Conducts research and development in the conservation of oil. Holds annual meeting, generally in April.

Atomic Industrial Forum—850 Third Ave., New York, N.Y. 10022

Industrial firms, research and service organizations, educational institutions, labor groups, and governmental agencies engaged in development and utilization of nuclear energy for constructive purposes. Publication: *AIF Memo to Members*, monthly. Holds annual meeting.

Combustion Institute—936-B Union Trust Bldg., Pittsburgh, Pa. 15219

Sponsors biennial symposia on combustion research. Publications: *Combustion & Flame*, quarterly; *Symposia Proceedings*, biennially.

Cooling Tower Institute—1120 W. 43 rd St., Houston, Tex. 77018

Conducts research to improve the design and performance of industrial water-cooling equipment, evaporative condensers, etc. Holds semiannual meetings, generally in January and June.

Coordinating Research Council—30 Rockefeller Plaza, New York, N.Y. 10020

Coordinates research activities between petroleum, equipment, and transportation industries.

Heat Exchange Institute—122 E. 42nd St., New York, N.Y. 10017

Manufacturers of steam condensers and heat-exchange equipment of powerhouse type.

Highway Research Board—2101 Constitution Ave. NW, Washington, D.C.

Encourages research and provides a national clearinghouse and correlation service for research activities and information on highway administration and technology. Sponsors research on finance, design, materials, maintenance, and construction. Publications: *Highway Research Abstracts*, monthly; material on accidents, administration, concrete, economics, planning, soils, etc. Holds annual meeting.

Independent Natural Gas Association of America—918 16th St., Washington, D.C. 20006

Producers and pipeline transporters of natural gas.

Independent Oil Compounders Association—4638 J. C. Nichols Parkway, Kansas City, Mo. 64112

Publication: *IOCA Compounding*, bimonthly.

Independent Petroleum Association of America—1430 S. Boulder, Tulsa, Okla.

Publications: *The Independent Petroleum*, monthly; *Current Oil Facts*, monthly; *U.S. Wholesale Prices of Crude Petroleum and Principal Products*, monthly.

Institute of Fuel—London

Promotes general advancement of the various branches of fuel technology and research. Publishes *Journal* and *Fuel Abstracts and Current Titles*.

Institute of Gas Technology—70 W. 34th St., Chicago, Ill. 60616

Educational and research facility sponsored by companies engaged in the production, processing, transmission, and distribution of utility gas and related fuels. Conducts research for the utility-gas industry. Maintains library. Publications: *Gas Abstracts*, monthly; *Research Bulletins*, irregularly. Holds annual meetings, generally in November.

Institute of Petroleum—London

Promotes study of petroleum and accumulates and disseminates information. Publishes *Journal, Review, Abstracts, IP Standards for Petroleum and Its Products*, and *Safety Codes*.

Institution of Chemical Engineers—London

Publishes *Transactions* and *The Chemical Engineer*.

Institution of Mechanical Engineers—London

Publishes *Proceedings* and *Chartered Mechanical Engineer*.

Instrument Society of America—530 Penn Place, Pittsburgh, Pa. 15279

Publications: *ISA Journal*, monthly; *Transactions*, annual; *ISA Proceedings*, irregular; *Recommended Practices*, irregular; *Electrical Safety Abstracts; Measurement Techniques*, monthly; *Instruments and Experimental Techniques*, bimonthly; *Automation and Remote Control*, monthly; *Industrial Laboratory*, monthly. Holds annual meeting, generally in September.

Liquefied Petroleum Gas Association—1111 S. LaSalle St., Chicago, Ill. 60603

Develops technical processes; provides safety regulations; sponsors service and management schools. Publications: *LPGA Times*, monthly; *LPGA Manual*, looseleaf format with periodic revisions; *Reports of Proceedings*, annual. Holds annual meetings, generally in spring.

National Association of Corrosion Engineers—1061 M & M Bldg., Houston, Tex.
Conducts research on corrosion control; sponsors short courses. Publications: *Corrosion*, monthly; *NACE Abstract Punch Card Service*, annual. Holds annual meeting, generally in March.

National Fire Protection Association—60 Batterymarch St., Boston, Mass. 02110
Organized in 1896 to promote the science and improve the methods of fire protection and prevention, to obtain and circulate information on these subjects, and to establish proper safeguards against loss of life and property by fire. Publishes the National Fire Codes, which are compilations of standards developed by NFPA.

National Lubricating Grease Institute—4638 J. C. Nichols Parkway, Kansas City, Mo. 64112
Publishes *NLGI Spokesman*, monthly.

National Petroleum Association—Merged with Western Petroleum Refiners Association to form National Petroleum Refiners Association, which see.

National Petroleum Council— 1625 K St. NW, Suite 601, Washington, D.C. 20006
Industry advisory council to the Secretary of Interior on matters relating to petroleum. Coordinates governmental-industry relationships.

National Petroleum Refiners Association—Suite 502, 1725 DeSales St., NW, Washington, D.C. 20036
Publishes *Saturday News Roundup* weekly. Meets annually, usually in April; numerous regional meetings.

Natural Gas Processors Association—429 Kennedy Bldg., Tulsa, Okla. 74103
Develops technical standards and specifications for products; also basic data on hydrocarbon behavior, testing procedures, laboratory analysis, plant safety practices. Publication: *Proceedings*, annual. Holds annual meeting.

Oil-Heat Institute of America—60 E. 42nd St., New York, N.Y.
Maintains testing laboratories, market research and technical services; works to improve burner standards, safety codes, and other industry regulations. Publications: *Dealer Digest*, monthly; *Public Relations Clip Sheet*, monthly; *Bulletin*.

Petroleum Electric Supply Association—1334 Fidelity Union Life Bldg., Dallas, Tex. 75201
Suppliers to firms interested in the construction, maintenance, and operation of electrical power and communications systems. Publication: *Electrical News*, monthly. Holds annual meeting, generally in April, in conjunction with the Petroleum Industry Electrical Association.

Petroleum Industry Electrical Association—1525 Fairfield Ave., Shreveport, La.
Engineers and technicians interested in the construction, maintenance, and operation of electrical power and communicating systems. Publication: *Electrical News*, monthly. Holds annual meeting, generally in April, in conjunction with the Petroleum Electric Supply Association.

Society for Nondestructive Testing—914 Chicago Ave., Evanston, Ill. 60201
Disseminates information on the various techniques of nondestructive testing. Publication: *Nondestructive Testing*.

Society of Automotive Engineers—485 Lexington Ave., New York, N.Y.
Promotes the arts, sciences, standards, and engineering practices of the design, construction, and utilization of self-propelled mechanisms. Publications: *SAE Journal*, monthly; *SAE Transactions*, annual. Holds annual meeting, generally in January.

Water Pollution Control Federation—4435 Wisconsin Ave. NW, Washington, D.C. 20016
Aim is to advance fundamental and practical knowledge concerning the nature, collection, treatment, and disposal of sewage and industrial wastes. Publishes *Journal of WPCF* monthly. Holds annual meeting, generally in October.

Western Oil and Gas Association—609 S. Grand Ave., Los Angeles, Calif. 90005
Participates in studies to determine the relationship between gasoline composition and smog; aids in establishing adequate smog-control regulations.

Western Petroleum Refiners Association—Merged with National Petroleum Association to form National Petroleum Refiners Association, which see.

PART 3. ABSTRACTING AND INDEXING SERVICES

Analytical Abstracts—1954
   Published monthly by the Society for Analytical Chemistry, London.
APCA Abstracts—1955
   Published monthly by the Air Pollution Control Assoc., Pittsburgh, Pa.
API Technical Abstracts Bulletins—1954—American Petroleum Institute, New York, N.Y. 10020
   The only abstract service which covers the literature and patents of particular interest to the refining industry.
Applied Mechanics Reviews: A Critical Review of the World Literature in Applied Mechanics—1948
   Published monthly by the ASME, New York. $25/year.
Applied Science and Technology Index
   A cumulative subject index to scientific periodicals. Published monthly by H. W. Wilson Co., 950 University Ave., New York, N.Y. 10052, on a service basis.
British Abstracts—1926–1953
   This abstracts journal at one time made longer abstracts than appeared in *Chemical Abstracts*, but after World War II the abstracts tended to become more descriptive.
British Technology Index—1962
   A monthly subject index to the contents of about 400 journals and covers engineering and chemical technology. Published by The Library Association, London, WC 1. $50/year.
Chemical Abstracts—1907
   Published by the American Chemical Society. This is the most comprehensive abstracting journal covering the field of chemistry. It attempts to cover the chemical literature completely and is currently abstracting the chemical contents of over 9,000 journals. Patents are also abstracted, and this coverage is expanding constantly. The indexes are very detailed, and in many cases entries are made from the original article, even though the information may not be contained in the abstract.
Chemical Market Abstracts—1950
   Published monthly by Foster D. Snell, Inc., 29 W. 15 St., New York, N.Y. 10011, $180/year.
Chemical Titles—1961
   Published semimonthly by the American Chemical Society. It is a concordance of titles of current chemical research papers published in three parts. The first is an index arranged alphabetically by key words from each title. The second is a bibliographic listing of titles of current papers from selected journals arranged in the form of tables of contents. The third is an author index.
Chemisches Zentralblatt—1830
   Currently being published weekly jointly by several scientific societies in Germany. Since 1919, it has attempted to cover the world's periodical literature thoroughly for both pure and applied chemistry. Patents have also been abstracted. Abstracts of journals behind the Iron Curtain sometimes appear sooner in *Chemisches Zentralblatt* than in *Chemical Abstracts*.
Chimie et Industrie—1918
   This is aimed particularly at good coverage of the literature of applied chemistry and was good up to 1939. It was useful for abstracts of French patents.
Current Chemical Papers—1954
   Chemical Society, Burlington House, Piccadilly, London, W 1. Monthly. 100s/year.
Current Contents—1959
   Reports tables of contents of more than 600 foreign and domestic journals in chemistry, space, electronic, and physical sciences. Published weekly by Institute for Scientific Information, 33 So. 17th Street, Philadelphia, Pa.
Dissertation Abstracts: Abstracts of Dissertations and Monographs in Microfilm—1938
   Published monthly by University Microfilms, 313 N. First St., Ann Arbor, Mich. $25.50.

Engineering Index—1913
Covers the entire field of engineering. Currently being issued monthly, in card form, and annually. Engineering Index, New York, N.Y.

Environmental Effects on Materials and Equipment Abstracts
Formerly *Prevention of Deterioration Abstracts*, which see.

Fire Research Abstracts and Reviews—1958
National Academy of Sciences, National Research Council, Washington, D.C. 3 times/year.

Fuel Abstracts—1947–1958
These were issued by the Department of Scientific and Industrial Research and were discontinued in 1958. It was resumed under the guidance of The Institute of Fuel in 1960 as *Fuel Abstracts and Current Titles*.

Fuel Abstracts and Current Titles—1960
A monthly publication, successor to *Fuel Abstracts*, published since May, 1960, by The Institute of Fuel.

Gas Abstracts—1945
Published monthly by Institute of Gas Technology, Chicago, Ill. 60616, and contains abstracts which are comprehensive for gas and partially for petroleum and coal. It covers about 175 journals and U.S. patents.

Highway Research Abstracts—1931
Published monthly (except August) by the National Research Council, Highway Research Board, 2101 Constitution Ave., Washington, D.C. 20025. $4/year.

Institute of Petroleum Journal—1914
Contains a very comprehensive abstracts section and is a very valuable source of information. The abstracts section includes the following sections: refinery operations, products, corrosion, engines and automotive equipment, safety precautions, transport and storage, economics and marketing.

Nuclear Science Abstracts—1948
Superintendent of Documents, Washington, D.C. 20402. Bimonthly. $6/year.

Petroleum. Zeitschrift für die gesamte Interessen der Erdöl—Industrie und des Mineralöl-Handels—1905–1939
Published a large number of abstracts, but was discontinued in 1939.

Prevention of Deterioration Abstracts—1946–1962
Published monthly by the Prevention of Deterioration Center, National Research Council, Washington, D.C. Unpublished reports and published articles are abstracted. Beginning with January, 1962, it combined with *Environmental Effects on Materials and Equipment Abstracts* under the latter title. The two journals are published as Parts I and II. The scope and coverage of each are unchanged.

Public Health Engineering Abstracts—1920
Published monthly by Public Health Service, Washington, D.C. $2/year.

Review of Metal Literature
Published monthly by the American Society for Metals, Metals Park, Ohio. $20/year to nonmembers.

Road Abstracts—1934
These are issued monthly by the Department of Scientific and Industrial Research (Great Britain).

Science Abstracts. Section B. Electrical Engineering Abstracts—1898
Published monthly by Institution of Electrical Engineers, Savoy Place, London, WC 2.

Synthetic Liquid Fuel Abstracts—1944–1951
These were issued by the U.S. Bureau of Mines and covered gas generation, Fischer-Tropsch process, Bergius process, coal, crude refining, analysis, organic and physical chemistry, and flames, explosions, electronics, and static electricity.

U.S. Government Research and Development Reports—1946
Superintendent of Documents, Washington, D.C. 20402. Monthly. $6/year.

Water Pollution Abstracts—1927
Published monthly by Department of Scientific and Industrial Research, Great Britain. 44s/year.

PART 4. BOOKS AND MONOGRAPHS

General

American Chemical Society
   Progress in Petroleum Technology (Advances in Chemistry Series No. 5). The
Society, Washington, D.C. 1951.
American Petroleum Institute
   Glossary of Terms Used in Petroleum Refining. A.P.I., New York, 1962, 2d ed.
American Society for Testing and Materials
   A.S.T.M. Standards on Petroleum Products and Lubricants. The Society, Phila-
delphia, 1961, 38th ed.
Bell, H. S.
   American Petroleum Refining. Van Nostrand, New York, 1959, 4th ed.
Brooks, B. T. et al.
   The Chemistry of Petroleum Hydrocarbons. Reinhold Publ. Co., New York, 1954–55,
3 vol.
Dunstan, A. E. et al., eds.
   The Science of Petroleum. Oxford Univ. Press, London, 1938–53, Vols. 1–6.
Gruse, W. A., and D. R. Stevens
   The Chemical Technology of Petroleum, McGraw-Hill Book Company, New York,
1960, 3d ed.
Institute of Petroleum
   IP Standards for Petroleum and Its Products, The Institute of Petroleum, London,
1962, 21st ed.
Institute of Petroleum
   Modern Petroleum Technology, The Institute of Petroleum, London, 1962, 3d ed.
Kirk, R. E., and D. F. Othmer, eds.
   Encyclopedia of Chemical Technology, Interscience Encyclopedia, Inc., New York,
1947–1960, vols. 1–15 and 2 supplements.
Kobe, K. A., and J. J. McKetta, eds.
   Advances in Petroleum Chemistry and Refining, Interscience Publishers, Inc., New
York, 1958–1965, vols. 1–10.
Nelson, W. L.
   Petroleum Refinery Engineering, McGraw-Hill Book Company, New York, 1958,
4th ed.
Petrie, D.
   Petroleum, Oxford University Press, Fair Lawn, N.J., 1961.
Purdy, G. A.
   Petroleum: Prehistoric to Petrochemicals, McGraw-Hill Book Company, New York,
1958.
Royal Dutch/Shell Group
   The Petroleum Handbook, Shell International Petroleum Co., Ltd., London, 1959,
4th ed.
Sell, G., ed.
   A Glossary of Petroleum Terms, Institute of Petroleum, London, 1961, 3d ed.
Sell, G., and H. A. Dossett, eds.
   Handbook of the Petroleum Industry, George Newnes, Ltd., London, 1958.
Society of Chemical Industry
   Reports on the Progress of Applied Chemistry, The Society of Chemical Industry,
London, Annual.

Crudes

American Petroleum Institute
   Measuring, Sampling, and Testing Crude Oil (API Standard 2500), The American
Petroleum Institute, New York, 1961, 2d ed.

American Society for Testing and Materials
  Chemical Composition of Petroleum Oils, STP 224, The American Society for Testing and Materials, Philadelphia, 1958.
Chantler, H. McD., et al.
  Analyses of Canadian Crude Oils, Canada, Department of Mines & Technical Survey, Mines Branch, Ottawa, 1951.
Henshaw, R. C., Jr.
  A Selected and Annotated Bibliography of Natural Gas, Bibliography No. 11, University of Texas, Bureau of Business Research, 1954.
McCoy, J. W.
  The Inorganic Analysis of Petroleum, Chemical Publishing Company, Inc., New York, 1962.
Nelson, W. L., et al.
  Crude Oils of Venezuela and Other Countries, Venezuela, Ministry of Mines and Hydrocarbons, Caracas, 1952.
Sachanan, A. N.
  The Chemical Constituents of Petroleum, Reinhold Publishing Corporation, New York, 1945.
U.S. Bureau of Mines
  Bibliography of Reports Containing Analyses of Crude Oils by the Bureau of Mines Routine Method, *Information Circular* 7921, by O. C. Blade, Superintendent of Documents, Washington, D.C., 1958.
U.S. Bureau of Mines
  An Index of Oil Shale Patents, *Bulletin* 468, compiled by S. Klosky, Superintendent of Documents, Washington, D.C., 1949.
van Nes, K., and H. A. van Westen
  Aspects of the Constitution of Mineral Oils, American Elsevier Publishing Company, New York, 1951.

Processes

Alders, L.
  Liquid-Liquid Extraction Theory and Laboratory Practice, American Elsevier Publishing Company, New York, 1959, 2d ed.
Brown, G. G., et al.
  Unit Operations, John Wiley & Sons, Inc., New York, 1950.
Cambel, A. B., and B. H. Jennings
  Gas Dynamics, McGraw-Hill Book Company, New York, 1958.
Chu, J. C.
  Distillation Equilibrium Data, Reinhold Publishing Corporation, New York, 1950.
Coulson, J. M., and J. F. Richardson
  Chemical Engineering, vol. 2, *Unit Operations*, Pergamon Press, London, 1955.
Cremer, H. W., ed.
  Chemical Engineering Practice, Academic Press Inc., New York, vols. 1–11, 1956–1959.
Enos, J. L.
  Petroleum Progress and Profits, The M.I.T. Press, Cambridge, Mass., 1962.
Groggins, P. H.
  Unit Processes in Organic Synthesis, McGraw-Hill Book Company, New York, 1958, 5th ed.
Hanson, D. N., et al.
  Computation of Multistage Separation Processes, Reinhold Publishing Corporation, New York, 1962.
Hengstebeck, R. J.
  Distillation: Principles and Design Procedures, Reinhold Publishing Corporation, New York, 1961.
Hengstebeck, R. J.
  Petroleum Processing: Principles and Applications, McGraw-Hill Book Company, New York, 1959.
Huntington, R. L.
  Natural Gas and Natural Gasoline, McGraw-Hill Book Company, New York, 1950.

Industrial & Engineering Chemistry
  Modern Chemical Processes, Reinhold Publishing Corporation, New York, vols. 1–5, 1950–1958.
Kalichevsky, V. A., and K. A. Kobe
  Petroleum Refining with Chemicals, American Elsevier Publishing Corporation, New York, 1956.
Katz, D. L., et al.
  Handbook of Natural Gas Engineering, McGraw-Hill Book Company, New York, 1959.
Kirschbaum, E.
  Destillier- und rektifiziertechnik, Springer-Verlag OHG Berlin, 1960, 3d ed.
Kohl, A. L., and F. C. Riesenfeld
  Gas Purification, McGraw-Hill Book Company, New York, 1960.
Leva, M.
  Fluidization, McGraw-Hill Book Company, New York, 1959.
McAdams, W. H.
  Heat Transmission, McGraw-Hill Book Company, New York, 1954, 3d ed.
Mantell, C. L.
  Adsorption, McGraw-Hill Book Company, New York, 1951, 2d ed.
Moulin, M.
  Petroleum Properties and Uses, vol. 4, Petroleum, Source of Chemical Products, Presses Documentaires, Paris, 1953.
Nielsen, C. H.
  Distillation in Practice, Reinhold Publishing Corporation, New York, 1956.
Noel, H. M.
  Petroleum Refinery Manual, Reinhold Publishing Corporation, New York, 1959.
Othmer, D. F., ed.
  Fluidization, Reinhold Publishing Corporation, New York, 1956.
Robinson, C. S., and E. R. Gilliland
  Elements of Fractional Distillation, McGraw-Hill Book Company, New York, 1950, 4th ed.
Rose, A., and E. Rose
  Distillation Literature, Index and Abstracts, 1946–1952, inclusive; 1953–1954, Applied Science Laboratories, Inc., State College, Pa.
Sherwood, T. K., and R. L. Pigford
  Absorption and Extraction, McGraw-Hill Book Company, New York, 1952, 2d ed.
Shreve, R. N.
  The Chemical Process Industries, McGraw-Hill Book Company, New York, 1956, 2d ed.
Smith, J. M.
  Chemical Engineering Kinetics, McGraw-Hill Book Company, New York, 1956.
Spausta, F.
  Triebstoffe für Verbrennungsmotoren, Band 1, *Flüssige Triebstoffe und ihre Herstellung*, Springer-Verlag OHG, Vienna, 1953.
Stewart, J. R., and F. E. Spicer
  An Encyclopedia of the Chemical Process Industries, Chemical Publishing Company, Inc., New York, 1956.
Storch, H. H., et al.
  The Fischer Tropsch and Related Processes, John Wiley & Sons, Inc., New York, 1951.
Streeter, V. L., ed.
  Handbook of Fluid Dynamics, McGraw-Hill Book Company, New York, 1961, 4th ed.
Treybal, R. E.
  Liquid Extraction, McGraw-Hill Book Company, New York, 1963, 2d ed.
Treybal, R. E.
  Mass-transfer Operation, McGraw-Hill Book Company, New York, 1955.
U.S. Bureau of Mines
  Bibliography of Pressure Hydrogenation, Part I, Review and Compilation of the Literature on Pressure Hydrogenation of Liquid and Solid Carbonaceous Materials; Part II, Patents; Part III, Subject Index and Numerical Patent List, *Bulletin* 485,

by J. L. Wiley and H. C. Anderson, Superintendent of Documents, Washington, D.C., 1950.

U.S. Bureau of Mines
  Bibliography of the Fischer-Tropsch Synthesis and Related Processes, Part I, Review and Compilation of the Literature on the Production of Synthetic Liquid Fuels and Chemicals by the Hydrogenation of Carbon Monoxide; Part II, Patents, *Bulletin* 544, by H. C. Anderson, J. L. Wiley, and A. Newell, Superintendent of Documents, Washington, D.C., 1954–1955.

Waterman, H. I., C. Boelhouwer, and D. Th. A. Huibers
  Process Characterization, American Elsevier Publishing Company, New York, 1960.

Winnacker, K., and E. Weingaertner
  Chemische Technologie organische Technologie, Carl Hanser Verlag, München, 1952.

Zenz, F. A., and D. F. Othmer
  Fluidization and Fluid-particle Systems, Reinhold Publishing Corporation, New York, 1960.

Equipment

A. I. Ch. E.
  Pump manual, The American Institute of Chemical Engineers, New York, 1960.

A. I. Ch. E. Research Committee, Distillation Subcommittee
  Bubble-tray Design Manual, Prediction of Fractionation Efficiency, The American Institute of Chemical Engineers, New York, 1958.

American Petroleum Institute
  Recommended Practice for the Design and Construction of Pressure-relieving Systems in Refineries, The American Petroleum Institute, New York, 1955.

American Petroleum Institute
  Centrifugal Pumps for General Refinery Services, The American Petroleum Institute, New York, 1960, 3d ed.

American Society for Metals
  American Standard Code for Pressure Piping, B31.1, The American Society for Metals, Cleveland, 1955.

American Society of Mechanical Engineers
  Boiler and Pressure Code, Unfired Pressure Vessels, Sec. VIII, The American Society of Mechanical Engineers, New York, 1959.

American Society of Mechanical Engineers
  Mechanical Engineers Catalog, The American Society of Mechanical Engineers, New York, 1962.

Binder, R. C.
  Fluid Mechanics, Prentice-Hall, Inc., Englewood Cliffs, N.J., 1955, 3d ed.

Brown, A. I., and S. M. Marco
  Introduction to Heat Transfer, McGraw-Hill Book Company, New York, 1958, 3d ed.

Brownell, L. E., and E. H. Young
  Process Equipment Design, John Wiley & Sons, Inc., New York, 1959.

Caddell, J. R., ed.
  Fluid Flow in Practice, Reinhold Publishing Corporation, New York, 1956.

Chuse, R.
  Unfired Pressure Vessels, The ASME Code Simplified, McGraw-Hill Book Company, New York, 1960, 4th ed.

Eckert, E. R. G.
  Heat and Mass Transfer, McGraw-Hill Book Company, New York, 1959, 2d ed.

Eckert, E. R. G., et al., eds.
  Recent Advances in Heat and Mass Transfer, McGraw-Hill Book Company, New York, 1961.

Giedt, W. H.
  Principles of Engineering Heat Transfer, D. Van Nostrand Company, Inc., Princeton, N.J., 1957.

Jackson, J.
  Cooling Towers—with Special Reference to Mechanical-draught Systems, Butterworth Scientific Publications, London, 1951.

Jakob, M., and S. P. Kezios
 Heat Transfer, John Wiley & Sons, Inc., New York, 1957, vol. 2.
Kjaer, J.
 Measurement and Calculation of Temperature and Conversion in Fixed-bed Catalytic Reactions, J. Gjellerups Forlag, Copenhagen, 1958.
Leva, M.
 Tower Packings and Packed Tower Design, U.S. Stoneware Co., Akron, Ohio, 1953, 2d ed.
McAdams, W. H.
 Heat Transmission, McGraw-Hill Book Company, New York, 1954, 3d ed.
McKelvey, K. K., and M. Brooke
 The Industrial Cooling Tower with Special Reference to the Design, Construction, Operation, and Maintenance of Water Cooling Towers, American Elsevier Publishing Company, New York, 1959.
Morris, G. A., and J. Jackson
 Absorption Towers, Butterworth Scientific Publications, London, 1953.
National Bureau of Casualty Underwriters
 Synopsis of Boiler and Pressure Vessel Laws, Rules and Regulations by States, Cities, Countries and Provinces (United States and Canada), The National Bureau of Casualty Underwriters, New York, 1960.
Norman, W. S.
 Absorption, Distillation and Cooling Towers, John Wiley & Sons, Inc., New York, 1962.
Petroleum Refiner
 The Refinery Catalog, Gulf Publishing Company, Houston, 1960, 27th ed.
Riegel, E. R.
 Chemical Process Machinery, Reinhold Publishing Corporation, New York, 1953, 2d ed.
Sieder, E. N.
 Heat Transfer Tables, American Locomotive Co., New York, 1952.
Smith, J. M.
 Chemical Engineering Kinetics, McGraw-Hill Book Company, New York, 1956.
Stepanoff, A. J.
 Centrifugal and Axial Pumps, John Wiley & Sons, Inc., New York, 1957, 2d ed.
Sweets' Catalog Service
 Plant Engineering Catalog File, McGraw-Hill Book Company, New York, 1963.
Tongue, H.
 The Design and Construction of High Pressure Chemical Plant. Chapman & Hall, Ltd., London, 1959, 2d ed.
Trinks, W.
 Industrial Furnaces, John Wiley & Sons, Inc., New York, vol. 1, 4th ed., 1951; vol. 2, 3d ed., 1955.
Wilkinson, W. L.
 Non-Newtonian Fluids; Fluid Mechanics, Mixing, and Heat Transfer, Pergamon Press, New York, 1960.
Zenz, F. A., and D. F. Othmer
 Fluidization and Fluid-particle Systems, Reinhold Publishing Corporation, New York, 1960.
Zimmerman, O. T., and I. Lavine
 Chemical Engineering Costs, Industrial Research Service, Dover, N.H., 1950.

Materials

American Society for Metals
 Metals Handbook, The American Society for Metals, Novelty, Ohio, 1961, 8th ed.
Brady, G. S.
 Materials Handbook, McGraw-Hill Book Company, New York, 1963, 9th ed.
Burns, R. M.
 Protective Coatings for Metals, Reinhold Publishing Corporation, New York, 1955, 2d ed.

Deitz, V. R.
Bibliography of Solid Adsorbents, Superintendent of Documents, Washington, D.C., 2 vols., 1944 and 1956.
Gackenbach, R. E.
Materials Selection for Process Plants, Reinhold Publishing Corporation, New York, 1960.
Goldman, J. E.
Science of Egineering Materials, John Wiley & Sons, Inc., New York, 1957.
Greathouse, G. A., and C. A. Wessel, eds.
Deterioration of Materials: Causes and Preventive Techniques, Reinhold Publishing Corporation, New York, 1954.
Jastrzebski, Z. D.
Nature and Properties of Engineering Materials, John Wiley & Sons, Inc., New York, 1959.
Lee, J. A.
Materials of Construction for Chemical Process Industries, McGraw-Hill Book Company, New York, 1950.
Mantell, C. L., ed.
Engineering Materials Handbook, McGraw-Hill Book Company, New York, 1958.
Miner, D. F., and J. B. Seastone
Handbook of Engineering Materials, John Wiley & Sons, Inc., New York, 1955.
Morley, A.
Strength of Materials, Longmans, Green & Co., Inc., New York, 1954, 11th ed.
Nelson, G. A., comp.
Corrosion Data Survey, Shell Development Co., Emeryville, Calif., 1960.
Rabald, E.
Corrosion Guide, D. Van Nostrand Company, Inc., Princeton, N.J., 1951.
Reinhold Publishing Corporation
Chemical Materials Catalog and Directory of Products, Reinhold Publishing Corporation, New York, 1963.
Reinhold Publishing Corporation
Materials in Design Engineering, Materials Selector (Reference Issue), Reinhold Publishing Corporation, New York, 1962.
Rumford, F.
Chemical Engineering Materials, Constable & Co., Ltd., London, 1960, 2d ed.
Smithells, C. J.
Metals Reference Book, Butterworths, Washington, D.C., 2 vols., 1962, 3d ed.
Society of Chemical Industry
Corrosion Problems of the Petroleum Industry, Monograph 10, The Society of Chemical Industry, London, 1960.
Uhlig, H. H.
The Corrosion Handbook, John Wiley & Sons, Inc., New York, 1948.
Young, J. F.
Materials and Processes, John Wiley & Sons, Inc., New York, 1954, 2d ed.

Process Control and Instrumentation

Ahrendt, W. R., and C. J. Savant
Servomechanism Practice, McGraw-Hill Book Company, New York, 1960, 2d ed.
American Petroleum Institute
Manual on Installation of Refinery Instruments and Control Systems, The American Petroleum Institute, New York, 1960.
Behar, M. F.
Handbook of Measurement and Control, Instruments Publishing Co., Pittsburgh, 1951.
Ceaglske, N. H.
Automatic Process Control for Chemical Engineers, John Wiley & Sons, Inc., New York, 1956.

Chestnut, H., and R. W. Mayer
  Servomechanisms and Regulating System Design, John Wiley & Sons, Inc., New York, vols. 1-2, 1951-1955.
Considine, D. M., ed.
  Process Instruments and Controls Handbook, McGraw-Hill Book Company, New York, 1957.
Coxon, W. F.
  Flow Measurement and Control, Heywood & Co., Ltd., London, 1959.
Coxon, W. F.
  Temperature Measurement and Control, The Macmillan Company, New York, vol. 5 of *Physical Processes in the Chemical Industry*, 1960.
Draper, C. S., et al.
  Instrument Engineering, McGraw-Hill Book Company, New York, vols. 1-3, 1952-1955.
Eckman, D. P.
  Automatic Process Control, John Wiley & Sons, Inc., New York, 1958.
Eckman, D. P.
  Industrial Instrumentation, John Wiley & Sons, Inc., New York, 1950.
Holzbock, W. G.
  Instruments for Measurement and Control, Reinhold Publishing Corporation, New York, 1955.
Instruments Publishing Co.
  Instruments and Automation, 1957 Handbook and Buyers Guide, The Instruments Publishing Co., Pittsburgh, 1956.
Jones, E. B.
  Instruments Technology, vol. 1, *Measurement of Pressure, Level, Flow, and Temperature*, Butterworth Scientific Publications, London, 1953.
Kallen, H. P., ed.
  Handbook of Instrumentation and Controls; a Practical Design and Applications Manual for the Mechanical Services Covering Steam Plants, Power Plants, Heating Systems, Air Conditioning Systems, Ventilation Systems, Diesel Plants, Refrigeration, and Water Treatment, McGraw-Hill Book Company, New York, 1961.
Kirk, R. E., and D. F. Othmer
  Encyclopedia of Chemical Technology, Interscience Encyclopedia Inc., New York, vols. 1-15 and 2 supplements, 1947-1960 (currently being revised).
La Joy, M. H.
  Industrial Automatic Control, Prentice-Hall, Inc., Englewood Cliffs, N.J., 1954.
Royds, R.
  The Measurement and Control of Temperatures in Industry, Chemical Publishing Company, Inc., New York, 1952.
Seifert, W. W., and C. W. Steeg, eds.
  Control Systems Engineering, McGraw-Hill Book Company, New York, 1960.
Siggia, S.
  Continuous Analysis of Chemical Process Systems, John Wiley & Sons, Inc., New York, 1959.
Spink, L. K.
  A.G.A. Flow Constants, Supplementary to Principles and Practices of Flow Meter Engineering, Foxboro Co., Foxboro, Mass., 1955.
Stearns, R. F., et al.
  Flow Measurement with Orifice Meters, D. Van Nostrand Company, Inc., New York, 1951.
Texas A. & M. College
  Instrumentation for the Process Industries, Instruments Publishing Co., Pittsburgh, 1953.
Truxal, J. G.
  Control Engineers' Handbook, McGraw-Hill Book Company, New York, 1958.
U.S. National Bureau of Standards
  Guide to Instrumentation Literature, Circular C-567, by W. G. Brombacher, J. F. Smith, and L. M. Van der Pyl, Superintendent of Documents, Washington, D.C., 1955.

Wade, W. F., and E. N. Kemler
   Automatic Control Bibliography, Summary Reports, Spring Park, Minn., 1955.
Williams, T. J., and V. A. Lauher
   Automatic Control of Chemical and Petroleum Processes, Gulf Publishing Company, Houston, 1961.
Young, A. J.
   An Introduction to Process Control System Design, Instruments Publishing Co., Pittsburgh, 1955.
Young, A. J.
   Process Control, Instruments Publishing Co., Pittsburgh, 1954.
Zoss, L. M., and B. C. Delahooke
   Theory and Application of Industrial Process Control, Delmar Publ., Albany, N.Y., 1961.

Maintenance and Construction

American Petroleum Institute
   API-ASME Code for the Design, Construction, Inspection, and Repair of Unfired Pressure Vessels for Petroleum Liquids and Gases, The American Petroleum Institute, New York, 1951, 5th ed.
American Petroleum Institute
   Guide for Inspection of Refinery Equipment, The American Petroleum Institute, New York, 1957–1961, Chaps. 1–20.
American Petroleum Institute
   Recommended Practices for Refinery Inspections, The American Petroleum Institute, New York, 1949.
National Petroleum Council
   Maintenance and Chemical Requirement for U.S. Petroleum Refineries and Natural Gasoline Plants, The National Petroleum Council, Washington, D.C., 1961.
Strachan, J. F.
   The Petroleum Refinery Engineers Handbook, E. & F. N. Spon, Ltd., London, 1955.

Offsite Facilities and Utilities

American Petroleum Institute
   Manual on Disposal of Refinery Wastes, vol. 1, *Waste Water Containing Oil,* 1959, 6th ed.   vol. 2, *Waste Gases and Particulate Matter,* 1957, 5th ed.   vol. 3, *Chemical Wastes,* 1958, 3d ed.   vol. 4, *Sampling and Analysis of Waste Water,* 1957, 2d ed. The American Petroleum Institute, New York.
American Petroleum Institute
   Recommended Practice for Electrical Installations in Petroleum Refineries, API Recommended Practice 540, The American Petroleum Institute, New York, 1959.
ASTM
   Manual on Industrial Water, STP 148-D, American Society for Testing and Materials, Philadelphia.
ASTM
   Manual on Measurement and Sampling of Petroleum and Petroleum Products, American Society for Testing and Materials, Philadelphia, 1950.
Besselievre, E. B.
   Industrial Waste Treatment, McGraw-Hill Book Company, New York, 1952.
Betz, W. H., and L. D. Betz
   Betz Handbook of Industrial Water Conditioning, Philadelphia, 1950.
Factory Insurance Association
   Recommended Good Practice for the Storage and Handling of Liquefied Petroleum Fuel Gases, The Factory Insurance Association, Hartford, Conn., 1950.
Fair, G. M., et al.
   Water Supply and Waste-water Disposal, John Wiley & Sons, Inc., New York, 1954.
Gurnham, C. F.
   Principles of Industrial Waste Treatment, John Wiley & Sons, Inc., New York, 1955.

Institute of Petroleum
  Petroleum Measurement Manual, The Institute of Petroleum, London, 1952.
Matthews, F. J.
  Boiler Feed Water Treatment, Chemical Publishing Company, Inc., New York, 1951,
3d ed.
National Tank Truck Carriers, Inc.
  Tank Truck Transportation of Chemicals and Other Bulk Liquids, The National Tank
Truck Carriers, Inc., Washington, D.C., 1956.
Nordell, E.
  Water Treatment for Industrial and Other Uses, Reinhold Publishing Corporation,
New York, 1961, 2d ed.
O'Brien, H. L.
  Petroleum Tankage and Transmission, Graver Tank & Mfg. Co., Chicago, 1951.
Powell, S. T.
  Water Conditioning for Industry, McGraw-Hill Book Company, New York, 1954.
Quinn, A. D.
  Design and Construction of Ports and Marine Structures, McGraw-Hill Book Company, New York, 1961.
Reeves, E. A., ed.
  Uses of Electricity in the Oil Industry, Ernest Benn, Ltd., London, 1960.
Rudolfs, W., ed.
  Industrial Wastes—Their Disposal and Treatment, Reinhold Publishing Corporation,
New York, 1953.
Society of Chemical Industries
  Disposal of Industrial Waste Materials, The Macmillan Company, New York, 1956.
Wooler, R. G.
  Tankerman's Handbook, E. W. Sweetman, New York, 1950, 2d ed.

Fire Protection

American Petroleum Institute
  Fire Protection in Refineries, RP-2001, The American Petroleum Institute, New
York, 1959, 4th ed.
American Petroleum Institute
  Protection Against Ignitions Arising out of Static, Lightning, and Stray Currents,
RP-2003, The American Petroleum Institute, New York, 1956.
Bahme, C. W.
  Fire Protection for Chemicals, National Fire Protection Association, Boston, 1956.
Institute of Petroleum
  Safety Codes, Part 1, *Electrical Code*, The Institute of Petroleum, London, 1956.
Klinkenberg, A., and J. L. van der Minne
  Electrostatics in the Petroleum Industry; the Prevention of Explosion Hazards, Elsevier
Publishing Company, Amsterdam, 1958.
National Fire Protection Association
  Flammable Liquids Code (NFPA No. 30) National Fire Protection Association,
Boston, 1959.
National Fire Protection Association
  Inspection Manual, The National Fire Protection Association, Boston, 1959.
National Fire Protection Association
  The National Fire Codes, vol. 1, *Flammable Liquids and Gases*, 1960; vol. 2, *Combustible Solids, Dusts, Chemicals and Explosives*, 1960; vol. 5, *Electrical*, 1962–1963,
vol. 6, *Transportation*, 1962, National Fire Protection Association, Boston.
National Fire Protection Association
  Standards for the Storage and Handling of Liquefied Petroleum Gases (NFPA No. 58),
National Fire Protection Association, Boston, 1955.
Stecher, G. E., and H. N. Lendall
  Fire Prevention and Protection Fundamentals, The Spectator, Philadelphia, 1953.
Tryon, G. H., ed.

Fire Protection Handbook, National Fire Protection Association, Boston, 1962, 12th ed.

U.S. National Bureau of Standards
Code for Protection against Lightning, Handbook 46, Superintendent of Documents, Washington, D.C., 1952.

Safety

American Petroleum Institute
Accident Prevention Manuals, The American Petroleum Institute, New York, Nos. 1–13, 1931–1959.
Armistead, G., comp.
Safety in Petroleum Refining and Related Industries, John G. Simmonds & Co., New York, 1959, 2d ed.
Association of Casualty & Surety Companies
Handbook of Industrial Safety Standards, The Association of Casualty & Surety Companies, New York, 1954, 9th ed.
Blake, R. P., ed.
Industrial Safety, Prentice-Hall, Inc., Englewood Cliffs, N.J., 1953, 2d ed.
Compressed Gas Manufacturers Association
Safe Handling of Compressed Gases, The Compressed Gas Manufacturers Association, New York.
De Reamer, R.
Modern Safety Practices, John Wiley & Sons, Inc., New York, 1958.
Heinrich, H. W.
Industrial Accident Prevention, McGraw-Hill Book Company, New York, 1959, 4th ed.
Institute of Petroleum
Safety Codes, Part II, *Marketing Safety Code*, 1954; Part III, *Refining Safety Code*, 1956; Part VI, *Code of Practice for Petroleum Pipelines*, 1962, The Institute of Petroleum, London.
Lamb, J.
Oil Tanker Cargoes. Their Safe and Efficient Handling, Charles Griffin & Company, Ltd., London, 1954.
National Safety Council
Accident Prevention Manual for Industrial Operations, The National Safety Council, Chicago, 1951, 2d ed.
Sax, N. I.
Dangerous Properties of Industrial Materials, Reinhold Publishing Corporation, New York, 1963, 2d ed.
Simonds, R. H., and J. V. Grimaldi
Safety Management, Accident Cost and Control, Richard D. Irwin, Inc., Homewood, Ill., 1956.

Air Pollution

Air Pollution Control Association
Directory of Government Air Pollution Agencies, The Air Pollution Control Association, Pittsburgh, 1956, 2d ed.
American Industrial Hygiene Association
Air Pollution Manual, Part I, Evaluation, The American Industrial Hygiene Association, Detroit, 1960.
Faith, W. L.
Air Pollution Control, John Wiley & Sons, Inc., New York, 1959.
Gibson, J. R., et al.
The Air Pollution Bibliography, vol. 1, Library of Congress, Washington, D.C., 1957.
Jacobius, A. J., et al.
The Air Pollution Bibliography, vol. 2, Library of Congress, Washington, D.C., 1959.

Magill, P. L., F. R. Holden, and C. Ackley, eds.
   Air Pollution Handbook, McGraw-Hill Book Company, New York, 1956.
Manufacturing Chemists' Association
   Air Pollution Abatement Manual, The Manufacturing Chemists' Association, Washington, D.C., 13 chapters and 3 bibliography supplements, 1952–1960.
Meetham, A. R.
   Atmospheric Pollution, Its Origins and Prevention, Pergamon Press, New York, 1956, 2d ed.
Stern, A. C., ed.
   Air Pollution, Academic Press Inc., New York, 2 vols., 1962.
U.S. Bureau of Mines
   Air Pollution, A Bibliography, Bulletin 537, by S. J. Davenport and G. G. Morgis, Superintendent of Documents, Washington, D.C., 1954.
U.S. Public Health Service
   Atmospheric Emissions from Petroleum Refineries; a Guide for Measurement and Control, Publication 763, Superintendent of Documents, Washington, D.C., 1960.
World Health Organization
   Air Pollution, Monograph Series 46, Columbia University Press, New York, 1961.

Products

General

Guthrie, V. B.
Petroleum Products Handbook, McGraw-Hill Book Company, New York, 1960.

Gases

Association of Casualty and Surety Companies
   LP-gas, Safe Handling and Use, The Association of Casualty and Surety Companies, New York, 1953.
Denny, L. C., et al., eds.
   Handbook, Butane-propane Gases, Chilton, Los Angeles, 1962, 4th ed.
Henshaw, R. C., Jr.
   A Selected and Annotated Bibliography of Natural Gas, University of Texas, 1954.
Huntington, R. L.
   Natural Gas and Natural Gasoline, McGraw-Hill Book Company, New York, 1950.
Lawrie, J.
   Natural Gas and Methane Sources, Chapman & Hall, Ltd., London, 1961.

Fuels and Lubricants

American Society for Testing and Materials
   ASTM Manual for Rating Motor Fuels by Motor and Research Methods, The American Society for Testing and Materials, Philadelphia, 1960.
American Society for Testing and Materials
   ASTM Standards on Petroleum Products and Lubricants, The American Society for Testing and Materials, Philadelphia, 1960.
Bestougeff, M. A.
   Separation et identification des hydrocarbures dans les essences de petrole, Tech. Appl. Petr., Paris, 1952.
Bondi, A.
   Physical Chemistry of Lubricating Oils, Reinhold Publishing Corporation, New York, 1951.
Boner, C. J.
   Manufacture and Application of Lubricating Greases, Reinhold Publishing Corporation, New York, 1954.
Boyd, J.
   Lubrication Science and Technology, Pergamon Press, New York, 1958.

Brame, J. S. S., and J. G. King
   Fuel—Solid, Liquid, and Gaseous, Edward Arnold (Publishers) Ltd., London, 1955.
Ellis, E. G.
   Lubricant Testing, Scientific Publ. Ltd., London, 1953.
Faust, F. H., and G. T. Kaufman
   Handbook of Oil Burning, Oil-Heat Institute of America, New York, 1951.
Georgi, C. W.
   Motor Oils and Engine Lubrication, Reinhold Publishing Corporation, New York, 1950.
Goodger, E. M.
   Petroleum and Performance in Internal Combustion Engineering, Butterworth & Co. (Publishers), Ltd., London, 1953.
Gunderson, R. C., and A. W. Hart, eds.
   Synthetic Lubricants, Reinhold Publishing Corporation, New York, 1962.
Henly, A. T.
   Oil Fuel Applications, Crosby Lockwood & Son, Ltd., London, 1956.
Institute of Petroleum and Royal Aeronautical Society
   Data Sheets on Fuels, Lubricants and Hydraulic Fluids, 1950.
Lamb, J.
   Petroleum and Its Combustion in Diesel Engines with Special Reference to the Use of Residual Fuels, Charles Griffin & Company, Ltd., London, 1955.
Popovich, M., and C. Hering
   Fuels and Lubricants, John Wiley & Sons, Inc., New York, 1959.
Schmidt, P. F.
   Fuel Oil Manual, The Industrial Press, New York, 1958, 2d ed.
Smith, M. L., and K. W. Stinson
   Fuels and Combustion, McGraw-Hill Book Company, New York, 1952.
Spausta, F.
   Treibstoffe für Verbrennungsmotoren, Springer-Verlag OHG, Vienna, 2 vols., 1953.
U.S. Bureau of Mines
   Burner Fuel Oils, 1960, Petroleum Products Survey 16, by O. C. Blade, Bureau of Mines, Washington, D.C., 1960.
Zuidema, H. H.
   The Performance of Lubricating Oils, Reinhold Publishing Corporation, New York, 1959, 2d ed.

Waxes and White Oils

Bennett, H.
   Commercial Waxes, Chemical Publishing Company, Inc., New York, 1956, 2d ed.
Ivanovszky, L.
   Wachs—enzyklopadie, Verlag für chem. Industrie, H. Ziolkowsky, K. G., Augsburg, vol. 1, 1954; vol. 2, 1960.
Meyer, E.
   White Mineral Oil and Petrolatum, Chemical Publishing Company, Inc., New York, 1950.
Warth, A. H.
   The Chemistry and Technology of Waxes, Reinhold Publishing Corporation, New York, 1956, 2d ed.

Asphalt and Bitumen

Abraham, H.
   Asphalts and Allied Substances, D. Van Nostrand Company, Inc., Princeton, N.J., 1960–1963, 5 vols, 6th ed.
American Society for Testing and Materials
   ASTM Standards on Bituminous Materials for Highway Construction, Waterproofing, and Roofing, The American Society for Testing and Materials, Philadelphia, 1960, 8th ed.

Asphalt Institute
 The Asphalt Handbook, The Asphalt Institute, College Park, Md.
Barth, E. J.
 Asphalt Science and Technology, Gordon and Breach, Science Publishers, Inc., New York, 1962.
Pfeiffer, J. P.
 The Properties of Asphaltic Bitumen, American Elsevier Publishing Company, New York, 1950.
Traxler, R. N.
 Asphalt: Composition, Properties, and Uses, Reinhold Publishing Corporation, New York, 1961.

Petrochemicals

Asinger, F.
 Chemie und Technologie der Monoolefine, Akademie-Verlag GmbH, Berlin, 1957.
Asinger, F.
 Chemie und Technologie der Paraffin-Kohlenwasserstoffe, Akademie-Verlag GmbH, Berlin, 1956.
Asinger, F.
 Einführung in die Petrolchemie, Akademie-Verlag GmbH, Berlin, 1959.
Astle, M. J.
 The Chemistry of Petrochemicals, Reinhold Publishing Corporation, New York, 1956.
Brooks, B. T.
 The Chemistry of the Nonbenzenoid Hydrocarbons, Reinhold Publishing Corporation, New York, 1956, 2d ed.
Davidson, J. G.
 Petrochemical Survey, Society of Chemical Industry, London, 1956.
Egloff, G., et al.
 Isomerization of Pure Hydrocarbons, Reinhold Publishing Corporation, New York, 1942.
Egloff, G., and G. Hulla
 Alkylation of Alkanes, vol. 1, *Patents*, Reinhold Publishing Corporation, New York, 1948.
Ellis, C.
 The Chemistry of Petroleum Derivatives, Reinhold Publishing Corporation, New York, vols. 1–2, 1934–1937.
Faith, W. L., et al.
 Industrial Chemicals, John Wiley & Sons, Inc., New York, 1965, 3d ed.
Goldstein, R. F.
 The Petroleum Chemicals Industry, John Wiley & Sons, Inc., New York, 1958, 2d ed.
Hatch, L. F.
 The Chemistry of Petrochemical Reactions, Gulf Publishing Company, Houston, 1955.
Littman, E. R., ed.
 Methods of Analysis for Petrochemicals, Chemical Publishing Company, New York, 1958.
Sittig, M.
 Organic Chemical Processes, Noyes Press, Inc., Pearl River, N.Y., 1962.
Steiner, H. M.
 Introduction to Petroleum Chemicals, Pergamon Press, New York, 1961.

Catalysis and Catalysts

Advances in Catalysis
 Academic Press Inc., New York, vols. 1–16, 1948–1966.
Berkman, S., et al.
 Catalysis, Reinhold Publishing Corporation, New York, 1940.
Bond, G. C.
 Catalysis by Metals, Academic Press Inc., New York, 1962.

Booth, H. S., and D. R. Martin
   Boron Trifluoride and Its Derivatives, John Wiley & Sons, Inc., New York, 1949.
Collier, C. H., ed.
   Catalysis in Practice, Reinhold Publishing Corporation, New York, 1957.
DeBoer, J. H., et al., eds.
   Symposium on the Mechanism of Heterogeneous Catalysis, Amsterdam, 1959, *Proceedings*, American Elsevier Publishing Company, New York, 1960.
Emmett, P. H., ed.
   Catalysis, Reinhold Publishing Corporation, New York, vols. 1–7, 1954–1960.
Griffith, R. H., and J. D. F. Marsh
   Contact Catalysis, Oxford University Press, Fair Lawn, N.J., 1957, 3d ed.
Komarewsky, V. I.
   Catalytic, Photochemical and Electrolytic Reactions, Interscience Publishers, Inc., New York, 1956, 2d ed.
Schwab, G. M.
   Handbuch der Katalyse, Springer-Verlag OHG, Vienna, vols. 1–7, 1941–1943.
Topchiev, A. V.
   Boron Fluoride and Its Compounds as Catalysts in Organic Chemistry, Pergamon Press, New York, 1959.

Physical Properties of Hydrocarbons

American Petroleum Institute
   Physical Constants of Hydrocarbons; C_1 to C_{10}, The American Petroleum Institute, New York, 1961.
American Society for Testing and Materials
   Knocking Characteristics of Pure Hydrocarbons, The American Society for Testing and Materials, Philadelphia, STP 225, 1958.
American Society for Testing and Materials
   Physical Constants of Hydrocarbons Boiling below 350°F, The American Society for Testing and Materials, Philadelphia, 1950.
California Natural Gasoline Association
   Physical Constants of the Components of Natural Gas and Natural Gasoline, TS 401, The California Natural Gasoline Association, Los Angeles, 1953.
California Research Corp.
   Raman Spectra of Hydrocarbons, The California Research Corp., Richmond, Calif., 1959.
Claxton, G., ed.
   Physical and Azeotropic Data; Hydrocarbons and Sulfur Compounds Boiling below 200°C, National Benzole & Allied Products Association, London, 1958.
Doss, M. P., comp.
   Physical Constants of the Principal Hydrocarbons, The Texas Co., New York, 1943.
Egloff, G.
   Physical Constants of Hydrocarbons, 5 vols., Reinhold Publishing Corporation, New York, 1939–1953.
Farkas, A.
   Physical Chemistry of the Hydrocarbons, 2 vols., Academic Press Inc., New York, 1950–1953.
Ferris, S. W.
   Handbook of Hydrocarbons, Academic Press Inc., New York, 1955.
Katz, D. L., and M. J. Rzasa
   Bibliography for Physical Behavior of Hydrocarbons under Pressure and Related Phenomena, J. W. Edwards Publisher, Incorporated, Ann Arbor, Mich., 1946.
Matheson Company
   Matheson Gas Data Book, The Matheson Company, East Rutherford, N.J., 1961, 3d ed.
Maxwell, J. B.
   Data Book on Hydrocarbons, D. Van Nostrand Company, Inc., New York, 1950.
Muckleroy, J. A.
   Bibliography on Hydrocarbons, 1946–1960, Natural Gas Processors Assoc., Tulsa, 1962.

Orlicek, A. F., and H. Pöll
Hilfsbuch für Mineralöl-techniker, Springer-Verlag OHG, Vienna, vol. 1, 1951.
Rossini, F. D., et al.
Data issued by Research Project 44, Tables of Physical and Thermodynamic Properties of Hydrocarbons; Infrared Spectral Data; Ultraviolet Spectral Data; Raman Spectral Data; Mass Spectral Data, American Petroleum Institute, New York, 1958.
Rossini, F. D., et al.
Hydrocarbons from Petroleum, Reinhold Publishing Corporation, New York, 1953.
Rossini, F. D., et al.
Selected Values of Physical and Thermodynamic Properties of Hydrocarbons and Related Compounds, Comprising the Tables of the A.P.I. Research Project 44, extant as of 12/1/52, Carnegie Press, Pittsburgh, 1953.
Rossini, F. D., et al., comp.
Selected Values of Properties of Hydrocarbons, U.S. National Bureau of Standards Circular 461, Superintendent of Documents, Washington, D.C., 1947.
Sage, B. H., and W. H. Lacey
Some Properties of the Lighter Hydrocarbons, Hydrogen Sulfide and Carbon Dioxide, American Petroleum Institute, New York, 1955.
Sage, B. H., and W. H. Lacey
Thermodynamic Properties of the Lighter Paraffin Hydrocarbons and Nitrogen, American Petroleum Institute, New York, 1950.

Handbooks and Tables of Data

American Chemical Society
Azeotropic Data I, II (Advances in Chemistry, Series, Nos. 6 and 35), The American Chemical Society, Washington, D.C., 1952.
American Gas Journal
Gas Handbook, The American Gas Journal, New York.
American Society for Testing and Materials
Manual on Measurement and Sampling of Petroleum and Petroleum Products, The American Society for Testing and Materials, Philadelphia, 1950.
American Society for Testing and Materials—Institute of Petroleum
Petroleum Measurement Tables, The American Society for Testing and Materials, Philadelphia, 1952.
Atack, F. W., and D. R. Atack, eds.
Handbook of Chemical Data, Reinhold Publishing Corporation, New York, 1957.
Baumeister, T., and L. S. Marks, eds.
Standard Handbook for Mechanical Engineers, McGraw-Hill Book Company, New York, 1966, 7th ed.
Bolz, H. A., and G. E. Hagemann, eds.
Materials Handling Handbook, The Ronald Press Company, New York, 1958.
Bruckner, H.
Gastafeln, Physikalische thermodynamische und brenntechnischer Eigenschaften der Gase und sonstigen Brennstoffen, R. Oldenbourg KG, Munchen, 1952, 2d ed.
Chemical Rubber Publishing Co.
Handbook of Chemistry and Physics, The Chemical Rubber Company, Cleveland, 1966, 47th ed.
Conway, B. E.
Electrochemical Data, American Elsevier Publishing Company, New York, 1952.
Dergazarian, T. E., et al., eds.
JANAF Interim Thermochemical Tables, Dow Chemical Co., Midland, Mich., vols. 1–2, 1960.
Din, F.
Thermodynamic Functions of Gases, vol. I, Ammonia, Carbon Dioxide, Carbon Monoxide; vol. II, Air, Acetylene, Ethylene, Propane, and Argon, Butterworth Scientific Publications, London, 1956.
Doss, M. P.
Properties of the Principal Fats, Fatty Oils, Waxes, Fatty Acids and Their Salts, Texas Co., New York, 1952.

Dreisbach, R. R.
    Physical Properties of Chemical Compounds, vols. 1–3, 1955–1961, American Chemical Society, Washington, D.C., 1955–1961.
Dreisbach, R. R.
    Pressure-volume-temperature Relationships of Organic Compounds, McGraw-Hill Book Company, New York, 1952, 3d ed.
Forsythe, W. E., comp.
    Smithsonian Physical Tables, Smithsonian Institution, Washington, D.C., 1954, 9th ed.
General Electric Co.
    Properties of Combustion Gases/Systems: C_nH_{2n}—air, vol. I, *Thermodynamic Properties;* vol. II, *Chemical Composition of Equilibrium Mixtures*, McGraw-Hill Book Company, New York, 1955.
Goldsmith, A., et al.
    Handbook of Thermophysical Properties of Solid Materials, The Macmillan Company, New York, 5 vols., 1961.
Hatt, H. H., et al., comp.
    Anti-composition Tables of Carbon Compounds, Cambridge University Press, New York, 1955.
Jordan, T. E.
    Vapor Pressure of Organic Compounds, Interscience Publishers, Inc., New York. 1954.
Kaye, G. W. C., and T. H. Laby
    Tables of Physical and Chemical Constants, John Wiley & Sons, Inc., New York, 1959, 12th ed.
Kent, W.
    Mechanical Engineers' Handbook, John Wiley & Sons, Inc., New York, 2 vols., 1950.
Kobe, K. A., et al.
  * Thermochemistry of Petrochemicals*, Texas University Bureau of Engineering Research, Austin, 1958.
Landolt-Börnstein's
    Zahlenwerte und Funktionen aus Physik, Chemie, Astronomie Geophysik und Technik, Springer-Verlag OHG Berlin, in several volumes, 1950–1959, 6th ed.
Manufacturing Chemists' Association
    Selected Values of Properties of Chemical Compounds, The Manufacturing Chemists' Association, Washington, D.C., 1955.
Menzel, D. H., ed.
    Fundamental Formulas of Physics, Prentice-Hall, Inc., Englewood Cliffs, N.J., 1955.
Office of Critical Tables
    Consolidated Index of Selected Property Values, Physical Chemistry and Thermodynamics, Publication 976, National Academy of Sciences—National Research Council, Washington 25, D.C., 1962.
Parsons, R.
    Handbook of Electrochemical Constants, Academic Press Inc., New York, 1959.
Perry, J. H., ed.
    Chemical Business Handbook, McGraw-Hill Book Company, New York, 1954.
Perry, J. H., ed.
    Chemical Engineers' Handbook, McGraw-Hill Book Company, New York, 1963, 4th ed.
Ross, T. K., and D. C. Freshwater
    Chemical Engineering Data Book, Leonard Hill Books Ltd., London, 1958.
Stanford Research Institute
    Chemical Economics Handbook, an annual publication, The Stanford Research Institute, Menlo Park, Calif.
Staniar, W.
    Plant Engineering Handbook, McGraw-Hill Book Company, New York, 1959, 2d ed.
Stull, D. R., and G. C. Sinke
    Thermodynamic Properties of the Elements (Advances in Chemistry Series, No. 18), American Chemical Society, Washington, D.C., 1956.

Timmermans, J.
  Physico-chemical Constants of Binary Systems in Concentrated Solutions, John Wiley & Sons, Inc., New York, 4 vols., 1959–1960.
Timmermans, J.
  Physico-chemical Constants of Pure Organic Compounds, American Elsevier Publishing Company, New York, 1950.
Touloukian, Y. S., ed.
  Retrieval Guide to Thermophysical Properties Research Literature, McGraw-Hill Book Company, New York, vol. 1 (in 3 parts), 1961.
U.S. National Bureau of Standards
  Selected Values of Chemical Thermodynamic Properties, Circular 500 by F. D. Rossini et al., Superintendent of Documents, Washington, D.C., 1952.
U.S. National Bureau of Standards
  Tables of Chemical Kinetics, Homogeneous Reactions, Circular 510, Superintendent of Documents, Washington, D.C., 1951.
U.S. National Bureau of Standards
  Tables of Thermal Properties of Gases, Circular 564, by J. Hilsenrath et al., Superintendent of Documents, Washington, D.C., 1955.
Woods, K. B., et al., eds.
  Highway Engineering Handbook, McGraw-Hill Book Company, New York, 1960.

Directories

Financial Post
  Survey of Oils, vol. 19, MacLean-Hunter Publ. Co., 481 University Ave., Toronto, Canada.
Guides Lesourd
  Guide du petrole et de l'equipment petrolier, O. Lesourd, Paris.
International Petroleum Register
  Mona Palmer, New York.
King, W., Ltd.
  Gas Journal Calendar and Directory, The W. King Company, Ltd., London.
Midwest Oil Register
  Directory of Oil Refineries; Construction, Petrochemical and Natural Gas Processing Plants, covers the world. The Midwest Oil Register Company, Tulsa.
Oil and Gas Journal
  Directory . . . oil refineries, field processing plants, and petrochemical plants . . . including company personnel, The Oil and Gas Journal, Tulsa.
Ozanne, H., comp.
  Oil Record, Petroleum Industry Projects, Washington, D.C., 3 vols., 1952–1954.
Ozanne, H., comp.
  The Gas Record, Petroleum Industry Projects, Washington, D.C., 2 vols., 1953.
Petroleum Engineer
  World-wide Directory of Petroleum Supply and Service Organizations, Petroleum Engineer Publ. Co., Dallas.
Petro/Chem Engineer
  Directory of Petrochemical Plants in the U.S. and Canada, Petroleum Engineer Publ. Co., Dallas.
Petroleum Publishing Company
  World-wide Personnel Directory; Refining and Gas Processing, The Petroleum Publishing Company, Tulsa.
Pipe Line News
  Annual Directory of Pipe Lines, Oildom Publishing Co., Bayonne, N.J.
Polk, R. L. & Co.
  Polk's Chemical and Oil Directory, R. L. Polk & Co., Nashville, Tenn.
Skinner, W. E., comp.
  Oil and Petroleum Yearbook, W. E. Skinner, 20 Copthall Ave., London, EC 2.
Stanford Research Institute
  Directory of Producers of Synthetic Organic Chemicals, The Stanford Research Institute, Menlo Park, Calif.

PART 5. STATISTICS

American Gas Association
  Gas facts.
  Gas requirements and supplies of the gas utility and pipeline industry. Annual.
  Historical statistics of the gas industry.
  Monthly bulletin of utility gas sales.
American Metal Market
  Metal statistics, 1961, The American Metal Market Company, 18 Cliff St., New York, 54th ed.
American Petroleum Institute
  Fire losses in the petroleum industry.
  Petroleum facts and figures.
  Semimonthly report covering production and inventories of LP and LR gas.
  Summary of injuries in the petroleum industry.
Andriot, J. L.
  Guide to United States government statistics, Documents Index, Arlington, Va.
Automobile Manufacturers Association
  Automobile facts and figures. Annual.
  Motor truck facts.
Automotive Industries
  Annual statistical issue.
Butane-Propane News
  Review and forecast number.
Chase Manhattan Bank
  Annual analysis of the petroleum industry.
  Monthly review of the petroleum situation.
Coman, E. T.
  Sources of business information, Prentice-Hall, Inc., Englewood Cliffs, N.J., 1949.
Ethyl Corp.
  Monthly report on retail gasoline sales.
  Yearly report on gasoline sales.
Frank, N. D., comp.
  Current sources of information for market research, American Marketing Assoc., Chicago, 1954.
Gas Age
  Annual review and forecast number.
Hauser, P. M., and W. R. Leonard, eds.
  Government Statistics for Business Use, John Wiley & Sons, Inc., New York, 1956, 2d ed.
Independent Petroleum Association of America
  Report on petroleum supply and demand. Annual.
  U.S. wholesale price of crude petroleum and principal products. Monthly.
Liquefied Petroleum Gas Association
  LP-Gas market facts, Chicago. Issued annually.
Manufacturing Chemist's Association
  The Chemical Industry Fact Book, The Manufacturing Chemist's Association, Washington, D.C.
National Association of Motor Bus Operators
  Bus facts. Biennial.
National Petroleum Refiners Association
  Digest of pipeline rates on crude petroleum oil. Annual.
  Digest of pipeline rates on gasoline and other petroleum products. Annual.
  Lubricating oil refineries, United States and Canada. Annual.
Oil & Gas Journal
  Review and forecast number.
  International number.
  Midyear review and outlook number.

Oil and Petroleum Yearbook, Incorporating the Oil and Petroleum Manual
  W. E. Skinner, London.
Oildom
  Weekly price supplement.
Platt's Oilgram Price Service
  McGraw-Hill. Daily.
  Crude oil supplement. Monthly.
  Price average supplement. Monthly.
  Tank wagon supplement. Weekly.
Platt's Oil Price Handbook and Oilmanac
  McGraw-Hill, New York. Annual.
Rocq, M. M.
  U.S. sources of petroleum and natural gas statistics, Special Libraries Association, New York, 1961.
Stanford Research Institute
  Chemical Economics Handbook, The Stanford Research Institute, Menlo Park, Calif.
U.S. Bureau of the Census
  Sales of lubricating oils and greases. Biennial.
U.S. Bureau of Labor Statistics
  Retail prices and indexes of fuel and electricity. Monthly.
U.S. Bureau of Mines
  Crude petroleum and petroleum products. Monthly petroleum statement.
  Crude petroleum and petroleum products. Annual statement number.
  District Five military and civilian petroleum demand.
  Forecast of market demand for crude oil.
  Injury experience in oil and gas industry of United States. Annual.
  International petroleum trade. Monthly.
  Liquefied petroleum gases. Monthly. LPG underground stocks.
  Marketed production of natural gas. Annual.
  Mineral industry surveys: Petroleum products surveys: Motor gasolines.
  Mineral facts and problems. 1960.
  Minerals yearbook. Annual.
  Natural gas monthly report.
  Natural gas quarterly report.
  Natural gas annual report.
  Natural gas liquids and liquefied refinery gases. Monthly.
  Natural gas statistics. 1936–1950.
  Natural gasoline and cycling plants in United States. Annual.
  Petroleum refineries, cracking plants, natural gasoline plants and cycling plants in District Five. Annual.
  Petroleum refineries including cracking plants in United States. Annual.
  Petroleum situation in District Five. Monthly.
  Sales of asphalt. Annual.
  Sales of fuel oil and kerosine. Annual.
  Sales of liquefied petroleum gases. Annual.
  Sales of liquefied petroleum gases (District Five). Annual.
  Weekly crude stocks report.
  World petroleum statistics. Monthly.
  World retail prices and taxes on gasoline, kerosine, and motor lubricating oils. Annual.
U.S. Department of Commerce, Bureau of the Census
  Current industrial reports, Series M29A.
U.S. Federal Power Commission
  Consumption of fuel for production of electric energy. Monthly.
  Statistics for natural gas companies. Annual.
U.S. Interstate Commerce Commission
  Distribution of petroleum products. Quarterly.

U.S. Tariff Commission
  Facts for industry.   Monthly.
  Synthetic organic chemicals.   U.S. production and sales.   Annual.
Wasserman, Paul, et al., eds.
  Statistics sources, Gale Research Co., Detroit, 1962.
World Petroleum
  Annual refinery number.
  Canadian review number.

Section 14

GLOSSARY OF PROCESSING TERMS

The following list represents a selection of definitions from the latest *Glossary of Terms Used in Petroleum Refining*, published by the American Petroleum Institute, Division of Refining.*

In general, the editors of this Handbook have included only those terms pertaining specifically to refining operations (processes, equipment, products) and have excluded many words dealing with test methods and specialty products, or which are of such wide usage as to be found in most current dictionaries.

GLOSSARY

A

absorber—See **absorption tower.**

absorption gasoline—Gasoline extracted from natural gas or refinery gas, e.g., by contacting the absorbed gas with an oil and subsequently distilling the gasoline from the heavier oil.

absorption oil—An oil used to separate the heavier components from a vapor mixture by absorption of the heavier components during intimate contacting of the oil and vapor. It is used in recovering natural gasoline from wet gas.

absorption plant—A plant for recovering the condensable portion of natural or refinery gas, by absorbing these heavier hydrocarbons in an absorption oil, followed by separation and fractionation of the absorbed material.

absorption system—A system of manufacturing casinghead gasoline, in which vapors of casinghead gas are absorbed in oil by being passed upward through a continuous spray of oil in a tower-like apparatus.

absorption tower—A tower or column which causes contact between a rising gas and a falling liquid so that part of the gas may be dissolved in the liquid.

accumulator—A vessel for the temporary storage of a gas or liquid; usually used for collecting sufficient material for a continuous charge to some refining process.

acetone-benzol process—A dewaxing process in which acetone and benzol are used as solvents.

acid blowcase—A small tank constructed of material to withstand corrosion and pressure, from which the sulfuric acid used in process treating is blown by compressed air to the agitator. When the tank is of cast iron, it is usually termed an "egg."

acid heat test—A test by which is measured the temperature rise resulting from the addition of commercial sulfuric acid to a petroleum distillate under controlled conditions.

acid sludge—The residue left after treating petroleum oil with sulfuric acid for the removal of impurities. It is a black, viscous substance containing the spent acid and impurities.

acid treating—A refining process in which unfinished petroleum products, such as gasoline, kerosine, and lubricating-oil stocks, are contacted with sulfuric acid to improve their color, odor, and other properties.

* 2d edition, 1962.

adsorption—The adhesion of the molecules of gases or dissolved substances to the surface of solid bodies, resulting in relatively high concentration of the gas or solution at the place of contact.

adsorption, chromatographic—See **chromatographic adsorption.**

adsorption gasoline—Natural gasoline obtained by the adsorption process (adsorbed from wet gas by activated carbon or charcoal). (See also **molecular sieves.**)

afterburning—In catalytic cracking, the combustion of carbon monoxide, entrained coke particles, or both (in the dilute phase) in the catalyst regenerator.

air-blown asphalt—Asphalt produced by blowing air through residual oils or similar mineral-oil products at moderately elevated temperatures.

air sweetening—A process in which air or oxygen is used to oxidize lead mercaptides to disulfides instead of using elemental sulfur.

Airco-Hoover sweetening—A mercaptan removal process in which gasoline charge stock is caustic-washed, water-washed, and dried. The stock is then heated and passed, with some oxygen, through a reactor containing a slurry of diatomaceous earth impregnated with copper chloride. The oxygen regenerates the reagent.

Airlift Thermofor catalytic cracking—A moving-bed, reactor-over-regenerator, continuous catalytic process for conversion of heavy gas oils into lighter products. The catalyst is moved by a stream of air.

albertite—A jet-black, brittle, natural hydrocarbon possessing a conchoidal fracture and a specific gravity of approximately 1.1.

aliphatic hydrocarbon—A hydrocarbon of open-chain structure, such as ethane, butane, octane, butene, and acetylene.

alkali treatment—See **caustic wash.**

alkali wash—See **caustic wash.**

alkylate bottoms—Residue from fractionation of total alkylate which boils higher than the aviation gasoline range; sometimes called heavy alkylate or alkylate polymer.

alkylation—In petroleum refining, usually the union of an olefin (ethylene through pentene) with isobutane to yield high-octane, branched-chain paraffinic hydrocarbons. Alkylation may be accomplished as a thermal or as a catalytic reaction. Alkylation of benzene and other aromatics with olefins is also done to yield alkyl aromatics. (See also specific alkylation processes under alphabetical listing.)

American melting point—An arbitrary temperature 3°F above the ASTM Method D 87 paraffin-wax melting point. The latter is also known as the English melting point (see **ASTM melting point**).

aniline point—The minimum temperature for complete miscibility of equal volumes of aniline and the sample under test. ASTM Method D 611 describes procedures for determining aniline point and mixed aniline point of petroleum products and hydrocarbon solvents. A product of high aniline point will be low in aromatics and naphthenes and, therefore, high in paraffins. Aniline point is often specified for spray oils, cleaning solvents, and thinners, where effectiveness depends upon aromatic content. In conjunction with API gravity, the aniline point may be used to calculate the net heat of combustion of aviation fuels. (See also **mixed aniline point.**)

antiknock—Resistance to detonation or pinging in spark-ignition engines.

antiknock agent—A chemical compound such as tetraethyl lead which, when added in small amount to the fuel charge of an internal-combustion engine, tends to lessen knocking.

antistripping agent—An additive used in an asphaltic binder to overcome the natural affinity of an aggregate for water instead of asphalt. It assists the asphalt to adhere to wet surfaces. Such agents may act either by affecting the interfacial tensions between bitumen-stone or bitumen-water or by forming a rigid interfacial film, which prevents establishment of the critical contact angle.

API Engine Service Classification System—Classifications and designations for lubricating oils for automotive engines, approved by the API Lubrication Committee and adopted by the API Division of Marketing. Service conditions are segregated into broad classifications for the purpose of guiding the choice of types of motor oils desirable for each service (see **Services DG, DM, DS, ML, MM, MS**).

API gravity—An arbitrary scale expressing the gravity or density of liquid petroleum products. The measuring scale is calibrated in terms of degrees API. It may be calculated in terms of the following formula:

$$\text{Deg API} = \frac{141.5}{\text{sp gr } 60°F/60°F} - 131.5$$

aromatic (adjective)—Derived from, or characterized by, the presence of the benzene ring.

aromatic (noun)—Benzene or a compound derived from benzene, with one or more rings of carbon atoms, as distinct from those of aliphatic or alicyclic character.

aromatization—The formation of aromatic hydrocarbons from nonaromatic hydrocarbons by: 1, rearrangement or combination into ring structures, sometimes with attendant dehydrogenation; and 2, dehydrogenation of naphthenes.

Arosorb process—Separation of aromatics from nonaromatics in refinery streams, such as catalytic reformate, by adsorption on a gel from which they are recovered by desorption.

asphalt—Black to dark-brown solid or semisolid cementitious material which gradually liquefies when heated and in which the predominating constituents are bitumens. These occur in the solid or semisolid form in nature; are obtained by refining petroleum; or are combinations with one another or with petroleum or derivatives thereof. (See also specific asphalts under alphabetical listing.)

asphalt cement—A fluxed or unfluxed asphalt especially prepared as to quality and consistency for direct use in the manufacture of bituminous pavements.

asphalt emulsion—Emulsion of asphalt cement in water containing a small amount of emulsifying agent.

asphalt flux—An oil used to reduce the consistency or viscosity of hard asphalt to the point required for use.

asphalt primer—A liquid asphaltic material of low viscosity which, upon application to a nonbituminous surface, is completely absorbed. Its purpose is to waterproof the existing surface and prepare it to serve as a base for further construction.

asphaltene—Any of the components of the bitumen in petroleums, petroleum products, asphalt cements, and solid native bitumens, which are soluble in carbon disulfide but insoluble in paraffin naphthas.

asphaltic road oil—A thick, fluid solution of asphalt. Sometimes heavy crude asphalt of petroleum is employed, but more often a residual oil, obtained by distilling off the more volatile constituents of crude asphalt petroleum, is used. (See also **nonasphaltic road oil**.)

asphaltum—See **asphalt**.

ASTM-CFR engine—A special engine developed by the Coordinating Fuel and Equipment Research Committee of the Coordinating Research Council, Inc., to determine the knock tendency of gasolines.

ASTM distillation—Any distillation made in accordance with an ASTM distillation procedure; and, especially, a distillation test made on such products as gasoline and kerosine to determine the initial and final boiling points and the boiling range (ASTM Method D 86).

ASTM gum test—1. An analytical method for determining the amount of existing gum in a gasoline by evaporating a sample from a glass dish on an elevated-temperature bath with the aid of circulating air. 2. Any gum test carried out in accordance with an ASTM gum test procedure. (ASTM Method D 381 and ASTM Method D 525 are generally used in the United States for the determination of gum in motor gasoline.)

ASTM melting point—The temperature at which wax first shows a minimum rate of temperature change; also known as the English melting point.

Attapulgus clay—See **fuller's earth**.

Autofining—A fixed-bed catalytic process for desulfurizing distillates. The pelleted catalyst is cobalt-molybdate on alumina and may be regenerated in place. Hydrogen for desulfurizing is produced in the process.

average boiling point—Unless otherwise indicated, the sum of the ASTM distillation temperatures from the 10 per cent point to the 90 per cent point, inclusive, divided by 9. Sometimes half the initial and half the maximum distillation temperatures are also added, and the sum then divided by 10.

aviation gasoline—Any of the special grades of gasoline suitable for use in certain airplane engines, as given in ASTM Method D 910.

Aviation Method—A method for determining the knock-limited power, under lean-mixture conditions, of fuels for use in spark-ignition aircraft engines. It is carried out as prescribed in ASTM Method D 614.

aviation mix—The antiknock fluid used in aviation gasoline, consisting of tetraethyl lead, ethylene dibromide, and dye.

aviation turbine fuel—See **jet fuel**.

B

Bari-Sol process—A dewaxing process which employs a mixture of ethylene dichloride and benzol as the solvent. (A mixture of ethylene dichloride and carbon tetrachloride has also been used.) The mixture of solvent and wax-containing hydrocarbon is amenable to centrifugal dewaxing.

barometric condenser—A device for condensing steam by direct contact with water. It produces a partial vacuum in refinery equipment such as a vacuum pipe still.

barrel—A common unit of measurement of liquids in the petroleum industry; it equals 42 U.S. standard gallons.

barrel, slack—See **slack barrel**.

basic sediment and water—The heavy material which collects in the bottom of storage tanks, usually composed of oil, water, and foreign matter. Also called bottoms, bottom settlings, etc.

batch oil—A pale, lemon-colored, low-viscosity mineral oil used particularly in the manufacture of cordage.

battery—A series of stills or other refinery equipment operated as a unit.

battery limits—A term used when a unit or a battery is to be built in a refinery by an outside contractor or construction company. It specifies the area within which the contractor shall supply all services, and defines the limits beyond which this shall be done by the refinery.

Baumé gravity—Specific gravity of liquids expressed as degrees on the Baumé scale. For liquids lighter than water, the formula is:

$$\text{Sp gr } 60°\text{F}/60°\text{F} = \frac{140}{130 + \text{deg Bé}}$$

For liquids heavier than water the formula is:

$$\text{Sp gr } 60°\text{F}/60°\text{F} = \frac{145}{145 - \text{deg Bé}}$$

bauxite—Mineral matter, essentially hydrated aluminum oxide, formed by the chemical weathering of igneous rocks; used as a treating agent.

B-B fraction—The butane-butene fraction. The boiling points of butane and butene (butylene) are so close that they are, as a rule, collected as one fraction.

belching—See **puking**.

bell cap—A hemispherical or triangular casting placed over the riser in a tower to direct the vapors through the liquid layer on the tray. (See also **bubble cap**.)

bend—A curved length of pipe struck to a larger radius than the elbow. Pipe bends of 45, 90, or 180° are often specified as one-eighth, one-quarter, or one-half bends. A slight bend is often called a spring. (See also **return bend**.)

Bender process—A continuous fixed-bed chemical treating process using lead sulfide catalyst for sweetening light distillates. The sweetening converts mercaptans to disulfides by oxidation.

bentonite—The mineral montmorillonite, a magnesium-aluminum silicate.   Used as a treating agent, as a component of drilling mud, and in greases.

benzene—Colorless liquid hydrocarbon, C_6H_6, with one ring of carbon atoms. Made from coal tar and by catalytic reforming of naphthenes, it is used in the manufacture of phenol, styrene, nylon, detergents, aniline, phthalic anhydride, diphenyl, nitrobenzene, chlorobenzene; as a solvent; and as a component of high-octane gasoline.

benzin—A refined light naphtha used for extraction purposes.   The term "petroleum benzin" has appeared in the U.S. Pharmacopoeia for many years; it should be used only for material meeting USP specifications.

benzine—An obsolete term for light petroleum distillates covering the gasoline and naphtha range.   (See **ligroine**.)

benzol—The general term which refers to commercial or technical (not necessarily pure) benzene.

biological oxidation—The consumption of organic matter by bacteria which live upon the organic matter and oxygen in the surrounding medium, converting the organic matter into nontoxic gases.

bitumen—A mixture of hydrocarbons of natural or pyrogenous origin, or both, frequently accompanied by their nonmetallic derivatives, and which are completely soluble in carbon disulfide.

bituminous—Containing bitumen or constituting the source of bitumen.

bituminous sand—Naturally occurring mixtures of asphalt and loose sand grains. The bituminous cementing material associated with the sand may run as high as 12 per cent.

BL Method—Primarily, a research test for refiners and automobile manufacturers. It provides five road octane numbers for a given fuel used in a car of a specified compression ratio.   These road octane numbers are obtained at five engine speeds from 750 to 2,500 rpm.

black acids—One of the sulfonates found in acid sludge which are insoluble in naphtha, benzene, and carbon tetrachloride.   These sulfonates are very soluble in water, but insoluble in 30 per cent sulfuric acid.   In the dry, oil-free state, the sodium soaps are black powders.

black oil—Any of the black-colored oils used for the lubrication of heavy, slow-moving, rough machinery where it would be impractical or uneconomical to use higher grade lubricants.

black soap—See **black acid.**

black strap—The black material (mainly lead sulfide) formed in the treatment of sour light oils with doctor solution (which see) and found at the interface between the oil and the solution.

blanket steam—See **top steam**.

blended fuel oil—A mixture of residual and distillate fuel oils.

blending naphtha—A distillate used to thin heavy stocks to facilitate processing, e.g., to thin lubricating oil in dewaxing processes.

blending value (antiknock)—Some antiknock blending agents possess the property of apparently increasing the rated octane number of certain gasoline base stocks to a higher octane number than their own value in terms of octane numbers.   This property is known as the blending value.

blending value (hydrocarbon)—In octane ratings of a hydrocarbon made on blends of 20 per cent hydrocarbon plus 80 per cent of a 60:40 mixture of isooctane and n-heptane, the "blending octane number" is a hypothetical value obtained by extrapolation to a rating of 100 per cent concentration of the hydrocarbon.   This value indicates the blending value of the hydrocarbon being examined.

blistering, hydrogen—See **hydrogen blistering.**

blocked operation—The use of a single process unit alternately in more than one operation.

bloom—Fluorescence; the color of an oil by reflected light when this differs from its color by transmitted light.   For certain purposes the trade has preferred oils of yellowish-green rather than bluish-green bloom.   This demand can be met by special processing.

blotter press—Plate and frame press with blotting paper for the filtering medium.

blowback—A system in which a liquid or a gas is continuously bled through the lead lines of an instrument meter into the main line. This prevents the main-line fluid from coming in contact with the meter body, thus eliminating vaporization, corrosion, or plugging.

blowdown stack—A stack into which the contents of a unit are emptied in an emergency.

blowing still—A still in which blown or oxidized asphalt is made.

blown asphalt—See **air-blown asphalt.**

body—A colloquial term referring to the viscosity of a lubricating oil.

boiler horsepower—A unit of rate of water evaporation. One boiler horsepower equals the evaporation, per hour, of $34\frac{1}{2}$ lb of water at a temperature of 212°F into steam at 212°F.

boiling range—The range of temperature, usually determined at atmospheric pressure in standard laboratory apparatus, over which the boiling or distillation of an oil commences, proceeds, and finishes.

boilup rate—See **throughput.**

Borderline Method—See **BL Method.**

bottled gas—Ordinarily, butane or propane, or butane-propane mixtures, liquefied and bottled under pressure for domestic use. (See also **liquefied petroleum gas.**)

bottoms—The liquid which collects in the bottom of a vessel (tower bottoms, tank bottoms) either during a fractionating process or while in storage. (See also **basic sediment and water; residue.**)

breathing—The movement of gas (oil vapors or air) in and out of the vent lines of storage tanks as a result of alternate heating and cooling.

bright stock—Refined, high-viscosity lubricating oils usually made from residual stocks by suitable treatment, such as a combination of acid treatment or solvent extraction with dewaxing or clay finishing.

bromine number—The number of grams of bromine absorbed by 100 g of oil; indicates the percentage of double bonds in the material. The content of olefins can thus be calculated if their type and molecular weight are known.

brown acid—One of the oil-soluble petroleum sulfonates found in acid sludge. These sulfonates can be recovered by extraction with naphtha solvent. Brown-acid sulfonates are somewhat similar to mahogany sulfonates but are more water-soluble. In the dry, oil-free state, the sodium soaps are light-colored powders.

brown soap—See **brown acid.**

BS doctor—See **black strap.**

BS&W—See **basic sediment and water.**

bubble cap—An inverted cup with a notched or slotted periphery to disperse the vapor in small bubbles beneath the surface of the liquid on the bubble plate in a distillation column.

bubble point—The temperature at which incipient vaporization of a liquid in a liquid mixture occurs. It corresponds with the equilibrium point of 0 per cent vaporization or 100 per cent condensation. The pressure should be specified if other than 1 atm.

bubble tower—A fractionating tower so constructed that the vapors rising pass up through layers of condensate on a series of plates or trays. The vapor passes from one plate to the next above by bubbling under one or more caps and out through the liquid on the plate. The less volatile portions of vapor condense in bubbling through the liquid on the plate, overflow to the next lower plate, and ultimately back into the reboiler. Fractionation is thereby effected.

bubble tray—One of the circular, perforated plates having the internal diameter of a tower, set at specified distances in a tower to collect the various fractions produced in fractional distillation.

bumping—The knocking against the walls of a still occurring during the boiling of a petroleum product containing water.

bunker "C" fuel oil—A heavy residual fuel oil used by ships, by industry, and for large-scale heating installations. The United States Navy calls it "Navy heavy"; in industry, it is often referred to as No. 6 fuel.

burning line—A line conveying refinery fuel gas, as distinguished from gas intended for subsequent processing.

burning oil—An illuminating oil, such as kerosine, mineral seal oil, etc., suitable for burning in a wick lamp.

burning point—See **fire point.**

burning-quality index—An empirical numerical indication of the likely burning performance of a furnace or heater oil; derived from certain ASTM distillation points and the API gravity, and generally recognizing the factors of paraffinicity and volatility.

Burton process—A thermal cracking process formerly used, in which oil was cracked in a pressure still and any condensation of the products of cracking also took place under pressure.

butane-butene fraction—See **B-B fraction.**

butane dehydrogenation—A process for splitting hydrogen from butane (see **Houdry butane dehydrogenation**).

butane vapor-phase isomerization—A process for isomerizing n-butane to isobutane using aluminum chloride catalyst on a granular alumina support and with hydrogen chloride as a promoter.

C

C_1, C_2, C_3, C_4, C_5 fractions—A common way of representing fractions containing a preponderance of hydrocarbons of 1, 2, 3, 4, or 5 carbon atoms, respectively, without reference to hydrocarbon type.

calendar day—1. The time from midnight of one day to midnight of the next day. 2. The $\frac{1}{365}$ part of a normal year; it relates to total lapsed time of a period measured in days.

California polymerization—A process developed by the Standard Oil Company of California for converting C_3-C_4 olefins to motor fuel. The catalyst is phosphoric acid on quartz chips.

carbene—Any of the components of bitumen which are soluble in carbon disulfide but insoluble in carbon tetrachloride. With the exception of solubility, they are similar to asphaltenes.

carbon burning rate—The weight of carbon burned from the cracking catalyst in the regenerator per unit of time.

carbon knock—An engine knock caused by carbon deposits.

carbon residue (Conradson; Ramsbottom)—The carbonaceous residue formed after evaporation and pyrolysis of a petroleum product. The residue is not entirely composed of carbon but is a coke which can be further changed by pyrolysis. Tests for the determination of carbon residue are intended to provide some indication of the relative coke-forming propensities of an oil (ASTM Methods D 189 and D 524).

carbonization—The conversion of an organic compound, such as oil, into char or coke by heat in the substantial absence of air.

carryover—1. Relatively nonvolatile contaminating material which is carried over by the overhead effluent from a fractionating column, absorber, or reaction vessel. It may be carried as liquid droplets or finely divided solids suspended in a gas, a vapor, or a discrete liquid. 2. That portion of a finely divided catalyst which escapes the cyclones of cracking units.

cascade tray—A fractionating device consisting of a series of parallel troughs arranged in stair-step fashion. Liquid from the tray above enters the uppermost trough. Liquid thrown from this trough by vapor rising from the tray below impinges against a plate and a perforated baffle. Liquid passing through the baffle enters the next lower of the troughs.

casinghead gas—The natural gas which issues from the casinghead (the mouth or opening) of an oil well.

casinghead gasoline—The liquid hydrocarbon product extracted from casinghead gas by one of three methods: compression, absorption, or refrigeration. (See also **natural gasoline.**)

cast—See **bloom.**

castor machine oil—A petroleum lubricating oil thickened with aluminum-base soap, such as aluminum oleate.

"cat" cracking—See **catalytic cracking.**

catalyst selectivity—The relative activity of a catalyst with respect to a particular compound in a mixture, or the relative rate in competing reactions of a single reactant.

catalyst stripping—The introduction of steam, at a point where spent catalyst leaves the reactor, in order to strip, i.e., remove, hydrocarbons retained on the catalyst.

catalytic activity—The ratio of the space velocity of the catalyst under test to the space velocity required for the standard catalyst to give the same conversion as the catalyst being tested. In most cases, the ratio is multiplied by 100 before being reported.

catalytic cracking—The conversion of high-boiling hydrocarbons into lower boiling substances by means of a catalyst which may be used in a fixed bed, moving bed, or fluid bed. Natural or synthetic catalysts are employed in bead, pellet, or powder form. Feedstocks may range from naphtha cuts to reduced crude oils. (See also specific catalytic cracking processes under alphabetical listing.)

catalytic reforming—The rearranging of hydrocarbon molecules in a gasoline-boiling-range feedstock to produce other hydrocarbons having a higher antiknock quality. Some coincident catalytic reforming reactions are isomerization of paraffins, cyclization to naphthenes, dehydrocyclization to aromatics, and hydrocracking to lower paraffin hydrocarbons. (See also specific catalytic reforming processes under alphabetical listing.)

Catforming—A naphtha-reforming process whose main feature is the catalyst. The catalyst is a platinum-silica-alumina composition which permits relatively high space velocities and results in very high hydrogen purity. Regeneration to prolong catalyst life is practiced on a blocked-out basis with a dilute air-in-steam mixture.

caustic wash—1. The process of treating a product with a solution of caustic soda to remove minor impurities. 2. The solution itself.

centrifuge refining—In this process, also called the centrifugal process, heated lubricating-oil stock is mixed with sulfuric acid and allowed to remain in contact with the acid for a certain time. The mixture is then centrifuged to separate the sour oil from the acid sludge. The sour oil is neutralized with alkali, filtered, etc.

ceresin—A hard, brittle wax obtained by purifying ozokerite (see **microcrystalline wax; ozokerite).**

cetane number (calculated)—The cetane number of distillate fuels as estimated from the API gravity and mid-boiling point by using a formula given in Appendix II of ASTM Method D 975. This estimate is used if a standard test engine is not available or if the sample is too small for an engine test.

cetane number (test method)—The percentage by volume of normal cetane, in a blend with heptamethylnonane (HMN), which matches the ignition quality of the fuel when compared with the procedure specified in ASTM Method D 613.

cetane number improver—A substance which, when added to a diesel fuel, has the effect of increasing its cetane number. In this class are nitro alkanes, nitrates, nitro carbonates, and peroxides.

cetene number—An obsolete designation for the starting and running quality of diesel fuel, using cetene as a reference fuel. It has been succeeded by the cetane number.

CFR—*Coordinating Fuel* and Equipment *Research* Committee, composed of engine-manufacturing, petroleum-refining, petroleum-consuming, university, government, and other technical men who supervise cooperative testing and study of engine fuels for the Coordinating Research Council, Inc.

CFR-ASTM engine—A special engine developed by the CRC Coordinating Fuel and Equipment Research Committee, and standardized by the American Society for Testing and Materials, to determine the knock tendency of gasolines.

CFR research test method—See **Research Method.**

channel process—Carbon-black process in which a system of iron channel beams is used as the depositing surface for the carbon black. These channel beams are set with the flat side downward over a horizontal row of burners and are given a slow, reciprocating motion. The black is scraped off these beams and is removed by spiral conveyors.

channeling—1. The phenomenon observed among gear lubricants and greases when they thicken, due to cold weather or other causes, to such an extent that a groove is formed through which the part to be lubricated moves without actually coming in full contact with the lubricant. 2. A term used in percolation filtration; may be defined as a preponderance of flow through certain portions of the clay bed.

characterization factor—The UOP characterization factor K. Variations in physical properties with change in character of the stock are quantitatively expressed by means of this factor, defined as the ratio of the cube root of the molal average boiling point, T_B, in degrees Rankine (deg R = deg F + 460), to the specific gravity at 60°F/60°F:

$$K = \frac{(T_B)^{1/3}}{\text{sp gr}}$$

This factor ranges from 12.5 for paraffinic stocks to 10.0 for the highly aromatic stocks. Also called Watson factor.

charcoal test—A test standardized by the American Gas Association and the Natural Gas Processors Association for determining the natural gasoline content of natural gas. The gasoline is adsorbed in activated charcoal and then recovered by distillation. The apparatus and procedures for this test, as well as specifications for the activated charcoal used, are described in Testing Code 101-43, a joint AGA-NGPA publication.

cheesebox still—An early type of vertical cylindrical still designed with a vapor dome.

chemical octane number—The octane number added to gasoline by refinery processes or by the use of octane number improvers such as tetraethyl lead. (Editor's note: Chemical octane numbers are frequently obtained by preparing blends which may contain such high-octane-number components as selected alkylate, isobutane, pure toluene, and others.)

chiller—Refining apparatus in which the temperature of paraffin distillates is lowered preparatory to filtering out the solid wax.

Chlorex process—A process for extracting lubricating-oil stocks in which the solvent used is Chlorex, a trade-mark for β,β-dichlorodiethyl ether.

chromatographic adsorption—Selective adsorption on materials (such as activated carbon, alumina, or silica gel). Liquid or gaseous mixtures of hydrocarbons are passed through the adsorbent in a stream of diluent, and certain components are preferentially adsorbed. As flow continues, heavier hydrocarbons replace or drive off the lighter ones that have been adsorbed, so that the effluent from the adsorbent contains the individual hydrocarbons in successive order of their molecular weights.

chromatography—A method of separation based on selective adsorption. A solution of the substance or substances desired is allowed to flow slowly through a column of adsorbent. Different substances will pass with different speeds down the column and will eventually be separated into zones. The column core can then be pushed out and the zones of material cut apart, or the zones can be eluted by passing more solvent down the column and collecting it in small fractions.

Partition chromatography involves the selective solution of the desired material between two solvents. The final solvent, usually water, is used to wet the solid material packed in the column, and the first solvent containing the desired material is poured into the column as described.

Paper chromatography is a micromethod. A drop of the liquid to be investigated is placed near one end of a strip of paper. This end is immersed in solvent which travels down the paper and distributes the materials present in the original drop selectively. Comparison with known substances makes identification possible.

Gas chromatography is an analytical technique for separating mixtures of volatile substances. The procedure consists of introducing the mixture to be examined into the chromatographic column and washing it down (eluting it) with an inert gas. The column is packed with adsorbent materials which selectively retard the components of the sample.

clarified oil—The heavy oil which has been taken from the bottom of a fractionator in a catalytic cracking process, and from which residual catalyst has been removed.

clarifier—An apparatus or device for removing the color or cloudiness of an oil or water by separating the foreign material through mechanical or chemical means. It may embody the principle of centrifugal action, filtration, simple heating, or treatment with acid or alkali.

clay—A filtering medium, especially fuller's earth, often used in refineries to absorb coloring matter from oil.

clay contact process—See **contact filtration.**

clay refining—A treating process in which vaporized gasoline or other light petroleum product is passed through a bed of granular clay such as fuller's earth. Certain olefinic compounds are polymerized to gums and are adsorbed by the clay.

clay regeneration—A process in which spent coarse-grained adsorbent clays from percolation processes are cleaned for reuse by deoiling them with naphtha, steaming out the excess naphtha, and then roasting in a stream of air to remove carbonaceous matter.

clay treating, Gray—See **Gray clay treating.**

clay wash—A light oil, such as kerosine or naphtha, used to clean fuller's earth after it has been used in a filter.

cleaner's naphtha—See **Stoddard solvent.**

cleaner's solvent—See **Stoddard solvent.**

clear gasoline—A gasoline which is free from antiknock additives such as tetraethyl lead. In making comparative engine tests between leaded and unleaded fuels, the clear, unleaded gasoline is sometimes referred to as straight gasoline base, base fuel, or as gasoline "neat."

closed steam—Steam used in a heating coil so that no direct contact takes place between the steam and the material to be heated. Contrast this with open steam, where intimate contact between the media is possible.

cloud point—With respect to a petroleum oil, the temperature at which paraffin wax or other solid substances begin to crystallize or separate from the solution, imparting a cloudy appearance to the oil when the oil is chilled under prescribed conditions. These conditions are described in ASTM Method D 97.

coal oil—1. Oil obtained by the destructive distillation of bituminous coal. 2. Archaic term for kerosine made from petroleum.

cobalt molybdate desulfurization—A desulfurization process using cobalt molybdate catalyst. (See also **Unifining.**)

codimer—The product obtained by the union of two dissimilar olefin molecules. In the petroleum refinery, it is generally understood to mean the product of polymerization of isobutylene with one of the two normal butylenes. The product can be hydrogenated and thus converted into a blending agent (hydrocodimer). (See also **polymerization.**)

coil—Any of the number of turns of piping, or one of a series of connected pipes in rows or layers, for the purpose of radiating or absorbing heat. (See also **cracking coil.**)

coke, petroleum—See **petroleum coke.**

coke drum—A vessel in which coke is formed and which can be cut off from the process for cleaning.

coke knocker—A mechanical device for breaking loose coke formations within a tower or drum.

coke number—Frequently used, particularly in Great Britain, to report the results of the Ramsbottom carbon residue test, which is also referred to as a coke test.

coker—The processing unit in which coking takes place.

coking—1. Any cracking process in which the time of cracking is so long that coke is produced as the bottom product. 2. Thermal cracking for conversion of heavy, low-grade oils into lighter products and a residue of coke.   3. The undesirable building up of coke or carbon deposits on refinery equipment.   (See also specific coking processes under alphabetical listing.)

coking still—A still, usually a batch still, in which coking is carried out.

cold pressing—The process of separating wax from oil by first chilling (to help form wax crystals) and then filtering under pressure in a plate and frame press.

cold settling—Processing for the removal of wax from high-viscosity stocks, wherein a naphtha solution of the waxy oil is chilled and the wax crystallizes out of the solution. Upon standing, the wax settles; it leaves a clear, oil-naphtha solution, nearly wax-free.

color stability—The resistance of oil to color change due to light, aging, etc.

column, fractionating—See **fractionating column.**

combination unit—A unit which combines more than one process, such as straight-run distillation and selective cracking, and which might also involve viscosity breaking, naphtha reforming, gas recovery, and coking.

combustion knock—See **diesel knock.**

compression process—The recovery of natural gasoline from "wet natural gas" by compression and condensation rather than by absorption.

Con Carbon (Conradson carbon residue)—See **Conradson carbon test.**

condensate—1. The liquid product coming from a condenser.   2. A light hydro-carbon mixture produced as a liquid product in a gas-recycling plant through expansion and cooling of the gas.

condenser—Ordinarily, a water-cooled heat exchanger used for cooling and liquefying oil vapors.   Where the cooling medium used is air, the condenser is called an air condenser.   (See also specific condensers under alphabetical listing.)

condenser box—A large box-shaped structure in which the condenser, which may consist of coils or "worms," is submerged in a heat-absorbing medium, usually water.

Conradson carbon test—A carbon residue test originated by Dr. Pontius H. Conradson and standardized by ASTM Method D 189.   (See also **carbon residue.**)

contact filtration—A process in which finely divided adsorbent clay is ultimately mixed with oil (and sometimes heated) to remove color bodies and to improve the stability of the oil.   The mixture is then filtered to remove the clay.   Also called contacting.

contact time—The time in which a substance is in direct, intimate relation with a treating agent.   For example, in fixed-bed polymerization, the volume of vaporized charge stock per unit time divided into the volume of catalyst will give the time of exposure.

continuous contact coking—A thermal conversion process employing the mass-flow lift principle to give continuous coke circulation.   Oil-wetted coke particles move downward into the reactor in which cracking, coking, and drying take place. Pelleted coke, gas, gasoline, and gas oil are products of this process.

continuous contact filtration—A process to finish lubricants, waxes, or special oils after acid treating, solvent extraction, or distillation.   The charge stock is slurried with finely divided adsorbent clay; the mixture is heated to hasten the reaction, stripped with steam, and cooled; and the clay is separated from the treated stock.

convection section—That portion of the furnace in which tubes receive heat by convection from the flue gases.

conversion per pass—1. In a once-through cracking operation, the yield of cracked gasoline per unit of fresh feed.   2. In a recycle cracking operation, the yield of gasoline per unit of total charge (fresh feed plus recycle).   3. The disappearance of a given component of the charge.

Coordinating Research Council, Inc.—An organization supported jointly by the American Petroleum Institute and the Society of Automotive Engineers, and which administers the work of CFR (which see) and other committees pertaining to correlation of test work on fuels, lubricants, engines, etc.

copolymer—A mixed polymer or heteropolymer, product of polymerization of two or more substances at the same time (copolymerization) to yield a product which is not a mixture of the separate polymers but a complex with properties different from those of each polymer alone.   For example, Buna S is a copolymer of butadiene and styrene.   (See also **polymerization.**)

copper dish gum—The milligrams of gum found in 100 ml of gasoline when evaporated under controlled conditions in a polished copper dish; indicates the potential gum content of a material.

copper number—Milligrams of mercaptan sulfur per 100 ml of sample.   A standard ammoniacal copper sulfate solution is used to titrate the mercaptans.

copper strip corrosion—A qualitative method of determining the corrosivity of a product by its effect on a small strip of polished copper suspended or placed in the product.   ASTM Method D 130 is generally used in the United States for testing copper corrosion by petroleum products.

copper sweetening—Any of the processes involving the use of cupric chloride as a clay slurry, or in the form of an aqueous solution, or deposited from an aqueous solution on a porous support such as fuller's earth, bauxite, etc.   Mercaptans are oxidized to disulfides by oxygen in the presence of cupric chloride.   (See also specific copper-sweetening processes under alphabetical listing.)

Cottrell precipitator—A piece of equipment designed for the removal of dusts or mists from gases by passing them through an electrostatic field.

cracker—See **cracking still.**

cracking—A phenomenon by which large oil molecules are decomposed into smaller, lower-boiling molecules; at the same time, certain of these molecules, which are reactive, combine with one another to give even larger molecules than those in the original stock.   The more stable molecules leave the system as cracked gasoline, but the reactive ones polymerize, forming tar and even coke.   (Editor's note: Cracking may be in either the liquid or vapor phase.   When a catalyst is used to bring about the desired chemical reaction, this is called catalytic cracking; otherwise, it is assumed to be thermal cracking.) (See also **reforming;** specific cracking processes under alphabetical listing.)

cracking activity—See **catalytic activity.**

cracking coil—A piece of refinery equipment used for cracking heavy petroleum products.   It consists of a coil of heavy pipe running through a furnace, so that the oil passing through it is subject to high temperature.

cracking severity—See **severity factor.**

cracking still—The combined equipment—furnace, reaction chamber, fractionator—for the thermal conversion of heavier charging stock to gasoline.

CRC—See **Coordinating Research Council, Inc.**

CRC method—Research technique formulated, approved, and published by the Coordinating Research Council, Inc.

crude assay—A procedure for determining the general distillation characteristics and other quality information of crude oil.   One well-known method is that of the Bureau of Mines, U.S. Department of the Interior, described in *Bulletin* 207: *The Analytical Distillation of Petroleum and Petroleum Products.*

crude naphtha—Light distillate made in the fractionation of crude oil.

crude scale wax—The wax product from the first sweating of the slack wax.   The melting point varies depending upon the ultimate wax product to be made.   It is resweated to yield waxes of higher melting points.

crude still—Distillation equipment in which crude oil is separated into various products.

cumene—A colorless liquid having the formula $C_6H_5CH(CH_3)_2$.   It is used as an aviation gasoline blending component and as an intermediate in the manufacture of chemicals.

cut—A fraction obtained by a separation process.   (See also **fractional distillation.**)

cut, heart—See **heart cut.**

cut oil—An oil which has been partially emulsified with water in the presence of air.

cut point—The boiling-temperature division between cuts of a crude oil or base stock.

cutback—The blending of heavier oils with lighter oils to bring the heavier oils to the desired specifications.

cutback asphalt—Asphalt liquefied by the addition of a volatile liquid such as naphtha or kerosine. Upon exposure to the atmosphere, the volatile product evaporates, leaving the asphalt.

cutting oil—An oil to lubricate and cool metal-cutting tools. Such oils may be water-soluble or water-insoluble. Usually, mineral oils are blended with lard oil or other oiliness or extreme-pressure agents to produce water-insoluble cutting oils, or with sulfonated products and other emulsifying agents to produce water-soluble cutting oils. Also called cutting fluid, cutting lubricant.

cycle plant—Similar to a natural gasoline plant in that the liquid hydrocarbons are removed from natural gas. In a cycle plant the gas is then put back into the ground to maintain pressure on the oil reservoir.

cycle stock—Product taken from some later stage of a process and recharged to the process at some earlier stage.

cyclization—The process of changing an open-chain hydrocarbon structure to a closed ring, e.g., hexane to benzene.

cyclone separator—A conical vessel provided with a tangential inlet for a gas stream containing finely divided solids or liquid droplets, normally designed with a centrally located overhead gas-withdrawal line. Powdered solids or coagulated liquids are separated by centrifugal force and pass downward along the incline (conical) to a centrally located outlet. Usually a pipe, known as a dip leg, is connected to this bottom outlet and, in catalytic cracking, serves to convey the solids back to the catalyst bed.

Cycloversion desulfurization—A continuous, fixed-bed catalytic process to desulfurize gasoline, kerosine, and other light fractions. Feedstocks are desulfurized by passing them in the vapor state over bauxite. The catalyst may be regenerated with air and steam.

cylinder oil—A viscous lubricating stock, used for lubricating the cylinders and valves of steam engines.

cylinder stock—1. The residuum remaining in a still after the lighter parts of certain crude oils have been vaporized. From it are made various steam-cylinder oils which, when filtered and processed, become bright stocks. 2. Compounded or straight oils, such as refined residuum or refined high-viscosity distillates, used for lubricating steam cylinders.

cymogene—A light, commercial petroleum product of gravity 100 to 110° API and boiling point at approximately 32°F: consisting largely of butane; used as a local anesthetic and in certain types of refrigerating machines. (Editor's note: According to ASTM D 288, this term is archaic and should not be used.)

D

1-D test—A modification of the L-1 test using a supercharged single-cylinder diesel engine (see **L-1 test**).

DC naphtha—See **Stoddard solvent.**

deactivation—Reduction in catalyst activity by coating of catalyst particles by contaminants, or by a change in the physical structure of the catalyst particles.

dead oil—An oil, with a density greater than water, which is distilled from tar.

dealkylation—The removal of an alkyl group from a compound. The dealkylation of higher aromatics is of interest in the production of increased quantities of benzene, toluene, naphthalene, etc.

deasphalted oil—See **propane deasphalting.**

deasphalting—See **solvent deasphalting.**

deblooming—A process by which the bloom of fluorescence is removed from oils by exposing the oil in shallow tanks to atmospheric conditions. A similar effect may be produced chemically or by masking. Chemicals, such as mononitronaphthalene and yellow coal-tar dyes, added to mineral oils to mask the fluorescence are called deblooming agents.

debutanization—Distillation to separate butane and lighter components from propane and heavier components.

decarbonizing—A thermal conversion process designed to maximize coker gas-oil production and minimize coke and gasoline yields. Flow is essentially the same as that for delayed coking. However, compared to conventional delayed coking, decarbonizing is operated at essentially lower temperatures and pressures. (Editor's note: Distinguish this process from solvent or propane decarbonizing, which see.)

decoking—The removal of petroleum coke from equipment such as coking drums. Hydraulic decoking uses high-velocity water streams as the cutting means.

decolorizing—The process of removing suspended, colloidal, and dissolved impurities from liquid petroleum products by filtering, adsorption, chemical treatment, distillation, bleaching, etc.

de-ethanization—Distillation to separate ethane and lighter components from propane and heavier components. Also called de-ethanation.

DeFlorez cracking—A process for the cokeless cracking of petroleum.

degree API—See **API gravity.**

degree Baume—See **Baume gravity.**

degree Engler—A measure of viscosity. The ratio of the time' of flow of 200 ml of the liquid tested, through the viscosimeter devised by Engler, to the time required for the flow of the same volume of water gives the number of degrees Engler.

dehydrocyclization—Any process involving both dehydrogenation and cyclization reactions. (See also **dehydrogenation; cyclization.**)

dehydrogenation—The removal of hydrogen from a chemical compound; for example, the removal of two hydrogen atoms from butane to make butylene, and the further removal of hydrogen to make butadiene. (See also specific dehydrogenation processes under alphabetical listing.)

delayed coking—A semicontinuous thermal process for the conversion of heavy stock to lighter material. Feedstock is preheated in a pipe still, discharged into large insulated coke drums, and held there for a length of time while cracking takes place. Provision is made for removing gas, gasoline, and gas oil as overhead products. Coke is recovered from the coke drums (see **decoking**).

delayed knock—See **diesel knock.**

demethanization—The process of distillation in which methane is separated from the heavier components. Also called demethanation.

deoiling—Reduction in quantity of liquid oil entrained in solid wax. The oil may be removed by draining (sweating) or by a selective solvent (see **MEK deoiling**).

depentanizer—A fractionating column for the removal of pentane and lighter fractions from a mixture of hydrocarbons.

depropanization—Distillation in which lighter components are separated from C_4's and heavier material. Also called depropanation.

deresining, solvent—See **solvent deasphalting.**

desalting—Removal of mineral salts (mostly chlorides) from crude oils.

desalting, electric—See **electric desalting.**

desorption—The reverse process of adsorption whereby adsorbed matter is removed from the adsorbent. The term is also used as the reverse of absorption.

desulfurization—The removal of sulfur or sulfur compounds from a charge stock. (See also specific desulfurization processes under alphabetical listing.)

detergent oil—A lubricating oil possessing special sludge-dispersing properties for use in internal-combustion engines. These properties are usually conferred on the oil by the incorporation of special additives. Detergent oils hold sludge particles in suspension and thus promote engine cleanliness.

dewaxing—See **solvent dewaxing.**

diesel fuel—Fuel used for internal combustion in diesel engines; usually that fraction which distills after kerosine; similar to gas oil.

diesel index—An expression for the ignitability of fuel relative to its aniline point:

$$\text{Diesel index} = \frac{\text{aniline point (deg F) times API gravity}}{100}$$

diesel knock (delayed knock, combustion knock, or lag knock)—When the delayed period of ignition is long, a large quantity of atomized fuel accumulates in the combustion chamber. When combustion does take place, the sudden high pressure due to the accumulated fuel causes diesel knock.

Diesulforming—A fixed-bed catalytic process, employing hydrogen, for desulfurizing naphtha, middle distillate, and gas oil. The catalyst is pelleted molybdenum oxide which may be regenerated *in situ* with an air-steam mixture.

dimer—The product formed when two identical olefin molecules are combined by polymerization to produce a single molecule having the same elements in the same proportions as in the original molecules. (See also **polymerization.**)

dimerization—The process of combining two identical olefinic molecules by polymerization. (See also **polymerization.**)

diolefin—An open-chain hydrocarbon having two double bonds per molecule. The empirical formula is C_nH_{2n-2}. The typical diolefin is butadiene, used to make synthetic rubber.

diolefin hydrogenation—A fixed-bed catalytic process to hydrogenate diolefins in C_4 and C_5 fractions to mono-olefins in alkylation feedstock. The catalyst is pelleted nickel sulfide on alumina and may be regenerated *in situ* with a mixture of air and steam, followed by sulfiding with hydrogen sulfide.

disk and doughnut—A type of fractionating tower construction consisting of alternating disks and doughnut-shaped plates which provide a mixing action.

distillate—That portion of a liquid which is removed as a vapor and condensed during a distillation process. (See also specific distillates under alphabetical listing.)

distillation—Vaporization of a liquid and its subsequent condensation in a different chamber. The separation of one group of petroleum constituents from another by means of volatilization in some form of closed apparatus, such as a still, by the aid of heat. (See also specific distillation processes under alphabetical listing.)

distillation curve—Curve made by plotting the percentage of gasoline (or other petroleum product) distilled versus the temperature. Such distillation curves are frequently made in connection with ASTM Method D 86.

distillation loss—The difference, in a laboratory distillation, between the volume of liquid originally introduced into the distilling flask and the sum of the residue and the condensate recovered.

distillation range—The difference between the temperature at the initial boiling point and at the end point, as obtained by the distillation test.

disulfide oil—One of the oils obtained by oxidizing the mercaptans extracted from light petroleum distillates to disulfides which separate from the extract as an oily layer.

doctor, BS—See **black strap.**

doctor solution—A solution (sodium plumbite) made from lead oxide and sodium hydroxide, used to treat gasoline or other light petroleum distillates to remove mercaptan sulfur. The "doctor test" is used for the detection of compounds in light petroleum distillates which react with sodium plumbite.

doctor sweetening—A refining process for sweetening gasoline, solvents, and kerosine which converts mercaptans to disulfides using sodium plumbite and sulfur.

doctor test—See **doctor solution.**

double-solvent extraction—A double-solvent process for the simultaneous deasphalting and solvent treating of a lubricating-oil stock or any residual or distillate base stock. Propane is the paraffinic solvent for the deasphalting operation: cresylic acid, usually containing from 20 to 40 per cent phenol, is the second solvent.

downcomer—A means of conveying liquid from one tray to the next below in a bubble tray column.

drawoff—A connection which allows liquid to flow from the side or bottom of a vessel.

drift—Water lost from a water-cooling tower as liquid droplets entrained in the exhaust air. Units: pounds per hour, or per cent of circulating water flow.

drip—1. Term applied to the oil which comes through the cloths of the paraffin-wax presses while pressing and "drips" into the trough below. Also applied to filter drainings too dark in color to be included in filter stock. They are set aside for charging into the next filter, together with the next lot of the same stock to be filtered. 2. A discharge mechanism installed at a low point in a gas transmission line to collect and remove liquid accumulations.

dropping point—The temperature at which grease passes from a semisolid to a liquid state under the conditions of ASTM Method D 566.

dry-cleaning fluid—See **Stoddard solvent.**

dry gas—A gas which does not contain fractions that may easily condense under normal atmospheric conditions.

dry point—In a distillation test, the temperature at which the last drop of petroleum fluid evaporates.

Dualayer distillate process—A process for removing mercaptans and oxygenated compounds from distillate fuel oils and similar products, using a combination of treatment with concentrated caustic solution and electrical precipitation of the impurities.

Dualayer gasoline process—A process for extracting mercaptans and other objectionable acidic compounds from petroleum distillates, using Dualayer solution, which consists of concentrated potassium or sodium hydroxide containing a solubilizer.

Dubbs cracking—A continuous, liquid-phase thermal cracking process formerly used. Fresh feed entered the dephlegmator and, together with the dephlegmator bottoms, was pumped to a pipe still and cracked under pressure. The cracked products passed successively through an expansion chamber and the dephlegmator. All components of the equipment were maintained under pressure.

Duo-Sol extraction—See **double-solvent extraction.**

E

Edeleanu process—A process for refining oils at low temperature with liquid sulfur dioxide (SO_2), or with liquid SO_2 and benzene. A modification is applicable to the recovery of aromatic concentrates from naphthas and heavier petroleum distillates.

effluent refrigeration alkylation—A modification of sulfuric acid alkylation in which reactor effluent is used as a refrigerant (by evaporation of a portion of the butane) to control reaction temperature and, at the same time, to separate isobutane for recycle.

electric desalting—A continuous process to remove inorganic salts and other impurities from crude oil by settling out in an electrostatic field.

electrical precipitation—A process using an electrical field to improve the separation of hydrocarbon reagent dispersions. May be used in chemical treating processes on a wide variety of refinery stocks.

Electrofining—A process for contacting a light hydrocarbon stream with a treating agent (acid, caustic, doctor, etc.), then assisting the action of separation of the chemical phase from the hydrocarbon phase by an electrostatic field.

electrolytic mercaptan process—A process in which aqueous caustic solution is used to extract mercaptans from refinery streams. An electrolytic process is used to regenerate the solution.

elektrion process—A process of polymerization and condensation in which a mixture of a relatively light mineral oil and a fatty oil is subjected to an electric discharge in an atmosphere of hydrogen. The product is a very viscous oil used for blending with lighter lubricating oils to improve the viscosity index.

emulsible oil—See **soluble oil.**

emulsified asphalt—See **asphalt emulsion.**

engine distillate—Petroleum distillate similar to naphtha but often of higher distillation range.

engine oil—A term applied to oils used for the bearing lubrication of all types of engines, machines, and shafting, and for cylinder lubrication in other than steam engines.

Engler degree—See **degree Engler.**

Engler distillation—A standard test for determining the volatility characteristics of a gasoline by measuring the per cent distilled at various specified temperatures. This test is described in ASTM Method D 86.

Engler viscosimeter—An instrument widely used in Europe to measure Engler viscosity (see **degree Engler**).

English melting point—See **ASTM melting point.**

EP additive—Extreme-pressure additive; any of the materials added to lubricants to improve their ability to adhere to metal surfaces and thus lubricate them under high bearing pressures.

EP lubricant—Extreme-pressure lubricant; any of the lubricating oils or greases which contain a substance or substances specifically introduced to prevent metal-to-metal contact in the operations of highly loaded gears. In some cases, this is accomplished by the substances reacting with the metal to form a protective film.

Ethyl fluid—A gasoline antiknock compound manufactured by Ethyl Corporation.

evaporator—A vessel which receives the hot discharge from a heating coil and, by a reduction in pressure, flashes off overhead the light products and allows the heavy residue to collect in the bottom. (See also **flash tower.**)

expansion drum—A vessel of varying size and construction, to which the vapor line of a still is connected, and in which vapors first expand on being driven over from the still.

explosive limits—The limits of percentage composition of mixtures of gases and air within which an explosion takes place when the mixture is ignited. The lower limit of flammability corresponds to the minimum amount of combustible gas and the upper limit to the maximum amount of combustible gas capable of conferring flammability on the mixture.

export kerosine—A grade of kerosine once used for exporting. It has the darkest shade of the standard colors of kerosine, namely, "standard white" (also called "export white").

extract—In solvent refining processes, that portion of the oil which is dissolved in, and removed by, the selective solvent used.

extractive crystallization—A process for separating components from eutectic mixtures. A solution of the mixture is cooled to cause one component to crystallize out, leaving the other component in solution. The crystalline product is then separated from the solution. An example is the separation of p-xylene and m-xylene, using normal pentane as the solvent.

extractive distillation—The separation of different components of mixtures which have similar vapor pressures by flowing a relatively high-boiling solvent, which is selective for one of the components in the feed, down a distillation column as the distillation proceeds. The relatively less soluble component passes overhead, while the selective solvent scrubs the soluble component from the vapor. The solvent containing the dissolved component is withdrawn from the bottom of the column, and the dissolved component and solvent may be separated in an auxiliary apparatus. Extractive distillation has been used for separating toluene from other hydrocarbons. Phenol may be used as the selective solvent.

extreme-pressure additive—See **EP additive.**

extreme-pressure lubricant—See **EP lubricant.**

F

F-1 method—Obsolete; see **Research Method.**

F-2 method—Obsolete; see **Motor Method.**

F-3 Method—Obsolete; see **Aviation Method.**

F-4 Method—Obsolete; see **Supercharge Method.**

farm tractor fuel—Any petroleum product, exclusive of gasoline, diesel fuel, and liquefied petroleum gas, which is used for the generation of power for the operation of farm implements and conforms to the specifications for the two grades in ASTM D 1215. Many tractors can operate on other fuels.

fat gas—See **wet gas.**

fat oil—The bottom or enriched oil drawn from the absorber, as opposed to lean oil. (See also **absorption tower; absorption plant.**)

ferrocyanide process—A regenerative chemical treatment for mercaptan removal using caustic-sodium ferrocyanide reagent.

filter aid—Finely divided, porous, absorbent solids, such as diatomaceous earth, used to increase the efficiency of filtering devices.

filter press—Apparatus employed to separate wax and oil in paraffin-wax distillates. It consists of a series of canvas-covered plates separated by narrow iron rings. The distillate is run into a narrow bore extending the entire length of the press, and is forced up into the spaces between the plates formed by the rings. The canvas covering the plates is easily penetrated by oil, but is impenetrable to the wax, so that the oil drips down from the canvas into a trough beneath the press. A mechanical separation of the oil and wax is thus effected on the principle of filtration.

filter wash—See **clay wash.**

filtrate—The liquid which has passed through a filter; the product from a filtration process.

fire point—The lowest temperature at which, under specified conditions in standardized apparatus, a petroleum product vaporizes sufficiently rapidly to form above its surface an air-vapor mixture which burns continuously when ignited by a small flame.

fire wall—An earth bank or cement wall built around an oil storage tank to prevent the spread of the oil in case of fire or bursting of the tank.

Fischer-Tropsch process—A process for synthesizing hydrocarbons and oxygenated chemicals from a mixture of hydrogen and carbon monoxide.

fixed-bed hydroforming—A reforming process using a regenerable catalyst consisting of molybdenum oxide deposited on activated alumina. The process is cyclic.

fixed-bed operation—A type of operation in which the catalyst remains stationary in the reactor. The catalyst may be regenerated in place periodically. To be contrasted with fluid-bed operation (which see).

flame arrestor—An assembly of perforated plates or screens enclosed in a case and attached to the breather vent on petroleum storage tanks.

flare—A device for disposing of gases by burning.

flash drum—See **flash tower.**

flash point—The lowest temperature at which vapors arising from the oil will ignite momentarily (i.e., flash) on application of a flame under specified conditions.

flash point test (Pensky-Martens closed tester)—A method of test for the determination of the flash point of fuel oils, cutback asphalts, and other viscous materials and suspensions of solids.

flash point test (Tag closed-cup tester)—A method of test for the determination of the flash point of mobile liquids flashing below 175°F, with the exception of fuel oils.

flash tower—The vessel used for the separation of oil fractions in a flash distillation process.

flash vaporization—An operation in which oil is heated to a high temperature, substantially in the liquid phase, and upon release at a lower pressure undergoes considerable vaporization (flashing).

Fleming cracking process—A liquid-phase thermal cracking process formerly used, in which the charge was heated under pressure in a vertical shell still. Fresh feed was introduced through a dephlegmator. Pressure distillate from the dephlegmator was released through a water-cooled needle valve in combination with a water-jet condenser.

floc point—The temperature at which wax or solids separate as a definite floc.

floc test—A quantitative test applied to kerosine and other illuminating oils to detect substances rendered insoluble by heat.

flooding—In a fractionating column, the filling up with a liquid.

Floyd tester—A portable instrument for evaluating the extreme-pressure properties of gear lubricants.

flue gas—Gas from the combustion of fuel, the heating value of which has been substantially spent and which is, therefore, discarded to the flue or stack.

fluid-bed operation—A type of operation based on the tendency of finely divided powders to develop high concentrations (i.e., settle) in a gas stream of low velocity. The fluidized-powder technique involves suspending the finely divided powder in an upwardly flowing stream of gas. Each particle is supported by the gas, and this suspension has most of the characteristics associated with a true liquid. Thus it can flow through pipes and valves, and it will develop hydrostatic heads which allow it to flow from one vessel to another.

fluid catalyst—The finely divided particles used as the catalyst in a fluid-bed catalytic process. Microspherical or powdered catalyst of either natural or synthetic origin may be used.

fluid catalytic cracking—A catalytic cracking process in which the oil is cracked in the presence of finely divided catalyst which is maintained in an aerated or fluidized state by the oil vapors. This powder, or fluid catalyst, is continuously circulated between the reactor and the regenerator, using air, oil vapor, and steam as the conveying media.

fluid coking—A thermal process utilizing the fluidized-solids technique for continuous conversion of heavy, low-grade oils into lighter products. Coking occurs in a thin film on circulating, fluidized, seed coke agitated by rising gaseous products in the reactor.

fluid hydroforming—A fluid-type process for upgrading low-octane-number stocks.

flushing oil—An oil or compound designed to remove used oil, decomposition products, and dirt from lubrication passages, crankcase surfaces, and moving parts of automotive engines accessible to the lubrication system.

flux oil—An oil of low volatility, suitable for blending with bitumen or with asphalt to yield a product of softer consistency or greater fluidity. Selected residual fuel oils may be used for this purpose.

foots oil—The oil sweated out of slack wax. It takes its name from the fact that the oil goes to the foot, or bottom, of the pan when sweated.

form oil—An oil used on wooden or metal concrete forms to keep concrete from sticking to them.

fouling factor—A term used in heat transfer to express the lowering of the clear-film transfer rates caused by corrosion, dirt, or roughness of the surface of tube walls of heat exchangers. This is usually referred to as the resistance to the conduction of the heat. The fouling factor is obtained by multiplying the fouling resistance by 1,000.

fractional condensation—A separation of the components of vaporized oil coming off during distillation by condensing the vapors in stages (partial condensation). The oil of highest boiling point will condense first and may be removed in the liquid state, allowing the portion still in the vapor state to pass on to the next-stage condenser.

fractional distillation—The separation of the components of a liquid mixture by vaporizing and collecting the fractions, or cuts, which condense in different temperature ranges.

fractionating column—A column arranged to separate various fractions of petroleum by a single distillation. The column may be tapped at different points along its length to separate various fractions in the order of their condensing temperatures (boiling points). (See also **bubble tower; tower.**)

fractionation—Separation in successive stages, each removing from a mixture some proportion of one of the substances. The operation may be precipitation, crystallization, distillation, etc. (See also specific fractionation processes under alphabetical listing.)

Frasch process—A process formerly used for removing sulfur by distilling oil in the presence of copper oxide.

front-end volatility—A term applied to the volatility of the lower-boiling fractions of gasoline. There is no general agreement at which a percentage should be distilled.

fuel oil—Any liquid or liquefiable petroleum product burned for the generation of heat in a furnace or firebox, or for the generation of power in an engine, exclusive of oils with a flash point below 100°F (Tag closed-cup tester) and oils burned in cotton- or wool-wick burners. (See also specific fuel oils under alphabetical listing.)

fuel-oil equivalent—The term used in indicating the No. 6 fuel-oil equivalent of a given fuel on a thermal value basis.

fuel sensitivity—The response of a motor fuel to the change in engine severity between the operating conditions of the ASTM Research Method (D 908) and ASTM Motor Method (D 357); numerically equal to the difference between the Research and Motor octane numbers.

fuller's earth—A clay which has high adsorptive capacity for removing color from oils. Floridin and Attapulgus clay are two widely used fuller's earths.

furfural extraction—A single-solvent process in which furfural is used to remove aromatic, naphthenic, olefinic, and unstable hydrocarbons from a lubricating-oil charge stock, thereby improving VI and stability characteristics.

furnace black—Carbon black made by incomplete combustion or thermal decomposition of natural gas or other hydrocarbons in a furnace.

furnace oil—A distillate fuel described in ASTM Specification D 396, primarily intended for use in domestic heating equipment.

G

gage table—One of the sheets prepared to show the contents of a tank for each $\frac{1}{8}$ in. or $\frac{1}{16}$ in. of oil contained in the tank. Tables of temperature corrections are often prepared, also, to serve as gage tables for reducing the contents of the tank to a standard volume at 60°F.

gas (gasoline)—Gasoline is commonly, but improperly, referred to simply as "gas."

gas absorber oil—See **absorption oil.**

gas black—Lampblack collected by introducing a cold iron surface into a luminous gas flame.

gas chromatograph—See **chromatography.**

gas oil—A petroleum distillate with a viscosity and boiling range between those of kerosine and lubricating oil.

gas reversion—A combination of thermal cracking or reforming of naphtha with thermal polymerization or alkylation of hydrocarbon gases carried out in the same reaction zone. (See also **polyforming.**)

gasoline pool—A planning concept which considers gasolines of various qualities as one group for the purpose of blending to meet final product specifications.

gear oil—A lubricating oil for use in standard transmissions, most types of differential gears, and gears contained in gear cases.

gilsonite—A black, lustrous natural asphalt, occurring on a large scale in Utah; it is 90 to 99 per cent bitumen.

Girbotol process—A continuous, regenerative process to separate hydrogen sulfide, carbon dioxide, and other acid impurities from natural gas, refinery gas, etc., using mono-, di-, or triethanolamine (MEA, DEA, or TEA) as the reagent.

glycol-amine gas treating—A continuous, regenerative process to simultaneously dehydrate and remove acid gases from natural gas or refinery gas. A mixture of an aqueous amine (generally monoethanolamine) and di- or triethylene glycol is the treating agent. Sour gas is contacted countercurrently with the solution in a bubble tower. Regeneration of the agent is effected by stripping or distillation in a second tower.

gravity—1. The ratio of the weight of a volume of any liquid to the weight of an equal volume of distilled water at 60°F. (See also **API gravity; Baume gravity.**) 2. The downward force which the earth exerts upon all objects.

Gray catalytic desulfurization—A fixed-bed process for vapor-phase desulfurization of light distillates, using a solid-absorbent-type catalyst such as fuller's earth. The process is continuous when two reactors are used in alternate regenerating and treating operations.

Gray clay treating—A fixed-bed vapor-phase treating process to selectively polymerize unsaturated gum-forming constituents (diolefins) in thermally cracked gasolines. A fixed bed of 30- to 60-mesh fuller's earth is used.

Gulf HDS process—A fixed-bed process for the catalytic hydrocracking of heavy stocks to lower-boiling distillates with accompanying desulfurization. The feedstock may be crude oils or residuals. The catalyst, supported on pelleted alumina, may be regenerated *in situ* with air and steam or flue gas.

Gulfining—A catalytic hydrogen treating process for cracked and straight-run distillates and fuel oils, to reduce sulfur content; improve carbon residue, color, and general stability; and effect a slight increase in gravity.

gum—Rosin-like insoluble deposits formed during the deterioration of petroleum and its products, particularly gasoline. The amount of gummy material in gasoline is known as its gum content. It is usually determined in the United States by ASTM Methods D 381 and D 525. (See also **ASTM gum test; copper dish gum.**)

gyro cracking—A vapor-phase thermal cracking process.

H

header—A common manifold in which a number of pipelines are united. It usually refers to the U-bend connection between two consecutive tubes in the coil.

heart cut—A narrow-range cut, usually taken near the middle portion of the stock being distilled or treated.

heat transfer oil—A medium used for the transfer of heat at temperature levels above that of steam. Probably the most widely used medium is a high-boiling petroleum fraction, usually in the gas-oil range.

heater coating index—A test designed to measure the tendency of fuel oil to form a deposit or to coat heater coils.

heating oil—See **furnace oil.**

heavy alkylate—See **alkylate bottoms.**

heavy-duty oil—An oil having proved oxidation stability, corrosion-preventive properties, and detergent-dispersant characteristics. Oils of this type are generally suitable for use in both high-speed diesel and gasoline engines under heavy-duty service conditions.

heavy ends—The highest boiling portion of a petroleum fraction.

heavy fuel—See **residual fuel oil.**

heel—Oil remaining in a tank after normal emptying.

HF alkylation—An alkylation process whereby olefins (C_3, C_4, C_5) are combined with isobutane in the presence of hydrofluoric acid catalyst.

high-solvency naphtha—One of the special naphthas, characterized by their high solvent power (or low precipitating tendency), in various resins, oils, and plastics used in paint and varnish manufacture. High aromatic content is often conducive to solvency. The kauri-butanol test is one of several used to evaluate solvency; the more naphtha which can be added to a standard kauri gum solution before a standardized degree of precipitation is encountered, the higher the solvency of the naphtha.

holdup—Any time period during which a certain proportion of the liquid introduced into the distillation apparatus is actually in the column of tower as reflux and rising vapor.

Hortonsphere—A spherical pressure-type tank used to store volatile liquids. Its purpose is to prevent excessive evaporation loss which occurs when such products are placed in conventional storage tanks.

hot spot—An area of a vessel or line wall appreciably above normal operating temperature, usually as a result of the deterioration of an internal insulating liner which exposes the line or vessel shell to the temperature of its contents.

Houdresid catalytic cracking—A continuous moving-bed process for catalytically cracking reduced crude oil to produce high-octane gasoline and light distillate fuels.

Houdriflow catalytic cracking—A continuous moving-bed catalytic cracking process employing an integrated single vessel for the reactor and regenerator kiln. Catalyst is transported from the bottom of the unit to the top by means of a gas lift employing steam and compressed flue gas.

Houdriforming—A continuous catalytic reforming process for producing aromatic concentrates and high-octane gasoline from low-octane straight naphthas.

Houdry butane dehydrogenation—A catalytic process for dehydrogenating light hydrocarbons to their corresponding mono- or diolefins. Chromia-alumina catalysts with inert material are used in pellet form.

Houdry fixed-bed catalytic cracking—A cyclic regenerable process for cracking of distillates. Synthetic or natural bead catalysts may be used.

Houdry hydrocracking—A catalytic process combining cracking and desulfurization in the presence of hydrogen. Catalysts include nickel oxide or nickel sulfide on silica alumina, and cobalt molybdate on alumina. (See also **hydrocracking**.)

hydraulic fluid—A fluid of petroleum or nonpetroleum origin used in hydraulic systems. Low viscosity, low rate of change of viscosity with temperature, and low pour point are desirable characteristics.

hydrocodimer—Saturated blending agent of high octane number, produced by the hydrogenation of codimer.

Hydrocol process—A catalytic process for synthesizing gasoline and other products from natural gas, crude oil, fuel oil, and coal.

Hydroconversion, Residuum—See **Residuum Hydroconversion—Humble.**

hydrocracking—A process combining cracking, or pyrolysis, with hydrogenation. Feedstocks can include crude oils, residua, petroleum tars, and asphalts.

hydrodesulfurization—A desulfurization process in which the oil is heated with hydrogen. (See also specific hydrodesulfurization processes under alphabetical listing.)

hydrofining—A fixed-bed catalytic process to desulfurize and hydrogenate a wide range of charge stocks from gases through waxes. The catalyst consists of cobalt and molybdenum oxides on an extruded alumina support and may be regenerated *in situ* with air and steam, or flue gas.

hydroforming—A process in which naphthas are passed over a catalyst at elevated temperatures and moderate pressures, in the presence of added hydrogen or hydrogen-containing gases, to form high-octane motor fuel or aromatics.

hydrogen blistering—Blistering of steel caused by trapped molecular hydrogen formed as atomic hydrogen during corrosion of steel by hydrogen sulfide.

hydrogen treating, Sinclair—See **Sinclair hydrogen treating.**

hydrogenation—The chemical addition of hydrogen to a material. In nondestructive hydrogenation, hydrogen is added to a molecule only if, and where, unsaturation with respect to hydrogen exists. In destructive hydrogenation, the operation is carried out under conditions which result in rupture of some of the hydrocarbon chains (cracking); hydrogen is added where the chain breaks have occurred.

hydrogenation, diolefin—See **diolefin hydrogenation.**

Hydrotreating—Texaco—A treating process using hydrogen for the desulfurization of cracked distillates.

Hygirtol process—A catalytic process for the production of hydrogen from steam and gaseous hydrocarbons.

Hyperforming—A catalytic hydrogenation process for improving the octane number of naphthas through removal of sulfur and nitrogen compounds. Light oil stocks may also be treated by this process for removal of sulfur and nitrogen to produce diesel fuels. The catalyst, consisting of cobalt molybdate on a silica-alumina base, moves by gravity down through the reactor and is returned to the top of the reactor by means of a solids-conveying technique called hyperflow.

hypersorption—A commercial process for the continuous separation of gases by selective adsorption on a moving bed of activated carbon. The carbon containing the adsorbed components is then contacted with heavy product to displace the lighter components, is stripped of heavy components, and is recycled to the primary adsorption step.

hypochlorite sweetening—The oxidation of mercaptans in a sour stock by agitation with aqueous, alkaline hypochlorite solution; used where avoidance of free-sulfur addition is desired, because of a stringent copper strip requirement, and minimum expense is not the primary object.

I

illuminating oil—Oil used for lighting purposes. Such oils are petroleum products heavier than gasoline.

inhibitor—A substance, the presence of which, in small amounts, in a petroleum product prevents or retards undesirable chemical changes from taking place in the product, or in the condition of the equipment in which the product is used. In general, the essential function of inhibitors is to prevent or retard oxidation or corrosion.

inhibitor, pour—See **pour depressor.**

inhibitor sweetening—A treating process to sweeten gasoline of low mercaptan content, using a phenylenediamine type of inhibitor, air, and caustic. It is generally used in connection with caustic washing to minimize disulfide and peroxide formation.

initial boiling point—According to ASTM Method D 86, the recorded temperature when the first drop of liquid falls from the end of the condenser.

initial vapor pressure—The vapor pressure of a liquid at a specified temperature and "zero per cent evaporated." It may be determined either by measuring the the vapor pressure in an apparatus with substantially zero vapor-to-liquid ratio, or by extrapolating to zero vapor space the experimental vapor pressures obtained at several low ratios of vapor-to-liquid volume. The initial vapor pressure (ivp) of ordinary wide petroleum cuts is somewhat higher than the standard Reid vapor pressure (Rvp), up to approximately 50 lb Rvp. With close-boiling cuts, the difference is less noticeable and with pure components, there is no difference between ivp and Rvp.

innage—Either the volume or the measured height of liquid in a tank or container, as measured from the bottom of the tank. (See also **outage.**)

insulating oil—An oil used in circuit breakers, switches, transformers, and other electrical apparatus for insulating, cooling, or both. In general, such oils are well-refined petroleum distillates of low volatility, with high resistance to oxidation and sludging.

iodine number—A measure of the iodine absorption by an oil under standard conditions; used to indicate the quantity of unsaturated compounds present. Also known as iodine value.

Isocrackate—The normally liquid product of Isocracking.

Isocracking—A hydrocracking process for conversion of hydrocarbons which operates at relatively low temperatures and pressures in the presence of hydrogen and a catalyst to produce more valuable, lower-boiling products.

Isoforming—A process in which olefinic naphtha is contacted with an alumina catalyst at high temperature and low pressure to produce isomers of higher octane number.

Iso-Kel process—A fixed-bed, vapor-phase isomerization process using a precious metal catalyst and external hydrogen. Feedstocks include natural gasoline, pentane, and hexane cuts. The product is high-octane blending stock.

Isomate process—A continuous, nonregenerative process for isomerizing C_5-C_8 normal paraffinic hydrocarbons, using aluminum chloride–hydrocarbon catalyst with anhydrous hydrochloric acid as a promoter.

Isomerate process—A fixed-bed isomerization process to convert pentane, hexane, and heptane to high-octane blending stocks. A rugged metal catalyst is used. Outside hydrogen is added to the feed along with recycle gas.

isomerization—A reaction which alters the fundamental arrangement of the atoms in the molecule without adding or removing anything from the original material. In the petroleum industry, straight-chain hydrocarbons are converted catalytically to branched-chain hydrocarbons of substantially higher octane number by isomerization. The process is important for the conversion of n-butane to provide increased supplies of isobutane for alkylation. (See also specific isomerization processes under alphabetical listing.)

Iso-plus Houdriforming—A combination process using a conventional Houdriformer (see **Houdriforming**) operated at moderate severity, in conjunction with one of three possible alternatives—including the use of an aromatic recovery unit or a thermal reformer. The raffinate from the aromatic extraction unit is either reformed separately or recycled. Feedstock is conventional naphtha reforming charge.

J

Jenkins cracking—A liquid-phase thermal cracking process formerly used, in which the charge was heated in a pipe still under pressure and circulated, also under pressure, to and from a bulk shell still by means of a pump. Cracked oil vapors left the shell still through a pressure-relief valve.

jet fuel—Fuel meeting the required properties for use in jet engines and aircraft turbine engines.

JP (jet propulsion)—See **jet fuel.**

K

kauri-butanol value—A measure of solvency used in a special test made on lacquers to show their solvent property. Derives its name from a solution of kauri resin in butyl alcohol. (See also **high-solvency naphtha.**)

Kellogg sulfuric acid alkylation—A process employing sulfuric acid catalyst to unite olefins with isobutane. It features a cascade reactor; autorefrigeration; series flow of acid, isobutane, and product; parallel-flow contact with olefins; and settling zones in the reaction vessel.

kerosine—A refined petroleum distillate suitable for use as an illuminant when burned in a wick lamp. (See also specific kerosines under alphabetical listing.)

K-factor—1. Term used in vapor-liquid equilibrium calculations to denote the ratio of mole fraction in vapor to mole fraction in liquid. 2. In a degree-day system for heating-oil deliveries, the factor which relates degree-day totals for the area to the individual consumer's requirements and controls deliveries into his tank. Each home or building has its own K-factor. 3. Another term for characterization factor (which see).

knock—The noise associated with self-ignition of a portion of the fuel-air mixture ahead of the advancing flame front. The flame front is presupposed to be moving at normal velocity. The source of the flame front is immaterial; it may be the result of surface ignition or spark ignition. (See also specific types of knock under alphabetical listing.)

knock intensity—The intensity of knock when testing gasoline for octane number or knock rating.

knock rating—See **octane number.**

knocker, coke—See **coke knocker.**

knockmeter—An instrument which is used to measure the output of the detonation meter in ASTM knock test methods for rating fuels

knockout—A drum or vessel, constructed with baffles, through which a mixture of gas and liquid is passed to disengage one from the other. As the mixture comes in contact with the baffles, the impact frees the gases and allows them to pass overhead; the heavier substance falls to the bottom of the drum.

Knox cracking—A vapor-phase thermal cracking process.

kogasin—German name for the oily fraction consisting of a mixture of straight- and slightly branched-chain saturated and unsaturated hydrocarbons, produced by the hydrogenation of carbon monoxide over iron, nickel, or cobalt catalyst at low pressure.

L

L-1 test—A full-scale 480-hr engine test made in single-cylinder Caterpillar diesel engine to determine the detergency of heavy-duty oil. This test is reported according to ring sticking, wear, and accumulation of engine deposits.

L-2 test—A test made in a single-cyclinder Caterpillar diesel engine to determine the oiliness of an engine oil. The test is reported according to the extent of scoring on cylinder and piston. Sometimes called the scoring test.

L-3 test—A test carried out in a special four-cylinder Caterpillar diesel engine to test the stability of crankcase oils at high temperatures under severe operating conditions. Used to determine the detergency, wear, ring sticking, engine deposits, and the oxidation stability of the oil, this test was formerly required for approval of oils by the military.

L-4 test—An engine test procedure conducted in a six-cylinder spark-ignition Chevrolet engine which is used for evaluating crankcase oil oxidation stability, bearing corrosion characteristics, and engine deposits.

L-5 test—A test procedure conducted in a General Motors diesel engine to determine the detergency, corrosiveness, ring sticking, and oxidation stability of lubricating oils.

lamp burning—A test of burning oils in which the oil is burned in a standard lamp under specified conditions in order to observe the steadiness of the flame, the degree of incrustation of the wick, and the rate of consumption of the kerosine.

lamp oil—See **kerosine.**

lampblack—A solid product, consisting largely of carbon, obtained by the incomplete combustion of hydrocarbon materials.

Lauson engine—A single-cylinder test engine used largely for making screening tests in connection with the L-series tests in the development of certain lubricating-oil products

lead—Industry parlance for the motor-fuel antiknock additive compound tetraethyl lead, or for other organometallic lead antiknock compounds.

lead susceptibility—Ability of gasolines to respond to the addition of tetraethyl lead, or other organometallic lead antiknock compounds, as reflected in the increase of antiknock quality per increment of lead.

leaded gasoline—Refers to gasoline containing tetraethyl lead or other organometallic lead antiknock compounds.

lean gas—The residual gas from the absorber after the condensable gasoline has been removed from the "wet" gas.

lean oil—1. Absorption oil from which gasoline fractions have been removed. 2. Oil leaving the stripper in a natural-gasoline plant.

life test—A test carried out to determine the probable useful life of a product. For example, ASTM Method D 943 is intended to determine the oxidation inhibitor life of inhibited turbine oils.

light ends—The lower-boiling components of a mixture of hydrocarbons.

light oil—Any of the products distilled or processed from crude oil up to, but not including, the first lubricating-oil distillate. Any of the products beginning with and following the first lubricating-oil distillate is referred to as a "heavy oil."

ligroine—A saturated petroleum naphtha boiling in the range of 20 to 135°C (68 to 275°F) and suitable for general laboratory use. (Editor's note: This term should be used in place of "benzine" or "petroleum ether.")

Linde copper sweetening—A process for treating gasolines and distillates with a slurry of clay and cupric chloride.

liquefied petroleum gas—Light hydrocarbon material, gaseous at atmospheric temperature and pressure, held in the liquid state by pressure to facilitate storage, transport, and handling. Commercial liquefied gas consists essentially of propane, butane, or mixtures thereof.

liquid chromatography—See **chomatography.**

liquid petrolatum—See **white oil.**

liquid seal—The depth of liquid above an orifice from which the vapor issues. For a bubble cap column, this is the depth of liquid above the tip of the slots; for a sieve plate, it is the total depth of liquid on the plate.

liquid SO$_2$-benzene process—A mixed-solvent process for treating lubricating-oil stocks to improve viscosity index (and to dewax also, if required).

litharge—Oxide of lead or plumbous oxide. It is combined with sodium hydroxide solution to form sodium plumbite or doctor solution, and is used to sweeten distillates (see **doctor solution; doctor sweetening**).

live steam—As contrasted to exhaust steam, steam coming directly from a boiler before being utilized for power or heat.

liver—The intermediate layer of dark-colored, oily material, insoluble in weak acid and in oil, which is formed when acid sludge is hydrolyzed.

long residuum—A residue from crude-oil distillation when a relatively small amount is taken overhead.

long-time burning oil—A carefully refined kerosine used in railway semaphore signal lamps. Heavy ends are completely removed so that the oil will burn in a lamp without charring the wick, thereby forming "wick crust." The longtime burning test is usually carried out in the United States as prescribed in ASTM Method D 219.

look box—A box with glass windows built in the rundown lines from stills, so arranged that the stream of oil coming from the condenser coils may be watched at all times and samples for tests drawn as desired.

lower toxic limit—The maximum permissible content of petroleum vapor (0.1 per cent by volume) in a tank or compartment which is to be entered by persons.

LPG—See **liquefied petroleum gas.**

lube—See **lubricating oil.**

lube cut—A fraction of crude oil of suitable boiling range and viscosity to yield lubricating oil when completely refined. Also referred to as lube-oil distillates or lube stock.

lubricated gasoline—A gasoline to which a lubricant has been added.

lubricating oil—A fluid lubricant used to reduce friction between bearing surfaces. Petroleum lubricating oils may be produced either from distillates or residues; amounts of other substances, known as additives, may be added to impart or improve certain required properties.

M

mahogany acid—One of the oil-soluble sulfonic acids formed by the action of sulfuric acid on petroleum distillates. These acids may be converted to their sodium soaps (mahogany soaps) and extracted from the oil with alcohol. The further refined soaps are used in the manufacture of soluble oils, rust preventives, and special greases. The calcium and barium soaps of these acids are used as detergent additives in motor oils. (See also **sulfonic acid.**)

manhead—An access opening into a tower, drum, or tank to allow entry of a man during shutdowns for inspection, cleaning, or repair. Also called manhole.

marine engine oil—A petroleum oil used as a crankcase oil in marine engines.

marine gasoline—Fuel for motors in marine service. Generally, but not necessarily, a regular-grade gasoline containing no organometallic antiknock compounds and compounded to provide protection against formation of gum.

mayonnaise—Low-temperature sludge; a black, brown, or gray deposit having a soft, mayonnaise-like consistency. It is formed in engines as the result of operation at low engine temperatures, such as exist during starting or warm-up and during intermittent start, stop, and low-speed driving.

mazut—A Russian name for distillation residues used largely as fuel oil; also spelled "masut" or "mazout."

mechanical octane number—Change in octane-number requirement of an engine due entirely to mechanical design. This may result from improvements in combustion-chamber design, or in changes made in manifolding, valve timing, cooling, etc.

medicinal oil—A highly refined, colorless, tasteless, and odorless petroleum oil used as a medicine in the nature of an internal lubricant.

MEK—Methyl ethyl ketone; a colorless liquid, $CH_3COCH_2CH_3$, made by dehydrogenation of secondary butyl alcohol, or from carbon monoxide and hydrogen by a modified Fischer-Tropsch process. Used as a solvent; as a chemical intermediate; and in the manufacture of lacquers, celluloid, and varnish removers.

MEK deoiling—A wax-deoiling process in which the solvent is generally a mixture of methyl ethyl ketone and toluene. (See also **deoiling.**)

MEK dewaxing—A continuous solvent dewaxing process in which the solvent is generally a mixture of methyl ethyl ketone and toluene.

Mercapsol process—A regenerative process for extracting mercaptans, utilizing aqueous sodium (or potassium) hydroxide containing mixed cresols as solubility promoters.

mercaptan—One of the organic compounds having the general formula R—SH, meaning that the thiol group, —SH, is attached to a radical such as CH_3, C_2H_5, etc. The simpler mercaptans have strong, repulsive, garlic-like odors which become less pronounced with increasing molecular weight and higher boiling points.

metal deactivator—A fuel or lubricant additive, usually a chelating agent, which converts into an inactive form the traces of metal (such as copper in fuels) and metal surfaces (such as copper in fuel lines) which, in the absence of the deactivator, would catalyze gum formation and other oxidation.

methyl ethyl ketone—See **MEK.**

microcrystalline wax—Wax extracted from certain petroleum residues and having a finer and less apparent crystalline structure than paraffin wax.

mid-boiling point—The temperature at which approximately 50 per cent of a material has distilled. There are, however, several specific definitions of the mid-boiling point, and the term should be used only with a clear understanding of its meaning in the context.

Mid-Continent oil—Petroleum oil obtained from the central regions of the United States (principally Oklahoma, Kansas, and North Texas), usually having characteristics between those of Pennsylvania and coastal oils.

middle distillate—One of the distillates obtained between kerosine and lubricating-oil fractions in the refining processes. These include light fuel oils and diesel fuel.

MIL Spec—Military specification; a guide in determining the quality requirements of products used by the miliary services, published by the United States Department of Defense.

mineral oil—Generally speaking, this term refers to a wide range of products derived from mineral substances.

mineral seal oil—A distillate cut between kerosine and gas oil. Used in signal lamps, as an absorption medium in gasoline-recovery processes, and as a carrier for insecticides and sprays.

mineral spirits—See **petroleum spirits.**

mineral turpentine—See **petroleum spirits.**

mix, motor—See **motor mix.**

mixed aniline point—The minimum solution temperature of a mixture of aniline, heptane, and a particular hydrocarbon; used to indicate the aromaticity of the latter. (See also **aniline point.**)

mixed-base crude—A crude oil which is a mixture of paraffin-base and naphthene-base crudes. (See also **paraffin-base crude; naphthenic crude.**)

mixed lead alkyl—One of the mixtures of organo-lead compounds used as antiknock agents.

mixed-phase cracking—The thermal decomposition of higher-boiling hydrocarbons to gasoline components. Moderate temperatures and high pressures are used. NOTE: The process was originally erroneously termed "liquid-phase cracking" to distinguish it from vapor-phase cracking. More proper even than "mixed-phase cracking" is the term "pressure cracking."

mixed polymerization—The polymerization of an olefin in the presence of another olefin, as of propylene and butylene.

mixed-solvent extraction—Solvent refining which uses a mixture of more than one solvent, e.g., the Duo-Sol extraction process. (See also **solvent extraction.**)

Modified Uniontown Method—See **Uniontown Method.**

mold lubricant—A lubricant, also called mold oil, used to insure easy parting of ceramic, glass, or other material from the mold in which it is cast. (See also **form oil.**)

molecular sieve—A synthetic zeolite mineral having pores of uniform size; it is capable of separating molecules, on the basis of their size, structure, or both, by absorption or sieving. (Editor's note: Molecular sieves are finding use in removing traces of water from jet fuel, and in separating *n*-paraffin hydrocarbons from admixture with isoparaffins, cyclo compounds, and aromatics.)

motor gasoline—A complex mixture of relatively volatile hydrocarbons, with or without small quantities of additives, which have been blended to form a fuel suitable for use in automotive internal-combustion engines.

Motor Method—A test for determining the knock rating, in terms of ASTM Motor octane numbers, of fuels for use in spark-ignition engines. The knocking tendency of the fuel is compared with those for blends of reference fuels of known octane number when run in the ASTM-CFR engine at 900 rpm, under standard operating conditions as prescribed in ASTM Method D 357.

motor mix—The antiknock fluid for gasoline used primarily in automotive engines.

Motor octane number—See **Motor Method.**

motor oil—An oil suitable for use in an engine crankcase; also applies to oils used to lubricate electric motors.

motor spirit—A term occasionally applied to gasoline (see **petrol**).

moving-bed catalytic cracking—A cracking process in which the catalyst, consisting of granules 4- to 12-mesh, is continuously cycled between the reactor and the regenerator. Examples: the Airlift Thermofor and the Houdriflow processes.

multigrade oil—One of the multi-viscosity number oils in which one oil combines three SAE viscosity number grades For example, multigrade SAE 10W-30 grade may be used where SAE 10W, SAE 20-20W, or SAE 30 grades are specified. Multigrade oils are usually made to meet the requirements of API Services MS, DG, and DM. They have been made possible by improved refining processes and the use of improved additives.

multipurpose grease—A lubricating grease suitable to meet the individual requirements for chassis lubricant, bearing lubricant, joint lubricant, water-pump lubricant, and cup grease.

N

napalm—A thickened gasoline used as an incendiary medium. It tends to adhere to the surface it strikes.

naphtha—See **petroleum naphtha;** see also specific naphthas under alphabetical listing.

naphtha scrubber—A tower in which gas is washed with mineral oil to absorb the naphtha. The "scrubbing" consists of spraying the mineral oil down through an ascending current of gas which may be under pressure.

naphthenic crude—Crude oil containing a relatively large percentage of naphthenes. An oil obtained from a naphthenic crude is said to be a naphthene-base oil. Lubricating oils made from such crudes are normally distinguished from similar oils made from paraffinic crudes (both oils equally well refined) by lower gravity, lower carbon content and pour point, and lower rating viscosity index.

natural gasoline—A mixture of liquid hydrocarbons extracted from natural gas by various methods and stabilized to obtain a liquid product suitable for blending with refinery gasoline.

natural gasoline plant—A plant for the extraction of fluid hydrocarbons, such as gasoline and liquefied petroleum gas, from natural gas.

neat gasoline—See **clear gasoline.**

neohexane alkylation—A noncatalytic alkylation process utilizing ethylene and isobutane as reactants to form neohexane.

neutral oil—A distillate lubricating oil with viscosity usually not above 200 sec at 100°F. (See also **nonviscous neutral oil; viscous neutral oil.**)

neutralization number—The weight, in milligrams, of potassium hydroxide needed to neutralize the acid in 1 g of oil. The neutralization number of an oil is an indication of its acidity.

NLGI number—One of a series of numbers classifying the consistency range of lubricating greases, based on the ASTM cone penetration number. The National Lubricating Grease Institute grades are in order of increasing consistency (hardness).

nonasphaltic road oil—Any of the nonhardening petroleum distillates or residual oils used as dust layers. They have sufficiently low viscosities to be applied without heating and, together with asphaltic road oils, are sometimes referred to as dust palliatives. (See also **asphaltic road oil.**)

nonviscous neutral oil—A neutral oil with a viscosity below 135 sec at 100°F.

normal benzin—A term used in Germany for a close-cut gasoline of specific gravity 0.695 to 0.705 at 15°C and boiling range of 60 to 95°C (140 to 203°F); used to detect and estimate asphalt in petroleum.

O

octane number—A term numerically indicating the relative antiknock value of a gasoline. For octane numbers 100 or below, it is based upon a comparison with the reference fuels isooctane (100 octane number) and n-heptane (0 octane number). The octane number of an unknown fuel is the per cent by volume of isooctane with n-heptane which matches the unknown fuel in knocking tendencies under a specified set of conditions. Above 100, the octane number of a fuel is based on the engine rating, in terms of milliliters of tetraethyl lead in isooctane, which matches that of the unknown fuel. (See also **chemical octane number; mechanical octane number; Motor Method; Research Method; road octane number.**)

octane scale—A series of arbitrary numbers, from 0 to 120.3, used to rate the octane number of gasoline. The scale is defined by three reference materials: n-heptane, assigned the number 0; 2, isooctane (2,2,4-trimethylpentane), assigned the number 100; 3, isooctane plus 6 ml tetraethyl lead, assigned the number 120.3.

OD (optical density) color—A term given to color values of oils calculated from the depth of oil (or depth of oil dissolved in a colorless solvent) having an optical density equal to that of the "neutral filter working standard" placed in the base of one of the metal cells of the Duboscq colorimeter. Neutral filters are pieces of gray-colored glass (disks) with an optical density of approximately 0.3. OD color is 1.57 times true color.

oil gas—A gas obtained by cracking gas oil.

oil sand—See **bituminous sand.**

oil separator—A tank into which waste oils and surface drainage are run for separation. Oil, called "separator slop," is recovered.

oil shale—A rock of sedimentary origin, with an ash content of more than 33 per cent; the contained organic matter yields oil when destructively distilled, but not appreciably when extracted with the ordinary solvents for petroleum.

onstream time—The length of time a unit is in actual production.

opalescence—Milkiness or cloudiness; the condition, when occurring in oil (particularly lubricating oil), may be attributable to wax.

outage—The difference between the full or the rated capacity and the actual contents of a barrel, tank, or tank car. The vertical distance between the surface or the liquid in a barrel, tank, or tank car, and the top of the container. (See also **innage.**)

overall plate efficiency—In a fractionating tower, the ratio of actual plates to theoretical plates required for a particular separation.

overhead—In a distilling operation, that portion of the charge which is vaporized and removed.

overlap—In adjacent fractions, the temperature interval between the initial boiling point of the higher-boiling fractions and the end point of the lower-boiling fractions.

overpoint—See **initial boiling point.**

oxidized asphalt—See **air-blown asphalt.**

ozokerite—A naturally occurring mineral wax, usually dark brown in color and containing mineral matter and oil. On purification, it yields a white-to-yellow microcrystalline wax. Fully refined ozokerite is a hard, white microcrystalline wax, substantially free from oil, also known as ceresin.

P

packed tower—A fractionating or absorber tower which is filled with small objects (packing) to effect an intimate contact between rising vapor and falling liquid.

painters' naphtha—See **varnish makers' and painters' naphtha.**

pale oil—A petroleum lubricating or process oil refined until its color, by transmitted light, is straw to pale yellow.

paper chromatograph—See **chromatography.**

paraffin—Any of the white, tasteless, odorless, and chemically inert waxy substances composed of saturated hydrocarbons obtained from petroleum.   (See also **paraffin series.**)

paraffin-base crude—Crude oil which contains paraffin hydrocarbons but no asphalt.

paraffin distillate—A refinery term for distillate oils containing crystalline wax before they are dewaxed to produce paraffin wax and paraffin oil.

paraffin oil—A light-colored, wax-free oil obtained by pressing paraffin distillate.

paraffin press—See **filter press.**

paraffin scale—Crude paraffin wax obtained from the sweating of slack wax.   (See also **sweating; slack wax.**)

paraffin series—A homologous series of open-chain saturated hydrocarbons of the general formula C_nH_{2n+2}, of which methane (CH_4) is the first member; sometimes referred to as the methane series.

paraffin wax—The wax removed from paraffin distillates by chilling and pressing. When separating from solutions, it is a colorless, more or less translucent, crystalline mass without odor and taste, slightly greasy to touch, and consisting of a mixture of solid hydrocarbons in which the paraffin series predominates.

paraffinicity—Describing the paraffinic nature or composition of crude oil or its products.   (See also **paraffin-base crude.**)

paraffinum liquidum—Another name for liquid petrolatum, white mineral oil, or heavy liquid petrolatum.

partition chromatography—See **chromatography.**

penetrating oil—A low-viscosity oil which can penetrate between closely fitted parts, such as the leaves of springs; also used to loosen rusted parts.

Penex process—A continuous, nonregenerative process for isomerization of C_5 and/or C_6 fractions in the presence of hydrogen (from reforming) and a platinum catalyst.   The system can operate in conjunction with reforming of the C_7+ naphtha fraction.

Pennsylvania crude—A type of oil produced in Pennsylvania, New York, West Virginia, and parts of Ohio.   It contains a high percentage of paraffin-base lubricating-oil stock.

Pensky-Martens closed tester—Apparatus used for the determination of the flash point of the fuel oils, cutback asphalts, and other viscous materials and suspensions of solids.   This apparatus, and the test method usually carried out in the United States, is described in ASTM Method D 93.

Pentafining —A pentane isomerization process using a regenerable platinum catalyst on a silica-alumina support and requiring outside hydrogen.

pentane insolubles—Usually called normal pentane insolubles; the insoluble matter which can be separated from a solution of used lubricating oil in normal pentane and, in addition to the benzene insolubles, may include resinous bitumens produced from the oxidation of oil and fuel.   Usually determined in the United States by ASTM Method D 893.

pepper sludge—The fine particles of sludge produced in acid treating.   Ordinarily, the main body of sludge settles out quickly; but in treating lubricating oils, considerable "pepper" may remain in suspension.

Perco copper sweetening—A process in which gasoline, stable to color and gum in the presence of air, is mixed with air and filtered through a bed of absorbent material impregnated with copper reagent; all other stocks are contacted with copper solution in the absence of air.   Reagent is made up of copper sulfate and sodium chloride.

Perco HF alkylation—A process using hydrofluoric acid as a catalyst to unite olefins with isobutane.   Bauxite-dried charge is intimately contacted with the acid at 60 to 100°F and optimum isobutane-to-olefin ratio of 6 to 1.   The mixture is separated in a settler, and acid is returned to the reactor.   An acid sidestream is continuously regenerated by fractionation.

percolation—The passing of a liquid through a bed of granules or powder, e.g., the slow flow of oil through a layer of decolorizing earth. Also, when gasoline boils up in a hot carburetor, sometimes the bubbles of gasoline vapor pass up (percolate) through the main discharge nozzle carrying liquid gasoline with them. (See also **Thermofor continuous percolation.**)

percolation filtration—A continuous, regenerative, liquid-phase clay-treating process to improve the color, odor, and stability of lubricating oils and waxes.

performance number—One of a series of numbers for converting fuel antiknock values in terms of a reference fuel into an index which is an approximate indication of relative engine performance. The PN scale is based on average engine performance. For supercharged engines, where the manifold pressure is varied to suit the knock value, the performance number is an index of relative engine output. In unsupercharged engines, or engines run at constant manifold pressure, the performance number is an index of relative, permissible compression ratio.

petrol—Term commonly used in Great Britain for motor spirit or gasoline.

petrolatum—A semisolid, unctuous product, ranging from white to yellow in color, produced during refining of residual stocks.

petroleum—A material occurring naturally in the earth, predominantly composed of mixtures of chemical compounds of carbon and hydrogen with or without other nonmetallic elements such as sulfur, oxygen, nitrogen, etc. Petroleum may contain, or be composed of, such compounds in the gaseous, liquid, and/or solid state, depending on the nature of these compounds and the existent conditions of temperature and pressure.

petroleum coke—A solid residue; the final product of the condensation processes in cracking. It consists probably of highly polycyclic aromatic hydrocarbons very poor in hydrogen. Calcination of petroleum coke can yield almost pure carbon or artificial graphite suitable for production of electrodes, motor brushes, dry cells, etc.

petroleum ether—A volatile fraction of petroleum consisting chiefly of pentanes and hexanes (see **ligroine**).

petroleum jelly—A colloquial expression for petrolatum.

petroleum naphtha—A generic term applied to refined, partly refined, or unrefined petroleum products and liquid products of natural gas, not less than 10 per cent of which distill below 347°F (175°C), and not less than 95 per cent of which distill below 464°F (240°C) when subjected to distillation in accordance with ASTM Method D 86. NOTE: The "naphthas" used for specific purposes, such as cleaning, manufacture of rubber, manufacture of paints and varnishes, etc., are made to conform to specifications which may require products of considerably greater volatility than that set by the limits of this generic definition.

petroleum product, solidified—See **solidified petroleum product.**

petroleum spirits—A refined petroleum distillate with volatility, flash point, and other properties making it suitable as a thinner and solvent in paints, varnishes, and similar products.

petroleum tar—A viscous black or dark-brown product obtained in petroleum refining which will yield a substantial quantity of solid residue when partly evaporated or fractionally distilled. The term "tar" should never be applied to materials derived from crude oil fractions thereof without the prefix "petroleum."

petroleum wax—See **microcrystalline wax; paraffin wax.**

phenol extraction—A liquid-liquid solvent extraction process for the removal of aromatic, unsaturated, and naphthenic constituents of lubricating-oil stocks. The solvent is phenol and is employed in either anhydrous or aqueous form.

Phillips catalytic isomerization—A fixed-bed, butane vapor-phase isomerization process using aluminum chloride (deposited on bauxite) catalyst with hydrogen chloride promotor.

phosphate desulfurization—A continuous, regenerative process for separating hydrogen sulfide from natural gas, refinery gas, or liquid hydrocarbons by means of tripotassium phosphate solution. The solution contains approximately 30 per cent tripotassium phosphate in water, and will selectively absorb hydrogen sulfide in the presence of carbon dioxide.

phosphoric acid polymerization—A process using a phosphoric acid catalyst to convert propylene, butylene, or both, to gasoline or petrochemical polymers.

pipe still—A still in which heat is applied to the oil while being pumped through a coil or pipe arranged in a suitable firebox. After leaving the heated zone, the pipe runs to a fractionator; a portion of oil is taken off overhead as vapor, and the liquid portion is removed continuously.

pitch—Black or dark-colored viscous substance obtained as a residue from the distillation of coal tar, wood tar, bone oil, etc. This term should not be applied to petroleum products.

plate—See **bubble tray; cascade tray;** specific plates under alphabetical listing.

Platforming—A reforming process using a platinum-containing catalyst which includes 0.1 to 8.0 per cent of fluorine or chlorine on an alumina base.

Platreating—A hydrogenation process for refining concentrates of aromatics to recover pure benzene and toluene.

POD (Podbielniak) analysis—A precision distillation procedure used to separate low-boiling hydrocarbon fractions quantitatively for analytical purposes.

pole height viscosity index—A system used in some European countries for expressing the viscosity index of lubricating oils. The pole height is the vertical height from a point on the zero line (the abscissa) to a point on a viscosity-temperature curve of an oil. A VI of 100 corresponds to a pole height of 1.9, and oils having a VI of zero will have a pole height of 3.7. (See also **VI.**)

Polyco catalytic polymerization—A process to convert propylene and butylene to motor fuel or special products (codimer, tetramer, etc.). The catalyst is pelleted copper pyrophosphate.

polyforming—A process charging both C_3 and C_4 gases with naphtha or gas oil under thermal conditions to produce gasoline. (See also **gas reversion.**)

polymer—A substance produced from another by polymerization (which see).

polymer gasoline—A product of polymerization of normally gaseous hydrocarbons to hydrocarbons boiling in the gasoline range.

polymerization—A reaction combining two or more molecules to form a single molecule having the same elements in the same proportions as in the original molecules; the union of light olefins to form hydrocarbons of higher molecular weight. The process may be thermal or catalytic. Of importance in the production of motor fuel and aviation fuel components from cracked gases. If unlike olefin molecules are combined, the process is referred to as "copolymerization" and produces "copolymers." High polymers are used as synthetic lubricating oils, plasticizers, etc. (See also **codimer; copolymer; dimer;** specific polymerization processes under alphabetical listing.)

polysulfide treating—A chemical treatment used to remove elemental sulfur from refinery liquids by contacting them with a nonregenerable solution of sodium polysulfide.

PONA analysis—Analysis for paraffins (P), olefins (O), naphthenes (N), and aromatics (A). Among methods used are ASTM Method D 1319 and "Proposed Method for Hydrocarbon Types in Low Olefinic Gasoline by Mass Spectrometry" (Appendix VII, 1950 *Report of Committee D-2*).

pour depressor—A lubricating-oil additive which lowers the pour point of an oil containing wax by reducing the tendency of the wax to form a solid mass in the oil. Also called pour-point depressor, pour depressant.

pour inhibitor—See **pour depressor.**

pour point—The lowest temperature at which oil will pour or flow when it is chilled without disturbance under definite conditions. In the United States these conditions are prescribed in ASTM Method D 97.

pour reversion—The difference between the original ASTM pour point of the oil and the relatively high solidification temperature observed in the field.

pour stability—The ability of a pour-depressant-treated oil to maintain its original ASTM pour point when subjected to storage at low temperature approximating winter conditions.

pour test—The chilling of a liquid under specified conditions to determine the pour point (which see). Observations are generally made over ranges of temperature equal to 5°F.

Powerforming—A fixed-bed naphtha-reforming process using a regenerable platinum catalyst.

precipitation number—The number of milliliters of precipitate formed when 10 ml of lubricating oil is mixed with 90 ml of petroleum naphtha of a definite quality and centrifuged under definitely prescribed conditions. The precipitation number should indicate the amount of the asphaltic bodies dissolved in the lubricating oil, although a certain amount of paraffin bodies may separate with the asphaltic bodies. A test for precipitation number of lubricating oils is described in ASTM Method D 91.

press drip—Oil which drips from the wax press after pressed distillate has been removed.

pressed distillate—The oil recovered when paraffin distillate is pressed to separate wax.

pressing, cold—See **cold pressing.**

pressure distillate—The light, gasoline-bearing distillate product from the pressure stills which has been produced by cracking, as contrasted with virgin or straight-run stock.

pressure still—A still in which the liquid oil and oil vapors are held by proper release valves under a pressure above atmospheric, under which condition the temperature is raised to a point where the oil will decompose (crack) at a satisfactory rate, giving lower-boiling products.

primary reference fuel (diesel)—Cetane, alpha-methylnaphthalene or mixtures thereof, used in the ASTM-CFR diesel test engine to rate the cetane number of commercial diesel fuels. Cetane (normal hexadecane) has been assigned the cetane rating of 100; alpha-methylnaphthalene, the rating of 0 on the cetane number scale. (See also **cetane number.**)

primary reference fuel (gasoline)—Isooctane, n-heptane, or mixtures thereof, used in the ASTM-CFR test engine to rate the octane number of commercial gasoline. Pure isooctane (2,2,4-trimethylpentane) has been assigned the rating of 100; n-heptane, the rating of 0 on the octane number scale. (See also **octane scale.**)

prime white kerosine—A kerosine having a color between that of water-white kerosine and standard white kerosine. The color is just off white (colorless). (See also **water-white kerosine; standard white kerosine.**)

propane deasphalting—A type of solvent deasphalting; an extraction process in which desirable oil in the charge is dissolved in the propane solvent, and undesirable heavy matter is precipitated and separated as asphalt. When vacuum-reduced crude oil is deasphalted for finishing into lubricating stock, the residue is contacted countercurrently with liquid propane in the deasphalting section. Deasphalted oil and propane are taken overhead; the bottoms are an asphalt-propane mixture. (See also **solvent deasphalting.**)

propane decarbonizing—A solvent extraction process used to recover catalytic cracking feed from heavy fuel residues. Process flow and equipment are essentially the same as those employed for the deasphalting of lubricating-oil stocks (see **propane deasphalting**). Because butane, alone or with propane, can be used as the solvent, the process is often referred to simply as solvent decarbonizing. Decarbonized and demetallized oil is recovered from topped or vacuum-reduced crude-oil charge.

propane dewaxing—A process for dewaxing lubricating oils in which propane serves as solvent, diluent, refrigerant, blanketing gas, and filter blowback.

propane fractionation—A continuous extraction process employing liquid propane as the solvent; a variant of the propane deasphalting process. Used for the segregation of long vacuum residuum into two or more grades of lubricating-oil stock. The products from the process are heavy neutral stock, bright stock, and asphalt.

puking—In a still or bubble tower, the foaming and rising of oil to the extent that part of the liquid is driven out of the vessel through the vapor line. (See also **surge.**)

pumpability test (fuel oil)—A test designed to ascertain the lowest temperature at which a fuel oil may be pumped, according to the procedure described in ASTM Method D 1659.

pumparound—A system on a distillation tower for withdrawing a liquid from a plate, cooling it, and returning it to another plate for the purpose of inducing the condensation of vapors.

pumpback—See **reflux.**

purging—The displacement of one material with another in process equipment; frequently, displacement of hydrocarbon vapor with steam or inert gas.

Q

quench—To suddenly cool hot material discharging from a cracking coil, usually by injecting cool oil into the discharge line; its purpose is to check the cracking reaction quickly.

R

raffinate—In solvent refining practice, that portion of the oil which remains undissolved and is not removed by the selective solvent. Also called "good oil" by operators.

Ramsbottom coke test—A carbon-residue test which originated in England (IP Method 14). The corresponding ASTM Method is D 524, which is widely used in the United States. (See also **carbon residue.**)

RE number—Resistance-to-emulsion number (see **steam emulsion test**).

reboiler—An auxiliary of a fractionating tower designed to supply additional heat to the lower portion. Liquid is usually withdrawn (or pumped) from the side or bottom of the tower and is reheated by means of heat exchange. The vapors and residual liquid, separately or together, are reintroduced to the tower.

receiving house—A building where the streams from the condensers are observed through a look box and samples taken for testing, and from where products are diverted to the proper storage tanks or to other parts of the refinery for further distillation or processing.

reclaimed oil—A lubricating oil which, after undergoing a period of service, is collected, reprocessed, and sold for reuse.

recovered oil—See **reclaimed oil.**

recycle ratio—The volume of recycle stock per volume of fresh feed. Sometimes expressed as the volume of recycle divided by the total charge.

recycle stock—That portion of a feedstock which has passed through a refining process and is recirculated through the process. It may include unwanted byproducts from the process.

red oil—Originally, the residual fraction obtained by fire-distilling pressed distillate. Such oils were acid-treated but not filtered. The term is now used to describe any oil of red color, regardless of the refining process by which it is made. The bulk of the so-called red engine oils, bearing oils, and machinery oils are of this class. They may be considered intermediate grades of lubricating oil for general purposes. (Editor's note: Term sometimes used for oleic acid.)

reduced crude—A residual product remaining after the removal, by distillation or other means, of an appreciable quantity of the more volatile components of crude oil.

Redwood viscosimeter—Standard British viscosimeter. The number of seconds required for 50 ml of an oil to flow out of a standard Redwood viscosimeter at a definite temperature is the Redwood viscosity prescribed by IP Method 70. Instrument is available in two sizes: Redwood No. I and Redwood No. II. When the flow time exceeds 2,000 sec, the No. II must be used.

reference fuel—See **primary reference fuel (diesel); primary reference fuel (gasoline); secondary reference fuel.**

refinery gas—Any form or mixture of gas gathered in a refinery from the various stills. (See also **still gas.**)

reflux—In fractional distillation, that part of the distillate which may be returned to the column to assist in making a better separation into desired fractions. This operation is called refluxing. Reflux may be either circulating or induced. Circulating reflux is liquid which is withdrawn hot, cooled, and pumped back to the tower. Induced reflux is liquid formed within a fractionation tower by condensation of vapors by means of an internal cooling coil.

reflux condenser—A condenser which constantly condenses vapors and returns liquid to the original distilling unit or to the lower levels of a fractionating tower.

reflux ratio—1. The quantity of reflux per unit quantity of distillate removed from the process as a product (forward flow). 2. For design purposes, the ratio of liquid reflux to vapor at any given point in a fractionating column. Values may range from zero to unity.

reformed gasoline—Gasoline made by a reforming process.

reforming—The thermal or catalytic conversion of naphtha into more volatile products of higher octane number. It represents the total effect of numerous reactions, such as cracking, polymerization, dehydrogenation, and isomerization, taking place simultaneously. (See also specific reforming processes under alphabetical listing.)

refrigeration—A process for making natural gasoline which employs low temperatures as a means of condensing (rather than absorbing) gasoline present in wet gas.

regeneration—In a catalytic process, the revivification or reactivation of the catalyst, sometimes done by burning off the coke deposits under carefully controlled conditions of temperature and oxygen content of the regeneration gas stream.

regenerator—In a catalytic process, that part of the system having as its primary function the revivification or reactivation of the catalyst (see **regeneration**).

Reid vapor pressure—An important test for gasolines. It is a measure of the vapor pressure of a sample at 100°F, and the test is commonly made in a bomb. The results are reported in pounds. This test is usually carried out in the United States in accordance with ASTM Method D 323.

re-refined oil—See **reclaimed oil**.

rerunning—The distillation of an oil which has already been distilled; it usually implies taking a large proportion of the charge overhead.

Research Method—A test for determining the knock rating, in terms of ASTM Research octane numbers, of fuels for use in spark-ignition engines. The knocking tendency of the fuel is compared with those for blends of reference fuels of known octane number when run in the ASTM-CFR engine at 600 rpm under standard operating conditions, as prescribed in ASTM Methods D 908 and D 1656.

Research octane number—See **Research Method**.

residence time—The average length of time a particle of reactant spends in contact with catalyst.

residual fuel oil—Topped crude oil or viscous residuum obtained in refinery operations. (See also **bunker "C" fuel oil**.)

residue—Heavy oil or bottoms left in the still after gasoline and other relatively low-boiling constituents have been distilled off, or the remaining fraction from crude oil after distilling off all the heaviest components. (See also specific residues under alphabetical listing.)

residuum—See **residue; long residuum; short residuum**.

Residuum Hydroconversion—**Humble**—A hydrocracking process to upgrade residual stocks (asphalt, visbreaker tar, crude-oil residuum, etc.). The cobalt molybdate catalyst supported on alumina is regenerable.

resin—A polymer of unsaturated hydrocarbons from petroleum processing, e.g., in the cracking of petroleum oils, propane deasphalting, clay treatment of thermally cracked naphthas. Chief uses include: rubber and plastics; impregnants; surface coatings. (See also **synthetic resin**.)

retort—A vessel in which substances are subjected to heat for the purpose of distillation or decomposition. A retort is distinguished from a still in that it is more often used for the treatment of solid or semisolid substances.

retrograde condensation and evaporation—The phenomena associated with the behavior of a hydrocarbon mixture in the critical region, wherein, at constant temperature, the liquid phase in contact with the vapor may be vaporized by an increase in pressure, or the vapor condensed by a decrease in pressure. Likewise, in the same region, at constant pressure, the liquid phase may be vaporized by a reduction of temperature, or the vapor condensed by an increase in temperature.

return bend—A pipe fitting, equal to two ells, used to connect parallel pipes so that fluid flowing into one will return in the opposite direction through the other.

Rexforming—A process combining Platforming (catalytic reforming) with aromatics extraction, wherein low-octane raffinate is recycled to the Platformer. Because of recycle, operating conditions in the reforming section can be milder than in conventional Platforming, and higher space velocities are used. Glycol is used to extract low-boiling, high-octane paraffins, as well as aromatics.

rhigolene—An archaic term denoting a petroleum product consisting principally of pentanes.

rich mixture—1. A fuel high in the combustible component. 2. The fuel mixture prescribed in ASTM Method D 909.

rich oil—Absorption oil containing dissolved natural gasoline fractions.

riser—That portion of the bubble-plate assembly which channels the vapor and causes it to flow downward to escape through the liquid.

road octane number—A numerical value based upon the relative antiknock performance in an automobile of a test gasoline as compared with specified reference fuels. Road octanes are determined by operating a car over a stretch of road or on a chassis dynamometer under conditions simulating those encountered on the highway. Several standardized procedures are in common use.

road oil—See **asphaltic road oil; nonasphaltic road oil.**

road test—A test conducted with motor vehicles on the highway or chassis dynamometer to determine the performance of fuels or lubricants. The octane number determined in this manner is known as the road octane number (which see).

roasting—The process of regenerating clay by heating or "burning" in some form of rotating cylinder which mechanically stirs the clay in contact with air. (See also **clay regeneration.**)

roily oil—Crude oil more or less emulsified with water.

run—The amount of stock processed by a particular unit in a given time. Often used colloquially in relation to the type of stock being processed.

runback—A pipe through which all or part of the condensate can be run back to a still instead of being run off.

rundown line—A line of pipe which connects the look box in the receiving house with the tank or receiver in which the distillate from the still is temporarily stored. The stream of oil coming from the condenser coils runs through the look box in the receiving house, and thence, by the rundown line, to the receiver. Each receiving house has one or more rundown lines for each kind of oil which is produced there, thus ensuring that the products from the stills will be kept free from contamination.

rundown tank—One of the tanks in which are received the condensates from the stills, agitators, or other refinery equipment, and from which the distillates are pumped to larger tanks known as work tanks or storage tanks. Rundown tanks are also known as "pans" or receiving tanks. If the condensates were received directly into the large storage tanks, the puking of a still would contaminate unnecessarily perhaps thousands of barrels of distillate.

Rvp—See **Reid vapor pressure.**

S

SAE EP lubricant tester—A machine designed to test the extreme-pressure properties of a lubricant under a combined rolling and sliding action. The revolving members are two bearing cups which rotate at different speeds.

SAE viscosity number—An arbitrary number; one of a system for classifying crankcase oils and automotive transmission and differential lubricants, according to their viscosities, established by the Society of Automotive Engineers. SAE numbers are used in connection with recommendations for crankcase oils to meet various design, service, and temperature requirements affecting viscosity only; they do not connote quality.

Saybolt color—A color standard for petroleum products. The procedure for determining Saybolt color and a description of the Saybolt chromometer are given in ASTM Method D 156.

Saybolt Furol viscosity—The time, in seconds, for 60 ml of fluid to flow through a capillary tube in a Saybolt Furol viscosimeter at specified temperatures between

70 and 210°F. This method is appropriate for high-viscosity oils such as transmission, gear, and heavy fuel oils. ASTM Method D 88 describes the equipment and procedure.

Saybolt Universal viscosity—The time, in seconds, for 60 ml of fluid to flow through a capillary tube in a Saybolt Universal viscosimeter at a given temperature, as described in ASTM Method D 88.

SBK catalytic reforming—A process which uses SBK (Sinclair-Baker-Kellogg) regenerable platinum catalyst. With multiple reactors, it is possible to regenerate the catalyst without interruption of the reforming. The process may employ a desulfurization reaction which also removes metal contaminants.

scale, paraffin—See **paraffin scale.**

scale wax—The paraffin derived by removing the greater part of the oil from slack wax by sweating or solvent deoiling. (See also **crude scale wax.**)

scrubbing—Purifying a gas by washing with water or chemical; less frequently, the removal of entrainment. The equipment used to give intimate contacting of the material to be purified is called a scrubber (see **naphtha scrubber; water scrubber.**)

secondary reference fuel—A commercially produced fuel which is acceptable for knock testing or cetane testing, and which has been calibrated by engine tests against primary reference fuels. Appropriate conversion tables are provided with each batch of such fuel supplied. Since calibrated data may change from batch to batch of a secondary reference fuel, it is essential that the calibrated data used match the fuel being used.

selective polymerization—The polymerization of a single type of molecule in a mixture; for example, the production of diisobutylene from a mixture of butylenes by sulfuric acid.

selective solvent—A solvent which, at certain temperatures and ratios, will preferentially dissolve more of one component of a mixture than of another and thereby permit partial separation. (Editor's note: Such solvents are widely used for refining lubricating-oil stocks.)

Separator-Nobel dewaxing—A solvent dewaxing process, the solvent being trichloroethylene. Since most chlorinated hydrocarbons have specific gravities greater than 1.0, centrifuging, rather than filtering, is used for separating the wax.

service classifications, API—See **API Engine Service Classification System.**

Service DG—Service typical of diesel engines in any operation where there are no severe requirements for wear or deposits control due to fuel, lubricating oil, or to engine design characteristics (see **API Engine Service Classification System**).

Service DM—Service typical of diesel engines operating under severe conditions or using fuel of a type normally tending to promote deposits and wear, but where there are design characteristics or operating conditions which may make the engine less sensitive to fuel effects or more sensitive to residues from lubricating oil (see **API Engine Service Classification System**).

Service DS—Service typical of diesel engines operating under very severe conditions, or having design characteristics or using fuel tending to produce excessive wear or deposits (see **API Engine Service Classification System**).

service factor—A measure of the continuity of an operation, generally expressed as the ratio of actual running time to calendar days. The factor is computed by dividing the time on-stream by total time.

Service ML—Service typical of gasoline and other spark-ignition engines used under the most favorable operating conditions, the engines having no special lubrication requirements and having no design characteristics sensitive to deposit formation (see **API Engine Service Classification System**).

Service MM—Service typical of gasoline and other spark-ignition engines used under moderate-to-severe operating conditions, but presenting problems of deposit or bearing corrosion control when crankcase-oil temperatures are high (see **API Engine Service Classification System**).

Service MS—Service typical of gasoline or other spark-ignition engines used under unfavorable or severe types of operating conditions, and where there are special lubrication requirements for deposit, wear, or bearing corrosion control due to operating conditions or to engine design or fuel characteristics (see **API Engine Service Classification System**).

settler—A separator; a tub, pan, vat, or tank in which the partial separation of a mixture is made due to difference in density. The operation may be continuous or batch. The separation may be solids from liquid or gas, liquid from liquid, liquid from gas.

severity factor—A measure of the severity of the overall reaction conditions. In dealing with reactor data and designs where both temperature and pressure vary, it is frequently convenient to correlate conversions and yields in terms of a severity factor. This approach is particularly desirable for complex systems, e.g., cracking where many reactions occur simultaneously and the complete compositions of both the charge and the products are generally unknown.

shale oil—An oil obtained from oil shale by destructive distillation (see **oil shale**).

Shell fluid catalytic cracking—A two-stage fluid catalytic cracking process in which the catalyst is regenerated.

shell still—A still formerly used in which the oil was charged into a closed, cylindrical shell and the heat required for distillation was applied to the outside of the bottom from a firebox.

short residuum—Low-gravity bottoms from distillation operations.

sidestream—A liquid stream taken from any one of the intermediate plates of a bubble tower.

sidestream stripper—A device used to perform further distillation on a liquid stream from any one of the plates of a bubble tower, usually by the use of steam.

sieve, molecular—See **molecular sieve.**

sieve plate—A fractionating tower tray which is perforated so that the vapor emerges vertically through the tray, making no turns. Some sieve trays are designed with no exit weirs. Sieve trays with perforations $\frac{1}{8}$ in. and $\frac{3}{16}$ in. in diameter have overall efficiencies equal to, or approximately 10 per cent higher than, those of bubble trays.

silica gel—A preparation consisting of incompletely dehydrated silicic acid; used in solid form as an adsorption agent.

Sinclair-Baker reforming—A naphtha-reforming process using platinum on alumina as catalyst.

Sinclair hydrogen treating—A hydrodesulfurization process using a catalyst of cobalt molybdenum on alumina.

skimming plant—An oil refinery designed to remove and finish only the lighter constituents of the crude oil, such as gasoline and kerosine. In such a plant the portion of the crude oil remaining after the aforementioned products are removed is usually sold as fuel oil.

slack barrel—A barrel for nonliquid materials. The construction is lighter than that of the usual barrel.

slack wax—The soft, oily crude wax obtained from the pressing of paraffin distillate or wax distillate.

slop—A term loosely used to denote odds and ends of oil produced in the plant, which must be rerun or further processed in order to make them suitable for use. Also called slop oil.

slow-curing liquid asphaltic material—An asphalt blended with gas oil. The gas oil does not volatilize readily; hence, "slow-curing."

slurry—A free-flowing mixture of solids and liquid; specifically, a suspension of cracking catalyst in cycle stock.

slushing grease—A grease used to provide a coating to prevent corrosion of metal or metal parts.

slushing oil—Oil used on metals to form a protective coating against rust and tarnish.

smoke point—A measure of the burning cleanliness of jet fuel and kerosine. The maximum flame height, in millimeters, at which a kerosine will burn without smoking when determined in the apparatus and under the conditions described in IP Method 57. The smoke point of jet fuels is determined by ASTM Method D 1322; this smoke test uses the same apparatus as that used by IP Method 57.

S-N dewaxing—See **Separator-Nobel dewaxing.**

SO$_2$ extraction—See **Edeleanu process.**

soaking drum—A high-pressure drum employed in connection with cracking coils to furnish the residence time necessary to complete the cracking reaction.

soaking factor—Important factor in designing cracking plants. The soaking factor, or time factor, is defined by the following equation:

$$\text{Soaking factor} = \left(\frac{a}{b}\right)\left(\frac{\text{mean } KT}{K_{800}}\right)$$

where a = cu ft coil volume above 800°F

b = BPD coil charge at 60°F

K = reaction velocity constant

$\dfrac{\text{mean } KT}{K_{800}}$ = average rate of reaction relative to the rate of reaction at 800°F, obtained by integration.

Values of K are obtained from a table of straight-line curves, where reaction velocity constants (K_1) are ordinates and temperatures (F) are the abscissas.

sodium hydroxide treatment—See **caustic wash.**

sodium plumbite—A solution prepared from a mixture of sodium hydroxide, lead oxide, and distilled water. Used in making the doctor test for light oils such as gasoline and kerosine.

solidified petroleum product—A combination of a petroleum product, e.g., gasoline, with some other substance to form a solid, highly flammable material (see **napalm**).

solubility test—As applied to asphalts and other bituminous materials, a test made to show the degree of their solubility in solvents such as carbon tetrachloride, carbon disulfide, or petroleum ether. One such test generally used in the United States is ASTM Method D 165.

soluble oil—Oil which readily forms stable emulsions or colloidal suspensions in water.

solutizer-air regenerative process—Identical with the solutizer-steam regenerative process (which see), except for the regeneration step. Newer units use uncatalyzed air regeneration.

solutizer-steam regenerative process—A chemical treating process for extracting mercaptans from gasoline or naphtha, using solutizers (potassium isobutyrate, potassium alkyl phenolate) in strong potassium hydroxide solution.

solutizer-tannin process—An early variation of the solutizer-air regenerative process for extracting mercaptans from gasoline, using a tannin-catalyzed oxidation regeneration step. Newer units eliminate tannin and use higher regeneration temperatures.

solvent deasphalting—A process for removing asphaltic and resinous materials from reduced crude oils, lubricating-oil stocks, gas oils, or middle distillates through the extraction or precipitant action of solvents. The principal deasphalting solvents are low-molecular-weight hydrocarbons, particularly liquid propane, and oxygenated compounds, such as alcohols and esters. (See also **propane deasphalting.**)

solvent decarbonizing—See **propane decarbonizing.**

solvent deresining—See **solvent deasphalting.**

solvent dewaxing—A process for removing wax from oils by means of solvents. The usual procedure is to chill a mixture of solvent and waxy oil, to separate by filtration or by centrifuging the wax which precipitates, and to recover the solvent from the wax and from the dewaxed oil for reuse. (See also specific solvent dewaxing processes under alphabetical listing.)

solvent extraction—The process of separating materials of different chemical type by means of selective solvent action. When a lubricating stock is treated, the purified product (raffinate) has improved VI and stability characteristics. Some solvents used for this purpose are furfural, phenol, Chlorex, liquid sulfur dioxide, and nitrobenzene. (See also **mixed-solvent extraction.**)

solvent naphtha—A refined naphtha of restricted boiling range used as a solvent; specifications are given in ASTM D 838, D 839, and D 840. (See also **petroleum naphtha; petroleum spirits.**)

solvent-refined oil—Oil which has been solvent-treated during the refining process.

solvent refining—See **solvent extraction.**

sour crude—Crude oil containing an abnormally large amount of sulfur compounds which, upon refining, liberate corrosive sulfur compounds. (See also **sweet crude.**)

Sovafining—A treating process using hydrogen. The operation includes pretreatment of straight-run and thermal naphtha for feed to the Sovaforming (catalytic reforming) unit, and the desulfurization and stabilization of sour No. 2 fuel-oil components.

Sovaforming—A multiple-reactor, fixed-bed reforming process using platinum catalyst.

space velocity—A convenient unit for expressing the relationship between feed rate and reactor volume in a flow process. It is defined as the volume or weight of feed (measured at standard conditions) per unit time per unit volume of reactor or per unit weight of catalyst.

splitter—A fractionating tower with an overhead stream and a bottoms stream only.

spray oil—A petroleum product of low viscosity, usually similar to lubricating oil, used to combat pests which attack trees, etc. ASTM Method D 447 gives some information on establishing the distillation range of such oils.

SR cylinder oil—Steam-refined cylinder oil; a viscous lubricating oil which has not been filtered. Such oils are generally reduced crudes which have had the lighter lubricant fractions removed by direct steam heating. They are used for the lubrication of steam engine cylinders and valves.

stability, color—See **color stability.**

stabilization—The process of separating light gases from petroleum or gasoline, thus leaving the liquid stable in the sense that it can be handled or stored with less liability to change in composition.

stabilized gasoline—Gasoline from which "wild" or low-boiling hydrocarbons (which have high vapor pressure) have been removed by stabilization.

stabilizer—A fractionating tower for removing light hydrocarbons from an oil to reduce vapor pressure particularly applied to gasoline.

standard white kerosine—A kerosine having the darkest of the regular color grades. The color is also called "export white" (see **export kerosine**). (See also **prime white kerosine; water-white kerosine.**)

standpipe—In fluid catalytic processes, a vertical tube filled with aerated, dense-phase, powdered catalyst, usually provided with a valved outlet at its lower end. This serves as a pressure leg for the injection of powdered catalyst into a zone wherein the pressure is lower than the hydrostatic head at the bottom of the standpipe.

steam cracking—High-temperature cracking of hydrocarbons in the presence of steam.

steam distillation—A distillation in which vaporization of the volatile constituents is effected at a lower temperature by introduction of steam directly into the charge. Steam used in this manner is termed open steam.

steam emulsion test—The method applied to lubricating oils whenever a knowledge of tendency to separate from water (resistance to emulsion) is desired. It is most significant in the case of turbine oils. In the United States, the test is carried out as prescribed in ASTM Method D 1401.

steam-refined cylinder oil—See **SR cylinder oil.**

steam refining—A type of refining in which the only heat used comes from steam in open and closed coils located near the bottom of the still. Such stills operate at low temperatures and are used in the production of gasoline and naphthas where odor and color are prime importance. Where open-steam coils are used, they perform a steam distillation (which see).

still gas—The mixture of extremely low-boiling hydrocarbons produced in the still on the distillation of crude oil.

stock—Petroleum, or products in storage, awaiting utilization or transfer to the ultimate point of utilization, whether with or without change of ownership. (See also specific stocks under alphabetical listing.)

Stoddard solvent—A refined petroleum distillate with volatility, flash point, and other properties making it suitable for use as a dry-cleaning solvent. Specifications are set forth in ASTM D 484.

stoke—A unit of viscosity.

straight-run product—A product produced by the primary distillation of crude oil.

strapping—The procedure whereby storage tanks are strapped (measured) on their outside with steel measuring tapes, and the measurements used to calculate the volumetric capacity of the tank for given increments in its height. Deductions are made for plates and interior fittings to determine the actual volume of the tank.

strapping table—See **gage table.**

straw oil—Pale paraffin oil of straw color, used for many process applications.

stream day—Denoting 24 hr of actual operation of a refinery unit; in contrast to a calendar day (which see).

stripping—Removal of the lightest fractions from a mixture. The process is usually carried out by passing the hot liquid from a flash drum or tower into a stripping vessel or section (stripper), through which open steam or inert gas is passed to remove the more volatile components of the cut. (See also **catalyst stripping.**)

sulfonic acid—An acid obtained by the addition of sulfur trioxide to a hydrocarbon, usually by treatment of an oil with strong sulfuric acid.

sulfur test (bomb method)—A test method to determine the sulfur content of a petroleum material by combustion in a bomb. This method is applicable to any petroleum product sufficiently low in volatility to be weighed accurately in an open sample boat, as described in ASTM Method D 129.

sulfur test (CO_2-O_2)—Test described in ASTM Method D 1266 for determining the amount of sulfur in petroleum products, including liquefied petroleum gas, by lamp combustion. Combustion is controlled by varying the flow of carbon dioxide and oxygen (CO_2-O_2) to the burner.

sulfuric acid alkylation—An alkylation process in which olefins (C_3, C_4, and C_5) combine with isobutane in the presence of a catalyst (sulfuric acid) to form branched-chain hydrocarbons used especially in gasoline blending stock.

sulfuric acid alkylation, Kellogg—See **Kellogg sulfuric acid alkylation.**

sulfuric acid treating—See **acid treating.**

sulfurized oil—A product formed from mineral oil combined with sulfur or certain sulfur compounds. It has far greater film strength and load-carrying ability than straight mineral oil and is used as cutting oil (which see).

Supercharge Method—A method for determining the knock-limited power, under supercharge rich-mixture conditions, of fuels for use in spark-ignition aircraft engines. It is carried out as prescribed in ASTM Method D 909. The knock characteristics of the fuels are expressed as octane numbers below 100 and as performance numbers above 100.

Supplement 1 oil—A high-detergency oil having an additive level of 7 per cent and intended for use in diesel engines burning fuels containing up to 0.75 per cent sulfur. The additive level is at least twice that of ordinary oil prescribed in military specifications (MIL-L-2104A).

Supplement 2 oil—A high-detergency oil having an additive level of 11 per cent and intended for use in diesel engines burning fuels containing over 0.75 per cent sulfur. The additive level is at least four times that of ordinary oil prescribed in military specifications (MIL-L-2104A).

Supplement 3 oil—A type of heavy-duty engine oil suitable for crankcase lubrication of high-output supercharged and unsupercharged diesel engines (MIL-L-45199).

surge—An upheaval of fluid in a system frequently causing a carryover of liquid through the vapor lines. (See also **puking.**)

Suspensoid catalytic cracking—A once-through, nonregenerative cracking process in which cracking stock is mixed with slurry of catalyst and cycle oil and passed through the coils of a heater. No other reactor is used; the catalyst is separated from the fractionator bottoms (heavy residual oil) by rotary filters. The catalyst is spent clay from the contact filtration of lubricating oils.

sweated wax—A crude wax freed from oil by having been passed through a sweater.

sweating—Separation of paraffin oil and low-melting wax from paraffin wax. This is done in a chamber (sweater) by first cooling the mixture until it has become a solid cake, then warming very gradually to cause partial fusion of the mixture and drainage of liquid from the cake.

sweet crude—Crude oil containing little sulfur. (See also **sour crude.**)

sweetening—The process by which petroleum products are improved in odor and color by oxidizing or removing the sulfur-containing and unsaturated compounds. The classic method is agitation with sodium plumbite (see **doctor sweetening**). Other processes use aluminum chloride, cuprous oxide, sodium hypochlorite, or ethanolamine. In some processes, the oil may be agitated with fuller's earth. (See also specific sweetening processes under alphabetical listing.)

synthetic crude—The total liquid, multicomponent hydrocarbon mixture resulting from a process involving molecular rearrangement of charge stock. Commonly applied to such a product from cracking, reforming, viscosity breaking, etc.

synthetic resin—Amorphous, organic, semisolid or solid material derived from certain petroleum oils among other sources; approximating natural resin in many qualities and used for similar purposes. (See also **resin.**)

T

Tag—Refers to the former C. J. Tagliabue Manufacturing Company, manufacturer of testing apparatus.

Tag closed-cup tester—An instrument used to determine the flash point of mobile liquids flashing below 175°F, as described in ASTM Method D 56.

Tag open-cup tester—An instrument used to determine the flash point of volatile flammable materials flashing below 175°F, as described in ASTM Method D 1310.

Tag-Robinson colorimeter—An instrument used for determining the color shades of lubricating and other oils. The color, reported as a number, is determined by varying the thickness of a column of oil until its color matches that of a standard color glass.

tail house—See **receiving house.**

tankage—The capacity of a tank, or of a series of tanks, in the same field.

tar sand—See **bituminous sand.**

technical oil—See **white oil.**

TEL—Tetraethyl lead; a lead compound, $Pb(C_2H_5)_4$, which, when added in small proportions to gasoline, increases the antiknock quality. (See also **lead susceptibility.**)

telltale—A marker on the outside of a tank which indicates on a scale the amount of fluid inside the tank.

temperature gradient—In a cracking coil, the rate of change of temperature based on temperatures taken at a number of points in the coil between the inlet and the outlet.

tetraethyl lead—See **TEL.**

tetramethyl lead—An organic compound of lead, $Pb(CH_3)_4$, which, when added in small amounts, increases the antiknock quality of gasoline.

thermal cracking—A refining process which decomposes, rearranges, or combines hydrocarbon molecules by the application of heat, without the aid of catalysts. The major variables involved are types of feed, time, pressure, and temperature. (See also specific thermal cracking processes under alphabetical listing.)

thermal polymerization—A thermal process to convert light hydrocarbon gases into liquid fuels. Paraffinic hydrocarbons are cracked to produce olefinic material, which is concurrently polymerized by heat and pressure to polymer gasoline. (See also **Unitary thermal polymerization.**)

thermal process—Any refining process which utilizes heat, without the aid of a catalyst.

thermal reforming—A process using heat (but no catalyst) to effect molecular rearrangement of low-octane naphtha into gasoline of higher antiknock quality.

Thermofor catalytic cracking—A continuous, moving-bed catalytic cracking process.

Thermofor catalytic reforming—A reforming process in which the synthetic, bead-type catalyst of coprecipitated chromia and alumina flows down through the reactor concurrent with the feed, which is a mixture of naphtha and recycle gas. The catalyst is transported from the bottom of the reactor to the top of the regenerator by bucket-type elevators and is thence recycled to the reactor.

Thermofor continuous percolation—A continuous clay treating process to stabilize and decolorize lubricants or waxes. The heated charge stock percolates upward through a bed of clay, is filtered in a blotter press, and is taken to storage. Spent clay is withdrawn continuously from the bottom of the percolator, and regenerated clay is added at the top to maintain a constant level of clay in the percolator.

thickener—Solid particles uniformly dispersed to form the structure of lubricating grease, in which the liquid is held by surface tension and other physical forces. (The solid particles may be fibers, as is the case with various metallic soaps; or plates or spheres, as is the case with some of the nonsoap thickeners.)

thief—A device which permits taking a sample from a definite predetermined location in the body of the liquid to be sampled. It consists of a metal box with a heavy lid, which opens very slowly on submerging the box in the liquid. The box is quickly lowered to a point in the tank from which the sample is to be taken, and is slowly filled at this location. Taking samples from a tank of oil in this manner is known as "thieving" a tank.

throughput—The volume of feedstock charged to process equipment in a specified time.

top steam—Steam admitted near the top of a shell still for purging the still, and to prevent vacuum when pumping out.

topped crude—See **reduced crude.**

topping—The distillation of crude oil to remove light fractions only.

tower—An apparatus for increasing the degree of separation obtained during the distillation of oil in a still. Towers may be divided into two general classes: those which secure separation by fractionation, and those which take advantage of partial condensation only. Towers of the first class are used when accurate work is necessary, as in the production of naphthas and gasoline to meet rigid distillation specifications. Towers operating by partial condensation are used to divide roughly the vapors from a still into several liquid portions. Commonly used in the former sense to mean bubble tower. (See also specific towers under alphabetical listing.)

tractor fuel—See **farm tractor fuel.**

transformer oil—See **insulating oil.**

trickle hydrodesulfurization—A fixed-bed process for desulfurizing middle distillates, gas oil, etc. The catalyst is pelleted cobalt molybdenum on alumina and can be regenerated *in situ.*

Trinidad asphalt—Natural asphalt from a deposit in Trinidad.

true-vapor-phase process—A thermal cracking process in which reaction occurs in the vapor phase.

tube-and-tank cracking—A liquid-phase thermal cracking process formerly used, in which the charge was heated to cracking temperature in a pipe still (the tube) under pressure. The hot, partly cracked oil passed to an insulated reaction chamber (the tank), and was maintained under cracking conditions of temperature and pressure for a period of time. The volatile products left this chamber through a pressure-relief valve.

tube still—See **pipe still.**

turbogrid plate—A tray for a fractionating column which consists of a flat grid of parallel slots extending over the entire cross-sectional area of the column. The slots can be stamped perforations in a flat metal plate, or can consist of the spaces between horizontal bars. The liquid level on each tray is maintained by a dynamic balance of liquid and vapor rates.

turnaround—Time necessary to clean and make repairs on refinery equipment after a normal run. It is the elapsed time between drawing the fires (shutting the unit down) and putting the unit on-stream again.

turnkey contract—A contract in which an independent agent undertakes to furnish all materials and labor, and to do all the work required to complete a project, for a fixed price.

U

Ultrafining—A fixed-bed catalytic hydrogenation process to desulfurize naphthas and upgrade distillates by essentially removing sulfur, nitrogen, and other materials.

Ultraforming—A low-pressure naphtha-reforming process employing onstream regeneration of a platinum-on-alumina catalyst and producing high yields of hydrogen and high-octane-number reformate.

undercutting—The practice of distilling a stock to a lower final boiling point than is usual for that stock.

Unifining (Union and UOP)—A fixed-bed catalytic process to desulfurize and hydrogenate refinery distillates. Contaminating metals, nitrogen compounds, and oxygen compounds also are essentially eliminated. The catalyst is cobalt-molybdenum-alumina and may be regenerated *in situ* with steam and air.

Uniontown Method—A type of road testing for determining the knocking characteristics of motor fuels. (Refer to *CRC Handbook*, 1946 edition; *CRC 259: Road Rating Techniques*, Appendix A, published under the auspices of the Coordinating Fuel and Equipment Research Committee, January 1951. Modified Uniontown Method also discussed.)

Unisol process—A chemical process for extracting mercaptan sulfur and certain nitrogen compounds from sour gasolines or distillates using regenerable aqueous solutions of sodium or potassium hydroxide containing methanol.

Unitary thermal polymerization—A thermal polymerization process for converting C_4 and lighter gases into liquid products. (See also **thermal polymerization.**)

Universal viscosity, Saybolt—See **Saybolt Universal viscosity.**

UOP alkylation—A process using hydrofluoric acid (which can be regenerated) as a catalyst to unite olefins with isobutane.

UOP copper sweetening—A fixed-bed process for sweetening gasolines by converting mercaptans to disulfides by contact with ammonium chloride and copper sulfate in a bed of Italian pumice. Charge stock is caustic-washed, then washed with hydrochloric acid to neutralize alkalinity and remove organic base compounds before processing.

UOP fluid catalytic cracking—A fluid process of a unitary reactor-over-regenerator design. It is adaptable to the needs of large or small refinery operations. The capacity of operating units varies from 1,400 BPD to 35,000 BPD.

urea dewaxing—A continuous dewaxing process for producing low-pour-point oils, and using urea which forms a solid complex (adduct) with the straight-chain wax paraffins in the stock. This complex is readily separated by filtration.

V

vacuum distillation—Distillation under reduced pressure. The boiling temperature is thereby reduced sufficiently to prevent decomposition or cracking of the material being distilled.

vapor phase—The term describing a substance in the gaseous state, under conditions in which it is capable of being liquefied either by pressure or cooling alone.

vapor-phase cracking—A high-temperature, low-pressure conversion process. Under these conditions, dehydrogenation reaction rates increase, with resultant higher yields of olefins and aromatics. Feedstocks can vary from gaseous hydrocarbons to gas oil.

vapor-phase hydrodesulfurization—A fixed-bed process which desulfurizes and hydrogenates napthas boiling up to approximately 500°F. Olefins may be saturated without hydrogenating aromatics. Tungsten-nickel-sulfide catalyst can be regenerated.

vapor recovery unit—A unit in which absorption; depropanization, debutanization, or both; stabilization; and rerunning are combined to separate a mixed charge of miscellaneous gases and gasolines into desired qualities for further processing.

varnish makers' and painters' naphtha—A petroleum naphtha used mainly as a thinner in paint and varnish. Distillation range is usually between 195 and 330°F, often narrower.

VGC—The viscosity-gravity constant; an index of the chemical composition of crude oil. It is defined as the general relation between specific gravity, g, at 60°F and Saybolt Universal viscosity, V, at 100°F, shown by the equation:

$$a = \frac{10g - 1.0752 \log (V - 38)}{10g - \log (V - 38)}$$

The constant (a) is low for the paraffinic crude oils and high for the naphthenic crude oils.

VI—Viscosity index; an arbitrary scale used to show the magnitude of viscosity changes in lubricating oils with changes in temperature. In the United States, the tables found in ASTM Method D 567 are widely used. (See also **pole height viscosity index.**)

virgin stock—Oil derived directly from crude oil which contains no cracked material. Also called straight-run stock.

visbreaking—Viscosity breaking; lowering or "breaking" the viscosity of residuum by cracking at relatively low temperatures.

viscosity—The measure of the internal friction or the resistivity to flow of a liquid. In measuring viscosities of petroleum products, the values of the viscosity are usually expressed as the number of seconds required for a certain volume of the oil to pass through a standard orifice under specified conditions. (See also specific types of viscosity under alphabetical listing.)

viscosity blending chart—A graphical means for estimating the viscosity, at a given temperature, of blends of petroleum liquids.

viscosity breaking—See **visbreaking.**

viscosity conversion table—A table or chart by means of which kinematic viscosity, in centistokes, can be converted to Saybolt viscosity, in seconds, at the same temperature. Conversion to Saybolt Universal viscosities may be done by the procedures and tables of ASTM Method D 446, and for Saybolt-Furol viscosities by ASTM Method D 666. Augmented tables are published in ASTM Special Technical Publication 43B.

viscosity curve—A graph showing the viscosity of an oil as a function of temperature.

viscosity-gravity constant—See **VGC.**

viscosity index—See **VI.**

viscosity number, SAE—See **SAE viscosity number.**

viscosity-temperature chart—A chart by means of which the viscosity of a petroleum oil at any temperature within a limited range may be ascertained, provided viscosities at two temperatures are known. Two charts for Saybolt viscosities and three for kinematic viscosities are available, as described in ASTM Method D 341.

viscous neutral oil—A term originating from the practice of reducing, by distillation, the neutral oil fraction to obtain an overhead and a bottom, the bottom being designated as the viscous neutral. Such oils are frequently blended with bright stock to make finished oils of various viscosities.

VM&P naphtha—See **varnish makers' and painters' naphtha.**

voltol process—See **elektrion process.**

W

wash—In petroleum refining, to cleanse or purify by intimate contact with water or chemicals. (See also **scrubbing;** specific washes under alphabetical listing.)

water scrubber—Apparatus in which gases are bubbled through water in order to wash out the last traces of any gas soluble in water.

water-white—A grade of color in oil, defined as 21 on the Saybolt chromometer scale (see **Saybolt color**).

water-white kerosine—A kerosine having the whitest (nearest colorless) of the three standard colors (see **water-white**). (See also **prime white kerosine; standard white kerosine.**)

Watson factor—See **characterization factor.**

wax distillate—A neutral distillate containing a high percentage of crystallizable paraffin wax, obtained on the distillation of paraffin or mixed-base crude, and on reducing neutral lubricating stocks. It is a primary base for paraffin wax and neutral lubricating oils.

wax fractionation—A continuous process for producing waxes of low oil content from wax concentrates, using a relatively large proportion of solvent at temperatures ranging from 5 to 60°F. The principle is repeated crystallization of a selected portion of the wax from a concentrate. (See also **MEK deoiling.**)

wax manufacturing—A process for producing oil-free waxes. Wax or a wax-concentrate feed is chilled and then crystallized from solvent. Operation and flow of the process after the chilling operation are essentially the same as for wax fractionation. Various crystalline (paraffin) products can be made having an oil content below 0.3 per cent. The process can also produce microcrystalline waxes of low oil content.

weathered crude—Crude oil which, due to natural causes during storage and handling, has lost an appreciable quantity of its more volatile components.

weep hole—A small hole in the bubble plate which serves to drain liquid from the plate when the column is not in use.

wet gas—A gas containing a relatively high proportion of hydrocarbons which are recoverable as liquids.

white oil—Generic name applied to highly refined, colorless hydrocarbon oils of low volatility, and covering a wide range of viscosities. They are widely used for the lubrication of food and textile machinery. (See also **medicinal oil; mineral oil.**)

wild gasoline—See **casinghead gasoline.**

Y

yield—The amount of a desired product or products obtained in a given process, expressed as a percentage of the feedstock. There are many yields, each of which should be specifically defined when used.

INDEX

1

8   PETROLEUM PROCESSING HANDBOOK